# Index of Tables

EXTENDED VERSION WITH MODERN PHYSICS
UNIVERSITY
Physics
EIGHTH EDITION

## CONTRIBUTORS

T. R. Sandin, *North Carolina A&T State University*
Principal editorial consultant; first draft author for modern physics chapters and modern physics case studies

A. Lewis Ford, *Texas A&M University*
Author for end-of-chapter problem sets

Craig Watkins, *Massachusetts Institute of Technology*
Principal proofreader and problem checker

Marvin DeJong, *College of the Ozarks*
First draft author for computational physics case studies

Michael Simon, *Housatonic Community College*
First draft author for extended applications case studies

William Cloud, *Eastern Illinois University*
Proofreader



## CONTRIBUTORS OF NEW PROBLEMS

Marie Baehr, *Elmhurst College*
Robert Cunningham, *Tri-State University*
David Groh, *Morningside College*
Virgil Highland, *Temple University*
David Smith, *Virginia Highlands Community College*
Jay Strieb, *Villanova University*
David Willey, *University of Pittsburgh–Johnstown*
Mark Williams, *Salisbury State University*

EXTENDED VERSION WITH MODERN PHYSICS

HUGH D. YOUNG
*Carnegie-Mellon University*

**Addison-Wesley Publishing Company**

*Reading, Massachusetts • Menlo Park, California • New York*
*Don Mills, Ontario • Wokingham, England • Amsterdam • Bonn • Sydney*
*Singapore • Tokyo • Madrid • San Juan • Milan • Paris*

The motion of a bicycle and its rider can be analyzed by considering the forces acting on the system. The downward force due to gravity (the weight, $\boldsymbol{w}$) is balanced by the upward normal forces ($\boldsymbol{n}_1$ and $\boldsymbol{n}_2$) applied to the tires by the road. There is no net vertical force, so the vertical motion does not change. The forward force $\mathcal{F}_{2\|}$ acting on the rear wheel, related to the forces the rider applies to the pedals, is partially offset by the drag force ($\boldsymbol{F}_D$) due to air resistance. If $\mathcal{F}_{2\|}$ is greater in magnitude than $\boldsymbol{F}_D$, the system has a forward acceleration. As the rider leans into the turn, the road exerts sideways frictional forces ($\mathcal{F}_{1\perp}$ and $\mathcal{F}_{2\perp}$) on the tires, directed toward the center of the curve. These forces cause the direction of the velocity to change as the system rounds the turn.



ISBN 0-201-52981-5
1 2 3 4 5 6 7 8 9 10-VH-95949392

*Sponsoring Editor:* Stuart Johnson
*Development Editor:* David M. Chelton
*Senior Production Supervisor:* Kazia Navas
*Managing Editor:* Mary Clare McEwing
*Art Development Editor:* Mary Hill
*Freelance Editorial/Production Services:*
Jane E. Hoover, Lifland et al., Bookmakers
*Design Director:* Marshall Henrichs
*Text Designer:* Carl Smizer
*Art Consultant:* Meredith Nightingale
*Technical Art Consultant:* Loretta Bailey
*Technical Art Supervisor:* Joe Vetere
*Illustrators:* James A. Bryant, George V. Kelvin,
Gary Torrisi, Darwen and Vally Hennings
*Electronic Art Source:* ST Associates
*Copyeditor:* Barbara Willette
*Layout Artist:* Lorraine Hodsdon
*Photo Researchers:* Laurie PiSierra, Mitchell Hammond
*Editorial Assistant:* Mitchell Hammond
*Permissions Editor:* Mary Dyer
*Senior Marketing Manager:* Susan Howell
*Composition and Separations Buyer:* Sarah McCracken
*Manufacturing Manager:* Roy Logan
*Compositor:* Typo-Graphics, Inc.
*Separator:* Lehigh Colortronics
*Printer:* Von Hoffmann Press, Inc.

*Text and photo credits appear on page 1355, which constitutes a continuation of the copyright page*

# Dedication

*Hugh D. Young, Mark W. Zemansky, Francis W. Sears—1972*

I would like to dedicate this new edition to the memory of Francis Sears and Mark Zemansky. Both were shining examples of excellence in physics teaching, and both were stalwart leaders in the physics teaching community throughout their professional lives. The very first edition of this book provided a rock-solid foundation on which all the later editions rest. When I became a co-author, beginning with the fifth edition, they firmly but kindly guided my hand, teaching me how to improve and update the book without ruining what was already good about it. I will never forget their kindness or their helpfulness. Thank you, Francis and Mark; I love you both.

H. D. Y.

# Preface

This new edition of *University Physics* represents a significant new stage in the evolution of a book that has played a prominent role in introductory physics education for over 40 years. This comprehensive revision reflects (a) the major changes in the philosophy of introductory physics courses that have occurred over these years, and (b) the changing backgrounds and needs of the students who take these courses.

I have tried very hard to preserve the qualities and features that users of earlier editions have found valuable. The basic goal is unchanged: to provide a broad, rigorous introduction to physics at the beginning college level for students who are currently learning elementary calculus. The most important objectives are the development of physical intuition and problem-solving skills.

The prose style is considerably more relaxed than in earlier editions. Without being colloquial or excessively familiar, I try to talk to the student as a partner in learning, not as an audience to be lectured to from atop a platform. I have found that this new style makes it a lot easier for me to convey my own excitement and enthusiasm about the beauty and intellectual challenge of physics than a more formal style would allow.

Physics educators face several major challenges today. One is to persuade students that the study of physics is meaningful and relevant to their professional preparation and to their lives. Another is to help them develop physical intuition and modeling skills along with problem-solving ability. While addressing these challenges, I would like to show the student as clearly as possible the beauty and fundamental unity of all branches of physics. The ways I have addressed these problems are shown in the following features of the new edition.

## Chapter Organization

There is always danger that the student will get lost in what may seem like a blizzard of detail. To help avoid this, each chapter begins with a list of Key Concepts to give the student perspective and guideposts through the chapter. Then, an opening paragraph gives specific examples of the chapter's content and connects it with what has come before. At the end of each chapter is a Summary, including a list of Key Terms that the student should have learned to use, and a synopsis of the most important principles introduced in the chapter, with the associated Key Equations.

### Contents

Here are the most significant changes in the organization of the new edition:

• The material on mechanical energy is divided into two chapters, the first on the work-energy theorem and applications, the second on potential energy and conservation of energy.

• The momentum chapter has been reorganized. It now begins with conservation of momentum and its applications to collision problems. The concept of impulse is introduced later.

• The treatment of rigid body motion is divided into two chapters. The first includes rotational kinematics and kinetic energy; the second includes the relation of torque to angular acceleration and angular momentum.

• The treatment of elasticity has been shortened and combined with the material on equilibrium.

• Gravitation and the motions of planets and satellites have been placed in a separate chapter, which also includes an expanded treatment of Kepler's laws and gravitational potential energy.

• The material on thermal phenomena and thermodynamics has been completely rewritten and considerably shortened, now comprising four chapters rather than the previous seven. The microscopic viewpoint, including kinetic theory of gases, is woven through the entire discussion rather than appearing as a separate chapter.

• The treatment of interference and diffraction has been expanded somewhat and divided into two chapters.

### Problems

The end-of-chapter problem collections have been extensively revised and enlarged. The problems are again grouped into three categories: *Exercises*, single-concept problems that can be keyed to specific sections of the text; *Problems*, usually requiring two or more nontrivial steps; and *Challenge Problems*, intended to challenge the strongest students. The number of problems has grown by 480; the total number is now 2300. The revision of the problem collections was carried out by Prof. A. Lewis Ford (Texas A&M University) with the assistance of Mr. Craig Watkins (M.I.T.).

### Problem-Solving Strategy

Problem-Solving Strategy sections originated with the seventh edition (1987) of this book and have since been imitated in several other books. They have proved to be a very substantial help, especially to the many earnest but bewildered students who "understood the material but couldn't do the problems."

### Examples

Each Problem-Solving Strategy section is followed immediately by one or more worked-out examples that illustrate the strategy. The number of examples has grown by 195 to a total of 486. Many are drawn from familiar, real-life situations related to the student's own experience. I have tried to emphasize the relevance of physics in a broad range of familiar situations, including applications to engineering, biology, astronomy, medicine, and many other areas.

Example solutions always begin with a statement of the general principles to be used and, when necessary, a discussion of the reason for choosing them. Heavy emphasis is placed on *modeling*—showing the student how to begin with a seemingly complex situation, make simplifying assumptions when needed, apply the appropriate physical principles, and evaluate the final result. Does it make sense? Is it what you expected? How can you check it?

### Illustrations

All the illustrations have been completely redone to take advantage of the new four-color format. In addition to giving the book greatly enhanced visual appeal for the student, the full-color treatment serves several vital pedagogical functions.

• Vectors are color-coded; each vector quantity has a characteristic color. Multiple graphs drawn on the same axis system are color-coded, and various materials in a diagram can be distinguished by color.

• Applications to a wide variety of real-world situations are made much more vivid by the use of full-color photographs and illustrations, often appearing side-by-side with analytic diagrams.

• The various pedagogical features of the book, such as introductions, examples, strategies, and summaries are easily recognizable by their distinctive color treatment.

### Mathematical Level

The mathematical level has not changed substantially from previous editions, but I have increased somewhat the use of unit vectors and calculus in worked-out examples. There are also more problems involving unit vectors and calculus. Some of the challenge problems invite students to stretch their mathematical ability, especially in situations that require numerical approximations.

### Units and Notation

In this edition SI units are used exclusively. English unit conversions are included where appropriate. The joule is used as the standard unit of energy of all forms, including heat. In examples, units and correct significant figures are always carried through all stages of numerical calculations. As usual, boldface symbols are used for vector quantities, and in addition boldface +, –, and = signs are used to remind the student at every opportunity of the crucial distinctions between operations with vectors and those with numbers.

### Enrichment

• Case Studies: I have resisted the temptation to include lengthy essays on topics unrelated to the chapters in which they appear. Instead, I have included 13 optional sections called Case Studies, each building on the material of its chapter. Some (Neutrinos, Black Holes, Photons) emphasize connections between classical and modern physics. Others (Automotive Power, Energy Resources, Power Distribution Systems) have an engineering flavor; still others (Baseball Trajectories, Chaos, Field Maps) emphasize computer simulations and include computer exercises for the student. All case studies have corresponding end-of-chapter problems.

• Building Physical Intuition: Twelve full-page layouts combine photographs, diagrams, and text to help build intuition for individual key concepts such as free-body diagrams, energy diagrams, entropy, and electromagnetic waves. These pages will help students build conceptual bridges between basic principles of physics and the world around them—to help them learn to think like physicists.

### Flexibility

This book is adaptable to a wide variety of course outlines. There is plenty of material for an intensive three-semester course. Most instructors will find that there is too much material for a one-year course, but the format makes it easy to tailor the book to any of a variety of one-year course plans by omitting certain chapters or sections. For example, any or all of the chapters on relativity, fluid mechanics, acoustics, electromagnetic waves, optical instruments, and several others can be omitted without loss of continuity. In addition, some sections that are unusually challenging or somewhat out of the mainstream have been identified with an asterisk preceding the section title. These too may be omitted without loss of continuity.

In any case, I hope no one will feel constrained to work straight through the book from cover to cover. Many chapters are of course inherently sequential in nature, but within this general limitation I would encourage instructors to select the chapters that fit their needs, omitting material that is not appropriate for the objectives of a particular course.

### Extended Version

This new edition is published in two versions. The regular version includes 39 chapters, ending with Relativity, and omits the modern physics material that was in Chapters 41 through 44 of the seventh edition. The Extended Version adds seven completely new chapters on modern physics. They include all the topics in the old Chapters 41 through 44 and considerable additional material, concluding with a chapter on particle physics and cosmology.

### Supplements

A textbook should stand on its own feet. Yet many students benefit from supplementary materials designed to be used with the text. With this thought in mind, we offer a Study Guide and a Solutions Manual.

- The Study Guide, prepared by Profs. James R. Gaines and William F. Palmer, includes for each chapter a statement of objectives, a review of central concepts, problem-solving hints, additional worked-out examples and a short quiz.
- The Solutions Manual, prepared by Prof. A. Lewis Ford, includes completely worked-out solutions for about two-thirds of the odd-numbered problems in the book. (Answers to all odd-numbered problems are listed at the end of the book.)

We also offer several aids for the instructor:

- The Instructor's Manual contains answers to all even-numbered problems and class discussion questions. This is available only to instructors.
- OmniTest is an algorithm-driven computerized testing system that enables instructors to generate tests effortlessly. It also allows instructors to add their own test items and edit existing items with an easy to use, on screen "What-You-See-Is-What-You-Get" text editor.

### ACKNOWLEDGMENTS

I want to extend my heartfelt thanks to my colleagues at Carnegie-Mellon, especially Profs. Robert Kraemer, Bruce Sherwood, Helmut Vogel, and Brian Quinn, for many stimulating discussions about physics pedagogy and for their support and encouragement during the writing of this new edition. I am equally indebted to the many generations of Carnegie-Mellon students who have helped me learn what good teaching and good writing are, by showing me what works and what doesn't. It is always a joy and a privilege to express my gratitude to my wife Alice and our children Gretchen and Rebecca for their love, support, and emotional sustenance during the writing of the new edition. May all men and women be blessed with love such as theirs.

As always, I welcome communications from students and professors, especially when they concern errors or deficiencies that you find in this edition. I have spent a lot of time and effort writing the best book I know how to write. It has been a labor of love; undergraduate teaching is the main focus of my professional life, and this edition is a direct outgrowth of my love of teaching. I hope it will help you to teach and learn physics. In turn, you can help me by letting me know what still needs to be improved!

*Pittsburgh, Pennsylvania* H. D. Y.
*November, 1991*

# Reviewers

Edward Adelson, Ohio State University
Harold Bale, University of North Dakota
John Barach, Vanderbilt University
Paul Baum, CUNY–Queens College
Robert Boeke, William Rainey Harper College
Tony Buffa, Cal Poly–San Luis Obispo
Roger Clapp, University of South Florida
Leonard Cohen, Drexel University
W. R. Coker, University of Texas–Austin
David Cook, Lawrence University
Larry Curtis, University of Toledo
George Dixon, Oklahoma State University
Boyd Edwards, West Virginia University
William Fasnacht, US Naval Academy
Neil Fletcher, Florida State University
Solomon Gartenhaus, Purdue University
Ron Gautreau, New Jersey Institute of Technology
John Gruber, San Jose State University
Graham Gutsche, US Naval Academy
Keith Honey, West Virginia Institute of Technology
Gregory Hood, Tidewater Community College
Alvin Jenkins, North Carolina State University
John Karchek, GMI Engineering & Management Institute
Elihu Lubkin, University of Wisconsin–Milwaukee
Jeffrey Mallow, Loyola University
Robert Marcini, Memphis State University
David Markowitz, University of Connecticut
Lawrence McIntyre, University of Arizona
Arnold Perlmutter, University of Miami
Ronald Poling, University of Minnesota
R. J. Peterson, University of Colorado–Boulder
Michael Rapport, Anne Arundel Community College
Ross Spencer, Brigham Young University
Edward Strother, Florida Institute of Technology
Albert Stwertka, US Merchant Marine Academy
Martin Tiersten, CUNY–City College
Somdev Tyagi, Drexel University
Thomas Wiggins, Pennsylvania State University
Stanley Williams, Iowa State University
Suzanne Willis, Northern Illinois University
James Wood, Palm Beach Junior College

# Contents

## 5
## Applications of Newton's laws 107



## 6
## Work and kinetic energy 145

## 7
## Conservation of energy 169



## 8
## Momentum and impulse 196



## 9
## Rotational motion 232

2

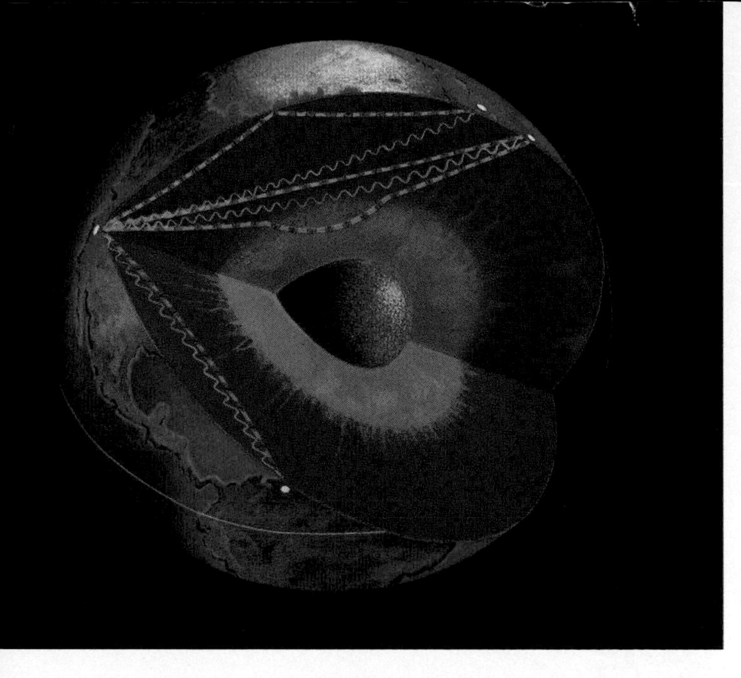

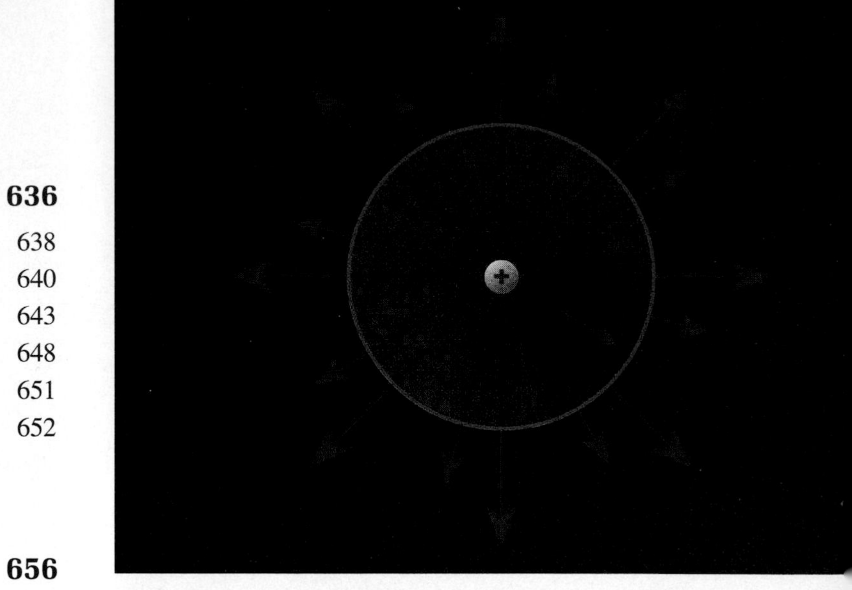

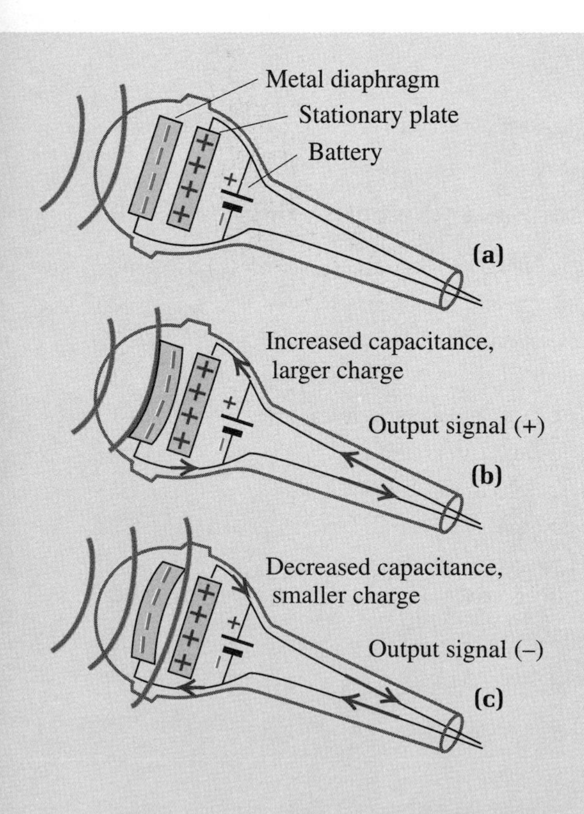

Metal diaphragm
Stationary plate
Battery
(a)
Increased capacitance,
larger charge
Output signal (+)
(b)
Decreased capacitance,
smaller charge
Output signal (–)
(c)

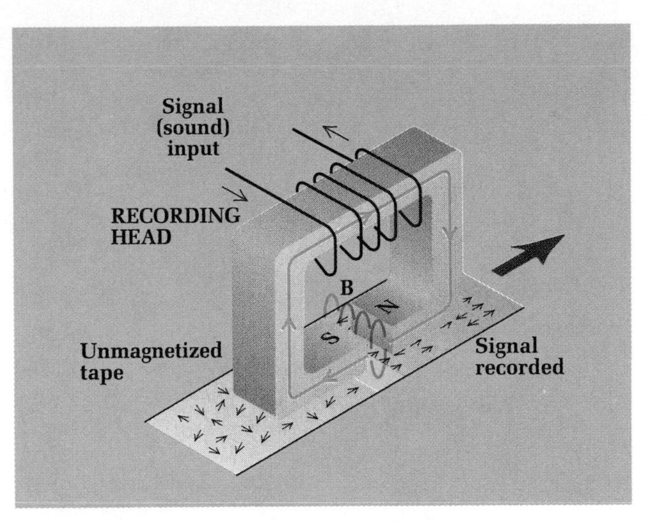

Signal (sound) input
RECORDING HEAD
B
S
N
Unmagnetized tape
Signal recorded

## 31 Inductance 866



## 32 Alternating currents 891



## 33 Electromagnetic waves 917



## 34 The nature and propagation of light 944



## 35 Geometric optics 974



## 36 Optical instruments 1003

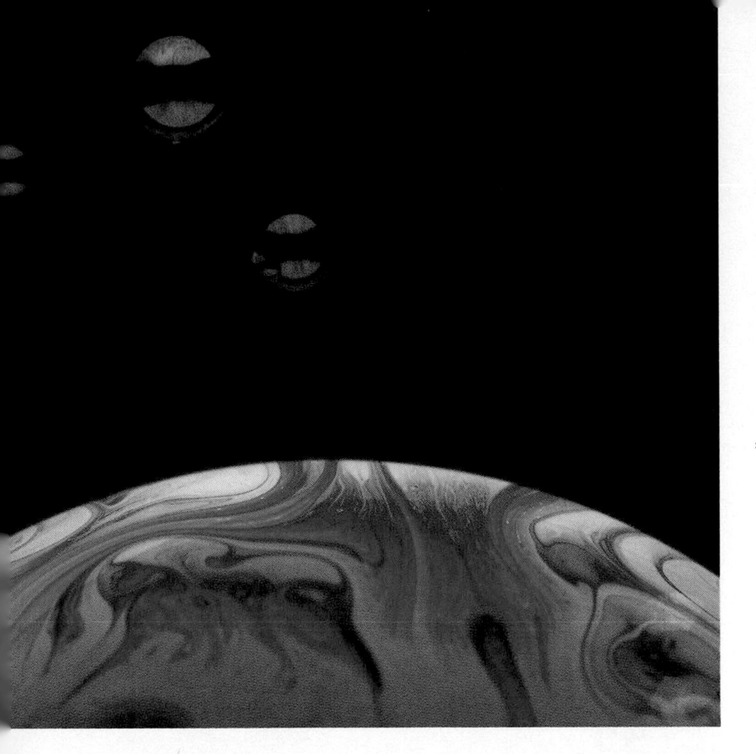

## 37

## Interference 1022



## 38

## Diffraction 1045



## 39

## Relativistic mechanics 1071



## 40

## Photons, electrons, and atoms 1101



## 41

## The wave nature of particles 1140

CHAPTER 1

# Units, Physical Quantities, and Vectors

This illustration from Sir Isaac Newton's *Principia* (1687) shows that free fall on the earth and orbital motion around the earth are both caused by the earth's gravitational attraction.

The space probe Galileo, launched October 18, 1989, is due to arrive at Jupiter in December, 1995, to study the planet and its moons. The technology that created this space explorer had its beginnings in the work of Newton and grew with the contributions of the physicists who came after him.

## KEY CONCEPTS

**• Physics is an experimental science; it relies on idealized models of complex physical situations. Physical theories and models evolve to include new ideas and observations.**

**• To make precise measurements, we need to define units of measurement that do not change and that can be duplicated easily.**

**• The precision of a calculated result is usually no greater than the precision of the input data. We indicate the precision of a measurement by the number of significant figures.**

**• A vector quantity has a direction in space as well as a magnitude; a scalar quantity has only a magnitude. Addition of vectors is a geometric process. There are two kinds of products of vectors; the dot product of two vectors is a scalar quantity, but the cross product is another vector.**

## INTRODUCTION

The achievement of space travel represents one of the greatest triumphs of modern technology. It has required pioneering work in many different areas, from aerodynamic design to materials science to sophisticated control systems. All these have required applications of basic physical laws, so the space program also represents many great achievements in basic science.

Why study physics? For two reasons. First, as we've just pointed out, basic science, including physics, is the foundation of all engineering and technology. No engineer could design any kind of practical device without first understanding the basic principles involved. To design a satellite or even a better mousetrap, you have to understand the basic laws of physics.

But there's another reason. The study of physics is an adventure. You will find it challenging, exhilarating, sometimes frustrating, occasionally painful, and often richly rewarding and satisfying. It will appeal to your sense of beauty as well as to your rational intelligence. Our present understanding of the physical world has been built on foundations laid by scientific giants such as Galileo, Newton, Maxwell, and Einstein, and their influence has extended far beyond science to affect profoundly the ways in which we live and think. You can share some of the excitement of their discoveries when you learn to use physics to solve practical problems and to gain insight into everyday phenomena. If you've ever wondered why the sky is blue, how radio waves can travel through empty space, or how a satellite maintains its orbital motion, you can find the answers as part of fundamental physics. Above all, you will come to see physics as a towering achievement of the human intellect in its quest for understanding of the world we live in.

In this opening chapter we go over some important preliminaries that we'll need throughout our study. We need to think about the philosophical framework in which we will operate — the nature of physical theory and the role of idealized models in representing physical systems. We discuss systems of units that are used to describe physical quantities and the precision of a number, which is often described by means of significant figures. We look at examples of problems in which we can't or don't want to make precise calculations but in which rough estimates can be interesting and useful. Finally, we study several aspects of vector algebra. We use vectors to describe and analyze many physical quantities, such as velocity and force, that have direction as well as magnitude.

## 1–1 INTRODUCTION

Physics is an *experimental* science. Physicists observe the phenomena of nature and try to find patterns and principles that relate these phenomena. These patterns are called physical theories or, when they are very broad and well established, physical laws. The development of physical theory requires creativity at every stage. The physicist has to learn to ask appropriate questions, design experiments to try to answer the questions, and draw appropriate conclusions from the results. Figure 1–1 shows two famous experimental facilities.

According to legend, Galileo Galilei (1564–1642) dropped light bodies and heavy bodies from the top of the Leaning Tower of Pisa to find out whether they fell at the same rate or at different rates. Galileo recognized that only an experimental investigation could answer this question. From the results of the experiments, he had to make the inductive leap to the principle, or theory, that the acceleration of a falling body is independent of its weight.

The development of physical theory is always a two-way process that starts and ends with observations or experiments. This development often takes an indirect path, with blind alleys, wrong guesses, and the discarding of unsuccessful theories in favor of

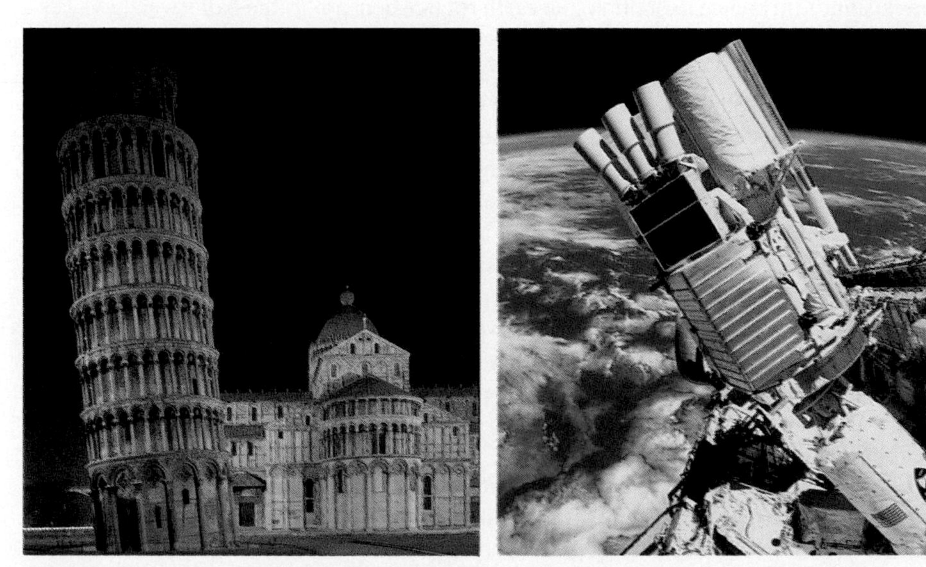

(a) (b)

**1–1** Two research laboratories. (a) The Leaning Tower of Pisa (Italy). According to legend, Galileo studied the motion of freely falling bodies by dropping them from the tower. He is also said to have gained insights into pendulum motion by observing the swinging of the hanging chandeliers in the cathedral behind the tower. (b) The Astro Observatory, operating from the space shuttle *Columbia* in December 1990. Its instruments measured x-ray and ultraviolet radiation from astronomical objects, finding signs of possible black holes at the centers of distant galaxies.

more promising ones. Physics is not a collection of facts and principles; it is the *process* by which we arrive at general principles that describe the behavior of the physical universe.

No theory is ever regarded as the final or ultimate truth; there is always the possibility that new observations will require revision of a theory. It is in the nature of physical theory that we can *disprove* a theory by finding behavior that is inconsistent with it, but we can never prove that a theory is always correct.

Getting back to Galileo, suppose we drop a feather and a cannonball. They certainly *do not* fall at the same rate. This does not mean that Galileo was wrong; it means that his theory was incomplete. If we drop the feather and the cannonball *in vacuum,* to eliminate air resistance, then they do fall at the same rate. Galileo's theory has a range of validity — specifically, those bodies for which the air resistance force is much less than the weight. A feather or a parachute is clearly not within this range.

Every physical theory has a range of validity and a boundary outside of which it is not applicable. Often a new development in physics has the effect of extending the range of validity of a principle. Galileo's analysis of falling bodies was greatly extended half a century later by Newton's laws of motion and law of gravitation.

An essential part of the interplay of theory and experiment is learning how to apply physical principles to a variety of practical problems. At various points in our study we will discuss systematic problem-solving procedures that will help you to set up problems and solve them efficiently and accurately. Learning to solve problems is absolutely essential; you don't *know* physics unless you can *do* physics. This means not only learning the general principles, but also learning how to apply them in specific situations.

## 1–2 IDEALIZED MODELS

In everyday conversation we often use the word "model" to mean either a small-scale replica, such as a model railroad, or a human body that displays articles of clothing (or the absence thereof). In physics a **model** is a simplified version of a physical system that would be too complicated to analyze in full without the simplifications.

Here's an example. Suppose we want to analyze the motion of a baseball thrown through the air. How complicated is this problem? The ball is neither perfectly spherical nor perfectly rigid; it has raised seams, and it spins as it moves through the air. Wind and air resistance influence its motion, the earth rotates beneath it, the ball's weight varies a little as its distance from the center of the earth changes, and so on. If we try to include all these things, the analysis gets pretty hopeless. Instead, we invent a simplified version of the problem. We neglect the size and shape of the ball, representing it as a point object, or *particle*. We neglect air resistance and make the ball move in vacuum; we forget about the earth's rotation, and we make the weight exactly constant. Now we have a problem that is simple enough to deal with. The ball is now a particle moving along a simple parabolic path; we will analyze this model in detail in Chapter 3.

The point is that we have to overlook quite a few minor effects in order to concentrate on the most important features of the motion. That's what we mean by making an idealized model of the system. Of course, we have to be careful not to neglect too much. If we ignore the effect of gravity completely, then when we throw the ball up, it goes in a straight line and disappears into space, never to be seen again. We need to use some judgment and creativity to construct a model that simplifies a problem enough to make it manageable, yet keeps its essential features.

When we analyze a system or predict its behavior on the basis of a model, the validity of our predictions is limited by the validity of the model (Fig. 1–2). Going back to Galileo once more, we see that his prediction about falling bodies corresponds to an ide-

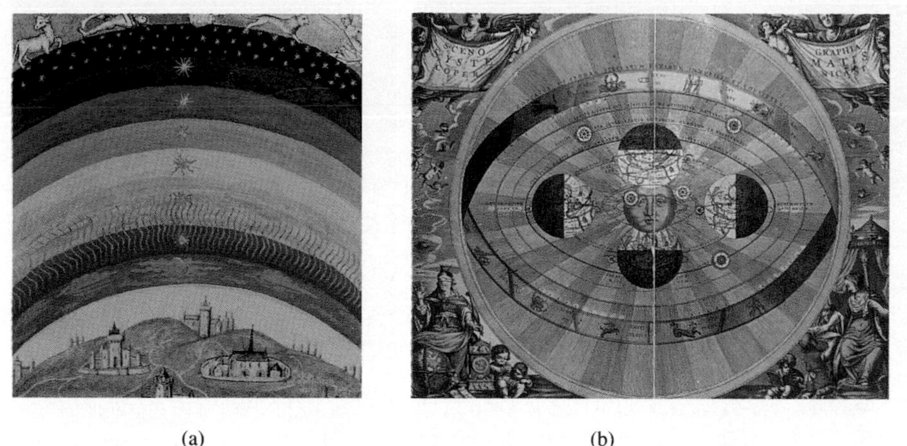

(a) (b)

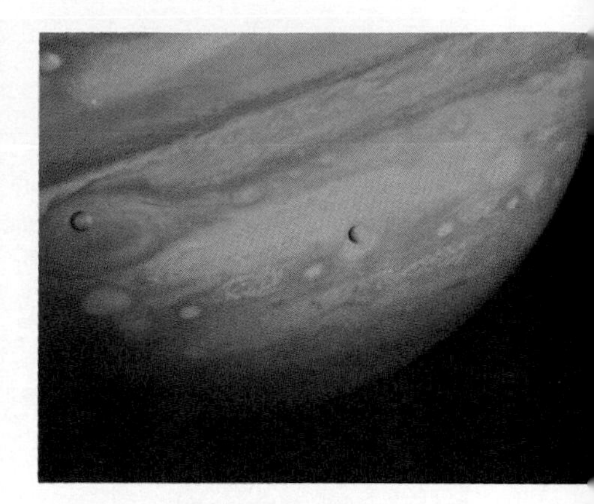

(c)

**1–2** (a) A sixteenth-century model of the universe placed the earth at the center, with planets and the stars circling around it. (b) Later observations of planetary motion and eclipses led to revision of this model. In 1543, Copernicus published *De Revolutionibus,* suggesting that the earth and the planets orbited around the sun. (c) Copernicus's model of the solar system was confirmed in 1610 when Galileo observed moons orbiting about the planet Jupiter, proving that the earth was not the center of all celestial motion. This photo from the Voyager 2 spacecraft shows the moons Io (left) and Europa (right) in front of Jupiter.

alized model that does not include the effects of air resistance. This model works fairly well for a bullet or a cannonball, not so well for a feather.

The concept of idealized models is of the utmost importance in all of physical science and technology. When we apply physical principles to complex systems, we always make extensive use of idealized models, and we have to be aware of the assumptions we are making. Indeed, the principles of physics themselves are stated in terms of idealized models; we speak about point masses, rigid bodies, ideal insulators, and so on. Idealized models play a crucial role throughout this book. Watch for them in discussions of physical theories and their applications to specific problems.

## 1–3 STANDARDS AND UNITS

Physics is an experimental science, and we usually use numbers to describe the results of measurements. Any number that is used to describe a physical phenomenon quantitatively is called a **physical quantity.** Some physical quantities are so fundamental that we can define them only by describing how to measure them. Such a definition is called an **operational definition.** In other cases we define a physical quantity by describing a way to calculate it from other quantities that we can measure. In the first case we might use a ruler to measure a distance or a stopwatch to measure a time interval. In the second case we might define the average speed of a moving object as the distance traveled (measured with a ruler) divided by the time of travel (measured with a stopwatch).

When we measure a quantity, we always compare it with some reference standard. When we say that a Porsche 944 is 4.29 meters long, we mean that it is 4.29 times as long as a meter stick, which we *define* to be 1 meter long. Such a standard defines a **unit** of the quantity. The meter is a unit of distance, and the second is a unit of time. When we use a number to describe a physical quantity, we must specify the unit that we are using; to describe a distance simply as "4.29" wouldn't mean anything.

To make precise measurements, we need units of measurement that do not change and that can be duplicated by observers in various locations. When the metric system was established in 1791 by the Paris Academy of Sciences, the **meter** was defined as one ten-millionth of the distance from the equator to the North Pole (Fig. 1–3). The **second** was defined as the time for a pendulum one meter long to swing from one side to the other.

These definitions were cumbersome and hard to duplicate precisely, and they have been replaced by more refined definitions. Since 1889 the definitions of the basic units have been established by an international organization, the General Conference on Weights and Measures. The system of units defined by this organization is based on the metric system, and since 1960 it has been known officially as the **International System,** or SI (the abbreviation for the French equivalent, *Système International*).

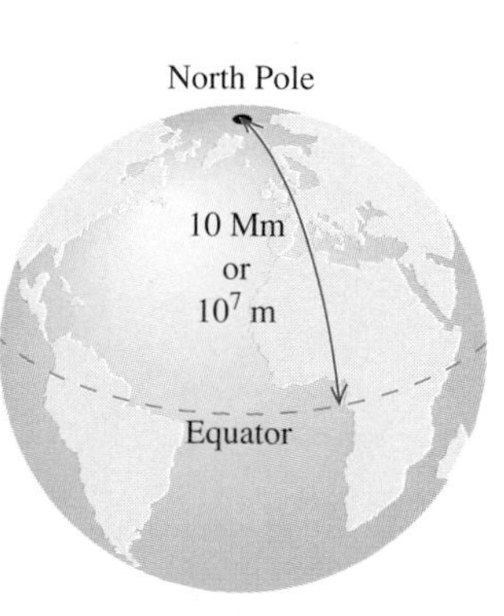


**1–3** The distance from the North Pole to the equator is about $10^7$ m.

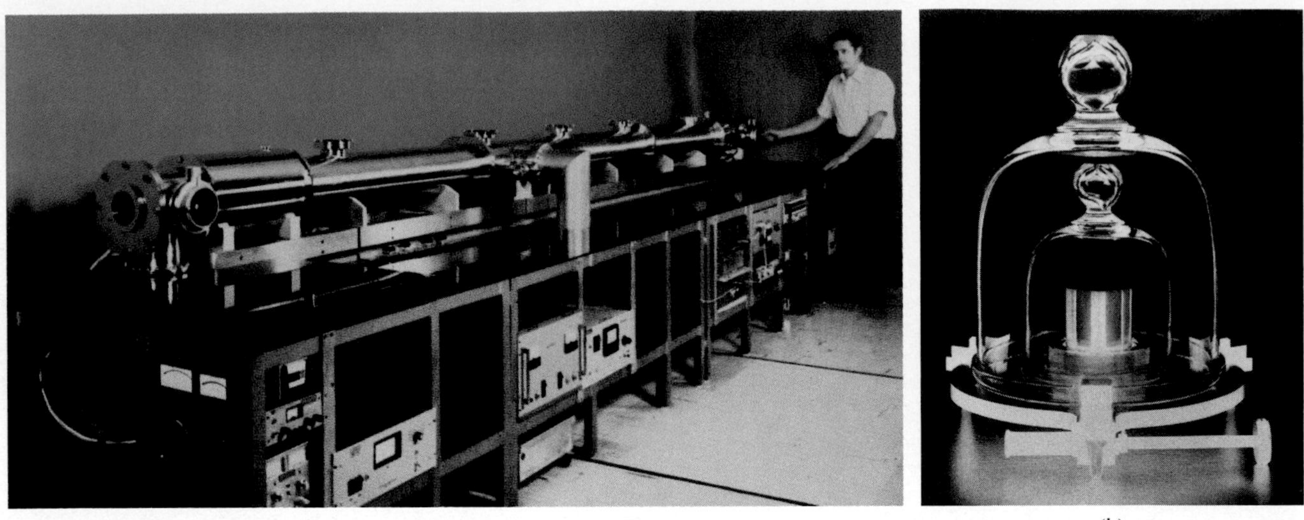

(a) (b)

**1–4** (a) NBS-6 is the primary atomic frequency standard developed by the National Institute of Standards and Technology. Consisting of a 6-m cesium beam tube, NBS-6 achieves an accuracy of better than one part in $10^{13}$ and, when operated as a clock, can keep time to within 3 millionths of a second per year. Clocks utilizing a *single* trapped ionized atom may be able to increase the accuracy to one part in $10^{17}$. (b) National Standard Kilogram No. 20 at the National Institute of Standards and Technology is an accurate copy of the international standard kilogram.

## Time

From 1889 until 1967 the unit of time was defined as a certain fraction of the mean solar day, the average time interval between successive arrivals of the sun at its highest point in the sky. The present standard, adopted in 1967, is much more precise. It is based on an atomic clock (Fig. 1–4a), which uses the energy difference between the two lowest energy states of the cesium atom. Electromagnetic radiation (microwaves) of precisely the proper frequency causes transitions from one of these states to the other. One second is defined as the time required for 9,192,631,770 cycles of this radiation.

## Length

In 1960 an atomic standard for the meter was also established, using the wavelength of the orange-red light emitted by atoms of krypton ($^{86}Kr$) in a glow discharge tube. In November 1983 the length standard was changed again, in a more radical way. The new definition of the meter is the distance that light travels (in vacuum) in 1/299,792,458 second. This has the effect of *defining* the speed of light to be precisely 299,792,458 m/s; we then define the meter to be consistent with this number and with the above definition of the second. This provides a much more precise standard of length than the one based on a wavelength of light.

## Mass

The standard of mass is the mass of a particular cylinder of platinum-iridium alloy. Its mass is defined to be one **kilogram,** and it is kept at the International Bureau of Weights and Measures at Sèvres, near Paris (Fig. 1–4b). An atomic standard of mass would be more fundamental, but at present we cannot measure masses on an atomic scale with as great precision as on a macroscopic scale.

## Prefixes

Once we have defined the fundamental units, it is easy to introduce larger and smaller units for the same physical quantities. In the metric system these other units are always related to the fundamental units by multiples of 10 or $\frac{1}{10}$. Thus one kilometer (1 km) is

1000 meters, one centimeter (1 cm) is $\frac{1}{100}$ meter, and so on. We usually express the multiplicative factors in exponential notation: $1000 = 10^3$, $\frac{1}{1000} = 10^{-3}$, and so on. With this notation,

$$1\ \text{km} = 10^3\ \text{m} \qquad \text{and} \qquad 1\ \text{cm} = 10^{-2}\ \text{m}.$$

The names of the additional units are always derived by adding a **prefix** to the name of the fundamental unit. For example, the prefix "kilo-," abbreviated k, always means a unit larger by a factor of 1000; thus

1 kilometer = 1 km = $10^3$ meters = $10^3$ m
1 kilogram = 1 kg = $10^3$ grams = $10^3$ g
1 kilowatt = 1 kW = $10^3$ watts = $10^3$ W

Table 1–1 lists the standard SI prefixes with their meanings and abbreviations. We note that most of these are multiples of $10^3$.

Here are several examples of the use of multiples of 10 and their prefixes with the units of length, mass, and time. Some additional time units are also included.

### Length

1 nanometer = 1 nm = $10^{-9}$ m (a few times the size of an atom)
1 micrometer = 1 μm = $10^{-6}$ m (size of some bacteria and cells)
1 millimeter = 1 mm = $10^{-3}$ m (point of a ballpoint pen)
1 centimeter = 1 cm = $10^{-2}$ m (diameter of your little finger)
1 kilometer = 1 km = $10^3$ m (a ten-minute walk)

### Mass

1 microgram = 1 μg = $10^{-9}$ kg (mass of a very small dust particle)
1 milligram = 1 mg = $10^{-6}$ kg (mass of a grain of salt)
1 gram = 1 g = $10^{-3}$ kg (mass of a paper clip)

**TABLE 1–1 Prefixes for Powers of 10**

| Power of ten | Prefix | Abbreviation | Pronunciation |
|---|---|---|---|
| $10^{-18}$ | atto- | a | *at*-toe |
| $10^{-15}$ | femto- | f | *fem*-toe |
| $10^{-12}$ | pico- | p | *pee*-koe |
| $10^{-9}$ | nano- | n | *nan*-oe |
| $10^{-6}$ | micro- | μ | *my*-kroe |
| $10^{-3}$ | milli- | m | *mil*-i |
| $10^{-2}$ | centi- | c | *cen*-ti |
| $10^{3}$ | kilo- | k | *kil*-oe |
| $10^{6}$ | mega- | M | *meg*-a |
| $10^{9}$ | giga- | G | *jig*-a |
| $10^{12}$ | tera- | T | *ter*-a |
| $10^{15}$ | peta- | P | *pet*-a |
| $10^{18}$ | exa- | E | *ex*-a |

# Relative Sizes

As technology has developed, the range of distances that we know about, from the very small to the very large, has become truly astonishing. Pictured on these pages are some of the objects that are central to various areas of physics. So that you can appreciate the different distance scales in each picture, we have expressed the distance along each side of the picture in terms of meters, using powers-of-ten notation.

# BUILDING PHYSICAL INTUITION

**$10^{21}$ m** A galaxy. Our own Milky Way galaxy has a spiral structure containing some 100 billion stars. The sun travels around the galaxy in a clockwise path, completing one orbit every 300 million years.

**$10^{9}$ m** The sun. The heat of the sun is generated by the fusion of hydrogen nuclei (one proton each) to form helium nuclei (two protons and two neutrons each). As in all stars, this fusion process is driven by the mutual gravitational attraction of the nuclei, which overwhelms all other forces in the sun by sheer numbers. The mass of a proton is $10^{-27}$ kg and the mass of the sun is $10^{30}$ kg, so there must be roughly $10^{57}$ protons in the sun—a truly enormous number.

**$10^{7}$ m** The earth. The average radius of the earth is about 6 x $10^{6}$ m. The atmosphere is 20-30 km thick, the land's average height above sea level is about 1 km, and the oceans' average depth is about 4 km. We live within a very thin shell above and below the surface of our planet.

**$10^{2}$ m** The Statue of Liberty. Structures built by humans often reach heights of several hundred meters, especially in cities. However, no building reaches as high as 500 m. On the other hand, horizontal structures such as bridges can exceed $10^{3}$ m in length.

**$10^0$ m** You. The scale of distance that we know best.

**$10^{-3}$ m** A computer chip. Modern microelectronics production can pack a device with over 6200 transistors onto a chip of silicon measuring about $5 \times 10^{-3}$ m on a side. Texas Instruments' memory chip shown here can store 74,000 bits of information, and is used in products ranging from computers to home video games.

**$10^{-5}$ m** A red blood cell. Mammalian red blood cells measure about $8 \times 10^{-6}$m in diameter. Their life span is only about 120 days. A human body contains about $25 \times 10^{12}$ red blood cells; new ones are formed at a rate of about 2 million per second to replace dying cells and maintain a constant number.

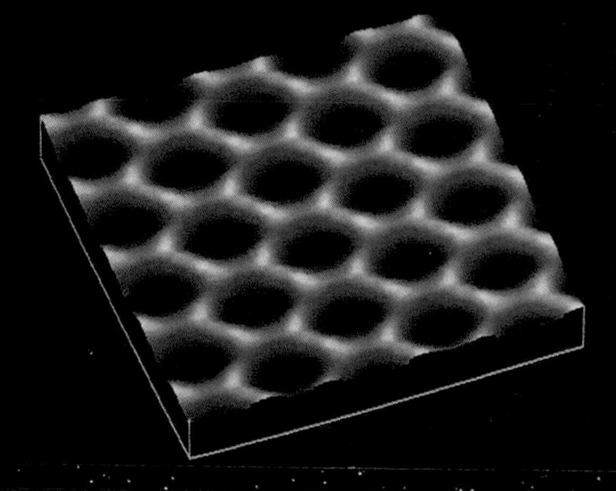

**$10^{-10}$ m** An atom. Assuming spherical atoms, atomic diameters range from about $0.5 \times 10^{-10}$ m for helium to $5 \times 10^{-10}$ m for cesium. Most of the atom is empty space, with the tiny nucleus in the center, but the atom's electrons may be found anywhere within this space. This atomic force microscope photograph shows carbon atoms in graphite, about $1.5 \times 10^{-10}$ m apart.

**$10^{-14}$ m** An atomic nucleus. As many as 209 total protons and neutrons can fit into this tiny volume and maintain stability. Protons repel one another; the particles are held together by the short-range but very strong nuclear force.

### Time

1 nanosecond = 1 ns = $10^{-9}$ s (time for light to travel 0.3 m)
1 microsecond = 1 $\mu$s = $10^{-6}$ s (time for a personal computer to do an addition)
1 millisecond = 1 ms = $10^{-3}$ s (time for sound to travel 0.35 m)
1 minute = 1 min = 60 s
1 hour = 1 h = 3600 s
1 day = 1 d = 86,400 s

## The British System

Finally, we mention the British system of units. These units are used only in the United States and a few other countries, and in most of these they are being replaced by SI units. British units are now officially defined in terms of SI units, as follows:

*Length:* 1 inch = 2.54 cm (exactly)

*Force:* 1 pound = 4.448221615260 newtons (exactly)

The newton, abbreviated N, is the SI unit of force. The British unit of time is the second, defined the same way as in SI. In physics, British units are used only in mechanics and thermodynamics; there is no British system of electrical units.

In this book we use SI units for all examples and problems, but in the early chapters we occasionally give approximate equivalents in British units. As you do problems using SI units, try to think of the approximate equivalents in British units, but also try to think directly in SI as much as you can. Much of U.S. industry has already converted to metric standards. We speak of 4-liter engines, 50-mm lenses, 35-mm film, 750-mL wine bottles, and so on. The use of SI units in everyday life is not far off.

# 1–4 UNIT CONSISTENCY AND CONVERSIONS

We use equations to express relations among physical quantities that are represented by algebraic symbols. Each algebraic symbol always denotes both a number and a unit. For example, $d$ might represent a distance of 10 m, $t$ a time of 5 s, and $v$ a speed of 2 m/s.

An equation must always be **dimensionally consistent.** You can't add apples and pomegranates; two terms may be added or equated only if they have the same units. For example, if a body moving with constant speed $v$ travels a distance $d$ in a time $t$, these quantities are related by the equation

$$d = vt. \tag{1-1}$$

If $d$ is measured in meters, then the product $vt$ must also be expressed in meters. Using the above numbers as an example, we may write

$$10\text{ m} = (2\text{ m}/\cancel{\text{s}})(5\ \cancel{\text{s}}).$$

Because the unit 1/s on the right side cancels the unit s, the product $vt$ has units of meters, as it must. In calculations, units are treated just like algebraic symbols with respect to multiplication and division.

When a problem requires calculations using numbers with units, always write the numbers with the correct units and carry the units through the calculation as in the example above. This provides a very useful check for calculations. If at some stage in a calculation you find that an equation or an expression has inconsistent units, you know you have made an error somewhere. In this book we will always carry units through all calculations, and we strongly urge you to follow this practice when you solve problems.

## PROBLEM-SOLVING STRATEGY

### Unit conversions

Units are multiplied and divided just like ordinary algebraic symbols. This fact gives us an easy way to convert a quantity from one set of units to another. The key idea is that we can express the same physical quantity in two different units and form an equality. For example, when we say that 1 min = 60 s, we don't mean that the number 1 is equal to the number 60; we mean that 1 min represents the same physical time interval as 60 s. We may multiply a quantity by the factor (1 min)/(60 s) without changing its physical meaning. To find the number of seconds in 3 min, we write

$$3\ \text{min} = (3\ \cancel{\text{min}})\left(\frac{60\ \text{s}}{1\ \cancel{\text{min}}}\right) = 180\ \text{s}.$$

This makes sense; the second is a smaller unit than the minute, so there are more seconds than minutes in the same time interval.

### EXAMPLE 1–1

The official world land speed record is 1019.5 km/h, set on October 4, 1983 by Richard Noble in the jet-engine car Thrust 2. Express this speed in m/s.

**SOLUTION** The prefix k means $10^3$. We multiply by the factor $(10^3\ \text{m})/(1\ \text{km})$ to change km to m and by the factor (1 h)/(3600 s) to change 1/h to 1/s.

$$1019.5\,\frac{\text{km}}{\text{h}} = \left(1019.5\,\frac{\cancel{\text{km}}}{\cancel{\text{h}}}\right)\left(\frac{10^3\ \text{m}}{1\ \cancel{\text{km}}}\right)\left(\frac{1\ \cancel{\text{h}}}{3600\ \text{s}}\right)$$
$$= 283.2\ \text{m/s}.$$

### EXAMPLE 1–2

**Converting volume units** The world's largest cut diamond is the Star of Africa No. 1 (mounted in the British Royal Scepter and kept in the Tower of London). Its volume is 1.84 cubic inches. What is its volume in cubic centimeters? In cubic meters?

**SOLUTION** To convert $\text{in.}^3$ to $\text{cm}^3$ we have to multiply by $[(2.54\ \text{cm})/(1\ \text{in.})]^3$, not just (2.54 cm)/(1 in.). We find

$$1.84\ \text{in.}^3 = (1.84\ \cancel{\text{in.}}^3)\left(\frac{2.54\ \text{cm}}{1\ \cancel{\text{in.}}}\right)^3 = 30.2\ \text{cm}^3.$$

Also, $1\ \text{cm} = 10^{-2}\ \text{m}$, and

$$30.2\ \text{cm}^3 = (30.2\ \cancel{\text{cm}}^3)\left(\frac{10^{-2}\ \text{m}}{1\ \cancel{\text{cm}}}\right)^3 = 3.02 \times 10^{-5}\ \text{m}^3.$$

## 1–5 PRECISION AND SIGNIFICANT FIGURES

Measurements always have uncertainties. When we measure a distance with an ordinary ruler, it is usually reliable only to the nearest millimeter, but with a precision micrometer caliper we can measure distances with an uncertainty of less than 0.01 mm. We often indicate the precision of a number by writing the number, the symbol ±, and a second number indicating the likely uncertainty. If the diameter of a steel rod is given as 56.47 ± 0.02 mm, this means that the true value is unlikely to be less than 56.45 mm or greater than 56.49 mm.

We can also express precision in terms of the maximum likely fractional or percent uncertainty. A resistor labeled "47 ohms ± 10%" probably has a true resistance differing from 47 ohms by no more than 10% of 47 ohms, that is, about 5 ohms. The resistance is probably between 42 and 52 ohms. For the diameter of the steel rod given above, the fractional uncertainty is (0.02 mm)/(56.47 mm), or about 0.0004; the percent uncertainty is (0.0004) (100%), or about 0.04%. Even small percent uncertainties can sometimes be very significant (see Fig. 1–5 on the next page).

**1–5** A very small percent error—say, traveling a few meters too far in a journey of hundreds of thousands of meters—can have spectacular consequences.

In a commonly used shorthand notation, the number 1.6454(21) means 1.6454 ± 0.0021. The numbers in parentheses show the uncertainty in the final digits of the main number.

When you use numbers having uncertainties to compute other numbers, the computed numbers are also uncertain. It is especially important to understand this when you compare a number obtained from measurements with a value obtained from a theoretical prediction. Suppose you want to verify the value of $\pi$, the ratio of the circumference to the diameter of a circle. The true value of this ratio, to ten digits, is 3.141592654. To make your own calculation, you draw a large circle and measure its diameter and circumference to the nearest millimeter, obtaining the values 135 mm and 424 mm. You punch these into your calculator and obtain the quotient 3.140740741. Does this agree with the true value or not?

First, the last seven digits in this answer are meaningless because they imply greater precision than is possible with your measurements. The number of meaningful digits in a number is called the number of **significant figures.** When numbers are multiplied or divided, the number of significant figures in the result is no greater than in the factor with the least number of significant figures. For example, $3.1416 \times 2.34 \times 0.58 = 4.3$. Therefore your measured value of $\pi$ has only three significant figures and should be stated simply as 3.14. Within the limit of three significant figures, your value does agree with the true value.

When we add and subtract numbers, it's the location of the decimal point that matters. For example, $123.62 + 8.9 = 132.5$. Although 123.62 is precise to the nearest hundredth (0.01), 8.9 is precise only to the nearest tenth (0.1). So their sum is precise only to the nearest tenth and should be written as 132.5, not 132.52.

In the examples and problems in this book we usually give numerical values with three significant figures, so your answers should usually have no more than three significant figures. (Many numbers that you encounter in the real world have only one- or two-figure precision. An automobile speedometer, for example, usually gives only two significant figures.) You may do the arithmetic with a calculator having a display with five to ten digits. But to give a ten-digit answer is not only unnecessary; it is genuinely wrong because it misrepresents the precision of the results. Always round your answer to keep only the correct number of significant figures or, in doubtful cases, one more at most. In Example 1–1 it would have been wrong to state the answer as 283.19444 m/s or to truncate it (without rounding) to 283.1 m/s.

Significant figures give us only a crude representation of the reliability of a number. The fractional uncertainty of 104 isn't much less than that of 96, despite the difference in the number of significant figures. When a better representation of uncertainty is needed, more sophisticated statistical methods are used.

When we calculate with very large or very small numbers, we can show significant figures much more easily by using powers-of-10 notation, sometimes called **scientific notation.** The distance from the earth to the sun is about 149,000,000,000 m, but writing the number in this form gives no indication of the number of significant figures. Certainly not all twelve are significant! Instead, we move the decimal point eleven places to the left (corresponding to dividing by $10^{11}$) and multiply by $10^{11}$. That is,

$$149{,}000{,}000{,}000 \text{ m} = 1.49 \times 10^{11} \text{ m}.$$

In this form it is clear that we have three significant figures. In scientific notation the usual practice is to express the quantity as a number between 1 and 10 multiplied by the appropriate power of 10. We can use the same technique when we multiply or divide very large or very small numbers.

## EXAMPLE 1–3

The rest energy $E$ of an electron with rest mass $m$ is given by Einstein's equation

$$E = mc^2, \tag{1-2}$$

where $c$ is the speed of light. Find $E$. The appropriate numbers are $m = 9.11 \times 10^{-31}$ kg and $c = 3.00 \times 10^8$ m/s.

**SOLUTION** Substituting the values of $m$ and $c$ into Eq. (1–2), we find

$$\begin{aligned} E &= (9.11 \times 10^{-31}\ \text{kg})(3.00 \times 10^8\ \text{m/s})^2 \\ &= (9.11)(3.00)^2(10^{-31})(10^8)^2\ \text{kg}\cdot\text{m}^2/\text{s}^2 \\ &= (82.0)(10^{[-31+(2\times 8)]})\ \text{kg}\cdot\text{m}^2/\text{s}^2 \\ &= 8.20 \times 10^{-14}\ \text{kg}\cdot\text{m}^2/\text{s}^2. \end{aligned}$$

Most calculators use scientific notation and add exponents automatically, but you should be able to do such calculations by hand when necessary. Incidentally, the value used for $c$ in Example 1–3 has three significant figures even though two of them are zeros. To greater accuracy, $c = 2.997925 \times 10^8$ m/s. Do you see why it would *not* be correct to write $c = 3.000 \times 10^8$ m/s?

When an integer or a fraction occurs in a general equation, we treat that number as having infinitely great precision. For example, in the equation $v^2 = v_0^2 + 2a(x - x_0)$, which is Eq. (2–14) in Chapter 2, the coefficient 2 is precisely 2, with no uncertainty at all, and we can consider it to have an infinite number of significant figures.

# 1–6 ESTIMATES AND ORDER OF MAGNITUDE

We have stressed the importance of knowing the precision of numbers that represent physical quantities. But even a very crude estimate of a quantity often gives us useful information. Sometimes we know how to calculate a certain quantity but have to guess at the data we need for the calculation. Or the calculation might be too complicated to carry out exactly, so we make some crude approximations. In either case our result is also a guess, but such a guess can be useful even if it is uncertain by a factor of two, ten, or more. Such calculations are often called **order-of-magnitude estimates.** The great nuclear physicist Enrico Fermi (1901–1954) called them "back-of-the-envelope calculations."

## EXAMPLE 1–4

You are writing an international espionage novel in which the hero escapes across the border with a billion dollars worth of gold in his suitcase. Is this possible? Would that amount of gold fit in a suitcase? Would it be too heavy to carry?

**SOLUTION** Gold sells for around \$400 an ounce. On a particular day the price might be \$200 or \$600, but never mind. An ounce is about 30 grams. Actually, an ordinary (avoirdupois) ounce is 28.35 g; an ounce of gold is a troy ounce, which is 9.45% more. Again, never mind. Ten dollars worth of gold has a mass somewhere around 1 gram, so a billion ($10^9$) dollars worth of gold is a hundred million ($10^8$) grams, or a hundred thousand ($10^5$) kilograms. This corresponds to a weight in British units of around 200,000 lb, or a hundred tons. Whether the precise number is 50 tons or 200 doesn't matter. Either way, the hero is not about to carry it across the border in a suitcase.

We can also estimate the *volume* of this gold. If its density were the same as that of water (1 g/cm$^3$), the volume would be $10^8$ cm$^3$, or 100 m$^3$. But gold is a heavy metal; we might guess its density to be ten times that of water. Gold is actually 19.3 times as dense as water. But guessing ten, we find a volume of 10 m$^3$. Visualize ten cubical stacks of gold bricks, each 1 meter on a side, and ask whether they would fit in a suitcase!

Exercises 1–12 through 1–20 at the end of this chapter are of the estimating, or "order-of-magnitude," variety. Some are silly, and most require guesswork for the needed input data. Don't try to look up a lot of data; make the best guesses you can. Even when they are off by a factor of ten, the results can be useful and interesting.

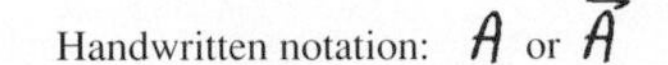


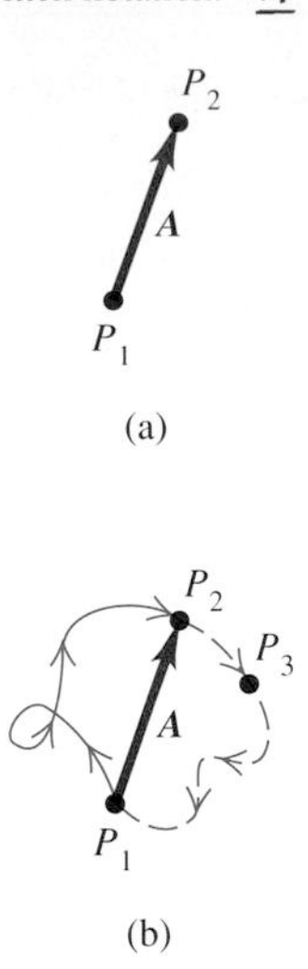


**1–6** (a) Vector $\boldsymbol{A}$ is the displacement from point $P_1$ to point $P_2$. (b) A displacement is always a straight-line segment, directed from the starting point to the end point, even if the actual path is curved. When a path ends at the same place it started, the displacement is zero.

## 1–7 VECTORS AND VECTOR ADDITION

Some physical quantities, such as time, temperature, mass, density, and electric charge, can be described completely by a single number with a unit. Many other quantities, however, have a *directional* quality and cannot be described by a single number. A familiar example is velocity. To describe the motion of an airplane, we must say not only how fast it is moving, but also in what direction. To fly from Chicago to New York, a plane has to head east, not south. Force is another example. When we push or pull on a body, we exert a force on it. To describe a force, we need to describe the direction in which it acts as well as its magnitude, or "how hard" the force pushes or pulls.

When a physical quantity is described by a single number, we call it a **scalar quantity.** In contrast, a **vector quantity** has both a *magnitude* (the "how much" or "how big" part) and a *direction* in space. Calculations with scalar quantities use the operations of ordinary arithmetic. For example, 6 kg + 3 kg = 9 kg, or 7 × 2 s = 14 s. However, combining vectors requires a different set of operations. Vector quantities play an essential role in all areas of physics, so let's talk next about what vectors are and how they combine.

We start with the simplest vector quantity, **displacement.** Displacement is simply a change in position of a point. (The point may be a model to represent a particle or a small body.) In Fig. 1–6a we represent the change of position from point $P_1$ to point $P_2$ by a line from $P_1$ to $P_2$ with an arrowhead at $P_2$ to represent the direction of motion. Displacement is a vector quantity because we must state not only how far the particle moves, but also in what direction. Walking 3 km north from your front door doesn't get you to the same place as walking 3 km southeast.

We usually represent a displacement by a single letter, such as $\boldsymbol{A}$ in Fig. 1–6a. In this book we always print vector symbols in boldface italic type to remind you that vector quantities have different properties from scalar quantities. In handwriting, vector symbols are usually underlined or written with an arrow above (Fig. 1–6), to indicate that they represent vector quantities. When you write a symbol for a vector quantity, *always* write it in one of these ways. If you don't distinguish between scalar and vector quantities in your notation, you probably won't make the distinction in your thinking either, and hopeless confusion will result.

Displacement is always a straight line segment, directed from the starting point to the end point, even though the actual path of the particle may be curved. In Fig. 1–6b, when the particle moves along the curved path shown from $P_1$ to $P_2$, the displacement is still the vector $\boldsymbol{A}$. If the particle were to continue on to $P_3$ and then return to $P_1$, the displacement for the entire trip would be zero. Displacement is not related directly to the total *distance* traveled.

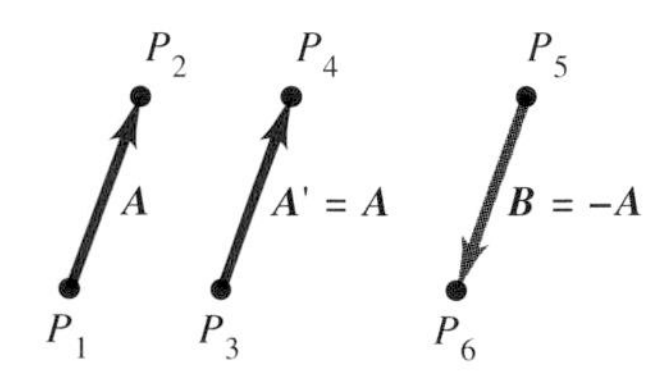


**1–7** The displacement from $P_3$ to $P_4$ is equal to that from $P_1$ to $P_2$. The displacement from $P_5$ to $P_6$ has the same magnitude as $\boldsymbol{A}$ and $\boldsymbol{A}'$, but opposite direction; displacement $\boldsymbol{B}$ is the negative of displacement $\boldsymbol{A}$.

The vector from point $P_3$ to point $P_4$ in Fig. 1–7 has the same length and direction as the one from $P_1$ to $P_2$. These two displacements are equal, even though they start at different points. By definition, two vector quantities are equal if they have the same magnitude (length) and direction, no matter where they are located in space. The vector $\boldsymbol{B}$, however, is not equal to $\boldsymbol{A}$ because its direction is *opposite* to that of $\boldsymbol{A}$. We define the *negative of a vector* as a vector having the same magnitude as the original vector but opposite direction. The negative of vector quantity $\boldsymbol{A}$ is denoted as $-\boldsymbol{A}$, and we use a boldface minus to emphasize the vector nature of the quantities. If $\boldsymbol{A}$ is 87 m south, then $-\boldsymbol{A}$ is 87 m north.

Thus the relation between $\boldsymbol{A}$ and $\boldsymbol{B}$ of Fig. 1–7 may be written as $\boldsymbol{A} = -\boldsymbol{B}$ or $\boldsymbol{B} = -\boldsymbol{A}$. When two vectors $\boldsymbol{A}$ and $\boldsymbol{B}$ have opposite directions, we say that they are *antiparallel*. Note that we also use a boldface equals sign to emphasize that equality of two vector quantities is not the same relationship as equality of scalar quantities. Two vector quantities are equal only when they have the same magnitude *and* the same direction.

We represent the **magnitude** of a vector quantity (its length in the case of a displacement vector) by the same letter used for the vector but in light italic type rather than boldface italic. An alternative notation is the vector symbol with vertical bars on both sides:

$$(\text{Magnitude of } \mathbf{A}) = A = |\mathbf{A}|. \qquad (1\text{–}3)$$

By definition the magnitude of a vector quantity is a scalar quantity (a number) and is *always positive*. We also note that a vector can never be equal to a scalar because they are different kinds of quantities. The expression "$\mathbf{A}$ = 6 m" is just as wrong as "2 oranges = 3 apples" or "6 lb = 7 km"!

Now suppose a particle undergoes a displacement $\mathbf{A}$, followed by a second displacement $\mathbf{B}$ (Fig. 1–8a). The final result is the same as though the particle had started at the same initial point and undergone a single displacement $\mathbf{C}$, as shown. We call displacement $\mathbf{C}$ the **vector sum,** or **resultant,** of displacements $\mathbf{A}$ and $\mathbf{B}$; we express this relationship symbolically as

$$\mathbf{C} = \mathbf{A} + \mathbf{B}. \qquad (1\text{–}4)$$

The boldface plus sign emphasizes that adding two vector quantities requires a geometrical process and is not the same operation as adding two scalar quantities such as 2 + 3 = 5. In vector addition we place the tail of the second vector at the tip, or point, of the first vector.

If we make the displacements $\mathbf{A}$ and $\mathbf{B}$ in reverse order, as in Fig. 1–8b, with $\mathbf{B}$ first and $\mathbf{A}$ second, the result is the same, as the figure shows. Thus

$$\mathbf{C} = \mathbf{B} + \mathbf{A} \quad \text{and} \quad \mathbf{A} + \mathbf{B} = \mathbf{B} + \mathbf{A}. \qquad (1\text{–}5)$$

This shows that the order of terms in a vector sum doesn't matter; vector addition obeys the *commutative law.*

Figure 1–8c also suggests an alternative graphical representation of the vector sum. When vectors $\mathbf{A}$ and $\mathbf{B}$ are both drawn with their tails at the same point, vector $\mathbf{C}$ is the diagonal of a parallelogram constructed with $\mathbf{A}$ and $\mathbf{B}$ as two adjacent sides.

Figure 1–9 shows special cases in which two vectors $\mathbf{A}$ and $\mathbf{B}$ are parallel (Fig. 1–9a) or antiparallel (Fig. 1–9b). When the vectors are parallel, the magnitude of the vector sum equals the *sum* of the magnitudes of $\mathbf{A}$ and $\mathbf{B}$. When they are antiparallel, it equals the *difference* of their magnitudes.

When we need to add more than two vectors, we may first find the vector sum of any two, add this vectorially to the third, and so on. Figure 1–10a shows three vectors $\mathbf{A}$, $\mathbf{B}$, and $\mathbf{C}$. In Fig. 1–10b vectors $\mathbf{A}$ and $\mathbf{B}$ are first added, giving a vector sum $\mathbf{D}$; vectors $\mathbf{C}$ and $\mathbf{D}$ are then added by the same process to obtain the vector sum $\mathbf{R}$:

$$\mathbf{R} = (\mathbf{A} + \mathbf{B}) + \mathbf{C} = \mathbf{D} + \mathbf{C}.$$

Alternatively, we can first add $\mathbf{B}$ and $\mathbf{C}$ (Fig. 1–10c) to obtain vector $\mathbf{E}$ and then add $\mathbf{A}$ and $\mathbf{E}$ to obtain $\mathbf{R}$:

$$\mathbf{R} = \mathbf{A} + (\mathbf{B} + \mathbf{C}) = \mathbf{A} + \mathbf{E}.$$

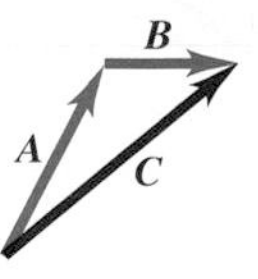


(a)

(b)

(c)

**1–8** Vector $\mathbf{C}$ is the vector sum of vectors $\mathbf{A}$ and $\mathbf{B}$. The order in vector addition doesn't matter; vector addition is commutative.

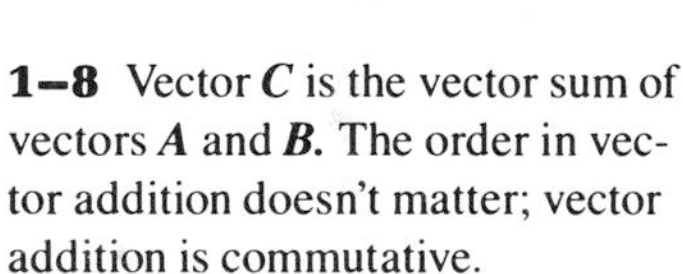


(a)

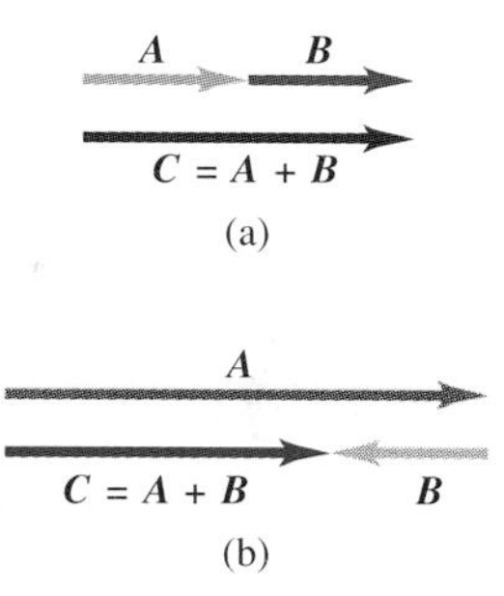


(b)

**1–9** The vector sum of (a) two parallel vectors and (b) two antiparallel vectors. Note that the vectors $\mathbf{A}$, $\mathbf{B}$, and $\mathbf{C}$ in (a) are not the same as the vectors $\mathbf{A}$, $\mathbf{B}$, and $\mathbf{C}$ in (b).

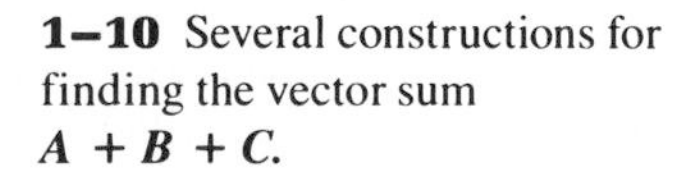

**1–10** Several constructions for finding the vector sum $\mathbf{A} + \mathbf{B} + \mathbf{C}$.

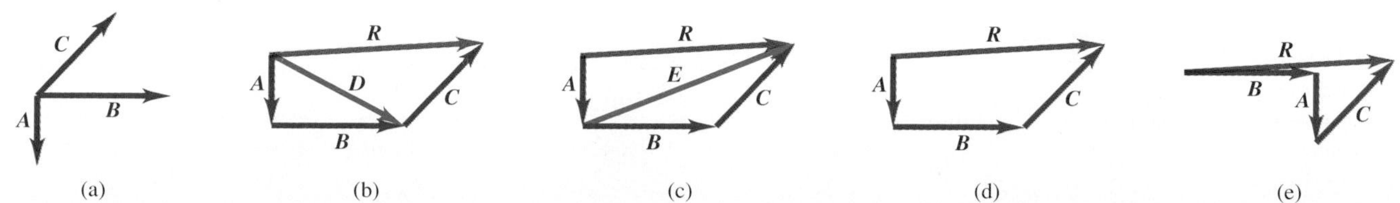

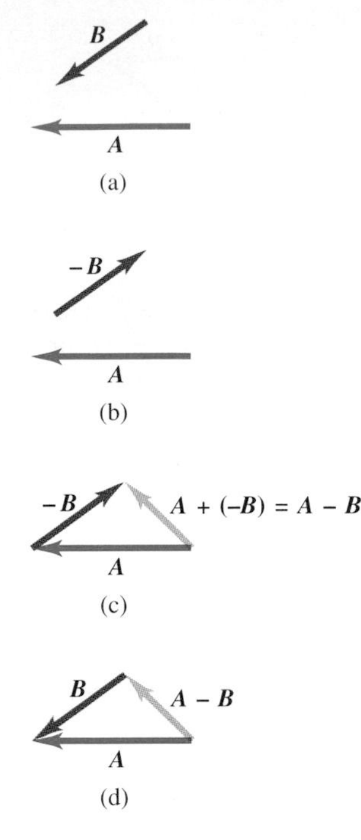


**1–11** (a) Vector $\boldsymbol{A}$ and vector $\boldsymbol{B}$. (b) Vector $\boldsymbol{A}$ and vector $-\boldsymbol{B}$. (c) The vector difference $\boldsymbol{A} - \boldsymbol{B}$ is the sum of the vectors $\boldsymbol{A}$ and $-\boldsymbol{B}$. The tail of $-\boldsymbol{B}$ is placed at the head of $\boldsymbol{A}$. (d) To check: $(\boldsymbol{A} - \boldsymbol{B}) + \boldsymbol{B} = \boldsymbol{A}$.

We don't even need to draw vectors $\boldsymbol{D}$ and $\boldsymbol{E}$; all we need to do is draw the given vectors in succession, with the tail of each at the head of the one preceding it, and complete the polygon by a vector $\boldsymbol{R}$ from the tail of the first vector to the head of the last vector (Fig. 1–10d). The order makes no difference; Fig. 1–10e shows a different order, and we invite you to try others. We see that vector addition obeys the *associative law.*

Diagrams for addition of displacement vectors don't have to be drawn actual size. We'll often use a scale similar to those used for maps, in which the distance on the diagram is proportional to the actual distance, such as 1 cm for 5 km. When we work with other vector quantities with different units, such as force or velocity, we *must* use a scale. In a diagram for force vectors we might use a scale in which a vector that is 1 cm long represents a force of magnitude 5 N. A 20-N force would then be represented by a vector 4 cm long with the appropriate direction.

A vector quantity such as a displacement can be multiplied by a scalar quantity (an ordinary number). The displacement $2\boldsymbol{A}$ is a displacement (vector quantity) in the same direction as the vector $\boldsymbol{A}$, but twice as long. The scalar quantity used to multiply a vector may also be a physical quantity having units. For example, you may be familiar with the relationship $\boldsymbol{F} = m\boldsymbol{a}$; force $\boldsymbol{F}$ (a vector quantity) is equal to the product of mass $m$ (a scalar quantity) and acceleration $\boldsymbol{a}$ (a vector quantity). The magnitude of the force is equal to the mass multiplied by the magnitude of the acceleration, and the unit of the magnitude of force is the unit of mass multiplied by the unit of the magnitude of acceleration. The direction of $\boldsymbol{F}$ is the same as that of $\boldsymbol{a}$.

We have already mentioned the special case of multiplication by $-1$: $(-1)\boldsymbol{A} = -\boldsymbol{A}$ is a vector having the same magnitude as $\boldsymbol{A}$ but opposite direction. This provides the basis for defining vector subtraction. We define the difference $\boldsymbol{A} - \boldsymbol{B}$ of two vectors $\boldsymbol{A}$ and $\boldsymbol{B}$ to be the vector sum of $\boldsymbol{A}$ and $-\boldsymbol{B}$:

$$\boldsymbol{A} - \boldsymbol{B} = \boldsymbol{A} + (-\boldsymbol{B}). \tag{1–6}$$

Figure 1–11 shows an example of vector subtraction.

## EXAMPLE 1–5

A cross-country skier skis 1.00 km north and then 2.00 km east. a) How far and in what direction is she from the starting point? b) What are the magnitude and direction of her resultant displacement?

**SOLUTION** a) Figure 1–12 is a scale diagram of the skier's displacements. By careful measurement we find that the distance from the starting point is about 2.2 km and the angle $\phi$ is about 63°. But it is much more accurate to *calculate* the result. The vectors in the diagram form a right triangle, and we can find the length of the hypotenuse by using the theorem of Pythagoras:

$$\sqrt{(1.00\text{ km})^2 + (2.00\text{ km})^2} = 2.24\text{ km}.$$

The angle $\phi$ can be found with a little simple trigonometry. By definition of the tangent function,

$$\tan\phi = \frac{\text{opposite side}}{\text{adjacent side}} = \frac{2.00\text{ km}}{1.00\text{ km}},$$

$$\phi = 63.4°.$$

b) The magnitude of the resultant displacement is just the distance that we found in part (a), 2.24 km. We can describe the direction as 63.4° east of north, or 26.6° north of east. Take your choice!

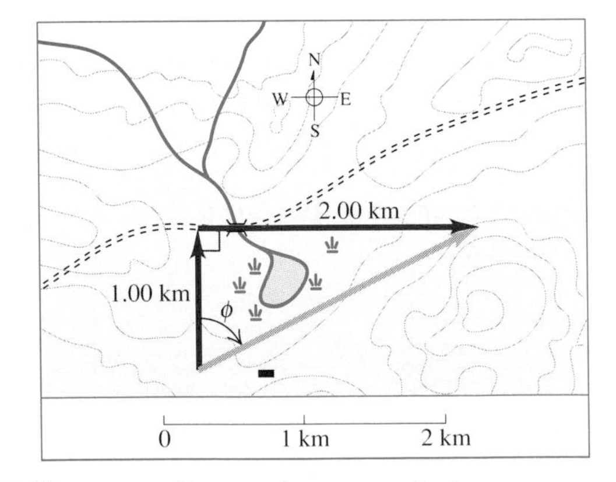


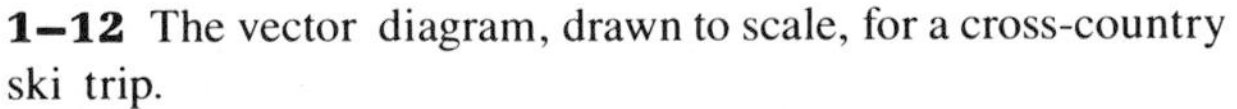


**1–12** The vector diagram, drawn to scale, for a cross-country ski trip.

## 1–8 COMPONENTS OF VECTORS

In Section 1–7 we added vectors by using a scale diagram or by using properties of right triangles. Measuring a diagram offers only very limited precision, and the calculation with right triangles works only when the two vectors are perpendicular. With an angle other than 90° or with more than two vectors, we would get into repeated trigonometric solutions of oblique triangles, which can get horribly complicated. So we need a simple but more general method for adding vectors.

Addition and subtraction of vectors are usually carried out by the use of **components.** To define what we mean by components, we begin with a rectangular (cartesian) coordinate system of axes (Fig. 1–13). We can represent any vector lying in the $xy$-plane as the sum of a vector parallel to the $x$-axis and a vector parallel to the $y$-axis. These two vectors are labeled $\boldsymbol{A}_x$ and $\boldsymbol{A}_y$ in the figure; they are called the component vectors of vector $\boldsymbol{A}$, and their vector sum is equal to $\boldsymbol{A}$. In symbols,

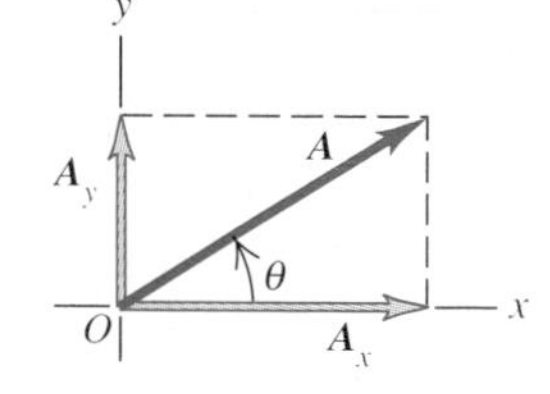


**1–13** Vectors $\boldsymbol{A}_x$ and $\boldsymbol{A}_y$ are the rectangular component vectors of $\boldsymbol{A}$ in the directions of the $x$- and $y$-axes.

$$\boldsymbol{A} = \boldsymbol{A}_x + \boldsymbol{A}_y. \qquad (1\text{–}7)$$

By definition, each component vector lies along a coordinate-axis direction. Thus we need only a single number to describe each one.

When the component vector $\boldsymbol{A}_x$ points in the $+x$-direction, we define the number $A_x$ to be the magnitude of $\boldsymbol{A}_x$. When $\boldsymbol{A}_x$ points in the $-x$-direction, we define the number $A_x$ to be the negative of that magnitude, keeping in mind that the magnitude of a vector quantity is always positive. We define the number $A_y$ the same way. The two numbers $A_x$ and $A_y$ are called the *components* of $\boldsymbol{A}$.

If we know the magnitude $A$ of the vector $\boldsymbol{A}$ and its direction, given by angle $\theta$ in Fig. 1–13, we can calculate the components. From the definitions of the trigonometric functions,

$$\frac{A_x}{A} = \cos\theta \quad \text{and} \quad \frac{A_y}{A} = \sin\theta; \qquad (1\text{–}8)$$
$$A_x = A\cos\theta \quad \text{and} \quad A_y = A\sin\theta.$$

These equations are correct when the angle $\theta$ is measured counterclockwise from the positive $x$-axis, a common convention. If $\theta$ is measured from a different reference direction, the relationships are different. Be careful!

In Fig. 1–14 the component $B_x$ is negative because its direction is opposite to that of the positive $x$-axis. This is consistent with Eqs. (1–8); the cosine of an angle in the second quadrant is negative. The component $B_y$ is positive, but both $C_x$ and $C_y$ are negative.

We can describe a vector completely by giving either its magnitude and direction or its $x$- and $y$-components. Equations (1–8) show how to find the components if we know the magnitude and direction. Or if we are given the components, we can find the magnitude and direction. Applying the Pythagorean theorem to Fig. 1–13, we find

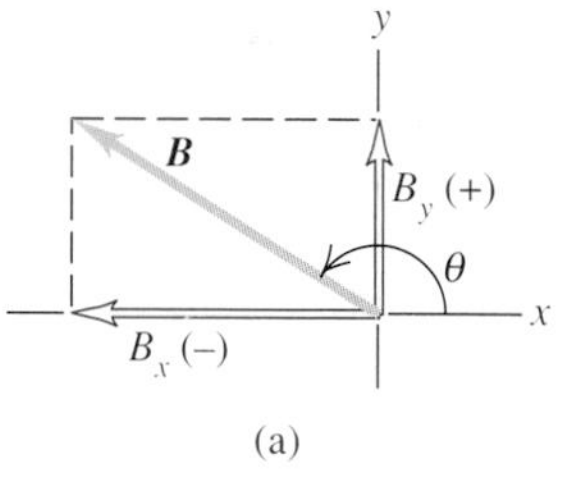


$$A = \sqrt{A_x^{\,2} + A_y^{\,2}}. \qquad (1\text{–}9)$$

Also, from the definition of the tangent of an angle,

$$\tan\theta = \frac{A_y}{A_x} \quad \text{and} \quad \theta = \arctan\frac{A_y}{A_x}. \qquad (1\text{–}10)$$

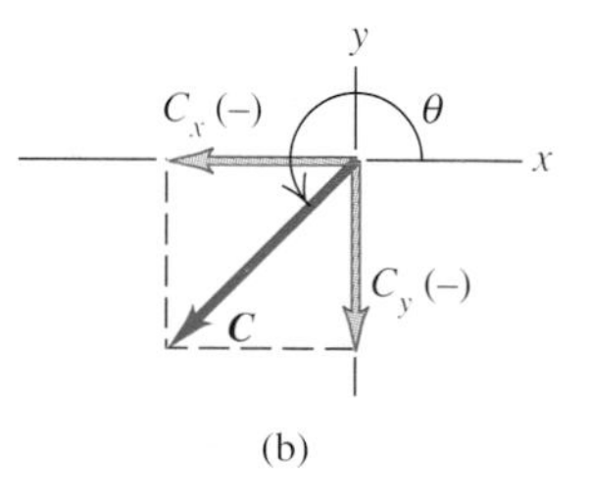


**1–14** The components of a vector may be positive or negative numbers.

We will always use the notation arctan for the inverse tangent function. The notation $\tan^{-1}$ is also commonly used, and your calculator may have an INV button to be used with the TAN button. Some computer languages use atan (FORTRAN) or atn (BASIC); Pascal uses ArcTan.

There is one slight complication in using Eq. (1–10) to find $\theta$. Suppose $A_x = 2$ m and $A_y = -2$ m; then $\tan\theta = -1$. But there are two angles having tangents of $-1$,

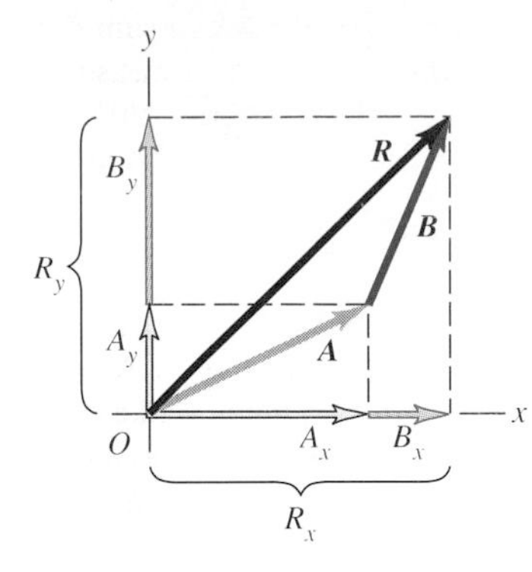


**1–15** Vector $\mathbf{R}$ is the vector sum (resultant) of $\mathbf{A}$ and $\mathbf{B}$. Its $x$-component, $R_x$, equals the sum of the $x$-components of $\mathbf{A}$ and $\mathbf{B}$. The $y$-components are similarly related.

namely, 135° and 315° (or −45°). To decide which is correct, we have to look at the individual components. Because $A_x$ is positive and $A_y$ is negative, the angle must be in the fourth quadrant; thus $\theta = 315°$ (or $-45°$) is the correct value. Most pocket calculators give $\arctan(-1) = -45°$. In this case that is correct; but if instead we have $A_x = -2$ m and $A_y = 2$ m, then the correct angle is 135°. You should always draw a sketch to check which of the two possibilities is the correct one. Similarly, when $A_x$ and $A_y$ are both negative, the tangent is positive, but the angle is in the third quadrant.

Here's how we use components to calculate the vector sum (resultant) of several vectors. Figure 1–15 shows two vectors $\mathbf{A}$ and $\mathbf{B}$ and their vector sum $\mathbf{R}$, along with $x$- and $y$-components of all three vectors. You can see from the diagram that the $x$-component $R_x$ of the vector sum is simply the sum $(A_x + B_x)$ of the $x$-components of the vectors being added. The same is true for the $y$-components. In symbols,

$$R_x = A_x + B_x, \qquad R_y = A_y + B_y. \tag{1–11}$$

Once we know the components of $\mathbf{A}$ and $\mathbf{B}$, perhaps by using Eqs. (1–8), we can compute the components of the vector sum $\mathbf{R}$. Then if the magnitude and direction of $\mathbf{R}$ are needed, we can obtain them from Eqs. (1–9) and (1–10), with the $A$'s replaced by $R$'s.

This procedure for finding the sum of two vectors can easily be extended to any number of vectors. Let $\mathbf{R}$ be the vector sum of $\mathbf{A}, \mathbf{B}, \mathbf{C}, \mathbf{D}, \mathbf{E}, \ldots$. Then

$$\begin{aligned} R_x &= A_x + B_x + C_x + D_x + E_x + \cdots, \\ R_y &= A_y + B_y + C_y + D_y + E_y + \cdots. \end{aligned} \tag{1–12}$$

## PROBLEM-SOLVING STRATEGY

### Vector addition

**1.** First find the $x$- and $y$-components of each individual vector. If a vector is described by its magnitude $A$ and its angle $\theta$, measured counterclockwise from the $+x$-axis, then the components are given by

$$A_x = A\cos\theta, \qquad A_y = A\sin\theta.$$

In general, some components are positive and some are negative. You can use this sign table as a check:

| Quadrant | I | II | III | IV |
|---|---|---|---|---|
| $A_x$ | + | − | − | + |
| $A_y$ | + | + | − | − |

If the vectors are given in some other way, perhaps by using a different angle, you will have to use the definitions of the sine and cosine functions; be particularly careful with signs. Make a table showing the components of each vector.

**2.** Add the individual $x$-components algebraically, including signs, to find $R_x$, the $x$-component of the vector sum. Do the same for the $y$-components.

**3.** Then the magnitude $R$ and direction $\theta$ of the vector sum are given by

$$R = \sqrt{R_x^2 + R_y^2},$$

$$\theta = \arctan\frac{R_y}{R_x}.$$

Remember that the magnitude $R$ is *always* positive.

## EXAMPLE 1–6

The three finalists in a contest are brought to the center of a large, flat field. Each is given a meter stick, a compass, a calculator, a shovel, and (in a different order for each) the following three displacements:

72.4 m, 32° east of north,

57.3 m, 36° south of west,

17.8 m straight south.

The three displacements lead to the point where the keys to a new Porsche are buried. Two contestants start measuring immediately, but the winner first *calculates* where to go. What does he calculate?

**SOLUTION** The situation is shown in Fig. 1–16. We have chosen the $x$-axis as east and the $y$-axis as north, the usual choice for maps. Let $\boldsymbol{A}$ be the first displacement, $\boldsymbol{B}$ the second, and $\boldsymbol{C}$ the third. We can estimate from the diagram that the vector sum $\boldsymbol{R}$ is about 10 m, 40° west of north. The angles of the vectors, measured counterclockwise from the $+x$-axis, are 58°, 216°, and 270°. We have to find the components of each. The components of $\boldsymbol{A}$ are

$$A_x = A\cos\theta = (72.4\text{ m})(\cos 58°) = 38.37\text{ m},$$
$$A_y = A\sin\theta = (72.4\text{ m})(\sin 58°) = 61.40\text{ m}.$$

The table below shows the components of all the displacements, the addition of components, and the other calculations. Always arrange your component calculations systematically like this. Note that we have kept one too many significant figures in the components; we wait until the end to round to the correct number of significant figures.

| Distance | Angle | $x$-component | $y$-component |
|---|---|---|---|
| $A = 72.4$ m | 58° | 38.37 m | 61.40 m |
| $B = 57.3$ m | 216° | −46.36 m | −33.68 m |
| $C = 17.8$ m | 270° | 0.00 m | −17.80 m |
| | | $R_x = -7.99$ m | $R_y = 9.92$ m |

$$R = \sqrt{(-7.99\text{ m})^2 + (9.92\text{ m})^2} = 12.7\text{ m},$$
$$\theta = \arctan\frac{9.92\text{ m}}{-7.99\text{ m}} = 129° = 39° \text{ west of north.}$$

The losers try to measure three angles and three distances totaling 147.5 m, one meter at a time. The winner measures only one angle and one much shorter distance.

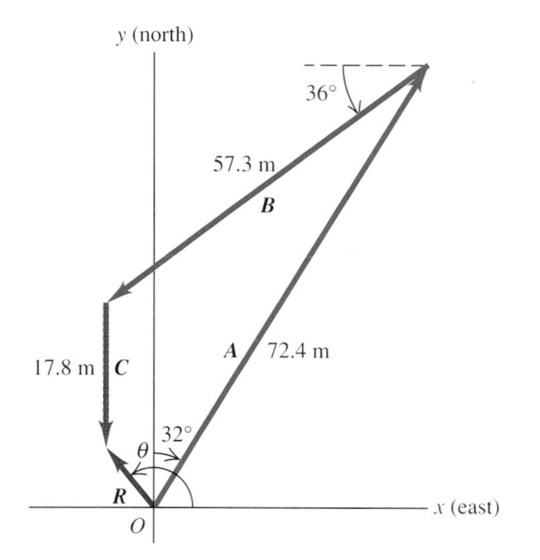


**1–16** Three successive displacements $\boldsymbol{A}$, $\boldsymbol{B}$, and $\boldsymbol{C}$ and the resultant (vector sum) displacement $\boldsymbol{R} = \boldsymbol{A} + \boldsymbol{B} + \boldsymbol{C}$.

We have talked only about vectors that lie in the $xy$-plane, but the component method works just as well for vectors having any direction in space. We introduce a $z$-axis perpendicular to the $xy$-plane; then in general a vector $\boldsymbol{A}$ has components $A_x$, $A_y$, and $A_z$ in the three coordinate directions. The magnitude $A$ is given by

$$A = \sqrt{A_x^2 + A_y^2 + A_z^2}, \qquad (1\text{–}13)$$

and Eqs. (1–12) for the components of the vector have an additional member:

$$R_z = A_z + B_z + C_z + D_z + E_z + \cdots.$$

## EXAMPLE 1–7

After an airplane takes off, it travels 10.4 km west, 8.7 km north, and 2.1 km up. How far is it from the takeoff point?

**SOLUTION** Let the $x$-axis be east, the $y$-axis north, and the $z$-axis up. Then $A_x = -10.4$ km, $A_y = 8.7$ km, $A_z = 2.1$ km, and

$$A = \sqrt{(-10.4\text{ km})^2 + (8.7\text{ km})^2 + (2.1\text{ km})^2} = 13.7\text{ km}.$$

Our discussion of vector addition has centered on combining *displacement* vectors, but the method is applicable to all other vector quantities as well. In Chapter 4, when we study the concept of force, we will make extensive use of the fact that forces can be combined according to the same methods of vector addition that we have used with displacement.

## 1–9 UNIT VECTORS

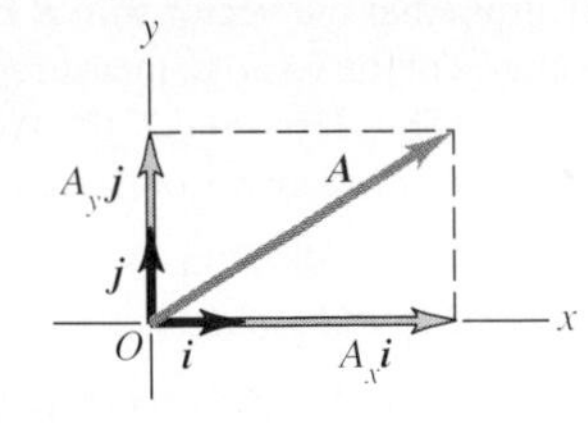


**1–17** Using unit vectors, we can express a vector $\boldsymbol{A}$ in terms of its components $A_x$ and $A_y$ as $\boldsymbol{A} = A_x\boldsymbol{i} + A_y\boldsymbol{j}$.

A **unit vector** is a vector having a magnitude of unity, with no units. Its only purpose is to describe a direction in space. Unit vectors provide a convenient notation for many expressions involving components of vectors.

In an $xy$-coordinate system we can define a unit vector $\boldsymbol{i}$ that points in the direction of the positive $x$-axis and a unit vector $\boldsymbol{j}$ that points in the direction of the positive $y$-axis. Then we can express the relationships between component vectors and components, described at the beginning of Section 1–8, as follows:

$$\boldsymbol{A}_x = A_x\boldsymbol{i}, \qquad \boldsymbol{A}_y = A_y\boldsymbol{j}. \tag{1–14}$$

Similarly, we can write vector $\boldsymbol{A}$ in terms of its components as

$$\boldsymbol{A} = A_x\boldsymbol{i} + A_y\boldsymbol{j}. \tag{1–15}$$

Equations (1–14) and (1–15) are vector equations; each term, such as $A_x\boldsymbol{i}$, is a vector quantity (Fig. 1–17). The boldface equals and plus signs denote vector equality and addition.

When two vectors $\boldsymbol{A}$ and $\boldsymbol{B}$ are represented in terms of their components, we can express the vector sum $\boldsymbol{R}$ using unit vectors as follows:

$$\begin{aligned} \boldsymbol{A} &= A_x\boldsymbol{i} + A_y\boldsymbol{j}, \\ \boldsymbol{B} &= B_x\boldsymbol{i} + B_y\boldsymbol{j}, \\ \boldsymbol{R} &= \boldsymbol{A} + \boldsymbol{B} \\ &= (A_x\boldsymbol{i} + A_y\boldsymbol{j}) + (B_x\boldsymbol{i} + B_y\boldsymbol{j}) \\ &= (A_x + B_x)\boldsymbol{i} + (A_y + B_y)\boldsymbol{j} \\ &= R_x\boldsymbol{i} + R_y\boldsymbol{j}. \end{aligned} \tag{1–16}$$

Equation (1–16) restates the content of Eqs. (1–11) in the form of a single vector equation rather than two component equations.

If the vectors do not all lie in the $xy$-plane, then we need a third component. We introduce a third unit vector $\boldsymbol{k}$ that points in the direction of the positive $z$-axis. The generalized forms of Eqs. (1–15) and (1–16) are

$$\boldsymbol{A} = A_x\boldsymbol{i} + A_y\boldsymbol{j} + A_z\boldsymbol{k}, \qquad \boldsymbol{B} = B_x\boldsymbol{i} + B_y\boldsymbol{j} + B_z\boldsymbol{k}, \tag{1–17}$$

$$\begin{aligned} \boldsymbol{R} &= (A_x + B_x)\boldsymbol{i} + (A_y + B_y)\boldsymbol{j} + (A_z + B_z)\boldsymbol{k} \\ &= R_x\boldsymbol{i} + R_y\boldsymbol{j} + R_z\boldsymbol{k}. \end{aligned} \tag{1–18}$$

### EXAMPLE 1–8

Given the two vectors

$$\boldsymbol{D} = 6\boldsymbol{i} + 3\boldsymbol{j} - \boldsymbol{k}$$

and

$$\boldsymbol{E} = 4\boldsymbol{i} - 5\boldsymbol{j} + 8\boldsymbol{k},$$

find the magnitude of the vector $2\boldsymbol{D} - \boldsymbol{E}$.

**SOLUTION** Letting $\boldsymbol{F} = 2\boldsymbol{D} - \boldsymbol{E}$, we have

$$\begin{aligned} \boldsymbol{F} &= 2(6\boldsymbol{i} + 3\boldsymbol{j} - \boldsymbol{k}) - (4\boldsymbol{i} - 5\boldsymbol{j} + 8\boldsymbol{k}) \\ &= (12 - 4)\boldsymbol{i} + (6 + 5)\boldsymbol{j} + (-2 - 8)\boldsymbol{k} \\ &= 8\boldsymbol{i} + 11\boldsymbol{j} - 10\boldsymbol{k}. \end{aligned}$$

Then, from Eq. (1–13),

$$F = \sqrt{F_x^2 + F_y^2 + F_z^2} = \sqrt{(8)^2 + (11)^2 + (-10)^2} = 17.$$

# 1–10 PRODUCTS OF VECTORS

We have seen how addition of vectors develops naturally from the problem of combining displacements, and we will use vector addition for many other vector quantities later. We will also find that we can express many physical relationships concisely by the use of *products* of vectors. Vectors are not ordinary numbers, so ordinary multiplication is not directly applicable to vectors. We will define two different kinds of products of vectors. The first, called the scalar product, yields a result that is a scalar quantity. The second, the vector product, yields another vector.

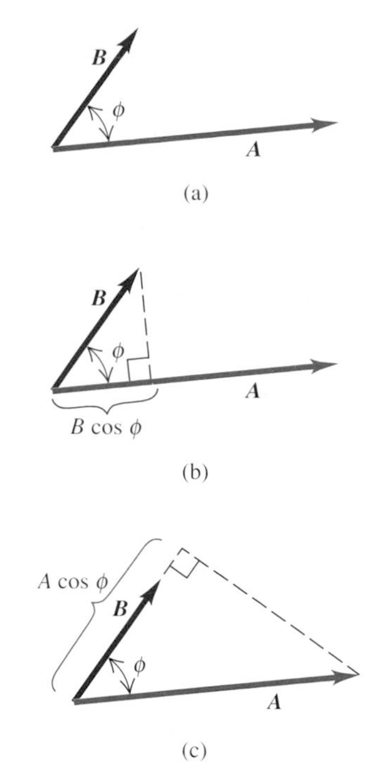


**1–18** (a) Two vectors drawn from a common starting point to define their scalar product. (b) $B \cos \phi$ is the component of $\boldsymbol{B}$ in the direction of $\boldsymbol{A}$, and $\boldsymbol{A} \cdot \boldsymbol{B}$ is the product of this component with the magnitude of $\boldsymbol{A}$. (c) $\boldsymbol{A} \cdot \boldsymbol{B}$ is also the product of the component of $\boldsymbol{A}$ in the direction of $\boldsymbol{B}$ with the magnitude of $\boldsymbol{B}$.

## Scalar Product

The **scalar product** of two vectors $\boldsymbol{A}$ and $\boldsymbol{B}$ is denoted by $\boldsymbol{A} \cdot \boldsymbol{B}$. Because of this notation, the scalar product is also called the *dot product*. We will use this product in Chapter 6 to describe work done by a force. When a constant force $\boldsymbol{F}$ is applied to a body that undergoes a displacement $\boldsymbol{s}$, the work $W$ (a scalar quantity) done by the force is given by

$$W = \boldsymbol{F} \cdot \boldsymbol{s}. \tag{1–19}$$

To define the scalar product $\boldsymbol{A} \cdot \boldsymbol{B}$ of two vectors $\boldsymbol{A}$ and $\boldsymbol{B}$, we draw the two vectors from a common point (Fig. 1–18a). The angle between their directions is $\phi$, as shown. (As usual, we use Greek letters for angles.) We define $\boldsymbol{A} \cdot \boldsymbol{B}$ as

$$\boldsymbol{A} \cdot \boldsymbol{B} = AB \cos \phi = |\boldsymbol{A}||\boldsymbol{B}| \cos \phi. \tag{1–20}$$

The scalar product is a scalar quantity, not a vector, and it may be positive, negative, or zero. When $\phi$ is between zero and 90°, the scalar product is positive; when $\phi$ is between 90° and 180°, it is negative; and when $\phi = 90°$, $\boldsymbol{A} \cdot \boldsymbol{B} = 0$. *The scalar product of two perpendicular vectors is always zero.*

For any two vectors $\boldsymbol{A}$ and $\boldsymbol{B}$, $AB \cos \phi = BA \cos \phi$. This means that $\boldsymbol{A} \cdot \boldsymbol{B} = \boldsymbol{B} \cdot \boldsymbol{A}$. The scalar product obeys the *commutative* law of multiplication; the order of the two vectors does not matter.

We can represent vector $\boldsymbol{B}$ in terms of a component parallel to $\boldsymbol{A}$ and a component perpendicular to $\boldsymbol{A}$ (Fig. 1–18b); the component parallel to $\boldsymbol{A}$ is $(B \cos \phi)$. Thus, from Eq. (1–20), $\boldsymbol{A} \cdot \boldsymbol{B}$ is equal to the component of $\boldsymbol{B}$ parallel to $\boldsymbol{A}$ multiplied by the magnitude of $\boldsymbol{A}$. Alternatively, $\boldsymbol{A} \cdot \boldsymbol{B}$ is also the component of $\boldsymbol{A}$ parallel to $\boldsymbol{B}$ multiplied by the magnitude of $\boldsymbol{B}$ (Fig. 1–18c).

If we know the components of $\boldsymbol{A}$ and $\boldsymbol{B}$, we can calculate the scalar product. The easiest procedure is to use unit vectors. First, let's work out the scalar products of the unit vectors themselves. This is easy; using Eq. (1–20), we find

$$\boldsymbol{i} \cdot \boldsymbol{i} = \boldsymbol{j} \cdot \boldsymbol{j} = \boldsymbol{k} \cdot \boldsymbol{k} = (1)(1) \cos 0 = 1,$$

and

$$\boldsymbol{i} \cdot \boldsymbol{j} = \boldsymbol{i} \cdot \boldsymbol{k} = \boldsymbol{j} \cdot \boldsymbol{k} = (1)(1) \cos 90° = 0. \tag{1–21}$$

Now we express $\boldsymbol{A}$ and $\boldsymbol{B}$ in terms of their components, expand the product, and use these products of unit vectors:

$$\begin{aligned} \boldsymbol{A} \cdot \boldsymbol{B} &= (A_x\boldsymbol{i} + A_y\boldsymbol{j} + A_z\boldsymbol{k}) \cdot (B_x\boldsymbol{i} + B_y\boldsymbol{j} + B_z\boldsymbol{k}) \\ &= A_x\boldsymbol{i} \cdot B_x\boldsymbol{i} + A_x\boldsymbol{i} \cdot B_y\boldsymbol{j} + A_x\boldsymbol{i} \cdot B_z\boldsymbol{k} \\ &\quad + A_y\boldsymbol{j} \cdot B_x\boldsymbol{i} + A_y\boldsymbol{j} \cdot B_y\boldsymbol{j} + A_y\boldsymbol{j} \cdot B_z\boldsymbol{k} \\ &\quad + A_z\boldsymbol{k} \cdot B_x\boldsymbol{i} + A_z\boldsymbol{k} \cdot B_y\boldsymbol{j} + A_z\boldsymbol{k} \cdot B_z\boldsymbol{k} \\ &= A_xB_x\boldsymbol{i} \cdot \boldsymbol{i} + A_xB_y\boldsymbol{i} \cdot \boldsymbol{j} + A_xB_z\boldsymbol{i} \cdot \boldsymbol{k} \\ &\quad + A_yB_x\boldsymbol{j} \cdot \boldsymbol{i} + A_yB_y\boldsymbol{j} \cdot \boldsymbol{j} + A_yB_z\boldsymbol{j} \cdot \boldsymbol{k} \\ &\quad + A_zB_x\boldsymbol{k} \cdot \boldsymbol{i} + A_zB_y\boldsymbol{k} \cdot \boldsymbol{j} + A_zB_z\boldsymbol{k} \cdot \boldsymbol{k}. \end{aligned} \tag{1–22}$$

From Eq. (1–21) we see that six of these nine terms are zero, and the three that survive give simply

$$\boldsymbol{A}\cdot\boldsymbol{B} = A_xB_x + A_yB_y + A_zB_z. \quad (1\text{–}23)$$

## EXAMPLE 1–9

Find the angle between the two vectors

$$\boldsymbol{A} = 2\boldsymbol{i} + 3\boldsymbol{j} + \boldsymbol{k} \quad \text{and} \quad \boldsymbol{B} = -4\boldsymbol{i} + 2\boldsymbol{j} - \boldsymbol{k}.$$

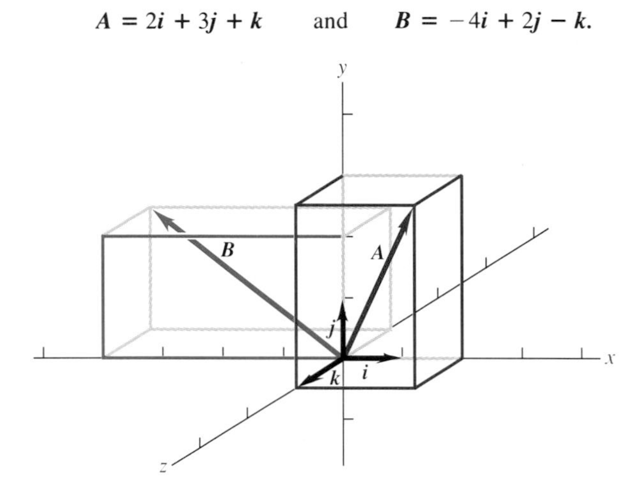


**1–19** Two vectors in three dimensions.

**SOLUTION** The vectors are shown in Fig. 1–19. We have

$$A_x = 2, \quad B_x = -4;$$
$$A_y = 3, \quad B_y = 2;$$
$$A_z = 1, \quad B_z = -1.$$

The scalar product is given by either Eq. (1–20) or Eq. (1–23). Equating these two and rearranging, we obtain

$$\cos\phi = \frac{A_xB_x + A_yB_y + A_zB_z}{AB}. \quad (1\text{–}24)$$

For our example,

$$A_xB_x + A_yB_y + A_zB_z = (2)(-4) + (3)(2) + (1)(-1) = -3,$$
$$A = \sqrt{2^2 + 3^2 + 1^2} = \sqrt{14},$$
$$B = \sqrt{(-4)^2 + 2^2 + (-1)^2} = \sqrt{21},$$
$$\cos\phi = \frac{-3}{\sqrt{14}\sqrt{21}} = -0.175,$$
$$\phi = 100°.$$

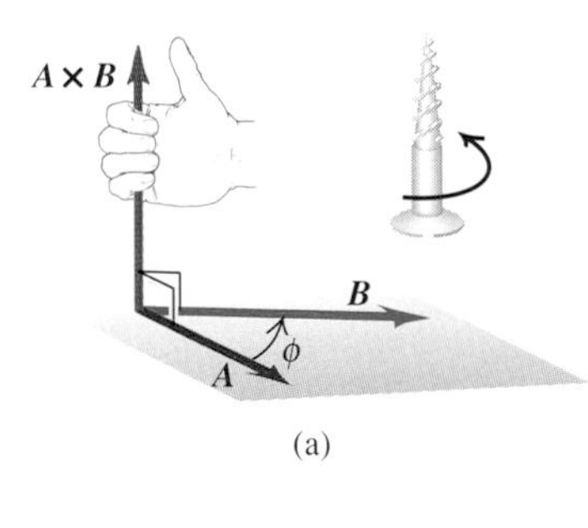


(a)

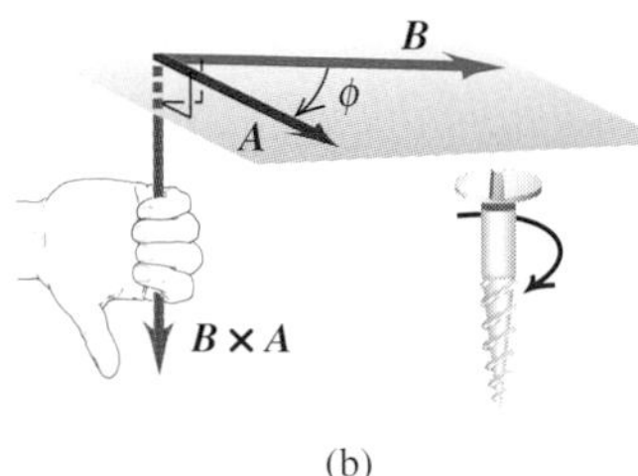


(b)

**1–20** (a) Vectors $\boldsymbol{A}$ and $\boldsymbol{B}$ lie in a plane; the vector product $\boldsymbol{A} \times \boldsymbol{B}$ is perpendicular to this plane in a direction determined by the right-hand rule. (b) $\boldsymbol{B} \times \boldsymbol{A} = -\boldsymbol{A} \times \boldsymbol{B}$; the vector product is anticommutative.

## Vector Product

The **vector product** of two vectors $\boldsymbol{A}$ and $\boldsymbol{B}$, also called the *cross product,* is denoted by $\boldsymbol{A} \times \boldsymbol{B}$. We will use this product in Chapter 10 to describe the quantities torque and angular momentum. Later we will also use it extensively for magnetic fields, where it will help us describe the relations among the directions of the various vector quantities involved.

To define the vector product $\boldsymbol{A} \times \boldsymbol{B}$ of two vectors $\boldsymbol{A}$ and $\boldsymbol{B}$, we again draw $\boldsymbol{A}$ and $\boldsymbol{B}$ from a common point (Fig. 1–20a). The two vectors then lie in a plane. We define the vector product to be a vector quantity with a direction perpendicular to this plane (that is, perpendicular to both $\boldsymbol{A}$ and $\boldsymbol{B}$) and a magnitude given by $AB \sin\phi$. That is, if $\boldsymbol{C} = \boldsymbol{A} \times \boldsymbol{B}$, then

$$C = AB\sin\phi. \quad (1\text{–}25)$$

We measure the angle $\phi$ *from* $\boldsymbol{A}$ *toward* $\boldsymbol{B}$ and take it always to be the smaller of the two possible angles, the one between 0 and 180°. Thus $C$ in Eq. (1–25) is always positive, as a vector magnitude must be. Note also that when $\boldsymbol{A}$ and $\boldsymbol{B}$ are parallel or antiparallel, $\phi = 0$ or 180° and $C = 0$. That is, *the vector product of two parallel or antiparallel vectors is always zero.* In particular, *the vector product of any vector with itself is zero.*

There are always *two* directions perpendicular to a given plane, one on each side of the plane. To distinguish between these, we imagine rotating vector $\boldsymbol{A}$ about the perpendicular line until it is aligned with $\boldsymbol{B}$ (choosing the smaller of the two possible angles). We then curl the fingers of the right hand around this perpendicular line so that the fingertips point in the direction of rotation; the thumb then gives the direction of the vector product.

This **right-hand rule** is shown in Fig. 1–20a. Alternatively, the direction of the vector product is the direction a right-hand-thread screw advances if turned in the sense from $\boldsymbol{A}$ toward $\boldsymbol{B}$, as shown.

Similarly, we determine the direction of the vector product $\boldsymbol{B} \times \boldsymbol{A}$ by rotating $\boldsymbol{B}$ into $\boldsymbol{A}$ in Fig. 1–20b. This yields a result *opposite* to that for $\boldsymbol{A} \times \boldsymbol{B}$. The vector product is not commutative! In fact, for any two vectors $\boldsymbol{A}$ and $\boldsymbol{B}$,

$$\boldsymbol{A} \times \boldsymbol{B} = -\boldsymbol{B} \times \boldsymbol{A} \tag{1–26}$$

If we know the components of $\boldsymbol{A}$ and $\boldsymbol{B}$, we can calculate the components of the vector product using a procedure similar to that for the scalar product. First we work out the multiplication table for the unit vectors $\boldsymbol{i}, \boldsymbol{j}$, and $\boldsymbol{k}$. The vector product of any vector with itself is zero, so

$$\boldsymbol{i} \times \boldsymbol{i} = \boldsymbol{j} \times \boldsymbol{j} = \boldsymbol{k} \times \boldsymbol{k} = \boldsymbol{0}.$$

Using Eq. (1–25) and the right-hand rule, we find

$$\begin{aligned} \boldsymbol{i} \times \boldsymbol{j} &= -\boldsymbol{j} \times \boldsymbol{i} = \boldsymbol{k}, \\ \boldsymbol{j} \times \boldsymbol{k} &= -\boldsymbol{k} \times \boldsymbol{j} = \boldsymbol{i}, \\ \boldsymbol{k} \times \boldsymbol{i} &= -\boldsymbol{i} \times \boldsymbol{k} = \boldsymbol{j}. \end{aligned} \tag{1–27}$$

Next we express $\boldsymbol{A}$ and $\boldsymbol{B}$ in terms of their components and the corresponding unit vectors, and we expand the expression for the vector product:

$$\begin{aligned} \boldsymbol{A} \times \boldsymbol{B} &= (A_x\boldsymbol{i} + A_y\boldsymbol{j} + A_z\boldsymbol{k}) \times (B_x\boldsymbol{i} + B_y\boldsymbol{j} + B_z\boldsymbol{k}) \\ &= A_x\boldsymbol{i} \times B_x\boldsymbol{i} + A_x\boldsymbol{i} \times B_y\boldsymbol{j} + A_x\boldsymbol{i} \times B_z\boldsymbol{k} \\ &\quad + A_y\boldsymbol{j} \times B_x\boldsymbol{i} + A_y\boldsymbol{j} \times B_y\boldsymbol{j} + A_y\boldsymbol{j} \times B_z\boldsymbol{k} \\ &\quad + A_z\boldsymbol{k} \times B_x\boldsymbol{i} + A_z\boldsymbol{k} \times B_y\boldsymbol{j} + A_z\boldsymbol{k} \times B_z\boldsymbol{k}. \end{aligned} \tag{1–28}$$

We can also rewrite the individual terms as $A_x\boldsymbol{i} \times B_y\boldsymbol{j} = (A_xB_y)\boldsymbol{i} \times \boldsymbol{j}$, and so on. Evaluating these by using the multiplication table for the unit vectors and grouping the terms, we find

$$\boldsymbol{A} \times \boldsymbol{B} = (A_yB_z - A_zB_y)\boldsymbol{i} + (A_zB_x - A_xB_z)\boldsymbol{j} + (A_xB_y - A_yB_x)\boldsymbol{k}. \tag{1–29}$$

If $\boldsymbol{C} = \boldsymbol{A} \times \boldsymbol{B}$, the components of $\boldsymbol{C}$ are given by

$$\begin{aligned} C_x &= A_yB_z - A_zB_y, \\ C_y &= A_zB_x - A_xB_z, \\ C_z &= A_xB_y - A_yB_x. \end{aligned} \tag{1–30}$$

The vector product can also be expressed in determinant form as

$$\boldsymbol{A} \times \boldsymbol{B} = \begin{vmatrix} \boldsymbol{i} & \boldsymbol{j} & \boldsymbol{k} \\ A_x & A_y & A_z \\ B_x & B_y & B_z \end{vmatrix}.$$

If you aren't familiar with determinants, don't worry about this form.

If with the axis system of Fig. 1–21a we reverse the direction of the $z$-axis, we get the system shown in Fig. 1–21b. Then, as you may verify, the definition of the vector product gives $\boldsymbol{i} \times \boldsymbol{j} = -\boldsymbol{k}$ instead of $\boldsymbol{i} \times \boldsymbol{j} = \boldsymbol{k}$. All these products have signs opposite to those in Eqs. (1–27). We see that there are two kinds of coordinate-axis systems, differing in the signs of the products of unit vectors. An axis system in which $\boldsymbol{i} \times \boldsymbol{j} = \boldsymbol{k}$ is called a **right-handed system.** The usual practice is to use *only* right-handed systems; we will follow that practice throughout this book.

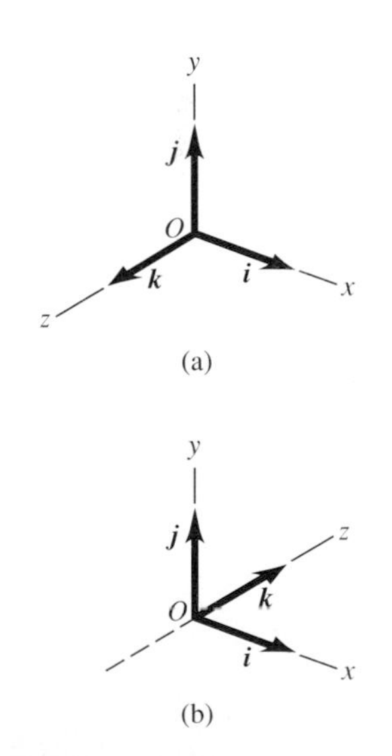


**1–21** (a) A right-handed coordinate system, in which $\boldsymbol{i} \times \boldsymbol{j} = \boldsymbol{k}$, $\boldsymbol{j} \times \boldsymbol{k} = \boldsymbol{i}$, and $\boldsymbol{k} \times \boldsymbol{i} = \boldsymbol{j}$. (b) A left-handed coordinate system, in which $\boldsymbol{i} \times \boldsymbol{j} = -\boldsymbol{k}$, and so on. We'll use only right-handed systems.

## EXAMPLE 1–10

Vector $\boldsymbol{A}$ has magnitude 6 units and is in the direction of the $+x$-axis; vector $\boldsymbol{B}$ has magnitude 4 units and lies in the $xy$-plane, making an angle of 30° with the $+x$-axis (Fig. 1–22). Find the vector product $\boldsymbol{A} \times \boldsymbol{B}$.

**SOLUTION** From Eq. (1–25) the magnitude of the vector product is

$$AB \sin \phi = (6)(4)(\sin 30°) = 12.$$

From the right-hand rule the direction of $\boldsymbol{A} \times \boldsymbol{B}$ is that of the $+z$-axis, so we have $\boldsymbol{A} \times \boldsymbol{B} = 12\boldsymbol{k}$.

Alternatively, we may write the components of $\boldsymbol{A}$ and $\boldsymbol{B}$ and use Eqs. (1–30):

$$A_x = 6, \quad A_y = 0, \quad A_z = 0;$$

$$B_x = 4 \cos 30° = 2\sqrt{3}, \quad B_y = 4 \sin 30° = 2, \quad B_z = 0.$$

If $\boldsymbol{C} = \boldsymbol{A} \times \boldsymbol{B}$, then

$$C_x = (0)(0) - (0)(2) = 0,$$

$$C_y = (0)(2\sqrt{3}) - (6)(0) = 0,$$

$$C_z = (6)(2) - (0)(2\sqrt{3}) = 12.$$

The vector product $\boldsymbol{C}$ has only a $z$-component, and it lies along the $z$-axis. The magnitude agrees with the above result.

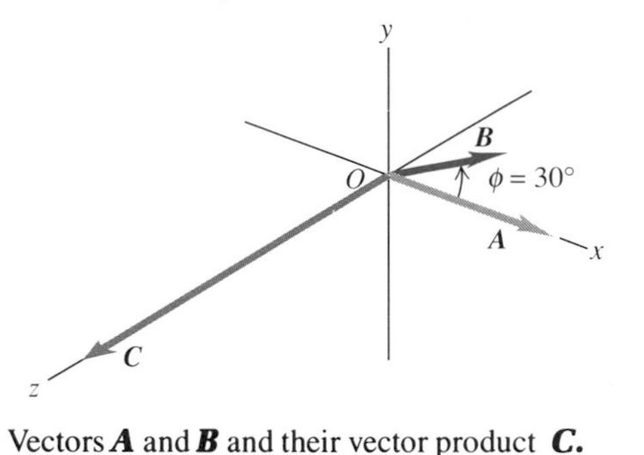


**1–22** Vectors $\boldsymbol{A}$ and $\boldsymbol{B}$ and their vector product $\boldsymbol{C}$.

# SUMMARY

### KEY TERMS

- model
- physical quantity
- operational definition
- unit
- meter
- second
- International System
- kilogram
- prefix
- dimensional consistency
- significant figures
- scientific notation
- order-of-magnitude estimates
- scalar quantity
- vector quantity
- displacement
- magnitude of a vector
- vector sum

• The fundamental physical quantities of mechanics are mass, length, and time; the corresponding SI units are the kilogram, the meter, and the second, respectively. Other units, related by powers of 10, are identified by adding prefixes to the basic unit. Derived units for other physical quantities are products or quotients of the basic units. Equations must be dimensionally consistent; two terms can be added or equated only when they have the same units.

• The uncertainty of a measurement can be indicated by the number of significant figures or by a stated uncertainty. The result of a calculation usually has no more significant figures than the input data. When only crude estimates are available for input data, we can often make useful order-of-magnitude estimates.

• Scalar quantities are numbers and are combined with the usual rules of arithmetic. Vector quantities have direction as well as magnitude and are combined according to the rules of vector addition.

• Vector addition can be carried out using components of vectors. If $A_x$ and $A_y$ are the components of vector $\boldsymbol{A}$ and $B_x$ and $B_y$ are the components of vector $\boldsymbol{B}$, the components of the vector sum $\boldsymbol{R} = \boldsymbol{A} + \boldsymbol{B}$ are given by

$$R_x = A_x + B_x, \qquad R_y = A_y + B_y. \tag{1–11}$$

• Unit vectors describe directions in space. A unit vector has a magnitude of unity, with no units. The unit vectors $\boldsymbol{i}, \boldsymbol{j}$, and $\boldsymbol{k}$, aligned with the coordinate axes of a rectangular coordinate system, are especially useful.

• The scalar product $C = \boldsymbol{A} \cdot \boldsymbol{B}$ of two vectors $\boldsymbol{A}$ and $\boldsymbol{B}$ is a scalar quantity, defined as

$$\boldsymbol{A} \cdot \boldsymbol{B} = AB \cos \phi = |\boldsymbol{A}|\,|\boldsymbol{B}| \cos \phi. \tag{1–20}$$

The scalar product can also be expressed in terms of components:

$$\boldsymbol{A} \cdot \boldsymbol{B} = A_xB_x + A_yB_y + A_zB_z. \tag{1–23}$$

The scalar product is commutative; for any two vectors $\boldsymbol{A}$ and $\boldsymbol{B}$, $\boldsymbol{A} \cdot \boldsymbol{B} = \boldsymbol{B} \cdot \boldsymbol{A}$. The scalar product of two perpendicular vectors is zero.

• The vector product $C = A \times B$ of two vectors $A$ and $B$ is another vector $C$, with magnitude given by

$$C = AB \sin \phi. \tag{1-25}$$

Its direction is perpendicular to the plane of the two vectors, as given by the right-hand rule. The components of the vector product can be expressed in terms of the components of the two vectors being multiplied:

$$C_x = A_y B_z - A_z B_y, \qquad C_y = A_z B_x - A_x B_z, \qquad C_z = A_x B_y - A_y B_x. \tag{1-30}$$

The vector product is not commutative; the order of the factors must not be interchanged. For any two vectors $A$ and $B$, $A \times B = -B \times A$. The vector product of two parallel or antiparallel vectors is zero.

**resultant**
**component**
**unit vector**
**scalar product**
**vector product**
**right-hand rule**
**right-handed system**

# EXERCISES

## Section 1–3 Standards and Units

## Section 1–4 Unit Consistency and Conversions

**1–1** Starting with the definition 1.00 in. = 2.54 cm, compute the number of miles in 1.00 kilometer.

**1–2** The density of water is 1.00 g/cm$^3$. What is this value in kilograms per cubic meter?

**1–3** The Concorde is the fastest airliner used for commercial service and can cruise at 1450 mi/h (about two times the speed of sound, or in other words Mach 2). a) What is the cruise speed of the Concorde in mi/s? b) What is it in m/s?

**1–4** The piston displacement of a car engine is given as 1.80 liters (L). Using only the facts that 1.00 L = 1000 cm$^3$ and 1.00 in. = 2.54 cm, express this volume in cubic inches.

**1–5** The gasoline consumption of a small car is 19.0 km/L (1 L = 1 liter). How many miles per gallon is this? Use the conversion factors in Appendix E.

**1–6** The speed limit on a highway in Lower Slobbovia is 135,000 furlongs per fortnight. How many miles per hour is this? (One furlong is $\frac{1}{8}$ mile, and a fortnight is 14 days. A furlong originally referred to the length of a plowed furrow.)

## Section 1–5 Precision and Significant Figures

**1–7** Figure 1–5 shows the result of unacceptable error in the stopping position of a train. If a train travels 270 km from Brussels to Paris and then overshoots the end of the track by 15 m, what is the percent error in the total distance covered?

**1–8** What is the percent error in each of the following approximations to $\pi$? a) $\frac{22}{7}$ b) $\frac{355}{113}$

**1–9** Estimate the percent error in measuring a) a distance of about 40 cm with a meter stick; b) a mass of about 8 g with a chemical balance; c) a time interval of about 5 min with a stopwatch.

**1–10** An angle is given, to one significant figure, as 7°, meaning that its value is between 6.5° and 7.5°. Find the corresponding range of possible values of the cosine of the angle. Is this a case in which there are more significant figures in the result than in the input data?

**1–11** The mass of the earth is $5.98 \times 10^{24}$ kg, and its radius is $6.38 \times 10^6$ m. Compute the density of the earth, using powers-of-10 notation and the correct number of significant figures. (The density of an object is defined as its mass divided by its volume. The formula for the volume of a sphere is given in Appendix B.)

## Section 1–6 Estimates and Order of Magnitude

**1–12** How many kernels of corn does it take to fill a 1-L soft drink bottle?

**1–13** A box of typewriter paper is 11″ × 17″ × 9″; it is marked "10 M." Does that mean that it contains ten thousand sheets or ten million?

**1–14** What total volume of air does a person breathe in a lifetime? How does that compare with the volume of the Houston Astrodome? (Estimate that a person breathes about 500 cm$^3$ of air with each breath.)

**1–15** How many hairs do you have on your head?

**1–16** How many times does a human heart beat during a lifetime? How many gallons of blood does it pump? (Estimate that the heart pumps 50 cm$^3$ of blood with each beat.)

**1–17** In Wagner's opera *Ring of the Nibelung,* the goddess Freia is ransomed by a pile of gold just tall enough and wide enough to hide her from sight. Estimate the monetary value of this pile. (Refer to Example 1–4 for information on the price per ounce and density of gold.)

**1–18** How many drops of water are in all the oceans?

**1–19** If you filled your room with rocks, what would they weigh? What if you filled it with crumpled newspapers?

**1–20** How many dollar bills would have to be stacked up to reach the moon? Would that be cheaper than building and launching a spacecraft?

## Section 1–7 Vectors and Vector Addition

**1–21** A postal employee drives a delivery truck along the route shown in Fig. 1–23. Determine the magnitude and direction of the resultant displacement by drawing a scale diagram.

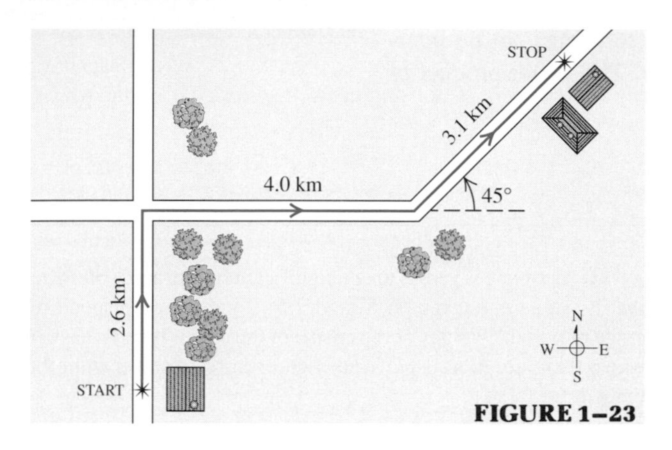


**FIGURE 1–23**

**1–22** For the vectors $\boldsymbol{A}$ and $\boldsymbol{B}$ in Fig. 1–24, use a scale drawing to find the magnitude and direction of a) the vector sum $\boldsymbol{A} + \boldsymbol{B}$; b) The vector difference $\boldsymbol{A} - \boldsymbol{B}$.

**1–23** A spelunker is surveying a cave. He follows a passage 180 m straight east, then 80 m in a direction 60.0° west of north, then 150 m at 45.0° west of south. After a fourth unmeasured displacement he finds himself back where he started. Use a scale drawing to determine the fourth displacement (magnitude and direction).

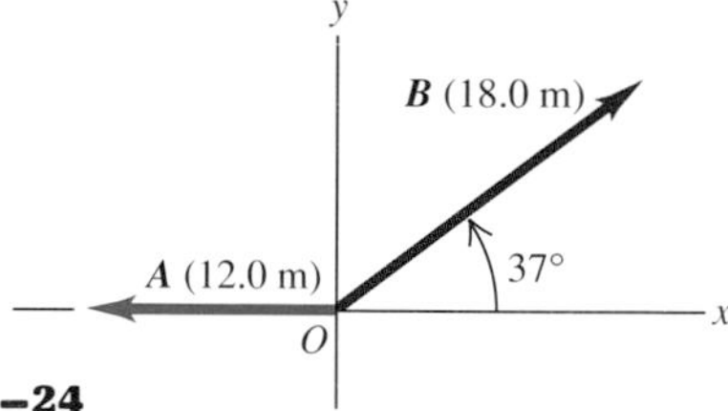


**FIGURE 1–24**

## Section 1–8 Components of Vectors

**1–24** Use a scale drawing to find the $x$- and $y$-components of the following vectors. In each case the magnitude of the vector and the angle, measured counterclockwise, that it makes with the $+x$-axis are given. a) magnitude 8.90 m, angle 60.0°; b) magnitude 12.0 km, angle 315°; c) magnitude 5.20 cm, angle 143°.

**1–25** Compute the $x$- and $y$- components of each of the vectors $\boldsymbol{A}$, $\boldsymbol{B}$, and $\boldsymbol{C}$ in Fig. 1–25.

**1–26** Find the magnitude and direction of the vector represented by each of the following pairs of components: a) $A_x = 2.80$ cm, $A_y = -4.10$ cm; b) $A_x = -5.40$ m, $A_y = -9.45$ m; c) $A_x = -2.25$ km, $A_y = 3.60$ km.

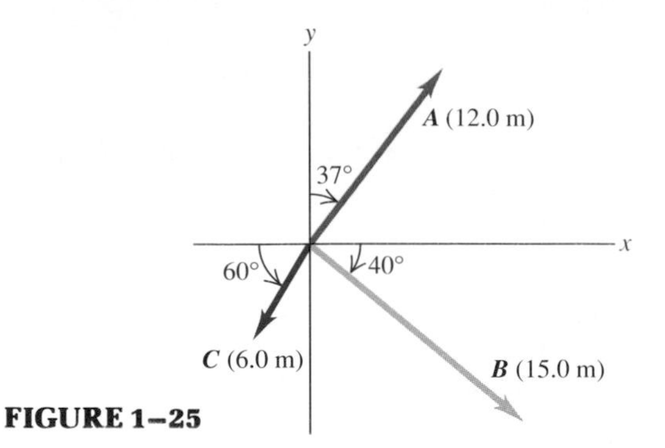


**FIGURE 1–25**

**1–27** For the vectors $\boldsymbol{A}$ and $\boldsymbol{B}$ in Fig. 1–24, use the method of components to find the magnitude and direction of a) the vector sum $\boldsymbol{A} + \boldsymbol{B}$; b) the vector difference $\boldsymbol{A} - \boldsymbol{B}$.

**1–28** Vector $\boldsymbol{A}$ has components $A_x = 2.90$ cm, $A_y = 3.45$ cm; vector $\boldsymbol{B}$ has components $B_x = 4.10$ cm, $B_y = -2.25$ cm. Find a) the components of the vector sum $\boldsymbol{A} + \boldsymbol{B}$; b) the magnitude and direction of $\boldsymbol{A} + \boldsymbol{B}$; c) the components of the vector difference $\boldsymbol{A} - \boldsymbol{B}$; d) the magnitude and direction of $\boldsymbol{A} - \boldsymbol{B}$.

**1–29** A disoriented physics professor drives 4.92 km east, then 3.95 km south, then 1.80 km west. Find the magnitude and direction of the resultant displacement, using the method of components.

**1–30** A postal employee drives a delivery truck over the route shown in Fig. 1–23. Use the method of components to determine the magnitude and direction of her resultant displacement.

**1–31** Vector $\boldsymbol{A}$ is 2.80 cm long and is 60.0° above the $x$-axis in the first quadrant. Vector $\boldsymbol{B}$ is 1.90 cm long and is 60.0° below the $x$-axis in the fourth quadrant (Fig. 1–26). Find the magnitude and direction of a) $\boldsymbol{A} + \boldsymbol{B}$; b) $\boldsymbol{A} - \boldsymbol{B}$; c) $\boldsymbol{B} - \boldsymbol{A}$.

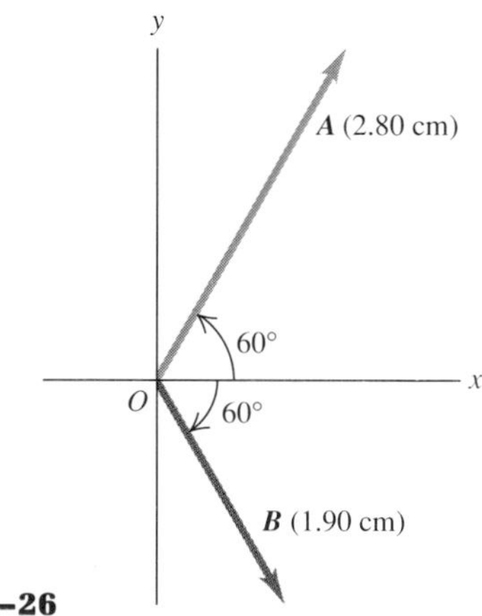


**FIGURE 1–26**

## Section 1–9 Unit Vectors

**1–32** Write each of the vectors in Fig. 1–24 in terms of the unit vectors $\boldsymbol{i}$ and $\boldsymbol{j}$.

**1–33** Given two vectors $\boldsymbol{A} = 4\boldsymbol{i} + 3\boldsymbol{j}$ and $\boldsymbol{B} = \boldsymbol{i} - 5\boldsymbol{j}$, a) find the magnitude of each vector; b) write an expression for the vector difference $\boldsymbol{A} - \boldsymbol{B}$, using unit vectors; c) find the magnitude and direction of the vector difference $\boldsymbol{A} - \boldsymbol{B}$.

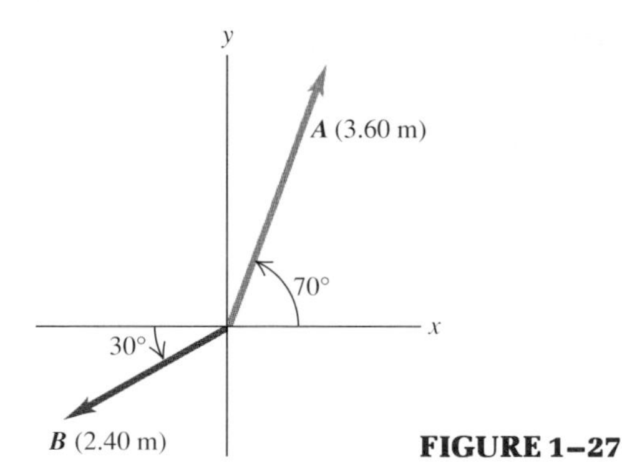


FIGURE 1–27

**1–34** a) Write each vector in Fig. 1–27 in terms of the unit vectors $\boldsymbol{i}$ and $\boldsymbol{j}$. b) Use unit vectors to express the vector $\boldsymbol{C}$, where $\boldsymbol{C} = 3\boldsymbol{A} - 2\boldsymbol{B}$. c) Find the magnitude and direction of $\boldsymbol{C}$.

## Section 1–10 Products of Vectors

**1–35** Write out a multiplication table for the scalar products of all possible pairs of unit vectors, such as $\boldsymbol{i} \cdot \boldsymbol{i} = ?$, $\boldsymbol{i} \cdot \boldsymbol{j} = ?$, and so on.

**1–36** Find the scalar product $\boldsymbol{A} \cdot \boldsymbol{B}$ of the two vectors in Fig. 1–24.

**1–37** Find the scalar product of the two vectors given in Exercise 1–33.

**1–38** Find the angle between the vectors $\boldsymbol{A} = -2\boldsymbol{i} + 7\boldsymbol{j}$ and $\boldsymbol{B} = 3\boldsymbol{i} - \boldsymbol{j}$.

**1–39** Write out a multiplication table for the vector products of all possible pairs of unit vectors, such as $\boldsymbol{i} \times \boldsymbol{i} = ?$, $\boldsymbol{i} \times \boldsymbol{j} = ?$, and so on, using a right-handed coordinate system.

**1–40** For the two vectors in Fig. 1–24, a) find the magnitude and direction of the vector product $\boldsymbol{A} \times \boldsymbol{B}$; b) find the magnitude and direction of $\boldsymbol{B} \times \boldsymbol{A}$.

**1–41** Find the vector product (expressed in unit vectors) of the two vectors given in Exercise 1–33. What is its magnitude?

**1–42** Which of the following are legitimate mathematical operations? a) $\boldsymbol{A} \cdot (\boldsymbol{B} - \boldsymbol{C})$ b) $(\boldsymbol{A} - \boldsymbol{B}) \times \boldsymbol{C}$ c) $\boldsymbol{A} \cdot (\boldsymbol{B} \times \boldsymbol{C})$ d) $\boldsymbol{A} \times (\boldsymbol{B} \times \boldsymbol{C})$ e) $\boldsymbol{A} \times (\boldsymbol{B} \cdot \boldsymbol{C})$

# PROBLEMS

**1–43 The Hydrogen Maser.** One standard of frequency is the radio waves generated by the hydrogen maser. These waves have a frequency of 1,420,405,751.786 hertz. (A hertz is a special name for cycles per second.) A clock controlled by a hydrogen maser is off by only 1 s in 100,000 years. For the following questions, use only three significant figures. (The large number of significant figures given for the frequency were to illustrate the remarkable precision to which it has been measured.) a) What is the time for one cycle of the radio wave? b) How many cycles occur in 1 h? c) How many cycles would have occurred during the age of the earth, which is estimated to be 4600 million years? d) By how many seconds would a hydrogen maser clock be off during the lifetime of the earth?

**1–44** The following is taken from a recent magazine article: "The most expensive land in Japan as of January 1, 1990 was located in Tokyo's downtown areas of Ginza and Marunouchi — worth 37.7 million Yen per square meter." The article then quoted the value, in U.S. dollars, of a piece of this land the size of a postage stamp. Assuming a postage stamp size of $\frac{7}{8}$ in. × 1.0 in., what value should the article have stated? At the time of the article, 1 dollar was equivalent to 136 Yen.

**1–45** Vector $\boldsymbol{M}$, with magnitude 5.20 cm, is at 49.0° counterclockwise from the $+x$-axis. It is added to vector $\boldsymbol{N}$, and the resultant is a vector with magnitude 5.20 cm, at 32.0° counterclockwise from the $+x$-axis. Find a) the components of $\boldsymbol{N}$; b) the magnitude and direction of $\boldsymbol{N}$.

**1–46** Find the magnitude and direction of the vector $\boldsymbol{R}$ that is the sum of the three vectors $\boldsymbol{A}$, $\boldsymbol{B}$, and $\boldsymbol{C}$ in Fig. 1–25.

**1–47** A spelunker is surveying a cave. He follows a passage 180 m straight east, then 80 m in a direction 30° west of north, then 150 m at 45° west of south. After a fourth unmeasured displacement he finds himself back where he started. Use the method of components to determine the fourth displacement (magnitude and direction).

**1–48** A sailor in a small sailboat encounters shifting winds. She sails 2.00 km east, then 3.50 km southeast, then an additional distance in an unknown direction. Her final position is 5.80 km directly east of the starting point (Fig. 1–28). Find the magnitude and direction of the third leg of the journey.

**1–49** A cross-country skier skis 7.40 km in the direction of 30.0° west of south, then 2.80 km in the direction 45.0° north of west, and finally 5.20 km in the direction 22.0° east of north. a) Show these displacements on a diagram. b) How far is he from his starting point?

**1–50** For the two vectors $\boldsymbol{A}$ and $\boldsymbol{B}$ in Fig. 1–27, a) find the scalar product $\boldsymbol{A} \cdot \boldsymbol{B}$; b) find the magnitude and direction of the vector product $\boldsymbol{A} \times \boldsymbol{B}$.

**1–51** Given two vectors $\boldsymbol{A} = -\boldsymbol{i} + 2\boldsymbol{j} - 5\boldsymbol{k}$ and $\boldsymbol{B} = 3\boldsymbol{i} + 2\boldsymbol{j} - 2\boldsymbol{k}$, obtain the following. a) Find the magnitude of each vector. b) Write an expression for the vector difference

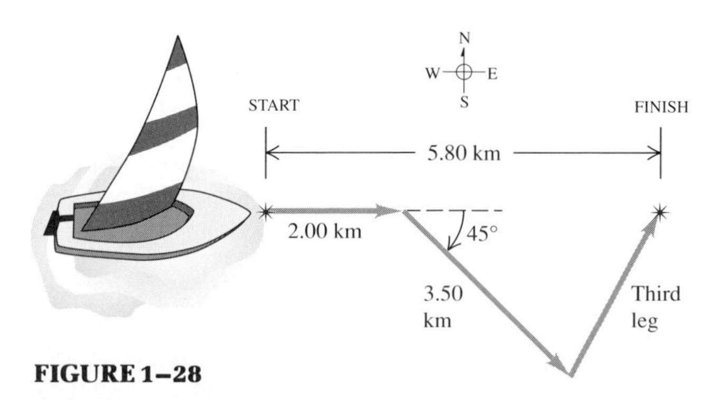


FIGURE 1–28

$\boldsymbol{A} - \boldsymbol{B}$, using unit vectors. c) Find the magnitude of the vector difference $\boldsymbol{A} - \boldsymbol{B}$. Is this the same as the magnitude of $\boldsymbol{B} - \boldsymbol{A}$? Explain.

**1–52 Bond Angle in Methane.** In methane, $CH_4$, each hydrogen atom is at a corner of a regular tetrahedron with the carbon atom at the center. In coordinates where one of the C — H bonds is in the direction of $\boldsymbol{i} + \boldsymbol{j} + \boldsymbol{k}$, then an adjacent C — H bond is in the $\boldsymbol{i} - \boldsymbol{j} - \boldsymbol{k}$ direction. Calculate the angle between these two bonds.

**1–53** A cube is placed so that one corner is at the origin and three edges are along the $x$-, $y$-, and $z$-axes of a coordinate system (Fig. 1–29). Compute a) the angle between the edge along the $z$-axis (line $ab$) and the diagonal from the origin to the opposite corner (line $ad$); b) the angle between line $ad$ and line $ac$ (the diagonal of a face).

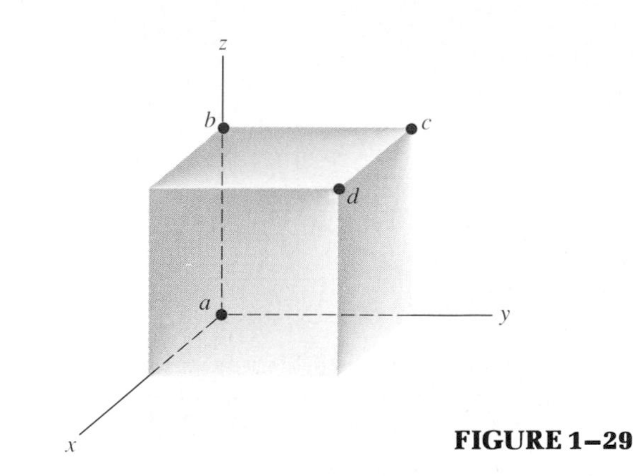


**FIGURE 1–29**

**1–54** When two vectors $\boldsymbol{A}$ and $\boldsymbol{B}$ are drawn from a common point, the angle between them is $\theta$. a) Show that the magnitude of their vector sum is given by

$$\sqrt{A^2 + B^2 + 2AB\cos\theta}.$$

b) If $\boldsymbol{A}$ and $\boldsymbol{B}$ have the same magnitude, under what circumstances will their vector sum have the same magnitude as $\boldsymbol{A}$ or $\boldsymbol{B}$? c) Derive a result analogous to that in part (a) for the magnitude of the vector difference $\boldsymbol{A} - \boldsymbol{B}$. d) If $\boldsymbol{A}$ and $\boldsymbol{B}$ have the same magnitude, under what circumstance will $\boldsymbol{A} - \boldsymbol{B}$ have this same magnitude?

## CHALLENGE PROBLEMS

**1–55** At Enormous State University (ESU) the football team records its plays using vector displacements, with the position of the ball before the play starts as the origin. In a certain pass play, the receiver starts at $+\boldsymbol{i} - 5\boldsymbol{j}$, where the units are yards, $\boldsymbol{i}$ is to the right, and $\boldsymbol{j}$ is downfield. Subsequent displacements of the receiver are $8\boldsymbol{i}$ (in motion before the snap), $+15\boldsymbol{j}$ (breaks downfield), $-6\boldsymbol{i} + 4\boldsymbol{j}$ (zigs), and $+12\boldsymbol{i} + 20\boldsymbol{j}$ (zags). Meanwhile, the quarterback has dropped straight back, with displacement $-8\boldsymbol{j}$. How far and in what direction must the quarterback throw the ball? (Like the coach, you will be well advised to diagram the situation before solving it numerically.)

**1–56** Physicists, mathematicians, and others often deal with large numbers. The number $10^{100}$ has been given the whimsical name *googol* by mathematicians. Let us compare some large numbers in physics with the googol. (*Note:* This problem requires numerical values that can be found in the appendices of this book, with which you should become familiar.) a) Approximately how many atoms make up the earth? For simplicity, take the average atomic mass of the atoms to be 14 g/mol. Avogadro's number gives the number of atoms in a mole. b) Approximately how many neutrons are in a neutron star? Neutron stars are made up of neutrons and have approximately twice the mass of the sun. c) In one theory of the origin of the universe, the universe at a very early time had a density (mass divided by volume) of $10^{15}$ g/cm$^3$, and its radius was approximately the present distance of the earth to the sun. Assuming that $\frac{1}{3}$ of the particles were protons, $\frac{1}{3}$ of the particles were neutrons, and the remaining $\frac{1}{3}$ were electrons, how many particles then made up the universe?

**1–57** The dot and cross product will appear later in many physical applications; for now, consider the following applications. a) Let $\boldsymbol{A}$ and $\boldsymbol{B}$ be vectors along the sides of a parallelogram, where the sides have lengths $A$ and $B$ (Fig. 1–30a). Show that $C$ is the area of the parallelogram, where $\boldsymbol{C} = \boldsymbol{A} \times \boldsymbol{B}$. b) The volume of a parallelepiped with sides of length $A$, $B$, and $C$ can be expressed in an elegant way as a product of three vectors $\boldsymbol{A}$, $\boldsymbol{B}$, and $\boldsymbol{C}$, where $\boldsymbol{A}$ is along the edge of length $A$, and so on (Fig. 1–30b). Express the volume of the parallelepiped as the magnitude of the appropriate product of the three vectors $\boldsymbol{A}$, $\boldsymbol{B}$, and $\boldsymbol{C}$.

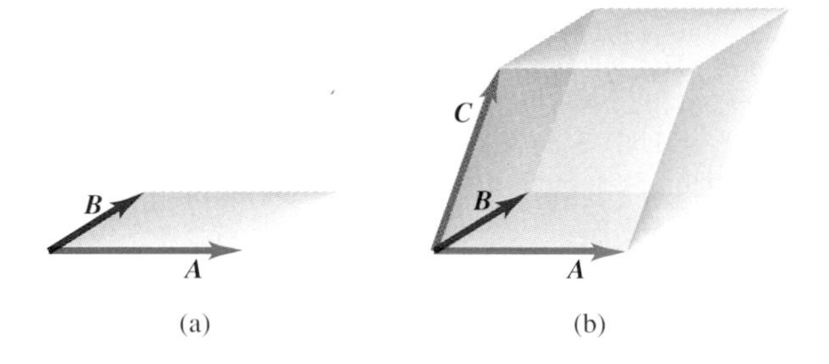


**FIGURE 1–30**

**1–58** Obtain a *unit vector* perpendicular to the two vectors given in Problem 1–51.

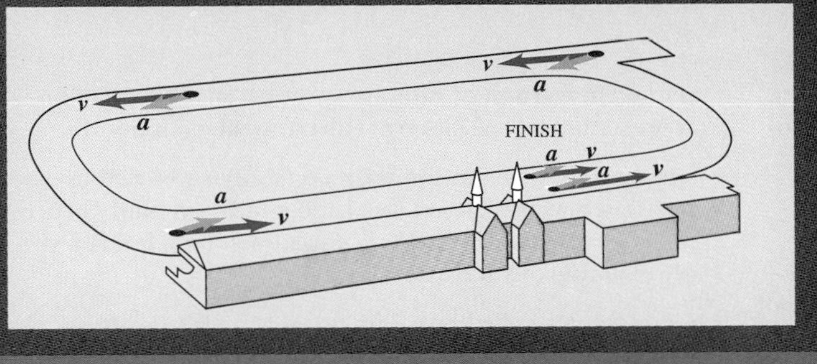


CHAPTER 2

# Motion Along a Straight Line

**In the 1973 Kentucky Derby, Secretariat ran the quarter-mile distances in 25.2 s, 24.0 s, 23.8 s, 23.4 s, and 23.0 s, running each quarter mile faster than the preceding one. In other words, he maintained a steady average acceleration throughout the 1 1/4 mile race—an incredible achievement that propelled him from last to first place with a winning time of 1:59.4 (a track record).**

## KEY CONCEPTS

**• The average velocity of a body is its displacement divided by the time interval during which the displacement occurs. Instantaneous velocity is the limit of this quantity as the time interval approaches zero.**

**• The average acceleration of a body is its change in velocity divided by the time interval during which the change occurs. Instantaneous acceleration is the limit of this quantity as the time interval approaches zero.**

**• On a graph of position versus time, velocity is represented graphically by the slope of the curve at each point. On a graph of velocity versus time, acceleration is represented by the slope of the curve at each point.**

**• In straight-line motion with constant acceleration, the position and velocity at any time and the velocity at any position are given by simple equations. Free fall is a simple example of such motion.**

**• The velocity of a body depends on the frame of reference in which it is observed. Different observers measure different velocities if they are moving relative to each other.**

## INTRODUCTION

How do you describe the motion of a horse coming into the home stretch, galloping toward the finish line in a race? When you throw a baseball straight up in the air, how high does it go? How fast do you have to throw it to reach that height? When a glass slips from your hand, how much time do you have to catch it before it hits the floor? These are the kinds of questions you will learn to answer in this chapter. We are beginning our study of physics with *mechanics,* the study of the relationships among force, matter, and motion. Our first goal is to develop general methods for *describing* motion. Later we will study the relation of motion to its causes.

In this chapter we limit our discussion to the simplest case, a single particle moving along a straight line. We will often use a particle as a model for a moving body in situations in which effects such as rotation and change of shape are not important. The problem is particularly simple when the acceleration of the particle is constant, and we will develop equations to fit this special case. We also need to understand how two observers who are moving relative to each other describe the same motion differently. Describing motion in a moving frame of reference contains the seeds of the special theory of relativity, although we won't get to that until later in the book. ■

## 2–1 AVERAGE VELOCITY

Suppose a drag racer drives his AA-fuel dragster in a straight line down the track. We'll use this line as the $x$-axis of our coordinate system (Fig. 2–1). The origin $O$ is at the starting line. The distance of the front of the dragster from the origin is given by the coordinate $x$, which varies with time. Suppose that 1.0 s after the start the car is at point $P_1$, 19 m from the origin, and that 4.0 s after the start it is at point $P_2$, 277 m from the origin. Then it has traveled a distance of (277 m − 19 m) = 258 m in a time of (4.0 s − 1.0 s) = 3.0 s. We define the car's **average velocity** during this interval to be a vector quantity whose $x$-component is the change in $x$ divided by the time interval, that is, (258 m)/(3.0 s) = 86 m/s.

Let's generalize this. At time $t_1$ a point at the front of the car is at point $P_1$, with coordinate $x_1$, and at time $t_2$ it is at point $P_2$, with coordinate $x_2$. The displacement during the time interval from $t_1$ to $t_2$ is the vector from $P_1$ to $P_2$; the $x$-component of the displacement vector is $(x_2 - x_1)$, and all other components are zero. We define the average velocity as a vector in the direction from $P_1$ to $P_2$, with an $x$-component $v_{av}$ given by

$$v_{av} = \frac{x_2 - x_1}{t_2 - t_1}. \tag{2–1}$$

Changes in quantities, such as $(x_2 - x_1)$ and $(t_2 - t_1)$, occur so often in physics that it is worthwhile to use a special notation. From now on we will use the Greek letter $\Delta$ to represent a *change* in any quantity. Thus we can write the $x$-component of the car's displacement as

$$\Delta x = x_2 - x_1. \tag{2–2}$$

Be sure you understand that $\Delta x$ is *not* the product of $\Delta$ and $x$. It is a *single symbol*, and it means "the change in the quantity $x$." With this same notation we denote the time interval from $t_1$ to $t_2$ as $\Delta t = t_2 - t_1$. Note that $\Delta x$ or $\Delta t$ always denotes the final value minus the initial value, never the reverse.

We can now define the $x$-component of average velocity as the ratio of the $x$-component of displacement $\Delta x$ to the time interval $\Delta t$. As above, we represent this quantity by the letter $v$ with a subscript "av" to signify average value. Thus

$$v_{av} = \frac{x_2 - x_1}{t_2 - t_1} = \frac{\Delta x}{\Delta t}. \tag{2–3}$$

From the above example we have $x_1 = 19$ m, $x_2 = 277$ m, $t_1 = 1.0$ s, and $t_2 = 4.0$ s, and Eq. (2–3) gives

$$v_{av} = \frac{277\text{ m} - 19\text{ m}}{4.0\text{ s} - 1.0\text{ s}} = 86\text{ m/s}.$$

In some cases, $v_{av}$ is negative. Suppose an official's truck is moving *toward* the origin, with $x_1 = 277$ m and $x_2 = 19$ m. If $t_1 = 6.0$ s and $t_2 = 16.0$ s, then $\Delta t = 10.0$ s, and the $x$-component of average velocity is −26 m/s. Whenever $x$ is positive and decreasing, or negative and becoming more negative, the particle is moving toward the left in Fig. 2–1, and $v_{av}$ is negative.

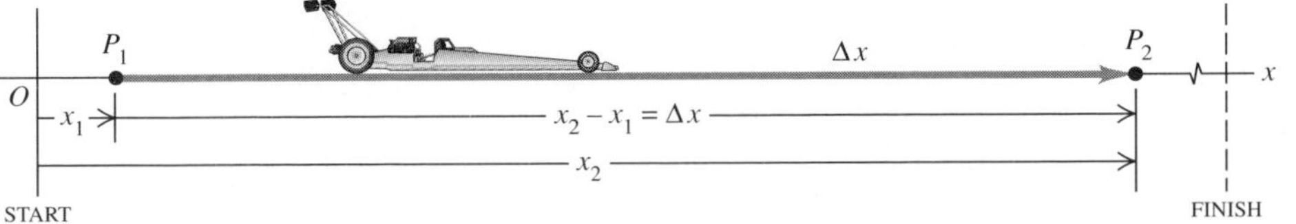


**2–1** Positions of an AA-fuel dragster at two times during its run.

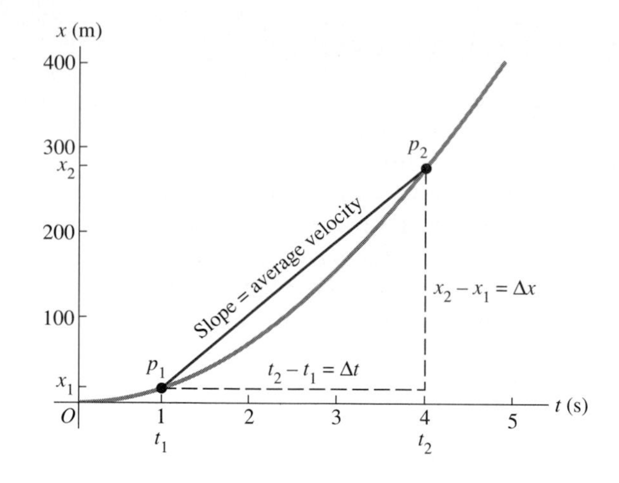


**2–2** The position of an AA-fuel dragster as a function of time. The average velocity between points $P_1$ and $P_2$ is the slope of the line $p_1p_2$.

Average velocity is a vector quantity, and Eq. (2–3) defines the $x$-component of this vector. With straight-line motion, when we call $\Delta x$ the displacement and $v_{av}$ the average velocity, we must remember that these are really the $x$-components of vector quantities that in this special case have *only* $x$-components. In Chapter 3, displacement, velocity, and acceleration vectors will have two or three nonzero components.

Sometimes, especially when the motion is along a vertical line, it is more natural to call that line the $y$-axis instead of the $x$-axis. Then Eq. (2–3), with the $x$'s replaced by $y$'s, gives the $y$-component of average velocity.

Figure 2–2 is a graph of the position of the dragster as a function of time. The curve in the figure *does not* represent the car's path in space; as Fig. 2–1 shows, the path is a straight line. Rather, the graph is a pictorial way to represent the car's change of position with time. The points on the graph corresponding to points $P_1$ and $P_2$ are labeled $p_1$ and $p_2$. The average velocity $v_{av}$ of the car is represented by the *slope* of the line $p_1p_2$, that is, the ratio of vertical to horizontal sides of the triangle. Thus $v_{av}$ is the ratio of $x_2 - x_1$, or $\Delta x$, to $t_2 - t_1$, or $\Delta t$.

It doesn't matter whether the car's velocity is or is not constant during the time interval $\Delta t = t_2 - t_1$. Another dragster might have started from rest, reached a maximum velocity, blown its engine, and then slowed down. To calculate its average velocity, we need only the total displacement $\Delta x = x_2 - x_1$ and the total time interval $\Delta t = t_2 - t_1$.

Table 2–1 shows a few typical velocity magnitudes.

**TABLE 2–1 Typical Velocity Magnitudes**

| | |
|---|---|
| A snail's pace | $10^{-3}$ m/s |
| A brisk walk | 2 m/s |
| Fastest human | 11 m/s |
| Running cheetah | 35 m/s |
| Fastest car | 283 m/s |
| Random speed of air molecules | 500 m/s |
| Fastest plane | 1000 m/s |
| Speed of communications satellite | 3000 m/s |
| Speed of electron in a hydrogen atom | $2 \times 10^6$ m/s |
| Speed of light | $3 \times 10^8$ m/s |

## 2–2 INSTANTANEOUS VELOCITY

Even when a particle speeds up or slows down, we can still define a velocity at any one specific instant of time or at one specific point in the path. Such a velocity is called **instantaneous velocity,** and it needs to be defined carefully.

To find the instantaneous velocity of the dragster in Fig. 2–1 at the point $P_1$, we imagine taking the second point $P_2$ closer and closer to the first point $P_1$. We compute the average velocity over these shorter and shorter displacements and time intervals. Both $\Delta x$ and $\Delta t$ become very small, but their ratio does not necessarily become small. We define the $x$-component of instantaneous velocity at $P_1$ as the limiting value that this series of average velocities approaches as the time interval becomes very small and $P_2$ gets closer and closer to $P_1$.

In the language of calculus the limit of $\Delta x/\Delta t$ as $\Delta t$ approaches zero is written $dx/dt$ and is called the **derivative** of $x$ with respect to $t$. We use the symbol $v$ with no subscript for instantaneous velocity, so

$$v = \lim_{\Delta t \to 0} \frac{\Delta x}{\Delta t} = \frac{dx}{dt}. \quad (2\text{–}4)$$

We always assume that $t_2$ is a *later* time than $t_1$; we always grow older, not younger. Then $\Delta t$ is always positive, and $v$ has the same algebraic sign as $\Delta x$. If the positive $x$-axis points to the right, as in Fig. 2–1, a positive value of $v$ means motion toward the right, and a negative value means motion toward the left. But a body can have a positive $x$ and a negative $v$ or the reverse; $x$ tells us where it is, and $v$ tells us how it's moving.

Instantaneous velocity, like average velocity, is a vector quantity. Equation (2–4) defines its $x$-component, which can be positive or negative. In straight-line motion all other components are zero, and in this case we will often call $v$ the instantaneous velocity. When we use the term "velocity," we always mean instantaneous rather than average velocity unless we state otherwise.

As point $P_2$ in Fig. 2–1 approaches point $P_1$, point $p_2$ in Fig. 2–2 approaches point $p_1$. In the limit, the slope of the line $p_1p_2$ equals the slope of the tangent to the curve at point $p_1$ (Fig. 2–3), which we obtain by dividing a vertical interval (with distance units) by a horizontal interval (with time units). *On a graph of a coordinate as a function of time, the instantaneous velocity at any point is equal to the slope of the tangent to the curve at that point.* If the tangent slopes upward to the right, its slope is positive, the

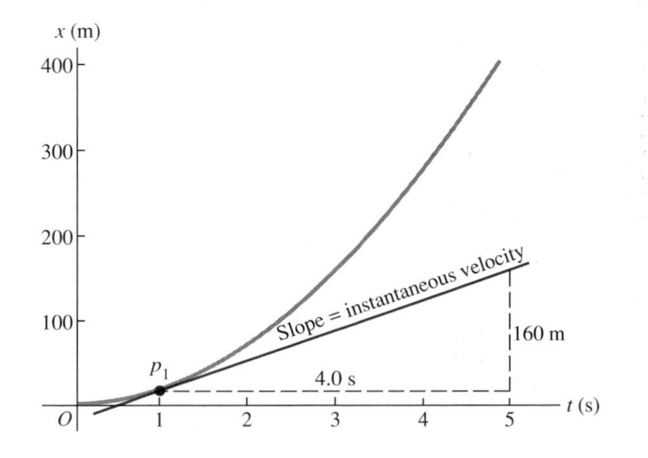


**2–3** The instantaneous velocity at point $P_1$ equals the slope of the tangent to the $x$-$t$ curve at point $p_1$. Here $v$ at $P_1$ equals 160 m/4.0 s = 40 m/s.

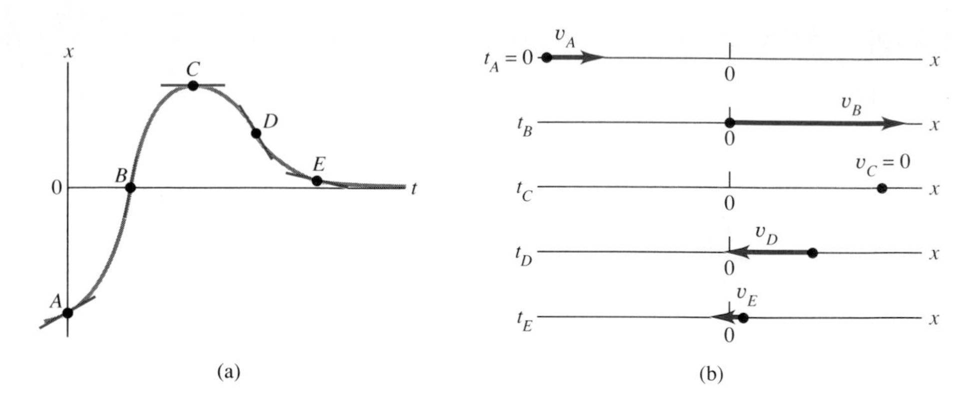


**2–4** (a) Between points $A$ and $B$, $x$ is negative but increasing, and the particle is moving with increasing velocity in the positive $x$-direction. At $B$ the slope and velocity are greatest. Between $B$ and $C$ the particle is still moving in the positive $x$-direction but with decreasing velocity, until at point $C$ the slope and velocity are zero. From $C$ to $E$, $x$ is positive but decreasing, so the velocity is negative; its most negative value occurs at $D$. At point $E$ the slope and velocity are approaching zero as $x$ approaches a constant value. (b) The positions and velocities of the particle at the five times shown on the $x$-$t$ graph.

velocity is positive, and the motion is in the positive $x$-direction. If the tangent slopes downward to the right, the velocity is negative, and the motion is in the negative $x$-direction. Where the tangent is horizontal, the slope is zero, and the velocity is zero. Figure 2–4 illustrates these three possibilities.

If we express distance in meters and time in seconds, velocity is expressed in meters per second (m/s). Other common units of velocity are kilometers per hour (km/h), feet per second (ft/s), miles per hour (mi/h), and knots (1 knot = 1 nautical mile/h or 6080 ft/h).

## EXAMPLE 2–1

**Average and instantaneous velocities** A cheetah is crouched in ambush 20 m to the east of an observer's blind (Fig. 2–5). At time $t = 0$ the cheetah charges an antelope in a clearing 50 m east of the observer. The cheetah runs along a straight line; later analysis of a videotape shows that during the first 2.0 s of the attack, the cheetah's coordinate $x$ varies with time according to the equation $x = 20\text{ m} + (5.0\text{ m/s}^2)t^2$. (Note that the numbers 20 and 5 in this expression *must* have the units shown in order to make the expression dimensionally consistent.)

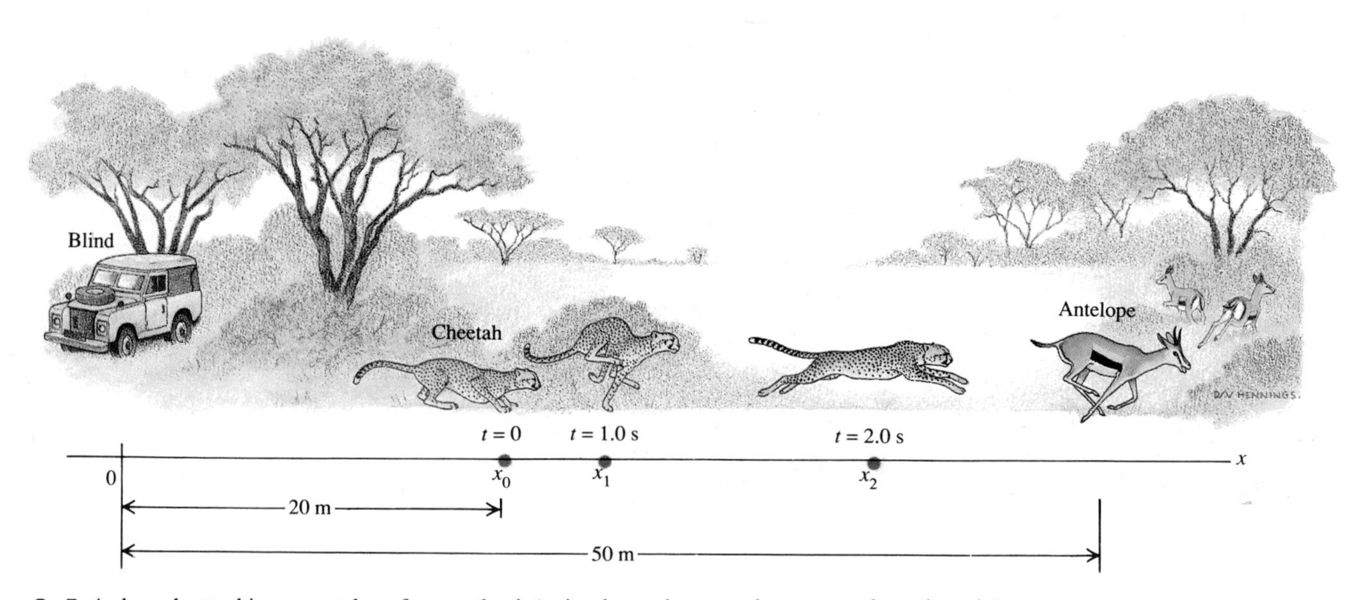


**2–5** A cheetah attacking an antelope from ambush (animals not drawn to the same scale as the axis).

a) Find the displacement of the cheetah during the interval between $t_1 = 1.0$ s and $t_2 = 2.0$ s.

At time $t_1 = 1.0$ s the cheetah's position $x_1$ is

$$x_1 = 20\ \text{m} + (5.0\ \text{m/s}^2)(1.0\ \text{s})^2 = 25\ \text{m}.$$

At time $t_2 = 2.0$ s its position $x_2$ is

$$x_2 = 20\ \text{m} + (5.0\ \text{m/s}^2)(2.0\ \text{s})^2 = 40\ \text{m}.$$

The displacement during this interval is

$$\Delta x = x_2 - x_1 = 40\ \text{m} - 25\ \text{m} = 15\ \text{m}.$$

b) Find the average velocity during the same time interval.

$$v_{\text{av}} = \frac{x_2 - x_1}{t_2 - t_1} = \frac{40\ \text{m} - 25\ \text{m}}{2.0\ \text{s} - 1.0\ \text{s}} = \frac{15\ \text{m}}{1.0\ \text{s}} = 15\ \text{m/s}.$$

c) Find the instantaneous velocity at time $t_1 = 1.0$ s by taking first $\Delta t = 0.1$ s, then $\Delta t = 0.01$ s, then $\Delta t = 0.001$ s.

When $\Delta t = 0.1$ s, $t_2 = 1.1$ s and

$$x_2 = 20\ \text{m} + (5.0\ \text{m/s}^2)(1.1\ \text{s})^2 = 26.05\ \text{m}.$$

The average velocity during this interval is

$$v_{\text{av}} = \frac{26.05\ \text{m} - 25\ \text{m}}{1.1\ \text{s} - 1.0\ \text{s}} = 10.5\ \text{m/s}.$$

We invite you to follow this same pattern to work out the average velocities for the 0.01-s and 0.001-s intervals. The results are 10.05 m/s and 10.005 m/s. As $\Delta t$ gets smaller and smaller, the average velocity gets closer and closer to 10.0 m/s, and we conclude that the instantaneous velocity at time $t = 1.0$ s is 10.0 m/s.

d) Derive a general expression for the instantaneous velocity as a function of time, and from it find $v$ at $t = 1.0$ s and $t = 2.0$ s.

We take the derivative of the expression for $x$. For any $n$ the derivative of $t^n$ is $nt^{n-1}$, so the derivative of $t^2$ is $2t$, and

$$v = \frac{dx}{dt} = (5.0\ \text{m/s}^2)(2t) = (10\ \text{m/s}^2)t.$$

At time $t = 1.0$ s, $v = 10$ m/s, as we found in part (c). At time $t = 2.0$ s, $v = 20$ m/s.

---

We use the term **speed** to denote distance traveled divided by time, on either an average or an instantaneous basis. Speed is a scalar quantity, not a vector; it has no direction. Instantaneous speed is the magnitude of instantaneous velocity, but average speed is *not* the magnitude of average velocity. When Matthew Biondi set his third world record of 1986 by swimming 100.0 m in 48.74 s, his average speed was (100.0 m)/(48.74 s) = 2.052 m/s. But because he swam two lengths in a 50-m pool, he started and ended at the same point, giving him zero total displacement and zero average *velocity* for his effort! The speedometer in a car indicates the car's instantaneous speed; the speedometer and a compass together indicate instantaneous velocity.

## 2–3 AVERAGE AND INSTANTANEOUS ACCELERATION

When the velocity of a moving body changes with time, we say that the body has an *acceleration*. Just as velocity describes the rate of change of position with time, acceleration describes the rate of change of velocity with time. Like velocity, acceleration is a vector quantity. In straight-line motion its only nonzero component is along the coordinate axis.

Let's consider again the motion of a particle (or a dragster) along the $x$-axis. Suppose at time $t_1$ the particle is at point $P_1$ and has $x$-component of (instantaneous) velocity $v_1$, and at a later time $t_2$ it is at point $P_2$ and has $x$-component of velocity $v_2$.

We define the **average acceleration** $\boldsymbol{a}_{\text{av}}$ of the particle, as it moves from $P_1$ to $P_2$, to be a vector quantity whose $x$-component is the ratio of the change in the $x$-component of velocity $\Delta v = v_2 - v_1$ to the time interval $\Delta t = t_2 - t_1$:

$$a_{\text{av}} = \frac{v_2 - v_1}{t_2 - t_1} = \frac{\Delta v}{\Delta t}. \qquad (2\text{–}5)$$

For straight-line motion we will usually call $a_{\text{av}}$ the average acceleration rather than the $x$-component of average acceleration.

## EXAMPLE 2–2

An astronaut has left the space shuttle to test a new personal maneuvering device (Fig. 2–6). As she moves along a straight line, her on-board partner measures her velocity before and after certain maneuvers, as follows: a) $v_1 = 0.8$ m/s, $v_2 = 1.2$ m/s (speed increases); b) $v_1 = 1.6$ m/s, $v_2 = 1.2$ m/s (speed decreases); c) $v_1 = -0.4$ m/s, $v_2 = -1.0$ m/s (speed increases); d) $v_1 = -1.6$ m/s, $v_2 = -0.8$ m/s (speed decreases). If the time interval for each maneuver is 2.0 s, find the average acceleration for each maneuver.

**SOLUTION** The various velocity changes are graphed in Fig. 2–7, starting at time $t = 1.0$ s and assuming 2.0-s intervals between successive maneuvers. The accelerations are as follows:

$$\text{a) } a_{av} = \frac{1.2 \text{ m/s} - 0.8 \text{ m/s}}{3.0 \text{ s} - 1.0 \text{ s}} = 0.2 \text{ m/s}^2;$$

$$\text{b) } a_{av} = \frac{1.2 \text{ m/s} - 1.6 \text{ m/s}}{7.0 \text{ s} - 5.0 \text{ s}} = -0.2 \text{ m/s}^2;$$

$$\text{c) } a_{av} = \frac{-1.0 \text{ m/s} - (-0.4 \text{ m/s})}{11.0 \text{ s} - 9.0 \text{ s}} = -0.3 \text{ m/s}^2;$$

$$\text{d) } a_{av} = \frac{-0.8 \text{ m/s} - (-1.6 \text{ m/s})}{15.0 \text{ s} - 13.0 \text{ s}} = 0.4 \text{ m/s}^2.$$

When the acceleration has the *same* direction as the initial velocity (a and c), the astronaut goes faster; when it is in the *opposite* direction (b and d), she slows down. When she moves in the negative direction with increasing speed (c), her velocity is algebraically decreasing (becoming more negative), and her acceleration is negative. But when she moves in the negative direction with decreasing speed (d), her velocity is increasing algebraically (becoming less negative), and her acceleration is positive. Be careful not to confuse acceleration with velocity. Velocity describes the astronaut's motion at any time; acceleration describes the change in her motion.

2–6 An astronaut testing a personal maneuvering device.

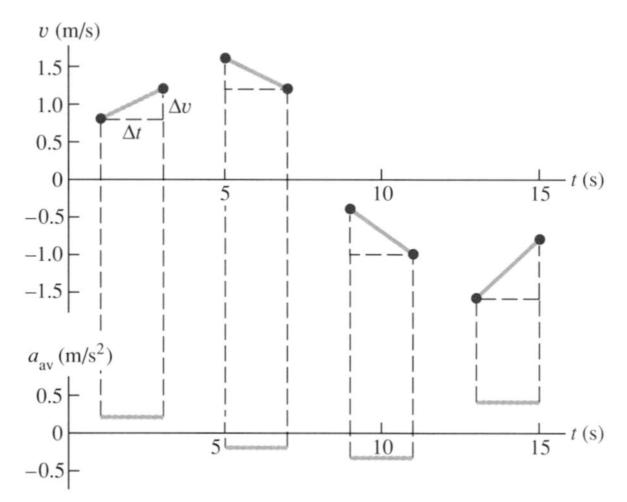


2–7 The slope of the line connecting two points on a graph of velocity versus time gives the average acceleration between those two points.

We can now define **instantaneous acceleration,** following the same procedure that we used for instantaneous velocity. Consider this situation: A sports car driver has just entered the final straightaway at the Grand Prix. He reaches point $P_1$ at time $t_1$, moving with velocity $v_1$. He passes point $P_2$, closer to the finish line, at time $t_2$ with velocity $v_2$ (Fig. 2–8).

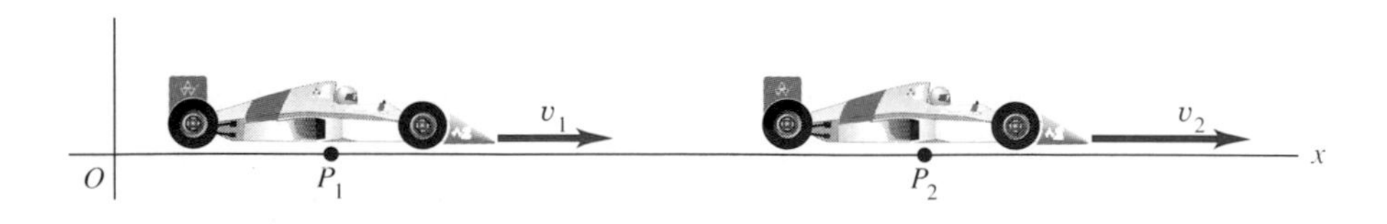


2–8 A Grand Prix car at two points on the straightaway.

To define instantaneous acceleration at point $P_1$, we take the second point $P_2$ in Fig. 2–8 to be closer and closer to the first point $P_1$, so that the average acceleration is computed over shorter and shorter time intervals. The instantaneous acceleration at point $P_1$ is the limit approached by the average acceleration as the time interval approaches zero. In the language of calculus the instantaneous acceleration is the *derivative* of the instantaneous velocity:

$$a = \lim_{\Delta t \to 0} \frac{\Delta v}{\Delta t} = \frac{dv}{dt}. \tag{2–6}$$

Instantaneous acceleration plays an essential role in the laws of mechanics. From now on, when we use the term "acceleration," we will always mean instantaneous acceleration, not average acceleration.

Instantaneous acceleration $a = dv/dt$ can be expressed in various ways. Because $v = dx/dt$, we can write

$$a = \frac{dv}{dt} = \frac{d}{dt}\left(\frac{dx}{dt}\right) = \frac{d^2x}{dt^2}. \tag{2–7}$$

That is, $a$ is the second derivative of $x$ with respect to $t$.

If we express velocity in meters per second and time in seconds, then acceleration is in meters per second per second, or (m/s)/s. This is usually written as $\text{m/s}^2$ and is read "meters per second squared."

## EXAMPLE 2–3

**Average and instantaneous accelerations** Suppose the velocity $v$ of the car in Fig. 2–8, at any time $t$, is given by the equation

$$v = 60\ \text{m/s} + (0.50\ \text{m/s}^3)t^2.$$

(Note that the numbers 60 and 0.50 *must* have the units shown for this equation to be dimensionally consistent.)

a) Find the change in velocity of the car in the time interval between $t_1 = 1.0$ s and $t_2 = 3.0$ s.

We first find the velocity at each time by substituting each value of $t$ into the equation. At time $t_1 = 1.0$ s,

$$v_1 = 60\ \text{m/s} + (0.50\ \text{m/s}^3)(1.0\ \text{s})^2 = 60.5\ \text{m/s}.$$

At time $t_2 = 3.0$ s,

$$v_2 = 60\ \text{m/s} + (0.50\ \text{m/s}^3)(3.0\ \text{s})^2 = 64.5\ \text{m/s}.$$

The change in velocity $\Delta v$ is

$$\Delta v = v_2 - v_1 = 64.5\ \text{m/s} - 60.5\ \text{m/s} = 4.0\ \text{m/s}.$$

The time interval is $\Delta t = 3.0\ \text{s} - 1.0\ \text{s} = 2.0$ s.

b) Find the average acceleration in this time interval.

$$a_{av} = \frac{v_2 - v_1}{t_2 - t_1} = \frac{4.0\ \text{m/s}}{2.0\ \text{s}} = 2.0\ \text{m/s}^2.$$

c) Find the instantaneous acceleration at time $t_1 = 1.0$ s by taking $\Delta t$ to be first 0.1 s, then 0.01 s, then 0.001 s.

When $\Delta t = 0.1$ s, $t_2 = 1.1$ s and

$$v_2 = 60\ \text{m/s} + (0.50\ \text{m/s}^3)(1.1\ \text{s})^2 = 60.605\ \text{m/s},$$

$$\Delta v = 0.105\ \text{m/s},$$

$$a_{av} = \frac{\Delta v}{\Delta t} = \frac{0.105\ \text{m/s}}{0.1\ \text{s}} = 1.05\ \text{m/s}^2.$$

We invite you to repeat this pattern for $\Delta t = 0.01$ s and $\Delta t = 0.001$ s; the results are $a_{av} = 1.005\ \text{m/s}^2$ and $a_{av} = 1.0005\ \text{m/s}^2$, respectively. As $\Delta t$ gets smaller and smaller, the average acceleration gets closer and closer to $1.0\ \text{m/s}^2$. We conclude that the instantaneous acceleration at $t = 2.0$ s is $1.0\ \text{m/s}^2$.

d) Derive an expression for the instantaneous acceleration at any time, and use it to find the acceleration at $t = 1.0$ s and $t = 3.0$ s.

We use the fact that $a = dv/dt$, along with the fact that the derivative of $t^2$ is $2t$, to obtain

$$a = \frac{dv}{dt} = \frac{d}{dt}[60\ \text{m/s} + (0.50\ \text{m/s}^3)t^2]$$
$$= (0.50\ \text{m/s}^3)(2t) = (1.0\ \text{m/s}^3)t.$$

When $t = 1.0$ s,

$$a = (1.0\ \text{m/s}^3)(1.0\ \text{s}) = 1.0\ \text{m/s}^2.$$

When $t = 3.0$ s,

$$a = (1.0\ \text{m/s}^3)(3.0\ \text{s}) = 3.0\ \text{m/s}^2.$$

Note that neither of these values is equal to the average acceleration found in part (b). The instantaneous acceleration varies with time.

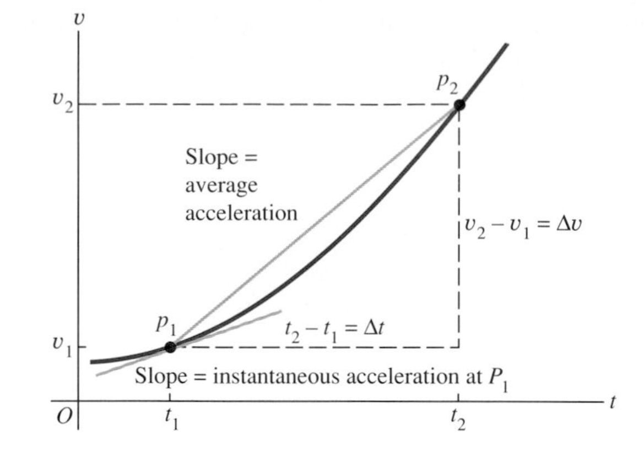


**2–9** Velocity-time graph of the motion in Fig. 2–8. The average acceleration between $t_1$ and $t_2$ equals the slope of the line $p_1p_2$. The instantaneous acceleration at $P_1$ equals the slope of the tangent at $p_1$.

We can get additional insight into the concepts of average and instantaneous acceleration by plotting a graph with velocity $v$ on the vertical axis and time $t$ on the horizontal axis (Fig. 2–9). The points on the graph corresponding to points $P_1$ and $P_2$ in Fig. 2–8 are labeled $p_1$ and $p_2$. The average acceleration during this interval is represented by the slope of the line $p_1p_2$, computed by using the appropriate scales and units of the graph. As point $P_2$ in Fig. 2–8 approaches point $P_1$, point $p_2$ in Fig. 2–9 approaches point $p_1$, and the slope of the line $p_1p_2$ approaches the slope of the line tangent to the curve at point $p_1$. Thus *the instantaneous acceleration at any point on the graph equals the slope of the line tangent to the curve at that point.* In Fig. 2–9 the instantaneous acceleration varies with time. Automotive engineers sometimes call the time rate of change of acceleration the "jerk."

A few remarks about the *sign* of acceleration may be helpful. First, the velocity and acceleration of a body don't necessarily have the same sign; they are independent quantities. When they do have the same sign, the body is speeding up. If both are positive, the body moves in the positive direction with increasing speed. If both are negative, the body moves in the negative direction with a velocity that is becoming more and more negative, again with increasing speed.

When $v$ and $a$ have opposite signs, the body is slowing down. If $v$ is positive and $a$ is negative, the body moves in the positive direction with decreasing speed. If $v$ is negative and $a$ is positive, the body moves in the negative direction with a velocity that is becoming less negative, and again the body slows down. Figure 2–10 illustrates these various possibilities.

The term *deceleration* is sometimes used for a decrease in speed. Because this may correspond to either positive or negative $a$, depending on the sign of $v$, we choose to avoid this term.

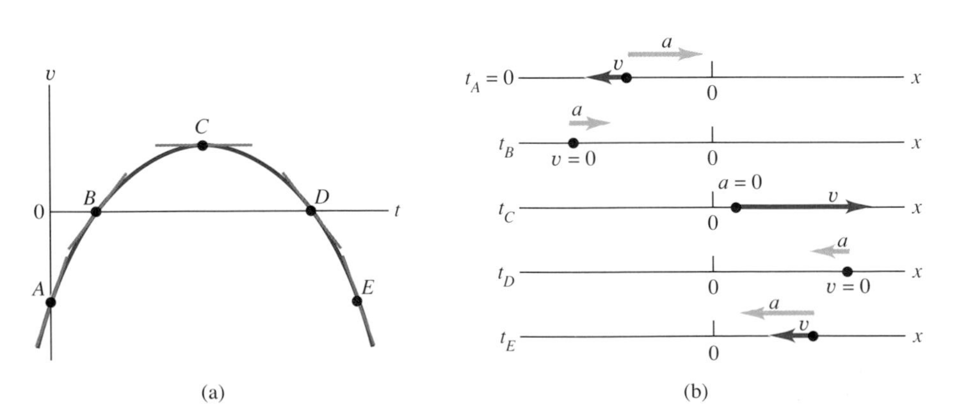


**2–10** (a) From $A$ to $B$, $v$ is negative but increasing, and the slope and acceleration are positive. The body slows down until $v = 0$ at time $B$. From $B$ to $C$, $v$ is positive and increasing, and the acceleration is positive. The body is speeding up. At point $C$ the velocity has its maximum value, and the acceleration is instantaneously zero. From $C$ to $D$ the velocity is positive but decreasing, and the acceleration is negative. The body slows down again until $v = 0$. From $D$ to $E$, $v$ is negative and decreasing, and the acceleration is negative. Once again, the body speeds up. At points $B$ and $D$ the velocity is zero, but the acceleration is not zero. (b) Position of the body at the times indicated on the $v$-$t$ graph.

## 2–4 MOTION WITH CONSTANT ACCELERATION

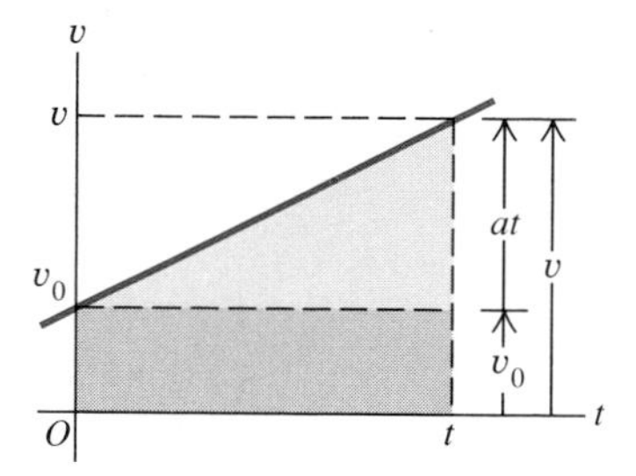


**2–11** A velocity-time graph for straight-line motion with constant positive acceleration.

The simplest accelerated motion is straight-line motion with *constant* acceleration; in this case the velocity changes at the same rate throughout the motion. The slope of the graph of velocity as a function of time is constant, and the curve is a straight line (Fig. 2–11). The velocity changes by equal amounts in equal time intervals. In this case it is easy to derive equations for position $x$ and velocity $v$ as functions of time. Let's start with velocity. In Eq. (2–5) we can replace the average acceleration $a_{av}$ by the constant (instantaneous) acceleration $a$. We then have

$$a = \frac{v_2 - v_1}{t_2 - t_1}. \tag{2–8}$$

Now we let $t_1 = 0$ and $t_2$ be any arbitrary later time $t$. We use the symbol $v_0$ for the velocity at the initial time $t = 0$; the velocity at the later time $t$ is $v$. Then Eq. (2–8) becomes

$$a = \frac{v - v_0}{t - 0}, \quad \text{or} \quad v = v_0 + at. \tag{2–9}$$

We can interpret this equation as follows: The acceleration $a$ is the constant rate of change of velocity, that is, the change in velocity per unit time. The term $at$ is the product of the change in velocity per unit time, $a$, and the time interval $t$. Therefore it equals the *total* change in velocity from the initial time $t = 0$ to the later time $t$. The velocity $v$ at any time $t$ then equals the initial velocity $v_0$ (at $t = 0$) plus the change in velocity $at$. Graphically, we can consider the height $v$ of the graph in Fig. 2–11 at any time $t$ as the sum of two segments: one with length $v_0$ equal to the initial velocity, the other with length $at$ equal to the change in velocity during time $t$.

Next we want to derive an equation for the position $x$ of a particle moving with constant acceleration. To do this we make use of two different expressions for the average velocity $v_{av}$ during the interval from $t = 0$ to any later time $t$. First, when the acceleration is constant and the velocity-time graph is a straight line, as in Fig. 2–11, the velocity changes at a uniform rate. In this particular case the average velocity throughout any time interval is the simple arithmetic average of the velocities at the beginning and the end of the interval. For the time interval 0 to $t$,

$$v_{av} = \frac{v_0 + v}{2}. \tag{2–10}$$

(In general, this is *not* true when the acceleration varies and the velocity-time graph is a curve, as in Fig. 2–10.) We also know that $v$, the velocity at any time $t$, is given by Eq. (2–9). Substituting that expression for $v$ into Eq. (2–10), we find

$$\begin{aligned} v_{av} &= \tfrac{1}{2}(v_0 + v_0 + at) \\ &= v_0 + \tfrac{1}{2}at. \end{aligned} \tag{2–11}$$

We can also get an expression for $v_{av}$ from its definition, Eq. (2–1). We call the position at time $t = 0$ the *initial position*, denoted by $x_0$. The position at the later time $t$ is simply $x$. Thus for the time interval $\Delta t = t - 0$ and the corresponding displacement $x - x_0$, Eq. (2–1) gives

$$v_{av} = \frac{x - x_0}{t}. \tag{2–12}$$

Finally, we equate Eqs. (2–11) and (2–12) and simplify the result:

$$v_0 + \frac{1}{2}at = \frac{x - x_0}{t}, \qquad \text{or}$$

$$x = x_0 + v_0 t + \tfrac{1}{2}at^2. \tag{2–13}$$

This equation states that if at the initial time $t = 0$ a particle is at position $x_0$ and has velocity $v_0$, its new position $x$ at any later time $t$ is the sum of three terms: its initial position $x_0$ plus the distance $v_0 t$ that it would move if its velocity were constant plus an additional distance $\frac{1}{2}at^2$ caused by the changing velocity.

We can check whether Eqs. (2–9) and (2–13) are consistent with the assumption of constant acceleration by taking the derivative of Eq. (2–13). We find

$$v = \frac{dx}{dt} = v_0 + at,$$

which is Eq. (2–9). Differentiating again, we find simply

$$\frac{dv}{dt} = a,$$

as we should expect.

We may also combine Eqs. (2–9) and (2–13) to obtain a relation for $x$, $v$, and $a$ that does not contain $t$. We first solve Eq. (2–9) for $t$ and then substitute the resulting expression into Eq. (2–13) and simplify:

$$t = \frac{v - v_0}{a},$$

$$x = x_0 + v_0\left(\frac{v - v_0}{a}\right) + \frac{1}{2}a\left(\frac{v - v_0}{a}\right)^2.$$

We transfer the term $x_0$ to the left side and multiply through by $2a$:

$$2a(x - x_0) = 2v_0 v - 2v_0^2 + v^2 - 2v_0 v + v_0^2.$$

Finally, simplifying gives us

$$v^2 = v_0^2 + 2a(x - x_0). \tag{2–14}$$

We can get one more useful relationship by equating the two expressions for $v_{av}$, Eqs. (2–10) and (2–12), and multiplying through by $t$. Doing this, we obtain

$$x - x_0 = \frac{v_0 + v}{2}t. \tag{2–15}$$

Note that Eq. (2–15) does not contain the acceleration $a$. This is sometimes a useful equation when $a$ is not given in the problem.

Equations (2–9), (2–13), (2–14), and (2–15) are the *equations of motion with constant acceleration*. Any kinematic problem involving motion of a particle on a straight line with constant acceleration can be solved by using these equations.

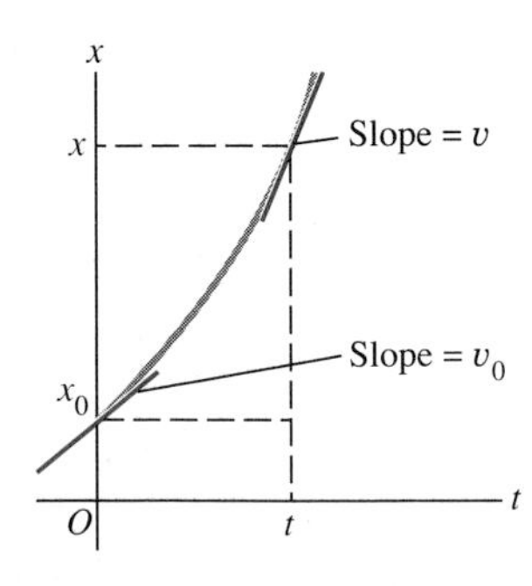


**2–12** Graph of coordinate $x$ as a function of time for straight-line motion with constant positive acceleration.

Figure 2–12 shows a graph of the coordinate $x$ as a function of time for motion with constant acceleration for a case in which $v_0$ and $a$ are positive. That is, it is a graph of Eq. (2–13); the curve is a *parabola*. The slope of the tangent at $t = 0$ equals the initial velocity $v_0$, and the slope of the tangent at any time $t$ equals the velocity $v$ at that time. The slope continuously increases with $t$ (because $a$ is positive), and measurements would show that the *rate* of increase with $t$ is constant.

## PROBLEM-SOLVING STRATEGY

### Motion with constant acceleration

**1.** You *must* decide at the beginning of a problem where the origin of coordinates is and which axis direction is positive. The choices are usually a matter of convenience; it is often easiest to place the particle at the origin at time $t = 0$; then $x_0 = 0$. A diagram showing these choices is always helpful. Then sketch some later positions of the particle on the same diagram.

**2.** Once you have chosen the positive axis direction, the positive directions for velocity and acceleration are also determined. It would be wrong to define $x$ as positive to the right of the origin and velocities as positive toward the left.

**3.** It often helps to restate the problem in prose first and then translate the prose description into symbols and equations. *When* does the particle arrive at a certain point (that is, at what value of $t$)? *Where* is the particle when its velocity has a certain value? (That is, what is the value of $x$ when $v$ has the specified value?) The following example asks "Where is the motorcyclist when his velocity is 25 m/s?" Translated into symbols, this becomes "What is the value of $x$ when $v = 25$ m/s?"

**4.** It always helps to make a list of quantities such as $x$, $x_0$, $v$, $v_0$, and $a$. In general, some of these will be known and some unknown. Write down those that are known. Be on the lookout for implicit information. "A car sits at a stoplight" usually means $v_0 = 0$, and so on. Once you have identified the unknowns, you may be able to choose an equation from Eqs. (2–9), (2–13), (2–14), and (2–15) that contains only one of the unknowns. Solve for the unknown; then substitute the known values and compute the value of the unknown. Sometimes you will have to solve two simultaneous equations for two unknowns.

**5.** Take a hard look at your results to see whether they make sense. Are they within the general range of magnitudes you expected?

## EXAMPLE 2–4

**Constant-acceleration calculations** A motorcyclist heading east through a small Iowa town accelerates after he passes the signpost at $x = 0$ marking the city limits (Fig. 2–13). His acceleration is constant, $a = 4.0\ \text{m/s}^2$. At time $t = 0$ he is 5.0 m east of the signpost and has a velocity of $v = 15$ m/s. a) Find his position and velocity at time $t = 2.0$ s. b) Where is the motorcyclist when his velocity is 25 m/s?

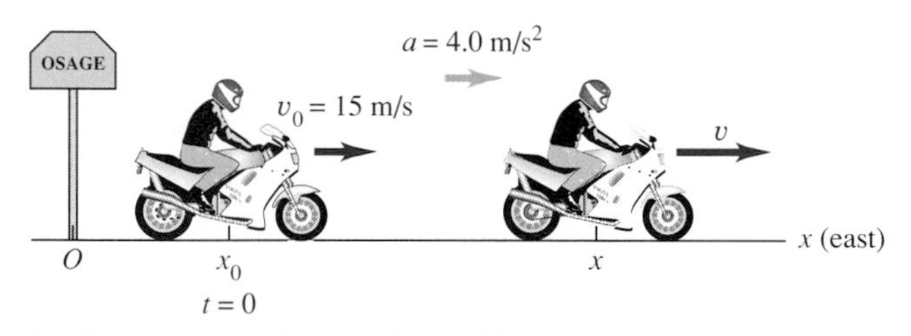


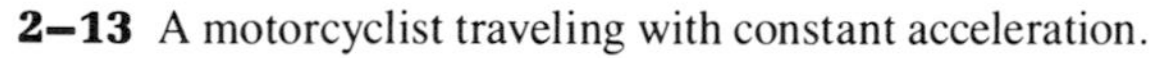
**2–13** A motorcyclist traveling with constant acceleration.

**SOLUTION** We take the signpost as the origin of coordinates, and the positive $x$-axis points east (Fig. 2–13). At the initial time $t = 0$ the initial position is $x_0 = 5.0$ m, the initial velocity is $v_0 = 15$ m/s, and the acceleration is $a = 4.0\ \text{m/s}^2$.

a) We want to know the position and velocity (the values of $x$ and $v$) at the later time $t = 2.0$ s. From Eq. (2–13), which gives position $x$ as a function of time $t$, we have

$$\begin{aligned} x &= x_0 + v_0 t + \tfrac{1}{2}at^2 \\ &= 5.0\ \text{m} + (15\ \text{m/s})(2.0\ \text{s}) + \tfrac{1}{2}(4.0\ \text{m/s}^2)(2.0\ \text{s})^2 \\ &= 43\ \text{m}. \end{aligned}$$

From Eq. (2–9), which gives velocity $v$ as a function of time $t$, we have

$$\begin{aligned} v &= v_0 + at \\ &= 15\ \text{m/s} + (4.0\ \text{m/s}^2)(2.0\ \text{s}) = 23\ \text{m/s}. \end{aligned}$$

b) We want to know the value of $x$ when $v = 25$ m/s. Note that this occurs at a time later than 2.0 s and at a point farther than 43 m from the signpost. From Eq. (2–14) we have

$$\begin{aligned} v^2 &= v_0^2 + 2a(x - x_0), \\ (25\ \text{m/s})^2 &= (15\ \text{m/s})^2 + 2(4.0\ \text{m/s}^2)(x - 5.0\ \text{m}), \\ x &= 55\ \text{m}. \end{aligned}$$

Alternatively, we may use Eq. (2–9) to find first the time when $v = 25$ m/s:

$$\begin{aligned} v &= v_0 + at, \\ 25\ \text{m/s} &= 15\ \text{m/s} + (4.0\ \text{m/s}^2)(t), \\ t &= 2.5\ \text{s}. \end{aligned}$$

Then from Eq. (2–13) we have

$$\begin{aligned} x &= x_0 + v_0 t + \tfrac{1}{2}at^2, \\ &= 5.0\ \text{m} + (15\ \text{m/s})(2.5\ \text{s}) + \tfrac{1}{2}(4.0\ \text{m/s}^2)(2.5\ \text{s})^2 \\ &= 55\ \text{m}. \end{aligned}$$

Finally, do the results make sense? The cyclist accelerates from 15 m/s (about 34 mi/h) to 23 m/s (about 51 mi/h) in 2.0 s while traveling a distance of 38 m (about 125 ft). This is pretty brisk acceleration, but well within the realm of possibility for a high-performance bike.

A special case of motion with constant acceleration occurs when the acceleration is zero. The *velocity* is then constant, and the equations of motion become simply

$$v = v_0 = \text{constant},$$
$$x = x_0 + vt.$$

## EXAMPLE 2–5

A motorist passes a school-crossing corner, where the speed limit is 10 m/s (about 22 mi/h), traveling with constant velocity of 15 m/s. A police officer on a motorcycle stopped at the corner starts off in pursuit with constant acceleration of 3.0 m/s$^2$ (Fig. 2–14a). a) How much time elapses before the officer catches up with the car? b) What is the officer's speed at that point? c) What is the total distance the officer has traveled at that point?

**SOLUTION** The motorcycle and the car both move with constant acceleration, so we can use the formulas we have developed. Take the origin at the corner, so $x_0 = 0$ for both. Let $x_P$ be the officer's position and $x_M$ the car's position at any time. Then, applying Eq. (2–13) to each, we find

$$x = x_0 + v_0 t + \tfrac{1}{2}at^2,$$
$$x_M = (15 \text{ m/s})t,$$
$$x_P = \tfrac{1}{2}(3.0 \text{ m/s}^2)t^2.$$

a) At the time the officer catches the car, they must be at the same position, so at this time, $x_M = x_P$. Equating the two expressions above, we have

$$(15 \text{ m/s})t = \tfrac{1}{2}(3.0 \text{ m/s}^2)t^2,$$

or

$$t = 0, 10 \text{ s}.$$

There are *two* times when the two vehicles have the same $x$-coordinate; the first is the time ($t = 0$) when the car passes the parked motorcycle at the corner, and the second is the time when the officer catches up.

b) From Eq. (2–9) we know that the officer's speed at any time is given by

$$v = v_0 + at = (3.0 \text{ m/s}^2)t,$$

so when $t = 10$ s, we find $v = 30$ m/s. When the officer overtakes the car, she is traveling twice as fast as the motorist is.

c) In 10 s the distance the car travels is

$$x_M = (15 \text{ m/s})(10 \text{ s}) = 150 \text{ m},$$

and the officer's distance traveled is

$$x_P = \tfrac{1}{2}(3.0 \text{ m/s}^2)(10 \text{ s})^2 = 150 \text{ m}.$$

This verifies that at the time the officer catches the car they have gone equal distances.

Figure 2–14b shows graphs of $x$ as a function of $t$ for each vehicle. We see again that there are two times when the two positions are the same. At neither of these times do the two vehicles have the same speed.

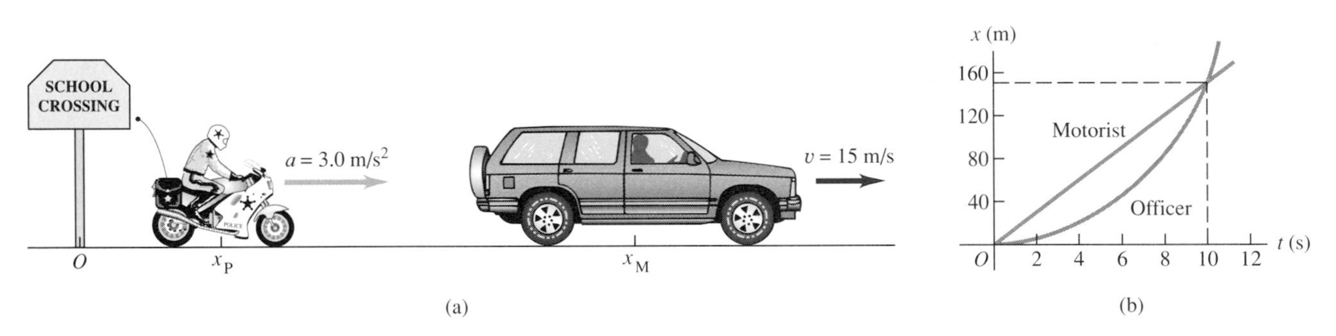


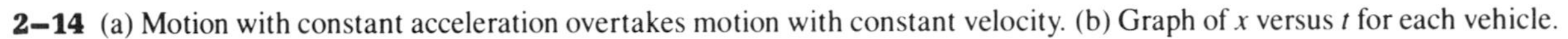
**2–14** (a) Motion with constant acceleration overtakes motion with constant velocity. (b) Graph of $x$ versus $t$ for each vehicle.

## 2–5 FREELY FALLING BODIES

The most familiar example of motion with (nearly) constant acceleration is that of a body falling under the influence of the earth's gravitational attraction. Such motion has held the attention of philosophers and scientists since ancient times. Aristotle thought (erroneously) that heavy objects fall faster than light objects, in proportion to their weight. Galileo argued that a body should fall with an acceleration that is constant and *independent* of its weight. We mentioned in Section 1–1 that, according to legend, Galileo experimented by dropping bullets and cannonballs from the Leaning Tower of Pisa.

The motion of falling bodies has since been studied with great precision. When air resistance can be neglected, Galileo is right; all bodies at a particular location fall with the same acceleration, regardless of their size or weight. If the distance of the fall is small compared to the radius of the earth, the acceleration is constant. In the following discussion we use an idealized model in which we neglect air resistance, the earth's rotation, and the decrease of acceleration with increasing altitude. We call this idealized motion **free fall,** although it includes rising as well as falling motion.

The constant acceleration of a freely falling body is called the **acceleration due to gravity,** or the *acceleration of free fall,* and we denote its magnitude with the letter $g$. At or near the earth's surface the value of $g$ is approximately 9.8 m/s$^2$, 980 cm/s$^2$, or 32 ft/s$^2$. Because $g$ is the magnitude of a vector quantity, it is always a *positive* number. On the surface of the moon the acceleration due to gravity is caused by the attractive force of the moon rather than the earth, and $g = 1.62$ m/s$^2$. Near the surface of the sun, $g = 274$ m/s$^2$.

In the following examples we use the constant-acceleration equations developed in Section 2–4. We suggest that you review the problem-solving strategies discussed in that section before you study these examples.

## EXAMPLE 2–6

A 500-lira coin is dropped from the Leaning Tower of Pisa; it starts from rest and falls freely. Compute its position and velocity after 1.0, 2.0, and 3.0 s.

**SOLUTION** Take the origin $O$ at the starting point, the coordinate axis vertical, and the upward direction as positive (Fig. 2–15). Because the coordinate axis is vertical, let's call the coordinate $y$ instead of $x$. We replace all the $x$'s in the constant-acceleration equations by $y$'s. The initial coordinate $y_0$ and the initial velocity $v_0$ are both zero. The acceleration is downward (in the negative $y$-direction), so $a = -g = -9.8$ m/s$^2$. (Remember that, by definition, $g$ itself is *always* positive.)

From Eqs. (2–13) and (2–9), with $x$ replaced by $y$, we get

$$y = v_0 t + \tfrac{1}{2}at^2 = 0 + \tfrac{1}{2}(-g)t^2 = (-4.9 \text{ m/s}^2)t^2,$$

$$v = v_0 + at = 0 + (-g)t = (-9.8 \text{ m/s}^2)t.$$

When $t = 1.0$ s, $y = (-4.9 \text{ m/s}^2)(1.0 \text{ s})^2 = -4.9$ m, and $v = (-9.8 \text{ m/s}^2)(1.0 \text{ s}) = -9.8$ m/s. The coin is therefore 4.9 m below the origin ($y$ is negative) and has a downward velocity ($v$ is negative) with magnitude 9.8 m/s.

The position and velocity at 2.0 and 3.0 s are found in the same way. The results are shown in Fig. 2–15; check the numerical values for yourself.

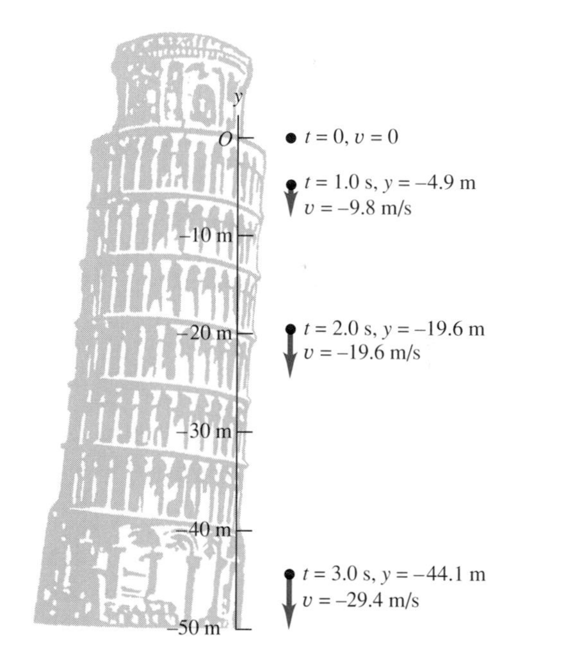


**2–15** Position and velocity of a coin freely falling from rest.

## EXAMPLE 2–7

Suppose you throw a ball vertically upward from the roof of a tall building. The ball leaves your hand at a point even with the roof railing, with an upward speed of 15.0 m/s. On its way back down, it just misses the railing. Find a) the position and velocity of the ball 1.00 s and 4.00 s after leaving your hand; b) the velocity when the ball is 5.00 m above the railing; c) the maximum height reached and the time at which it is reached.

**SOLUTION** In Fig. 2–16 the downward path is displaced a little to the right of its actual position for clarity. Take the origin at the roof railing, at the point where the ball leaves your hand, and take the positive direction to be upward. First, let's collect our data. The initial position $y_0$ is zero. The initial velocity $v_0$ is $+15.0$ m/s, and the acceleration is $a = -9.80\ \text{m/s}^2$.

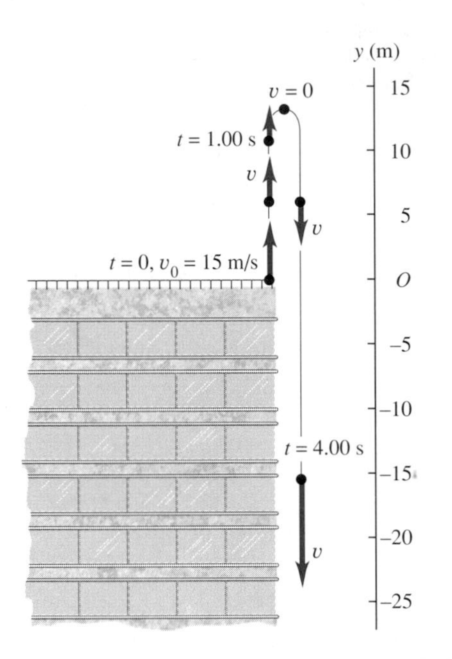


**2–16** The position and velocity of a ball thrown vertically upward.

Next, what equations do we have to work with? The velocity $v$ at any time $t$ is

$$v = v_0 + at = 15.0\ \text{m/s} + (-9.80\ \text{m/s}^2)t. \quad (2\text{–}16)$$

The position $y$ at any time $t$ is

$$y = v_0t + \tfrac{1}{2}at^2 = (15.0\ \text{m/s})t + \tfrac{1}{2}(-9.80\ \text{m/s}^2)t^2. \quad (2\text{–}17)$$

The velocity $v$ at any position $y$ is given by

$$v^2 = v_0^2 + 2ay = (15.0\ \text{m/s})^2 + 2(-9.80\ \text{m/s}^2)y. \quad (2\text{–}18)$$

a) When t = 1.00 s, Eqs. (2–16) and (2–17) give

$$y = +10.1\ \text{m}, \qquad v = +5.2\ \text{m/s}.$$

The ball is 10.1 m above the origin ($y$ is positive), and it has an upward velocity ($v$ is positive) of 5.2 m/s (less than the initial velocity, as expected).

When $t = 4.00$ s, again from Eqs. (2–16) and (2–17),

$$y = -18.4\ \text{m}, \qquad v = -24.2\ \text{m/s}.$$

The ball has passed its highest point and is 18.4 m *below* the origin ($y$ is negative). It has a *downward* velocity ($v$ is negative) with magnitude 24.2 m/s (greater than the initial velocity, as we should expect). Note that to get these results, we don't need to find the highest point reached or the time at which it was reached. The equations of motion give the position and velocity at *any* time, whether the ball is on the way up or the way down.

b) When the ball is 5.0 m above the origin,

$$y = +5.0\ \text{m},$$

and, from Eq. (2–18),

$$v^2 = (15.0\ \text{m/s})^2 + 2(-9.80\ \text{m/s}^2)(5.00\ \text{m}) = 127\ \text{m}^2/\text{s}^2,$$

$$v = \pm 11.3\ \text{m/s}.$$

We get two values of $v$, positive and negative. The ball passes this point *twice*, once on the way up and again on the way down. The velocity on the way up is $+11.3$ m/s, and on the way down it is $-11.3$ m/s.

c) At the highest point the ball stops going up (positive $v$) and starts going down (negative $v$). Just at the instant it reaches the highest point, $v = 0$. From Eq. (2–18),

$$0 = (15.0\ \text{m/s})^2 - (19.6\ \text{m/s}^2)y,$$

and the maximum height (where $v = 0$) is given by

$$y = 11.5\ \text{m}.$$

We can now find the time $t_1$ at the highest point, from Eq. (2–16), setting $v = 0$:

$$0 = 15.0\ \text{m/s} + (-9.80\ \text{m/s}^2)t_1,$$

$$t_1 = 1.53\ \text{s}.$$

Alternatively, to find the maximum height, we may ask first *when* the maximum height is reached. That is, at what $t$ is $v = 0$? As we just found, $v = 0$ when $t = 1.53$ s. Substituting this value of $t$ back into Eq. (2–17), we find

$$y = (15\ \text{m/s})(1.53\ \text{s}) + \tfrac{1}{2}(-9.8\ \text{m/s}^2)(1.53\ \text{s})^2 = 11.5\ \text{m}.$$

Although at the highest point the velocity is instantaneously zero, the *acceleration* at this point is still $-9.8\ \text{m/s}^2$. The ball stops for an instant, but its velocity is continuously changing, from positive values through zero to negative values. The acceleration is constant throughout.

Figure 2–17 shows graphs of position and velocity as functions of time for this problem. Note that the graph of $v$ versus $t$ has a constant negative slope. The acceleration is negative on the way up, at the highest point, and on the way down.

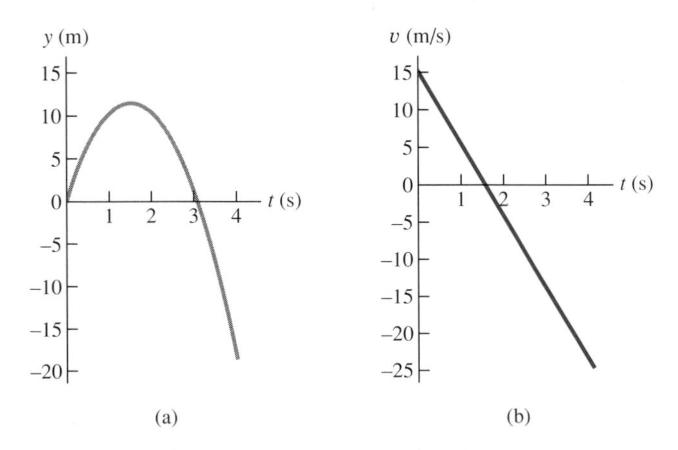


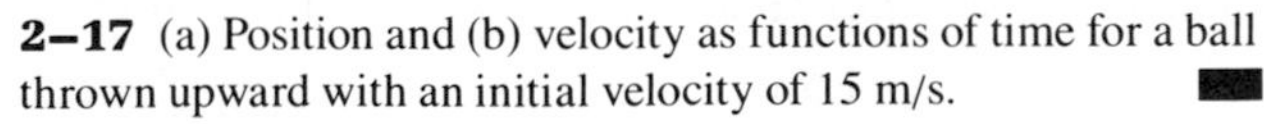
**2–17** (a) Position and (b) velocity as functions of time for a ball thrown upward with an initial velocity of 15 m/s.

Figure 2–18 is a multiflash photograph of a falling ball taken with a stroboscopic light source that produces a series of intense flashes at equal time intervals. Each flash is so short (a few millionths of a second) that there is no blur in the image of even a rapidly moving body. As each flash occurs, the position of the ball at that instant is recorded on the film. Because of the equal time intervals between flashes, the average velocity of the ball between any two flashes is proportional to the distance between corresponding images in the photograph. The increasing distances between images show that the velocity is continuously increasing; the ball accelerates. Careful measurement shows that the velocity change is the same in each time interval; the acceleration is constant.

**2–18** Multiflash photo of a freely falling ball.

## *2–6 VELOCITY AND COORDINATE BY INTEGRATION

This optional section is intended for students who have already learned a little integral calculus. In Section 2–4 we analyzed the special case of straight-line motion with constant acceleration. When $a$ is *not* constant, the equations that we derived in that section are no longer valid. But even when $a$ varies with time, we can still use the relation $v = dx/dt$ to find the velocity $v$ as a function of time if the position $x$ is a given function of time. And we can still use $a = dv/dt$ to find the acceleration $a$ as a function of time if the velocity $v$ is a given function of time.

We can also reverse this process. Suppose we know $v$ as a function of time. How can we find $x$ as a function of time? Let's first consider a graphical approach. Figure 2–19 shows a velocity-versus-time curve for a body whose acceleration (the slope of the curve) is not constant but increases with time. We can divide the time interval between times $t_1$ and $t_2$ into many smaller intervals, calling a typical one $\Delta t$. Let the average velocity during $\Delta t$ be $v_{av}$. From Eq. (2–3) the displacement $\Delta x$ during $\Delta t$ is

$$\Delta x = v_{av}\Delta t.$$

This corresponds graphically to the area of the shaded strip with height $v_{av}$ and width $\Delta t$, that is, the area under the curve between the two sides of $\Delta t$. The total displacement during any interval (say, $t_1$ to $t_2$) is the sum of the displacements $\Delta x$ in the small subintervals, so the total displacement is represented graphically by the *total* area under the curve between the vertical lines $t_1$ and $t_2$. In the limit, when all the $\Delta t$'s become very small and their number very large, each $v_{av}$ approaches the instantaneous velocity $v$ at each point. This limit, the area under the curve, is the *integral* of $v$ (which is in general a function of $t$) from $t_1$ to $t_2$. If $x_1$ is the position of the body at time $t_1$ and $x_2$ the position at time $t_2$, then

$$x_2 - x_1 = \int_{x_1}^{x_2} dx = \int_{t_1}^{t_2} v\,dt. \qquad (2\text{–}19)$$

We can carry out exactly the same procedure with the acceleration-versus-time curve, where $a$ is in general a function of $t$. If $v_1$ is a body's velocity at time $t_1$ and $v_2$ is its velocity at time $t_2$, the change in velocity $\Delta v$ during a small time interval $\Delta t$ is approximately equal to $a_{av}\,\Delta t$, and the total change in velocity $(v_2 - v_1)$ during the interval $t_2 - t_1$ is given by

$$v_2 - v_1 = \int_{v_1}^{v_2} dv = \int_{t_1}^{t_2} a\,dt. \qquad (2\text{–}20)$$

If $t_1 = 0$ and $t_2$ is any later time $t$, and if $x_0$ and $v_0$ are the position and velocity, respectively, at time $t = 0$, then we can rewrite Eqs. (2–19) and (2–20) as follows:

$$x = x_0 + \int_0^t v\,dt \qquad (2\text{–}21)$$

$$v = v_0 + \int_0^t a\,dt. \qquad (2\text{–}22)$$

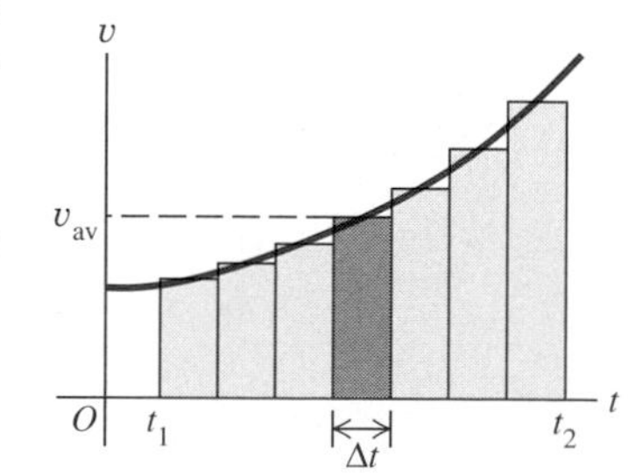


**2–19** The area under a velocity-time graph between times $t_1$ and $t_2$ equals the displacement $x_2 - x_1$ between those times.

### EXAMPLE 2–8

You are driving along a straight highway in your new Mustang. At time $t = 0$, when you are moving at 10 m/s, you pass a signpost at $x = 50$ m. Your acceleration is a function of time:

$$a = 2.0\ \text{m/s}^2 - (0.10\ \text{m/s}^3)t.$$

a) Derive expressions for your velocity and position as functions of time. b) At what time is your velocity greatest? c) What is the maximum velocity? d) Where is the car when it reaches maximum velocity?

**SOLUTION** a) The position at time $t = 0$ is $x_0 = 50$ m, and the velocity at time $t = 0$ is $v_0 = 10$ m/s. First we use Eq. (2–22) to find the velocity $v$ at time $t$:

$$\begin{aligned} v &= 10\ \text{m/s} + \int_0^t [2.0\ \text{m/s}^2 - (0.10\ \text{m/s}^3)t]\,dt \\ &= 10\ \text{m/s} + (2.0\ \text{m/s}^2)t - \tfrac{1}{2}(0.10\ \text{m/s}^3)t^2. \end{aligned}$$

Then we use Eq. (2–21) to find $x$ as a function of $t$:

$$\begin{aligned} x &= 50\ \text{m} + \int_0^t [10\ \text{m/s} + (2.0\ \text{m/s}^2)t - \tfrac{1}{2}(0.10\ \text{m/s}^3)t^2]dt \\ &= 50\ \text{m} + (10\ \text{m/s})t + \tfrac{1}{2}(2.0\ \text{m/s}^2)t^2 - \tfrac{1}{6}(0.10\ \text{m/s}^3)t^3. \end{aligned}$$

b) The maximum value of $v$ occurs when $v$ stops increasing and begins to decrease. At this instant, $dv/dt = 0$. Taking the derivative of the above expression for $v$ and setting it equal to zero, we obtain

$$0 = 2.0\ \text{m/s}^2 - (0.10\ \text{m/s}^3)t,$$

$$t = \frac{2.0\ \text{m/s}^2}{0.10\ \text{m/s}^3} = 20\ \text{s}.$$

We can also get this result directly from the expression for $a$ by noting that $a$ is positive between $t = 0$ and $t = 20$ s and negative after that. It is zero at $t = 20$ s, corresponding to the fact that $dv/dt = 0$ at this time. Thus the car speeds up until $t = 20$ s and then begins to slow down again.

c) We find the maximum velocity by substituting $t = 20$ s (the time when the maximum velocity is attained) into the general velocity equation:

$$\begin{aligned} v_{max} &= 10\ \text{m/s} + (2.0\ \text{m/s}^2)(20\ \text{s}) - \tfrac{1}{2}(0.10\ \text{m/s}^3)(20\ \text{s})^2 \\ &= 30\ \text{m/s}. \end{aligned}$$

d) The maximum value of $v$ occurs at time $t = 20$ s. We obtain the position of the car (value of $x$) at that time by substituting $t = 20$ s into the general expression for $x$:

$$\begin{aligned} x &= 50\ \text{m} + (10\ \text{m/s})(20\ \text{s}) + \tfrac{1}{2}(2.0\ \text{m/s}^2)(20\ \text{s})^2 \\ &\quad - \tfrac{1}{6}(0.10\ \text{m/s}^3)(20\ \text{s})^3 \\ &= 517\ \text{m}. \end{aligned}$$

Figure 2–20 shows graphs of $x$, $v$, and $a$ as functions of time.

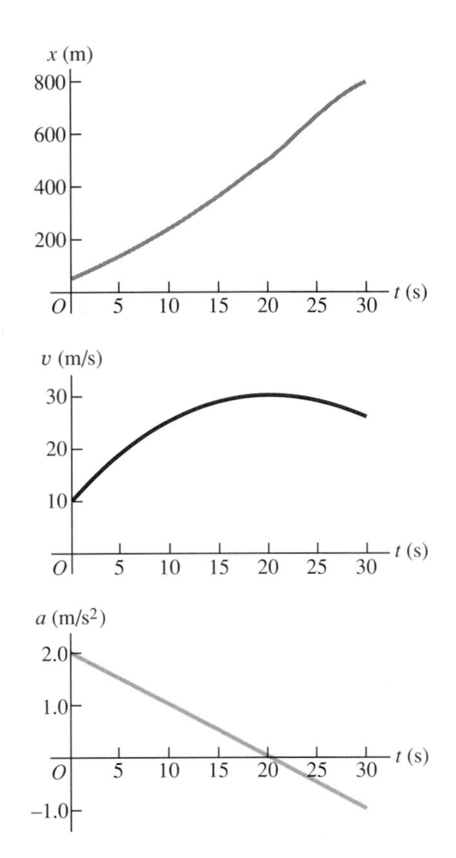


**2–20** The position, velocity, and acceleration of the car in Example 2–8 as functions of time. If this motion continues, can you show that the car will stop at $t = 44.5$ s?

## 2–7 RELATIVE VELOCITY ALONG A STRAIGHT LINE

When two observers measure the velocity of a moving body, they get different results if one observer is moving relative to the other. The velocity seen by a particular observer is called the velocity *relative* to that observer, or simply **relative velocity.** Here is an example of this concept in action. A woman walks with a velocity of 1.0 m/s along the aisle of a train that is moving with a velocity of 3.0 m/s (Fig. 2–21a). What is the woman's velocity?

It's a simple enough question, but it has no single answer. As seen by a passenger sitting in the train, she is moving at 1.0 m/s. A person on a bicycle standing beside the train sees the woman moving at 1.0 m/s + 3.0 m/s = 4.0 m/s. An observer in another

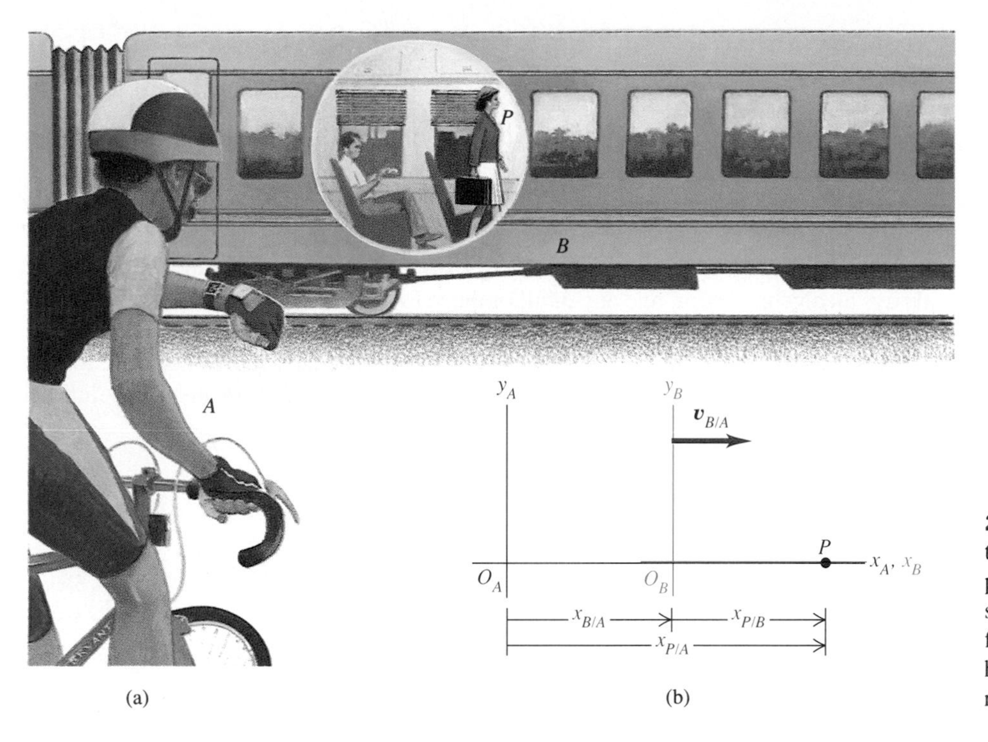


**2–21** (a) A woman walking in a train. (b) At the instant shown, the position of the woman, represented as particle $P$, relative to frame of reference $A$ differs from her position relative to frame of reference $B$.

train going in the opposite direction would give still another answer. The velocity is different for different observers. We have to specify which observer we mean, and we speak of the velocity *relative* to a particular observer. The woman's velocity relative to the train is 1.0 m/s, her velocity relative to the cyclist is 4.0 m/s, and so on. Each observer, equipped in principle with a meter stick and a stopwatch, forms what we call a **frame of reference.** Thus a frame of reference is a coordinate system plus a timer.

Let's call the cyclist's frame of reference (at rest with respect to the ground) $A$ and the frame of reference of the moving train $B$ (Fig. 2–21b). In straight-line motion the position of a point $P$ relative to frame of reference $A$ is given by the distance $x_{P/A}$ (the position of $P$ with respect to $A$), and the position relative to frame $B$ is given by $x_{P/B}$. The distance from the origin of $A$ to the origin of $B$ (position of $B$ with respect to $A$) is $x_{B/A}$. We can see from the figure that

$$x_{P/A} = x_{P/B} + x_{B/A}. \qquad (2\text{–}23)$$

This says that the total distance from the origin of $A$ to point $P$ is the distance from the origin of $B$ to point $P$ plus the distance from the origin of $A$ to the origin of $B$.

The velocity of $P$ relative to frame $A$, denoted by $v_{P/A}$, is the derivative of $x_{P/A}$ with respect to time, and the other velocities are similarly obtained. So the time derivative of Eq. (2–23) gives us a relationship among the various velocities:

$$\frac{dx_{P/A}}{dt} = \frac{dx_{P/B}}{dt} + \frac{dx_{B/A}}{dt},$$

or

$$v_{P/A} = v_{P/B} + v_{B/A}. \qquad (2\text{–}24)$$

Getting back to the woman on the train, $A$ is the cyclist's frame of reference, $B$ is the frame of reference of the train, and point $P$ represents the woman. Using the above notation, we have

$$v_{P/B} = 1.0\text{ m/s}, \qquad v_{B/A} = 3.0\text{ m/s}.$$

From Eq. (2–24) the woman's velocity $v_{P/A}$ relative to the cyclist is

$$v_{P/A} = 1.0\text{ m/s} + 3.0\text{ m/s} = 4.0\text{ m/s},$$

as we already knew.

In this example both velocities are toward the right, and we have implicitly taken this as the positive direction. If the woman walks toward the *left* relative to the train, then $v_{P/B} = -1.0$ m/s, and her velocity relative to the cyclist is 2.0 m/s. The sum in Eq. (2–24) is always an algebraic sum, and any or all of the velocities may be negative.

When the woman looks out the window, the stationary cyclist on the ground appears to her to be moving backward; we can call the cyclist's velocity relative to her $v_{A/P}$. Clearly, this is just the negative of $v_{P/A}$. In general, if $A$ and $B$ are any two points or frames of reference,

$$v_{A/B} = -v_{B/A}. \tag{2–25}$$

## PROBLEM-SOLVING STRATEGY

### Relative velocity

Note the order of the double subscripts on the velocities above; $v_{A/B}$ always means "velocity of $A$ relative to $B$." These subscripts obey an interesting kind of algebra, as Eq. (2–24) shows. If we regard each one as a fraction, then the fraction on the left side is the *product* of the fractions on the right sides: $P/A = (P/B)(B/A)$. This is a handy rule to use when you apply Eq. (2–24). If there are *three* different frames of reference $A$, $B$, and $C$, we can write immediately

$$v_{P/A} = v_{P/C} + v_{C/B} + v_{B/A},$$

and so on.

## EXAMPLE 2–9

You are driving north on a straight two-lane road at a constant 88 km/h. A truck traveling at a constant 104 km/h approaches you (in the other lane, fortunately). a) What is the truck's velocity relative to you? b) What is your velocity with respect to the truck?

**SOLUTION** Let you be $Y$, the truck be $T$, the earth be $E$, and let the positive direction be north (Fig. 2–22). Then $v_{Y/E} = +88$ km/h.

a) The truck is approaching you, so it must be moving south, giving $v_{T/E} = -104$ km/h. We want to find $v_{T/Y}$. Transcribing Eq. (2–24), we have

$$v_{T/E} = v_{T/Y} + v_{Y/E},$$
$$v_{T/Y} = v_{T/E} - v_{Y/E}$$
$$= -104\text{ km/h} - 88\text{ km/h} = -192\text{ km/h}.$$

The truck is moving 192 km/h south relative to you.

b) From Eq. (2–25),

$$v_{Y/T} = -v_{T/Y} = -(-192\text{ km/h}) = +192\text{ km/h}.$$

You are moving 192 km/h north relative to the truck.

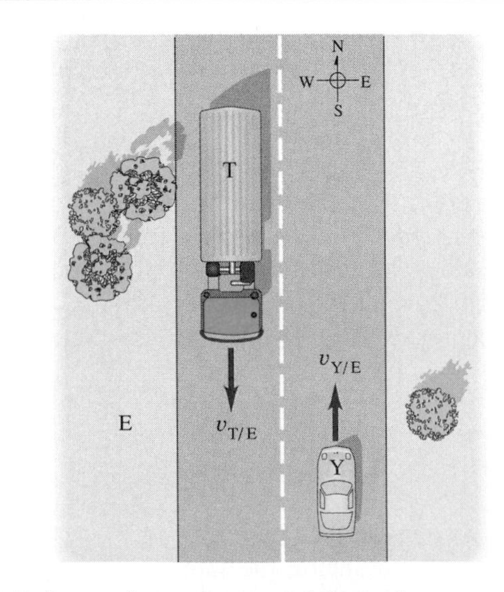


**2–22** Reference frames for you and the truck.

## EXAMPLE 2–10

How do the relative velocities of Example 2–9 change after you and the truck have passed?

**SOLUTION** They don't change at all. The relative positions of the bodies don't matter. The velocity of the truck relative to you is still $-192$ km/h, but it is now moving away from you instead of toward you.

When we derived the relative-velocity relations, we assumed that all the observers use the same time scale. This may seem so obvious that it isn't even worth mentioning, but this is precisely the point at which Einstein's special theory of relativity departs from the physics of Galileo and Newton. When any of the speeds approach the speed of light, denoted by $c$, the velocity-addition equation has to be modified. It turns out that if the woman could walk down the aisle at $0.30c$ and the train could move at $0.90c$, then her speed relative to the ground would be not $1.20c$ but $0.94c$. Nothing can travel faster than light; we'll return to the special theory of relativity later in the book.

## SUMMARY

- When a particle moves along a straight line, we describe its position with respect to an origin $O$ by means of a coordinate such as $x$.
- The particle's average velocity during a time interval $\Delta t = t_2 - t_1$ is defined as

$$v_{av} = \frac{x_2 - x_1}{t_2 - t_1} = \frac{\Delta x}{\Delta t}. \quad (2\text{–}3)$$

The instantaneous velocity at any time $t$ is defined as

$$v = \lim_{\Delta t \to 0} \frac{\Delta x}{\Delta t} = \frac{dx}{dt}. \quad (2\text{–}4)$$

- The average acceleration during a time interval $\Delta t = t_2 - t_1$ is defined as

$$a_{av} = \frac{v_2 - v_1}{t_2 - t_1} = \frac{\Delta v}{\Delta t}. \quad (2\text{–}5)$$

The instantaneous acceleration at any time $t$ is defined as

$$a = \lim_{\Delta t \to 0} \frac{\Delta v}{\Delta t} = \frac{dv}{dt}. \quad (2\text{–}6)$$

- When the acceleration is constant, the position $x$ and velocity $v$ at any time $t$ are related to the acceleration $a$, the initial position $x_0$, and the initial velocity $v_0$ (both at time $t = 0$) by the following equations:

$$x = x_0 + v_0 t + \tfrac{1}{2}at^2, \quad (2\text{–}13)$$

$$v = v_0 + at, \quad (2\text{–}9)$$

$$v^2 = v_0^2 + 2a(x - x_0), \quad (2\text{–}14)$$

$$x - x_0 = \frac{v_0 + v}{2}t. \quad (2\text{–}15)$$

- When the acceleration is not constant but is a known function of time, we can find the velocity and coordinate as functions of time by integrating the acceleration function.
- When a body $P$ moves relative to a body (or reference frame) $B$, and $B$ moves relative to $A$, we denote the velocity of $P$ relative to $B$ by $v_{P/B}$, the velocity of $P$ relative to $A$ by $v_{P/A}$, and the velocity of $B$ relative to $A$ by $v_{B/A}$. These velocities are related by

$$v_{P/A} = v_{P/B} + v_{B/A} \quad (2\text{–}24)$$

### KEY TERMS

average velocity
instantaneous velocity
derivative
speed
average acceleration
instantaneous acceleration
free fall
acceleration due to gravity
relative velocity
frame of reference

# EXERCISES

## Section 2–1 Average Velocity

**2–1** A rocket carrying a satellite is accelerating straight up from the earth's surface. The rocket clears the top of its launch platform, 47 m above ground, 1.35 s after lift-off. After an additional 4.45 s it is 1.00 km above the ground. Calculate the magnitude of the average velocity of the rocket for a) the 4.45-s part of its flight; b) the first 5.80 s of its flight.

**2–2** A hiker travels in a straight line for 40 min with an average velocity of magnitude 1.4 m/s. What distance does she cover in this time?

**2–3** You normally drive on the freeway between San Francisco and Sacramento at an average speed of 96 km/h (60 mi/h), and the trip takes 2 h and 10 min. On a rainy day you slow down and drive the same distance at an average speed of 72 km/h (45 mi/h). How much longer does the trip take?

**2–4 The World's Fastest Man.** On August 30, 1987, in Rome, Ben Johnson ran the 100-m dash in 9.83 s. He ran the 50.0-m to 70.0-m stretch in 1.70 s. What was the magnitude of his average velocity for a) the whole race? b) the 50.0- to 70.0-m segment?

**2–5** A car is traveling in a straight line along a road. Its distance $x$ from a stop sign is given as a function of time $t$ by the equation $x = \alpha t^2 + \beta t^3$, where $\alpha = 1.80\ \text{m/s}^2$ and $\beta = 0.250\ \text{m/s}^3$. Calculate the average velocity of the car for the following time intervals: a) $t = 0$ to $t = 2.00$ s; b) $t = 0$ to $t = 4.00$ s; c) $t = 2.00$ s to $t = 4.00$ s.

## Section 2–2 Instantaneous Velocity

**2–6** A physics professor leaves her house and walks along the sidewalk toward campus. After 5 min it starts to rain, and she returns home. Her distance from her house as a function of time is shown in Fig. 2–23. At which of the labeled points is her velocity a) zero? b) constant and positive? c) constant and negative? d) increasing in magnitude? e) decreasing in magnitude?

**2–7** A hobbyist is testing a new model rocket engine by using it to propel a cart along a model railroad track. He determines that its motion along the $x$-axis is described by the equation $x = bt^2$, where $b = 12.0\ \text{cm/s}^2$. Compute the instantaneous velocity of the cart at time $t = 3.00$ s.

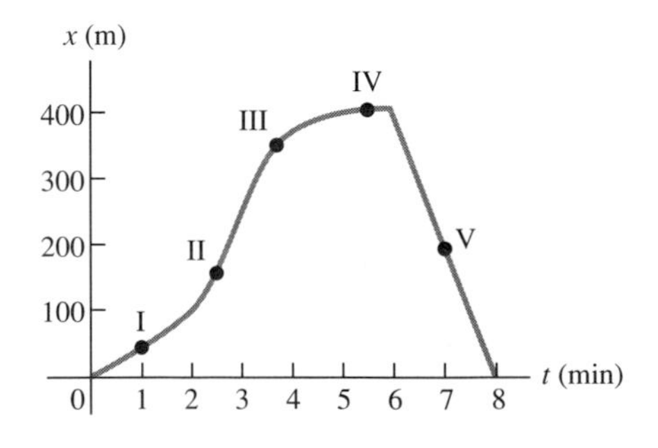

**FIGURE 2–23**

**2–8** A car is stopped at a traffic light. It then travels along a straight road so that its distance from the light is given by $x = bt^2 + ct^3$, where $b = 1.40\ \text{m/s}^2$ and $c = 0.100\ \text{m/s}^3$. a) Calculate the average velocity of the car for the time interval $t = 0$ to $t = 10.0$ s. b) Calculate the instantaneous velocity of the car at i) $t = 0$; ii) $t = 5.0$ s; iii) $t = 10.0$ s.

## Section 2–3 Average and Instantaneous Acceleration

**2–9** A truck is traveling on a straight stretch of highway at 120 km/h (75 mi/h). The driver spots a highway patrol car ahead. If the truck slows with an average acceleration of $-4.0\ \text{m/s}^2$, how much time does it take to reduce its speed to the legal limit of 96 km/h (60 mi/h)?

**2–10** Figure 2–24 shows the velocity of a solar-powered car as a function of time. The driver accelerates from a stop sign, cruises for 20 s at a constant speed of 60 km/h, and then brakes to come to a stop 40 s after leaving the stop sign. a) Compute the average acceleration for each of the following time intervals: i) $t = 0$ to $t = 10$ s; ii) $t = 30$ s to $t = 40$ s; iii) $t = 10$ s to $t = 30$ s; iv) $t = 0$ to $t = 40$ s. b) At what time in the motion does the instantaneous acceleration $a$ have its largest value? c) What is the instantaneous acceleration at $t = 20$ s? d) What is the instantaneous acceleration at $t = 35$ s?

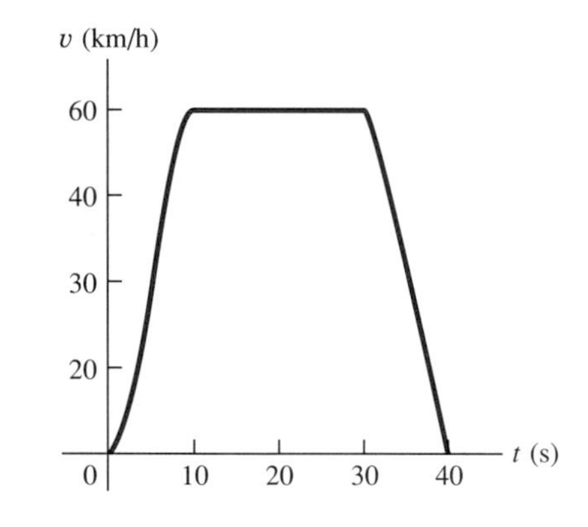

**FIGURE 2–24**

**2–11** A test driver at Incredible Motors, Inc. is testing a new model of car having a speedometer calibrated to read m/s rather than mi/h. The following series of speedometer readings was obtained during a test run.

| **Time (s)** | 0 | 2 | 4 | 6 | 8 | 10 | 12 | 14 | 16 |
|---|---|---|---|---|---|---|---|---|---|
| **Velocity (m/s)** | 0 | 0 | 2 | 5 | 10 | 15 | 20 | 22 | 22 |

a) Compute the average acceleration during each 2-s interval. Is the acceleration constant? Is it constant during any part of the time? b) Make a velocity-time graph of the data, using scales of 1 cm = 1 s horizontally and 1 cm = 2 m/s vertically. Draw a smooth curve through the plotted points. By measuring the slope of your curve, find the instantaneous acceleration at $t = 8$ s, 13 s, and 15 s.

**2–12** An astronaut has left Spacelab V to test a new space scooter for use in constructing Space Habitat I. Her partner measures the following velocity changes, each taking place in a 10-s interval. What are the magnitude, the algebraic sign, and the direction of the average acceleration in each interval? Assume that the positive direction is to the right. a) At the beginning of the interval the astronaut is moving toward the right along the $x$-axis at 25.0 m/s, and at the end of the interval she is moving toward the right at 5.0 m/s. b) At the beginning she is moving toward the left at 5.0 m/s, and at the end she is moving toward the left at 25.0 m/s. c) At the beginning she is moving toward the right at 25.0 m/s, and at the end she is moving toward the left at 25.0 m/s.

**2–13** Figure 2–25 is a graph of the coordinate of a spider crawling along the $x$-axis. Sketch the graphs of its velocity and acceleration as functions of time.

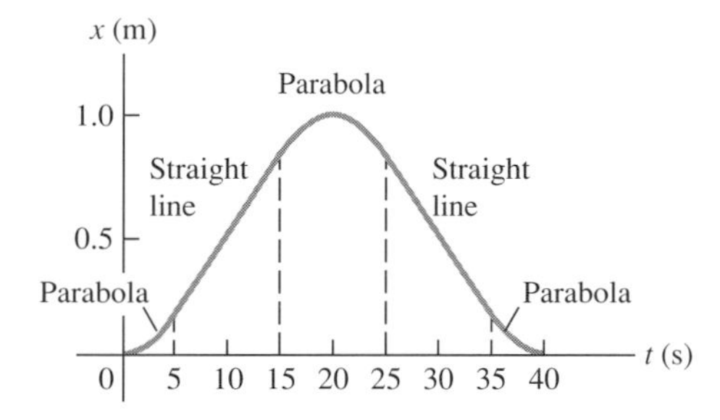


**FIGURE 2–25**

**2–14** A car's velocity as a function of time is given by $v(t) = \alpha + \beta t^2$, where $\alpha = 5.00$ m/s and $\beta = 0.200$ m/s$^3$. a) Calculate the average acceleration for the time interval $t = 0$ to $t = 5.00$ s. b) Calculate the instantaneous acceleration for i) $t = 0$; ii) $t = 5.00$ s.

## Section 2–4 Motion with Constant Acceleration

**2–15** An antelope moving with constant acceleration covers the distance between two points 80 m apart in 6.00 s. Its speed as it passes the second point is 15.0 m/s. a) What is its speed at the first point? b) What is the acceleration?

**2–16** An airplane travels 480 m down the runway before taking off. If it starts from rest, moves with constant acceleration, and becomes airborne in 16.0 s, what is its speed, in m/s, when it takes off?

**2–17** The graph in Fig. 2–26 shows the velocity of a motorcycle police officer plotted as a function of time. a) Find the instantaneous acceleration at $t = 3$ s, at $t = 7$ s, and at $t = 11$ s. b) How far does the officer go in the first 5 s? The first 9 s? The first 13 s?

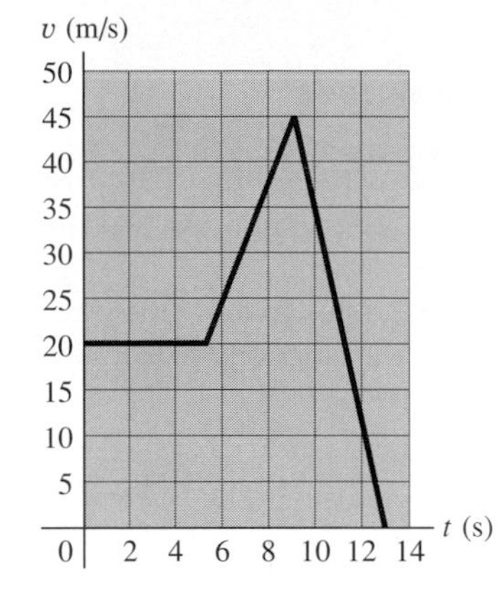


**FIGURE 2–26**

**2–18** Figure 2–27 is a graph of the acceleration of a model railroad locomotive moving on the $x$-axis. Sketch the graphs of its velocity and coordinate as functions of time if $x = 0$ and $v = 0$ when $t = 0$.

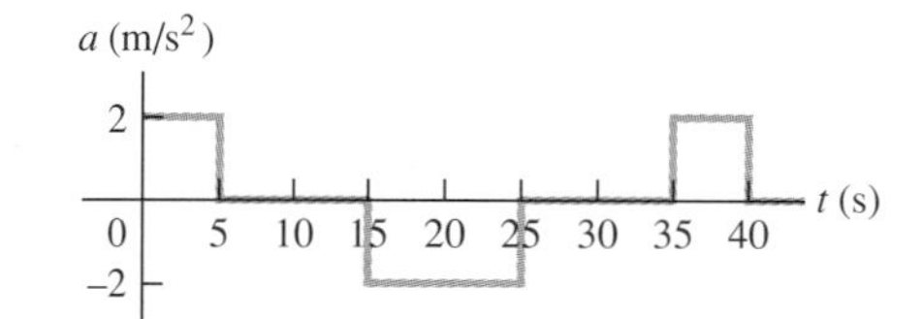

**FIGURE 2–27**

**2–19** The reaction time of the average automobile driver is about 0.7 s. (The reaction time is the interval between the perception of a signal to stop and the application of the brakes.) If an automobile can slow down with an acceleration of $-4.0$ m/s$^2$, compute the total distance covered in coming to a stop after a signal is observed a) from an initial speed of 6.7 m/s (15 mi/h) in a school zone; b) from an initial speed of 24.6 m/s (55 mi/h).

**2–20** A subway train starts from rest at a station and accelerates at a rate of 1.80 m/s$^2$ for 12.0 s. It runs at constant speed for 30.0 s and slows down at $-3.50$ m/s$^2$ until it stops at the next station. Find the *total* distance covered.

**2–21** A spaceship ferrying workers to Moon Base I takes a straight-line path from the earth to the moon, a distance of about 400,000 km. Suppose it accelerates at 15.0 m/s$^2$ for the first 10.0 min of the trip, then travels at constant speed until the last 10.0 min, when it accelerates at $-15.0$ m/s$^2$, just coming to rest as it reaches the moon. a) What is the maximum speed attained? b) What fraction of the total distance is traveled at constant speed? c) What total time is required for the trip?

## Section 2–5 Freely Falling Bodies

**2–22 Touch-down on the Moon.** A lunar lander is making its descent to Moon Base I (Fig. 2–28). The lander descends slowly under the retro-thrust of its descent engine. The engine is cut off when the lander is 5.0 m above the landing pad and has a downward speed of 2.0 m/s. What is the speed of the lander just before it touches the surface? The acceleration due to gravity on the moon is 1.6 $m/s^2$.

**2–23** a) If a flea can jump to a height of 0.640 m, what is its initial speed as it leaves the ground? b) For how much time is it in the air?

**2–24** A brick is dropped (zero initial speed) from the roof of a building. The brick strikes the ground in 4.90 s. a) How tall, in meters, is the building? b) What is the magnitude of the brick's velocity just before it reaches the ground?

**2–25** A student throws a water balloon vertically downward from the top of a building. The balloon leaves the thrower's hand with a speed of 12.0 m/s. a) What is its speed after falling for 2.00 s? b) How far does it fall in 2.00 s? c) What is the magnitude of its velocity after falling 10.0 m?

**2–26** A rock is thrown vertically upward with a speed of 12.0 m/s from the roof of a building that is 40.0 m above the ground. a) In how many seconds after being thrown does the rock strike the ground? b) What is the speed of the rock just before it strikes the ground?

**2–27** A rock is thrown straight up with an initial speed of 8.00 m/s from the roof of a building, 12.0 m above the ground. For the motion from the roof to the ground, what are the magnitude and direction of a) the average velocity of the rock? b) the average acceleration of the rock?

**2–28** A hot-air balloonist, rising vertically with a constant velocity of magnitude 5.00 m/s, releases a sandbag at an instant when the balloon is 40.0 m above the ground (Fig. 2–29). a) Compute the position and velocity of the sandbag at 0.500 s and at 2.00 s after its release. b) How many seconds after its release will the bag strike the ground? c) With what magnitude of velocity does it strike?

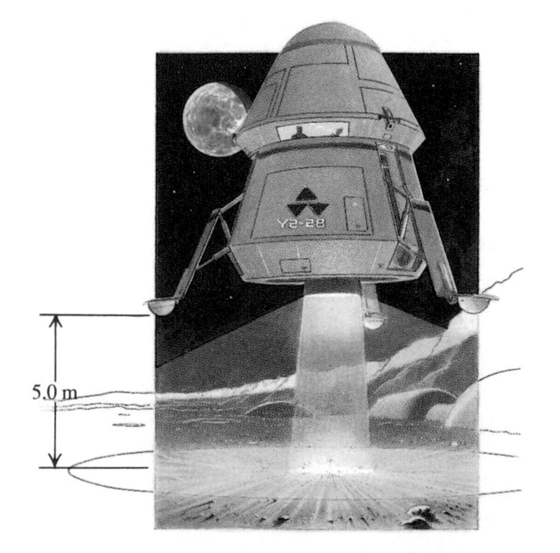


**FIGURE 2–28**

**FIGURE 2–29**

**2–29** An egg is thrown nearly vertically upward from a point near the cornice of a tall building. It just misses the cornice on the way down and passes a point 50.0 m below its starting point 7.00 s after it leaves the thrower's hand. a) What is the initial speed of the egg? b) How high does it rise above its starting point? c) What is the magnitude of its velocity at the highest point? d) What are the magnitude and direction of its acceleration at the highest point?

**2–30** A ball is thrown vertically upward from the ground with a velocity of magnitude 45.0 m/s. a) At what time after being thrown does the ball have a velocity of 15.0 m/s upward? b) At what time does it have a velocity of 15.0 m/s downward? c) When is the displacement of the ball zero? d) When is the velocity of the ball zero? e) What are the magnitude and direction of the acceleration while the ball is i) moving upward? ii) moving downward? iii) at the highest point?

**2–31** The rocket-driven sled Sonic Wind No. 2, used for investigating the physiological effects of large accelerations, runs on a straight, level track 1070 m (3500 ft) long. Starting from rest, it can reach a speed of 447 m/s (1000 mi/h) in 1.80 s. a) Compute the acceleration in $m/s^2$, assuming that it is constant. b) What is the ratio of this acceleration to that of a freely falling body ($g$)? c) What is the distance covered in 1.80 s? d) A magazine article states that at the end of a certain run, the speed of the sled was decreased from 283 m/s (632 mi/h) to zero in 1.40 s and that during this time its passenger was subjected to more than 40 times the pull of gravity (that is, the acceleration's magnitude was greater than $40g$). Are these figures consistent?

**2–32** Suppose the acceleration of gravity were only 2.0 $m/s^2$, instead of 9.8 $m/s^2$. a) Estimate the height to which you could jump vertically from a standing start if you can jump to 0.75 m when $g = 9.8\ m/s^2$. b) How high could you throw a baseball if you can throw it 18 m up when $g = 9.8\ m/s^2$? c) Estimate the maximum height of a window from which you would care to jump to a concrete sidewalk below if for $g = 9.8\ m/s^2$ you are willing to jump from a height of 2.0 m, which is the typical height of a first-story window.

## Section 2–6 Velocity and Coordinate by Integration

***2–33** The acceleration of a motorcycle is given by $a = At - Bt^2$, where $A = 1.90\ m/s^3$ and $B = 0.120\ m/s^4$. It is at rest at the origin at time $t = 0$. a) Find its position and velocity as functions of time. b) Calculate the maximum velocity it attains.

***2–34** The acceleration of a bus is given by $a = \alpha t$, where $\alpha = 3.0\ m/s^3$. a) If the bus's velocity at time $t = 1.0$ s is 5.0 m/s, what is its velocity at time $t = 2.0$ s? b) If the bus's position at time $t = 1.0$ s is 6.0 m, what is its position at time $t = 2.0$ s?

***2–35** Equations (2–21) and (2–22) can be applied to the case in which the acceleration $a$ is constant. For constant $a$, use Eq. (2–22) to calculate $v(t)$. Then use this $v(t)$ in Eq. (2–21) to calculate $x(t)$. How do your results compare to Eqs. (2–9) and (2–13)?

## Section 2–7 Relative Velocity along a Straight Line

**2–36** A railroad flatcar is traveling to the right at a speed of 13.0 m/s relative to an observer standing on the ground. A motor scooter is being ridden on the flatcar (Fig. 2–30). What is the velocity (magnitude and direction) of the motor scooter relative to the flatcar if its velocity relative to the observer on the ground is a) 18.0 m/s to the right? b) 1.5 m/s to the left? c) zero?

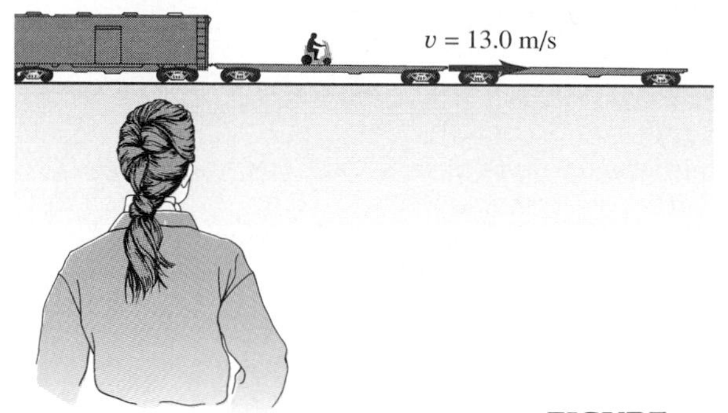


FIGURE 2–30

**2–37** A "moving sidewalk" in an airport terminal building moves 1.0 m/s and is 80.0 m long. If a woman steps on at one end and walks 2.0 m/s relative to the moving sidewalk, how much time does she require to reach the opposite end if she walks a) in the same direction the sidewalk is moving? b) in the opposite direction?

**2–38** Two piers $A$ and $B$ are located on a river: $B$ is 1500 m downstream from $A$ (Fig. 2–31). Two friends must make round trips from pier $A$ to pier $B$ and return. One rows a boat at a constant speed of 6.00 km/h relative to the water; the other walks on the shore at a constant speed of 6.00 km/h. The velocity of the river is 3.20 km/h in the direction from $A$ to $B$. How much time does it take each person to make the round trip?

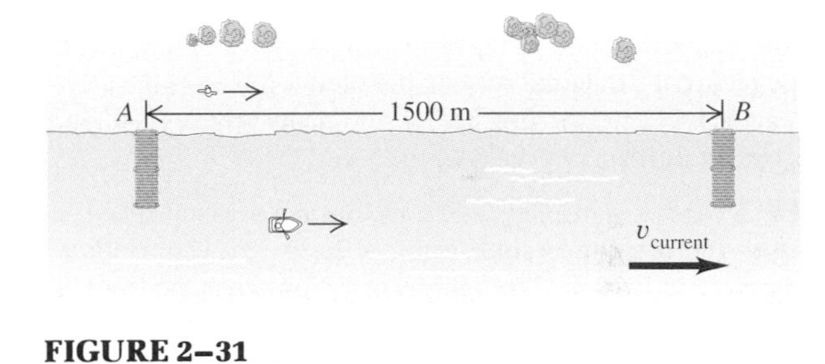


FIGURE 2–31

## PROBLEMS

**2–39** The motion of a particle along a straight line is described by the function $x = \alpha + \beta t^2 - \gamma t^4$, where $\alpha = 6.00$ m, $\beta = 4.80\ m/s^2$, and $\gamma = 1.30\ m/s^4$. a) Find the position, velocity, and acceleration at time $t = 2.00$ s. b) Sketch graphs of the position, velocity, and acceleration of the particle as functions of time, for $t = 0$ to $t = 2.00$ s. c) What is the maximum positive velocity attained by the particle? At what time is it reached?

**2–40** In a relay race each contestant runs 20.0 m while carrying an egg balanced on a spoon, turns around, and comes back to the starting point. Edith runs the first 20.0 m in 12.0 s. On the return trip she is more confident and takes only 8.0 s. What is the magnitude of her average velocity for a) the first 20.0 m? b) the return trip? c) What is her average velocity for the entire round trip? d) What is her average speed for the entire round trip?

**2–41** A typical world-class sprinter accelerates to maximum speed in 4.0 s. If such a runner finishes a 100-m race in 9.1 s, what is his average acceleration during the first 4.0 s of the race?

**2–42 Automobile Airbags.** The human body can survive a negative acceleration trauma incident (sudden stop) if the magnitude of the acceleration is less than 250 $m/s^2$ (approximately $25g$). If you are in an automobile accident at an initial speed of 96 km/h (60 mi/h) and are stopped by an airbag that inflates from the dashboard, over what distance must the airbag stop you for you to survive the crash?

**2–43** Dan gets on Interstate Highway I-80 at Seward, Nebraska and drives due west in a straight line and at an average velocity of magnitude 88 km/h. After traveling 76 km he reaches the Aurora

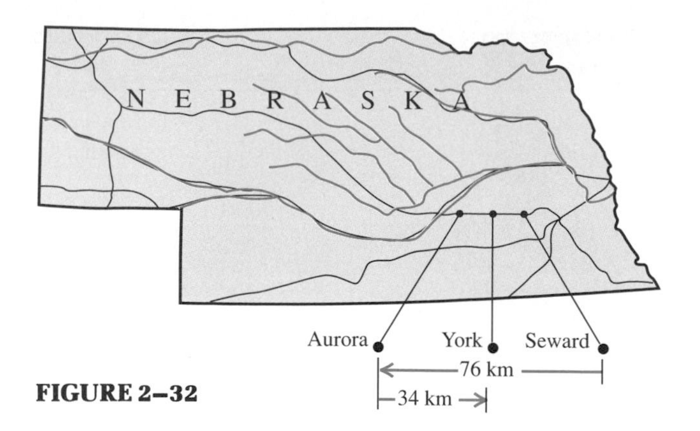


**FIGURE 2–32**

exit (Fig. 2–32). Realizing that he has gone too far, he turns around and drives due east 34 km back to the York exit at an average velocity of magnitude 79 km/h. For his whole trip from Seward to the York exit, what is a) his average speed? b) the magnitude of his average velocity?

**2–44 Freeway Traffic.** According to a *Scientific American* article (May 1990), current freeways can sustain about 2400 vehicles per lane per hour in smooth traffic flow at 96 km/h (60 mi/h). Above that figure the traffic flow becomes "turbulent" (stop-and-go). a) If a vehicle is 4.6 m (15 ft) long on the average, what is the average spacing between vehicles at the above traffic density? b) Collision-avoidance automated control systems, which operate by bouncing radar or sonar signals off surrounding vehicles and then accelerate or brake the car when necessary, could greatly reduce the required spacing between vehicles. If the average spacing is 9.2 m (two car lengths), how many vehicles per hour can a lane of traffic carry at 96 km/h?

**2–45** A car 3.5 m in length and traveling at a constant speed of 20 m/s is approaching an intersection (Fig. 2–33). The width of the intersection is 20 m. The light turns yellow when the front of the car is 50 m from the beginning of the intersection. If the driver steps on the brake, the car will slow at $-4.2\ \text{m/s}^2$. If the driver instead steps on the gas pedal, the car will accelerate at $1.5\ \text{m/s}^2$. The light will be yellow for 3.0 s. Ignore the reaction time of the driver. To avoid being in the intersection while the light is red, should the driver hit the brake pedal or the gas pedal?

**2–46** A sled starts from rest at the top of a hill and slides down with a constant acceleration. At some later time it is 32.0 m from the top; 2.00 s after that it is 50.0 m from the top; 2.00 s later it is 72.0 m from the top; and 2.00 s later it is 98.0 m from the top. a) What is the magnitude of the average velocity of the sled during each of the 2-s intervals after passing the 32.0-m point? b) What is the acceleration of the sled? c) What is the speed of the sled when it passes the 32.0-m point? d) How much time did it take to go from the top to the 32.0-m point? e) How far did the sled go during the first second after passing the 32.0-m point?

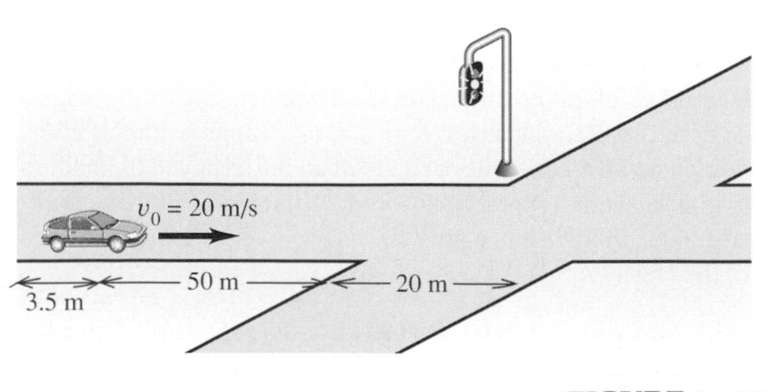


**FIGURE 2–33**

**2–47** At the instant the traffic light turns green, an automobile that has been waiting at an intersection starts ahead with a constant acceleration of $2.00\ \text{m/s}^2$. At the same instant a truck, traveling with a constant speed of 14.0 m/s, overtakes and passes the automobile. a) How far beyond its starting point does the automobile overtake the truck? b) How fast is the automobile traveling when it overtakes the truck? c) On the same graph, sketch the position of each vehicle as a function of time. Take $x = 0$ at the intersection.

**2–48** The engineer of a passenger train traveling at 25.0 m/s sights a freight train whose caboose is 200 m ahead on the same track (Fig. 2–34). The freight train is traveling in the same direction as the passenger train with a speed of 15.0 m/s. The engineer of the passenger train immediately applies the brakes, causing a constant acceleration of $-0.100\ \text{m/s}^2$, while the freight train continues with constant speed. a) Will the cows witness a collision? b) If so, where will it take place? c) On the same graph, sketch the position of the front of each train as a function of time. Take $x = 0$ at the initial location of the passenger train.

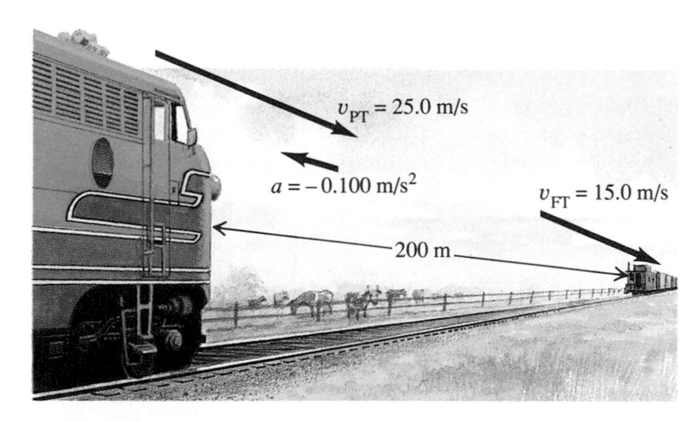


**FIGURE 2–34**

**2–49** An automobile and a truck start from rest at the same instant, with the automobile initially at some distance behind the truck. The truck has a constant acceleration of $2.20\ \text{m/s}^2$, and the automobile an acceleration of $3.50\ \text{m/s}^2$. The automobile overtakes the truck after the truck has moved 75.0 m. a) How much time does it take the automobile to overtake the truck? b) How far was the automobile behind the truck initially? c) What is the speed of each when they are abreast? d) Sketch on the same graph the position of each vehicle as a function of time. Take $x = 0$ at the initial location of the truck.

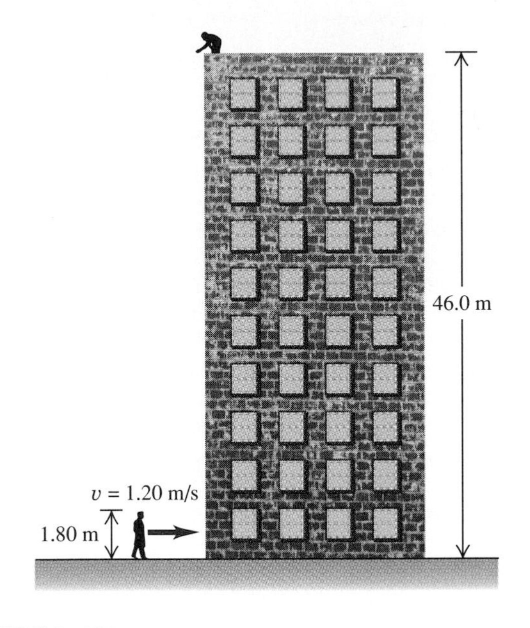

FIGURE 2–35

✲**2–50** An object's velocity is measured to be $v(t) = \alpha - \beta t^2$, where $\alpha = 6.00$ m/s and $\beta = 2.00$ m/s$^3$. At $t = 0$ the object is at $x = 0$. a) Calculate the object's position and acceleration as functions of time. b) What is the object's maximum *positive* displacement from the origin?

**2–51** You are on the roof of the physics building, 46.0 m above the ground (Fig. 2–35). Your physics professor, who is 1.80 m tall, is walking toward the building at a constant speed of 1.20 m/s. If you wish to drop an egg on your professor's head, where should the professor be when you release the egg?

**2–52 A Simple Reaction-Time Test.** A meter stick is held vertically above your hand, with the lower end between your thumb and first finger. On seeing the meter stick released, you grab it with these two fingers. Your reaction time can be calculated from the distance the meter stick falls, read directly from the point where your fingers grabbed it. a) Derive a relationship for your reaction time in terms of this measured distance, $d$. b) If the measured distance is 22.8 cm, what is the reaction time?

**2–53** Visitors at Great Adventure, an amusement park, are able to watch divers step off a platform 24.4 m (80 ft) above a pool of water. According to the announcer, the divers enter the water at a speed of 65 mi/h (105 km/h). a) Is the announcer correct in this claim? b) Is it possible for a diver to leap directly upward off the board so that, missing the board on the way down, she enters the water at 105 km/h? If so, what initial upward velocity is required? Is the required initial speed physically attainable?

**2–54** A student determined to test the law of gravity for himself walks off a skyscraper 250 m high, stopwatch in hand, and starts his free fall (zero initial velocity). Five seconds later, Superman arrives at the scene and dives off the roof to save the student. a) What must the magnitude of Superman's initial velocity be so that he catches the student just before the ground is reached? (Assume that Superman's acceleration is that of any freely falling body.) b) If the height of the skyscraper is less than some minimum value, even Superman can't reach the student before he hits the ground. What is this minimum height?

**2–55** A football is kicked vertically upward from the ground, and a student gazing out of the window sees it moving upward past her at 5.00 m/s. The window is 12.0 m above the ground. a) How high does the football go above ground? b) How much time does it take to go from the ground to its highest point?

**2–56** A flowerpot falls off a windowsill and falls past the window below. It takes 0.420 s to pass this window, which is 1.90 m high. How far is the top of this window below the windowsill above?

**2–57** The driver of a car wishes to pass a truck that is traveling at a constant speed of 20.0 m/s (about 45 mi/h). The car initially is also traveling at 20.0 m/s. The car's maximum acceleration in this speed range is 0.800 m/s$^2$. Initially, the vehicles are separated by 25.0 m, and the car pulls back into the truck's lane after it is 25.0 m ahead of the truck. The car is 5.0 m long, and the truck is 20.0 m long. The car's acceleration is a constant 0.800 m/s$^2$. a) How much time is required for the car to pass the truck? b) What distance does the car travel during this time? c) What is the final speed of the car?

**2–58** A juggler performs in a room whose ceiling is 4.50 m above the level of his hands. He throws an apple vertically upward so that it just reaches the ceiling. a) With what initial speed does he throw the apple? b) What time is required for the apple to reach the ceiling? He throws an orange upward with the same initial speed at the instant that the apple is at the ceiling. c) How long after the orange is thrown do the two objects pass each other? d) When the objects pass each other, how far are they above the juggler's hands?

**2–59 The Carrier Pigeon Problem.** Larry is driving east at 40 km/h. His twin brother Harry is driving west at 30 km/h toward Larry in an identical car on the same straight road. When they are 35 km apart, Larry sends out a carrier pigeon, which flies at a constant speed of 50 km/h. (All speeds are relative to the earth.) The pigeon flies to Harry, gets confused and immediately returns to Larry, gets more confused and immediately flies back to Harry. This continues until the twins meet, at which time the dazed pigeon drops to the ground in exhaustion. Ignoring turnaround time, how far did the pigeon fly?

**2–60** Two cars, $A$ and $B$, travel in a straight line. The distance of $A$ from the starting point is given as a function of time by $x_A = \alpha t + \beta t^2$, with $\alpha = 4.00$ m/s and $\beta = 1.20$ m/s$^2$. The distance of $B$ from the starting point is $x_B = \gamma t^2 + \delta t^3$, with $\gamma = 1.60$ m/s$^2$ and $\delta = 1.20$ m/s$^3$. a) Which car is ahead just after they leave the starting point? b) At what values of $t$ are the cars at the same point? c) At what value of $t$ is the velocity of $B$ relative to $A$ zero? d) At what value of $t$ is the distance from $A$ to $B$ neither increasing nor decreasing?

## CHALLENGE PROBLEMS

**2–61** As is implied by Eqs. (2–4) and (2–6), the definition of the derivative $df/dt$ of the function $f(t)$ is

$$\frac{df}{dt} = \lim_{\Delta t \to 0} \frac{\Delta f}{\Delta t},$$

where $\Delta f = f(t + \Delta t) - f(t)$. Use this definition to prove the following: a) $(d/dt)t^n = nt^{n-1}$. (*Hint:* Use the binomial theorem, which may be found in Appendix B.) b) $(d/dt)[af(t)] = a(df/dt)$. c) $(d/dt)[f(t) + g(t)] = (df/dt) + (dg/dt)$.

**2–62** An alert hiker sees a boulder fall from the top of a distant cliff and notes that it takes 1.50 s for the boulder to fall the last third of the way to the ground. a) What is the height of the cliff in meters? b) If in part (a) you get two roots to a quadratic equation and use one for your answer, what does the other root represent?

**2–63** A student is running to catch the campus shuttle bus, which is stopped at the bus stop. The student is running at a constant speed of 6.0 m/s; she can't run any faster. When the student is still 60.0 m from the bus, it starts to pull away. The bus moves with a constant acceleration of 0.180 $m/s^2$. a) For how much time and what distance does the student have to run before she overtakes the bus? b) When she reaches the bus, how fast is the bus traveling? c) Sketch a graph showing $x(t)$ for both the student and the bus. Take $x = 0$ as the initial position of the student. d) The equations that you used in part (a) to find the time have a second solution, corresponding to a later time for which the student and bus are again at the same place if they continue their specified motions. Explain the significance of this second solution. How fast is the bus traveling at this point? e) If the student's constant speed is 4.0 m/s, will she catch the bus? f) What is the *minimum* speed the student must have to just catch up with the bus? For what time and what distance does she have to run in that case?

**2–64** A ball is thrown straight up from the edge of the roof of a building. A second ball is dropped from the roof 2.00 s later. a) If the height of the building is 60.0 m, what must be the initial speed of the first ball if both are to hit the ground at the same time? Consider the same situation, but now let the initial speed $v_0$ of the first ball be given and treat the height $h$ of the building as an unknown. b) What must the height of the building be for both balls to reach the ground at the same time for each of the following values of $v_0$: i) 13.0 m/s; ii) 19.2 m/s? c) If $v_0$ is greater than some value $v_{max}$, a value of $h$ does not exist that allows both balls to hit the ground at the same time. Solve for $v_{max}$. The value $v_{max}$ has a simple physical interpretation. What is it? d) If $v_0$ is less than some value $v_{min}$, a value of $h$ does not exist that allows both balls to hit the ground at the same time. Solve for $v_{min}$. The value $v_{min}$ also has a simple physical interpretation. What is it?

CHAPTER 3

# Motion in a Plane

Aircraft carrier catapults accelerate jets at take-off from rest to speeds of 90 m/s (200 mi/h) within a very short distance.

When a plane lands, its acceleration is opposite in direction to its velocity, and it slows down.

## KEY CONCEPTS

**• Motion in two or three dimensions is described by displacement, velocity, and acceleration vectors.**

**• Projectile motion is a combination of two independent motions: horizontal motion with constant velocity and vertical motion with constant acceleration. The path of the projectile is a parabola.**

**• In uniform circular motion a particle moves in a circular path with constant speed (magnitude of velocity) but continuously changing direction of velocity. The particle's acceleration at each point is directed toward the center of the circle. Its magnitude depends on the speed and the radius of the circle.**

**• Determining relative velocity in two or three dimensions requires combining the velocities using vector addition.**

## INTRODUCTION

A Navy F14 Tomcat jet fighter roars off from the deck of an aircraft carrier, moving 55 m/s relative to the deck. It quickly accelerates into the sky. The instruments in the cockpit tell the pilot his speed and altitude and warn him of other planes nearby. He must constantly be aware of the three-dimensional nature of his motion.

In the straight-line problems of Chapter 2, we could describe where a particle is with a single coordinate. But the real world is three-dimensional. If someone placed a narrow steel beam between the fiftieth floors of two skyscrapers, would you volunteer to walk the beam? Probably not; you would doubt your ability to maintain straight-line motion. You'd be afraid of adding a sideways dimension of motion followed by a third dimension—downward.

To understand the curved flight of a baseball, the orbital motion of a satellite, or the path of a projectile shot from a gun, we need to extend our descriptions of motion to two- and three-dimensional situations. The vector quantities displacement, velocity, and acceleration now have two or three components and no longer lie along a single line. Several important kinds of motion take place in a *plane* and can be described with two coordinates and two components of velocity and acceleration. We will also generalize the concept of relative velocity to motion in space, such as a jet plane moving relative to an aircraft carrier.

This chapter represents a merging of the vector language we learned in Chapter 1 with the kinematic language of Chapter 2. As before, we are concerned with *describing* motion, not with analyzing its causes. But the language you learn here will be an essential tool in later chapters when you use Newton's laws of motion to study the relation between force and motion.

# 3–1 THE VELOCITY VECTOR

To describe the *motion* of a particle at point $P$ in space, we first need to be able to describe the *position* of the particle. The **position vector $\boldsymbol{r}$** of point $P$ is a vector that goes from the origin of the coordinate system to the point (Fig. 3–1). The figure also shows that the cartesian coordinates $x$, $y$, and $z$ of point $P$ are the $x$-, $y$-, and $z$-components of vector $\boldsymbol{r}$. Using the unit vectors introduced in Section 1–9, we can write

$$\boldsymbol{r} = x\boldsymbol{i} + y\boldsymbol{j} + z\boldsymbol{k}. \tag{3–1}$$

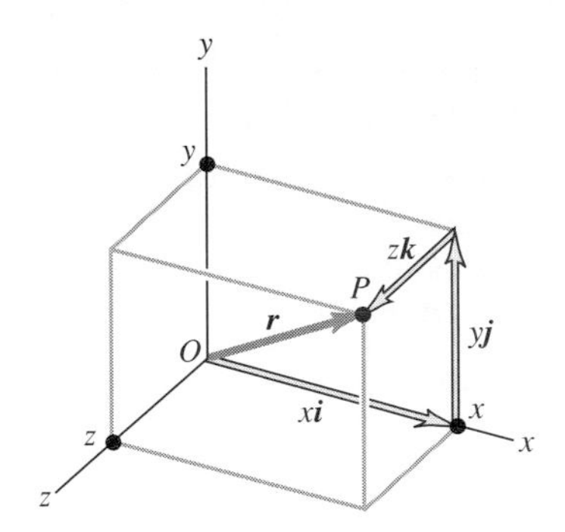


**3–1** The position vector $\boldsymbol{r}$ from the origin to point $P$ has components $x$, $y$, and $z$.

As the particle moves, its path is in general a curve (Fig. 3–2). During a time interval $\Delta t$ the particle moves from $P_1$, where its position vector is $\boldsymbol{r}_1$, to $P_2$, where its position vector is $\boldsymbol{r}_2$. The change in position (the displacement) during this interval is $\Delta\boldsymbol{r} = \boldsymbol{r}_2 - \boldsymbol{r}_1$. We define the ***average velocity*** $\boldsymbol{v}_{av}$ during this interval in the same way we did in Chapter 2 for straight-line motion, as the displacement divided by the time interval:

$$\boldsymbol{v}_{av} = \frac{\boldsymbol{r}_2 - \boldsymbol{r}_1}{t_2 - t_1} = \frac{\Delta\boldsymbol{r}}{\Delta t}. \tag{3–2}$$

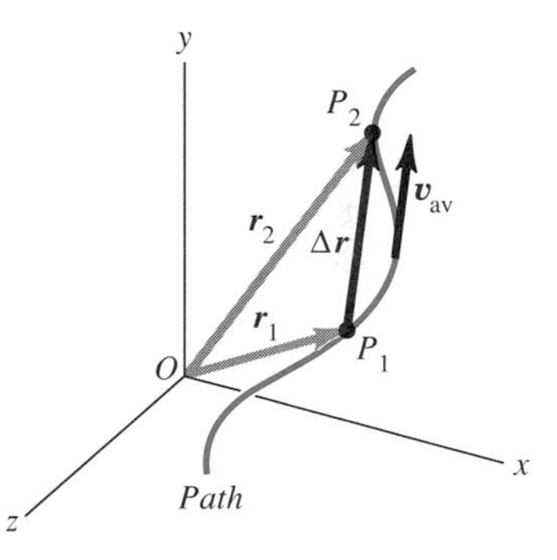


**3–2** The average velocity $\boldsymbol{v}_{av}$ between points $P_1$ and $P_2$ has the same direction as the displacement $\Delta\boldsymbol{r}$.

We define the ***instantaneous velocity*** as the limit of the average velocity as the time interval $\Delta t$ approaches zero:

$$\boldsymbol{v} = \lim_{\Delta t \to 0} \frac{\Delta\boldsymbol{r}}{\Delta t} = \frac{d\boldsymbol{r}}{dt}. \tag{3–3}$$

As $\Delta t \to 0$, points $P_1$ and $P_2$ move closer and closer together. In this limit, the vector $\Delta\boldsymbol{r}$ becomes tangent to the curve. The direction of $\Delta\boldsymbol{r}$ in the limit is also the direction of the instantaneous velocity $\boldsymbol{v}$. This leads to an important conclusion: **At every point along the path, the instantaneous velocity vector is tangent to the path** (Fig. 3–3).

During any displacement $\Delta\boldsymbol{r}$ the changes $\Delta x$, $\Delta y$, and $\Delta z$ in the three coordinates are the *components* of $\Delta\boldsymbol{r}$. It follows that the components $v_x$, $v_y$, and $v_z$ of the instantaneous velocity $\boldsymbol{v}$ are simply the time derivatives of the coordinates $x$, $y$, and $z$. That is,

$$v_x = \frac{dx}{dt}, \qquad v_y = \frac{dy}{dt}, \qquad v_z = \frac{dz}{dt}. \tag{3–4}$$

We can also get this result by taking the derivative of Eq. (3–1). The cartesian unit vectors $\boldsymbol{i}, \boldsymbol{j}$, and $\boldsymbol{k}$ are constant in magnitude and direction, so their derivatives are zero, and we find

$$\boldsymbol{v} = \frac{d\boldsymbol{r}}{dt} = \frac{dx}{dt}\boldsymbol{i} + \frac{dy}{dt}\boldsymbol{j} + \frac{dz}{dt}\boldsymbol{k}. \tag{3–5}$$

This shows again that the components of $\boldsymbol{v}$ are $dx/dt$, $dy/dt$, and $dz/dt$.

The *speed* $v$ of the particle is the *magnitude* $v$ of the instantaneous velocity vector $\boldsymbol{v}$. This is given by the Pythagorean relation

$$v = \sqrt{v_x^2 + v_y^2 + v_z^2}. \tag{3–6}$$

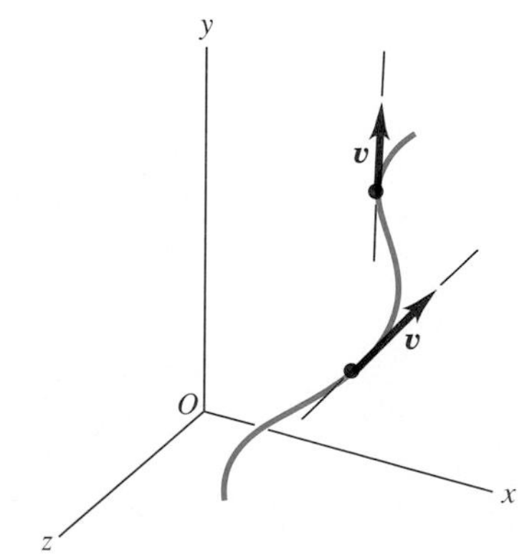


**3–3** The instantaneous velocity at any point is tangent to the path at that point.

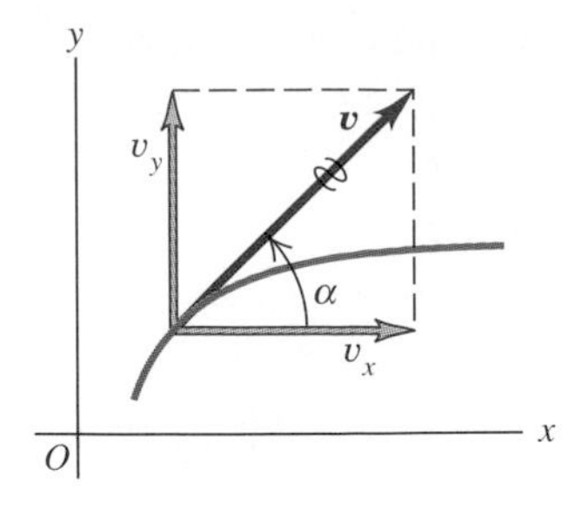


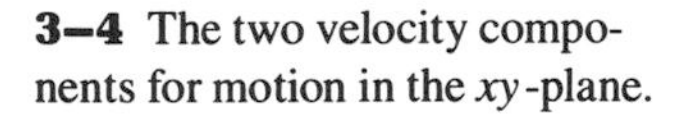
**3–4** The two velocity components for motion in the $xy$-plane.

Figure 3–4 shows the situation when the particle moves in the $xy$-plane. In this case, $z$ and $v_z$ are zero. Then the magnitude of $\boldsymbol{v}$ is

$$|\boldsymbol{v}| = v = \sqrt{{v_x}^2 + {v_y}^2},$$

and its direction is given by the angle $\alpha$ in the figure. We see that

$$\tan\alpha = \frac{v_y}{v_x}. \tag{3–7}$$

(We always use Greek letters for angles; we use $\alpha$ for the direction of the velocity vector to avoid confusion with the direction $\theta$ of the *position* vector of the particle.)

We may represent velocity, a vector quantity, in terms of its components or in terms of its magnitude and direction, just as we do with other vector quantities such as displacement and force. The *direction* of the instantaneous velocity of a particle at any point is *always* tangent to the path at that point.

## EXAMPLE 3–1

**Calculating average and instantaneous velocity** You are operating a radio-controlled model car on a vacant tennis court. Your position is the origin of coordinates, and the surface of the court lies in the $xy$-plane. The car, which we represent as a point, has $x$- and $y$-coordinates that vary with time according to

$$x = 3.0\text{ m} + (2.0\text{ m/s}^2)t^2,$$
$$y = (10.0\text{ m/s})t + (0.25\text{ m/s}^3)t^3.$$

a) Find the car's coordinates and its distance from you at time $t = 2.0$ s. b) Find the car's displacement and average velocity during the interval from $t = 0$ to $t = 2.0$ s. c) Derive a general expression for the car's instantaneous velocity, and find the instantaneous velocity at time $t = 2.0$ s.

**SOLUTION** a) The car's path is shown in Fig. 3–5. At time $t = 2.0$ s,

$$x = 3.0\text{ m} + (2.0\text{ m/s}^2)(2.0\text{ s})^2 = 11.0\text{ m},$$
$$y = (10.0\text{ m/s})(2.0\text{ s}) + (0.25\text{ m/s}^3)(2.0\text{ s})^3 = 22.0\text{ m}.$$

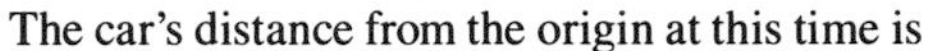

The car's distance from the origin at this time is

$$r = \sqrt{x^2 + y^2} = \sqrt{(11.0\text{ m})^2 + (22.0\text{ m})^2} = 24.6\text{ m}.$$

b) The position vector $\boldsymbol{r}$ at any time $t$ is given by

$$\boldsymbol{r} = [3.0\text{ m} + (2.0\text{ m/s}^2)t^2]\boldsymbol{i} + [(10.0\text{ m/s})t + (0.25\text{ m/s}^3)t^3]\boldsymbol{j}.$$

At time $t = 0$,

$$\boldsymbol{r}_0 = (3.0\text{ m})\boldsymbol{i} + (0)\boldsymbol{j}.$$

From part (a) we find that at time $t = 2.0$ s the position vector $\boldsymbol{r}_2$ is

$$\boldsymbol{r}_2 = (11.0\text{ m})\boldsymbol{i} + (22.0\text{ m})\boldsymbol{j}.$$

Therefore the displacement from $t = 0$ to $t = 2.0$ s is

$$\Delta\boldsymbol{r} = \boldsymbol{r}_2 - \boldsymbol{r}_0 = (11.0\text{ m})\boldsymbol{i} + (22.0\text{ m})\boldsymbol{j} - (3.0\text{ m})\boldsymbol{i}$$
$$= (8.0\text{ m})\boldsymbol{i} + (22.0\text{ m})\boldsymbol{j}.$$

The average velocity during the interval is

$$\boldsymbol{v}_{\text{av}} = \frac{\Delta\boldsymbol{r}}{\Delta t} = \frac{(8.0\text{ m})\boldsymbol{i} + (22.0\text{ m})\boldsymbol{j}}{2.0\text{ s} - 0\text{ s}}$$
$$= (4.0\text{ m/s})\boldsymbol{i} + (11.0\text{ m/s})\boldsymbol{j}.$$

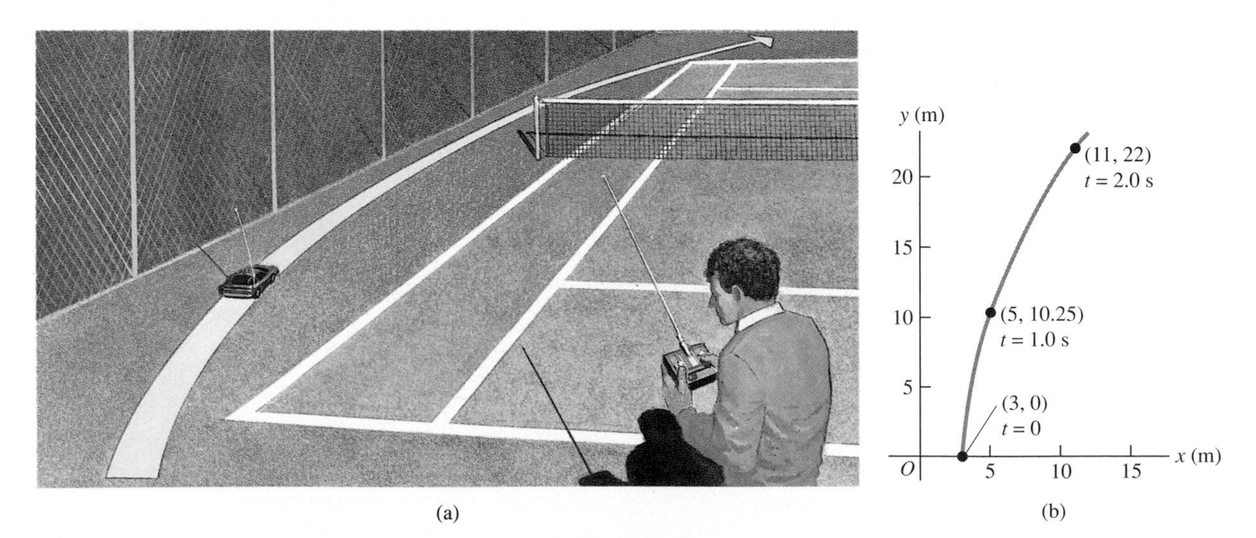


**3–5** (a) Motion of a radio-controlled model car. (b) Path of the car from $t = 0$ to $t = 2.0$ s.

The components of average velocity are

$$(v_x)_{av} = 4.0 \text{ m/s}, \qquad (v_y)_{av} = 11.0 \text{ m/s}.$$

c) The components of instantaneous velocity are the time derivatives of the coordinates:

$$v_x = \frac{dx}{dt} = (2.0 \text{ m/s}^2)(2t),$$

$$v_y = \frac{dy}{dt} = 10.0 \text{ m/s} + (0.25 \text{ m/s}^3)(3t^2).$$

We can write the instantaneous velocity vector $\boldsymbol{v}$ as

$$\boldsymbol{v} = (4.0 \text{ m/s}^2)t\boldsymbol{i} + [10.0 \text{ m/s} + (0.75 \text{ m/s}^3)t^2]\boldsymbol{j}.$$

At time $t = 2.0$ s the components of instantaneous velocity are

$$v_x = (4.0 \text{ m/s}^2)(2.0 \text{ s}) = 8.0 \text{ m/s},$$
$$v_y = 10.0 \text{ m/s} + (0.75 \text{ m/s}^3)(2.0 \text{ s})^2 = 13.0 \text{ m/s}.$$

The magnitude of the velocity at this time is

$$v = \sqrt{v_x{}^2 + v_y{}^2} = \sqrt{(8.0 \text{ m/s})^2 + (13.0 \text{ m/s})^2} = 15 \text{ m/s}.$$

Its direction with respect to the positive $x$-axis is given by the angle $\alpha$, where

$$\tan\alpha = \frac{v_y}{v_x} = \frac{13.0 \text{ m/s}}{8.0 \text{ m/s}}, \qquad \alpha = 58°.$$

## 3–2 THE ACCELERATION VECTOR

Now let's consider the acceleration of a particle moving in space. In Figure 3–6a the vectors $\boldsymbol{v}_1$ and $\boldsymbol{v}_2$ represent the particle's instantaneous velocities at points $P_1$ and $P_2$ and at times $t_1$ and $t_2$. The two velocities may differ in both magnitude and direction.

We define the ***average acceleration*** $\boldsymbol{a}_{av}$ of the particle as it moves from $P_1$ to $P_2$ as the *vector change in velocity,* $\boldsymbol{v}_2 - \boldsymbol{v}_1 = \Delta\boldsymbol{v}$, divided by the time interval $t_2 - t_1 = \Delta t$:

$$\text{Average acceleration} = \boldsymbol{a}_{av} = \frac{\boldsymbol{v}_2 - \boldsymbol{v}_1}{t_2 - t_1} = \frac{\Delta\boldsymbol{v}}{\Delta t}. \tag{3–8}$$

Average acceleration is a vector quantity in the same direction as the vector $\Delta\boldsymbol{v}$ (Fig. 3–6a). Note that $\boldsymbol{v}_2$ is the vector sum of the original velocity $\boldsymbol{v}_1$ and the change $\Delta\boldsymbol{v}$ (Fig. 3–6b).

We define the ***instantaneous acceleration a*** at point $P_1$ as the limit approached by the average acceleration when point $P_2$ approaches point $P_1$ and $\Delta\boldsymbol{v}$ and $\Delta t$ approach zero:

$$\boldsymbol{a} = \lim_{\Delta t \to 0} \frac{\Delta\boldsymbol{v}}{\Delta t} = \frac{d\boldsymbol{v}}{dt}. \tag{3–9}$$

The instantaneous acceleration vector at point $P_1$ (Fig. 3–6c) does *not* have the same direction as the velocity vector at that point; in general there is no reason it should. The construction in Fig. 3–6c shows that the acceleration vector must always point toward the *concave* side of the curved path. When a particle moves in a curved path, it *always* has nonzero acceleration, even when it moves with constant speed. More generally, acceleration is associated with change of speed, change of direction of velocity, or both.

Each component of the acceleration vector is the derivative of the corresponding component of velocity:

$$a_x = \frac{dv_x}{dt}, \qquad a_y = \frac{dv_y}{dt}, \qquad a_z = \frac{dv_z}{dt}. \tag{3–10}$$

In terms of unit vectors,

$$\boldsymbol{a} = \frac{dv_x}{dt}\boldsymbol{i} + \frac{dv_y}{dt}\boldsymbol{j} + \frac{dv_z}{dt}\boldsymbol{k}. \tag{3–11}$$

Also, because each component of velocity is the derivative of the corresponding coordinate, we can express the components $a_x$, $a_y$, and $a_z$ of the acceleration vector $\boldsymbol{a}$ as

$$a_x = \frac{d^2x}{dt^2}, \qquad a_y = \frac{d^2y}{dt^2}, \qquad a_z = \frac{d^2z}{dt^2}, \tag{3–12}$$

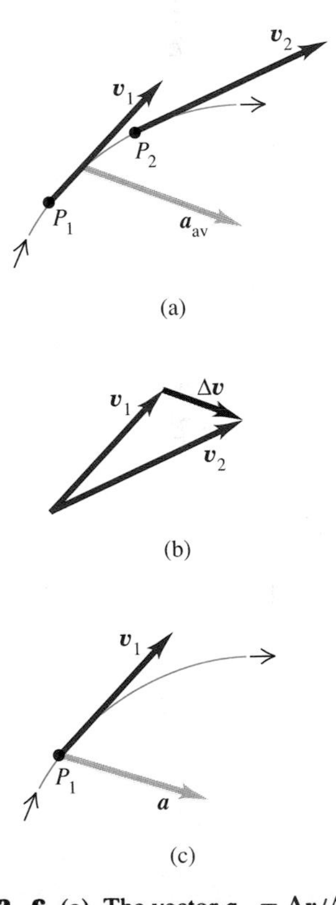


**3–6** (a) The vector $\boldsymbol{a}_{av} = \Delta\boldsymbol{v}/\Delta t$ represents the average acceleration between $P_1$ and $P_2$. (b) Construction for obtaining $\Delta\boldsymbol{v} = \boldsymbol{v}_2 - \boldsymbol{v}_1$. (c) Instantaneous acceleration $\boldsymbol{a}$ at point $P_1$. Vector $\boldsymbol{v}$ is tangent to the path; vector $\boldsymbol{a}$ points toward the concave side of the path.

and the acceleration vector $\boldsymbol{a}$ as

$$\boldsymbol{a} = \frac{d^2x}{dt^2}\boldsymbol{i} + \frac{d^2y}{dt^2}\boldsymbol{j} + \frac{d^2z}{dt^2}\boldsymbol{k}. \tag{3-13}$$

## EXAMPLE 3–2

**Calculating average and instantaneous acceleration** Let's look again at the radio-controlled model car in Example 3–1. We found that the components of instantaneous velocity at any time $t$ are

$$v_x = \frac{dx}{dt} = (2.0 \text{ m/s}^2)(2t),$$

$$v_y = \frac{dy}{dt} = 10.0 \text{ m/s} + (0.25 \text{ m/s}^3)(3t^2),$$

and that the velocity vector is

$$\boldsymbol{v} = (4.0 \text{ m/s}^2)t\boldsymbol{i} + [10.0 \text{ m/s} + (0.75 \text{ m/s}^3)t^2]\boldsymbol{j}.$$

a) Find the average acceleration in the interval from $t = 0$ to $t = 2.0$ s. b) Find the instantaneous acceleration at $t = 2.0$ s.

**SOLUTION** a) We find the components of instantaneous velocity at time $t = 0$ by substituting this value into the above expressions for $v_x$ and $v_y$. We find that at time $t = 0$,

$$v_x = 0, \qquad v_y = 10.0 \text{ m/s}.$$

We found in Example 3–1 that at time $t = 2.0$ s the values of these components are

$$v_x = 8.0 \text{ m/s}, \qquad v_y = 13.0 \text{ m/s}.$$

Thus the components of average acceleration in this interval are

$$(a_x)_{\text{av}} = \frac{\Delta v_x}{\Delta t} = \frac{8.0 \text{ m/s} - 0}{2.0 \text{ s} - 0} = 4.0 \text{ m/s}^2,$$

$$(a_y)_{\text{av}} = \frac{\Delta v_y}{\Delta t} = \frac{13.0 \text{ m/s} - 10.0 \text{ m/s}}{2.0 \text{ s} - 0} = 1.5 \text{ m/s}^2.$$

b) The components of instantaneous acceleration are the time derivatives of the components of instantaneous velocity. We find

$$a_x = \frac{dv_x}{dt} = 4.0 \text{ m/s}^2, \qquad a_y = \frac{dv_y}{dt} = (0.75 \text{ m/s}^3)(2t)$$

We can write the instantaneous acceleration vector $\boldsymbol{a}$ as

$$\boldsymbol{a} = (4.0 \text{ m/s}^2)\boldsymbol{i} + (1.5 \text{ m/s}^3)t\boldsymbol{j}.$$

At time $t = 2.0$ s the components of instantaneous acceleration are

$$a_x = 4.0 \text{ m/s}^2, \qquad a_y = (1.5 \text{ m/s}^3)(2.0 \text{ s}) = 3.0 \text{ m/s}^2.$$

The acceleration vector at this time is

$$\boldsymbol{a} = (4.0 \text{ m/s}^2)\boldsymbol{i} + (3.0 \text{ m/s}^2)\boldsymbol{j}.$$

The magnitude of acceleration at this time is

$$a = \sqrt{a_x^2 + a_y^2}$$

$$= \sqrt{(4.0 \text{ m/s}^2)^2 + (3.0 \text{ m/s}^2)^2} = 5.0 \text{ m/s}^2.$$

Its direction with respect to the positive $x$-axis is given by the angle $\beta$, where

$$\tan\beta = \frac{3.0 \text{ m/s}^2}{4.0 \text{ m/s}^2},$$

$$\beta = 37°.$$

Figure 3–7 shows the positions of the car at times from $t = 0$ to $t = 2.0$ s and the velocity and acceleration vectors at $t = 2.0$ s. Note that these two vectors *do not* have the same direction. The velocity vector is tangent to the path at each point, and the acceleration vector points toward the concave side of the path.

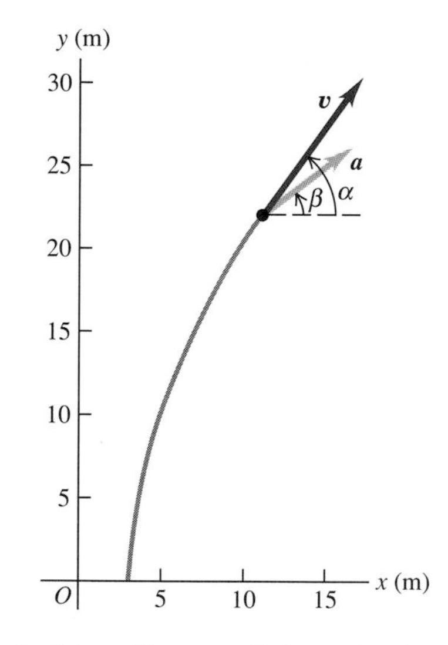


**3–7** The path of the radio-controlled car, showing the velocity and acceleration at $t = 2.0$ s.

We can also represent the acceleration of a particle moving in a curved path in terms of components parallel and perpendicular to the velocity at each point (Fig. 3–8). In the figure these components are labeled $a_\parallel$ and $a_\perp$. To see why these components are useful, we'll consider two special cases. In Fig. 3–9a the acceleration vector is *parallel* to the velocity $\boldsymbol{v}_1$. Then because $\boldsymbol{a}$ gives the rate of change of velocity, the change in $\boldsymbol{v}$ during a small time interval $\Delta t$ is a vector $\Delta \boldsymbol{v}$ having the same direction as $\boldsymbol{a}$ and hence the same direction as $\boldsymbol{v}_1$. The velocity $\boldsymbol{v}_2$ at the end of $\Delta t$, given by $\boldsymbol{v}_2 = \boldsymbol{v}_1 + \Delta \boldsymbol{v}$, is a vector having the same direction as $\boldsymbol{v}_1$ but somewhat greater magnitude.

In Fig. 3–9b the acceleration $\boldsymbol{a}$ is *perpendicular* to the velocity $\boldsymbol{v}$. In a small time interval $\Delta t$ the change $\Delta \boldsymbol{v}$ is a vector very nearly perpendicular to $\boldsymbol{v}_1$, as shown. Again $\boldsymbol{v}_2 = \boldsymbol{v}_1 + \Delta \boldsymbol{v}$, but in this case $\boldsymbol{v}_1$ and $\boldsymbol{v}_2$ have different directions. As the time interval $\Delta t$ approaches zero, the angle $\phi$ in the figure also approaches zero, $\Delta \boldsymbol{v}$ becomes perpendicular to *both* $\boldsymbol{v}_1$ and $\boldsymbol{v}_2$, and $\boldsymbol{v}_1$ and $\boldsymbol{v}_2$ have the same magnitude.

Thus when $\boldsymbol{a}$ is *parallel* (or antiparallel) to $\boldsymbol{v}$, its effect is to change the magnitude of $\boldsymbol{v}$ but not its direction; when $\boldsymbol{a}$ is *perpendicular* to $\boldsymbol{v}$, its effect is to change the direction of $\boldsymbol{v}$ but not its magnitude. In general, $\boldsymbol{a}$ may have components *both* parallel and perpendicular to $\boldsymbol{v}$, but the above statements are still valid for the individual components. In particular, when a particle travels along a curved path with constant speed, its acceleration is not zero, even though the magnitude of $\boldsymbol{v}$ does not change. In this case the acceleration is always perpendicular to $\boldsymbol{v}$ at each point. For example, when a particle moves in a circle with constant speed, the acceleration at each instant is directed toward the center of the circle. We will consider this special case in detail in Section 3–4.

More generally, the two quantities

$$\frac{d|\boldsymbol{v}|}{dt} \quad \text{and} \quad \left|\frac{d\boldsymbol{v}}{dt}\right|$$

are not the same. The first is the rate of change of *speed;* it is zero whenever a particle moves with constant speed, even when its direction of motion changes. The second is the magnitude of the vector acceleration; it is zero only when the particle's acceleration is zero, that is, when the particle moves in a straight line with constant velocity.

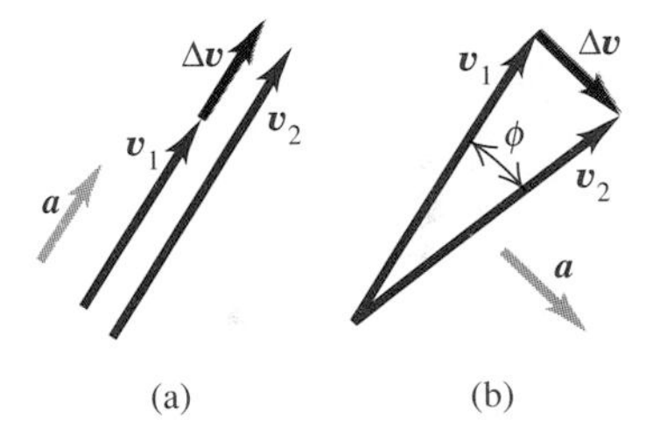


**3–8** The acceleration can be resolved into a component $a_{\|}$ parallel to the path (and to the velocity) and $a_{\perp}$ perpendicular to the path.

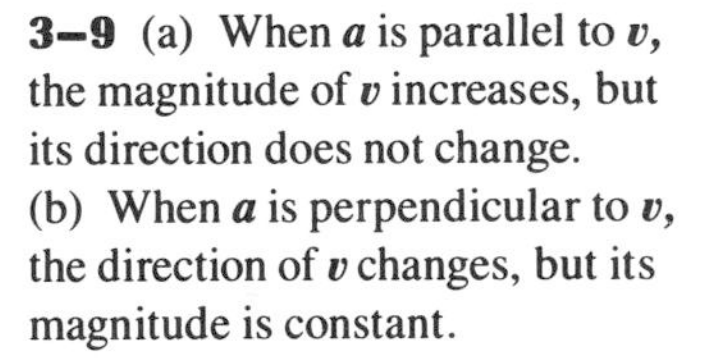


**3–9** (a) When $\boldsymbol{a}$ is parallel to $\boldsymbol{v}$, the magnitude of $\boldsymbol{v}$ increases, but its direction does not change. (b) When $\boldsymbol{a}$ is perpendicular to $\boldsymbol{v}$, the direction of $\boldsymbol{v}$ changes, but its magnitude is constant.

## 3–3 PROJECTILE MOTION

A **projectile** is any body that is given an initial velocity and then follows a path determined entirely by the effects of gravitational acceleration and air resistance. A batted baseball, a thrown football, a package dropped from an airplane, and a bullet shot from a rifle are all projectiles. The path followed by a projectile is called its *trajectory*.

To analyze this common type of motion, we'll start with an idealized model, representing the projectile as a single particle with an acceleration (due to gravity) that is constant in both magnitude and direction. We'll neglect the effects of air resistance and the curvature and rotation of the earth. Like all models, this one has limitations. Curvature of the earth has to be considered in the flight of ICBMs, and air resistance is of crucial importance to a sky diver. Nevertheless, we can learn a lot from analysis of this simple model.

We first notice that projectile motion is always confined to a vertical plane determined by the direction of the initial velocity (Fig. 3–10). We will call this plane the $xy$-coordinate plane, with the $x$-axis horizontal and the $y$-axis vertically upward.

The key to analysis of projectile motion is the fact that *we can treat the x- and y-coordinates separately.* The $x$-component of acceleration is zero, and the $y$-component is constant and equal to $-g$. (Remember that, by definition, $g$ is always positive, and with our choice of coordinate directions, $a_y$ is negative.) So we can think of projectile motion as a combination of horizontal motion with constant velocity and vertical motion with constant acceleration. We can then express all the vector relationships in terms of separate equations for the horizontal and vertical components. The actual motion is the superposition of these separate motions. The components of $\boldsymbol{a}$ are

$$a_x = 0, \qquad a_y = -g. \tag{3–14}$$

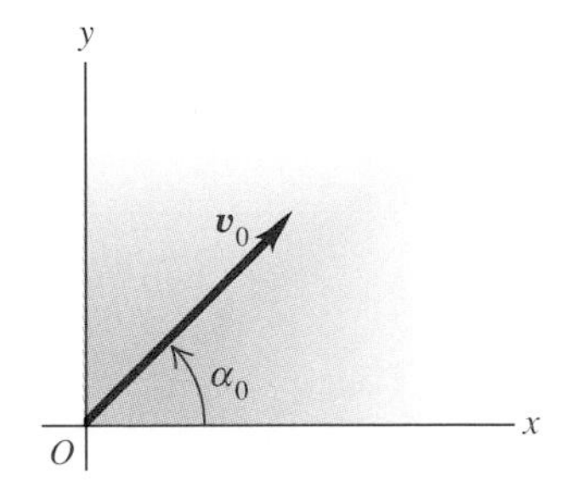


**3–10** All projectile motion occurs in the vertical plane containing $\boldsymbol{v}_0$.

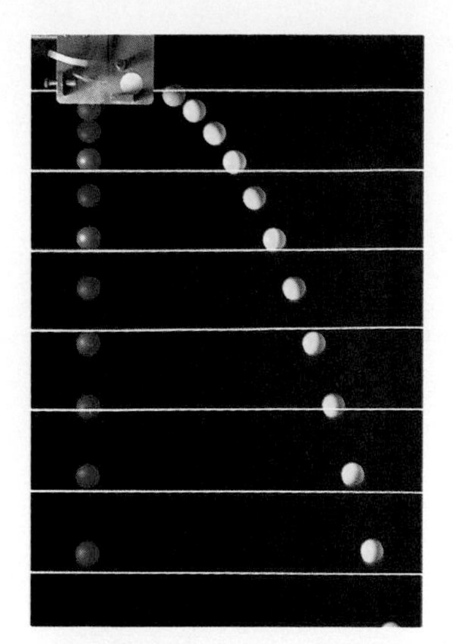

**3–11** Independence of horizontal and vertical motion: At any given time both balls have the same $y$-position, velocity, and acceleration despite having different $x$-positions and velocities.

We will usually use $g = 9.80\ \text{m/s}^2$. Figure 3–11 shows two projectiles with different $x$-motion but identical $y$-motion.

In projectile motion each coordinate varies with constant-acceleration motion, and we can use Eqs. (2–9), (2–13), and (2–14) directly. For example, suppose that at time $t = 0$ the particle is at the point $(x_0, y_0)$ and its velocity components have the initial values $v_{0x}$ and $v_{0y}$. The components of acceleration are $a_x = 0$, $a_y = -g$. Considering the $x$-motion, we substitute $v_x$ for $v$, $v_{0x}$ for $v_0$, and 0 for $a$ in Eqs. (2–9) and (2–13). We find

$$v_x = v_{0x}, \tag{3–15}$$

$$x = x_0 + v_{0x}t. \tag{3–16}$$

For the $y$-motion we substitute $y$ for $x$, $v_y$ for $v$, $v_{0y}$ for $v_0$, and $-g$ for $a$:

$$v_y = v_{0y} - gt, \tag{3–17}$$

$$y = y_0 + v_{0y}t - \tfrac{1}{2}gt^2. \tag{3–18}$$

Usually, it is simplest to take the initial position (at time $t = 0$) as the origin; in this case, $x_0 = y_0 = 0$. This point might be, for example, the position of a ball at the instant it leaves the thrower's hand or the position of a bullet at the instant it leaves the gun barrel.

Figure 3–12 shows the path of a projectile that starts at (or passes through) the origin at time $t = 0$. The position, velocity, and velocity components are shown at a series of times separated by equal intervals. The $x$-component of acceleration is zero, so $v_x$ is constant, but $v_y$ changes by equal amounts in equal times, corresponding to constant $y$-acceleration. At the highest point in the trajectory, $v_y = 0$.

We can also represent the initial velocity $\boldsymbol{v}_0$ by its magnitude $v_0$ (the initial speed) and its angle $\alpha_0$ with the positive $x$-axis. In terms of these quantities the components $v_{0x}$ and $v_{0y}$ of initial velocity are

$$v_{0x} = v_0 \cos \alpha_0, \qquad v_{0y} = v_0 \sin \alpha_0. \tag{3–19}$$

Using these relations in Eqs. (3–15) through (3–18) and setting $x_0 = y_0 = 0$, we find

$$x = (v_0 \cos \alpha_0)t, \tag{3–20}$$

$$y = (v_0 \sin \alpha_0)t - \tfrac{1}{2}gt^2, \tag{3–21}$$

$$v_x = v_0 \cos \alpha_0, \tag{3–22}$$

$$v_y = v_0 \sin \alpha_0 - gt. \tag{3–23}$$

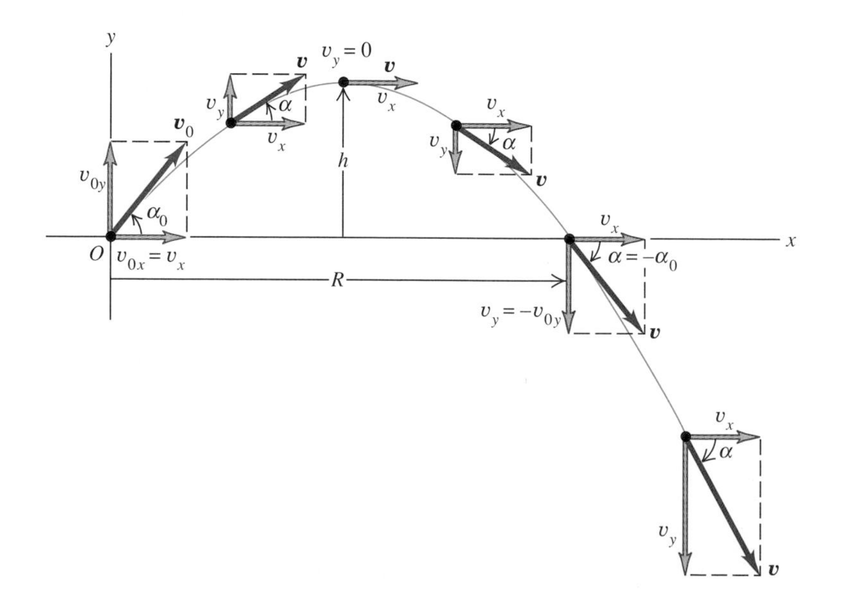


**3–12** The trajectory of a body projected with an initial velocity $\boldsymbol{v}_0$ at an angle above the horizontal $\alpha_0$ with negligible air resistance. The distance $R$ is the horizontal range, and $h$ is the maximum height.

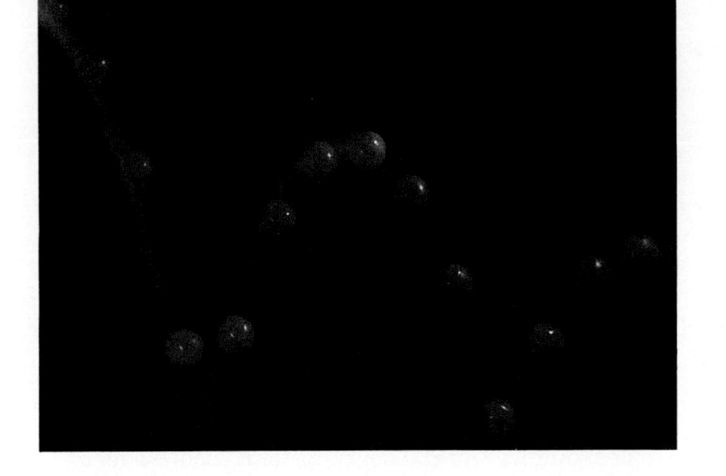

**3–13** Stroboscopic photograph of a bouncing ball, showing parabolic trajectories after each bounce. Successive images are separated by equal time intervals, as in Fig. 3–11. Each peak in the trajectories is lower than the preceding one because of energy loss during the "bounce," or collision with the horizontal surface.

These equations describe the position and velocity of the projectile in Fig. 3–12 at any time $t$.

We can get a lot of information from these equations. For example, the distance $r$ of the projectile from the origin at any time (the magnitude of the position vector $\boldsymbol{r}$) is given by

$$r = \sqrt{x^2 + y^2}. \tag{3–24}$$

The projectile's speed (the magnitude of its velocity) at any time is

$$v = \sqrt{v_x^2 + v_y^2}. \tag{3–25}$$

The *direction* of the velocity, in terms of the angle $\alpha$ it makes with the positive $x$-axis, is given by

$$\tan \alpha = \frac{v_y}{v_x}. \tag{3–26}$$

The velocity vector $\boldsymbol{v}$ is tangent to the trajectory at each point.

We can derive an equation for the shape of the trajectory in terms of $x$ and $y$ by eliminating $t$. We find $t = x/(v_0 \cos \alpha_0)$ and

$$y = (\tan \alpha_0)x - \frac{g}{2v_0^2 \cos^2 \alpha_0}x^2. \tag{3–27}$$

Don't worry about the details of this equation; the important point is its general form. The quantities $v_0$, $\tan \alpha_0$, $\cos \alpha_0$, and $g$ are constants, so the equation has the form

$$y = bx - cx^2,$$

where $b$ and $c$ are positive constants. This is the equation of a *parabola*. With projectile motion, for our simple model, the trajectory is always a parabola (Fig. 3–13).

## PROBLEM-SOLVING STRATEGY

### Projectile motion

The strategies that we used in Sections 2–4 and 2–5 for straight-line, constant-acceleration problems are also useful here.

**1.** Define your coordinate system and make a sketch showing your axes. Usually, it is easiest to place the origin at the initial ($t = 0$) position of the projectile, with the $x$-axis horizontal and the $y$-axis upward. Then $x_0 = 0$, $y_0 = 0$, $a_x = 0$, and $a_y = -g$.

**2.** List the known and unknown quantities. In some problems the components (or magnitude and direction) of initial velocity will be given, and you can use Eqs. (3–20) through (3–23) to find the coordinates and velocity components at some later time. In other problems you might be given two points on the trajectory and be asked to find the initial velocity. Be sure you know which quantities are given and which are to be found.

**3.** It often helps to state the problem in prose and then translate into symbols. For example, *when* does the particle arrive at a certain point (that is, at what value of $t$)? *Where* is the particle when its velocity has a certain value? (That is, what are the values of $x$ and $y$ when $v_x$ or $v_y$ has the specified value?)

**4.** At the highest point in a trajectory, $v_y = 0$. So the question "When does the projectile reach its highest point?" translates into "What is the value of $t$ when $v_y = 0$?" Similarly, if $y_0 = 0$, then "When does the projectile return to its initial elevation?" translates into "What is the value of $t$ when $y = 0$?" And so on.

**5.** Resist the temptation to break the trajectory into segments and analyze each one separately. You don't have to start all over, with a new axis system and a new time scale, when the projectile reaches its highest point. It is usually easier to set up Eqs. (3–20) through (3–23) at the start and use the same axes and time scale throughout the problem.

## EXAMPLE 3–3

**A body projected horizontally.** A motorcycle stunt rider rides off the edge of a cliff. Just at the edge his velocity is horizontal, with magnitude 5.0 m/s. Find the rider's position and velocity after $\frac{1}{4}$ s.

**SOLUTION** The coordinate system is shown in Fig. 3–14. We know the following: $x_0 = 0$, $y_0 = 0$, $v_{0x} = 5.0$ m/s, $v_{0y} = 0$, $a_x = 0$, $a_y = -g = -9.8\ \text{m/s}^2$.

Where is the motorcycle at $t = \frac{1}{4}$ s? The answer is contained in Eqs. (3–20) and (3–21), which give $x$ and $y$ as functions of time. When $t = \frac{1}{4}$ s, the $x$- and $y$-coordinates are

$$x = v_{0x}t = (5.0\ \text{m/s})(\tfrac{1}{4}\ \text{s}) = 1.2\ \text{m},$$

$$y = -\tfrac{1}{2}gt^2 = -\tfrac{1}{2}(9.8\ \text{m/s}^2)(\tfrac{1}{4}\ \text{s})^2 = -0.31\ \text{m}.$$

The negative value of $y$ shows that at this time the motorcycle is below its starting point.

What is the rider's distance from the origin at this time? From Eq. (3–24),

$$r = \sqrt{x^2 + y^2} = \sqrt{(1.2\ \text{m})^2 + (-0.31\ \text{m})^2} = 1.2\ \text{m}.$$

What is the velocity at time $t = \frac{1}{4}$ s? From Eqs. (3–22) and (3–23), which give $v_x$ and $v_y$ as functions of time, the components of velocity at time $t = \frac{1}{4}$ s are

$$v_x = v_{0x} = 5.0\ \text{m/s},$$

$$v_y = -gt = (-9.8\ \text{m/s}^2)(\tfrac{1}{4}\ \text{s}) = -2.4\ \text{m/s}.$$

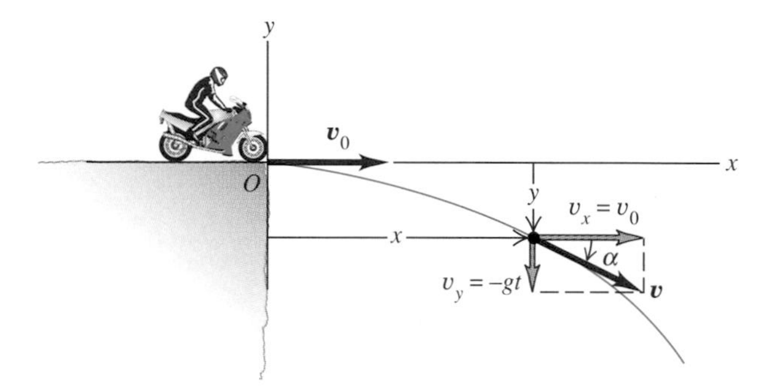


**3–14** Trajectory of a body projected horizontally.

From Eq. (3–25) the speed (magnitude of the velocity) at this time is

$$v = \sqrt{v_x^2 + v_y^2}$$
$$= 5.5\ \text{m/s}.$$

From Eq. (3–26) the angle $\alpha$ of the velocity vector is

$$\alpha = \arctan\frac{v_y}{v_x}$$
$$= \arctan\frac{-2.4\ \text{m/s}}{5.0\ \text{m/s}} = -26°.$$

At this time the velocity is 26° below the horizontal.

## EXAMPLE 3–4

**Height and range of a batted baseball** In Fig. 3–12, suppose the projectile is a home-run ball hit with an initial speed $v_0 = 37.0$ m/s at an initial angle $\alpha_0 = 53.1°$. Let's see how we can predict where it is and how it is moving at a certain time and how we can find its maximum height and the distance from home plate when it comes down.

The ball is probably struck a meter or so above ground level, but we neglect this distance and assume that it starts at ground level ($y_0 = 0$). Then

$$v_{0x} = v_0 \cos\alpha_0 = (37.0\ \text{m/s})(0.600) = 22.2\ \text{m/s},$$

$$v_{0y} = v_0 \sin\alpha_0 = (37.0\ \text{m/s})(0.800) = 29.6\ \text{m/s}.$$

a) Find the position of the ball and the magnitude and direction of its velocity when $t = 2.00$ s.

Using the coordinate system shown in Fig. 3–12, we want to find $x$, $y$, $v_x$, and $v_y$ at time $t = 2.00$ s. From Eqs. (3–20) through (3–23),

$$x = v_{0x}t = (22.2\ \text{m/s})(2.00\ \text{s}) = 44.4\ \text{m},$$

$$y = v_{0y}t - \tfrac{1}{2}gt^2$$
$$= (29.6\ \text{m/s})(2.00\ \text{s}) - \tfrac{1}{2}(9.80\ \text{m/s}^2)(2.00\ \text{s})^2$$
$$= 39.6\ \text{m},$$

$$v_x = v_{0x} = 22.2\ \text{m/s},$$

$$v_y = v_{0y} - gt = 29.6\ \text{m/s} - (9.80\ \text{m/s}^2)(2.00\ \text{s}) = 10.0\ \text{m/s},$$

$$v = \sqrt{v_x{}^2 + v_y{}^2} = \sqrt{(22.2\ \text{m/s})^2 + (10.0\ \text{m/s})^2} = 24.3\ \text{m/s},$$

$$\alpha = \arctan\frac{10.0\ \text{m/s}}{22.2\ \text{m/s}} = \arctan 0.450 = 24.2°.$$

b) Find the time when the ball reaches the highest point of its flight, and find its height $h$ at this point.

At the highest point the vertical velocity $v_y$ is zero. When does this happen? Call the time $t_1$; then

$$v_y = 0 = 29.6\ \text{m/s} - (9.80\ \text{m/s}^2)t_1,$$

$$t_1 = 3.02\ \text{s}.$$

The height $h$ at this time is the value of $y$ when $t = 3.02$ s:

$$h = (29.6\ \text{m/s})(3.02\ \text{s}) - \tfrac{1}{2}(9.80\ \text{m/s}^2)(3.02\ \text{s})^2 = 44.7\ \text{m}.$$

Alternatively, we can use the constant-acceleration formula:

$$v_y{}^2 = v_{0y}{}^2 + 2a(y - y_0) = v_{0y}{}^2 - 2g(y - y_0).$$

At the highest point, $v_y = 0$ and $y = h$. Substituting these in, along with $y_0 = 0$, we find

$$0 = (29.6\ \text{m/s})^2 - 2(9.80\ \text{m/s}^2)h,$$

$$h = 44.7\ \text{m}.$$

This is roughly half the height above the playing field of the Toronto Skydome.

c) Find the *horizontal range R*, that is, the horizontal distance from the starting point to the point at which the ball hits the ground.

First, *when* does the ball hit the ground? This occurs when $y = 0$. Call this time $t_2$; then

$$y = 0 = (29.6\ \text{m/s})t_2 - \tfrac{1}{2}(9.80\ \text{m/s}^2)t_2{}^2.$$

This is a quadratic equation for $t_2$; it has two roots,

$$t_2 = 0 \quad \text{and} \quad t_2 = 6.04\ \text{s}.$$

There are two times at which $y = 0$; $t_2 = 0$ is the time the ball *leaves* the ground, and $t_2 = 6.04$ s is the time of its return. This is exactly twice the time to reach the highest point, so the time of descent equals the time of ascent. (This is *always* true if the starting and end points are at the same elevation and air resistance is neglected. We will prove this in Example 3–6.)

The horizontal range $R$ is the value of $x$ when the ball returns to the ground, that is, when $t = 6.04$ s:

$$R = v_{0x}t_2 = (22.2\ \text{m/s})(6.04\ \text{s}) = 134\ \text{m}.$$

For comparison, the distance from home plate to center field at Pittsburgh's Three Rivers Stadium is 125 m (Fig. 3–15), so the ball really *is* a home run! The vertical component of velocity at this point is

$$v_y = 29.6\ \text{m/s} - (9.80\ \text{m/s}^2)(6.04\ \text{s}) = -29.6\ \text{m/s}.$$

That is, $v_y$ has the same magnitude as the initial vertical velocity $v_{0y}$ but the opposite direction. Since $v_x$ is constant, the angle $\alpha = -53.1°$ (below the horizontal) at this point is the negative of the initial angle $\alpha_0 = 53.1°$.

d) If the ball could continue to travel *below* its original level (through an appropriately shaped hole in the ground), then negative values of $y$ corresponding to times greater than 6.04 s would be possible. Can you compute the position and velocity at a time 8.00 s after the start, corresponding to the last position shown in Fig. 3–12? The results are

$$x = 178\ \text{m}, \qquad y = -76.8\ \text{m},$$

$$v_x = 22.2\ \text{m/s}, \qquad v_y = -48.6\ \text{m/s}.$$

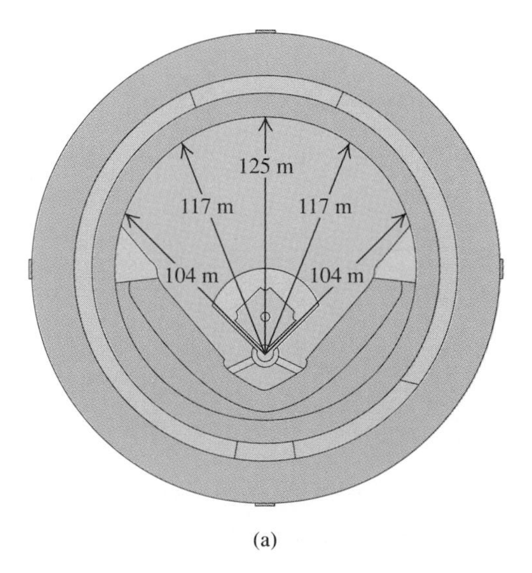


(a)

(b)

**3–15** (a) Distances for hitting a home run at Three Rivers Stadium, Pittsburgh. (b) Can the ball clear the fence?

## EXAMPLE 3–5

**The hunter and the monkey: A classic demonstration** A clever monkey escapes from the zoo. The zoo keeper finds him in a tree. Failing to entice the monkey down, the zoo keeper points her tranquilizer gun directly at the monkey and shoots (Fig. 3–16). The clever monkey lets go of the tree at the same instant the dart leaves the gun barrel, intending to land on the ground and escape. Show that the dart always hits the monkey, regardless of the dart's muzzle velocity (provided that it gets to the monkey before he hits the ground).

**SOLUTION** To show that the dart hits the monkey, we have to prove that there is some time when the monkey and the dart have the same $x$- and $y$-coordinates. First, let's ask when the $x$-coordinates are the same. The monkey drops straight down, so for him, $x = d$ at *all* times. For the dart, $x$ is given by Eq. (3–20): $x = (v_0 \cos \alpha_0)t$. When these are equal, $d = (v_0 \cos \alpha_0)t$, or

$$t = \frac{d}{v_0 \cos \alpha_0}.$$

Now we ask whether $y_{\text{monkey}}$ and $y_{\text{dart}}$ are also equal at this time; if they are, we have a hit. The monkey is in one-dimensional free fall, and his position at any time is given by Eq. (2–13), with appropriate symbol changes. The initial height is $d \tan \alpha_0$, and we find

$$y_{\text{monkey}} = d \tan \alpha_0 - \tfrac{1}{2}gt^2.$$

For the dart we use Eq. (3–21):

$$y_{\text{dart}} = (v_0 \sin \alpha_0)t - \tfrac{1}{2}gt^2.$$

So we see that if $d \tan \alpha_0 = (v_0 \sin \alpha_0)t$ at the time when the two $x$'s are equal, we have a hit. To prove that this happens, we replace $t$ with the expression above. The right side of the equation then becomes

$$(v_0 \sin \alpha_0)t = (v_0 \sin \alpha_0)\frac{d}{v_0 \cos \alpha_0} = d \tan \alpha_0.$$

We have proved that at the time the $x$-coordinates are equal, the $y$-coordinates are also equal; a dart aimed at the initial position of the monkey always hits him, no matter what $v_0$ is. With no gravity ($g = 0$) the monkey would remain motionless, and the dart would travel in a straight line to hit him. With gravity, both "fall" the same distance ($\tfrac{1}{2}gt^2$) below their $g = 0$ positions, and the dart still hits the monkey.

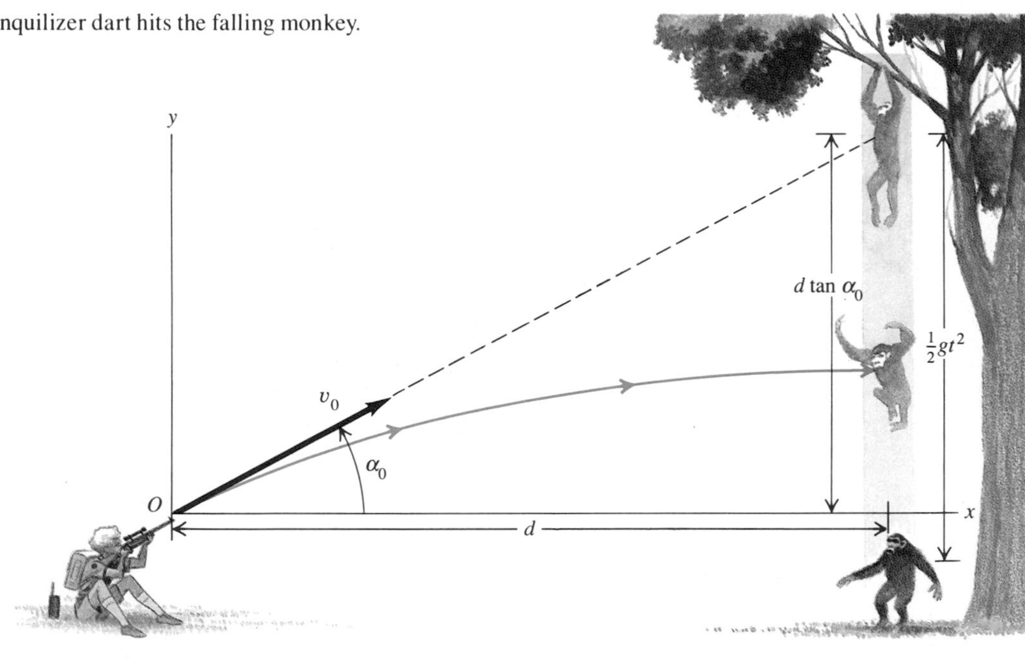


**3–16** The tranquilizer dart hits the falling monkey.

## EXAMPLE 3–6

**Height and range of a projectile** For a projectile launched with speed $v_0$ at an initial angle $\alpha_0$ (between 0 and 90°), derive general expressions for the maximum height $h$ and horizontal range $R$ (Fig. 3–12). For a given $v_0$, what value of $\alpha_0$ gives maximum height? Maximum horizontal range?

**SOLUTION** We follow the same pattern as in Example 3–4, part (b). First, for a given $\alpha_0$, when does the projectile reach its highest point? At this point, $v_y = 0$, so the time $t_1$ at the highest point ($y = h$) is given by

$$v_y = v_0 \sin \alpha_0 - gt_1 = 0, \qquad t_1 = \frac{v_0 \sin \alpha_0}{g}.$$

Next, in terms of $v_0$ and $\alpha_0$, what is the value of $y$ at this time? From Eq. (3–21),

$$h = v_0 \sin\alpha_0\left(\frac{v_0 \sin\alpha_0}{g}\right) - \frac{1}{2}g\left(\frac{v_0 \sin\alpha_0}{g}\right)^2$$
$$= \frac{v_0^2 \sin^2\alpha_0}{2g}. \qquad (3\text{–}28)$$

If we vary $\alpha_0$, the maximum value of $h$ occurs when $\sin\alpha_0 = 1$ and $\alpha_0 = 90°$, that is, when the projectile is launched straight up. That's what we should expect. If it is launched horizontally, $\alpha_0 = 0$, and the maximum height is zero!

To derive a general expression for the horizontal range $R$, we first find an expression for the time $t_2$ when the projectile returns to its initial elevation. At that time, $y = 0$, and from Eq. (3–21),

$$t_2(v_0 \sin\alpha_0 - \tfrac{1}{2}gt_2) = 0.$$

The two roots of this quadratic equation for $t_2$ are $t_2 = 0$ (launch time) and $t_2 = 2v_0 \sin\alpha_0/g$. The horizontal range $R$ is the value of $x$ at the second time. From Eq. (3–20),

$$R = (v_0 \cos\alpha_0)\frac{2v_0 \sin\alpha_0}{g}.$$

We can now use the trigonometric identity $2 \sin\alpha_0 \cos\alpha_0 = \sin 2\alpha_0$ to rewrite this as

$$R = \frac{v_0^2 \sin 2\alpha_0}{g}. \qquad (3\text{–}29)$$

**3–17** A firing angle of 45° gives the maximum horizontal range (based on a multiflash photograph).

The maximum value of $\sin 2\alpha_0$—namely, unity—occurs when $2\alpha_0 = 90°$, or $\alpha_0 = 45°$. This angle gives the maximum range for a given initial speed.

Figure 3–17 is based on a composite photograph of three trajectories of a ball projected from a spring gun at angles of 30°, 45°, and 60°. The initial speed $v_0$ is approximately the same in all three cases. The horizontal ranges are nearly the same for the 30° and 60° angles, and the range for 45° is greater than either.

Comparing the expressions for $t_1$ and $t_2$, we see that $t_2 = 2t_1$; that is, the total flight time is twice the time to reach the highest point. It follows that the time to reach the highest point equals the time to fall from there back to the initial elevation, as we asserted in Example 3–4.

*Note:* We don't recommend memorizing Eqs. (3–28) and (3–29). They are applicable only in the special circumstances that we have described. In particular, Eq. (3–29) can be used *only* when launch and landing heights are equal. There are many end-of-chapter problems to which these equations are *not* applicable.

## EXAMPLE 3–7

**A touchdown pass** A quarterback wants to throw a football at 20 m/s to a receiver 30 m away. At what angle should he throw it? Assume that the receiver catches the ball at the same height as the quarterback released it, and ignore air resistance.

**SOLUTION** We want the range to be 30 m. Solving Eq. (3–29) for $\alpha_0$, we find

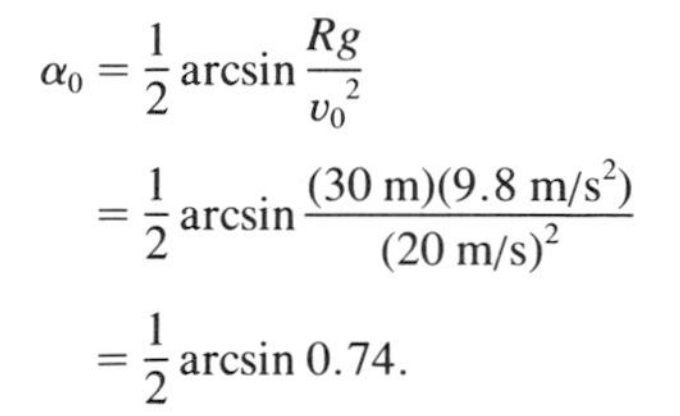

$$\alpha_0 = \frac{1}{2}\arcsin\frac{Rg}{v_0^2}$$
$$= \frac{1}{2}\arcsin\frac{(30\ \text{m})(9.8\ \text{m/s}^2)}{(20\ \text{m/s})^2}$$
$$= \frac{1}{2}\arcsin 0.74.$$

There are two values of $\alpha_0$ between 0 and 90° that satisfy this equation: arcsin 0.74 = 47° or 133°, giving $\alpha_0 = 24°$ or 66°. Both of these angles give the same range, although the time of flight and the maximum height are greater for the higher-angle trajectory. Incidentally, the sum of these two values of $\alpha_0$ is exactly 90°. This is *not* a coincidence; can you prove this? (See Problem 3–47.)

We mentioned at the beginning of this section that air resistance isn't always negligible. When it has to be included, the calculations become a lot more complicated because the air-resistance forces depend on velocity and the acceleration is no longer constant. Figure 3–18 shows a computer simulation of the trajectory of the baseball in Example 3–4, for 10 s of flight, with an air-resistance force proportional to the square of the particle's speed. We see that air resistance decreases the maximum height and range, and the trajectory becomes asymmetric. Computer simulations of projectile motion are discussed in greater detail in Section 3–6.

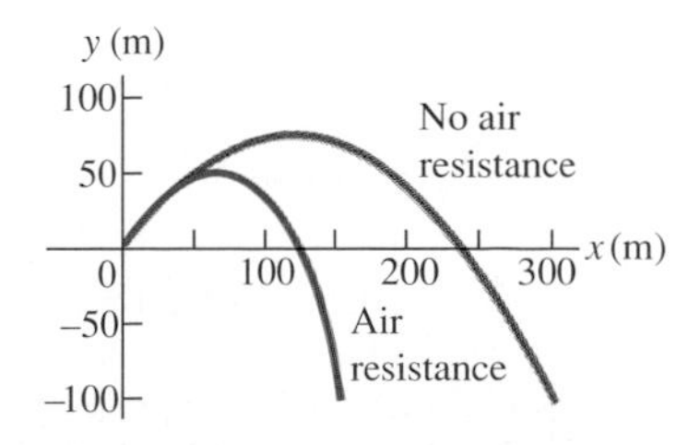


**3–18** Air resistance has a large cumulative effect on the motion of a baseball.

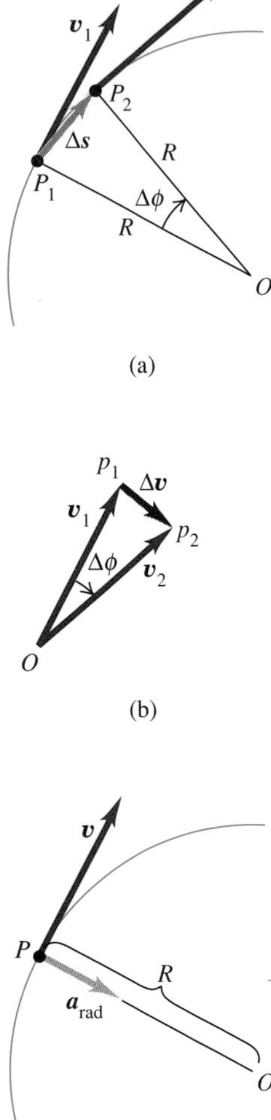


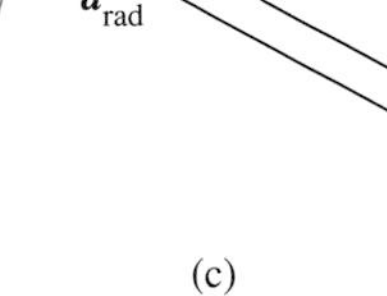




**3–19** Finding the change in velocity, $\Delta \boldsymbol{v}$, of a particle moving in a circle with constant speed.

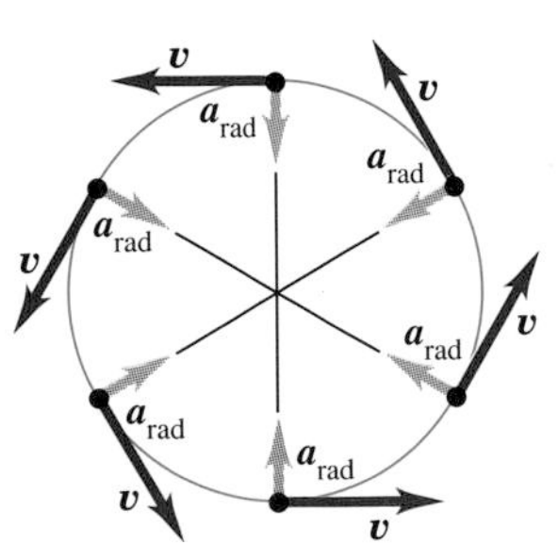


**3–20** For a particle in uniform circular motion the velocity at each point is tangent to the circle, and the acceleration is directed toward the center.

## 3–4 UNIFORM CIRCULAR MOTION

We talked about components of acceleration in Section 3–2. When a particle moves along a curved path, the direction of its velocity changes. Thus it *must* have a component of acceleration perpendicular to the path even if its speed is constant.

When a particle moves in a circle with constant speed, the motion is called **uniform circular motion.** A car rounding a curve with constant radius at constant speed, a satellite moving in a circular orbit, and an ice skater skating in a circle with constant speed are all examples of uniform circular motion. There is no component of acceleration parallel (tangent) to the path; otherwise, the speed would change. The component of acceleration normal to the path, which causes the direction of the velocity to change, is related in a simple way to the speed of the particle and the radius of the circle. Our next project is to derive this relation.

First we note that this is a different problem from the projectile-motion situation in Section 3–3, in which the acceleration was always straight down and was constant in both magnitude and direction. Here the acceleration is perpendicular to the velocity at each instant; as the direction of the velocity changes, the direction of the acceleration also changes. As we will see, the acceleration vector at each point in the circular path is directed toward the *center* of the circle.

Figure 3–19a shows a particle moving with constant speed in a circular path of radius $R$ with center at $O$. The particle moves from $P_1$ to $P_2$ in a time $\Delta t$. The vector change in velocity $\Delta \boldsymbol{v}$ during this time is shown in Fig. 3–19b.

The triangles $OP_1P_2$ and $Op_1p_2$ in Fig. 3–19 are *similar* because both are isosceles triangles and the angles labeled $\Delta\phi$ are the same. Ratios of corresponding sides are equal, so

$$\frac{|\Delta \boldsymbol{v}|}{v_1} = \frac{\Delta s}{R}, \qquad \text{or} \qquad |\Delta \boldsymbol{v}| = \frac{v_1}{R}\Delta s.$$

The magnitude $a_{av}$ of the average acceleration $\boldsymbol{a}_{av}$ during $\Delta t$ is therefore

$$a_{av} = \frac{|\Delta \boldsymbol{v}|}{\Delta t} = \frac{v_1}{R}\frac{\Delta s}{\Delta t}.$$

The magnitude $a$ of the instantaneous acceleration $\boldsymbol{a}$ at point $P_1$ is the limit of this expression as we take point $P_2$ closer and closer to point $P_1$:

$$a = \lim_{\Delta t \to 0} \frac{v_1}{R}\frac{\Delta s}{\Delta t} = \frac{v_1}{R} \lim_{\Delta t \to 0} \frac{\Delta s}{\Delta t}.$$

But the limit of $\Delta s / \Delta t$ is the speed $v_1$ at point $P_1$. Also, $P_1$ can be any point on the path, so we can drop the subscript and let $v$ represent the speed at any point. Then

$$a_{rad} = \frac{v^2}{R}. \qquad (3\text{–}30)$$

We have added the subscript "rad" as a reminder that the direction of the instantaneous acceleration at each point is always along a radius of the circle, toward its center, and perpendicular to the instantaneous velocity at the point (Fig. 3–19c).

We conclude: **In uniform circular motion the magnitude $a$ of the instantaneous acceleration is equal to the square of the speed $v$ divided by the radius $R$ of the circle. Its direction is perpendicular to $v$ and inward along the radius.** Because the acceleration is always directed toward the center of the circle, it is sometimes called **centripetal acceleration.** The word *centripetal* is derived from two Greek words meaning "seeking the center." Figure 3–20 shows the directions of the velocity and acceleration vectors at several points for a particle moving with uniform circular motion. Compare this with the projectile motion shown in Fig. 3–12, in which the acceleration is always directed straight down and is *not* perpendicular to the path, except at one point.

We can also express the magnitude of the acceleration in uniform circular motion in terms of the **period** $T$ of the motion, the time for one revolution. If a particle travels once around the circle, a distance of $2\pi R$, in a time $T$, its speed $v$ is

$$v = \frac{2\pi R}{T}. \tag{3–31}$$

When we substitute this into Eq. (3–30), we obtain the alternative expression

$$a_{\text{rad}} = \frac{4\pi^2 R}{T^2}. \tag{3–32}$$

## EXAMPLE 3–8

The 1990 Corvette claims a "lateral acceleration" of $0.91g$, which is $(0.91)(9.8\ \text{m/s}^2) = 8.9\ \text{m/s}^2$. If this represents the maximum centripetal acceleration that can be attained without skidding out of the circular path, and if the car is traveling at a constant 45 m/s (about 101 mi/h), what is the minimum radius of curve it can negotiate? (Assume that the curve is unbanked.)

**SOLUTION** We are given $a_{\text{rad}}$ and $v$, so we first solve Eq. (3–30) for $R$:

$$R = \frac{v^2}{a_{\text{rad}}} = \frac{(45\ \text{m/s})^2}{8.9\ \text{m/s}^2} = 230\ \text{m} \quad (\text{about } \tfrac{1}{7}\ \text{mi}).$$

If the curve is banked, the radius can be smaller, as we will see in Chapter 5.

## EXAMPLE 3–9

In a carnival ride the passengers travel in a circle of radius 5.0 m (Fig. 3–21). They make one complete circle in 4.0 s. What is their acceleration?

**SOLUTION** The speed is the circumference of the circle divided by the period $T$ (the time for one revolution):

$$v = \frac{2\pi R}{T} = \frac{2\pi(5.0\ \text{m})}{4.0\ \text{s}} = 7.9\ \text{m/s}.$$

The centripetal acceleration is

$$a_{\text{rad}} = \frac{v^2}{R} = \frac{(7.9\ \text{m/s})^2}{5.0\ \text{m}} = 12\ \text{m/s}^2,$$

or, from Eq. (3–32),

$$a_{\text{rad}} = \frac{4\pi^2(5.0\ \text{m})}{(4.0\ \text{s})^2} = 12\ \text{m/s}^2.$$

As in the preceding example, the direction of $\boldsymbol{a}$ is always toward the center of the circle. The magnitude of $\boldsymbol{a}$ is greater than $g$, the acceleration due to gravity, so this is not a ride for the faint-hearted. (But some roller coasters subject their passengers to accelerations as great as $4g$.)

**3–21** A carnival ride.

We have assumed throughout this discussion that the particle's speed is constant. If the speed varies, Eq. (3–30) still gives the *radial* component of acceleration $a_{\text{rad}} = v^2/R$, which is always perpendicular to the instantaneous velocity and directed toward the center of the circle. In that case, however, there is also a component of acceleration $a_{\text{tan}}$ that is tangent to the circle, parallel to the instantaneous velocity. From the discussion at the end of Section 3–2, we see that the tangential component of acceleration $a_{\text{tan}}$ is equal to

## EXAMPLE 3–11

In what direction should the pilot in Example 3–10 head to travel due north? What will be her velocity relative to the earth then? (Assume that the wind velocity and the magnitude of her airspeed are the same as in Example 3–10.)

**SOLUTION** Now the information given is

$$v_{P/A} = 240\ \text{km/h} \quad \text{direction unknown},$$

$$\boldsymbol{v}_{A/E} = 100\ \text{km/h} \quad \text{due east}.$$

We want to find $\boldsymbol{v}_{P/E}$; its magnitude is unknown, but we know that its direction is due north. Note that both this and the preceding example require us to determine two unknown quantities. In Example 3–10 these were the magnitude and direction of $\boldsymbol{v}_{P/E}$; in this example the unknowns are the direction of $\boldsymbol{v}_{P/A}$ and the *magnitude* of $\boldsymbol{v}_{P/E}$.

The three relative velocities must still satisfy the vector equation

$$\boldsymbol{v}_{P/E} = \boldsymbol{v}_{P/A} + \boldsymbol{v}_{A/E}.$$

The appropriate vector diagram is shown in Fig. 3–24. We find

$$v_{P/E} = \sqrt{(240\ \text{km/h})^2 - (100\ \text{km/h})^2} = 218\ \text{km/h},$$

$$\phi = \arcsin\frac{100\ \text{km/h}}{240\ \text{km/h}} = 25°.$$

The pilot should head 25° west of north, and her ground speed is then 218 km/h.

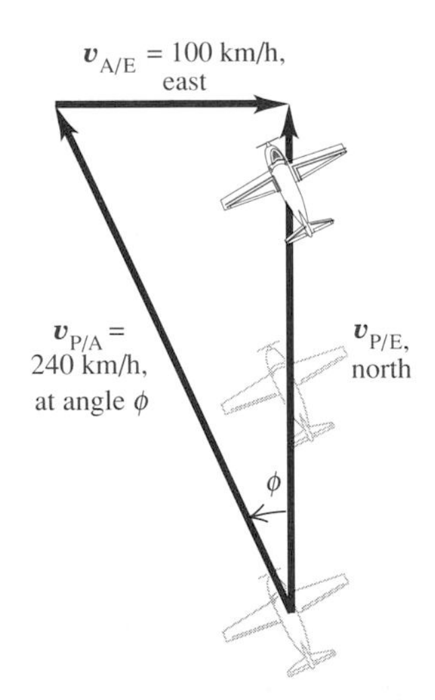


**3–24** The pilot must point the plane in the direction of the vector $\boldsymbol{v}_{P/A}$ to travel due north relative to the earth.

# 3–6 BASEBALL TRAJECTORIES: A Case Study in Computer Modeling

Throughout our study of projectile motion in this chapter we have assumed that air-resistance forces are negligibly small. In fact, though, air resistance (often called air drag, or simply drag) has a major effect on many forms of motion, including the motion of bicycle riders, tennis balls, airplanes, and everything in between. It's not difficult to include an air-resistance force in the equations for a projectile, but solving the equations to find the position and velocity as functions of time, or the shape of the path, can get quite complex. Fortunately, by using a computer we can obtain close numerical approximations to these solutions. That is what this section is about.

When we omit air drag, the components of acceleration $\boldsymbol{a}$ of a projectile are simply

$$a_x = 0, \qquad a_y = -g.$$

The effect of air drag is represented by additional terms in these expressions. The magnitude of the acceleration $\boldsymbol{a}_{\text{drag}}$ caused by air drag alone is approximately proportional to the square of the body's speed and can be represented with a proportionality constant $k$ as

$$a_{\text{drag}} = kv^2, \tag{3–37}$$

where $v^2 = v_x{}^2 + v_y{}^2$. The corresponding acceleration components are

$$(a_x)_{\text{drag}} = -kvv_x, \qquad (a_y)_{\text{drag}} = -kvv_y. \tag{3–38}$$

Note that each component is opposite in direction to the corresponding component of velocity and that $a_{\text{drag}}{}^2 = (a_x)_{\text{drag}}{}^2 + (a_y)_{\text{drag}}{}^2$.

The constant $k$ depends on the density $\rho$ of air, the silhouette area $A$ of the body, its mass $m$, and a dimensionless constant $C$, called the drag coefficient, that depends on the shape of the body. Typical values of $C$ range from 0.3 to 0.6. In terms of these quantities, $k$ is given by

$$k = \frac{\rho CA}{2m} \tag{3–39}$$

The acceleration components, including the effects of both gravity and air drag, are

$$a_x = -kvv_x, \qquad a_y = -g - kvv_y. \tag{3–40}$$

Now comes the basic idea of our numerical calculation. If we know the coordinates and the velocity components at some time $t$, we can find these quantities at a slightly later time $t + \Delta t$. Here's how we do it. The average $x$-component of velocity is $v_x = \Delta x/\Delta t$. During time $\Delta t$ the coordinate $x$ changes by an amount $\Delta x = v_x\,\Delta t$. Similarly, $y$ changes by an amount $v_y\,\Delta t$. So the values of $x$ and $y$ at the end of the interval are

$$x + \Delta x = x + v_x\,\Delta t, \qquad y + \Delta y = v_y + \Delta t. \tag{3–41}$$

While this is happening, the velocity components are also changing. The average component of acceleration $a_x$ is $a_x = \Delta v_x/\Delta t$, so during $\Delta t$, $v_x$ changes by an amount $\Delta v_x = a_x\,\Delta t$, and similarly for $v_y$. So the values of $v_x$ and $v_y$ at the end of the interval are

$$v_x + \Delta v_x = v_x + a_x\,\Delta t, \qquad v_y + \Delta v_y = v_y + a_y\,\Delta t. \tag{3–42}$$

We have to specify the starting conditions, that is, the initial values of $x$, $y$, $v_x$, and $v_y$. Then we can step through the calculation to find the position and velocity at the end of each interval in terms of their values at the beginning and thus find the values at the end of any number of intervals. It would be a lot of work to do all the calculations for 100 or more steps with a hand calculator, but the computer does it for us quickly and accurately.

Exactly how you implement this plan depends on whether you use a computer language, such as BASIC, FORTRAN, or Pascal, or a spreadsheet or numerical-analysis software package, such as MathCAD. Here's a sketch of a general algorithm using a programming language.

*Step 1:* Identify the parameters of the problem, $m$, $A$, $C$, and $\rho$, and evaluate $k$.

*Step 2:* Choose the time interval $\Delta t$ (which we'll abbreviate below as $h$) and the initial values of $x$, $y$, $v_x$, $v_y$, and $t$. You may want to express the initial velocity components in terms of the magnitude $v_0$ and direction $\alpha_0$ of the initial velocity.

*Step 3:* Choose the maximum number of intervals $N$ or the maximum time $t_{max} = Nh$ for which you want to get the numerical solution.

*Step 4:* Loop (or iterate) Steps 5–9 while $n < N$ or $t < t_{max}$:

*Step 5:* Calculate the acceleration components:

$$a_x = -kvv_x, \qquad a_y = -g - kvv_y.$$

*Step 6:* Print or plot $x$, $y$, $v_x$, $v_y$, $a_x$, and $a_y$.

*Step 7:* Calculate the new coordinates from Eqs. (3–41):

$$x = x + v_x h, \qquad y = y + v_y h.$$

*Step 8:* Calculate the new velocity components from Eqs. (3–42):

$$v_x = v_x + a_x h, \qquad v_y = v_y + a_y h.$$

*Step 9:* Increment the time by $h$:

$$t = t + h.$$

*Step 10:* Stop.

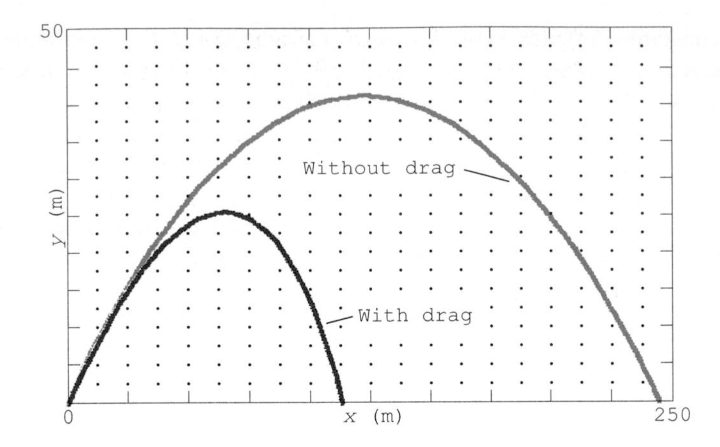


**3–25** Computer-generated trajectories of a baseball with and without drag.

If you are using a spreadsheet, the following notation may be useful. Let $x(n)$, $y(n)$, $v_x(n)$, and $v_y(n)$ be the values at the end of the $n$th interval. These values appear across the $n$th line of the spreadsheet. The initial values are $x(1)$, $y(1)$, $v_x(1)$, and $v_y(1)$. Then Eqs. (3–41) and (3–42) become

$$x(n+1) = x(n) + v_x(n)h, \qquad y(n+1) = y(n) + v_y(n)h, \tag{3–43}$$

$$v_x(n+1) = v_x(n) + a_x(n)h, \qquad v_y(n+1) = v_y(n) + a_y(n)h. \tag{3–44}$$

In Eqs. (3–44) we compute $a_x(n)$ and $a_y(n)$ by substituting the values of $v_x(n)$ and $v_y(n)$ into Eqs. (3–40).

Figure 3–25 shows the trajectory of a baseball, with and without air drag. The radius of the baseball is $r = 0.0366$ m, and $A = \pi r^2$. The mass is $m = 0.145$ kg, the drag coefficient is about $C = 0.5$, and the density of air is about $\rho = 1.2\ \text{kg/m}^3$. In this example the baseball was given an initial velocity of 50 m/s at an angle of 35° from the $+x$-axis. We see that air drag is an important part of the game of baseball.

## SUMMARY

### KEY TERMS

- position vector
- average velocity
- instantaneous velocity
- average acceleration
- instantaneous acceleration
- projectile
- uniform circular motion

• The position vector $\boldsymbol{r}$ of a point $P$ in a plane is the displacement vector from the origin to $P$. Its components are the coordinates $x$ and $y$.

• The average velocity $\boldsymbol{v}_{av}$ during the time interval $\Delta t$ is the displacement $\Delta \boldsymbol{r}$ (the change in the position vector $\boldsymbol{r}$) divided by $\Delta t$:

$$\boldsymbol{v}_{av} = \frac{\boldsymbol{r}_2 - \boldsymbol{r}_1}{t_2 - t_1} = \frac{\Delta \boldsymbol{r}}{\Delta t}. \tag{3–2}$$

Instantaneous velocity is $\boldsymbol{v} = d\boldsymbol{r}/dt$; its components are

$$v_x = \frac{dx}{dt}, \qquad v_y = \frac{dy}{dt}, \qquad v_z = \frac{dz}{dt}. \tag{3–4}$$

• The average acceleration $\boldsymbol{a}_{av}$ during the time interval $\Delta t$ is the velocity change $\Delta \boldsymbol{v}$ divided by $\Delta t$:

$$\boldsymbol{a}_{av} = \frac{\boldsymbol{v}_2 - \boldsymbol{v}_1}{t_2 - t_1} = \frac{\Delta \boldsymbol{v}}{\Delta t}. \tag{3–8}$$

Instantaneous acceleration is $\boldsymbol{a} = d\boldsymbol{v}/dt$. Its components are

$$a_x = \frac{dv_x}{dt}, \qquad a_y = \frac{dv_y}{dt}, \qquad a_z = \frac{dv_z}{dt}. \tag{3–10}$$

Acceleration can also be represented in terms of its components parallel and perpendicular to the direction of the instantaneous velocity.

- In projectile motion, $a_x = 0$ and $a_y = -g$. The coordinates and velocity components, as functions of time, are

$$x = (v_0 \cos \alpha_0)t, \tag{3–20}$$

$$y = (v_0 \sin \alpha_0)t - \tfrac{1}{2}gt^2, \tag{3–21}$$

$$v_x = v_0 \cos \alpha_0, \tag{3–22}$$

$$v_y = v_0 \sin \alpha_0 - gt. \tag{3–23}$$

The shape of the path in projectile motion is always a parabola.

- When a particle moves in a circular path of radius $R$ with speed $v$, it has an acceleration with magnitude

$$a_{\text{rad}} = \frac{v^2}{R}, \tag{3–30}$$

always directed toward the center of the circle and perpendicular to $\boldsymbol{v}$. The period $T$ of a circular motion is the time for one revolution. If the speed is constant, then $v = 2\pi R/T$, and

$$a_{\text{rad}} = \frac{4\pi^2 R}{T^2}. \tag{3–32}$$

When the speed is not constant, there is also a component of $\boldsymbol{a}$ parallel to the path; this is equal to the rate of change of speed, $dv/dt$.

- When a body $P$ moves relative to a body (or reference frame) $B$, and $B$ moves relative to $A$, we denote the velocity of $P$ relative to $B$ by $\boldsymbol{v}_{P/B}$, the velocity of $P$ relative to $A$ by $\boldsymbol{v}_{P/A}$, and the velocity of $B$ relative to $A$ by $\boldsymbol{v}_{B/A}$. These velocities are related by

$$\boldsymbol{v}_{P/A} = \boldsymbol{v}_{P/B} + \boldsymbol{v}_{B/A}. \tag{3–35}$$

**centripetal acceleration**
**period**
**relative velocity**

# EXERCISES

## Section 3–1 The Velocity Vector

**3–1** A squirrel has $x$- and $y$-coordinates (2.7 m, 3.8 m) at time $t_1 = 0$ and coordinates $(-4.5\text{ m}, 8.1\text{ m})$ at time $t_2 = 4.0$ s. For this time interval, find a) the components of the average velocity; b) the magnitude and direction of the average velocity.

**3–2** An elephant is at the origin of coordinates at time $t_1 = 0$. For the time interval from $t_1 = 0$ to $t_2 = 20.0$ s, the average velocity of the elephant has components $(v_x)_{\text{av}} = 3.6$ m/s and $(v_y)_{\text{av}} = -5.2$ m/s. At time $t_2 = 20.0$ s, a) what are the $x$- and $y$-coordinates of the elephant? b) how far is the elephant from the origin?

## Section 3–2 The Acceleration Vector

**3–3** A jet plane at time $t_1 = 0$ has components of velocity $v_x = 190$ m/s, $v_y = -120$ m/s. At time $t_2 = 20.0$ s the velocity components are $v_x = 110$ m/s, $v_y = 60$ m/s. For this time interval, calculate a) the components of the average acceleration; b) the magnitude and direction of the average acceleration.

**3–4** A dog running in an open field has components of velocity $v_x = 4.5$ m/s and $v_y = 3.2$ m/s at $t_1 = 10.0$ s. For the time interval from $t_1 = 10.0$ s to $t_2 = 20.0$ s, the average acceleration of the dog has magnitude 0.55 m/s$^2$ and direction 52.0° counterclockwise from the $+x$-axis. At $t = 20.0$ s, a) what are the $x$- and $y$-components of the dog's velocity? b) what are the magnitude and direction of the dog's velocity?

**3–5** The coordinates of a bird flying in the $xy$-plane are given as functions of time by $x = 2.0\text{ m} - \alpha t$ and $y = \beta t^2$, where $\alpha = 3.6$ m/s and $\beta = 2.8$ m/s$^2$. a) Calculate the velocity and acceleration vectors of the bird as functions of time. b) Calculate the magnitude and direction of the bird's velocity and acceleration at $t = 3.0$ s.

**3–6** A motorcycle moves in the $xy$-plane with acceleration $\boldsymbol{a} = \alpha t^2\boldsymbol{i} + \beta t\boldsymbol{j}$, where $\alpha = 1.2$ m/s$^4$ and $\beta = 3.5$ m/s$^3$. a) Assuming that the motorcycle is at rest at the origin at time $t = 0$, derive expressions for the velocity and position vectors as functions of time. b) Sketch the path of the motorcycle. c) Find the magnitude and direction of the velocity at $t = 3.0$ s.

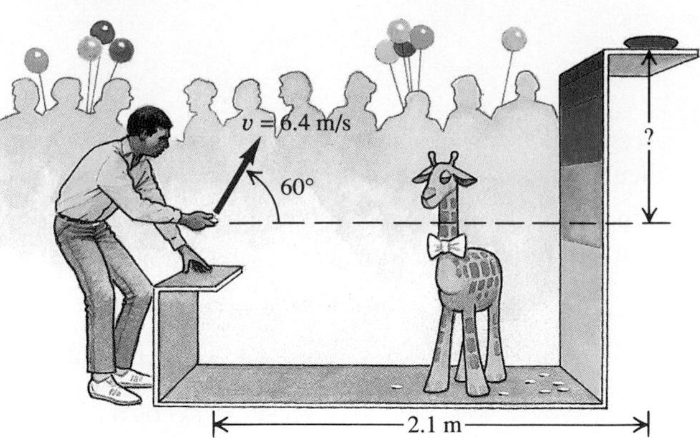


**FIGURE 3–26**

## Section 3–3 Projectile Motion

**3–7** A physics book slides off a horizontal table top with a speed of 3.60 m/s. It strikes the floor in 0.500 s. Find a) the height of the table top above the floor; b) the horizontal distance from the edge of the table to the point where the book strikes the floor; c) the horizontal and vertical components of the book's velocity and the magnitude and direction of its velocity just before it reaches the floor.

**3–8** A tennis ball rolls off the edge of a table top 1.00 m above the floor and strikes the floor at a point 2.20 m horizontally from the edge of the table. a) Find the time of flight. b) Find the magnitude of the initial velocity. c) Find the magnitude and direction of the velocity of the ball just before it strikes the floor. Draw a diagram to scale.

**3–9** A sharpshooter fires a .22-caliber rifle horizontally at a target. The bullet has a muzzle velocity with magnitude 275 m/s. a) How much does the bullet drop in flight if the target is 50 m away? b) Sketch a graph of the vertical drop of the bullet as a function of the distance to the target.

**3–10 Dropping a Bomb.** A military airplane on a routine training mission is flying horizontally at a speed of 100 m/s and accidentally drops a bomb (fortunately not armed) at an elevation of 2000 m. a) How much time is required for the bomb to reach the earth? b) How far does it travel horizontally while falling? c) Find the horizontal and vertical components of its velocity just before it strikes the earth. d) Where is the airplane when the bomb strikes the earth if the velocity of the airplane remains constant?

**3–11** Jose Canseco throws a baseball at an angle of 53.1° above the horizontal with an initial speed of 45.0 m/s. a) At what *two* times is the baseball at a height of 25.0 m above the point from which it was thrown? b) Calculate the horizontal and vertical components of the baseball's velocity at each of the two times calculated in part (a). c) What are the magnitude and direction of the baseball's velocity when it returns to the level from which it was thrown?

**3–12** John Elway throws a football with an initial upward velocity component of 18.0 m/s and a horizontal velocity component of 25.0 m/s. a) How much time is required for the football to reach the highest point of the trajectory? b) How high is this point? c) How much time (after being thrown) is required for the football to return to its original level? How does this compare with the time calculated in part (a)? d) How far has it traveled horizontally during this time?

**3–13** A pistol that fires a signal flare gives the flare an initial speed (muzzle speed) of 240 m/s. a) If the flare is fired at an angle of 55° above the horizontal on the level salt flats of Utah, what is its horizontal range? b) If the flare is fired at the same angle over the flat Sea of Tranquility on the moon, where $g = 1.6$ m/s$^2$, what is its horizontal range?

**3–14** A batted baseball leaves the bat at an angle of 30° above the horizontal and is caught by an outfielder 122 m (400 ft) from the plate. Assume that the height of the point where it was struck by the bat equals the height of the point where it was caught. a) What was the initial speed of the ball? b) How high did it rise above the point where it struck the bat?

**3–15** A man stands on the roof of a building that is 30.0 m tall and throws a rock with a velocity of magnitude 60.0 m/s at an angle of 33.0° above the horizontal. Calculate a) the maximum height above the roof reached by the rock; b) the magnitude of the velocity of the rock just before it strikes the ground; c) the horizontal distance from the base of the building to the point where the rock strikes the ground.

**3–16** A Civil War mortar called the Dictator fired its 90.7-kg (200-lb) shell a maximum horizontal distance of 4345 m (4752 yd) when the shell was projected at an angle of 45° above the horizontal. a) What was the muzzle speed of the shell (the speed of the shell as it left the barrel of the mortar)? b) What maximum height above the ground did the shell reach?

**3–17** In a carnival booth you win a stuffed giraffe if you toss a quarter into a small dish. The dish is on a shelf above the point where the quarter leaves your hand and is a horizontal distance of 2.1 m from this point (Fig. 3–26). If you toss the coin with a velocity of 6.4 m/s at an angle of 60° above the horizontal, the coin lands in the dish. a) What is the height of the shelf above the point where the quarter leaves your hand? b) What is the vertical component of the velocity of the quarter just before it lands in the dish?

**3–18** Suppose the departure angle $\alpha_0$ in Fig. 3–16 is 58.0° and the distance $d$ is 5.00 m. Where will the dart and monkey meet if the initial speed of the dart is a) 22.0 m/s; b) 14.0 m/s? c) What will happen if the initial speed of the dart is 6.0 m/s? Sketch the trajectory in each case.

## Section 3–4 Uniform Circular Motion

**3–19** The earth has a radius of $6.38 \times 10^6$ m and turns around once on its axis in 24 h. What is the radial acceleration of an object at the earth's equator in units of m/s$^2$?

**3–20** The radius of the earth's orbit around the sun (assumed to be circular) is $1.49 \times 10^{11}$ m, and the earth travels around this orbit in 365 days. a) What is the magnitude of the orbital velocity of the earth in m/s? b) What is the radial acceleration of the earth toward the sun in m/s$^2$?

FIGURE 3–27

**3–21** A Ferris wheel with radius 14.0 m is turning about a horizontal axis through its center (Fig. 3–27). The linear speed of a passenger on the rim is constant and equal to 9.00 m/s. a) What are the magnitude and direction of the passenger's acceleration as she passes through the lowest point in her circular motion? b) How much time does it take the Ferris wheel to make one revolution?

**3–22** A model of a helicopter rotor has four blades, each 3.20 m in length from the central shaft to the blade tip. The model is rotated in a wind tunnel at 1500 rev/min. a) What is the linear speed of the blade tip in m/s? b) What is the radial acceleration of the blade tip expressed as a multiple of the acceleration of gravity, $g$?

## Section 3–5 Relative Velocity

**3–23** An airplane pilot wishes to fly due north. A wind of 80.0 km/h (about 50 mi/h) is blowing toward the west. a) If the flying speed of the plane (its speed in still air) is 290.0 km/h (about 180 mi/h), in what direction should the pilot head? b) What is the speed of the plane over the ground? Illustrate with a vector diagram.

**3–24** A passenger on a ship traveling due east with a speed of 26.0 knots observes that the stream of smoke from the ship's funnels makes an angle of 20° with the ship's wake (Fig. 3–28). The wind is blowing from south to north. Assume that the smoke acquires a velocity (with respect to the earth) equal to the velocity of the wind as soon as it leaves the funnels. Find the magnitude of the velocity of the wind in knots. (A knot is a unit of speed used by sailors; 1 knot = 1.852 km/h.)

**3–25** A river flows due north with a speed of 2.4 m/s. A man rows a boat across the river; his velocity relative to the water is 3.5 m/s due east. The river is 1000 m wide. a) What is his velocity relative to the earth? b) How much time is required to cross the river? c) How far north of his starting point will he reach the opposite bank?

**3–26** a) In what direction should the rowboat in Exercise 3–25 be headed to reach a point on the opposite bank directly east from the starting point? b) What is the velocity of the boat relative to the earth? c) How much time is required to cross the river?

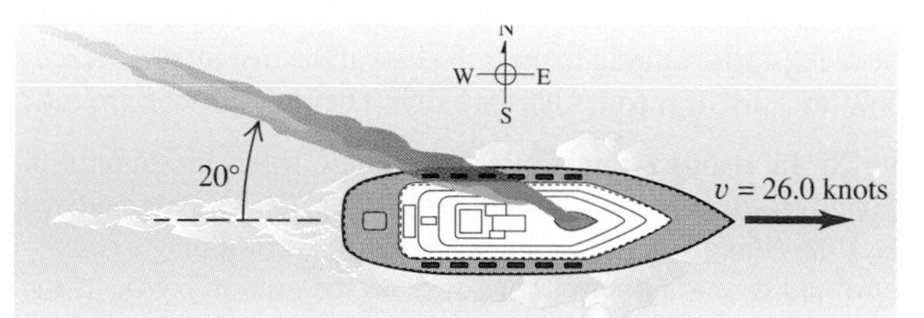


FIGURE 3–28

## Section 3–6 Baseball Trajectories: A Case Study in Computer Modeling

**3–27** Implement the algorithm in Section 3–6 on a computer. Assume that the object is a baseball with $m = 0.145$ kg, $r = 0.0366$ m, and $C = 0.5$. At what angle should the ball be hit or thrown to maximize the range? Assume that $v_0 = 50$ m/s.

**3–28** Bo Jackson made a flat-footed 300-ft throw (no bounces) from left field to home plate to put out a tenth-inning tying run. Assuming that $\theta = 40$ degrees, with what speed did he throw the ball? Is this more or less than a good fastball pitcher?

**3–29** In tennis, 100 mph is a very good serving speed. How fast is the ball moving when it crosses the baseline in the opposite court (24 m distant)? Assume that $m = 0.055$ kg, $r = 0.031$ m, $C = 0.75$.

**3–30** Estimate the maximum distance a human being can throw a ping-pong ball. A ping-pong ball has $r = 0.019$ m and $m = 0.0024$ kg. Why can a baseball be thrown much farther than a ping-pong ball?

# PROBLEMS

**3–31** The coordinates of a particle moving in the $xy$-plane are given as functions of time by $x = \alpha t$ and $y = 19.0\text{ m} - \beta t^2$, where $\alpha = 1.40$ m/s and $\beta = 0.800$ m/s$^2$. a) What is the particle's distance from the origin at time $t = 2.00$ s? b) What is the particle's velocity (magnitude and direction) at time $t = 2.00$ s? c) What is the particle's acceleration (magnitude and direction) at time $t = 2.00$ s? d) At what times is the particle's velocity perpendicular to its acceleration? e) At what times is the particle's velocity perpendicular to its position vector? What are the locations of the particle at these times? f) What is the particle's minimum distance from the origin? At what times does this minimum occur? g) Sketch the path of the particle.

***3–32** A faulty model rocket moves in the $xy$-plane in a coordinate system in which the positive $y$-direction is vertically upward. The rocket's acceleration has components given by $a_x = \alpha t^2$ and $a_y = \beta - \gamma t$, where $\alpha = 2.50$ m/s$^4$, $\beta = 12.0$ m/s$^2$, and $\gamma = 2.00$ m/s$^3$. At $t = 0$ the rocket is at the origin and has an initial velocity

$v_0 = v_{0x}i + v_{0y}j$ with $v_{0x} = 2.00$ m/s and $v_{0y} = 6.00$ m/s. a) Calculate the velocity and position vectors as functions of time. b) What is the maximum height reached by the rocket? c) What is the horizontal displacement of the rocket when it returns to $y = 0$?

***3–33** A bird flies in the $xy$-plane with a velocity vector given by $v = (\alpha - \beta t^2)i + \gamma t j$, where $\alpha = 2.1$ m/s, $\beta = 3.6$ m/s$^3$, and $\gamma = 5.0$ m/s$^2$ and where the positive $y$-direction is vertically upward. At $t = 0$ the bird is at the origin. a) Calculate the position and acceleration vectors of the bird as functions of time. b) What is the bird's altitude ($y$-coordinate) as it flies over $x = 0$ for the first time after $t = 0$?

**3–34** A player kicks a football at an angle of 40.0° above the horizontal with an initial speed of 14.0 m/s. A second player standing at a distance of 30.0 m from the first (in the direction of the kick) starts running to meet the ball at the instant it is kicked. How fast must he run to catch the ball just before it hits the ground?

**3–35** In fighting forest fires, airplanes work in support of ground crews by dropping water on the fires. (This was depicted in the 1989 film *Always.*) A pilot is practicing by dropping a cannister of red dye, hoping to hit a target on the ground below. If the plane is flying in a horizontal path 70.0 m above the ground and with a speed of 54.0 m/s (120 mi/h), at what horizontal distance from the target should the pilot release the cannister?

**3–36** A girl throws a water-filled balloon at an angle of 50.0° with a speed of 12.0 m/s. A car is advancing toward the girl at a constant speed of 8.00 m/s (Fig. 3–29). If the balloon is to hit the car, how far away should the car be when the balloon is thrown?

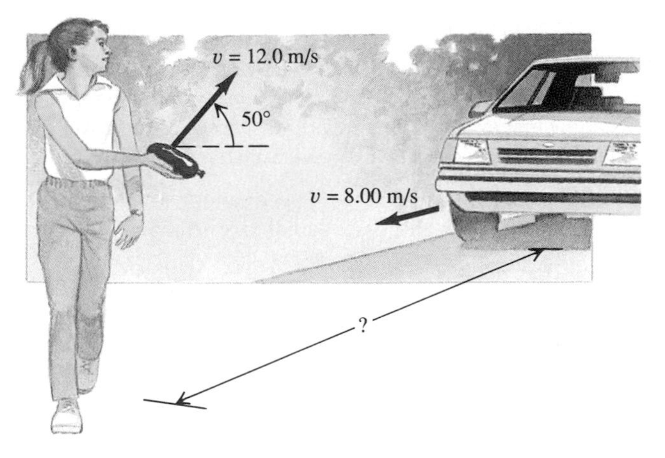


**FIGURE 3–29**

**3–37 The Longest Home Run.** According to the *Guinness Book of World Records,* the longest home run ever measured was hit by Roy "Dizzy" Carlyle in a minor league game and traveled 188 m (618 ft) before landing on the ground outside the ballpark. a) Assuming that the ball's initial velocity was 45° above the horizontal and neglecting air resistance, what was the initial speed of the ball if it was hit at a point 0.9 m (3.0 ft) above ground level? Assume that the ground was perfectly flat. b) How far would the ball be above a fence 3.0 m (10 ft) in height and 116 m (380 ft) from home plate?

**3–38** A baseball thrown at an angle of 60.0° above the horizontal strikes a building 36.0 m away at a point 15.0 m above the point from which it is thrown. a) Find the magnitude of the initial velocity of the baseball (the velocity with which it is thrown). b) Find the magnitude and direction of the velocity of the baseball just before it strikes the building.

**3–39** An airplane diving at an angle of 40.9° below the horizontal drops a mailbag from an altitude of 900 m. The bag strikes the ground 6.00 s after its release. a) What is the speed of the plane? b) How far does the bag travel horizontally during its fall? c) What are the horizontal and vertical components of its velocity just before it strikes the ground?

**3–40** A snowball rolls off a barn roof that slopes downward at an angle of 40° (Fig. 3–30). The edge of the roof is 14.0 m above the ground, and the snowball has a speed of 7.00 m/s as it rolls off

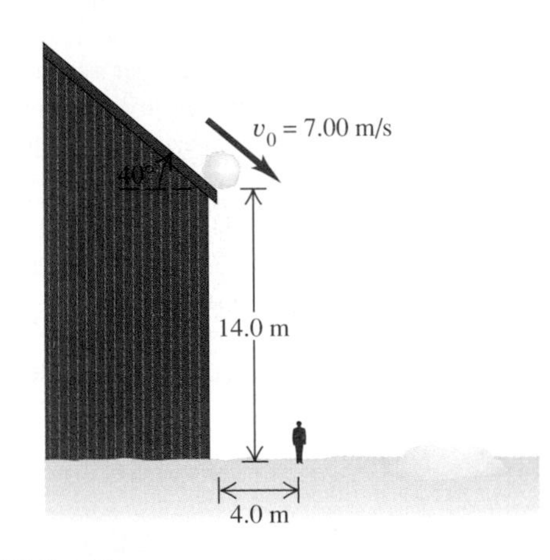

**FIGURE 3–30**

the roof. a) How far from the edge of the barn does the snowball strike the ground if it doesn't strike anything else while falling? b) A man 1.9 m tall is standing 4.0 m from the edge of the barn. Will he be hit by the snowball?

**3–41** A baseball thrown by a centerfielder toward home plate reaches a maximum height above the point where it was thrown of 24.4 m (80 ft). The baseball was thrown at an angle of 40.0° above the horizontal and is caught by an infielder. a) How far does it travel horizontally? b) For how much time is it in the air? c) What is the magnitude of its velocity just before it is caught?

**3–42 On the Flying Trapeze.** A new circus act is called the Texas Tumblers. Lovely Mary Belle swings from a trapeze, projects herself at an angle of 53°, and is supposed to be caught by Joe Bob, whose hands are 6.1 m above and 8.2 m horizontally from her launch point (Fig. 3–31). a) What initial speed $v_0$ must Mary Belle have to just reach Joe Bob? b) For the initial speed calculated in part (a), what are the magnitude and the direction of her velocity when Mary Belle reaches Joe Bob? c) The night of their debut performance, Joe Bob misses her completely as she flies past. How far horizontally does Mary Belle travel, from her

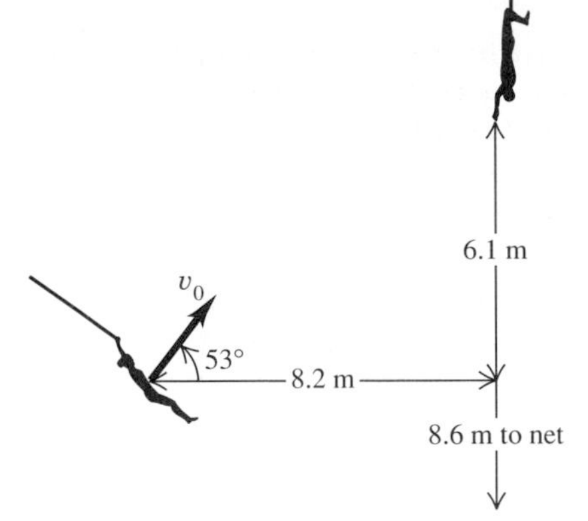


**FIGURE 3–31**

initial launch point, before landing in the safety net 8.6 m below her starting point?

**3–43** A physics professor did daredevil stunts in his spare time. His last stunt was to attempt to jump across a river on a motorcycle (Fig. 3–32). The takeoff ramp was inclined at 53.0°, the river was 40.0 m wide, and the far bank was 15.0 m lower than the top of the ramp. The river itself was 100 m below the ramp. What should his speed have been at the top of the ramp to have just made it to the edge of the far bank?

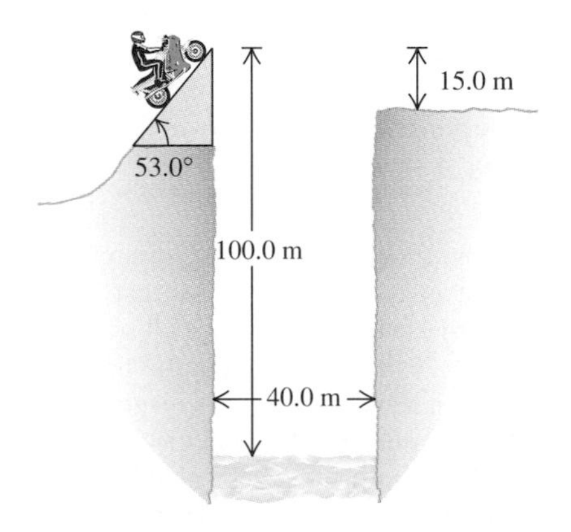

**FIGURE 3–32**

**3–44** A movie stuntwoman drops from a helicopter 30.0 m above ground and moving with a constant velocity whose components are 10.0 m/s upward and 20.0 m/s horizontal and toward the east. Where on the ground (relative to the position of the helicopter when she drops) should the stuntwoman have placed her foam mats to break her fall?

**3–45** A basketball player is fouled and knocked to the floor during a layup attempt. He is awarded two free throws. The center of the basket is a horizontal distance of 4.21 m (13.8 ft) from the foul line and is a height of 3.05 m (10.0 ft) above the floor (Fig. 3–33). On the first free-throw attempt he shoots the ball at an angle of 35° above the horizontal and with a speed of $v_0 = 4.88$ m/s (16.0 ft/s). The ball is released 1.83 m (6.0 ft) above the floor. This shot misses badly. a) What is the maximum height reached by the ball? b) At what distance along the floor from the free-throw line does the ball land? For the second throw the ball goes through the center of the basket. For this second free throw the player again shot the ball at 35° above the horizontal and released it 1.83 m above the floor. c) What initial speed did the player give the ball on this second attempt? d) For the second throw, what is the maximum height reached by the ball? At this point, how far horizontally is the ball from the basket?

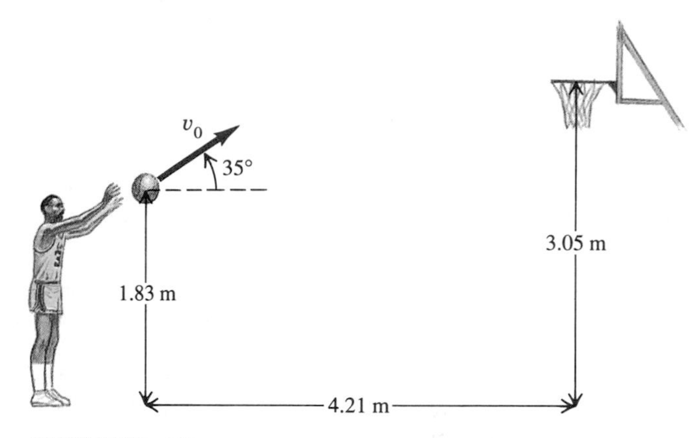

**FIGURE 3–33**

**3–46** A rock is thrown from the roof of a building with a velocity $v_0$ at an angle $\theta$ from the horizontal. The building has height $h$. Calculate the magnitude of the velocity of the rock just before it strikes the ground, and show that this speed is independent of $\theta$.

**3–47** Prove that a projectile launched at angle $\alpha_0$ has the same range as one launched with the same speed at angle $(90° - \alpha_0)$.

**3–48** In an action-adventure film the hero is supposed to throw a grenade from his car, which is going 70.0 km/h, to his enemy's car, which is going 110 km/h. The enemy's car is 14.6 m in front of the hero's when he lets go of the grenade. If the hero throws the grenade so that its initial velocity relative to him is at an angle of 45° above the horizontal, what should be the magnitude of the velocity? The cars are both traveling in the same direction on a level road. Find the magnitude of the velocity both relative to the hero and relative to the earth.

**3–49** A rock tied to a rope moves in the $xy$-plane; its coordinates are given as functions of time by

$$x = R\cos\omega t, \qquad y = R\sin\omega t,$$

where $R$ and $\omega$ are constants. a) Show that the rock's distance from the origin is constant and equal to $R$, that is, that its path is a circle of radius $R$. b) Show that at every point the rock's velocity is perpendicular to its position vector. c) Show that the rock's acceleration is always opposite in direction to its position vector and has magnitude $\omega^2 R$. d) Show that the magnitude of the rock's velocity is constant and equal to $\omega R$. e) Combine the results of parts (c) and (d) to show that the rock's acceleration has constant magnitude $v^2/R$.

**3–50** When a train's velocity is 12.0 m/s eastward, raindrops that are falling vertically with respect to the earth make traces that are inclined 30° to the vertical on the windows of the train. a) What is the horizontal component of a drop's velocity with respect to the earth? With respect to the train? b) What is the magnitude of the velocity of the raindrop with respect to the earth? With respect to the train?

**3–51** An airplane pilot sets a compass course due west and maintains an airspeed of 220 km/h. After flying for 0.500 h she finds herself over a town that is 150 km west and 40 km south of her starting point. a) Find the wind velocity (magnitude and direction). b) If the wind velocity is 120 km/h due south, in what direction should the pilot set her course to travel due west? Take the same airspeed of 220 km/h.

**3–52** A motorboat is traveling 18.0 km/h relative to the earth in the direction 37.0° north of east. If the velocity of the boat due to the wind is 3.20 km/h eastward and that due to the current is 6.40 km/h southward, what are the magnitude and direction of the velocity of the boat due to its own power?

## CHALLENGE PROBLEMS

**3–53** A shotgun fires a large number of pellets upward, with some pellets traveling very nearly vertically and others as much as 1.0° from the vertical. Assume that the initial speed of the pellets is uniformly 200 m/s and ignore air resistance. a) Within what radius from the point of firing will the pellets land? b) If there are 1000 pellets, and they fall in a uniform distribution over a circle with the radius calculated in part (a), what is the probability that at least one pellet will fall on the head of the person who fires the shotgun? Assume that his head has a radius of 10 cm. c) Air resistance in fact has several effects. It slows down the rising pellets, decreases their horizontal component of velocity, and limits the speed with which they fall. Which of these effects will tend to make the radius greater than calculated in part (a), and which will tend to make it less? What do you think the overall effect of air resistance will be? (The effect of air resistance on a velocity component increases as the magnitude of the component increases.)

**3–54** A man is riding on a flatcar traveling with a constant speed of 9.10 m/s (Fig. 3–34). He wishes to throw a ball through a stationary hoop 4.90 m above the height of his hands in such a manner that the ball will be moving horizontally as it passes through the hoop. He throws the ball with a speed of 13.4 m/s with respect to himself. a) What must be the vertical component of the initial velocity of the ball? b) How many seconds after he releases the ball will it pass through the hoop? c) At what horizontal distance in front of the hoop must he release the ball?

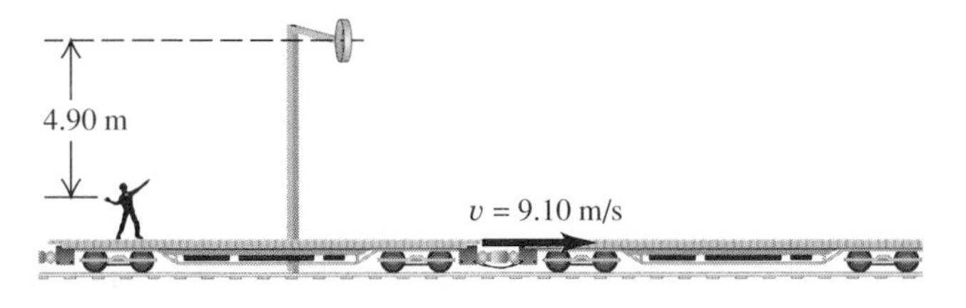


**FIGURE 3–34**

**3–55** A projectile is given an initial velocity with magnitude $v_0$ at an angle $\phi$ above the surface of an incline, which is in turn inclined at an angle $\theta$ above the horizontal (Fig. 3–35). a) Calculate the distance, measured along the incline, from the launch point to where the object strikes the incline. Your answer will be in terms of $v_0$, $g$, $\theta$, and $\phi$. b) What angle $\phi$ gives the maximum range, measured along the incline? (*Note:* You might be interested in the three different methods of solution presented by I. R. Lapidus in *Amer. Jour. of Phys.*, Vol. 51 (1983), pp. 806 and 847. See also H. A. Buckmaster in *Amer. Jour. of Phys.*, Vol. 53 (1985), pp. 638–641, for a thorough study of this and some similar problems.)

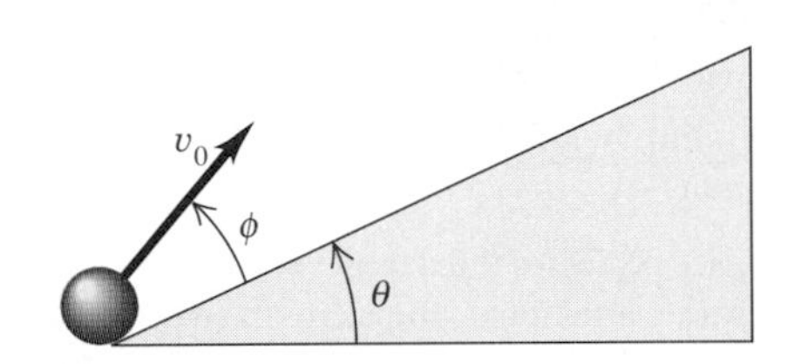


**FIGURE 3–35**

**3–56** Refer to the Problem 3–55. a) An archer on ground of constant upward slope of 30.0° aims at a target 50.0 m farther up the incline. The arrow in the bow and the bull's-eye at the center of the target are each 1.50 m above the ground. The initial velocity of the arrow just after it leaves the bow has magnitude 32.0 m/s. At what angle above the *horizontal* should the archer aim to hit the bull's-eye? If there are two such angles, calculate the smaller of the two. You might have to solve the equation for the angle by iteration, that is, by trial and error. How does the angle compare to that required when the ground is level, with zero slope? b) Repeat the above for ground of constant *downward* slope of 30.0°.

**3–57** An object is traveling in a circle with radius $R = 3.00$ m with a constant speed of $v = 6.00$ m/s. Let $\boldsymbol{v}_1$ be the velocity vector at time $t_1$ and $\boldsymbol{v}_2$ be the velocity vector at time $t_2$. Consider $\Delta\boldsymbol{v} = \boldsymbol{v}_2 - \boldsymbol{v}_1$ and $\Delta t = t_2 - t_1$. Recall that $\boldsymbol{a}_{av} = \Delta\boldsymbol{v}/\Delta t$. For $\Delta t = 0.5$ s, 0.1 s, and 0.05 s, calculate the magnitude (to four significant figures) and the direction (relative to $\boldsymbol{v}_1$) of the average acceleration $\boldsymbol{a}_{av}$. Compare your results to the general expression for the instantaneous acceleration $\boldsymbol{a}$ for uniform circular motion that is derived in the text.

C H A P T E R 4

# Newton's Laws of Motion

Force is a physical quantity that describes the interaction between two bodies. When a train starts to move after stopping at a station, the locomotive exerts a forward force $F$ on the first car, making the cars accelerate forward. The locomotive also accelerates forward under the combined action of the forward forces $F_W$ applied to its wheels by the rails and the backward force $-F$ applied by the first car.

## KEY CONCEPTS

**• Force is a vector quantity; forces combine by vector addition, which is often carried out by use of components.**

**• When the vector sum of forces on a body is zero, the body is at rest or moving with constant velocity. This condition is called equilibrium.**

**• The mass of a body describes its inertial properties. The vector sum of forces on a body equals its mass times its acceleration.**

**• When two bodies interact, the forces they exert on each other are equal in magnitude and opposite in direction.**

**• Free-body diagrams are helpful in determining the forces that act on a body.**

## INTRODUCTION

How can a tugboat push a cruise ship that's much heavier than the tug? Why does it take a long distance to stop the ship once it is in motion? Why does your foot hurt more when you kick a big rock than when you kick an empty cardboard box? Why is it harder to control a car on wet ice than on dry concrete? The answers to these and similar questions take us into the subject of **dynamics,** the relationship of motion to the forces that cause it. In the two preceding chapters we studied *kinematics,* the language for describing motion. Now we are ready to think about what makes bodies move the way they do. In this chapter we will use the kinematic quantities displacement, velocity, and acceleration, along with two new concepts, *force* and *mass.*

All the principles of dynamics can be wrapped up in a neat package containing three statements called **Newton's laws of motion.** The first law states that when the net force on a body is zero, its motion doesn't change. The second law relates force to acceleration when the net force is *not* zero. The third law gives a relation between the forces that two interacting bodies exert on each other.

Newton's laws, the cornerstone of mechanics, are based on experimental studies of moving bodies. They are fundamental laws of nature; they cannot be deduced or proved from any other principles. They were clearly stated for the first time by Sir Isaac Newton (1642–1727), who published them in 1686 in his *Philosophiae Naturalis Principia Mathematica* ("Mathematical Principles of Natural Philosophy"). Many other scientists before Newton contributed to the foundations of mechanics, including Copernicus, Brahe, Kepler, and especially Galileo Galilei (1564–1642), who died the same year Newton was born. Indeed, Newton himself said, "If I have been able to see a little farther than other men, it is because I have stood on the shoulders of giants."

# 4–1 FORCE

The concept of **force** gives us a quantitative description of the interaction between two bodies or between a body and its environment. When you push on a car that is stuck in the snow, you exert a force on it. A locomotive exerts a force on the train it is pulling or pushing, a steel cable exerts a force on the beam it is hoisting at a construction site, and so on. When a force involves direct contact between two bodies, we call it a *contact force*. There are also forces, including gravitational and electrical forces, that act even when the bodies are separated by empty space. Viewed on an atomic scale, contact forces are the electrical attractions and repulsions of the electrons and nuclei in the atoms of materials. The force of gravitational attraction that the earth exerts on a body is called the *weight* of the body.

Force is a vector quantity; to describe a force we need to describe the direction in which it acts (Fig. 4–1) as well as its magnitude, the quantity that describes "how much" or "how hard" the force pushes or pulls. The SI unit of the magnitude of force is the *newton*, abbreviated N. (We haven't yet given a precise definition of the newton. The definition is based on the standard kilogram; we'll get into this in Section 4–3.) Table 4–1 lists some typical force magnitudes.

A common instrument for measuring forces is the spring balance. It consists of a coil spring, enclosed in a case for protection, with a pointer attached to one end. When forces are applied to the ends of the spring, it stretches; the amount of stretch depends on the force. We can make a scale for the pointer and calibrate it by using a number of identical bodies with weights of exactly 1 N each. When two, three, or more of these are suspended simultaneously from the balance, the total force stretching the spring is 2 N, 3 N, and so on, and we can label the corresponding positions of the pointer 2 N, 3 N, and so

**TABLE 4–1 Some Typical Force Magnitudes**

| | |
|---|---|
| Sun's gravitational force on earth | $3.5 \times 10^{22}$ N |
| Thrust of the *Energia* rocket | $3.9 \times 10^{7}$ N |
| Weight of a large blue whale | $1.9 \times 10^{6}$ N |
| Maximum pulling force of a locomotive | $8.9 \times 10^{5}$ N |
| Weight of a 250-lb linebacker | $1.1 \times 10^{3}$ N |
| Weight of a medium apple | 1 N |
| Weight of the smallest insect eggs | $2 \times 10^{-6}$ N |
| Force on an electron in a hydrogen atom | $8.2 \times 10^{-8}$ N |
| Weight of a very small bacterium | $1 \times 10^{-18}$ N |
| Weight of a hydrogen atom | $1.6 \times 10^{-26}$ N |
| Weight of an electron | $8.9 \times 10^{-30}$ N |
| Gravitational attraction of the proton and the electron in a hydrogen atom | $3.6 \times 10^{-47}$ N |

(a)

(b)

(c)

**4–1** (a) The motion of a freight train is an example of one-dimensional motion. Each car in the line moves in the direction of the force exerted by the locomotive. (b) The forces between these two ice skaters act along the line of their arms. (c) The forces between atoms in a crystal, such as diamond, act in a straight line between neighboring atoms. The geometric planes of the crystal reflect the orderly arrangement of the atoms.

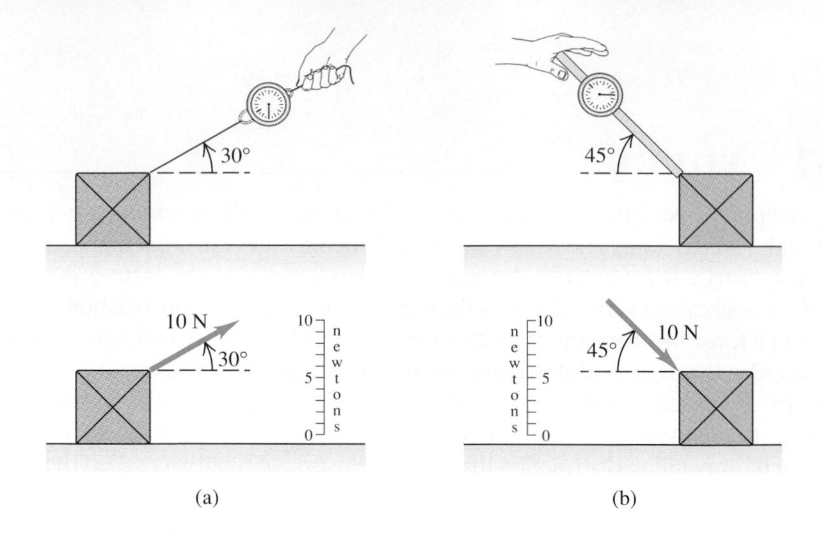


**4–2** Force may be exerted on the box by either (a) pulling it or (b) pushing it. A force diagram illustrates each case.

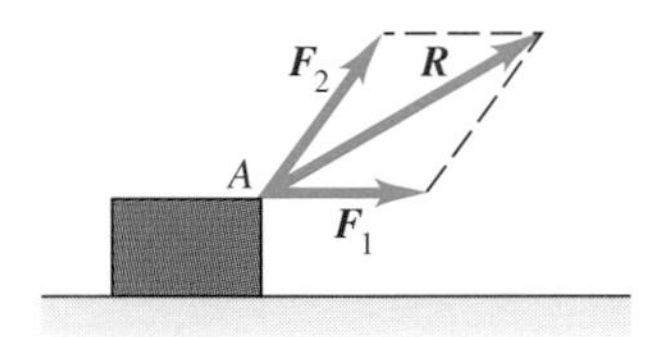

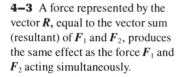

**4–3** A force represented by the vector $\boldsymbol{R}$, equal to the vector sum (resultant) of $\boldsymbol{F}_1$ and $\boldsymbol{F}_2$, produces the same effect as the force $\boldsymbol{F}_1$ and $\boldsymbol{F}_2$ acting simultaneously.

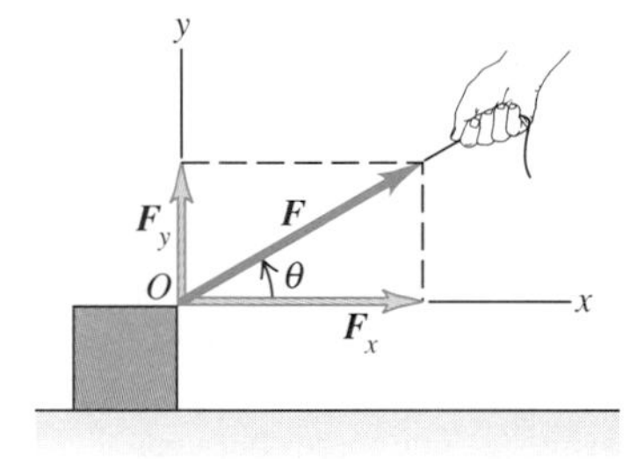

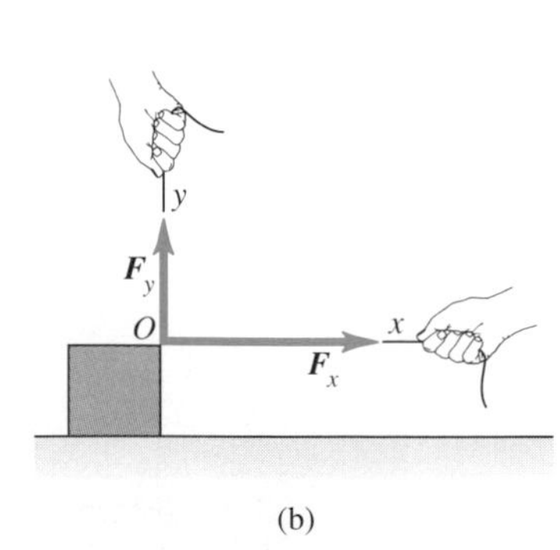


**4–4** The inclined force $\boldsymbol{F}$ may be replaced by its rectangular components $\boldsymbol{F}_x$ and $\boldsymbol{F}_y$. $F_x = F \cos\theta$ and $F_y = \sin\theta$.

on. Then we can use this instrument to measure the magnitude of an unknown force. We can also make a similar instrument that measures pushes instead of pulls.

Suppose we slide a box along the floor, applying a force to it by pulling it with a string or pushing it with a stick (Fig. 4–2) . In each case the force applied can be represented by a vector. The labels indicate the magnitude and direction of the force, and the length of the arrow (drawn to some scale such as 1 cm = 5 N) also shows the magnitude.

When two forces $\boldsymbol{F}_1$ and $\boldsymbol{F}_2$ act at the same time on a point $A$ of a body (Fig. 4–3), experiment shows that the effect is the same as the effect of a single force equal to the *vector sum,* or **resultant,** $\boldsymbol{R} = \boldsymbol{F}_1 + \boldsymbol{F}_2$ of the original forces. More generally, the effect of any number of forces applied at a point on a body is the same as the effect of a single force equal to the vector sum of the forces. This important principle goes by the name **superposition of forces.**

The discovery that forces combine according to vector addition is of the utmost importance, and we will use this fact many times throughout our study of physics. In particular, it allows us to represent a force by means of *components,* as we did with displacements in Section 1–8. For example, in Fig. 4–4a force $\boldsymbol{F}$ acts on a body at point $O$. The component vectors of $\boldsymbol{F}$ in the directions $Ox$ and $Oy$ are $\boldsymbol{F}_x$ and $\boldsymbol{F}_y$. When $\boldsymbol{F}_x$ and $\boldsymbol{F}_y$ are applied simultaneously, as in Fig. 4–4b, the effect is exactly the same as the effect of the original force. **Any force can be replaced by its components, acting at the same point.**

There is no law that says our coordinate axes have to be vertical and horizontal. Figure 4–5 shows a stone block being pulled up a ramp by a force $\boldsymbol{F}$, represented by its components $F_x$ and $F_y$, parallel and perpendicular to the sloping surface of the ramp. We draw a wiggly line through the force vector $\boldsymbol{F}$ to show that we have replaced it by its $x$- and $y$-components. Otherwise, the diagram would include the same force twice.

We will often need to find the vector sum (resultant) of several forces acting on a body. We will use the Greek letter $\Sigma$ (capital sigma, equivalent to the Roman S) as a shorthand notation for a sum. If the forces are labeled $\boldsymbol{F}_1$, $\boldsymbol{F}_2$, $\boldsymbol{F}_3$, and so on, we abbreviate the sum operation as

$$\boldsymbol{R} = \boldsymbol{F}_1 + \boldsymbol{F}_2 + \boldsymbol{F}_3 + \cdots = \Sigma \boldsymbol{F}, \qquad (4\text{–}1)$$

where $\Sigma \boldsymbol{F}$ is read as "the vector sum of the forces." The component version of Eq. (4–1) is the pair of component equations

$$R_x = \Sigma F_x, \qquad R_y = \Sigma F_y, \qquad (4\text{–}2)$$

where $\Sigma F_x$ is the sum of the $x$-components, and so on. Each component may be positive or negative; be careful with signs when you evaluate the sums in Eqs. (4–2). Equation

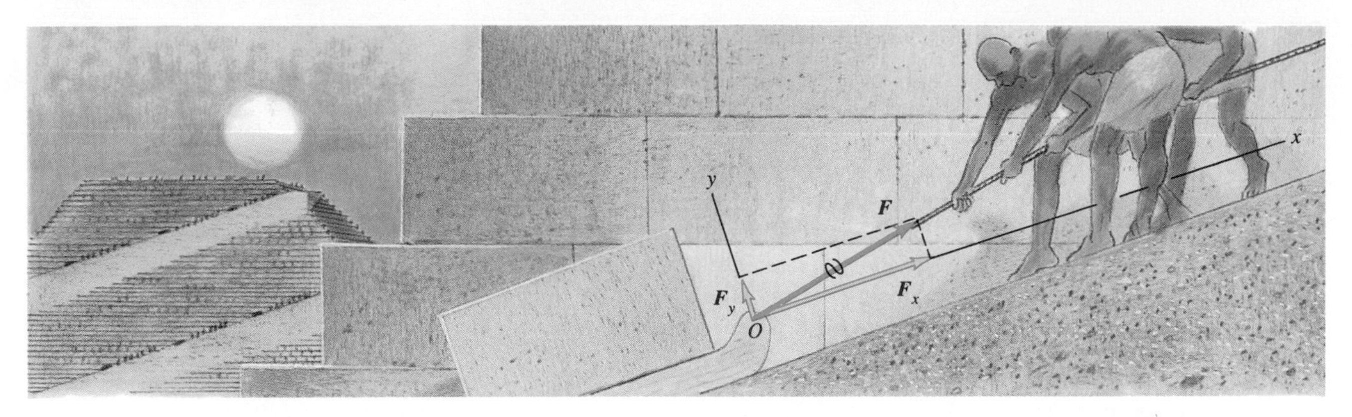


**4–5** $\boldsymbol{F}_x$ and $\boldsymbol{F}_y$ are the rectangular components of $\boldsymbol{F}$ parallel and perpendicular to the sloping surface of the inclined plane.

(4–1) uses a boldface sigma ($\boldsymbol{\Sigma}$) as a reminder that the sum is a vector sum, and Eqs. (4–2) use lightface sigmas ($\Sigma$) because components are ordinary numbers.

Once we have $R_x$ and $R_y$, we can find the magnitude and direction of $\boldsymbol{R}$. The magnitude is

$$R = \sqrt{R_x^{\,2} + R_y^{\,2}},$$

and the angle $\theta$ between $\boldsymbol{R}$ and the $x$-axis can be found from the relation $\tan\theta = R_y/R_x$. The components $R_x$ and $R_y$ may be positive or negative, and the angle $\theta$ may be in any of the four quadrants.

In three-dimensional problems forces may also have $z$-components; then we add the equation $R_z = \Sigma F_z$ to Eqs. (4–2). The magnitude of the vector sum of forces is then

$$R = \sqrt{R_x^{\,2} + R_y^{\,2} + R_z^{\,2}}.$$

## 4–2 NEWTON'S FIRST LAW

The fundamental role of force is to change the state of motion of the body on which the force acts. The key word is "change." Newton's first law of motion, translated from the Latin of the *Principia,* states:

> **Every body continues in its state of rest, or of uniform motion in a straight line, unless it is compelled to change that state by forces impressed on it.**

When no net force acts on a body, the body either remains at rest or moves with constant velocity in a straight line. Once a body has been set in motion, no net force is needed to keep it moving. In other words, **a body acted on by no net force moves with constant velocity (which may be zero) and zero acceleration.**

Everyday experience may seem to contradict this statement. Suppose you slide a hockey puck along a horizontal tabletop, applying a horizontal force to it with your hand (Fig. 4–6a). After you stop pushing, the puck *does not* continue to move indefinitely; it slows down and stops. To keep it moving, you have to keep pushing. This is because as the puck slides, the tabletop applies a frictional force to it in a direction *opposite* to the puck's motion.

But now imagine pushing the puck across the smooth ice of a skating rink (Fig. 4–6b). It will move a lot farther after you quit pushing before it stops. Put it on an air-hockey table, where it floats on a thin cushion of air, and it slides still farther (Fig. 4–6c). The more slippery the surface, the less friction. The first law states that if we could eliminate friction completely, we would need *no forward force at all* to keep the puck moving once it had been started.

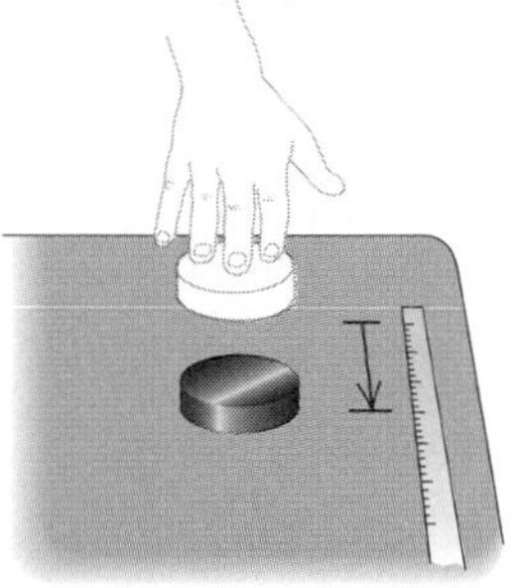

(a)

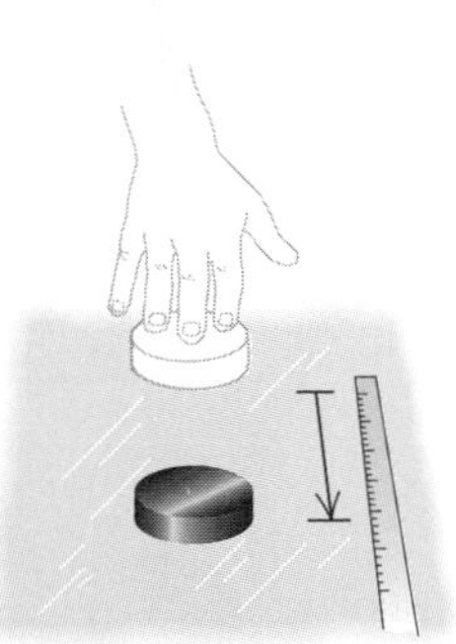
(b)

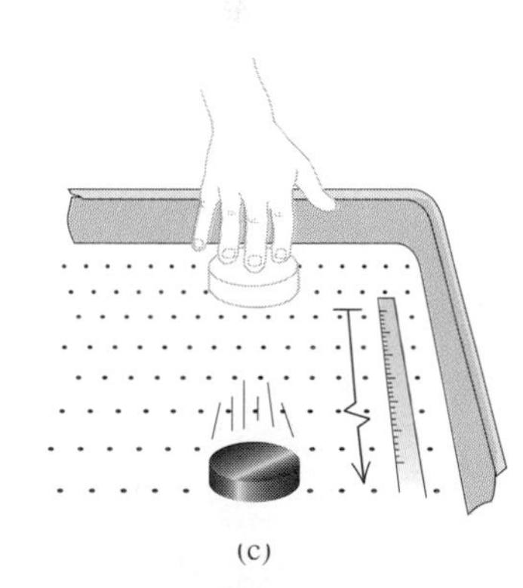
(c)

**4–6** A hockey puck is given an initial velocity. (a) It stops in a short distance on a tabletop. (b) An ice surface decreases the friction force, and the puck slides farther. (c) On an air-hockey table the friction force is practically zero, so the puck continues with almost constant velocity.

## Inertia

The tendency of a body to remain at rest, or to keep moving once it is set in motion, results from a property called *inertia*. That's what you feel when you're behind home plate trying to catch a fastball pitch. That baseball really wants to keep moving. The tendency of a body at rest to remain at rest is also due to inertia. You may have seen a tablecloth yanked out from under the china without anything being broken. The force on the china isn't great enough to make it move appreciably.

### EXAMPLE 4–1

In a TV science fiction show, the hero is in the vacuum of outer space when the engine of his spaceship suddenly dies. He slows down and stops, hoping for a rescue before his air runs out. What does Newton's first law say about this event?

**SOLUTION** In this situation there are no external forces acting on the spaceship, so according to Newton's first law, it doesn't stop. It continues to move in a straight line with constant speed. Science fiction sometimes contains more fiction than science.

When a single force acts on a body, it changes the state of motion of the body. A body that is initially at rest starts to move. If the body is initially moving, a force in the direction opposite to the motion causes the body to slow down or stop. Suppose a hockey puck rests on a horizontal surface with negligible friction, such as an air-hockey table or a slab of wet ice. If the puck is initially at rest and a single force $\boldsymbol{F}_1$ acts on it (Fig. 4–7a), the puck starts to move. If the puck is in motion at the start, the force makes it speed up, slow down, or change direction, depending on the direction of the force.

Now suppose we apply a second force $\boldsymbol{F}_2$ (Fig. 4–7b), equal in magnitude to $\boldsymbol{F}_1$ but opposite in direction. The two forces are negatives of each other, $\boldsymbol{F}_2 = -\boldsymbol{F}_1$, and their vector sum is zero:

$$\boldsymbol{R} = \boldsymbol{F}_1 + \boldsymbol{F}_2 = \boldsymbol{0}.$$

We find that if the body is at rest at the start, it remains at rest. If it is initially moving, it continues to move in the same direction with constant speed. These results show that in Newton's first law, *zero resultant force is equivalent to no force at all.*

When a body is acted on by no forces or by several forces such that their vector sum (resultant) is zero, we say that the body is in **equilibrium.** For a body in equilibrium,

$$\boldsymbol{R} = \Sigma \boldsymbol{F} = \boldsymbol{0}. \tag{4–3}$$

For this to be true, each component of $\boldsymbol{R}$ must be zero, so

$$\Sigma F_x = 0, \qquad \Sigma F_y = 0. \tag{4–4}$$

When Eqs. (4–4) are satisfied, the body is in equilibrium. (We are assuming that the body can be represented adequately as a point. When the body has finite size, we also have to consider *where* on the body the forces are applied. We will return to this point in Chapter 11.)

$F_1$

(a)

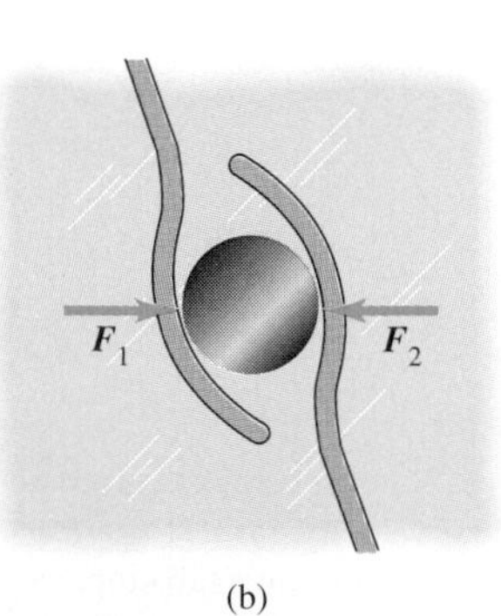


(b)

**4–7** (a) A hockey puck accelerates in the direction of an applied force $\boldsymbol{F}_1$. (b) When the net force is zero, the acceleration is zero, and the puck is in equilibrium.

## Inertial Frames of Reference

In our discussions of relative velocity at the ends of Chapters 2 and 3, we stressed the concept of *frame of reference*. This concept also plays a central role in Newton's laws of motion. Suppose you are sitting in an airplane that is accelerating down the runway during takeoff. You feel a forward force pushing on your back, but you do not start moving forward relative to the airplane. If you could stand in the aisle on roller skates, you would start moving *backward* relative to the plane as the pilot cracks the throttle. In either case it looks as though Newton's first law is not obeyed. Forward force but no acceleration, or no force and backward acceleration. What's wrong?

The point is that the plane, accelerating with respect to the earth, is not a suitable frame of reference for Newton's first law. This law is valid in some frames of reference and not in others. A frame of reference in which Newton's first law *is* valid is called an **inertial frame of reference.** The earth is at least approximately an inertial frame of reference, but the airplane is not.

This may sound as though there's only one inertial frame of reference in the universe. On the contrary, if we have an inertial frame of reference $A$, in which Newton's first law is obeyed, then this law is also obeyed in any second frame of reference $B$ that moves relative to the first with constant velocity $\boldsymbol{v}_{B/A}$. In that case, $B$ is also inertial. To prove this, we use the relative-velocity relation from Section 3–5:

$$\boldsymbol{v}_{P/A} = \boldsymbol{v}_{P/B} + \boldsymbol{v}_{B/A}$$

The time derivative of this equation is

$$\frac{d\boldsymbol{v}_{P/A}}{dt} = \frac{d\boldsymbol{v}_{P/B}}{dt} + \frac{d\boldsymbol{v}_{B/A}}{dt}. \tag{4–5}$$

Now $d\boldsymbol{v}_{P/A}/dt$ is the acceleration of point $P$ with respect to frame $A$, and $d\boldsymbol{v}_{P/B}/dt$ is its acceleration with respect to frame $B$. When the relative velocity $\boldsymbol{v}_{B/A}$ is constant, its derivative is zero, and these two accelerations are equal. Thus if $P$ has constant velocity (zero acceleration) with respect to frame $A$, it must also have constant velocity with respect to frame $B$. If $A$ is inertial, $B$ must also be inertial. Equation (4–5) also shows that, in this case, if $P$ has an acceleration with respect to $A$, its acceleration with respect to $B$ is the same. We will use this fact later in connection with Newton's second law.

There is no single inertial frame of reference that is preferred over all others for formulating Newton's laws. If one frame is inertial, then every other frame moving relative to it with constant velocity is also inertial. Viewed in this light, the state of rest and the state of uniform motion (with constant velocity) are not very different; both occur when the vector sum of forces acting on the body is zero. Because Newton's first law can be used to define what we mean by an inertial frame of reference, it is sometimes called the *law of inertia.*

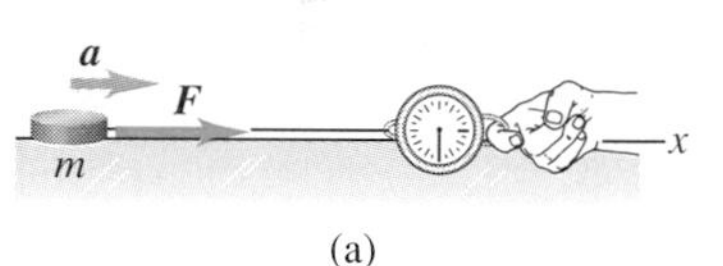


## 4–3 MASS AND NEWTON'S SECOND LAW

When a body is acted on by no force or zero resultant force, it moves with constant velocity and zero acceleration. But what happens when the resultant force is *not* zero? If the body is initially at rest, it starts to move. If it is initially moving, the force may speed it up, slow it down, or change the direction of its velocity. In each case the body has an *acceleration*. We want to know the relation of the acceleration to the force, and this is what Newton's second law of motion is all about.

Let's look at several fundamental experiments. Consider a small body moving on a flat, level, frictionless surface. The body could be a puck on an air-hockey table, as we described in Section 4–2. Initially, it is moving to the right along the $x$-axis of a coordinate system (Fig. 4–8a). We apply a constant horizontal force $\boldsymbol{F}$ to the body, using the spring balance we described in Section 4–1, with the spring stretched a constant amount. We find that during the time the force is acting, the velocity of the body changes at a

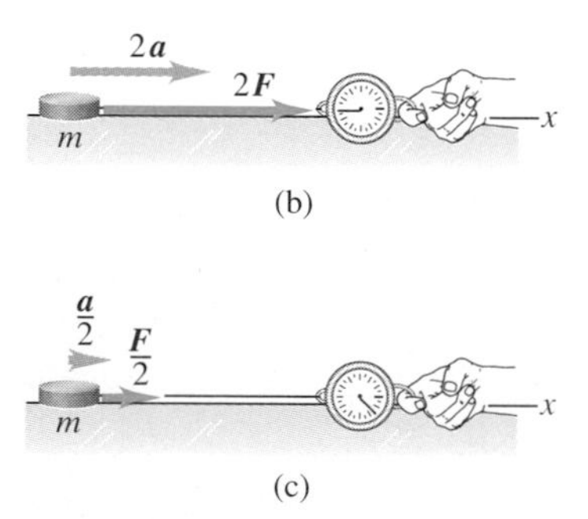


**4–8** (a) Acceleration $\boldsymbol{a}$ is proportional to the applied force $\boldsymbol{F}$. (b) Doubling the force doubles the acceleration. (c) Halving the force halves the acceleration.

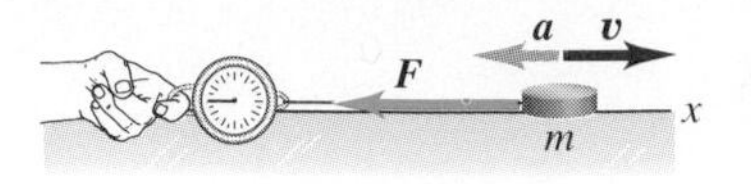


**4–9** The acceleration $\boldsymbol{a}$ is in the same direction as $\boldsymbol{F}$, independent of the direction of $\boldsymbol{v}$.

constant rate; that is, the body moves with *constant acceleration*. If we change the magnitude of the force, the acceleration changes in the same proportion. Doubling the force doubles the acceleration, halving the force halves the acceleration, and so on (Fig. 4–8b and 4–8c). When we take away the force, the acceleration becomes zero, and the body moves with constant velocity. We conclude that for any given body the acceleration is directly proportional to the force acting on the body.

In another experiment we give the body the same initial velocity as before, but we reverse the direction of the force (Fig. 4–9). We find that the body moves more and more slowly to the right, stops, and begins to move more and more rapidly toward the left. The direction of the acceleration is toward the *left,* in the same direction as the force $\boldsymbol{F}$. We conclude that the *magnitude* of the acceleration is proportional to that of the force, and the *direction* of the acceleration is the same as that of the force, regardless of the direction of the velocity.

For a given body the ratio of the force to the acceleration is constant, regardless of the magnitude of the force. We call this ratio the inertial mass, or simply the **mass,** of the body, denoted by $m$. That is, $m = F/a$, or

$$F = ma. \tag{4–6}$$

The concept of mass is a familiar one. If you hit a table-tennis ball with a paddle and then hit a basketball with the same force, the basketball has much smaller acceleration because it has much greater mass. When a large force is needed to give a body a certain acceleration, the mass of the body is large. When only a small force is needed for the same acceleration, the mass is small. The greater the mass, the more the body "resists" being accelerated. Thus mass is a quantitative measure of inertia, which we discussed in Section 4–2.

The SI unit of mass is the **kilogram.** We mentioned in Section 1–3 that the kilogram is officially defined to be the mass of a chunk of platinum-iridium alloy kept in a vault near Paris. We can use this standard kilogram, along with Eq. (4–6), to define the **newton:**

> **One newton is the amount of force that gives an acceleration of one meter per second squared to a body with mass of one kilogram.**

We can use this definition to calibrate the spring balances and other instruments that we use to measure forces.

We can also use Eq. (4–6) to compare a mass with the standard mass and thus to *measure* masses. Suppose we apply a constant force $F$ to a body having a known mass $m_1$ and we find an acceleration $a_1$. We then apply the *same* force to another body having an unknown mass $m_2$, and we find an acceleration $a_2$. Then, according to Eq. (4–6),

$$m_1 a_1 = m_2 a_2,$$

or

$$\frac{m_2}{m_1} = \frac{a_1}{a_2}. \tag{4–7}$$

The ratio of the masses is the inverse of the ratio of the accelerations. In principle, we could use this relation to measure an unknown mass $m_2$, but it is usually easier to determine mass indirectly by measuring the body's weight. We'll return to this point in Section 4–4.

When two bodies with masses $m_1$ and $m_2$ are fastened together, we find that the mass of the composite body is always $m_1 + m_2$ (Fig. 4–10). This additive property of mass may seem obvious, but it has to be verified experimentally. Ultimately, the mass of a body is related to the numbers of protons, electrons, and neutrons it contains. This wouldn't be a good way to *define* mass because there is no practical way to count these particles. But the concept of mass is the most fundamental way to characterize the quantity of matter in a body. Figure 4–10 shows the properties of mass that we have described.

We need to generalize Eq. (4–6) in two ways. First, the particle doesn't necessarily have to move along a straight line; its path may be a curve in space. Its velocity, its acceleration, and the force acting on it then have to be treated as vector quantities. Second, there may be several forces acting on the particle. We can combine the forces using vector addition; the effect of the forces is the same as the effect of a single force equal to the vector sum of the individual forces. We have discussed this principle (superposition of forces) in connection with Newton's first law, and it also turns out to be valid even when the vector sum of forces is not zero.

Newton wrapped up all these relationships and experimental results in a single concise statement that we now call *Newton's second law of motion:*

$$\Sigma \boldsymbol{F} = m\boldsymbol{a}. \tag{4–8}$$

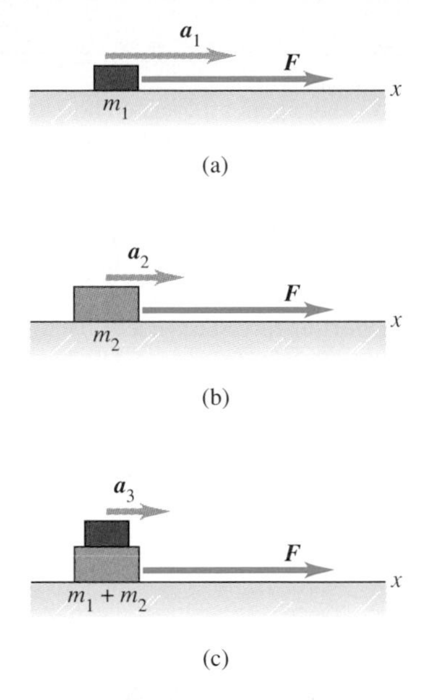


**4–10** For a constant force, acceleration is inversely proportional to the mass of a body, and masses are additive as scalars. (a) $\boldsymbol{a}_1 = \boldsymbol{F}/m_1$; (b) $\boldsymbol{a}_2 = \boldsymbol{F}/m_2$; (c) $\boldsymbol{a}_3 = \boldsymbol{F}/(m_1 + m_2)$

This is a fundamental law of nature, the basic relation between force and motion. Most of the remainder of this chapter and all of the next are devoted to learning how to apply this principle in various situations.

Equation (4–8) is a *vector* equation. Usually, we will use it in component form, a separate equation for each component of force and the corresponding acceleration:

$$\Sigma F_x = ma_x, \qquad \Sigma F_y = ma_y, \qquad \Sigma F_z = ma_z. \tag{4–9}$$

This set of component equations is equivalent to the single vector equation, Eq. (4–8). The acceleration of a body (the rate of change of its velocity) is equal to the vector sum (resultant) of all forces acting on the body, divided by its mass. The acceleration has the same direction as the resultant force. Each component of the total force equals the mass times the corresponding component of acceleration.

Equations (4–8) and (4–9) are valid only when the mass $m$ is *constant.* It's easy to think of systems whose masses change, such as a leaking tank truck, a rocket ship, or a moving railroad car being loaded with coal. But such systems are better handled by using the concept of momentum; we'll get to that in Chapter 8.

Like the first law, Newton's second law is valid only in inertial frames of reference. We will usually assume that the earth is an adequate approximation to an inertial frame, even though because of its rotation and orbital motion, it is not precisely inertial.

Because of the way we have defined the newton, it is related to the units of mass, length, and time. For $\Sigma \boldsymbol{F} = m\boldsymbol{a}$ to be dimensionally consistent, it must be true that

$$1 \text{ newton} = (1 \text{ kilogram})(1 \text{ meter per second squared}),$$

or

$$1 \text{ N} = 1 \text{ kg} \cdot \text{m/s}^2.$$

We will use this relation many times in the next few chapters, so keep it in mind.

In learning how to use Newton's second law, we will begin in this chapter with examples of straight-line motion. Then in Chapter 5 we will consider more general cases and develop more detailed problem-solving strategies for applying Newton's laws of motion.

## EXAMPLE 4–2

A worker applies a constant horizontal force with magnitude 20 N to a box with mass 40 kg resting on a level frictionless surface. What is the acceleration of the box?

**SOLUTION** We take the $+x$-axis in the direction of the horizontal force (Fig. 4–11). The forces acting on the box are this force and two vertical forces: the weight $w$ of the box and the upward supporting force $\mathcal{n}$ exerted on it by the surface. (We'll discuss this supporting force in more detail in Chapter 5.) The acceleration is given by Newton's second law, Eq. (4–9); there is only one horizontal force, and we have

$$\Sigma F_x = ma_x,$$

$$a_x = \frac{F_x}{m} = \frac{20\ \text{N}}{40\ \text{kg}} = \frac{20\ \text{kg}\cdot\text{m/s}^2}{40\ \text{kg}} = 0.5\ \text{m/s}^2.$$

The force is constant, so the acceleration is also constant. If we are given the initial position and velocity of the box, we can find the position and velocity at any later time from the equations of motion with constant acceleration. There is no vertical acceleration, so we know that the two vertical forces must sum to zero. The upward force $\mathcal{n}$ is called a *normal* force because it is normal (perpendicular) to the surface of contact.

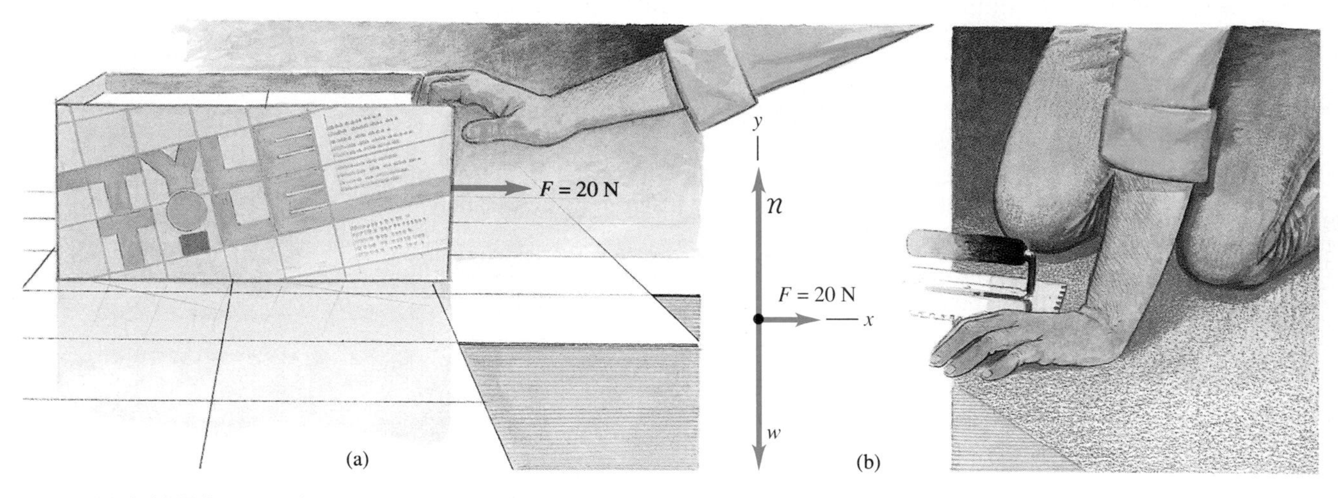


**4–11** (a) A 20-N force accelerates the box to the right. (b) The free-body diagram for the box, considered as a particle.

## EXAMPLE 4–3

A waitress shoves a ketchup bottle with mass 0.2 kg toward the right along a smooth, level lunch counter. As the bottle leaves her hand, it has an initial velocity of 2.8 m/s. As it slides, it slows down because of the horizontal friction force exerted on it by the countertop. It slides a distance of 1.0 m before coming to rest. What are the magnitude and direction of the friction force acting on it?

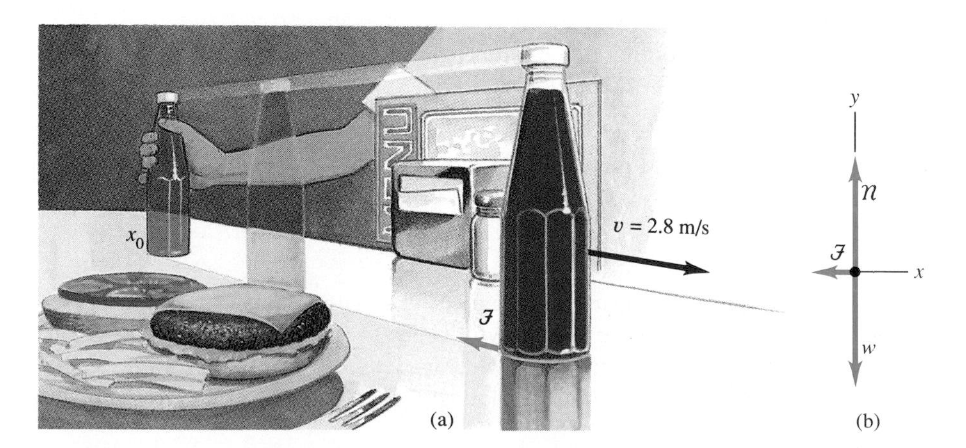


**4–12** (a) As the bottle slides to the right, the friction force $\mathcal{F}$ slows it down. (b) Diagram of forces acting on the bottle, considered as a particle.

**SOLUTION** Suppose the bottle slides in the $+x$-direction, starting at the point $x_0 = 0$ with the given initial velocity (Fig. 4–12a). The forces acting on the bottle are shown in Fig. 4–12b. We assume that the friction force $\mathcal{F}$ is constant. The acceleration is then constant also; from Eq. (2–14) we have

$$v^2 = v_0^2 + 2a(x - x_0),$$
$$0 = (2.8\text{ m/s})^2 + (2a)(1.0\text{ m}),$$
$$a = -3.9\text{ m/s}^2.$$

The negative sign means that the acceleration is toward the *left* (although the velocity is toward the right). The $x$-component $F_x$ of the friction force on the bottle is

$$F_x = ma = (0.2\text{ kg})(-3.9\text{ m/s}^2) = -0.8\text{ kg}\cdot\text{m/s}^2 = -0.8\text{ N}.$$

Again the negative sign shows that the force on the bottle is directed toward the left.

## EXAMPLE 4–4

In a color TV picture tube an electric field exerts a net force of $1.60 \times 10^{-13}$ N on an electron ($m = 9.11 \times 10^{-31}$ kg). What is the electron's acceleration?

**SOLUTION** Using Newton's second law, we get

$$a = \frac{F}{m} = \frac{1.60 \times 10^{-13}\text{ N}}{9.11 \times 10^{-31}\text{ kg}} = \frac{1.60 \times 10^{-13}\text{ kg m/s}^2}{9.11 \times 10^{-31}\text{ kg}}$$
$$= 1.76 \times 10^{17}\text{ m/s}^2.$$

Even though a TV set is a very familiar device, the magnitudes of these quantities are far outside the range of everyday experience.

A few words about units are in order. In the cgs metric system (not used in this book), the unit of mass is the gram, equal to $10^{-3}$ kg, and the unit of distance is the centimeter, equal to $10^{-2}$ m. The corresponding unit of force is called the *dyne:*

$$1\text{ dyne} = 1\text{ g}\cdot\text{cm/s}^2.$$

One dyne is equal to $10^{-5}$ N. In the British system the unit of force is the *pound* (or pound-force), and the unit of mass is the *slug.* The unit of acceleration is one foot per second squared, so

$$1\text{ pound} = 1\text{ slug}\cdot\text{ft/s}^2.$$

One pound is defined officially as

$$1\text{ pound} = 4.448221615260\text{ newtons}.$$

It is handy to remember that a pound is about 4.4 N and a newton is about 0.22 pound. Next time you want to order a "quarter-pounder," try asking for a "one-newtoner" and see what happens. Another useful fact: A body with a mass of 1 kg has a weight of about 2.2 lb at the earth's surface.

The units of force, mass, and acceleration in the three systems are summarized in Table 4–2.

**TABLE 4–2 Units of Force, Mass, and Acceleration**

| System of units | Force | Mass | Acceleration |
|---|---|---|---|
| SI | newton (N) | kilogram (kg) | $\text{m/s}^2$ |
| cgs | dyne (dyn) | gram (g) | $\text{cm/s}^2$ |
| British | pound (lb) | slug | $\text{ft/s}^2$ |

## 4–4 MASS AND WEIGHT

The **weight** of a body is a familiar force. It is the force of the earth's gravitational attraction for the body. We will study gravitational interactions in detail in Chapter 12, but we need some preliminary discussion now. The terms *mass* and *weight* are often misused and interchanged in everyday conversation. It is absolutely essential for you to understand clearly the distinctions between these two physical quantities.

Mass characterizes the *inertial* properties of a body. Mass is what keeps the china on the table when you yank the tablecloth out from under it. The greater the mass, the greater the force needed to cause a given acceleration; this meaning is reflected in Newton's second law, $\Sigma\mathbf{F} = m\mathbf{a}$. Weight, on the other hand, is a *force* exerted on a body by the pull of the earth or some other large body. Everyday experience shows us that bodies having large mass also have large weight. A cart loaded with bricks is hard to get started rolling because of its large *mass,* and it is also hard to lift off the ground because of its large *weight.* On the moon the cart would be just as hard to get rolling, but it would be easier to lift. So what exactly *is* the relationship between mass and weight?

The answer to this question, according to legend, came to Newton as he sat under an apple tree watching the apples fall. A freely falling body has an acceleration equal to $g$, and because of Newton's second law, this acceleration requires a force. If a 1-kg body falls with an acceleration of 9.8 m/s$^2$, the required force has magnitude

$$F = ma = (1\text{ kg})(9.8\text{ m/s}^2) = 9.8\text{ kg}\cdot\text{m/s}^2 = 9.8\text{ N}.$$

But the force that makes the body accelerate downward is the gravitational pull of the earth, that is, the *weight* of the body. Any body near the surface of the earth that has a mass of 1 kg *must* have a weight of 9.8 N to give it the acceleration we observe when it is in free fall. More generally, a body with mass $m$ must have weight with magnitude $w$ given by

$$w = mg. \tag{4–10}$$

The weight of a body is a force, a vector quantity, and we can write Eq. (4–10) as a vector equation:

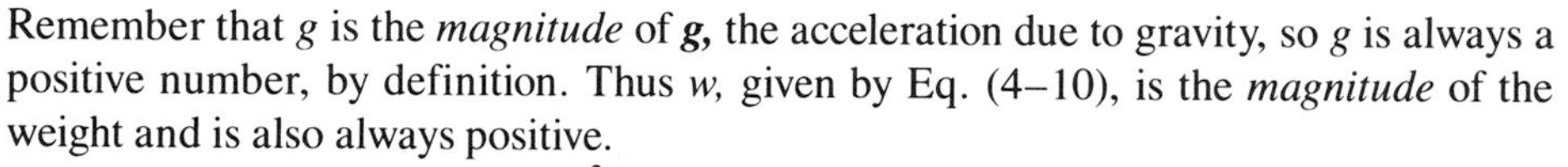

$$\mathbf{w} = m\mathbf{g}. \tag{4–11}$$

Remember that $g$ is the *magnitude* of $\mathbf{g}$, the acceleration due to gravity, so $g$ is always a positive number, by definition. Thus $w$, given by Eq. (4–10), is the *magnitude* of the weight and is also always positive.

We will use $g = 9.80\text{ m/s}^2$ for problems on earth. In fact, the value of $g$ varies somewhat from point to point on the earth's surface, from about 9.78 to 9.82 m/s$^2$, because the earth is not perfectly spherical and because of effects due to its rotation and orbital motion. At a point where $g = 9.80\text{ m/s}^2$, the weight of a standard kilogram is $w = 9.80$ N. At a different point, where $g = 9.78\text{ m/s}^2$, the weight is $w = 9.78$ N but the mass is still 1 kg. The weight of a body varies from one location to another; the mass does not. If we take a standard kilogram to the surface of the moon, where the acceleration of free fall (equal to the gravitational field at the moon's surface) is 1.62 m/s$^2$, its weight is 1.62 N but its mass is still 1 kg (Fig. 4–13). An 80.0-kg man has a weight on earth of $(80.0\text{ kg})(9.80\text{ m/s}^2) = 784$ N, but on the moon his weight is only $(80.0\text{ kg})\cdot(1.62\text{ m/s}^2) = 130$ N.

(a)

(b)

**4–13** (a) A standard kilogram weighs about 9.8 N on earth. (b) The same kilogram weighs only about 1.6 N on the moon.

The following brief excerpt from Astronaut Buzz Aldrin's book *Men from Earth* offers some interesting insights into the distinction between mass and weight (Fig. 4–14):

> Our portable life-support system backpacks looked simple, but they were hard to put on and tricky to operate. On Earth the portable life-support sys-

tem and space suit combination weighed 190 pounds, but on the moon it was only 30. Combined with my own body weight, that brought me to a total lunar-gravity weight of around 60 pounds.

One of my tests was to jog away from the lunar module to see how maneuverable an astronaut was on the surface. I remembered what Isaac Newton had taught us two centuries before: Mass and weight are not the same. I weighed only 60 pounds, but my *mass* was the same as it was on Earth. Inertia was a problem. I had to plan ahead several steps to bring myself to a stop or to turn without falling.

**4–14** Astronaut Buzz Aldrin on the surface of the moon, trying to cope with his own inertia.

It is important to understand that the weight of a body, as given by Eq. (4–11), acts on the body *all the time,* whether it is in free fall or not. When a 10-kg flowerpot hangs suspended from a chain, it is in equilibrium, and its acceleration is zero. But its weight, given by Eq. (4–11), is still pulling down on it. In this case the chain pulls up on the pot, applying an upward force. The *vector sum* of the forces is zero, and the pot is in equilibrium.

## EXAMPLE 4–5

A $1.96 \times 10^4$ N Lincoln Town Car traveling in the $+x$-direction makes a fast stop; the $x$-component of the net force acting on it is $-1.50 \times 10^4$ N. What is its acceleration?

**SOLUTION** Because the newton is a unit of force, we know that $1.96 \times 10^4$ N is the weight, not the mass, of the car. Its mass $m$ is

$$m = \frac{w}{g} = \frac{1.96 \times 10^4\ \text{N}}{9.80\ \text{m/s}^2} = \frac{1.96 \times 10^4\ \text{kg} \cdot \text{m/s}^2}{9.80\ \text{m/s}^2} = 2000\ \text{kg}.$$

Then $\Sigma F_x = ma_x$ gives

$$a_x = \frac{F_x}{m} = \frac{-1.50 \times 10^4\ \text{N}}{2000\ \text{kg}} = \frac{-1.50 \times 10^4\ \text{kg} \cdot \text{m/s}^2}{2000\ \text{kg}}$$
$$= -7.5\ \text{m/s}^2.$$

This acceleration can be written as $-0.77g$. Note that $-0.77$ is also the ratio of $-1.50 \times 10^4$ N to $1.96 \times 10^4$ N.

---

Usually, the easiest way to measure the mass of a body is to measure its weight, often by comparing with a standard. Because of Eq. (4–10), two bodies that have the same weight at a particular location also have the same mass. We can compare weights very precisely; the familiar equal-arm balance (Fig. 4–15) can determine with great precision (to one part in $10^6$) when the weights of two bodies are equal and hence when their masses are equal. This method doesn't work in the "zero-gravity" environment of outer space. Instead, we have to use Newton's second law directly. We apply a known force to the body, measure its acceleration, and compute the mass as the ratio of force to acceleration. This method, or a variation of it, is used to measure the masses of astronauts in orbiting space stations and also the masses of atomic and subatomic particles.

The concept of mass plays two rather different roles in mechanics. The weight of a body (the gravitational force acting on it) is proportional to its mass; we may call the property related to gravitational interactions *gravitational mass.* On the other hand, we can call the inertial property that appears in Newton's second law the *inertial mass.* If these two quantities were different, the acceleration due to gravity might well be different for different bodies. However, extraordinarily precise experiments have established with a precision of better than one part in $10^{12}$ that in fact the two *are* the same. Recent efforts to find departures from this equivalence, possibly showing evidence for a previously unknown force, have been inconclusive.

Finally, we remark that the SI units for mass and weight are often misused in everyday life. Incorrect expressions such as "This box weighs 6 kg" are nearly universal. What is meant is that the *mass* of the box, probably determined indirectly by weighing, is 6 kg.

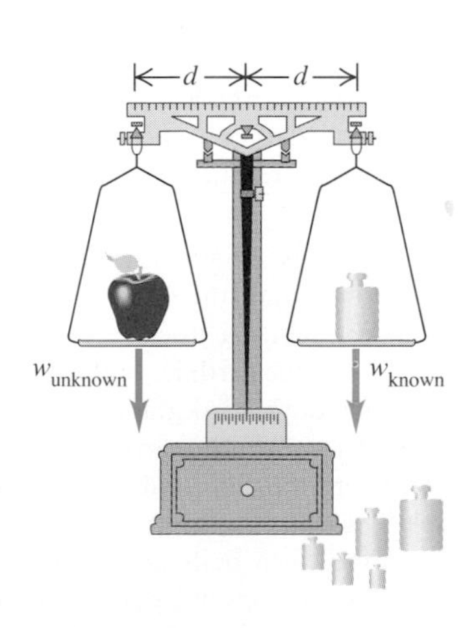


**4–15** An equal-arm balance determines the mass of a substance by comparing its weight to a known weight.

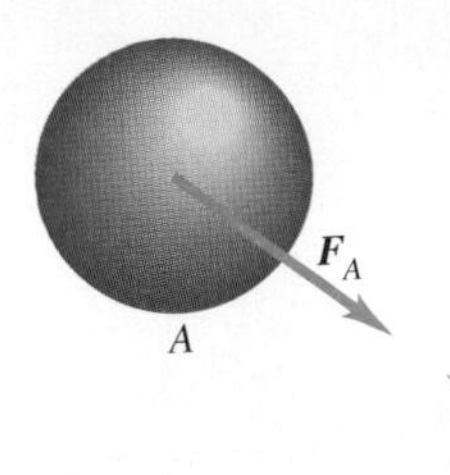
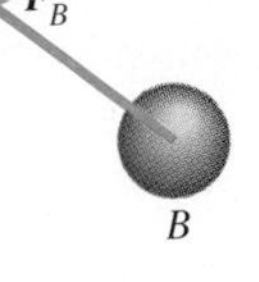


**4–16** If object $A$ exerts a force $\boldsymbol{F}_B$ on object $B$, then object $B$ exerts a force $\boldsymbol{F}_A$ on object $A$ that is equal in magnitude and opposite in direction, whether they are touching or not. $\boldsymbol{F}_A = -\boldsymbol{F}_B$.

This usage is so common that there is probably no hope of straightening things out, but be sure you recognize that the term *weight* is often used when *mass* is meant. To keep your own thinking clear, be careful to avoid this kind of mistake! In SI units, weight (a force) is measured in newtons, and mass is measured in kilograms.

## 4–5 NEWTON'S THIRD LAW

A force acting on a body is always the result of its interaction with another body, so forces always come in pairs. I can't push on you unless you push back on me at the same time. When you kick a football, the force your foot exerts on the ball launches it into its trajectory, but you also feel the force the ball exerts on your foot.

Experiments show that whenever two bodies interact, the two forces they exert on each other are equal in magnitude and opposite in direction. This fact is called *Newton's third law.* In Fig. 4–16, $\boldsymbol{F}_A$ is the force applied to body $A$ by body $B$, and $\boldsymbol{F}_B$ is the force applied to $B$ by $A$. The directions of the forces in this example correspond to an *attractive* interaction, such as the gravitational attraction of two masses or the electrical attraction of two particles with opposite charges. The formal statement of Newton's third law is

$$\boldsymbol{F}_B = -\boldsymbol{F}_A, \quad \text{or} \quad \boldsymbol{F}_A + \boldsymbol{F}_B = \boldsymbol{0}. \tag{4–12}$$

Newton's own statement, translated from the Latin of the *Principia,* is

> **To every action there is always opposed an equal reaction; or, the mutual actions of two bodies upon each other are always equal, and directed to contrary parts.**

In this statement "action" and "reaction" are the two opposite forces. This is not meant to imply any cause-and-effect relationship; we can consider either force as the "action" and the other the "reaction." We often say simply that the forces are "equal and opposite," meaning that they have equal magnitudes and opposite directions; their vector sum is then always zero.

### EXAMPLE 4–6

An apple sits on a table in equilibrium. What forces act on it? What is the reaction force to each of the forces acting on the apple? What are the action-reaction pairs?

**SOLUTION** Figure 4–17a shows the apple on the table, and Fig. 4–17b shows the forces acting on the apple. We have elaborated our notation a little to help explain things. The apple is A, the table T, and the earth E. In the diagram, $\boldsymbol{F}_{\text{AE}}$ is the weight of the apple, the downward gravitational force exerted *on* the apple A (first subscript) *by* the earth E (second subscript). Similarly, $\boldsymbol{F}_{\text{AT}}$ is the upward force exerted *on* the apple A (first subscript) *by* the table T (second subscript).

As the earth pulls down on the apple, the apple exerts an equally strong upward pull $\boldsymbol{F}_{\text{EA}}$ on the earth, as shown in Fig. 4–17d. $\boldsymbol{F}_{\text{EA}}$ and $\boldsymbol{F}_{\text{AE}}$ are an action-reaction pair, representing the mutual interaction of the apple and the earth, so

$$\boldsymbol{F}_{\text{EA}} = -\boldsymbol{F}_{\text{AE}}.$$

Also, as the table pushes up on the apple with force $\boldsymbol{F}_{\text{AT}}$, the corresponding reaction is the downward force $\boldsymbol{F}_{\text{TA}}$ exerted on the table by the apple (Fig. 4–17c), and we have

$$\boldsymbol{F}_{\text{AT}} = -\boldsymbol{F}_{\text{TA}}.$$

The two forces acting on the apple are $\boldsymbol{F}_{\text{AT}}$ and $\boldsymbol{F}_{\text{AE}}$. Are they an action-reaction pair? No, they aren't, despite the fact that they are equal and opposite. They do not represent the mutual interaction of two bodies; they are two different forces acting on the same body. *The two forces in an action-reaction pair* ***never*** *act on the same body.* Here's another way to look at it. Suppose we suddenly yank the table out from under the apple (Fig. 4–17e). The two forces $\boldsymbol{F}_{\text{AT}}$ and $\boldsymbol{F}_{\text{TA}}$ then become zero. But $\boldsymbol{F}_{\text{AE}}$ and $\boldsymbol{F}_{\text{EA}}$ are still there, and they *do* form an action-reaction pair. In that case, $\boldsymbol{F}_{\text{AT}}$ is zero, and it can't be the negative of $\boldsymbol{F}_{\text{AE}}$.

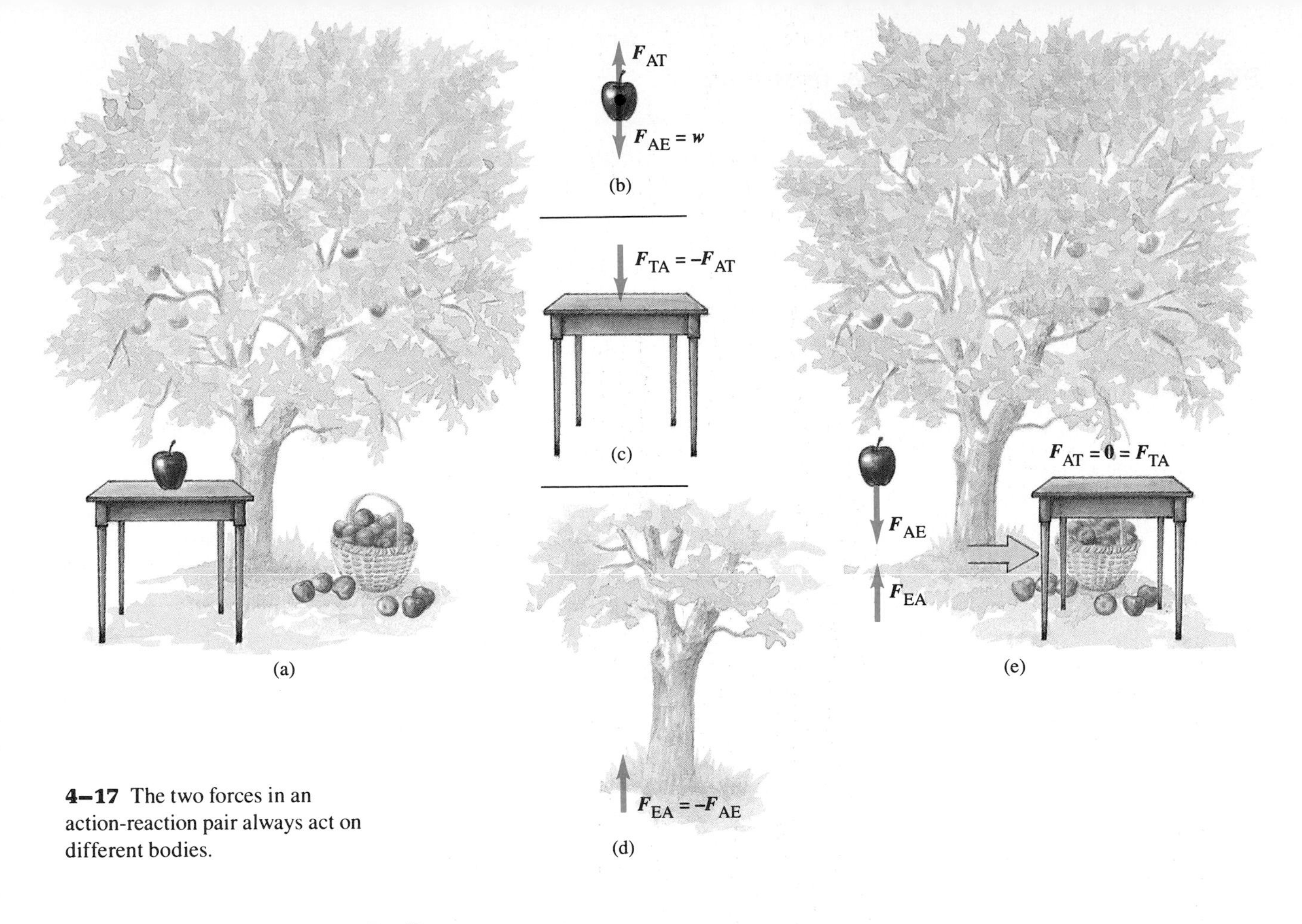


**4–17** The two forces in an action-reaction pair always act on different bodies.

## EXAMPLE 4–7

A stonemason drags a marble block across a floor by pulling on a rope attached to the block (Fig. 4–18a). The block may or may not be in equilibrium. How are the various forces related? What are the action-reaction pairs?

**SOLUTION** Figure 4–18b shows the horizontal forces acting on each body, the block B, the rope R, and the mason M. Vector $\boldsymbol{F}_{RM}$ represents the force exerted *on* the rope *by* the mason. Its reaction is the equal and opposite force $\boldsymbol{F}_{MR}$ exerted *on* the mason *by*

**4–18** (a) A mason pulls on a rope attached to a block. (b) Separate diagrams showing the force on the block by the rope, the forces on the rope by the block and the mason, and the force on the mason by the rope. (c) If the rope is not accelerating or if its mass can be neglected, it can be considered to transmit a force from the mason to the block, and vice versa.

the rope. Vector $\boldsymbol{F}_{BR}$ represents the force exerted on the block by the rope. The reaction to it is the equal and opposite force $\boldsymbol{F}_{RB}$ exerted on the rope by the block. Thus

$$\boldsymbol{F}_{MR} = -\boldsymbol{F}_{RM} \quad \text{and} \quad \boldsymbol{F}_{RB} = -\boldsymbol{F}_{BR} \tag{4-13}$$

Be sure you understand that the forces $\boldsymbol{F}_{RM}$ and $\boldsymbol{F}_{RB}$ are *not* an action-reaction pair. Both of these forces act on the *same* body (the rope); an action and its reaction *must* always act on *different* bodies. Furthermore, the forces $\boldsymbol{F}_{RM}$ and $\boldsymbol{F}_{RB}$ are not necessarily equal in magnitude. Applying Newton's second law to the rope, we get

$$\boldsymbol{F}_{RM} + \boldsymbol{F}_{RB} = m_{rope}\boldsymbol{a}_{rope}.$$

If block and rope are accelerating to the right, the rope is not in equilibrium, and $\boldsymbol{F}_{RM}$ must have greater magnitude than $\boldsymbol{F}_{RB}$. In the case in which the rope is in equilibrium, the forces $\boldsymbol{F}_{RM}$ and $\boldsymbol{F}_{RB}$ are equal in magnitude, but this is an example of Newton's *first* law, not his *third*. Newton's third law holds whether the rope is accelerating or not. The action-reaction forces $\boldsymbol{F}_{RM}$ and $\boldsymbol{F}_{MR}$ are always equal in magnitude to *each other,* as are $\boldsymbol{F}_{BR}$ and $\boldsymbol{F}_{RB}$. But $\boldsymbol{F}_{RM}$ is *not* equal in magnitude to $\boldsymbol{F}_{RB}$ when the rope is accelerating.

If you feel as though you're drowning in subscripts at this point, take heart. Go over this discussion again, comparing the symbols with the vector diagrams, until you're sure you see what's going on.

In the special case in which the rope is in equilibrium, or when we can consider it as massless, then $\boldsymbol{F}_{RB}$ equals $-\boldsymbol{F}_{RM}$ because of Newton's *first* or *second* law. Also, $\boldsymbol{F}_{RB}$ always equals $-\boldsymbol{F}_{BR}$ by Newton's *third* law, so in this special case, $\boldsymbol{F}_{BR}$ also equals $\boldsymbol{F}_{RM}$. We can then think of the rope as "transmitting" to the block, without change, the force that the stonemason exerts on it (Fig. 4–18c). This is a useful point of view, but you have to remember that it is valid only when the rope has negligibly small mass or is in equilibrium. ■

---

A body such as the rope in Fig. 4–18 that has pulling forces applied at its ends is said to be in **tension.** The tension at any point is the magnitude of force acting at that point. In Fig. 4–18b the tension at the right-hand end of the rope is the magnitude of $\boldsymbol{F}_{RM}$ (or of $\boldsymbol{F}_{MR}$), and the tension at the left-hand end equals the magnitude of $\boldsymbol{F}_{RB}$ (or of $\boldsymbol{F}_{BR}$). If the rope is in equilibrium and if no forces act except at its ends, the tension is the same at both ends and throughout the rope. If the magnitudes of $\boldsymbol{F}_{RB}$ and $\boldsymbol{F}_{RM}$ are 50 N each, the tension in the rope is 50 N (*not* 100 N). Resist the temptation to add the two forces; remember that the total force $\boldsymbol{F}_{RB} + \boldsymbol{F}_{RM}$ on the rope in this case is zero!

Finally, we emphasize once more a fundamental truth: The two forces in an action-reaction pair *never* act on the same body. Remembering this simple fact can often help you avoid confusion about action-reaction pairs and Newton's third law.

## 4–6 USING NEWTON'S LAWS

Newton's three laws of motion are a beautifully wrapped package containing the basic principles we need to solve a wide variety of problems in mechanics. They are very simple in form, but the process of applying them to specific situations can pose real challenges.

Let's talk about some useful techniques. When you use Newton's first or second law, $\Sigma\boldsymbol{F} = \boldsymbol{0}$ for an equilibrium situation or $\Sigma\boldsymbol{F} = m\boldsymbol{a}$ for a nonequilibrium situation, you must apply it to some specific body. It is absolutely essential to decide at the beginning what body you are talking about. This may sound trivial, but it isn't. Once you have chosen a body, then you have to identify all the forces acting on it. These are the forces that are included in $\Sigma\boldsymbol{F}$. As you may have noticed in Section 4–5, it is easy to get confused between the forces acting *on* a body and the forces exerted *by* that body *on* some other body. Only the forces acting *on* the body go into $\Sigma\boldsymbol{F}$.

To help identify the relevant forces, draw a **free-body diagram.** What's that? It is a diagram showing the chosen body by itself, "free" of its surroundings, and with vectors drawn to show the forces applied to it by the various other bodies that interact with it. We have already shown some free-body diagrams in Figs. 4–17 and 4–18. Be careful to include *all* the forces acting on the body, but be equally careful *not* to include any forces that the body exerts on any other body. In particular, the two forces in an action-reaction pair must *never* appear in the same free-body diagram because they never act on the same body.

# Free-BodyDiagrams

Drawing a correct free-body diagram is the first step in analyzing almost any situation in physics or engineering. The complexity of the diagram reflects the complexity of the mathematical model you use to represent the situation and does not necessarily reflect other physical characteristics. Note that the following figures are all in very different surroundings, but the free-body diagrams are remarkably similar.

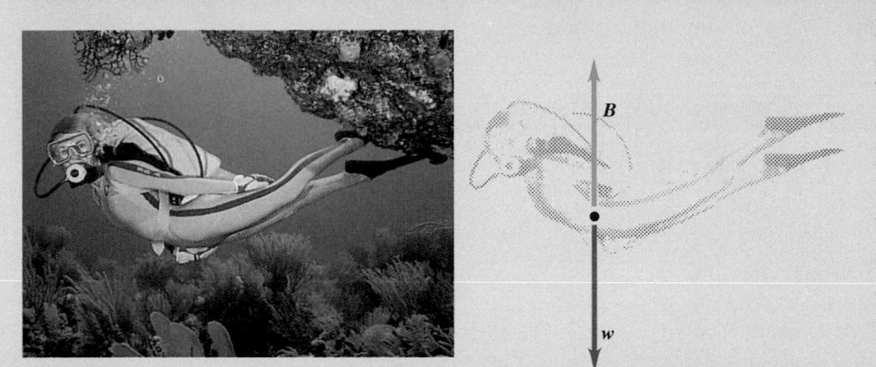


Immersed in water, a person's body experiences an upward force due to buoyancy. This is balanced by the downward force of the diver's weight. In this situation, the swimmer's motion depends on the force with which the water presses down on him or her, either due to the water currents or to reaction forces to the swimmer's arm and leg movements.

# BUILDING PHYSICAL INTUITION

A person jumps by pushing with his feet against the ground. The forces acting on him are his weight and the reaction force with which the ground pushes back up. Once the player is in the air, the only force acting on him is his weight. Even as he rises up, his acceleration is directed down.

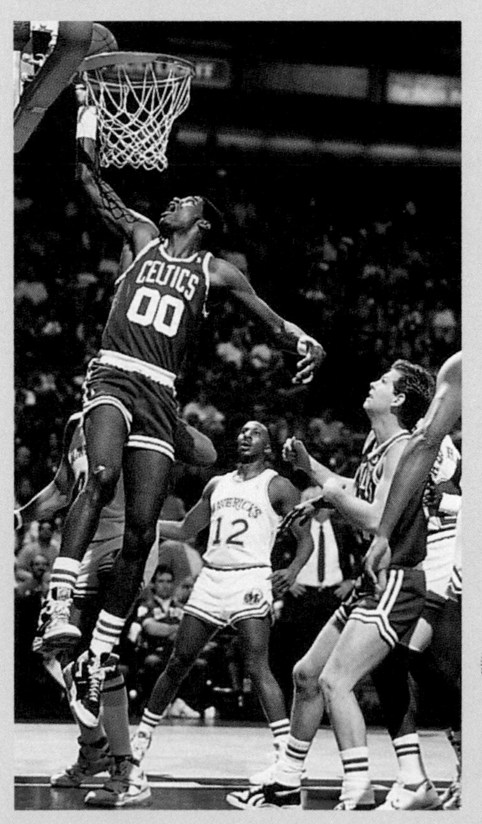

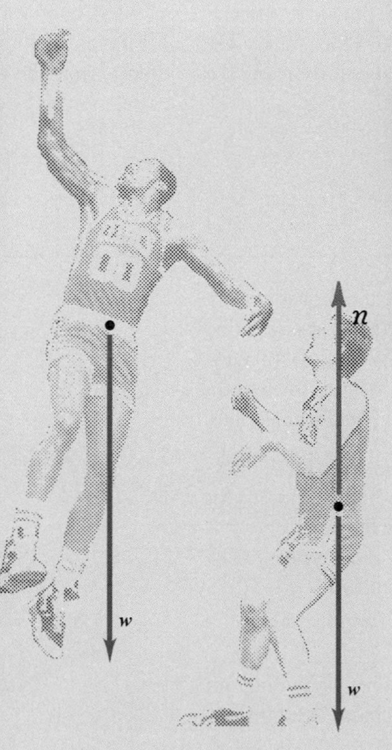


A sprinter gains a large forward acceleration at the start of a race by kicking back hard against the angled starting blocks. The reaction force on the runner has a small upward component and a large forward component that springs her into motion.

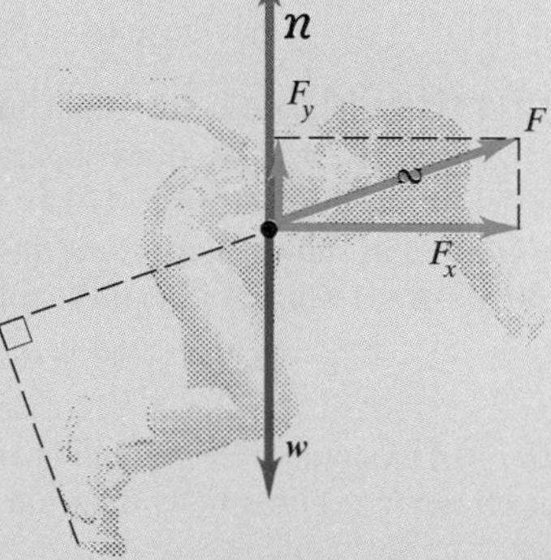

When you have a complete free-body diagram, you should be able to answer for each force the question "What other body is applying this force?" If you can't answer that question, you may be dealing with a nonexistent force. Sometimes you will have to take the problem apart and draw a separate free-body diagram for each part. That's what we did in Fig. 4–18b, which has separate diagrams for the block, the rope (considered to be massless), and the mason. Of these, only the diagram for the rope is a complete free-body diagram. What's missing in the others?

## PROBLEM-SOLVING STRATEGY

### Newton's laws

**1.** Always define your coordinate system. A diagram showing the location of the origin and the positive axis direction is always helpful. If you know the direction of the acceleration, it is often convenient to take that as your positive direction.

**2.** Be consistent with signs. Once you define the $x$-axis and its positive direction, then velocity, acceleration, and force components in that direction are also positive.

**3.** In applying Newton's laws, always concentrate on a specific body. Draw a free-body diagram showing all the forces (magnitudes and directions) acting *on* this body, but *do not* include forces that the body exerts on any other body. The acceleration of the body is determined by the forces acting on it, not by the forces it exerts on something else. Represent the body as a particle; you don't have to be an artist. Using a colored pencil or pen for the force vectors may help. In this chapter and the next we will use different colors to distinguish certain force vectors in free-body diagrams: dark blue for weight, purple for normal forces, and dark green for friction forces. All other forces will appear in the bright blue used for force vectors throughout the remainder of this book.

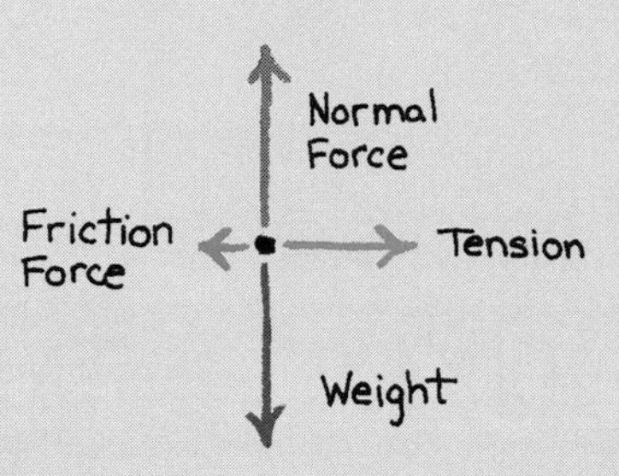


**4.** Identify the known and unknown quantities, and give each unknown quantity an algebraic symbol. If you know the direction of a force at the start, use a symbol to represent the *magnitude* of the force, always a positive quantity. Keep in mind that the *component* of this force along a particular axis direction may still be either positive or negative.

**5.** Always check for unit consistency. When appropriate, use the conversion $1\ \text{N} = 1\ \text{kg} \cdot \text{m/s}^2$.

In Chapter 5 we will expand this strategy to deal with more complex problems, but it is important for you to use it consistently from the very start, to develop good habits in the systematic analysis of problems.

## EXAMPLE 4–8

**Tension in a massless chain** To improve the acoustics in an auditorium, a sound reflector with a mass of 200 kg is suspended by a chain from the ceiling. What is its weight? What force (magnitude and direction) does the chain exert on it? What is the tension in the chain? Assume that the mass of the chain itself is negligible.

**SOLUTION** The reflector is in equilibrium, so we use Newton's first law, $\Sigma \mathbf{F} = \mathbf{0}$. We draw separate free-body diagrams for the reflector and the chain (Fig. 4–19). We take the positive $y$-axis to be upward, as shown. Each force has only a $y$-component. The magnitude of the weight of the reflector is given by Eq. (4–10):

$$w = mg = (200\ \text{kg})(9.80\ \text{m/s}^2) = 1960\ \text{N}.$$

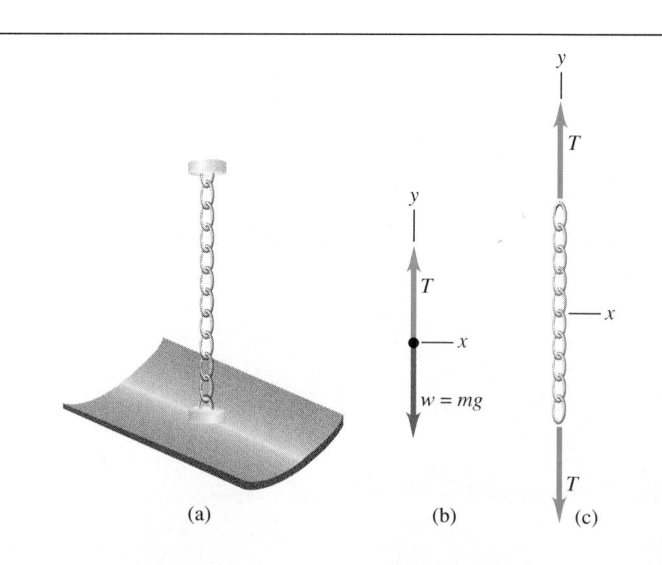


**4–19** (a) The sound reflector and chain. (b) Free-body diagram of the reflector. (c) Free-body diagram of the chain, assuming its weight to be negligible.

This force points in the negative $y$-direction, so its $y$-component is $-1960$ N. The upward force exerted by the chain has unknown magnitude $T$. Because the reflector is in equilibrium, the sum of the $y$-components of force on it must be zero:

$$\Sigma F_y = T + (-1960 \text{ N}) = 0, \qquad \text{so} \qquad T = 1960 \text{ N}.$$

The chain pulls *up* on the reflector with a force $T$ of magnitude 1960 N. By Newton's third law the reflector pulls *down* on the chain with a force of magnitude 1960 N.

The chain is also in equilibrium. We have assumed that it is weightless, so an upward force of 1960 N must act on it at its top end to make the vector sum of forces on it equal to zero. The tension in the chain is 1960 N.

Note that we have defined $T$ to be the *magnitude* of a force, so it is always positive. But the $y$-component of force acting on the chain at its lower end is $-T = -1960$ N.

## EXAMPLE 4–9

**Tension in a chain with mass** Suppose the mass of the chain in Example 4–8 is not negligible but is 10.0 kg. Find the forces at the ends of the chain.

**SOLUTION** The weight of the chain is $(10.0 \text{ kg})(9.8 \text{ m/s}^2) =$ 98 N. Again we draw separate free-body diagrams for the reflector and the chain; each is in equilibrium. The free-body diagrams, shown in Fig. 4–20, differ from those in Fig. 4–19 because the magnitudes of the forces at the two ends of the chain are no longer equal. We label them $T_1$ and $T_2$, as shown. Note that the two forces labeled $T_2$ form an action-reaction pair; that's how we know that they have the same magnitude. The equilibrium condition $\Sigma F_y =$ 0 for the reflector is

$$T_2 + (-1960 \text{ N}) = 0, \qquad \text{so} \qquad T_2 = 1960 \text{ N}.$$

There are now three forces acting on the chain: its weight and the forces at the two ends. The equilibrium condition $\Sigma F_y = 0$ for the chain is

$$T_1 + (-T_2) + (-98 \text{ N}) = 0.$$

Note that the $y$-component of $T_1$ is positive because it points in the $+y$-direction, but the $y$-components of both $T_2$ and 98 N are negative. When we substitute the value $T_2 = 1960$ N and solve for $T_1$, we find

$$T_1 = 2058 \text{ N}.$$

Alternatively, we could draw a free-body diagram for the composite body consisting of the reflector and the chain together (Fig. 4–20c). The two forces on this composite object are the upward force $T_1$ at the top of the chain and the total weight, with magnitude $1960 \text{ N} + 98 \text{ N} = 2058$ N. Again we find $T_1 = 2058$ N. Note that we cannot find $T_2$ by this method.

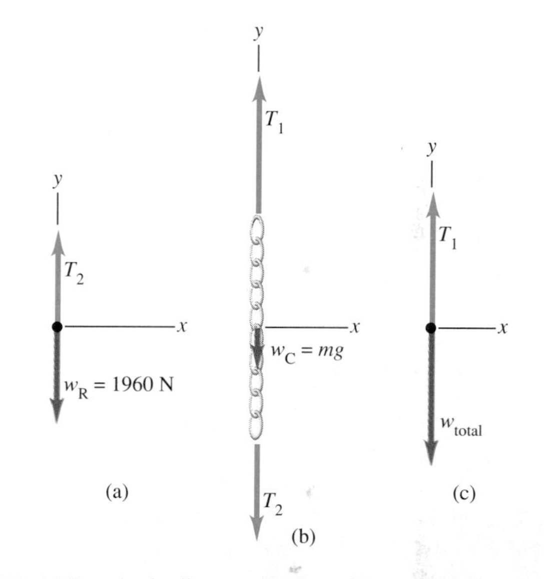


**4–20** (a) Free-body diagram for the reflector (weight $w_R$). (b) Free-body diagram for the chain (weight $w_C$). (c) Free-body diagram for the reflector and chain, considered as a single particle.

## EXAMPLE 4–10

The cutting blade assembly on a radial-arm saw has a mass of 5.0 kg. It is pulled along a pair of frictionless horizontal rails aligned with the $x$-axis by a force $F_x$. Its position is given as a function of time by

$$x = (0.18 \text{ m/s}^2)t^2 - (0.030 \text{ m/s}^3)t^3.$$

Find the force acting on the assembly as a function of time. What is the force at time $t = 5.0$ s? For what times is the force positive? Negative? Zero?

**SOLUTION** Figure 4–21 shows the free-body diagram and coordinate axes. The forces are the horizontal force $F_x$, the weight $w$, and the upward force $n$ that the rails exert to support the blade assembly. There is no acceleration in the vertical direction, so the sum of the vertical components of force must be zero. We need to

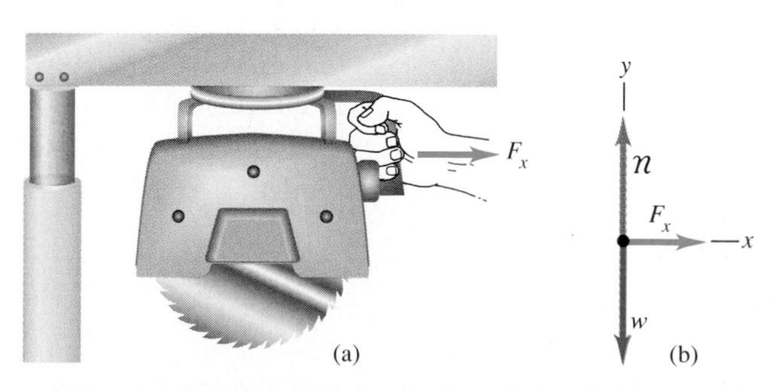


**4–21** (a) A radial-arm saw cutting blade. (b) Free-body diagram of blade assembly. Both the magnitude and direction of $\boldsymbol{F}_x$ are functions of time.

be concerned only with horizontal components. We first find the acceleration $a_x$ by taking the second derivative of $x$:

$$a_x = \frac{d^2x}{dt^2} = 0.36\ \text{m/s}^2 - (0.18\ \text{m/s}^3)t.$$

Then, from Newton's second law,

$$\Sigma F_x = ma_x = (5.0\ \text{kg})[0.36\ \text{m/s}^2 - (0.18\ \text{m/s}^3)t] = 1.80\ \text{N} - (0.90\ \text{N/s})t.$$

At time $t = 5.0$ s the force is $1.80\ \text{N} - (0.90\ \text{N/s})(5.0\ \text{s}) = -2.70$ N. The force is zero when $a = 0$, that is, when $0.36\ \text{m/s}^2 - (0.18\ \text{m/s}^3)t = 0$. This happens when $t = 2.0$ s. When $t < 2.0$ s, $F$ is positive, and when $t > 2.0$ s, $F$ is negative. Figure 4–22 shows graphs of $F_x$ and $a_x$ as functions of time.

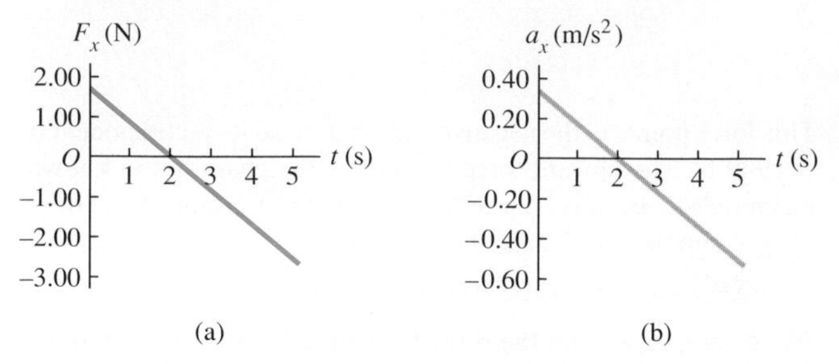


**4–22** (a) The net force on the cutting blade assembly is directly proportional to (b) its acceleration.

## SUMMARY

**dynamics**
**Newton's laws of motion**
**force**
**resultant**
**superposition of forces**
**equilibrium**
**inertial frame of reference**
**mass**
**kilogram**
**newton**
**weight**
**tension**
**free-body diagram**

- Force, a vector quantity, is a quantitative measure of the mechanical interaction between two bodies. When several forces act on a body, the effect is the same as when a single force, equal to the vector sum, or resultant, of the forces, acts on the body.
- Newton's first law states that when no force acts on a body, or when the vector sum of all forces acting on it is zero, the body is in equilibrium. If the body is initially at rest, it remains at rest; if it is initially in motion, it continues to move with constant velocity. This law is valid only in inertial frames of reference.
- The inertial properties of a body are characterized by its *mass*. The acceleration of a body under the action of a given set of forces is directly proportional to the vector sum of the forces and inversely proportional to the mass of the body. This relationship is Newton's second law:

$$\Sigma \boldsymbol{F} = m\boldsymbol{a}. \qquad (4\text{–}8)$$

Like the first law, this is valid only in inertial frames of reference.

- The unit of force is defined in terms of the units of mass and acceleration. In SI units the unit of force is the newton (N), equal to $1\ \text{kg} \cdot \text{m/s}^2$.
- The weight of a body is the gravitational force exerted on it by the earth (or whatever other body exerts the gravitational force). Weight is a force and is therefore a vector quantity. The magnitude of the weight of a body at any specific location is equal to the product of its mass $m$ and the magnitude of the acceleration due to gravity $g$ at that location:

$$w = mg. \qquad (4\text{–}10)$$

The weight of a body depends on its location, but the mass is independent of location.

- Newton's third law states that "action equals reaction"; when two bodies interact, they exert forces on each other that at each instant are equal in magnitude and opposite in direction. The two forces in an action-reaction pair always act on two different bodies; they never act on the same body.

## EXERCISES

### Section 4–1 Force

**4–1** A warehouse worker pushes a crate along the floor as in Fig. 4–2b by a force of 80.0 N that points downward at an angle of 30.0° below the horizontal. Find the horizontal and vertical components of the force.

**4–2** A man is dragging a trunk up the loading ramp of a mover's truck. The ramp has a slope angle of 20.0°, and the man pulls upward with force $\boldsymbol{F}$ whose direction makes an angle of 30.0° with the ramp (Fig. 4–23). a) How large a force $\boldsymbol{F}$ is necessary for the

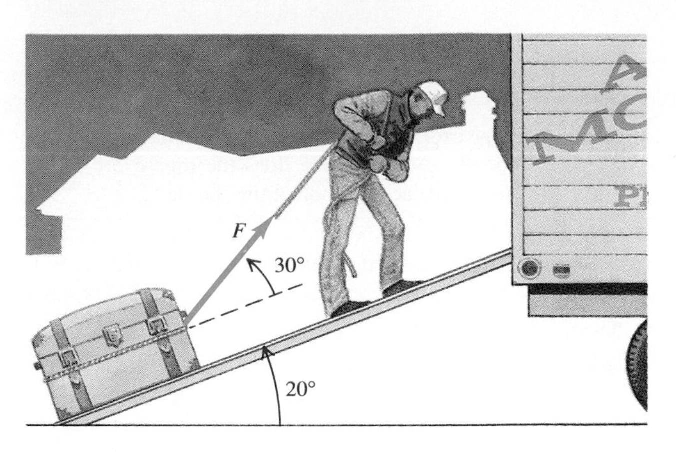


**FIGURE 4–23**

component $F_x$ parallel to the ramp to be 60.0 N? b) How large will the component $F_y$ then be?

**4–3** Two dogs pull horizontally on ropes attached to a post; the angle between the ropes is 50.0°. If dog $A$ exerts a force of 310 N and dog $B$ exerts a force of 220 N, find the magnitude of the resultant force and angle it makes with dog $A$'s rope.

**4–4** Two forces, $\boldsymbol{F}_1$ and $\boldsymbol{F}_2$, act at a point. The magnitude of $\boldsymbol{F}_1$ is 8.00 N, and its direction is 60.0° above the $x$-axis in the first quadrant. The magnitude of $\boldsymbol{F}_2$ is 6.00 N, and its direction is 53.1° below the $x$-axis in the fourth quadrant. a) What are the $x$- and $y$-components of the resultant force? b) What is the magnitude of the resultant? c) What is the magnitude of the vector difference $\boldsymbol{F}_1 - \boldsymbol{F}_2$?

## Section 4–3 Mass and Newton's Second Law

**4–5** A box rests on a frozen pond, which serves as a frictionless horizontal surface. If a fisherman applies a horizontal force with magnitude 48.0 N to the box and produces an acceleration of magnitude 8.50 m/s$^2$, what is the mass of the box?

**4–6** What magnitude of force is required to give a 125-kg refrigerator an acceleration of magnitude 1.60 m/s$^2$?

**4–7** If a horizontal force of magnitude 158.0 N is applied to a crate with mass 120 kg that is resting on a horizontal frictionless surface, what magnitude of acceleration is produced?

**4–8** A crate with mass 37.5 kg initially at rest on a frictionless horizontal plane is acted on by a horizontal force of 160 N. a) What acceleration is produced? b) How far does the crate travel in 10.0 s? c) What is its speed at the end of 10.0 s?

**4–9** World-class sprinters can accelerate out of the starting blocks with an acceleration that is nearly horizontal and has magnitude 15 m/s$^2$. How much horizontal force does an 80-kg sprinter apply during a start to produce this acceleration?

**4–10** A dockworker applies a constant horizontal force of 90.0 N to a block of ice on a smooth horizontal floor. The block starts from rest and moves 16.0 m in 5.00 s. a) What is the mass of the block of ice? b) If the worker stops pushing at the end of 5.00 s, how far does the block move in the next 5.00 s?

**4–11** A hockey puck with mass 0.160 kg is at rest at the origin ($x = 0$) on the horizontal frictionless surface of the rink. At time $t = 0$ a player applies a force of 0.400 N to the puck, parallel to the $x$-axis; the player continues to apply this force until $t = 2.00$ s. a) What are the position and speed of the puck at $t = 2.00$ s? b) If the same force is again applied at $t = 5.00$ s, what are the position and speed of the puck at $t = 7.00$ s?

**4–12** An electron (mass = $9.11 \times 10^{-31}$ kg) leaves one end of a TV picture tube with zero initial speed and travels in a straight line to the accelerating grid, which is 1.50 cm away. It reaches the grid with a speed of $6.00 \times 10^6$ m/s. If the accelerating force is constant, compute a) the acceleration; b) the time to reach the grid; c) the accelerating force in newtons. (The gravitational force on the electron may be neglected.)

## Section 4–4 Mass and Weight

**4–13** At the surface of Mars the acceleration due to gravity is $g = 3.7$ m/s$^2$. A watermelon weighs 64.0 N at the surface of the earth. What are its mass and weight on the surface of Mars?

**4–14** a) What is the mass of a book that weighs 4.50 N at a point where $g = 9.80$ m/s$^2$? b) At the same location, what is the weight of a dog whose mass is 12.0 kg?

**4–15** Superman throws a 2800-N boulder at an adversary. What horizontal force must Superman apply to the boulder to give it a horizontal acceleration of 24.0 m/s$^2$?

**4–16** A bowling ball weighs 71.2 N (16 lb). The bowler applies a horizontal force of 214 N (48 lb) to the ball. What is the magnitude of the acceleration of the ball?

## Section 4–5 Newton's Third Law

**4–17** A bottle is given a push along a tabletop and slides off the edge of the table. Neglect air resistance. a) What forces are exerted on the bottle while it is falling from the table to the floor? b) What is the reaction to each force; that is, on what body and by what body is the reaction exerted?

**4–18** Imagine that you are holding a book weighing 4 N at rest on the palm of your hand. Complete the following sentences: a) A downward force of magnitude 4 N is exerted on the book by ________. b) An upward force of magnitude ________ is exerted on ________ by the hand. c) Is the upward force in part (b) the reaction to the downward force in part (a)? d) The reaction to the force in part (a) is a force of magnitude ________, exerted on ________ by ________. Its direction is ________. e) The reaction to the force in part (b) is a force of magnitude ________, exerted on ________ by ________. Its direction is ________. f) The forces in parts (a) and (b) are equal and opposite because of Newton's ________ law. g) The forces in parts (b) and (e) are equal and opposite because of Newton's ________ law. Now suppose you exert an upward force of magnitude 5 N on the book. h) Does the book remain in equilibrium? i) Is the force exerted on the book by your hand equal and opposite to the force exerted on the book by the earth? j) Is the force exerted on the book by the earth equal and opposite to the force exerted on the earth by the book? k) Is

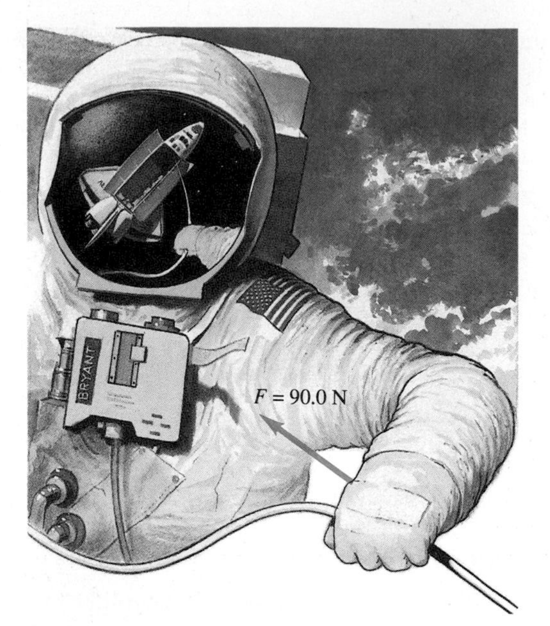


**FIGURE 4–24**

the force exerted on the book by your hand equal and opposite to the force exerted on your hand by the book? Finally, suppose that you snatch your hand away while the book is moving upward. l) How many forces then act on the book? m) Is the book in equilibrium?

**4–19** A 4.40-kg bucket of water is accelerated upward by a cord whose breaking strength is 60.0 N. Find the maximum upward acceleration that can be given to the bucket without breaking the cord.

**4–20** An astronaut with mass 80.0 kg is tethered by a strong rope to a space shuttle (Fig. 4–24). The mass of the shuttle is $8.55 \times 10^4$ kg, and the mass of the rope can be neglected. The shuttle is far from both the moon and the earth, so the gravitational forces on it and the astronaut are negligible. Both the shuttle and the astronaut are initially at rest in an inertial reference frame. The astronaut then pulls on the rope with a force of 90.0 N. a) What force does the rope exert on the astronaut? b) What is the astronaut's acceleration? c) What force does the rope exert on the shuttle? d) What is the acceleration of the shuttle?

**4–21** A parachutist relies on the drag force of her parachute to reduce her acceleration toward the earth. If she has a mass of 50.0 kg and her parachute drag supplies an upward force of 340 N, what is her acceleration?

**4–22** An elevator with mass 2000 kg rises with an acceleration of 1.50 $m/s^2$. What is the tension in the supporting cable?

### Section 4–6 Using Newton's Laws

**4–23** Two crates, one with mass 4.00 kg and the other with mass 6.00 kg, sit on the frictionless surface of a frozen pond, connected by a light rope (Fig. 4–25). A woman wearing golf shoes (so that she can get traction on the ice) pulls horizontally on the 6.00-kg crate with a force $\boldsymbol{F}$ that gives the crate an acceleration of 2.50 $m/s^2$. a) What is the magnitude of the force $\boldsymbol{F}$? b) What is the tension $T$ in the rope connecting the two crates?

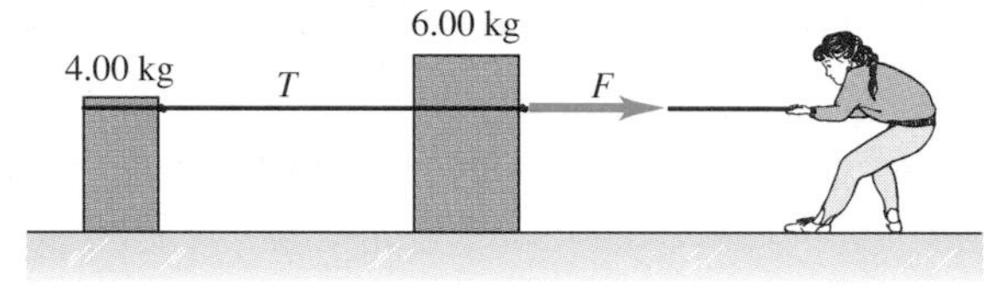


**FIGURE 4–25**

**4–24** Refer to Fig. 4–25. The crates are on a horizontal frictionless surface. The woman applies a horizontal force $F$ = 60.0 N to the 6.00-kg crate. a) What is the magnitude of the acceleration of the 6.00-kg crate? b) What is the tension $T$ in the rope connecting the two crates?

**4–25** An object with mass $m$ moves along the $x$-axis. Its position as a function of time is given by $x(t) = At + Bt^3$, where $A$ and $B$ are constants. Calculate the resultant force on the object as a function of time.

## PROBLEMS

**4–26** Three customers are fighting over the same bargain basement coat. They apply the three horizontal forces to the coat that are shown in Fig. 4–26, where the coat is located at the origin. a) Find the $x$- and $y$-components of each of the three forces. b) Find the magnitude and direction of the resultant of these forces. c) Find the magnitude and direction of the horizontal force that the sales clerk must apply to make the resultant horizontal force zero. Indicate this fourth force by a diagram.

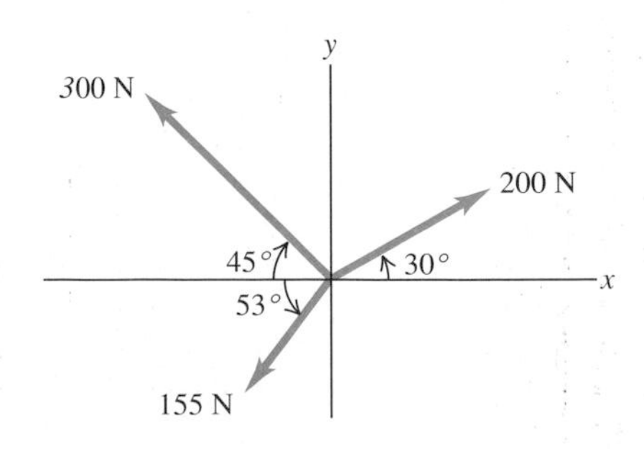


**FIGURE 4–26**

**4–27** Two horses pull horizontally on ropes attached to a stump. The two forces $\boldsymbol{F}_1$ and $\boldsymbol{F}_2$ that they apply to the stump are such that the resultant force $\boldsymbol{R}$ has a magnitude equal to that of $\boldsymbol{F}_1$ and makes an angle of 90° with $\boldsymbol{F}_1$ (Fig. 4–27). Let $F_1$ = 1200 N and $R$ = 1200 N also. Find the magnitude of $\boldsymbol{F}_2$ and its direction (relative to $\boldsymbol{F}_1$).

**4–28** The resultant of four forces is 1400 N in the direction 30.0° west of north. Three of the forces are 400 N, 60.0° north of east; 200 N, south; and 400 N, 53.1° west of south. Find the magnitude and direction of the fourth force.

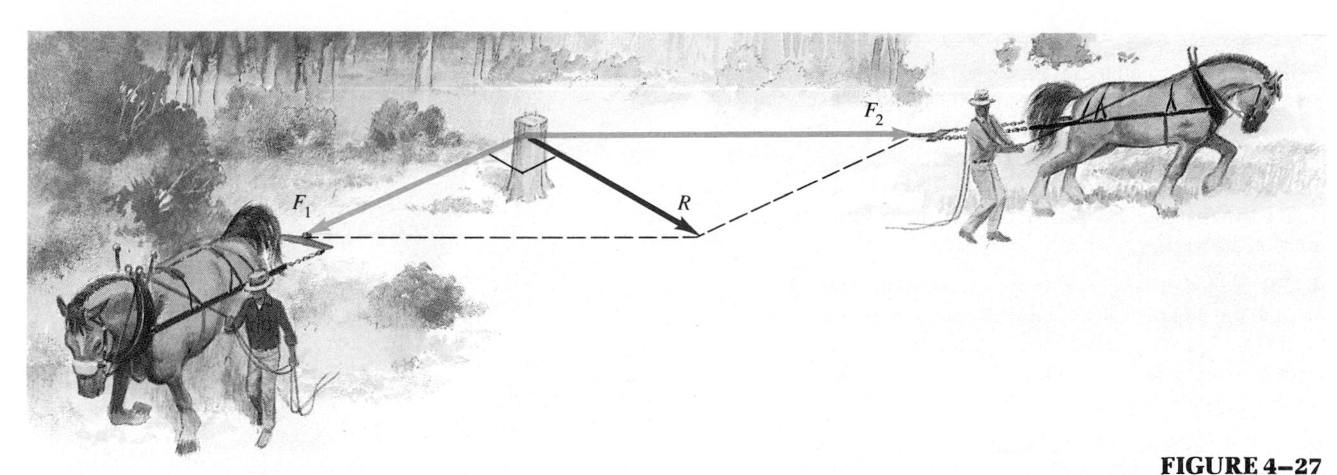


FIGURE 4–27

**4–29** Two adults and a child want to push a crate in the direction marked $x$ in Fig. 4–28. The two adults push with horizontal forces $\boldsymbol{F}_1$ and $\boldsymbol{F}_2$, whose magnitudes and directions are indicated in the figure. Find the magnitude and direction of the *smallest* force that the child should exert.

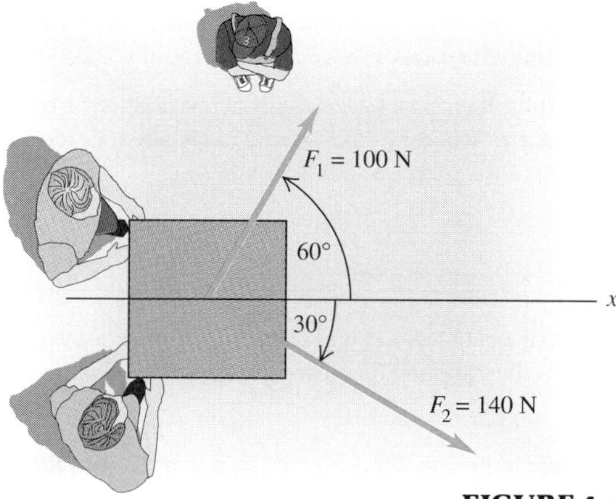


**FIGURE 4–28**

**4–30** A .22-caliber bullet, traveling at 360 m/s, strikes a block of soft wood, which it penetrates to a depth of 0.140 m. The mass of the bullet is 1.80 g. Assume a constant retarding force. a) How much time is required for the bullet to stop? b) What force, in newtons, does the wood exert on the bullet?

**4–31 A Standing Vertical Jump.** According to the *Guinness Book of World Records,* basketball player Darrell Griffith holds the record for the standing vertical jump, with a jump of 1.2 m (4 ft). If Griffith weighs 890 N (200 lb) and the time of the jump before his feet leave the ground is 0.400 s, what is the average force he applies to the ground?

**4–32** An oil tanker's engines have broken down, and the wind has accelerated the tanker to a speed of 1.5 m/s straight toward a reef (Fig. 4–29). When the tanker is 500 m from the reef, the wind dies down just as the engineer gets the engines going again. The rudder is stuck, so the only choice is to try to accelerate straight backward away from the reef. The mass of the tanker and cargo is $3.6 \times 10^7$ kg, and the engines produce a resultant horizontal force of $8.0 \times 10^4$ N on the tanker. Will the ship hit the reef? If it does, will the oil be safe? The hull can withstand an impact at a speed of 0.2 m/s or less.

**4–33 Stopping on a Dime.** An advertisement asserts that a particular automobile can "stop on a dime." What force would be necessary to stop a 950-kg automobile traveling initially at 13.4 m/s (30 mi/h) in a distance equal to the diameter of a dime, which is 1.8 cm?

**4–34** A large fish hangs from a spring balance supported from the roof of an elevator. a) If the elevator has an upward acceleration of 2.45 m/s$^2$ and the balance reads 55.0 N, what is the true weight of the fish? b) Under what circumstances will the balance read 25.0 N? c) What will the balance read if the elevator cable breaks?

**4–35** A 560-N physics student stands on a bathroom scale in an elevator. As the elevator starts moving, the scale reads 800 N. a) Find the acceleration of the elevator (magnitude and direction). b) What is the acceleration if the scale reads 450 N? c) If the scale reads zero, should the student worry? Explain.

**4–36** A 65.0-kg gymnast climbs a vertical rope attached to the ceiling. The weight of the rope can be neglected. Calculate the tension in the rope if the gymnast a) climbs at a constant rate; b) hangs motionless on the rope; c) accelerates up the rope at 0.40 m/s$^2$; d) slides down the rope with a downward acceleration of 0.40 m/s$^2$.

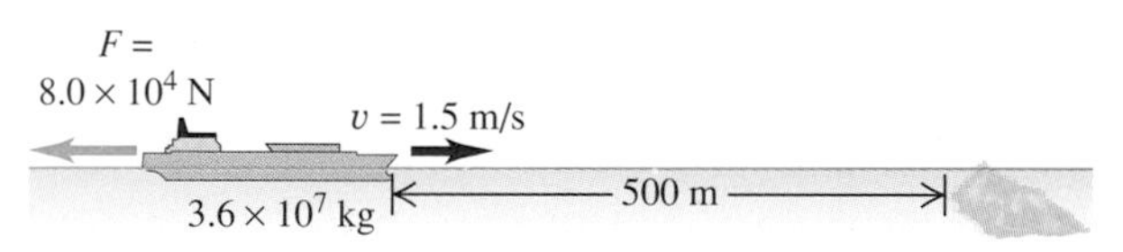


**FIGURE 4–29**

**4–37** A train (an engine plus four cars) is accelerating at 0.800 m/s$^2$. If each car has a mass of $3.90 \times 10^4$ kg and if each car has negligible friction forces acting on it, what is a) the force of the engine on the first car? b) the force of the first car on the second car? c) the force of the second car on the third car? d) the force of the fourth car on the third car? e) What would these same four forces be if the train were slowing down with an acceleration of $-0.800$ m/s$^2$? (When solving, note the importance of selecting the correct set of cars as your object.)

**4–38** A short commuter train consists of a locomotive and two cars. The mass of the locomotive is 6000 kg, and the mass of each car is 2000 kg. The train pulls away from a station with an acceleration of $0.600\ \text{m/s}^2$. a) Find the tension in the coupler joining the locomotive to the first car and in the coupler joining the two cars. b) What total horizontal force must the locomotive wheels exert on the track?

**4–39** A loaded elevator with very worn cables has a total mass of 1800 kg, and the cables can withstand a maximum tension of 24,000 N. a) What is the maximum upward acceleration for the elevator if the cables are not to break? b) What is the answer to part (a) if the elevator is taken to the moon, where $g = 1.62\ \text{m/s}^2$?

**4–40 Jumping to the Ground.** An 80.0-kg man steps off a platform 2.50 m above the ground. He keeps his legs straight as he falls, but at the moment his feet touch the ground his knees begin to bend, and his torso moves an additional 0.60 m before coming to rest. a) What is his speed at the instant his feet touch the ground? b) What is the acceleration of his torso as he slows down if the acceleration is assumed to be constant? c) What force do his feet exert on the ground while he slows down? Express this force in newtons and also as a multiple of his weight.

**4–41** The two blocks in Fig. 4–30 are connected by a heavy uniform rope with a mass of 4.00 kg. An upward force of 200 N is applied as shown. a) What is the acceleration of the system? b) What is the tension at the top of the heavy rope? c) What is the tension at the midpoint of the rope?

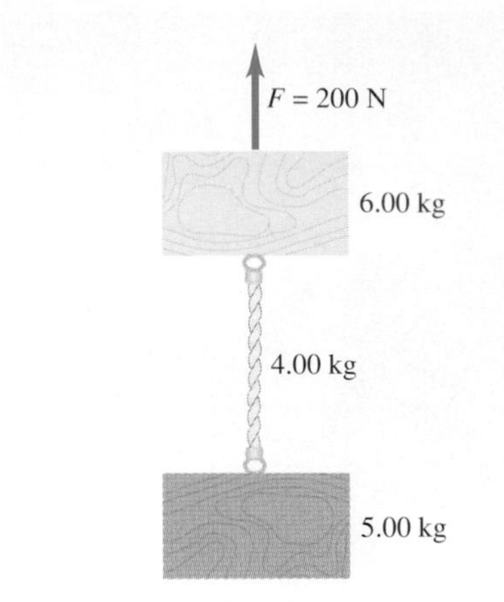


**FIGURE 4–30**

**4–42** A model airplane with mass 1.40 kg moves in the $xy$-plane such that its $x$- and $y$-coordinates vary in time according to $x(t) = 2.00\ \text{m} - \alpha t^3$ and $y(t) = \beta t^2$, where $\alpha = 0.500\ \text{m/s}^3$ and $\beta = 1.60\ \text{m/s}^2$. a) Calculate the $x$- and $y$-components of the resultant force on the plane as functions of time. b) What are the magnitude and direction of the resultant force at $t = 3.00$ s?

*__4–43__ An object with mass $m$ initially at rest is acted on by a force $\boldsymbol{F} = k_1\boldsymbol{i} + k_2t^2\boldsymbol{j}$, where $k_1$ and $k_2$ are constants. Calculate the velocity $\boldsymbol{v}(t)$ of the object as a function of time.

## CHALLENGE PROBLEMS

**4–44** If a beach ball with mass 0.0900 kg is thrown vertically upward in a vacuum, so there is no drag force on it, it reaches a height of 10.0 m. If the ball is thrown upward with the same initial velocity but in air instead of vacuum, its maximum height is 8.4 m. What is the drag force on the ball, assuming that it is constant during the upward motion of the ball?

**4–45** An object of mass $m$ is at rest at the origin at time $t = 0$. A force $\boldsymbol{F}(t)$ is then applied that has components

$$F_x(t) = k_1 + k_2 y, \qquad F_y(t) = k_3 t,$$

where $k_1$, $k_2$, and $k_3$ are constants. Calculate the position $\boldsymbol{r}(t)$ and velocity $\boldsymbol{v}(t)$ vectors as functions of time.

CHAPTER 5

# Applications of Newton's Laws

An ice boat, like a sailboat, travels into the wind by a process called "tacking." When the sail is held parallel to the direction of the wind, it produces a force perpendicular to the sail. This force has components in the forward direction of the boat and sideways to the boat. Friction from the ice boat's runners or the sailboat's keel prevents sideways motion, and the boat moves forward.

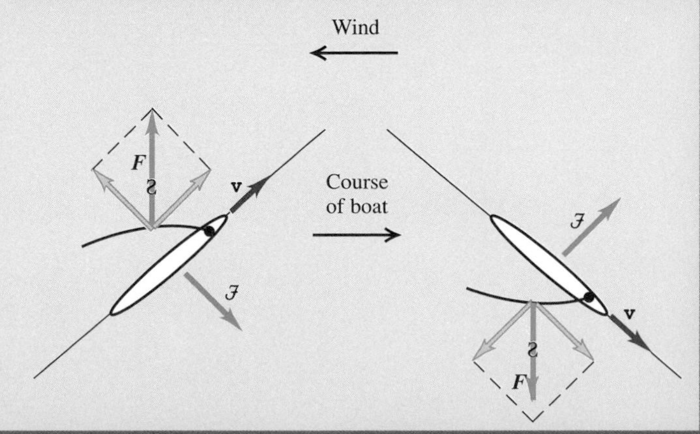

## KEY CONCEPTS

**• When a body is in equilibrium, the vector sum of the forces acting on it must be zero. Alternatively, the sum of components in each coordinate direction must be zero.**

**• The vector sum of forces acting on a body equals its mass times its acceleration. Alternatively, the sum of components of force in each coordinate direction equals the mass times the corresponding component of acceleration.**

**• When a body is in contact with a surface, the force on the body can always be represented in terms of a normal component (perpendicular to the surface) and a friction component (parallel to the surface). The magnitude of the friction force is approximately proportional to the magnitude of the normal force; the ratio of the two is called the coefficient of friction.**

**• Newton's laws are applicable to the special case of circular motion, in which the acceleration is related to the radius of the circle and the body's speed in the circular path.**

**• Good systematic problem-solving technique is of the utmost importance for solving the problems of this chapter.**

## INTRODUCTION

Newton's three laws of motion, the foundation of classical mechanics, can be stated very simply, as we have seen. But applying these laws to situations such as an iceboat, a suspension bridge, a car rounding a banked curve, or a toboggan sliding down a hill requires some analytical skills and some problem-solving technique. In this chapter we introduce no new principles, but we will try to help you develop some of the problem-solving skills you will need to analyze such situations.

We begin with equilibrium problems, concentrating on systems at rest. Then we generalize our problem-solving techniques to include systems that are not in equilibrium, for which we need to deal precisely with the relationships between forces and motion. We will learn how to describe and analyze the contact force acting on a body when it rests or slides on a surface. Next we study the important case of uniform circular motion, when a body moves in a circle with constant speed. Finally, we take a brief look at the fundamental nature of force and the classes of forces found in the physical universe.

# 5–1 EQUILIBRIUM OF A PARTICLE

We learned in Chapter 4 that a body is in **equilibrium** when it is at rest or moving with constant velocity in an inertial frame of reference. A hanging lamp, a rope and pulley set-up for hoisting heavy loads, a suspension bridge—all are examples of equilibrium situations. In this section we consider only equilibrium of a body that can be modeled as a particle. (Later, in Chapter 11, we'll consider the additional principles needed when a body can't be represented adequately as a particle.) The essential physical principle is Newton's first law: When a particle is at rest or is moving with constant velocity in an inertial frame of reference, the vector sum of all the forces acting on it must be zero. That is,

$$\Sigma \mathbf{F} = \mathbf{0}. \qquad (5\text{–}1)$$

We will usually use this in component form:

$$\Sigma F_x = 0, \qquad \Sigma F_y = 0. \qquad (5\text{–}2)$$

This chapter is about problem solving. We strongly recommend that you study the following strategy carefully, look for its applications in the worked-out examples, and try to apply it when you solve assigned problems.

## PROBLEM-SOLVING STRATEGY

### Equilibrium of a particle

**1.** Draw a simple sketch of the apparatus or structure, showing dimensions and angles.

**2.** Choose some body that is in equilibrium and draw a diagram of this body. For the present we will consider it as a particle, so a large dot will do to represent the body. *Do not* include the other bodies that interact with it, such as a surface it may be resting on or a rope pulling on it.

**3.** Now ask yourself what is interacting with the body by touching it or in any other way. Draw a force vector (using a colored pencil or pen) for each of the interactions. If you know the direction of a force, draw it accurately and label it. A surface in contact with the body exerts a normal force perpendicular to the surface and possibly a friction force parallel to the surface. Remember that a rope or chain can't push on a body, it can only pull in a direction along its length. Don't forget the body's weight. If the mass is given, use $w = mg$ to find the weight. Label each force with a symbol representing the *magnitude* of the force; indicate the direction with appropriate angles. For each force, make sure that you can answer the question "What other body causes that force?" If you can't answer that question, you may be imagining a force that isn't there.

**4.** *Do not* show in the free-body diagram any of the forces exerted *by* the body on any other body.

**5.** Choose a set of coordinate axes and represent each force acting on the body in terms of its components along these axes. Cross out lightly each force that has been replaced by its components so that you don't count it twice. You can often simplify the problem by using a particular choice of coordinate axes. For example, when a body rests or slides on a plane surface, it is usually simplest to take the axes in the directions parallel and perpendicular to this surface, even when the plane is tilted.

**6.** Set the algebraic sum of all $x$-components of force equal to zero. In a separate equation, set the algebraic sum of all $y$-components equal to zero. (Never add $x$- and $y$-components in a single equation.) You can then solve these equations for up to two unknown quantities, which may be force magnitudes, components, or angles.

**7.** If there are two or more bodies, repeat Steps 2 through 6 for each body. If the bodies interact with each other, use Newton's third law to relate the forces they exert on each other. You need to find as many independent equations as the number of unknown quantities. Then solve these equations to obtain the unknowns. This part is algebra, not physics, but it's an essential step.

**8.** Whenever possible, look at your results and ask whether they make sense. When the result is a symbolic expression or formula, try to think of special cases (particular values or extreme cases for the various quantities) for which you can guess what the results ought to be. Check to see that your formula works in these particular cases.

## EXAMPLE 5–1

**One-dimensional equilibrium** A gymnast has just begun climbing up a rope hanging from a gymnasium ceiling (Fig. 5–1a). She stops, suspended from the lower end of the rope by her hands. Her weight is 500 N, and the weight of the rope is 100 N. Analyze the forces on the gymnast and on the rope.

**SOLUTION** Figure 5–1b is a free-body diagram for the gymnast. The forces acting on her are her weight (magnitude 500 N) and the upward force (magnitude $T_{GR}$) exerted on her by the rope. We *don't* include the downward force she exerts on the rope because it isn't a force that acts *on* her. We take the $y$-axis vertically upward and the $x$-axis horizontal; there are no $x$-components of force. The rope pulls upward (in the positive $y$-direction), and the corresponding $y$-component of force is just the magnitude $T_{GR}$, a positive (scalar) quantity. But the weight acts in the negative $y$-direction, and its $y$-component is the *negative* of the magnitude, that is, $-500$ N. The algebraic sum of $y$-components is $T_{GR} + (-500\text{ N})$, and from the equilibrium condition, $\Sigma F_y = 0$, we have

$$\Sigma F_y = T_{GR} + (-500\text{ N}) = 0,$$

$$T_{GR} = 500\text{ N}.$$

The tension at the bottom of the rope equals the gymnast's weight, as you probably expected.

The two forces acting on the gymnast, both with magnitude 500 N, are *not* an action-reaction pair, even though they are equal in magnitude and opposite in direction. The weight is the attractive (downward) force the earth exerts on the gymnast. Its reaction is the equal and opposite (upward) attractive force exerted *on* the earth *by* the gymnast. This force acts on the earth, not on the gymnast, so it does not appear in the free-body diagram for the gymnast.

Figure 5–1c shows a free-body diagram for the rope. The reaction to the upward force of magnitude 500 N on the gymnast is a downward force acting on the rope, as shown. According to Newton's third law, the magnitude of this downward force is also 500 N. The other forces on the rope are its own weight (magnitude 100 N) and the upward force (magnitude $T_{RC}$) exerted on the upper end of the rope by the ceiling.

Figure 5–1c, the free-body diagram for the rope, shows that the $y$-component of the tension at the top end of the rope is $+T_{RC}$, the $y$-component at the bottom end is $-500$ N, and the $y$-component of the weight is $-100$ N. The equilibrium condition $\Sigma F_y = 0$ for the rope gives

$$\Sigma F_y = T_{RC} + (-100\text{ N}) + (-500\text{ N}) = 0,$$

$$T_{RC} = 600\text{ N}.$$

The tension is 100 N greater at the top of the rope (where it must support the weights of both the rope and the gymnast) than at the bottom (where it supports only the gymnast).

Figure 5–1d is a partial free-body diagram for the ceiling, showing that the rope exerts a downward force $F_{CR}$ on the ceiling. From Newton's third law the magnitude of this force is also 600 N.

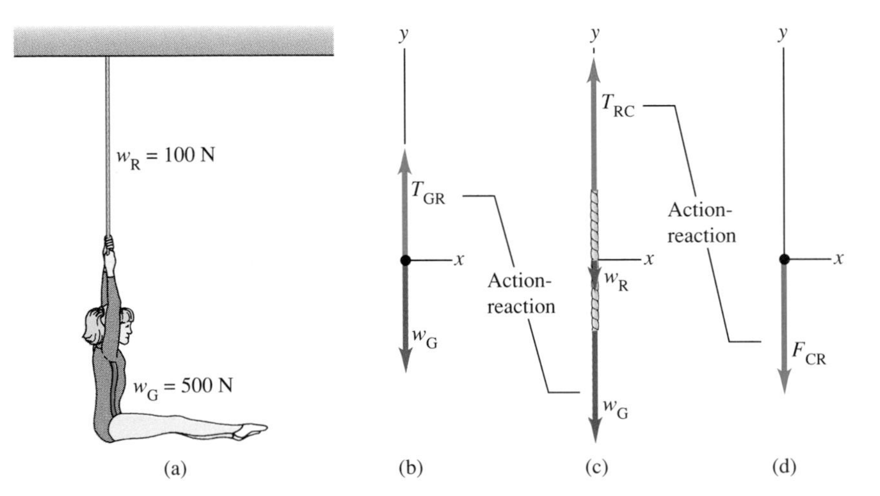


**5–1** (a) A gymnast hanging at rest from the end of a vertical rope. (b) Free-body diagram for the gymnast. (c) Free-body diagram for the rope. (d) The force exerted on the ceiling by the rope (not a complete free-body diagram).

The strength of the rope probably isn't a major concern in Example 5–1. The breaking strength of a string, rope, or cable is described by the maximum tension it can withstand without breaking. A few typical breaking strengths are shown in Table 5–1.

**TABLE 5–1 Approximate Breaking Strengths**

| | |
|---|---|
| Thin white string | 50 N |
| $\frac{1}{4}''$ nylon clothesline rope | 4000 N |
| 11 mm Perlon mountaineering rope | $3 \times 10^4$ N |
| $1\frac{1}{4}''$ manila climbing rope | $6 \times 10^4$ N |
| $\frac{1}{4}''$ steel cable | $6 \times 10^4$ N |

## EXAMPLE 5–2

**Two-dimensional equilibrium** In Fig. 5–2a a car engine with weight $w$ hangs from a chain that is linked at point $O$ to two other chains, one fastened to the ceiling and the other to the wall. Find the tensions in these three chains, assuming that $w$ is given and the weights of the chains themselves are negligible.

**SOLUTION** Figure 5–2b is a free-body diagram for the engine. Without further ceremony we can conclude that $T_1 = w$. The horizontal and slanted chains do not exert forces on the engine itself because they are not attached to it, but they do exert forces on the ring where three chains join. So let's consider the *ring* as a particle in equilibrium; the weight of the ring itself is negligible.

In the free-body diagram for the ring (Fig. 5–2c), remember that $T_1$, $T_2$, and $T_3$ are the *magnitudes* of the forces; their directions are shown by the vectors on the diagram. An $x$-$y$ coordinate axis system is also shown, and the force with magnitude $T_3$ has been resolved into its $x$- and $y$-components. Note that the downward force with magnitude $T_1$ acting on the ring and the upward force with magnitude $T_1$ acting on the engine are an action-reaction pair. That's how we know their magnitudes are equal.

We now apply the equilibrium conditions *for the ring,* writing separate equations for the $x$- and $y$-components. (Note that $x$- and $y$-components are *never* added together in a single equation.) We find

$$\Sigma F_x = 0, \qquad T_3 \cos 60^\circ - T_2 = 0;$$

$$\Sigma F_y = 0, \qquad T_3 \sin 60^\circ - T_1 = 0.$$

Because $T_1 = w$, we can rewrite the second equation as

$$T_3 = \frac{T_1}{\sin 60^\circ} = \frac{w}{\sin 60^\circ} = 1.155w.$$

We can now use this result in the first equation:

$$T_2 = T_3 \cos 60^\circ = (1.155w) \cos 60^\circ = 0.577w.$$

So we can express all three tensions as multiples of the weight $w$ of the engine, which we assume is known. To summarize,

$$T_1 = w,$$

$$T_2 = 0.577w,$$

$$T_3 = 1.155w.$$

If the engine's weight is $w = 2200$ N (about 500 lb), then

$$T_1 = 2200 \text{ N},$$

$$T_2 = (0.577)(2200 \text{ N}) = 1270 \text{ N},$$

$$T_3 = (1.155)(2200 \text{ N}) = 2540 \text{ N}.$$

We find that $T_3$ is greater than the weight of the engine. If this seems strange, note that $T_3$ must be large enough for its vertical component to be equal in magnitude to $w$, so $T_3$ itself must have somewhat *larger* magnitude than $w$.

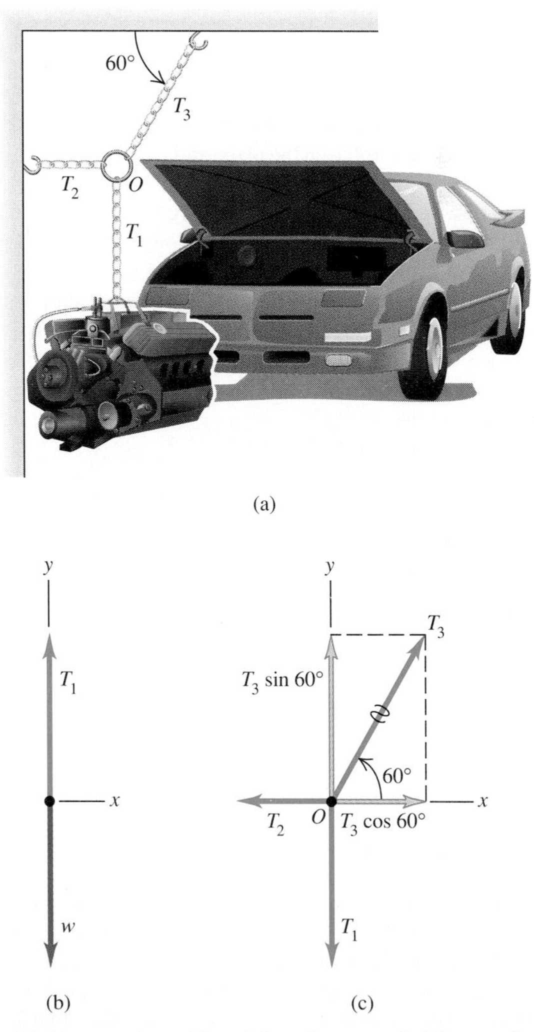


**5–2** (a) A car engine with weight $w$ is suspended from a chain linked at $O$ to two other chains. The chains are considered to be massless. (b) Free-body diagram for the engine. (c) Free-body diagram for the ring with $T_3$ replaced by its components.

## EXAMPLE 5–3

**An inclined plane** A car rests on the slanted tracks of the top level of a car-transporter trailer (Fig. 5–3a). The car's brakes and transmission lock are released; only a cable attached to the car and to the frame of the trailer prevents the car from rolling backward off the trailer. If the car's weight is $w$, find the tension in the cable and the force with which the tracks push on the car tires.

**SOLUTION** Figure 5–3b shows a free-body diagram for the car. The three forces acting on the car are its weight (magnitude $w$), the tension in the cable (magnitude $T$), and the forces exerted on the wheels by the tracks; we have lumped the forces on the wheels together as a single force with magnitude $\mathcal{n}$. Note that this force must be perpendicular to the tracks, and for that reason we call it a *normal force*. We use a special script letter $\mathcal{n}$ to avoid confusion with the abbreviation N for newton. (If this force did have a component *along* the track, it would tend to prevent the car from moving along the track, and we have assumed that there is no such effect.)

We take our coordinate axes parallel and perpendicular to the track, as shown. To find the components of the weight, we first note that the angle $\alpha$ between the plane and the horizontal is equal to the angle $\alpha$ between the weight vector and the normal to the plane, as shown. The angle $\alpha$ is *not* measured counterclockwise from the $+x$-axis, so we can't use Eqs. (1–8) to find the components. Instead, consider the right triangles in Fig. 5–3c. The magnitude of the $y$-component of $\mathbf{w}$, divided by the magnitude $w$, is the cosine of $\alpha$. Similarly, the magnitude of the $x$-component divided by $w$ is the sine of $\alpha$. Both components are negative. We cross out the original vector representing the weight. The equilibrium conditions then give us

$$\Sigma F_x = 0, \qquad T + (-w \sin \alpha) = 0;$$

$$\Sigma F_y = 0, \qquad \mathcal{n} + (-w \cos \alpha) = 0.$$

Be sure you understand how the signs are related to our choice of coordinates, recalling that, by definition, $T$, $w$, and $\mathcal{n}$ are all *magnitudes* of vectors and are therefore all positive.

Solving these equations for $T$ and $\mathcal{n}$, we find

$$T = w \sin \alpha,$$

$$\mathcal{n} = w \cos \alpha.$$

To check some special cases, note that if the angle $\alpha$ is zero, then $\sin \alpha = 0$ and $\cos \alpha = 1$. In this case the tracks are horizontal; no cable tension $T$ is needed to hold the car, and the total normal force $\mathcal{n}$ is equal in magnitude to the weight. If the angle is 90°, then $\sin \alpha = 1$ and $\cos \alpha = 0$. Then the cable tension $T$ equals the weight $w$, and the normal force $\mathcal{n}$ is zero. Are these the results you would expect for these particular cases?

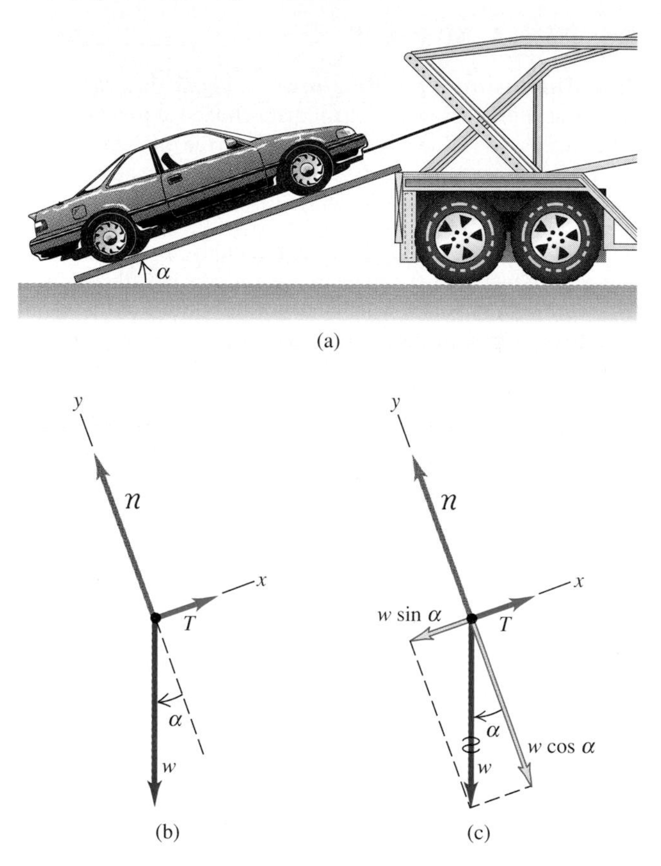


**5–3** (a) A cable holds the car on the trailer. (b) Free-body diagram for the car. (c) The weight component $w \sin \alpha$ acts down the plane, and $w \cos \alpha$ is perpendicular to the plane.

## EXAMPLE 5–4

**Tension over a frictionless pulley** Blocks of granite are being hauled up a 15° slope out of a quarry (Fig. 5–4a). For environmental reasons, dirt is also being dumped into the quarry to fill up old holes. You have been asked to find a way to use this dirt to move the granite out more easily. You design a system that lets the dirt (weight $w_2$, including the bucket) dropping vertically into the quarry pull out a granite block on a cart with steel wheels rolling on steel rails (Fig. 5–4a). Ignoring friction in the pulley and wheels and the weight of the cable, determine how the weights $w_1$ and $w_2$ must be related for the system to move with constant speed.

**SOLUTION** Our idealized model for the system is shown in Fig. 5–4b. The free-body diagram for the dirt and bucket is Fig. 5–4c, and the free-body diagram for the granite block on its cart is Fig. 5–4d. We have drawn an axis system for each body. Note that we are at liberty to orient the axes differently for each body; the choices shown are the most convenient ones. We have represented the weight of the granite block in terms of its components in the chosen axis system. The tension $T$ in the cable is the same throughout because we are assuming the weight of the cable to be negligible.

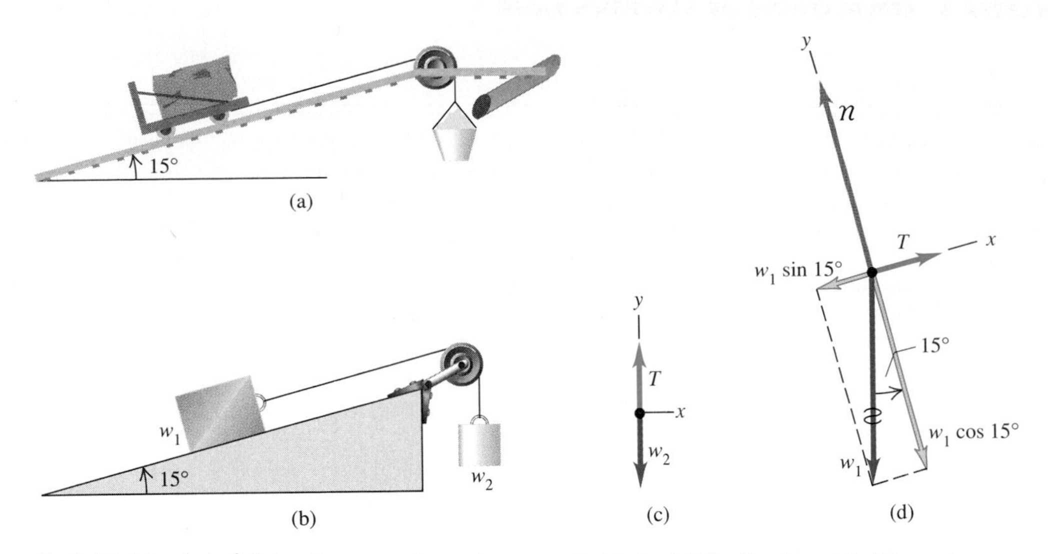


**5–4** (a) A bucket of dirt pulls up a cart carrying a granite block. (b) Idealized model of the system. (c) Free-body diagram for the dirt and bucket. (d) Free-body diagram for the cart and block.

Applying $\Sigma F_y = 0$ to Fig. 5–4c, we find

$$T - w_2 = 0, \qquad T = w_2.$$

Applying $\Sigma F_x$ to Fig. 5–4d, we get

$$T - w_1 \sin 15° = 0, \qquad T = w_1 \sin 15°.$$

Equating the two expressions for $T$, we find

$$w_2 = w_1 \sin 15° = 0.26 w_1.$$

If the weight of dirt and bucket totals 26% of the weight of the granite block and cart, the system can move in either direction with constant speed. ■

## 5–2 APPLICATIONS OF NEWTON'S SECOND LAW

We are now ready to discuss problems in **dynamics,** showing applications of Newton's second law to systems that are *not* in equilibrium. In this case,

$$\Sigma \mathbf{F} = m\mathbf{a}. \tag{5–3}$$

We will usually use this relation in component form:

$$\Sigma F_x = ma_x, \qquad \Sigma F_y = ma_y. \tag{5–4}$$

The following problem-solving strategy is very similar to our strategy for equilibrium problems in Section 5–1. We urge you to study this strategy carefully, watch for its applications in our examples, and use it when you tackle the end-of-chapter problems.

### PROBLEM-SOLVING STRATEGY

#### Newton's second law

**1.** Draw a sketch of the physical situation and identify one or more moving bodies to which you will apply Newton's second law.

**2.** Draw a free-body diagram for the chosen body. Be sure to include all the forces acting *on* the body, but be equally careful *not* to include any force exerted *by* the body on some other body. Some of the forces may be unknown; label their magnitudes with algebraic symbols. Usually, one of the forces will be the body's weight; it is usually best to label this as $mg$. If a numerical value of mass is given, you can compute the corresponding weight.

**3.** Show your coordinate axes explicitly in the free-body diagram and then determine components of forces with reference to these axes. If you know the direction of the acceleration, it is usually best to take that direction as one of the axes. When you represent a force in terms of its components, cross out the original force so as not to include it twice. When there are two or more bodies, you can use a separate axis system for each body; you don't have to use the same axis system for all the bodies. But in the equations for each body the signs of the components *must* be consistent with the axes you have chosen for that body.

**4.** Write the equations for Newton's second law, Eqs. (5–4), a separate equation for each component.

**5.** If more than one body is involved, repeat Steps 2 through 4 for each body. In addition, the motions of the bodies may be related. For example, they may be connected by a rope. Express these relations in algebraic form as relations between the accelerations of the various bodies. Then solve the equations to find the required unknowns.

**6.** Check particular cases or extreme values of quantities, when possible, and compare the results for these particular cases with your intuitive expectations. Ask yourself: "Does this result make sense?"

## EXAMPLE 5–5

**Acceleration in one dimension** An iceboat is at rest on a perfectly frictionless horizontal surface (Fig. 5–5a). What constant horizontal force $F$ do we need to apply (along the direction of the runners) to give the iceboat a velocity of 4.0 m/s at the end of 2.0 s? The mass of iceboat and rider is 200 kg.

**SOLUTION** Figure 5–5b shows a free-body diagram and a coordinate system. We can find the unknown force by using Eqs. (5–4) if we can find the acceleration, so let's start there. The $y$-component of acceleration is zero, and we can get the $x$-component from the velocity data. (The forces are all constant, so $a_x$ is also constant.) We find

$$a_x = \frac{v - v_0}{t} = \frac{4.0\ \text{m/s} - 0}{2.0\ \text{s}} = 2.0\ \text{m/s}^2.$$

The sum of the $x$-components of force is simply

$$\Sigma F_x = F,$$

and Newton's second law gives

$$\Sigma F_x = F = ma_x,$$

$$F = (200\ \text{kg})(2.0\ \text{m/s}^2) = 400\ \text{kg}\cdot\text{m/s}^2 = 400\ \text{N}.$$

Note that we did not need the $y$-components at all in this problem. Here they are anyway:

$$a_y = 0,$$

$$\Sigma F_y = n + (-mg) = ma_y = 0,$$

$$n = mg = (200\ \text{kg})(9.8\ \text{m/s}^2) = 1960\ \text{N}.$$

The upward normal force $n$ that the surface exerts on the iceboat is equal to the weight of the iceboat and rider.

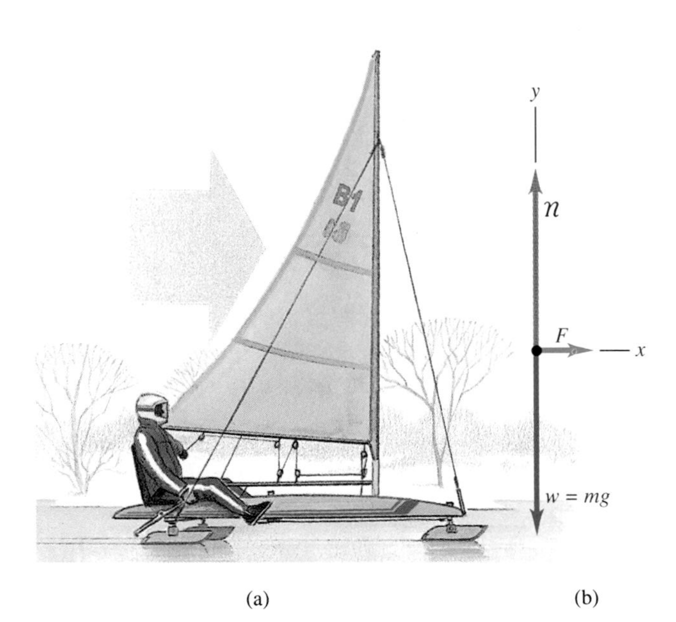


**5–5** (a) An iceboat starting from rest. (b) Free-body diagram for the iceboat with no friction.

## EXAMPLE 5–6

Consider the same situation as in Example 5–5, but now the motion of the iceboat is opposed by a constant horizontal friction force with magnitude 100 N. Now what force $F$ must we apply to give the iceboat a velocity of 4.0 m/s at the end of 2.0 s?

**SOLUTION** The acceleration is the same as before, $a_x = 2.0$ m/s$^2$. The new free-body diagram is shown in Fig. 5–6; the difference between it and Fig. 5–5 is the addition of the friction force $\mathcal{F}$. (Note that its *magnitude*, $\mathcal{F} = 100$ N, is a positive quantity but that its *component* in the $x$-direction is negative, equal to $-\mathcal{F}$, or $-100$ N.) Now Newton's second law gives

$$\Sigma F_x = T + (-\mathcal{F}) = ma_x,$$
$$T + (-100\text{ N}) = (200\text{ kg})(2.0\text{ m/s}^2) = 400\text{ N},$$
$$T = 500\text{ N}.$$

We need 100 N to overcome friction and 400 N more to give the iceboat the necessary acceleration. Does this seem reasonable?

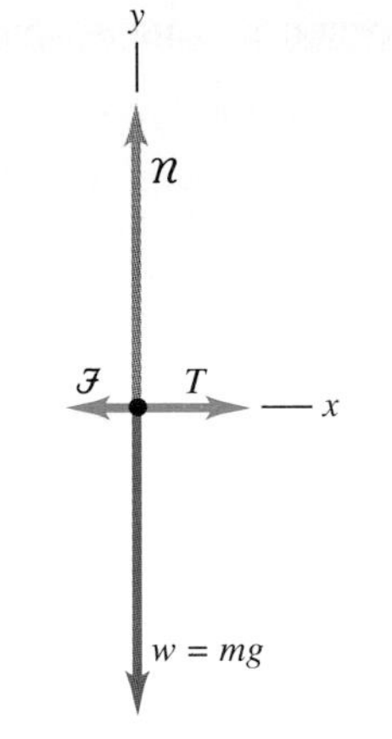


**5–6** Free-body diagram for the iceboat with a frictional force $\mathcal{F}$ opposing the motion.

## EXAMPLE 5–7

**Tension in an elevator cable** An elevator and its load have a total mass of 800 kg (Fig. 5–7). The elevator is originally moving downward at 10.0 m/s, and it is brought to rest with constant acceleration in a distance of 25.0 m. Find the tension $T$ in the supporting cable.

**SOLUTION** We can find the forces on the elevator if we can find its acceleration. The simplest way to do this is to use the constant-acceleration formula $v^2 = v_0^2 + 2a(y - y_0)$. Taking the positive $y$-axis upward, we have $v_0 = -10$ m/s, $v = 0$, and $y - y_0 = -25.0$ m. Then

$$(0)^2 = (-10\text{ m/s})^2 + 2a_y(-25\text{ m}),$$
$$a_y = +2.00\text{ m/s}^2.$$

Note that even though the velocity is downward, the acceleration is upward, corresponding to downward motion with decreasing speed.

Now we are ready to use Newton's second law:

$$\Sigma F_y = T + (-w) = ma_y,$$
$$T = w + ma_y = mg + ma_y = m(g + a_y)$$
$$= (800\text{ kg})(9.80\text{ m/s}^2 + 2.00\text{ m/s}^2) = 9440\text{ N}.$$

The tension must be 1600 N *greater* than the weight ($w = mg = 7840$ N) to stop the elevator in the required distance.

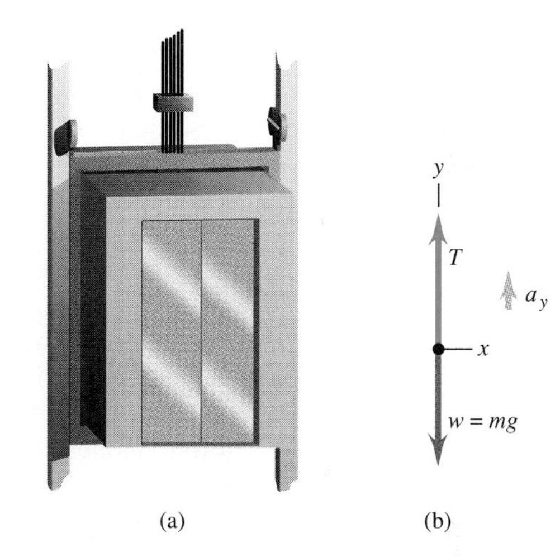


**5–7** (a) A loaded elevator moving downward being brought to rest. (b) Free-body diagram for the elevator.

## EXAMPLE 5–8

In Example 5–7, what force do a passenger's feet exert downward on the floor if her mass is 50.0 kg?

**SOLUTION** We first find the force the floor exerts *on* the passenger's feet, which is the reaction force to the force asked for. Figure 5–8 shows a free-body diagram for the passenger. The forces on her are her weight, (50.0 kg)(9.80 m/s²) = 490 N, and the upward force $F$ applied by the floor. Her acceleration is the same as the elevator's, and Newton's second law gives

$$F + (-mg) = ma_y,$$
$$F = m(g + a_y) = (50.0\text{ kg})(9.80\text{ m/s}^2 + 2.00\text{ m/s}^2)$$
$$= 590\text{ N}.$$

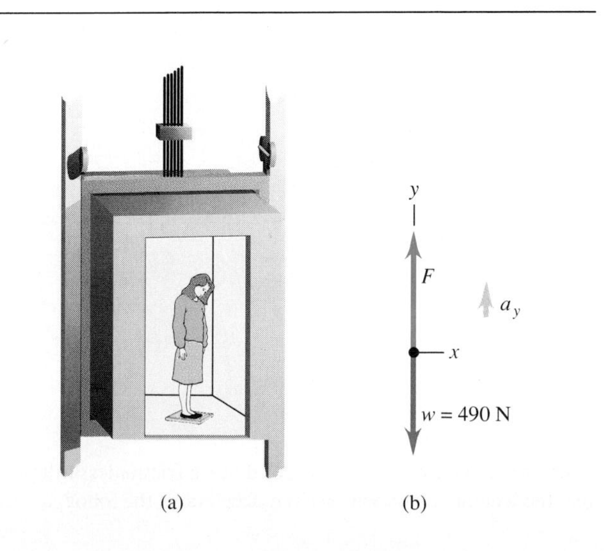


**5–8** (a) A passenger in the elevator as it slows down. (b) Free-body diagram for the passenger.

While the elevator is stopping, the passenger pushes down on the floor with a force of 590 N, 100 N more than her weight. The passenger feels a greater strain in her legs and feet than when the elevator is stationary or moving with constant velocity because the floor is pushing up harder on her feet. You can feel this yourself; try taking a few steps in an elevator that is coming to a stop after descending.

Let's generalize the result of Example 5–8. A passenger with mass $m$ rides in an elevator with acceleration $a_y$. What force $F$ does the passenger exert on the floor? We have shown above that

$$F = m(g + a_y).$$

When $a_y$ is positive, the elevator is accelerating upward, and $F$ is greater than the passenger's weight $w = mg$. When the elevator is accelerating downward, $a_y$ is negative, and $F$ is less than the weight. If the passenger doesn't know the elevator is accelerating, she may feel as though her weight is changing; indeed, if she is standing on a bathroom scale, it reads $F$, not $w$. (Why?) With upward acceleration the apparent weight is greater than the true weight, and the passenger feels heavier; with downward acceleration she feels lighter.

The extreme case occurs when the elevator has a downward acceleration $a_y = -g$, that is, when it is in free fall. In that case, $F = 0$, and the passenger *seems* to be weightless. Similarly, an astronaut orbiting the earth in a space capsule experiences apparent weightlessness. In each case the person is not really weightless because there is still a gravitational force acting. But the effect of the free-fall condition is exactly the same as though the body were in outer space with no gravitational force at all.

The physiological effect of prolonged apparent weightlessness is an interesting medical problem that is being actively explored. Gravity plays an important role in blood distribution in the body; one reaction to apparent weightlessness is a decrease in blood volume through increased excretion of water. In some cases astronauts returning to earth have experienced temporary impairment of their sense of balance and a greater tendency toward motion sickness.

## EXAMPLE 5–9

**Acceleration down a hill** A toboggan loaded with vacationing students (total weight $w$) slides down a long, snow-covered slope (Fig. 5–9a). The hill slopes at a constant angle $\alpha$, and the toboggan is so well waxed that there is virtually no friction at all. What is its acceleration?

**5–9** (a) A loaded toboggan slides down a frictionless hill. (b) The free-body diagram shows that the weight component $w \sin \alpha$ accelerates the toboggan down the hill.

**SOLUTION** The only forces acting on the toboggan are its weight $w$ and the normal force $\mathcal{N}$ exerted by the hill (Fig. 5–9b). We take axes parallel and perpendicular to the surface of the hill and resolve the weight into $x$- and $y$-components. Then

$$\Sigma F_x = w \sin \alpha .$$

From $\Sigma F_x = ma_x$ we have

$$w \sin \alpha = ma_x ,$$

and since $w = mg$,

$$a_x = g \sin \alpha .$$

The mass does not appear in the final result, which means that *any* toboggan, regardless of its mass or number of passengers, slides down a frictionless hill with an acceleration of $g \sin \alpha$. In particular, when $\alpha = 0$, $a_x = 0$, and when the surface is vertical, $\alpha = 90°$ and $a_x = g$, as we should expect.

Note again that we did not need the $y$-components; we know that $a_y = 0$, so $\Sigma F_y = 0$, and the normal force $\mathcal{N}$ exerted on the toboggan by the surface of the hill is

$$\mathcal{N} = mg \cos \alpha .$$

This force is *not* equal to the toboggan's weight. We don't need this result here, but it will be useful in a later example.

## EXAMPLE 5–10

A robot arm pulls a 4.0-kg cart along a horizontal frictionless track by applying a horizontal force with magnitude $F = 9.0$ N to a 0.50-kg rope (Fig. 5–10a). Find the acceleration of the system and the tension at the point where the rope is fastened to the cart. (On earth the rope would sag a little; to avoid this complication, suppose the robot arm is operating in a zero-gravity space station.)

**SOLUTION** Once again, only $x$-components are relevant. We can proceed in either of two ways.

*Method 1:* Take the cart and rope as a composite body with a total mass of 4.50 kg. A 9.0-N force gives this body an acceleration

$$a = \frac{F}{m} = \frac{9.0\ \text{N}}{4.5\ \text{kg}} = 2.0\ \text{m/s}^2.$$

Then, looking at the 4.0-kg cart by itself, we see that to give it an acceleration of 2.0 m/s$^2$ requires a force

$$T = ma = (4.0\ \text{kg})(2.0\ \text{m/s}^2) = 8.0\ \text{N}.$$

Then the net forward force on the *rope* is 9.0 N − 8.0 N = 1.0 N. This corresponds to a mass of 0.50 kg accelerating at 2.0 m/s$^2$, the same acceleration as that of the cart.

This method may seem slightly acrobatic. Let's go a little more slowly.

*Method 2:* Write an equation of motion for the cart and another for the rope, using $a$ and $T$ for the unknown acceleration and tension (Fig. 5–10b). For the cart,

$$T = m_{\text{cart}}a = (4.0\ \text{kg})a,$$

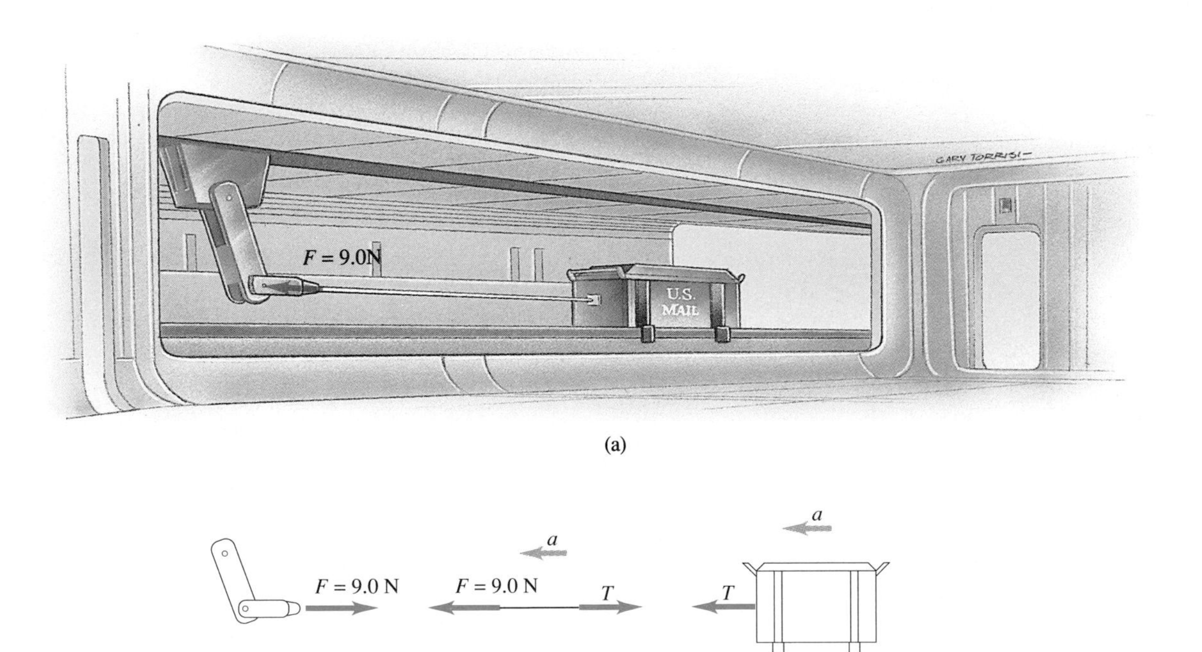


**5–10** (a) A robot arm pulling a cart along a frictionless track in a zero-gravity space station. (b) Free-body diagrams for the rope and the cart.

and for the rope,

$$9.0\text{ N} - T = m_{\text{rope}}a = (0.5\text{ kg})a.$$

(Note the role of Newton's third law in equating the magnitudes of the two forces labeled $T$. Also, the accelerations of cart and rope are equal; we represent both with the symbol $a$.) We now have two simultaneous equations for $T$ and $a$. An easy way to solve them is to substitute the first equation for $T$ into the second and then solve for $a$. We leave the details for you to work out; the results are again $a = 2.0\text{ m/s}^2$ and $T = 8.0\text{ N}$.

---

An essential element in both solution methods in Example 5–10 is the fact that the two bodies in the system have the same acceleration. We can apply Newton's second law to either body separately or to the composite system as a whole. In the next example the two bodies have different accelerations, and we *must* treat them separately.

## EXAMPLE 5–11

In Fig. 5–11 an air-track glider with mass $m_1$ moves on a level, frictionless air track in the physics lab. It is connected by a light, flexible, nonstretching string passing over a small frictionless pulley to a weight hanger with mass $m_2$. Find the acceleration of each body and the tension in the string.

**SOLUTION** The two bodies have different motions, so we can't consider them together as we did the bodies in Example 5–10. We need a separate free-body diagram and $\Sigma F = ma$ equations for each. Figure 5–11 shows a free-body diagram and a coordinate system for each body. There is no friction in the pulley, and we consider the string to be massless, so the tension $T$ in the string is the same throughout; it applies a force of magnitude $T$ to each body. The weights are $m_1 g$ and $m_2 g$. For the glider on the track, Newton's second law gives

$$\Sigma F_x = T = m_1 a_{1x},$$

$$\Sigma F_y = n + (-m_1 g) = m_1 a_{1y} = 0,$$

and for the weight hanger,

$$\Sigma F_y = m_2 g + (-T) = m_2 a_{2y}.$$

We note again that it is perfectly all right to use different coordinate axes for the two bodies. Here it is convenient to take the $+y$-direction as downward for $m_2$, so both bodies move in positive axis directions.

Now if the string doesn't stretch, the two bodies must move equal distances in equal times, so their speeds at any instant must be equal. When the speeds change, they change by equal amounts in a given time, so the accelerations of the two bodies must have the same magnitude $a$. We can express this relation as

$$a_{1x} = a_{2y} = a.$$

(The directions of the two accelerations are different, of course.) The two Newton's-second-law equations are then

$$T = m_1 a,$$

$$m_2 g + (-T) = m_2 a.$$

These are two simultaneous equations for the unknowns $T$ and $a$. An easy way to solve them is to add the two equations; this eliminates $T$, giving

$$m_2 g = m_1 a + m_2 a = (m_1 + m_2)a,$$

and

$$a = \frac{m_2}{m_1 + m_2} g.$$

Substituting this back into the first equation, we get

$$T = \frac{m_1 m_2}{m_1 + m_2} g.$$

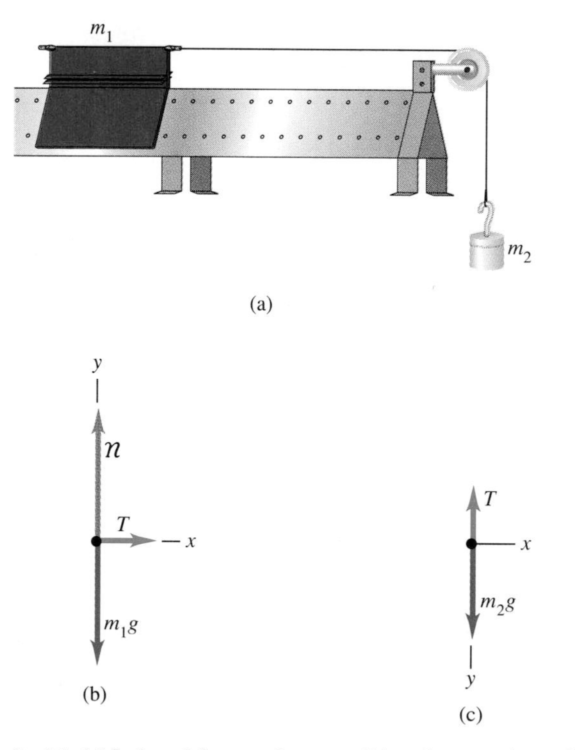


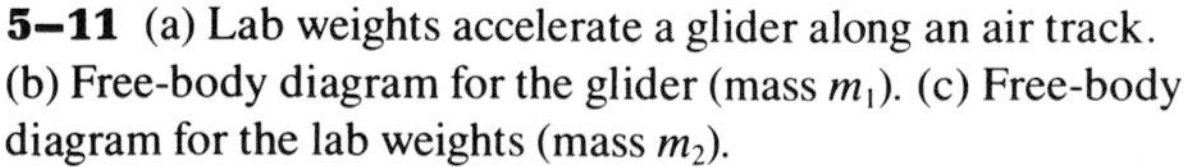
**5–11** (a) Lab weights accelerate a glider along an air track. (b) Free-body diagram for the glider (mass $m_1$). (c) Free-body diagram for the lab weights (mass $m_2$).

We see that the tension $T$ is *not* equal to the weight $m_2 g$ of mass $m_2$ but is *less* by a factor of $m_1/(m_1 + m_2)$. If $T$ were equal to $m_2 g$, then $m_2$ would be in equilibrium, and it isn't.

Now let's check some special cases. If $m_1 = 0$, we expect that $m_2$ would fall freely and there would be no tension in the string. The equations do give $T = 0$ and $a = g$ when $m_1 = 0$. Also, if $m_2 = 0$, we expect no tension and no acceleration; for this case the equations give $T = 0$ and $a = 0$. Thus in these two special cases the results agree with our intuitive expectations.

### EXAMPLE 5-12

**A simple accelerometer** Figure 5-12a shows a lead fish-line sinker hanging from a string attached to point $P$ on the ceiling of a car. When the system has an acceleration $a$ toward the right, the string makes an angle $\beta$ with the vertical. In a practical instrument some form of damping would be needed to keep the string from swinging when the acceleration changes. For example, the system might hang in a tank of oil. (You can make a primitive version of this device by taping a thread to the ceiling light in a car, tying a key or other small object to the other end, and using a protractor. Let someone else drive the car while you are doing the experiment.) The problem is this: Given $m$ and $\beta$, what is the acceleration $a$?

**SOLUTION** As is shown in the free-body diagram, Fig. 5-12b, two forces act on the body: its weight $w = mg$ and the tension $T$ in the string. In finding the components of the tension, note that the angle $\beta$ is not measured counterclockwise from the $+x$-axis. Referring to the right triangles in Fig. 5-12b and recalling the definitions of the sine and cosine functions, we find the components shown. The sum of the horizontal components of force is

$$\Sigma F_x = T \sin \beta,$$

and the sum of the vertical components is

$$\Sigma F_y = T \cos \beta + (-mg).$$

The $x$-acceleration is the acceleration $a$ of the system, and the $y$-acceleration is zero, so

$$T \sin \beta = ma, \qquad T \cos \beta = mg.$$

When we divide the first equation by the second, we get

$$a = g \tan \beta.$$

The acceleration $a$ is proportional to the tangent of the angle $\beta$.

When $\beta = 0$, the sinker hangs vertically, and the acceleration is zero; when $\beta = 45°$, $a = g$, and so on. We note that $\beta$ can never be 90° because that would require an infinite acceleration. Also note that this result doesn't depend on the mass of the sinker.

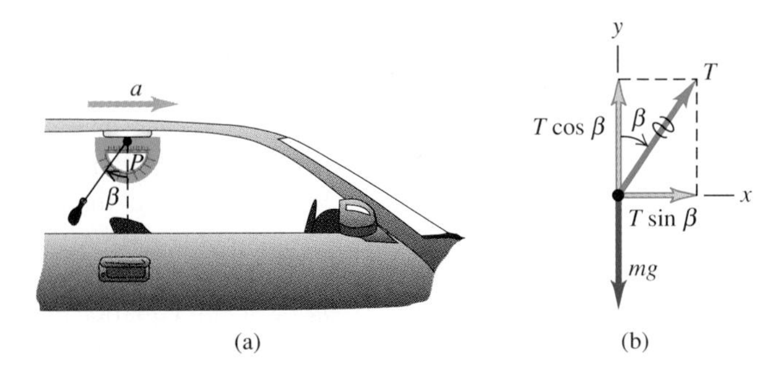


**5-12** (a) A simple accelerometer. (b) Free-body diagram for the lead sinker.

## 5-3 CONTACT FORCES AND FRICTION

We have seen several problems in which a body rests or slides on a surface that exerts forces on the body, and we have used the terms *normal force* and *friction force* to describe these forces. Whenever two bodies interact by direct contact (touching) of their surfaces, we call the interaction forces **contact forces.** Normal and friction forces are both contact forces. Friction is important in many aspects of everyday life. The oil in a car engine minimizes friction between moving parts, but without friction between the tires and the road we couldn't drive or turn the car. Air drag decreases automotive fuel economy but makes parachutes work. Without friction, nails would pull out, light bulbs and bottle caps would unscrew effortlessly, and riding a bicycle would be hopeless.

Let's consider a body sliding across a surface. When you try to slide a heavy box of books across the floor, the box doesn't move at all until you reach a certain critical force. Then the box starts moving, and you can usually keep it moving with less force than you needed to get it started. If you take some of the books out, you need less force than before to get it started or keep it moving. What general statements can we make about this behavior?

First, when a body rests or slides on a surface, we can always represent the contact force exerted by the surface on the body in terms of components of force perpendicular and parallel to the surface. We call the perpendicular component the **normal force,** denoted by $\mathcal{n}$. (*Normal* is a synonym for *perpendicular.*) The component parallel to the surface is the **friction force,** denoted by $\mathcal{f}$. By definition, $\mathcal{n}$ and $\mathcal{f}$ are always perpendicular to each other. We use special script symbols for these quantities to emphasize their special role in representing the contact forces. The script letter $\mathcal{n}$ also avoids confusion with the abbreviation N for newton. If the surface is frictionless (an unattainable idealization), then the contact force has *only* a normal component, and $\mathcal{f}$ is zero. The direction of the friction force is always such as to oppose relative motion of the two surfaces.

The *magnitude* of the friction force usually increases when the normal force increases. It takes more force to slide a box full of books across the floor than to slide the same box when it is empty. This principle is also used in automotive braking systems; the harder the brake pads are squeezed against the rotating brake disks, the greater the braking effect. In some cases the magnitude of the sliding friction force $\mathcal{f}_k$ is found to be approximately *proportional* to the magnitude $\mathcal{n}$ of the normal force. In such cases we represent the relation by the equation

$$\mathcal{f}_k = \mu_k \mathcal{n}, \tag{5-5}$$

where $\mu_k$ is a constant called the **coefficient of kinetic friction.** The more slippery the surface, the smaller the coefficient of friction. The adjective "kinetic" and the subscript k refer to the fact that the two surfaces are moving relative to each other. Because it is a quotient of two force magnitudes, $\mu_k$ is a pure number, without units.

The friction force and the normal force are always perpendicular, so Eq. (5–5) is *not* a vector equation but a relation between the *magnitudes* of the two forces.

Table 5–2 shows a few representative values of $\mu_k$. Although these are given with two significant figures, they are only approximate values; indeed, Eq. (5–5) is only an approximate representation of a complex phenomenon. On a microscopic level, friction and normal forces result from the intermolecular forces (fundamentally electrical in nature) between two rough surfaces at high points where they come into contact. The actual area of contact is usually much smaller than the total surface area. When two smooth surfaces of the same metal are brought together, these forces cause a "cold weld"; a similar effect occurs with aluminum pistons in an unlubricated aluminum cylinder.

**TABLE 5–2 Approximate Coefficients of Friction**

| Materials | Static, $\mu_s$ | Kinetic, $\mu_k$ |
|---|---|---|
| Steel on steel | 0.74 | 0.57 |
| Aluminum on steel | 0.61 | 0.47 |
| Copper on steel | 0.53 | 0.36 |
| Brass on steel | 0.51 | 0.44 |
| Zinc on cast iron | 0.85 | 0.21 |
| Copper on cast iron | 1.05 | 0.29 |
| Glass on glass | 0.94 | 0.40 |
| Copper on glass | 0.68 | 0.53 |
| Teflon on Teflon | 0.04 | 0.04 |
| Teflon on steel | 0.04 | 0.04 |
| Rubber on concrete (dry) | 1.0 | 0.8 |
| Rubber on concrete (wet) | 0.30 | 0.25 |

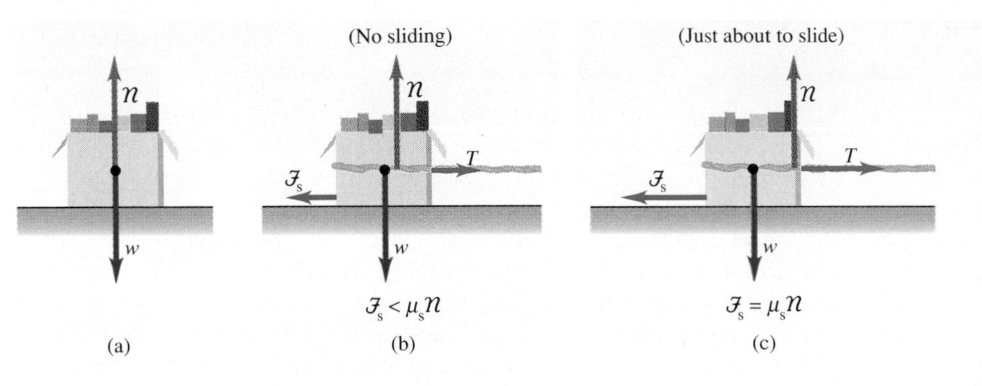


**5–13** When there is no relative motion of the surfaces, the magnitude of the friction force $\mathcal{F}_s$ is less than or equal to $\mu_s n$. When there is relative motion, the friction force $\mathcal{F}_k$ equals $\mu_k n$.

Indeed, the success of lubricating oils depends on the fact that they maintain a film between the surfaces that prevents them from coming into actual contact.

Friction forces can also depend on the *velocity* of the body relative to the surface. We will ignore that complication for now to concentrate on the simplest cases.

Friction forces may also act when there is *no* relative motion. If you try to slide that box of books across the floor, the box may not move at all because the floor exerts an equal and opposite friction force on the box. This is called a **static friction force $\mathcal{F}_s$**. In Fig. 5–13a the box is at rest, in equilibrium, under the action of its weight $\boldsymbol{w}$ and the upward normal force $\boldsymbol{n}$, which is equal in magnitude to the weight and exerted on the box by the floor. Now we tie a rope to the box (Fig. 5–13b) and gradually increase the tension $T$ in the rope. At first the box remains at rest because, as $T$ increases, the force of static friction $\mathcal{F}_s$ also increases (staying equal in magnitude to $T$).

At some point, $T$ becomes greater than the maximum friction force $\mathcal{F}_s$ that the surface can exert; the box "breaks loose" and starts to slide. Figure 5–13c is the force diagram when $T$ is at this critical value. If $T$ exceeds this value, the box is no longer in equilibrium. For a given pair of surfaces the maximum value of $\mathcal{F}_s$ depends on the normal force. In some cases the maximum value of $\mathcal{F}_s$ is approximately *proportional* to $n$; we call the proportionality factor $\mu_s$ the **coefficient of static friction.** In a particular situation the actual force of static friction can have any magnitude between zero (when there is no other force parallel to the surface) and a maximum value given by $\mu_s n$. In symbols,

$$\mathcal{F}_s \leq \mu_s n. \tag{5–6}$$

The equality sign holds only when the applied force $T$, parallel to the surface, has reached the critical value at which motion is about to start (Fig. 5–13c). When $T$ is less than this value (Fig. 5–13b), the inequality sign holds. In that case we have to use the equilibrium conditions ($\Sigma \boldsymbol{F} = \boldsymbol{0}$) to find $\mathcal{F}_s$. As soon as sliding starts, the friction force usually *decreases* because, as Table 5–2 shows, the coefficient of kinetic friction is usually *less* than the coefficient of static friction for any given pair of surfaces (Fig. 5–13d). If we start with no applied force ($T = 0$) at time $t = 0$ and gradually increase the force, the friction force varies somewhat as shown in Fig. 5–14.

In some situations the surfaces will alternately stick (static friction) and slip (kinetic friction); this is what causes the horrible squeak made by chalk held at the wrong

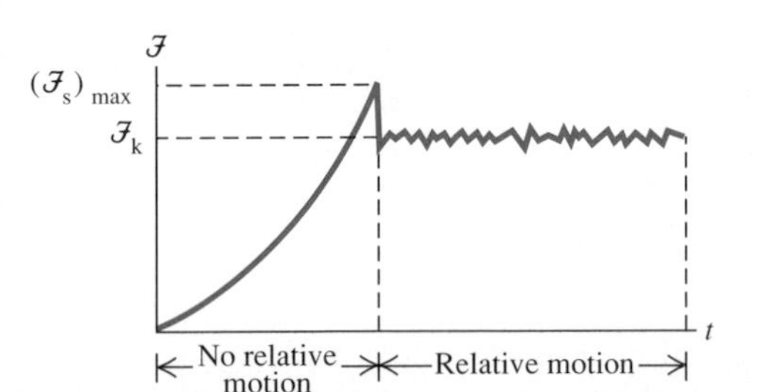


**5–14** In response to an externally applied force, the friction force increases to $(\mathcal{F}_s)_{max}$. Then the surfaces begin to slide across one another, and the frictional force drops back to a nearly constant value $\mathcal{F}_k$.

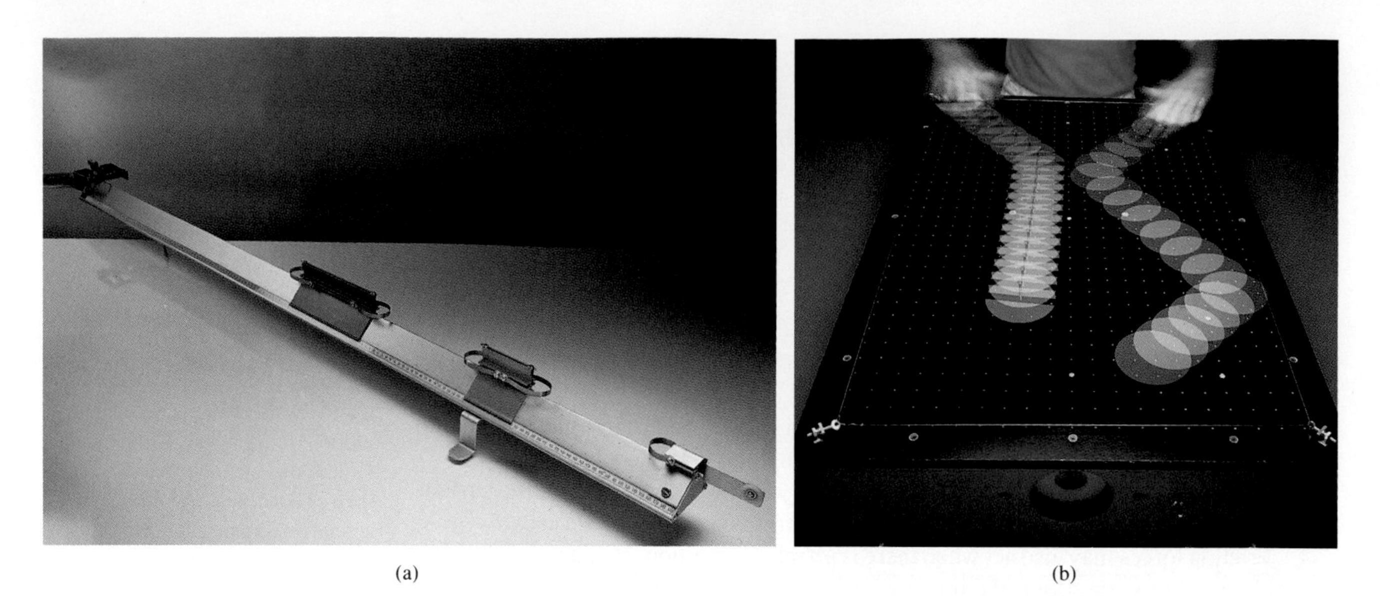

(a) (b)

**5–15** (a) The Ealing-Stull linear air track. Inverted Y-shaped sliders ride on a layer of air streaming through many fine holes in the inverted V-shaped surface. (b) The Ealing-Daw two-dimensional air table. Plastic pucks slide on a cushion of air issuing from more than 1000 minute holes in the tabletop.

angle while writing on the blackboard. Another slip-stick phenomenon is the squeaky noise your windshield-wiper blades make when the glass is nearly dry; still another is the outraged shriek of tires sliding on asphalt pavement. A more positive example is the sound produced by the motion of a violin bow against the string.

Liquids and gases also show frictional effects. Friction between two solid surfaces separated by a layer of liquid or gas is determined by the *viscosity* of the fluid. In a car engine the pistons are separated from the cylinder walls by a thin layer of oil. When a body slides on a layer of gas, friction can be made very small. In the familiar linear air track (Fig. 5–15a) the gliders are supported on a layer of air. The frictional force is velocity-dependent, but at typical speeds the effective coefficient of friction is of the order of 0.001. A similar device is the frictionless air table (Fig. 5–15b), where the pucks are supported by an array of small air jets about 2 cm apart.

## EXAMPLE 5–13

**Friction in horizontal motion** A delivery company has just unloaded a 500-N crate full of home exercise equipment in your driveway (Fig. 5–16a). You find that to get it started moving toward your garage, you have to pull with a horizontal force of magnitude 230 N. Once it "breaks loose" and starts to move, you can keep it moving at constant velocity with only 200 N of force. What are the coefficients of static and kinetic friction?

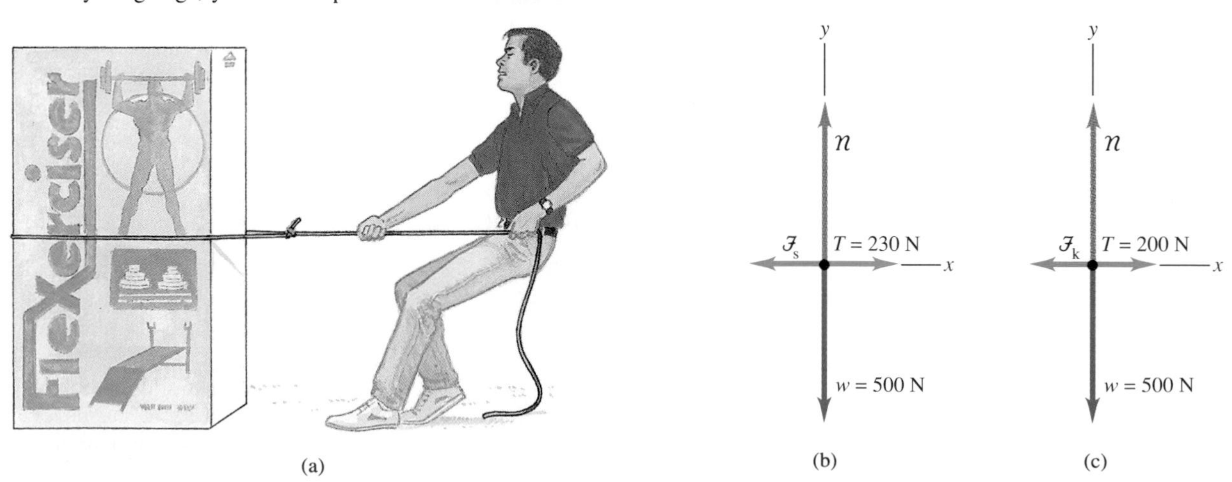


**5–16** (a) Pulling a crate with a horizontal force. (b) Free-body diagram for the crate as it starts to move. (c) Free-body diagram for the crate moving at constant velocity.

**SOLUTION** An instant before the crate starts to move, the static friction force has its maximum possible value, $\mathcal{F}_s = \mu_s \mathcal{N}$. The appropriate force diagram is Fig. 5–16b. The state of rest and the state of motion with constant velocity are both equilibrium conditions; using Eqs. (5–2), we find

$$\Sigma F_y = \mathcal{N} + (-w) = \mathcal{N} - 500\ \text{N} = 0, \qquad \mathcal{N} = 500\ \text{N},$$

$$\Sigma F_x = T + (-\mathcal{F}_s) = 230\ \text{N} - \mathcal{F}_s = 0, \qquad \mathcal{F}_s = 230\ \text{N},$$

$$\mathcal{F}_s = \mu_s \mathcal{N} \qquad \text{(motion about to start)},$$

$$\mu_s = \frac{\mathcal{F}_s}{\mathcal{N}} = \frac{230\ \text{N}}{500\ \text{N}} = 0.46.$$

After the crate starts to move, the forces are as shown in Fig. 5–16c, and we have

$$\Sigma F_y = \mathcal{N} + (-w) = \mathcal{N} - 500\ \text{N} = 0, \qquad \mathcal{N} = 500\ \text{N},$$

$$\Sigma F_x = T + (-\mathcal{F}_k) = 200\ \text{N} - \mathcal{F}_k = 0, \qquad \mathcal{F}_k = 200\ \text{N},$$

$$\mathcal{F}_k = \mu_k \mathcal{N} \qquad \text{(motion occurs)}.$$

The coefficient of kinetic friction is

$$\mu_k = \frac{200\ \text{N}}{500\ \text{N}} = 0.40.$$

## EXAMPLE 5–14

What is the friction force if the crate in Example 5–13 is at rest on the surface and a horizontal force of 50 N is applied to it?

**SOLUTION** From the equilibrium conditions we have

$$\Sigma F_x = T + (-\mathcal{F}_s) = 50\ \text{N} - \mathcal{F}_s = 0,$$

$$\mathcal{F}_s = 50\ \text{N}.$$

In this case $\mathcal{F}_s < \mu_s \mathcal{N}$. The frictional force can prevent motion for any horizontal force less than 230 N. ■

It is a lot easier to move a loaded filing cabinet across a horizontal floor using rollers or a cart with wheels than to slide it. How much easier? For a wheeled vehicle we can define a **coefficient of rolling friction** $\mu_r$, which is the horizontal force needed for constant speed on a flat surface divided by the upward normal force exerted by the surface. Transportation engineers call $\mu_r$ the *tractive resistance*. Typical values of $\mu_r$ are 0.002 to 0.003 for steel wheels on steel rails and 0.01 to 0.02 for rubber tires on concrete. These values show one reason why railroad trains are in general much more fuel-efficient than highway trucks. Rolling vehicles also experience air drag, which usually increases proportionally to the square of the speed. It is often negligible at low speeds but comparable to or greater than rolling friction at highway speeds.

## EXAMPLE 5–15

A 1200-kg car weighs about 12,000 N (about 2700 lb). If the coefficient of rolling friction is $\mu_r = 0.01$, what horizontal force is needed to make the car move with constant speed?

**SOLUTION** The normal force $\mathcal{N}$ is equal to the weight $w$. From the definition of $\mu_r$ the friction force $\mathcal{F}_r$ is

$$\mathcal{F}_r = (0.01)(12{,}000\ \text{N}) = 120\ \text{N} \qquad \text{(about 27 lb)}.$$

From Newton's first law, a forward force with this magnitude is needed to keep the car moving.

We invite you to apply this analysis to the crate of exercise equipment in Example 5–13. If the delivery company brings it on a rubber-wheel dolly with $\mu_r = 0.02$ (Fig. 5–17), only a 10-N force is needed to keep it moving at constant velocity. Can you verify this?

**5–17** Pushing a crate on a wheeled dolly.

## EXAMPLE 5–16

Suppose you try to move the crate full of exercise equipment in Example 5–13 by tying a rope around it and pulling upward on the rope at an angle of 30° above the horizontal (Fig. 5–18a). How hard do you have to pull to keep the crate moving with constant velocity? Is this easier or harder than pulling horizontally? Recall that $w = 500\ \text{N}$ and $\mu_k = 0.40$.

**SOLUTION** Figure 5–18b is a free-body diagram showing the forces on the crate. The friction force $\mathcal{F}_k$ is still equal to $\mu_k \mathcal{N}$, but now the normal force $\mathcal{N}$ is *not* equal in magnitude to the weight of the crate. The force exerted by the rope has an additional vertical component that tends to lift the crate off the floor. From the equilibrium conditions,

$$\Sigma F_x = T\cos 30° - \mathcal{F}_k = T\cos 30° - 0.40\mathcal{N} = 0,$$

$$\Sigma F_y = T\sin 30° + \mathcal{N} + (-500\ \text{N}) = 0.$$

These are two simultaneous equations for the two unknown quantities $T$ and $\mathcal{N}$. To solve them we can eliminate one unknown and solve for the other. There are many ways to do this. Here is one way: Rearrange the second equation to the form

$$\mathcal{N} = 500\ \text{N} - T\sin 30°.$$

Substitute this expression for $\mathcal{N}$ back into the first equation:

$$T\cos 30° - 0.40(500\ \text{N} - T\sin 30°) = 0.$$

Finally, solve this equation for $T$, then substitute the result back into either of the original equations to obtain $\mathcal{N}$. The results are

$$T = 188\ \text{N}, \qquad \mathcal{N} = 406\ \text{N}.$$

Note that the normal force is *less* than the weight ($w = 500$ N) of the box and that the tension required is a little less than the force needed (200 N) when you pulled horizontally. Try pulling at 22°; you'll find that you need even less force.

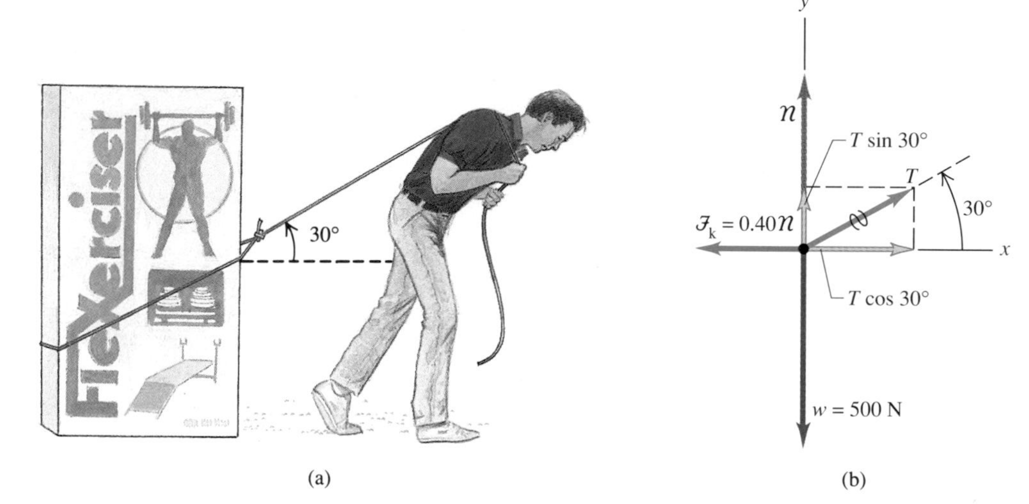


**5–18** (a) Pulling a crate with a force applied at an upward angle. (b) Free-body diagram for the crate moving at constant velocity.

## EXAMPLE 5–17

**Toboggan ride with friction** Let's go back to the toboggan we studied in Example 5–9 (Section 5–2). The wax has worn off, and there is now a coefficient of kinetic friction $\mu_k$. The slope has just the right angle to make the toboggan slide with constant speed. Derive an expression for the slope angle in terms of $w$ and $\mu_k$.

**SOLUTION** Figure 5–19 shows the free-body diagram. The slope angle is $\alpha$. The forces on the toboggan are its weight $w$ and the normal and frictional components of the contact force exerted on it by the sloping surface. We take axes perpendicular and parallel to the surface and represent the weight in terms of its components in these two directions, as shown. The equilibrium conditions are

$$\Sigma F_x = w\sin\alpha + (-\mathcal{F}_k) = w\sin\alpha - \mu_k \mathcal{N} = 0,$$

$$\Sigma F_y = \mathcal{N} + (-w\cos\alpha) = 0.$$

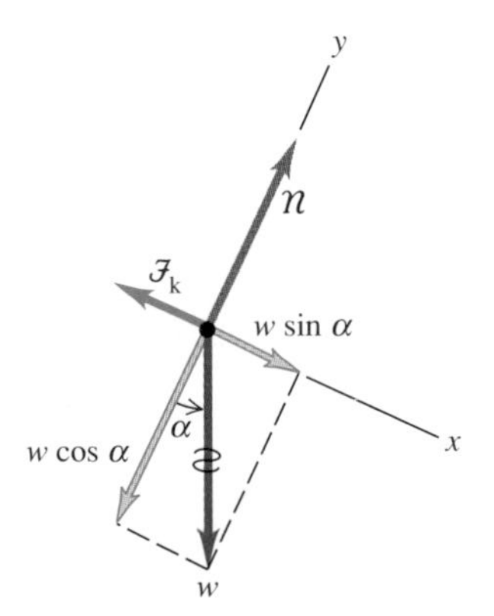


**5–19** Free-body diagram for the toboggan with friction.

Rearranging, we get

$$\mu_k \mathcal{N} = w \sin\alpha, \qquad \mathcal{N} = w\cos\alpha.$$

Note that the normal force $\mathcal{N}$ is *not* equal to the weight $w$. When we divide the first of these equations by the second, we find

$$\mu_k = \frac{\sin\alpha}{\cos\alpha} = \tan\alpha.$$

The weight $w$ doesn't appear in this expression, so *any* toboggan, regardless of its weight, slides down an incline with constant speed if the tangent of the slope angle of the hill equals the coefficient of kinetic friction. The greater the coefficient of friction, the steeper the slope has to be for the toboggan to slide with constant velocity. This is just what we should expect.

## EXAMPLE 5–18

What if we have the same toboggan and coefficient of friction as in Example 5–17 but a steeper hill? This time the toboggan accelerates, although not as much as in Example 5–9, when there was no friction. Derive an expression for the acceleration in terms of $g$, $\alpha$, $\mu_k$, and $w$.

**SOLUTION** The free-body diagram (Fig. 5–20) is almost the same as for Example 5–17, but the body is no longer in equilibrium; $a_y$ is still zero, but $a_x$ is not. Newton's second law gives us the two equations

$$\Sigma F_x = mg\sin\alpha + (-\mathcal{F}_k) = ma_x,$$

$$\Sigma F_y = \mathcal{N} + (-mg\cos\alpha) = 0.$$

From the second equation and Eq. (5–5) we get an expression for $\mathcal{F}_k$:

$$\mathcal{N} = mg\cos\alpha,$$

$$\mathcal{F}_k = \mu_k \mathcal{N} = \mu_k\, mg\cos\alpha.$$

We substitute this back into the $x$-component equation. The result is

$$mg\sin\alpha + (-\mu_k\, mg\cos\alpha) = ma_x,$$

$$a_x = g(\sin\alpha - \mu_k\cos\alpha).$$

Does this result make sense? Here are some special cases we can check. First, if the hill is *vertical*, $\alpha = 90°$; then $\sin\alpha = 1$, $\cos\alpha = 0$, and $a_x = g$. This is free fall, just what we would expect. Second, on a hill at angle $\alpha$ with *no* friction, $\mu_k = 0$. Then $a_x = g\sin\alpha$. The situation is the same as in Example 5–9, and we get the same result; that's encouraging! Next, suppose that $a_x = 0$. In that case there is just enough friction to make the toboggan move with constant velocity. In that case, $a_x = 0$, and our result gives

$$\sin\alpha = \mu_k\cos\alpha \qquad \text{and} \qquad \mu_k = \tan\alpha.$$

This agrees with our result from Example 5–17; good! Finally, note that there may be so much friction that $\mu_k\cos\alpha$ is actually greater than $\sin\alpha$. In that case, $a_x$ is negative. If we give the toboggan an initial downhill push, it will slow down and eventually stop.

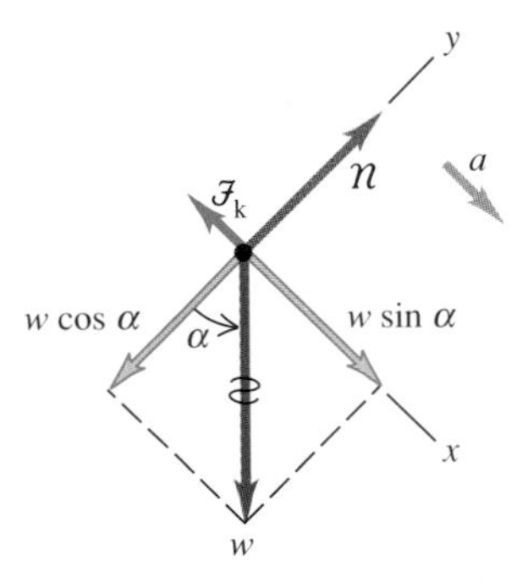


**5–20** Free-body diagram for the toboggan, with friction, going down a steeper hill.

We have pretty much beaten the toboggan problem to death, but there is an important lesson to be learned. We started out with a simple problem and then extended it to more and more general situations. Our most general result includes all the previous ones as special cases, and that's a nice, neat package! Don't memorize this package; it is useful only for this one set of problems. But do try to understand how we obtained it and what it means.

One final variation that you may want to try out is the case in which we give the toboggan an initial push *up* the hill. The direction of the friction force is now reversed, so the acceleration is different from the downhill value. It turns out that the expression for $a_x$ is the same as before except that the minus sign becomes plus. Can you prove this?

## EXAMPLE 5–19

Friction forces resulting from motion through a fluid (such as air or water) usually increase with the body's speed through the fluid. Suppose you release a rock at the surface of a deep pond, and it falls to the bottom. For small speeds the resisting force $\mathcal{F}$ of the liquid is approximately proportional to the rock's speed $v$:

$$\mathcal{F} = kv, \tag{5–7}$$

where $k$ is a proportionality constant that depends on the shape and size of the rock and the properties of the fluid. Taking the positive direction to be downward and neglecting any force associated with buoyancy in the fluid, we find that the net vertical component of force is $mg - kv$. Newton's second law gives

$$mg - kv = ma.$$

When the rock first starts to move, $v = 0$, the resisting force is zero, and the initial acceleration is $a = g$. As its speed increases, the resisting force also increases until finally it equals the weight in magnitude. At this time $mg - kv = 0$, the acceleration becomes zero, and there is no further increase in speed. The final speed $v_t$, called the **terminal speed,** is given by

$$mg - kv_t = 0,$$

or

$$v_t = \frac{mg}{k}. \tag{5-8}$$

Figure 5–21 shows how the acceleration, velocity, and position vary with time.

To find the relation between speed and time during the interval before the terminal speed is reached, we go back to Newton's second law, which we rewrite as

$$m\frac{dv}{dt} = mg - kv.$$

After rearranging terms and replacing $mg/k$ by $v_t$, we find

$$\frac{dv}{v - v_t} = -\frac{k}{m}dt.$$

We now integrate both sides, noting that $v = 0$ when $t = 0$:

$$\int_0^v \frac{dv}{v - v_t} = -\frac{k}{m}\int_0^t dt,$$

which integrates to

$$\ln\frac{v_t - v}{v_t} = -\frac{k}{m}t, \qquad \text{or} \qquad 1 - \frac{v}{v_t} = e^{-(k/m)t},$$

and finally

$$v = v_t[1 - e^{-(k/m)t}]. \tag{5-9}$$

From this result we can derive expressions for the acceleration and position as functions of time. We leave the derivations as an exercise; the results are

$$a = ge^{-(k/m)t}, \tag{5-10}$$

$$y = v_t\left[t - \frac{m}{k}\left(1 - e^{-(k/m)t}\right)\right]. \tag{5-11}$$

Now look again at Fig. 5–21, which shows graphs of these three relations.

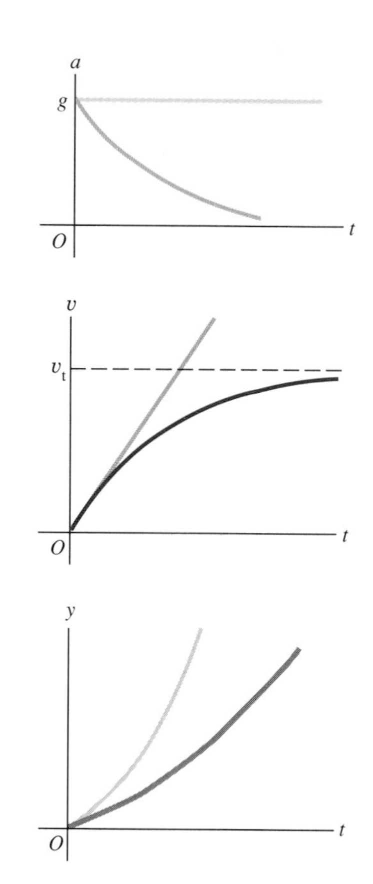


**5–21** Graphs of acceleration, velocity, and position versus time for a body falling in a viscous fluid, shown as dark color curves. The light color curves show the corresponding relations if there is *no* viscous friction.

In high-speed motion through air the resisting force is approximately proportional to $v^2$ rather than to $v$; it is then called *air drag,* or simply *drag.* Falling raindrops, airplanes, and cars moving at high speed all experience air drag. In this case we replace Eq. (5–7) by $\mathcal{F} = Cv^2$. We invite you to show that the terminal speed $v_t$ is then given by

$$v_t = \sqrt{\frac{mg}{C}}, \tag{5-12}$$

and that the units of the constant $C$ are $N \cdot s^2/m^2$ or kg/m.

## EXAMPLE 5–20

**Terminal speed of a sky diver** For a human body falling through air in a spread-eagle position (Fig. 5–22), the numerical value of the constant $C$ in Eq. (5–12) is about 0.25 kg/m. If an 80-kg sky diver jumps out of an airplane and reaches terminal speed before opening the parachute, the terminal speed is, from Eq. (5–12),

$$v_t = \sqrt{\frac{mg}{C}}$$
$$= \sqrt{\frac{(80\ \text{kg})(9.8\ \text{m/s}^2)}{0.25\ \text{kg/m}}}$$
$$= 56\ \text{m/s} \quad \text{(about 125 mi/h)}.$$

**5–22** How fast does a sky diver fall?

# 5–4 DYNAMICS OF CIRCULAR MOTION

We talked about uniform circular motion in Section 3–4. We showed that when a particle moves in a circular path with constant speed, the particle's acceleration is directed always toward the center of the circle (perpendicular to the instantaneous velocity). The magnitude $a_{\text{rad}}$ of the acceleration is constant and is given in terms of the speed $v$ and the radius $R$ of the circle by

$$a_{\text{rad}} = \frac{v^2}{R}. \tag{5–13}$$

The subscript "rad" is a reminder that at each point the acceleration is radially inward toward the center of the circle, perpendicular to the instantaneous velocity. We explained in Section 3–4 why this acceleration is sometimes called *centripetal acceleration*.

We can also express the centripetal acceleration $a_{\text{rad}}$ in terms of the **period** $T$, the time for one revolution:

$$T = \frac{2\pi R}{v}. \tag{5–14}$$

In terms of the period, $a_{\text{rad}}$ is

$$a_{\text{rad}} = \frac{4\pi^2 R}{T^2}. \tag{5–15}$$

Circular motion, like all other motion of a particle, is governed by Newton's second law. The particle's acceleration toward the center of the circle must be caused by a force, or several forces, such that the vector sum $\Sigma \boldsymbol{F}$ is a vector directed always toward the center with constant magnitude. If we tie a small object, such as a key, to a string and whirl it in a circle, the string has to pull constantly toward the center of the circle. If the string breaks, then the inward force no longer acts, and the key flies off along a tangent to the circle.

The magnitude of the radial acceleration is given by $a_{\text{rad}} = v^2/R$, so the magnitude of the net inward radial force $F_{\text{net}}$ on a particle of mass $m$ must be

$$F_{\text{net}} = ma_{\text{rad}} = m\frac{v^2}{R}. \tag{5–16}$$

## PROBLEM-SOLVING STRATEGY

### Circular motion

The strategies for dynamics problems outlined at the beginning of Section 5–2 are equally applicable here, and we suggest that you reread them before studying the examples below.

A serious peril in circular-motion problems is the temptation to regard $m(v^2/R)$ as a *force,* as though the body's circular motion somehow generates an extra force in addition to the real physical forces exerted by strings, contact with other bodies, gravitation, or whatever. *Resist this temptation!* The quantity $m(v^2/R)$ is not a force; it corresponds to the $ma$ side of $\Sigma F = ma$ and *does not* appear in $\Sigma F$. It may help to draw the free-body diagram with colored pencils, using one color for the forces and a different color for the acceleration. Then in the $\Sigma F = ma$ equations, write the force terms in the force color and the $m(v^2/R)$ term in the acceleration color.

## EXAMPLE 5–21

A small plastic box with a mass of 0.200 kg revolves uniformly in a circle on a horizontal frictionless surface such as an air-hockey table (Fig. 5–23). The box is attached by a horizontal cord 0.200 m long to a pin set in the surface. If the box makes two complete revolutions per second, find the force $F$ exerted on it by the cord.

**SOLUTION** The period is $T = (1.00 \text{ s})/(2 \text{ rev}) = 0.500$ s. From Eq. (5–14) the speed $v$ is

$$v = \frac{2\pi R}{T} = \frac{2\pi(0.200 \text{ m})}{0.500 \text{ s}} = 2.513 \text{ m/s}.$$

The magnitude of the centripetal acceleration is

$$a_{\text{rad}} = \frac{v^2}{R} = \frac{(2.513 \text{ m/s})^2}{0.200 \text{ m}} = 31.6 \text{ m/s}^2.$$

Alternatively, we may use Eq. (5–15):

$$a_{\text{rad}} = \frac{4\pi^2 R}{T^2} = \frac{4\pi^2(0.200 \text{ m})}{(0.500 \text{ s})^2} = 31.6 \text{ m/s}^2.$$

The box has no vertical acceleration, so the forces $\boldsymbol{n}$ and $\boldsymbol{w}$ are equal and opposite. The only force toward the center of the circle is the tension $F$ in the cord, and

$$F = ma = (0.200 \text{ kg})(31.6 \text{ m/s}^2)$$
$$= 6.32 \text{ kg} \cdot \text{m/s}^2 = 6.32 \text{ N}.$$

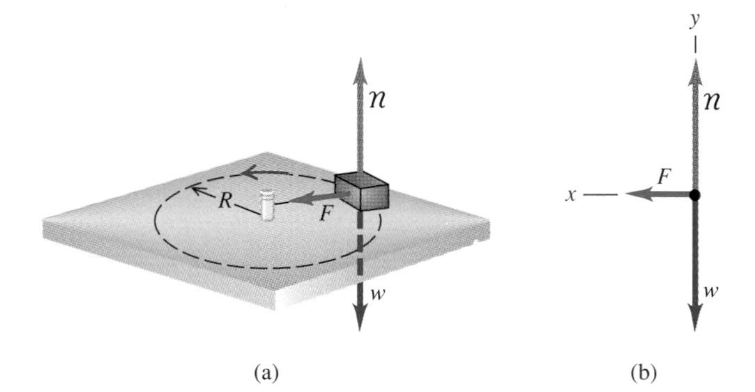


**5–23** (a) A box moving in uniform circular motion on a horizontal frictionless surface. (b) Free-body diagram for the box.

## EXAMPLE 5–22

**The conical pendulum** An inventor who dares to be different proposes to make a pendulum clock with a pendulum bob of mass $m$ at the end of a thin wire of length $L$. Instead of swinging back and forth, the bob moves in a horizontal circle with constant speed $v$, with the wire making a constant angle $\beta$ with the vertical direction (Fig. 5–24). Assuming that the time $T$ for one revolution (that is, the period) is known, find the tension $F$ in the wire and the angle $\beta$.

**SOLUTION** A free-body diagram for the bob and a coordinate system are shown in Fig. 5–24b. The center of the circular path is at the center of the shaded area, *not* at the top end of the

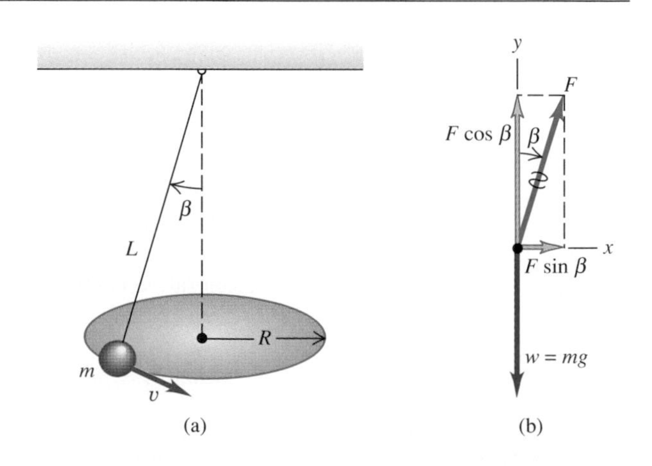


**5–24** (a) The bob on the end of the wire moves in uniform circular motion. (b) Free-body diagram for the bob.

wire. The forces on the bob in the position shown are the weight $mg$ and the tension $F$ in the wire. The tension has a horizontal component $F \sin \beta$ and a vertical component $F \cos \beta$. (This free-body diagram is exactly like the one in Fig. 5–12b, but in this case the acceleration $a$ is the *radial* acceleration, $v^2/R$.) The system has no vertical acceleration, so the sum of the vertical components of force is zero. The horizontal force must equal the mass $m$ times the radial (centripetal) acceleration. Thus the $\Sigma F = ma$ equations are

$$F \sin \beta = m\frac{v^2}{R}, \qquad F \cos \beta - mg = 0.$$

These are two simultaneous equations for the unknowns $F$ and $\beta$. An easy way to eliminate $F$ is to divide the first of these equations by the second; the result is

$$\tan \beta = \frac{v^2}{gR}. \tag{5–17}$$

Also, the radius $R$ of the circle is given by $R = L \sin \beta$, and the speed is the circumference $2\pi L \sin \beta$ divided by the period $T$:

$$v = \frac{2\pi L \sin \beta}{T}.$$

We use these relations to eliminate $R$ and $v$ from Eq. (5–17), obtaining

$$\cos \beta = \frac{gT^2}{4\pi^2 L},$$

or

$$T = 2\pi\sqrt{\frac{L \cos \beta}{g}}. \tag{5–18}$$

Once we know $\beta$, we can find the tension $F$ from $F = mg/\cos \beta$. For a given length $L$, $\cos \beta$ decreases and $\beta$ increases as $T$ becomes smaller. The angle can never be 90°, however; this would require that $T = 0$, $F = \infty$, and $v = \infty$. This system is called a *conical pendulum* because it is constructed like a pendulum but the suspending wire traces out a cone. It would not make a very good clock because the period depends on $\beta$ in such a direct way.

You may be tempted to add to the forces in Fig. 5–24 an extra outward force to "keep the body out there" at angle $\beta$ or to "keep it in equilibrium." Perhaps you have heard the term *centrifugal force;* centrifugal means "fleeing from the center." Resist that temptation! First, the body *doesn't* stay "out there"; it is in constant motion around its circular path. Its velocity is constantly changing in direction, it accelerates, and it is *not* in equilibrium. Second, if there *were* an additional outward ("centrifugal") force to balance the inward force, there would then be *no* net inward force to cause the circular motion, and the body would move in a straight line, not a circle. If you were rotating with the body, you might *think* there was an extra force and that the body was in equilibrium, but this is because you would then be in a noninertial frame of reference. In an inertial frame of reference there is no such thing as centrifugal force, and we promise not to mention this term again.

## EXAMPLE 5–23

**Rounding a flat curve** The Corvette in Example 3–8 (Section 3–4) is rounding a flat, unbanked curve with radius $R$. If the coefficient of friction between tires and road is $\mu_s$, what is the maximum speed $v$ at which the driver can take the curve without sliding?

**SOLUTION** Figure 5–25 shows a free-body diagram for the car. The acceleration $v^2/R$ toward the center of the curve must be caused by the friction force $\mathcal{F}$, and there is no vertical acceleration. Thus we have

$$\mathcal{F} = m\frac{v^2}{R}, \qquad \mathcal{N} - mg = 0.$$

The first equation shows that the friction force *needed* to keep the car moving in its circular path increases with the car's speed. But the maximum friction force *available* is $\mathcal{F} = \mu_s \mathcal{N} = \mu_s mg$, which is constant, and this determines the car's maximum speed.

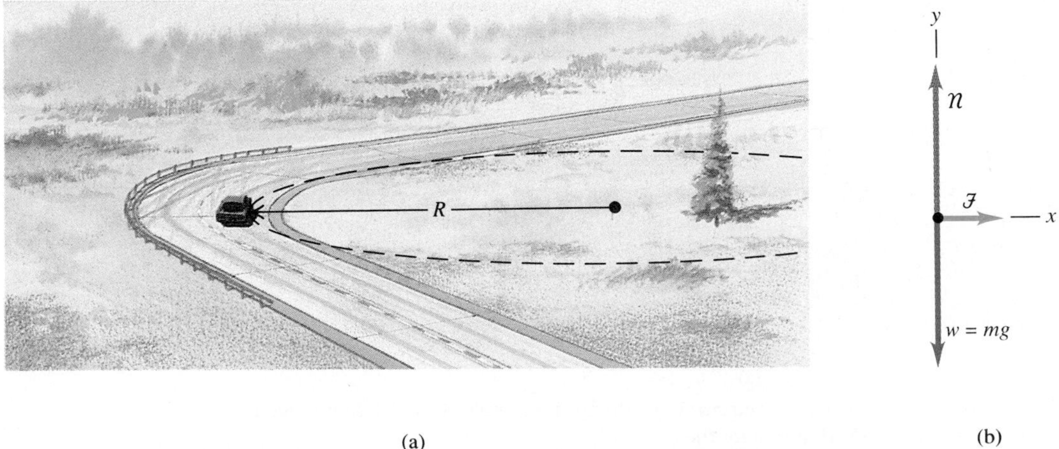


**5–25** (a) A sports car rounding a curve on a level road. (b) Free-body diagram for the car.

Combining these relations to eliminate $\mathcal{F}$, we find

$$\mu_s mg = m\frac{v^2}{R},$$

or

$$v = \sqrt{\mu_s gR}. \tag{5-19}$$

If $\mu_s = 0.91$ and $R = 230$ m, then

$$v = \sqrt{(0.91)(9.8\ \text{m/s}^2)(230\ \text{m})} = 45\ \text{m/s},$$

or about 100 mi/h. This is the maximum speed for this radius, typical in modern highway design. If we take the curve more slowly than this, the friction force is *less than* $\mu_s mg$. Note that the maximum centripetal acceleration (called the "lateral acceleration" in Example 3–8 in Section 3–4) is equal to $\mu_s g$. Why do we use the coefficient of *static* friction for a moving car? What happens when we try to go *faster* than this maximum speed? ■

---

It is possible to bank a curve so that at one certain speed, no friction at all is needed. Then a car can round the curve even on wet ice with Teflon tires. Airplanes *always* bank at this angle when making turns, and bobsled racing depends on this idea.

## EXAMPLE 5–24

**Rounding a banked curve** An engineer proposes to rebuild the curve in Example 5–23, banking it so that at a certain speed $v$, *no* friction at all is needed for the car to make the curve (Fig. 5–26). At what angle $\beta$ should it be banked?

**SOLUTION** The free-body diagram is shown in Fig. 5–26b. The normal force $\mathcal{N}$ is no longer vertical but is perpendicular to the roadway at an angle $\beta$ with the vertical. Thus it has a vertical component $\mathcal{N}\cos\beta$ and a horizontal component $\mathcal{N}\sin\beta$, as shown. We want to get around the curve without relying on friction, so no friction force is included. The horizontal component of $\mathcal{N}$ must now cause the acceleration $v^2/R$; there is no vertical acceleration. Thus

$$\mathcal{N}\sin\beta = \frac{mv^2}{R},$$

$$\mathcal{N}\cos\beta - mg = 0.$$

Dividing the first of these equations by the second, we find

$$\tan\beta = \frac{v^2}{gR}. \tag{5-20}$$

If $R = 230$ m and $v = 45$ m/s, as in Example 5–23, then

$$\beta = \arctan\frac{(45\ \text{m/s})^2}{(9.8\ \text{m/s}^2)(230\ \text{m})} = 42°.$$

The banking angle depends on the speed and the radius. For a given radius, no one angle is correct for all speeds. In the design of highways and railroads, curves are often banked for the *average speed* of the traffic over them. The same considerations apply to the correct banking angle of an airplane when it makes a turn in level flight. We also note that the banking angle is given by the same expression as that for the angle of a conical pendulum, Eq. (5–17) in Example 5–22. In fact, the free-body diagrams of Figs. 5–24b and 5–26b are very similar, with the normal force playing the role of the tension in the pendulum.

**5–26** (a) The sports car rounding a banked curve. At the banking angle $\beta$, no friction is needed to round the curve. (b) Free-body diagram for the car.

# 5–5 MOTION IN A VERTICAL CIRCLE

In all the examples in Section 5–4 the body moved in a horizontal circle. Motion in a vertical circle is no different in principle, but the weight of the body has to be treated carefully. The following example shows what we mean.

## EXAMPLE 5–25

In a carnival ride such as a Ferris wheel, a passenger moves in a vertical circle of radius $R$ with constant speed $v$. Assuming that the seat remains upright during the motion, derive expressions for the force the seat exerts on the passenger at the top of the circle and at the bottom.

**SOLUTION** Figure 5–27 is a diagram of the situation, with free-body diagrams for the two positions. We take the positive direction as upward in both cases. Let $F_T$ be the force the seat applies to the passenger at the top of the circle, and let $F_B$ be the force at the bottom. At the top the acceleration has magnitude $v^2/R$, but its vertical component is negative because its direction is downward, toward the center of the circle. So the $\Sigma F = ma$ equation at the top is

$$F_T + (-mg) = -m\frac{v^2}{R}, \quad \text{or}$$

$$F_T = m\left(g - \frac{v^2}{R}\right).$$

This means that at the top of the Ferris wheel the upward force the seat applies to the passenger is *smaller* in magnitude than the passenger's weight. If the ride goes fast enough that $g - v^2/R$ becomes zero, the seat applies *no* force, and the passenger is about to become airborne. If $v$ becomes still larger, $F_T$ becomes negative; this means that a *downward* force is needed to keep the passenger in the seat. We could supply a seat belt, or we could glue the passenger to the seat.

At the bottom the acceleration is upward, and the equation of motion is

$$F_B + (-mg) = +m\frac{v^2}{R}, \quad \text{or}$$

$$F_B = m\left(g + \frac{v^2}{R}\right).$$

At this point the upward force provided by the seat is always *greater than* the passenger's weight. You feel the seat pushing up on you more firmly than when you are at rest.

As a specific example, suppose the passenger's mass is 60.0 kg, the radius of the circle is $R = 8.00$ m, and the wheel makes one revolution in 10.0 s. The speed is the circumference divided by the time for one revolution, $v = (2\pi)(8.00\text{ m})/(10.0\text{ s}) = 5.03\text{ m/s}$, and

$$a_{\text{rad}} = v^2/R = (5.03\text{ m/s})^2/(8.00\text{ m}) = 3.16\text{ m/s}^2.$$

The two forces are

$$F_T = (60.0\text{ kg})(9.80\text{ m/s}^2 - 3.16\text{ m/s}^2) = 398\text{ N},$$

$$F_B = (60.0\text{ kg})(9.80\text{ m/s}^2 + 3.16\text{ m/s}^2 = 778\text{ N}.$$

The passenger's weight is 588 N, and $F_T$ and $F_B$ are about 30% less than and greater than the weight, respectively.

Some military aircraft are capable of $9g$ acceleration when pulling out of a dive (that is, $g + v^2/R = 9g$ in the above analysis). This would eliminate blood supply to the brain of the pilot, who would black out under this acceleration.

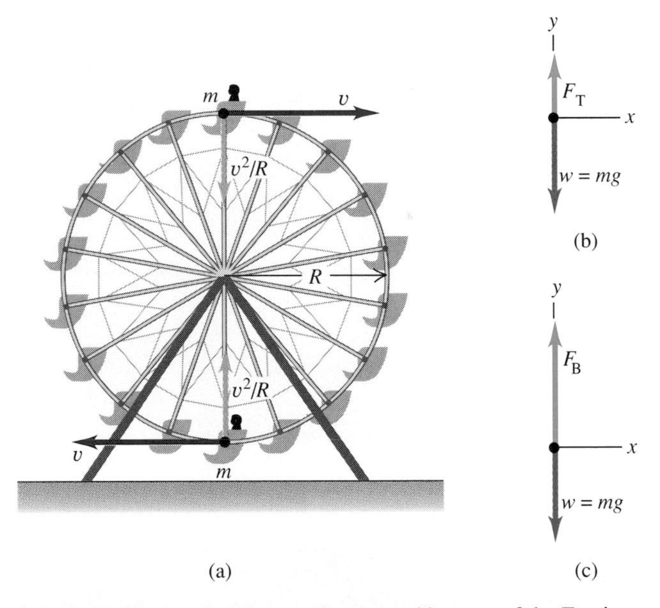


**5–27** (a) The accelerations at the top and bottom of the Ferris wheel are equal in magnitude but opposite in direction. (b) Free-body diagram for a passenger at the top of the circle. (c) Free-body diagram for a passenger at the bottom of the circle.

When we tie a string to an object and whirl it in a vertical circle, the preceding analysis isn't directly applicable; $v$ isn't constant in this case because at every point except the top and bottom there is a component of force (and therefore of acceleration) tangent to the circle. Even worse, we can't use the constant-acceleration formulas to relate the speeds at various points because *neither* the magnitude nor the direction of the acceleration is constant. The speed relations we need are best obtained by using energy relations. We will consider such problems in Chapter 7.

## *5–6 FORCES IN NATURE

The historical development of our understanding of the forces (or interactions) found in nature has traditionally placed them in four distinct classes (Fig. 5–28). Two are familiar in everyday experience. The other two involve fundamental particle interactions that we cannot observe with the unaided senses.

Of the two familiar classes, **gravitational interactions** were the first to be studied in detail. The *weight* of a body results from the earth's gravitational attraction acting on it. The sun's gravitational attraction for the earth keeps the earth in its nearly circular orbit around the sun. Newton recognized that the motions of the planets around the sun and the free fall of objects on earth are both the result of gravitational forces. In Chapter 12 we will study gravitational interactions in greater detail, and we will analyze their vital role in the motions of planets and satellites.

The second familiar class of forces, **electromagnetic interactions,** includes electric and magnetic forces. When you run a comb through your hair, you can then use it to pick up bits of paper or fluff; this interaction is the result of electric charge on the comb, which also causes the infamous "static cling." *Magnetic* forces occur in interactions between magnets or between a magnet and a piece of iron. These may seem to form a different category, but magnetic interactions are actually the result of electric charges in

(a)

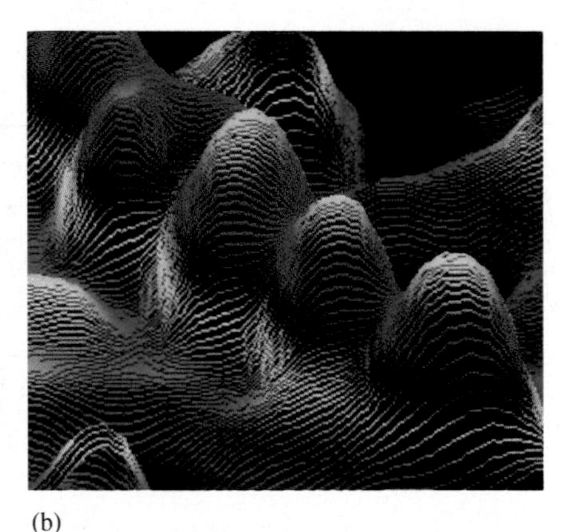

(b)

(c)

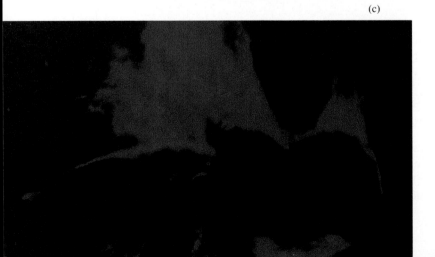

(d)

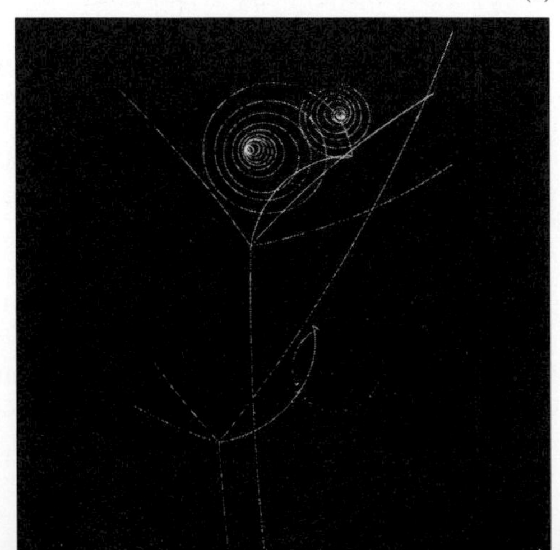

**5–28** Examples of the fundamental forces in nature. (a) Gravitational forces cause the orbital motions of the earth around the sun and the moon around the earth. (b) Electromagnetic forces act between atoms to form molecules, as in this false-color scanning tunneling micrograph of DNA. (The orange and yellow peaks represent the ridges of the double helix.) (c) Strong forces between nuclear particles are associated with the thermonuclear reactions that cause solar flares. (d) Weak forces, characteristic of interactions involving neutrinos, may be observed in high-energy experiments with sub-atomic particles.

motion. In an electromagnet an electric current in a coil of wire causes magnetic interactions. We will study electric and magnetic interactions in detail in the second half of this book.

These two interactions differ enormously in their strength. The electrical repulsion between two protons at a given distance is stronger than their gravitational attraction by a factor of the order of $10^{35}$. Gravitational forces play no significant role in atomic or molecular structure. But in bodies of astronomical size, positive and negative charge are usually present in nearly equal amounts; the resulting electrical interactions nearly cancel out. Gravitational interactions are the dominant influence in the motion of planets and also in the internal structure of stars.

The other two classes of interactions are less familiar. One, the **strong interaction,** is responsible for holding the nucleus of an atom together. Nuclei contain electrically neutral and positively charged particles. The charged particles repel each other, and a nucleus could not be stable if it were not for the presence of an attractive force of a different kind that counteracts the repulsive electrical interactions. In this context the strong interaction is also called the *nuclear force*. It has much shorter range than electrical interactions, but within its range it is much stronger. The strong interaction is also responsible for the creation of unstable particles in high-energy particle collisions.

Finally, there is the **weak interaction.** It plays no direct role in the behavior of ordinary matter, but it is of vital importance in interactions among fundamental particles. The weak interaction is responsible for beta decay; this is the emission of an electron (beta particle) from a radioactive nucleus by conversion of a neutron into a proton, an electron, and an antineutrino. The weak interaction is also responsible for the decay of many unstable particles produced in high-energy collisions of fundamental particles.

During the past decade a unified theory of the electromagnetic and weak interactions has been developed. This interaction is called the *electroweak interaction,* and in a sense this reduces the number of classes of force from four to three. Similar attempts have been made to understand all strong, electromagnetic, and weak interactions on the basis of a single unified theory called a *grand unified theory* (GUT). Such theories are still speculative, and there are many unanswered questions. The entire area is a very active field of current research.

## SUMMARY

- When a body is in equilibrium in an inertial frame of reference, the vector sum of forces acting on it must be zero. In component form,

$$\Sigma F_x = 0, \qquad \Sigma F_y = 0. \tag{5–2}$$

Free-body diagrams are useful in identifying the forces acting on the body being considered. Newton's third law is also frequently needed in equilibrium problems. The two forces in an action-reaction pair never act on the same body.

- When the vector sum of forces on a body is not zero, the body has an acceleration determined by Newton's second law:

$$\Sigma \mathbf{F} = m\mathbf{a}. \tag{5–3}$$

In component form,

$$\Sigma F_x = ma_x, \qquad \Sigma F_y = ma_y. \tag{5–4}$$

- The contact force between two bodies can always be represented in terms of a normal component $\mathcal{n}$ perpendicular to the interaction surface and a frictional component $\mathcal{f}$ parallel to the surface. When sliding occurs, $\mathcal{f}_k$ is approximately proportional to $\mathcal{n}$,

### KEY TERMS

equilibrium

dynamics

contact force

normal force

friction force

coefficient of kinetic friction

static friction force

coefficient of static friction

**coefficient of rolling friction**

**terminal speed**

**period**

**gravitational interaction**

**electromagnetic interaction**

**strong interaction**

**weak interaction**

and the proportionality constant is $\mu_k$, the coefficient of kinetic friction,

$$\mathcal{F}_k = \mu_k n. \tag{5–5}$$

When there is no relative motion, the maximum possible friction force is approximately proportional to the normal force, and the proportionality constant is $\mu_s$, the coefficient of static friction,

$$\mathcal{F}_s \leq \mu_s n. \tag{5–6}$$

The actual static friction force may be anything from zero to this maximum value, depending on the situation. Usually, $\mu_k$ is less than $\mu_s$ for a given pair of surfaces.

• In circular motion the acceleration vector is directed toward the center of the circle and has magnitude $v^2/R$. The motion is governed by $\Sigma \mathbf{F} = m\mathbf{a}$, just as for any other dynamics problem.

• The classes of forces found in nature are the gravitational, electromagnetic, strong, and weak interactions. The electromagnetic and weak interactions have been unified into a single interaction, the electroweak interaction.

## EXERCISES

### Section 5–1 Equilibrium of a Particle

**5–1** Two 15.0-N weights are suspended at opposite ends of a rope that passes over a light, frictionless pulley. The pulley is attached to a chain that goes to the ceiling. a) What is the tension in the rope? b) What is the tension in the chain?

**5–2** In each of the situations in Fig. 5–29 the blocks suspended from the rope have weight $w$. The pulleys are frictionless. In each case, calculate the tension $T$ in the rope in terms of the weight $w$.

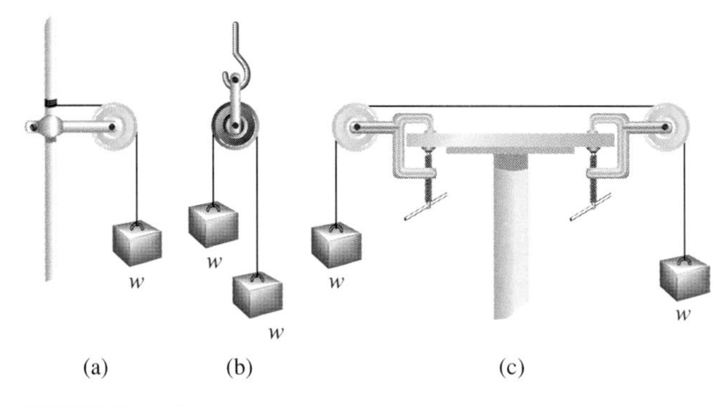


**FIGURE 5–29**

**5–3** An adventurous archaeologist crosses between two rock cliffs, with a raging river far below, by slowly going hand-over-hand along a rope stretched between the cliffs. He stops to rest at the middle of the rope (Fig. 5–30). The breaking strength of the rope is 24,000 N, and our hero's mass is 81.6 kg. a) If the angle $\theta$ is 15.0°, find the tension in the rope. b) What is the smallest value the angle $\theta$ can have if the rope is not to break?

**FIGURE 5–30**

**5–4** A picture frame hung against a wall is suspended by two wires attached to its upper corners. If the two wires make the same angle with the vertical, what must this angle be if the tension in each wire is equal to the weight of the frame? (Neglect any friction between the wall and the picture frame.)

**5–5** A large wrecking ball is held in place by two light steel cables (Fig. 5–31). If the tension $T_A$ in the horizontal cable is 460

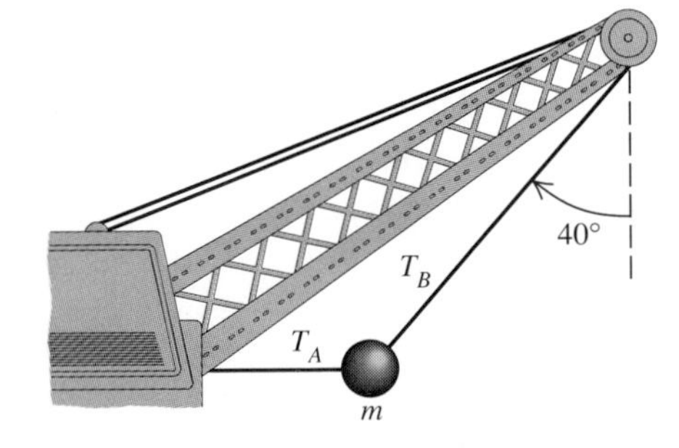

**FIGURE 5–31**

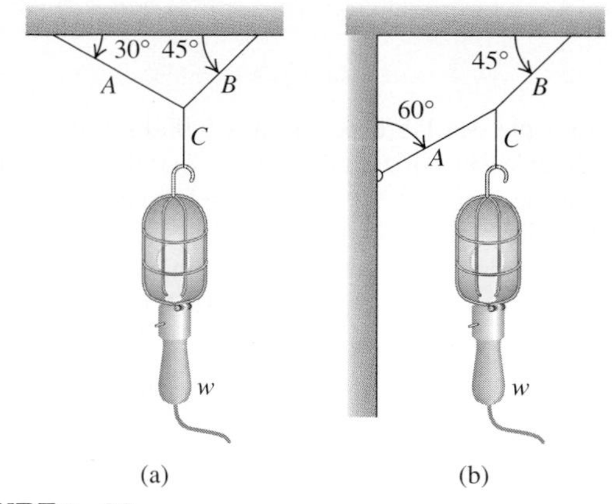


**FIGURE 5–32**

N, what is a) the tension $T_B$ in the other cable that makes an angle of 40° with the vertical? b) the mass $m$ of the wrecking ball?

**5–6** Find the tension in each cord in Fig. 5–32 if the weight of the suspended lamp is $w$.

**5–7** In Fig. 5–33 the tension in the diagonal string is 40.0 N. a) Find the magnitudes of the horizontal forces $\boldsymbol{F}_1$ and $\boldsymbol{F}_2$ that must be applied to hold the system in the position shown. b) What is the weight of the suspended block?

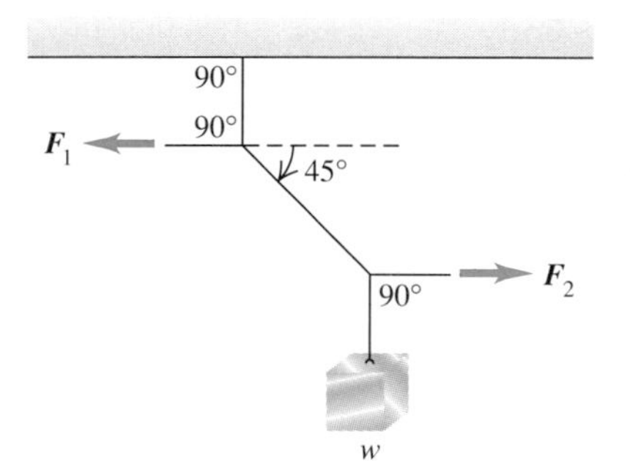


**FIGURE 5–33**

**5–8** A tether ball leans against the post to which it is attached (Fig. 5–34). If the string to which the ball is attached is 1.80 m long, the ball has a radius of 0.200 m, and the ball has a mass of 0.400 kg, what are the tension in the rope and the force the pole exerts on the ball? Assume that there is so little friction between the ball and the pole that it can be neglected. (The string is attached to the ball such that a line along the string passes through the center of the ball.)

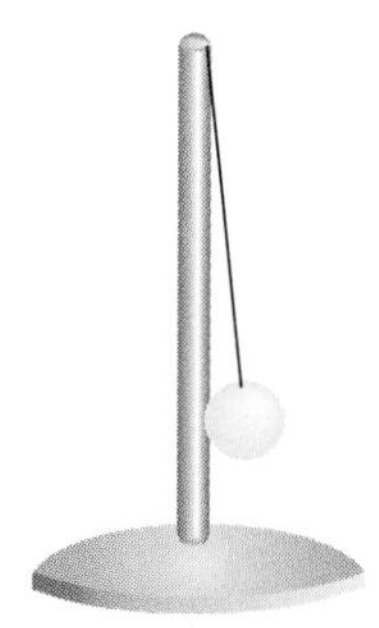

**FIGURE 5–34**

**5–9** Two blocks, each with weight $w$, are held in place on a frictionless incline (Fig. 5–35). In terms of $w$ and the angle $\theta$ of the incline, calculate the tension in a) the rope connecting the blocks; b) the rope that connects block $A$ to the wall.

**5–10** A man is pushing a piano with mass 140 kg at constant velocity up a ramp that is inclined at 36.9° above the horizontal. Neglect friction. If the force applied by the man is parallel to the incline, calculate the magnitude of this force.

**5–11** At a certain part of your drive from home to school, your car will coast in neutral at a constant speed of 92 km/h if there is no wind. Examination of a topographical map shows that for this stretch of straight highway the elevation decreases 300 m for each 6000 m of road. What is the total resistive force (friction plus air resistance) that acts on your car when it is traveling at 92 km/h? Assume a total mass of 1200 kg.

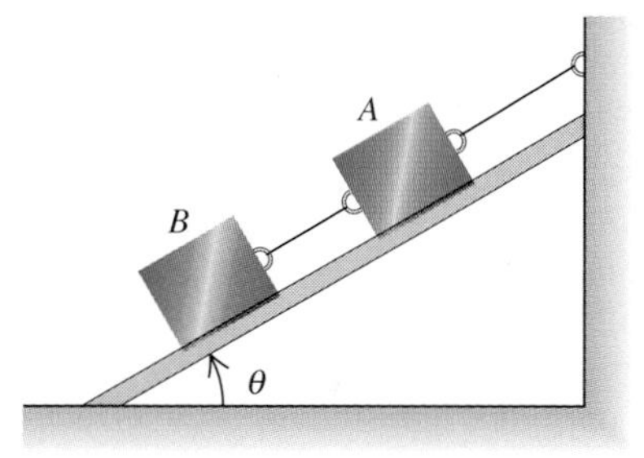


**FIGURE 5–35**

## Section 5–2
## Applications of Newton's Second Law

**5–12** A block of ice released from rest at the top of a 4.00-m long ramp slides to the bottom in 2.20 s. What is the angle between the ramp and the horizontal?

**5–13** A light rope is attached to a block with a mass of 6.00 kg that rests on a horizontal, frictionless surface. The rope passes over a frictionless, massless pulley, and a block of mass $m$ is suspended from the other end. When the blocks are released, the tension in the rope is 12.0 N. What is a) the acceleration of the 6.00-kg block? b) the mass $m$ of the hanging block?

**5–14** A physics student playing with an air-hockey table (just a frictionless surface) finds that if she gives the puck a velocity of 4.60 m/s along the length (1.75 m) of the table at one end, by the time it has reached the other end the puck has drifted 2.80 cm to the right but still has a velocity along the length of 4.60 m/s. She concludes correctly that the table is not level and correctly calculates its inclination from the above information. What is the angle of inclination?

**5–15 Atwood's Machine.** A 15.0-kg load of bricks hangs from one end of a rope that passes over a small, frictionless pulley. A 28.0-kg counterweight is suspended from the other end of the rope (Fig. 5–36). The system is released from rest. a) What is the magnitude of the upward acceleration of the load of bricks? b) What is the tension in the rope while the load is moving?

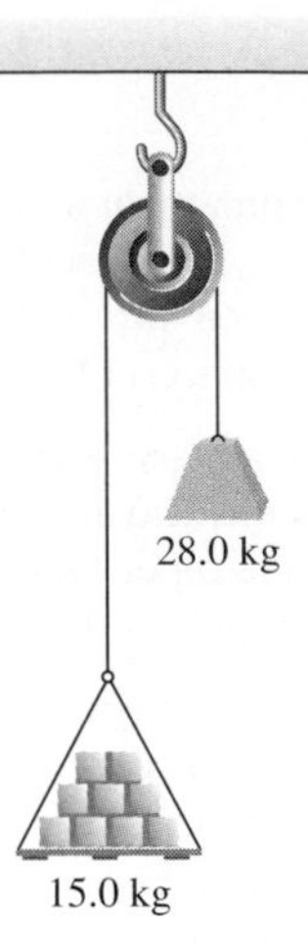


**FIGURE 5–36**

**5–16** Which way will the accelerometer in Fig. 5–12 deflect under the following conditions? a) The car is moving toward the right and traveling faster. b) The car is moving toward the right and traveling slower. c) The car is moving toward the left and traveling faster.

**5–17** Which way will the accelerometer in Fig. 5–12 deflect under the following conditions? a) The car is at rest on a sloping surface. b) The car has an upward velocity on a frictionless inclined plane. It first moves up, then stops, and then moves down. What is the deflection in each stage of the motion?

**5–18** A transport plane takes off from a level landing field with two gliders in tow, one behind the other. The mass of each glider is 400 kg, and the friction force (or drag) on each may be assumed constant and equal to 2000 N. The tension in the towrope between the transport plane and the first glider is not to exceed 10,000 N. a) If a speed of 40 m/s is required for takeoff, what minimum length of runway is needed? b) What is the tension in the towrope between the two gliders while they are accelerating for the takeoff?

## Section 5–3 Contact Forces and Friction

**5–19** A stockroom worker pushes a small crate with mass 8.75 kg on a horizontal surface with a constant speed of 4.50 m/s. The coefficient of kinetic friction between the crate and the surface is 0.20. a) What horizontal force must be applied by the worker to maintain the motion? b) If the force calculated in part (a) is removed, how soon does the crate come to rest?

**5–20** A box of bananas weighing 25.0 N rests on a horizontal surface. The coefficient of static friction between the box and the surface is 0.40, and the coefficient of kinetic friction is 0.20. a) If no horizontal force is applied to the box and the box is at rest, how large is the friction force exerted on the box? b) What is the magnitude of the friction force if a monkey applies a horizontal force of 7.0 N to the box and the box is initially at rest? c) What minimum horizontal force must the monkey apply to start the box in motion? d) What minimum horizontal force must the monkey apply to keep the box moving at constant velocity once it has been started? e) If the monkey applies a horizontal force of 15.0 N, what is the magnitude of the friction force?

**5–21** In a physics lab experiment a 6.00-kg box is pushed across a flat table by a horizontal force $\boldsymbol{F}$. a) If the box is moving at a constant speed of 0.350 m/s and the coefficient of kinetic friction is 0.14, what is the magnitude of $\boldsymbol{F}$? b) What is the magnitude of $\boldsymbol{F}$ if the box has a constant acceleration of 0.220 m/s$^2$? c) How would your answers to parts (a) and (b) change if the experiments were performed on the moon, where $g = 1.62$ m/s$^2$?

**5–22** a) A large rock rests on a rough horizontal surface. A bulldozer pushes on the rock with a horizontal force $T$ that is slowly increased, starting from zero. Draw a graph with $T$ along the $x$-axis and the friction force $\mathcal{F}$ along the $y$-axis, starting at $T = 0$ and showing the region of no motion, the point at which the rock is ready to move, and the region where the rock is moving. b) A block with weight $w$ rests on a rough horizontal plank. The slope angle of the plank $\theta$ is gradually increased until the block starts to slip. Draw two graphs, both with $\theta$ along the $x$-axis. In one graph, show the ratio of the normal force to the weight, $\mathcal{N}/w$, as a function of $\theta$. In the second graph, show the ratio of the friction force to the weight, $\mathcal{F}/w$. Indicate the region of no motion, the point at which the block is ready to move, and the region where the block is moving.

**5–23 Free-Body Diagrams.** The first two steps in the solution of problems involving Newton's second law are to select an object for analysis and then to draw free-body diagrams for that object. Draw a free-body diagram for each of the following situations. a) A mass $M$ sliding down a frictionless inclined plane of angle $\theta$. b) A mass $M$ sliding up a frictionless inclined plane of angle $\theta$. c) A mass $M$ sliding up an inclined plane of angle $\theta$ with kinetic friction present. d) Masses $M$ and $m$ sliding down an inclined plane of angle $\theta$ with friction present (Fig. 5–37a). Draw free-body diagrams for both $m$ and $M$. Identify the forces that are action-reaction pairs. e) Draw free-body diagrams for masses $m$ and $M$ in Fig. 5–37b. Identify all action-reaction pairs. There is a frictional force between all surfaces in contact. The pulley is frictionless and massless. In all cases, be sure you have the correct direction of the forces and are completely clear as to what object is causing each force in your free-body diagram.

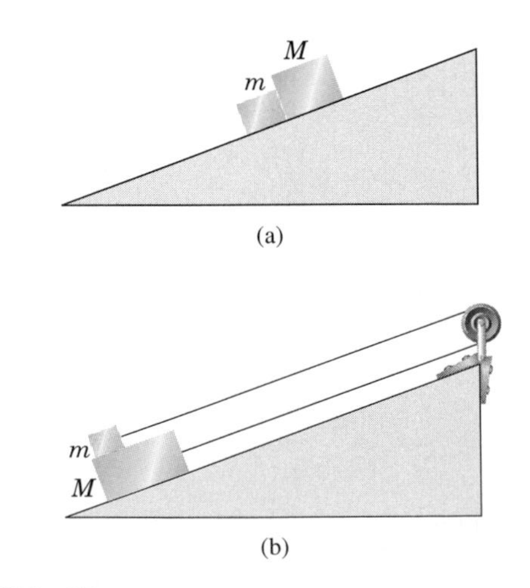

**FIGURE 5–37**

**5–24** a) If the coefficient of kinetic friction between tires and dry pavement is 0.80, what is the shortest distance in which an automobile can be stopped by locking the brakes when traveling at 20.1 m/s (about 45 mi/h)? b) On wet pavement the coefficient of kinetic friction may be only 0.25. How fast should you drive on wet pavement in order to be able to stop in the same distance as in part (a)? (*Note:* Locking the brakes is *not* the safest way to stop.)

**5–25** A safe with mass 215 kg is lowered at constant speed down skids 4.00 m long from a truck 2.00 m high. a) If the coefficient of kinetic friction between safe and skids is 0.30, does the safe need to be pulled down or held back? b) How great a force parallel to the skids is needed?

**5–26 Effect of Air Pressure on Rolling Friction of Bicycle Tires.** Two bicycle tires are set rolling with the same initial speed of 5.40 m/s along a long, straight road, and the distance that each travels before its speed is reduced by half is measured. One tire is inflated to a pressure of 40 psi and goes 45.6 m; the other is at 105 psi and goes 213 m. What is the coefficient of rolling friction $\mu_r$ for each? Assume that the net horizontal force is due to rolling friction only.

**5–27** In emergency situations with major blood loss, the doctor will order the patient placed in the Trendelberg position, in which the foot of the bed is raised to get maximum blood flow to the brain. If the coefficient of static friction between the typical patient and the bed sheets is 0.80, what is the maximum angle the bed can be tilted with respect to the floor before the patient begins to slide?

**5–28** Consider the system shown in Fig. 5–38. The coefficient of kinetic friction between block $A$ (weight $w_A$) and the tabletop is $\mu_k$. Calculate the weight $w_B$ of the hanging block required if this block is to descend at constant speed once it has been set into motion.

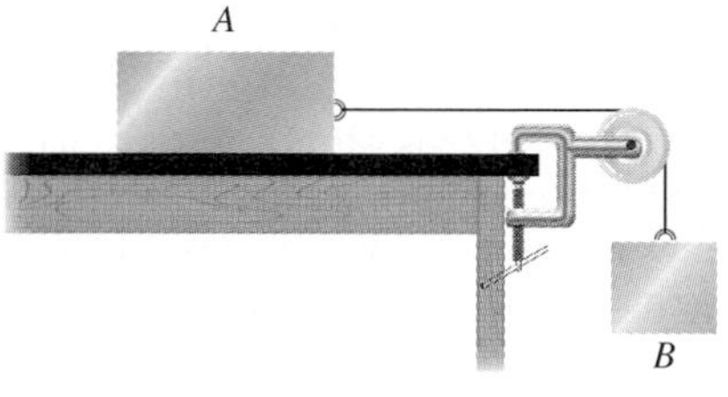


**FIGURE 5–38**

**5–29** Two crates connected by a rope lie on a horizontal surface (Fig. 5–39). Crate $A$ has mass $m_A$, and crate $B$ has mass $m_B$. The coefficient of kinetic friction between the crates and the surface is $\mu_k$. The crates are pulled to the right at constant velocity by a horizontal force $\boldsymbol{F}$. In terms of $m_A$, $m_B$, and $\mu_k$, calculate a) the magnitude of the force $\boldsymbol{F}$; b) the tension in the rope connecting the blocks.

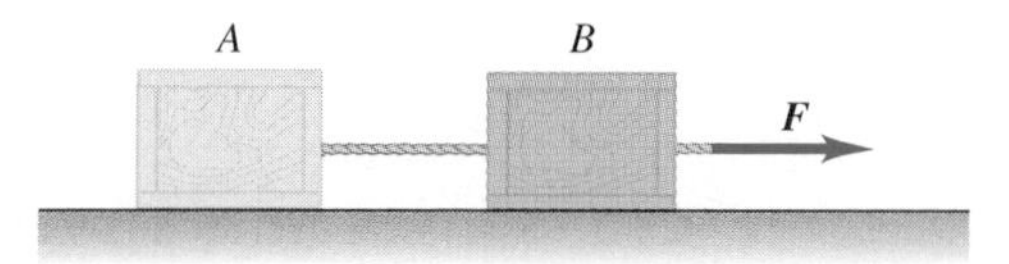


**FIGURE 5–39**

**5–30** A child pushes on a box resting on the floor. The box weighs 260 N, and the child is pushing down on the box with a force that is directed at 40.0° below the horizontal. If the smallest force the child can apply that gets the box moving is 150 N, what is the coefficient of static friction between the box and the floor?

**5–31** A block with mass 5.00 kg resting on a horizontal surface is connected by a horizontal cord passing over a light, frictionless pulley to a hanging block with mass 4.00 kg. The coefficient of kinetic friction between the block and the horizontal surface is 0.50. After the blocks are released, find a) the acceleration of each block; b) the tension in the cord.

**5–32** A packing crate rests on a loading ramp that makes an angle $\theta$ with the horizontal. The coefficient of kinetic friction is 0.30, and the coefficient of static friction is 0.40. a) As the angle $\theta$ is increased, find the minimum angle at which the crate starts to slip. b) At this angle, find the acceleration once the crate has begun to move. c) How much time is required for the crate to slip 8.0 m along the inclined plane?

**5–33** A large crate with mass $m$ rests on a horizontal floor. The coefficients of friction between the crate and the floor are $\mu_s$ and $\mu_k$. A woman pushes downward at an angle $\theta$ below the horizontal on the crate with a force $\boldsymbol{F}$. a) What magnitude of force $\boldsymbol{F}$ is required to keep the crate moving at constant velocity? b) If $\mu_s$ is larger than some critical value, the woman cannot start the crate moving no matter how hard she pushes. Calculate this critical value of $\mu_s$.

**5–34** A box with mass $m$ is dragged across a rough, level floor having a coefficient of kinetic friction $\mu_k$ by a rope that is pulled upward at an angle $\theta$ above the horizontal with a force of magnitude $F$. In terms of $m$, $\mu_k$, $\theta$, $F$, and $g$, obtain expressions for a) the normal force; b) the frictional force; c) the acceleration of the box; d) the magnitude of force required to move the box with constant speed.

**5–35** Two blocks, $A$ and $B$, are placed as in Fig. 5–40 and connected by ropes to block $C$. Blocks $A$ and $B$ weigh 20.0 N each, and the coefficient of kinetic friction between each block and the surface is 0.40. Block $C$ descends with constant velocity. a) Draw two separate free-body diagrams showing the forces acting on $A$ and on $B$. b) Find the tension in the rope connecting blocks $A$ and $B$. c) What is the weight of block $C$?

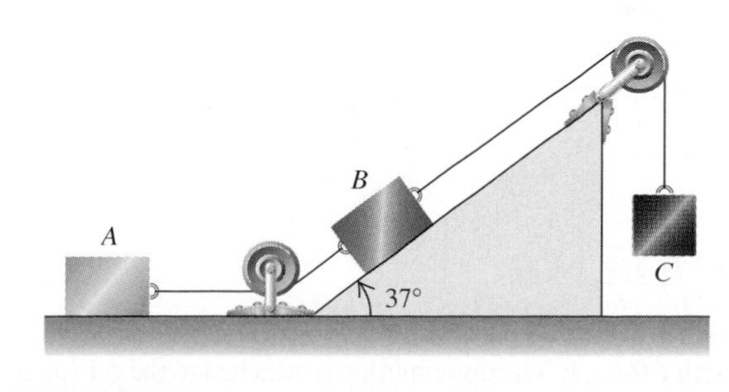


**FIGURE 5–40**

## Section 5–4 Dynamics of Circular Motion

**5–36** A flat (unbanked) curve on a highway has a radius of 240 m. A car rounds the curve at a speed of 26.0 m/s. What is the minimum coefficient of friction that will prevent sliding?

**5–37** A stone with a mass of 1.50 kg is attached to one end of a string 0.80 m long with a breaking strength of 500 N. The stone is whirled in a horizontal circle on a frictionless tabletop; the other end of the string is kept fixed. Find the maximum speed the stone can attain without breaking the string.

**5–38** A small button placed on a record with diameter 0.305 m will revolve with the record when it is brought up to a speed of $33\frac{1}{3}$ rev/min, provided that the button is no more than 0.120 m from the axis. a) What is the coefficient of static friction between the button and the record? b) How far from the axis can the button be placed, without slipping, if the turntable rotates at 45.0 rev/min?

**5–39** A highway curve with radius 350 m is to be banked so that a car traveling 24.6 m/s (55 mi/h) will not skid sideways even in the absence of friction. At what angle should the curve be banked?

**5–40** The "Giant Swing" at a county fair consists of a vertical central shaft with a number of horizontal arms attached at its upper end (Fig. 5–41). Each arm supports a seat suspended from a cable 5.00 m long, the upper end of the cable being fastened to the arm at a point 3.00 m from the central shaft. a) Find the time of one revolution of the swing if the cable supporting a seat makes an angle of 30.0° with the vertical. b) Does the angle depend on the weight of the passenger for a given rate of revolution?

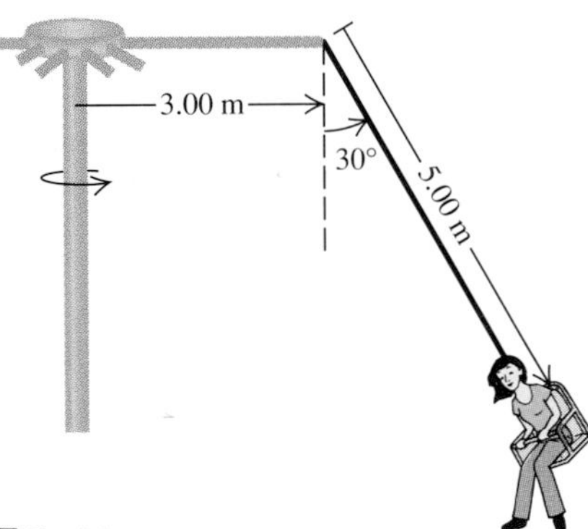


FIGURE 5–41

**5–41 Rotating Space Stations.** One of the problems for humans living in outer space is that they are weightless. One way around this problem is to design space stations as shown in Fig. 5–42 and have them spin about their center at a constant rate. This creates artificial gravity. If the diameter of the space station is 1200 m, how many revolutions per minute are needed for the artificial-gravity acceleration to be 9.8 m/s$^2$?

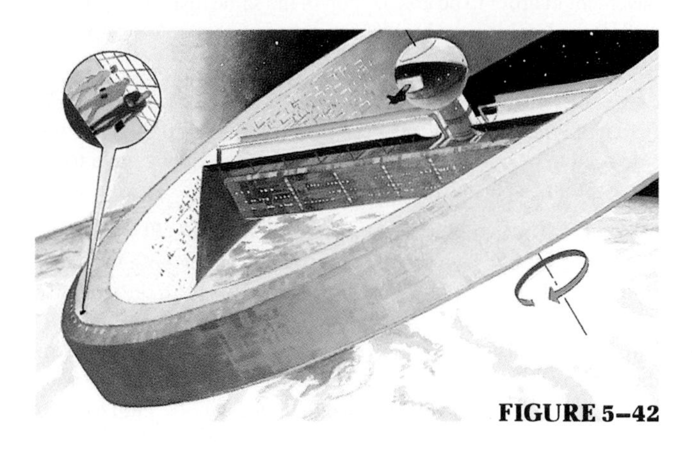

FIGURE 5–42

## Section 5–5 Motion in a Vertical Circle

**5–42** A bowling ball weighing 71.2 N (16 lb) is attached to the ceiling by a 4.20-m rope. The ball is pulled to one side and released; it then swings back and forth as a pendulum. As the rope swings through the vertical, the speed of the bowling ball is 3.60 m/s. a) What is the acceleration of the bowling ball, in magnitude and direction, at this instant? b) What is the tension in the rope at this instant?

**5–43** A cord is tied to a pail of water, and the pail is swung in a vertical circle of radius 0.800 m. What must be the minimum speed of the pail at the highest point of the circle if no water is to spill from it?

**5–44** The radius of a Ferris wheel is 9.00 m, and it makes one revolution in 12.0 s. a) Find the apparent weight of an 80.0-kg passenger at the highest and lowest points. b) What would the time for one revolution be if the passenger's apparent weight at the highest point were zero? c) What would then be the passenger's apparent weight at the lowest point?

**5–45** A stunt pilot who has been diving vertically at a speed of 75.0 m/s pulls out of the dive by changing his course to a circle in a vertical plane. a) What is the minimum radius of the circle for the acceleration at the lowest point not to exceed $7g$? (The plane's speed at this point is 75.0 m/s.) b) What is the apparent weight of an 80.0-kg pilot at the lowest point of the pullout?

# PROBLEMS

**5–46** In Fig. 5–43 a man lifts a weight $w$ by pulling down on a rope with a force $\boldsymbol{F}$. The upper pulley is attached to the ceiling by a chain, and the lower pulley is attached to the weight by another chain. In terms of $w$, find the tension in each chain and the force $\boldsymbol{F}$ if the weight is lifted at constant speed. Assume that the weights of the rope, pulleys, and chains are negligible.

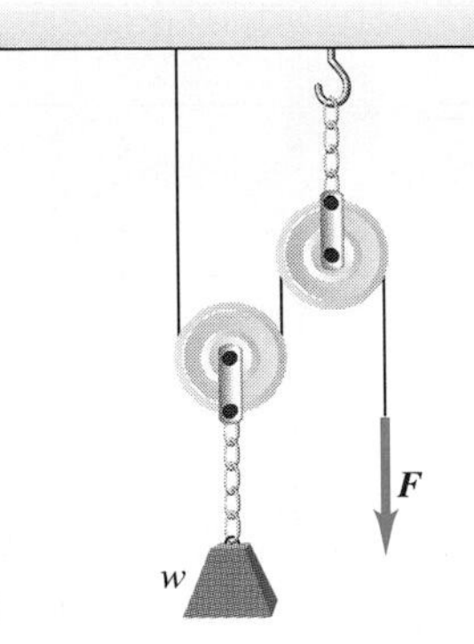


**FIGURE 5–43**

**5–47** A refrigerator with mass $m$ is pushed up a ramp at constant speed by a man applying a force $\boldsymbol{F}$. The ramp is at an angle $\theta$ above the horizontal. Neglect friction for the refrigerator. If the force $\boldsymbol{F}$ is *horizontal,* calculate its magnitude in terms of $m$ and $\theta$.

**5–48 A Rope with Mass.** Most problems in this book do not give the mass of ropes, cords, or cables, and therefore we do not consider their mass or weight. In some problems the ropes are said to be massless. It is understood that the ropes, cords, and cables, compared to the other objects in the problem, have so little mass that their mass can be neglected. But if the rope is the *only* object in the problem, then its mass clearly cannot be neglected. For example, suppose we have a clothesline attached to two poles (Fig. 5–44). The clothesline has a mass $M$, and each end makes an angle $\theta$ with the horizontal. What is a) the tension at the ends of the clothesline? b) the tension at the lowest point? The curve of the clothesline, or of any flexible cable hanging under its own weight, is called a catenary. For a more advanced treatment of this curve, see J. L. Synge and B. A. Griffith, *Principles of Mechanics* (New York: McGraw-Hill, 1959), pp. 94–97.

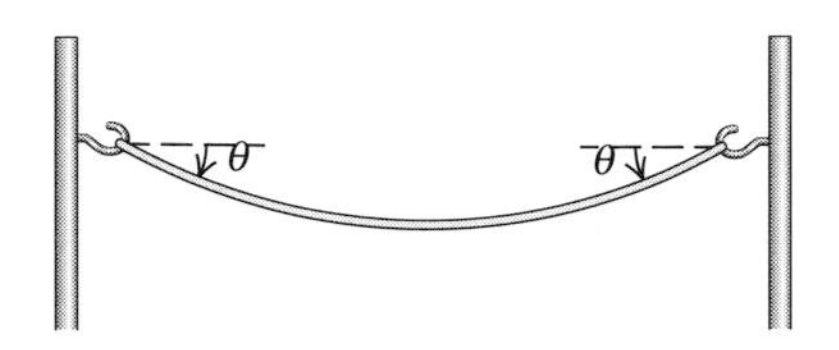


**FIGURE 5–44**

**5–49** If the coefficient of static friction between a table and a rope is $\mu_s$, what fraction of the rope can hang over the edge of a table without the rope sliding?

**5–50** A block with mass 10.0 kg is placed on an inclined plane with slope angle 30° and is connected to a second hanging block that has mass $m$ by a cord passing over a small, frictionless pulley (Fig. 5–45). The coefficient of static friction is 0.45, and the coefficient of kinetic friction is 0.35. a) Find the mass $m$ for which the 10.0-kg block moves up the plane at constant speed once it has been set in motion. b) Find the mass $m$ for which it moves down the plane at constant speed once it has been set in motion. c) For what range of values of $m$ will the block remain at rest if it is released from rest?

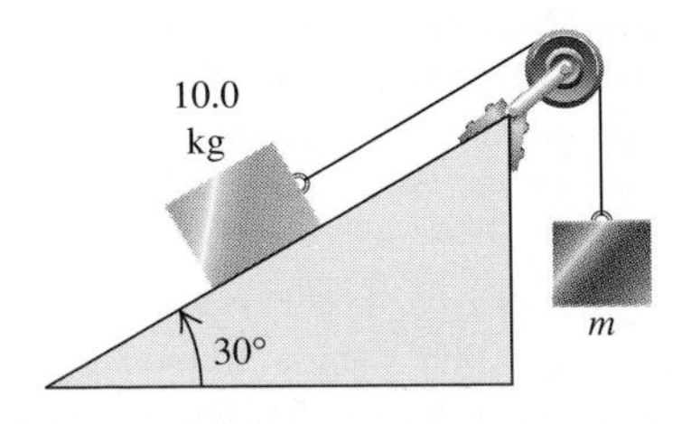

**FIGURE 5–45**

**5–51** a) Block $A$ in Fig. 5–46 weighs 80.0 N. The coefficient of static friction between the block and the surface on which it rests is 0.30. The weight $w$ is 20.0 N, and the system is in equilibrium. Find the friction force exerted on block $A$. b) Find the maximum weight $w$ for which the system will remain in equilibrium.

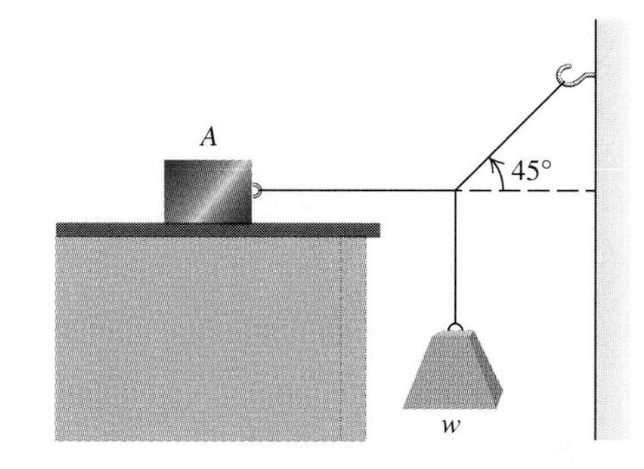


**FIGURE 5–46**

**5–52** Block $A$ in Fig. 5–47 weighs 3.60 N, and block $B$ weighs 5.40 N. The coefficient of kinetic friction between all surfaces is 0.25. Find the magnitude of the horizontal force $\boldsymbol{F}$ necessary to drag block $B$ to the left at constant speed a) if $A$ rests on $B$ and moves with it (Fig. 5–47a); b) if $A$ is held at rest (Fig. 5–47b).

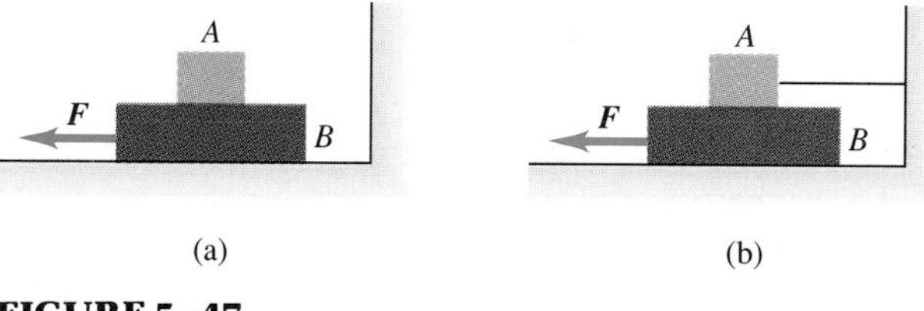

**FIGURE 5–47**

**5–53** A window washer pushes his scrub brush up a vertical window at constant speed by applying a force $\boldsymbol{F}$ as shown in Fig. 5–48. The brush weighs 8.00 N, and the coefficient of kinetic

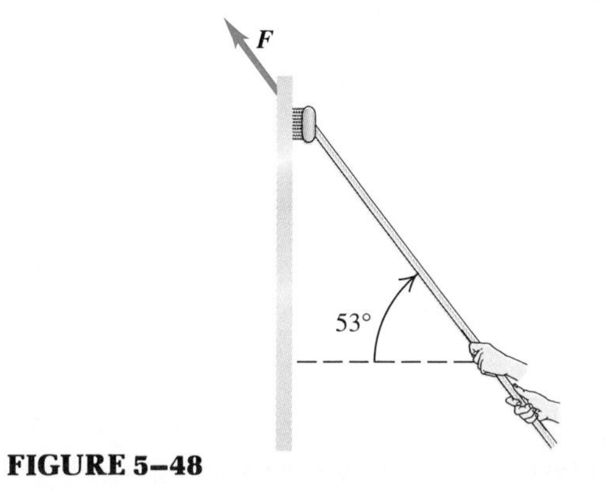


**FIGURE 5–48**

friction is $\mu_k = 0.40$. Calculate a) the magnitude of the force $\boldsymbol{F}$; b) the normal force exerted by the window on the brush.

**5–54** A woman attempts to push a box of books that has mass $m$ up a ramp inclined at an angle $\theta$ above the horizontal. The coefficients of friction between the ramp and the box are $\mu_s$ and $\mu_k$. The force $\boldsymbol{F}$ applied by the woman is *horizontal*. a) If $\mu_s$ is greater than some critical value, the woman cannot start the box moving up the ramp no matter how hard she pushes. Calculate this critical value of $\mu_s$. b) Assume that $\mu_s$ is less than this critical value. What magnitude of force must the woman apply to keep the box moving up the ramp at constant speed?

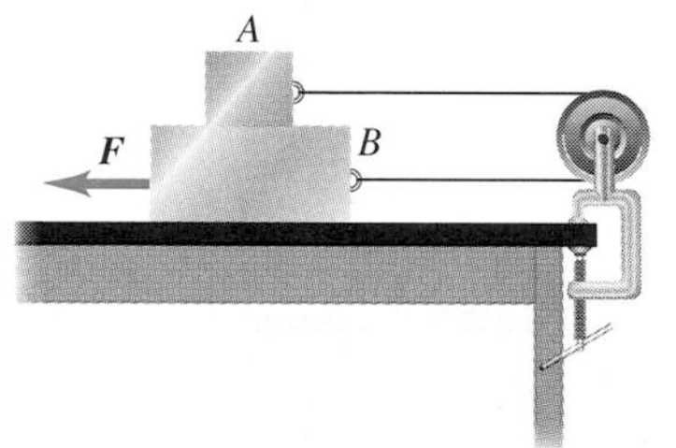


FIGURE 5–49

**5–55** Block $A$ in Fig. 5–49 weighs 3.60 N, and block $B$ weighs 5.40 N. The coefficient of kinetic friction between all surfaces is 0.25. Find the magnitude of the horizontal force $\boldsymbol{F}$ necessary to drag block $B$ to the left at constant speed if $A$ and $B$ are connected by a light, flexible cord passing around a fixed, frictionless pulley.

**5–56** A 20.0-kg box rests on the flat floor of a truck. The coefficients of friction between box and floor are $\mu_s = 0.15$ and $\mu_k = 0.10$. The truck stops at a stop sign and then starts to move with an acceleration of 2.00 m/s$^2$. If the box is 2.80 m from the rear of the truck when the truck starts, how much time elapses before the box falls off the rear of the truck? How far does the truck travel in this time?

**5–57** A 30.0-kg packing case is initially at rest on the floor of a truck. The coefficient of static friction between the case and the floor is 0.30, and the coefficient of kinetic friction is 0.20. Before each acceleration given below, the truck is traveling due east at constant speed. Find the magnitude and direction of the friction force acting on the case a) when the truck accelerates at 2.20 m/s$^2$ eastward; b) when it accelerates at 3.50 m/s$^2$ westward.

**5–58** A 150-kg crate is released from an airplane traveling due east at an altitude of 8500 m with a ground speed of 120 m/s. The wind applies a constant force on the crate of 250 N directed horizontally in the opposite direction to the plane's flight path. Where and when (with respect to the release location and time) does the crate hit the ground?

**5–59** Block $A$ in Fig. 5–50 has a mass of 4.00 kg, and block $B$ has a mass of 25.0 kg. The coefficient of kinetic friction between $B$ and the horizontal surface is 0.20. a) What is the mass of block $C$ if block $B$ is moving to the right with an acceleration of 2.00 m/s$^2$? b) What is the tension in each cord when $B$ has this acceleration?

**5–60** Two blocks connected by a cord passing over a small, frictionless pulley rest on frictionless planes (Fig. 5–51). a) Which way will the system move when the blocks are released from rest? b) What is the acceleration of the blocks? c) What is the tension in the cord?

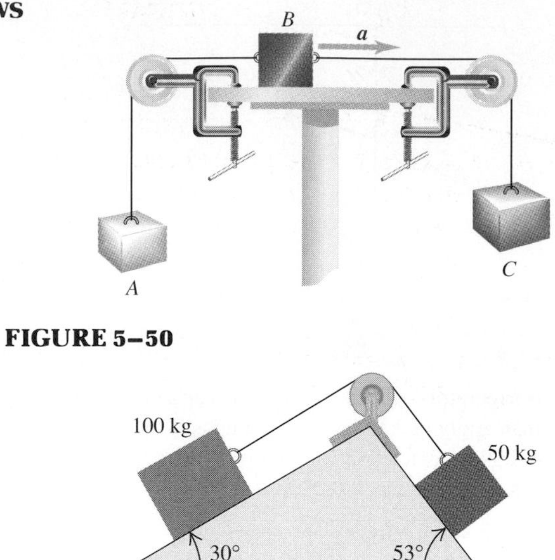


FIGURE 5–50



FIGURE 5–51

**5–61** In terms of $m_1$, $m_2$, and $g$, find the accelerations of each block in Fig. 5–52. There is no friction anywhere in the system.

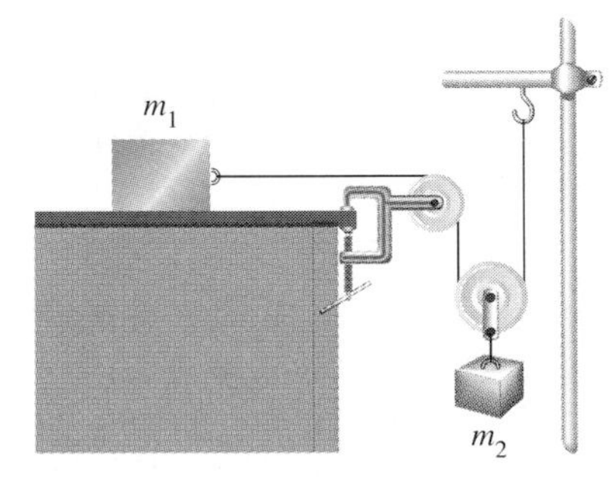


FIGURE 5–52

**5–62 Friction in an Elevator.** You are riding in an elevator on the way to your apartment on the eighteenth floor of your dormitory. The elevator is accelerating upward with $a = 5.40$ m/s$^2$. Beside you is the box containing your new TV set; box and contents have a total mass of 36.0 kg. While the elevator is accelerating upward, you push horizontally on the box to slide it at constant speed toward the elevator door. If the coefficient of kinetic friction between the box and elevator floor is $\mu_k = 0.34$, what magnitude of force must you apply?

**5–63 Atwood's Machine.** Two objects with masses of 5.00 kg and 3.00 kg, respectively, hang 0.500 m above the floor from the ends of a cord 4.00 m long passing over a frictionless pulley. Both objects start from rest. Find the maximum height reached by the 3.00-kg object.

**5–64** Two blocks with masses of 4.00 kg and 8.00 kg are connected by a string and slide down a 30.0° inclined plane (Fig. 5–53). The coefficient of kinetic friction between the 4.00-kg block and the plane is 0.20; that between the 8.00-kg block and the

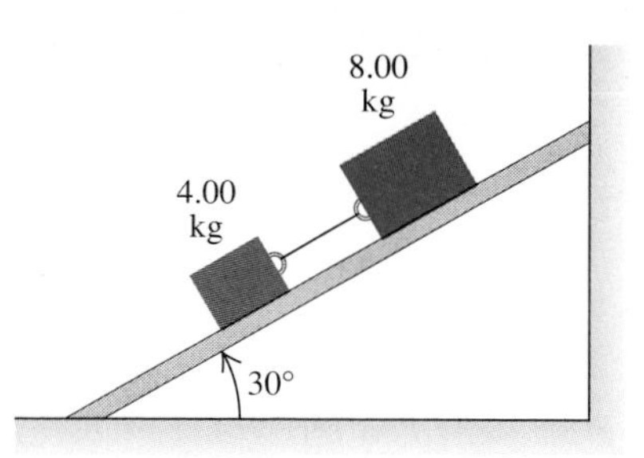

**FIGURE 5–53**

plane is 0.30. a) Calculate the acceleration of each block. b) Calculate the tension in the string.

**5–65** What acceleration must the cart in Fig. 5–54 have in order that block $A$ does not fall? The coefficient of static friction between the block and the cart is $\mu_s$. How would the behavior of the block be described by an observer on the cart?

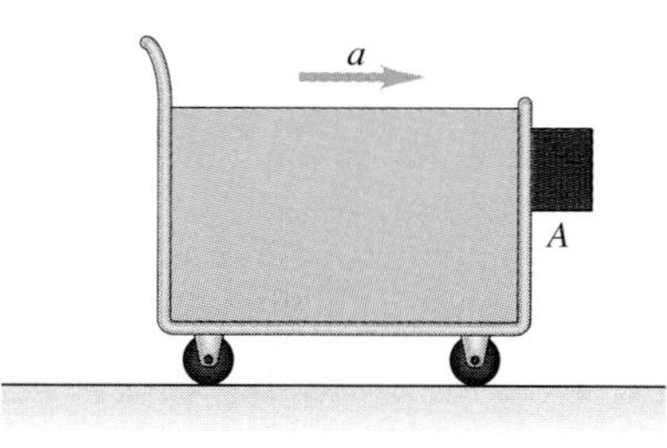


**FIGURE 5–54**

**5–66** A man with mass 85.0 kg stands on a platform with mass 25.0 kg. He pulls on the free end of a rope that runs over a pulley on the ceiling and has its other end fastened to the platform. With what force does he have to pull in order to give himself and the platform an upward acceleration of 1.20 m/s$^2$?

**5–67 The Monkey and Bananas Problem.** A 20-kg monkey has a firm hold on a light rope that passes over a frictionless pulley and is attached to a 20-kg bunch of bananas (Fig. 5–55). The monkey looks upward, sees the bananas, and starts to climb the rope to get them. a) As the monkey climbs, do the bananas move up, move down, or remain at rest? b) As the monkey climbs, does the distance between the monkey and the bananas decrease, increase, or remain constant? c) The monkey releases her hold on the rope. What about the distance between the monkey and the bananas while she is falling? d) Before reaching the ground, the monkey grabs the rope to stop her fall. What do the bananas do?

**5–68** You are riding in a school bus. As the bus rounds a flat curve, a lunch box with a mass of 0.500 kg suspended from the ceiling of the bus by a string 1.80 m long is found to be in equilibrium when the string makes an angle of 37.0° with the vertical. In this position the lunch box is 50.4 m from the center of curvature of the curve. What is the speed $v$ of the bus?

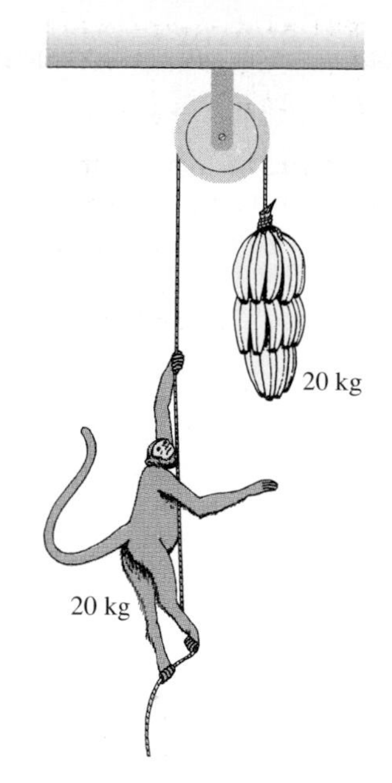


**FIGURE 5–55**

**5–69** Consider a roadway banked as in Example 5–24 (Section 5–4), where there is a coefficient of static friction of 0.35 and a coefficient of kinetic friction of 0.25 between the tires and the roadway. The radius of the curve is $R = 36$ m. a) If the banking angle is $\theta = 25°$, what is the *maximum* speed the automobile can have before sliding *up* the banking? b) What is the *minimum* speed the automobile can have before sliding *down* the banking?

**5–70** A curve with a 175-m radius on a level road is banked at the correct angle for a speed of 15 m/s. If an automobile rounds this curve at 30 m/s, what is the minimum coefficient of friction between tires and road needed so that the automobile will not skid?

**5–71** In the ride "Spindletop" at the amusement park Six Flags Over Texas, people stand against the inner wall of a hollow vertical cylinder with radius 3.0 m. The cylinder starts to rotate, and when it reaches a constant rotation rate of 0.60 rev/s, the floor on which people are standing drops about 0.5 m. The people remain pinned against the wall. a) Draw a force diagram for a person in this ride after the floor has dropped. b) What minimum coefficient of static friction is required if the person in the ride is not to slide downward to the new position of the floor? c) Does your answer in part (b) depend on the mass of the passenger? (*Note:* When the ride is over, the cylinder is slowly brought to rest. As it slows down, people slide down the walls to the floor.)

**5–72** A physics major is working to pay his college tuition by performing in a traveling carnival. He rides a motorcycle inside a hollow steel sphere. The surface of the sphere is full of holes so the audience can see in. After gaining sufficient speed, he travels in a vertical circle of radius 20.0 m. The physics major has a mass of 60.0 kg, and his motorcycle has a mass of 40.0 kg. a) What minimum speed must he have at the top of the circle if the tires of the motorcycle are not to lose contact with the steel sphere? b) At the bottom of the circle his speed is twice the value calculated in part (a). What is the magnitude of the normal force exerted on the motorcycle by the steel sphere at this point?

**5–73** You are driving with a friend who is sitting to your right on the passenger side of the front seat. You would like to be closer to your friend and decide to use physics to achieve your romantic goal by making a quick turn. a) Which way (to the left or to the right) should you turn the car to get your friend to slide closer to you? b) If the coefficient of static friction between your friend and the car seat is 0.40 and you keep driving at a constant speed of 18 m/s, what is the maximum radius you could make your turn and still have your friend slide your way?

**5–74** The 4.00-kg block in Fig. 5–56 is attached to a vertical rod by means of two strings. When the system rotates about the axis of the rod, the strings are extended as shown in the diagram, and the tension in the upper string is 85.0 N. a) What is the tension in the lower cord? b) How many revolutions per minute does the system make?

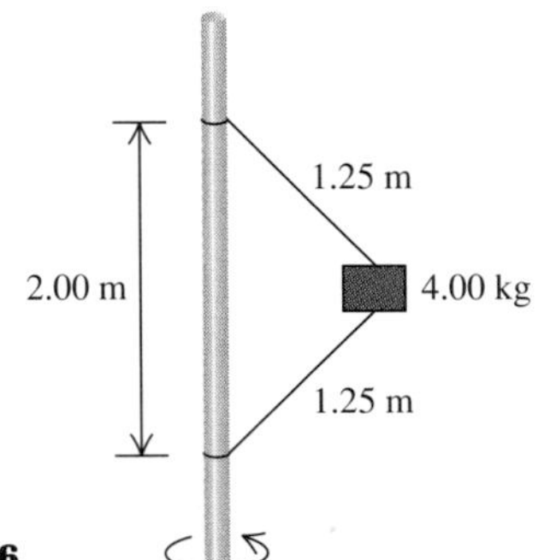


**FIGURE 5–56**

**5–75** Identical twins, Kathy and Karen, are playing one December on a large merry-go-round (a disk mounted parallel to the ground on a vertical axle through its center) in their school playground in northern Minnesota. The merry-go-round surface is coated with ice and therefore is frictionless. The merry-go-round is turning at a constant rate of revolution as the twins ride on it. The mass of each twin is 47.6 kg. Kathy, sitting 2.00 m from the center of the merry-go-round, must hold onto one of the metal posts attached to the merry-go-round with a horizontal force of 80.0 N to keep from sliding off. Karen is sitting at the edge, 4.00 m from the center. With what horizontal force must she hold on to keep from falling off?

**5–76** Aircraft experience a lift force, due to the air, that is perpendicular to the plane of the wings. A small airplane is flying at a constant speed of 350 km/h. At what angle from the horizontal must the wings of the airplane be tilted for the plane to execute a horizontal turn from east to north with a turning radius of 1200 m?

**5–77** A small bead can slide without friction on a circular hoop that is in a vertical plane and has a radius of 0.100 m. The hoop rotates at a constant rate of 4.00 rev/s about a vertical diameter (Fig. 5–57). a) Find the angle $\theta$ at which the bead is in vertical equilibrium. (Of course, it has a radial acceleration toward the axis.) b) Is it possible for the bead to "ride" at the same elevation as the center of the hoop? c) What will happen if the hoop rotates at 1.00 rev/s?

**5–78** A small block with mass $m = 0.100$ kg rests on a frictionless horizontal tabletop a distance $r = 0.900$ m from a hole in the center of the table (Fig. 5–58). A string tied to the small block passes down through the hole, and a larger block with mass $M = 1.60$ kg is suspended from the free end of the string. The small block is set into uniform circular motion with radius 0.900 m and speed $v$. What must $v$ be if the large block is to remain motionless when released?

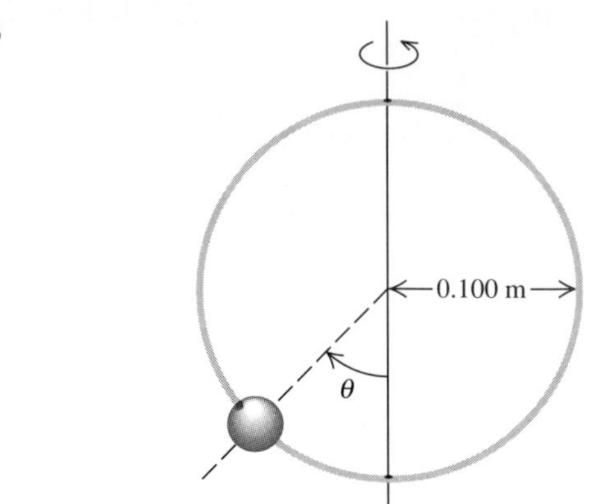


**FIGURE 5–57**

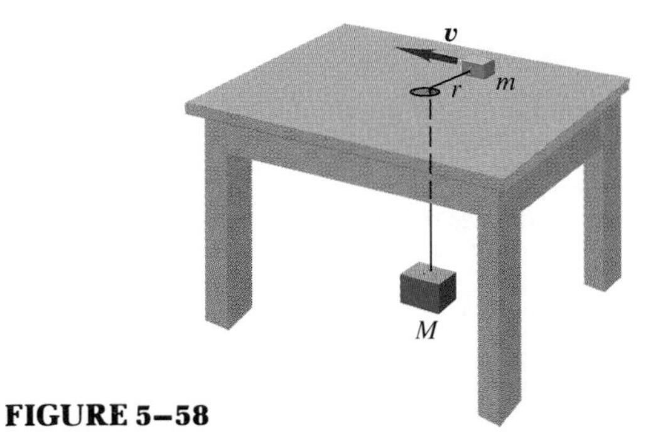


**FIGURE 5–58**

**5–79** A small block with mass $m$ is placed inside an inverted cone that is rotating about a vertical axis such that the time for one revolution of the cone is $T$ (Fig. 5–59). The walls of the cone make an angle $\theta$ with the vertical. The coefficient of static friction between the block and the cone is $\mu_s$. If the block is to remain at a constant height $h$ above the apex of the cone, what are the maximum and minimum values of $T$?

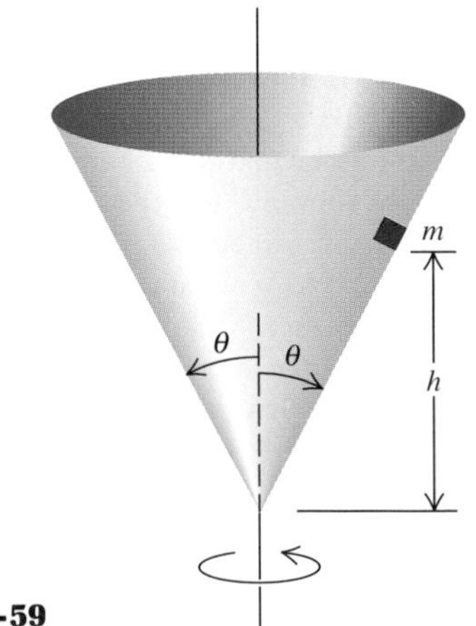


**FIGURE 5–59**

**5–80** A small remote-control car with a mass of 1.50 kg moves at a constant speed of $v = 12.0$ m/s in a vertical circle inside a hollow metal cylinder that has a radius of 5.00 m (Fig. 5–60). What is the magnitude of the normal force exerted on the car by the walls of the cylinder at a) point $A$ (at the bottom of the vertical circle)? b) point $B$ (at the top of the vertical circle)?

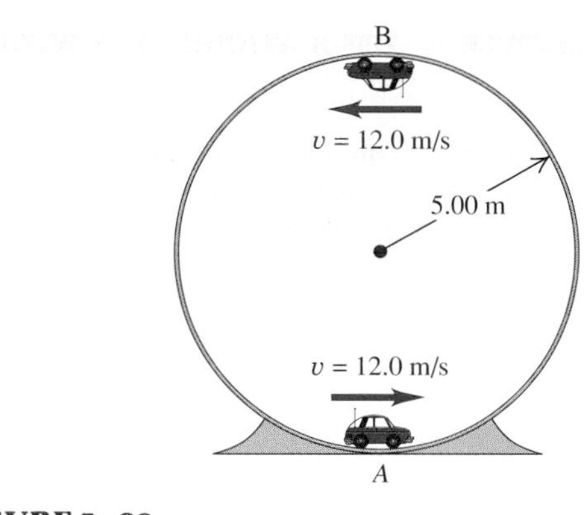

**FIGURE 5–60**

## CHALLENGE PROBLEMS

**5–81** A box with weight $w$ is pulled at constant speed along a level floor by a force $\boldsymbol{F}$ that is at an angle $\phi$ above the horizontal. The coefficient of kinetic friction between the floor and box is $\mu_k$. a) In terms of $\phi$, $\mu_k$, and $w$, calculate $F$. b) For $w = 500$ N and $\mu_k = 0.30$, calculate $F$ for $\phi$ ranging from 0° to 90° in increments of 10°. Sketch a graph of $F$ versus $\phi$. c) From the general expression in part (a), calculate the value of $\phi$ for which the $F$ required to maintain constant speed is a minimum. For the special case of $w = 500$ N and $\mu_k = 0.30$, evaluate this optimal $\phi$ and compare your result to the graph you constructed in part (b). [*Note:* At the value of $\phi$ where the function $F(\phi)$ has a minimum, $dF(\phi)/d\phi = 0$.]

**5–82** Block $A$, with weight $2w$, slides down an inclined plane $S$ of slope angle 37.0° at a constant speed while the plank $B$, with weight $w$, rests on top of $A$. The plank is attached by a cord to the top of the plane (Fig. 5–61). a) Draw a diagram of all the forces acting on block $A$. b) If the coefficient of friction is the same between $A$ and $B$ and between $S$ and $A$, determine its value.

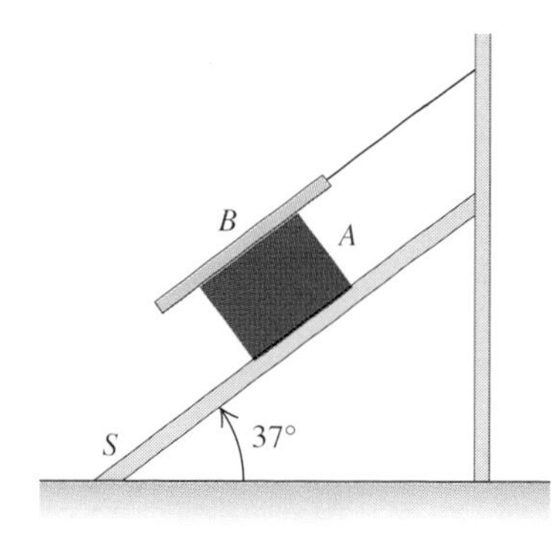


**FIGURE 5–61**

**✲5–83 Variable Coefficient of Kinetic Friction.** The assumption that the coefficient of kinetic friction is independent of speed when an object is sliding is true only over a suitably small range of speeds. For example, in some cases for a speed of 130 km/h (about 80 mi/h) $\mu_k$ may be only half of what it is for speeds of 8 to 16 km/h. For a car sliding with its brakes locked, take $\mu_k = 0.80$ at 8 km/h and $\mu_k = 0.40$ at 130 km/h, and assume that $\mu_k$ varies linearly with speed. Then how far would a car slide after the brakes are applied at 130 km/h before coming to rest? [For a still more realistic characterization of the friction forces involved in sliding, see J. D. Edmonds, Jr., *Amer. Jour. of Phys.*, Vol. 48 (1980), pp. 253–254.]

**5–84** a) A wedge with mass $M$ rests on a frictionless horizontal tabletop. A block with mass $m$ is placed on the wedge (Fig. 5–62a). There is no friction between the block and the wedge. The system is released from rest. Calculate i) the acceleration of the wedge; ii) the horizontal and vertical components of the acceleration of the block. Do your answers reduce to the correct results when $M$ is very large? b) The wedge and block are as in part (a). Now a horizontal force $\boldsymbol{F}$ is applied to the wedge (Fig. 5–62b). What magnitude must $\boldsymbol{F}$ have if the block is to remain at constant height above the tabletop?

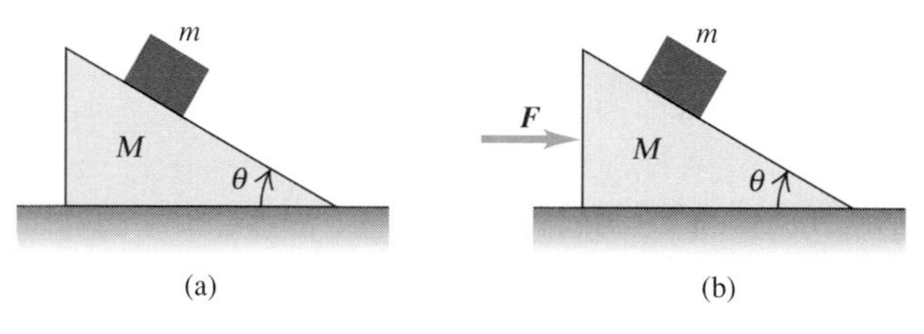


**FIGURE 5–62**

**✲5–85** A rock with mass $m = 3.00$ kg falls from rest in a viscous medium. The rock is acted on by a net constant downward force of 20.0 N (a combination of gravity and the buoyant force exerted by the medium) and by a viscous retarding force $R = \alpha v$, where $v$ is the speed in m/s and $\alpha = 4.00\ \text{N}\cdot\text{s/m}$. (See Example 5–19 in Section 5–3.) a) Find the initial acceleration, $a_0$. b) Find the acceleration when the speed is 3.00 m/s. c) Find the speed when the acceleration equals $0.1a_0$. d) Find the terminal speed, $v_t$. e) Find the coordinate, speed, and acceleration 2.00 s after the start of the motion. f) Find the time required to reach a speed of $0.9v_t$. g) Construct a graph of $v$ versus $t$ for the first 3.00 s of the motion.

**✲5–86** A marble falls from rest through a medium that exerts a resisting force that varies directly with the square of the speed

## KEY CONCEPTS

**• When a force acts on a body that moves, the force does work on the body. Work is a scalar quantity, computed from the force and the displacement.**

**• Kinetic energy is a scalar quantity associated with the motion of a body. It is defined as one half of the product of the body's mass and the square of its speed.**

**• In any displacement of a body the change in its kinetic energy equals the total work done by all the forces acting on the body.**

**• Power is time rate of doing work, that is, work per unit time.**

## INTRODUCTION

Some problems aren't as simple as they look. Suppose you try to find the speed of a toy glider that has been launched by a compressed spring that applies a varying force. You apply Newton's laws and all the problem-solving techniques that we've learned, but you find that the acceleration isn't constant and the simple methods we've learned can't be implemented directly. Never fear; we aren't by any means finished with the subject of mechanics. There are other methods for dealing with such problems. The method we'll consider next involves the concept of energy and one of the great conservation laws of physics.

Energy — its supply, distribution, conversion, and use — has vast economic, political, social, and even moral consequences in our world. Energy is one of the most important unifying concepts in all of physical science. Its importance stems from the principle of conservation of energy. Energy can be converted from one form to another, but it cannot be created or destroyed. In this chapter and the next we concentrate on mechanical energy, which is associated with the motion and position of mechanical systems. We begin with the concepts of work and kinetic energy (energy of motion) and the relationship between them. We also consider power, which is the time rate of transfer of energy into or out of a system.

## 6–1 CONSERVATION OF ENERGY

The concept of **energy** appears throughout every area of physics, yet it's not easy to define in a general way just what energy is. This concept is the cornerstone of a fundamental law of nature called **conservation of energy,** which states that the total energy in any isolated system is constant, no matter what happens within the system. Conservation laws, including conservation of mass, energy, momentum, angular momentum, electric charge, and others, play vital roles in every area of physics and provide unifying threads that run through the whole fabric of the subject. Every conservation law says that the total amount of some physical quantity in every isolated (closed) system is constant.

A familiar example is conservation of mass in chemical reactions. The total mass of the material participating in a chemical reaction is always the same (within the limits of experimental precision) before and after the reaction. This generalization is called the principle of *conservation of mass,* and it is obeyed in all chemical reactions.

Something analogous happens in collisions between bodies. For a particle with mass $m$ moving with speed $v$, we define a quantity $\frac{1}{2}mv^2$ called the *kinetic energy* of the particle. When two highly elastic, or "springy," bodies (such as two hard steel ball bearings) collide, we find that the individual speeds change but that the total kinetic energy (the sum of the $\frac{1}{2}mv^2$ quantities) is the same after the collision as before. We say that kinetic energy is *conserved* in such collisions. We haven't really said what kinetic energy is, only that this product of half the mass and the square of the speed is useful in representing a conservation principle in a particular class of interaction.

When two soft, deformable bodies (such as two balls of putty or chewing gum) collide, we find that kinetic energy is *not* conserved, but the bodies become warmer. It turns out that there is a definite relationship between the rise in temperature and the loss of kinetic energy. We can define a new quantity, *internal energy,* that increases with temperature in a definite way such that the *sum* of kinetic energy and internal energy *is* conserved in these collisions.

We can also define additional forms of energy that enable us to extend the principle of conservation of energy to broader classes of phenomena. It's like studying a snowflake under a microscope, or the finer and finer details of a fractal pattern, or the structure of a Bach fugue. The more closely you look, the richer the subject becomes.

The astonishing thing is that in every situation in which it seems that the total energy in all known forms is not conserved, it has been found possible to identify a new form of energy such that the *total* energy, including the new form, *is* conserved. Energy is transformed from one form to another, but it is not created or destroyed. There is energy associated with heat, elastic deformations, electric and magnetic fields, and, according to the theory of relativity, even with mass itself (Fig. 6–1). Newtonian

(a)

(b)

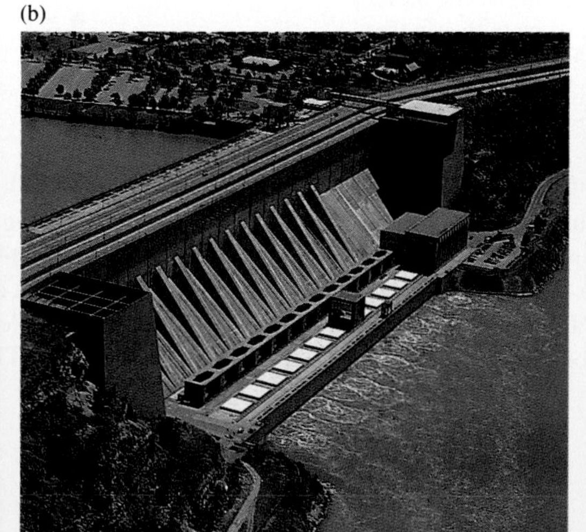

(c)

**6–1** Different forms of energy. (a) An explosion is a release of chemical energy. (b) A hydroelectric generating station uses the gravitational potential energy released by falling water to generate electricity. (c) At a solar-energy installation in the Mojave Desert, near Barstow, California, an array of mirrors concentrates heat energy from the sun onto a boiler (central tower) to generate steam for driving turbines. The mirrors move continually, controlled by a computer that tracks the sun across the sky.

mechanics has its limitations; it doesn't work for very fast motion, for which we have to use the theory of relativity, or for atomic or subatomic systems, for which we need quantum mechanics. But conservation of energy holds even in these realms. It is a *universal* conservation principle; no exception has ever been found.

In this chapter and the next we are concerned only with *mechanical* energy, the energy associated with motion, position, and deformation of bodies. We will find that in some interactions mechanical energy is conserved; in others there is a conversion of mechanical energy to or from other forms. In Chapters 15 and 17 we will study in detail the relation of mechanical energy to heat, and in later chapters we will define still other forms of energy.

## 6–2 WORK

The English playwright Jerome K. Jerome wrote: "I like work; it fascinates me. I can sit and look at it for hours." He was using the everyday meaning of the word *work* — any activity that requires muscular or mental exertion. Physicists have a much more specific definition involving a force acting on a body while the body undergoes a displacement. The importance of the concept of work stems from the fact that in any motion the change in *kinetic energy* of a body equals the total work done on it. We'll develop this relationship in Section 6–4; meanwhile, let's learn how to calculate work in a variety of situations.

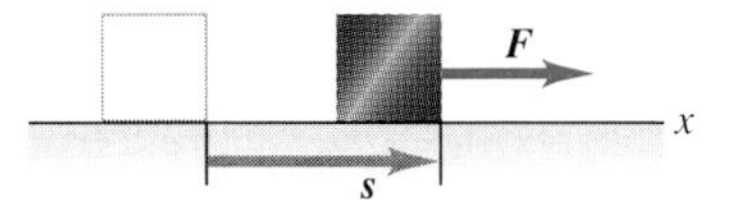


**6–2** Work is the product of a displacement $\boldsymbol{s}$ and the constant force $\boldsymbol{F}$ in the direction of $\boldsymbol{s}$.

In the simplest case a body undergoes a displacement with magnitude $s$ along a straight line while a constant force with magnitude $F$, directed along the same line, acts on it (Fig. 6–2). We define the **work** $W$ done by the force on the body as

$$W = Fs. \tag{6–1}$$

From this definition we see that the unit of work in any system is the unit of force multiplied by the unit of distance. In SI units the unit of force is the newton, the unit of distance is the meter, and the unit of work is the *newton-meter* (N · m). This combination appears so often in mechanics that we give it a special name, the **joule** (abbreviated J, pronounced "jewel," named in honor of the English physicist James Prescott Joule):

$$1 \text{ joule} = (1 \text{ newton})(1 \text{ meter}) \qquad \text{or} \qquad 1 \text{ J} = 1 \text{ N} \cdot \text{m}.$$

In the British system the unit of force is the pound (lb), the unit of distance is the foot, and the unit of work is the *foot-pound* (ft-lb). The following conversions are useful:

$$1 \text{ J} = 0.7376 \text{ ft} \cdot \text{lb}, \qquad 1 \text{ ft} \cdot \text{lb} = 1.356 \text{ J}.$$

### EXAMPLE 6–1

Steve is trying to impress Elaine with his new car, but the engine dies in the middle of an intersection. While Elaine steers, Steve pushes the car 19 m to clear the intersection. If he pushes in the direction of motion with a constant force of 210 N (about 47 lb), how much work does he do on the car?

**SOLUTION** From Eq. (6–1),

$$W = Fs = (210 \text{ N})(19 \text{ m}) = 4.0 \times 10^3 \text{ J}.$$

In Example 6–1, Steve pushed the car in the direction he wanted it to go. What if he had pushed at an angle $\phi$ with the car's displacement (Fig. 6–3)? Only the component in the direction of the car's motion, (210 N) cos $\phi$, would be effective in moving the car. When the force $\boldsymbol{F}$ and the displacement $\boldsymbol{s}$ have different directions, we take the component of $\boldsymbol{F}$ in the direction of the displacement $\boldsymbol{s}$, and we define the work as the product of this component and the magnitude of the displacement. The component of $\boldsymbol{F}$ in the direc-

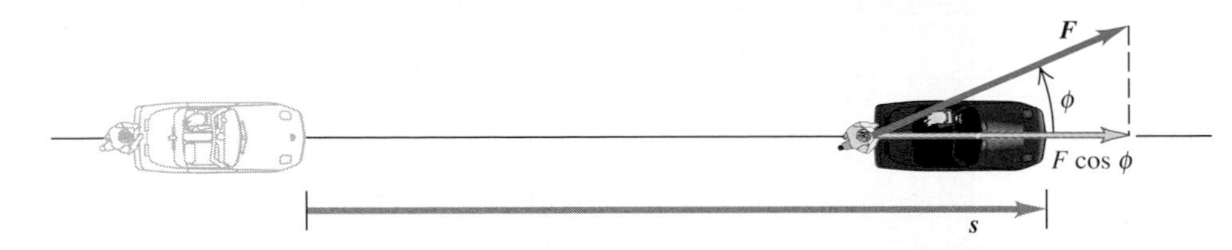


**6–3** The work done by a constant force $\boldsymbol{F}$ during a straight-line displacement $\boldsymbol{s}$ is $(F \cos \phi)\, s$.

tion of $\boldsymbol{s}$ is $F \cos \phi$, so

$$W = (F \cos \phi)s. \tag{6–2}$$

We are assuming that $F$ and $\phi$ are constant during the displacement. If $\phi$ happens to be zero, then $\boldsymbol{F}$ and $\boldsymbol{s}$ have the same direction, $\cos \phi = 1$, and we are back to Eq. (6–1). By definition the component of the force $\boldsymbol{F}$ *perpendicular* to the displacement $\boldsymbol{s}$ does no work.

Equation (6–2) has the form of the *scalar product* of two vectors, introduced in Section 1–10: $\boldsymbol{A} \cdot \boldsymbol{B} = AB \cos \phi$. You may want to review that definition. Using this, we can write Eq. (6–2) more compactly as

$$W = \boldsymbol{F} \cdot \boldsymbol{s}. \tag{6–3}$$

It is important to understand that work is a *scalar* quantity, even though it is calculated by using two vector quantities (force and displacement). A 5-N force toward the east acting on a body that moves 6 m to the east does exactly the same work as a 5-N force toward the north acting on a body that moves 6 m to the north. Work can be positive or negative. When the force has a component in the *same direction* as the displacement ($\phi$ between zero and 90°), $\cos \phi$ in Eq. (6–2) is positive, and the work $W$ is *positive* (Fig. 6–4a). When the force has a component *opposite* to the displacement ($\phi$ between 90° and 180°), $\cos \phi$ is negative, and the work is *negative* (Fig. 6–4b). When the force is *perpendicular* to the displacement, $\phi = 90°$, and the work done by the force is *zero* (Fig. 6–4c).

We always speak of work done *by* a specific force *on* a particular body. Always be sure to specify exactly what force is doing the work you are talking about. When a body is lifted, the work done on the body by the lifting force is positive; when a spring is stretched, the work done by the stretching force is positive. But the work done by the *gravitational* force on a body being lifted is *negative* because the (downward) gravitational force is opposite to the (upward) displacement. You might think it is "hard work" to hold this book out at arm's length for five minutes, but you aren't actually doing any work at all on the book because there is no displacement. Even when you walk with constant velocity on a level floor while carrying the book, you do no work on it because the (vertical) supporting force you exert on the book has no component in the direction of the

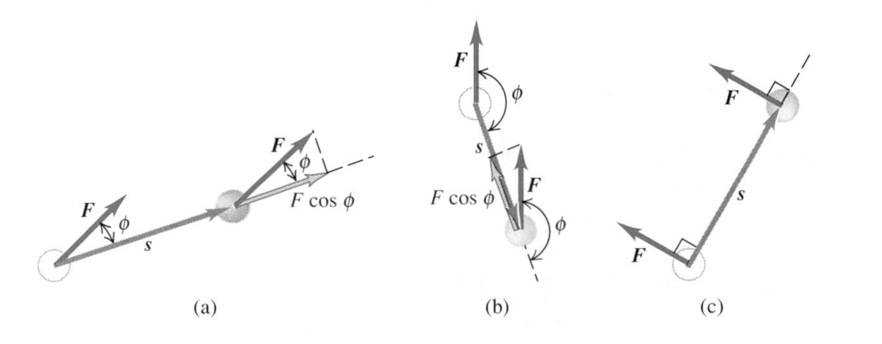


**6–4** (a) $W$ is positive because $\boldsymbol{F}$ has a component in the direction of $\boldsymbol{s}$. (b) $W$ is negative because $\boldsymbol{F}$ has a component opposite the direction of $\boldsymbol{s}$. (c) $W$ is zero because $\boldsymbol{F}$ has no component in the direction of $\boldsymbol{s}$.

**6–5** These circus performers do no work while walking the tightrope when their supporting forces are perpendicular to their displacement.

horizontal) motion. In this case, in Eq. (6–2), $\phi = 90^\circ$ and $\cos\phi = 0$. When a body slides along a surface, the work done by the normal force acting on the body is zero; and when a body moves in a circle, the work done by the centripetal force on the body is also zero. Figure 6–5 shows another example of zero work.

How do we calculate work when several forces act on a body? One way is to use Eq. (6–2) or (6–3) to compute the work done by each separate force. Then, because work is a scalar quantity, the *total* work $W_{tot}$ done on the body by all the forces is the algebraic sum of the quantities of work done by the individual forces. An alternative route to finding the total work $W_{tot}$ is to compute the vector sum (resultant) of the forces and then use this vector sum as $\boldsymbol{F}$ in Eq. (6–2) or (6–3).

## EXAMPLE 6–2

**Work done by several forces** Farmer Johnson hitches his tractor to a sled loaded with firewood and pulls it a distance of 20 m along level frozen ground (Fig. 6–6). The total weight of sled and load is 14,700 N. The tractor exerts a constant 5000-N force at an angle of 36.9° above the horizontal, as shown. There is a 3500-N friction force opposing the motion. Find the work done by each force acting on the sled and the total work done by all the forces.

**6–6** (a) A tractor pulling a sled of firewood. (b) Free-body diagram for the sled, treated as a particle.

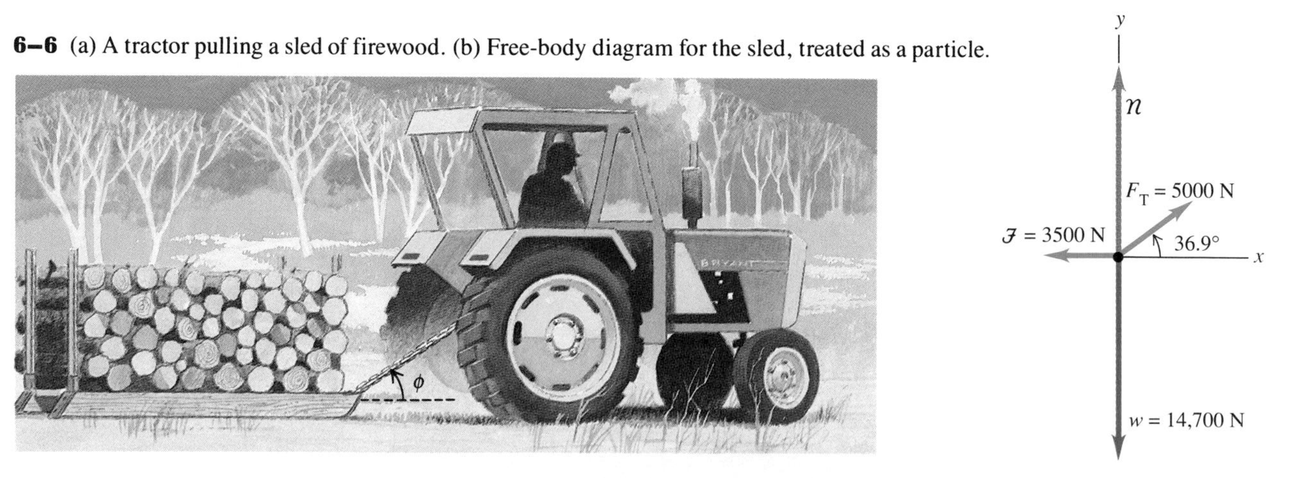

**SOLUTION** Let's do the easy parts first. The work $W_w$ done by the weight is zero because its direction is perpendicular to the displacement. (The angle between them is 90°, and the cosine of the angle is zero.) For the same reason the work $W_{\mathcal{n}}$ done by the normal force (which, incidentally, is *not* equal in magnitude to the weight) is also zero. So $W_w = W_{\mathcal{n}} = 0$.

That leaves $\boldsymbol{F}_T$ and $\boldsymbol{\mathcal{J}}$. From Eq. (6–2) the work $W_T$ done by the tractor is

$$W_T = (F_T \cos \phi)s$$
$$= (5000\text{ N})(0.800)(20\text{ m}) = 80{,}000\text{ N}\cdot\text{m} = 80\text{ kJ}.$$

The friction force $\boldsymbol{\mathcal{J}}$ is opposite to the displacement, so for this force, $\phi = 180°$ and $\cos \phi = -1$. The work $W_{\mathcal{J}}$ done by the friction force is

$$W_{\mathcal{J}} = \mathcal{J}s \cos 180° = -(3500\text{ N})(20\text{ m})$$
$$= -70{,}000\text{ N}\cdot\text{m} = -70\text{ kJ}.$$

The total work $W_{tot}$ done by all forces on the sled is the algebraic sum (*not* the vector sum) of the work done by the individual forces:

$$W_{tot} = W_T + W_W + W_{\mathcal{n}} + W_{\mathcal{J}}$$
$$= 80\text{ kJ} + 0 + 0 + (-70\text{ kJ}) = 10\text{ kJ}.$$

In the alternative approach we first find the vector sum (resultant) of the forces and then use it to compute the total work. The vector sum is best found by using components. From Fig. 6–6,

$$\Sigma F_x = (5000\text{ N}) \cos 36.9° - 3500\text{ N} = 500\text{ N},$$

$$\Sigma F_y = (5000\text{ N}) \sin 36.9° + \mathcal{n} + (-14{,}700\text{ N}).$$

We don't really need the second equation; we know that the $y$-component of force is perpendicular to the displacement, so it does no work. Besides, there is no $y$-component of acceleration, so $\Sigma F_y$ has to be zero anyway. The work done by the total $x$-component is therefore the total work:

$$W_{tot} = (500\text{ N})(20\text{ m}) = 10{,}000\text{ J} = 10\text{ kJ}.$$

This is the same result that we found by computing the work of each force separately.

## EXAMPLE 6–3

An electron moves toward the east with constant velocity (in a straight line) of magnitude $8 \times 10^7$ m/s. It has electric, magnetic, and gravitational forces acting on it. Calculate the total work done on the electron during a 1-m displacement.

**SOLUTION** The electron moves with constant velocity, so its acceleration is zero. For that reason the vector sum of forces is also zero. Therefore the total work done by all the forces (equal to the work done by the vector sum of the forces) must be zero. Individual forces may do nonzero work, but that's not what the problem asks for. ■

Once more, for emphasis: When several forces act on a body, there are always two equivalent ways to calculate the total work. We may calculate the work done by each force separately and take the algebraic sum, or we may compute the vector sum (resultant) of the forces and then find the work done by the resultant.

# 6–3 WORK DONE BY A VARYING FORCE

In Section 6–2 we defined work done by a *constant* force. But what happens when you stretch a spring? The more you stretch it, the harder you have to pull, so the force is *not* constant as the string is stretched. You can think of many other situations in which a force that varies in both magnitude and direction acts on a body moving along a curved path. We need to be able to compute the work done by the force in these more general situations.

To add one complication at a time, let's consider a straight-line motion with a force that is directed along the line but may change in magnitude as the body moves. For example, imagine a train on a straight track, with the engineer constantly changing the locomotive's throttle setting or applying the brakes. Suppose a particle moves along the $x$-axis

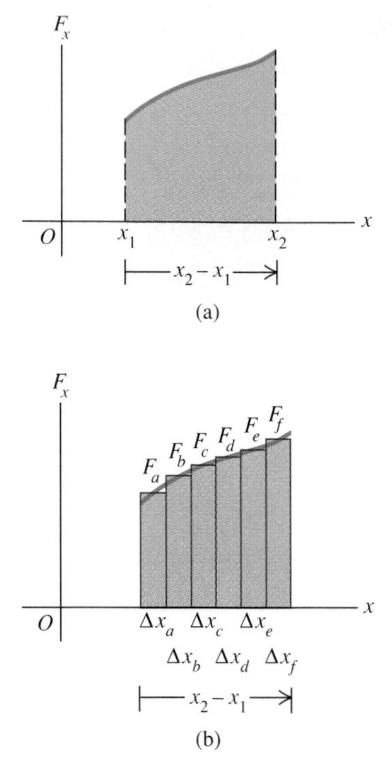


**6–7** (a) Curve showing how $F$ varies with $x$. (b) If the area is partitioned into small rectangles, the sum of their areas approximates the total work done during the displacement; the greater the number of rectangles used, the closer the approximation.

from point $x_1$ to $x_2$. Figure 6–7a is a graph of the $x$-component of force as a function of the particle's coordinate $x$. To find the work done by this force, we divide the total displacement into small segments $\Delta x_a$, $\Delta x_b$, and so on (Fig. 6–7b). We approximate the work done by the force during segment $\Delta x_a$ as the average force $F_a$ in that segment multiplied by its length $\Delta x_a$. We do this for each segment and then add the results for all the segments. The work done by the force in the total displacement from $x_1$ to $x_2$ is approximately

$$W = F_a \Delta x_a + F_b \Delta x_b + \cdots .$$

As the number of segments becomes very large and the size of each becomes very small, this sum becomes (in the limit) the *integral* of $F$ from $x_1$ to $x_2$:

$$W = \int_{x_1}^{x_2} F\, dx. \tag{6–4}$$

Note that $F_a \Delta x_a$ represents the *area* of the first vertical strip in Fig. 6–7b and that the integral in Eq. (6–4) represents the area under the curve of Fig. 6–7a between $x_1$ and $x_2$. On a graph of force as a function of position, the total work done by the force is represented by the area under the curve between the initial and final positions.

Not let's get back to the stretched spring. To keep a spring stretched an amount $x$ beyond its unstretched length, we have to apply a force with magnitude $F$ at each end (Fig. 6–8). If the elongation is not too great, we find that $F$ is directly proportional to $x$:

$$F = kx, \tag{6–5}$$

where $k$ is a constant called the **force constant** (or spring constant) of the spring. Equation (6–5) shows that the units of $k$ are force divided by distance, N/m in SI units and lb/ft in British units. The fact that elongation is directly proportional to force for elongations that are not too great was discovered by Robert Hooke in 1678 and is known as **Hooke's law.** We will discuss it more fully in Chapter 13. It really shouldn't be called a *law;* it is a statement about a specific device, not a fundamental law of nature. Real springs don't always obey Eq. (6–5) precisely, but it's still a useful idealized model.

Suppose we apply equal and opposite forces to the ends of the spring and gradually increase the forces, starting from zero. We hold the left end stationary; the force at this end does no work. The force at the moving end *does* do work. Figure 6–9 is a graph of $F$ as a function of $x$ (the elongation of the spring). The work done by $F$ when the elongation

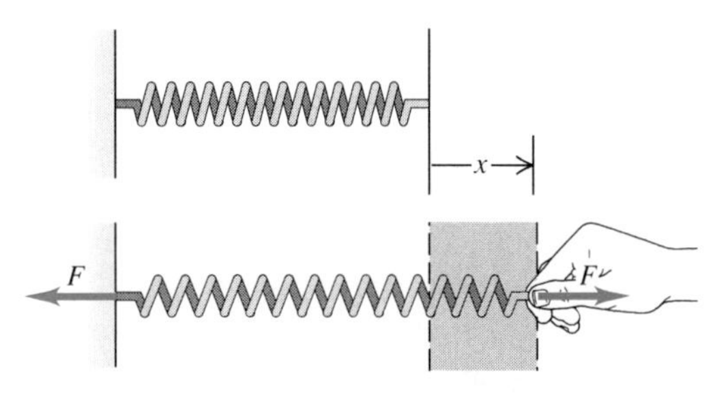


**6–8** The force needed to stretch an ideal spring is proportional to its elongation: $F = kx$.

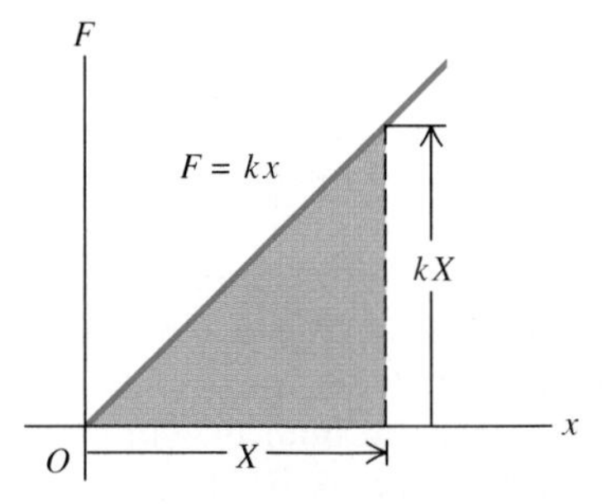


**6–9** The work done in stretching a spring is equal to the area of the shaded triangle.

goes from zero to a maximum value $X$ is

$$W = \int_0^X F\,dx = \int_0^X kx\,dx = \tfrac{1}{2}kX^2. \qquad (6\text{–}6)$$

We can also obtain this result graphically. The area of the shaded triangle in Fig. 6–9, representing the total work, done by the force, is equal to half the product of the base and altitude, or

$$W = \tfrac{1}{2}(X)(kX) = \tfrac{1}{2}kX^2.$$

This also says that the work is the *average* force $kX/2$ multiplied by the total displacement $X$.

We see that the total work is proportional to the *square* of the final elongation $X$. When the elongation is doubled, the total work necessary to produce the new elongation increases by a factor of four.

The spring also exerts a force on the hand, which moves during the stretching process. The displacement of the hand is the same as that of the moving end of the spring, but the force on it is opposite in direction to the force on the spring because the two forces form an action-reaction pair. Thus the work done *on* the hand *by* the spring is the negative of the work done on the spring, namely, $-\tfrac{1}{2}kX^2$. In problems involving work and its relation to kinetic energy, we will nearly always want to find the work done *by* a force *on* the body under study, and we have to be careful to write work quantities with the correct sign.

Suppose the spring is stretched a distance $x_1$ at the start. Then the work we have to do on it to stretch it to a greater elongation $x_2$ is

$$W = \int_{x_1}^{x_2} F\,dx = \int_{x_1}^{x_2} kx\,dx = \tfrac{1}{2}kx_2^2 - \tfrac{1}{2}kx_1^2. \qquad (6\text{–}7)$$

If the spring has spaces between the coils when it is unstretched, then it can also be compressed, and Hooke's law holds for compression as well as stretching. In this case, $F$ and $x$ in Eq. (6–5) are both negative. The force again has the same direction as the displacement, and the work done by $F$ is again positive. So the total work is still given by Eq. (6–6) or (6–7), even when $X$ or either or both of $x_1$ or $x_2$ are negative.

## EXAMPLE 6–4

**Work done on a spring scale** A woman weighing 600 N steps on a bathroom scale containing a heavy spring (Fig. 6–10). The spring is compressed 1.0 cm under her weight. Find the force constant of the spring and the total work done on it during the compression.

**SOLUTION** If positive values of $x$ correspond to elongation, then $x = -0.010$ m when $F = -600$ N. From Eq. (6–5) the force constant $k$ is

$$k = \frac{F}{x} = \frac{-600\text{ N}}{-0.010\text{ m}} = 60{,}000\text{ N/m}.$$

Then, from Eq. (6–6),

$$W = \tfrac{1}{2}kX^2 = \tfrac{1}{2}(60{,}000\text{ N/m})(-0.010\text{ m})^2 = 3.0\text{ N}\cdot\text{m} = 3.0\text{ J}.$$

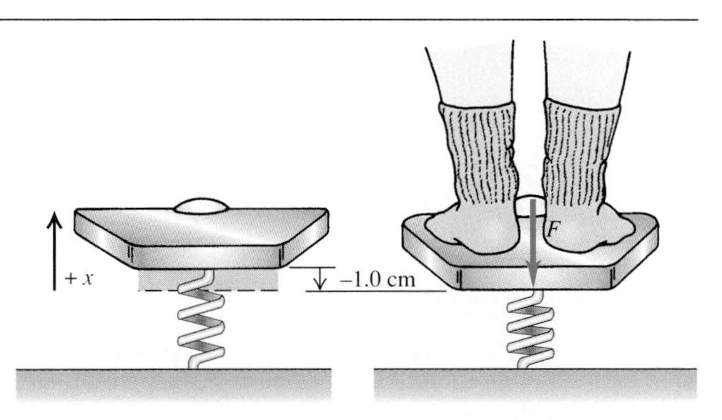


**6–10** Compressing a spring in a bathroom scale results in negative displacement by negative force but a positive amount of work.

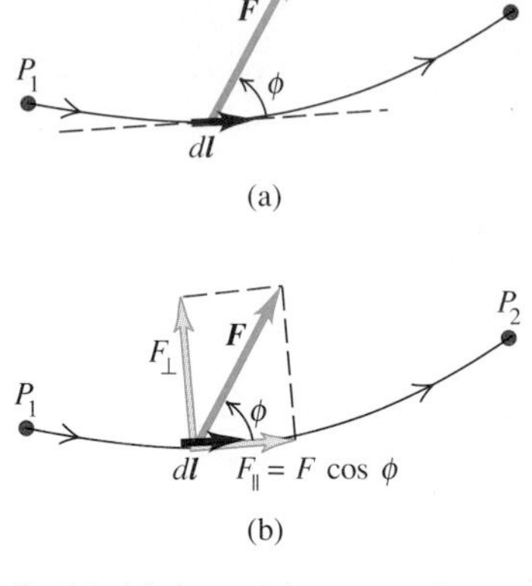


**6–11** (a) A particle moves along a curved path from point $P_1$ to $P_2$, acted on by a force $\boldsymbol{F}$ that varies in magnitude and direction. During an infinitesimal displacement $d\boldsymbol{l}$ (a small segment of the path), the work $dW$ done by the force is given by $dW = \boldsymbol{F}\cdot d\boldsymbol{l} = F\cos\phi\, dl$. (b) The force contributing to the work is the component of force parallel to the displacement, $F_\parallel = F\cos\phi$.

We can generalize our definition of work further to include a force that varies in direction as well as magnitude and a displacement that may occur along a curved path. Suppose a particle moves from point $P_1$ to $P_2$ along a curve, as shown in Fig. 6–11. We imagine dividing the portion of the curve between these points into many infinitesimal vector displacements, and we call a typical one of these $d\boldsymbol{l}$. Each $d\boldsymbol{l}$ is tangent to the path at its position. Let $\boldsymbol{F}$ be the force at a typical point along the path, and let $\phi$ be the angle between $\boldsymbol{F}$ and $d\boldsymbol{l}$ at this point. Then the small element of work $dW$ done on the particle during the displacement $d\boldsymbol{l}$ may be written as

$$dW = F\cos\phi\, dl = F_\parallel\, dl = \boldsymbol{F}\cdot d\boldsymbol{l},$$

where $F_\parallel = F\cos\phi$ is the component of $\boldsymbol{F}$ in the direction parallel to $d\boldsymbol{l}$ (Fig. 6–11b). The total work done by $\boldsymbol{F}$ on the particle as it moves from $P_1$ to $P_2$ is then represented symbolically as

$$W = \int_{P_1}^{P_2} F\cos\phi\, dl = \int_{P_1}^{P_2} F_\parallel\, dl = \int_{P_1}^{P_2} \boldsymbol{F}\cdot d\boldsymbol{l}. \tag{6–8}$$

This integral is called a *line integral*. To evaluate Eq. (6–8) in a specific problem, we need some sort of detailed description of the path and of the variation of $\boldsymbol{F}$ along the path. We usually express the integral in terms of some scalar variable, as in the following example.

## EXAMPLE 6–5

At a family picnic you are appointed to push your obnoxious cousin Throckmorton in a swing (Fig. 6–12a). His weight is $w$, the length of the chains is $R$, and you push Throcky up until the chains make an angle $\theta_0$ with the vertical. To do this, you exert a varying horizontal force $\boldsymbol{F}$ that starts at zero and gradually increases just enough so that Throcky and the swing move very slowly and remain very nearly in equilibrium. What is the total work done by the tension $T$ in the chains? What is the total work done by the force $\boldsymbol{F}$?

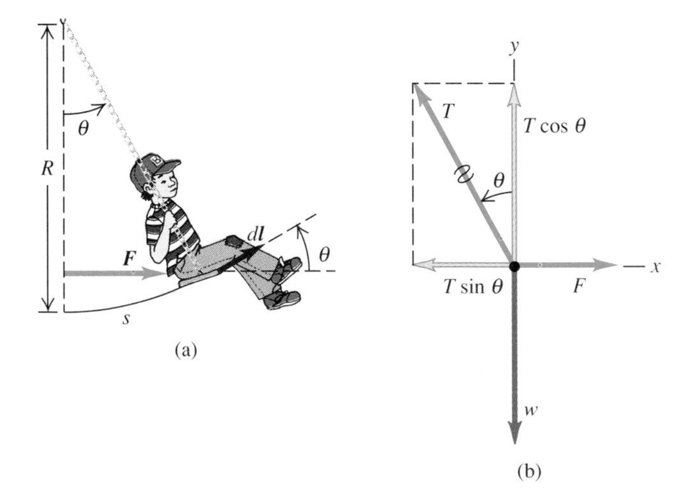


**6–12** (a) Pushing cousin Throckmorton in a swing. (b) Free-body diagram for Throcky, considered as a particle and neglecting the small weight of chains and seat.

**SOLUTION** The free-body diagram is shown in Fig. 6–12b. We have replaced the tensions in the two chains with a single tension $T$. At any point during the motion the chain force on Throcky is perpendicular to each $d\boldsymbol{l}$, so the angle between the chain force and the displacement is always 90°. Therefore, the total work done by the chain tension is zero.

To compute the work done by $\boldsymbol{F}$, we have to find how it varies with the angle $\theta$. Throcky is in equilibrium at every point; from $\Sigma F_x = 0$ we get

$$F - T\sin\theta = 0,$$

and from $\Sigma F_y = 0$ we find

$$T\cos\theta - w = 0.$$

Dividing these two equations, we have

$$F = w\tan\theta.$$

The point where $\boldsymbol{F}$ is applied swings through the arc $s$. The arc length $s$ equals the radius $R$ of the circular path multiplied by the angle $\theta$ (in radians), so $s = R\theta$. Therefore the displacement $d\boldsymbol{l}$ corresponding to a small change of angle $d\theta$ has a magnitude $ds = R\, d\theta$. The work done by $\boldsymbol{F}$ is

$$W = \int \boldsymbol{F}\cdot d\boldsymbol{l} = \int F\cos\theta\, ds.$$

Now we express everything in terms of the varying angle $\theta$:

$$W = \int_0^{\theta_0} (w \tan\theta) \cos\theta \, R \, d\theta$$
$$= wR \int_0^{\theta_0} \sin\theta \, d\theta = wR(1 - \cos\theta_0). \qquad (6\text{–}9)$$

If $\theta_0 = 0$, there is no displacement; in that case, $\cos\theta_0 = 1$ and $W = 0$, as we should expect. If $\theta_0 = 90°$, then $\cos\theta_0 = 0$ and $W = wR$. In that case the total work you do is the same as if you had lifted Throcky straight up a distance $R$ with a force equal to his weight $w$. In fact, the quantity $R(1 - \cos\theta_0)$ is the increase in his height above the ground during the displacement, so for any value of $\theta_0$ the work done by force $\boldsymbol{F}$ is the change in height multiplied by the weight. We will prove this result more generally in Section 7–1.

## 6–4 WORK AND KINETIC ENERGY

In Section 6–2 we promised to work out the relation between work and kinetic energy. This important relationship is called the *work-energy theorem*; without it the concept of work wouldn't be of much use. So now it's time to make good on our promise.

Let's consider a particle with mass $m$ moving along the $x$-axis under the action of a constant resultant force with magnitude $F$ directed along the axis. The particle's acceleration is constant and is given by Newton's second law, $F = ma$. Suppose the speed increases from $v_1$ to $v_2$ while the particle undergoes a displacement $s$ from point $x_1$ to $x_2$. Using the constant-acceleration equation, Eq. (2–14), and replacing $v_0$ by $v_1$, $v$ by $v_2$, and $(x - x_0)$ by $s$, we have

$$v_2^2 = v_1^2 + 2as,$$
$$a = \frac{v_2^2 - v_1^2}{2s}.$$

When we multiply this equation by $m$ and replace $ma$ with $F$, we find

$$F = ma = m\frac{v_2^2 - v_1^2}{2s}$$

and

$$Fs = \tfrac{1}{2}mv_2^2 - \tfrac{1}{2}mv_1^2. \qquad (6\text{–}10)$$

The product $Fs$ is the work done by the resultant force $F$ and thus is equal to the total work $W_{tot}$ done by all the forces. The quantity $\frac{1}{2}mv^2$ is the **kinetic energy** $K$ of the particle, which we defined in Section 6–1:

$$K = \tfrac{1}{2}mv^2. \qquad (6\text{–}11)$$

The first term on the right side of Eq. (6–10) is the final kinetic energy of the particle, $K_2 = \frac{1}{2}mv_2^2$ (after the displacement). The second term is the initial kinetic energy, $K_1 = \frac{1}{2}mv_1^2$, and the difference between these terms is the *change* in kinetic energy. So Eq. (6–10) says that **the work done by the resultant external force on a particle is equal to the change in kinetic energy of the particle.** In symbols,

$$W_{tot} = K_2 - K_1 = \Delta K. \qquad (6\text{–}12)$$

This result is called the **work-energy theorem;** it is the foundation for most of what follows in this chapter.

Kinetic energy, like work, is a scalar quantity. It can never be negative, although work can be either positive or negative. The kinetic energy of a moving particle depends only on its speed (the magnitude of its velocity), not on the direction of its motion. A car (viewed as a particle) has the same kinetic energy when going north at 5 m/s as when going east at 5 m/s. The *change* in kinetic energy during any displacement is determined by the total work $W_{tot}$ done by all the forces acting on the particle. When this work $W_{tot}$ is *positive,* the particle speeds up during the displacement, $K_2$ is greater than $K_1$, and the kinetic energy *increases.* When $W_{tot}$ is *negative,* the kinetic energy and speed *decrease;* and when $W_{tot} = 0$, $K$ is *constant.*

In SI units, $m$ is measured in kilograms and $v$ in meters per second, so kinetic energy has the units $\text{kg} \cdot \text{m}^2/\text{s}^2$. From Eq. (6–10) or (6–12) kinetic energy must have the same units as work. To verify this, we recall that $1 \text{ N} = 1 \text{ kg} \cdot \text{m/s}^2$, so

$$1 \text{ J} = 1 \text{ N} \cdot \text{m} = 1 \text{ (kg} \cdot \text{m/s}^2) \cdot \text{m} = 1 \text{ kg} \cdot \text{m}^2/\text{s}^2.$$

The joule is the SI unit of both work and kinetic energy and, as we will see later, of all kinds of energy. In the British system,

$$1 \text{ ft} \cdot \text{lb} = 1 \text{ ft} \cdot \text{slug} \cdot \text{ft/s}^2 = 1 \text{ slug} \cdot \text{ft}^2/\text{s}^2.$$

A reminder: In Eq. (6–12), $W_{tot}$ is the work done by the *resultant* force, the *vector sum* of *all* the forces acting on the particle. Alternatively, we can calculate the work done by each separate force; $W_{tot}$ is then the *algebraic sum* of all these quantities of work. Example 6–2 (Section 6–2) illustrates these two alternatives.

Although we derived Eq. (6–12) for the special case of a constant resultant force, it is true even when the force varies with position. To prove this, we divide the total displacement $x$ into a large number of small segments $\Delta x$, just as in the calculation of work done by a varying force. The change of kinetic energy in segment $\Delta x_a$ is equal to the work $F_a \, \Delta x_a$, and so on. The total change of kinetic energy is the sum of the changes in the individual segments and is thus equal to the total work done during the entire displacement.

Here is an alternative derivation of Eq. (6–12) for a force that varies with position. It involves making a change of variable from $x$ to $v$ in the work integral. As a preliminary, we note that the acceleration $a$ of the particle can be expressed in various ways, using $a = dv/dt$, $v = dx/dt$, and the chain rule for derivatives:

$$a = \frac{dv}{dt} = \frac{dv}{dx}\frac{dx}{dt} = v\frac{dv}{dx}. \tag{6–13}$$

Using this result, we can rewrite the work integral as

$$W = \int_{x_1}^{x_2} F \, dx = \int_{x_1}^{x_2} ma \, dx = \int_{x_1}^{x_2} mv \frac{dv}{dx} \, dx. \tag{6–14}$$

Now $(dv/dx)\,dx$ is the change in velocity $dv$ during the displacement $dx$, so in Eq. (6–14) we can substitute $dv$ for $(dv/dx)\,dx$. This changes the integration variable from $x$ to $v$, so we change the limits from $x_1$ and $x_2$ to the corresponding velocities $v_1$ and $v_2$ at these points. This gives us

$$W = \int_{v_1}^{v_2} mv \, dv. \tag{6–15}$$

The integral of $v \, dv$ is just $v^2/2$; substituting the upper and lower limits, we finally find

$$W = \tfrac{1}{2}mv_2^2 - \tfrac{1}{2}mv_1^2. \tag{6–16}$$

This is the same result as Eq. (6–10). But in this derivation we have not assumed that $F$ is constant, so we have proved that the work-energy theorem is valid even when $F$ varies during the displacement. We could also extend the proof further to motion along a curved path, but that is beyond our scope.

## PROBLEM-SOLVING STRATEGY

### Work and kinetic energy

**1.** Draw a free-body diagram; make sure you show all the forces that act on the body. List the forces, and calculate the work done by each force. In some cases, one or more forces may be unknown; represent the unknowns by algebraic symbols. Be sure to check signs. When a force has a component in the same direction as the displacement, its work is positive; when the direction is opposite to the displacement, the work is negative. When force and displacement are perpendicular, the work is zero.

**2.** Add the amounts of work done by the separate forces to find the total work. Again be careful with signs. Sometimes, though not often, it may be easier to calculate the resultant of the forces first and then find the work done by the resultant force.

**3.** List the initial and final kinetic energies $K_1$ and $K_2$. If a quantity such as $v_1$ or $v_2$ is unknown, express it in terms of the corresponding algebraic symbol. When you calculate kinetic energies, make sure you use the *mass* of the body, not its *weight.*

**4.** Use the relationship $W_{tot} = K_2 - K_1 = \Delta K$; insert the results from the above steps and solve for whatever unknown is required. Remember that kinetic energy can never be negative. If you come up with a negative $K$, you've made a mistake. Maybe you interchanged subscripts 1 and 2 or made a sign error in one of the work calculations.

## EXAMPLE 6–6

**Using work and energy to calculate speed** Let's look again at the sled in Fig. 6–6 and the numbers at the end of Example 6–2. The free-body diagram is shown again in Fig. 6–13. We found that the total work done by all the forces is 10,000 N · m = 10 kJ, so the kinetic energy of the sled must increase by 10 kJ. The mass of the sled is $m = (14{,}700\ \text{N})/(9.8\ \text{m/s}^2) = 1500\ \text{kg}$. Suppose the initial speed $v_1$ is 2.0 m/s. What is the final speed?

**SOLUTION** Steps 1 and 2 of the problem-solving strategy were done in Example 6–2, where we found $W_{tot} = 10\ \text{kJ}$. The initial kinetic energy $K_1$ is

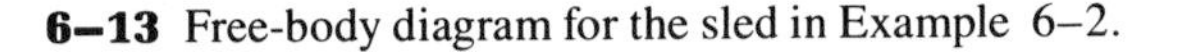


**6–13** Free-body diagram for the sled in Example 6–2.

$$K_1 = \tfrac{1}{2}mv_1^2 = \tfrac{1}{2}(1500\ \text{kg})(2.0\ \text{m/s})^2 = 3000\ \text{kg}\cdot\text{m}^2/\text{s}^2 = 3000\ \text{J}.$$

The final kinetic energy $K_2$ is

$$K_2 = \tfrac{1}{2}(1500\ \text{kg})v_2^2,$$

where $v_2$ is the unknown final speed we want to find. Equation (6–12) gives

$$K_2 = K_1 + W_{tot}.$$

$$\tfrac{1}{2}(1500\ \text{kg})v_2^2 = 3000\ \text{J} + 10{,}000\ \text{J} = 13{,}000\ \text{J}.$$

Solving for $v_2$, we find

$$v_2 = 4.2\ \text{m/s}.$$

This problem can also be done without the work-energy theorem. We can find the acceleration from $F = ma$ and then use the equations of motion with constant acceleration to find $v_2$:

$$a = \frac{F}{m} = \frac{4000\ \text{N} - 3500\ \text{N}}{1500\ \text{kg}} = 0.333\ \text{m/s}^2;$$

then

$$v_2^2 = v_1^2 + 2as = (2.0\ \text{m/s})^2 + 2(0.333\ \text{m/s}^2)(20\ \text{m}) = 17.3\ \text{m}^2/\text{s}^2,$$

$$v_2 = 4.2\ \text{m/s}.$$

This is the same result we obtained with the work-energy approach, but there we avoided the intermediate step of finding the acceleration. You will find several other examples and problems in this chapter and the next that *can* be done without using energy considerations but that are easier when energy methods are used. Also, when a problem can be done by two different methods, doing it both ways is always a good way to check your work.

## EXAMPLE 6–7

**Forces on a hammerhead** In a pile driver a steel hammerhead with mass 200 kg is lifted 3.00 m above the top of a vertical I-beam being driven into the ground (Fig. 6–14a). The hammer is then dropped, driving the I-beam 7.4 cm farther into the ground. The vertical rails that guide the hammerhead exert a constant 60-N friction force on the hammerhead. Use the work-energy theorem to find a) the speed of the hammerhead just as it hits the I-beam; b) the average force the hammerhead exerts on the I-beam.

**SOLUTION** Figure 6–14b is a free-body diagram showing the vertical forces on the falling hammerhead. Because the displacement is vertical, any horizontal forces that may be present do no work. The vertical forces are the downward weight $w = mg = (200 \text{ kg})(9.8 \text{ m/s}^2) = 1960 \text{ N}$ and the upward friction force of 60 N. Thus the net downward force is 1900 N, and the total work done on the hammerhead during the 3.00-m drop is

$$W_{tot} = (1900 \text{ N})(3.00 \text{ m}) = 5700 \text{ J}.$$

a) The hammerhead is initially at rest (point 1), so its initial kinetic energy $K_1$ is zero. Equation (6–12) gives

$$W_{tot} = K_2 - K_1,$$

$$5700 \text{ J} = \tfrac{1}{2}(200 \text{ kg})v_2^2 - 0,$$

$$v_2 = 7.55 \text{ m/s}.$$

This is the hammerhead's speed at point 2, just as it hits the beam.

b) Let point 3 be the place the hammerhead finally comes to rest; then $K_3 = 0$. As Fig. 6–14c shows, there is now an additional force, the upward normal force $\boldsymbol{\mathcal{n}}$ (assumed to be constant) that the beam exerts on the hammerhead during the additional downward displacement of 7.4 cm. The total work done on the hammerhead during this displacement is

$$W_{tot} = (1900 \text{ N})(0.074 \text{ m}) + \mathcal{n}(0.074 \text{ m})(-1).$$

This is equal to the change in kinetic energy $K_3 - K_2$, so we have

$$(1900 \text{ N} - \mathcal{n})(0.074 \text{ m}) = 0 - 5700 \text{ J},$$

$$\mathcal{n} = 79{,}000 \text{ N}.$$

The force *on* the I-beam is the equal but opposite reaction force of 79,000 N (about 9 tons) downward. The total change in the hammerhead's kinetic energy during the whole process is zero; a relatively small force does positive work over a large distance, and then a much larger force does negative work over a much smaller distance. The same thing happens if you speed your car up gradually and then drive it into a brick wall. The very large force needed to reduce the kinetic energy to zero over a short distance is what does the damage to your car—and possibly to you.

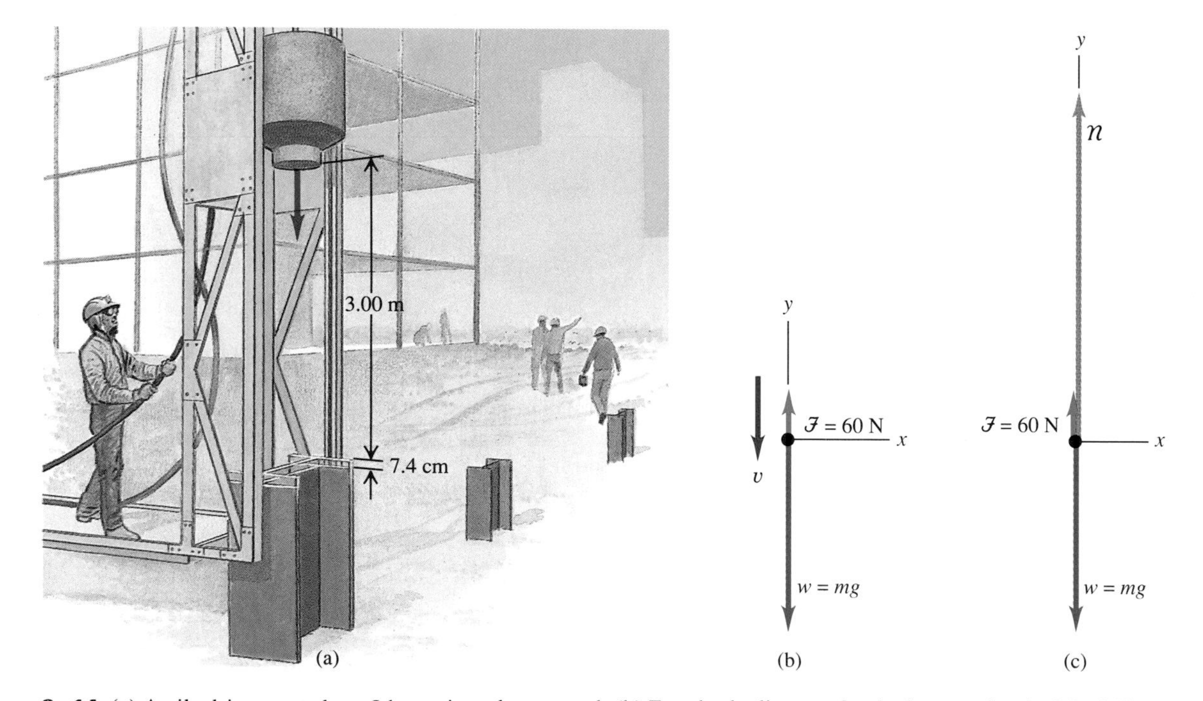


**6–14** (a) A pile driver pounds an I-beam into the ground. (b) Free-body diagram for the hammerhead while falling. (c) Free-body diagram for the hammerhead while driving the I-beam. Vector lengths are not to scale.

Two final comments: First, because of the role of Newton's laws in deriving the work-energy theorem, we must use it only in an inertial frame of reference. The speeds we use to compute the kinetic energies and the distances we use to compute work *must* be measured in an inertial frame. Second, the work-energy theorem is valid in *all* inertial frames. For a given situation the work and kinetic energies will be different in different frames because the speed of a body is different in different frames of reference.

## 6–5 POWER

Time considerations are not involved in the definition of work. If you lift a barbell weighing 400 N through a vertical distance of 0.5 m at constant velocity, you do 200 J of work whether it takes you 1 second, 1 hour, or 1 year to do it. Often, though, we need to know how quickly work is done. The time rate at which work is done or energy is transferred is called **power.** Like work and energy, power is a scalar quantity.

When a quantity of work $\Delta W$ is done during a time interval $\Delta t$, the **average power** $P_{\text{av}}$, or work per unit time, is defined as

$$P_{\text{av}} = \frac{\Delta W}{\Delta t}. \tag{6–17}$$

The rate at which work is done might not be constant. Even when it varies, we can define **instantaneous power** $P$ as the limit of this quotient as $\Delta t$ approaches zero:

$$P = \lim_{\Delta t \to 0} \frac{\Delta W}{\Delta t} = \frac{dW}{dt}. \tag{6–18}$$

The SI unit of power, the joule per second (J/s), is called the **watt** (W). The kilowatt (1 kW $= 10^3$ W) and the megawatt (1 MW $= 10^6$ W) are also commonly used. In the British system, in which work is expressed in foot-pounds, the unit of power is the foot-pound per second. A larger unit called the *horsepower* (hp) is also used:

$$1 \text{ hp} = 550 \text{ ft} \cdot \text{lb/s} = 33{,}000 \text{ ft} \cdot \text{lb/min}.$$

That is, a 1-hp motor running at full load does 33,000 ft · lb of work every minute. A useful conversion factor is

$$1 \text{ hp} = 746 \text{ W} = 0.746 \text{ kW},$$

or 1 horsepower equals about $\frac{3}{4}$ of a kilowatt, a useful figure to remember.

The watt is a familiar unit of *electrical* power; a 100-W light bulb converts 100 J of electrical energy into light and heat each second. But there is nothing inherently electrical about the watt; a light bulb could be rated in horsepower, and some automobile manufacturers rate their engines in kilowatts rather than horsepower.

Power units can be used to define new units of work or energy. The *kilowatt-hour* (kWh) is the usual commercial unit of electrical energy. One kilowatt-hour is the total work done in 1 hour (3600 s) when the power is 1 kilowatt ($10^3$ J/s), so

$$1 \text{ kWh} = (10^3 \text{ J/s})(3600 \text{ s}) = 3.6 \times 10^6 \text{ J} = 3.6 \text{ MJ}.$$

The kilowatt-hour is a unit of *work* or *energy,* not power.

It is a curious fact of modern life that although energy is an abstract physical quantity, it is bought and sold. We don't buy a newton of force or a meter per second of velocity, but a kilowatt-hour of electrical energy usually costs from two to ten cents, depending on location and amount purchased.

When a force acts on a moving body, it does work on the body (unless the force and velocity are always perpendicular). The corresponding power can be expressed in terms of force and velocity. Suppose a force $\boldsymbol{F}$ acts on a body while it undergoes a vector displacement $\Delta\boldsymbol{s}$. If $F_{\parallel}$ is the component of $\boldsymbol{F}$ tangent to the path (parallel to $\Delta\boldsymbol{s}$), then the

the following results:

| $v$ (m/s) | $F_{roll}$ (N) | $F_{air}$ (N) | $F_{tot}$ (N) | $P$ (kW) | $P$ (hp) |
|---|---|---|---|---|---|
| 10 | 184 | 40 | 224 | 2.24 | 3.00 |
| 15 | 184 | 90 | 274 | 4.11 | 5.51 |
| 30 | 184 | 360 | 544 | 16.3 | 21.9 |

How is all this related to fuel consumption? Let's look at the 15-m/s case. The power required is 4.11 kW = 4110 J/s. In 1 hour (3600 s) the total energy required is

$$(4110\ \text{J/s})(3600\ \text{s}) = 1.48 \times 10^7\ \text{J}.$$

During that hour the car travels a distance of

$$(15\ \text{m/s})(3600\ \text{s}) = 5.4 \times 10^4\text{m} = 54\ \text{km}.$$

As we explained above, only 15% of the $3.5 \times 10^7$ J of energy obtained by burning 1 liter (1 L) of gasoline is available to propel the car. The available energy per liter is

$$(0.15)(3.5 \times 10^7\ \text{J/L}) = 5.25 \times 10^6\ \text{J/L}.$$

So the amount of fuel consumed in 1 hour, traveling 54 km at 15 m/s, is actually

$$(1.48 \times 10^7\ \text{J})/(5.25 \times 10^6\ \text{J/L}) = 2.82\ \text{L}.$$

That amount of gasoline gets the car 54 km, so the fuel consumption per unit distance is (54 km)/(2.82 L) = 19.1 km/L, or about 44.9 miles per gallon. (Figure 6–16 shows some of the features of contemporary car design that have improved fuel efficiency.)

The power required for a steady 15 m/s on level ground is 4.11 kW, but the power required for acceleration and hill climbing may be much greater. The 911 Carrera is advertised as going from zero to 60 mi/h (26.8 m/s) in 6.1 s. The final kinetic energy is then

$$K = \tfrac{1}{2}mv^2 = \tfrac{1}{2}(1251\ \text{kg})(26.8\ \text{m/s})^2 = 4.5 \times 10^5\ \text{J}.$$

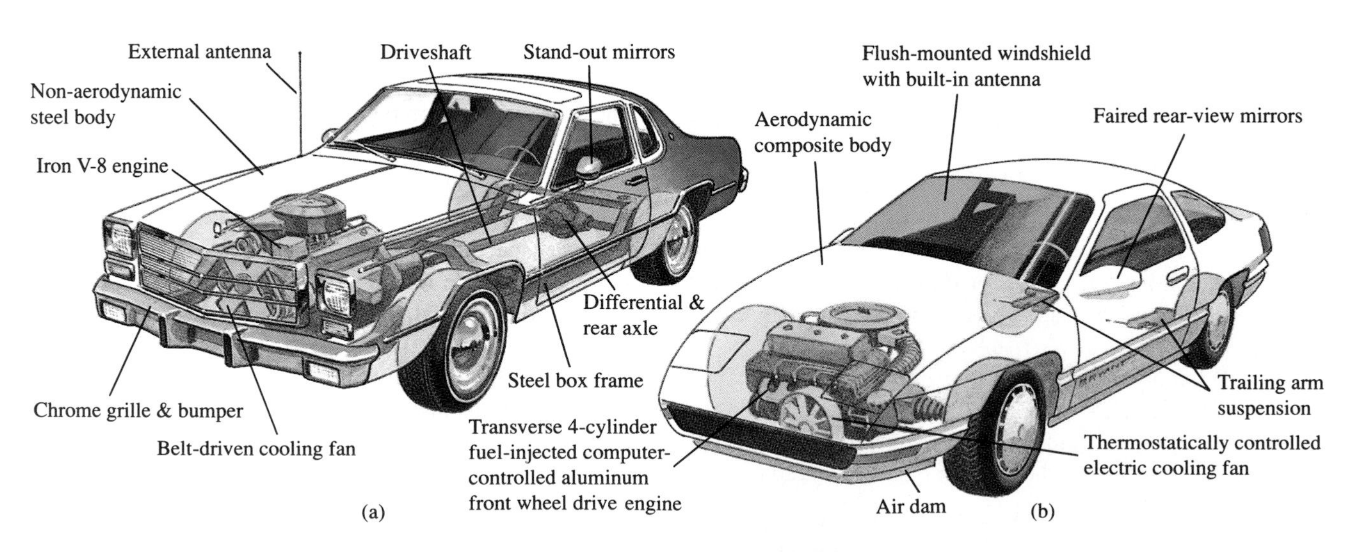


**6–16** (a) Cars designed in the early 1970s did not feature aerodynamic styling and used heavy materials such as iron for engines and steel for body panels. Rear-wheel drive required a heavy drive train as well. (b) By the 1990s economics forced new ideas in car designs. Body shapes have lower drag coefficients, engines are built of aluminum, and body panels are often made of plastic. These changes have doubled the typical fuel efficiency of cars.

The average additional power required for the acceleration is

$$P_{av} = \frac{4.5 \times 10^5 \text{ J}}{6.1 \text{ s}} = 7.38 \times 10^4 \text{ W} = 73.8 \text{ kW} = 98.9 \text{ hp}.$$

Thus this rapid acceleration requires about 18 times as much power (not including the power to overcome friction) as cruising at a steady 15 m/s.

What about hill climbing? A 5% grade, which is about the maximum found on most interstate highways, rises 5 meters for every 100 meters of horizontal distance. A car moving at 30 m/s up a 5% grade is gaining elevation at the rate of (0.05) · (30 m/s) = 1.50 m/s. To lift the 911, weighing 12,260 N, at this rate requires a power of

$$\begin{aligned} P &= Fv = (12{,}260 \text{ N})(1.50 \text{ m/s}) \\ &= 18{,}390 \text{ J/s} = 18.4 \text{ kW} = 24.7 \text{ hp}. \end{aligned}$$

The *total* power required is this amount plus the 16.3 kW needed to maintain 30 m/s on a level road, that is,

$$P_{tot} = 18.4 \text{ kW} + 16.3 \text{ kW} = 34.7 \text{ kW} = 46.5 \text{ hp}.$$

For the record, the 911 advertises a maximum horsepower of 214 hp at an engine speed of 5900 rpm.

Finally, let's compare these energy and power quantities with some purely *thermal* considerations. This is a little premature; we will study the relationship of heat to mechanical energy in detail in later chapters, but here's a little preview. We will learn that to heat 1 kg of water from 0°C to 100°C requires an energy input to the water of $4.19 \times 10^5$ J. Thus the $3.5 \times 10^7$ J obtained from 1 L of gasoline is enough to heat $(3.5 \times 10^7 \text{ J})/(4.19 \times 10^5 \text{ J/kg}) = 84$ kg of water from freezing to boiling. That's not a whole lot of water, only about 23 gallons. So the amount of energy needed to heat 23 gallons of water from freezing to boiling is enough to push our car more than 19 km! Does that surprise you?

# SUMMARY

- The concepts of work and kinetic energy play central roles in one of the universal conservation laws of physics, the law of conservation of energy.
- When a force $\boldsymbol{F}$ acts on a particle that undergoes a displacement $\boldsymbol{s}$, the work $W$ done by the force is defined as

$$W = Fs \cos \phi = \boldsymbol{F} \cdot \boldsymbol{s}, \qquad (6\text{–}3)$$

where $\phi$ is the angle between the directions of $\boldsymbol{F}$ and $\boldsymbol{s}$. The unit of work in SI units is 1 newton · meter = 1 joule (1 N · m = 1 J). Work is a scalar quantity; it has an algebraic sign (positive or negative) but no direction in space.

- When the force varies during the displacement, the work done by the force is given by

$$W = \int_{x_1}^{x_2} F\, dx \qquad (6\text{–}4)$$

if the force has the same direction as the displacement, or by

$$W = \int_{P_1}^{P_2} F \cos \phi\, dl = \int_{P_1}^{P_2} F_{\parallel}\, dl = \int_{P_1}^{P_2} \boldsymbol{F} \cdot d\boldsymbol{l} \qquad (6\text{–}8)$$

if it makes an angle $\phi$ with the displacement. The path may be curved, and the angle $\phi$ may vary during the displacement.

### KEY TERMS

conservation of energy
energy
work
joule
force constant
Hooke's law
kinetic energy
work-energy theorem
power
average power
instantaneous power
watt

• The kinetic energy $K$ of a particle with mass $m$ and speed $v$ is

$$K = \tfrac{1}{2}mv^2. \qquad (6\text{–}11)$$

Kinetic energy is a scalar quantity; it has a magnitude (always positive) but no direction in space. Its units, $kg \cdot m^2/s^2$, are the same as the units of work: $1\ kg \cdot m^2/s^2 = 1\ N \cdot m = 1\ J$.

• When forces act on a particle while it undergoes a displacement, the total work $W_{tot}$ done on the particle by all the forces equals the change in the particle's kinetic energy:

$$W_{tot} = K_2 - K_1 = \Delta K. \qquad (6\text{–}12)$$

This relation is called the work-energy theorem.

• Power is the time rate of doing work. If an amount of work $\Delta W$ is done in a time $\Delta t$, the average power $P_{av}$ is

$$P_{av} = \frac{\Delta W}{\Delta t}. \qquad (6\text{–}17)$$

The instantaneous power $P$ is defined as

$$P = \lim_{\Delta t \to 0} \frac{\Delta W}{\Delta t} = \frac{dW}{dt}. \qquad (6\text{–}18)$$

When a force $\boldsymbol{F}$ acts on a particle moving with velocity $\boldsymbol{v}$, the instantaneous power, or rate at which the force does work, is

$$P = \boldsymbol{F} \cdot \boldsymbol{v}. \qquad (6\text{–}21)$$

Power, like work and kinetic energy, is a scalar quantity. Its SI unit, one joule per second (1 J/s), is called one watt (1 W).

# EXERCISES

## Section 6–2 Work

**6–1** A physics book is pushed 1.50 m along a horizontal tabletop by a horizontal force of 2.00 N. The opposing force of friction is 0.400 N. a) How much work is done on the book by the 2.00-N force? b) What is the work done on the book by the friction force?

**6–2** A factory worker pushes a 35.0-kg crate a distance of 5.0 m along a level floor at constant speed by pushing horizontally on it. The coefficient of kinetic friction between the crate and floor is 0.25. a) What magnitude of force must the worker apply? b) How much work is done on the crate by this force? c) How much work is done on the crate by friction? d) How much work is done by the normal force? By gravity?

**6–3** A fisherman reels in 20.0 m of line while pulling in a fish that exerts a constant resisting force of 18.0 N. If the fish is pulled in at constant speed, how much work is done on it by the tension in the line?

**6–4** A water skier is pulled by a tow rope behind a boat. She skis off to the side, and the rope makes an angle of 20.0° with her direction of motion. The tension in the rope is 120 N. How much work is done on the skier by the rope during a displacement of 150 m?

**6–5** Suppose the worker in Exercise 6–2 pushes downward at an angle of 30° below the horizontal. a) What magnitude of force must the worker apply to move the crate at constant speed? b) How much work is done on the crate by this force when the crate is pushed a distance of 5.0 m? c) How much work is done on the crate by friction during this displacement? d) How much work is done by the normal force? By gravity?

**6–6** The old oaken bucket that hangs in a well has a mass of 6.75 kg. You slowly raise it a distance of 7.00 m by pulling horizontally on a rope passing over a pulley at the top of the well. a) How much work do you do in pulling the bucket up? b) How much work is done by the gravitational force acting on the bucket?

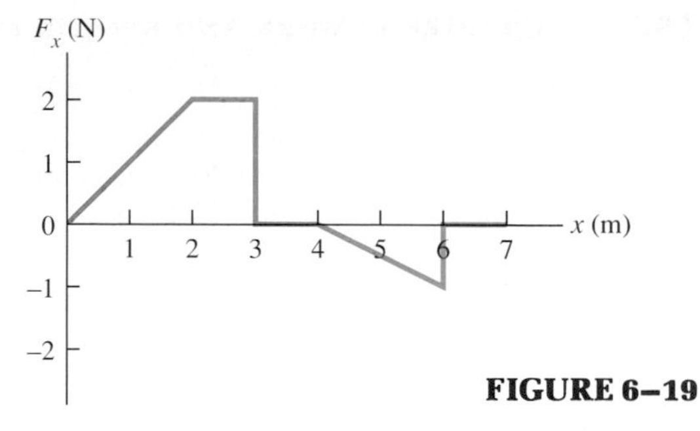

FIGURE 6–19

## Section 6–3 Work Done by a Varying Force

**6–7** A force of 90.0 N stretches a spring 0.400 m beyond its unstretched length. a) What magnitude of force is required to stretch the spring 0.100 m beyond its unstretched length? To compress the spring 0.200 m? b) How much work must be done to stretch the spring 0.100 m beyond its unstretched length? How much work to compress the spring 0.200 m from its unstretched length?

**6–8** If 8.00 J of work must be done to stretch a spring 5.00 cm beyond its unstretched length, how much work must be done to compress the same spring 10.0 cm from its unstretched length?

**6–9** An exercise machine for doing arm curls has a handle attached to a stiff spring. A woman does 20.0 J of work to move the handle 0.150 m and stretch the spring this same distance from its unstretched length. What magnitude of force must she apply to hold the handle in this position?

**6–10 Leg Presses.** As part of your daily workout, you lie on your back and push with your feet against a platform attached to two stiff springs (Fig. 6–17). When you push the platform, you compress the springs. You do 40.0 J of work when you compress the springs 0.200 m from their uncompressed length. How much *additional* work must you do to move the platform 0.200 m *farther?*

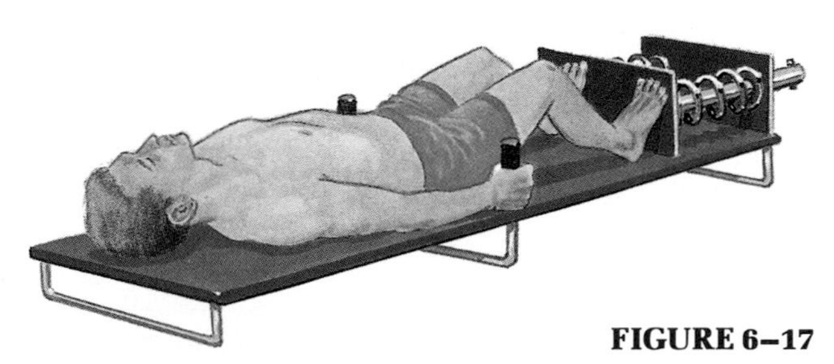
FIGURE 6–17

**6–11** A force $\boldsymbol{F}$ that is parallel to the $x$-axis is applied to an object. The $x$-component of the force varies with the $x$-coordinate of the object (Fig. 6–18). Calculate the work done by the force $\boldsymbol{F}$ when the object moves from a) $x = 0$ to $x = 12.0$ m; b) $x = 12.0$ m to $x = 8.0$ m.

**6–12** A child applies a force $\boldsymbol{F}$ parallel to the $x$-axis to a sled moving on the frozen surface of a small pond. As the child controls the speed of the sled, the $x$-component of the force that she applies varies with the $x$-coordinate of the sled as shown in Fig. 6–19. Calculate the work done by the force $\boldsymbol{F}$ when the sled moves from a) $x = 0$ to $x = 3.0$ m; b) $x = 3.0$ m to $x = 4.0$ m; c) $x = 4.0$ m to $x = 7.0$ m; d) $x = 0$ to $x = 7.0$ m.

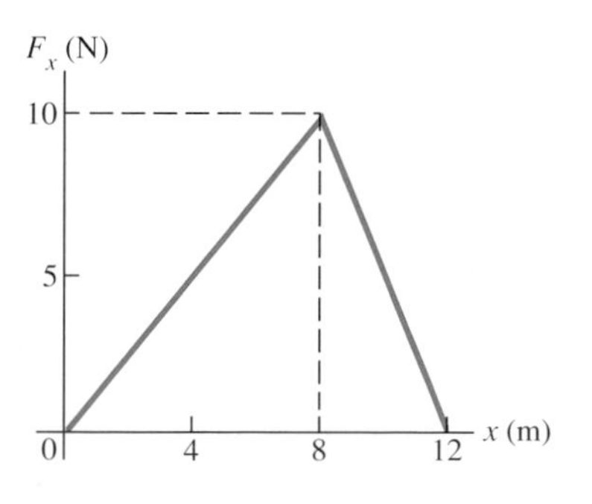

FIGURE 6–18

## Section 6–4 Work and Kinetic Energy

**6–13** Compute the kinetic energy, in joules, of a 5.00-g rifle bullet traveling at 500 m/s.

**6–14** a) Compute the kinetic energy of a 1200-kg automobile traveling at 25.0 km/h. b) By what factor does the kinetic energy increase if the speed is doubled?

**6–15** Baseball pitcher Dwight Gooden throws a fastball that has a speed (leaving his hand) of 42.0 m/s. The mass of the baseball is 0.145 kg. How much work has the pitcher done on the ball in throwing it?

**6–16** A television set with a mass of 18.2 kg is initially at rest on a horizontal frictionless surface. It is then pulled 2.50 m by a horizontal force with magnitude 112 N. Use the work-energy theorem to find its final speed.

**6–17** A little red wagon with a mass of 2.50 kg moves in a straight line on a frictionless horizontal surface. It has an initial speed of 3.00 m/s and then is pushed 4.0 m by a force with a magnitude of 2.50 N and in the direction of the initial velocity. a) Use the work-energy theorem (Eq. 6–12) to calculate the wagon's final speed. b) Calculate the acceleration produced by the force. Use this acceleration in the kinematic relations of Chapter 2 to calculate the wagon's final speed. Compare this result to that calculated in part (a).

**6–18** A sled with a mass of 8.00 kg moves in a straight line on a frictionless horizontal surface. At one point in its path its speed is 4.00 m/s; after it has traveled 3.00 m, its speed is 9.00 m/s in the same direction. Use the work-energy theorem to find the force acting on the sled, assuming that this force is constant and that it acts in the direction of the sled's motion.

**6–19** A force with magnitude 30.0 N acts on a 0.420-kg soccer ball moving initially in the direction of the force with a speed of 4.00 m/s. Over what distance must the force act to increase the ball's speed to 6.00 m/s?

**6–20** A 1.20-kg rock is dropped (zero initial speed) from the roof of a 20.0-m tall building. a) Calculate the work done by gravity on the rock during the rock's displacement from the roof to the ground. b) What is the kinetic energy of the rock just before it strikes the ground?

**6–21** A baseball with mass 0.145 kg is thrown straight upward with an initial speed of 27.0 m/s. a) How much work has gravity done on the baseball when it reaches a height of 20.0 m? b) Use the work-energy theorem to calculate the speed of the baseball at a height of 20.0 m.

**6–22** An ice cube with mass 0.050 kg slides 0.40 m down an inclined plane that slopes downward at an angle of 60° below the horizontal. If the ice cube starts from rest, what is its final speed? Friction can be neglected.

**6–23** A car is traveling on a level road with speed $v_0$ at the instant when the brakes are locked. a) Use the work-energy theorem (Eq. 6–12) to calculate the minimum stopping distance of the car in terms of $v_0$, the coefficient of kinetic friction $\mu_k$ between the tires and the road, and the acceleration $g$ due to gravity. b) The car stops in a distance of 42.7 m if $v_0 = 17.9$ m/s (40 mi/h). What is the stopping distance if $v_0 = 26.8$ m/s (60 mi/h)? Assume that $\mu_k$ and thus the friction force remain the same.

**6–24** A 2.00-kg block of ice is placed against a horizontal spring that has a force constant $k = 150$ N/m and is compressed 0.040 m. The spring is released and accelerates the block along a horizontal surface. Friction can be neglected. a) Calculate the work done on the block by the spring during the motion of the block from its initial position to where the spring has returned to its uncompressed length. b) What is the speed of the block after it leaves the spring?

**6–25** A spring with a force constant $k = 300$ N/m rests on a frictionless horizontal surface. One end is in contact with a stationary wall, and a 1.60-kg can of beans is pushed against the other end, compressing the spring 0.100 m. The can is then released with no initial velocity. What is the can's speed when the spring a) returns to its uncompressed length? b) is still compressed 0.060 m?

**6–26** An ingenious bricklayer builds a device for shooting bricks up to the top of the wall where he is working. He places a brick on a compressed spring with force constant $k = 250$ N/m. When the spring is released, the brick is propelled upward. If the brick has mass 1.50 kg and is to reach a maximum height of 3.8 m above its initial position on the compressed spring, what distance must the spring be compressed initially?

**6–27** A small glider with mass 0.0500 kg is placed against a compressed spring that has $k = 150$ N/m at the bottom of an air track that slopes upward at an angle of 40.0° above the horizontal. When the spring is released, the glider leaves the spring and travels a maximum distance of 1.80 m along the air track before sliding back down. a) What distance was the spring originally compressed? b) What is the kinetic energy of the glider when it has traveled along the air track 0.80 m from its initial position against the compressed spring?

## Section 6–5 Power

**6–28** At 7.50 cents per kWh, what does it cost to operate a 4000-W motor for 8.00 h?

**6–29** A tandem (two-person) bicycle team must overcome a force of 150 N to maintain a speed of 9.20 m/s. Find the power required per rider, assuming that each contributes equally. Express your answer in horsepower.

**6–30** The total consumption of electrical energy in the United States is about $1.0 \times 10^{19}$ joules per year. a) What is the average rate of electrical energy consumption in watts? b) If the population of the United States is 240 million, what is the average rate of electrical energy consumption per person? c) The sun transfers energy to the earth by radiation at a rate of approximately 1.4 kW per square meter of surface. If this energy could be collected and converted to electrical energy with 100% efficiency, how great an area (in square kilometers) would be required to collect the electrical energy used by the United States?

**6–31** A man whose mass is 80.0 kg walks up to the third floor of a building. This is a vertical height of 12.0 m above the street level. If he climbs the stairs in 20.0 s, what was his rate of working, in watts?

**6–32** The hammer of a pile driver has a weight of 4800 N and must be lifted at constant speed a vertical distance of 1.80 m in 3.00 s. What horsepower engine is required?

**6–33** The engine of a motorboat delivers 30.0 kW to the propeller while the boat is moving at 12.0 m/s. If the boat is towed at the same speed, what is the tension in the towline?

**6–34** A ski tow is operated on a 37.0° slope 300 m long. The rope moves at 12.0 km/h, and power is provided for 80 riders at one time, with an average mass per rider of 65.0 kg. Estimate the power required to operate the tow.

## Section 6–6 Automotive Power: A Case Study in Energy Relations

**6–35** Consider the car of Section 6–6. a) Verify that the power needed for a constant speed of 30.0 m/s on a level road is 16.3 kW. b) What volume of gasoline is consumed in 1.0 h at this speed if 15% of the $3.5 \times 10^7$ J of energy obtained by burning 1.0 L of gasoline is available to propel the car? c) Calculate the fuel consumption per unit distance in L/km and in gal/mi.

**6–36** A truck engine develops 20.0 kW (26.8 hp) when the truck is traveling at 50.0 km/h. a) What is the resisting force acting on the truck? b) If the resisting force is proportional to the speed, what power will drive the truck at 25.0 km/h? At 100.0 km/h? Give your answers in kilowatts and in horsepower.

**6–37** a) If 6.00 hp are required to drive a 1200-kg automobile at 50.0 km/h on a level road, what is the total retarding force due to friction, air resistance, etc.? b) What power is necessary to drive the car at 50.0 km/h up a 10% grade (a hill rising 10.0 m vertically in 100 m horizontally)? c) What power is necessary to drive the car at 50.0 km/h *down* a 2% grade? d) Down what percent grade would the car coast at 50.0 km/h?

## PROBLEMS

**6–38** A 6.00-kg package slides 4.00 m down a ramp that is inclined at 53.1° below the horizontal. The coefficient of kinetic friction between the package and the ramp is $\mu_k = 0.40$. Calculate a) the work done by friction on the package; b) the work done by gravity; c) the work done by the normal force; d) the total work done on the package.

**6–39** A 90.0-N suitcase is pulled up a frictionless ramp inclined at 30° above the horizontal by a force $\boldsymbol{F}$ with a magnitude of 80.0 N and acting parallel to the ramp. If the suitcase travels 4.60 m along the ramp, calculate a) the work done on the suitcase by the force $\boldsymbol{F}$; b) the work done by the gravity force; c) the work done by the normal force; d) the total work done on the suitcase.

**6–40** An object is attracted toward the origin with a force given by $F_x = -k/x^2$. (Gravitational and electrical forces have this distance dependence.) Calculate the work done by the force $F_x$ when the object moves in the $x$-direction from $x_1$ to $x_2$. If $x_2 > x_1$, is the work done by $F_x$ positive or negative?

**6–41** An object is attracted toward the origin with a force given by $F = \alpha x^3$, where $\alpha = 8.00\ \mathrm{N/m^3}$. What is the force $F$ when the object is a) at point $a$, 1.00 m from the origin? b) at point $b$, 2.00 m from the origin? c) How much work is done by the force $F$ when the object moves from $a$ to $b$? Is this work positive or negative?

**6–42** A 4.00-kg package slides 2.00 m down a ramp that is inclined at 50.0° below the horizontal. The coefficient of kinetic friction between the package and the ramp is $\mu_k = 0.30$. a) Calculate the work done on the package by gravity. b) Calculate the work done on the package by friction. c) If the package has a speed of 1.2 m/s at the top of the ramp, what is its speed after sliding 2.00 m down the ramp?

**6–43** A small block with a mass of 0.0500 kg is attached to a cord passing through a hole in a horizontal frictionless surface (Fig. 6–20). The block is originally revolving at a distance of 0.40 m from the hole with a speed of 0.80 m/s. The cord is then pulled from below, shortening the radius of the circle in which the block revolves to 0.20 m. At this new distance the speed of the block is observed to be 1.60 m/s. a) What is the tension in the cord when the block has speed $v = 0.80$ m/s? b) What is the tension in the cord when the block has speed $v = 1.60$ m/s? c) How much work was done by the person who pulled on the cord?

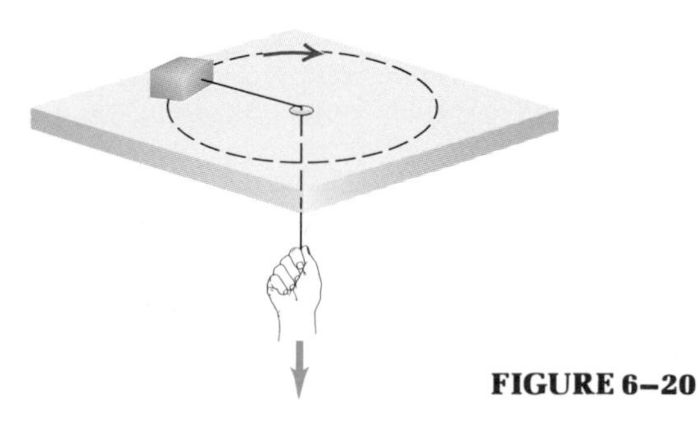

**FIGURE 6–20**

**6–44 Proton Bombardment.** A proton with a mass of $1.67 \times 10^{-27}$ kg is propelled at an initial speed of $8.00 \times 10^4$ m/s directly toward a gold nucleus that is 4.00 m away. The proton is repelled by the gold nucleus with a force of magnitude $F = \alpha/x^2$, where $x$ is the separation between the two objects and $\alpha = 1.82 \times 10^{-26}\ \mathrm{N \cdot m^2}$. Assume that the gold nucleus remains at rest. a) What is the speed of the proton when it is $8.00 \times 10^{-9}$ m from the gold nucleus? b) How close to the gold nucleus does the proton get? (Calculate the position of the proton for which the work done by the repulsive force has reduced the velocity of the proton momentarily to zero.)

**6–45** A block of ice with mass 5.00 kg is initially at rest on a frictionless horizontal surface. A worker then applies a horizontal force $F$ to it. As a result, the block moves along the $x$-axis such that its position as a function of time is given by $x(t) = \alpha t^2 + \beta t^3$, where $\alpha = 3.00\ \mathrm{m/s^2}$ and $\beta = 0.500\ \mathrm{m/s^3}$. a) Calculate the velocity of the object when $t = 4.00$ s. b) Calculate the work done by the force $F$ during the first 4.00 s of the motion.

**6–46** The spring of a spring gun has force constant $k = 500$ N/m. The spring is compressed 0.0500 m, and a ball with mass 0.0300 kg is placed in the horizontal barrel against the compressed spring. The spring is then released, and the ball is propelled out the barrel. a) Compute the speed with which the ball leaves the gun if friction forces are negligible. b) Determine the speed of the ball as it leaves the end of the barrel if a constant resisting force of 10.0 N acts on the ball while it travels the 0.0500 m along the barrel. Note that in this case the maximum speed does not occur at the end of the barrel.

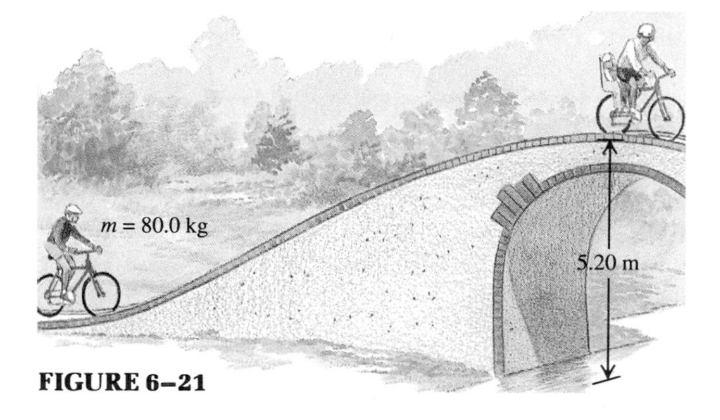


**FIGURE 6–21**

**6–47** You and your bicycle have a combined mass of 80.0 kg. When you reach the base of an overpass, you are traveling along the road at 3.00 m/s (Fig. 6–21). The vertical height of the overpass is 5.20 m, and your speed at the top is 1.50 m/s. Ignore work done by friction and any inefficiency in the bike or your legs. a) What is the total work done on you and your bicycle when you go from the base to the top of the overpass? b) How much work have you done with the force you apply to the pedals?

**6–48** A 6.00-kg dictionary is pushed up a frictionless ramp inclined upward at 30.0° above the horizontal. It is pushed 2.00 m

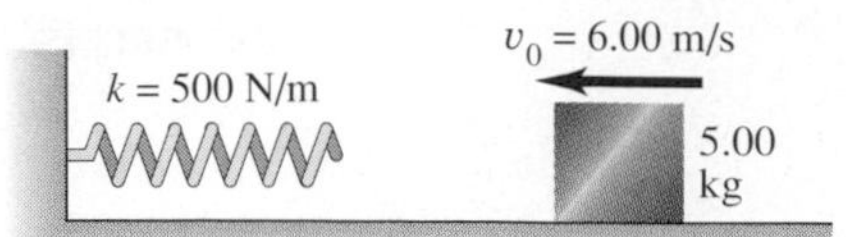


**FIGURE 6–22**

along the incline by a constant 100-N force parallel to the ramp. If the dictionary's speed at the bottom is 2.00 m/s, what is its speed at the top? Use energy methods.

**6–49** You are asked to design spring bumpers for the walls of a parking garage. A freely rolling 1200-kg car moving at 0.50 m/s is to compress the spring no more than 0.075 m before stopping. What should be the force constant of the spring?

**6–50** A 5.00-kg block is moving at 6.00 m/s along a frictionless horizontal surface toward a spring with force constant $k =$ 500 N/m that is attached to a wall (Fig. 6–22). Find the maximum distance the spring will be compressed. The spring has negligible mass.

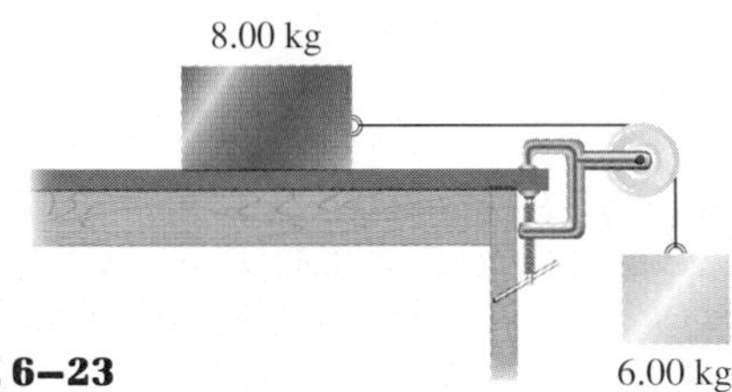


**FIGURE 6–23**

**6–51 Power of Human Heart.** The human heart is a powerful and extremely reliable pump. Each day it takes in and discharges about 7500 L of blood. If the work done by the heart is equal to the work required to lift this amount of blood a height equal to that of the average American female (1.63 m), and if the density (mass per unit volume) of blood is the same as that of water ($1.00 \times 10^3$ kg/m$^3$), a) how much work does the heart do in a day? b) what is its power output in watts?

**6–52** Consider the system shown in Fig. 6–23. The coefficient of kinetic friction between the 8.00-kg block and the tabletop is $\mu_k = 0.30$. Neglect the mass of the rope and of the pulley, and assume that the pulley is frictionless. Use energy methods to calculate the speed of the 6.00-kg block after it has descended 1.50 m, starting from rest.

**6–53** A pump is required to lift 800 kg (about 200 gallons) of water per minute from a well 15.0 m deep and eject it with a speed of 20.0 m/s. a) How much work is done per minute in lifting the water? b) How much in giving it kinetic energy? c) What engine power is needed?

***6–54** An automobile with mass $m$ accelerates, starting from rest, while the engine supplies constant power $P$. a) Show that the speed is given as a function of time by $v = (2Pt/m)^{1/2}$. b) Show that the acceleration is given as a function of time by $a = (P/2mt)^{1/2}$ and thus is not constant. c) Show that the displacement is given as a function of time by $x - x_0 = (8P/9m)^{1/2}t^{3/2}$.

**6–55** The Grand Coulee Dam is 1270 m long and 170 m high. The electrical power output from generators at its base is approximately 2000 MW. How many cubic meters of water must flow over the dam per second to produce this amount of power? (The density, or mass per unit volume, of water is $1.00 \times 10^3$ kg/m$^3$.)

**6–56** Figure 6–24 shows how the force exerted by the string of a compound bow on an arrow varies as a function of how far back the arrow is pulled. Assume that the same force is exerted on the arrow as it moves forward after being released. Full draw for this bow is at a draw length of 75.0 cm. If the bow shoots a 0.0425-kg arrow from full draw, what is the speed of the arrow as it leaves the bow?

**6–57** For a touring bicyclist the drag coefficient is 1.00, the frontal area is 0.463 m$^2$, and the coefficient of rolling friction is 0.0045. The rider weighs 540 N, and her bike weighs 111 N. a) To maintain a speed of 14.0 m/s (about 31 mi/h) on a level road, what must the rider's power output be? b) For racing, the same rider uses a different bike, one with coefficient of rolling friction 0.0030 and a weight of 89 N. She also crouches down, reducing her drag coefficient to 0.88 and reducing her frontal area to 0.366 m$^2$. What then must her power output be to maintain a speed of 14.0 m/s? c) For the situation in part (b), what power output is required to maintain a speed of 7.0 m/s? Note the great drop in power requirement when the speed is only halved. (See "The Aerodynamics of Human-Powered Land Vehicles," *Scientific American*, December 1983. The article discusses aerodynamic speed limitations for a wide variety of human-powered vehicles.)

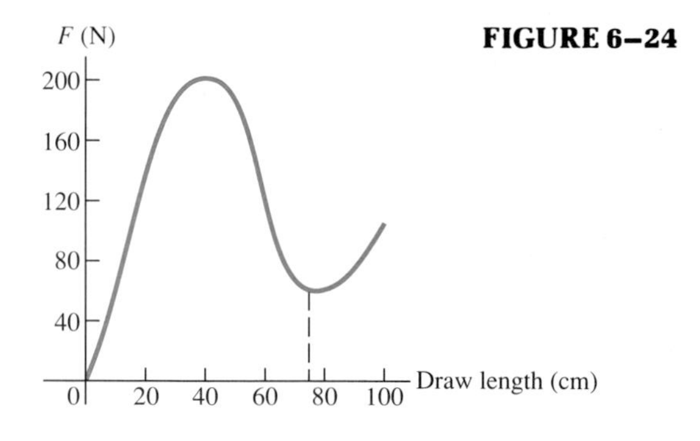


**FIGURE 6–24**

## CHALLENGE PROBLEM

**6–58** An object has several forces acting on it. One of these forces is $\boldsymbol{F} = \alpha xy\boldsymbol{i}$, a force in the $+x$-direction whose magnitude depends on the position of the object and where the constant $\alpha = 5.00$ N/m$^2$. Calculate the work done on the object by this force for each of the following displacements of the object: a) The object starts at the point $x = 0$, $y = 4.00$ m and moves parallel to the $x$-axis to the point $x = 2.00$ m, $y = 4.00$ m. b) The object starts at the point $x = 2.00$ m, $y = 0$ and moves in the $y$-direction to the point $x = 2.00$ m, $y = 4.00$ m. c) The object starts at the origin and moves on the line $y = 2x$ to the point $x = 2.00$ m, $y = 4.00$ m.

# CHAPTER 7

# Conservation of Energy

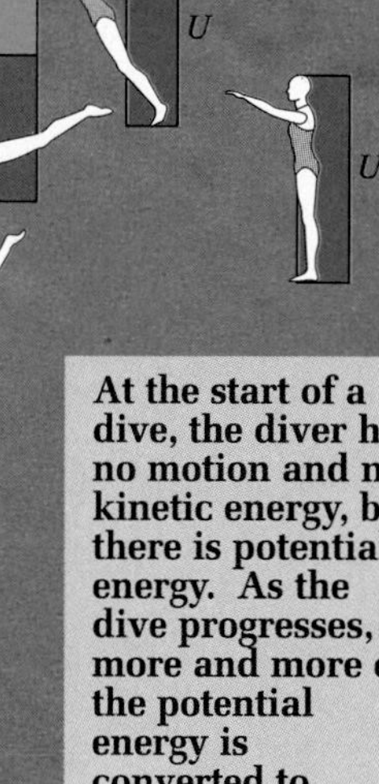


At the start of a dive, the diver has no motion and no kinetic energy, but there is potential energy. As the dive progresses, more and more of the potential energy is converted to kinetic energy. The total energy—potential plus kinetic—is constant throughout the dive.

## KEY CONCEPTS

**• Potential energy is energy associated with the position of a system rather than its motion. Examples are a body acted on by the earth's gravitational field and a body acted on by a stretched spring.**

**• Potential energy can be used to calculate work done by forces acting on a body. When the work of all the forces can be represented in terms of potential energy changes, the total mechanical energy of the system (kinetic plus potential) is constant. Otherwise, the change in the total mechanical energy equals the work done by the forces not included in the potential energy.**

**• If the potential energy of a system is known as a function of its coordinates, the components of force acting on a body can be expressed as the negative derivatives of the potential-energy function.**

## INTRODUCTION

Remember the first time you jumped off the high board into a swimming pool? Maybe you only thought about jumping, knowing that when you hit the water, you'd be moving pretty fast, with a lot of kinetic energy. Where did that energy come from? The answer we learned in Chapter 6 was that the gravitational force (your weight) did *work* on you as you dropped. Your kinetic energy increased by an amount equal to the work done. We're now going to learn an alternative and very useful way to think about these situations. Our new approach is based on a new concept called *potential energy,* which is energy associated with the *position* of a system rather than its *motion*. The gravitational potential energy is greater when you're on the high board than when you hit the water; as you drop, potential energy is converted to kinetic energy.

If you bounce on the end of the board before you jump, the bent board is storing another kind of energy called *elastic potential energy.* We'll discuss elastic potential energy of simple systems such as a stretched or compressed spring. We will prove that in some cases the *total* energy of a system, kinetic plus potential, is constant during the motion of the system. This will give us the beginning of an important conservation principle called *conservation of energy.* That's what this chapter is about.

## 7–1 POTENTIAL ENERGY AND CONSERVATIVE FORCES

A body gains or loses kinetic energy because it interacts with other bodies that exert forces on it. We learned in Chapter 6 that during any interaction the change in a body's kinetic energy is equal to the total work done on the body by the forces that act on it during the interaction.

In many situations it seems as though kinetic energy has been *stored* in a system, to be recovered later. It's like a savings account in a bank: You deposit money, and then you can draw it out later. For example, the pile-driver of Example 6–7 raises the 200-kg steel hammerhead 3.00 m in the air with a winch mechanism. The hammerhead is then released; it falls freely and strikes the top of the vertical steel I-beam that is being driven into the ground. It seems reasonable that in hoisting the hammer into the air we are storing energy in the system, energy that is later converted into kinetic energy as the hammer falls.

Or consider your cousin Throckmorton on the swing (Example 6–5, Section 6–3). You give him a shove, which gives him some kinetic energy, and then you let him swing freely back and forth. He stops when he reaches the front and back end points of his arc, so at these points he has no kinetic energy. But he regains his kinetic energy as he passes through the low point in the arc. It seems as though at the high points the energy is stored in some other form, related to his height above the ground, and is converted back to kinetic energy as he swings toward the low point.

We can cite many other examples: The humble pogo stick, in which a spring is compressed as you jump on the stick and then rebounds, launching you into the air. A bow and arrow, where you do work to pull the string back, storing energy that is later converted partly to kinetic energy of the arrow. A mousetrap, in which the spring stores energy that is released when the trigger is tripped. A wind-up alarm clock. And on and on.

All these examples point to the idea of an energy associated with the *position* of bodies in a system. Changes in this energy may accompany opposite changes in kinetic energy in such a way that the *total* energy remains constant, or is *conserved.* Energy associated with position is called **potential energy,** and forces that can be associated with a potential energy are called **conservative forces.** A system in which the total mechanical energy, kinetic and potential, is constant, is called a **conservative system.** As we will see, there is potential energy associated with a body's weight and its height above the ground. There is also potential energy associated with *elastic* deformations of a body, such as stretching or compressing a spring (Fig. 7–1). Later, in Chapter 24, we will deal with potential energy associated with the positions of electrically charged particles that exert forces on each other.

**7–1** Stroboscopic photograph of a pole-vaulter. The athlete's initial kinetic energy is partly stored as elastic potential energy in the flexed pole. Most of this elastic potential energy and some of his internal energy are then used to give the increase in gravitational potential energy needed to clear the bar.

Another way to look at potential energy is that "potential" refers to the potential for a force to do work on a body. When the pile-driver hammer is raised, the earth's gravitational force (the hammer's weight) gains the opportunity to do work on the hammer as it falls. When we set a mousetrap, we give it the opportunity to do work on whatever is in its way when the spring snaps back. And so on. So potential energy is associated with the opportunity, or potential, to do work.

The kinetic energy of a falling body increases as it drops because a force (the earth's gravity, its weight) is doing work on it. Remember, $W_{\text{tot}} = \Delta K$. Another way to say this is that the kinetic energy increases because the potential energy decreases. Putting these two pieces together, we see that there has to be a direct relation between work and potential energy. When the force we are considering (such as the pile-driver's weight) does positive work, the potential energy must decrease.

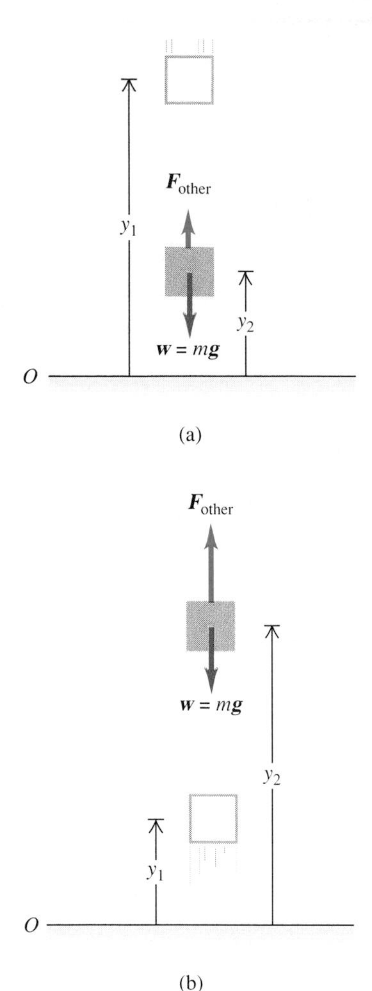


**7–2** Work done by the gravitational force $w$ during the motion of a body from one point to another. The vertical displacement is (a) downward or (b) upward with initial height $y_1$ and final height $y_2$.

So we can think of potential energy as a shorthand way to calculate work done by some of the forces in a system.

All this may sound pretty abstract. Never fear; we are now going to look in detail at several particular examples of potential energy and their uses. Our first example is gravitational potential energy, the subject of the next section.

## 7–2 GRAVITATIONAL POTENTIAL ENERGY

Let's apply the work-energy theorem to the work done on a body when it moves under the action of the earth's gravitational force (the body's weight). This will lead directly to the concept of *gravitational potential energy,* energy associated with the *position* of the body relative to the earth.

Let's consider a body with mass $m$ that moves along the (vertical) $y$-axis (Fig. 7–2). The forces acting on it are its weight, with magnitude $w = mg$, and possibly some other forces; we call the resultant of all the other forces $\boldsymbol{F}_{other}$. We'll assume that the body stays close enough to earth's surface that the weight is constant. We want to find the work done by the weight when the body drops from a height $y_1$ above the origin to a smaller height $y_2$ (Fig. 7–2a). The weight and displacement are in the same direction, so the work $W_{grav}$ done on the body by its weight is positive,

$$W_{grav} = Fs = w(y_1 - y_2) = mgy_1 - mgy_2. \tag{7–1}$$

This expression also gives the correct work when the body moves upward and $y_2$ is greater than $y_1$ (Fig.7–2b). In that case the quantity $(y_1 - y_2)$ is negative, and $W_{grav}$ is negative because the weight and displacement are opposite in direction.

Equation (7–1) shows that we can express $W_{grav}$ in terms of the values of the quantity $mgy$ at the beginning and at the end of the displacement. This quantity, the product of the weight $mg$ and the height $y$ above the origin of coordinates, is called the **gravitational potential energy,** $U$:

$$U = mgy. \tag{7–2}$$

Its initial value is $U_1 = mgy_1$, and its final value is $U_2 = mgy_2$. We can express the work $W_{grav}$ done by the gravitational force during the displacement from $y_1$ to $y_2$ as

$$W_{grav} = U_1 - U_2 = -\Delta U. \tag{7–3}$$

The negative sign in front of $\Delta U$ is essential. Remember that $\Delta U$ always means the final value minus the initial value. When the body moves down, $y$ decreases, the gravitational force does *positive* work, and the potential energy *decreases* (Fig. 7–3a). It's like drawing some money out of the bank (decreasing $U$) and spending it (doing positive work). When the body moves *up* (Fig. 7–3b), the work done by the gravitational force is *negative,* and the potential energy *increases.*

To see what all this is good for, suppose the body's weight is the *only* force acting on it, so $\boldsymbol{F}_{other} = \boldsymbol{0}$. The body may be in free fall, or perhaps we threw it up in the air. Let its speed at point $y_1$ be $v_1$ and its speed at $y_2$ be $v_2$. The work-energy theorem, Eq. (6–12), says $W_{tot} = K_2 - K_1$, and in our case, $W_{tot} = W_{grav} = U_1 - U_2$. Putting these together, we get

$$U_1 - U_2 = K_2 - K_1,$$

which we can rewrite as

$$K_1 + U_1 = K_2 + U_2, \tag{7–4}$$

**7–3** (a) Gravitational potential energy decreases as a sky diver falls toward earth. (b) Gravitational potential energy increases as a rock climber ascends a cliff.

or

$$\tfrac{1}{2}mv_1^2 + mgy_1 = \tfrac{1}{2}mv_2^2 + mgy_2. \qquad (7\text{–}5)$$

We now define $K + U$ to be the **total mechanical energy** (kinetic plus potential) of the system; let's call it $E$. Then $E_1$ $(=K_1 + U_1)$ is the total energy at $y_1$, and $E_2$ $(=K_2 + U_2)$ is the total energy at $y_2$. Equations (7–4) and (7–5) say that when the body's weight is the only force doing work on it, $E_1 = E_2$. That is, $E$ is constant; it has the same value at $y_1$ and $y_2$ and at all points during the motion. **When $F_{\text{other}} = 0$, the total mechanical energy is constant or conserved.** This is our first example of the principle of **conservation of energy.**

When we throw a ball into the air, it slows down on the way up as kinetic energy is converted to potential energy. On the way back down, potential energy is converted back to kinetic energy, and the ball speeds up. But the total energy, kinetic plus potential, is the same at every point in the motion.

## EXAMPLE 7–1

**Height of a baseball from energy conservation** You throw a 0.150-kg baseball straight up in the air, giving it an initial upward velocity of magnitude 20.0 m/s. Use conservation of energy to find how high it goes, ignoring air resistance.

**SOLUTION** The only force doing work on the ball after it leaves your hand is its weight, and we can use Eq. (7–5). Let's take the origin at the starting point (point 1), where the ball leaves your hand; then $y_1 = 0$ (Fig. 7–4). At this point, $v_1 = 20.0$ m/s. We want to find the height at point 2, where it stops and begins to fall back to earth. At this point, $v_2 = 0$ and $y_2$ is unknown. Eq. (7–5) says $K_1 + U_1 = K_2 + U_2$, or

$$\tfrac{1}{2}mv_1^2 + mgy_1 = \tfrac{1}{2}mv_2^2 + mvy_2,$$

$$\tfrac{1}{2}(0.150\text{ kg})(20.0\text{ m/s})^2 + (0.150\text{ kg})(9.80\text{ m/s}^2)(0)$$
$$= \tfrac{1}{2}(0.150\text{ kg})(0)^2 + (0.150\text{ kg})(9.80\text{ m/s}^2)y_2,$$

$$y_2 = 20.4\text{ m}.$$

The mass divides out, as we should expect; we learned in Chapter 2 that the motion of a body in free fall doesn't depend on its mass.

We could have substituted the values $y_1 = 0$ and $v_2 = 0$ in Eq. (7–5) and then solved algebraically for $y_2$ to get

$$y_2 = \frac{v_1^2}{2g}.$$

This gives us a result that we could have derived using Eq. (2–14).

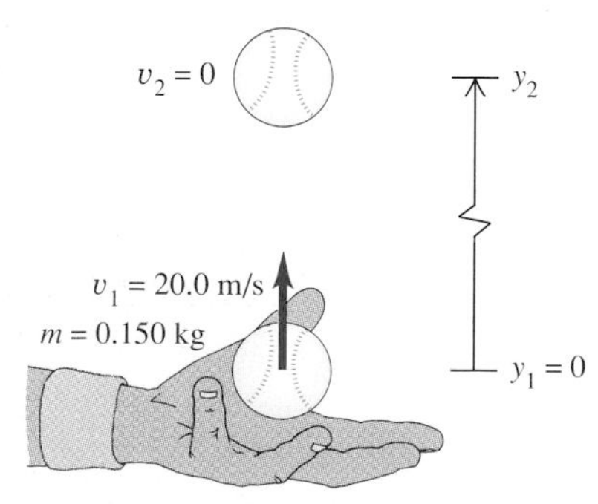


**7–4** After a baseball leaves your hand, the only force acting on it is its weight, so energy is conserved.

An important point about potential energy is that it doesn't matter where we put the origin of coordinates. If we shift the origin for $y$, both $y_1$ and $y_2$ change, but the *difference* $(y_2 - y_1)$ does not. It follows that although $U_1$ and $U_2$ depend on where we place the origin, the difference $(U_2 - U_1)$ does not. The physically significant quantity is not the value of $U$ at a particular point, but only the *difference* in $U$ between two points. We can define $U$ to be zero at any point we choose (in this case, at the origin); the difference in values of $U$ between any two points is independent of this choice.

When other forces act on the body in addition to its weight, then $\boldsymbol{F}_{\text{other}}$ in Fig. 7–2 is *not* zero. In the pile-driver example that we mentioned in Section 7–1 the force applied by the hoisting cable and the friction with the vertical guide rails are examples of forces that might be included in $\boldsymbol{F}_{\text{other}}$. The gravitational work $W_{\text{grav}}$ is still given by Eq. (7–3), but the total work $W_{\text{tot}}$ is then the sum of $W_{\text{grav}}$ and the work done by $\boldsymbol{F}_{\text{other}}$. We will call this additional work $W_{\text{other}}$, so $W_{\text{tot}} = W_{\text{grav}} + W_{\text{other}}$. Equating this to the change in kinetic energy, we have

$$W_{\text{other}} + W_{\text{grav}} = K_2 - K_1. \tag{7–6}$$

Also, from Eq. (7–3), $W_{\text{grav}} = U_1 - U_2$, so

$$W_{\text{other}} + U_1 - U_2 = K_2 - K_1,$$

which we can rearrange in the form

$$K_1 + U_1 + W_{\text{other}} = K_2 + U_2. \tag{7–7}$$

Finally, using the appropriate expressions for the various energy terms, we obtain

$$\tfrac{1}{2}mv_1^2 + mgy_1 + W_{\text{other}} = \tfrac{1}{2}mv_2^2 + mgy_2. \tag{7–8}$$

We can translate Eqs. (7–7) and (7–8): **The work done by all the forces *except the gravitational force* equals the change in the total mechanical energy $E = K + U$ of the system.** When $W_{\text{other}}$ is positive, $E$ increases, and $K_2 + U_2$ is greater than $K_1 + U_1$. When $W_{\text{other}}$ is negative, $E$ decreases. In the special case in which no forces other than the body's weight do work, $W_{\text{other}} = 0$. The total mechanical energy is then constant, and we are back to Eq. (7–4) or (7–5).

## PROBLEM-SOLVING STRATEGY

### Conservation of mechanical energy I

**1.** Decide what the initial and final states (the positions and velocities) of the system are; use the subscript 1 for the initial state and 2 for the final state. It helps to draw sketches showing the initial and final states.

**2.** Define your coordinate system, particularly the level at which $y = 0$. You will use this to compute potential energies. Equation (7–2) assumes that the positive direction for $y$ is upward; we suggest that you use this choice consistently.

**3.** List the initial and final kinetic and potential energies, that is, $K_1$, $K_2$, $U_1$, and $U_2$. In general, some of these will be known and some unknown. Use algebraic symbols for any unknown coordinates or velocities.

**4.** Identify all nongravitational forces that do work. A free-body diagram is always helpful. Calculate the work $W_{\text{other}}$ done by all these forces. If some of the quantities that you need are unknown, represent them by algebraic symbols.

**5.** Use Eq. (7–7) to relate these quantities. If there is no non-gravitational work $W_{\text{other}}$, this becomes Eq. (7–4). Then solve to find whatever unknown quantity is required.

**6.** Keep in mind, here and in later sections, that the work done by each force must be represented either in $U_1 - U_2 = -\Delta U$ or as $W_{\text{other}}$ but *never* in both places. If the gravitational work is included in $\Delta U$, do not include it again in $W_{\text{other}}$.

## EXAMPLE 7–2

**Work and energy in throwing a baseball** Referring to Example 7–1, suppose your hand moves up 0.50 m while you are throwing the ball, which leaves your hand with an upward velocity of 20 m/s. Again ignore air resistance. a) Assuming that your hand exerts a constant upward force on the ball, find the magnitude of that force. b) Find the speed of the ball at a point 15 m above the point where it leaves your hand.

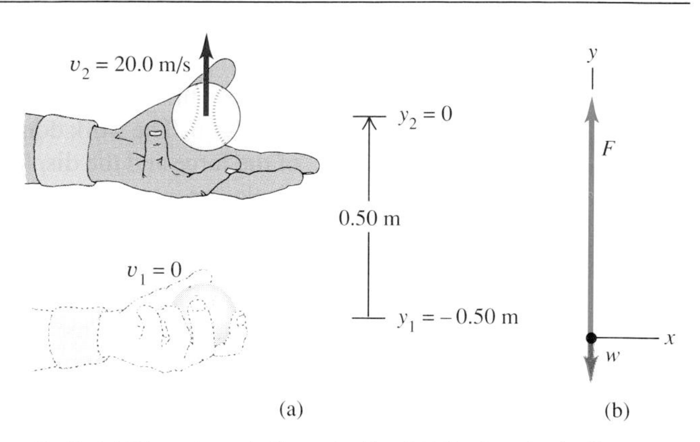


**7–5** (a) Throwing a ball vertically. (b) The free-body diagram for the ball as the force $\boldsymbol{F}$ applied by the hand does the work $W_{\text{other}}$ on the ball.

**SOLUTION** Figure 7–5 shows a diagram of the situation, including a free-body diagram for the ball while it is being thrown.

a) Let point 1 be the point where your hand first starts to move, and let point 2 be the point where the ball leaves your hand. With the same coordinate system as before, we have $y_1 = -0.50$ m and $y_2 = 0$. Then

$$K_1 = 0,$$

$$U_1 = mgy_1 = (0.150\ \text{kg})(9.8\ \text{m/s}^2)(-0.50\ \text{m}) = -0.74\ \text{J},$$

$$U_2 = mgy_2 = (0.150\ \text{kg})(9.8\ \text{m/s}^2)(0) = 0,$$

$$K_2 = \tfrac{1}{2}mv_2^{\,2} = \tfrac{1}{2}(0.150\ \text{kg})(20\ \text{m/s})^2 = 30\ \text{J}.$$

According to Eq. (7–7), $K_1 + U_1 + W_{\text{other}} = K_2 + U_2$, so

$$0 + (-0.74\ \text{J}) + W_{\text{other}} = 30\ \text{J} + 0,$$

$$W_{\text{other}} = 31\ \text{J}.$$

The kinetic energy of the ball increases by 30 J, and the potential energy increases by less than 1 J; the sum is equal to $W_{\text{other}}$.

The work $W_{\text{other}}$ is the work done by the upward force $\boldsymbol{F}$ that your hand applies to the ball as you throw it. Assuming that this force is constant, we have

$$W_{\text{other}} = F(y_2 - y_1) = (0.50\ \text{m})F,$$

$$F = \frac{W_{\text{other}}}{y_2 - y_1} = \frac{31\ \text{J}}{0.50\ \text{m}} = 62\ \text{N}.$$

This is about 40 times as great as the weight of the ball.

b) Let point 3 be at the 15-m height. Then $y_3 = 15$ m, and we want to find the speed $v_3$ at this point. Between points 2 and 3 the force of your hand no longer acts, and $W_{\text{other}} = 0$. We have

$$U_3 = mgy_3 = (0.150\ \text{kg})(9.8\ \text{m/s}^2)(15\ \text{m}) = 22\ \text{J},$$

$$K_3 = \tfrac{1}{2}mv_3^{\,2} = \tfrac{1}{2}(0.150\ \text{kg})v_3^{\,2}.$$

Conservation of energy, Eq. (7–4), gives

$$K_2 + U_2 = K_3 + U_3,$$

$$30\ \text{J} + 0 = \tfrac{1}{2}(0.150\ \text{kg})v_3^{\,2} + 22\ \text{J},$$

$$v_3 = \pm 10\ \text{m/s}.$$

We can interpret $v_3$ as the $y$-component of the ball's velocity. Then the significance of the plus-or-minus sign is that the ball passes this point *twice*, once on the way up and again on the way down. The *potential* energy at this point is the same whether the ball is moving up or down, so its kinetic energy and its *speed* are the same. The velocity is positive (plus) when the ball is moving up and negative (minus) when it is moving down.

---

In our first two examples the body moved along a straight vertical line. What happens when the path is a slanted or curved path (Fig. 7–6a)? The forces include the gravitational force $w = mg$ and possibly an additional force $\boldsymbol{F}_{\text{other}}$. The work done by the

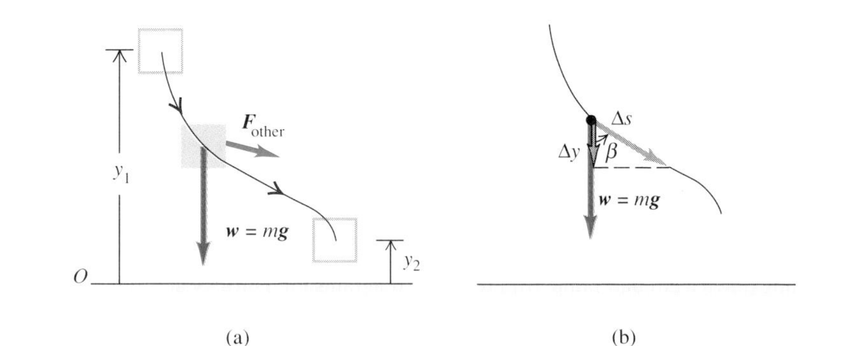


**7–6** (a) A displacement along a curved path. (b) The work done by the gravitational force $w$ depends only on the vertical component of displacement $\Delta y$.

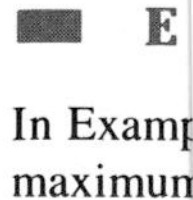

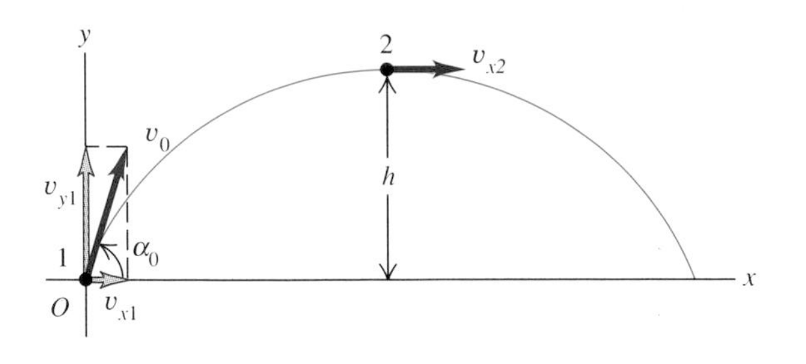


**7–10** Trajectory of a projectile.

Conservation of energy then gives $K_1 + U_1 = K_2 + U_2$ and

$$\tfrac{1}{2}m(v_{x1}^2 + v_{y1}^2) + 0 = \tfrac{1}{2}m(v_{x2}^2 + v_{y2}^2) + mgh.$$

To simplify this, we multiply through by $2/m$ to obtain

$$v_{x1}^2 + v_{y1}^2 = v_{x2}^2 + v_{y2}^2 + 2gh.$$

Now for the *coup de grace:* We recall that in projectile motion the $x$-component of acceleration is zero, so the $x$-component of velocity is constant, and $v_{x1} = v_{x2}$. Also, because point 2 is the highest point, the vertical component of velocity is zero at that point: $v_{y2} = 0$. Subtracting the $v_x^2$ terms from both sides, we get

$$v_{y1}^2 = 2gh.$$

But $v_{y1}$ is just the $y$-component of initial velocity, which is equal to $v_0 \sin\alpha_0$. Making this substitution and solving for $h$, we find

$$h = \frac{v_0^2 \sin^2\alpha_0}{2g},$$

which agrees with Eq. (3–28).

## EXAMPLE 7–7

A rebuilt transmission in a crate sits on the floor; the total mass is 80 kg. The crate must be brought to the top of a loading dock by sliding it up a ramp 2.5 m long, inclined at 30°. The shop foreman, giving no thought to the force of friction, calculates that he can get the crate up the ramp by giving it an initial speed of 5.0 m/s at the bottom and letting it go. Unfortunately, friction is *not* negligible; the crate slides 1.6 m up the ramp, stops, and slides back down. Figure 7–11 shows the situation. a) Assuming that the friction force acting on the crate is constant, find its magnitude. b) How fast is the crate moving when it reaches the bottom of the ramp?

**SOLUTION** a) Let point 1 be the bottom of the ramp and point 2 the point where the crate stops (Fig. 7–11a). This point is (1.6 m)(sin 30°) = 0.80 m above the floor. If we take $U = 0$ at floor level, we have $y_1 = 0$, $y_2 = 0.80$ m. The energy quantities are

$$K_1 = \tfrac{1}{2}(80\text{ kg})(5.0\text{ m/s})^2 = 1000\text{ J}, \qquad K_2 = 0,$$

$$U_1 = 0, \qquad U_2 = (80\text{ kg})(9.8\text{ m/s}^2)(0.80\text{ m}) = 627\text{ J},$$

$$W_{\text{other}} = -\mathcal{F}s = -\mathcal{F}(1.6\text{ m}),$$

where $\mathcal{F}$ is the magnitude of the unknown friction force. Using Eq. (7–7), we find

$$K_1 + U_1 + W_{\text{other}} = K_2 + U_2$$

$$1000\text{ J} + 0 - \mathcal{F}(1.6\text{ m}) = 0 + 627\text{ J},$$

$$\mathcal{F} = 233\text{ N}.$$

b) Let point 3 be the bottom of the ramp where the crate arrives after its little journey (Fig. 7–11b). The friction force reverses direction on the way down, so the frictional work is negative for both halves of the trip. The total gravitational work for the round trip is zero, so the total work is

$$W_{\text{other}} = W_{\mathcal{F}} = -2(1.6\text{ m})(233\text{ N}) = -746\text{ J}.$$

The other energy quantities are

$$K_1 = 1000\text{ J}, \qquad U_1 = U_3 = 0, \qquad K_3 = \tfrac{1}{2}(80\text{ kg})v_3^2,$$

where $v_3$ is the unknown final speed. Substituting these values into Eq. (7–7), we get

$$1000\text{ J} + 0 + (-746\text{ J}) = \tfrac{1}{2}(80\text{ kg})v_3^2 + 0,$$

$$v_3 = 2.5\text{ m/s}.$$

Alternatively, we could have applied Eq. (7–7) to points 2 and 3, considering the second half of the trip by itself. Try it and see whether you get the same result.

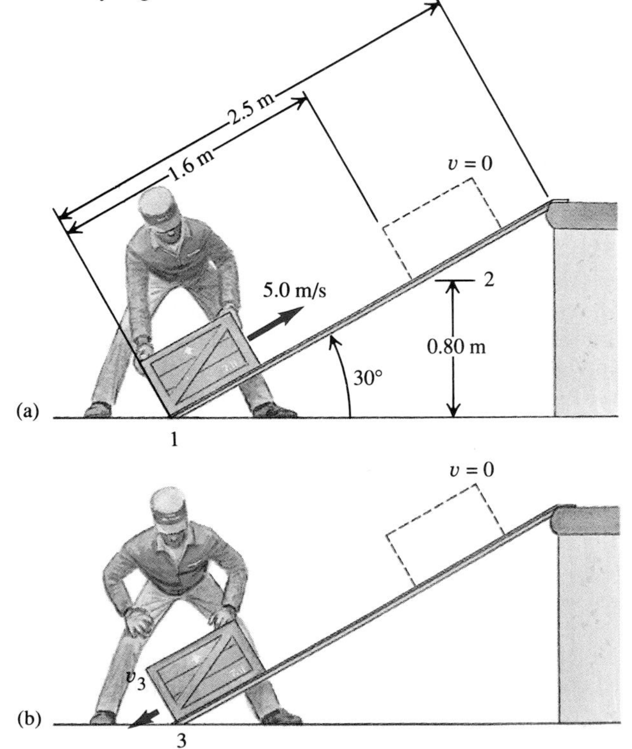


**7–11** (a) A crate slides part way up the ramp, stops, and (b) slides back down. Can you show that the crate would easily make it to the top with an initial speed of 6.3 m/s?

## 7–3 ELASTIC POTENTIAL ENERGY

When a railroad car runs into a spring bumper at the end of the track, the spring is compressed as the car is brought to a stop. If there is no friction, the bumper springs back, and the car rolls away in the opposite direction. During the interaction with the spring, the car's kinetic energy has been "stored" in the elastic deformation of the spring. Something very similar happens in a spring gun. Work is done on the spring by the force that compresses it; that work is stored in the spring until you pull the trigger and the spring gives kinetic energy to the projectile.

This is the same pattern that we saw with the pile-driver: Do work on the system to store energy; then this energy can later be converted to kinetic energy. We can apply the concept of elastic potential energy to this storage process. We proceed just as we did for gravitational potential energy. We begin with the work done by the spring force and then combine this with the work-energy theorem.

In Section 6–3 we discussed work associated with stretching or compressing a spring. Figure 7–12 shows the spring from Fig. 6–8, with its left end held stationary and its right end attached to a body with mass $m$ that can move along the $x$-axis. The body is at $x = 0$ when the spring is neither stretched nor compressed (Fig. 7–12a). An applied force $\boldsymbol{F}$ acts on the mass to make it undergo a displacement (Fig. 7–12b). How much work does the spring force do on the body?

We found in Section 6–3 that the work we must do *on* the spring to stretch it from an elongation $x_1$ to a greater elongation $x_2$ is $\frac{1}{2}kx_2^2 - \frac{1}{2}kx_1^2$, where $k$ is the force constant of the spring. Now we need to find the work done *by* the spring *on* the body. From Newton's third law the two quantities of work are just negatives of each other. Changing the signs in the expression in Eq. (6–7), we find that in a displacement from $x_1$ to $x_2$ the spring does an amount of work $W_{el}$ given by

$$W_{el} = \tfrac{1}{2}kx_1^2 - \tfrac{1}{2}kx_2^2,$$

where the subscript "el" stands for *elastic*. When $x_2$ is greater than $x_1$ (Fig. 7–12c), this quantity is negative because the body moves in the $+x$-direction while the spring pulls

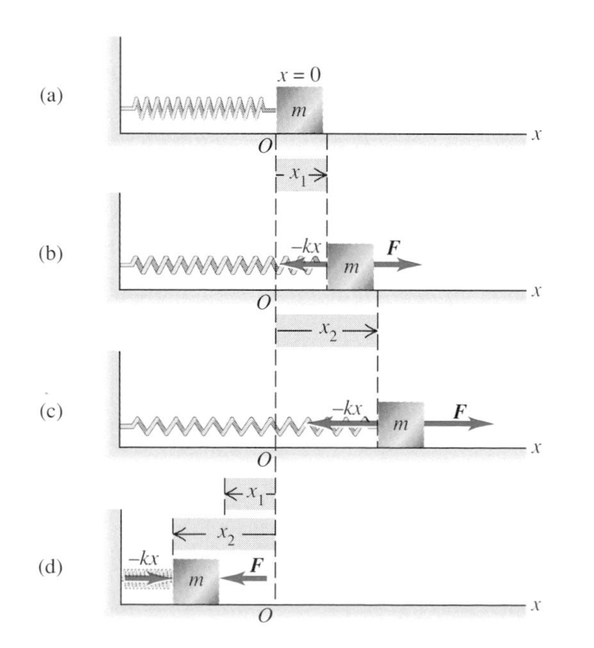


**7–12** (a) A block attached to a spring in equilibrium ($x = 0$) on a horizontal surface. (b) An applied force $\boldsymbol{F}$ to the right produces an extension (positive $x_1$) of the spring. (c) More force applied to the right produces greater spring extension ($x_2 > x_1$) and positive work. (d) An applied force to the left produces a compression ($x_2 < x_1$) of the spring, but the work is still positive.

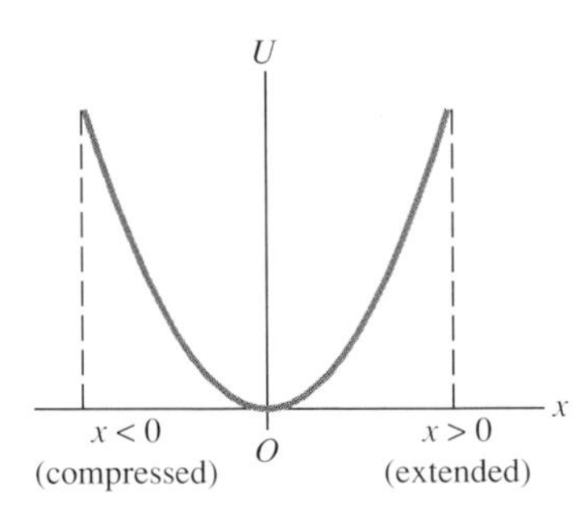


**7–13** The potential-energy curve for an ideal spring is a parabola, $U = \frac{1}{2}kx^2$, where $x$ is the change in length of the spring. For stretching, $x$ is positive. For compression (when that is possible), $x$ is negative; but $U$ is never negative.

on it in the $-x$-direction. If the spring can be compressed as well as stretched, $x_1$ or $x_2$ or both may be negative (Fig. 7–12d), and the expression for $W_{el}$ is still valid.

Following the same procedure as for gravitational work, we define **elastic potential energy** as

$$U = \tfrac{1}{2}kx^2. \tag{7–9}$$

Figure 7–13 is a graph of Eq. (7–9). The units of $k$ are N/m, so the units of $U$ are N · m. We have defined 1 N · m to be 1 J, the unit used for *all* energy and work quantities.

We can use Eq. (7–9) to express the work $W_{el}$ done on the body by the spring force in terms of the change in potential energy:

$$W_{el} = \tfrac{1}{2}kx_1^2 - \tfrac{1}{2}kx_2^2 = U_1 - U_2 = -\Delta U. \tag{7–10}$$

When $x$ increases, $W_{el}$ is *negative* and $U$ *increases;* when $x$ decreases, $W_{el}$ is positive and $U$ *decreases.* If the spring can be compressed as well as stretched, then $x$ is negative for compression. But, as Fig. 7–13 shows, $U$ is positive for both positive and negative $x$, and Eqs. (7–9) and (7–10) are valid for both cases. When $x = 0$, $U = 0$, and the spring is neither stretched nor compressed.

The work-energy theorem says that $W_{tot} = K_2 - K_1$, no matter what kind of forces are acting. If the spring force is the *only* force that does work on the body, then

$$W_{tot} = W_{el} = U_1 - U_2.$$

The work-energy theorem then gives us

$$K_1 + U_1 = K_2 + U_2 \tag{7–11}$$

and

$$\tfrac{1}{2}mv_1^2 + \tfrac{1}{2}kx_1^2 = \tfrac{1}{2}mv_2^2 + \tfrac{1}{2}kx_2^2. \tag{7–12}$$

In this case the total energy $E = K + U$ is *conserved.*

If other forces also do work on the body, we call their work $W_{other}$, as before. Then the total work is $W_{tot} = W_{el} + W_{other}$, and the work-energy theorem gives

$$W_{el} + W_{other} = K_2 - K_1.$$

The work done by the spring is still $W_{el} = U_1 - U_2$, so again

$$K_1 + U_1 + W_{other} = K_2 + U_2 \tag{7–13}$$

and

$$\tfrac{1}{2}mv_1^2 + \tfrac{1}{2}kx_1^2 + W_{other} = \tfrac{1}{2}mv_2^2 + \tfrac{1}{2}kx_2^2. \tag{7–14}$$

This equation shows that **the work done by all the forces *except the elastic force* equals the change in the total mechanical energy $E = K + U$ of the body.** When $W_{other}$ is positive, $E$ increases; when $W_{other}$ is negative, $E$ decreases.

What happens when we have *both* gravitational and elastic forces, such as for a body hanging at the end of a spring in a gravitational field? We can still use the relationship

$$K_2 + U_2 = K_1 + U_1 + W_{other},$$

but now $U_1$ and $U_2$ are the initial and final values of the *total* potential energy, including both gravitational and elastic potential energies. That is, $U = U_{grav} + U_{el}$. If the gravitational and elastic forces are the *only* forces that do work on the body, then $W_{other} = 0$. Figure 7–14 shows a familiar example of transformations between elastic potential energy, kinetic energy, and gravitational potential energy.

**7–14** Elastic potential energy in the bow is converted to kinetic energy of the arrow and then partly to gravitational potential energy.

## PROBLEM-SOLVING STRATEGY

### Conservation of mechanical energy II

The strategy outlined in Section 7–2 is equally useful here. In the list of kinetic and potential energies in Step 3, you should include both gravitational and elastic potential energies when appropriate. Remember that every force that does work must be represented *either* in $U$ *or* in $W_{\text{other}}$ but *never* in both places. The work done by the gravitational and elastic forces is accounted for by the potential energies; the work of the other forces, $W_{\text{other}}$, has to be included separately. In some cases there will be "other" forces that do no work, as we mentioned in Section 7–2. Remember that in the expression $U = \frac{1}{2}kx^2$, $x$ must be the displacement of the spring from its unstretched length (because we have used $F = kx$ to derive it).

### EXAMPLE 7–8

In Fig. 7–15 a glider with mass $m = 0.200$ kg sits on a frictionless, horizontal air track, connected to a spring with force constant $k = 5.00$ N/m. You pull on the glider, stretching the spring 0.100 m, and then release it with no initial velocity. The glider begins to move back toward its equilibrium position ($x = 0$). What is its velocity when $x = 0.080$ m?

**SOLUTION** As the body starts to move, potential energy is converted into kinetic energy. The spring force is the only force doing work on the body, so $W_{\text{other}} = 0$, and we may use Eq. (7–11). The energy quantities are

$$K_1 = \tfrac{1}{2}(0.200\text{ kg})(0)^2 = 0,$$

$$U_1 = \tfrac{1}{2}(5.00\text{ N/m})(0.100\text{ m})^2 = 0.0250\text{ J},$$

$$K_2 = \tfrac{1}{2}(0.200\text{ kg})v_2{}^2,$$

$$U_2 = \tfrac{1}{2}(5.00\text{ N/m})(0.080\text{ m})^2 = 0.0160\text{ J}.$$

Then, from Eq. (7–11),

$$0 + 0.0250\text{ J} = \tfrac{1}{2}(0.200\text{ kg})v_2{}^2 + 0.0160\text{ J},$$

$$v_2 = \pm 0.30\text{ m/s}.$$

What is the physical significance of the plus-or-minus sign?

Note that this problem *cannot* be done by using the equations of motion with constant acceleration because the spring force varies with position. The energy method, in contrast, offers a simple and elegant solution.

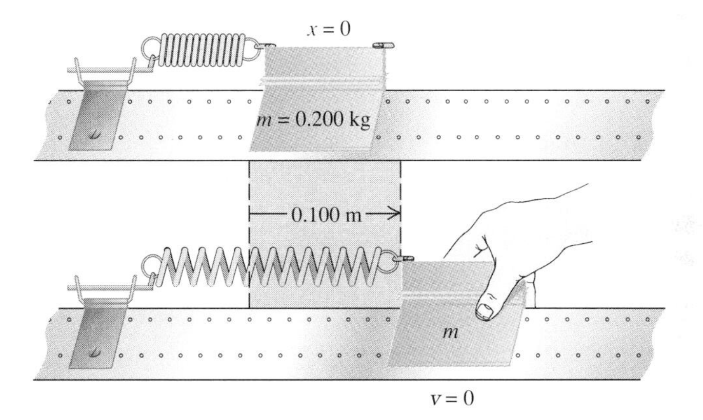


**7–15** Elastic potential energy is transformed to kinetic energy as the glider moves back toward its equilibrium position.

### EXAMPLE 7–9

For the system of Example 7–8, suppose the glider is initially at rest at $x = 0$, with the spring unstretched. Then you apply a constant force $\boldsymbol{F}$ with magnitude 0.610 N to the glider. What is the glider's speed when it has moved to $x = 0.100$ m?

**SOLUTION** Again, we don't have constant acceleration. (Do you see why not?) Mechanical energy is not conserved because of the work done by the force $\boldsymbol{F}$, but we can still use the energy relation of Eq. (7–13). Let point 1 be at $x = 0$ and point 2 at $x = 0.100$ m. The energy quantities are

$$K_1 = 0, \qquad K_2 = \tfrac{1}{2}(0.200\text{ kg})v_2{}^2,$$

$$U_1 = 0, \qquad U_2 = \tfrac{1}{2}(5.00\text{ N/m})(0.100\text{ m})^2 = 0.0250\text{ J},$$

$$W_{\text{other}} = (0.610\text{ N})(0.100\text{ m}) = 0.0610\text{ J}.$$

Putting the pieces into Eq. (7–13), we find

$$K_1 + U_1 + W_{\text{other}} = K_2 + U_2,$$

$$0 + 0 + 0.0610\text{ J} = \tfrac{1}{2}(0.200\text{ kg})v_2{}^2 + 0.0250\text{ J},$$

$$v_2 = 0.60\text{ m/s}.$$

## EXAMPLE 7–10

Suppose the force $\boldsymbol{F}$ in Example 7–9 is removed when the glider reaches the point $x = 0.100$ m. How much farther does the glider move before coming to rest?

**SOLUTION** The elastic force is now the only force doing work, and total mechanical energy $(K + U)$ is conserved. The kinetic energy at the 0.100-m point (point 2) is $K_2 = \frac{1}{2}mv_2^2 = \frac{1}{2}(0.200 \text{ kg})(0.60 \text{ m/s})^2 = 0.036$ J, and the potential energy at this point is $U_2 = \frac{1}{2}kx^2 = 0.025$ J. The total energy is therefore 0.061 J (equal to the work done by the force $\boldsymbol{F}$). When the body comes to rest at $x_{max}$ (point 3), its kinetic energy $K_3$ is zero, and its potential energy $U_3$ is

$$U_3 = \tfrac{1}{2}kx_{max}^2 = \tfrac{1}{2}(5.00 \text{ N/m})x_{max}^2.$$

From $K_2 + U_2 = K_3 + U_3$ we find

$$0.036 \text{ J} + 0.025 \text{ J} = 0 + \tfrac{1}{2}(5.00 \text{ N/m})x_{max}^2,$$

$$x_{max} = 0.156 \text{ m}.$$

So the body moves an additional 0.056 m after the force is removed at $x = 0.100$ m.

## EXAMPLE 7–11

In a "worst-case" design scenario a 2000-kg elevator with broken cables is falling at 25 m/s when it first contacts a cushioning spring at the bottom of the shaft. The spring is supposed to stop the elevator, compressing 3.0 m as it does so (Fig. 7–16). During the motion a safety clamp applies a constant 17,000-N frictional force to the elevator. As an energy consultant, you are asked to determine what the force constant of the spring should be.

**SOLUTION** Take point 1 as the elevator's position when it initially contacts the spring ($y_1 = 3.00$ m) and point 2 as its position when the spring is fully compressed ($y_2 = 0$). The elevator's initial speed is $v_1 = 25$ m/s, so

$$K_1 = \tfrac{1}{2}mv_1^2 = \tfrac{1}{2}(2000 \text{ kg})(25.0 \text{ m/s})^2 = 625{,}000 \text{ J}.$$

The elevator stops at point 2, so $K_2 = 0$. The elastic potential energy at point 1 is zero because the spring isn't yet compressed. There *is* gravitational potential energy:

$$U_1 = mgy_1 = (2000 \text{ kg})(9.80 \text{ m/s}^2)(3.00 \text{ m}) = 58{,}800 \text{ J}.$$

At point 2 the gravitational potential energy is zero because $y = 0$, and the elastic potential energy is

$$U_2 = \tfrac{1}{2}kx^2 = \tfrac{1}{2}k(3.00 \text{ m})^2,$$

where the force constant $k$ is to be determined.

The other force is the 17,000-N friction force, acting opposite to the 3.00-m displacement, and

$$W_{other} = -(17{,}000 \text{ N})(3.00 \text{ m}) = -51{,}000 \text{ J}.$$

Inserting all these values in Eq. (7–13), you find

$$K_1 + U_1 + W_{other} = K_2 + U_2,$$

$$625{,}000 \text{ J} + 58{,}800 \text{ J} + (-51{,}000 \text{ J}) = 0 + \tfrac{1}{2}k(3.00 \text{ m})^2,$$

$$k = 1.41 \times 10^5 \text{ N/m}.$$

You then have to explain to the client that the elevator won't stay at the bottom of the shaft but will bounce back up and then return to hit the spring again and again until enough energy has been removed by friction for it to stop.

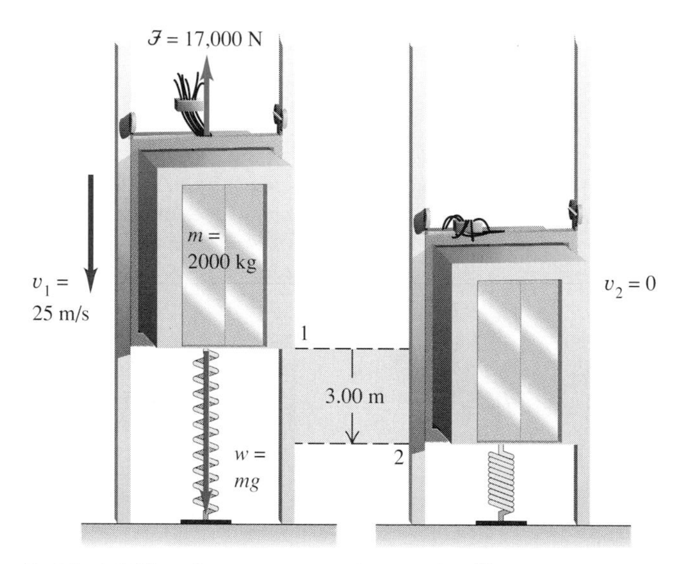


**7–16** A falling elevator compressing a spring illustrates conversion of kinetic energy and gravitational and elastic potential energy.

# 7–4 CONSERVATIVE AND NONCONSERVATIVE FORCES

In our intuitive discussions of potential energy we have talked about "storing" kinetic energy by converting it to potential energy. We always have in mind that later we may retrieve it again as kinetic energy. When you throw a ball up in the air, it slows down as kinetic energy is converted into potential energy. But on the way

**7–17** The work done by a conservative force is independent of the path; it depends only on the end points.

down, the conversion is reversed, and the ball speeds up as potential energy is converted back to kinetic energy. If there is no air resistance, the ball is moving just as fast when you catch it as when you threw it.

If a glider moving on a frictionless, horizontal air track runs into a spring bumper at the end of the track, the spring compresses and the glider stops. But then the glider bounces back, and if there is no friction, it has the same speed and kinetic energy that it had before the collision. Again, there is a two-way conversion from kinetic to potential energy and back. In both cases we find that we can define a potential-energy function such that the total mechanical energy, kinetic plus potential, is constant or *conserved* during the motion.

A force that offers this opportunity of two-way conversion between kinetic and potential energies is called a **conservative force.** An essential feature of conservative forces is that their work is always *reversible.* Anything that we deposit in the energy "bank" can later be withdrawn without loss. Gravitational forces show another important aspect of conservative forces. A body may move from point 1 to point 2 by various paths (Fig. 7–17), but the work done by the gravitational force is the same for all of these paths. In a uniform gravitational field the work depends only on the change in height. If a body moves around a closed path, ending at the same point where it started, the *total* work done by the gravitational field is always zero.

The work done by a conservative force always has these properties:

1. It can always be expressed as the difference between the initial and final values of a *potential energy* function.
2. It is reversible.
3. It is independent of the path of the body and depends only on the starting and ending points.
4. When the starting and ending points are the same, the total work is zero.

When all the work done on a body is done by conservative forces, the total mechanical energy $E = K + U$ is constant.

Not all forces are conservative. Consider the friction force acting on the crated transmission sliding on a ramp in Example 7–7 in Section 7–2. There is no potential-energy function for the friction force. When the body slides up and then back down to the starting point, the total work done on it by the friction force is *not* zero. When the direction of motion reverses, so does the friction force, and it does *negative* work in *both* directions. When a car with its brakes locked skids across the pavement with decreasing speed

**7–18** A motorcycle skidding sideways. Mechanical energy is not conserved during a skid; the kinetic energy decreases as a result of negative work done by the nonconservative friction forces. There is no corresponding potential energy, and the kinetic energy cannot be recovered.

(and decreasing kinetic energy), the lost kinetic energy cannot be recovered by reversing the motion or in any other way, and mechanical energy is *not* conserved (Fig. 7–18).

Such a force is called a **nonconservative force,** or a **dissipative force.** To describe the associated energy relations we have to introduce additional kinds of energy and a more general energy-conservation principle. When a body slides on a rough surface, such as tires on pavement or a box dropped on a conveyer belt, the surfaces become hotter; the energy associated with this change in the state of the materials is called *internal energy.* In later chapters we will study the relation of internal energy to temperature changes, heat, and work. This is the heart of the area of physics called *thermodynamics.*

### EXAMPLE 7–12

**Work done by friction** Let's look again at Example 7–4 in Section 7–2, in which your cousin Throcky skateboards down a curved ramp. He starts with 735 J of potential energy, and at the bottom he has 613 J of kinetic energy. The work $W_{\mathcal{F}}$ done by the nonconservative friction forces is $-122$ J. As he rolls down, the wheels, the bearings, and the ramp all get a little warmer. The same temperature changes could have been produced by adding 122 J of heat to these bodies. Their *internal* energy increases by 122 J; the sum of this energy and the final mechanical energy equals the initial mechanical energy, and the total energy of the system (including nonmechanical forms of energy) is conserved.

## *7–5 FORCE AND POTENTIAL ENERGY

For the two kinds of conservative forces (gravitational and elastic) that we have studied, we started with a description of the behavior of the force and derived from that an expression for the potential energy. For a body with mass $m$ in a uniform gravitational field, the gravitational force is $F_y = -mg$. We found that the corresponding potential energy is $U(y) = mgy$. When an ideal spring exerts a force $F_x = -kx$ on a body, the corresponding potential energy function is $U(x) = \frac{1}{2}kx^2$.

We can reverse this procedure. If we are given a potential-energy expression, we can find the corresponding force. Here's how we do it. First, let's consider motion along a straight line, with coordinate $x$. We denote the $x$-component of force, a function of $x$, by $F_x(x)$ and the potential energy by $U(x)$. This notation reminds us that both $F_x$ and $U$ are *functions* of $x$. Now we recall that in any displacement the work $W$ done by a conservative force equals the negative of the change $\Delta U$ in potential energy:

$$W = -\Delta U.$$

Let's apply this to a small displacement $\Delta x$. The work done by the force $F_x(x)$ during this displacement is approximately equal to $F_x(x)\ \Delta x$. We have to say "approximately" because $F_x(x)$ may vary a little over the interval $\Delta x$. But it is at least approximately true that

$$F_x(x)\Delta x = -\Delta U, \qquad \text{and} \qquad F_x(x) = -\frac{\Delta U}{\Delta x}.$$

You can probably see what's coming. We take the limit as $\Delta x \rightarrow 0$. In this limit the variation of $F_x$ becomes negligible, and we have the exact relation

$$F_x(x) = -\frac{dU}{dx}. \tag{7–15}$$

This result makes sense; in regions where $U(x)$ changes most rapidly with $x$ (that is, where $dU/dx$ is large), the greatest amount of work is done during a given displacement, and this corresponds to a large force magnitude. Also, when $F$ is in the positive $x$-direction, $U$ *decreases* with increasing $x$. So $F$ and $dU/dx$ should have opposite signs.

## EXAMPLE 7–13

**An electrical force and its potential energy** An electrically charged particle is held at rest at the point $x = 0$, and a second particle with equal charge is free to move along the positive $x$-axis. The potential energy of the system is

$$U(x) = \frac{k}{x},$$

where $k$ is a constant that depends on the magnitude of the charges. Derive an expression for the $x$-component of force acting on the moving charge as a function of its position.

**SOLUTION** We are given the potential-energy function $U(x)$, so we can use Eq. (7–15). The derivative with respect to $x$ of the function $1/x$ is $-1/x^2$, and we find

$$F_x(x) = -\frac{dU}{dx} = -k\left(-\frac{1}{x^2}\right) = \frac{k}{x^2}.$$

The $x$-component of force is positive, corresponding to a repulsive interaction between like electric charges. The force varies as $1/x^2$; it is very small when the particles are very far apart (large $x$), and the force and potential energy both become very large when $x$ is very small and the particles are very close together. This is an example of Coulomb's law for electrical interactions; we will study it in greater detail in Chapter 22.

We can extend this analysis to three dimensions, where the particle may move in the $x$-, $y$-, or $z$-direction, or all at once, under the action of a force that has components $F_x$, $F_y$, and $F_z$, each a function of the coordinates $x$, $y$, and $z$. The potential-energy function $U$ is also a function of all three space coordinates, and we write it as $U(x, y, z)$. We can now use Eq. (7–15) to find each component of force. The change in potential energy when the particle moves a small distance $\Delta x$ in the $x$-direction is again given by $-F_x\Delta x$; it doesn't depend on $F_y$ and $F_z$, which represent forces that are perpendicular to the displacement and do no work. So we again have

$$F_x = -\frac{dU}{dx}.$$

The $y$- and $z$-components of force are determined in exactly the same way:

$$F_y = -\frac{dU}{dy}, \qquad F_z = -\frac{dU}{dz}.$$

Because $U$ may be a function of all three coordinates, we need to remember that when we calculate each of these derivatives, only one coordinate changes at a time. We compute $dU/dx$ by assuming that $y$ and $z$ are constant and only $x$ varies, and so on. Such a derivative is called a *partial derivative,* and the usual notation is $\partial U/\partial x$, and so on. The symbol $\partial$ is a modified $d$ to remind us of the nature of this operation. So we write

$$F_x = -\frac{\partial U}{\partial x}, \qquad F_y = -\frac{\partial U}{\partial y}, \qquad F_z = -\frac{\partial U}{\partial z}. \tag{7–16}$$

We can use unit vectors to write a single compact vector expression for the force $\boldsymbol{F}$:

$$\boldsymbol{F} = -\left(\frac{\partial U}{\partial x}\boldsymbol{i} + \frac{\partial U}{\partial y}\boldsymbol{j} + \frac{\partial U}{\partial z}\boldsymbol{k}\right). \tag{7–17}$$

The expression inside the parentheses represents a particular operation on the function $U$, in which we take the partial derivative of $U$ with respect to each coordinate, multiply by the corresponding unit vector, and take the vector sum. This operation is called the **gradient** of $U$ and is often abbreviated as $\nabla U$. Thus the force is the negative of the gradient of the potential-energy function:

$$\boldsymbol{F} = -\nabla U. \tag{7–18}$$

## EXAMPLE 7–14

**Force and potential energy in two dimensions** A puck slides on a frictionless, level air table; its coordinates are $x$ and $y$. It is acted on by a conservative force described by the potential-energy function

$$U(x,y) = \tfrac{1}{2}k(x^2 + y^2).$$

Derive an expression for the force acting on the puck.

**SOLUTION** We can use Eqs. (7–16) to find the components of force:

$$F_x = -\frac{\partial U}{\partial x} = -kx, \qquad F_y = -\frac{\partial U}{\partial y} = -ky.$$

These correspond to the following vector expression, from Eq. (7–17):

$$\boldsymbol{F} = -k(x\boldsymbol{i} + y\boldsymbol{j}).$$

Now $x\boldsymbol{i} + y\boldsymbol{j}$ is just the position vector $\boldsymbol{r}$ of the particle, so we can rewrite the vector expression as $\boldsymbol{F} = -k\boldsymbol{r}$. This represents a force that at each point is opposite in direction to the position vector of the point, that is, a force that at each point is directed toward the origin. Furthermore, the *magnitude* of the force at any point is

$$F = k\sqrt{x^2 + y^2} = kr,$$

Where $r$ is the particle's distance from the origin. This is the force exerted by a spring that obeys Hooke's law and has a negligibly small length (compared to the other distances in the problem) when it is not stretched. The potential-energy function can also be expressed as $U = \frac{1}{2}kr^2$, so the radial force component $F_r$ can be expressed as

$$F_r = -\frac{\partial U}{\partial r}. \tag{7–19}$$

## *7–6 ENERGY DIAGRAMS

When a particle moves along a straight line under the action of a conservative force, we can get a lot of insight into the possible motions by looking at the graph of the potential-energy function $U(x)$. Figure 7–19a shows a glider with mass $m$ that moves

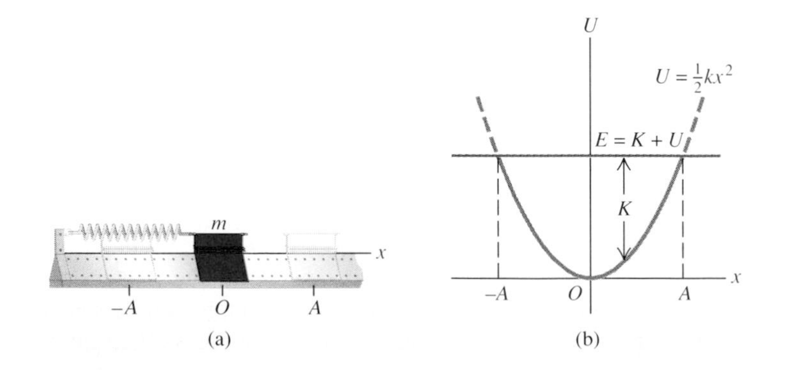


**7–19** (a) A glider on an air track. The spring exerts a force $F_x = -kx$. (b) The potential-energy function. The limits of the motion are the points where the $U$ curve intersects the horizontal line representing the total energy $E$.

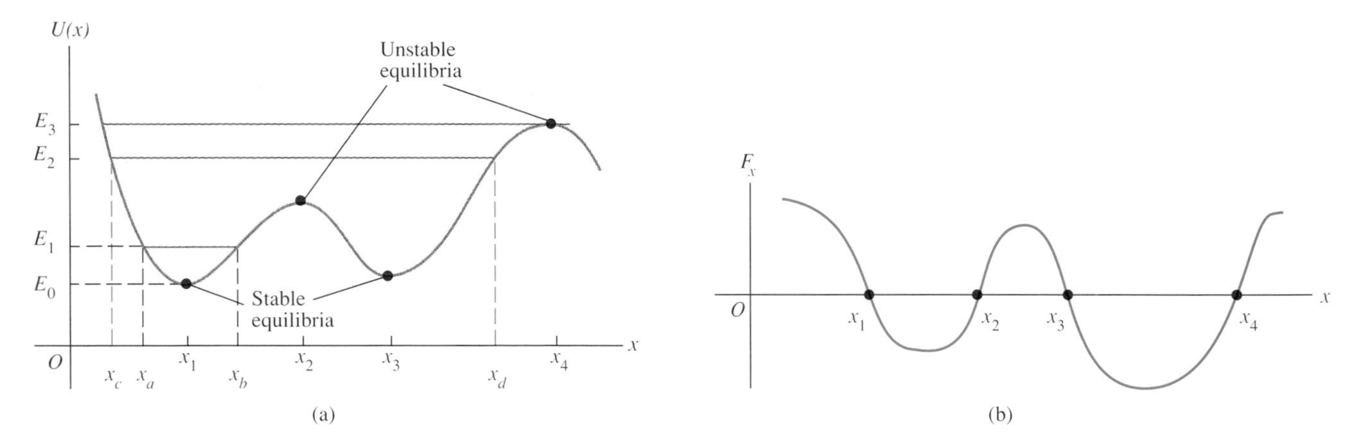


**7–20** (a) A hypothetical potential-energy function $U(x)$. (b) The corresponding force $F_x = -dU/dx$. Maxima and minima of $U(x)$ correspond to points where $F_x = 0$.

along the $x$-axis on an air track. The spring exerts on the glider a force with $x$-component $F_x = -kx$. Figure 7–19b is a graph of the corresponding potential-energy function $U(x) = \frac{1}{2}kx^2$. If the spring force is the *only* horizontal force acting on the glider, the total energy $E = K + U$ is constant, independent of $x$. A graph of $E$ as a function of $x$ is thus a straight horizontal line.

The vertical distance between the $U$ and $E$ graphs at each point represents the difference $E - U$, equal to the kinetic energy $K$ at that point. We see that $K$ is greatest at $x = 0$ and is zero at the values of $x$ where the two graphs cross, labeled $A$ and $-A$ in the diagram. Thus the speed $v$ is greatest at $x = 0$ and is zero at $x = \pm A$, the points of *maximum* possible displacement from $x = 0$ for a given value of the total energy $E$. The potential energy $U$ can never be greater than the total energy $E$; if it were, $K$ would be negative, and that's impossible. The motion is a back-and-forth oscillation between the points $x = A$ and $x = -A$.

At each point the force $F_x$ on the glider is equal to the negative of the slope of the $U(x)$ curve: $F_x = -dU/dx$. When the particle is at $x = 0$, the slope and the force are zero, so this is an *equilibrium* position. When $x$ is positive, the slope of the $U$ curve is positive, and the force $F_x$ is negative, directed toward the origin. When $x$ is negative, the slope is negative, and $F_x$ is positive, again toward the origin. Such a force is sometimes called a *restoring force* because when the glider is displaced to either side of $x = 0$, the resulting force tends to "restore" it back to $x = 0$. An analogous situation is a marble rolling around in a round-bottomed salad bowl. We say that $x = 0$ is a point of stable *equilibrium*. More generally, **any minimum in a potential-energy curve is a stable equilibrium position.**

Figure 7–20a shows a hypothetical but more general potential-energy function $U(x)$, and Fig. 7–20b is the corresponding force $F_x = -dU/dx$. Points $x_1$ and $x_3$ are stable equilibrium points. At each, $F_x$ is zero because the slope of the $U$ curve is zero. When the particle is displaced to either side, the force pushes back toward the equilibrium point. The slope of the $U$ curve is also zero at points $x_2$ and $x_4$, and these are also equilibrium points. But when the particle is displaced a little to the right of either of these points, the slope of the $U$ curve becomes negative, corresponding to a positive $F_x$ that tends to push the particle still farther from the point. When the particle is displaced a little to the left, $F_x$ is negative, again pushing away from equilibrium. This is analogous to a marble rolling on the top of a bowling ball. We conclude that **any maximum in a potential-energy curve is an unstable equilibrium position.**

If the total energy is $E_1$ and the particle is initially near $x_1$, it can move only in the region between $x_a$ and $x_b$ determined by the intersection of the $E_1$ and $U$ graphs. Again, $U$ cannot be greater than $E$ because $K$ can't be negative. We speak of the particle as moving in a *potential well*. If we increase the total energy to the level $E_2$, the particle can move

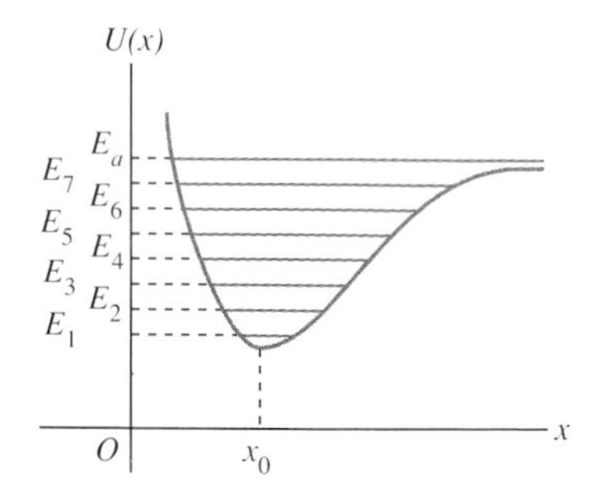


**7–21** A potential-energy function for a diatomic molecule. The equilibrium distance between atoms is $x_0$. Several energy levels, each corresponding to a particular quantum state of the molecule, are shown. In general, the lowest energy state $E_1$ is somewhat higher than the value of $U(x)$ at the minimum point of the curve.

over a wider range, from $x_c$ to $x_d$. If the total energy is $E_3$ or greater, the particle can "escape" and move to indefinitely large values of $x$. At the other extreme, $E_0$ represents the least possible total energy that the system can have.

Energy diagrams similar to these are often used to represent the interaction forces between atoms or molecules. Detailed analysis of such systems requires quantum mechanics, not Newtonian mechanics. One of the basic concepts of quantum mechanics is that the total energy of a system cannot change by an arbitrary amount. The possible amounts of energy that a system can have form a discrete set of *energy levels.* Figure 7–21 shows a possible potential-energy function for the interaction force in a diatomic molecule. Several energy levels are shown, and $E_a$ represents the minimum total energy needed for the molecule to dissociate into two separate atoms. The existence of these *quantized* energy levels is an essential ingredient in understanding atomic and molecular spectra, specific heat capacities of materials, and many other phenomena that depend on intermolecular interactions.

## *7–7 INTERNAL WORK AND ENERGY

We have talked about mechanical energy mostly with reference to bodies that we can represent as *particles.* For more complex systems that have to be represented in terms of many particles with different motions, new subtleties in energy considerations appear. We can't go into these in detail, but here are three examples.

First, consider a man standing on frictionless roller skates on a level surface, facing a rigid wall (Fig. 7–22). He pushes against the wall, setting himself in motion backward (to the right). The forces acting on him are his weight $\boldsymbol{w}$, the upward normal forces $\boldsymbol{n}_1$ and $\boldsymbol{n}_2$ exerted by the ground on his skates, and the horizontal force $\boldsymbol{F}$ exerted on him by the wall. There is no vertical displacement, so $\boldsymbol{w}$, $\boldsymbol{n}_1$, and $\boldsymbol{n}_2$ do no work. The force $\boldsymbol{F}$ is the horizontal force that accelerates him to the right, but the point where that force is applied (the man's hands) does not move, so the force $\boldsymbol{F}$ also does no work. So where does the man's kinetic energy come from?

**7–22** The external forces acting on a skater pushing off a wall. The work done by these forces is zero, but his kinetic energy changes.

The difficulty is that representing the man as a single point is not an adequate model. For the motion to occur as we have described it, different parts of the man's body must have different motions. His hands are stationary against the wall while his torso is moving away from it. In the interactions between various parts of his body, one part can exert forces and do work on another part. Therefore the *total* kinetic energy of this composite system can change, even though no work is done by forces applied by bodies outside the system. This would not be possible with a system that can be represented as a single point. In Chapter 8 we will consider the motion of a collection of moving particles that interact with each other. We will discover that the total kinetic energy of such a system can change even when no work is done on any part of the system by anything outside it.

As a second example, consider a car skidding on horizontal pavement with its brakes locked. The friction forces on the car's tires act in the direction opposite to the car's motion relative to the pavement, so these forces do negative work on the car. If they are the only horizontal forces on the car, its kinetic energy decreases. At the same time the tires exert an equal and opposite reaction force, in the direction of motion, on the pavement. But because the pavement doesn't move, those forces do no work. The work done on the pavement is *not* the negative of the work done on the car. Thus there is a decrease in the car's kinetic energy but no corresponding increase in the kinetic energy of the pavement. So where does the energy go?

We have already pointed out in Section 7–4 that in this situation, mechanical energy is not conserved. The surfaces become warmer, and the corresponding energy

# Potential-Energy Diagrams

Graphs of potential energy as a function of distance can give us very useful information about a physical situation. The key to the analysis is that the derivative of the potential-energy function (the slope of the potential-energy curve) at any point is the negative of the force acting on the body at that point.

## BUILDING PHYSICAL INTUITION

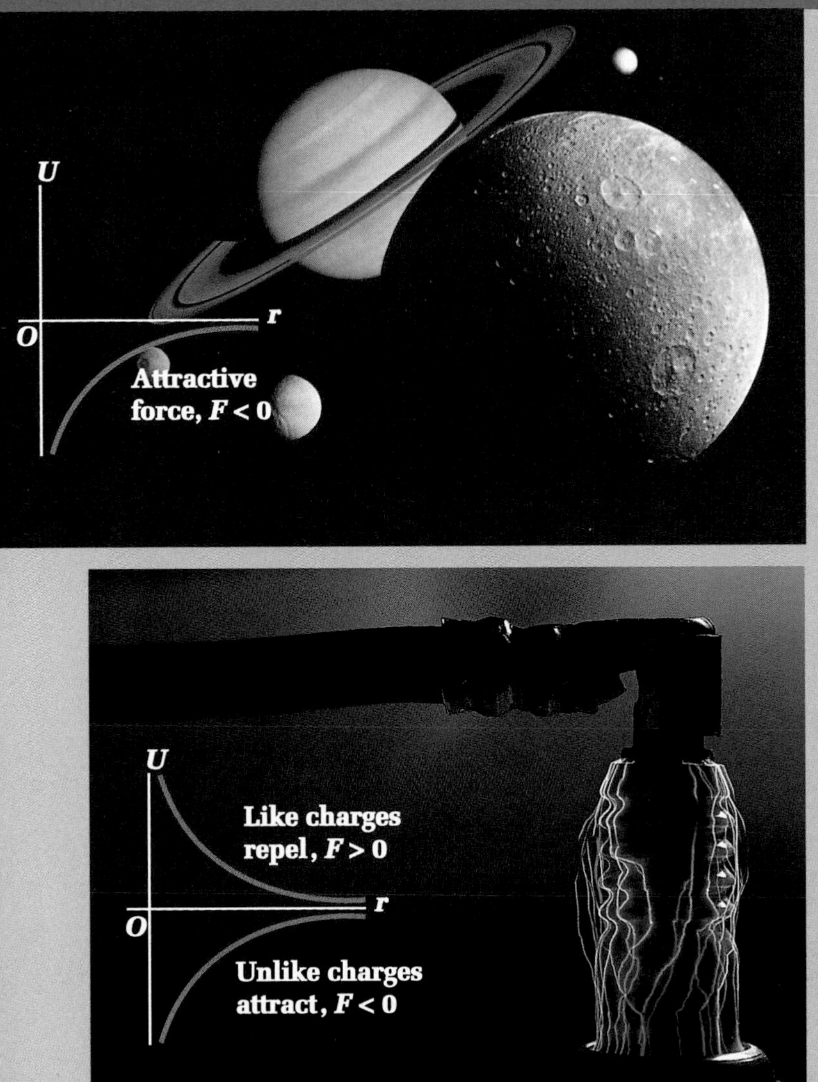


For a body in a gravitational field created by a point mass at the origin of coordinates, the graph of potential energy versus distance always has a positive slope. The force exerted on any other body is always directed toward the origin. Close to the origin, the slope of the curve is steep and the force is large. Far from the origin, the slope and the force are small.

For a charged particle in the electric field created by a positive point charge at the origin, the graph of electrical potential energy as a function of distance depends on the sign of the charge. For a negative charge, the slope is always positive and the force is always negative (directed toward the origin), corresponding to an *attractive* force. For a positive charge, the slope is always negative and the force is always positive (directed away from the origin), corresponding to a *repulsive* force. The electric discharge shown here is from an automobile spark plug.

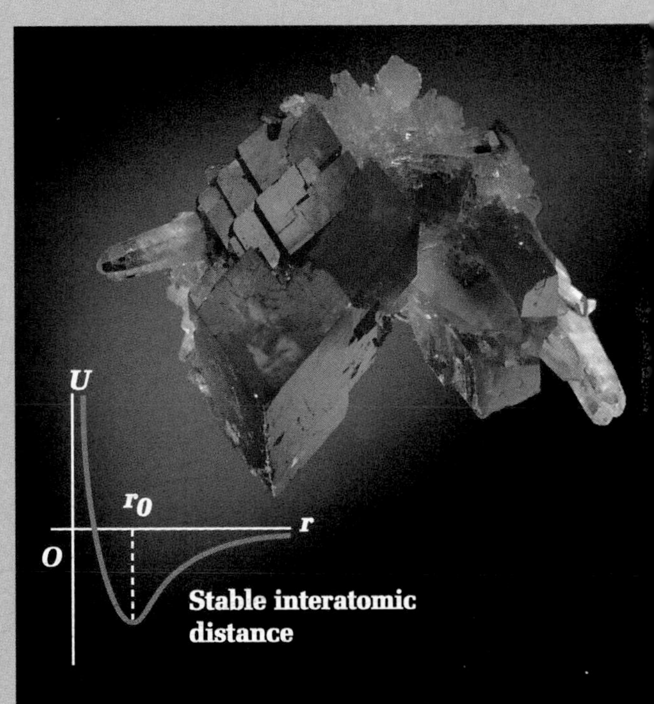


For an atom in a crystalline solid, the potential-energy function has a minimum at a distance from neighboring atoms corresponding to the spacing of atoms in the crystal lattice. At greater distances the potential-energy function has a positive slope and the force is negative, tending to pull the atoms together. At smaller distances the potential-energy function has a negative slope and the force is positive, tending to push the atoms apart. A stable equilibrium position always corresponds to a minimum in the potential-energy function.

transferred to the materials is called *internal energy.* In later chapters we will study internal energy and its relation to temperature changes, heat, and work in considerable detail; this is the heart of the area of physics called *thermodynamics.*

As a final example, suppose we drop a box onto a moving conveyor belt. If the box is dropped from a very small initial height, it is essentially at rest at the instant it contacts the belt. It slips at first but eventually acquires the same velocity as the belt. In doing so the box acquires kinetic energy, and some force has to act on it, doing positive work on it, to give it this kinetic energy. What is the force?

Clearly, the answer has to be "the friction force exerted on the box by the belt"; that's the *only* horizontal force acting on the box. So here is a case in which a friction force on a moving body does *positive* work on the body and *increases* its kinetic energy. The same thing happens when a truck starts from rest and accelerates. If there is a box sitting in the truck, the force that gives the box its forward acceleration and adds to its kinetic energy is again the friction force exerted on the box, this time by the moving floor of the truck. We're accustomed to thinking of moving bodies slowing down and stopping as a result of friction forces, but here are two examples in which friction forces do positive work and increase the kinetic energy of the bodies on which they act.

Still, we are left with the nagging feeling that somehow friction forces ought to dissipate or waste mechanical energy. After all, don't we oil bearings and sliding surfaces to decrease friction and make them more efficient? True enough! Going back to the conveyor belt, let's consider the work done *on the belt* by the friction force exerted on the belt by the box. This is negative, because the force on the belt has the opposite direction to its motion. By Newton's third law the two forces have equal magnitude. Furthermore, during the sliding of the box, the belt moves *farther* than the box; thus the negative work done on the belt by the box has greater magnitude than the positive work done on the box by the belt. Thus the *total* work done by these two forces is negative, and in this energy-exchange process there is indeed a net loss of mechanical energy. As was mentioned above, there is a corresponding increase of internal energy of the materials, showing itself as a rise in temperature.

Despite these complications, it remains true that for any system that can be adequately represented as a moving *point* mass, the total change in kinetic energy in any process is *always* equal to the total work done by all the forces acting on the system. Only when a system consists of several interacting masses do complications develop.

## SUMMARY

- The work done on a particle in a uniform gravitational field can be represented in terms of a potential energy $U = mgy$ as

$$W_{\text{grav}} = mgy_1 - mgy_2 = U_1 - U_2. \tag{7–1}$$

- When gravitational and other forces act on a body, the work $W_{\text{other}}$ done by the forces other than gravitational equals the total change in kinetic energy plus gravitational potential energy:

$$K_1 + U_1 + W_{\text{other}} = K_2 + U_2. \tag{7–7}$$

- The sum $K + U = E$ is called the total mechanical energy.
- The work done by a stretched or compressed spring that exerts a force $F_x = -kx$ on a particle, where $x$ is the amount of stretch or compression, can be represented in terms of a potential-energy function $U = \frac{1}{2}kx^2$:

$$W_{\text{el}} = \tfrac{1}{2}kx_1^2 - \tfrac{1}{2}kx_2^2 = U_1 - U_2 = -\Delta U. \tag{7–10}$$

If other forces also act on the body, then the work of the other forces equals the total change in kinetic energy plus elastic potential energy:

$$K_1 + U_1 + W_{\text{other}} = K_2 + U_2. \tag{7-13}$$

• Equation (7–13) is valid when both gravitational and elastic forces are present. In such cases, $U$ is the total potential energy, both gravitational and elastic. When no forces other than elastic or gravitational do work on the body, $W_{\text{other}} = 0$ and the total energy $E = K + U$ is conserved or constant:

$$K_1 + U_1 = K_2 + U_2, \tag{7-11}$$

where $U$ in general includes both gravitational and elastic potential energies.

• A conservative force is one for which the work-kinetic energy relation is completely reversible. The work of a conservative force can always be represented in terms of a potential energy, but the work of a nonconservative force cannot.

• For motion along a straight line a conservative force $F_x$ and its associated potential energy $U(x)$ are related by

$$F_x(x) = -\frac{dU}{dx}. \tag{7-15}$$

In three dimensions, where $U$ is a function of $x$, $y$, and $z$, the components of force are

$$F_x = -\frac{\partial U}{\partial x}, \qquad F_y = -\frac{\partial U}{\partial y}, \qquad F_z = -\frac{\partial U}{\partial z}, \tag{7-16}$$

or, in vector form,

$$\boldsymbol{F} = -\left(\frac{\partial U}{\partial x}\boldsymbol{i} + \frac{\partial U}{\partial y}\boldsymbol{j} + \frac{\partial U}{\partial z}\boldsymbol{k}\right) = -\boldsymbol{\nabla} U. \tag{7-17}$$

**KEY TERMS**

- potential energy
- conservative force
- conservative system
- gravitational potential energy
- total mechanical energy
- conservation of energy
- elastic potential energy
- nonconservative (dissipative) force
- gradient

# EXERCISES

## Section 7–2 Gravitational Potential Energy

**7–1** What is the potential energy for a 600-kg elevator at the top of the Empire State Building, 380 m above street level? Assume that the potential energy at street level is zero.

**7–2** A 4.00-kg sack of flour is lifted vertically at a constant speed of 4.00 m/s through a height of 12.0 m. a) How great a force is required? b) How much work is done on the sack by the lifting force? What becomes of this work?

**7–3** A baseball is thrown from the roof of a 27.5-m tall building with an initial velocity of magnitude 18.5 m/s and directed at an angle of 37.0° above the horizontal. a) What is the speed of the ball just before it strikes the ground? Use energy methods. b) What is the answer for part (a) if the initial velocity is at an angle of 37.0° *below* the horizontal?

**7–4** A mail bag with a mass of 120 kg is suspended by a vertical rope 8.0 m long. a) What horizontal force is necessary to hold the bag in a position displaced sideways 4.0 m from its initial position (Fig. 7–23)? b) How much work is done by the worker in moving the bag to this position?

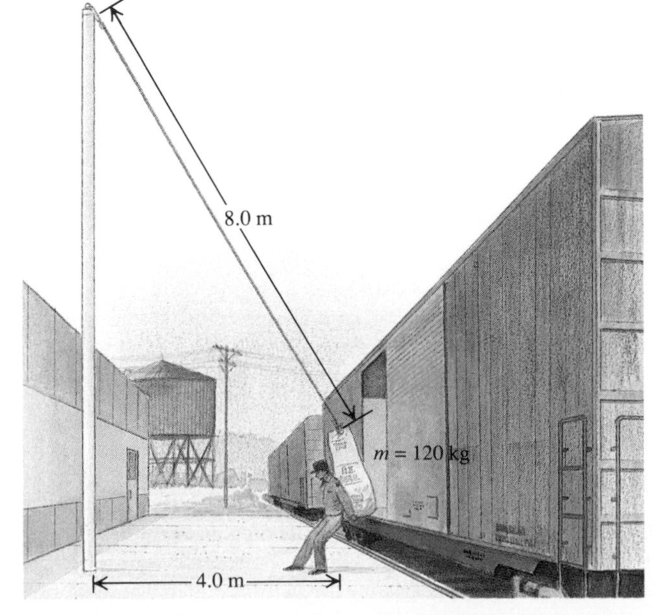


FIGURE 7–23

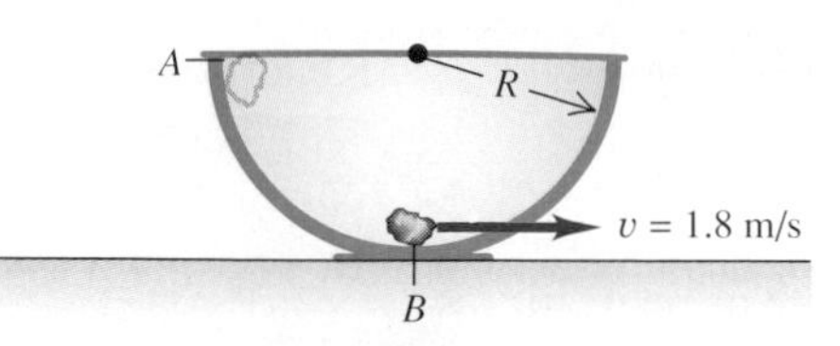


**FIGURE 7–24**

**7–5** A small rock with a mass of 0.10 kg is released from rest at point $A$, which is at the top edge of a hemispherical bowl with radius $R = 0.50$ m (Fig. 7–24). When it reaches point $B$ at the bottom of the bowl, the rock is observed to have speed $v = 1.8$ m/s. Calculate the *work* done by friction on the rock when it moves from $A$ to $B$. (*Note*: The friction force is not constant. You can easily solve for the friction *work* but not for the friction *force*.)

**7–6** A small rock with mass $m$ is fastened to a massless string with length 0.60 m to form a pendulum. The pendulum is swinging so as to make a maximum angle of 60° with the vertical. What is the speed of the rock when the string passes through the vertical position?

**7–7** A 12.0-kg microwave oven is pushed 16.0 m up the sloping surface of a loading ramp inclined at an angle of 37° above the horizontal, by a constant force $\boldsymbol{F}$ with a magnitude of 120 N and acting parallel to the ramp. The coefficient of kinetic friction between the oven and the ramp is 0.25. a) What is the work done by the force $\boldsymbol{F}$ on the oven? b) What is the work done by the friction force? c) Compute the increase in potential energy. d) Use your answers to parts (a), (b), and (c) to calculate the increase in the oven's kinetic energy. e) Use $\Sigma \boldsymbol{F} = m\boldsymbol{a}$ to calculate the acceleration of the oven. Assuming that the oven is initially at rest, use the acceleration to calculate the oven's speed after traveling 16.0 m. Compute from this the increase in the oven's kinetic energy, and compare with the answer you got in part (d).

## Section 7–3 Elastic Potential Energy

**7–8** A spring has a force constant $k = 300$ N/m. How far must the spring be stretched for 80.0 J of potential energy to be stored in it?

**7–9** A force of 1400 N stretches a certain spring a distance of 0.100 m. a) What is the potential energy of the spring when it is stretched 0.100 m? b) What is its potential energy when it is compressed 0.050 m?

**7–10** A force of 820 N stretches a certain spring a distance of 0.150 m. What is the potential energy of the spring when a 60.0-kg mass hangs vertically from it?

**7–11** A 1.20-kg book is dropped from a height of 0.40 m onto a spring with force constant $k = 1960$ N/m. Find the maximum distance the spring will be compressed.

**7–12** A slingshot will shoot a 10-g pebble 35.0 m straight up. a) How much potential energy is stored in the slingshot's rubber band? b) With the same potential energy stored in the rubber band, how high can the slingshot shoot a 20-g pebble?

**7–13** A brick with mass 0.600 kg is placed on a vertical spring with force constant $k = 500$ N/m that is compressed 0.20 m. When the brick is released, how high does it rise from this initial position? (The brick and the spring are *not* attached. The spring has negligible mass.)

## Section 7–5 Force and Potential Energy

***7–14** A force parallel to the $x$-axis acts on a particle moving along the $x$-axis. This force produces a potential energy $U(x)$ given by $U(x) = \alpha x^3$, where $\alpha = 2.5$ J/m$^3$. What is the force (magnitude and direction) when the particle is at $x = 1.20$ m?

***7–15** The potential energy of a pair of hydrogen atoms separated by a large distance $x$ is given by $U(x) = -C_6/x^6$, where $C_6$ is a positive constant. What is the force that one atom exerts on the other? Is this force attractive or repulsive?

***7–16** An object moving in the $xy$-plane is acted on by a conservative force described by the potential-energy function $U(x, y) = \alpha(1/x + 1/y)$. Derive an expression for the force expressed in terms of the unit vectors $\boldsymbol{i}$ and $\boldsymbol{j}$.

***7–17** An object moving in the $xy$-plane is acted on by a conservative force described by the potential-energy function $U(x, y) = k(x^2 + y^2) + k'xy$. Derive an expression for the force expressed in terms of the unit vectors $\boldsymbol{i}$ and $\boldsymbol{j}$.

***7–18** A marble moves along the $x$-axis in the potential-energy function sketched in Fig. 7–25. a) At which of the labeled $x$-coordinates is the force on the marble zero? b) Which of the labeled $x$-coordinates is a position of stable equilibrium? c) Which of the labeled $x$-coordinates is a position of unstable equilibrium?

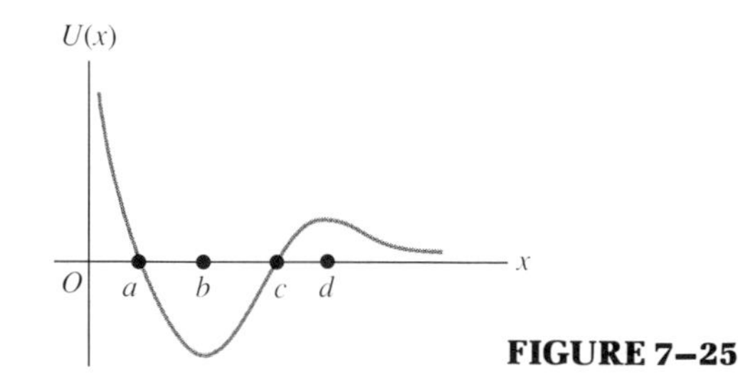


**FIGURE 7–25**

# PROBLEMS

**7–19** A man with mass 65.0 kg sits on a platform suspended from a movable pulley as shown in Fig. 7–26 and raises himself at constant speed by a rope passing over a fixed pulley. The platform and the pulleys have negligible mass. Assume that there are no friction losses. a) Find the force he must exert. b) Find the increase in his energy when he raises himself 1.50 m. (Answer by

FIGURE 7–26

calculating his increase in potential energy and also by computing the product of the force on the rope and the length of the rope passing through his hands.)

**7–20** A 2.00-kg block is pushed against a spring with force constant $k = 400$ N/m, compressing it 0.220 m. When the block is released, it moves along a frictionless, horizontal surface and then up a frictionless incline with slope 37.0° (Fig. 7–27). a) What is the speed of the block as it slides along the horizontal surface after having left the spring? b) How far does the block travel up the incline before starting to slide back down?

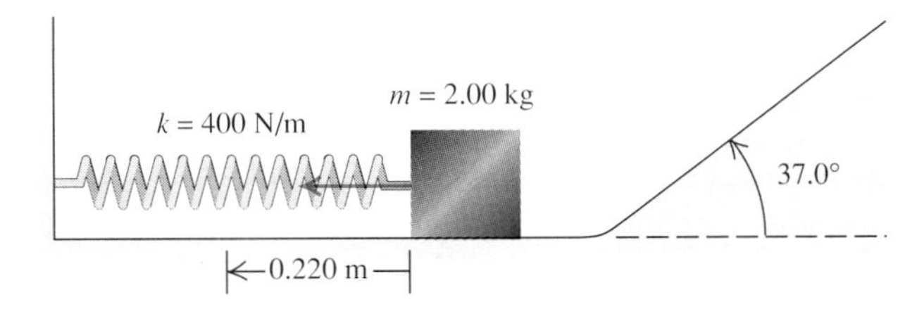


FIGURE 7–27

**7–21** A block with mass 0.50 kg is forced against a horizontal spring of negligible mass, compressing the spring a distance of 0.20 m (Fig. 7–28). When released, the block moves on a horizontal tabletop for 1.00 m before coming to rest. The spring constant $k$ is 100 N/m. What is the coefficient of kinetic friction, $\mu_k$, between the block and the table?

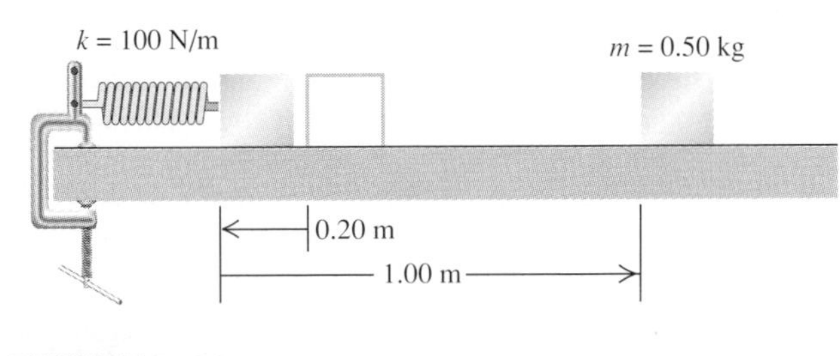


FIGURE 7–28

**7–22 Riding a Loop-the-loop.** A car in an amusement park ride rolls without friction around the track shown in Fig. 7–29. It starts from rest at point $A$ at a height $h$ above the bottom of the loop. a) What is the minimum value of $h$ (in terms of $R$) such that the car moves around the loop without falling off at the top (point $B$)? b) If $h = 3.50R$ and $R = 30.0$ m, compute the speed, radial acceleration, and tangential acceleration of the passengers when the car is at point $C$, which is at the end of a horizontal diameter. Show these acceleration components in a diagram, approximately to scale.

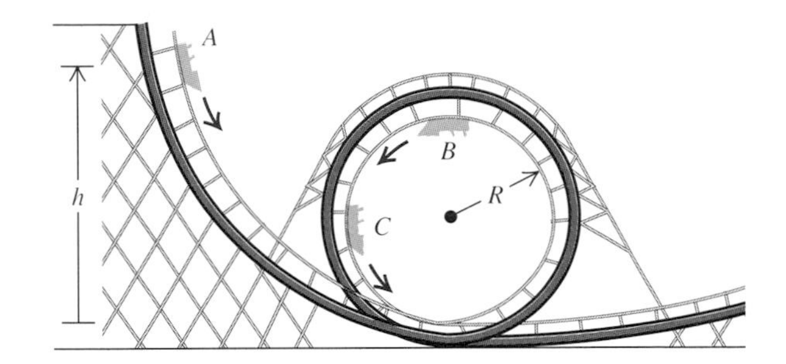

FIGURE 7–29

**7–23** The system of Fig. 7–30 is released from rest with the 12.0-kg block 2.00 m above the floor. Use the principle of conservation of energy to find the speed with which the block strikes the floor. Neglect friction and the inertia of the pulley.

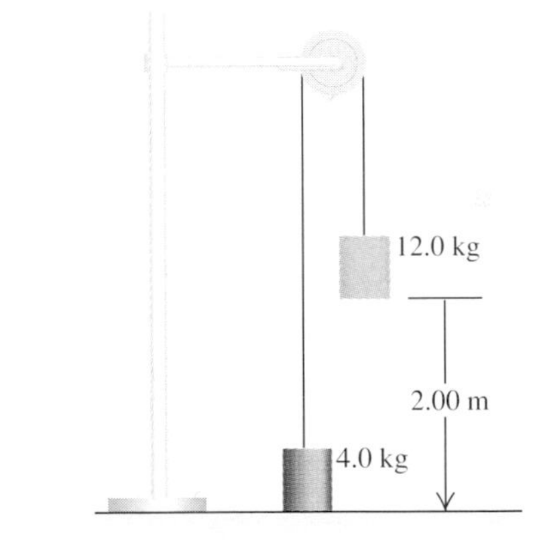


FIGURE 7–30

**7–24** a) A skier with mass 80.0 kg starts from rest at the top of a ski slope 75.0 m high. Assuming negligible friction between the skis and the snow, how fast is she going at the bottom of the slope? b) Now moving horizontally, the skier crosses a patch of rough snow, where $\mu_k = 0.20$. If the patch is 225 m wide, how fast is she going after crossing the patch? c) The skier hits a snowdrift and penetrates 2.5 m into it before coming to a stop. What is the average force exerted on her by the snowdrift as it stops her?

**7–25** A skier starts at the top of a very large frictionless snowball, with a very small initial speed, and skis straight down the

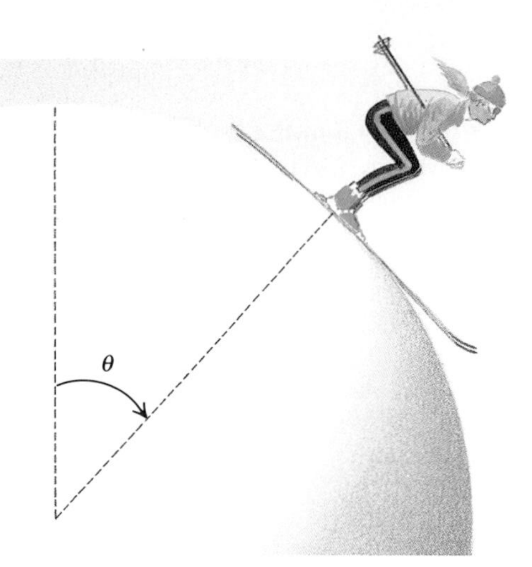


**FIGURE 7–31**

side (Fig. 7–31). At what point does she lose contact with the snowball and fly off at a tangent? That is, at the instant she loses contact with the snowball, what angle $\theta$ does a radial line from the center of the snowball to the skier make with the vertical?

**7–26** A meter stick, pivoted about a horizontal axis through its center, has a metal clamp with mass 3.00 kg attached to one end and a second clamp with mass 2.00 kg attached to the other. The mass of the meter stick can be neglected. The system is released from rest with the stick horizontal. What is the speed of each clamp as the stick swings through a vertical position?

**7–27** A 0.500-kg ball is tied to a string 2.00 m long, and the other end of the string is tied to a rigid support. The ball is held straight out horizontally from the point of support, with the string pulled taut, and is then released. a) What is the speed of the ball at the lowest point of its motion? b) What is the tension in the string at this point?

**7–28** A rock is tied to a cord, and the other end of the cord is held fixed. The rock is given an initial tangential velocity that causes it to rotate in a vertical circle. Prove that the tension in the cord at the lowest point exceeds that at the highest point by six times the weight of the rock.

**7–29** In a truck-loading station at a post office a 2.00-kg package is released from rest at point $A$ on a track that is one quarter of a circle with radius 1.60 m (Fig. 7–32). It slides down the track and reaches point $B$ with a speed of 4.00 m/s. From point $B$ it slides on a level surface a distance of 3.00 m to point $C$, where it comes to rest. a) What is the coefficient of kinetic friction on the horizontal surface? b) How much work is done on the package by friction as it slides down the circular arc from $A$ to $B$?

**FIGURE 7–32**

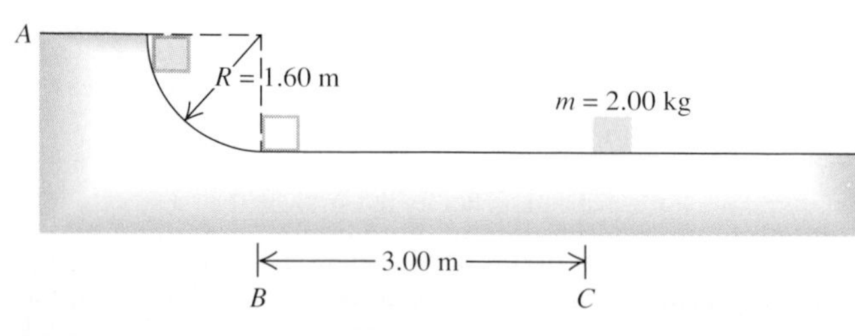


**7–30** A variable force $\boldsymbol{F}$ is maintained tangent to a frictionless surface of radius $a$ (Fig. 7–33). By slowly varying the force, a block with weight $w$ is moved, and the spring to which it is attached is stretched from position 1 to position 2. Calculate the work done by the force $\boldsymbol{F}$.

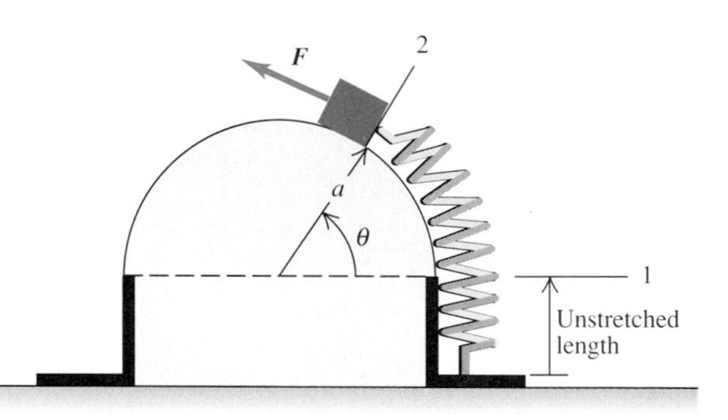


**FIGURE 7–33**

**7–31** A small block of ice with mass 0.120 kg is placed against a horizontal compressed spring mounted on a horizontal tabletop that is 1.90 m above the floor. The spring has a force constant $k = 2940$ N/m and is initially compressed 0.045 m. The spring is released, and the block slides along the table, goes off the edge, and travels to the floor. If there is negligible friction between the ice and the table, what is the speed of the block of ice when it reaches the floor?

**7–32** An 80.0-kg man jumps from a height of 2.50 m onto a platform mounted on springs. As the springs compress, the platform is pushed down a maximum distance of 0.200 m below its initial position, and then it rebounds. The platform and springs have negligible mass. a) What is the man's speed at the instant the platform is depressed 0.100 m? b) If the man had just stepped gently onto the platform, what maximum distance would it have been pushed down?

**7–33** A certain spring is found *not* to obey Hooke's law; it exerts a restoring force $F_x(x) = -\alpha x - \beta x^2$ if it is stretched or compressed a distance $x$, where $\alpha = 70.0$ N/m and $\beta = 12.0$ N/m$^2$. a) Calculate the potential energy function $U(x)$ for this spring. Let $U = 0$ when $x = 0$. b) An object with a mass of 2.00 kg is attached to this spring, pulled 1.00 m to the right on a frictionless, horizontal surface, and released. What is the speed of the object when it is 0.50 m to the right of the $x = 0$ equilibrium position?

**7–34** If a fish is attached to a vertical spring and slowly lowered to its equilibrium position, the spring stretches by an amount $d$. If the same fish is attached to the end of the unstretched spring and then allowed to fall from rest, through what maximum distance does it stretch the spring? (*Hint:* Calculate the force constant of the spring in terms of the distance $d$ and the mass $m$ of the fish.)

**7–35** A wooden block with mass 2.00 kg is placed against a compressed spring at the bottom of an incline of slope 37° (point

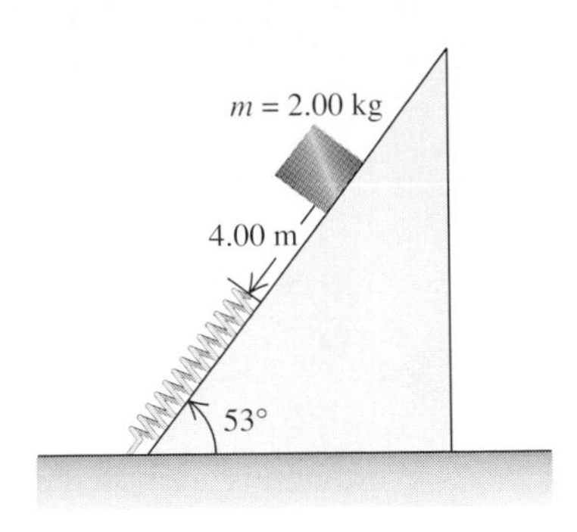


FIGURE 7–34

$A$). When the spring is released, it projects the block up the incline. At point $B$, a distance of 6.00 m up the incline from $A$, the block has a velocity with magnitude 4.00 m/s, directed up the incline. The coefficient of kinetic friction between the block and incline is $\mu_k = 0.50$. Calculate the amount of potential energy that was initially stored in the spring.

**7–36** A 2.00-kg block is released on a 53° incline, 4.00 m from a long spring with force constant $k = 70.0$ N/m that is attached at the bottom of the incline (Fig. 7–34). The coefficients of friction between the block and the incline are $\mu_s = 0.40$ and $\mu_k = 0.20$. a) What is the speed of the block just before it reaches the spring? b) What is the maximum compression of the spring? c) The block rebounds back up the incline. How close does it get to its initial position?

**7–37** A 0.500-kg block attached to a spring with length 0.60 m and force constant $k = 40.0$ N/m is at rest at point $A$ on a frictionless, horizontal surface (Fig. 7–35). A constant horizontal force $F = 20.0$ N is applied to the block and moves the block to the right along the surface. a) What is the speed of the block when it reaches point $B$, which is 0.25 m to the right of point $A$? b) When the block reaches point $B$, the force $F$ is suddenly removed. In the subsequent motion, how close does the block get to the wall?

*__7–38__ An object moves along the $x$-axis while acted on by a single conservative force parallel to the $x$- axis. The force corresponds to the potential-energy function graphed in Fig. 7–36. The object is released from rest at point $A$. a) What is the direction of the force on the object when it is at point $A$? b) At point $B$? c) At what value of $x$ is the kinetic energy of the object a maximum? d) What is the force on the particle when it is at point $C$? e) What is the largest value of $x$ reached by the particle during its motion? f) What value or values of $x$ correspond to points of stable equilibrium? g) Of unstable equilibrium?

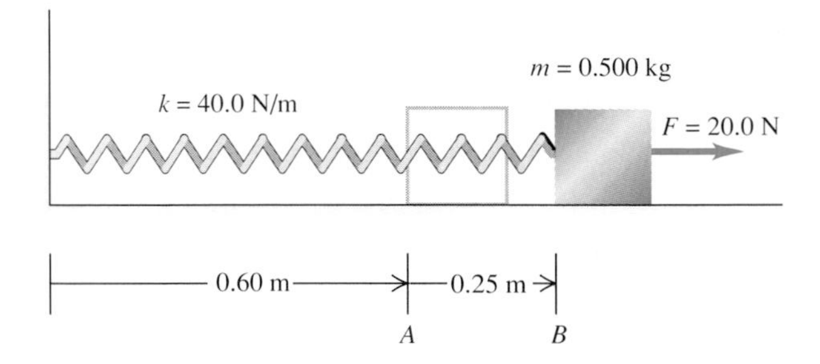


FIGURE 7–35

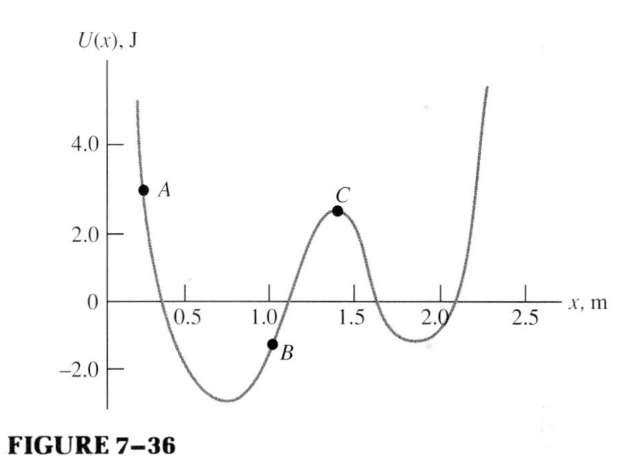


FIGURE 7–36

## CHALLENGE PROBLEMS

**7–39** An object has several forces acting on it. One of these forces is $\boldsymbol{F} = -\alpha x y^2 \boldsymbol{j}$, a force in the negative $y$-direction whose magnitude depends on the position of the object, where the constant $\alpha = 3.00\ \text{N/m}^3$. Consider the displacement of the object from the origin to the point $x = 2.00$ m, $y = 2.00$ m. a) Calculate the work done on the object by $\boldsymbol{F}$ if this displacement is along the straight line $y = x$ that connects these two points. b) Calculate the work done on the object by $\boldsymbol{F}$ if this displacement is carried out by the object first moving out along the $x$-axis to the point $x = 2.00$ m, $y = 0$ and then moving parallel to the $y$-axis to $x = 2.00$ m, $y = 2.00$ m. c) Compare the work done by $\boldsymbol{F}$ along these two paths. Is the force $\boldsymbol{F}$ conservative or nonconservative?

*__7–40__ A particle with mass $m$ moves in a one-dimensional potential-energy function

$$U(x) = \frac{\alpha}{x^2} - \frac{\beta}{x}.$$

The particle is released from rest at $x_0 = \alpha/\beta$. a) Show that $U(x)$ can be written as

$$U(x) = \frac{\alpha}{x_0^2}\left[\left(\frac{x_0}{x}\right)^2 - \left(\frac{x_0}{x}\right)\right].$$

Sketch $U(x)$. Calculate $U(x_0)$ and thereby locate the point $x_0$ on the sketch. b) Calculate $v(x)$, the speed of the particle as a function of position. Draw a sketch of $v$ as a function of $x$, and give a qualitative description of the motion. c) For what value of $x$ is the speed of the particle a maximum? What is the value of that maximum speed? d) What is the force on the particle at the point where the speed is maximum? e) Now let the particle be released instead at $x_1 = 3(\alpha/\beta)$. Locate the point $x_1$ on the sketch of $U(x)$. Calculate $v(x)$, and give a qualitative description of the motion. f) In each case, when the particle is released at $x = x_0$ and at $x = x_1$, what are the maximum and minimum values of $x$ reached during the motion?

CHAPTER 8

# Momentum and Impulse

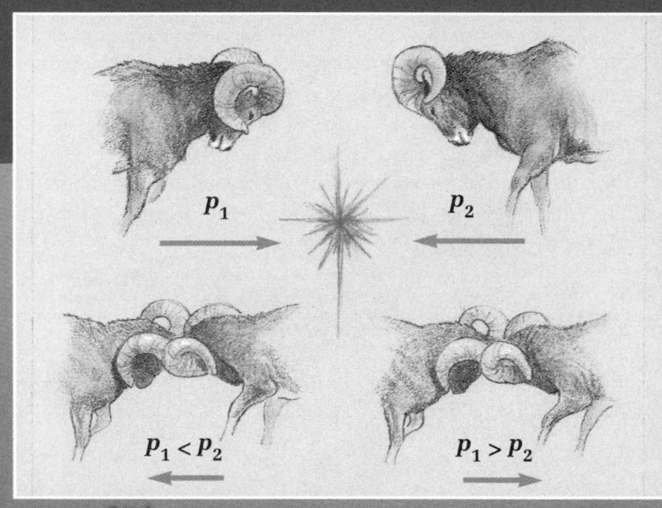


The result of a collision between two bodies depends on their momenta. (Momentum is the product of a body's mass and velocity.) When two bighorn sheep collide and lock horns, their resulting combined motion is in the direction of the ram with the larger initial magnitude of momentum.

## KEY CONCEPTS

**• The momentum of a particle, a vector quantity, is defined as the product of the particle's mass and velocity.**

**• In any interaction of two or more particles in which there are no forces except those the particles exert on each other, the total momentum (vector sum of the momenta of the particles) is constant or conserved.**

**• A collision in which total kinetic energy is conserved is called an elastic collision. When kinetic energy is not conserved, the collision is inelastic.**

**• When a constant force acts for a certain time, the impulse of the force is the product of force and time. The change of momentum of a body or system equals the total impulse of all the forces acting on it.**

**• The center of mass of a system is the average position of the mass of the system. Its motion under given forces is the same as though all the mass were concentrated at the center of mass.**

## INTRODUCTION

A .22 rifle bullet and a pitched baseball have roughly the same kinetic energy—a hundred or so joules. Which would you rather catch with a catcher's mitt? Why? When an eighteen-wheeler collides head-on with a compact car, why are the occupants of the car much more likely to be injured than those of the truck? How do you decide how to aim the cue ball in pool to knock the eight ball into the pocket? How can a rocket engine accelerate a space shuttle in outer space where there's nothing to push against?

To answer these and similar questions we need two new concepts, *momentum* and *impulse,* and a new conservation law, *conservation of momentum.* The position of this conservation law among the laws of physics is every bit as exalted as that of conservation of energy. Also like that of conservation of energy, its validity extends far beyond the bounds of classical mechanics to include relativistic mechanics (the mechanics of the very fast) and quantum mechanics (the mechanics of the very small). Within classical mechanics it enables us to analyze many situations that would be very difficult if we tried to use Newton's laws directly. Among these are *collision* problems, in which two bodies collide and exert very large forces on each other for a short time. ■

## 8–1 MOMENTUM

Let's come straight to the point. When a particle with mass $m$ moves with velocity $\boldsymbol{v}$, we define its **momentum** as the product of its mass and velocity. Using the symbol $\boldsymbol{p}$ for momentum, we have

$$\boldsymbol{p} = m\boldsymbol{v}. \tag{8–1}$$

Because velocity is a vector quantity, momentum is also a vector quantity. It has a magnitude ($mv$) and a direction (the same as the velocity vector). The momentum of a car driving north at 20 m/s is different from the momentum of the same car driving east at the same speed. A fast ball thrown by a major-league pitcher has greater momentum than the same ball thrown by a child because the velocity is greater. An eighteen-wheeler going 55 mi/h has greater momentum than a Saturn automobile going the same speed because the truck's mass is greater. The units of the magnitude of momentum are units of mass times speed; in SI units the units of momentum are kg · m/s.

We will often express the momentum of a particle in terms of its components. It follows from Eq. (8–1) that if the particle has velocity components $v_x$, $v_y$, and $v_z$, its components of momentum $p_x$, $p_y$, and $p_z$ are given by

$$p_x = mv_x, \qquad p_y = mv_y, \qquad p_z = mv_z. \tag{8–2}$$

These three component equations are equivalent to Eq. (8–1).

The importance of momentum can be traced to its close relation to Newton's second law, $\Sigma \boldsymbol{F} = m\boldsymbol{a}$. We assume for now that the mass $m$ of the particle is constant. (Later we will learn how to deal with systems in which the mass of a body changes.) We can re-express Newton's second law in terms of momentum. First, $\boldsymbol{a} = d\boldsymbol{v}/dt$, so

$$\Sigma \boldsymbol{F} = m\frac{d\boldsymbol{v}}{dt}.$$

Then, because $m$ is constant,

$$m\frac{d\boldsymbol{v}}{dt} = \frac{d}{dt}(m\boldsymbol{v}) = \frac{d\boldsymbol{p}}{dt}.$$

Putting the pieces together, we restate Newton's second law as

$$\Sigma \boldsymbol{F} = \frac{d\boldsymbol{p}}{dt}. \tag{8–3}$$

**The vector sum of forces acting on a particle equals the time rate of change of momentum of the particle.** This, not $\Sigma \boldsymbol{F} = m\boldsymbol{a}$, is the modern equivalent of Newton's original statement of his second law. It is valid only in inertial frames of reference. A large ocean vessel such as a supertanker takes a long time and requires large forces to come to a stop from cruising speed. We can think of this effect in terms of the large *mass* of the vessel, but an alternative viewpoint is that the vessel at cruising speed has an enormous amount of *momentum*.

What do we mean by the *total momentum* of two particles? We define it as simply the vector sum of the momenta (plural of momentum) of the two particles. Let $\boldsymbol{p}_1$ and $\boldsymbol{p}_2$ be the momenta of two bodies, and denote their total momentum as $\boldsymbol{P}$. Then, by definition,

$$\boldsymbol{P} = \boldsymbol{p}_1 + \boldsymbol{p}_2. \tag{8–4}$$

We can extend this definition directly to any number of particles. **The total momentum**

***P* of any number of particles is equal to the vector sum of momenta $p_i$ of the individual particles.**

We will often use Eq. (8–4) in component form. If $p_{x1}$, $p_{y1}$, and $p_{z1}$ are the components of momentum of the first particle, and similar notation is used for the second, third, and further particles, then Eq. (8–4) is equivalent to the component equations

$$
\begin{aligned}
P_x &= p_{x1} + p_{x2} + \cdots, \\
P_y &= p_{y1} + p_{y2} + \cdots, \\
P_z &= p_{z1} + p_{z2} + \cdots.
\end{aligned}
\tag{8–5}
$$

It is absolutely essential to remember that momentum is a vector quantity; Eqs. (8–1), (8–3), and (8–4) are *vector* equations. We will often add momentum vectors by use of components, so everything that you have learned about components of vectors will be useful with momentum.

## EXAMPLE 8–1

**Preliminary analysis of a collision** A small compact car with a mass of 1000 kg is traveling north on Morewood Avenue with a speed of 15 m/s. At the intersection of Morewood and Fifth Avenue it collides with an enormous luxury car with a mass of 2000 kg, traveling east on Fifth Avenue at 10 m/s (Fig. 8–1). Treating each car as a particle, find the total momentum just before the collision.

**SOLUTION** We have drawn coordinate axes in the figure and have labeled the small car 1 and the large car 2. The components of momentum of the two cars are

$$
\begin{aligned}
p_{x1} &= m_1 v_{x1} = (1000\text{ kg})(0) = 0, \\
p_{y1} &= m_1 v_{y1} = (1000\text{ kg})(15\text{ m/s}) = 1.5 \times 10^4\text{ kg}\cdot\text{m/s}, \\
p_{x2} &= m_2 v_{x2} = (2000\text{ kg})(10\text{ m/s}) = 2.0 \times 10^4\text{ kg}\cdot\text{m/s}, \\
p_{y2} &= m_2 v_{y2} = (2000\text{ kg})(0) = 0.
\end{aligned}
$$

From Eqs. (8–5) the components of the total momentum $\boldsymbol{P}$ are

$$
\begin{aligned}
P_x &= 0 + 2.0 \times 10^4\text{ kg}\cdot\text{m/s} = 2.0 \times 10^4\text{ kg}\cdot\text{m/s}, \\
P_y &= 1.5 \times 10^4\text{ kg}\cdot\text{m/s} + 0 = 1.5 \times 10^4\text{ kg}\cdot\text{m/s}.
\end{aligned}
$$

The total momentum $\boldsymbol{P}$ is a vector quantity with these components. Its magnitude is

$$
\begin{aligned}
P &= \sqrt{(2.0 \times 10^4\text{ kg}\cdot\text{m/s})^2 + (1.5 \times 10^4\text{ kg}\cdot\text{m/s})^2} \\
&= 2.5 \times 10^4\text{ kg}\cdot\text{m/s}.
\end{aligned}
$$

Its direction is given by the angle $\theta$ in Fig. 8–1b, where

$$
\tan\theta = \frac{1.5 \times 10^4\text{ kg}\cdot\text{m/s}}{2.0 \times 10^4\text{ kg}\cdot\text{m/s}} = \frac{3}{4}, \qquad \theta = 36.9^\circ.
$$

Remember this problem. Later in the chapter an insurance adjustor will want to know the direction in which the wreckage moved and how fast it moved.

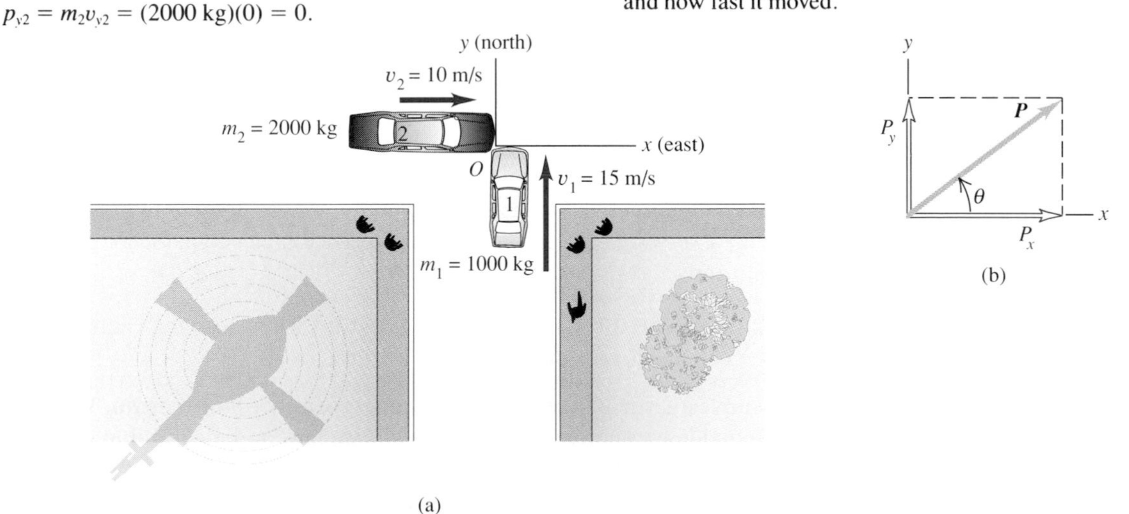


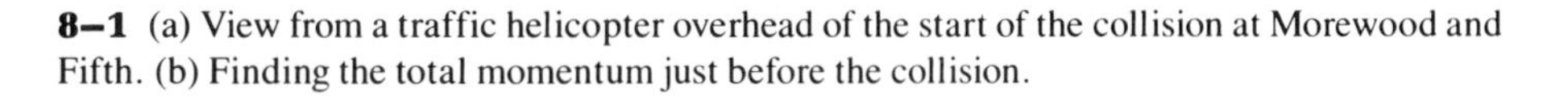

**8–1** (a) View from a traffic helicopter overhead of the start of the collision at Morewood and Fifth. (b) Finding the total momentum just before the collision.

**8–2** Two astronauts push each other as they float freely in the apparent zero-gravity environment of space. These are internal forces; they don't change the total momentum of the two astronauts.

## 8–2 CONSERVATION OF MOMENTUM

The concept of momentum is particularly important in situations in which we have two or more interacting bodies. Let's consider first an idealized system consisting of two bodies that interact with each other but not with anything else — for example, two astronauts who touch each other as they float freely in the zero-gravity environment of outer space (Fig. 8–2). Think of the astronauts as particles. Each particle exerts a force on the other; according to Newton's third law, the two forces are always equal in magnitude and opposite in direction. From Newton's *second* law, as reformulated in Section 8–1, the force on each particle equals the rate of change of its momentum. This means that the two rates of change of momentum are equal and opposite, so the rate of change of the *total* momentum is zero, and the total momentum is constant.

Let's go over that again with some new terminology. For any system the forces that the various particles exert on each other are called **internal forces.** Forces exerted on any part of the system by some agency outside it are called **external forces.** A system such as we have described, with *no* external forces, is called an **isolated system.** We have defined the **total momentum** of the system, $\boldsymbol{P}$, as the vector sum of momenta of the individual bodies,

$$\boldsymbol{P} = \boldsymbol{p}_1 + \boldsymbol{p}_2.$$

The time rate of change of the total momentum is the time derivative of this:

$$\frac{d\boldsymbol{P}}{dt} = \frac{d\boldsymbol{p}_1}{dt} + \frac{d\boldsymbol{p}_2}{dt}. \tag{8–6}$$

Let $\boldsymbol{F}_1$ and $\boldsymbol{F}_2$ be the internal forces on the two bodies due to their mutual interaction. From Newton's second law,

$$\boldsymbol{F}_1 = \frac{d\boldsymbol{p}_1}{dt}, \qquad \boldsymbol{F}_2 = \frac{d\boldsymbol{p}_2}{dt}.$$

From Newton's third law, $\boldsymbol{F}_1 + \boldsymbol{F}_2 = \boldsymbol{0}$. Combining these with Eq. (8–6), we find

$$\frac{d\boldsymbol{P}}{dt} = \boldsymbol{F}_1 + \boldsymbol{F}_2 = \boldsymbol{0}. \tag{8–7}$$

The time rate of change of the total momentum $\boldsymbol{P}$ is zero.

If external forces are also present, the right side of Eq. (8–7) must include them as well as the internal forces. Then the total momentum is, in general, not constant. But if the vector sum of the external forces is zero, these forces don't contribute to the sum, and $d\boldsymbol{P}/dt$ is again zero.

**The total momentum of a system is constant whenever the vector sum of the external forces on the system is zero.**

This is the simplest form of the principle of **conservation of momentum.** We have used Newton's second law to derive this principle, so we have to be careful to use it only in inertial frames of reference.

We can generalize this principle for a system containing any number of bodies interacting only with each other. We make the same argument as before; the total rate of change of momentum of the system due to each action-reaction pair of internal forces is zero. Thus the total rate of change of momentum of the entire system is zero whenever the vector sum of the external forces acting on it is zero. The internal forces can change the momenta of individual particles in the system but not the *total* momentum of the system.

In some ways the principle of conservation of momentum is more general than the principle of conservation of mechanical energy. For example, it is valid *even when the internal forces are not conservative,* but mechanical energy is conserved only when the internal forces are *conservative.* In this chapter we will analyze situations in which both momentum and mechanical energy are conserved and others in which only momentum is conserved. These two principles play a fundamental role in all areas of physics, and we will encounter them throughout our study.

## PROBLEM-SOLVING STRATEGY

### Conservation of momentum

Momentum is a vector quantity; you *must* use vector addition to compute the total momentum of a system. Using components is usually the simplest method.

**1.** Define a coordinate system. Make a sketch showing the coordinate axes, including the positive direction for each. Often it is easiest to choose the $x$-axis as having the direction of one of the initial velocities. Make sure you are using an inertial frame of reference. Most of the problems in this chapter deal with two-dimensional situations, in which the vectors have only $x$- and $y$-components; all the following statements can be generalized to include $z$-components when necessary.

**2.** Treat each body as a particle. Draw "before" and "after" sketches, and include vectors on each to represent all known velocities. Label the vectors with magnitudes, angles, components, or whatever information is given, and give each unknown magnitude, angle, or component an algebraic symbol.

**3.** Compute the $x$- and $y$-components of momentum of each particle, both before and after the interaction, using the relations $p_x = mv_x$ and $p_y = mv_y$. Even when all the velocities lie along a line (such as the $x$-axis), the components of velocity along this line can be positive or negative; be careful with signs!

**4.** Write an equation equating the total *initial* $x$-component of momentum to the total *final* $x$-component of momentum. Write another equation for the $y$-components. These two equations express conservation of momentum in component form. Some of the components will be expressed in terms of symbols representing unknown quantities.

**5.** Solve these equations to determine whatever results are required. In some problems you will have to convert from the $x$- and $y$-components of a velocity to its magnitude and direction, or the reverse.

**6.** Remember that the $x$- and $y$-components of velocity or momentum are *never* added together in the same equation.

**7.** In some problems energy considerations give additional relationships among the various velocities, as we will see later in this chapter.

## EXAMPLE 8–2

**Recoil of a rifle** A marksman holds a 3.00-kg rifle loosely, so as to let it recoil freely when fired, and fires a bullet of mass 5.00 g horizontally with a muzzle velocity $v_B = 300$ m/s (Fig. 8–3). What is the recoil velocity $v_R$ of the rifle? What is the final kinetic energy of the bullet? Of the rifle?

**SOLUTION** We consider an idealized model in which the horizontal forces the marksman exerts on the rifle are negligible. Then with respect to horizontal forces the rifle and bullet can be considered an isolated system. Then the total momentum $\boldsymbol{P}$ of the system is zero both before and after firing. Take the positive $x$-axis to be the direction in which the rifle is aimed. The initial $x$-component of total momentum is zero. After the bullet is fired, the $x$-component of its momentum is (0.00500 kg) (300 m/s), and that of the rifle is $(3.00\text{ kg})v_R$. Conservation of the $x$-component of total momentum gives

$$P_x = 0 = (0.00500\text{ kg})(300\text{ m/s}) + (3.00\text{ kg})v_R,$$
$$v_R = -0.500\text{ m/s}.$$

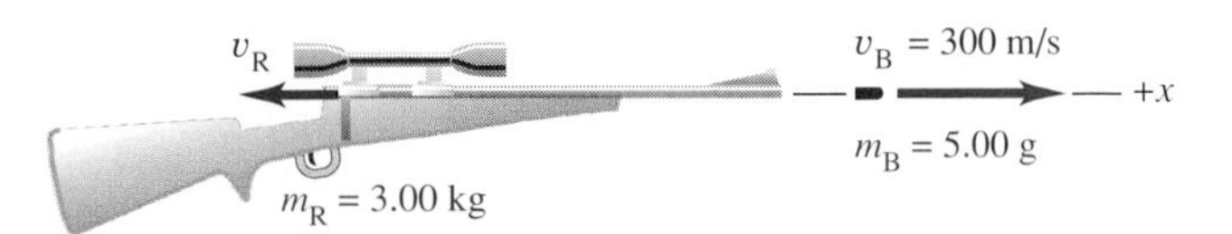


**8–3** The ratio of bullet speed to recoil speed is the inverse of the ratio of bullet mass to rifle mass.

The negative sign means that the recoil is in the direction opposite to that of the bullet. If the butt of a rifle were to hit your shoulder traveling this speed, you would probably feel it. That's what the "kick" of a rifle is all about, and that's why it is more comfortable to hold the rifle tightly against your shoulder when you fire it.

The kinetic energy of the bullet is

$$K_B = \tfrac{1}{2}(0.00500\text{ kg})(300\text{ m/s})^2 = 225\text{ J},$$

and the kinetic energy of the rifle is

$$K_R = \tfrac{1}{2}(3.00\text{ kg})(0.500\text{ m/s})^2 = 0.375\text{ J}.$$

The bullet acquires much greater kinetic energy than the rifle. The reason is that the bullet moves much farther than the rifle during the interaction, so the interaction force on the bullet does more work than the force on the rifle (even though the two forces are equal and opposite). In fact, the ratio of the two kinetic energies, 600:1, is equal to the inverse ratio of the masses. It can be shown that this always happens in recoil situations; we leave the proof as a problem (see Exercise 8–9).

## EXAMPLE 8–3

**A straight-line collision** Two gliders move toward each other on a linear air track (Fig. 8–4). After they collide, $B$ moves away with a final velocity of $+2.0$ m/s. What is the final velocity of $A$?

**SOLUTION** The total vertical force on each glider is zero; the resultant force on each glider is the horizontal force exerted on it by the other glider. The combination is an isolated system, so its total momentum is constant. Take the $x$-axis as lying along the air track, with the positive direction to the right. The masses of the gliders and their initial velocities are shown in Fig. 8–4. Let the final velocity of $A$ be $v_{A2}$. Then we write an equation showing the equality of the total $x$-component of momentum before and after the collision:

$$\begin{aligned} P_x = m_A v_{A1} + m_B v_{B1} &= (0.50\text{ kg})(2.0\text{ m/s}) \\ &\quad + (0.30\text{ kg})(-2.0\text{ m/s}) \\ = m_A v_{A2} + m_B v_{B2} &= (0.50\text{ kg})v_{A2} + (0.30\text{ kg})(+2.0\text{ m/s}). \end{aligned}$$

Solving this equation for $v_{A2}$, we find

$$v_{A2} = -0.40\text{ m/s}.$$

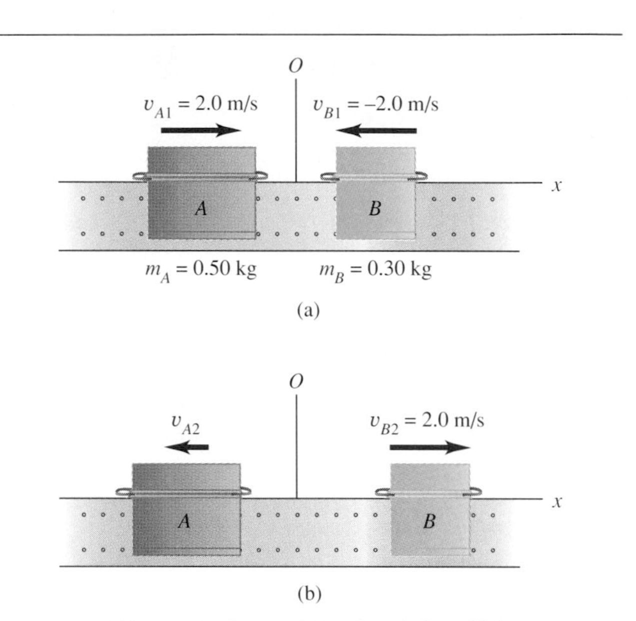


**8–4** Two gliders (a) before and (b) after their collision.

## EXAMPLE 8–4

**Collision in a horizontal plane** Figure 8–5 shows two chunks of ice sliding on a frictionless frozen pond. Chunk $A$, with mass $m_A = 5.0$ kg, moves with initial velocity $v_{A1} = 2.0$ m/s parallel to the $x$-axis. It collides with chunk $B$, which has mass $m_B = 3.0$ kg and is initially at rest. After the collision the velocity of $A$ is found to be $v_{A2} = 1.0$ m/s in a direction at an angle $\alpha = 30°$ with the initial direction. What is the final velocity of $B$?

**SOLUTION** There are no horizontal external forces, so the total horizontal momentum of the system is the same before and after the collision. In this case the velocities are not all along a single line. We have to treat momentum as a *vector* quantity, using components of each momentum in the $x$- and $y$-directions. Then conservation of momentum requires that the sum of the $x$-components before the collision must equal the sum after the collision, and similarly for the $y$-components. Just as with force equilibrium problems, we write a separate equation for each component. For the $x$-components (before and after) we have

$$\begin{aligned} m_A v_{A1x} + m_B v_{B1x} &= (5.0\text{ kg})(2.0\text{ m/s}) + (3.0\text{ kg})(0) \\ &= m_A v_{A2x} + m_B v_{B2x} \\ &= (5.0\text{ kg})(1.0\text{ m/s})(\cos 30°) + (3.0\text{ kg})v_{B2x}, \\ v_{B2x} &= 1.89\text{ m/s}. \end{aligned}$$

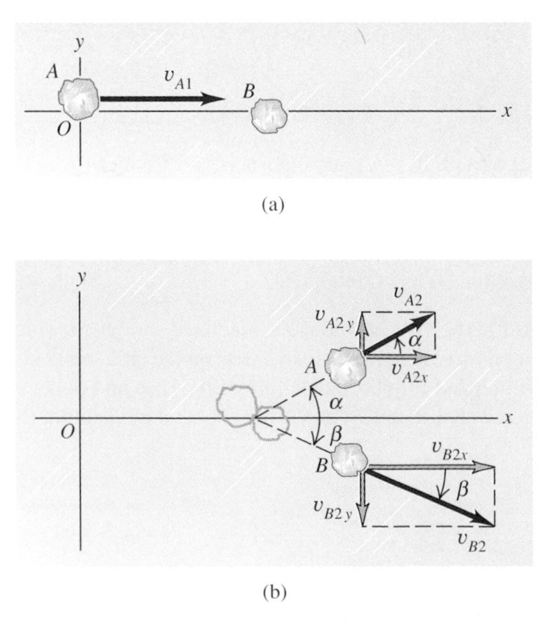


**8–5** Velocities, before and after the collision, and components.

Conservation of the $y$-component of total momentum gives

$$m_A v_{A1y} + m_B v_{B1y} = (5.0\text{ kg})(0) + (3.0\text{ kg})(0)$$
$$= m_A v_{A2y} + m_B v_{B2y}$$
$$= (5.0\text{ kg})(1.0\text{ m/s})(\sin 30°) + (3.0\text{kg})v_{B2y}.$$

Solving this for $v_{B2y}$, we find

$$v_{B2y} = -0.83\text{ m/s}.$$

The magnitude of $v_{B2}$ is

$$v_{B2} = \sqrt{(1.89\text{ m/s})^2 + (-0.83\text{ m/s})^2}$$
$$= 2.1\text{ m/s},$$

and the angle of its direction from the positive $x$-axis is

$$\beta = \arctan\frac{-0.83\text{ m/s}}{1.9\text{ m/s}} = -24°.$$

## 8–3 INELASTIC COLLISIONS

To the average person on the street the term *collision* is likely to mean some sort of automotive disaster. We'll use it in that sense, but we'll also broaden the meaning to include any strong interaction between two bodies that lasts a relatively short time. So we include not only car accidents but also balls hitting on a billiard table, the slowing down of neutrons in a nuclear reactor, a bowling ball striking the pins, the impact of a meteor on the Arizona desert, and a close encounter of a space probe with the planet Saturn.

If the interaction forces are much larger than any external forces, we can model the system as an *isolated* system, neglecting the external forces entirely. Two cars colliding at an icy intersection provide a good example. Even two cars colliding on dry pavement can be treated as an isolated system during the collision if, as happens all too often, the interaction forces between the cars are much larger than the friction forces of pavement against tires.

If the interaction forces between the bodies are *conservative,* the total *kinetic energy* of the system is the same after the collision as before. Such a collision is called an **elastic collision.** A collision between two hard steel balls or two billiard balls is almost completely elastic, and collisions between subatomic particles are often, though not always, elastic. Figure 8–6 shows a model for an elastic collision. When the bodies collide, the springs are momentarily compressed, and some of the original kinetic energy is momentarily converted to elastic potential energy. Then the bodies bounce apart, the springs expand, and this potential energy is reconverted to kinetic energy.

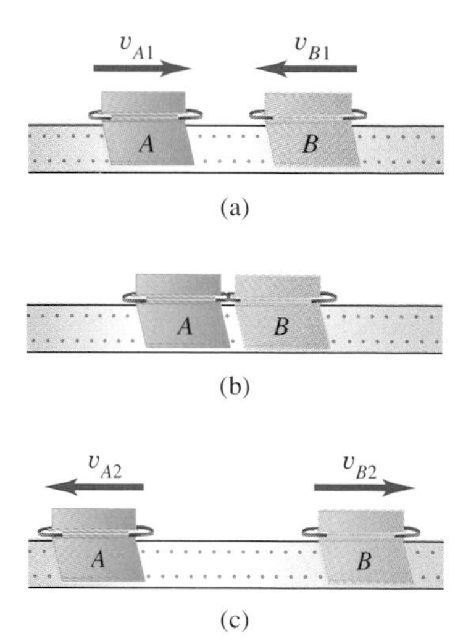


**8–6** Model of an elastic collision: Body $A$ and body $B$ approach each other on a frictionless surface. Each glider has a steel spring bumper on the end to ensure an elastic collision.

A collision in which the total kinetic energy after the collision is *less* than that before the collision is called an **inelastic collision.** In one kind of inelastic collision the colliding bodies stick together and move as one body after the collision; this is often called a **completely inelastic collision.** In Fig. 8–7 we have replaced the spring in Fig. 8–6 with a ball of chewing gum or putty that squashes and sticks the two bodies together. A spitball striking a window shade, a bullet embedding itself in a block of wood, and two cars colliding and locking their fenders are examples of inelastic collisions.

Let's do a little general analysis of a completely inelastic collision of two bodies ($A$ and $B$). Because they stick together after the collision, their final velocities must be equal:

$$\boldsymbol{v}_{A2} = \boldsymbol{v}_{B2} = \boldsymbol{v}_2.$$

Conservation of momentum gives the relation

$$m_A \boldsymbol{v}_{A1} + m_B \boldsymbol{v}_{B1} = (m_A + m_B)\boldsymbol{v}_2. \tag{8–8}$$

If we know the masses and initial velocities, we can compute the common final velocity $v_2$.

Suppose, for example, that a body with mass $m_A$ and initial velocity $v_1$ along the $+x$-axis collides inelastically with a body with mass $m_B$ that is initially at rest ($v_{B1} = 0$). From Eq. (8–8) the common $x$-component of velocity $v_2$ of both bodies after the collision

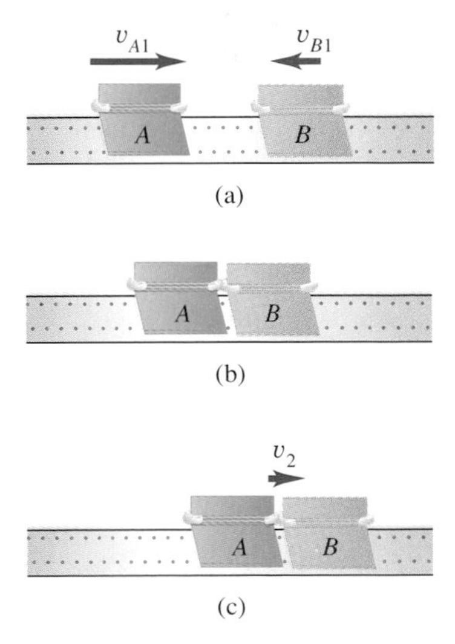


**8–7** Model of a completely inelastic collision. The spring bumpers on the gliders are replaced by putty, so the bodies stick together after collision.

is

$$v_2 = \frac{m_A}{m_A + m_B} v_1. \tag{8–9}$$

Let's verify that the total kinetic energy after this completely inelastic collision is less than before the collision. The kinetic energies $K_1$ and $K_2$ before and after the collision, respectively, are

$$K_1 = \tfrac{1}{2} m_A v_1^2,$$

$$K_2 = \tfrac{1}{2}(m_A + m_B) v_2^2 = \tfrac{1}{2}(m_A + m_B)\left(\frac{m_A}{m_A + m_B}\right)^2 v_1^2.$$

The ratio of final to initial kinetic energy is

$$\frac{K_2}{K_1} = \frac{m_A}{m_A + m_B}. \tag{8–10}$$

The right side of this equation is always less than unity because the denominator is always greater than the numerator. Even when the initial velocity of $m_B$ is not zero, it is not hard to verify that the kinetic energy after a completely inelastic collision is always less than before.

*Please note:* We don't recommend memorizing these equations. We derived them only to prove that kinetic energy is always lost in a completely inelastic collision.

## EXAMPLE 8–5

Suppose that in the collision in Fig. 8–4 the two gliders stick together after the collision; the masses and initial velocities are as shown. Find the common final velocity, and compare the initial and final kinetic energies.

**SOLUTION** Let the $x$-axis lie along the direction of motion, as before. From conservation of the $x$-component of momentum,

$$(0.50\text{ kg})(2.0\text{ m/s}) + (0.30\text{ kg})(-2.0\text{ m/s}) = (0.50\text{ kg} + 0.30\text{ kg})v_2,$$

$$v_2 = 0.50\text{ m/s}.$$

Because $v_2$ is positive, the gliders move together to the right (the $+x$-direction) after the collision. Before the collision the kinetic energy of glider $A$ is

$$\tfrac{1}{2} m_A v_{A1}^2 = \tfrac{1}{2}(0.50\text{ kg})(2.0\text{ m/s})^2 = 1.0\text{ J},$$

and that of glider $B$ is

$$\tfrac{1}{2} m_B v_{B1}^2 = \tfrac{1}{2}(0.30\text{ kg})(-2.0\text{ m/s})^2 = 0.60\text{ J}.$$

Note that the kinetic energy of glider $B$ is positive, even though the $x$-components of its velocity $v_{B1}$ and momentum $mv_{B1}$ are both negative.

The *total* kinetic energy before the collision is 1.6 J. The kinetic energy after the collision is

$$\tfrac{1}{2}(m_A + m_B)v_2^2 = \tfrac{1}{2}(0.50\text{ kg} + 0.30\text{ kg})(0.50\text{ m/s})^2 = 0.10\text{ J}.$$

The final kinetic energy is only $\frac{1}{16}$ of the original, and $\frac{15}{16}$ is "lost" in the collision.

The energy isn't really lost; it is converted from mechanical energy to various other forms. If there is a ball of chewing gum between the gliders, it squashes irreversibly and becomes warmer. If the gliders couple together like two freight cars, the energy goes into elastic waves that are eventually dissipated. If there is a spring between the gliders that is compressed as they lock together, then the energy is stored as potential energy in the spring. In all of these cases the *total* energy of the system is conserved, although *kinetic* energy is not. However, in an isolated system, momentum is *always* conserved, whether the collision is elastic or not.

## EXAMPLE 8–6

**The ballistic pendulum** Figure 8–8 shows a ballistic pendulum, a system for measuring the speed of a bullet. The bullet, with mass $m$, is fired into a block of wood with mass $M$, suspended like a pendulum, and makes a completely inelastic collision with it. After the impact of the bullet the block swings up to a maximum height $y$. Given values of $y$, $m$, and $M$, how can we find the initial speed $v$ of the bullet?

**SOLUTION** We analyze this event in two stages; the first is the embedding of the bullet in the block, and the second is the subsequent swinging of the block on its strings. As we will see, momentum is conserved in the first stage, energy in the second.

During the first stage, the bullet embeds itself in the block so quickly that the block doesn't have time to swing appreciably away from its initial position. So during the impact the supporting

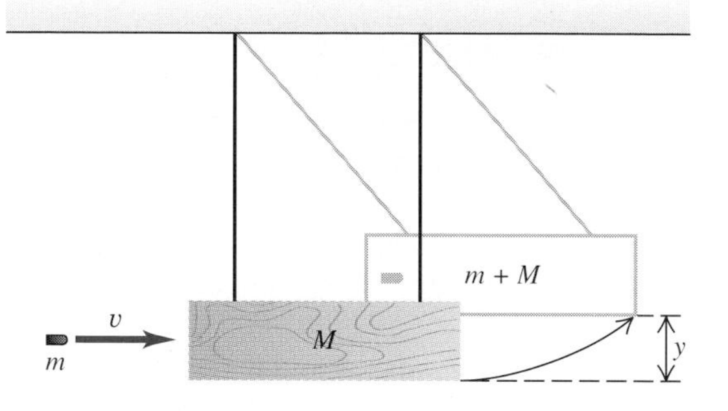


**8–8** A ballistic pendulum.

strings remain very nearly vertical, there is negligible external horizontal force acting on the system, and the horizontal component of momentum is conserved. Let $V$ be the speed of bullet and block just after the collision; momentum conservation gives

$$mv = (m + M)V, \qquad v = \frac{m + M}{m}V.$$

The kinetic energy of the system just after the collision is $K = \frac{1}{2}(m + M)V^2$.

In the second stage, after the collision, the block and bullet move as a unit. The only forces are the weight (a conservative force) and the string tensions (which do no work). So as the pendulum swings upward and to the right, mechanical energy is conserved. The pendulum comes to rest (for an instant) at a height $y$, where its kinetic energy $\frac{1}{2}(m + M)V^2$ has all become potential energy $(m + M)gy$; then it swings back down. At the highest point energy conservation gives

$$\tfrac{1}{2}(m + M)V^2 = (m + M)gy,$$

from which we get

$$V = \sqrt{2gy}.$$

Now we substitute this into the momentum equation to find $v$:

$$v = \frac{m + M}{m}\sqrt{2gy}.$$

By measuring $m$, $M$, and $y$ we can compute the original velocity $v$ of the bullet. For example, if $m = 5.00$ g $= 0.00500$ kg, $M = 2.00$ kg, and $y = 3.00$ cm $= 0.0300$ m,

$$v = \frac{2.00 \text{ kg} + 0.00500 \text{ kg}}{0.00500 \text{ kg}}\sqrt{2(9.80 \text{ m/s}^2)(0.0300 \text{ m})}$$
$$= 307 \text{ m/s}.$$

The velocity $V$ of the block just after impact is

$$V = \sqrt{2gy} = \sqrt{2(9.80 \text{ m/s}^2)(0.0300 \text{ m})}$$
$$= 0.767 \text{ m/s}.$$

The kinetic energy just before impact is $\frac{1}{2}(0.00500 \text{ kg}) \cdot (307 \text{ m/s})^2 = 236$ J. Just after impact it is $\frac{1}{2}(2.005 \text{ kg}) \cdot (0.767 \text{ m/s})^2 = 0.590$ J. Nearly all the kinetic energy disappears as the wood splinters and the bullet and wood become hotter.

## EXAMPLE 8–7

**Collision analysis continued** In Example 8–1 we considered the impending collision of a small compact car with an enormous luxury car. It is now two seconds later, and the collision has occurred. Fortunately, all occupants were wearing seat belts, and there were no injuries, but the two cars became thoroughly tangled and moved away from the impact point as one mass. The insurance adjustor has asked you to help find the velocity of the wreckage just after impact.

**SOLUTION** We worked out the initial momentum components in Example 8–1. To find out what happens *after* the collision, we'll assume that we can treat the cars as an isolated system during the collision. This may seem implausible, so let's look at some numbers. The mass of the large car is 2000 kg, its weight is about 20,000 N, and if the coefficient of friction is 0.5, the friction force when it slides across the pavement is about 10,000 N. That sounds like a large force. But suppose the car runs into a brick wall while going 10 m/s. Its kinetic energy just before impact is $\frac{1}{2}(2000 \text{ kg})(10 \text{ m/s})^2 = 1.0 \times 10^5$ J, so the force applied by the wall must do $-1.0 \times 10^5$ J of work on the car to stop it. The car may crumple 0.2 m or so; to do $-1.0 \times 10^5$ J of work in 0.2 m requires a force with magnitude $5.0 \times 10^5$ N, or 50 times as great as the friction force.

So it's not so unreasonable to neglect the friction forces on both cars and assume that during the collision they form an isolated system. In that case the total momentum just after the collision is the same as that just before, which we found to be $2.5 \times 10^4$ kg·m/s in a direction 36.9° north from straight east. Assuming that no parts fall off, the total mass of wreckage is $M = 3000$ kg. Calling the final velocity $\boldsymbol{V}$, we have $\boldsymbol{P} = M\boldsymbol{V}$. The direction of the velocity is the same as that of the momentum, and its magnitude is

$$V = \frac{P}{M} = \frac{2.5 \times 10^4 \text{ kg}\cdot\text{m/s}}{3000 \text{ kg}} = 8.3 \text{ m/s}.$$

This is an inelastic collision, so we expect the total kinetic energy to be less after the collision than before. We invite you to carry out the calculations; you will find that the initial kinetic energy is $2.1 \times 10^5$ J and the final kinetic energy is $1.0 \times 10^5$ J. Over half of the initial kinetic energy is converted to other forms of energy.

If you were tempted to try to find the final velocity by taking the vector sum of the initial velocities, ask yourself why in the world you expected that to work. Is there a law of conservation of velocities? Absolutely not; the only conserved quantity is the total momentum of the system, and the above analysis is the only correct way to do the problem.

## 8–4 ELASTIC COLLISIONS

In Section 8–3 we defined an *elastic collision* in an isolated system to be a collision in which kinetic energy (as well as momentum) is conserved. Elastic collisions occur when the interaction forces between the bodies are *conservative*. When two steel balls collide, they squash a little near the surface of contact, but then they spring back. Some of the kinetic energy is stored temporarily as elastic potential energy, but at the end it is reconverted to kinetic energy.

Let's look at an elastic collision between two bodies $A$ and $B$. We start with a head-on collision, in which all the velocities lie along the same line; we choose this line to be the $x$-axis. Each momentum and velocity then has only an $x$-component. Later we will consider more general collisions. We call the $x$-components of velocity before the collision $v_{A1}$ and $v_{B1}$ and those after the collision $v_{A2}$ and $v_{B2}$. From conservation of kinetic energy we have

$$\tfrac{1}{2}m_A v_{A1}{}^2 + \tfrac{1}{2}m_B v_{B1}{}^2 = \tfrac{1}{2}m_A v_{A2}{}^2 + \tfrac{1}{2}m_B v_{B2}{}^2,$$

and conservation of momentum gives

$$m_A v_{A1} + m_B v_{B1} = m_A v_{A2} + m_B v_{B2}.$$

If the masses $m_A$ and $m_B$ and the initial velocities $v_{A1}$ and $v_{B1}$ are known, these two equations can be solved simultaneously to find the two final velocities $v_{A2}$ and $v_{B2}$.

The general solution is a little complicated, so we will concentrate on a particular case, in which body $B$ is at rest before the collision. Think of it as a target for $m_A$ to hit. We can then simplify the velocity notation; we let $v$ be the initial $x$-component of velocity of $A$, and $v_A$ and $v_B$ the final $x$-components for $A$ and $B$. Then the kinetic energy and momentum conservation equations are, respectively,

$$\tfrac{1}{2}m_A v^2 = \tfrac{1}{2}m_A v_A{}^2 + \tfrac{1}{2}m_B v_B{}^2, \qquad (8\text{–}11)$$

$$m_A v = m_A v_A + m_B v_B. \qquad (8\text{–}12)$$

We may solve for $v_A$ and $v_B$ in terms of the masses and the initial velocity $v$. This involves some fairly strenuous algebra, but it's worth it. No pain, no gain! The simplest approach is somewhat indirect, but it shows another interesting feature of elastic collisions.

First we rearrange Eqs. (8–11) and (8–12) as follows:

$$m_B v_B{}^2 = m_A(v^2 - v_A{}^2) = m_A(v - v_A)(v + v_A), \qquad (8\text{–}13)$$

$$m_B v_B = m_A(v - v_A). \qquad (8\text{–}14)$$

Now we divide Eq. (8–13) by Eq. (8–14) to obtain

$$v_B = v + v_A. \qquad (8\text{–}15)$$

We substitute this back into Eq. (8–14) to eliminate $v_B$ and then solve for $v_A$:

$$m_B(v + v_A) = m_A(v - v_A),$$

$$v_A = \frac{m_A - m_B}{m_A + m_B}v. \qquad (8\text{–}16)$$

Finally, we substitute this result back into Eq. (8–14) to obtain

$$v_B = \frac{2m_A}{m_A + m_B}v. \qquad (8\text{–}17)$$

Now we can interpret the results. Suppose body $A$ is a ping-pong ball and $B$ is a bowling ball. Then we expect $A$ to bounce off after the collision with a velocity nearly equal to its original value but in the opposite direction (Fig. 8–9a), and we expect $B$'s

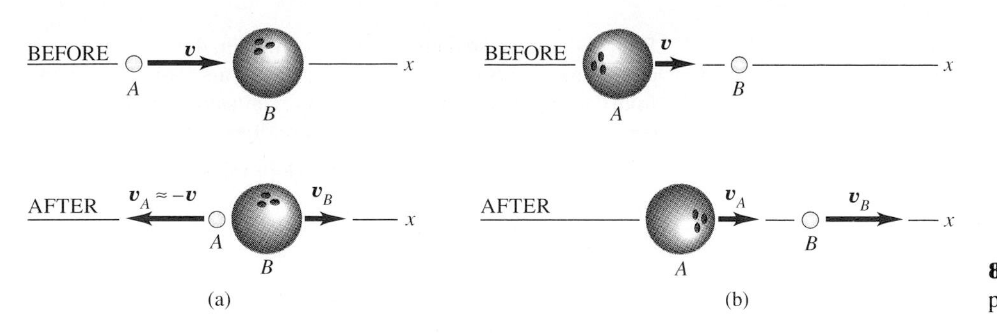


**8–9** Collisions between a ping-pong ball and a bowling ball.

velocity to be much smaller. That's just what the equations predict. When $m_A$ is much smaller than $m_B$, the fraction in Eq. (8–16) is approximately equal to $(-1)$; so $v_A$ is approximately equal to $-v$. When $m_A$ is much smaller than $m_B$, the fraction in Eq. (8–17) is much smaller than unity; so $v_B$ is much less than $v$. Figure 8–9b shows the opposite case, in which $A$ is the bowling ball and $B$ is the ping-pong ball, and $m_A$ is much larger than $m_B$. What do you expect to happen then? Check your predictions against Eqs. (8–16) and (8–17).

Another interesting case occurs when the masses are equal (Fig. 8–10). If $m_A = m_B$, then Eqs. (8–16) and (8–17) give $v_A = 0$ and $v_B = v$. That is, the body that was moving stops dead; it gives all its momentum and kinetic energy to the body that was at rest. This behavior is familiar to all pool players and marble shooters.

Now comes the surprise bonus. Equation (8–15) can be rewritten as

$$v = v_B - v_A. \tag{8–18}$$

Now $v_B - v_A$ is just the velocity of $B$ relative to $A$ after the collision, and $v$, apart from sign, is the velocity of $B$ relative to $A$ before the collision. *The relative velocity has the same magnitude before and after the collision.* If we view this collision from a second coordinate system moving with constant velocity relative to the first, the velocities of the bodies are different but the *relative* velocities are the same. So we can generalize Eq. (8–18) to include the case in which neither body is at rest initially. Going back to our original notation, we state

$$v_{B2} - v_{A2} = -(v_{B1} - v_{A1}). \tag{8–19}$$

We have proved this only for straight-line collisions, but it turns out that a similar *vector* relationship is a general property of *all* elastic collisions, even when both bodies are moving initially and the velocities do not all lie along the same line. This result provides an alternative and equivalent definition of an elastic collision: **In an elastic collision the relative velocity of the two bodies has the same magnitude before and after the collision.** Whenever this condition is satisfied, the total kinetic energy is also conserved.

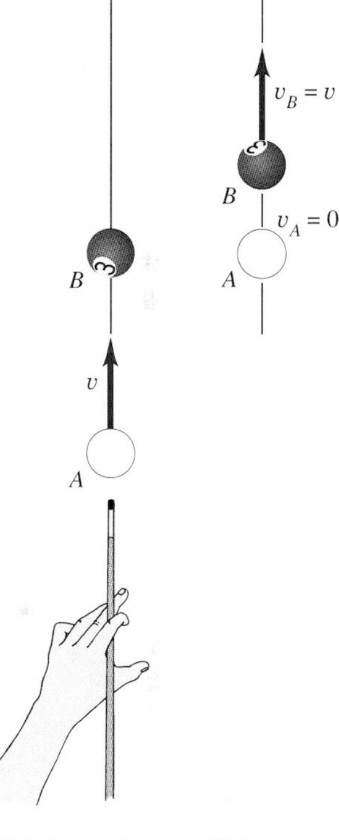


**8–10** In an elastic collision between bodies of equal mass, if one body is at rest, it receives all the momentum and kinetic energy of the moving body.

## EXAMPLE 8–8

The situation in Fig. 8–11 (on page 208) is the same as for Example 8–3 in Section 8–2, but now we've added springs to the gliders. Suppose the collision is elastic. What are the velocities of $A$ and $B$ after the collision?

**SOLUTION** From conservation of momentum,

$$m_A v_{A1} + m_B v_{B1} = m_A v_{A2} + m_B v_{B2},$$

$$(0.50\text{ kg})(2.0\text{ m/s}) + (0.30\text{ kg})(-2.0\text{ m/s}) = (0.50\text{ kg})v_{A2} + (0.30\text{ kg})v_{B2},$$

$$0.50 v_{A2} + 0.30 v_{B2} = 0.40\text{ m/s}.$$

(In the last equation we have divided through by the unit "kg.") From the relative velocity relation for an elastic collision,

$$\begin{aligned} v_{B2} - v_{A2} &= -(v_{B1} - v_{A1}) \\ &= -(-2.0\text{ m/s} - 2.0\text{ m/s}) = 4.0\text{ m/s}. \end{aligned}$$

Solving these equations simultaneously, we find

$$v_{A2} = -1.0\text{ m/s}, \qquad v_{B2} = 3.0\text{ m/s}.$$

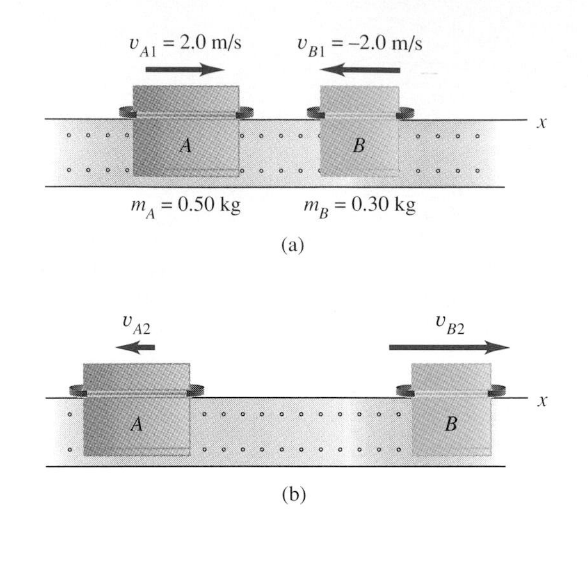


Both bodies reverse their directions of motion; $A$ moves to the left at 1.0 m/s, and $B$ moves to the right at 3.0 m/s. This is different from the result of Example 8–3, but that collision was *not* an elastic one. The total kinetic energy after the collision is

$$\tfrac{1}{2}(0.50\text{ kg})(-1.0\text{ m/s})^2 + \tfrac{1}{2}(0.30\text{ kg})(3.0\text{ m/s})^2 = 1.6\text{ J}.$$

This equals the total kinetic energy before the collision, as expected.

**8–11** Air-track gliders (a) before and (b) after an elastic collision.

## EXAMPLE 8–9

**Moderator in a nuclear reactor** High-speed neutrons are produced in a nuclear reactor during nuclear fission processes. Before a neutron can trigger additional fissions, it has to be slowed down by collisions with nuclei in the *moderator* of the reactor. The first nuclear reactor and the reactor involved in the Chernobyl accident used carbon (graphite) as the moderator material. A neutron (mass 1.0 u) traveling at $2.6 \times 10^7$ m/s makes an elastic head-on collision with a carbon nucleus (mass 12 u) that is initially at rest. What are the velocities after the collision? (1 u is the *atomic mass unit,* equal to $1.66 \times 10^{-27}$ kg.)

**SOLUTION** Because one body is initially at rest, we can use Eqs. (8–16) and (8–17), with

$$m_A = 1.0\text{ u}, \quad m_B = 12\text{ u}, \quad \text{and} \quad v = 2.6 \times 10^7\text{ m/s}.$$

We'll let you do the arithmetic; the results are

$$v_A = -2.2 \times 10^7\text{ m/s},$$
$$v_B = 0.4 \times 10^7\text{ m/s}.$$

The neutron ends up with $\frac{11}{13}$ of its original speed, and the speed of the recoiling carbon nucleus is $\frac{2}{13}$ of the neutron's original speed. Kinetic energy is proportional to speed squared, so the neutron's final kinetic energy is $(\frac{11}{13})^2$ or about 0.72 of its original value. If it makes a second such collision, its kinetic energy is $(0.72)^2$ or about half its original value, and so on.

## EXAMPLE 8–10

**The slingshot effect** Figure 8–12 shows the planet Saturn moving in the negative $x$-direction at its orbital speed (with respect to the sun) of 9.6 km/s. The mass of Saturn is $5.63 \times 10^{26}$ kg. A space probe with mass 150 kg approaches Saturn, moving initially in the $+x$-direction with a speed of 10.4 km/s. The gravitational attraction of Saturn (a conservative force) causes the probe to swing around it and head off in the opposite direction. Find its final speed $v$ after it is far enough away to be nearly free of Saturn's gravitational pull.

**SOLUTION** Here the "collision" is not an impact but a gravitational interaction. Let the probe be body $A$ and Saturn be body $B$. Saturn's mass is so much greater than that of the probe that we can assume its speed to be essentially constant during this interaction. So we have a one-dimensional elastic collision in which $v_{B1} = v_{B2} = -9.6$ km/s and $v_{A1} = 10.4$ km/s. The relative velocity of the two bodies has the same magnitude before and after the interaction, as is shown by Eq. (8–19). Being very careful with signs, we find

$$\begin{aligned} v_{A2} &= v_{B2} + v_{B1} - v_{A1} \\ &= (-9.6\text{ km/s}) + (-9.6\text{ km/s}) - 10.4\text{ km/s} \\ &= -29.6\text{ km/s}. \end{aligned}$$

The space probe's final speed (relative to the sun) is nearly three times as great as before the encounter with Saturn. This is a simplified version of the *slingshot effect,* which is used to give an extra boost to space probes. Not only does the probe get a close look at Saturn, but its kinetic energy increases by a factor of $(29.6/10.4)^2 = 8.1$ in the process. In this bit of celestial gymnastics, does the space probe get something for nothing? Where does its extra energy come from?

**8–12** The slingshot effect: A simplified model of a space probe making a nonimpact "collision" with the planet Saturn.

When an elastic two-body collision isn't head-on, the velocities don't all lie along a single line. If they all lie in a plane, each final velocity has two unknown components, and there are four unknowns in all. Conservation of energy and conservation of the $x$- and $y$-components of momentum give only three equations. To determine the final velocities uniquely we need additional information, such as the direction or magnitude of one of the final velocities.

## EXAMPLE 8–11

**An off-center elastic collision** Figure 8–13 shows an elastic collision of two pucks on a frictionless air table. Puck $A$ has mass $m_A = 0.50$ kg, and puck $B$ has mass $m_B = 0.30$ kg. Puck $A$ has an initial velocity of 4.0 m/s in the positive $x$-direction and a final velocity of 2.0 m/s in an unknown direction. Find the final speed $v_{B2}$ of puck $B$ and the angles $\alpha$ and $\beta$ in the figure, assuming that the collision is elastic.

**SOLUTION** Because the collision is elastic, the initial and final kinetic energies are equal:

$$\tfrac{1}{2}m_A v_{A1}{}^2 = \tfrac{1}{2}m_A v_{A2}{}^2 + \tfrac{1}{2}m_B v_{B2}{}^2,$$

$$\tfrac{1}{2}(0.50\text{ kg})(4.0\text{ m/s})^2 = \tfrac{1}{2}(0.50\text{ kg})(2.0\text{ m/s})^2 + \tfrac{1}{2}(0.30\text{ kg})v_{B2}{}^2,$$

$$v_{B2} = 4.47\text{ m/s}.$$

Conservation of the $x$- and $y$-components of total momentum gives, respectively,

$$m_A v_{A1x} = m_A v_{A2x} + m_B v_{B2x},$$

$$(0.50\text{ kg})(4.0\text{ m/s}) = (0.50\text{ kg})(2.0\text{ m/s})(\cos\alpha) + (0.30\text{ kg})(4.47\text{ m/s})(\cos\beta),$$

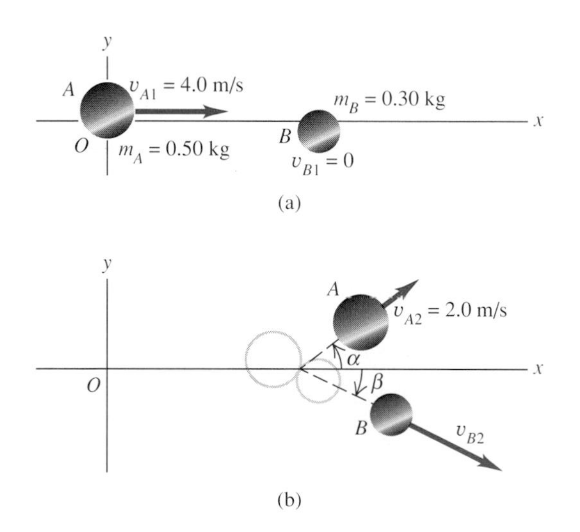


**8–13** (a) Before and (b) after an elastic collision that isn't head-on.

$$0 = m_A v_{A2y} + m_B v_{B2y},$$

$$0 = (0.50 \text{ kg})(2.0 \text{ m/s})(\sin \alpha) - (0.30 \text{ kg})(4.47 \text{ m/s})(\sin \beta).$$

These are two simultaneous equations for $\alpha$ and $\beta$. The simplest solution is to eliminate $\beta$ as follows: We solve the first equation for $\cos \beta$ and the second for $\sin \beta$; we then square each equation and add. Since $\sin^2 \beta + \cos^2 \beta = 1$, this eliminates $\beta$ and leaves an equation that we can solve for $\cos \alpha$ and thus for $\alpha$. We can then substitute this value back into either of the two equations and solve the result for $\beta$. We leave the details for you to work out as a problem (see Exercise 8–24); the results are

$$\alpha = 36.9°, \qquad \beta = 26.6°.$$

The examples in this and the preceding sections show that we can classify collisions according to energy considerations. A collision in which kinetic energy is conserved is called *elastic*, or sometimes *completely elastic* or *perfectly elastic*. A collision in which the total kinetic energy decreases is called *inelastic*. When the two bodies have a common final velocity, we say that the collision is *completely inelastic* or *perfectly inelastic*. In other books you may see the term *semielastic* used to describe a collision in which kinetic energy is not conserved but the two bodies do not have the same final velocity. There are also cases in which the final kinetic energy is *greater* than the initial value. Rifle recoil, discussed in Example 8–2 (Section 8–2) is an example.

Finally, we emphasize again that we can sometimes use conservation of momentum even when there are external forces acting on the system. If the vector sum of the external forces is zero, momentum is conserved. Also, it may happen that the net external force acting on the colliding bodies is small in comparison with the internal forces during the collision. Then we may use an idealized model that neglects the external forces during the actual collision.

## 8–5 IMPULSE

Impulse is a quantity closely related to momentum. Let's begin by defining it. When a constant force $\boldsymbol{F}$ acts on a body during a time interval $\Delta t$, from $t_1$ to $t_2$, the **impulse** of the force, denoted by $\boldsymbol{J}$, is defined to be

$$\boldsymbol{J} = \boldsymbol{F}(t_2 - t_1) = \boldsymbol{F}\,\Delta t. \tag{8–20}$$

Impulse is a vector quantity. The SI unit of the magnitude of impulse is the newton · second (N · s). Because 1 N = 1 kg · m/s$^2$, an alternative set of units for impulse is kg · m/s. So the units of impulse are the same as the units of momentum.

To see what impulse is good for, let's go back to Newton's second law, as restated in terms of momentum: $\boldsymbol{F} = d\boldsymbol{p}/dt$, where $\boldsymbol{F}$ represents the vector sum (resultant) of forces acting on a body. We can combine this with Eq. (8–20):

$$\boldsymbol{F}(t_2 - t_1) = \frac{d\boldsymbol{p}}{dt}(t_2 - t_1).$$

If $\boldsymbol{F}$ is constant, then $d\boldsymbol{p}/dt$ is also constant. In that case, $d\boldsymbol{p}/dt$ is equal to the *total* change in momentum $\boldsymbol{p}_2 - \boldsymbol{p}_1$ during the time interval $t_2 - t_1$, divided by the interval:

$$\frac{d\boldsymbol{p}}{dt} = \frac{\boldsymbol{p}_2 - \boldsymbol{p}_1}{t_2 - t_1}.$$

The impulse $\boldsymbol{J}$ of the force is then

$$\boldsymbol{J} = \boldsymbol{F}(t_2 - t_1) = \boldsymbol{p}_2 - \boldsymbol{p}_1. \tag{8–21}$$

That is, the impulse of the force acting on a body equals the total change of momentum of the body. This relation is called the **impulse-momentum theorem.**

We can generalize the concept of impulse to include a force that varies with time. Again writing Newton's second law in the form $\boldsymbol{F} = d\boldsymbol{p}/dt$, we integrate both sides of this equation over time, between the limits $t_1$ and $t_2$:

$$\int_{t_1}^{t_2} \boldsymbol{F}\,dt = \int_{t_1}^{t_2} \frac{d\boldsymbol{p}}{dt}dt = \int_{\boldsymbol{p}_1}^{\boldsymbol{p}_2} d\boldsymbol{p} = \boldsymbol{p}_2 - \boldsymbol{p}_1.$$

The integral on the left is defined to be the impulse $\boldsymbol{J}$ of the force during this interval:

$$\boldsymbol{J} = \int_{t_1}^{t_2} \boldsymbol{F}\,dt. \tag{8–22}$$

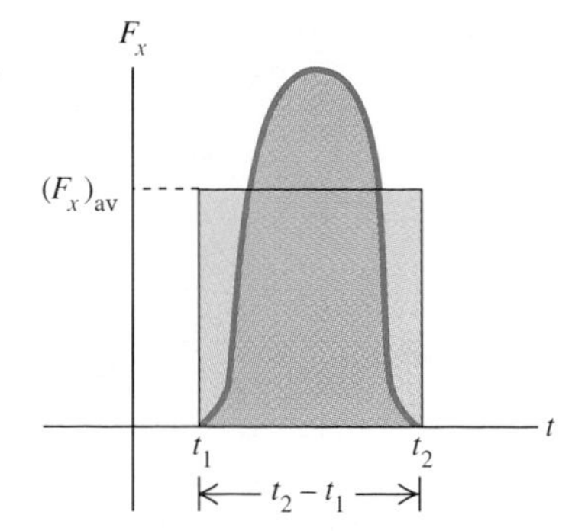


**8–14** The $x$-component of the impulse of $F_x$ between $t_1$ and $t_2$ is the area under the $F$-$t$ curve as well as the area under the rectangle with height $(F_x)_{av}$.

With this definition the impulse-momentum theorem ($\boldsymbol{J} = \boldsymbol{p}_2 - \boldsymbol{p}_1$) is valid even when the force $\boldsymbol{F}$ varies with time.

We can define an average force $\boldsymbol{F}_{av}$ such that even when $\boldsymbol{F}$ is not constant, its impulse $\boldsymbol{J}$ is given by

$$\boldsymbol{J} = \boldsymbol{F}_{av}(t_2 - t_1). \tag{8–23}$$

When $\boldsymbol{F}$ is constant, $\boldsymbol{F} = \boldsymbol{F}_{av}$, and Eq. (8–23) reduces to Eq. (8–20).

Figure 8–14 shows a graph of the $x$-component of force $F_x$ in a collision as a function of time. The $x$-component of total impulse from time $t_1$ to $t_2$ is represented by the area under the curve between $t_1$ and $t_2$. This area is equal to the rectangular area bounded by $t_1$, $t_2$, and $(F_x)_{av}$, and $(F_x)_{av}(t_2 - t_1)$ is equal to the impulse of the actual time-varying force during the same interval.

Impulse and momentum are both vector quantities, and Eqs. (8–20) through (8–23) are all vector equations. In specific problems it is often easiest to use them in component form:

$$\int_{t_1}^{t_2} F_x\,dt = p_{x2} - p_{x1} = mv_{x2} - mv_{x1},$$
$$\int_{t_1}^{t_2} F_y\,dt = p_{y2} - p_{y1} = mv_{y2} - mv_{y1}. \tag{8–24}$$

## EXAMPLE 8–12

**A ball hits a wall** Suppose you throw a ball of mass 0.40 kg against a brick wall. It hits the wall moving horizontally to the left at 30 m/s and rebounds horizontally to the right at 20 m/s. a) Find the impulse of the force exerted on the ball by the wall. b) If the ball is in contact with the wall for 0.010 s, find the average force on the ball during the impact.

**SOLUTION** a) Take the $x$-axis as horizontal and the positive direction to the right (Fig. 8–15). Then the initial $x$-component of momentum of the ball is

$$p_1 = mv_1 = (0.40\text{ kg})(-30\text{ m/s}) = -12\text{ kg}\cdot\text{m/s}.$$

The final $x$-component of momentum is

$$p_2 = mv_2 = +8.0\text{ kg}\cdot\text{m/s}.$$

The *change* in the $x$-component of momentum is

$$p_2 - p_1 = mv_2 - mv_1 = 8.0\text{ kg}\cdot\text{m/s} - (-12\text{ kg}\cdot\text{m/s})$$
$$= 20\text{ kg}\cdot\text{m/s}$$

According to Eq. (8–21), this equals the $x$-component of impulse of the force acting on the ball, so $J_x = 20\text{ kg}\cdot\text{m/s} = 20\text{ N}\cdot\text{s}$.

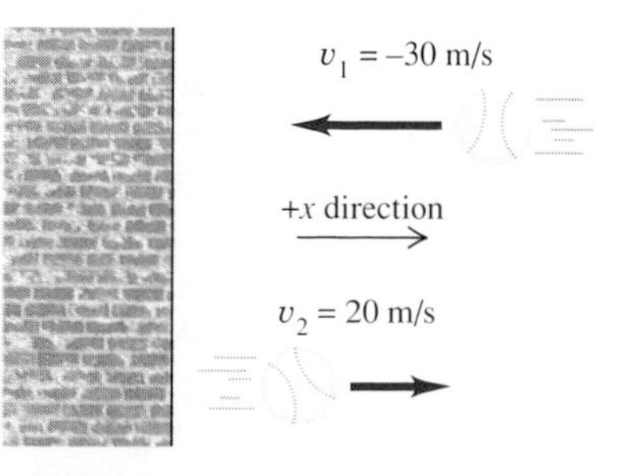


**8–15** Impulse of a ball moving to the left, hitting a wall, and rebounding to the right.

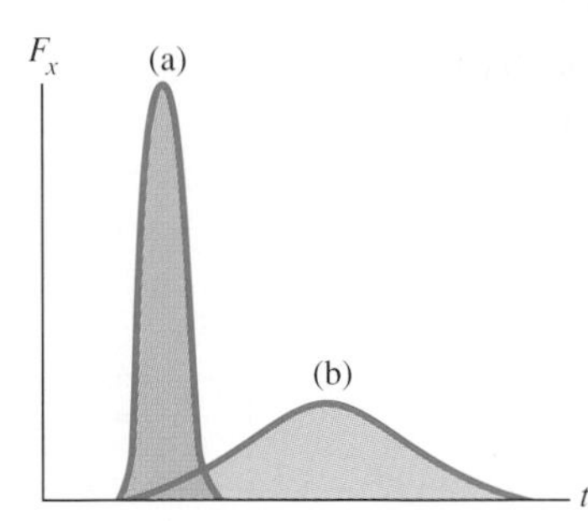


**8–16** (a) A "hard" collision, such as that of a golf ball. (b) A "soft" collision, such as that of a tennis ball. For a given impulse the maximum force is less for the longer-lasting collision. This is the basic principle behind cushioning objects to prevent breakage.

The time variation of the force may be similar to one of the curves in Fig. 8–16. The force is zero before impact, rises to a maximum, and then decreases to zero when the ball loses contact with the wall. If the ball is relatively rigid, like a baseball or a golf ball, the time of collision is small and the maximum force is large, as in curve (a). If the ball is softer, like a tennis ball, the collision time is larger and the maximum force is less, as in curve (b). In any case the *area* under the curve represents the impulse $J = 20\ \text{N} \cdot \text{s}$.

b) If the collision time is 0.010 s, then from Eq. (8–23),

$$F_{\text{av}}(0.01\ \text{s}) = 20\ \text{N} \cdot \text{s}, \qquad F_{\text{av}} = 2000\ \text{N}.$$

This average force is represented by the horizontal line $(F_x)_{\text{av}}$ in Fig. 8–14.

Figure 8–17 is a photograph showing the impact of a tennis ball and racket.

**8–17** Boris Becker hitting a tennis ball. Typically, the ball is in contact with the racket for approximately 0.01 s. The ball flattens noticeably on both sides, and the frame of the racket vibrates during and after the impact.

## EXAMPLE 8–13

**Kicking a soccer ball** Suppose the ball in Example 8–12 is a soccer ball of mass 0.40 kg. Initially it is moving to the left at 20 m/s, but then it is kicked and given a velocity at 45° upward and to the right, with a magnitude of 30 m/s (Fig. 8–18a). Find the impulse of the force and the average force, assuming a collision time $\Delta t = 0.010$ s.

**SOLUTION** The velocities are not along the same line, and we have to be careful to treat momentum and impulse as vector quantities, using their $x$- and $y$-components. Taking the $x$-axis horizontally to the right and the $y$-axis vertically upward, we find the following velocity components:

$$v_{x1} = -20\ \text{m/s}, \qquad v_{y1} = 0,$$

$$v_{x2} = v_{y2} = (0.707)(30\ \text{m/s}) = 21\ \text{m/s}.$$

The $x$-component of impulse is equal to the $x$-component of momentum change, and the same is true for the $y$-components:

$$\begin{aligned} J_x &= m(v_{x2} - v_{x1}) \\ &= (0.40\ \text{kg})[21\ \text{m/s} - (-20\ \text{m/s})] \\ &= 16.4\ \text{kg} \cdot \text{m/s}, \end{aligned}$$

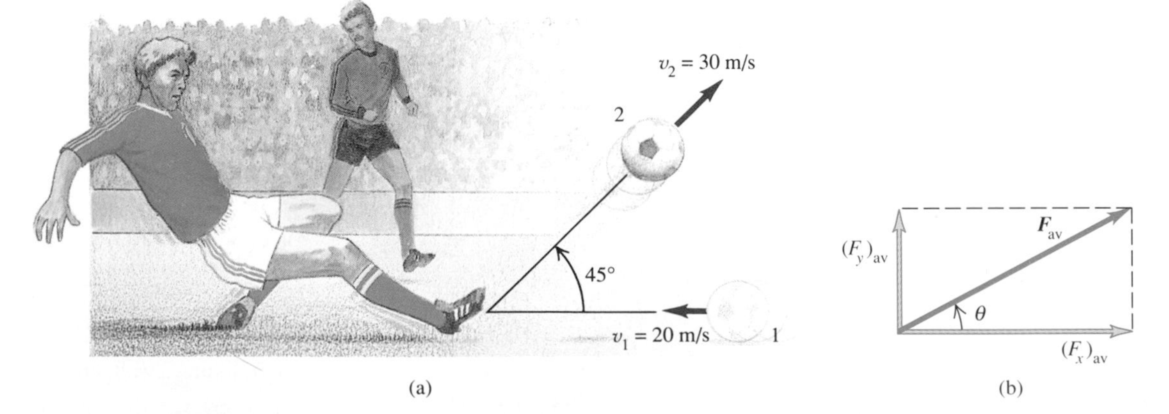


**8–18** (a) A soccer ball (1) before and (2) after being kicked. (b) Finding the average force from its components.

$$J_y = m(v_{y2} - v_{y1}) = (0.40\ \text{kg})(21\ \text{m/s} - 0) = 8.4\ \text{kg}\cdot\text{m/s}.$$

The components of average force on the ball (Fig. 8–18b) are

$$(F_x)_{av} = \frac{J_x}{\Delta t} = 1640\ \text{N}, \qquad (F_y)_{av} = \frac{J_y}{\Delta t} = 840\ \text{N}.$$

The magnitude and direction of the average force are

$$F_{av} = \sqrt{(1640\ \text{N})^2 + (840\ \text{N})^2} = 1.8 \times 10^3\ \text{N},$$

$$\theta = \arctan\frac{840\ \text{N}}{1640\ \text{N}} = 27°,$$

where $\theta$ is measured upward from the $+x$-axis. Note that the ball's final velocity does not have the same direction as the average force acting on it.

The impulse-momentum theorem has a superficial resemblance to the work-energy theorem that we developed in Chapter 6, but there are important differences. First, impulse is a product of a force and a *time* interval, but work is a product of a force and a *distance* and depends on the angle between force and displacement. Second, impulse and momentum are *vector* quantities, and work and kinetic energy are *scalars*. Even in straight-line motion, in which only one component of a vector is involved, force and velocity may have components along this line that are either positive or negative.

Both the impulse-momentum and work-energy theorems are relationships between force and motion, and both rest on the foundation of Newton's laws. They are *integral* principles, relating the motion at two different times separated by a finite interval. Newton's second law, by contrast, is a *differential* principle, relating the forces to the rate of change of velocity at each instant.

## 8–6 CENTER OF MASS

We can restate the principle of conservation of momentum in a useful way by using the concept of **center of mass.** Suppose we have several particles, with masses $m_1$, $m_2$, and so on. Let the coordinates of $m_1$ be $(x_1, y_1)$, those of $m_2$ be $(x_2, y_2)$, and so on. We define the center of mass of the system as the point having coordinates $(x_{cm}, y_{cm})$ given by

$$x_{cm} = \frac{m_1x_1 + m_2x_2 + m_3x_3 + \cdots}{m_1 + m_2 + m_3 + \cdots} = \frac{\sum_i m_ix_i}{\sum_i m_i},$$

$$y_{cm} = \frac{m_1y_1 + m_2y_2 + m_3y_3 + \cdots}{m_1 + m_2 + m_3 + \cdots} = \frac{\sum_i m_iy_i}{\sum_i m_i}. \tag{8–25}$$

The position vector $\boldsymbol{r}_{cm}$ of the center of mass can be expressed in terms of the position vectors $\boldsymbol{r}_1, \boldsymbol{r}_2, \ldots$, of the particles, as

$$\boldsymbol{r}_{cm} = \frac{m_1\boldsymbol{r}_1 + m_2\boldsymbol{r}_2 + m_3\boldsymbol{r}_3 + \cdots}{m_1 + m_2 + m_3 + \cdots} = \frac{\sum_i m_i\boldsymbol{r}_i}{\sum_i m_i}. \tag{8–26}$$

In statistical language the center of mass is a *mass-weighted average* position of the particles.

### EXAMPLE 8–14

**The water molecule** Figure 8–19 (on page 214) shows a simple model of the structure of a water molecule. We represent each atom as a point because nearly all the mass of each atom is concentrated in its nucleus, which is only about $10^{-5}$ times the overall size of the atom. Using the coordinate system shown, find the position of the center of mass.

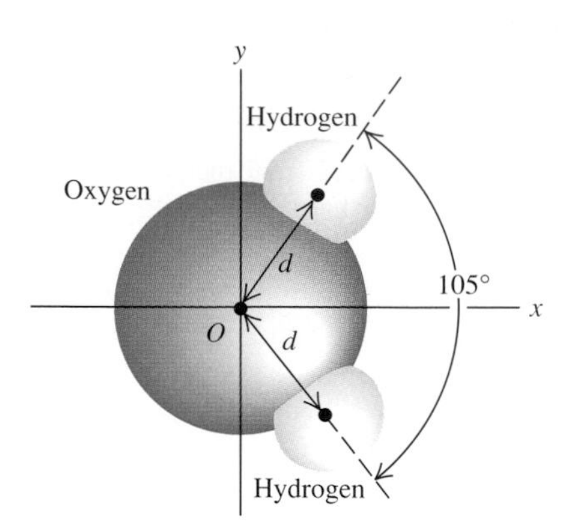


**8–19** Where is the center of mass of a water molecule?

**SOLUTION** From symmetry we see that the center of mass is somewhere along the $x$-axis, so we need to find only its $x$-coordinate. The $x$-coordinate of each hydrogen atom is $d \cos(105°/2)$. The mass of each hydrogen atom is 1.0 u, and the mass of the oxygen atom is 16.0 u. From Eqs. (8–25) the $x$-coordinate of the center of mass is

$$x_{cm} = \frac{(1.0\text{ u})(d\cos 52°) + (1.0\text{ u})(d\cos 52°) + (16.0\text{ u})(0)}{1.0\text{ u} + 1.0\text{ u} + 16.0\text{ u}}$$
$$= 0.068d.$$

The actual value of $d$ for the water molecule is $0.957 \times 10^{-10}$ m $= 95.7$ pm, so we find

$$x_{cm} = (0.068)(95.7\text{ pm}) = 6.5\text{ pm}.$$

The center of mass is much closer to the oxygen atom than to either hydrogen atom because the oxygen atom is much more massive.

## EXAMPLE 8–15

Cyrano is gazing lovingly at Roxanne, who is 12.0 m west of him, when he notices Christian, 12.0 m from each of them and to their north, also gazing lovingly at Roxanne. If Cyrano's mass is 80 kg, Roxanne's is 50 kg, and Christian's is 70 kg, where is the center of mass of this famous triangle relative to Roxanne?

**SOLUTION** Treating the three as particles, we set up the coordinate system shown in Fig. 8–20, with Roxanne at the origin. Then we have

$m_1 = 50$ kg, $\quad x_1 = y_1 = 0,$

$m_2 = 80$ kg, $\quad x_2 = 12.0$ m, $\quad y_2 = 0,$

$m_3 = 70$ kg, $\quad x_3 = 6.0$ m, $\quad y_3 = 12.0\text{ m}\sin 60° = 10.4$ m.

When these values are substituted into Eqs. (8–25), which we leave as an exercise, the results are $x_{cm} = 6.9$ m and $y_{cm} = 3.6$ m. The center of mass of this eternal triangle is 6.9 m east and 3.6 m north of Roxanne.

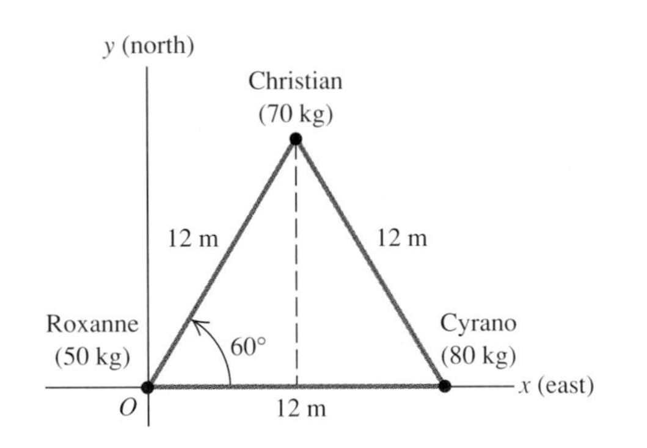


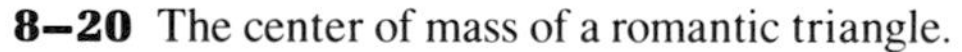

**8–20** The center of mass of a romantic triangle.

For solid bodies, which we think of (at least on a macroscopic level) as having a continuous distribution of matter, the sums in Eqs. (8–25) have to be replaced by integrals. The calculations can get quite involved, but we can say some general things about such problems. First, whenever a homogeneous body has a geometric center, such as a billiard ball, an ice cube, or a can of frozen orange juice, the center of mass is at the geometric center. Whenever a body has an axis of symmetry, such as a wheel or a pulley, the center of mass always lies on that axis. Second, there is no law that says the center of mass has to be within the body. For example, the center of mass of a doughnut is right in the middle of the hole. You can probably think of other examples.

We'll talk a little more about locating the center of mass in Chapter 11 in connection with the related concept *center of gravity.*

Getting back to the center of mass for a collection of particles, we can ask what happens to the center of mass when the particles move. The $x$- and $y$-components of velocity of the center of mass, $v_{xcm}$ and $v_{ycm}$, are the time derivatives of $x_{cm}$ and $y_{cm}$. Also, $dx_1/dt$ is the $x$-component of velocity of particle 1, and so on, so $dx_1/dt = v_{x1}$, and so on.

Taking time derivatives of Eqs. (8–25), we get

$$v_{x\mathrm{cm}} = \frac{m_1 v_{x1} + m_2 v_{x2} + m_3 v_{x3} + \cdots}{m_1 + m_2 + m_3 + \cdots},$$

$$v_{y\mathrm{cm}} = \frac{m_1 v_{y1} + m_2 v_{y2} + m_3 v_{y3} + \cdots}{m_1 + m_2 + m_3 + \cdots}. \qquad (8\text{–}27)$$

These equations are equivalent to the single vector equation

$$\boldsymbol{v}_{\mathrm{cm}} = \frac{m_1 \boldsymbol{v}_1 + m_2 \boldsymbol{v}_2 + m_3 \boldsymbol{v}_3 + \cdots}{m_1 + m_2 + m_3 + \cdots}. \qquad (8\text{–}28)$$

We denote the *total* mass $m_1 + m_2 + \cdots$ by $M$; we can then rewrite Eq. (8–28) as

$$M\boldsymbol{v}_{\mathrm{cm}} = m_1 \boldsymbol{v}_1 + m_2 \boldsymbol{v}_2 + m_3 \boldsymbol{v}_3 + \cdots = \boldsymbol{P}. \qquad (8\text{–}29)$$

The right side is simply the total momentum $\boldsymbol{P}$ of the system. Thus we have proved that *the total momentum is equal to the total mass times the velocity of the center of mass.* It follows that for an isolated system, in which the total momentum is constant, the velocity of the center of mass is also constant.

## EXAMPLE 8–16

A 2.0-kg cat and a 3.0-kg dog are moving along the $x$-axis, heading for a fight. At a particular instant the 2.0-kg cat is 1.0 m to the right of the origin and has a velocity (that is, $x$-component) of 3.0 m/s, and the 3.0-kg dog is 2.0 m to the right of the origin and has a velocity of $-1.0$ m/s. Find the position and velocity of the center of mass, and also find the total momentum.

**SOLUTION** From Eqs. (8–25),

$$x_{\mathrm{cm}} = \frac{(2.0\ \mathrm{kg})(1.0\ \mathrm{m}) + (3.0\ \mathrm{kg})(2.0\ \mathrm{m})}{2.0\ \mathrm{kg} + 3.0\ \mathrm{kg}} = 1.6\ \mathrm{m},$$

and from Eqs. (8–27) the $x$-component of the center-of-mass velocity is

$$v_{x\mathrm{cm}} = \frac{(2.0\ \mathrm{kg})(3.0\ \mathrm{m/s}) + (3.0\ \mathrm{kg})(-1.0\ \mathrm{m/s})}{2.0\ \mathrm{kg} + 3.0\ \mathrm{kg}} = 0.60\ \mathrm{m/s}.$$

The total $x$-component of momentum is

$$P_x = (2.0\ \mathrm{kg})(3.0\ \mathrm{m/s}) + (3.0\ \mathrm{kg})(-1.0\ \mathrm{m/s}) = 3.0\ \mathrm{kg\cdot m/s}.$$

Alternatively, from Eq. (8–29),

$$P_x = (5.0\ \mathrm{kg})(0.60\ \mathrm{m/s}) = 3.0\ \mathrm{kg\cdot m/s}.$$

## 8–7 MOTION OF THE CENTER OF MASS

In Section 8–6 we defined the center of mass of a system of particles and developed the relation between its velocity and the total momentum of the system. Let's now continue that analysis and look at the relation between the motion of the center of mass and the forces acting on the system.

Equations (8–27) and (8–28) give the *velocity* of the center of mass in terms of the velocities of the individual particles. Proceeding one additional step, we take the time derivatives of these equations to show that the accelerations are related in the same way. Let $\boldsymbol{a}_{\mathrm{cm}} = d\boldsymbol{v}_{\mathrm{cm}}/dt$ be the acceleration of the center of mass; then we find

$$M\boldsymbol{a}_{\mathrm{cm}} = m_1 \boldsymbol{a}_1 + m_2 \boldsymbol{a}_2 + m_3 \boldsymbol{a}_3 + \cdots. \qquad (8\text{–}30)$$

Now $m_1\boldsymbol{a}_1$ is equal to the vector sum of forces on the first particle, and so on, so the right side of Eq. (8–30) is equal to the vector sum $\Sigma \boldsymbol{F}$ of *all* the forces on *all* the particles. Just as we did in Section 8–2, we may classify each force as *internal* or *external*. The sum of all forces on all the particles is then

$$\Sigma \boldsymbol{F} = \Sigma \boldsymbol{F}_{\mathrm{ext}} + \Sigma \boldsymbol{F}_{\mathrm{int}} = M\boldsymbol{a}_{\mathrm{cm}}.$$

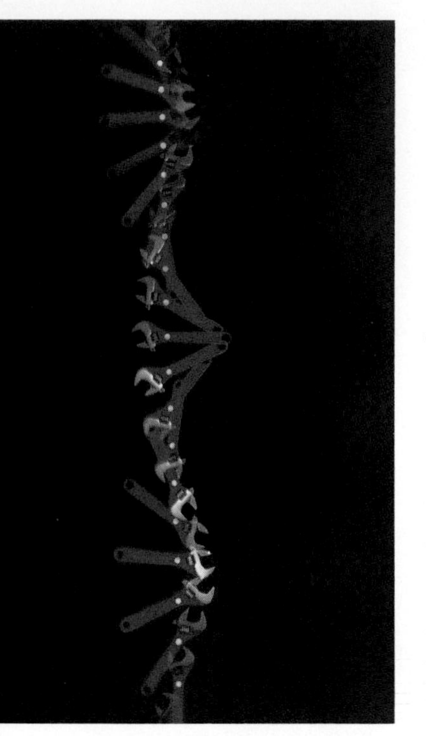

**8–21** The center of mass of this wrench is marked with an X. The total external force acting on the wrench is almost zero. As it spins on a smooth horizontal surface, the center of mass moves in a straight line with constant velocity.

Because of Newton's third law, the internal forces all cancel in pairs, and $\Sigma \boldsymbol{F}_{\text{int}} = \mathbf{0}$. Only the sum of the *external* forces is left, and we have

$$\Sigma \boldsymbol{F}_{\text{ext}} = M\boldsymbol{a}_{\text{cm}}. \tag{8–31}$$

**When a body or a collection of particles is acted on by external forces, the center of mass moves just as though all the mass were concentrated at that point and it were acted on by a resultant force equal to the sum of the external forces on the system.**

This result may not sound very impressive, but in fact it is central to the whole subject of mechanics. Without it we would not be able to represent an extended body as a point when we apply Newton's laws, and in fact we have been using it all along.

Let's look at some additional implications. Suppose we mark the center of mass of an adjustable wrench, which lies at a point partway down the handle. We then slide the wrench with a spinning motion across a smooth tabletop (Fig. 8–21). The overall motion appears complicated, but the center of mass follows a straight line, as though all the mass were concentrated at that point. As another example, suppose a cannon shell traveling in a parabolic trajectory (neglecting air resistance) explodes in flight, splitting into two fragments with equal mass (Fig. 8–22a). The fragments follow new parabolic paths, but the center of mass continues on the original parabolic trajectory, just as though all the mass were still concentrated at that point. A skyrocket exploding in air (Fig. 8–22b) is a spectacular example.

This property of the center of mass is important when we analyze the motion of rigid bodies. We describe the motion of an extended body as a combination of translational motion of the center of mass and rotational motion about an axis through the center of mass. We will return to this topic in Chapter 10. This property also plays an important role in the motion of astronomical objects. The earth and the moon revolve in orbits centered on their center of mass, as do the two stars in a binary system.

Finally, we note again that for an *isolated* system, $\Sigma \boldsymbol{F}_{\text{ext}} = \mathbf{0}$. In this case Eq. (8–31) shows that the acceleration $\boldsymbol{a}_{\text{cm}}$ of the center of mass is zero, so its velocity $\boldsymbol{v}_{\text{cm}}$ is constant, and from Eq. (8–29) the total momentum $\boldsymbol{P}$ is also constant. This reaffirms our statement in Section 8–2 that the total momentum of an isolated system is constant.

Because the center of mass of an isolated system moves with constant velocity in an inertial frame of reference, this point itself may be used as the origin of an inertial frame

**8–22** (a) A shell explodes into two parts in flight. If air resistance is ignored, the fragments follow individual parabolic paths, but the center of mass continues on the same trajectory as the shell's path before exploding—until a fragment hits the ground. (b) The same effect occurs with exploding fireworks.

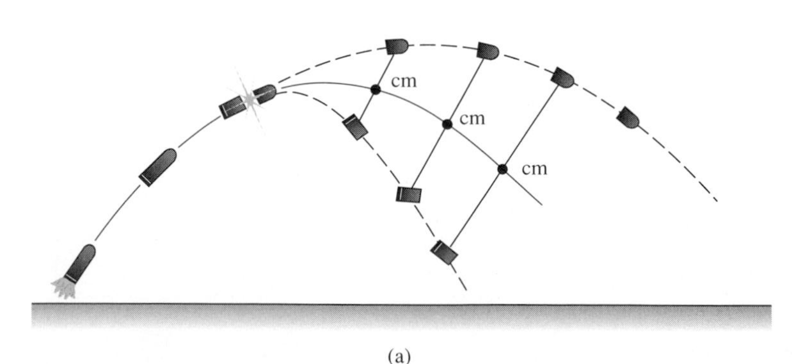


(a)

(b)

of reference. Such a frame, in which the center of mass of a system is at rest, is called the **center-of-mass coordinate system.** Because the center of mass is at rest in this system, the total momentum with respect to this system is zero. This fact leads to significant simplifications in the analysis of collisions.

Here's an example. We have learned that a perfectly elastic collision can be defined in two equivalent ways: as one in which kinetic energy is conserved and as one in which the relative velocity of the two bodies has the same magnitude before and after the collision. It is almost trivial to prove this equivalence by using the center-of-mass coordinate system. First, the total momentum can be zero only if the initial velocities are along a single line, and to satisfy $m_1\mathbf{v}_1 + m_2\mathbf{v}_2 = \mathbf{0}$, the speeds must be inversely proportional to the masses of the particles. This must be true both before and after the collision. The total momentum of the system would still be conserved if each particle had twice the speed after the collision as before, but this would violate conservation of kinetic energy. So we are forced to the conclusion that in the center-of-mass system, energy and momentum can be conserved only if each particle has the same speed after the collision as before. It follows that the relative velocity after the collision (the velocity with which they separate) must have the same magnitude as before the collision (the "closing velocity").

## *8–8 ROCKET PROPULSION

Momentum considerations are particularly useful when we have to analyze a system in which the masses of parts of the system change with time. In such cases we can't use Newton's second law ($\sum \mathbf{F} = m\mathbf{a}$) directly because $m$ changes. Rocket propulsion offers a typical and interesting example of this kind of analysis. A rocket is propelled forward by rearward ejection of burned fuel that initially was in the rocket. The forward force on the rocket is the reaction to the backward force on the ejected material. The total mass of the system is constant, but the mass of the rocket itself decreases as material is ejected. As a simple example, we consider a rocket fired in outer space, where there is no gravitational field and no air resistance. We choose our $x$-axis to be along the rocket's direction of motion.

Figure 8–23a shows the rocket at a time $t$ after firing, when its mass is $m$ and the $x$-component of its velocity relative to our coordinate system is $v$. The $x$-component of total momentum $P$ at this instant is $P = mv$. In a short time interval $dt$ the mass of the rocket changes by an amount $dm$. This is an inherently negative quantity because the rocket's mass $m$ *decreases* with time. During $dt$ a *positive* mass $-dm$ of burned fuel is ejected from the rocket. Let $v_{ex}$ be the exhaust speed of this material *relative to the rocket.*

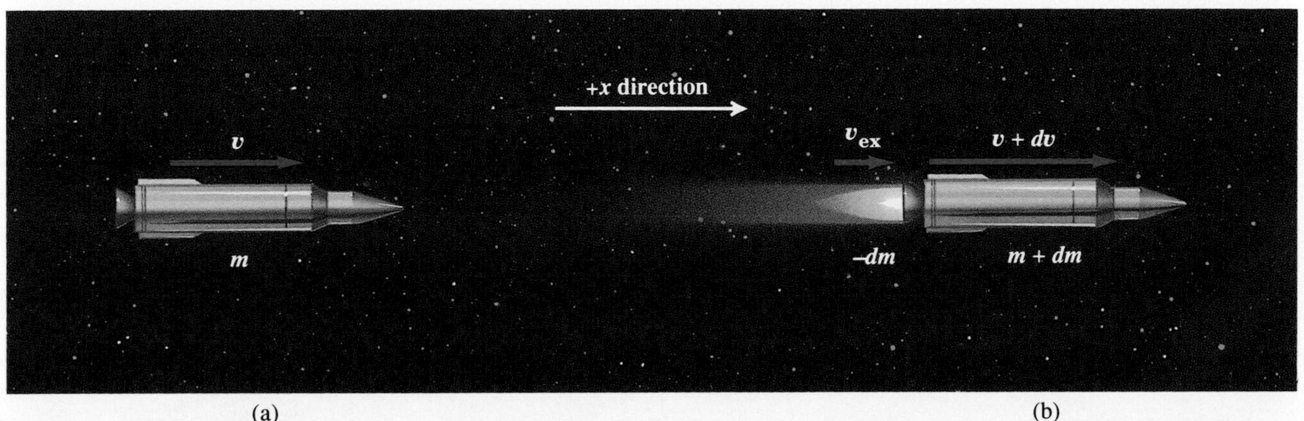


**8–23** (a) A rocket moving in gravity-free outer space with mass $m$ and $x$-component of velocity $v$ at time $t$. (b) At time $t + dt$ the rocket's mass has decreased, so $dm$ is negative. The mass of the rocket (including the unburned fuel) is $m + dm$, and its velocity component is $v + dv$. The mass of burned fuel in the exhaust is $-dm$, and its velocity relative to the coordinate system shown is $v'$.

The $x$-component of velocity $v'$ of the fuel relative to our coordinate system is then

$$v' = v - v_{ex},$$

and the $x$-component of momentum of the mass $(-dm)$ is

$$(-dm)v' = (-dm)(v - v_{ex}).$$

At the end of the time interval $dt$ the mass of rocket and unburned fuel has decreased to $m + dm$, and its velocity has increased to $v + dv$ (Fig. 8–23b). (Remember that $dm$ is negative.) The rocket's momentum at this time is

$$(m + dm)(v + dv).$$

Thus the *total* momentum $P$ of rocket plus ejected fuel at time $t + dt$ is

$$P = (m + dm)(v + dv) + (-dm)(v - v_{ex}).$$

According to our initial assumption, the rocket and fuel are an isolated system. Thus momentum is conserved; the total momentum $P$ of the system must be the same at time $t$ and at time $t + dt$:

$$mv = (m + dm)(v + dv) + (-dm)(v - v_{ex}).$$

This can be simplified to

$$m\,dv = -dm\,v_{ex} - dm\,dv.$$

We can neglect the term $dm\,dv$ because it is a product of two small quantities and thus is much smaller than the other terms. Dropping this term, dividing by $dt$, and rearranging, we find

$$m\frac{dv}{dt} = -v_{ex}\frac{dm}{dt}. \tag{8–32}$$

Since $dv/dt$ is the acceleration of the rocket, the left side of this equation (mass times acceleration) equals the resultant force $F$, or *thrust*, on the rocket:

$$F = -v_{ex}\frac{dm}{dt}. \tag{8–33}$$

The thrust is proportional both to the relative speed $v_{ex}$ of the ejected fuel and to the mass of fuel ejected per unit time, $-dm/dt$. (Remember that $dm/dt$ is negative because it is the rate of change of the rocket's mass.)

The acceleration of the rocket is

$$a = \frac{dv}{dt} = -\frac{v_{ex}}{m}\frac{dm}{dt}. \tag{8–34}$$

The rocket's mass $m$ decreases continuously while the fuel is being consumed. If $v_{ex}$ and $dm/dt$ are constant, the acceleration increases until all the fuel is gone.

We can integrate Eq. (8–34) to find a relation between the velocity $v$ at any time and the remaining mass $m$. Let the mass be $m_0$ and the velocity $v_0$ at time $t = 0$. Then we rewrite Eq. (8–34) as

$$dv = -v_{ex}\frac{dm}{m}.$$

We change the integration variables to $v'$ and $m'$ so that we can use $v$ and $m$ as the upper limits (the final speed and mass). Then we integrate both sides, using limits $v_0$ to $v$ and $m_0$ to $m$:

$$\int_{v_0}^{v} dv' = -\int_{m_0}^{m} v_{ex}\frac{dm'}{m'}, \qquad v - v_0 = -v_{ex}\ln\frac{m}{m_0} = v_{ex}\ln\frac{m_0}{m}. \tag{8–35}$$

## EXAMPLE 8–17

In the first second of its flight a rocket ejects $\frac{1}{60}$ of its mass with a relative velocity of 2400 m/s. What is the rocket's acceleration?

**SOLUTION** We are given that

$$\frac{dm}{dt} = -\frac{m/60}{1\text{ s}} = -\frac{m}{60\text{ s}}.$$

From Eq. (8–34),

$$a = -\frac{v_{ex}}{m}\frac{dm}{dt} = -\frac{2400\text{ m/s}}{m}\left(-\frac{m}{60\text{ s}}\right) = 40\text{ m/s}^2.$$

## EXAMPLE 8–18

Suppose that $\frac{3}{4}$ of the initial mass $m_0$ of the rocket in Example 8–17 is fuel, so the final mass is $m = m_0/4$, and that the fuel is consumed at a constant rate in a total time $t = 60$ s. If the rocket starts from rest in our coordinate system, find its speed at the end of this time.

**SOLUTION** We have $m_0/m = 4$, so from Eq. (8–35),

$$v = v_0 + v_{ex}\ln\frac{m_0}{m}$$
$$= 0 + (2400\text{ m/s})(\ln 4) = 3327\text{ m/s}.$$

At the start of the flight, when the velocity of the rocket is zero, the ejected fuel is moving to the left, relative to our coordinate system, at 2400 m/s.

At the end of the first second ($t = 1$ s) the rocket is moving at 30 m/s, and the fuel's speed relative to our system is 2370 m/s. During the next second the acceleration, given by Eq. (8–34), is a little greater. At $t = 2$ s the rocket is moving at a little over 60 m/s, and the fuel's speed is a little less than 2340 m/s. And so on.

Detailed calculation shows that at about $t = 50.6$ s the rocket's velocity $v$ in our coordinate system passes 2400 m/s. The burned fuel ejected after this time moves *forward*, not backward, in our system. The final velocity acquired by the rocket can be greater in magnitude (and is often *much* greater) than the relative speed $v_{ex}$. In this example, in which the final velocity of the rocket is 3327 m/s and the relative velocity is 2400 m/s, the last portion of the ejected fuel has a *forward* velocity (relative to our frame) of 3327 m/s − 2400 m/s = 927 m/s. We are using too many figures, but you can see the point.

In the early days of rocket propulsion, people who didn't understand conservation of momentum thought that a rocket couldn't function in outer space because "it doesn't have anything to push against." On the contrary, rockets work *best* in outer space! Figure 8–24 shows a dramatic example of rocket propulsion. The rocket is *not* "pushing against the ground" to get into the air.

**8–24** Launch of a space shuttle, a dramatic example of rocket propulsion. Successful launch and landing of space shuttles are the first steps toward feasibility of a manned space station in orbit around the earth.

# 8–9 THE NEUTRINO: A Case Study in Modern Physics

The laws of conservation of energy and momentum, which are vitally important in all areas of physical science, rest on a very solid foundation of experimental evidence. Nevertheless, the decay of radioactive nuclei by a process called *beta decay* led the great physicist Neils Bohr to suggest seriously in 1930 that these laws might not be obeyed in nuclear processes. Most physicists disagreed; they believed so strongly in these conservation laws that they accepted instead an alternative hypothesis proposed by Wolfgang Pauli that an unseen "ghost particle" is emitted in beta decay. Not until 25 years later was the particle actually observed experimentally, but so great was the faith in the conservation laws that during those 25 years no one seriously doubted the existence of the particle.

To see how this challenge to the conservation laws arose, let's consider an isolated system consisting of two particles, with masses $m_1$ and $m_2$, initially at rest. The system might be the bullet and rifle of Example 8–2, two balls tied together with a compressed spring between them, or an unstable nucleus that comes apart into two fragments. When the two bodies fly apart, with speeds $v_1$ and $v_2$, some total quantity of energy $Q$ is divided between them:

$$Q = K_1 + K_2 = \tfrac{1}{2}m_1 v_1{}^2 + \tfrac{1}{2}m_2 v_2{}^2. \tag{8–36}$$

In Example 8–2, $Q$ was 225 J. We noted in that example that the ratio of the kinetic energies of the two particles equals the inverse ratio of their masses. With a 600:1 mass ratio the recoiling rifle and the bullet get kinetic energies of $\frac{1}{601}Q$ and $\frac{600}{601}Q$, respectively. The energy always divides this way, the lighter particle always getting more kinetic energy than the heavier one.

To prove this we note that the system is isolated; the total momentum is zero both before and after it flies apart. Thus the momenta of the two final particles must have equal magnitude:

$$m_1 v_1 = m_2 v_2. \tag{8–37}$$

When we square this equation and divide by 2, we get

$$\tfrac{1}{2}m_1{}^2 v_1{}^2 = \tfrac{1}{2}m_2{}^2 v_2{}^2,$$

$$m_1 K_1 = m_2 K_2, \qquad \text{or} \qquad \frac{K_2}{K_1} = \frac{m_1}{m_2}. \tag{8–38}$$

When we combine this with Eq. (8–36) to get expressions for $K_1$ and $K_2$, we find

$$K_1 = \frac{m_2}{m_1 + m_2}Q \qquad \text{and} \qquad K_2 = \frac{m_1}{m_1 + m_2}Q. \tag{8–39}$$

The essential point is that when there are only two final particles, the total available energy $Q$ *must* divide this way every time. If there are three or more final particles, the energy can be divided up in many different ways, depending on the directions of the particles.

## EXAMPLE 8–19

The radioactive isotope $^{222}$Rn, an inert gas, is naturally present in the air. It undergoes a process called *alpha decay*; the end products are a $^{218}$Po nucleus (mass $= 3.62 \times 10^{-25}$ kg) and an alpha particle (mass $= 6.65 \times 10^{-27}$ kg). (This radiation, along with that produced by further decay events, is responsible for the health hazards of radon.) The quantity of energy released in the decay of $^{222}$Rn is $Q = 9.0 \times 10^{-13}$ J. What are the kinetic energies of the emitted alpha particle and the recoiling $^{218}$Po nucleus?

**SOLUTION** Let $m_1$ be the alpha particle and $m_2$ the $^{218}$Po nucleus. We can use Eqs. (8–39) with $m_1 = 6.65 \times 10^{-27}$ kg and $m_2 = 3.62 \times 10^{-25}$ kg:

$$K_1 = \frac{m_2}{m_1 + m_2}Q = \frac{3.62 \times 10^{-25}\ \text{kg}}{3.69 \times 10^{-25}\ \text{kg}}(9.0 \times 10^{-13}\ \text{J})$$
$$= 8.8 \times 10^{-13}\ \text{J}.$$

The alpha particle takes $8.8 \times 10^{-13}$ J, or about $0.98Q$, leaving

$$K_2 = Q - K_1 = 0.2 \times 10^{-13}\ \text{J}$$

for the nucleus.

---

Experimental measurements of $^{222}$Rn decay give data on the number of alpha particles emitted with various kinetic energies (Fig. 8–25). The graph shows that *every* alpha particle emitted from *every* $^{222}$Rn nucleus at rest has the same kinetic energy, within experimental error. (The alpha particles are said to be *monoenergetic*.) If there are only two final particles, then this is to be expected. If all $^{222}$Rn nuclei are identical, then $Q$ is the same for all, and $K_1$ must be the same. Thus these measurements confirm the two-particle picture of the decay process.

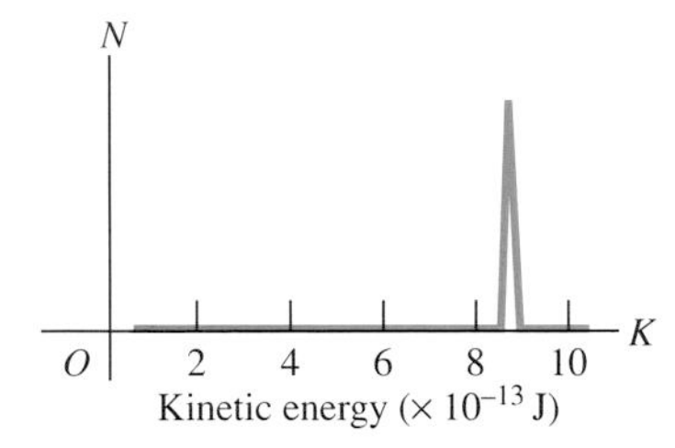


**8–25** A graph of the numbers of alpha particles ($N$) emitted with various energies $K$ in the decay of $^{222}$Rn to $^{218}$Po. The graph consists of a single sharp peak at $K = 8.8 \times 10^{-13}$ J.

Another type of radioactive decay process, called beta decay, involves an electron or its antiparticle, the positron. Electrons are sometimes called beta-minus ($\beta^-$) particles and positrons are beta-plus ($\beta^+$) particles.

An example of $\beta^-$ decay is the unstable nucleus $^{210}$Bi, which decays into an electron and a $^{210}$Po nucleus. Suppose this is a two-particle decay, like alpha emission. The rest mass of the $^{210}$Po nucleus is 383,000 times the rest mass of the electron. A calculation similar to the one in Example 8–19 predicts that virtually all the $Q$ energy of $1.86 \times 10^{-13}$ J should go to the electron and that the electrons from all decays should be monoenergetic. (The calculation requires relativistic expressions for kinetic energy and momentum, developed in Chapter 39, because the electrons that are emitted in $^{210}$Bi decay move at about 97% of the speed of light.)

A. H. Becquerel, who discovered radioactivity in 1896, found indications by 1900 that the electrons emitted in $\beta^-$ decay were *not* monoenergetic. Measurements made in 1914 by James Chadwick, the discoverer of the neutron, suggested this even more strongly. In the case of $^{210}$Bi decay later experimenters found a whole range of electron energies, with a maximum value of $1.86 \times 10^{-13}$ J (Fig. 8–26). Since conservation of energy and momentum predicted that all the electrons should have the *same* energy, these results cast doubt on the validity of these conservation laws.

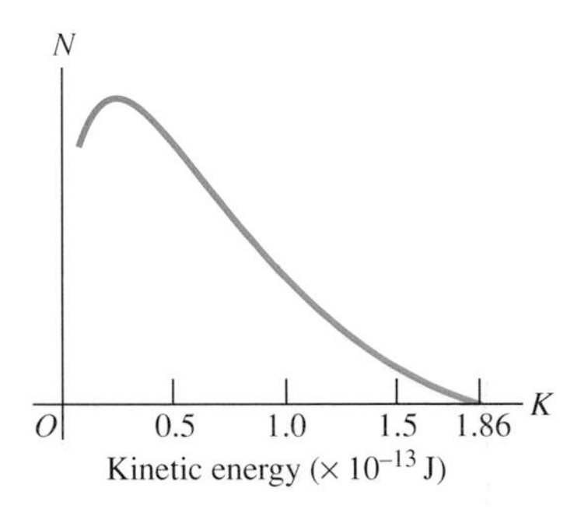


**8–26** A graph of $N_K$, the number $N$ of electrons emitted with various energies $K$, in the $\beta^-$ decay of $^{210}$Bi to $^{210}$Po. The graph shows a continuous distribution of electron energies, with a maximum value of $1.86 \times 10^{-13}$ J.

Looking for a way to save the conservation laws, Wolfgang Pauli suggested in 1931 that an unseen, electrically neutral third particle might be emitted during $\beta^-$ decay. If so, it could carry off some of the energy and momentum. With three particles the energy and momentum can be shared in many different ways that are consistent with the conservation laws. With two particles (Fig. 8–27a) the momenta must be equal and opposite, and this requirement determines the energy of each particle uniquely. But with three particles zero total momentum is represented by a triangle of momentum vectors (Figs. 8–27b,

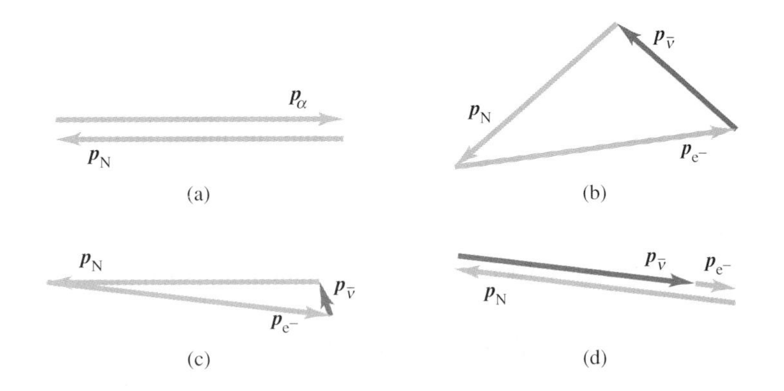


**8–27** (a) Two-particle decay: The momenta are equal in magnitude and opposite in direction. (b) In the case of three particles the electron's kinetic energy will be somewhere between zero and $Q$. (c) When the momentum and energy of the antineutrino are small, the electron's kinetic energy is nearly $Q$. (d) When the antineutrino's energy is nearly $Q$, the electron's kinetic energy is nearly zero.

**8–4** A golf ball of mass 0.045 kg is moving in the $+y$-direction with a speed of 5.00 m/s, and a baseball of mass 0.145 kg is moving in the $-x$-direction with a speed of 2.00 m/s. What are the magnitude and direction of the total momentum of the system?

## Section 8–2 Conservation of Momentum

**8–5 Energy Change during a Hip Check.** Hockey star Wayne Gretzky is skating at 13.0 m/s toward a defender, who is in turn skating at 5.0 m/s toward Gretzky (Fig. 8–29). Gretzky's weight is 756 N; that of the defender is 900 N. Immediately after the collision Gretzky is moving at 2.50 m/s in his original direction. Neglect external horizontal forces applied by the ice to the skaters during the collision. a) What is the velocity of the defender immediately after the collision? b) Calculate the change in total kinetic energy of the two players.

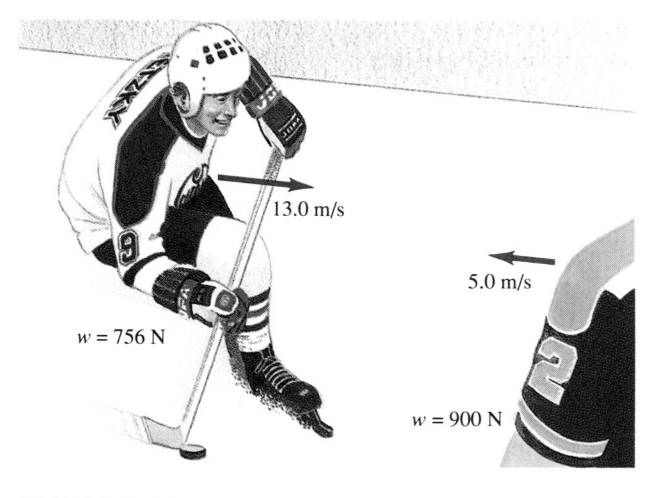


**FIGURE 8–29**

**8–6** On a frictionless horizontal surface, block $A$ (mass 3.00 kg) is moving toward block $B$ (mass 5.00 kg), which is initially at rest. After the collision block $A$ has a velocity of 1.20 m/s to the left, and block $B$ has velocity 6.50 m/s to the right. a) What was the speed of block $A$ initially, before the collision? b) Calculate the change in the total kinetic energy of the system that occurs during the collision.

**8–7** You are standing on a sheet of ice that covers a parking lot; there is negligible friction between your feet and the ice. A friend throws you a 0.400-kg ball that is traveling horizontally at 12.0 m/s. Your mass is 80.0 kg. a) If you catch the ball, with what speed do you and the ball move afterward? b) If the ball hits you and bounces off your chest, so that afterward it is moving horizontally at 12.0 m/s in the opposite direction, what is your speed after the collision?

**8–8** An 80.0-kg man standing on ice throws a 0.300-kg ball horizontally with a speed of 30.0 m/s. With what speed and in what direction will the man begin to move if there is no friction between his feet and the ice?

**8–9** Consider the following recoil situation. Initially, a combined object with mass $m_A + m_B$ is at rest at the origin. Then, owing to some internal force, the object breaks into two pieces. One piece, mass $m_A$, travels off to the left with speed $v_A$. The other piece, mass $m_B$, travels off to the right with speed $v_B$. a) Use conservation of momentum to solve for $v_B$ in terms of $m_A$, $m_B$, and $v_A$. b) Use the results of part (a) to show that $K_A/K_B = m_B/m_A$, where $K_A$ and $K_B$ are the kinetic energies of the two pieces.

**8–10** One of James Bond's adversaries is standing on a frozen lake; there is no friction between his feet and the ice. He throws his steel-lined hat with a velocity of 25.0 m/s at 36.9° above the horizontal, hoping to hit James. If his mass is 160 kg and that of his hat is 9.00 kg, what is the magnitude of his horizontal recoil velocity?

**8–11** Block $A$ in Fig. 8–30 has a mass of 1.00 kg, and block $B$ has a mass of 3.00 kg. The blocks are forced together, compressing a spring $S$ between them; then the system is released from rest on a level, frictionless surface. The spring is not fastened to either block and drops to the surface after it has expanded. Block $B$ acquires a speed of 0.500 m/s. a) What is the final speed of block $A$? b) How much potential energy was stored in the compressed spring?

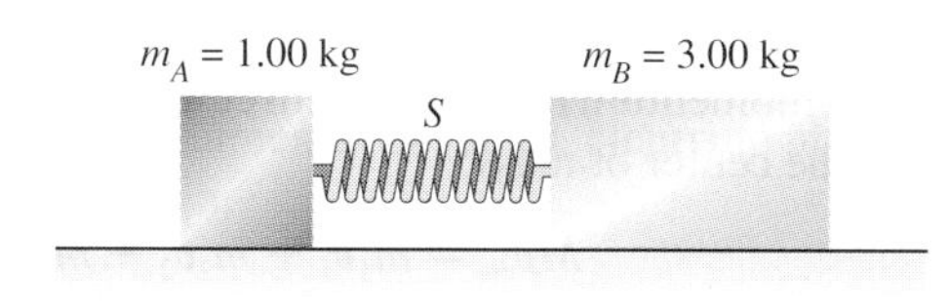

**FIGURE 8–30**

**8–12** An open-topped freight car of mass 10,000 kg is coasting without friction along a level track. It is raining very hard, and the rain is falling vertically downward. The car is originally empty and moving with a speed of 2.00 m/s. What is the speed of the car after it has collected 1000 kg of rainwater?

**8–13** A hockey puck $B$ rests on a smooth ice surface and is struck by a second puck $A$, which was originally traveling at 40.0 m/s and which is deflected 30.0° from its original direction (Fig. 8–31). Puck $B$ acquires a velocity at 45.0° with the original velocity of $A$. The pucks have the same mass. a) Compute the speed of each puck after the collision. b) What fraction of the original kinetic energy of puck $A$ is dissipated during the collision?

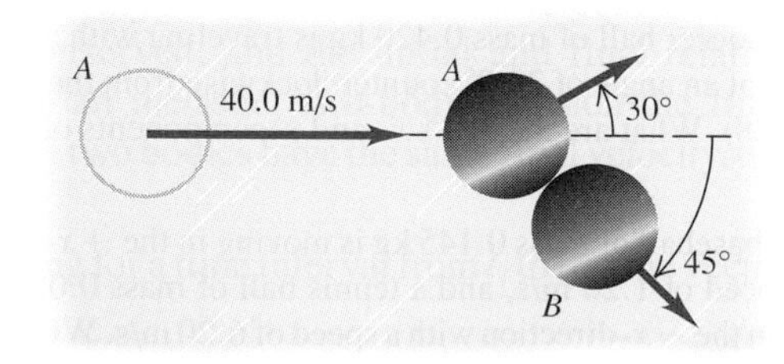

**FIGURE 8–31**

**8–14** Two ice skaters, Daniel and Rebecca, are practicing. Daniel's mass is 60.0 kg, and Rebecca's mass is 40.0 kg. Daniel stops to tie his shoelace and, while at rest, is struck by Rebecca, who is moving at 12.5 m/s before she collides with him. After the collision Rebecca has a velocity of magnitude 10.0 m/s at an angle of 37.0° from her initial direction. Both skaters move on the frictionless horizontal surface of the rink. a) What are the magnitude and direction of Daniel's velocity after the collision? b) What is the change in total kinetic energy of the two skaters as a result of the collision?

## Section 8–3 Inelastic Collisions

**8–15** a) An empty freight car of mass 10,000 kg rolls at 4.00 m/s along a level track and collides with a loaded car of mass 20,000 kg, standing at rest with brakes released. Friction can be neglected. If the cars couple together, find their speed after the collision. b) Find the change in kinetic energy of the two freight cars as a result of the collision. c) With what speed should the loaded car be rolling toward the empty car for both to be brought to rest by the collision?

**8–16** On a frictionless horizontal air track a 0.300-kg glider moving 6.00 m/s to the right collides with an 0.800-kg glider moving 1.50 m/s to the left. a) If the two gliders stick together, what is the final velocity? b) How much mechanical energy is dissipated in the collision?

**8–17** An 18.0-kg fish moving horizontally to the right at 3.20 m/s swallows a 2.0-kg fish that is swimming to the left at 6.80 m/s. What is the speed of the large fish immediately after its lunch if the forces exerted on the fishes by the water can be neglected?

**8–18** At the intersection of Texas Avenue and University Drive a small subcompact car of mass 900 kg traveling east on University collides with a maroon pickup truck of mass 1500 kg that is traveling north on Texas and ran a red light (Fig. 8–32). The two vehicles stick together as a result of the collision, and immediately after the collision the wreckage is sliding at 16.0 m/s in the direction 24.0° east of north. Calculate the speed of each vehicle just before the collision. (Friction forces between the vehicles and the road can be neglected during the collision.)

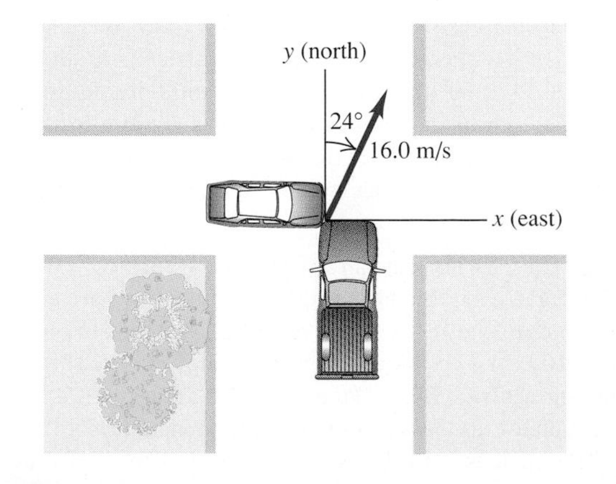


**FIGURE 8–32**

**8–19** A 2000-kg automobile going eastward on Chestnut Street at 40.0 km/h collides with a 4000-kg truck that is going southward *across* Chestnut Street at 20.0 km/h. If they become coupled on collision, what are the magnitude and direction of their velocity immediately after colliding? (Friction forces between the vehicles and the road can be neglected during the collision.)

**8–20 A Ballistic Pendulum.** A 10.0-g rifle bullet is fired with a speed of 500 m/s into a ballistic pendulum of mass 5.00 kg, suspended from a cord 0.600 m long. Compute a) the vertical height through which the pendulum rises; b) the initial kinetic energy of the bullet; c) the kinetic energy of the bullet and pendulum immediately after the bullet becomes embedded in the pendulum.

**8–21** A 5.00-g bullet is fired horizontally into a 2.50-kg wooden block resting on a horizontal surface. The coefficient of kinetic friction between block and surface is 0.20. The bullet remains embedded in the block, which is observed to slide 0.250 m along the surface before stopping. What was the initial speed of the bullet?

## Section 8–4 Elastic Collisions

**8–22** A 10.0-g marble travels to the left with a velocity of magnitude 0.400 m/s on a smooth, level surface and makes a head-on collision with a larger 30.0-g marble moving to the right with a velocity of magnitude 0.200 m/s (Fig. 8–33). If the collision is perfectly elastic, find the velocity of each marble (magnitude and direction) after the collision. (Since the collision is head-on, all the motion is along a line.)

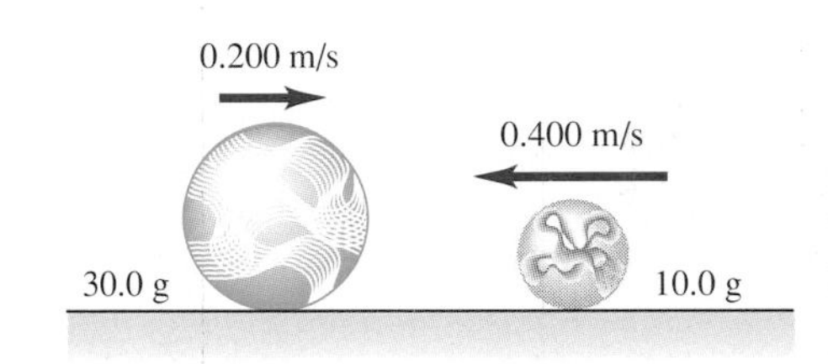


**FIGURE 8–33**

**8–23** A 0.300-kg block is moving to the right on a horizontal, frictionless surface with a speed of 0.60 m/s. It makes a head-on collision with a 0.200-kg block that is moving to the left with a speed of 1.50 m/s. Find the final velocity (magnitude and direction) of each block if the collision is elastic. (Since the collision is head-on, all motion is along a line.)

**8–24** Supply the details of the calculation of $\alpha$ and $\beta$ in Example 8–11.

## Section 8–5 Impulse

**8–25** A 1.50-kg block of ice is moving on a frictionless horizontal surface. At $t = 0$ the block is moving to the right with a velocity of magnitude 5.00 m/s. Calculate the velocity of the block (magnitude and direction) after each of the following forces has been applied for 5.00 s: a) a force of 5.00 N directed to the right; b) a force of 7.00 N directed to the left.

**8–26 Force of a Baseball Swing.** A baseball has a mass of 0.145 kg. a) If the velocity of a pitched ball has a magnitude of 30.0 m/s, and after the ball is batted the velocity is 50.0 m/s in the opposite direction, find the magnitude of the change in momentum of the ball and the impulse applied to it by the bat. b) If the ball remains in contact with the bat for $2.00 \times 10^{-3}$ s, find the magnitude of the average force applied by the bat.

**8–27 Force of a Golf Swing.** A 0.0450-kg golf ball initially at rest is given a speed of 50.0 m/s when struck by a club. If the club and ball are in contact for $2.00 \times 10^{-3}$ s, what average force acts on the ball? Is the effect of the ball's weight during the time of contact significant?

**8–28** A 0.145-kg baseball is struck by a bat. Just before impact, the ball is traveling horizontally toward the right at 40.0 m/s, and it leaves the bat traveling to the left at an angle of 30° above horizontal with a speed of 60.0 m/s. If the ball and bat are in contact for $5.00 \times 10^{-3}$ s, find the horizontal and vertical components of the average force on the ball.

**8–29** A resultant force with magnitude $F(t) = A + Bt^2$ and directed to the right is applied to a girl on roller skates. The girl has mass $m$. The force starts at $t_1 = 0$ and continues until $t = t_2$. a) What is the impulse $J$ of the force? b) If the girl is initially at rest, what is her speed at time $t_2$?

## Section 8–6 Center of Mass

**8–30** A 1000-kg automobile is moving along a straight highway at 12.0 m/s. Another car, with mass 2000 kg and speed 20.0 m/s, is 40.0 m ahead of the first (Fig. 8–34). a) Find the position of the center of mass of the two automobiles. b) Find the magnitude of the total momentum from the above data. c) Find the speed of the center of mass. d) Find the total momentum, using the speed of the center of mass. Compare the result with part (b).

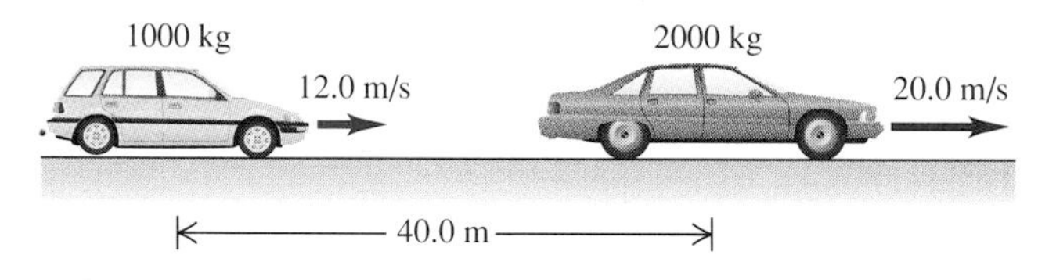

**FIGURE 8–34**

**8–31** Find the position of the center of mass of the earth-moon system. Use the data in Appendix F.

**8–32** Three particles have the following masses and coordinates: (1) 2.00 kg, (4.00 m, 2.00 m); (2) 3.00 kg, (1.00 m, −4.00 m); (3) 4.00 kg, (−3.00 m, 6.00 m). Find the coordinates of the center of mass of the system.

## Section 8–8 Rocket Propulsion

***8–33** A rocket is fired from an orbiting space station, where the gravitational force is very small. If the rocket has an initial mass of 5000 kg and ejects gas at a relative velocity of magnitude 2000 m/s, how much gas must it eject in the first second to have an initial acceleration of 20.0 m/s$^2$?

***8–34** A rocket is fired in deep space, where there is no gravitational field. In the first second it ejects $\frac{1}{80}$ of its mass as exhaust gas and has an acceleration of 40.0 m/s$^2$. What is the speed of the exhaust gas relative to the rocket?

***8–35** A small rocket burns 0.0500 kg of fuel per second, ejecting it as a gas with a velocity relative to the rocket of magnitude 800 m/s. a) What force does this gas exert on the rocket? b) Would the rocket operate in free space? c) If it would operate in free space, how would you steer it? Could you brake it?

***8–36** A single-stage rocket is fired from rest from a space platform, where the gravitational force is negligible. If the rocket burns its fuel in a time of 40.0 s and the relative speed of the exhaust gas is $v_{ex} = 3000$ m/s, what must be the mass ratio $m_0/m$ for a final speed $v$ of 8.00 km/s (about equal to the orbital speed of an earth satellite)?

***8–37** Obviously rockets can be made to go very fast, but what is a reasonable top speed? Assume that a rocket is fired from rest at a space station, where there is no gravitational force. a) If the rocket ejects gas at a relative speed of 2000 m/s and you want the rocket's speed eventually to be $1.00 \times 10^{-3}\,c$, where $c$ is the speed of light, what fraction of the initial mass of the rocket and fuel is *not* fuel? b) What is this fraction if the final speed is to be 3000 m/s?

## Section 8–9 The Neutrino: A Case Study in Modern Physics

**8–38** A $^{190}$Pt nucleus decays to an $^{186}$Os nucleus with the emission of an alpha particle. The total kinetic energy of the decay fragments is $5.20 \times 10^{-13}$ J. An alpha particle has 2.15% of the mass of an $^{186}$Os nucleus. Calculate the kinetic energy of a) the recoiling $^{186}$Os nucleus; b) the alpha particle.

**8–39** In a certain alpha decay the kinetic energy of the alpha particle is $1.070 \times 10^{-12}$ J and the $Q$ value for the decay is $1.090 \times 10^{-12}$ J. What is the mass of the recoiling nucleus?

**8–40** The $^{210}$Bi nucleus can undergo beta decay to $^{210}$Po. Suppose the emitted electron moves to the right with a momentum, calculated relativistically, of $6.20 \times 10^{-22}$ kg · m/s. The $^{210}$Po nucleus, with a mass of $3.50 \times 10^{-25}$ kg, recoils to the left at a speed of $1.14 \times 10^3$ m/s. Calculate the magnitude and direction of the momentum of the antineutrino that is emitted in the decay. [The $^{210}$Po nucleus moves at much less than the speed of light, so its momentum can be calculated from the nonrelativistic expression, Eq. (8–1).]

**8–41** The $^{210}$Bi nucleus can undergo $\beta^-$ decay to $^{210}$Po. In a particular decay event the electron and antineutrino are emitted at right angles to one another. The magnitudes of their momenta are $3.60 \times 10^{-22}$ kg · m/s for the electron and $4.40 \times 10^{-22}$ kg · m/s for the antineutrino. The $^{210}$Po nucleus has a mass of $3.50 \times 10^{-25}$ kg. Calculate a) the magnitude of the momentum of the recoiling $^{210}$Po nucleus; b) the kinetic energy of the recoiling $^{210}$Po nucleus.

## PROBLEMS

**8–42** A 1500-kg yellow convertible is traveling south, and a 2000-kg red station wagon is traveling west. If the total momentum of the system consisting of the two cars is 9000 kg · m/s directed at 30.0° west of south, what is the speed of each car?

**8–43** Two railroad cars roll along and couple with a third car, which is initially at rest. These three roll along and couple with a fourth. This process continues until the speed of the final collection of railroad cars is $\frac{1}{10}$ of the speed of the initial two railroad cars. All the cars are identical. Ignoring friction, how many cars are in the final collection?

**8–44** Objects $A$ (mass 2.00 kg), $B$ (mass 3.00 kg), and $C$ (mass 4.00 kg) are each approaching the origin (Fig. 8–35). The initial velocities of $A$ and $B$ are given in the figure. All three objects arrive at the origin at the same time. What must be the $x$- and $y$-components of the initial velocity of $C$ if all three objects are to end up at rest after the collision?

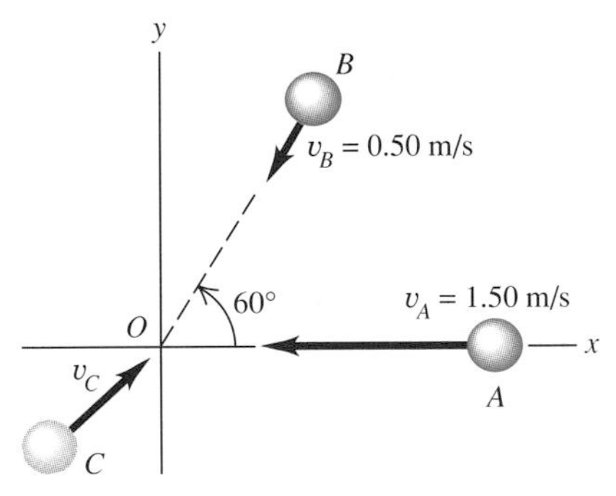


**FIGURE 8–35**

**8–45** A railroad handcar is moving along straight frictionless tracks. In each of the following cases the car initially has a total mass (car and contents) of 200 kg and is traveling east with a velocity of magnitude 5.00 m/s. Find the *final velocity* of the car in each case. a) A 20.0-kg mass is thrown sideways out of the car with a velocity of magnitude 2.00 m/s relative to the initial velocity of the car. b) A 20.0-kg mass is thrown backward out of the car with a velocity of 5.00 m/s relative to the initial motion of the car. c) A 20.0-kg mass is thrown into the car with a velocity of 6.00 m/s relative to the ground and opposite in direction to the initial velocity of the car.

**8–46 A Railroad Car Leaking Sand.** A railroad car filled with sand is rolling with an initial speed of 12.0 m/s on straight horizontal tracks. The total mass of the car plus sand is 1200 kg. The floor of the railroad car has a hole through which the sand leaks out. After 20 min, 200 kg of sand has leaked out. What then is the speed of the railroad car? (Compare your analysis to that used to solve Exercise 8–12.)

**8–47** A 2.00-g bullet traveling horizontally with a velocity of magnitude 500 m/s is fired into a wooden block of mass 1.00 kg, initially at rest on a level surface. The bullet passes through the block and emerges with its speed reduced to 100 m/s. The block slides a distance of 0.30 m along the surface from its initial position. a) What is the coefficient of kinetic friction between block and surface? b) What is the decrease in kinetic energy of the bullet? c) What is the kinetic energy of the block at the instant after the bullet passed through it?

**8–48** A 5.00-g bullet is shot *through* a 1.00-kg wood block suspended on a string 2.000 m long. The center of mass of the block is observed to rise a distance of 0.45 cm. Find the speed of the bullet as it emerges from the block if its initial speed is 300 m/s.

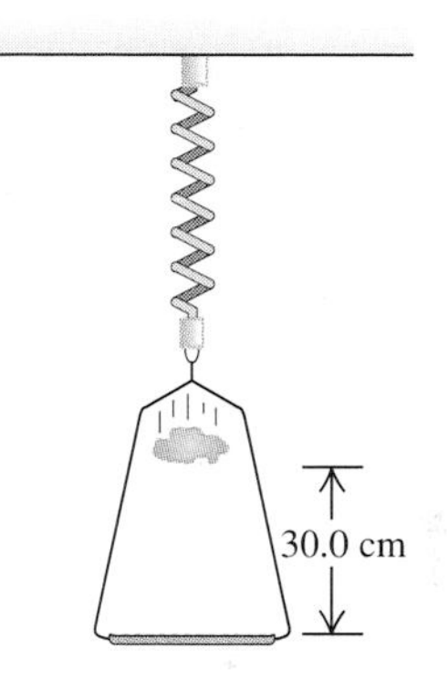


**FIGURE 8–36**

**8–49** A 0.200-kg frame, when suspended from a coil spring, is found to stretch the spring 0.050 m. A 0.200-kg lump of putty is dropped from rest onto the frame from a height of 0.300 m (Fig. 8–36). Find the maximum distance the frame moves downward.

**8–50** A rifle bullet of mass 10.0 g strikes and embeds itself in a block of mass 0.990 kg, which rests on a frictionless horizontal surface and is attached to a coil spring (Fig. 8–37). The impact compresses the spring 10.0 cm. Calibration of the spring shows that a force of 2.00 N is required to compress the spring 1.00 cm. a) Find the magnitude of the velocity of the block just after impact. b) What was the initial speed of the bullet?

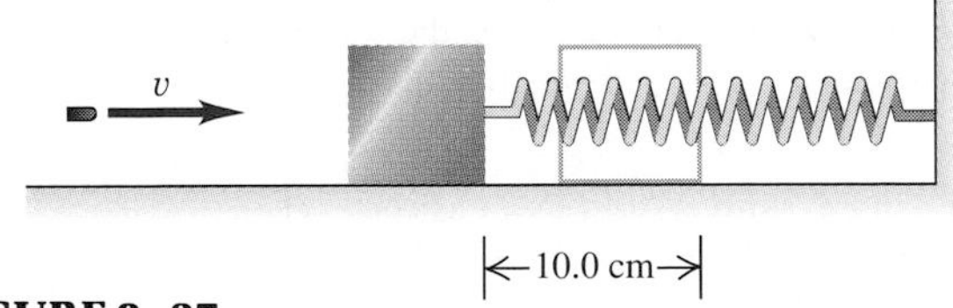

**FIGURE 8–37**

**8–51 A Ricocheting Bullet.** A 0.100-kg stone rests on a frictionless horizontal surface. A bullet of mass 2.50 g, traveling horizontally at 450 m/s, strikes the stone and rebounds horizontally at right angles to its original direction with a speed of 300 m/s. a) Compute the magnitude and direction of the velocity of the stone after it is struck. b) Is the collision perfectly elastic?

**8–52** A movie stuntman (mass 80.0 kg) stands on a window ledge 5.0 m above the floor (Fig. 8–38). Grabbing a rope attached to a chandelier, he swings down to grapple with the movie's villain

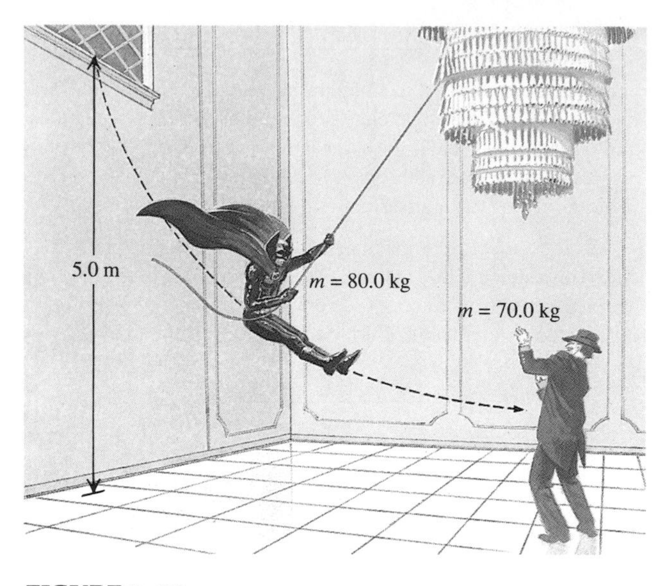


**FIGURE 8–38**

(mass 70.0 kg), who is standing directly under the chandelier. (Assume that the stuntman's center of mass moves downward 5.0 m. He releases the rope just as he reaches the villain.) a) With what speed do the entwined foes start to slide across the floor? b) If the coefficient of kinetic friction of their bodies with the floor is $\mu_k = 0.300$, how far do they slide along the floor?

**8–53** A neutron of mass $m$ collides elastically with a nucleus of mass $M$, which is initially at rest. Show that if the neutron's initial kinetic energy is $K_0$, the maximum kinetic energy that it can *lose* during the collision is

$$4mMK_0/(M + m)^2.$$

(*Hint:* The maximum energy loss occurs in a head-on collision.)

**8–54** A blue block of mass 0.400 kg, sliding with a velocity of magnitude 0.120 m/s on a smooth level surface, makes a perfectly elastic head-on collision with a red block of mass $m$, initially at rest. After the collision the velocity of the 0.400-kg block is 0.040 m/s in the same direction as its initial velocity. Find a) the velocity (magnitude and direction) of the red block after the collision; b) the mass $m$ of the red block.

**8–55** Two asteroids of masses $m_A$ and $m_B$ are moving with velocities $\boldsymbol{v}_A$ and $\boldsymbol{v}_B$ with respect to an astronomer in a space vehicle. a) Show that the total kinetic energy as measured by the astronomer is

$$K = \tfrac{1}{2}Mv_{\text{cm}}{}^2 + \tfrac{1}{2}(m_A v_A{}'^2 + m_B v_B{}'^2)$$

with $\boldsymbol{v}_{\text{cm}}$ and $M$ defined as in Section 8–6, $\boldsymbol{v}_A{}' = \boldsymbol{v}_A - \boldsymbol{v}_{\text{cm}}$, and $\boldsymbol{v}_B{}' = \boldsymbol{v}_B - \boldsymbol{v}_{\text{cm}}$. In this expression the total kinetic energy of the two asteroids is the energy associated with their center of mass plus that associated with the internal motion relative to the center of mass. b) If the asteroids collide, what is the *minimum* possible kinetic energy they can have after the collision as measured by the astronomer?

**8–56** A small steel ball moving with speed $v_0$ in the positive $x$-direction makes a perfectly elastic noncentral collision with an identical ball that is originally at rest. After impact the first ball moves with speed $v_1$ in the first quadrant at an angle $\theta_1$ with the $x$-axis, and the second moves with speed $v_2$ in the fourth quadrant at an angle $\theta_2$ with the $x$-axis. a) Write the equations expressing conservation of linear momentum in the $x$- and $y$-directions. b) Square the equations from part (a) and add them. c) At this point, introduce the fact that the collision is perfectly elastic. d) Prove that $\theta_1 + \theta_2 = \pi/2$. (You have shown that this equation is obeyed in any elastic collision between objects of equal mass when one object is initially at rest.)

**8–57** Hockey puck $B$ rests on a smooth ice surface and is struck by a second puck $A$, which has the same mass. Puck $A$ is initially traveling at 30.0 m/s and is deflected 30.0° from its initial direction. Assume that the collision is perfectly elastic. Find the final speed of each puck and the direction of $B$'s velocity after the collision. [*Hint:* Use the relation derived in part (d) of Problem 8–56.]

**8–58** A man and a woman are sitting in a sleigh that is at rest on frictionless ice. The weight of the man is 800 N, the weight of the woman is 600 N, and that of the sleigh is 1200 N. The people suddenly see a poisonous spider on the floor of the sleigh and jump out. The man jumps to the left with a velocity of 6.00 m/s at 30.0° above the horizontal, and the woman jumps to the right at 9.00 m/s at 36.9° above the horizontal. Calculate the horizontal velocity (magnitude and direction) that the sleigh has after they jump out.

**8–59** Jack and Jill are standing on a crate at rest on the frictionless horizontal surface of a frozen pond. Jack has mass 80.0 kg, Jill has mass 50.0 kg, and the crate has mass 20.0 kg. They remember that they must go and fetch a pail of water, so each jumps horizontally from the top of the crate with a speed of 5.00 m/s *relative to the crate.* a) What is the final speed of the crate if both Jack and Jill jump simultaneously and in the same direction? b) What is the final speed of the crate if Jack jumps first and then a few seconds later Jill jumps in the same direction? (*Hint:* Use an inertial coordinate system attached to the ground.) c) What is the final speed of the crate if Jill jumps first and then Jack, again in the same direction?

**8–60** A uniform steel rod 0.800 m in length is bent in a 90° angle at its midpoint. Determine the position of its center of mass. (*Hint:* The mass of each side of the angle may be assumed to be concentrated at its center.)

**8–61** The objects in Fig. 8–39 are constructed of uniform wire bent into the shapes shown. Find the position of the center of mass of each.

**8–62** James and Ramon are standing on the smooth surface of a frozen pond. Ramon has mass 60.0 kg, and James has mass 90.0 kg. They are 20.0 m apart and hold the ends of a light rope that is stretched between them. Midway between the two men a mug of beer sits on the ice (Fig. 8–40). James pulls on the rope and, as a result, slides toward the mug. When James has moved 4.0 m, how far and in what direction has Ramon moved?

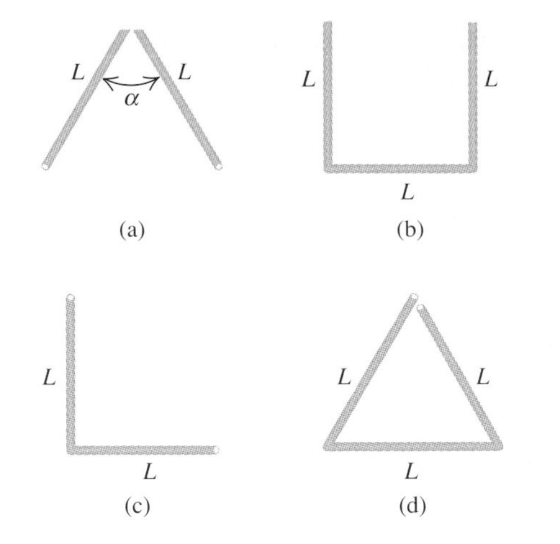


**FIGURE 8–39**

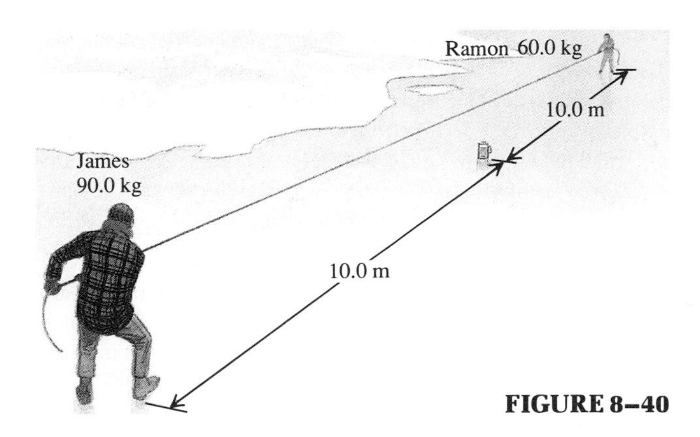


**FIGURE 8–40**

**8–63** You are standing on a concrete slab, which in turn is resting on a frozen lake. Assume that there is no friction between the slab and the ice. The slab has a weight five times your weight. If you begin walking forward at 3.00 m/s relative to the ice, with what speed relative to the ice does the slab move?

**8–64** A 70.0-kg woman stands up in a 30.0-kg canoe of length 5.00 m. She walks from a point 1.00 m from one end to a point 1.00 m from the other end. If resistance to motion of the canoe in the water can be neglected, how far does the canoe move during this process?

**8–65 A Nuclear Reaction.** Fission, the process that supplies energy in nuclear power plants, occurs when a heavy nucleus is split into two medium-sized nuclei. One such reaction occurs when a neutron colliding with a $^{235}$U nucleus splits that nucleus into a $^{141}$Ba nucleus and a $^{92}$Kr nucleus. In this reaction two neutrons are also split off from the original $^{235}$U nucleus. Before the collision we have the arrangement in Fig. 8–41a. After the collision we have the Ba nucleus moving in the $+z$-direction and the Kr nucleus in the $-z$-direction. The three neutrons are moving in the $xy$-plane as shown in Fig. 8–41b. If the incoming neutron has an initial velocity of magnitude $4.0 \times 10^6$ m/s and a final velocity of magnitude $2.0 \times 10^6$ m/s in the directions shown, what are the speeds of the other two neutrons, and what can you say about the speeds of the Ba and Kr nuclei? (The mass of the Ba nucleus is approximately $2.3 \times 10^{-25}$ kg, and that of Kr is about $1.5 \times 10^{-25}$ kg.)

**8–66** A 0.200-kg steel ball is dropped from a height of 4.00 m onto a horizontal steel slab. The collision is elastic, and the ball rebounds to its original height. a) Calculate the impulse delivered to the ball during impact. b) If the ball is in contact with the slab for $2.00 \times 10^{-3}$ s, find the average force on the ball during impact.

**8–67** Find the average recoil force on a machine gun firing 120 shots per minute. The mass of each bullet is 10.0 g, and the muzzle speed is 500 m/s.

**8–68** A tennis ball weighing 0.560 N has $\boldsymbol{v}_1 = (22.0\text{ m/s})\boldsymbol{i} - (4.0\text{ m/s})\boldsymbol{j}$ before being struck by a racket. The racket applies a force $\boldsymbol{F} = -(400\text{ N})\boldsymbol{i} + (120\text{ N})\boldsymbol{j}$ that we will assume to be constant during the $4.00 \times 10^{-3}$ s that the racket and ball are in contact. a) What are the $x$- and $y$-components of the impulse of the force applied to the ball? b) What are the $x$- and $y$-components of the final velocity of the ball?

**8–69** A force $\boldsymbol{F} = (\alpha t^2)\boldsymbol{i} - (\beta - \gamma t)\boldsymbol{j}$, where $\alpha = 15.0\text{ N/s}^2$, $\beta = 12.0$ N, and $\gamma = 20.0$ N/s is applied to an object with mass 2.00 kg. If the object was originally at rest, what is its velocity after the force has acted for 0.500 s? Express your answer in terms of the $\boldsymbol{i}$ and $\boldsymbol{j}$ unit vectors.

**8–70 Center-of-Mass Coordinate System.** Block $A$ (mass $m_A$) is moving on a frictionless horizontal surface in the $+x$-direction with velocity $v_{A1}$ and makes an elastic collision with block $B$ (mass $m_B$) that is initially at rest. a) Calculate the velocity of the center of mass of the two-block system before the collision. b) Consider a coordinate system whose origin is at the center of mass and moves with it. Is this an inertial reference

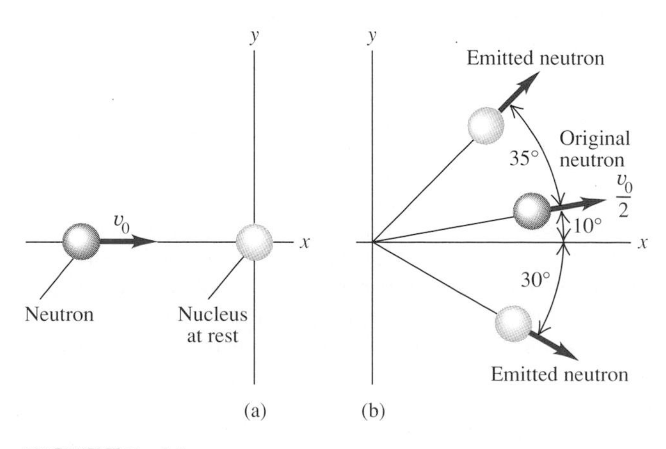

**FIGURE 8–41**

frame? c) What are the initial velocities $\boldsymbol{u}_{A1}$ and $\boldsymbol{u}_{B1}$ of the two blocks in this center-of-mass reference frame? What is the total momentum in this frame? d) Use conservation of momentum and energy, applied in the center-of-mass reference frame, to relate the final momentum of each block to its initial momentum and thus the final velocity of each block to its initial velocity. Your results should show that a one-dimensional elastic collision has a very simple description in center-of-mass coordinates, as was discussed in Section 8–7. e) Let $m_A = 2.00$ kg, $m_B = 4.00$ kg, and $v_{A1} = 5.00$ m/s. Find the center-of-mass velocities $\boldsymbol{u}_{A1}$ and $\boldsymbol{u}_{B1}$, apply the simple result found in part (d), and transform back to velocities in a stationary frame to find the final velocities of the blocks. Does your result agree with Eqs. (8–16) and (8–17)?

***8–71 A Multistage Rocket.** Suppose the first stage of a two-stage rocket has a total mass of 12,000 kg, of which 9000 kg is fuel. The total mass of the second stage is 1000 kg, of which 600 kg is fuel. Assume that the relative speed $v_{ex}$ of ejected material is constant, and neglect any effect of gravity. (The latter effect is small during the firing period if the rate of fuel consumption is large.) a) Suppose the entire fuel supply carried by the two-stage rocket is utilized in a single-stage rocket of the same total mass of 13,000 kg. What is the speed of the rocket, starting from rest, when its fuel is exhausted? b) What is the speed when the fuel of the first stage is exhausted if the first stage carries the second stage with it to this point? This speed then becomes the initial speed of the second stage. At this point the second stage separates from the first stage. c) What is the final speed of the second stage? d) What value of $v_{ex}$ is required to give the second stage of the rocket a speed of 8.00 km/s?

## CHALLENGE PROBLEMS

***8–72 A Variable-Mass Raindrop.** In a rocket-propulsion problem the mass is variable. Another such problem is a raindrop falling through a cloud of small water droplets. Some of these small droplets adhere to the raindrop, thereby *increasing* its mass as it falls The force on the raindrop is

$$F_{ext} = \frac{dp}{dt} = m\frac{dv}{dt} + v\frac{dm}{dt}.$$

Suppose the mass of the raindrop depends on how far it has fallen. Then $m = kx$, where $k$ is a constant, and $dm/dt = kv$. This gives, since $F_{ext} = mg$,

$$mg = m\frac{dv}{dt} + v(kv),$$

or, dividing by $k$,

$$xg = x\frac{dv}{dt} + v^2.$$

This is a differential equation that has a solution of the form $v = at$, where $a$ is the acceleration. Take the initial velocity of the raindrop to be zero. a) Using the proposed solution for $v$, find the acceleration $a$. b) Find the distance the raindrop has fallen in $t = 0.500$ s. c) Given that $k = 2.00$ g/m, find the mass of the raindrop at $t = 0.500$ s. For many more intriguing aspects of this problem, see K. S. Krane, *Amer. Jour. of Phys.*, Vol. 49 (1981), pp. 113–117.

**8–73** In Section 8–6 we calculated the center of mass by considering objects composed of a *finite* number of point masses or objects that by symmetry could be represented by a finite number of point masses. For a solid object whose mass distribution does not allow for a simple determination of the center of mass by symmetry, the sums of Eqs. (8–25) must be generalized to integrals:

$$x_{cm} = \frac{1}{M}\int x\,dm, \qquad y_{cm} = \frac{1}{M}\int y\,dm,$$

where $x$ and $y$ are the coordinates of the small piece of the object that has mass $dm$. The integration is over the whole of the object. Consider a thin rod of length $L$, mass $M$, and cross-sectional area $A$. Let the origin of coordinates be at the left end of the rod and the positive $x$-axis lie along the rod. a) If the density $\rho = M/V$ of the object is *constant,* perform the integration described above to show that the $x$-coordinate of the center of mass of the rod is at its geometrical center. b) If the density of the object varies linearly with $x$, that is, $\rho = \alpha x$, where $\alpha$ is a constant, calculate the $x$-coordinate of the rod's center of mass.

**8–74** Use the methods of Problem 8–73 to calculate the $x$- and $y$-coordinates of the center of mass of a semicircular metal plate with uniform density $\rho$ and thickness $t$. Let the radius of the plate be $a$. The mass of the plate is thus $M = \frac{1}{2}\rho\pi a^2 t$. Use the coordinate system indicated in Fig. 8–42.

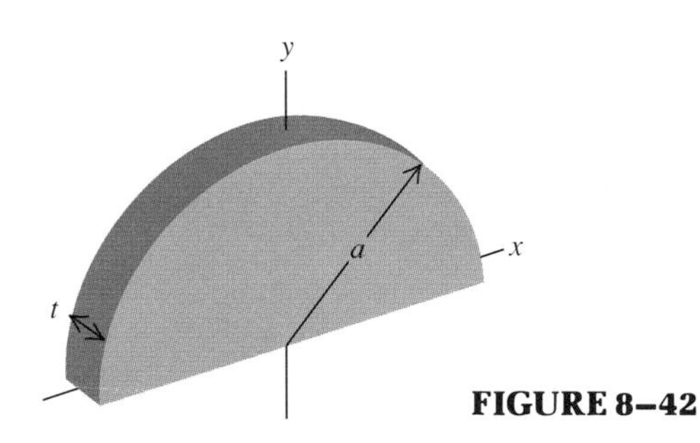


FIGURE 8–42

**8–75 An Application of Center of Mass.** Suppose one fourth of a rope of length $l$ is hanging down over the edge of a frictionless table. The rope has a linear density (mass per unit length) $\lambda$, and the end already on the table is held by a person. How much work is done by the person when she pulls on the rope to slowly raise the rest of the rope onto the table? Do the problem in two ways as follows. a) Find the force that the person must exert to raise the rope and from this the work done. Note this is a variable force because at different times, different amounts of rope are hanging over the edge. b) Suppose the segment of the rope initially hanging over the edge of the table has all of its mass concentrated at the center of mass. Find the work necessary to raise this to table height. You will probably find this approach simpler than the approach of part (a). How do the answers compare, and why is this so?

**8–76** A 20.0-kg projectile is fired at an angle of 60.0° above the horizontal and with a muzzle velocity of magnitude 300 m/s. At the highest point of its trajectory the projectile explodes into two fragments with equal mass, one of which falls vertically with zero initial speed. a) How far from the point of firing does the other fragment strike if the terrain is level? b) How much energy is released during the explosion?

**8–77** A wagon containing two boxes of gold and having mass 300 kg altogether has been cut loose from the horses by an outlaw when the wagon is 50 m up a 6.0° slope (Fig. 8–43). The outlaw plans to have the wagon roll back down the slope and across the level ground and then crash into a canyon where his confederates wait. But in a tree 40 m from the canyon edge wait the Lone Ranger (mass 80.0 kg) and Tonto (mass 70.0 kg). They drop vertically into the wagon as it passes beneath them. If they require 5.0 s to grab the gold and jump out, will they make it before the wagon goes over the edge? Assume that the wagon rolls with negligible friction.

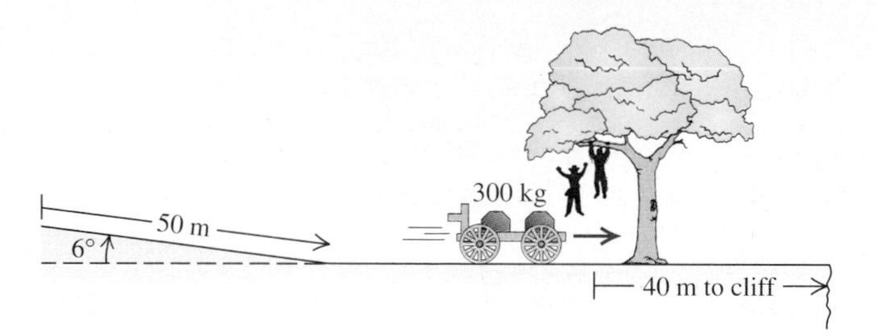

**FIGURE 8–43**

**8–78** Your favorite after-dinner trick is to pull the tablecloth out from under the dishes on the table. You are going to perform this trick at your Aunt Emma's birthday party. The birthday cake is resting on a tablecloth at the center of a round table of radius $r = 1.2$ m. The tablecloth is the same size as the tabletop. You grab the edge of the tablecloth and pull sharply. The tablecloth and cake are in contact for time $t$ after you start pulling. Then the sliding cake is stopped (you hope) by the friction between the cake and tabletop. The coefficient of kinetic friction between the cake and tablecloth is $\mu_{k1} = 0.30$, and that between the cake and tabletop is $\mu_{k2} = 0.40$. Apply the impulse-momentum theorem (Eq. 8–24) and the work-energy theorem (Eq. 6–12) to calculate the maximum value of $t$ if the cake is not to end up on the floor. (*Hint:* Assume that the cake moves a distance $d$ while still on the tablecloth and therefore a distance $r - d$ while sliding on the tabletop. Assume that the friction forces are independent of the relative speed of the sliding surfaces. You can easily try this trick yourself by pulling a sheet of paper out from under a glass of water, but have a mop handy just in case!)

CHAPTER 9

# Rotational Motion

Hurricane Gilbert, shown here centered in the Gulf of Mexico in September, 1988, achieved top wind speeds of 120 mi/h. In a rigid body, the highest linear speeds occur farthest from the axis of rotation. Although a hurricane is not a rigid body, lowest wind speeds do occur in the eye of the storm. The highest winds occur near the outer regions.

## KEY CONCEPTS

**• A rigid body is a body with a definite and unchanging shape and size.**

**• When a rigid body rotates about a stationary axis, its motion is described by its angular position, angular velocity (the time derivative of angular position), and angular acceleration (the time derivative of angular velocity).**

**• The velocity and acceleration of any point in a rotating rigid body can be expressed in terms of the point's distance from the axis and the body's angular velocity and acceleration.**

**• The kinetic energy of a rotating rigid body can be expressed in terms of the body's angular velocity and its moment of inertia, a quantity that depends on the distribution of mass of the body and the location of the axis of rotation. Several specialized methods are available for computing moments of inertia.**

## INTRODUCTION

What do a Ferris wheel, a circular saw blade, and a ceiling fan have in common? None of these can be represented adequately as a moving *point;* each involves a rigid body that *rotates* about an axis that is stationary in some inertial frame of reference. Rotation occurs on all scales, from the motion of electrons in atoms to the motion of hurricanes, to the motions of entire galaxies. We need to develop some general methods for analyzing the motion of a rotating body. In this chapter and the next we consider bodies that have definite size and definite shape and that in general can have rotational as well as translational motion.

Real-world bodies can be even more complicated; the forces that act on them can deform them — stretching, twisting, and squeezing them. We'll neglect these deformations for now and assume that the body has a perfectly definite and unchanging shape and size. We call this idealized model a **rigid body.** This chapter and the next are mostly about rotational motion of a rigid body. We begin with kinematic language for *describing* rotational motion. Next we look at the kinetic energy of rotation, the key to using energy methods for rotational motion. Then in Chapter 10 we'll develop dynamic principles that relate the forces on a body to its rotational motion. ■

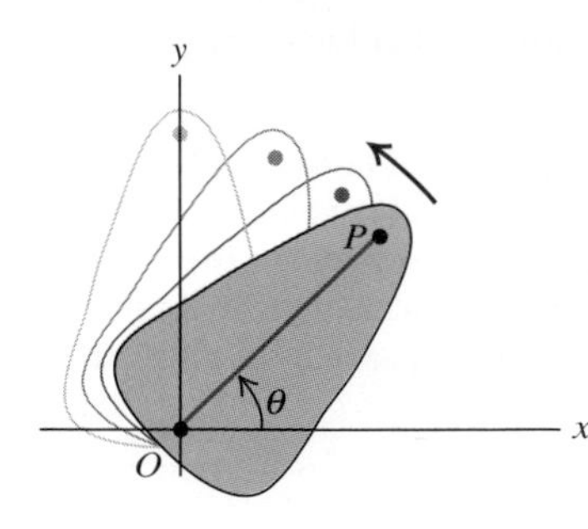


**9–1** A rigid body rotating counterclockwise about a fixed axis out of the page and passing through $O$.

## 9–1 ANGULAR VELOCITY AND ACCELERATION

In analyzing rotational motion, let's think first about a rigid body that rotates about a fixed axis. By fixed axis we mean an axis that is at rest in some inertial frame of reference and does not change direction relative to that frame. The body might be a motor shaft, a chunk of roast beef on a barbeque skewer, or a merry-go-round. Figure 9–1 shows a rigid body rotating about a stationary axis passing through point $O$ and perpendicular to the plane of the diagram. Line $OP$ is fixed in the body and rotates with it. The angle $\theta$ that this line makes with the $+x$-axis describes the rotational position of the body and serves as a *coordinate* for rotational position.

We usually measure the angle $\theta$ in **radians.** As is shown in Fig. 9–2a, one radian (1 rad) is the angle subtended at the center of a circle by an arc with a length equal to the radius of the circle. In Fig. 9–2b an angle $\theta$ is subtended by an arc of length $s$ on a circle of radius $r$; $\theta$ (in radians) is equal to $s$ divided by $r$:

$$\theta = \frac{s}{r}, \qquad \text{or} \qquad s = r\theta. \tag{9–1}$$

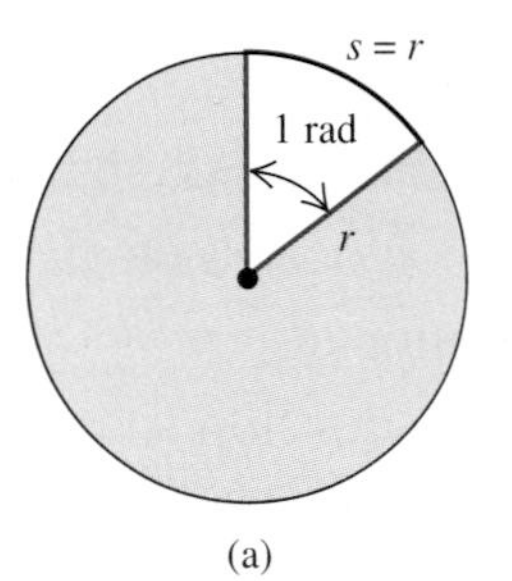


(a)

An angle in radians is the quotient of two lengths, so it is a pure number, without dimensions. If $s = 1.5$ m and $r = 1.0$ m, then $\theta = 1.5$, but we will often write this as 1.5 rad to distinguish it from an angle measured in degrees or revolutions.

The circumference of a circle is $2\pi$ times the radius, so there are $2\pi$ (about 6.283) radians in one complete revolution (360°). Therefore

$$1 \text{ rad} = \frac{360°}{2\pi} = 57.3°.$$

Similarly, $180° = \pi$ rad, $90° = \pi/2$ rad, and so on.

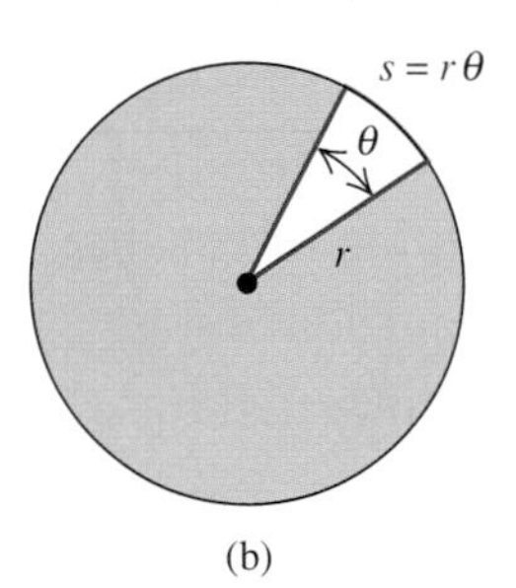


(b)

**9–2** An angle $\theta$ in radians is defined as the ratio of the arc length $s$ to the radius $r$.

### Angular Velocity

We can describe rotational *motion* of a rigid body in terms of the rate of change of $\theta$. In Fig. 9–3a a reference line $OP$ in a rotating body makes an angle $\theta_1$ with the $+x$-axis at time $t_1$. At a later time $t_2$ the angle has changed to $\theta_2$. We define the **average angular velocity** of the body $\omega_{av}$ in the time interval $\Delta t = t_2 - t_1$ as the ratio of the angular displacement $\Delta\theta = \theta_2 - \theta_1$ to $\Delta t$:

$$\omega_{av} = \frac{\theta_2 - \theta_1}{t_2 - t_1} = \frac{\Delta\theta}{\Delta t}. \tag{9–2}$$

The **instantaneous angular velocity** $\omega$ is the limit of $\omega_{av}$ as $\Delta t$ approaches zero, that is,

**9–3** (a) Angular displacement $\Delta\theta$ of a rotating body. (b) Every part of a rigid body has the same angular velocity.

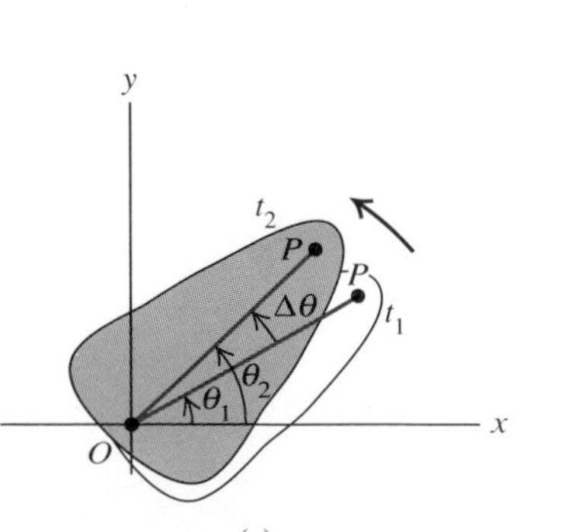


(a)

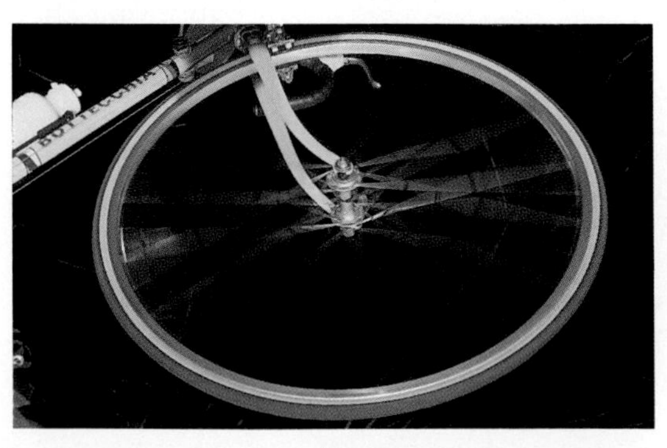

(b)

the derivative of $\theta$ with respect to $t$:

$$\omega = \lim_{\Delta t \to 0} \frac{\Delta \theta}{\Delta t} = \frac{d\theta}{dt}. \qquad (9\text{–}3)$$

Because the body is rigid, *all* lines in it rotate through the same angle in the same time (Fig. 9–3b). **At any instant, every part of a rigid body has the same angular velocity.** If the angle $\theta$ is in radians, the unit of angular velocity is one radian per second (1 rad/s). Other units, such as the revolution per minute (rev/min or rpm), are often used. Two useful conversions are

$$1 \text{ rev/s} = 2\pi \text{ rad/s}$$

and

$$1 \text{ rev/min} = 1 \text{ rpm} = \frac{2\pi}{60} \text{ rad/s}.$$

## EXAMPLE 9–1

**Calculating angular velocity** Figure 9–4 shows the flywheel in a car engine under test. The angular position $\theta$ of the flywheel is given by

$$\theta = (2.0 \text{ rad/s}^3)t^3.$$

The diameter of the flywheel is 0.36 m. a) Find the angle $\theta$, in radians and in degrees, at times $t_1 = 2.0$ s and $t_2 = 5.0$ s. b) Find the distance that a particle on the rim moves during that time interval. c) Find the average angular velocity, in rad/s and in rev/min (rpm), between $t_1 = 2.0$ s and $t_2 = 5.0$ s. d) Find the instantaneous angular velocity at time $t = 2.0$ s.

**SOLUTION** a) We substitute the values of $t$ into the given equation:

$$\theta_1 = (2.0 \text{ rad/s}^3)(2.0 \text{ s})^3 = 16 \text{ rad}$$
$$= (16 \text{ rad})\frac{360°}{2\pi \text{ rad}} = 917°,$$
$$\theta_2 = (2.0 \text{ rad/s}^3)(5.0 \text{ s})^3 = 250 \text{ rad}$$
$$= (250 \text{ rad})\frac{360°}{2\pi \text{ rad}} = 1.4 \times 10^{4}°.$$

b) The flywheel turns through an angle of $\theta_2 - \theta_1 = 250$ rad $- 16$ rad $= 234$ rad. The radius $r$ is half the diameter, or 0.18 m. Equation (9–1) gives

$$s = r\theta = (0.18 \text{ m})(234 \text{ rad}) = 42 \text{ m}.$$

Notice that when we use Eq. (9–1), the angle *must* be expressed in radians, but then we can drop it from the units for $s$ because $\theta$ is really a dimensionless number.

c) Using Eq. (9–2), we have

$$\omega_{av} = \frac{\theta_2 - \theta_1}{t_2 - t_1} = \frac{250 \text{ rad} - 16 \text{ rad}}{5.0 \text{ s} - 2.0 \text{ s}} = 78 \text{ rad/s}$$
$$= \left(78 \frac{\text{rad}}{\text{s}}\right)\left(\frac{1 \text{ rev}}{2\pi \text{ rad}}\right)\left(\frac{60 \text{ s}}{1 \text{ min}}\right) = 740 \text{ rev/min} = 740 \text{ rpm}.$$

d) We use Eq. (9–3):

$$\omega = \frac{d\theta}{dt} = \frac{d}{dt}[(2.0 \text{ rad/s}^3)t^3] = (2.0 \text{ rad/s}^3)(3t^2)$$
$$= (6.0 \text{ rad/s}^3)t^2.$$

At time $t = 2.0$ s,

$$\omega = (6.0 \text{ rad/s}^3)(2.0 \text{ s})^2 = 24 \text{ rad/s}.$$

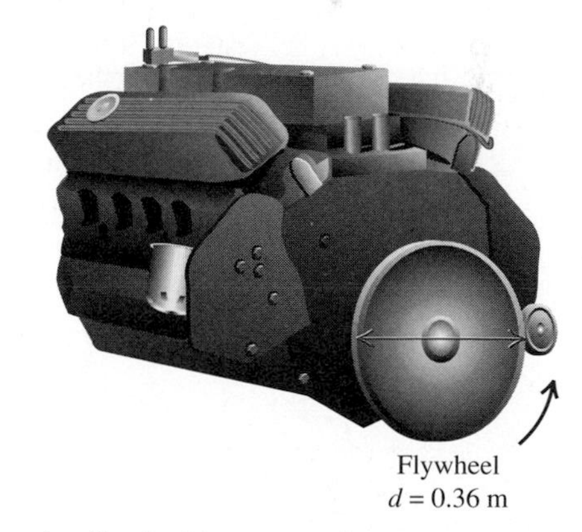


**9–4** Rotating flywheel in a car engine.

## Angular Acceleration

When the angular velocity of a rigid body changes, it has an angular acceleration. If $\omega_1$ and $\omega_2$ are the instantaneous angular velocities at times $t_1$ and $t_2$, we define the **average angular acceleration** $\alpha_{av}$ as

$$\alpha_{av} = \frac{\omega_2 - \omega_1}{t_2 - t_1} = \frac{\Delta\omega}{\Delta t}. \qquad (9\text{–}4)$$

The **instantaneous angular acceleration** $\alpha$ is the limit of $\alpha_{av}$ as $\Delta t \to 0$:

$$\alpha = \lim_{\Delta t \to 0} \frac{\Delta\omega}{\Delta t} = \frac{d\omega}{dt}. \tag{9–5}$$

The usual unit of angular acceleration is 1 rad/s$^2$. Because $\omega = d\theta/dt$, we can also express angular acceleration as

$$\alpha = \frac{d}{dt}\frac{d\theta}{dt} = \frac{d^2\theta}{dt^2}. \tag{9–6}$$

## EXAMPLE 9–2

**Calculating angular acceleration** In Example 9–1 we found that the instantaneous angular velocity $\omega$ of the flywheel at any time $t$ is given by

$$\omega = (6.0\ \text{rad/s}^3)t^2.$$

a) Find the average angular acceleration between $t_1 = 2.0$ s and $t_2 = 5.0$ s. b) Find the instantaneous angular acceleration at time $t = 2.0$ s.

**SOLUTION** a) The values of $\omega$ at the two times are

$$\omega_1 = (6.0\ \text{rad/s}^3)(2.0\ \text{s})^2 = 24\ \text{rad/s},$$

$$\omega_2 = (6.0\ \text{rad/s}^3)(5.0\ \text{s})^2 = 150\ \text{rad/s}.$$

From Eq. (9–4) the average angular acceleration is

$$\alpha_{av} = \frac{150\ \text{rad/s} - 24\ \text{rad/s}}{5.0\ \text{s} - 2.0\ \text{s}} = 42\ \text{rad/s}^2.$$

b) From Eq. (9–5) the instantaneous angular acceleration at any time $t$ is

$$\alpha = \frac{d\omega}{dt} = \frac{d}{dt}[(6.0\ \text{rad/s}^3)(t^2)] = (6.0\ \text{rad/s}^3)(2t)$$
$$= (12\ \text{rad/s}^3)t.$$

At time $t = 2.0$ s,

$$\alpha = (12\ \text{rad/s}^3)(2.0\ \text{s}) = 24\ \text{rad/s}^2.$$

You have probably noticed that we are using Greek letters for angular kinematic quantities: $\theta$ for angular position, $\omega$ for angular velocity, and $\alpha$ for angular acceleration. These are analogous to $x$ for position, $v$ for velocity, and $a$ for acceleration, respectively, in straight-line motion. In each case velocity is the rate of change of position, and acceleration is the rate of change of velocity.

In some problems it is useful to think of $\omega$ and $\alpha$ as vector quantities. Then $\boldsymbol{\omega}$ is a vector quantity directed along the axis of rotation. As Fig. 9–5 shows, the direction is given by the right-hand rule that we used to define the vector product in Section 1–10. We define the vector angular acceleration as the time derivative of the vector angular velocity. This formulation is useful in problems in which the direction of the axis changes and the direction of the vector $\boldsymbol{\alpha}$ may be different from that of $\boldsymbol{\omega}$. We'll return to these matters briefly at the end of Chapter 10.

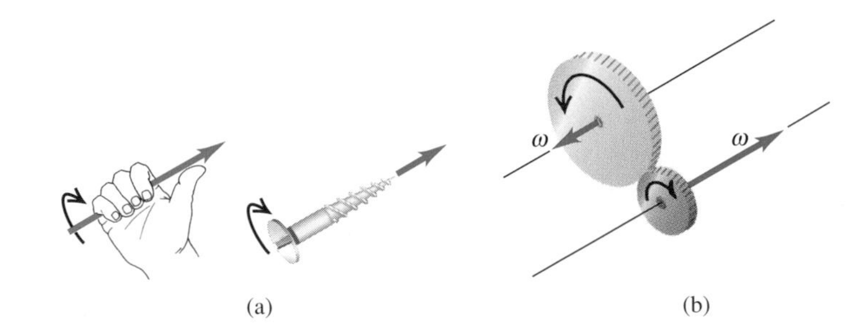


**9–5** (a) When the fingers of the right hand curl in the direction of rotation, the thumb points in the direction of the angular velocity vector. The direction is also the direction of advance of a screw with a right-hand thread. (b) When the two gears rotate, their angular velocity vectors point in opposite directions.

## 9–2 ROTATION WITH CONSTANT ANGULAR ACCELERATION

In Chapter 2 we found that straight-line motion is particularly simple when the acceleration is constant. This is also true of rotational motion; when the angular acceleration is constant, we can derive equations for angular velocity and angular position using exactly the same procedure that we used in Section 2–4. In fact, the equations that we are about to derive are identical to Eqs. (2–9), (2–13), and (2–14) if we replace $x$ with $\theta$, $v$ with $\omega$, and $a$ with $\alpha$. We suggest that you review Section 2–4 before continuing.

Let $\omega_0$ be the angular velocity of a rigid body at time $t = 0$, and let $\omega$ be its angular velocity at any later time $t$. The angular acceleration $\alpha$ is constant and equal to the average value for any interval. Using Eq. (9–4) with the interval from 0 to $t$, we find

$$\alpha = \frac{\omega - \omega_0}{t - 0}, \qquad \text{or}$$

$$\omega = \omega_0 + \alpha t. \tag{9–7}$$

The product $\alpha t$ is the total change in $\omega$ between $t = 0$ and the later time $t$; the angular velocity $\omega$ at time $t$ is the sum of the initial value $\omega_0$ and this total change.

With constant angular acceleration the angular velocity changes at a uniform rate, so its average value between 0 and $t$ is the average of the initial and final values:

$$\omega_{av} = \frac{\omega_0 + \omega}{2}. \tag{9–8}$$

We also know that $\omega_{av}$ is the total angular displacement $(\theta - \theta_0)$ divided by the time interval $(t - 0)$:

$$\omega_{av} = \frac{\theta - \theta_0}{t - 0}. \tag{9–9}$$

When we equate Eqs. (9–8) and (9–9) and multiply the result by $t$, we get

$$\theta - \theta_0 = \tfrac{1}{2}(\omega_0 + \omega)t. \tag{9–10}$$

To obtain a relation between $\theta$ and $t$ that doesn't contain $\omega$, we subsitute Eq. (9–7) into Eq. (9–10):

$$\theta - \theta_0 = \tfrac{1}{2}(\omega_0 + \omega_0 + \alpha t)t,$$

$$\theta = \theta_0 + \omega_0 t + \tfrac{1}{2}\alpha t^2. \tag{9–11}$$

That is, if at the initial time $t = 0$ the body is at angular position $\theta_0$ and has angular velocity $\omega_0$, then its angular position $\theta$ at any later time $t$ is the sum of three terms: its initial angular position $\theta_0$, plus the rotation $\omega_0 t$ it would have if the angular velocity were constant, plus an additional rotation $\frac{1}{2}\alpha t^2$ caused by the changing angular velocity.

Following the same procedure as in Section 2–4, we can combine Eqs. (9–7) and (9–11) to obtain a relation between $\theta$ and $\omega$ that does not contain $t$. We invite you to work out the details, following the same procedure we used to get Eq. (2–14). (See Exercise 9–6.) In fact, because of the perfect analogy between straight-line and rotational quanti-

**TABLE 9–1 Comparison of Linear and Angular Motion with Constant Acceleration**

| Straight-line motion with constant linear acceleration | Fixed-axis rotation with constant angular acceleration |
|---|---|
| $a = \text{constant}$ | $\alpha = \text{constant}$ |
| $v = v_0 + at$ | $\omega = \omega_0 + \alpha t$ |
| $x = x_0 + v_0 t + \frac{1}{2}at^2$ | $\theta = \theta_0 + \omega_0 t + \frac{1}{2}\alpha t^2$ |
| $v^2 = v_0^2 + 2a(x - x_0)$ | $\omega^2 = \omega_0^2 + 2\alpha(\theta - \theta_0)$ |
| $x - x_0 = \frac{1}{2}(v + v_0)t$ | $\theta - \theta_0 = \frac{1}{2}(\omega + \omega_0)t$ |

ties, we can simply take Eq. (2–14) and replace each straight-line quantity by its rotational analog; we get

$$\omega^2 = \omega_0^2 + 2\alpha(\theta - \theta_0). \tag{9–12}$$

Keep in mind that all of these results are valid *only* when the angular acceleration $\alpha$ is *constant*; be careful not to try to apply them to problems in which $\alpha$ is *not* constant. Table 9–1 shows Eqs. (9–7), (9–10), (9–11), and (9–12) for fixed-axis rotation with constant angular acceleration and the analogous equations for straight-line motion with constant linear acceleration.

## EXAMPLE 9–3

The angular velocity of the front wheel of an exercise bike is 4.00 rad/s at time $t = 0$, and its angular acceleration is constant and equal to 2.00 rad/s$^2$. A spoke $OP$ on the wheel coincides with the $+x$-axis at time $t = 0$ (Fig. 9–6a). a) What angle does this spoke make with the $+x$-axis at time $t = 3.00$ s (Fig. 9–6b)? b) What is the wheel's angular velocity at this time?

**SOLUTION** We can use Eqs. (9–7) and (9–11) to find $\theta$ and $\omega$ at any time in terms of the given initial conditions.

a) The angle $\theta$ is given as a function of time by Eq. (9–11):

$$\begin{aligned}\theta &= \theta_0 + \omega_0 t + \tfrac{1}{2}\alpha t^2 \\ &= 0 + (4.00 \text{ rad/s})(3.00 \text{ s}) + \tfrac{1}{2}(2.00 \text{ rad/s}^2)(3.00 \text{ s})^2 \\ &= 21.0 \text{ rad} = \frac{1 \text{ rev}}{2\pi \text{ rad}} 21.0 \text{ rad} = 3.34 \text{ rev}.\end{aligned}$$

The wheel has turned through three complete revolutions plus an additional 0.34 rev, or (0.34 rev)($2\pi$ rad/rev) = 2.14 rad = 123°. The spoke $OP$ is at an angle of 123° with the $+x$-axis.

b) From Eq. (9–7), $\omega = \omega_0 + \alpha t$. At time $t = 3.0$ s,

$$\omega = 4.00 \text{ rad/s} + (2.00 \text{ rad/s}^2)(3.00 \text{ s}) = 10.0 \text{ rad/s}.$$

Alternatively, from Eq. (9–12),

$$\begin{aligned}\omega^2 &= \omega_0^2 + 2\alpha(\theta - \theta_0) \\ &= (4.00 \text{ rad/s})^2 + 2(2.00 \text{ rad/s}^2)(21.0 \text{ rad}) \\ &= 100 \text{ rad}^2/\text{s}^2, \\ \omega &= 10.0 \text{ rad/s} \\ &= 1.59 \text{ rev/s}.\end{aligned}$$



**9–6** The spoke $OP$ of a rotating wheel.

## 9–3 VELOCITY AND ACCELERATION RELATIONS

How do we find the velocity and acceleration of a particular point in a rotating rigid body? We will need to answer this question to find the kinetic energy of a rotating body and to work out general relations between force and motion for rotation. To find the kinetic energy of a rotating body, we have to start from $K = \frac{1}{2}mv^2$ for a particle, and this requires knowing $v$ for a particle in a rigid body. In Chapter 10 we will develop a general relation between force and motion for rotation, starting with $\Sigma \boldsymbol{F} = m\boldsymbol{a}$. So it's worthwhile to develop general relations between the *angular* velocity and acceleration of a rigid body rotating about a fixed axis and the velocity and acceleration of a specific point in the body. We will sometimes use the terms *linear* velocity and *linear* acceleration for the familiar quantities that we defined in Chapters 2 and 3, to make sure we distinguish clearly between these and the *angular* quantities introduced in this chapter.

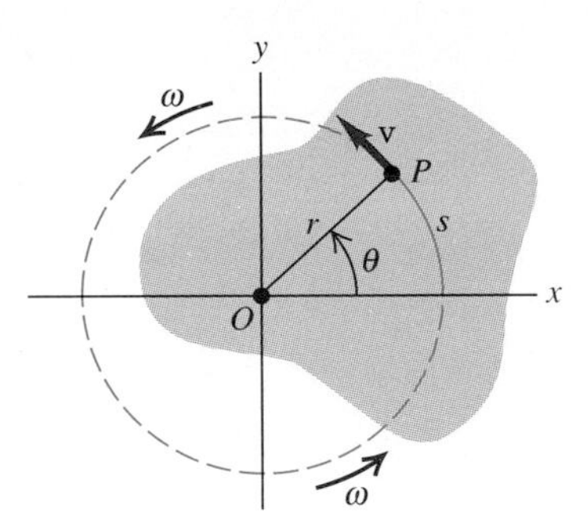


**9–7** The distance $s$ moved through by point $P$ on a rigid body equals $r\theta$ if $\theta$ is measured in radians.

When a rigid body rotates about a fixed axis, every particle in the body moves in a circular path. The circle lies in a plane perpendicular to the axis and is centered on the axis. The speed of a particle is directly proportional to the body's angular velocity; the faster the body rotates, the greater the speed of each particle. In Fig. 9–7 point $P$ is a constant distance $r$ away from the axis of rotation, and it moves in a circle of radius $r$. At any time the angle $\theta$ and the arc length $s$ are related by

$$s = r\theta.$$

We take the time derivative of this, noting that for any specific particle, $r$ is constant:

$$\frac{ds}{dt} = r\frac{d\theta}{dt}.$$

Now $ds/dt$ is the rate of change of arc length, which is equal to the instantaneous speed $v$ (magnitude of the instantaneous linear velocity) of the particle, and $d\theta/dt$ is the angular velocity $\omega$. Thus

$$v = r\omega. \tag{9–13}$$

The *direction* of the particle's linear velocity is tangent to its circular path at each point.

We can represent the acceleration of a particle in terms of its radial and tangential components, $a_{\text{rad}}$ and $a_{\text{tan}}$ (Fig. 9–8), as we did in Section 3–4. It would be a good idea to review that section now. We found that the **tangential component of acceleration** $a_{\text{tan}}$, the component parallel to the instantaneous velocity, acts to change the *magnitude* of velocity of the particle (that is, its speed) and is equal to the rate of change of speed. Taking the derivative of Eq. (9–13), we find

$$a_{\text{tan}} = \frac{dv}{dt} = r\frac{d\omega}{dt} = r\alpha. \tag{9–14}$$

This component of a particle's acceleration is always tangent to the circular path of the particle.

The **radial component of acceleration** $a_{\text{rad}}$ of the particle is associated with change of *direction* of the particle's velocity. For a particle moving in a circle of radius $r$ we worked out in Section 3–4 the relation $a_{\text{rad}} = v^2/r$. We can also express this in terms of $\omega$ by using Eq. (9–13):

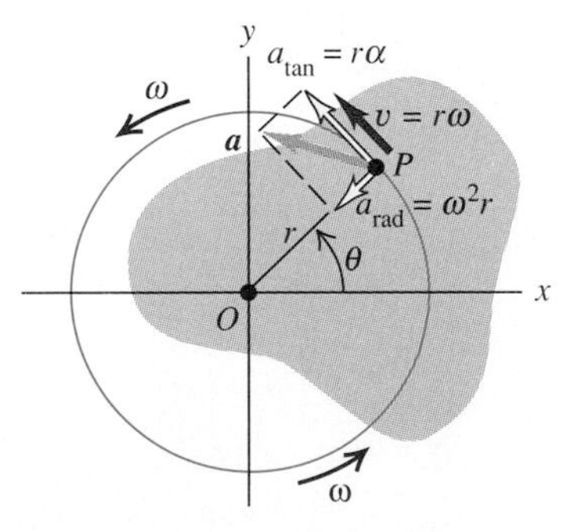


**9–8** Nonuniform rotation about a fixed axis through point $O$. The component of acceleration of point $P$ parallel to $\boldsymbol{v}$ equals $r\alpha$; the component perpendicular to $\boldsymbol{v}$ equals $\omega^2 r$. The rigid body is speeding up.

$$a_{\text{rad}} = \frac{v^2}{r} = \omega^2 r. \tag{9–15}$$

This is true at each instant *even when $\omega$ and $v$ are not constant.*

The vector sum of the radial and tangential components of acceleration of a particle in a rotating body is the linear acceleration $\boldsymbol{a}$ (Fig. 9–8).

It's important to remember that Eq. (9–1) ($s = r\theta$) is valid *only* when $\theta$ is measured in radians. The same is true of any equation derived from this, including Eqs.

(9–13), (9–14), and (9–15). *When you use these equations, you **must** express the angular quantities in radians, not revolutions or degrees.*

Equations (9–1), (9–13), and (9–14) are also correct for any particle that has the same tangential velocity as a point in a rotating rigid body. For example, when a rope wound around a circular cylinder unwraps without stretching or slipping, its velocity and tangential acceleration at any instant are the same as those for the point at which it is tangent to the cylinder. The same principle holds for similar situations such as bicycle chains and sprockets, V-belts and pulleys that turn without slipping, and so on. We will have several opportunities to use these relations in Chapter 10. We note that Eq. (9–15) for the radial component $a_{rad}$ is applicable to the rope or chain only at points that are in contact with the cylinder or sprocket; other points do not have the same acceleration toward the center of the circle that points on the cylinder or sprocket have.

## EXAMPLE 9–4

**Throwing a discus** A discus thrower turns with angular acceleration $\alpha = 50\ \text{rad/s}^2$, moving the discus in a circle of radius 0.80 m. We model the thrower's arm as a rigid body, so $r$ is constant. (This may not be completely realistic.) Find the radial and tangential components of acceleration of the discus and the magnitude of its acceleration at the instant when the angular velocity is 10 rad/s.

**SOLUTION** We model the discus as a particle moving in a circular path (Fig. 9–9). The acceleration components are given by Eqs. (9–14) and (9–15):

$$a_{rad} = \omega^2 r = (10\ \text{rad/s})^2(0.80\ \text{m}) = 80\ \text{m/s}^2,$$

$$a_{tan} = r\alpha = (0.80\ \text{m})(50\ \text{rad/s}^2) = 40\ \text{m/s}^2.$$

The magnitude of the acceleration vector is

$$a = \sqrt{a_{rad}^2 + a_{tan}^2} = 89\ \text{m/s}^2,$$

or about nine times the acceleration due to gravity. Note that we have dropped the unit "radian" from our final results; we can do this because a radian is a dimensionless quantity.

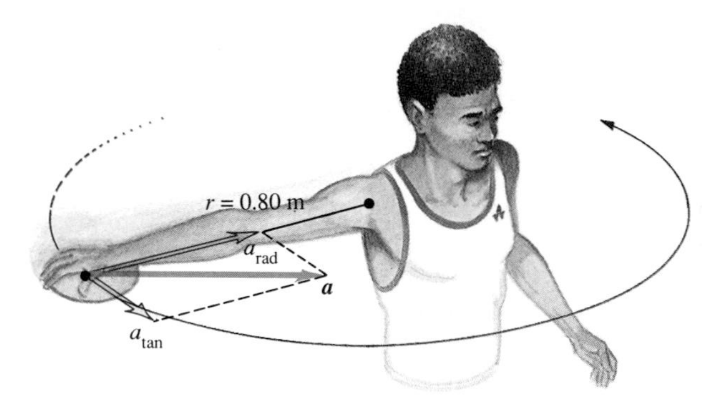


**9–9** Treating a discus thrower's arm as a rigid body.

## EXAMPLE 9–5

**Designing a propeller** You are asked to design an airplane propeller to turn at 3000 rpm. The forward airspeed of the plane is to be 180 m/s, and the speed of the tips of the propeller blades through the air must not exceed 300 m/s, the speed of sound in very cold air. What is the maximum radius the propeller can have?

**9–10** Designing an airplane propeller.

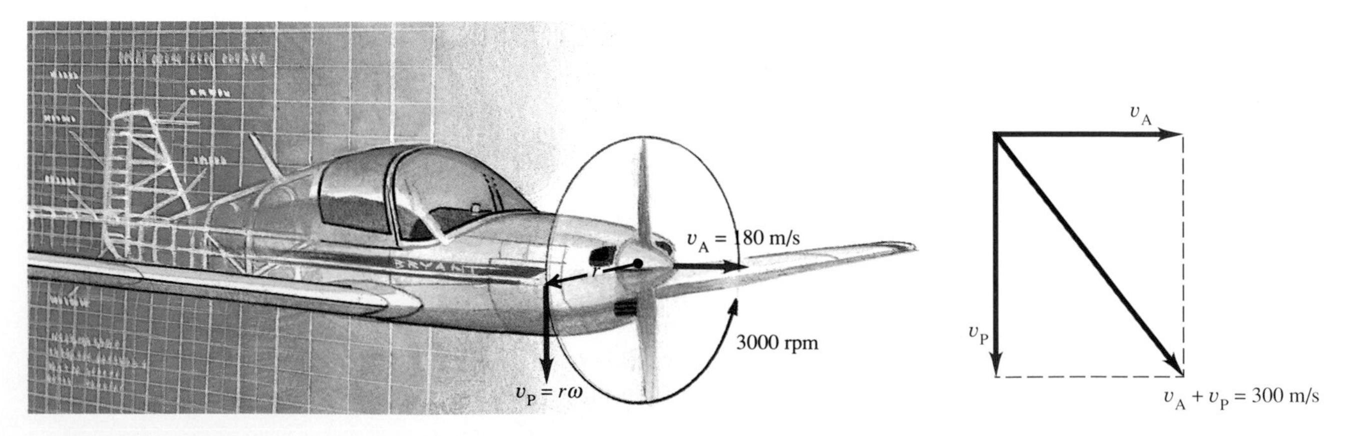

**SOLUTION** First we convert $\omega$ to rad/s:

$$\omega = 3000 \text{ rpm} = \left(3000 \frac{\text{rev}}{\text{min}}\right)\left(\frac{2\pi \text{ rad}}{1 \text{ rev}}\right)\left(\frac{1 \text{ min}}{60 \text{ s}}\right) = 314 \text{ rad/s}.$$

The tangential velocity $v_P = r\omega$ of the propeller tip is perpendicular to the forward velocity $v_A$ through the air (Fig. 9–10). The magnitude of the vector sum of these velocities is equal to 300 m/s when

$$(180 \text{ m/s})^2 + (r\omega)^2 = (300 \text{ m/s})^2,$$

$$r\omega = r(314 \text{ rad/s})$$

$$r\omega = \sqrt{(300 \text{ m/s})^2 - (180 \text{ m/s})^2} = 240 \text{ m/s},$$

$$r = 0.764 \text{ m}.$$

Just for fun, let's calculate the acceleration of the propeller tip with this radius:

$$a_{rad} = \omega^2 r$$
$$= (314 \text{ rad/s})^2(0.764 \text{ m}) = 7.53 \times 10^4 \text{ m/s}^2.$$

From $\Sigma F = ma$ the propeller must provide a centripetal force of $7.53 \times 10^4$ N for each kilogram of tip material! The propeller has to be made out of tough material.

## EXAMPLE 9–6

**Bicycle gears** How are the angular speeds of the two bicycle sprockets in Fig. 9–11 related to their numbers of teeth, $N_1$ and $N_2$?

**SOLUTION** The chain does not slip or stretch but moves at the same tangential speed $v$ on both sprockets:

$$v = r_1\omega_1 = r_2\omega_2, \quad \text{or} \quad \frac{\omega_2}{\omega_1} = \frac{r_1}{r_2}.$$

The angular speed is inversely proportional to the radius. This relationship also holds for pulleys connected by a belt, provided that the belt doesn't slip. Also, for chain sprockets the teeth must be equally spaced on the circumferences of both sprockets in order for the chain to mesh properly with both. Let $N_1$ and $N_2$ be the numbers of teeth; then

$$\frac{2\pi r_1}{N_1} = \frac{2\pi r_2}{N_2}, \quad \text{or} \quad \frac{r_1}{r_2} = \frac{N_1}{N_2}.$$

Combining this with the other equation, we get

$$\frac{\omega_2}{\omega_1} = \frac{N_1}{N_2}.$$

The angular speed of each sprocket is inversely proportional to the number of teeth. On a ten-speed bike you get the highest angular speed $\omega_2$ of the rear wheel for a given pedaling rate $\omega_1$ using the larger-radius front sprocket and the smallest-radius rear sprocket, giving the maximum front-to-back-teeth ratio $N_1/N_2$.

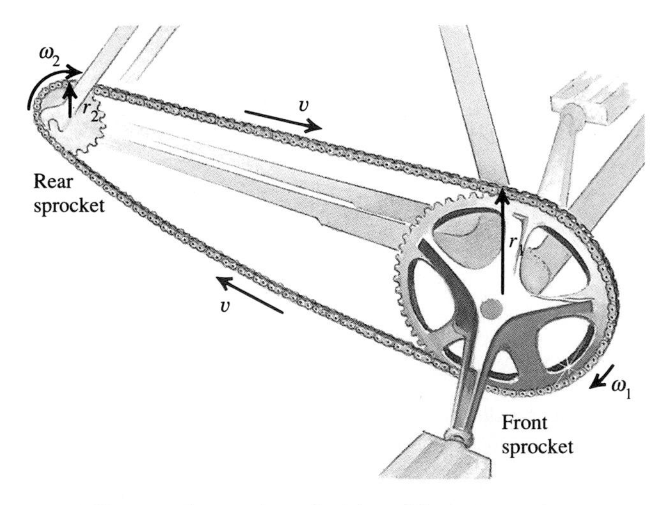


**9–11** How are the angular velocities of the two sprockets related to their number of teeth?

# 9–4 KINETIC ENERGY OF ROTATION

A rotating rigid body consists of mass in motion, so it has kinetic energy. We can express this kinetic energy in terms of the body's angular velocity and a new quantity called *moment of inertia* that we will define below. To develop this relationship, we think of the body as made up of a lot of particles, with masses $m_1, m_2, \ldots$, at distances $r_1, r_2, \ldots$ from the axis of rotation. We label the particles with the index $i$: The mass of a typical particle is $m_i$, and its distance from the axis of rotation is $r_i$. The particles don't necessarily all lie in the same plane, so we specify that $r_i$ is the *perpendicular* distance from the particle to the axis.

When a rigid body rotates about a fixed axis, the speed $v_i$ of a typical particle is given by Eq. (9–13), $v_i = r_i\omega$, where $\omega$ is the body's angular velocity. Different particles have different values of $r$, but $\omega$ is the same for all (otherwise the body wouldn't be rigid). The kinetic energy of a typical particle can be expressed as

$$\tfrac{1}{2}m_i v_i^2 = \tfrac{1}{2}m_i r_i^2 \omega^2.$$

The *total* kinetic energy of the body is the sum of the kinetic energies of all its particles:

$$K = \tfrac{1}{2}m_1 r_1{}^2 \omega^2 + \tfrac{1}{2}m_2 r_2{}^2 \omega^2 + \cdots$$
$$= \sum_i \tfrac{1}{2}m_i r_i{}^2\, \omega^2$$

Taking the common factor $\omega^2/2$ out of this expression, we get

$$K = \tfrac{1}{2}(m_1 r_1{}^2 + m_2 r_2{}^2 + \cdots)\omega^2$$
$$= \tfrac{1}{2}\left(\sum_i m_i r_i{}^2\right)\omega^2.$$

The quantity in parentheses, obtained by multiplying the mass of each particle by the square of its distance from the axis of rotation and adding these products, is called the **moment of inertia** of the body, denoted by $I$:

$$I = m_1 r_1{}^2 + m_2 r_2{}^2 + \cdots = \sum_i m_i r_i{}^2. \tag{9–16}$$

In terms of moment of inertia $I$ the rotational kinetic energy $K$ of a rigid body is

$$K = \tfrac{1}{2}I\omega^2. \tag{9–17}$$

This is analogous to the expression $K = \frac{1}{2}mv^2$ for kinetic energy of a particle. Moment of inertia is analogous to mass, and angular velocity $\omega$ is analogous to velocity $v$. This kinetic energy is not a new form of energy; it's the same physical quantity that we expressed as $\frac{1}{2}mv^2$ for a single particle. But Eq. (9–17) is much easier to use when we have to find the kinetic energy of a rotating body.

The SI unit of moment of inertia is one kilogram-meter$^2$ ($1\ \text{kg}\cdot\text{m}^2$). As in our previous discussion, $\omega$ *must* be measured in radians per unit time, not revolutions or degrees.

## EXAMPLE 9–7

An engineer is designing a one-piece machine part consisting of three heavy connectors linked by light molded struts (Fig. 9–12). The connectors can be considered massive particles connected by massless rods. a) What is the moment of inertia of this machine part about an axis through point $A$ and perpendicular to the plane of the diagram? b) What is the moment of inertia about an axis coinciding with rod $BC$? c) If the body rotates about an axis through $A$ and perpendicular to the plane of the diagram, with angular velocity $\omega = 4.0$ rad/s, what is its kinetic energy?

**SOLUTION** a) The particle at point $A$ lies *on* the axis. Its distance $r$ *from* the axis is zero, so it contributes nothing to the moment of inertia. Equation (9–16) gives

$$I = \Sigma m_i r_i{}^2 = (0.10\ \text{kg})(0.50\ \text{m})^2 + (0.20\ \text{kg})(0.40\ \text{m})^2$$
$$= 0.057\ \text{kg}\cdot\text{m}^2.$$

b) The particles at $B$ and $C$ both lie *on* the axis, so for them $r = 0$, and neither contributes to the moment of inertia. Only $A$ contributes, and we have

$$I = \Sigma m_i r_i{}^2 = (0.30\ \text{kg})(0.40\ \text{m})^2 = 0.048\ \text{kg}\cdot\text{m}^2.$$

c) From Eq. (9–17),

$$K = \tfrac{1}{2}I\omega^2 = \tfrac{1}{2}(0.057\ \text{kg}\cdot\text{m}^2)(4.0\ \text{rad/s})^2 = 0.46\ \text{J}.$$

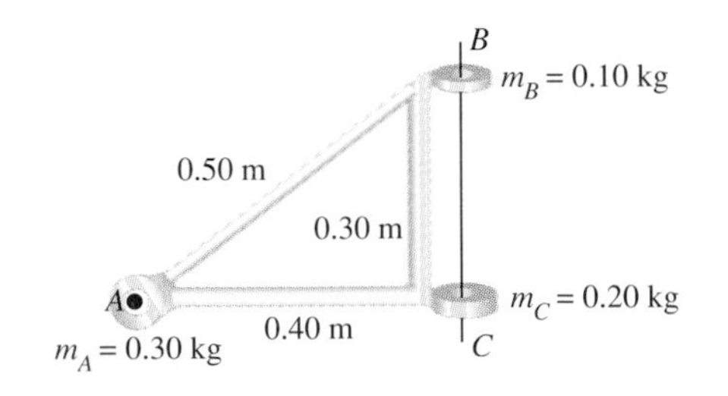


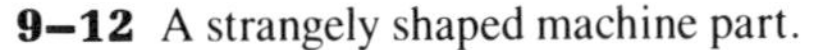
**9–12** A strangely shaped machine part.

The results of parts (a) and (b) of Example 9–7 show that the moment of inertia of a body depends on the location of the axis. We can't just say, "The moment of inertia of this body is $0.048\ \text{kg}\cdot\text{m}^2$." We have to say, "The moment of inertia of this body *with respect to axis BC* is $0.048\ \text{kg}\cdot\text{m}^2$."

In Example 9–7 we represented the body as several point masses, and we evaluated the sum in Eq. (9–16) directly. When the body is a *continuous* distribution of matter,

**TABLE 9–2 Moments of Inertia for Various Bodies**

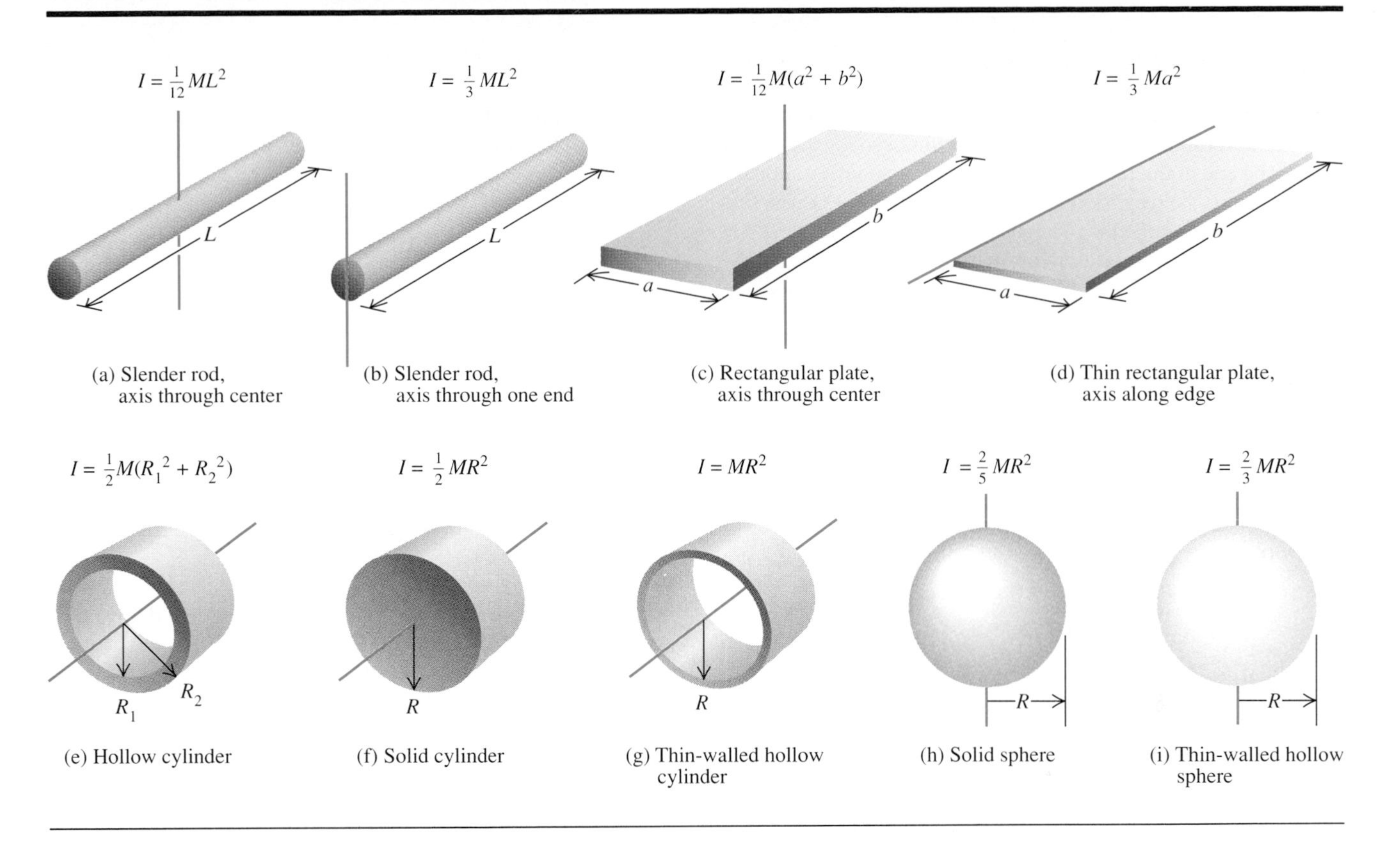


(a) Slender rod, axis through center

(b) Slender rod, axis through one end

(c) Rectangular plate, axis through center

(d) Thin rectangular plate, axis along edge

(e) Hollow cylinder

(f) Solid cylinder

(g) Thin-walled hollow cylinder

(h) Solid sphere

(i) Thin-walled hollow sphere

such as a solid cylinder or plate, we need methods of integral calculus to evaluate the sum. We will give several examples of moment-of-inertia calculations in Section 9–5; meanwhile, Table 9–2 gives moments of inertia for several familiar shapes in terms of the masses and dimensions.

You may be tempted to try to compute the moment of inertia of a body by assuming that all the mass is concentrated at the center of mass and then multiplying the total mass by the square of the distance from the center of mass to the axis. Resist that temptation; it doesn't work! For example, when a uniform thin rod of length $L$ and mass $M$ is pivoted about an axis through one end and perpendicular to the rod, the moment of inertia is $I = ML^2/3$. If we took the mass as concentrated at the center, a distance $L/2$ from the axis, we would obtain the *incorrect* result $I = M(L/2)^2 = ML^2/4$.

Now that we know how to calculate the kinetic energy of a rotating rigid body, we can apply the energy principles of Chapter 7 to rotational motion. Here are some points of strategy and some examples.

## PROBLEM-SOLVING STRATEGY

### Rotational energy

**1.** We suggest that you review the problem-solving strategies outlined in Sections 7–2 and 7–3; they are equally useful here. The only new idea here is that kinetic energy $K$ is expressed in terms of the moment of inertia $I$ and angular velocity $\omega$ of the body instead of its mass $m$ and speed $v$. You can use work-energy relations and conservation of energy to find relations involving position and motion of a rotating body.

**2.** The kinematic relations of Section 9–3, particularly Eqs. (9–13) and (9–14), are often useful, especially when a rotating cylindrical body functions as any sort of pulley. Examples 9–8 and 9–9 illustrate this point.

## EXAMPLE 9–8

**An unwinding cable** A light, flexible, nonstretching cable is wrapped several times around a winch drum, a solid cylinder of mass 50 kg and diameter 0.12 m, which rotates about a stationary horizontal axis, held by frictionless bearings (Fig. 9–13). The free end of the cable is pulled with a constant force of magnitude 9.0 N for a distance of 2.0 m. It unwinds without slipping, turning the cylinder as it does so. If the cylinder is initially at rest, find its final angular velocity and the final speed of the cable.

**SOLUTION** Because no energy is lost in friction, the final kinetic energy $K = \frac{1}{2}I\omega^2$ of the cylinder is equal to the work $W = Fd$ done by the force, which is (9.0 N)(2.0 m) = 18 J. From Table 9–2 the moment of inertia is

$$I = \tfrac{1}{2}MR^2 = \tfrac{1}{2}(50\text{ kg})(0.060\text{ m})^2 = 0.090\text{ kg}\cdot\text{m}^2.$$

The work-energy theorem then gives

$$18\text{ J} = \tfrac{1}{2}(0.090\text{ kg}\cdot\text{m}^2)\omega^2,$$

$$\omega = 20\text{ rad/s}.$$

The final speed of the cable is equal to the final tangential speed $v$ of the cylinder, which is given by Eq. (9–13):

$$\begin{aligned} v &= r\omega \\ &= (0.060\text{ m})(20\text{ rad/s}) \\ &= 1.2\text{ m/s}. \end{aligned}$$

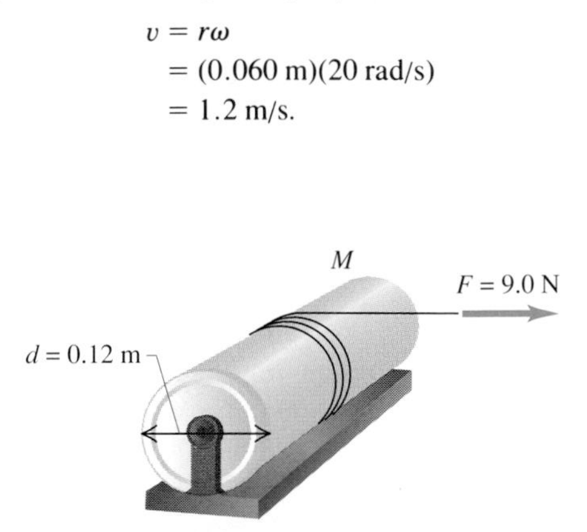


**9–13** A cable unwinds from a cylinder. The force $F$ does work and increases the kinetic energy of the cylinder.

## EXAMPLE 9–9

**Conservation of energy** In a lab experiment to test conservation of energy in rotational motion we wrap a light, flexible cable around a solid cylinder with mass $M$ and radius $R$. The cylinder rotates with negligible friction about a stationary horizontal axis (Fig. 9–14). We tie the free end of the cable to a mass $m$ and release the mass with no initial velocity at a distance $h$ above the floor. As the mass drops, the cable unwinds without stretching or slipping, turning the cylinder. Find the speed of mass $m$ and the angular velocity of the cylinder just as mass $m$ strikes the floor.

**SOLUTION** Initially, the system has no kinetic energy ($K_1 = 0$). We take the potential energy to be zero when $m$ is at floor level; then $U_1 = mgh$ and $U_2 = 0$. There are no nonconservative forces, so $W_{\text{other}} = 0$ and $K_1 + U_1 = K_2 + U_2$. (The cable does no net work; at one end the force and displacement are in the same direction, and at the other end they are in opposite directions. So the total work done by both ends of the cable is zero.) Just before $m$ hits the floor, both this mass and the cylinder have kinetic energy. The total kinetic energy $K_2$ at that time is

$$K_2 = \tfrac{1}{2}mv^2 + \tfrac{1}{2}I\omega^2.$$

From Table 9–2 the moment of inertia of the cylinder is $I = \frac{1}{2}MR^2$. Also, $v$ and $\omega$ are related by $v = R\omega$, since the speed of mass $m$ must be equal to the tangential speed at the outer surface of the cylinder. Using these relations and equating the initial and final total energies, we find

$$0 + mgh = \tfrac{1}{2}mv^2 + \tfrac{1}{2}(\tfrac{1}{2}MR^2)\left(\frac{v}{R}\right)^2 = \tfrac{1}{2}(m + \tfrac{1}{2}M)v^2,$$

$$v = \sqrt{\frac{2gh}{1 + M/2m}}.$$

The final angular velocity $\omega$ is obtained from $\omega = v/R$. Let's check some particular cases. When $M$ is much larger than $m$, $v$ is very small, as we would expect. When $M$ is much smaller than $m$, $v$ is nearly equal to the speed of a body in free fall with initial height $h$, namely, $\sqrt{2gh}$. Does it surprise you that $v$ doesn't depend on the radius of the cylinder?

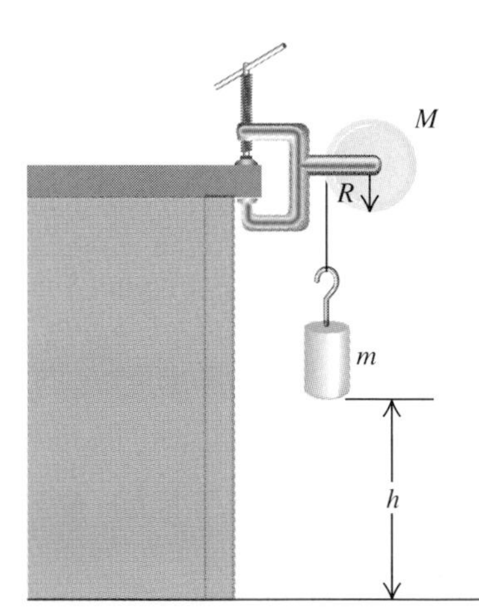


**9–14** As the cylinder rotates counterclockwise, the cable unwinds, mass $m$ drops, and gravitational potential energy is transformed into kinetic energy.

Later we will need to know how to calculate the *potential* energy for a rigid body. In a uniform gravitational field $g$ the potential energy is the same as though all the mass were concentrated at the center of mass of the body. Suppose we take the $y$-axis vertically upward. Then for a body with total mass $M$ the potential energy $U$ is simply

$$U = Mgy_{\text{cm}}, \tag{9–18}$$

where $y_{\text{cm}}$ is the $y$-coordinate of the center of mass.

To prove Eq. (9–18) we again represent the body as a collection of mass elements $m_i$. The potential energy for element $m_i$ is $m_i g y_i$, and the total potential energy is

$$U = m_1 g y_1 + m_2 g y_2 + \cdots = (m_1 y_1 + m_2 y_2 + \cdots)g.$$

But from Eqs. (8–25), which define the coordinates of the center of mass,

$$m_1 y_1 + m_2 y_2 + \cdots = (m_1 + m_2 + \cdots)y_{\text{cm}} = My_{\text{cm}},$$

where $M = m_1 + m_2 + \cdots$ is the total mass. Combining this with the above expression for $U$, we find

$$U = Mgy_{\text{cm}},$$

in agreement with Eq. (9–18).

We will need this relationship in Chapter 10 in the analysis of rigid-body problems in which the axis of rotation moves.

## *9–5 MOMENT-OF-INERTIA CALCULATIONS

When a rigid body cannot be represented by a few point masses but is a continuous distribution of mass, the sum in Eq. (9–16) that defines moment of inertia becomes an integral. We imagine dividing the body into small mass elements $dm$ so that all points in a particular element are at very nearly the same perpendicular distance from the axis of rotation; we call this distance $r$, as before. Then the moment of inertia is

$$I = \int r^2\, dm. \tag{9–19}$$

To evaluate the integral, we have to represent $r$ and $dm$ in terms of the same integration variable. When we have a one-dimensional object, such as rods (a) and (b) in Table 9–2, we can use a coordinate $x$ along the length and relate $dm$ to an increment $dx$. For a three-dimensional object it is usually easiest to express $dm$ in terms of an element of volume $dV$ and the density $\rho$ of the body. Density is mass per unit volume, $\rho = dm/dV$, so we may also write

$$I = \int r^2 \rho\, dV.$$

If the body is homogeneous (uniform in density), then we may take $\rho$ outside the integral:

$$I = \rho \int r^2\, dV. \tag{9–20}$$

To use this equation, we have to express the volume element $dV$ in terms of the differentials of the integration variables, such as $dV = dx\, dy\, dz$. The element $dV$ must always be chosen so that all points within it are at very nearly the same distance from the axis of rotation. The limits on the integral are determined by the shape and dimensions of the body. For regularly shaped bodies this integration can often be carried out quite easily.

## EXAMPLE 9–10

**Uniform thin rod, axis perpendicular to length** Figure 9–15 shows a slender uniform rod with mass $M$ and length $L$. It might be a baton held by a twirler in a marching band. We want to compute its moment of inertia about an axis through $O$ at an arbitrary distance $h$ from one end. We choose as an element of mass a short section of rod with length $dx$ at a distance $x$ from point $O$. The ratio of the mass $dm$ of this element to the total mass $M$ is equal to the ratio of its length $dx$ to the total length $L$:

$$\frac{dm}{M} = \frac{dx}{L}.$$

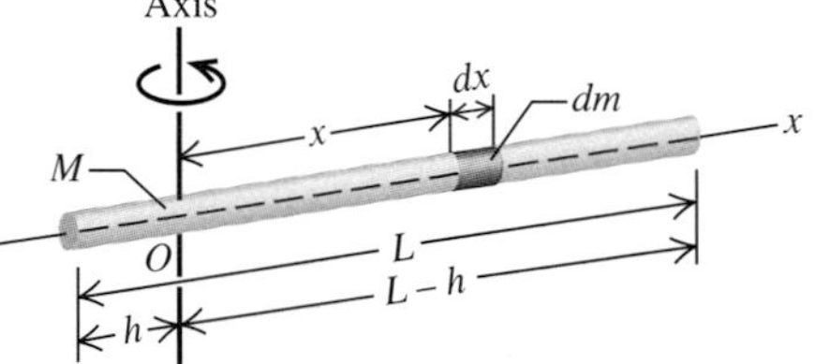


**9–15** Moment of inertia of a thin rod. The mass element is a segment with length $dx$.

We solve this for $dm$, substitute into Eq. (9–19), and add the appropriate integration limits on $x$ to obtain

$$I = \int x^2\,dm = \frac{M}{L}\int_{-h}^{L-h} x^2\,dx$$
$$= \left[\frac{M}{L}\left(\frac{x^3}{3}\right)\right]_{-h}^{L-h} = \tfrac{1}{3}M(L^2 - 3Lh + 3h^2).$$

From this general expression we can find the moment of inertia about an axis through any point on the rod. For example, if the axis is at the left end, $h = 0$ and

$$I = \tfrac{1}{3}ML^2.$$

If the axis is at the right end, we should get the same result. With $h = L$, we again get

$$I = \tfrac{1}{3}ML^2.$$

If the axis passes through the center, the usual place for a twirled baton, then $h = L/2$ and

$$I = \tfrac{1}{12}ML^2.$$

These results agree with the expressions in Table 9–2.

## EXAMPLE 9–11

**Hollow or solid cylinder, rotating about axis of symmetry** Figure 9–16 shows a hollow cylinder with length $L$ and inner and outer radii $R_1$ and $R_2$, perhaps a steel cylinder in a printing press or a sheet-steel rolling mill. We choose as a volume element a thin cylindrical shell of radius $r$, thickness $dr$, and length $L$. All parts of this element are at very nearly the same distance from the axis. The volume of the element is very nearly equal to that of a flat sheet with thickness $dr$, length $L$, and width $2\pi r$ (the circumference of the shell). Then

$$dm = \rho\,dV = 2\pi\rho Lr\,dr.$$

The moment of inertia is given by

$$I = \int r^2\,dm$$
$$= 2\pi\rho L\int_{R_1}^{R_2} r^3\,dr$$
$$= \frac{2\pi\rho L}{4}(R_2^4 - R_1^4)$$
$$= \frac{\pi\rho L}{2}(R_2^2 - R_1^2)(R_2^2 + R_1^2).$$

It is usually more convenient to express the moment of inertia in terms of the total mass $M$ of the body, which is its density $\rho$ multiplied by the total volume $V$. The volume is

$$V = \pi L(R_2^2 - R_1^2),$$

so the total mass $M$ is

$$M = \pi L\rho(R_2^2 - R_1^2),$$

and the moment of inertia is

$$I = \tfrac{1}{2}M(R_1^2 + R_2^2),$$

as is also shown in Table 9–2.

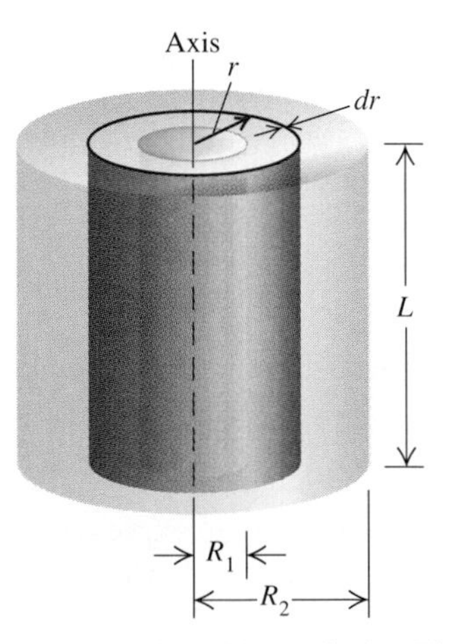


**9–16** Moment of inertia of a hollow cylinder. The mass element is a cylindrical shell with radius $r$ and thickness $dr$.

If the cylinder is solid, such as a lawn roller, $R_1 = 0$. Calling the outer radius $R_2$ simply $R$, we find that the moment of inertia of a solid cylinder of radius $R$ is

$$I = \tfrac{1}{2}MR^2.$$

If the cylinder has a very thin wall (like a stovepipe), $R_1$ and $R_2$ are very nearly equal; if $R$ represents this common radius,

$$I = MR^2.$$

Note that the moment of inertia of a cylinder about an axis coinciding with its axis of symmetry does not depend on the length $L$. Two hollow cylinders with the same inner and outer radii, one of wood and one of brass, but having the same mass $M$, have equal moments of inertia even though the length of the wood cylinder is much greater. Moment of inertia depends only on the *radial* distribution of mass, not on its distribution along the axis. Thus the above results also hold for a very short cylinder, such as a washer, and for a thin disk, such as a compact disk.

## EXAMPLE 9–12

**Uniform sphere with radius *R*, axis through center** The object might be a billiard ball, a steel ball bearing, or the earth. Divide the sphere into thin disks (Fig. 9–17). The radius $r$ of the disk shown is

$$r = \sqrt{R^2 - x^2}.$$

Its volume is

$$dV = \pi r^2\,dx = \pi(R^2 - x^2)\,dx,$$

and its mass is

$$dm = \rho\,dV = \pi\rho(R^2 - x^2)\,dx.$$

From Example 9–11 the moment of inertia of this disk is

$$dI = \tfrac{1}{2}r^2\,dm = \frac{\pi\rho}{2}(R^2 - x^2)^2\,dx.$$

Integrating this expression from 0 to $R$ gives the moment of inertia of the right hemisphere; from symmetry the total $I$ for the entire sphere is just twice this:

$$I = (2)\frac{\pi\rho}{2}\int_0^R (R^2 - x^2)^2\,dx.$$

Carrying out the integration, we obtain

$$I = \frac{8\pi\rho}{15}R^5.$$

The mass $M$ of the sphere is

$$M = \rho V = \frac{4\pi\rho R^3}{3}.$$

Thus

$$I = \tfrac{2}{5}MR^2.$$

This agrees with the expression in Table 9–2.

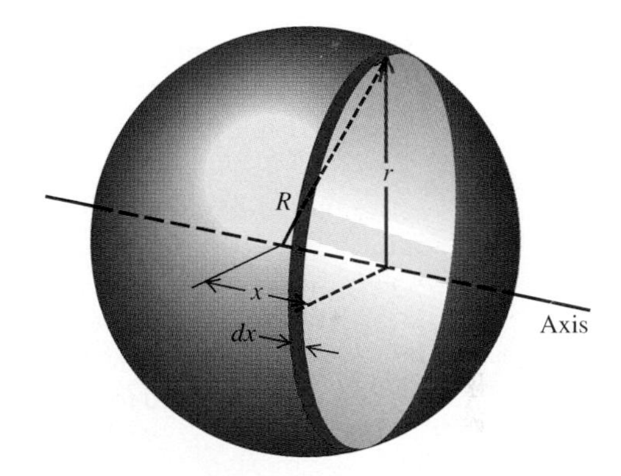


**9–17** Moment of inertia of a sphere. The mass element is a disk with thickness $dx$.

# *9–6 PARALLEL-AXIS THEOREM

We pointed out in Section 9–4 that a body doesn't have just one moment of inertia; it has infinitely many because there are infinitely many axes about which it might rotate. It turns out that there is a simple relationship between the moment of inertia $I_{\text{cm}}$ of a body about an axis through its center of mass and the moment of inertia $I_P$ about any other axis parallel to the original one but displaced from it by a distance $d$. This relationship, called the **parallel-axis theorem,** states that

$$I_P = I_{\text{cm}} + Md^2. \qquad (9\text{–}21)$$

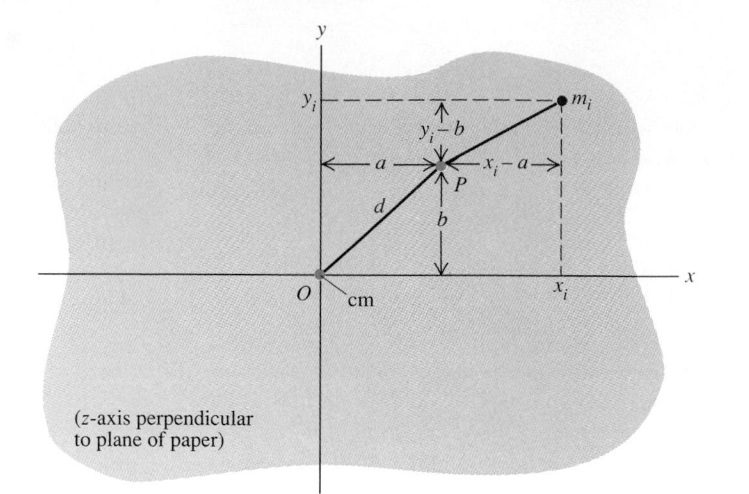


**9–18** The mass element $m_i$ has coordinates $(x_i, y_i)$ with respect to the axis of rotation through the center of mass and coordinates $(x_i - a, y_i - b)$ with respect to the parallel axis through point $P$.

To prove this theorem, we consider two axes, both parallel to the $z$-axis, one through the center of mass and the other through point $P$ (Fig. 9–18). First we take a very thin slice of the body, parallel to the $xy$-plane and perpendicular to the $z$-axis. We take the origin of our coordinate system to be at the center of mass of the body; the coordinates of the center of mass are then $x_{\text{cm}} = y_{\text{cm}} = z_{\text{cm}} = 0$. The axis through the center of mass passes through the figure at point $O$, and the parallel axis passes through point $P$, whose $x$- and $y$-coordinates are $(a, b)$. The distance of this axis from the axis through the center of mass is $d$, where $d^2 = a^2 + b^2$.

We can write an expression for the moment of inertia $I_P$ about the axis through point $P$. Let $m_i$ be a mass element in our slice, with coordinates $(x_i, y_i, z_i)$. Then the moment of inertia $I_{\text{cm}}$ of the slice about the axis through the center of mass (at $O$) is

$$I_{\text{cm}} = \Sigma m_i (x_i^2 + y_i^2).$$

The moment of inertia of the slice, about the axis through $P$, is

$$I_P = \Sigma m_i [(x_i - a)^2 + (y_i - b)^2].$$

These expressions don't contain the coordinates $z_i$ measured parallel to the slices, so we can extend the sums to include all the particles in all the slices. We then expand the squared terms and regroup, obtaining

$$I_P = \Sigma m_i (x_i^2 + y_i^2) - 2a\Sigma m_i x_i - 2b\Sigma m_i y_i + (a^2 + b^2)\,\Sigma m_i.$$

The first sum is $I_{\text{cm}}$. The second and third sums are zero; from Eqs. (8–25) they are $M$ times the $x$- and $y$-coordinates of the center of mass, which are zero because we have taken our origin to be the center of mass. The final term is $d^2$ multiplied by the total mass. This completes our proof.

## EXAMPLE 9–13

**Using the parallel-axis theorem** You measure the moment of inertia of an irregularly shaped machine part (Fig. 9–19) of mass 3.6 kg about an axis 0.15 m from its center of mass, using a method (which we'll work out later) that can't be used if the axis passes through the center of mass. Your result is $I_P = 0.132\ \text{kg}\cdot\text{m}^2$. What is the moment of inertia $I_{\text{cm}}$ about a parallel axis through the center of mass?

**SOLUTION** We rearrange Eq. (9–21) and substitute in the given values:

$$I_{\text{cm}} = I_P - Md^2 = 0.132\ \text{kg}\cdot\text{m}^2 - (3.6\ \text{kg})(0.15\ \text{m})^2$$
$$= 0.051\ \text{kg}\cdot\text{m}^2.$$

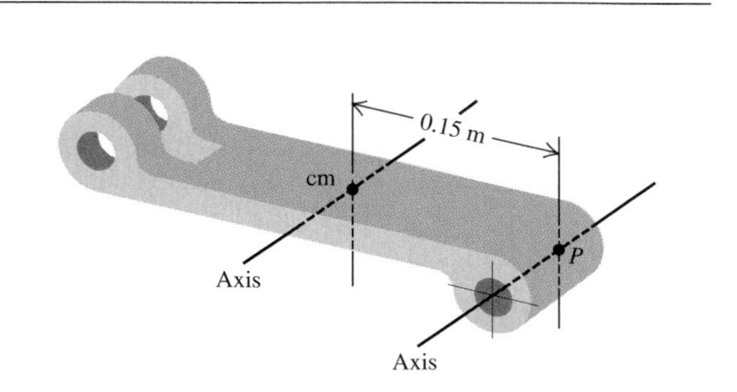


**9–19** Calculating $I_{\text{cm}}$ from a measurement of $I_P$.

## EXAMPLE 9–14

Find the moment of inertia of a thin, uniform disk about an axis perpendicular to its plane at the edge.

**SOLUTION** From Table 9–2, $I_{\text{cm}} = MR^2/2$, and in this case, $d = R$. The parallel-axis theorem gives

$$I_P = I_{\text{cm}} + Md^2 = \frac{MR^2}{2} + MR^2 = \frac{3MR^2}{2}.$$

This is a very simple solution of a problem that would be fairly complicated if we tried to solve it by direct integration.

Another moment-of-inertia theorem that's sometimes useful is the **perpendicular-axis theorem.** It is valid only for a body in the form of a thin plane sheet, which we will take to lie in the $xy$-plane (Fig. 9–20). Let $I_x$ and $I_y$ be the moments of inertia about the $x$- and $y$-axes and $I_O$ the moment of inertia about an axis through $O$ perpendicular to the plane. For this theorem, $O$ doesn't have to be the center of mass. The theorem states that

$$I_O = I_x + I_y. \tag{9–22}$$

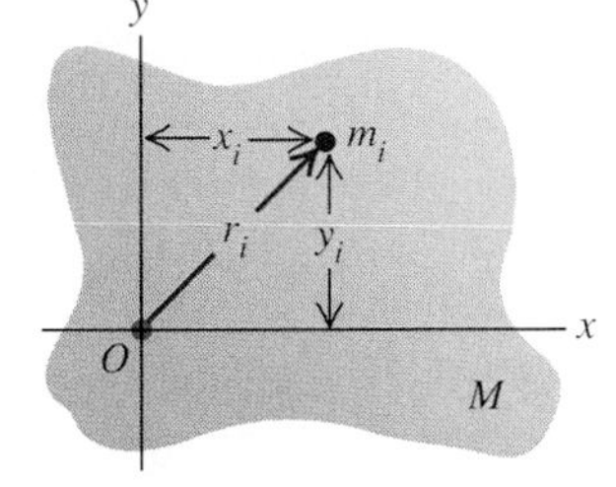


($z$-axis perpendicular to plane of paper)

**9–20** For a body in the form of a thin plane sheet in the $xy$-plane the moment of inertia about the $x$-axis plus the moment of inertia about the $y$-axis equals the moment of inertia about the perpendicular axis through $O$.

The proof is almost trivial. Take a mass element $m_i$ with coordinates $(x_i, y_i)$. The moment of inertia of the body about the $x$-axis is

$$I_x = \Sigma m_i y_i^2,$$

and its moment of inertia about the $y$-axis is

$$I_y = \Sigma m_i x_i^2.$$

The distance $r_i$ of $m_i$ from $O$ is $r_i = (x_i^2 + y_i^2)^{1/2}$, so the moment of inertia about the axis through $O$ is

$$I_O = \Sigma m_i (x_i^2 + y_i^2) = \Sigma m_i x_i^2 + \Sigma m_i y_i^2 = I_y + I_x.$$

This proves the theorem.

## EXAMPLE 9–15

For a thin washer with inner and outer radii $R_1$ and $R_2$, find the moment of inertia about an axis in the plane of the washer passing through its center (Fig. 9–21).

**SOLUTION** We choose the coordinate system shown in the figure; the washer lies in the $xy$-plane with its center at the origin. From Table 9–2, $I_O = \frac{1}{2}M(R_1^2 + R_2^2)$. From symmetry, $I_x = I_y$, and each is equal to the moment of inertia that we are asked to find. The perpendicular-axis theorem gives

$$I_O = I_x + I_y = 2I_x,$$

$$I_x = \frac{I_O}{2} = \frac{1}{4}M(R_1^2 + R_2^2).$$

This is the moment of inertia for *any* axis in the plane of the washer passing through its center.

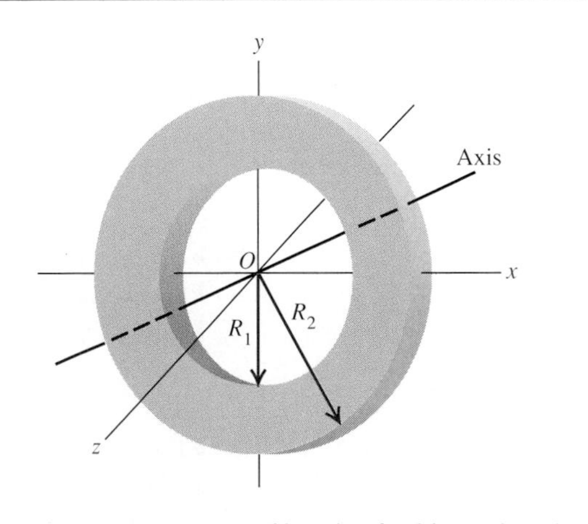


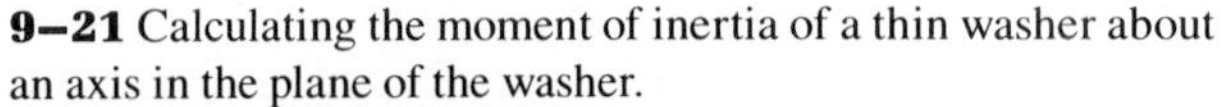
**9–21** Calculating the moment of inertia of a thin washer about an axis in the plane of the washer.

## SUMMARY

• When a rigid body rotates about a stationary axis, its position is described by an angular coordinate $\theta$. The angular velocity $\omega$ is defined as the derivative of the angular coordinate $\theta$:

$$\omega = \lim_{\Delta t \to 0} \frac{\Delta\theta}{\Delta t} = \frac{d\theta}{dt}. \tag{9-3}$$

The angular acceleration $\alpha$ is defined as the derivative of the angular velocity $\omega$ or the second derivative of the angular coordinate $\theta$:

$$\alpha = \frac{d\omega}{dt} = \frac{d^2\theta}{dt^2}. \tag{9-5}$$

• When a body rotates with constant angular acceleration, the angular position, velocity, and acceleration are related by

$$\theta = \theta_0 + \omega_0 t + \tfrac{1}{2}\alpha t^2, \tag{9-11}$$

$$\omega = \omega_0 + \alpha t, \tag{9-7}$$

$$\omega^2 = \omega_0{}^2 + 2\alpha(\theta - \theta_0), \tag{9-12}$$

where $\theta_0$ and $\omega_0$ are the initial values of angular position and angular velocity, respectively.

• A particle in a rotating rigid body has a speed $v$ given by

$$v = r\omega \tag{9-13}$$

and an acceleration $\boldsymbol{a}$ with tangential component

$$a_{\text{tan}} = \frac{dv}{dt} = r\frac{d\omega}{dt} = r\alpha \tag{9-14}$$

and radial component

$$a_{\text{rad}} = \frac{v^2}{r} = \omega^2 r. \tag{9-15}$$

• The moment of inertia $I$ of a body with respect to a given axis is defined as

$$I = m_1 r_1{}^2 + m_2 r_2{}^2 + \cdots = \sum_i m_i r_i{}^2, \tag{9-16}$$

where $r_i$ is the distance of mass $m_i$ from the axis of rotation.

• The kinetic energy of a rotating rigid body is given by

$$K = \tfrac{1}{2}I\omega^2. \tag{9-17}$$

• When the moment of inertia $I_{\text{cm}}$ of a body about an axis through the center of mass is known, the moment of inertia $I_P$ about a parallel axis at a distance $d$ from the first axis is given by

$$I_P = I_{\text{cm}} + Md^2. \tag{9-21}$$

• For a thin sheet lying in the $xy$-plane the moment of inertia about an axis through $O$ and perpendicular to the plane is

$$I_O = I_x + I_y. \tag{9-22}$$

### KEY TERMS

rigid body
radian
average angular velocity
instantaneous angular velocity
average angular acceleration
instantaneous angular acceleration
tangential component of acceleration
radial component of acceleration
moment of inertia
parallel-axis theorem
perpendicular-axis theorem

# EXERCISES

## Section 9–1 Angular Velocity and Acceleration

**9–1** a) What angle in radians is subtended by an arc 3.00 m in length on the circumference of a circle whose radius is 1.50 m? What is this angle in degrees? b) The angle between two radii of a circle with radius 2.00 m is 0.600 rad. What length of arc is intercepted on the circumference of the circle by the two radii? c) An arc 35.0 cm in length on the circumference of a circle subtends an angle of 42.0°. What is the radius of the circle?

**9–2** Compute the angular velocity in rad/s of the crankshaft of an automobile engine that is rotating at 4200 rev/min.

**9–3** A child is pushing a merry-go-round. The angle the merry-go-round has turned through varies with time according to $\theta(t) = \gamma t + \beta t^3$, where $\gamma = 2.50$ rad/s and $\beta = 0.0400$ rad/s$^3$. a) Calculate the angular velocity of the merry-go-round as a function of time. b) What is the initial value of the angular velocity? c) Calculate the instantaneous value of the angular velocity $\omega$ at $t = 5.00$ s and the average angular velocity $\omega_{av}$ for the time interval $t = 0$ to $t = 5.00$ s. How do these two quantities compare?

**9–4** A fan blade rotates with angular velocity given by $\omega(t) = \gamma - \beta t^2$, where $\gamma = 3.00$ rad/s and $\beta = 0.800$ rad/s$^3$. a) Calculate the angular acceleration as a function of time. b) Calculate the instantaneous angular acceleration $\alpha$ at $t = 2.00$ s and the average angular acceleration $\alpha_{av}$ for the time interval $t = 0$ to $t = 2.00$ s. How do these two quantities compare?

**9–5** The angle $\theta$ through which a bicycle wheel turns is given by $\theta(t) = a + bt^2 + ct^3$, where $a$, $b$, and $c$ are constants such that for $t$ in seconds, $\theta$ will be in radians. Calculate the angular acceleration of the wheel as a function of time.

## Section 9–2 Rotation with Constant Angular Acceleration

**9–6** Derive Eq. (9–12) by combining Eqs. (9–7) and (9–11) to eliminate $t$.

**9–7** A wheel turns with constant angular acceleration 0.520 rad/s$^2$. a) How much time does it take to reach an angular velocity of 8.00 rad/s, starting from rest? b) Through how many revolutions does the wheel turn in this time interval?

**9–8** A bicycle wheel has an initial angular velocity of 1.50 rad/s. If its angular acceleration is constant and equal to 0.200 rad/s$^2$, what is its angular velocity after it has turned through 3.50 revolutions?

**9–9** An electric motor is turned off, and its angular velocity decreases uniformly from 900 rev/min to 400 rev/min in 5.00 s. a) Find the angular acceleration in rev/s$^2$ and the number of revolutions made by the motor in the 5.00-s interval. b) How many more seconds are required for the motor to come to rest if the angular acceleration remains constant at the value calculated in part (a)?

**9–10** A circular saw blade 0.600 m in diameter starts from rest and accelerates with constant angular acceleration to an angular velocity of 140 rad/s in 8.00 s. Find the angular acceleration and the angle through which the blade has turned.

**9–11** A flywheel whose angular acceleration is constant and equal to 2.50 rad/s$^2$ rotates through an angle of 100 rad in 5.00 s. What was the angular velocity of the flywheel at the beginning of the 5.00-s interval?

**9–12** A flywheel requires 3.00 s to rotate through 212 rad. Its angular velocity at the end of this time is 108 rad/s. Find a) the angular velocity at the beginning of the 3.00-s interval; b) the constant angular acceleration.

## Section 9–3 Velocity and Acceleration Relations

**9–13** a) A cylinder 0.150 m in diameter rotates in a lathe at 700 rev/min. What is the tangential speed of the surface of the cylinder? b) The proper tangential speed for machining cast iron is about 0.60 m/s. At how many rev/min should a piece of stock 0.090 m in diameter be rotated in a lathe to produce this tangential speed?

**9–14** Find the required angular velocity of an ultracentrifuge in rev/min for the radial acceleration of a point 1.00 cm from the axis to equal 400,000$g$ (that is, 400,000 times the acceleration of gravity).

**9–15** A helicopter is rising vertically at 20.0 m/s while the main rotor is turning in a horizontal plane at 400 rev/min (Fig. 9–22). The rotor has a length (tip to tip) of 10.0 m, so the distance from the rotor shaft to each blade tip is 5.0 m. Calculate the magnitude of the resultant velocity of the blade tip through the air.

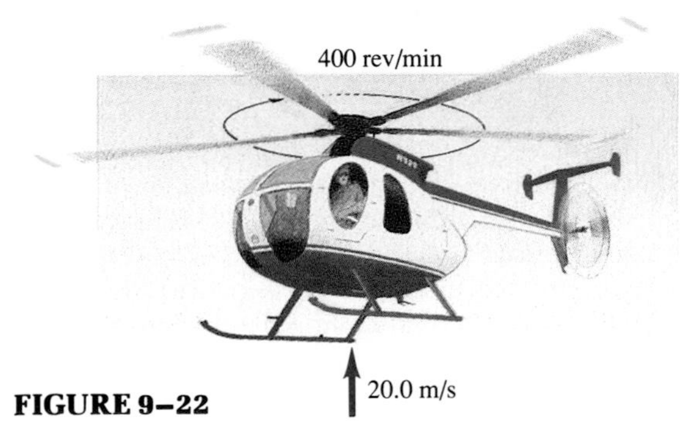


**FIGURE 9–22**

**9–16** A machine part has a disk of radius 4.50 cm fixed to the end of a shaft that has radius 0.25 cm. If the tangential speed of a point on the surface of the shaft is 1.50 m/s, what is the tangential speed of a point on the rim of the disk?

**9–17** A wheel rotates with a constant angular velocity of 12.0 rad/s. a) Compute the radial acceleration of a point 0.500 m from the axis from the relation $a_{rad} = \omega^2 r$. b) Find the tangential speed of the point, and compute its radial acceleration from the relation $a_{rad} = v^2/r$.

**9–18** An electric fan blade 0.850 m in diameter is rotating about a fixed axis with an initial angular velocity of 3.00 rev/s. The angular acceleration is 2.00 rev/$s^2$. a) Compute the angular velocity after 1.00 s. b) Through how many revolutions has the blade turned in this time interval? c) What is the tangential speed of a point on the tip of the blade at $t = 1.00$ s? d) What is the magnitude of the resultant acceleration of a point on the tip of the blade at $t = 1.00$ s?

**9–19** A flywheel with a radius of 0.300 m starts from rest and accelerates with a constant angular acceleration of 0.600 rad/$s^2$. Compute the magnitude of the tangential acceleration, the radial acceleration and the resultant acceleration of a point on its rim a) at the start; b) after it has turned through 120°; c) after it has turned through 240°.

## Section 9–4 Kinetic Energy of Rotation

**9–20** Small blocks, each with mass $m$, are clamped at the ends and at the center of a light rod of length $L$. Compute the moment of inertia of the system about an axis perpendicular to the rod and passing through a point one-third of the length from one end. Neglect the moment of inertia of the light rod.

**9–21** Four small spheres, each of mass 0.200 kg, are arranged in a square 0.400 m on a side and connected by light rods (Fig. 9–23). Find the moment of inertia of the system about an axis a) through the center of the square, perpendicular to its plane (an axis through point $O$ in the figure); b) bisecting two opposite sides of the square (an axis along line $AB$ in the figure).

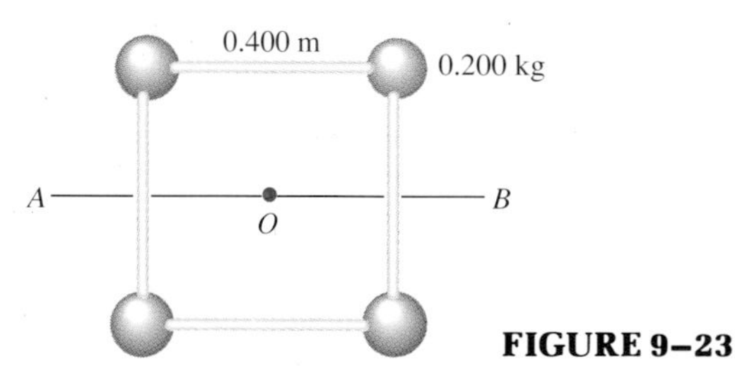


FIGURE 9–23

**9–22** Find the moment of inertia about each of the following axes for a rod that is 4.00 cm in diameter and 2.00 m long and has a mass of 6.00 kg. Use the formulas of Table 9–2. a) An axis perpendicular to the rod and passing through its center. b) An axis perpendicular to the rod and passing through one end. c) A longitudinal axis passing through the center of the rod.

**9–23** A wagon wheel is constructed as in Fig. 9–24. The radius of the wheel is 0.300 m, and the rim has a mass of 1.20 kg. Each of the eight spokes, which lie along a diameter and are 0.300 m long, has a mass of 0.375 kg. What is the moment of inertia of the wheel about an axis through its center and perpendicular to the plane of the wheel? (Use the formulas given in Table 9–2.)

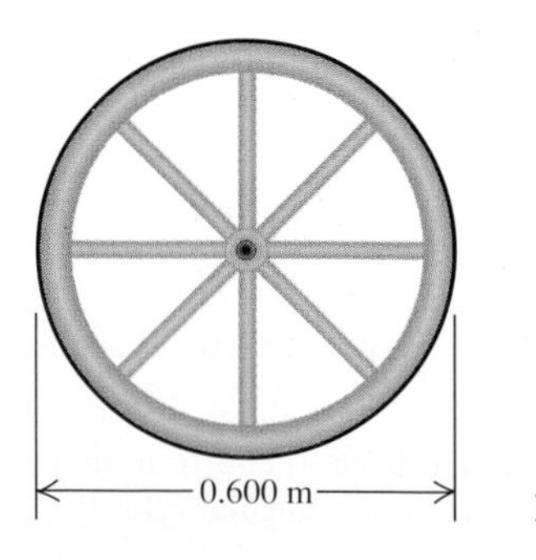


FIGURE 9–24

**9–24** A grinding wheel in the shape of a solid disk is 0.200 m in diameter and has a mass of 3.00 kg. The wheel is rotating at 2800 rev/min about an axis through its center. a) What is its kinetic energy? b) How far would it have to drop in free fall to acquire the same kinetic energy?

**9–25** The flywheel of a gasoline engine is required to give up 300 J of kinetic energy while its angular velocity decreases from 660 rev/min to 540 rev/min. What moment of inertia is required?

**9–26** A phonograph turntable has a kinetic energy of 0.0700 J when turning at 78.0 rev/min. What is the moment of inertia of the turntable about the rotation axis?

**9–27** Energy is to be stored in a large flywheel in the shape of a disk with radius $R = 1.20$ m and mass 80.0 kg. To prevent structural failure of the flywheel, the maximum allowed radial acceleration of a point on its rim is 5000 m/$s^2$. What is the maximum kinetic energy that can be stored in the flywheel?

**9–28** A light, flexible rope is wrapped several times around a solid cylinder with a weight of 40.0 N and a radius of 0.25 m, which rotates without friction about a fixed horizontal axis. The free end of the rope is pulled with a constant force $P$ for a distance of 5.00 m. What must $P$ be for the final speed of the end of the rope to be 4.00 m/s?

**9–29 Center of Mass of an Extended Object.** How much work must a wrestler do to raise the center of mass of his 120-kg opponent a vertical distance of 0.400 m?

**9–30** A uniform rope 12.0 m long with mass 2.50 kg is hanging with one end attached to a gymnasium ceiling and the other end just touching the floor. The upper end of the rope is released, and the rope falls to the floor. What is the change in the gravitational potential energy if the rope ends up flat on the floor (not coiled up)?

## Section 9–5 Moment-of-Inertia Calculations

***9–31** Use Eq. (9–19) to calculate the moment of inertia of a slender, uniform rod with mass $M$ and length $L$ about an axis at one end, perpendicular to the rod.

***9–32** Use Eq. (9–19) to calculate the moment of inertia of a uniform disk with mass $M$ and radius $R$ about an axis perpendicular to the plane of the disk and passing through its center.

***9–33** A slender rod with length $L$ has a mass per unit length that varies with distance from the left end according to $dm/dx = \gamma x$, where $\gamma$ has units of kg/$m^2$. a) Calculate the total mass of the rod in terms of $\gamma$ and $L$. b) Use Eq. (9–19) to calculate the moment of inertia of the rod for an axis at one end, perpendicular to the rod. Use the expression that you derived in part (a) to express $I$ in terms of $M$ and $L$. How does your result compare with that for a uniform rod?

## Section 9–6 Parallel-Axis Theorem

***9–34** The moment of inertia of a slender, uniform rod with mass $M$ and length $L$ for an axis through its center is $I_{cm} = \frac{1}{12}ML^2$. Use the parallel-axis theorem to calculate the moment of inertia for an axis perpendicular to the rod at one end. Compare your result with that given in Table 9–2.

***9–35** Use the parallel-axis theorem to calculate the moment of inertia of a square sheet of metal of side length $a$ and mass $M$ for an axis perpendicular to the sheet and passing through one corner.

***9–36** Find the moment of inertia of a thin-walled hollow cylinder (hoop) of mass $M$ and radius $R$ about an axis perpendicular to its plane at an edge.

***9–37** A thin, rectangular sheet of steel is 0.30 m by 0.40 m and has mass 16.0 kg. a) Find the moment of inertia about an axis in the plane of the sheet and through the center of the sheet and parallel to the long sides. b) Find the moment of inertia about an axis through the center of the sheet and parallel to the short sides. c) Use the perpendicular-axis theorem (Eq. 9–22) to calculate the moment of inertia about an axis perpendicular to the sheet and through its center. Compare your result to that calculated from part (c) of Table 9–2.

# PROBLEMS

**9–38** a) Prove that when an object starts from rest and rotates about a fixed axis with constant angular acceleration, the radial acceleration of a point in the object is directly proportional to its angular displacement. b) Through what angle has the object turned at the instant when the resultant acceleration of a point makes an angle of 36.9° with the radial direction?

**9–39** A roller in a printing press turns through an angle $\theta(t)$ given by $\theta(t) = \gamma t^2 - \beta t^3$, where $\gamma = 2.50 \text{ rad/s}^2$ and $\beta = 0.400 \text{ rad/s}^3$. a) Calculate the angular velocity of the roller as a function of time. b) Calculate the angular acceleration of the roller as a function of time. c) What is the maximum positive angular velocity, and at what value of $t$ does it occur?

**9–40** A 20.0-g weight is attached to the free end of a 1.60-m piece of string that is attached to the ceiling. The weight is pulled to one side so that the string makes a 53.1° angle with the vertical and is then released. What is the angular velocity of the weight when its angular acceleration is zero?

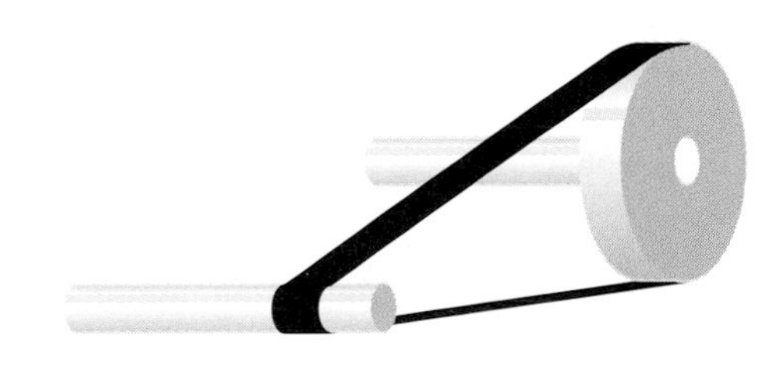

**FIGURE 9–25**

**9–41** A vacuum cleaner belt is looped over a shaft of radius 0.40 cm and a wheel of radius 2.00 cm (Fig. 9–25). The motor turns the shaft at 60.0 rev/s, and the moving belt turns the wheel, which is in turn connected by another shaft to a roller that beats the dirt out of the rug being vacuumed. Assume that the belt doesn't slip on either the shaft or the wheel. a) What is the speed of a point on the belt? b) What is the angular velocity of the wheel in rad/s?

***9–42** A bicycle wheel with a radius of 0.33 m turns with angular acceleration $\alpha(t) = \gamma - \beta t$, where $\gamma = 1.20 \text{ rad/s}^2$ and $\beta = 0.30 \text{ rad/s}^3$. It is at rest at $t = 0$. a) Calculate the angular velocity and angular displacement as functions of time. b) Calculate the maximum positive angular velocity and maximum positive angular displacement of the wheel.

**9–43 A Toy Car.** When a toy car is rapidly scooted across the floor, it stores energy in a flywheel. The car has mass 0.120 kg, and its flywheel has a moment of inertia of $4.00 \times 10^{-5} \text{ kg}\cdot\text{m}^2$. The car is 10.0 cm long. An advertisement claims that the car can be made to travel at a scale speed of up to 800 km/h (500 mi/h). a) For a scale speed of 800 km/h, what is the actual translational speed of the car? (The speed scales according to the length of the toy compared to that of a real car. Assume a length of 3.0 m for a real car.) b) If all the kinetic energy that is initially in the flywheel is converted to the translational kinetic energy of the toy, how much energy is originally stored in the flywheel? c) What is the initial angular velocity of the flywheel to store the amount of energy calculated in part (b)?

**9–44** The three objects shown in Fig. 9–26 have equal masses, $m$. Object $A$ is a solid cylinder with radius $R$. Object $B$ is a hollow, thin cylinder with radius $R$. Object $C$ is a solid square whose length of side equals $2R$. The objects have axes of rotation as shown in the figure, through the center of mass of each object. a) Which object has the smallest moment of inertia? b) Which object has the largest moment of inertia?

**9–45** The flywheel of a punch press has a moment of inertia of $16.0 \text{ kg}\cdot\text{m}^2$ and runs at 300 rev/min. The flywheel supplies all the

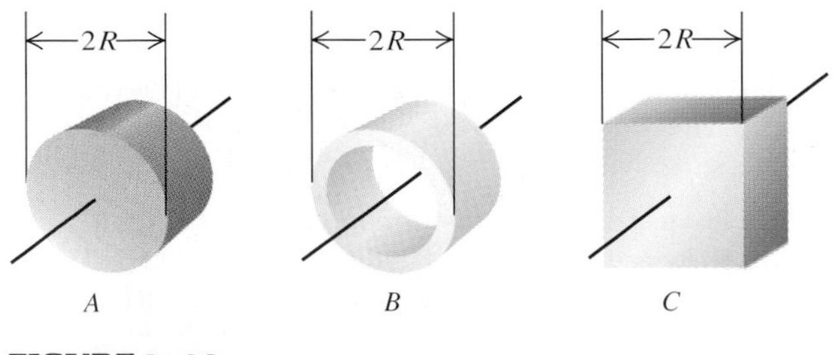


**FIGURE 9–26**

energy needed in a quick punching operation. a) Find the speed in rev/min to which the flywheel will be reduced by a sudden punching operation requiring 4000 J of work. b) What must be the constant power supply to the flywheel (in watts) to bring it back to its initial speed in a time of 5.00 s?

**9–46** A uniform solid disk with mass $m$ and radius $R$ is pivoted about a horizontal axis through its center, and a small object of mass $m$ is attached to the rim of the disk. If the disk is released from rest with the small object at the end of a horizontal radius, find the angular velocity when the small object is at the bottom of the disk.

**9–47** A meter stick with a mass of 0.300 kg is pivoted about one end so that it can rotate without friction about a horizontal axis. The meter stick is held in a horizontal position and released. As it swings through the vertical, calculate a) the change in gravitational potential energy that has occurred; b) the angular velocity of the stick; c) the linear velocity of the end of the stick opposite the axis. d) Compare the answer in part (c) to the speed of a particle that has fallen 1.00 m, starting from rest.

**9–48 The Swiss Bus.** A magazine article described a passenger bus in Zurich, Switzerland that derived its motive power from the energy stored in a large flywheel. The wheel was brought up to speed periodically, when the bus stopped at a station, by an electric motor, which could then be attached to the electric power lines. The flywheel was a solid cylinder of mass 1000 kg and diameter 1.80 m; its top angular velocity was 3000 rev/min. a) At this angular velocity, what is the kinetic energy of the flywheel? b) If the average power required to operate the bus is $1.86 \times 10^4$ W, how long could it operate between stops?

**9–49** The pulley in Fig. 9–27 has radius $R$ and moment of inertia $I$. The rope does not slip over the pulley. The coefficient of friction between block $A$ and the tabletop is $\mu_k$. The system is released from rest, and block $B$ descends. Use energy methods to calculate the speed of block $B$ as a function of the distance $d$ that it has descended.

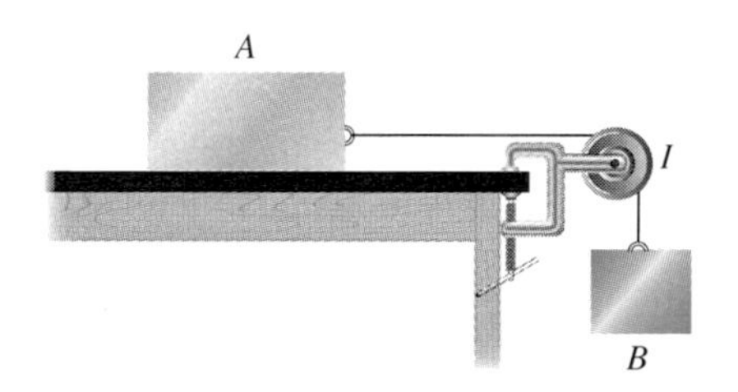


**FIGURE 9–27**

**9–50** The pulley in Fig. 9–28 has radius 0.200 m and moment of inertia 0.420 kg $\cdot$ m$^2$. The rope does not slip on the pulley rim. Use energy methods to calculate the speed of the 4.00-kg block just before it strikes the floor.

**9–51** Two metal disks, one with radius $R_1 = 4.00$ cm and mass $M_1 = 0.80$ kg and the other with radius $R_2 = 8.00$ cm and mass $M_2 = 1.60$ kg, are welded together and mounted on a frictionless axis through their common center (Fig. 9–29). a) What is the total moment of inertia of the two disks? b) A light string is wrapped around the edge of the smaller disk, and a 1.50-kg block is suspended from the free end of the string. If the block is released from rest a distance of 2.00 m above the floor, what is its speed just before it strikes the floor? c) Repeat the calculation of part (b), this time with the string wrapped around the edge of the larger disk. In which case is the final speed of the block the greatest? Does your answer make sense?

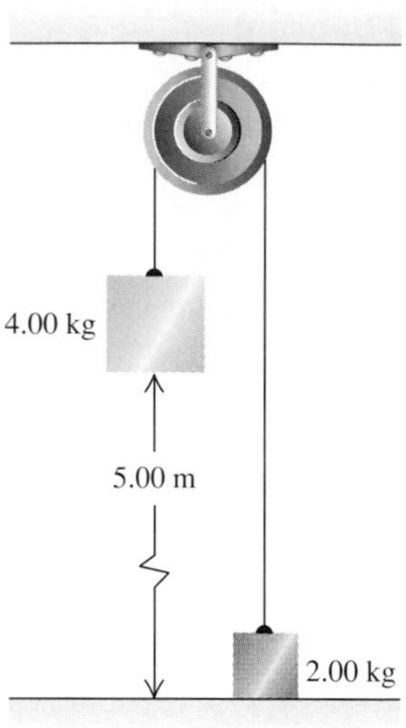

4.00 kg

5.00 m

2.00 kg

**FIGURE 9–28**

**9–52** A uniform, thin rod is bent into a square of side length $a$. If the total mass is $M$, find the moment of inertia about an axis through the center and perpendicular to the plane of the square. (*Hint:* Use the parallel-axis theorem.)

**9–53** A cylinder of radius $R$ and mass $M$ has density that increases linearly with distance $r$ from the cylinder axis, $\rho = \alpha r$. Calculate the moment of inertia of the cylinder about a longitudinal axis through its center in terms of $M$ and $R$. Compare your result with that for a cylinder of uniform density. Is the proper qualitative relation obtained?

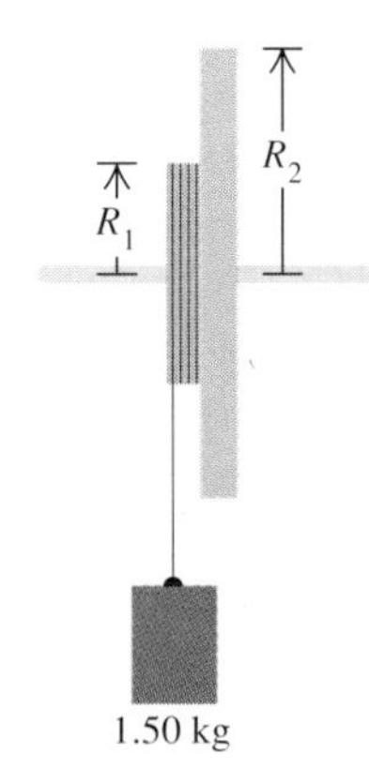


**FIGURE 9–29**

## CHALLENGE PROBLEMS

**9–54** Calculate the moment of inertia of a solid cone of uniform density about an axis through the center of the cone (Fig. 9–30). The cone has mass $M$ and altitude $h$. The radius of its circular base is $R$.

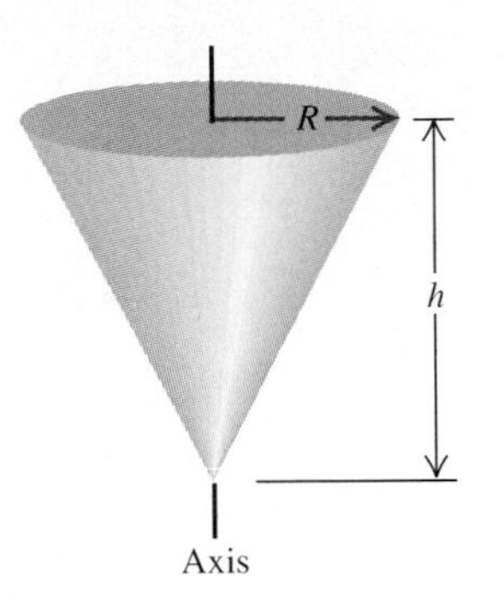


**FIGURE 9–30**

**9–55 The Earth's Moment of Inertia.** The moment of inertia of a sphere with uniform density about an axis through its center is $\frac{2}{5}MR^2 = 0.400\,MR^2$ (Table 9–2). Recent satellite observations show that the earth's moment of inertia is $0.3308MR^2$. The earth's interior consists of a mantle and a core (Fig. 9–31). The mantle occurs from the earth's surface to a depth of 3600 km, and the remainder is core. The average density (mass per unit volume) of the core is approximately $11.0 \times 10^3\ \text{kg/m}^3$; the mantle's average density is $5.0 \times 10^3\ \text{kg/m}^3$. For this two-part model of the earth, what is the earth's moment of inertia? Express your answer in terms of $M$ and $R$ as above.

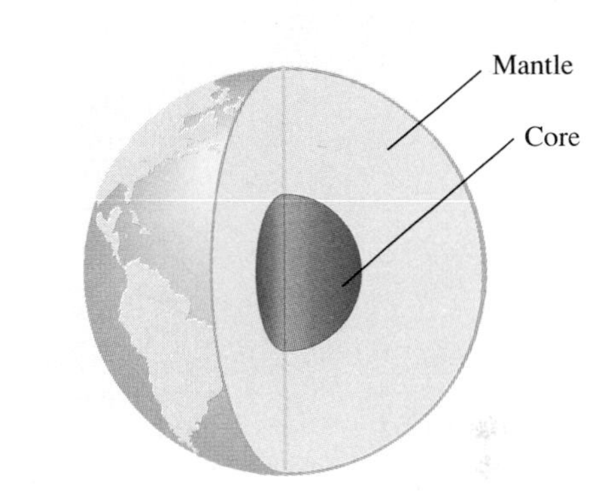

**FIGURE 9–31**

The spiral galaxy NGC 5236 is approximately 10 million light years distant from earth. As stars and other matter move away from the rapidly rotating center of the galaxy, the angular momentum of this material is roughly constant. Thus as the distance from the center increases, the angular velocity must decrease, forming the spiral pattern.

CHAPTER 10

# Dynamics of Rotational Motion

In a rotating solid body, the outer edge has a greater linear speed than parts rotating near the center.

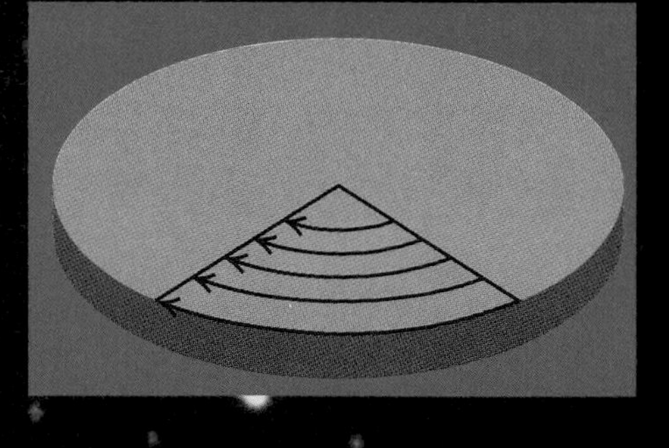

In a rotating galaxy, stars and gas have a slower linear speed the farther they are from the center of the galaxy.

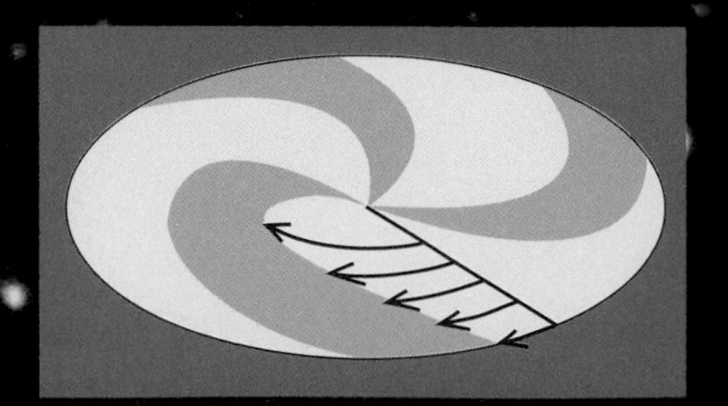

## KEY CONCEPTS

**• The torque of a force describes the tendency of the force to cause a rotation of the body on which the force acts.**

**• The sum of the torques acting on a body is proportional to its angular acceleration; the proportionality constant is the moment of inertia.**

**• The motion of a body can always be described as a combination of translational motion of a point and rotational motion about an axis through the point.**

**• When a force acts on a rotating body, the work done during an angular displacement is the product of torque and angular displacement. Power is the product of torque and angular velocity.**

**• Angular momentum is the rotational analog of linear momentum. When two or more bodies form an isolated system, the total angular momentum of the system is constant (conserved).**

**• When the direction of the axis of rotation changes, torque and angular momentum are represented as vector quantities. The time rate of change of angular momentum of a body equals the vector sum of torques acting on it.**

## INTRODUCTION

When you use a wrench to loosen the lug nuts when you change a flat tire, the whole wheel starts to turn unless you figure out a way to hold it. What is it about the force you apply to the wrench handle that makes the wheel turn? More generally, what is it that gives a rotating body an *angular* acceleration? A force is a push or a pull, but to produce rotational motion, we need a twisting or turning action.

In this chapter we will define a physical quantity, torque, that describes the twisting or turning effort of a force. We will incorporate this into a fundamental dynamic relationship for rotational motion. We also need to look at work and power in rotational motion in order to understand such problems as how energy is transmitted by a rotating shaft such as a drive shaft or axle in a car or the shaft of a rotating turbine in a jet engine. Finally, we will develop a new conservation principle, conservation of angular momentum, that is useful for rotational motion but also has much broader significance. Rotational motion can have some interesting surprises, as you may know if you've ever played with a toy gyroscope. We'll finish this chapter by studying how gyroscopes work. ■

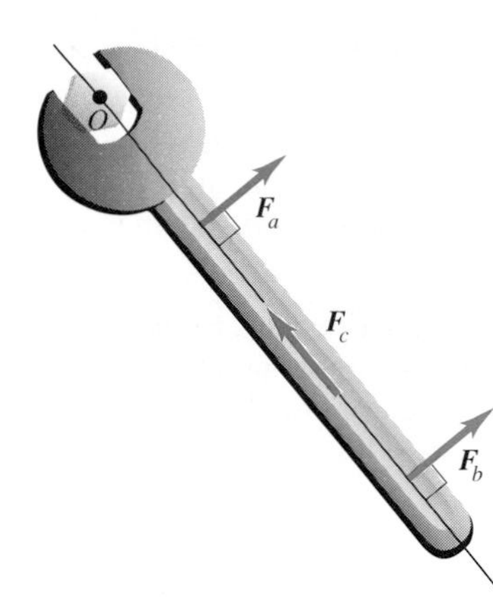


**10–1** Which of these three forces is most likely to loosen the tight bolt?

## 10–1 TORQUE

What is it about a force that determines how effective it is in causing or changing a rotational motion? The magnitude and direction of the force certainly play a role, but the position of the point where the force is applied is also important. When you try to swing a heavy door open, it's a lot easier if you push near the knob than if you push close to the hinges. The farther away from the axis of rotation you push, the more effective the push is. In Fig. 10–1, where a wrench is being used to loosen a tight bolt, force $\boldsymbol{F}_b$, applied at the end of the handle, is more effective than an equal force $\boldsymbol{F}_a$ applied near the bolt. Force $\boldsymbol{F}_c$, directed along the length of the handle, doesn't do any good at all.

The quantitative measure of the tendency of a force to cause or change rotational motion of a body is called **torque.** Figure 10–2 shows a body that can rotate about an axis that passes through point $O$ and is perpendicular to the plane of the figure. The body is acted on by three forces, $\boldsymbol{F}_1$, $\boldsymbol{F}_2$, and $\boldsymbol{F}_3$, in the plane of the figure. The tendency of force $\boldsymbol{F}_1$ to cause a rotation about point $O$ depends on the magnitude $F_1$ of the force. It also depends on the perpendicular distance $l_1$ between the line of action of the force (that is, the line along which the force vector lies) and point $O$. The distance $l_1$ plays the role of the wrench handle, and we call it the **moment arm** (or *lever arm*) of force $\boldsymbol{F}_1$ about $O$. The twisting effort is proportional to both $F_1$ and $l_1$. We define the **torque** or **moment** of the force $\boldsymbol{F}_1$ with respect to point $O$ as the produce $F_1 l_1$. We will use the Greek letter $\tau$ for torque:

$$\tau = Fl. \qquad (10\text{–}1)$$

We will use the terms *torque* and *moment* interchangeably. Physicists usually use "torque"; engineers usually use "moment" unless they are talking about a rotating shaft. Both groups use the term "moment arm" for the distance $l$.

The moment arm of $\boldsymbol{F}_1$ is the perpendicular distance $OA$ or $l_1$, and the moment arm of $\boldsymbol{F}_2$ is the perpendicular distance $OB$ or $l_2$. Force $\boldsymbol{F}_3$ has a line of action that passes through the reference point $O$. The moment arm for $\boldsymbol{F}_3$ is zero, and its torque with respect to point $O$ is zero. Note that torque is always defined with reference to a specific point, often (but not always) the origin of a coordinate system. If we shift the position of this point, the torque of each force may also change.

Force $\boldsymbol{F}_1$ tends to cause *counterclockwise* rotation about $O$, and $\boldsymbol{F}_2$ tends to cause *clockwise* rotation. To distinguish between these two possibilities, we will usually use the convention that *counterclockwise torques are positive and clockwise torques are negative.*

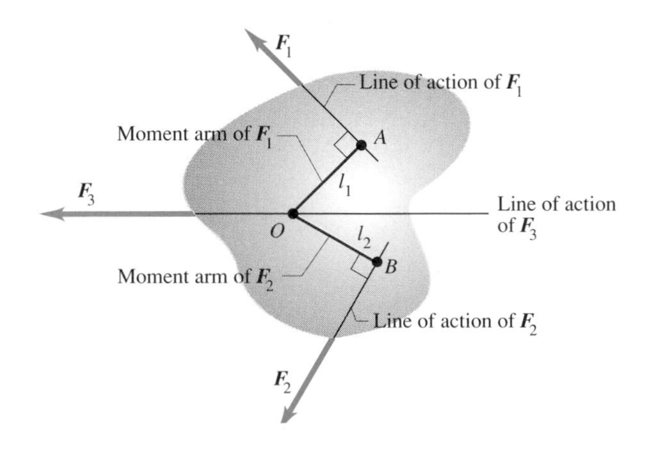


**10–2** The torque or moment of a force about a point is the product of the force magnitude and the moment arm.

With this convention the torque $\tau_1$ of $\boldsymbol{F}_1$ about $O$ is

$$\tau_1 = +F_1 l_1,$$

and the torque $\tau_2$ of $\boldsymbol{F}_2$ is

$$\tau_2 = -F_2 l_2.$$

We will often use the symbol

$$\circlearrowright +$$

to indicate our choise of positive sense of rotation.

The SI unit of torque is the newton-meter. In our discussion of work and energy we called this combination the *joule*. But torque is *not* work or energy, and torque should be expressed in newton-meters, not joules.

Often, one of the important forces acting on a body is its *weight.* This force is not concentrated at a single point but is distributed over the entire body. Nevertheless, it turns out that in a uniform gravitational field we always get the correct torque (about any specified point) if we assume that all the weight is concentrated at the *center of mass* of the body. We will prove this statement in Chapter 11, but meanwhile we will use it for some of the problems in this chapter.

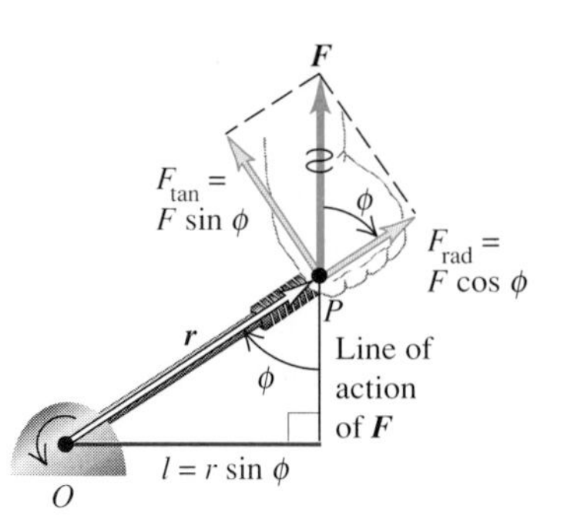


**10–3** The torque of the force $\boldsymbol{F}$ about the point $O$ is defined by $\boldsymbol{\tau} = \boldsymbol{r} \times \boldsymbol{F}$. Therefore the magnitude of $\boldsymbol{\tau}$ is $rF \sin \phi$. In this figure, $\boldsymbol{r}$ and $\boldsymbol{F}$ are in the plane of the paper, so $\boldsymbol{\tau}$ points out of the page.

Figure 10–3 shows a force $\boldsymbol{F}$ applied at a point $P$ described by a position vector $\boldsymbol{r}$ (with respect to the chosen point $O$). There are several alternative ways to calculate the torque of this force. We can find the moment arm $l$ and use $\tau = Fl$. Or we can determine the angle $\phi$; the moment arm is $r \sin \phi$, so $\tau = rF \sin \phi$. Finally, we can represent $\boldsymbol{F}$ in terms of a radial component $F_{rad}$ along the direction of $\boldsymbol{r}$ and a tangential component $F_{tan}$ at right angles, perpendicular to $\boldsymbol{r}$. (We call this a tangential component because if the body rotates, the point where the force acts moves in a circle, and this component is tangent to that circle.) Then $F_{tan} = F \sin \phi$, and

$$\tau = F_{tan} r = r(F \sin \phi) = Fl. \tag{10–2}$$

The component $F_{rad}$ has no torque with respect to $O$ because its moment arm with respect to that point is zero.

In Eq. (10–2) the quantity $rF \sin \phi$ is the magnitude of the *vector product* $\boldsymbol{r} \times \boldsymbol{F}$ that we defined in Section 1–10, Eq. (1–25). You may want to review that definition. This suggests that we generalize the definition of torque as follows. When a force $\boldsymbol{F}$ acts at a point having a position vector $\boldsymbol{r}$ with respect to an origin $O$, as in Fig. 10–3, the torque $\boldsymbol{\tau}$ of the force with respect to point $O$ is defined to be the vector quantity

$$\boldsymbol{\tau} = \boldsymbol{r} \times \boldsymbol{F}. \tag{10–3}$$

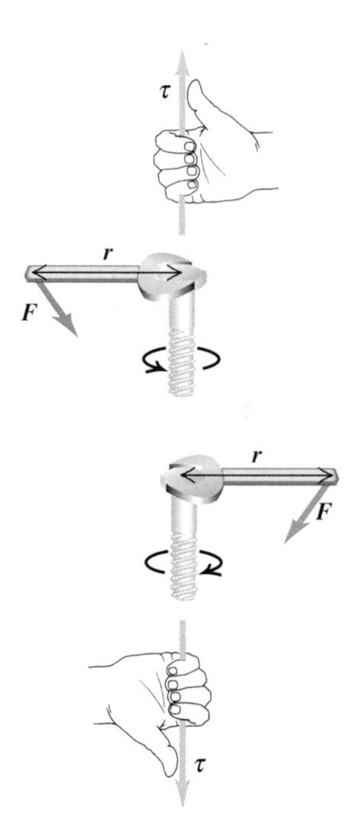


**10–4** The torque, $\boldsymbol{\tau} = \boldsymbol{r} \times \boldsymbol{F}$, is directed along the axis of the bolt. When the fingers of the right hand curl from the direction of $\boldsymbol{r}$ into the direction of $\boldsymbol{F}$, the outstretched right thumb points in the direction of $\boldsymbol{\tau}$ (perpendicular to both $\boldsymbol{r}$ and $\boldsymbol{F}$).

The magnitude of $\boldsymbol{r} \times \boldsymbol{F}$ is the torque that we have just defined. What about its direction? If $\boldsymbol{F}$ lies in the plane of Fig. 10–3 and the axis of rotation is perpendicular to that plane, then the direction of the torque vector $\boldsymbol{\tau} = \boldsymbol{r} \times \boldsymbol{F}$ is along the axis of rotation, with a sense given by the right-hand rule shown in Fig. 1–20. The direction relationships are shown in Fig. 10–4.

In the following sections we will usually be concerned with rotation of a body about an axis with a specified direction. In that case only the component of torque along that axis is of interest, and we often call that component the torque with respect to the specified axis.

## EXAMPLE 10–1

A weekend plumber, unable to loosen a pipe fitting, slips a piece of scrap pipe (a "cheater") over his wrench handle. He then applies his full weight of 900 N to the end of the cheater by standing on it. The distance from the center of the fitting to the point where the weight acts is 0.80 m, and the wrench handle and cheater make an angle of 19° with the horizontal (Fig. 10–5a). Find the magnitude and direction of the torque of his weight about the center of the pipe fitting.

**SOLUTION** From Fig. 10–5b the angle $\phi$ between $\boldsymbol{r}$ and $\boldsymbol{F}$ is 109°, and the moment arm $l$ is

$$l = (0.80\text{ m})(\sin 109°) = (0.80\text{ m})(\sin 71°) = 0.76\text{ m}.$$

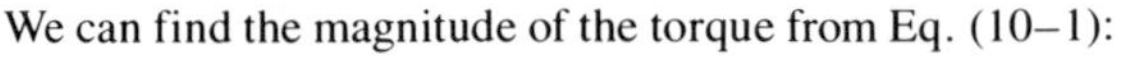

We can find the magnitude of the torque from Eq. (10–1):

$$\tau = Fl = (900\text{ N})(0.76\text{ m}) = 680\text{ N}\cdot\text{m}.$$

Or, from Eq. (10–2),

$$\tau = rF\sin\phi = (0.80\text{ m})(900\text{ N})(\sin 109°) = 680\text{ N}\cdot\text{m}.$$

Or we can find the components of $\boldsymbol{F}$. From the diagram the component perpendicular to $\boldsymbol{r}$ (which we call $F_{\tan}$, as explained above) is $F_{\tan} = F(\cos 19°) = (900\text{ N})(\cos 19°) = 850\text{ N}$. Then the torque is

$$\tau = F_{\tan}r = (850\text{ N})(0.80\text{ m}) = 680\text{ N}\cdot\text{m}.$$

The force tends to produce a clockwise rotation about $O$, and the direction of the vector torque $\boldsymbol{\tau}$ is *into* the plane of the figure.

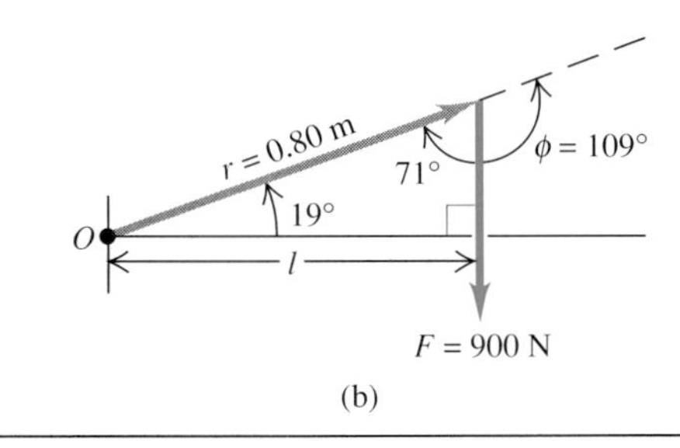


**10–5** (a) A weekend plumber tries to loosen a pipe fitting by standing on the "cheater." (b) Free-body diagram showing torque about $O$.

## 10–2 TORQUE AND ANGULAR ACCELERATION

We are now ready to develop the fundamental dynamical relation for rotational motion of a rigid body. We will show that the angular acceleration of a rotating body is directly proportional to the sum of the torque components along the axis of rotation. We will call these components "torques with respect to the axis of rotation." The proportionality factor is the moment of inertia.

To develop this relationship, we again imagine the body as being made up of a large number of particles. The first particle on our list has mass $m_1$ and distance $r_1$ from the axis (Fig. 10–6). We represent the *total force* acting on this particle in terms of a component $F_{1\text{rad}}$ that acts along the radial direction, a component $F_{1\tan}$ that is tangent to the circle of radius $r_1$ in which the particle moves as the body rotates, and a component $F_{1y}$ along the axis of rotation. Newton's second law for the tangential components is

$$F_{1\tan} = m_1 a_{1\tan}. \tag{10–4}$$

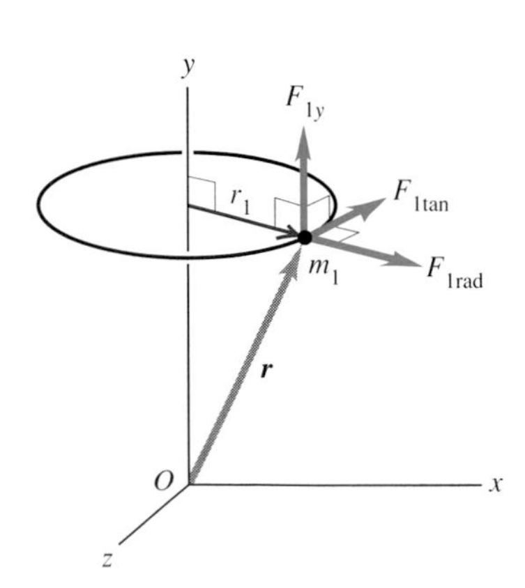


**10–6** Three components of the net force acting on one of the particles of a rigid body. Only $F_{1\tan}$ has a $y$-component of torque about $O$.

We can express $a_{1\tan}$ in terms of the angular acceleration $\alpha$, using Eq. (9–14): $a_{1\tan} = r_1\alpha$. Using this relation and multiplying both sides of Eq. (10–4) by $r_1$, we obtain

$$F_{1\tan}r_1 = m_1 r_1^2\alpha. \tag{10–5}$$

Now $F_{1\tan}r_1$ is just the *torque* of the force with respect to the axis through point $O$ (equal to the component of vector torque, with respect to point $O$, along the axis). Also, $m_1r_1^2$ is the moment of inertia of the particle. The component $F_{1\text{rad}}$ acts along a line passing through the axis and so has no torque with respect to the axis. The force component $F_{1y}$ along the axis of rotation also has no torque with respect to the axis. So we can rewrite Eq. (10–5) as

$$\tau_1 = I_1\alpha .$$

We may write an equation like this for every particle in the body and add all these equations:

$$\tau_1 + \tau_2 + \cdots = m_1r_1^2\alpha + m_2r_2^2\alpha + \cdots,$$

or

$$\Sigma\tau_i = (\Sigma m_ir_i^2)\alpha .$$

The left side of this equation is the sum of all the torques acting on all the particles. The right side is the total moment of inertia $I = \Sigma\, m_ir_i^2$ multiplied by the angular acceleration $\alpha$, which is the same for every particle. Thus for the entire body,

$$\Sigma\tau = I\alpha . \qquad (10\text{–}6)$$

Now let's think a little more about the forces. The force $\boldsymbol{F}$ on each particle is the vector sum of external and internal forces, as we defined these in Section 8–2. The two internal forces that any pair of particles exert on each other are an action-reaction pair; according to Newton's third law, they have equal magnitude and opposite direction. Assuming that they act along the line joining the two particles, their moment arms with respect to any axis are also equal, so the torques for each such pair add to zero. Indeed, *all* the internal torques add to zero, so the sum $\Sigma\tau$ includes only the torques of the *external* forces.

Equation (10–6) is the rotational analog of Newton's second law, $\Sigma\,\boldsymbol{F} = m\boldsymbol{a}$. It gives us the dynamic relationship we need to relate the rotational motion of a rigid body to the forces acting on it.

## PROBLEM-SOLVING STRATEGY

### Rotational dynamics

Our strategy for solving problems in rotational dynamics is very similar to that used in Section 5–2 for applications of Newton's laws:

**1.** Select a body for analysis. You will apply $\Sigma\,\boldsymbol{F} = m\boldsymbol{a}$ or $\Sigma\,\tau = I\alpha$, or sometimes both, to this body.

**2.** Draw a free-body diagram, as you have done before, isolating the body and including all the forces (and *only* those) that act on the body, including its weight. Label unknown quantities with algebraic symbols. A new consideration is that you must show the *shape* of the body accurately, with all dimensions and angles that you will need for torque calculations.

**3.** Choose coordinate axes for each body, and also indicate a positive sense of rotation for each rotating body. If there is a linear acceleration, it's usually simplest to pick a positive axis in its direction. If you know the sense of $\alpha$ in advance, picking that as the positive sense of rotation simplifies the calculations. When you represent a force in terms of its components, cross out the original force so that you don't include it twice.

**4.** If more than one body is involved, carry out the above steps for each body. Some problems will include bodies that have translational motion and others that have rotational motion. There may also be *geometrical* relations between the motions of two or more bodies. Express these in algebraic form, usually as relations between two linear accelerations or a linear acceleration and an angular acceleration.

**5.** Write the appropriate dynamical equations, mentioned in Step 1 above. Write a separate equation of motion for each body, and then solve the equations to find the unknown quantities. Often this involves solving a set of simultaneous equations.

**6.** Check special cases or extreme values of quantities when possible, and compare the results for these particular cases with your intuitive expectations. Ask yourself: "Does this result make sense?"

## EXAMPLE 10–2

Figure 10–7a shows the same situation that we analyzed in Example 9–8 (Section 9–4). A cable is wrapped several times around a uniform solid cylinder with diameter 0.12 m and mass 50 kg, pivoted so that it can rotate about its axis. The cable is pulled with a force of 9.0 N. Assuming that it unwinds without stretching or slipping, what is its acceleration?

**SOLUTION** Figure 10–7b shows the free-body diagram. The torque is $\tau = Fl = (0.060\ \text{m})(9.0\ \text{N}) = 0.54\ \text{N}\cdot\text{m}$. The moment of inertia, from Example 9–8, is $I = \frac{1}{2}mR^2 = \frac{1}{2}(50\ \text{kg})\cdot(0.060\ \text{m})^2 = 0.090\ \text{kg}\cdot\text{m}^2$. The angular acceleration $\alpha$ is given by Eq. (10–6):

$$\alpha = \frac{\tau}{I} = \frac{0.54\ \text{N}\cdot\text{m}}{0.090\ \text{kg}\cdot\text{m}^2} = 6.0\ \text{rad/s}^2.$$

We invite you to check the units in this equation and make sure they come out right. The acceleration of the cable is given by Eq. (9–14):

$$a = r\alpha = (0.060\ \text{m})(6.0\ \text{rad/s}^2) = 0.36\ \text{m/s}^2.$$

Can you use this result, together with an equation from Chapter 2, to determine the speed of the cable after it has been pulled 2.0 m? Try it, and compare your result with Example 9–8, in which we found this speed using work and energy considerations.

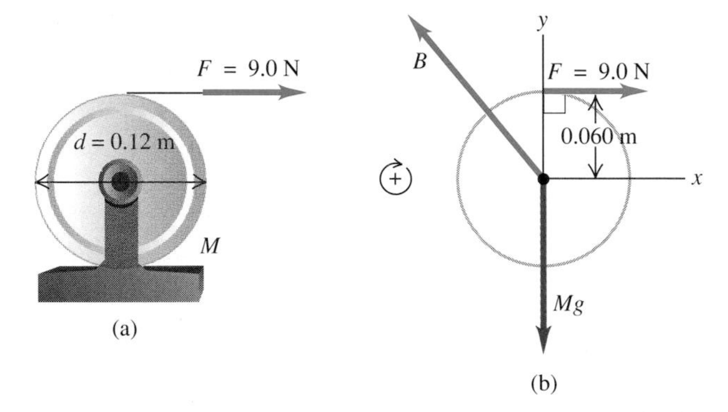


**10–7** (a) Cylinder and cable. (b) The free-body diagram for the cylinder. The bearing force $B$ and the weight $Mg$ give no torque about the center of the cylinder.

## EXAMPLE 10–3

Figure 10–8a shows the same situation as in Example 9–9. Find the acceleration of mass $m$ and the angular acceleration of the cylinder.

**SOLUTION** We have to treat the two bodies separately. Figure 10–8b shows a free-body diagram for each body. We take the positive sense of rotation for the cylinder to be counterclockwise and the positive direction of the $y$-coordinate for $m$ to be downward. Newton's second law applied to $m$ gives

$$\Sigma F_y = mg - T = ma.$$

The forces $Mg$ and $n$ (exerted by the bearing) act along lines through the axis of rotation and thus have no torque with respect to that axis. Applying Eq. (10–6) to the cylinder gives

$$\Sigma\tau = RT = I\alpha = \tfrac{1}{2}MR^2\alpha.$$

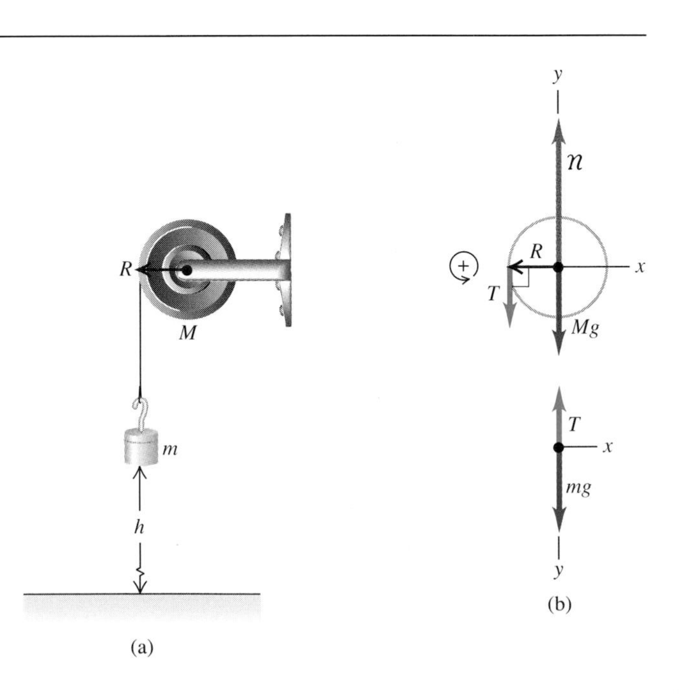


**10–8** (a) Cylinder, mass, and rope. (b) Free-body diagrams for the cylinder and for the mass hanging from the rope. The rope is assumed to have negligible mass.

Now for the kinematic relation. We remarked in Section 9–3 that the acceleration of a cable unwinding from a cylinder is the same as the tangential component of acceleration of a point on the surface of the cylinder where the cable is tangent to it. According to Eq. (9–14), this is given by $a_{tan} = R\alpha$. We use this to replace $(R\alpha)$ with $a$ in the cylinder equation above and then divide by $R$; the result is

$$T = \tfrac{1}{2}Ma.$$

Now we substitute this expression for $T$ into the equation of motion for $m$:

$$mg - \tfrac{1}{2}Ma = ma.$$

When we solve this for $a$, we get

$$a = \frac{g}{1 + M/2m}.$$

Finally, we can substitute this back into the $\Sigma F = ma$ equation for $m$ to find $T$:

$$T = mg - ma = mg - m\frac{g}{1 + M/2m} = \frac{mg}{1 + 2m/M}.$$

We note that the tension in the cable is *not* equal to the weight $mg$ of mass $m$; if it were, $m$ could not accelerate.

Let's check two particular cases. When $M$ is much larger than $m$, the tension is nearly equal to $mg$, and the acceleration is correspondingly much less than $g$. When $M$ is zero, $T = 0$ and $a = g$; the mass then falls freely. If mass $m$ starts at a height $h$ above the floor with an initial speed $v_0$, its speed $v$ when it strikes the gound is given by $v^2 = v_0^2 + 2ah$. If it starts from rest, $v_0 = 0$ and

$$v = \sqrt{2ah} = \sqrt{\frac{2gh}{1 + M/2m}}.$$

This is the same result that we obtained from energy considerations in Section 9–4.

## EXAMPLE 10–4

In Fig. 10–9 mass $m_1$ is a glider that slides without friction on a horizontal air track. The pulley is in the form of a thin cylindrical shell (with massless spokes) with mass $M$ and radius $R$, and the string turns the pulley without slipping or stretching. Find the acceleration of each body, the angular acceleration of the pulley, and the tension in each part of the string.

**SOLUTION** Figure 10–9b shows the free-body diagrams and coordinate systems for the three bodies. Note that the mass of the pulley plays an essential role. The two tensions $T_1$ and $T_2$ *cannot* be equal; if they were, the pulley could not have an angular acceleration. Thus labeling the tension in both parts of the string as simply $T$ would be a serious error.

The equations of motion for masses $m_1$ and $m_2$ are

$$\Sigma F_x = T_1 = m_1 a_1, \qquad (10\text{–}7)$$

$$\Sigma F_y = m_2 g - T_2 = m_2 a_2. \qquad (10\text{–}8)$$

The unknown normal force $n_2$ and the pulley's weight both act on a line through the pulley's axis of rotation, so they have no moment arm or torque with respect to that axis. From Table 9–2 the moment of inertia of the pulley is $I = MR^2$. We take the positive sense of rotation as clockwise (the sense of the pulley's actual angular acceleration). Then the pulley's equation of motion is

$$\Sigma\tau = T_2 R - T_1 R = I\alpha = (MR^2)\alpha. \qquad (10\text{–}9)$$

Because the string does not stretch or slip, we have the additional *kinematic* relations

$$a_1 = a_2 = R\alpha. \qquad (10\text{–}10)$$

(The accelerations of $m_1$ and $m_2$ have different directions but the same magnitude.)

Equations (10–7) through (10–10) are five simultaneous equations for the five unknowns $a_1$, $a_2$, $\alpha$, $T_1$, and $T_2$. [Equation (10–10) is actually two equations.] This may sound like an awesome problem, but we have as many equations as unknowns, and solving them is a lot easier than you might imagine. First we use Eqs. (10–10) to eliminate $a_2$ and $\alpha$ from Eqs. (10–7) through (10–9). We then have three equations for the three unkowns $T_1$, $T_2$, and $a_1$:

$$T_1 = m_1 a_1,$$

$$m_2 g - T_2 = m_2 a_1,$$

$$T_2 - T_1 = Ma_1.$$

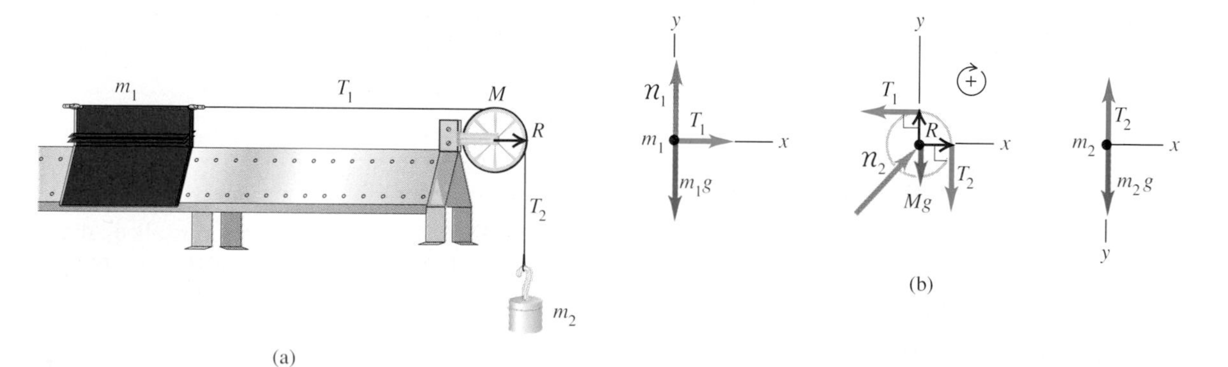


**10–9** (a) An air-track glider pulled by a mass hanging over a pulley. (b) Free-body diagrams for $m_1$, $m_2$, and $M$.

The easiest way to solve these is simply to *add* the three equations, which eliminates $T_1$ and $T_2$, and then to solve for $a_1$:

$$a_1 = \frac{m_2 g}{m_1 + m_2 + M}.$$

We can then substitute this back into Eqs. (10–7) and (10–8) to find the tensions. The results are

$$T_1 = \frac{m_1 m_2 g}{m_1 + m_2 + M}, \qquad T_2 = \frac{(m_1 + M) m_2 g}{m_1 + m_2 + M}.$$

Let's check some special cases to see whether these results make sense. First, if either $m_1$ or $M$ is much larger than $m_2$, the accelerations are very small, and $T_2$ is approximately $m_2 g$. But if $m_2$ is much *larger* than either $m_1$ or $M$, the acceleration is approximately $g$. Both of these results are what we should expect. When $M = 0$, we have the same situation as in Example 5–11 (Section 5–2). Do we get the same result? Can you think of other special cases to check?

## 10–3 ROTATION ABOUT A MOVING AXIS

We can extend our analysis of the dynamics of rotational motion to some cases in which the axis of rotation moves. When that happens, the body has both translational and rotational motion. In fact, every possible motion of a rigid body can be represented as a combination of translational motion of the center of mass and rotation about an axis through the center of mass. A ball rolling down a hill and a yo-yo unwinding at the end of a string are familiar examples of such motion.

### Dynamic Relations

The key to this more general analysis, which we will not derive in detail, is that Eq. (10–6), $\Sigma\tau = I\alpha$, is valid even when the axis of rotation moves if the following two conditions are met: (1) The axis must be an axis of symmetry and must pass through the center of mass of the body; (2) the axis must not change direction. Note that, in general, this moving axis is *not* at rest in an inertial frame of reference.

### PROBLEM-SOLVING STRATEGY

**Rotation about a moving axis**

The strategy outlined in Section 10–2 is equally useful here. There is one new wrinkle: When a body undergoes translational and rotational motion at the same time, we need two separate equations of motion *for the same body.* One of these is based on $\Sigma \boldsymbol{F} = M\boldsymbol{a}_{cm}$ for the translational motion of the center of mass. We showed in Section 8–7 that for a body with total mass $M$ the acceleration $\boldsymbol{a}_{cm}$ of the center of mass is the same as that of a point mass $M$, acted on by all the forces on the actual body. The other equation of motion is based on $\Sigma\tau = I_{cm}\alpha$ for the rotational motion about the axis through the center of mass, with moment of inertia $I_{cm}$. In addition, there is often a geometric relation between the linear and angular accelerations, as with a wheel that rolls without slipping or a string that unwinds from a pulley while turning it.

### EXAMPLE 10–5

**A primitive yo-yo** In Fig. 10–10a a string is wrapped several times around a solid cylinder with mass $M$ and radius $R$. We hold the end of the string stationary while releasing the cylinder with no initial motion. The string unwinds but does not slip or stretch as the cylinder drops and rotates. Find the downward acceleration of the cylinder and the tension in the string.

**SOLUTION** Figure 10–10b shows a free-body diagram for the situation, including the choice of positive coordinate directions. The equation for the translational motion of the center of mass is

$$\Sigma F_y = Mg - T = Ma_{cm}. \qquad (10\text{–}11)$$

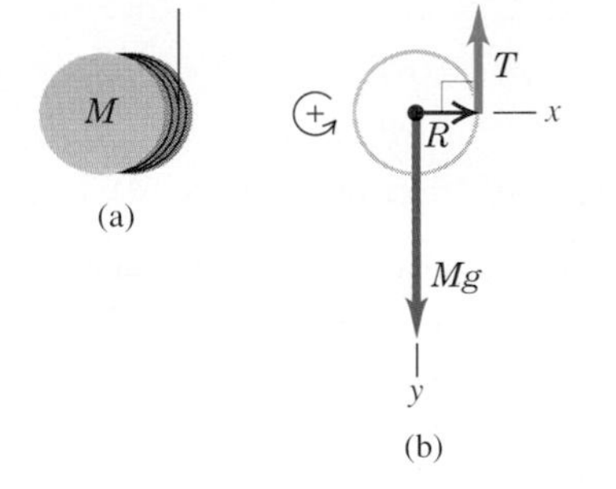


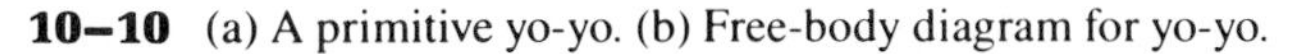

**10–10** (a) A primitive yo-yo. (b) Free-body diagram for yo-yo.

The moment of inertia for an axis through the center of mass is $I_{cm} = \frac{1}{2}MR^2$, and the equation for rotational motion about the axis through the center of mass is

$$\Sigma\tau = TR = I_{cm}\alpha = \tfrac{1}{2}MR^2\alpha. \qquad (10\text{–}12)$$

If the string unwinds without slipping, the linear displacement of the center of mass in any time interval equals $R$ times the angular displacement. This gives us the kinematic relations

$$v_{cm} = R\omega, \qquad (10\text{–}13)$$

$$a_{cm} = R\alpha. \qquad (10\text{–}14)$$

(If this isn't obvious, imagine moving along with the center of mass of the cylinder and watching the string unwind. From that point of view the kinematic situation is just the same as in Examples 10–2 and 10–3.)

We now use Eq. (10–14) to eliminate $\alpha$ from Eq. (10–12) and then solve Eqs. (10–11) and (10–12) simultaneously for $T$ and $a_{cm}$. The results are amazingly simple:

$$a_{cm} = \tfrac{2}{3}g, \qquad T = \tfrac{1}{3}Mg.$$

## EXAMPLE 10–6

A bowling ball rolls without slipping down the return ramp at the side of the alley (Fig. 10–11a). The ramp is inclined at an angle $\theta$ to the horizontal. What is the ball's acceleration? Treat the ball as a solid sphere, ignoring the finger holes.

**SOLUTION** Figure 10–11b shows a free-body diagram, with positive coordinate directions indicated. From Table 9–2 the moment of inertia of a solid sphere is $I = \frac{2}{5}MR^2$. The equations of motion for translational motion and for rotation about the axis through the center of mass, respectively, are

$$\Sigma F_x = Mg\sin\theta - \mathcal{F} = Ma_{cm}, \qquad (10\text{–}15)$$

$$\Sigma\tau = \mathcal{F}R = I\alpha = (\tfrac{2}{5}MR^2)\alpha. \qquad (10\text{–}16)$$

If the ball rolls without slipping, we have the same kinematic relation as in Example 10–5:

$$a_{cm} = R\alpha.$$

We use this to eliminate $\alpha$ from Eq. (10–16):

$$\mathcal{F}R = \tfrac{2}{5}MRa_{cm}.$$

Then we solve Eq. (10–15) for $\mathcal{F}$, substitute the expression into this equation to eliminate $\mathcal{F}$, and then solve for $a_{cm}$ to obtain

$$a_{cm} = \tfrac{5}{7}g\sin\theta.$$

Finally, we substitute this back into Eq. (10–15) and solve for  to obtain

$$\mathcal{F} = \tfrac{2}{7}Mg\sin\theta.$$

The acceleration is just $\frac{5}{7}$ as large as it would be if the ball could *slide* without friction down the slope, like the toboggan in Example 5–9 (Section 5–2).

The friction force $\mathcal{F}$ is a *static* friction force; it is needed to prevent slipping and to give the ball its angular acceleration. We can derive an expression for the minimum coefficient of static friction $\mu_s$ needed to prevent slipping. The normal force is $n = Mg\cos\theta$. To prevent slipping, the coefficient of friction must be at least as great as

$$\mu_s = - = \frac{\frac{2}{7}Mg\sin\theta}{Mg\cos\theta} = \tfrac{2}{7}\tan\theta.$$

If the plane is tilted only a little, $\theta$ is small, and only a small value of $\mu_s$ is needed to prevent slipping. But as the angle increases, the required value of $\mu_s$ increases, as we would expect intuitively.

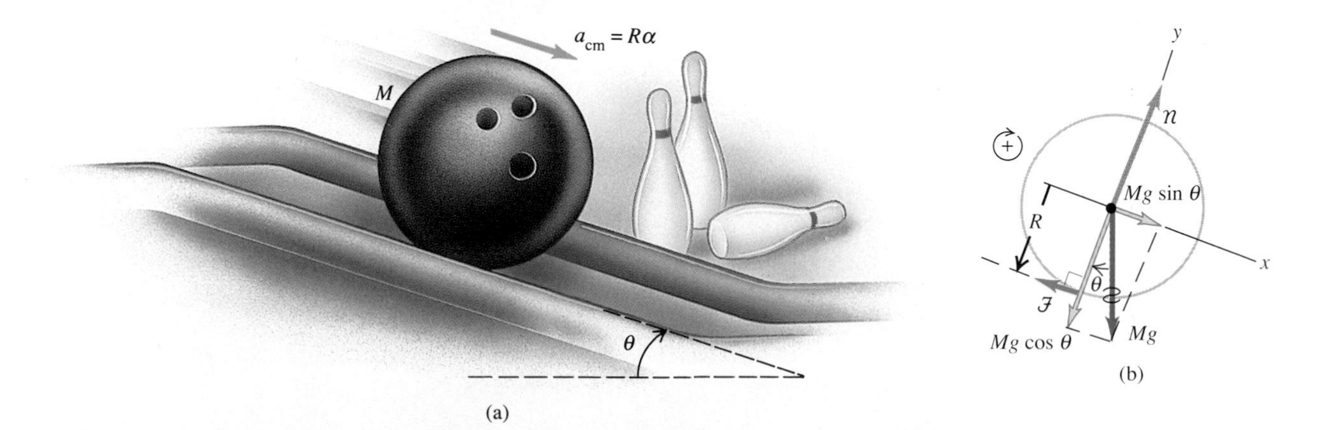


**10–11** (a) A bowling ball rolling down a ramp. (b) The free-body diagram for the bowling ball.

Finally, suppose $\mu_s$ is *not* large enough to prevent slipping. Then we have a whole new ball game, so to speak. Equations (10–15) and (10–16) are still valid, but now the ball slides and rotates at the same time. Thus there is no longer a definite relation between $a_{cm}$ and $\alpha$. In Eq. (10–16), $\mathcal{F}$ is now a *kinetic* friction force, given by $\mathcal{F} = \mu_k \mathcal{N} = \mu_k Mg \cos\theta$, and the center-of-mass acceleration is

$$a_{cm} = g(\sin\theta - \mu_k \cos\theta).$$

This is the same result that we found in Example 5–18 (Section 5–3), in which we hoped that the toboggan slides, not rolls, down the hill.

As we stated at the beginning of this section, every motion of a rigid body can be described as a combination of motion of the center of mass plus rotation about an axis through the center of mass. For example, Fig. 10–12 shows a wheel that rolls without slipping. The center-of-mass velocity is $\boldsymbol{v}_{cm}$, and the vectors around the rim, labeled $\boldsymbol{v}'$, are the velocities of points on the rim *relative to the center,* that is, the velocities associated with rotation about an axis through the center. The velocity of each point in the inertial frame in which we view this motion is the vector sum of these two velocities, as shown. Note that the point of the wheel in contact with the surface is instantaneously *at rest,* the point at the top of the wheel is moving forward *twice as fast* as the center of mass, and the points at the sides have velocities at 45° to the horizontal. At any instant we can think of the wheel as rotating about the point of contact with the ground, an "instantaneous axis" of rotation.

## Kinetic Energy of a Rigid Body

When a rigid body has both translational and rotational motion, the total kinetic energy is the sum of a part $\frac{1}{2}Mv_{cm}^2$ associated with motion of the center of mass, plus a part $\frac{1}{2}I_{cm}\omega^2$ associated with rotation about an axis through the center of mass:

$$K = \tfrac{1}{2}Mv_{cm}^2 + \tfrac{1}{2}I_{cm}\omega^2. \tag{10–17}$$

To prove this relation, we consider the velocity relations shown in Fig. 10–13. For a typical point with mass $m_i$ the velocity $\boldsymbol{v}_i$ relative to our inertial frame is the vector sum of the

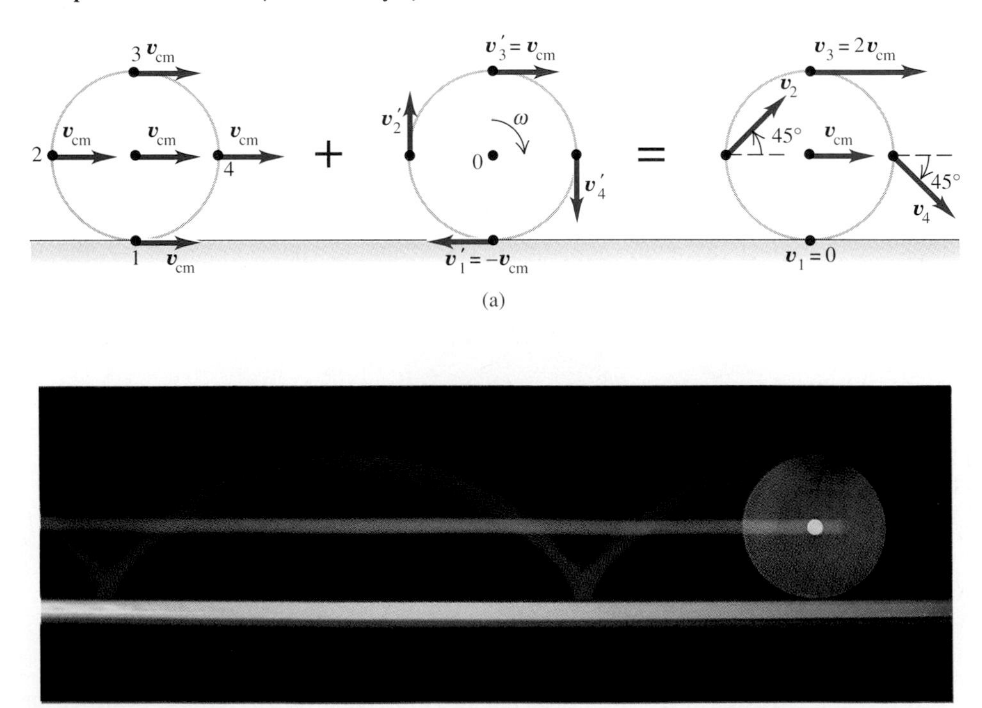


**10–12** (a) The velocity $\boldsymbol{v}$ of a point on the rim of a rolling wheel is the vector sum of the velocity $\boldsymbol{v}_{cm}$ of the center of mass and the velocity $\boldsymbol{v}'$ of the point relative to the center of mass due to the rotation of the wheel about the center of mass. The magnitude of $\boldsymbol{v}'$ is equal to $R\omega$. When the wheel rolls without slipping, $v_{cm} = R\omega$ and $v' = v_{cm}$. (b) The path traced out by a point on the rim of a wheel that rolls without slipping. This curve is called a cycloid.

of the velocity $\boldsymbol{v}_{cm}$ of the center of mass and the velocity $\boldsymbol{v}_i'$ of the particle *relative* to the center of mass:

$$\boldsymbol{v}_i = \boldsymbol{v}_{cm} + \boldsymbol{v}_i'. \tag{10–18}$$

The kinetic energy $K_i$ of this particle in the inertial frame is $\frac{1}{2}m_i v_i^2$, which we can also express as $\frac{1}{2}m_i(\boldsymbol{v}_i \cdot \boldsymbol{v}_i)$. Substituting Eq. (10–18) into this, we get

$$\begin{aligned} K_i &= \tfrac{1}{2}m_i(\boldsymbol{v}_{cm} + \boldsymbol{v}_i') \cdot (\boldsymbol{v}_{cm} + \boldsymbol{v}_i') \\ &= \tfrac{1}{2}m_i(\boldsymbol{v}_{cm} \cdot \boldsymbol{v}_{cm} + 2\boldsymbol{v}_{cm} \cdot \boldsymbol{v}_i' + \boldsymbol{v}_i' \cdot \boldsymbol{v}_i') \\ &= \tfrac{1}{2}m_i(v_{cm}^2 + 2\boldsymbol{v}_{cm} \cdot \boldsymbol{v}_i' + v_i'^2). \end{aligned}$$

The total kinetic energy is the sum $\Sigma K_i$ for all the particles making up the body. Expressing the three terms in this equation as separate sums, we get

$$K = \Sigma\tfrac{1}{2}m_i v_{cm}^2 + \Sigma m_i \boldsymbol{v}_{cm} \cdot \boldsymbol{v}_i' + \Sigma\tfrac{1}{2}m_i v_i'^2.$$

Each term has common factors that can be taken outside the sum:

$$K = \tfrac{1}{2}(\Sigma m_i)v_{cm}^2 + \boldsymbol{v}_{cm} \cdot (\Sigma m_i \boldsymbol{v}_i') + \Sigma\tfrac{1}{2}m_i v_i'^2. \tag{10–19}$$

Now comes the reward for our effort. In the first term, $\Sigma m_i$ is the total mass $M$. The second term is zero because $\Sigma m_i \boldsymbol{v}_i'$ is $M$ times the velocity of the center of mass *relative to the center of mass,* and this is zero by definition. The last term is the sum of kinetic energies of the particles computed by using their speeds with respect to the center of mass, and this is just $\frac{1}{2}I_{cm}\omega^2$. So Eq. (10–19) reduces to

$$K = \tfrac{1}{2}Mv_{cm}^2 + \tfrac{1}{2}I_{cm}\omega^2. \tag{10–20}$$

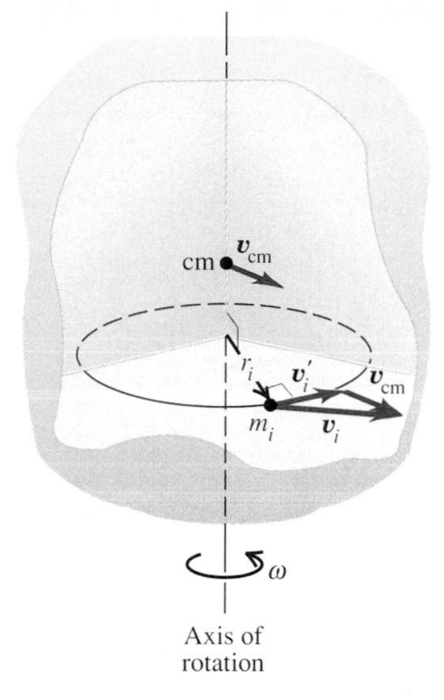


**10–13** A rigid body is rotating about an axis through its center of mass. For a typical particle, $v_i' = r_i\omega$, so $\frac{1}{2}m_i v_i'^2 = \frac{1}{2}(m_i r_i^2)\omega^2$.

## EXAMPLE 10–7

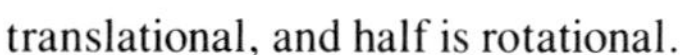

**A rolling cylindrical shell** A hollow cylindrical shell with mass $M$ and radius $R$ rolls without slipping with speed $v_{cm}$ on a flat surface. What is its kinetic energy?

**SOLUTION** The angular velocity $\omega$ is equal to $v_{cm}/R$, and the moment of inertia is $I = MR^2$. Substituting these expressions into Eq. (10–20), we get

$$\begin{aligned} K &= \tfrac{1}{2}Mv_{cm}^2 + \tfrac{1}{2}(MR^2)\left(\frac{v_{cm}}{R}\right)^2 \\ &= Mv_{cm}^2. \end{aligned}$$

The kinetic energy is twice as great as it would be if the cylinder were sliding without rolling; half of the total kinetic energy is translational, and half is rotational.

We can get this same result by considering rotation about the instantaneous axis at the line of contact of the cylinder with the surface. We use the parallel-axis theorem, Eq. (9–20), to find the moment of inertia about this axis:

$$I_O = I_{cm} + MR^2 = MR^2 + MR^2 = 2MR^2.$$

The angular velocity about this axis is $\omega = v_{cm}/R$, so the kinetic energy is

$$K = \tfrac{1}{2}I_O\omega^2 = \tfrac{1}{2}(2MR^2)\left(\frac{v_{cm}}{R}\right)^2 = Mv_{cm}^2,$$

which agrees with our other result.

## EXAMPLE 10–8

For the primitive yo-yo in Example 10–5, use energy considerations to find the speed $v_{cm}$ of the center of mass of the solid cylinder after it has dropped a distance $h$ (Fig. 10–14).

**SOLUTION** The potential energies are $U_1 = Mgh$ and $U_2 = 0$. The initial kinetic energy $K_1$ is zero, and the final kinetic energy $K_2$ is given by Eq. (10–20):

$$K_2 = \tfrac{1}{2}Mv_{cm}^2 + \tfrac{1}{2}I_{cm}\omega^2.$$

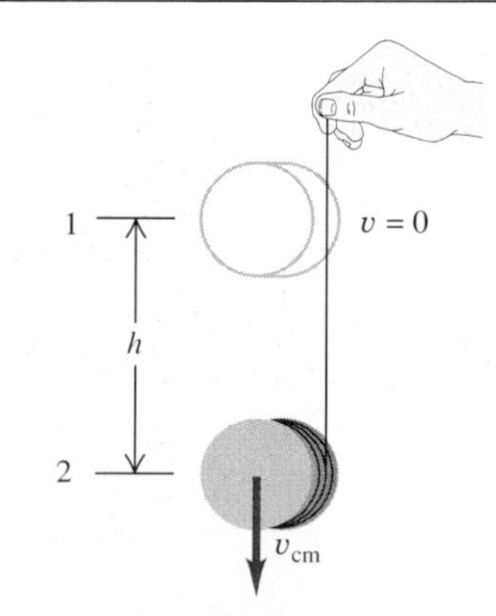


**10–14** Calculating the velocity of a primitive yo-yo.

But $\omega = v_{cm}/R$ from Eq. (10–13), and $I_{cm} = \frac{1}{2}MR^2$, so

$$K_2 = \tfrac{1}{2}(\tfrac{1}{2}MR^2)\left(\frac{v_{cm}}{R}\right)^2 + \tfrac{1}{2}Mv_{cm}{}^2$$
$$= \tfrac{3}{4}Mv_{cm}{}^2.$$

Finally, conservation of energy gives

$$K_1 + U_1 = K_2 + U_2,$$
$$0 + Mgh = \tfrac{3}{4}Mv_{cm}{}^2 + 0,$$

and

$$v_{cm} = \sqrt{\tfrac{4}{3}gh}.$$

We can also get this same result by using the constant-acceleration formula $v_{cm}{}^2 = v_0{}^2 + 2a_{cm}h$, with the acceleration from Example 10–5. Try it!

## EXAMPLE 10–9

**Race of the rolling bodies** In a physics lecture demonstration an instructor "races" various round objects by releasing them from rest at the top of an inclined plane (Fig. 10–15). The students guess (and sometimes bet on) which body will win. How can you predict the results and make your bets a sure thing?

**SOLUTION** Let's use conservation of energy, ignoring rolling friction and air drag. If the bodies roll without slipping, then no work is done by friction. Each body starts from rest at the top of an incline with height $h$, so $K_1 = 0$, $U_1 = Mgh$, and $U_2 = 0$. From Eq. (10–20),

$$K_2 = \tfrac{1}{2}Mv_{cm}{}^2 + \tfrac{1}{2}I_{cm}\omega^2.$$

If the bodies roll without slipping, $\omega = v_{cm}/R$. The moments of inertia of all the round bodies in Table 9–2 (about axes through their centers of mass) can be expressed as $I_{cm} = fMR^2$, where $f$ is a pure number between 0 and 1 that depends on the shape of the body. Then, from conservation of energy,

$$K_1 + U_1 = K_2 + U_2,$$
$$0 + Mgh = \tfrac{1}{2}Mv_{cm}{}^2 + \tfrac{1}{2}fMR^2(v_{cm}/R)^2$$
$$= \tfrac{1}{2}(1 + f)Mv_{cm}{}^2,$$

and we find

$$v_{cm} = \sqrt{\frac{2gh}{1 + f}}.$$

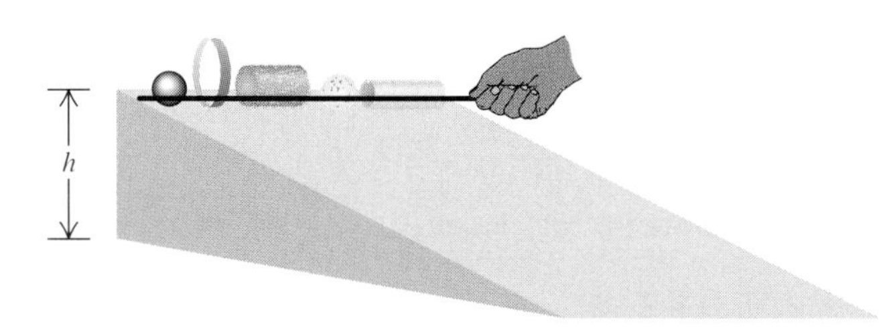


**10–15** Which object rolls down the plane fastest, and why?

This is a fairly amazing result; the velocity doesn't depend on either the mass $M$ of the body or its radius $R$. All uniform solid cylinders have the same speed at the bottom, even if their masses and radii are different, because they have the same $f$. All solid spheres have the same speed, and so on. The smaller the value of $f$, the faster the body is moving at the bottom (and at any point on the way down). Small-$f$ bodies always beat large-$f$ bodies because they have less kinetic energy tied up in rotation and have more available for translation. Reading the values of $f$ from Table 9–2, we see that the order of finish is as follows: any solid sphere, any hollow sphere, any solid cylinder, any hollow cylinder, and any thin cylindrical shell.

# 10–4 WORK AND POWER IN ROTATIONAL MOTION

When you pedal a bicycle, you apply forces to a rotating body and do work on it. Similar things happen in many real-life situations, such as a rotating motor shaft driving a power tool or a car engine propelling the vehicle. We can express this work in terms of torque and angular displacement.

Suppose a tangential force $\boldsymbol{F}_{tan}$ acts at the rim of a pivoted wheel of radius $R$ (Fig. 10–16), while the wheel rotates through an infinitesimal angle $d\theta$ about a fixed axis during an infinitesimal time interval $dt$. By definition, the work $dW$ done by the force $F_{tan}$ is

$dW = F_{\tan}\, ds$. If $\theta$ is measured in radians, $ds = R\, d\theta$, so

$$dW = F_{\tan} R\, d\theta.$$

Now $F_{\tan}R$ is the *torque* $\tau$ due to the force $F_{\tan}$, so

$$dW = \tau\, d\theta. \tag{10–21}$$

The total work $W$ done by the torque during an angular displacement from $\theta_1$ to $\theta_2$ is

$$W = \int_{\theta_1}^{\theta_2} \tau\, d\theta. \tag{10–22}$$

If the torque is *constant* while the angle changes by a finite amount $\Delta\theta = \theta_2 - \theta_1$, then

$$W = \tau(\theta_2 - \theta_1) = \tau\, \Delta\theta. \tag{10–23}$$

If $\tau$ is expressed in newton-meters (N · m) and $\theta$ in radians, the work is in joules. Equation (10–23) is the rotational analog of Eq. (6–1), $W = Fs$, and Eq. (10–22) is the analog of Eq. (6–4), $W = \int F\, dx$, for the work done by a force in a straight-line displacement.

The force in Fig. 10–16 has no axial or radial component. If there were such a component, it would do no work because the displacement of the point of application has no radial component. A radial or axial component of force would also make no contribution to the torque about the axis of rotation, so Eqs. (10–22) and (10–23) are correct for any force, even if it does have a radial or axial component.

When a torque does work on a rotating body, the kinetic energy changes by an amount equal to the work done. We can prove this by using exactly the same procedure that we used in Eqs. (6–13) through (6–16) for a particle. We transform the integrand in Eq. (10–22) into an integral on $\omega$ as follows:

$$\tau\, d\theta = (I\alpha)\, d\theta = I\frac{d\omega}{dt}\, d\theta = I\frac{d\theta}{dt}\, d\omega = I\omega\, d\omega.$$

The work integral, Eq. (10–22), then becomes

$$W = \int_{\omega_1}^{\omega_2} I\omega\, d\omega = \tfrac{1}{2}I\omega_2^{\,2} - \tfrac{1}{2}I\omega_1^{\,2}. \tag{10–24}$$

This equation is completely analogous to Eq. (6–16), the work-energy theorem for a particle.

What about the *power* associated with work done by a torque acting on a rotating body? When we divide both sides of Eq. (10–21) by the time interval $dt$ during which the angular displacement occurs, we find

$$\frac{dW}{dt} = \tau\frac{d\theta}{dt}.$$

But $dW/dt$ is the rate of doing work, or *power* $P$, and $d\theta/dt$ is angular velocity $\omega$, so

$$P = \tau\omega. \tag{10–25}$$

When a torque $\tau$ (with respect to the axis of rotation) acts on a body that rotates with angular velocity $\omega$, its power (rate of doing work) is the product of torque and angular velocity. This is the analog of the relation $P = \boldsymbol{F} \cdot \boldsymbol{v}$ that we developed in Section 6–5 for particle motion.

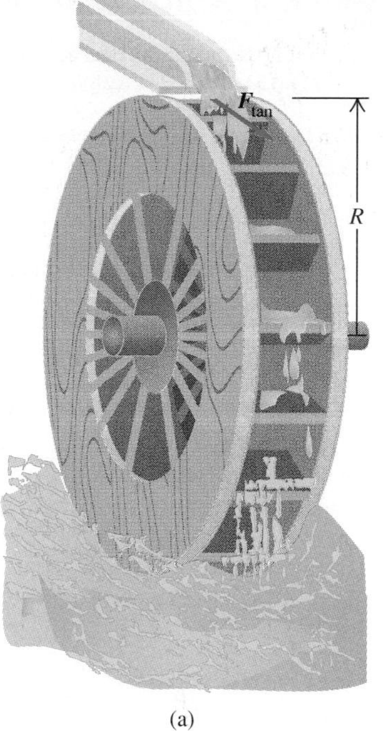


(a)

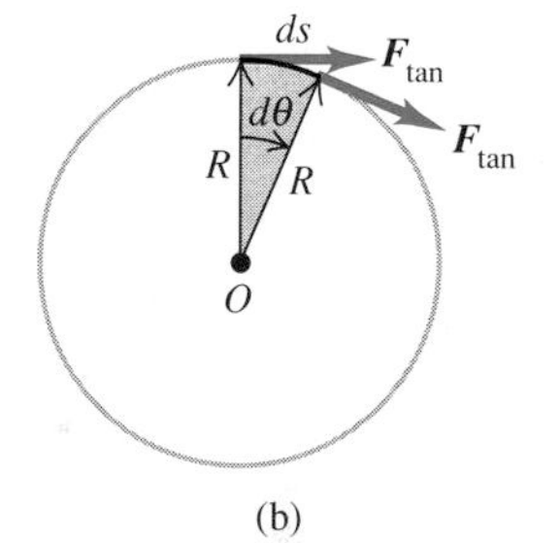


(b)

**10–16** A force acting on a rotating body does work.

### EXAMPLE 10–10

The power output of the engine of a Toyota Supra 6 is advertised to be 200 hp at 6000 rpm. What is the corresponding torque?

**SOLUTION** First we have to convert the power to watts and the angular velocity to rad/s:

$$P = 200 \text{ hp} = 200 \text{ hp}\left(\frac{746 \text{ W}}{1 \text{ hp}}\right) = 1.49 \times 10^5 \text{ W},$$

$$\omega = 6000 \text{ rev/min} = \left(\frac{6000 \text{ rev}}{1 \text{ min}}\right)\left(\frac{2\pi \text{ rad}}{1 \text{ rev}}\right)\left(\frac{1 \text{ min}}{60 \text{ s}}\right) = 200\pi \text{ rad/s}.$$

From Eq. (10–25),

$$\tau = \frac{P}{\omega} = \frac{1.49 \times 10^5 \text{ N}\cdot\text{m/s}}{200\pi \text{ rad/s}} = 237 \text{ N}\cdot\text{m}.$$

We could apply this much torque by using a wrench 1 m long and applying a force of 237 N (about 53 lb) to the end of its handle.

### EXAMPLE 10–11

An electric motor exerts a constant torque of $\tau = 10 \text{ N}\cdot\text{m}$ on a grindstone mounted on its shaft; the moment of inertia of the grindstone is $I = 2.0 \text{ kg}\cdot\text{m}^2$. If the system starts from rest, find the work done by the motor in 8.0 s and the kinetic energy at the end of this time. What was the average power delivered by the motor?

**SOLUTION** From $\tau = I\alpha$ the angular acceleration is 5.0 $\text{rad/s}^2$. The angular velocity after 8.0 s is

$$\omega = \alpha t = (5.0 \text{ rad/s}^2)(8.0 \text{ s}) = 40 \text{ rad/s}.$$

The kinetic energy at this time is

$$K = \tfrac{1}{2}I\omega^2 = \tfrac{1}{2}(2.0 \text{ kg}\cdot\text{m}^2)(40 \text{ rad/s})^2 = 1600 \text{ J}.$$

The total angle through which the system turns in 8.0 s is

$$\theta = \tfrac{1}{2}\alpha t^2 = \tfrac{1}{2}(5.0 \text{ rad/s}^2)(8.0 \text{ s})^2 = 160 \text{ rad},$$

and the total work done by the torque is

$$W = \tau\theta = (10 \text{ N}\cdot\text{m})(160 \text{ rad}) = 1600 \text{ J}.$$

This equals the total kinetic energy, as it must.

The average power is the total work divided by the time interval:

$$P_{\text{av}} = \frac{1600 \text{ J}}{8.0 \text{ s}} = 200 \text{ J/s} = 200 \text{ W}.$$

The instantaneous power, given by $P = \tau\omega$, is not constant because $\omega$ increases continuously. But we can compute the total work by taking the time integral of $P$, as follows:

$$\begin{aligned} W &= \int P\,dt = \int \tau\omega\,dt = \int \tau(\alpha t)dt \\ &= \int_0^{8.0 \text{ s}} (10 \text{ N}\cdot\text{m})(5 \text{ rad/s}^2)t\,dt = \left[(50 \text{ J/s}^2)\frac{t^2}{2}\right]_0^{8.0 \text{ s}} \\ &= 1600 \text{ J}, \end{aligned}$$

as we found previously. The instantaneous power $P$ increases from zero at the start to $(10 \text{ N}\cdot\text{m})(40 \text{ rad/s}) = 400 \text{ W}$ at time $t = 8.0$ s. The angular velocity and the power increase uniformly with time, so the *average* power is just half this maximum value, or 200 W.

## 10–5 ANGULAR MOMENTUM

Every rotational quantity that we have seen in Chapters 9 and 10 is the analog of some quantity of motion of a particle. The analog of *momentum* of a particle is **angular momentum,** a vector quantity denoted as $\boldsymbol{L}$. Its relation to momentum $\boldsymbol{p}$ (which we will sometimes call *linear momentum* for clarity) is exactly the same as the relation of torque $\boldsymbol{\tau} = \boldsymbol{r} \times \boldsymbol{F}$ to force $\boldsymbol{F}$. For a particle with constant mass $m$, velocity $\boldsymbol{v}$, momentum $\boldsymbol{p} = m\boldsymbol{v}$, and position vector $\boldsymbol{r}$ relative to the origin $O$ of an inertial frame, we define angular momentum $\boldsymbol{L}$ as

$$\boldsymbol{L} = \boldsymbol{r} \times \boldsymbol{p} = \boldsymbol{r} \times m\boldsymbol{v}. \tag{10–26}$$

The units of angular momentum are $\text{kg}\cdot\text{m}^2/\text{s}$.

When a force $\boldsymbol{F}$ acts on a particle, its velocity changes, so its angular momentum also changes. We can show that the *rate of change* of angular momentum is equal to the torque of the force. We take the time derivative of Eq. (10–26), using the rule for the derivative of a product:

$$\frac{d\boldsymbol{L}}{dt} = \frac{d\boldsymbol{r}}{dt} \times m\boldsymbol{v} + \boldsymbol{r} \times m\frac{d\boldsymbol{v}}{dt} = \boldsymbol{v} \times m\boldsymbol{v} + \boldsymbol{r} \times m\boldsymbol{a}.$$

The first term is zero because it contains the vector product of the vector $\boldsymbol{v}$ with itself. In the second term we replace $m\boldsymbol{a}$ with $\boldsymbol{F}$, obtaining

$$\frac{d\boldsymbol{L}}{dt} = \boldsymbol{r} \times \boldsymbol{F} = \boldsymbol{\tau}. \qquad (10\text{–}27)$$

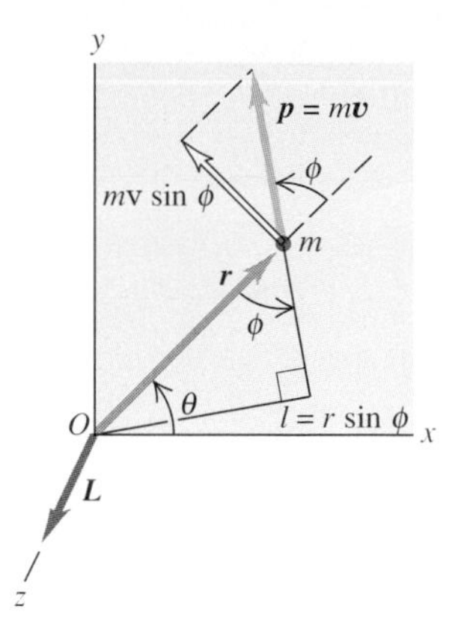


**10–17** A particle with mass $m$ moving in the $xy$-plane. The angular momentum of the particle is $\boldsymbol{L} = \boldsymbol{r} \times m\boldsymbol{v} = \boldsymbol{r} \times \boldsymbol{p}$, and its magnitude is $L = mvl$.

**The rate of change of angular momentum of a particle equals the torque of the net force acting on it.**

In Fig. 10–17 a particle moves in the $xy$-plane; its position vector $\boldsymbol{r}$ and momentum $\boldsymbol{p} = m\boldsymbol{v}$ are shown. The angular momentum vector $\boldsymbol{L}$ is perpendicular to the $xy$-plane. The right-hand rule for vector products shows that its direction is that of the $+z$-axis, and its magnitude is

$$L = mvr \sin\phi = mvl, \qquad (10\text{–}28)$$

where $l$ is the perpendicular distance from the line of $\boldsymbol{v}$ to $O$. This distance plays the role of "moment arm" for the momentum vector.

We can use Eq. (10–28) to find the total angular momentum of a rigid body rotating about the $z$-axis with angular velocity $\omega$. First consider a thin slice of the body lying in the $xy$-plane (Fig. 10–18). Each particle in the slice moves in a circle centered at the origin, and at each instant its velocity $\boldsymbol{v}_i$ is perpendicular to its position vector $\boldsymbol{r}_i$, as shown. In Eq. (10–28), $\phi = 90°$ for every particle. A particle with mass $m_i$, at a distance $r_i$ from $O$, has a speed $v_i$ equal to $r_i\omega$. From Eq. (10–28) the magnitude of its angular momentum $L_i$ is

$$L_i = m_i(r_i\omega)r_i = m_i r_i^2\omega. \qquad (10\text{–}29)$$

The direction of each particle's angular momentum, as given by the right-hand rule for the vector product, is along the $+z$-axis.

The *total* angular momentum of the slice of the body lying in the $xy$-plane is the sum $\Sigma L_i$ of the angular momenta $L_i$ of the particles. Summing Eq. (10–29), we have

$$L = \Sigma L_i = (\Sigma m_i r_i^2)\omega = I\omega.$$

We can do this same calculation for the other slices of the body, all parallel to the $xy$-plane. For points that do not lie in the $xy$-plane a complication arises because the $\boldsymbol{r}$ vectors have components in the $z$-direction as well as the $x$- and $y$-directions; this gives the angular momentum of each particle a component perpendicular to the $z$-axis. But *if the z-axis is an axis of symmetry,* these perpendicular components must add up to zero. Otherwise, we would have a symmetric body with a lopsided angular-momentum vector, and that wouldn't make physical sense. When a body rotates about an axis of symmetry, its angular momentum vector $\boldsymbol{L}$ has the direction of the symmetry axis, and its magnitude $L$ is given by

$$L = I\omega. \qquad (10\text{–}30)$$

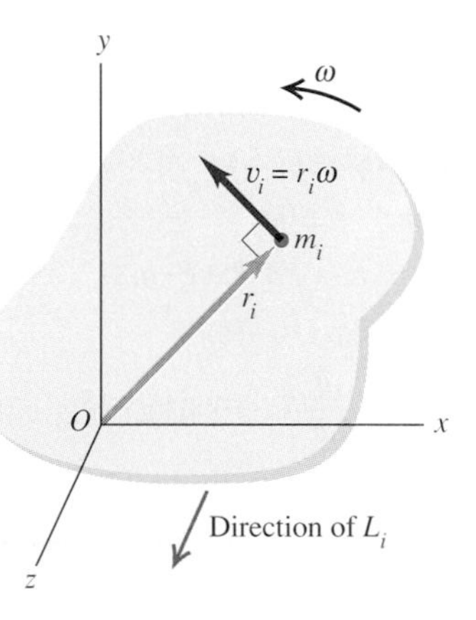


**10–18** Every particle in a rigid body rotates in a circle about the $z$-axis with the same angular velocity.

*Caution:* When the axis of rotation is *not* a symmetry axis, the angular momentum is in general *not* parallel to the axis. This is one of the reasons an unbalanced car wheel tends to wobble and vibrate. We won't consider this more general case further, but it is important in more advanced work in mechanics.

From Eq. (10–27) the rate of change of angular momentum of each particle equals the torque of the resultant force acting on the particle. For a collection of particles or a rigid body, the rate of change of the *total* angular momentum equals the sum of the torques of all forces acting on all the particles. The torques of the *internal* forces add to zero because of Newton's third law, so the sum of torques includes only the torques of the *external* forces. (A similar cancellation occurred in our discussion of center-of-mass motion in Section 8–7.) If the total angular momentum of the body is $\mathbf{L}$ and the sum of external torques is $\Sigma\,\boldsymbol{\tau}$, then

$$\Sigma\boldsymbol{\tau} = \frac{d\mathbf{L}}{dt}. \tag{10–31}$$

Finally, for a rigid body rotating about a symmetry axis, $L = I\omega$, and $I$ is constant. If the axis has a fixed direction in space, then the vector $\boldsymbol{\omega}$ changes only in magnitude, not in direction. In that case, $dL/dt = I\,d\omega/dt = I\alpha$, or

$$\Sigma\tau = I\alpha,$$

which is again our basic dynamical relation for rigid-body rotation. Note, however, that if the body is not rigid, $I$ may change, and in that case, $L$ changes even when $\omega$ is constant. In this case Eq. (10–31) is still valid, even when Eq. (10–6) is not.

In fixed-axis rotation we often call $L$ the angular momentum of the body, with a sign to indicate the sense of rotation, just as with angular velocity. This quantity is really the *component* of the vector angular momentum along the axis of rotation.

## EXAMPLE 10–12

A turbine fan (Fig. 10–19) has a moment of inertia of 2.5 kg · m² about its axis of rotation. Its angular velocity, a function of time, is

$$\omega = (400\ \text{rad/s}^3)t^2.$$

a) Find the fan's angular momentum as a function of time, and find its value at time $t = 2.0$ s. b) Find the net torque acting on the fan, as a function of time, and find the torque at time $t = 2.0$ s.

**10–19** A turbine fan is used in turbojet engines to force air through the engine at high speeds.

**SOLUTION** a) From Eq. (10–30),

$$L = I\omega = (2.5\ \text{kg}\cdot\text{m}^2)(400\ \text{rad/s}^3)t^2 = (1000\ \text{kg}\cdot\text{m}^2/\text{s}^3)t^2.$$

At time $t = 2.0$ s,

$$L = (1000\ \text{kg}\cdot\text{m}^2/\text{s}^3)(2.0\ \text{s})^2 = 4000\ \text{kg}\cdot\text{m}^2/\text{s}.$$

b) From Eq. (10–31),

$$\tau = \frac{dL}{dt} = (1000\ \text{kg}\cdot\text{m}^2/\text{s}^3)(2t) = (2000\ \text{kg}\cdot\text{m}^2/\text{s}^3)t.$$

At time $t = 2.0$ s,

$$\tau = (2000\ \text{kg}\cdot\text{m}^2/\text{s}^3)(2.0\ \text{s}) = 4000\ \text{kg}\cdot\text{m}^2/\text{s}^2 = 4000\ \text{N}\cdot\text{m}.$$

We have discussed angular momentum in the context of macroscopic bodies, but it is vitally important in physics at all scales, from nuclear to galactic. Angular momentum plays an essential role in classifying the quantum states of nuclei, atoms, and molecules and in describing their interactions. Angular-momentum considerations are also an important tool in analyzing the motions of stars, galaxies, and nebula. Along with mass, energy, and momentum, angular momentum is one of the unifying threads woven through the entire fabric of physical science.

## 10–6 CONSERVATION OF ANGULAR MOMENTUM

We have just seen that angular momentum can be used for an alternative statement of the basic dynamic principle for rotational motion. Angular momentum also forms the basis for a very important conservation principle, the principle of **conservation of angular momentum.** Like conservation of energy and of linear momentum, this principle appears to be a universal conservation law, valid at all scales from atomic and nuclear systems to the motions of galaxies.

This principle follows directly from Eq. (10–31): $\Sigma \boldsymbol{\tau} = d\boldsymbol{L}/dt$. **When the sum of torques of all the external forces acting on a system is zero, the total angular momentum of the system is constant (conserved).** We can generalize the concept of an **isolated system,** introduced in Section 8–2, to mean a system in which the total external force is zero *and* the total external torque is zero. Then we can restate the principle of conservation of angular momentum more simply: **The total angular momentum of an isolated system is constant.**

When a system has several parts, the internal forces the parts exert on each other cause changes in the angular momenta of the parts, but the *total* angular momentum doesn't change. Here's an example. When two bodies $A$ and $B$ interact with each other but not with anything else, $A$ and $B$ together form an *isolated system.* Suppose body $A$ exerts a force $\boldsymbol{F}_B$ on body $B$; the corresponding torque (with respect to whatever point we choose) is $\boldsymbol{\tau}_B$. According to Eq. (10–31), this torque is equal to the rate of change of angular momentum of $B$:

$$\boldsymbol{\tau}_B = \frac{d\boldsymbol{L}_B}{dt}.$$

At the same time, body $B$ exerts a force $\boldsymbol{F}_A$ on body $A$, with a corresponding torque $\boldsymbol{\tau}_A$, and

$$\boldsymbol{\tau}_A = \frac{d\boldsymbol{L}_A}{dt}.$$

From Newton's third law, $\boldsymbol{F}_A = -\boldsymbol{F}_B$. Furthermore, if the forces act along the same line, their moment arms with respect to the chosen axis are equal. Thus the *torques* of these two forces are equal and opposite, and $\boldsymbol{\tau}_A = -\boldsymbol{\tau}_B$. So if we add the two preceding equations, we find

$$\frac{d\boldsymbol{L}_A}{dt} + \frac{d\boldsymbol{L}_B}{dt} = \mathbf{0},$$

or, because $\boldsymbol{L}_A + \boldsymbol{L}_B$ is the *total* angular momentum $\boldsymbol{L}$ of the system,

$$\frac{d\boldsymbol{L}}{dt} = \mathbf{0}. \tag{10–32}$$

That is, the total angular momentum of the system is constant. The torques of the internal forces can transfer angular momentum from one body to the other, but they can't change the *total* angular momentum of the system.

**10–20** Conservation of angular momentum. No external torque acts on the acrobat. When he decreases his moment of inertia $I$ by going into a tuck position, his angular velocity $\omega$ increases, keeping his angular momentum $L$ constant.

A circus acrobat, a diver, and an ice skater performing a pirouette on the toe of one skate all take advantage of this principle. Suppose an acrobat has just left a swing (Fig. 10–20), with arms and legs extended and with a counterclockwise angular momentum. When he pulls his arms and legs in, his moment of inertia $I$ becomes much smaller. His angular momentum $L = I\omega$ remains constant and $I$ decreases, so his angular velocity $\omega$ increases. That is,

$$I_1\omega_1 = I_2\omega_2. \tag{10–33}$$

When a skater or ballerina spins with arms outstretched and then pulls them in, her angular velocity increases as her moment of inertia decreases. In each case we are seeing conservation of angular momentum in an isolated system.

## EXAMPLE 10–13

**Rotation of a neutron star** Astronomers believe that under some circumstances a star can collapse into an extremely dense structure made mostly of neutrons and called a *neutron star.* The density of neutron stars is roughly $10^{15}$ times as great as that of ordinary solid matter. Suppose we represent the star as a rigid sphere, both before and after the collapse. Initially, the star has a radius $R_1$ and is rotating with an angular velocity $\omega_1$. After the collapse it has a smaller radius $R_2$. For a star comparable to our sun, suppose $R_1 = 7.0 \times 10^8$ m and $R_2 = 1.6 \times 10^4$ m. If the original star rotates once per month, find the angular velocity of the neutron star.

**SOLUTION** There is no external torque acting on the star during this process. (What would exert it?) The angular momentum is therefore constant, and

$$I_1\omega_1 = I_2\omega_2.$$

For a sphere, $I = \frac{2}{5}MR^2$, so

$$\tfrac{2}{5}MR_1{}^2\omega_1 = \tfrac{2}{5}MR_2{}^2\omega_2,$$

$$\omega_2 = \left(\frac{R_1}{R_2}\right)^2 \omega_1.$$

One month is (30 d)(24 h/d)(3600 s/h) $= 2.59 \times 10^6$ s, so $\omega_1 =$ (1 rev)/($2.59 \times 10^6$ s) $= 3.9 \times 10^{-7}$ rev/s. Substituting this and the given values of $R_1$ and $R_2$, we find that the angular velocity $\omega_2$ after collapse is

$$\omega_2 = \left(\frac{7.0 \times 10^8 \text{ m}}{1.6 \times 10^4 \text{ m}}\right)^2 (3.9 \times 10^{-7} \text{ rev/s})$$
$$= 750 \text{ rev/s}.$$

Astronomers have discovered objects called *pulsars* that seem to spin as fast as this. Note that we didn't have to change "revolutions" to "radians" in this calculation. Why not?

## EXAMPLE 10–14

Figure 10–21 shows two disks, one an engine flywheel, the other a clutch plate attached to a transmission shaft. Their moments of inertia are $I_A$ and $I_B$; initially, they are rotating with constant angular velocities $\omega_A$ and $\omega_B$, respectively. We then push the disks together with forces acting along the axis, so as not to apply any torque on either disk. The disks rub against each other and eventually reach a common final angular velocity $\omega$. Derive an expression for $\omega$.

**SOLUTION** The two disks are an isolated system. The only torque acting on either disk is the torque applied by the other disk; there are no external torques. Thus the total angular momentum of

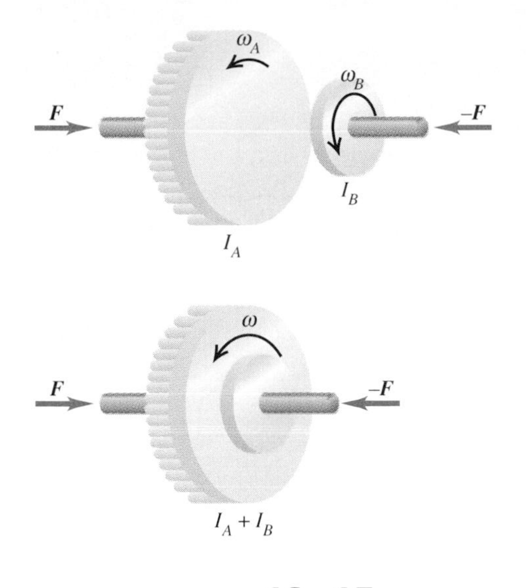


the system is the same before and after they are pushed together. At the end they rotate together as one body with total moment of inertia $I = I_A + I_B$ and angular velocity $\omega$. Conservation of angular momentum gives

$$I_A\omega_A + I_B\omega_B = (I_A + I_B)\omega,$$

$$\omega = \frac{I_A\omega_A + I_B\omega_B}{I_A + I_B}.$$

**10–21** When the net external torque is zero, angular momentum is conserved. The forces shown are along the axis of rotation and thus exert no torque about any point on it.

## EXAMPLE 10–15

Suppose that flywheel $A$ in Example 10–14 has a mass of 2.0 kg, a radius of 0.20 m, and an initial angular velocity of 50 rad/s (about 500 rpm) and that clutch plate $B$ has a mass of 4.0 kg, a radius of 0.10 m, and an initial angular velocity of 200 rad/s. Find the common final angular velocity $\omega$ after the disks are pushed into contact. Is kinetic energy conserved during this process?

**SOLUTION** The moments of inertia of the two disks are

$$I_A = \tfrac{1}{2}m_A r_A{}^2 = \tfrac{1}{2}(2.0\ \text{kg})(0.20\ \text{m})^2 = 0.040\ \text{kg}\cdot\text{m}^2,$$

$$I_B = \tfrac{1}{2}m_B r_B{}^2 = \tfrac{1}{2}(4.0\ \text{kg})(0.10\ \text{m})^2 = 0.020\ \text{kg}\cdot\text{m}^2.$$

From conservation of angular momentum we have

$$I_A\omega_A + I_B\omega_B = (I_A + I_B)\omega,$$

$$(0.040\ \text{kg}\cdot\text{m}^2)(50\ \text{rad/s}) + (0.020\ \text{kg}\cdot\text{m}^2)(200\ \text{rad/s}) = (0.040\ \text{kg}\cdot\text{m}^2 + 0.020\ \text{kg}\cdot\text{m}^2)\omega,$$

$$\omega = 100\ \text{rad/s}.$$

The initial kinetic energy is

$$K_1 = \tfrac{1}{2}I_A\omega_A{}^2 + \tfrac{1}{2}I_B\omega_B{}^2$$
$$= \tfrac{1}{2}(0.040\ \text{kg}\cdot\text{m}^2)(50\ \text{rad/s})^2 + \tfrac{1}{2}(0.020\ \text{kg}\cdot\text{m}^2)(200\ \text{rad/s})^2$$
$$= 450\ \text{J}.$$

The final kinetic energy is

$$K_2 = \tfrac{1}{2}(I_A + I_B)\omega^2$$
$$= \tfrac{1}{2}(0.040\ \text{kg}\cdot\text{m}^2 + 0.020\ \text{kg}\cdot\text{m}^2)(100\ \text{rad/s})^2 = 300\ \text{J}.$$

One third of the initial kinetic energy was lost during this "angular collision," the rotational analog of an inelastic collision. We shouldn't expect kinetic energy to be conserved, even though the resultant external force and torque are zero, because nonconservative (frictional) internal forces act while the two disks rub together as they gradually approach a common angular velocity.

## EXAMPLE 10–16

**Anyone can be a ballerina** An acrobatic physics professor stands at the center of a turntable, holding his arms extended horizontally, with a 5.0-kg dumbbell in each hand (Fig. 10–22). He is set rotating about a vertical axis, making one revolution in 2.0 s. Find the prof's new angular velocity if he drops his hands to his sides. His moment of inertia (without the dumbbells) is 3.0 kg · m² when his arms are outstretched, dropping to 2.2 kg · m² when his arms are at his sides. The dumbbells are 1.0 m from the axis initially and 0.20 m from it at the end.

**SOLUTION** If we neglect friction in the turntable, no external torques act about the vertical axis, and the angular momentum about this axis is constant:

$$I_1\omega_1 = I_2\omega_2,$$

**10–22** Fun with conservation of angular momentum—if you don't get dizzy.

where $I_1$ and $\omega_1$ are the initial moment of inertia and angular velocity and $I_2$ and $\omega_2$ are the final values.

In each case, $I = I_{\text{prof}} + I_{\text{dumb}}$.

$$I_1 = 3.0\ \text{kg}\cdot\text{m}^2 + 2(5.0\ \text{kg})(1.0\ \text{m})^2 = 13\ \text{kg}\cdot\text{m}^2,$$

$$I_2 = 2.2\ \text{kg}\cdot\text{m}^2 + 2(5.0\ \text{kg})(0.20\ \text{m})^2 = 2.6\ \text{kg}\cdot\text{m}^2,$$

$$\omega_1 = 2\pi\frac{1\ \text{rev}}{2.0\ \text{s}} = \pi\ \text{rad/s}.$$

From conservation of angular momentum,

$$(13\ \text{kg}\cdot\text{m}^2)(\pi\ \text{rad/s}) = (2.6\ \text{kg}\cdot\text{m}^2)\omega_2,$$

$$\omega_2 = 5.0\pi\ \text{rad/s} = 2.5\ \text{rev/s}.$$

That is, the angular velocity increases by a factor of five.

The initial kinetic energy is

$$K_1 = \tfrac{1}{2}I_1\omega_1^2 = \tfrac{1}{2}(13\ \text{kg}\cdot\text{m}^2)(\pi\ \text{rad/s})^2 = 64\ \text{J}.$$

The final kinetic energy is

$$K_2 = \tfrac{1}{2}I_2\omega_2^2 = \tfrac{1}{2}(2.6\ \text{kg}\cdot\text{m}^2)(5.0\ \pi\ \text{rad/s})^2 = 320\ \text{J}.$$

Where did the extra energy come from?

## EXAMPLE 10–17

**Angular momentum in a crime bust** A door 1.0 m wide, with a mass of 15 kg, is hinged at one side so that it can rotate without friction about a vertical axis. It is unlatched. A police detective fires a bullet with a mass of 10 g and a speed of 400 m/s into the exact center of the door in a direction perpendicular to the plane of the door (Fig. 10–23). Find the angular velocity of the door just after the bullet embeds itself in it. Is kinetic energy conserved?

**SOLUTION** We consider the door and bullet together as a system. There is no external torque about the axis defined by the hinges, so angular momentum about this axis is conserved. The initial angular momentum of the bullet is given by Eq. (10–28):

$$L = mvl = (0.010\ \text{kg})(400\ \text{m/s})(0.50\ \text{m}) = 2.0\ \text{kg}\cdot\text{m}^2/\text{s}.$$

This is equal to the final angular momentum $I\omega$, where $I = I_{\text{door}} + I_{\text{bullet}}$. From Table 9–2,

$$I_{\text{door}} = \frac{ML^2}{3} = \frac{(15\ \text{kg})(1.0\ \text{m})^2}{3} = 5.0\ \text{kg}\cdot\text{m}^2.$$

The moment of inertia of the bullet (with respect to the axis along the hinges) is

$$I_{\text{bullet}} = ml^2 = (0.010\ \text{kg})(0.50\ \text{m})^2 = 0.0025\ \text{kg}\cdot\text{m}^2.$$

Conservation of angular momentum requires that $mvl = I\omega$, or

$$2.0\ \text{kg}\cdot\text{m}^2/\text{s} = (5.0\ \text{kg}\cdot\text{m}^2 + 0.0025\ \text{kg}\cdot\text{m}^2)\omega,$$

$$\omega = 0.40\ \text{rad/s}.$$

The collision of bullet and door is inelastic because nonconservative forces act during the impact. Thus we do not expect kinetic energy to be conserved. To check, we calculate initial and final kinetic energies:

$$K_1 = \tfrac{1}{2}mv^2 = \tfrac{1}{2}(0.010\ \text{kg})(400\ \text{m/s})^2 = 800\ \text{J},$$

$$K_2 = \tfrac{1}{2}I\omega^2 = \tfrac{1}{2}(5.0025\ \text{kg}\cdot\text{m}^2)(0.40\ \text{rad/s})^2 = 0.40\ \text{J}.$$

The final kinetic energy is only $\frac{1}{2000}$ of the initial value!

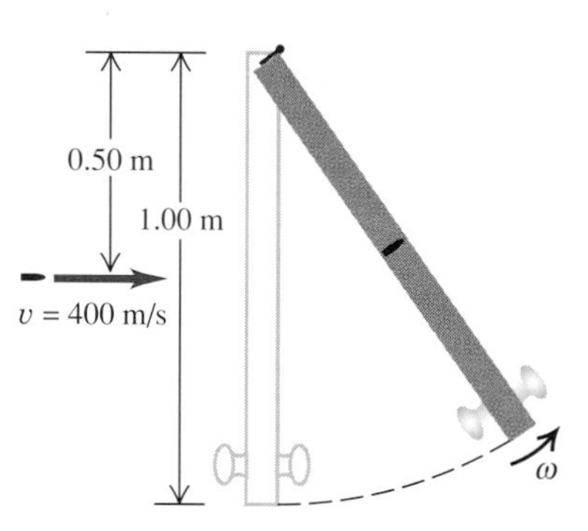


**10–23** Shooting open a door (top view). The bullet embeds itself in the center of the door.

## 10–7 GYROSCOPES

Angular velocity, angular momentum, and torque are all *vector* quantities. We discussed the vector nature of angular velocity $\boldsymbol{\omega}$ at the end of Section 9–1 and the vector nature of angular momentum $\boldsymbol{L}$ in Section 10–5. For a body rotating about a fixed axis that is an axis of symmetry the directions of these vector quantities are along the axis of rotation (Fig. 10–24). For this case we can take the $z$-axis to be the rotation axis; $\boldsymbol{\tau}$ and $\boldsymbol{L}$ have only $z$-components, and we have a rotational situation that is comparable to straight-line motion of a particle.

When the axis of rotation can change direction, the situation becomes much richer in the variety of physical phenomena (some quite unexpected) that can occur. For example, how can a spinning top stand up straight on its point and not fall over? How can the axle of a spinning gyroscope stay horizontal when it is supported only at one end?

General answers to these questions are quite complex, but we can go a little way into understanding this behavior.

A symmetric body turning about its axis of symmetry has a vector angular momentum $\boldsymbol{L}$ directed along its axis, parallel to the angular velocity vector. For such a body we derived in Section 10–5 the *vector* relationship

$$\boldsymbol{L} = I\boldsymbol{\omega}, \tag{10–34}$$

where $I$ is the moment of inertia about the axis of rotation. We also have the general relation between torque $\boldsymbol{\tau}$ and rate of change of angular momentum, given by Eq. (10–31):

$$\Sigma\,\boldsymbol{\tau} = \frac{d\boldsymbol{L}}{dt} \quad \text{or} \quad d\boldsymbol{L} = \Sigma\,\boldsymbol{\tau}\,dt. \tag{10–31}$$

In the problems we have worked out so far, $\Sigma\,\boldsymbol{\tau}$ and $\boldsymbol{L}$ have always had the same direction, so only one component of each vector quantity is different from zero. But Eq. (10–31) is valid even when $\Sigma\,\boldsymbol{\tau}$ and $\boldsymbol{L}$ have different directions. Then the *direction* of the axis of rotation may change, as with a spinning top or gyroscope, and we have to consider the vector nature of the various angular quantities very carefully.

An example of a three-dimensional application of Eq. (10–31) is the familiar toy gyroscope (Fig. 10–25). We set the flywheel spinning by wrapping a string around its shaft and pulling. When the shaft is supported at only one end, as shown in the figure, one possible motion is a steady circular motion of the axis in a horizontal plane, combined with the spin motion about the axis. This phenomenon is quite unexpected when you see it for the first time. Intuition suggests that the free end of the axis should simply drop if it isn't supported. If the flywheel isn't spinning, that *is* what happens. The vector nature of angular momentum is the key to understanding the horizontal motion of the axis when the wheel is spinning.

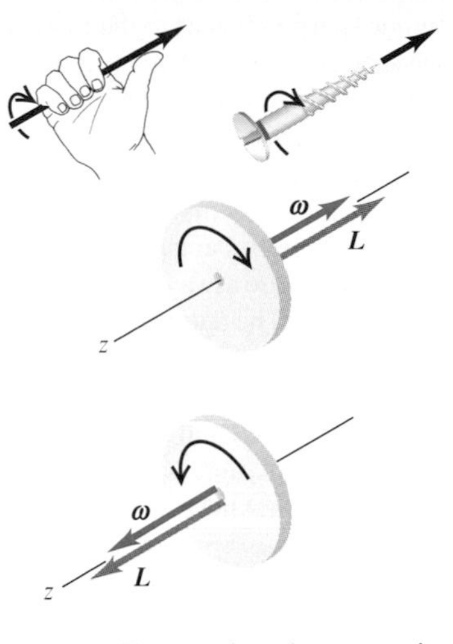


**10–24** For rotation about an axis of symmetry, $\boldsymbol{\omega}$ and $\boldsymbol{L}$ are along the axis; the directions are given by the same right-hand rule that we use to define the vector product.

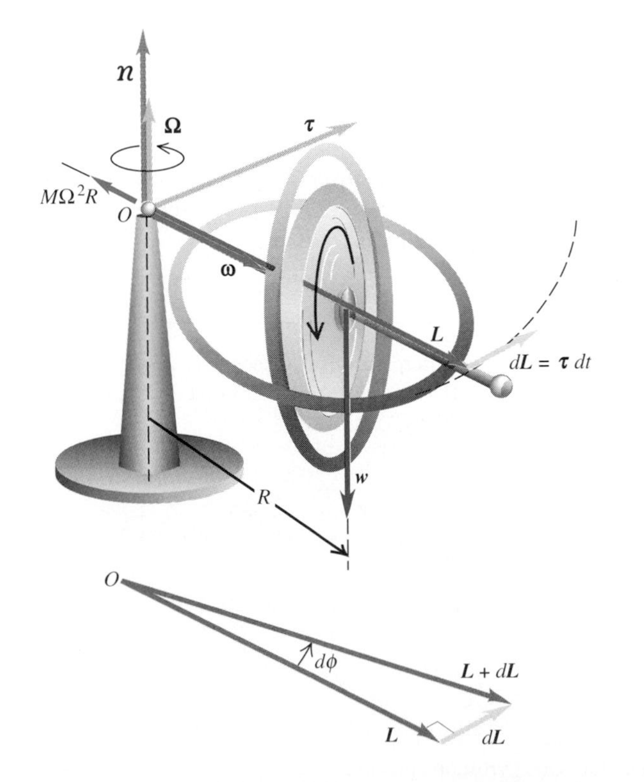


**10–25** Vector $d\boldsymbol{L}$ is the change in the angular momentum produced in the time $dt$ by the torque $\boldsymbol{\tau}$ of the weight $\boldsymbol{w}$. Vectors $d\boldsymbol{L}$ and $\boldsymbol{\tau}$ are in the same direction. Adding $d\boldsymbol{L}$ to $\boldsymbol{L}$ changes the direction, but not the magnitude, of $\boldsymbol{L}$ and gives precession about the vertical axis.

10–26 Looking down from directly over the gyroscope. In each successive time interval $dt$ the torque (perpendicular to the flywheel axis) provides a $d\mathbf{L}$. (a) If the flywheel has zero initial angular momentum, all the $d\mathbf{L}$'s are parallel, and the final angular momentum $\mathbf{L}_f$ has that direction. (b) With an initial angular momentum the $d\mathbf{L}$'s are always perpendicular to $\mathbf{L}$. The addition of the $d\mathbf{L}$'s changes the direction of $\mathbf{L}$, causing precession of the gyroscope. The initial and final angular momenta, $\mathbf{L}_i$ and $\mathbf{L}_f$, have the same magnitude but different directions.

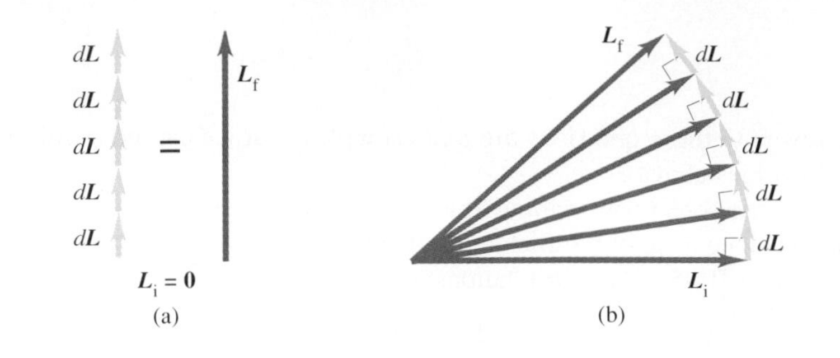


Two of the forces acting on the gyroscope are its weight $\mathbf{w}$, acting downward at the center of mass, and the normal force $\mathbf{n}$ at the pivot point $O$. If the flywheel is initially at rest, all the $d\mathbf{L}$'s caused by the constant-direction torque $\boldsymbol{\tau}$, as given by Eq. (10–31), are parallel (Fig. 10–26a). These $d\mathbf{L}$'s add to give an ever increasing horizontal angular momentum as the free end drops until the gyroscope hits the stand or table. But if the flywheel is initially rotating, its initial $\mathbf{L}$ is *along* the axis, while each $d\mathbf{L}$ is perpendicular to the axis (Fig. 10–26b). This causes the *direction* of $\mathbf{L}$ to change but not its *magnitude*. The situation is analogous to a particle moving in uniform circular motion. The centripetal force gives changes $d\mathbf{p}$ in linear momentum that are always perpendicular to the linear momentum $\mathbf{p}$ that already exists, changing its direction but not its magnitude.

In Fig. 10–25, at the instant shown, the gyroscope has a definite, nonzero angular momentum described by the vector $\mathbf{L}$. The *change* $d\mathbf{L}$ in angular momentum in a short time interval $dt$ following this is given by Eq. (10–31):

$$d\mathbf{L} = \boldsymbol{\tau}\, dt.$$

The direction of $d\mathbf{L}$ is the same as that of $\boldsymbol{\tau}$. After a time $dt$ the angular momentum is $\mathbf{L} + d\mathbf{L}$. As the vector diagram shows, this means that the gyroscope axis has turned through a small angle $d\phi$ given by $d\phi = |d\mathbf{L}|/|\mathbf{L}|$. We see that this motion of the axis is required by the torque–angular momentum relationship. This motion of the axis is called **precession.**

The rate at which the axis moves, $d\phi/dt$, is called the *precession angular velocity;* denoting this quantity by $\Omega$, we find

$$\Omega = \frac{d\phi}{dt} = \frac{|d\mathbf{L}|/|\mathbf{L}|}{dt} = \frac{\tau}{L} = \frac{wR}{I\omega}. \tag{10–35}$$

Thus the precession angular velocity is *inversely* proportional to the angular velocity of spin about the axis. A rapidly spinning gyroscope precesses slowly; as the flywheel slows down, the precession angular velocity *increases!*

## EXAMPLE 10–18

Figure 10–27 shows a top view of a cylindrical gyroscope wheel driven by an electric motor. The pivot is at $O$, and the mass of the axle and motor are negligible. a) Is the precession clockwise or counterclockwise, as seen from above? b) If the gyroscope takes 4.0 s for one revolution of precession, what is the wheel's angular velocity?

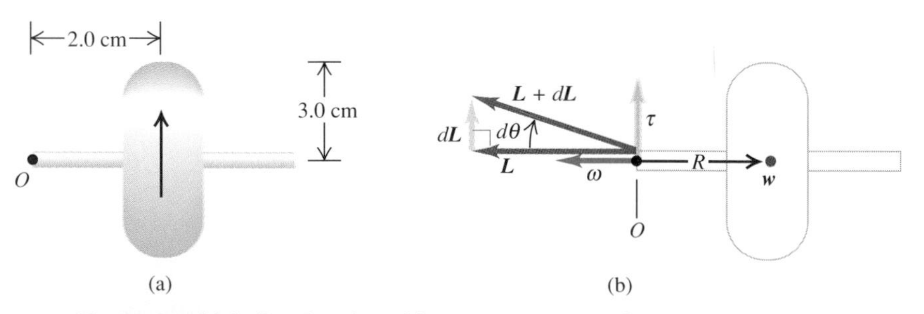


10–27 In which direction does this gyroscope precess?

**SOLUTION** a) The right-hand rule shows that $\boldsymbol{\omega}$ and $\boldsymbol{L}$ are to the left (Fig. 10–27b). The weight $\boldsymbol{w}$ points into the page, the torque $\boldsymbol{\tau} = \boldsymbol{R} \times \boldsymbol{w}$ is toward the top of the page, and $d\boldsymbol{L}/dt$ is also toward the top of the page. Adding a small $d\boldsymbol{L}$ to the $\boldsymbol{L}$ that we have at the start changes the direction of $\boldsymbol{L}$ as shown, so the precession is clockwise, as seen from above.

b) Be careful not to mix up $\omega$ and $\Omega$. We are given $\Omega = (1 \text{ rev})/(4.0 \text{ s}) = (2\pi \text{ rad})/(4.0 \text{ s}) = \pi/2$ rad/s. The weight is equal to $mg$, and the moment of inertia of a solid cylinder with radius $r$ is $I = \frac{1}{2}mr^2$. Solving Eq. (10–35) for $\omega$, we find

$$\omega = \frac{wR}{I\Omega} = \frac{mgR}{(mr^2/2)\Omega}$$

$$= \frac{2(9.8 \text{ m/s}^2)(2.0 \times 10^{-2}\text{m})}{(3.0 \times 10^{-2} \text{ m})^2(\pi/2 \text{ rad/s})} = 280 \text{ rad/s} = 2600 \text{ rev/min}.$$

Note that the mass of the flywheel divides out.

As the gyroscope precesses, the center of mass moves in a circle with radius $R$ in a horizontal plane. Its vertical component of acceleration is zero, so the pivot must exert an upward normal force $\boldsymbol{n}$ just equal in magnitude to the weight. The circular motion of the center of mass with angular velocity $\Omega$ requires a force $\boldsymbol{F}$ directed toward the center of the circle, with magnitude $F = M\Omega^2 R$. This force must also be supplied by the pivot (Fig. 10–25).

This analysis of the gyroscope gives us a glimpse of the richness of the dynamics of rotational motion, which can involve some very complex phenomena.

## SUMMARY

- When a force $\boldsymbol{F}$ acts on a body, the torque $\tau$ of that force with respect to a point $O$ is given by

$$\tau = Fl. \quad (10\text{–}1)$$

A common sign convention is "counterclockwise positive, clockwise negative." A generalized definition of torque as a vector quantity $\boldsymbol{\tau}$ is

$$\boldsymbol{\tau} = \boldsymbol{r} \times \boldsymbol{F}, \quad (10\text{–}3)$$

where $\boldsymbol{r}$ is the position vector of the point at which the force acts.

- The angular acceleration $\alpha$ of a rigid body rotating about a stationary axis is related to the total torque $\Sigma\,\tau$ and the moment of inertia $I$ of the body by

$$\Sigma\,\tau = I\alpha. \quad (10\text{–}6)$$

This relation is also valid for a moving axis, provided that it is an axis of symmetry that passes through the center of mass and does not change direction.

- When a torque $\tau$ acts on a rigid body that undergoes an angular displacement from $\theta_1$ to $\theta_2$, the work $W$ done by the torque is

$$W = \int_{\theta_1}^{\theta_2} \tau\, d\theta. \quad (10\text{–}22)$$

If the torque is *constant,* then

$$W = \tau(\theta_2 - \theta_1) = \tau\,\Delta\theta. \quad (10\text{–}23)$$

The work-energy theorem for rotational motion is

$$W = \tfrac{1}{2}I\omega_2^2 - \tfrac{1}{2}I\omega_1^2. \quad (10\text{–}24)$$

When the body rotates with angular velocity $\omega$, the power $P$ (rate at which the torque does work) is

$$P = \tau\omega. \quad (10\text{–}25)$$

### KEY TERMS

torque
moment arm
moment
angular momentum
conservation of angular momentum
isolated system
precession

• The angular momentum $\boldsymbol{L}$, with respect to point $O$, of a particle with mass $m$ and velocity $\boldsymbol{v}$ is

$$\boldsymbol{L} = \boldsymbol{r} \times \boldsymbol{p} = \boldsymbol{r} \times m\boldsymbol{v}, \qquad (10\text{–}26)$$

where $\boldsymbol{r}$ is the particle's position vector with respect to point $O$. When a rigid body with moment of inertia $I$ rotates with angular velocity $\omega$ about a stationary axis, its component of angular momentum along that axis is given by

$$L = I\omega. \qquad (10\text{–}30)$$

In terms of vector torque $\boldsymbol{\tau}$ and angular momentum $\boldsymbol{L}$ the basic dynamic relation for rotational motion can be restated as

$$\Sigma \boldsymbol{\tau} = \frac{d\boldsymbol{L}}{dt}. \qquad (10\text{–}31)$$

• If a system consists of bodies that interact with each other but not with anything else, or if the total torque associated with the external forces is zero, the total angular momentum of the system is constant (conserved).

# EXERCISES

## Section 10–1 Torque

**10–1** Calculate the torque (magnitude and direction) about point $O$ due to the force $\boldsymbol{F}$ in each of the situations sketched in Fig. 10–28. In each case the object to which the force is applied has length 4.00 m, and $F = 20.0$ N.

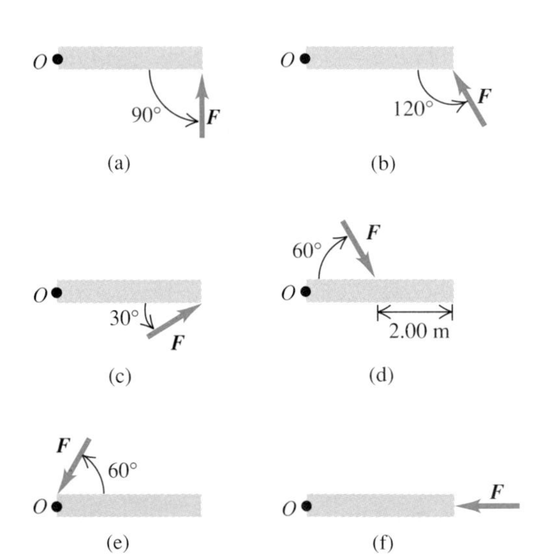


FIGURE 10–28

**10–2** Calculate the resultant torque about point $O$ for the two forces applied as in Fig. 10–29.

**10–3** A square metal plate 0.180 m on each side is pivoted about an axis through point $O$ at its center and perpendicular to the plate

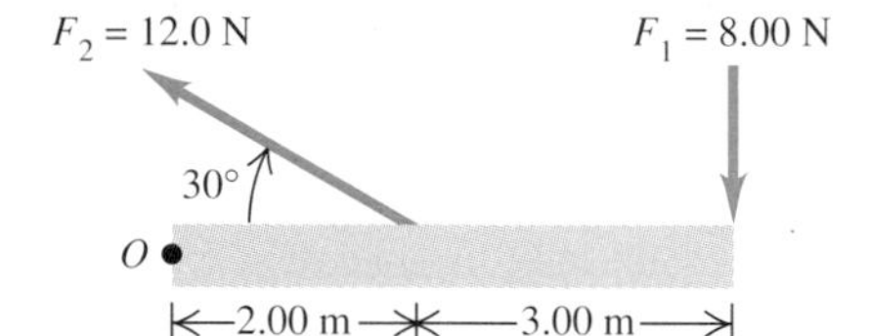

FIGURE 10–29

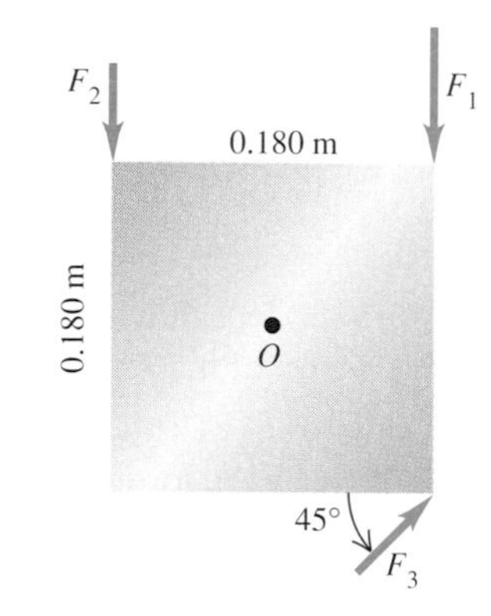


FIGURE 10–30

(Fig. 10–30). Calculate the net torque about this axis due to the three forces shown in the figure if the magnitudes of the forces are $F_1 = 24.0$ N, $F_2 = 16.0$ N, and $F_3 = 18.0$ N.

**10–4** Forces $F_1 = 8.60$ N and $F_2 = 3.80$ N are applied tangentially to a wheel with radius 0.330 m, as shown in Fig. 10–31. What is the net torque on the wheel due to these two forces for an axis perpendicular to the wheel and passing through its center?

**10–5** A force $\boldsymbol{F}$ is given by $\boldsymbol{F} = (4.00 \text{ N})\boldsymbol{i} - (5.00 \text{ N})\boldsymbol{j}$, and the vector $\boldsymbol{r}$ from the axis to point of application of the force is

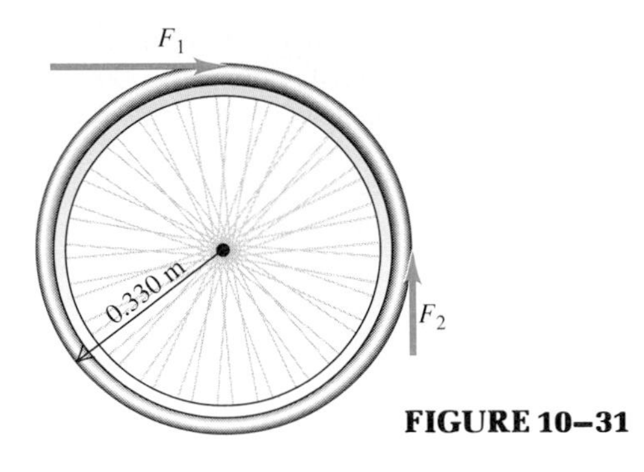


FIGURE 10–31

$\boldsymbol{r} = (-0.300\ \text{m})\boldsymbol{i} + (0.500\ \text{m})\boldsymbol{j}$. Calculate the vector torque produced by this force.

## Section 10–2
## Torque and Angular Acceleration

**10–6** The flywheel of an engine has moment of inertia 3.50 $\text{kg}\cdot\text{m}^2$. a) What constant torque is required to bring it up to an angular velocity of 900 rev/min in 10.0 s, starting from rest? b) What is its final kinetic energy?

**10–7** A 5.00-kg block rests on a frictionless horizontal surface. A cord attached to the block passes over a pulley whose diameter is 0.120 m to a hanging block also of mass 5.00 kg. The system is released from rest, and the blocks are observed to move 3.00 m in 2.00 s. a) What is the tension in each part of the cord? b) What is the moment of inertia of the pulley?

**10–8** A cord is wrapped around the rim of a flywheel 0.400 m in radius, and a steady pull of 50.0 N is exerted on the cord. The wheel is mounted in frictionless bearings on a horizontal shaft through its center. The moment of inertia of the wheel is 4.00 $\text{kg}\cdot\text{m}^2$. Compute the angular acceleration of the wheel.

**10–9** A grindstone in the shape of a solid disk with a diameter of 0.600 m and a mass of 50.0 kg is rotating at 900 rev/min. You press an ax against the rim with a normal force of 160 N (Fig. 10–32), and the grindstone comes to rest in 10.0 s. Find the coefficient of friction between the ax and the grindstone. Neglect friction in the bearings.

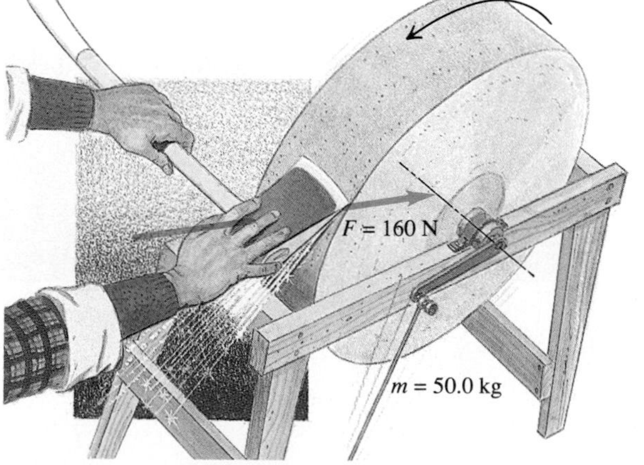

FIGURE 10–32

**10–10** A thin rod with length $l$ and mass $M$ is pivoted about a vertical axis at one end. A force with constant magnitude $F$ is applied to the other end, causing the rod to rotate. The force is maintained perpendicular to the rod and to the axis of rotation. Calculate the angular acceleration $\alpha$ of the rod.

**10–11 A Dropping Bucket.** A bucket of water of mass 20.0 kg is suspended by a rope wrapped around a windlass in the form of a solid cylinder 0.300 m in diameter, also of mass 20.0 kg. The cylinder is pivoted on a frictionless axle through its center. The bucket is released from rest at the top of a well and falls 12.0 m to the water. Neglect the weight of the rope. a) What is the tension in the rope while the bucket is falling? b) With what speed does the bucket strike the water? c) What is the time of fall? d) While the bucket is falling, what is the force exerted on the cylinder by the axle?

## Section 10–3
## Rotation about a Moving Axis

**10–12** A solid cylinder of mass 4.00 kg rolls, without slipping, down a 34.0° slope. Find the acceleration, the friction force, and the minimum coefficient of friction needed to prevent slipping.

**10–13** A string is wrapped several times around the rim of a small hoop of radius 0.0800 m and mass 0.140 kg. If the free end of the string is held in place and the hoop is released from rest (Fig. 10–33), calculate a) the tension in the string while the hoop descends as the string unwinds; b) the time it takes the hoop to descend 0.600 m; c) the angular velocity of the rotating hoop after it has descended 0.600 m.

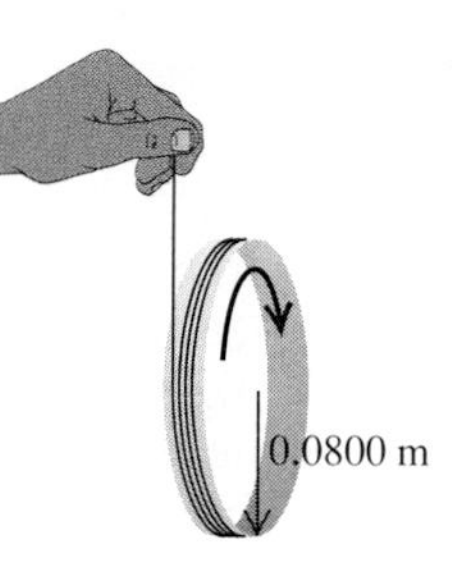


FIGURE 10–33

**10–14** Repeat part (c) of Exercise 10–13, this time using energy considerations.

## Section 10–4
## Work and Power in Rotational Motion

**10–15** A grindstone in the form of a solid cylinder has a radius of 0.200 m and a mass of 30.0 kg. a) What torque will bring it from rest to an angular velocity of 300 rev/min in 10.0 s? b) Through what angle has it turned during that time? c) Use Eq. (10–23) to calculate the work done by the torque. d) What is the grindstone's kinetic energy when it is rotating at 300 rev/min? Compare your answer to the result in part (c).

**10–16** What is the power output in horsepower of an electric motor turning at 3600 rev/min and developing a torque of 3.25 N · m?

**10–17** A playground merry-go-round has radius 4.40 m and moment of inertia 160 kg · $m^2$ and turns with negligible friction about a vertical axle through its center. a) A child applies a 25.0-N force tangentially to the edge of the merry-go-round for 20.0 s. If the merry-go-round is initially at rest, what is its angular velocity after this 20.0-s interval? b) How much work did the child do on the merry-go-round? c) What is the average power supplied by the child?

**10–18** The flywheel of a motor has mass 30.0 kg and moment of inertia 67.5 kg · $m^2$. The motor develops a constant torque of 150 N · m, and the flywheel starts from rest. a) What is the angular acceleration of the flywheel? b) What is its angular velocity after making 4.00 revolutions? c) How much work is done by the motor during the first 4.00 revolutions?

**10–19** a) Compute the torque developed by an airplane engine whose output is $1.80 \times 10^6$ W at an angular velocity of 2400 rev/min. b) If a drum of negligible mass, 0.500 m in diameter, is attached to the motor shaft and the power output of the motor is used to raise a weight hanging from a rope wrapped around the drum, how large a weight can be lifted? (Assume that the weight is lifted at constant speed.) c) With what speed will it rise?

## Section 10–5 Angular Momentum

**10–20** Calculate the angular momentum of a uniform sphere of radius 0.160 m and mass 4.00 kg if it is rotating about an axis along a diameter at 6.00 rad/s.

**10–21** What is the angular momentum of the second hand on a clock about an axis through the center of the clock face if the clock hand has a length of 25.0 cm and a mass of 20.0 g? Take the second hand to be a slender rod rotating with constant angular velocity about one end.

**10–22** A woman of mass 70.0 kg is standing on the rim of a large disk that is rotating at 0.600 rev/s about an axis through its center. The disk has a mass of 120 kg and a radius of 4.00 m. Calculate the total angular momentum of the woman-plus-disk system. (Assume that the woman can be treated as a point.)

**10–23** A 0.600-kg rock is thrown with speed $v = 12.0$ m/s. When it is at point $P$ in Fig. 10–34, what is its angular momentum relative to point $O$? Assume that the rock travels in a straight line with constant speed.

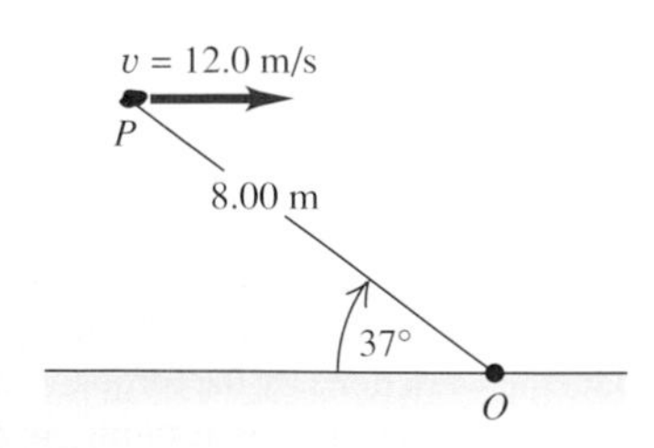


FIGURE 10–34

**10–24** A merry-go-round starts from rest. The merry-go-round is in the shape of a disk of radius 3.00 m, and the moment of inertia about its rotation axis is 600 kg · $m^2$. a) What resultant torque must be applied to increase its angular velocity at a constant rate of $d\omega/dt = 0.300$ rad/$s^2$? b) If friction at the axis can be neglected, what magnitude of force applied tangentially produces this torque?

## Section 10–6 Conservation of Angular Momentum

**10–25** A small block on a frictionless horizontal surface has a mass of 0.0500 kg and is attached to a cord passing through a hole in the surface (Fig. 10–35). The block is originally revolving at a distance of 0.200 m from the hole with an angular velocity of 1.50 rad/s. The cord is then pulled from below, shortening the radius of the circle in which the block revolves to 0.100 m. The block may be considered a point mass. a) What is the new angular velocity? b) Find the change in kinetic energy of the block. c) How much work was done in pulling the cord?

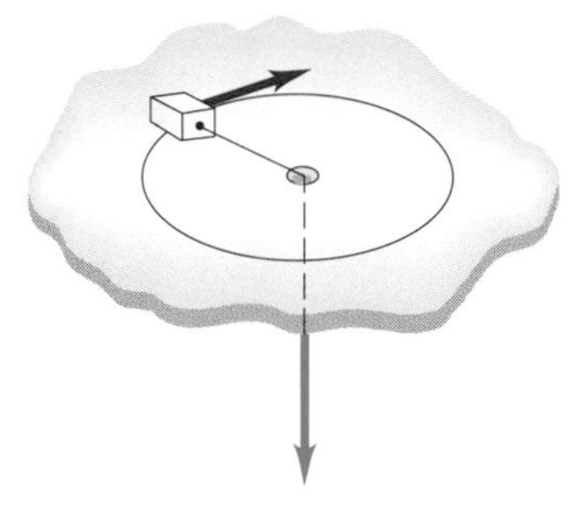

FIGURE 10–35

**10–26 The Spinning Figure Skater.** The outstretched arms of a figure skater preparing for a spin can be considered to be a slender rod pivoting about an axis through its center (Fig. 10–36). When her arms are brought in and wrapped around her body to execute the spin, they can be considered to be a thin-walled hollow cylinder. Her arms have a combined mass of 8.00

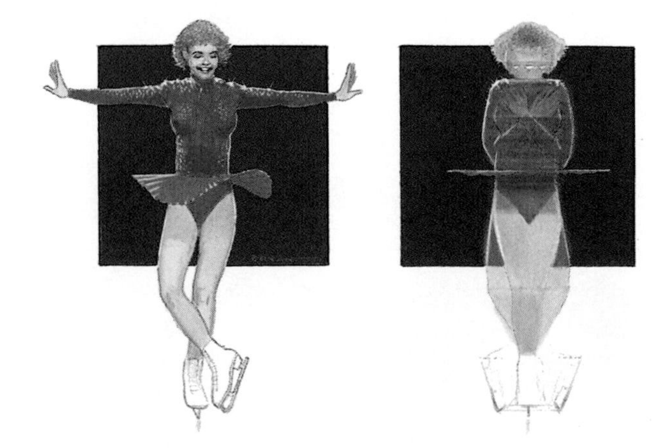

FIGURE 10–36

kg. When outstretched, they span 1.80 m; when wrapped, they form a cylinder of radius 25.0 cm. The moment of inertia of the remainder of her body is constant and equal to $3.00 \text{ kg} \cdot \text{m}^2$. If her original angular velocity is 0.600 rev/s, what is her final angular velocity?

**10–27** A solid wood door 1.00 m wide and 2.00 m high is hinged along one side and has a total mass of 50.0 kg. Initially open and at rest, the door is struck at its center by a handful of sticky mud of mass 0.500 kg, traveling 8.00 m/s just before impact. Find the final angular velocity of the door. Is the moment of inertia of the mud significant?

**10–28** A large turntable rotates about a fixed vertical axis, making one revolution in 8.00 s. The moment of inertia of the turntable about this axis is $1200 \text{ kg} \cdot \text{m}^2$. A man of mass 80.0 kg, initially standing at the center of the turntable, runs out along a radius. What is the angular velocity of the turntable when the man is 2.00 m from the center? (Assume that the man can be treated as a point.)

**10–29** A large wooden turntable in the shape of a flat disk has a radius of 2.00 m and a total mass of 120 kg. The turntable is initially rotating about a vertical axis through its center with an angular velocity of 1.50 rad/s. From a very small height a 100-kg bag of sand is dropped vertically onto the turntable at a point near the outer edge. a) Find the angular velocity of the turntable after the bag is dropped. (Assume that the bag of sand can be treated as a point.) b) Compute the kinetic energy of the system before and after the bag is dropped. Why are these kinetic energies not equal?

### Section 10–7 Gyroscopes

**10–30** The rotor of a toy gyroscope has a mass of 0.150 kg, and its moment of inertia about its axis is $1.50 \times 10^{-4} \text{ kg} \cdot \text{m}^2$. The mass of the frame is 0.0300 kg. The gyroscope is supported on a single pivot (Fig. 10–37) with its center of mass a horizontal distance of 4.00 cm from the pivot. The gyroscope is precessing in a horizontal plane at the rate of one revolution in 4.00 s. a) Find the upward force exerted by the pivot. b) Find the angular velocity with which the rotor is spinning about its axis, expressed in rev/min. c) Copy the diagram, and show by vectors the angular momentum of the rotor and the torque acting on it.

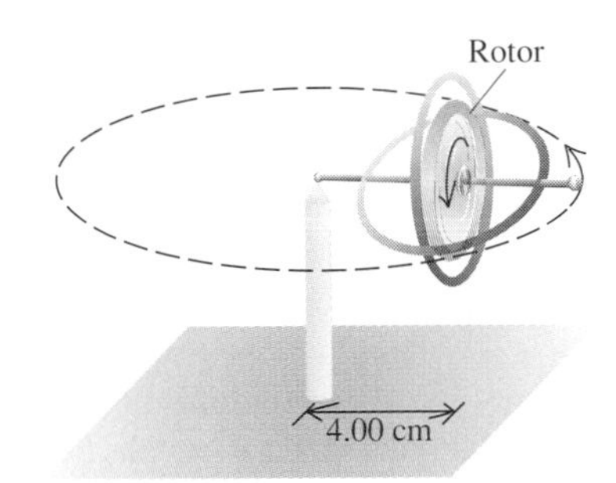


**FIGURE 10–37**

**10–31 A Gyro Stabilizer.** The stabilizing gyroscope of a ship is a solid disk of mass 50,000 kg; its radius is 2.00 m, and it rotates about a vertical axis with an angular velocity of 800 rev/min. a) How much time is required to bring it up to speed, starting from rest, with a constant power input of $7.46 \times 10^4$ W? b) Find the torque needed to cause the axis to precess in a vertical fore-and-aft plane at the rate of 1.00°/s.

## PROBLEMS

**10–32** A 60.0-kg grindstone is 0.600 m in diameter and has a moment of inertia of $4.50 \text{ kg} \cdot \text{m}^2$. You press a knife down on the rim with a normal force of 50.0 N. The coefficient of kinetic friction between the blade and the stone is 0.60, and there is a constant friction torque of $5.00 \text{ N} \cdot \text{m}$ between the axle of the stone and its bearings. a) How much force must be applied tangentially at the end of a crank handle 0.500 m long to bring the stone from rest to 120 rev/min in 9.00 s? b) After the grindstone attains an angular velocity of 120 rev/min, what tangential force at the end of the handle is needed to maintain a constant angular velocity of 120 rev/min? c) How much time does it take the grindstone to go from 120 rev/min to rest if it is acted on by the axle friction alone?

**10–33** A constant torque equal to $20.0 \text{ N} \cdot \text{m}$ is exerted on a pivoted wheel for 8.00 s, during which time the angular velocity of the wheel increases from zero to 100 rev/min. The external torque is then removed, and the wheel is brought to rest by friction in its bearings in 100 s. Compute a) the moment of inertia of the wheel; b) the friction torque; c) the total number of revolutions made by the wheel in the 108-s time interval.

**10–34** A flywheel 1.00 m in diameter is pivoted on a horizontal axis. A rope is wrapped around the outside of the flywheel, and a steady pull of 60.0 N is exerted on the rope. The flywheel starts from rest, and 8.00 m of rope are unwound in 4.00 s. a) What is the angular acceleration of the flywheel? b) What is its final angular velocity? c) What is its final kinetic energy? d) What is its moment of inertia?

**10–35** Dirk the Dragonslayer is exploring a castle. He is spotted by a dragon, which chases him down a hallway. Dirk runs into a room and attempts to swing the heavy door shut before the dragon gets him. The door is initially perpendicular to the wall, so it must be turned through 90° to close. The door is 3.00 m tall and 1.00 m wide, and it weighs 700 N. The friction at the hinges can be neglected. If Dirk applies a force of 180 N at the edge of the door and perpendicular to it, how much time does it take him to close the door?

**10–36** A large 15.0-kg roll of wrapping paper of radius $R = 12.0$ cm rests against the wall and is held in place by a bracket

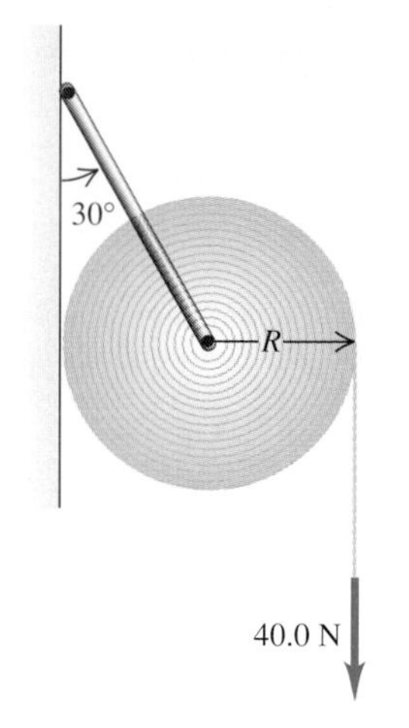


**FIGURE 10–38**

attached to a rod through the center of the roll (Fig. 10–38). The rod turns without friction in the bracket, and the moment of inertia of the paper and rod about the axis is $0.120\ \text{kg}\cdot\text{m}^2$. The coefficient of kinetic friction between the paper and the wall is $\mu_k = 0.20$. A constant vertical force $F = 40.0$ N is applied to the paper, and the paper unrolls. a) What is the magnitude of the force that the rod exerts on the paper as it unrolls? b) What is the angular acceleration of the roll?

**10–37** The mechanism shown in Fig. 10–39 is used to raise a crate of supplies from a ship's hold. The crate has a total mass of 50.0 kg. A rope is wrapped around a wooden cylinder that turns on a metal axle. The cylinder has radius 0.250 m and moment of inertia $I = 0.920\ \text{kg}\cdot\text{m}^2$ about the axle. The crate is suspended from the free end of the rope. One end of the axle is pivoted on frictionless bearings; a crank handle is attached to the other end. When the crank is turned, the end of the handle rotates about the axle in a vertical circle of radius 0.12 m, the cylinder turns, and the crate is raised. What magnitude of the force $\boldsymbol{F}$ applied tangentially to the rotating crank is required to raise the crate with an acceleration of $1.20\ \text{m/s}^2$? (The moment of inertia of the axle and of the crank and the mass of the rope can be neglected.)

**10–38** Two metal disks, one with radius $R_1 = 4.00$ cm and mass $M_1 = 0.80$ kg and the other with radius $R_2 = 8.00$ cm and mass $M_2 = 1.60$ kg, are welded together and mounted on a frictionless axis through their common center, as in Problem 9–51. a) A light string is wrapped around the edge of the smaller disk, and a 1.50-kg block is suspended from the free end of the string. What is the magnitude of the downward acceleration of the block after it is released? b) Repeat the calculation of part (a), this time with the string wrapped around the edge of the larger disk. In which case is the acceleration of the block the greatest, and does your answer make sense?

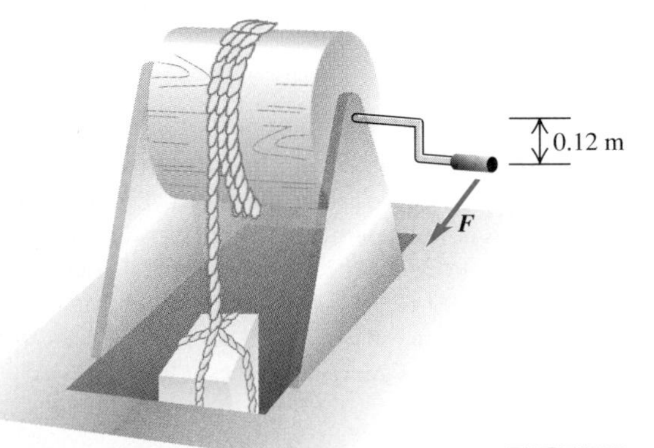


**FIGURE 10–39**

**10–39** A block with mass $m = 5.00$ kg slides down a surface inclined 37.0° to the horizontal (Fig. 10–40). The coefficient of kinetic friction is 0.20. A string attached to the block is wrapped around a flywheel on a fixed axis at $O$. The flywheel has a mass $M = 20.0$ kg, an outer radius $R = 0.200$ m, and a moment of inertia with respect to the axis of $0.300\ \text{kg}\cdot\text{m}^2$. a) What is the acceleration of the block down the plane? b) What is the tension in the string?

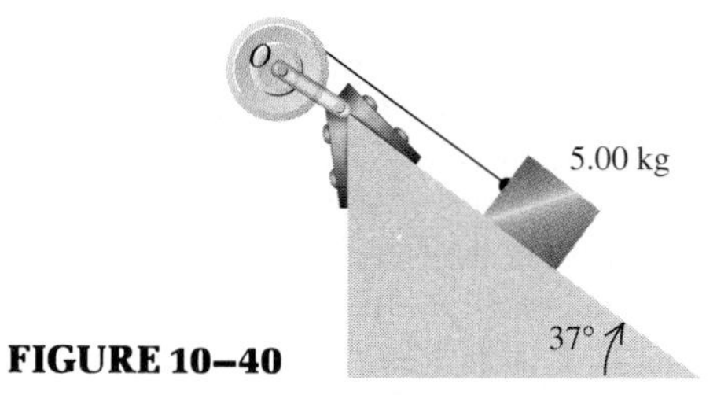


**FIGURE 10–40**

**10–40 Atwood's Machine.** Figure 10–41 represents an Atwood's machine. Find the linear accelerations of blocks $A$ and $B$, the angular acceleration of the wheel $C$, and the tension in each side of the cord if there is no slipping between the cord and the surface of the wheel. Let the masses of blocks $A$ and $B$ be 5.00 kg and 2.00 kg, respectively, the moment of inertia of the wheel about its axis be $0.200\ \text{kg}\cdot\text{m}^2$, and the radius of the wheel be 0.100 m.

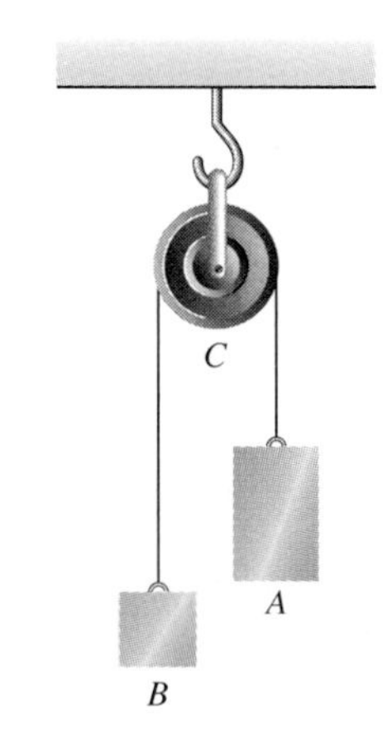


**FIGURE 10–41**

**10–41** A lawn roller in the form of a thin-walled hollow cylinder of mass $M$ is pulled horizontally with a constant force $F$ applied by a handle attached to the axle. If it rolls without slipping, find the acceleration and the friction force.

**10–42 The Yo-yo.** A yo-yo is made from two uniform disks, each with mass $m$ and radius $R$, connected by a light axle of radius $b$. A string is wound several times around the axle and then held stationary while the yo-yo is released from rest, dropping as the string unwinds. Find the acceleration of the yo-yo and the tension in the string.

**10–43** A solid disk is rolling without slipping on a level surface at a constant speed of 4.00 m/s. How far can it roll up a 30.0° ramp before it stops?

**10–44** A solid wood door 1.00 m wide and 2.00 m high is hinged along one side and has a total mass of 40.0 kg. Initially open and at rest, the door is struck with a hammer at its center. During the blow an average force of 2000 N acts for $1.00 \times 10^{-2}$ s. Find the angular velocity of the door after the impact. [*Hint:* Integrating Eq. (10–31) yields $\Delta L = \int_{t_1}^{t_2} (\Sigma\tau)\, dt = (\Sigma\tau)_{av}\, \Delta t$. The quantity $\int_{t_1}^{t_2} (\Sigma\tau)\, dt$ is called the angular impulse.]

**10–45** A uniform solid cylinder of mass $M$ and radius $2R$ rests on a horizontal tabletop. A string is attached by a yoke to a frictionless axle through the center of the cylinder so that the cylinder can rotate about the axle. The string runs over a pulley in the shape of a disk of mass $M$ and radius $R$ that is mounted on a frictionless axle through its center. A block of mass $M$ is suspended from the free end of the string (Fig. 10–42). The string doesn't slip over the pulley surface, and the cylinder rolls without slipping on the tabletop. After the system is released from rest, what is the magnitude of the downward acceleration of the block?

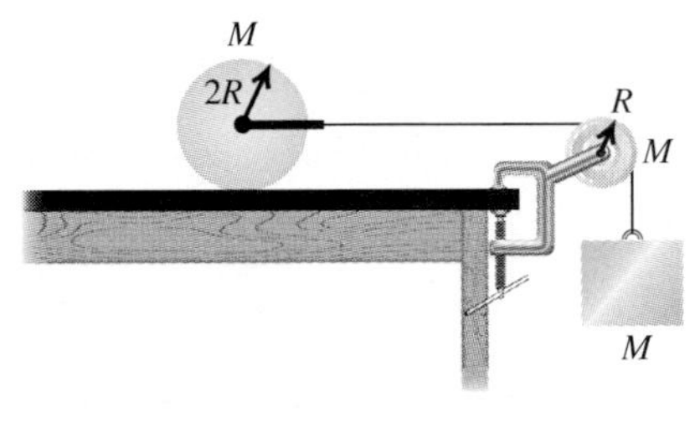


**FIGURE 10–42**

**10–46** A uniform rod of mass 0.0300 kg and length 0.400 m rotates in a horizontal plane about a fixed axis through its center and perpendicular to the rod. Two small rings, each of mass 0.0200 kg, are mounted so that they can slide along the rod. They are initially held by catches at positions 0.0500 m on each side of the center of the rod, and the system is rotating at 30.0 rev/min. Without otherwise changing the system the catches are released, and the rings slide outward along the rod and fly off at the ends. a) What is the angular velocity of the system at the instant when the rings reach the ends of the rod? b) What is the angular velocity of the rod after the rings leave it?

**10–47** Disks $A$ and $B$ are mounted on a shaft $SS$ and may be connected or disconnected by a clutch $C$ (Fig. 10–43). The moment of inertia of disk $A$ is half that of disk $B$. With the clutch disconnected, $A$ is brought up to an angular velocity $\omega_0$. The accelerating torque is then removed from $A$, and $A$ is coupled to disk $B$ by the clutch. (Bearing friction may be neglected.) It is found that 4000 J of heat are developed in the clutch when the connection is made. What was the original kinetic energy of disk $A$?

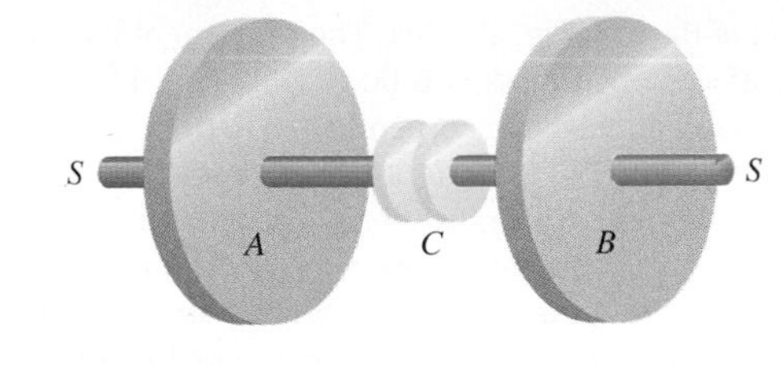

**FIGURE 10–43**

**10–48** A small block of mass 0.400 kg is attached to a cord passing through a hole in a frictionless horizontal surface (see Fig. 10–35). The block is originally revolving in a circle with a radius of 0.500 m about the hole with a tangential speed of 4.00 m/s. The cord is then pulled slowly from below, shortening the radius of the circle in which the block revolves. The breaking strength of the cord is 60.0 N. What is the radius of the circle when the cord breaks?

**10–49 Hitting a Bull's-eye.** A target in a shooting gallery consists of a vertical square wooden board, 0.200 m on a side and of mass 2.20 kg, that is pivoted on an axis along its top edge. It is struck at the center by a bullet of mass 5.00 g that is traveling at 300 m/s. a) What is the angular velocity of the board just after the bullet's impact? b) What maximum height above the equilibrium position does the center of the board reach before starting to swing down again? c) What bullet speed would be required for the board to swing all the way over after impact?

**10–50** A 60.0-kg runner runs around the edge of a horizontal turntable mounted on a vertical frictionless axis through its center. The runner's velocity relative to the earth has magnitude 4.00 m/s. The turntable is rotating in the opposite direction with an angular velocity of magnitude 0.200 rad/s relative to the earth. The radius of the turntable is 2.00 m, and its moment of inertia about the axis of rotation is 400 kg · m$^2$. Find the final angular velocity of the system if the runner comes to rest relative to the turntable. (The runner may be treated as a point.)

**10–51 A falling bicycle.** The moment of inertia of the front wheel of a bicycle about its axle is 0.340 kg · m$^2$, the wheel's radius is 0.360 m, and the forward speed of the bicycle is 5.00 m/s. With what angular velocity must the wheel be turned about a vertical axis to counteract the capsizing torque due to a mass of 60.0 kg located 0.040 m horizontally to the right or left of the line of contact of wheels and ground? (Bicycle riders, compare your own experience and decide whether your answer seems reasonable.)

**10–52 Center of Percussion.** A baseball bat rests on a horizontal frictionless surface. The bat has a length of 0.900 m and a mass of 0.900 kg, and its center of mass is 0.600 m from the handle end of the bat (Fig. 10–44). The moment of inertia of the bat about its center of mass is $6.00 \times 10^{-2}\ \text{kg} \cdot \text{m}^2$. The bat is struck by a baseball traveling perpendicular to the bat. The impact applies an impulse $J = \int_{t_1}^{t_2} F\,dt$ at a point a distance $x$ from the handle end of the bat. What must $x$ be so that the handle end of the bat remains at rest as the bat begins to move? [*Hint:* Consider the motion of the center of mass and the rotation about the center of mass. Find $x$ so that these two motions combine to give $v = 0$ for the end of the bat just after the collision. Also, note that integration of Eq. (10–31) gives $\Delta L = \int_{t_1}^{t_2} (\Sigma\tau)\,dt$ (Problem 10–44). The point on the bat that you are locating is called the center of percussion. Hitting a pitched ball at the center of percussion of the bat minimizes the "sting" the batter experiences on the hands.]

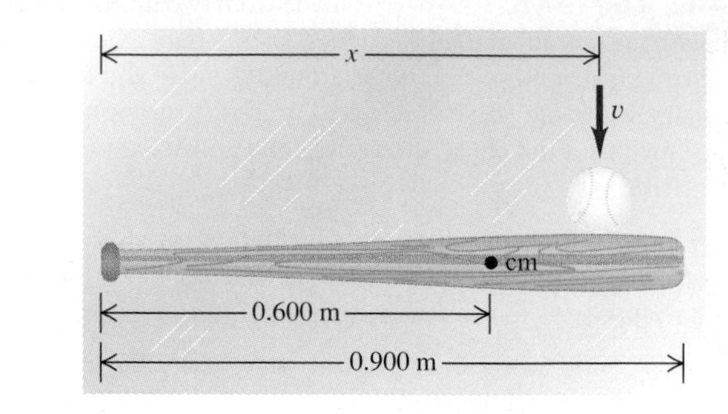


**FIGURE 10–44**

## CHALLENGE PROBLEMS

**10–53** A block with mass $m$ is revolving with linear speed $v_1$ in a circle of radius $r_1$ on a frictionless horizontal surface (see Fig. 10–35). The string is slowly pulled from below until the radius of the circle in which the block is revolving is reduced to $r_2$. a) Calculate the tension $T$ in the string as a function of $r$, the distance of the block from the hole. Your answer will be in terms of the initial velocity $v_1$ and the radius $r_1$. b) Use $W = \int_{r_1}^{r_2} \mathbf{T}(r) \cdot d\mathbf{r}$ to calculate the work done by $T$ when $r$ changes from $r_1$ to $r_2$. c) Compare the results of part (b) to the change in the kinetic energy of the block.

**10–54 The Bicycle Wheel Gyroscope.** A demonstration gyroscope wheel is constructed by removing the tire from a bicycle wheel 0.700 m in diameter, wrapping lead wire around the rim, and taping it in place. The shaft projects 0.200 m at each side of the wheel, and a man holds the ends of the shaft in his hands. The mass of the system is 6.00 kg; its entire mass may be assumed to be located at its rim. The shaft is horizontal, and the wheel is spinning about the shaft at 5.00 rev/s. Find the magnitude and direction of the force each hand exerts on the shaft a) when the shaft is at rest; b) when the shaft is rotating in a horizontal plane about its center at 0.040 rev/s; c) when the shaft is rotating in a horizontal plane about its center at 0.200 rev/s. d) At what rate must the shaft rotate in order that it may be supported at one end only?

**10–55** When an object is rolling without slipping, the rolling friction force is much less than the friction force when the object is sliding; a silver dollar will roll on its edge much farther than it will slide on its flat side. (See Section 5–3.) When an object is rolling without slipping on a horizontal surface, we can take the friction force to be zero, so $a$ and $\alpha$ are approximately zero and $v$ and $\omega$ are approximately constant. Rolling without slipping means $v = r\omega$ and $a = r\alpha$. If an object is set into motion on a surface *without* these equalities, sliding (kinetic) friction will act on the object as it slips until rolling without slipping is established. A solid cylinder with mass $M$ and radius $R$, rotating with angular velocity $\omega_0$ about an axis through its center, is set on a horizontal surface for which the kinetic friction coefficient is $\mu_k$. a) Draw a free-body diagram for the cylinder on the surface. Think carefully about the direction of the kinetic friction force on the cylinder. Calculate the accelerations $a$ of the center of mass and $\alpha$ of rotation about the center of mass. b) The cylinder is initially slipping completely, as initially $\omega = \omega_0$ but $v = 0$. Rolling without slipping sets in when $v = R\omega$. Calculate the *distance* the cylinder rolls before slipping stops. c) Calculate the work done by the friction force on the cylinder as it moves from where it was set down to where it begins to roll without slipping.

CHAPTER 11

# Equilibrium and Elasticity

Gothic cathedrals, such as this one (Notre Dame de Paris), use delicate-looking flying buttresses to support the upper walls and roof. They provide stability against compressive forces due to the downward weight and outward pressure of the walls. However, they are vulnerable to tensile (stretching) forces caused by strong winds.

## KEY CONCEPTS

**• For a rigid body to be in equilibrium, the vector sum of forces acting on it must be zero, and the sum of the torques of these forces with respect to any specified point must be zero.**

**• The torque due to the weight of a body can be computed by assuming that the entire weight acts at the center of mass (center of gravity).**

**• The deformation of an elastic material is often approximately proportional to the forces causing the deformation. The concepts of stress and strain are used to describe the force and deformation in a general way. The ratio of stress to strain is called an elastic modulus. The three kinds of deformation are tensile (or compressive), volume change, and shear.**

## INTRODUCTION

Try standing facing a wall, with your toes directly against the wall. Now try standing on tiptoe. What happens? You start falling over backward. Why? The vector sum of forces on you is zero, yet you're not in equilibrium. Similar problems occur with a suspension bridge, a ladder leaning against a wall, the walls of a building, or a crane hoisting a bucket full of concrete.

A body that can be modeled as a *particle* is in equilibrium whenever the vector sum of the forces acting on it is zero. But for the problems that we've just described, that condition isn't enough. Many bodies can't be adequately modeled as particles. The vector sum of the forces on such a body must still be zero, but an additional requirement must be satisfied to ensure that the body has no tendency to *rotate*. This requirement is based on the principles of rotational dynamics developed in Chapter 10. For a rigid body in equilibrium, the sum of the *torques* about any point must be zero. We can compute the torque due to the weight of a body using the concept of center of mass from Section 8–6 and the related concept of center of gravity, which we introduce in this chapter.

Rigid bodies don't bend, stretch, or squash when forces act on them. But the rigid body is an idealized model; all real materials do deform to some extent. In this chapter we will introduce the concepts of stress, strain, and elastic modulus and a simple principle that helps us predict what deformations will occur when forces are applied to a real (not perfectly rigid) body. Elastic properties of materials are tremendously important. You want the wings of an airplane to be able to bend a little, but you'd rather not have them break off. The steel frame of an earthquake-resistant building has to be able to flex, but not too much. Many of the necessities of everyday life, from rubber bands to suspension bridges, depend on the elastic properties of materials.

## 11–1 CONDITIONS FOR EQUILIBRIUM

We learned in Section 5–1 that a particle acted on by several forces is in equilibrium (in an inertial frame of reference) if the vector sum of the forces acting on the particle is zero, $\Sigma \boldsymbol{F} = \boldsymbol{0}$. This is often called the **first condition for equilibrium.** In terms of components,

$$\Sigma F_x = 0, \qquad \Sigma F_y = 0, \qquad \Sigma F_z = 0. \tag{11–1}$$

For a rigid body there is a **second condition for equilibrium:**

**The sum of the torques due to all forces acting on the body, with respect to any specified point, must be zero.**

This condition is needed so that the body has no tendency to begin to *rotate*.

The second condition for equilibrium is based on the dynamics of rotational motion in exactly the same way that the first condition is based on Newton's first law. A rigid body at rest has zero angular momentum, $\boldsymbol{L} = \boldsymbol{0}$. For it to *remain* at rest, the rate of change of angular momentum $d\boldsymbol{L}/dt$ must also be zero. From the discussion in Section 10–5, particularly Eq. (10–31), this means that the sum of torques $\Sigma \tau$ of all the forces acting on the body must be zero. Furthermore, because every point in the body is at rest in an inertial frame, any point can be used as an origin. Therefore the sum of torques about *any* point must be zero. The body doesn't actually have to be pivoted about an axis through the chosen point. If a rigid body is in equilibrium, it can't have any tendency to begin to rotate about *any* axis. Thus the sum of torques must be zero, *no matter what point is chosen*. That is,

$$\Sigma \tau = 0 \quad \textit{about any point}. \tag{11–2}$$

Although the choice of reference point is arbitrary, once we choose it we must use the *same* point to calculate *all* of the torques. An important element of problem-solving strategy is to pick the point or points so as to simplify the calculations as much as possible.

When a rigid body is in equilibrium, *every point* in the body must be in equilibrium. A particle in uniform motion (constant velocity) in an inertial frame is in equilibrium. Similarly, a rigid body in uniform *translational* motion (without rotation) is in equilibrium. Uniform *rotational* motion of a rigid body is *not* an equilibrium state, even though the sum of forces and sum of torques are zero, because the individual particles in the body have accelerations.

## 11–2 CENTER OF GRAVITY

In most equilibrium problems one of the forces acting on the body is its weight. We need to be able to calculate the *torque* of this force. The weight doesn't act at a single point; it is distributed over the entire body. But when the acceleration due to gravity is the same at every point in the body, we can always calculate the torque due to the body's weight by assuming that the entire force of gravity (weight) is concentrated at a point called the **center of gravity.** This point is identical with the **center of mass** of the body, which we defined in Section 8–6. We mentioned this result without proof in Section 10–1, and now we'll prove it.

First, let's review the definition of the center of mass. For a collection of particles with masses $m_1, m_2, \ldots$ and coordinates $(x_1, y_1, z_1), (x_2, y_2, z_2), \ldots$, the coordinates $x_{\mathrm{cm}}$,

$y_{cm}$, and $z_{cm}$ of the center of mass are given by

$$x_{cm} = \frac{m_1x_1 + m_2x_2 + \cdots}{m_1 + m_2 + \cdots} = \frac{\Sigma m_i x_i}{\Sigma m_i},$$
$$y_{cm} = \frac{m_1y_1 + m_2y_2 + \cdots}{m_1 + m_2 + \cdots} = \frac{\Sigma m_i y_i}{\Sigma m_i}, \qquad (11\text{–}3)$$
$$z_{cm} = \frac{m_1z_1 + m_2z_2 + \cdots}{m_1 + m_2 + \cdots} = \frac{\Sigma m_i z_i}{\Sigma m_i}.$$

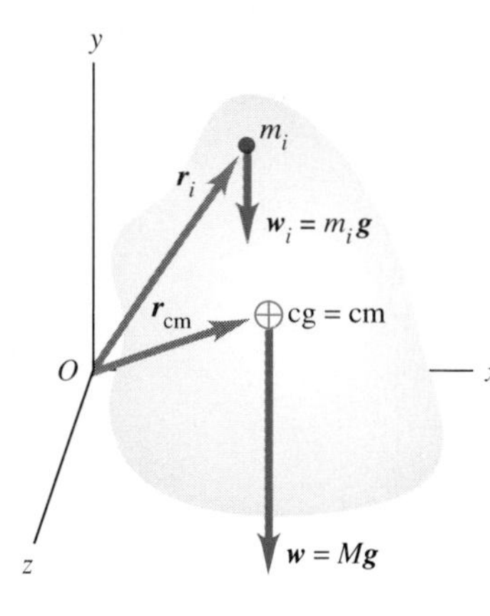


**11–1** The torque about any point can be obtained by assuming that all the weight of the body acts at its center of gravity (center of mass).

Also, $x_{cm}$, $y_{cm}$, and $z_{cm}$ are the components of the position vector $\boldsymbol{r}_{cm}$ of the center of mass, so Eqs. (11–3) are equivalent to the vector equation

$$\boldsymbol{r}_{cm} = \frac{m_1\boldsymbol{r}_1 + m_2\boldsymbol{r}_2 + \cdots}{m_1 + m_2 + \cdots} = \frac{\Sigma\, m_i\boldsymbol{r}_i}{\Sigma\, m_i}. \qquad (11\text{–}4)$$

Now let's consider the gravitational torque on a body of arbitrary shape (Fig. 11–1). We assume that the acceleration due to gravity $\boldsymbol{g}$ has the same magnitude and direction at every point in the body. Every particle in the body experiences a gravitational force, and the total weight of the body is the vector sum of a large number of parallel forces. A typical particle has mass $m_i$ and position vector $\boldsymbol{r}_i$, and the total mass of the body is $M = \Sigma\, m_i$. The weight $\boldsymbol{w}_i$ of a particle with mass $m_i$ is given by $\boldsymbol{w}_i = m_i\boldsymbol{g}$. If $\boldsymbol{r}_i$ is the position vector of this particle with respect to an arbitrary origin $O$, the torque $\boldsymbol{\tau}_i$ of the weight $\boldsymbol{w}_i$ with respect to $O$ is, from Eq. (10–3),

$$\boldsymbol{\tau}_i = \boldsymbol{r}_i \times \boldsymbol{w}_i = \boldsymbol{r}_i \times m_i\boldsymbol{g}.$$

The *total* torque due to the gravitational forces on all the particles is

$$\begin{aligned}\boldsymbol{\tau} = \Sigma\boldsymbol{\tau}_i &= \boldsymbol{r}_1 \times m_1\boldsymbol{g} + \boldsymbol{r}_2 \times m_2\boldsymbol{g} + \cdots \\ &= (m_1\boldsymbol{r}_1 + m_2\boldsymbol{r}_2 + \cdots) \times \boldsymbol{g} \\ &= (\Sigma m_i\boldsymbol{r}_i) \times \boldsymbol{g}.\end{aligned} \qquad (11\text{–}5)$$

When we multiply and divide this by the total mass

$$M = m_1 + m_2 + \cdots = \Sigma m_i,$$

we get

$$\boldsymbol{\tau} = \frac{m_1\boldsymbol{r}_1 + m_2\boldsymbol{r}_2 + \cdots}{m_1 + m_2 + \cdots} \times M\boldsymbol{g} = \frac{\Sigma\, m_i\boldsymbol{r}_i}{\Sigma\, m_i} \times M\boldsymbol{g}. \qquad (11\text{–}6)$$

The fraction in this equation is simply the position vector $\boldsymbol{r}_{cm}$ of the center of mass, with components $x_{cm}$, $y_{cm}$, and $z_{cm}$, as given by Eq. (11–4), and $M\boldsymbol{g}$ is equal to the total weight $\boldsymbol{w}$ of the body. Thus

$$\boldsymbol{\tau} = \boldsymbol{r}_{cm} \times M\boldsymbol{g} = \boldsymbol{r}_{cm} \times \boldsymbol{w}. \qquad (11\text{–}7)$$

The total torque is the same as though the total weight $\boldsymbol{w}$ were acting at the position $\boldsymbol{r}_{cm}$ of the center of mass, which we also call the *center of gravity*. In a uniform gravitational field $\boldsymbol{g}$, **the center of gravity of any body is identical to its center of mass.** Note, however, that the center of mass is defined independently of any gravitational effect.

We can often use symmetry considerations to locate the center of gravity of a body. The center of gravity of a homogeneous sphere, cube, circular disk, or rectangular plate is at its geometric center. The center of gravity of a right circular cylinder or cone is on the axis of symmetry, and so on.

In some cases we can think of a body as being made up of several pieces that individually have some symmetry that lets us determine their individual centers of gravity.

For example, we could approximate the human body as a collection of solid cylinders, with a sphere for the head. Then we can compute the coordinates of the center of gravity of the combination from Eqs. (11–3), letting $m_1, m_2, \ldots$ be the masses of the individual pieces and $(x_1, y_1, z_1), (x_2, y_2, z_2), \ldots$ be the coordinates of their centers of gravity.

The center of gravity (center of mass) has several other important properties. First, when a body is in equilibrium in a gravitational field, supported or suspended at a single point, the center of gravity is always at or directly above or below the point of suspension. If it were anywhere else, the weight would have a torque with respect to this point, and the body could not be in rotational equilibrium. This fact can be used to determine experimentally the location of the center of gravity of an irregular body (Fig. 11–2).

Second, a force applied to a body at its center of gravity does not tend to cause the body to rotate about an axis through this point. A force applied at any other point can cause both rotational and translational motion. We used this fact implicitly in the examples of Section 10–2.

Third, as we discussed in Section 8–6, the total (linear) momentum of a body equals the product of its total mass and its center-of-mass velocity. The acceleration of the center of mass is the same as that of a point mass equal to the total mass of the body, acted on by the resultant external force.

**11–2** Locating the center of gravity of a body by balancing. The meter stick balances at its midpoint, but the hammer has a heavy head at one end.

## EXAMPLE 11–1

**A center of gravity demonstration** In Fig. 11–3a we show the obvious, that a piece of steel won't balance with one end resting on the edge of the table. But suppose we fasten on two other pieces (Fig. 11–3b). All pieces have the same square cross section and density. The system can now balance if we find the right length $l$. What is $l$?

**SOLUTION** For balance, the center of gravity must be directly below the table edge. The mass of each piece is proportional to its length; using a proportionality constant $A$, we can write $m_1 = A(10.0\text{ cm})$, $m_2 = A(5.0\text{ cm})$, and $m_3 = Al$. We neglect the masses of the fasteners that hold the pieces together.

We place the origin at the edge of the table. The center of mass of each piece is at its center, so the $x$-coordinate of the center of mass of the system is

$$x_{cm} = \frac{A(10.0\,\text{cm})(5.0\,\text{cm}) + A(5.0\,\text{cm})(9.5\,\text{cm}) + Al(10.0\,\text{cm} - l/2)}{A(10.0\,\text{cm}) + A(5.0\,\text{cm}) + Al}.$$

The proportionality constant $A$ divides out, as you should expect. We want $x_{cm}$ to be zero; setting the numerator, with $A$ divided out, equal to zero gives us a quadratic equation for $l$, which we invite you to solve. One root is negative and not physically meaningful; the root we want is

$$l = (10 + \sqrt{295})\text{ cm} = 27.2\text{ cm}.$$

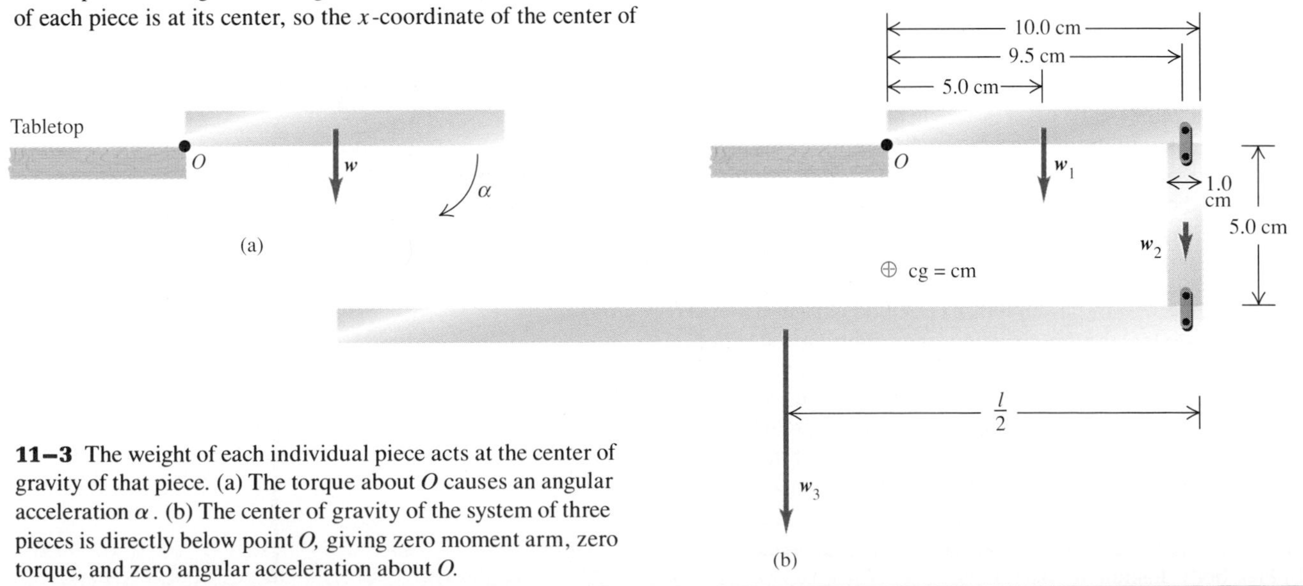


**11–3** The weight of each individual piece acts at the center of gravity of that piece. (a) The torque about $O$ causes an angular acceleration $\alpha$. (b) The center of gravity of the system of three pieces is directly below point $O$, giving zero moment arm, zero torque, and zero angular acceleration about $O$.

## 11–3 SOLVING EQUILIBRIUM PROBLEMS

The principles of rigid-body equilibrium are few and simple: The vector sum of the forces on the body must be zero, and the sum of torques about any point must be zero. The hard part is applying these principles to specific problems. Careful and systematic problem-solving methods always pay off. The following strategy is very similar to the suggestions in Section 5–1 for equilibrium of a particle.

### PROBLEM-SOLVING STRATEGY

#### Equilibrium of a rigid body

**1.** Sketch the physical situation, including dimensions.

**2.** Choose some appropriate body as the body in equilibrium, and draw a free-body diagram showing the forces acting *on* this body and no others. *Do not* include forces exerted *by* this body on other bodies. Be careful to show correctly the point at which each force acts; this is crucial for correct torque calculations. You can't represent a rigid body as a point.

**3.** Draw coordinate axes and specify a positive sense of rotation for torques. Represent forces in terms of their components with respect to the axes you have chosen.

**4.** Write equations expressing the equilibrium conditions. Remember that $\Sigma F_x = 0$, $\Sigma F_y = 0$, and $\Sigma \tau = 0$ are always separate equations; *never* add $x$- and $y$-components in a single equation.

**5.** In choosing a point to compute torques, note that if a force has a line of action that goes *through* a particular point, the torque of the force with respect to that point is zero. You can often eliminate unknown forces or components from the torque equation by a clever choice of point for your calculation. Also remember that when a force is represented in terms of its components, you can compute the torque of that force by finding the torque of each component separately, each with its appropriate moment arm and sign, and adding the results. This is often easier than determining the moment arm of the original force.

**6.** You always need as many equations as you have unknowns. Depending on the number of unknowns, you may need to compute torques with respect to two or more axes to obtain enough equations. There often are several equally good sets of force and torque equations for a particular problem; there is usually no single "right" combination of equations. When you have as many independent equations as unknowns, you can solve the equations simultaneously. We will illustrate this point in some of the following examples by using various sets of equations.

### EXAMPLE 11–2

**Weight distribution for a car** An auto magazine reports, "The Nissan 240SX has 53% of its weight on its front wheels and 47% on its rear wheels, with a 2.46-m wheelbase" (Fig. 11–4a). This means that the total normal force on the front wheels is $0.53w$ and that on the rear wheels is $0.47w$, where $w$ is the total weight. The wheelbase is the distance between front and rear axles. How far in front of the rear axle is the 240SX's center of gravity?

**SOLUTION** Figure 11–4b shows a free-body diagram for the car. We can see that the first condition for equilibrium is satisfied; $\Sigma F_x = 0$ because there aren't any $x$-components of force, and $\Sigma F_y = 0$ because $0.47w + 0.53w + (-w) = 0$. So far, so good. We have drawn the weight $w$ as acting at the center of gravity, and the distance we want is $L_{cg}$. This is the moment arm of the weight with respect to the rear axle $R$, so it is reasonable to take torques

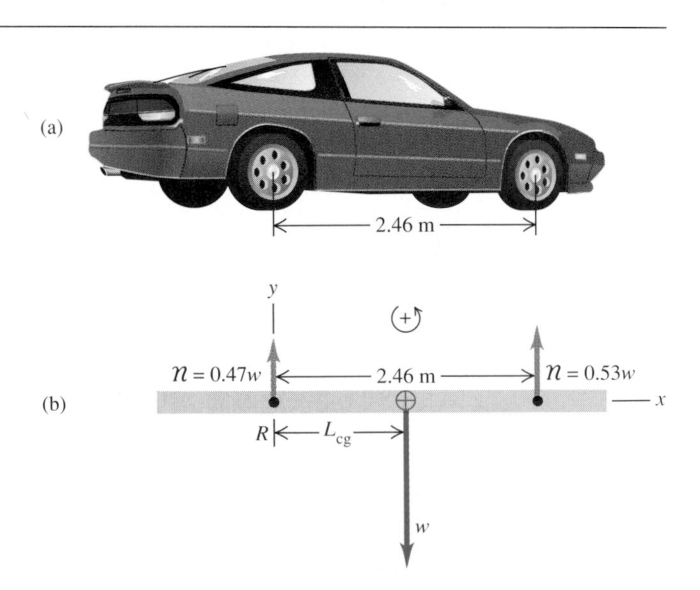


11–4 (a) The Nissan 240SX at rest on a horizontal surface. (b) Free-body diagram for the car.

with respect to $R$. Note that the torque due to the weight is negative because it tends to cause a clockwise rotation about $R$. The torque due to the upward normal force at the front wheels is positive because it tends to cause a counterclockwise rotation about $R$. The torque equation is

$$\Sigma\tau_R = 0.47w(0) - w(L_{cg}) + 0.53w(2.46\text{ m}) = 0,$$

$$L_{cg} = 1.30\text{ m}.$$

Can you show that if $f$ is the fraction of the weight on the front wheels and $d$ is the wheelbase, the center of gravity is a distance $fd$ in front of the rear wheels? In terms of the center of gravity, why do owners of rear-wheel-drive vehicles put bags of sand in their trunks to improve traction on snow and ice? Would this help with a front-wheel-drive car? With a four-wheel-drive car?

## EXAMPLE 11–3

**A heroic rescue** Sir Lancelot is trying to rescue the Lady Elayne from the Black Castle by climbing a uniform ladder that is 5.0 m long and weighs 180 N. Lancelot, who weighs 800 N, stops a third of the way up the ladder (Fig. 11–5a). The bottom of the ladder rests on a horizontal stone ledge and leans across the moat in equilibrium against a vertical wall that is frictionless because of a thick layer of moss. The ladder makes an angle of 53.1° with the horizontal, conveniently forming a 3-4-5 right triangle. a) Find the normal and friction forces on the ladder at its base. b) Find the minimum coefficient of static friction needed to prevent slipping. c) Find the magnitude and direction of the contact force on the ladder at the base.

**SOLUTION** a) Figure 11–5b shows the free-body diagram for the ladder. The ladder is described as "uniform," so we assume that its center of gravity is at its center, halfway between the base and the wall. Lancelot's weight pushes down on the ladder at a point one-third of the way from the base toward the wall. The forces at the base are the upward normal force and the friction force, which must point to the right to prevent slipping. The first condition for equilibrium, in component form, gives

$$\Sigma F_x = \mathcal{F}_s + (-n_1) = 0,$$

$$\Sigma F_y = n_2 + (-800\text{ N}) + (-180\text{ N}) = 0.$$

These are two equations for the three unknowns $n_1$, $n_2$, and $\mathcal{F}_s$. The first equation tells us that the two horizontal forces must be equal and opposite, and the second equation gives $n_2 = 980$ N. The ledge pushes up with a force of 980 N to balance the total (downward) weight (800 N + 180 N).

We don't yet have enough equations, but now we can use the second condition for equilibrium. We can take torques about any point we choose. The smart choice is the point that will give us the fewest terms and fewest unknowns in the torque equation. With this thought in mind we choose point $B$, at the base of the ladder. The two forces $n_2$ and $\mathcal{F}_s$ have no torque about that point. From Fig. 11–5b we see that the moment arm for the ladder's weight is 1.5 m, and the moment arm for Lancelot's weight is 1.0 m. The torque equation for point $B$ is

$$\Sigma\tau_B = n_1(4.0\text{ m}) - (180\text{ N})(1.5\text{ m}) - (800\text{ N})(1.0\text{ m}) + n_2(0) + \mathcal{F}_s(0) = 0.$$

Solving for $n_1$, we get

$$n_1 = 268\text{ N}.$$

Now we substitute this back into the $\Sigma F_x = 0$ equation to get

$$\mathcal{F}_s = 268\text{ N}.$$

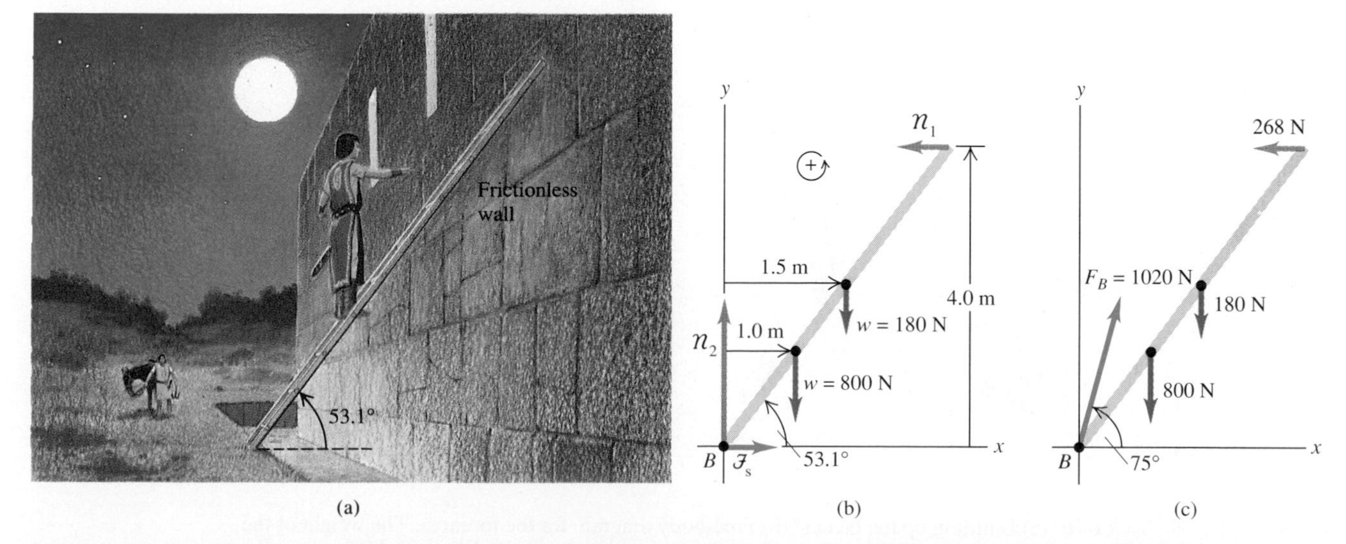


**11–5** (a) Sir Lancelot pauses a third of the way up the ladder, fearing that it will slip. (b) Free-body diagram for the ladder.

b) The static friction force $\mathcal{F}_s$ cannot exceed $\mu_s n_2$, so the *minimum* coefficient of static friction to prevent slipping is

$$(\mu_s)_{min} = \frac{\mathcal{F}_s}{n_2} = \frac{268\text{ N}}{980\text{ N}} = 0.27$$

c) The components of the contact force $\boldsymbol{F}_B$ at the base are the static friction force $\mathcal{F}_s$ and the normal force $n_2$, so

$$\boldsymbol{F}_B = (268\text{ N})\boldsymbol{i} + (980\text{ N})\boldsymbol{j}.$$

The magnitude of $\boldsymbol{F}_B$ is

$$F_B = \sqrt{(268\text{ N})^2 + (980\text{ N})^2} = 1020\text{ N},$$

and its direction (Fig. 11–5c) is

$$\theta = \arctan\frac{980\text{ N}}{268\text{ N}} = 75°.$$

Here are a few final comments. First, the contact force $\boldsymbol{F}_B$, at an angle of 75°, is *not* directed along the length of the ladder, which is at an angle of 53.1°. If it were, the sum of torques with respect to the point where Lancelot stands couldn't possibly be zero.

Second, as Lancelot climbs higher on the ladder, the moment arm and torque of his weight about $B$ increase; this increases the values of $n_1$, $\mathcal{F}_s$, and $(\mu_s)_{min}$. At the top his moment arm would be nearly 3 m, giving a minimum coefficient of static friction of nearly 0.7. A present-day aluminum ladder on a wood floor would not have a value of $\mu_s$ this large; such ladders are usually equipped with nonslip rubber pads.

Third, a larger ladder angle would decrease the moment arms with respect to $B$ of the weights of the ladder and the man and increase the moment arm of $n_1$, all of which would decrease the required friction force. The R. D. Werner Ladder Co. recommends that its ladders be used at an angle of 75°. (Why not 90°?)

Finally, if we had assumed friction on the wall as well as on the ledge, there would be no way to separate the effects of the two friction forces, and the problem would be impossible to solve. (Try it!) Such a problem is said to be *statically indeterminate*. Another simple example of such a problem is a four-legged table; there is no way to use the equilibrium conditions to find the force on each separate leg. In a sense, four legs are one too many; three, properly located, are sufficient for stability.

## EXAMPLE 11–4

**Equilibrium and pumping iron** Figure 11–6a shows a human arm lifting a dumbbell, and Fig. 11–6b is a free-body diagram for the forearm, showing the forces involved. The forearm is in equilibrium under the action of the weight $w$ of the dumbbell, the tension $T$ in the tendon connected to the biceps muscle, and the forces exerted on the forearm by the upper arm at the elbow joint. For clarity the tendon force has been displaced away from the elbow farther than its actual position. The weight $w$ and the angle $\theta$ are given; we want to find the tendon tension and the two components of force at the elbow (three unknown scalar quantities in all). We neglect the weight of the forearm itself.

**SOLUTION** First we represent the tendon force in terms of its components $T_x$ and $T_y$, using the given angle $\theta$ and the unknown magnitude $T$:

$$T_x = T\cos\theta, \qquad T_y = T\sin\theta.$$

We also represent the force at the elbow in terms of its components $E_x$ and $E_y$. We don't need to know at the beginning whether these are positive or negative. They could come out either way. Next we note that if we take torques about the elbow joint, the resulting torque equation does not contain $E_x$, $E_y$, or $T_x$ because the lines of action of all these forces pass through this point. The torque equation is then simply

$$\Sigma\tau_E = lw - dT_y = 0.$$

From this we find

$$T_y = \frac{lw}{d} \qquad \text{and} \qquad T = \frac{lw}{d\sin\theta}.$$

(a)

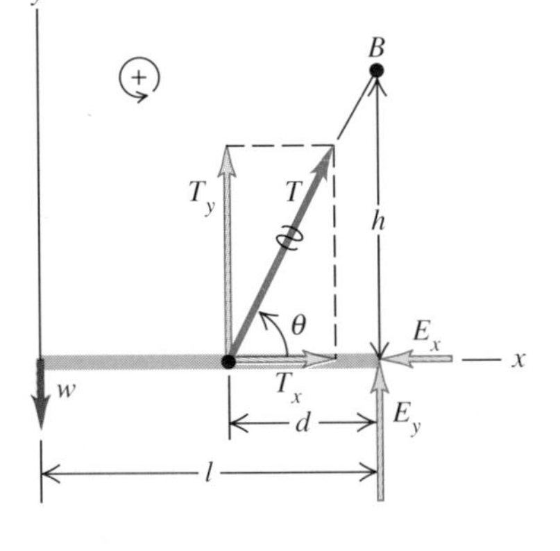


(b)

**11–6** (a) Building up the biceps. (b) Free-body diagram for the forearm. The weight of the forearm is neglected, and the distance $d$ is exaggerated for clarity.

To find $E_x$ and $E_y$ we use the first conditions for equilibrium, $\sum F_x = 0$ and $\sum F_y = 0$:

$$\sum F_x = T_x + (-E_x) = 0,$$

$$E_x = T_x = T\cos\theta = \frac{lw}{d\sin\theta}\cos\theta = \frac{lw}{d}\cot\theta = \frac{lw}{d}\frac{d}{h} = \frac{lw}{h};$$

$$\sum F_y = T_y + E_y + (-w) = 0,$$

$$E_y = w - \frac{lw}{d} = -\frac{(l-d)w}{d}.$$

The negative sign shows that our initial guess for the direction of $E_y$ was wrong; it is actually vertically *downward*.

An alternative way to find $E_x$ and $E_y$ is to use two more torque equations. We take torques about the point $A$ where the tendon is attached:

$$\sum \tau_A = (l-d)w + dE_y = 0 \quad \text{and} \quad E_y = -\frac{(l-d)w}{d}.$$

Finally, we take torques about point $B$ in the figure:

$$\sum \tau_B = lw - hE_x = 0 \quad \text{and} \quad E_x = \frac{lw}{h}.$$

Notice how much we have simplified the calculations by choosing the point for calculating torques so as to eliminate one or more of the unknown quantities. In the last step the force $T$ has no torque about point $B$; thus when the torques of $T_x$ and $T_y$ are computed separately, they must add to zero. We invite you to check out this statement (see Exercise 11–9).

As a specific example, suppose $w = 200$ N, $d = 0.050$ m, $l = 0.30$ m, and $\theta = 80°$. Then from $\tan\theta = h/d$ we find

$$h = d\tan\theta = (0.050\text{ m})(5.67) = 0.28\text{ m}.$$

From the previous general results we find

$$T = \frac{lw}{d\sin\theta} = \frac{(0.30\text{ m})(200\text{ N})}{(0.050\text{ m})(0.98)} = 1220\text{ N},$$

$$E_y = -\frac{(l-d)w}{d} = -\frac{(0.30\text{ m} - 0.050\text{ m})(200\text{ N})}{0.050\text{ m}} = -1000\text{ N},$$

$$E_x = \frac{lw}{h} = \frac{(0.30\text{ m})(200\text{ N})}{0.28\text{ m}} = 210\text{ N}.$$

The magnitude of the force at the elbow is

$$E = \sqrt{E_x^2 + E_y^2} = 1020\text{ N}.$$

In our alternative determination of $E_x$ and $E_y$ we didn't explicitly use the first condition for equilibrium, that the vector sum of the forces be zero. As a check, compute $\sum F_x$ and $\sum F_y$ to verify that they really *are* zero. Consistency checks are always a good idea!

In view of the magnitudes of our results, neglecting the weight of the forearm itself, which may be 20 N or so, will cause relatively small errors.

## EXAMPLE 11–5

**Another rescue attempt** After Sir Lancelot falls into the moat, Sir Gawain tosses a grappling hook through an open upstairs window and is resting partway up the rope (Fig. 11–7a). Inside the window are Lady Elayne and five armed guards. Sir Gawain weighs 700 N; his body makes an angle of 60° with the wall, and his center of gravity is 0.85 m from his feet. The rope force acts 1.30 m from his feet, and the rope makes an angle of 20° with the vertical. Find the tension in the rope and the forces exerted on his feet by the moss-free region of the wall.

**SOLUTION** Figure 11–7b shows the free-body diagram for Sir Gawain. The forces on his feet include a (horizontal) normal component $\mathcal{n}$ and a (vertical) frictional component $\mathcal{F}$. The rope tension is $T$. The angles needed to find the torques are shown.

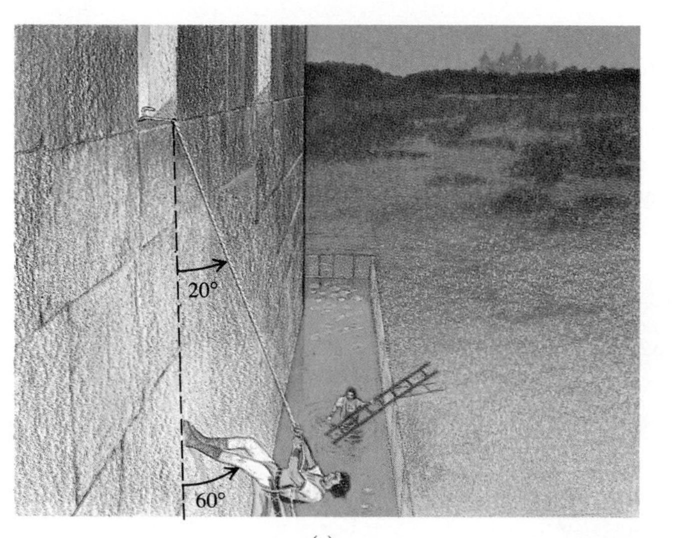


(a)

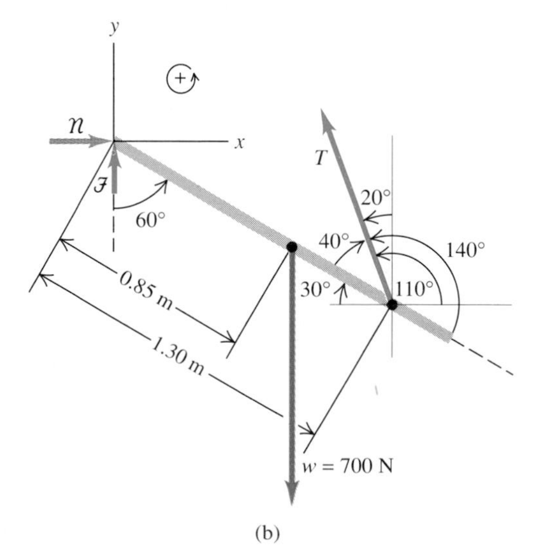


(b)

**11–7** (a) Sir Gawain pausing to catch his breath. (b) Free-body diagram for Sir Gawain.

Because two unknown forces ($\mathcal{N}$ and $\mathcal{F}$) act at Gawain's feet, we take torques there, obtaining a torque equation in which $T$ is the only unknown. The easiest way to find the torques is to use $\tau = rF \sin \phi$ for each one. Taking counterclockwise torques as positive, we obtain the torque equation

$$\begin{aligned}\Sigma \tau_F &= (1.30 \text{ m})T(\sin 140°) - (0.85 \text{ m})(700 \text{ N})(\sin 60°) \\ &\quad + (0)\mathcal{N} + (0)\mathcal{F} \\ &= 0.\end{aligned}$$

We solve this equation for $T$; the result is

$$T = 617 \text{ N}.$$

We note that the rope tension is *less than* Gawain's weight.

To find the components of force $\mathcal{N}$ and $\mathcal{F}$ at Gawain's feet, we use the first equilibrium condition:

$$\Sigma F_x = \mathcal{N} + T \cos 110° = 0,$$

$$\Sigma F_y = \mathcal{F} + T \sin 110° + (-700 \text{ N}) = 0.$$

When we substitute the value of $T$ into these equations, we get

$$\mathcal{N} = 211 \text{ N} \quad \text{and} \quad \mathcal{F} = 120 \text{ N}.$$

The sum of the vertical component of the tension and the vertical friction force must equal Gawain's weight. ■

**11–8** Applying forces with equal magnitude and opposite direction on the wrench and the pipe.

## 11–4 COUPLES

Suppose you stand on a ladder to tighten a threaded pipe connection. You pull on the wrench handle with one hand, brace your other hand beside the connection, and push with an equal and opposite force (Fig. 11–8). If you didn't do this, you would just pull yourself toward the pipe; this would risk bending the pipe and tipping the ladder over. You want to exert a torque but no net force on the pipe.

When two forces with equal magnitude and opposite direction and different lines of action act on a body, the pair of forces is called a **couple.** Another familiar example is a compass needle in the earth's magnetic field. The north and south poles of the needle are acted on by equal forces, one toward the north and the other toward the south. Except when the needle points in the N-S direction, the two forces do not have the same line of action.

Figure 11–9 shows a couple consisting of two forces, each with magnitude $F$, separated by a perpendicular distance $l$. The vector sum of the forces is zero, so they have no effect on the translational motion of the body on which they act. The only effect of a couple is to apply a torque to the body. The total torque of the couple in Fig. 11–9 about an arbitrary point $O$ is

$$\begin{aligned}\Sigma \tau_O &= x_2F - x_1F = (x_1 + l)F - x_1F \\ &= lF.\end{aligned} \tag{11–8}$$

The distances $x_1$ and $x_2$ do not appear in the result, only the distance $l$. **The torque of a couple is the same about every point and is equal to the product of the magnitude of either force and the perpendicular distance between their lines of action.**

We will find several uses for the concept of a couple, particularly in studying the interaction of an electric dipole with an electric field and of a magnetic dipole (such as a compass needle) in a magnetic field.

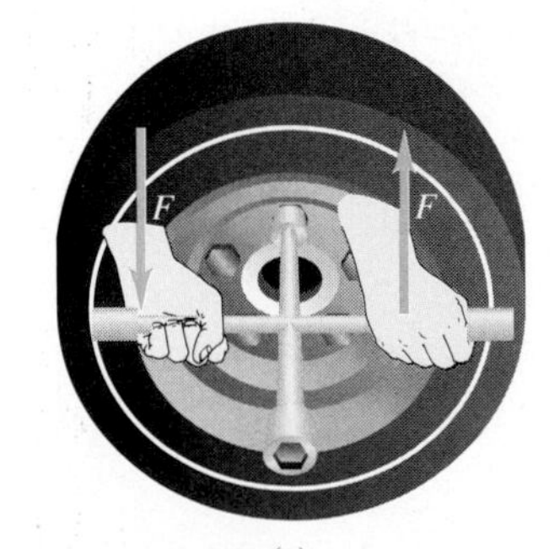


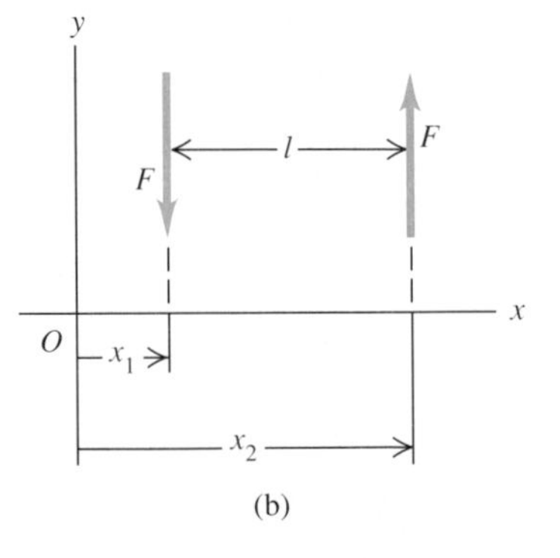


(a) (b)

**11–9** Two equal and opposite forces having different lines of action are called a *couple*. The torque of the couple is the same about all points and is equal to $Fl$ in magnitude.

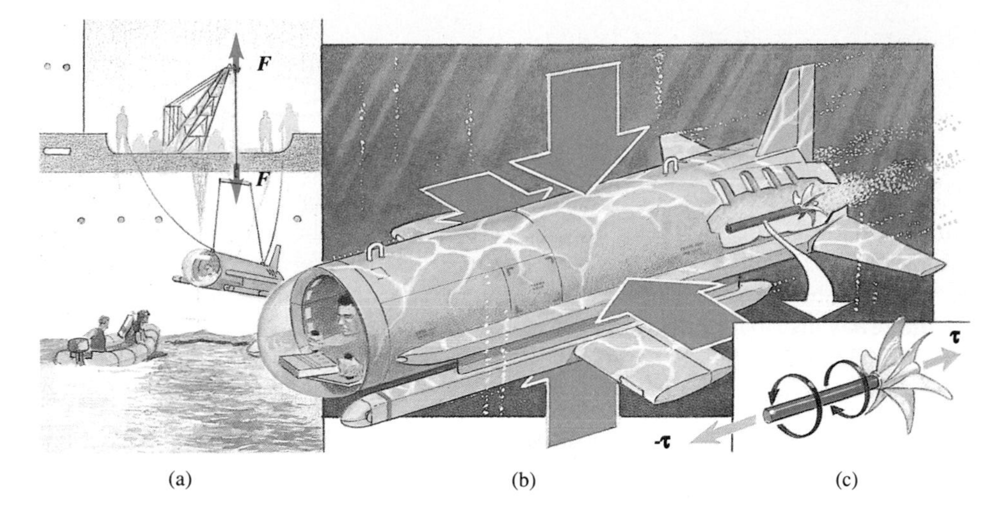


**11–10** (a) A cable is stretched by the weight of the object hanging from it. (b) The aircraft-inspired submersible is squeezed from all sides by the force of the water pressure. (c) A torsion bar or drive shaft is twisted by forces at its ends that cause torques about its axis.

## 11–5 TENSILE STRESS AND STRAIN

The rigid body is a useful idealized model, but the stretching, squeezing, and twisting of real bodies when forces are applied are often too important to ignore. Figure 11–10 shows three examples. We want to study the relation between the forces and deformations for each case.

For each kind of deformation we will introduce a quantity called **stress** that characterizes the strength of the forces causing the stretch, squeeze, or twist, usually on a "force per unit area" basis. Another quantity, **strain,** describes the resulting deformation. When the stress and strain are small enough, we often find that the two are directly proportional, and we call the proportionality constant an **elastic modulus.** The harder you pull on something, the more it stretches; and the more you squeeze it, the more it compresses. The general pattern that emerges can be formulated as

$$\frac{\text{Stress}}{\text{Strain}} = \text{Elastic modulus.} \tag{11–9}$$

The proportionality of stress and strain (under certain conditions) is called **Hooke's law,** after Robert Hooke (1635–1703), a contemporary of Newton. We saw a special case of Hooke's law in Sections 6–3 and 7–3: the elongation of a spring is proportional to the stretching forces.

The simplest elastic behavior to understand is the stretching of a bar, rod, or wire when its ends are pulled. Figure 11–11a shows a bar with uniform cross-section area $A$, with equal and opposite forces $F$ pulling at its ends. We say that the bar is in **tension.** We have already talked a lot about tensions in ropes and strings; it's the same concept here. Figure 11–11b shows a cross section through the bar; the part to the right of the section pulls on the part to the left with a force $F_\perp$, and vice versa. We use the notation $F_\perp$ as a

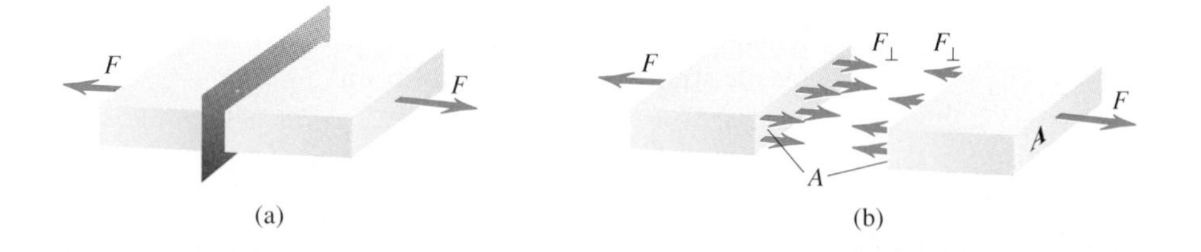


**11–11** (a) A bar in tension. (b) The stress at a particular section equals $F_\perp/A$.

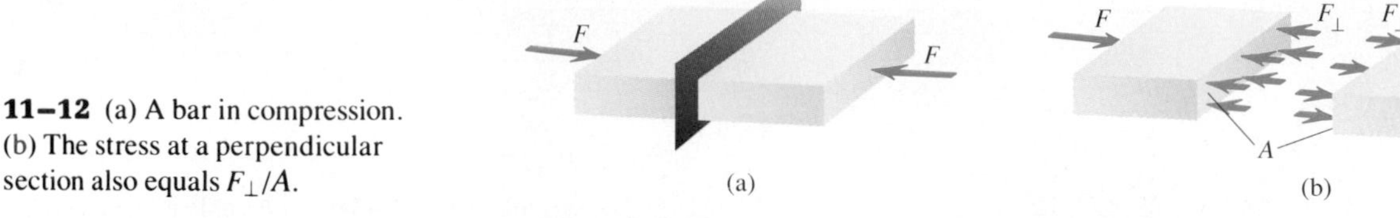


**11–12** (a) A bar in compression. (b) The stress at a perpendicular section also equals $F_\perp/A$.

reminder that the force acts in a direction perpendicular to the cross section. We will assume that the forces at every cross section are distributed uniformly over the section, as shown by the short arrows in Fig. 11–11b. (This is always the case if the forces at the *ends* are uniformly distributed.)

We define the **tensile stress** at the cross section as the ratio of the force $F_\perp$ to the cross-section area $A$:

$$\text{Tensile stress} = \frac{F_\perp}{A}. \tag{11–10}$$

The SI unit of stress is the newton per square meter ($N/m^2$). This unit is also given a special name, the **pascal** (abbreviated Pa):

$$1 \text{ pascal} = 1 \text{ Pa} = 1 \text{ N/m}^2.$$

In the British system the logical unit of stress would be the pound per square foot, but the pound per square inch ($lb/in.^2$, or psi) is more commonly used. The conversion factors are

$$1 \text{ psi} = 6891 \text{ Pa} \qquad \text{and} \qquad 1 \text{ Pa} = 1.451 \times 10^{-4} \text{ psi}.$$

The units of stress are the same as those of *pressure,* which we will encounter often in later chapters. Air pressure in automobile tires is typically of the order of $3 \times 10^5$ Pa = 300 kPa, and steel cables are commonly required to withstand tensile stresses of the order of $10^8$ Pa.

When the forces on the ends of a bar are pushes rather than pulls (Fig. 11–12a), the bar is in **compression,** and the stress is a **compressive stress.** At the cross section shown, each side pushes rather than pulls on the other.

The fractional change of length (stretch) of a body under a tensile stress is called the **tensile strain.** Figure 11–13 shows a bar with unstretched length $l_0$ that stretches to a length $l = l_0 + \Delta l$ when equal and opposite forces $F$ are applied to its ends. The elongation $\Delta l$ does not occur only at the ends; every part of the bar stretches in the same proportion. The tensile strain is defined as the ratio of the elongation $\Delta l$ to the original length $l_0$:

$$\text{Tensile strain} = \frac{l - l_0}{l_0} = \frac{\Delta l}{l_0}. \tag{11–11}$$

Tensile strain is amount of stretch per unit length. It is a ratio of two lengths, always measured in the same units, so strain is a pure (dimensionless) number with no units. The **compressive strain** of a bar in compression is defined in the same way as tensile strain, but $\Delta l$ has the opposite direction. In this case it is often convenient to treat $\Delta l$ as a negative quantity.

Experiments have shown that for a sufficiently small tensile or compressive stress, stress and strain are proportional, as stated by Hooke's law, Eq. (11–9). The corresponding elastic modulus is called **Young's modulus,** denoted by $Y$:

$$Y = \frac{\text{Tensile stress}}{\text{Tensile strain}} = \frac{\text{Compressive stress}}{\text{Compressive strain}},$$

$$Y = \frac{F_\perp/A}{\Delta l/l_0} = \frac{l_0}{A}\frac{F_\perp}{\Delta l}. \tag{11–12}$$

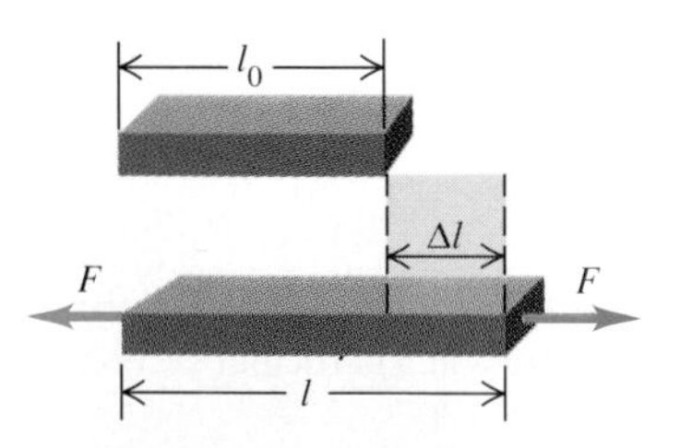


**11–13** The tensile strain is defined as $\Delta l/l_0$. The elongation $\Delta l$ is exaggerated for clarity.

## Equilibrium and Stress

A large static structure, such as a bridge or a building, experiences large tensile and compressive stresses due to the weight it is designed to support. The construction materials used must be able to withstand these stresses to maintain the structure's equilibrium. Often an engineer must choose from among many different design ideas and construction materials to achieve a practical, cost-efficient, and attractive result.

# BUILDING PHYSICAL INTUITION

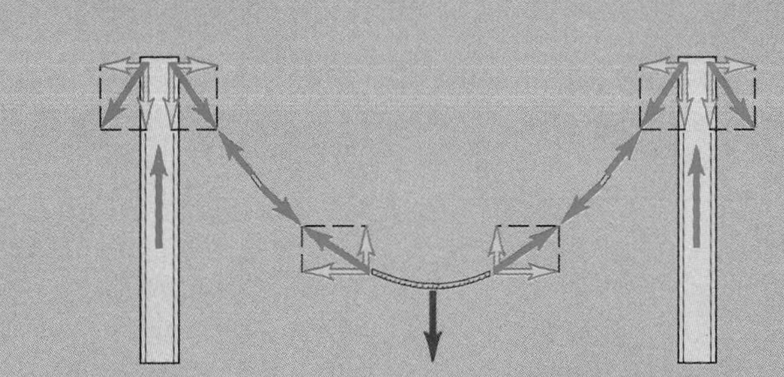

A suspension bridge supports its load primarily through tension in the cables and compression in the towers. Tension structures are most efficient for supporting light or moderate loads over long distances. The reaction force to the tension in the supporting cables of a suspension bridge is transmitted to the ground under the bridge towers. The ground must be strong enough to bear this stress without fracturing.

An arch supports its load primarily through compression. Compression structures are very efficient for supporting large loads or for transmitting stresses over short distances. In an arched bridge, the compressive stress is transmitted directly to the ground through the ends of the arch.

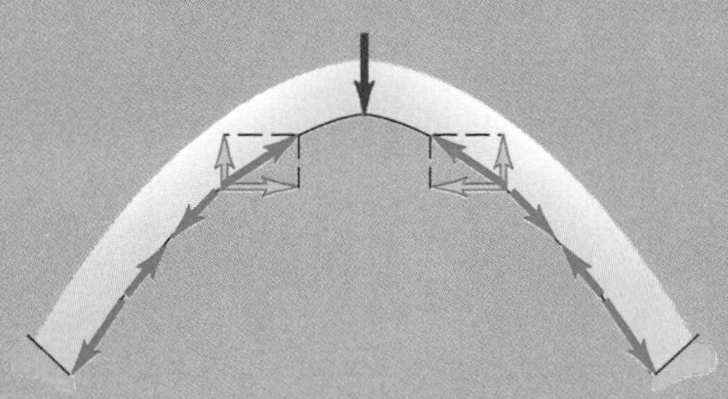

**TABLE 11–1 Approximate Elastic Constants**

| Material | Young's modulus, $Y$ (Pa) | Shear modulus, $S$ (Pa) | Bulk modulus, $B$ (Pa) | Poisson's ratio, $\sigma$ |
|---|---|---|---|---|
| Aluminum | $0.70 \times 10^{11}$ | $0.30 \times 10^{11}$ | $0.70 \times 10^{11}$ | 0.16 |
| Brass | $0.91 \times 10^{11}$ | $0.36 \times 10^{11}$ | $0.61 \times 10^{11}$ | 0.26 |
| Copper | $1.1 \times 10^{11}$ | $0.42 \times 10^{11}$ | $1.4 \times 10^{11}$ | 0.32 |
| Glass | $0.55 \times 10^{11}$ | $0.23 \times 10^{11}$ | $0.37 \times 10^{11}$ | 0.19 |
| Iron | $1.9 \times 10^{11}$ | $0.70 \times 10^{11}$ | $1.0 \times 10^{11}$ | 0.27 |
| Lead | $0.16 \times 10^{11}$ | $0.056 \times 10^{11}$ | $0.077 \times 10^{11}$ | 0.43 |
| Nickel | $2.1 \times 10^{11}$ | $0.77 \times 10^{11}$ | $2.6 \times 10^{11}$ | 0.36 |
| Steel | $2.0 \times 10^{11}$ | $0.84 \times 10^{11}$ | $1.6 \times 10^{11}$ | 0.19 |
| Tungsten | $3.6 \times 10^{11}$ | $1.5 \times 10^{11}$ | $2.0 \times 10^{11}$ | 0.20 |

Strain is a pure number, so the units of Young's modulus are the same as those of stress, force per unit area. Some typical values are listed in Table 11–1. (This table also gives values of three other elastic constants we will discuss shortly.) A material with a large value of $Y$ is relatively *unstretchable;* a large stress is required for a given strain. For example, steel has a much larger value of $Y$ than rubber.

When you stretch a wire or a rubber band, it gets *thinner* as well as longer. When Hooke's law holds, the fractional decrease in width is proportional to the tensile strain. If $w_0$ is the original width and $\Delta w$ is the change in width, then

$$\frac{\Delta w}{w_0} = -\sigma \frac{\Delta l}{l_0}, \tag{11–13}$$

where $\sigma$ is a dimensionless constant, different for different materials and called **Poisson's ratio.** For many common materials, $\sigma$ has a value between 0.1 and 0.4; several representative values are listed in Table 11–1. Similarly, a material under compressive stress bulges at the sides, and again the fractional change in width is given by Eq. (11–13). For many materials, though not all, Young's modulus and Poisson's ratio have the same values for both tensile and compressive stresses.

## EXAMPLE 11–6

A small elevator of mass 550 kg hangs from a steel cable 3.0 m long. The wires making up the cable have a total cross-section area of 0.20 cm$^2$, and with this load the cable stretches 0.40 cm beyond its no-load length. Determine the stress, the strain, and the value of Young's modulus for the steel in the cable. Assume that the cable behaves like a solid rod with the same cross-section area.

**SOLUTION** We use the definitions of stress, strain, and Young's modulus given by Eqs. (11–10), (11–11), and (11–12):

$$\text{Stress} = \frac{F_\perp}{A} = \frac{(550\text{ kg})(9.8\text{ m/s}^2)}{2.0 \times 10^{-5}\text{ m}^2} = 2.7 \times 10^8\text{ Pa};$$

$$\text{Strain} = \frac{\Delta l}{l_0} = \frac{0.0040\text{ m}}{3.0\text{ m}} = 0.00133;$$

$$Y = \frac{\text{Stress}}{\text{Strain}} = \frac{2.7 \times 10^8\text{ Pa}}{0.00133} = 2.0 \times 10^{11}\text{ Pa}.$$

As the cable stretches, it becomes a little narrower; according to Eq. (11–13) and Table 11–1,

$$\frac{\Delta w}{w_0} = -\sigma \frac{\Delta l}{l_0} = -(0.19)(0.00133) = -0.00025.$$

If the cable has a circular cross section, its diameter is about 0.5 cm, and the change in diameter when it stretches is about

$$\Delta w = (-0.00025)(0.5\text{ cm}) = -0.0001\text{ cm}.$$

This is not quite within the precision of ordinary micrometer calipers.

## 11–6 BULK STRESS AND STRAIN

When a submersible plunges deep into the ocean, the water exerts nearly uniform pressure everywhere on its surface and squeezes it to a slightly smaller volume (Fig. 11–10b). This situation is different from the tensile and compressive stresses and strains that we have discussed. The stress is now a uniform pressure on all sides, and the resulting deformation is a volume change. We use the terms **bulk stress** and **bulk strain** to describe these quantities. Another familiar example is the compression of a gas under pressure, such as the air in a car's tires.

If we choose an arbitrary cross section within a fluid (liquid or gas) at rest, the force acting on each side of the section is always *perpendicular* to it. If we tried to exert a force parallel to a section, the fluid would slip sideways to counteract the effort. When a solid is immersed in a fluid and both are at rest, the forces that the fluid exerts on the surface of the solid are always perpendicular to the surface at each point. The force $F_\perp$ per unit area $A$ on such a surface is called the **pressure** $p$ in the fluid,

$$p = \frac{F_\perp}{A}. \tag{11–14}$$

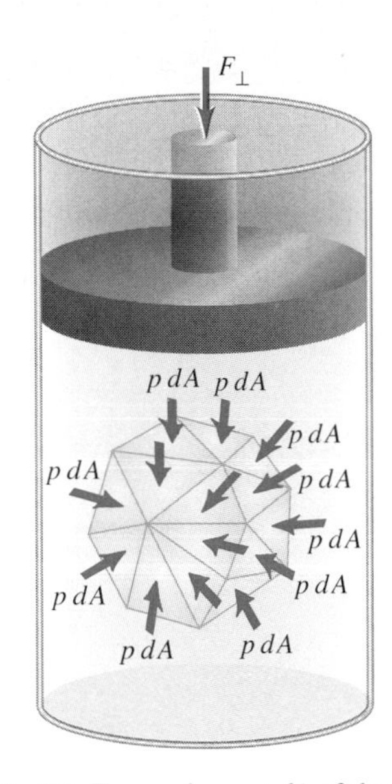

**11–14** Every element $dA$ of the surface of a submerged body experiences an inwardly normal force $dF_\perp = p\,dA$ from the surrounding fluid.

When we apply pressure to the surface of a fluid in a container, such as the cylinder and piston shown in Fig. 11–14, the pressure is transmitted through the fluid and also acts on the surface of any body immersed in the fluid. This principle is called *Pascal's law.* If pressure differences due to differences in depth within the fluid can be neglected, the pressure is the same at every point in the fluid and at every point on the surface of any submerged body.

Pressure has the same units as stress; commonly used units include 1 Pa (or 1 $N/m^2$) and 1 $lb/in.^2$ (or 1 psi). Also in common use is the **atmosphere,** abbreviated atm. One atmosphere is defined to be the average pressure of the earth's atmosphere at sea level:

$$1 \text{ atmosphere} = 1 \text{ atm} = 1.013 \times 10^5 \text{ Pa} = 14.7 \text{ lb/in.}^2$$

Pressure is a scalar quantity, not a vector quantity; it has no direction.

Pressure plays the role of *stress* (also called bulk stress) in a volume deformation. The corresponding strain is **volume strain,** defined as the fractional change in volume (Fig. 11–15), that is, the ratio of the volume change $\Delta V$ to the original volume $V_0$:

$$\text{Volume strain} = \frac{\Delta V}{V_0}. \tag{11–15}$$

Volume strain is change in volume per unit volume. Like tensile or compressive strain, it is a pure number, without units.

When Hooke's law is obeyed, the volume strain is *proportional* to the volume stress (pressure change). The corresponding elastic modulus (ratio of stress to strain) is called the **bulk modulus,** denoted by $B$. When the pressure on a body changes by a small amount $\Delta p$, from $p_0$ to $p_0 + \Delta p$, and the resulting volume strain (fractional change in volume) is $\Delta V/V_0$, Hooke's law takes the form

$$B = -\frac{\Delta p}{\Delta V/V_0}. \tag{11–16}$$

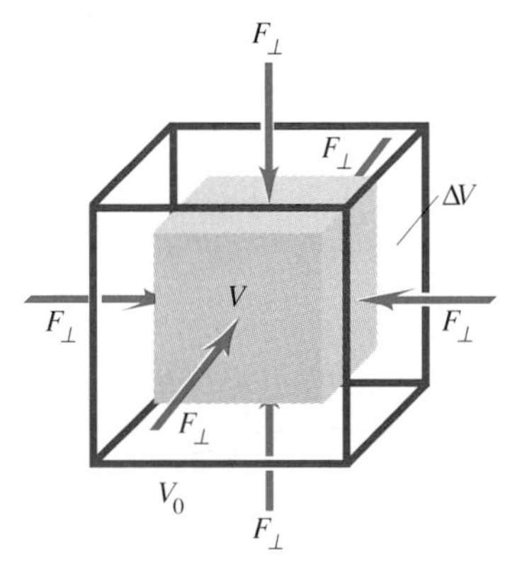

**11–15** Volume strain is fractional change in volume, $\Delta V/V_0$.

We include a minus sign in this equation because an *increase* of pressure always causes a *decrease* in volume. In other words, if $\Delta p$ is positive, $\Delta V$ is negative. The negative sign in Eq. (11–16) makes $B$ itself a positive quantity.

**TABLE 11–2 Compressibilities of Liquids**

| Liquid | Compressibility, $k$ | |
|---|---|---|
| | $Pa^{-1}$ | $atm^{-1}$ |
| Carbon disulfide | $93 \times 10^{-11}$ | $94 \times 10^{-6}$ |
| Ethyl alcohol | $110 \times 10^{-11}$ | $111 \times 10^{-6}$ |
| Glycerine | $21 \times 10^{-11}$ | $21 \times 10^{-6}$ |
| Mercury | $3.7 \times 10^{-11}$ | $3.8 \times 10^{-6}$ |
| Water | $45.8 \times 10^{-11}$ | $46.4 \times 10^{-6}$ |

For small pressure changes in a solid or a liquid we consider $B$ to be constant. The bulk modulus of a *gas*, however, depends on the initial pressure $p_0$. Table 11–1 includes values of the bulk modulus for several solid materials. Its units, force per unit area, are the same as those of pressure (and of tensile or compressive stress).

The reciprocal of the bulk modulus is called the **compressibility** $k$. From Eq. (11–16),

$$k = \frac{1}{B} = -\frac{\Delta V/V_0}{\Delta p} = -\frac{1}{V_0}\frac{\Delta V}{\Delta p}. \qquad (11\text{–}17)$$

Compressibility is the *fractional decrease in volume,* $-\Delta V/V_0$, *per unit increase* $\Delta p$ *in pressure*. The units of compressibility are those of *reciprocal pressure,* area per unit force.

Values of compressibility $k$ for several liquids are listed in Table 11–2. From this table the compressibility of water is $46.4 \times 10^{-6}\ \text{atm}^{-1}$. This means that for each atmosphere of increase in pressure the volume decreases by 46.4 parts per million. Materials with a small bulk modulus and a large compressibility are easy to compress; those with a larger bulk modulus and a smaller compressibility compress less with the same pressure increase.

## EXAMPLE 11–7

The volume of oil contained in a certain hydraulic press is 0.25 $m^3$ = 250 L. Find the decrease in the volume of the oil when it is subjected to a pressure increase $\Delta p = 1.6 \times 10^7$ Pa (about 160 atm or 2300 psi). The bulk modulus of the oil is $B = 5.0 \times 10^9$ Pa (about $5.0 \times 10^4$ atm), and its compressibility is $k = 1/B = 20 \times 10^{-6}\ \text{atm}^{-1}$.

**SOLUTION** To find the volume change $\Delta V$, we solve Eq. (11–16) for $\Delta V$ and then substitute in the given values:

$$\Delta V = -\frac{V_0\,\Delta p}{B}$$

$$\Delta V = -\frac{(0.25\ \text{m}^3)(1.6 \times 10^7\ \text{Pa})}{5.0 \times 10^9\ \text{Pa}} = -8.0 \times 10^{-4}\ \text{m}^3 = -0.80\ \text{L}.$$

Alternatively, we can use Eq. (11–17). Solving for $\Delta V$ and using the approximate unit conversions given above, we get

$$\Delta V = -kV_0\,\Delta p = -(20 \times 10^{-6}\ \text{atm}^{-1})(0.25\ \text{m}^3)(160\ \text{atm}) = -8.0 \times 10^{-4}\ \text{m}^3.$$

This represents a substantial compression of the oil under the action of a very large pressure.

# 11–7 SHEAR STRESS AND STRAIN

The third kind of stress-strain situation, shown in Fig. 11–10, is called *shear.* The body in Fig. 11–16 is under **shear stress,** which we define as the force $F_{\parallel}$, *tangent* to a material surface, divided by the area $A$ on which it acts:

$$\text{Shear stress} = \frac{F_{\parallel}}{A}. \qquad (11\text{–}18)$$

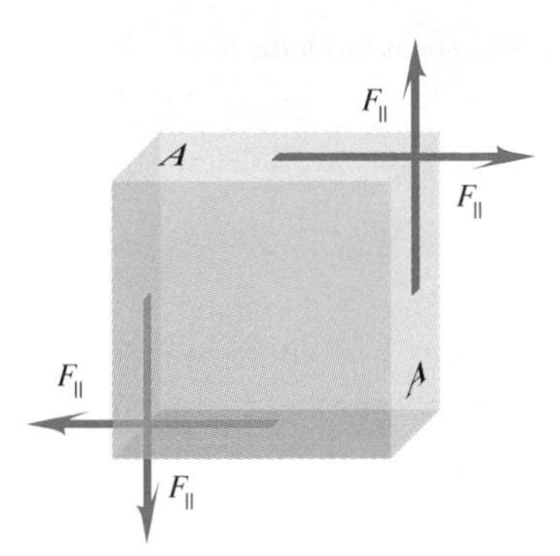


**11–16** A body under shear stress. The area $A$ is the edge area on which each force $F_{\parallel}$ acts.

Note that the forces acting on opposite sides of the body are *couples* (Section 11–4). Shear stress, like the other two types of stress, is a force per unit area. For systems in equilibrium, shear stress can exist only in *solid* materials.

A shear deformation is shown in Fig. 11–17. The black outline *abcd* represents an unstressed block of material, such as a book or a chunk of metal. The area $a'b'c'd'$ shaded in color shows the same block under shear stress. In part (a) of the figure the centers of the stressed and unstressed block coincide. The deformation in part (b) is the same as that in part (a), but the block has been shifted to make the edges *ad* and $a'd'$ coincide. In shear stress the lengths of the faces remain nearly constant; all dimensions parallel to the diagonal *ac* increase in length, and those parallel to the diagonal *bd* decrease in length. We define **shear strain** as the ratio of the displacement $x$ of corner $b$ to the transverse dimension $h$:

$$\text{Shear strain} = \frac{x}{h} = \tan\phi. \tag{11–19}$$

In real-life situations, $x$ is nearly always much smaller than $h$, $\tan\phi$ is very nearly equal to $\phi$, and the strain is simply the angle $\phi$, measured in radians. Like all strains, shear strain is a dimensionless number because it is a ratio of two lengths.

If the forces are small enough that Hooke's law is obeyed, the shear strain is *proportional* to the shear stress. The corresponding elastic modulus (ratio of shear stress to shear strain) is called the **shear modulus,** denoted by $S$.

$$S = \frac{\text{Shear stress}}{\text{Shear strain}} = \frac{F_{\parallel}/A}{x/h} = \frac{h}{A}\frac{F_{\parallel}}{x} = \frac{F_{\parallel}/A}{\phi}, \tag{11–20}$$

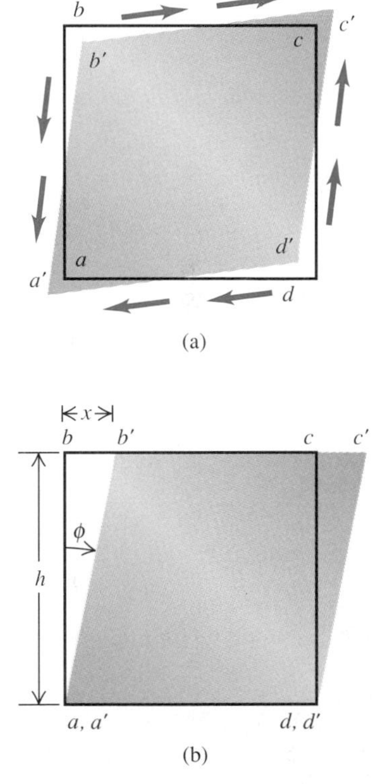


**11–17** The change in the shape of a block under shear stress. The shear strain, defined as $x/h$, is very nearly equal to $\phi$ in radians. In part (b) the deformed block has been shifted and rotated so that its bottom edge coincides with that of the undeformed block in order to show the strain $\phi$ clearly.

with $x$ and $h$ defined as in Fig. 11–17.

The shear modulus is also called the *modulus of rigidity,* or the *torsion modulus.* Several values of shear modulus are given in Table 11–1. For a given material, $S$ is usually a third to a half as large as $Y$. Keep in mind that only a *solid* material has a shear modulus. A liquid or gas flows freely under the action of a shear stress, so a fluid at rest cannot sustain such a stress.

## EXAMPLE 11–8

Suppose the body in Fig. 11–17 is the brass base plate of an outdoor sculpture; it experiences shear forces as a result of a mild earthquake. The plate is 0.80 m square and 0.50 cm thick. How large a force $F$ must be exerted on each of its edges if the displacement $x$ (see Fig. 11–17b) is 0.016 cm? The shear modulus of brass is $0.36 \times 10^{11}$ Pa.

**SOLUTION** The shear stress at each edge of the plate is the force divided by the area of the edge:

$$\text{Shear stress} = \frac{F_{\parallel}}{A} = \frac{F_{\parallel}}{(0.80\text{ m})(0.0050\text{ m})} = (250\text{ m}^{-2})F_{\parallel}.$$

The shear strain is

$$\text{Shear strain} = \frac{x}{h} = \frac{1.6 \times 10^{-4}\text{ m}}{0.80\text{ m}} = 2.0 \times 10^{-4}.$$

The shear modulus $S$ is

$$S = \frac{\text{Stress}}{\text{Strain}} = 0.36 \times 10^{11}\text{ Pa} = \frac{(250\text{ m}^{-2})F_{\parallel}}{2.0 \times 10^{-4}},$$

and

$$F_{\parallel} = 2.9 \times 10^{4}\text{ N}.$$

## 11–8 ELASTICITY AND PLASTICITY

Hooke's law, the proportionality of stress and strain in elastic deformations, has a limited range of validity. In the past three sections we have used phrases such as "provided that the forces are small enough that Hooke's law is obeyed." Just what *are* the limitations of Hooke's law? We know that if you pull, squeeze, or twist *anything* hard enough, it will bend or break. Can we be more precise than that?

Let's look at tensile stress and strain again. Suppose we plot a graph of stress as a function of strain. If Hooke's law is obeyed, the graph is a straight line with a slope equal to Young's modulus. Figure 11–18 shows a typical stress-strain graph for a metal such as copper or soft iron. The strain is shown as the *percent* elongation, and the horizontal scale is not uniform beyond the first portion of the curve, up to a strain of less than 1%. The first portion is a straight line, indicating Hooke's-law behavior with stress directly proportional to strain. This straight-line portion ends at point $a$; the stress at this point is called the **proportional limit.**

From $a$ to $b$ stress and strain are no longer proportional, and *Hooke's law is not obeyed.* If the load is gradually removed, starting at any point between $O$ and $b$, the curve is retraced until the material returns to its original length. The deformation is *reversible,* and the forces are conservative; the energy put into the material to cause the deformation is recovered when the stress is removed. In region $Ob$ we say that the material shows *elastic behavior.* Point $b$, the end of this region, is called the **yield point;** the stress at the yield point is called the **elastic limit.**

When we increase the stress beyond point $b$, the strain continues to increase. But now when we remove the load at some point beyond $b$, say $c$, the material does not come back to its original length. Instead, it follows the red line in Fig. 11–18. The length at

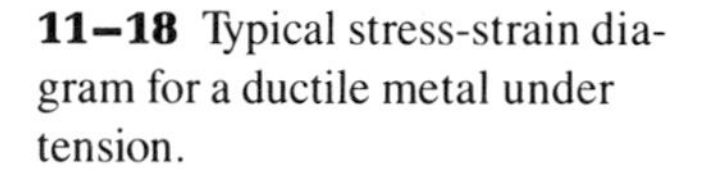

**11–18** Typical stress-strain diagram for a ductile metal under tension.

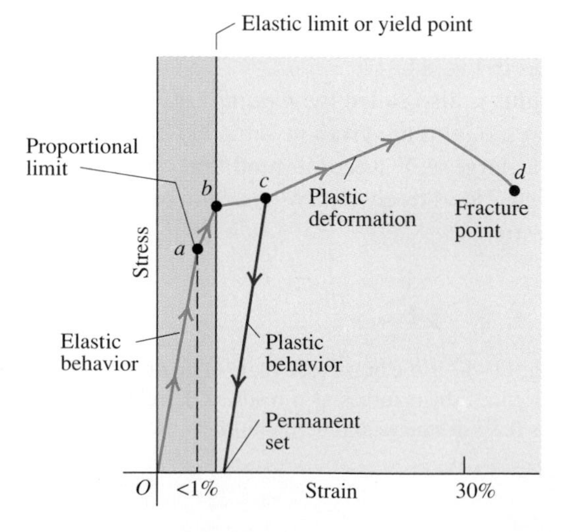

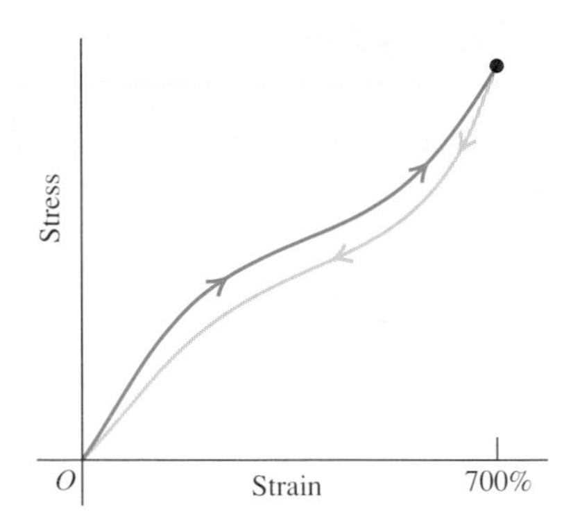


**11–19** Typical stress-strain diagram for vulcanized rubber, showing elastic hysteresis.

zero stress is now *greater* than the original length; the material has undergone an *irreversible deformation* and has acquired what we call a permanent set. Further increase of load beyond $c$ produces a large increase in strain with a relatively small increase in stress until a point $d$ is reached at which *fracture* takes place. The behavior of the material from $b$ to $d$ is called *plastic flow,* or *plastic deformation*. A plastic deformation is irreversible; *when the stress is removed, the material does not return to its original state.*

For some materials a large amount of plastic deformation takes place between the elastic limit and the fracture point. Such a material is said to be *ductile*. But if fracture occurs soon after the elastic limit is passed, the material is said to be *brittle*. A soft iron wire that can have considerable permanent stretch without breaking is ductile; a steel piano string that breaks soon after its elastic limit is reached is brittle.

Figure 11–19 shows a stress-strain curve for vulcanized rubber that has been stretched to over seven times its original length. The stress is not proportional to the strain, but the behavior is elastic because when the load is removed the material returns to its original length. The fact that the material follows different curves for increasing and decreasing stress is called *elastic hysteresis*. The work done by the material when it returns to its original shape is less than the work required to deform it; there are nonconservative forces associated with internal friction. Rubber with large elastic hysteresis is very useful for absorbing vibrations, as in engine mounts and shock-absorber bushings in cars.

The stress required to cause actual fracture of a material is called the **breaking stress,** the *ultimate strength,* or (for tensile stress) the *tensile strength*. Two materials, such as two types of steel, may have very similar elastic constants but vastly different breaking stresses. Table 11–3 gives a few typical values of breaking stress for several materials in tension.

The conversion factor $6.9 \times 10^8$ Pa $= 100{,}000$ psi will help put these numbers in perspective. For example, if the breaking stress of a particular steel is $6.9 \times 10^8$ Pa, then a bar with a 1-in.$^2$ cross section has a breaking strength of 100,000 lb.

**TABLE 11–3 Breaking Stresses of Materials**

| Material | Breaking stress (Pa or N/m$^2$) |
|---|---|
| Aluminum | $2.2 \times 10^8$ |
| Brass | $4.7 \times 10^8$ |
| Glass | $10 \times 10^8$ |
| Iron | $3.0 \times 10^8$ |
| Phosphor bronze | $5.6 \times 10^8$ |
| Steel | $5\text{–}20 \times 10^8$ |

## SUMMARY

### KEY TERMS

first condition for equilibrium
second condition for equilibrium
center of gravity
center of mass
couple
stress
strain
elastic modulus
Hooke's law
tension
tensile stress
pascal
compression
compressive stress
tensile strain
compressive strain
Young's modulus
Poisson's ratio
bulk stress
bulk strain
pressure
atmosphere
volume strain
bulk modulus
compressibility
shear stress
shear strain
shear modulus
proportional limit
yield point
elastic limit
breaking stress

• For a rigid body to be in equilibrium, two conditions must be satisfied. First, the vector sum of forces must be zero. Second, the sum of torques about any point must be zero. The first condition, in terms of components, is

$$\Sigma F_x = 0, \qquad \Sigma F_y = 0, \qquad \Sigma F_z = 0. \tag{11-1}$$

The second condition is

$$\Sigma \tau = 0 \quad \textit{about any point.} \tag{11-2}$$

• The torque due to the weight of a body can be obtained by assuming the entire weight to be concentrated at the center of mass, in this context also called the center of gravity. The coordinates of the center of gravity are given by

$$x_{\text{cm}} = \frac{m_1 x_1 + m_2 x_2 + \cdots}{m_1 + m_2 + \cdots} = \frac{\Sigma m_i x_i}{\Sigma m_i},$$
$$y_{\text{cm}} = \frac{m_1 y_1 + m_2 y_2 + \cdots}{m_1 + m_2 + \cdots} = \frac{\Sigma m_i y_i}{\Sigma m_i}, \tag{11-3}$$
$$z_{\text{cm}} = \frac{m_1 z_1 + m_2 z_2 + \cdots}{m_1 + m_2 + \cdots} = \frac{\Sigma m_i z_i}{\Sigma m_i}.$$

• The torque of a force can be computed by finding the torque of each component, using its appropriate moment arm and sign, and then adding these component torques.

• A couple consists of two forces with equal magnitude $F$ and opposite direction, acting on a body. If the lines of action of the forces are separated by a distance $l$, the magnitude of the total torque due to the couple is

$$\tau = lF. \tag{11-8}$$

This torque is the same for all points.

• Hooke's law states that in elastic deformations stress is proportional to strain:

$$\frac{\text{Stress}}{\text{Strain}} = \text{Elastic modulus.} \tag{11-9}$$

• Tensile stress is tensile force per unit area, $F_\perp/A$. Tensile strain is fractional change in length, $\Delta l/l_0$. Young's modulus $Y$ is the ratio of tensile stress to tensile strain:

$$Y = \frac{F_\perp/A}{\Delta l/l_0} = \frac{l_0}{A}\frac{F_\perp}{\Delta l}. \tag{11-12}$$

Compressive stress and strain are defined in the same way as tensile stress and strain; for many materials Young's modulus has the same value for both tensile and compressive stresses and strains.

• Poisson's ratio $\sigma$ is the negative of the ratio of the fractional change of width $\Delta w/w_0$ of a bar under tension or compression and the fractional change of length $\Delta l/l_0$:

$$\frac{\Delta w}{w_0} = -\sigma\frac{\Delta l}{l_0}, \tag{11-13}$$

• Pressure is force per unit area:

$$p = \frac{F_\perp}{A}. \tag{11-14}$$

The bulk modulus $B$ is the negative of the ratio of pressure change $\Delta p$ (bulk stress) to fractional volume change $\Delta V/V_0$:

$$B = -\frac{\Delta p}{\Delta V/V_0}. \qquad (11\text{–}16)$$

Compressibility $k$ is the reciprocal of bulk modulus:

$$k = \frac{1}{B} = -\frac{\Delta V/V_0}{\Delta p} = -\frac{1}{V_0}\frac{\Delta V}{\Delta p}. \qquad (11\text{–}17)$$

• Shear stress is force per unit area $F_{\parallel}/A$ for a force applied parallel to a surface. Shear strain is the angle $\phi$ shown in Fig. 11–17. The shear modulus $S$ is the ratio of shear stress to shear strain:

$$S = \frac{F_{\parallel}/A}{x/h} = \frac{h}{A}\frac{F_{\parallel}}{x} = \frac{F_{\parallel}/A}{\phi}. \qquad (11\text{–}20)$$

• The proportional limit is the maximum stress for which stress and strain are proportional. Beyond the proportional limit Hooke's law is not valid. The elastic limit is the stress beyond which irreversible deformation occurs. The breaking stress, or ultimate strength, is the stress at which the material breaks.

# EXERCISES

## Section 11–2 Center of Gravity

**11–1** A ball with radius $r_1 = 0.060$ m and mass 1.00 kg is attached by a light rod 0.400 m long to a second ball with radius $r_2 = 0.080$ m and mass 4.00 kg (Fig. 11–20). Where is the center of gravity of this system?

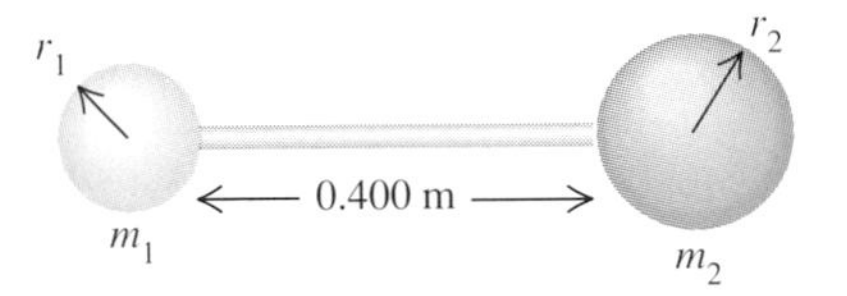


**FIGURE 11–20**

**11–2** Suppose the rod in Exercise 11–1 is uniform and has a mass of 2.00 kg. Where is the system's center of gravity?

## Section 11–3 Solving Equilibrium Problems

**11–3** Two people carry a heavy electric motor by placing it on a light board 2.00 m long. One person lifts at one end with a force of 700 N, and the other lifts the opposite end with a force of 400 N. What is the weight of the motor, and where is its center of gravity located along the board?

**11–4** Suppose the board in Exercise 11–3 is not light but weighs 200 N, with its center of gravity at its center. The two people exert the same forces as before. What is the weight of the motor in this case, and where is its center of gravity located?

**11–5 Raising a Ladder.** An 1800-N ladder carried by a fire truck is 20.0 m long, and its center of gravity is at its center.

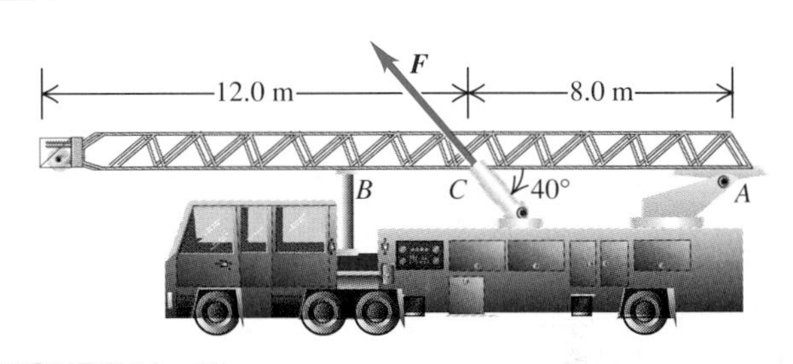


**FIGURE 11–21**

The ladder is pivoted at one end ($A$) about a pin (Fig. 11–21); the friction torque at the pin can be neglected. The ladder is raised into position by a force applied by a hydraulic piston at $C$. Point $C$ is 8.0 m from $A$, and the force $\boldsymbol{F}$ exerted by the piston makes an angle of 40° with the ladder. What magnitude must $\boldsymbol{F}$ have to just lift the ladder off the support bracket at $B$?

**11–6** Two people are carrying a uniform ladder that is 6.00 m long and weighs 500 N. If one applies an upward force equal to 180 N at one end, at what point does the other person lift?

**11–7** A diving board 3.00 m long is supported at a point 1.00 m from the end, and a diver weighing 580 N stands at the free end (Fig. 11–22). The diving board is of uniform cross section and

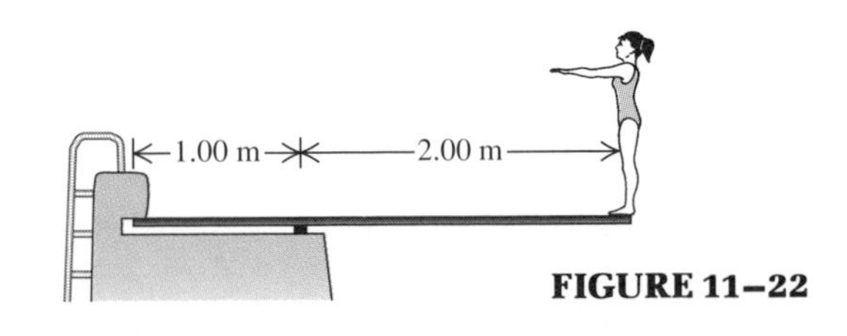


**FIGURE 11–22**

weighs 400 N. Find a) the force at the support point; b) the force at the end that is held down.

**11–8** A uniform 350-N trapdoor in a ceiling is hinged at one side. Find the net upward force needed to begin to open it and the total force exerted on the door by the hinges, a) if the upward force is applied at the center; b) if the upward force is applied at the center of the edge opposite the hinges.

**11–9** Show that the torques about point $B$ due to $T_x$ and $T_y$ in Example 11–4 sum to zero. Do this for the general case, rather than for the specific numerical values given at the end of the example.

**11–10 A Ladder and a Frictionless Wall.** A uniform ladder 10.0 m long rests against a frictionless vertical wall with its lower end 6.0 m from the wall. The ladder weighs 400 N. The coefficient of static friction between the foot of the ladder and the ground is 0.40. A man weighing 860 N climbs slowly up the ladder. a) What is the maximum frictional force that the ground can exert on the ladder at its lower end? b) What is the actual frictional force when the man has climbed 3.0 m along the ladder? c) How far along the ladder can the man climb before the ladder starts to slip?

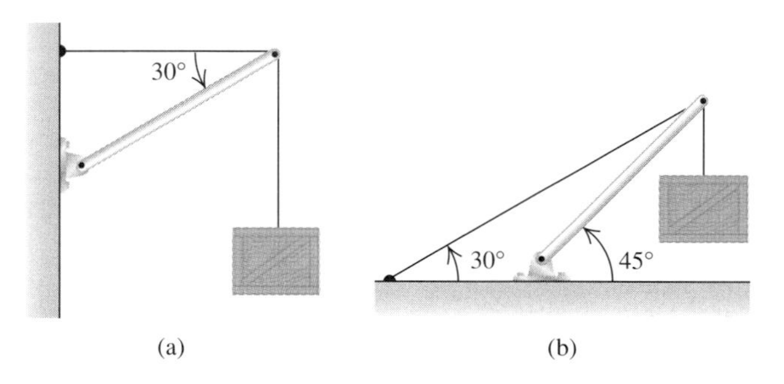


**FIGURE 11–23**

**11–11** Find the tension $T$ in each cable and the magnitude and direction of the force exerted on the strut by the pivot in each of the arrangements in Fig. 11–23. Let the weight of the suspended object in each case be $w$. The strut is uniform and also has weight $w$.

**11–12** The horizontal beam in Fig. 11–24 weighs 150 N, and its center of gravity is at its center. Find a) the tension in the upper cable; b) the horizontal and vertical components of the force exerted on the beam at the wall.

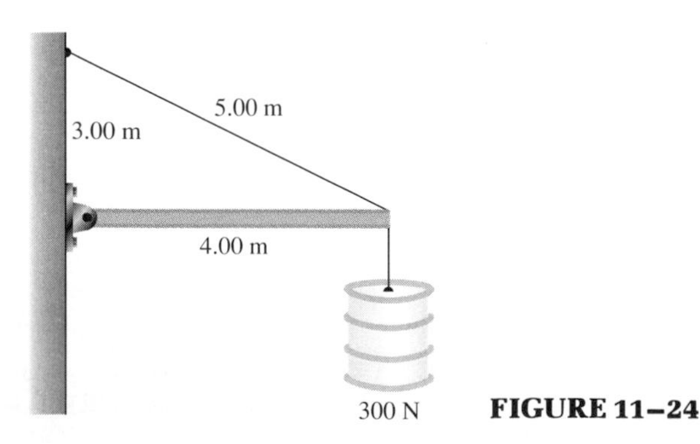


**FIGURE 11–24**

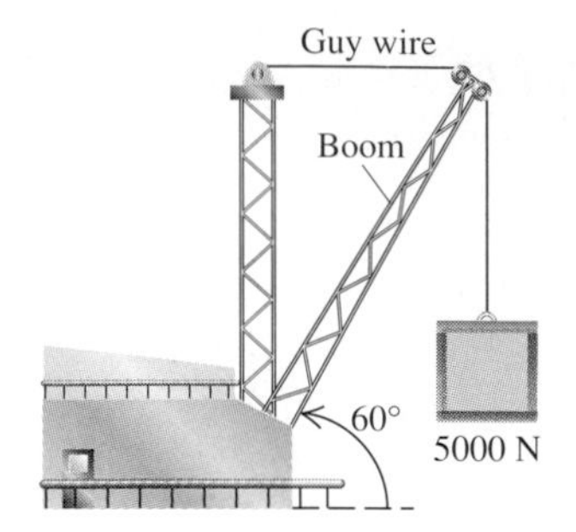


**FIGURE 11–25**

**11–13** A door 1.00 m wide and 2.50 m high weighs 250 N and is supported by two hinges, one 0.50 m from the top and the other 0.50 m from the bottom. Each hinge supports half the total weight of the door. Assuming that the door's center of gravity is at its center, find the horizontal components of force exerted on the door by each hinge.

**11–14** The boom in Fig. 11–25 is uniform and weighs 3200 N. The boom is attached with a frictionless pivot at its lower end. a) Find the tension in the guy wire and the horizontal and vertical components of the force exerted on the boom at its lower end. b) Does the line of action of the force exerted on the boom at its lower end lie along the boom?

**11–15** Two forces with magnitude $F_1 = F_2 = 6.00$ N are applied to a rod as shown in Fig. 11–26. a) Calculate the net torque about point $O$ due to these two forces by calculating the torque due to each separate force. b) Calculate the net torque about point $P$ due to these two forces by calculating the torque due to each separate force. c) Compare your results to Eq. (11–8).

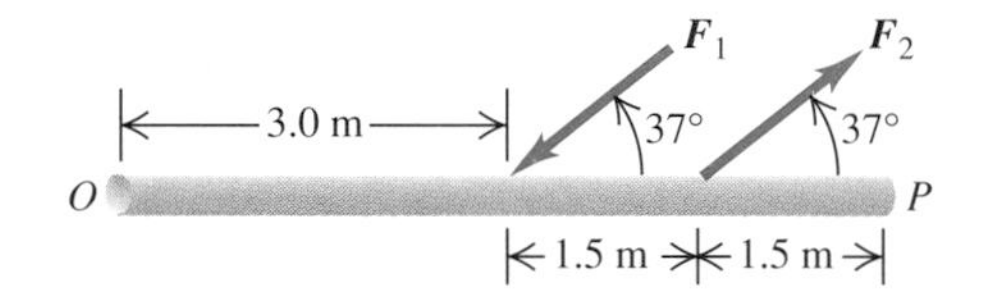


**FIGURE 11–26**

**11–16** Two equal parallel forces of magnitude $F_1 = F_2 = 8.00$ N are applied to a rod as shown in Fig. 11–27. a) What should the distance $l$ between the forces be if they are to provide a net

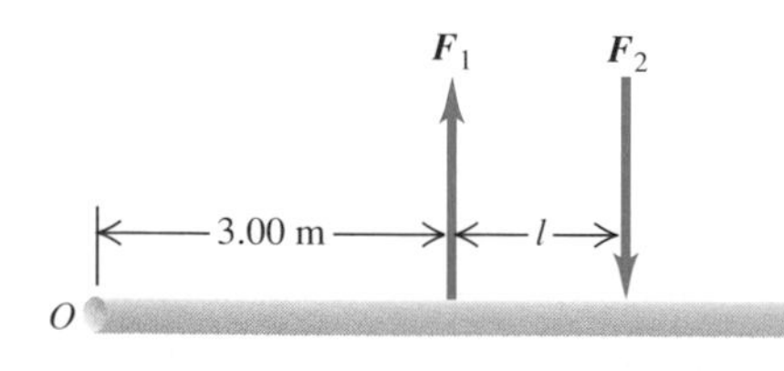


**FIGURE 11–27**

torque of 9.60 N · m about the left-hand end of the rod? b) Is this torque clockwise or counterclockwise?

### Section 11–5 Tensile Stress and Strain

**11–17** A metal rod that is 4.00 m long and 0.50 $cm^2$ in cross-section area is found to stretch 0.20 cm under a tension of 4,000 N. What is Young's modulus for this metal?

**11–18 Stress on a Mountaineer's Rope.** A nylon rope used by mountaineers elongates 1.50 m under the weight of an 80.0-kg climber. a) If the rope is 50.0 m in length and 7.0 mm in diameter, what is Young's modulus for this material? b) If Poisson's ratio for nylon is 0.20, find the change in diameter under this stress.

**11–19** A circular steel wire 3.00 m long is to stretch no more than 0.20 cm when a tensile force of 300 N is applied to each end of the wire. What minimum diameter is required for the wire?

**11–20 The Biceps Curl.** A relaxed biceps muscle requires a force of 25.0 N for an elongation of 3.0 cm; the same muscle under maximum tension requires a force of 500 N for the same elongation. Find Young's modulus for the muscle tissue under each of these conditions if the muscle is assumed to be a uniform cylinder with length 0.200 m and cross-section area 50.0 $cm^2$.

**11–21** A vertical steel post 15 cm in diameter and 3.00 m long is required to support a load of 6000 kg. What is a) the stress in the post? b) the strain in the post? c) the change in the post's length?

**11–22** Two round rods, one steel and the other brass, are joined end to end. Each rod is 0.500 m long and 2.00 cm in diameter. The combination is subjected to a tensile force with magnitude 3000 N. For each rod, what is a) the strain? b) the elongation? c) the change in diameter?

**11–23** A 5.0-kg mass hangs on a vertical steel wire 0.50 m long and $3.0 \times 10^{-3}$ $cm^2$ in cross-section area. Hanging from the bottom of this mass is a similar steel wire from which is hung a 10.0-kg mass. For each wire, compute a) the tensile strain; b) the elongation.

### Section 11–6 Bulk Stress and Strain

**11–24** In the Challenger Deep of the Marianas Trench the depth of seawater is 10.9 km, and the pressure is $1.10 \times 10^8$ Pa (about $1.09 \times 10^3$ atm). a) If a cubic meter of water is taken from the surface to this depth, what is the change in its volume? (Normal atmospheric pressure is about $1.0 \times 10^5$ Pa. Assume that $k$ for seawater is the same as the freshwater value given in Table 11–2.) b) What is the density of seawater at this depth? (At the surface seawater has a density of $1.03 \times 10^3$ $kg/m^3$.)

**11–25** A specimen of oil having an initial volume of 1000 $cm^3$ is subjected to a pressure increase of $1.8 \times 10^6$ Pa, and the volume is found to decrease by 0.30 $cm^3$. What is the bulk modulus of the material? The compressibility?

### Section 11–7 Shear Stress and Strain

**11–26** Suppose the object in Fig. 11–16 is a square steel plate, 10.0 cm on a side and 1.00 cm thick. Find the magnitude of force required on each of the four sides to cause a shear strain of 0.0300.

**11–27** Two strips of metal are riveted together at their ends by four rivets, each with diameter 0.300 cm. What is the maximum tension that can be exerted by the riveted strip if the shearing stress on each rivet is not to exceed $6.00 \times 10^8$ Pa? Assume that each rivet carries one quarter of the load.

### Section 11–8 Elasticity and Plasticity

**11–28** A steel wire has the following properties:

$$\begin{aligned} \text{Length} &= 5.00 \text{ m} \\ \text{Cross-section area} &= 0.050 \text{ cm}^2 \\ \text{Young's modulus} &= 2.0 \times 10^{11} \text{ Pa} \\ \text{Shear modulus} &= 0.84 \times 10^{11} \text{ Pa} \\ \text{Proportional limit} &= 3.6 \times 10^8 \text{ Pa} \\ \text{Breaking stress} &= 11.0 \times 10^8 \text{ Pa} \end{aligned}$$

The wire is fastened at its upper end and hangs vertically. a) How great a weight can be hung from the wire without exceeding the proportional limit? b) How much will the wire stretch under this load? c) What is the maximum weight that can be supported?

**11–29** The elastic limit of a steel elevator cable is $2.75 \times 10^8$ Pa, and the cross-section area is 4.00 $cm^2$. Find the maximum upward acceleration that can be given a 900-kg elevator supported by the cable if the stress is not to exceed one-third of the elastic limit.

## PROBLEMS

**11–30** A station wagon weighing $1.96 \times 10^4$ N has a wheelbase of 3.00 m. Ordinarily, 10,780 N rests on the front wheels and 8820 N on the rear wheels. A box weighing 2200 N is now placed on the tailgate, 1.00 m behind the rear axle. How much total weight now rests on the front wheels? On the rear wheels?

**11–31 Center of Gravity of a Car.** For an automobile with weight $w$ the front wheels support a fraction $f$ of the weight, so the normal force exerted at the front wheels is $fw$, and at the rear wheels the normal force is $(1 - f)w$. The distance between the front and rear axles is $d$. a) Show that the center of gravity of the car is a distance $fd$ in front of the rear wheels. b) Show that the general result derived in part (a) reproduces the numerical answer of Example 11–2.

**11–32** Sir Lancelot slowly rides out of the castle at Camelot and onto the 12.0-m-long drawbridge that passes over the moat

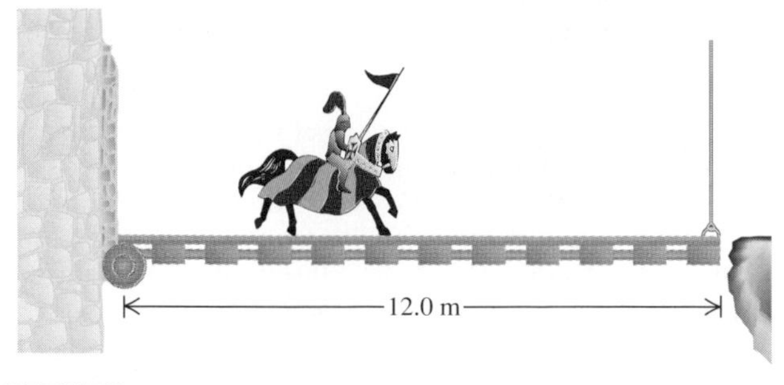


**FIGURE 11–28**

(Fig. 11–28). Unbeknownst to him, his enemies have partially severed the vertical cable holding up the front end of the bridge so that it will break under a tension of $5.00 \times 10^3$ N. The bridge has a mass of 200 kg and its center of gravity is at its center. Lancelot, his lance and armor, and his horse have a combined mass of 600 kg. Will the cable break before Lancelot reaches the end of the drawbridge? If so, how far from the castle end of the bridge will the center of gravity of the horse plus rider be when the cable breaks?

**11–33 Walking the Plank.** A uniform plank 15.0 m long, weighing 400 N, rests symmetrically on two supports 8.00 m apart (Fig. 11–29). A boy weighing 690 N starts at point $A$ and walks toward the right. a) Construct in the same diagram two graphs showing the upward forces $F_A$ and $F_B$ exerted on the plank at points $A$ and $B$, as functions of the coordinate $x$ of the boy. Let 1 cm = 100 N vertically and 1 cm = 1.00 m horizontally. b) From your diagram, how far beyond point $B$ can the boy walk before the plank tips? c) How far from the right end of the plank should support $B$ be placed so that the boy can walk just to the end of the plank without causing it to tip?

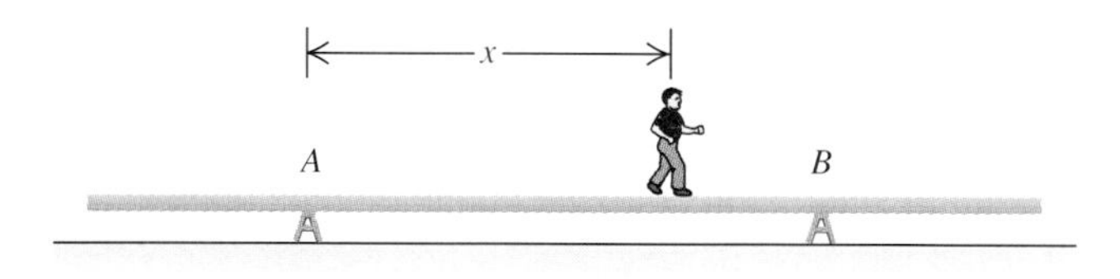


**FIGURE 11–29**

**11–34** A claw hammer is used to pull a nail out of a board (Fig. 11–30). The nail is at an angle of 60° to the board, and a force $\boldsymbol{F}_1$ of magnitude 400 N applied to the nail is required to pull it from the board. The hammer head contacts the board at point $A$, which is 0.080 m from where the nail enters the board. A horizontal force $\boldsymbol{F}_2$ is applied to the hammer handle at a distance of 0.300 m above the board. What magnitude of force $\boldsymbol{F}_2$ is required to apply the required 400-N force ($\boldsymbol{F}_1$) to the nail? (Neglect the weight of the hammer.)

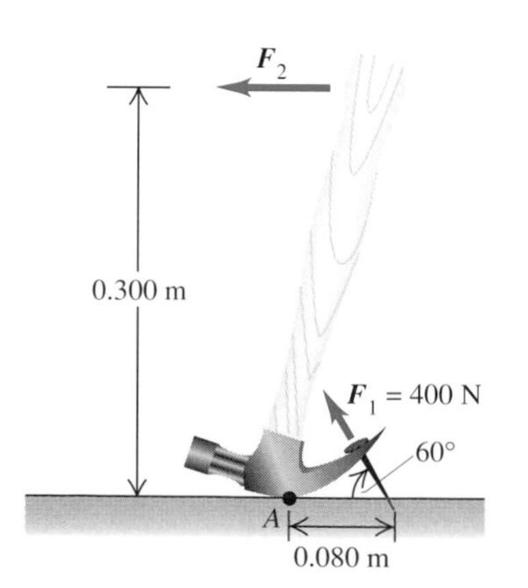


**FIGURE 11–30**

**11–35** Your dog Trixie is 0.740 m long (nose to hind legs). Her fore (front) legs are 0.15 m horizontally from her nose, her center of gravity is 0.15 m horizontally from her hind legs, and she weighs 140 N. a) How much force does a level floor exert on each of Trixie's fore feet and on each of her hind feet? b) If Trixie picks up a 25-N bone and holds it in her mouth (directly under her nose), what is the force exerted by the floor on each of her fore feet and each of her hind feet?

**11–36 A Truck on a Drawbridge.** A loaded cement mixer drives onto an old drawbridge, where it stalls with its center of gravity three-quarters of the way across the span. The truck driver radios for help, sets the hand-brake, and waits. Meanwhile, a boat approaches, so the drawbridge is raised by means of a cable attached to its end opposite the hinge (Fig. 11–31). The drawbridge span is 40.0 m long and has a mass of 3000 kg; its center of gravity is at its midpoint. The cement mixer, with driver, has a mass of 6000 kg. When the drawbridge has been raised to an angle of 30° above the horizontal, the cable makes an angle of 70° with the surface of the bridge. a) What is the tension $T$ in the cable when the drawbridge is held in this position? b) What are the horizontal and vertical components of the force the hinge exerts on the span?

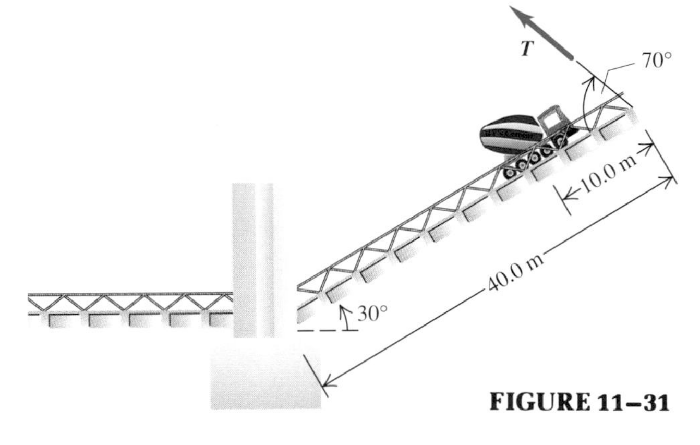


**FIGURE 11–31**

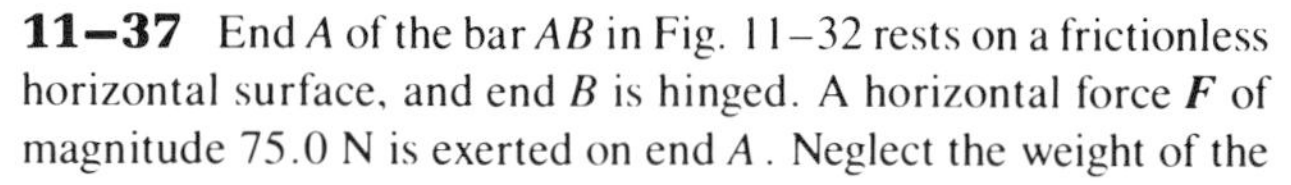
**11–37** End $A$ of the bar $AB$ in Fig. 11–32 rests on a frictionless horizontal surface, and end $B$ is hinged. A horizontal force $\boldsymbol{F}$ of magnitude 75.0 N is exerted on end $A$. Neglect the weight of the

**FIGURE 11–32**

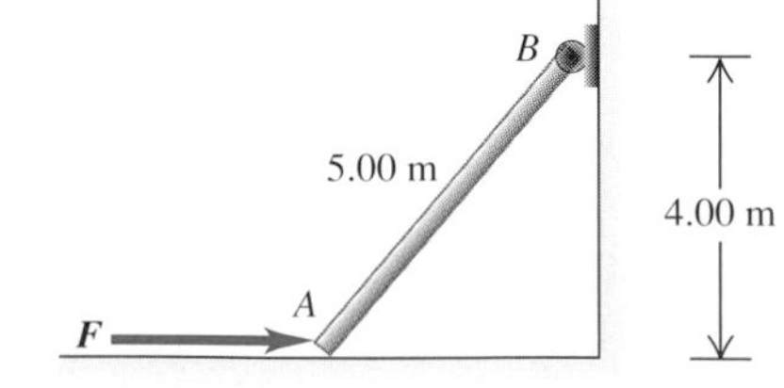

bar. What are the horizontal and vertical components of the force exerted by the bar on the hinge at $B$?

**11–38** A single additional force is to be applied to the bar in Fig. 11–33 to maintain it in equilibrium in the position shown. The weight of the bar can be neglected. a) What are the horizontal and vertical components of the required force? b) What is the angle the force must make with the bar? c) What is the magnitude of the required force? d) Where should the force be applied?

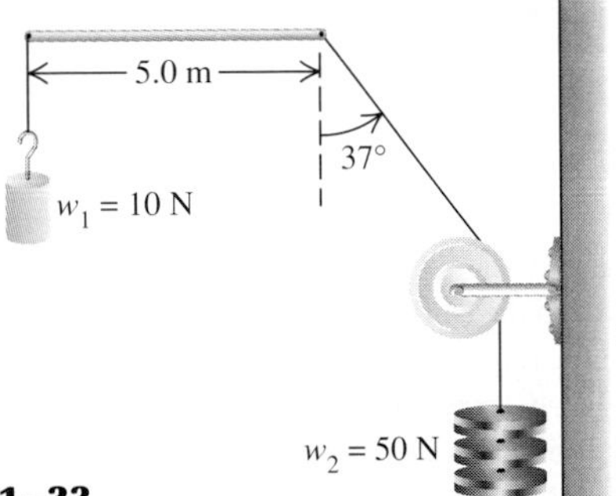


**FIGURE 11–33**

**11–39** A horizontal boom 6.00 m long is hinged to a vertical wall at one end, and a 500-N object hangs from its other end. The boom is supported by a guy wire running from its outer end to a point on the wall directly above the boom. The weight of the boom can be neglected. a) If the tension in this wire is not to exceed 1000 N, what is the minimum height above the boom at which it may be fastened to the wall? b) If the boom remains horizontal, by how many newtons would the tension be increased if the wire were fastened 0.50 m below this point?

**11–40** a) In Fig. 11–34 a 6.00-m-long uniform beam is hanging from a point 1.00 m to the right of its center. The beam weighs 80.0 N and makes an angle of 30.0° with the vertical. At the right-hand end of the beam a 100.0-N weight is hung; an unknown weight $w$ hangs at the other end. If the system is in equilibrium, what is $w$? b) If the beam makes instead an angle of 45.0° with the vertical, what is $w$?

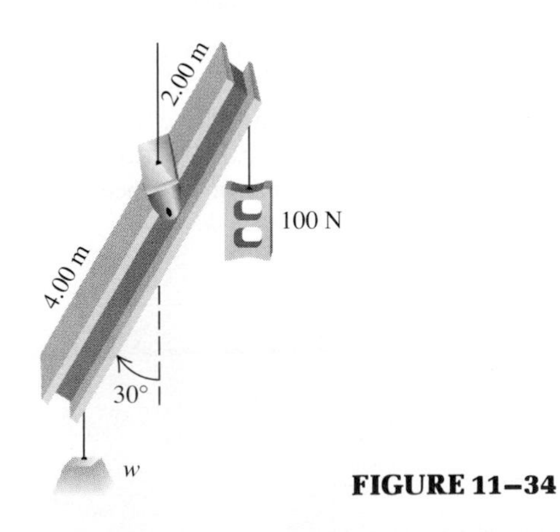


**FIGURE 11–34**

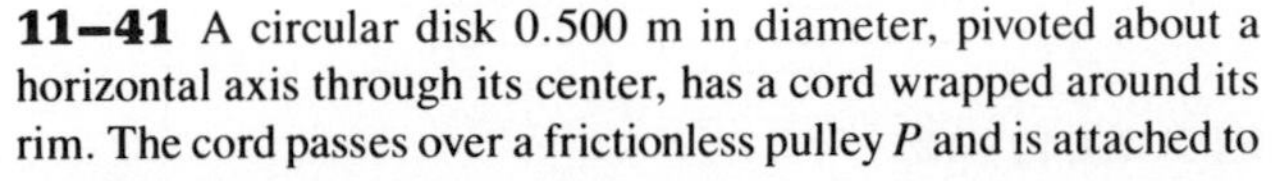

**11–41** A circular disk 0.500 m in diameter, pivoted about a horizontal axis through its center, has a cord wrapped around its rim. The cord passes over a frictionless pulley $P$ and is attached to

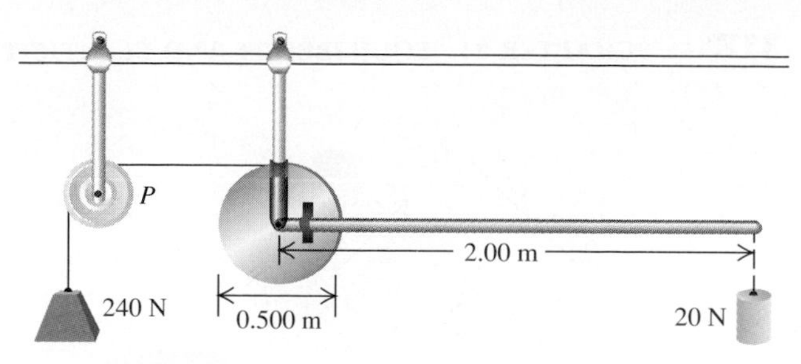


**FIGURE 11–35**

an object that weighs 240 N. A uniform rod 2.00 m long is fastened to the disk, with one end at the center of the disk. The apparatus is in equilibrium with the rod horizontal (Fig. 11–35). a) What is the weight of the rod? b) What is the new equilibrium direction of the rod when a second object weighing 20.0 N is suspended from the other end of the rod, as shown by the broken line? That is, what angle does the rod then make with the horizontal?

**11–42** A holiday decoration consists of two shiny glass spheres with masses 0.0240 kg and 0.0360 kg suspended as shown in Fig. 11–36 from a uniform rod with mass 0.200 kg and length 1.00 m. The rod is suspended from the ceiling by a vertical cord at each end so that it is horizontal. Calculate the tension in each of the cords $A$ through $F$.

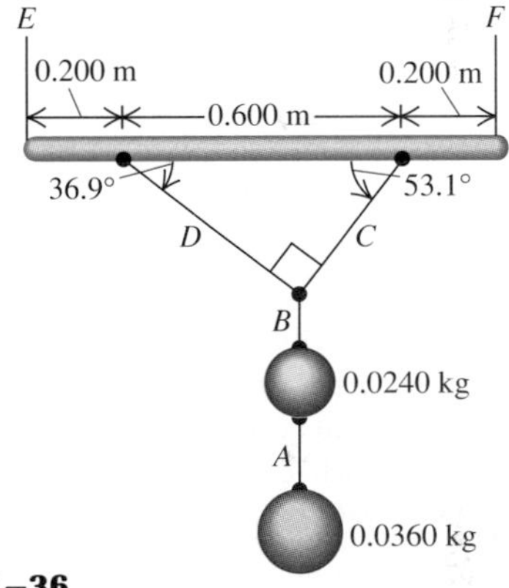


**FIGURE 11–36**

**11–43 The Farmyard Gate.** A gate 4.00 m long and 2.00 m high weighs 500 N. Its center of gravity is at its center, and it is hinged at $A$ and $B$. To relieve the strain on the top hinge, a wire $CD$ is connected as shown in Fig. 11–37. The tension in $CD$ is

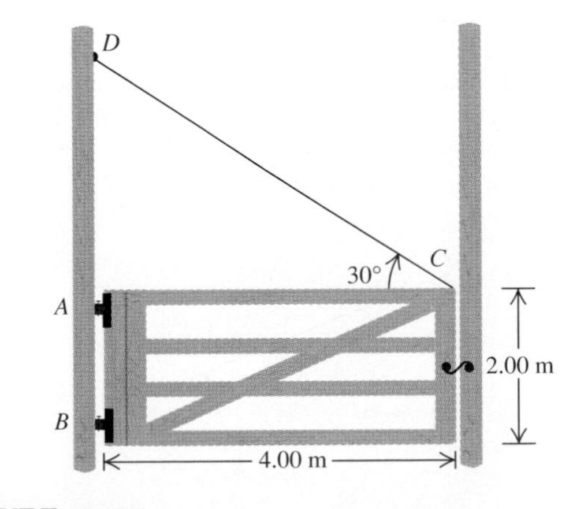


**FIGURE 11–37**

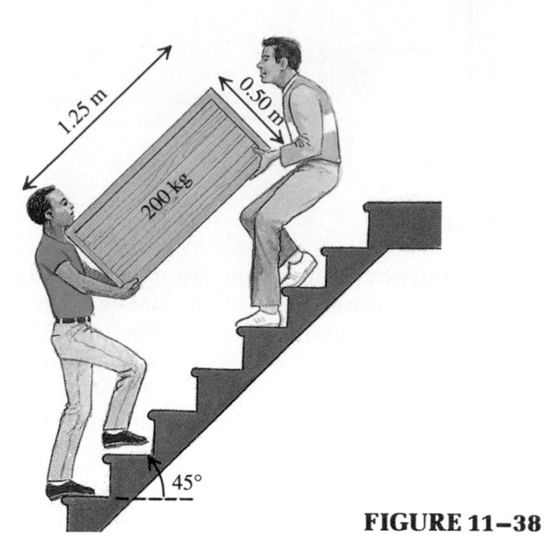

FIGURE 11–38

increased until the horizontal force at hinge $A$ is zero. a) What is the tension in the wire $CD$? b) What is the magnitude of the horizontal component of the force at hinge $B$? c) What is the combined vertical force exerted by hinges $A$ and $B$?

**11–44** A roller whose diameter is 1.00 m weighs 450 N. What is the minimum horizontal force necessary to pull the roller over a 0.100-m-high brick when the force is applied a) at the center of the roller? b) at the top of the roller?

**11–45** You and a friend are carrying a 200-kg crate up a flight of stairs, with you at the lower end. The crate is 1.25 m long and 0.50 m high, and its center of gravity is at its center. The stairs make a 45.0° angle with respect to the floor. The crate is carried also at a 45.0° angle, so its bottom side is parallel to the slope of the stairs (Fig. 11–38). If the force each of you applies is vertical, what is the magnitude of each of these forces? Is it best to be the person above or below on the stairs?

**11–46** One end of a meter stick is placed against a vertical wall (Fig. 11–39). The other end is held by a light cord making an angle $\theta$ with the stick. The coefficient of static friction between the end of the meter stick and the wall is 0.40. a) What is the maximum value the angle $\theta$ can have if the stick is to remain in equilibrium? b) Let the angle $\theta$ be 10°. A block of the same weight as the meter stick is suspended from the stick as shown, at a distance $x$ from the wall. What is the minimum value of $x$ for which the stick will remain in equilibrium? c) When $\theta = 10°$, how large must the coefficient of static friction be so that the block can be attached at the left end of the stick without causing it to slip?

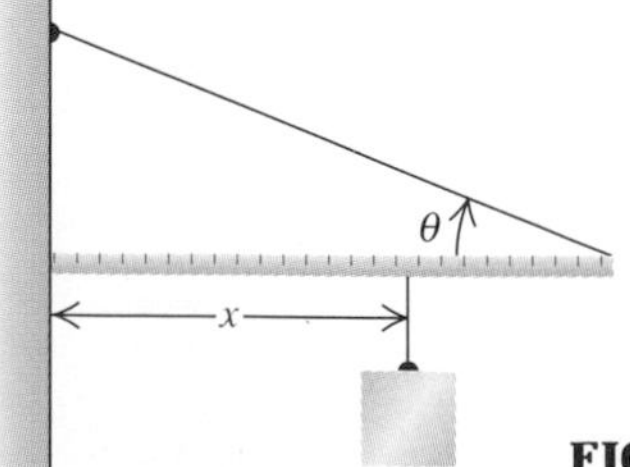

FIGURE 11–39

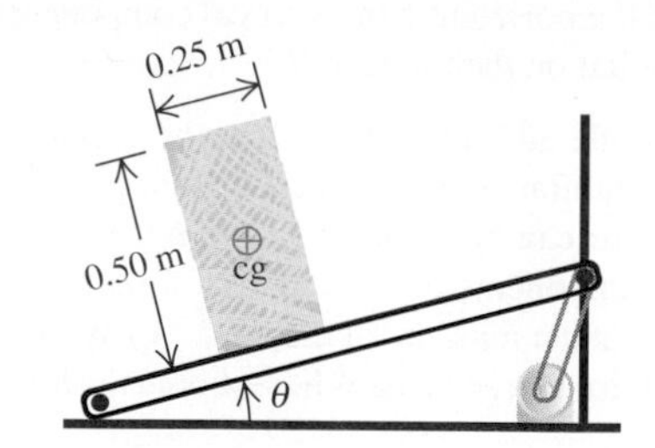

FIGURE 11–40

**11–47** An engineer is designing a conveyor system for loading hay bales into a wagon. Each bale has a mass of 25.0 kg and is 0.75 m long, 0.25 m wide, and 0.50 m high. The center of gravity of each bale is at its geometrical center. The coefficient of static friction between a bale and the conveyor belt is 0.30, and the belt moves with constant speed (Fig. 11–40). a) The angle $\theta$ of the conveyor is slowly increased. At some critical angle a bale will tip (if it doesn't slip first), and at some different critical angle it will slip (if it doesn't tip first). Find the two critical angles, and determine which happens first. b) Would the outcome of part (a) be different if the coefficient of friction were 0.75?

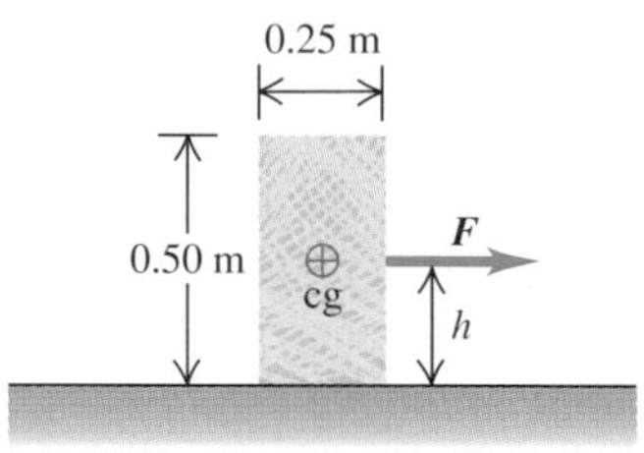

FIGURE 11–41

**11–48** The hay bale of Problem 11–47 is dragged along a horizontal surface with constant speed by a force $\boldsymbol{F}$ (Fig. 11–41). The coefficient of kinetic friction is 0.35. a) Find the magnitude of the force $\boldsymbol{F}$. b) Find the value of $h$ for which the bale just begins to tip.

**11–49** A garage door is mounted on an overhead rail (Fig. 11–42). The wheels at $A$ and $B$ have rusted so that they do not roll, but rather slide along the track. The coefficient of kinetic friction is 0.55. The distance between the wheels is 2.00 m, and each is

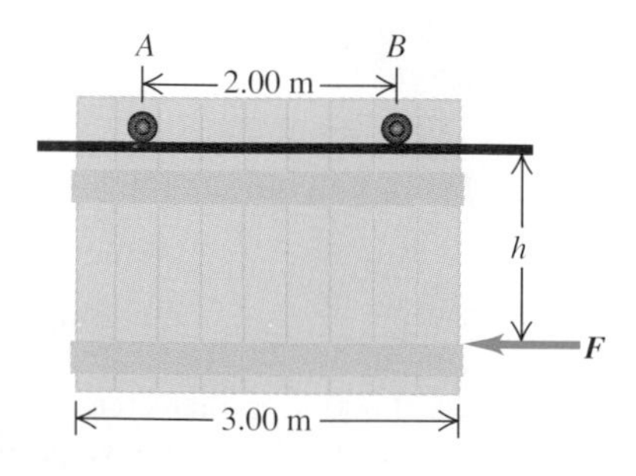

FIGURE 11–42

0.50 m from the vertical sides of the door. The door is uniform and weighs 800 N. It is pushed to the left at constant speed by a horizontal force $\boldsymbol{F}$. a) If the distance $h$ is 1.50 m, what is the vertical component of the force exerted on each wheel by the track? b) Find the maximum value $h$ can have without causing one wheel to leave the track.

**11–50** A copper wire 4.40 m long and 1.00 mm in diameter was given the test below. A load weighing 20.0 N was originally hung from the wire to keep it taut. The position of the lower end of the wire was read on a scale.

| Added load (N) | 0 | 10 | 20 | 30 | 40 | 50 | 60 | 70 |
|---|---|---|---|---|---|---|---|---|
| Scale reading (cm) | 3.02 | 3.07 | 3.12 | 3.17 | 3.22 | 3.27 | 3.32 | 4.27 |

a) Graph these values, plotting the increase in length horizontally and the added load vertically. b) Calculate the value of Young's modulus. c) What was the stress at the proportional limit?

**11–51** A 15.0-kg mass, fastened to the end of a steel wire with an unstretched length of 0.50 m, is whirled in a vertical circle with an angular velocity of 2.00 rev/s at the bottom of the circle. The cross-section area of the wire is 0.014 $cm^2$. Calculate the elongation of the wire when the mass is at the lowest point of the path.

**11–52 Stress on the Shin Bone.** Compressive strength of bones is important in everyday life. Young's modulus for bone is about $1.0 \times 10^{10}$ Pa. Bone can take only about a 1.0% change in its length before fracturing. a) What is the maximum force that can be applied to a bone whose minimum cross-section area is 3.0 $cm^2$? (This is approximately the cross-section area of a tibia, or shin bone, at its narrowest point.) b) Estimate from what maximum height a 70-kg person (one weighing about 150 lb) could jump and not fracture the tibia. Take the time between when the person first touches the floor and when he or she has stopped to be 0.030 s, and assume that the stress is distributed equally between the person's two legs.

**11–53** A 1.05-m-long rod whose weight is negligible is supported at its ends by wires $A$ and $B$ of equal length (Fig. 11–43). The cross-section area of $A$ is 1.00 $mm^2$; that of $B$ is 4.00 $mm^2$. Young's modulus for wire $A$ is $2.40 \times 10^{11}$ Pa; that for $B$ is $1.60 \times 10^{11}$ Pa. At what point along the rod should a weight $w$ be suspended to produce a) equal stresses in $A$ and $B$? b) equal strains in $A$ and $B$?

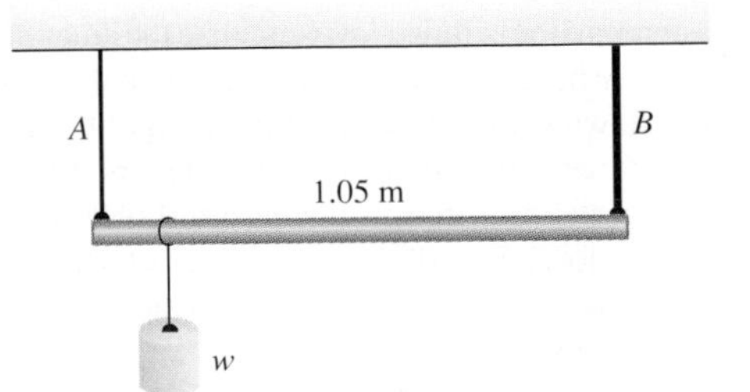


**FIGURE 11–43**

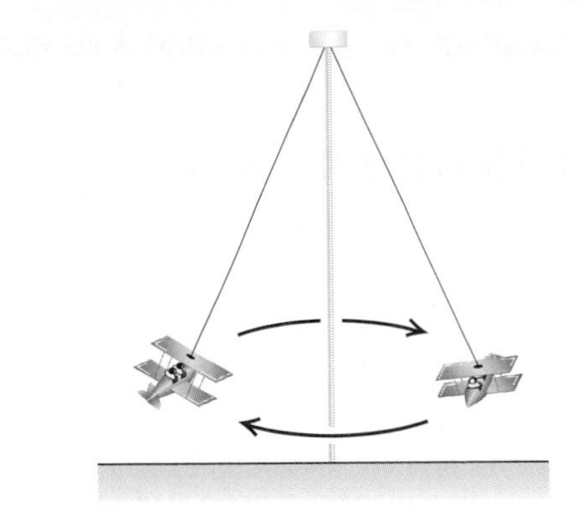

**FIGURE 11–44**

**11–54** An amusement park ride consists of seats attached to cables as shown in Fig. 11–44. Each steel cable has a length of 20.0 m and a cross-section area of 7.00 $cm^2$. a) What amount is the cable stretched when the ride is at rest? (Assume that each seat plus two people seated in it has a total weight of 2500 N.) b) When turned on, the ride has a maximum angular velocity of 0.90 rad/s. How much is the cable stretched then?

**11–55** A copper rod with a length of 1.40 m and a cross-section area of 2.00 $cm^2$ is fastened end to end to a steel rod with length $L$ and cross-section area 1.00 $cm^2$. The compound rod is subjected to equal and opposite pulls of magnitude $4.00 \times 10^4$ N at its ends. a) Find the length $L$ of the steel rod if the elongations of the two rods are equal. b) What is the stress in each rod? c) What is the strain in each rod?

**11–56** A moonshiner produces pure ethanol (ethyl alcohol) late at night and stores it in a stainless steel tank that is a cylinder 0.300 m in diameter with a tight-fitting piston at the top. The total volume of the tank is 200 L (0.200 $m^3$). In an attempt to squeeze a little more into the tank, the moonshiner piles lead bricks on the piston so that the total mass of bricks is 1420 kg. What additional volume of ethanol can the moonshiner squeeze into the tank? (Assume that the wall of the tank is perfectly rigid.)

**11–57** A bar with cross-section area $A$ is subjected to equal and opposite tensile forces $\boldsymbol{F}$ at its ends. Consider a plane through the bar making an angle $\theta$ with a plane at right angles to the bar (Fig. 11–45). a) What is the tensile (normal) stress at this plane in terms of $F$, $A$, and $\theta$? b) What is the shear (tangential) stress at the plane in terms of $F$, $A$, and $\theta$? c) For what value of $\theta$ is the tensile stress a maximum? d) For what value of $\theta$ is the shear stress a maximum?

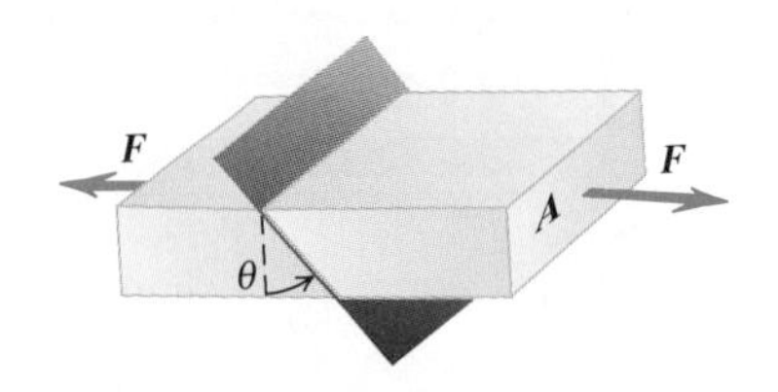


**FIGURE 11–45**

## CHALLENGE PROBLEMS

**11–58 Knocking over a Post.** One end of a post weighing 500 N and with height $h$ rests on a rough horizontal surface with $\mu_s = 0.35$. The upper end is held by a rope fastened to the surface and making an angle of 37.0° with the post (Fig. 11–46). A horizontal force $\boldsymbol{F}$ is exerted on the post as shown. a) If the force $\boldsymbol{F}$ is applied at the midpoint of the post, what is the largest value it can have without causing the post to slip? b) How large can the force be without causing the post to slip if its point of application is six-tenths of the way from the ground to the top of the post? c) Show that if the point of application of the force is too high, the post cannot be made to slip, no matter how great the force. Find the critical height for the point of application.

**FIGURE 11–46**

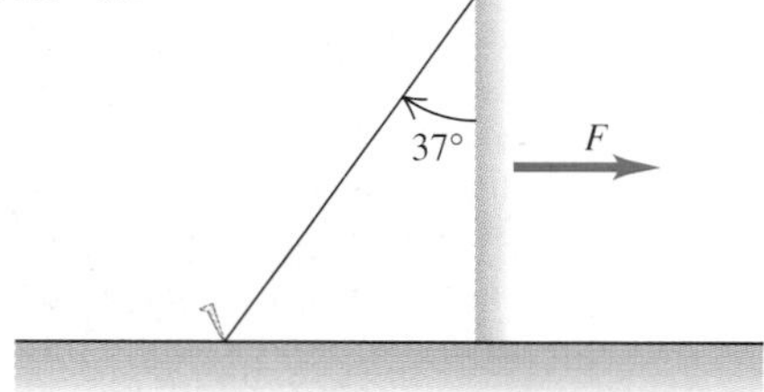


**11–59** A bookcase weighing 1500 N rests on a horizontal surface for which the coefficient of static friction is $\mu_s = 0.30$. The bookcase is 1.80 m tall and 2.00 m wide; its center of gravity is at its geometrical center. The bookcase rests on four short legs that are each 0.10 m from the edge of the bookcase. A person pulls on a rope attached to an upper corner of the bookcase with a force $\boldsymbol{F}$ that makes an angle $\theta$ with the bookcase (Fig. 11–47). a) If $\theta = 90°$, so $\boldsymbol{F}$ is horizontal, show that as $F$ is increased from zero, the bookcase will start to slide before it tips, and calculate the magnitude of $\boldsymbol{F}$ that will start the bookcase sliding. b) If $\theta = 0°$, so $\boldsymbol{F}$ is vertical, show that the bookcase will tip over rather than slide, and calculate the magnitude of $\boldsymbol{F}$ that will cause the bookcase to start to tip. c) Calculate as a function of $\theta$ the magnitude of $\boldsymbol{F}$ that will cause the bookcase to start to slide and the magnitude that will cause it to start to tip. What is the smallest value that $\theta$ can have and the bookcase still start to slide before it starts to tip?

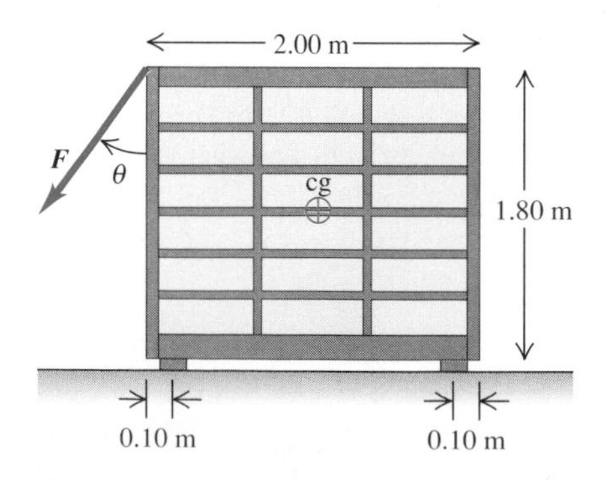

**FIGURE 11–47**

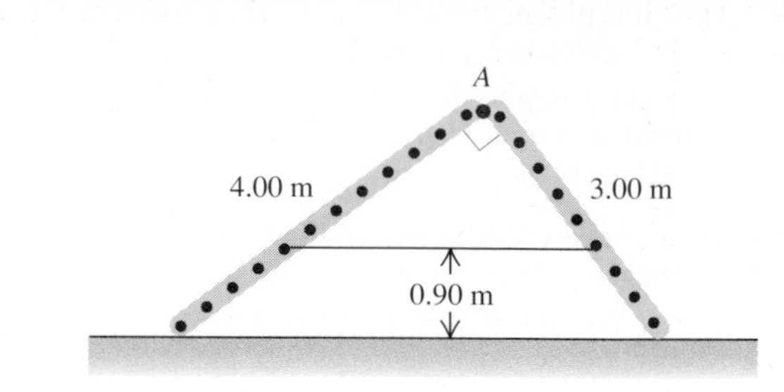


**FIGURE 11–48**

**11–60** Two ladders, 4.00 m and 3.00 m long, are hinged at point $A$ and tied together by a horizontal rope 0.90 m above the floor (Fig. 11–48). The ladders weigh 700 N and 450 N, respectively, and the center of gravity of each is at its center. Assume that the floor is frictionless. a) Find the upward force at the bottom of each ladder. b) Find the tension in the rope. c) Find the magnitude of the force one ladder exerts on the other at point $A$. d) If a load of 1000 N is suspended from point $A$, find the tension in the horizontal rope.

**11–61** The compressibility of sodium is to be measured by observing the displacement of the piston in Fig. 11–14 when a force is applied. The sodium is immersed in an oil that fills the cylinder below the piston. Assume that the piston and walls of the cylinder are perfectly rigid and that there is no friction and no oil leak. Compute the compressibility of the sodium in terms of the applied force $F$, the piston displacement $x$, the piston area $A$, the initial volume of the oil $V_0$, the initial volume of the sodium $v_0$, and the compressibility of the oil $k_0$.

**11–62 Bulk Modulus of an Ideal Gas.** The equation of state (the equation relating pressure, volume, and temperature) for an ideal gas is $pV = nRT$, where $n$ and $R$ are constants. a) Show that if the gas is compressed while the temperature $T$ is held constant, the bulk modulus is equal to the pressure. b) When an ideal gas is compressed without the transfer of any heat into or out of it, the pressure and volume are related by $pV^{\gamma} =$ constant, where $\gamma$ is a constant having different values for different gases. Show that in this case the bulk modulus is given by $B = \gamma p$.

**11–63** A 5.00-kg mass is hung from a vertical steel wire 2.00 m long and $5.00 \times 10^{-3}$ cm$^2$ in cross-section area. The wire is securely fastened to the ceiling. a) Calculate the amount the wire is stretched by the hanging mass. Now assume that the mass is very slowly pulled downward 0.0600 cm from its equilibrium position by an external force of magnitude $F$. Calculate b) the work done by gravity when the mass moves downward 0.0600 cm; c) the work done by the force $\boldsymbol{F}$; d) the work done by the force the wire exerts on the mass; e) the change in the elastic potential energy (the potential energy associated with the tensile stress in the wire) when the mass moves downward 0.0600 cm. Compare the answers in parts (d) and (e).

The planet Neptune is more than four billion kilometers distant from Earth and is almost four times larger in diameter than Earth. The Voyager 2 space probe revealed the presence of a large ($10^4$ km) cyclonic storm system on the planet, similar to Jupiter's Great Red Spot.

CHAPTER 12

# Gravitation

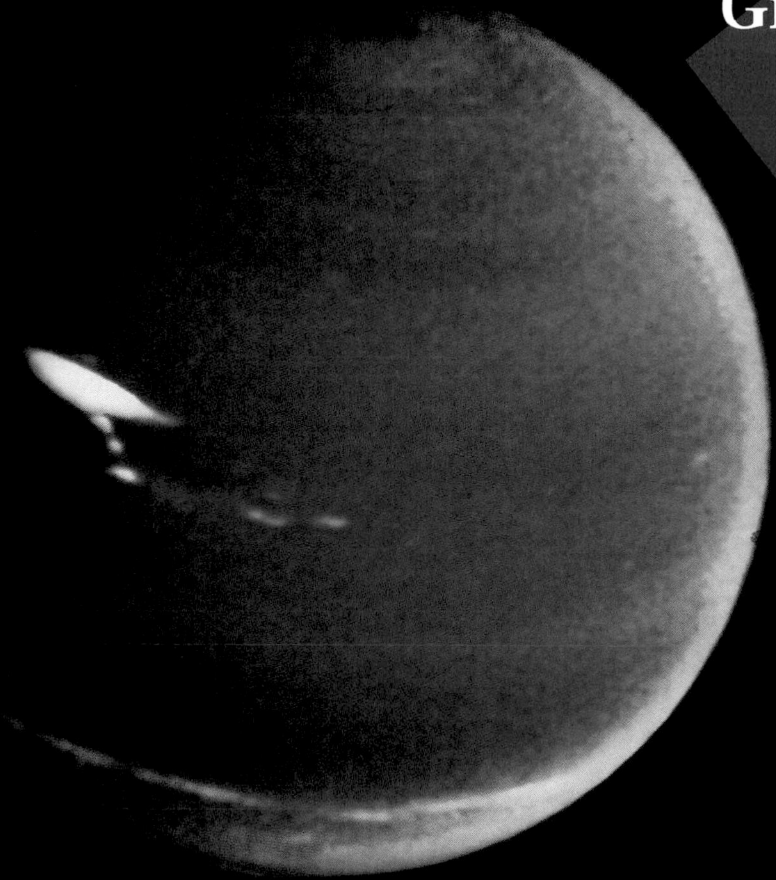

How do we know the size of the solar system? Kepler's third law relates the period of a planet's orbit around the sun to the planet's mean distance from the sun. By observing the planet's motion, we can calculate its distance. Neptune's orbital period is 165 years and the semimajor axis of its orbit is 30.1 astronomical units (AU).

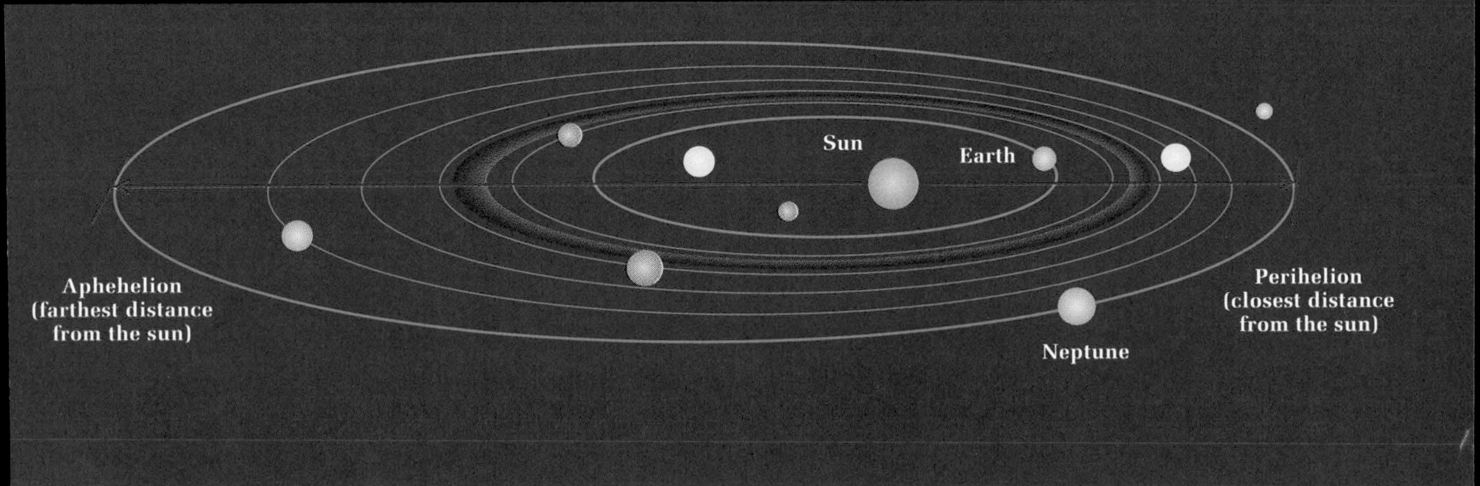

Sun

Earth

Aphehelion (farthest distance from the sun)

Perihelion (closest distance from the sun)

Neptune

## KEY CONCEPTS

- **The gravitational attraction between two particles is proportional to the product of their masses and inversely proportional to the square of the distance between them. Weight is a particular case of gravitational attraction.**
- **Gravitational field is gravitational force per unit mass.**
- **Gravitational potential energy is inversely proportional to the distance between two particles.**
- **When a planet or satellite moves in a circular orbit, the centripetal force is supplied to the body around which the planet or satellite orbits.**
- **Kepler's laws describe the shapes of orbits of planets and satellites and give relations between their speed and the sizes and shapes of the orbits.**
- **The gravitational interaction outside of any spherically symmetric body is the same as though all its mass were concentrated at its center.**
- **The rotation of the earth causes small changes in the apparent acceleration of a falling body near its surface.**

## INTRODUCTION

Why do some earth satellites circle the earth in 90 minutes, while the moon (our only natural earth satellite) takes 28 days for the trip? Why is the value of $g$, the acceleration of gravity, different at the equator from its value at the North Pole? What forces were responsible for forming the earth, the moon, and the sun in their present shapes? The study of gravitation provides the answers for these and many related questions.

Newton discovered 300 years ago that the same interaction that makes an apple fall out of a tree also keeps the planets in their nearly circular orbits around the sun. He could hardly have anticipated the many artificial satellites that circle the earth today, made possible by our knowledge of gravitational interactions. Satellites also give us a new tool for investigating gravitational interactions. As we remarked in Chapter 5, gravitation is one of the four classes of interactions found in nature, and it was the earliest one of the four to be studied extensively.

In this chapter you will learn the basic law that governs gravitational interactions. Then you will learn to apply it to phenomena such as the variation of weight with altitude, the motion of planets and satellites, and the effect of the earth's rotation on the acceleration of a falling body.

## 12–1 NEWTON'S LAW OF GRAVITATION

The *weight* of a body, the force that attracts it toward the earth (or whatever celestial body it is near) is the most familiar example of gravitational attraction. Newton discovered the **law of gravitation** during his study of the motions of planets around the sun, and he published it in 1686. It may be stated as follows:

> **Every particle of matter in the universe attracts every other particle with a force that is directly proportional to the product of the masses of the particles and inversely proportional to the square of the distance between them.**

Translating this into an equation, we have

$$F_g = G\frac{m_1 m_2}{r^2}, \tag{12–1}$$

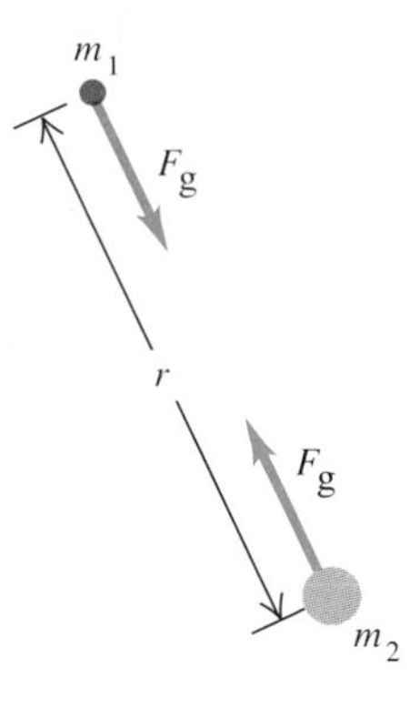


**12–1** Two particles, separated by a distance $r$, exert attractive gravitational forces of equal magnitude on one another even if their masses are quite different.

where $F_g$ is the magnitude of the gravitational force on either particle, $m_1$ and $m_2$ are their masses, $r$ is the distance between them (Fig. 12–1), and $G$ is a fundamental physical constant called the **gravitational constant.** The numerical value of $G$ depends on the system of units used.

Gravitational attraction forces always act along the line joining the two particles, and they form an action-reaction pair. Even when the masses of the particles are different, the two interaction forces have equal magnitude. The attractive force that your body exerts on the earth has the same magnitude as the force it exerts on you.

We have stated the law of gravitation in terms of the interaction forces between two *particles.* It turns out that the interaction of any two bodies having *spherically symmetric* mass distributions (such as solid spheres or spherical shells) is the same as though all the mass of each were concentrated at its center, as in Fig. 12–2, in effect replacing each spherically symmetric body with a particle at the center. Thus if we model the earth as a solid sphere with mass $m_E$, the force exerted by it on a particle or a spherically symmetric body with mass $m$, at a distance $r$ between centers, is

$$F_g = G\frac{m m_E}{r^2}, \tag{12–2}$$

provided that the body lies outside the earth. A force of the same magnitude is exerted *on* the earth by the body. We will prove these statements about spherically symmetric mass distributions in Section 12–7, but meanwhile we will use them without proof.

At points *inside* the earth the situation is different. If we could drill a hole to the center of the earth and measure the gravitational force on a particle at various depths, we would find that the force *decreases* as we approach the center, rather than increasing as $1/r^2$. As the body entered the interior of the earth (or other spherical body), some of the earth's mass would be on the side of the body opposite from the center and would pull in the opposite direction. Exactly at the center, the gravitational force on the body would be zero.

To determine the value of the gravitational constant $G$, we have to *measure* the gravitational force between two bodies of known masses $m_1$ and $m_2$ at a known distance $r$. For bodies of reasonable size the force is extremely small, but it can be measured with an instrument called a *torsion balance,* used by Sir Henry Cavendish in 1798 to determine $G$.

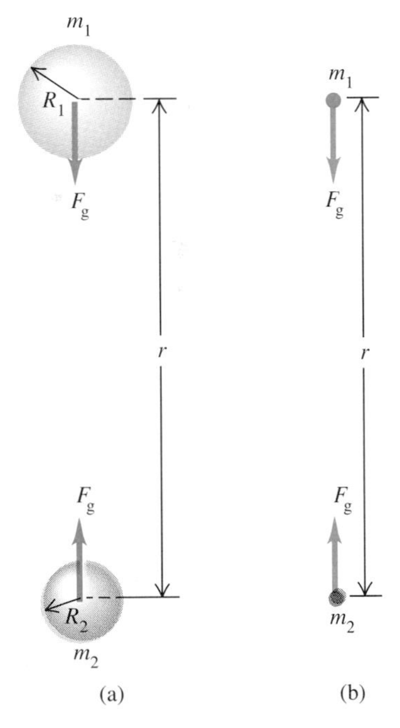


**12–2** The gravitational effect *outside* any spherically symmetric mass distribution is the same as though all the mass of the sphere were concentrated at its center.

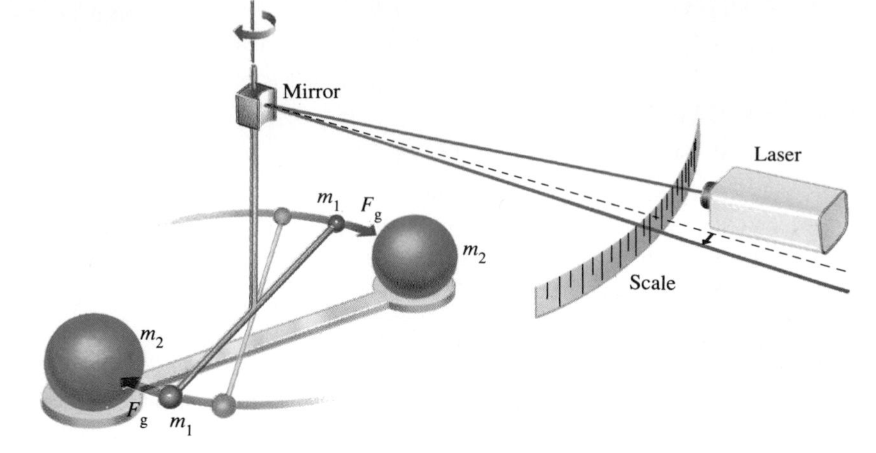


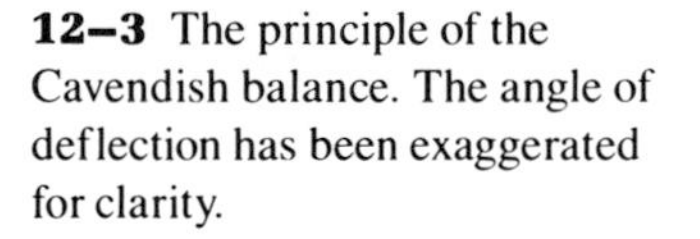
**12–3** The principle of the Cavendish balance. The angle of deflection has been exaggerated for clarity.

The Cavendish balance consists of a light, rigid rod shaped like an inverted T (Fig. 12–3) and supported by a very thin vertical quartz fiber. Two small spheres, each of mass $m_1$, are mounted at the ends of the horizontal arms of the T. When we bring two large spheres, each of mass $m_2$, to the positions shown, the attractive gravitational forces twist the T through a small angle. To measure this angle we shine a beam of light on a mirror fastened to the T; the reflected beam strikes a scale, and as the T twists, the reflected beam moves along the scale.

After calibrating the instrument we can measure gravitational forces and thus determine $G$. The currently accepted value (in SI units) is

$$G = 6.673 \times 10^{-11}\ \text{N} \cdot \text{m}^2/\text{kg}^2.$$

Because $1\ \text{N} = 1\ \text{kg} \cdot \text{m/s}^2$, the units of $G$ can also be expressed (in fundamental SI units) as $\text{m}^3/(\text{kg} \cdot \text{s}^2)$.

Gravitational forces combine vectorially. If each of two masses exerts a force on a third, the *total* force on the third mass is the vector sum of the individual forces due to the first two. Example 12–3 makes use of this property, which is often called *superposition of forces*.

## EXAMPLE 12–1

The mass $m_1$ of one of the small spheres of a Cavendish balance is 0.0100 kg, the mass $m_2$ of one of the large spheres is 0.500 kg, and the center-to-center distance between each large sphere and the nearer small one is 0.0500 m. Find the gravitational force $F_g$ on each sphere due to the nearest other sphere.

**SOLUTION** The magnitude of each force is

$$F_g = (6.67 \times 10^{-11}\ \text{N} \cdot \text{m}^2/\text{kg}^2)\frac{(0.0100\ \text{kg})(0.500\ \text{kg})}{(0.0500\ \text{m})^2}$$
$$= 1.33 \times 10^{-10}\ \text{N}.$$

This is a very small force. *Reminder:* The two bodies experience the *same* magnitude of force, even though their masses are very different.

## EXAMPLE 12–2

Suppose one large and one small sphere are detached from the apparatus in Example 12–1 and placed 0.0500 m (between centers) from each other at a point in space far removed from all other bodies. What is the acceleration of each, relative to an inertial system?

**SOLUTION** The force on each sphere has the same magnitude we found in Example 12–1. The acceleration $a_1$ of the smaller sphere is

$$a_1 = \frac{F_g}{m_1} = \frac{1.33 \times 10^{-10}\ \text{N}}{1.00 \times 10^{-2}\ \text{kg}} = 1.33 \times 10^{-8}\ \text{m/s}^2.$$

The acceleration $a_2$ of the larger sphere is

$$a_2 = \frac{F_g}{m_2} = \frac{1.33 \times 10^{-10}\ \text{N}}{0.500\ \text{kg}} = 2.67 \times 10^{-10}\ \text{m/s}^2.$$

Although the forces on the bodies are equal in magnitude, the two accelerations are *not* equal. Also, they are not constant because the gravitational forces increase as the spheres start to move toward each other.

## EXAMPLE 12–3

**Superposition of gravitational forces** Three spheres are arranged as shown in Fig. 12–4. Find the magnitude and direction of the total gravitational force exerted on the small sphere by both large ones.

**SOLUTION** We use the principle of superposition: The total force on the small sphere is the vector sum of the two forces due to the two individual large spheres. We can compute the vector sum by using components, but first we find the magnitudes of the forces. The magnitude $F_1$ of the force on the small mass due to the upper large one is

$$F_1 = \frac{(6.67 \times 10^{-11}\ \mathrm{N \cdot m^2/kg^2})(0.500\ \mathrm{kg})(0.0100\ \mathrm{kg})}{(0.200\ \mathrm{m})^2 + (0.200\ \mathrm{m})^2} = 4.17 \times 10^{-12}\ \mathrm{N}.$$

The magnitude $F_2$ of the force due to the lower large mass is

$$F_2 = \frac{(6.67 \times 10^{-11}\ \mathrm{N \cdot m^2/kg^2})(0.500\ \mathrm{kg})(0.0100\ \mathrm{kg})}{(0.200\ \mathrm{m})^2} = 8.34 \times 10^{-12}\ \mathrm{N}.$$

The $x$- and $y$-components of these forces are

$$F_{1x} = (4.17 \times 10^{-12}\ \mathrm{N})(\cos 45°) = 2.95 \times 10^{-12}\ \mathrm{N},$$
$$F_{1y} = (4.17 \times 10^{-12}\ \mathrm{N})(\sin 45°) = 2.95 \times 10^{-12}\ \mathrm{N},$$
$$F_{2x} = 8.34 \times 10^{-12}\ \mathrm{N},$$
$$F_{2y} = 0.$$

The components of the total force on the small mass are

$$F_x = F_{1x} + F_{2x} = 11.3 \times 10^{-12}\ \mathrm{N},$$
$$F_y = F_{1y} + F_{2y} = 2.95 \times 10^{-12}\ \mathrm{N}.$$

The magnitude of this total force is

$$F = \sqrt{(11.3 \times 10^{-12}\ \mathrm{N})^2 + (2.95 \times 10^{-12}\ \mathrm{N})^2} = 1.17 \times 10^{-11}\ \mathrm{N},$$

and its direction is

$$\theta = \arctan \frac{2.95 \times 10^{-12}\ \mathrm{N}}{11.3 \times 10^{-12}\ \mathrm{N}} = 14.6°.$$

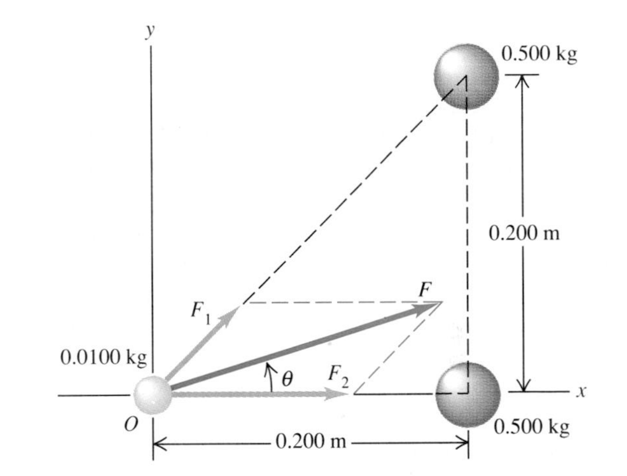


**12–4** The total gravitational force on the 0.0100-kg sphere is the vector sum of the forces exerted on it by the two 0.500-kg spheres.

## 12–2 WEIGHT

We defined the *weight* of a body tentatively in Section 4–4 as the attractive gravitational force exerted on it by the earth. We can now broaden our definition. **The weight of a body is the total gravitational force exerted on the body by all other bodies in the universe.** When the body is near the surface of the earth, we can neglect all other gravitational forces and consider the weight as just the earth's gravitational attraction. At the surface of the *moon* we consider a body's weight to be the gravitational attraction of the moon, and so on.

If we again model the earth as a spherically symmetric body with radius $R_E$ and mass $m_E$, the weight $w$ of a small body of mass $m$ at the earth's surface (distance $R_E$ from its center) is

$$w = F_g = \frac{Gmm_E}{R_E^2}. \qquad (12\text{–}3)$$

But we also know from Section 4–4 that the weight $w$ of a body is its mass $m$ times the acceleration $g$ of free fall: $w = mg$. Weight is the force that causes the acceleration of free fall. Equating this with Eq. (12–3) and dividing by $m$, we find

$$g = \frac{Gm_E}{R_E^2}. \qquad (12\text{–}4)$$

Because this result does not contain $m$, the acceleration due to gravity $g$ is independent of the mass $m$ of the body. We already knew that, but we can now see how it follows from the law of gravitation.

Furthermore, we can *measure* all the quantities in Eq. (12–4) except for $m_E$, so this relation allows us to compute the mass of the earth. We first solve Eq. (12–4) for $m_E$, obtaining

$$m_E = \frac{gR_E{}^2}{G}.$$

Taking $R_E = 6380\ \text{km} = 6.38 \times 10^6$ m and $g = 9.80\ \text{m/s}^2$, we find

$$m_E = 5.98 \times 10^{24}\ \text{kg}.$$

Once Cavendish had measured $G$, he could compute the mass of the earth. He described his measurement of $G$ with the grandiose phrase "weighing the earth." That's not really what he did; he certainly didn't hang our planet from a spring balance. But after he had determined $G$, he carried out the above calculation and determined the mass (not the weight) of the earth.

At a distance $r$ from the center of the earth (distance $r - R_E$ above the surface), the weight of a body is given by Eq. (12–3) with $R_E$ replaced by $r$:

$$w = F_g = \frac{Gmm_E}{r^2}. \qquad (12\text{–}5)$$

The weight of a body decreases inversely with the square of its distance from the earth's center. At a radial distance $r = 2R_E$ from the center, the weight is one quarter of its value at the earth's surface. Figure 12–5 shows how the weight varies with height above the earth for an astronaut who weighs 700 N on earth.

Once we know the mass and radius of the earth, we can compute its average density. The volume, assuming a spherical earth, is

$$V = \tfrac{4}{3}\pi R_E{}^3 = \tfrac{4}{3}\pi(6.38 \times 10^6\ \text{m})^3 = 1.09 \times 10^{21}\ \text{m}^3.$$

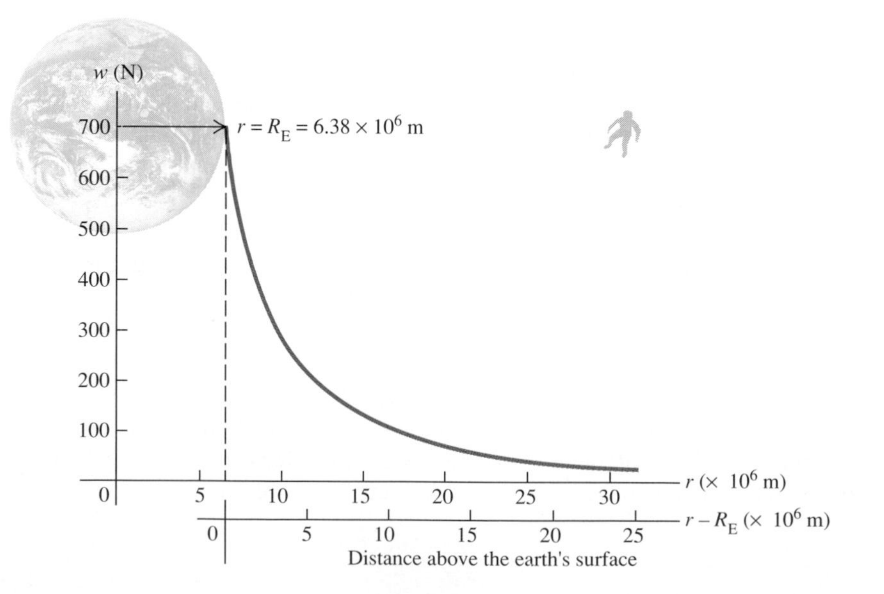


**12–5** An astronaut weighing 700 N at the earth's surface experiences less gravitational attraction when above that surface ($10^6$ m = 621 mi). The astronaut's distance from the *center* of the earth is $r$; her distance from the *surface* of the earth is $r - R_E = r - 6.38 \times 10^6$ m.

The average density $\rho$ is the total mass divided by the total volume:

$$\rho = \frac{M}{V} = \frac{5.98 \times 10^{24}\ \text{kg}}{1.09 \times 10^{21}\ \text{m}^3} = 5500\ \text{kg/m}^3 = 5.50\ \text{g/cm}^3.$$

For comparison, the density of water is $1.00\ \text{g/cm}^3 = 1000\ \text{kg/m}^3$. The density of igneous rock near the earth's surface, such as granite or gneiss, is about $3\ \text{g/cm}^3 = 3000\ \text{kg/m}^3$, although some basaltic rock has a density of about $5\ \text{g/cm}^3 = 5000\ \text{kg/m}^3$. So the interior of the earth must be much more dense than the surface. In fact, the maximum density at the center, where the pressure is enormous, is thought to be nearly $15\ \text{g/cm}^3$. Figure 12–6 is a graph of density as a function of distance from the earth's center.

The *apparent* weight of a body on earth differs slightly from the earth's gravitational force because the earth rotates and is therefore not precisely an inertial frame of reference. We have ignored this effect in the above discussion and have assumed that the earth *is* an inertial system. We will return to the effect of the earth's rotation in Section 12–8.

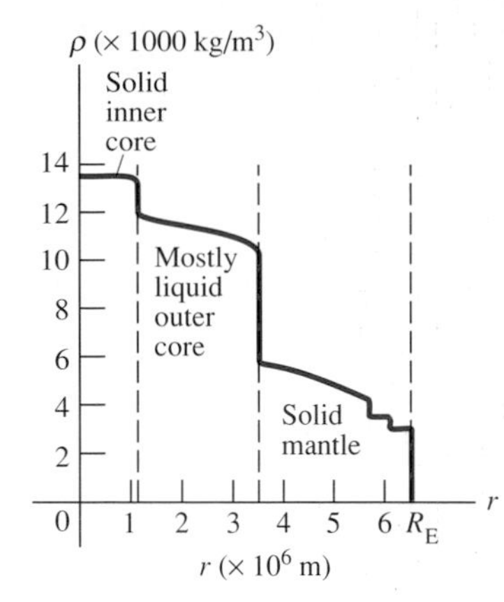


**12–6** The density of the earth decreases with increasing distance from its center.

## EXAMPLE 12–4

**Gravity on Mars** You're involved in the design of a mission carrying humans to the surface of the planet Mars, which has a radius $R_M = 3.38 \times 10^6$ m and a mass $m_M = 6.42 \times 10^{23}$ kg (Fig. 12–7). The earth weight of the Mars lander is 39,200 N. Calculate its weight $F_g$ and the acceleration due to Mars' gravity $g_M$: a) at $6.0 \times 10^6$ m above the surface of Mars (corresponding to the orbit of the Martian moon Phobos); b) at the surface of Mars. Neglect the gravitational effects of the moons of Mars.

**SOLUTION** a) In Eq. (12–5) we replace $m_E$ with $m_M$. The value of $G$ is the same everywhere; it is a fundamental physical constant. The distance $r$ from the *center* of Mars is

$$r = 6.0 \times 10^6\text{m} + 3.38 \times 10^6\text{m} = 9.4 \times 10^6\text{m}.$$

The mass $m$ of the lander is its earth weight $w$ divided by the acceleration of gravity $g$ on the earth:

$$m = \frac{w}{g} = \frac{39{,}200\ \text{N}}{9.8\ \text{m/s}^2} = 4000\ \text{kg}.$$

The mass is the same whether the lander is on the earth, on Mars, or in between. From Eq. (12–5),

$$F_g = \frac{Gmm_M}{r^2}$$
$$= \frac{(6.67 \times 10^{-11}\text{N}\cdot\text{m}^2/\text{kg}^2)(4000\,\text{kg})(6.42 \times 10^{23}\text{kg})}{(9.4 \times 10^6\text{m})^2}$$
$$= 1940\text{N}.$$

The acceleration due to the gravity of Mars at this point is

$$g_M = \frac{F_g}{m} = \frac{1940\ \text{N}}{4000\ \text{kg}} = 0.48\ \text{m/s}^2.$$

This is also the acceleration experienced by Phobos in its orbit, $6.0 \times 10^6$ m above the surface of Mars.

b) To find $F_g$ and $g_M$ at the surface, we repeat the above calculations, replacing $r = 9.4 \times 10^6$ m with $R_M = 3.38 \times 10^6$ m. Or, because $F_g$ and $g_M$ are inversely proportional to $1/r^2$ (at any point outside the planet), we can multiply the results of part (a) by the factor

$$\left(\frac{9.4 \times 10^6\ \text{m}}{3.38 \times 10^6\ \text{m}}\right)^2.$$

We invite you to complete the calculation by both methods and show that at the surface $F_g = 15{,}000$ N and $g_M = 3.7\ \text{m/s}^2$. That is, on the surface of Mars, $F_g$ and $g$ are roughly 40% as large as on the surface of the earth.

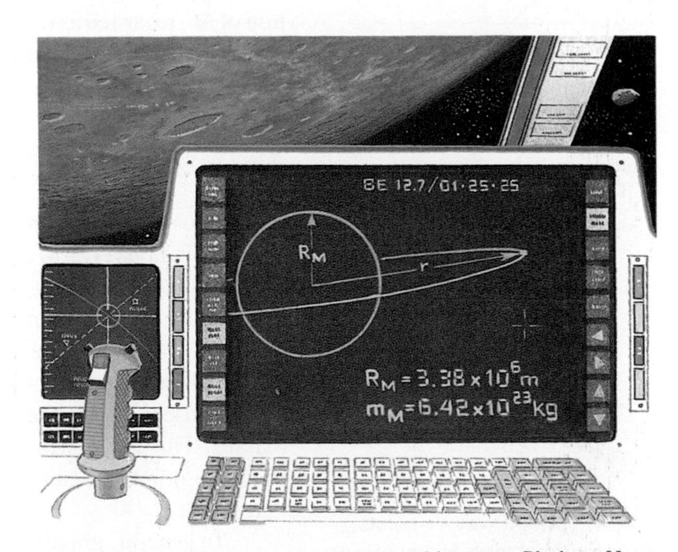


**12–7** Approaching the planet Mars and its moon Phobos. How does Martian gravity compare to that of the earth?

## 12–3 GRAVITATIONAL FIELD

We can restate Newton's law of gravitation in a useful way using the concept of **gravitational field.** Instead of calculating the interaction forces between two masses by using Eq. (12–1) directly, we consider a two-stage process. A mass creates a *gravitational field* in the space around it.

But what *is* a gravitational field, and how do we know when one is present? We can use a test mass $m$ as a detector of gravitational field. We take the mass to various points; at each point we measure the gravitational force that acts on it. We then *define* the gravitational field $\boldsymbol{g}$ at each point as the gravitational force $\boldsymbol{F}_g$ experienced by the test mass $m$ at that point, divided by the mass. That is,

$$\boldsymbol{g} = \frac{\boldsymbol{F}_g}{m}. \qquad (12\text{–}6)$$

To say it another way, at a point where the gravitational field is $\boldsymbol{g}$, the gravitational force $\boldsymbol{F}_g$ on a mass $m$ is

$$\boldsymbol{F}_g = m\boldsymbol{g}. \qquad (12\text{–}7)$$

Force is a vector quantity, so gravitational field is also a vector quantity; Eqs. (12–6) and (12–7) are vector equations.

In Example 12–3, instead of calculating directly the forces on the 0.0100-kg body by the large bodies, we may take the point of view that the large bodies create a gravitational field at the location of the small body and that the field exerts a force on any mass at that location.

The gravitational field $\boldsymbol{g}$ caused by a point mass $M$, at a distance $r$ away from it, has magnitude

$$g = \frac{GM}{r^2} \qquad (12\text{–}8)$$

and is directed toward the point mass $M$. We can express both the magnitude and direction of $\boldsymbol{g}$ by using a unit vector $\hat{\boldsymbol{r}}$ (Fig. 12–8) that points from the "source point" (the location of $M$, regarded as the source of the field) to the "field point" $P$ (the location of the test mass $m$). The vector expression is

$$\boldsymbol{g} = -\frac{GM}{r^2}\hat{\boldsymbol{r}}. \qquad (12\text{–}9)$$

The minus sign shows that the direction of $\boldsymbol{g}$ is always opposite that of $\hat{\boldsymbol{r}}$.

Equations (12–8) and (12–9) are also valid for points outside any spherically symmetric mass distribution; we will prove in Section 12–7 that the gravitational field caused by any spherically symmetric mass distribution is the same, at all points outside the distribution, as the field caused by an equal point mass at the center.

When several masses are located in the vicinity of a point such as $P$, the *total* gravitational field at $P$ is the *vector* sum of the gravitational fields caused by the separate masses. This is the principle of superposition of forces once again. When we know the field at a point, we can quickly calculate the gravitational force on any body located at that point.

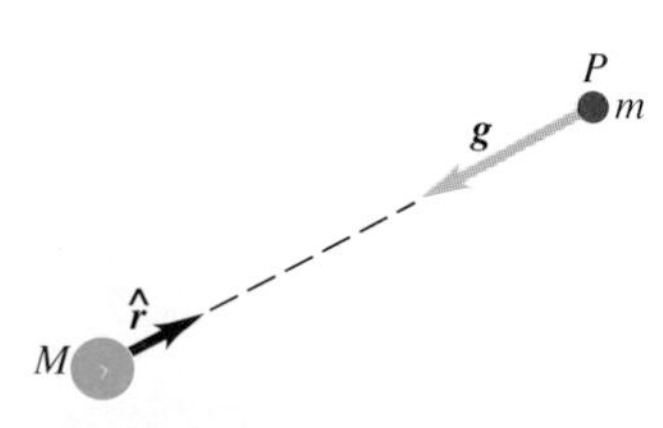


**12–8** The unit vector $\hat{\boldsymbol{r}}$ points from the point mass $M$ that sets up the gravitational field to the point $P$ (the location of the test mass $m$). The direction of the gravitational field is always opposite that of $\hat{\boldsymbol{r}}$.

In general, gravitational field varies from point to point. Thus it is not a single vector quantity but rather a whole collection of vector quantities, one vector associated with each point in space. It is an example of a **vector field.** Another familiar example of a vector field is the velocity in a flowing fluid; different parts of the fluid have different velocities, and we speak of the *velocity field* in the fluid. When we study electricity and magnetism in the second half of this book, we will work a lot with electric and magnetic fields, which are both vector fields.

We have used the same symbol $g$ for gravitational field magnitude that we used earlier for the acceleration of free fall. This is not an accident. You may have noticed that the units of gravitational field are m/s$^2$. In an inertial frame of reference, the acceleration of free fall of a body at any point is *equal* to the gravitational field at that point. When the force $\mathbf{F}_g = m\mathbf{g}$ is the *only* force acting on a body with mass $m$, then from Newton's second law ($\mathbf{F} = m\mathbf{a}$) the acceleration is $\mathbf{a} = \mathbf{g}$. In particular, when $\mathbf{g}$ is the gravitational field caused by the earth, the gravitational force $\mathbf{F}_g$ is simply the *weight* $\mathbf{w}$ of the body, and we obtain the familiar relation $\mathbf{w} = m\mathbf{g}$, which is Eq. (4–11).

## EXAMPLE 12–5

Figure 12–9 shows the same situation as in Example 12–3. Find the magnitude of the gravitational field at the location of the 0.0100-kg mass.

**SOLUTION** We found in Example 12–3 that the total force on this mass has magnitude $F_g = 1.17 \times 10^{-11}$ N. From Eq. (12–6) the gravitational field at this point has magnitude

$$g = \frac{F_g}{m} = \frac{1.17 \times 10^{-11}\ \text{N}}{0.0100\ \text{kg}} = 1.17 \times 10^{-9}\ \text{m/s}^2.$$

The direction of the field is the same as that of the force on the small body.

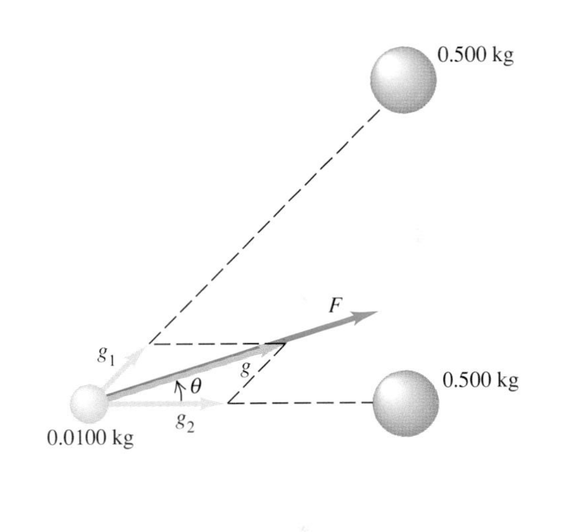


**12–9** The gravitational field at the center of the 0.0100-kg sphere has the same direction as the gravitational force on the sphere. Its magnitude is the magnitude of that gravitational force divided by the mass of the sphere.

## EXAMPLE 12–6

A ring-shaped body with radius $a$ has total mass $M$. Find the gravitational field at point $P$, at a distance $x$ from the center of the ring, along the line through the center and perpendicular to the plane of the ring (Fig. 12–10).

**SOLUTION** We imagine the ring as being divided into small segments $\Delta s$, each with mass $\Delta M$. At point $P$ each segment produces a gravitational field $\Delta \mathbf{g}$ with magnitude

$$\Delta g = \frac{G\,\Delta M}{r^2} = \frac{G\,\Delta M}{x^2 + a^2}.$$

The component of this field along the $x$-axis is given by

$$\Delta g_x = -\Delta g \cos\phi = -\frac{G\,\Delta M}{x^2 + a^2}\frac{x}{\sqrt{x^2 + a^2}} = -\frac{G\,\Delta M\, x}{(x^2 + a^2)^{3/2}}. \tag{12–10}$$

To find the *total* $x$-component of the gravitational field from all the mass elements $\Delta M$, we sum all the $\Delta g_x$'s. This is easy; $x$ and $a$ are the same for every segment around the ring, so we simply sum all the $\Delta M$'s. This sum is equal to the total mass $M$. The result is

$$g_x = -\frac{GMx}{(x^2 + a^2)^{3/2}}. \tag{12–11}$$

The $y$-component of gravitational field at point $P$ is zero because the $y$-components of field produced by two segments at opposite sides of the ring cancel each other, and the contributions of all such pairs added over the entire ring sum to zero.

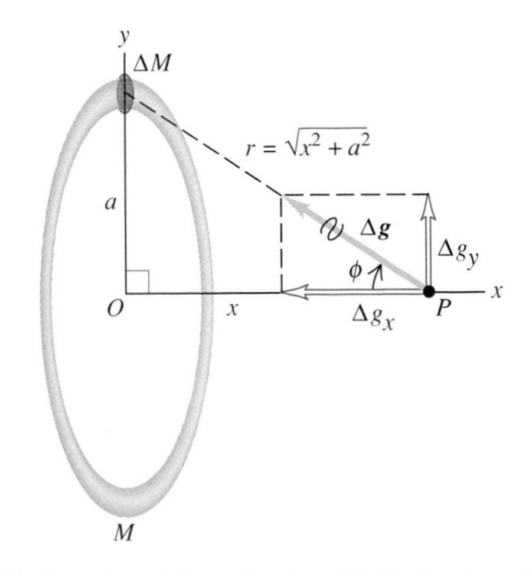


**12–10** Finding the axial gravitational field of a ring with mass $M$ and radius $a$.

Equation (12–11) shows that at the center of the ring ($x = 0$) the total gravitational field is zero. We should expect this; mass elements on opposite sides pull in opposite directions, and their fields cancel. Also, when $x$ is much larger than $a$, we can neglect $a$ in the denominator of Eq. (12–11); the expression then becomes approximately equal to $-GM/x^2$. This shows that at distances ($x$) much greater than the radius ($a$) of the ring, the ring appears as a point mass.

## 12–4 GRAVITATIONAL POTENTIAL ENERGY

When we first developed the concept of gravitational potential energy in Section 7–2, we assumed that the gravitational force on a body is constant in magnitude and direction. But we now know that the earth's gravitational force on a body of mass $m$, at any point outside the earth, is given more generally by

$$w = F_g = \frac{Gmm_E}{r^2}, \tag{12–12}$$

where $m_E$ is the mass of the earth and $r$ is the distance of the body from the earth's center. In problems in which $r$ changes enough that the gravitational force can't be considered as constant, we have to revise our concept of gravitational potential energy.

Equation (12–12) shows that the weight $w$ of a body at a distance $r$ from the center of the earth is proportional to $1/r^2$. To find the corresponding generalization of gravitational potential energy, we have to compute the work $W_{grav}$ done by the gravitational force when $r$ changes from $r_1$ to $r_2$. This is given by

$$W_{grav} = \int_{r_1}^{r_2} F_r \, dr,$$

where $F_r$ is the *component* of the gravitational force in the direction of increasing $r$, that is, the direction outward from the center of the earth. Because the gravitational force actually points in the direction of *decreasing* $r$, $F_r$ differs from the expression in Eq. (12–12) by a minus sign. That is, the magnitude of the gravitational force is positive, but its component along the direction of increasing $r$ (the radial direction) is negative.

Thus $W_{grav}$ is given by

$$W_{grav} = -Gmm_E \int_{r_1}^{r_2} \frac{dr}{r^2} = \frac{Gmm_E}{r_2} - \frac{Gmm_E}{r_1}. \tag{12–13}$$

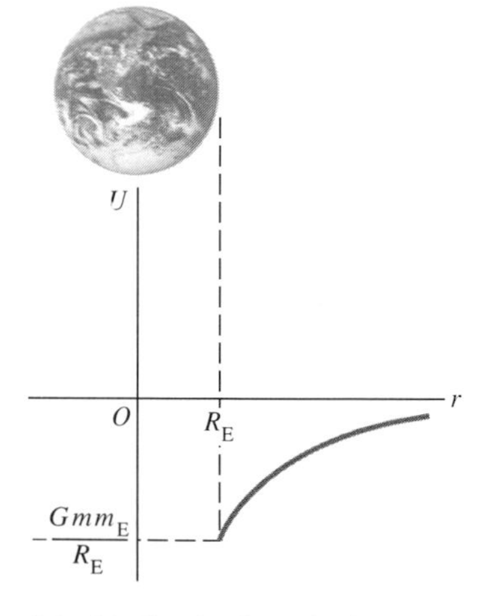


**12–11** Graph of gravitational potential energy $U$ versus distance from the center of the earth $r$. Note that $U$ is always negative, but $U$ becomes less negative with increasing $r$.

The path doesn't have to be a straight line; we could make an argument similar to that in Section 7–2 to show that this work depends only on the initial and final values of $r$. This also proves that the gravitational force is always *conservative*.

We want to define the corresponding potential energy $U$ so that $W_{grav} = U_1 - U_2$, as before. Comparing this with Eq. (12–13), we see that the appropriate definition is

$$U = -\frac{Gmm_E}{r}. \tag{12–14}$$

When $r$ increases, the gravitational force does negative work and $U$ increases (that is, becomes less negative), as Fig. 12–11 shows. When $r$ decreases, the body "falls" toward earth, the gravitational work is positive, and the potential energy decreases (that is, becomes more negative).

One slightly peculiar feature of Eq. (12–14) is that $U$ is always negative. The reason for this is that, according to Eq. (12–14), $U$ is zero when the mass $m$ is infinitely far away from the earth ($r = \infty$). This is quite different from making $U = 0$ at some arbitrary height, as we did before. But the choice of zero point is arbitrary. Only *differences* of $U$

from one point to another are significant, so negative values of $U$ should not be too alarming. If we wanted, we could make $U = 0$ at the surface of the earth, where $r = R_E$, by simply adding the quantity $Gmm_E/R_E$ to Eq. (12–14). This would make $U$ zero when $r = R_E$ and positive when $r > R_E$, at the price of making the expression for $U$ more complicated. This added term would not affect the *difference* in $U$ between any two points, which is the only physically significant quantity, so it's usually easier to omit the term.

Armed with Eq. (12–14), we can now use general energy relations for problems in which the $1/r^2$ behavior of the earth's gravitational force has to be included. If the gravitational force on the body is the only force, the total mechanical energy of the system is constant, or *conserved*.

## EXAMPLE 12–7

**"From the Earth to the Moon"** In Jules Verne's story with this title, written in 1865, three men were shot to the moon in a shell fired from a giant cannon sunk in the earth in Florida. a) What muzzle velocity would be needed to shoot the shell straight up to a height above the earth equal to the earth's radius? b) What muzzle velocity would allow the shell to escape from the earth completely? To simplify the calculation, neglect the gravitational pull of the moon and all air-resistance effects.

**SOLUTION** a) We use the general energy relation from Chapter 7:

$$K_1 + U_1 + W_{\text{other}} = K_2 + U_2.$$

The only force doing work is the gravitational force, so $W_{\text{other}} = 0$. Let point 1 be the starting point and point 2 be the point of maximum height, where $v = 0$ (Fig. 12–12). If the radius of the earth is $R_E$, then $r_1 = R_E$ and $r_2 = 2R_E$. Also, $v_2 = 0$, and $v_1$ is to be determined. The mass of the shell (with passengers) is $m$. The energy-conservation equation $K_1 + U_1 = K_2 + U_2$ gives

$$\tfrac{1}{2}mv_1^2 + \left(-G\frac{mm_E}{R_E}\right) = 0 + \left(-G\frac{mm_E}{2R_E}\right).$$

Rearranging this, we find

$$v_1 = \sqrt{\frac{Gm_E}{R_E}} = \sqrt{\frac{(6.67 \times 10^{-11}\ \text{N}\cdot\text{m}^2/\text{kg}^2)(5.98 \times 10^{24}\ \text{kg})}{6.38 \times 10^6\ \text{m}}}$$
$$= 7910\ \text{m/s} \quad (= 17{,}690\ \text{mi/h}).$$

b) When $v_1$ is the magnitude of the escape velocity, $r_1 = R_E$, $r_2 = \infty$, and $v_2 = 0$. We want the body barely to be able to "reach" $r_2 = \infty$, with no kinetic energy left over. Then $K_1 + U_1 = K_2 + U_2$ gives

$$\tfrac{1}{2}mv_1^2 + \left(-G\frac{mm_E}{R_E}\right) = 0 + 0,$$

$$v_1 = \sqrt{\frac{2Gm_E}{R_E}}$$
$$= \sqrt{\frac{2(6.67 \times 10^{-11}\ \text{N}\cdot\text{m}^2/\text{kg}^2)(5.98 \times 10^{24}\ \text{kg})}{6.38 \times 10^6\ \text{m}}}$$
$$= 1.12 \times 10^4\ \text{m/s} \quad (= 25{,}000\ \text{mi/h} = 6.95\ \text{mi/s}).$$

Generalizing this result, we may say that the initial speed $v_1$ needed for a body to escape from the surface of a spherical mass $M$ with radius $R$ (ignoring air resistance) is

$$v_1 = \sqrt{\frac{2GM}{R}}. \tag{12–15}$$

We invite you to use this result to compute the escape velocity for other bodies. You will find $0.497 \times 10^4$ m/s for Mars, $5.91 \times 10^4$ m/s for Jupiter, and $61.8 \times 10^4$ m/s for the sun.

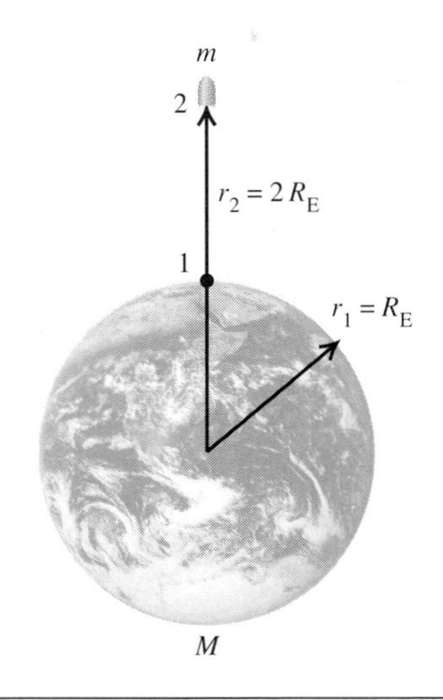


**12–12** A projectile fired from the surface of the earth to a height equal to the earth's radius.

When we are close to the earth's surface, Eq. (12–13) reduces to the familiar $U = mgy$ from Chapter 7. To show this we rewrite Eq. (12–13) as

$$W_{\text{grav}} = Gmm_E\frac{r_1 - r_2}{r_1 r_2}.$$

If the body stays close to the earth, then in the denominator we may replace $r_1$ and $r_2$ by $R_E$, the earth's radius, obtaining

$$W_{grav} = Gmm_E \frac{r_1 - r_2}{R_E^{\ 2}}.$$

According to Eq. (12–4), $g = Gm_E/R_E^{\ 2}$, so

$$W_{grav} = mg(r_1 - r_2).$$

This agrees with Eq. (7–1), with the $y$'s replaced by $r$'s, so we may consider Eq. (7–1) as a special case of the more general Eq. (12–13).

The relationship between the gravitational force on a body, given by Eq. (12–12), and the gravitational potential energy, given by Eq. (12–14), can be expressed in a different way, using the general analysis of Section 7–5. Equation (7–15) showed that the component of force in a given direction equals the negative of the derivative of $U$ with respect to the corresponding coordinate. For motion along the $x$-axis,

$$F_x = -\frac{dU}{dx}. \tag{12–16}$$

In our case we can replace $x$ by $r$ in Eq. (12–16); we find

$$F_r = -\frac{dU}{dr} = -\frac{d}{dr}\left(-\frac{Gmm_E}{r}\right) = -\frac{Gmm_E}{r^2}. \tag{12–17}$$

This agrees with the $F_r$ that we started with. As we remarked before, $F_r$ is negative, showing that the force points in the direction opposite to the direction of increasing $r$.

## 12–5 SATELLITE MOTION

Artificial satellites orbiting the earth are a familiar fact of contemporary life. We know that they differ only in scale, not in principle, from the motion of our moon around the earth or the motion of the moons of Jupiter. But we still have to deal with questions such as "What holds that thing up there, anyway?" So let's see how we can use Newton's laws and the law of gravitation for a detailed analysis of satellite motion.

To begin, think back to the discussion of projectile motion in Section 3–3. In Example 3–3 a motorcycle rider rides horizontally off the edge of a cliff, launching himself into a parabolic path that ends on the flat ground at the base of the cliff. If he survives and repeats the experiment with increased launch speed, he will land farther from the starting point. We can imagine him launching himself with great enough speed that the earth's curvature becomes significant. As he falls, the earth curves away beneath him. If he is going fast enough, and if his launch point is high enough that he clears the mountaintops, he may be able to go right on around the earth without ever landing.

Figure 12–13a shows a variation on this theme. We launch a projectile from point $A$ in the direction $AB$, tangent to the earth's surface. Trajectories (1) through (7) show the effect of increasing the initial speed. Trajectory (3) just misses the earth, and the projectile has become an earth satellite. Its speed when it returns to point $A$ is the same as its initial speed; if there is no retarding force (such as air resistance), it repeats its motion indefinitely.

Trajectories (1) through (5) are ellipses or segments of ellipses; trajectory (4) is a circle (a special case of an ellipse). Trajectories (6) and (7) are not closed orbits; for these the projectile never returns to its starting point but travels farther and farther away from the earth.

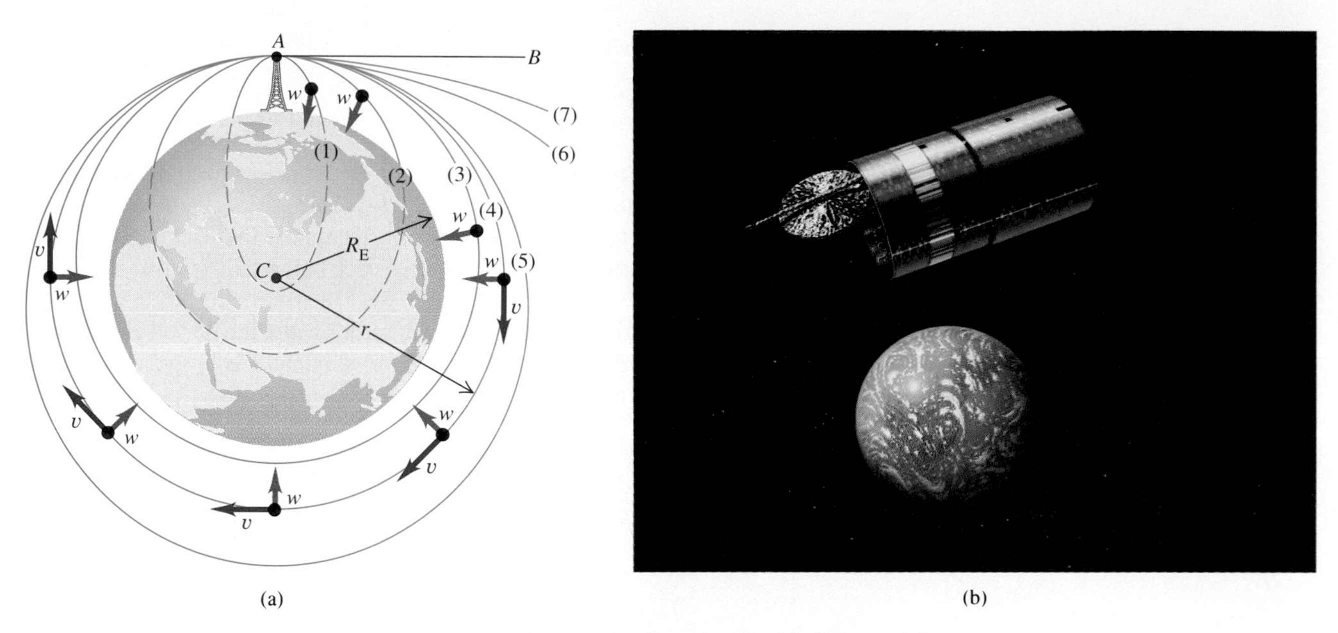


**12–13** (a) Trajectories of a body projected from point *A* in the direction *AB* with different initial velocities. Orbits (1) and (2) would be completed as shown if the earth were a point mass at *C*. (b) A modern communications satellite. Solar energy collector panels can be seen on the sides and a parabolic antenna at the end.

A circular orbit is the simplest case; it is the only one that we will analyze in detail. Artificial satellites often have nearly circular orbits, and the orbits of the planets around the sun are also nearly circular. As we learned in Section 3–4, a particle in uniform circular motion with speed $v$ and radius $r$ has an acceleration, with magnitude $a_{rad} = v^2/r$, directed always toward the center of the circle. For a satellite the *force* that provides this acceleration is the gravitational attraction of the earth (mass $m_E$) for the satellite (mass $m$), as shown in Fig. 12–14. If the radius of the circular orbit, measured from the *center* of the earth, is $r$, then the gravitational force is given by the law of gravitation: $F_g = Gmm_E/r^2$.

The principle governing the motion is Newton's second law; the force is $F_g$, and the acceleration is $v^2/r$, so the $\Sigma \boldsymbol{F} = m\boldsymbol{a}$ equation is

$$\frac{Gmm_E}{r^2} = \frac{mv^2}{r}.$$

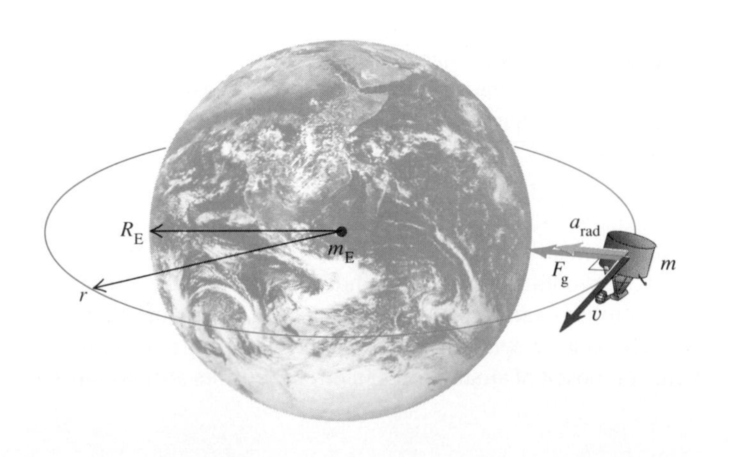


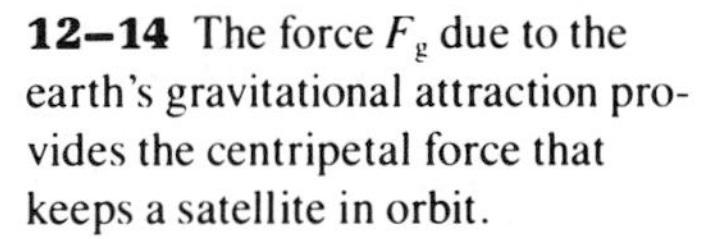

**12–14** The force $F_g$ due to the earth's gravitational attraction provides the centripetal force that keeps a satellite in orbit.

Solving this for $v$, we find

$$v = \sqrt{\frac{Gm_E}{r}}. \tag{12–18}$$

This relation shows that we can't choose the orbit radius $r$ and the speed $v$ independently; if we choose a value of $r$, $v$ is determined. Equation (12–18) also shows that the satellite's motion does not depend on its mass $m$, because $m$ doesn't appear in the equation. If we could cut a satellite in half without changing its speed, each half would continue on with the original motion.

We can also derive a relation between the orbit radius $r$ and the period $T$, the time for one revolution. The speed $v$ is the distance $2\pi r$ traveled in one revolution, divided by the time $T$ for one revolution. Thus,

$$v = \frac{2\pi r}{T}. \tag{12–19}$$

To get an expression for $T$, we solve Eq. (12–19) for $T$ and combine it with Eq. (12–18):

$$T = \frac{2\pi r}{v} = 2\pi r\sqrt{\frac{r}{Gm_E}} = \frac{2\pi r^{3/2}}{\sqrt{Gm_E}}. \tag{12–20}$$

Equations (12–18) and (12–20) show that larger orbits correspond to slower speeds and longer periods.

An interesting by-product of this analysis emerges when we compare Eq. (12–18) with Eq. (12–15). We see that the escape speed from a spherical body with radius $R$ is just $\sqrt{2}$ times the speed of a satellite in a circular orbit at that radius. If our spaceship is in circular orbit around the planet Vulcan, we have to multiply our speed by a factor of $\sqrt{2}$ to escape from Vulcan's gravity, regardless of the planet's mass.

## EXAMPLE 12–8

Suppose we want to place a weather satellite into a circular orbit 300 km above the earth's surface. What speed, period, and radial acceleration must it have? The earth's radius is 6380 km = $6.38 \times 10^6$ m, and its mass is $5.98 \times 10^{24}$ kg. (See Appendix F.)

**SOLUTION** The radius of the satellite's orbit is

$$r = 6380\text{ km} + 300\text{ km} = 6680\text{ km} = 6.68 \times 10^6\text{ m}.$$

From Eq. (12–18),

$$v = \sqrt{\frac{Gm_E}{r}} = \sqrt{\frac{(6.67 \times 10^{-11}\text{ N}\cdot\text{m}^2/\text{kg}^2)(5.98 \times 10^{24}\text{ kg})}{6.68 \times 10^6\text{ m}}}$$
$$= 7730\text{ m/s}.$$

From Eq. (12–20),

$$T = \frac{2\pi r}{v} = \frac{2\pi(6.68 \times 10^6\text{ m})}{7730\text{ m/s}}$$
$$= 5430\text{ s} = 90.5\text{ min}.$$

The radial acceleration is

$$a_{rad} = \frac{v^2}{r} = \frac{(7730\text{ m/s})^2}{6.68 \times 10^6\text{ m}}$$
$$= 8.94\text{ m/s}^2.$$

This is somewhat less than the value of the free-fall acceleration $g$ at the earth's surface; it is equal to the value of $g$ at a height of 300 km above the earth's surface.

We have talked mostly about artificial earth satellites, but we can apply the same analysis to the circular motion of *any* body under its gravitational attraction to a stationary body. Other familiar examples are our moon, the moons of other planets, and the planets orbiting the sun in nearly circular paths. The rings of Saturn and Neptune (Fig. 12–15) are composed of small pieces in circular orbits around those planets.

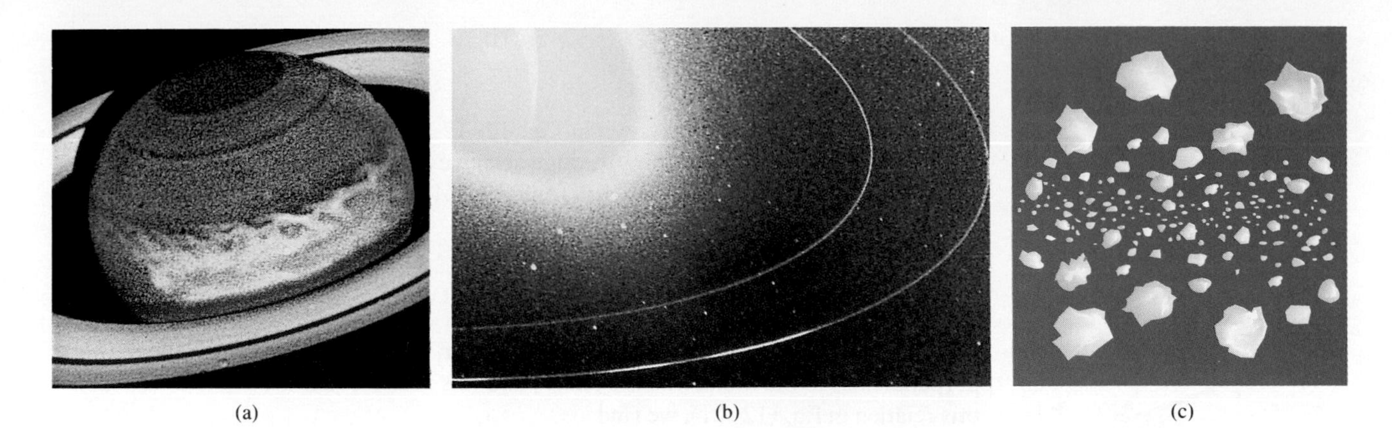

(a) (b) (c)

**12–15** (a) Saturn's rings, photographed from the Hubble Space Telescope in November 1990. The outer portions take more time for one revolution than the inner. (The white spot is an atmospheric storm.) (b) Neptune's rings, photographed from Voyager 1. (c) Planetary rings consist mainly of ice chunks, each moving in a circular orbit with period given by Eq. (12–20), using the planet's mass.

## *12–6 KEPLER'S LAWS

The motions of the planets, as they travel slowly across the sky, have fascinated humankind at least since the beginning of recorded history. The discovery of Kepler's laws of planetary motion forms a remarkable chapter in the history of science. Johannes Kepler (1571–1630) was a German astronomer. Like all astronomers of his day, he was deeply involved in trying to understand the motions of the planets, an endeavor with theological as well as astronomical significance at the time. Kepler based his analysis on observations made by his predecessor and teacher, the Danish astronomer Tycho Brahe (1546–1601). Brahe's measurements were remarkabley precise, considering that he worked only with direct-vision sighting devices. (The first telescope was invented in 1608, seven years after Brahe's death.)

Without any of the theoretical underpinnings supplied by Newton (1642–1727) nearly a century later, Kepler discovered by trial and error three empirical laws that described and correlated the motions of the five planets then known:

1. **Each planet moves in an elliptical orbit, with the sun at one focus of the ellipse.**
2. **A line from the sun to each planet sweeps out equal areas in equal times.**
3. **The periods of the planets are proportional to the $\frac{3}{2}$ powers of the major axis lengths of their orbits.**

Newton's laws of motion and law of gravitation can be used to derive these laws.

First, let's consider the elliptical orbits. Figure 12–16 shows the geometry of the ellipse. The $x$-axis is the *major axis,* with half-length $a$. The sum of the distances from the two points $F$ and $F'$ to $P$ is the same for all points on the curve; that's one way to *define* an ellipse. $F$ and $F'$ are the *foci* (plural of focus). The sun is at $F$, the planet at $P$; we think of them both as points because the size of each is very small in comparison to the distance between them.

The distance of each focus from the center of the ellipse is $ea$, where $e$ is a dimensionless number between 0 and 1 called the *eccentricity.* If $e = 0$, the ellipse is a circle. The actual orbits of the planets are nearly circular; the eccentricity of earth's orbit is about 0.017. As Fig. 12–13 shows, the orbit of a body moving under a $1/r^2$ force *must* be a circle, an ellipse, a parabola, or a hyperbola. This result can be derived by a straightforward application of Newton's laws and the law of gravitation, together with a lot more differential equations than we're ready for.

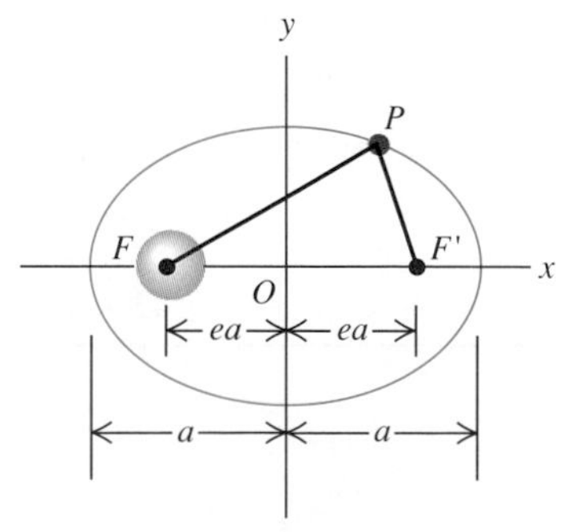


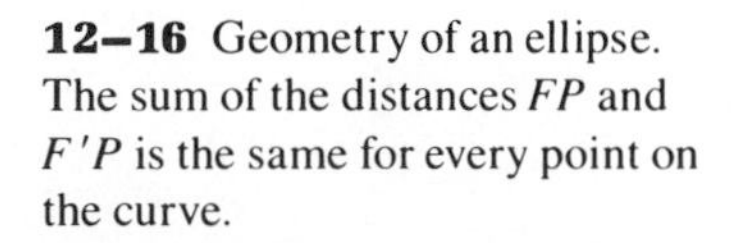

**12–16** Geometry of an ellipse. The sum of the distances $FP$ and $F'P$ is the same for every point on the curve.

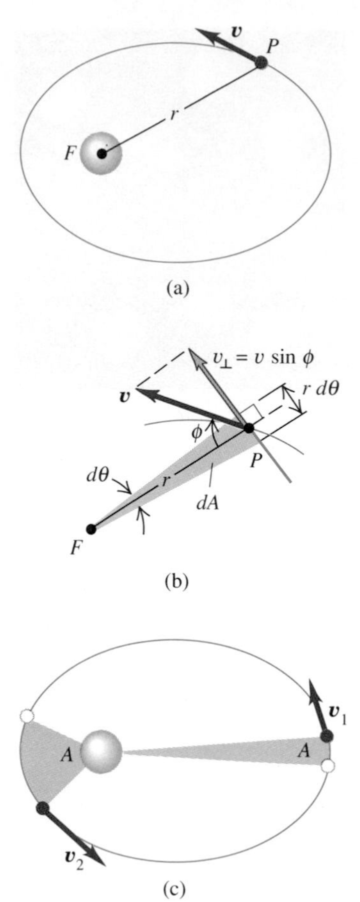


**12–17** (a) The planet $P$ moves about the sun (at $F$) in an elliptical orbit. (b) In a time $dt$ the line from the sun to the planet sweeps out a triangular area $dA = \frac{1}{2}(r\,d\theta)r = \frac{1}{2}r^2\,d\theta$. (c) The planet's velocity varies so that the ratio of area $dA$ to time interval $dt$ is constant, regardless of the planet's position in orbit.

Kepler's second law is shown in Fig. 12–17. In a small time interval $dt$ the line from the sun $F$ to the planet $P$ turns through an angle $d\theta$. The area swept out is the colored triangle with height $r$, base length $r\,d\theta$, and area $dA = \frac{1}{2}r^2\,d\theta$. The rate of sweeping out area, $dA/dt$, is called the *sector velocity:*

$$\frac{dA}{dt} = \frac{1}{2}r^2\frac{d\theta}{dt}. \tag{12–21}$$

We can express this rate in terms of the velocity $\boldsymbol{v}$ of the planet $P$; the component of $\boldsymbol{v}$ perpendicular to the radial line is $v \sin\phi$, and this is also equal to $r\omega = r\,d\theta/dt$. Using this relation in Eq. (12–21), we find

$$\frac{dA}{dt} = \tfrac{1}{2}rv\sin\phi.$$

But $rv\sin\phi$ is the magnitude of the vector product $\boldsymbol{r} \times \boldsymbol{v}$, which in turn is $1/m$ times the angular momentum ($\boldsymbol{L} = \boldsymbol{r} \times m\boldsymbol{v}$) of the planet. So we have

$$\frac{dA}{dt} = \frac{1}{2m}|\boldsymbol{r} \times m\boldsymbol{v}| = \frac{L}{2m}. \tag{12–22}$$

Constant sector velocity is equivalent to constant angular momentum!

In fact, it is easy to see why the angular momentum of the planet *must* be constant. According to Eq. (10–27),

$$\frac{d\boldsymbol{L}}{dt} = \boldsymbol{\tau} = \boldsymbol{r} \times \boldsymbol{F}.$$

In our situation, $\boldsymbol{r}$ and $\boldsymbol{F}$ always lie along the same line; their vector product $\boldsymbol{r} \times \boldsymbol{F}$ is zero, so $d\boldsymbol{L}/dt = \boldsymbol{0}$. Thus Kepler's second law is equivalent to conservation of angular momentum. This conclusion *does not* depend on the $1/r^2$ behavior of the force; angular momentum is conserved for *any* force that always acts along the line joining the particle to a fixed point. Such a force is called a *central force*. (The first and third laws are valid *only* for a $1/r^2$ force.)

We have already derived Kepler's third law for the particular case of circular orbits. Equation (12–20) shows that the period of a satellite or planet in a circular orbit is proportional to the $\frac{3}{2}$ power of the orbit radius.

## EXAMPLE 12–9

Using the data in Appendix F, check whether the orbit radii and periods of Uranus and Saturn obey Kepler's third law.

**SOLUTION** The ratio of the orbit radius of Uranus to that of Saturn is

$$\frac{r_U}{r_S} = \frac{2.87 \times 10^{12}\ \text{m}}{1.43 \times 10^{12}\ \text{m}} = 2.01.$$

According to Kepler's third law, the ratio of the periods should be the $\frac{3}{2}$ power of this, or 2.85. The actual ratio is

$$\frac{T_U}{T_S} = \frac{84.02\ \text{y}}{29.46\ \text{y}} = 2.85.$$

Voila!

---

The study of planetary motion played a pivotal role in the early development of physics. Kepler spent upwards of 15 years analyzing Brahe's observations and searching for patterns that would correlate them. He published his first two laws in 1609 and the third in 1619. But it remained for Newton to show, nearly a hundred years later, that this behavior of the planets could be understood by using the very same laws of motion that he had developed to analyze *terrestrial* motion.

Newton recognized that a falling apple (Section 4–4) was pulled to the earth by the same gravitational attraction that kept the planets in orbit. But to arrive at the $1/r^2$ form of the law of gravitation, he had to do something much subtler; he had to take Brahe's observations of planetary motions, as systematized by Kepler, and show that Kepler's laws demanded a $1/r^2$ force law.

From our historical perspective 300 years later, there is absolutely no doubt that this **Newtonian synthesis,** as it has come to be called, is one of the greatest achievements in the entire history of science, certainly comparable in significance to the development of quantum mechanics, the theory of relativity, and the understanding of DNA in our own century. It was an astonishing leap in understanding, made by a giant intellect!

## *12–7 SPHERICAL MASS DISTRIBUTIONS

We have used without proof the fact that the gravitational interaction between two spherically symmetric mass distributions is the same as though all the mass of each were concentrated at its center. Now we're ready to prove this statement. Newton searched for a proof for several years, and he delayed publication of the law of gravitation until he found one.

Here's our program. Rather than starting with two spherically symmetric masses, we'll tackle the simpler problem of a point mass $m$ interacting with a thin spherical shell with total mass $M$. We will show that when $m$ is outside the sphere, the *potential energy* associated with this gravitational interaction is the same as though $M$ were all concentrated at the center of the sphere. According to Eq. (12–16), the force is the negative derivative of the potential energy, so the *force* on $m$ is also the same as for a point mass $M$. Any spherically symmetric mass distribution can be thought of as being made up of many concentric spherical shells, so our result will also hold for *any* spherically symmetric $M$.

We start by considering a ring on the surface of the shell (Fig. 12–18a), centered on the line from the center of the shell to $m$. The radius of the shell is $R$; in terms of the angle $\phi$ shown in the figure, the radius of the ring is $R \sin \phi$, and its circumference is $2\pi R \sin \phi$. The width of the ring is $R\,d\phi$, and its area $dA$ is approximately equal to its width times its circumference:

$$dA = 2\pi R^2 \sin \phi \, d\phi .$$

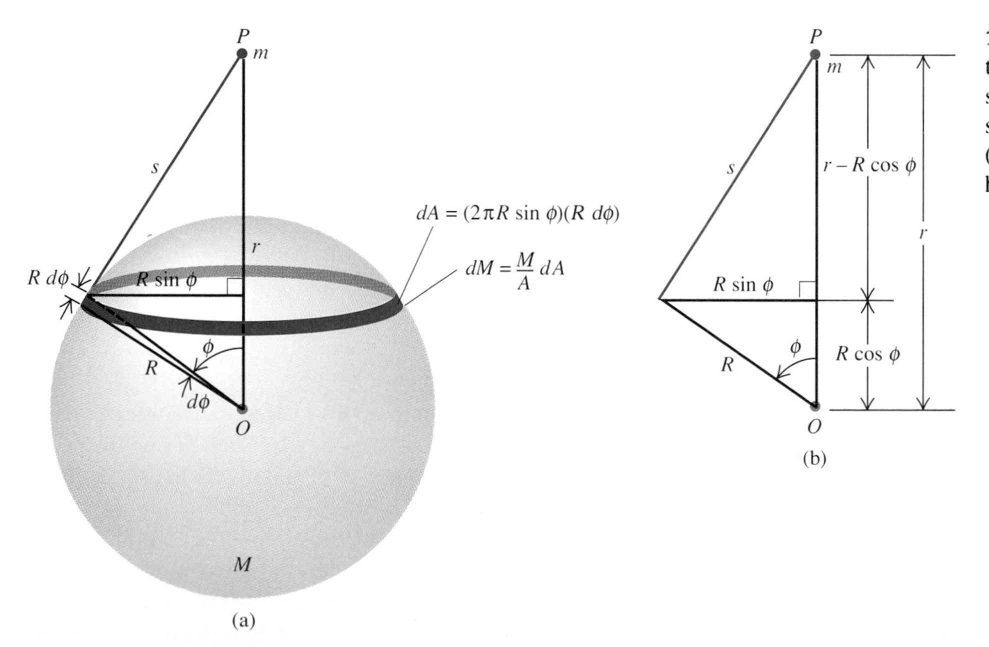


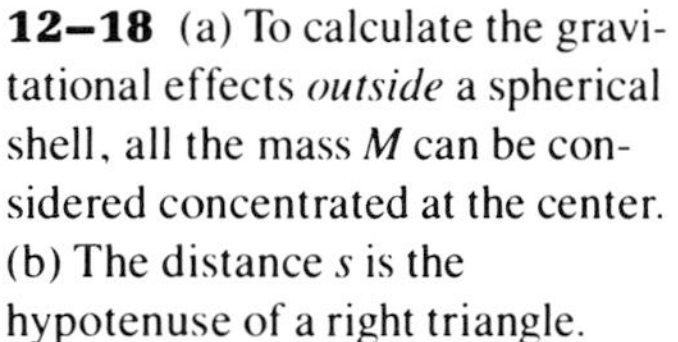
**12–18** (a) To calculate the gravitational effects *outside* a spherical shell, all the mass $M$ can be considered concentrated at the center. (b) The distance $s$ is the hypotenuse of a right triangle.

We need to know the mass $dM$ of this ring. The ratio of $dM$ to the total mass $M$ of the shell is equal to the ratio of the area $dA$ of the ring to the total area $A = 4\pi R^2$ of the shell:

$$\frac{dM}{M} = \frac{2\pi R^2 \sin\phi \, d\phi}{4\pi R^2} = \tfrac{1}{2}\sin\phi \, d\phi. \qquad (12\text{–}23)$$

The potential energy of interaction of $m$ and $dM$ can be obtained from Eq. (12–14), $U = -Gmm_E/r$, with the necessary changes of notation. In our case the masses are $m$ and $dM$ instead of $m$ and $m_E$, and the distance between the masses is the distance $s$ in the figure. Calling this potential energy $dU$, we have

$$dU = -\frac{Gm\,dM}{s}.$$

Now we solve Eq. (12–23) for $dM$ and substitute the result into the expression for $dU$:

$$dU = -\frac{GMm \sin\phi \, d\phi}{2s}. \qquad (12\text{–}24)$$

The total potential energy is the integral of Eq. (12–24) over the whole sphere as $\phi$ varies from 0 to $\pi$ (*not* $2\pi$!) and $s$ varies from $r - R$ to $r + R$. To carry out the integration, we have to express the integrand in terms of a single variable; we choose $s$. To express $\phi$ and $d\phi$ in terms of $s$, we have to do a little geometry. Figure 12–18b shows that $s$ is the hypotenuse of a right triangle with sides

$$(r - R\cos\phi) \qquad \text{and} \qquad (R\sin\phi),$$

so the Pythagorean theorem gives

$$\begin{aligned} s^2 &= (r - R\cos\phi)^2 + (R\sin\phi)^2 \\ &= r^2 - 2rR\cos\phi + R^2. \end{aligned} \qquad (12\text{–}25)$$

We take differentials of both sides:

$$2s\,ds = 2rR\sin\phi\,d\phi.$$

Next we divide this by $2rR$ and substitute the result into Eq. (12–24):

$$dU = -\frac{GMm}{2s}\frac{s\,ds}{rR} = -\frac{GMm}{2rR}\,ds. \qquad (12\text{–}26)$$

To find the total potential energy, we have to add the effects of all the rings that make up the spherical shell. We do this by integrating Eq. (12–26), recalling that $s$ varies from $r - R$ to $r + R$:

$$U = -\frac{GMm}{2rR}\int_{r-R}^{r+R} ds = -\frac{GMm}{2rR}[(r + R) - (r - R)]. \qquad (12\text{–}27)$$

Finally, we have

$$U = -\frac{GmM}{r}. \qquad (12\text{–}28)$$

This is equal to the potential energy of two point masses $m$ and $M$ at a distance $r$. So we have proved that the gravitational potential energy of the spherical shell $M$ and the point mass $m$ at any distance $r$ is the same as though they were point masses. Because the force is given by $F_r = -dU/dr$, the force is also the same.

Any spherically symmetric mass distribution can be thought of as a combination of concentric spherical shells. Because of the principle of superposition of forces, what is true of one shell is also true of the combination. So we have proved half of what we set out to prove, namely, that the gravitational interaction between any spherically symmetric

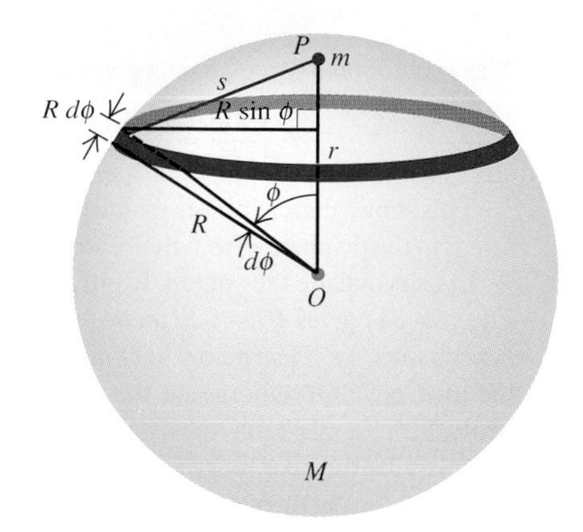


**12–19** When $m$ is *inside* a uniform spherical shell of mass $M$, the potential energy is constant, and the force from the masses' mutual gravitational interaction is zero.

mass distribution and a point mass is the same as though all the mass of the spherically symmetric distribution were concentrated at its center.

The other half is to prove that *two* spherically symmetric mass distributions interact as though they were both points. That's easier. In Fig. 12–18 the forces that the two bodies exert on each other are an action-reaction pair, and they obey Newton's third law. So we have also proved that the force that $m$ exerts *on* the sphere $M$ is the same as though $M$ were a point. But now if we replace $m$ with a spherically symmetric mass distribution centered at $m$'s location, the resulting gravitational field at any point on $M$ is the same as before, and so is the total force. This completes our proof.

We assumed at the beginning that $m$ was outside the spherical shell, so our proof is valid only when $m$ is outside a spherically symmetric mass distribution. When $m$ is *inside* a spherical shell, the geometry is as shown in Fig. 12–19. The entire analysis goes just as before; Eqs. (12–23) through (12–26) are still valid. But when we get to Eq. (12–27), the limits of integration have to be changed to $R - r$ and $R + r$. We then have

$$U = -\frac{GMm}{2rR}\int_{R-r}^{R+r} ds = -\frac{GMm}{2rR}[(R+r) - (R-r)], \tag{12–29}$$

and the final result is

$$U = -\frac{GmM}{R}. \tag{12–30}$$

Instead of having $r$, the distance between $m$ and the center of $M$, in the denominator, we have $R$, the radius of the shell. This means that $U$ doesn't depend on $r$ and thus has the same value everywhere inside the shell. When $m$ moves around inside the shell, no work is done on it, so the force on $m$ at any point inside the shell must be zero.

More generally, at any point in the interior of any spherically symmetric mass distribution (not necessarily a shell), at a distance $r$ from its center, the gravitational force on a point mass $m$ is the same as though we removed all the mass at points farther than $r$ from the center and concentrated all the remaining mass at the center.

## EXAMPLE 12–10

**"Journey to the Center of the Earth"** Suppose we drill a hole through the earth (radius $R_E$, mass $M_E$) from one side to the other along a diameter and drop a mail pouch down the hole. Derive an expression for the gravitational force on the pouch (mass $m$) as a function of its distance $r$ from the center. Assume that the density of the earth is uniform (not a very realistic model).

**SOLUTION** The gravitational force at a distance $r$ from the center is determined only by the mass $M$ within a sphere of radius $r$ (Fig. 12–20, on the next page). With uniform density the mass is proportional to the volume of the sphere, which is $\frac{4}{3}\pi r^3$ for this sphere and $\frac{4}{3}\pi R_E^3$ for the entire earth. So we have

$$\frac{M}{M_E} = \frac{\frac{4}{3}\pi r^3}{\frac{4}{3}\pi R_E^3} = \frac{r^3}{R_E^3}.$$

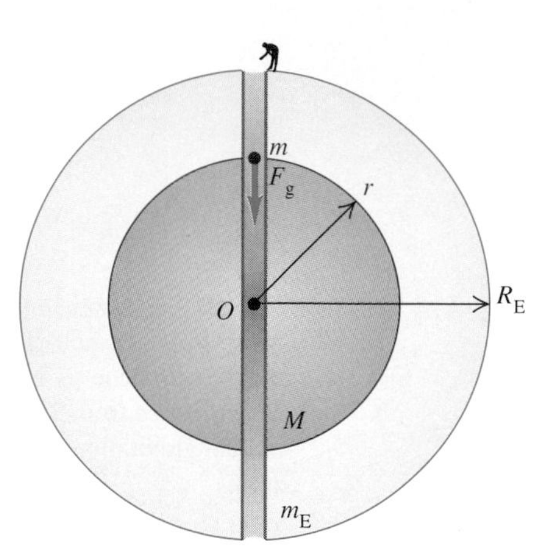


**12–20** A hole through the earth's center. When an object is at a distance $r$ from the center, only the mass inside a sphere of radius $r$ exerts a net gravitational force on it.

The gravitational force on $m$ is given by

$$F_g = \frac{GMm}{r^2} = \frac{GM_Em}{r^2}\frac{r^3}{R_E{}^3} = \frac{GM_Em}{R_E{}^3}r. \tag{12–31}$$

This shows that, at points inside the earth, $F_g$ is *directly proportional* to the distance $r$ from the center, rather than proportional to $1/r^2$ as it is outside the sphere. Right at the surface, where $r = R_E$, Eq. (12–31) gives $F_g = GM_Em/R_E{}^2$, as we should expect. In the next chapter we'll learn how to compute the time it would take for the mail pouch to emerge on the other side of the earth. Meanwhile, keep in mind that the assumption of uniform density of the earth is, like the hole, pure fantasy. We discussed the variation of density in the earth in Section 12–2.

## *12–8 EFFECT OF THE EARTH'S ROTATION ON THE ACCELERATION DUE TO GRAVITY

Because the earth rotates on its axis, it is not precisely an inertial frame of reference. For this reason the apparent weight of a body on the earth is not precisely equal to the earth's gravitational attraction, which we will call the **true weight** $\boldsymbol{w}_0$ of the body. Figure 12–21 is a cutaway view of the earth, showing three observers. Each one holds a body of mass $m$ hanging from a spring balance. Each balance applies a tension force $\boldsymbol{F}$ to the body hanging from it, and the reading on each balance is the magnitude $F$ of this force. If the observers are unaware of the earth's rotation, each one *thinks* that the scale reading equals the weight of the body because he thinks the body on his spring balance is in equilibrium. If it is, he thinks, the tension $\boldsymbol{F}$ must be opposed by an equal and opposite force $\boldsymbol{w}$, which we call the **apparent weight.** Because of rotation, the bodies are *not* precisely in equilibrium. Our problem is to find the relation between the apparent weight $\boldsymbol{w}$ and the true weight $\boldsymbol{w}_0$.

If we assume that the earth is spherically symmetric, then the true weight $\boldsymbol{w}_0$ has magnitude $Gmm_E/R_E{}^2$, where $m_E$ and $R_E$ are the mass and radius of the earth. This value is the same for all points on the earth's surface. If the center of the earth can be taken as the origin of an inertial coordinate system, then the body at the North Pole really *is* in equilibrium in an inertial system, and the reading on that observer's spring balance is equal to $w_0$. But the body at the equator is moving in a circle of radius $R_E$ with speed $v$, and there must be a net inward force equal to the mass times the centripetal acceleration:

$$w_0 - F = \frac{mv^2}{R_E}. \tag{12–32}$$

So the magnitude of the apparent weight (equal to $F$ at this location) is

$$w = w_0 - \frac{mv^2}{R_E}. \tag{12–33}$$

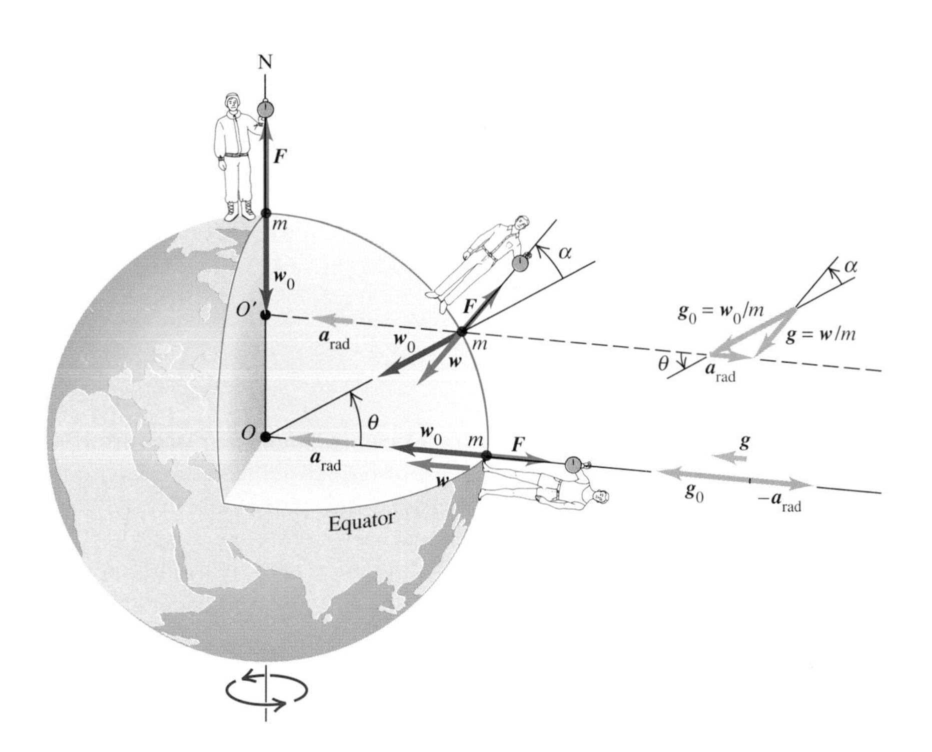


**12–21** Except at the poles, the scale reads less than the gravitational force of attraction because a net force is needed to provide the centripetal acceleration. The sketch exaggerates the difference for clarity. On the earth the maximum value of $\alpha$ is 0.1°, and the magnitude of the apparent weight is smaller than that of the true weight by less than 0.35%.

If the earth were not rotating, the body when released would have a free-fall acceleration $g_0$ given by $g_0 = w_0/m$, but its actual acceleration relative to the observer at the equator is given by $g = w/m$. Dividing Eq. (12–33) by $m$ and using these relations, we find

$$g = g_0 - \frac{v^2}{R_E}. \quad (12\text{–}34)$$

To evaluate $v^2/R_E$ we note that in 86,164 s a point on the equator moves a distance equal to the earth's circumference, $2\pi R_E = 2\pi(6.38 \times 10^6 \text{ m})$. (The solar day, 86,400 s, is $\frac{1}{365}$ longer than this because in one day the earth also completes $\frac{1}{365}$ of its orbit around the sun.) Thus we find

$$v = \frac{2\pi(6.38 \times 10^6 \text{ m})}{86{,}164 \text{ s}} = 465 \text{ m/s},$$

$$\frac{v^2}{R_E} = \frac{(465 \text{ m/s})^2}{6.38 \times 10^6 \text{ m}} = 0.0339 \text{ m/s}^2.$$

So for a spherically symmetric earth the acceleration due to gravity should be about 0.03 m/s$^2$ less at the equator than at the poles.

At locations intermediate between the equator and the poles, the true weight $\mathbf{w}_0$ and the centripetal acceleration are not along the same line, and we need to write a vector equation corresponding to Eq. (12–33). From Fig. 12–21 we see that the appropriate equation is

$$\mathbf{w} = \mathbf{w}_0 - m\mathbf{a}_{rad} = m\mathbf{g}_0 - m\mathbf{a}_{rad}. \quad (12\text{–}35)$$

The difference in the magnitudes $g$ and $g_0$ is somewhere between zero and 0.0339 m/s$^2$, and the *direction* of the apparent weight differs from the radial direction by a small angle $\alpha$, 0.1° or less, as shown.

**TABLE 12–1 Variations of $g$ with Latitude and Elevation**

| Station | North latitude | Elevation (m) | $g$ (m/s$^2$) | $g$ (ft/s$^2$) |
|---|---|---|---|---|
| Canal Zone | 9° | 0 | 9.78243 | 32.0944 |
| Jamaica | 18° | 0 | 9.78591 | 32.1059 |
| Bermuda | 32° | 0 | 9.79806 | 32.1548 |
| Denver | 40° | 1638 | 9.79609 | 32.1393 |
| Cambridge, Mass. | 42° | 0 | 9.80398 | 32.1652 |
| Pittsburgh, Pa. | 40.5° | 235 | 9.80118 | 32.1561 |
| Greenland | 70° | 0 | 9.82534 | 32.2353 |

Table 12–1 gives the values of $g$ at several locations, showing variations with latitude of approximately this magnitude. There are also small additional variations due to the lack of perfect spherical symmetry of the earth, local variations in density (used for mineral prospecting), and differences in elevation.

Our discussion of apparent weight can also be applied to the phenomenon of apparent weightlessness in orbiting spacecraft. Bodies in an orbiting vehicle are *not* weightless; the earth's gravitational attraction continues to act on them just as though they were at rest relative to the earth. A space vehicle in orbit has an acceleration $\boldsymbol{a}_{\text{rad}}$ toward the earth's center that is equal to the value of the acceleration of gravity $\boldsymbol{g}_0$ at its orbit radius. The apparent weight $\boldsymbol{w}$ is given as before by

$$\boldsymbol{w} = \boldsymbol{w}_0 - m\boldsymbol{a}_{\text{rad}} = m\boldsymbol{g}_0 - m\boldsymbol{a}_{\text{rad}}.$$

But in this case,

$$\boldsymbol{g}_0 = \boldsymbol{a}_{\text{rad}},$$

so

$$\boldsymbol{w} = \boldsymbol{0}.$$

This is what we mean when we say that an astronaut or other body in a spacecraft is apparently weightless (Fig. 12–22). The earth's gravity acts on both the spacecraft and the bodies in it, giving each the same acceleration. Thus a body released inside the spacecraft does not fall relative to it and *appears* to be weightless.

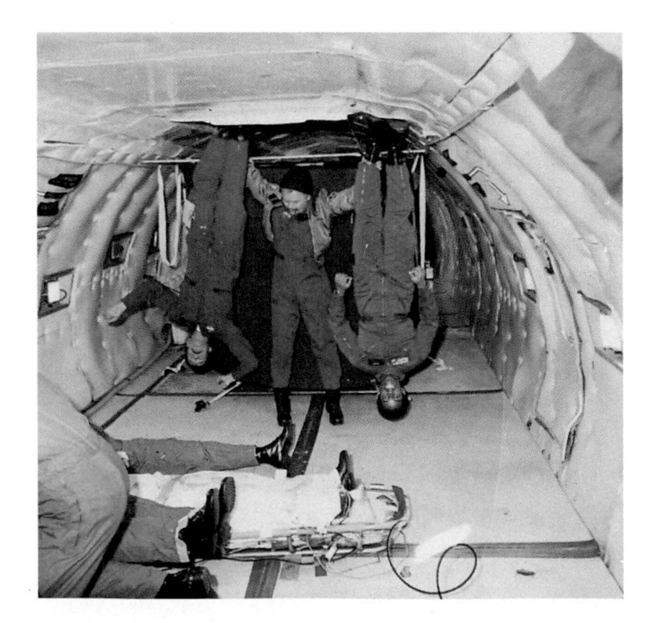

**12–22** Two astronauts having fun with apparent weightlessness.

Something similar happens in an accelerating elevator. Suppose a person with mass $m$ stands on the floor of an elevator that has an upward acceleration $a_y$, relative to an inertial frame of reference. We showed in Example 5–8 that the upward normal force on the passenger's feet has magnitude $m(g + a_y)$, and if she stands on a spring scale calibrated to read weight, it will read this value rather than her weight $mg$. If the acceleration is downward, $a_y$ is negative and the normal force is smaller in magnitude than $mg$. In particular, if the elevator is, unfortunately, in free fall, $a_y = -g$, the normal force is zero, and the person seems to be weightless, although the earth's gravitational attraction still acts on her.

## 12–9 BLACK HOLES: A Case Study in Modern Physics

The concept of a black hole is one of the most interesting and startling products of modern gravitational theory, yet the basic idea can be understood on the basis of Newtonian principles. Think first about the properties of our sun. Its mass $M = 1.99 \times 10^{30}$ kg and radius $R = 6.96 \times 10^{8}$ m are much larger than those of any planet; but compared to other stars, our sun is not exceptionally massive.

What's the average density $\rho$ of the sun? You can find it the same way we found the density of the earth in Section 12–2:

$$\rho = \frac{M}{V} = \frac{M}{\frac{4}{3}\pi R^3} = \frac{1.99 \times 10^{30}\ \text{kg}}{\frac{4}{3}\pi(6.96 \times 10^{8}\ \text{m})^3}$$
$$= 1410\ \text{kg/m}^3.$$

The sun's temperatures range from 5800 K (about 10,000 °F) at the surface up to $1.5 \times 10^{7}$ K (about $2.7 \times 10^{7}$ °F) in the interior, so it surely contains no solids or liquids. Yet gravitational attraction pulls the sun's gas atoms together until the sun is, on average, 41% denser than water and about 1200 times as dense as the air we breathe.

Now think about the escape speed for a body at the surface of the sun; this is given by Eq. (12–15). We can relate this to the average density. Substituting $M = \rho V = \rho \frac{4}{3}\pi R^3$ into Eq. (12–15) gives

$$v = \sqrt{\frac{2GM}{R}} = \sqrt{\frac{8\pi G\rho}{3}}\,R. \qquad (12\text{–}36)$$

Using either form of this equation, you can show that the escape speed for a body at the surface of our sun is $v = 6.18 \times 10^{5}$ m/s (about 1.4 million mi/h). This value, roughly $\frac{1}{500}$ the speed of light, is independent of the mass of the escaping body; it depends only on the mass and radius (or average density and radius) of the sun.

Now consider various stars with the same average density $\rho$ and different radii $R$. Equation (12–36) shows that for a given value of $\rho$ the escape speed $v$ is directly proportional to $R$. In 1783 the Rev. John Mitchell, an amateur astronomer, noted that if a body with the same average density as the sun had about 500 times the radius of the sun, the magnitude of its escape velocity would be greater than the speed of light $c$. With his statement that "all light emitted from such a body would be made to return toward it," Mitchell became the first person to suggest the existence of what we now call a **black hole.**

What's the critical radius of a body if it is to act as a black hole? You might think that you can find the answer by simply setting $v = c$ in Eq. (12–36). As a matter of fact, this

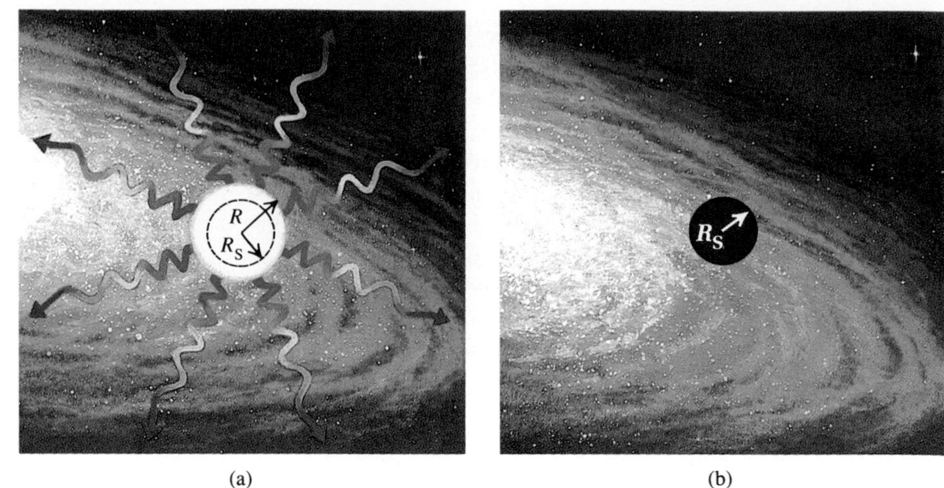


**12–23** (a) When the radius $R$ of a body is greater than the Schwarzschild radius $R_S$, light can escape from the surface of the body. As it travels away, it is "red-shifted" to longer wavelengths. (b) When the body is within the event horizon (radius $R_S$), it is a black hole with an escape speed greater than the speed of light. In that case we can obtain very little information about it.

does give the correct result, but only because of two compensating errors. The kinetic energy of light is *not* $mc^2/2$, and the gravitational potential energy near a black hole is *not* given by Eq. (12–14). In 1916 Karl Schwarzschild used Einstein's general theory of relativity (in part a generalization and extension of Newtonian gravitation theory) to derive an expression for the critical radius $R_S$, now called the **Schwarzschild radius.** The result turns out to be the same as though we had set $v = c$ in Eq. (12–36); the Schwarzschild radius is

$$c = \sqrt{\frac{2GM}{R_S}} \quad \text{or} \quad R_S = \frac{2GM}{c^2}. \tag{12–37}$$

If a spherical, nonrotating body with mass $M$ has a radius less than $R_S$, then *nothing* (not even light) can escape from the surface of the body, and the body functions as a black hole (Fig. 12–23). In this case any other body within a distance $R_S$ of the center of the body is "trapped" by the gravitational field and cannot escape from it.

The surface of the sphere with radius $R_S$ surrounding a black hole is called the **event horizon** because, since light can't escape from within that sphere, we can't see events occurring inside. All we can know about a black hole is its mass (from its gravitational field), its charge (from its electric field, expected to be zero), and its rotation (because a rotating black hole tends to drag space — and everything in that space — along with it.

## EXAMPLE 12–11

Current astrophysical theory suggests that a burned-out star can collapse under its own gravity to form a black hole when its mass is as small as two solar masses. If it does, what is the radius of its event horizon?

**SOLUTION** The radius is $R_S$, and "two solar masses" means $M = 2(1.99 \times 10^{30}\ \text{kg}) = 4.0 \times 10^{30}\ \text{kg}$. From Eq. (12–37),

$$R_S = \frac{2GM}{c^2} = \frac{2(6.67 \times 10^{-11}\ \text{N}\cdot\text{m}^2/\text{kg}^2)(4.0 \times 10^{30}\ \text{kg})}{(3.00 \times 10^8\ \text{m/s})^2}$$
$$= 5.9 \times 10^3\ \text{m} = 5.9\ \text{km},$$

or less than 4 miles.

The average density of such an object has the incredibly large *minimum* value

$$\rho = \frac{M}{\frac{4}{3}\pi R^3} = \frac{4.0 \times 10^{30}\ \text{kg}}{\frac{4}{3}\pi(5.9 \times 10^3\ \text{m})^3}$$
$$= 4.6 \times 10^{18}\ \text{kg/m}^3.$$

This is of the order of $10^{15}$ times as great as the density of familiar matter on earth and is comparable to the densities of atomic nuclei. In fact, contemporary theory suggests that in the final stage of collapse of a dying star its mass is crushed down to a single point called a *singularity*, with infinite density.

At points far from a black hole its gravitational effects are the same as those of any normal body with the same mass. However, any body moving close to a black hole experiences huge and strongly position-dependent forces. These cause the body to become so hot that it emits not just visible light (as in "red-hot" or "white-hot") but x rays. Astronomers look for these x rays (emitted *before* the body crosses the event horizon) to signal the presence of a black hole (Fig. 12–24). Several promising candidates have been found, and many astronomers now express considerable confidence in the existence of black holes.

Don't think about jumping into a black hole unless you really want to be a martyr for science. Those you left behind would notice several odd effects as you moved toward the event horizon, most of them associated with effects of general relativity. If you carried a radio transmitter to send back your comments on what was happening, they would have to retune the receiver continuously to lower and lower frequencies, an effect called the *gravitational red shift.* Consistent with this shift, they would observe that your clocks (electronic or biological) would appear to run more and more slowly, an effect called *time dilation*. In fact, you would never make it to the event horizon during their lifetimes, even though your own clocks wouldn't show this effect. But you'd never live long enough to reach the event horizon anyway; the nonuniform gravitational field over your body would stretch you along the direction of the field and compress you perpendicular to it. These effects (called *tidal forces*) would finish you off long before your bodily atoms got hot enough to emit x rays.

Theorist Stephen Hawking believes that black holes aren't completely black. He has suggested a way in which black holes can emit energy (and thus lose mass) through a quantum-mechanical process called *tunneling,* which is somewhat analogous to emission of an alpha particle from an unstable nucleus. But, he estimates, a two-solar-mass black hole would have a temperature of the order of $10^{-6}$ K and would need about $10^{67}$ years to "evaporate" by this process. So don't hold your breath! Meanwhile, experimental and theoretical studies of black holes continue to be a vital and exciting area of research in contemporary physics.

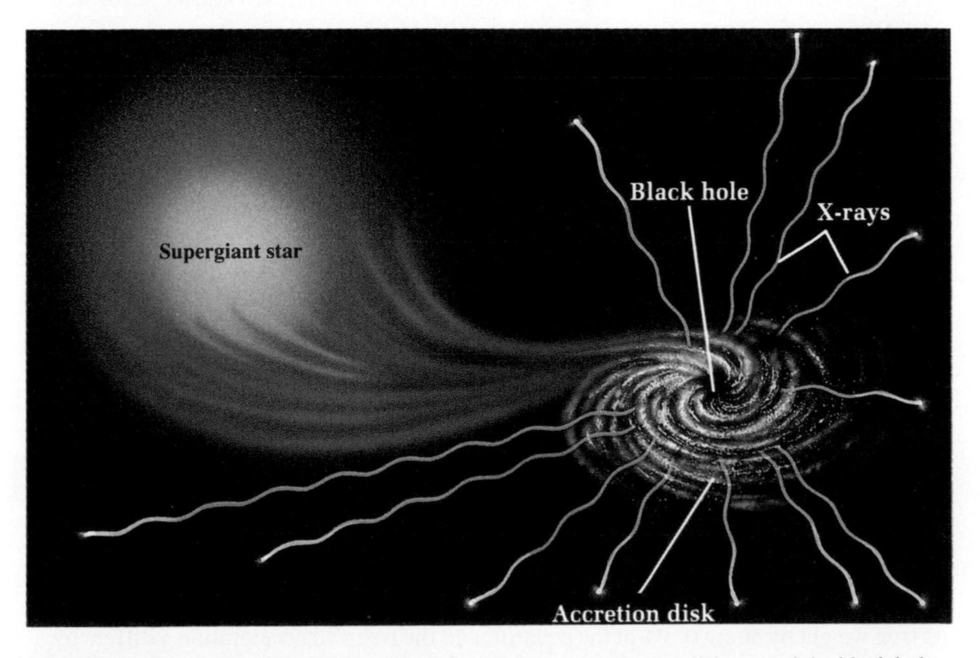


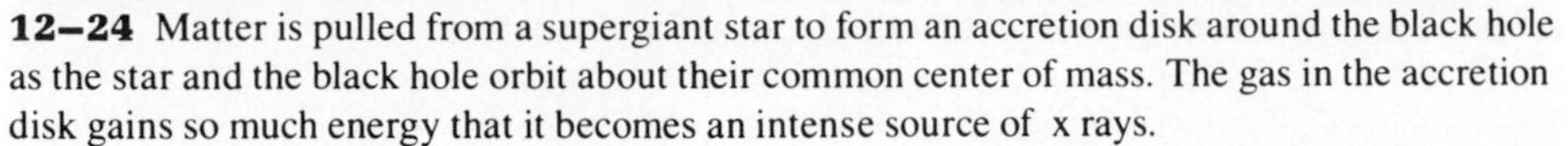
**12–24** Matter is pulled from a supergiant star to form an accretion disk around the black hole as the star and the black hole orbit about their common center of mass. The gas in the accretion disk gains so much energy that it becomes an intense source of x rays.

## SUMMARY

### KEY TERMS

law of gravitation
gravitational constant
weight
gravitational field
vector field
Newtonian synthesis
true weight
apparent weight
black hole
Schwarzschild radius
event horizon

• Newton's law of gravitation: Two particles with masses $m_1$ and $m_2$, a distance $r$ apart, attract each other with forces of magnitude

$$F_g = G\frac{m_1 m_2}{r^2}. \tag{12-1}$$

These forces form an action-reaction pair and obey Newton's third law. When two or more bodies exert gravitational forces on a particular body, the total gravitational force on it is the vector sum of the forces exerted by the individual bodies.

• The weight $w$ of a body is the total gravitational force exerted on it by all other bodies in the universe. Near the surface of the earth (mass $m_E$) this is essentially equal to the gravitational force of the earth alone. The weight $w$ of a body with mass $m$ is then

$$w = F_g = \frac{Gmm_E}{R_E^2}, \tag{12-3}$$

and the acceleration due to gravity $g$ is

$$g = \frac{Gm_E}{R_E^2}. \tag{12-4}$$

• The gravitational field $\boldsymbol{g}$ at a point in space is the gravitational force on a mass $m$ at that point, divided by $m$. The gravitational field caused by a particle of mass $M$ at a distance $r$ from it is

$$\boldsymbol{g} = -\frac{GM}{r^2}\hat{\boldsymbol{r}}. \tag{12-9}$$

The total gravitational field at any point due to several masses is the vector sum of gravitational fields due to the individual masses.

• The gravitational potential energy $U$ of two particles of masses $m$ and $m_E$ separated by a distance $r$ is

$$U = -\frac{Gmm_E}{r}. \tag{12-14}$$

• When a satellite moves in a circular orbit, the centripetal acceleration is provided by the gravitational attraction of the earth. The speed $v$ and period $T$ of a satellite in an orbit with radius $r$ are

$$v = \sqrt{\frac{Gm_E}{r}}, \tag{12-18}$$

$$T = \frac{2\pi r}{v} = 2\pi r\sqrt{\frac{r}{Gm_E}} = \frac{2\pi r^{3/2}}{\sqrt{Gm_E}}. \tag{12-20}$$

• Kepler's three laws describe characteristics of the elliptical orbits of planets around the sun or of satellites around the parent planet.

• The gravitational interaction of any spherically symmetric mass distribution, at points outside the distribution, is the same as though all the mass were concentrated at the center.

• Because of the earth's rotation, the apparent weight of a body on earth differs from its true weight by about 0.3% at the equator, and the free-fall acceleration $g$ differs by the same amount from the value it would have without rotation.

- If a nonrotating spherical mass distribution with total mass $M$ has a radius less than $R_S = 2GM/c^2$, then the gravitational interaction prevents anything, including light, from escaping from within a sphere with radius $R_S$, called the Schwarzschild radius. Such a body is called a black hole.

## EXERCISES

### Section 12–1 Newton's Law of Gravitation

**12–1** What is the ratio of the gravitational pull of the sun on the moon to that of the earth on the moon? Use the data in Appendix F. What is the significance of the result to the motion of the moon?

**12–2** A 200-kg communications satellite is in a circular orbit with a radius of 30,000 km, measured from the center of the earth. What is the gravitational force on the satellite? What fraction is this of its weight at the surface of the earth?

**12–3** A spaceship travels from the earth directly toward the sun. At what distance from the center of the earth do the gravitational forces of the sun and the earth on the ship exactly cancel? Use the data in Appendix F.

**12–4** A small, uniform sphere of mass 0.100 kg is located on the line between a 5.00-kg and a 10.0-kg uniform sphere (Fig. 12–25). The small sphere is 4.00 m to the right of the center of the 5.00-kg sphere and 6.00 m to the left of the center of the 10.0-kg sphere. What are the magnitude and direction of the force on the 0.100-kg sphere?

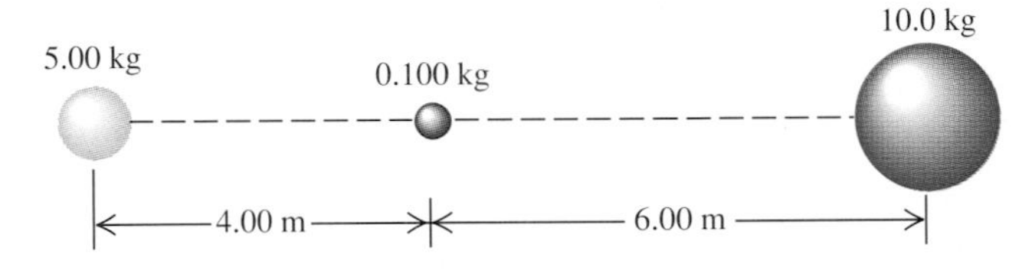


FIGURE 12–25

**12–5** Two spheres, each of mass 6.40 kg, are fixed at points $A$ and $B$ (Fig. 12–26). Find the magnitude and direction of the initial acceleration of a sphere with mass 0.010 kg if it is released from rest at point $P$ and acted on only by forces of gravitational attraction of the spheres at $A$ and $B$.

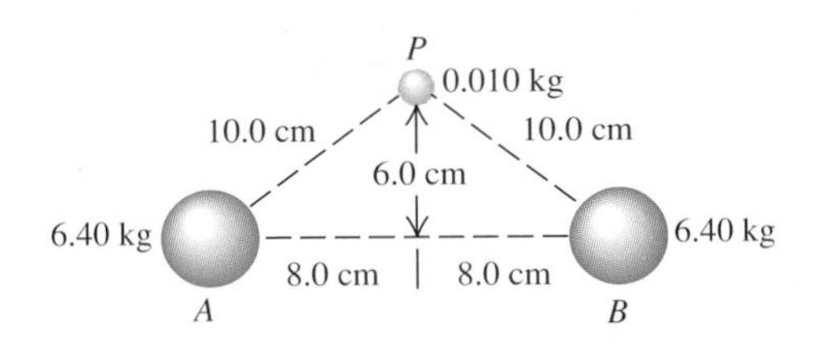

FIGURE 12–26

### Section 12–2 Weight

**12–6 Weight on the Moon.** The mass of the moon is about $\frac{1}{81}$ that of the earth and its radius is $\frac{1}{4}$ that of the earth. Compute the acceleration due to gravity on the moon's surface from these data.

**12–7** An experiment using the Cavendish balance to measure the gravitational constant $G$ found that a mass of 0.800 kg attracts another sphere of mass $4.00 \times 10^{-3}$ kg with a force of $1.30 \times 10^{-10}$ N when the distance between the centers of the spheres is 0.0400 m. The acceleration of gravity at the earth's surface is 9.80 m/s$^2$, and the radius of the earth is 6380 km. Compute the mass of the earth from these data.

**12–8** Use the mass and radius of the planet Mercury given in Appendix F to calculate the acceleration due to gravity at the surface of Mercury.

### Section 12–3 Gravitational Field

**12–9** What is the magnitude of the gravitational field 2.00 m from a 5.00-kg point mass?

**12–10** What are the magnitude and direction of the gravitational force on a 0.100-kg point mass placed at a point where the gravitational field has components $g_x = 4.00$ m/s$^2$ and $g_y = -6.00$ m/s$^2$?

**12–11** The gravitational force $\boldsymbol{F}$ on a 0.0100-kg test mass is found to be $\boldsymbol{F} = (-0.225\ \text{N})\boldsymbol{i} + (0.540\ \text{N})\boldsymbol{j}$ at a certain point. What are the components of the gravitational field vector at that point?

**12–12** What are the magnitude and direction of the gravitational field at a point midway between two point masses, one of mass $m_1$ and the other of mass $m_2$, where $m_2 > m_1$? The two point masses are a distance $d$ apart.

**12–13** At what distance above the surface of the earth is the magnitude of its gravitational field 4.90 m/s$^2$ if the gravitational field at the surface has magnitude 9.80 m/s$^2$?

**12–14** An object in the shape of a thin ring has radius $a = 0.800$ m and mass $M = 2.50$ kg. A small, uniform sphere with a mass $m = 0.0200$ kg is placed at a distance $x = 3.00$ m to the right of the center of the ring along a line through the center of

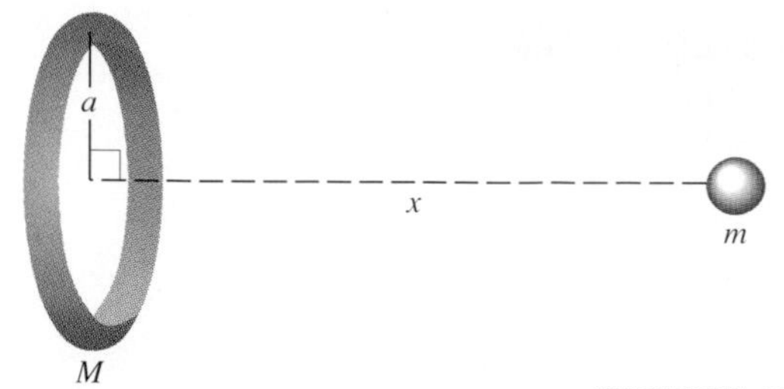


FIGURE 12–27

the ring and perpendicular to its plane (Fig. 12–27). What is the gravitational force that the sphere exerts on the ring-shaped object?

## Section 12–4 Gravitational Potential Energy

**12–15** Use the results of Example 12–7 to calculate the magnitude of the escape velocity for an object a) from the surface of the moon; b) from the surface of Saturn. c) Why is the escape velocity for an object independent of the object's mass?

**12–16** An artillery shell with mass $m$ is shot vertically upward from the surface of the earth. If the shell's initial speed is $8.00 \times 10^3$ m/s, to what height above the surface of the earth will it rise? (Neglect the effect of the drag force exerted by the air, so the only force on the shell is assumed to be gravity.)

**12–17** The asteroid Toro, discovered in 1964, has a radius of about 5.0 km and a mass of about $2.0 \times 10^{15}$ kg. Use the results of Example 12–7 to calculate the escape speed for an object at the surface of Toro. Could a person reach this speed just by running?

## Section 12–5 Satellite Motion

## Section 12–6 Kepler's Laws

**12–18** Pluto orbits the sun in a nearly circular orbit with radius $5.90 \times 10^{12}$ m and an orbital period of 247.7 years. Use these data to calculate the mass of the sun.

**12–19** To launch a satellite in a circular orbit 1200 km above the surface of the earth, what orbital speed must be given to the satellite?

**12–20** What is the period of revolution of an artificial satellite of mass $m$ that is orbiting the earth in a circular path of radius 8380 km (about 2000 km above the surface of the earth)?

**12–21** An earth satellite moves in a circular orbit with an orbital speed of 6400 m/s. a) Find the time of one revolution. b) Find the radial acceleration of the satellite in its orbit.

**12–22** A science fiction writer speculates that a small, as yet unobserved, planet circles the sun inside the orbit of Mercury. The orbit radius of Mercury is $5.8 \times 10^{10}$ m, and its orbital period is 88.0 days. If the hypothetical planet has an orbital period of 60.0 days, what is the radius of its orbit?

## Section 12–7 Spherical Mass Distributions

***12–23** Consider the ring-shaped body of Example 12–6. A point mass $m$ is placed a distance $x$ from the center of the ring, along the line through the center of the ring and perpendicular to its plane (Fig. 12–27). a) Use Eq. (12–14) to calculate the gravitational potential energy of this system. b) Show that your answer reduces to the expected result when $x$ is much larger than the radius $a$ of the ring. c) Use $F_x = -dU/dx$ to calculate the force between the objects. Compare your result to that calculated from Eq. (12–11).

***12–24** A thin, uniform rod has length $L$ and mass $M$. A small, uniform sphere of mass $m$ is placed a distance $x$ from one end of the rod along the axis of the rod (Fig. 12–28). Calculate the gravitational potential energy of this pair of masses. Show that your answer reduces to the expected result when $x$ is much larger than $L$.

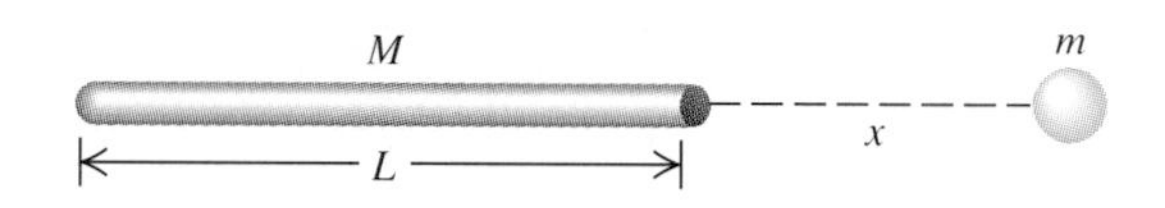


FIGURE 12–28

## Section 12–8 Effect of the Earth's Rotation on the Acceleration due to Gravity

***12–25 Weight on Jupiter.** The acceleration due to gravity at the north pole of Jupiter is approximately 25 m/s$^2$. Jupiter has mass $1.9 \times 10^{27}$ kg and radius $7.1 \times 10^7$ m and makes one revolution about its axis in about 10 h. a) What is the gravitational force on a 5.0-kg object at the north pole of Jupiter? b) What is the apparent weight of this same object at the equator of Jupiter?

***12–26** The weight of a man as determined by a spring balance at the equator is 650 N. By how much does this differ from the true force of gravitational attraction at the same point?

## Section 12–9 Black Holes: A Case Study in Modern Physics

**12–27** More than 200 asteroids have a diameter greater than 100 km. What is the magnitude of the escape velocity from a 100-km diameter asteroid with a density of $5.0 \times 10^3$ kg/m$^3$?

**12–28** Derive the relation $R_S \cong$ (3.0 km/solar mass)$M$, where $M$ is in solar masses. (One solar mass is the mass of the sun.) How accurate is this relation?

**12–29** To what fraction of its current radius would the sun have to be compressed to become a black hole?

**12–30** Calculate the magnitude of the gravitational field strength 6380 km from a black hole of Schwarzschild radius 8.88 mm.

**12–31** Show that a black hole attracts a distant mass $m$ with a force of $mc^2R_S/(2r^2)$.

**12–32 Mini Black Holes.** It is possible that mini black holes may have been created in the Big Bang. The evaporation time $t_{evap}$ is $1.0 \times 10^{66}$ years for a one-solar-mass black hole and is proportional to the cube of the mass. a) Show that $t_{evap} = (1.3 \times 10^{-25} \text{y/kg}^3)M^3$. b) If the Big Bang occurred $2 \times 10^{10}$ years ago, all black holes of less than a certain mass and Schwarzschild radius must have evaporated by now. Calculate that mass and Schwarzschild radius.

## PROBLEMS

**12–33** Two point masses are located at $(x, y)$ coordinates as follows: 5.00 kg at (1.00 m, 0) and 3.00 kg at (0, −0.500 m). a) What are the components of the gravitational field at the origin due to these two point masses? b) What will be the magnitude and direction of the gravitational force on a 0.0100-kg test mass placed at the origin?

**12–34** Three point masses are fixed at positions shown in Fig. 12–29. a) Calculate the $x$- and $y$-components of the gravitational field at point $P$, which is at the origin of coordinates, due to these three point masses. b) What would be the magnitude and direction of the force on a 0.0200-kg mass placed at $P$?

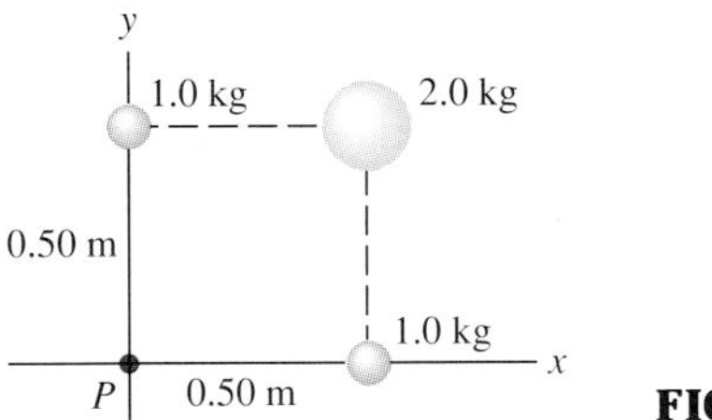


**FIGURE 12–29**

**12–35 Geosynchronous Satellites.** Several satellites are moving in a circle in the earth's equatorial plane. They are at such a height above the earth's surface that they remain always above the same point. Find the altitude of these satellites above the earth's surface. (Such an orbit is said to be *geosynchronous.*)

**12–36** The asteroid Toro was discovered in 1964. Its radius is about 5.0 km, and its mass is about $2.0 \times 10^{15}$ kg. Suppose an object is to be placed in a circular orbit around Toro, with the radius of its orbit just slightly larger than the asteroid's radius. What is the speed of the object? Could you launch yourself into orbit around Toro by running?

**12–37** Assume that the moon orbits the earth in a circular orbit. From the observed orbital period of 27.3 days, calculate the distance of the moon from the center of the earth. Assume that the motion of the moon is determined solely by the gravitational force exerted on it by the earth, and use the mass of the earth given in Appendix F.

**12–38** What would be the length of a day (that is, the time required for one revolution of the earth on its axis) if the rate of revolution of the earth were such that $g = 0$ at the equator?

**12–39** A uniform sphere with mass 0.600 kg is held with its center at the origin, and a second uniform sphere is held with its center at the point $x = 0.300$ m, $y = 0$. What are the magnitude and direction of the resultant gravitational force due to these objects on a third uniform sphere with mass 0.050 kg placed at the point $x = 0$, $y = 0.400$ m?

**12–40** A uniform sphere with mass 0.200 kg is 6.00 m to the left of a second uniform sphere with mass 0.300 kg. Where, in addition to infinitely far away, is the resultant gravitational field due to these masses equal to zero?

**12–41** There are two equations from which a change in the gravitational potential energy $U$ for a mass $m$ can be calculated. One is $U = mgy$ (Eq. 7–2). The other is $U = -Gmm_E/r$ (Eq. 12–14). As is shown in Section 12–4, the first equation is correct only if the gravitational force is constant over the change in height $\Delta y$. The second equation is always correct. Actually, the gravitational force is never exactly constant over any change in height, but the variation may be so small that we ignore it. Consider the difference in $U$ between a mass at the earth's surface and a distance $h$ above it using both equations, and find the height $h$ for which Eq. (7–1) is in error by 1%. Express this $h$ as a fraction of the earth's radius, and also obtain a numerical value for it.

**12–42** A satellite moving at 2000 m/s just outside the earth's atmosphere would very quickly fall to earth because its speed is far less than that required to maintain its orbit. A rocket plane flying at 2000 m/s at nearly the same altitude but within the earth's atmosphere can maintain its altitude by virtue of the lift provided by its wings. What lift force is required to maintain a rocket plane with weight 50,000 N moving at a speed of 2000 m/s at an altitude of 50.0 km above the surface of the earth?

**12–43** A hammer with mass $m$ is dropped from a height $h$ above the earth's surface. This height is not necessarily small in comparison with the radius $R_E$ of the earth. If air resistance is neglected, derive an expression for the speed $v$ of the hammer when it reaches the surface of the earth. Your expression should involve $h$, $R_E$, and $m_E$, the mass of the earth.

**12–44** An object is thrown straight up from the surface of the asteroid Toro (see Problem 12–36) with a speed 25.0 m/s, which is greater than the escape speed. What is the speed of the object when it is far from Toro? Neglect the gravitational forces due to all other astronomical objects.

**12–45** Comets travel around the sun in elliptical orbits with large eccentricities. If a comet has speed $2.0 \times 10^4$ m/s when at a distance of $2.0 \times 10^{11}$ m from the center of the sun, what is its speed when at a distance of $4.0 \times 10^{10}$ m?

**12–46** As the earth orbits the sun in its elliptical orbit, its distance of closest approach to the center of the sun (at the perihelion) is $1.471 \times 10^{11}$ m, and its maximum distance from the center of the sun (at aphelion) is $1.521 \times 10^{11}$ m. If the earth's orbital speed at perihelion is $3.027 \times 10^{4}$ m/s, what is its orbital speed at aphelion? (Neglect the effect of the moon and other planets.)

**12–47** An experiment is performed in deep space with two uniform spheres, one with mass 50.0 kg and the other with mass 100.0 kg. The spheres are released from rest with their centers 40.0 m apart. They accelerate toward each other because of their mutual gravitational attraction. (Neglect all gravitational forces other than that between the two spheres. Note that linear momentum is conserved.) When their centers are 20.0 m apart, what is a) the speed of each sphere? b) the magnitude of the relative velocity with which one sphere is approaching the other?

**12–48** A 400-kg lunar lander satellite is in a circular orbit 2000 km above the surface of the moon. How much work must the lander engines perform to move the lander to a circular orbit with radius 4000 km?

***12–49** A thin, uniform rod has length $L$ and mass $M$. Calculate the magnitude of the gravitational field produced by the rod at a point along the axis of the rod at distance $d$ from one end. Show that your result reduces to the expected result when $d$ is much larger than $L$.

***12–50** A shaft is drilled from the surface to the center of the earth (Fig. 12–20). The gravitational force on an object of mass $m$ that is inside the earth at a distance $r$ from the center has magnitude $F_g = GM_E mr/R_E^3$ (Eq. 12–31) and points toward the center of the earth. a) Derive an expression for the gravitational potential energy $U(r)$ of the object as a function of its distance from the center of the earth. b) If an object is released in the shaft at the surface of the earth, what speed will it have when it reaches the center of the earth?

## CHALLENGE PROBLEMS

**12–51** Mass $M$ is distributed uniformly along a line of length $2L$. Calculate the gravitational field components perpendicular and parallel to the line at a point that is a distance $a$ above the center of the line on its perpendicular bisector (Fig. 12–30). Does your result reduce to the correct expression as $a$ becomes very large?

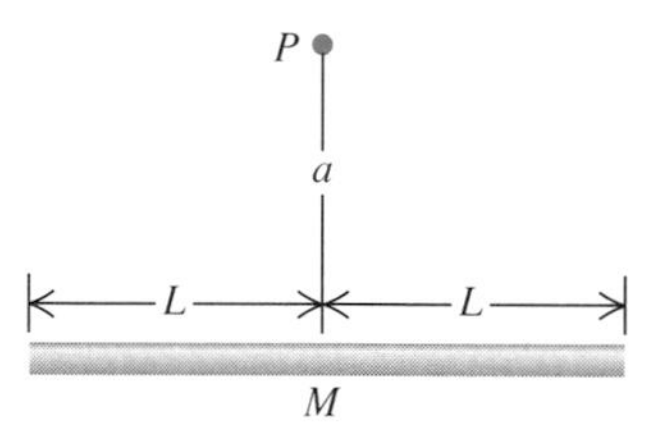


**FIGURE 12–30**

**12–52** Mass $M$ is distributed uniformly over a disk of radius $a$. Find the gravitational force (magnitude and direction) between this disk-shaped mass and a point mass $m$ located a distance $x$ above the center of the disk (Fig. 12–31). Does your result reduce to the correct expression as $x$ becomes very large? (*Hint:* Divide the disk into infinitesimally thin concentric rings, use the expression derived in Section 12–3 for the gravitational field due to each ring, and integrate to find the total field.)

**12–53 Gravitational Field of a Black Hole.** An astronaut inside a spacecraft that protects her from harmful radiation is orbiting a black hole at a distance of 150 km from its center. The mass of the black hole is 10.0 times the mass of the sun (10.0 solar masses), and the Schwarzchild radius is 30 km. She is positioned inside the spaceship such that one of her 0.030-kg ears is 6.0 cm farther from the black hole than the center of mass of the spacecraft and the other ear is 6.0 cm closer. What is the tension between her ears? Would the astronaut find it difficult to keep from being torn apart by the gravitational forces? (Since her whole body orbits with the same angular velocity, one ear is moving too slowly for the radius of its orbit and the other is moving too fast. Thus her head must exert forces on her ears to keep them in their orbits.)

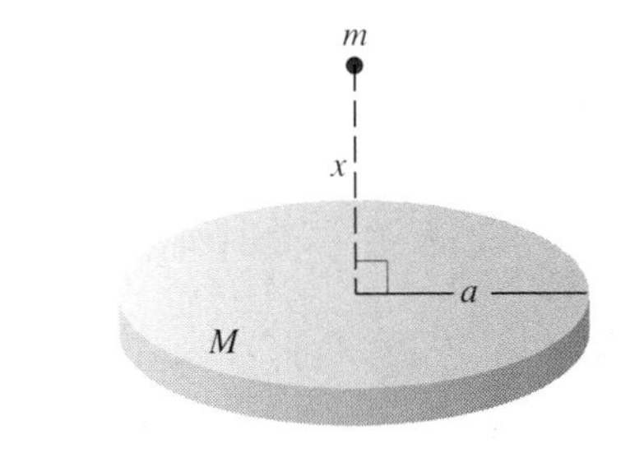


**FIGURE 12–31**

CHAPTER 13

# Periodic Motion

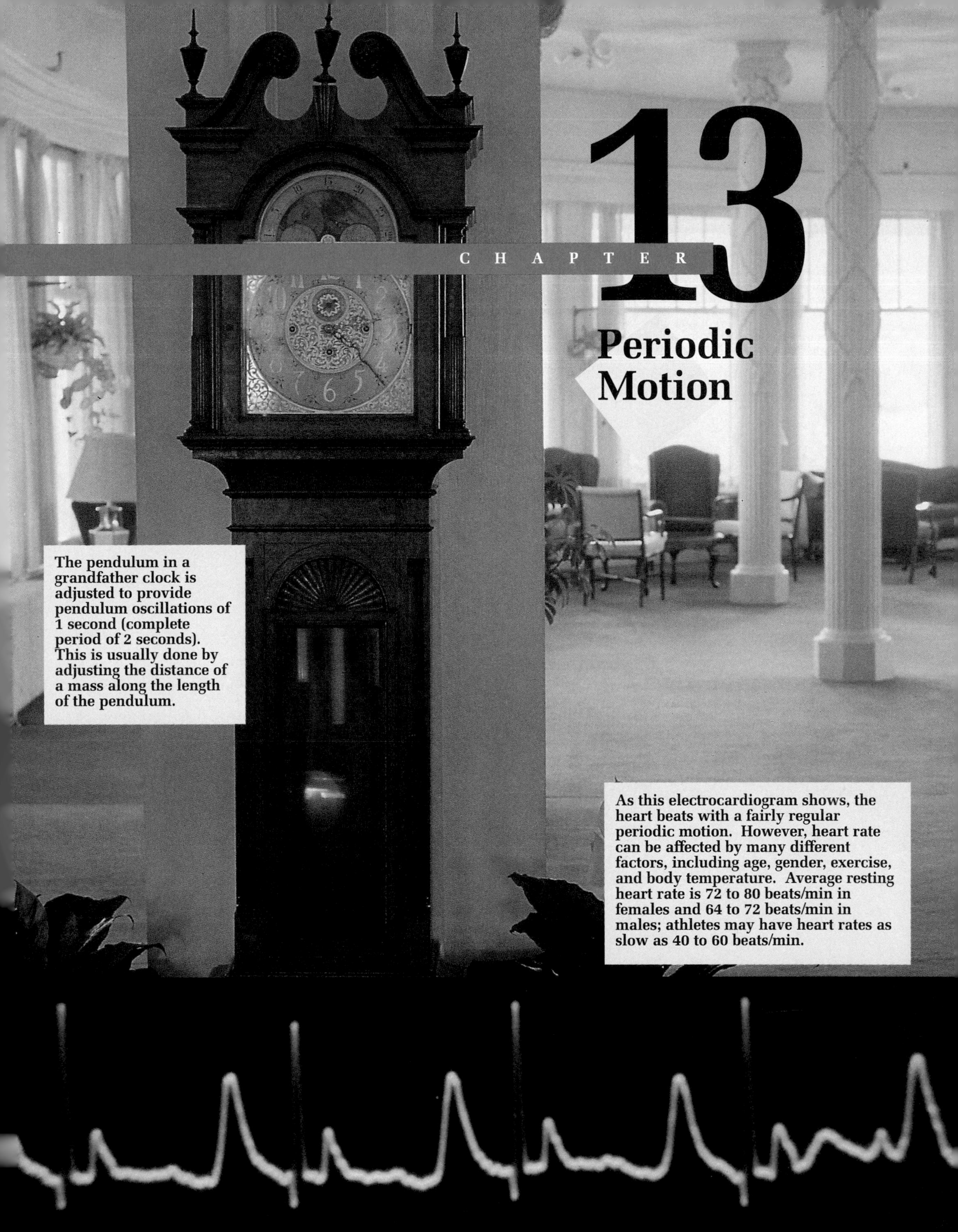

The pendulum in a grandfather clock is adjusted to provide pendulum oscillations of 1 second (complete period of 2 seconds). This is usually done by adjusting the distance of a mass along the length of the pendulum.

As this electrocardiogram shows, the heart beats with a fairly regular periodic motion. However, heart rate can be affected by many different factors, including age, gender, exercise, and body temperature. Average resting heart rate is 72 to 80 beats/min in females and 64 to 72 beats/min in males; athletes may have heart rates as slow as 40 to 60 beats/min.

## KEY CONCEPTS

- **Periodic motion is motion that repeats itself in a regular cycle. A periodic motion is described by its frequency, period, and amplitude.**
- **The simplest periodic motion is simple harmonic motion (SHM). It occurs when the restoring force on a body is directly proportional to the displacement from its equilibrium position. In SHM the period and frequency depend on the mass of the body and the strength of the restoring force but not on amplitude. The motion can be analyzed in detail by using energy considerations or by application of Newton's second law.**
- **Analysis of SHM is aided by use of the circle of reference, which makes use of a simple relation between SHM and uniform circular motion.**
- **Many systems undergo motion that is approximately simple harmonic, including pendulums and other systems that have rotational oscillations.**
- **When damping forces are present, the amplitude of oscillation decreases with time. When there is an additional force that varies sinusoidally with time, the system oscillates with the same frequency as the additional force; this is called forced oscillation.**

## INTRODUCTION

What is periodic motion? The vibration of a quartz crystal in a watch, the pendulum of a grandfather clock, the sound vibrations produced by a clarinet or an organ pipe, the back-and-forth motion of the pistons in a car engine — all these are examples of motion that repeats itself over and over. We call this *periodic motion,* or *oscillation,* and it is the subject of this chapter.

A body that undergoes periodic motion always has a stable equilibrium position. When it is moved away from this position and released, a force comes into play to pull it back toward equilibrium. But by the time it gets there, it has picked up some kinetic energy, so it overshoots, stopping somewhere on the other side, and is again pulled back toward equilibrium. Picture a ball rolling back and forth in a round bowl or a pendulum that swings back and forth past its straight-down position.

In this chapter we will concentrate on two simple examples of systems that can undergo periodic motions: spring-mass systems and pendulums. We will also study why oscillations often tend to die out with time and why some oscillations can build up to greater and greater displacements from equilibrium when periodically varying forces act. ■

## 13–1 BASIC CONCEPTS

Motion that repeats itself in a regular cycle is called **periodic motion,** or **oscillation.** One of the simplest systems that can have periodic motion is shown in Fig. 13–1a. A body with mass $m$ moves on a frictionless horizontal guide system, such as a linear air track, so it moves only along the $x$-axis. It is attached to a spring that can be either stretched or compressed. The spring obeys Hooke's law and has force constant $k$. (You may want to review the definition of the force constant in Section 6–3.) The left end of the spring is held fixed, and the right end is attached to the body. The spring force is the only horizontal force acting on the body. Because the two vertical forces, $\mathcal{n}$ and $mg$, always add to zero, the motion is one-dimensional; $x$, $v$, $a$, and $F$ refer to the $x$-components of the position, velocity, acceleration, and force vectors, respectively. Like all components, they can be positive or negative.

It is simplest to define our coordinate system so that the origin $O$ is at the equilibrium position, where the spring is neither stretched nor compressed. Then $x$ is the displacement of the body from equilibrium and is also the change in length of the spring. The acceleration $a$ is given by $a = F/m$. Figure 13–1b shows complete free-body diagrams for three different positions of the body.

1. $x$ is negative: The body is to the left of $O$, the spring is compressed, and it pushes on the body to the right (toward the equilibrium point). The acceleration is also toward the right; $F$ and $a$ are both positive.
2. $x = 0$: The body is at $O$, the equilibrium position. The spring is neither stretched nor compressed; it exerts no force on the body, and the acceleration is zero.
3. $x$ is positive: The body is to the right of $O$, the spring is stretched, and it pulls on the body to the left (toward the equilibrium point). The acceleration is also toward the left; $F$ and $a$ are both negative.

The force acting on the body tends always to restore it to the equilibrium position, and we call it a **restoring force.** On either side of the equilibrium position, $F$ and $x$ always have opposite signs. In Section 6–3 we represented the force acting *on* a stretched spring

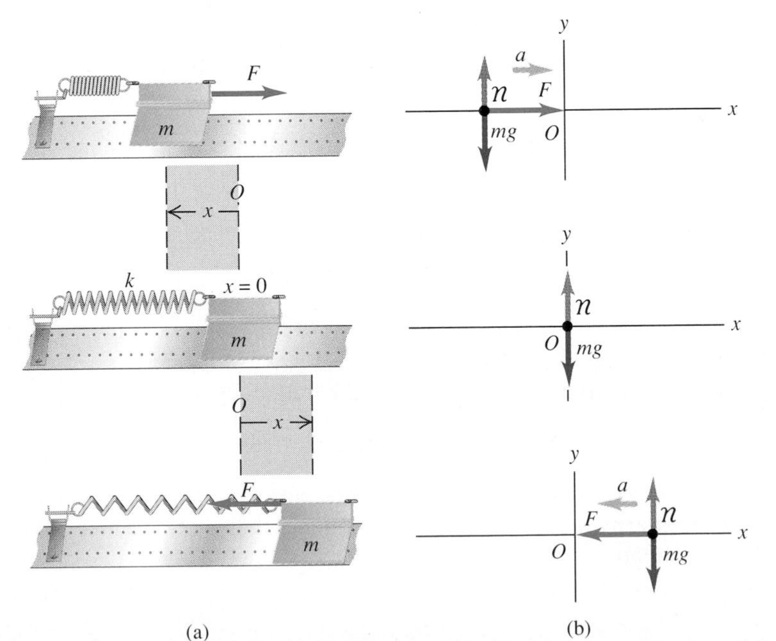


**13–1** Model for periodic motion. If the spring obeys Hooke's law, the motion is simple harmonic motion. (a) When the body is to the left of the equilibrium position, the spring exerts a force to the right. When the body is at the equilibrium position, it experiences no horizontal force. When it is to the right of the equilibrium position, the spring exerts a force to the left. (b) Free-body diagrams for the three positions.

as $F = kx$. The force the spring exerts *on the body* is the negative of this, so the $x$-component of force $F$ on the body is

$$F = -kx. \tag{13–1}$$

This equation gives the correct magnitude and sign of the force, whether $x$ is positive, negative, or zero. The force constant $k$ is always positive and has units of N/m. A useful alternative set of units for $k$ is kg/s$^2$.

The acceleration $a$ of the body at any point is determined by $\Sigma \mathbf{F} = m\mathbf{a}$, with the $x$-component of force given by Eq. (13–1):

$$a = \frac{F}{m} = -\frac{k}{m}x. \tag{13–2}$$

This acceleration is *not* constant, so don't even think of using the constant-acceleration equations from Chapter 2. If we displace the body to the right a distance $A$ (so $x = A$) and release it, its initial acceleration is, from Eq. (13–2), $-kA/m$. As it approaches the equilibrium position, its acceleration decreases in magnitude but its speed increases. It overshoots the equilibrium position and comes to a stop on the other side; we will show later that the stopping point is at $x = -A$. Then it accelerates to the right, overshoots equilibrium again, and stops at the starting point $x = A$, ready to repeat the whole process. If there is no friction or other force to remove mechanical energy from the system, this motion repeats itself forever.

The motion that we have described, under a restoring force that is directly proportional to the displacement from equilibrium, as given by Eq. (13–1), is called **simple harmonic motion,** abbreviated **SHM.** Many periodic motions, such as the vibration of the quartz crystal in a watch, vibrations of molecules, the motion of a tuning fork, and the electric current in an alternating-current circuit, are approximately simple harmonic. However, not *all* periodic motions are simple harmonic.

Here are some terms that we'll use in discussing periodic motions:

The **amplitude** of the motion, denoted by $A$, is the maximum magnitude of displacement from equilibrium, that is, the maximum value of $|x|$. It is always positive. The total overall range of the motion is $2A$. The SI unit of $A$ is the meter. A complete vibration, or **cycle,** is one complete round trip, say, from $A$ to $-A$ and back to $A$, or from $O$ to $A$, back through $O$ to $-A$, and back to $O$. Note that motion from one side to the other (say, from $-A$ to $A$) is a half-cycle, not a whole cycle.

The **period,** $T$, is the time for one cycle. It is always positive; the SI unit is the second, but it is sometimes expressed as "seconds per cycle."

The **frequency,** $f$, is the number of cycles in a unit of time. It is always positive; the SI unit of frequency is the hertz:

$$1 \text{ hertz} = 1 \text{ Hz} = 1 \text{ cycle/s} = 1 \text{ s}^{-1}.$$

This unit is named in honor of Heinrich Hertz, one of the pioneers in investigating electromagnetic waves during the late nineteenth century.

The **angular frequency,** $\omega$, is $2\pi$ times the frequency:

$$\omega = 2\pi f.$$

We'll learn later why $\omega$ is a useful quantity. It represents the rate of change of an angular quantity (not necessarily related to a rotational motion) that is always measured in radians, so its units are rad/s.

From the definitions of period $T$ and frequency $f$ we see that each is the reciprocal of the other:

$$f = \frac{1}{T}, \qquad T = \frac{1}{f}. \tag{13–3}$$

Also, from the definition of $\omega$,

$$\omega = 2\pi f = \frac{2\pi}{T}. \tag{13–4}$$

### EXAMPLE 13–1

An ultrasonic transducer (a kind of ultrasonic loudspeaker) used for medical diagnosis is oscillating at a frequency of 6.7 MHz (that is, $6.7 \times 10^6$ Hz) (Fig. 13–2). How much time does each oscillation take, and what is the angular frequency?

**SOLUTION** The period $T$ is given by Eq. (13–3):

$$T = \frac{1}{f} = \frac{1}{6.7 \times 10^6\ \text{Hz}} = 1.5 \times 10^{-7}\ \text{s}$$
$$= 0.15\ \mu\text{s}.$$

We get $\omega$ from Eq. (13–4):

$$\omega = 2\pi f = 2\pi(6.7 \times 10^6\ \text{Hz})$$
$$= 4.2 \times 10^7\ \text{rad/s}.$$

A very rapid vibration corresponds to large $f$ and $\omega$ and small $T$, and conversely.

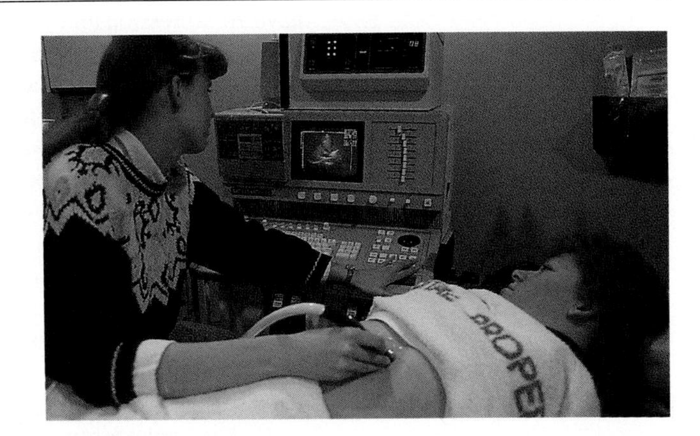

**13–2** Ultrasonic vibrations for medical diagnosis are produced by an electronic transducer.

Simple harmonic motion is the simplest of all periodic motions to analyze; that's why we call it "simple harmonic." In more complex periodic motion the force may depend on displacement in a more complicated way, but only when it is *directly proportional* to displacement do we use the term "simple harmonic motion." However, many periodic motions are *approximately* simple harmonic if the amplitude is small enough that the force is approximately proportional to displacement. Thus we can use SHM as an approximate model for many different periodic motions, such as pendulums and vibrations of atoms in molecules and solids. Also, we can represent *any* periodic motion, no matter how complicated, as a combination of simple harmonic motions with frequencies that are integer multiples of the frequency of the complex motion. The branch of mathematics concerned with this representation is called *Fourier analysis.*

## 13–2 ENERGY IN SIMPLE HARMONIC MOTION

In this section and the next we'll analyze the spring-mass model of SHM that we've just described. What questions shall we ask? Given the mass $m$ of the object and the force constant $k$ of the spring, we'd like to be able to predict the period $T$ and the frequency $f$. It would also be of interest to know how fast the body is moving at each point in its motion, that is, to know its velocity $v$ as a function of position $x$. Best of all would be to

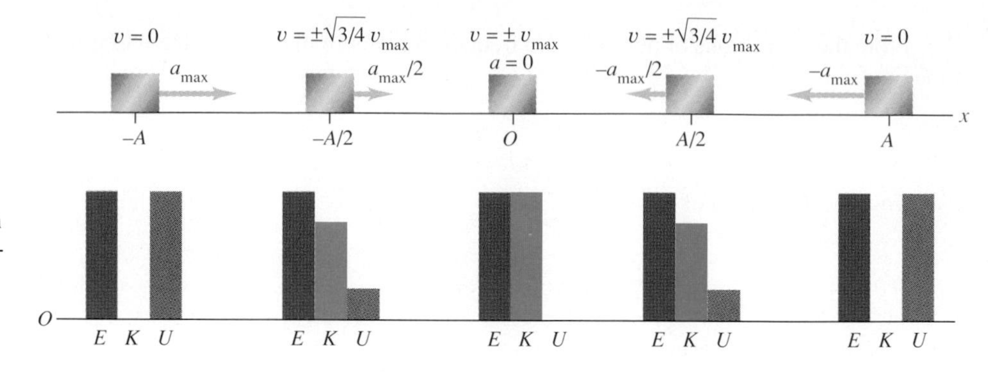


**13–3** In simple harmonic motion the total energy $E$ is constant, continuously transforming from potential energy $U$ to kinetic energy $K$, and back again.

have an expression for the position $x$ as a function of *time*. With that, not only would we be able to find where the body is at any time, but by taking derivatives we could also find its velocity and acceleration at any time. In the next few pages we'll work all these things out.

We will use two different starting points, the first using energy considerations and the second using $\Sigma \boldsymbol{F} = m\boldsymbol{a}$ directly. We've already noted that the spring force is the only horizontal force on the body. It is a *conservative* force, and the vertical forces do no work; so the total energy of the system is conserved. We will also assume that the mass of the spring itself is negligible.

The kinetic energy of the body is $K = \frac{1}{2}mv^2$, and the potential energy is $U = \frac{1}{2}kx^2$, just as in Examples 7–8 and 7–10 (Section 7–3). (It would be helpful to review those examples.) The energy equation is $K + U = E =$ constant, or

$$E = \tfrac{1}{2}mv^2 + \tfrac{1}{2}kx^2 = \text{constant}. \tag{13–5}$$

The total energy $E$ is also directly related to the amplitude $A$ of the motion. When the body reaches the point $x = A$, its maximum displacement from equilibrium, it momentarily stops as it turns back toward the equilibrium position. That is, when $x = A$ (or $-A$), $v = 0$. At this point the energy is entirely potential, and $E = \frac{1}{2}kA^2$. Because $E$ is constant, this quantity equals $E$ at any other point; combining this expression with Eq. (13–5), we get

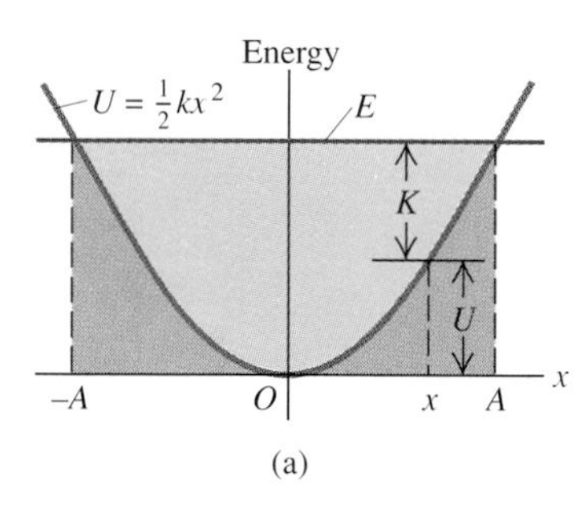


$$E = \tfrac{1}{2}kA^2 = \tfrac{1}{2}mv^2 + \tfrac{1}{2}kx^2. \tag{13–6}$$

Solving this for $v$, we find

$$v = \pm\sqrt{\frac{k}{m}}\sqrt{A^2 - x^2}. \tag{13–7}$$

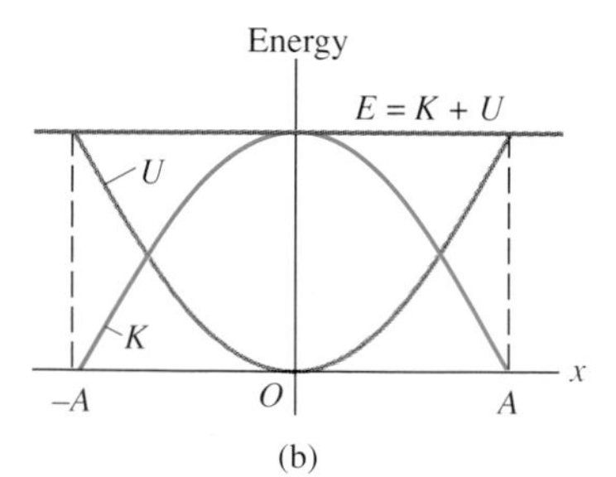


**13–4** Kinetic, potential, and total energies as functions of position for SHM. At each value of $x$ the sum of $K$ and $U$ equals the constant total energy $E$.

We can use this equation to find the velocity (but not its sign) for any given position. For example, when $x = \pm A/2$,

$$v = \pm\sqrt{\frac{k}{m}}\sqrt{A^2 - \left(\pm\frac{A}{2}\right)^2} = \pm\sqrt{\frac{3}{4}}\sqrt{\frac{k}{m}}A.$$

Equation (13–6) also shows that the *maximum* speed $v_{max}$ occurs at $x = 0$ and is given by

$$v_{max} = \sqrt{\frac{k}{m}}A. \tag{13–8}$$

Figure 13–3 is a pictorial display of the energy quantities at these points, and Figure 13–4 is a graphical display of Eq. (13–5). Energy (kinetic, potential, and total) is plotted vertically, and the coordinate $x$ is plotted horizontally. The parabolic curve in Fig.

13–4a represents the potential energy $U = \frac{1}{2}kx^2$. The horizontal line represents the total constant energy (which does not vary with $x$). This line intersects the potential-energy curve at $x = -A$ and $x = A$, where the energy is entirely potential. Suppose we start on the horizontal axis at some value of $x$ between $-A$ and $A$. The vertical distance from the $x$-axis to the parabola is $U$, and since $E = K + U$, the remaining vertical distance up to the horizontal line is $K$. Figure 13–4b shows both $K$ and $U$ as functions of $x$. As the body oscillates between $-A$ and $A$, the energy is continuously transformed from potential to kinetic to potential. At $x = \pm A$, $E = U$ and $K = 0$.

We see again why $A$ is the maximum displacement. If we tried to make $x$ greater than $A$ (or less than $-A$), $U$ would be greater than $E$, and $K$ would have to be negative. Because $K$ can never be negative, $x$ can't be greater than $A$ or less than $-A$. (In quantum mechanics it *is* possible in some circumstances for a particle to traverse regions where classically it would have negative kinetic energy. This concept is important in alpha decay of nuclei and in the tunneling electron microscope. But it plays no role in the mechanics of macroscopic bodies.)

## EXAMPLE 13–2

A spring is mounted horizontally (Fig. 13–5a), with the left end held stationary. By attaching a spring balance to the free end and pulling toward the right, we determine that the stretching force is proportional to the displacement and that a force of 6.0 N causes a displacement of 0.030 m. We remove the spring balance and attach a 0.50-kg body to the end, pull it a distance of 0.040 m, release it, and watch it oscillate in SHM.

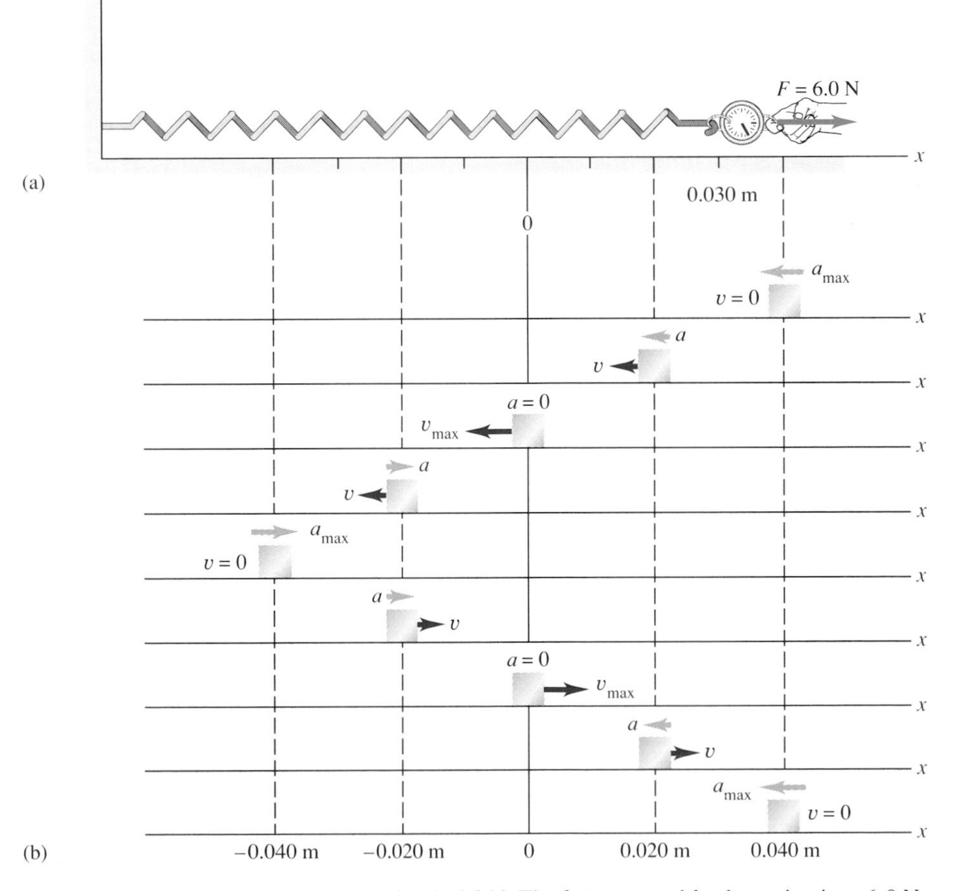


**13–5** (a) The force exerted *on* the spring is 6.0 N. The force exerted *by* the spring is $-6.0$ N. (b) The body is attached to the spring, pulled 0.040 m from the equilibrium position, and released.

a) Find the force constant of the spring.

When $x = 0.030$ m, the force the spring exerts on the body is $F = -6.0$ N. From Eq. (13–1),

$$k = -\frac{F}{x} = -\frac{-6.0\ \text{N}}{0.030\ \text{m}} = 200\ \text{N/m}.$$

b) Find the maximum and minimum velocities attained by the vibrating body.

The velocity $v$ at any position $x$ is given by Eq. (13–7):

$$v = \pm\sqrt{\frac{k}{m}}\sqrt{A^2 - x^2}.$$

The maximum velocity occurs at the equilibrium position, where $x = 0$:

$$v = v_{\text{max}} = \sqrt{\frac{k}{m}}A = \sqrt{\frac{200\ \text{N/m}}{0.50\ \text{kg}}}(0.040\ \text{m}) = 0.80\ \text{m/s}.$$

The minimum (most negative) velocity also occurs at $x = 0$ and is the negative of this value.

c) Compute the maximum acceleration.

From Eq. (13–2),

$$a = -\frac{k}{m}x.$$

The maximum (most positive) acceleration occurs at the most negative value of $x$, that is, $x = -A$; therefore,

$$a_{\text{max}} = -\frac{k}{m}A = -\frac{200\ \text{N/m}}{0.50\ \text{kg}}(-0.040\ \text{m}) = 16.0\ \text{m/s}^2.$$

The minimum (most negative) acceleration is $-16.0\ \text{m/s}^2$, occurring at $x = +A = +0.040$ m.

d) Determine the velocity and acceleration when the body has moved halfway to the center from its initial position.

At this point, $x = A/2 = 0.020$ m. From Eq. (13–7),

$$v = -\sqrt{\frac{200\ \text{N/m}}{0.50\ \text{kg}}}\sqrt{(0.040\ \text{m})^2 - (0.020\ \text{m})^2} = -0.69\ \text{m/s}.$$

We have chosen the negative square root because the body is moving from $x = A$ toward $x = 0$. From Eq. (13–2),

$$a = -\frac{200\ \text{N/m}}{0.50\ \text{kg}}(0.020\ \text{m}) = -8.0\ \text{m/s}^2.$$

The conditions at $x = 0$, $\pm A/2$, and $\pm A$ are shown in Fig. 13–5b.

---

The position-velocity relation given by Eq. (13–6) is very useful, but expressions that give position, velocity, and acceleration as functions of time would be even more useful. We can obtain these relations either by integrating the energy equation, Eq. (13–7), or by working directly with Newton's second law, Eq. (13–2). Let's continue our energy approach first.

In Eq. (13–7) we replace $v$ by $dx/dt$:

$$\frac{dx}{dt} = \pm\sqrt{\frac{k}{m}}\sqrt{A^2 - x^2}.$$

The next step is called *separation of variables;* we write all factors containing $x$ on one side and all those containing $t$ on the other so that each side can be integrated:

$$\int \frac{dx}{\sqrt{A^2 - x^2}} = \pm\sqrt{\frac{k}{m}}\int dt,$$

$$-\arccos\frac{x}{A} = \pm\sqrt{\frac{k}{m}}\,t + \phi,$$

where $\phi$ is an integration constant. [Your integral table may give $\arcsin(x/A)$ instead of $-\arccos(x/A)$; both are correct.] Solving this equation for $x$ and using the fact that for any angle $\alpha$, $\cos(-\alpha) = \cos\alpha$, we obtain

$$x = A\cos\left(\sqrt{\frac{k}{m}}\,t + \phi\right). \qquad (13\text{–}9)$$

Once we have $x$ as a function of $t$, we can get the velocity $v$ and acceleration $a$ as functions of $t$ by taking derivatives of Eq. (13–9). We'll do that in the next section.

Because the cosine function is *periodic,* Eq. (13–9) shows that the position $x$ is a periodic function of time, as we expected for SHM. We could also have written Eq. (13–9) in terms of a sine function rather than a cosine by using the identity $\cos\alpha =$

sin $(\alpha + \pi/2)$. **In simple harmonic motion the position is a sinusoidal function of time.** The value of the cosine function is always between $-1$ and 1, so in Eq. (13–9), $x$ is always between $-A$ and $A$. This confirms that $A$ is the amplitude of the motion. Figure 13–6a shows a graph of Eq. (13–9) for the particular case $\phi = 0$. Figures 13–6b and 13–6c are graphs of velocity $v$ and acceleration $a$ for this same motion. We will explore the significance of $\phi$ later.

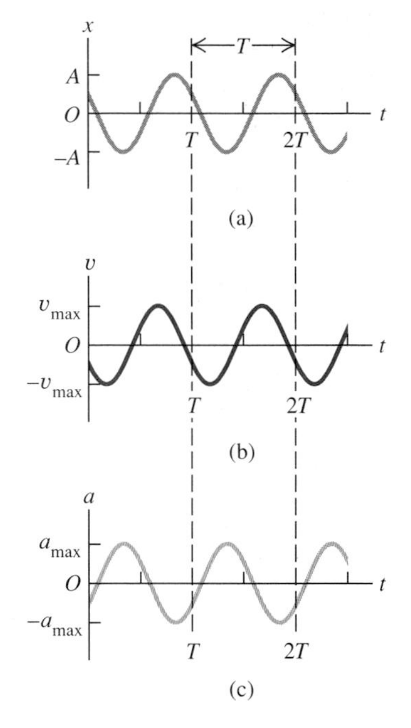


**13–6** (a) Graph of $x$ versus $t$ (Eq. 13–9) for simple harmonic motion. In this graph, $\phi = 0$. (b) Graph of $v$ versus $t$ for the same motion. (c) Graph of $a$ versus $t$ for the same motion.

The period $T$ is the time for one complete cycle, or oscillation. The cosine function repeats itself whenever the quantity in parentheses in Eq. (13–9) increases by $2\pi$ radians. Thus if we start at time $t = 0$, the time $T$ at which one cycle has been completed is given by

$$\sqrt{\frac{k}{m}}T = 2\pi, \qquad \text{or}$$

$$T = 2\pi\sqrt{\frac{m}{k}}. \tag{13–10}$$

The period of the motion is determined by the mass $m$ and the force constant $k$. It *does not* depend on the amplitude or on the total energy. For given values of $m$ and $k$ the time of one complete oscillation is the same whether the amplitude is large or small. Equation (13–7) shows us why we should expect this; larger $A$ means larger speed at any given $x$. For a complete cycle a larger distance is traveled at a proportionately greater average speed, so the same total time is involved.

With everything else constant we see from Eq. (13–10) that a larger mass $m$, with its greater inertia, will have less acceleration, move more slowly, and take a longer time for a complete cycle. In contrast, a stiffer spring (one with a larger force constant $k$) exerts a greater force at a given deformation $x$, causing greater acceleration, higher speeds, and a shorter time $T$ per cycle. This is what Eq. (13–10) and Fig. 13–7 show.

We can easily check the units of Eq. (13–10). The units of $k$ are N/m or $(\text{kg}\cdot\text{m/s}^2)/\text{m} = \text{kg/s}^2$, so the units of $m/k$ are $\text{kg}/(\text{kg/s}^2) = \text{s}^2$. When we take the square root, we get seconds, as expected. Strictly speaking, the quantity $(k/m)^{1/2}$ should have units of rad/s in order for the angular quantity in Eq. (13–9) to be in radians, so the numerical factor $2\pi$ in Eq. (13–10) has units rad/cycle.

The frequency $f$, the number of complete oscillations per unit time, is the reciprocal of the period $T$:

$$f = \frac{1}{T} = \frac{1}{2\pi}\sqrt{\frac{k}{m}}. \tag{13–11}$$

The SI unit for $f$ is cycles per second, or hertz (Hz). The angular frequency $\omega$ defined in Section 13–1 is given by

$$\omega = 2\pi f = \sqrt{\frac{k}{m}}. \tag{13–12}$$

As we pointed out above, the units of this quantity are rad/s. Now we see the motivation for defining the angular frequency in this way. Most of the equations of simple harmonic

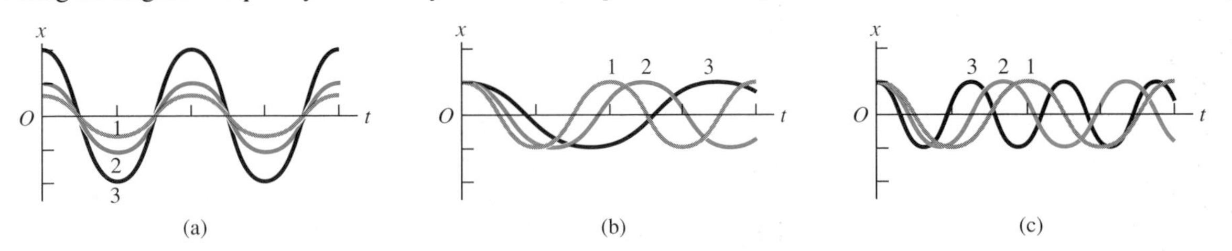


**13–7** Variations of simple harmonic motion. (a) Amplitude $A$ increases from curve 1 to 2 to 3, while mass $m$ and force constant $k$ are unchanged. (b) Mass $m$ increases from 1 to 2 to 3, while $A$ and $k$ are unchanged. (c) Force constant $k$ increases from 1 to 2 to 3, while $A$ and $m$ are unchanged.

motion are simplified by using Eq. (13–12). For example, we can rewrite Eq. (13–7) in the form

$$v = \pm\omega\sqrt{A^2 - x^2}. \tag{13–13}$$

Equation (13–8) for $v_{max}$ can be written

$$v_{max} = \omega A, \tag{13–14}$$

and Eq. (13–2), relating $a$ to $x$, becomes

$$a = -\omega^2 x. \tag{13–15}$$

## EXAMPLE 13–3

Find the angular frequency of oscillation, the frequency, and the period for the system described in Example 13–2.

**SOLUTION** We are given $m = 0.50$ kg and $k = 200$ N/m $= 200$ kg/s$^2$. Using Eq. (13–12), we find

$$\omega = \sqrt{\frac{k}{m}} = \sqrt{\frac{200\ \text{kg/s}^2}{0.50\ \text{kg}}} = 20\ \text{rad/s}.$$

The frequency $f$ is

$$f = \frac{\omega}{2\pi} = \frac{20\ \text{rad/s}}{2\pi\ \text{rad/cycle}} = 3.2\ \text{cycle/s} = 3.2\ \text{Hz}.$$

The period $T$ is the reciprocal of the frequency $f$:

$$T = \frac{1}{f} = \frac{1}{3.2\ \text{cycle/s}} = 0.31\ \text{s/cycle}.$$

Periods are usually given simply in "seconds" rather than "seconds per cycle."

---

We would like to stress once more the fact that the expressions for the frequency and period of SHM *do not* contain the amplitude $A$. This is one of the most important characteristics of SHM: **In simple harmonic motion the frequency does not depend on the amplitude.** This is why a tuning fork can be used as a pitch standard; it always vibrates with the same frequency, independent of amplitude. If it were not for simple harmonic motion, it would be impossible to make familiar types of mechanical and electronic clocks run accurately or to play most musical instruments in tune.

## 13–3 EQUATIONS OF SIMPLE HARMONIC MOTION

Let's explore further the significance of Eq. (13–9). Restated below as Eq. (13–16), this equation gives the position $x$ of a body in SHM as a function of time:

$$x = A\cos\left(\sqrt{\frac{k}{m}}\,t + \phi\right). \tag{13–16}$$

We would like to know how to determine the constants $A$ and $\phi$ for specific problems and how to use this equation to get useful information.

First, let's backtrack a little to see how Eq. (13–16) fits in with Newton's second law, $\Sigma \boldsymbol{F} = m\boldsymbol{a}$, expressed by Eq. (13–2). We can rewrite that equation as

$$\frac{d^2x}{dt^2} = -\frac{k}{m}x. \tag{13–17}$$

Newton's second law doesn't tell us directly what $x$ is, as a function of time. It tells us only that when we take the second time derivative of $x$, the result must equal $-k/m$ times the original function $x$.

What function do you know whose second derivative is a negative constant times the function itself? The sine and cosine are the *only* real (not complex) functions with this property. In fact, when we take the second derivative of Eq. (13–16), we get

$$\frac{d^2x}{dt^2} = -\frac{k}{m}A\cos\left(\sqrt{\frac{k}{m}}\,t + \phi\right) = -\frac{k}{m}x. \tag{13–18}$$

So this function *does* have the property described by Eq. (13–17). This means that the motion described by Eq. (13–16) is consistent with (or allowed by) Newton's second law. If we wanted to use fancier language, we could say that Eq. (13–17) is a *differential equation* and that Eq. (13–16) is a *solution* of this equation. This is a common pattern, in mechanics as well as many other areas of physics: The laws of motion give us differential equations that we have to solve to obtain detailed descriptions of the motion.

Next, we note that Eq. (13–16) can be written more simply in terms of the angular frequency $\omega = (k/m)^{1/2}$:

$$x = A\cos(\omega t + \phi). \tag{13–19}$$

Again, $A$ is the amplitude of the motion, representing the maximum displacement from equilibrium.

The constant $\phi$ is called the **phase angle;** it tells us at what point in the cycle the motion started at the initial time $t = 0$. We denote the initial position (the position at time $t = 0$) by $x_0$. Putting $t = 0$ and $x = x_0$ in Eq. (13–19), we get

$$x_0 = A\cos\phi. \tag{13–20}$$

If $\phi = 0$, then $x_0 = A\cos 0 = A$, and the body starts at its maximum positive displacement (Fig. 13–8). If $\phi = \pi$, then $x_0 = A\cos\pi = -A$, and the particle starts at its maximum *negative* displacement. If $\phi = \pi/2$, then $x_0 = A\cos(\pi/2) = 0$, and the particle starts at the origin.

From Eq. (13–19) we can find the velocity $v$ and acceleration $a$ as functions of time by taking derivatives with respect to time:

$$v = \frac{dx}{dt} = -\omega A\sin(\omega t + \phi), \tag{13–21}$$

$$a = \frac{dv}{dt} = \frac{d^2x}{dt^2} = -\omega^2 A\cos(\omega t + \phi). \tag{13–22}$$

The velocity $v$ oscillates between $-\omega A$ and $+\omega A$; this is consistent with Eq. (13–13). The initial velocity $v_0$ is the velocity at time $t = 0$. Putting $v = v_0$ and $t = 0$ in Eq. (13–21), we find

$$v_0 = -\omega A\sin\phi. \tag{13–23}$$

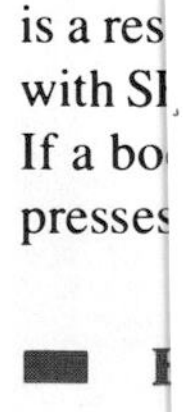

If we are given the initial position $x_0$ and the initial velocity $v_0$, we can use Eqs. (13–20) and (13–23) to determine the amplitude $A$ and the phase angle $\phi$. Here's how we do it. To find $\phi$, divide Eq. (13–23) by Eq. (13–20). This eliminates $A$ and gives an equation that we can solve for $\phi$:

$$\frac{v_0}{x_0} = \frac{-\omega A\sin\phi}{A\cos\phi},$$

$$\phi = \arctan\left(-\frac{v_0}{\omega x_0}\right). \tag{13–24}$$

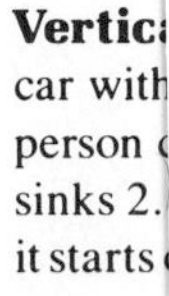

**13–8** Position as a function of time for simple harmonic motion with the same frequency and amplitude but with three different phase angles.

It is also easy to find the amplitude $A$ if we are given $x_0$ and $v_0$. We'll sketch the derivation, and you can fill in the details. Square Eq. (13–20); divide Eq. (13–23) by $\omega$, square it,

the effect of all four springs) is

$$k = -\frac{F}{x} = -\frac{980\ \text{N}}{-0.028\ \text{m}} = 3.5 \times 10^4\ \text{N/m} = 3.5 \times 10^4\ \text{kg/s}^2.$$

The person's mass is $m = w/g = (980\ \text{N})/(9.8\ \text{m/s}^2) = 100$ kg. The *total* oscillating mass is 1000 kg + 100 kg = 1100 kg. The period $T$ is

$$T = 2\pi\sqrt{\frac{m}{k}} = 2\pi\sqrt{\frac{1100\ \text{kg}}{3.5 \times 10^4\ \text{kg/s}^2}} = 1.11\ \text{s},$$

and the frequency is

$$f = \frac{1}{T} = \frac{1}{1.11\ \text{s}} = 0.90\ \text{Hz}.$$

If you have ridden in a car suffering from this affliction, the persistent oscillation may have left your stomach in an altered state.

## Angular SHM

As a final example, here is a brief discussion of *angular* simple harmonic motion. Figure 13–11 shows the balance wheel of a mechanical watch or alarm clock. The wheel has a moment of inertia $I$ about its axis. A coil spring (called the *hairspring*) exerts a restoring torque $\tau$ that is proportional to the angular displacement $\theta$ from the equilibrium position, where $\theta = 0$. We write $\tau = -\kappa\theta$, where $\kappa$ is a constant called the *torsion constant.* The equation of motion, from $\Sigma\,\tau = I\alpha = I\,d^2\theta/dt^2$, is

$$-\kappa\theta = I\alpha, \qquad \text{or} \qquad \frac{d^2\theta}{dt^2} = -\frac{\kappa}{I}\theta.$$

The form of this equation is exactly the same as Eq. (13–17), with $x$ replaced by $\theta$ and $k/m$ replaced by $\kappa/I$. We conclude that we are dealing with a form of simple harmonic motion. The frequency and angular frequency are given by Eqs. (13–11) and (13–12), respectively, with the same replacement:

$$f = \frac{1}{2\pi}\sqrt{\frac{\kappa}{I}} \qquad \text{and} \qquad \omega = \sqrt{\frac{\kappa}{I}}. \tag{13–26}$$

The motion is described by the function

$$\theta = \Theta \cos(\omega t + \phi),$$

where $\Theta$ plays the role of an angular amplitude.

It's a good thing the motion of a balance wheel *is* simple harmonic. If it weren't, the frequency might depend on the amplitude, and the clock would run too fast or too slow as the spring ran down.

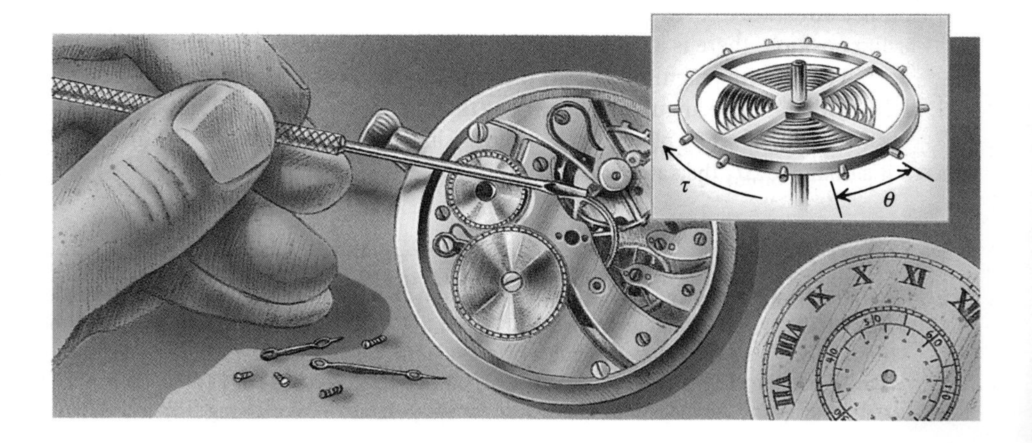


**13–11** The balance wheel of a mechanical watch. The spring exerts a restoring torque that is proportional to the angular displacement from the equilibrium position. Therefore the motion is angular SHM.

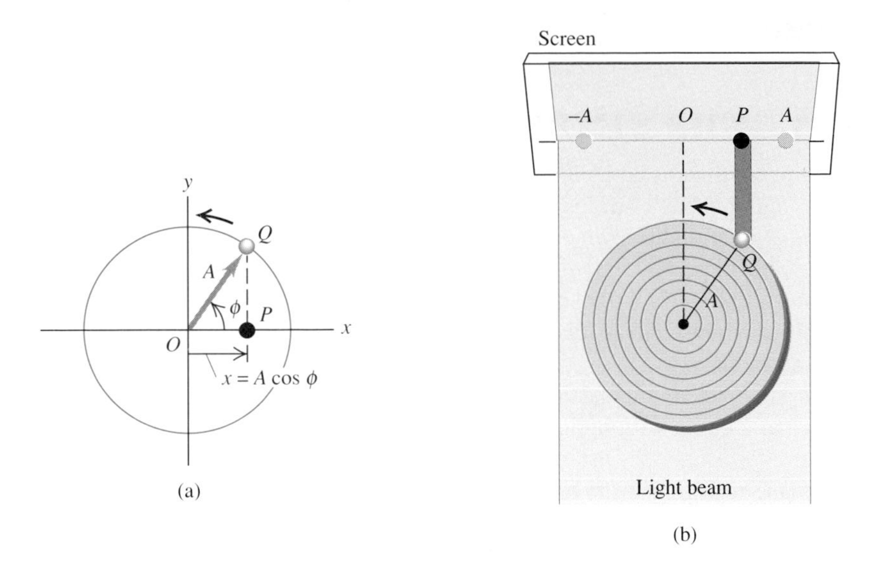


**13–12** (a) The $x$-coordinate of point $Q$'s position changes with time as point $Q$ rotates counterclockwise in uniform circular motion. (b) The location of $P$ is that of a shadow of $Q$ projected onto the $x$-axis by a beam of light parallel to the $y$-axis.

## 13–4 CIRCLE OF REFERENCE

Simple harmonic motion is closely related to uniform circular motion, which we have studied several times. By examining this relationship we can gain additional insight into both kinds of motion. For this purpose we use a geometric representation called the **circle of reference.** The basic idea is shown in Fig. 13–12. Point $Q$ moves counterclockwise around a circle with a radius $A$ equal to the amplitude of the actual simple harmonic motion, with constant angular velocity $\omega$ (measured in rad/s) equal to the angular frequency of the simple harmonic motion that we want to represent.

The vector from $O$ to $Q$ is the position vector of point $Q$ relative to $O$; this vector has constant magnitude $A$. If at time $t = 0$, vector $OQ$ makes an angle $\phi$ with the horizontal axis, then at a later time $t$ the angle is $\phi + \omega t$. As $Q$ moves, the angle changes at a constant rate $\omega$, and the vector rotates counterclockwise with constant angular velocity $\omega$. Such a rotating vector is called a **phasor.** (This term was in use long before the invention of the Star Trek stun-gun with the similar name.)

The horizontal component of the phasor $OQ$ represents the actual motion of the body under study. In Fig. 13–12a, point $P$ lies on the horizontal diameter of the circle, directly below point $Q$. We call $Q$ the *reference point,* the circle itself the *circle of reference,* and $P$ the *projection* of $Q$ onto the diameter. The location of $P$ is that of a *shadow* of $Q$ on the $x$-axis, cast by a spotlight beam parallel to the $y$-axis (Fig. 13–12b). As $Q$ revolves, $P$ moves back and forth along the diameter, staying always directly below (or above) $Q$. The position of $P$ with respect to the origin $O$ at any time $t$ is the coordinate $x$; from Fig. 13–13a,

$$x = A\cos(\omega t + \phi).$$

But this is identical to Eq. (13–19); this shows that the motion of point $P$ is simple harmonic motion. That is, **simple harmonic motion is the projection of uniform circular motion onto a diameter.** It is also clear from this discussion that the *angular velocity* of point $Q$ in its circular motion is identical with the *angular frequency* of the SHM of point $P$.

The horizontal components of the velocity and acceleration of point $Q$ are also identical to the velocity and acceleration, respectively, of $P$. The speed of $Q$ is equal to $\omega A$, and the magnitude of its acceleration is $\omega^2 A$. By taking the horizontal components of the velocity and acceleration vectors of $Q$, we can derive Eqs. (13–21) and (13–22) geometrically, as shown in Figure 13–13. We'll leave the details as problems.

The phasor representation of quantities that vary sinusoidally with time is also useful in many other areas of physics, including ac circuit analysis (Chapter 32) and interference phenomena in optics (Chapters 37 and 38).

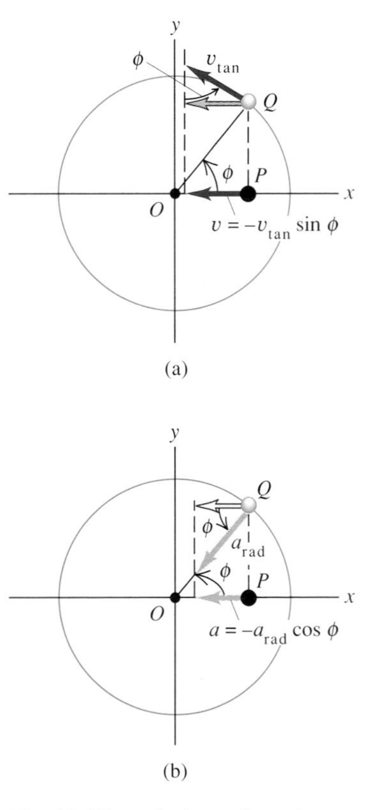


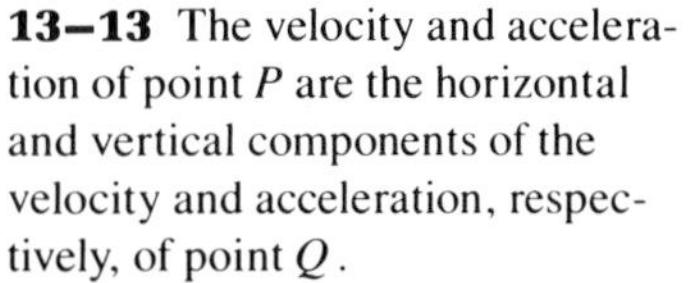

**13–13** The velocity and acceleration of point $P$ are the horizontal and vertical components of the velocity and acceleration, respectively, of point $Q$.

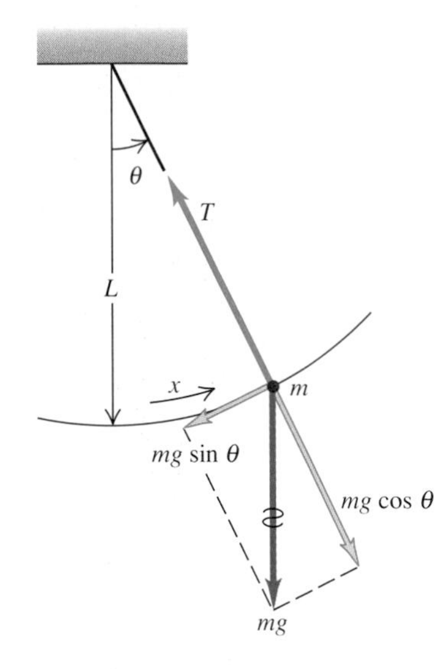


**13–14** The forces on the bob of a simple pendulum.

## 13–5 THE SIMPLE PENDULUM

A **simple pendulum** is an idealized model consisting of a point mass suspended by a weightless, unstretchable string in a uniform gravitational field. When the mass is pulled to one side of its straight-down equilibrium position and released, it oscillates about the equilibrium position. Familiar situations such as a wrecking ball on a crane's cable, the plumb bob on a surveyor's transit, and a child on a swing can be modeled as simple pendulums. We can now analyze the motion of such systems, asking in particular whether it is simple harmonic.

The path of the point mass (sometimes called a pendulum bob) is not a straight line but the arc of a circle with radius $L$ equal to the length of the string (Fig. 13–14). We use as our coordinate the distance $x$ measured along the arc. If the motion is simple harmonic, the restoring force must be directly proportional to $x$ or (because $x = L\theta$) to $\theta$. Is it?

In Fig. 13–14 we represent the forces on the mass in terms of tangential and radial components. The restoring force $F$ is

$$F = -mg \sin \theta. \tag{13–27}$$

The restoring force is therefore proportional *not* to $\theta$ but to $\sin \theta$, so the motion is *not* simple harmonic. However, *if the angle $\theta$ is small,* $\sin \theta$ is very nearly equal to $\theta$. For example, when $\theta = 0.1$ rad (about 6°), $\sin \theta = 0.0998$, a difference of only 0.2%. With this approximation, Eq. (13–27) becomes

$$F = -mg\theta = -mg\frac{x}{L}, \qquad \text{or}$$

$$F = -\frac{mg}{L}x. \tag{13–28}$$

The restoring force is then proportional to the coordinate *for small displacements,* and the constant $mg/L$ represents the force constant $k$. From Eq. (13–12) the angular frequency $\omega$ of a simple pendulum with small amplitude is

$$\omega = \sqrt{\frac{k}{m}} = \sqrt{\frac{mg/L}{m}} = \sqrt{\frac{g}{L}}. \tag{13–29}$$

The corresponding frequency and period relations are

$$f = \frac{\omega}{2\pi} = \frac{1}{2\pi}\sqrt{\frac{g}{L}}, \tag{13–30}$$

$$T = \frac{2\pi}{\omega} = \frac{1}{f} = 2\pi\sqrt{\frac{L}{g}}. \tag{13–31}$$

Note that these expressions do not contain the *mass* of the particle. This is because the restoring force, a component of the particle's weight, is proportional to $m$. Thus the mass appears on *both* sides of $F = ma$ and cancels out. For small oscillations the period of a pendulum for a given value of $g$ is determined entirely by its length.

Galileo invented an elegant argument four centuries ago to show that the period should be independent of the mass. Make a simple pendulum, said Galileo, measure its period, and then split the pendulum down the middle, string and all. Because of the symmetry, neither half could have been pushing or pulling on the other before it was split. So splitting it can't change the motion, and each half must swing with the same period as the original! Therefore the period cannot depend on the mass.

The dependence on $L$ and $g$ in Eqs. (13–29) through (13–31) is just what we should expect on the basis of everyday experience. A long pendulum has a longer period than a shorter one. Increasing $g$ increases the restoring force, causing the frequency to increase and the period to decrease.

We emphasize again that the motion of a pendulum is only *approximately* simple harmonic. When the amplitude is not small, the departures from simple harmonic motion can be substantial. But how small is "small"? The period can be expressed by an infinite series; when the maximum angular displacement is $\Theta$, the period $T$ is given by

$$T = 2\pi\sqrt{\frac{L}{g}}\left(1 + \frac{1^2}{2^2}\sin^2\frac{\Theta}{2} + \frac{1^2 \cdot 3^2}{2^2 \cdot 4^2}\sin^4\frac{\Theta}{2} + \cdots\right). \tag{13–32}$$

We can compute the period to any desired degree of precision by taking enough terms in the series. We invite you to check that when $\Theta = 15°$ (on either side of the central position), the true period differs from the approximate value given by Eq. (13–31) by less than 0.5%.

The usefulness of the pendulum as a timekeeper depends on the fact that for small amplitudes the motion is *very nearly* isochronous (the period is independent of amplitude). Thus, as a pendulum clock runs down and the amplitude of the swings decreases a little, the clock still keeps very nearly correct time.

The simple pendulum, or a variation, is also a precise and convenient method for measuring the acceleration of gravity $g$, since $L$ and $T$ can be measured easily and precisely. Such measurements are often used in geophysics. Local deposits of ore or oil affect the local value of $g$ because their density differs from that of their surroundings. Precise measurements of this quantity over an area being surveyed often furnish valuable information about the nature of underlying deposits.

## EXAMPLE 13–6

Find the period and frequency of a simple pendulum 1.000 m long at a location where $g = 9.800\ \text{m/s}^2$.

**SOLUTION** From Eq. (13–31),

$$T = 2\pi\sqrt{\frac{L}{g}} = 2\pi\sqrt{\frac{1.000\ \text{m}}{9.800\ \text{m/s}^2}} = 2.007\ \text{s}.$$

Then

$$f = \frac{1}{T} = 0.4983\ \text{Hz}.$$

The period is almost exactly 2 s. In fact, when the metric system was first established, the second was defined as half the period of a one-meter pendulum.

# 13–6 THE PHYSICAL PENDULUM

A **physical pendulum** is any *real* pendulum, using a body of finite size, compared to the idealized model of the *simple* pendulum, with all the mass concentrated at a point. For small oscillations, analyzing the motion of a real pendulum is almost as easy as for a simple pendulum. Figure 13–15 shows a body with irregular shape, pivoted so that it can turn without friction about an axis through point $O$. In the equilibrium position the center of gravity is directly below the pivot; in the position shown in the figure the body is displaced from equilibrium by an angle $\theta$, which we use as a coordinate for the system. The distance from $O$ to the center of gravity is $d$, the moment of inertia of the body about the axis of rotation through $O$ is $I$, and the total mass is $m$. When the body is displaced as shown, the weight $mg$ causes a restoring torque

$$\tau = -(mg)(d \sin\theta). \tag{13–33}$$

The negative sign shows that the restoring torque is clockwise when the displacement is counterclockwise, and vice versa.

When the body is released, it oscillates about its equilibrium position. The motion is not simple harmonic because the torque $\tau$ is proportional to $\sin\theta$ rather than to $\theta$ itself.

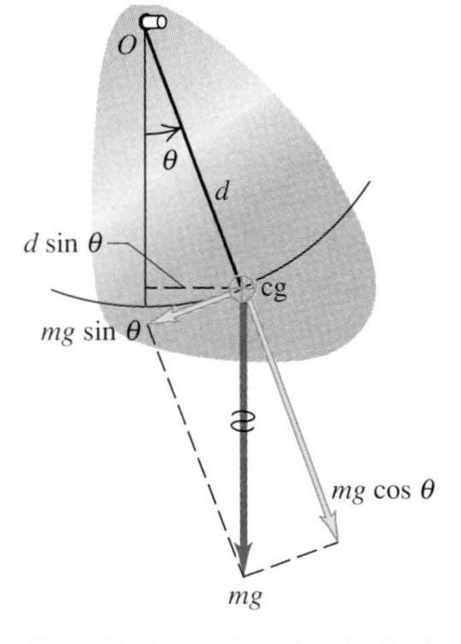


**13–15** Dynamics of a physical pendulum.

However, if $\theta$ is small, we can again approximate $\sin\theta$ by $\theta$, and the motion is *approximately* harmonic. With this approximation,

$$\tau = -(mgd)\theta.$$

The equation of motion is $\Sigma\tau = I\alpha$, where $I$ is the body's moment of inertia about the axis of rotation. We find

$$-(mgd)\theta = I\alpha = I\frac{d^2\theta}{dt^2},$$

$$\frac{d^2\theta}{dt^2} = -\frac{mgd}{I}\theta. \tag{13–34}$$

Comparing this with Eq. (13–17), we see that the role of $k/m$ for the mass-spring system is played here by the quantity $mgd/I$. Thus the angular frequency is given by

$$\omega = \sqrt{\frac{mgd}{I}}. \tag{13–35}$$

The frequency $f$ is $1/2\pi$ times this, and the period $T$ is

$$T = 2\pi\sqrt{\frac{I}{mgd}}. \tag{13–36}$$

## EXAMPLE 13–7

**Physical pendulum versus simple pendulum** Suppose the body in Fig. 13–15 is a uniform rod with length $L$, pivoted at one end. Find the period of its motion.

**SOLUTION** From Chapter 9 the moment of inertia of a uniform rod about an axis through one end is $I = \frac{1}{3}ML^2$. The distance from the pivot to the center of gravity is $d = L/2$. From Eq. (13–36),

$$T = 2\pi\sqrt{\frac{\frac{1}{3}ML^2}{MgL/2}} = 2\pi\sqrt{\frac{2L}{3g}}.$$

If the rod is a meter stick ($L = 1.00$ m) and $g = 9.80\ \text{m/s}^2$, then

$$T = 2\pi\sqrt{(\tfrac{2}{3})(1.00\ \text{m})/9.80\ \text{m/s}^2} = 1.64\ \text{s}.$$

The period is smaller by a factor of $(\frac{2}{3})^{1/2} = 0.816$ than the period of a simple pendulum with the same length.

## EXAMPLE 13–8

**Brisk walking and the physical pendulum** Many people find that they have a natural walking pace, a number of steps per minute that is more comfortable than a faster or slower pace. Suppose this natural pace is equal to the period of the leg, viewed as a uniform rod pivoted at the hip joint (Fig. 13–16). For a person with legs 1.0 m long, what is the natural pace in steps per minute?

**SOLUTION** The analysis is exactly the same as in Example 13–7. Each period (a complete back-and-forth swing of the leg) corresponds to two steps, so the pace is $(1.64\ \text{s})/2 = 0.82$ s per step. This corresponds to $1/(0.82\ \text{s}) = 1.2$ steps per second, or about 72 steps per minute.

In fact, people with 1-m legs (a little taller than average) typically have a natural pace that is closer to 120 steps per minute, corresponding to $T = 1.0$ s. This pace is often used by marching

**13–16** Modeling a human leg as a uniform rod pivoted at the hip joint.

bands and military organizations. The discrepancy in our calculation shows that a uniform rod isn't a very good model for the human leg. Most people's legs are tapered; there is a lot more mass between the knee and the hip than between the knee and the toes. Thus the center of mass is less than $L/2$ from the hip; a reasonable guess would be about $L/4$. The moment of inertia is *considerably* less than $ML^2/3$, probably somewhere around $ML^2/15$. Try these numbers out with the analysis of Example 13–7 and see whether you come closer to $T = 1.0$ s.

## EXAMPLE 13–9

**Measuring moments of inertia** How can the period of a physical pendulum be used to determine the moment of inertia of the body about the rotational axis through the pivot?

**SOLUTION** Assuming that $I$ is the unknown quantity to be determined, we solve Eq. (13–36) for $I$:

$$I = \frac{T^2 mgd}{4\pi^2}.$$

All the quantities on the right can be measured directly, so we can determine the moment of inertia of a body with a complicated shape by suspending it as a physical pendulum and measuring its period of vibration. We can locate the center of gravity by balancing. For example, Fig. 13–17 shows a connecting rod from a car engine, pivoted about a horizontal knife edge. Suppose that the mass of the rod is 2.00 kg and that when it is set into oscillation, it makes 100 complete swings in 120 s, so $T = (120\text{ s})/100 = 1.20$ s.

From the above relation,

$$I = \frac{(1.20\text{ s})^2(2.00\text{ kg})(9.80\text{ m/s}^2)(0.200\text{ m})}{4\pi^2} = 0.143\text{ kg}\cdot\text{m}^2.$$

$d = 0.200$ m

cg

**13–17** The moment of inertia of this connecting rod about the rotational axis through the pivot can be determined by measuring its small-angle period of oscillation.

# 13–7 DAMPED OSCILLATIONS

The idealized oscillating systems that we have discussed thus far are frictionless. There are no nonconservative forces, the total mechanical energy is constant, and a system set into motion continues oscillating forever with no decrease in amplitude.

Real-world systems always have some friction, however, and oscillations do die out with time unless we provide some means for replacing the mechanical energy lost to friction. A mechanical pendulum clock continues to run because potential energy stored in the spring or a hanging weight system replaces the mechanical energy lost because of friction in the pendulum and the gears. But the spring eventually runs down, or the weights reach the bottom of their travel. Then no more energy is available, and the pendulum swings decrease in amplitude and stop.

The decrease in amplitude caused by dissipative forces is called **damping,** and the corresponding motion is called **damped oscillation.** The simplest case to analyze in detail is that of a frictional damping force that is directly proportional to the *velocity* of the oscillating body. This behavior occurs in friction involving viscous fluid flow, such as in shock absorbers or sliding between oil-lubricated surfaces. We then have an additional force on the body due to friction, $F = -bv$, where $v = dx/dt$ is the velocity and $b$ is a constant that describes the strength of the damping force. The negative sign shows that the force is always opposite in direction to the velocity. The *total* force on the body is then

$$F = -kx - bv, \qquad (13\text{–}37)$$

and Newton's second law for the system is

$$-kx - bv = ma, \qquad \text{or} \qquad -kx - b\frac{dx}{dt} = m\frac{d^2x}{dt^2}. \qquad (13\text{–}38)$$

Equation (13–38) is a differential equation for $x$; it would be same as Eq. (13–17) except for the added term $-b\,dx/dt$. Solving this equation is a straightforward problem in differential equations, but we won't go into the details here. If the damping force is relatively small and the body is given an initial displacement $A$, the motion is described by

$$x = A\,e^{-(b/2m)t}\cos\omega' t. \tag{13–39}$$

The angular frequency of oscillation $\omega'$ is given by

$$\omega' = \sqrt{\frac{k}{m} - \frac{b^2}{4m^2}}. \tag{13–40}$$

You can verify that Eq. (13–39) is a solution of Eq. (13–38) by calculating the first and second derivatives of $x$, substituting them into Eq. (13–38), and checking whether the left and right sides are equal. This is a straightforward but slightly tedious procedure.

The motion described by Eq. (13–39) differs from the undamped case in two ways. First, the amplitude $Ae^{-(b/2m)t}$ is not constant but decreases with time because of the decreasing exponential factor $e^{-(b/2m)t}$. The larger the value of $b$, the more quickly the amplitude decreases. Figure 13–18 shows graphs of Eq. (13–39) for two different values of the constant $b$.

Second, the angular frequency $\omega'$, given by Eq. (13–40), is no longer equal to $\omega = (k/m)^{1/2}$ but is somewhat smaller. It becomes zero when $b$ becomes so large that

$$\frac{k}{m} - \frac{b^2}{4m^2} = 0, \qquad \text{or}$$

$$b = 2\sqrt{km}. \tag{13–41}$$

When $b$ exceeds this value, the system no longer oscillates but returns to its equilibrium position without oscillation when it is displaced and released. When Eq. (13–41) is satisfied, the condition is called **critical damping.** The nonoscillating motion that occurs when $b$ is even larger is called **overdamping.** For the overdamped case the solutions of Eq. (13–38) have the form

$$x = C_1 e^{-a_1 t} + C_2 e^{-a_2 t},$$

where $C_1$ and $C_2$ are constants that depend on the initial conditions and $a_1$ and $a_2$ are positive constants determined by $m$, $k$, and $b$. When $b$ is less than the critical value, the situation is called **underdamping.**

Similar behavior occurs in electric circuits containing inductance, capacitance, and resistance. There is a natural frequency of oscillation, and the resistance plays the

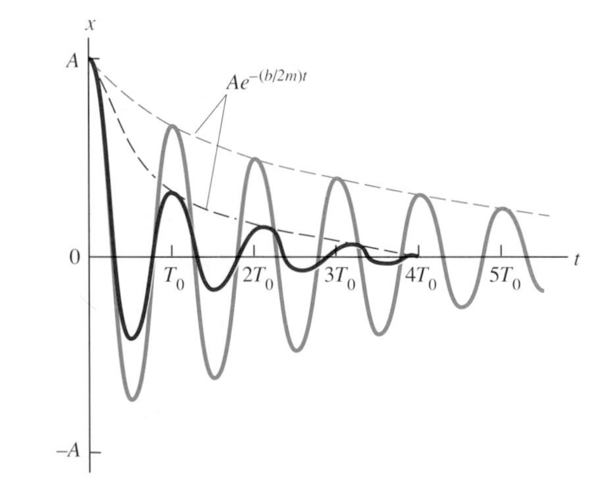


**13–18** Graphs of Eq. (13–39), showing damped harmonic motion. The period when there is no damping ($b = 0$) is $T_0$. The blue curve shows the motion when $b = 0.1\sqrt{km}$ and the red curve when $b = 0.4\sqrt{km}$. The amplitude decreases more rapidly for the larger value of $b$. Close inspection of the points where the curves cross the $t$-axis also reveals that the period increases slightly with increasing $b$. The critical-damping condition is $b = 2\sqrt{km}$.

role of the damping constant $b$. We will study these circuits in detail in Chapters 31 and 32.

In damped oscillations the damping force is nonconservative; the mechanical energy of the system is not constant but decreases continuously, approaching zero after a long time. To derive an expression for the rate of change of energy, we first write an expression for the total mechanical energy $E$ at any instant:

$$E = \tfrac{1}{2}mv^2 + \tfrac{1}{2}kx^2.$$

To find the rate of change of this quantity, we take its time derivative:

$$\frac{dE}{dt} = mv\frac{dv}{dt} + kx\frac{dx}{dt}.$$

But $dv/dt = a$ and $dx/dt = v$, so

$$\frac{dE}{dt} = v(ma + kx).$$

From Eq. (13–38), $ma + kx = -b\,dx/dt = -bv$, so

$$\frac{dE}{dt} = v(-bv) = -bv^2. \qquad (13\text{–}42)$$

The right side of Eq. (13–42) is always negative, whether $v$ is positive or negative. This shows that $E$ continuously decreases, though not at a uniform rate. The term $-bv^2 = (-bv)v$ (force times velocity) is power, the rate at which the damping force does (negative) work on the system, and is equal to the rate of change of the total mechanical energy of the system.

The suspension system of an automobile is a familiar example of damped oscillations. The shock absorbers provide a velocity-dependent damping force so that when the car goes over a bump, it doesn't continue bouncing forever (Fig. 13–19). For optimal passenger comfort the system should be critically damped or slightly underdamped. As the shocks get old and worn, the value of $b$ decreases, and the bouncing persists longer. Not only is this nauseating, but it is bad for steering because the front wheels have less positive contact with the ground. Thus damping is an advantage in this system. Conversely, in a system such as a clock or an electrical oscillating system of the type found in radio transmitters it is usualy desirable to minimize damping.

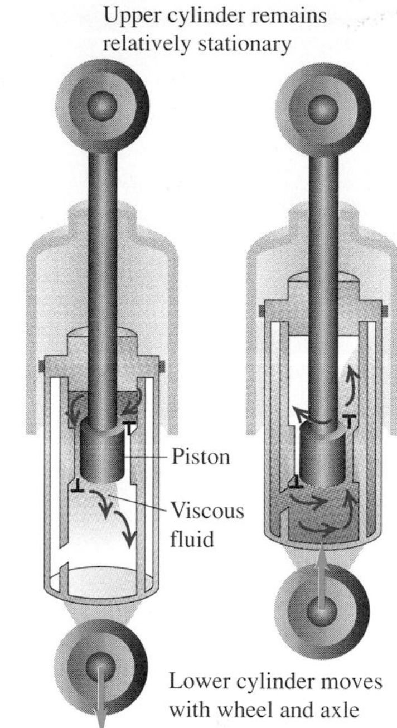


**13–19** A car shock absorber. The top part, connected to the piston, is attached to the car's frame; the bottom part, connected to the lower cylinder, is connected to the axle. The viscous fluid causes a damping force that depends on the relative velocity of the two ends of the unit. This helps control wheel bounce and jounce.

## *13–8 FORCED OSCILLATIONS AND RESONANCE

What happens when a periodically varying force is applied to a vibrating system? A factory floor supported by a slightly flexible steel framework may begin to vibrate when an oscillating machine rests on it. Or consider your cousin Throckmorton on a playground swing. You can keep him swinging with constant amplitude by giving him a little push once each cycle. More generally, we can maintain a constant-amplitude oscillation in a damped harmonic oscillator by applying a force that varies with time in a periodic or cyclic way, with a definite period and frequency. We call this additional force a **driving force.**

We denote the angular frequency of the driving force by $\omega_d$. It is not necessarily equal to the angular frequency $\omega'$ with which the system would naturally oscillate without a driving force. When a periodically varying driving force is present, the mass can undergo a periodic motion *with the same angular frequency $\omega_d$ as that of the driving force*. We call this motion a **forced oscillation,** or a *driven oscillation;* it is different from the motion that occurs when the system is simply set into motion and then left alone to oscillate with a natural frequency $\omega'$ determined by $m$, $k$, and $b$.

When the angular frequency of the driving force is nearly *equal* to the natural angular frequency $\omega'$ of the system, we are forcing it to vibrate with a frequency that is close to the frequency it would have even with no driving force. In this case we expect the amplitude of the resulting oscillation to be larger than when the two frequencies are very different. This expectation is borne out by more detailed analysis and experiment. The easiest case to analyze is that of a *sinusoidally* varying force, say, $F_x = F_{max} \sin \omega_d t$, where $\omega_d$ is not necessarily equal to the natural angular frequency $\omega'$ of the system, given by Eq. (13–40). If we vary the frequency $\omega_d$ of the driving force, the amplitude of the resulting forced oscillation varies in an interesting way (Fig. 13–20). When there is very little damping (small $b$), the amplitude goes through a sharp peak as the driving angular frequency $\omega_d$ nears the natural oscillation angular frequency $\omega'$. When the damping is increased (larger $b$), the peak becomes broader and smaller in height and shifts toward lower frequencies.

We could work out an expression that shows how the amplitude $A$ of the forced oscillation depends on the frequency of a sinusoidal driving force, with maximum value $F_{max}$. That would involve more differential equations than we're ready for, but here is the result:

$$A = \frac{F_{max}}{\sqrt{(k - m\omega_d^2)^2 + b^2\omega_d^2}}. \tag{13–43}$$

When $k - m\omega_d^2 = 0$, the first term under the radical is zero, so $A$ has a maximum near $\omega_d = (k/m)^{1/2}$. The height of the curve at this point is proportional to $1/b$; the less damping, the higher the peak. At the low-frequency extreme, when $\omega_d = 0$, we get $A = F_{max}/k$. This corresponds to a *constant* force $F_{max}$ and a constant displacement $A = F_{max}/k$ from equilibrium, as we might expect.

The fact that there is an amplitude peak at driving frequencies close to the natural frequency of the system is called **resonance.** Physics is full of examples of resonance; building up the oscillations of a child on a swing by pushing with a frequency equal to the swing's natural frequency is one. A vibrating rattle in a car that occurs only at a certain engine speed or wheel-rotation speed is an all too familiar example. You may have heard of the dangers of a band marching in step across a bridge; if the frequency of their steps is close to a natural vibration frequency of the bridge, dangerously large oscillations can build up. Inexpensive loudspeakers often have an annoying buzz or boom when a musical note happens to coincide with the resonant frequency of the speaker cone. A few years

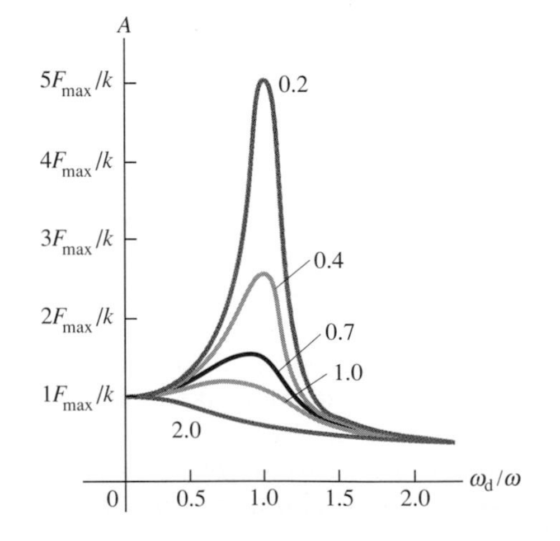


**13–20** Graph of the amplitude $A$ of forced oscillation of a damped harmonic oscillator as a function of the frequency $\omega_d$ of the driving force, plotted on the horizontal axis as the ratio of $\omega_d$ to the angular frequency $\omega = \sqrt{k/m}$ of an undamped oscillator. Each curve is labeled with the value of the dimensionless quantity $b/\sqrt{km}$, which characterizes the amount of damping. The highest curve has $b = 0.2\sqrt{km}$, the next has $b = 0.4\sqrt{km}$, and so on. As $b$ increases, the peak becomes broader and less sharp and shifts toward lower frequencies. When $b$ is larger than $\sqrt{2km}$, the peak disappears completely.

13–21 The Tacoma Narrows Bridge collapsed four months and six days after it was opened for traffic. The main span was 2800 ft long and 39 ft wide, with 8-ft-high steel stiffening girders on both sides. The maximum amplitude of the torsional vibrations was 35°; the frequency was about 0.2 Hz.

ago vibrations from the engines of a particular airplane had just the right frequency to resonate with the natural frequencies of its wings. Dangerously large oscillations built up, and the wings occasionally fell off. And the list goes on. A tuned circuit in a radio or television receiver responds strongly to waves having frequencies near its resonant frequency, and this fact is used to select a particular station and reject the others. We will study resonance in electric circuits in detail in Chapter 32.

Finally, we mention the famous collapse of the Tacoma Narrows suspension bridge in 1940; nearly everyone has seen the film of this catastrophe (Fig. 13–21). This is usually cited as an example of resonance, but there's some doubt as to whether it should be called that. The wind didn't have to vary *periodically* with a frequency close to a natural frequency of the bridge. The air flow past the bridge is turbulent, and vortices are formed with a regular frequency that depends on the flow speed. It is conceivable that this frequency may have resonated with a natural frequency of the bridge. But the cause may well have been something more subtle called a *self-excited oscillation,* in which the aerodynamic forces caused by a *steady* wind blowing on the bridge tended to displace it farther from equilibrium at times when it was already moving away from equilibrium. It is as though we had a damping force such as the $-bv$ term in Eq. (13–37) but with the sign reversed. Instead of draining mechanical energy away from the system, this antidamping force pumps energy into the system, building up the oscillations to destructive amplitudes. The approximate differential equation is Eq. (13–38) with the sign of the $b$ term reversed, and the oscillating solution is Eq. (13–39) with a *positive* sign in the exponent. You can see that this means trouble. Engineers have now learned how to stabilize suspension bridges, both structurally and aerodynamically, to prevent such disasters.

## 13–9 CHAOS: A Case Study in Dynamic Analysis

The world of Newtonian mechanics has often been called "the clockwork universe." This nickname refers to the fact that when a mechanical system is governed by Newton's laws, the motion is completely predictable, just as a mechanical clock runs in a definite way once it is wound up and started.

Here are two examples: If you know the initial position and velocity of a baseball, you can predict the shape of its entire trajectory and the time when it arrives at any given point. When a harmonic oscillator is given an initial displacement $x_0$ and an initial velocity $v_0$, Eqs. (13–19), (13–21), and (13–22) give the position, velocity, and acceleration at any later time. Any system for which the entire course of the motion can be predicted when we know the appropriate initial conditions is called a *deterministic system.*

During the last half of this century, physicists and other scientists have begun to study dynamical systems that are deterministic and yet have behavior that is unpredictable or *chaotic.* One aspect of chaos in mechanical systems is sensitivity to initial conditions. For a baseball trajectory, if you were to find that changing the initial velocity by a tiny amount caused a very large change in the subsequent motion, then in a practical sense you couldn't predict the long-term behavior precisely, no matter how precisely your initial conditions were specified.

Another aspect of chaos appears in forced oscillations (Section 13–8). If the spring force is not a linear function of displacement (that is, if its behavior is more complicated than $F = -kx$), then the motion that results from a sinusoidal driving force may not have the same frequency as the force. Indeed, it may not be periodic at all but may appear to be random and unpredictable. We'll look at a simple example that shows both of these features of chaotic behavior.

Graphical representations are a great help in our study of chaos. We humans are most directly aware of the *time* evolution of motion, so we often plot graphs of position, velocity, and acceleration as functions of time. We did this in Chapter 2 for straight-line motion of a particle and again in Section 13–3 for SHM. But for the study of chaos it is often more useful to plot velocity $v$ as a function of position $x$. A coordinate system with $x$ on the horizontal axis and $v$ on the vertical axis is called a **phase space.** A point in phase space represents the instantaneous position and velocity of the particle under study.

Here's an example: For undamped SHM the total energy $E$ is constant, and $x$ and $v$ are related by the energy-conservation equation:

$$E = \tfrac{1}{2}mv^2 + \tfrac{1}{2}kx^2 = \text{constant}. \qquad (13\text{–}44)$$

In the $x$-$v$ coordinate system this is the equation of an ellipse (Fig. 13–22). The half-axis lengths are the maximum displacement $A$ and the maximum velocity $v_{\max}$, related to $E$ by

$$E = \tfrac{1}{2}kA^2 = \tfrac{1}{2}mv_{\max}{}^2. \qquad (13\text{–}45)$$

Each point on the ellipse represents the position and velocity of the body at one instant of time, and as time goes on, the point moves clockwise around the curve. If there is no damping, the curve is traced out over and over, with a frequency $f$ equal to the frequency of the motion. Every cyclic motion corresponds to a closed curve in phase space.

When there is damping, the total energy $E$ decreases continuously, and the point in phase space *does not* trace the same path over and over. Instead, as the amplitude, maximum velocity, and total energy decrease with time, the point spirals in toward the origin

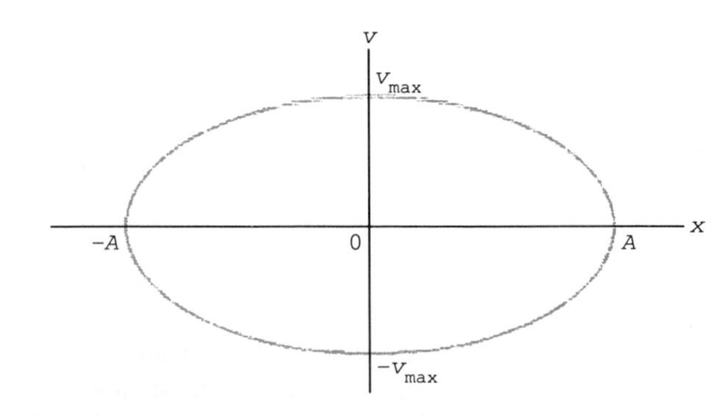


**13–22** Phase-space plot for an undamped harmonic oscillator.

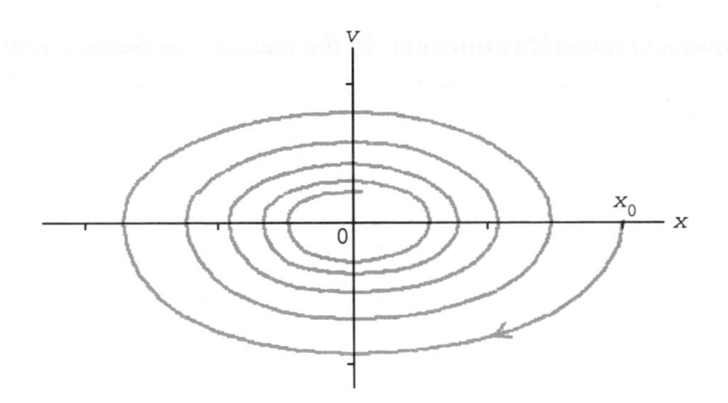


**13–23** Phase-space plot for damped harmonic motion.

(Fig. 13–23). To plot this curve, we could use Eq. (13–39) and its time derivative to find $x$ and $v$ for each of a succession of times and then plot the resulting points $(x, v)$.

How do *forced* oscillations look in phase space? Suppose we apply a sinusoidal driving force $F = F_{\max} \cos \omega_d t$ (just as in Section 13–8) to an underdamped ($b^2 < 4km$) system. The $\Sigma F = ma$ equation is the same as Eq. (13–38) except for the additional sinusoidal force term. The complete equation is

$$-kx - b\frac{dx}{dt} + F_{\max} \cos \omega_d t = m\frac{d^2x}{dt^2}. \tag{13–46}$$

This differential equation can be solved exactly. We won't go into the details, but the complete solution has two parts. There is a sinusoidal term with the same frequency $\omega_d$ as the driving force and an amplitude $A$ given by Eq. (13–43). This is the forced-oscillation part, also called the *steady-state* motion; it persists as long as the driving force is applied. In addition, there is a *transient* part, representing a damped oscillation (which could also occur with *no* driving force) that depends on the initial conditions. This second part always has an exponential factor $e^{-(b/2m)t}$, just as in Eq. (13–39), so it always dies out after a long time.

Thus the motion of this system may begin with some irregular-looking motion, but it eventually settles down to a sinusoidal motion with angular frequency $\omega_d$, represented again by an ellipse in phase space (Fig. 13–24a). If we choose different initial conditions $x_0$ and $v_0$ (Fig. 13–24b), the initial irregularity is different, but the final sinusoidal motion is the same. We call this final motion a **limit cycle.** For each case the effect of the initial conditions is a transient effect that dies out after a long time, leaving a final steady-state motion that is independent of the initial conditions.

Now we're ready for the really interesting part of this discussion. The potential-energy function $U(x)$ corresponding to our spring force $F = -kx$ is $U(x) = \frac{1}{2}kx^2$;

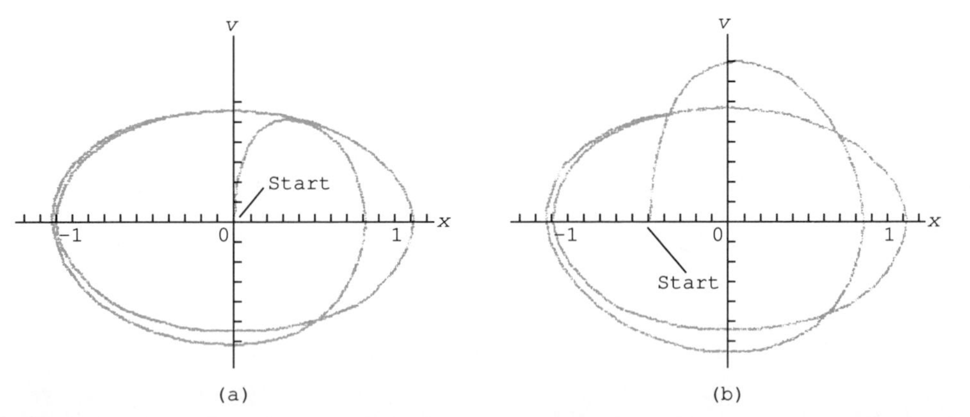


**13–24** Phase-space plot of forced oscillation, showing approach to steady-state forced oscillation (limit cycle). (a) Initial conditions $x_0 = 0$, $v_0 = 0$; (b) initial conditions $x_0 = -0.5$, $v_0 = 0$. The limit cycle is the same for both.

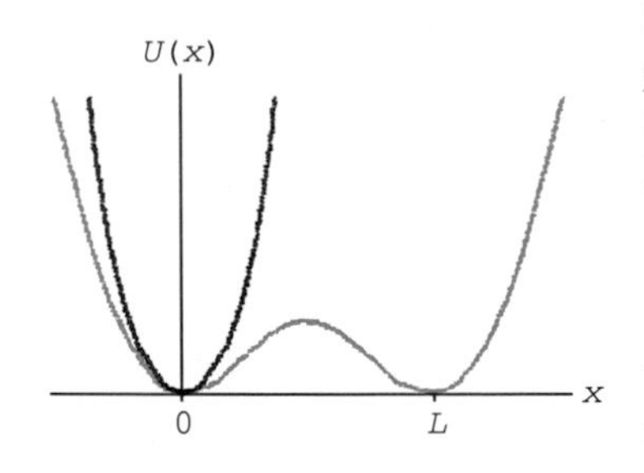


**13–25** Graphs of the functions $U(x) = \frac{1}{2}kx^2$ and $U(x) = \frac{1}{2}kx^2(1 - x/L)^2$. The two curves coincide near $x = 0$ but become very different for larger $x$.

Fig. 13–4 shows a graph of this function. In the language of Section 7–6 we call this a *potential well;* its minimum is a stable equilibrium position. Now suppose we modify this function by multiplying it by the factor $(1 - x/L)^2$, where $L$ is a constant with units of length. Then we have

$$U(x) = \frac{1}{2}kx^2\left(1 - \frac{x}{L}\right)^2. \tag{13–47}$$

When the particle's coordinate $x$ is much smaller than $L$, the quantity $x/L$ is much smaller than unity and can be neglected. Then Eq. (13–47) is equal to the original function, $U = \frac{1}{2}kx^2$. But for larger $x$ the two functions are quite different. We leave it as an exercise to show that Eq. (13–47) has *two* minima, one at $x = 0$ and the other at $x = L$, and thus has *two* potential wells. Figure 13–25 shows graphs of both functions; they are nearly identical near $x = 0$ but quite different when $x$ is comparable to $L$.

Suppose the body is initially near $x = 0$ and we apply a driving force with $F_{max}$ small enough that the amplitude $A$ of the forced oscillations is much smaller than $L$. Then we expect the same behavior as for $U = \frac{1}{2}kx^2$, a sinusoidal forced oscillation with angular frequency $\omega_d$. Figures 13–26a and 13–26b show phase-space plots of this motion for two different initial conditions. The effect of the initial conditions dies out after a long time, and the limit cycle is the same in both cases. When we increase $F_{max}$ (Figs. 13–26c and 13–26d), the limit cycle looks less like an ellipse (because the force is no longer a linear function of $x$), but it is still independent of initial conditions.

When we increase $F_{max}$ still more, we find that at a certain critical value the behavior suddenly becomes much more complex. For some values of $F_{max}$ the particle oscillates between the two wells (Fig. 13–27a). At still larger values there is no closed curve in phase space (which would represent a regular cycle) but instead a random and chaotic curve (Fig. 13–27b). At times the particle seems to be oscillating within one potential well; then it hops over to the other or wanders between the two with no apparent pattern.

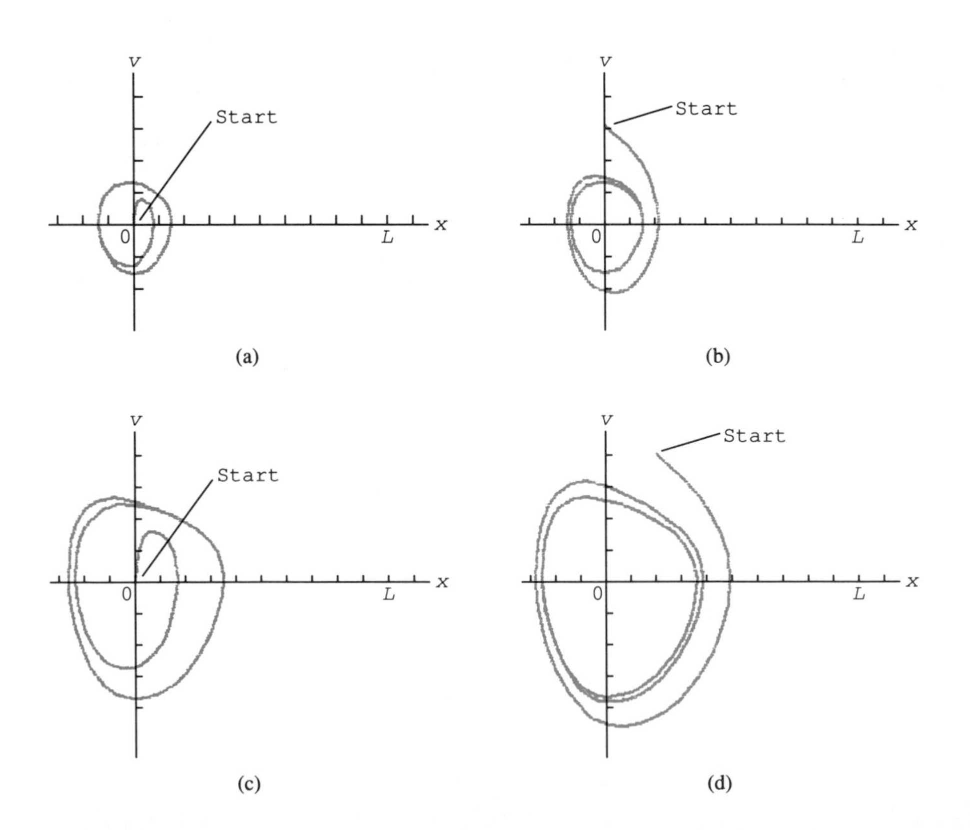


**13–26** Phase-space plots of forced oscillations with two-well potential energy. (a) Small-amplitude oscillation with small driving force. (b) Same driving force as part (a) but different initial conditions; same limit cycle as part (a). (c) Larger driving-force amplitude. (d) Same driving force as part (c) but different initial conditions; same limit cycle as part (c).

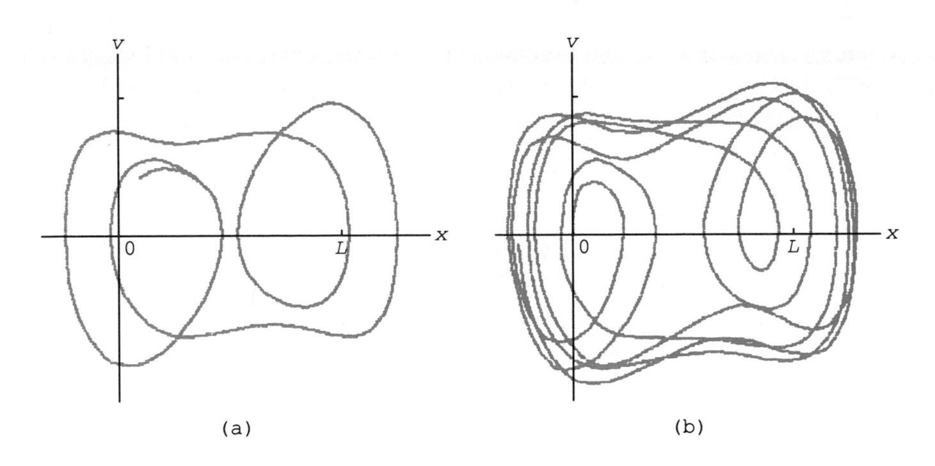


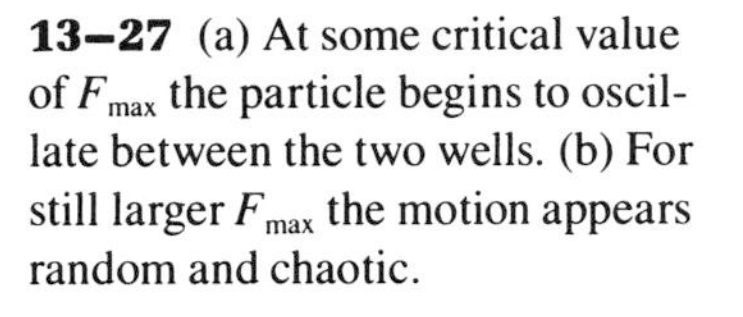
**13–27** (a) At some critical value of $F_{max}$ the particle begins to oscillate between the two wells. (b) For still larger $F_{max}$ the motion appears random and chaotic.

Another new feature that sets in when we reach the critical value of $F_{max}$ is sensitivity to initial conditions. When we start the system with two very slightly different initial conditions, represented by two neighboring points in phase space (Fig. 13–28), the two resulting motions very quickly diverge from each other. This is in contrast to the complete *lack* of dependence on initial conditions that we found with the simpler potential-energy function. It is as though the transient part of the motion, instead of decaying exponentially, is *growing* exponentially.

The $\Sigma F = ma$ equation for this system is obtained by replacing the spring force $F(x) = -kx$ in Eq. (13–46) by the more general force given by $F = -dU/dx$ for the function $U$ of Eq. (13–47). We leave it as an exercise to show that

$$F(x) = -\frac{dU}{dx} = -kx\left(1 - \frac{x}{L}\right)\left(1 - \frac{2x}{L}\right) \tag{13–48}$$

and that the resulting differential equation is

$$-kx\left(1 - \frac{x}{L}\right)\left(1 - \frac{2x}{L}\right) - b\frac{dx}{dt} + F_{max}\cos\omega_d t = m\frac{d^2x}{dt^2}. \tag{13–49}$$

This equation *cannot* be solved in terms of familiar functions. But we can carry out a numerical calculation just as we did for the baseball trajectory with air resistance in Section 3–6. We choose values for the constants $k$, $b$, $m$, $L$, $F_{max}$, and $\omega_d$ and for the initial conditions $x_0$ and $v_0$. Using these initial values, we compute the values of $v$ and $x$ at a slightly later time. We use these for another time step and so on. That's how we obtained the curves in Figs. 13–26 and 13–27.

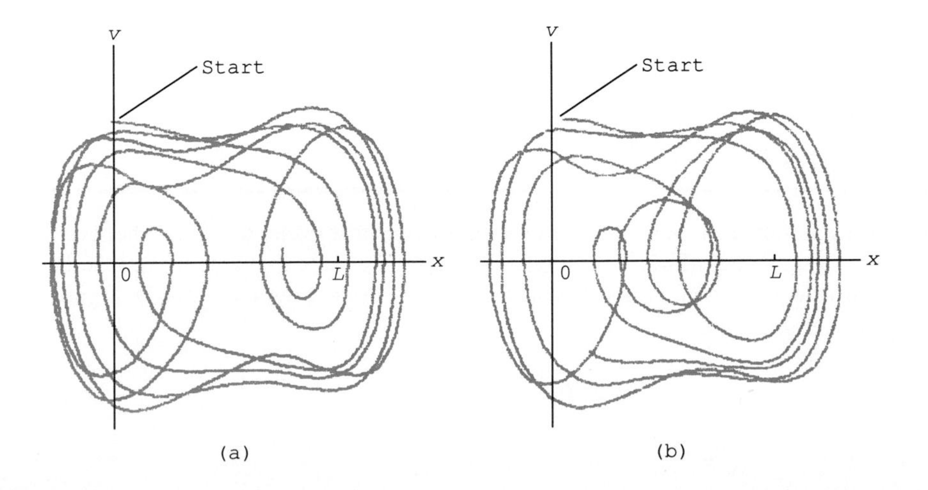


**13–28** Phase-space plot for two forced oscillations with the same driving force but different initial conditions. (a) $x = 0$, $v = 1$; (b) $x = 0.1L$, $v = 1$. After a few cycles of the driving force, the two motions have become very different.

Here's a sketch of a program to compute phase-space curves such as Figs. 13–26 and 13–27). Just as in Section 3–6, you can implement this scheme either with a programming language such as BASIC or with a spreadsheet.

*Step 1:* Specify the parameters $m$, $k$, $b$, $L$, $\omega_d$, and $F_{max}$. For simplicity we suggest $m = 1$, $k = 1$, and $L = \frac{1}{2}$.

*Step 2:* Specify the time interval $\Delta t$, which we'll abbreviate as $h$, and the initial conditions $x_0$ and $v_0$.

*Step 3:* Choose the maximum number of intervals $N$ or the maximum time $t_{max}$ for which you want to compute the solution.

*Step 4:* Loop (or iterate) Steps 5 through 9 while $n < N$ or $t < t_{max}$.

*Step 5:* Print or plot $(x, v)$.

*Step 6:* Calculate the acceleration $a$ from Eq. (13–49):

$$a = \left(\frac{1}{m}\right)\left[-kx\left(1 - \frac{x}{L}\right)\left(1 - \frac{2x}{L}\right) - bv + F_{max}\cos\omega_d t\right].$$

*Step 7:* Calculate the new velocity:

$$v = v + ah.$$

*Step 8:* Calculate the new position:

$$x = x + vh.$$

*Step 9:* Calculate the new time or update the counter $n$:

$$t_{n+1} = t_n + h \qquad \text{or} \qquad n = n + 1.$$

*Step 10:* Loop until $t_{max}$ or $N$ is reached.

*Step 11:* End.

If you are using a spreadsheet, the following notation may be helpful. Let $x(n)$, $v(n)$, and $a(n)$ be the values at the end of the $n$th interval. These values appear across the $n$th line of the spreadsheet. The initial values are $x(1) = x_0$ and $v(1) = v_0$. The equations in Steps 7 and 8 become

$$v(n + 1) = v(n) + a(n)h,$$

$$x(n + 1) = x(n) + v(n + 1)h.$$

The value of $a(n)$ is computed by substituting the values $x(n)$ and $v(n)$ into the equation in Step 6.

## SUMMARY

• Periodic motion is motion that repeats itself in a definite cycle. It occurs whenever a body has a stable equilibrium position and a restoring force that acts when it is displaced from equilibrium. Period $T$ is the time for one cycle. Frequency $f$ is the number of cycles per unit time:

$$f = \frac{1}{T}, \qquad T = \frac{1}{f}. \tag{13–3}$$

Angular frequency $\omega$ is

$$\omega = 2\pi f = \frac{2\pi}{T}. \tag{13–4}$$

• If the restoring force $F$ is directly proportional to the displacement $x$, the motion is called simple harmonic motion (SHM). In that case,

$$F = -kx, \tag{13–1}$$

$$a = \frac{F}{m} = -\frac{k}{m}x. \tag{13–2}$$

• Conservation of energy in SHM leads to the relations

$$E = \tfrac{1}{2}mv^2 + \tfrac{1}{2}kx^2 = \text{constant}, \tag{13–5}$$

$$v = \pm\sqrt{\frac{k}{m}}\sqrt{A^2 - x^2}. \tag{13–7}$$

• The position $x$ is given as a function of time $t$ by

$$x = A\cos\left(\sqrt{\frac{k}{m}}t + \phi\right). \tag{13–9}$$

The constant $\phi$ is called the phase angle. When $\phi = 0$, the body starts from its point of maximum positive displacement. When $\phi = \pi/2$, the body starts at the origin with an initial velocity. The period, frequency, and angular frequency in SHM are

$$T = 2\pi\sqrt{\frac{m}{k}}, \tag{13–10}$$

$$f = \frac{1}{T} = \frac{1}{2\pi}\sqrt{\frac{k}{m}}, \tag{13–11}$$

$$\omega = 2\pi f = \sqrt{\frac{k}{m}}. \tag{13–12}$$

• In angular simple harmonic motion the frequency and angular frequency are related to the moment of inertia $I$ and the torsion constant $\kappa$ by

$$f = \frac{1}{2\pi}\sqrt{\frac{\kappa}{I}} \quad \text{and} \quad \omega = \sqrt{\frac{\kappa}{I}}. \tag{13–26}$$

• The circle of reference construction uses a rotating vector called a phasor, having a length equal to the amplitude of the motion. Its projection on the horizontal axis represents the actual motion of the body.

• A simple pendulum consists of a point mass $m$ at the end of a string of length $L$. Its motion is approximately simple harmonic for sufficiently small amplitude; the angular frequency, frequency, and period are then given by

$$\omega = \sqrt{\frac{g}{L}}, \tag{13–29}$$

$$f = \frac{\omega}{2\pi} = \frac{1}{2\pi}\sqrt{\frac{g}{L}}, \tag{13–30}$$

$$T = \frac{2\pi}{\omega} = \frac{1}{f} = 2\pi\sqrt{\frac{L}{g}}. \tag{13–31}$$

These quantities are independent of $m$.

## KEY TERMS

periodic motion
oscillation
restoring force
simple harmonic motion (SHM)
amplitude
cycle
period
frequency
angular frequency
phase angle
circle of reference
phasor
simple pendulum
physical pendulum
damping
damped oscillation
critical damping
overdamping
underdamping
driving force
forced oscillation
resonance
phase space

• A physical pendulum is a body suspended from an axis of rotation a distance $d$ from its center of gravity. If the moment of inertia about the axis of rotation is $I$, the angular frequency and period are

$$\omega = \sqrt{\frac{mgd}{I}}, \tag{13-35}$$

$$T = 2\pi\sqrt{\frac{I}{mgd}}. \tag{13-36}$$

• When a damping force $F = -bv$ proportional to velocity is added to a simple harmonic oscillator, the motion is called a damped oscillation:

$$x = Ae^{-(b/2m)t} \cos \omega' t. \tag{13-39}$$

The angular frequency of oscillation $\omega'$ is given by

$$\omega' = \sqrt{\frac{k}{m} - \frac{b^2}{4m^2}}. \tag{13-40}$$

This motion occurs when $b^2 < 4km$, a condition called underdamping. When $b^2 = 4km$, the system is critically damped and no longer oscillates. When $b$ is still larger, the system is overdamped.

• When a sinusoidally varying driving force is added to a damped harmonic oscillator, the resulting motion is called a forced oscillation. The amplitude is given as a function of driving frequency $\omega_d$ by

$$A = \frac{F_{max}}{\sqrt{(k - m\omega_d^2)^2 + b^2\omega_d^2}}. \tag{13-43}$$

The amplitude reaches a peak at a driving frequency close to the natural oscillation frequency of the system. This behavior is called resonance.

## EXERCISES

### Section 13–1 Basic Concepts

**13–1** A vibrating object goes through four complete vibrations in 1.00 s. Find the angular frequency and the period of the motion.

**13–2** The glider in Fig. 13–1 is displaced 0.160 m from its equilibrium position and released with zero initial speed. After 1.10 s its displacement is found to be 0.160 m on the opposite side, and it has passed the equilibrium position once during this interval. Find a) the amplitude; b) the period; c) the frequency.

**13–3** a) A force of 36.0 N is required to displace the end of a spring 0.120 m. What is the force constant $k$ of the spring? b) A spring has force constant $k = 1500$ N/m. What force is required to displace the end of the spring 0.060 m?

### Section 13–2 Energy in Simple Harmonic Motion

**13–4** A spring with force constant $k = 600$ N/m is mounted as in Fig. 13–1. A 0.400-kg glider attached to the end is undergoing simple harmonic motion with an amplitude 0.075 m. There is no friction force on the glider. Compute a) the maximum speed of the glider; b) the speed of the glider when it is at $x = 0.030$ m; c) the magnitude of the maximum acceleration of the glider; d) the acceleration of the glider when it is at $x = 0.030$ m; e) the total mechanical energy of the glider at any point in its motion.

**13–5** A 0.500-kg object is undergoing simple harmonic motion on the end of a horizontal spring with force constant $k = 300$ N/m. When the object is 0.012 m from its equilibrium position, it is observed to have a speed of 0.300 m/s. What is a) the total energy of the object at any point of its motion? b) the amplitude of the motion? c) the maximum speed attained by the object during its motion?

**13–6** A 0.400-kg object is undergoing simple harmonic motion with amplitude 0.025 m on the end of a horizontal spring. The maximum acceleration of the object is observed to have magnitude 6.00 m/s$^2$. What is a) the force constant of the spring? b) the maximum speed of the object? c) the acceleration (magnitude and direction) of the object when it is displaced 0.012 m to the left of its equilibrium position?

**13–7** A harmonic oscillator has a mass of 0.800 kg and a spring with force constant 140 N/m. Find a) the period; b) the frequency; c) the angular frequency.

**13–8** A harmonic oscillator is made by using a 0.400-kg block and a spring of unknown force constant. The oscillator is found to have a period of 0.200 s. Find the force constant of the spring.

**13–9** A glider of unknown mass is attached to a spring with force constant 200 N/m in the arrangement shown in Fig. 13–1. It is found to vibrate with a frequency of 4.00 Hz. Find a) the period; b) the angular frequency; c) the mass of the glider.

**13–10** A tuning fork labeled 440 Hz has the tip of each of its two prongs vibrating with an amplitude of 0.90 mm. a) What is the maximum speed of the tip of a prong? b) What is the maximum acceleration of the tip of a prong?

## Section 13–3 Equations of Simple Harmonic Motion

**13–11** Derive Eq. (13–25) a) from Eqs. (13–20) and (13–23); b) from conservation of energy, Eq. (13–6).

**13–12** A 3.00-kg block is attached to a spring with force constant $k = 150$ N/m. The block is given an initial velocity in the positive direction of magnitude $v_0 = 12.0$ m/s and no initial displacement ($x_0 = 0$). Find a) the amplitude; b) the phase angle; c) the total energy of the motion. d) Write an equation for the position as a function of time.

**13–13** Repeat Exercise 13–12, but assume that the block is given an initial velocity of $-6.00$ m/s and an initial displacement of $x_0 = +0.200$ m.

**13–14** The scale of a spring balance reading from 0 to 180 N is 9.00 cm long. A fish suspended from the balance is observed to oscillate vertically at 2.20 Hz. What is the mass of the fish? Neglect the mass of the spring.

**13–15** An object is vibrating with simple harmonic motion that has an amplitude of 18.0 cm and a frequency of 4.00 Hz. Compute a) the maximum magnitude of the acceleration and of the velocity; b) the acceleration and speed when the object's coordinate is $x = +9.0$ cm; c) the time required to move from the equilibrium position directly to a point 12.0 cm distant from it.

**13–16** A block with a mass of 1.50 kg is suspended from a spring having negligible mass and stretches the spring 0.200 m. a) What is the force constant of the spring? b) What is the period of oscillation of the block if it is pulled down and released?

**13–17** A block with a mass of 5.00 kg hangs from a spring. When displaced from equilibrium and released, the block oscillates with a period of 0.400 s. How much is the spring stretched when the block hangs in equilibrium (at rest)?

**13–18** A block with a mass of 4.00 kg is attached to a coil spring and oscillates vertically in simple harmonic motion. The amplitude is 0.250 m, and at the highest point of the motion the spring has its natural unstretched length. Calculate the elastic potential energy of the spring (take it to be zero for the unstretched spring), the kinetic energy of the block, the block's gravitational potential energy relative to the lowest point of the motion, and the sum of these three energies when the body is a) at its lowest point; b) at its equilibrium position; c) at its highest point.

**13–19** An alarm clock ticks four times each second, each tick representing half a period. The balance wheel consists of a thin rim with radius 0.65 cm, connected to the balance staff by thin spokes of negligible mass. The total mass of the balance wheel is 0.800 g. a) What is the moment of inertia of the balance wheel about its shaft? b) What is the torsion constant of the hairspring?

**13–20** The balance wheel of a watch vibrates with an angular amplitude $\pi/4$ rad and with a period of 0.500 s. a) Find its maximum angular velocity. b) Find its angular velocity when its displacement is one half its angular amplitude. c) Find its angular acceleration when its angular displacement is $\pi/8$ rad.

## Section 13–4 Circle of Reference

**13–21** An object is undergoing simple harmonic motion with period $\pi/2$ s and amplitude $A = 0.300$ m. At $t = 0$ the object is at $x = 0$. How far is the object from the equilibrium position when $t = \pi/10$ s?

**13–22** An object is undergoing simple harmonic motion with period $T = 0.800$ s and amplitude $A$. The object is initially at $x = 0$ and has velocity in the positive direction. Use the circle of reference to calculate the time it takes the object to go from $x = 0$ to $x = A/4$.

## Section 13–5 The Simple Pendulum

**13–23** Find the length of a simple pendulum that makes 100 complete swings in 75.0 s at a point where $g = 9.80$ m/s$^2$.

**13–24** A simple pendulum 4.00 m long swings with a maximum angular displacement of 0.400 rad. a) Compute the linear speed $v$ at the pendulum's lowest point. b) Compute its linear acceleration $a$ at the ends of its path.

**13–25 A Pendulum on the Moon.** A certain simple pendulum has a period on the earth of 1.20 s. What is its period on the surface of the moon, where $g = 1.62$ m/s$^2$?

## Section 13–6 The Physical Pendulum

**13–26** A 1.50-kg monkey wrench is pivoted at one end and allowed to swing as a physical pendulum. The period is 0.820 s, and the pivot is 0.200 m from the center of mass. a) What is the moment of inertia of the wrench about an axis through the pivot? b) If the wrench is initially displaced 0.600 rad from its equilibrium position, what is its angular velocity as it passes through the equilibrium position?

**13–27** A Christmas ornament in the shape of a solid sphere with mass $M = 0.015$ kg and radius $R = 0.050$ m is hung from a tree limb by a small loop of wire attached to the surface of the

FIGURE 13–29

sphere (Fig. 13–29). If the ornament is displaced a small distance and released, it swings back and forth as a physical pendulum. Calculate its period. (Neglect friction at the pivot. The moment of inertia of the sphere about the pivot at the tree limb is $7MR^2/5$.)

## Section 13–7 Damped Oscillations

**13–28** A 0.200-kg mass moving on the end of a spring with force constant $k = 320$ N/m has an initial displacement of 0.300 m. A damping force $F_x = -bv$ acts on the mass, and the amplitude of the motion decreases to 0.100 m in 5.00 s. Calculate the magnitude of the damping constant $b$.

**13–29** A 0.400-kg mass is moving on the end of a spring with force constant $k = 300$ N/m and is acted on by a damping force $F_x = -bv$. a) If the constant $b$ has the value 6.00 kg/s, what is the frequency of oscillation of the mass? b) For what value of the constant $b$ will the motion be critically damped?

## Section 13–9 Chaos: A Case Study in Dynamic Analysis

**13–30** a) For damped harmonic motion, use $v(t) = dx/dt$ and $x(t)$ given in Eq. (13–39) to calculate $v(t)$. b) Calculate $x$ and $v$ for $t = 0$, $\pi/2\omega'$, $\pi/\omega'$, $3\pi/2\omega'$, $2\pi/\omega'$, $3\pi/\omega'$, and $4\pi/\omega'$. Put these points on a phase-space plot, and show that the plot has the form of Fig. 13–23.

**13–31** Show that $U(x)$ given by Eq. (13–47) has two minima, one at $x = 0$ and the other at $x = L$.

**13–32** Derive Eq. (13–48) from Eq. (13–47). Show that when $x << L$, Eq. (13–48) reduces to $F = -kx$, and that for $x$ near $L$, it reduces to $F = -k(x - L)$.

**13–33** For $U(x)$ given by Eq. (13–47), show that $x = L/2$ is a point of unstable equilibrium.

# PROBLEMS

**13–34** a) A block suspended from a spring vibrates with simple harmonic motion. At an instant when the displacement of the block is equal to one-half the amplitude, what fraction of the total energy of the system is kinetic and what fraction is potential? Assume that $U = 0$ at equilibrium. b) When the block is in equilibrium, the length of the spring is an amount $s$ greater than in the unstretched state. Prove that $T = 2\pi\sqrt{s/g}$.

**13–35 SHM in a Car Engine.** The motion of the piston of an automobile engine (Fig. 13–30) is approximately simple harmonic. a) If the stroke of an engine (twice the amplitude) is 0.100 m and the engine runs at 2500 rev/min, compute the acceleration of the piston at the end point of its stroke. b) If the piston has a mass of 0.400 kg, what resultant force must be exerted on it at this point? c) What is the speed of the piston, in meters per second, at the mid point of its stroke?

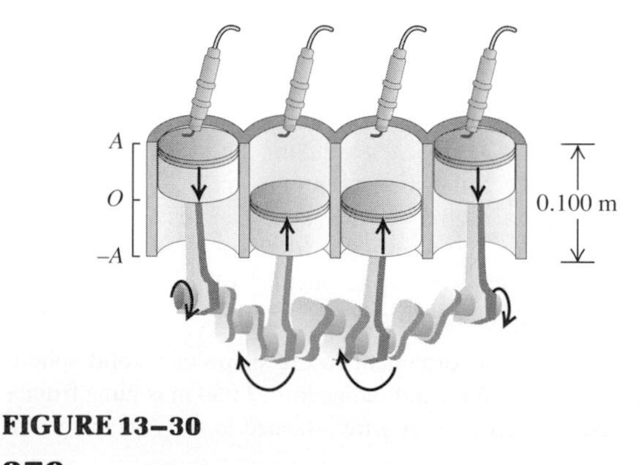


FIGURE 13–30

**13–36** Four passengers whose combined mass is 350 kg compress the springs of a car with worn-out shock absorbers by 5.00 cm when they enter it. If the total load supported by the springs is 1200 kg, find the period of vibration of the loaded car.

**13–37** A block is executing simple harmonic motion on a frictionless horizontal surface with an amplitude of 0.100 m. At a point 0.060 m away from equilibrium the speed of the block is 0.320 m/s. a) What is the period? b) What is the displacement when the speed is 0.120 m/s? c) A small object whose mass is much less than the mass of the block is placed on the oscillating block. If the small object is just on the verge of slipping at the end point of the path, what is the coefficient of static friction between the small object and the block?

**13–38** An object with a mass of 0.200 kg is acted on by an elastic restoring force with force constant $k = 25.0$ N/m. a) Construct the graph of elastic potential energy $U$ as a function of displacement $x$ over a range of $x$ from $-0.300$ m to $+0.300$ m. Let 1 cm = 0.1 J vertically and 1 cm = 0.05 m horizontally. The object is set into oscillation with an initial potential energy of 0.500 J and an initial kinetic energy of 0.200 J. Answer the following questions by referring to the graph. b) What is the amplitude of oscillation? c) What is the potential energy when the displacement is one-half the amplitude? d) At what displacement are the kinetic and potential energies equal? e) What is the phase angle $\phi$ if the initial velocity $v_0$ is negative and the initial displacement $x_0$ is positive?

**13–39** As we discussed in Chapter 11, a wire stretches when a tensile stress is applied to it. a) Compare Eqs. (13–1) and

(11–12) to derive an expression for the force constant $k$ of the wire in terms of the length $l_0$, cross-section area $A$, and Young's modulus $Y$. b) A copper wire has length 2.00 m and diameter 1.50 mm. What is the force constant $k$ for this wire?

**13–40** A 100-kg mass suspended from a wire whose unstretched length $l_0$ is 2.00 m is found to stretch the wire by $4.00 \times 10^{-3}$ m. The cross-section area of the wire, which can be assumed to be constant, is 0.100 $cm^2$. If the mass is pulled down a small additional distance and released, find the frequency at which it will vibrate.

**13–41** A rubber raft bobs up and down, executing simple harmonic motion due to the waves on a lake. The amplitude of the motion is 0.600 m, and the period is 3.50 s. A stable dock is next to the raft at a level equal to the highest level of the raft. People wish to step off the raft onto the dock but can do so comfortably only if the level of the raft is within 0.300 m of the dock level. How much time do the people have to get off comfortably during each period of the simple harmonic motion?

**13–42** A 0.0100-kg object moves with simple harmonic motion that has an amplitude of 0.240 m and a period of 3.00 s. The $x$-coordinate of the object is +0.240 m when $t = 0$. Compute a) the $x$-coordinate of the object when $t = 0.500$ s; b) the magnitude and direction of the force acting on the object when $t = 0.500$ s; c) the minimum time required for the object to move from its initial position to the point where $x = -0.120$ m; d) the speed of the object when $x = -0.120$ m.

**13–43** A 0.100-kg object hangs from a long spiral spring. When the object is pulled down 0.100 m below its equilibrium position and released, it vibrates with a period of 1.80 s. a) What is its speed as it passes through the equilibrium position? b) What is its acceleration when it is 0.050 m above the equilibrium position? c) When it is moving upward, how much time is required for it to move from a point 0.050 m below its equilibrium position to a point 0.050 m above it? d) The motion of the object is stopped, and then the object is removed from the spring. How much does the spring shorten?

**13–44** A 30.0-N force stretches a vertical spring 0.200 m. a) What mass must be suspended from the spring so that the system will oscillate with a period of $\pi/4$ s? b) If the amplitude of the motion is 0.050 m and the period is that specified in part (a), where is the object and in what direction is it moving $\pi/12$ s after it has passed the equilibrium position, moving downward? c) What force (magnitude and direction) does the spring exert on the object when it is 0.030 m below the equilibrium position, moving upward?

**13–45 SHM of a Butcher's Scale.** A spring with force constant $k = 400$ N/m is hung vertically, and a 0.200-kg pan is suspended from its lower end. A butcher drops a 1.8-kg steak onto the pan from a height of 0.40 m; the steak makes a totally inelastic collision with the pan and sets the system into vertical simple harmonic motion. What is a) the speed of the pan and steak immediately after the collision? b) the amplitude of the subsequent motion? c) the period of that motion?

**13–46 Journey to the Center of the Earth, continued.** A very interesting though impractical example of simple harmonic motion occurs in the motion of an object dropped down a hole that extends from one side of the earth, through its center, to the other side. Prove that the motion is simple harmonic and find the period. [*Note:* The gravitational force on the object as a function of the object's distance $r$ from the center of the earth was derived in Section 12–7, and the result is given in Eq. (12–31). The motion is simple harmonic if the acceleration $a$ and the displacement from equilibrium $x$ are related by Eq. (13–15), and the period is then $T = 2\pi/\omega$.]

**13–47** A 0.50-kg block sits on top of a 5.00-kg block that rests on the floor. The larger block is attached to a horizontal spring that has a force constant of 15.0 N/m. It is displaced and undergoes simple harmonic motion. What is the largest amplitude the 5.00-kg mass can have for the smaller mass to remain at rest relative to the larger block? The coefficient of static friction between the two blocks is 0.25. There is no friction between the larger block and the floor.

**13–48** Consider the circle of reference shown in Fig. 13–13. a) The horizontal component of the velocity of $Q$ is the velocity of $P$. Compute this component and show that the velocity of $P$ is as given by Eq. (13–21). b) The horizontal component of the acceleration of $Q$ is the acceleration of $P$. Compute this component and show that the acceleration of $P$ is as given by Eq. (13–22).

**13–49** To measure $g$ in an unorthodox manner, a student places a ball bearing on the concave side of a lens (Fig. 13–31). She attaches the lens to a simple harmonic oscillator (actually a small stereo speaker) whose amplitude is $A$ and whose frequency $f$ can be varied. She can measure both $A$ and $f$ with a strobe light. a) If the ball bearing has a mass $m$, find the normal force exerted by the lens on the ball bearing as a function of time. Your result should be in terms of $A$, $f$, $m$, $g$, and a phase angle $\phi$. b) The frequency is slowly increased. When it reaches a value $f_b$, the ball bearing is heard to bounce. What is $g$ in terms of $A$ and $f_b$?

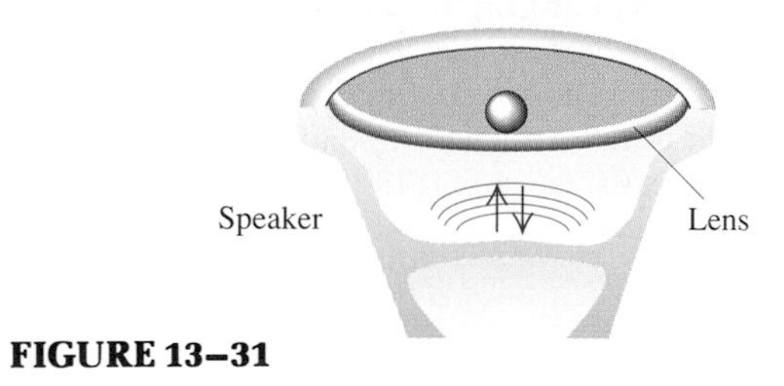


**FIGURE 13–31**

**13–50** A block with mass $m_1$ attached to a horizontal spring with force constant $k$ is moving with simple harmonic motion having amplitude $A$. At the instant when the block passes through its equilibrium position, a lump of putty with mass $m_2$ is dropped vertically onto the block from a very small height and sticks to it. a) Find the new period and amplitude. b) Was there a loss of mechanical energy? If so, where did it go? Calculate the ratio between the final and the initial mechanical energy. c) What are the answers to part (a) if the putty is dropped on the block when it is at one end of its path?

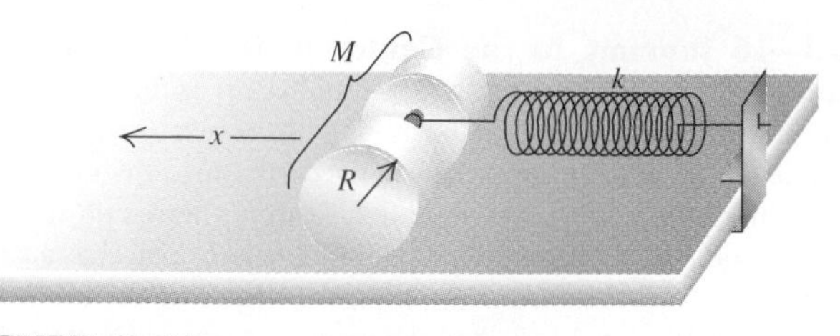


**FIGURE 13–32**

**13–51** A solid cylinder with mass $M$ and radius $R$ rests on a horizontal tabletop. A spring with force constant $k$ has one end attached to a clamp and the other end attached to a frictionless pivot at the center of mass of the cylinder (Fig. 13–32). (The spring is attached through a narrow slot cut in the cylinder, so the cylinder can still turn freely about its axis.) The cylinder is pulled to the left a distance $x$, which stretches the spring, and released. There is sufficient friction between the tabletop and the cylinder for the cylinder to roll without slipping as it moves back and forth on the end of the spring. Show that the motion of the center of mass of the cylinder is simple harmonic, and calculate its period in terms of $M$ and $k$. [*Hint:* The motion is simple harmonic if $a$ and $x$ are related by Eq. (13–15), and the period then is $T = 2\pi/\omega$. Apply $\Sigma\tau = I_{cm}\alpha$ and $\Sigma F = Ma_{cm}$ to the cylinder in order to relate $a_{cm}$ and the displacement $x$ of the cylinder from its equilibrium position.]

**13–52** A slender, uniform metal rod with mass $M$ is pivoted without friction about an axis through its midpoint and perpendicular to the rod. A horizontal spring with force constant $k$ is attached to the lower end of the rod, with the other end of the spring attached to a rigid support. If the rod is displaced by a small angle $\theta$ from the vertical (Fig. 13–33) and released, show that it moves in angular simple harmonic motion and calculate the period. (*Hint:* Assume that the angle $\theta$ is small enough for the approximations $\sin\theta \approx \theta$ and $\cos\theta \approx 1$ to be valid. The motion is simple harmonic if $d^2\theta/dt^2 = -\omega^2\theta$, and the period is then $T = 2\pi/\omega$.)

**13–53 The Silently Ringing Bell Problem.** A large bell is hung from a wooden beam so that it can swing back and forth with negligible friction. The center of mass of the bell is 0.60 m below the pivot, the bell's mass is 34.0 kg, and the moment of inertia of the bell for an axis at the pivot is 18.0 kg · m$^2$. The clapper is a small 1.8-kg mass attached to one end of a slender rod that has length $L$ and negligible mass. The other end of the rod is attached to the inside of the bell so that it can swing freely about the same axis as the bell (Fig. 13–34). What should be the length $L$ of the clapper rod for the bell to ring silently, that is, for the period of oscillation for the bell to equal that for the clapper?

**FIGURE 13–33**

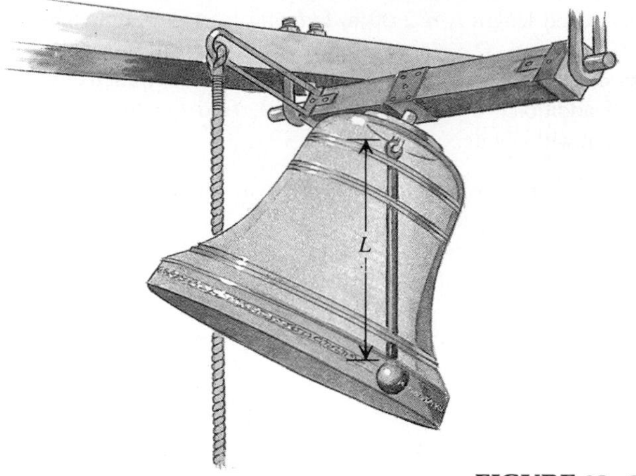

**FIGURE 13–34**

**13–54** Show that $x(t)$ as given in Eq. (13–39) is a solution of Newton's second law for damped oscillations (Eq. 13–38) if $\omega'$ is as defined in Eq. (13–40).

**13–55 A Three-second Pendulum.** You want to construct a pendulum with a period of 3.00 s. a) What is the length of a *simple* pendulum having this period? b) Suppose the pendulum must be mounted in a case that is not over 0.50 m high. Can you devise a pendulum having a period of 3.00 s that will satisfy this requirement?

***13–56** A solid disk with radius $R = 16.0$ cm oscillates as a physical pendulum about an axis perpendicular to the plane of the disk at a distance $r$ from its center (Fig. 13–35). a) Calculate the period of oscillation (for small amplitudes) for the following values of $r$: $R/4$, $R/2$, $3R/4$, $R$, and $3R/2$. (*Hint:* Use the parallel-axis theorem of Section 9–6.) b) Let $T_0$ represent the period when $r = R$ and $T$ the period at any other value of $r$. Construct a graph of the dimensionless ratio $T/T_0$ as a function of the dimensionless ratio $r/R$. (Note that the graph then describes the behavior of *any* solid disk, whatever its radius.) c) Find the value of $r/R$ that minimizes the period, and calculate the minimum value of the period.

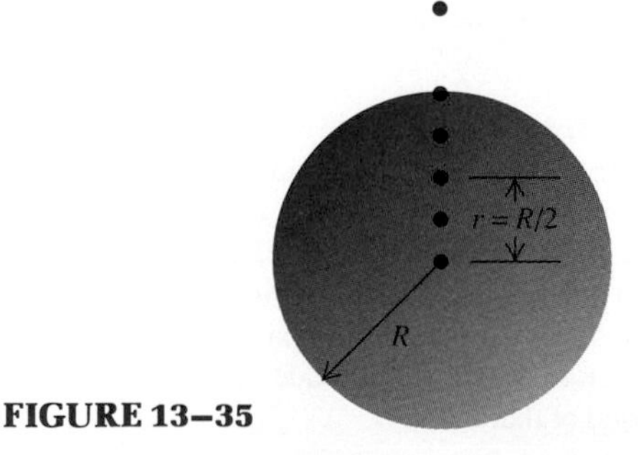


**FIGURE 13–35**

## CHALLENGE PROBLEMS

**13–57** a) What is the change $\Delta T$ in the period of a simple pendulum when the acceleration of gravity $g$ changes by $\Delta g$? [*Hint:* The new period $T + \Delta T$ is obtained by substituting $g + \Delta g$ for $g$:

$$T + \Delta T = 2\pi\sqrt{L/(g + \Delta g)}.$$

To obtain an approximate expression, expand the factor $(g + \Delta g)^{-1/2}$, using the binomial theorem (Appendix B) and keeping only the first two terms:

$$(g + \Delta g)^{-1/2} = g^{-1/2} - \tfrac{1}{2}g^{-3/2}\,\Delta g + \cdots.$$

The other terms contain higher powers of $\Delta g$ and are very small if $\Delta g$ is small.] b) Find the *fractional* change in period $\Delta T/T$ in terms of the fractional change $\Delta g/g$. c) A pendulum clock, which keeps correct time at a point where $g = 9.800\ \text{m/s}^2$, is found to lose 10 s each day at a higher elevation. Use the result of part (b) to find approximately the value of $g$ at this new location.

**13–58 The Effective Force Constant of Two Springs.** Two springs with the same unstretched length but different force constants $k_1$ and $k_2$ are attached to a block of mass $m$ on a level, frictionless surface. Calculate the effective force constant $k_{\text{eff}}$ in each of the three cases (a), (b), and (c) depicted in Fig. 13–36. (The effective force constant is defined by $\Sigma F = -k_{\text{eff}}x$.) d) An object with mass $m$, suspended from a spring with a force constant $k$, vibrates with a frequency $f_1$. When the spring is cut in half and the same object is suspended from one of the halves, the frequency is $f_2$. What is the ratio $f_2/f_1$?

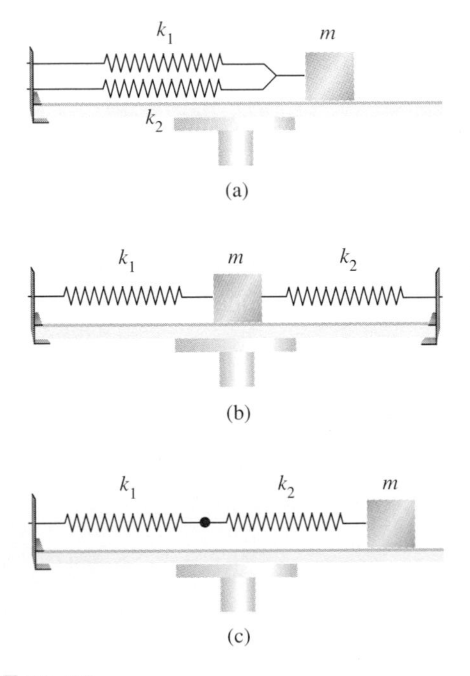


**FIGURE 13–36**

**13–59** Two springs, each with unstretched length 0.200 m but having different force constants $k_1$ and $k_2$, are attached to opposite ends of a block with mass $m$ on a level, frictionless surface. The outer ends of the springs are now attached to two pins $P_1$ and $P_2$, 0.100 m from the original positions of the ends of the springs. Let

$$k_1 = 1.00\ \text{N/m}, \qquad k_2 = 3.00\ \text{N/m}, \qquad m = 0.100\ \text{kg}.$$

(See Fig. 13–37.) a) Find the length of each spring when the block is in its new equilibrium position after the springs have been attached to the pins. b) Find the period of vibration of the block if it is slightly displaced from its new equilibrium position and released.

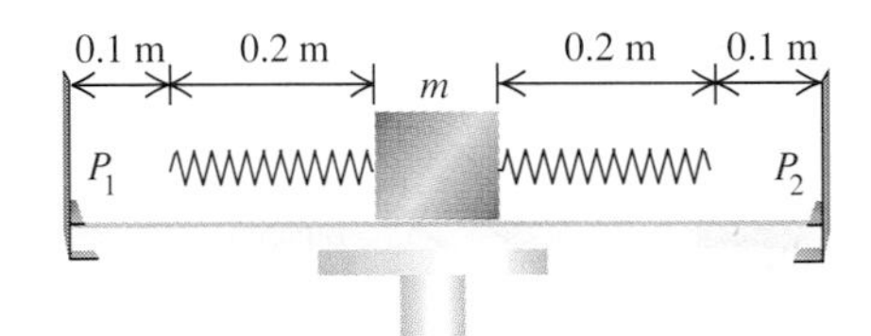


**FIGURE 13–37**

**13–60 A Spring with Mass.** All the previous problems in this chapter have assumed that the springs had negligible mass. But of course no spring is completely massless. To find the effect of the spring's mass, consider a spring of mass $M$, equilibrium length $L_0$, and spring constant $k$. When stretched or compressed to a length $L$, the potential energy is $\frac{1}{2}kx^2$, where $x = L - L_0$. a) Consider a spring as described above that has one end fixed and the other end moving with speed $v$. Assume that the speed of points along the length of the spring varies linearly with distance $l$ from the fixed end. Assume also that the mass $M$ of the spring is distributed uniformly along the length of the spring. Calculate the kinetic energy of the spring in terms of $M$ and $v$. (*Hint:* Divide the spring into pieces of length $dl$; find the speed of each piece in terms of $l$, $v$, and $L$; find the mass of each piece in terms of $dl$, $M$, and $L$; and integrate from 0 to $L$. The result is *not* $\frac{1}{2}Mv^2$, since not all of the spring moves with the same speed.) b) Take the time derivative of the conservation of energy equation, Eq. (13–6), for a mass $m$ moving on the end of a *massless* spring. By comparing your results to Eq. (13–15), which defines $\omega$, show that the angular frequency of oscillation is $\omega = \sqrt{k/m}$. c) Apply the procedure of part (b) to obtain the angular frequency of oscillation $\omega$ of the spring considered in part (a). If the *effective mass* $M'$ of the spring is defined by $\omega = \sqrt{k/M'}$, what is $M'$ in terms of $M$?

**13–61** A meterstick hangs from a horizontal axis at one end and oscillates as a physical pendulum. An object of small dimensions and of mass equal to that of the meterstick can be clamped to the stick at a distance $y$ below the axis. Let $T$ represent the period of the system with the body attached and $T_0$ the period of the meterstick alone. a) Find the ratio $T/T_0$. Evaluate your expression for $y$ ranging from 0 to 1.0 m in steps of 0.1 m, and sketch a graph of $T/T_0$ versus $y$. b) Is there any value of $y$ for which $T = T_0$? If so, find it and explain why the period is unchanged when $y$ has this value.

## 14–1 DENSITY

The **density** of a material is defined as its mass per unit volume. A homogeneous material such as ice or iron has the same density throughout. The SI unit of density is the kilogram per cubic meter (1 $kg/m^3$). The cgs unit, the gram per cubic centimeter (1 $g/cm^3$), is also widely used. We use the Greek letter $\rho$ (rho) for density. If a mass $m$ of material has volume $V$, the density $\rho$ is

$$\rho = \frac{m}{V}. \tag{14–1}$$

Densities of several common solids and liquids at ordinary temperatures are given in Table 14–1. The conversion factor

$$1\ g/cm^3 = 1000\ kg/m^3$$

is useful. Note the wide range of magnitudes. The densest material found on earth is the metal osmium (22,500 $kg/m^3$, or 22.5 $g/cm^3$). The density of air is about 1.2 $kg/m^3$ (0.0012 $g/cm^3$), but the density of white-dwarf stars is of the order of $10^{10}$ $kg/m^3$, and that of neutron stars is of the order of $10^{18}$ $kg/m^3$!

The **specific gravity** of a material is the ratio of its density to the density of water; it is a pure (unitless) number. For example, the specific gravity of aluminum is 2.7. "Specific gravity" is a poor term, since it has nothing to do with gravity; "relative density" would be better.

Density measurements are an important analytical technique. For example, we can determine the charge condition of a storage battery by measuring the density of its electrolyte, a sulfuric acid solution. As the battery discharges, the sulfuric acid ($H_2SO_4$) combines with lead in the battery plates to form insoluble lead sulfate ($PbSO_4$), decreasing the concentration of the solution. The density decreases from about $1.30 \times 10^3$ $kg/m^3$ for a fully charged battery to $1.15 \times 10^3$ $kg/m^3$ for a discharged battery.

Similarly, permanent-type antifreeze is usually a solution of ethylene glycol (density $1.12 \times 10^3$ $kg/m^3$) in water. The glycol concentration determines the freezing point of the solution; it can be found from a simple density measurement. Both these measurements are performed routinely in service stations with the aid of a hydrometer, which measures density by observation of the level at which a calibrated body floats in a sample of the solution. The hydrometer is discussed in Section 14–3.

**TABLE 14–1 Densities of Some Common Substances**

| Material | Density ($kg/m^3$) | Material | Density ($kg/m^3$) |
|---|---|---|---|
| Air (1 atm, 20°C) | 1.20 | Iron | $7.8 \times 10^3$ |
| Aluminum | $2.7 \times 10^3$ | Lead | $11.3 \times 10^3$ |
| Benzene | $0.90 \times 10^3$ | Mercury | $13.6 \times 10^3$ |
| Blood | $1.06 \times 10^3$ | Neutron star | $10^{18}$ |
| Brass | $8.6 \times 10^3$ | Platinum | $21.4 \times 10^3$ |
| Concrete | $2 \times 10^3$ | Seawater | $1.03 \times 10^3$ |
| Copper | $8.9 \times 10^3$ | Silver | $10.5 \times 10^3$ |
| Ethanol | $0.81 \times 10^3$ | Steel | $7.8 \times 10^3$ |
| Glycerin | $1.26 \times 10^3$ | Water | $1.00 \times 10^3$ |
| Gold | $19.3 \times 10^3$ | White-dwarf star | $10^{10}$ |
| Ice | $0.92 \times 10^3$ | | |

To obtain the densities in grams per cubic centimeter, simply divide by $10^3$.

## EXAMPLE 14–1

**The weight of a roomful of air** Find the mass of air, and its weight, in a living room with a 4.0 m × 5.0 m floor and a ceiling 3.0 m high. What would be the mass and weight of an equal volume of water?

**SOLUTION** The volume of the room is $V = (3.0\text{ m})(4.0\text{ m}) \cdot (5.0\text{ m}) = 60\text{ m}^3$. The mass $m$ is given by Eq. (14–1):

$$m = \rho V = (1.2\text{ kg/m}^3)(60\text{ m}^3) = 72\text{ kg}.$$

The weight of the air is

$$w = mg = (72\text{ kg})(9.8\text{ m/s}^2) = 706\text{ N} = 160\text{ lb}.$$

Does it surprise you that a roomful of air weighs this much? The mass of an equal volume of water is

$$m = \rho V = (1000\text{ kg/m}^3)(60\text{ m}^3) = 6.0 \times 10^4\text{ kg}.$$

The weight is

$$\begin{aligned} w = mg &= (6.0 \times 10^4\text{ kg})(9.8\text{ m/s}^2) = 5.9 \times 10^5\text{ N} \\ &= 1.33 \times 10^5\text{ lb} = 66\text{ tons}. \end{aligned}$$

This much weight would certainly collapse the floor of an ordinary house.

## 14–2 PRESSURE IN A FLUID

When a fluid (either liquid or gas) is at rest, it exerts a force perpendicular to any surface in contact with it, such as a container wall or the surface of a body immersed in it. If we think of an imaginary surface *within* the fluid, the fluid on the two sides of the surface exerts equal and opposite forces on the surface. (Otherwise, the surface would accelerate, and the fluid would not remain at rest.)

Pressure is force per unit area. We define the **pressure** $p$ at a point in a fluid as the ratio of the normal force $dF_{\perp}$ on a small area $dA$ around that point to the area:

$$p = \frac{dF_{\perp}}{dA}. \tag{14–2}$$

If the pressure is the same at all points of a finite plane surface with area $A$, then

$$p = \frac{F_{\perp}}{A}. \tag{14–3}$$

The SI unit of pressure is 1 N/m$^2$; as we stated in Chapter 11, this is given a special name, the **pascal:**

$$1\text{ pascal} = 1\text{ Pa} = 1\text{ N/m}^2.$$

Two related units, used principally in meteorology, are the *bar,* equal to $10^5$ Pa, and the *millibar,* equal to 100 Pa.

**Atmospheric pressure** $p_a$ is the pressure of the earth's atmosphere, the pressure at the bottom of the sea of air in which we live. This pressure varies with weather changes and with elevation. *Normal* atmospheric pressure at sea level (an average value) is

$$\begin{aligned} (p_a)_{av} &= 1.013 \times 10^5\text{ Pa} = 14.7\text{ lb/in.}^2 = 1\text{ atm} \\ &= 1.013\text{ bar} = 1013\text{ millibar}. \end{aligned}$$

## EXAMPLE 14–2

In the room described in Example 14–1, what is the total force on the floor surface due to air pressure of 1.00 atm?

**SOLUTION** The floor area is $A = (4.0\text{ m})(5.0\text{ m}) = 20\text{ m}^2$. From Eq. (14–3) the total force is

$$\begin{aligned} F = pA &= (1.013 \times 10^5\text{ N/m}^2)(20\text{ m}^2) \\ &= 2.0 \times 10^6\text{ N} = 4.5 \times 10^5\text{ lb} = 225\text{ tons}. \end{aligned}$$

As in Example 14–1, this is more than enough force to collapse the floor. So why doesn't it collapse? Because there is an upward force on the underside of the floor. If we neglect the thickness of the floor, this upward force is exactly equal to the downward force on the top surface, and the total force due to air pressure is zero.

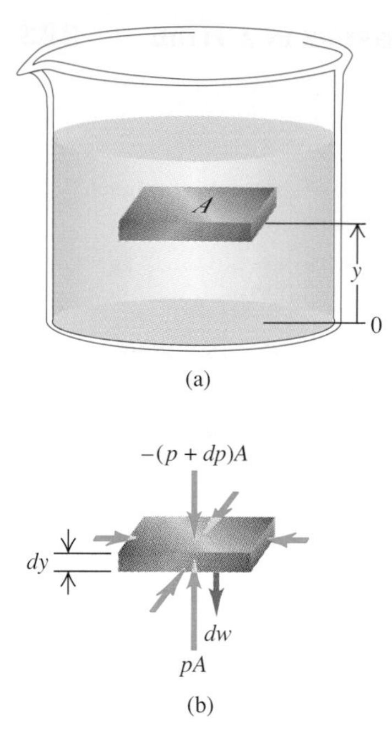


**14–1** The forces on an element of fluid in equilibrium.

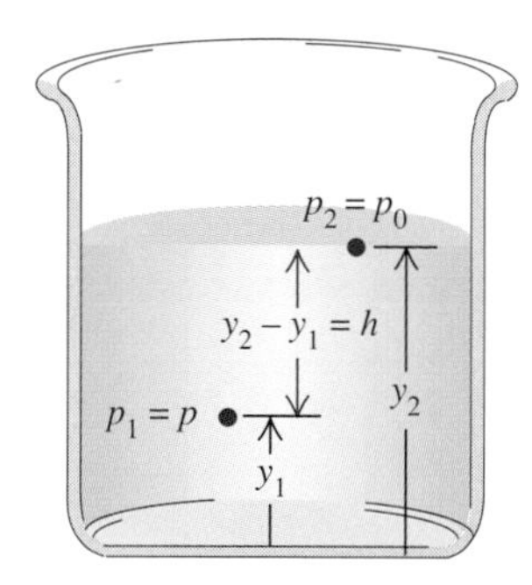


**14–2** The pressure at a depth $h$ in a liquid is greater than the surface pressure $p_0$ by $\rho gh$.

**14–3** The pressure in a fluid is the same at all points having the same elevation. The shape of the container does not affect the pressure.

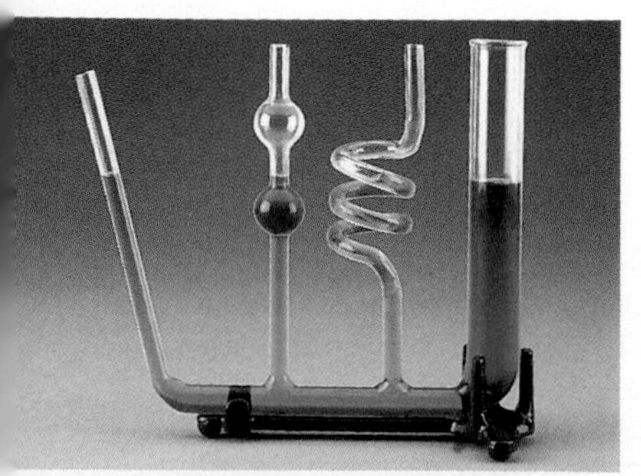

## Pascal's Law

If the weight of the fluid can be neglected, the pressure in a fluid is the same throughout. We used that approximation in our discussion of bulk stress and strain in Section 11–6. But often the fluid's weight is *not* negligible. Atmospheric pressure is greater at sea level than in high mountains, and an airplane cabin has to be pressurized when the plane flies at 35,000 feet. When you dive into deep water, your ears tell you that the pressure increases rapidly with increasing depth below the surface.

We can derive a general relation between the pressure $p$ at any point in a fluid and the elevation $y$ of the point. If the fluid is in equilibrium, every volume element is in equilibrium. For the present, let's assume that the density $\rho$ and the acceleration due to gravity $g$ are the same throughout the fluid. Consider a thin element of fluid, with height $dy$ (Fig. 14–1). The bottom and top surfaces each have area $A$, and they are at elevations $y$ and $y + dy$ above some reference level where $y = 0$. The volume of the fluid is $dV = A\,dy$, its mass is $dm = \rho\,dV = \rho A\,dy$, and its weight is $w = dm\,g = \rho g A\,dy$.

What are the other forces on this fluid element? Call the pressure at the bottom surface $p$; then the total $y$-component of upward force on this surface is $pA$. The pressure at the top surface is $p + dp$, and the total $y$-component of (downward) force on the top surface is $-(p + dp)A$. The fluid in this volume is in equilibrium, so the total $y$-component of force, including the weight and the forces at the bottom and top surfaces, must be zero:

$$\Sigma F_y = 0, \qquad pA - (p + dp)A - \rho g A\,dy = 0.$$

When we divide out the area $A$ and rearrange, we get

$$\frac{dp}{dy} = -\rho g. \tag{14–4}$$

This equation shows that when $y$ increases, $p$ decreases, as we expect. If $p_1$ and $p_2$ are the pressures at elevations $y_1$ and $y_2$ and if $\rho$ and $g$ are constant, then

$$p_2 - p_1 = -\rho g(y_2 - y_1). \tag{14–5}$$

Let's apply this equation to a liquid in a container (Fig. 14–2). Take point 1 at any level and let $p$ represent the pressure at this point. Take point 2 at the *surface* of the liquid, where the pressure is $p_0$ (subscript zero for zero depth). Then

$$p_0 - p = -\rho g(y_2 - y_1),$$
$$p = p_0 + \rho g(y_2 - y_1) = p_0 + \rho g h, \tag{14–6}$$

where $h$ is the height of the surface above point 1 or the depth of point 1 below the surface. The pressure $p$ at a depth $h$ is greater than the pressure $p_0$ at the surface by an amount $\rho g h$.

Equation (14–6) also shows that if we increase the pressure $p_0$ at the top surface, possibly by inserting a piston that fits tightly inside the container and using the piston to push down on the fluid surface, the pressure $p$ at any depth increases by exactly the same amount. This fact was recognized in 1653 by the French scientist Blaise Pascal (1623–1662) and is called **Pascal's law: Pressure applied to an enclosed fluid is transmitted undiminished to every portion of the fluid and the walls of the containing vessel.** The pressure is the same in every direction, and it depends only on depth. The *shape* of the container does not matter (Fig. 14–3).

The hydraulic lift shown schematically in Fig. 14–4 illustrates Pascal's law. A piston with small cross-section area $A_1$ exerts a force $F_1$ on the surface of a liquid, such as oil. The applied pressure $p = F_1/A_1$ is transmitted through the connecting pipe to a larger

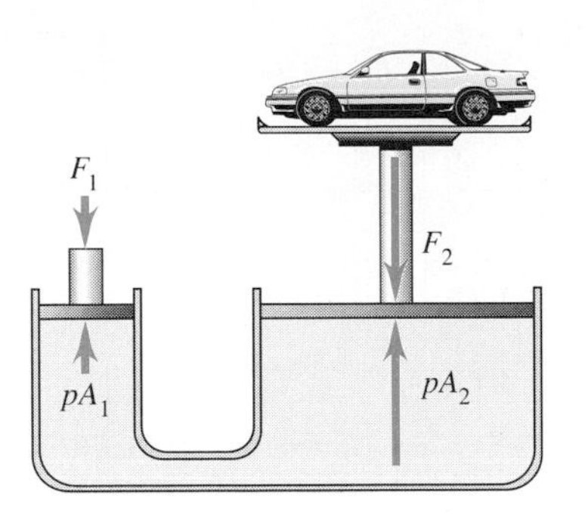


**14–4** Principle of the hydraulic lift, an application of Pascal's law.

piston of area $A_2$. The applied pressure is the same in both cylinders, so

$$p = \frac{F_1}{A_1} = \frac{F_2}{A_2} \quad \text{and} \quad F_2 = \frac{A_2}{A_1}F_1. \tag{14–7}$$

The hydraulic lift is a force-multiplying device with a multiplication factor equal to the ratio of the areas of the two pistons. Dentist chairs, hydraulic jacks, many elevators, and hydraulic brakes all use this principle.

The assumption that the density $\rho$ of a fluid is constant is often reasonable for liquids, which are relatively incompressible, but it is realistic for gases only over short vertical distances. In a room with a ceiling height of 3.0 m filled with air of uniform density 1.2 kg/m$^3$, the difference in pressure between floor and ceiling, given by Eq. (14–6), is

$$\rho gh = (1.2\ \text{kg/m}^3)(9.8\ \text{m/s}^2)(3.0\ \text{m}) = 35\ \text{Pa},$$

or about 0.00035 atm, a very small difference. But between sea level and the summit of Mount Everest (8882 m) the density of air changes by a factor of nearly three, and in this case we cannot use Eq. (14–6). At great enough pressures, even liquids are compressed appreciably. The deepest point in the oceans is the Marianas Trench, which is 10,920 m deep. If the density of seawater were constant, the pressure at this depth would be $1.10 \times 10^8$ Pa. But because seawater is compressed to greater density under high pressures, the actual pressure at this depth is $1.16 \times 10^8$ Pa (nearly 17,000 lb/in.$^2$).

If the pressure inside a car tire is equal to atmospheric pressure, the tire is flat. The pressure has to be *greater than* atmospheric to support the car, and the significant quantity is the *difference* between the inside and outside pressures. When we say that the pressure in a car tire is "32 pounds" (actually 32 lb/in.$^2$, equal to 220 kPa or $2.2 \times 10^5$ Pa), we mean that it is *greater* than atmospheric pressure (14.7 lb/in.$^2$ or $1.01 \times 10^5$ Pa) by this amount. The *total* pressure in the tire is 47 lb/in.$^2$, or 320 kPa. The excess pressure above atmospheric pressure is usually called **gauge pressure,** and the total pressure is called **absolute pressure.** Engineers use the abbreviations psig and psia for "pounds per square inch gauge" and "pounds per square inch absolute," respectively. If the pressure is *less than* atmospheric, as in a partial vacuum, the gauge pressure is negative.

## EXAMPLE 14–3

**Finding absolute and gauge pressures** A solar water-heating system uses solar panels on the roof, 12.0 m above the storage tank. The pressure at the level of the panels is one atmosphere. What is the absolute pressure in the tank? The gauge pressure?

**SOLUTION** From Eq. (14–6) the absolute pressure is

$$\begin{aligned} p &= p_0 + \rho gh \\ &= (1.01 \times 10^5\ \text{Pa}) + (1000\ \text{kg/m}^3)(9.8\ \text{m/s}^2)(12\ \text{m}) \\ &= 2.19 \times 10^5\ \text{Pa} = 2.16\ \text{atm} = 31.8\ \text{lb/in.}^2. \end{aligned}$$

The gauge pressure is

$$\begin{aligned} p - p_0 &= (2.19 - 1.01) \times 10^5\ \text{Pa} = 1.18 \times 10^5\ \text{Pa} \\ &= 1.16\ \text{atm} = 17.1\ \text{lb/in.}^2. \end{aligned}$$

If the tank has a pressure gauge, it is usually calibrated to read gauge pressure rather than absolute pressure. As we have mentioned, the variation in *atmospheric* pressure over this height is negligibly small.

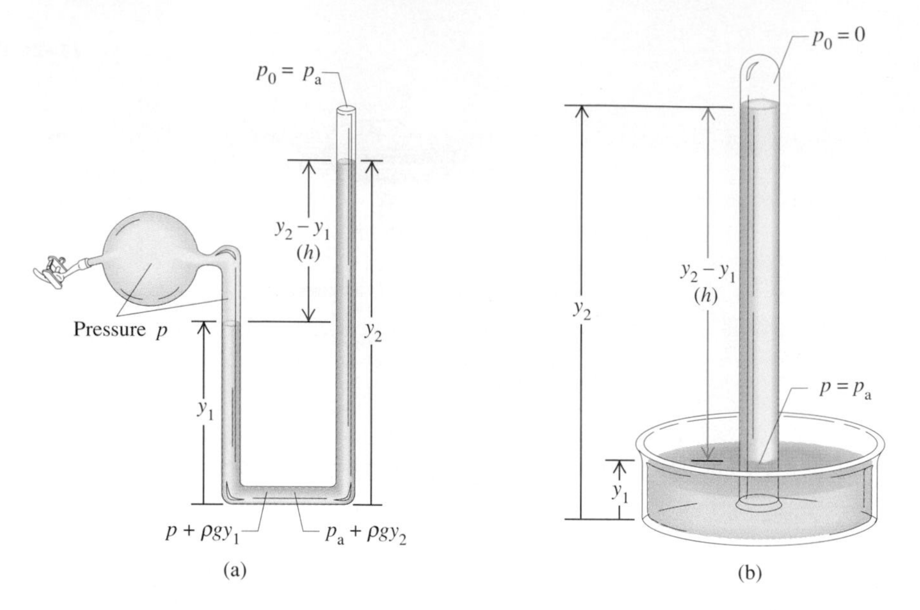


**14–5** Pressure gauges: (a) The open-tube manometer. (b) The mercury barometer.

## Pressure Gauges

The simplest pressure gauge is the open-tube manometer (Fig. 14–5a). The U-shaped tube contains a liquid, often mercury or water. One end of the tube is connected to the container where the pressure $p$ is to be measured, and the other end is open to the atmosphere at pressure $p_0 = p_a$. The pressure at the bottom of the left column is $p + \rho g y_1$, and the pressure at the bottom of the right column is $p_a + \rho g y_2$, where $\rho$ is the density of the liquid in the manometer. These are the same point, so these pressures must be equal:

$$p + \rho g y_1 = p_a + \rho g y_2, \qquad \text{and}$$

$$p - p_a = \rho g (y_2 - y_1) = \rho g h. \tag{14–8}$$

In Eq. (14–8), $p$ is the *absolute pressure,* and the difference $p - p_a$ between absolute and atmospheric pressure is the gauge pressure. Thus the gauge pressure is proportional to the difference in height $(y_2 - y_1)$ of the liquid columns.

Another common pressure gauge is the **mercury barometer.** It consists of a long glass tube, closed at one end, that has been filled with mercury and then inverted in a dish of mercury (Fig. 14–5b). The space above the mercury column contains only mercury vapor; its pressure is negligibly small, so the pressure $p_0$ at the top of the mercury column is practically zero. From Eq. (14–6),

$$p_a = p = 0 + \rho g (y_2 - y_1) = \rho g h. \tag{14–9}$$

Thus the mercury barometer reads the atmospheric pressure $p_a$ directly from the height of the mercury column.

In the past, pressures were commonly described in terms of the height of the corresponding mercury column, as so many "inches of mercury" or "millimeters of mercury" (abbreviated mm Hg). A pressure of 1 mm Hg was called *one torr,* after Evangelista Torricelli, inventor of the mercury barometer. These units depend on the density of mercury, which varies with temperature, and on the value of $g$, which varies with location, so the pascal is the preferred unit of pressure.

### EXAMPLE 14–4

Compute the atmospheric pressure $p_a$ on a day when the height of mercury in a barometer is 76.0 cm.

**SOLUTION** We use Eq. (14–9). Assuming that $g = 9.80\ \text{m/s}^2$ and $\rho = 13.6 \times 10^3\ \text{kg/m}^3$ for mercury, we find

$$p_a = \rho g h = (13.6 \times 10^3\ \text{kg/m}^3)(9.80\ \text{m/s}^2)(0.760\ \text{m})$$
$$= 101{,}300\ \text{N/m}^2 = 1.013 \times 10^5\ \text{Pa} = 1.013\ \text{bar}.$$

One common type of blood-pressure gauge, called a *sphygmomanometer,* uses a mercury-filled manometer. Blood-pressure readings, such as 130/80, refer to the maximum and minimum gauge pressures in the arteries, measured in mm Hg or torrs. Blood pressure varies with height; the standard reference point is the upper arm, level with the heart.

Many types of pressure gauges use a flexible sealed vessel (Fig. 14–6). A change in the pressure either inside or outside the vessel causes a change in its dimensions. This change is detected optically, electrically, or mechanically.

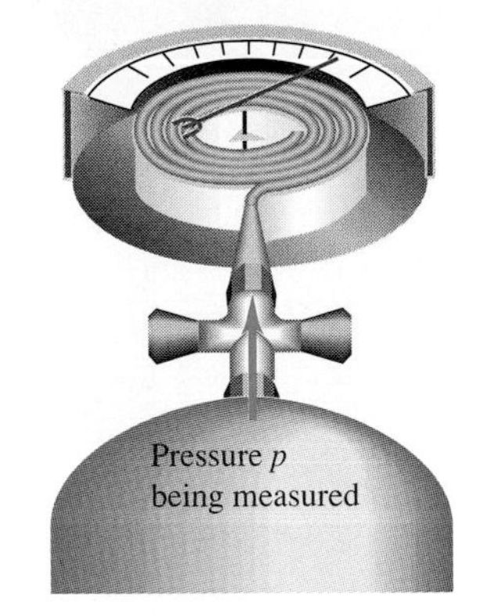


**14–6** A Bourdon pressure gauge. The spiral metal tube is attached to a pointer. When the pressure inside the tube decreases, the tube straightens out a little, deflecting the pointer on the scale.

## 14–3 BUOYANCY

**Buoyancy** is a familiar phenomenon: A body immersed in water seems to weigh less than when it is in air. When the body is less dense than the fluid, it floats. The human body usually floats in water, and a helium-filled balloon floats in air (Fig. 14–7).

**Archimedes' principle** states: **When a body is completely or partially immersed in a fluid, the fluid exerts an upward force on the body equal to the weight of the fluid displaced by the body.** To prove this principle, we consider an arbitrary portion of fluid at rest. In Fig. 14–8a the irregular outline is the surface bounding this portion of fluid. The arrows represent the forces exerted on the boundary surface by the surrounding fluid.

The entire fluid is in equilibrium. The sum of all the $y$-components of force on this portion of fluid is zero, so the sum of the $y$-components of the surface forces must be an upward force equal in magnitude to the weight $mg$ of the fluid inside the surface. Also, the sum of the torques must be zero, so the line of action of the resultant $y$-component of surface force must pass through the center of gravity of this fluid (which doesn't necessarily coincide with the center of gravity of the body).

Now we remove the fluid inside the surface and replace it with a solid body having exactly the same shape (Fig. 14–8b). The pressure at every point is exactly the same as before, so the total upward force exerted on the body by the fluid is also the same, again equal in magnitude to the weight $mg$ of fluid displaced. We call this upward force the **buoyant force** on the solid body. The line of action of this force again passes through the center of gravity of the displaced fluid.

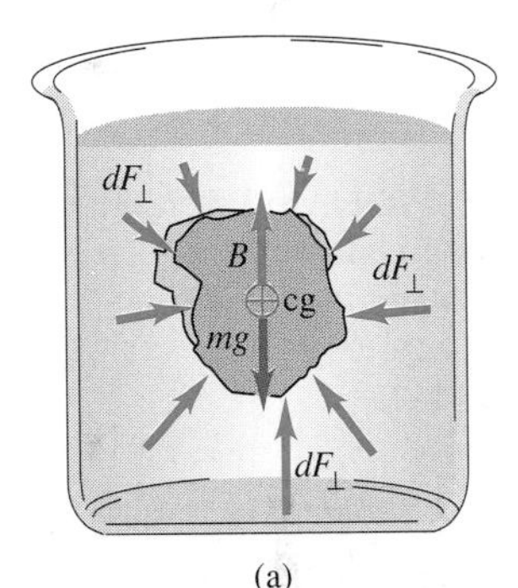


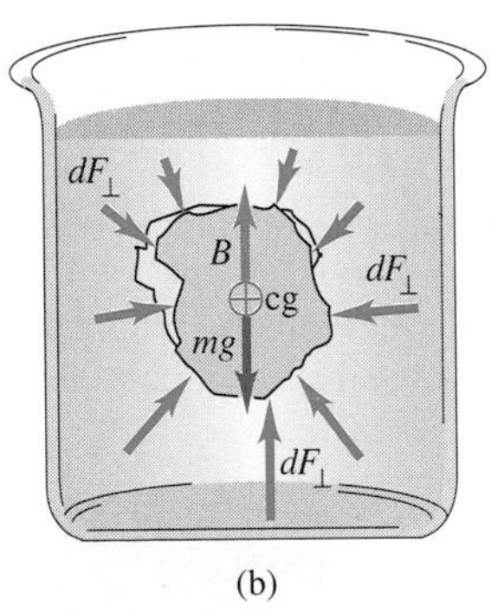


**14–8** (a) An element of fluid in equilibrium under its weight and the buoyant force of the surrounding fluid. (b) When the element is replaced by a body with identical shape, the buoyant force is the same and is equal to the weight of the displaced fluid.

**14–7** A body floats in a fluid if its average density is less than that of the fluid. The average density of a boat is less than that of water; the average density of an airship is less than that of air.

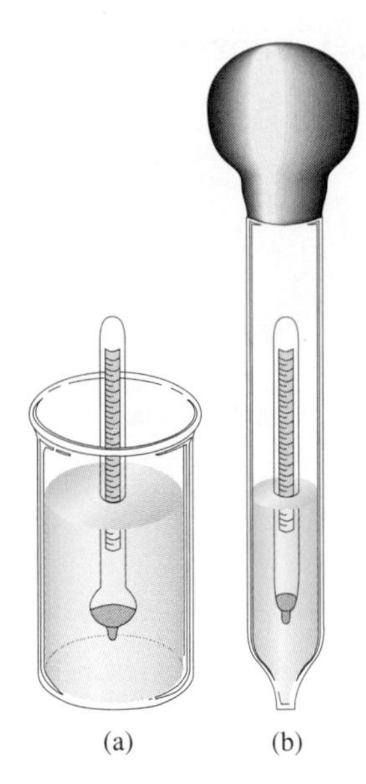


**14–9** (a) A simple hydrometer. (b) Use of a hydrometer as a tester for battery acid or antifreeze.

When a balloon floats in equilibrium in air, its weight (including the gas inside it) must be the same as the weight of the air displaced by the balloon. When a submerged submarine is in equilibrium, its weight must equal the weight of the water it displaces. A body whose average density is less than that of a liquid can float partially submerged at the free upper surface of the liquid. The greater the density of the liquid, the less of the body is submerged. When you swim in seawater (density 1030 $kg/m^3$), your body floats higher than it does in fresh water (1000 $kg/m^3$). Unlikely as it may seem, lead floats in mercury. Very flat-surfaced "float glass" is made by floating glass on molten tin and then letting it cool.

Another familiar example of buoyancy is the hydrometer (Fig. 14–9a). The calibrated float sinks into the fluid until the weight of the fluid it displaces is exactly equal to its own weight. The hydrometer floats *higher* in denser liquids than in less dense liquids. It is weighted at its bottom end so that the upright position is stable, and a scale in the top stem permits direct density readings. Figure 14–9b shows a type of hydrometer that is commonly used to measure the density of battery acid or antifreeze. The bottom of the large tube is immersed in the liquid; the bulb is squeezed to expel air and then released, like a giant medicine dropper. The liquid rises into the outer tube, and the hydrometer floats in this sample of the liquid.

## EXAMPLE 14–5

A 15.0-kg solid gold statue is being raised from a sunken treasure ship. What is the tension in the hoisting cable a) when the statue is completely immersed; b) when it is out of the water?

**SOLUTION** a) When the statue is immersed, it experiences an upward buoyant force that is equal in magnitude to the weight of the water displaced (Fig. 14–10). To find this force, we first find the volume of the statue, using the density of gold from Table 14–1:

$$V = \frac{m}{\rho} = \frac{15.0\ \text{kg}}{19.3 \times 10^3\ \text{kg/m}^3} = 7.77 \times 10^{-4}\ \text{m}^3.$$

Using Table 14–1 again, we find the weight of this volume of seawater:

$$\begin{aligned} w_{sw} &= m_{sw}g = \rho_{sw}Vg \\ &= (1.03 \times 10^3\ \text{kg/m}^3)(7.77 \times 10^{-4}\ \text{m}^3)(9.80\ \text{m/s}^2) \\ &= 7.84\ \text{N}. \end{aligned}$$

This equals the buoyant force $B$.

To find the cable tension $T$, we draw a free-body diagram for the statue (Figure 14–10b). We use this to identify the forces to be included in the equilibrium condition:

$$\Sigma F_y = B + T - mg = 0,$$

$$\begin{aligned} T &= mg - B = (15.0\ \text{kg})(9.80\ \text{m/s}^2) - 7.84\ \text{N} \\ &= 139\ \text{N}. \end{aligned}$$

This is about 5% less than the statue's weight in vacuum, $mg = (15.0\ \text{kg})(9.80\ \text{m/s}^2) = 147\ \text{N}$.

b) The density of air is about 1.2 $kg/m^3$, so the buoyant force of air on the statue is

$$\begin{aligned} B &= \rho_{air}Vg = (1.2\ \text{kg/m}^3)(7.77 \times 10^{-4}\ \text{m}^3)(9.80\ \text{m/s}^2) \\ &= 9.1 \times 10^{-3}\ \text{N}. \end{aligned}$$

This is only 62 parts per million of the statue's weight in vacuum. This effect is not within the precision of our data, and we ignore it. Thus the tension in the cable with the statue in air is equal to the statue's weight, 147 N.

**14–10** (a) The completely immersed statue at rest. (b) Free-body diagram for the immersed statue.

## EXAMPLE 14–6

Suppose you place a container of seawater on a scale and suspend the statue of Example 14–5 in it (Fig. 14–11). How does the scale reading differ from the reading when the statue is in air?

**SOLUTION** Consider the water, the statue, and the container together as a system; the total weight of the system does not depend on whether or not the statue is immersed. The total supporting force, including the tension $T$ and the upward force $F$ of the scale on the container (equal to the scale reading), is the same in both cases. But $T$ decreases by 7.8 N when the statue is immersed, so the scale reading $F$ must *increase* by 7.8 N. An alternative viewpoint is that the water exerts an upward buoyant force of 7.8 N on the statue, so the statue must exert an equal downward force on the water, making the scale reading 7.8 N greater than the weight of water and container.

**14–11** How does the scale reading change when the statue is immersed in the water?

## 14–4 SURFACE TENSION

What do the following observations have in common? A liquid emerges from the tip of a medicine dropper as a succession of drops, not a continuous stream. A paper clip rests on a water surface (Fig. 14–12), even though its density is several times that of water. Some insects can walk on the surface of water; their feet make indentations in the surface but do not penetrate it. When a small clean glass tube is dipped into water, the water rises in the tube, but when the tube is dipped in mercury, the mercury is depressed.

In all these situations the surface of the liquid seems to be under *tension*. Here's another example. We attach a loop of thread to a wire ring, dip the ring and thread into a soap solution, and remove them, forming a thin film of liquid in which the thread floats (Fig. 14–13a). When we puncture the film inside the loop, the thread springs out into a circular shape (Fig. 14–13b) as the tension of the liquid surface pulls radially outward on it.

**14–12** A paper clip "floating" on water. The clip is supported not by buoyant forces (Section 14–3) but by surface tension of the water.

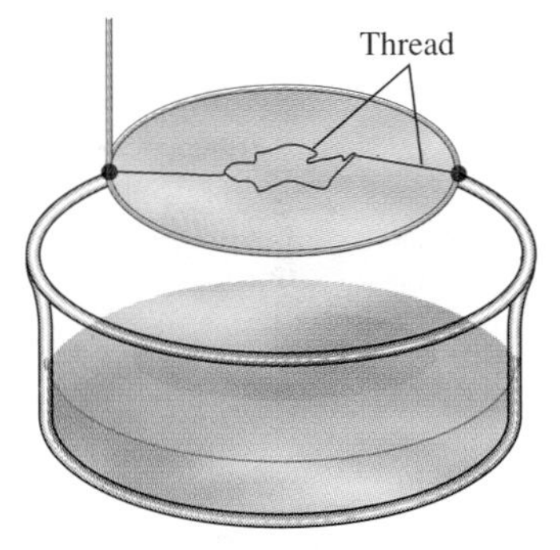


(a)

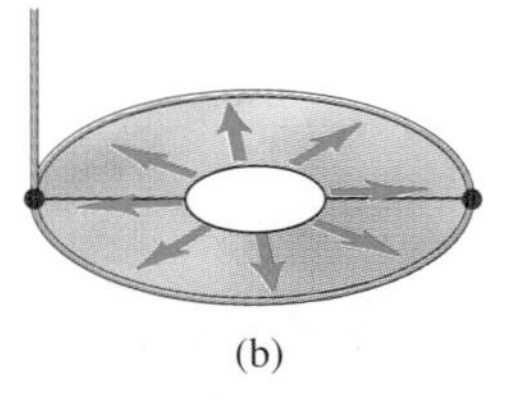

(b)

**14–13** A wire ring with a flexible loop of thread, dipped in a soap solution (a) before and (b) after puncturing the surface films inside the loop.

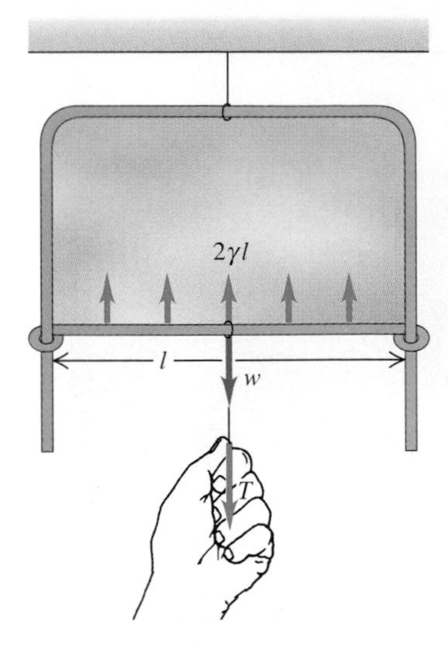


**14–14** The horizontal slide wire is in equilibrium under the action of the upward surface force $2\gamma l$ and downward pull $w + T$.

Figure 14–14 shows another example of surface tension. A piece of wire is bent into a U shape, and a second piece of wire slides on the arms of the U. When the apparatus is dipped into a soap solution and removed, the slider (if its weight $w$ is not too great) is quickly pulled up toward the top of the inverted U. To hold the slider in equilibrium, we need a total downward force $F = w + T$. When we pull the slider down, increasing the area of the film, molecules move from the interior of the liquid (which is many molecules thick, even in a thin film) into the surface layers. The surface layers are not stretched like a rubber sheet. Instead, more surface is created by molecules moving from the bulk liquid.

Let $l$ be the length of the wire slider. The film has two surfaces, so the surface force acts along a total length $2l$. The **surface tension** $\gamma$ in the film is defined as *the ratio of the surface force F to the length d along which the force acts:*

$$\gamma = \frac{F}{d}. \tag{14–10}$$

In this case, $d = 2l$ and

$$\gamma = \frac{F}{2l}.$$

Surface tension is a *force per unit length*. The SI unit is the newton per meter (N/m), but the cgs unit, the dyne per centimeter (dyn/cm), is more commonly used:

$$1 \text{ dyn/cm} = 10^{-3} \text{ N/m} = 1 \text{ mN/m}.$$

Table 14–2 shows some typical values of surface tension. The value for a particular liquid usually decreases as temperature increases; the table shows this behavior for water.

A drop of liquid in free fall in vacuum is always spherical because a surface under tension tends to have the minimum possible area. A sphere has smaller surface area for a given volume than any other shape. Figure 14–15 is a beautiful example of the formation of a spherical droplet.

**TABLE 14–2 Experimental Values of Surface Tension**

| Liquid in contact with air | Temperature (°C) | Surface tension (mN/m, or dyn/cm) |
|---|---|---|
| Benzene | 20 | 28.9 |
| Carbon tetrachloride | 20 | 26.8 |
| Ethanol | 20 | 22.3 |
| Glycerin | 20 | 63.1 |
| Mercury | 20 | 465.0 |
| Olive oil | 20 | 32.0 |
| Soap solution | 20 | 25.0 |
| Water | 0 | 75.6 |
| Water | 20 | 72.8 |
| Water | 60 | 66.2 |
| Water | 100 | 58.9 |
| Oxygen | −193 | 15.7 |
| Neon | −247 | 5.15 |
| Helium | −269 | 0.12 |

**14–15** Aftermath of a drop of water splashing on a liquid surface.

## Pressure inside a Bubble

Surface tension causes a pressure difference between the inside and outside of a soap bubble or a liquid drop. A soap bubble consists of two spherical surface films with a thin layer of liquid between. The films tend to contract, but as the bubble contracts, it compresses the inside air, eventually increasing the interior pressure to a point that prevents further contraction.

We can derive an expression for the excess pressure inside a bubble in terms of its radius $R$ and the surface tension $\gamma$ of the liquid. Assume first that there is no external pressure. Each half of the soap bubble is in equilibrium; the lower half is shown in Fig. 14–16. The forces at the flat circular surface where this half joins the upper half are the upward force of surface tension and the downward force due to the pressure of air in the upper half. The circumference of the circle along which the surface tension acts is $2\pi R$. (We neglect the small difference between inner and outer radii.) The total surface-tension force for each surface (inner and outer) is $\gamma(2\pi R)$, for a total of $(2\gamma)(2\pi R)$. The force due to air pressure is the pressure $p$ times the area $\pi R^2$ of the circle over which it acts. For the sum of these forces to be zero we must have

$$(2\gamma)(2\pi R) = p(\pi R^2),$$

$$p = \frac{4\gamma}{R}. \tag{14–11}$$

In general, the pressure outside the bubble is *not* zero. But Eq. (14–11) still gives the *difference* between inside and outside pressure. If the outside pressure is atmospheric pressure $p_a$, then

$$p - p_a = \frac{4\gamma}{R} \quad \text{(soap bubble)}. \tag{14–12}$$

A liquid drop has only *one* surface film, and the difference between pressure in the liquid and that in the outside air is half that for a soap bubble:

$$p - p_a = \frac{2\gamma}{R} \quad \text{(liquid drop)}. \tag{14–13}$$

Liquid between surfaces
R
O
Air fills interior
(a)

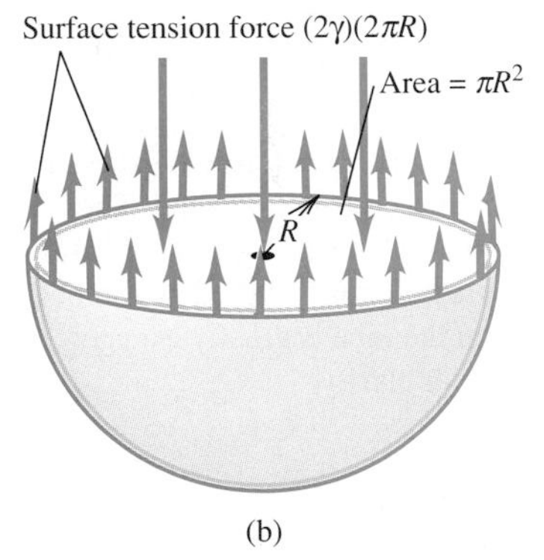


(b)

**14–16** (a) Cross section of a soap bubble showing the two surfaces, the thin layer of liquid between them, and the air inside the bubble. (b) Equilibrium of half a soap bubble. The surface-tension force exerted by the other half is $(2\gamma)(2\pi R)$, and the force exerted by the air inside the bubble is the pressure $p$ times the area $\pi R^2$.

## EXAMPLE 14–7

Calculate the excess pressure inside a drop of mercury 4.00 mm in diameter at 20°C.

**SOLUTION** From Table 14–2,

$$\gamma = 465 \text{ mN/m} = 465 \times 10^{-3} \text{ N/m}.$$

From Eq. (14–13),

$$p - p_a = \frac{2\gamma}{R} = \frac{2(465 \times 10^{-3} \text{ N/m})}{0.00200 \text{ m}}$$
$$= 465 \text{ N/m}^2 = 465 \text{ Pa} = 0.00459 \text{ atm}.$$

You can see that the pressure difference between inside and outside the drop is a very small fraction of atmospheric pressure.

## Capillarity

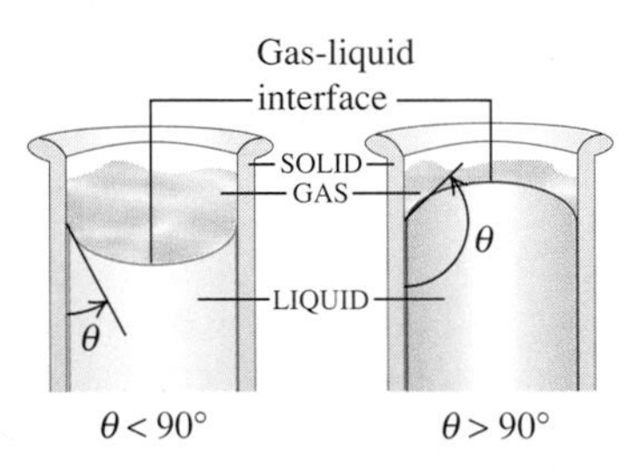


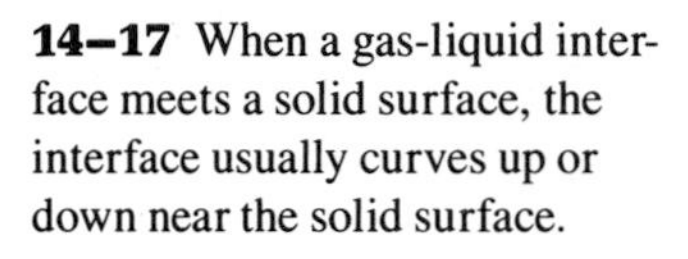


**14–17** When a gas-liquid interface meets a solid surface, the interface usually curves up or down near the solid surface.

When a gas-liquid interface meets a solid surface, such as the wall of a container (Fig. 14–17), the gas-liquid interface usually curves up or down near the solid surface. The angle $\theta$ at which it meets the surface is called the **contact angle.** When the surface curves up, as with water and glass, $\theta$ is less than 90°; when it curves down, as with mercury and glass, $\theta$ is greater than 90°. The first case occurs with a liquid that "wets," or adheres to, the solid surface; the second occurs with a nonwetting liquid.

Surface tension causes elevation or depression of the liquid in a narrow tube (Fig. 14–18). This effect is called **capillarity.** When the contact angle is less than 90° (Fig. 14–18a), the liquid *rises* until it reaches an equilibrium height $y$ determined by the requirement that the total surface-tension force along the line of contact with the tube wall must balance the extra weight of the liquid in the tube. The curved liquid surface is called a *meniscus*. For a nonwetting liquid such as mercury (Figure 14–18b) the contact angle is greater than 90°. The meniscus curves down, and the surface is depressed, pulled *down* by the surface-tension forces.

Capillarity is responsible for the absorption of water by paper towels, the rise of melted wax in a candle wick, and many other everyday phenomena. In higher animals, including humans, blood is *pumped* through the arteries and veins, but capillarity is still important in the smallest blood vessels, which indeed are called capillaries.

Related to surface tension is the phenomenon of *negative pressure*. The stress in a liquid is ordinarily *compressive,* but in some circumstances, liquids can sustain *tensile* stresses. Consider a cylindrical tube that is closed at one end and has a tight-fitting piston

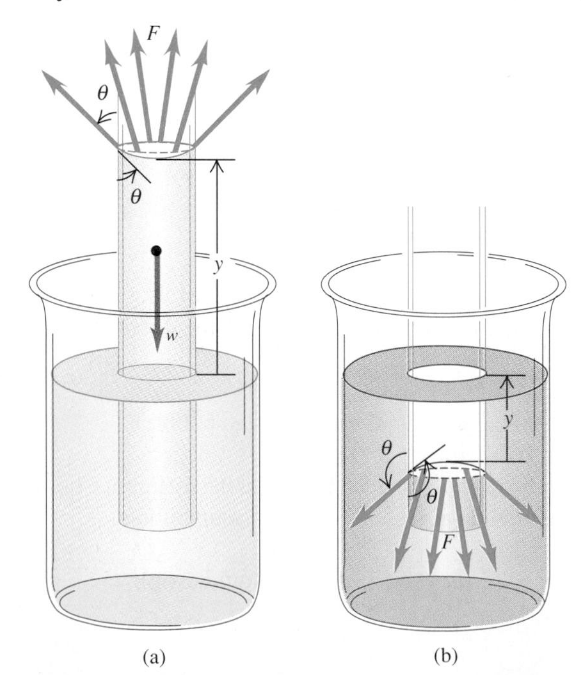


**14–18** Surface-tension forces on a liquid in a capillary tube. The liquid (a) rises if $\theta < 90°$ or (b) is depressed if $\theta > 90°$. The diameter of the tube is greatly exaggerated for clarity.

at the other. We fill the tube completely with a liquid that wets both the inner surface of the tube and the piston face; the molecules of liquid adhere to all the surfaces. If the surfaces are very clean and the liquid very pure, then when we pull the piston face, we observe a *tensile* stress and a slight *increase* in volume; we are *stretching* the liquid. Adhesive forces prevent it from pulling away from the walls of the container.

With water, tensile stresses as large as 300 atm have been observed in the laboratory. This situation is highly unstable; a liquid under tension tends to break up into many small droplets. In tall trees, however, negative pressures are a regular occurrence. Negative pressure is believed to be an important mechanism for the transport of water and nutrients from the roots to the leaves in the small xylem tubes (diameter of the order of 0.1 mm) in the growing layers of the tree.

## 14–5 FLUID FLOW

We are now ready to consider *motion* of a fluid. Fluid flow can be extremely complex, as is shown by the currents in a river in flood or the swirling flames of a campfire. Despite this, some situations can be represented by relatively simple idealized models. An **ideal fluid** is a fluid that is *incompressible* and has no internal friction or viscosity. The assumption of incompressibility is usually a good approximation for liquids, and we may also treat a gas as incompressible whenever the pressure differences from one region to another are not too great. Internal friction in a fluid causes shear stresses when two adjacent layers of fluid move relative to each other, as when fluid flows inside a tube or around an obstacle. In some cases we can neglect these shear forces in comparison with forces arising from gravitation and pressure differences.

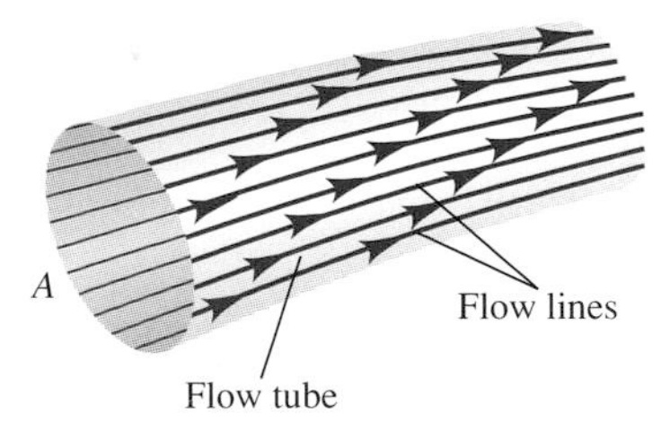

**14–19** A flow tube bounded by flow lines. In steady flow, fluid cannot cross the walls of a flow tube.

The path of an individual particle in a moving fluid is called a **flow line.** If the overall flow pattern does not change with time, the flow is called **steady flow.** In steady flow, every element passing through a given point follows the same flow line. In this case the "map" of the fluid velocities at various points in space remains constant, although the velocity of a particular particle may change in both magnitude and direction during its motion. A **streamline** is a curve whose tangent at any point is in the direction of the fluid velocity at that point. When the flow pattern changes with time, the streamlines do not coincide with the flow lines. We will consider only steady-flow situations, in which flow lines and streamlines are identical.

The flow lines passing through the edge of an imaginary element of area, such as the area $A$ in Fig. 14–19, form a tube called a **flow tube.** From the definition of a flow line, no fluid can cross the side walls of a flow tube; there can be no mixing of the fluids in different flow tubes.

Figure 14–20 shows patterns of fluid flow from left to right around a number of obstacles and in a channel of varying cross section. The photographs were made by

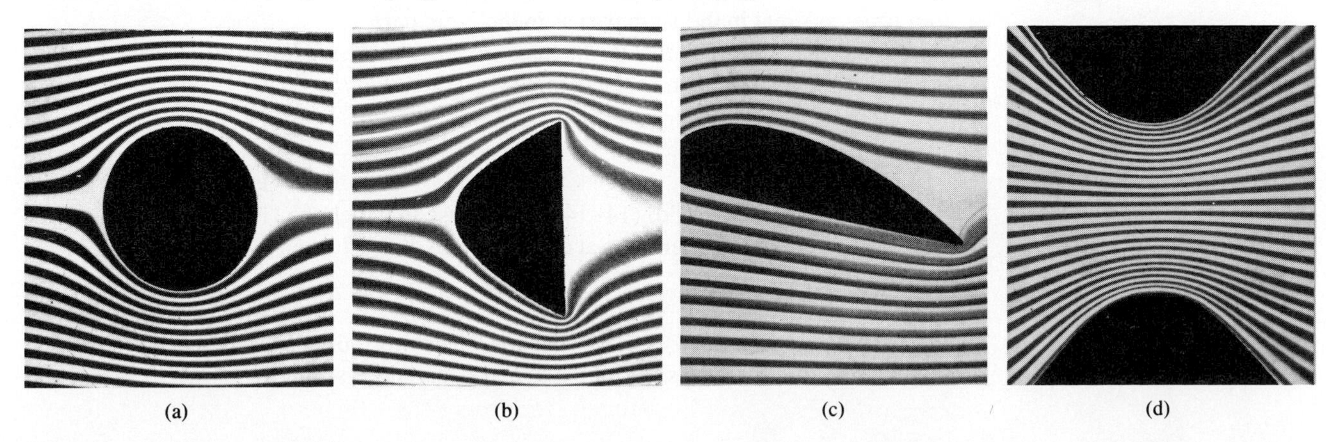

**14–20** (a), (b), (c) Laminar flow around obstacles of different shapes. (d) Flow through a channel of varying cross-section area.

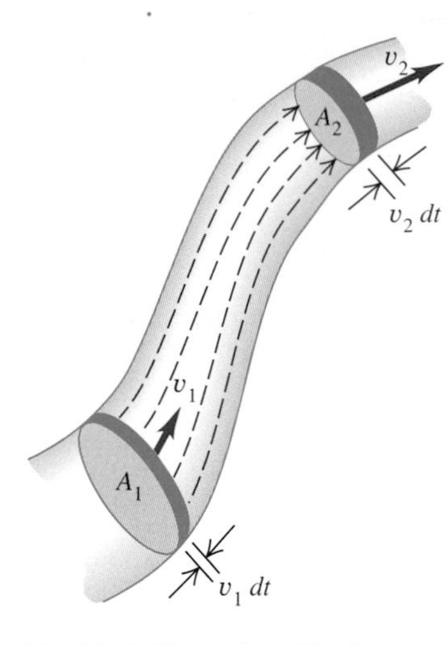


**14–21** A flow tube with changing cross-section area. The product $Av$ is constant for an incompressible fluid.

injecting dye into water flowing between two closely spaced glass plates. These patterns are typical of **laminar flow,** in which adjacent layers of fluid slide smoothly past each other. At sufficiently high flow rates, or when boundary surfaces cause abrupt changes in velocity, the flow becomes irregular and chaotic; this is called **turbulent flow.** In turbulent flow there is no steady-state pattern; the flow pattern continuously changes.

Conservation of mass in fluid flow gives us an important relationship called the **continuity equation.** The essence of this relation is contained in the familiar maxim "Still waters run deep." If the flow rate of a river with constant width is the same everywhere along a certain length, the water must run faster where it is shallow than where it is deep. As another example, the stream of water from a faucet tapers down in width because the water speeds up as it falls and a smaller diameter is needed to carry the same volume flow rate as at the top.

We can turn these observations into a quantitative relation. We'll consider only incompressible fluids, for which the density is constant. Figure 14–21 shows a portion of a flow tube between two stationary cross sections with areas $A_1$ and $A_2$. The fluid speeds at these sections are $v_1$ and $v_2$. No fluid flows across the side wall of the tube because the fluid velocity is tangent to the wall at every point on the wall. During a small time interval $dt$ the fluid at $A_1$ moves a distance $v_1\,dt$. The volume of fluid $dV_1$ that flows into the tube across $A_1$ during this interval is the fluid in the cylindrical element with base $A_1$ and height $v_1\,dt$: $dV_1 = A_1 v_1\,dt$. If the density of the fluid is $\rho$, the mass $dm_1$ flowing into the tube is $dm_1 = \rho A_1 v_1\,dt$. Similarly, the mass $dm_2$ that flows out across $A_2$ in the same time is $dm_2 = \rho A_2 v_2\,dt$. In steady flow the total mass in the tube is constant, so $dm_1 = dm_2$, and

$$\rho A_1 v_1\,dt = \rho A_2 v_2\,dt, \qquad \text{or}$$

$$A_1 v_1 = A_2 v_2. \tag{14–14}$$

The product $Av$ is the *volume flow rate, $dV/dt$,* the rate at which volume crosses a section of the tube:

$$\frac{dV}{dt} = Av. \tag{14–15}$$

Equation (14–14) shows that the volume flow rate is constant along any flow tube. When the cross section of a flow tube decreases (Fig. 14–20d), the speed increases. The shallow part of the river has smaller cross section and faster current than the deep part, but the volume flow rates are the same in both. The stream from the faucet narrows as it gains speed during its fall, but $dV/dt$ is the same everywhere along the stream. If a water pipe with 2-cm diameter is connected to a pipe with 1-cm diameter, the flow speed is four times as great in the 1-cm part as in the 2-cm part.

The *mass* flow rate $dm/dt$, the mass flow per unit time through any cross section, is equal to the density $\rho$ times the volume flow rate $dV/dt$:

$$\frac{dm}{dt} = \rho\frac{dV}{dt}. \tag{14–16}$$

We can generalize Eq. (14–14) for the case in which the fluid is *not* incompressible. If $\rho_1$ and $\rho_2$ are the densities at cross sections 1 and 2, then

$$\rho_1 A_1 v_1 = \rho_2 A_2 v_2. \tag{14–17}$$

We leave the details for an exercise. We don't need this more general form when we deal with incompressible fluids, for which $\rho_1$ and $\rho_2$ are always equal.

# Shape and Drag in Fluid Flow

A major factor affecting the motion of bodies in a fluid is the amount of drag force between body and fluid. One measure of this drag force is the drag coefficient $C_D$, which depends to a large degree on the shape of the body. Other factors affect the drag coefficient to significant extents, factors such as surface roughness, relative flow velocity, and cross-sectional area in the direction perpendicular to the flow. But a major consideration for objects designed to move through air or water is the amount of "streamlining" given to the object's shape.

# BUILDING PHYSICAL INTUITION

Girth measurements of trout or tuna match almost exactly the shape of a modern, low-drag airfoil. Similarly, submarine designs closely resemble the shapes of marine swimmers such as whales; boat hulls are shaped much like duck bodies; and small airplane bodies mimic the shapes of large birds.

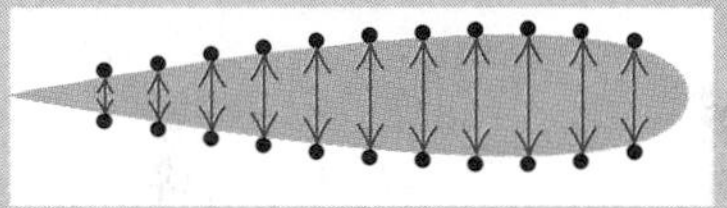

Wind tunnels are the major experimental tool for testing the shapes of cars, planes, or other vehicles for minimal drag. With the help of wind tunnels, airfoils have been developed with drag coefficients actually less than that of a thin plate parallel to the flow; however, these shapes cause turbulence unless the flow is very smooth and precisely oriented relative to the object.

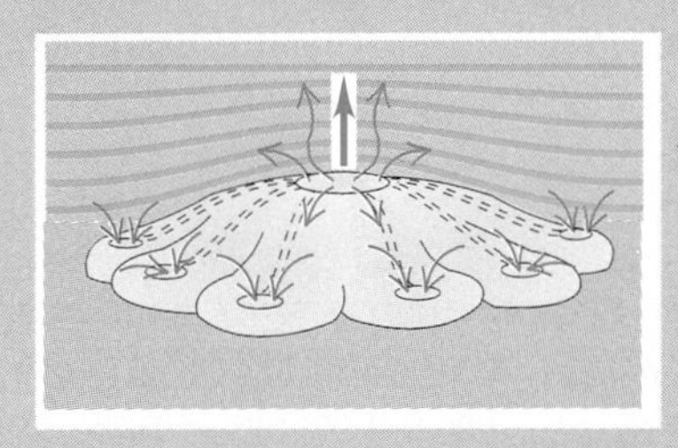

The shape of a Frisbee® or boomerang gives small drag and large lift, enabling it to glide long distances through the air. This same shape occurs in colonies of stationary sea creatures, minimizing the drag of ocean currents tending to dislodge them and also circulating water through them for food intake and waste discharge.

## 14–6 BERNOULLI'S EQUATION

According to the continuity equation, the speed of fluid flow can vary along the paths of the fluid. The pressure can also vary; it depends on height as in the static situation, and it also depends on the speed of flow. We can derive an important relationship called *Bernoulli's equation* that relates the pressure, flow speed, and height for flow of an ideal fluid.

The dependence of pressure on speed follows from Eq. (14–14). When an incompressible fluid flows along a flow tube with varying cross section, its speed *must* change. Therefore each element of fluid must have an acceleration, and the force that causes this acceleration has to be applied by the surrounding fluid. This means that the pressure *must* be different in different regions; if it were the same everywhere, the net force on every fluid element would be zero. When a fluid element speeds up, it must move from a region of larger to smaller pressure in order to have a net forward force to accelerate it. When the cross section of a flow tube varies, the pressure *must* vary along the tube, even when there is no difference of elevation. If the elevation also changes, this causes an additional pressure difference.

To derive the Bernoulli equation, we apply the work-energy theorem to the fluid in a section of a flow tube. In Fig. 14–22 we consider the fluid that at some initial time lies between the two cross sections $a$ and $c$. The speed at the lower end is $v_1$, and the pressure is $p_1$. In a small time interval $dt$ the fluid that is initially at $a$ moves to $b$, a distance $ds_1 = v_1\,dt$. In the same time interval the fluid that is initially at $c$ moves to $d$, a distance $ds_2 = v_2\,dt$. The cross-section areas at the two ends are $A_1$ and $A_2$, as shown. Because of the continuity relation, the volume of fluid $dV$ passing any cross section during time $dt$ is $dV = A_1\,ds_1 = A_2\,ds_2$.

Let's compute the *work* done on this fluid during $dt$. The total force on the cross section at $a$ is $p_1A_1$, and the force at $c$ is $p_2A_2$, where $p_1$ and $p_2$ are the pressures at the two ends. The net work $dW$ done on the element during this displacement is therefore

$$dW = p_1A_1\,ds_1 - p_2A_2\,ds_2 = (p_1 - p_2)\,dV. \tag{14–18}$$

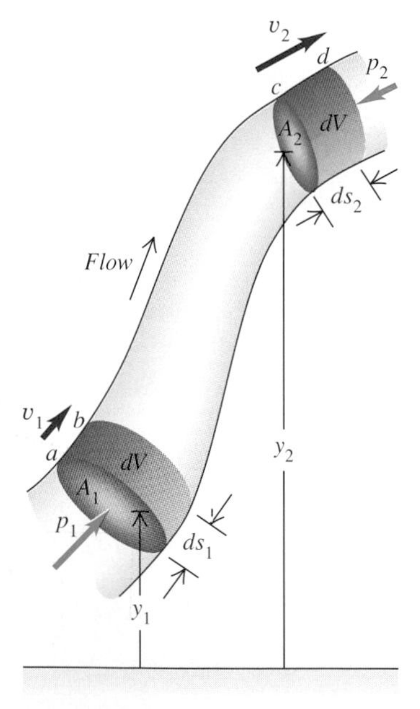


**14–22** The net work done by the pressure equals the change in the kinetic energy plus the change in the gravitational potential energy.

We need the negative sign for the second term because the force at $c$ is opposite in direction to the displacement.

We now equate this work to the total change in energy, kinetic and potential, for the fluid element. The energy for the fluid between sections $b$ and $c$ does not change. At the beginning of $dt$ the fluid between $a$ and $b$ has volume $A_1\,ds_1$, mass $\rho A_1\,ds_1$, and kinetic energy $\frac{1}{2}\rho(A_1\,ds_1)v_1^2$. Similarly, at the end of $dt$ the fluid between $c$ and $d$ has kinetic energy $\frac{1}{2}\rho(A_2\,ds_2)v_2^2$. The net change in kinetic energy $dK$ during time $dt$ is

$$dK = \tfrac{1}{2}\rho\,dV(v_2^2 - v_1^2). \tag{14–19}$$

What about the change in potential energy? The potential energy for the mass between $a$ and $b$ at the beginning of $dt$ is $dm\,gy_1 = \rho\,dV\,gy_1$. The potential energy for the mass between $c$ and $d$ at the end of $dt$ is $dm\,gy_2 = \rho\,dV\,gy_2$. The net change in potential energy $dU$ during $dt$ is

$$dU = \rho\,dV\,g(y_2 - y_1). \tag{14–20}$$

Combining Eqs. (14–18), (14–19), and (14–20) in the work-energy theorem $dW = dK + dU$, we obtain

$$(p_1 - p_2)\,dV = \tfrac{1}{2}\rho\,dV\,(v_2^2 - v_1^2) + \rho\,dV\,g(y_2 - y_1),$$

$$p_1 - p_2 = \tfrac{1}{2}\rho(v_2^2 - v_1^2) + \rho g(y_2 - y_1). \tag{14–21}$$

This is **Bernoulli's equation.** In this form, it states that the work per unit volume of fluid $(p_1 - p_2)$ is equal to the sum of the changes in kinetic and potential energies per unit volume that occur during the flow. Or we may interpret Eq. (14–21) in terms of pressures. The second term on the right is the pressure difference caused by the weight of the fluid and the difference in elevation of the two ends. The first term on the right is the additional pressure difference associated with the change of speed of the fluid.

We can also express Eq. (14–21) as

$$p_1 + \rho g y_1 + \tfrac{1}{2}\rho v_1^2 = p_2 + \rho g y_2 + \tfrac{1}{2}\rho v_2^2. \qquad (14\text{–}22)$$

The subscripts 1 and 2 refer to *any* two points along the flow tube, so we can also write

$$p + \rho g y + \tfrac{1}{2}\rho v^2 = \text{constant}. \qquad (14\text{–}23)$$

## PROBLEM-SOLVING STRATEGY

### Bernoulli's equation

Bernoulli's equation is derived from the work-energy relationship, so it isn't surprising that much of the problem-solving strategy suggested in Section 7–2 is equally applicable here. In particular:

**1.** Always begin by identifying clearly points 1 and 2, with reference to Bernoulli's equation.

**2.** Make lists of the known and unknown quantities in Eq. (14–22). The variables are $p_1, p_2, v_1, v_2, y_1$, and $y_2$, and the constants are $\rho$ and $g$. What is given? What do you need to determine?

**3.** In some problems you will need to use the continuity equation, Eq. (14–14), to get a relation between the two speeds in terms of cross-section areas of pipes or containers. Or perhaps you will know both speeds and will need to determine one of the areas. You may also need to use the flow-rate expressions given by Eqs. (14–15) and (14–16) to find $dV/dt$ or $dm/dt$.

**4.** As always, consistent units are essential. In SI units, pressure is in pascals, density in kilograms per cubic meter, and speed in meters per second. Also note that the pressures must be either all absolute pressures or all gauge pressures.

## EXAMPLE 14–8

**Water pressure in the home** Water enters a house through a pipe with an inside diameter of 2.0 cm at an absolute pressure of $4.0 \times 10^5$ Pa (about 4 atm). The pipe leading to the second-floor bathroom 5.0 m above is 1.0 cm in diameter (Fig. 14–23). When the flow speed at the inlet pipe is 2.0 m/s, find the flow speed, pressure, and volume flow rate in the bathroom.

**SOLUTION** Let point 1 be at the inlet pipe and point 2 at the bathroom. The speed $v_2$ at the bathroom is obtained from the continuity equation, Eq. (14–14):

$$v_2 = \frac{A_1}{A_2}v_1 = \frac{\pi(1.0\text{ cm})^2}{\pi(0.50\text{ cm})^2}(2.0\text{ m/s}) = 8.0\text{ m/s}.$$

We take $y_1 = 0$ (at the inlet) and $y_2 = 5.0$ m (at the bathroom). We are given $p_1$ and $v_1$; we can find $p_2$ from Bernoulli's equation:

$$\begin{aligned} p_2 &= p_1 - \tfrac{1}{2}\rho(v_2^2 - v_1^2) - \rho g(y_2 - y_1) \\ &= 4.0 \times 10^5\text{ Pa} - \tfrac{1}{2}(1.0 \times 10^3\text{ kg/m}^3)(64\text{ m}^2/\text{s}^2 - 4.0\text{ m}^2/\text{s}^2) \\ &\quad - (1.0 \times 10^3\text{ kg/m}^3)(9.8\text{ m/s}^2)(5.0\text{ m}) \end{aligned}$$

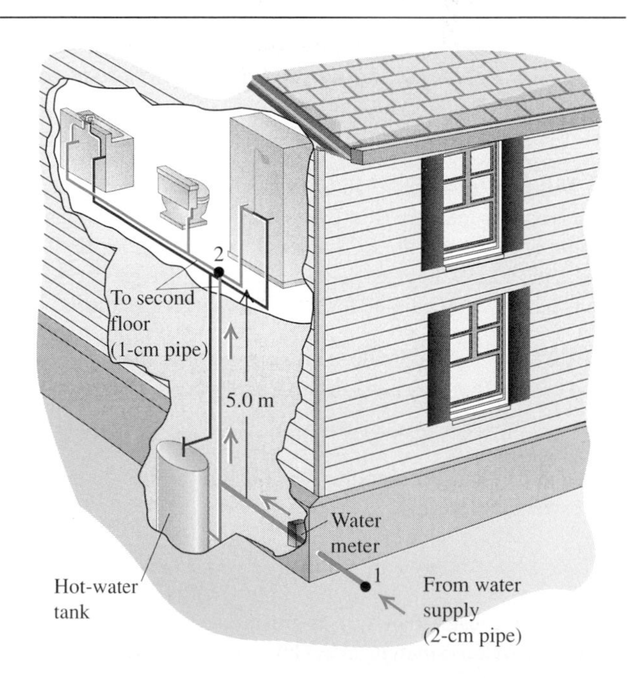


**14–23** What is the water pressure in the second-story bathroom of this house?

$$= 4.0 \times 10^5\ \text{Pa} - 0.30 \times 10^5\ \text{Pa} - 0.49 \times 10^5\ \text{Pa}$$
$$= 3.2 \times 10^5\ \text{Pa} = 3.2\ \text{atm} = 47\ \text{psia}.$$

The volume flow rate is

$$\frac{dV}{dt} = A_2 v_2$$
$$= \pi(0.50 \times 10^{-2}\ \text{m})^2(8.0\ \text{m/s})$$
$$= 6.3 \times 10^{-4}\ \text{m}^3/\text{s} = 0.63\ \text{L/s}.$$

Note that after the water is turned off, the second term on the right side of the equation for pressure vanishes, and the pressure rises to $3.5 \times 10^5$ Pa. In fact, when the fluid is not moving, Bernoulli's equation, Eq. (14–21), reduces to the pressure relation that we derived for a fluid at rest, Eq. (14–5).

## EXAMPLE 14–9

**Speed of efflux** Figure 14–24 shows a gasoline storage tank with cross-section area $A_1$, filled to a depth $h$. The space above the gasoline contains air at pressure $p_0$, and the gasoline flows out through a short pipe with area $A_2$. Derive expressions for the flow speed in the pipe and the volume flow rate.

**SOLUTION** We can consider the entire volume of moving liquid as a single flow tube; $v_1$ and $v_2$ are the speeds at points 1 and 2 in Fig. 14–24. The pressure at point 2 is atmospheric pressure, $p_a$. Applying Bernoulli's equation to points 1 and 2 and taking $y = 0$ at the bottom of the tank, we find

$$p_0 + \tfrac{1}{2}\rho v_1^2 + \rho g h = p_a + \tfrac{1}{2}\rho v_2^2,$$

$$v_2^2 = v_1^2 + 2\frac{p_0 - p_a}{\rho} + 2gh.$$

Because $A_2$ is much smaller than $A_1$, $v_1^2$ is much smaller than $v_2^2$ and can be neglected. We then find

$$v_2^2 = 2\frac{p_0 - p_a}{\rho} + 2gh. \qquad (14\text{–}24)$$

The speed $v_2$, sometimes called the *speed of efflux,* depends on both the pressure difference $(p_0 - p_a)$ and the height $h$ of the liquid level in the tank. If the top of the tank is vented to the atmosphere, there is no excess pressure, $p_0 = p_a$, and $p_0 - p_a = 0$. In that case,

$$v_2 = \sqrt{2gh}. \qquad (14\text{–}25)$$

That is, *the speed of efflux from an opening at a distance h below the top surface of the liquid is the same as the speed a body would acquire in falling freely through a height h.* This result is called *Torricelli's theorem.* It is valid not only for an opening in the bottom of a container, but also for a hole in a side wall at a depth $h$ below the surface. We find the volume flow rate from Eq. (14–15):

$$\frac{dV}{dt} = A_2\sqrt{2gh}.$$

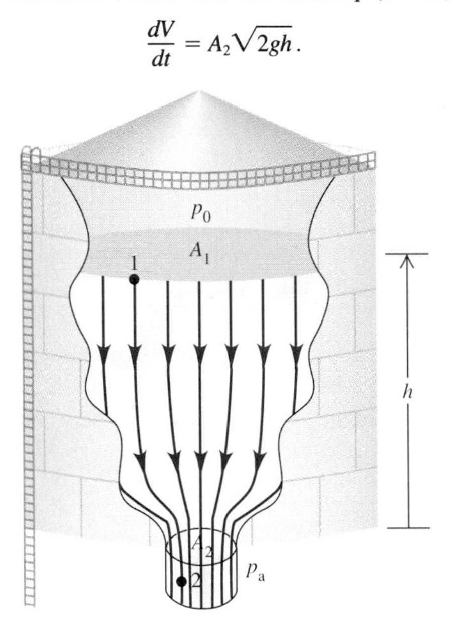


**14–24** Calculating the speed of efflux for gasoline flowing out the bottom of a storage tank.

## EXAMPLE 14–10

**The Venturi meter** Figure 14–25 shows a *Venturi meter,* which is used to measure flow speed in a pipe. The narrow part of the pipe is called the *throat.* Derive an expression for the flow speed $v_1$ in terms of the cross-section areas $A_1$ and $A_2$ and the difference in height $h$ of the liquid levels in the two vertical tubes.

**SOLUTION** We apply Bernoulli's equation to the wide (point 1) and narrow (point 2) parts of the pipe, with $y_1 = y_2$:

$$p_1 + \tfrac{1}{2}\rho v_1^2 = p_2 + \tfrac{1}{2}\rho v_2^2.$$

From the continuity equation, $v_2 = (A_1/A_2)v_1$. Substituting this and rearranging, we get

$$p_1 - p_2 = \tfrac{1}{2}\rho v_1^2\left(\frac{A_1^2}{A_2^2} - 1\right).$$

Because $A_1$ is greater than $A_2$, $v_2$ is greater than $v_1$ and the pressure $p_2$ in the throat is *less* than $p_1$. A net force to the right accelerates the fluid as it enters the throat, and a net force to the left slows it as it leaves. The pressure difference $p_1 - p_2$ is also equal to $\rho g h$, where $h$ is the difference in liquid level in the two tubes. Combining this with the above result and solving for $v_1$, we get

$$v_1 = \sqrt{\frac{2gh}{(A_1/A_2)^2 - 1}}.$$

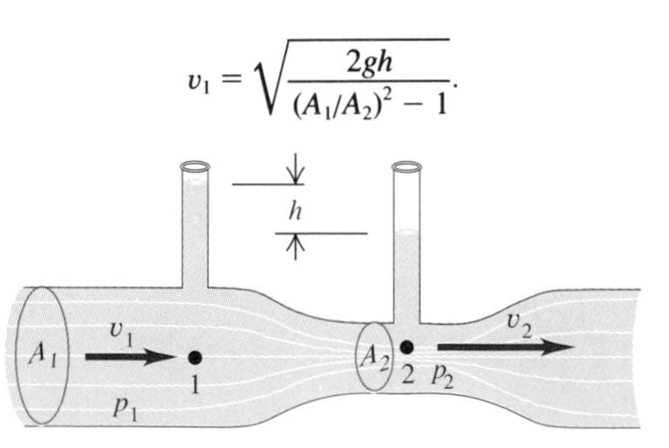


**14–25** The Venturi meter.

## EXAMPLE 14–11

**Lift on an airplane wing** Figure 14–26 shows flow lines around a cross section of an airplane wing. The lines crowd together above the wing, corresponding to increased flow speed and reduced pressure in this region, just as in the Venturi throat. Because the upward force on the underside of the wing is greater than the downward force on the top side, there is a net upward force, or *lift.* (This highly simplified discussion ignores the effect of turbulent flow and the formation of vortices; a more complete discussion would take these into account.)

We can also understand the lift force on the basis of momentum changes. Figure 14–26 shows that there is a net *downward* change in the vertical component of momentum of the air flowing past the wing, corresponding to the downward force the wing exerts on it. The reaction force *on* the wing is *upward*, as we concluded above.

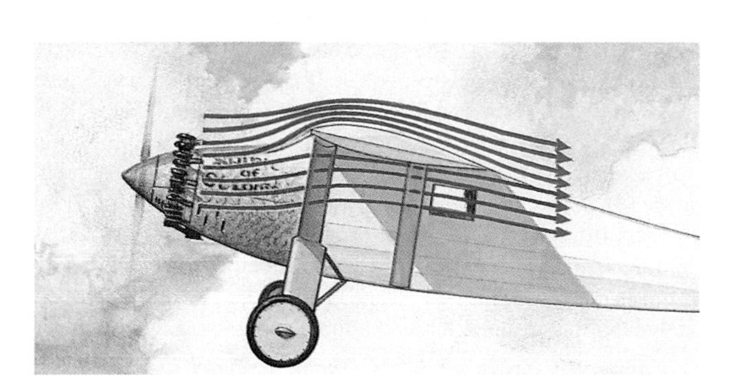

**14–26** Flow lines around an airplane wing.

## EXAMPLE 14–12

**The curve ball** Does a curve ball *really* curve? Yes, it certainly does. Figure 14–27a shows a ball moving through the air from left to right. To an observer moving with the center of the ball the air stream appears to move from right to left, as shown by the flow lines in the figure. Because of the large speeds that are ordinarily involved, there is a region of turbulent flow behind the ball.

When the ball is spinning (Fig. 14–27b), the viscosity of air causes layers of air near the ball's surface to be pulled around in the direction of spin. The speed of air relative to the ball's surface becomes smaller at the top of the ball than at the bottom. The region of turbulence becomes asymmetric; turbulence occurs farther forward on the top side than on the bottom. This asymmetry causes a pressure difference; the average pressure at the top of the ball becomes greater than that at the bottom. The corresponding net downward force deflects the ball as shown. This could be a side view of a tennis ball (Fig. 14–27d) moving from left to right with "top spin," the only way to keep a very fast return in the court. In a baseball curve pitch (Fig. 14–27e) the ball spins about a

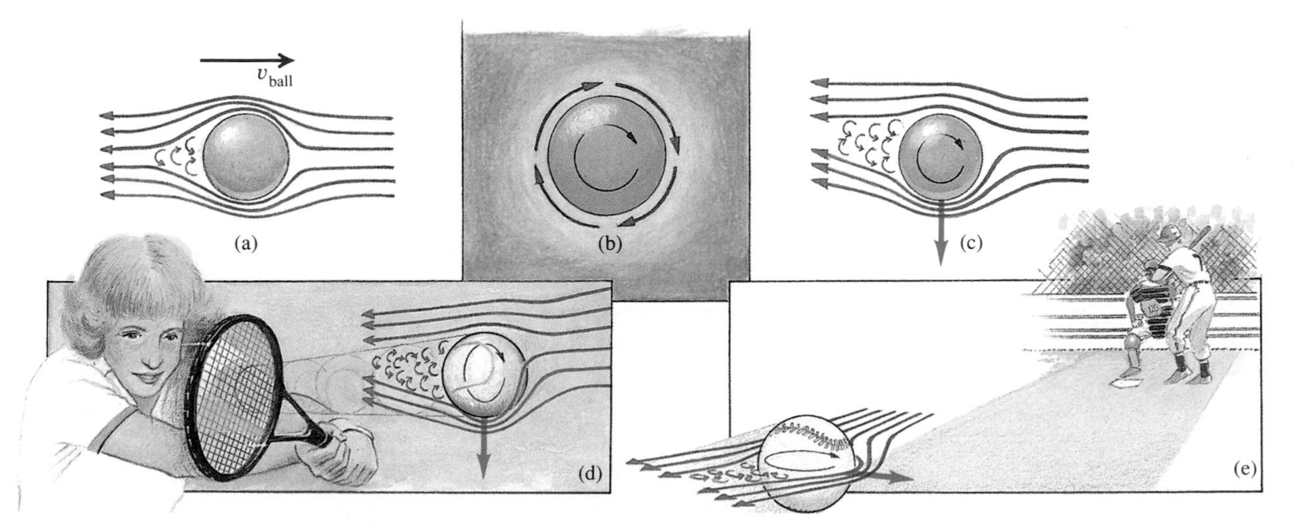


**14–27** Motion of the air from right to left relative to the ball corresponds to motion of a ball through still air from left to right. (a) A nonspinning ball has a symmetric region of turbulence behind it. (b) A spinning ball drags along layers of air near its surface. (c) The resulting asymmetric region of turbulence and the deflection of the air stream by the spinning ball. The force shown is force exerted on the ball by the air stream.

a nearly vertical axis, and the actual deflection is sideways. In that case, Fig. 14–27c is a *top* view of the situation. Note that a curve ball thrown by a right-handed pitcher curves *away from* a right-handed batter, making it harder to hit.

A similar effect occurs with golf balls, which always have "backspin" from impact with the slanted face of the golf club. The resulting pressure difference between the top and bottom of the ball causes a "lift" force that keeps the ball in the air considerably longer than would be possible without spin. A well-hit drive appears from the tee to "float" or even curve *upward* during the initial portion of its flight. This is a real effect, not an illusion. The dimples on the ball play an essential role; because of effects associated with viscosity of air, an undimpled ball has a much shorter trajectory than a dimpled one, given the same initial velocity and spin. One manufacturer even claims that the *shape* of the dimples is significant; polygonal dimples are alleged to be better than round ones. Figure 14–28 shows the backspin of a golf ball just after it is struck by a club.

**14–28** Stroboscopic photograph of a golf ball being struck by a club. The picture was taken at 1000 flashes per second. The ball rotates about once in eight pictures, corresponding to an angular velocity of 125 rev/s, or 7500 rpm.

## 14–7 REAL FLUIDS

We call this section "Real Fluids" because we want to talk briefly about two important aspects of fluid flow, viscosity and turbulence, that are not present in the idealized models we have been using up to now.

### Viscosity

*Viscosity* is internal friction in a fluid; viscous forces oppose the motion of one portion of a fluid relative to another. Viscosity is the reason it takes effort to paddle a canoe through calm water, but it is also the reason the paddle works. Viscosity plays a vital role in the flow of fluids in pipes, the flow of blood, the lubrication of engine parts, and many other areas of practical importance.

A viscous fluid always tends to cling to a solid surface that is in contact with it. There is always a thin *boundary layer* of fluid near the surface in which the fluid is nearly at rest with respect to the surface. That's why dust particles can cling to a fan blade even when it is rotating rapidly and why you can't get all the mud off your car by just squirting a hose at it.

The simplest example of viscous flow is motion of a fluid between two parallel plates (Fig. 14–29). The bottom plate is stationary, and the top plate moves with constant velocity $\boldsymbol{v}$. The fluid in contact with each surface has the same velocity as that surface. The flow speeds of intermediate layers of fluid increase uniformly from one surface to the other, as shown by the arrows.

Flow of this type is called *laminar.* (A lamina is a thin sheet.) In laminar flow, the layers of fluid slide smoothly over one another. A portion of the fluid that has the shape *abcd* at some instant has the shape *abc'd'* a moment later and becomes more and more

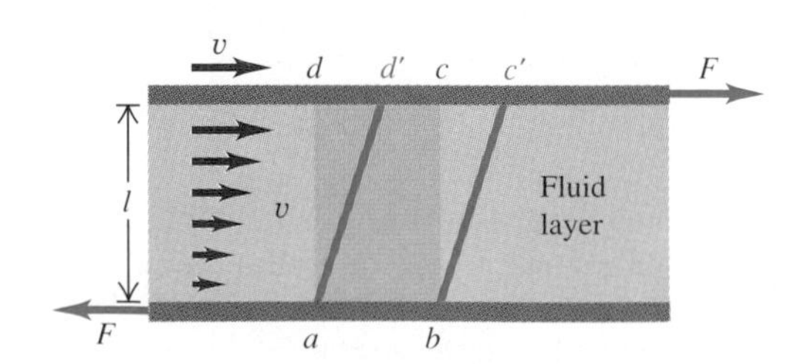


**14–29** Laminar flow of a viscous fluid. Viscosity is the ratio of shear stress $F/A$ to strain rate $v/l$.

distorted as the motion continues. That is, the fluid is in a state of continuously increasing *shear strain*. To maintain this motion, a constant force $F$ must be applied toward the right on the upper plate to keep it moving and a force of equal magnitude toward the left on the lower plate to hold it stationary. If $A$ is the surface area of each plate, the ratio $F/A$ is the *shear stress* exerted on the fluid.

In Section 11–7 we defined shear strain as the ratio of the displacement $dd'$ to the length $l$. In a solid, shear strain is proportional to shear stress. In a fluid we have a *continuously increasing* shear strain that increases without limit as long as the stress is applied. The stress depends not on the shear strain, but on its *rate of change*. The strain shown in Fig. 14–29 (at the instant when the volume of fluid has the shape $abc'd'$) is $dd'/ad$, or $dd'/l$. The *rate of change* of strain, also called the *strain rate*, equals the rate of change of $dd'$ (the speed $v$ of the moving surface) divided by $l$. That is,

$$\text{Rate of change of shear strain} = \frac{v}{l}.$$

We define the **viscosity** of the fluid, denoted by $\eta$, as the ratio of the shear stress, $F/A$, to the strain rate:

$$\eta = \frac{\text{Shear stress}}{\text{Strain rate}} = \frac{F/A}{v/l}, \quad \text{or} \tag{14–26}$$

$$F = \eta A \frac{v}{l}. \tag{14–27}$$

Fluids that flow readily, such as water or gasoline, have smaller viscosities than "thick" liquids such as honey or motor oil. Viscosities of all fluids are strongly temperature-dependent, increasing for gases and decreasing for liquids as the temperature increases. An important goal in the design of oils for engine lubrication is to *reduce* the temperature variation of viscosity as much as possible.

From Eq. (14–26) the unit of viscosity is that of force times distance, divided by area times speed. The SI unit is

$$1\ \text{N}\cdot\text{m}/[\text{m}^2\cdot(\text{m/s})] = 1\ \text{N}\cdot\text{s/m}^2.$$

The corresponding cgs unit, $1\ \text{dyn}\cdot\text{s/cm}^2$, is the only viscosity unit in common use; it is called a **poise,** in honor of the French scientist Jean Louis Marie Poiseuille:

$$1\ \text{poise} = 1\ \text{dyn}\cdot\text{s/cm}^2 = 10^{-1}\ \text{N}\cdot\text{s/m}^2.$$

The centipoise and the micropoise are also used.

The viscosity of water is 1.79 centipoise at 0°C and 0.28 centipoise at 100°C. Viscosities of lubricating oils are typically 1 to 10 poise, and the viscosity of air at 20°C is 181 micropoise.

A fluid that behaves according to Eq. (14–27) is called a *Newtonian fluid*. Not all fluids show this direct proportionality of force and speed. Fluids that are suspensions or dispersions are often non-Newtonian in their viscous behavior. An interesting example is blood, for which force increases more slowly than speed. Blood is a suspension of solid particles in a liquid. As the strain rate increases, these particles deform and become preferentially oriented to facilitate flow. The fluids that lubricate human joints show similar behavior. Some paints cling to the brush but flow more readily when brushed on.

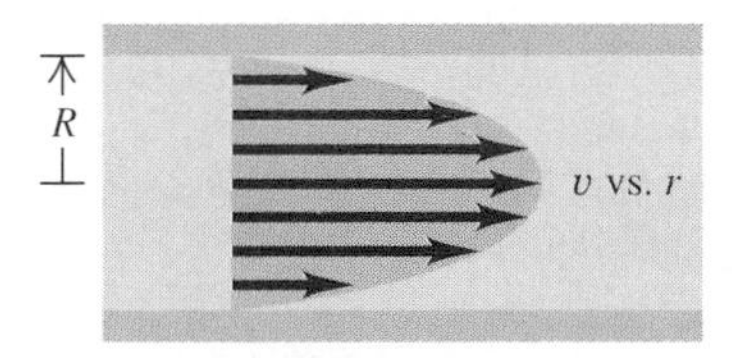

**14–30** Velocity profile for a viscous fluid in a cylindrical pipe.

Figure 14–30 shows the flow speed profile for flow of a viscous fluid in a cylindrical pipe. The speed is greatest along the axis and zero at the pipe walls, and the motion is like a lot of telescoping tubes sliding relative to one another, the central tube moving fastest and the outermost tube at rest. By applying Eq. (14–26) to a cylindrical fluid element we could derive an equation describing the speed profile. We'll omit the details; the flow speed $v$ at a distance $r$ from the axis of a pipe with radius $R$ is

$$v = \frac{p_1 - p_2}{4\eta L}(R^2 - r^2), \tag{14–28}$$

where $p_1$ and $p_2$ are the pressures at the two ends of a pipe with length $L$. The speed at any point is proportional to the change of pressure per unit length, $(p_2 - p_1)/L$ or $dp/dx$, called the *pressure gradient.* (The flow direction is always opposite to $dp/dx$.) We can also integrate this expression over the cross section, using an element of cross section area $dA = 2\pi r\,dr$ and integrating from $r = 0$ to $r = R$, to find the *total* volume flow rate. The result is

$$\frac{dV}{dt} = \frac{\pi}{8}\left(\frac{R^4}{\eta}\right)\left(\frac{p_1 - p_2}{L}\right). \tag{14–29}$$

This relation was first derived by Poiseuille and is called **Poiseuille's law.** It shows that the volume rate of flow is inversely proportional to viscosity, as we would expect. The volume flow rate is also proportional to the pressure gradient $(p_1 - p_2)/L$, and it varies as the *fourth power* of the radius $R$. If we double $R$, the flow rate increases by a factor of 16. This relation is important for design of plumbing systems and also for hypodermic needles. Needle size is much more important than thumb pressure in determining the flow rate from the needle; doubling the needle diameter has the same effect as increasing the thumb force sixteenfold. Similarly, blood flow in arteries and veins can be controlled over a wide range by relatively small changes in diameter, an important temperature-control mechanism in warm-blooded animals. Relatively slight narrowing of arteries due to arteriosclerosis can result in elevated blood pressure and added strain on the heart muscle.

One more useful relation in viscous fluid flow is the expression for the force $F$ exerted on a sphere of radius $r$ moving with speed $v$ through a fluid with viscosity $\eta$. When the flow is laminar, the relationship is simple:

$$F = 6\pi\eta r v. \tag{14–30}$$

We have encountered this kind of velocity-proportional force before, in Example 5–19 (Section 5–3). Equation (14–30) is called **Stokes' law.**

## EXAMPLE 14–13

**Terminal speed in a viscous fluid** Derive an expression for the terminal speed $v_t$ of a sphere falling in a viscous fluid in terms of the sphere's radius $r$ and density $\rho$ and the viscosity $\eta$ of the fluid, assuming that Stokes' law is valid.

**SOLUTION** The terminal speed is reached when the total force, including the weight of the sphere, the viscous retarding force, and the buoyant force, is zero (Fig. 14–31). Let $\rho$ be the density of the sphere and $\rho'$ the density of the fluid. The weight of the sphere is then $\frac{4}{3}\pi r^3 \rho g$, and the buoyant force is $\frac{4}{3}\pi r^3 \rho' g$. At terminal speed,

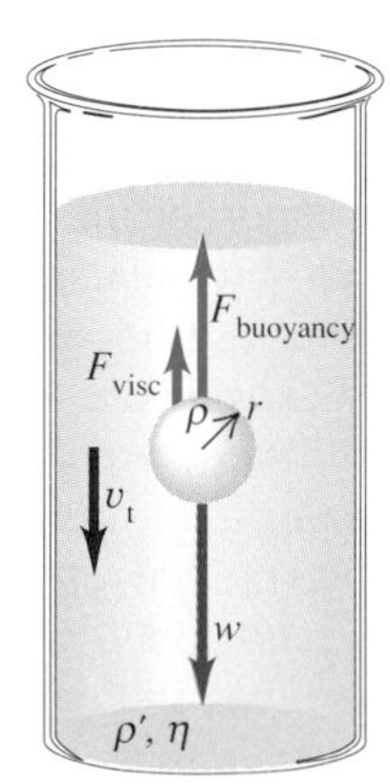


**14–31** A sphere falling in a viscous fluid reaches a terminal speed when the sum of the forces acting on it is zero.

$$\Sigma F_y = \tfrac{4}{3}\pi r^3 \rho' g + 6\pi\eta r v_t - \tfrac{4}{3}\pi r^3 \rho g = 0,$$

$$v_t = \frac{2}{9}\frac{r^2 g}{\eta}(\rho - \rho'). \qquad (14\text{–}31)$$

We can use this formula to measure the viscosity of a fluid. Or, if we know the viscosity, we can determine the radius of the sphere (or the approximate sizes of other small particles) by measuring the terminal speed. Robert Millikan used this method to determine the radii of very small electrically charged oil drops by observing their terminal speeds in air. He used these drops to measure the charge of an individual electron; we will discuss this landmark experiment further in Section 24–6.

The Stokes'-law force is what keeps clouds in the air. The terminal speed for a water droplet of radius $10^{-5}$ m is of the order of 1 mm/s. At higher speeds the flow often becomes turbulent, and Stokes' law is no longer valid. In air the drag force at highway speeds is approximately proportional to $v^2$. We have discussed the applications of this relation to skydiving (Example 5–20 in Section 5–3) and to air drag of a moving automobile (Section 6–6). Raindrops have terminal speeds of a few meters per second because of air drag; otherwise, they would smash everything in their path!

At temperatures below 2.17 K the common isotope of helium ($^4$He) acts as though it were made up of two fluids, one that has viscosity and one with zero viscosity. The zero-viscosity, or *superfluid,* component can flow through tiny channels, resulting in superleaks. It creeps up the sides of an open container and shows a variety of other unexpected behavior.

## Turbulence

When the speed of a flowing fluid exceeds a certain critical value, the flow is no longer laminar. For a fluid in a pipe the flow pattern that we described above as telescoping tubes breaks down. The pattern becomes extremely irregular and complex, and it changes continuously with time; there is no steady-state pattern. This irregular, chaotic flow is called **turbulence.** Figure 14–32 shows the contrast between laminar and

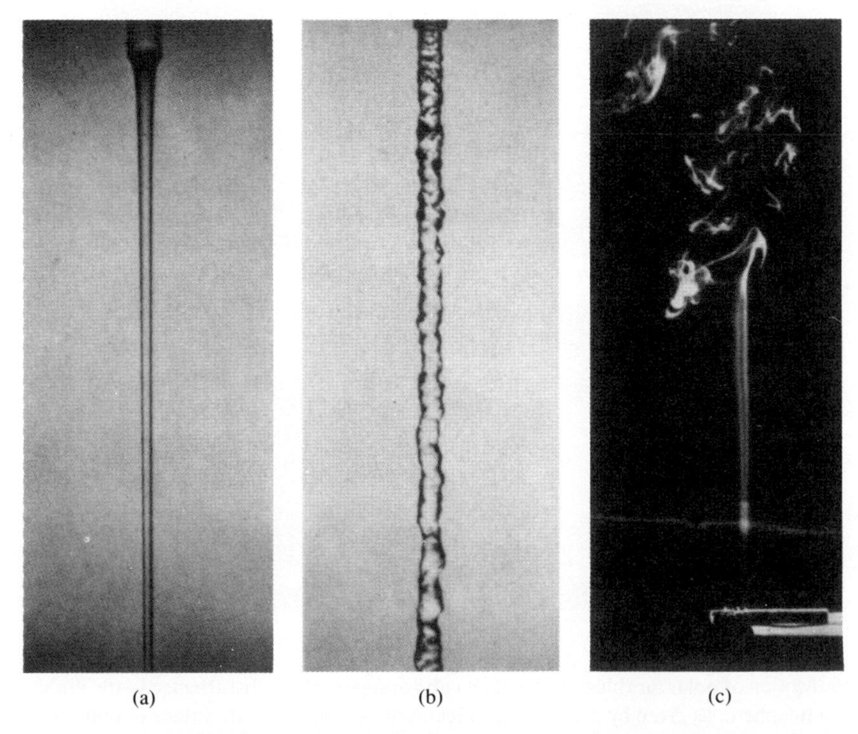


**14–32** (a) Laminar flow. (b) Turbulent flow. (c) First laminar, then turbulent flow.

**14–33** An example of the onset of turbulence produced by a cylindrical obstacle. Upstream flow is laminar, but downstream flow shows the formation of eddies, or vortices. The alternating series of vortices is called a Karman vortex street. The pattern is made visible by injecting smoke filaments into the air upstream from the obstacle.

turbulent flow in two familiar systems: a stream of water and rising smoke in air. Figure 14–33 shows turbulence caused by an obstacle in a flowing fluid.

The transition from laminar to turbulent flow is often very sudden. A flow pattern that is stable at low speeds suddenly becomes unstable when a critical speed is reached. Irregularities in the flow pattern can be caused by roughness in the pipe wall, variations in density of the fluid, and many other factors. At small flow speeds these disturbances damp out; the flow pattern is *stable* and tends to maintain its laminar nature. But when the critical speed is reached, the flow pattern becomes unstable. The disturbances no longer damp out; instead, they grow until they destroy the entire laminar-flow pattern.

Turbulence poses some profound questions for theoretical physics. The motion of any particular particle in a fluid is presumably determined by Newton's laws. If we know the motion of the fluid at some initial time, shouldn't we be able to predict the motion at any later time? After all, when we launch a projectile, we can compute the entire trajectory if we know the initial position and velocity. How is it that a system that obeys well-defined physical laws can have unpredictable, chaotic behavior?

There is no simple answer to these questions. Indeed, the study of chaotic behavior in deterministic systems is a very active field of research in theoretical physics. We had a glimpse in Section 13–9 of the richness of this area of research. The development of supercomputers in recent years has enabled scientists to carry out computer simulations of the behavior of physical systems on a scale that would have been impossible only a few years ago (Fig. 14–34). The results of such simulations have often guided new developments in the theory of chaotic behavior. A particularly significant result of these studies has been the discovery of common characteristics in the behavior of widely divergent kinds of systems, including turbulent fluid flow, population fluctuations in ecosystems, the growth of crystals, the shapes of coastlines, and many others.

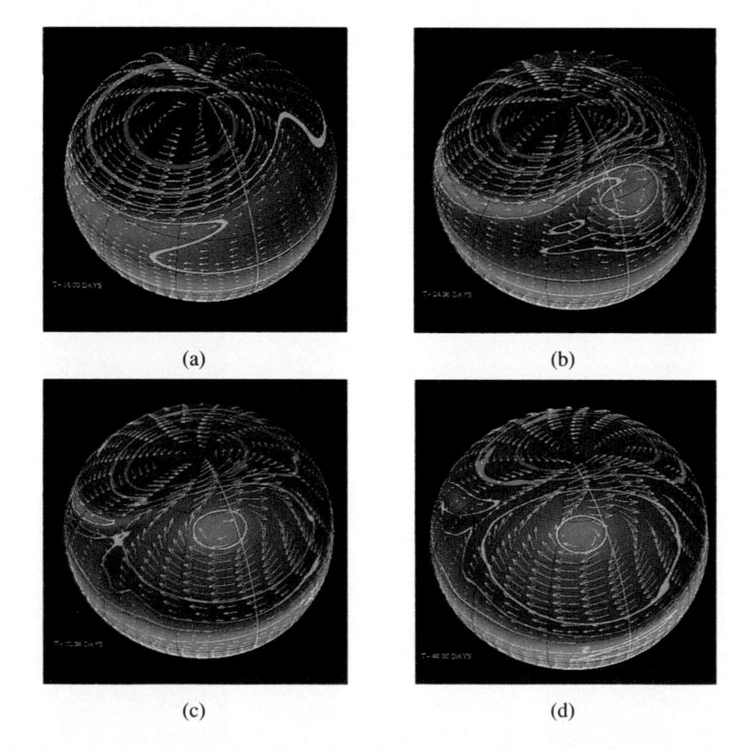


**14–34** Motion of polar air (blue) and tropical air (orange) after a disturbance to the circulation of the stratosphere, as given by a numerical calculation. Colors indicate values of potential vorticity, which is conserved for individual bodies of air and therefore traces out their motion. The sequence shows tropical air being drawn poleward in the form of low potential vorticity.

## SUMMARY

• Density is mass per unit volume. If a mass $m$ of material has volume $V$, its density $\rho$ is

$$\rho = \frac{m}{V}. \tag{14–1}$$

Specific gravity is the ratio of the density of a material to the density of water.

• Pressure is force per unit area. Pascal's law: Pressure applied to the surface of an enclosed fluid is transmitted undiminished to every portion of the fluid. Absolute pressure is the total pressure in a fluid; gauge pressure is the difference between absolute pressure and atmospheric pressure. The SI unit of pressure is the pascal (Pa).

$$1 \text{ Pa} = 1 \text{ N/m}^2,$$

$$1 \text{ bar} = 10^5 \text{ Pa},$$

$$1 \text{ torr} = 1 \text{ millimeter of mercury (mm Hg)},$$

$$1 \text{ atmosphere} = 1.013 \times 10^5 \text{ Pa} = 760 \text{ torr} = 1.013 \text{ bar}.$$

• Archimedes' principle: When a body is immersed in a fluid, the fluid exerts an upward buoyant force on the body equal to the weight of the fluid the body displaces.

• A boundary surface between a liquid and a gas behaves like a surface under tension; the force per unit length across a line on the surface is called the surface tension, denoted by $\gamma$.

• An ideal fluid is an incompressible fluid with no viscosity. A flow line is the path of a fluid particle; a streamline is a curve that is tangent at each point to the velocity vector at that point. A flow tube is a tube bounded at its sides by flow lines. In laminar flow, layers of fluid slide smoothly past each other. In turbulent flow, there is great disorder and a constantly changing flow pattern.

• Conservation of mass for an incompressible fluid is expressed by the equation of continuity: For two cross sections $A_1$ and $A_2$ in a flow tube the flow speeds $v_1$ and $v_2$ are related by

$$A_1 v_1 = A_2 v_2. \tag{14–14}$$

The product $Av$ is the *volume flow rate, $dV/dt$*, the rate at which volume crosses a section of the tube:

$$\frac{dV}{dt} = Av. \tag{14–15}$$

• Bernoulli's equation relates the pressure $p$, flow speed $v$, and elevation $y$ in an ideal fluid. For any two points, indicated by subscripts 1 and 2,

$$p_1 + \rho g y_1 + \tfrac{1}{2}\rho v_1{}^2 = p_2 + \rho g y_2 + \tfrac{1}{2}\rho v_2{}^2. \tag{14–22}$$

• The viscosity of a fluid characterizes its resistance to shear strain. In a Newtonian fluid the viscous force is proportional to strain rate. When such a fluid flows in a cylindrical pipe of inner radius $R$, the flow speed $v$ at a distance $r$ from the axis is given by

$$v = \frac{p_1 - p_2}{4\eta L}(R^2 - r^2), \tag{14–28}$$

### KEY TERMS

**density**
**specific gravity**
**pressure**
**pascal**
**atmospheric pressure**
**Pascal's law**
**gauge pressure**
**absolute pressure**
**mercury barometer**
**buoyancy**
**Archimedes' principle**
**buoyant force**
**surface tension**
**contact angle**
**capillarity**
**ideal fluid**
**flow line**
**steady flow**
**streamline**
**flow tube**
**laminar flow**
**turbulent flow**
**continuity equation**
**Bernoulli's equation**
**viscosity**
**poise**
**Poiseuille's law**
**Stokes' law**
**turbulence**

where $L$ is the length of pipe, $p_1$ and $p_2$ are the pressures at the two ends, and $\eta$ is the viscosity. The total volume rate is given by Poiseuille's equation:

$$\frac{dV}{dt} = \frac{\pi}{8}\left(\frac{R^4}{\eta}\right)\left(\frac{p_1 - p_2}{L}\right). \qquad (14\text{–}29)$$

A sphere of radius $r$ moving with speed $v$ through a fluid having viscosity $\eta$ experiences a viscous resisting force $F$ given by Stokes' law:

$$F = 6\pi\eta r v. \qquad (14\text{–}30)$$

## EXERCISES

### Section 14–1 Density

**14–1** A rectangular block of an unidentified material has dimensions $5.0 \times 15.0 \times 30.0$ cm and a mass of 1.80 kg. What is its density?

**14–2** A lead cube has a total mass of 30.0 kg. What is the length of a side?

**14–3** A cylindrical aluminum rod has a length of 62.0 cm and a diameter of 0.60 cm. What is its mass?

**14–4** A uniform solid metal sphere has a radius of 4.50 cm and a mass of 1.60 kg. What is its density?

### Section 14–2 Pressure in a Fluid

**14–5** You are designing a diving bell to withstand the pressure of seawater at a depth of 800 m. a) What is the gauge pressure at this depth? (Neglect changes in the density of the water with depth.) b) At this depth, what is the net force on a circular glass window 15.0 cm in diameter if the pressure inside the diving bell equals the pressure at the surface of the water? (Neglect the small variation of pressure over the surface of the window.)

**14–6** What gauge pressure must a pump produce to pump water from the bottom of the Grand Canyon (730-m elevation) to Indian Gardens (1370 m)? Express your results in pascals and in atmospheres.

**14–7** A barrel contains a 0.180-m layer of oil floating on water that is 0.300 m deep. The density of the oil is 600 kg/m³. a) What is the gauge pressure at the oil-water interface? b) What is the gauge pressure at the bottom of the barrel?

**14–8 A Hydraulic Lift.** The piston of a hydraulic automobile lift is 0.30 m in diameter. What gauge pressure, in pascals, is required to lift a car with a mass of 900 kg? Also express this pressure in atmospheres.

**14–9** The liquid in the open-tube manometer in Fig. 14–5a is mercury, $y_1 = 3.00$ cm, and $y_2 = 7.00$ cm. Atmospheric pressure is 970 millibars. a) What is the absolute pressure at the bottom of the U-shaped tube? b) What is the absolute pressure in the open tube at a depth of 4.00 cm below the free surface? c) What is the absolute pressure of the gas in the container? d) What is the gauge pressure of the gas in pascals?

**14–10 Maximum Snorkeling Depth.** There is a maximum depth at which a diver can breathe through a snorkel tube (Fig. 14–35) because as the depth increases, so does the pressure difference, tending to collapse the diver's lungs. Since the snorkel connects the air in the lungs to the atmosphere at the surface, the pressure inside the lungs is atmospheric pressure. What is the external-internal pressure difference when the diver's lungs are at a depth of 6.1 m (about 20 ft)? Assume that the diver is in seawater. (A scuba diver breathing from compressed air tanks can operate at greater depths than can a snorkeler, since the pressure of the air inside the scuba diver's lungs increases to match the external pressure of the water.)

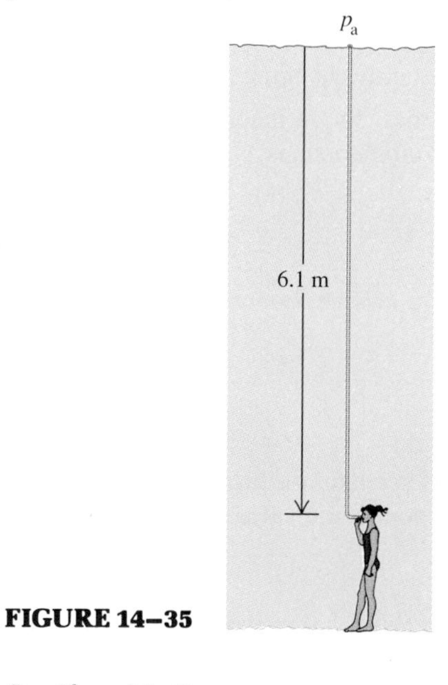


**FIGURE 14–35**

### Section 14–3 Buoyancy

**14–11 The Tip of the Iceberg.** When an iceberg floats in seawater, what fraction of its volume is submerged? (The ice, being of glacial origin, is frozen fresh water.)

**14–12** A solid brass statue weighs 145 N in air. a) What is its volume? b) The statue is suspended from a rope and totally immersed in water. What is the tension in the rope (the *apparent* weight of the statue in water)?

**14–13** A slab of ice floats on a fresh water lake. What minimum volume must the slab have for a 48.0-kg woman to be able to stand on it without getting her feet wet?

**14–14** An ore sample weighs 15.00 N in air. When the sample is suspended by a light cord and totally immersed in water, the tension in the cord is 9.00 N. Find the total volume and the density of the sample.

**14–15** A cubical block of wood 10.0 cm on a side floats at the interface between oil and water with its lower surface 2.00 cm below the interface (Fig. 14–36). The density of the oil is 650 $kg/m^3$. a) What is the gauge pressure at the upper face of the block? b) What is the gauge pressure at the lower face of the block? c) What is the mass of the block?

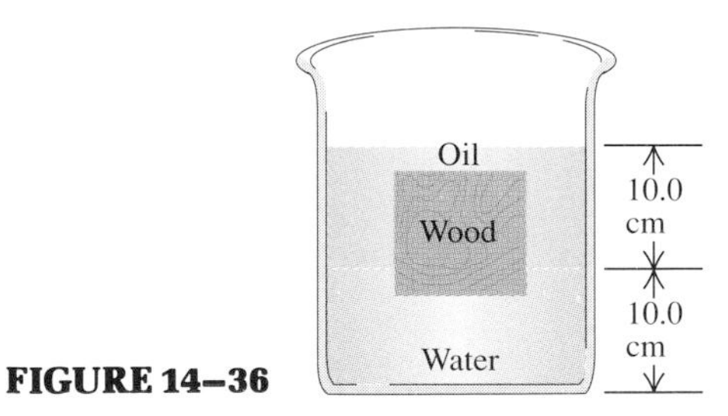


**FIGURE 14–36**

**14–16** A large, hollow plastic sphere is held below the surface of a fresh water lake by a cable anchored to the bottom of the lake. The sphere has a volume of 0.300 $m^3$, and the tension in the cable is 700 N. a) Calculate the buoyant force exerted by the water on the sphere. b) What is the mass of the sphere? c) The cable breaks, and the sphere rises to the surface of the lake. When the sphere comes to rest again, what fraction of its volume will be submerged?

## Section 14–4 Surface Tension

**14–17** Find the gauge pressure in pascals in a soap bubble 3.00 cm in diameter. The surface tension is $25.0 \times 10^{-3}$ N/m.

**14–18** Assume that $T = 20°C$. Find the excess pressure a) inside a water drop with a radius of 1.00 mm; b) inside a water drop with a radius of 0.0100 mm (typical of water droplets in fog).

## Section 14–5 Fluid Flow

**14–19** Water is flowing in a circular pipe of varying cross-section area, and at all points the water completely fills the pipe. a) At one place in the pipe the radius is 0.200 m. What is the magnitude of the water velocity at this point if the volume flow rate in the pipe is 0.600 $m^3/s$? b) At a second point in the pipe the water velocity has a magnitude of 3.80 m/s. What is the radius of the pipe at this point?

**14–20** Water is flowing in a pipe with varying cross-section area, and at all points the water completely fills the pipe. At point 1 the cross-section area of the pipe is 0.080 $m^2$, and the magnitude of the fluid velocity is 2.50 m/s. a) What is the fluid speed at points in the pipe where the cross-section area is i) 0.060 $m^2$; ii) 0.112 $m^2$? b) Calculate the volume of water discharged from the open end of the pipe in 1.00 s.

## Section 14–6 Bernoulli's Equation

**14–21** A sealed tank containing seawater to a height of 12.0 m also contains air above the water at a gauge pressure of 4.00 atm. Water flows out from the bottom through a hole whose cross-section area is 10.0 $cm^2$. Calculate the efflux speed of the water.

**14–22** A circular hole 2.00 cm in diameter is cut in the side of a large water tank, 16.0 m below the water level in the tank. The top of the tank is open to the air. Find a) the speed of efflux; b) the volume discharged per unit time.

**14–23** Water flowing in a horizontal pipe discharges at the rate of $4.00 \times 10^{-3}$ $m^3/s$. At a point in the pipe where the cross-section area is $1.00 \times 10^{-3}$ $m^2$ the absolute pressure is $1.60 \times 10^5$ Pa. What is the cross-section area of a constriction in the pipe if the pressure there is reduced to $1.20 \times 10^5$ Pa?

**14–24** At a certain point in a horizontal water pipeline the water's speed is 2.00 m/s, and the gauge pressure is $2.00 \times 10^4$ Pa. Find the gauge pressure at a second point in the line if the cross-section area at the second point is one half that at the first.

**14–25 Pressure in a Fire Hose.** What gauge pressure is required in the city mains for a stream from a fire hose connected to the mains to reach a vertical height of 25.0 m? (Assume that the mains have a much larger diameter than the fire hose.)

**14–26** At a certain point in a pipeline the water's speed is 4.00 m/s, and the gauge pressure is $5.00 \times 10^4$ Pa. Find the gauge pressure at a second point in the line, 12.0 m lower than the first, if the cross-section area at the second point is twice that at the first.

**14–27 Lift on an Airplane.** Air is streaming horizontally past a small airplane's wings such that the speed is 50.0 m/s over the top surface and 30.0 m/s past the bottom surface. If the plane has a mass of 700 kg and a wing area of 9.00 $m^2$, what is the net force (including the effects of gravity) on the airplane? The density of the air is 1.20 $kg/m^3$.

**14–28** An airplane with a mass of 6000 kg has a wing area of 30.0 $m^2$. If the pressure on the lower wing surface is $6.00 \times 10^4$ Pa during level flight at an elevation of 4000 m, what is the pressure on the upper wing surface?

## Section 14–7 Real Fluids

**14–29** Water at 20°C is flowing in a pipe with a radius of 20.0 cm. The viscosity of water at 20°C is 1.005 centipoise. If the water's speed in the center of the pipe is 2.60 m/s, what is the water's speed a) 10.0 cm from the center of the pipe (halfway between the center and the walls); b) at the walls of the pipe?

**14–30** Water at 20°C flows through a pipe with a radius of 1.00 cm. The viscosity of water at 20°C is 1.005 centipoise. If the flow speed at the center is 0.200 m/s and the flow is laminar, find the pressure drop due to viscosity along a 2.00-m section of pipe.

**14–31** Water at 20°C is flowing in a horizontal pipe that is 15.0 m long; the flow is laminar, and the water completely fills the pipe. A pump maintains a gauge pressure of 1400 Pa at a large tank at one end of the pipe. The other end of the pipe is open to the air. The viscosity of water at 20°C is 1.005 centipoise. a) If the pipe has diameter 8.00 cm, what is the volume flow rate? b) What gauge pressure must the pump provide to achieve the same volume flow rate for a pipe with a diameter of 4.00 cm? c) For the pipe in part (a) and the same gauge pressure maintained by the pump, what does the volume flow rate become if the water is at a temperature of 60°C? (The viscosity of water at 60°C is 0.469 centipoise.)

**14–32** A viscous liquid flows through a tube with laminar flow, as in Fig. 14–30. Prove that the volume rate of flow is the same as if the speed were uniform at all points of a cross section and equal to half the speed at the axis.

**14–33** What speed must a gold sphere with a radius of 6.00 mm have in castor oil at 20°C for the viscous drag force to be one-fourth the weight of the sphere? (The viscosity of castor oil at this temperature is 9.86 poise.)

**14–34 Measuring Viscosity.** A copper sphere with a mass of 0.20 g falls with a terminal speed of 5.0 cm/s in an unknown liquid. if the density of copper is 8900 kg/m$^3$ and that of the liquid is 2800 kg/m$^3$, what is the viscosity of the liquid?

## PROBLEMS

**14–35 Pressure in the Ocean Deep.** The deepest point known in any of the earth's oceans is in the Marianas Trench, 10.92 km deep. a) Assuming water to be incompressible, what is the pressure at this depth? Use the density of seawater. b) The actual pressure is $1.16 \times 10^8$ Pa. Your calculated value of the pressure will be less because the density actually varies with depth. Take your value for the pressure, and using the compressibility of water, find its density at the bottom of the Marianas Trench. What is the percent change in the density of the water?

**14–36** A swimming pool measures 25.0 m long × 8.0 m wide × 4.0 m deep. Compute the force exerted by the water against a) the bottom; b) either end. (*Hint:* Calculate the force on a thin horizontal strip at a depth $h$, and integrate this over the end of the pool. Do not include the force due to air pressure.)

**14–37** The upper edge of a gate in a dam runs along the water surface. The gate is 2.00 m high and 4.00 m wide and is hinged along a horizontal line through its center (Fig. 14–37). Calculate the torque about the hinge arising from the force due to the water. (*Hint:* Use a procedure similar to that used in Problem 14–36; calculate the torque of a thin horizontal strip at a depth $h$ and integrate this over the gate.)

**FIGURE 14–37**

**FIGURE 14–38**

**14–38 The Great Molasses Flood.** On the afternoon of January 15, 1919, which was an unusually warm day in Boston, a 27.4-m (90-ft) high, 27.4-m diameter cylindrical metal tank used for storing molasses ruptured. Molasses flooded into the streets in a 9-m deep stream, killing pedestrians and horses and knocking down buildings. The molasses had a density of 1600 kg/m$^3$. If the tank was full before the accident, what was the total outward force the molasses exerted on its sides? [*Hint:* Consider the outward force on a circular ring of width $dy$ at a depth $y$ below the surface (Fig. 14–38). Integrate to find the total outward force.]

**14–39** A U-shaped tube open to the air at both ends contains some mercury. A quantity of water is carefully poured into the left arm of the U-shaped tube until the vertical height of the water column is 15.0 cm (Fig. 14–39). a) What is the gauge pressure at the water-mercury interface? b) Calculate the vertical distance $h$ from the top of the mercury in the right-hand arm of the tube to the top of the water in the left-hand arm.

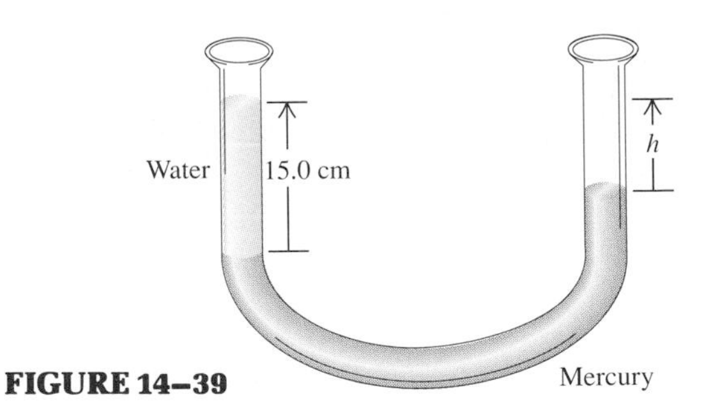


**FIGURE 14–39**

**14–40** According to an old advertising claim, a certain small car would float in water. a) If the car's mass is 800 kg and its interior volume is 4.0 $m^3$, what fraction of the car is immersed when it floats? The volume of steel and other materials may be neglected. b) As water gradually leaks in and displaces the air in the car, what fraction of the interior volume is filled with water when the car sinks?

**14–41 Remember the Hindenburg.** The densities of air, helium, and hydrogen (at $p = 1.0$ atm and $T = 20°C$) are 1.20 $kg/m^3$, 0.166 $kg/m^3$, and 0.0899 $kg/m^3$, respectively. a) What is the volume in cubic meters displaced by a hydrogen-filled dirigible that has a total "lift" of 12,000 kg? (The "lift" is the total mass of the dirigible that can be supported by the buoyant force in addition to the mass of the gas with which it is filled.) b) What would be the "lift" if helium were used instead of hydrogen? In view of your answer, why is helium the gas actually used?

**14–42 A Hydrometer.** A hydrometer consists of a spherical bulb and a cylindrical stem with a cross-section area of 0.400 $cm^2$ (Fig. 14–9a). The total volume of bulb and stem is 13.2 $cm^3$. When immersed in water, the hydrometer floats with 8.00 cm of the stem above the water surface. In alcohol, 1.20 cm of the stem is above the surface. Find the density of the alcohol. (*Note:* This illustrates the accuracy of such a hydrometer. Relatively small density differences give rise to relatively large differences in hydrometer readings.)

**14–43** An open barge is 22 m wide, 40 m long, and 12 m deep (Fig. 14–40). If the barge is made out of 7.5-cm thick steel plate on each of its four sides and its bottom, what mass of coal can the barge carry without sinking? Is there enough room in the barge to hold this amount of coal? (The density of coal is about 1500 $kg/m^3$.)

**14–44 SHM of a Floating Object.** Consider an object with height $h$, mass $M$, and uniform cross-section area $A$ floating upright in a liquid with density $\rho$. a) Calculate the vertical distance from the surface of the liquid to the bottom of the floating object at equilibrium. b) A downward force with magnitude $F$ is applied to the top of the object. At the new equilibrium position, how much farther below the surface of the liquid is the bottom of the object than it was in part (a)? (Assume that $F$ is small enough for some of the object to remain above the surface of the liquid.) c) Your result in part (b) shows that if the force is suddenly removed, the object will oscillate up and down in simple harmonic motion. Calculate the period of this motion in terms of the density $\rho$ of the liquid and the mass $M$ and cross-section area $A$ of the object.

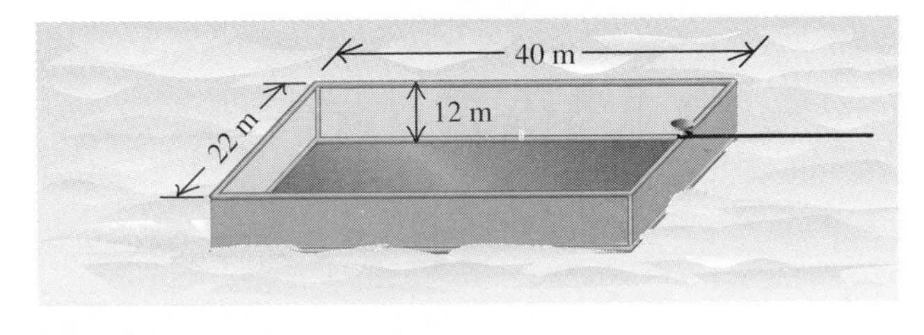


**FIGURE 14–40**

**14–45** A 1500-kg cylindrical can buoy floats vertically in salt water. The diameter of the buoy is 0.900 m. a) Calculate the additional distance the buoy will sink when a 100-kg man stands on top. [Use the expression derived in part (b) of Problem 14–44.] b) Calculate the period of the resulting vertical simple harmonic motion when the man dives off. [Use the expression derived in part (c) of Problem 14–44.]

**14–46** Calculate the period of oscillation of an ice cube 3.50 cm on a side floating in water if it is pushed down and released. [Use the expression derived in part (c) of Problem 14–44.]

**14–47** A piece of wood is 0.500 m long, 0.200 m wide, and 0.030 m thick. Its density is 600 $kg/m^3$. What volume of lead must be fastened underneath to sink the wood in calm water so that its top is just even with the water level? What is the mass of this volume of lead?

**14–48** In seawater a life preserver with a volume of 0.0400 $m^3$ will support an 80.0-kg person (average density 920 $kg/m^3$) with 20% of the person's volume above water when the life preserver is fully submerged. What is the density of the material composing the life preserver?

**14–49** A block of balsa wood placed in one scale pan of an equal-arm balance is found to be exactly balanced by a 0.0750-kg brass mass in the other scale pan. Find the true mass of the balsa wood if its density is 150 $kg/m^3$. Neglect the buoyancy of the brass in air.

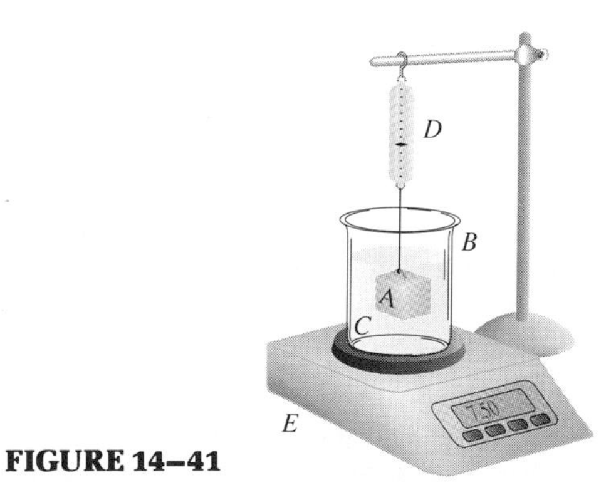


**FIGURE 14–41**

**14–50** Block $A$ in Fig. 14–41 hangs by a cord from spring balance $D$ and is submerged in a liquid $C$ contained in beaker $B$. The mass of the beaker is 1.00 kg; the mass of the liquid is 1.50 kg. Balance $D$ reads 2.50 kg, and balance $E$ reads 7.50 kg. The volume of block $A$ is $3.80 \times 10^{-3}$ $m^3$. a) What is the density of the liquid? b) What will each balance read if block $A$ is pulled up out of the liquid?

**14–51** A piece of gold-aluminum alloy weighs 45.0 N. When the alloy is suspended from a spring balance and submerged in water, the balance reads 32.0 N. What is the weight of gold in the alloy?

**14–52** A cubical block of wood 0.100 m on a side and with a density of 500 kg/m$^3$ floats in a jar of water. Oil with a density of 700 kg/m$^3$ is poured on the water until the top of the oil layer is 0.040 m below the top of the block. a) How deep is the oil layer? b) What is the gauge pressure at the block's lower face?

**14–53 Dropping Anchor.** An iron anchor with mass 25.0 kg and density 786 kg/m$^3$ lies on the deck of a small barge that has vertical sides and is floating in a fresh water river. The area of the bottom of the barge is 6.00 m$^2$. The anchor is thrown overboard but is suspended above the bottom of the river by a rope; the mass and volume of the rope are small enough to ignore. When the anchor is dropped overboard, does the barge rise or sink down in the water, and by what vertical distance?

**14–54** Assume that crude oil from the tanker *Exxon Valdez* has density 730 kg/m$^3$. The tanker ran aground on a reef. To refloat the tanker, its oil cargo is pumped out into steel barrels, each of which has a mass of 20.0 kg when empty and holds 0.120 m$^3$ of oil. Neglect the volume occupied by the steel from which the barrel is made. a) If a salvage worker accidently drops a filled, sealed barrel overboard, will it float or sink in the seawater? b) If the barrel floats, what fraction of its volume will be above the water surface? If it sinks, what minimum tension would have to be exerted by a rope to haul the barrel up from the ocean floor?

**14–55** Repeat Problem 14–54, but let the oil's density be 890 kg/m$^3$ and let the mass of each empty barrel be 35.0 kg.

**14–56** A barge is in a rectangular lock on a fresh water river. The lock is 50.0 m long and 20.0 m wide, and the steel doors on each end are closed. With the barge floating in the lock, a $2.00 \times 10^6$ N load of scrap metal is put onto the barge. The metal has density 9000 kg/m$^3$. a) When the load of scrap metal, initially on the bank, is placed on the barge, what vertical distance does the water in the lock rise? b) The scrap metal is now pushed overboard into the water. Does the water level in the lock rise, fall, or remain the same? If it rises or falls, by what vertical distance does it change?

**14–57** A cubical brass block with sides of length $L$ floats in mercury. a) What fraction of the block is above the mercury surface? b) If water is poured on the mercury surface, how deep must the water layer be so that the water surface just rises to the top of the brass block? (Express your answer in terms of $L$.)

**14–58** What radius must a water drop have for the difference between inside and outside pressures to be 0.0200 atm? Assume that $T = 20°C$.

**14–59** A cubical block of wood 0.30 m on a side is weighted so that its center of gravity is at the point shown in Fig. 14–42a and it floats in water with one half its volume submerged. If the block is "heeled" at an angle of 45.0°, as in Fig. 14–42b, compute the restoring torque (the net torque about a horizontal axis perpendicular to the block and passing through its geometrical center).

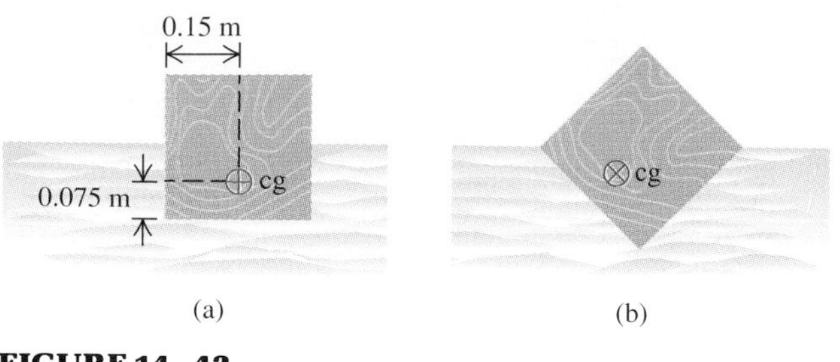

**FIGURE 14–42**

**14–60** Water stands at a depth $H$ in a large open tank whose side walls are vertical (Fig. 14–43). A hole is made in one of the walls at a depth $h$ below the water surface. a) At what distance $R$ from the foot of the wall does the emerging stream of water strike the floor? b) Let $H = 14.0$ m and $h = 3.0$ m. At what height below the water surface could a second hole be cut so that the stream emerging from it would have the same range as that from the first hole?

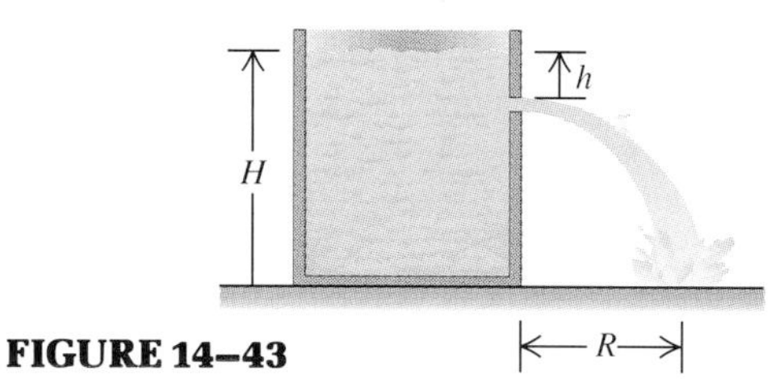


**FIGURE 14–43**

**14–61** A cylindrical bucket, open at the top, is 0.200 m high and 0.100 m in diameter. A circular hole with cross-section area 1.00 cm$^2$ is cut in the center of the bottom of the bucket. Water flows into the bucket from a tube above it at the rate of $1.50 \times 10^{-4}$ m$^3$/s. How high will the water in the bucket rise?

**14–62** Water flows steadily from an open tank, as in Fig. 14–44. The elevation of point 1 is 10.0 m, and the elevation of points 2 and 3 is 2.00 m. The cross-section area at point 2 is 0.0400 m$^2$; at point 3 it is 0.0200 m$^2$. The area of the tank is very large in comparison with the cross-section area of the pipe. Assuming that Bernoulli's equation applies, compute a) the discharge rate in cubic meters per second; b) the gauge pressure at point 2.

**14–63** Modern airplane design calls for a lift due to the net force of the moving air on the wing of about 1000 N per square meter of wing area. Assume that air flows past the wing of an aircraft with streamline flow. If the speed of flow past the lower wing surface is

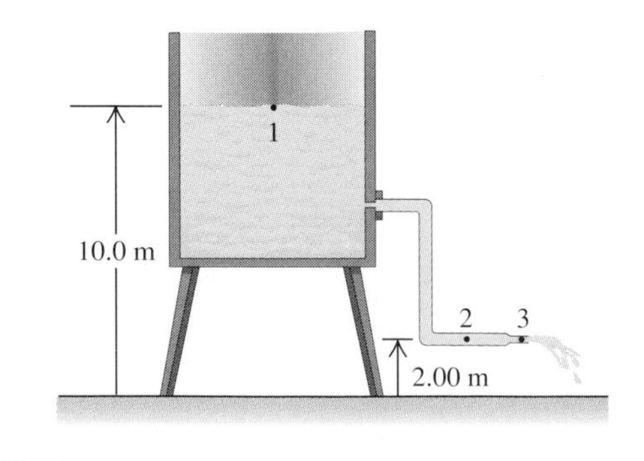

**FIGURE 14–44**

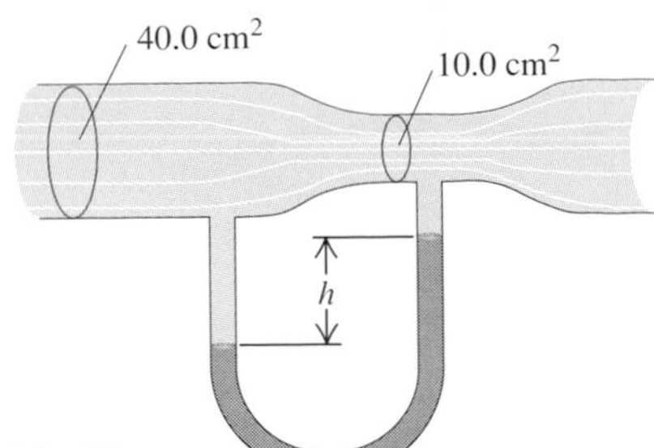


FIGURE 14–45

140 m/s, what is the required speed over the upper surface to give a lift of 1000 N/m$^2$? The density of the air is 1.20 kg/m$^3$.

**14–64** The horizontal section of pipe shown in Fig. 14–45 has a cross-section area of 40.0 cm$^2$ at the wider portions and 10.0 cm$^2$ at the constriction. Water is flowing in the pipe, and the discharge from the pipe is $4.00 \times 10^{-3}$ m$^3$/s (3.00 L/s). Find a) the flow speeds at the wide and the narrow portions; b) the pressure difference between these portions; c) the difference in height between the mercury columns in the U-shaped tube.

**14–65** Two very large open tanks, $A$ and $F$ (Fig. 14–46), contain the same liquid. A horizontal pipe $BCD$, having a constriction at $C$, leads out of the bottom of tank $A$, and a vertical pipe $E$ opens into the constriction at $C$ and dips into the liquid in tank $F$. Assume streamline flow and no viscosity. If the cross-section area at $C$ is half that at $D$, and if $D$ is a distance $h_1$ below the level of the liquid in $A$, to what height $h_2$ will liquid rise in pipe $E$? Express your answer in terms of $h_1$.

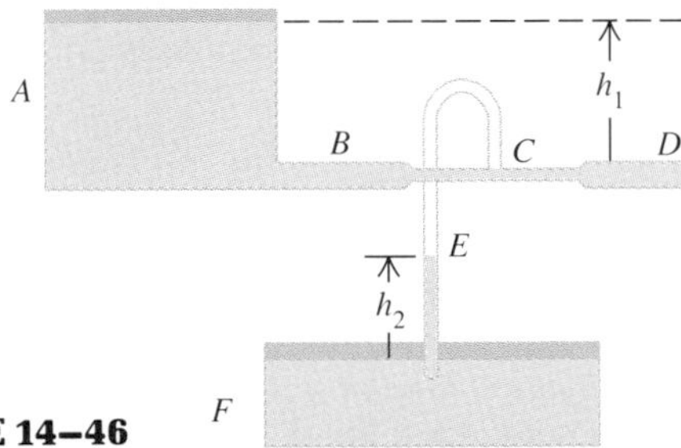


FIGURE 14–46

**14–66** a) With what speed is a steel ball 2.00 mm in radius falling in a tank of glycerin at an instant when its acceleration is half that of a freely falling body? The viscosity of the glycerin is 8.30 poise. b) What is the terminal speed of the ball?

**14–67** Figure 14–32a shows a liquid flowing from a vertical pipe. Note that the vertical stream of the liquid has a very definite shape as it flows from the pipe. a) To get the equation for this shape, assume that the liquid is in free fall once it leaves the pipe and then find an equation for the speed of the liquid as a function of the distance it has fallen. Combining this with the equation of continuity, find an expression for the radius of the stream of liquid. b) If water flows out of a vertical pipe with a speed of 1.60 m/s as it exits from the pipe, how far below the outlet will the radius be half of the original radius of the stream?

**14–68** Oil with a viscosity of 3.00 poise and a density of 860 kg/m$^3$ is to be pumped from one large open tank to another through 1.00 km of smooth steel pipe that is 0.150 m in diameter. The line discharges into the air at a point 30.0 m above the level of the oil in the supply tank. a) What gauge pressure, in pascals and in atmospheres, must the pump exert to maintain a flow of 0.0600 m$^3$/s? b) What is the power consumed by the pump?

**14–69 Speed of a Bubble in Liquid.** a) With what terminal speed does an air bubble 2.00 mm in diameter rise in a liquid with viscosity 1.50 poise and density 800 kg/m$^3$? (Assume that the density of the air is 1.20 kg/m$^3$.) b) What is the terminal speed of the same bubble in water at 20°C, which has viscosity 1.005 centipoise?

**14–70** The tank at the left in Fig. 14–47a has a very large cross-section area and is open to the atmosphere. The depth is $y = 0.500$ m. The cross-section areas of the horizontal tubes leading out of the tank are 1.00 cm$^2$, 0.50 cm$^2$, and 0.20 cm$^2$, respectively. The liquid is ideal, having zero viscosity. a) What is the volume rate of flow out of the tank? b) What is the speed in each portion of the horizontal tube? c) What is the height of the liquid in each of the five vertical side tubes? d) Suppose the liquid in Fig. 14–47b has a viscosity of 0.500 poise and a density of 800 kg/m$^3$ and that the depth of liquid in the large tank is such that the volume rate of flow is the same as that in part (a). The distance between the side tubes at $c$ and $d$, and between those at $e$ and $f$, is 0.200 m. The cross-section areas of the horizontal tubes are the same in both diagrams. What is the difference in level between the tops of the liquid columns in tubes $c$ and $d$? e) In tubes $e$ and $f$? f) What is the flow speed on the axis of each part of the horizontal tube?

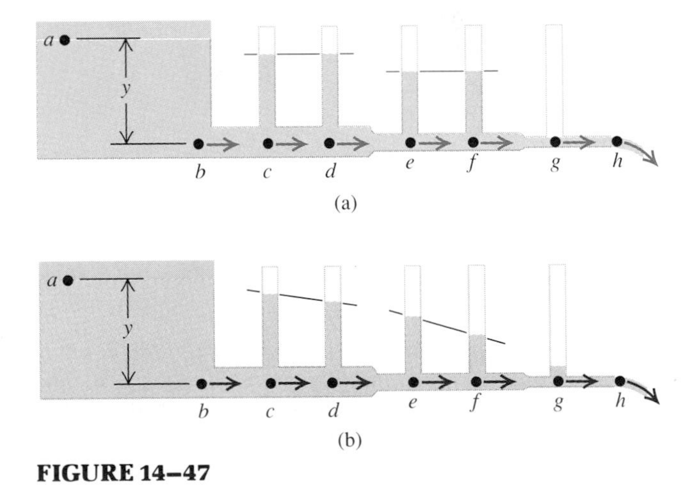


FIGURE 14–47

## CHALLENGE PROBLEMS

**14–71** Suppose a piece of styrofoam, $\rho = 150$ kg/m$^3$, is held completely submerged in water (see Fig. 14–48 on page 412). a) What is the tension in the cord? Find this using Archimedes' principle. b) Use $p = p_0 + \rho g h$ to directly calculate the force on the two sloped sides and the bottom of the styrofoam; then show that the vector sum of these forces is the buoyant force.

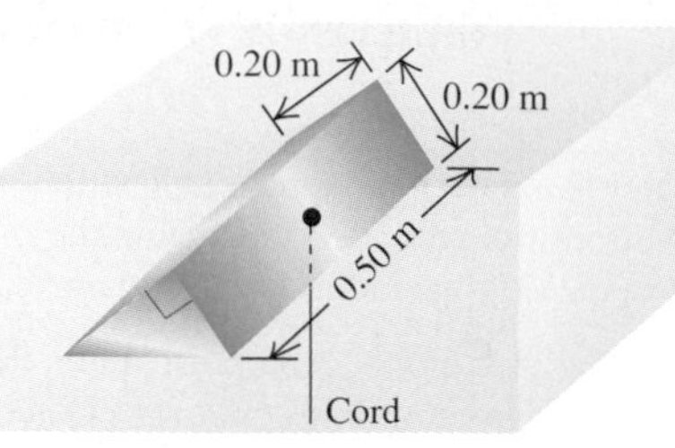


**FIGURE 14–48**

**14–72** A rock with mass $m = 5.00$ kg is suspended from the roof of an elevator by a light cord. The rock is totally immersed in a bucket of water that sits on the floor of the elevator, but the rock doesn't touch the bottom or sides of the bucket. a) When the elevator is at rest, the tension in the cord is 38.0 N. Calculate the volume of the rock. b) Derive an expression for the tension in the cord when the elevator is accelerating *upward* with an acceleration $a$. Calculate the tension when $a = 2.50\ \mathrm{m/s^2}$ upward. c) Derive an expression for the tension in the cord when the elevator is accelerating *downward* with an acceleration $a$. Calculate the tension when $a = 2.50\ \mathrm{m/s^2}$ downward. d) What is the tension when the elevator is in free fall with a downward acceleration equal to $g$?

**14–73** A U-shaped tube with a horizontal portion of length $l$ (see Fig. 14–49) contains a liquid. What is the difference in height between the liquid columns in the vertical arms a) if the tube has an acceleration $a$ toward the right? b) If the tube is mounted on a horizontal turntable rotating with an angular velocity $\omega$ with one of the vertical arms on the axis of rotation? c) Explain why the difference in height does not depend on the density of the liquid or on the cross-section area of the tube. Would it be the same if the vertical tubes did not have equal cross-section areas? Would it be the same if the horizontal portion were tapered from one end to the other?

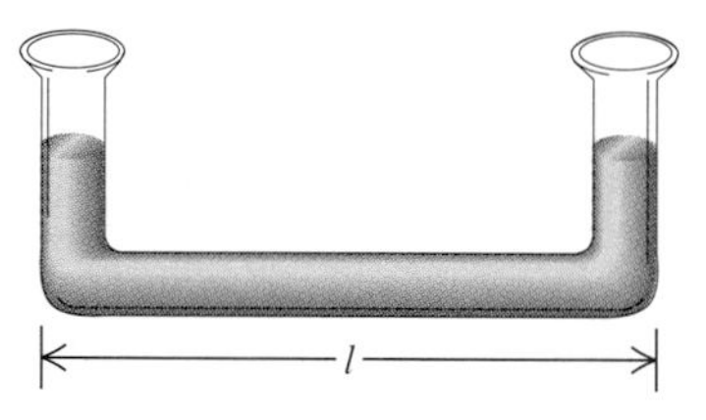


**FIGURE 14–49**

**14–74** A large tank with diameter $D$, open to the air, contains water to a height $H$. A small hole with diameter $d$ ($d \ll D$) is made at the base of the tank. Neglecting any effects of viscosity, calculate the time it takes for the tank to drain completely.

**14–75** A *siphon,* as depicted in Fig. 14–50, is a convenient device for removing liquids from containers. To establish the flow, the tube must be initially filled with fluid. Let the fluid have density $\rho$ and let the atmospheric pressure be $p_a$. Assume that the cross-section area of the tube is the same at all points along it. a) If the lower end of the siphon is at a distance $h$ below the surface of the liquid in the container, what is the speed of the fluid as it flows out the lower end of the siphon? (Assume that the container has a very large diameter, and neglect any effects of viscosity.) b) A curious feature of a siphon is that the fluid initially flows "uphill." What is the greatest height $H$ that the high point of the tube can have and the flow will still occur?

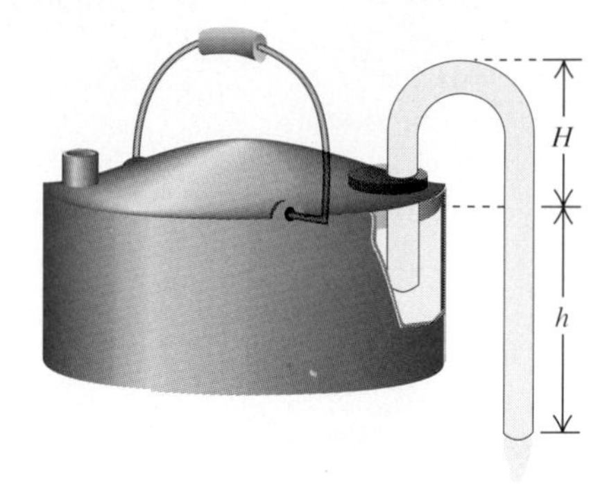


**FIGURE 14–50**

**14–76** For an object falling in air, Stokes' law is obeyed only for small speeds. The air drag on an object falling through the earth's atmosphere is more accurately given by $\frac{1}{2}CA\rho v^2$, where $A$ is the area of the object perpendicular to the motion, $\rho$ is the density of the air, and $C$ is the drag coefficient. (Refer to Section 6–6.) $C$ is determined experimentally; for a spherical object having the speeds of this problem it is approximately 0.5. [*Note:* For a graph of the values of $C$ over a wide range of values of speed, see J. E. A. John and W. Haberman, *Introduction to Fluid Mechanics* (Englewood Cliffs, N.J.: Prentice-Hall), p. 198.] A ball bearing with radius 0.500 cm and mass 4.00 g is dropped from the top of the Sears Tower in Chicago, which is 443 m tall. a) Calculate the speed the ball bearing would have at the ground if air resistance were negligible. b) Taking into account air resistance, what is the terminal speed of the ball bearing? c) What is its actual speed just before reaching the ground? (*Note:* Refer to Problem 5–86.)

**14–77** The following is quoted from a letter. How would you reply?

*It is the practice of carpenters hereabouts, when laying out and leveling up the foundations of relatively long buildings, to use a garden hose filled with water, into the ends of the hose being thrust glass tubes 10 to 12 inches long.*

*The theory is that water, seeking a common level, will be of the same height in both the tubes and thus effect a level. Now the question rises as to what happens if a bubble of air is left in the hose. Our greybeards contend the air will not affect the reading from one end to the other. Others say that it will cause important inaccuracies.*

Can you give a relatively simple answer to this question, together with an explanation? Fig. 14–51 gives a rough sketch of the situation that caused the dispute.

**FIGURE 14–51**

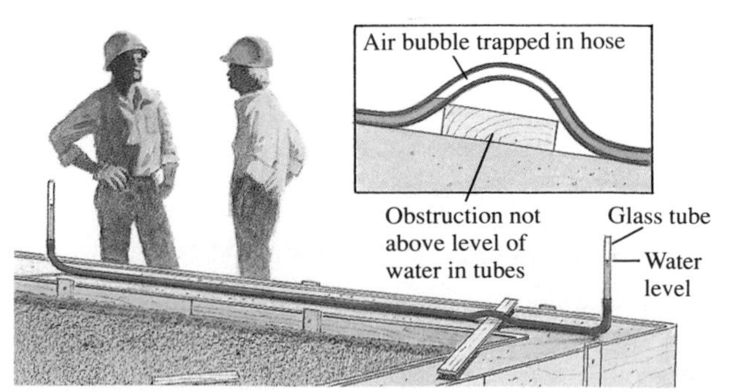

CHAPTER

# 15 Temperature and Heat

Steel is an alloy; its principal constituents are iron and carbon. It can have several solid phases with differing crystal structure, depending on temperature and amount of carbon. The phase diagram shows this relationship, and it also shows how the melting temperature depends on the amount of carbon. Cast iron usually has 2 to 3.5% carbon, while steels intended for forging usually have less than 1.2% carbon. Specialty steels have small amounts of other metals, such as chromium, molybdenum, and vanadium.

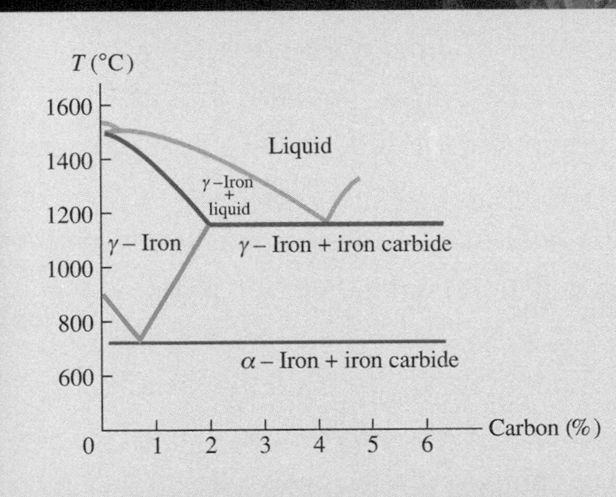


Metals used for jewelry and fine craft work are often worked at room temperature or softened by heating in a small flame.

## KEY CONCEPTS

**• Temperature is a quantitative description of hotness and coldness. A thermometer is an instrument that measures temperature. Two bodies are in thermal equilibrium if their temperatures are the same. Heat flow occurs between bodies only when their temperatures are different.**

**• The gas-thermometer temperature scale is nearly independent of the gas used. This thermometer suggests the existence of an absolute zero point for temperature.**

**• Materials usually expand when the temperature increases. The changes in volume and linear dimensions are approximately proportional to the temperature change.**

**• The amount of heat needed to increase the temperature of a quantity of matter depends on the mass of material, the temperature change, and a property of the material called the specific heat capacity. Heat transfer can also occur during a change of phase when there is no temperature change.**

**• The three mechanisms for heat transfer are conduction, convection, and radiation. The rate of heat transfer by conduction depends on the temperature difference and the cross-section area across which heat flows. The rate of radiative heat transfer depends on the fourth power of the absolute temperature.**

## INTRODUCTION

Which season do you like better, summer or winter? Whatever your choice, you use a lot of physics to keep comfortable in changing weather conditions. Your body needs to be kept at constant temperature. It has effective temperature-control mechanisms, but sometimes it needs help. On a hot day you wear less clothing to improve heat transfer from your body to the air and to allow better cooling by evaporation of perspiration. You probably drink cold beverages, possibly with ice in them, and sit near a fan or in an air-conditioned room. On a cold day you wear more clothes or stay indoors where it's warm. When you're outside, you keep active and drink hot liquids to stay warm. The concepts in this chapter will help you understand the basic physics of keeping warm or cool.

First we need to define temperature, including temperature scales and ways to measure temperature. Next we discuss changes of dimensions and volumes caused by temperature changes. We'll encounter the concept of *heat,* which describes energy transfer caused by temperature differences, and we'll learn to calculate the *rates* of such energy transfers.

This chapter lays the groundwork for the subject of *thermodynamics,* the study of energy transformations and their relationships to the properties of matter. Thermodynamics forms an indispensable part of the foundation of physics, chemistry, and the life sciences, and its applications turn up in such places as car engines, refrigerators, biochemical processes, and the structure of stars.

## 15–1 TEMPERATURE AND THERMAL EQUILIBRIUM

The concept of **temperature** is rooted in qualitative ideas of "hot" and "cold" based on our sense of touch. A body that feels hot usually has a higher temperature than the same body when it feels cold. That's pretty vague, and the senses can be deceived. But many properties of matter that we can *measure* depend on temperature. The length of a metal rod, steam pressure in a boiler, the ability of a wire to conduct an electric current, and the color of a very hot glowing object—all these depend on temperature.

Temperature is also related to the kinetic energies of the molecules of a material. In general, this relationship is fairly complex, so it's not a good place to start in defining temperature. In Chapter 16 we'll look at the relationship between temperature and the energy of molecular motion for an ideal gas. It is important to understand, however, that temperature and heat are inherently *macroscopic* concepts. They can and must be defined independently of any detailed molecular picture. In this section we'll develop a macroscopic definition of temperature.

Temperature is a measure of hotness or coldness on some scale. But *what* scale? We can use any measurable property of a system that varies with its "hotness" or "coldness" to assign numbers to various states of hotness and coldness. Figure 15–1a shows a familiar type of thermometer. When the system becomes hotter, the liquid (usually mercury or ethanol) expands and rises in the tube, and the value of $L$ increases. Another simple system is a quantity of gas in a constant-volume container (Fig. 15–1b). The pressure $p$, measured by the gauge, increases or decreases as the gas becomes hotter or colder. A third example is the electrical resistance $R$ of a conducting wire, which also varies when the wire becomes hotter or colder. Each of these properties gives us a number ($L$, $p$, or $R$) that varies with hotness and coldness, so each property can be used to make a **thermometer.**

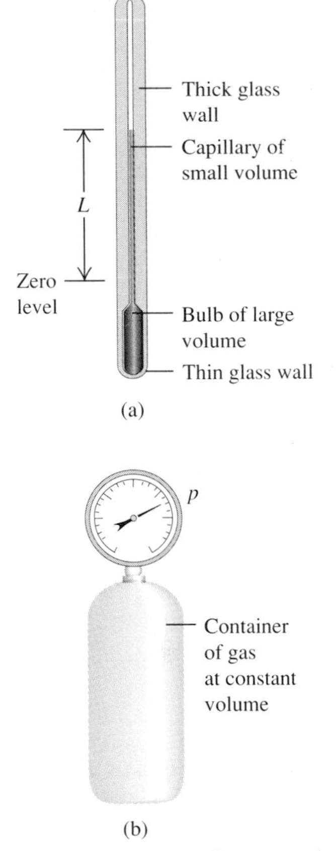


**15–1** (a) A system whose state is specified by the value of length $L$. (b) A system whose state is given by the value of the pressure $p$.

To measure the temperature of a body, you place the thermometer in contact with the body. If you want to know the temperature of a cup of hot coffee, you stick the thermometer in the coffee; as the two interact, the thermometer becomes hotter and the coffee cools off a little. After the thermometer settles down to a steady value, you read the temperature. The system has reached an *equilibrium* condition, in which the interaction between the thermometer and the coffee causes no further change in the system. We call this a state of **thermal equilibrium.**

In a state of thermal equilibrium, the thermometer and the coffee have the same temperature. How do we know this? Consider the interaction of any two systems, $A$ and $B$. System $A$ might be the tube-and-liquid system of Fig. 15–1a, and system $B$ the container of gas and pressure gauge of Fig. 15–1b. When the two systems come into contact, in general the values of $L$ and $p$ change as the systems move toward thermal equilibrium. Initially, one system is hotter than the other, and during the interaction each system changes the state of the other.

If the systems are separated by an insulating material, or **insulator,** such as wood, plastic foam, or fiberglass, they influence each other more slowly. An *ideal insulator* is a material that permits *no* interactions at all between the two systems. It *prevents* the systems from attaining thermal equilibrium if they aren't in thermal equilibrium at the start. This is why camping coolers are made with insulating materials. We don't want the ice and cold food inside to warm up and attain thermal equilibrium with the hot summer air outside; the insulation delays this, although not forever.

Now consider three systems, $A$, $B$, and $C$, that are initially *not* in thermal equilibrium (Fig. 15–2). We surround them with an ideal insulating box so that

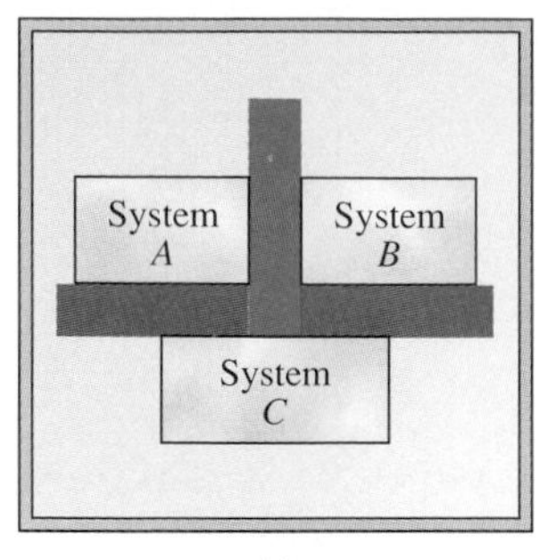


(a)

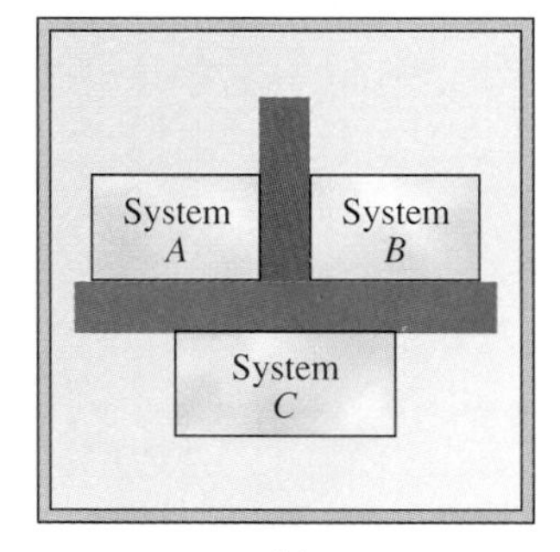


(b)

**15–2** The zeroth law of thermodynamics. (a) If $A$ and $B$ are each in thermal equilibrium with $C$, then (b) $A$ and $B$ are in thermal equilibrium with each other. Blue layers represent insulating walls; orange layers represent conducting walls.

they cannot interact with anything except each other. We separate $A$ and $B$ with an ideal insulating wall, the blue slab in Fig. 15–2a, but we let system $C$ interact with both $A$ and $B$. This interaction is shown in the figure by an orange slab representing a thermal **conductor,** a material that *permits* thermal interactions through it. We wait until thermal equilibrium is attained; then $A$ and $B$ are each in thermal equilibrium with $C$. But are they in thermal equilibrium *with each other?*

To find out, we separate $C$ from $A$ and $B$ with an ideal insulating wall (Fig. 15–2b), and then we replace the insulating wall between $A$ and $B$ with a *conducting* wall that lets $A$ and $B$ interact. What happens? Experiment shows that *nothing* happens; the states of $A$ and $B$ do not change. Conclusion: If $C$ is initially in thermal equilibrium with both $A$ and $B$, then $A$ and $B$ are also in thermal equilibrium with each other. This result may seem trivial and obvious, but even so it needs to be verified by experiment.

**Two systems each in thermal equilibrium with a third system are in thermal equilibrium with each other.** This principle is called the **zeroth law of thermodynamics.** The importance of this law was recognized only after the first, second, and third laws of thermodynamics had been named. Since it is fundamental to all of them, the label "zeroth" seemed appropriate.

The temperature of a system determines whether or not the system is in thermal equilibrium with another system. **Two systems are in thermal equilibrium if and only if they have the same temperature.** When the temperatures of two systems are different, the systems *cannot* be in thermal equilibrium. A thermometer actually measures *its own* temperature, but when a thermometer is in thermal equilibrium with another body, the temperatures must be equal. A temperature scale, such as the Celsius or Fahrenheit scale, is just a particular scheme for assigning numbers to temperatures. We will study temperature scales in the next section.

## 15–2 THERMOMETERS AND TEMPERATURE SCALES

To make the tube-and-liquid device shown in Fig. 15–1a into a thermometer, we need a scale with numbers on it. These numbers are arbitrary; historically, many different schemes have been used. Suppose we label the thermometer's liquid level at the freezing temperature of pure water "zero" and the level at the boiling temperature "100," and divide the distance between these two points into 100 equal intervals called *degrees.* The result is the **Celsius temperature scale** (formerly called the *centigrade scale* in English-speaking countries). We interpolate between these reference temperatures 0°C and 100°C, or extrapolate beyond them, by using the liquid-in-tube thermometer or one of the other thermometers that we will describe. The temperature for a state colder than freezing water is a negative number. The Celsius scale is used, both in everyday life and in science and industry, everywhere in the world except for a few English-speaking countries.

Another common type of thermometer contains a *bimetallic strip,* made by welding strips of two different metals together (Fig. 15–3a). When the system gets hotter, one metal expands more than the other, so the composite strip bends when the temperature changes. This strip is usually formed into a spiral, with the outer end anchored to the thermometer case and the inner end attached to a pointer (Fig. 15–3b). The pointer rotates in response to temperature changes.

In a *resistance thermometer* the changing electrical resistance of a coil of fine wire, a carbon cylinder, or a germanium crystal is measured. Because resistance can be measured very precisely, resistance thermometers are usually more precise than most other types.

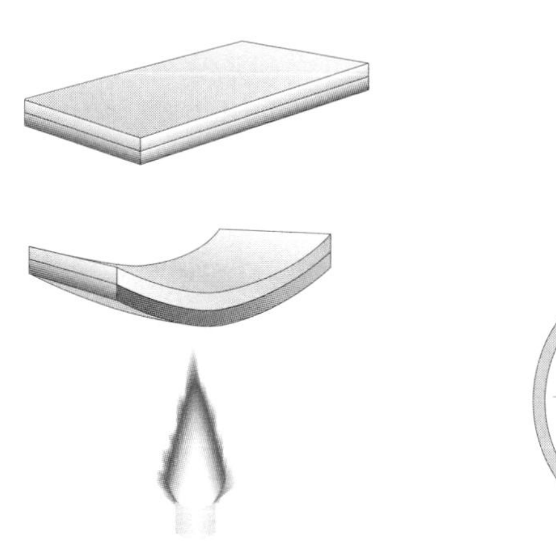


**15–3** (a) A bimetallic strip bends when heated because one metal expands more than the other. (b) A bimetallic strip, usually in the form of a spiral, may be used as a thermometer.

To measure very high temperatures, an *optical pyrometer* can be used. It measures the intensity of radiation emitted by a red-hot or white-hot substance (Fig. 15–4). The instrument does not touch the hot substance, so the optical pyrometer can be used at temperatures that would destroy most other thermometers.

With the **Fahrenheit temperature scale,** still used in everyday life in the United States, the freezing temperature of water is 32°F (thirty-two degrees Fahrenheit), and the boiling temperature is 212°F, both at normal atmospheric pressure. There are 180 degrees between freezing and boiling, compared to 100 on the Celsius scale, so one Fahrenheit degree represents only $\frac{100}{180}$, or $\frac{5}{9}$, as great a temperature change as one Celsius degree.

To convert temperatures from Celsius to Fahrenheit, note that a Celsius temperature $T_C$ is the number of Celsius degrees above freezing; the number of Fahrenheit degrees above freezing is $\frac{9}{5}$ of this. But freezing on the Fahrenheit scale is at 32°F, so to obtain the actual Fahrenheit temperature, multiply the Celsius value by $\frac{9}{5}$ and then add 32°. Symbolically,

$$T_F = \tfrac{9}{5}T_C + 32°. \tag{15–1}$$

To convert Fahrenheit to Celsius, solve this equation for $T_C$:

$$T_C = \tfrac{5}{9}(T_F - 32°). \tag{15–2}$$

That is, subtract 32° to get the number of Fahrenheit degrees above freezing, and then multiply by $\frac{5}{9}$ to obtain the number of Celsius degrees above freezing, that is, the Celsius temperature.

We don't recommend memorizing Eqs. (15–1) and (15–2). Instead, try to understand the reasoning that led to them so that you can derive them on the spot when you need them, checking your reasoning with the relation 100°C = 212°F.

It is useful to distinguish between an actual temperature and a temperature *interval* (a difference or change in temperature). An actual temperature of 20° is stated as 20°C (twenty degrees Celsius), and a temperature *interval* of 10° is 10 C° (ten Celsius degrees). A beaker of water heated from 20°C to 30°C undergoes a temperature change of 10 C°.

When we calibrate two thermometers, such as a liquid-in-tube system and a resistance thermometer, so that they agree at 0°C and 100°C, they may not agree exactly at intermediate temperatures. Any temperature scale defined in this way depends somewhat on the specific properties of the material used. Ideally, we would like to define a temperature scale that *doesn't* depend on the properties of a particular material. The gas thermometer, discussed in the next section, comes close. But to establish a truly material-independent scale, we first need to develop some principles of thermodynamics. We'll return to this fundamental problem in Chapter 18.

**15–4** An optical pyrometer measures high temperature by measuring the intensity of radiation emitted by the hot object.

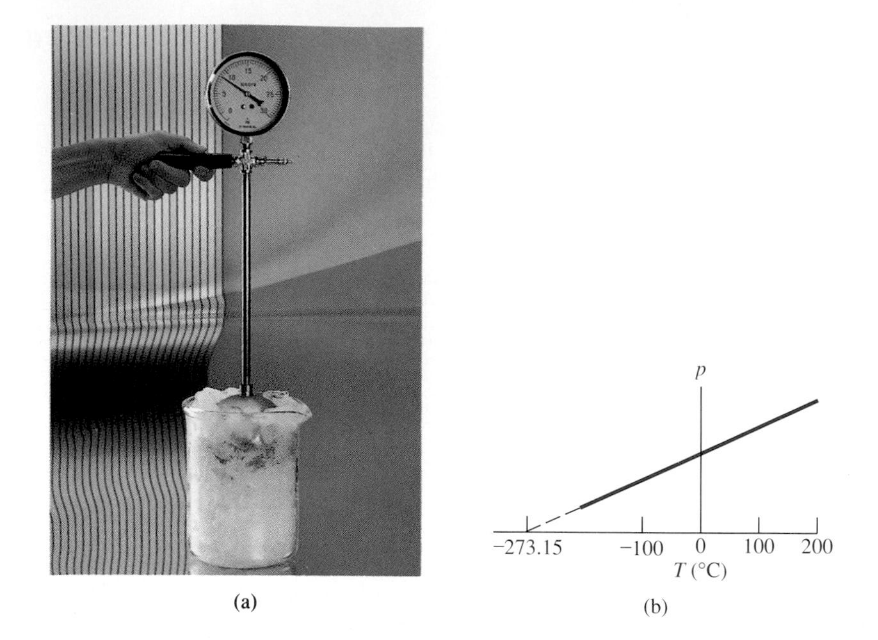


(a)

(b)

**15–5** (a) A constant-volume gas thermometer. (b) Graph of pressure versus temperature for a constant-volume low-density gas thermometer. A straight-line extrapolation predicts that the pressure would become zero at −273.15°C if that temperature could be reached and the proportionality of pressure to temperature held exactly.

## 15–3 GAS THERMOMETERS AND THE KELVIN SCALE

The pressure of a gas increases with temperature; we can make a thermometer based on this behavior. The gas, often helium, is placed in a constant-volume container (Fig. 15–5a), and its pressure is measured by one of the devices described in Section 15–2. To calibrate a constant-volume gas thermometer, we measure the pressure at two temperatures, say 0°C and 100°C, plot these points on a graph, and draw a straight line between them (Fig. 15–5b). Then we can read from the graph the temperature corresponding to any other pressure.

By extrapolating this graph we see that there is a hypothetical temperature at which the gas pressure would become zero. We might expect that this temperature would be different for different gases, but it turns out to be the same for many different gases (at least in the limit of very low gas concentrations), namely, −273.15°C. We can't actually observe this zero-pressure condition. Gases liquefy and solidify at very low temperatures, and the proportionality of pressure to temperature no longer holds.

We can use this extrapolated zero-pressure temperature as the basis for a new temperature scale with its zero at this temperature. This is the **Kelvin temperature scale,** named for Lord Kelvin (1824–1907). The units are the same size as those of the Celsius scale, but the zero is shifted so that 0 K = −273.15°C and 273.15 K = 0°C; that is,

$$T_K = T_C + 273.15. \tag{15–3}$$

A common room temperature, 20°C ( = 68°F), is 20 + 273.15, or about 293 K.

In SI nomenclature, "degree" is not used with the Kelvin scale; the temperature mentioned above is read "293 kelvins," not "degrees Kelvin." We capitalize Kelvin when it refers to the temperature scale, but the *unit* of temperature is the *kelvin,* not capitalized but abbreviated K.

### EXAMPLE 15–1

**Body temperature** You place a small piece of melting ice in your mouth. Eventually, the water all converts from ice at $T_1 = 32.00°F$ to body temperature, $T_2 = 98.60°F$. Express these temperatures in degrees Celsius and kelvins, and find $\Delta T = T_2 - T_1$ in both cases.

**SOLUTION** First we find the Celsius temperatures. We know that $T_1 = 32.00°F = 0.00°C$, and 98.60°F is 98.60 − 32.00 = 66.60 F° above freezing; multiply this by (5 C°/9 F°) to find 37.00 C° above freezing, or $T_2 = 37.00°C$.

To get the Kelvin temperatures, we just add 273.15 to each

Celsius temperature: $T_1 = 273.15$ K and $T_2 = 310.15$ K.

Many hospitals now use Celsius thermometers; "normal" body temperature is 37.0°C. On the other hand, if your doctor says your temperature is 310 K, don't be alarmed.

The temperature *difference* $\Delta T = T_2 - T_1$ is 37.00 C° = 37.00 K. The Celsius and Kelvin scales have different zero points but the same size degrees. Therefore *any temperature difference is the same on the two scales.*

---

The Celsius scale has two fixed points: the normal freezing and boiling temperatures of water. But we can define the Kelvin scale using a gas thermometer with only a single reference temperature. We define the ratio of any two temperatures $T_1$ and $T_2$ on the Kelvin scale as the ratio of the corresponding gas-thermometer pressures $p_1$ and $p_2$:

$$\frac{T_2}{T_1} = \frac{p_2}{p_1}. \tag{15–4}$$

To complete the definition, we need only specify the Kelvin temperature of a single specific state. For reasons of precision and reproducibility the state chosen is the *triple point* of water. This is the unique condition of temperature and pressure at which solid water (ice), liquid water, and water vapor can all coexist. This occurs at a temperature of 0.01°C and a water-vapor pressure of 610 Pa (about 0.006 atm). (This is the pressure of the *water;* it has nothing to do directly with the gas pressure in the *thermometer.*) The triple-point temperature $T_{\text{triple}}$ of water is *defined* to have the value $T_{\text{triple}} = 273.16$ K. Thus with reference to Eq. (15–4), if $p_{\text{triple}}$ is the pressure in a gas thermometer at temperature $T_{\text{triple}}$ and $p$ is the pressure at some other temperature $T$, then $T$ is given on the Kelvin scale by

$$T = T_{\text{triple}}\frac{p}{p_{\text{triple}}} = 273.16\frac{p}{p_{\text{triple}}}. \tag{15–5}$$

## EXAMPLE 15–2

Suppose a constant-volume gas thermometer has a pressure of $1.50 \times 10^4$ Pa at temperature $T_{\text{triple}}$ and a pressure of $1.95 \times 10^4$ Pa at some unknown temperature $T$. What is $T$?

**SOLUTION** From Eq. (15–5) the Kelvin temperature is

$$T = (273.16\ \text{K})\frac{1.95 \times 10^4\ \text{Pa}}{1.50 \times 10^4\ \text{Pa}} = 355\ \text{K} \quad (= 82°\text{C}).$$

---

The relationships of the three temperature scales we have discussed are shown graphically in Fig. 15–6. The Kelvin scale is called an **absolute temperature scale,** and its zero point is called **absolute zero.** To define more completely what we mean by absolute zero, we need to use the thermodynamic principles developed in the next several chapters. We will return to this concept in Chapter 18.

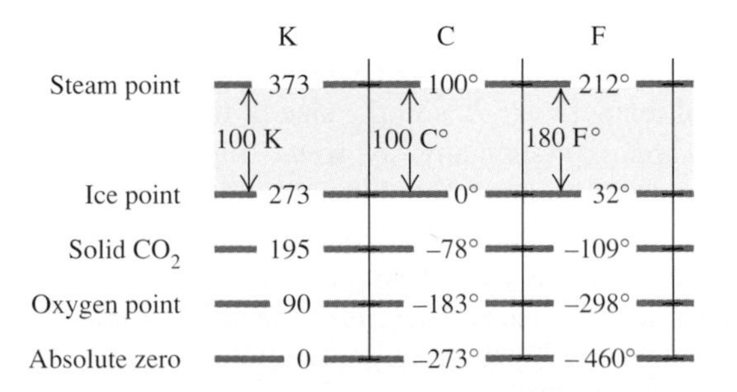


**15–6** Relations among Kelvin, Celsius, and Fahrenheit temperature scales. Temperatures have been rounded off to the nearest degree.

Low-pressure gas thermometers using various gases are found to agree very closely. They are used principally to establish high-precision standards and to calibrate other thermometers. They are large, bulky, and very slow to come to thermal equilibrium.

## 15–4 THERMAL EXPANSION

Most materials expand when their temperatures increase. Bridge decks need special joints and supports to allow for expansion. A completely filled and tightly capped bottle of water cracks when it is heated, but you can loosen a metal jar lid by running hot water on it. These are all examples of *thermal expansion*.

### Linear Expansion

Suppose a rod of material has a length $L_0$ at some initial temperature $T_0$. When the temperature changes by $\Delta T$, the length changes by $\Delta L$. Experiment shows that if $\Delta T$ is not too large (say, less than 100 C° or so), $\Delta L$ is *directly proportional* to $\Delta T$. If two rods made of the same material have the same temperature change, but one is twice as long as the other, then the *change* in its length is also twice as great. Therefore $\Delta L$ must also be proportional to $L_0$. Introducing a proportionality constant $\alpha$ (which is different for different materials), we may express these relations in an equation:

$$\Delta L = \alpha L_0 \, \Delta T. \tag{15–6}$$

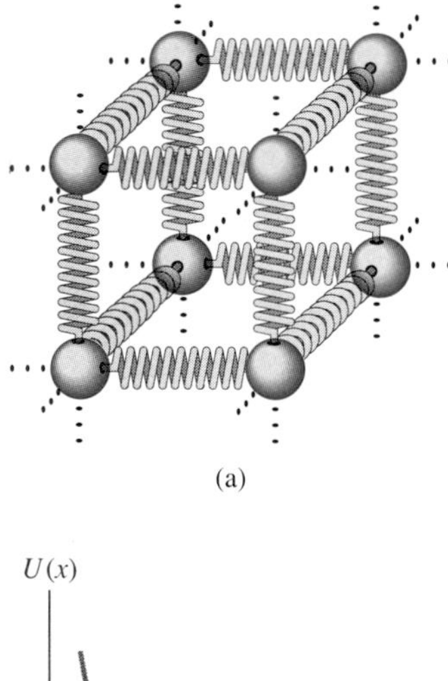


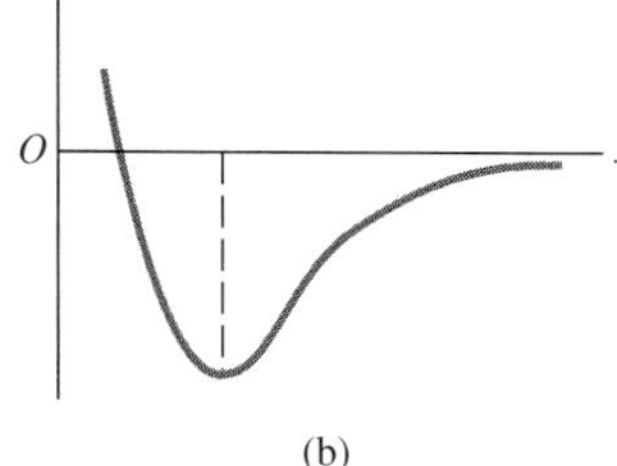


**15–7** (a) We can visualize the forces between neighboring atoms in a solid by imagining them to be connected by springs that are easier to stretch than compress. (b) A graph of potential energy versus distance between neighboring atoms shows that the forces are not symmetric.

If a body has length $L_0$ at temperature $T_0$, then its length $L$ at a temperature $T = T_0 + \Delta T$ is

$$L = L_0 + \Delta L = L_0(1 + \alpha \, \Delta T). \tag{15–7}$$

The constant $\alpha$, which describes the thermal expansion properties of a particular material, is called the **coefficient of linear expansion.** The units of $\alpha$ are $K^{-1}$ or $(C°)^{-1}$. (Remember that a temperature *interval* is the same in Celsius and Kelvin.) For many materials every linear dimension changes according to Eq. (15–6) or (15–7). Thus $L$ could be the thickness of a rod, the side length of a square sheet, or the diameter of a hole. Some materials, such as wood or single crystals, expand differently in different directions. We won't consider this complication.

We can understand thermal expansion qualitatively on a molecular basis. Picture the interatomic forces in a solid as springs (Fig. 15–7). Each atom vibrates about its equilibrium position. When the temperature increases, the amplitude and associated energy of the vibration also increase. The interatomic spring forces and their associated potential energies are not symmetric about the equilibrium position; they usually behave like a spring that is easier to stretch than to compress. When the amplitude of vibration increases, the *average* distance between molecules also increases. As the atoms get farther apart, every dimension increases, including the sizes of holes.

The direct proportionality expressed by Eq. (15–6) is not exact; it is *approximately* correct for sufficiently small temperature changes. For a given material, $\alpha$ varies somewhat with the initial temperature $T_0$ and the size of the temperature interval. But Eq. (15–6) is at best an approximation anyway, so we'll ignore this complication. Average values of $\alpha$ for several materials are listed in Table 15–1. Within the precision of these values we don't need to worry about whether $T_0$ is 0°C or 20°C or some other temperature. We also note that $\alpha$ is usually a very small number; even for a temperature change of 100 C°, the fractional length change $\Delta L / L_0$ is of the order of $\frac{1}{1000}$.

**TABLE 15–1 Coefficients of Linear Expansion**

| Material | $\alpha$ [$K^{-1}$ or $(C°)^{-1}$] |
|---|---|
| Aluminum | $2.4 \times 10^{-5}$ |
| Brass | $2.0 \times 10^{-5}$ |
| Copper | $1.7 \times 10^{-5}$ |
| Glass | $0.4–0.9 \times 10^{-5}$ |
| Invar (nickel-iron alloy) | $0.09 \times 10^{-5}$ |
| Quartz (fused) | $0.04 \times 10^{-5}$ |
| Steel | $1.2 \times 10^{-5}$ |

## Volume Expansion

Temperature increases usually cause increases in *volume* for both solid and liquid materials. Experiments show that if the temperature change $\Delta T$ is not too great (less than 100 C° or so), the increase in volume $\Delta V$ is approximately proportional to the temperature change. The volume change is also proportional to the initial volume $V_0$, just as with linear expansion, so

$$\Delta V = \beta V_0 \, \Delta T. \qquad (15–8)$$

The constant $\beta$ characterizes the volume expansion properties of a particular material; it is called the **coefficient of volume expansion.** The units of $\beta$ are $K^{-1}$ or $(C°)^{-1}$. As with linear expansion, $\beta$ varies somewhat with temperature, and Eq. (15–8) is an approximate relation for small temperature changes. For many substances, $\beta$ decreases at low temperatures. Several values of $\beta$ in the neighborhood of room temperature are listed in Table 15–2. Note that the values for liquids are much larger than those for solids.

For solid materials there is a simple relation between the volume expansion coefficient $\beta$ and the linear expansion coefficient $\alpha$. To derive this relation, we consider a cube of material with side length $L$ and volume $V = L^3$. At the initial temperature the values are $L_0$ and $V_0$. When the temperature increases by $dT$, the side length increases by $dL$ and the volume increases by an amount $dV$ given by

$$dV = \frac{dV}{dL} dL = 3L^2 \, dL.$$

Now we replace $L$ and $V$ by the initial values $L_0$ and $V_0$. From Eq. (15–6), $dL$ is

$$dL = \alpha L_0 \, dT.$$

**TABLE 15–2 Coefficients of Volume Expansion**

| Solids | $\beta$ ($K^{-1}$) | Liquids | $\beta$ ($K^{-1}$) |
|---|---|---|---|
| Aluminum | $7.2 \times 10^{-5}$ | Carbon disulfide | $115 \times 10^{-5}$ |
| Brass | $6.0 \times 10^{-5}$ | Ethanol | $75 \times 10^{-5}$ |
| Copper | $5.1 \times 10^{-5}$ | Glycerin | $49 \times 10^{-5}$ |
| Glass | $1.2–2.7 \times 10^{-5}$ | Mercury | $18 \times 10^{-5}$ |
| Invar | $0.27 \times 10^{-5}$ | | |
| Quartz (fused) | $0.12 \times 10^{-5}$ | | |
| Steel | $3.6 \times 10^{-5}$ | | |

Thus $dV$ can also be expressed as

$$dV = 3L_0^2 \alpha L_0 \, dT = 3\alpha V_0 \, dT.$$

Comparing this with $dV = \beta V_0 \, dT$, we see that the two are consistent only if

$$\beta = 3\alpha. \tag{15–9}$$

We invite you to check this relation for some of the materials listed in Tables 15–1 and 15–2.

## PROBLEM-SOLVING STRATEGY

### Thermal expansion

**1.** Identify which quantities in Eq. (15–6) or (15–8) are known and which are unknown. Often you will be given two temperatures and will have to compute $\Delta T$. Or you may be given an initial temperature $T_0$ and have to find a final temperature corresponding to a given length or volume change. In this case, find $\Delta T$ first, then the final temperature is $T_0 + \Delta T$.

**2.** Unit consistency is crucial, as always. $L_0$ and $\Delta L$ (or $V_0$ and $\Delta V$) must have the same units, and if you use a value of $\alpha$ or $\beta$ in $K^{-1}$ or $(C°)^{-1}$, then $\Delta T$ must be in kelvins or Celsius degrees (C°). But you can use K and C° interchangeably.

**3.** Remember that the sizes of holes in a material expand with temperature just the same way as any other linear dimension, and the volume of a hole (such as the volume of a container) expands the same way as the corresponding solid shape.

## EXAMPLE 15–3

A surveyor uses a steel measuring tape that is exactly 50.000 m long at a temperature of 20°C. What is its length on a hot summer day when the temperature is 35°C?

**SOLUTION** We have $L_0 = 50.000$ m, $T_0 = 20°C$, $T = 35°C$, and $\Delta T = 15$ C° $= 15$ K. From Eq. (15–6),

$$\begin{aligned}\Delta L &= \alpha L_0 \, \Delta T = (1.2 \times 10^{-5}\,\text{K}^{-1})(50\,\text{m})(15\,\text{K}) \\ &= 9.0 \times 10^{-3}\,\text{m} = 9.0\,\text{mm},\end{aligned}$$

$$L = L_0 + \Delta L = 50.000\,\text{m} + 0.009\,\text{m} = 50.009\,\text{m}.$$

Thus the length at 35°C is 50.009 m. Note that $L_0$ is given to five significant figures but that we need only two of them to compute $\Delta L$.

## EXAMPLE 15–4

The surveyor of Example 15–3 measures a distance when the temperature is 35°C and obtains the result 35.794 m. What is the actual distance?

**SOLUTION** At 35°C the distance between two successive meter marks on the tape is a little more than a meter, by the ratio (50.009 m)/(50.000 m). Thus the true distance is

$$\frac{50.009\,\text{m}}{50.000\,\text{m}}(35.794\,\text{m}) = 35.800\,\text{m}.$$

## EXAMPLE 15–5

A glass flask with a volume of 200 cm$^3$ is filled to the brim with mercury at 20°C. How much mercury overflows when the temperature of the system is raised to 100°C? The coefficient of volume expansion of the glass is $1.2 \times 10^{-5}\,\text{K}^{-1}$.

**SOLUTION** Mercury overflows because $\beta$ is much larger for mercury than for glass. The increase in volume of the flask is

$$\begin{aligned}\Delta V &= \beta V_0 \, \Delta T \\ &= (1.2 \times 10^{-5}\,\text{K}^{-1})(200\,\text{cm}^3)(100°\text{C} - 20°\text{C}) \\ &= 0.19\,\text{cm}^3.\end{aligned}$$

The increase in volume of the mercury is

$$\begin{aligned}\Delta V &= (18 \times 10^{-5}\,\text{K}^{-1})(200\,\text{cm}^3)(100°\text{C} - 20°\text{C}) \\ &= 2.9\,\text{cm}^3.\end{aligned}$$

The volume of mercury that overflows is

$$2.9\ \text{cm}^3 - 0.19\ \text{cm}^3 = 2.7\ \text{cm}^3.$$

This is basically how a mercury-in-glass thermometer works, except that instead of letting the mercury overflow and run all over the place, the thermometer has it rise inside a sealed tube as $T$ increases.

## Thermal Expansion of Water

Water, in the temperature range from 0°C to 4°C, *decreases* in volume with increasing temperature; in this range its coefficient of volume expansion is *negative*. Above 4°C, water expands when heated (Fig. 15–8); thus water has its greatest density at 4°C. Water also expands when it freezes; this is why ice humps up in the middle of the compartments in an ice cube tray. Most materials contract when they freeze.

This anomalous behavior of water has an important effect on plant and animal life in lakes. When a lake cools, the cooled water at the surface flows to the bottom because of its greater density. But when the temperature reaches 4°C, this flow ceases and the water near the surface remains colder (and less dense) than that at the bottom. As the surface freezes, the ice floats because it is less dense than water. The water at the bottom remains at 4°C until nearly the entire lake is frozen. If water behaved like most substances, contracting continuously on cooling and freezing, lakes would freeze from the bottom up. Circulation due to density differences would continuously carry warmer water to the surface for efficient cooling, and lakes would freeze solid much more easily. This would destroy all plant and animal life that can withstand cold water but not freezing. All life forms known on earth depend on chemical systems based on aqueous solutions; if water did not have this special property, the evolution of various life forms would have taken a very different course.

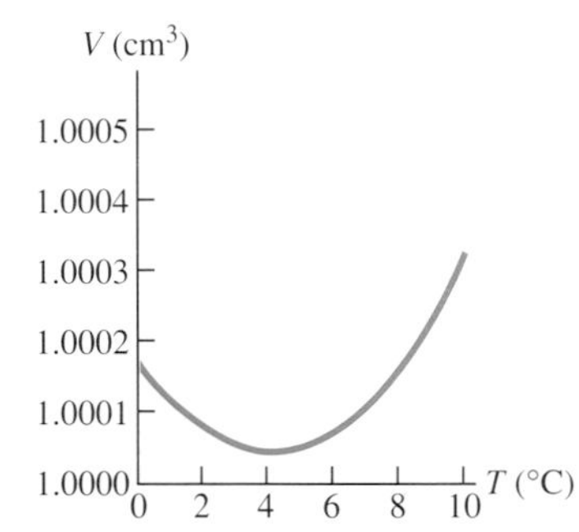


**15–8** The volume of one gram of water in the temperature range from 0°C to 10°C. By 100°C the volume has increased to 1.0434 $\text{cm}^3$. If the coefficient of volume expansion were constant, the curve would be a straight line.

## Thermal Stress

If we clamp the ends of a rod rigidly to prevent expansion or contraction and then change the temperature, tensile or compressive stresses called **thermal stresses** develop. The rod would like to expand or contract, but the clamps won't let it. The resulting stresses may become large enough to strain the rod irreversibly or even break it. Concrete highways and bridge decks (Fig. 15–9) usually have gaps between sections, filled with a flexible material or bridged by interlocking teeth, to permit expansion and contraction of the concrete. Long steam pipes have expansion joints or U-shaped sections to prevent buckling or stretching with temperature changes. If one end of a steel bridge is rigidly fastened to its abutment, the other end usually rests on rollers.

To calculate the thermal stress in a clamped rod, we compute the amount the rod *would* expand (or contract) if not held and then find the stress needed to compress (or stretch) it back to its original length. Suppose a rod with length $L_0$ and cross-section area $A$ is held at constant length while the temperature is reduced (negative $\Delta T$), causing a tensile stress. The fractional change in length if the rod were free to contract would be

$$\frac{\Delta L}{L_0} = \alpha\,\Delta T. \tag{15–10}$$

Both $\Delta L$ and $\Delta T$ are negative. The tension $F$ must increase enough to produce an equal and opposite fractional change in length. From the definition of Young's modulus, Eq. (11–12),

$$Y = \frac{F/A}{\Delta L/L_0}, \qquad \frac{\Delta L}{L_0} = \frac{F}{AY}. \tag{15–11}$$

**15–9** The interlocking teeth of an expansion joint on a bridge; these joints are needed to accommodate changes in length that result from thermal expansion.

If the *total* fractional change in length is zero, then

$$\alpha\,\Delta T + \frac{F}{AY} = 0, \qquad \text{or} \qquad F = -AY\alpha\,\Delta T. \tag{15–12}$$

For a decrease in temperature, $\Delta T$ is negative, so $F$ is positive. The tensile *stress* $F/A$ in the rod is

$$\frac{F}{A} = -Y\alpha\,\Delta T. \tag{15–13}$$

If $\Delta T$ is positive, $F$ and $F/A$ are negative, corresponding to *compressive* force and stress.

## EXAMPLE 15–6

An aluminum cylinder 10 cm long, with a cross-section area of 20 cm$^2$, is to be used as a spacer between two steel walls. At 17.2°C it just slips in between the walls. Then it warms to 22.3°C. Calculate the stress in the cylinder and the total force it exerts on each wall, assuming that the walls are perfectly rigid and a constant distance apart.

**SOLUTION** Equation (15–13) relates the stress to the temperature change. From Table 11–1 we find $Y = 0.70 \times 10^{11}$ Pa, and from Table 15–1, $\alpha = 2.4 \times 10^{-5}\,\text{K}^{-1}$. The temperature change is 22.3°C − 17.2°C = 5.1 C° = 5.1 K. The stress is $F/A$. From Eq. (15–13),

$$\frac{F}{A} = -Y\alpha\,\Delta T = -(0.70 \times 10^{11}\,\text{Pa})(2.4 \times 10^{-5}\,\text{K}^{-1})(5.1\,\text{K})$$
$$= -8.6 \times 10^{6}\,\text{Pa} \quad (\text{or } -1240\,\text{lb/in.}^2).$$

(The negative sign indicates compressive rather than tensile stress.) This stress is independent of the length and cross-section area of the cylinder. The total force $F$ is

$$F = -AY\alpha\,\Delta T = (20 \times 10^{-4}\,\text{m}^2)(-8.6 \times 10^{6}\,\text{Pa})$$
$$= -1.7 \times 10^{4}\,\text{N},$$

or nearly two tons. The negative sign indicates compression.

Thermal stresses can be induced by nonuniform expansion due to temperature differences. Have you ever broken a thick glass vase by pouring very hot water into it? Thermal stress caused by temperature differences exceeded the breaking stress of the material, and it cracked. The same phenomenon makes ice cubes crack when they're dropped into a warm liquid. Heat-resistant glasses such as Pyrex have exceptionally low expansion coefficients and high strength.

## 15–5 QUANTITY OF HEAT

When you put a cold spoon into a cup of hot coffee, the spoon warms up and the coffee cools off as they approach thermal equilibrium. The interaction that causes these temperature changes is fundamentally a transfer of *energy* from one substance to another. Energy transfer that takes place solely because of a temperature difference is called *heat flow,* or *heat transfer,* and energy transferred in this way is called **heat.**

An understanding of the relation between heat and other forms of energy emerged gradually during the eighteenth and nineteenth centuries. Sir James Joule (1818–1889) discovered that water can be warmed by vigorous stirring with a paddle wheel (Fig. 15–10a). The paddle wheel adds energy to the water by doing *work* on it, and Joule found that *the temperature rise is directly proportional to the amount of work done.* The same temperature change could have been caused by putting the water in contact with some hotter body (Fig. 15–10b), so that interaction must also involve an energy exchange. Similarly, when water is boiled by the heat from burning coal, it acquires a greater ability to do *work* (by pushing against a turbine blade or the piston in a steam engine) than an equal mass of cold liquid water. We will explore the relation between heat and mechanical energy in greater detail in Chapters 17 and 18.

The term *heat* always refers to *transfer* of energy from one body or system to another, never to the amount of energy *contained within* a particular system. We can define a *unit* of quantity of heat based on temperature changes of some specific material. The **calorie** (abbreviated cal) is defined as *the amount of heat required to raise the temperature of one gram of water from 14.5°C to 15.5°C.* The kilocalorie (kcal), equal to 1000 cal, is also used; a food-value calorie is actually a kilocalorie. A corresponding unit of heat using Fahrenheit degrees and British units is the **British thermal unit,** or Btu. One Btu is the quantity of heat required to raise the temperature of one pound (weight) of water 1 F° from 63°F to 64°F.

Because heat is energy in transit, there must be a definite relation between these units and the familiar mechanical energy units, such as the joule. Experiments similar in concept to Joule's have shown that

$$1 \text{ cal} = 4.186 \text{ J},$$

$$1 \text{ kcal} = 1000 \text{ cal} = 4186 \text{ J},$$

$$1 \text{ Btu} = 778 \text{ ft} \cdot \text{lb} = 252 \text{ cal} = 1055 \text{ J}.$$

Figure 15–11 shows an example of this equivalence.

The calorie is not a fundamental SI unit. The International Committee on Weights and Measures recommends use of the joule as the basic unit of energy in all forms, including heat. We will follow that recommendation in this book.

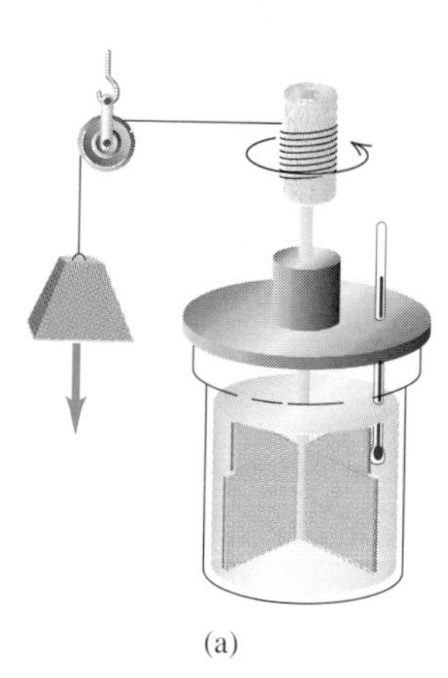

(a)

(b)

**15–10** The same temperature change of the same system may be accomplished by (a) doing work on it or (b) adding heat to it.

## Specific Heat Capacity

We use the symbol $Q$ for quantity of heat. When it is associated with an infinitesimal temperature change $dT$, we call it $dQ$. The quantity of heat $Q$ required to increase the temperature of a mass $m$ of a certain material from $T_1$ to $T_2$ is found to be approximately proportional to the temperature change $\Delta T = T_2 - T_1$. It is also proportional to the mass $m$ of substance. When you heat water to make tea, you need twice as much heat for two cups as for one if the temperature interval is the same. The quantity of heat needed also depends on the nature of the material; to raise the temperature of one kilogram of water by 1 C° requires over five times as much heat as for one kilogram of aluminum and the same temperature change.

Putting all these relationships together, we have

$$Q = mc\,\Delta T, \tag{15–14}$$

where $c$ is a quantity, different for different materials, called the **specific heat capacity** (or sometimes *specific heat*) for the material. For an infinitesimal temperature change $dT$ and corresponding quantity of heat $dQ$,

$$dQ = mc\,dT, \tag{15–15}$$

$$c = \frac{1}{m}\frac{dQ}{dT}. \tag{15–16}$$

In Eqs. (15–14) and (15–15), $Q$ (or $dQ$) and $\Delta T$ (or $dT$) can be either positive or negative. When they are positive, heat enters the body and its temperature increases; when they are negative, heat leaves the body and its temperature decreases. Remember that $dQ$ does not represent a change in the amount of heat *contained* in a body; this is a meaningless concept. Heat is always energy *in transit* as a result of a temperature difference. There is no such thing as "the amount of heat in a body." The term *heat capacity* is unfortunate because it tends to suggest the erroneous idea that a body contains a certain amount of heat.

The specific heat capacity of water is approximately

$$4190 \text{ J/kg} \cdot \text{C}°, \qquad 1 \text{ cal/g} \cdot \text{C}°, \qquad \text{or} \qquad 1 \text{ Btu/1b} \cdot \text{F}°.$$

**15–11** A soft drink sold in Australia lists energy content in joules instead of calories.

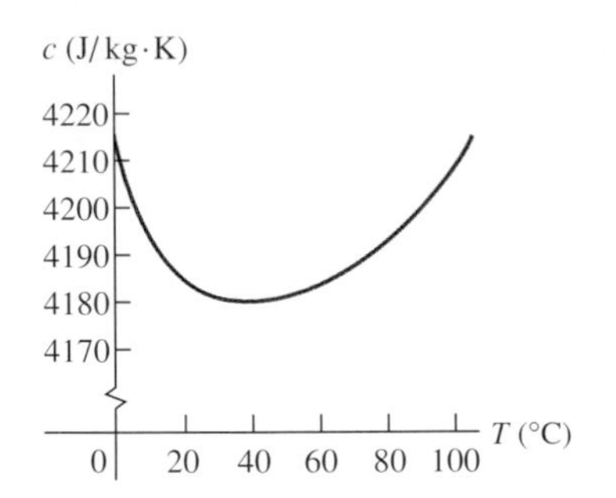


**15–12** Specific heat capacity of water as a function of temperature.

The specific heat capacity of a material always depends somewhat on the initial temperature and the temperature interval. Figure 15–12 shows this variation for water. In the problems and examples in this chapter we will usually ignore this small variation.

### EXAMPLE 15–7

During a bout with the flu an 80-kg man ran a fever of 2.0 C° above normal; that is, his body temperature was 39.0°C (or 102.2°F) instead of the normal 37.0°C (or 98.6°F). Assuming that the human body is mostly water, how much heat is required to raise his temperature by that amount?

**SOLUTION** The temperature change is $\Delta T = 39.0°\text{C} - 37.0°\text{C} = 2.0\ \text{C}° = 2.0\ \text{K}$. From Eq. (15–14),

$$Q = mc\,\Delta T = (80\ \text{kg})(4190\ \text{J/kg}\cdot\text{K})(2.0\ \text{K}) = 6.7 \times 10^5\ \text{J}.$$

This corresponds to 160 kcal (160 food-value calories).

### EXAMPLE 15–8

You are designing an electronic circuit element made of 23 mg of silicon. The electric current through it adds energy at the rate of $7.4\ \text{mW} = 7.4 \times 10^{-3}\ \text{J/s}$. If your design doesn't allow any heat transfer out of the element, at what rate does its temperature increase? The specific heat capacity of silicon is 705 J/kg · K.

**SOLUTION** In one second, $Q = (7.4 \times 10^{-3}\ \text{J/s})(1\ \text{s}) = 7.4 \times 10^{-3}\ \text{J}$. From Eq. (15–14) the temperature change in one second is

$$\Delta T = \frac{Q}{mc} = \frac{7.4 \times 10^{-3}\ \text{J}}{(23 \times 10^{-6}\ \text{kg})(705\ \text{J/kg}\cdot\text{K})} = 0.46\ \text{K}.$$

Alternatively, we can divide both sides of Eq. (15–15) by $dt$ and rearrange:

$$\frac{dT}{dt} = \frac{dQ/dt}{mc} = \frac{7.4 \times 10^{-3}\ \text{J/s}}{(23 \times 10^{-6}\ \text{kg})(705\ \text{J/kg}\cdot\text{K})} = 0.46\ \text{K/s}.$$

At this rate of temperature rise (27 K every minute) the circuit element would soon self-destruct. Heat transfer is an important design consideration in electronic circuit elements.

## Molar Heat Capacity

Sometimes it's more convenient to describe a quantity of a substance in terms of the number of *moles* $n$ rather than the *mass* $m$ of material. Recall from your study of chemistry that the *molecular mass* of any substance, denoted by $M$, is the mass per mole. For example, the molecular mass of water is $18.0\ \text{g/mol} = 18.0 \times 10^{-3}\ \text{kg/mol}$; one mole of water has a mass of $18.0\ \text{g} = 0.0180\ \text{kg}$. The total mass $m$ of material is equal to the mass per mole $M$ times the number of moles $n$:

$$m = nM. \tag{15–17}$$

(The quantity $M$ is sometimes called *molecular weight,* but *molecular mass* is preferable because the quantity depends on the mass of a molecule, not its weight. A mole of any pure substance always contains the same number of molecules; we will discuss this point in detail in Chapter 16.)

Replacing the mass $m$ in Eq. (15–14) by the product $nM$, we find

$$Q = nMc\,\Delta T. \tag{15–18}$$

The product $Mc$ is called the **molar heat capacity** (or *molar specific heat*) and is denoted by $C$. With this notation we rewrite Eq. (15–14) as

$$Q = nC\,\Delta T \tag{15–19}$$

and Eq. (15–16) as

$$C = \frac{1}{n}\frac{dQ}{dT}. \tag{15–20}$$

For example, the molar heat capacity of water is

$$C = Mc = (0.0180\text{ kg/mol})(4190\text{ J/kg}\cdot\text{K}) = 75.4\text{ J/mol}\cdot\text{K}.$$

Values of specific and molar heat capacities for several substances are given in Table 15–3.

Precise measurements of heat capacities require great experimental skill. Usually, a measured quantity of energy is supplied by an electric current in a heater wire wound around the specimen. The temperature change $\Delta T$ is measured with a resistance thermometer or thermocouple embedded in the specimen. This sounds simple, but great care is needed to avoid or compensate for unwanted heat transfer between the sample and its surroundings. Measurements for solid materials are usually made at constant atmospheric pressure; the corresponding values are called *heat capacities at constant pressure,* denoted by $c_p$ and $C_p$. For a gas it is usually easier to keep the substance in a container with constant *volume;* the corresponding values are called *heat capacities at constant volume,* denoted by $c_V$ and $C_V$. For a given substance, $C_V$ and $C_p$ are different. If the system can expand while heat is added, there is additional energy exchange through the performance of *work* by the system on its surroundings. If the volume is constant, the system does no work. For gases the difference between $C_p$ and $C_V$ is substantial. We will study heat capacities of gases in detail in Section 17–7.

The last column of Table 15–3 shows something interesting. The molar heat capacities for most elemental solids are about the same, about 25 J/mol · K. This correlation, named the **rule of Dulong and Petit** (for its discoverers), forms the basis for a very

**TABLE 15–3 Approximate Specific and Molar Heat Capacities (Constant Pressure)**

| Substance | Specific heat capacity, $c$ (J/kg · K) | $M$ (kg/mol) | Molar heat capacity, $C$ (J/mol · K) |
|---|---|---|---|
| Aluminum | 910 | 0.0270 | 24.6 |
| Beryllium | 1970 | 0.00901 | 17.7 |
| Copper | 390 | 0.0635 | 24.8 |
| Ethanol | 2428 | 0.0460 | 112.0 |
| Ethylene glycol | 2386 | 0.0620 | 148.0 |
| Ice | 2000 | 0.0180 | 36.5 |
| Iron | 470 | 0.0559 | 26.3 |
| Lead | 130 | 0.207 | 26.9 |
| Marble ($CaCO_3$) | 879 | 0.100 | 87.9 |
| Mercury | 138 | 0.201 | 27.7 |
| Salt (NaCl) | 879 | 0.0585 | 51.4 |
| Silver | 234 | 0.108 | 25.3 |
| Water | 4190 | 0.0180 | 75.4 |

important idea. The number of atoms in one mole is the same for all elemental substances. This means that on a *per-atom* basis, about the same amount of heat is required to raise the temperature of each of these elements by a given amount, even though the *masses* of the atoms are very different. The heat required for a given temperature increase depends only on *how many* atoms the sample contains and not on the mass of an individual atom. We will study the molecular basis of heat capacities in greater detail in Chapter 16.

We conclude this section with a short sermon. It is absolutely essential for you to keep clearly in mind the distinction between *temperature* and *heat.* Temperature depends on the physical state of a material and is a quantitative description of its hotness or coldness. Heat is energy transferred from one body to another because of a temperature difference. We can change the temperature of a body by adding heat to it or taking heat away or by adding or subtracting energy in other ways, such as mechanical work. If we cut a body in half, each half has the same temperature as the whole; but to raise the temperature of each half by a given interval, we add *half* as much heat as for the whole.

## 15–6 CALORIMETRY

Calorimetry means "measuring heat." We have discussed energy transfer (heat) involved in temperature changes. Heat is also involved in *phase changes,* such as melting of ice or boiling of water. Once we understand these additional heat relationships, we can analyze a variety of problems involving quantity of heat.

### Phase Changes

We use the term **phase** to describe a specific state of matter, such as solid, liquid, or gas. The compound $H_2O$ exists in the *solid phase* as ice, in the *liquid phase* as water, and in the *gaseous phase* as steam. A transition from one phase to another is called a **phase change,** or phase transition. For any given pressure a phase change takes place at a definite temperature, usually accompanied by absorption or emission of heat and a change of volume and density.

A familiar example of a phase change is the melting of ice. When we add heat to ice at 0°C and normal atmospheric pressure, the temperature of the ice *does not* increase. Instead, some of it melts to form liquid water. If we add the heat slowly, to maintain the system very close to thermal equilibrium, the temperature remains at 0°C until all the ice is melted. The effect of adding heat to this system is not to raise its temperature but to change its *phase* from solid to liquid.

To change 1 kg of ice at 0°C to 1 kg of liquid water at 0°C and normal atmospheric pressure requires $3.34 \times 10^5$ J of heat. The heat required per unit mass is called the **heat of fusion** (or sometimes *latent* heat of fusion), denoted by $L_f$. The heat of fusion $L_f$ of water, at normal atmospheric pressure, is

$$L_f = 3.34 \times 10^5 \text{ J/kg} = 79.6 \text{ cal/g} = 143 \text{ Btu/lb}.$$

More generally, to melt a mass $m$ of material that has a heat of fusion $L_f$ requires a quantity of heat $Q$ given by

$$Q = mL_f.$$

This process is *reversible.* To freeze liquid water to ice at 0°C, we have to *remove* heat. The magnitude is the same, but in this case $Q$ is negative because heat is removed rather than added. To cover both possibilities and to include other kinds of phase changes,

**TABLE 15–4 Heats of Fusion and Vaporization**

| Substance | Normal melting point K | Normal melting point °C | Heat of fusion, $L_f$ (J/kg) | Normal boiling point K | Normal boiling point °C | Heat of vaporization, $L_v$ (J/kg) |
|---|---|---|---|---|---|---|
| Helium | 3.5 | −269.65 | $5.23 \times 10^3$ | 4.216 | −268.93 | $20.9 \times 10^3$ |
| Hydrogen | 13.84 | −259.31 | $58.6 \times 10^3$ | 20.26 | −252.89 | $452 \times 10^3$ |
| Nitrogen | 63.18 | −209.97 | $25.5 \times 10^3$ | 77.34 | −195.81 | $201 \times 10^3$ |
| Oxygen | 54.36 | −218.79 | $13.8 \times 10^3$ | 90.18 | −182.97 | $213 \times 10^3$ |
| Ethyl alcohol | 159 | −114 | $104.2 \times 10^3$ | 351 | 78 | $854 \times 10^3$ |
| Mercury | 234 | −39 | $11.8 \times 10^3$ | 630 | 357 | $272 \times 10^3$ |
| Water | 273.15 | 0.00 | $334 \times 10^3$ | 373.15 | 100.00 | $2256 \times 10^3$ |
| Sulfur | 392 | 119 | $38.1 \times 10^3$ | 717.75 | 444.60 | $326 \times 10^3$ |
| Lead | 600.5 | 327.3 | $24.5 \times 10^3$ | 2023 | 1750 | $871 \times 10^3$ |
| Antimony | 903.65 | 630.50 | $165 \times 10^3$ | 1713 | 1440 | $561 \times 10^3$ |
| Silver | 1233.95 | 960.80 | $88.3 \times 10^3$ | 2466 | 2193 | $2336 \times 10^3$ |
| Gold | 1336.15 | 1063.00 | $64.5 \times 10^3$ | 2933 | 2660 | $1578 \times 10^3$ |
| Copper | 1356 | 1083 | $134 \times 10^3$ | 1460 | 1187 | $5069 \times 10^3$ |

we write

$$Q = \pm mL. \tag{15–21}$$

The plus sign (heat entering) is used when the material melts, the minus sign (heat leaving) when it freezes. The heat of fusion is different for different materials, and it also varies somewhat with pressure.

At any given pressure there is a unique temperature at which liquid water and ice can coexist in a condition called **phase equilibrium.** The freezing temperature is always the same as the melting temperature.

We can go through this whole story again, changing the names of the characters, for *boiling* or *evaporation,* a phase transition between liquid and gaseous phases. The corresponding heat (per unit mass) is called the **heat of vaporization** $L_v$. Both $L_v$ and the boiling temperature of a material depend on pressure. Water boils at a lower temperature (about 95°C) in Denver than in Pittsburgh because Denver is at higher elevation and the average atmospheric pressure is less. The heat of vaporization is somewhat greater at this lower pressure, about $2.27 \times 10^6$ J/kg. At normal atmospheric pressure the heat of vaporization $L_v$ for water is

$$L_v = 2.26 \times 10^6 \text{ J/kg} = 539 \text{ cal/g} = 970 \text{ Btu/lb}.$$

Note that over five times as much heat is required to vaporize a quantity of water at 100°C as to raise its temperature from 0°C to 100°C.

Like melting, boiling is a reversible transition. When heat is removed from a gas at the boiling temperature, the gas returns to the liquid phase, or *condenses,* giving up to its surroundings the same quantity of heat (heat of vaporization) as was needed to vaporize it. At a given pressure the boiling and condensation temperatures are always the same.

Table 15–4 lists heats of fusion and vaporization for several materials and their melting and boiling temperatures at normal atmospheric pressure. Very few *elements* have melting temperatures in the vicinity of ordinary room temperatures; one of the few is the metal gallium, shown in Fig. 15–13.

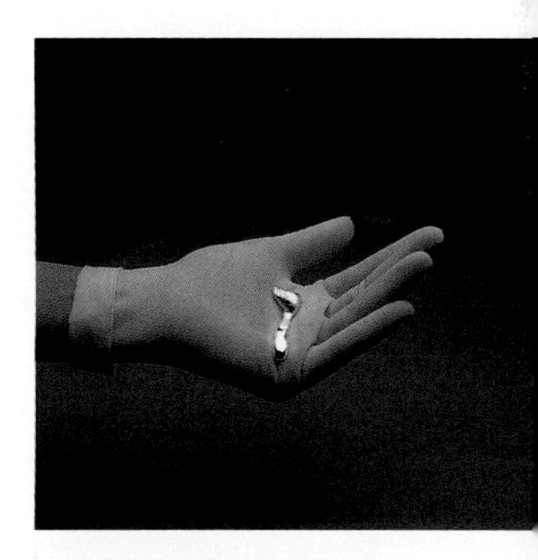

**15–13** The metal gallium, shown here melting in a person's hand, is one of the few elements that melt in the vicinity of room temperature. Its melting temperature is 29.8°C, and its heat of fusion is $8.04 \times 10^4$ J/kg.

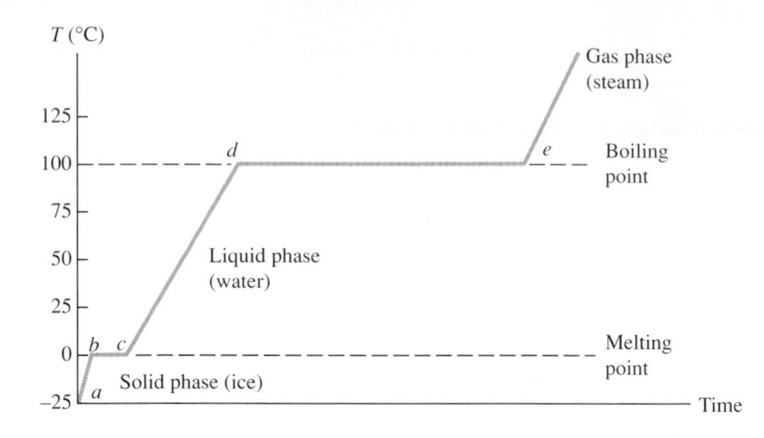


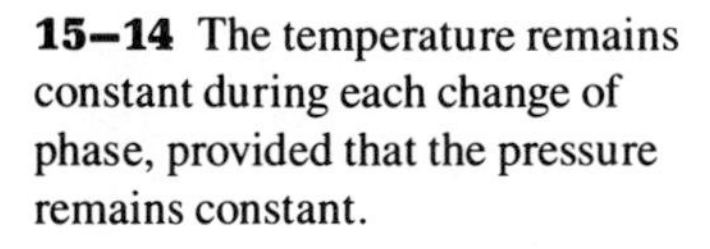
**15–14** The temperature remains constant during each change of phase, provided that the pressure remains constant.

Figure 15–14 shows how the temperature varies when heat is added continuously to a specimen of ice with an initial temperature below 0°C. The temperature rises (from *a* to *b*) until the melting point is reached. Then as more heat is added (*b* to *c*), the temperature remains constant until all the ice has melted. Then the temperature starts to rise again (*c* to *d*) until the boiling temperature is reached. At that point the temperature again is constant (*d* to *e*) until all the water is transformed into the vapor phase. If the rate of heat input is constant, the slope of the line for the liquid phase is smaller than that of the solid phase. Do you see why?

A substance can sometimes change directly from the solid to the gaseous phase; this process is called **sublimation,** and the solid is said to *sublime*. The corresponding heat is called the **heat of sublimation,** $L_s$. Liquid carbon dioxide cannot exist at a pressure lower than about $5 \times 10^5$ Pa (about 5 atm), and "dry ice" (solid carbon dioxide) sublimes at atmospheric pressure. Sublimation of water from frozen food causes freezer burn. The reverse process, a phase change from gas to solid, occurs when frost forms on cold bodies such as refrigerator cooling coils.

Very pure water can be cooled several degrees below the freezing temperature without freezing; the resulting unstable state is described as *supercooled.* When a small ice crystal is dropped in or the water is agitated, it crystallizes within a second or less. Supercooled water *vapor* condenses quickly into fog droplets when a disturbance, such as dust particles or ionizing radiation, is introduced. This principle is used in "seeding" clouds, which often contain supercooled water vapor, to cause condensation and rain.

A liquid can sometimes be *superheated* above its normal boiling temperature. Again, any small disturbance such as agitation or the passage of a charged particle through the material causes local boiling with bubble formation. This is the basis of operation of the *bubble chamber,* historically an important instrument used to observe tracks of high-energy particles by the lines of bubbles they leave in a superheated liquid.

Steam heating systems for buildings use a boiling-condensing process to transfer heat from the furnace to the radiators. Each kilogram of water that is turned to steam in the boiler absorbs over $2 \times 10^6$ J (the heat of vaporization $L_v$ of water) from the boiler and gives it up when it condenses in the radiators. Boiling-condensing processes are also used in refrigerators, air conditioners, and heat pumps. We will discuss these systems in Chapter 18.

The temperature-control mechanisms of many warm-blooded animals make use of heat of vaporization, removing heat from the body by using it to evaporate sweat (mostly water). Evaporative cooling enables humans to maintain normal body temperature in hot, dry desert climates where the air temperature may reach 55°C (about 130°F). The skin temperature may be as much as 30 C° cooler than the surrounding air. Under these conditions a normal person may perspire several liters per day. Unless this lost water is replaced, dehydration, heat stroke, and death result. Old-time desert rats (such as the author) state that in the desert any canteen that holds less than a gallon should be viewed as a toy!

Evaporative cooling is also used to cool buildings in hot, dry climates and to condense and recirculate "used" steam in coal-fired or nuclear-powered electric-generating plants. That's what is going on in the large cone-shaped concrete towers that you see at such plants.

Chemical reactions such as combustion involve definite quantities of heat. Complete combustion of one gram of gasoline produces about 46,000 J or about 11,000 cal, so the **heat of combustion** $L_c$ of gasoline is

$$L_c = 46{,}000 \text{ J/g} = 46 \times 10^6 \text{ J/kg}.$$

Energy values of foods are defined similarly; the unit of food energy, although called a calorie, is a *kilocalorie,* equal to 1000 cal or 4186 J. When we say that a gram of peanut butter "contains 6 calories," we mean that, when it reacts with oxygen, with the help of enzymes, to convert the carbon and hydrogen completely to $CO_2$ and $H_2O$, the total energy liberated as heat is 6 kcal (6000 cal or 25,000 J). Not all of this energy is directly useful for mechanical work. We will study the *efficiency* of energy utilization in Chapter 18.

## Heat Calculations

Let's look at some examples of calorimetry calculations (calculations with heat). The basic principle is very simple: When heat flow occurs between two bodies that are isolated from their surroundings, the amount of heat lost by one body must equal the amount gained by the other. Heat is energy in transit, so this principle is really just conservation of energy. We take each quantity of heat *added to* a body as *positive* and each quantity *leaving* a body as *negative*. When several bodies interact, the *algebraic sum* of the quantities of heat transferred to all the bodies must be zero. Calorimetry, dealing entirely with one conserved quantity, is in many ways the simplest of all physical theories!

## PROBLEM-SOLVING STRATEGY

### Calorimetry problems

**1.** To avoid confusion about algebraic signs when calculating quantities of heat, use Eqs. (15–14) and (15–21) consistently for each body, noting that each $Q$ is positive when heat enters a body and negative when it leaves. The algebraic sum of all the $Q$'s must be zero.

**2.** Often you will need to find an unknown temperature. Represent it by an algebraic symbol such as $T$. Then if a body has an initial temperature of 20°C and an unknown final temperature $T$, the temperature change for the body is $\Delta T = T_{\text{final}} - T_{\text{initial}} = T - 20°\text{C}$ (*not* $20°\text{C} - T$). And so on.

**3.** In problems in which a phase change takes place, as when ice melts, you might not know in advance whether *all* the material undergoes a phase change or only part of it. You can always assume one or the other, and if the resulting calculation gives an absurd result (such as a final temperature that is higher or lower than *any* of the initial temperatures), you know that the initial assumption was wrong. Back up and try again!

## EXAMPLE 15–9

An ethnic restaurant down by the docks serves coffee in copper mugs (with the inside tin-plated to avoid copper poisoning). A waiter fills a cup having a mass of 0.100 kg, initially at 20.0°C, with 0.300 kg of coffee that is initially at 70.0°C. What is the final temperature after the coffee and the cup attain thermal equilibrium? (Assume that coffee has the same specific heat capacity as water and that there is no heat exchange with the surroundings.)

**SOLUTION** Call the final temperature $T$. The (negative) heat gained by the coffee is

$$Q_{coffee} = mc_{water}\,\Delta T_{coffee} = (0.300\text{ kg})(4190\text{ J/kg}\cdot\text{K})(T - 70.0°\text{C}).$$

The (positive) heat gained by the copper cup is

$$Q_{copper} = mc_{copper}\,\Delta T_{copper} = (0.100\text{ kg})(390\text{ J/kg}\cdot\text{K})(T - 20.0°\text{C}).$$

We equate the sum of these two quantities of heat to zero, obtaining an algebraic equation for $T$:

$$Q_{coffee} + Q_{copper} = 0, \quad \text{or}$$

$$(0.300\text{ kg})(4190\text{ J/kg}\cdot\text{K})(T - 70.0°\text{C}) + (0.100\text{ kg})(390\text{ J/kg}\cdot\text{K})(T - 20.0°\text{C}) = 0.$$

Solution of this equation gives $T = 68.5°\text{C}$. The final temperature is much closer to the initial temperature of the coffee than to that of the cup; water has a much larger specific heat capacity than copper, and we have three times as much mass of water.

We can also find the quantities of heat by substituting this value for $T$ back into the original equations. We find $Q_{coffee} = -1890$ J and $Q_{copper} = +1890$ J; $Q_{coffee}$ is negative, as expected.

## EXAMPLE 15–10

A physics student wants to cool 0.25 kg of Diet Omni-Cola (mostly water), initially at 20°C, by adding ice initially at −20°C. How much ice should she add so that the final temperature will be 0°C with all the ice melted if the heat capacity of the container may be neglected?

**SOLUTION** The (negative) heat added to the water is

$$Q = mc_{water}\,\Delta T_{Omni} = (0.25\text{ kg})(4190\text{ J/kg}\cdot\text{K})(0°\text{C} - 20°\text{C}) = -21{,}000\text{ J}.$$

The specific heat capacity of ice (not the same as for liquid water) is about $2.0 \times 10^3$ J/kg · K. Let the mass of ice be $m$; then the heat needed to warm it from −20°C to 0°C is

$$Q = mc_{ice}\,\Delta T_{ice} = m(2.0 \times 10^3\text{ J/kg}\cdot\text{K})[0°\text{C} - (-20°\text{C})] = m(4.0 \times 10^4\text{ J/kg}).$$

The additional heat needed to melt the ice is the mass times the heat of fusion:

$$Q = mL_f = m(3.34 \times 10^5\text{ J/kg}).$$

The sum of these three quantities must equal zero:

$$-21{,}000\text{ J} + m(40{,}000\text{ J/kg}) + m(334{,}000\text{ J/kg}) = 0.$$

Solving this for $m$, we get $m = 0.056$ kg $= 56$ g (two or three medium-size ice cubes).

## EXAMPLE 15–11

In a particular gasoline camp stove, 30% of the energy released in burning the fuel actually goes to heating the pot on the stove. If we heat 1.00 L (1.00 kg) of water from 20°C to 100°C and boil 0.25 kg of it away, how much gasoline do we burn in the process?

**SOLUTION** The heat required to raise the temperature of the water from 20°C to 100°C is

$$Q = mc\,\Delta T = (1.00\text{ kg})(4190\text{ J/kg}\cdot\text{K})(80\text{ K}) = 3.35 \times 10^5\text{ J}.$$

To boil 0.25 kg of water at this temperature requires

$$Q = mL_v = (0.25\text{ kg})(2.26 \times 10^6\text{ J/kg}) = 5.65 \times 10^5\text{ J}.$$

The total energy needed is the sum of these, or $9.00 \times 10^5$ J. This is only 0.30 of the total heat of combustion, so that energy is $(9.00 \times 10^5\text{ J})/0.30 = 3.00 \times 10^6$ J. As we mentioned above, each gram of gasoline releases 46,000 J, so the mass of gasoline required is

$$\frac{3.00 \times 10^6\text{ J}}{46{,}000\text{ J/g}} = 65\text{ g},$$

or about 0.09 L of gasoline.

# 15–7 MECHANISMS OF HEAT TRANSFER

We have talked about *conductors* and *insulators,* materials that permit or prevent heat transfer between bodies. Now let's look in more detail at *rates* of energy transfer. In the kitchen you use an aluminum pot for good heat transfer from the stove to whatever you're cooking, but your refrigerator is insulated with styrofoam to *prevent* conduction of heat. How do we describe the difference between these two materials?

The three mechanisms of heat transfer are conduction, convection, and radiation. *Conduction* occurs within a body or between two bodies in contact. *Convection* depends

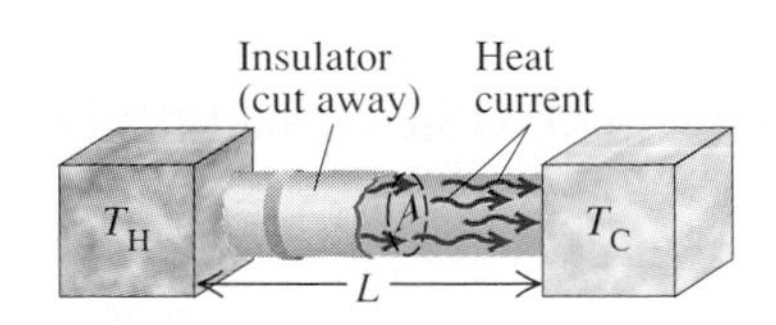


**15–15** Steady-state heat flow due to conduction in a uniform rod.

on motion of mass from one region of space to another. *Radiation* is heat transfer by electromagnetic radiation, such as sunshine, with no need for matter to be present in the space between bodies.

## Conduction

If you hold one end of a copper rod and place the other end in a flame, the end you are holding gets hotter and hotter, even though it is not in direct contact with the flame. Heat reaches the cooler end by **conduction** through the material. On the atomic level, the atoms in the hotter regions have more kinetic energy, on the average, than their cooler neighbors. They jostle their neighbors, giving them some of their energy. The neighbors jostle *their* neighbors, and so it goes through the material. The atoms themselves do not move from one region of material to another, but their energy does.

In metals, electron motion provides another mechanism for heat transfer. Most metals are good conductors of electricity because some electrons can leave their parent atoms and wander through the crystal lattice. These "free" electrons can carry energy from the hotter to the cooler regions of the metal. Good thermal conductors such as silver, copper, aluminum, and gold are also good electrical conductors.

Heat transfer occurs only between regions at different temperatures, and the direction of heat flow is always from higher to lower temperature. Figure 15–15 shows a rod of conducting material with cross-section area $A$ and length $L$. The left end of the rod is kept at a temperature $T_H$ and the right end at a lower temperature $T_C$, and heat flows from left to right. The sides of the rod are covered by an insulator, so no heat transfer occurs at the sides.

When a quantity of heat $dQ$ is transferred in a time $dt$, the rate of heat flow is $dQ/dt$. We call this rate the **heat current,** denoted by $H$. That is, $H = dQ/dt$. Experiments show that the heat current is proportional to the cross-section area $A$ and to the temperature difference $(T_H - T_C)$ and is inversely proportional to the length $L$. Introducing a proportionality constant $k$ called the **thermal conductivity** of the material, we have

$$H = \frac{dQ}{dt} = kA\frac{T_H - T_C}{L}. \qquad (15\text{–}22)$$

The quantity $(T_H - T_C)/L$ is the temperature difference *per unit length;* it is called the **temperature gradient.** The numerical value of $k$ depends on the material of the rod. Materials with large $k$ are good conductors of heat; materials with small $k$ are poor conductors or insulators. Equation (15–22) also gives the heat current through a slab, or *any* homogeneous body with uniform cross-section area $A$ perpendicular to the direction of flow; $L$ is the length of the heat-flow path.

The units of heat current $H$ are units of energy per unit time, or power; the SI unit of heat current is the joule per second, or watt (W). We can find the units of $k$ by solving Eq. (15–22) for $k$. We invite you to verify that the SI units of $k$ are W/m · K. Some numerical values of $k$ are given in Table 15–5.

The thermal conductivity of "dead" (nonmoving) air is very small. A wool sweater is warm because it traps air between the fibers. In fact, many insulating materials such as

**TABLE 15–5 Thermal Conductivities**

| | **k (W/m · K)** |
|---|---|
| *Metals* | |
| Aluminum | 205.0 |
| Brass | 109.0 |
| Copper | 385.0 |
| Lead | 34.7 |
| Mercury | 8.3 |
| Silver | 406.0 |
| Steel | 50.2 |
| *Various solids (representative values)* | |
| Brick, insulating | 0.15 |
| Brick, red | 0.6 |
| Concrete | 0.8 |
| Cork | 0.04 |
| Felt | 0.04 |
| Fiberglass | 0.04 |
| Glass | 0.8 |
| Ice | 1.6 |
| Rock wool | 0.04 |
| Styrofoam | 0.01 |
| Wood | 0.12–0.04 |
| *Gases* | |
| Air | 0.024 |
| Argon | 0.016 |
| Helium | 0.14 |
| Hydrogen | 0.14 |
| Oxygen | 0.023 |

**15–16** This protective tile, developed for use in the space shuttle, has extraordinary thermal properties. The extremely small thermal conductivity and small heat capacity of the material make it possible to hold the tile by its edges, even though it is hot enough to emit the light for this photograph.

styrofoam and fiberglass are mostly dead air. Figure 15–16 shows a "space-age" ceramic material with very unusual thermal properties, including very small conductivity.

If the temperature varies in a non-uniform way along the length of the conducting rod, we can introduce a coordinate $x$ along the length and generalize the temperature gradient as $dT/dx$. The corresponding generalization of Eq. (15–22) is

$$H = \frac{dQ}{dt} = -kA\frac{dT}{dx}. \tag{15–23}$$

The negative sign shows that heat always flows in the direction of *decreasing* temperature.

For thermal insulation in buildings, engineers use the concept of **thermal resistance,** denoted by $R$. The thermal resistance $R$ of a slab of material with surface area $A$ is defined so that the heat current $H$ through the slab is

$$H = \frac{A(T_H - T_C)}{R}, \tag{15–24}$$

where $T_H$ and $T_C$ are the temperatures on the two sides of the slab. Comparing this with Eq. (15–22), we see that $R$ is given by

$$R = \frac{L}{k}, \tag{15–25}$$

where $L$ is the thickness of the slab. The SI unit of $R$ is $1\ m^2 \cdot K/W$. In the units used for commercial insulating materials, $H$ is expressed in Btu/h, $A$ is in $ft^2$, and $T_H - T_C$ is in F° (1 Btu/h = 0.293 W). The units of $R$ are then $ft^2 \cdot F° \cdot h/Btu$. Values of $R$ are usually quoted without units: A 6-inch-thick layer of fiberglass has an $R$ value of 19, a two-inch slab of polyurethane foam has a value of 12, and so on. Doubling the thickness doubles the $R$ value. Common practice in new construction in severe northern climates is to specify $R$ values of around 30 for exterior walls and ceilings. When the insulating material is in layers, such as a plastered wall, fiberglass insulation, and wood exterior siding, the $R$ values are additive. Do you see why? (See Problem 15–80.)

Electronics engineers concerned with cooling microprocessor chips use a different definition of thermal resistance and denote the quantity by $r$. This subject is discussed in Section 15–8.

## PROBLEM-SOLVING STRATEGY

### Heat conduction

**1.** Identify the direction of heat flow in the problem (from hot to cold). In Eq. (15–22), $L$ is always measured along this direction, and $A$ is always an area perpendicular to this direction. Often when a box or other container has an irregular shape but uniform wall thickness, you can approximate it as a flat slab with the same thickness and total wall area.

**2.** In some problems the heat flows through two different materials in succession. The temperature at the interface between the two materials is then intermediate between $T_H$ and $T_C$; represent it by a symbol such as $T$. The temperature differences for the two materials are then $(T_H - T)$ and $(T - T_C)$. In steady-state heat flow the same heat has to pass through both materials in succession, so the heat current $H$ must be *the same* in both materials. It's like a series electrical circuit.

**3.** If there are two *parallel* heat flow paths, so that some heat flows through each, then the total $H$ is the sum of the quantities $H_1$ and $H_2$ for the separate paths. An example is heat flow from inside to outside a house, both through the glass in a window and through the surrounding frame. In this case the temperature difference is the same for the two paths, but $L$, $A$, and $k$ may be different for the two paths.

**4.** As always, it is essential to use a consistent set of units. If you use a value of $k$ expressed in $W/m \cdot K$, don't use distances in cm, heat in calories, or $T$ in °F!

## EXAMPLE 15–12

**Conduction through a picnic cooler** A styrofoam box used to keep drinks cold at a picnic has total wall area (including the lid) of 0.80 $m^2$ and wall thickness 2.0 cm. It is filled with ice, water, and cans of Omni-Cola at 0°C. What is the rate of heat flow into the box if the temperature of the outside wall is 30°C? How much ice melts in one day?

**SOLUTION** We assume that the total heat flow is approximately the same as it would be through a flat slab of area 0.80 $m^2$ and thickness 2.0 cm = 0.020 m (Fig. 15–17). We find $k$ from Table 15–5. From Eq. (15–22) the heat current (rate of heat flow) is

$$H = kA\frac{T_H - T_C}{L}$$

$$= (0.010\ \text{W/m}\cdot\text{K})(0.80\ \text{m}^2)\frac{30°\text{C} - 0°\text{C}}{0.020\ \text{m}}$$

$$= 12\ \text{W} = 12\ \text{J/s}.$$

There are 86,400 s in one day, so the total heat flow $Q$ in one day is

$$(12\ \text{J/s})(86{,}400\ \text{s}) = 1.04 \times 10^6\ \text{J}.$$

The heat of fusion of ice is $3.33 \times 10^5$ J/kg, so the quantity of ice melted by this quantity of heat is

$$m = \frac{Q}{L_F}$$

$$= \frac{1.04 \times 10^6\ \text{J}}{3.33 \times 10^5\ \text{J/kg}} = 3.1\ \text{kg}.$$

**15–17** Conduction: We can approximate heat flow through a picnic cooler by heat flow through a flat slab of styrofoam.

## EXAMPLE 15–13

A steel bar 10.0 cm long is welded end-to-end to a copper bar 20.0 cm long (Fig. 15–18). Both bars are insulated at their sides. Each bar has a square cross section, 2.00 cm on a side. The free end of the steel bar is in contact with steam at 100°C, and the free end of the copper bar is in contact with ice at 0°C. Find the temperature at the junction of the two bars and the total rate of heat flow.

**SOLUTION** As we discussed in the Problem-Solving Strategy, the heat currents in the two bars must be equal; this is the key to the solution. Let $T$ be the unknown junction temperature. We use Eq. (15–22) for each bar and equate the two expressions:

$$\frac{k_s A(100°\text{C} - T)}{L_s} = \frac{k_c A(T - 0°\text{C})}{L_c}.$$

The areas $A$ are equal and may be divided out. After substituting numerical values, we have

$$\frac{(50.2\ \text{W/m}\cdot\text{K})(100°\text{C} - T)}{0.100\ \text{m}} = \frac{(385\ \text{W/m}\cdot\text{K})(T - 0°\text{C})}{0.200\ \text{m}}.$$

Rearranging and solving for $T$, we find

$$T = 20.7°\text{C}.$$

Even though the steel bar is shorter, the temperature drop across it is much greater than that across the copper bar because steel is a much poorer conductor.

We can find the total heat current by substituting this value for $T$ back into either of the above expressions:

$$H = \frac{(50.2\ \text{W/m}\cdot\text{K})(0.0200\ \text{m})^2(100°\text{C} - 20.7°\text{C})}{0.100\ \text{m}}$$

$$= 15.9\ \text{W},$$

or

$$H = \frac{(385\ \text{W/m}\cdot\text{K})(0.0200\ \text{m})^2(20.7°\text{C} - 0°\text{C})}{0.200\ \text{m}}$$

$$= 15.9\ \text{W}.$$

Insulator (cutaway)
2.00 cm
$T_H = 100$ °C
STEEL
COPPER
$T_C = 0$ °C
10.0 cm
20.0 cm

**15–18** Heat flow along two metal bars, one of steel and one of copper, connected end-to-end.

## EXAMPLE 15–14

Suppose the two bars in Example 15–13 are separated. One end of each bar is placed in contact with steam at 100°C, and the other end of each bar contacts ice at 0°C (Fig. 15–19). What is the *total* rate of heat flow in the two bars?

**SOLUTION** In this case the bars are in parallel rather than in series. The total heat current is now the *sum* of the currents in the two bars, and for each bar, $T_H - T_C = 100°C - 0°C = 100\ K$:

$$H = \frac{(50.2\ W/m\cdot K)(0.0200\ m)^2(100\ K)}{0.100\ m} + \frac{(385\ W/m\cdot K)(0.0200\ m)^2(100\ K)}{0.200\ m}$$

$$= 20.1\ W + 77.0\ W = 97.1\ W.$$

The heat flow in the copper bar is much greater than that in the steel bar, even though it is longer, because its thermal conductivity is much larger. The total heat flow is much greater than that in Example 15–13, partly because the cross section for heat flow is greater and partly because the full 100 C° temperature difference appears across each bar.

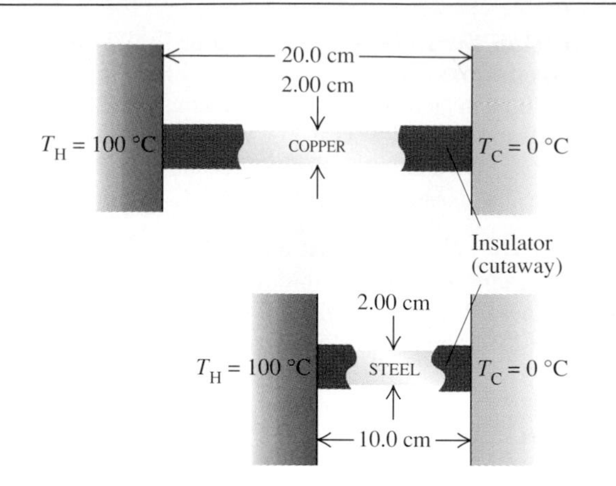


**15–19** Heat flow along two metal bars, one of steel and one of copper, parallel to and separated from each other.

## Convection

**Convection** is transfer of heat by mass motion of a fluid from one region of space to another. Familiar examples include hot-air and hot-water home heating systems, the cooling system of an automobile engine, and the heating of the body by the flow of blood. If the fluid is circulated by a blower or pump, the process is called *forced convection;* if the flow is caused by differences in density due to thermal expansion, such as hot air rising, the process is called *natural convection,* or *free convection.*

Convection in the atmosphere plays a dominant role in determining weather patterns (Fig. 15–20), and convection in the oceans is an important global heat-transfer mechanism. On a smaller scale, glider pilots make use of thermal updrafts from the warm earth. But sometimes the air cools enough as it rises to form thunderheads; these contain convection currents that can tear a glider to pieces.

The most important mechanism for heat transfer within the human body (needed to maintain nearly constant temperature in various environments) is forced convection of blood, with the heart serving as the pump. The total rate of heat loss from the body is of the order of 100 to 200 W (2000 to 4000 kcal per day). A dry, unclothed body in still air

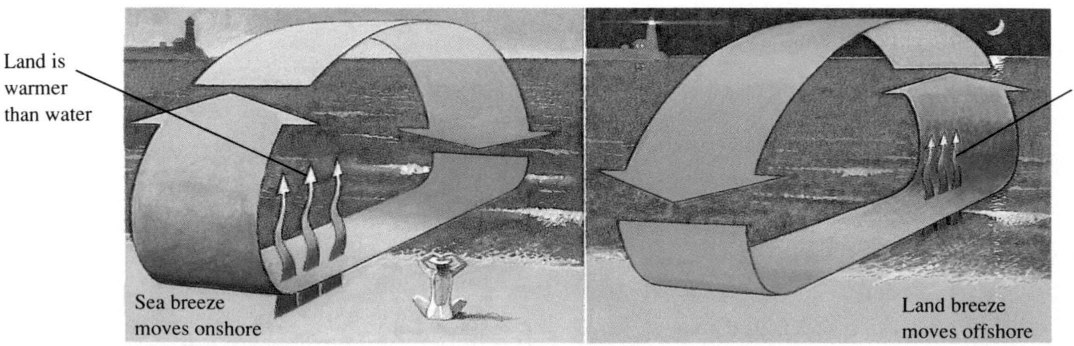


**15–20** Convection currents in the air determine daytime sea breezes and nighttime land breezes near shore.

loses about half this heat by radiation, but during vigorous exercise and plentiful perspiration, evaporative cooling is the dominant mechanism.

Convective heat transfer is a very complex process, and there is no simple equation to describe it. Here are a few experimental facts:

The heat current due to convection is directly proportional to the surface area. This is the reason for the large surface areas of radiators and cooling fins.

The viscosity of fluids slows natural convection near a stationary surface, giving a surface film that on a vertical surface typically has about the same insulating value as 1.3 cm of plywood (*R* value of 0.7). Forced convection decreases the thickness of this film, increasing the rate of heat transfer. This is the reason for the "wind-chill factor"; you get cold faster in a cold wind than in still air with the same temperature.

The heat current due to convection is found to be approximately proportional to the $\frac{5}{4}$ power of the temperature difference between the surface and the main body of fluid.

## Radiation

Heat transfer by **radiation** depends on electromagnetic waves such as visible light, infrared, and ultraviolet radiation. Everyone has felt the warmth of the sun's radiation (Fig. 15–21) and the intense heat from a charcoal grill or the glowing coals in a fireplace. Heat from these very hot bodies reaches you not by conduction or convection in the intervening air but by *radiation*. This heat transfer would occur even if there were nothing but vacuum between you and the source of heat.

*Every* body, even at ordinary temperatures, emits energy in the form of electromagnetic radiation. At ordinary temperatures, say 20°C, nearly all the energy is carried by infrared waves with wavelengths much longer than those of visible light. As the temperature rises, the wavelengths shift to shorter values. At 800°C a body emits enough visible radiation to be self-luminous and appears "red-hot," although even at this temperature most of the energy is carried by infrared waves. At 3000°C, the temperature of an incandescent lamp filament, the radiation contains enough visible light that the body appears "white-hot."

The rate of energy radiation from a surface is proportional to the surface area $A$. The rate increases very rapidly with temperature in proportion to the fourth power of the absolute temperature. The rate also depends on the nature of the surface; this dependence is described by a quantity $e$ called the **emissivity.** This is a dimensionless number between 0 and 1 representing the ratio of the rate of radiation from a particular surface to that of an equal area of an ideal radiating surface at the same temperature. Emissivity depends somewhat on temperature. The heat current $H = dQ/dt$ due to radiation from a surface area $A$ with emissivity $e$ at absolute temperature $T$ can be expressed as

$$H = Ae\sigma T^4, \qquad (15\text{–}26)$$

where $\sigma$ is a fundamental physical constant called the **Stefan-Boltzmann constant.** This relation is called the **Stefan-Boltzmann law** in honor of its late nineteenth-century discoverers. The numerical value of $\sigma$ is found experimentally to be

$$\sigma = 5.6705 \times 10^{-8}\ \text{W/m}^2 \cdot \text{K}^4.$$

We invite you to check unit consistency in Eq. (15–26). Emissivity ($e$) is often larger for dark surfaces than for light ones. The emissivity of a smooth copper surface is about 0.3, but $e$ for a dull black surface can be close to unity.

**15–21** Heat transfer by radiation from the sun. The heat is not dangerous to human skin, but the ultraviolet radiation is.

## EXAMPLE 15–15

A thin, square steel plate, 10 cm on a side, is heated in a blacksmith's forge to a temperature of 800°C. If the emissivity is 0.60, what is the total rate of radiation of energy?

**SOLUTION** The total surface area, including both sides, is $2(0.10\ \text{m})^2 = 0.020\ \text{m}^2$. We must convert the temperature to the Kelvin scale: 800°C = 1073 K. Then Eq. (15–26) gives

$$H = Ae\sigma T^4$$
$$= (0.020\ \text{m}^2)(0.60)(5.67 \times 10^{-8}\ \text{W/m}^2 \cdot \text{K}^4)(1073\ \text{K})^4$$
$$= 900\ \text{W}.$$

While a body at absolute temperature $T$ is radiating, its surroundings at temperature $T_s$ are also radiating, and the body *absorbs* some of this radiation. If it is in thermal equilibrium with its surroundings, the rates of radiation and absorption must be equal. For this to be true, the rate of absorption must be given in general by $H = Ae\sigma T_s^4$. Then the *net* rate of radiation from a body at temperature $T$ with surroundings at temperature $T_s$ is

$$H_{\text{net}} = Ae\sigma T^4 - Ae\sigma T_s^4 = Ae\sigma(T^4 - T_s^4). \tag{15–27}$$

In this equation a positive value of $H$ means a net heat flow *out of* the body. Equation (15–27) shows that for radiation, as for conduction and convection, the heat current depends on the temperature *difference* between two bodies.

## EXAMPLE 15–16

**Radiation from the human body** If the total surface of the human body is 1.2 m² and the surface temperature is 30°C = 303 K, find the total rate of radiation of energy from the body. If the surroundings are at a temperature of 20°C, what is the *net* rate of heat loss from the body by radiation? The emissivity of the body is very close to unity, irrespective of skin pigmentation.

**SOLUTION** The rate of radiation of energy per unit area is given by Eq. (15–26). Taking $e = 1$, we find

$$H = Ae\sigma T^4$$
$$= (1.2\ \text{m}^2)(1)(5.67 \times 10^{-8}\ \text{W/m}^2 \cdot \text{K}^4)(303\ \text{K})^4$$
$$= 574\ \text{W}.$$

This loss is partly offset by *absorption* of radiation, which depends on the temperature of the surroundings. The *net* rate of radiative energy transfer is given by Eq. (15–27):

$$H = Ae\sigma(T^4 - T_s^4)$$
$$= (1.2\ \text{m}^2)(1)(5.67 \times 10^{-8}\ \text{W/m}^2 \cdot \text{K}^4)[(303\ \text{K})^4 - (293\ \text{K})^4]$$
$$= 72\ \text{W}.$$

Heat transfer by radiation is important in some surprising places. A premature baby in an incubator can be cooled dangerously by radiation if the walls of the incubator happen to be cold, even when the *air* in the incubator is warm. Some incubators regulate the air temperature by measuring the baby's skin temperature.

A body that is a good absorber must also be a good emitter. An ideal radiator, with an emissivity of unity, is also an ideal absorber, absorbing *all* of the radiation that strikes it. Such an ideal surface is called an ideal black body, or simply a **blackbody.** Conversely, an ideal *reflector,* which absorbs *no* radiation at all, is a very ineffective radiator.

This is the reason for the silver coatings on vacuum ("Thermos") bottles, which were invented by Sir James Dewar (1842–1923). A vacuum bottle has double glass walls. The air is pumped out of the spaces between the walls; this eliminates nearly all heat transfer by conduction and convection. The silver coating on the walls reflects most of the radiation from the contents back into the container, and the wall itself is a very poor emitter. Thus a vacuum bottle can keep coffee or soup hot for several hours. The Dewar flask, used to store very cold liquefied gases, is exactly the same in principle.

# 15–8 INTEGRATED CIRCUITS: A Case Study in Heat Transfer

Integrated circuits (ic's) and VLSI (very large-scale integration) chips are at the heart of most modern electronic devices, including computers, stereo systems, fuel-injection systems in car engines, and many other applications. The widespread use of these devices has posed new and interesting problems in heat transfer. When a chip gets too hot, it doesn't work right and can be permanently ruined. Keeping chips cool during operation is vitally important.

To introduce the problem, let's think about what happens when you leave your car out in the sun on a bright summer day. The car gets pretty hot; it eventually reaches a temperature at which the rate of transfer of energy from the car to its surroundings just balances the rate of energy input from sunlight. This rate isn't particularly large, typically about 1000 W per square meter of surface area, or 0.1 $W/cm^2$. Something similar happens with light bulbs. A standard 60-W light bulb has a power input of 60 W and a surface area of about 120 $cm^2$. Its power dissipation per unit surface area (by radiation and by conduction and convection to the surrounding air) is 0.5 $W/cm^2$. After a light bulb has been turned on for a few minutes, it gets too hot to touch.

Now consider the *much larger* power dissipation values that occur in microelectronics. Thousands of electronic elements are packed tightly together on a silicon chip only a few millimeters on a side. Power densities for present-day VLSI chips range up to 40 $W/cm^2$. For comparison, heat shields that protect space vehicles when they reenter the earth's atmosphere typically have to dissipate 100 $W/cm^2$.

Modern electronic circuits are made by depositing and etching layers of metals and oxides onto a silicon base chip. If an ic chip gets too hot, the circuits become unreliable or are permanently destroyed. For an ic chip in a plastic package, the maximum safe temperature is about 100°C. A chip in a ceramic package can operate reliably at about 120°C (Fig. 15–22).

The temperature of an operating chip is determined the same way as for your car sitting in the hot sun: Power input equals power output. The power output $H$ (the rate of

(a) (b)

**15–22** (a) Modern ic chips are usually placed in plastic or ceramic packages for protection. The packages contain conducting paths that lead from the chip to outside pins, which connect the circuit to the overall circuit board in the final product. (b) This ceramic package was specifically designed to ensure $r_{th}$ of less than 1 K/W.

heat transfer out of the chip) is approximately proportional to the difference $T_{ic} - T_{amb}$ between the temperature $T_{ic}$ of the chip and the temperature $T_{amb}$ of the surroundings (the *ambient* temperature). Using a proportionality constant $r_{th}$ that depends on the shape and size of the ic, we express the rate of heat loss $H$ as

$$H = \frac{T_{ic} - T_{amb}}{r_{th}}. \tag{15-28}$$

When the ic reaches its final operating temperature, this must equal the electrical power $P$ dissipated in the device. Equating $H$ and $P$ and solving for $T_{ic}$, we find

$$T_{ic} = T_{amb} + r_{th}P. \tag{15-29}$$

Values of $r_{th}$ for common ic packages in still air vary from 30 to 70 K/W (30 to 70 C°/W). For example, one watt of electrical power ($P = 1$ W) into a 40-pin plastic package ic will raise its temperature about 62 C° = 62 K above the surrounding air temperature. For this unit, $r_{th} = 62$ K/W.

## EXAMPLE 15–17

A 40-pin ceramic-package ic has a thermal resistance of 40 K/W. If the maximum temperature the circuit can safely reach is 120°C, what is the highest power level at which the circuit can safely operate in an ambient temperature of 75°C?

**SOLUTION** We use Eq. (15–28), replacing $H$ by $P$:

$$P = \frac{T_{ic} - T_{amb}}{r_{th}},$$

$$P = \frac{120°\text{C} - 75°\text{C}}{40 \text{ K/W}} = 1.1 \text{ W}.$$

This power level is adequate in many typical ic applications, but chips used in high-speed computing applications often require considerably higher power levels.

To remove heat from a chip more efficiently, we can force air past the circuit, improving the heat transfer by convection. Blowing air through the system at a rate of about 20 $m^3$/min can reduce the effective thermal resistance of the chip and package by about 10 to 15 K/W. This is an improvement, but in many cases it is still not enough cooling for a high-performance ic.

One method of cooling that is now being used is direct immersion of the ic package in a fluorocarbon fluid. These fluids are electrical insulators and are chemically inert, making them compatible with operation of electronic components. (One such liquid was first developed as artificial blood!) Their thermal transport properties are not very favorable because of small values of thermal conductivity and heat of vaporization. To improve heat transfer in these fluids, design engineers have tried forced convection, structural modification of packages, and boiling.

In forced convection, microscopic channels are cut into the back of the silicon base of the ic chip. Typical channel dimensions are 50 $\mu$m wide and 300 $\mu$m deep. Experiments with water (not suitable for cooling an actual ic) have achieved heat transfers as large as 790 W per square centimeter of surface area.

Adding fins to the basic cylindrical pin form of an ic package can increase the heated surface area of the package by a factor of 8 to 12. The larger surface area increases heat transfer by both conduction and radiation [see Eqs. (15–24) and (15–26)], reducing the effective thermal resistance by as much as a factor of 20.

Values of heat flow of 45 W/$cm^2$ have been obtained by using boiling techniques, which use heat transfer by convection involving both liquid and vapor at the same time (two-phase convective heat transfer). However, the sudden drop in temperature at the surface of the ic package when boiling begins results in thermal stress on the package. Further work in package design is underway to try to minimize this problem.

Chip packaging has become both an art and a science. Careful layout of electronic components on the circuit board can minimize the overall thermal resistance of the chip, allowing the heat generated to dissipate more readily. Much recent attention has been devoted to research and development of new packages, so there will be rapid changes over the next few years in the heat limitations of ic technology.

## SUMMARY

• A thermometer measures temperature. Two bodies in thermal equilibrium must have the same temperature. A conducting material between two bodies permits interaction leading to thermal equilibrium; an insulating material prevents or impedes this interaction.

• The Celsius and Fahrenheit temperature scales are based on the freezing (0°C = 32°F) and boiling (100°C = 212°F) temperatures of water. They are related by

$$T_F = \tfrac{9}{5}T_C + 32°, \qquad T_C = \tfrac{5}{9}(T_F - 32°), \tag{15–1}$$

and 1 C° $= \frac{9}{5}$ F°.

• The Kelvin scale has its zero at the extrapolated zero-pressure temperature for a gas thermometer, which is $-273.15$°C. Thus 0 K = $-273.15$°C, and

$$T_K = T_C + 273.15. \tag{15–3}$$

In the gas-thermometer scale the ratio of two temperatures is defined to be equal to the ratio of the two corresponding gas-thermometer pressures:

$$\frac{T_2}{T_1} = \frac{p_2}{p_1}. \tag{15–4}$$

The triple-point temperature of water (0.01°C) is defined to be 273.16 K.

• The change $\Delta L$ of any linear dimension $L_0$ of a solid body with a temperature change $\Delta T$ is approximately

$$\Delta L = \alpha L_0 \, \Delta T, \tag{15–6}$$

where $\alpha$ is the coefficient of linear expansion. The change $\Delta V$ in the volume $V_0$ of any solid or liquid material with a temperature change $\Delta T$ is approximately

$$\Delta V = \beta V_0 \, \Delta T, \tag{15–8}$$

where $\beta$ is the coefficient of volume expansion. For solids,

$$\beta = 3\alpha. \tag{15–9}$$

• When a material is cooled and held so that it cannot contract, the tensile stress $F/A$ is

$$\frac{F}{A} = -Y\alpha \, \Delta T. \tag{15–13}$$

• Heat is energy transferred from one body to another as a result of a temperature difference. The quantity of heat $Q$ required to raise the temperature of a mass $m$ of material by a small amount $\Delta T$ is

$$Q = mc \, \Delta T, \tag{15–14}$$

where $c$ is the specific heat capacity of the material. When the quantity of material is

### KEY TERMS

temperature
thermometer
thermal equilibrium
insulator
conductor
zeroth law of thermodynamics
Celsius scale
Fahrenheit scale
Kelvin scale
absolute temperature scale
absolute zero
coefficient of linear expansion
coefficient of volume expansion
thermal stress
heat
calorie
British thermal unit
specific heat capacity
molar heat capacity
rule of Dulong and Petit
phase
phase change
heat of fusion
phase equilibrium
heat of vaporization
sublimation

**heat of sublimation**
**heat of combustion**
**conduction**
**heat current**
**thermal conductivity**
**temperature gradient**
**thermal resistance**
**convection**
**radiation**
**emissivity**
**Stefan-Boltzmann constant**
**Stefan-Boltzmann law**
**blackbody**

given by the number of moles $n$, the corresponding relation is

$$Q = nC\,\Delta T, \tag{15-19}$$

where $C$ is the molar heat capacity ($C = Mc$, and $M$ is the molecular mass). The number of moles $n$ and the mass $m$ of material are related by $m = nM$.

• The molar heat capacities of many solid elements are approximately 25 J/mol · K; this is the rule of Dulong and Petit.

• To change a mass $m$ of a material to a different phase (such as liquid to solid or liquid to vapor) at the same temperature requires the addition or subtraction of a quantity of heat $Q$ given by

$$Q = \pm mL, \tag{15-21}$$

where $L$ the heat of fusion, vaporization, or sublimation.

• When heat is added to a body, the corresponding $Q$ is positive; when it is removed, $Q$ is negative. In a system whose parts interact by heat exchange, the algebraic sum of the $Q$'s for all parts of the system must be zero.

• The three mechanisms of heat transfer are conduction, convection, and radiation. Conduction is transfer of energy of molecular motion within materials without bulk motion of the materials. Convection involves mass motion from one region to another. Radiation is energy transfer through electromagnetic radiation.

• The heat current $H$ for conduction depends on the area $A$ through which the heat flows, the length $L$ of the heat path, the temperature difference $(T_H - T_C)$, and the thermal conductivity $k$ of the material, according to

$$H = \frac{dQ}{dt} = kA\frac{T_H - T_C}{L}. \tag{15-22}$$

• Heat transfer by convection is a complex process, depending on surface area and orientation and the temperature difference between a body and its surroundings.

• The heat current $H$ due to radiation is given by

$$H = Ae\sigma T^4, \tag{15-26}$$

where $A$ is the surface area, $e$ the emissivity of the surface (a pure number between 0 and 1), $T$ the absolute temperature, and $\sigma$ a fundamental constant called the Stefan-Boltzmann constant. When a body at temperature $T$ is surrounded by material at temperature $T_s$, the net heat current from the body to its surroundings is

$$H_{net} = Ae\sigma T^4 - Ae\sigma T_s^4 = Ae\sigma(T^4 - T_s^4). \tag{15-27}$$

# EXERCISES

## Section 15–2 Thermometers and Temperature Scales

**15–1** a) If you feel sick and are told that you have a temperature of 40.0°C, should you be concerned? b) What is normal body temperature on the Celsius scale?

**15–2** When the United States finally converts officially to metric units, the Celsius temperature scale will replace the Fahrenheit scale for everyday use. As a familiarization exercise, find the Celsius temperatures corresponding to a) a cool room (64.0°F); b) a hot summer day in Texas (99.0°F); c) a cold winter day in Chicago (12.0°F).

**15–3** At what temperature do the Fahrenheit and Celsius scales coincide?

## Section 15–3 Gas Thermometers and the Kelvin Scale

**15–4** The ratio of the pressure of a gas at the melting point of lead and its pressure at the triple point of water, when the gas is kept at constant volume, is found to be 2.1982. What is the Kelvin temperature of the melting point of lead?

**15–5** The normal boiling point of liquid nitrogen is $-195.81°C$. What is this temperature on the Kelvin scale?

**15–6** A gas thermometer registered an absolute pressure corresponding to 12.0 cm of mercury when in contact with water at the triple point. What pressure does it read when in contact with water at the normal boiling point?

**15–7 An Experiment with a Constant-Volume Gas Thermometer.** An experimenter found the pressure at the triple point of water (0.01°C) to be $4.00 \times 10^4$ Pa and the pressure at the normal boiling point (100°C) to be $5.40 \times 10^4$ Pa. Assuming that the pressure varies linearly with temperature, use these two data points to find the Celsius temperature (absolute zero) at which the gas pressure would be zero.

## Section 15–4 Thermal Expansion

**15–8** The pendulum shaft of a clock is made of aluminum. What is the fractional change in length of the shaft when it is cooled from 31.0°C to 9.0°C?

**15–9** A steel bridge is built in the summer when the temperature is 32.0°C. At the time of construction its length is 80.00 m. What is the length of the bridge on a cold winter day when the temperaure is $-8.0°C$?

**15–10 Ensuring a Tight Fit.** Aluminum rivets used in airplane construction are made slightly larger than the rivet holes and cooled by "dry ice" (solid $CO_2$) before being driven. If the diameter of a hole is 0.3000 cm, what should be the diameter of a rivet at 20.0°C if its diameter is to equal that of the hole when the rivet is cooled to $-78.0°C$, the temperature of dry ice? Assume that the expansion coefficient remains constant at the value given in Table 15–1.

**15–11** A metal rod is 80.000 cm long at 20.0°C and 80.028 cm long at 40.0°C. Calculate the average coefficient of linear expansion of the rod for this temperature range.

**15–12** If an area measured on the surface of a solid body is $A_0$ at some initial temperature and then changes by $\Delta A$ when the temperature changes by $\Delta T$, show that

$$\Delta A = (2\alpha)A_0\,\Delta T.$$

**15–13** A machinist bores a hole of 1.400 cm in diameter in a brass plate at a temperature of 20°C. What is the diameter of the hole when the temperature of the plate is increased to 200°C? Assume that the expansion coefficient remains constant over this temperature range.

**15–14** A brass cylinder is initially at 20.0°C. At what temperature will its volume be 0.300% larger than it is at 20.0°C?

**15–15** A glass flask whose volume is exactly 1000 $cm^3$ at 0°C is completely filled with mercury at this temperature. When flask and mercury are heated to 100°C, 15.4 $cm^3$ of mercury overflow. If the coefficient of volume expansion of mercury is $18.0 \times 10^{-5}$ $(C°)^{-1}$, compute the coefficient of volume expansion of the glass.

**15–16** The cross-section area of a steel rod is 4.20 $cm^2$. What force is needed to prevent the rod from contracting when it is cooled from 520°C to 20°C?

**15–17** An underground tank with a capacity of 1800 L (1.80 $m^3$) is filled with ethanol that has an initial temperature of 25.0°C. After the ethanol has cooled off to the temperature of the tank and ground, which is 10.0°C, how much air space will there be above the ethanol in the tank? (Assume that the volume of the tank doesn't change.)

**15–18** Steel train rails 12.0 m long are laid on a winter day when the temperature is $-8.0°C$. a) How much space must be left between adjacent rails if they are to just touch on a summer day when the temperature is 39.0°C? b) If the rails are originally laid in contact, what is the stress in them on a summer day when the temperature is 39.0°C?

**15–19** a) A wire that is 3.00 m long at 20°C is found to increase in length by 1.9 cm when heated to 520°C. Compute its average coefficient of linear expansion for this temperature range. b) Find the stress in the wire if it is stretched just taut (zero tension) at 520°C and cooled to 20°C without being allowed to contract. Young's modulus for the wire is $2.0 \times 10^{11}$ Pa.

## Section 15–5 Quantity of Heat

**15–20** A fruit crate with a mass of 50.0 kg slides down a ramp inclined at 53.1° below the horizontal. The ramp is 12.0 m long. If the crate was at rest at the top of the incline and has a speed of 8.00 m/s at the bottom, how much heat is generated by friction? Express your answer in joules, calories, and Btu.

**15–21** An engineer is working on an innovative engine design. One of the moving parts, designed to operate at 150°C, contains 1.20 kg of aluminum and 0.60 kg of iron. How much heat is required to raise its temperature from 20°C to 150°C?

**15–22** A nail being driven into a board heats up. If we assume that all the kinetic energy delivered by a 1.60-kg hammer with a speed of 8.40 m/s is converted into heat that stays in the nail, what is the temperature increase of a 10.0-g aluminum nail after it is struck five times?

**15–23 Heat Loss during Breathing.** A significant mechanism for heat loss in very cold weather is energy expended in warming the air taken into the lungs with each breath. Assume that the specific heat capacity of air is 1020 J/kg · K and that 1.0 L of air has mass $1.3 \times 10^{-3}$ kg. a) On an arctic winter day when the temperature is $-30°C$, calculate the amount of heat needed to warm to body temperature (37°C) the 0.50 L of air exchanged with each breath. b) How much heat is lost per hour if the respiration rate is 12 breaths per minute?

**15–24** While jogging, the average 65-kg student generates heat at a rate of 300 W. If this heat is not disposed of (by perspiration and other mechanisms) but remains in the student's body, how much would the body temperature rise after jogging for an hour? As in Example 15–7, assume that the student's specific heat capacity is 4190 J/kg · K.

**15–25** A copper tea kettle with a mass of 2.00 kg, containing 3.00 kg of water, is placed on a stove. How much heat must be added to raise the temperature from 20.0°C to 80.0°C?

**15–26** A student uses a 200-W electric immersion heater to heat 0.200 kg of water from 20.0°C to 100.0°C to make tea. a) How much heat must be added to the water? b) How much time is required?

**15–27** A technician measures the specific heat capacity of an unidentified liquid by immersing an electrical resistor in it. Electrical energy is converted to heat and transferred to the liquid for 100 s at a constant rate of 65.0 W. The mass of the liquid is 0.530 kg, and its temperature increases from 17.64°C to 20.77°C. Find the average specific heat capacity of the liquid in this temperature range. Assume that negligible heat is transferred to the container that holds the liquid and that no heat is lost to the surroundings.

## Section 15–6 Calorimetry

**15–28** An ice-cube tray of negligible mass contains 0.400 kg of water at 20.0°C. How much heat must be removed to cool the water to 0.0°C and freeze it? Express your answer in joules, calories, and Btu.

**15–29** How much heat is required to convert 6.00 g of ice at −10.0°C to steam at 100.0°C? Express your answer in joules, calories, and Btu.

**15–30 Steam Burns versus Water Burns.** A skin burn caused by steam at 100°C is much worse than a burn caused by water at 100°C. To see why, calculate the amount of heat input to your skin when it receives the heat released a) by 10.0 g of steam initially at 100.0°C when it is cooled to skin temperature (34.0°C); b) by 10.0 g of water initially at 100.0°C when it is cooled to 34.0°C.

**15–31** What must the initial speed of a lead bullet at a temperature of 25°C be so that the heat developed when it is brought to rest will be just sufficient to melt it? Assume that all the initial mechanical energy of the bullet is converted to heat and that all this heat energy stays in the bullet without any being lost to the surroundings.

**15–32** An open container has in it 0.600 kg of ice at −20.0°C. The mass of the container can be neglected. Heat is supplied to the container at the constant rate of 800 J/min for 500 min. a) After how many minutes does the ice *start* to melt? b) After how many minutes, from the time when the heating is first started, does the temperature start to rise above 0°C? c) Plot a curve showing the elapsed time horizontally and the temperature vertically.

**15–33** The capacity of air conditioners is sometimes expressed in "tons," the number of tons of ice (1 ton = 2000 lb) that can be frozen from water at 0°C in 24 hours by the unit. Express the capacity of a one-ton air conditioner in watts and Btu/h.

**15–34** Evaporation of sweat is an important mechanism for temperature control in warm-blooded animals. What mass of water must evaporate from the surface of a 75.0-kg human body to cool it 1.00 C°? The heat of vaporization of water at body temperature (37°C) is $2.42 \times 10^6$ J/kg. As in Example 15–7, assume that the specific heat capacity of the human body is approximately 4190 J/kg · K.

**15–35** An aluminum can with a mass of 0.500 kg contains 0.130 kg of water at a temperature of 20.0°C. A 0.200-kg block of iron at 75.0°C is dropped into the can. Find the final temperature, assuming no heat loss to the surroundings.

**15–36** In a physics lab experiment a student immersed 100 copper pennies (having a mass of 3.00 g each) in boiling water. After they reached thermal equilibrium, she fished them out and dropped them into 0.120 kg of water at 20.0°C in a container of negligible mass. What was the final temperature?

**15–37** A 3.40-kg iron block is taken from a furnace where its temperature is 650°C and placed on a large block of ice at 0°C. Assuming that all the heat given up by the iron is used to melt the ice, how much ice is melted?

**15–38** A laboratory technician drops a 0.050-kg sample of unknown material, at a temperature of 100.0°C, into a calorimeter containing 0.200 kg of water that is initially at 20.0°C. The calorimeter is made of copper, and its mass is 0.100 kg. The final temperature of the calorimeter is 23.6°C. Compute the specific heat capacity of the sample.

**15–39** A beaker with negligible mass contains 0.500 kg of water at a temperature of 80.0°C. How many kilograms of ice at a temperature of −20.0°C must be added to the water so that the final temperature of the system will be 40.0°C?

**15–40** A copper calorimeter with a mass of 0.100 kg contains 0.150 kg of water and 0.012 kg of ice in thermal equilibrium at atmospheric pressure. If 0.600 kg of lead at a temperature of 200°C is dropped into the calorimeter, what is the final temperature, assuming that no heat is lost to the surroundings?

**15–41** A vessel whose walls are thermally insulated contains 2.10 kg of water and 0.200 kg of ice, all at a temperature of 0.0°C. The outlet of a tube leading from a boiler in which water is boiling at atmospheric pressure is inserted into the water. How many grams of steam must condense to raise the temperature of the system to 34.0°C? Neglect the heat transferred to the container.

**15–42** A glass vial containing a 14.0-g sample of an enzyme must be cooled in an ice bath. The bath contains water and 0.120 kg of ice. The sample has specific heat capacity 2450 J/kg · K; the glass vial has mass 6.0 g and specific heat capacity 2800 J/kg · K. How much ice melts in cooling the enzyme sample from room temperature (24.2°C) to the temperature of the ice bath?

## Section 15–7 Mechanisms of Heat Transfer

**15–43** Use Eq. (15–22) to show that the SI units of thermal conductivity are W/m · K.

**15–44 Heat Leakage from an Oven.** The electric oven in a kitchen range has a total wall area of 1.50 $m^2$ and is insulated with a layer of fiberglass 4.0 cm thick. The inside surface has a temperature of 200°C, and the outside surface is at room temperature, 20°C. The fiberglass has a thermal conductivity of 0.040 W/m · K. a) What is the heat current through the insulation, assuming that it may be treated as a flat slab with an area of 1.50 $m^2$? b) What electric-power input to the heating element is required to maintain this temperature?

**15–45 House Insulation.** A carpenter builds an outer house wall with a layer of wood 2.0 cm thick on the outside and a layer of styrofoam insulation 3.0 cm thick as the inside wall surface. The wood has $k = 0.080$ W/m · K, and the styrofoam has $k = 0.010$ W/m · K. The interior surface temperature is 20.0°C, and the exterior surface temperature is $-10.0$°C. a) What is the temperature at the plane where the wood meets the styrofoam? b) What is the rate of heat flow per square meter through this wall?

**15–46** One end of an insulated metal rod is maintained at 100°C, and the other end is placed in an ice-water mixture. The bar has a length of 50.0 cm and a cross-section area of 0.800 $cm^2$. The heat conducted by the rod melts 3.00 g of ice in 5.00 min. Calculate the thermal conductivity $k$ of the metal.

**15–47** The ceiling of a room has an area of 150 $ft^2$. The ceiling is insulated to an $R$ value of 30 (in commercial units of $ft^2$ · F° · h/Btu). The surface in the room is maintained at 72°F, and the surface in the attic has a temperature of 105°F. How many Btu of heat flow through the ceiling into the room in 6.0 h? Also express your answer in joules.

**15–48** A long rod, insulated to prevent heat loss along its sides, is in contact with boiling water (at atmospheric pressure) at one end and with an ice-water mixture at the other (Fig. 15–23). The rod consists of a 1.00-m section of copper (one end in steam) joined end-to-end to a length $L_2$ of steel (one end in ice). Both sections of the rod have cross-section areas of 4.00 $cm^2$. The temperature of the copper-steel junction is 65.0°C after a steady state has been set up. a) How much heat per second flows from the steam bath to the ice-water mixture? b) What is the length $L_2$ of the steel section?

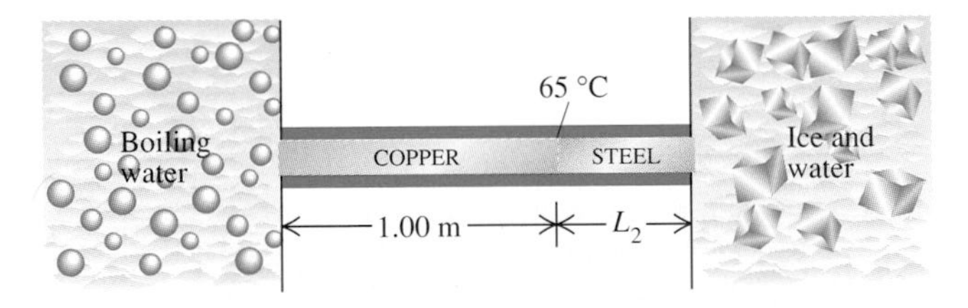


**FIGURE 15–23**

**15–49** Suppose the rod in Fig. 15–15 is made of copper, has a length of 20.0 cm, and has a cross-section area of 1.20 $cm^2$. Let $T_H = 100.0$°C and $T_C = 0.0$°C. a) What is the final steady-state temperature gradient along the rod? b) What is the heat current in the rod in the final steady state? c) What is the final steady-state temperature at a point in the rod 6.00 cm from its left end?

**15–50** A pot with a steel bottom 1.50 cm thick rests on a hot stove. The area of the bottom of the pot is 0.150 $m^2$. The water inside the pot is at 100°C, and 0.840 kg are evaporated every 5.00 min. Find the temperature of the lower surface of the pot, which is in contact with the stove.

**15–51** What is the net rate of heat loss by radiation in Example 15–16 if the temperature of the surroundings is 12.0°C?

**15–52** What is the rate of energy radiation per unit area of a blackbody at a temperature of a) 300 K; b) 3000 K?

**15–53 Size of a Light-Bulb Filament.** The operating temperature of a tungsten filament in an incandescent lamp is 2450 K, and its emissivity is 0.35. Find the surface area of the filament of a 60-W lamp if all the electrical energy consumed by the lamp is radiated by the filament as electromagnetic waves.

**15–54** The emissivity of tungsten is 0.35. A tungsten sphere with a radius of 2.00 cm is suspended within a large evacuated enclosure whose walls are at 300 K. What power input is required to maintain the sphere at a temperature of 3000 K if heat conduction along the supports is neglected?

## Section 15–8 Integrated Circuits: A Case Study in Heat Transfer

**15–55** A night light has a 12-W bulb with a surface area of the glass envelope of 15 $cm^2$. For this bulb, what is the power dissipation per unit surface area (by radiation, conduction, and convection to the surrounding air)?

**15–56** The thermal resistance $r_{th}$ of the light bulb in Exercise 15–55 is 5.0 K/W. What is the operating temperature of the bulb surface if the ambient temperature is 20°C?

**15–57 Maximum Temperature for an Operating Chip.** One method for controlling the temperature of an operating VLSI chip is to decrease the ambient temperature by air conditioning or by immersing the chip in a cold liquid. A VLSI chip has a heat power output of 36 W and a thermal resistance of $r_{th} = 3.0$ K/W. If the chip is to operate at a temperature of 120°C, what must be the ambient temperature?

**15–58** The thermal resistance $r_{th}$ of an ic chip is measured to determine the effectiveness of forced air cooling. Without the forced air, when the electrical power dissipation (by conversion to heat) in the chip is 1.5 W, the chip temperature rises to 78 C° above the ambient temperature. With the forced air system operating, the chip temperature rises 54 C° above the ambient temperature for the same 1.5-W power dissipation. What is the thermal resistance $r_{th}$ of the chip a) without the forced air cooling; b) with the forced air cooling?

## PROBLEMS

**15–59** An aluminum cube 0.100 m on a side is heated from 10.0°C to 60.0°C. a) What is the change in its volume; b) in its density?

**15–60** Suppose a steel hoop could be constructed around the earth's equator, just fitting it at a temperature of 20.0°C. What would be the thickness of space between the hoop and the earth if the temperature of the hoop were increased by 1.0 C°?

**15–61** A steel ring with a 3.0000-in. inside diameter at 20.0°C is to be heated and slipped over a brass shaft measuring 3.0050 in. in diameter at 20.0°C. a) To what temperature should the ring be heated? b) If the ring and shaft together are cooled by some means such as liquid air, at what temperature will the ring just slip off the shaft?

**15–62** At a temperature of 20.0°C the volume of a certain glass flask, up to a reference mark on the long stem of the flask, is exactly 100 cm$^3$. The flask is filled to this point with a liquid whose coefficient of volume expansion is $4.00 \times 10^{-4}\,K^{-1}$, with both flask and liquid at 20.0°C. The coefficient of volume expansion of the glass is $2.00 \times 10^{-5}\,K^{-1}$. The cross-section area of the stem is 50.0 mm$^2$ and can be considered constant. (Do you see why this is a good approximation?) How far will the liquid rise or fall in the stem when the temperature is raised to 40.0°C?

**15–63** A metal rod that is 30.0 cm long expands by 0.0750 cm when its temperature is raised from 0°C to 100°C. A rod of a different metal and of the same length expands by 0.0400 cm for the same rise in temperature. A third rod, also 30.0 cm long, is made up of pieces of each of the above metals placed end-to-end and expands 0.0650 cm between 0°C and 100°C. Find the length of each portion of the composite bar.

**15–64** A steel rod with a length of 0.400 m and a copper rod with a length of 0.300 m, both with the same diameter, are placed end-to-end between rigid supports with no initial stress in the rods. The temperature of the rods is now raised by 50.0 C°. What is the stress in each rod?

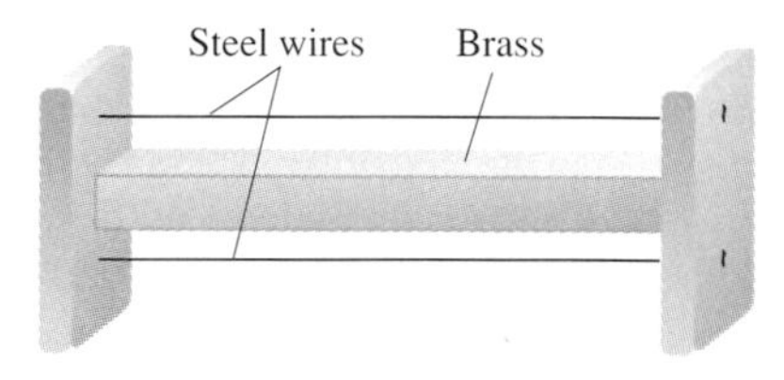


**FIGURE 15–24**

**15–65** A heavy brass bar has projections at its ends, as in Fig. 15–24. Two fine steel wires fastened between the projections are just taut (zero tension) when the whole system is at 0°C. What is the tensile stress in the steel wires when the temperature of the system is raised to 220°C? Make any simplifying assumptions you think are justified, but state what they are.

**15–66 Bulk Stress due to Heating.** a) Prove that if a body under hydrostatic pressure is raised in temperature but not allowed to expand, the increase in pressure is

$$\Delta p = B\beta\,\Delta T,$$

where the bulk modulus $B$ and the average coefficient of volume expansion $\beta$ are both assumed to be positive and constant. b) What hydrostatic pressure is necessary to prevent a copper block from expanding when its temperature is increased from 20.0°C to 35.0°C?

**15–67** A liquid is enclosed in a metal cylinder with a piston of the same metal. The system is originally at atmospheric pressure (1.00 atm) and at a temperature of 60.0°C. The piston is forced down until the pressure on the liquid is increased by 100 atm ($1.013 \times 10^7$ Pa), and it is then clamped in this position. Find the new temperature at which the pressure of the liquid is again 1.00 atm ($1.013 \times 10^5$ Pa). Assume that the cylinder is sufficiently strong that its volume is not altered by changes in pressure, but only by changes in temperature. Use the result derived in Problem 15–66.

Compressibility of liquid: $k = 7.00 \times 10^{-10}\,Pa^{-1}$.
Coefficient of volume expansion of liquid: $\beta = 5.30 \times 10^{-4}\,K^{-1}$.
Coefficient of volume expansion of metal: $\beta = 4.00 \times 10^{-5}\,K^{-1}$.

**15–68** A thirsty farmer cools a 1.00-L bottle of a soft drink (mostly water) by pouring it into a large tin-plated copper mug with a mass of 0.278 kg and adding 0.054 kg of ice initially at −16.0°C. If soft drink and mug are initially at 20.0°C, what is the final temperature of the system, assuming no heat losses?

**15–69 Satellite Reentry.** An artificial satellite made of aluminum circles the earth at a speed of 7000 m/s. a) Find the ratio of its kinetic energy to the energy required to raise its temperature by 600 C°. (The melting point of aluminum is 660°C. Assume a constant specific heat capacity of 910 J/kg·K.) b) Discuss the bearing of your answer on the problem of the reentry of a satellite into the earth's atmosphere.

**15–70** A capstan is a rotating drum or cylinder over which a rope or cord slides to provide a great amplification of the rope's tension while keeping both ends free (see Fig. 15–25). Since the added tension in the rope is due to friction, the capstan generates

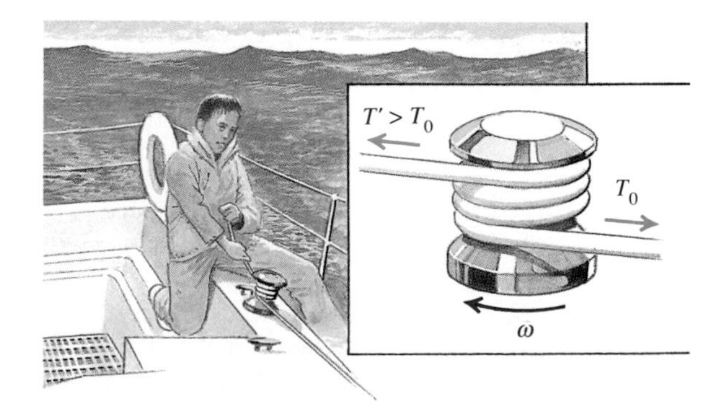


**FIGURE 15–25**

heat. a) If the difference in tension between the two parts of the rope is 540 N, and the capstan has a diameter of 12.0 cm and turns once in 1.00 s, find the rate at which heat is being generated. Why does the number of turns not matter? b) If the capstan is made of iron and has a mass of 5.00 kg, at what rate does its temperature rise? Assume that the temperature in the capstan is uniform and that all the heat generated remains in it.

**15–71 Debye's $T^3$ Law.** At very low temperatures the molar heat capacity of rock salt varies with temperature according to Debye's $T^3$ law:

$$C = k\frac{T^3}{\Theta^3},$$

where $k = 1940$ J/mol · K and $\Theta = 281$ K. a) How much heat is required to raise the temperature of 2.00 mol of rock salt from 10.0 K to 60.0 K? [*Hint:* Use Eq. (15–15) for $dQ$ and integrate.] b) What is the average molar heat capacity in this range? c) What is the true molar heat capacity at 60.0 K?

**15–72** The molar heat capacity of a certain substance varies with temperature according to the empirical equation

$$C = 27.2 \text{ J/mol} \cdot \text{K} + (4.00 \times 10^{-3} \text{ J/mol} \cdot \text{K}^2)T.$$

How much heat is necessary to change the temperature of 2.00 mol of this substance from 27°C to 527°C? [*Hint:* Use Eq. (15–15) for $dQ$ and integrate.]

**15–73 Hot Air in a Physics Lecture.** a) A typical student listening attentively to a physics lecture has a heat output of 200 W. How much heat energy does a class of 140 physics students release into a lecture hall over the course of a 50-min lecture? b) Assume that all the heat energy in part (a) is transferred to the 3200 $m^3$ of air in the room. The air has specific heat capacity 1020 J/kg · K and density 1.20 kg/$m^3$. If none of the heat escapes and the air conditioning system is off, how much will the temperature of the air in the room rise during the 50-min lecture? c) If the class is taking an exam, the heat output per student rises to 350 W. What is the temperature rise during 50 min in this case?

**15–74 Hot Water versus Steam Heating.** In a household hot-water heating system, water is delivered to the radiators at 60.0°C (140°F) and leaves at 27.0°C (81°F). The system is to be replaced by a steam system in which steam at atmospheric pressure condenses in the radiators and the condensed steam leaves the radiators at 34.0°C (93°F). How many kilograms of steam will supply the same heat as was supplied by 1.00 kg of hot water in the first system?

**15–75** An ice cube with a mass of 0.070 kg is taken from a freezer, where the cube's temperature was − 10.0°C, and dropped into a glass of water at 0.0°C. If no heat is gained or lost from outside, how much water will freeze onto the cube?

**15–76** A tube leads from a flask in which water is boiling under atmospheric pressure to a calorimeter. The mass of the calorimeter is 0.150 kg, its specific heat capacity is 420 J/kg · K, and it originally contains 0.340 kg of water at 15.0°C. Steam is allowed to condense in the calorimeter until its temperature increases to 71.0°C, after which the total mass of calorimeter and contents is found to be 0.525 kg. Compute the heat of vaporization of water from these data.

**15–77** A copper calorimeter can with a mass of 0.322 kg contains 0.046 kg of ice. The system is initially at 0.0°C. If 0.012 kg of steam at 100.0°C and 1.00 atm pressure is admitted into the calorimeter, what is the final temperature of the calorimeter and its contents?

**15–78** One experimental method of measuring an insulating material's thermal conductivity is to construct a box of the material and measure the power input to an electric heater inside the box that maintains the interior at a measured temperature above the outside surface. Suppose that in such an apparatus a power input of 180 W is required to keep the interior surface of the box 65.0 C° (about 120 F°) above the temperature of the outer surface. The total area of the box is 2.32 $m^2$, and the wall thickness is 3.8 cm. Find the thermal conductivity of the material in SI units.

**15–79 Effect of a Window in a Door.** A carpenter builds a solid wood door with dimensions 2.00 m × 0.80 m × 4.0 cm. Its thermal conductivity is $k = 0.0500$ W/m · K. The inside air temperature is 20.0°C, and the outside air temperature is − 10.0°C. a) What is the rate of heat flow through the door, assuming that the surface temperatures are those of the surrounding air? b) By what factor is the heat flow increased if a window 0.50 m on a side is inserted in the door, assuming that the glass is 0.40 cm thick and the surface temperatures are again those of the surrounding air? The glass has thermal conductivity 0.80 W/m · K.

**15–80** A wood ceiling with thermal resistance $R_1$ is covered with a layer of insulation with thermal resistance $R_2$. Prove that the effective thermal resistance of the combination is $R = R_1 + R_2$.

**15–81** Compute the ratio of the rate of heat loss through a single-pane window with area 0.15 $m^2$ to that for a double-pane window with the same area. The glass of the single pane is 2.5 mm thick, and the air space between the two panes of the double-pane window is 5.0 mm thick. The glass has thermal conductivity 0.80 W/m · K.

**15–82** Rods of copper, brass, and steel are welded together to form a Y-shaped figure. The cross-section area of each rod is 2.00 $cm^2$. The free end of the copper rod is maintained at 100.0°C, and the free ends of the brass and steel rods at 0.0°C. Assume that there is no heat loss from the surfaces of the rods. The lengths of the rods are as follows: copper, 24.0 cm; brass, 13.0 cm; steel, 18.0 cm. a) What is the temperature of the junction point? b) What is the heat current in each of the three rods?

**15–83 Time Needed for a Lake to Freeze Over.** a) Show that the thickness of the ice sheet formed on the surface of a lake is proportional to the square root of the time if the heat of fusion of the water freezing on the underside of the ice sheet is conducted through the sheet. b) Assuming that the upper surface of the ice sheet is at − 10°C and that the bottom surface is at 0°C, calculate the time it will take to form an ice sheet 30 cm thick.

**15–84** A rod is initially at a uniform temperature of 0°C throughout. One end is kept at 0°C, and the other is brought into contact with a steam bath at 100°C. The surface of the rod is insulated so that heat can flow only lengthwise along the rod. The cross-section area of the rod is 2.00 $cm^2$, its length is 100 cm, its thermal conductivity is 335 W/m · K, its density is $1.00 \times 10^4$ $kg/m^3$, and its specific heat capacity is 419 J/kg · K. Consider a short cylindrical element of the rod 1.00 cm in length. a) If the temperature gradient at the cooler end of this element is 160 C°/m, how many joules of heat energy flow across this end per second? b) If the average temperature of the element is increasing at the rate of 0.300 C°/s, what is the temperature gradient at the other end of the element?

**15–85** If the solar radiation energy incident per second on the frozen surface of a lake is 600 $W/m^2$ and 70% of this energy is absorbed by the ice, how much time will it take for a 1.40-cm-thick layer of ice to melt? The ice and the water beneath it are at a temperature of 0°C.

**15–86 Temperature of the Sun.** The rate at which radiant energy reaches the upper atmosphere of the earth from the sun is about 1.40 $kW/m^2$. The distance from the earth to the sun is $1.49 \times 10^{11}$ m, and the radius of the sun is $6.95 \times 10^8$ m. a) What is the rate of radiation of energy per unit area from the sun's surface? b) If the sun radiates as an ideal blackbody, what is the temperature of its surface?

**15–87** A physicist uses a cylindrical metal can 0.100 m high and 0.060 m in diameter to store liquid helium at 4.22 K; at that temperature its heat of vaporization is $2.09 \times 10^4$ J/kg. Completely surrounding the metal can are walls maintained at the temperature of liquid nitrogen, 77.3 K, the intervening space being evacuated. How much helium is lost per hour? The emissivity of the metal can is 0.200. The only heat transfer between the metal can and the surrounding walls is by radiation.

**15–88 Thermal Expansion of an Ideal Gas.** a) The pressure $p$, volume $V$, number of moles $n$, and Kelvin temperature $T$ of an ideal gas are related by the equation $pV = nRT$, where $R$ is a constant. Prove that the coefficient of volume expansion for an ideal gas is equal to the reciprocal of the Kelvin temperature if the expansion occurs at constant pressure. b) Compare the coefficients of volume expansion of copper and air at a temperature of 20°C. Assume that air may be treated as an ideal gas and that the pressure remains constant.

## CHALLENGE PROBLEMS

**15–89 Temperature Change in a Clock.** A clock whose pendulum makes one oscillation in 2.00 s is correct at 25.0°C. The pendulum shaft is of steel, and its mass may be neglected in comparison with that of the bob. a) What is the fractional change in length of the shaft when it is cooled to 5.0°C? b) How many seconds per day will the clock gain or lose at 5.0°C? c) How closely must the temperature be controlled if the clock is not to gain or lose more than 1.00 s a day? Does the answer depend on the period of the pendulum?

**15–90** An engineer is developing an electric water heater to provide a continuous supply of hot water. One trial design is shown in Fig. 15–26. Water is flowing at the rate of 0.300 kg/min, the inlet thermometer registers 18.0°C, the voltmeter reads 120 V, and the ammeter reads 10.0 A [corresponding to a power input of (120 V)(10.0 A) = 1200 W]. a) When a steady state is finally reached, what is the reading of the outlet thermometer? b) Why is it unnecessary to take into account the heat capacity $mc$ of the apparatus itself?

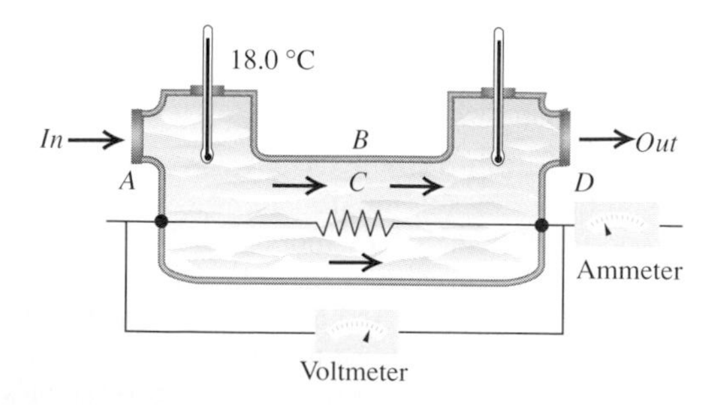

**FIGURE 15–26**

**15–91** a) A spherical shell has inner and outer radii $a$ and $b$, respectively, and the temperatures at the inner and outer surfaces are $T_2$ and $T_1$. The thermal conductivity of the material of which the shell is made is $k$. Derive an equation for the total heat current through the shell. b) Derive an equation for the temperature variation within the shell in part (a); that is, calculate $T$ as a function of $r$, the distance from the center of the shell. c) A hollow cylinder has length $L$, inner radius $a$, and outer radius $b$, and the temperatures at the inner and outer surfaces are $T_2$ and $T_1$. (The cylinder could represent an insulated hot-water pipe, for example.) The thermal conductivity of the material of which the cylinder is made is $k$. Derive an equation for the total heat current through the walls of the cylinder. d) For the cylinder of part (c), derive an equation for the temperature variation inside the cylinder walls. e) For the spherical shell of part (a) and the hollow cylinder of part (c), show that the equation for the total heat current in each case reduces to Eq. (15–22) for linear heat flow when the shell or cylinder is very thin.

**15–92 Food Intake of a Hamster.** The energy output of an animal engaged in an activity is called the basal metabolic rate (BMR) and is a measure of the conversion of food energy into other forms of energy. A simple calorimeter to measure the BMR consists of an insulated box with a thermometer to measure the temperature of the air. The air has density 1.20 $kg/m^3$ and specific heat capacity 1020 J/kg · K. A 50.0-g hamster is placed in a calorimeter that contains 0.0500 $m^3$ of air at room temperature. a) When the hamster is running in a wheel, the temperature of the air in the calorimeter rises 1.60 C° per hour. How much heat does the running hamster generate in an hour? Assume that all this heat goes into the air in the calorimeter. Neglect the heat that goes into the walls of the box and into the thermometer, and assume that no heat is lost to the surroundings. b) Assuming that the hamster converts seed into heat with an efficiency of 10% and that hamster seed has a food energy value of 24 J/g, how many grams of seed must the hamster eat per hour to supply this energy?

**15–93** Suppose that both ends of the rod in Fig. 15–15 are kept at a temperature of 0°C and that the initial temperature distribution along the rod is given by $T = (100°C) \sin \pi x/L$, where $x$ is measured from the left end of the rod. Let the rod be copper, with length $L = 0.100$ m and cross-section area $1.00$ cm$^2$. a) Show the initial temperature distribution in a diagram. b) What is the final temperature distribution after a very long time has elapsed? c) Sketch curves that you think would represent the temperature distribution at intermediate times. d) What is the initial temperature gradient at the ends of the rod? e) What is the initial heat current from the ends of the rod into the bodies making contact with its ends? f) What is the initial heat current at the center of the rod? Explain. What is the heat current at this point at any later time? g) What is the value of the *thermal diffusivity* $k/\rho c$ for copper, and in what unit is it expressed? (Here $k$ is the thermal conductivity, $\rho$ is the density, and $c$ is the specific heat capacity.) h) What is the initial rate of change of temperature at the center of the rod? i) How much time would be required for the center of the rod to reach its final temperature if the temperature continued to decrease at this rate? (This time is called the *relaxation time* of the rod.) j) From the graphs in part (c), would you expect the rate of change of temperature at the midpoint to remain constant, increase, or decrease as a function of time? k) What is the initial rate of change of temperature at a point in the rod 2.5 cm from its left end?

**15–94** A steam pipe with a radius of 2.00 cm, carrying steam at 130°C, is surrounded by a cylindrical jacket of cork with inner and outer radii 2.00 cm and 4.00 cm; this in turn is surrounded by a cylindrical jacket of styrofoam having inner and outer radii 4.00 cm and 6.00 cm (Fig. 15–27). The outer surface of the styrofoam is in contact with air at 20°C. Assume that this outer surface has a temperature of 20°C. a) What is the temperature at a radius of 4.00 cm, where the two insulating layers meet? b) What is the total rate of transfer of heat out of a 2.00-m length of pipe? [*Hint:* Use the expression derived in part (c) of Problem 15–91.]

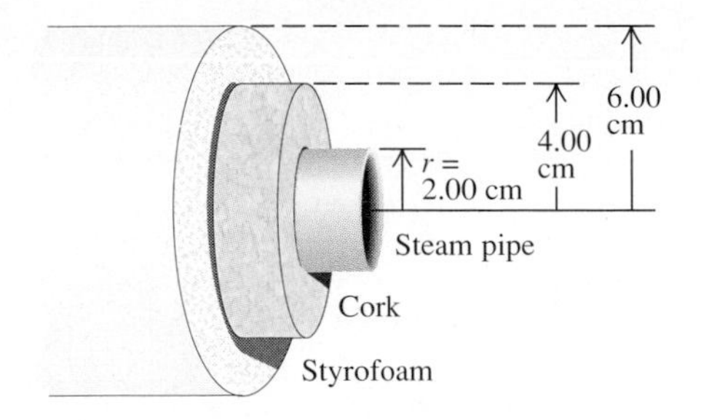


**FIGURE 15–27**

**15–95** A solid cylindrical copper rod 0.100 m long has one end maintained at a temperature of 20.00 K. The other end is blackened and exposed to thermal radiation from surrounding walls at 400 K. The sides of the rod are insulated, so no energy is lost or gained except at the ends of the rod. When equilibrium is reached, what is the temperature of the blackened end? (*Hint:* Since copper is a very good conductor of heat at low temperatures, with $k = 1670$ W/m·K at 20 K, the temperature of the blackened end is only slightly greater than 20.00 K.)

CHAPTER 16

# Thermal Properties of Matter

Time-delay fuse

Colored stars and bursting mixture

Fast fuse

Gunpowder propels shell from launch tube

Fireworks consist of a chemical fuel and a chemical oxidizer (oxygen source). When the mixture is ignited, it vaporizes in an intense energy-releasing reaction. The sudden increase in volume bursts the card - board shell with great force. The shape and colors of the explosion depend on the type of shell and the composition of the chemicals.

## KEY CONCEPTS

**• The equation of state for a substance relates the pressure, volume, and temperature. The ideal-gas equation is an approximate equation of state for gases.**

**• Molecular mass is the mass of a mole of substance. Avogadro's number is the number of molecules in a mole.**

**• The ideal-gas equation can be derived from a molecular model in which moving gas molecules collide elastically with the walls of a container. The average molecular speed depends on temperature and molecular mass.**

**• Heat capacities are related to changes in molecular kinetic and potential energies brought about by additions of heat. The molar heat capacity of a gas depends on its molecular structure.**

**• Speeds of gas molecules are distributed according to the Maxwell-Boltzmann distribution law.**

**• The equation of state for a material, and the conditions under which various phases can occur, can be represented graphically by means of $pV$-graphs, phase diagrams, or thermodynamic surfaces. At the triple point, solid, liquid, and vapor phases can all coexist. At the critical point, the distinction between liquid and vapor disappears.**

## INTRODUCTION

You can't see air molecules, but they're moving all around you. Did you know that in one day you are hit $10^{32}$ times by air molecules? And did you know that the average speed of an air molecule is about 1000 mi/h? Or that the impacts of these molecules on a surface are responsible for the pressure that air exerts on a surface?

To understand and relate such facts, we need to be able to use both a macroscopic (bulk) and microscopic (molecular) perspective on the properties of matter. The *macroscopic* perspective considers such large-scale variables as pressure, volume, temperature, and mass of substance. In contrast, the *microscopic* point of view investigates small-scale quantities such as the speeds, kinetic energies, momenta, and masses of the individual molecules that make up a substance. Where these two perspectives overlap, they must be consistent. For example, the (microscopic) collision forces that occur when air molecules strike a solid surface (such as your skin) cause (macroscopic) atmospheric pressure.

In this chapter we'll use both macroscopic and microscopic approaches to gain an understanding of thermal properties and behavior of matter. One of the simplest kinds of matter to understand is the *ideal gas*. For this class of materials we'll relate pressure, volume, temperature, and amount of substance to one another and to the speeds and masses of individual molecules. We will explore the molecular basis of heat capacities of both gases and solids, and we will take a look at the various *phases* of matter — gas, liquid, and solid — and the conditions under which each occurs.

## 16–1 EQUATIONS OF STATE

The conditions in which a particular material exists are described by physical quantities such as pressure, volume, temperature, and amount of the substance. For example, a tank of oxygen in a welding outfit has a pressure gauge and a label stating its volume. We could add a thermometer and place the tank on a scale. These variables describe the *state* of the material, and they are called *state variables,* or **state coordinates.**

The volume $V$ of a substance is usually determined by its pressure $p$, temperature $T$, and amount of substance, mass $m$ or number of moles $n$. Ordinarily, we can't change one of these variables without causing a change in another. When the tank of oxygen gets hotter, the pressure increases. When the tank gets too hot, it explodes; this happens occasionally with overheated steam boilers (Fig. 16–1).

**16–1** An overheated steam boiler can explode with great violence.

In a few cases the relationship among $p$, $V$, $T$, and $m$ (or $n$) is simple enough that we can express it as an equation called the **equation of state.** When it's too complicated for that, we can use graphs or numerical tables. Even then, the relation among the variables still exists; we call it an equation of state even when we don't know the actual equation.

Here's a simple (though approximate) equation of state for a solid material. The temperature coefficient of volume expansion $\beta$ is the fractional volume change $dV/V_0$ per unit temperature change, and the compressibility $k$ is the fractional volume change $dV/V_0$ per unit pressure change. If a certain amount of material has volume $V_0$ when the pressure is $p_0$ and the temperature is $T_0$, the volume $V$ at slightly differing pressure $p$ and temperature $T$ is approximately

$$V = V_0[1 + \beta(T - T_0) - k(p - p_0)]. \tag{16–1}$$

Equation (16–1) is called an *equation of state* for the material.

### The Ideal Gas Equation

Another simple equation of state is the one for an ideal gas. Figure 16–2 shows an experimental setup to study the behavior of a gas. The cylinder has a movable piston and is equipped with a pressure gauge and a thermometer. We can vary the pressure, volume, and temperature, and we can pump any desired mass of any gas into the cylinder. It is usually easiest to describe the amount of gas in terms of the number of moles $n$, rather than the mass. We did this when we defined molar heat capacity in Section 15–5 (you may want to review that section). Molecular mass $M$ is mass per mole, and the total mass $m_{tot}$ is the number of moles $n$ times the mass per mole $M$:

$$m_{tot} = nM. \tag{16–2}$$

We are calling the total mass $m_{tot}$ because later in the chapter we will use $m$ for the mass of one molecule.

Measurements of the behavior of various gases lead to several conclusions. First, the volume $V$ is proportional to the number of moles $n$. If we double the number of moles, keeping pressure and temperature constant, the volume doubles.

Second, the volume varies *inversely* with pressure $p$. If we double the pressure, holding the temperature $T$ and amount of substance $n$ constant, we compress the gas to one-half of its initial volume. This relation is called Boyle's law, after Robert Boyle (1627–1691), a contemporary of Newton. It states that $pV = \text{constant}$ when $n$ and $T$ are constant.

Third, the pressure is proportional to the absolute temperature. If we double the absolute temperature, keeping the volume and quantity of material constant, the pressure doubles. This is called Charles' law, after Jacques Charles (1746–1823). It states that $p = (\text{constant})\,T$ when $n$ and $V$ are constant.

These three relationships can be combined neatly into a single equation:

$$pV = nRT, \tag{16–3}$$

which we call the **ideal-gas equation.** We might expect the constant $R$ in this equation to have different values for different gases, but it turns out to have the same value for *all* gases, at least at sufficiently high temperatures and low pressures. It is called the **ideal gas constant.**

The numerical value of $R$ depends on the units of $p$, $V$, and $T$. In SI units, in which the unit of $p$ is Pa or $N/m^2$ and the unit of $V$ is $m^3$, the numerical value of $R$ is

$$R = 8.3145 \text{ J/mol}\cdot\text{K}.$$

Note that the units of pressure times volume are the same as units of work or energy (for example, $N/m^2$ times $m^3$); that's why $R$ has units of energy per mole per unit of absolute temperature. In chemical calculations, volumes are often expressed in liters (L) and pressures in atmospheres (atm). In this system,

$$R = 0.08206 \text{ L}\cdot\text{atm/mol}\cdot\text{K}.$$

We can express Eq. (16–3) in terms of the mass $m_{tot}$ of material, using Eq. (16–2):

$$pV = \frac{m_{tot}}{M}RT. \tag{16–4}$$

From this we can get an expression for the density $\rho = m_{tot}/V$ of the gas:

$$\rho = \frac{pM}{RT}. \tag{16–5}$$

An **ideal gas** is one for which Eq. (16–3) holds precisely for *all* pressures and temperatures. This is an idealized model; it works best at very low pressures and high temperatures, when the gas molecules are far apart and in rapid motion. It is reasonably good (within a few percent) at moderate pressures (such as a few atmospheres) and at temperatures well above those at which the gas liquefies.

For a *constant mass* (or constant number of moles) of an ideal gas the product $nR$ is constant, so the quantity $pV/T$ is also constant. Thus, using the subscripts 1 and 2 to refer to any two states of the same mass of a gas, we have

$$\frac{p_1V_1}{T_1} = \frac{p_2V_2}{T_2} = \text{constant}. \tag{16–6}$$

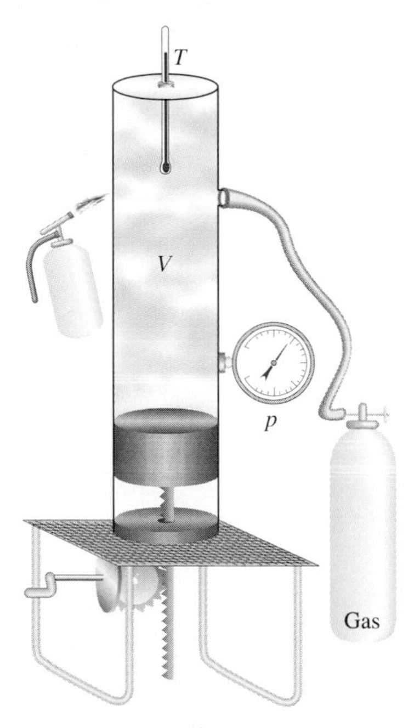


**16–2** A hypothetical setup for studying the behavior of gases. The pressure, volume, temperature, and number of moles of a gas can be controlled and measured.

The proportionality of pressure to absolute temperature is familiar; in fact, in Chapter 15 we *defined* a temperature scale in terms of pressure in a constant-volume gas thermometer. That may make it seem that the pressure-temperature relation in the ideal-gas equation is just a result of the way we define temperature. But the equation also tells us what happens when we change the volume or the amount of substance. Also, the gas-thermometer scale turns out to correspond closely to a scale that we will define in Chapter 18 that doesn't depend on properties of any particular material. For now, consider the ideal-gas equation as being based on this genuinely material-independent temperature scale, even though we haven't defined it yet.

## PROBLEM-SOLVING STRATEGY

### Ideal gases

**1.** In some problems you will be concerned with only one state of the system; some of the quantities in Eq. (16–3) will be known, some unknown. Make a list of what you know and what you have to find. For example, $p = 1.0 \times 10^6$ Pa, $V = 4\ \text{m}^3$, $T = ?$, $n = 2$ mol, or something comparable.

**2.** In other problems there will be two different states of the same amount of gas. Decide which is state 1 and which is state 2, and make a list of the quantities for each: $p_1$, $p_2$, $V_1$, $V_2$, $T_1$, $T_2$. If all but one of these quantities are known, you can use Eq. (16–6). Otherwise, you have to use Eq. (16–3). For example, if $p_1$, $V_1$, and $n$ are given, you can't use Eq. (16–6) because you don't know $T_1$.

**3.** As always, be sure to use a consistent set of units. First decide which value of the gas constant $R$ you are going to use, and then convert the units of the other quantities accordingly. You may have to convert atmospheres to pascals or liters to cubic meters ($1\ \text{m}^3 = 10^3\ \text{L} = 10^6\ \text{cm}^3$). Sometimes the problem statement will make one system of units clearly more convenient than others. Decide on your system, and stick to it.

**4.** Don't forget that $T$ must always be an *absolute* temperature. If you are given temperatures in °C, be sure to add 273 to convert to kelvins.

**5.** You may sometimes have to convert between mass $m_{tot}$ and number of moles $n$. The relationship is $m_{tot} = Mn$, where $M$ is the molecular mass. Here's a tricky point: If you use Eq. (16–4), you *must* use the same mass units for $m_{tot}$ and $M$. So if $M$ is in grams per mole (the usual units for molecular mass), then $m_{tot}$ must be in grams. If you want to use $m_{tot}$ in kg, then you must convert $M$ to kg/mol. For example, the molecular mass of oxygen is 32 g/mol, or $32 \times 10^{-3}$ kg/mol. Be careful!

## EXAMPLE 16–1

**Volume of a gas at STP** The condition called **standard temperature and pressure** (STP) for a gas is defined to be a temperature of 0°C = 273 K and a pressure of 1 atm = $1.013 \times 10^5$ Pa. If you want to keep a mole of an ideal gas in your room at STP, how big a container do you need?

**SOLUTION** From Eq. (16–3), using $R$ in J/mol · K,

$$V = \frac{nRT}{p} = \frac{(1\ \text{mol})(8.315\ \text{J/mol}\cdot\text{K})(273\ \text{K})}{1.013 \times 10^5\ \text{Pa}}$$
$$= 0.0224\ \text{m}^3 = 22.4\ \text{L}.$$

A cube 0.282 m on a side would do the job (Fig. 16–3).

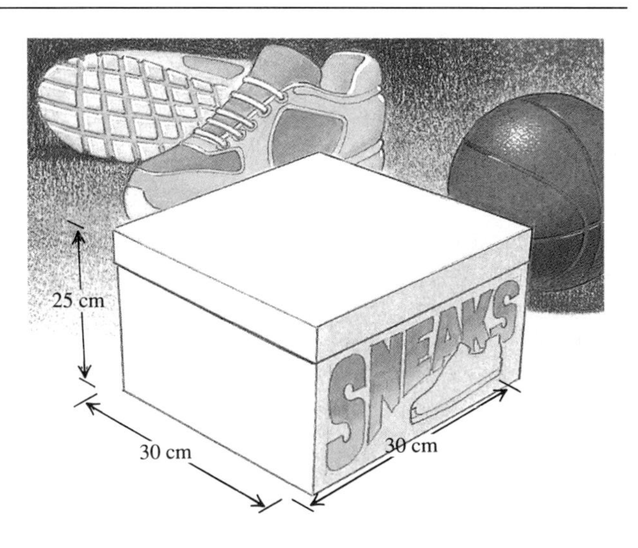


**16–3** At STP this box contains one mole of air molecules. For comparison, a basketball has almost exactly one-third of this volume.

## EXAMPLE 16–2

**Mass of air in a scuba tank** A typical tank used for scuba diving (Fig. 16–4) has a volume of 11.0 L (about 0.4 ft$^3$) and a gauge pressure, when full, of $2.10 \times 10^7$ Pa (about 3000 psig). The "empty" tank contains 11.0 L of air at 21°C and 1 atm ($1.013 \times 10^5$ Pa). The air is hot when it comes out of the compressor; when the tank is filled, the temperature is 42°C and the gauge pressure is $2.10 \times 10^7$ Pa. What mass of air was added? (Air is a mixture of gases, about 78% nitrogen, 21% oxygen, and 1% miscellaneous; its average molecular mass is 28.8 g/mol = $28.8 \times 10^{-3}$ kg/mol.)

**SOLUTION** Let's first find the number of moles in the tank at the beginning and at the end. We must remember to convert the temperatures to the Kelvin scale by adding 273 and to convert the pressure to absolute by adding $1.013 \times 10^5$ Pa. From Eq. (16–3) the number of moles $n_1$ in the "empty" tank is

$$n_1 = \frac{pV}{RT} = \frac{(1.013 \times 10^5 \text{ Pa})(11.0 \times 10^{-3} \text{ m}^3)}{(8.315 \text{ J/mol}\cdot\text{K})(294 \text{ K})} = 0.456 \text{ mol}.$$

The number of moles in the full tank is

$$n_2 = \frac{(21.1 \times 10^6 \text{ Pa})(11.0 \times 10^{-3} \text{ m}^3)}{(8.315 \text{ J/mol}\cdot\text{K})(315 \text{ K})} = 88.6 \text{ mol}.$$

We added 88.6 mol − 0.456 mol = 88.1 mol to the tank. The added mass is (88.1 mol)($28.8 \times 10^{-3}$ kg/mol) = 2.54 kg.

**16–4** An air tank for scuba diving. Does the tank weigh measurably more when full?

## EXAMPLE 16–3

Find the density of air at 20°C and normal atmospheric pressure.

**SOLUTION** We use Eq. (16–5), taking care to convert $T$ to kelvins:

$$\rho = \frac{pM}{RT} = \frac{(1.013 \times 10^5 \text{ Pa})(28.8 \times 10^{-3} \text{ kg/mol})}{(8.315 \text{ J/mol}\cdot\text{K})(293 \text{ K})}$$
$$= 1.20 \text{ kg/m}^3.$$

## EXAMPLE 16–4

**Variation of atmospheric pressure with elevation** Find the variation of atmospheric pressure with elevation in the earth's atmosphere, assuming that the temperature is 0°C throughout.

**SOLUTION** We begin with the pressure relation that we derived in Section 14–2, Eq. (14–4): $dp/dy = -\rho g$. The density $\rho$ is given by Eq. (16–5): $\rho = pM/RT$. We substitute this into Eq. (14–4), separate variables, and integrate, letting $p_1$ be the pressure at elevation $y_1$ and $p_2$ the pressure at $y_2$:

$$\frac{dp}{dy} = -\frac{pM}{RT}g,$$

$$\int_{p_1}^{p_2} \frac{dp}{p} = -\frac{Mg}{RT}\int_{y_1}^{y_2} dy,$$

$$\ln\frac{p_2}{p_1} = -\frac{Mg}{RT}(y_2 - y_1).$$

Now let $y_1 = 0$ and let the pressure at that point be $p_0$. Then the pressure $p$ at any height $y$ is

$$p = p_0 e^{-Mgy/RT}. \qquad (16\text{–}7)$$

At the summit of Mt. Everest, where $y$ = 8863 m,

$$\frac{Mgy}{RT} = \frac{(28.8 \times 10^{-3} \text{ kg/mol})(9.80 \text{ m/s}^2)(8863 \text{ m})}{(8.315 \text{ J/mol}\cdot\text{K})(273 \text{ K})} = 1.10;$$

$$p = (1.01 \times 10^5 \text{ Pa})e^{-1.10} = 0.336 \times 10^5 \text{ Pa}$$
$$= 0.333 \text{ atm}.$$

The assumption of constant temperature isn't realistic, and $g$ decreases a little with increasing elevation. Even so, this example shows why mountaineers need to carry oxygen on Mt. Everest (see Fig. 16–5 on page 456). Air pressure at the summit is only one-third of its sea-level value.

The ability of the human body to absorb oxygen from the atmosphere depends critically on atmospheric pressure. Absorption drops sharply when the pressure is less than about $0.65 \times 10^5$ Pa, corresponding to an elevation above sea level of about 4700 m (15,000 ft). At pressures less than $0.55 \times 10^5$ Pa, oxygen absorption is not sufficient to maintain life. There is no permanent human habitation on earth above 5000 m (about 16,000 ft), although survival for short periods of time is possible at higher elevations. Jet airplanes, which typically fly at altitudes of 8000 to 12,000 m, *must* have pressurized cabins for passenger comfort and health.

**16–5** Air pressure at the summit of Mt. Everest (8863 m above sea level) is only one-third of its value at sea level.

## The van der Waals Equation

The ideal-gas equation can be derived from a simple molecular model that ignores the volumes of the molecules themselves and the attractive forces between them. We'll do that derivation in Section 16–3. Meanwhile, we mention another equation of state, the **van der Waals equation,** which makes approximate corrections for these two omissions. The van der Waals equation is

$$\left(p + \frac{an^2}{V^2}\right)(V - nb) = nRT. \qquad (16\text{–}8)$$

The constants $a$ and $b$ are empirical constants, different for different gases. Roughly speaking, $b$ represents the volume of a mole of molecules; the total volume of the molecules is then $nb$, and the net volume available for the molecules to move around in is $V - nb$. The constant $a$ depends on the attractive intermolecular forces, which contribute an additional effective pressure by *pulling* the molecules together as they are *pushed* together by the walls of the container. The net force on each molecule near the wall is proportional to the number of molecules per unit volume and thus to $n/V$, and the *total* increase in effective pressure is proportional to $n^2/V^2$. When $n/V$ is very small, the molecules are so far apart, on average, that these corrections become insignificant, and in that limit, Eq. (16–8) reduces to the ideal-gas equation.

## *pV*-Diagrams

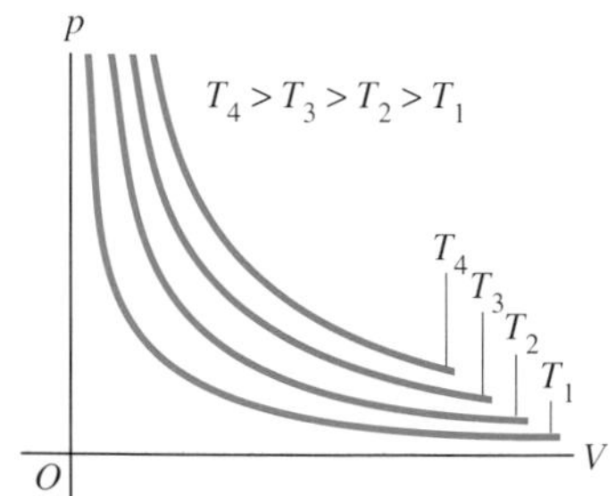


**16–6** Isotherms, or constant-temperature curves, for a constant amount of an ideal gas. This is a graphical representation of Boyle's law: For each curve, the product $pV$ is constant, and the value of that constant is directly proportional to the absolute temperature $T$.

We could in principle represent the $p$-$V$-$T$ relationship graphically as a *surface* in a three-dimensional space with coordinates $p$, $V$, and $T$. This representation sometimes helps in grasping the overall behavior of the material, but ordinary two-dimensional graphs are usually more convenient. One of the most useful of these is a set of graphs of pressure as a function of volume, each for a particular constant temperature. Such a diagram is called a ***pV*-diagram.** Each curve, representing behavior at a specific temperature, is called an **isotherm,** or a *pV-isotherm.*

Figure 16–6 shows $pV$-isotherms for a constant amount of an ideal gas. The highest temperature is $T_4$; the lowest is $T_1$. This is a graphical representation of the ideal-gas equation of state. We can read off the volume $V$ corresponding to any given pressure $p$ and temperature $T$ in the range shown.

Figure 16–7 shows a $pV$-diagram for a material that *does not* obey the ideal-gas equation. At high temperatures such as $T_4$ the curves resemble the ideal-gas curves of Fig. 16–6. However, at temperatures below $T_c$ the isotherms develop flat parts, regions where the material can be compressed without an increase in pressure. Observation of the gas shows that it is *condensing* from the vapor (gas) to the liquid phase. The flat parts of the isotherms in the shaded area of Fig. 16–7 represent conditions of liquid-vapor *phase equilibrium.* As the volume decreases, more and more material goes from vapor to liquid, but the pressure does not change. To keep the temperature constant, we have to remove heat, the heat of vaporization that we discussed in Section 15–6.

When we compress such a gas at a constant temperature $T_2$ in Fig. 16–7, it is vapor until point $a$ is reached. There it begins to liquefy; as the volume decreases further, more material liquefies, with *both* pressure and temperature remaining constant. At point $b$, *all* the material is in the liquid state. After this, any further compression results in a very rapid rise of pressure because in general liquids are much less compressible than gases. At a lower constant temperature $T_1$, similar behavior occurs, but the onset of condensation occurs at lower pressure and greater volume than at the constant temperature $T_2$. At temperatures greater than $T_c$, *no* phase transition occurs as the material is compressed. We call $T_c$ the *critical temperature* for this material.

We will use $pV$-diagrams often in the next two chapters. We will show that the area under a $pV$-curve (regardless of whether or not it is an isotherm) represents the *work* done by the system during a volume change. This work, in turn, is directly related to heat transfer and changes in the *internal energy* of the system, which we'll get to in Chapter 17.

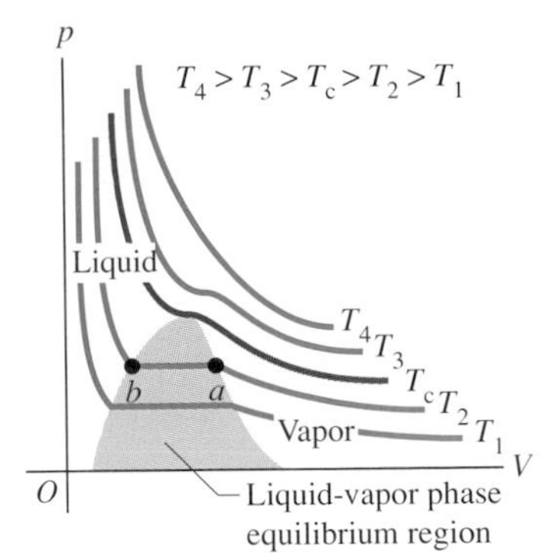


**16–7** A $pV$-diagram for a non-ideal gas, showing $pV$-isotherms for temperatures above and below the critical temperature $T_c$. The liquid-vapor equilibrium region is shown as a shaded area. At still lower temperature the material might undergo phase transitions from solid to liquid or from gas to solid; these are not shown on this diagram.

## 16–2 MOLECULAR PROPERTIES OF MATTER

We have studied several properties of matter in bulk, including elasticity, density, surface tension, equations of state, and heat capacities, with only passing references to molecular structure. Now we want to look in more detail at the relation of bulk behavior to microscopic structure. We begin with a general discussion of the molecular structure of matter. Then in the next two sections we develop the kinetic-molecular model of an ideal gas, deriving from this molecular model the equation of state and an expression for heat capacity.

All familiar matter is made up of **molecules.** For any specific chemical compound, all the molecules are identical. The smallest molecules contain one atom each and are of the order of $10^{-10}$ m in size; the largest contain many atoms and are at least 10,000 times that large. In gases the molecules move nearly independently; in liquids and solids they are held together by intermolecular forces that are electrical in nature, arising from interactions of the electrically charged particles that make up the molecules. Gravitational forces between molecules are negligible in comparison with electrical forces.

The interaction of two *point* electric charges is described by a force (repulsive for like charges, attractive for unlike charges) with a magnitude proportional to $1/r^2$, where $r$ is the distance between the points. This relationship is called *Coulomb's law;* we will study it in detail in Chapter 22. Molecules are *not* point charges but complex structures containing both positive and negative charge, and their interactions are more complex. For molecules of a gas the intermolecular force varies with the distance $r$ between molecules somewhat as shown in Fig. 16–8, where a positive $F$ corresponds to a repulsive force and a negative $F$ to an attractive force. When molecules are far apart, the intermolecular forces are very small and usually attractive. As a gas is compressed and its molecules are brought closer together, the attractive forces increase. In liquids and

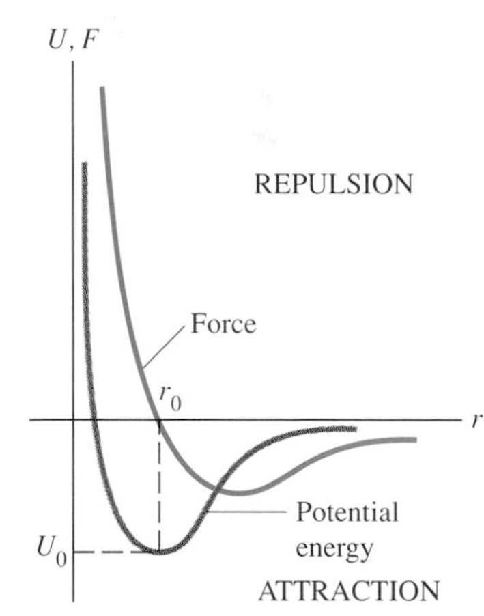


**16–8** The force between two molecules (blue curve) changes from an attraction when the separation is large to a repulsion when the separation is small. The potential energy (dark-red curve) is minimum at $r_0$, where the force is zero.

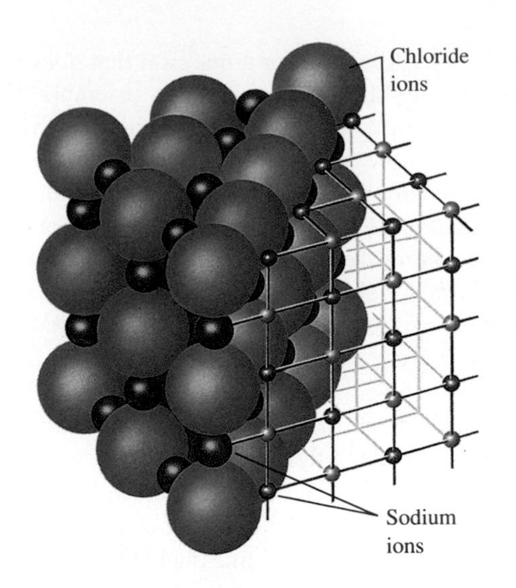


**16–9** Schematic representation of the cubic crystal structure of sodium chloride.

solids, relatively large pressures are needed to compress the substance appreciably. This shows that at molecular distances that are slightly *less* than the equilibrium spacing, the forces become *repulsive* and relatively large. The force is zero at an equilibrium separation $r_0$.

Figure 16–8 also shows the potential energy $U$ as a function of $r$. This function has a *minimum* at $r_0$, where the force is zero. The two curves are related by $F(r) = -dU/dr$, as we showed in Section 7–5. Such a potential energy function is often called a **potential well.** A molecule at rest at $r_0$ would need an additional energy $U_0$, the "depth" of the potential well, to "escape" to an indefinitely large value of $r$.

Molecules are always in motion; their kinetic energies usually increase with temperature. At very low temperatures the average kinetic energy of a molecule may be much *less* than the depth of the potential well. The molecules then condense into the liquid or solid phase with average intermolecular spacings of about $r_0$. But at higher temperatures the average kinetic energy becomes larger than the depth $U_0$ of the potential well. Molecules can then escape the intermolecular force and become free to move independently, as in the gaseous phase of matter.

In *solids,* molecules vibrate about more or less fixed centers. In a crystalline solid these centers are arranged in a recurring *crystal lattice*. Figure 16–9 shows the cubic crystal structure of sodium chloride (ordinary salt). A scanning tunneling microscope photograph of the individual molecules of a fatty acid bi-layer deposited on graphite is shown in Fig. 16–10.

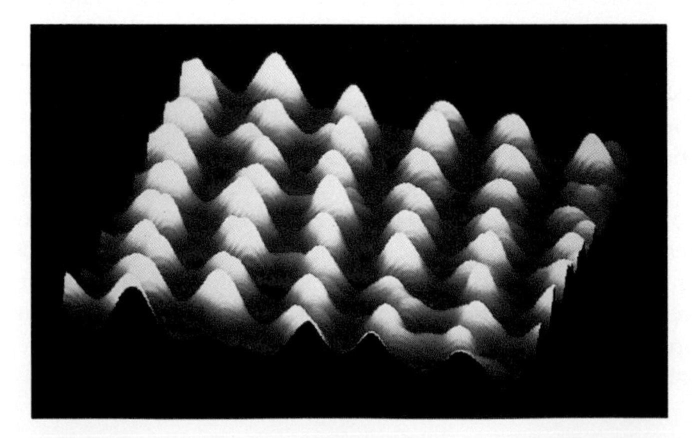

**16–10** A scanning tunneling microscope photo of a fatty acid bi-layer. Looking like a bunch of stacked oranges, each atom has a diameter of about $10^{-10}$ m.

At distances close to $r_0$ the potential well may be approximately parabolic in shape; the resulting molecular motion is nearly simple harmonic. If the potential-energy curve rises more gradually for $r > r_0$ than for $r < r_0$, the average position shifts to larger $r$ with increasing amplitude. As we pointed out in Section 15–4, this is the basis of thermal expansion.

In a *liquid,* the intermolecular distances are usually only slightly greater than in the solid phase of the same substance, but the molecules have much greater freedom of movement. Liquids show regularity of structure only in the immediate neighborhood of a few molecules. This is called *short-range order,* in contrast with the *long-range order* of a solid crystal.

The molecules of a *gas* are usually widely separated and have only very small attractive forces. A gas molecule moves in a straight line until it collides with another molecule or with a wall of the container. In molecular terms an *ideal gas* is a gas whose molecules exert *no* attractive forces on each other and therefore have no *potential* energy.

At low temperatures, most common substances are in the solid phase. As the temperature rises, a substance melts and then vaporizes. From a *molecular* point of view these transitions are in the direction of increasing molecular kinetic energy. Thus temperature and molecular kinetic energy are closely related.

We have used the mole as a measure of quantity of substance. One **mole** of any pure chemical element or compound contains a definite number of molecules, the same number for all elements and compounds. The official SI definition of the mole states:

**The mole is the amount of substance that contains as many elementary entities as there are atoms in 0.012 kilogram of carbon 12.**

In our discussion the "elementary entities" are molecules. (In a monatomic substance such as carbon or helium, each molecule is a single atom, but in this discussion we'll still call it a molecule.)

The number of molecules in a mole is called **Avogadro's number,** denoted by $N_A$. The numerical value is

$$N_A = 6.022 \times 10^{23} \text{ molecules/mol}.$$

Figure 16–11 shows one mole of each of several familiar materials.

The **molecular mass** $M$ of a compound (which would be better called the *molar mass*) is the mass of one mole; this is equal to the mass $m$ of a single molecule, multiplied by Avogadro's number:

$$M = N_A m. \tag{16–9}$$

When the molecule consists of a single atom, the term *atomic mass* is often used instead of molecular mass.

**16–11** How much is one mole? The photograph shows one mole each of (left to right) lead shot, copper pennies, sucrose (ordinary sugar, top dish), potassium dichromate, salt (bottom dish), mercury (in beaker), water (in flask), and sulfur.

### EXAMPLE 16–5

Find the mass of a single hydrogen atom and the mass of an oxygen molecule.

**SOLUTION** We use Eq. (16–9). From the periodic table of the elements (Appendix D) the mass per mole of atomic hydrogen (that is, the atomic mass) is 1.008 g/mol. Therefore the mass $m_H$ of a single hydrogen atom is

$$m_H = \frac{1.008 \text{ g/mol}}{6.022 \times 10^{23} \text{ molecules/mol}} = 1.674 \times 10^{-24} \text{ g/molecule}.$$

The atomic mass of oxygen (again from Appendix D) is 16.0 g/mol, so the molecular mass of oxygen, which has diatomic (two-atom) molecules, is 32.0 g/mol. The mass of a single molecule of $O_2$ is

$$m_{O_2} = \frac{32.0 \text{ g/mol}}{6.022 \times 10^{23} \text{ molecules/mol}} = 53.1 \times 10^{-24} \text{ g/molecule}.$$

---

The goal of any molecular theory of matter is to understand the *macroscopic* properties of matter in terms of its atomic or molecular structure and behavior. Such theories are of tremendous practical importance; once we have this understanding, we can *design* materials to have specific desired properties. Such analysis has led to the development of high-strength steels, glasses with special optical properties, semiconductor materials for electronic devices, and countless other materials that are essential to contemporary technology.

In the following sections we will consider a simple molecular model of a gas, representing it as a large number of particles bouncing around in a container. We can relate this model to the ideal-gas equation of state and use it to predict the molar heat capacity of such a gas. This model can then be elaborated, using "particles" that are not points but have a finite size. We will be able to see why polyatomic gases have larger heat capacities than monatomic gases.

## 16–3 KINETIC-MOLECULAR MODEL OF AN IDEAL GAS

One of the simplest examples of a molecular theory is the kinetic-molecular model of an ideal gas. In this section we use the kinetic-molecular model to understand how the ideal-gas equation of state, Eq. (16–3), is related to Newton's laws. The following development has several steps, and you may need to go over it several times to grasp how it all goes together. Don't get discouraged.

Here are the assumptions of the kinetic-molecular model:

1. A container with volume $V$ contains a very large number $N$ of identical molecules, each with mass $m$.
2. The molecules behave as point particles; their size is small in comparison to the average distance between particles and to the dimensions of the container.
3. The molecules are in constant motion; they obey Newton's laws of motion. Each molecule collides occasionally with a wall of the container. These collisions are perfectly elastic.
4. The container walls are perfectly rigid and infinitely massive and do not move.

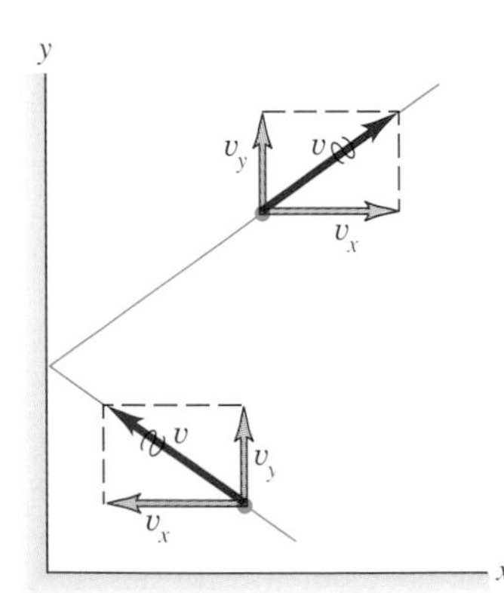


**16–12** Elastic collision of a molecule with an idealized container wall. The component $v_y$ parallel to the wall does not change; the component $v_x$ perpendicular to the wall reverses direction. The speed $v$ does not change.

During collisions the molecules exert *forces* on the walls of the container; this is the origin of the *pressure* that the gas exerts. In a typical collision (Fig. 16–12) the velocity component $v_y$ parallel to the wall is unchanged, and the component $v_x$ perpendicular to the wall changes direction but not magnitude.

Here is our program. First we determine the *number* of collisions per unit time for a certain wall area $A$. Then we find the total momentum change associated with these collisions and the force needed to cause this momentum change. Then we can determine the pressure, which is force per unit area, and compare that result to the ideal-gas equation. Let $v_x$ be the magnitude of the $x$-component of velocity of a molecule of the gas. For now we assume that all molecules have the same $v_x$. This isn't right, but making this temporary assumption helps to clarify the basic ideas. We will show later that this assumption isn't really necessary.

For each collision the $x$-component of momentum changes from $-mv_x$ to $+mv_x$, so the *change* in the $x$-component of momentum is $mv_x - (-mv_x) = 2mv_x$. If a molecule is going to collide with a given wall area $A$ during a small time interval $dt$, then at the beginning of $dt$ it must be within a distance $v_x\,dt$ from the wall (Fig. 16–13), and it must be headed toward the wall. So the number of molecules that collide with $A$ during $dt$ is equal to the number of molecules that are within a cylinder with base area $A$ and length $v_x\,dt$ and have their $v_x$ aimed toward the wall. The volume of such a cylinder is $Av_x\,dt$.

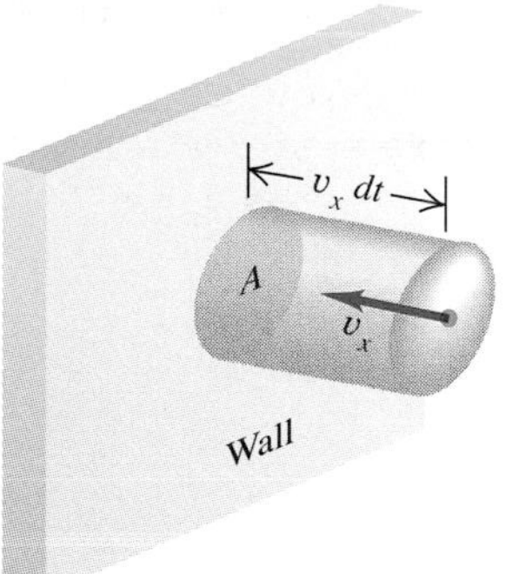


**16–13** A molecule moving toward the wall with speed $v_x$ collides with the area $A$ during the time interval $dt$ only if it is within a distance $v_x\,dt$ of the wall at the beginning of the interval. All such molecules are contained within a volume $Av_x\,dt$.

Assuming that the number of molecules per unit volume ($N/V$) is uniform, the *number* of molecules in this cylinder is $(N/V)(Av_x\,dt)$. On the average, half of these molecules are moving toward the wall and half away from it. So the number of collisions with $A$ during $dt$ is

$$\frac{1}{2}\left(\frac{N}{V}\right)(Av_x dt).$$

The total momentum change $dP_x$ due to all these collisions is the *number* of collisions multiplied by $2mv_x$:

$$dP_x = \frac{1}{2}\left(\frac{N}{V}\right)(Av_x\,dt)(2mv_x) = \frac{NAmv_x^{\,2}\,dt}{V}. \tag{16–10}$$

(We are using capital $P$ for momentum and small $p$ for pressure. Be careful!) The *rate* of change of momentum component $P_x$ is

$$\frac{dP_x}{dt} = \frac{NAmv_x^{\,2}}{V}. \tag{16–11}$$

According to Newton's second law, this rate of change of momentum equals the force exerted by the wall area $A$ on the molecules. From Newton's *third* law this is equal and opposite to the force exerted *on* the wall *by* the molecules. Finally, pressure $p$ is the magnitude of the force per unit area, and we obtain

$$p = \frac{F}{A} = \frac{Nmv_x^{\,2}}{V}. \tag{16–12}$$

We have mentioned that $v_x$ is *not* really the same for all the molecules. But we could have sorted the molecules into groups having the same $v_x$ within each group and added up the resulting contributions to the pressure. The net effect of all this is just to replace $v_x^{\,2}$ in Eq. (16–12) by the *average* value of $v_x^{\,2}$, which we denote by $(v_x^{\,2})_{\text{av}}$. Furthermore, $(v_x^{\,2})_{\text{av}}$ is related simply to the *speeds* of the molecules. The speed $v$ of any molecule is related to the velocity components $v_x$, $v_y$, and $v_z$ by

$$v^2 = v_x^{\,2} + v_y^{\,2} + v_z^{\,2}.$$

We can average this relation over all molecules:

$$(v^2)_{\text{av}} = (v_x^{\,2})_{\text{av}} + (v_y^{\,2})_{\text{av}} + (v_z^{\,2})_{\text{av}}.$$

But the $x$-, $y$-, and $z$-directions are all equivalent, so

$$(v_x^{\,2})_{\text{av}} = (v_y^{\,2})_{\text{av}} = (v_z^{\,2})_{\text{av}}.$$

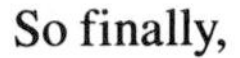
So finally,

$$(v_x^{\,2})_{\text{av}} = \tfrac{1}{3}(v^2)_{\text{av}},$$

and Eq. (16–12) becomes

$$pV = \tfrac{1}{3}Nm(v^2)_{\text{av}} = \tfrac{2}{3}N[\tfrac{1}{2}m(v^2)_{\text{av}}]. \tag{16–13}$$

We notice that $\frac{1}{2}m(v^2)_{\text{av}}$ is the average translational kinetic energy of a single molecule. The product of this and the total number of molecules $N$ equals the total random kinetic energy $K_{\text{tr}}$ of translational motion of all the molecules. (The notation $K_{\text{tr}}$ reminds us that this energy is associated with *translational* motion and anticipates the possibility of additional energies associated with rotational and vibrational motion of molecules.) The product $pV$ equals two-thirds of the total translational kinetic energy:

$$pV = \tfrac{2}{3}K_{\text{tr}}. \tag{16–14}$$

Now we compare this with the ideal-gas equation,

$$pV = nRT,$$

which is based on experimental studies of gas behavior. For the two equations to agree we must have

$$K_{\text{tr}} = \tfrac{3}{2}nRT. \tag{16–15}$$

This remarkably simple result shows that $K_{\text{tr}}$ is *directly proportional* to the absolute temperature $T$. We will use this important result several times in the following discussion.

The average translational kinetic energy of a single molecule is the total translational kinetic energy $K_{\text{tr}}$ of all molecules divided by the number $N$ of molecules:

$$\frac{K_{\text{tr}}}{N} = \tfrac{1}{2}m(v^2)_{\text{av}} = \frac{3nRT}{2N}.$$

Also, the total number of molecules $N$ is the number of moles $n$ multiplied by Avogadro's number $N_A$, so

$$n = \frac{N}{N_A}, \qquad \frac{n}{N} = \frac{1}{N_A},$$

$$\frac{1}{2}m(v^2)_{\text{av}} = \frac{3}{2}\left(\frac{R}{N_A}\right)T. \tag{16–16}$$

The ratio $R/N_A$ occurs frequently in molecular theory. It is called the **Boltzmann constant,** $k$:

$$k = \frac{R}{N_A} = \frac{8.315\ \text{J/mol}\cdot\text{K}}{6.022 \times 10^{23}\ \text{molecules/mol}}$$
$$= 1.381 \times 10^{-23}\ \text{J/molecule}\cdot\text{K}.$$

In terms of $k$ we can rewrite Eq. (16–16) as

$$\tfrac{1}{2}m(v^2)_{\text{av}} = \tfrac{3}{2}kT. \tag{16–17}$$

This shows that the average translational kinetic energy *per molecule* depends only on the temperature, not on the pressure, volume, or kind of molecule. We can obtain an equivalent statement from Eq. (16–16) by using the relation $M = N_A m$:

$$N_A\tfrac{1}{2}m(v^2)_{\text{av}} = \tfrac{1}{2}M(v^2)_{\text{av}} = \tfrac{3}{2}RT. \tag{16–18}$$

That is, the translational kinetic energy of a *mole* of ideal-gas molecules depends only on $T$.

Finally, it is sometimes convenient to rewrite the ideal-gas equation on a molecular basis. We use $N = N_A n$ and $R = N_A k$ to obtain the following alternative form:

$$pV = NkT. \qquad (16\text{–}19)$$

This shows that we can think of $k$ as a gas constant on a "per-molecule" basis instead of the usual "per-mole" basis for $R$.

## Molecular Speeds

From Eqs. (16–17) and (16–18) we can obtain expressions for the square root of $(v^2)_{av}$, called the *root-mean-square speed* $v_{rms}$:

$$v_{rms} = \sqrt{(v^2)_{av}} = \sqrt{\frac{3kT}{m}} = \sqrt{\frac{3RT}{M}}. \qquad (16\text{–}20)$$

It might seem more natural to characterize molecular speeds by their *average* value rather than with $v_{rms}$, but we see that $v_{rms}$ evolves more simply from our analysis. To compute the rms speed, we square each molecular speed, add, divide by the number of molecules, and take the square root; $v_{rms}$ is the *root* of the *mean* of the *squares.* Example 16–9 illustrates this procedure.

Gaseous-diffusion isotope-separation plants use the fact that $v_{rms}$ is different for different molecular masses to separate isotopes, such as the two uranium isotopes $^{235}U$ and $^{238}U$. Another interesting application of Eq. (16–20) has to do with the fact that there's practically no hydrogen or helium in the earth's atmosphere, despite the fact that about 98% of all the atoms in the universe are one of these. A sizable fraction of any hydrogen or helium molecules present in the atmosphere have speeds greater than the "escape speed" needed to escape from the earth's gravity. We found in Section 12–4 that this is about $1.12 \times 10^4$ m/s.

## PROBLEM-SOLVING STRATEGY

### Kinetic-molecular theory

As usual, using a consistent set of units is essential. Following are several places where caution is needed.

**1.** The usual units for molecular mass $M$ are grams per mole; the molecular mass of oxygen is 32 g/mol, for example. These units are often omitted in tables. When you use SI units in equations such as Eq. (16–20), you *must* express $M$ in kilograms per mole by multiplying the table value by (1 kg/$10^3$ g). Thus in SI units, $M$ for oxygen is $32 \times 10^{-3}$ kg/mol.

**2.** Are you working on a "per-molecule" basis or a "per-mole" basis? Remember that $m$ is the mass of a molecule and $M$ is the mass of a mole; $N$ is the number of molecules, $n$ is the number of moles; $k$ is the gas constant per molecule, and $R$ is the gas constant per mole. Although $N$, the number of molecules, is in one sense a dimensionless number, you can do a complete unit check if you think of $N$ as having the unit "molecules"; then $m$ has units "mass per molecule," and $k$ has units "joules per molecule per kelvin."

**3.** Remember that $T$ is always *absolute* (Kelvin) temperature.

## EXAMPLE 16–6

**Kinetic energy of a molecule** What is the average translational kinetic energy of a molecule of an ideal gas at a temperature of 27°C?

**SOLUTION** To use Eq. (16–17) we first convert the temperature to the Kelvin scale: 27°C = 300 K.

$$\begin{aligned} \tfrac{1}{2}m(v^2)_{av} &= \tfrac{3}{2}kT \\ &= \tfrac{3}{2}(1.38 \times 10^{-23}\ \text{J/K})(300\ \text{K}) \\ &= 6.21 \times 10^{-21}\ \text{J}. \end{aligned}$$

## EXAMPLE 16–7

What is the total random translational kinetic energy of the molecules in one mole of an ideal gas at a temperature of 300 K?

**SOLUTION** From Eq. (16–15),

$$K_{tr} = \tfrac{3}{2}nRT = \tfrac{3}{2}(1 \text{ mol})(8.315 \text{ J/mol} \cdot \text{K})(300 \text{ K})$$
$$= 3740 \text{ J}.$$

This is about the same kinetic energy as that of a sprinter in a 100-m dash. This result can also be obtained by multiplying the result in Example 16–6 by Avogadro's number. Do you see why?

## EXAMPLE 16–8

**Calculating $v_{rms}$** What is the root-mean-square speed of a diatomic hydrogen molecule at 300 K?

**SOLUTION** From Example 16–5 (Section 16–2) the mass of a hydrogen molecule is

$$m_{H_2} = (2)(1.674 \times 10^{-27} \text{ kg}) = 3.348 \times 10^{-27} \text{ kg}.$$

From Eq. (16–20),

$$v_{rms} = \sqrt{\frac{3kT}{m}} = \sqrt{\frac{3(1.38 \times 10^{-23} \text{ J/K})(300 \text{ K})}{3.348 \times 10^{-27} \text{ kg}}}$$
$$= 1.93 \times 10^3 \text{ m/s} = 1.93 \text{ km/s}.$$

This is over 4300 mi/h! Alternatively,

$$v_{rms} = \sqrt{\frac{3RT}{M}} = \sqrt{\frac{3(8.315 \text{ J/mol} \cdot \text{K})(300 \text{ K})}{2(1.008 \times 10^{-3} \text{ kg/mol})}}$$
$$= 1.93 \times 10^3 \text{ m/s}.$$

Note again that when we use Eq. (16–20) with $R$ in SI units, we have to express $M$ in *kilograms* per mole, not grams per mole. In this example, $M = 2.016 \times 10^{-3}$ kg/mol, *not* 2.016 g/mol.

## EXAMPLE 16–9

Five gas molecules chosen at random are found to have speeds of 500, 600, 700, 800, and 900 m/s. Find the rms speed. Is it the same as the *average* speed?

**SOLUTION** The average value of $v^2$ for the five molecules is

$$(v^2)_{av} = \frac{(500 \text{ m/s})^2 + (600 \text{ m/s})^2 + (700 \text{ m/s})^2 + (800 \text{ m/s})^2 + (900 \text{ m/s})^2}{5}$$
$$= 51 \times 10^4 \text{ m}^2/\text{s}^2.$$

The square root of this is $v_{rms}$:

$$v_{rms} = 714 \text{ m/s}.$$

The *average* speed $v_{av}$ is given by

$$v_{av} = \frac{500 \text{ m/s} + 600 \text{ m/s} + 700 \text{ m/s} + 800 \text{ m/s} + 900 \text{ m/s}}{5}$$
$$= 700 \text{ m/s}.$$

We see that in general $v_{rms}$ and $v_{av}$ are not the same. Roughly speaking, $v_{rms}$ gives greater weight to the larger speeds than does $v_{av}$.

## EXAMPLE 16–10

**Volume of a gas at STP, revisited** Find the number of molecules and the number of moles in one cubic meter of air at atmospheric pressure and 0°C.

**SOLUTION** From Eq. (16–19),

$$N = \frac{pV}{kT} = \frac{(1.013 \times 10^5 \text{ Pa})(1 \text{ m}^3)}{(1.38 \times 10^{-23} \text{ J/K})(273 \text{ K})}$$
$$= 2.69 \times 10^{25}.$$

The number of moles $n$ is

$$n = \frac{N}{N_A} = \frac{2.69 \times 10^{25} \text{ molecules}}{6.022 \times 10^{23} \text{ molecules/mol}} = 44.7 \text{ mol}.$$

*To check:* The total volume is 1.00 m³, so the volume of one mole is

$$\frac{1.00 \text{ m}^3}{44.7 \text{ mol}} = 0.0224 \text{ m}^3/\text{mol} = 22.4 \text{ L/mol}.$$

This is the same result that we obtained in Example 16–1 (Section 16–1), using only macroscopic quantities.

## Molecular Collisions

When a molecule collides with a stationary surface such as a wall, it exerts a momentary force on the wall but does no work because the wall does not move. But if the surface is moving, such as a piston in a gasoline engine, the molecule *does* do work on the surface during the collision. If the piston in Fig. 16–14 is moving to the left, the molecules that strike it exert a force through a distance and do work; their speeds and kinetic energies after colliding are smaller than they were before. If the piston is moving toward the right, the kinetic energy of a colliding molecule *increases* by an amount equal to the work done on the molecule.

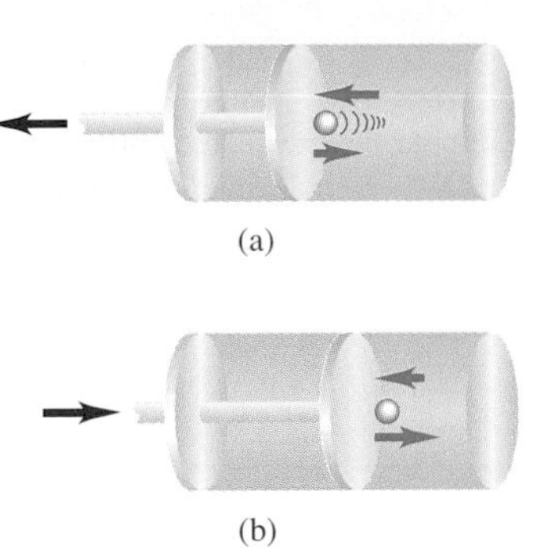


16–14 (a) When a molecule strikes a wall moving away from it, the molecule does work on the wall; the molecule's speed and kinetic energy decrease. (b) When a molecule strikes a wall moving toward it, the wall does work on the molecule; its speed and kinetic energy increase.

The assumption that individual molecules undergo elastic collisions with the wall of a container is actually a little too simple. More detailed investigation has shown that in most cases, molecules actually adhere to the wall for a short time and then leave again with speeds characteristic of the temperature *of the wall*. However, the gas and the wall are ordinarily in thermal equilibrium, with no net energy transfer between gas and wall, and this discovery does not alter the validity of our conclusions.

We have ignored the possibility that two gas molecules might collide. If they are really points, they *never* collide. But consider a more realistic model in which the molecules are rigid spheres with radius $r$. How often do they collide with other molecules? How far do they travel, on the average, between collisions? We can get approximate answers from the following rather primitive model.

Consider $N$ spherical molecules of radius $r$ in a volume $V$. Suppose only one molecule is moving. It collides with another molecule whenever the distance between centers is $2r$. Suppose we draw a cylinder with radius $2r$, with its axis parallel to the velocity of the molecule (Fig. 16–15). The moving molecule will collide with any other molecule whose center is inside this cylinder. In a short time $dt$ a molecule with speed $v$ travels a distance $v\,dt$; during this time it collides with any molecule that is in the cylindrical volume of radius $2r$ and length $v\,dt$. The volume of the cylinder is $4\pi r^2 v\,dt$. There are $N/V$ molecules per unit volume, so the number $dN$ with centers in this cylinder is

$$dN = 4\pi r^2 v\,dt\,N/V.$$

Thus the number of collisions *per unit time* is

$$\frac{dN}{dt} = \frac{4\pi r^2 vN}{V}.$$

This result assumes that only one molecule is moving. The analysis is quite a bit more involved when all the molecules move at once. It turns out that in this case the collisions are more frequent by a factor of $\sqrt{2}$, and the above equation has to be multiplied by that factor:

$$\frac{dN}{dt} = \frac{4\pi\sqrt{2}r^2\,vN}{V}.$$

The average time $t_{mean}$ between collisions, called the *mean free time,* is the reciprocal of this expression:

$$t_{mean} = \frac{V}{4\pi\sqrt{2}r^2 vN}. \tag{16–21}$$

The average distance traveled between collisions is called the **mean free path,** denoted by $\lambda$. In our simple model this is just the molecule's speed $v$ multiplied by $t_{mean}$:

$$\lambda = vt_{mean} = \frac{V}{4\pi\sqrt{2}r^2 N}. \tag{16–22}$$

This result shows that the mean free path is inversely proportional to the number of molecules per unit volume ($N/V$) and that it is directly proportional to the cross-section area $\pi r^2$ of a molecule. Note that this result *does not* depend on the speed of the molecule.

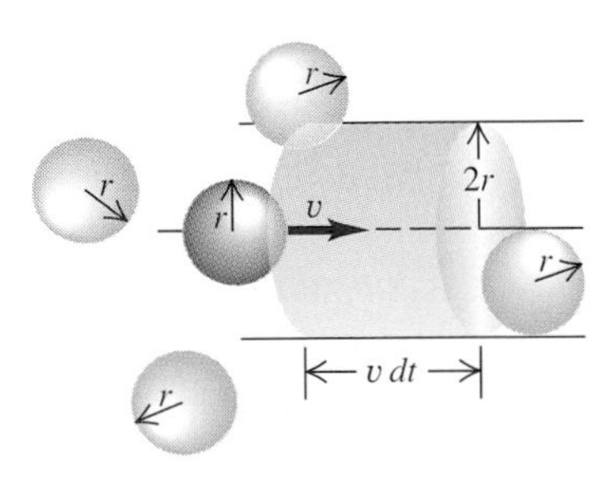


16–15 In a time $dt$ a molecule with radius $r$ will collide with any other molecule within a cylindrical volume of radius $2r$ and length $v\,dt$.

We can express Eq. (16–22) in terms of macroscopic properties of the gas, using the ideal-gas equation in the form of Eq. (16–19), $pV = NkT$. We find

$$\lambda = \frac{kT}{4\pi\sqrt{2}r^2 p}. \tag{16–23}$$

**EXAMPLE 16–11**

Estimate the mean free path of air molecules at 20°C and 1 atm. Model each molecule as a sphere with radius $r = 2.0 \times 10^{-10}$ m.

**SOLUTION** From Eq. (16–23),

$$\lambda = \frac{(1.38 \times 10^{-23}\ \text{J/K})(293\ \text{K})}{4\pi\sqrt{2}(2.0 \times 10^{-10}\ \text{m})^2(1.01 \times 10^5\ \text{Pa})} = 5.6 \times 10^{-8}\ \text{m}.$$

The molecule doesn't get very far between collisions, but the distance is still several hundred times the size of the molecule. To have a mean free path equal to a meter, the pressure must be about 1/18,000,000 atm. Pressures of this magnitude are found 115 km or so above the earth's surface, at the outer fringe of our atmosphere.

## 16–4 HEAT CAPACITIES

When we introduced the concept of heat capacity in Section 15–5, we talked about ways to *measure* the specific or molar heat capacity of a particular material. Now we'll see how these numbers can be *predicted* on theoretical grounds. That's a significant step forward.

### Heat Capacities of Gases

The basis of our analysis is the fact that heat is energy in transit. When we add heat to a substance, we are increasing its molecular energy. In this discussion we will keep the volume constant so that we don't have to worry about energy transfer through mechanical work. If we were to let the gas expand, it would do work by pushing on moving walls of its container, and this additional energy transfer would have to be included in our calculations. We'll return to this more general case in Section 17–7. For now, with the volume held constant, we are concerned with $C_V$, the molar heat capacity *at constant volume*.

In the simple kinetic-molecular model presented in the last section the molecular energy consists only of the kinetic energy $K_{tr}$ associated with the translational motion of the pointlike molecules. This energy is directly proportional to the absolute temperature $T$, as shown by Eq. (16–15), $K_{tr} = \frac{3}{2}nRT$. When the temperature changes by a small amount $dT$, the corresponding change in kinetic energy is

$$dK_{tr} = \tfrac{3}{2}nR\,dT. \tag{16–24}$$

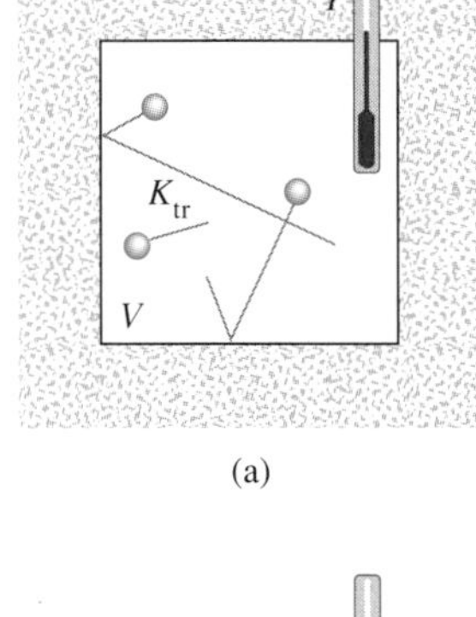

(a)

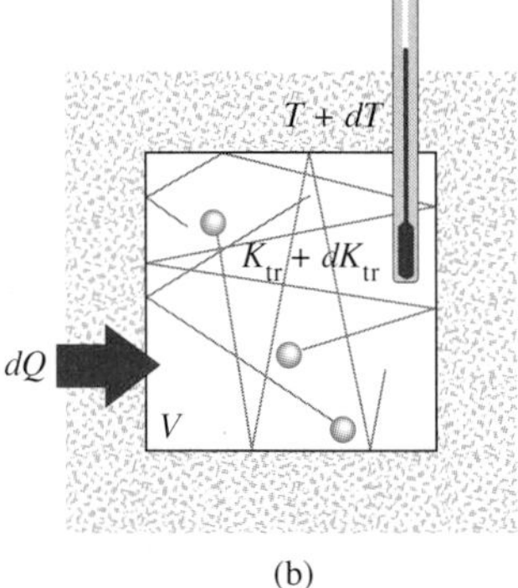

(b)

**16–16** When an amount of heat $dQ$ is added to (a) a constant volume of monatomic ideal gas molecules, (b) the total translational kinetic energy increases by $dK_{tr} = dQ$ and the temperature increases by $dT = dQ/nC_V$.

From the definition of molar heat capacity at constant volume, $C_V$, we also have

$$dQ = nC_V\,dT, \tag{16–25}$$

where $dQ$ is the heat input needed for a temperature change $dT$. Now if $K_{tr}$ represents the total molecular energy, as we have assumed, then $dQ$ and $dK_{tr}$ must be *equal* (Fig. 16–16). Equating the expressions given by Eqs. (16–24) and (16–25), we get

$$nC_V\,dT = \tfrac{3}{2}nR\,dT,$$

$$C_V = \tfrac{3}{2}R. \tag{16–26}$$

This surprisingly simple result says that the molar heat capacity (at constant volume) of *every* gas whose molecules can be represented as points is equal to $3R/2$.

To see whether this makes sense, let's first check the units. The gas constant *does* have units of energy per mole per kelvin, the correct units for a molar heat capacity. But more important is whether Eq. (16–26) agrees with *measured* values of molar heat capacities. In SI units, Eq. (16–26) gives

$$C_V = \tfrac{3}{2}(8.315\ \text{J/mol}\cdot\text{K}) = 12.47\ \text{J/mol}\cdot\text{K}.$$

For comparison, Table 16–1 gives measured values of $C_V$ for several gases. We see that for monatomic gases our prediction is right on the money, but that it is way off for diatomic and polyatomic gases.

**TABLE 16–1 Molar Heat Capacities of Gases**

| Type of gas | Gas | $C_V$ (J/mol · K) |
|---|---|---|
| Monatomic | He | 12.47 |
| | A | 12.47 |
| Diatomic | $H_2$ | 20.42 |
| | $N_2$ | 20.76 |
| | $O_2$ | 21.10 |
| | CO | 20.85 |
| Polyatomic | $CO_2$ | 28.46 |
| | $SO_2$ | 31.39 |
| | $H_2S$ | 25.95 |

This comparison tells us that our point-molecule model is good enough for monatomic gases but that for diatomic and polyatomic molecules we need something more sophisticated. For example, we can picture a diatomic molecule as *two* point masses, like a little elastic dumbbell, with an interaction force of the kind shown in Fig. 16–8. Such a molecule can have additional kinetic energy associated with *rotation* about axes through its center of mass, and the atoms may also have a back-and-forth *vibrating* motion along the line joining them, with additional kinetic and potential energies. These possibilities are shown in Fig. 16–17.

Equation (16–17) shows that the average random *translational* kinetic energy of the molecules of an ideal gas is proportional to the absolute temperature $T$ of the gas. When heat flows into a *monatomic* gas at constant volume, *all* of this energy goes into an increase in random *translational* molecular kinetic energy. But when the temperature of a *diatomic* or *polyatomic* gas is increased by the same amount, additional energy is needed for the increased rotational and vibrational motion. Thus polyatomic gases have *larger* molar heat capacities than monatomic gases, as Table 16–1 shows.

But how do we know how much energy is associated with each additional kind of motion of a complex molecule compared to the translational kinetic energy? The new principle that we need is called the principle of **equipartition of energy.** It can be derived from sophisticated statistical-mechanics considerations; that derivation is beyond our scope, and we will treat the principle as an axiom.

The principle of equipartition of energy states that each velocity component (either linear or angular) has, on the average, an associated kinetic energy per molecule of $\frac{1}{2}kT$. The number of velocity components needed to describe the motion of a molecule completely is called the number of **degrees of freedom.** For a monatomic gas the number is three; this gives a total average kinetic energy per molecule of $3(\frac{1}{2}kT)$, consistent with Eq. (16–17). For a *diatomic* molecule there are two possible axes of rotation, perpendicular to each other and to the molecule's axis. (We don't include rotation about the molecule's own axis because in ordinary collisions there is no way for this rotational motion to change.) If we assign five degrees of freedom to a diatomic molecule, the average total kinetic energy per molecule is $\frac{5}{2}kT$ instead of $\frac{3}{2}kT$. The total kinetic energy of $n$ moles is $K_{\text{tot}} = \frac{5}{2}nRT$, and the molar heat capacity (at constant volume) is

$$C_V = \tfrac{5}{2}R. \qquad (16\text{–}27)$$

In SI units,

$$C_V = \tfrac{5}{2}(8.315 \text{ J/mol} \cdot \text{K}) = 20.79 \text{ J/mol} \cdot \text{K}.$$

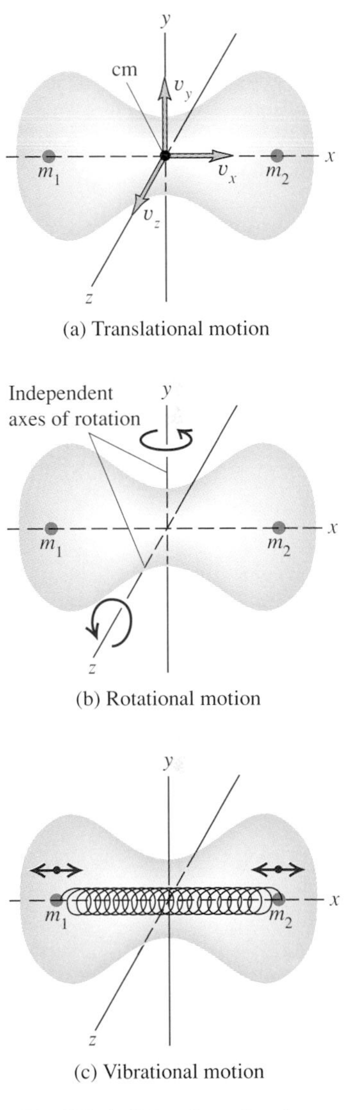


**16–17** A diatomic molecule: Almost all the mass of each atom is in its tiny nucleus. (a) The center of mass has three independent velocity components. (b) The molecule has two independent axes of rotation through its center of mass. (c) The atoms and "spring" have kinetic and potential energies for vibration.

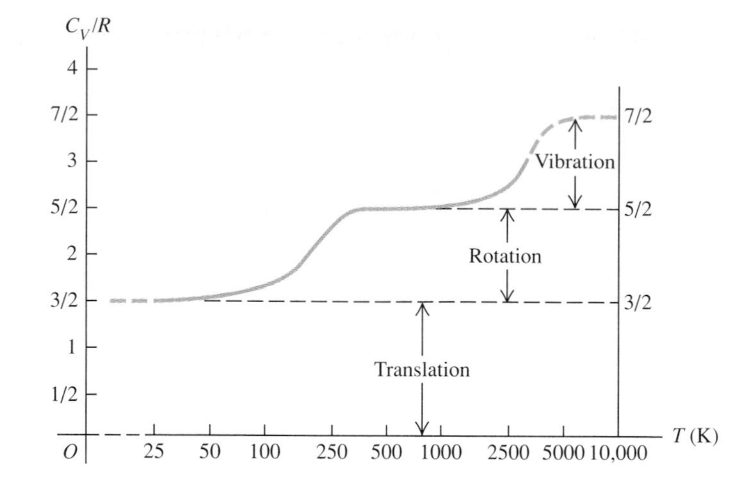


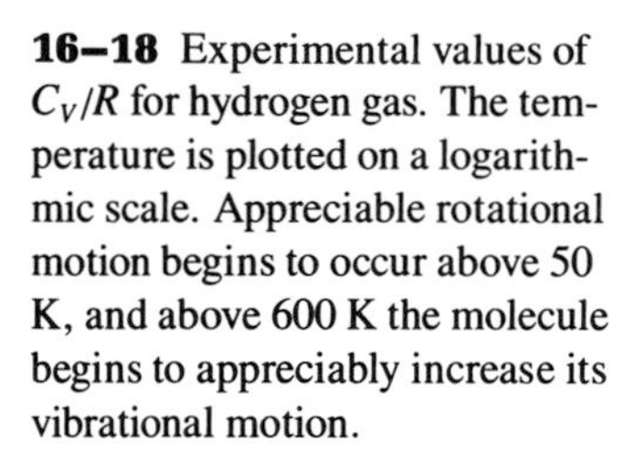

**16–18** Experimental values of $C_V/R$ for hydrogen gas. The temperature is plotted on a logarithmic scale. Appreciable rotational motion begins to occur above 50 K, and above 600 K the molecule begins to appreciably increase its vibrational motion.

This agrees within a few percent with the measured values for diatomic gases given in Table 16–1.

*Vibrational* motion can also contribute to the heat capacities of gases. Molecular bonds are not rigid; they can stretch and bend, and the resulting vibrations lead to additional degrees of freedom and additional energies. For most diatomic gases, however, vibrational motion does *not* contribute appreciably to heat capacity. The reason for this is a little subtle and involves some concepts of quantum mechanics. Briefly, vibrational energy can change only in finite steps. If the energy change of the first step is much larger than the energy possessed by most molecules, then nearly all the molecules remain in the minimum-energy state of motion. In that case, changing the temperature does not change their average vibrational energy appreciably, and the vibrational degrees of freedom are said to be "frozen out." In more complex molecules the gaps between permitted energy levels are sometimes much smaller, and then vibration *does* contribute to heat capacity. The rotational energy of a molecule also changes by finite steps, but they are usually much smaller, and the freezing out of rotational degrees of freedom occurs only in rare instances, such as for the hydrogen molecule.

In Table 16–1 the large values of $C_V$ for some polyatomic molecules show the contributions of vibrational energy. In addition, a molecule with three or more atoms that are not in a straight line has three, not two, rotational degrees of freedom.

From this discussion we expect heat capacities to be temperature-dependent, generally increasing with increasing temperature. Figure 16–18 shows the temperature dependence of $C_V$ for hydrogen, showing the temperatures at which the rotational and vibrational energies begin to contribute to the heat capacity.

## Heat Capacities of Solids

We can carry out a similar heat-capacity analysis for a crystalline solid. Consider a crystal consisting of $N$ identical atoms. Each atom is bound to an equilibrium position by interatomic forces. The macroscopic elasticity of solid materials shows us that these forces must permit stretching and bending of the bonds. We can think of a crystal lattice as an array of atoms connected by little springs (Fig. 16–19). Each atom can *vibrate* about its equilibrium position.

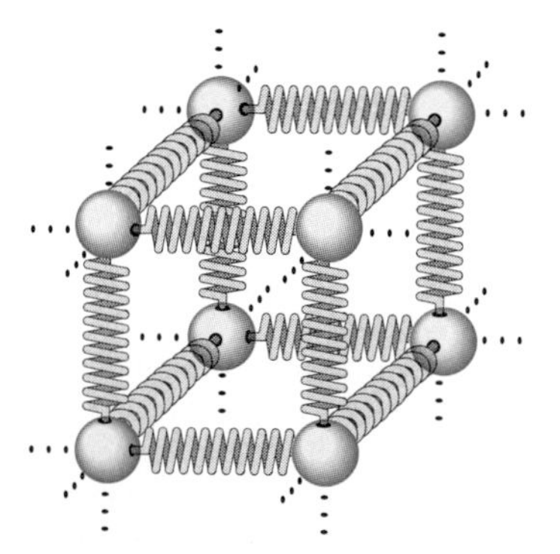

**16–19** The forces between neighboring particles in a crystal may be visualized by imagining every particle as being connected to its neighbors by springs.

Each atom has three degrees of freedom, corresponding to its three components of velocity. According to the equipartition principle, each atom has an average kinetic energy of $\frac{1}{2}kT$ for each degree of freedom. In addition, each atom has *potential* energy associated with the elastic deformation. For a simple harmonic oscillator (discussed in Chapter 13) it is not hard to show that the average kinetic energy of an atom is *equal* to its average potential energy. In our crystal lattice model, each atom is essentially a three-dimensional harmonic oscillator; it can be shown that the equality of average kinetic and potential energies also holds here, provided that the "spring" forces obey Hooke's law.

Thus we expect each atom to have an average kinetic energy of $\frac{3}{2}kT$ and an average potential energy of $\frac{3}{2}kT$, or an average total energy of $3kT$ per molecule. If the crystal contains $N$ atoms, or $n$ moles, its total energy is

$$K_{\text{tot}} = 3NkT, \qquad (16\text{–}28)$$

or, in molar terms,

$$K_{\text{tot}} = 3nRT. \qquad (16\text{–}29)$$

From this we conclude that the molar heat capacity of a crystal should be

$$C_V = 3R. \qquad (16\text{–}30)$$

In SI units,

$$C_V = (3)(8.315\ \text{J/mol}\cdot\text{K}) = 24.9\ \text{J/mol}\cdot\text{K}.$$

This is the **rule of Dulong and Petit,** which we encountered in Section 15–5. There we regarded it as an *empirical* rule, but now we have *derived* it from kinetic theory. The agreement is only approximate, to be sure, but considering the very simple nature of our model, it is quite significant.

At low temperatures the heat capacities of most solids *decrease* with decreasing temperature for the same reason that vibrational degrees of freedom of molecules are frozen out at low temperatures. At very low temperatures the quantity $kT$ is much *smaller* than the smallest energy step the vibrating atoms can take. At low $T$, most of the atoms remain in their lowest energy states because the next higher energy level is out of reach. The average vibrational energy per atom is then *less* than $3kT$, and the heat capacity per molecule is *less* than $3k$. At higher temperatures, when $kT$ is *large* in comparison to the minimum energy step, the classical equipartition principle holds, and the total heat capacity is $3k$ per molecule, or $3R$ per mole, as the Dulong and Petit rule predicts. Quantitative understanding of the temperature variation of heat capacities was one of the triumphs of quantum mechanics during its initial development in the 1920s.

## *16–5 MOLECULAR SPEEDS

As we mentioned in Section 16–3, the molecules in a gas don't all have the same speed. We can measure the distribution of molecular speeds directly; one experimental scheme is shown in Fig. 16–20. A substance is vaporized in a hot oven; molecules of the vapor escape through an aperture in the oven wall and into a vacuum chamber. A series of slits

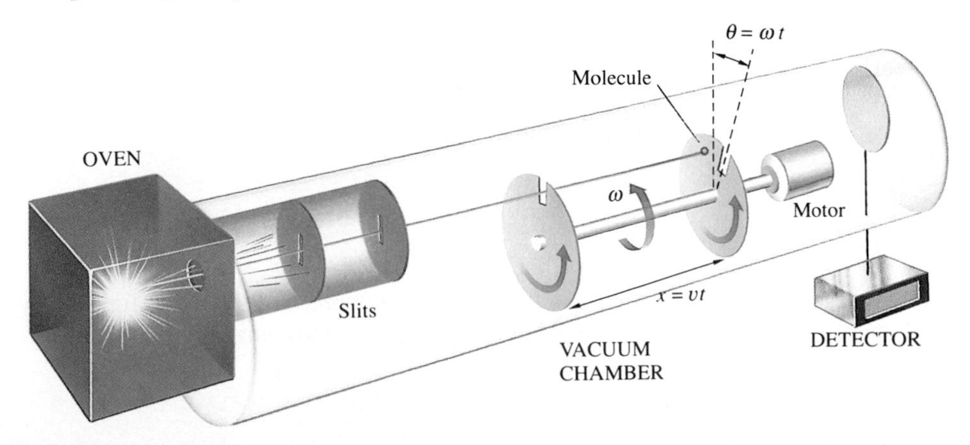


**16–20** A molecule with a speed $v$ is passing through the first slit. When it reaches the second slit, the slits have rotated through the offset angle $\theta$. If $v = \omega x/\theta$, the molecule passes through the second slit and reaches the detector.

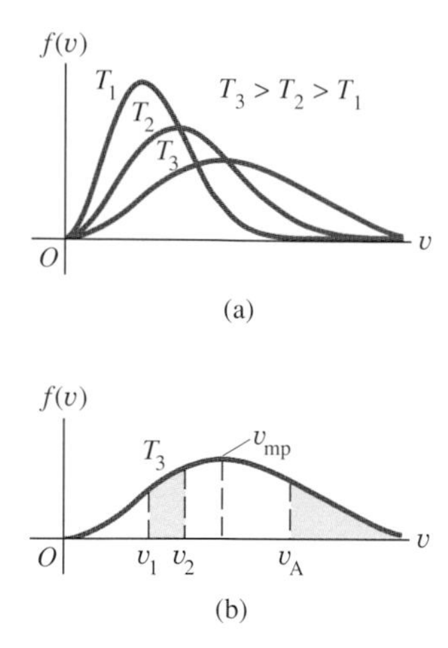


**16–21** (a) Maxwell-Boltzmann distribution curves for various temperatures. As the temperature increases, the curve becomes flatter, and its maximum shifts to higher speeds. (b) At temperature $T_3$ the fraction of molecules having speeds in the range $v_1$ to $v_2$ is shown by the shaded area under the $T_3$ curve. The fraction with speeds greater than $v_A$ would be the area from $v_A$ to infinity.

blocks all molecules except those in a narrow beam; the beam is aimed at a pair of rotating disks. A molecule passing through the slit in the first disk is blocked by the second disk unless it arrives just as the slit in the second disk is lined up with the beam. The disks function as a speed selector that passes only molecules within a certain narrow speed range. This speed can be varied by changing the disk rotation speed, and we can measure how many molecules have each of various speeds.

The results of such measurements can be represented graphically (Fig. 16–21a). We define a function $f(v)$ called a *distribution function,* such that if $N$ molecules are observed, the number $dN$ having speeds in any range between $v$ and $v + dv$ is given by

$$dN = Nf(v)\,dv. \tag{16–31}$$

We can also say that the *probability* that a randomly chosen molecule will have a speed in the interval $v$ to $v + dv$ is $f(v)\,dv$. The figure shows distribution functions for several different temperatures. At each temperature the height of the curve for any value of $v$ is proportional to the number of molecules with speeds near $v$. The peak of the curve represents the *most probable speed* $v_{mp}$ for the corresponding temperature. As the temperature increases, the peak shifts to higher and higher speeds, corresponding to the increase in average molecular kinetic energy with temperature.

Figure 16–21b shows that the area under a curve between any two values of $v$ represents the fraction of all the molecules having speeds in that range. Every molecule must have *some* value of $v$, so the integral of $f(v)$ over all $v$ must be unity for any $T$.

If we know $f(v)$, we can calculate the most probable speed $v_{mp}$ (corresponding to the peak of the curve), the average speed $v_{av}$, and the rms speed $v_{rms}$. To find $v_{mp}$, we simply find the point where $df/dv = 0$; this gives the maximum point on the curve. To find $v_{av}$, we take the number $Nf(v)\,dv$ of molecules having speeds in each interval $dv$, multiply each number by the corresponding speed $v$, add all these products (by integrating over $v$), and finally divide by $N$. That is,

$$v_{av} = \int_0^\infty vf(v)\,dv. \tag{16–32}$$

The rms speed is obtained similarly; the average of $v^2$ is given by

$$(v^2)_{av} = \int_0^\infty v^2 f(v)\,dv, \tag{16–33}$$

and $v_{rms}$ is the square root of this.

The function $f(v)$ describing the actual distribution of molecular speeds is called the **Maxwell-Boltzmann distribution.** It can be derived from statistical-mechanics considerations, but that derivation is beyond our scope. Here is the result:

$$f(v) = 4\pi\left(\frac{m}{2\pi kT}\right)^{3/2} v^2 e^{-mv^2/2kT}. \tag{16–34}$$

We can also express this function in terms of the translational kinetic energy of a molecule, which we denote by $\epsilon$. That is, $\epsilon = \frac{1}{2}mv^2$. We invite you to verify that when this is substituted into Eq. (16–34), the result is

$$f(v) = \frac{8\pi}{m}\left(\frac{m}{2\pi kT}\right)^{3/2} \epsilon e^{-\epsilon/kT}. \tag{16–35}$$

This form shows that the exponent in the Maxwell-Boltzmann distribution function is $\epsilon/kT$ and that the shape of the curve is determined by the relative magnitudes of $\epsilon$ and $kT$

at any point. We leave it as an exercise (see Exercise 16–29) to prove that the *peak* of each curve occurs where $\epsilon = kT$, corresponding to a speed $v_{mp}$ given by

$$v_{mp} = \sqrt{\frac{2kT}{m}}. \qquad (16\text{–}36)$$

This is therefore the *most probable* speed.

To find the average speed, we substitute Eq. (16–34) into Eq. (16–32) and carry out the integration, making a change of variable $v^2 = x$ and then integrating by parts. We leave the details as an exercise; the result is

$$v_{av} = \sqrt{\frac{8kT}{\pi m}}. \qquad (16\text{–}37)$$

To find the rms speed, we substitute Eq. (16–34) into Eq. (16–33). Evaluating the resulting integral takes some mathematical acrobatics, but we can find it in a table of integrals. The result is

$$v_{rms} = \sqrt{\frac{3kT}{m}}. \qquad (16\text{–}38)$$

This result agrees with Eq. (16–20); it *must* agree if the Maxwell-Boltzmann distribution is to be consistent with the equipartition principle and our other kinetic-theory calculations.

Table 16–2 shows fractions of all the molecules in an ideal gas that have speeds *less than* various multiples of $v_{rms}$. These numbers were obtained by numerical integration; they are the same for all ideal gases.

**TABLE 16–2 Fractions of Molecules in an Ideal Gas with Speeds Less Than $v/v_{rms}$**

| $v/v_{rms}$ | Fraction |
|---|---|
| 0.20 | 0.011 |
| 0.40 | 0.077 |
| 0.60 | 0.218 |
| 0.80 | 0.411 |
| 1.00 | 0.608 |
| 1.20 | 0.771 |
| 1.40 | 0.882 |
| 1.60 | 0.947 |
| 1.80 | 0.979 |
| 2.00 | 0.993 |

The distribution of molecular speeds in liquids is similar, although not identical, to that for gases. We can understand the vapor pressure of a liquid and the phenomenon of boiling on this basis. Suppose a molecule must have a speed at least as great as $v_A$ in Fig. 16–21b to escape from the surface of a liquid into the adjacent vapor. The number of such molecules, represented by the area under the "tail" of each curve (to the right of $v_A$), increases rapidly with temperature. Thus the rate at which molecules can escape is strongly temperature-dependent. This process is balanced by another one in which molecules in the vapor phase collide inelastically with the surface and are trapped back into the liquid phase. The number of molecules suffering this fate, per unit time, is proportional to the pressure in the vapor phase. Phase equilibrium between liquid and vapor occurs when these two competing processes proceed at exactly the same rate. So if the molecular speed distributions are known for various temperatures, we can make a theoretical prediction of vapor pressure as a function of temperature. When liquid evaporates, it's the high-speed molecules that escape from the surface. The ones that are left have less energy, on the average, than the liquid molecules before evaporation; this gives us a molecular view of evaporative cooling.

Rates of chemical reactions are often strongly temperature-dependent, and the Maxwell-Boltzmann distribution contains the reason for this dependence. When two reacting molecules collide, the reaction can occur only when the molecules are close enough for the electric-charge distributions of their electrons to interact strongly. This requires a minimum energy, called the *activation energy,* and thus a certain minimum molecular speed. Figure 16–21 shows that the number of molecules in the high-speed tail of the curve increases rapidly with temperature. Thus we expect the rate of any reaction that depends on an activation energy to increase rapidly with temperature. Similarly, many plant growth processes have strongly temperature-dependent rates, as can be seen by the rapid and diverse growth in tropical rain forests.

## 16–6 PHASES OF MATTER

We've talked a lot about ideal gases in the last few sections. An ideal gas is the simplest system to analyze from a molecular viewpoint because we ignore the interactions between molecules. But those interactions are the very thing that makes matter condense into the liquid and solid phases under some conditions. So it's not surprising that theoretical analyses of liquid and solid structure and behavior is a lot more complicated than those for gases. We won't try to go far here with a microscopic picture, but we can talk in general about phases of matter, phase equilibrium, and phase transitions.

In Section 15–5 we learned that each phase is stable only in certain ranges of temperature and pressure. A transition from one phase to another takes place ordinarily under conditions of **phase equilibrium** between the two phases, and for a given pressure this occurs at only one specific temperature. We can represent these conditions on a graph with axes $p$ and $T$, called a **phase diagram**; Fig. 16–22 shows an example. Each point on the diagram represents a pair of values of $p$ and $T$. At each point, only a single phase can exist, except for points on the solid lines, where two phases can coexist in phase equilibrium.

These lines separate the diagram into solid, liquid, and vapor regions. For example, the fusion curve separates the solid and liquid areas and represents possible conditions of solid-liquid phase equilibrium. Similarly, the vaporization curve separates the liquid and vapor areas, and the sublimation curve separates the solid and vapor areas. The three curves meet at the **triple point,** the only condition under which all three phases can coexist. In Section 15–3 we used the triple-point temperature of water to define the Kelvin temperature scale. Triple-point data for several substances are given in Table 16–3.

If we add heat to a substance at a constant pressure $p_a$, it goes through a series of states represented by the horizontal line (a) in Fig. 16–22. The melting and boiling temperatures at this pressure are the temperatures at which the line intersects the fusion and vaporization curves, respectively. When the pressure is $p_s$, constant-pressure heating transforms a substance from solid directly to vapor. This process is called *sublimation;* the intersection of line (s) with the sublimation curve gives the temperature $T_s$ at which sublimation occurs for a pressure $p_s$. At any pressure less than the triple-point pressure, no liquid phase is possible. The triple-point pressure for carbon dioxide is 5.1 atm. At normal atmospheric pressure, solid carbon dioxide ("dry ice") undergoes sublimation; there is no liquid phase at this pressure.

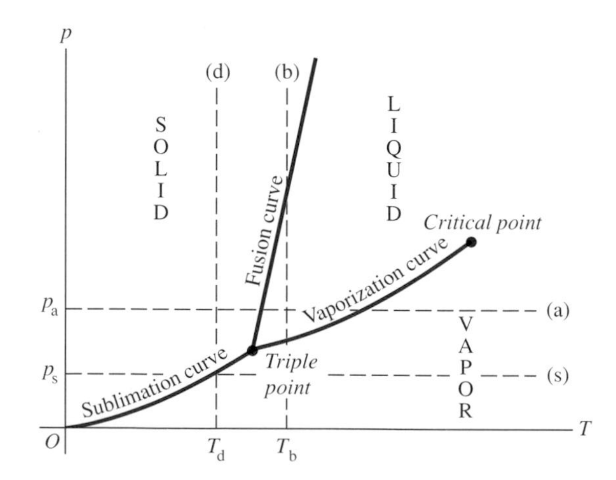


**16–22** A typical $pT$ phase diagram, showing regions of temperature and pressure at which the various phases exist and where phase changes occur.

**TABLE 16–3 Triple-Point Data**

| Substance | Temperature (K) | Pressure (Pa) |
|---|---|---|
| Hydrogen | 13.84 | $0.0704 \times 10^5$ |
| Deuterium | 18.63 | $0.171 \times 10^5$ |
| Neon | 24.57 | $0.432 \times 10^5$ |
| Oxygen | 54.36 | $0.00152 \times 10^5$ |
| Nitrogen | 63.18 | $0.125 \times 10^5$ |
| Ammonia | 195.40 | $0.0607 \times 10^5$ |
| Sulfur dioxide | 197.68 | $0.00167 \times 10^5$ |
| Carbon dioxide | 216.55 | $5.17 \times 10^5$ |
| Water | 273.16 | $0.00610 \times 10^5$ |

Line (b) in Fig. 16–22 represents constant-temperature compression at temperature $T_b$; the material passes from vapor to liquid and then to solid at the points where the broken line crosses the vaporization curve and fusion curve, respectively. Line (d) shows constant-temperature compression at a lower temperature $T_d$; the material passes from vapor to solid at the point where the line crosses the sublimation curve.

We saw in the $pV$-diagram of Fig. 16–7 that a liquid-vapor phase transition occurs only when the temperature and pressure are less than those at the point lying at the top of the tongue-shaped shaded area labeled "liquid-vapor phase equilibrium region." This point corresponds to the end point at the top of the vaporization curve in Fig. 16–22. It is called the **critical point,** and the corresponding values of $p$ and $T$ are called the critical pressure and temperature, $p_c$ and $T_c$. A gas at a pressure *above* the critical pressure does not separate into two phases when it is cooled at constant pressure (along a horizontal line above the critical point in Fig. 16–22). Instead, its properties change gradually and continuously from those we ordinarily associate with a gas (low density, large compressibility) to those of a liquid (high density, small compressibility) *without a phase transition*.

If this stretches credibility, think about liquid-phase transitions at successively higher points on the vaporization curve. As we approach the critical point, the *differences* in physical properties, such as density, bulk modulus, index of refraction, and viscosity, between the liquid and vapor phases become smaller and smaller. Exactly *at* the critical point these differences all become zero, and at this point the distinction between liquid and vapor disappears. The heat of vaporization also grows smaller and smaller as we approach the critical point, and it too becomes zero at the critical point.

For nearly all familiar materials the critical pressures are much greater than atmospheric pressure, so we don't observe this behavior in everyday life. For example, the critical point for water is at 647.4 K and $221.2 \times 10^5$ Pa (about 218 atm or 3210 psi). But high-pressure steam boilers in electric-generating plants regularly run at pressures and temperatures well above this critical point.

Many substances can exist in more than one solid phase. A familiar example is carbon, which exists as noncrystalline soot and crystalline graphite and diamond. Water is another example; at least eight types of ice, differing in crystal structure and physical properties, have been observed at very high pressures.

We remarked at the end of Section 16–1 that the equation of state for any material can be represented graphically as a surface in a three-dimensional space with coordinates $p$, $V$, and $T$. Such a surface is seldom useful in representing detailed quantitative information, but it can add to our general understanding of the behavior of materials at

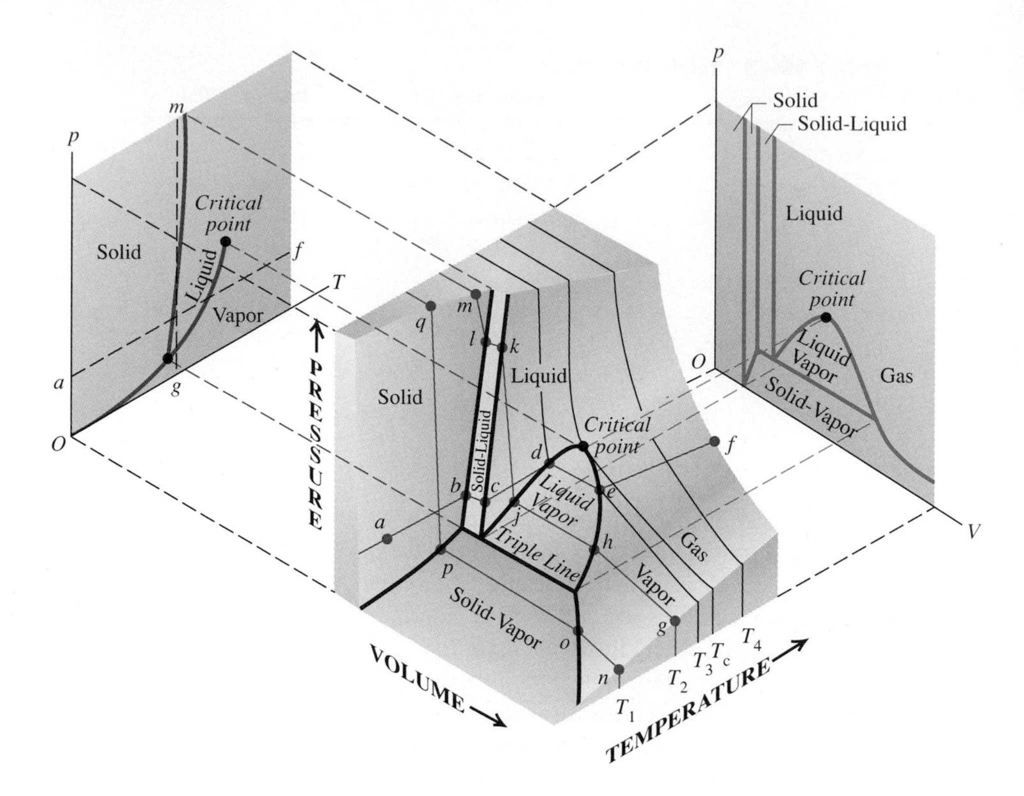


**16–23** *pVT*-surface for a substance that expands on melting. Projections of the boundaries on the surface on the *pT*- and *pV*-planes are also shown.

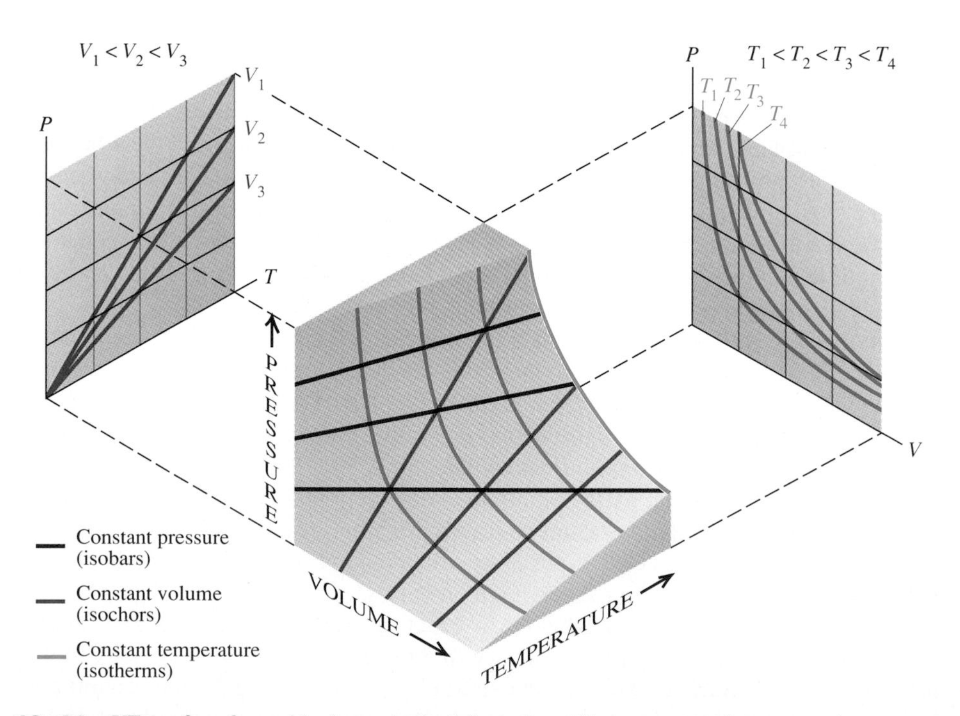


**16–24** *pVT*-surface for an ideal gas. At the left, each red line corresponds to a certain constant volume; at the right, each blue line corresponds to a certain constant temperature.

various temperatures and pressure. Figure 16–23 shows a typical $pVT$-surface. The fine red and black lines represent $pV$-isotherms; projecting them onto the $pV$-plane would give a diagram similar to Fig. 16–7. The $pV$-isotherms represent contour lines on the $pVT$-surface, just as contour lines on a topographic map represent the elevation (the third dimension) at each point. The projections of the edges of the surface onto the $pT$-plane gives the $pT$-phase diagram of Fig. 16–22.

Line *abcdef* represents constant-pressure heating, with melting along *bc* and vaporization along *de*. Note the volume changes that occur as $T$ increases along this line. Line *ghjklm* corresponds to an isothermal (constant temperature) compression, with liquefaction along *hj* and solidification along *kl*. Between these, segments *gh* and *jk* represent isothermal compression with increase in pressure; the pressure increases are much greater in the liquid region *jk* and the solid region *lm* than in the vapor region *gh*. Finally, *nopq* represents isothermal solidification directly from the vapor phase; this is the process involved in growth of crystals directly from vapor, as in the formation of snowflakes or frost and in the fabrication of some solid-state electronic devices. These three lines on the $pVT$-surface are worth careful study.

For contrast, Fig. 16–24 shows the much simpler $pVT$-surface for a substance that obeys the ideal-gas equation of state under all conditions. The projections of the constant-temperature curves onto the $pV$-plane correspond to the curves of Fig. 16–6 (Boyle's law), and the projections of the constant-volume curves onto the $pT$-plane show the direct proportionality of pressure to absolute temperature (Charles' law).

## SUMMARY

- The pressure, volume, and temperature of a given quantity of a substance are called state coordinates. They are related by an equation of state. This relation pertains only to equilibrium states, where $p$ and $T$ are uniform throughout the system. The ideal-gas equation is the equation of state for an ideal gas:

$$pV = nRT. \qquad (16\text{–}3)$$

The constant $R$ is the same for all gases for conditions in which this equation is applicable.

- A $pV$-diagram is a set of graphs, called isotherms, each showing pressure as a function of volume for a constant temperature.
- The molecular mass $M$ of a pure substance is the mass per mole. The total mass $m_{tot}$ is related to the number of moles $n$ by

$$m_{tot} = nM. \qquad (16\text{–}2)$$

- Avogadro's number $N_A$ is the number of molecules in a mole. The mass $m$ of an individual molecule is related to $M$ and $N_A$ by

$$M = N_A m. \qquad (16\text{–}9)$$

- The average translational kinetic energy of molecules of an ideal gas is directly proportional to absolute temperature:

$$K_{tr} = \tfrac{3}{2}nRT. \qquad (16\text{–}15)$$

This can be expressed on a molecular basis by using Boltzmann's constant, $k = R/N_A$:

$$\tfrac{1}{2}m(v^2)_{av} = \tfrac{3}{2}kT. \qquad (16\text{–}17)$$

### KEY TERMS

**state coordinates**
**equation of state**
**ideal-gas equation**
**ideal gas constant**
**ideal gas**
**van der Waals equation**
**$pV$-diagram**
**isotherm**
**molecule**
**potential well**
**mole**

**Avogadro's number**
**molecular mass**
**Boltzmann constant**
**mean free path**
**equipartition of energy**
**degrees of freedom**
**rule of Dulong and Petit**
**Maxwell-Boltzmann distribution**
**phase equilibrium**
**phase diagram**
**triple point**
**critical point**

The root-mean-square speed of molecules in an ideal gas is

$$v_{\text{rms}} = \sqrt{(v^2)_{\text{av}}} = \sqrt{\frac{3kT}{m}} = \sqrt{\frac{3RT}{M}}. \qquad (16\text{–}20)$$

• The mean free path $\lambda$ of molecules in an ideal gas is

$$\lambda = vt_{\text{mean}} = \frac{V}{4\pi\sqrt{2}r^2N}. \qquad (16\text{–}22)$$

• The molar heat capacity $C_V$ at constant volume for an ideal monatomic gas is

$$C_V = \tfrac{3}{2}R. \qquad (16\text{–}26)$$

For an ideal diatomic gas, including rotational but not vibrational kinetic energy,

$$C_V = \tfrac{5}{2}R. \qquad (16\text{–}27)$$

For a monatomic solid,

$$C_V = 3R. \qquad (16\text{–}30)$$

• The speeds of molecules in an ideal gas are distributed according to the Maxwell-Boltzmann distribution:

$$f(v) = 4\pi\left(\frac{m}{2\pi kT}\right)^{3/2} v^2 e^{-mv^2/2kT}. \qquad (16\text{–}34)$$

• Ordinary matter exists in the solid, liquid, and gas phases. A phase diagram shows conditions under which two phases can coexist in phase equilibrium. All three phases can coexist at the triple point. The vaporization curve ends at the critical point, above which the distinction between liquid and gas phases disappears.

# EXERCISES

## Section 16–1 Equations of State

**16–1** A cylindrical tank has a tight-fitting piston that allows the volume of the tank to be changed. The tank originally contains 0.120 $m^3$ of air at a pressure of 1.00 atm. The piston is slowly pushed in until the volume of the gas is decreased to 0.050 $m^3$. If the temperature remains constant, what is the final value of the pressure?

**16–2** A 2.00-L tank contains air at 1.00 atm and 20.0°C. The tank is sealed and heated until the pressure is 5.00 atm. a) What is the temperature then in degrees Celsius? Assume that the volume of the tank is constant. b) If the temperature is kept at the value found in part (a) and the gas is permitted to expand, what is the volume when the pressure again becomes 1.00 atm?

**16–3** A 20.0-L tank contains 0.280 kg of helium at 27.0°C. The atomic mass of helium is 4.00 g/mol. a) What is the number of moles of helium? b) What is the pressure in pascals and in atmospheres?

**16–4** Helium gas with a volume of 1.80 L, under a pressure of 3.00 atm and at a temperature of 57.0°C, is heated until both pressure and volume are doubled. a) What is the final temperature? b) How many grams of helium are there? The atomic mass of helium is 4.00 g/mol.

**16–5** A large cylindrical tank with a tight-fitting piston that allows the volume to be changed contains 0.500 $m^3$ of nitrogen gas at 27°C and $1.50 \times 10^5$ Pa (absolute pressure). What is the pressure if the volume is increased to 4.00 $m^3$ and the temperature is increased to 327°C?

**16–6** A room with dimensions 4.00 m × 5.00 m × 3.00 m is filled with pure oxygen at 20.0°C and 1.00 atm. The molecular mass of oxygen is 32.0 g/mol. a) What is the number of moles of oxygen? b) What is the mass of oxygen in kilograms?

**16–7 Change of State in a Diesel Engine.** At the beginning of the compression stroke, a cylinder of a diesel engine contains 800 $cm^3$ of air at atmospheric pressure ($1.01 \times 10^5$ Pa) and a temperature of 27.0°C. At the end of the stroke the air has been compressed to a volume of 75.0 $cm^3$, and the gauge pressure has increased to $2.25 \times 10^6$ Pa. Compute the final temperature.

**16–8** A welder using a tank having a volume of 0.0800 $m^3$ fills it with oxygen (whose molecular mass is 32.0 g/mol) at a gauge pressure of $4.00 \times 10^5$ Pa and temperature of 47.0°C. The tank has a small leak, and in time some of the gas leaks out. On a day when the temperature is 27.0°C the gauge pressure of the gas in the tank is $3.00 \times 10^5$ Pa. Find a) the initial mass of oxygen; b) the mass that has leaked out.

**16–9** The total lung volume for a typical physics student is 6.0 L. A physics student fills her lungs with air at an absolute pressure of 1.0 atm. Then, holding her breath, she compresses her chest cavity, decreasing her lung volume to 5.5 L. What is then the pressure of the air in her lungs? Assume that the temperature of the air remains constant.

**16–10** A diver observes a bubble of air rising from the bottom of a lake, where the pressure is 2.50 atm, to the surface, where the pressure is 1.00 atm. The temperature at the bottom is 7.0°C, and the temperature at the surface is 27.0°C. What is the ratio of the volume of the bubble as it reaches the surface to its volume at the bottom?

**16–11** The gas inside a balloon will always have a pressure nearly equal to atmospheric pressure, since that is the pressure applied to the outside of the balloon. You fill a balloon with an ideal gas to a volume of 0.800 L at a temperature of 27.0°C. What is the volume of the balloon if you cool it to the boiling point of liquid nitrogen (77.3 K)?

**16–12** What is the atmospheric pressure at an altitude of 5000 m if the air temperature is a uniform 0°C? This is roughly the maximum altitude usually attained by aircraft with non-pressurized cabins.

**16–13** A tank with volume 2.50 L will burst if the absolute pressure of the gas it contains exceeds 100 atm. If 8.0 mol of an ideal gas is put into the tank at a temperature of 23.0°C, to what temperature can the gas be heated before the tank ruptures? Neglect the thermal expansion of the tank.

## Section 16–2 Molecular Properties of Matter

**16–14** Consider an ideal gas at 27°C and 1.00 atm pressure. Imagine each molecule to be, on average, at the center of a small cube. a) What is the length of an edge of this small cube? b) How does this distance compare with the diameter of a molecule?

**16–15** How many moles are there in a glass of water (0.250 kg)? How many molecules? The molecular mass of water is 18.0 g/mol.

**16–16** The lowest pressures that are readily attainable in the laboratory are of the order of $10^{-13}$ atm. At a pressure of $1.00 \times 10^{-13}$ atm and an ordinary temperature (say, $T = 300$ K), how many molecules are present in a volume of 1.00 cm$^3$?

**16–17** Consider 1.00 mol of liquid water. a) What volume is occupied by this amount of water? The molecular mass of water is 18.0 g/mol. b) Imagine each molecule to be, on the average, at the center of a small cube. What is the length of an edge of this small cube? c) How does this distance compare with the diameter of a molecule?

**16–18** What is the length of the side of a cube, in a gas at standard conditions, that contains a number of molecules equal to the population of the United States (about 250 million)?

## Section 16–3 Kinetic-Molecular Model of an Ideal Gas

**16–19** At what temperature is the root-mean-square speed of nitrogen molecules equal to the root-mean-square speed of hydrogen molecules at 27.0°C? The molecular mass of hydrogen ($H_2$) is 2.02 g/mol, and that of nitrogen ($N_2$) is 28.01 g/mol.

**16–20** A flask contains a mixture of krypton, neon, and helium gases. The atomic masses are: helium, 4.00 g/mol; neon, 20.18 g/mol; krypton, 83.80 g/mol. Compare a) the average kinetic energies of the three types of atoms; b) the root-mean-square speeds.

**16–21 Gaseous Diffusion of Uranium.** Isotopes of uranium are sometimes separated by gaseous diffusion, using the fact that the root-mean-square speeds of the molecules in vapor are slightly different, so the vapors diffuse at slightly different rates. The atomic masses for $^{235}$U and $^{238}$U are 0.235 kg/mol and 0.238 kg/mol, respectively. What is the ratio of the root-mean-square speed of the $^{235}$U atoms in the vapor to that of the $^{238}$U atoms if the temperature is uniform?

**16–22** Calculate the mean free path of air molecules at a pressure of $1.00 \times 10^{-13}$ atm and a temperature of 300 K. (This is of the order of the lowest pressure that is readily attainable in the laboratory; see Exercise 16–16.)

**16–23** a) What is the average translational kinetic energy of a helium atom (with an atomic mass of 4.00 g/mol) at a temperature of 300 K? b) What is the average value of the square of its speed? c) What is the root-mean-square speed? d) What is the momentum of a helium atom traveling at this speed? e) Suppose an atom traveling at this speed bounces back and forth between opposite sides of a cubical vessel 0.10 m on a side. What is the average force that it exerts on one of the walls of the container? (Assume that the atom's velocity is perpendicular to the two sides that it strikes.) f) What is the average force per unit area? g) How many molecules traveling at this speed are necessary to produce an average pressure of 1 atm? h) Compute the number of helium atoms that are actually contained in a vessel of this size at 300 K and atmospheric pressure. i) Your answer for part (h) should be three times as large as the answer for part (g). Where does this discrepancy arise?

**16–24** Smoke particles in the air typically have masses of the order of $10^{-16}$ kg. The Brownian motion (rapid, irregular movement) of these particles, resulting from collisions with air molecules, can be observed with a microscope. a) Find the root-mean-square speed of Brownian motion for a particle with a mass of $1.00 \times 10^{-16}$ kg in air at 300 K. b) Would the rms speed be different if the particle were in hydrogen gas at 300 K? Explain.

## Section 16–4 Heat Capacities

**16–25** a) How much heat does it take to increase the temperature of 4.00 mol of a diatomic ideal gas by 25.0 K near room temperature if the gas is held at constant volume? b) What is the answer to the question in part (a) if the gas is monatomic rather than diatomic?

**16–26** Compute the specific heat capacity at constant volume of hydrogen gas, and compare it with the specific heat capacity of liquid water. The molecular mass of hydrogen ($H_2$) is 2.02 g/mol.

**16–27** Calculate the molar heat capacity at constant volume of water vapor, assuming that the nonlinear triatomic molecule has three translational and three rotational degrees of freedom and that vibrational motion does not contribute. The actual specific heat capacity of water vapor at low pressures is about 2000 J/kg · K. Compare this with your calculation and comment on the actual role of vibrational motion. The molecular mass of water is 18.0 g/mol.

## Section 16–5 Molecular Speeds

***16–28** Derive Eq. (16–35) from Eq. (16–34).

***16–29** Prove that $f(v)$ as given by Eq. (16–35) is maximum for $\epsilon = kT$.

## Section 16–6 Phases of Matter

**16–30** Solid water (ice) is slowly heated from a very low temperature. a) What minimum external pressure $p_1$ must be applied to the solid if a melting phase transition is to be observed? Describe the sequence of phase transitions that occur if the applied pressure $p$ is such that $p < p_1$. b) Above a certain maximum pressure $p_2$, no melting transition is observed. What is this pressure? Describe the sequence of phase transitions that occur if $p_1 < p < p_2$.

**16–31** A physicist places a piece of ice at 0.00°C alongside a beaker of water at 0.00°C in a glass vessel from which all the air has been removed. If the ice, water, and vessel are all maintained at a temperature of 0.00°C, describe the final equilibrium state inside the vessel.

# PROBLEMS

**16–32** Partial vacuums in which the residual gas pressure is about $1.00 \times 10^{-13}$ atm are not difficult to obtain with modern technology. Calculate the mass of nitrogen present in a volume of 4000 $cm^3$ at this pressure if the temperature of the gas is 27.0°C. The molecular mass of nitrogen ($N_2$) is 28.0 g/mol.

**16–33** An automobile tire has a volume of 0.0150 $m^3$ on a cold day when the temperature of the air in the tire is 5.0°C and atmospheric pressure is 1.03 atm. Under these conditions the gauge pressure is measured to be 2.00 atm (about 30 lb/in.$^2$). After the car has been driven on the highway for 30 min, the temperature of the air in the tires has risen to 47°C and the volume to 0.0160 $m^3$. What then is the gauge pressure?

**16–34** The submarine *Squalus* sank at a point where the depth of water was 73.0 m. The temperature at the surface was 27.0°C, and at the bottom it was 7.0°C. The density of seawater is 1030 kg/$m^3$. a) If a diving bell in the form of a circular cylinder 2.60 m high, open at the bottom and closed at the top, is lowered to this depth, to what height will water rise within the diving bell when it reaches the bottom? b) At what gauge pressure must compressed air be supplied to the bell while it is on the bottom to expel all the water from it?

**16–35** A glassblower makes a barometer using a tube 0.900 m long and with a cross-section area of 0.620 $cm^2$. Mercury stands in this tube to a height of 0.750 m. The room temperature is 27.0°C. A small amount of nitrogen is introduced into the evacuated space above the mercury, and the column drops to a height of 0.700 m. How many grams of nitrogen were introduced? The molecular mass of nitrogen ($N_2$) is 28.0 g/mol.

**16–36** A hot-air balloon makes use of the fact that hot air at atmospheric pressure is less dense than cooler air at the same pressure; the buoyant force is calculated as discussed in Chapter 14. If the volume of the balloon is 500 $m^3$ and the surrounding air is at 0°C, what must be the temperature of the air in the balloon for it to lift a total load of 250 kg (in addition to the mass of the hot air)? The density of air at 0°C and atmospheric pressure is 1.29 kg/$m^3$.

**16–37** A flask with a volume of 1.20 L, provided with a stopcock, contains oxygen (whose molecular mass is 32.0 g/mol) at 300 K and atmospheric pressure ($1.013 \times 10^5$ Pa). The system is heated to a temperature of 400 K, with the stopcock open to the atmosphere. The stopcock is then closed, and the flask is cooled to its original temperature. a) What is the final pressure of the oxygen in the flask? b) How many grams of oxygen remain in the flask?

**16–38** A vertical cylindrical tank 0.900 m high has its top end closed by a tightly fitting frictionless piston of negligible weight. The air inside the cylinder is at an absolute pressure of 1.00 atm. The piston is depressed by pouring mercury on it slowly (Fig. 16–25). How far will the piston descend before mercury spills over the top of the cylinder? The temperature of the air is kept constant.

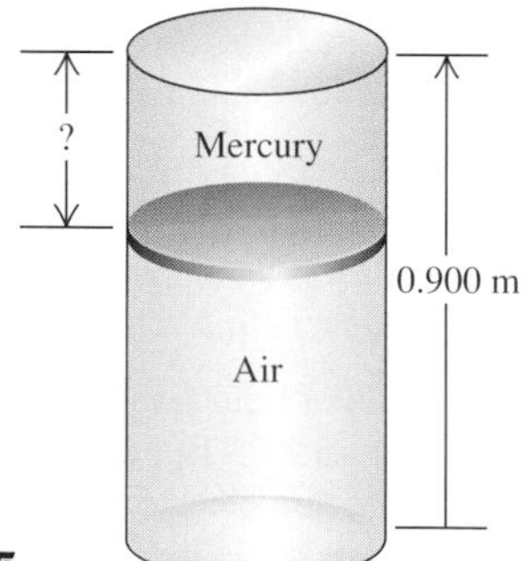


**FIGURE 16–25**

**16–39** A balloon whose volume is 500 $m^3$ is to be filled with hydrogen at atmospheric pressure ($1.01 \times 10^5$ Pa). a) If the hydrogen is stored in cylinders with a volume of 2.50 $m^3$ at an absolute pressure of $2.50 \times 10^6$ Pa, how many cylinders are required? Assume that the temperature of the hydrogen remains constant. b) What is the total weight (in addition to the weight of the gas) that can be supported by the balloon if the gas in the balloon and the surrounding air are both at 0°C? The molecular mass of hydrogen ($H_2$) is 2.02 g/mol. The density of air at 0°C and atmospheric pressure is 1.29 kg/m$^3$. c) What weight could be supported if the balloon were filled with helium (with an atomic mass of 4.00 g/mol) instead of hydrogen, again at 0°C?

**16–40** Experiments show that the size of an oxygen molecule is of the order of $2.0 \times 10^{-10}$ m. Make a rough estimate of the pressure at which the finite volume of the molecules should cause noticeable deviations from ideal-gas behavior at ordinary temperatures ($T = 300$ K).

**16–41 How Many Atoms Are You?** Estimate the number of atoms in the body of a 60-kg physics student. Base the estimate on the fact that the human body is mostly water. The molecular mass of water ($H_2O$) is 18.0 g/mol, and each water molecule contains three atoms.

**16–42** a) Compute the increase in gravitational potential energy of an oxygen molecule (molecular mass 32.0 g/mol) for an increase in elevation of 500 m near the earth's surface. b) At what temperature is this equal to the average kinetic energy of oxygen molecules?

**16–43** The speed of propagation of a sound wave in air at 27°C is about 350 m/s. Calculate, for comparison, a) $v_{rms}$ for nitrogen molecules; b) the root-mean-square value of $v_x$ at this temperature. The molecular mass of nitrogen ($N_2$) is 28.0 g/mol. [If sound propagation were an isothermal process, which it ordinarily is not, the speed of sound would be equal to $(v_x)_{rms}$. See Section 19–6.]

**16–44** a) What is the total random translational kinetic energy of 8.00 L of helium gas, with pressure $1.01 \times 10^5$ Pa and temperature 300 K? b) If the tank containing the gas is moved with a speed of 20.0 m/s, by what percentage is the total kinetic energy of the gas increased? The atomic mass of helium is 4.00 g/mol.

**16–45** a) For what mass of molecule or particle is $v_{rms}$ equal to 0.100 m/s at a temperature of 300 K? b) If the particle is an ice crystal, how many molecules does it contain? (The molecular mass of water is 18.0 g/mol.) c) Calculate the diameter of the particle if it is a spherical piece of ice. Would it be visible to the naked eye?

**16–46 Hydrogen in the Sun.** The surface of the sun has a temperature of about 6000 K and consists largely of hydrogen atoms. (In fact, most of the atoms will be ionized, so the hydrogen "atom" is actually a proton.) a) Find the rms speed of a hydrogen atom at this temperature. (The mass of a single hydrogen atom is $1.67 \times 10^{-27}$ kg.) b) The escape speed for a particle to leave the gravitational influence of the sun is given by $(2GM/R)^{1/2}$, where $M$ is the sun's mass, $R$ its radius, and $G$ the gravitational constant (Example 12–7). Use the data in Appendix F to calculate this escape speed. c) Can appreciable quantities of hydrogen escape from the sun's gravitational field?

**16–47 "Escape Temperature."** a) Show that a projectile with mass $m$ can "escape" from the earth's gravitational field if it is launched vertically upward with a kinetic energy greater than $mgR$, where $g$ is the acceleration due to gravity at the earth's surface and $R$ is the earth's radius. (See the preceding problem.) b) At what temperature does the average translational kinetic energy of an oxygen molecule equal that required to escape? What about a hydrogen molecule?

**16–48** For each of the polyatomic gases in Table 16–1, compute the value of $C_V$ on the assumption that there is no vibrational energy. Compare with the measured values in the table, and compute the fraction of the total heat capacity that is due to vibration for each of the three gases. (*Note:* $CO_2$ is linear; $SO_2$ and $H_2S$ are not. Recall that a linear polyatomic molecule has two rotational degrees of freedom, and a nonlinear molecule has three.)

**16–49** a) Show that

$$\int_0^\infty f(v)\,dv = 1,$$

where $f(v)$ is the Maxwell-Boltzmann distribution of Eq. (16–34). b) Calculate

$$\int_0^\infty v^2 f(v)\,dv$$

and compare this result to $(v^2)_{av}$ as given by Eq. (16–17). [*Hint:* You may use the tabulated integral

$$\int_0^\infty x^{2n} e^{-ax^2}\,dx = \frac{1 \cdot 3 \cdot 5 \cdots (2n-1)}{2^{n+1} a^n}\sqrt{\frac{\pi}{a}},$$

where $n$ is an integer and $a$ is a positive constant.]

**16–50 Vapor Pressure and Relative Humidity.** The vapor pressure is the pressure of the vapor phase of a substance when it is in equilibrium with the solid or liquid phase of the substance. The relative humidity is the partial pressure of water vapor in the air divided by the vapor pressure of water at the same temperature, expressed as a percentage. The air is saturated when the humidity is 100%. a) What is the partial pressure of water vapor in the atmosphere when the air temperature is 15°C and the humidity is 35%? The vapor pressure of water at 15°C is $1.69 \times 10^3$ Pa. b) Under the conditions of part (a), what is the mass of water in 1.00 m$^3$ of air? (The molecular mass of water is $18.0 \times 10^{-3}$ kg/mol. Assume that water vapor can be treated as an ideal gas.)

**16–51 The Dew Point.** The vapor pressure of water (Problem 16–50) decreases as the temperature decreases. If the partial pressure of water vapor in the air is kept constant as the air is cooled, a temperature is reached, called the dew point, at which the partial pressure and vapor pressure coincide and the vapor is saturated. If the air is cooled further, vapor condenses to liquid until the partial pressure again equals the vapor pressure at that temperature. The temperature in a room is 40°C. A meteorologist cools a metal can by gradually adding cold water. At 15°C, water droplets form on the outside surface of the can. What is the relative humidity in the 40°C air in the room? (The vapor pressure of water is $7.34 \times 10^3$ Pa at 40°C and $1.69 \times 10^3$ Pa at 15°C.)

## CHALLENGE PROBLEMS

**16–52 Atmospheric Pressure at the Top of Mt. Everest.** In the lower part of the atmosphere (the troposphere) the temperature is not uniform but decreases with increasing elevation. a) Show that if the temperature variation is approximated by the linear relation

$$T = T_0 - \alpha y,$$

where $T_0$ is the temperature at the earth's surface and $T$ is the temperature at height $y$, then the pressure is given by

$$\ln\left(\frac{p}{p_0}\right) = \frac{Mg}{R\alpha}\ln\left(\frac{T_0 - \alpha y}{T_0}\right)$$

where $p_0$ is the pressure at the earth's surface and $M$ is the molecular mass for air. The coefficient $\alpha$ is called the lapse rate of temperature. It varies with atmospheric conditions, but an average value is about 0.6 C°/100 m. b) Show that the above result reduces to Eq. (16–7) in the limit $\alpha \rightarrow 0$. c) With $\alpha = 0.6$ C°/100 m, calculate $p$ for $y = 8863$ m and compare your answer to the result of Example 16–4. Take $T_0 = 273$ K and $p_0 = 1.00$ atm.

**16–53** A large tank of water has a hose connected to it, as shown in Fig. 16–26. The tank is sealed at the top and has compressed air between the water surface and the top. When the water height $h$ has the value 3.00 m, the absolute pressure $p$ is $2.50 \times 10^5$ Pa. Assume that the air above the water expands isothermally (at constant $T$). Take the atmospheric pressure to be $1.00 \times 10^5$ Pa. a) What is the speed of flow out of the hose when $h = 3.00$ m? b) As water flows out of the tank, $h$ decreases. Calculate the speed of flow for $h = 2.00$ m and for $h = 1.60$ m. c) At what value of $h$ does the flow stop?

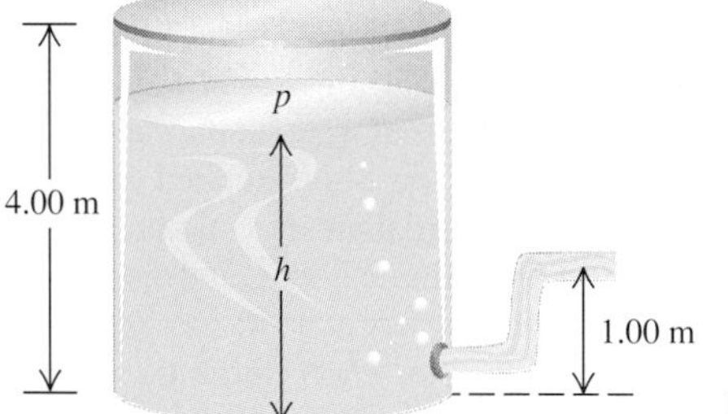

**FIGURE 16–26**

**16–54** The position and velocity of a simple harmonic oscillator are given by Eqs. (13–19) and (13–21). For simplicity, assume that the initial position and velocity make the phase angle $\phi$ equal to zero. Use those equations to show that the potential energy of the oscillator averaged over one period of the motion is equal to the kinetic energy averaged over one period. This result was used in Section 16–4. (*Hint:* Use the trigonometric identities $\cos^2\theta = [1 + \cos(2\theta)]/2$ and $\sin^2\theta = [1 - \cos(2\theta)]/2$. What is the average value of $\cos(2\omega t)$ averaged over one period?)

**16–55 Van der Waals Equation and Critical Points.** In $pV$-diagrams the slope $\partial p/\partial V$ along an isotherm is never positive. (You should be able to explain why.) Regions where $\partial p/\partial V = 0$ represent equilibrium between two phases; volume can change with no change in pressure. We can use this fact to determine critical constants if we have an equation of state. With $p = p(V, T, n)$, if $T > T_c$, then $p(V)$ has no maximum along an isotherm, but if $T < T_c$, then $p(V)$ has a maximum. This leads to the following condition for determining the critical point:

$$\frac{\partial p}{\partial V} = 0 \quad \text{and} \quad \frac{\partial^2 p}{\partial V^2} = 0 \quad \text{at the critical point.}$$

a) Solve the van der Waals equation (Eq. 16–8) for $p$. That is, find $p(V, T, n)$. Find $\partial p/\partial V$ and $\partial^2 p/\partial V^2$. Set these equal to zero to obtain two equations for $V$, $T$, and $n$. b) Simultaneous solution of the two equations obtained in part (a) gives the critical values $T_c$ and $(V/n)_c$. Find these constants in terms of $a$ and $b$. (*Hint:* Divide one equation by the other to eliminate $T$.) c) Substitute these values into the equation of state to find $p_c$, the critical pressure. d) Find the ratio

$$\frac{RT_c}{p_c(V/n)_c}.$$

This should not contain either $a$ or $b$. e) Compute the ratio in part (d) for the gases $H_2$, $N_2$, and $H_2O$ using the critical point data given below. Compare the results to the prediction you made based on the van der Waals equation.

| Gas | $T_c$ (K) | $p_c$ (Pa) | $V_c/n$ (m³/mol) |
|---|---|---|---|
| $H_2$ | 33.3 | $13.0 \times 10^5$ | $65.0 \times 10^{-6}$ |
| $N_2$ | 126.2 | $33.9 \times 10^5$ | $90.1 \times 10^{-6}$ |
| $H_2O$ | 647.4 | $221.2 \times 10^5$ | $56.0 \times 10^{-6}$ |

**16–56** In Example 16–9 we saw that $v_{rms} > v_{av}$. It is not difficult to show that this is *always* the case. (Unless all the particles have the same speed, in which case $v_{rms} = v_{av}$.) a) For two particles with speeds $v_1$ and $v_2$, show that $v_{rms} \geq v_{av}$, regardless of the numerical values of $v_1$ and $v_2$. Then show that $v_{rms} > v_{av}$ if $v_1 \neq v_2$. b) Suppose that for a collection of $N$ particles you know that $v_{rms} > v_{av}$. Another particle, with speed $u$, is added to the collection of particles. If the new rms and average speeds are denoted as $v'_{rms}$ and $v'_{av}$, show that

$$v'_{rms} = \sqrt{\frac{Nv^2_{rms} + u^2}{N+1}}, \qquad v'_{av} = \frac{Nv_{av} + u}{N+1}.$$

c) Use the expressions in part (b) to show that $v'_{rms} > v'_{av}$, regardless of the numerical value of $u$. d) Explain why your results for (a) and (c) together show that $v_{rms} > v_{av}$ for any collection of particles if the particles do not all have the same speed.

**16–57 The Lennard-Jones Potential.** A commonly used potential-energy function for the interaction of two molecules (Fig. 16–8) is the Lennard-Jones potential

$$U(r) = U_0\left[\left(\frac{\sigma}{r}\right)^{12} - \left(\frac{\sigma}{r}\right)^6\right].$$

a) What is the force $F(r)$ that corresponds to this potential? Sketch $U(r)$ and $F(r)$. b) In terms of $\sigma$, what are the values of $r_1$ defined by $U(r_1) = 0$ and the values of $r_2$ defined by $F(r_2) = 0$? Also, what is the ratio $r_1/r_2$? Show the location of $r_1$ and $r_2$ on your sketch of $U(r)$. c) If the molecules are located a distance $r_2$ apart [as calculated in part (b)], how much work must be done to pull them apart such that $r \rightarrow \infty$?

The types of engines that power cars in the Indianapolis 500 car race are turbocharged; extra air is forced into the cylinders under pressure for a bigger explosion when the spark plugs ignite the fuel. The fuel used is methanol, which is less dense than gasoline, and typical engine output is about 720 horsepower at 12,000 rpm. (Common car engines typically produce 170 horsepower at 4200 rpm.) The waste heat produced by an Indy-type engine is so great that it usually overheats at speeds below 90 mi/h; common life expectancy of an engine is only 2500 mi.

CHAPTER 17

# First Law of Thermodynamics

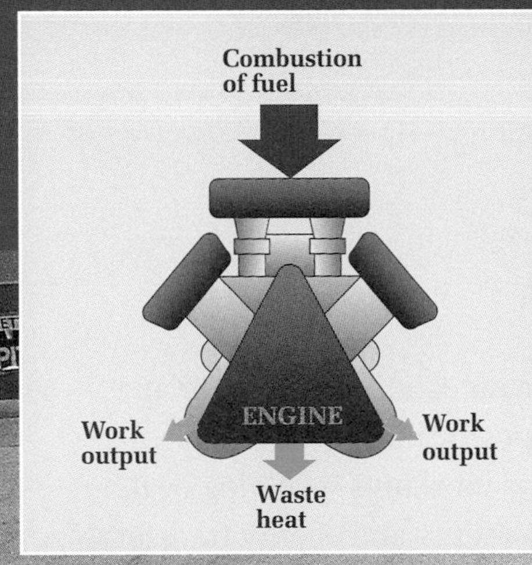

## KEY CONCEPTS

- **A thermodynamic system can exchange energy with its surroundings in two ways: heat transfer and mechanical work. The system does work when its volume changes.**
- **The internal energy of a system depends only on the state of the system. Conversely, there is no such thing as the quantity of heat or work in a system. The internal energy changes when either heat transfer or work occurs.**
- **The first law of thermodynamics is an expression of conservation of energy (including internal energy, heat, and work) together with the concept that, in any particular state, a system has a definite amount of internal energy.**
- **Several kinds of thermodynamic processes are of special importance, including adiabatic, isochoric, isobaric, and isothermal processes.**
- **The internal energy of an ideal gas depends only on its temperature. This fact can be used to derive a relation between the molar heat capacity at constant volume and that at constant pressure for an ideal gas.**
- **When an ideal gas undergoes an adiabatic process (no heat transfer), the changes in pressure, volume, and temperature are related by simple equations.**

## INTRODUCTION

Every time you drive a car, turn on an air conditioner, or use an electrical appliance, you reap the practical benefits of *thermodynamics,* the study of energy relationships involving heat, mechanical work, and other aspects of energy and energy transfer. If you've seen demonstrations that use liquid nitrogen, you may have wondered how gases are liquefied. One method consists of first compressing the gas to very high pressure while keeping the temperature constant, then insulating it and allowing it to expand. The gas cools so much during the expansion that it liquefies. This is an example of a *thermodynamic process.*

The first law of thermodynamics, which is central to the understanding of such processes, is an extension of the principle of conservation of energy. It broadens this principle to include energy exchange by both heat transfer and mechanical work and introduces the concept of *internal energy* of a system. Conservation of energy plays a vital role in every area of physical science, and the first law has extremely broad usefulness. To state energy relationships precisely, we need the concept of a *thermodynamic system,* and we discuss *heat* and *work* as two means of transferring energy into or out of such a system.

## 17–1 ENERGY, HEAT, AND WORK

We have studied energy transfer through mechanical work (Chapter 6) and through heat transfer (Chapters 15 and 16). Now we are ready to combine and generalize these principles. We will always talk about some specific *system;* it might be a quantity of expanding steam in a turbine, the Freon in an air conditioner, some other specific quantity of material, or sometimes a particular device or an organism. A **thermodynamic system** (Fig. 17–1) is a system that can interact (and exchange energy) with its surroundings, or environment, in at least two ways, one of which is heat transfer. A familiar example of a thermodynamic system is a quantity of gas confined in a cylinder with a piston, similar to the cylinders in internal-combustion engines. Energy can be added to the system by conduction of heat, and the system can also do *work* as the gas exerts a force on the piston and moves it through a displacement.

We have often used the concept *system* in mechanics in connection with free-body diagrams and conservation of energy and momentum. With thermodynamic systems, as with all others, it is essential to define clearly at the start exactly what is and is not included in the system. Only then can we describe unambiguously the energy transfers into and out of that system.

As we have mentioned, thermodynamics has its roots in practical problems. The engine in an automobile and the jet engines in an airplane use the heat of combustion of their fuel to perform mechanical work in propelling the vehicle (Fig. 17–2). Muscle tissue in living organisms metabolizes chemical energy in food and performs mechanical work on the organism's surroundings. A steam engine or steam turbine uses the heat of combustion of coal or other fuel to perform mechanical work such as driving an electric generator or pulling a train.

In all these situations we describe the energy relations in terms of the quantity of heat $Q$ added *to* the system and the work $W$ done *by* the system. Both $Q$ and $W$ may be positive or negative. A positive value of $Q$ represents heat flow *into* the system, with a corresponding input of energy to it; negative $Q$ represents heat flow *out of* the system. A positive value of $W$ represents work done *by* the system against its surroundings, such as work done by an expanding gas, and thus corresponds to energy *leaving* the system. Negative $W$, such as work done during compression

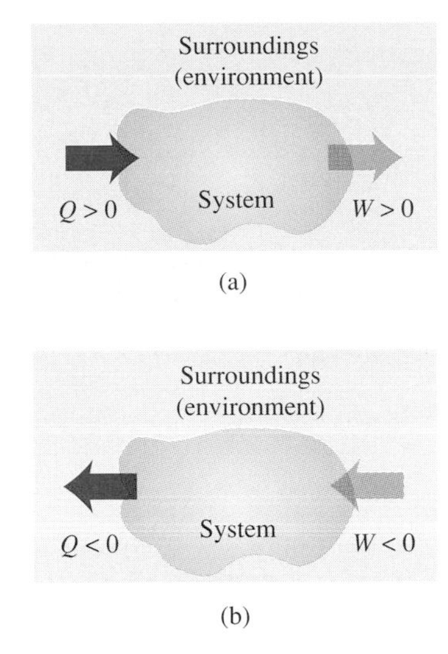


**17–1** A thermodynamic system may exchange energy with its surroundings (environment) by means of heat and work. (a) When heat is added *to* the system and work is done *by* it, $Q$ and $W$ are both positive. (b) When heat is transferred *out* of the system and work is done *on* it, $Q$ and $W$ are both negative. Either $Q$ or $W$ or both may also be zero.

(a)

(b)

**17–2** (a) A jet engine uses the heat of combustion of its fuel to do work, propelling the plane. (b) Humans and other biological organisms are more complicated systems than we can analyze in this book, but the same basic principles of thermodynamics apply to them.

**17–3** The catastrophic eruption of Mt. St. Helens on May 18, 1980, a dramatic illustration of the enormous pressure developed in a confined hot gas. As the gas expanded, it did enough work on the overlying rock and soil to lift roughly a cubic mile of material several thousand feet into the air.

of a gas in which work is done *on the gas* by its surroundings, represents energy entering the system. We will use these conventions consistently in the examples in this chapter and the next. Note that our sign rule for work is *opposite* to the one we used in mechanics, in which we always spoke of the work done *by* the forces acting on a body. In thermodynamics it is usually more convenient to call $W$ the work done *by* the system so that when a system expands, the pressure, volume change, and work are all positive.

The microscopic viewpoint, looking at the kinetic and potential energies of individual molecules in a material, is very helpful in developing intuition about thermodynamic quantities. But it is also important to understand that the central principles and concepts of thermodynamics can be treated in a completely *macroscopic* way, without reference to microscopic models. Indeed, part of the great power and generality of thermodynamics springs from the fact that it *does not* depend on details of the structure of matter.

## 17–2 WORK DONE DURING VOLUME CHANGES

A gas in a cylinder with a movable piston is a simple example of a thermodynamic system. Internal-combustion engines, steam engines, and compressors in refrigerators and air conditioners all use some version of such a system. In the next several sections we will use the gas-in-cylinder system to explore several kinds of processes involving energy transformations. First we consider the *work* done by the system during a volume change. When a gas expands, it pushes out on its boundary surfaces as they move outward; an expanding gas always does positive work. The same thing is true of any solid or fluid material confined under pressure. Figure 17–3 shows an impressive example of the work done by expanding gases.

Figure 17–4 shows a solid or fluid in a cylinder with a movable piston. Suppose that the cylinder has a cross-section area $A$ and that the pressure exerted by the system at the piston face is $p$. The total force $F$ exerted by the system is $F = pA$. When the piston moves out an infinitesimal distance $dx$, the work $dW$ done by this force is

$$dW = F\,dx = pA\,dx.$$

But

$$A\,dx = dV,$$

where $dV$ is the infinitesimal change of volume of the system. Thus we can express the work done by the system as

$$dW = p\,dV, \tag{17–1}$$

and for a finite change of volume from $V_1$ to $V_2$,

$$W = \int_{V_1}^{V_2} p\,dV. \tag{17–2}$$

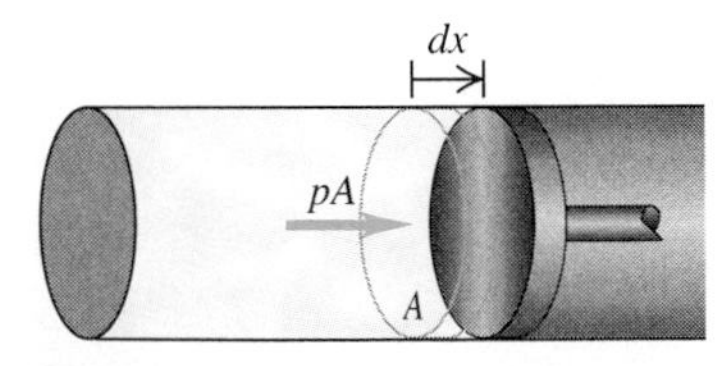


**17–4** The infinitesimal work done by the system during the small expansion $dx$ is $dW = pA\,dx$.

In general the pressure of the system may vary during the volume change, and to evaluate this integral, we have to know how the pressure varies as a function of volume. We can represent this relationship as a graph of $p$ as a function of $V$ (a $pV$-diagram, described at

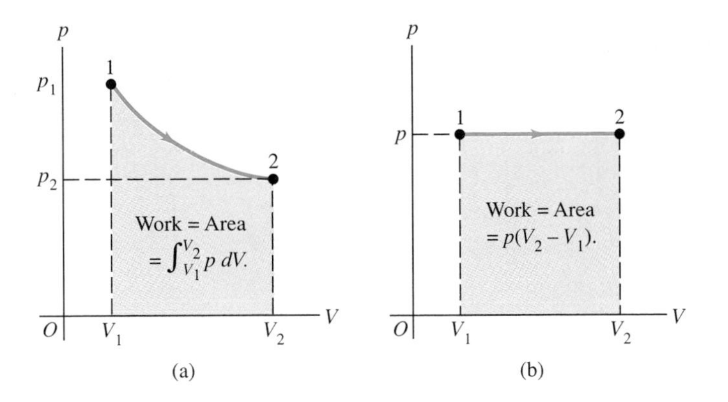


**17–5** The work done equals the area under the curve on a $pV$-diagram. (a) In a change from state 1 to state 2 the work and area are positive, but in a change from state 2 to state 1 the area is taken to be negative to agree with the sign of $W$ for decreasing volume. (b) The area of the rectangle gives the work done for a constant-pressure process.

the end of Section 16–1). Figure 17–5a shows a simple example; Eq. (17–2) is represented graphically as the *area* under the curve between the limits $V_1$ and $V_2$.

According to the rule we stated in Section 17–1, work is *positive* when a system *expands.* For an expansion from state 1 to state 2 in Fig. 17–5a the area and the work are positive. A *compression* from 2 to 1 gives a *negative* area; when a system is compressed, its volume decreases, and it does *negative* work on its surroundings.

If the pressure $p$ remains constant while the volume changes from $V_1$ to $V_2$ (Fig. 17–5b), the work done by the system is

$$W = p(V_2 - V_1) \quad \text{(constant pressure only).} \qquad (17\text{–}3)$$

In any process in which the volume is *constant,* the system does no work. In Eq. (17–1), $dV = 0$, and in Eq. (17–2), $V_1 = V_2$; in both cases, $W = 0$.

## EXAMPLE 17–1

**Isothermal expansion of an ideal gas** An ideal gas undergoes an *isothermal* (constant-temperature) *expansion* at temperature $T$, during which its volume changes from $V_1$ to $V_2$. How much work does the gas do?

**SOLUTION** From Eq. (17–2),

$$W = \int_{V_1}^{V_2} p\,dV.$$

For an ideal gas,

$$p = \frac{nRT}{V}.$$

We substitute this into the integral, take the constants $n$, $R$, and $T$ outside, and evaluate the integral:

$$W = nRT \int_{V_1}^{V_2} \frac{dV}{V} = nRT \ln \frac{V_2}{V_1}. \qquad (17\text{–}4)$$

In an expansion, $V_2 > V_1$, and $W$ is positive. Also, when $T$ is constant,

$$p_1V_1 = p_2V_2,$$

or

$$\frac{V_2}{V_1} = \frac{p_1}{p_2},$$

so the isothermal work may also be expressed as

$$W = nRT \ln \frac{p_1}{p_2}. \qquad (17\text{–}5)$$

In an isothermal expansion the pressure drops, so $p_1 > p_2$, and again the work is positive.

When a system changes from an initial state to a final state, it passes through a series of intermediate states; we call this series of states a **path.** There are always infinitely many different possibilities for these intermediate states. When they are all

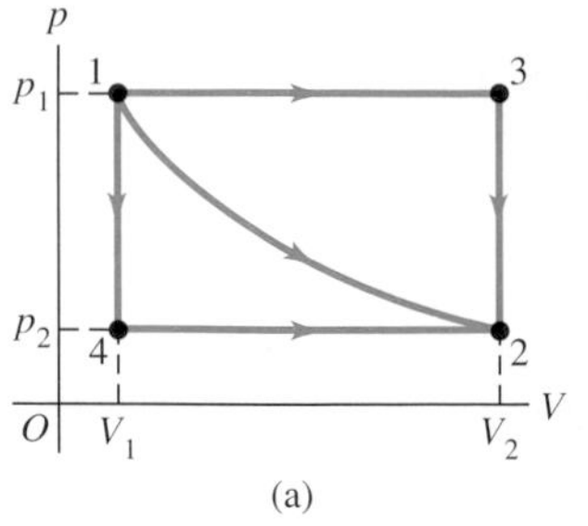


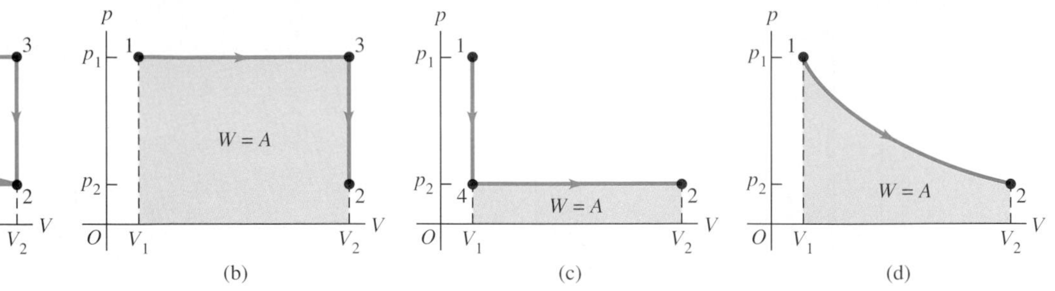


**17–6** (a) Three different paths between state 1 and state 2. (b)–(d) The work done by the system during a transition between two states depends on the path chosen.

equilibrium states, the path can be plotted on a $pV$-diagram (Fig. 17–6a). Point 1 represents an initial state with pressure $p_1$ and volume $V_1$, and point 2 represents a final state with pressure $p_2$ and volume $V_2$. For example, we could keep the pressure constant at $p_1$ while the system expands to volume $V_2$ (point 3 in Fig. 17–6b) and then reduce the pressure to $p_2$ (probably by decreasing the temperature) while keeping the volume constant at $V_2$ (to point 2 on the diagram). The work done by the system during this process is the area under the line $1 \rightarrow 3$; no work is done during the constant-volume process $3 \rightarrow 2$. Or the system might traverse the path $1 \rightarrow 4 \rightarrow 2$ (Fig. 17–6c); in that case the work is the area under the line $4 \rightarrow 2$. The smooth curve from 1 to 2 is another possibility (Fig. 17–6d), and the work is different from either of the other paths.

We conclude that the *work done by the system depends not only on the initial and final states, but also on the intermediate states, that is, on the path*. Furthermore, we can take the system through a series of states forming a closed loop, such as $1 \rightarrow 3 \rightarrow 2 \rightarrow 4 \rightarrow 1$. In this case the final state is the same as the initial state, but the total work done by the system is *not* zero. (In fact, it is represented on the graph by the area enclosed by the loop; can you prove that? See Exercise 17–6.) It follows that it doesn't make sense to talk about the amount of work *contained in* a system. In a particular state, a system may have definite values of the state coordinates $p$, $V$, and $T$, but it wouldn't make sense to say that it has a definite value of $W$.

## 17–3 HEAT TRANSFER DURING VOLUME CHANGES

We have learned that *heat* is energy transferred into or out of a system as a result of a temperature difference between the system and its surroundings. Heat transfer and work are two means by which the energy of a thermodynamic system may increase or decrease.

We have just seen in Section 17–2 that when a thermodynamic system undergoes a change of state, the *work* done by the system depends not only on the initial and final states but also on the series of intermediate states through which it passes, that is, on the *path* from the initial state to the final state. We will now show that this is also true of the *heat* added to the system.

Here is an example. Suppose we want to change the volume of a certain quantity of an ideal gas from 2.0 L to 5.0 L while keeping the temperature constant at $T = 300$ K. Figure 17–7 shows two different ways we can do this. In Fig. 17–7a the gas is contained in a cylinder with a piston and has initial volume of 2.0 L. We let the gas expand slowly, supplying heat from the electric heater to keep the temperature at 300 K. After expanding in this slow, controlled, isothermal manner the gas reaches its final volume of 5.0 L; it absorbs a definite amount of heat in the process.

Figure 17–7b shows a different process leading to the same final state. The container is surrounded by insulating walls and is divided by a thin, breakable partition into two compartments. The lower part has volume 2.0 L, and the upper part has volume

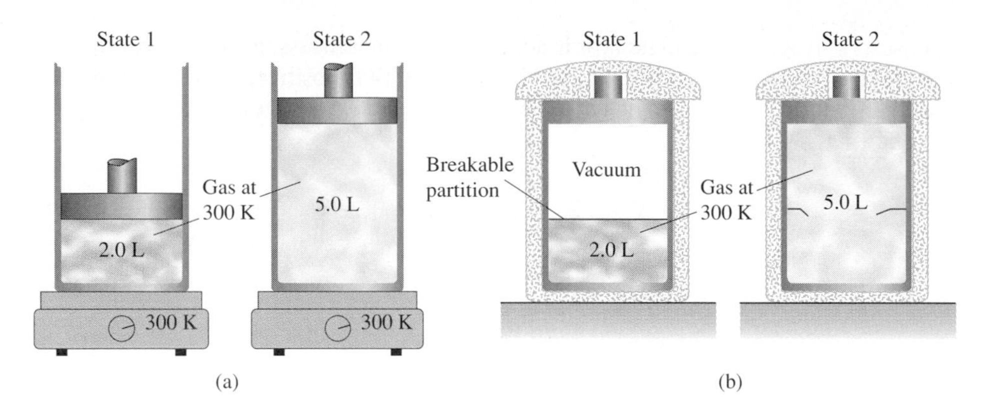


**17–7** (a) Slow, controlled isothermal expansion of a gas from an initial state 1 to a final state 2 with the same temperature but lower pressure. (b) Rapid, uncontrolled expansion of the same gas starting at the same state 1 and ending at the same state 2.

3.0 L. In the lower compartment we place the same amount of the same gas as in Fig. 17–7a, again at $T = 300$ K. The initial state is the same as before. Now we break the partition; the gas undergoes a rapid, uncontrolled expansion, with no heat passing through the insulating walls. The final volume is 5.0 L, the same as in Fig. 17–7a. The gas does no work during this expansion because it doesn't push against anything that moves. This uncontrolled expansion of a gas into vacuum is called a **free expansion;** we will discuss it further in Section 17–6.

Experiments have shown that when an ideal gas undergoes a free expansion, there is no temperature change. Therefore the final state of the gas in Fig. 17–7b is the same as in Fig. 17–7a. The intermediate states (pressures and volumes) during the transition from state 1 to state 2 are entirely different in the two cases; Figs. 17–7a and 17–7b represent *two different paths* connecting the *same states* 1 and 2. For path (b), *no* heat is transferred into the system, and it does no work. Like work, *heat depends not only on the initial and final states but also on the path*.

Because of this path dependence, it would not make sense to say that a system "contains" a certain quantity of heat. Consider what happens if we try to develop such an idea. Suppose we assign an arbitrary value to "the heat in a body" in some reference state. Then presumably the "heat in the body" in some other state would equal the heat in the reference state plus the heat added when the body goes to the second state. But that's ambiguous, as we have just seen; the heat added depends on the *path* we take from the reference state to the second state. We are forced to conclude that there is no consistent way to define "heat in a body"; it is not a useful concept. It *does* make sense, however, to speak of the amount of *internal energy* in a body; this important concept is our next topic.

## 17–4 INTERNAL ENERGY AND THE FIRST LAW

The concept of internal energy is one of the most important concepts in thermodynamics. We can look at it in various ways, some simple, some subtle. Let's start with the simple. Matter consists of atoms and molecules, and these are made up of particles having kinetic and potential energies. We tentatively define the **internal energy** of a system as the sum of all the kinetic and potential energies of its constituent particles. We use the symbol $U$ for internal energy. During a change of state of the system, the internal energy may change from an initial value $U_1$ to a final value $U_2$; we denote the change in internal energy as $\Delta U = U_2 - U_1$.

We also know that heat transfer is energy transfer. When we add a quantity of heat $Q$ to a system and it does no work during the process, the internal energy should increase by an amount equal to $Q$; that is, $\Delta U = Q$. When a system does work $W$ by expanding against its surroundings and no heat is added during the process, energy leaves the system and the internal energy decreases. That is, when $W$ is positive, $\Delta U$ is negative, and conversely; so $\Delta U = -W$. When *both* heat transfer and work occur, the *total* change in internal energy is

$$U_2 - U_1 = \Delta U = Q - W. \tag{17-6}$$

We can rearrange this to the form

$$Q = \Delta U + W. \tag{17-7}$$

*Translation:* When heat $Q$ is added to a system, some of this added energy remains within the system, increasing its internal energy by an amount $\Delta U$; the remainder leaves the system again as the system does work $W$ against its surroundings. Because $W$ and $Q$ may be either positive or negative, we also expect $\Delta U$ to be positive for some processes and negative for others.

Equation (17–6) or (17–7) is the **first law of thermodynamics.** It represents a generalization of the principle of conservation of energy to include energy transfer through heat as well as mechanical work. We studied conservation of energy in Chapter 7 in a purely *mechanical* context, and it seems reasonable that it should have this more general validity. Nevertheless, it is worth a closer look.

One problem is that defining internal energy in terms of microscopic kinetic and potential energies isn't entirely convincing. Actually *calculating* this total energy for any real system would be hopelessly complicated. We can't use this as an *operational* definition because it doesn't describe how to determine internal energy from physical quantities that we can measure directly. So we need to backtrack a little to show clearly the empirical and experimental basis of the first law of thermodynamics.

Starting over, we *define* the change in internal energy $\Delta U$ during any change of a system as the quantity given by Eq. (17–6). This *is* an operational definition because we can measure $Q$ and $W$. It does not define $U$ itself, only $\Delta U$. This is not a serious shortcoming; we can *define* the internal energy of a system to have a specified value in some reference state and then use Eq. (17–6) to define the internal energy in any other state. This is analogous to our treatment of potential energy in Chapter 7, in which we arbitrarily defined the potential energy of a mechanical system to be zero at a certain position.

This new definition trades one difficulty for another. If we define $\Delta U$ by Eq. (17–6), then when the system goes from state 1 to state 2 by two different paths, how do we know that $\Delta U$ is the same for the two paths? We have already seen that $Q$ and $W$ are, in general, *not* the same for different paths. If $\Delta U$, which equals $Q - W$, is also path-dependent, then $\Delta U$ is ambiguous. If so, the concept of internal energy of a system is subject to the same criticism as the erroneous concept of quantity of heat in a system, as we discussed at the end of Section 17–3.

The only way to answer this question is through *experiment.* We study the properties of various materials; in particular, we measure $Q$ and $W$ for various changes of state and various paths to learn whether $\Delta U$ is or is not path-dependent. The results of many such investigations are clear and unambiguous: *$\Delta U$ is independent of path*. The change in internal energy of a system during any thermodynamic process depends only on the initial and final states, *not* on the path leading from one to the other.

Experiment, then, is the ultimate justification for believing that a thermodynamic system in a specific state has a unique internal energy (relative to a given reference) that depends only on that state. An equivalent statement is that the internal energy $U$ of a system is a function of the state coordinates $p$, $V$, and $T$ (actually, any two of these, since the three variables are related by the equation of state).

Now let's return to the first law of thermodynamics. To say that the first law, given by Eq. (17–6) or (17–7), represents conservation of energy for thermodynamic processes is correct, as far as it goes. But an important *additional* aspect of the first law is the fact that internal energy depends only on the state of a system. In changes of state, the change in internal energy is path-independent.

All this may seem a little abstract if you are satisfied to think of internal energy as microscopic mechanical energy. There's nothing wrong with that view. But in the interest of precise *operational* definitions, internal energy, like heat, can and must be defined in a way that is independent of the detailed microscopic structure of the material.

If we take a system through a process that eventually returns it to its initial state (a *cyclic* process), the *total* internal energy change must be zero. Then

$$U_2 = U_1 \qquad \text{and} \qquad Q = W.$$

If a net quantity of work $W$ is done by the system during this process, an equal amount of energy must have flowed into the system as heat $Q$. But there is no reason why either $Q$ or $W$ individually has to be zero.

An *isolated* system is one that does no work on its surroundings and has no heat flow to or from its surroundings. For any process taking place in an isolated system,

$$W = Q = 0;$$

therefore,

$$U_2 - U_1 = \Delta U = 0.$$

In other words, *the internal energy of an isolated system is constant.*

## PROBLEM-SOLVING STRATEGY

### First law of thermodynamics

**1.** The internal energy change $\Delta U$ in any thermodynamic process or series of processes is independent of the path, no matter whether the substance is an ideal gas or not. This is of the utmost importance in the problems in this chapter and the next. Sometimes you will be given enough information about one path between given initial and final states to calculate $\Delta U$ for that path. Then you can use the fact that $\Delta U$ is the same for every other path between the same two states to relate the various energy quantities for other paths.

**2.** As usual, consistent units are essential. If $p$ is in Pa and $V$ in $m^3$, then $W$ is in joules. Otherwise, you may want to convert the pressure and volume units into Pa and $m^3$. If a heat capacity is given in terms of calories, the simplest procedure is usually to convert it to joules. Be especially careful with moles. When you use $n = m_{tot}/M$ to convert between total mass and number of moles, remember that if $m_{tot}$ is in kilograms, $M$ must be in *kilograms* per mole. The usual units for $M$ are *grams* per mole; be careful!

**3.** When a process consists of several distinct steps, it often helps to make a chart showing $Q$, $W$, and $\Delta U$ for each step. Put these quantities for each step on a different line, and arrange them so that the $Q$'s, $W$'s, and $\Delta U$'s form columns. Then you can apply the first law to each line; in addition, you can add each column and apply the first law to the sums. Do you see why?

## EXAMPLE 17–2

**Working off your dessert** You propose to eat a 900-calorie hot fudge sundae (with whipped cream) and then run up several flights of stairs to work off the energy that you have taken in. How high do you have to climb? Assume that your mass is 60 kg.

**SOLUTION** The system consists of you and the earth. Remember that one food-value calorie is 1 kcal = 1000 cal = 4190 J. The energy intake is

$$Q = 900\ \text{kcal}\ (4190\ \text{J}/1\ \text{kcal}) = 3.77 \times 10^6\ \text{J}.$$

The energy output is

$$W = mgh = (60\ \text{kg})(9.8\ \text{m/s}^2)h = (588\ \text{N})h.$$

If the final state of the system is the same as the initial state (that is, no fatter, no leaner), these two energy quantities must be equal, and we find

$$3.77 \times 10^6\ \text{J} = (588\ \text{N})h,$$

$$h = 6410\ \text{m} \quad (\text{or } 21{,}000\ \text{ft}).$$

Good luck! We have assumed 100% efficiency in the conversion of food energy into mechanical work; this isn't completely realistic. We'll talk more about efficiency later.

## EXAMPLE 17–3

Figure 17–8 shows a *cyclic* process (one in which the initial and final states are the same). It starts at point $a$ and proceeds counterclockwise to point $b$, then back to $a$, and the total work is $W = -500\ \text{J}$. a) Why is the work negative? b) Find the change in internal energy and the heat added during this process.

**SOLUTION** a) The work done equals the area under the curve, with the area taken as positive for increasing volume and negative for decreasing volume. From $a$ to $b$ the area is positive, but it is smaller than the absolute value of the negative area from $b$ back to $a$. Therefore the net area (the area enclosed by the path) and the work are negative. In other words, 500 more joules of work are done *on* the system than *by* the system.

b) For this and any other cyclic process (for which the beginning and end points are the same), $\Delta U = 0$, so $Q = W = -500\ \text{J}$. That is, 500 joules of heat must come *out of* the system.

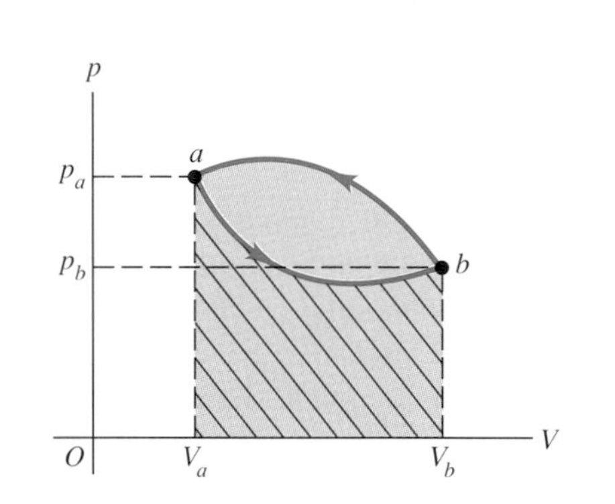


**17–8** The net work done by the system in the process *aba* is $-500$ J. What would it have been if the process had proceeded clockwise?

## EXAMPLE 17–4

A series of thermodynamic processes is shown in the $pV$-diagram of Fig. 17–9. In process $ab$, 150 J of heat are added to the system, and in process $bd$, 600 J of heat are added. Find a) the internal energy change in process $ab$; b) the internal energy change in process $abd$; c) the total heat added in process $acd$.

**SOLUTION** a) No volume change occurs during process $ab$, so $W_{ab} = 0$ and $\Delta U_{ab} = Q_{ab} = 150\ \text{J}$.

b) Process $bd$ occurs at constant pressure, so the work done by the system during this expansion is

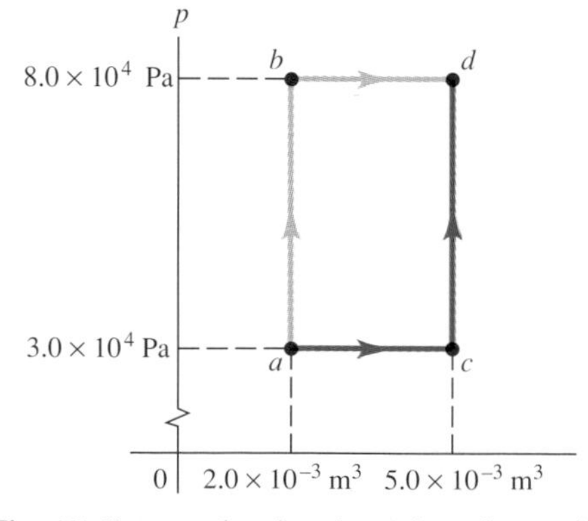


**17–9** The $pV$-diagram showing the various thermodynamic processes.

$$\begin{aligned} W_{bd} &= p(V_2 - V_1) \\ &= (8.0 \times 10^4\ \text{Pa})(5.0 \times 10^{-3}\ \text{m}^3 - 2.0 \times 10^{-3}\ \text{m}^3) \\ &= 240\ \text{J}. \end{aligned}$$

The total work for process $abd$ is

$$W_{abd} = W_{ab} + W_{bd} = 0 + 240\ \text{J} = 240\ \text{J},$$

and the total heat added is

$$Q_{abd} = Q_{ab} + Q_{bd} = 150\ \text{J} + 600\ \text{J} = 750\ \text{J}.$$

Applying Eq. (17–6) to $abd$, we find

$$\Delta U_{abd} = Q_{abd} - W_{abd} = 750\ \text{J} - 240\ \text{J} = 510\ \text{J}.$$

c) Because $\Delta U$ is independent of path, the internal energy change is the same for path $acd$ as for $abd$; that is,

$$\Delta U_{acd} = \Delta U_{abd} = 510\ \text{J}.$$

The total work for the path $acd$ is

$$\begin{aligned} W_{acd} &= W_{ac} + W_{cd} = p(V_2 - V_1) + 0 \\ &= (3.0 \times 10^4\ \text{Pa})(5.0 \times 10^{-3}\ \text{m}^3 - 2.0 \times 10^{-3}\ \text{m}^3) \\ &= 90\ \text{J}. \end{aligned}$$

Now we apply Eq. (17–7) to process $acd$:

$$Q_{acd} = \Delta U_{acd} + W_{acd} = 510\ \text{J} + 90\ \text{J} = 600\ \text{J}.$$

We see that although $\Delta U$ is the same (510 J) for $abd$ and $acd$, $W$ (240 J versus 90 J) and $Q$ (750 J versus 600 J) are quite different for the two processes.

Here is a tabulation of the various quantities:

| Step | $Q$ | $W$ | $\Delta U = Q - W$ |
|---|---|---|---|
| $ab$ | 150 J | 0 | 150 J |
| $bd$ | 600 J | 240 J | 360 J |
| $abd$ | 750 J | 240 J | 510 J |

| Step | $Q$ | $W$ | $\Delta U = Q - W$ |
|---|---|---|---|
| $ac$ | ? | 90 J | ? |
| $cd$ | ? | 0 | ? |
| $acd$ | 600 J | 90 J | 510 J |

## EXAMPLE 17–5

**Thermodynamics of boiling water** One gram of water (1 $cm^3$) becomes 1671 $cm^3$ of steam when boiled at a constant pressure of 1 atm ($1.013 \times 10^5$ Pa). The heat of vaporization at this pressure is $L_v = 2.256 \times 10^6$ J/kg = 2256 J/g. Compute a) the work done by the water when it vaporizes; b) its increase in internal energy.

**SOLUTION** a) For a constant-pressure process we may use Eq. (17–3) to compute the work done by the vaporizing water:

$$\begin{aligned} W &= p(V_2 - V_1) \\ &= (1.013 \times 10^5 \text{ Pa})(1671 \times 10^{-6} \text{ m}^3 - 1 \times 10^{-6} \text{ m}^3) \\ &= 169 \text{ J}. \end{aligned}$$

b) The heat added to the water is the heat of vaporization:

$$\begin{aligned} Q &= mL_v \\ &= (1 \text{ g})(2256 \text{ J/g}) \\ &= 2256 \text{ J}. \end{aligned}$$

From the first law of thermodynamics, Eq. (17–6), the change in internal energy is

$$\begin{aligned} \Delta U &= Q - W \\ &= 2256 \text{ J} - 169 \text{ J} \\ &= 2087 \text{ J}. \end{aligned}$$

To vaporize one gram of water, we have to add 2256 J of heat. Most (2087 J) of this added energy remains in the system as an increase in internal energy. The remaining 169 J leaves the system again as it does work against the surroundings while expanding from liquid to vapor. The increase in internal energy is associated mostly with the intermolecular forces that hold the molecules together in the liquid state. Because these forces are attractive, the associated potential energies are greater after work has been done on the molecules to pull them apart in forming the vapor state. It's like increasing the potential energy by pulling an elevator farther from the center of the earth.

In the preceding examples the initial and final states differ by a finite amount. Later we will consider *infinitesimal* changes of state in which a small amount of heat $dQ$ is added to the system, the system does a small amount of work $dW$, and its internal energy changes by an amount $dU$. For such a process we state the first law in differential form as

$$dU = dQ - dW. \qquad (17\text{–}8)$$

For the systems that we'll discuss, the work $dW$ is given by $dW = p\,dV$, so we can also state the first law as

$$dU = dQ - p\,dV. \qquad (17\text{–}9)$$

# 17–5 THERMODYNAMIC PROCESSES

Here are four specific kinds of thermodynamic processes that occur often enough in practical situations to be worth some discussion. For some of these we can use a simplified form of the first law of thermodynamics. The processes can be summarized briefly as "no heat transfer," "constant volume," "constant pressure," and "constant temperature," respectively.

## Adiabatic Process

An **adiabatic process** is defined as one with no heat transfer into or out of a system; $Q = 0$ or $dQ = 0$. We can prevent heat flow either by surrounding the system with thermally insulating material or by carrying out the process so quickly that there is not enough time for appreciable heat flow. From the first law we find that, for every adiabatic process,

$$U_2 - U_1 = \Delta U = -W \quad \text{(adiabatic process).} \tag{17–10}$$

When a system expands under adiabatic conditions, $W$ is positive, $\Delta U$ is negative, and the internal energy decreases. When a system is *compressed* adiabatically, $W$ is negative and $U$ increases. An increase of internal energy is often, though not always, accompanied by a rise in temperature.

The compression stroke in an internal-combustion engine is approximately an adiabatic process. The temperature rises as the air-fuel mixture in the cylinder is compressed. The expansion of the burned fuel during the power stroke is approximately an adiabatic expansion with a drop in temperature.

## Isochoric Process

An **isochoric process** is a *constant-volume* process. When the volume of a thermodynamic system is constant, it does no work on its surroundings. Then $W = 0$, and

$$U_2 - U_1 = \Delta U = Q \quad \text{(isochoric process).} \tag{17–11}$$

In an isochoric process, all the energy added as heat remains in the system as an increase in internal energy. Heating a gas in a closed, constant-volume container is an example of an isochoric process. (There are exceptional cases; we can do work on a fluid without changing its volume by stirring it. For such cases, "isochoric" is used in some literature to mean "zero-work," irrespective of volume changes.)

## Isobaric Process

An **isobaric process** is a *constant-pressure* process. In general, for an isobaric process, *none* of the three quantities $\Delta U$, $Q$, and $W$ in the first law is zero, but calculating $W$ is easy. From Eq. (17–3),

$$W = p(V_2 - V_1) \quad \text{(isobaric process).} \tag{17–12}$$

Example 17–5 concerned an example of an isobaric process; Example 17–4 includes both isobaric and isochoric processes.

## Isothermal Process

An **isothermal process** is a *constant-temperature* process. For a process to be isothermal, any heat flow into or out of the system must occur slowly enough that thermal equilibrium is maintained. In general, *none* of the quantities $\Delta U$, $Q$, and $W$ is zero.

In some special cases the internal energy of a system depends *only* on its temperature, not on its pressure or volume. The most familiar system having this special property is an ideal gas; we'll discuss this point in the next section. For such systems, if the temperature is constant, the internal energy is also constant; $\Delta U = 0$ and $Q = W$. That is, any energy entering the system as heat $Q$ must leave it again as work $W$ done by the system. Example 17–1, involving an ideal gas, is an example of an isothermal process in which $U$ is also constant. For most systems other than ideal gases the internal energy depends on pressure as well as temperature, so $U$ may vary even when $T$ is constant.

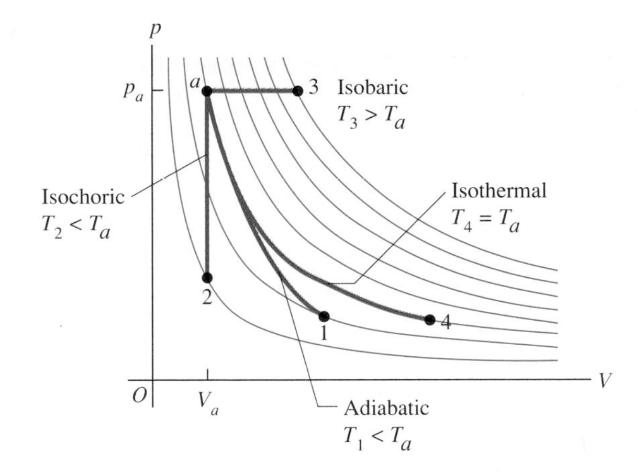


**17–10** Four different processes for a constant amount of an ideal gas, all starting at state $a$. For the adiabatic process, $Q = 0$; for the isochoric process, $W = 0$; and for the isothermal process, $\Delta U = 0$. The temperature increases only during the isobaric expansion.

Figure 17–10 shows a $pV$-diagram for each of these four processes for a constant amount of an ideal gas.

## 17–6 INTERNAL ENERGY OF AN IDEAL GAS

We stated in Section 17–5 that for an ideal gas the internal energy $U$ depends only on temperature, not on pressure or volume. How do we know this? Let's think again about the free-expansion experiment described in Section 17–3. A thermally insulated container with rigid walls is divided into two compartments by a partition (Fig. 17–11). One compartment contains a quantity of an ideal gas, and the other is evacuated.

When the partition is removed or broken, the gas expands to fill both parts of the container. This process is called a *free expansion*. The gas does no work on its surroundings because the walls of the container don't move. Both $Q$ and $W$ are zero, so the internal energy $U$ is constant. This is true of any substance, whether it is an ideal gas or not.

Does the *temperature* change during a free expansion? Suppose it *does* change, while the internal energy stays the same. In that case we have to conclude that the internal energy depends on both the temperature and the volume, or both the temperature and the pressure, but certainly not the temperature alone. But if $T$ is constant during a free expansion, for which we know that $U$ is constant even though both $p$ and $V$ change, then we have to conclude that $U$ depends only on $T$, not on $p$ or $V$.

Many experiments have shown that when an ideal gas undergoes a free expansion, its temperature *does not* change. The conclusion is: **The internal energy of an ideal gas depends only on its temperature, not on its pressure or volume.** This property, in addition to the ideal-gas equation of state, is part of the ideal-gas model. In the following sections we will use several times the fact that for an ideal gas, $U$ depends only on $T$, so make sure you understand it.

For nonideal gases, some temperature change occurs during free expansions, even though the internal energy is constant. This shows that the internal energy cannot depend *only* on temperature; it must depend on pressure as well. From a microscopic viewpoint this is not surprising. Nonideal gases usually have attractive intermolecular forces. When molecules move farther apart, the associated potential energies increase. If the total internal energy is constant, the kinetic energies must decrease. Temperature is directly related to molecular *kinetic* energy, and for such a gas a free expansion is usually accompanied by a *drop* in temperature.

Insulation
Vacuum
Breakable partition
Gas at temperature $T$

**17–11** The partition is broken (or removed) to start the free expansion of gas into the vacuum region.

## 17–7 HEAT CAPACITIES OF AN IDEAL GAS

We defined specific heat capacity and molar heat capacity in Section 15–5. We also remarked at the end of that section that the specific or molar heat capacity of a substance depends on the conditions under which the heat is added. It is usually easiest to measure the heat capacity of a gas in a closed container under constant-volume conditions. The corresponding heat capacity is the **molar heat capacity at constant volume,** denoted by $C_V$. Heat capacity measurements for solids and liquids are usually carried out in the atmosphere under constant atmospheric pressure, and we call the resulting heat capacity the **molar heat capacity at constant pressure,** $C_p$. If neither $p$ nor $V$ is constant, we have an infinite number of possible heat capacities.

Let's consider $C_V$ and $C_p$ for gases. To measure $C_V$, we raise the temperature of a gas in a rigid container with constant volume (neglecting its thermal expansion). To measure $C_p$, we let the gas expand just enough to keep the pressure constant as the temperature rises.

Why should these two molar heat capacities be different? The answer lies in the first law of thermodynamics. In a constant-volume temperature change, the system does no work, and the change in internal energy $\Delta U$ equals the heat added $Q$. In a constant-pressure temperature change, on the other hand, the volume *must* increase; otherwise, the pressure could not remain constant. As the material expands, it does an amount of work $W$. According to the first law,

$$Q = \Delta U + W. \tag{17–13}$$

For a given temperature change, the heat input for a constant-pressure process must be *greater* than that for a constant-volume process because additional energy must be supplied to account for the work $W$ done during the expansion, so $C_p$ is greater than $C_V$. The $pV$-diagram in Fig. 17–12 shows this relationship. For air, $C_p$ is 40% greater than $C_V$. If the volume were to *decrease* during heating, $W$ would be negative, the heat input would be *less* than that in the constant-volume case, and $C_p$ would be *less than* $C_V$. That is an exceptional case; water between 0°C and 4°C is one of the few examples.

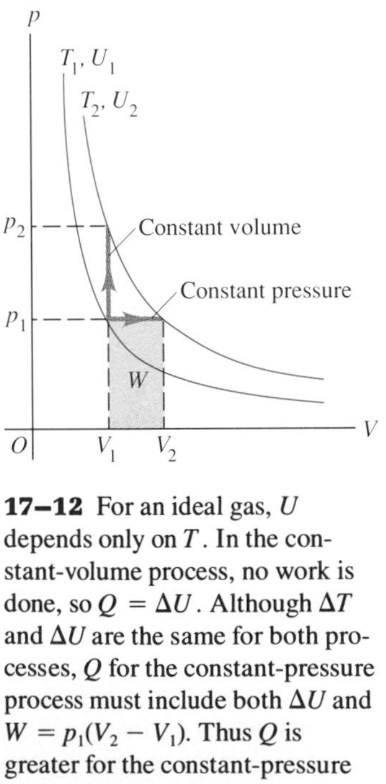

17–12 For an ideal gas, $U$ depends only on $T$. In the constant-volume process, no work is done, so $Q = \Delta U$. Although $\Delta T$ and $\Delta U$ are the same for both processes, $Q$ for the constant-pressure process must include both $\Delta U$ and $W = p_1(V_2 - V_1)$. Thus $Q$ is greater for the constant-pressure process than for the constant-volume process, and $C_p > C_V$.

We can derive a simple relation between $C_V$ and $C_p$ for an ideal gas. First consider the *constant-volume* process. We place $n$ moles of an ideal gas at temperature $T$ in a constant-volume container. We place it in thermal contact with a hotter body; an infinitesimal quantity of heat $dQ$ flows into the gas, and its temperature increases by an infinitesimal amount $dT$. By definition of $C_V$, the molar heat capacity at constant volume,

$$dQ = nC_V\,dT. \tag{17–14}$$

The pressure increases during this process, but the gas does no *work* ($dW = 0$) because the volume is constant. The first law in differential form, Eq. (17–9), is $dQ = dU + p\,dV$. When $dW = 0$, $dQ = dU$, so Eq. (17–14) can also be written as

$$dU = nC_V\,dT. \tag{17–15}$$

Now consider a *constant-pressure* process. We place the same gas in a cylinder with a piston that we can allow to move just enough to maintain constant pressure. Again we bring the system into contact with a hotter body. As heat flows into the gas, it expands at constant pressure and does work. By definition of $C_p$, the molar heat capacity at constant pressure, the amount of heat $dQ$ entering the gas is

$$dQ = nC_p\,dT. \tag{17–16}$$

The work $dW$ done by the gas in this constant-pressure process is

$$dW = p\,dV.$$

We can also express $dW$ in terms of the temperature change $dT$ by using the ideal-gas equation of state $pV = nRT$. Because $p$ is constant, the change in $V$ is proportional to the change in $T$:

$$p\,dV = nR\,dT$$

and

$$dW = nR\,dT. \tag{17–17}$$

Now we substitute Eqs. (17–16) and (17–17) into the first law, $dQ = dU + dW$. We obtain

$$nC_p\,dT = dU + nR\,dT. \tag{17–18}$$

Now, here comes the crux of the calculation. The internal energy change $dU$ is again given by Eq. (17–15), $dU = nC_V\,dT$, *even though now the volume is not constant.* Why is this so? Recall the discussion of Section 17–6; one of the special properties of an ideal gas is that its internal energy depends *only* on temperature. Thus the *change* in internal energy during any process must be determined only by the temperature change. If Eq. (17–15) is valid for an ideal gas for one particular kind of process, it must be valid for an ideal gas for *every* kind of process with the same $dT$.

To complete our derivation, we replace $dU$ in Eq. (17–18) by $nC_V\,dT$:

$$nC_p\,dT = nC_V\,dT + nR\,dT.$$

When we divide each term by the common factor $n\,dT$, we get

$$C_p = C_V + R. \tag{17–19}$$

As we predicted, the molar heat capacity of an ideal gas at constant pressure is *greater* than the molar heat capacity at constant volume; the difference is the gas constant $R$. (Of course, $R$ must be expressed in the same units as $C_p$ and $C_V$, such as J/mol · K.)

We have used the ideal-gas model to derive Eq. (17–19), but it turns out to be obeyed quite well (within a few percent) for many real gases at moderate pressures. Measured values of $C_p$ and $C_V$ are given in Table 17–1 for several real gases at low pressures; the difference in most cases is approximately 8.31 J/mol · K.

The table also shows that the molar heat capacity of a gas is related to its molecular structure, as we discussed in Section 16–4. In fact, the first two columns of Table 17–1 are the same as those in Table 16–1.

**TABLE 17–1 Molar Heat Capacities of Gases at Low Pressure**

| Type of gas | Gas | $C_V$ (J/mol · K) | $C_p$ (J/mol · K) | $C_p - C_V$ (J/mol · K) | $\gamma = C_p/C_V$ |
|---|---|---|---|---|---|
| Monatomic | He | 12.47 | 20.78 | 8.31 | 1.67 |
| | A | 12.47 | 20.78 | 8.31 | 1.67 |
| Diatomic | $H_2$ | 20.42 | 28.74 | 8.32 | 1.41 |
| | $N_2$ | 20.76 | 29.07 | 8.31 | 1.40 |
| | $O_2$ | 20.85 | 29.17 | 8.31 | 1.40 |
| | CO | 20.85 | 29.16 | 8.31 | 1.40 |
| Polyatomic | $CO_2$ | 28.46 | 36.94 | 8.48 | 1.30 |
| | $SO_2$ | 31.39 | 40.37 | 8.98 | 1.29 |
| | $H_2S$ | 25.95 | 34.60 | 8.65 | 1.33 |

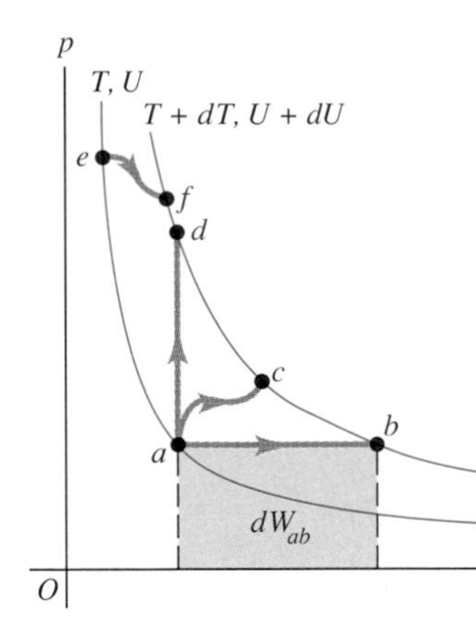


**17–13** The change in internal energy of an ideal gas is the same for all processes between two given temperatures.

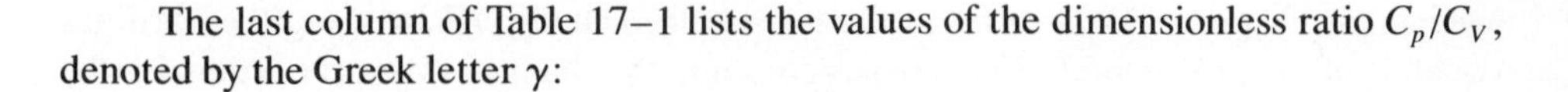

The last column of Table 17–1 lists the values of the dimensionless ratio $C_p/C_V$, denoted by the Greek letter $\gamma$:

$$\gamma = \frac{C_p}{C_V}. \tag{17–20}$$

Because $C_p$ is always greater than $C_V$ for gases, $\gamma$ is always greater than unity. This quantity plays an important role in *adiabatic* processes for an ideal gas, which we will study in the next section.

We can use our kinetic-theory discussion of heat capacity of an ideal gas (Section 16–4) to predict values of $\gamma$ from kinetic theory. An ideal monatomic gas has $C_V = \frac{3}{2}R$. From Eq. (17–19),

$$C_p = C_V + R = \tfrac{3}{2}R + R = \tfrac{5}{2}R,$$

so

$$\gamma = \frac{C_p}{C_V} = \frac{\frac{3}{2}R + R}{\frac{3}{2}R} = \frac{5}{3} = 1.67.$$

As Table 17–1 shows, this agrees well with values of $\gamma$ computed from measured heat capacities. For a diatomic gas, $C_V = \frac{5}{2}R$, $C_p = \frac{7}{2}R$, and

$$\gamma = \frac{C_p}{C_V} = \frac{\frac{5}{2}R + R}{\frac{5}{2}R} = \frac{7}{5} = 1.40,$$

also in good agreement with measured values.

Here's a graphical approach to molar heat capacities of an ideal gas. Figure 17–13 is a $pV$-diagram showing two isotherms for an ideal gas, one for temperature $T$ and the other for $T + dT$. The internal energy $U$ of an ideal gas depends only on the temperature $T$, so the gas has the same value $U$ at every point on the isotherm for temperature $T$ and the same value $U + dU$ at every point on the isotherm for $T + dT$. The *change* in internal energy, $dU$, is the same in *every* process that takes the gas from *any* point on one isotherm to *any* point on the other. In Fig. 17–13, $dU$ is the same for processes $ab$, $ac$, $ad$, and $ef$. In particular, $ad$ represents a constant-volume process from temperature $T$ to $T + dT$, and $ab$ is a constant-pressure process through the same temperature interval. No work is done in process $ad$; the work in process $ab$ is the area under the curve.

Here's a final reminder: For an ideal gas the internal energy change for *any* infinitesimal process is given by $dU = nC_V\,dT$, *whether the volume is constant or not.* This relation, which comes in handy in the following example, holds for other substances *only* when the volume is constant.

## EXAMPLE 17–6

**Cooling your room** A typical dorm room contains about 2500 mol of air. Find the change in the internal energy of this much air when it is cooled from 23.9°C to 11.6°C at a constant pressure of 1.00 atm. Treat the air as an ideal gas with $\gamma = 1.400$.

**SOLUTION** This is a constant-pressure process. Your first impulse may be to find $C_p$, then calculate $Q$ from $Q = nC_p\,\Delta T$; find the volume change and find the work done by the gas from $W = p\,\Delta V$; then finally use the first law to find $U$. This would be perfectly correct, but there's a much easier way. For an ideal gas the internal-energy change is $\Delta U = nC_V\,\Delta T$ for *every* process, *whether the volume is constant or not.* So all we have to do is find $C_V$ and use this expression for $\Delta U$. From Eq. (17–20),

$$\gamma = \frac{C_p}{C_V} = \frac{C_V + R}{C_V} = 1 + \frac{R}{C_V},$$

$$C_V = \frac{R}{\gamma - 1} = \frac{8.315\ \text{J/mol}\cdot\text{K}}{1.400 - 1} = 20.79\ \text{J/mol}\cdot\text{K}.$$

Then

$$\begin{aligned} U &= nC_V\,\Delta T \\ &= (2500\ \text{mol})(20.79\ \text{J/mol}\cdot\text{K})(11.6 - 23.9)\ \text{K} \\ &= -6.39 \times 10^5\ \text{J}. \end{aligned}$$

# 17–8 ADIABATIC PROCESSES FOR AN IDEAL GAS

An adiabatic process, defined in Section 17–5, is a process in which no heat transfer takes place between a system and its surroundings. In an infinitesimal adiabatic process, $dQ = 0$ and (from the first law) $dU = -dW$. Zero heat transfer is an idealization, but a process is approximately adiabatic if the system is well insulated or if the process takes place so quickly that there is not enough time for appreciable heat flow to occur.

An adiabatic process for an ideal gas is shown on the $pV$-diagram of Fig. 17–14. As the gas expands from volume $V_a$ to $V_b$, its temperature drops because it does positive work, decreasing its internal energy. If point $a$, representing the initial state, lies on an isotherm at temperature $T + dT$, then point $b$ for the final state is on a different isotherm at a lower temperature $T$. An adiabatic curve at any point is always *steeper* than the isotherm passing through the same point. For an adiabatic *compression* from $V_b$ to $V_a$ the situation is reversed, and the temperature rises.

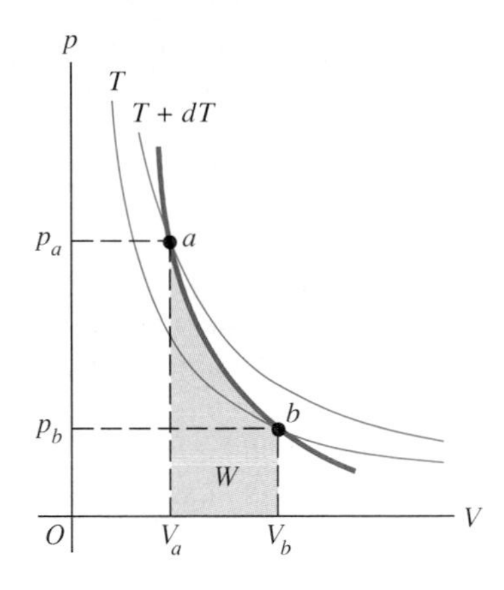


**17–14** A $pV$-diagram of an adiabatic process for an ideal gas. As the gas expands from $V_a$ to $V_b$, its temperature drops from $T + dT$ to $T$, corresponding to the decrease in internal energy due to the work $W$ done by the gas (indicated by shaded area). For an ideal gas, when an isotherm and an adiabat pass through the same point, the adiabat is always steeper.

The air in the output pipes of air compressors used in gasoline stations and paint-spraying equipment and to fill scuba tanks is always warmer than the air entering the compressor because of the approximately adiabatic compression. When air is compressed in the cylinders of a diesel engine during the compression stroke, it gets so hot that fuel injected into the cylinders ignites spontaneously.

We can derive a relation between volume and temperature changes for an infinitesimal adiabatic process. Equation (17–15) gives the internal energy change $dU$ for *any* process for an ideal gas, adiabatic or not, so we have $dU = nC_V\,dT$. Also, the work done by the gas during the process is given by $dW = p\,dV$. Then, since $dU = -dW$ for an adiabatic process, we have

$$nC_V\,dT = -p\,dV. \qquad (17\text{–}21)$$

To obtain a relation containing only the volume $V$ and temperature $T$, we eliminate $p$ using the ideal-gas equation in the form $p = nRT/V$. Substituting this into Eq. (17–21) and rearranging, we get

$$nC_V\,dT = -\frac{nRT}{V}\,dV,$$

$$\frac{dT}{T} + \frac{R}{C_V}\frac{dV}{V} = 0.$$

The coefficient $R/C_V$ can be expressed in terms of $\gamma$. We have

$$\frac{R}{C_V} = \frac{C_p - C_V}{C_V} = \frac{C_p}{C_V} - 1 = \gamma - 1,$$

$$\frac{dT}{T} + (\gamma - 1)\frac{dV}{V} = 0. \qquad (17\text{–}22)$$

Because $\gamma$ is always greater than unity for a gas, $(\gamma - 1)$ is always positive. This means that in Eq. (17–22), $dV$ and $dT$ always have opposite signs. An adiabatic *expansion* of an ideal gas always occurs with a *drop* in temperature, and an adiabatic *compression* occurs with a *rise* in temperature; this confirms our earlier prediction.

For finite changes in temperature and volume we integrate Eq. (17–22), obtaining

$$\ln T + (\gamma - 1)\ln V = \text{constant},$$

$$\ln T + \ln V^{\gamma - 1} = \text{constant},$$

$$\ln (TV^{\gamma - 1}) = \text{constant},$$

and, finally,

$$TV^{\gamma - 1} = \text{constant}. \qquad (17\text{–}23)$$

Thus for an initial state $(T_1, V_1)$ and a final state $(T_2, V_2)$,

$$T_1V_1^{\gamma-1} = T_2V_2^{\gamma-1}. \tag{17–24}$$

Because we have used the ideal-gas equation in our derivation of Eqs. (17–23) and (17–24), the $T$'s must always be *absolute* (Kelvin) temperatures.

We can also convert Eq. (17–23) into a relation between pressure and volume by eliminating $T$, using the ideal-gas equation in the form $T = pV/nR$. Substituting this into Eq. (17–23), we find

$$\frac{pV}{nR}V^{\gamma-1} = \text{constant},$$

or, because $n$ and $R$ are constant,

$$pV^{\gamma} = \text{constant}. \tag{17–25}$$

For an initial state $(p_1, V_1)$ and a final state $(p_2, V_2)$,

$$p_1V_1^{\gamma} = p_2V_2^{\gamma}. \tag{17–26}$$

## EXAMPLE 17–7

**Adiabatic compression in a diesel engine** The compression ratio of a diesel engine is 15 to 1; this means that air in the cylinders is compressed to $\frac{1}{15}$ of its initial volume (Fig. 17–15). If the initial pressure is $1.01 \times 10^5$ Pa and the initial temperature is 27°C (300 K), find the final pressure and the temperature after compression. Air is mostly a mixture of diatomic oxygen and nitrogen; treat it as an ideal gas with $\gamma = 1.40$.

**SOLUTION** We have $p_1 = 1.01 \times 10^5$ Pa, $T_1 = 300$ K, and $V_1/V_2 = 15$. From Eq. (17–24),

$$T_2 = T_1\left(\frac{V_1}{V_2}\right)^{\gamma-1} = (300\text{ K})(15)^{0.40} = 886\text{ K} = 613°\text{C}.$$

From Eq. (17–26),

$$p_2 = p_1\left(\frac{V_1}{V_2}\right)^{\gamma} = (1.01 \times 10^5\text{ Pa})(15)^{1.40}$$
$$= 44.8 \times 10^5\text{ Pa} = 44\text{ atm}.$$

If the compression had been isothermal, the final pressure would have been 15 atm, but because the temperature also increases during an adiabatic compression, the final pressure is much greater. The high temperature attained during compression causes the fuel to ignite spontaneously, without the need for spark plugs, when it is injected into the cylinders near the end of the compression stroke.

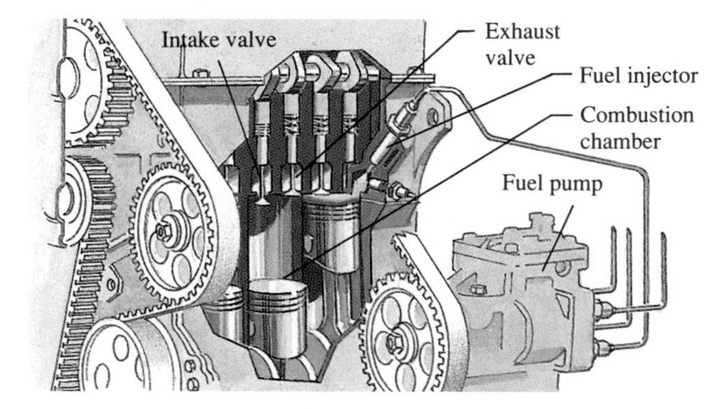


**17–15** The high compression ratio in the cylinders of a diesel engine causes high enough temperatures to ignite the fuel without spark plugs.

We can calculate the *work* done by an ideal gas during an adiabatic process. We know that $Q = 0$ and $W = -\Delta U$ for *any* adiabatic process. For an ideal gas, $\Delta U = nC_V(T_2 - T_1)$. If the number of moles $n$ and the initial and final temperatures $T_1$ and $T_2$ are known, we have simply

$$W = nC_V(T_1 - T_2). \tag{17–27}$$

We may also use $pV = nRT$ in this equation to obtain

$$W = \frac{C_V}{R}(p_1V_1 - p_2V_2) = \frac{1}{\gamma - 1}(p_1V_1 - p_2V_2). \quad (17\text{–}28)$$

Note that if the process is an expansion, the temperature drops, $T_1$ is greater than $T_2$, $p_1V_1$ is greater than $p_2V_2$, and the work is *positive*, as we should expect. If the process is a compression, the work is negative.

## EXAMPLE 17–8

In Example 17–7, how much work does the gas do during the compression if the initial volume of the cylinder is 1.00 L = $1.00 \times 10^{-3}\,\text{m}^3$? Assume that $C_V$ for air is 20.8 J/mol · K and $\gamma = 1.40$.

**SOLUTION** We may determine the number of moles and then use Eq. (17–27), or we may use Eq. (17–28). With the first method we have

$$n = \frac{p_1V_1}{RT_1} = \frac{(1.01 \times 10^5\,\text{Pa})(1.00 \times 10^{-3}\,\text{m}^3)}{(8.315\,\text{J/mol}\cdot\text{K})(300\,\text{K})} = 0.0405\,\text{mol},$$

and Eq. (17–27) gives

$$\begin{aligned} W &= nC_V(T_1 - T_2) \\ &= (0.0405\,\text{mol})(20.8\,\text{J/mol}\cdot\text{K})(300\,\text{K} - 886\,\text{K}) \\ &= -494\,\text{J}. \end{aligned}$$

With the second method,

$$\begin{aligned} W &= \frac{1}{\gamma - 1}(p_1V_1 - p_2V_2) \\ &= \frac{1}{1.40 - 1}[(1.01 \times 10^5\,\text{Pa})(1.00 \times 10^{-3}\,\text{m}^3) \\ &\quad - (44.8 \times 10^5\,\text{Pa})(\tfrac{1}{15} \times 10^{-3}\,\text{m}^3)] \\ &= -494\,\text{J}. \end{aligned}$$

The work is negative because the gas is compressed.

---

Throughout this analysis we have used the ideal-gas equation of state, which is valid only for *equilibrium* states. Strictly speaking, our results are valid only for a process that is fast enough to prevent appreciable heat exchange with the surroundings yet slow enough that the system does not depart very much from thermal and mechanical equilibrium. Even when these conditions are not strictly satisfied, though, these equations give useful approximate results.

## SUMMARY

- A thermodynamic system can exchange energy with its surroundings by heat transfer or mechanical work and in some cases by other mechanisms. When a system at pressure $p$ expands from volume $V_1$ to $V_2$, it does an amount of work $W$ given by

$$W = \int_{V_1}^{V_2} p\,dV. \quad (17\text{–}2)$$

If the pressure $p$ is constant during the expansion,

$$W = p(V_2 - V_1) \quad \text{(constant pressure only).} \quad (17\text{–}3)$$

- In any thermodynamic process the heat added to the system and the work done by the system depend not only on the initial and final states, but also on the path (the series of intermediate states through which the system passes).
- The first law of thermodynamics states that when heat $Q$ is added to a system while it does work $W$, the internal energy $U$ changes by an amount

$$U_2 - U_1 = \Delta U = Q - W. \quad (17\text{–}6)$$

## KEY TERMS

thermodynamic system

path

free expansion

internal energy

first law of thermodynamics

adiabatic process

isochoric process

isobaric process

isothermal process

molar heat capacity at constant volume

molar heat capacity at constant pressure

In an infinitesimal process,

$$dU = dQ - dW = dQ - p\,dV. \tag{17-8}$$

The internal energy of any thermodynamic system depends only on its state. The change in internal energy in any process depends only on the initial and final states, not on the path. The internal energy of an isolated system is constant.

- Adiabatic process: No heat transfer in or out of a system; $Q = 0$.
- Isochoric process: Constant volume; $W = 0$.
- Isobaric process: Constant pressure; $W = p(V_2 - V_1)$.
- Isothermal process: Constant temperature.
- The internal energy of an ideal gas depends only on its temperature, not its pressure or volume. For other substances the internal energy generally depends on both pressure and temperature.
- The molar heat capacities $C_V$ and $C_p$ of an ideal gas are related by

$$C_p = C_V + R. \tag{17-19}$$

The ratio of $C_p$ to $C_V$ is denoted by $\gamma$:

$$\gamma = \frac{C_p}{C_V}. \tag{17-20}$$

- For an adiabatic process for an ideal gas the quantities $TV^{\gamma-1}$ and $pV^{\gamma}$ are constant. For an initial state $(p_1, V_1, T_1)$ and a final state $(p_2, V_2, T_2)$,

$$T_1V_1^{\gamma-1} = T_2V_2^{\gamma-1}, \tag{17-24}$$

$$p_1V_1^{\gamma} = p_2V_2^{\gamma}. \tag{17-26}$$

The work done during an adiabatic expansion of an ideal gas is

$$W = \frac{C_V}{R}(p_1V_1 - p_2V_2) = \frac{1}{\gamma - 1}(p_1V_1 - p_2V_2). \tag{17-28}$$

# EXERCISES

## Section 17–2 Work Done during Volume Changes

**17–1** Two moles of oxygen are in a container with rigid walls. The gas is heated until the pressure doubles. Neglecting the thermal expansion of the container, calculate the work done by the gas.

**17–2** A gas under a constant pressure of $4.00 \times 10^5$ Pa and with an initial volume of $0.0500 \text{ m}^3$ is cooled until its volume becomes $0.0400 \text{ m}^3$. Calculate the work done by the gas.

**17–3** Three moles of an ideal gas are heated at constant pressure from $T = 27°\text{C}$ to $177°\text{C}$. Calculate the work done by the gas.

**17–4** Two moles of an ideal gas have an initial temperature of 27.0°C. While the temperature is kept constant, the volume is decreased until the pressure doubles. Calculate the work done by the gas.

**17–5** A system is taken from state $a$ to state $b$ along the three paths shown in Fig. 17–16. a) Along which path is the work done by the system the greatest? The least? b) If $U_b > U_a$, along which path is the magnitude of the heat transfer $Q$ the greatest? For this path, is heat absorbed or liberated by the system?

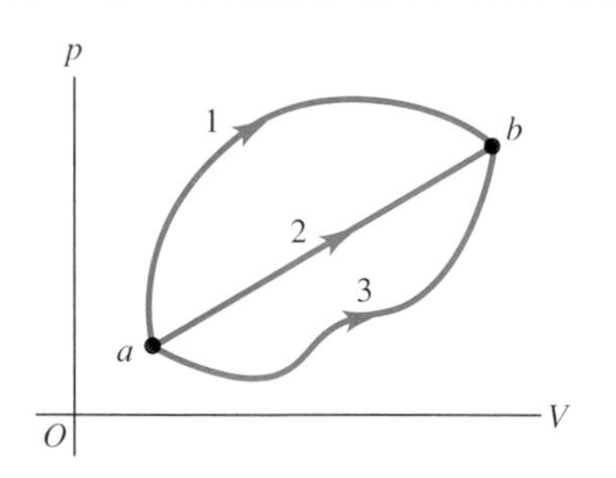


**FIGURE 17–16**

**17–6 Work and *pV*-Diagrams.** a) In Fig. 17–6a, consider the closed loop $1 \rightarrow 3 \rightarrow 2 \rightarrow 4 \rightarrow 1$. Prove that the total work done by the system is equal to the area enclosed by the loop.

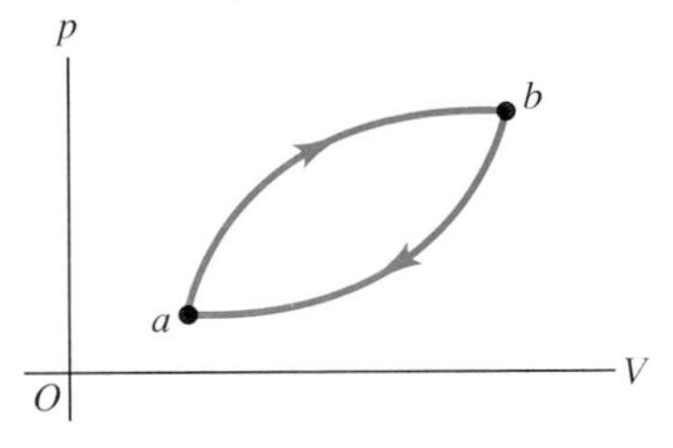


**FIGURE 17–17**

b) How is the work done for the process in part (a) related to the work done if the loop is traversed in the opposite direction, $1 \rightarrow 4 \rightarrow 2 \rightarrow 3 \rightarrow 1$?

**17–7** A system is taken around the cycle shown in Fig. 17–17, from state $a$ to state $b$ and then back to state $a$. The magnitude of the heat transfer during one cycle is 400 J. a) Does the system absorb or liberate heat when it goes around the cycle in the direction shown in the figure? b) What is the work $W$ done by the system in one cycle? c) If the system goes around the cycle in a counterclockwise direction, does it absorb or liberate heat in one cycle? What is the magnitude of the heat exchanged in one cycle?

## Section 17–4 Internal Energy and the First Law

**17–8** A liquid is irregularly stirred in a well-insulated container and thereby undergoes a rise in temperature. Regard the liquid as a system. a) Has heat been transferred? b) Has work been done? c) What is the sign of $\Delta U$?

**17–9** A student performs a combustion experiment by burning a mixture of fuel and oxygen in a constant-volume metal can surrounded by a water bath. During the experiment the temperature of the water is observed to rise. Regard the mixture of fuel and oxygen as the system. a) Has heat been transferred? b) Has work been done? c) What is the sign of $\Delta U$?

**17–10 Energy from Butter.** The nominal food-energy value of butter is about 6.0 kcal/g. If all this energy could be converted completely to mechanical energy, how much butter would be required to power an 80-kg mountaineer on her journey from Lupine Meadows (elevation 2070 m) to the summit of Grand Teton (4196 m)?

**17–11** In running a certain chemical process a lab technician supplies 140 J of heat to a system, and at the same time 100 J of work are done on the system by its surroundings. What is the increase in the internal energy of the system?

**17–12** A gas in a cylinder expands from a volume of 0.400 $m^3$ to 0.600 $m^3$. Heat is added just rapidly enough to keep the pressure constant at $1.50 \times 10^5$ Pa during the expansion. The total heat added is $1.20 \times 10^5$ J. a) Find the work done by the gas. b) Find the change in internal energy of the gas. c) Does it matter whether or not the gas is ideal?

**17–13** A gas in a cylinder is cooled and compressed at a constant pressure of $2.00 \times 10^5$ Pa, from 1.20 $m^3$ to 0.80 $m^3$. A quantity of heat of magnitude $2.80 \times 10^5$ J is removed from the gas. a) Find the work done by the gas. b) Find the change in internal energy of the gas. c) Does it matter whether or not the gas is ideal?

**17–14 Boiling Water at High Pressure.** When water is boiled under a pressure of 2.00 atm, the heat of vaporization is $2.20 \times 10^6$ J/kg and the boiling point is 120°C. At this pressure, 1.00 kg of water has a volume of $1.00 \times 10^{-3}$ $m^3$, and 1.00 kg of steam has a volume of 0.824 $m^3$. a) Compute the work done when 1.00 kg of steam is formed at this temperature. b) Compute the increase in internal energy of the water.

## Section 17–7 Heat Capacities of an Ideal Gas

**17–15** Consider the isothermal compression of 0.100 mol of an ideal gas at $T = 27.0°C$. The initial pressure is 1.00 atm, and the final volume is one-eighth the initial volume. a) Determine the work required. b) What is the change in internal energy? c) Does the gas exchange heat with its surroundings? If so, how much? Does the gas absorb or liberate heat?

**17–16** During an isothermal compression of an ideal gas, 135 J of heat must be removed from the gas to maintain constant temperature. How much work is done by the gas during the process?

**17–17** A certain ideal gas has $\gamma = 1.333$. Determine the molar heat capacity at constant volume and the molar heat capacity at constant pressure.

**17–18** A cylinder contains 1.00 mol of oxygen gas at a temperature of 27.0°C. The cylinder is provided with a frictionless piston, which maintains a constant pressure of 1.00 atm on the gas. The gas is heated until its temperature increases to 177.0°C. Assume that the oxygen can be treated as an ideal gas. a) Draw a $pV$-diagram representing the process. b) How much work is done by the gas in this process? c) On what is this work done? d) What is the change in internal energy of the gas? e) How much heat was supplied to the gas? f) How much work would have been done if the pressure had been 0.50 atm?

## Section 17–8 Adiabatic Processes for an Ideal Gas

**17–19** A gasoline engine takes in air at 20.0°C and 1.00 atm and compresses it adiabatically to one-third the original volume. Find the final temperature and pressure.

**17–20** An ideal gas initially at 5.00 atm and 400 K is permitted to expand adiabatically until its volume doubles. Find the final pressure and temperature if the gas is a) monatomic; b) diatomic.

**17–21** During an adiabatic expansion the temperature of 0.600 mol of oxygen drops from 30.0°C to 10.0°C. a) How much work does the gas do? b) How much heat is added to the gas?

**17–22** A monatomic ideal gas initially at a pressure of $4.00 \times 10^5$ Pa and with a volume of 0.0800 $m^3$ is compressed adiabatically to a volume of 0.0300 $m^3$. a) What is the final pressure? b) How much work is done by the gas?

## PROBLEMS

**17–23** When a system is taken from state $a$ to state $b$ in Fig. 17–18, along the path $acb$, 90.0 J of heat flow into the system and 70.0 J of work are done by the system. a) How much heat flows into the system along path $adb$ if the work done by the system is 10.0 J? b) When the system is returned from $b$ to $a$ along the curved path, the magnitude of the work done by the system is 45.0 J. Does the system absorb or liberate heat? How much heat? c) If $U_a = 0$ and $U_d = 6.0$ J, find the heat absorbed in the processes $ad$ and $db$.

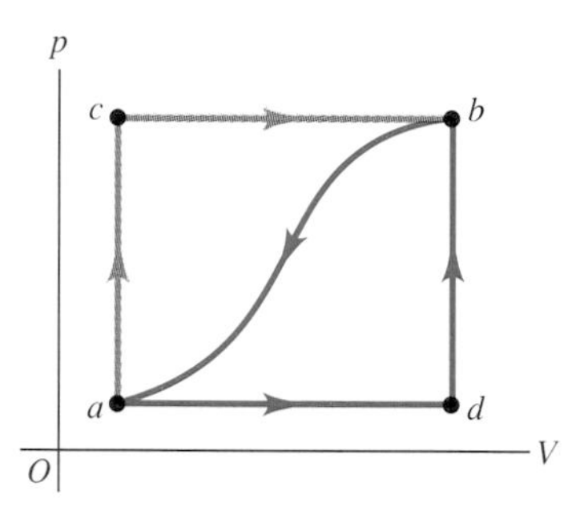


**FIGURE 17–18**

**17–24** A thermodynamic system is taken from state $a$ to state $c$ in Fig. 17–19, along either path $abc$ or path $adc$. Along path $abc$ the work $W$ done by the system is 500 J. Along path $adc$, $W$ is 200 J. The internal energies of each of the four states shown in the figure are $U_a = 100$ J, $U_b = 500$ J, $U_c = 800$ J, and $U_d = 600$ J. Calculate the heat flow $Q$ for each of the four processes $ab$, $bc$, $ad$, and $dc$. In each process, does the system absorb or liberate heat?

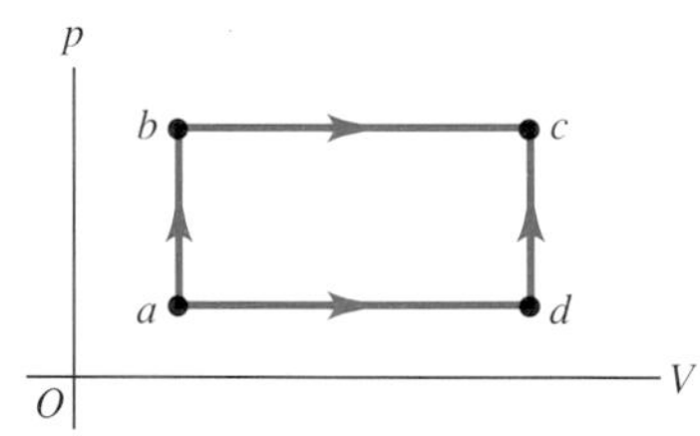


**FIGURE 17–19**

**17–25** In a certain process, $2.65 \times 10^5$ J of heat is supplied to a system, and at the same time the system expands against a constant external pressure of $6.90 \times 10^5$ Pa. The internal energy of the system is the same at the beginning and end of the process. Find the increase in volume of the system. (The system is *not* an ideal gas.)

**17–26** Nitrogen gas in an expandable container is raised from 0.0°C to 50.0°C, with the pressure held constant at $4.00 \times 10^5$ Pa. The total heat added is $6.0 \times 10^4$ J. a) Find the number of moles of gas. b) Find the change in internal energy of the gas. c) Find the work done by the gas. d) How much heat would be needed to cause the same temperature change if the volume were constant?

**17–27** A chemical engineer studying the properties of glycerin uses a steel cylinder having a cross-section area of 0.0200 m$^2$ and containing $1.50 \times 10^{-2}$ m$^3$ of glycerin. The cylinder is equipped with a tightly fitting piston that supports a load of $3.00 \times 10^4$ N. The temperature of the system is increased from 20.0°C to 70.0°C. The coefficient of volume expansion of glycerin is given in Table 15–2. Neglect the expansion of the steel cylinder. Find a) the increase in volume of the glycerin; b) the mechanical work of the $3.00 \times 10^4$ N force; c) the amount of heat added to the glycerin ($c_p$ for glycerin is $2.43 \times 10^3$ J/kg · K); d) the change in internal energy of the glycerin.

**17–28 A Compressed-Air Engine.** You are designing an engine that runs on compressed air. Air enters the engine at a pressure of $2.00 \times 10^6$ Pa and leaves at a pressure of $3.00 \times 10^5$ Pa. What must the temperature of the compressed air be for there to be no possibility of frost forming in the exhaust ports of the engine? Assume that the expansion is adiabatic. (*Note:* Frost frequently forms in the exhaust ports of an air-driven engine. This happens when the moist air is cooled below 0°C by the expansion that takes place in the engine.)

**17–29** A pump compressing air from atmospheric pressure ($1.01 \times 10^5$ Pa) into a very large tank at a gauge pressure of $4.40 \times 10^5$ Pa has a cylinder 0.220 m long. ($C_V$ for air is 20.8 J/mol · K.) a) At what position in the stroke will air begin to enter the tank? Assume that the compression is adiabatic. (You are being asked to calculate the distance the piston has moved in the cylinder.) b) If the air is taken into the pump at 27.0°C, what is the temperature of the compressed air? c) How much work does the pump do in putting 30.0 mol of air into the tank?

**17–30** Initially at a temperature of 80.0°C, 0.28 m$^3$ of air expands at a constant gauge pressure of $1.38 \times 10^5$ Pa to a volume of 1.42 m$^3$ and then expands further adiabatically to a final volume of 2.27 m$^3$ and a final gauge pressure of $2.29 \times 10^4$ Pa. Draw a $pV$-diagram for this sequence of processes and compute the total work done by the air. (Take atmospheric pressure to be $1.01 \times 10^5$ Pa. $C_V$ for air is 20.8 J/mol · K.)

**17–31** An ideal gas expands slowly to twice its original volume, doing 600 J of work in the process. Find the heat added to the gas and the change in internal energy of the gas if the process is a) isothermal; b) adiabatic.

**17–32 Comparing Thermodynamic Processes.** In a cylinder, 3.00 mol of an ideal monatomic gas initially at $1.00 \times 10^6$ Pa and 300 K expands until its volume doubles. Compute the work done by the gas if the expansion is a) isothermal; b) adiabatic; c) isobaric. d) Show each process on a $pV$-diagram. In which case is the magnitude of the work done by the gas greatest? Least? e) In which case is the magnitude of the heat transfer greatest? Least? f) In which case is the magnitude of the change in internal energy of the gas greatest? Least?

**17–33** Two moles of helium initially at a temperature of 27.0°C occupy a volume of 0.0400 m$^3$. The helium first expands at constant pressure until the volume has doubled. Then it expands adiabatically until the temperature returns to its initial value. Assume that the helium can be treated as an ideal gas. a) Draw a diagram of the process in the $pV$-plane. b) What is the total heat supplied to the helium in the process? c) What is the total change in the internal energy of the helium? d) What is the total work done by the helium? e) What is the final volume?

**17–34** A cylinder with a piston contains 0.500 mol of oxygen at $5.00 \times 10^5$ Pa and 300 K. The oxygen may be treated as an ideal gas. The gas first expands at constant pressure to twice its original volume. It is then compressed isothermally back to its original volume, and finally it is cooled at constant volume to its original pressure. a) Show the series of processes on a $pV$-diagram. b) Compute the temperature during the isothermal compression. c) Compute the maximum pressure.

**17–35** Use the conditions and processes of Problem 17–34 to compute a) the work done by the gas, the heat added to it, and its internal-energy change during the initial expansion; b) the work done, the heat added, and the internal-energy change during the final cooling; c) the internal-energy change during the isothermal compression.

**17–36** A quantity of air is taken from state $a$ to state $b$ along a path that is a straight line in the $pV$-diagram (Fig. 17–20). If $V_a = 0.0400\ \text{m}^3$, $V_b = 0.0900\ \text{m}^3$, $p_a = 1.00 \times 10^5$ Pa, and $p_b = 1.60 \times 10^5$ Pa, what is the work $W$ done by the gas in this process?

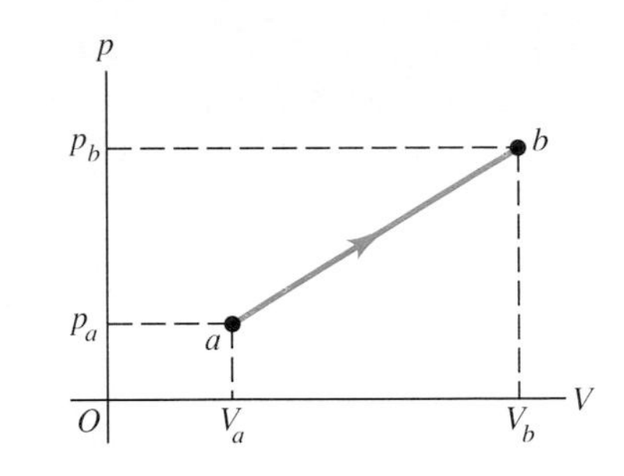


**FIGURE 17–20**

## CHALLENGE PROBLEM

**17–37** The van der Waals equation, an approximate representation of the behavior of gases at high pressure, is (Eq. 16–8),

$$\left(p + \frac{an^2}{V^2}\right)(V - nb) = nRT,$$

where $a$ and $b$ are constants having different values for different gases. (In the special case when $a = b = 0$ this is the ideal-gas equation.) Calculate the work done by a gas with this equation of state in an isothermal expansion from $V_1$ to $V_2$. Show that your answer agrees with the ideal-gas result, Eq. (17–4), when you set $a = b = 0$.

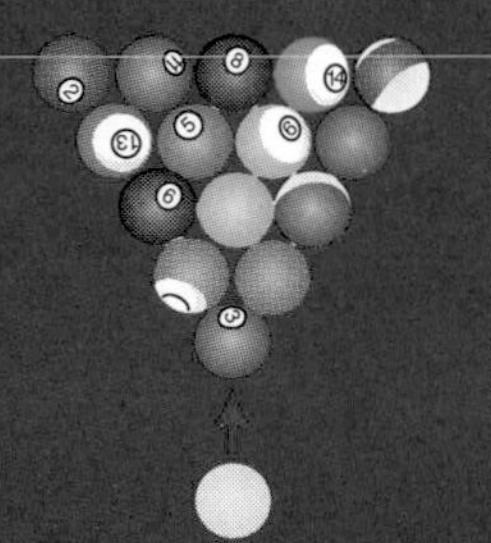

In the starting position of a game of billiards, the 15 billiard balls are arranged in one particular stationary pattern.

CHAPTER 18

# Second Law of Thermodynamics

After the first play, the positions of the balls are more disordered, and the entropy of the system has increased.

The thermodynamic property called entropy measures the disorder or randomness in any thermodynamic system.

## KEY CONCEPTS

- **A heat engine converts heat partly into mechanical work; the fraction converted is called the thermal efficiency. Gasoline and diesel engines are examples of heat engines.**
- **A refrigerator takes heat from a cold place to a hotter place. The heat removed from the cold place divided by the work needed is called the performance coefficient.**
- **The second law of thermodynamics states that it is impossible to make a cyclic device that converts heat completely into work or a cyclic device that transfers heat from a cooler to a hotter body without requiring work input.**
- **The Carnot engine is an idealized engine that uses only reversible processes and has the greatest possible thermal efficiency for given input and output temperatures. It uses two isothermal and two adiabatic processes.**
- **Entropy is a measure of the disorder of a system in a given state. Increases in disorder are accompanied by increases in entropy. The total entropy of the universe can never decrease.**
- **The Carnot cycle can be used to define a temperature scale that does not depend on the physical properties of any specific substance.**

## INTRODUCTION

When you put 0.1 kg of boiling water and 0.1 kg of ice in an insulated cup, you end up with 0.2 kg of water at about 10°C. That's not surprising. But you'd be very surprised if you came back later and found that the water had turned back to 0.1 kg of ice and 0.1 kg of boiling water. That wouldn't violate the first law of thermodynamics; energy would be conserved. But it doesn't happen in nature. Why not? Why does a power plant convert less than half of the heat from burning coal into electrical energy, discarding the remainder of the heat? Why does heat always flow from hotter places to cooler places, never the reverse? When you drop ink into water, it mixes spontaneously, coloring the water, but it never spontaneously unmixes. Why not? What do all these things have in common?

The first law of thermodynamics expresses conservation of energy in thermodynamic processes. But there is a whole category of questions that the first law cannot answer, having to do with the *directions* of thermodynamic processes. A study of inherently one-way processes, such as the flow of heat from hotter to colder regions and the irreversible conversion of work into heat by friction, leads to the *second law of thermodynamics.* This law places fundamental limitations on the efficiency of an engine or a power plant; it also places limitations on the minimum energy input needed to operate a refrigerator. So the second law is directly relevant for many important practical problems. We can also state the second law in terms of the concept of *entropy,* a quantitative measure of the degree of disorder or randomness of a system.

## 18–1 DIRECTIONS OF THERMODYNAMIC PROCESSES

Many thermodynamic processes proceed naturally in one direction but not the opposite. For example, heat always flows from a hot body to a cooler body, never the reverse. Heat flow from a cool body to a hot body would not violate the first law, but it doesn't happen in nature. Or suppose all the air in a box could rush to one side, leaving vacuum in the other side, the reverse of the free expansion described in Section 17–3. This does not occur in nature either, although the first law does not forbid it. It is easy to convert mechanical energy completely into heat; we do this every time we use a car's brakes to stop it. It is not so easy to convert heat into mechanical energy. Many would-be inventors have proposed cooling some air, to extract heat from it, and converting that heat to mechanical energy to propel a car or an airplane. None has ever succeeded; no one has ever built a machine that converts heat *completely* into mechanical energy.

What these examples have in common is a preferred *direction*. In each case a process proceeds spontaneously in one direction but not in the other. The law that determines the preferred direction for a given process is the *second law of thermodynamics,* our main topic in this chapter.

Despite there being a preferred direction for every natural process, we can think of a class of idealized processes that are *reversible*. We say that a system undergoes a **reversible process** if the system is always very close to being in thermodynamic equilibrium, within itself and with its surroundings. When this is the case, any change of state that takes place can be reversed (that is, made to go the other way) by making only an infinitesimal change in the conditions of the system. For example, heat flow between two bodies whose temperatures differ only infinitesimally can be reversed by making only a very small change in one temperature or the other. A gas expanding slowly and adiabatically can be compressed slowly and adiabatically by an infinitesimal increase in pressure.

Reversible processes are thus **equilibrium processes.** In contrast, heat flow with finite temperature difference, free expansion of a gas, and conversion of work to heat by friction are all **irreversible processes;** no small change in conditions could make any of them go the other way. They are also all *nonequilibrium* processes.

You may notice an apparent contradiction in this discussion. If a system is *really* in thermodynamic equilibrium, how can any change of state at all take place? How can heat flow into or out of a system if the temperature is uniform throughout? How can it start to move to expand and do work against its surroundings if it is in mechanical equilibrium?

The answer to all these questions is that a reversible process is an idealization that can never be precisely attained in the real world. But by making the temperature gradients and the pressure differences in the substance very small, we can keep the system very close to equilibrium states. We use the term *quasi-equilibrium process* to emphasize the idealized nature of a reversible process.

Finally, we will find that there is a relation between the direction of a process and the *disorder,* or *randomness,* of the resulting state. For example, imagine a tedious sorting job, such as alphabetizing a thousand book titles written on file cards. Throw the alphabetized stack of cards into the air. Do they come down in alphabetical order? No, their tendency is to come down in a random or disordered state. In the free expansion example the air is more disordered after it has expanded into the entire box than when it was confined in one side, because the molecules are scattered over more space.

Similarly, macroscopic kinetic energy is energy associated with organized, coordinated motions of many molecules, but heat transfer involves changes in energy of random, disordered molecular motion. Therefore conversion of mechanical energy into heat involves an increase of randomness, or disorder.

In the following sections we will introduce the second law of thermodynamics by considering two broad classes of devices, *heat engines,* which are partly successful in converting heat into work, and *refrigerators,* which are partly successful in transporting heat from cooler to hotter bodies.

## 18–2 HEAT ENGINES

The essence of a technological society is the ability to use sources of energy other than muscle power. Sometimes mechanical energy is directly available; water power is an example. But most of our energy comes from the burning of fossil fuels (coal, oil, and gas) and from nuclear reactions. These supply energy that is transferred as *heat.* This is directly useful for heating buildings, for cooking, and for chemical and metallurgical processing, but to operate a machine or propel a vehicle, we need *mechanical* energy.

Thus it's important to know how to take heat from a source and convert as much of it as possible into mechanical energy or work. This is what happens in gasoline engines in automobiles, jet engines in airplanes, steam turbines in electric power plants, and many other systems. Closely related processes occur in the animal kingdom when food energy is "burned" (that is, carbohydrates combine with oxygen to yield water, carbon dioxide, and energy) and partly converted to mechanical energy as the animal's muscles do work on its surroundings.

Any device that transforms heat partly into work or mechanical energy is called a **heat engine.** Usually a quantity of matter inside the engine undergoes addition and subtraction of heat, expansion and compression, and sometimes a phase change. We call this matter the **working substance** of the engine. In internal-combustion engines the working substance is a mixture of air and fuel; in a steam turbine it is water.

The simplest kind of engine to analyze is one in which the working substance undergoes a **cyclic process,** a sequence of processes that eventually leaves the substance in the same state as that in which it started. In a steam turbine the water is recycled and used over and over. Internal-combustion engines do not use the same air over and over, but we can still analyze them in terms of cyclic processes that approximate their actual operation.

All the heat engines mentioned absorb heat from a source at a relatively high temperature, perform some mechanical work, and discard some heat at a lower temperature. As far as the engine is concerned, the discarded heat is wasted. In internal-combustion engines the waste heat is discarded in the hot exhaust gases and the cooling system; in a steam turbine it is the heat that must be taken out of the used steam in order to condense and recycle the water.

When a system is carried through a cyclic process, its initial and final internal energies are equal. For any cyclic process the first law of thermodynamics requires that

$$U_2 - U_1 = 0 = Q - W \qquad \text{and} \qquad Q = W.$$

That is, the net heat flowing into the engine in a cyclic process equals the net work done by the engine.

When we analyze heat engines, it helps to think of two bodies with which the working substance of the engine can interact. One of these, called the *hot reservoir,* can give the working substance large amounts of heat at a constant temperature $T_H$ without appreciably changing its own temperature. The other body, called the *cold reservoir,* can

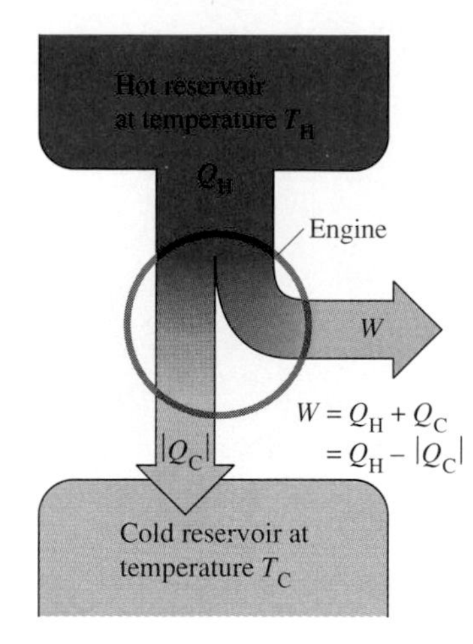


**18–1** Schematic energy-flow diagram for a heat engine.

absorb large amounts of discarded heat from the engine at a constant lower temperature $T_C$. In a steam-turbine system the flames and hot gases in the boiler are the hot reservoir, and the cold water and air used to condense and cool the used steam are the cold reservoir.

We denote the quantities of heat transferred from the hot and cold reservoirs as $Q_H$ and $Q_C$, respectively. A quantity of heat $Q$ is positive when heat is transferred *from* a reservoir *into* the working substance, negative when heat leaves the working substance. Thus in a heat engine, $Q_H$ is positive but $Q_C$ is negative, representing heat *leaving* the working substance. This sign convention is consistent with the rules that we stated in Chapter 17; we will continue to use those rules here. Sometimes it clarifies the relationships to state them in terms of the absolute values of the $Q$'s and $W$'s because absolute values are always positive. When we do this, our notation will show it explicitly.

We can represent the energy transformations in a heat engine by the *energy-flow diagram* of Fig. 18–1. The engine itself is represented by the circle. The amount of heat $Q_H$ supplied to the engine by the hot reservoir is proportional to the width of the incoming "pipeline" at the top of the diagram. The width of the outgoing pipeline at the bottom is proportional to the magnitude $|Q_C|$ of the heat discarded in the exhaust. The branch line to the right represents that portion of the heat supplied that the engine converts to mechanical work, $W$.

When an engine repeats the same cycle over and over, $Q_H$ and $Q_C$ represent the quantities of heat absorbed and rejected by the engine *during one cycle;* $Q_H$ is positive, $Q_C$ is negative. The *net* heat $Q$ absorbed per cycle is

$$Q = Q_H + Q_C = |Q_H| - |Q_C|. \tag{18–1}$$

The useful output of the engine is the net work $W$ done by the working substance; from the first law,

$$W = Q = Q_H + Q_C = |Q_H| - |Q_C|. \tag{18–2}$$

Ideally, we would like to convert *all* the heat $Q_H$ into work; in that case we would have $Q_H = W$ and $Q_C = 0$. Experience shows that this is impossible; there is always some heat wasted, and $Q_C$ is never zero. We define the **thermal efficiency** of an engine, denoted by $e$, as the quotient

$$e = \frac{W}{Q_H}. \tag{18–3}$$

The thermal efficiency $e$ represents the fraction of $Q_H$ that *is* converted to work. To put it another way, $e$ is what you get divided by what you pay for. This is always less than unity, an all-too-familiar experience! In terms of the flow diagram of Fig. 18–1 the most efficient engine is the one for which the branch pipeline representing the work output is as wide as possible and the exhaust pipeline representing the heat thrown away is as narrow as possible.

When we substitute the two expressions for $W$ given by Eq. (18–1) into Eq. (18–3), we get the following equivalent expressions for $e$:

$$e = \frac{W}{Q_H} = 1 + \frac{Q_C}{Q_H} = 1 - \frac{|Q_C|}{|Q_H|}. \tag{18–4}$$

Note that $e$ is a quotient of two energy quantities and thus is a pure number, without units. Of course, we must always express $W$, $Q_H$, and $Q_C$ in the same units.

## PROBLEM-SOLVING STRATEGY

### Heat engines

We suggest that you reread the strategy in Section 17–4; those suggestions are equally useful throughout the present chapter. The following points may need additional emphasis.

**1.** Be very careful with the sign conventions for $W$ and the various $Q$'s. $W$ is positive when the system expands and does work, negative when it is compressed. Each $Q$ is positive if it represents heat entering the working substance of the engine or other system, negative for heat leaving the system. When in doubt, use the first law if possible, to check consistency. When you know that a quantity is negative, such as $Q_C$ in the above discussion, it sometimes helps to write it as $Q_C = -|Q_C|$.

**2.** Some problems deal with power rather than energy quantities. Power is work per unit time ($P = W/t$), and rate of heat transfer (heat current) $H$ is heat transfer per unit time ($H = Q/t$). Sometimes it helps to ask, "What is $W$ or $Q$ in one second (or one hour)?"

### EXAMPLE 18–1

A gasoline engine in a large truck takes in 2500 J of heat and delivers 500 J of mechanical work per cycle. The heat is obtained by burning gasoline with heat of combustion $L_c = 5.0 \times 10^4$ J/g. a) What is the thermal efficiency of this engine? b) How much heat is discarded in each cycle? c) How much gasoline is burned in each cycle? d) If the engine goes through 100 cycles per second, what is its power output in watts? In horsepower? e) How much gasoline is burned per second? Per hour?

**SOLUTION** We have $Q_H = 2500$ J and $W = 500$ J.

a) From Eq. (18–3) the thermal efficiency is

$$e = \frac{W}{Q_H} = \frac{500\text{ J}}{2500\text{ J}} = 0.20 = 20\%.$$

This is a fairly typical figure for cars and trucks if $W$ includes only the work actually delivered to the wheels.

b) From Eq. (18–2),

$$W = Q_H + Q_C$$

$$500\text{ J} = 2500\text{ J} + Q_C$$

$$Q_C = -2000\text{ J}.$$

That is, 2000 J of heat leave the engine during each cycle.

c) Let $m$ be the mass of gasoline burned during each cycle; then

$$Q = mL_c$$

$$2500\text{ J} = m(5.0 \times 10^4\text{ J/g})$$

$$m = 0.050\text{ g}.$$

d) The power $P$ (rate of doing work) is the work per cycle multiplied by the number of cycles per second:

$$P = (500\text{ J/cycle})(100\text{ cycles/s}) = 50{,}000\text{ W} = 50\text{ kW},$$

$$P = (50{,}000\text{ W})\frac{1\text{ hp}}{746\text{ W}} = 67\text{ hp}.$$

e) The mass of gasoline burned per second is the mass per cycle multiplied by the number of cycles per second:

$$(0.050\text{ g/cycle})(100\text{ cycles/s}) = 5.0\text{ g/s}.$$

The mass burned per hour is

$$(5.0\text{ g/s})\frac{3600\text{ s}}{1\text{ h}} = 18{,}000\text{ g/h} = 18\text{ kg/h}.$$

The density of gasoline is about 0.70 g/cm$^3$, so this is about 25,700 cm$^3$, 25.7 L, or 6.8 gallons of gasoline per hour. If the truck is traveling at 55 mi/h, this represents fuel consumption of 8.1 miles/gallon.

## 18–3 INTERNAL-COMBUSTION ENGINES

The gasoline engine, used in automobiles and many other types of machinery, is a familiar example of a heat engine. Let's look at its thermal efficiency. Figure 18–2 (page 510) shows the sequence of processes. First a mixture of air and gasoline vapor flows into a cylinder through an open intake valve while the piston descends, increasing the volume of the cylinder from a minimum of $V$ (when the piston is all the way up) to a maximum of

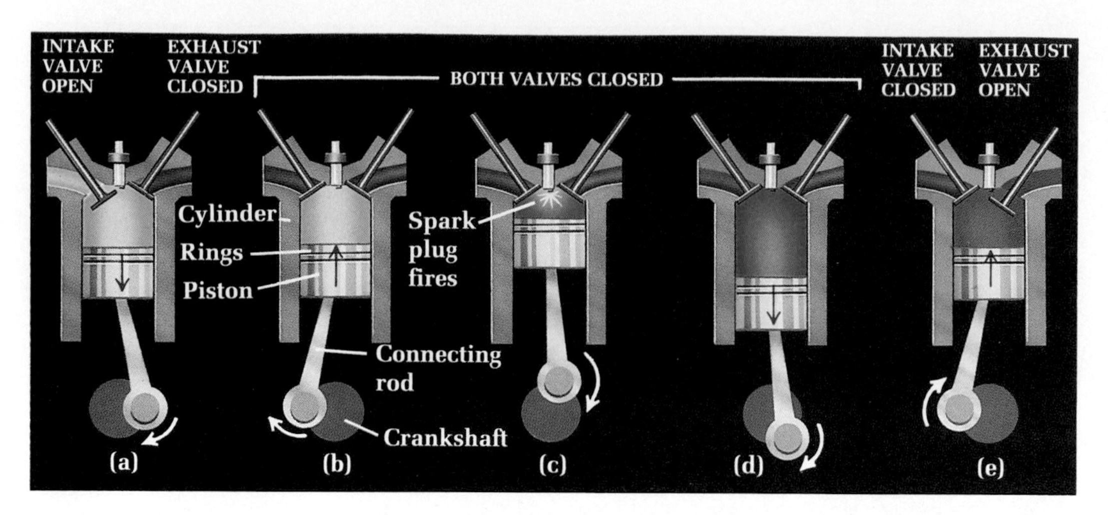


**18–2** Cycle of a four-stroke internal-combustion engine. (a) *Intake stroke:* Piston moves down, causing a partial vacuum in cylinder; gasoline-air mixture flows through open intake valve into cylinder. (b) *Compression stroke:* Intake valve closes, and mixture is compressed as piston moves up. (c) *Ignition:* Spark plug ignites mixture. (d) *Power stroke:* Hot burned mixture pushes piston down, doing work. (e) *Exhaust stroke:* Exhaust valve opens and piston moves up, pushing burned mixture out of cylinder. Engine is now ready for next intake stroke, and the cycle repeats.

$rV$ (when it is all the way down). The quantity $r$ is called the **compression ratio;** for present-day automobile engines it is typically about 8. At the end of this *intake stroke* the intake valve closes and the mixture is compressed, approximately adiabatically, to volume $V$ during the *compression stroke*. The mixture is then ignited by the spark plug, and the heated gas expands, approximately adiabatically, back to volume $rV$, pushing on the piston and doing work; this is the *power stroke*. Finally, the exhaust valve opens and the combustion products are pushed out (during the *exhaust stroke*), leaving the cylinder ready for the next intake stroke.

## The Otto Cycle

Figure 18–3 is a $pV$-diagram showing an idealized model of the thermodynamic processes in a gasoline engine. This model is called the **Otto cycle.** At point $a$ the gasoline-air mixture has entered the cylinder. The mixture is compressed adiabatically (line $ab$) and is then ignited. Heat $Q_H$ is added to the system by the burning gasoline (line $bc$), and the power stroke is the adiabatic expansion $cd$. The gas is cooled to the temperature of the outside air ($da$); during this process, heat $Q_C$ is released. In practice, this same air does not enter the engine again, but since an equivalent amount does enter, we may consider the process to be cyclic.

We can calculate the efficiency of this idealized cycle. Processes $bc$ and $da$ are constant-volume, so the heats $Q_H$ and $Q_C$ are related simply to the temperatures:

$$Q_H = nC_V(T_c - T_b),$$

$$Q_C = nC_V(T_a - T_d).$$

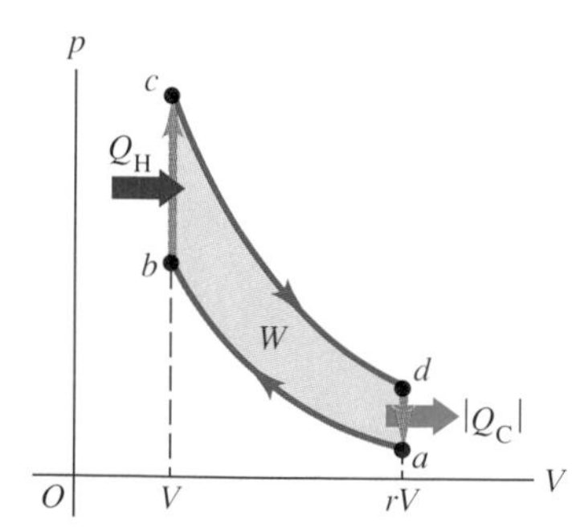


**18–3** The $pV$-diagram for the Otto cycle, an idealized model of the thermodynamic processes in a gasoline engine.

The thermal efficiency is given by Eq. (18–4); inserting the above expressions and canceling out the common factor $nC_V$, we find

$$e = \frac{Q_H + Q_C}{Q_H} = \frac{T_c - T_b + T_a - T_d}{T_c - T_b}. \tag{18–5}$$

To simplify this further, we use the temperature-volume relation for adiabatic processes for an ideal gas, Eq. (17–24). For the two adiabatic processes $ab$ and $cd$,

$$T_a(rV)^{\gamma-1} = T_bV^{\gamma-1},$$

$$T_d(rV)^{\gamma-1} = T_cV^{\gamma-1}.$$

We divide each of these equations by the common factor $V^{\gamma-1}$ and substitute the resulting expressions for $T_b$ and $T_c$ back into Eq. (18–5). The result is

$$e = \frac{T_d r^{\gamma-1} - T_a r^{\gamma-1} + T_a - T_d}{T_d r^{\gamma-1} - T_a r^{\gamma-1}} = \frac{(T_d - T_a)(r^{\gamma-1} - 1)}{(T_d - T_a)r^{\gamma-1}}.$$

Dividing out the common factor $(T_d - T_a)$, we get

$$e = 1 - \frac{1}{r^{\gamma-1}}. \qquad (18\text{–}6)$$

The thermal efficiency given by Eq. (18–6) is always less than unity, even for this idealized model. With $r = 8$ and $\gamma = 1.4$ (the value for air) the theoretical efficiency is $e = 0.56$, or 56%. The efficiency can be increased by increasing $r$. However, this also increases the temperature at the end of the adiabatic compression of the air-fuel mixture. If the temperature is too high, the mixture explodes spontaneously during compression instead of burning evenly after the spark plug ignites it. This is called *pre-ignition,* or *detonation;* it causes a knocking sound and can damage the engine. The octane rating of a gasoline is a measure of its antiknock qualities. The maximum practical compression ratio for high-octane, or "premium," gasoline is about 10. Higher ratios can be used with more exotic fuels.

The Otto cycle, which we have just described, is a highly idealized model. It assumes that the air-fuel mixture behaves as an ideal gas; it neglects friction, turbulence, loss of heat to cylinder walls, and many other effects that combine to reduce the efficiency of a real engine. Another source of inefficiency is incomplete combustion. A mixture of gasoline vapor with just enough air for complete combustion of the hydrocarbons to water and carbon dioxide ($CO_2$) does not ignite readily. Reliable ignition requires a mixture that is "richer" in gasoline; the resulting incomplete combustion leads to carbon monoxide (CO) and unburned hydrocarbons in the exhaust. The heat obtained from the gasoline is then less than the total heat of combustion; the difference is wasted, and the exhaust products contribute to air pollution. Efficiencies of real gasoline engines are typically around 20%.

## The Diesel Cycle

The Diesel engine is similar in operation to the gasoline engine. The most important difference is that there is no fuel in the cylinder at the beginning of the compression stroke. A little before the beginning of the power stroke, the injectors start to inject fuel directly into the cylinder, just fast enough to keep the pressure approximately constant during the first part of the power stroke. Because of the high temperature developed during the adiabatic compression, the fuel ignites spontaneously as it is injected; no spark plugs are needed.

The idealized **Diesel cycle** is shown in Fig. 18–4. Starting at point $a$, air is compressed adiabatically to point $b$, heated at constant pressure to point $c$, expanded adiabatically to point $d$, and cooled at constant volume to point $a$. Because there is no fuel in the cylinder during most of the compression stroke, pre-ignition cannot occur, and the compression ratio $r$ can be much higher than for a gasoline engine. This improves efficiency and ensures reliable ignition when the fuel is injected (because of the high temperature reached during the adiabatic compression). Values of $r$ of 15 to 20 are typical; with these values and $\gamma = 1.4$ the theoretical efficiency of the idealized Diesel cycle is about 0.65 to 0.70. As with the Otto cycle, the efficiency of any actual engine is substantially less than this. Diesel engines are usually more efficient than gasoline engines. They are

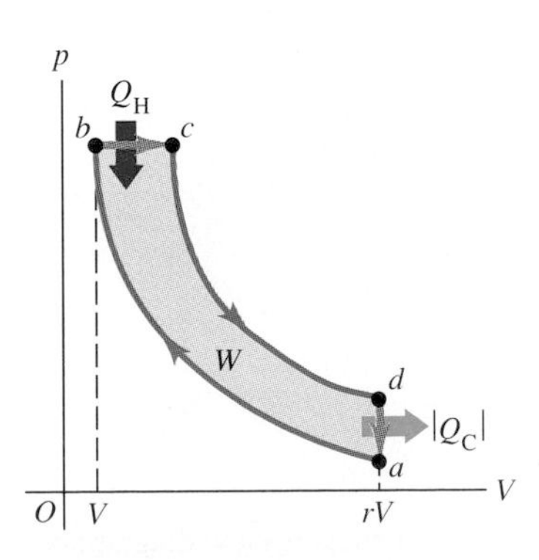

**18–4** The $pV$-diagram for the idealized Diesel cycle.

also heavier (per unit power output) and often harder to start. They need no carburetor or ignition system, but the fuel-injection system requires expensive high-precision machining.

## 18–4 REFRIGERATORS

We can think of a **refrigerator** as a heat engine operating in reverse. A heat engine takes heat from a hot place and gives off heat to a colder place. A refrigerator does the opposite; it takes heat from a cold place (the inside of the refrigerator) and gives it off to a warmer place (usually the air in the room where the refrigerator is located). A heat engine has a net *output* of mechanical work; the refrigerator requires a net *input* of mechanical work. With the symbols we used in Section 18–2, $Q_C$ is positive for a refrigerator, but both $W$ and $Q_H$ are negative, so $|W| = -W$ and $|Q_H| = -Q_H$.

A flow diagram for a refrigerator is shown in Fig. 18–5. From the first law for a cyclic process,

$$Q_H + Q_C - W = 0, \qquad \text{or} \qquad -Q_H = Q_C - W,$$

or, because both $Q_H$ and $W$ are negative,

$$|Q_H| = Q_C + |W|. \tag{18–7}$$

Thus, as the diagram shows, the heat $|Q_H|$ leaving the working substance and given to the hot reservoir is always *greater* than the heat $Q_C$ taken from the cold reservoir. Note that the absolute-value relation

$$|Q_H| = |Q_C| + |W| \tag{18–8}$$

is valid for both heat engines and refrigerators.

From an economic point of view the best refrigeration cycle is one that removes the greatest amount of heat $Q_C$ from the refrigerator for the least expenditure of mechanical work, $W$. The relevant ratio is therefore $|Q_C/W| = -Q_C/W$. We call this the **performance coefficient,** denoted by $K$. Also, we have $W = Q_H + Q_C = -|Q_H| + Q_C$, so

$$K = \frac{Q_C}{|W|} = \frac{|Q_C|}{|Q_H| - |Q_C|}. \tag{18–9}$$

As always, we measure $Q_H$, $Q_C$, and $W$ all in the same energy units; $K$ is then a dimensionless number.

The principles of the common refrigeration cycle are shown schematically in Fig. 18–6. The fluid "circuit" contains a refrigerant (the working substance). In the past this has usually been $CCl_2F_2$ or another member of the Freon family; because halogens in the upper atmosphere deplete the ozone layer, alternative refrigerants are being developed. The left side of the circuit (including the cooling coils inside the refrigerator) is at low temperature and low pressure; the right side (including the condenser coils outside the refrigerator) is at high temperature and high pressure. Ordinarily, both sides contain liquid and vapor in phase equilibrium.

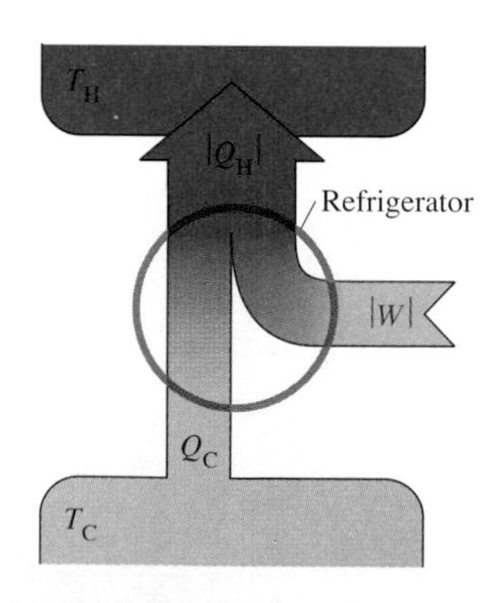

**18–5** Schematic flow diagram of a refrigerator.

The compressor takes in fluid, compresses it adiabatically, and delivers it to the condenser coil at high pressure. The fluid temperature is then higher than that of the air surrounding the condenser, so the refrigerant gives off heat $|Q_H|$ and partially condenses to liquid. The fluid then expands adiabatically, at a rate controlled by the expansion valve, into the evaporator. As it does so, it cools considerably, enough so that the fluid in the evaporator coil is colder than its surroundings. It absorbs heat $|Q_C|$ from its surroundings, cooling them and partially vaporizing. The fluid then enters the compressor to

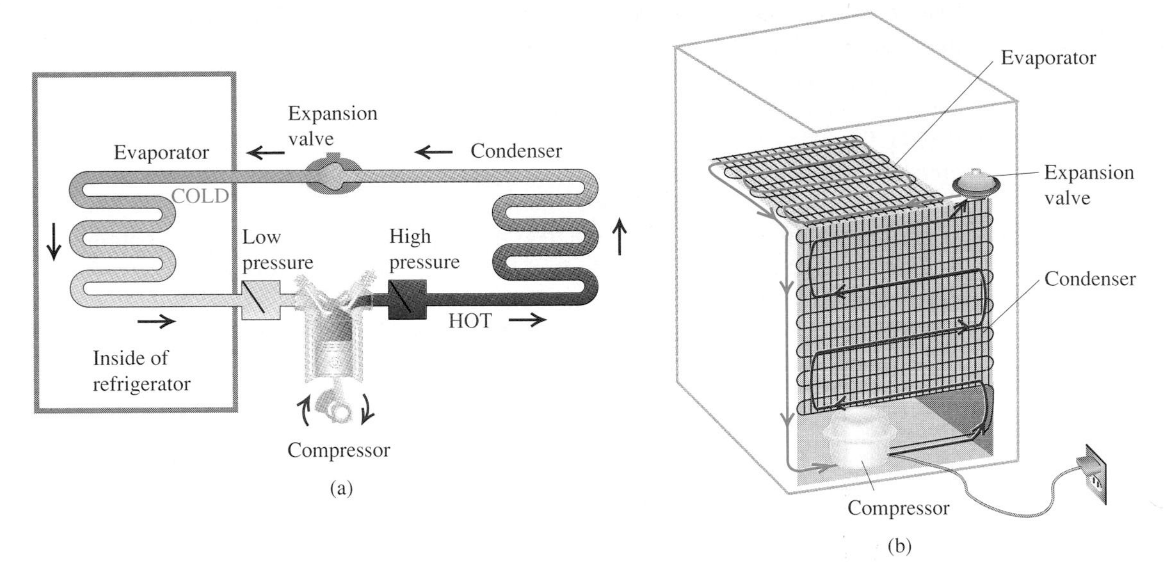


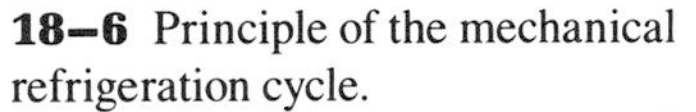
**18–6** Principle of the mechanical refrigeration cycle.

begin another cycle. The compressor, usually driven by an electric motor, requires energy input and does work $|W|$ *on* the working substance during each cycle.

An air conditioner operates on exactly the same principle. In this case the refrigerator box becomes a room or an entire building. The evaporator coils are inside, the condenser is outside, and fans circulate air through these (Fig. 18–7). In large installations the condenser coils are often cooled by water. For air conditioners the quantities of greatest practical importance are the *rate* of heat removal (the heat current $H$ from the region being cooled) and the *power* input $P = W/t$ to the compressor. If heat $Q_C$ is removed in time $t$, then $H = Q_C/t$. Then we can express the performance coefficient as

$$K = \frac{Q_C}{|W|} = \frac{Ht}{Pt} = \frac{H}{P}.$$

Typical room air conditioners have heat removal rates $H$ of 5000 to 10,000 Btu/h, or about 1500 to 3000 W, and require electric power input of about 600 to 1200 W. Typical performance coefficients are about 2.5, with somewhat larger values for larger-capacity units.

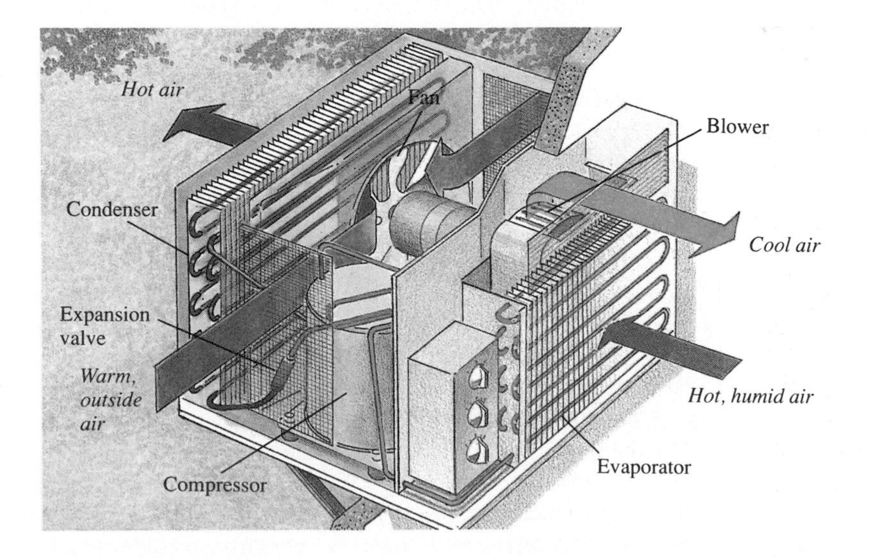


**18–7** An air conditioner works on the same principle as a refrigerator.

Values of $K$ are measured under standardized conditions of inside and outside temperatures; actual values of $K$ vary with operating conditions.

Unfortunately, $K$ is often expressed commercially in mixed units, with $H$ in Btus per hour and $P$ in watts. In these units, $H/P$ is called the **energy efficiency rating (EER).** Because 1 W = 3.413 Btu/h, the EER is numerically 3.413 times as large as the dimensionless $K$. Room air conditioners typically have an EER of 7 to 10. The units, customarily omitted, are (Btu/h)/W.

A variation on this theme is the **heat pump,** used to heat buildings by cooling the outside air. It functions like a refrigerator turned inside out. The evaporator coils are outside, where they take heat from cold air, and the condenser coils are inside, where they give off heat to the warmer air. With proper design the heat $|Q_H|$ delivered to the inside per cycle can be considerably greater than the work $|W|$ required to get it there.

Work is *always* needed to transfer heat from a colder to a hotter body. Heat flows spontaneously from hotter to colder, and to reverse this flow requires the addition of work from the outside. Experience shows that it is impossible to make a refrigerator that transports heat from a cold body to a hotter body without the addition of work. If no work were needed, the performance coefficient would be infinite. We call such a device a *workless refrigerator;* it is a mythical beast, like the unicorn and the free lunch.

## 18–5 THE SECOND LAW OF THERMODYNAMICS

Experimental evidence strongly suggests that it is impossible to build a heat engine that converts heat completely to work, that is, an engine with 100% thermal efficiency. This impossibility is the basis of one form of the **second law of thermodynamics,** as follows:

> **It is impossible for any system to undergo a process in which it absorbs heat from a reservoir at a single temperature and converts the heat completely into mechanical work, with the system ending in the same state in which it began.**

We will call this the "engine" statement of the second law.

The basis of the second law of thermodynamics lies in the difference between the nature of internal energy and that of macroscopic mechanical energy. In a moving body the molecules have random motion, but superimposed on this is a coordinated motion of every molecule in the direction of the body's velocity. The kinetic energy associated with this *coordinated* macroscopic motion is what we call the kinetic energy of the moving body. The kinetic and potential energies associated with the *random* motion of the body's molecules constitute the internal energy.

When a moving body comes to rest as a result of friction or an inelastic collision, the organized part of the motion is converted to random motion. Since we cannot control the motions of individual molecules, we cannot convert this random motion completely back to organized motion. We can convert *part* of it, and this is what a heat engine does.

Think what we could do if the second law were *not* true. We could power an automobile or run a power plant by cooling the surrounding air. Neither of these impossibilities violates the *first* law of thermodynamics. The second law, therefore, is not a deduction from the first but stands by itself as a separate law of nature. The first law denies the possibility of creating or destroying energy; the second law limits the *availability* of energy and the ways in which it can be used and converted.

Our analysis of refrigerators in Section 18–4 forms the basis for an alternative statement of the second law of thermodynamics. Heat flows spontaneously from hotter to colder bodies, never the reverse. A refrigerator does take heat from a colder to a hotter body, but its operation depends on input of mechanical energy or work.

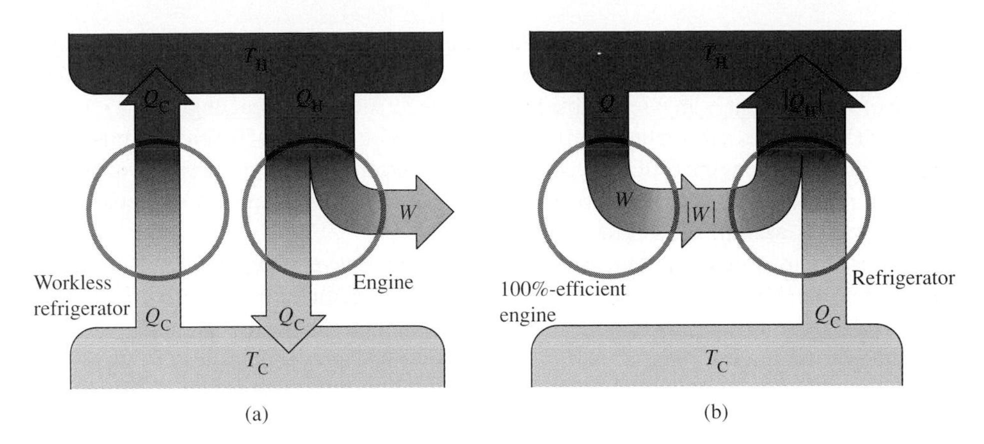


**18–8** Energy-flow diagrams for equivalent forms of the second law. (a) A workless refrigerator (left), if it existed, could be used in combination with an ordinary heat engine (right) to form a composite device that functions as an engine with 100% efficiency, converting heat $Q_H - |Q_C|$ completely to work. (b) An engine with 100% efficiency (left), if it existed, could be used in combination with an ordinary refrigerator (right) to form a workless refrigerator, transferring heat $Q_C$ from the cold reservoir to the hot with no net input of work. Since either of these is impossible, the other must be also.

Generalizing this observation, we state:

**It is impossible for any process to have as its sole result the transfer of heat from a cooler to a hotter body.**

We'll call this the "refrigerator" statement of the second law. It may not seem to be very closely related to the "engine" statement. In fact, though, the two statements are completely equivalent. For example, if we could build a workless refrigerator, violating the "refrigerator" statement of the second law, we could use it in conjunction with a heat engine, pumping the heat rejected by the engine back to the hot reservoir to be reused. This composite machine (Fig. 18–8a) would violate the "engine" statement of the second law because its net effect would be to take a net quantity of heat $Q_H - |Q_C|$ from the hot reservoir and convert it completely to work $W$.

Alternatively, if we could make an engine with 100% thermal efficiency, in violation of the first statement, we could run it using heat from the hot reservoir and use the work output to drive a refrigerator that pumps heat from the cold reservoir to the hot (Fig. 18–8b). This composite device would violate the "refrigerator" statement because its net effect would be to take heat $Q_C$ from the cold reservoir and deliver it to the hot reservoir without requiring any input of work. Thus any device that violates one form of the second law can also be used to make a device that violates the other form. If violations of the first form are impossible, so are violations of the second!

The conversion of work to heat, as in friction or viscous fluid flow, and heat flow from hot to cold across a finite temperature gradient are *irreversible* processes. The "engine" and "refrigerator" statements of the second law state that these processes can be only partially reversed. We could cite other examples. Gases always seep through an opening spontaneously from a region of high pressure to a region of low pressure; gases and miscible liquids left by themselves always tend to mix, not to unmix. The second law of thermodynamics is an expression of the inherent one-way aspect of these and many other irreversible processes. Energy conversion is an essential aspect of all plant and animal life and of many mechanical devices, so the second law of thermodynamics is of the utmost fundamental importance in the world we live in.

## 18–6 THE CARNOT CYCLE

According to the second law, no heat engine can have 100% efficiency. How great an efficiency *can* an engine have, given two heat reservoirs at temperatures $T_H$ and $T_C$? This question was answered in 1824 by the French engineer Sadi Carnot (1796–1832), who developed a hypothetical, idealized heat engine that has the maximum possible efficiency consistent with the second law. The cycle of this engine is called the **Carnot cycle.**

To understand the rationale of the Carnot cycle, we return to a recurrent theme in this chapter, *reversibility* and its relation to directions of thermodynamic processes. Conversion of work to heat is an irreversible process; the purpose of a heat engine is a *partial* reversal of this process, the conversion of heat to work with as great efficiency as possible. For maximum heat-engine efficiency, therefore, *we must avoid all irreversible processes.* This requirement turns out to be enough to determine the basic sequence of steps in the Carnot cycle, as we will show next.

*Heat flow* through a finite temperature drop is an irreversible process. Therefore, during heat transfer in the Carnot cycle there must be *no* finite temperature difference. When the engine takes heat from the hot reservoir at temperature $T_H$, the working substance of the engine must also be at $T_H$; otherwise, irreversible heat flow would occur. Similarly, when the engine discards heat to the cold reservoir at $T_C$, the engine itself must be at $T_C$. That is, every process that involves heat transfer must be *isothermal* at either $T_H$ or $T_C$.

Conversely, in any process in which the temperature of the working substance of the engine is intermediate between $T_H$ and $T_C$, there must be *no* heat transfer between the engine and either reservoir because such heat transfer could not be reversible. Therefore, any process in which the temperature $T$ of the working substance changes must be adiabatic. The bottom line is that *every process* in our idealized cycle must be either *isothermal* or *adiabatic*. In addition, thermal and mechanical equilibrium must be maintained at all times so that each process is completely reversible.

## The Carnot Cycle

The Carnot cycle consists of two isothermal and two adiabatic processes. Figure 18–9 shows a Carnot cycle using as its working substance an ideal gas in a cylinder with a

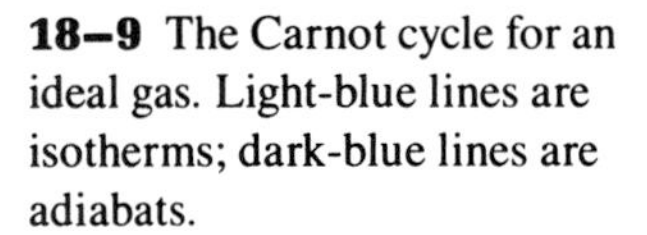

**18–9** The Carnot cycle for an ideal gas. Light-blue lines are isotherms; dark-blue lines are adiabats.

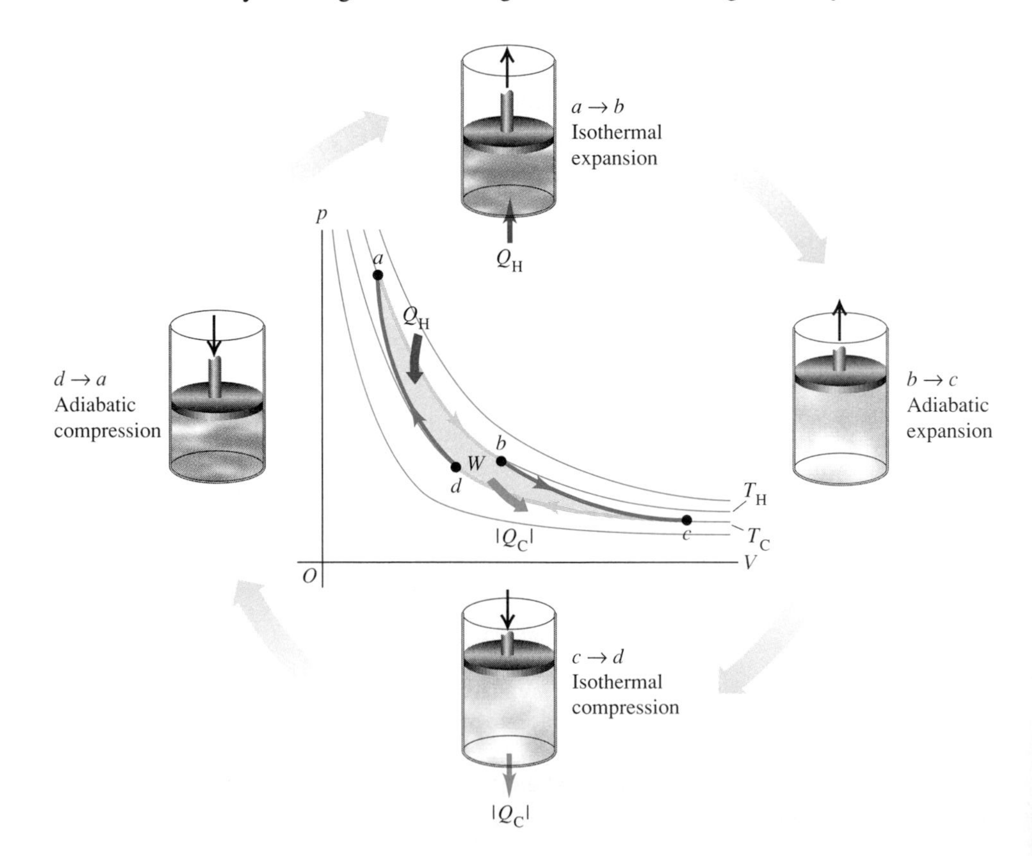

piston. It consists of the following steps:

1. The gas expands isothermally at temperature $T_H$, absorbing heat $Q_H$ ($ab$).
2. It expands adiabatically until its temperature drops to $T_C$ ($bc$).
3. It is compressed isothermally at $T_C$, rejecting heat $|Q_C|$ ($cd$).
4. It is compressed adiabatically back to its initial state at temperature $T_H$ ($da$).

When the working substance in a Carnot engine is an ideal gas, we can calculate the thermal efficiency $e$. To carry out this calculation, we will first find the ratio $Q_C/Q_H$ of the quantities of heat transferred in the two isothermal processes, and then use Eq. (18–4) to find $e$.

For an ideal gas the internal energy $U$ depends only on temperature and is thus constant in any isothermal process. For the isothermal expansion $ab$, $\Delta U_{ab} = 0$, and $Q_H$ is equal to the work $W_{ab}$ done by the gas during its isothermal expansion at temperature $T_H$. This in turn is given by Eq. (17–4):

$$W_{ab} = Q_H = nRT_H \ln \frac{V_b}{V_a}. \tag{18–10}$$

Similarly,

$$W_{cd} = Q_C = nRT_C \ln \frac{V_d}{V_c} = -nRT_C \ln \frac{V_c}{V_d}. \tag{18–11}$$

This quantity is negative because $V_d$ is less than $V_c$.

The ratio of the two quantities of heat is thus

$$\frac{Q_C}{Q_H} = \left(-\frac{T_C}{T_H}\right)\frac{\ln (V_c/V_d)}{\ln (V_b/V_a)}. \tag{18–12}$$

This can be simplified further by use of the temperature-volume relation for an adiabatic process, Eq. (17–24). We find for the two adiabatic processes:

$$T_H V_b^{\gamma-1} = T_C V_c^{\gamma-1},$$

$$T_H V_a^{\gamma-1} = T_C V_d^{\gamma-1}.$$

Dividing the first of these by the second, we find

$$\frac{V_b^{\gamma-1}}{V_a^{\gamma-1}} = \frac{V_c^{\gamma-1}}{V_d^{\gamma-1}}, \quad \text{and} \quad \frac{V_b}{V_a} = \frac{V_c}{V_d}.$$

Thus the two logarithms in Eq. (18–12) are equal, and that equation reduces to

$$\frac{Q_C}{Q_H} = -\frac{T_C}{T_H}, \quad \text{or} \quad \frac{|Q_C|}{|Q_H|} = \frac{T_C}{T_H}. \tag{18–13}$$

Then from Eq. (18–4) the efficiency of the Carnot engine is

$$e_{\text{Carnot}} = 1 - \frac{T_C}{T_H} = \frac{T_H - T_C}{T_H}. \tag{18–14}$$

This surprisingly simple result says that the efficiency of a Carnot engine depends only on the temperatures of the two heat reservoirs. The efficiency is large when the temperature *difference* is large, and it is very small when the temperatures are nearly equal. The efficiency can never be exactly unity unless $T_C = 0$; we'll see later that this too is impossible.

## EXAMPLE 18–2

A Carnot engine takes 2000 J of heat from a reservoir at 500 K, does some work, and discards some heat to a reservoir at 350 K. How much work does it do, how much heat is discarded, and what is the efficiency?

**SOLUTION** From Eq. (18–13) the heat $Q_C$ discarded by the engine is

$$Q_C = -Q_H \frac{T_C}{T_H} = -(2000\text{ J})\frac{350\text{ K}}{500\text{ K}} = -1400\text{ J}.$$

Then from the first law, the work $W$ done by the engine is

$$W = Q_H + Q_C = 2000\text{ J} + (-1400\text{ J}) = 600\text{ J}.$$

From Eq. (18–14) the thermal efficiency is

$$e = 1 - \frac{T_C}{T_H} = 1 - \frac{350\text{ K}}{500\text{ K}} = 0.30 = 30\%.$$

Alternatively, from the basic definition of thermal efficiency,

$$e = \frac{W}{Q_H} = \frac{600\text{ J}}{2000\text{ J}} = 0.30 = 30\%.$$

## EXAMPLE 18–3

Suppose 0.200 mol of an ideal diatomic gas ($\gamma = 1.40$) undergoes a Carnot cycle with temperatures $T_H = 400$ K and $T_C = 300$ K. The initial pressure is $p_a = 10.0 \times 10^5$ Pa, and during the isothermal expansion at temperature $T_H$ the volume doubles. a) Find the pressure and volume at each of points $a$, $b$, $c$, and $d$ in Fig. 18–9. b) Find $Q$, $W$, and $\Delta U$ for each step and for the entire cycle. c) Determine the efficiency directly from the results of part (b), and compare it with the result from Eq. (18–14).

**SOLUTION** a) We use the ideal-gas equation to find $V_a$:

$$V_a = \frac{nRT_H}{p_a} = \frac{(0.200\text{ mol})(8.315\text{ J/mol}\cdot\text{K})(400\text{ K})}{10.0 \times 10^5\text{ Pa}} = 6.65 \times 10^{-4}\text{ m}^3.$$

The volume doubles during the isothermal expansion $a \rightarrow b$, so

$$V_b = 2V_a = 2(6.65 \times 10^{-4}\text{ m}^3) = 13.3 \times 10^{-4}\text{ m}^3.$$

Also, during the isothermal expansion $a \rightarrow b$, $p_aV_a = p_bV_b$, so

$$p_b = \frac{p_aV_a}{V_b} = 5.00 \times 10^5\text{ Pa}.$$

For the adiabatic expansion $b \rightarrow c$, $T_HV_b^{\gamma-1} = T_CV_c^{\gamma-1}$, so

$$V_c = V_b\left(\frac{T_H}{T_C}\right)^{1/(\gamma-1)} = (13.3 \times 10^{-4}\text{ m}^3)\left(\frac{400\text{ K}}{300\text{ K}}\right)^{2.5} = 27.3 \times 10^{-4}\text{ m}^3.$$

Using the ideal-gas equation again for point $c$, we find

$$p_c = \frac{nRT_C}{V_c} = \frac{(0.200\text{ mol})(8.315\text{ J/mol}\cdot\text{K})(300\text{ K})}{27.3 \times 10^{-4}\text{ m}^3} = 1.83 \times 10^5\text{ Pa}.$$

For the adiabatic compression $d \rightarrow a$, $T_CV_d^{\gamma-1} = T_HV_a^{\gamma-1}$, and

$$V_d = V_a\left(\frac{T_H}{T_C}\right)^{1/(\gamma-1)} = (6.65 \times 10^{-4}\text{m}^3)\left(\frac{400\text{ K}}{300\text{ K}}\right)^{2.5} = 13.6 \times 10^{-4}\text{m}^3,$$

$$p_d = \frac{nRT_C}{V_d} = \frac{(0.200\text{mol})(8.315\text{ J/mol}\cdot\text{K})(300\text{ K})}{13.6 \times 10^{-4}\text{m}^3} = 3.67 \times 10^5\text{ Pa}.$$

b) For the isothermal expansion $a \rightarrow b$, $\Delta U_{ab} = 0$. To find $W_{ab}$ (which is equal to $Q_H$), we use Eq. (17–4) in Example 17–1 for the work done by an ideal gas in an isothermal expansion:

$$W_{ab} = Q_H = nRT_H \ln\frac{V_b}{V_a} = (0.200\text{ mol})(8.315\text{ J/mol}\cdot\text{K})(400\text{ K})(\ln 2) = 461\text{ J}.$$

For the adiabatic expansion $b \rightarrow c$, $Q_{bc} = 0$. We find the work $W_{bc}$ from Eq. (17–27), using $C_V = 20.8$ J/mol · K for an ideal diatomic gas:

$$W_{bc} = -\Delta U_{bc} = nC_V(T_H - T_C) = (0.200\text{ mol})(20.8\text{ J/mol}\cdot\text{K})(400\text{ K} - 300\text{ K}) = 416\text{ J}.$$

For the isothermal compression $c \rightarrow d$, $\Delta U_{cd} = 0$; Eq. (17–4) gives

$$W_{cd} = Q_C = nRT_C \ln\frac{V_d}{V_c} = (0.200\text{ mol})(8.315\text{ J/mol}\cdot\text{K})(300\text{ K})\left(\ln\frac{13.6 \times 10^{-4}\text{ m}^3}{27.3 \times 10^{-4}\text{ m}^3}\right) = -348\text{ J}.$$

For the adiabatic compression $d \rightarrow a$, $Q_{da} = 0$, and

$$W_{da} = -\Delta U_{da} = nC_V(T_C - T_H) = (0.200\text{ mol})(20.8\text{ J/mol}\cdot\text{K})(300\text{ K} - 400\text{ K}) = -416\text{ J}.$$

We can tabulate the results as follows:

| Process | $Q$ | $W$ | $\Delta U$ |
|---|---|---|---|
| $a \rightarrow b$ | 461 J | 461 J | 0 |
| $b \rightarrow c$ | 0 | 416 J | −416 J |
| $c \rightarrow d$ | −348 J | −348 J | 0 |
| $d \rightarrow a$ | 0 | −416 J | 416 J |
| Total | 113 J | 113 J | 0 |

Note that for the entire cycle, $Q = W$ and $\Delta U = 0$. Also note that the quantities of work in the two adiabatic processes are negatives of each other; it is easy to prove from the analysis leading to Eq. (18–13) that this must *always* be the case.

c) From the table the total work is 113 J and $Q_H = 461$ J. Thus

$$e = \frac{W}{Q_H} = \frac{113\text{ J}}{461\text{ J}} = 0.25.$$

We can compare this with the result from Eq. (18–14):

$$e = \frac{T_H - T_C}{T_H} = \frac{400\text{ K} - 300\text{ K}}{400\text{ K}} = 0.25.$$

## The Carnot Refrigerator

Because each step in the Carnot cycle is reversible, the *entire cycle* may be reversed, converting the engine into a refrigerator. The performance coefficient of the Carnot refrigerator is obtained by combining the general definition of $K$, Eq. (18–9), with Eq. (18–13) for the Carnot cycle. We first rewrite Eq. (18–9) as

$$K = \frac{|Q_C|}{|Q_H| - |Q_C|} = \frac{|Q_C|/|Q_H|}{1 - |Q_C|/|Q_H|}.$$

Then we substitute Eq. (18–13), $|Q_C|/|Q_H| = T_C/T_H$, into this. The result is

$$K_{\text{Carnot}} = \frac{T_C}{T_H - T_C}. \qquad (18\text{–}15)$$

When the temperature difference $T_H - T_C$ is small, $K$ is much larger than unity; in this case a lot of heat can be "pumped" from the lower to the higher temperature with only a little expenditure of work. But the greater the temperature difference, the smaller $K$ is and the more work is required to transfer a given quantity of heat.

### EXAMPLE 18–4

If the cycle described in Example 18–3 is run backwards as a refrigerator, the performance coefficient is given by Eq. (18–9):

$$K = \frac{Q_C}{|W|} = \frac{348\text{ J}}{113\text{ J}} = 3.0.$$

Because the cycle is a Carnot cycle, we may also use Eq. (18–15):

$$K = \frac{T_C}{T_H - T_C} = \frac{300\text{ K}}{400\text{ K} - 300\text{ K}} = 3.0.$$

For a Carnot cycle, $e$ and $K$ depend only on the temperatures, as shown by Eqs. (18–14) and (18–15), and we don't need to calculate $Q$ and $W$. For cycles containing irreversible processes, however, these equations are not valid, and more detailed calculations are necessary.

## The Carnot Cycle and the Second Law

We can prove that **no engine can be more efficient than a Carnot engine operating between the same two temperatures.** The key to the proof is the above observation that since each step in the Carnot cycle is reversible, the *entire cycle* may be reversed. Run backward, the engine becomes a refrigerator. Suppose we have an engine that is more efficient than a Carnot engine (Fig. 18–10 on page 520). Let the Carnot engine, run backward as a refrigerator by negative work $-W$, take in heat $Q_C$ from the cold reservoir and expel heat $|Q_H|$ to the hot reservoir. The superefficient engine expels heat $|Q_C|$, but to do this, it takes in a greater amount of heat $Q_H + \Delta$. Its work output is then $W + \Delta$, and the net effect of the two machines together is to take a quantity of heat $\Delta$ and convert it completely into work. This violates the "engine" statement of the second law. We could construct a similar argument that a superefficient engine could be used to violate the "refrigerator" statement of the second law. Note that we don't have to assume that the superefficient engine is reversible.

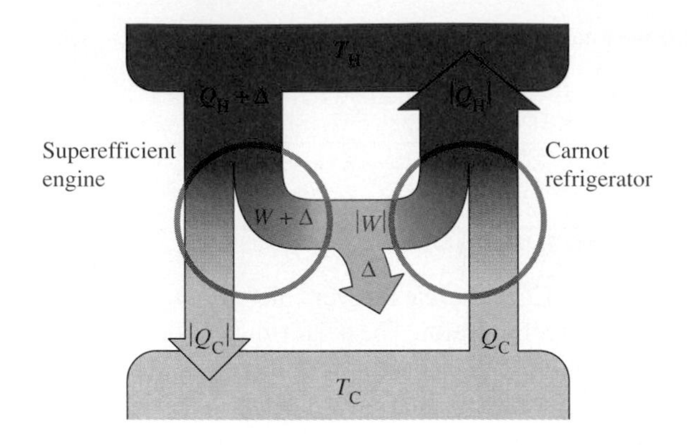


**18–10** If there were a more efficient engine than a Carnot engine, it could be used in conjunction with a Carnot refrigerator to convert the heat $\Delta$ completely into work, with no net heat transfer to the cold reservoir. By the second law of thermodynamics this cannot happen.

Thus the statement that no engine can be more efficient than a Carnot engine is yet another equivalent statement of the second law of thermodynamics. It also follows directly that **all Carnot engines operating between the same two temperatures have the same efficiency, irrespective of the nature of the working substance.** Although we derived Eq. (18–14) for a Carnot engine using an ideal gas as its working substance, it is in fact valid for *any* Carnot engine, no matter what its working substance.

Equation (18–14) shows how to maximize the efficiency of a real engine, such as a steam turbine. The intake temperature $T_H$ must be made as high as possible and the exhaust temperature $T_C$ as low as possible.

The exhaust temperature cannot be lower than the lowest temperature available for cooling the exhaust. This is usually the temperature of the air, or perhaps of river or lake water if this is available at the plant. Then we want the boiler temperature $T_H$ to be as high as possible. The vapor pressures of all liquids increase rapidly with increasing temperature, so we are limited by the mechanical strength of the boiler. At 500°C the vapor pressure of water is about $240 \times 10^5$ Pa (235 atm or 3450 psi), and this is about the maximum practical pressure in large present-day steam boilers.

The unavoidable exhaust heat loss in electric power plants creates a serious environmental problem. When a lake or river is used for cooling, the temperature of the body of water may be raised several degrees. Such a temperature change has a severely disruptive effect on the overall ecological balance, since relatively small temperature changes can have significant effects on metabolic rates in plants and animals. Because **thermal pollution,** as this effect is called, is an inevitable consequence of the second law of thermodynamics, careful planning is essential to minimize the ecological impact of new power plants.

## *18–7 ENTROPY

The second law of thermodynamics, as we have stated it, is rather different in form from many familiar physical laws. It is not an equation or a quantitative relationship but rather a statement of *impossibility.* However, the second law *can* be stated as a quantitative relation using the concept of *entropy,* the subject of this section.

We have talked about several processes that proceed naturally in the direction of increasing disorder. Irreversible heat flow increases disorder because the molecules are initially sorted into hotter and cooler regions; this sorting is lost when the system comes to thermal equilibrium. Adding heat to a body increases its disorder because it increases average molecular speeds and therefore the randomness of molecular motion. Free expansion of a gas increases its disorder because the molecules have greater randomness of position after the expansion than before.

## Entropy and Disorder

**Entropy** provides a *quantitative* measure of disorder. To introduce this concept, let's consider an infinitesimal isothermal expansion of an ideal gas. We add heat $dQ$ and let the gas expand just enough so that the temperature remains constant. Because the internal energy of an ideal gas depends only on its temperature, the internal energy is also constant; thus, from the first law, the work $dW$ done by the gas is equal to the heat $dQ$ added. That is,

$$dQ = dW = p\,dV = \frac{nRT}{V}\,dV.$$

The gas is in a more disordered state after the expansion than before because the molecules are moving in a larger volume and have more randomness of position. Thus the fractional volume change $dV/V$ is a measure of the increase in disorder, and it is proportional to the quantity $dQ/T$. We introduce the symbol $S$ for the entropy of the system, and we define the infinitesimal entropy change $dS$ during an infinitesimal reversible process at absolute temperature $T$ as

$$dS = \frac{dQ}{T} \quad \text{(reversible isothermal process).} \tag{18–16}$$

If a total amount of heat $Q$ is added during a reversible isothermal process at absolute temperature $T$, the total entropy change $\Delta S = S_2 - S_1$ is given by

$$\Delta S = S_2 - S_1 = \frac{Q}{T}. \tag{18–17}$$

Entropy has units of energy divided by temperature; the SI unit of entropy is 1 J/K.

### EXAMPLE 18–5

One kilogram of ice at 0°C is melted and converted to water at 0°C. Compute its change in entropy. The heat of fusion of water is $L_f = 3.34 \times 10^5$ J/kg.

**SOLUTION** The temperature $T$ is constant at 273 K. The heat needed to melt the ice is $Q = mL_f = 3.34 \times 10^5$ J. From Eq. (18–17) the increase in entropy of the system is

$$\Delta S = S_2 - S_1 = \frac{Q}{T} = \frac{3.34 \times 10^5\ \text{J}}{273\ \text{K}} = 1.22 \times 10^3\ \text{J/K}.$$

This increase corresponds to the increase in disorder when the water molecules go from the highly ordered state of a crystalline solid to the much more disordered state of a liquid.

In any *isothermal* reversible process the entropy change equals the heat transferred divided by the absolute temperature. When we refreeze the water, its entropy change is $\Delta S = -1.22 \times 10^3$ J/K.

---

We can generalize the definition of entropy change to include *any* reversible process leading from one state to another, whether it is isothermal or not. We represent the process as a series of infinitesimal reversible steps. During a typical step an infinitesimal quantity of heat $dQ$ is added to the system at absolute temperature $T$. Then we sum (integrate) the quotients $dQ/T$ for the entire process; that is,

$$\Delta S = \int_1^2 \frac{dQ}{T} \quad \text{(reversible process).} \tag{18–18}$$

The limits 1 and 2 refer to the initial and final states.

In such a process we can again see the relation of entropy increase to increase in disorder. Higher temperature means greater randomness of motion. When the substance is initially cold, with little molecular motion, a given amount of heat $Q$ causes a substantial increase in molecular motion and randomness. But when the substance is already hot, the same quantity of heat adds relatively little to the greater molecular motion that is already present. So the quotient $Q/T$ is an appropriate characterization of the increase in randomness, or disorder.

Because entropy is a measure of the disorder of a system in any specific state, it must depend only on the current state of the system, not on its past history. We will show later, by using the second law of thermodynamics, that this is indeed the case. When a system proceeds from an initial state with entropy $S_1$ to a final state with entropy $S_2$, the change in entropy $\Delta S = S_2 - S_1$, defined by Eq. (18–18), does not depend on the path leading from the initial to the final state but is the same for *all possible* processes leading from state 1 to state 2. Thus the entropy of a system must also have a definite value for any given state of the system. We recall that *internal energy,* which was introduced in Chapter 17, also has this property, although entropy and internal energy are very different quantities.

The fact that entropy is a function of only the state of a system shows us how to compute entropy changes in *irreversible* (nonequilibrium) processes, for which Eq. (18–16) is not applicable. We simply invent a path connecting the given initial and final states that *does* consist entirely of reversible, equilibrium processes and compute the total entropy change for that path. It is not the actual path, but the entropy change must be the same as for the actual path.

As with internal energy, the above discussion does not define entropy itself, but only the change in entropy in any given process. To complete the definition, we may arbitrarily assign a value to the entropy of a system in a specified reference state and then calculate the entropy of any other state with reference to this.

## EXAMPLE 18–6

One kilogram of water at 0°C is heated to 100°C. Compute its change in entropy.

**SOLUTION** The temperature is not constant; to carry out the integral in Eq. (18–18), we use Eq. (15–16) to replace $dQ$ by $mc\,dT$, obtaining

$$\Delta S = S_2 - S_1 = \int_1^2 \frac{dQ}{T} = \int_{T_1}^{T_2} mc\frac{dT}{T} = mc \ln\frac{T_2}{T_1}$$

$$= (1.00\text{ kg})(4190\text{ J/kg}\cdot\text{K})\left(\ln\frac{373\text{ K}}{273\text{ K}}\right)$$

$$= 1310\text{ J/K}.$$

## EXAMPLE 18–7

A gas expands adiabatically and reversibly. What is its change in entropy?

**SOLUTION** In an adiabatic process, no heat enters or leaves the system; $dQ = 0$, and there is *no* change in entropy: $\Delta S = 0$. Every *reversible* adiabatic process is a constant-entropy process. The increase in disorder resulting from the gas occupying a greater volume after the expansion is exactly balanced by the decrease in disorder associated with the lowered temperature and reduced molecular speeds.

## EXAMPLE 18–8

A thermally insulated box is divided by a partition into two compartments, each having volume $V$ (Fig. 18–11). Initially, one compartment contains $n$ moles of an ideal gas at temperature $T$, and the other compartment is evacuated. We then break the partition,

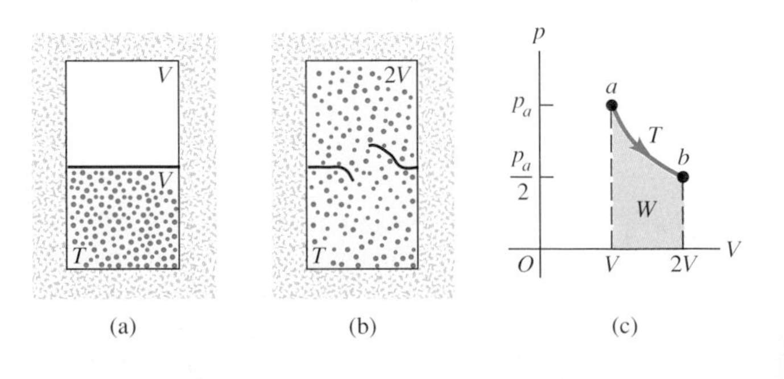


**18–11** (a, b) Free expansion of an insulated ideal gas. (c) The free-expansion process doesn't pass through equilibrium states from $a$ to $b$. However, the entropy change $S_b - S_a$ can be calculated by using the isothermal path shown or *any* reversible path from $a$ to $b$.

## Entropy and Disorder

**Entropy is a quantitative measure of disorder, and the total entropy of the universe never decreases. From these two statements we can infer that all natural processes for any system end up with the system in a more disordered state than that in which it began. Consider the following physical processes.**

# BUILDING PHYSICAL INTUITION

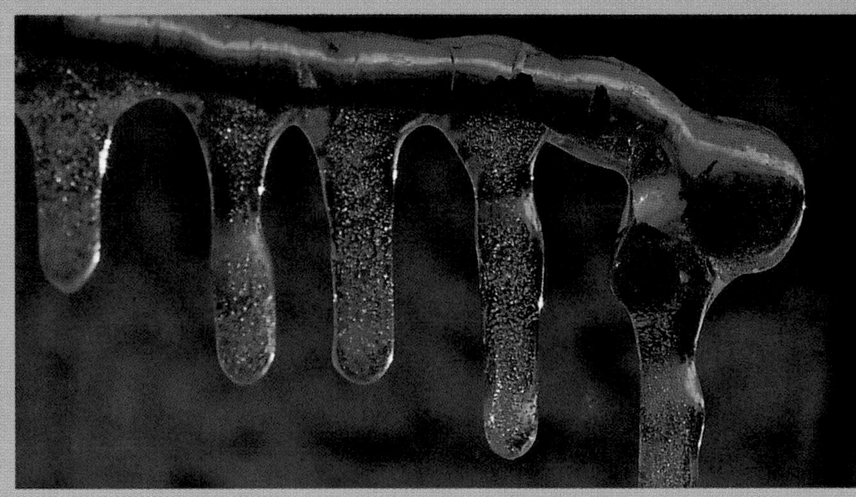

**A new deck of playing cards is sorted out by suit (hearts, diamonds, clubs, spades) and by number. Shuffling a deck of cards increases its disorder into a random arrangement. Shuffling a deck of cards back into its original order is highly unlikely.**

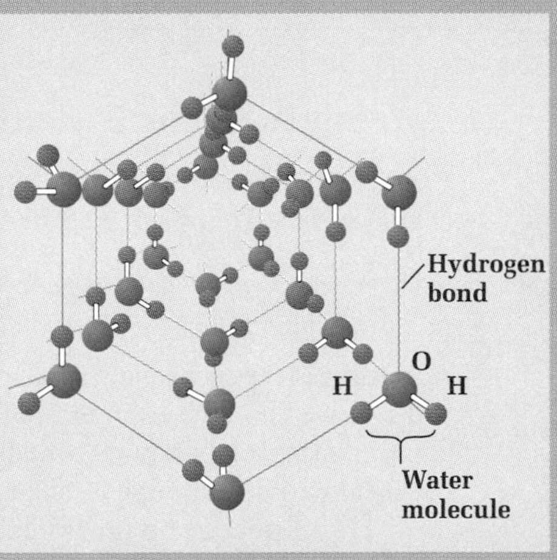


**Heat flows from a warm body to a cold one. As a result, the internal energy of the colder body, such as ice, increases and its molecules gain increased random motion. The increase of disorder in the ice is greater than the decrease of disorder in the surroundings, and there is a net increase of disorder in the system. When 1 mol of ice melts at $0^{o}C$, the water has an increase in entropy of 22 J/K.**

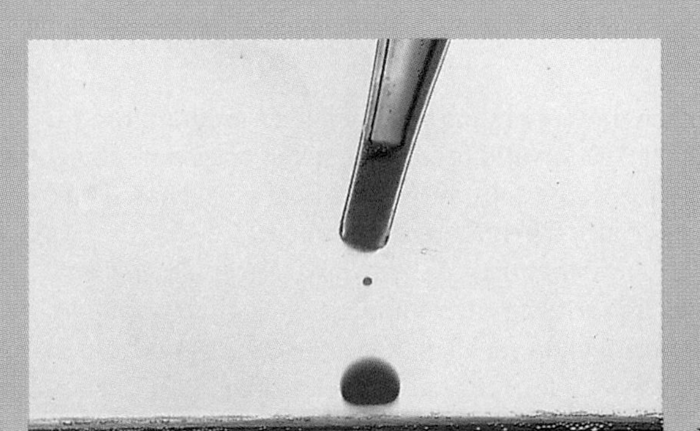

**Physical mixing of two fluids starts from a state of relative order in which each fluid is separate and distinct from the other one. The final state after mixing is more disordered because particles of each fluid are randomly surrounded by particles of both fluids. The spontaneous unmixing of fluids is highly unlikely.**

and the gas expands to fill both compartments. What is the entropy change in this free-expansion process?

**SOLUTION** For this process, $Q = 0$, $W = 0$, $\Delta U = 0$, and therefore (because the system is an ideal gas) $\Delta T = 0$. We might think that the entropy change is zero because there is no heat exchange. But Eq. (18–18) is valid only for *reversible* processes; this free expansion is *not* reversible, and there *is* an entropy change. To calculate $\Delta S$, we use the fact that the entropy change depends only on the initial and final states. We can devise a *reversible* process having the same end points, use Eq. (18–18) to calculate its entropy change, and thus determine the entropy change in the original process. The appropriate reversible process in this case is an isothermal expansion from $V$ to $2V$ at temperature $T$. The gas does work during this substitute expansion, so heat must be supplied to keep the internal energy constant. The total heat equals the total work, which is given by Eq. (17–4):

$$W = Q = nRT \ln 2.$$

Thus the entropy change is

$$\Delta S = \frac{Q}{T} = nR \ln 2,$$

and this is also the entropy change for the free expansion. For one mole,

$$\begin{aligned}\Delta S &= (1 \text{ mol})(8.315 \text{ J/mol} \cdot \text{K})(0.693) \\ &= 5.76 \text{ J/K}.\end{aligned}$$

## EXAMPLE 18–9

**Entropy and the Carnot cycle** For the Carnot engine in Example 18–2 (Section 18–6), find the total entropy change in the engine during one cycle.

**SOLUTION** During the isothermal expansion at 500 K the engine takes in 2000 J, and its entropy change is

$$\Delta S = \frac{Q}{T} = \frac{2000 \text{ J}}{500 \text{ K}} = 4.0 \text{ J/K}.$$

During the isothermal compression at 350 K the engine gives off 1400 J of heat, and its entropy change is

$$\Delta S = \frac{-1400 \text{ J}}{350 \text{ K}} = -4.0 \text{ J/K}.$$

The total entropy change in the engine during one cycle is $4.0 \text{ J/K} - 4.0 \text{ J/K} = 0$. The total entropy change of the two heat reservoirs is also zero, although each individual reservoir has an entropy change. This cycle contains no irreversible processes, and the total entropy change of the system and its surroundings is zero.

## Entropy in Cyclic Processes

Example 18–9 showed that the total entropy change for a cycle of a particular Carnot engine is zero. This result is related directly to Eq. (18–13), which we can rewrite as

$$\frac{Q_{\text{H}}}{T_{\text{H}}} + \frac{Q_{\text{C}}}{T_{\text{C}}} = 0. \tag{18–19}$$

According to the second law, all Carnot engines operating between given temperature $T_{\text{H}}$ and $T_{\text{C}}$ have the same efficiency, so this result is valid for every Carnot engine working between these temperatures, whether its working substance is an ideal gas or not. Thus **the total entropy change in one cycle of any Carnot engine is zero.**

This result can be generalized to show that the total entropy change during *any* reversible cyclic process is zero. A reversible cyclic process appears on a $pV$-diagram as a closed path (Fig. 18–12a). We can approximate such a path as closely as we like as a sequence of isothermal and adiabatic processes forming parts of many long, thin Carnot

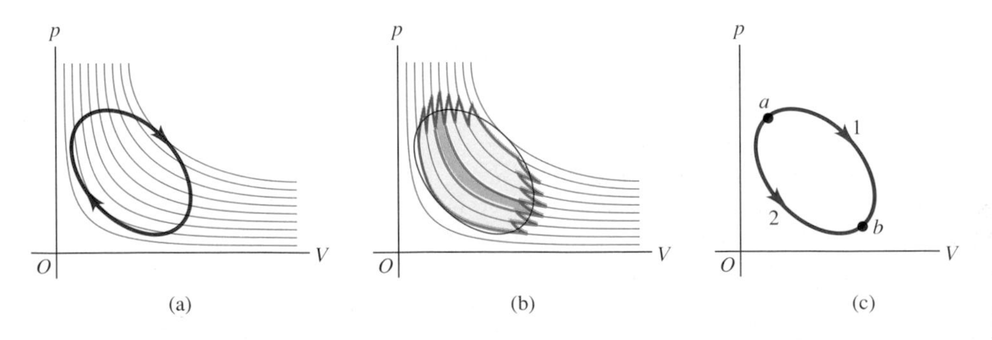


**18–12** (a) A reversible cyclic process for an ideal gas, shown as a closed path on a $pV$-diagram. Several ideal-gas isotherms are also shown. (b) The path can be approximated by a series of long, thin Carnot cycles; one cycle is highlighted. The work and heat along the isotherm between each pair of adjacent cycles cancel out because the two cycles traverse the isotherm in opposite directions. The outer sections of isotherms and adiabats approximate the actual cyclic process; the total entropy change is zero for both. (c) The entropy change $S_b - S_a$ is the same for all paths, such as paths 1 and 2, connecting points $a$ and $b$.

cycles (Fig. 18–12b). Any two adjacent cycles have a common isotherm, but each common isotherm is traversed in opposite directions. Thus the total $Q$ and $W$ for each interior isotherm are both zero. The total entropy change for each small Carnot cycle is zero, so **the total entropy change during *any* reversible cycle is zero:**

$$\int \frac{dQ}{T} = 0 \quad \text{(reversible cyclic process).} \tag{18–20}$$

From this result it follows that when a system undergoes a reversible transition leading from any state $a$ to any other state $b$, **the entropy change is independent of the path** (Fig. 18–12c). If the entropy change for path 1 were different from the change for path 2, the system could be taken along path 1 and then backward along path 2 to the starting point, with a nonzero net change in entropy. This would violate the previous conclusion that the total entropy change in such a process must be zero. Because the entropy change in such processes is independent of path, we conclude that in any given state the system has a definite value of entropy (relative to a given reference) that depends only on the state, not on the processes that led to that state.

## Entropy in Irreversible Processes

In an idealized, reversible process involving only equilibrium states, the total entropy change of the system and its surroundings is zero. But all *irreversible* processes involve an increase in entropy. Unlike energy, entropy is *not* a conserved quantity. The entropy of an isolated system can change, but, as we shall see, it can never decrease. An entropy increase occurs in every natural (irreversible) process, if all systems taking part in the process are included. Example 18–8 showed an entropy increase during an irreversible process in an isolated system.

### EXAMPLE 18–10

Suppose 1.00 kg of water at 100°C is placed in thermal contact with 1.00 kg of water at 0°C. What is the total change in entropy? Assume that the specific heat capacity of water is constant at 4190 J/kg · K over this temperature range.

**SOLUTION** This process involves irreversible heat flow because of the temperature differences. The final temperature is 50°C = 323 K. The first 4190 J of heat transferred cools the hot water to 99°C and warms the cold water from 0°C to 1°C. The net change of entropy is approximately

$$\Delta S = \frac{-4190\ \text{J}}{373\ \text{K}} + \frac{4190\ \text{J}}{273\ \text{K}} = 4.1\ \text{J/K}.$$

Further increases in entropy occur as the system approaches thermal equilibrium at 50°C (323 K). The *total* increase in entropy can be calculated by using the same method as in Example 18–6. The entropy change of the hot water is

$$\Delta S_{\text{hot}} = mc\int_{T_1}^{T_2}\frac{dT}{T} = (1.00\ \text{kg})(4190\ \text{J/kg}\cdot\text{K})\int_{373\ \text{K}}^{323\ \text{K}}\frac{dT}{T}$$
$$= (4190\ \text{J/K})\left(\ln\frac{323\ \text{K}}{373\ \text{K}}\right) = -603\ \text{J/K}.$$

The entropy change of the cold water is

$$\Delta S_{\text{cold}} = (4190\ \text{J/K})\left(\ln\frac{323\ \text{K}}{273\ \text{K}}\right) = +705\ \text{J/K}.$$

The *total* entropy change of the system is

$$\Delta S_{\text{total}} = +705\ \text{J/K} - 603\ \text{J/K} = +102\ \text{J/K}.$$

An irreversible heat flow in an isolated system is accompanied by an increase in entropy. We could have reached the same end state by simply mixing the two quantities of water. This too is an irreversible process; because the entropy depends only on the state of the system, the total entropy change would be the same, 102 J/K.

## Entropy and the Second Law

These examples of the mixing of substances at different temperatures, or the flow of heat from a higher to a lower temperature, are characteristic of *all* natural (that is, irreversible) processes. When all the entropy changes in the process are included, the increases in entropy are always greater than the decreases. In the special case of a *reversible* process the increases and decreases are equal. Thus we can state the following general principle: **When all systems taking part in a process are included, the entropy either remains constant or increases.** Or: **No process is possible in which the total entropy decreases when all systems taking part in the process are included.** This alternative statement of the second law of thermodynamics, in terms of entropy, is equivalent to the "engine" and "refrigerator" statements discussed earlier.

The increase of entropy in every natural (irreversible) process measures the increase of disorder, or randomness, in the universe associated with that process. Consider again the example of mixing hot and cold water. We *might* have used the hot and cold water as the high- and low-temperature reservoirs of a heat engine. While removing heat from the hot water and giving heat to the cold water, we could have obtained some mechanical work. But once the hot and cold water have been mixed and have come to a uniform temperature, this opportunity to convert heat to mechanical work is lost irretrievably. The lukewarm water will never *unmix* itself and separate into hotter and colder portions. No decrease in *energy* occurs when the hot and cold water are mixed. What has been lost is not *energy*, but *opportunity,* the opportunity to convert part of the heat from the hot water into mechanical work. When entropy increases, energy becomes less *available,* and the universe has become more random, or "run down."

## *18–8 THE KELVIN TEMPERATURE SCALE

When we studied temperature scales in Chapter 15, we expressed the need for a temperature scale that doesn't depend on the properties or behavior of any particular material. We can now use the Carnot cycle to define such a scale. The efficiency of a Carnot engine operating between two heat reservoirs at temperatures $T_H$ and $T_C$ is independent of the nature of the working substance and depends only on the temperatures. If several Carnot engines with different working substances operate between the same two heat reservoirs, their thermal efficiencies are all the same:

$$e = \frac{Q_H + Q_C}{Q_H} = 1 + \frac{Q_C}{Q_H}.$$

Therefore the ratio $Q_C/Q_H$ is the same for *all* Carnot engines operating between two given temperatures $T_H$ and $T_C$.

Kelvin proposed that we *define* the ratio of the temperatures, $T_C/T_H$, to be equal to the magnitude of the ratio $Q_C/Q_H$ of the quantities of heat absorbed and rejected:

$$\frac{T_C}{T_H} = \frac{|Q_C|}{|Q_H|} = -\frac{Q_C}{Q_H}. \qquad (18\text{–}21)$$

Equation (18–21) looks identical to Eq. (18–13), but there is a subtle and crucial difference. The temperatures in Eq. (18–13) are based on an ideal-gas thermometer, as defined in Section 15–3, but Eq. (18–21) *defines* a temperature scale based on the Carnot cycle and the second law of thermodynamics and independent of the behavior of any particular substance. Thus the **Kelvin temperature scale** is truly *absolute*. To complete the definition of the Kelvin scale, we proceed, as in Section 15–3, to assign the arbitrary value of 273.16 K to the temperature of the triple point of water. When a substance is taken around a Carnot cycle, the ratio of the heats absorbed and rejected, $|Q_H|/|Q_C|$, is equal to the ratio of the temperatures of the reservoirs *as expressed on the gas scale,* defined in Section 15–3. Since, in both scales, the triple point of water is chosen to be 273.16 K, it follows that *the Kelvin and the ideal-gas scales are identical*.

The zero point on the Kelvin scale is called **absolute zero.** There are theoretical reasons for believing that absolute zero cannot be attained experimentally, although temperatures below $10^{-7}$ K have been achieved. The more closely we approach absolute zero, the more difficult it is to get closer. Absolute zero can also be interpreted on a molecular level, although this must be done with some care. Because of quantum effects, it would *not* be correct to say that at $T = 0$, all molecular motion ceases. At absolute zero the system has its *minimum* possible total energy (kinetic plus potential). One statement of the *third law of thermodynamics* is that it is impossible to reach absolute zero in a finite number of thermodynamic steps.

## 18–9 ENERGY RESOURCES: A Case Study in Thermodynamics

The laws of thermodynamics place very general limitations on conversion of energy from one form to another. In these times of increasing energy demand and diminishing resources these matters are of the utmost practical importance. We conclude this chapter with a brief discussion of a few energy-conversion systems, present and proposed.

Over half of the electric power generated in the United States is obtained from coal-fired steam-turbine generating plants. Modern boilers can transfer about 80% to 90% of the heat of combustion of coal into steam. The theoretical maximum thermal efficiency of the turbine, given by Eq. (18–14), is usually limited to about 0.55, and the actual efficiency is typically 90% of this value, or about 0.50. The efficiency of large generators in converting mechanical power to electrical is very high, typically 99%. Thus the overall thermal efficiency of such a plant is roughly (0.85) (0.50) (0.99), or about 40%.

The Tennessee Valley Authority's Paradise power plant has a generating capacity of about 1 gigawatt (1000 MW, or $10^9$ W) of electrical power. The steam is heated to 540°C (1004°F) at a pressure of 248 atm ($2.51 \times 10^7$ Pa or 3650 psi). The plant burns 10,500 tons of coal per day, and the overall thermal efficiency is 39.3%.

In a nuclear power plant the heat to generate steam is supplied by a nuclear reaction rather than the chemical reaction of burning coal. The steam turbines in nuclear power plants have the same theoretical efficiency limit as those in coal-fired plants. At present it is not practical to run nuclear reactors at temperatures and pressures as high as those in coal boilers, so the thermal efficiency of a nuclear plant is somewhat lower, typically 30%.

In the year 1990, 16% of the world's electric power supply came from nuclear power plants (19% in the United States). High construction costs, questions of public safety, and the problem of disposal of radioactive waste have slowed the development of additional nuclear power. However, there is also increasing concern about the environmental impact of coal-burning power plants. The health hazards of coal smoke are serious and well documented, and the burning of coal contributes to global warming through the greenhouse effect. It is important to manage our development of energy resources so as to minimize the risks to human life and health, and nuclear power continues to receive attention as a future energy resource. We noted at the end of Section 18–6 that thermal pollution from both coal-fired and nuclear plants poses serious environmental problems.

Solar energy is an inviting possibility. The power per unit surface area in the sun's radiation (before it passes through the earth's atmosphere) is about 1.4 kW/m$^2$. A maximum of about 1.0 kW/m$^2$ reaches the surface of the earth on a clear day, and the time

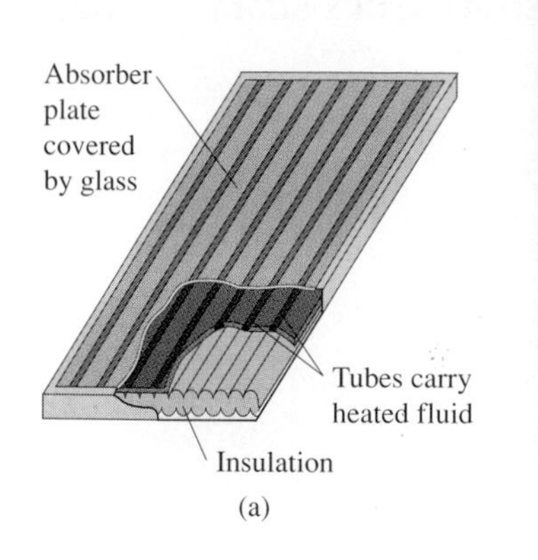

**18–13** (a) A passive solar collecting panel for home water heating. (b) An array of photovoltaic cells for direct conversion of sun power to electric power. (c) Solar-powered cars presently under development use photovoltaic cells to generate electric power for motion.

average, over a 24-hour period, is about 0.2 kW/m$^2$. This radiation can be collected and used to heat hot water for home use (Fig. 18–13a).

A different solar-energy scheme is to use large banks of photovoltaic (PV) cells (Fig. 18–13b) for direct conversion of solar energy to electricity. Such a system is not a heat engine in the usual sense and is not limited by the Carnot efficiency. There are other fundamental limitations on photocell efficiency, but 50% seems attainable in multilayer semiconductor photocells. Because of high capital costs, the price of PV-produced energy is about 25 cents per kilowatt-hour (compared with about 5 cents for conventional coal-fired steam-turbine plants). But such a system offers many potential advantages, such as lack of noise, lack of moving parts, minimal maintenance requirements, and freedom from pollution. Capital costs are likely to drop with improved technology, and PV systems continue to be studied and developed (Fig. 18–13c).

The energy of wind, fundamentally solar in origin, can be gathered and converted by "forests" of windmills (Fig. 18–14). About 17,000 wind turbines (85% of the U.S. total) are located in California, where the state government encourages their use. As with PV systems, the capital costs of such systems are higher than for coal-fired generating plants of equal capacity, although some installations are producing energy at a cost of about 7 cents per kilowatt-hour.

**18–14** An array of windmills to collect and convert wind and energy. The propellerlike blades turn electric generators, converting the kinetic energy of moving air into electrical energy.

Biomass is yet another mechanism for using solar energy. Plants absorb solar energy and store it; the energy can be utilized by burning the plant matter or by using fermentation and distillation to make alcohols. Ethanol made from fermented corn is used in the United States as a gasoline additive, although its cost is not quite competitive at present crude-oil price levels. An important feature of biomass materials is that they take as much carbon dioxide out of the atmosphere when they grow as is released when they burn; thus they don't contribute to the greenhouse effect.

An indirect scheme for collection and conversion of solar energy would use the temperature gradient in the ocean. In the Caribbean, for example, the water temperature near the surface is about 25°C, and at a depth of a few hundred meters it is about 10°C. The second law of thermodynamics forbids cooling the ocean and converting the extracted heat completely into work, but there is nothing to forbid running a heat engine between these two temperatures. The thermodynamic efficiency would be very low (about 5%), but a vast supply of energy would be available.

The present level of activity in energy-conversion research in the United States is rather low in comparison to the period of the 1970s, when we experienced a serious energy shortage. Political unrest in the Middle East has underlined once again the dangers of excessive reliance on nonrenewable resources such as oil, especially imported oil. Critics have accused the U.S. government of not having a coherent and effective energy policy compared with other countries. India, for example, has recently proposed an ambitious plan to develop, within the next decade, generating capacity for 15,000 MW of electric power from a combination of nonconventional sources (including all those mentioned above) and another 30,000 MW from nuclear energy. As we use up our available fossil-fuel resources, we will certainly have to become more active in developing alternative energy sources. The principles of thermodynamics will play a central role in this development.

## SUMMARY

- A reversible process is one whose direction can be reversed by an infinitesimal change in the conditions of the process. A reversible process is also an equilibrium process.
- A heat engine takes heat $Q_H$ from a source, converts part of it to work $W$, and discards the remainder $|Q_C|$ at a lower temperature. The thermal efficiency $e$ of a heat engine is

$$e = \frac{W}{Q_H} = 1 + \frac{Q_C}{Q_H} = 1 - \frac{|Q_C|}{|Q_H|}. \qquad (18\text{–}4)$$

- A gasoline engine operating on the Otto cycle with compression ratio $r$ has a theoretical maximum thermal efficiency $e$ given by

$$e = 1 - \frac{1}{r^{\gamma - 1}}. \qquad (18\text{–}6)$$

- A refrigerator takes heat $Q_C$ from a cold place, has a work input $W$, and discards heat $|Q_H|$ at a warmer place. The performance coefficient $K$ is defined as

$$K = \frac{Q_C}{|W|} = \frac{|Q_C|}{|Q_H| - |Q_C|}. \qquad (18\text{–}9)$$

- The second law of thermodynamics describes the directionality of natural thermodynamic processes. It can be stated in several equivalent forms. The "engine" statement: No cyclic process can convert heat completely into work. The "refrigerator"

**KEY TERMS**

reversible process
equilibrium process
irreversible process
heat engine
working substance
cyclic process
thermal efficiency
compression ratio
Otto cycle
Diesel cycle
refrigerator
performance coefficient
energy efficiency rating (EER)
heat pump
second law of thermodynamics
Carnot cycle
thermal pollution
entropy
Kelvin temperature scale
absolute zero

statement: No cyclic process can transfer heat from a cold place to a hotter place with no input of mechanical work.

• The Carnot cycle operates between two heat reservoirs at temperatures $T_H$ and $T_C$ and uses only reversible processes. Its thermal efficiency is

$$e_{\text{Carnot}} = 1 - \frac{T_C}{T_H} = \frac{T_H - T_C}{T_H}. \tag{18–14}$$

Two additional equivalent statements of the second law: No engine operating between the same two temperatures can be more efficient than a Carnot engine. All Carnot engines operating between the same two temperatures have the same efficiency.

• A Carnot engine run backward makes a Carnot refrigerator. Its performance coefficient is

$$K_{\text{Carnot}} = \frac{T_C}{T_H - T_C}. \tag{18–15}$$

Another form of the second law: No refrigerator operating between the same two temperatures can have a larger performance coefficient than a Carnot refrigerator. All Carnot refrigerators operating between the same two temperatures have the same performance coefficient.

• Entropy is a quantitative measure of the disorder of a system. The entropy change in any reversible process is

$$\Delta S = \int_1^2 \frac{dQ}{T} \quad \text{(reversible process)}, \tag{18–18}$$

where $T$ is absolute temperature. Entropy depends only on the state of the system, and the change in entropy between given initial and final states is the same for all processes leading from one to the other. This fact can be used to find the entropy change in an irreversible process, where Eq. (18–18) is not applicable.

• The second law can be stated in terms of entropy: The entropy of an isolated system may increase but can never decrease. When a system interacts with its surroundings, the total entropy of system and surroundings can never decrease. When the interaction involves only reversible processes, the total entropy is constant; when there is any irreversible process, the total entropy increases.

• The Kelvin temperature scale is based on the efficiency of the Carnot cycle and is independent of the properties of any specific material. The zero point on the Kelvin scale is called absolute zero.

# EXERCISES

## Section 18–2 Heat Engines

**18–1 A Diesel Engine.** A large diesel engine takes in 8000 J of heat and delivers 2000 J of work per cycle. The heat is obtained by burning diesel fuel with a heat of combustion of $5.00 \times 10^4$ J/g. a) What is the thermal efficiency? b) How much heat is discarded in each cycle? c) What mass of fuel is burned in each cycle? d) If the engine goes through 40.0 cycles per second, what is its power output in watts? In horsepower?

**18–2** A gasoline engine has a power output of 20.0 kW (about 27 hp). Its thermal efficiency is 30.0%. a) How much heat must be supplied to the engine per second? b) How much heat is discarded by the engine per second?

**18–3** A nuclear-power plant has a mechanical-power output (used to drive an electric generator) of 200 MW. Its rate of heat input from the nuclear reactor is 700 MW. a) What is the thermal efficiency of the system? b) At what rate is heat discarded by the system?

**18–4** An engine takes in 8000 J of heat and discards 5000 J each cycle. a) What is the mechanical work output of the engine during one cycle? b) What is the thermal efficiency of the engine?

**18–5** A gasoline engine performs 9000 J of mechanical work and discards 6000 J of heat each cycle. a) How much heat must be supplied to the engine in each cycle? b) What is the thermal efficiency of the engine?

## Section 18–3 Internal-Combustion Engines

**18–6** For a gas with $\gamma = 1.40$, what compression ratio $r$ must an Otto cycle have to achieve an ideal efficiency of 65.0%?

**18–7** For an Otto cycle with $\gamma = 1.40$ and $r = 7.00$ the temperature of the gasoline-air mixture when it enters the cylinder is 22.0°C (point $a$ of Fig. 18–3). What is the temperature at the end of the compression stroke (point $b$)?

## Section 18–4 Refrigerators

**18–8** A window air-conditioner unit absorbs $9.00 \times 10^4$ J of heat per minute from the room being cooled and in the same time period deposits $1.30 \times 10^5$ J of heat into the outside air. a) What is the power consumption of the unit in watts? b) What is the performance coefficient of the unit?

**18–9** A refrigerator has a performance coefficient of 2.00. During each cycle it absorbs $3.00 \times 10^4$ J of heat from the cold reservoir. a) How much mechanical energy is required each cycle to operate the refrigerator? b) During each cycle, how much heat is discarded to the high-temperature reservoir?

**18–10** A freezer has performance coefficient $K = 4.00$. The freezer is to convert 1.50 kg of water at $T = 20.0$°C to 1.50 kg of ice at $T = -10.0$°C in an hour. a) What amount of heat must be removed from the water at 20.0°C to convert it to ice at −10.0°C? b) How much electrical energy is consumed by the freezer during this hour? c) How much wasted heat is rejected to the room in which the freezer sits?

## Section 18–6 The Carnot Cycle

**18–11** Show that the efficiency $e$ of a Carnot engine and the performance coefficient $K$ of a Carnot refrigerator are related by $K = (1 - e)/e$. The engine and refrigerator operate between the same hot and cold reservoirs.

**18–12** A Carnot engine is operated between two heat reservoirs at temperatures of 500 K and 300 K. a) If the engine receives 5000 J of heat energy from the reservoir at 500 K in each cycle, how many joules per cycle does it reject to the reservoir at 300 K? b) How much mechanical work is performed by the engine during each cycle? c) What is the thermal efficiency of the engine?

**18–13** A Carnot engine whose high-temperature reservoir is at 400 K takes in 480 J of heat at this temperature in each cycle and gives up 335 J to the low-temperature reservoir. a) What is the temperature of the low-temperature reservoir? b) What is the thermal efficiency of the cycle? c) How much mechanical work does the engine perform during each cycle?

**18–14** A Carnot refrigerator is operated between two heat reservoirs at temperatures of 350 K and 250 K. a) If in each cycle the refrigerator receives 300 J of heat energy from the reservoir at 250 K, how many joules of heat energy does it deliver to the reservoir at 350 K? b) If the refrigerator goes through 3.0 cycles each second, what power input is required to operate the refrigerator? c) What is the performance coefficient of the refrigerator?

**18–15** An ice-making machine operates in a Carnot cycle; it takes heat from water at 0.0°C and rejects heat to a room at 27.0°C. Suppose 40.0 kg of water at 0.0°C are converted to ice at 0.0°C. a) How much heat is rejected to the room? b) How much energy must be supplied to the refrigerator?

## Section 18–7 Entropy

***18–16** Two moles of an ideal gas undergo a reversible isothermal expansion at a temperature of 300 K. During this expansion the gas does 1800 J of work. What is the change of entropy of the gas?

***18–17 Entropy and Condensation.** What is the entropy change of 0.800 kg of steam at 1.00 atm pressure and 100°C when it condenses to 0.800 kg of water at 100°C?

***18–18** A sophomore with nothing better to do adds heat to 0.300 kg of ice at 0.0°C until it is all melted. a) What is the change in entropy of the water? b) The source of heat is a very massive body at a temperature of 20.0°C. What is the change in entropy of this body? c) What is the total change in entropy of the water and the heat source?

***18–19** Calculate the entropy change that occurs when 1.00 kg of water at 20.0°C is mixed with 2.00 kg of water at 60.0°C.

***18–20** A block of aluminum with a mass of 1.00 kg, initially at 100.0°C, is dropped into 0.500 kg of water initially at 0.0°C. a) What is the final temperature? b) What is the total change in entropy of the system?

***18–21** Two moles of an ideal gas undergo a reversible isothermal expansion from 0.0200 $m^3$ to 0.0500 $m^3$ at a temperature of 300 K. What is the change in entropy of the gas?

## Section 18–9 Energy Resources: A Case Study in Thermodynamics

**18–22 A Coal-Burning Power Plant.** A coal-fired steam-turbine power plant has a mechanical-power output of 500 MW and a thermal efficiency of 35.0%. a) At what rate must heat be supplied by burning coal? b) If the heat of combustion of coal is $2.50 \times 10^4$ J/g, what mass of coal is burned per second? Per day? c) At what rate is heat discarded by the system? d) If the discarded heat is given to water in a river and its temperature is to rise by 5.0 C°, what volume of water is needed per second? e) In part (d), if the river is 100 m wide and 5.0 m deep, what must be the speed of flow of the water?

**18–23 Solar Heating.** A well-insulated house of moderate size in a temperate climate requires a maximum heat input rate of 20.0 kW. If this heat is to be supplied by a solar collector with an average (night and day) energy input of 200 $W/m^2$ and a collection efficiency of 60.0%, what area of solar collector is required?

**18–24** An engine is to be built to extract power from the temperature gradient of the ocean. If the surface and deep-water temperatures are 25.0°C and 10.0°C, respectively, what is the maximum theoretical efficiency of such an engine?

**18–25** A solar water heater for domestic hot-water supply uses solar collecting panels with collection efficiency 50% in a location where the average solar-energy input is 200 $W/m^2$. If the water comes into the house at 15.0°C and is to be heated to 60.0°C, what volume of water can be heated per hour if the collector area is 30.0 $m^2$?

**18–26** a) A homeowner in a cold climate has a coal-burning furnace that burns 9000 kg (about 10 tons) of coal during a winter. The heat of combustion of coal is $2.50 \times 10^7$ J/kg. If stack losses (the amount of heat energy lost up the chimney) are 20%, how many joules were actually used to heat the house? b) The homeowner proposes to install a solar heating system, heating large tanks of water by solar radiation during the summer and using the stored energy for heating during the winter. Find the required dimensions of the storage tank, assuming it to be a cube, to store a quantity of energy equal to that computed in part (a). Assume that the water is raised to 49.0°C (120°F) in the summer and cooled to 27.0°C (81°F) in the winter.

**18–27** A "solar house" has storage facilities for $5.25 \times 10^9$ J (about 5 million Btu). Compare the space requirements (in $m^3$) for this storage on the assumption a) that the energy is stored in water heated from a minimum temperature of 21.0°C (70°F) to a maximum of 49.0°C (120°F); b) that the energy is stored in Glauber salt heated in the same temperature range.

Properties of Glauber Salt ($Na_2SO_4 \cdot 10\,H_20$):

| | |
|---|---|
| Specific heat capacity | |
| Solid | 1930 J/kg · K |
| Liquid | 2850 J/kg · K |
| Density | 1600 $kg/m^3$ |
| Melting point | 32.0°C |
| Heat of fusion | $2.42 \times 10^5$ J/kg |

## PROBLEMS

**18–28** A Carnot engine whose low-temperature reservoir is at 250 K has an efficiency of 40.0%. An engineer is assigned the problem of increasing this to 50.0%. a) By how many degrees must the temperature of the high-temperature reservoir be increased if the temperature of the low-temperature reservoir remains constant? b) By how many degrees must the temperature of the low-temperature reservoir be decreased if that of the high-temperature reservoir remains constant?

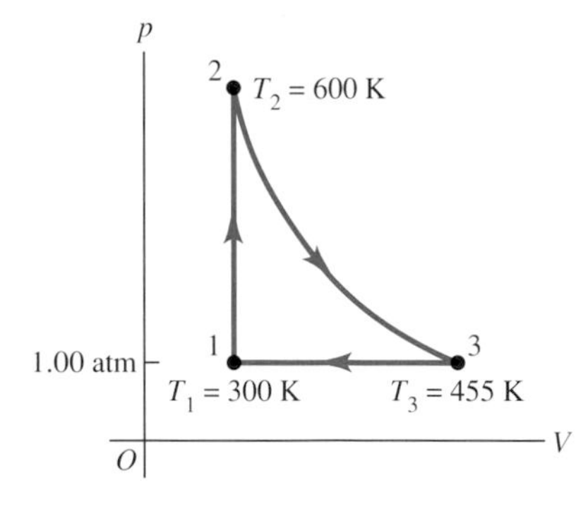


**FIGURE 18–15**

**18–29** A heat engine takes 0.300 mol of an ideal gas around the cycle shown in the $pV$-diagram of Fig. 18–15. Process $1 \rightarrow 2$ is at constant volume, process $2 \rightarrow 3$ is adiabatic, and process $3 \rightarrow 1$ is at a constant pressure of 1.00 atm. The value of $\gamma$ for this gas is 1.67. a) Find the pressure and volume at points 1, 2, and 3. b) Calculate $Q$, $W$, and $\Delta U$ for each of the three processes. c) Find the net work done by the gas in the cycle. d) Find the net heat flow into the engine in one cycle. e) What is the thermal efficiency of the engine?

**18–30** A cylinder contains oxygen at a pressure of 2.00 atm. The volume is 5.00 L, and the temperature is 300 K. The oxygen is carried through the following processes:

1. Heated at constant pressure from the initial state (state 1) to state 2, which has $T = 500$ K.
2. Cooled at constant volume to 250 K (state 3).
3. Cooled at constant pressure to 150 K (state 4).
4. Heated at constant volume to 300 K, which takes the system back to state 1.

a) Show these four processes in a $pV$-diagram, giving the numerical values of $p$ and $V$ for each of the four states. b) Calculate $Q$ and $W$ for each of the four processes. c) Calculate the net work done by the oxygen. d) What is the efficiency of this device as a heat engine?

**18–31** What is the thermal efficiency of an engine that operates by taking $n$ moles of an ideal gas through the following cycle (Fig. 18–16)? Let $C_V = 20.5$ J/mol · K.

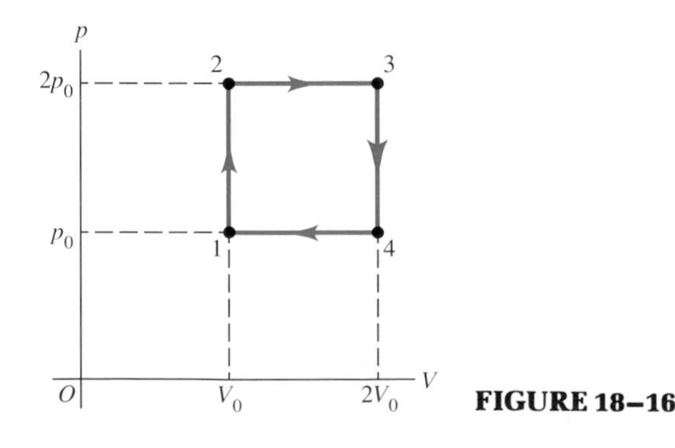


**FIGURE 18–16**

1. Start with $n$ moles at $p_0$, $V_0$, $T_0$.
2. Change to $2p_0$, $V_0$, at constant volume.
3. Change to $2p_0$, $2V_0$, at constant pressure.
4. Change to $p_0$, $2V_0$, at constant volume.
5. Change to $p_0$, $V_0$, at constant pressure.

**18–32** A monatomic ideal gas is taken around the cycle shown in Fig. 18–17 in the direction shown in the figure. The path for process $ca$ is a straight line in the $pV$-diagram. a) Calculate $Q$, $W$, and $\Delta U$ for each process $ab$, $bc$, and $ca$. b) What are $Q$, $W$, and $\Delta U$ for one complete cycle?

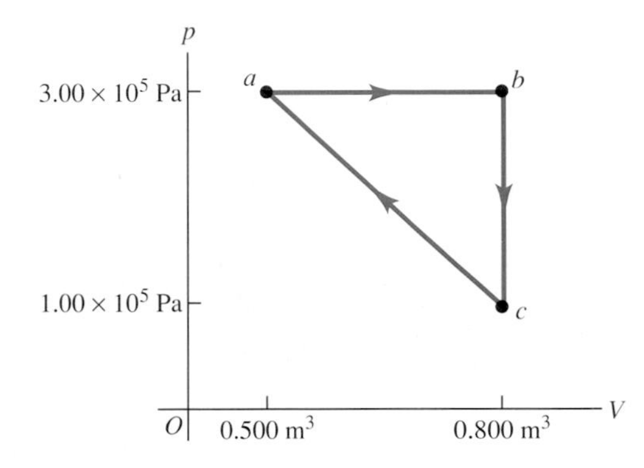


**FIGURE 18–17**

**18–33 Thermodynamic Processes for a Refrigerator.** A refrigerator operates on the cycle shown in Fig. 18–18. The compression ($da$) and expansion ($bc$) steps are adiabatic.

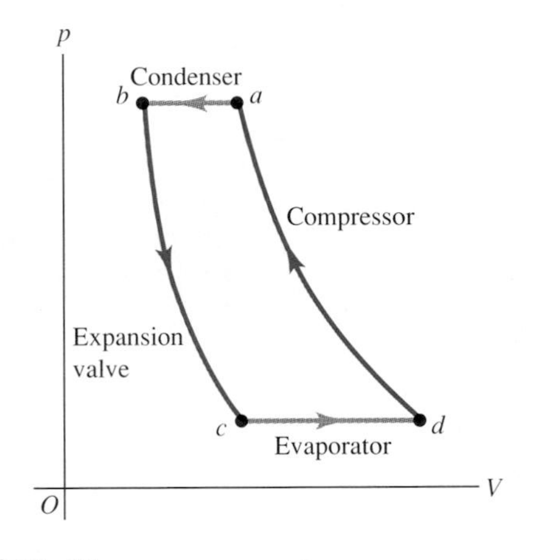


**FIGURE 18–18**

The temperature, pressure, and volume of the coolant in each of the four states $a$, $b$, $c$, and $d$ are given in the table below.

| State | $T$ (°C) | $P$ (kPa) | $V$ (m³) | $U$ (kJ) | Percentage that is liquid |
|---|---|---|---|---|---|
| $a$ | 80 | 2305 | 0.0682 | 1969 | 0 |
| $b$ | 80 | 2305 | 0.00946 | 1171 | 100 |
| $c$ | 5 | 363 | 0.2202 | 1005 | 54 |
| $d$ | 5 | 363 | 0.4513 | 1657 | 5 |

a) In each cycle, how much heat is taken from inside the refrigerator into the coolant while the coolant is in the evaporator? b) In each cycle, how much heat is exhausted from the coolant into the air outside the refrigerator while the coolant is in the condenser? c) In each cycle, how much work is done by the motor that operates the compressor? d) Calculate the performance coefficient of the refrigerator.

**18–34** A Carnot engine operates between two heat reservoirs at temperatures $T_H$ and $T_C$. An inventor proposes to increase the efficiency by running one engine between $T_H$ and an intermediate temperature $T'$ and a second engine between $T'$ and $T_C$ using the heat expelled by the first engine. Compute the efficiency of this composite system and compare it to that of the original engine.

***18–35** A 0.0600-kg cube of ice at an initial temperature of $-15.0$°C is placed in 0.500 kg of water at $T = 60.0$°C in an insulated container of negligible mass. Calculate the entropy change of the system.

***18–36** A physics student performing a heat-conduction experiment immerses one end of a copper rod in boiling water at 100°C and the other end in an ice-water mixture at 0°C. The sides of the rod are insulated. After steady-state conditions have been achieved in the rod, 0.250 kg of ice melts in a certain time interval. Find a) the entropy change of the boiling water; b) the entropy change of the ice-water mixture; c) the entropy change of the copper rod; d) the total entropy change of the entire system.

***18–37 A *TS*-Diagram.** a) Draw a graph of a Carnot cycle, plotting Kelvin temperature vertically and entropy horizontally (a temperature-entropy diagram, or $TS$-diagram). b) Show that the area under any curve in a temperature-entropy diagram represents the heat absorbed by the system. c) Derive from your diagram the expression for the thermal efficiency of a Carnot cycle.

## CHALLENGE PROBLEM

**18–38** Consider a Diesel cycle that starts (at point $a$ in Fig. 18–4) with 1.20 L of air at a temperature of 300 K and a pressure of $1.00 \times 10^5$ Pa. The air may be treated as an ideal gas. If the temperature at point $c$ is $T_c = 1200$ K, derive an expression for the efficiency of the cycle in terms of the compression ratio $r$. What is the efficiency when $r = 20.0$?

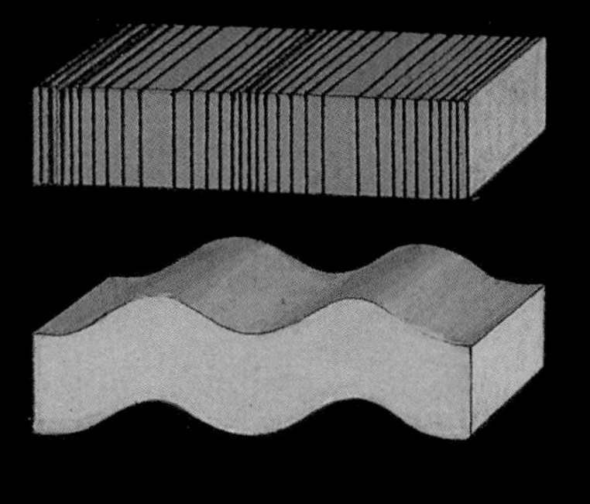

An earthquake generates two kinds of waves: primary waves (P-waves) and secondary waves (S-waves). P-waves are longitudinal waves (compression waves) and travel through solid or liquid rock. S-waves are transverse waves (shear waves) and do not pass through the molten rock in the earth's core. The speeds of both waves depend on the density of the rock through which they pass.

# CHAPTER 19

# Mechanical Waves

Seismic waves from the San Francisco earthquake centered at Loma Prieta, CA, at 5:04 PM local time on October 17, 1989 (0h 4m 15s on October 18, 1989, Universal Time Coordinated) were detected by seismographs around the world. Deviations from expected detection times were generally less than 10 seconds.

Kevo, Finland, detected P-waves at 0h 15m 35s UTC, reflected P-waves at 0h 18m 15s UTC, and S-waves at 0h 24m 50s UTC.

The San Francisco earthquake began Oct. 18, 1989, at 0h 4m 15s

Vienna, Austria, detected P-waves at 0h 17m 02 s UTC, S-waves at 0h 27m 44s UTC, and refracted P-waves at 0h 28m 56s UTC.

Inner core

Outer core

Mantle

Crust

Caracas, Venezuela, detected P-waves at 0h 13m 54s UTC and S-waves at 0h 21m 45s UTC

The main shock of the Loma Prieta earthquake registered 7.1 on the Richter scale. Numerous landslides occurred in the area, and 62 people were killed.

## KEY CONCEPTS

**• A wave is a disturbance from an equilibrium condition that travels, or propagates, from one region of space to another. The speed of propagation is called the wave speed. Waves can be transverse, longitudinal, or a combination.**

**• In a periodic wave the disturbance at each point is a periodic function of time, and the pattern of the disturbance is a periodic function of distance. A periodic wave has definite frequency and wavelength. The simplest periodic waves are sinusoidal waves.**

**• A wave function describes the position of each point in a wave medium at any time.**

**• Wave speed in any wave medium, such as a stretched rope or an elastic solid, is determined by the elastic and inertial properties of the medium.**

**• A sound wave in a gas is a longitudinal wave. The wave speed is determined by the temperature and molecular mass of the gas.**

**• Waves transport energy from one region of space to another. For sinusoidal waves the rate of energy transport is proportional to the square of the frequency and to the square of the amplitude.**

## INTRODUCTION

When you go to the beach to enjoy the ocean surf, you're experiencing a wave motion. Ripples on a pond, musical sounds, sounds we *can't* hear, the wiggles of a Slinky stretched out on the floor—all these are *wave* phenomena. Waves can occur whenever a system is disturbed from its equilibrium position and when the disturbance can travel, or *propagate,* from one region of the system to another. Sound, light, ocean waves, radio and television transmission, and earthquakes are all wave phenomena. Waves occur in all branches of physical and biological science, and the concept of waves is one of the most important unifying threads running through the entire fabric of the natural sciences.

This chapter and the next two are about mechanical waves. Every such wave travels within some material called a *medium.* The speed of travel depends on the mechanical properties of the medium. Some waves are *periodic,* and the particles of the medium undergo periodic motion during wave propagation. If the motion of each particle is simple harmonic (sinusoidal), the wave is called a *sinusoidal* wave. The concepts of this chapter form a foundation for our later study of other kinds of waves, including electromagnetic waves.

## 19–1 TYPES OF MECHANICAL WAVES

Every type of mechanical wave is associated with some material or substance called the **medium** for that type. As the wave travels through the medium, the particles that make up the medium undergo displacements of various kinds, depending on the nature of the wave. Figure 19–1 shows waves produced by drops falling on a water surface.

Here are a few other examples of mechanical waves. In Fig. 19–2a the medium is a wire or rope under tension; it could also be a long spring, or Slinky, lying on a smooth floor. If we give the left end a small upward shake or wiggle, the wiggle travels down the length of the rope. Successive sections of spring go through the same up-and-down motion that we gave to the end but at successively later times. Because the displacements of the medium are perpendicular (transverse) to the direction of travel of the wave along the medium, this is called a **transverse wave.**

In Fig. 19–2b the medium is a liquid or gas in a tube with a rigid wall at the right end and a movable piston at the left end. If we give the piston a single back-and-forth motion, displacement and pressure fluctuations travel down the length of the medium. This time the motions of the particles of the medium are back and forth along the *same* direction in which the wave travels, and we call this a **longitudinal wave.**

In Fig. 19–2c the medium is water in a trough, such as an irrigation ditch or canal. When we move the flat board at the left end back and forth once, a wave disturbance travels down the length of the trough. In this case the displacements of the water have *both* longitudinal and transverse components.

Each of these systems has an equilibrium state. For the Slinky or stretched rope it is the state in which the system is at rest, stretched out along a straight line. For the fluid in a tube it is a state in which the fluid is at rest with uniform pressure, and for the water in a trough it is a smooth, level water surface. In each case the wave motion is a disturbance from the equilibrium state that travels from one region of the medium to another. In each case there are forces that tend to restore the system to its equilibrium position when it is displaced, just as the force of gravity tends to pull a pendulum toward its straight-down equilibrium position when it is displaced.

These examples have three things in common. First, in each case the disturbance travels, or *propagates,* with a definite speed through the medium. This speed is called the speed of propagation, or simply the **wave speed.** It is determined in each case by the mechanical properties of the medium. We will use the symbol $v$ for wave speed. Second,

**19–1** A series of drops falling vertically into water produces a wave pattern that moves radially outward from its source. The wave crests and troughs are concentric circles.

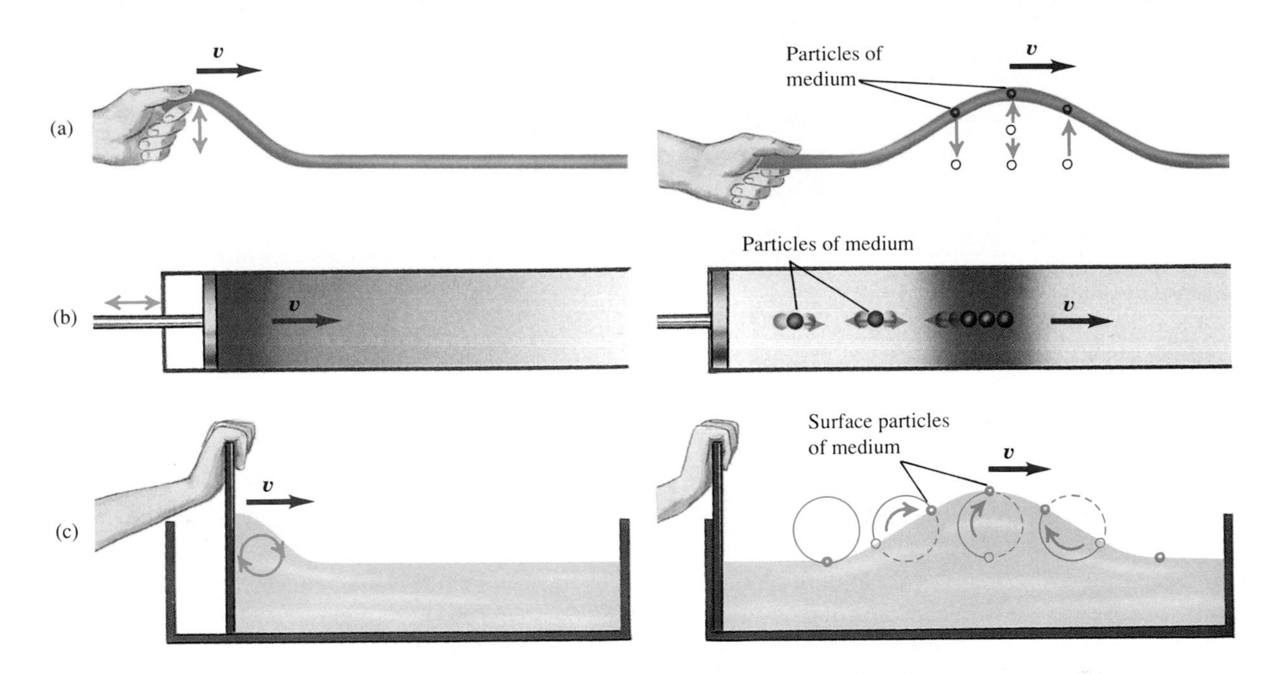


**19–2** (a) The hand moves the spring to the side, then returns, producing a transverse wave. (b) The piston compresses the liquid or gas to the right, then returns, producing a longitudinal wave. (c) The board pushes to the right, then returns, producing a sum of transverse and longitudinal waves. In all three cases the solitary wave propagates to the right.

the medium itself does not travel through space; its individual particles undergo back-and-forth or up-and-down motions around their equilibrium positions. The pattern of the wave disturbance is what travels. Third, to set any of these systems into motion, we have to put in energy by doing mechanical work on the system. The wave motion transports this energy from one region of the medium to another. *Waves transport energy, but not matter, from one region to another.*

Not all waves are mechanical in nature. Another broad class is *electromagnetic waves,* including light, radio waves, infrared and ultraviolet radiation, X-rays, and gamma rays. There is *no* medium for electromagnetic waves; they can travel through empty space. Yet another class of wave phenomena is the wavelike behavior of atomic and subatomic particles. This behavior forms part of the foundation of quantum mechanics, the basic theory used for the analysis of atomic and molecular structure. We will return to electromagnetic waves in later chapters. Meanwhile, we can learn the essential language of waves in the context of mechanical waves.

## 19–2 PERIODIC WAVES

One of the easiest kinds of wave to demonstrate is a transverse wave on a stretched string (which may also be a rope or a flexible wire or cable). Suppose we tie one end of a long flexible string to a stationary object and pull on the other end, stretching the string out horizontally. We then give this end an up-and-down shake, exerting a transverse force on it as we do so. The result is a "wiggle," or *wave pulse,* that travels down the length of the string. The tension in the string restores its straight-line shape once the wiggle has passed.

A more interesting situation develops when we give the free end of the string a repetitive, or *periodic,* motion. (You may want to review the discussion of periodic motion in Chapter 13 before going ahead.) In particular, suppose we move the string up and down with *simple harmonic motion* with amplitude $A$, frequency $f$, angular frequency $\omega = 2\pi f$, and period $T = 1/f = 2\pi/\omega$. A possible experimental setup is shown in Fig. 19–3. As we will see, waves with simple harmonic motion are particularly easy to analyze; we call them **sinusoidal waves.** It also turns out that *any* periodic wave can be

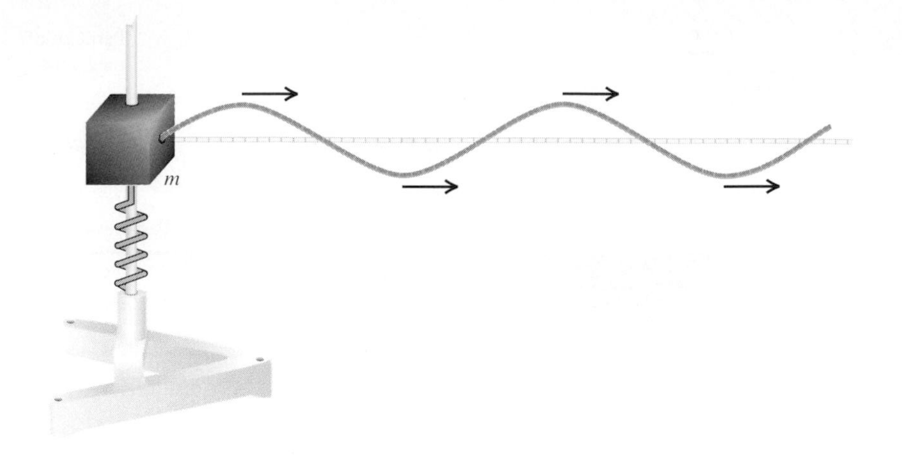


**19–3** The spring-mass system undergoes simple harmonic motion, producing a sinusoidal wave that travels to the right on the string. In a real-life system a driving force would have to be applied to the mass $m$ to replace the energy carried away by the wave.

represented as a combination of sinusoidal waves. So this particular kind of wave motion is worth special attention.

In Fig. 19–3 a *continuous succession* of transverse sinusoidal waves advances along the string. Figure 19–4 shows the shape of a part of the string near the left end, at intervals of $\frac{1}{8}$ period, for a total time of one period. The waveform advances steadily toward the right, as indicated by the short red arrow pointing to a particular wave crest, while any one point on the string (the blue dot, for example) oscillates up and down about its equilibrium position with simple harmonic motion. Be very careful to distinguish between the motion of a *waveform,* which moves with constant speed $v$ *along* the string, and the motion of a *particle of the string,* which is simple harmonic and *transverse* (perpendicular) to the length of the string.

The shape of the string at any instant is a repeating pattern, a series of identical shapes. For a periodic wave the length of one complete wave pattern is the distance between any two points at corresponding positions on successive repetitions in the wave shape. We call this the **wavelength** of the wave, denoted by $\lambda$. The waveform travels with constant speed $v$ and advances a distance of one wavelength $\lambda$ in a time interval of one period $T$, so the wave speed $v$ is given by $v = \lambda/T$, or, because $f = 1/T$,

$$v = \lambda f. \tag{19–1}$$

The speed of propagation equals the product of wavelength and frequency.

To understand the mechanics of a *longitudinal* wave, we consider a long tube filled with a fluid, with a plunger at the left end (Fig. 19–5). If we push the plunger in, we compress the fluid near the plunger, increasing the pressure in this region. This region then pushes against the neighboring region of fluid, and so on, as a wave pulse moves along the tube. The pressure fluctuations play the role of the restoring force, trying to restore the fluid to equilibrium and uniform pressure.

Now suppose we move the plunger back and forth with simple harmonic motion along a line parallel to the axis of the tube. This motion forms regions in the fluid in which the pressure and density are greater or less than the equilibrium values. We call a region of increased pressure a *compression*. In the figure, compressions are represented by darkly shaded areas. A region of reduced pressure is an *expansion;* in the figure, expansions are represented by lightly shaded areas. The compressions and expansions move to the right with constant speed $v$, as indicated by successive positions of the small vertical arrow.

The motion of a single particle of the medium, shown by a black dot, is simple harmonic, parallel to the direction of propagation. The wavelength is the distance

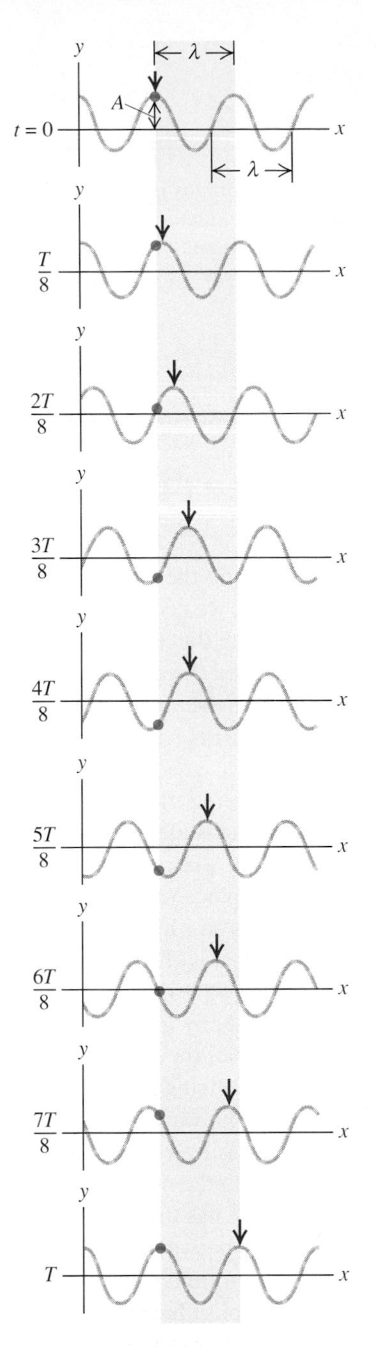


**19–4** Sinusoidal transverse wave traveling toward the right. The shape of the string is shown at intervals of $\frac{1}{8}$ of a period. The vertical scale is exaggerated.

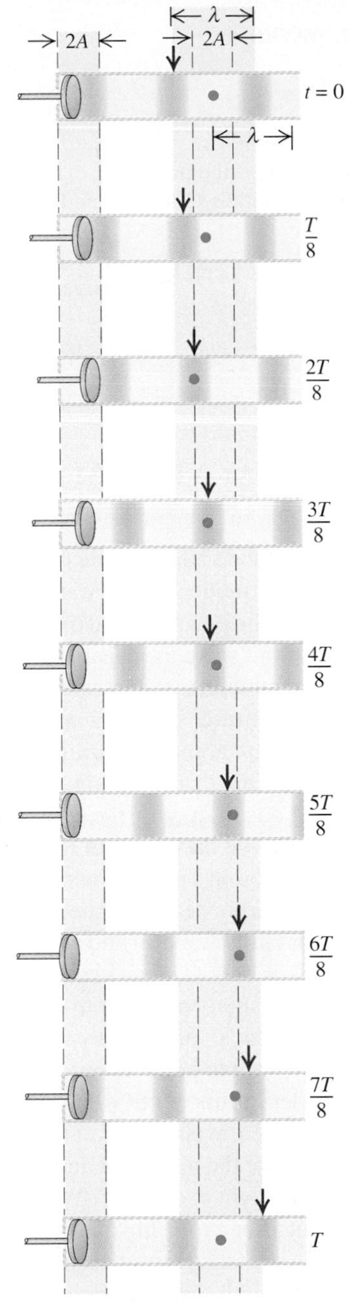


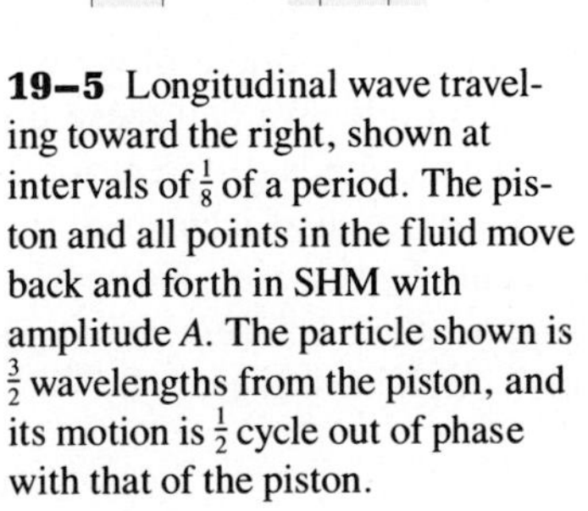
**19–5** Longitudinal wave traveling toward the right, shown at intervals of $\frac{1}{8}$ of a period. The piston and all points in the fluid move back and forth in SHM with amplitude $A$. The particle shown is $\frac{3}{2}$ wavelengths from the piston, and its motion is $\frac{1}{2}$ cycle out of phase with that of the piston.

between two successive compressions or two successive expansions. The same fundamental equation, $v = \lambda f$, holds for this example, as for transverse waves and indeed for *all* types of periodic waves. The speed of longitudinal waves in air (sound waves) depends on temperature. At 20°C it is 344 m/s, or 1130 ft/s.

### EXAMPLE 19–1

What is the wavelength of a sound wave in air at 20°C if the frequency is $f = 262$ Hz (the approximate frequency of middle C on the piano)?

**SOLUTION** At 20°C the speed of sound in air is $v = 344$ m/s, as we mentioned above. From Eq. (19–1),

$$\lambda = \frac{v}{f} = \frac{344 \text{ m/s}}{262 \text{ s}^{-1}} = 1.31 \text{ m}.$$

The high C sung by coloratura sopranos is two octaves above middle C. Each octave corresponds to a factor of two in frequency, so the frequency of high C is four times that of middle C, or $f = 4(262 \text{ Hz}) = 1048$ Hz, and the wavelength is one-fourth as large, $\lambda = (1.31 \text{ m})/4 = 0.328$ m.

## 19–3 MATHEMATICAL DESCRIPTION OF A WAVE

We have described some characteristics of periodic waves using the concepts of wave speed, period, frequency, and wavelength. Sometimes, though, we need a more detailed description of the positions and motions of individual particles of the medium at particular times during wave propagation. For this description we need the concept of a **wave function,** a function that describes the position of any particle in the medium at any time. We will concentrate on *sinusoidal waves,* in which each particle undergoes *simple harmonic motion* about its equilibrium position.

As a specific example, let's look at waves on a stretched string. If we ignore the sag of the string due to gravity, the equilibrium position of the string is along a straight line. We take this to be the $x$-axis of a coordinate system. Waves on a string are *transverse;* during wave motion a particle with equilibrium position $x$ is displaced some distance $y$ in the direction perpendicular to the $x$-axis. The value of $y$ depends on which particle we are talking about (that is, on $x$) and also on the time $t$ when we look at it. Thus $y$ is a *function* of $x$ and $t$: $y = f(x, t)$. If we know this function for a particular wave motion, we can use it to find the displacement (from equilibrium) of any particle at any time. From this we can find the velocity and acceleration of any particle, the shape of the string, and in fact anything we want to know about the position and motion of the string at any time.

Now let's think about what a wave function for a sinusoidal wave might look like. Suppose a wave travels from left to right (the direction of increasing $x$) along the string. We can compare the motion of any one particle of the string with the motion of a second particle to the right of the first. We find that the second particle has the same motion as the first, but after a *time lag* that is proportional to the distance between the particles. If one end of a stretched string oscillates with simple harmonic motion, every other point also oscillates with simple harmonic motion with the same amplitude and frequency.

But the cyclic motions of various points are out of step with each other by various fractions of a cycle. We call these differences *phase differences,* and we say that the **phase** of the motion is different for different points. For example, if one point has its maximum positive displacement at the same time as another has its maximum negative displacement, the two are a half-cycle out of phase.

Suppose the displacement of a particle at the left end of the string (at $x = 0$), where the motion originates, is given by

$$y = A \sin \omega t = A \sin 2\pi f t. \tag{19–2}$$

The wave disturbance travels from $x = 0$ to some point $x$ to the right of the origin in an amount of time given by $x/v$, where $v$ is the wave speed. So the motion of point $x$ at time $t$ is the same as the motion of point $x = 0$ at the earlier time $t - x/v$, and we can find the displacement of point $x$ at time $t$ by simply replacing $t$ in Eq. (19–2) by $(t - x/v)$. When we do that, we find

$$y(x, t) = A \sin \omega\left(t - \frac{x}{v}\right) = A \sin 2\pi f\left(t - \frac{x}{v}\right). \tag{19–3}$$

The notation $y(x, t)$ is a reminder that the displacement $y$ is a function of both the location $x$ of the point and the time $t$. We could make Eq. (19–3) more general by including a phase angle, as we did for simple harmonic motion in Section 13–3, but for now we omit this.

We can rewrite Eq. (19–3) in several useful ways, presenting the same relationship in different forms. We can express it in terms of the period $T = 1/f$ and the wavelength $\lambda = v/f$:

$$y(x, t) = A \sin 2\pi\left(\frac{t}{T} - \frac{x}{\lambda}\right). \tag{19–4}$$

We get another convenient form if we define a quantity $k$, called the **wave number,** or the **propagation constant:**

$$k = \frac{2\pi}{\lambda}. \tag{19–5}$$

In terms of $k$ and the angular frequency $\omega$ the wavelength-frequency relation $v = \lambda f$ becomes

$$\omega = vk, \tag{19–6}$$

and we can rewrite Eq. (19–4) as

$$y(x, t) = A \sin(\omega t - kx). \tag{19–7}$$

Which of these various forms we use in any specific problem is a matter of convenience. They all say the same thing, but they express it in terms of different quantities. Note that $\omega$ has units rad/s, so for unit consistency in Eqs. (19–6) and (19–7) the wave number $k$ must have the units rad/m. (A few authors define the wave number as $1/\lambda$ rather than $2\pi/\lambda$. In that case our $k$ is still called the propagation constant.)

At any specific time $t$, Eq. (19–3), (19–4), or (19–7) gives the displacement $y$ of a particle from its equilibrium position as a function of the *coordinate* $x$ of the particle. If the wave is a transverse wave on a string, the equation represents the *shape* of the string at that instant, as if we had taken a stop-action photograph of the string. In particular, at time $t = 0$,

$$y = A \sin(-kx) = -A \sin kx = -A \sin 2\pi\frac{x}{\lambda}.$$

This curve is plotted in Fig. 19–6.

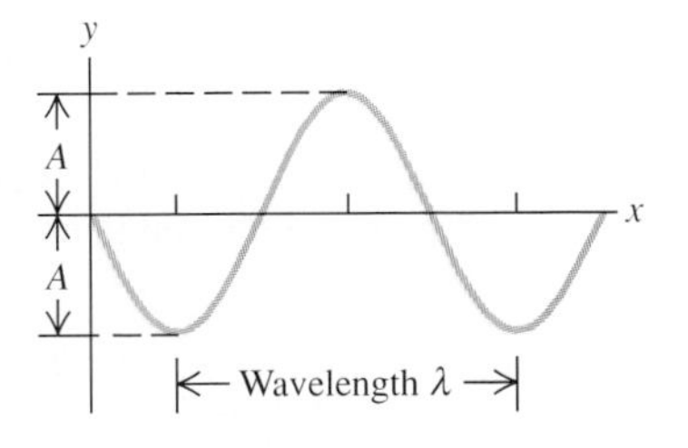


**19–6** Waveform at $t = 0$. The vertical scale is exaggerated.

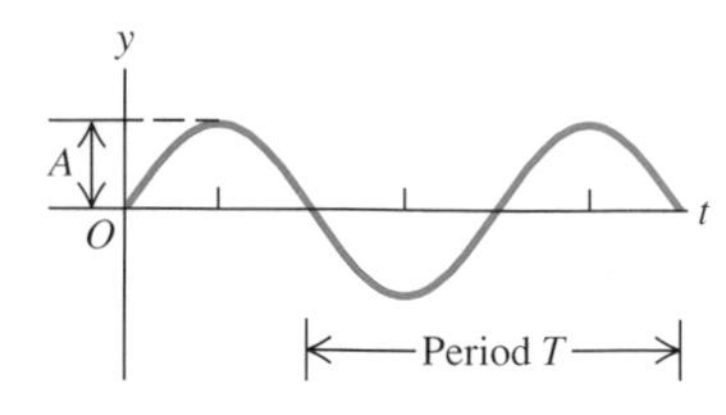


**19–7** Motion of a point at $x = 0$ as a function of time. The vertical scale is exaggerated.

At any specific *coordinate* $x$, Eq. (19–3), (19–4), or (19–7) gives the displacement $y$ of the particle at that coordinate as a function of *time*. That is, it describes the motion of that particle. In particular, at the position $x = 0$,

$$y = A \sin \omega t = A \sin 2\pi \frac{t}{T}.$$

This is consistent with our original assumption about the motion at $x = 0$. This curve is plotted in Fig. 19–7, which is *not* a picture of the shape of the string but rather a graph of the position $y$ of a particle as a function of time.

## PROBLEM-SOLVING STRATEGY

### Mechanical waves

**1.** It helps to make a distinction between *kinematics* problems and *dynamics* problems. In kinematics problems, we are concerned only with *describing* motion; the relevant quantities are wave speed, wavelength (or wave number), frequency (or angular frequency), amplitude, and the position, velocity, and acceleration of individual particles. In dynamics problems, concepts such as force and mass enter; the relation of wave speed to the mechanical properties of a system is an example. We'll get into these relations in the next few sections.

**2.** If $f$ is given, you can find $T = 1/f$, and vice versa. If $\lambda$ is given, you can find $k = 2\pi/\lambda$, and vice versa. If any two of the quantities $v$, $\lambda$, and $f$ (or $v$, $k$, and $\omega$) are known, you can find the third, using $v = \lambda f$ (or $\omega = vk$). In some problems that's all you need. To determine the wave function completely, you need to know $A$ and any two of $v$, $\lambda$, and $f$ (or $v$, $k$, and $\omega$). Once you have this information, you can use it in Eq. (19–3), (19–4), or (19–7) to get the specific wave function for the problem at hand. Once you have that, you can find the value of $y$ at any point (any value of $x$) and at any time by substituting into the wave function.

**3.** If the wave speed $v$ is not given, you may be able to find it in one of two ways: Use the frequency-wavelength relation $v = \lambda f$, or use relations between $v$ and the mechanical properties of the system, such as tension and mass per unit length for a string. We will develop these relations in the next three sections. Which method you use to find $v$ will depend on what information you are given.

## EXAMPLE 19–2

**Wave on a clothesline rope** Your cousin Throckmorton is playing with the clothesline (Fig. 19–8). He unties one end, holds it taut, and wiggles the end up and down sinusoidally with frequency 5.0 Hz and amplitude 0.010 m. The wave speed is $v = 10.0$ m/s. At time $t = 0$, the end has zero displacement and is moving in the $+y$-direction. Assume that no wave bounces back from the far end to muddle up the pattern. a) Find the amplitude, angular frequency, period, wavelength, and wave number of the wave. b) Write a wave function describing the wave. c) Find the position of the point at $x = 0.25$ m at time $t = 0.10$ s.

**SOLUTION** a) The amplitude $A$ of the wave is just the amplitude of the motion of the end point, $A = 0.010$ m. The angular frequency is

$$\begin{aligned}\omega &= 2\pi f = (2\pi \text{ rad/cycle})(5.0 \text{ cycles/s}) \\ &= 31.4 \text{ rad/s}.\end{aligned}$$

The period is $T = 1/f = 0.20$ s. We get the wavelength from Eq. (19–1):

$$\lambda = \frac{v}{f} = \frac{10.0 \text{ m/s}}{5.0 \text{ s}^{-1}} = 2.0 \text{ m}.$$

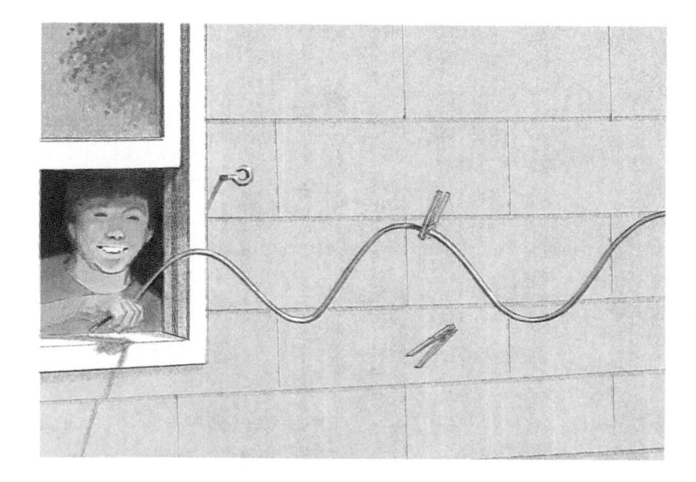

**19–8** Throcky making waves on a clothesline.

We find the wave number from Eq. (19–5) or (19–6):

$$k = \frac{2\pi}{\lambda} = \frac{2\pi \text{ rad}}{2.0 \text{ m}} = 3.14 \text{ rad/m}, \qquad \text{or}$$

$$k = \frac{\omega}{v} = \frac{31.4 \text{ rad/s}}{10.0 \text{ m/s}} = 3.14 \text{ rad/m}.$$

b) The wave function, in the form of Eq. (19–4), is

$$y = A \sin 2\pi\left(\frac{t}{T} - \frac{x}{\lambda}\right) = (0.010 \text{ m}) \sin 2\pi\left(\frac{t}{0.20 \text{ s}} - \frac{x}{2.0 \text{ m}}\right)$$
$$= (0.010 \text{ m}) \sin [(31.4 \text{ rad/s})t - (3.14 \text{ rad/m})x].$$

We can also get this same equation from Eq. (19–7), using the values of $\omega$ and $k$ obtained above.

c) We find the displacement of the point $x = 0.25$ m at time $t = 0.10$ s by substituting these values into the above wave equation:

$$y = (0.010 \text{ m}) \sin 2\pi\left(\frac{0.10 \text{ s}}{0.20 \text{ s}} - \frac{0.25 \text{ m}}{2.0 \text{ m}}\right)$$
$$= (0.010 \text{ m}) \sin 2\pi(0.375) = 0.0071 \text{ m}.$$

Make sure you understand that the quantity $2\pi(0.375)$ represents an angle measured in *radians*.

---

We can modify Eqs. (19–3) through (19–7) to represent a wave traveling in the *negative x*-direction. In this case the displacement of point $x$ at time $t$ is the same as the motion of point $x = 0$ at the *later* time $(t + x/v)$, so in Eq. (19–2) we replace $t$ by $(t + x/v)$. For a wave traveling in the negative $x$-direction,

$$y = A \sin 2\pi f\left(t + \frac{x}{v}\right) = A \sin 2\pi\left(\frac{t}{T} + \frac{x}{\lambda}\right)$$
$$= A \sin (\omega t + kx). \qquad (19\text{–}8)$$

In the expression $y = A \sin (\omega t \pm kx)$ the quantity $(\omega t \pm kx)$ is called the *phase*. It plays the role of an angular quantity (always measured in radians) in Eq. (19–7) or (19–8), and its value for any values of $x$ and $t$ determines what part of the sinusoidal cycle is occurring at a particular point and time. For a positive crest (where $y = A$ and the sine function has the value 1), the phase could be $\pi/2$, $5\pi/2$, and so on; for a point of zero displacement, it could be 0, $\pi$, $2\pi$, and so on. The wave speed is the speed with which we have to move along with the wave to keep abreast of a point with *constant phase*. For a wave traveling in the $+x$-direction, that means $\omega t - kx =$ constant. Taking the derivative with respect to $t$, we find $\omega = k\, dx/dt$, or

$$\frac{dx}{dt} = \frac{\omega}{k}.$$

Comparing this with Eq. (19–6), we see that $dx/dt$ is equal to the speed $v$ of the wave. Because of this relationship, $v$ is sometimes called the *phase velocity* of the wave. (*Phase speed* would be a better term.)

From the wave function we can get an expression for the transverse velocity of any *particle* in a transverse wave; we call this $v_y$ to distinguish it from the wave propagation speed $v$. To find the transverse velocity $v_y$ at a particular point $x$, we take the derivative of the wave equation with respect to $t$, keeping $x$ constant. If the wave function is

$$y = A \sin (\omega t - kx),$$

then

$$v_y = \frac{\partial y}{\partial t} = \omega A \cos (\omega t - kx). \qquad (19\text{–}9)$$

The $\partial$ in this expression is a modified $d$, used to remind us that $y$ is really a function of *two* variables and that we are allowing only one ($t$) to vary while the other ($x$) is constant (because we are looking at a particular point on the string). This derivative is called a *partial derivative*. If you haven't reached this point yet in your study of calculus, don't fret; it's a simple idea.

Going on, we find the *acceleration* of any particle as the *second* partial derivative of $y$ with respect to $t$:

$$a_y = \frac{\partial^2 y}{\partial t^2} = -\omega^2 A \sin(\omega t - kx). \tag{19–10}$$

We can also compute partial derivatives of $y$ with respect to $x$, holding $t$ constant. This corresponds to studying the shape of the string at one instant of time, like a flash photo. The first derivative $\partial y/\partial x$ is the *slope* of the string at any point. The second partial derivative with respect to $x$ is

$$\frac{\partial^2 y}{\partial x^2} = -k^2 A \sin(\omega t - kx). \tag{19–11}$$

From Eqs. (19–10) and (19–11) and the relation $\omega = vk$ we see that

$$\frac{\partial^2 y/\partial t^2}{\partial^2 y/\partial x^2} = \frac{\omega^2}{k^2} = v^2,$$

$$\frac{\partial^2 y}{\partial x^2} = \frac{1}{v^2}\frac{\partial^2 y}{\partial t^2}. \tag{19–12}$$

The wave function $y = A \sin(\omega t + kx)$ also satisfies this relationship (verify this).

Equation (19–12), called the **wave equation,** is one of the most important equations in all of physics. Whenever it occurs, we know that the disturbance described by the function $y$ propagates as a wave along the $x$-axis with wave speed $v$.

The concept of wave function is equally useful with *longitudinal* waves, and everything that we have said about wave functions can be adapted to this case. The quantity $y$ still measures the displacement of a particle of the medium from its equilibrium position; the difference is that for a longitudinal wave this displacement is *parallel* to the $x$-axis instead of perpendicular to it. We'll discuss longitudinal waves in detail in Section 19–5.

## 19–4 SPEED OF A TRANSVERSE WAVE

How is the speed of propagation $v$ of a transverse wave on a string related to the *mechanical* properties of the system? The relevant physical quantities are the *tension* in the string and its *mass per unit length*. We might guess that increasing the tension should increase the restoring forces that tend to straighten the string when it is disturbed, thus increasing the wave speed. We might also guess that increasing the mass should make the motion more sluggish and decrease the speed. Both these guesses turn out to be right. We'll develop the exact relationship by two different methods. The first is simple in concept and considers a specific type of waveform; the second is more general but also more formal. Choose whichever you like better.

### Wave Speed on a String: First Method

We consider a perfectly flexible string (Fig. 19–9). In the equilibrium position the tension is $F$, and the linear mass density (mass per unit length) is $\mu$. (When portions of the string are displaced from equilibrium, the mass per unit length decreases a little, and the tension increases a little.) In Fig. 19–9a the string is at rest. We ignore the weight of the string; in the equilibrium position it forms a perfectly straight line.

Starting at time $t = 0$, we apply a constant transverse force $F_y$ at the left end of the string. We might expect that the end would move with constant acceleration; that would

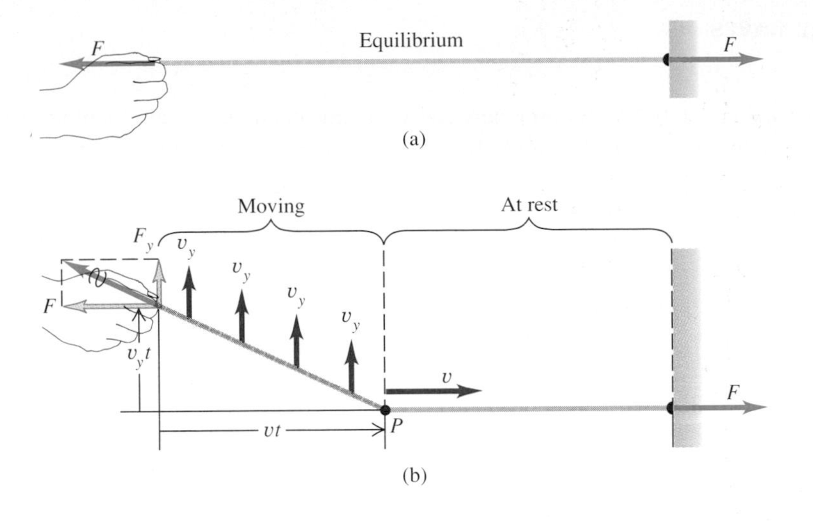


**19–9** Propagation of a transverse wave on a string. (a) String in equilibrium; (b) part of the string in motion.

happen if the force were applied to a *point* mass. But here the effect of the force $F_y$ is to set successively more and more mass in motion. The wave travels with constant speed $v$, so the division point between moving and nonmoving portions moves with constant speed $v$.

The total mass in motion is proportional to the time $t$ the force has been acting and thus to the *impulse* of the force up to time $t$. According to the impulse-momentum theorem (Section 8–5), the impulse is equal to the change in the total transverse component of momentum $(mv_y - 0)$ of the moving part of the string. Because the system started with *no* transverse momentum, this is equal to the total momentum at time $t$. The total momentum thus must increase proportionately with time, so the *change* of momentum must be associated entirely with the increasing amount of mass in motion, not with an increasing velocity of an individual mass element. That is, $mv_y$ changes because $m$ changes, not $v_y$. Thus the end of the string moves upward with constant *velocity* $v_y$.

Figure 19–9b shows the shape of the string at time $t$. All particles of the string to the left of point $P$ are moving upward with speed $v_y$, and all particles to the right of $P$ are still at rest. The boundary point $P$ between the moving and the stationary portions is traveling to the right along the string with speed equal to the speed of propagation or wave speed $v$. At time $t$ the left end of the string has moved up a distance $v_y t$, and the boundary point $P$ has advanced a distance $vt$.

The total force at the left end of the string has components $F$ and $F_y$. Why $F$? There is no motion in the direction along the length of the string, so there is no unbalanced horizontal force. Therefore $F$, the magnitude of the horizontal component, does not change when the string is displaced. In the displaced position the tension is $(F^2 + F_y{}^2)^{1/2}$ (greater than $F$), and the string stretches somewhat.

To derive an expression for the wave speed $v$, we apply the impulse-momentum theorem to the portion of the string that is in motion at time $t$, that is, the darkly shaded portion in Fig. 19–9b. We set the transverse *impulse* (transverse force times time) equal to the change of transverse *momentum* of the moving portion (mass times transverse component of velocity). The impulse of the transverse force $F_y$ in time $t$ is $F_y t$. By similar triangles,

$$\frac{F_y}{F} = \frac{v_y t}{vt}, \qquad F_y = F\frac{v_y}{v},$$

and

$$\text{Transverse impulse} = F_y t = F\frac{v_y}{v}t.$$

The mass of the moving portion of the string is the product of the mass per unit length $\mu$ and the length $vt$, or $\mu vt$. The transverse momentum is the product of this mass and the transverse velocity $v_y$:

$$\text{Transverse momentum} = (\mu vt)v_y.$$

We note again that the momentum increases with time *not* because mass is moving faster, as was usually the case in Chapter 8, but because *more mass* is brought into motion. But the impulse of the force $F_y$ is still equal to the total change in momentum of the system. Applying this relation, we obtain

$$F\frac{v_y}{v}t = \mu v t v_y.$$

Solving this for $v$, we find

$$v = \sqrt{\frac{F}{\mu}} \quad \text{(transverse wave).} \tag{19–13}$$

This confirms our prediction that the wave speed $v$ should increase when the tension $F$ increases but decrease when the mass per unit length $\mu$ increases.

Note that $v_y$ does not appear in Eq. (19–13). The wave speed doesn't depend on $v_y$. Our calculation considered only a very special kind of pulse, but we can consider *any* shape of wave disturbance as a series of pulses with different values of $v_y$. So even though we derived Eq. (19–13) for a special case, it is valid for *any* transverse wave motion on a string, including, in particular, the sinusoidal and other periodic waves that we discussed in Section 19–2.

## Wave Speed on a String: Second Method

Here is an alternative derivation of Eq. (19–13). If you aren't comfortable with partial derivatives, it can be omitted. We apply Newton's second law, $\Sigma \mathbf{F} = m\mathbf{a}$, to a small segment of string whose length in the equilibrium position is $\Delta x$ (Fig. 19–10). The mass of the segment is $m = \mu\ \Delta x$; the forces at the ends are represented in terms of their $x$- and $y$-components. The $x$-components have equal magnitude $F$ and add to zero because the motion is transverse and there is no component of acceleration in the $x$-direction. To obtain $F_{1y}$ and $F_{2y}$, we note that the ratio $F_{1y}/F$ is equal in magnitude to the slope of the string at point $x$ and that $F_{2y}/F$ is equal to the slope at point $x + \Delta x$. Taking proper account of signs, we find

$$\frac{F_{1y}}{F} = -\left(\frac{\partial y}{\partial x}\right)_x, \qquad \frac{F_{2y}}{F} = \left(\frac{\partial y}{\partial x}\right)_{x+\Delta x}. \tag{19–14}$$

The notation reminds us that the derivatives are evaluated at points $x$ and $x + \Delta x$, respectively. The net $y$-component of force is

$$F_y = F_{1y} + F_{2y} = F\left[\left(\frac{\partial y}{\partial x}\right)_{x+\Delta x} - \left(\frac{\partial y}{\partial x}\right)_x\right]. \tag{19–15}$$

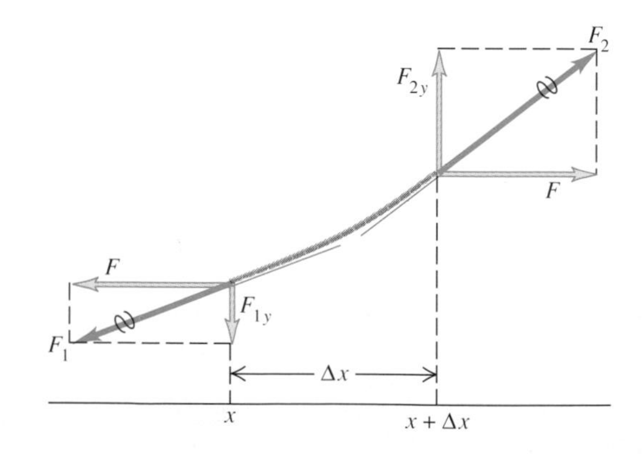


**19–10** Free-body diagram for a segment of string whose length in its equilibrium position is $\Delta x$. The force at each end of the string is tangent to the string at the point of application; each force is represented in terms of its $x$- and $y$-components.

We now equate this to the mass $\mu\,\Delta x$ times the $y$-component of acceleration $\partial^2 y/\partial t^2$ to obtain

$$F\left[\left(\frac{\partial y}{\partial x}\right)_{x+\Delta x} - \left(\frac{\partial y}{\partial x}\right)_x\right] = \mu\,\Delta x\,\frac{\partial^2 y}{\partial t^2}, \tag{19–16}$$

or, dividing by $F\,\Delta x$,

$$\frac{\left(\frac{\partial y}{\partial x}\right)_{x+\Delta x} - \left(\frac{\partial y}{\partial x}\right)_x}{\Delta x} = \frac{\mu}{F}\frac{\partial^2 y}{\partial t^2}. \tag{19–17}$$

We now take the limit as $\Delta x \rightarrow 0$. In this limit the left side of Eq. (19–17) becomes the derivative of $\partial y/\partial x$ with respect to $x$ (at constant $t$), that is, the *second* (partial) derivative of $y$ with respect to $x$:

$$\frac{\partial^2 y}{\partial x^2} = \frac{\mu}{F}\frac{\partial^2 y}{\partial t^2}. \tag{19–18}$$

Now, finally, comes the punch line of our story. Equation (19–18) has exactly the same form as the general differential equation that we derived at the end of Section 19–3, Eq. (19–12). That equation and Eq. (19–18) describe the very same wave motion, so they must be identical. Comparing the two equations, we see that for this to be so we must have

$$v = \sqrt{\frac{F}{\mu}}, \tag{19–19}$$

which is the same expression as Eq. (19–13).

## EXAMPLE 19–3

The linear mass density of the clothesline in Example 19–2 is 0.250 kg/m. How much tension does Throcky have to apply to produce the observed wave speed of 10.0 m/s?

**SOLUTION** We use Eq. (19–13). Solving for $F$, we find

$$F = \mu v^2 = (0.250\ \text{kg/m})(10.0\ \text{m/s})^2 = 25.0\ \text{kg}\cdot\text{m/s}^2$$
$$= 25.0\ \text{N} \quad (5.62\ \text{lb}).$$

This much force is probably within Throcky's capability.

## EXAMPLE 19–4

One end of a nylon rope is tied to a stationary support at the top of a vertical mineshaft that is 80.0 m deep (Fig. 19–11). The rope is stretched taut by a box of mineral samples with mass 20.0 kg attached at the lower end. The mass of the rope is 2.0 kg. The geologist at the bottom of the mine signals to his colleague at the top by jerking the rope sideways. a) What is the speed of a transverse wave on the rope? b) If a point on the rope is given a transverse simple harmonic motion with a frequency of 20 Hz, what is the wavelength of the wave?

**SOLUTION** a) The tension at the bottom of the rope is the weight of the 20.0-kg load:

$$T = (20.0\ \text{kg})(9.8\ \text{m/s}^2)$$
$$= 196\ \text{N}.$$

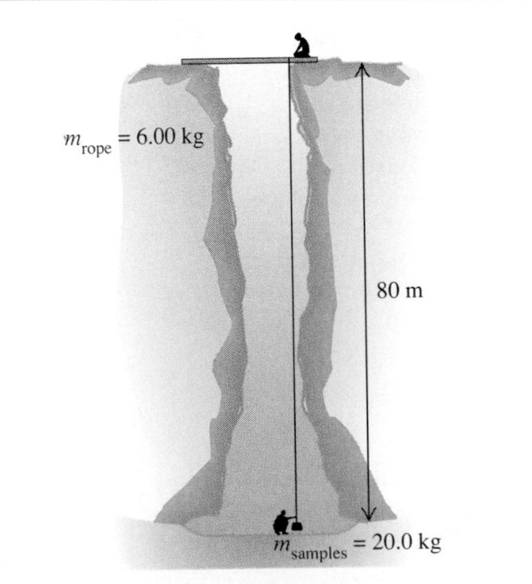


**19–11** Sending signals by way of transverse waves on a vertical rope.

The mass per unit length is

$$\mu = \frac{m}{L} = \frac{2.00\ \text{kg}}{80.0\ \text{m}} = 0.0250\ \text{kg/m}.$$

The wave speed is given by Eq. (19–13):

$$v = \sqrt{\frac{F}{\mu}} = \sqrt{\frac{196\ \text{N}}{0.0250\ \text{kg/m}}} = 88.5\ \text{m/s}.$$

b) From Eq. (19–1),

$$\lambda = \frac{v}{f} = \frac{88.5\ \text{m/s}}{20\ \text{s}^{-1}} = 4.43\ \text{m}.$$

We have neglected the 10% increase in tension in the rope between bottom and top due to the rope's own weight. Can you verify that the wave speed at the top is 92.9 m/s?

## Polarization

An important property of transverse waves is **polarization.** When we produce a transverse wave on a horizontal string, we can move the end either up and down or sideways; in either case the wave displacements are perpendicular, or *transverse,* to the length of the string. If the end moves up and down, the motion of the entire string is confined to a vertical plane. If the end moves sideways, the wave moves in a horizontal plane. In either case the wave is said to be *linearly polarized* because the individual particles move back and forth along straight lines perpendicular to the string.

The motion may also be more complex, containing both vertical and horizontal components. If we combine two perpendicular sinusoidal motions that have equal amplitude but are out of step by a quarter-cycle, the result is a wave in which each particle moves in a *circular* path perpendicular to the string. Such a wave is said to be *circularly polarized.* Figure 19–12 shows these various polarizations.

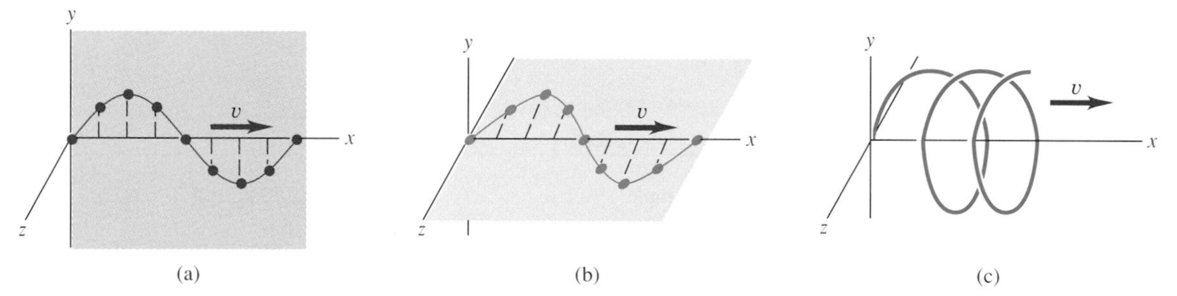


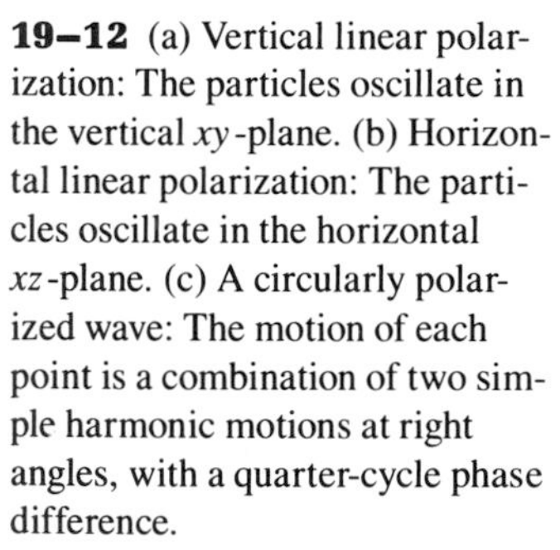

**19–12** (a) Vertical linear polarization: The particles oscillate in the vertical $xy$-plane. (b) Horizontal linear polarization: The particles oscillate in the horizontal $xz$-plane. (c) A circularly polarized wave: The motion of each point is a combination of two simple harmonic motions at right angles, with a quarter-cycle phase difference.

We can make a device to separate the various components of motion. We cut a thin slot in a flat board, thread the string through it, and orient the board with its plane perpendicular to the string (Fig. 19–13). Then any transverse motion parallel to the slot passes through unimpeded, while any motion perpendicular to the slot is blocked. Such a device is called a *polarizing filter.* Analogous optical devices for polarized light form the basis of some kinds of sunglasses and also of polarizing filters used in photography. We will discuss polarization of light waves in Chapter 34.

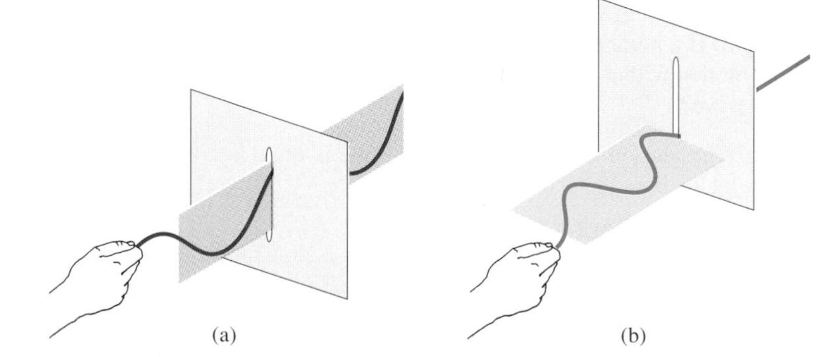


**19–13** (a) The slot passes a parallel polarization, but (b) it blocks a perpendicular polarization.

## 19–5 SPEED OF A LONGITUDINAL WAVE

Propagation speeds of longitudinal as well as transverse waves are determined by the mechanical properties of the medium. We can derive relations for longitudinal waves that are analogous to Eq. (19–13) for transverse waves on a rope. As in the wave-function discussion in Section 19–3, $x$ is still the coordinate measured along the length of the wave medium, but for a longitudinal wave the displacement $y$ is along that same direction rather than perpendicular to it as in a transverse wave.

Here is a derivation for the speed of longitudinal waves in a fluid in a pipe. The steps are completely parallel to the derivation of Eq. (19–13), and we invite you to compare the two developments.

Figure 19–14 shows a fluid (either liquid or gas) with density $\rho$ in a pipe with cross-section area $A$. In the equilibrium state the fluid is under a uniform pressure $p$. In Fig. 19–14a the fluid is at rest. At time $t = 0$ we start the piston at the left end moving toward the right with constant speed $v_y$. This initiates a wave motion that travels to the right along the length of the pipe, in which successive sections of fluid begin to move and become compressed at successively later times.

Figure 19–14b shows the fluid at time $t$. All portions of fluid to the left of point $P$ are moving to the right with speed $v_y$, and all portions to the right of $P$ are still at rest. The boundary between the moving and stationary portions travels to the right with a speed equal to the speed of propagation or wave speed $v$. At time $t$ the piston has moved a distance $v_y t$, and the boundary has advanced a distance $vt$. As with a transverse disturbance in a string, we can compute the speed of propagation from the impulse-momentum theorem.

The quantity of fluid set in motion in time $t$ is the amount that originally occupied a section of the cylinder with length $vt$, cross-section area $A$, and volume $vtA$. The mass of this fluid is $\rho vtA$, and its longitudinal momentum is

$$\text{Longitudinal momentum} = (\rho vtA)v_y.$$

Next we compute the increase of pressure, $\Delta p$, in the moving fluid. The original volume of the moving fluid, $Avt$, has decreased by an amount $Av_y t$. From the definition of the bulk modulus $B$, Eq. (11–16) in Section 11–6,

$$B = \frac{-\text{Pressure change}}{\text{Fractional volume change}} = \frac{-\Delta p}{-Av_y t/Avt},$$

$$\Delta p = B\frac{v_y}{v}.$$

The pressure in the moving fluid is $p + \Delta p$, and the force exerted on it by the piston is $(p + \Delta p)A$. The net force on the moving fluid (see Fig. 19–14b) is $\Delta pA$, and the longitudinal impulse is

$$\text{Longitudinal impulse} = \Delta p\, At = B\frac{v_y}{v}At.$$

Because the fluid was at rest at time $t = 0$, the change in momentum up to time $t$ is equal to the momentum at that time. Applying the impulse-momentum theorem to the mass of the fluid $\rho vtA$, we find

$$B\frac{v_y}{v}At = \rho vtAv_y. \tag{19–20}$$

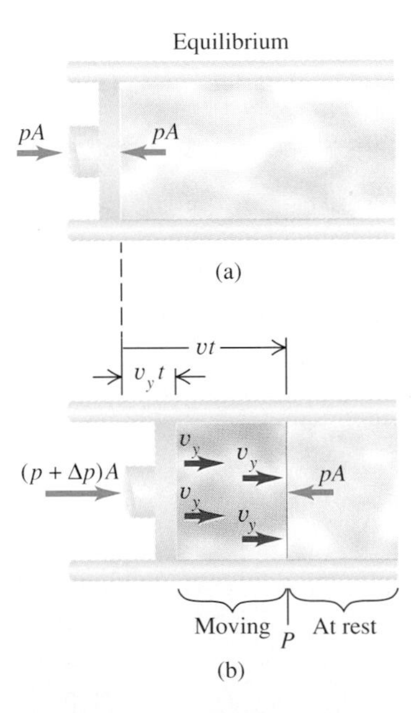


**19–14** Propagation of a longitudinal wave in a fluid confined in a tube. (a) Fluid in equilibrium; (b) part of the fluid in motion.

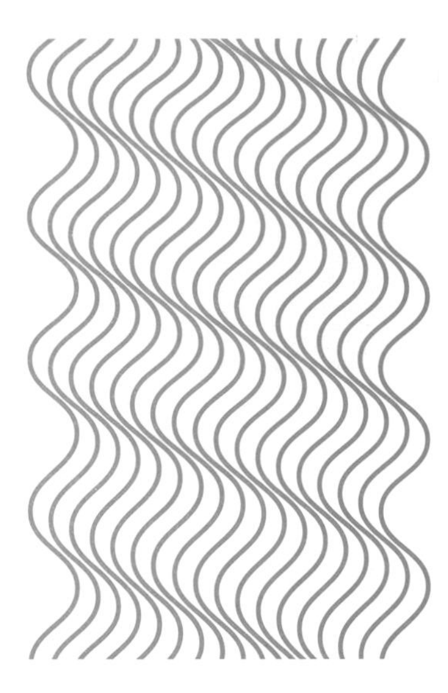

**19–15** Diagram for illustrating longitudinal traveling waves.

When we solve this for $v$, we get

$$v = \sqrt{\frac{B}{\rho}} \quad \text{(longitudinal wave).} \tag{19–21}$$

The speed of propagation of a longitudinal pulse in a fluid depends only on the bulk modulus $B$ and the density $\rho$ of the medium. The *form* of this relation is pleasingly similar to that of Eq. (19–13). In both cases the numerator is an elastic property describing the restoring force, and the denominator is an inertial property of the medium.

When a longitudinal wave propagates in a *solid* bar, the situation is somewhat different. The bar expands sidewise slightly when it is compressed longitudinally, but a fluid in a pipe with constant cross section cannot move sideways. Using the same kind of reasoning that led us to Eq. (19–21), we can show that the speed of a longitudinal pulse in the bar is given by

$$v = \sqrt{\frac{Y}{\rho}} \quad \text{(longitudinal wave),} \tag{19–22}$$

where $Y$ is Young's modulus, defined in Section 11–5. Note the similarity of form of Eqs. (19–19), (19–21), and (19–22).

Equation (19–22) applies only to a bar or rod whose sides are free to bulge and shrink a little as the wave travels. It does not apply to longitudinal waves in a bulk material, either solid or fluid, because sideways motion of any element in such a material is prevented by the surrounding material. The speed of longitudinal waves in bulk matter is given by Eq. (19–21), not Eq. (19–22). When the frequency of a longitudinal wave is within the range of human hearing, we call it **sound.** The speed of sound in air or water is determined by Eq. (19–21). We'll work out the details of this relationship in the next section.

**TABLE 19–1 Speed of Sound in Various Materials**

| Material | Speed of sound (m/s) |
|---|---|
| *Gases* | |
| Air (20°C) | 344 |
| Helium (20°C) | 999 |
| Hydrogen (20°C) | 1330 |
| *Liquids* | |
| Liquid helium (4 K) | 211 |
| Mercury (20°C) | 1451 |
| Water (0°C) | 1402 |
| Water (20°C) | 1482 |
| Water (100°C) | 1543 |
| *Solids* | |
| Beryllium | 12870 |
| Bone | 3445 |
| Brass | 3480 |
| Pyrex glass | 5170 |
| Polystyrene | 1840 |
| Steel | 5000 |

As with the derivation for a transverse wave on a string, Eqs. (19–21) and (19–22) are valid for sinusoidal and other periodic waves, not just for the special case discussed here.

Longitudinal waves have no component of displacement perpendicular to the direction of propagation, so the concept of polarization has no meaning for a longitudinal wave.

Visualizing the relation between particle motion and wave motion is not as easy for longitudinal waves as for transverse waves on a rope. Figure 19–15 will help you understand these motions. To use this figure, tape two index cards edge-to-edge with a gap of 1 mm or so between them, forming a thin slit. Place the cards over the figure with the slit horizontal at the top of the diagram, and move them downward with constant speed. The portions of the sine curves that are visible through the slit correspond to a row of particles in a medium in which a longitudinal sinusoidal wave is traveling. Each particle undergoes simple harmonic motion about its equilibrium position, with delays or phase shifts that increase continuously along the slit. The regions of maximum compression and expansion move from left to right with constant speed. Moving the card upward simulates a wave traveling from right to left.

Table 19–1 lists the speed of sound in several materials. Note that sound waves travel faster in those materials that have large values for the elastic constants, such as steel and beryllium.

## EXAMPLE 19–5

**Wavelength of sonar waves** A ship uses a sonar system to detect underwater objects (Fig. 19–16). The system emits underwater sound waves and measures the time interval for the reflected wave (echo) to return to the detector. Determine the speed of sound waves in water, and find the wavelength of a wave having a frequency of 262 Hz.

**19–16** A sonar system uses underwater sound waves to detect and locate underwater objects.

**SOLUTION** We use Eq. (19–21) to find the wave speed. From Table 11–2 we find that the compressibility of water, which is the reciprocal of its bulk modulus, is $k = 45.8 \times 10^{-11}\ \text{Pa}^{-1}$. Thus, $B = (1/45.8) \times 10^{11}$ Pa. The density of water is $\rho = 1.00 \times 10^3$ kg/m$^3$. We obtain

$$v = \sqrt{\frac{B}{\rho}} = \sqrt{\frac{(1/45.8) \times 10^{11}\ \text{Pa}}{1.00 \times 10^3\ \text{kg/m}^3}}$$
$$= 1480\ \text{m/s}.$$

This is over four times the speed of sound in air at ordinary temperatures. The wavelength is given by

$$\lambda = \frac{v}{f} = \frac{1480\ \text{m/s}}{262\ \text{s}^{-1}}$$
$$= 5.65\ \text{m}.$$

A wave of this frequency in air has a wavelength of 1.31 m, as we found in Example 19–1 (Section 19–2).

Dolphins emit high-frequency sound waves (typically 100,000 Hz) and use the echoes for guidance and for hunting. The corresponding wavelength in water is 1.48 cm. With this high-frequency "sonar" system they can sense objects of about the size of the wavelength but not much smaller.

## EXAMPLE 19–6

What is the speed of longitudinal sound waves in a steel rod?

**SOLUTION** This is the situation for which Eq. (19–22) is applicable. From Table 11–1, $Y = 2.0 \times 10^{11}$ Pa, and from Table 14–1, $\rho = 7.8 \times 10^3\ \text{kg/m}^3$. We find

$$v = \sqrt{\frac{Y}{\rho}} = \sqrt{\frac{2.0 \times 10^{11}\ \text{Pa}}{7.8 \times 10^3\ \text{kg/m}^3}} = 5.1 \times 10^3\ \text{m/s},$$

nearly 15 times the speed of sound in air at ordinary temperatures.

# 19–6 SOUND WAVES IN GASES

In the preceding section we derived an expression for the speed of sound in a fluid in terms of its density $\rho$ and bulk modulus $B$:

$$v = \sqrt{\frac{B}{\rho}} \quad \text{(longitudinal wave).}$$

We can use this to find the speed of sound in an ideal gas.

The bulk modulus is defined in general as in Eq. (11–16); for infinitesimal pressure and volume changes, $B = -V\,dp/dV$. So we need to know how $p$ varies with $V$ for an ideal gas. If the temperature is constant, then, according to Boyle's law, the product $pV$ is constant, and we can use this to evaluate $dp/dV$. But when a gas is compressed *adiabatically,* its temperature rises, and when it expands adiabatically, its temperature drops. In an adiabatic process for an ideal gas, Eq. (17–25) states that $pV^\gamma$ is constant, and we get a different result for $B$. Does this happen when a wave travels through a gas, or is there enough heat conduction between adjacent layers of the gas to maintain a nearly constant temperature throughout?

Thermal conductivities of gases are very small, and it turns out that for ordinary sound frequencies, from 20 to 20,000 Hz, propagation of sound is very nearly *adiabatic.*

Thus in Eq. (19–21) we use the **adiabatic bulk modulus** $B_{ad}$, derived from the assumption that

$$pV^{\gamma} = \text{constant}. \tag{19–23}$$

We take the derivative of Eq. (19–23) with respect to $V$:

$$\frac{dp}{dV}V^{\gamma} + \gamma p V^{\gamma-1} = 0.$$

Dividing by $V^{\gamma-1}$ and rearranging, we find

$$B_{ad} = -V\frac{dp}{dV} = \gamma p. \tag{19–24}$$

For an *isothermal* process, $pV$ = constant; we invite you to prove (Exercise 19–18) that the *isothermal* bulk modulus is

$$B_{iso} = p. \tag{19–25}$$

The adiabatic modulus is *larger* than the isothermal modulus by a factor of $\gamma$.

Combining Eqs. (19–21) and (19–24), we find

$$v = \sqrt{\frac{\gamma p}{\rho}} \quad \text{(ideal gas)}. \tag{19–26}$$

We can get a useful alternative form by using Eq. (16–5) for the density $\rho$ of an ideal gas:

$$\rho = \frac{pM}{RT},$$

where $R$ is the gas constant, $M$ is the molecular mass, and $T$ is the absolute temperature. Combining this with Eq. (19–26), we find

$$v = \sqrt{\frac{\gamma RT}{M}} \quad \text{(ideal gas)}. \tag{19–27}$$

For any particular gas, $\gamma$, $R$, and $M$ are constants, and the wave speed is proportional to the square root of the absolute temperature. Except for the numerical factor of 3 in one and $\gamma$ in the other, this expression is identical to Eq. (16–20), which gives the rms speed of molecules in an ideal gas. This shows that sound speeds and molecular speeds are closely related; exploring that relationship in detail is beyond our scope.

## EXAMPLE 19–7

Compute the speed of longitudinal waves in air at room temperature ($T$ = 20°C).

**SOLUTION** From Example 16–2 (Section 16–1) the mean molecular mass of air is $28.8 \times 10^{-3}$ kg/mol. Also, $\gamma = 1.40$ for air, and $R = 8.315$ J/mol · K. At $T$ = 20°C = 293 K, we find

$$v = \sqrt{\frac{\gamma RT}{M}} = \sqrt{\frac{(1.40)(8.315\ \text{J/mol}\cdot\text{K})(293\ \text{K})}{28.8 \times 10^{-3}\ \text{kg/mol}}} = 344\ \text{m/s}.$$

This agrees with the measured speed of sound at this temperature to within 0.3%.

---

The human ear is sensitive to a range of sound frequencies from about 20 Hz to about 20,000 Hz. From the relation $v = \lambda f$ the corresponding wavelength range is from about 17 m, corresponding to a 20-Hz note, to about 1.7 cm, corresponding to a 20,000-Hz note.

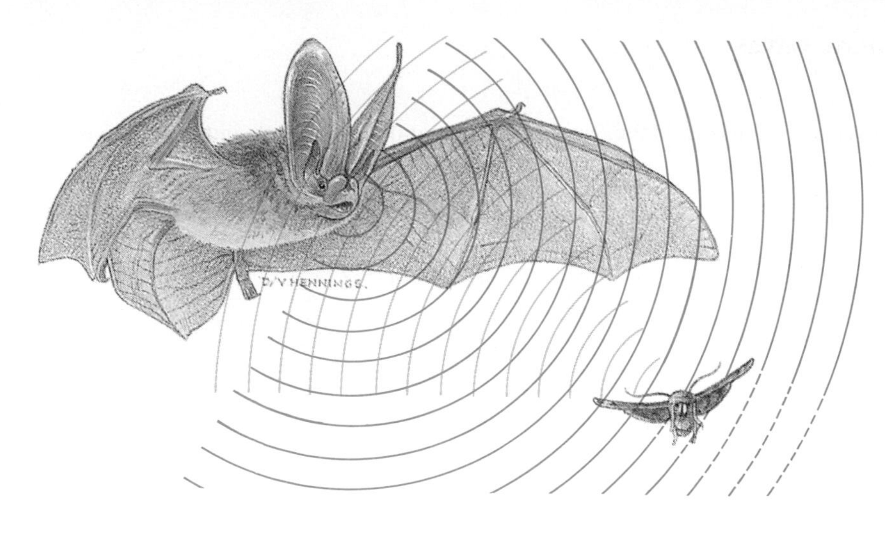


19–17 Bats emit ultrasonic (about 100 kHz) clicks or cries that reflect from surrounding objects. Detection of these reflections enables bats to avoid objects at night and locate insects for food.

Bats can hear much higher frequencies. Like dolphins, bats use high-frequency sound waves for navigation. A typical frequency is 100 kHz; the corresponding wavelength in air is about 3.4 mm, small enough for the bats to detect the flying insects they eat (Fig. 19–17).

In this discussion we have ignored the *molecular* nature of a gas and have treated it as a continuous medium. We know that a gas is actually composed of molecules in random motion, separated by distances that are large in comparison to their diameters. The vibrations that constitute a wave in a gas are superposed on the random thermal motion. At atmospheric pressure a molecule travels an average distance (the mean free path) of about $10^{-7}$ m between collisions, and the displacement amplitude of a faint sound may be only $10^{-9}$ m. We can think of a gas with a sound wave passing through it as being comparable to a swarm of bees; the swarm as a whole oscillates slightly while individual insects move about through the swarm, apparently at random.

## 19–7 ENERGY IN WAVE MOTION

19–18 An ocean wave crashing onto the shore has an enormous amount of energy, which can be very destructive during large storms.

Every wave motion has *energy* associated with it; the energy that we receive from sunlight and the destructive effects of ocean surf and earthquakes bear this out (Fig. 19–18). To produce any of the wave motions that we have discussed in this chapter, we have to apply a force to a portion of the wave medium; the point where the force is applied moves, so we do *work* on the system. In the same way a wave can transport energy from one region of space to another.

As an example of energy considerations in wave motion, let's look again at transverse waves on a string. How is energy transferred from one portion of string to another? Picture a wave traveling from left to right (the positive $x$-direction) along the string, and consider a particular point $a$ on the string (Fig. 19–19a). The string to the left of $a$ exerts a force on the string to the right of it, and vice versa. In Fig. 19–19b the string to the left of $a$ has been removed, and the force it exerts at $a$ is represented by the components $F$ and $F_y$, as we have done before. We note again that $F_y/F$ is equal to the negative of the *slope* of the string at $a$, which is also given by $\partial y/\partial x$. Putting these together, we have

$$F_y = -F\frac{\partial y}{\partial x}. \tag{19–28}$$

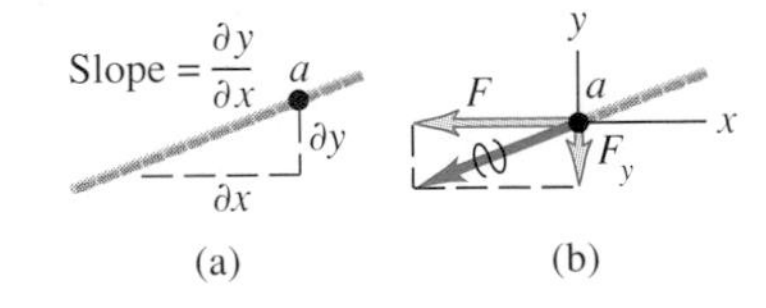


19–19 (a) Point $a$ on a string carrying a wave from left to right. (b) The components of the force exerted on the right part of the string by the left part at point $a$.

We need the negative sign because $F_y$ is negative when the slope is positive.

When point $a$ moves, $F_y$ does *work* on this point and therefore transfers energy into the part of the string to the right of $a$. The corresponding power $P$ (rate of doing work) is the transverse force $F_y$ times the transverse velocity $v_y = \partial y/\partial t$:

$$P = F_y v_y = -F\frac{\partial y}{\partial x}\frac{\partial y}{\partial t}. \tag{19–29}$$

This expression is valid for *any* wave, sinusoidal or not. For a sinusoidal wave with wave function

$$y(x, t) = A \sin(\omega t - kx), \tag{19–30}$$

$$\frac{\partial y}{\partial x} = -kA \cos(\omega t - kx),$$

$$\frac{\partial y}{\partial t} = \omega A \cos(\omega t - kx),$$

$$P = Fk\omega A^2 \cos^2(\omega t - kx). \tag{19–31}$$

By using the relations $\omega = vk$ and $v^2 = F/\mu$ we can also express Eq. (19–31) in the alternative form

$$P = \sqrt{\mu F}\,\omega^2 A^2 \cos^2(\omega t - kx). \tag{19–32}$$

This equation gives the *instantaneous* rate of energy transmission along the string; it depends on both $x$ and $t$. Its maximum value occurs when the $\cos^2$ function has the value unity:

$$P_{\text{max}} = \sqrt{\mu F}\,\omega^2 A^2. \tag{19–33}$$

We can obtain the *average* power using the fact that the *average* value of the $\cos^2$ function, averaged over any whole number of cycles, is $\frac{1}{2}$. Thus the average power is

$$P_{\text{av}} = \tfrac{1}{2}\sqrt{\mu F}\omega^2 A^2. \tag{19–34}$$

The rate of energy transfer is proportional to the *square* of the amplitude and to the square of the frequency.

## EXAMPLE 19–8

a) In Examples 19–2 and 19–3, at what maximum rate does Throcky put energy into the rope? That is, what is his maximum instantaneous power? b) What is his average power? c) As Throcky tires, the amplitude decreases. What is his average power when the amplitude has dropped to 2.00 mm?

**SOLUTION** a) From Eq. (19–33),

$$\begin{aligned} P_{\text{max}} &= \sqrt{\mu F}\omega^2 A^2 \\ &= \sqrt{(0.25\ \text{kg/m})(25\ \text{N})}(31.4\ \text{rad/s})^2(0.010\ \text{m})^2 \\ &= 0.25\ \text{W}. \end{aligned}$$

b) From Eqs. (19–33) and (19–34), $P_{\text{av}} = \frac{1}{2}P_{\text{max}}$, so

$$\begin{aligned} P_{\text{av}} &= \tfrac{1}{2}(0.25\ \text{W}) \\ &= 0.12\ \text{W}. \end{aligned}$$

c) This amplitude is $\frac{1}{5}$ of the value we used in parts (a) and (b). The average power is proportional to the *square* of the amplitude, so now the average power is

$$\begin{aligned} P_{\text{av}} &= \frac{1}{5^2}(0.12\ \text{W}) = 0.0048\ \text{W} \\ &= 4.8\ \text{mW}. \end{aligned}$$

Power relations analogous to Eq. (19–33) can be worked out for longitudinal waves. We won't go into detail; the results are most easily stated in terms of the average power $P$ *per unit cross-section area* $A$ in the wave motion. The quantity $P_{\text{av}}/A$ is called the *intensity,* denoted by $I$. For fluids in a pipe it is given by

$$I = \tfrac{1}{2}\sqrt{\rho B}\omega^2 A^2, \tag{19–35}$$

and for a solid rod by

$$I = \tfrac{1}{2}\sqrt{\rho Y}\omega^2 A^2. \tag{19–36}$$

Again the power is proportional to $A^2$ and to $\omega^2$.

# 19–8 SOUND WAVES IN CRYSTALS: A Case Study in Modern Physics

In our discussion of sound we've treated the wave medium (a fluid or solid material) as a smooth, continuous distribution of matter. But we know that matter on a microscopic scale is lumpy because it is made of atoms. When the wavelength of the sound is comparable to the interatomic spacing, new aspects of sound emerge. We'll explore two of these briefly, one associated with the effect of the interatomic distances, the other with the fact that the energy of a sound wave is *quantized*. We'll find that the speed of sound becomes frequency-dependent and that there is a maximum frequency for sound propagation in a crystal.

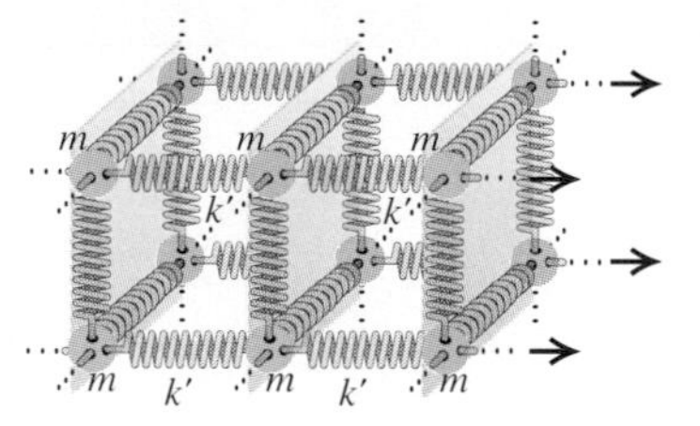


**19–20** Model of a simple cubic crystal lattice. The atoms are represented as identical point masses $m$, and the interatomic forces are represented as identical springs with force constant $k'$, acting only between nearest-neighbor atoms.

To analyze wave propagation in a crystal, we'll use the simplest possible model, a simple cubic lattice of identical atoms (Fig. 19–20), with the sound traveling parallel to one of the lattice directions. We model the lattice as an array of identical point masses $m$, connected by identical springs with spring constants $k'$. (We use this symbol because we'll use $k$ for the wave number, $k = 2\pi/\lambda$.) With the direction of propagation shown, the atoms lie in planes perpendicular to the direction of propagation. All the atoms in a particular plane move in unison, so all we really have to do is analyze the motion of a single row of atoms parallel to the propagation direction.

Figure 19–21a shows a few atoms in such a row. The atoms, numbered 0, 1, 2, 3, and 4, lie along the $x$-axis; their equilibrium positions are $x_0$, $x_1$, $x_2$, and so on. In these positions they are spaced a distance $a$ apart, and atom 0 is at the origin. Thus $x_0 = 0$, $x_1 = a$, $x_2 = 2a$, and so on. Figure 19–21b shows three atoms farther down the row, numbered $n - 1$, $n$, and $n + 1$. Their equilibrium positions are $x_{n-1} = (n - 1)a$, $x_n = na$, and $x_{n+1} = (n + 1)a$. During wave motion, each atom is displaced along the $x$-axis; let $y_n$ be the displacement of the $n$th atom from its equilibrium position. All the $y_n$'s are functions of time.

Let's apply $\Sigma F_x = ma_x$ to the $n$th atom. The spring to its right is stretched by an amount $y_{n+1} - y_n$. It exerts a force on atom $n$ with $x$-component equal to $k'(y_{n+1} - y_n)$. (When $y_{n+1} < y_n$, the spring is compressed and the force component is negative.) Similarly, the spring to the left of this atom exerts a force with $x$-component $-k'(y_n - y_{n-1})$. The total $x$-component of force on the $n$th atom is

$$\Sigma F_x = k'(y_{n+1} - y_n) - k'(y_n - y_{n-1}) = k'(y_{n+1} - 2y_n + y_{n-1}),$$

and the $\Sigma F_x = ma_x$ equation for the $n$th atom is

$$k'(y_{n+1} - 2y_n + y_{n-1}) = m\frac{d^2y_n}{dt^2}. \tag{19–37}$$

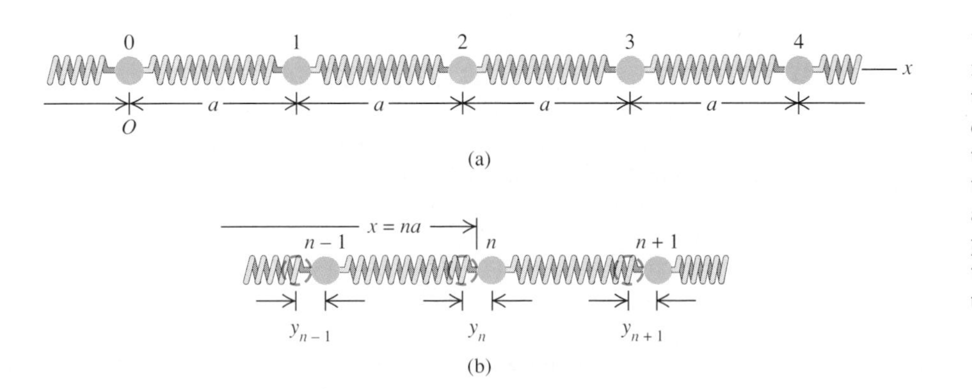


**19–21** (a) Part of a line of atoms in a cubic crystal lattice, shown at their equilibrium positions. The distance $x_n$ of the $n$th atom from the origin is $x_n = na$, where $a$ is the lattice spacing. (b) Three adjacent atoms in the row, displaced from their equilibrium positions as a longitudinal wave moves to the right.

Now suppose a wave with wave function $y = A \sin(\omega t - kx)$ travels through the crystal. Here $x$ represents the equilibrium position of a particular atom. For atom $n$, $x = na$, and so on, and we can write

$$y_n = A \sin(\omega t - kna),$$

$$y_{n-1} = A \sin[\omega t - k(n-1)a] = A \sin(\omega t - kna + ka),$$

$$y_{n+1} = A \sin[\omega t - k(n+1)a] = A \sin(\omega t - kna - ka),$$

$$\frac{d^2 y_n}{dt^2} = -\omega^2 A \sin(\omega t - kna).$$

To find out when such a wave is physically possible, we substitute these expressions into Eq. (19–37) to see the conditions under which it is satisfied. This requires some computation, and we need to use the trigonometric identities

$$\sin(\alpha \pm \beta) = \sin\alpha\cos\beta \pm \cos\alpha\sin\beta,$$

where $\alpha = \omega t - kna$ and $\beta = ka$ for our problem. We'll leave the details as an exercise. The common factor $A \sin(\omega t - kna)$ can be divided out, leaving the simple result

$$2k'(1 - \cos ka) = \omega^2 m,$$

$$\omega^2 = \frac{k'}{m} 2(1 - \cos ka).$$

We can simplify this further by using the identity $1 - \cos\beta = 2\sin^2(\beta/2)$; we find

$$\omega = 2\sqrt{\frac{k'}{m}} \sin\frac{ka}{2}. \tag{19–38}$$

Compared to the simple relationship $\omega = vk$ that we used in Section 19–3 for sinusoidal waves with constant speed $v$, this is quite an unexpected result. It shows that the ratio $\omega/k$, which equals the wave speed $v$, is not constant but depends on $k$ (and therefore on frequency). Such a wave is said to be *dispersive*. However, if the wavelength $\lambda$ is much larger than the lattice spacing $a$, then the wave number $k = 2\pi/\lambda$ is much *smaller* than $a$, and the product $ka$ is much smaller than unity. In that case we can approximate $\sin(ka/2)$ in Eq. (19–38) by the quantity $ka/2$ itself. (We used this same approximation in Section 13–5 when we analyzed small-amplitude motion of a simple pendulum.)

Replacing $\sin(ka/2)$ by $ka/2$ in Eq. (19–38), we find

$$\omega = 2\sqrt{\frac{k'}{m}}\frac{ka}{2} = a\sqrt{\frac{k'}{m}}k, \qquad \frac{\omega}{k} = v = a\sqrt{\frac{k'}{m}}. \tag{19–39}$$

This shows that when the wavelength is much larger than the lattice spacing (typically a few tenths of a nanometer), the wave speed $v$ becomes constant and that it is determined by the microscopic constants $a$, $k'$, and $m$. But when the wavelength gets down to the same range as the lattice spacing, the waves begin to "see" that matter is not continuous but that its distribution in space is *quantized* into individual atoms.

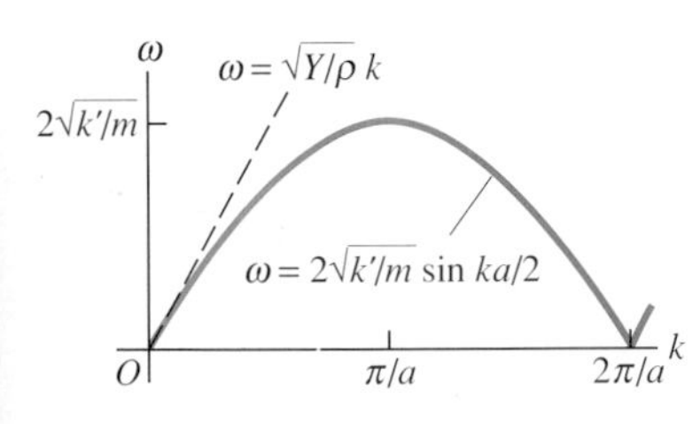


**19–22** The angular frequency $\omega$ is approximately proportional to the wave number $k$ only when $ka \ll 1$. No plane longitudinal waves can move through the crystal with an angular frequency greater than $\omega_{max} = 2\sqrt{k'/m}$.

Figure 19–22 is a graph of Eq. (19–38), showing the relationship between angular frequency $\omega$ and wave number $k$. At small $k$ (long wavelength) the curve is nearly a straight line, showing constant wave speed. As $k$ increases, the curve bends over and reaches a maximum value

$$\omega_{max} = 2\sqrt{\frac{k'}{m}}. \tag{19–40}$$

For $\omega$ greater than $\omega_{max}$ there is *no* $k$ for which Eq. (19–38) is satisfied. Therefore $\omega_{max}$ is the maximum possible angular frequency of waves in our simple model. If you try to make a wave with $\omega > \omega_{max}$ travel through the crystal, it won't go; it will bounce back out. Numerical values of $\omega_{max}$ are typically of the order of $10^{12}$ to $10^{14}$ rad/s.

Here's another unexpected feature of wave motion in a crystal. Experimental studies show that the energy carried by sound waves in crystals is quantized; that is, the energy comes in packages called *phonons.* The energy $E$ of each phonon is proportional to the frequency $f$ of the wave;

$$E = hf, \tag{19–41}$$

where $h$ is a fundamental physical constant called *Planck's constant.* Its numerical value in SI units is

$$h = 6.626 \times 10^{-34}\ \text{J}\cdot\text{s}.$$

We don't notice this quantization of energy for audible sound because the phonon energies are of the order of $10^{-29}$ J. However, many aspects of the behavior of solid materials show the existence of these bundles of energy.

## EXAMPLE 19–9

Copper has a Young's modulus $Y = 1.1 \times 10^{11}$ Pa (Table 11–1), a density $\rho = 8.9 \times 10^3$ kg/m$^3$ (Table 14–1), an atomic mass $M = 63.5 \times 10^{-3}$ kg/mol, and a lattice spacing $a = 0.255$ nm. Assume that the atoms are arranged in a simple cubic lattice. (The actual lattice is somewhat more complex than this.) a) Find the force constant $k'$ of the interatomic "springs." b) Find the maximum frequency for longitudinal mechanical waves. c) Find the phonon energy corresponding to a wave number $k = 2\pi \times 10^9$ rad/m.

**SOLUTION** a) For long wavelengths the wave speed $v$ given by Eq. (19–39) must agree with Eq. (19–22), $v = (Y/\rho)^{1/2}$, derived in Section 19–5. Using that expression, we find

$$v = \sqrt{\frac{Y}{\rho}} = \sqrt{\frac{1.1 \times 10^{11}\ \text{Pa}}{8.9 \times 10^3\ \text{kg/m}^3}} = 3.52 \times 10^3\ \text{m/s}.$$

The mass $m$ of an atom is the molecular mass $M$ divided by Avogadro's number $N_A$:

$$m = \frac{M}{N_A} = \frac{63.5 \times 10^{-3}\ \text{kg/mol}}{6.02 \times 10^{23}\ \text{atoms/mol}} = 1.05 \times 10^{-25}\ \text{kg}.$$

Solving Eq. (19–39) for $k'$ and substituting in the numbers, we get

$$k' = \frac{mv^2}{a^2} = \frac{(1.05 \times 10^{-25}\ \text{kg})(3.52 \times 10^3\ \text{m/s})^2}{(0.255 \times 10^{-9}\ \text{m})^2} = 20.0\ \text{N/m}.$$

b) From Eq. (19–40) the maximum angular frequency is

$$\omega_{max} = 2\sqrt{\frac{k'}{m}} = 2\sqrt{\frac{20.0\ \text{N/m}}{1.05 \times 10^{-25}\ \text{kg}}} = 2.76 \times 10^{13}\ \text{rad/s}.$$

The maximum frequency is $f_{max} = \omega_{max}/2\pi = 4.4 \times 10^{12}$ Hz. Thus even ultrasonic waves with $f = 10^7$ Hz propagate through copper with no difficulty. The wavelengths of such waves are much larger than $a$.

c) We need to find the angular frequency $\omega$. Note that $ka = (2\pi \times 10^9\ \text{rad/m})(0.255 \times 10^{-9}\ \text{m}) = 1.60$ rad, $ka/2 = 0.80$ rad, and we can't use the small-$k$ approximation. We can rewrite Eq. (19–38) as

$$\omega = \omega_{max} \sin\frac{ka}{2}.$$

Substituting numerical values, we get

$$\omega = (2.76 \times 10^{13}\ \text{rad/s}) \sin(0.80\ \text{rad}) = 1.98 \times 10^{13}\ \text{rad/s}.$$

The corresponding frequency is $f = \omega/2\pi = 3.15 \times 10^{12}$ Hz, and the energy $E$ of one phonon is

$$E = hf = (6.63 \times 10^{-34}\ \text{J}\cdot\text{s})(3.15 \times 10^{12}\ \text{s}^{-1}) = 2.1 \times 10^{-21}\ \text{J}.$$

The behavior of mechanical waves in crystals has a particlelike aspect in that the energy is divided into bundles called phonons. Also, mechanical waves have only a certain allowed range of frequency; therefore only certain ranges of phonon energy are possible. Conversely, quantum mechanics also predicts that particles have wavelike properties. As a result, electrons in a crystal have allowed ranges of energy, called *energy bands,* separated by forbidden ranges. The existence of these energy bands helps us to understand why some solids are conductors, some are insulators, and some are semiconductors.

## SUMMARY

### KEY TERMS

medium
transverse wave
longitudinal wave
wave speed
sinusoidal wave
wavelength
wave function
phase
wave number
propagation constant
wave equation
polarization
sound
adiabatic bulk modulus

• A wave is any disturbance from an equilibrium condition that propagates from one region to another. A mechanical wave always travels within some material called the medium. In a periodic wave the motion of each point of the medium is periodic. When the motion is sinusoidal, the wave is called a sinusoidal wave. The frequency $f$ of a periodic wave is the number of repetitions per unit time, and the period $T$ is the time for one cycle. The wavelength $\lambda$ is the distance over which the spatial pattern repeats. The speed of propagation $v$ is the speed with which the wave disturbance travels. For any periodic wave these quantities are related by

$$v = \lambda f. \tag{19–1}$$

• A wave function $y(x, t)$ describes the displacements of individual particles in the medium. For a sinusoidal wave,

$$y(x, t) = A \sin \omega \left(t - \frac{x}{v}\right) = A \sin 2\pi f\left(t - \frac{x}{v}\right), \tag{19–3}$$

$$y(x, t) = A \sin 2\pi\left(\frac{t}{T} - \frac{x}{\lambda}\right), \tag{19–4}$$

or

$$y(x, t) = A \sin (\omega t - kx), \tag{19–7}$$

where the wave number $k$ is defined as $k = 2\pi/\lambda$. In all three forms, $A$ is the amplitude, the maximum displacement of a particle from its equilibrium position.

• The wave function must obey a partial differential equation called the wave equation,

$$\frac{\partial^2 y}{\partial x^2} = \frac{1}{v^2}\frac{\partial^2 y}{\partial t^2}. \tag{19–12}$$

• The speed of a transverse wave on a string with tension $F$ and mass per unit length $\mu$ is

$$v = \sqrt{\frac{F}{\mu}} \quad \text{(transverse wave).} \tag{19–13}$$

Transverse waves have the property of polarization; longitudinal waves do not.

• The speed of a longitudinal wave in a fluid with bulk modulus $B$ and density $\rho$ is

$$v = \sqrt{\frac{B}{\rho}} \quad \text{(longitudinal wave).} \tag{19–21}$$

The speed of a longitudinal wave in a solid rod with Young's modulus $Y$ and density $\rho$ is

$$v = \sqrt{\frac{Y}{\rho}} \quad \text{(longitudinal wave).} \tag{19–22}$$

• Sound propagation in gases is ordinarily an adiabatic process; the adiabatic bulk modulus for an ideal gas is

$$B_{\text{ad}} = -V\frac{dp}{dV} = \gamma p. \tag{19–24}$$

The speed of sound in an ideal gas is

$$v = \sqrt{\frac{\gamma p}{\rho}} \quad \text{(ideal gas).} \qquad (19\text{–}26)$$

or

$$v = \sqrt{\frac{\gamma RT}{M}} \quad \text{(ideal gas).} \qquad (19\text{–}27)$$

• Wave motion conveys energy from one region to another. For a transverse wave on a stretched string the rate of energy transfer (power) is

$$P = F_y v_y = -F\frac{\partial y}{\partial x}\frac{\partial y}{\partial t}. \qquad (19\text{–}29)$$

For a sinusoidal wave the average power is

$$P_{av} = \tfrac{1}{2}\sqrt{\mu F}\,\omega^2 A^2. \qquad (19\text{–}34)$$

For a longitudinal wave the average power $P_{av}$ per unit cross-section area $A$ in the wave motion is called the intensity, $I = P_{av}/A$. For a fluid in a pipe,

$$I = \tfrac{1}{2}\sqrt{\rho B}\,\omega^2 A^2. \qquad (19\text{–}35)$$

For a solid rod,

$$I = \tfrac{1}{2}\sqrt{\rho Y}\,\omega^2 A^2. \qquad (19\text{–}36)$$

# EXERCISES

## Section 19–2 Periodic Waves

**19–1** The speed of sound in air at 20°C is 344 m/s. a) What is the wavelength of a sound wave with a frequency of 32.0 Hz, corresponding to the note produced by the lowest pedal on a medium-sized pipe organ? b) What is the frequency of a wave with a wavelength of 1.22 m (4 ft), corresponding approximately to the note D above middle C on the piano?

**19–2** The speed of radio waves in vacuum (equal to the speed of light) is $3.00 \times 10^8$ m/s. Find the wavelength for a) an AM radio station with frequency 1240 kHz; b) an FM radio station with frequency 90.0 MHz.

**19–3** Provided that the amplitude is sufficiently great, the human ear can respond to longitudinal waves over a range of frequencies from about 20.0 Hz to about 20,000 Hz. Compute the wavelengths corresponding to these frequencies a) for waves in air ($v$ = 344 m/s); b) for waves in water ($v$ = 1480 m/s).

**19–4** The sound waves from a loudspeaker spread out nearly uniformly in all directions when their wavelengths are large in comparison with the diameter of the speaker. When the wavelength is small in comparison with the diameter of the speaker, much of the sound energy is concentrated forward. For a speaker with a diameter of 20.0 cm, compute the frequency for which the wavelength of the sound waves in air ($v$ = 344 m/s) is a) ten times the diameter of the speaker; b) equal to the diameter of the speaker; c) one-tenth of the diameter of the speaker.

**19–5 Ocean Waves.** A fisherman notices that his boat is moving up and down periodically owing to waves on the surface of the water. It takes 3.0 s for the boat to travel from its highest point to its lowest, a total distance of 0.800 m. The fisherman sees that the wave crests are spaced 8.0 m apart. a) How fast are the waves traveling? b) What is the amplitude of each wave?

## Section 19–3 Mathematical Description of a Wave

**19–6** The equation for a certain transverse wave is

$$y = (3.00\text{ cm})\sin 2\pi\left(\frac{t}{0.0200\text{ s}} - \frac{x}{40.0\text{ cm}}\right).$$

Determine the wave's a) amplitude; b) wavelength; c) frequency; d) speed of propagation.

**19–7** Transverse waves on a string have wave speed 15.0 m/s, amplitude 0.0800 m, and wavelength 0.300 m. The waves travel in the $+x$-direction, and at $t = 0$ the $x = 0$ end of the string has zero displacement and is moving upward. a) Find the frequency, period, and wave number of these waves. b) Write a wave function describing the wave. c) Find the transverse displacement of a point at $x$ = 0.200 m at time $t$ = 0.200 s.

**19–8** Show that Eq. (19–3) may be written

$$y = -A\sin\frac{2\pi}{\lambda}(x - vt).$$

**19–9** A transverse wave traveling on a string is represented by the equation in Exercise 19–8. Let $A = 8.0$ cm, $\lambda = 12.0$ cm, and $v = 2.0$ cm/s. a) At time $t = 0$, compute the transverse displacement $y$ at 2-cm intervals of $x$ (that is, at $x = 0$, $x = 2$ cm, $x = 4$ cm, and so on) from $x = 0$ to $x = 24$ cm. Show the results on a graph. This is the shape of the string at time $t = 0$. b) Repeat the calculations for the same values of $x$ at times $t = 2$ s and $t = 4$ s. Show on the same graph the shape of the string at these instants. In what direction is the wave traveling?

## Section 19–4 Speed of a Transverse Wave

**19–10** A steel wire 5.00 m long has a mass of 0.0600 kg and is stretched with a tension of 800 N. What is the speed of propagation of a transverse wave on the wire?

**19–11** One end of a horizontal string is attached to a prong of an electrically driven tuning fork whose frequency of vibration is 240 Hz. The other end passes over a pulley and supports a 5.00-kg mass. The linear mass density of the string is 0.0180 kg/m. a) What is the speed of a transverse wave on the string? b) What is the wavelength?

**19–12** With what tension must a rope of length 5.00 m and mass 0.160 kg be stretched for transverse waves of frequency 60.0 Hz to have a wavelength of 0.600 m?

## Section 19–5 Speed of a Longitudinal Wave

## Section 19–6 Sound Waves in Gases

**19–13** A 120-m-long steel pipe is struck at one end. A person at the other end hears two sounds as a result of two longitudinal waves, one traveling in the metal pipe and the other traveling in the air. What is the time interval between the two sounds? (Young's modulus for steel is $2.00 \times 10^{11}$ Pa, the density of steel is 7800 kg/m$^3$, and the speed of sound in air is 344 m/s.)

**19–14** A metal bar with a length of 20.0 m has density 6000 kg/m$^3$. Longitudinal sound waves take $5.00 \times 10^{-3}$ s to travel from one end of the bar to the other. What is Young's modulus for this metal?

**19–15** Longitudinal waves with frequency 150 Hz in a liquid with density 800 kg/m$^3$ are found to have wavelength 8.00 m. Calculate the bulk modulus of the liquid.

**19–16** At a temperature of 27.0°C, what is the speed of longitudinal waves in a) argon (atomic mass 39.9 g/mol); b) hydrogen (molecular mass 2.02 g/mol)? c) Compare your answers for parts (a) and (b) with the speed in air at the same temperature.

**19–17 Speed of Sound in Water.** A scuba diver below the surface of a lake hears the sound of a boat horn on the surface directly above her at the same time as a friend standing on dry land 22.0 m from the boat (Fig. 19–23). At what depth is the diver?

**19–18** Use the definition $B = -V\, dp/dV$ and the relation between $p$ and $V$ for an isothermal process to derive Eq. (19–25).

**19–19** What is the difference between the speed of longitudinal waves in air at 7.0°C and their speed at 67.0°C?

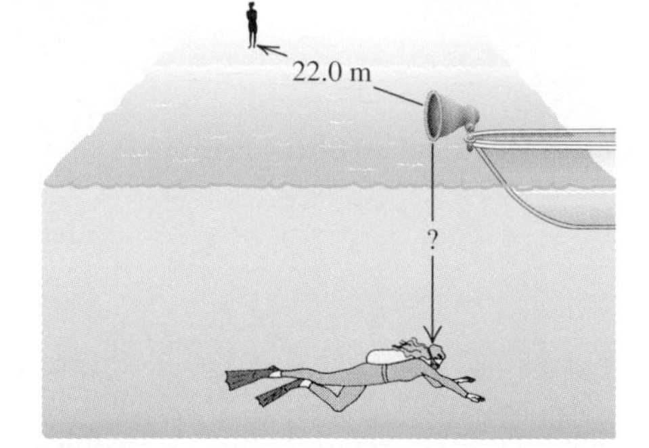


**FIGURE 19–23**

## Section 19–7 Energy in Wave Motion

**19–20** A piano wire with mass 4.00 g and length 0.800 m is stretched with a tension of 30.0 N. Waves with frequency $f = 60.0$ Hz and amplitude 1.5 mm are traveling along the wire. a) Calculate the average power carried by these waves. b) What happens to the average power if the amplitude of the waves is doubled?

**19–21** Show that Eq. (19–34) can also be written $P_{av} = \frac{1}{2}Fk\omega A^2$, where $k$ is the wave number of the wave.

## Section 19–8 Sound Waves in Crystals: A Case Study in Modern Physics

**19–22** In the simple model of a crystal introduced in Section 19–8, how is the force constant related to the atomic mass $M$, the distance $a$, the density $\rho$, and Young's modulus $Y$?

**19–23** The magnitude of the phase velocity (wave speed) is $\omega/k$. The magnitude of the group velocity is defined as $d\omega/dk$. a) How is the group velocity related to Fig. 19–22? b) What is the expression for the group velocity in our model? c) For wave numbers from $k = 0$ to $k = \pi/a$, at what wavelength does the group velocity equal zero? d) For wave numbers from $k = 0$ to $k = \pi/a$, for what wavelengths does the magnitude of the group velocity approximately equal that of the phase velocity?

**19–24 Inelastic Scattering in Sodium.** Physicists have collided neutrons with sodium and observed the creation or absorption of phonons (called inelastic scattering). They find a maximum frequency of $3.8 \times 10^{12}$ Hz for plane longitudinal waves in one particular crystal direction. Sodium has an atomic mass of 23.0 g/mol. Use our simple crystal model. a) What is the force constant in this direction? b) What is the maximum phonon energy for these waves? c) What is Young's modulus if the atoms are 0.30 nm apart and the density of sodium is 540 kg/m$^3$? d) What is the largest wavelength for which the phonon energy is half its maximum value?

**19–25** The energy of an electron in germanium is increased by $1.230 \times 10^{-19}$ J by the simultaneous absorption of $1.189 \times 10^{-19}$ J of light energy and a single phonon. What are the frequency and the angular frequency for this phonon?

**19–26** A sound wave is sent down a copper rod with an average power of 3.7 $\mu$W and a frequency of 264 Hz. a) How many phonons per second are sent down the rod? b) Would the answer be different for aluminum?

## PROBLEMS

**19–27** A transverse sine wave with an amplitude of 0.0800 m and a wavelength of 1.60 m travels from left to right along a long horizontal stretched string with a speed of 100 m/s. Take the origin at the left end of the undisturbed string. At time $t = 0$ the left end of the string is at the origin and is moving downward. a) What are the frequency, angular frequency, and propagation constant of the wave? b) What is the equation of the wave? c) What is the equation of motion of the left end of the string? d) What is the equation of motion of a particle 1.20 m to the right of the origin? e) What is the maximum magnitude of the transverse velocity of any particle of the string? f) Find the transverse displacement and the transverse velocity of a particle 1.20 m to the right of the origin at time $t = 0.0240$ s.

**19–28** The equation of a transverse wave traveling on a string is

$$y = (1.75 \text{ cm}) \sin \pi [(250/\text{s})t + (0.400/\text{cm})x].$$

a) Find the amplitude, wavelength, frequency, period, and speed of propagation. b) Sketch the shape of the string at the following values of $t$: 0, 0.0020 s, and 0.0040 s. c) Is the wave traveling in the positive or negative $x$-direction? d) If the mass per unit length of the string is 0.50 kg/m, find the tension.

**19–29** Show that $y(x, t) = A \cos (\omega t + kx)$ satisfies the wave equation, Eq. (19–12).

**19–30 Locating Lightning by Radio.** A student is listening in her dorm room to the radio broadcast of a Red Sox–Yankees baseball game at Fenway Park while doing her physics homework. In the bottom of the fourth inning a thunderstorm approaching from a generally westward direction makes its presence known in three ways: (1) The student sees a lightning flash (and hears the electromagnetic pulse on her radio receiver); (2) 3.00 s later, she hears the thunder over the radio; (3) 4.43 s after the lightning flash, the thunder rattles her window. By a previous careful measurement she knows that she is 1.12 km due north of the broadcast booth at the ballpark. The speed of sound is 344 m/s. Where did the lightning flash occur in relation to the ballpark?

**19–31** One end of a rubber tube 12.0 m long, with a total mass of 0.800 kg, is fastened to a fixed support. A cord attached to the other end passes over a pulley and supports an object with a mass of 15.0 kg. The tube is struck a transverse blow at one end. Find the time required for the pulse to reach the other end.

**19–32** A cowgirl ties one end of a 20.0-m long rope to a fence post and pulls on the other end so that the rope is stretched horizontally with a tension of 70.0 N. The mass of the rope is 0.200 kg. a) If the cowgirl moves the free end up and down with a frequency of 6.00 Hz, what is the wavelength of the transverse waves on the rope? b) The cowgirl pulls harder on the rope so that the tension is doubled, to 140.0 N. With what frequency must she move the free end of the rope up and down to produce transverse waves of the same wavelength as in part (a)?

**19–33** What must be the stress ($F/A$) in a stretched wire of a material whose Young's modulus is $Y$ for the speed of longitudinal waves to equal fifty times the speed of transverse waves?

**19–34** A metal wire, with a density of $4.00 \times 10^3$ kg/m$^3$ and a Young's modulus of $2.00 \times 10^{11}$ Pa, is stretched between rigid supports. At one temperature the speed of a transverse wave is found to be 200 m/s. When the temperature is raised 30.0 C°, the speed decreases to 160 m/s. Determine the coefficient of linear expansion of the wire.

## CHALLENGE PROBLEM

**19–35** A deep-sea diver is suspended beneath the surface of Loch Ness by a 100-m-long cable that is attached to a boat on the surface (Fig. 19–24). The diver and his suit have a total mass of 120 kg and a volume of 0.0800 m$^3$. The cable has a diameter of 2.00 cm and a linear mass density of $\mu = 1.30$ kg/m. The diver thinks he sees something moving in the murky depths and jerks the end of the cable back and forth to send transverse waves up the cable as a signal to his companions in the boat. a) What is the tension in the cable at its lower end, where it is attached to the diver? Do not forget to include the buoyant force that the seawater (density 1030 kg/m$^3$) exerts on him. b) Calculate the tension in the cable a distance $x$ above the diver. The buoyant force on the cable must be included in this calculation. c) The speed of transverse waves on the cable is given by $v = \sqrt{F/\mu}$ (Eq. 19–13). The speed therefore varies along the cable, since the tension is not constant. (This expression neglects the damping force that the water exerts on the moving cable.) Integrate to find the time required for the first signal to reach the surface.

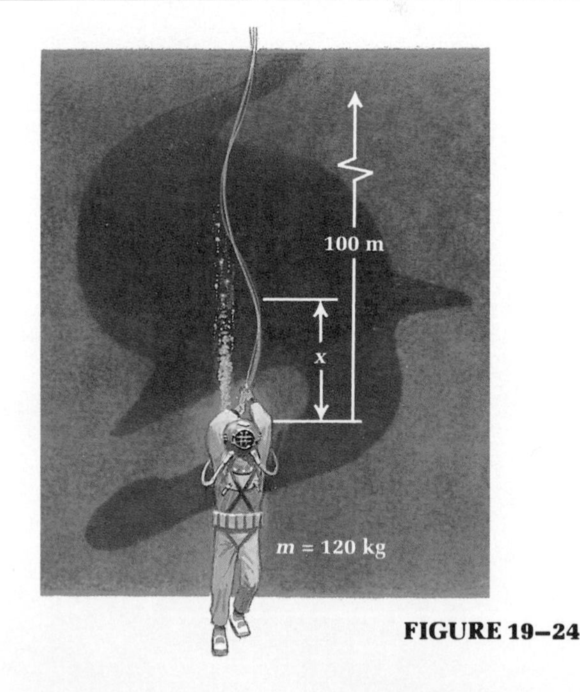


FIGURE 19–24

The organ at Wellesley College, Massachusetts, was built in 1977-1981 by Charles Fisk, according to the methods used in the 17th century. The air supply comes from a bellows operated by human power.

CHAPTER 20

# Superposition and Normal Modes

Organs usually have several sets of pipes available for sound, which can be brought into play or stopped as the player wishes. Most organs have at least two keyboards, often more, and can control as many as several thousand pipes.

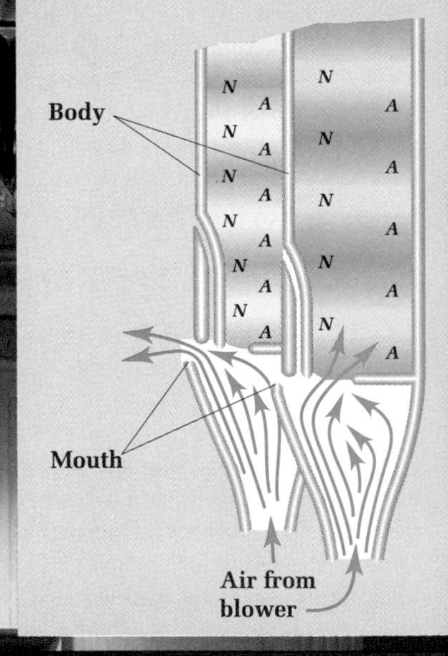


Air blown into the bottom of an organ pipe flows across the mouth of the pipe and sets up a steady wave oscillation in the air column, causing the pipe to "speak." The length and diameter of the pipe determine the wavelengths of sound that can interfere constructively and form a standing wave in the pipe..

## KEY CONCEPTS

- **When a wave reaches a boundary of the medium, such as the end of a string or a pipe, a reflected wave traveling in the opposite direction is produced.**
- **Two waves can occupy the same region of space at the same time. The total wave disturbance at any point is the sum of the disturbances of the separate waves. Two waves with the same amplitude and frequency but opposite directions combine to form a standing wave, which has a series of alternating nodes and antinodes and does not appear to move in either direction.**
- **When both ends of a string are held, only certain frequencies of standing waves can occur. The frequencies and associated wave patterns are called normal modes. Similar things happen with a pipe of finite length. The pitches of many musical instruments are determined by normal-mode frequencies in the instruments.**
- **Interference effects occur in all kinds of wave motion, light as well as sound.**
- **When a periodically varying force is applied to a system having normal modes, large-amplitude vibrations are induced if the force frequency coincides with one of the normal-mode frequencies. This behavior is called resonance.**

## INTRODUCTION

Why do you get a musical note when you pluck a guitar string but not when you drop a bag full of empty tin cans on the ground? What determines the pitch of a piano string, an organ pipe, or a xylophone bar? Why do suspension bridges sometimes vibrate with dangerously large amplitudes? Each of these situations involves a wave that strikes the boundaries of its medium. In each case the wave is *reflected* at this boundary, and the resulting wave motion is a combination of the initial wave and the reflected wave.

It's easy to think of examples of wave reflection. When you yell at a building wall or a cliff face some distance away, the sound wave is reflected from the rigid surface and an *echo* comes back. When you flip the end of a rope whose far end is tied to a rigid support, a pulse travels the length of the rope and is reflected back to you. In both cases the initial and reflected waves overlap in the same region of the medium. This overlapping of waves is called *interference*. When there are *two* boundary points or surfaces, we get repeated reflections. In such situations we find that sinusoidal waves can occur only for particular frequencies, which are determined by the properties and dimensions of the medium. These special frequencies and their associated wave patterns are called *normal modes*. The pitches of most musical instruments are determined by normal-mode frequencies, and many other mechanical vibrations (including some that can destroy buildings and make airplane wings fall off) involve normal-mode motion. Interference is also the basic principle behind holograms and the use of x rays to explore crystal structure. Interference and normal modes also appear in determining the energy levels of atoms.

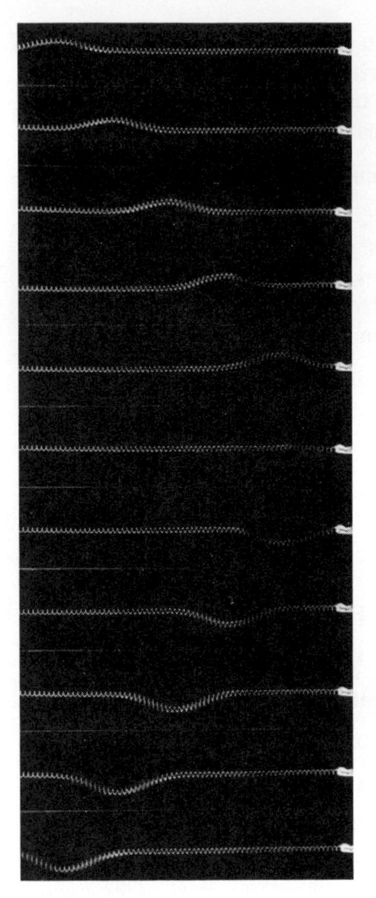

**20–1** A pulse starts at the left in the top image, travels to the right, and is reflected from the fixed end of the "string" at the right.

## 20–1 BOUNDARY CONDITIONS FOR A STRING

As a simple example of wave reflections and the role of the boundary of a wave medium, let's look again at transverse waves on a stretched string. What happens when a wave pulse or a sinusoidal wave arrives at the end of the string? If the end is fastened to a rigid support, the end cannot move. The arriving wave exerts a force on the support; the reaction to this force, exerted *by* the support *on* the string, "kicks back" on the string and sets up a *reflected* pulse or wave traveling in the reverse direction.

The opposite situation from an end that is held stationary is one that is perfectly free to move in the direction perpendicular to the length of the string. For example, the string might be tied to a light ring that slides on a frictionless rod perpendicular to the string. The ring and rod maintain the tension but exert no transverse force. When a wave arrives at such a free end, the end "overshoots," and again a reflected wave is set up. The physical conditions at the end of the string, such as the presence of a rigid support or the complete absence of transverse force, are called **boundary conditions.**

Figure 20–1 is a multiflash photograph showing the reflection of a pulse at the stationary end of a rubber tube. (The camera was tilted upward while the photographs were being taken, in order to spread the successive images out vertically.) The reflected pulse moves in the opposite direction from the initial pulse, and its displacement is also opposite. This situation is illustrated by a wave pulse in a string in Fig. 20–2a. When reflection takes place at a *free* end, as shown in Figure 20–2b, the direction of propagation is again reversed, but the direction of the displacement is the same as for the initial pulse.

The formation of the reflected pulse is similar to the overlap of two pulses traveling in opposite directions. Figure 20–3 shows two pulses with the same shape, one inverted with respect to the other, traveling in opposite directions. As the pulses overlap and pass each other, the total displacement of any point of the string is the *algebraic sum* of the displacements at that point in the individual pulses. At points along the vertical line in the middle of the figure, the total displacement is zero at all times. Thus the motion of the left half of the string would be the same if we cut the string at point $O$, threw away the right side, and held the end at $O$ stationary. The two pulses on the left side then correspond to the incident and reflected pulses, combining so that the total displacement at the end of the string is *always* zero. For this to occur, the reflected pulse must be inverted relative to the incident pulse.

Figure 20–4 shows two pulses with the same shape, traveling in opposite directions but not inverted relative to each other. The displacement at the midpoint is not zero, but the slope of the string at this point is always zero. According to Eq. (19–28), this corresponds to the absence of any transverse force at this point. In this case the motion of the left half of the string would be the same if we cut the string at point $O$ and anchored the end with a sliding ring (Fig. 20–2b), to maintain tension without exerting any transverse force. This situation corresponds to reflection of a pulse at a free end of a string at this point. In this case the reflected pulse is *not* inverted.

Combining the displacements of the separate pulses at each point to obtain the actual displacement is an example of the *principle of superposition*, to be discussed in detail in the next section. This principle plays a central role in most of this chapter, as well as in later discussions involving wave motion.

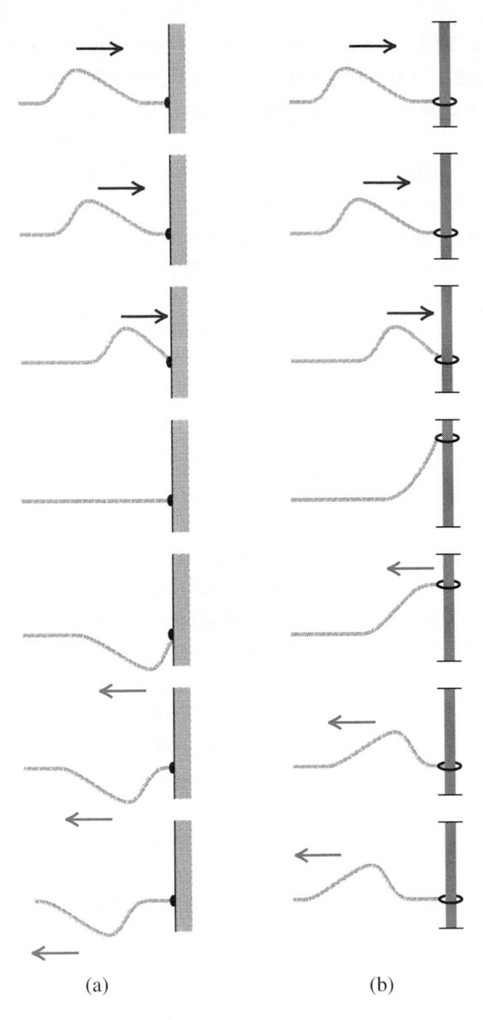


**20–2** Reflection of a wave pulse (a) at a fixed end of a string and (b) at a free end.

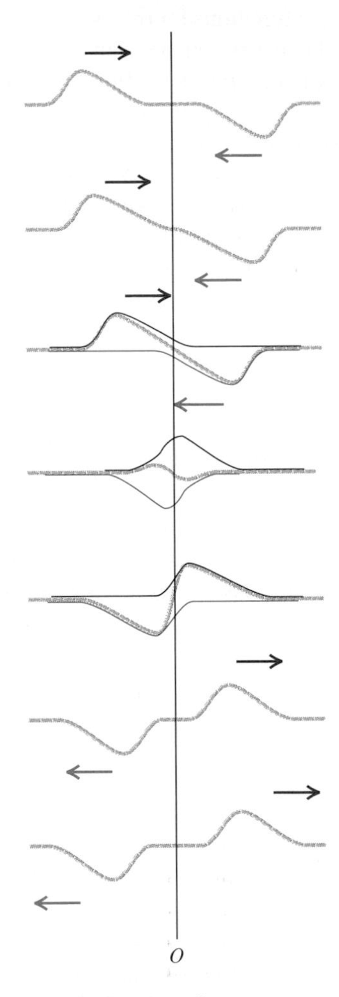


**20–3** Overlap of two wave pulses traveling in opposite directions with one pulse inverted with respect to the other.

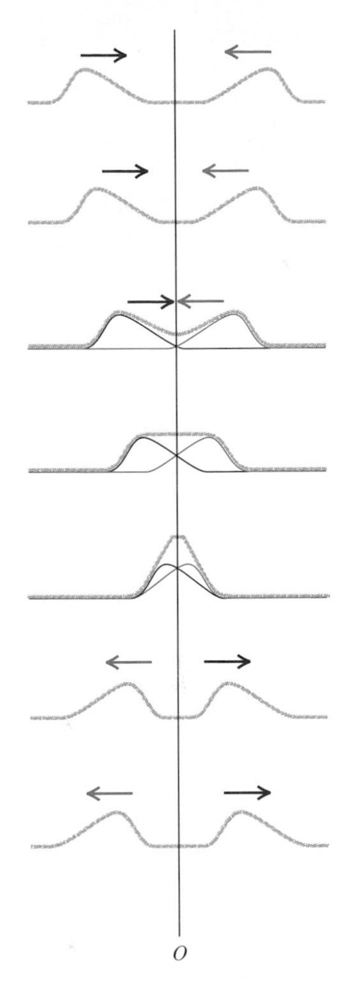


**20–4** Overlap of two wave pulses traveling in opposite directions with no inversion of one pulse.

## 20–2 SUPERPOSITION AND STANDING WAVES

We have talked about reflection of a wave pulse on a rope or string when it arrives at a boundary point (either a stationary point or an end that is free to move transversely). The same thing happens when a *sinusoidal* wave reaches such a boundary point. A reflected wave originates at this point and travels in the opposite direction. The resulting motion of the string is determined by the **principle of superposition.** This principle states that when two waves overlap, the actual displacement of any point on the string, at any time, is obtained by adding two displacements: the displacement the point would have if only the first wave were present and the displacement it would have with only the second wave.

Mathematically speaking, the principle of superposition states that the wave function $y(x, t)$ that describes the resulting motion in this situation is obtained by *adding* the two wave functions for the two separate waves. This additive property of wave functions depends, in turn, on the form of the wave equation, Eq. (19–12) or (19–18), which every physically possible wave function must satisfy. Specifically, the wave equation is *linear;* that is, it contains derivatives of the function $y(x, t)$ only to the first power. As a result, if any two functions $y_1(x, t)$ and $y_2(x, t)$ each satisfy the wave equation separately, their sum $y_1 + y_2$ also satisfies it and is therefore a physically possible motion. Because this principle depends on the linearity of the wave equation and the corresponding linear-combination property of its solutions, it is also called the *principle of linear superposition*. For some physical phenomena, such as nonlinear elastic behavior (behavior not obeying Hooke's law), the wave equation is *not* linear; this principle does not hold for such systems.

The principle of superposition is of central importance in all types of wave motion. It applies not only to waves on a string, but also to sound waves, electromagnetic waves (such as light), and many other types of waves. The general term **interference** is used to describe the result of two or more waves passing through the same region at the same time.

Now let's look in more detail at what happens when a *sinusoidal* wave is reflected by a stationary point on a string. When the two waves combine, the resulting motion no longer looks like two waves traveling in opposite directions. The string appears to be subdivided into a number of segments, as in the time-exposure photograph of Fig. 20–5. Figure 20–5e shows a few instantaneous shapes of the string in Figure 20–5b. Let's compare this behavior with the traveling waves that we studied in Chapter 19. In a traveling wave the amplitude is constant, and the waveform moves with a speed equal to the wave speed. Here, instead, the waveform remains in the same position along the string and its amplitude fluctuates. There are particular points called **nodes** (labeled $N$ in Fig. 20–5e that never move at all. Midway between the nodes are points called **antinodes** (labeled $A$ in Fig. 20–5e) where the amplitude of motion is greatest. Because the wave pattern doesn't appear to be moving in either direction along the string, it is called a **standing wave.**

Figure 20–6 shows how we can use the superposition principle to understand the formation of a standing wave. The figure shows separate graphs of the waveforms at four

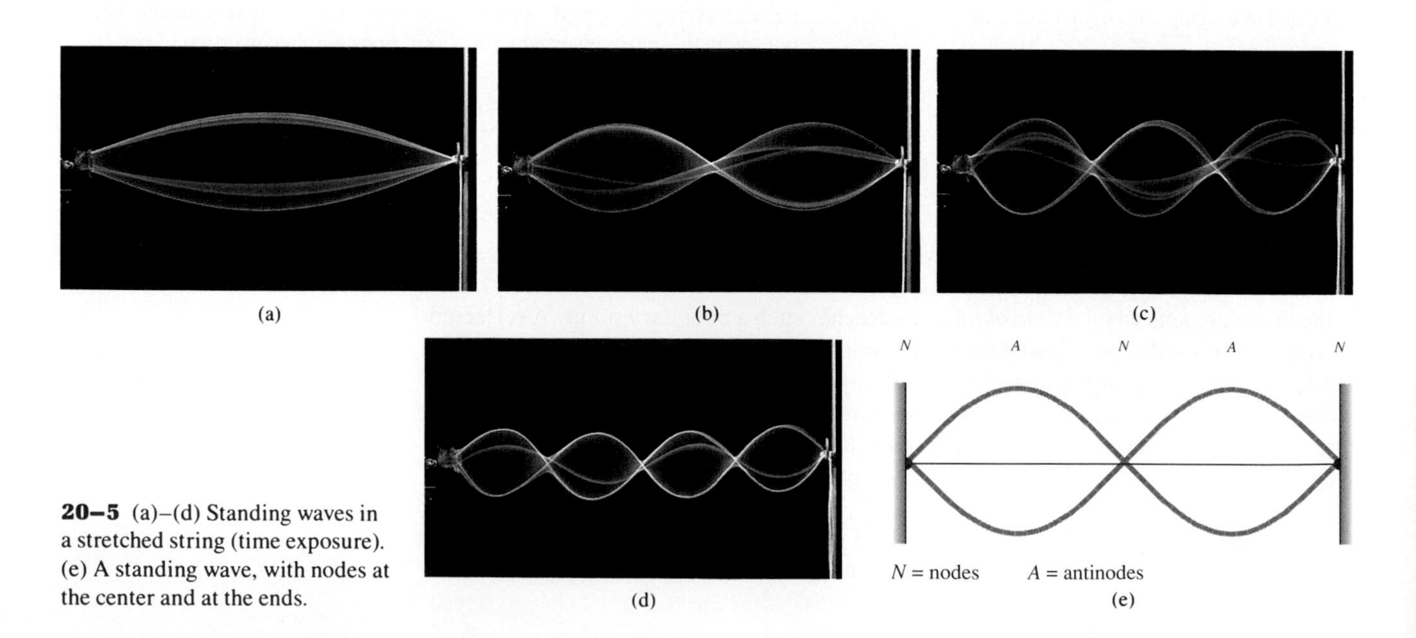


**20–5** (a)–(d) Standing waves in a stretched string (time exposure). (e) A standing wave, with nodes at the center and at the ends.

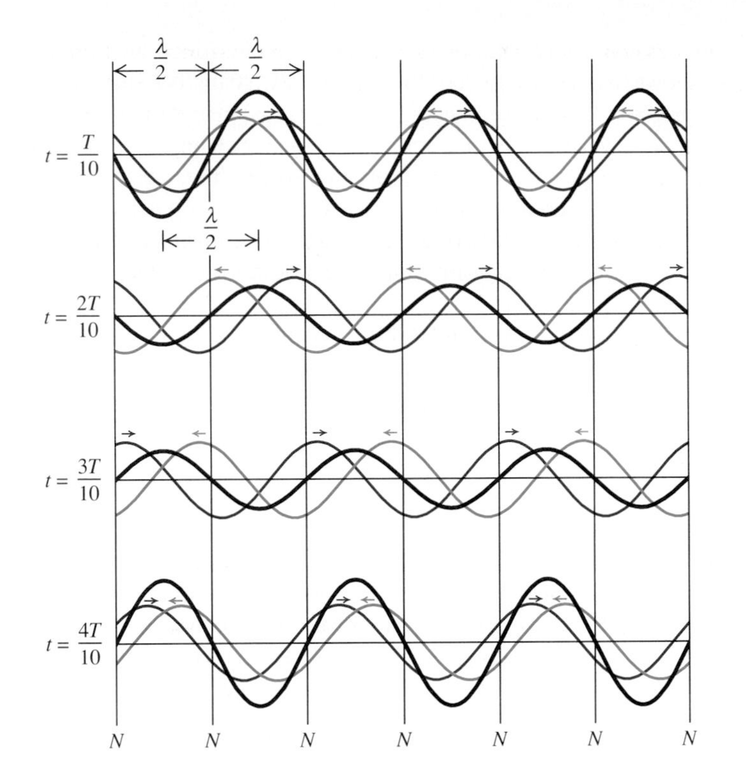


**20–6** Formation of a standing wave. A wave traveling to the right (red lines) combines with a wave traveling to the left (blue lines) to form a standing wave (black lines). The thin horizontal line in each part shows the equilibrium position of the string.

instants, one-tenth of a period apart. The red curves show a wave traveling to the right. The blue curves show a wave with the same propagation speed, wavelength, and amplitude traveling to the left. The black curves are the resultant waveform, obtained by applying the principle of superposition. At each point we add the displacements (the values of $y$) for the two separate waves. At those places marked $N$ at the bottom of Fig. 20–6 the resultant displacements are *always* zero. These are the *nodes.* Midway between the nodes are the points of *greatest* amplitude; these are the *antinodes.* We can see from the figure that *the distance between successive nodes or between successive antinodes is one half-wavelength,* or $\lambda/2$.

We can derive a wave function for the standing wave of Fig. 20–6 by adding the wave functions for two waves with equal amplitude, period, wavelength, and polarization, traveling in opposite directions. Call these two wave functions $y_1$ and $y_2$; we can think of $y_1$ as an *incident wave* traveling to the left, arriving at the point $x = 0$ and being reflected, and $y_2$ as the *reflected wave* proceeding to the right from $x = 0$. We noted in Section 20–1 that the reflected wave is inverted, so we give its amplitude a negative sign:

$$y_1 = A \sin(\omega t + kx) \quad \text{(traveling to the left)},$$

$$y_2 = -A \sin(\omega t - kx) \quad \text{(traveling to the right)}.$$

The wave function for the standing wave is the sum of these:

$$y = y_1 + y_2 = A[\sin(\omega t + kx) - \sin(\omega t - kx)].$$

We can rearrange this by using the identities for the sine of the sum or the difference of two angles: $\sin(a \pm b) = \sin a \cos b \pm \cos a \sin b$. Using these and combining terms, we obtain

$$y = y_1 + y_2 = (2A \cos \omega t) \sin kx. \tag{20–1}$$

This expression has two factors: a function of $t$ multiplied by a function of $x$. The factor $\sin kx$ shows that at each instant the shape of the string is a sine curve. The amplitude is the expression in the parentheses, and it varies sinusoidally with time from $2A$ to $-2A$ and back. The wave shape does not move along the string; it stays in the same position, but the amplitude grows large and smaller with time. This behavior is shown graphically in Fig. 20–6. Each point in the string still undergoes simple harmonic motion, but all the points between any successive pair of nodes move *in phase*. This is in contrast to the phase differences between motions of adjacent points that we see with a wave traveling in one direction.

We can use Eq. (20–1) to find the positions of the nodes. At any point for which $\sin kx = 0$ the displacement is *always* zero. This occurs when $kx = 0, \pi, 2\pi, 3\pi, \ldots$, or

$$\begin{aligned} x &= 0, \quad \frac{\pi}{k}, \quad \frac{2\pi}{k}, \quad \frac{3\pi}{k}, \quad \ldots, \\ &= 0, \quad \frac{\lambda}{2}, \quad \frac{2\lambda}{2}, \quad \frac{3\lambda}{2}, \quad \ldots. \end{aligned} \tag{20–2}$$

In particular, there is a node at $x = 0$. This standing wave could correspond to a wave traveling in the $-x$-direction, reflected from a stationary point at $x = 0$. The reflected wave is inverted; that's why we chose the two wave functions $y_1$ and $y_2$ to have opposite amplitudes $A$ and $-A$. Note also that the change of sign corresponds to a shift in *phase* of 180° or $\pi$ radians. At $x = 0$ the motion from the incident wave is $A \cos \omega t$, and the motion from the reflected wave is $-A \cos \omega t$, which we can also write as $A \cos (\omega t + \pi)$.

A standing wave, unlike a traveling wave, *does not* transfer energy from one end to the other. The two waves that form a standing wave would individually carry equal amounts of power in opposite directions. There is a local flow of energy from each node to the adjacent antinodes and back, but the *average* rate of energy transfer is zero at every point. We invite you to evaluate Eq. (19–29) with the wave function of Eq. (20–1) and show that the average power is zero.

## EXAMPLE 20–1

Two waves are traveling on a string. One is moving in the $+x$-direction at 84.0 m/s with an amplitude of 15 mm and a frequency of 120 Hz. The other wave is identical except that it is traveling in the $-x$-direction. The resulting wave is the linear superposition of these two waves. a) Find the points on the string that don't move at all, assuming that one of them is at $x = 0$. b) Find the equation giving the displacement of a point on the string as a function of position and time. c) Find the amplitude at points of maximum oscillation.

**SOLUTION** The superposition of these two traveling waves with equal frequency and amplitude is a standing wave.

a) The nodes are points on the string that don't move; their positions are given by Eq. (20–2). The wavelength is $\lambda = v/f = (84.0 \text{ m/s})/(120 \text{ s}^{-1}) = 0.700$ m, and Eq. (20–2) gives

$$x = 0, \quad 0.350 \text{ m}, \quad 0.700 \text{ m}, \quad 1.050 \text{ m}, \quad \ldots.$$

b) We are given the amplitude $A = 15 \text{ mm} = 0.015$ m. The wave function is Eq. (20–1); we need to find $\omega$ and $k$. We have

$$\omega = 2\pi f = (2\pi \text{rad})(120 \text{ s}^{-1}) = 754 \text{ rad/s},$$

$$k = \frac{2\pi}{\lambda} = \frac{2\pi \text{ rad}}{0.700 \text{ m}} = 8.98 \text{ rad/m}.$$

Then Eq. (20–1) gives

$$y = [(0.030 \text{ m}) \cos (754 \text{ rad/s})t] \sin (8.98 \text{ rad/m})x.$$

We could now take first and second partial derivatives of $y$ with respect to time to find the transverse velocity and acceleration of any particle on the string as functions of time. Or we could take the partial derivative of $y$ with respect to $x$ to find the slope of the string at any point on the string as a function of time.

c) From the expression that we just found for $y$ we see that the maximum displacement from equilibrium is 0.030 m. This maximum occurs at the antinodes, where $\sin (8.98 \text{ rad/m})x = \pm 1$ and $x = 0.175$ m, 0.525 m, 0.875 m, ….

## 20–3 NORMAL MODES OF A STRING

We have mentioned the reflection or echo of a sound wave from a rigid wall. Now suppose we have two parallel walls. If we produce a sharp sound pulse, such as a hand clap, at some point between the walls, the result is a series of regularly spaced echoes caused by repeated back-and-forth reflection between the walls. In room acoustics this phenomenon is called "flutter echo"; it is the bane of acoustical engineers.

The analogous situation with transverse waves on a string is a string with some definite length $L$, rigidly held at *both* ends. If we produce a sinusoidal wave on a guitar string, the wave is reflected and re-reflected from the ends. The waves combine to form a standing wave. Both ends must be nodes, and adjacent nodes are one half-wavelength $(\lambda/2)$ apart. The length of the string therefore has to be $(\lambda/2)$, $2(\lambda/2)$, $3(\lambda/2)$, or in general some integer number of half-wavelengths:

$$L = n\frac{\lambda}{2} \quad (n = 1, 2, 3, \ldots). \tag{20–3}$$

That is, a standing wave can exist in a string with length $L$, held at both ends, only when its wavelength satisfies Eq. (20–3).

Solving this equation for $\lambda$ and labeling the possible values of $\lambda$ as $\lambda_n$, we find

$$\lambda_n = \frac{2L}{n} \quad (n = 1, 2, 3, \ldots). \tag{20–4}$$

When the wavelength is *not* equal to one of these values, no standing wave is possible.

Corresponding to this series of possible wavelengths is a series of possible frequencies $f_n$, each related to its corresponding wavelength by $f_n = v/\lambda_n$. The smallest frequency $f_1$ corresponds to the largest wavelength (the $n = 1$ case), $\lambda = 2L$:

$$f_1 = \frac{v}{2L} \quad \text{(string, held at both ends).} \tag{20–5}$$

This is called the **fundamental frequency.** The other frequencies are $f_2 = 2v/2L$, $f_3 = 3v/2L$, and so on. These are all integer multiples of $f_1$, such as $2f_1$, $3f_1$, $4f_1$, and so on, and we can express *all* the frequencies as

$$f_n = n\frac{v}{2L} = nf_1 \quad (n = 1, 2, 3, \ldots). \tag{20–6}$$

These frequencies, all integer multiples of $f_1$, are called **harmonics,** and the series is called a **harmonic series.** Musicians sometimes call $f_2$, $f_3$, and so on **overtones;** $f_2$ is the second harmonic or the first overtone, $f_3$ is the third harmonic or the second overtone, and so on.

In deriving Eq. (20–2) we found that there are nodes at $x = 0$ and at $kx = n\pi$, where $n = 1, 2, 3, \ldots$. To have a node at each end of a string whose ends are at $x = 0$ and $x = L$, we must have

$$kL = n\pi \quad (n = 1, 2, 3, \ldots).$$

Replacing $k$ by $2\pi/\lambda_n$, we obtain

$$\frac{2\pi L}{\lambda_n} = n\pi \quad \text{or} \quad \lambda_n = \frac{2L}{n}, \tag{20–7}$$

in agreement with Eq. (20–4).

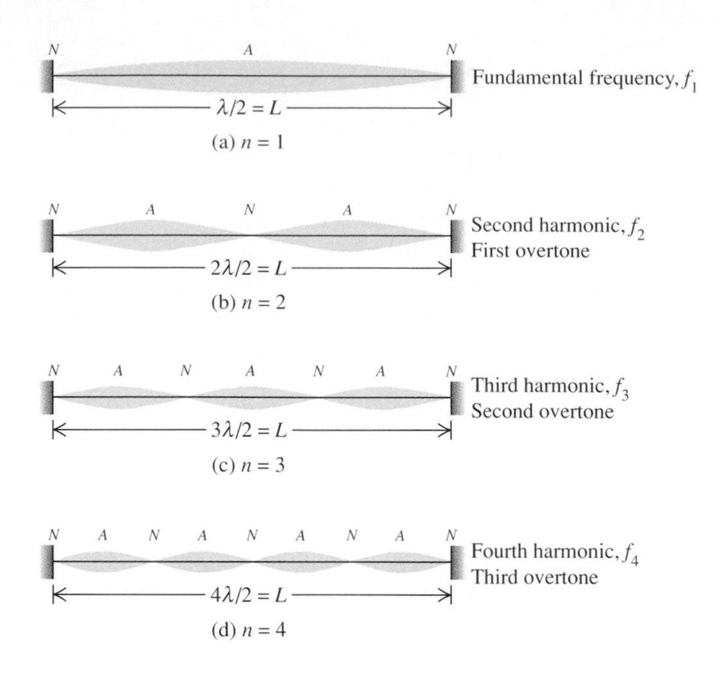


**20–7** The first four normal modes of a string fixed at both ends.

A **normal mode** is a motion in which all particles of the string move sinusoidally with the same frequency. Each of the wavelengths given by Eq. (20–7) corresponds to a possible normal mode pattern and frequency. There are infinitely many normal modes, each with its characteristic frequency and vibration pattern. Figure 20–7 shows the first four normal-mode patterns and their associated frequencies and wavelengths. We can contrast this situation with a simpler vibrating system, the harmonic oscillator, which has only one normal mode and one characteristic frequency.

If we displace a string so that its shape is the same as one of the normal-mode patterns, and then release it, it vibrates with the frequency of that mode. But when a piano string is struck or a guitar string is plucked, not only the fundamental but many overtones are present in the resulting vibration. This motion is therefore a combination or *superposition* of many normal modes. Several frequencies and motions are present simultaneously, and the displacement of any point on the string is the sum (or superposition) of displacements associated with the individual modes.

Indeed, it is possible to represent *every possible* motion of the string as some superposition of normal-mode motions. Finding this representation for a given vibration pattern is called *harmonic analysis.* The sum of sinusoidal functions that represents a complex wave is called a *Fourier series.* Figure 20–8 shows an example of representing a triangular wave shape, perhaps produced by plucking a guitar string, as a combination of sinusoidal functions.

As we have seen, the fundamental frequency of a vibrating string is $f_1 = v/2L$. The wave speed $v$ is determined by Eq. (19–13), $v = \sqrt{F/\mu}$. Combining these equations, we have

$$f_1 = \frac{1}{2L}\sqrt{\frac{F}{\mu}}. \tag{20–8}$$

All string instruments are "tuned" to the correct frequencies by varying the tension $F$. An increase of tension increases the wave speed $v$ and thus increases the frequency (and the

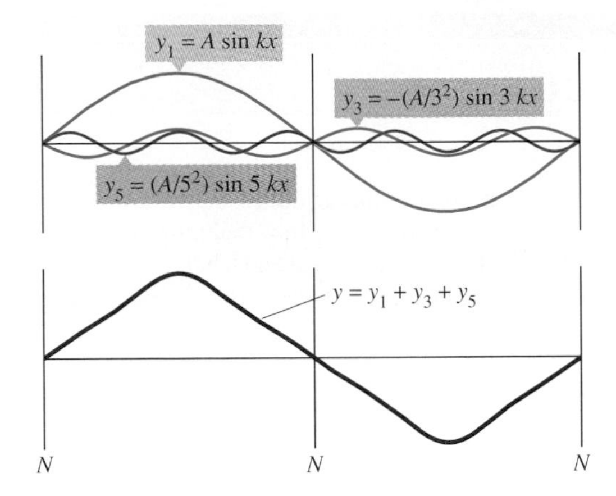


**20–8** A triangular wave is well represented (except at the sharp maximum and minimum points) by the sum of only three sinusoidal functions.

pitch) for a string of any fixed length $L$. The inverse dependence of frequency on length $L$ is illustrated by the long strings of the bass (low-frequency) section of the piano or the string bass compared with the shorter strings of the piano treble or the violin (Fig. 20–9). One reason for winding the bass strings of a piano with wire is to increase the mass per unit length $\mu$ so as to obtain the desired low frequency without resorting to a string that is inconveniently long or inflexible. In playing the violin or guitar the usual means of varying the pitch is to press the strings against the fingerboard with the fingers to change the length $L$ of the vibrating portion of the string. As Eq. (20–8) shows, decreasing $L$ increases $f_1$.

**20–9** Comparison of ranges of a concert grand piano, a violin, a viola, a cello, and a string bass. In all cases, longer strings provide bass notes and shorter strings produce treble notes.

## PROBLEM-SOLVING STRATEGY

### Standing waves

**1.** As with the problems in Chapter 19, with standing wave problems it is useful to distinguish between the purely kinematic quantities, such as wave speed $v$, wavelength $\lambda$, and frequency $f$, and the dynamic quantities involving the properties of the medium, including $F$, $\mu$, and (in the next section) $B$ and $\rho$. You can compute the wave speed if you know either $\lambda$ and $f$ or $F$ and $\mu$. Try to determine whether the properties of the medium are involved in the problem at hand or whether the problem is only kinematic in nature.

**2.** In visualizing nodes and antinodes in standing waves it is always helpful to draw diagrams. For a string you can draw the shape at one instant and label the nodes $N$ and the antinodes $A$. For longitudinal waves (discussed in the next section) it is not so easy to draw the shape, but you can still label the nodes and antinodes. The distance between two adjacent nodes or two adjacent antinodes is always $\lambda/2$, and the distance between a node and the adjacent antinode is always $\lambda/4$.

### EXAMPLE 20–2

**A giant bass viol** In an effort to get your name in the *Guinness Book of World Records* you set out to build a bass viol with strings that have a length of 5.00 m between fixed points. One string has a linear mass density of 40 g/m and a fundamental frequency of 20 Hz. a) Calculate the tension of this string. b) Calculate the frequency and wavelength of the second harmonic. c) Calculate the frequency and wavelength of the second overtone.

**SOLUTION** a) To find the tension $F$ in the string, we first solve Eq. (20–8) for $F$:

$$F = 4\mu L^2 f_1^2.$$

We have $\mu = 40 \times 10^{-3}$ kg/m, $L = 5.00$ m, and $f_1 = 20\ \text{s}^{-1}$; substituting these, we find

$$F = 4(40 \times 10^{-3}\ \text{kg/m})(5.00\ \text{m})^2(20\ \text{s}^{-1})^2 = 1600\ \text{N} \quad \text{(or 360 lb)}.$$

(In a real bass viol the string tension is typically a few hundred newtons.)

b) From Eq. (20–6) the second harmonic frequency ($n = 2$) is

$$f_2 = 2f_1 = 2(20\ \text{Hz}) = 40\ \text{Hz}.$$

From Eq. (20–4) the wavelength of the second harmonic is

$$\lambda_2 = \frac{2L}{2} = 5.00\ \text{m}.$$

c) The second overtone is the "second tone over" (above) the fundamental, that is, $n = 3$, and

$$f_3 = 3f_1 = 3(20\ \text{Hz}) = 60\ \text{Hz}.$$

The wavelength is

$$\lambda_3 = \frac{2L}{3} = 3.33\ \text{m}.$$

## 20–4 LONGITUDINAL STANDING WAVES

When longitudinal waves propagate in a fluid in a pipe with finite length, the waves are reflected from the ends in the same way that transverse waves on a string are reflected at its ends. The superposition of the waves traveling in opposite directions again forms a standing wave.

When reflection takes place at a *closed* end (an end with a rigid barrier or plug), the displacement of the particles at this end must always be zero. Just as with a stationary end of a string, there is no displacement at the end, and the end is a *node*. In the following discussion we will call a closed end of a pipe a *displacement node*. If an end of a pipe is *open* and the pipe is narrow in comparison with the wavelength (which is true for most musical instruments), the open end is a *displacement antinode*. We will explain below why this is the case. (A *free* end of a stretched string, mentioned in Section 20–1, is also a displacement antinode.) Thus longitudinal waves in a column of fluid are reflected at the closed and open ends of a pipe in the same way that transverse waves in a string are reflected at stationary and free ends, respectively.

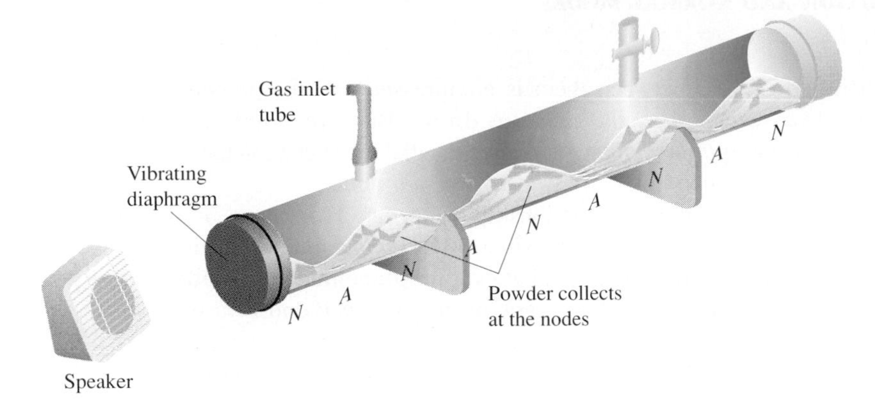


**20–10** Kundt's tube for determining the velocity of sound in a gas. The shading represents the density of the gas molecules at an instant when the pressure at the displacement nodes is a maximum or a minimum.

We can demonstrate longitudinal standing waves in a column of gas and also measure the wave speed, using an apparatus called Kundt's tube (Fig. 20–10). A horizontal glass tube a meter or so long is closed at one end and at the other end has a flexible diaphragm that can transmit vibrations. We use as our sound source a small loudspeaker driven by an audio oscillator and amplifier to vibrate the diaphragm sinusoidally with a frequency that we can vary. We place a small amount of light powder or cork dust inside the pipe and distribute it uniformly along the bottom side of the pipe. As we vary the frequency of the sound, we pass through frequencies where the amplitude of the standing waves becomes large enough for the moving gas to sweep the cork dust along the pipe at all points where the gas is in motion. The powder therefore collects at the displacement nodes (where the gas is not moving). Adjacent nodes are separated by a distance equal to $\lambda/2$, and we can measure this distance. We read the frequency $f$ from the oscillator dial, and we can then calculate the speed $v$ of the waves from the relation $v = \lambda f$.

Figure 20–11 will help you to visualize longitudinal standing waves; it is analogous to Fig. 19–15 for longitudinal traveling waves. Again tape two index cards together edge-to-edge, with a gap of a millimeter or two forming a thin slit. Place the card over the diagram with the slit horizontal and move it vertically with constant velocity. The portions of the sine curves that appear in the slit correspond to the oscillations of the particles in a longitudinal standing wave. Each particle moves with longitudinal simple harmonic motion about its equilibrium position. The particles at the nodes do not move, and the nodes are regions of maximum compression and expansion. Midway between the nodes are the antinodes, regions of maximum displacement but zero compression and expansion.

**20–11** Diagram for illustrating longitudinal standing waves. The straight vertical lines mark the locations of displacement nodes and pressure antinodes.

At a displacement node the pressure variations above and below the average have their *maximum* value; at a displacement antinode the pressure does not vary. To understand this, note that points on opposite sides of a displacement *node* vibrate in *opposite phase*. You can see this using Fig. 20–12. When the points approach

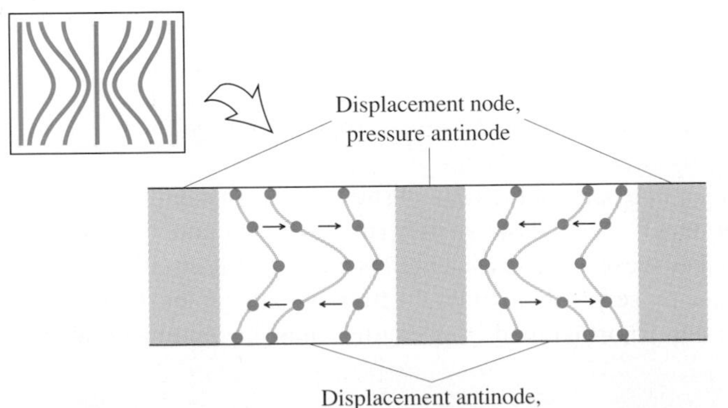


**20–12** Points on opposite sides of a displacement node vibrate in opposite phase, creating a pressure antinode. Points on opposite sides of a displacement antinode vibrate in phase, creating a constant pressure node.

each other, the gas between them is compressed and the pressure rises; when they recede from each other, the pressure drops. But two points on opposite sides of a displacement *antinode* vibrate *in phase;* the distance between them is nearly constant, and there is *no* pressure variation at the antinode.

We can describe this relationship in terms of **pressure nodes,** points where the pressure does not vary, and **pressure antinodes,** points where the pressure variation is greatest. **A pressure node is always a displacement antinode, and a pressure antinode is always a displacement node.** An *open* end of a narrow tube or pipe is a pressure node because it is open to the atmosphere, where the pressure is constant. Because of this, an open end is always a displacement *antinode.*

### EXAMPLE 20–3

**Sound speed in hydrogen** At a frequency of 25 kHz the distance from a closed end of a tube of hydrogen gas to the first pressure antinode of a standing wave is 0.026 m. Calculate the wave speed.

**SOLUTION** The closed end is a displacement node; a pressure antinode is a displacement node, and these two displacement nodes are 0.026 m apart. Adjacent nodes are a half-wavelength apart, so we have $\lambda/2 = 0.026$ m and $\lambda = 0.052$ m. Then

$$v = \lambda f = (0.052\ \text{m})(25{,}000\ \text{s}^{-1}) = 1300\ \text{m/s}.$$

From Table 19–1 the measured speed of sound in hydrogen at 20°C is 1330 m/s.

## 20–5 NORMAL MODES OF AIR COLUMNS

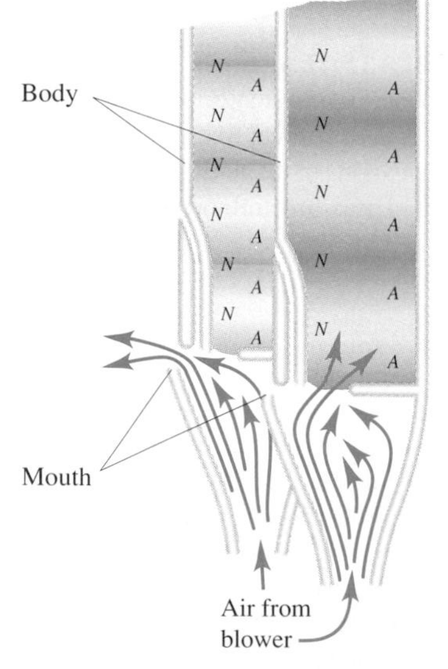


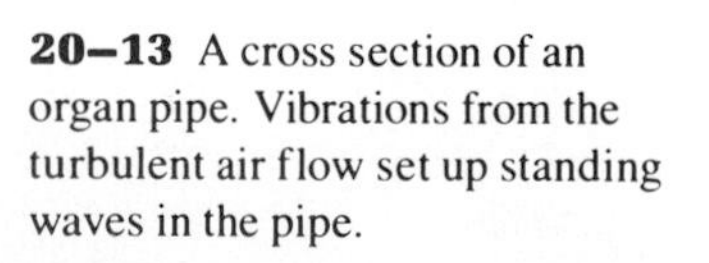
**20–13** A cross section of an organ pipe. Vibrations from the turbulent air flow set up standing waves in the pipe.

Many musical instruments, including organ pipes and all the woodwind and brass instruments, use longitudinal standing waves (normal modes) in vibrating air columns to produce musical tones. Organ pipes are one of the simplest examples. Air is supplied by a blower, at a gauge pressure typically of the order of $10^3$ Pa or $10^{-2}$ atm, to the bottom end of the pipe (Fig. 20–13). A stream of air emerges from the narrow slit to the left of the horizontal surface (the *languid*) and is directed against the top edge of the opening, which is called the *mouth* of the pipe. The column of air in the pipe is set into vibration, and there is a series of possible normal modes, just as with a stretched string. The mouth always acts as an open end; thus it is a pressure node and a displacement antinode. The top may be either open or closed.

Let's consider the general problem of normal modes in an air column. The pipe may be a flute, a clarinet, an organ pipe, or any other wind instrument. In Fig. 20–14 both ends of the pipe are open, so both are pressure nodes and displacement antinodes. An organ pipe that is open at both ends is called an *open pipe.* The fundamental frequency $f_1$ corresponds to a standing-wave pattern with a displacement antinode at each end and a displacement node in the middle (Fig. 20–14a). The distance between adjacent nodes is always equal to one half-wavelength, and in this case that is equal to the length $L$ of the pipe: $\lambda/2 = L$. The corresponding frequency, obtained from the relation $f = v/\lambda$, is

$$f_1 = \frac{v}{2L} \quad \text{(open pipe)}. \tag{20–9}$$

The other two parts of Fig. 20–14 show the second and third harmonics (first and second overtones); their vibration patterns have two and three displacement nodes, respectively. For these a half-wavelength is equal to $L/2$ and $L/3$, respectively, and the frequencies are twice and three times the fundamental, respectively. That is, $f_2 = 2f_1$ and $f_3 = 3f_1$. For *every* normal mode the length $L$ must be an integer number of half-wavelengths, and the possible wavelengths $\lambda_n$ are given by

$$L = n\frac{\lambda_n}{2}$$

or

$$\lambda_n = \frac{2L}{n} \quad (n = 1, 2, 3, \ldots). \tag{20–10}$$

The corresponding frequencies $f_n$ are given by $f_n = v/\lambda_n$, so all the normal-mode frequencies for a pipe open at both ends are given by

$$f_n = \frac{nv}{2L} \quad (n = 1, 2, 3, \ldots) \quad \text{(open pipe)}. \tag{20–11}$$

The value $n = 1$ corresponds to the fundamental frequency, $n = 2$ to the second harmonic (or first overtone), and so on. Alternatively, we can say

$$f_n = nf_1, \tag{20–12}$$

with $f_1$ given by Eq. (20–9).

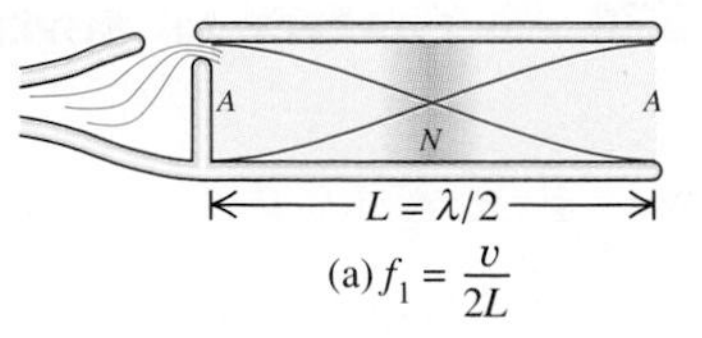


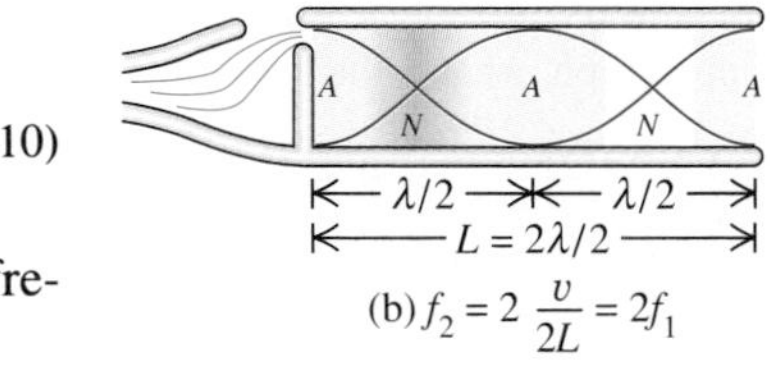


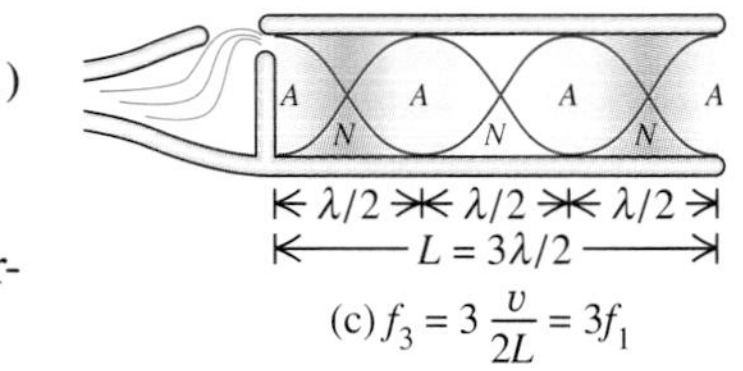


**20–14** A cross section of an open pipe showing the first three normal modes as well as the *displacement* nodes and antinodes. Interchange the *A*'s and *N*'s to show the *pressure* antinodes and nodes.

Figure 20–15 shows a pipe open at the left end but closed at the right end. An organ pipe closed at one end is called a *stopped pipe*. The left (open) end is a displacement antinode (pressure node), but the right (closed) end is a displacement node (pressure antinode). The distance between a node and the adjacent antinode is always one quarter-wavelength. Figure 20–15a shows the lowest-frequency mode; the length of the pipe is a quarter-wavelength ($L = \lambda_1/4$). The fundamental frequency is $f_1 = v/\lambda_1$, or

$$f_1 = \frac{v}{4L} \quad \text{(stopped pipe)}. \tag{20–13}$$

This is one-half the fundamental frequency for an *open* pipe of the same length. In musical language the *pitch* of a stopped pipe is one octave lower (a factor of two in frequency) than that of an open pipe of the same length. Figure 20–15b shows the next mode, for which the length of the pipe is *three-quarters* of a wavelength, corresponding to a frequency $3f_1$. For Fig. 20–15c, $L = 5\lambda/4$ and the frequency is $5f_1$. The possible wavelengths are given by

$$L = n\frac{\lambda_n}{4} \quad \text{or} \quad \lambda_n = \frac{4L}{n} \quad (n = 1, 3, 5, \ldots). \tag{20–14}$$

The normal-mode frequencies are given by $f_n = v/\lambda_n$, or

$$f_n = \frac{nv}{4L} \quad (n = 1, 3, 5, \ldots) \quad \text{(stopped pipe)}, \tag{20–15}$$

or

$$f_n = nf_1 \quad (n = 1, 3, 5, \ldots), \tag{20–16}$$

with $f_1$ given by Eq. (20–13). We see that the second, fourth, and all *even* harmonics are missing. In a pipe that is closed at one end the fundamental frequency is $f_1 = v/4L$, and only the odd harmonics in the series ($3f_1$, $5f_1$, ...) are possible.

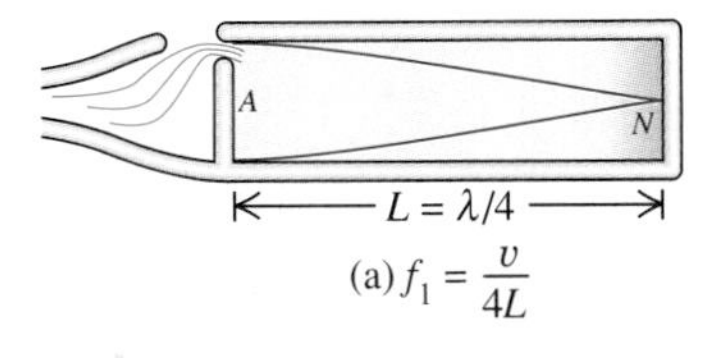


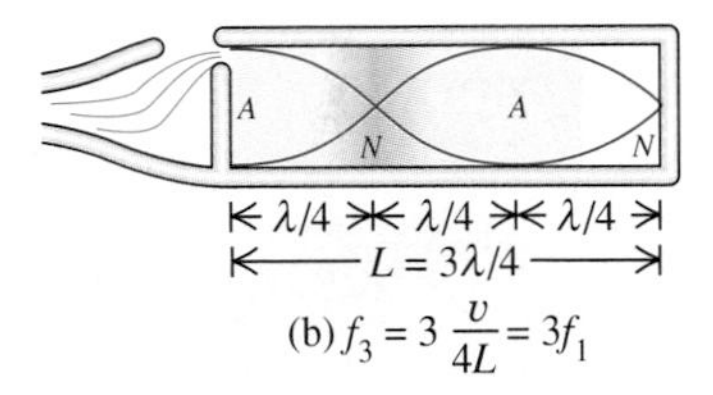


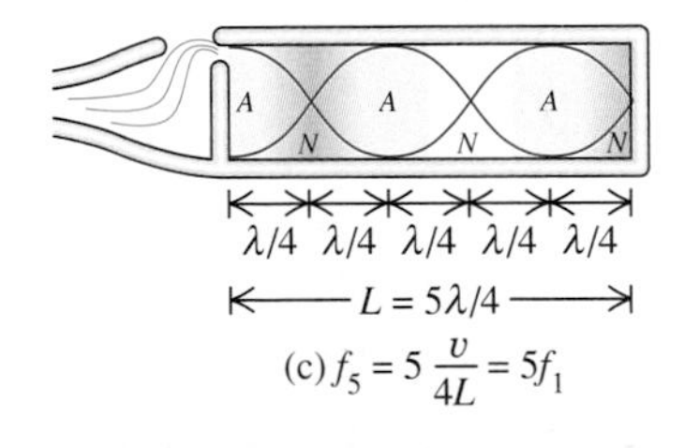


**20–15** A cross section of a stopped pipe showing the first three normal modes as well as the displacement nodes and antinodes. Only odd harmonics are possible.

## EXAMPLE 20–4

On a day when the speed of sound is 345 m/s, the fundamental frequency of an open organ pipe is 690 Hz. If the second *harmonic* of this pipe has the same wavelength as the second *overtone* of a stopped pipe, what is the length of each pipe?

**SOLUTION** For an open pipe, $f_1 = v/2L$, so the length of the open pipe is

$$L_{\text{open}} = \frac{v}{2f_1} = \frac{345\ \text{m/s}}{2(690\ \text{s}^{-1})} = 0.250\ \text{m}.$$

The second harmonic of the open pipe has a frequency of

$$f_2 = 2f_1 = 2(690\ \text{Hz}) = 1380\ \text{Hz}.$$

If the wavelengths are the same, the frequencies are the same, so the frequency of the second overtone of the stopped pipe is also 1380 Hz. The first overtone of a stopped pipe is at $3f_1$, and the second at $5f_1 = 5(v/4L)$. If this equals 1380 Hz, then

$$1380\ \text{Hz} = 5\frac{345\ \text{m/s}}{4L_{\text{stopped}}} \quad \text{and} \quad L_{\text{stopped}} = 0.313\ \text{m}.$$

A final possibility is a pipe closed at *both* ends, with displacement nodes and pressure antinodes at both ends. This wouldn't be of much use as a musical instrument because there would be no way for the vibrations to get out of the pipe.

In an organ pipe in actual use, several modes are always present at once. This situation is analogous to a string that is struck or plucked, producing several modes at the same time. In each case the motion is a *superposition* of various modes. The extent to which modes higher than the fundamental are present depends on the cross section of the pipe, the proportion of length to width, the shape of the mouth, and other more subtle factors. The harmonic content of the tone is an important factor in determining the tone quality, or timbre. A very narrow pipe produces a tone that is rich in higher harmonics, which we hear as a thin and "stringy" tone; a fatter pipe produces mostly the fundamental mode, heard as a softer, more flutelike tone.

We have talked about organ pipes, but this discussion is also applicable to other wind instruments. The flute and the recorder are directly analogous. The most significant difference is that those instruments have holes along the pipe. Opening and closing the holes with the fingers changes the effective length $L$ of the air column and thus changes the pitch. Any individual organ pipe, by comparison, plays only a single note. The flute and recorder behave as *open* pipes, but the clarinet acts as a *stopped* pipe (closed at the reed end, open at the bell).

Equations (20–11) and (20–15) show that the frequencies of any such instrument are proportional to the speed of sound $v$ in the air column inside the instrument. As Eq. (19–27) shows, $v$ depends on temperature; it increases when temperature increases. Thus the pitch of all wind instruments rises with increasing temperature. An organ that has some of its pipes at one temperature and others at a different temperature is bound to sound out of tune.



**20–16** Two speakers driven by the same amplifier. The waves emitted by the speakers are in phase; they arrive at point $P$ in phase because the two path lengths are the same. They arrive at point $Q$ a half-cycle out of phase because the path lengths differ by $\lambda/2$.

## 20–6 INTERFERENCE OF WAVES

Wave phenomena that occur when two or more waves overlap in the same region of space are grouped under the heading *interference*. As we have seen, standing waves are a simple example of an interference effect. Two waves traveling in opposite directions in a medium combine to produce a standing wave pattern with nodes and antinodes that do not move.

Figure 20–16 shows an example of interference involving waves that spread out in space. Two speakers, driven in phase by the same amplifier, emit identical sinusoidal sound waves with the same constant frequency. We place a microphone at point $P$ in the figure, equidistant from the speakers. Wave crests emitted from the two speakers at the same time travel equal distances and arrive at point $P$ at the same time. The amplitudes add, according to the principle of superposition. The total wave amplitude at $P$ is twice the amplitude from each individual wave, and we can measure this combined amplitude with the microphone.

Now let's move the microphone to point $Q$, where the distances from the two speakers to the microphone differ by a half-wavelength. Then the two waves arrive a half-cycle out of step, or *out of phase;* a positive crest from one speaker arrives at the same time as a negative crest from the other, and the amplitude measured by the microphone is much *smaller* than when only one speaker is present. If the amplitudes from the two speakers are equal, the two waves cancel each other out completely at point $Q$, and the total amplitude there is zero.

## EXAMPLE 20-5

Two small loudspeakers, $A$ and $B$ (Fig. 20-17), are driven by the same amplifier and emit pure sinusoidal waves in phase. If the speed of sound is 350 m/s, a) for what frequencies does reinforcement occur at point $P$; b) for what frequencies does destructive interference occur at point $P$?

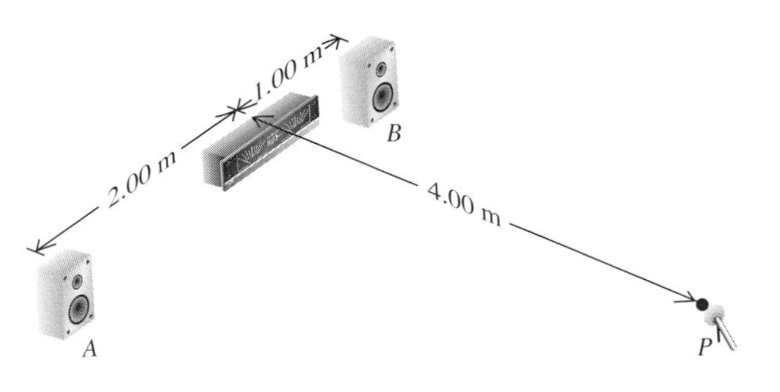


**20-17** The distances from $A$ and $B$ to $P$ are hypotenuses of right triangles.

**SOLUTION** First we need to find the difference in path lengths from points $A$ and $B$ to point $P$. The distance from speaker $A$ to point $P$ is $[(2.00\text{ m})^2 + (4.00\text{ m})^2]^{1/2} = 4.47$ m, and the distance from speaker $B$ to point $P$ is $[(1.00\text{ m})^2 + (4.00\text{ m})^2]^{1/2} =$ 4.12 m. The path difference is $d = 4.47\text{ m} - 4.12\text{ m} = 0.35$ m.

a) Reinforcement (constructive interference) occurs when the path difference $d$ is an integer number of wavelengths, $0, \lambda, 2\lambda, \ldots$, or $d = 0, v/f, 2v/f, \ldots = nv/f$. So the possible frequencies are

$$f_n = \frac{nv}{d} = n\frac{350\text{ m/s}}{0.35\text{ m}} \quad (n = 1, 2, 3, \ldots)$$
$$= 1000\text{ Hz}, 2000\text{ Hz}, 3000\text{ Hz}, \ldots.$$

b) Destructive interference (cancellation) occurs when the path difference is a half-integer number of wavelengths, $d = \lambda/2, 3\lambda/2, 5\lambda/2, \ldots$, or $d = v/2f, 3v/2f, 5v/2f, \ldots$. The possible frequencies are

$$f_n = \frac{nv}{2d} = n\frac{350\text{ m/s}}{2(0.35\text{ m})} \quad (n = 1, 3, 5, \ldots)$$
$$= 500\text{ Hz}, 1500\text{ Hz}, 2500\text{ Hz}, \ldots.$$

As we increase the frequency, the sound at point $P$ alternates between large and small amplitudes; the maxima and minima occur at the frequencies we have found. It would be hard to notice this effect in an ordinary room because of multiple reflections from the walls, floor, and ceiling. Such an experiment is best done in an anechoic chamber (Fig. 20-18).

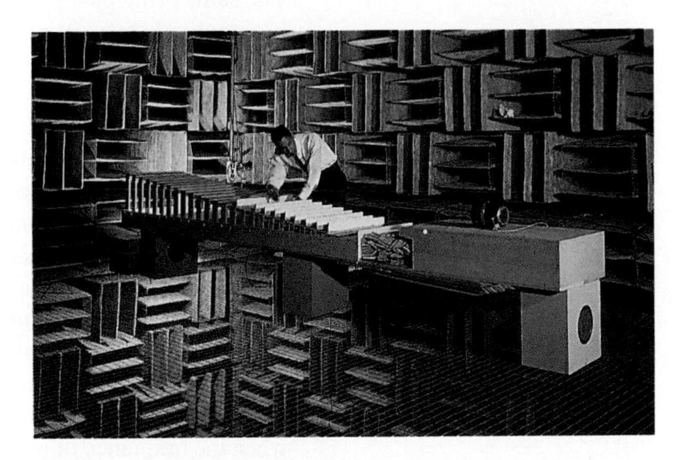

**20-18** An anechoic chamber.

Experiments closely analogous to this one, but using light, have provided strong evidence for the wave nature of light and a means of measuring its wavelengths. We will discuss these experiments in detail in Chapter 37.

Interference effects are used to control noise from very loud sound sources such as gas-turbine power plants and jet engine test cells. The idea is to use additional sound sources that in some regions of space interfere destructively with the unwanted sound and cancel it out. Microphones in the controlled area feed signals back to the sound sources, which are continuously adjusted for optimum cancellation of noise in the controlled area.

## 20–7 RESONANCE

We have discussed several mechanical systems that have normal modes of oscillation. In each mode, every particle of the system oscillates with simple harmonic motion with the same frequency as the frequency of this mode. The systems that we have discussed have an infinite series of normal modes, but the basic concept is closely related to the simple harmonic oscillator, discussed in Chapter 13, which has only a single normal mode.

Suppose we apply a periodically varying force to a system that has normal modes. The system is then forced to vibrate with a frequency equal to the frequency of the *force*. This motion is called a **forced oscillation.** In general, the amplitude of this motion is relatively small, but if the frequency of the force is close to one of the normal-mode frequencies, the amplitude can become quite large. We talked about forced oscillations of the harmonic oscillator in Section 13–8, and we suggest that you review that discussion.

If the frequency of the force is precisely *equal* to a normal-mode frequency, and if there is no friction or other energy-dissipating mechanism, then the force continues to add energy to the system, and the amplitude increases indefinitely. In any real system there is always some dissipation of energy, or damping, as we discussed in Section 13–8. Even so, the "response" of the system (that is, the amplitude of the forced oscillation) is greatest when the force frequency is equal to one of the normal-mode frequencies. This behavior is called **resonance.**

In Section 13–8 we mentioned pushing Cousin Throckmorton on a swing as a familiar example of mechanical resonance. The swing is a pendulum; it has only a single normal mode, with a frequency determined by its length. If we push the swing periodically with this frequency, we can build up the amplitude of the motion. But if we push with a very different frequency, or if we push randomly, the swing hardly moves at all. The same principle applies to "pumping up" — when the person on the swing builds up his amplitude by shifting his weight in phase with the back-and-forth oscillations.

A stretched string (and the other systems discussed in this chapter) has not just one normal mode but an infinite number, each with its own frequency. Suppose one end of a stretched string is held stationary while the other is given a transverse sinusoidal motion with small amplitude, setting up standing waves. If the frequency of the driving mechanism is *not* equal to one of the normal-mode frequencies of the string, the amplitude at the antinodes is fairly small.

However, if the frequency is equal to any one of the natural frequencies, the string is in resonance, and the amplitude at the antinodes is very much *larger* than that at the driven end. The driven end is not precisely a node, but it lies much closer to a node than to an antinode when the string is in resonance. The photographs in Fig. 20–5 were made this way, with the right end of the string stationary and the left oscillating vertically with small amplitude. The photographs show the large-amplitude standing waves that result when the frequency of oscillation of the left end was equal to the fundamental frequency or to one of the first three overtones.

A steel bridge, like any elastic structure, has normal modes and can vibrate with certain natural frequencies. If the regular footsteps of a marching band have a frequency equal to one of the natural frequencies of a bridge that the band is crossing, a vibration of dangerously large amplitude may result. Therefore, when crossing a bridge, a marching group always "breaks step." Similarly, an unbalanced wheel on a car can cause large-amplitude vibrations of parts of the suspension system at certain speeds. There have been cases in which vibrations of aircraft engines happened to coincide with normal-mode frequencies of an airplane frame, resulting in catastrophic resonance-induced vibrations.

It is easy to demonstrate resonance with a piano. Try this: Push down the damper pedal (the right-hand pedal) so that the dampers are lifted and the strings are free to vibrate, and then sing a steady tone into the piano. When you stop singing, the piano

**20–19** A singer's amplified voice matches a normal-mode frequency of a wineglass — with shattering effect.

seems to continue to sing the same note. The sound waves from your voice excite vibrations in the strings that have natural frequencies close to the frequencies (fundamental and harmonics) present in the note you sang. A more spectacular example is a singer breaking a wine glass with her amplified voice (Fig. 20–19). A good-quality wine glass has normal-mode frequencies that you can hear by tapping it. If the singer emits a loud note with a frequency corresponding exactly to one of these normal-mode frequencies, large-amplitude oscillations can build up and break the glass.

Resonance is a very important concept, not only in mechanical systems but in all areas of physics. Later we will see examples of resonance in electric circuits.

## EXAMPLE 20–6

A stopped organ pipe is sounded near a guitar, causing one of the strings to vibrate with large amplitude. We vary the tension of the string until we find the maximum amplitude. The string is 80% as long as the stopped pipe. Assuming that both the pipe and the string vibrate at their fundamental frequency, calculate the ratio of the wave speed on the string to the speed of sound in air.

**SOLUTION** The large response of the string is an example of resonance; it occurs because the organ pipe and the guitar string have the same fundamental frequency. Letting the subscript letters s and a stand for string and air, respectively, we have $f_{1a} = f_{1s}$. We also know from Eq. (20–13) that $f_{1a} = v_a/4L_a$ and from Eq. (20–5) that $f_{1s} = v_s/2L_s$. Putting these together, we have

$$\frac{v_a}{4L_a} = \frac{v_s}{2L_s}.$$

Substituting $L_s = 0.80\,L_a$ and rearranging, we get

$$\frac{v_s}{v_a} = 0.40.$$

For example, if the speed of sound in air is 345 m/s, the wave speed on the string is (0.40)(345 m/s) = 138 m/s.

## SUMMARY

- A wave that reaches a boundary of the medium in which it propagates is reflected. The principle of superposition states that the total wave displacement at any point where two or more waves overlap is the sum of the displacements of the individual waves.
- When a wave is reflected from a fixed or free end of a stretched string, the incident and reflected waves combine to form a standing wave containing nodes and antinodes. Adjacent nodes are spaced a distance $\lambda/2$ apart, as are adjacent antinodes.

• When both ends of a string with length $L$ are held, standing waves can occur only when $L$ is an integer multiple of $\lambda/2$; the corresponding possible frequencies are

$$f_n = n\frac{v}{2L} = nf_1 \quad (n = 1, 2, 3, \ldots). \tag{20–6}$$

Each frequency with its associated vibration pattern is called a normal mode. The lowest frequency $f_1$ is called the fundamental frequency. In terms of the mechanical properties $F$ and $\mu$ of the string the fundamental frequency is

$$f_1 = \frac{1}{2L}\sqrt{\frac{F}{\mu}}. \tag{20–8}$$

• For sound waves in a pipe or tube a closed end is a displacement node and a pressure antinode; an open end is a displacement antinode and a pressure node. For a pipe open at both ends the normal-mode frequencies are

$$f_n = \frac{nv}{2L} \quad (n = 1, 2, 3, \ldots) \quad \text{(open pipe)}. \tag{20–11}$$

For a pipe open at one end and closed at the other the normal-mode frequencies are

$$f_n = \frac{nv}{4L} \quad (n = 1, 3, 5, \ldots) \quad \text{(stopped pipe)}. \tag{20–15}$$

• When two or more waves overlap in the same region of space, the resulting effects are called interference. The resulting amplitude can be either larger or smaller than the amplitude of each individual wave, depending on whether the waves are in phase or out of phase. When the waves are in phase, the result is called reinforcement or constructive interference; when they are out of phase, it is called cancellation or destructive interference.

• When a periodically varying force is applied to a system having normal modes, the system vibrates with the same frequency as that of the force; this is called a forced oscillation. If the force frequency is equal to or close to one of the normal-mode frequencies, the amplitude of the resulting forced oscillation can become very large; this phenomenon is called resonance.

## KEY TERMS

boundary condition
principle of superposition
interference
node
antinode
standing wave
fundamental frequency
harmonics
harmonic series
overtone
normal mode
pressure node
pressure antinode
forced oscillation
resonance

# EXERCISES

## Section 20–2 Superposition and Standing Waves

**20–1** Let $y_1(x, t) = A_1 \sin(\omega_1 t - k_1 x)$ and let $y_2(x, t) = A_2 \sin(\omega_2 t - k_2 x)$ be two solutions to the wave equation, Eq. (19–12), for the same $v$. Show that $y(x, t) = y_1(x, t) + y_2(x, t)$ is also a solution to the wave equation.

**20–2** Give the details of the derivation of Eq. (20–1) from $y_1 + y_2 = A[\sin(\omega t + kx) - \sin(\omega t - kx)]$.

**20–3** Standing waves on a wire with length 2.40 m are described by Eq. (20–1), with $A = 3.00$ cm, $\omega = 314$ rad/s, and $k = 1.67\pi$ rad/m and with the left-hand end of the wire at $x = 0$. At what distances from the left-hand end are a) the nodes of the standing wave; b) the antinodes of the standing wave?

**20–4** Prove by direct substitution that $y = [2A\cos\omega t]\sin kx$ is a solution of the wave equation, Eq. (19–12), for $v = \omega/k$.

## Section 20–3 Normal Modes of a String

**20–5** A physics student observes a stretched string vibrating with a frequency of 40.0 Hz in its fundamental mode when the supports to which the ends of the string are tied are 0.600 m apart. The amplitude at the antinode is 0.50 cm. The string has a mass of 0.0500 kg. a) What is the speed of propagation of a transverse wave in the string? b) Compute the tension in the string.

**20–6** A piano tuner stretches a steel piano wire with a tension of 400 N. The steel wire is 0.500 m long and has a mass of 6.00 g. a) What is the frequency of its fundamental mode of vibration? b) What is the number of the highest harmonic that could be heard by a person who is capable of hearing frequencies up to 10,000 Hz?

**20–7 Playing a Cello.** The portion of a cello string between the bridge and the upper end of the fingerboard (that part of the string that is free to vibrate) is 60.0 cm long, and this length of the string has a mass of 2.00 g. The string sounds an A note (440 Hz) when played. a) Where must the cellist put a finger (what distance $x$ from the bridge) to play a C note (528 Hz)? (See Fig. 20–20.) For both the A and C notes the string vibrates in its fundamental mode. b) Without retuning, is it possible to play a D note (294 Hz) on this string? Why or why not?

**FIGURE 20–20**

**20–8** A string is vibrating in its fundamental mode. The waves have velocity $v$, frequency $f$, amplitude $A$, and wavelength $\lambda$. a) Calculate the maximum velocity and acceleration of points located at (i) $x = \lambda/2$, (ii) $x = \lambda/4$, and (iii) $x = \lambda/8$ from the left-hand end of the string. b) At each of the points in part (a), what is the amplitude of the motion? c) At each of the points in part (a), how much time does it take the string to go from its largest upward displacement to its largest downward displacement?

**20–9** A rope with a length of 1.60 m is stretched between two supports with a tension that makes the speed of transverse waves 40.0 m/s. What are the wavelength and frequency of a) the fundamental; b) the first overtone; c) the third harmonic?

## Section 20–4 Longitudinal Standing Waves

## Section 20–5 Normal Modes of Air Columns

**20–10** Standing sound waves are produced in a pipe that is 1.80 m long and open at both ends. For the fundamental and first two overtones, where along the pipe (measured from one end) are the a) displacement nodes; b) pressure nodes?

**20–11** Standing sound waves are produced in a pipe that is 1.80 m long and open at one end and closed at the other. For the fundamental and first two overtones, where along the pipe (measured from the closed end) are the a) displacement antinodes; b) pressure antinodes?

**20–12** Find the fundamental frequency and the frequency of the first three overtones of a 28.0-cm pipe a) if the pipe is open at both ends; b) if the pipe is closed at one end. c) What is the number of the highest harmonic that may be heard by a person having normal hearing for each of the above cases? (A person with normal hearing can hear frequencies in the range 20–20,000 Hz.)

**20–13 The Bass Notes of an Organ.** The longest pipe found in most medium-sized pipe organs is 4.88 m (16 ft) long. What is the frequency of the note corresponding to the fundamental mode if the pipe is a) open at both ends; b) open at one end and closed at the other?

**20–14** A certain pipe produces a frequency of 520 Hz in air. If the pipe is filled with helium at the same temperature, what frequency does it produce? (The molecular mass of air is 28.8 g/mol, and the atomic mass of helium is 4.00 g/mol.)

## Section 20–6 Interference of Waves

**20–15** Figure 20–21 shows two rectangular wave pulses on a stretched string traveling toward each other. Each pulse is traveling with a speed of 1.00 mm/s and has the height and width shown in the figure. If the leading edges of the pulses are 8.00 mm apart at $t = 0$, sketch the shape of the string at $t = 4.00$ s, $t = 6.00$ s, and $t = 10.0$ s.

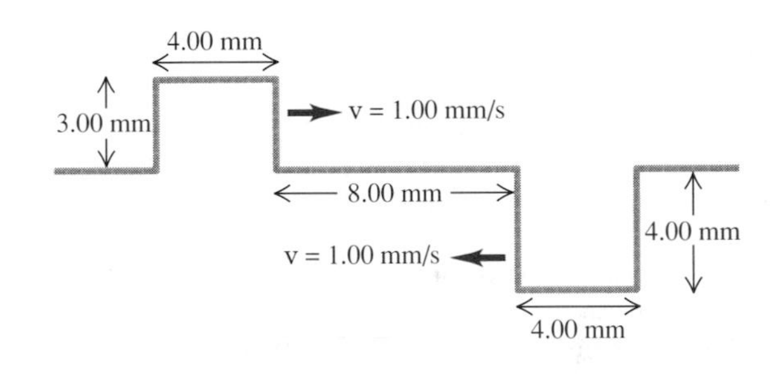


**FIGURE 20–21**

**20–16 Interference in a Stereo System.** Two loudspeakers, $A$ and $B$ (Fig. 20–22), are driven by the same amplifier and emit sinusoidal waves in phase. Speaker $B$ is 2.00 m to the right of speaker $A$. The frequency of the sound waves produced by the loudspeakers is 700 Hz, and their speed in air is 350 m/s. Consider point $P$ between the speakers and along the line connecting them, a distance $x$ to the right of speaker $A$. For what values of $x$ will destructive interference occur at point $P$?

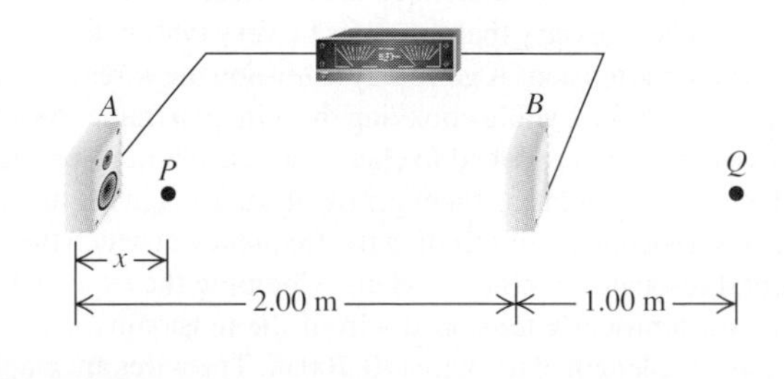

**FIGURE 20–22**

**20–17** Two loudspeakers, $A$ and $B$ (Fig. 20–22), are driven by the same amplifier and emit sinusoidal waves in phase. Speaker $B$ is 2.00 m to the right of speaker $A$. The speed of sound in the air is 350 m/s. Consider point $Q$ along the extension of the line connecting the speakers, 1.00 m to the right of speaker $B$. a) For what frequencies does reinforcement occur at point $P$? b) For what frequencies does destructive interference occur at point $P$?

## PROBLEMS

**20–18** A string with both ends held stationary is vibrating in its fundamental mode. The waves have a speed of 32.0 m/s and a frequency of 20.0 Hz. The amplitude of the standing wave at its antinode is 1.20 cm. a) Calculate the amplitude of the motion of points on the string a distance of (i) 80 cm, (ii) 40 cm, and (iii) 20 cm from the left-hand end of the string. b) At each of the points in part (a), how much time does it take the string to go from its largest upward displacement to its largest downward displacement?

**20–19 The Pit and the Plank.** A wooden plank is placed over a pit which is 10.0 m wide. A physics student stands in the middle of the plank and begins to jump up and down in such a fashion that she jumps upward from the plank two times each second. The plank oscillates with a large amplitude, with maximum amplitude at its center. a) What is the speed of transverse waves on the plank? b) At what rate does the student have to jump to produce large-amplitude oscillations if she is standing 2.5 m from the edge of the pit? (*Note:* The transverse standing waves of the plank have nodes at the two ends that rest on the ground on either side of the pit.)

**20–20** Your physics professor has invented a musical instrument. It consists of a metal can with length $L$ and diameter $L/10$. The top of the can is cut out, and a string is stretched across this open end of the can. a) The tension in the string is adjusted so that the fundamental frequency for longitudinal sound waves in the air column in the can equals the frequency of the third harmonic for transverse waves on the string. What is the relationship between the speed $v_t$ of transverse waves on the string and the speed $v_a$ of sound waves in the air? b) What happens to the sound produced by the instrument if the tension in the string is increased by a factor of four?

**20–21** An organ pipe open at both ends has two successive harmonics with frequencies 240 and 280 Hz. a) What is the length of the pipe? b) What two harmonics are these?

**20–22** To determine the paths of subnuclear particles, physicists arrange fine parallel wires into planes (each plane like a harp) and electrify them. In each plane the wire to which the particle passes closest generates a distinctive electrical signal that can be recorded. It is necessary that the wires be very taut, with all wires having the same tension. A good way to tension the wires is to hang a mass $m$ from a wire while epoxying the wire to a frame. As a student assistant, you are asked to check the tension in a completed panel of wires by vibrating them gently (using a magnetic method) at various frequencies and finding the frequency at which the fundamental resonance vibration occurs. Compute the expected frequency for a properly tensioned wire if the tensioning mass was 50.0 g and the length of the wire is 0.700 m. The wires are made of tungsten (density = 19.3 g/cm$^3$) and are 200 $\mu$m in diameter.

**20–23** Cellist Yo-Yo Ma tunes the A-string of his instrument to a fundamental frequency of 220 Hz. The vibrating portion of the string is 0.680 m long and has a mass of 1.42 g. a) With what tension must it be stretched? b) What percent increase in tension is needed to increase the frequency from 220 Hz to 233 Hz, corresponding to a rise in pitch from A to A-sharp?

**20–24** A solid aluminum sculpture is hung from a steel wire. The fundamental frequency for transverse standing waves on the wire is 240 Hz. The sculpture is then immersed in water so that one third of its volume is submerged. What is the new fundamental frequency?

**20–25** A long tube contains air at a pressure of 1.00 atm and a temperature of 77.0°C. The tube is open at one end and closed at the other by a movable piston. A tuning fork near the open end is vibrating with a frequency of 500 Hz. Resonance is produced when the piston is at distances 18.0, 55.5, and 93.0 cm from the open end. a) From these measurements, what is the speed of sound in air at 77.0°C? b) From the result of part (a), what is the ratio $\gamma$ of the specific heat capacities at constant pressure and constant volume for air at this temperature? (The molecular mass of air is 28.8 g/mol.)

**20–26** A standing wave with a frequency of 1100 Hz in a column of methane ($CH_4$) at 20.0°C produces nodes that are 0.200 m apart. What is the ratio $\gamma$ of the heat capacity at constant pressure to that at constant volume for methane? (The molecular mass of methane is 16.0 g/mol.)

**20–27** The frequency of middle C is 262 Hz. a) If an organ pipe is open at both ends, what length must it have for its fundamental mode to produce this note at 20.0°C? b) At what temperature will the frequency be 6.00% higher, corresponding to a rise in pitch from C to C-sharp?

**20–28** Two identical loudspeakers are located at points $A$ and $B$, 2.00 m apart. The loudspeakers are driven by the same amplifier and produce sound waves with a frequency of 480 Hz. Take the speed of sound in air to be 340 m/s. A small microphone is moved out from point $B$ along a line perpendicular to the line connecting $A$ and $B$ (line $BC$ in Fig. 20–23). At what distances from $B$ will there be *destructive* interference?

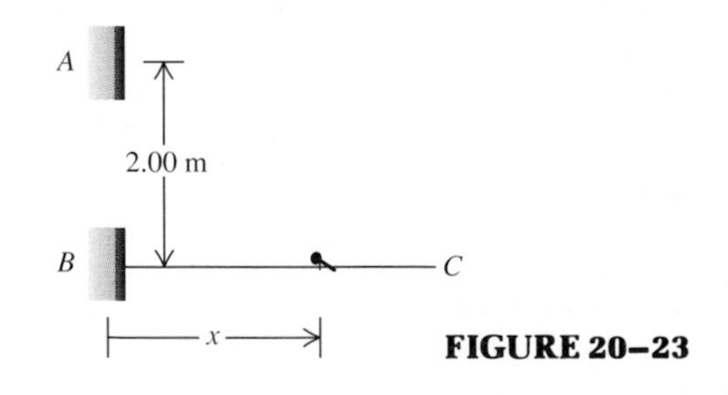


**FIGURE 20–23**

## CHALLENGE PROBLEMS

**20–29 Tuning a Guitar.** The steel B string of an acoustic guitar is 63.5 cm long and has a diameter of 0.406 mm (16 gauge). a) Under what tension must the string be placed to give a frequency for transverse waves of 247.5 Hz? The density of steel is 7800 kg/m$^3$. Assume that the string vibrates in its fundamental mode. b) If the tension is changed by a small amount $\Delta F$, the frequency changes by $\Delta f$. Show that

$$\frac{\Delta f}{f} = \frac{1}{2}\left(\frac{\Delta F}{F}\right).$$

c) If the string is tuned indoors as in part (a), where the temperature is 24.0°C, and the guitar is then taken to an outdoor stage where the temperature is 11.0°C, the frequency will change, with unpleasant results. Find $\Delta f$ if Young's modulus $Y$ of the steel string is $2.00 \times 10^{11}$ Pa and the coefficient of linear expansion $\alpha$ is $1.20 \times 10^{-5}\,(\text{C}°)^{-1}$. Will the pitch be raised or lowered?

**20–30 Resonance in a Mechanical System.** A mass $m$ is attached to one end of a massless spring with a force constant $k$ and an unstretched length $l_0$. The other end of the spring is free to turn about a nail driven into a frictionless horizontal surface (Fig. 20–24). The mass is made to revolve in a circle with an angular frequency of revolution $\omega'$. a) Calculate the length $l$ of the spring as a function of $\omega'$. b) What happens to the result in part (a) when $\omega'$ approaches the natural frequency $\omega = \sqrt{k/m}$ of the mass-spring system? (If your result bothers you, remember that massless springs and frictionless surfaces don't precisely exist but can only be approximated. Also, Hooke's law is itself only an approximation to the way real springs behave; the greater the elongation of the spring, the greater the deviation from Hooke's law.)

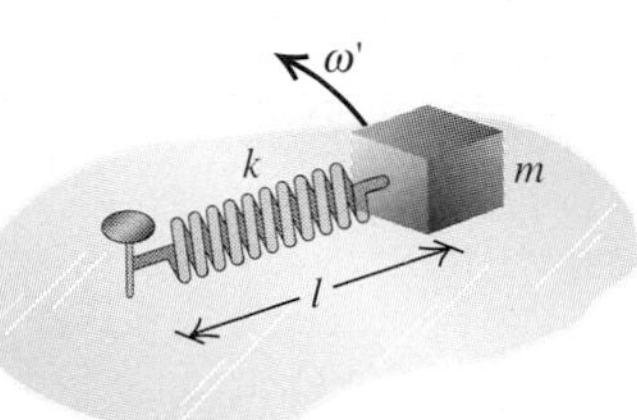


**FIGURE 20–24**

CHAPTER 21

# Sound

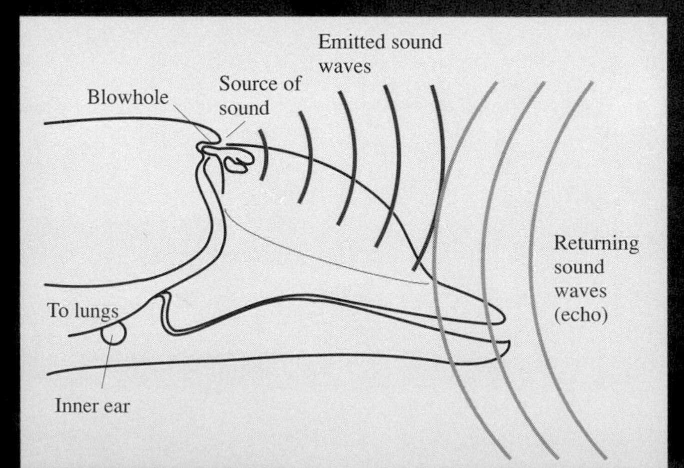


**Dolphins emit ultrasonic (about $10^6$ Hz) sound waves in the form of whistles. The returning echos of the sound waves give the dolphin information about its surroundings at a greater distance than underwater vision can provide. Dolphins use this information primarily to locate small fish for food.**

## KEY CONCEPTS

**• Sound consists of longitudinal waves in air. The human ear is sensitive to sound waves in a limited frequency range. A sound wave can be described in terms of displacements of air molecules or in terms of pressure fluctuations.**

**• The intensity of a sound wave is the average energy carried per unit area per unit time. Sound intensity can be described by using the decibel scale.**

**• Superposition of sound waves with slightly differing frequencies produces beats, in which the wave amplitude fluctuates at the difference frequency.**

**• The Doppler effect is a frequency shift caused by motion of a source of sound or a listener relative to the air.**

**• Shock waves are produced when a sound source moves through air with a speed greater than the speed of sound.**

**• Medical applications of sound include imaging techniques and destruction of kidney stones and gallstones. Geological applications include studies of the earth's structure, mineral prospecting, and analysis of earthquakes.**

**• Musical tones can be described in terms of the physical characteristics of sound waves.**

## INTRODUCTION

It can be gentle or painful. It can be infrasonic, audible, or ultrasonic, and its source can be supersonic. It shows interference effects, both in space and in time. It can be used to cause bodily harm or to find and cure bodily defects. What is it? *Sound,* longitudinal waves in air.

We will discuss several important properties of sound waves, including frequency, amplitude, and intensity. We'll study the relations among displacement, pressure variation, and intensity and the connections between these quantities and humans' perception of sound. We will find that interference of two sound waves that differ slightly in frequency causes a phenomenon called *beats.* When a source of sound or a listener is moving through the air, the listener may hear a different frequency than the one emitted by the source. This is the Doppler effect, another topic of this chapter. We'll also take a brief look at shock waves, hearing, and musical sounds. ■

## 21–1 SOUND WAVES

Our main subject in this chapter is longitudinal waves in air, that is, **sound.** The simplest sound waves are sinusoidal waves, which have definite frequency, amplitude, and wavelength. The human ear is sensitive to waves in the frequency range from about 20 to 20,000 Hz, but we also use the term *sound* for similar waves with frequencies above **(ultrasonic)** and below **(infrasonic)** the range of human hearing.

Sound waves may also be described in terms of variations of *air pressure* at various points. In a sinusoidal sound wave the pressure fluctuates above and below atmospheric pressure $p_a$ with a sinusoidal variation having the same frequency as the motions of the air particles.

In Section 19–3 we discussed wave functions that describe the displacements $y(x, t)$ of particles in a medium during wave propagation. One of the forms we used was Eq. (19–7):

$$y = A \sin(\omega t - kx). \tag{21–1}$$

Remember that in a longitudinal wave the displacements are *parallel* to the direction of travel of the wave, so distances $x$ and $y$ are measured parallel to each other, not perpendicular as in a transverse wave. This also means that sound waves cannot have polarization. The amplitude $A$ is the maximum displacement of a particle from its equilibrium position.

Equation (21–1) describes a sound wave that travels only in one direction; actual sound waves usually travel out in all directions from the source with an amplitude that depends on the direction and distance from the source. We'll come back to this point in the next section.

Microphones and similar devices usually sense pressure differences, not displacements, so it is very useful to develop a relation between the two. Let $p$ be the instantaneous pressure fluctuation at any point, that is, the amount by which the pressure *differs* from normal atmospheric pressure $p_a$. Think of $p$ as the *gauge pressure* defined in Section 14–2; it can be either positive or negative. The *absolute* pressure at a point is then $p_a + p$.

If the displacements of two neighboring points $x$ and $x + \Delta x$ are the same, the air between these points is neither compressed nor expanded, there is no volume change, and consequently $p = 0$. Only when $y$ varies from one point to a neighboring point is there a change of volume and therefore of pressure. The fractional volume change $dV/V$ in a volume element near point $x$ turns out to be given simply by $\partial y/\partial x$, which is the rate of change of $y$ with respect to $x$ as we go from one point to a neighboring point.

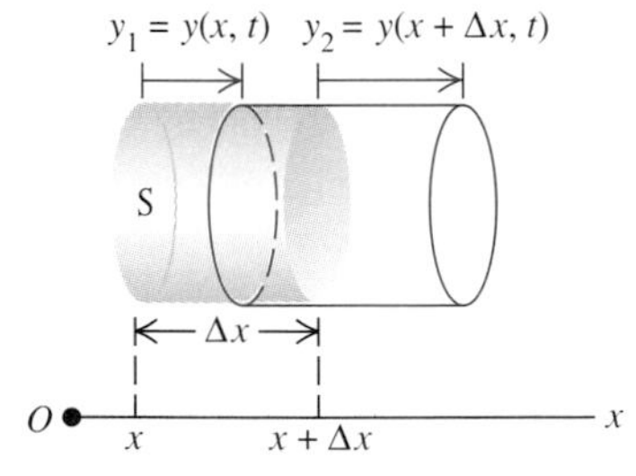


**21–1** A cylindrical volume of gas with cross-section area $S$. The length in the undisplaced position is $\Delta x$. During wave propagation along the axis the left end is displaced to the right a distance $y_1$, and the right end is displaced a different distance $y_2$. The resulting change in volume is $S(y_2 - y_1)$.

To see why this is so, consider an imaginary cylinder of air (Fig. 21–1), with cross-section area $S$ and axis along the direction of propagation. The cylinder shows the undisplaced position, and the red lines show the displaced position. When no sound disturbance is present, the cylinder's length is $\Delta x$, and its volume is $V = S\,\Delta x$. When a wave is present, the end of the cylinder that is initially at $x$ is displaced a distance $y_1 = y(x, t)$, and the end that is initially at $x + \Delta x$ is displaced a distance $y_2 = y(x + \Delta x, t)$. The change in volume $\Delta V$ of this element is

$$\Delta V = S(y_2 - y_1) = S[y(x + \Delta x, t) - y(x, t)].$$

In the limit as $\Delta x \to 0$ the fractional change in volume $dV/V$ is

$$\frac{dV}{V} = \lim_{\Delta x \to 0} \frac{y(x + \Delta x, t) - y(x, t)}{\Delta x} = \frac{\partial y}{\partial x}. \tag{21–2}$$

Now from the definition of the bulk modulus $B$, Eq. (11–16), $p = -B\,dV/V$, and we have

$$p = -B\frac{\partial y}{\partial x}. \tag{21–3}$$

The negative sign arises because, when $\partial y/\partial x$ is positive, the displacement is greater at $x + \Delta x$ than at $x$, corresponding to an increase in volume and a *decrease* in pressure.

When we evaluate $\partial y/\partial x$ for the sinusoidal wave of Eq. (21–1), we find

$$p = BkA \cos(\omega t - kx). \tag{21-4}$$

This expression shows that the quantity $BkA$ represents the maximum pressure variation. We call this the **pressure amplitude,** denoted by $p_{max}$:

$$p_{max} = BkA. \tag{21-5}$$

The pressure amplitude is directly proportional to the displacement amplitude $A$, as we might expect, and it also depends on wavelength. Waves of shorter wavelength (larger $k$) have greater pressure variations for a given amplitude because the maxima and minima are squeezed closer together. Media that have large values of bulk modulus $B$ require relatively greater pressures for a given displacement because large $B$ means a less compressible medium. That is, a greater pressure change is required for a given volume change.

## EXAMPLE 21–1

**Amplitude of a loud sound wave** Measurements of sound waves show that in the loudest sounds that the ear can tolerate without pain, the maximum pressure variations are of the order of 30 Pa above and below atmospheric pressure $p_a$ (nominally $1.013 \times 10^5$ Pa at sea level). Find the corresponding maximum displacement if the frequency is 1000 Hz and $v = 350$ m/s.

**SOLUTION** We have $\omega = 2\pi f = (2\pi \text{ rad})(1000 \text{ Hz}) = 6283$ rad/s, and

$$k = \frac{\omega}{v} = \frac{6283 \text{ rad/s}}{350 \text{ m/s}} = 18 \text{ rad/m}.$$

The adiabatic bulk modulus for air at normal atmospheric pressure is

$$B = \gamma p_a = (1.40)(1.013 \times 10^5 \text{ Pa}) = 1.42 \times 10^5 \text{ Pa}.$$

From Eq. (21–5),

$$A = \frac{p_{max}}{Bk} = \frac{30 \text{ Pa}}{(1.42 \times 10^5 \text{ Pa})(18 \text{ rad/m})} = 1.2 \times 10^{-5} \text{ m} = 0.012 \text{ mm}$$

This shows that the displacement amplitude of even the loudest sound is *extremely* small. The maximum pressure variation in the *faintest* audible sound of frequency 1000 Hz is only about $3 \times 10^{-5}$ Pa. The corresponding displacement amplitude is about $10^{-11}$ m. For comparison the wavelength of yellow light is $6 \times 10^{-7}$ m, and the diameter of a molecule is about $10^{-10}$ m. The ear is an extremely sensitive organ!

## 21–2 INTENSITY

Like all other waves, sound waves transfer energy from one region of space to another. We define the **intensity** of a wave (denoted by $I$) to be *the time average rate at which energy is transported by the wave per unit area* across a surface perpendicular to the direction of propagation. That is, intensity $I$ is average *power* per unit area.

Our study of power in Section 6–5 showed that the power developed by a force equals the product of force and velocity; see Eq. (6–20). So the power *per unit area* in a sound wave equals the product of the excess pressure $p$ (force per unit area) and the *particle* velocity $v_y$. For the sinusoidal wave described by Eq. (21–1), $p$ is given by Eq. (21–4), and $v_y$ is obtained by taking the time derivative of Eq. (21–1). We find

$$v_y = \omega A \cos(\omega t - kx), \tag{21-6}$$

$$pv_y = \omega BkA^2 \cos^2(\omega t - kx). \tag{21-7}$$

The intensity is, by definition, the average value of this quantity. The average value of the function $\cos^2(\omega t - kx)$ over one cycle is $\frac{1}{2}$, so

$$I = \tfrac{1}{2}\omega BkA^2. \tag{21–8}$$

By using the relations $\omega = vk$ and $v^2 = B/\rho$, we can transform this into the form

$$I = \tfrac{1}{2}\sqrt{\rho B}\,\omega^2 A^2, \tag{21–9}$$

which we cited at the end of Section 19–7. This equation shows why in a stereo system a low-frequency woofer has to vibrate with much larger amplitude than a high-frequency tweeter to produce the same sound intensity.

It is usually more useful to express $I$ in terms of the pressure amplitude $p_{max}$. Using Eq. (21–5) and the relation $\omega = vk$, we find

$$I = \frac{\omega p_{max}^2}{2Bk} = \frac{v p_{max}^2}{2B}. \tag{21–10}$$

By using the wave speed relation $v^2 = B/\rho$ we can also write this in the alternative forms

$$I = \frac{p_{max}^2}{2\rho v} = \frac{p_{max}^2}{2\sqrt{\rho B}}. \tag{21–11}$$

We invite you to verify these expressions. (See Exercise 21–5.)

## EXAMPLE 21–2

Find the intensity of the loud sound wave in Example 21–1, with $p_{max} = 30$ Pa, if the temperature is 20°C.

**SOLUTION** At 20°C the density of air is $\rho = 1.20\ \text{kg/m}^3$, and the speed of sound is $v = 344$ m/s. From Eq. (21–11),

$$I = \frac{p_{max}^2}{2\rho v} = \frac{(30\ \text{Pa})^2}{2(1.20\ \text{kg/m}^3)(344\ \text{m/s})}$$
$$= 1.1\ \text{J/s}\cdot\text{m}^2 = 1.1\ \text{W/m}^2.$$

The intensity of a sound wave at the pain threshold is about $1\ \text{W/m}^2$. The pressure amplitude of the *faintest* sound wave that can be heard is about $3 \times 10^{-5}$ Pa, and the corresponding intensity is about $10^{-12}\ \text{W/m}^2$. We invite you to verify this number.

## EXAMPLE 21–3

What amplitude at 20 Hz would give the same intensity as the 1000-Hz sound wave in Examples 21–1 and 21–2?

**SOLUTION** In Eq. (21–9), $\rho$ and $B$ depend on the medium, not the amplitude or frequency of the wave. For $I$ to be constant the product $\omega A$ must be constant. That is,

$$(20\ \text{Hz})\,A_{20} = (1000\ \text{Hz})\,(1.2 \times 10^{-5}\text{m}),$$
$$A_{20} = 6.0 \times 10^{-4}\ \text{m} = 0.60\ \text{mm}.$$

Do you understand why we didn't have to convert the frequencies to angular frequencies?

The *total* power carried across a surface by a sound wave equals the product of the intensity at the surface and the surface area if the intensity over the surface is uniform. The average total sound power emitted by a person speaking in an ordinary conversational tone is about $10^{-5}$ W, and a loud shout corresponds to about $3 \times 10^{-2}$ W. If all the residents of New York City were to talk at the same time, the total sound power would be about 100 W, equivalent to the electric power requirement of a medium-sized light bulb. On the other hand, the power required to fill a large auditorium with loud sound is considerable.

## EXAMPLE 21–4

For an auditorium sound system, suppose we want the sound intensity over the surface of a hemisphere 20 m in radius to be 1 W/m$^2$. What acoustic power output would be needed from a speaker array (Fig. 21–2) at the center of the sphere?

**SOLUTION** The area of the hemispherical surface is $\frac{1}{2}(4\pi)\cdot(20\text{ m})^2$, or about 2500 m$^2$. The acoustic power needed is

$$(1\text{ W/m}^2)(2500\text{ m}^2) = 2500\text{ W} = 2.5\text{ kW}.$$

The electrical power input to the speakers would need to be considerably larger because the efficiency of such devices is not very high (typically a few percent for ordinary speakers, up to 25% for horn-type speakers).

**21–2** Rock bands produce tremendous volumes of sound. On their 1989 tour the Rolling Stones utilized a half-million watts of power on 96 audio channels.

If a source of sound can be considered as a point, the intensity at a distance $r$ from the source is inversely proportional to $r^2$. This follows directly from energy conservation; if the power output of the source is $P$, then the average intensity $I_1$ through a sphere with radius $r_1$ and surface area $4\pi r_1^2$ is

$$I_1 = \frac{P}{4\pi r_1^2}.$$

The average intensity $I_2$ through a sphere with a different radius $r_2$ is given by a similar expression. If no energy is absorbed between the two spheres, the power $P$ must be the same for both, and

$$4\pi r_1^2 I_1 = 4\pi r_2^2 I_2,$$

$$\frac{I_1}{I_2} = \frac{r_2^2}{r_1^2}. \qquad (21\text{–}12)$$

The intensity $I$ at any distance $r$ is therefore inversely proportional to $r^2$. This "inverse-square" relationship also holds for various other energy-flow situations with a point source, such as light emitted by a point source.

## The Decibel Scale

Because the ear is sensitive over such a broad range of intensities, a *logarithmic* intensity scale is usually used. The **intensity level** $\beta$ of a sound wave (also called the *sound level*) is defined by the equation

$$\beta = (10\text{ dB})\log\frac{I}{I_0}. \qquad (21\text{–}13)$$

**TABLE 21–1 Sound Intensity Levels from Various Sources (Representative Values)**

| Source or description of sound | Intensity level (dB) | Intensity ($W/m^2$) |
|---|---|---|
| Threshold of pain | 120 | 1 |
| Riveter | 95 | $3.2 \times 10^{-3}$ |
| Elevated train | 90 | $10^{-3}$ |
| Busy street traffic | 70 | $10^{-5}$ |
| Ordinary conversation | 65 | $3.2 \times 10^{-6}$ |
| Quiet automobile | 50 | $10^{-7}$ |
| Quiet radio in home | 40 | $10^{-8}$ |
| Average whisper | 20 | $10^{-10}$ |
| Rustle of leaves | 10 | $10^{-11}$ |
| Threshold of hearing | 0 | $10^{-12}$ |

In Eq. (21–13), $I_0$ is a reference intensity, chosen to be $10^{-12}$ W/m$^2$, approximately the threshold of human hearing at 1 kHz. Intensity levels are expressed in **decibels,** abbreviated dB. A decibel is $\frac{1}{10}$ of a bel, a unit named for Alexander Graham Bell. The bel is inconveniently large for most purposes, and the decibel is the usual unit of sound intensity level.

If the intensity of a sound wave equals $I_0$, or $10^{-12}$ W/m$^2$, its intensity level is 0 dB. An intensity of 1 W/m$^2$ corresponds to 120 dB. Table 21–1 gives the intensity levels in decibels of several familiar sounds.

## Hearing

The normal human ear is sensitive to sounds with frequencies from about 20 to 20,000 Hz. Higher frequencies are called *ultrasonic*. Within the audible range, the sensitivity of the ear varies with frequency. A sound at one frequency may seem louder than one of equal intensity at a different frequency. At 1000 Hz the minimum intensity level that can be perceived is about 0 dB; at 200 or 15,000 Hz it is about 20 dB. Sensitivity at the high-frequency end usually falls off with age. An intensity level of 120 dB or above, within the audible range, causes pain.

Some sound-level meters take into account the ear's varying sensitivity by weighting the various frequencies unequally. One such scheme leads to the so-called dBA scale; this scale de-emphasizes the low and very high frequencies, at which the ear is less sensitive than at midrange frequencies.

## PROBLEM-SOLVING STRATEGY

### Sound intensity

**1.** Quite a few quantities are involved in characterizing the amplitude and intensity of a sound wave, and it's easy to get lost in the maze of relationships. It helps to put them in categories: the amplitude is described by $A$ or $p_{max}$, and the frequency $f$ can be determined from $\omega$, $k$, or $\lambda$. These quantities are related through the wave speed $v$, which in turn is determined by the properties of the medium, $B$ and $\rho$. Take a hard look at the problem at hand, identifying which of these quantities are given and which you have to find; then start looking for relationships that take you where you want to go.

**2.** In using Eq. (21–13) for the sound intensity level, remember that $I$ and $I_0$ must be in the same units, usually W/m$^2$. If they aren't, convert!

## EXAMPLE 21–5

**Temporary deafness** A ten-minute exposure to 120-dB sound will typically shift your threshold of hearing from 0 dB up to 28 dB for a while. Ten years of exposure to 92-dB sound will cause a *permanent* shift up to 28 dB. What intensities (sound levels) correspond to 28 dB and 92 dB?

**SOLUTION** We rearrange Eq. (21–13) by dividing both sides by 10 dB and then taking inverse logs of both sides:

$$I = I_0\, 10^{(\beta/10\ \text{dB})}.$$

When $\beta = 28$ dB,

$$I = (10^{-12}\ \text{W/m}^2)\, 10^{(28\ \text{dB}/10\ \text{dB})}$$
$$= (10^{-12}\ \text{W/m}^2)\, 10^{2.8} = 6.3 \times 10^{-10}\ \text{W/m}^2.$$

Similarly, for $\beta = 92$ dB,

$$I = (10^{-12}\ \text{W/m}^2)\, 10^{(92\ \text{dB}/10\ \text{dB})} = 1.6 \times 10^{-3}\ \text{W/m}^2.$$

If your answers are a factor of 10 too large, you may have entered $10 \times 10^{-12}$ in your calculator instead of $1 \times 10^{-12}$. Be careful!

## EXAMPLE 21–6

**A bird sings in a meadow** Consider an idealized model with a point source emitting constant sound power with intensity inversely proportional to the square of the distance from the bird. By how many dB does the sound level (intensity) drop when you move twice as far away from the bird?

**SOLUTION** We label the two points 1 and 2 (Fig. 21–3), and we use Eq. (21–13) twice. The difference in sound level, $\beta_2 - \beta_1$, is given by

$$\beta_2 - \beta_1 = (10\ \text{dB})\left(\log\frac{I_2}{I_0} - \log\frac{I_1}{I_0}\right)$$
$$= (10\ \text{dB})\,[(\log I_2 - \log I_0) - (\log I_1 - \log I_0)]$$
$$= (10\ \text{dB})\log\frac{I_2}{I_1}.$$

Now we use Eq. (21–12), inverted: $I_2/I_1 = r_1^2/r_2^2$, so

$$\beta_2 - \beta_1 = (10\ \text{dB})\log\frac{r_1^2}{r_2^2} = (10\ \text{dB})\log\frac{r_1^2}{(2r_1)^2}$$
$$= (10\ \text{dB})\log\frac{1}{4} = -6.0\ \text{dB}.$$

A decrease in intensity of a factor of four corresponds to a 6-dB decrease in sound intensity level.

We invite you to prove that an increase of a factor of two corresponds to a 3-dB increase. This change is barely perceptible to the human ear. An increase of 8 to 10 dB in intensity level is usually interpreted by the ear as a doubling of loudness.

**21–3** When you double your distance from a point source of sound, how much does the intensity decrease?

# 21–3 BEATS

In Section 20–6 we talked about *interference* effects that occur when two different waves with the same frequency overlap in the same region of space. Now let's look at what happens when we have two waves with equal amplitude but slightly different frequencies. This occurs, for example, when two tuning forks with slightly different frequencies are

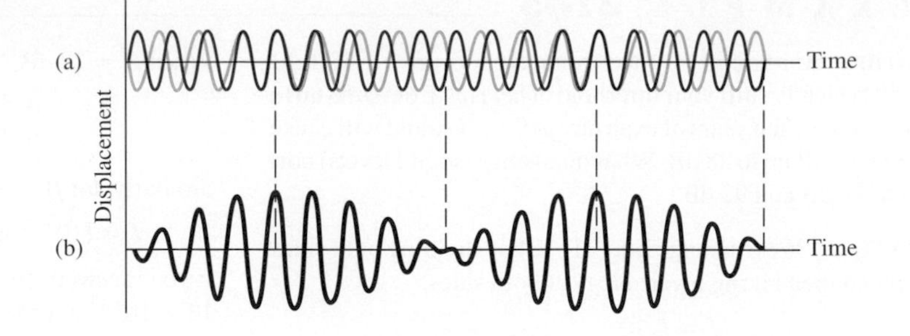


**21–4** Beats are fluctuations in amplitude produced by two sound waves of slightly different frequency. (a) Individual waves. (b) Pattern formed by superposition of the two waves.

sounded together or when two organ pipes that are supposed to have exactly the same frequency are slightly "out of tune."

Consider a particular point in space where the two waves overlap. The displacements of the individual waves at this point are plotted as functions of time in Fig. 21–4a. The total length of the time axis represents about one second, and the frequencies are 16 Hz (blue graph) and 18 Hz (red graph). Applying the principle of superposition, we add the two displacements at each instant of time to find the total displacement at that time. The result is the graph in Fig. 21–4b. At certain times the two waves are in phase; their maxima coincide, and their amplitudes add. But as time goes on, they become more and more out of phase because of their slightly different frequencies. Eventually, a positive peak of one wave coincides with a negative peak of the other wave. The two waves then cancel each other, and the total amplitude is zero.

The resulting wave looks like a single sinusoidal wave with a varying amplitude that goes from a maximum to zero and back, as Fig. 21–4b shows. In this example the amplitude goes through two maxima and two minima in one second; thus the frequency of this amplitude variation is 2 Hz. The amplitude variation causes variations of loudness called **beats,** and the frequency with which the amplitude varies is called the **beat frequency.** In this example the beat frequency is the *difference* of the two frequencies. If the beat frequency is a few hertz, we hear it as a waver or pulsation in the tone.

We can prove that the beat frequency is *always* the difference of the two frequencies $f_1$ and $f_2$. Suppose $f_1$ is larger than $f_2$; the corresponding periods are $T_1$ and $T_2$, with $T_1 < T_2$. If the two waves start out in phase at time $t = 0$, they will again be in phase at a time $T$ such that the first wave has gone through exactly one more cycle than the second. Let $n$ be the number of cycles of the first wave in time $T$; then the number of cycles of the second wave in the same time is $(n - 1)$, and we have the relations

$$T = nT_1 = (n - 1)T_2.$$

We solve the second equation for $n$ and substitute the result back into the first equation, obtaining

$$T = \frac{T_1 T_2}{T_2 - T_1}.$$

Now $T$ is the *period* of the beat, and its reciprocal is the beat *frequency*, $f_{\text{beat}} = 1/T$, so

$$f_{\text{beat}} = \frac{T_2 - T_1}{T_1 T_2} = \frac{1}{T_1} - \frac{1}{T_2},$$

and finally

$$f_{\text{beat}} = f_1 - f_2. \qquad (21\text{–}14)$$

As claimed, the beat frequency is the difference of the two frequencies.

An alternative way to derive Eq. (21–14) is to write functions to describe the curves in Fig. 21–4a and then add them, using a trigonometric identity. Suppose $y_1 = A \sin 2\pi f_1 t$ and $y_2 = -A \sin 2\pi f_2 t$. We use the identity

$$\sin a - \sin b = 2 \sin \tfrac{1}{2}(a - b) \cos \tfrac{1}{2}(a + b).$$

We can then express $y = y_1 + y_2$ as

$$y_1 + y_2 = [2A \sin \tfrac{1}{2}(2\pi)(f_1 - f_2)t] \cos \tfrac{1}{2}(2\pi)(f_1 + f_2)t.$$

The amplitude factor (the quantity in square brackets) varies slowly with the frequency $\frac{1}{2}(f_1 - f_2)$. The cosine factor varies with a frequency equal to the *average* frequency $\frac{1}{2}(f_1 + f_2)$. The *square* of the amplitude factor, which is what the ear hears, goes through two maxima and two minima per cycle, so the beat frequency $f_{beat}$ that is heard is twice the quantity $\frac{1}{2}(f_1 - f_2)$, or just $f_1 - f_2$, in agreement with Eq. (21–14).

Beats between two tones can be heard up to a beat frequency of 6 or 7 Hz. Two piano strings or two organ pipes differing in frequency by 2 or 3 Hz sound wavery and "out of tune," although some organ stops contain two sets of pipes deliberately tuned to beat frequencies of about 1 to 2 Hz for a gently undulating effect. Listening for beats is an important technique in tuning all musical instruments.

At higher frequency differences we no longer hear individual beats, and the sensation merges into one of *consonance* or *dissonance,* depending on the frequency ratio of the two tones. In some cases the ear perceives a tone called a *difference tone,* with a pitch equal to the beat frequency of the two tones.

For multiengine aircraft the engines have to be synchronized so that the sounds don't cause annoying beats, which are heard as loud throbbing sounds. On some planes this is done electronically; on others the pilot does it by ear, just like tuning a piano.

## 21–4 THE DOPPLER EFFECT

You've probably noticed that when a car approaches you with its horn sounding, the pitch seems to drop as the car passes. This phenomenon is called the **Doppler effect.** When a source of sound and a listener are in motion relative to each other, the frequency of the sound heard by the listener is not the same as the source frequency. We can work out a relation between the frequency shift and the source and listener velocities.

To keep things simple, we'll consider only the special case in which the velocities of both source and listener lie along the line joining them. Let $v_S$ and $v_L$ be the velocities of source and listener relative to the air. We will consider the direction from the listener L to the source S as the positive direction for both $v_S$ and $v_L$. The speed of sound $v$ is always considered positive.

### Moving Listener

Let's think first about a listener L moving with velocity $v_L$ toward a stationary source S (Fig. 21–5). The source emits a sound wave with frequency $f_S$ and wavelength $\lambda = v/f_S$. The figure shows several wave crests, separated by equal distances $\lambda$. The waves approaching the moving listener have a speed of propagation *relative to the listener* of $v + v_L$, so the frequency $f_L$ with which the wave crests arrive at the listener's position (the frequency the listener hears) is

$$f_L = \frac{v + v_L}{\lambda} = \frac{v + v_L}{v/f_S}, \qquad (21\text{–}15)$$

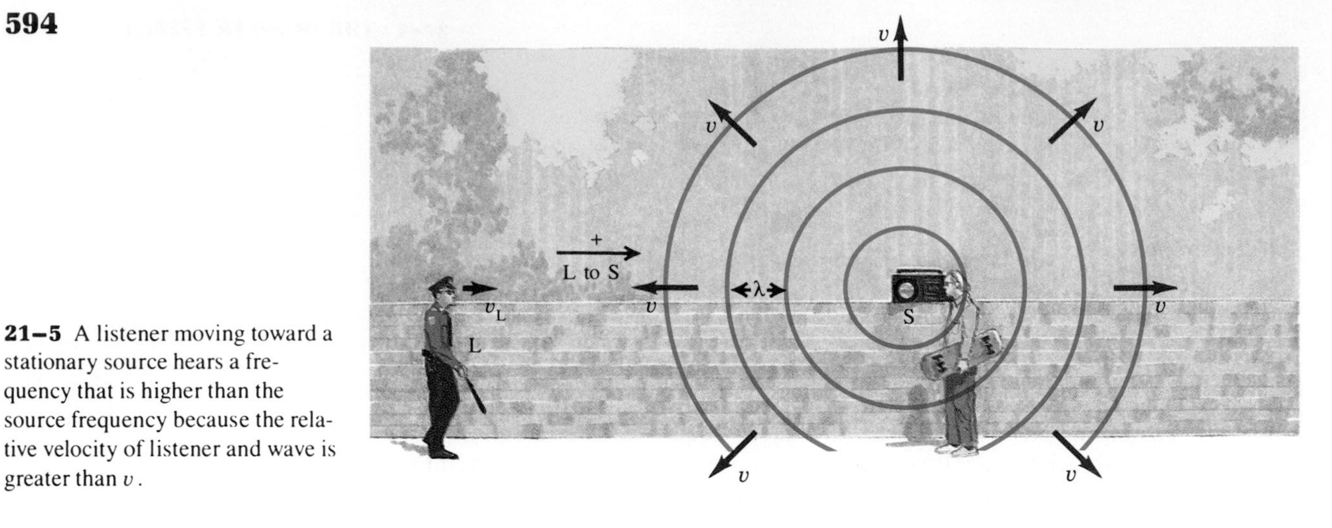


**21–5** A listener moving toward a stationary source hears a frequency that is higher than the source frequency because the relative velocity of listener and wave is greater than $v$.

or

$$f_L = \left(\frac{v + v_L}{v}\right) f_S = \left(1 + \frac{v_L}{v}\right) f_S. \tag{21–16}$$

So a listener moving toward a source ($v_L > 0$) hears a larger frequency (higher pitch) than a stationary listener hears. A listener moving away from the source ($v_L < 0$) hears a smaller frequency (lower pitch).

## Moving Source

Now suppose the source is also moving, with velocity $v_S$ (Fig. 21–6). The wave speed relative to the air is still $v$; it is determined by the properties of the wave medium and is not changed by the motion of the source. But the wavelength is no longer equal to $v/f_S$. Here's why: The time for emission of one cycle of the wave is the period $T = 1/f_S$. During this time the wave travels a distance $vT = v/f_S$, and the source moves a distance $v_S T = v_S/f_S$. The wavelength is the distance between successive wave crests, and this is determined by the *relative* displacement of source and wave. As Fig. 21–6 shows, this is different in front of and behind the source. In the region to the right of the source in Fig. 21–6 the wavelength is

$$\lambda = \frac{v}{f_S} - \frac{v_S}{f_S} = \frac{v - v_S}{f_S}. \tag{21–17}$$

In the region to the left of the source it is

$$\lambda = \frac{v + v_S}{f_S}. \tag{21–18}$$

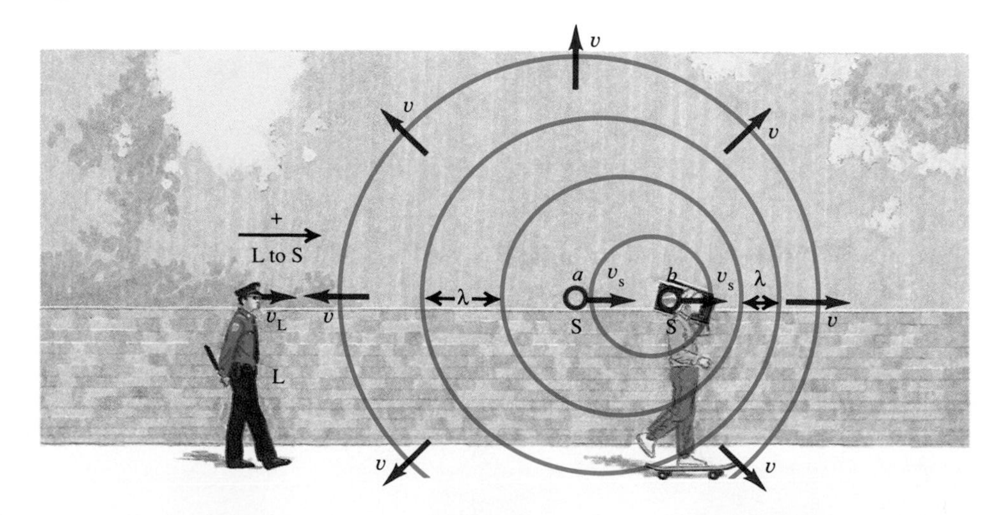


**21–6** Wave crests emitted by a moving source are crowded together in front of the source and stretched out behind it.

The waves are compressed and stretched out, respectively, by the motion of the source.

To find the frequency heard by the listener, we substitute Eq. (21–18) into the first form of Eq. (21–15):

$$f_L = \frac{v + v_L}{\lambda} = \frac{v + v_L}{(v + v_S)/f_S},$$

$$f_L = \frac{v + v_L}{v + v_S} f_S. \tag{21–19}$$

This expresses the frequency $f_L$ heard by the listener in terms of the frequency $f_S$ of the source.

Equation (21–19) includes all possibilities for motion of source and listener (relative to the medium) along the line joining them. If the listener happens to be at rest in the medium, $v_L$ is zero. When both source and listener are at rest or have the same velocity relative to the medium, then $v_L = v_S$ and $f_L = f_S$. Whenever the direction of the source or listener velocity is opposite to the direction from listener toward source (which we have defined as positive), the corresponding velocity to be used in Eq. (21–19) is negative.

## PROBLEM-SOLVING STRATEGY

### Doppler effect

**1.** Establish a coordinate system. Define the positive direction to be the direction from listener to source, and make sure you know the signs of all the relevant velocities. A velocity in the direction from listener source is positive; a velocity in the opposite direction is negative. Also, the velocities must all be measured relative to the air in which the sound is traveling.

**2.** Use consistent notation to identify the various quantities: subscript S for source, L for listener.

**3.** When a wave is reflected from a surface, either stationary or moving, the analysis can be carried out in two steps. In the first, the surface plays the role of listener; the frequency with which the wave crests arrive at the surface is $f_L$. Then think of the surface as a new source, emitting waves with this same frequency, $f_L$. Finally, determine what frequency is heard by a listener detecting this new wave.

**4.** Ask whether your final result makes sense. If the source and listener are moving toward each other, $f_L$ is always greater than $f_S$; if they are moving apart, $f_L$ is always less than $f_S$.

## EXAMPLE 21–7

A police siren emits a sinusoidal wave with frequency $f_S = 300$ Hz. The speed of sound is 340 m/s. a) Find the wavelength of the waves if the siren is at rest in the air. b) If the siren is moving with velocity $v_S = 30$ m/s, find the wavelengths of the waves ahead of and behind the source.

**SOLUTION** a) When the source is at rest,

$$\lambda = \frac{v}{f_S} = \frac{340 \text{ m/s}}{300 \text{ Hz}} = 1.13 \text{ m}.$$

b) The situation is shown in Fig. 21–7. We use Eqs. (21–17) and (21–18). In front of the siren,

$$\lambda = \frac{v - v_S}{f_S} = \frac{340 \text{ m/s} - 30 \text{ m/s}}{300 \text{ Hz}} = 1.03 \text{ m}.$$

Behind the siren,

$$\lambda = \frac{v + v_S}{f_S} = \frac{340 \text{ m/s} + 30 \text{ m/s}}{300 \text{ Hz}} = 1.23 \text{ m}.$$

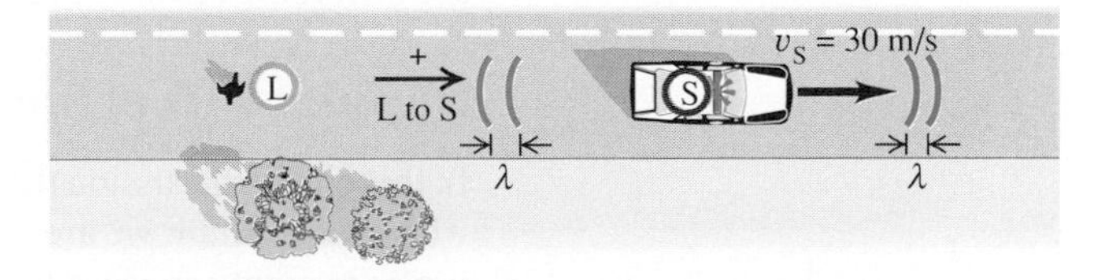


**21–7** Wavelengths ahead of and behind the police siren when it is moving through the air at 30 m/s.

## EXAMPLE 21–8

If the listener is at rest and the siren is moving away from the listener at 30 m/s (Fig. 21–8), what frequency does the listener hear?

**SOLUTION** We have $v_L = 0$ and $v_S = 30$ m/s. (The source velocity $v_S$ is positive because the siren is moving in the same direction as the direction from listener to source.) From Eq. (21–19),

$$f_L = \frac{v}{v + v_S} f_S = \frac{340 \text{ m/s}}{340 \text{ m/s} + 30 \text{ m/s}} (300 \text{ Hz}) = 276 \text{ Hz}.$$

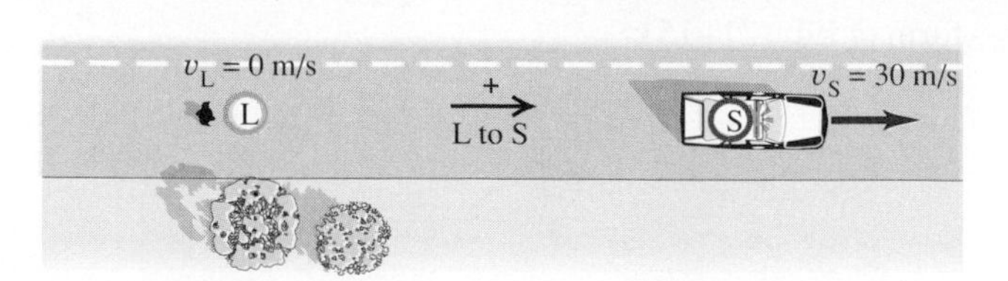


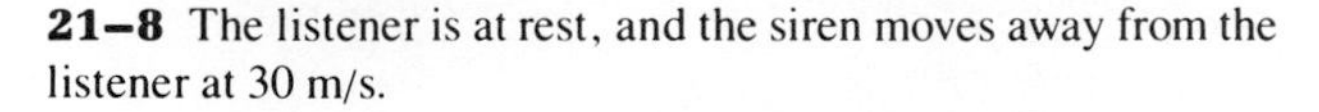

**21–8** The listener is at rest, and the siren moves away from the listener at 30 m/s.

## EXAMPLE 21–9

If the siren is at rest and the listener is moving toward the left at 30 m/s (Fig. 21–9), what frequency does the listener hear?

**SOLUTION** The positive direction (from listener to source) is still from left to right, so $v_L = -30$ m/s, $v_S = 0$, and

$$f_L = \frac{v + v_L}{v} f_S = \frac{340 \text{ m/s} + (-30 \text{ m/s})}{340 \text{ m/s}} (300 \text{ Hz}) = 274 \text{ Hz}.$$

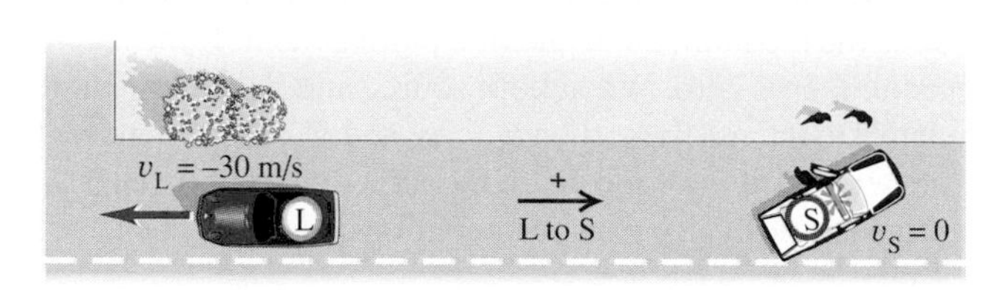


**21–9** The siren is at rest, and the listener moves away from it at 30 m/s.

## EXAMPLE 21–10

If the siren is moving away from the listener with a speed of 45 m/s relative to the air and the listener is moving toward the siren with a speed of 15 m/s relative to the air (Fig. 21–10), what frequency does the listener hear?

**SOLUTION** In this case, $v_L = 15$ m/s and $v_S = 45$ m/s. (Both velocities are positive because both velocity vectors point in the direction from listener to source.) From Eq. (21–19),

$$f_L = \frac{v + v_L}{v + v_S} f_S = \frac{340 \text{ m/s} + 15 \text{ m/s}}{340 \text{ m/s} + 45 \text{ m/s}} (300 \text{ Hz}) = 277 \text{ Hz}.$$

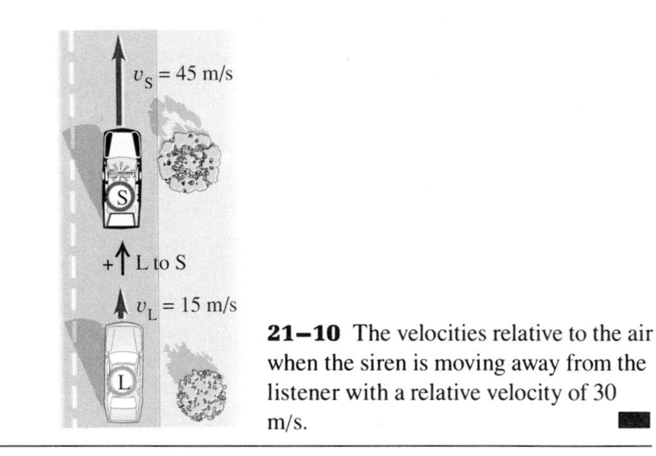


**21–10** The velocities relative to the air when the siren is moving away from the listener with a relative velocity of 30 m/s.

When the source and listener are moving apart, the frequency $f_L$ heard by the listener is always *less than* the frequency $f_S$ emitted by the source. This is the case in all of the last three examples. Note that the *relative velocity* of source and listener is the same in all three, but the Doppler shifts are all different because the velocities relative to the air are different.

## Doppler Effect for Light

In this entire discussion the velocities $v_L$ and $v_S$ are always measured *relative to the air* or whatever medium we are considering. There is also a Doppler effect for electromagnetic waves in empty space, such as light waves or radio waves. In this case there is no medium that we can use as a reference to measure velocities, and the only velocity that we can talk about is the *relative* velocity $v$ of source and receiver.

To derive the expression for the Doppler frequency shift for light, we have to use relativistic kinematic relations. We will derive these later, but meanwhile we quote the result without derivation. The wave speed is the speed of light, usually denoted by $c$, and it is the same for both source and listener. In the frame of reference in which the listener is at rest, the source is moving away from the listener with velocity $v$. (If the source is *approaching* the listener, $v$ is negative.) The source frequency is again $f_S$. The frequency $f_L$ measured by the listener (the frequency of the waves on arrival at the listener) is then given by

$$f_L = \sqrt{\frac{c - v}{c + v}} f_S. \tag{21–20}$$

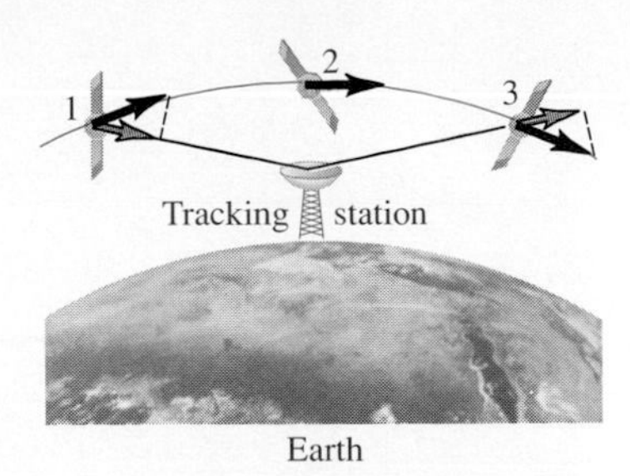


**21–11** Change of velocity component along the line of sight of a satellite passing a tracking station. The distance first decreases, then increases.

When $v$ is positive, the source is moving directly *away* from the listener and $f_L$ is always less than $f_S$; when $v$ is negative, the source is moving directly *toward* the listener and $f_L$ is *greater* than $f_S$. The qualitative effect is the same as for sound, but the quantitative relationship is different.

A familiar application of the Doppler effect with radio waves is the black radar device mounted on the side window of a police car to check other cars' speeds. The electromagnetic wave emitted by the device is reflected from a moving car, which acts as a moving source, and the wave reflected back to the device is Doppler-shifted in frequency. The transmitted and reflected signals are combined to produce beats, and the speed can be computed from the frequency of the beats. Similar techniques ("Doppler radar") are used to measure wind velocities in the atmosphere.

The Doppler effect is also used to track satellites and other space vehicles. In Fig. 21–11 a satellite emits a radio signal with constant frequency $f_S$. As the satellite orbits overhead, its distance from the listener first decreases, then increases, and the frequency $f_L$ of the signal received on earth decreases as the satellite passes overhead.

The Doppler effect for *light* is important in astronomy. Astronomers compare wavelengths of light from distant stars to those emitted by the same elements on earth. In a binary star system, in which two stars rotate about their common center of mass, the light is Doppler-shifted to higher frequencies when a star is moving toward an observer on earth and to lower frequencies when it's moving away.

Light from most galaxies is shifted toward the longer-wavelength, or red, end of the visible spectrum, an effect called the *red shift.* This is often described as a Doppler shift resulting from motion of these galaxies away from us. But from the point of view of the general theory of relativity it is something much more fundamental, associated with the expansion of space itself. The most distant galaxies have the greatest red shifts because their light has shared in the expansion of the space through which it moved. Extrapolating this expansion backward to over ten billion years ago leads to the "big bang" picture. From this point of view the big bang was not an explosion in space but the initial rapid expansion of space itself.

## *21–5 SHOCK WAVES AND APPLICATIONS

Everyone has experienced "sonic booms" caused by an airplane flying over at supersonic speed. We can see qualitatively what happens from Fig. 21–12 (on page 598), which is similar to Fig. 21–6. The waves in front of a moving source are crowded together with a wavelength given by Eq. (21–17):

$$\lambda = \frac{v - v_S}{f_S}.$$

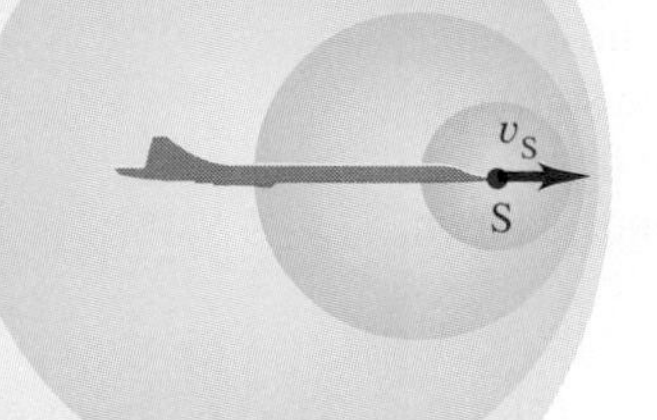

(c)

**21–12** (a) As the speed $v_S$ of the source of sound S approaches the speed of sound $v$, the wave crests begin to pile up in front of S. (b) Formation of the shock wave when the speed of the source $v_S$ is greater than the speed of sound $v$. (c) You can see the effect of the shock wave around this supersonic jet, causing compression of the air molecules.

As the speed $v_S$ of the source approaches the speed $v$ of sound, the wavelength approaches zero and the wave crests pile up on each other. In Eq. (21–19), when $v_S = -v$ (corresponding to the source moving toward the listener at the speed of sound $v$), $f_L$ becomes infinite. With all that energy piled up in front of the source you can see why going faster than the speed of sound is called "breaking the sound barrier."

## Shock Waves

When $v_S$ is greater in magnitude than $v$, our whole analysis of the Doppler effect breaks down. Figure 21–12b is a diagram of what happens. The source emits a series of wave crests; each spreads out in a circle centered at the position of the source when it emitted the crest. After a time $t$ the crest emitted from point $S_1$ has spread to a circle with radius $vt$, and the source has moved a greater distance $v_S t$, to position $S_2$. You can see that the circular crests interfere constructively at points along the green line making an angle $\alpha$ with the direction of the source velocity, leading to a very large-amplitude wave crest along this line. This crest is called a **shock wave.**

From the right triangle shown in the figure we can see that the angle $\alpha$ is given by

$$\sin\alpha = \frac{vt}{v_S t} = \frac{v}{v_S}. \quad (21\text{–}21)$$

In this relation, $v_S$ is the *speed* of the source (the magnitude of its velocity) and is always positive.

The actual situation is three-dimensional; the shock wave forms a *cone* around the direction of motion of the source (Fig. 21–12c). If the source (possibly a supersonic jet plane or a rifle bullet) moves with constant velocity, the angle $\alpha$ is constant, and the shock-wave cone moves along with the source. It's the arrival of this shock wave that causes the sonic boom that you hear after the plane has passed by.

The ratio $v_S/v$ is called the **Mach number;** it is greater than unity for all supersonic speeds, and $\sin\alpha$ in Eq. (21–21) is the reciprocal of the Mach number. The first person officially to break the sound barrier was Chuck Yeager, flying at Mach 1.015 on October 14, 1947.

## EXAMPLE 21–11

**Sonic boom of the Concorde** The Concorde is flying at Mach 1.75 at an altitude of 8000 m, where the speed of sound is 320 m/s. How long after the plane passes directly overhead will you hear the sonic boom?

**SOLUTION** Figure 21–13 shows the situation just as the shock wave reaches you. From Eq. (21–21) the angle $\alpha$ is

$$\alpha = \arcsin\frac{1}{1.75} = 34.8°.$$

The speed of the plane is the speed of sound multiplied by the Mach number:

$$v_S = (1.75)(320\ \text{m/s}) = 560\ \text{m/s}.$$

From the figure we have

$$\tan\alpha = \frac{8000\ \text{m}}{v_S t},$$

$$t = \frac{8000\ \text{m}}{(560\ \text{m/s})(\tan 34.8°)} = 20.5\ \text{s}.$$

You hear the boom 20.5 s after the Concorde passes overhead, and at that time it has traveled (560 m/s)(20.5 s), or 11.5 km, past the straight-overhead point.

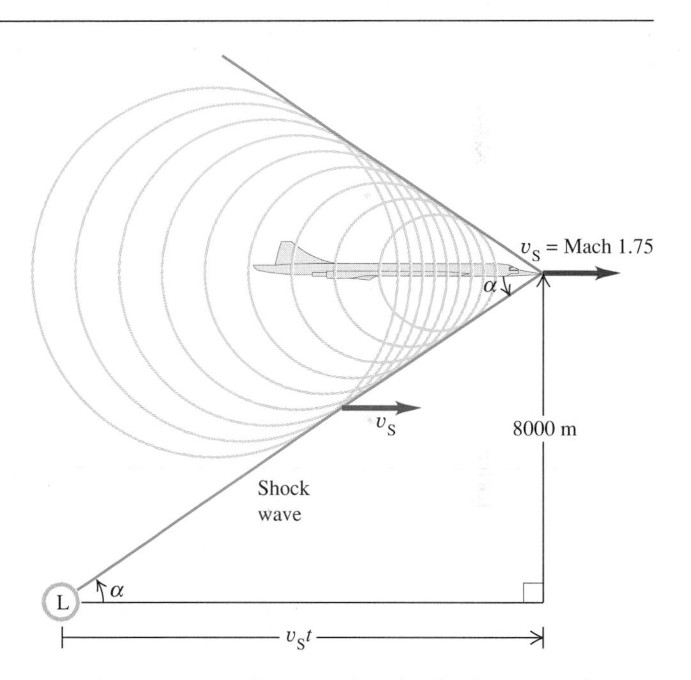

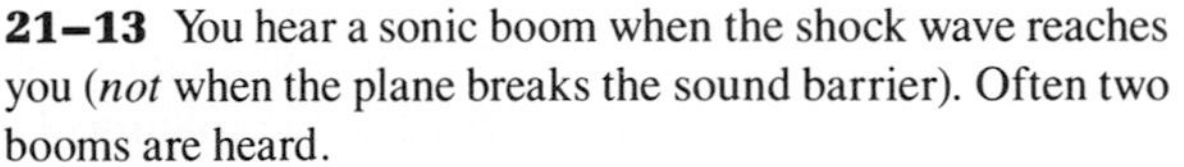


**21–13** You hear a sonic boom when the shock wave reaches you (*not* when the plane breaks the sound barrier). Often two booms are heard.

## Other Applications

Shock waves are used to break up kidney stones and gallstones without invasive surgery, using a technique with the impressive name *extracorporeal shock-wave lithotripsy.* A shock wave is produced outside the body and then focused by a reflector or acoustic lens so that as much of it as possible converges on the stone. When the resulting stresses in the stone exceed its tensile strength, it breaks into small pieces and can be eliminated. This technique requires accurate determination of the location of the stone, which is done by using ultrasonic imaging techniques.

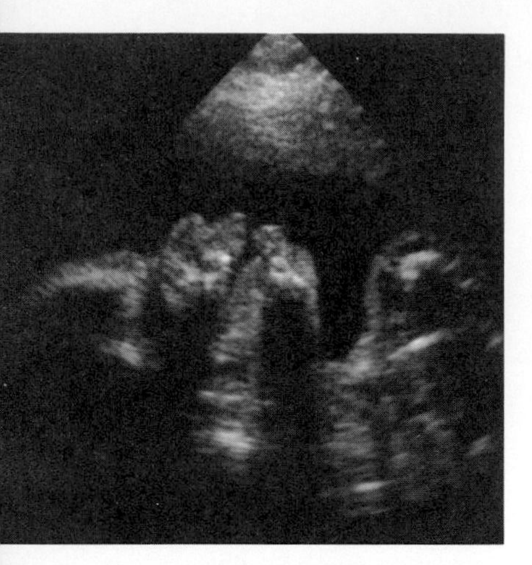

**21–14** Sonogram of a human fetus in the womb showing the facial profile of a nineteen-week fetus.

Acoustic effects have many other important medical applications. Ultrasonic imaging (reflection of ultrasonic waves from regions in the interior of the body) is used for prenatal examinations, detection of anomalous conditions such as tumors, and the study of heart-valve action, to name a few possibilities. Ultrasound is more sensitive than X-rays in distinguishing various kinds of tissues, and it does not have the radiation hazards associated with X-rays. At much higher power levels, ultrasound appears to have promise as a selective destroyer of pathological tissue in the treatment of arthritis and certain cancers. The sound is always produced by first generating an electrical wave, then using this to drive a loudspeakerlike device called a **transducer** that converts electrical waves to sound. Techniques have been developed in which transducers move over, or *scan,* the region of interest and a computer-reconstructed image is produced. An example of such an image is shown in Fig. 21–14.

Analysis of elastic waves in the earth provides important information about its structure. The interior of the earth may be pictured roughly as being made of concentric spherical shells around a fluid core. Mechanical properties such as density and elastic moduli are different in different shells. Waves produced by explosions or earthquakes are reflected and refracted at the interfaces between these shells, and analysis of these waves helps geologists to measure the dimensions and properties of the shells. Local anomalies such as oil deposits can also be detected by study of wave propagation in the earth.

Acoustic principles have many important applications to environmental problems such as noise control. The design of quiet mass-transit vehicles, for example, involves detailed study of sound generation and propagation in the motors, wheels, and supporting structures. Excessive noise levels often lead to permanent hearing impairment; studies have shown that many young rock musicians have hearing that is typical of persons 65 years of age. Prolonged listening to music at high intensity levels (100 to 120 dB) can lead to permanent hearing loss. Stereo headsets used at high volume levels pose similar threats to hearing. Be careful!

## *21–6 MUSICAL TONES

Several aspects of musical sound are directly related to the physical characteristics of sound waves. The **pitch** of a musical tone is the quality that lets us classify it as "high" or "low." Middle C on the piano has a frequency of 262 Hz. The musical interval of an *octave* corresponds to a factor of two in frequency. The C note an octave above middle C has a frequency of $2 \times 262$ Hz $= 524$ Hz, and the "high C" sung by coloratura sopranos, two octaves above middle C, has a frequency of $4 \times 262$ Hz, or 1048 Hz. The ear can be fooled. When a listener hears two sinusoidal tones with the same frequency but different intensities, the louder one usually seems to be slightly lower in pitch.

Musical tones usually contain many frequencies. A plucked, bowed, or struck string or the column of air in a wind instrument vibrates with a fundamental frequency and many harmonics at the same time. Two tones may have the same fundamental frequency (and thus the same pitch) but sound different because of the presence of different intensities of the various harmonics. The difference is called *tone color, quality,* or **timbre,** often described in subjective terms such as reedy, golden, round, mellow, or tinny. A tone that is rich in harmonics usually sounds thin and "stringy" or "reedy," and a tone containing mostly a fundamental is more mellow and flutelike.

Another factor in determining tone quality is the behavior at the beginning and end of a tone. A piano tone begins percussively with a thump and then dies away gradually. A harpsichord tone, in addition to having different harmonic content, begins much more quickly and incisively with a click, and the higher harmonics begin before the lower ones. The ending of the tone, when the key is released, is also much more incisive on the harpsichord than the piano. Similar effects are present with other musical instruments; with wind and string instruments the player has considerable control over the attack and decay of the tone, and these characteristics help to define the unique characteristics of each instrument.

When combinations of musical tones are heard together or in succession, even a listener with no musical training hears a relationship among them; they sound comfortable together, or they clash. Musicians use the terms *consonant* and *dissonant* to describe these effects. A *consonant interval* is a combination of two tones that sound comfortable or pleasing together. The most consonant interval is the *octave;* it consists of two tones with a frequency ratio of 2:1. An example is middle C on the piano and the next C above it. Another set of consonant tones is obtained by playing C, E, and G. These form what is called a *major triad,* and the frequencies are in the ratios 4:5:6. The interval from C to G, with a frequency ratio of 3:2, is a perfect fifth; C to E, with 5:4, is a major third; and so on. These combinations sound good together because they have many harmonics in common. For example, for the interval of the octave, every harmonic in the harmonic series of the lower-frequency tone is also present in the harmonic series of the upper tone. For other intervals the overlap of harmonics is only partial, but it is there.

If we want to be able to play major triads starting on various tones on a keyboard instrument such as the piano or organ, complications arise. Either we add a lot of extra keys in each octave to get the exact frequencies we need, or else we compromise a little on the frequencies. The most common compromise in piano tuning is called *equal temperament,* in which every pair of adjacent keys (white and black) is tuned to the same frequency ratio of $2^{1/12}$. This interval is called a *semitone* or a *halftone;* a succession of 12 of these intervals then forms an exact 2:1 octave. The perfect fifth is seven semitones; the corresponding frequency ratio is $2^{7/12} = 1.4983$. This is close to the ideal ratio of 3:2, or 1.5000, but a sensitive ear can hear the difference.

Thus a piano tuned in equal temperament is not quite in tune, in terms of the ideal ratios, in *any* key, but it is equally good (or bad) in all keys. An alternative would be to tune the white keys to form ideal intervals. This would sound better for music in the key of C, but music in some other keys would sound worse in comparison with equal temperament. So an instrument that is intended to be used for compositions in all keys is usually tuned to equal temperament.

In the Baroque period, however, keys with more than three sharps or flats were seldom used, and various compromise temperaments were invented to favor the commonly used keys. Organs and harpsichords that are intended primarily for music of this period are often tuned to one of these unequal temperaments. The great J. S. Bach favored these rather than equal temperament. His composition "The Well-Tempered Clavier" contains preludes and fugues in all the major and minor keys, but it is important not to misconstrue *well-tempered* as meaning *equally tempered!*

## SUMMARY

- Sound consists of longitudinal waves in air. A sinusoidal sound wave is characterized by its frequency $f$, wavelength $\lambda$ (or angular frequency $\omega$, wave number $k$), and amplitude $A$. The amplitude is also related to the pressure amplitude $p_{\max}$ by

$$p_{\max} = BkA. \qquad (21\text{–}5)$$

- The intensity $I$ of a wave is the time average rate at which energy is transported by the wave per unit area. In terms of the amplitude $A$,

$$I = \tfrac{1}{2}\omega BkA^2. \qquad (21\text{–}8)$$

In terms of the pressure amplitude,

$$I = \frac{p_{max}^2}{2\rho v} = \frac{p_{max}^2}{2\sqrt{\rho B}}. \qquad (21\text{–}11)$$

**KEY TERMS**

sound
ultrasonic
infrasonic
pressure amplitude
intensity
intensity level
decibel
beat
beat frequency
Doppler effect
shock wave
Mach number
transducer
pitch
timbre

• The intensity level $\beta$ of a sound wave is defined as

$$\beta = (10\text{ dB}) \log \frac{I}{I_0}, \qquad (21\text{–}13)$$

where $I_0$ is a reference intensity defined to be $10^{-12}$ W/m$^2$. Intensity levels are expressed in decibels (dB).

• Beats are heard when two tones with slightly different frequencies $f_1$ and $f_2$ are sounded together. The beat frequency $f_{beat}$ is

$$f_{beat} = f_1 - f_2. \qquad (21\text{–}14)$$

• The Doppler effect is the frequency shift that occurs when there is relative motion of a source of sound and a listener. The source and listener frequencies $f_S$ and $f_L$ and their velocities $v_S$ and $v_L$ are related by

$$f_L = \frac{v + v_L}{v + v_S} f_S. \qquad (21\text{–}19)$$

• A sound source moving with a speed $v_S$ greater than the speed of sound $v$ creates a shock wave. The wave front is a cone with angle $\alpha$ given by

$$\sin\alpha = \frac{vt}{v_S t} = \frac{v}{v_S}. \qquad (21\text{–}21)$$

• The pitch of a musical tone depends primarily on its frequency. The tone quality, or timbre, depends on the harmonic content and the attack and decay characteristics. Consonance and dissonance are determined by frequency ratios of tones.

## EXERCISES

Unless indicated otherwise, assume the speed of sound in air to be $v$ = 344 m/s.

### Section 21–1 Sound Waves

### Section 21–2 Intensity

**21–1** Consider a sound wave in air that has displacement amplitude 0.0125 mm. Calculate the pressure amplitude for frequencies of a) 500 Hz; b) 20,000 Hz (barely audible). In each case, compare the results to the pain threshold given in Example 21–1.

**21–2** a) If the pressure amplitude of a sound wave is tripled, by what factor is the intensity of the wave increased? b) By what factor must the pressure amplitude of a sound wave be increased to increase the intensity by a factor of 4?

**21–3** The pressure amplitude of the faintest sound wave that can be heard is about $3.0 \times 10^{-5}$ Pa. Calculate the corresponding intensity in W/m$^2$.

**21–4 Power of a Siren.** A tornado warning siren on top of a pole radiates sound waves uniformly in all directions. At a distance of 20.0 m the intensity of the sound is 0.400 W/m$^2$. a) At what distance from the siren is the intensity 0.100 W/m$^2$? b) What is the total acoustical power output of the siren?

**21–5** Derive Eq. (21–10) from the preceding equations.

**21–6** a) Relative to the arbitrary reference intensity of $1.00 \times 10^{-12}$ W/m$^2$, what is the intensity level in decibels of a sound wave whose intensity is $1.00 \times 10^{-7}$ W/m$^2$? b) What is the intensity level of a sound wave in air at 20°C whose pressure amplitude is 0.250 Pa?

**21–7** A sound wave in air has a frequency of 400 Hz, a wave speed of 344 m/s, and a displacement amplitude of $6.00 \times 10^{-3}$ mm. Calculate the intensity (in W/m$^2$) and intensity level (in decibels) for this sound wave.

**21–8** When a physics professor lectures, she produces sound with an intensity 500 times greater than when she whispers. What is the difference in intensity levels, in decibels?

**21–9 Cries and Whispers.** What is the difference in intensity level of a baby's cry heard by its father if the baby's mouth is 40 cm from his ear compared to the intensity level heard by the baby's mother, who is 3.00 m away from the baby?

**21–10** The intensity due to a number of independent sound sources is the sum of the individual intensities. a) How many decibels greater is the intensity level when four quadruplets cry simultaneously than when a single one cries? b) How many more crying babies are required to produce a further increase in the intensity level of an equal number of decibels?

### Section 21–3
### Beats

**21–11 Tuning Trumpets.** A trumpet player is tuning his instrument by playing an A note simultaneously with the first-chair trumpeter, who has perfect pitch. The first chair's note is exactly 440 Hz, and 3.2 beats per second are heard. What are the two possible frequencies of the other player's note?

**21–12** Two identical piano strings, when stretched with the same tension, have a fundamental frequency of 440 Hz. By what fractional amount must the tension in one string be increased so that 2.5 beats per second will occur when both strings vibrate simultaneously?

### Section 21–4
### The Doppler Effect

**21–13** A railroad train is traveling at 30.0 m/s in still air. The frequency of the note emitted by the locomotive whistle is 400 Hz. What is the wavelength of the sound waves a) in front of the locomotive; b) behind the locomotive? What is the frequency of the sound heard by a stationary listener c) in front of the locomotive; d) behind the locomotive?

**21–14** On the planet Vulcan a male Nameloc is flying toward his mate at 0.250 m/s while singing at a frequency of 1200 Hz. If the stationary female hears a tone of 1500 Hz, what is the speed of sound in the atmosphere of Vulcan?

**21–15** A railroad train is traveling at 30.0 m/s in still air. The frequency of the note emitted by the train whistle is 400 Hz. What frequency is heard by a passenger on a train moving in the direction opposite to the first at 15.0 m/s and a) approaching the first; b) receding from the first?

**21–16** Two train whistles, $A$ and $B$, each have a frequency of 500 Hz. $A$ is stationary, and $B$ is moving toward the right (away from $A$) at a speed of 35.0 m/s. A listener is between the two whistles and is moving toward the right with a speed of 15.0 m/s (Fig. 21–15). No wind is blowing. a) What is the frequency from $A$ as heard by the listener? b) What is the frequency from $B$ as heard by the listener? c) What is the beat frequency detected by the listener?

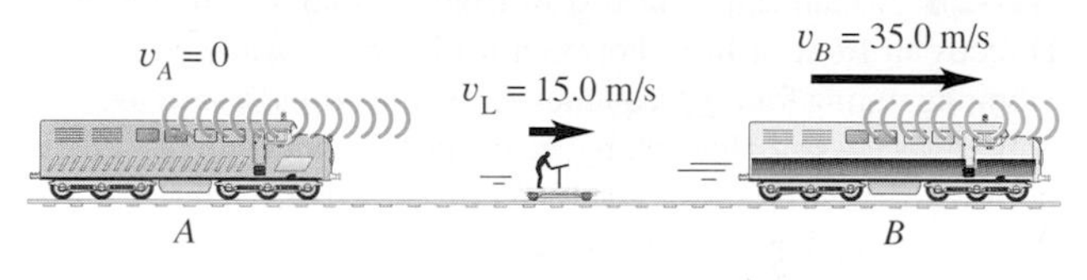


**FIGURE 21–15**

### Section 21–5
### Shock Waves and Applications

***21–17** A jet plane flies overhead at Mach 2.00 and at a constant altitude of 1200 m. The speed of sound in the air is 340 m/s. a) What is the angle $\alpha$ of the shock-wave cone? b) How much time after the plane passes directly overhead do you hear the sonic boom?

## PROBLEMS

**21–18** A window whose area is 1.20 $m^2$ opens on a street where the street noises result in an intensity level, at the window, of 60.0 dB. How much "acoustic power" enters the window via the sound waves?

**21–19** The sound from a trumpet radiates uniformly in all directions in air at 20°C. At a distance of 6.00 m from the trumpet the sound level is 50.0 dB. The frequency is 440 Hz. a) What is the pressure amplitude at this distance? b) What is the displacement amplitude? c) At what distance is the sound level 30.0 dB?

**21–20** A very noisy chain saw operated by a tree surgeon emits a total sound power of 16.0 W uniformly in all directions. At what distance from the source is the sound level a) 100 dB; b) 60 dB?

**21–21 Speed of a Duck.** A swimming duck paddles the water with its feet once every 2.0 s, producing surface waves with this frequency. The duck is moving at constant speed in a pond where the speed of surface waves is 0.40 m/s, and the crests of the waves ahead of the duck are spaced 0.20 m apart. a) What is the duck's speed? b) How far apart are the crests behind the duck?

**21–22 Medical Ultrasound.** A 2.00-MHz sound wave travels through a mother's abdomen and is reflected from the fetal heart wall of her unborn baby, which is moving toward the sound receiver as the heart beats. The reflected sound is then mixed with the transmitted sound, and 160 beats per second are detected. The speed of sound in body tissue is 1500 m/s. Calculate the speed of the fetal heart wall at the instant when this measurement is made.

**21–23** The frequency ratio of a half-tone interval on the equally tempered scale is 1.059. Find the speed of an automobile passing a listener at rest in still air if the pitch of the car's horn drops a half-tone between the times when the car is coming directly toward her and when it is moving directly away from her.

**21–24** Roger Rabbit is cruising through downtown Toontown on Friday night at 20.0 m/s on the way to a movie. As he approaches the corner of 9th Street and Maple Avenue, Roger sees Betty Boop standing there and waving at him. a) When Roger whistles at Betty with a frequency of 1000 Hz, what frequency does she hear? b) As Roger passes by her, Betty whistles back at him with a frequency of 1000 Hz. What frequency does he hear?

**21–25** a) Show that Eq. (21–20) can be written

$$f_L = f_S\left(1 - \frac{v}{c}\right)^{1/2}\left(1 + \frac{v}{c}\right)^{-1/2}.$$

b) Use the binomial theorem to show that if $v << c$, this is approximately equal to

$$f_L = f_S\left(1 - \frac{v}{c}\right).$$

c) An airplane starting its initial landing approach emits a radio signal with a frequency of $1.00 \times 10^8$ Hz. An air traffic controller in the control tower on the ground detects beats between the received signal and a local signal also of frequency $1.00 \times 10^8$ Hz (Fig. 21–16). At a particular moment the beat frequency is 22.0 Hz. What is the component of the airplane's velocity directed toward the control tower at this moment? (The speed of light is $c = 3.00 \times 10^8$ m/s.)

**FIGURE 21–16**

**21–26** A man stands at rest in front of a large, smooth wall. Directly in front of him, between him and the wall, he holds a vibrating tuning fork of frequency $f_0$. He now runs toward the wall with a speed $v_m$. How many beats per second will he detect between the sound waves reaching him directly from the fork and those reaching him after being reflected from the wall? (*Note:* If the beat frequency is too large, the man may have to use some instrumentation other than his ears to detect and count the beats.)

**21–27 Ship's Sonar.** The sound source of a ship's sonar system operates at a frequency of 30.0 kHz. The speed of sound in water is 1480 m/s. a) What is the wavelength of the waves emitted by the source? b) What is the difference in frequency between the directly radiated waves and the waves reflected from a whale traveling directly away from the ship at 6.95 m/s? The ship is at rest in the water.

***21–28** On a clear day you see a jet plane flying overhead. From the apparent size of the plane you determine that it is flying at a constant altitude of 2400 m. You hear the sonic boom 6.2 s after the plane passes directly overhead. The speed of sound is 340 m/s. What is the speed $v_S$ of the plane? (*Hint:* Consider the square of the trigonometric equations involving $\alpha$ and $v_S$, and use the identity $\sin^2\alpha + \cos^2\alpha = 1$ to write $\tan^2\alpha$ in terms of $\sin^2\alpha$.)

## CHALLENGE PROBLEMS

**21–29** Two loudspeakers, $A$ and $B$, radiate sound uniformly in all directions in air at 20°C. The output of acoustic power from $A$ is $8.00 \times 10^{-4}$ W, and from $B$ it is $12.0 \times 10^{-4}$ W. Both loudspeakers are vibrating in phase at a frequency of 172 Hz. a) Determine the difference in phase of the two signals at a point $C$ along the line joining $A$ and $B$, 3.00 m from $B$ and 4.00 m from $A$ (Fig. 21–17). b) Determine the intensity and intensity level at point $C$ from speaker $A$ if speaker $B$ is turned off and the intensity and intensity level at point $C$ from speaker $B$ if speaker $A$ is turned off. c) With both speakers on, what are the intensity and intensity level at point $C$?

**21–30** A sound wave with frequency $f_0$ and wavelength $\lambda_0$ travels horizontally toward the right. It strikes and is reflected from a large, rigid, vertical plane surface, perpendicular to the direction of propagation of the wave and moving toward the left with a speed $v_1$. a) How many positive wave crests strike the surface in a time interval $t$? b) At the end of this time interval, how far to the left of the surface is the wave that was reflected at the beginning of the time interval? c) What is the wavelength of the reflected waves in terms of $\lambda_0$? d) What is the frequency in terms of $f_0$? Is your result consistent with the assertion made in point 3 of the Problem-Solving Strategy in Section 21–4? e) A listener is at rest at the left of the moving surface. How many beats per second does she detect as a result of the combined effect of the incident and reflected waves?

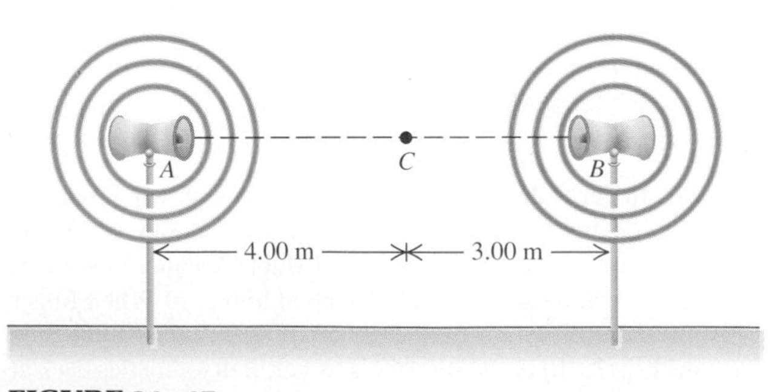


**FIGURE 21–17**

**21–31** A source of sound waves, $S$, emitting waves of frequency $f_0$, is traveling toward the right in still air with a speed $v_1$. At the right of the source is a large, smooth, reflecting surface moving toward the left with a speed $v_2$. The speed of the sound waves is $v$. a) How far does an emitted wave travel in time $t$? b) What is the wavelength of the emitted waves in front of (at the right of) the source? c) How many waves strike the reflecting surface in time $t$? d) What is the speed of the reflected waves? e) What is the wavelength of the reflected waves? f) What is the frequency of the reflected waves as heard by a stationary listener? Is your result consistent with the assertion made in point 3 of the Problem-Solving Strategy in Section 21–4? g) Calculate a numerical value for the frequency in part (f) for $v = 344$ m/s, $v_1 = 25.0$ m/s, $v_2 = 60.0$ m/s, and $f_0 = 1000$ Hz.

C H A P T E R 22

# Electric Charge and Electric Field

Electrostatic principles of a xerographic copier.

Original copy, face down

Movable light

Movable mirror

Lens assembly

Positively charged electrode

Heater assembly

Fixed mirror

Positively charged paper

Negatively charged toner brush

Selenium-coated drum

1) The imaging drum is a rotating aluminum cylinder with a thin coating of selenium, which is an insulator in the dark but becomes conductive when exposed to light. The drum is given a uniform surface charge in the dark.

2) An image of the original document is projected optically onto the drum. Light areas become discharged; dark areas retain their charge.

3) As the drum rotates further, negatively charged particles of a dark fusible powder called toner are applied to the drum. They stick only to the positively charged dark parts of the image.

4) The drum transfers the toner to a positively charged sheet of paper, forming a copy of the original document.

5) The sheet is heated; the toner particles fuse into the paper to form the final copy.

## KEY CONCEPTS

**• Electric charge is the basis of one of the fundamental interactions among particles. Charge is positive or negative. Like charges repel; unlike charges attract.**

**• Conductors are materials that permit charge to move within them; charge moves much less readily in insulators.**

**• The total electric charge in any isolated system is constant. All charges are integer multiples of the magnitude of the electron charge, the fundamental unit of charge.**

**• Coulomb's law describes how the interaction forces depend on the charges and the distance between them. The total force on a given charge due to two or more other charges is the vector sum of the individual forces.**

**• Electric field, a vector quantity, describes the electric force on a charge at a given point of space on a force-per-unit-charge basis. The total field at any point due to a collection of charges is the vector sum of the fields due to the individual charges.**

**• Electric field lines provide a graphical representation of electric-field patterns.**

**• An electric dipole consists of two charges with the same magnitude and opposite sign, separated in space.**

## INTRODUCTION

When you scuff your shoes across a carpet, you can get zapped by an annoying spark of static electricity. That same spark would totally destroy the function of a computer chip. Lightning, the same phenomenon on a vast scale, can destroy a lot more than chips. The clinging of newly laundered synthetic fabrics is related to these sparks. All these phenomena involve electric charge and electrical interactions, one of nature's fundamental classes of interactions.

In this chapter we'll study the interactions of electric charges that are at rest in our frame of reference; we call these *electrostatic* interactions. They are governed by Coulomb's law and are most conveniently described using the concept of *electric field*. We will find that charge is quantized and that it obeys a conservation principle. Electrostatic interactions hold atoms, molecules, and our bodies together, but they also are constantly trying to tear apart the nuclei of atoms. We'll explore all these concepts in this chapter and the following ones. ■

## 22–1 ELECTRIC CHARGE

The ancient Greeks discovered as early as 600 B.C. that when they rubbed amber with wool, the amber could then attract other objects. Today we say that the amber has acquired a net **electric charge,** or has become *charged.* The word *electric* is derived from the Greek word *elektron,* meaning amber. When you scuff your shoes across a nylon carpet, you become electrically charged, and you can charge a comb by passing it through dry hair.

Plastic rods and fur (real or fake) are particularly good for demonstrating electric-charge interactions. Figure 22–1a shows two plastic rods and two pieces of fur. After we charge them by rubbing each rod with its piece of fur, we find that the rods repel each other (Fig. 22–1b). When we rub glass rods (Fig. 22–1c) with silk, the glass rods also become charged and repel each other (Fig. 22–1d). But a charged plastic rod *attracts* a charged glass rod (Fig. 22–1e). Furthermore, the plastic rod and the fur attract each other, and the glass rod and the silk attract each other (Fig. 22–1f).

These experiments and many others like them have shown that there are exactly two (no more) kinds of electric charge, the kind on the plastic rod rubbed with fur and the kind on the glass rod rubbed with silk. Benjamin Franklin (1706–1790) suggested calling these two kinds of charge *negative* and *positive,* respectively, and these names are still used. **Two positive charges or two negative charges repel each other. A positive and a negative charge attract each other.** The plastic rod and the silk have negative charge; the glass rod and the fur have positive charge.

When we rub a plastic rod with fur (or a glass rod with silk), *both* objects acquire a net charge, and their net charges are always equal in magnitude and opposite in sign. These experiments show that in the charging process we are not *creating* electric charge but *transferring* it from one body to another. We now know that the plastic rod acquires extra electrons, which have negative charge. These electrons are taken from the fur, which is left with a deficiency of electrons and a net positive charge. The *total* electric charge on *both* bodies does not change. This is an example of *conservation of charge;* we'll come back to this important principle later.

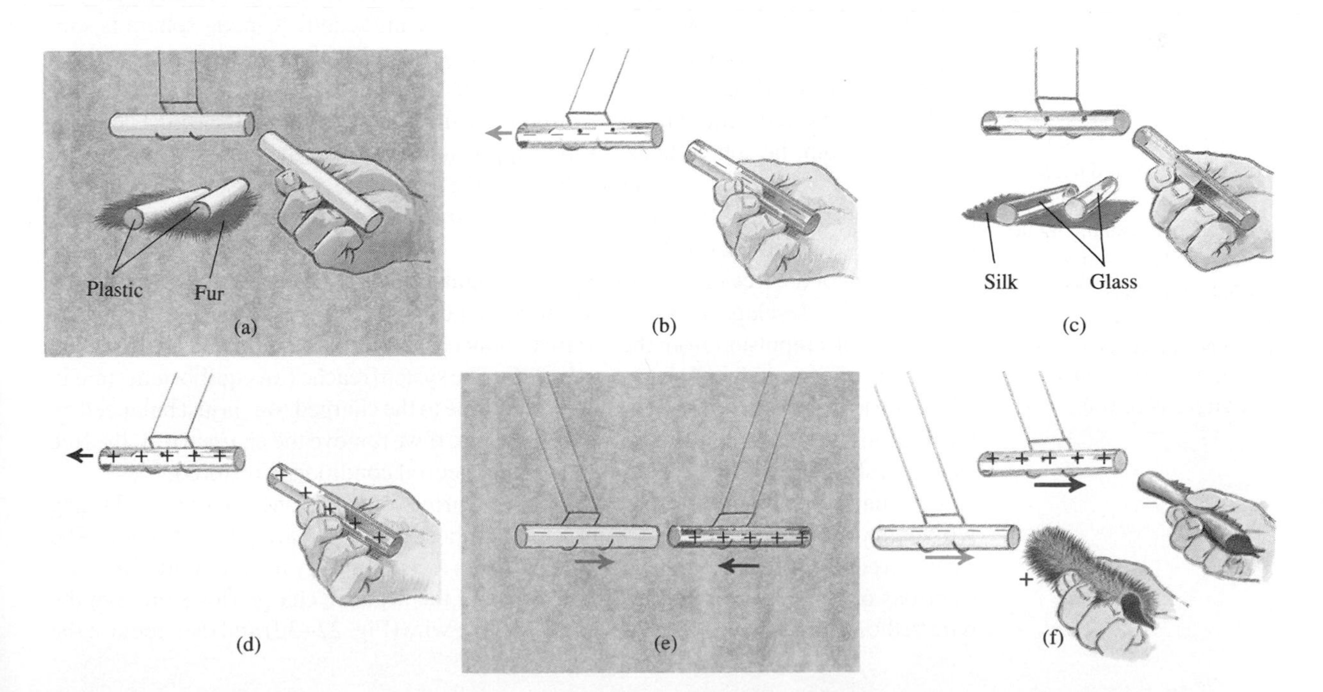


**22–1** (a, b) After being rubbed with fur, two plastic rods repel each other. (c, d) After being rubbed with silk, two glass rods repel each other. (e) The charged plastic rod attracts the charged glass rod. (f) The fur attracts the plastic rod, and the silk attracts the glass rod.

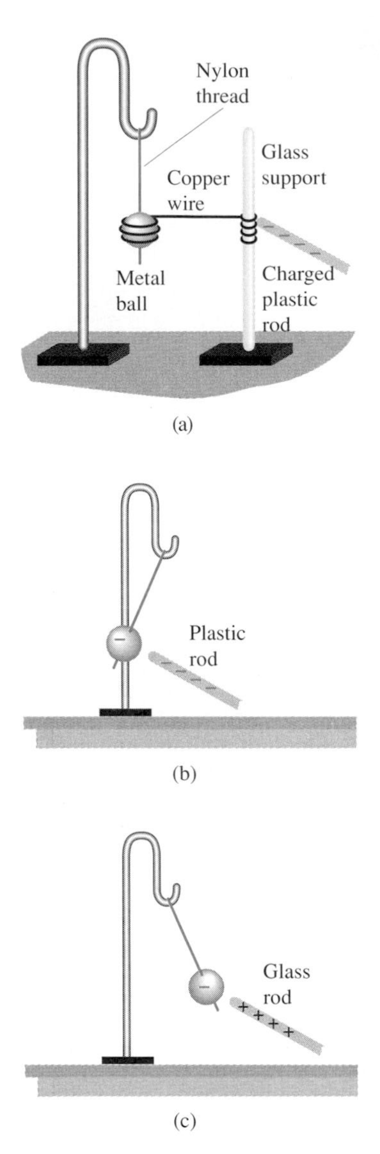


**22–2** Copper is a good conductor of electricity; glass and nylon are good insulators. (a) The wire conducts charge between the metal ball and charged plastic rod to charge the ball negatively. (b) Afterward, the metal ball is repelled by a negatively charged plastic rod, and (c) attracted to a positively charged glass rod.

## 22–2 CONDUCTORS AND INSULATORS

Some materials permit electric charge to move from one region of the material to another; others do not. For example, Fig. 22–2 shows a copper wire supported by a glass rod. Suppose you touch one end of the wire to a charged plastic rod and attach the other end to a metal ball that is initially uncharged. When you bring another charged body up close to the ball, the ball is attracted or repelled, showing that it has become electrically charged. Electric charge has been transferred through the copper wire between the ball and the surface of the plastic rod.

The wire is called a **conductor** of electricity. If you repeat the experiment using a rubber band or nylon thread in place of the wire, you find that *no* charge is transferred to the ball. These materials are called **insulators.** Conductors permit the movement of charge through them; insulators do not. Carpet fibers on a dry day are good insulators and allow charge to build up on us as we walk across the carpet. Coating the fibers with an antistatic layer that does not easily transfer electrons to or from our shoes is one solution to the charge buildup problem; another is to wind some of the fibers around conducting cores.

Most *metals* are good conductors, and most *nonmetals* are insulators. Within a solid metal such as copper, one or more outer electrons in each atom become detached and can move freely throughout the material, just as the molecules of a gas can move through the spaces between the grains in a bucket of sand. The other electrons remain bound to the positively charged nuclei, which themselves are bound in fixed positions within the material. In an insulator there are no, or at most very few, free electrons, and electric charge cannot move freely through the material. Some materials called *semiconductors* are intermediate in their properties between good conductors and good insulators.

We have talked about charging a metal ball by touching it with an electrically charged plastic rod. In this process, some of the excess electrons on the rod are transferred from it to the ball, leaving the rod with a smaller negative charge. There is a different technique in which the plastic rod can give another body a charge of *opposite* sign without losing any of its own charge. This process is called charging by **induction.**

Figure 22–3 shows an example of charging by induction. A metal sphere is supported on an insulating stand. When you bring a negatively charged rod near it, without actually touching it (Fig. 22–3b), the free electrons in the metal sphere are repelled by the excess electrons on the rod, and they shift toward the right, away from the rod. They cannot escape from the sphere because the supporting stand and the surrounding air are insulators. So we get excess negative charge at the right surface of the sphere and excess positive charge (or a deficiency of negative charge) at the left surface. These excess charges are called **induced charges.**

Not all of the free electrons move to the right surface of the sphere. As soon as any induced charge develops, it exerts forces toward the *left* on the other free electrons. These electrons feel a repulsion from the negative induced charge on the right and an attraction toward the positive induced charge on the left. The system reaches an equilibrium state in which the force toward the right on an electron, due to the charged rod, is just balanced by the force toward the left due to the induced charge. If we remove the charged rod, the free electrons shift back to the left, and the original neutral condition is restored.

What happens if, while the plastic rod is nearby, you touch one end of a conducting wire to the right surface of the sphere and the other end to the earth (Fig. 22–3c)? The earth is a conductor, and it is so large that it can act as a practically infinite source of extra electrons or sink of unwanted electrons. Some of the negative charge flows through the wire to the earth. Now suppose you disconnect the wire (Fig. 22–3d) and then remove the

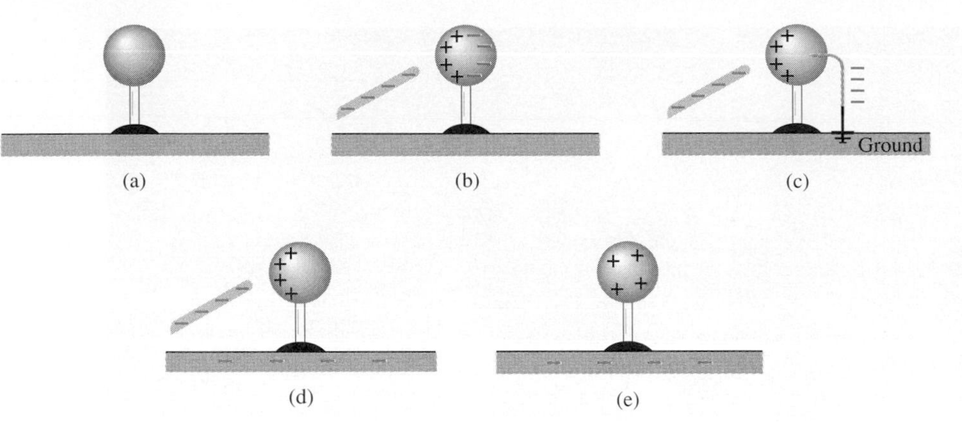


**22–3** Charging a metal sphere by induction.

rod (Fig. 22–3e); a net positive charge is left on the sphere. The charge on the negatively charged rod has not changed during this process. The earth acquires a negative charge that is equal in magnitude to the induced positive charge remaining on the sphere.

Charging by induction would work just as well if the mobile charges in the sphere were positive charges instead of (negatively charged) electrons or even if both positive and negative mobile charges were present. In a metallic conductor the mobile charges are always negative electrons, but it is often convenient to describe a process *as though* the moving charges were positive. In ionic solutions and ionized gases, both positive and negative charges do participate in conduction.

A charged body can exert forces even on objects that are *not* charged themselves. If you rub a balloon on the rug and then hold it against the ceiling, the balloon sticks, even though the ceiling has no net electric charge. After you electrify a comb by running it through your hair, you can pick up uncharged bits of paper on the comb. How is this possible?

This interaction is an induced-charge effect. In Fig. 22–3b the plastic rod exerts a net attractive force on the sphere, even though the total charge on the sphere is zero, because the positive charges are closer to the rod than the negative charges. Even in an insulator the electric charges can shift back and forth a little when there is charge nearby. As Fig. 22–4 shows, a negatively charged plastic comb causes a slight shifting of charge even in insulating materials; this effect is called *polarization*. The positive and negative charges in the material are present in equal amounts, but the positive charges are closer to the plastic rod and so feel a stronger (attractive) force than the (repulsive) forces felt by the negative charges, giving a net attractive force. (We will study the dependence of electric forces on distance in Section 22–4.)

Such forces have many important practical applications, including electrostatic dust precipitators, attracting droplets of sprayed paint to a car body, and attracting toner particles to charged regions of the imaging drum in a copying machine (see the chapter opening figure).

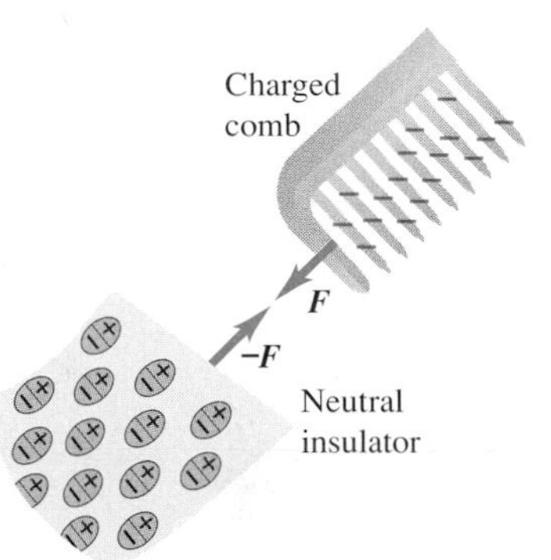


**22–4** The charges within the molecules of an insulating material can shift slightly. As a result, a charged rod or comb attracts the neutral insulator. By Newton's third law the neutral insulator exerts an equal-magnitude attractive force on the charged comb.

## 22–3 CONSERVATION AND QUANTIZATION OF CHARGE

When all is said and done, we can't say what electric charge *is;* we can only describe its properties and its behavior. We *can* say with certainty that electric charge is one of the fundamental attributes of the particles of which matter is made. The interactions responsible for the structure and properties of atoms and molecules, and indeed of all ordinary matter, are primarily *electrical* interactions between electrically charged particles.

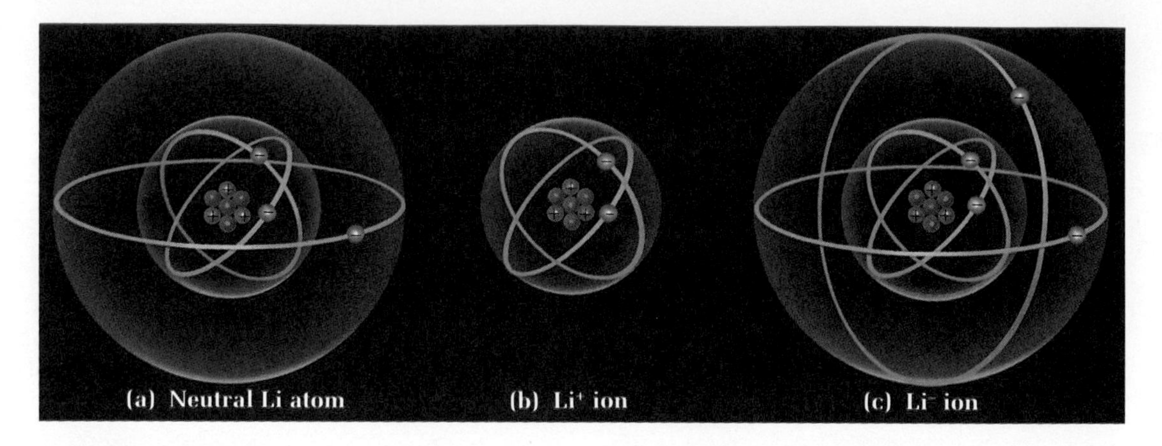


**22–5** (a) The neutral Li atom has three protons in its nucleus and three electrons. (b) A positive ion is made by removing an electron from a neutral atom. (c) A negative ion is made by adding an electron to a neutral atom. (The electron "orbits" are a schematic representation of the actual electron distribution, actually a diffuse cloud many times larger than the nucleus.)

The structure of ordinary matter can be described in terms of three particles: the negatively charged **electron,** the positively charged **proton,** and the uncharged **neutron.** The proton and neutron are combinations of other entities called *quarks,* which have charges of $\pm\frac{1}{3}$ and $\pm\frac{2}{3}$ of the electron charge. Individual quarks have not been observed, and there are theoretical reasons to believe that it is impossible in principle to observe an individual quark.

The protons and neutrons in an atom make up a small, very dense core called the **nucleus,** with dimensions of the order of $10^{-15}$ m. Surrounding the nucleus are the electrons, extending out to distances of the order of $10^{-10}$ m from the nucleus. If an atom were a few miles across, its nucleus would be the size of a tennis ball.

The negative charge of the electron has (within experimental error) *exactly* the same magnitude as the positive charge of the proton. In a neutral atom the number of electrons equals the number of protons in the nucleus, and the net electrical charge (the algebraic sum of all the charges) is exactly zero (Fig. 22–5a). The number of protons or electrons in neutral atoms of any element is called the **atomic number** of the element. If one or more electrons are removed, the remaining positively charged structure is called a **positive ion** (Fig. 22–5b). A **negative ion** is an atom that has *gained* one or more electrons (Fig. 22–5c). This gaining or losing of electrons is called **ionization.**

The masses of the individual particles, to the precision that they are currently known, are

$$\text{Mass of electron} = m_e = 9.1093897(54) \times 10^{-31}\ \text{kg},$$
$$\text{Mass of proton} = m_p = 1.6726231(10) \times 10^{-27}\ \text{kg},$$
$$\text{Mass of neutron} = m_n = 1.6749286(10) \times 10^{-27}\ \text{kg}.$$

The numbers in parentheses are the uncertainties in the last two digits. Note that the masses of the proton and neutron are nearly equal and that the mass of the proton is roughly 2000 times that of the electron. Over 99.9% of the mass of any atom is concentrated in its nucleus.

When the total number of protons in a body equals the total number of electrons, the total charge is zero, and the body as a whole is electrically neutral. To give a body an excess negative charge, we may either *add negative* charges to a neutral body or *remove positive* charges. Similarly, we can create an excess positive charge by either *adding positive* charge or *removing negative* charge. In most cases, negative charges (electrons) are added or removed, and a "positively charged body" is one that has lost some of its normal complement of electrons. When we speak of the charge of a body, we always mean its *net* charge. The net charge is always a very small fraction (typically $10^{-14}$) of the total positive or negative charge in the body.

Implicit in this discussion are two very important principles. First is the principle of **conservation of charge:**

**The algebraic sum of all the electric charges in any closed system is constant.**

Charge can be transferred from one body to another, but it cannot be created or destroyed. Conservation of charge is believed to be a *universal* conservation law; there is no experimental evidence for any violation of this principle. Even in high-energy interactions in which particles are created and destroyed, such as the creation of electron-positron pairs, the total charge of any closed system is exactly constant.

Second, the magnitude of charge of the electron or proton is a natural unit of charge. Every amount of observable electric charge is always an integer multiple of this basic unit. We say that charge is *quantized.* A more familiar example of quantization is money. When you pay cash for an item in a store, you have to do it in one-cent increments. If grapefruit are selling three for a dollar, you can't buy one for $33\frac{1}{3}$ cents; you have to pay 34 cents. Cash can't be divided into smaller amounts than one cent, and electric charge can't be divided into smaller amounts than the charge of one electron or proton. (The quark charges, $\pm\frac{1}{3}$ and $\pm\frac{2}{3}$ of the electron charge, are probably not observable as individual charges.)

The forces that hold atoms together in a molecule or a solid crystal lattice, the adhesive force of glue, the forces associated with surface tension — all of these are *electrical* in nature, arising from the electrical forces between charged particles in the interacting atoms. The structure of atomic *nuclei,* however, requires an attractive interaction that holds the protons and neutrons of stable nuclei together despite the electrical repulsion of the protons. This interaction is called the *nuclear force;* it is an example of a class of interactions called the *strong interaction*. The nuclear force has a short range, of the order of nuclear dimensions, and its effects do not extend far beyond the nucleus.

## 22–4 COULOMB'S LAW

Charles Augustin de Coulomb (1736–1806) studied the interaction forces of charged particles in detail in 1784. He used a torsion balance (Fig. 22–6a) similar to the one used 13 years later by Cavendish to study the (much weaker) gravitational interaction, as we discussed in Section 12–1. For *point charges* (charged bodies that are very small in comparison with the distance $r$ between them), Coulomb found that the electric force is proportional to $1/r^2$. That is, when the distance doubles, the force decreases to $\frac{1}{4}$ of its initial value.

The force also depends on the quantity of charge on each body, which we will denote by $q$ or $Q$. To explore this dependence, Coulomb divided a charge into two equal parts by placing a small charged spherical conductor into contact with an identical but uncharged sphere; by symmetry the charge is shared equally between the two spheres. (Note the essential role of the principle of conservation of charge in this procedure.) Thus he could obtain one-half, one-quarter, and so on, of any initial charge. He found that the forces that two point charges $q_1$ and $q_2$ exert on each other are proportional to each charge and therefore are proportional to the *product* $q_1q_2$ of the two charges.

The magnitude $F$ of the force that each of two point charges $q_1$ and $q_2$ a distance $r$ apart exerts on the other (Fig. 22–6) can be expressed as

$$F = k\frac{|q_1q_2|}{r^2}, \tag{22–1}$$

where $k$ is a proportionality constant whose numerical value depends on the system of units used. Equation (22–1) is the mathematical statement of what we now call

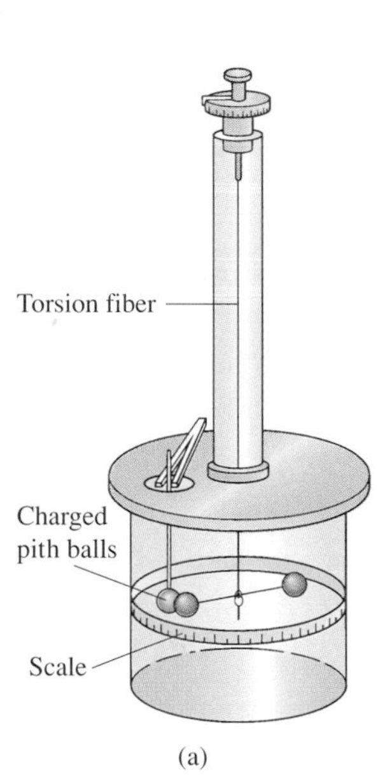


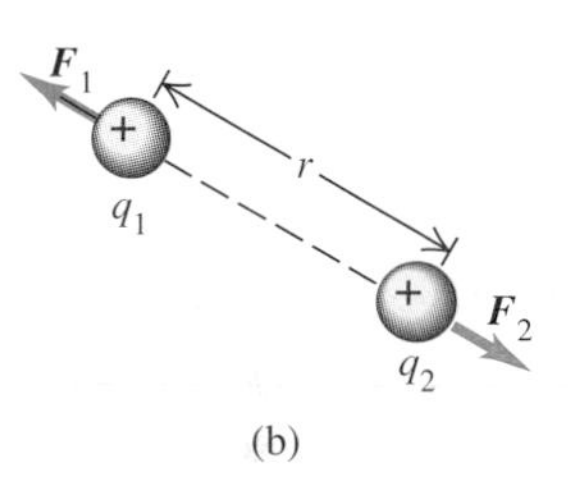


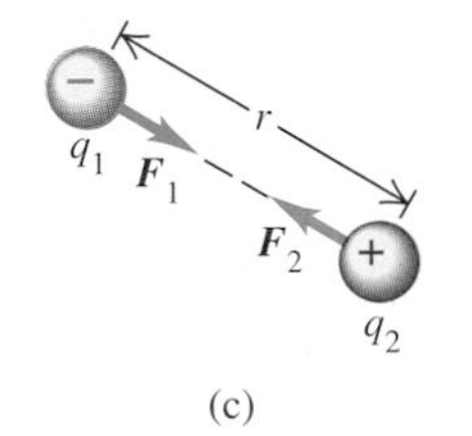


**22–6** (a) A torsion balance of the type used by Coulomb to measure the electric force. (b) Like electric charges repel each other. (c) Unlike electric charges attract each other. In each case the magnitude of the force on each charge is proportional to the product of the charges and is inversely proportional to the square of the distance between them. The forces obey Newton's third law: $\boldsymbol{F}_2 = -\boldsymbol{F}_1$.

**Coulomb's law:**

> **The magnitude of the force of interaction between two point charges is directly proportional to the product of the charges and inversely proportional to the square of the distance between them.**

The directions of the forces that the two charges exert on each other are always along the line joining them. The two forces are always equal in magnitude and opposite in direction, even when the charges are not equal. *The forces obey Newton's third law.*

Let's look at some other properties of the electric force. As we have seen, $q_1$ and $q_2$ can be either positive or negative quantities. When the charges have the same sign (both positive or both negative), the forces are repulsive; when they are unlike, the forces are attractive. We need the absolute value bars in Eq. (22–1) because $F$ is the magnitude of a vector quantity and by definition is always positive, but the product $q_1 q_2$ is negative whenever the two charges have opposite signs.

The proportionality of the electrical force to $1/r^2$ has been verified with great precision. There is no reason to suspect that the exponent is anything different from precisely 2. Thus the form of Eq. (22–1) is the same as that of the law of gravitation, but electrical and gravitational interactions are two distinct classes of phenomena. Electrical interactions depend on electric charges and can be either attractive or repulsive; gravitational interactions depend on mass and are always attractive (because there is no such thing as negative mass).

When two charges exert forces simultaneously on a third charge, the total force acting on that charge is the *vector sum* of the forces that the two charges would exert individually. This important property, called the **principle of superposition,** holds for any number of charges. Coulomb's law as we have stated it describes only the interaction of two *point* charges, but by using the superposition principle we can apply it to *any* collection of charge. Several of the following examples illustrate the superposition principle.

Strictly speaking, Coulomb's law as we have stated it should be used only for point charges *in vacuum.* If matter is present in the space between the charges, the net force acting on each charge is altered because charges are induced in the molecules of the intervening material. We will describe this effect later. As a practical matter, though, we can use Coulomb's law unaltered for point charges in air; at normal atmospheric pressure the presence of air changes the electrical force from its vacuum value by only about 1 part in 2000.

The value of the proportionality constant $k$ in Coulomb's law depends on the system of units used. In the electricity and magnetism chapters of this book we will use SI units exclusively. The SI electrical units include most of the familiar units such as the volt, the ampere, the ohm, and the watt. (There is *no* British system of electrical units.) The SI unit of electric charge is called one **coulomb** (1 C). In this system the constant $k$ in Eq. (22–1) is

$$k = 8.987551787 \times 10^9 \text{ N} \cdot \text{m}^2/\text{C}^2 \cong 8.988 \times 10^9 \text{ N} \cdot \text{m}^2/\text{C}^2.$$

In principle we can measure the interaction force $F$ between two equal charges $q$ at a measured distance $r$ and use Coulomb's law to determine the charge. Thus we could regard this value of $k$ as an operational definition of the coulomb. For reasons of experimental precision it is better to define the coulomb instead in terms of a unit of electric *current* (charge per unit time), the *ampere,* equal to one coulomb per second. We will return to this definition in Chapter 29.

When we study electromagnetic radiation in Chapter 33, we will show that the numerical value of $k$ is closely related to the speed of light. As we discussed in Section 1–3, this speed is *defined* to be exactly

$$c = 2.99792458 \times 10^8 \text{ m/s}.$$

The numerical value of $k$ is defined to be precisely

$$k = (10^{-7}\,\mathrm{N \cdot s^2/C^2})c^2.$$

That's why it is known to such a large number of significant figures. You may want to check this expression to see whether $k$ has the right units.

In SI units the constant $k$ in Eq. (22–1) is usually written not as $k$ but as $1/4\pi\epsilon_0$, where $\epsilon_0$ is another constant. This appears to complicate matters, but it actually simplifies many formulas that we will encounter later. From now on, we will usually write Coulomb's law as

$$F = \frac{1}{4\pi\epsilon_0}\frac{|q_1 q_2|}{r^2}. \qquad (22\text{–}2)$$

The constants in Eq. (22–2) are approximately

$$\frac{1}{4\pi\epsilon_0} = k = 8.988 \times 10^9\,\mathrm{N \cdot m^2/C^2},$$

$$\epsilon_0 = 8.854 \times 10^{-12}\,\mathrm{C^2/N \cdot m^2}.$$

In examples and problems we will often use the approximate value

$$\frac{1}{4\pi\epsilon_0} = 9.0 \times 10^9\,\mathrm{N \cdot m^2/C^2},$$

which is within about 0.1% of the correct value.

As we mentioned in Section 22–3, the most fundamental unit of charge is the magnitude of the charge of an electron or a proton, denoted by $e$. The most precise value available, as of 1990, is

$$e = 1.60217733(49) \times 10^{-19}\,\mathrm{C}.$$

One coulomb represents the negative of the total charge carried by about $6 \times 10^{18}$ electrons. For comparison, the population of the earth is about $5 \times 10^9$ persons, and a cube of copper 1 cm on a side contains about $2.4 \times 10^{24}$ electrons.

In electrostatics problems, charges as large as one coulomb are very unusual. Two charges with magnitude 1 C, at a distance 1 m apart, would exert forces of magnitude $9 \times 10^9$ N (about a million tons) on each other! A more typical range of magnitude is $10^{-9}$ to $10^{-6}$ C. The microcoulomb ($1\,\mu\mathrm{C} = 10^{-6}$ C) and the nanocoulomb ($1\,\mathrm{nC} = 10^{-9}$ C) are often used as practical units of charge. The total charge of all the electrons in a copper penny is about $1.4 \times 10^5$ C. This number shows that we can't disturb electrical neutrality very much without using enormous forces.

## PROBLEM-SOLVING STRATEGY

### Coulomb's law

**1.** As always, consistent units are essential. With the value of $k = 1/4\pi\epsilon_0$ given above, distances *must* be in meters, charge in coulombs, and force in newtons. If you are given distances in centimeters, inches, or furlongs, don't forget to convert! When a charge is given in microcoulombs, remember that $1\,\mu\mathrm{C} = 10^{-6}$ C.

**2.** When the forces acting on a charge are caused by two or more other charges, the total force on the charge is the *vector sum* of the individual forces. Don't forget what you have learned about vector addition; you may want to go back and review the vector algebra in Sections 1–7 and 1–8. It's often useful to use components in an $xy$-coordinate system. Be sure to use correct vector notation; if a symbol represents a vector quantity, underline it or put an arrow over it. If you get sloppy with your notation, you will also get sloppy with your thinking. It is absolutely essential to distinguish between vector quantities and scalar quantities and to treat vectors properly as vectors.

**3.** Some of the examples and problems in this and later chapters involve a continuous distribution of charge along a line or over a surface. In these cases the vector sum described in Step 2 becomes a vector integral, usually carried out by use of components. We divide the total charge distribution into infinitesimal pieces, use Coulomb's law for each piece, and then integrate to find the vector sum. Sometimes this can be done without explicit use of integration.

## EXAMPLE 22–1

If the current in a car's tail-light bulb is $2.8\ \text{A} = 2.8\ \text{C/s}$, how much charge passes through the bulb's filament each hour? How many electrons?

**SOLUTION** The total charge $q$ is the current (charge per unit time) multiplied by the time:

$$q = (2.8\ \text{C/s})(3600\ \text{s}) = 1.0 \times 10^4\ \text{C}.$$

The number of electrons $n$ is given by

$$n = \frac{1.0 \times 10^4\ \text{C}}{1.60 \times 10^{-19}\ \text{C/electron}} = 6.3 \times 10^{22}\ \text{electrons}.$$

## EXAMPLE 22–2

Two point charges are located on the positive $x$-axis of a coordinate system (Fig. 22–7). Charge $q_1 = 2.0\ \text{nC}$ is 2.0 cm from the origin, and charge $q_2 = -3.0\ \text{nC}$ is 4.0 cm from the origin. What is the total force exerted by these two charges on a charge $q_3 = 5.0\ \text{nC}$ located at the origin?

**SOLUTION** The total force on $q_3$ is the vector sum of the forces due to $q_1$ and $q_2$ individually. Converting distance to meters and charge to coulombs, we use Eq. (22–2) to find the magnitude $F_1$ of the force on $q_3$ due to $q_1$:

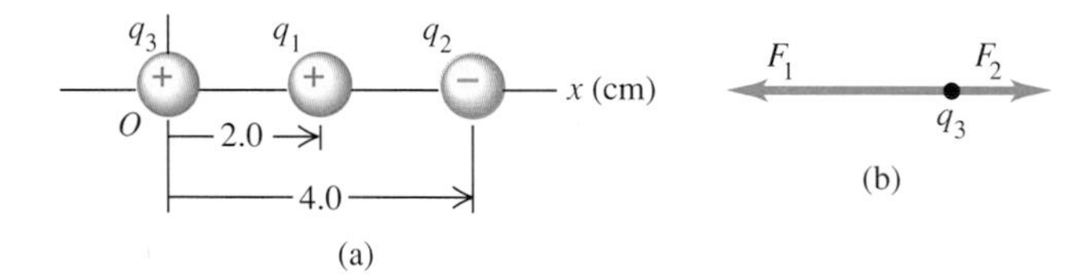


**22–7** What is the total force exerted on the point charge $q_3$ by the other two point charges? (a) The three charges, (b) Free-body diagram for charge $q_3$.

$$F_1 = \frac{1}{4\pi\epsilon_0}\frac{|q_1 q_3|}{r^2}$$

$$= (9.0 \times 10^9\ \text{N}\cdot\text{m}^2/\text{C}^2)\frac{(2.0 \times 10^{-9}\ \text{C})(5.0 \times 10^{-9}\ \text{C})}{(0.020\ \text{m})^2}$$

$$= 2.25 \times 10^{-4}\ \text{N}.$$

This force has a negative $x$-component because $q_3$ is repelled (pushed in the negative $x$-direction) by $q_1$, which has the same sign.

The magnitude $F_2$ of the force due to $q_2$ is

$$F_2 = \frac{1}{4\pi\epsilon_0}\frac{|q_2 q_3|}{r^2}$$

$$= (9.0 \times 10^9\ \text{N}\cdot\text{m}^2/\text{C}^2)\frac{(3.0 \times 10^{-9}\ \text{C})(5.0 \times 10^{-9}\ \text{C})}{(0.040\ \text{m})^2}$$

$$= 0.84 \times 10^{-4}\ \text{N}.$$

This force has a positive $x$-component because $q_3$ is attracted (pulled in the positive $x$-direction) by the opposite charge $q_2$. The sum of the $x$-components is

$$F_x = -2.25 \times 10^{-4}\ \text{N} + 0.84 \times 10^{-4}\ \text{N} = -1.41 \times 10^{-4}\ \text{N}.$$

There are no $y$- or $z$-components. Thus the total force on $q_3$ is directed to the left, with magnitude $1.41 \times 10^{-4}\ \text{N}$.

## EXAMPLE 22–3

An $\alpha$-particle is the nucleus of a helium atom. It has a mass $m$ of $6.64 \times 10^{-27}$ kg and a charge $q$ of $+2e$, or $3.2 \times 10^{-19}$ C. Compare the force of the electrostatic repulsion between two $\alpha$-particles with the force of gravitational attraction between them.

**SOLUTION** The magnitude $F_e$ of the electrostatic force is

$$F_e = \frac{1}{4\pi\epsilon_0}\frac{q^2}{r^2},$$

and the magnitude $F_g$ of the gravitational force is

$$F_g = G\frac{m^2}{r^2}.$$

The ratio of the two magnitudes is

$$\frac{F_e}{F_g} = \frac{1}{4\pi\epsilon_0 G}\frac{q^2}{m^2} = \frac{9.0 \times 10^9\ \text{N}\cdot\text{m}^2/\text{C}^2}{6.67 \times 10^{-11}\ \text{N}\cdot\text{m}^2/\text{kg}^2}\frac{(3.2 \times 10^{-19}\ \text{C})^2}{(6.64 \times 10^{-27}\ \text{kg})^2}$$

$$= 3.1 \times 10^{35}.$$

This astonishingly large number shows that the gravitational force is completely negligible in comparison to the electrostatic force on a charged body. This is always true for interactions of atomic and subatomic particles. But for objects the size of the earth, the positive and negative charges are nearly equal, and the net electrical interactions are usually much *smaller* than the gravitational.

### EXAMPLE 22–4

**Vector addition of electric forces** In Fig. 22–8, two equal positive point charges $q = 2.0\ \mu\text{C}$ interact with a third point charge $Q = 4.0\ \mu\text{C}$. Find the magnitude and direction of the total (resultant) force on $Q$.

**SOLUTION** The key word is *total*. We have to compute the force each charge exerts on $Q$ and then find the *vector sum* of the forces. The easiest way to do this is to use components. The figure shows the force on $Q$ due to the upper charge $q$. From Coulomb's law the magnitude $F$ of this force is

$$F = 9.0 \times 10^9\ \text{N}\cdot\text{m}^2/\text{C}^2 \frac{(4.0 \times 10^{-6}\ \text{C})(2.0 \times 10^{-6}\ \text{C})}{(0.50\ \text{m})^2}$$
$$= 0.29\ \text{N}.$$

The components of this force are given by

$$F_x = F\cos\alpha = (0.29\ \text{N})\frac{0.40\ \text{m}}{0.50\ \text{m}} = 0.23\ \text{N},$$

$$F_y = -F\sin\alpha = -(0.29\ \text{N})\frac{0.30\ \text{m}}{0.50\ \text{m}} = -0.17\ \text{N}.$$

The lower charge $q$ exerts a force with the same magnitude but a different direction. From symmetry we see that its $x$-component is the same as that due to the upper charge, but its $y$-component has the opposite direction. So we find

$$F_x = 0.23\ \text{N} + 0.23\ \text{N} = 0.46\ \text{N},$$
$$F_y = -0.17\ \text{N} + 0.17\ \text{N} = 0.$$

The total force on $Q$ is in the $+x$-direction, with magnitude 0.46 N. How would this solution differ if the lower charge were *negative?*

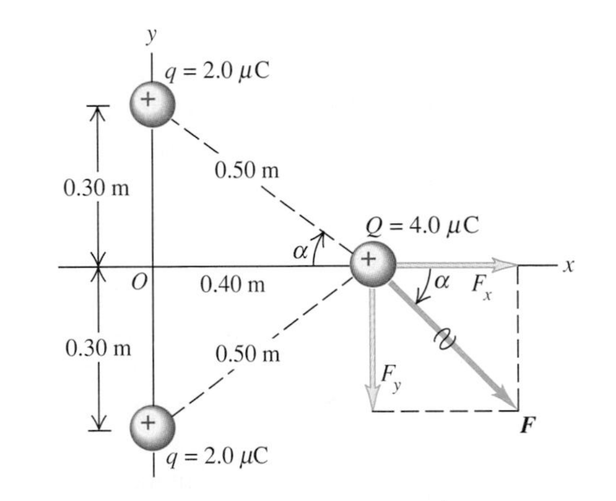


**22–8** $F$ is the force on $Q$ due to the upper charge $q$.

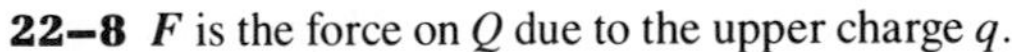

## 22–5 ELECTRIC FIELD AND ELECTRIC FORCES

When two electrically charged particles in empty space interact, how does each one know that the other is there? What goes on in the space between them to communicate the effect of each one to the other? We can begin to answer these questions, and at the same time reformulate Coulomb's law in a very useful way, by using the concept of *electric field*. This concept is directly analogous to the concept of *gravitational field,* which we introduced in Section 12–3. We suggest that you review that section now to help you understand what comes next.

To introduce the concept of electric field, let's look at the mutual repulsion of two positively charged bodies $A$ and $B$ (Fig. 22–9a). Suppose $B$ is a point charge $q'$, and let $\boldsymbol{F}'$ be the force on $B$, as shown in the figure. One way to think about this force is as an "action-at-a-distance" force, that is, as a force that acts across empty space without needing any matter (such as a push rod or a rope) to transmit it through the intervening space.

Now think of body $A$ as having the effect of somehow modifying the properties of the space around it. We remove body $B$ and label its former position as point $P$ (Fig. 22–9b). We say that the charged body $A$ produces or causes an **electric field** at point $P$ (and at all other points in the neighborhood). Then when point charge $B$ is placed at point $P$ and experiences the force $\boldsymbol{F}'$, we take the point of view that the force is exerted on $B$ *by the field* at $P$ (Fig. 22–9c). Because $B$ would experience a force at *any* point in the neighborhood of $A$, the electric field exists at all points in the region around $A$. (We can also say that point charge $B$ sets up an electric field, which in turn exerts a force on body $A$.)

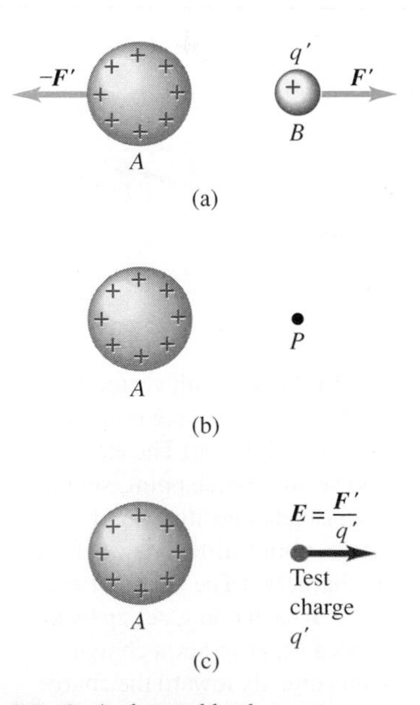


**22–9** A charged body creates an electric field in the space around it.

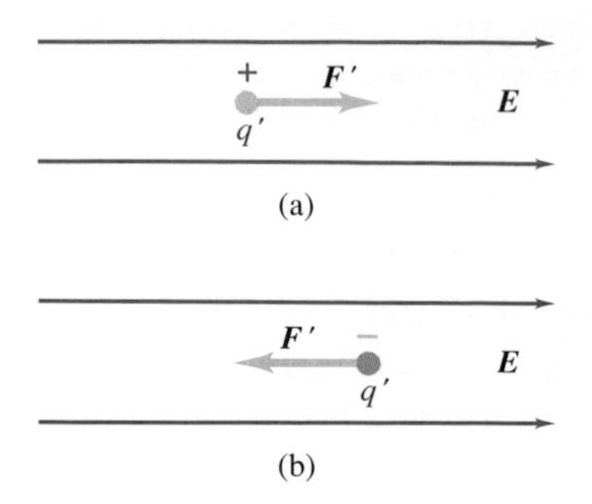


**22–10** The force $\boldsymbol{F}'$ exerted on a charge $q'$ by an electric field $\boldsymbol{E}$. (a) If $q'$ is positive, then $\boldsymbol{F}'$ and $\boldsymbol{E}$ are in the same direction. (b) If $q'$ is negative, $\boldsymbol{F}'$ and $\boldsymbol{E}$ are in opposite directions.

To find out experimentally whether there is an electric field at a particular point, we place a charged body, which we call a **test charge,** at the point. If the test charge experiences an electric force, then there is an electric field at that point.

Force is a vector quantity, so electric field is also a vector quantity. (Note the use of boldface letters and plus, minus, and equals signs in the following discussion.) To define the *electric field* **E** at any point, we place a test charge $q'$ at the point and measure the electrical force $\boldsymbol{F}'$ on it (Fig. 22–9c). We define $\boldsymbol{E}$ at this point to be equal to $\boldsymbol{F}'$ divided by $q'$:

$$\boldsymbol{E} = \frac{\boldsymbol{F}'}{q'}$$

or (22–3)

$$\boldsymbol{F}' = q'\boldsymbol{E}.$$

The test charge $q'$ can be either positive or negative. If it is positive, the directions of $\boldsymbol{E}$ and $\boldsymbol{F}'$ are the same; if it is *negative,* they are opposite (Fig. 22–10).

In SI units, in which the unit of force is 1 N and the unit of charge is 1 C, the unit of electric field magnitude is one newton per coulomb (1 N/C).

The force experienced by the test charge $q'$ varies from point to point, so the electric field is also different at different points. Be sure that you understand that $\boldsymbol{E}$ is not a single vector quantity but an infinite set of vector quantities, one associated with each point in space. This is an example of a **vector field,** as is the gravitational field mentioned above. If we use a rectangular ($xyz$) coordinate system, each component of $\boldsymbol{E}$ at any point is in general a function of the coordinates $(x, y, z)$ of the point. We can represent the functions as $E_x(x, y, z)$, $E_y(x, y, x)$, and $E_z(x, y, z)$. Vector fields are an important part of the language of physics, particularly in electricity and magnetism.

There is a slight difficulty with our definition of electric field: In Fig. 22–9 the force exerted by the test charge $q'$ on the charge distribution $A$ may cause this distribution to shift around, especially if the body is a conductor, where charge is free to move. So the electric field around $A$ when $q'$ is present may not be the same as when $q'$ is absent. But if $q'$ is very small, the redistribution of charge on body $A$ is also very small. So we refine our definition of electric field by taking the limit of Eq. (22–3) as the test charge $q'$ approaches zero and its disturbing effect on the charge distribution becomes negligible:

$$\boldsymbol{E} = \lim_{q'\to 0} \frac{\boldsymbol{F}'}{q'}. \tag{22–4}$$

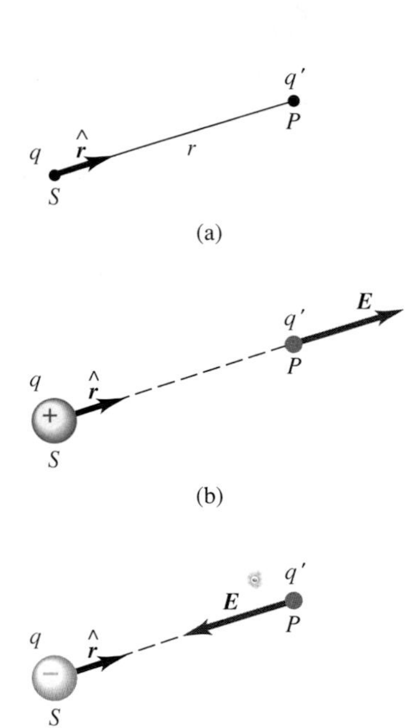


**22–11** (a) The unit vector $\hat{\boldsymbol{r}}$ points from the source point $S$ to the field point $P$. (b) The electric field vector at each point, set up by an isolated positive point charge, points directly away from the charge. (c) The electric field vector at each point, set up by an isolated negative point charge, points directly toward the charge.

If an electric field exists within a *conductor,* the field exerts a force on every charge in the conductor, causing the free charges to move. By definition an *electrostatic* situation is one in which the charges *do not* move. We conclude that *in electrostatics the electric field at every point within the material of a conductor must be zero.* (Note that we are *not* saying that the field is necessarily zero in a hole inside a conductor.)

In general the magnitude and direction of an electric field can vary from point to point. If in a particular situation the magnitude and direction of the field are *constant* throughout a certain region, we say that the field is *uniform* in this region.

If the source distribution is a point charge $q$, it is easy to find the electric field that it produces. We call the location of the charge the **source point** $S$ and the point $P$ where we are determining the field the **field point.** It is also useful to introduce a *unit vector* $\hat{\boldsymbol{r}}$ that points along the line from source point to field point (Fig. 22–11). If we place a small test charge $q'$ at the field point $P$, at a distance $r$ from the source point, the force $\boldsymbol{F}'$ is given by Coulomb's law, Eq. (22–1):

$$F' = \frac{1}{4\pi\epsilon_0}\frac{|qq'|}{r^2}.$$

From Eq. (22–4) the magnitude $E$ of the electric field at $P$ is

$$E = \frac{1}{4\pi\epsilon_0}\frac{|q|}{r^2}. \quad (22\text{–}5)$$

Using the unit vector $\hat{\mathbf{r}}$, we can write a *vector* equation that gives both the magnitude and direction of the electric field $\mathbf{E}$:

$$\mathbf{E} = \frac{1}{4\pi\epsilon_0}\frac{q}{r^2}\hat{\mathbf{r}}. \quad (22\text{–}6)$$

By definition the electric field of a point charge always points *away from* a positive charge but *toward* a negative charge.

## EXAMPLE 22–5

What is the electric field 30 cm from a charge $q = 4.0\ \text{nC}$?

**SOLUTION** From Eq. (22–5),

$$E = \frac{1}{4\pi\epsilon_0}\frac{q}{r^2} = (9.0 \times 10^9\ \text{N}\cdot\text{m}^2/\text{C}^2)\frac{4.0 \times 10^{-9}\ \text{C}}{(0.30\ \text{m})^2} = 400\ \text{N/C}.$$

Alternatively, we first use Coulomb's law to find the magnitude $F'$ of the force on test charge $q'$ 30 cm from $q$:

$$F' = (9.0 \times 10^9\ \text{N}\cdot\text{m}^2/\text{C}^2)\frac{(4.0 \times 10^{-9}\ \text{C})(q')}{(0.30\ \text{m})^2}.$$

$$= (400\ \text{N/C})q'.$$

Then from Eq. (22–4) the magnitude of $\mathbf{E}$ is

$$E = \frac{F'}{q'} = 400\ \text{N/C}.$$

The *direction* of $\mathbf{E}$ at this point is along the line from $q$ toward $q'$, as Fig. 22–11 shows. The sign of $q'$ doesn't matter. Do you see why?

With the concept of electric field our description of electrical interactions has two parts. First, a given charge distribution acts as a *source* of electric field. Second, the electric field exerts a force on any charge that is present in the field. Our analysis often has two corresponding steps: first, calculating the field caused by a source charge distribution, and second, looking at the effect of the field in terms of force and motion. The second step often involves Newton's laws as well as the electrostatic principles. In the next section we show how to calculate fields caused by various source distributions, but first here are some examples of the second-step calculations.

## EXAMPLE 22–6

When the terminals of a 100-V battery are connected to two large parallel horizontal plates 1.0 cm apart, the resulting charges on the plates cause an electric field $\mathbf{E}$ in the region between the plates that is very nearly uniform and has magnitude $E = 10^4\ \text{N/C}$. Suppose the direction of $\mathbf{E}$ is vertically upward, as shown by the vectors in Fig. 22–12 (on page 618). Compute the electrical force on an electron in this field and compare it with the gravitational force on the electron (its weight on earth).

**SOLUTION** We need the following data from this chapter:

Magnitude of electron charge $e = 1.60 \times 10^{-19}$ C,

Electron mass $m = 9.11 \times 10^{-31}$ kg.

From Eq. (22–3),

$$F_{\text{elec}} = eE = (1.60 \times 10^{-19}\ \text{C})(10^4\ \text{N/C}) = 1.60 \times 10^{-15}\ \text{N};$$

$$F_{\text{grav}} = mg = (9.11 \times 10^{-31}\ \text{kg})(9.8\ \text{m/s}^2) = 8.93 \times 10^{-30}\ \text{N}.$$

The ratio of the electrical to the gravitational force is

$$\frac{F_{\text{elec}}}{F_{\text{grav}}} = \frac{1.60 \times 10^{-15}\ \text{N}}{8.93 \times 10^{-30}\ \text{N}} = 1.8 \times 10^{14}.$$

The gravitational force is *extremely* small in comparison to the electrical force, and we neglect the electron's weight in the next two examples.

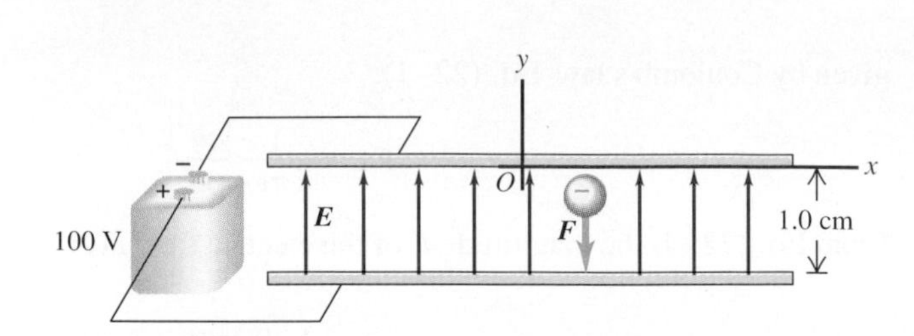


**22–12** A uniform electric field between two parallel conducting plates connected to a 100-V battery. (The separation of the plates is exaggerated in this figure relative to the dimensions of the plates.)

## EXAMPLE 22–7

**Electron in a uniform field** If the electron of Example 22–6 is released from rest at the upper plate, what speed does it acquire while traveling 1.0 cm to the lower plate? What is its kinetic energy after 1.0 cm? How much time is required for it to travel this distance?

**SOLUTION** Note that $\boldsymbol{E}$ is upward but $\boldsymbol{F}$ is downward because the charge is negative. Thus $F_y$ is negative. Because $F_y$ is constant, the electron moves with constant acceleration $a_y$ given by

$$a_y = \frac{F_y}{m} = \frac{-eE}{m} = \frac{-1.60 \times 10^{-15}\ \text{N}}{9.11 \times 10^{-31}\ \text{kg}}$$
$$= -1.76 \times 10^{15}\ \text{m/s}^2.$$

This is an enormous acceleration; to give a 1000-kg car this acceleration, we would need a force of about $2 \times 10^{18}$ N, or about $2 \times 10^{14}$ tons.

A coordinate system is shown in Fig. 22–12. We can find the electron's speed at any position from one of the constant-acceleration formulas, Eq. (2–14): $v^2 = v_0^2 + 2a_y(y - y_0)$. In our case, $v_0 = 0$ and $y_0 = 0$, so the speed $v$ when $y = -1.0$ cm $= -1.0 \times 10^{-2}$ m is

$$v = \sqrt{2a_y y} = \sqrt{2(-1.76 \times 10^{15}\ \text{m/s}^2)(-1.0 \times 10^{-2}\ \text{m})}$$
$$= 5.9 \times 10^6\ \text{m/s}.$$

The velocity is downward, so its $y$-component is $v_y = -5.9 \times 10^6$ m/s. The electron's kinetic energy is

$$\tfrac{1}{2}mv^2 = \tfrac{1}{2}(9.11 \times 10^{-31}\ \text{kg})(5.9 \times 10^6\ \text{m/s})^2$$
$$= 1.6 \times 10^{-17}\ \text{J}.$$

The time required is

$$t = \frac{v_y}{a_y} = \frac{-5.9 \times 10^6\ \text{m/s}}{-1.76 \times 10^{15}\ \text{m/s}^2}$$
$$= 3.4 \times 10^{-9}\ \text{s}.$$

## EXAMPLE 22–8

**An electron trajectory** If we launch an electron into the electric field of Example 22–6 with an initial horizontal velocity $v_0$ (Fig. 22–13), what is the equation of its trajectory?

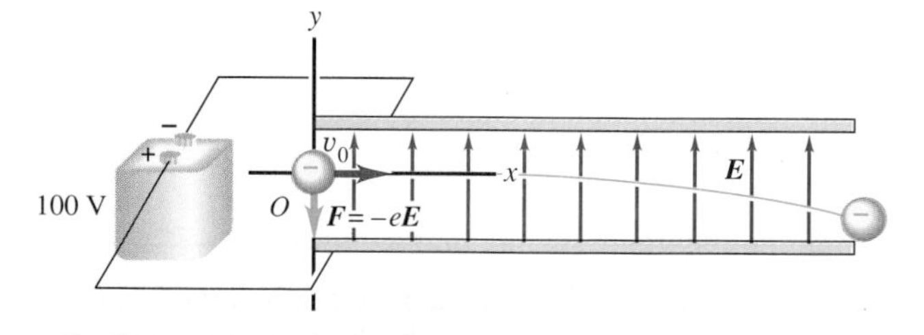


**22–13** The parabolic trajectory of an electron in a uniform electric field.

**SOLUTION** The force and acceleration are the same as in Example 22–7. The field is upward, but the force on the (negatively charged) electron is downward. The $x$-acceleration is zero; the $y$-acceleration is $-(eE/m)$. At time $t$,

$$x = v_0 t,$$

$$y = \tfrac{1}{2}a_y t^2 = -\tfrac{1}{2}\frac{eE}{m}t^2.$$

Eliminating $t$ between these equations, we get

$$y = -\tfrac{1}{2}\frac{eE}{mv_0^2}x^2.$$

This is the equation of a parabola, just like the trajectory of a projectile launched horizontally in the earth's gravitational field (discussed in Section 3–3). In Section 24–7 we will see how electric fields are used to control electron beams in cathode-ray oscilloscopes. A similar scheme is used in inkjet printers to control the direction of a stream of ink droplets from a nozzle (Fig. 22–14).

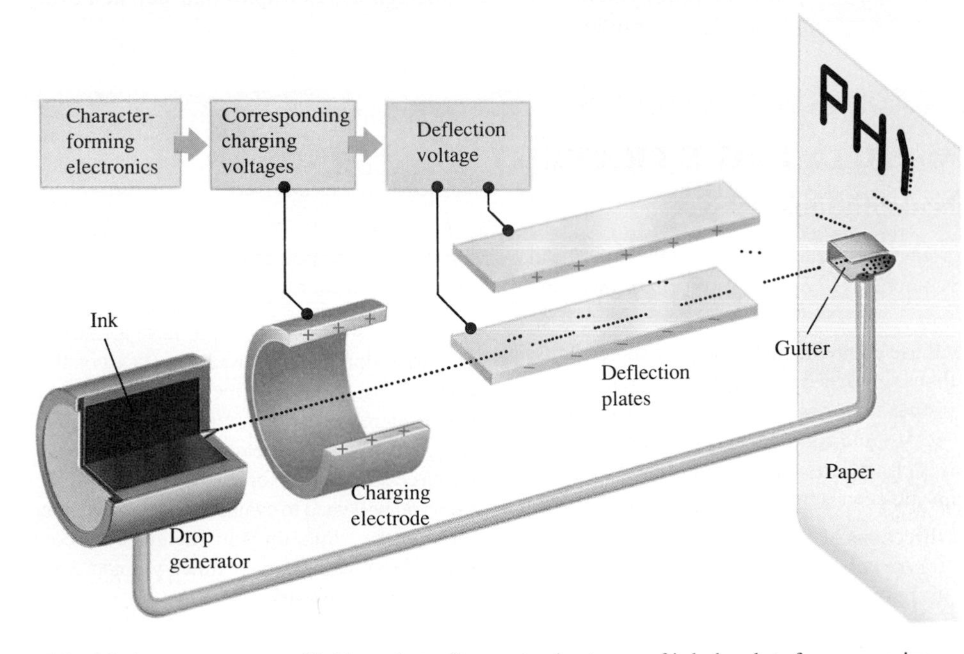


**22–14** A common type of inkjet printer fires a steady stream of ink droplets from a moving nozzle at a piece of paper. Each droplet (about $10^{-5}$ m across) passes through a computer-controlled charging electrode and a pair of deflecting plates. If a dot of ink is to be placed on the paper, the charging electrode applies the appropriate voltage, and the ink droplet is deflected onto the paper. If the paper is to remain blank, the charging electrode shuts off, and the ink droplet shoots straight through the deflection plates into a gutter and back to the ink supply. ■

## 22–6 ELECTRIC-FIELD CALCULATIONS

In this section we'll learn to calculate electric fields caused by various distributions of electric charge. Equation (22–6) gives the field caused by a *point-charge* source. To find the field caused by an *extended* source distribution, we imagine the source to be made up of many point charges $q_1, q_2, q_3, \ldots$. Then we calculate the fields $\boldsymbol{E}_1, \boldsymbol{E}_2, \boldsymbol{E}_3, \ldots$ caused by the individual point charges and take their vector sum (using the superposition principle) to find the total field $\boldsymbol{E}$ at a point $P$. That is,

$$\boldsymbol{E} = \boldsymbol{E}_1 + \boldsymbol{E}_2 + \boldsymbol{E}_3 + \cdots.$$

When charge is distributed along a line, over a surface, or through a volume, a few additional terms are useful. For a line charge distribution we use $\lambda$ to represent the **linear charge density** (charge per unit length, measured in C/m). When charge is distributed over a surface, $\sigma$ represents the **surface charge density** (charge per unit area, measured in $\mathrm{C/m^2}$); and when charge is distributed through a volume, $\rho$ is the **volume charge density** (charge per unit volume, measured in $\mathrm{C/m^3}$).

Some of the calculations in the following examples may look fairly intricate; in electric-field calculations a certain amount of mathematical complexity is in the nature of things. After you have worked through the examples one step at a time, the process will seem less formidable.

## PROBLEM-SOLVING STRATEGY

### Electric-field calculations

**1.** Be sure to use a consistent set of units. Distances must be in meters, charge in coulombs. If you are given cm or nC, don't forget to convert.

**2.** Usually, you will use components to compute vector sums. Use the methods that you learned in Chapter 1; review them if you need to. Use proper vector notation; distinguish carefully between scalars, vectors, and components of vectors. Indicate your coordinate axes clearly on your diagram, and be certain that the components are consistent with your choice of axes.

**3.** In working out directions of $\boldsymbol{E}$ vectors, be careful to distinguish between the *source point S* and the *field point P.* The field produced by a positive point charge always points in the direction from source point to field point; the opposite is true for a negative point charge.

**4.** In some situations you will have a continuous distribution of charge along a line, over a surface, or through a volume. Then you will have to define a small element of charge that can be considered as a point, find its electric field at point $P$, and find a way to add the fields of all the charge elements. Usually, it is easiest to do this for each component of $\boldsymbol{E}$ separately, and often you will need to evaluate one or more integrals. Make certain that the limits on your integrals are correct. Especially when the situation has symmetry, make sure that you don't count the charge twice.

## EXAMPLE 22–9

**Field of an electric dipole** Point charges $q_1$ and $q_2$ of $+12$ nC and $-12$ nC, respectively, are placed 0.10 m apart (Fig. 22–15). This combination of two charges with equal magnitude and opposite sign is called an *electric dipole;* we'll study dipoles in more detail in Section 22–8. Compute the electric fields caused by these charges at points *a*, *b*, and *c*.

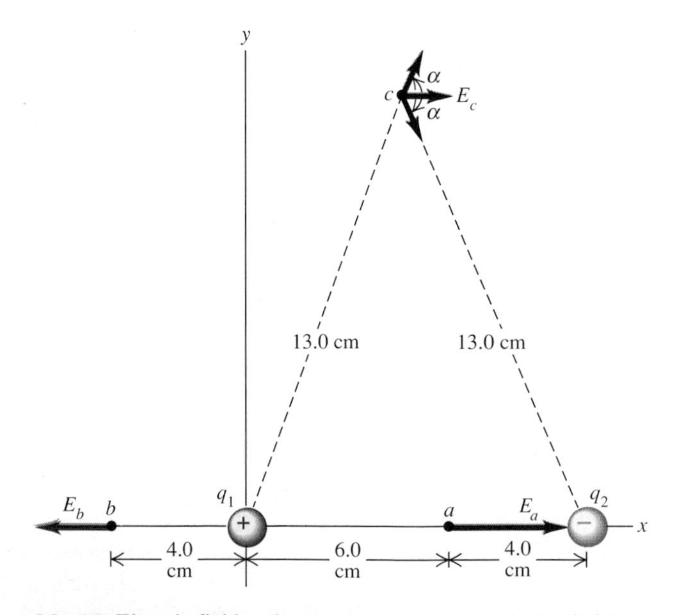


**22–15** Electric field at three points, *a*, *b*, and *c*, in the field set up by charges $q_1$ and $q_2$, which form an electric dipole.

**SOLUTION** At point *a* the field $\boldsymbol{E}_1$ caused by the positive charge $q_1$ is directed toward the right; its magnitude is

$$E_1 = \frac{1}{4\pi\epsilon_0}\frac{|q|}{r^2}$$
$$= (9.0 \times 10^9\ \mathrm{N\cdot m^2/C^2})\frac{12 \times 10^{-9}\ \mathrm{C}}{(0.060\ \mathrm{m})^2}$$
$$= 3.0 \times 10^4\ \mathrm{N/C}.$$

The components of $\boldsymbol{E}_1$ are

$$E_{1x} = 3.0 \times 10^4\ \mathrm{N/C}, \qquad E_{1y} = 0.$$

The field caused by the negative charge $q_2$ is also directed toward the right; its magnitude is

$$E_2 = (9.0 \times 10^9\ \mathrm{N\cdot m^2/C^2})\frac{12 \times 10^{-9}\ \mathrm{C}}{(0.040\ \mathrm{m})^2}$$
$$= 6.8 \times 10^4\ \mathrm{N/C}.$$

The components of $\boldsymbol{E}_2$ are

$$E_{2x} = 6.8 \times 10^4\ \mathrm{N/C}, \qquad E_{2y} = 0.$$

So, at point *a*,

$$(E_a)_x = (3.0 + 6.8) \times 10^4\ \mathrm{N/C}, \qquad (E_a)_y = 0,$$

and

$$E_a = 9.8 \times 10^4\ \mathrm{N/C} \quad \text{toward the right},$$

or

$$\boldsymbol{E}_a = (9.8 \times 10^4\ \mathrm{N/C})\boldsymbol{i}.$$

At point $b$ the field due to $q_1$ is directed toward the left, with magnitude

$$E_1 = \frac{1}{4\pi\epsilon_0}\frac{q}{r^2} = (9.0 \times 10^9\ \text{N}\cdot\text{m}^2/\text{C}^2)\frac{12 \times 10^{-9}\ \text{C}}{(0.040\ \text{m})^2}$$
$$= 6.8 \times 10^4\ \text{N/C}.$$

Its components are

$$E_{1x} = -6.8 \times 10^4\ \text{N/C}, \qquad E_{1y} = 0.$$

The field due to $q_2$ is directed toward the right, with magnitude

$$E_2 = (9.0 \times 10^9\ \text{N}\cdot\text{m}^2/\text{C}^2)\frac{12 \times 10^{-9}\ \text{C}}{(0.140\ \text{m})^2} = 0.55 \times 10^4\ \text{N/C}.$$

Its components are

$$E_{2x} = 0.55 \times 10^4\ \text{N/C}, \qquad E_{2y} = 0.$$

Thus, at point $b$,

$$(E_b)_x = (-6.8 + 0.55) \times 10^4\ \text{N/C}, \qquad (E_b)_y = 0,$$

and

$$E_b = 6.2 \times 10^4\ \text{N/C} \quad \text{toward the left,}$$

or

$$\mathbf{E}_b = (-6.2 \times 10^4\ \text{N/C})\mathbf{i}.$$

At point $c$ the magnitude of each vector is

$$E_1 = \frac{1}{4\pi\epsilon_0}\frac{q}{r^2} = (9.0 \times 10^9\ \text{N}\cdot\text{m}^2/\text{C}^2)\frac{12 \times 10^{-9}\ \text{C}}{(0.13\ \text{m})^2}$$
$$= 6.39 \times 10^3\ \text{N/C}.$$

The directions of these vectors are shown in the figure. The $x$-component of each is

$$E_x = E\cos\alpha = (6.39 \times 10^3\ \text{N/C})(\tfrac{5}{13}) = 2.46 \times 10^3\ \text{N/C}.$$

From symmetry the $y$-components add to zero. We find

$$(E_c)_x = 2(2.46 \times 10^3\ \text{N/C}) = 4.9 \times 10^3\ \text{N/C},$$
$$(E_c)_y = 0,$$

and

$$E_c = 4.9 \times 10^3\ \text{N/C} \quad \text{toward the right,}$$

or

$$\mathbf{E}_c = (4.9 \times 10^3\ \text{N/C})\mathbf{i}.$$

Does it surprise you that the field at this point is parallel to the line between the two charges?

## EXAMPLE 22–10

**Field of a ring of charge** A ring-shaped conductor with radius $a$ carries a total charge $Q$, uniformly distributed around it (Fig. 22–16). Find the electric field at a point $P$ that lies on the axis of the ring at a distance $x$ from its center.

**SOLUTION** Note the close resemblance of this problem to the gravitational-field calculation in Example 12–6 (Section 12–3). We imagine the ring divided into small segments $ds$, as shown. Let $dQ$ be the charge of segment $ds$, as shown in the figure. The square of the distance $r$ from this charge to the field point $P$ is $r^2 = x^2 + a^2$. The magnitude $dE$ of this charge's contribution $d\mathbf{E}$ to the electric field at the field point $P$ is given by

$$dE = \frac{1}{4\pi\epsilon_0}\frac{dQ}{x^2 + a^2}.$$

The component $dE_x$ of this field along the $x$-axis is

$$dE_x = dE\cos\alpha = \frac{1}{4\pi\epsilon_0}\frac{dQ}{x^2 + a^2}\frac{x}{\sqrt{x^2 + a^2}} = \frac{1}{4\pi\epsilon_0}\frac{x\,dQ}{(x^2 + a^2)^{3/2}}.$$

To find the *total* $x$-component $E_x$ of the field at $P$, we integrate this expression:

$$E_x = \int \frac{1}{4\pi\epsilon_0}\frac{x\,dQ}{(x^2 + a^2)^{3/2}}.$$

Since $x$ does not vary as we move from point to point around the ring, all the factors on the right side except $dQ$ are constant and can be taken outside the integral. The integral of $dQ$ is just the total charge $Q$, and we finally get

$$E_x = \frac{1}{4\pi\epsilon_0}\frac{Qx}{(x^2 + a^2)^{3/2}}. \qquad (22\text{–}7)$$

We can see from symmetry that there can't be any component of $\mathbf{E}$ perpendicular to the $x$-axis, so the field at $P$ is described completely by its $x$-component.

Equation (22–7) shows that at the center of the ring ($x = 0$) the field is zero. We should expect this; charges on opposite sides of the ring would push in opposite directions on a test charge at this point, and the forces would add to zero. When the field point $P$ is much farther from the ring than its size (that is, $x >> a$), the denominator in Eq. (22–7) becomes approximately equal to $x^3$, and the expression becomes approximately

$$E_x = \frac{1}{4\pi\epsilon_0}\frac{Q}{x^2}.$$

When we are so far from the ring that its size $a$ is negligible in comparison to the distance $x$, it looks like a point charge. Again, this is just what we should expect.

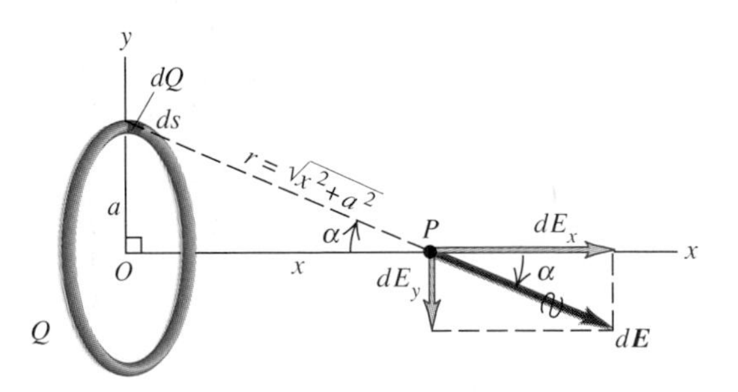


**22–16** Calculating the electric field on the axis of a ring of charge.

## EXAMPLE 22–11

**Field of a line of charge** Electric charge $Q$ is distributed uniformly along a line with length $2a$, lying along the $y$-axis (Fig. 22–17). Find the electric field at point $P$ on the $x$-axis at a distance $x$ from the origin.

**SOLUTION** We divide the line charge into infinitesimal segments; let the length of a typical segment at height $y$ be $dy$, as shown. To find the charge $dQ$ in this segment, note that if the charge is distributed uniformly, the ratio of $dQ$ to the total charge $Q$ is equal to the ratio of $dy$ to the total length $2a$. Thus

$$\frac{dQ}{Q} = \frac{dy}{2a}$$

and

$$dQ = \frac{Q\,dy}{2a}.$$

The distance $r$ from this segment to $P$ is $(x^2 + y^2)^{1/2}$, so the magnitude of field $dE$ at $P$ due to this segment is

$$dE = \frac{Q}{4\pi\epsilon_0}\frac{dy}{2a(x^2 + y^2)}.$$

We represent this field in terms of its $x$- and $y$-components:

$$dE_x = dE\cos\alpha, \qquad dE_y = -dE\sin\alpha.$$

We note that

$$\sin\alpha = \frac{y}{\sqrt{x^2 + y^2}}, \qquad \cos\alpha = \frac{x}{\sqrt{x^2 + y^2}}.$$

Combining these with the expression for $dE$, we find

$$dE_x = \frac{Q}{4\pi\epsilon_0}\frac{x\,dy}{2a(x^2 + y^2)^{3/2}},$$

$$dE_y = -\frac{Q}{4\pi\epsilon_0}\frac{y\,dy}{2a(x^2 + y^2)^{3/2}}.$$

To find the total field components $E_x$ and $E_y$, we integrate these expressions, noting that to include all of $Q$ we must integrate from $y = -a$ to $y = +a$. We invite you to work out the details of the integration; an integral table is helpful. The final results are

$$E_x = \frac{1}{4\pi\epsilon_0}\frac{Qx}{2a}\int_{-a}^{a}\frac{dy}{(x^2 + y^2)^{3/2}} = \frac{Q}{4\pi\epsilon_0}\frac{1}{x\sqrt{x^2 + a^2}},$$

$$E_y = -\frac{1}{4\pi\epsilon_0}\frac{Q}{2a}\int_{-a}^{a}\frac{y\,dy}{(x^2 + y^2)^{3/2}} = 0,$$

or, in vector form,

$$\boldsymbol{E} = \frac{Q}{4\pi\epsilon_0}\frac{1}{x\sqrt{x^2 + a^2}}\boldsymbol{i}. \qquad (22\text{–}8)$$

We could have guessed from symmetry that $E_y$ would be zero; if there is a positive charge at $P$, the upper half of $Q$ pushes downward on it, and the lower half pushes up with equal magnitude. We also note that when $x$ is much larger than $a$, we can neglect $a$ in the denominator, and our result becomes approximately

$$\boldsymbol{E} = \frac{Q}{4\pi\epsilon_0}\frac{1}{x^2}\boldsymbol{i}.$$

This means that if the line charge $Q$ is far away from point $P$ in comparison to its size, it looks like a point.

We get an added dividend from Eq. (22–8) if we express it in terms of the linear charge density $\lambda = Q/2a$. Substituting $Q = 2a\lambda$ into Eq. (22–8) and simplifying, we get

$$\boldsymbol{E} = \frac{1}{2\pi\epsilon_0}\frac{\lambda}{x\sqrt{(x^2/a^2) + 1}}\boldsymbol{i}. \qquad (22\text{–}9)$$

Now what happens if we make the line of charge longer and longer, adding charge in proportion to the total length so that $\lambda$, the charge per unit length, remains constant? What is $\boldsymbol{E}$ at a distance $x$ from a *very* long line of charge?

To answer the question, we take the *limit* of Eq. (22–9) as $a$ becomes very large. In this limit, the term $x^2/a^2$ in the denominator becomes much smaller than unity and can be thrown away. We are left with

$$\boldsymbol{E} = \frac{\lambda}{2\pi\epsilon_0 x}\boldsymbol{i}. \qquad (22\text{–}10)$$

The field magnitude depends only on the distance of point $P$ from the line of charge, so we can say that at any point $P$ at a distance $r$ out from the line in any direction, the field has magnitude

$$E = \frac{\lambda}{2\pi\epsilon_0 r}. \qquad (22\text{–}11)$$

The direction of $\boldsymbol{E}$ is radially outward from the line. Thus the electric field due to an infinitely long line of charge is proportional to $1/r$ rather than to $1/r^2$ as for a point charge.

Of course, there's really no such thing in nature as an infinite line of charge. But when the field point is close enough to the line, there's very little difference between the result for an infinite line and the real-life finite case. For example, if the distance $r$ of the field point from the center of the line is 1% of the length of the line, the value of $E$ differs from the infinite-length value by less than 2 parts in 10,000.

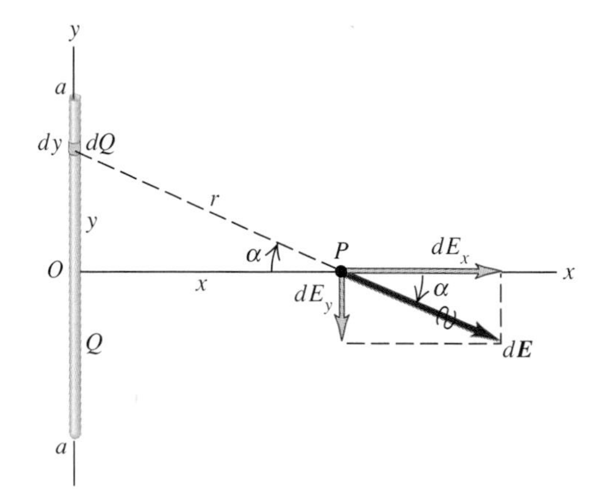


**22–17** A line of charge with length $2a$ and total charge $Q$ creates an electric field $\boldsymbol{E}$ at point $P$.

## EXAMPLE 22–12

**Field of a uniformly charged disk** Find the electric field caused by a uniform surface charge density (charge per unit area) $\sigma$ on a disk with radius $R$, at a point along the axis of the disk a distance $x$ from its center. Assume that $x$ is positive.

**SOLUTION** The situation is shown in Fig. 22–18. We can represent this charge distribution as a collection of concentric rings of charge. We already know from Example 22–10 how to find the field of a single ring, so all we have to do is add the contributions of all the rings.

A typical ring has inner radius $r$ and outer radius $r + dr$, as shown in the figure. Its area $dA$ is approximately equal to its width $dr$ times its circumference $2\pi r$, or $dA = 2\pi r\,dr$. The charge $dQ$ of this ring is given by $dQ = \sigma\,dA$, or

$$dQ = 2\pi\sigma r\,dr.$$

We use this in place of $Q$ in Eq. (22–7) and replace $a$ with $r$. The field component $dE_x$ at point $P$ due to charge $dQ$ is

$$dE_x = \frac{1}{4\pi\epsilon_0}\frac{(2\pi\sigma r\,dr)\,x}{(x^2 + r^2)^{3/2}}.$$

To find the total field due to all the rings, we integrate $dE_x$ over $r$. To include the whole disk, we must integrate from 0 to $R$ (*not* from $-R$ to $R$):

$$E_x = \int_0^R \frac{1}{4\pi\epsilon_0}\frac{(2\pi\sigma r\,dr)\,x}{(x^2 + r^2)^{3/2}} = \frac{\sigma x}{2\epsilon_0}\int_0^R \frac{r\,dr}{(x^2 + r^2)^{3/2}}.$$

Remember that $x$ is a constant during the integration and that the integration variable is $r$. The integral can be evaluated by use of the substitution $z = x^2 + r^2$. We'll let you work out the details; the result is

$$E_x = \frac{\sigma x}{2\epsilon_0}\left[-\frac{1}{\sqrt{x^2 + R^2}} + \frac{1}{x}\right]$$

$$= \frac{\sigma}{2\epsilon_0}\left[1 - \frac{1}{\sqrt{(R^2/x^2) + 1}}\right]. \qquad (22\text{–}12)$$

Again we can ask what happens if the charge distribution gets very large. Suppose we keep increasing the radius $R$ of the disk, adding charge so that the surface charge density $\sigma$ (charge per unit area) is constant. When we take the limit of Eq. (22–12) as $R$ gets much larger than the distance $x$ of the field point from the disk, the term with $R$ in the denominator becomes negligibly small, and we get

$$E = \frac{\sigma}{2\epsilon_0}. \qquad (22\text{–}13)$$

Our final result does not contain the distance $x$ from the plane. This correct but rather surprising result means that the electric field produced by an infinite plane sheet of charge is *independent of the distance from the charge*. Thus the field is *uniform;* its direction is everywhere perpendicular to the sheet, away from it. Again, there is no such thing as an infinite sheet of charge, but if the dimensions of the sheet are much larger than the distance $x$ of the field point $P$ from the sheet, the field is very nearly the same as for an infinite sheet.

If $P$ is to the left of the plane ($x < 0$) instead of to the right, the result is the same except that the direction of $\boldsymbol{E}$ is to the left instead of the right. Also, if $\sigma$ is negative, the directions of the fields on both sides of the plane are toward it rather than away from it.

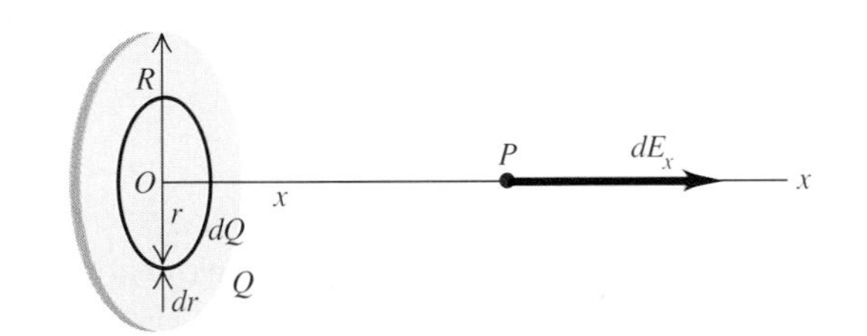


**22–18** Finding the axial electric field of a uniformly charged disk.

# *22–7 ELECTRIC FIELD LINES

The concept of an electric field can be a little elusive because you can't perceive an electric field directly with your senses. Electric field lines can be a big help for visualizing electric fields and making them seem more real. An **electric field line** is an imaginary line drawn through a region of space so that at every point it is tangent to the direction of the electric-field vector at that point. The basic idea is shown in Fig. 22–19. Michael Faraday (1791–1867) first introduced the concept of field lines. He called them "lines of force," but the term "field lines" is preferable.

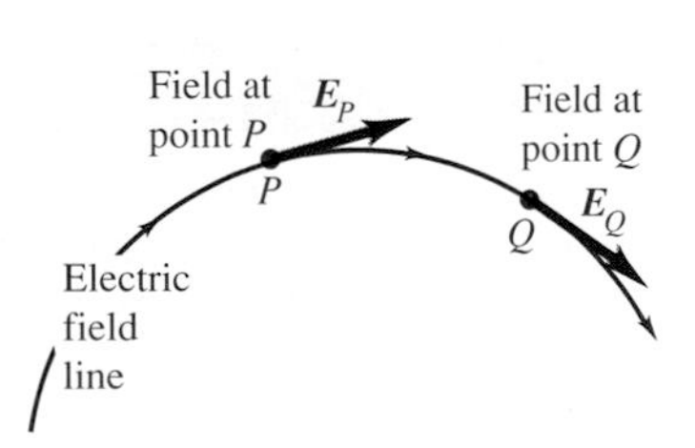


**22–19** The direction of the electric field at any point is tangent to the field line through that point.

Electric field lines show the direction of $\boldsymbol{E}$ at each point, and their spacing gives a general idea of the *magnitude* of $\boldsymbol{E}$ at each point. Where $\boldsymbol{E}$ is strong, we draw lines bunched closely together; where $\boldsymbol{E}$ is weaker, they are farther apart. At any particular point the electric field has a unique direction, so only one field line can pass through each point of the field. In other words, *field lines never intersect.*

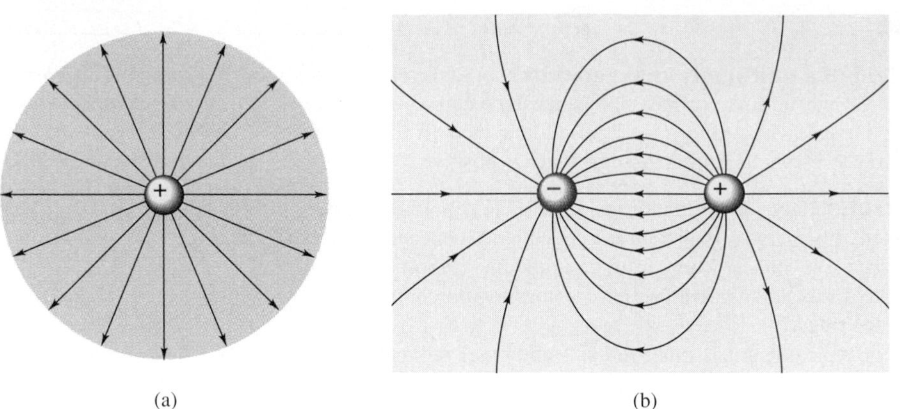


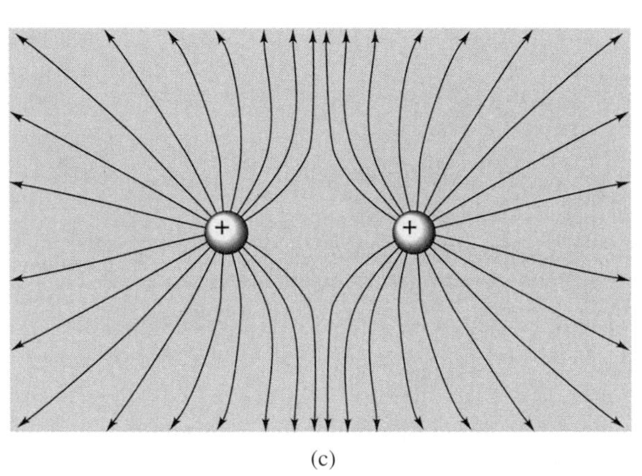


**22–20** Electric field lines for several charge distributions. (a) A single positive charge; (b) two equal and opposite charges (an electric dipole); (c) two equal positive charges.

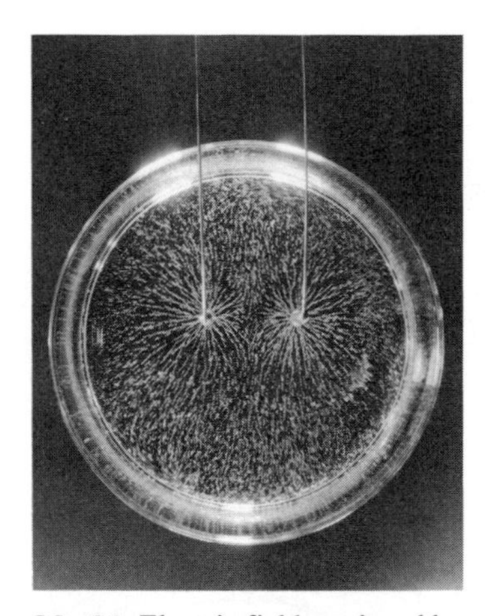

**22–21** Electric field produced by two point charges. The pattern is formed by grass seeds floating in a liquid above the charges.

Figure 22–20 shows some of the electric field lines in a plane containing (a) a single positive charge, (b) two equal charges, one positive and one negative (a dipole), and (c) two equal positive charges. These are cross sections of the actual three-dimensional patterns. The direction of the total electric field at every point in each diagram is along the tangent to the electric field line passing through the point. Arrowheads on the field lines indicate the sense of the $\boldsymbol{E}$-field vector along each line. In regions where the field magnitude is large, such as the space between the positive and negative charges in Fig. 22–20b, the field lines are drawn close together. In regions where it is small, such as between the two positive charges in Fig. 22–20c, the lines are widely separated. In a *uniform* field the field lines are straight, parallel, and uniformly spaced, as in Fig. 22–12.

Diagrams such as this are sometimes called *field maps*. Figure 22–21 shows a demonstration setup for visualizing electric field lines.

## *22–8 ELECTRIC DIPOLES

We mentioned the term *electric dipole* in Example 22–9 (Section 22–6). There are many physical systems, from molecules to TV antennas, that can be described as electric dipoles, so the concept is worth exploring further. We will also need this concept in our discussion of dielectrics in Chapter 25.

An **electric dipole** is a pair of point charges with equal magnitude and opposite sign, say, $q$ and $-q$, separated by a distance $l$. In the following discussion we'll assume that $q$ is positive. We will ask two questions. First, what forces and torques does an electric dipole experience when placed in an external electric field (that is, a field set up by charges outside the dipole)? Second, what electric field does an electric dipole itself produce?

To start with the first question, let's place an electric dipole in a *uniform* external electric field $\boldsymbol{E}$ (Fig. 22–22). The forces $\boldsymbol{F}_+$ and $\boldsymbol{F}_-$ on the two charges both have magnitude $qE$, but their directions are opposite and they add to zero. *The net force on an electric dipole in a uniform external electric field is zero.*

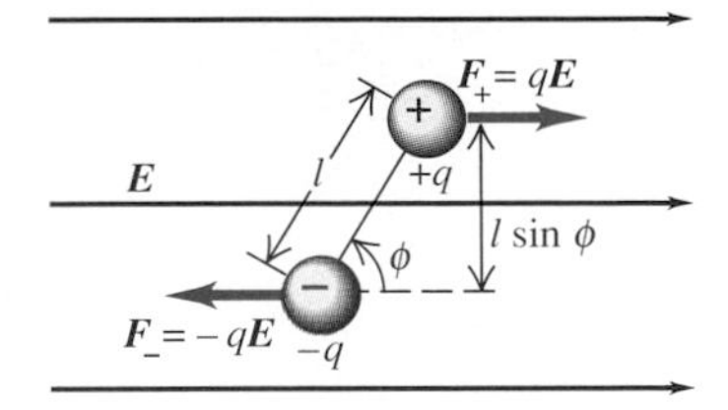


**22–22** The net force on this electric dipole is zero, but there is a torque directed into the page, tending to rotate the dipole clockwise.

However, the two forces don't act along the same line, so their *torques* don't add to zero. They form a *couple;* we defined that term in Section 11–4, and you may want to review it. Let the angle between the electric field $\boldsymbol{E}$ and the dipole axis be $\phi$; then the torque $\tau$ exerted by the couple is

$$\tau = (qE)(l \sin\phi), \tag{22–14}$$

where $l \sin\phi$ is the perpendicular distance between the lines of action of the two forces. Remember that in general we compute torque with respect to a specific point but that for a couple the torque is the same for *all* points.

The product of the charge $q$ and the separation $l$ is the magnitude of a quantity called the **electric dipole moment,** denoted by $p$. Its units are charge times distance (C · m). (Be careful not to confuse it with momentum or pressure. There aren't as many letters in the alphabet as physical quantities; some letters are used several times. Usually, the context makes it clear what is meant, but be careful.)

In terms of $p$ the torque $\tau$ exerted by the field has magnitude

$$\tau = pE \sin\phi. \tag{22–15}$$

We can write this more compactly in vector form. We define the electric dipole moment to be a vector quantity $\boldsymbol{p}$ with magnitude $ql$ and direction along the dipole axis, from $-$ to $+$ charge. Then $\phi$ is the angle between the directions of the vectors $\boldsymbol{p}$ and $\boldsymbol{E}$, and

$$\boldsymbol{\tau} = \boldsymbol{p} \times \boldsymbol{E}. \tag{22–16}$$

The torque is greatest when $\boldsymbol{p}$ and $\boldsymbol{E}$ are perpendicular and is zero when they are parallel or antiparallel. The torque always tends to turn $\boldsymbol{p}$ to line it up with $\boldsymbol{E}$. The position $\phi = 0$, with $\boldsymbol{p}$ *parallel* to $\boldsymbol{E}$, is a position of *stable* equilibrium, and the position $\phi = \pi$, with $\boldsymbol{p}$ and $\boldsymbol{E}$ *antiparallel,* is a position of *unstable* equilibrium.

When a dipole changes direction in a field, the electric-field torque does *work* on it, with a corresponding change in potential energy. The work $dW$ done by a torque $\tau$ during an infinitesimal displacement $d\phi$ is given by Eq. (10–21): $dW = \tau\, d\phi$. Because the torque is in the direction of *decreasing* $\phi$, $\tau = -pE \sin\phi$, and

$$dW = \tau\, d\phi = -pE \sin\phi\, d\phi.$$

In a finite displacement from $\phi_1$ to $\phi_2$ the total work done on the dipole is

$$W = \int_{\phi_1}^{\phi_2} -pE \sin\phi\, d\phi$$
$$= pE\cos\phi_2 - pE\cos\phi_1.$$

The work is the negative of the change of potential energy, just as in Chapter 7: $W = U_1 - U_2$. So we see that a suitable definition of potential energy $U$ for this system is

$$U(\phi) = -pE\cos\phi. \tag{22–17}$$

In Eq. (22–17) we recognize the *scalar product* $\boldsymbol{p} \cdot \boldsymbol{E} = pE \cos \phi$, so we can also write

$$U = -\boldsymbol{p} \cdot \boldsymbol{E}. \tag{22–18}$$

The potential energy is minimum (most negative) at the stable equilibrium position, where $\phi = 0$ and $\boldsymbol{p}$ is parallel to $\boldsymbol{E}$. It is maximum when $\phi = \pi$ and $\boldsymbol{p}$ is antiparallel to $\boldsymbol{E}$. At $\phi = \pi/2$, where $\boldsymbol{p}$ is perpendicular to $\boldsymbol{E}$, $U$ is zero. We could define $U$ differently so that it is zero at some other orientation of $\boldsymbol{p}$, but our definition is simplest.

## EXAMPLE 22–13

Figure 22–23a shows an electric dipole in a uniform electric field with magnitude $5.0 \times 10^5$ N/C directed parallel to the plane of the figure. The charges are $\pm 1.6 \times 10^{-19}$ C, both in the plane and separated by 0.125 nm $= 0.125 \times 10^{-9}$ m. (Note that both the charge magnitude and the distance are typical of molecular quantities.) Find a) the net force exerted by the field on the dipole; b) the magnitude and direction of the electric dipole moment; c) the magnitude and direction of the torque; d) the potential energy of the system in the position shown.

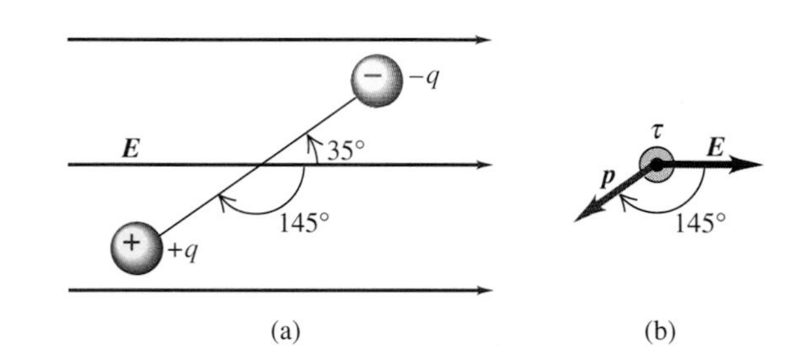


**22–23** (a) An electric dipole. (b) Directions of the electric dipole moment and torque.

**SOLUTION** a) In any uniform field the forces on the two charges are equal and opposite, and the total force is zero.

b) The magnitude $p$ of the electric dipole moment $\boldsymbol{p}$ is

$$p = ql = (1.60 \times 10^{-19}\ \text{C})(0.125 \times 10^{-9}\ \text{m}) = 2.0 \times 10^{-29}\ \text{C} \cdot \text{m}.$$

The direction of $\boldsymbol{p}$ is from the negative to the positive charge, 145° clockwise from the electric-field direction (Fig. 22–23b).

c) The magnitude of the torque is

$$\tau = pE \sin \phi = (2.0 \times 10^{-29}\ \text{C} \cdot \text{m})(5.0 \times 10^5\ \text{N/C})(\sin 145°) = 5.7 \times 10^{-24}\ \text{N} \cdot \text{m}.$$

From the right-hand rule for vector products (Section 1–10) the direction of the torque is out of the page. This corresponds to a counterclockwise torque that tends to align $\boldsymbol{p}$ with $\boldsymbol{E}$.

d) The potential energy is

$$U = -pE \cos \phi = -(2.0 \times 10^{-29}\ \text{C} \cdot \text{m})(5.0 \times 10^5\ \text{N/C})(\cos 145°) = 8.2 \times 10^{-24}\ \text{J}.$$

---

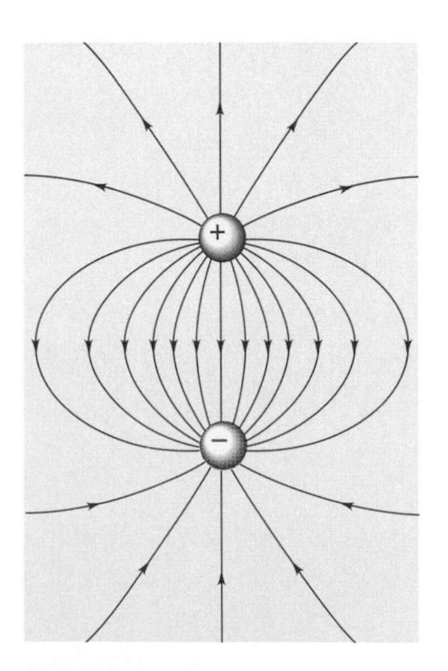


**22–24** Electric field lines in the plane of an electric dipole.

In this discussion we have always assumed that $\boldsymbol{E}$ is uniform, so there is no net force on the dipole. If $\boldsymbol{E}$ is not uniform, the forces at the ends may not cancel completely, and the net force may not be zero. Thus a body with zero net charge but an electric dipole moment can experience a net force in a non-uniform electric field. As we mentioned in Section 22–2, an electric dipole moment can be *induced* in an uncharged body by an electric field, and this is how uncharged bodies can experience electrostatic forces.

Now let's think of an electric dipole as a *source* of electric field. What does the field look like? The general shape is shown by the field map of Fig. 22–24. At each point in the pattern the total $\boldsymbol{E}$ field is the vector sum of the fields from the two individual charges, as in Example 22–9. We suggest that you try drawing diagrams showing this vector sum for several points.

To get quantitative information about a dipole field, we have to do some calculating. Here's an example. Notice the use of the superposition principle to add up the contributions to the field of the individual charges. Also notice that we need to use approximation techniques even for the relatively simple case of a field due to two charges. Field calculations often become very complicated. After the following example, we discuss a graphic technique using computer analysis to determine the field due to an electric charge distribution.

## EXAMPLE 22–14

In Fig. 22–25 an electric dipole is centered at the origin, with $\boldsymbol{p}$ in the direction of the $+y$-axis. Derive an approximate expression for the electric field at a point on the $y$-axis for which $y$ is much larger than $l$. Use the binomial expansion of $(1 + x)^n$, that is, $(1 + x)^n \cong 1 + nx + n(n - 1)x^2/2 + \cdots$, for the case $x << 1$. (This problem illustrates a useful calculational technique.)

**SOLUTION** The total $y$-component $E_y$ of electric field from the two charges is

$$E_y = \frac{q}{4\pi\epsilon_0}\left[\frac{1}{(y - l/2)^2} - \frac{1}{(y + l/2)^2}\right]$$
$$= \frac{q}{4\pi\epsilon_0 y^2}\left[(1 - l/2y)^{-2} - (1 + l/2y)^{-2}\right].$$

Now comes the approximation. When $y$ is much greater than $l$, that is, when we are far away from the dipole in comparison to its size, the quantity $l/2y$ is much smaller than 1. With $n = -2$ and $l/2y$ playing the role of $x$ in the binomial expansion we keep only the first two terms. The terms that we discard are much smaller than those that we keep, and we have

$$(1 - l/2y)^{-2} \cong 1 + l/y,$$
$$(1 + l/2y)^{-2} \cong 1 - l/y,$$

and

$$E \cong \frac{q}{4\pi\epsilon_0 y^2}[1 + l/y - (1 - l/y)]$$
$$= \frac{ql}{2\pi\epsilon_0 y^3} = \frac{p}{2\pi\epsilon_0 y^3}.$$

An alternative route to this expression is to put the fractions in the $E_y$ expression over a common denominator and combine, then approximate the denominator $(y - l/2)^2(y + l/2)^2$ as $y^4$. We leave the details as a problem.

For points $P$ off the coordinate axes the expressions are more complicated, but at *all* points far away from the dipole (in any direction) the field drops off as $1/r^3$. We can compare this with the $1/r^2$ behavior of a point charge, the $1/r$ behavior of a long line charge, and the independence of $r$ for a large sheet of charge. There are charge arrangements for which the field drops off even more quickly. A *quadrupole* consists of two equal dipoles with opposite orientation, separated by a small distance. The field of a quadrupole at large distances drops off as $1/r^4$.

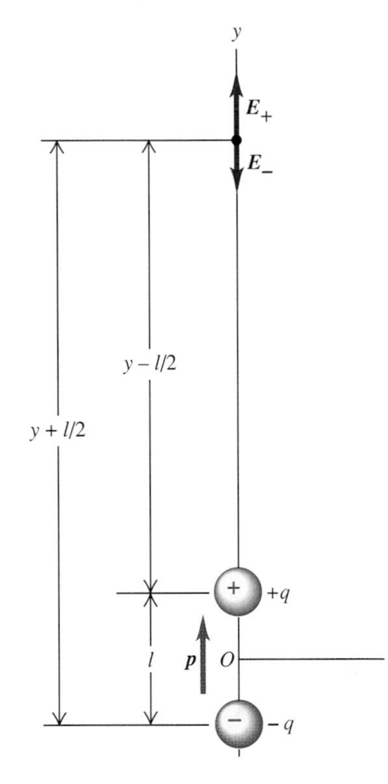


**22–25** Finding the electric field of an electric dipole at a point on its axis.

# 22–9 FIELD MAPS: A Case Study in Computer Analysis

In Section 22–7 we introduced electric field lines as an aid to understanding and visualizing the general properties of electric fields. Any such graphic representation of a field is called a *field map.* To make an accurate field map, we have to compute the electric field at many points in the region of interest. It would be enormously time-consuming to do these computations with a hand calculator, but the problem can be handled very nicely with a computer and a fairly simple program. In the following discussion we won't trace out actual field lines; we'll obtain a slightly different kind of field map that requires only simple concepts and computations.

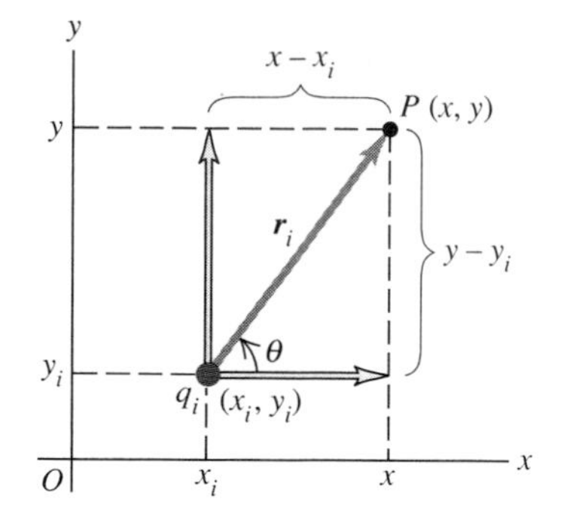


**22–26** The components of the vector $\boldsymbol{r}_i$ from charge $q_i$ to field point $P$ are $x - x_i$ and $y - y_i$.

Let's consider the following general problem. Suppose we have $N$ point charges $q_1$, $q_2, \ldots, q_N$, located in the $xy$-plane at points $(x_1, y_1)$, $(x_2 y_2), \ldots, (x_N, y_N)$. How can we find the direction of the resulting $\boldsymbol{E}$ field at any point in the plane and display it graphically?

Here's an outline of a possible approach. For a particular field point $P$ with coordinates $(x, y)$ we find the components $E_x$ and $E_y$ of the field caused by a single charge $q_i$ at point $(x_i, y_i)$. Then we sum these over all the charges to find the $x$- and $y$-components $E_{xr}$ and $E_{yr}$ and the magnitude $E_r$ of the *total* (resultant) field $\boldsymbol{E}_r$ at $(x, y)$. Finally, we draw a little line segment through point $P$ in the direction of the field. The result is not a set of continuous field lines but a set of lines showing the direction of the field at each point.

Now let's fill in some details. For a charge $q_i$ at source point $(x_i, y_i)$ the vector $\boldsymbol{r}_i$ from the source point to a field point $P$ with coordinates $(x, y)$ has components $(x - x_i$, $y - y_i)$ (Fig. 22–26). Its direction (measured counterclockwise from the $+x$-axis) is given by the angle $\theta$, where

$$\theta = \arctan\frac{y - y_i}{x - x_i}. \tag{22–19}$$

The distance $r_i$ from $q_i$ to point $P$ is

$$r_i = \sqrt{(x - x_i)^2 + (y - y_i)^2}. \tag{22–20}$$

The magnitude $E$ of the field $\boldsymbol{E}$ at $P$ due to $q_i$ is

$$E = \frac{1}{4\pi\epsilon_0}\frac{q_i}{r_i^{\,2}}. \tag{22–21}$$

The components of $\boldsymbol{E}$ are

$$\begin{aligned} E_x = E\cos\theta &= \frac{1}{4\pi\epsilon_0}\frac{q_i}{r_i^{\,2}}\cos\theta = \frac{1}{4\pi\epsilon_0}\frac{q_i}{r_i^{\,2}}\left(\frac{x - x_i}{r_i}\right) \\ &= \frac{1}{4\pi\epsilon_0}\frac{q_i(x - x_i)}{r_i^{\,3}}, \end{aligned} \tag{22–22}$$

$$\begin{aligned} E_y = E\sin\theta &= \frac{1}{4\pi\epsilon_0}\frac{q_i}{r_i^{\,2}}\sin\theta = \frac{1}{4\pi\epsilon_0}\frac{q_i}{r_i^{\,2}}\left(\frac{y - y_i}{r_i}\right) \\ &= \frac{1}{4\pi\epsilon_0}\frac{q_i(y - y_i)}{r_i^{\,3}}. \end{aligned} \tag{22–23}$$

The components $E_{xr}$ and $E_{yr}$ of the total (resultant) field $\boldsymbol{E}_r$ at $P$ are the sums of such expressions over all the charges, as $i$ ranges from 1 to $N$. In the program these sums will be done by using a loop on $i$.

The magnitude $E_r$ of the total field $\boldsymbol{E}_r$ is

$$E_r = \sqrt{E_{xr}^{\,2} + E_{yr}^{\,2}}. \tag{22–24}$$

The direction of $\boldsymbol{E}_r$ is described by the angle $\theta_r$ (measured counterclockwise from the $+x$-axis) given by

$$\theta_r = \arctan\frac{E_{yr}}{E_{xr}}.$$

To draw a line with length $c$ through point $P$ in this direction, we draw a line between the two points with coordinates

$$\left(x - \frac{cE_{xr}}{2E_r}, y - \frac{cE_{yr}}{2E_r}\right) \quad \text{and} \quad \left(x + \frac{cE_{xr}}{2E_r}, y + \frac{cE_{yr}}{2E_r}\right). \tag{22–25}$$

You can verify that the distance between these points is $c$ and that the direction of the vector is the same as that of $\boldsymbol{E}_r$ at the point.

We have to decide *how many* field points to use. Drawing a little line for *every* point in the plane would take an infinite amount of computation and would fill up the plane completely. Instead, let's divide the region of interest into a rectangular array of boxes with a field point at the center of each. We'll make the boxes small enough to give us a good idea of the shape of the field but not so small that we have a useless filigree of tiny lines. The computer program will loop through the above calculation as many times as the number of field points, incrementing the field-point coordinates appropriately.

One possible computational difficulty is that if a field point is too close to (or coincides with) a charge, we may get an overflow or divide-by-zero error. To avoid this, we skip all field points that are within some arbitrary distance (such as the width of one box) of any charge.

Here's a skeleton of a program to carry out this calculation.

*Step 1:* Choose values of the charges $q_i$ and their positions $(x_i, y_i)$.

*Step 2:* Choose the range of values of $x$ and $y$ for which the computation is to be carried out, that is, values of $x_{\mathrm{min}}$, $y_{\mathrm{min}}$, $x_{\mathrm{max}}$, and $y_{\mathrm{max}}$. If the $x$'s and $y$'s are screen coordinates (addresses of pixels on the screen), these values are determined by the characteristics of your computer. For example, a PC with a VGA video adapter has $x_{\mathrm{min}} = 0$, $x_{\mathrm{max}} = 639$, $y_{\mathrm{min}} = 0$, $y_{\mathrm{max}} = 479$. Some languages let you change the range of the screen coordinates. In BASIC, for example, the WINDOW command lets you choose any range of $x$ and $y$ you like; it then does the appropriate conversion to pixel coordinates for you.

*Step 3:* Choose the number $m$ of boxes across and down the region. The number of field points is then $m^2$. Values of $m$ between 15 and 40 work well. Large $m$'s produce lengthy calculation; small $m$'s result in a coarse pattern with low resolution.

*Step 4:* Calculate the width $dx$ and height $dy$ of a box:

$$dx = \frac{x_{\mathrm{max}} - x_{\mathrm{min}}}{m}, \qquad dy = \frac{y_{\mathrm{max}} - y_{\mathrm{min}}}{m}.$$

*Step 5:* Let $(j, k)$ be a pair of integer indexes that identify a particular box and its screen location (the $j$th column and $k$th row in the grid); each ranges from 1 to $m$.

*Step 6:* Begin a loop on $k$, from $k = 1$ to $k = m$ (successive rows of boxes).

*Step 7:* Calculate the $y$-coordinate of the center of the box:

$$y = y_{\mathrm{min}} + (k - \tfrac{1}{2})\,dy.$$

*Step 8:* Begin a loop on $j$, from $j = 1$ to $j = m$ (going across a row of boxes).

*Step 9:* Calculate the $x$-coordinate of the center of the box:

$$x = x_{\mathrm{min}} + (j - \tfrac{1}{2})\,dx.$$

*Step 10:* Begin a loop on $i$, from $i = 1$ to $i = N$ (including the charges one at a time).

*Step 11:* Calculate $r_i$, the distance of the field point $(x, y)$ from the charge $q_i$ at point $(x_i, y_i)$, using Eq. (22–20).

*Step 12:* If $r_i > dx$, proceed to calculate the components of $\boldsymbol{E}_r$ in Steps 13 through 16 and continue the program through Step 17. Otherwise, set a flag (FLAG = 1) to tell the program not to calculate the field or plot this point, and proceed directly to Step 17.

*Step 13:* Calculate $E_x$ from Eq. (22–22).

*Step 14:* Add this contribution to the sum of $x$-components: $E_{xr} = E_{xr} + E_x$.

*Step 15:* Calculate $E_y$ from Eq. (22–23).

*Step 16:* Add this contribution to the sum of $y$-components: $E_{yr} = E_{yr} + E_y$.

*Step 17:* End of the IF in Step 12.

*Step 18:* End of Step 10 loop over $i$.

*Step 19:* If FLAG = 0, compute the value of $E_r$ from Eq. (22–24). Set $c = dx$ (that is, a line with length equal to the width of a box), and draw a line between the points given by Eq. (22–25).

*Step 20:* If FLAG = 1, do not draw the line but reset the flag to FLAG = 0 and proceed.

*Step 21:* End of loops over $j$ and $k$.

*Step 22:* END.

Figure 22–27 shows the result of this calculation for the case of two charges with equal magnitude but opposite sign.

This procedure can be elaborated in several ways. You may want to draw at each grid point a vector with length proportional to the field magnitude to show the magnitude of the field as well as its direction. To do this, simply replace the parameter $c$ in Step 19 by a multiple of $E_r$. You may have to experiment to find the appropriate scale factor for the best picture.

Another possibility is to draw actual field lines. For this it is best to abandon the fixed grid of boxes. Instead, compute the field and its components at some starting point. Draw a line from that point to a neighboring point. Then compute the field at that point, draw another line, and so on. Repeat for other starting points. In this calculation the location of each field point is determined by the components of field at the preceding point. The procedure is similar to our calculations of baseball trajectories in Section 3–6 and phase-space paths in Section 13–9. The program is only a little more complex than the one outlined above, and it yields genuine field lines.

All the field maps that we have discussed are inherently two-dimensional representations. We don't want to create the impression that the electric field exists *only* in the $xy$-plane. Every electric field, even one resulting from charges that lie in a plane, is an inherently three-dimensional entity. When more detailed information about the field is needed, we can make field maps showing various cross sections of the field.

Finally, we remark that for complicated geometries it is sometimes easier to abandon the Coulomb's-law approach and instead work directly with the differential equations that the field must satisfy, including the conditions that are imposed by boundary surfaces. These and similar calculations are of tremendous practical importance in the design of electronic devices and electromagnets for particle accelerators and in many other applications.

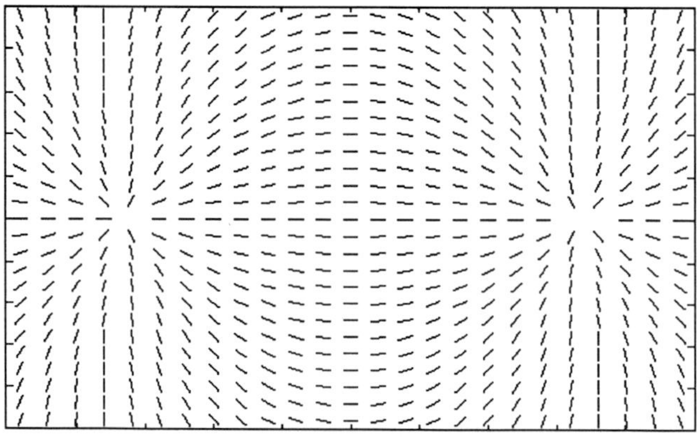

**22–27** Field map for two charges with equal magnitude and opposite sign. For this map, $m = 25$.

## SUMMARY

• The fundamental entity in electrostatics is electric charge. There are two kinds of charge, positive and negative. Like charges repel each other; unlike charges attract. Charge is conserved; the total charge in an isolated system is constant.

• Conductors are materials that permit electric charge to move within them. Insulators permit charge to move much less readily. Most metals are good conductors; most nonmetals are insulators.

• All ordinary matter is made of protons, neutrons, and electrons. The protons and neutrons in the nucleus of an atom are bound together by the nuclear force; the electrons surround the nucleus at distances much greater than its size. Electrical interactions are chiefly responsible for the structure of atoms, molecules, and solids.

• Coulomb's law is the basic law of interaction for point electric charges. For charges $q_1$ and $q_2$ separated by a distance $r$ the magnitude of the force is

$$F = \frac{1}{4\pi\epsilon_0}\frac{|q_1 q_2|}{r^2}. \qquad (22\text{–}2)$$

The force on each charge is along the line joining the two charges, repulsive if $q_1$ and $q_2$ have the same sign, attractive if they have opposite signs. The forces form an action-reaction pair and obey Newton's third law. In SI units the unit of electric charge is the coulomb, abbreviated C, and

$$\frac{1}{4\pi\epsilon_0} = 8.988 \times 10^9\ \mathrm{N \cdot m^2/C^2}.$$

• The principle of superposition states that when two or more charges each exert a force on another charge, the total force on that charge is the vector sum of the forces exerted by the individual charges.

• Electric field, a vector quantity, is the force per unit charge exerted on a test charge at any point, provided that the test charge is small enough that it does not disturb the charges that cause the field. From Coulomb's law the electric field produced by a point charge is

$$\boldsymbol{E} = \frac{1}{4\pi\epsilon_0}\frac{q}{r^2}\hat{\boldsymbol{r}}. \qquad (22\text{–}6)$$

• The principle of superposition states that the electric field of any combination of charges is the vector sum of the fields caused by the individual charges. To calculate the electric field caused by a continuous distribution of charge, divide the distribution into small elements, calculate the field caused by each element, and then carry out the vector sum or each component sum, usually by integrating. Charge distributions are described by linear charge density $\lambda$, surface charge density $\sigma$, and volume charge density $\rho$.

• Field lines provide a graphical representation of electric fields. A field line at any point in space is tangent to the direction of $\boldsymbol{E}$ at that point, and the number of lines per unit area (perpendicular to their direction) is proportional to the magnitude of $\boldsymbol{E}$ at the point.

• An electric dipole is a pair of electric charges of equal magnitude $q$ but opposite sign, separated by a distance $l$. The electric dipole moment $p$ is defined to be $p = ql$. The vector dipole moment is the vector $\boldsymbol{p}$ having this magnitude and a direction from

### KEY TERMS

electric charge
conductor
insulator
induction
induced charge
electron
proton
neutron
nucleus
atomic number
positive ion
negative ion
ionization
conservation of charge
Coulomb's law
principle of superposition
coulomb
electric field
test charge
vector field
source point
field point
linear charge density
surface charge density
volume charge density
electric field line
electric dipole
electric dipole moment

negative toward positive charge. An electric dipole in an electric field experiences a torque $\tau$ given by

$$\tau = pE \sin \phi, \quad (22\text{–}15)$$

where $\phi$ is the angle between the directions of $\boldsymbol{p}$ and $\boldsymbol{E}$. Vector torque $\boldsymbol{\tau}$ is

$$\boldsymbol{\tau} = \boldsymbol{p} \times \boldsymbol{E}. \quad (22\text{–}16)$$

# EXERCISES

## Section 22–3 Conservation and Quantization of Charge

**22–1** What is the total positive charge, in coulombs, of all the protons in 2.00 mol of hydrogen atoms?

**22–2 Particles in a Gold Ring.** You have a pure (24-karat) gold ring with a mass of 13.4 g. Gold has an atomic mass of 197 g/mol and an atomic number of 79. a) How many protons are in the ring and what is their total positive charge? b) How many electrons are in the ring if it carries no net charge?

## Section 22–4 Coulomb's Law

**22–3** Two equal point charges of $+3.00\ \mu\text{C}$ are placed 0.600 m apart. What is the magnitude of the force each exerts on the other? What are the directions of the forces?

**22–4** A negative charge $-0.500\ \mu\text{C}$ exerts an attractive force with a magnitude of 0.600 N on an unknown charge 0.200 m away. a) What is the unknown charge (magnitude and sign)? b) What is the magnitude of the force that the unknown charge exerts on the $-0.500\ \mu\text{C}$ charge?

**22–5** Two small plastic spheres are given positive electrical charges. When they are 40.0 cm apart, the repulsive force between them has magnitude 0.250 N. What is the charge on each sphere a) if the two charges are equal; b) if one sphere has twice the charge of the other?

**22–6** How many excess electrons must be present on each of two small spheres spaced 15.0 cm apart if the spheres have equal charge and if the magnitude of the force of repulsion between them is $5.00 \times 10^{-19}$ N?

**22–7** How far does the electron of a hydrogen atom have to be removed from the nucleus for the force of attraction to equal the weight of the electron at the surface of the earth?

**22–8** Two copper spheres, each having a mass of 0.400 kg, are separated by 2.00 m. a) How many electrons does each sphere contain? (The atomic mass of copper is 63.5 g/mol, and its atomic number is 29.) b) How many electrons would have to be removed from one sphere and added to the other to cause an attractive force with magnitude $1.00 \times 10^{4}$ N (roughly one ton)? c) What fraction of all the electrons in each sphere does this represent?

**22–9** Two point charges are located on the $y$-axis as follows: Charge $q_1 = +3.80$ nC is at $y = 0.600$ m, and charge $q_2 = -2.50$ nC is at the origin ($y = 0$). What is the total force (magnitude and direction) exerted by these two charges on a third point charge $q_3 = +5.00$ nC located at $y = -0.400$ m?

**22–10** Two point charges are placed on the $x$-axis as follows: Charge $q_1 = +3.00$ nC is located at $x = 0.400$ m, and charge $q_2 = +5.00$ nC is at $x = -0.200$ m. What are the magnitude and direction of the total force exerted by these two charges on a negative point charge $q_3 = -8.00$ nC at the origin?

**22–11** A ring-shaped conductor with radius $a = 0.250$ m carries a total positive charge $Q = +8.40\ \mu\text{C}$, uniformly distributed around it, as shown in Fig. 22–16. The center of the ring is at the origin of coordinates. A point charge $q = -2.50\ \mu\text{C}$ is located at point $P$, which is at $x = 0.500$ m. What are the magnitude and direction of the force exerted *by* the charge $q$ *on* the ring?

## Section 22–5 Electric Field and Electric Forces

**22–12** Find the magnitude and direction of the electric field at a point 0.500 m directly above a particle having an electric charge of $+4.00\ \mu\text{C}$.

**22–13** At what distance from a particle with a charge of 5.00 nC does the electric field of that charge have a magnitude of 6.00 N/C?

**22–14** a) What is the electric field of a gold nucleus at a distance of $6.00 \times 10^{-10}$ m from the nucleus? The atomic number of gold is 79. b) What is the electric field of a proton at a distance of $5.28 \times 10^{-11}$ m from the proton? (This is the radius of the electron orbit in the Bohr model for the ground state of the hydrogen atom.)

**22–15** A small object carrying a charge of $-5.00$ nC experiences a downward force of $3.00 \times 10^{-8}$ N when placed at a certain point in an electric field. a) What are the magnitude and direction of the electric field at this point? b) What would be the magnitude and direction of the force acting on a proton placed at this same point in the electric field?

**22–16** What must be the charge (sign and magnitude) of a particle with a mass of 5.60 g for it to remain stationary in the laboratory when placed in a downward-directed electric field of magnitude 5000 N/C?

**22–17** What is the magnitude of an electric field in which the electrical force on an electron is equal in magnitude to its weight?

**22–18 Electric Field of the Earth.** The earth has a net electric charge that causes a field at points near its surface equal to 150 N/C and directed in toward the center of the earth. a) What magnitude and sign of charge would a 60.0-kg human have to acquire to overcome his or her weight by the force exerted by the earth's electric field? b) What would be the force of repulsion between two people each with the charge calculated in part (a) and separated by a distance of 100 m? Is use of the earth's electric field a feasible means of flight?

**22–19 Electric Field between Parallel Plates.** An electron is projected with an initial speed $v_0 = 7.00 \times 10^6$ m/s into the uniform field between the parallel plates in Fig. 22–28. The direction of the field is vertically downward, and the field is zero except in the space between the plates. The electron enters the field at a point midway between the plates. If the electron just misses the upper plate as it emerges from the field, find the magnitude of the electric field.

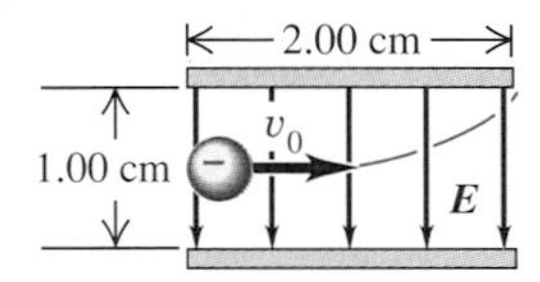


**FIGURE 22–28**

**22–20** A uniform electric field exists in the region between two oppositely charged plane parallel plates. An electron is released from rest at the surface of the negatively charged plate and strikes the surface of the opposite plate, 2.60 cm distant from the first, in a time interval of $1.50 \times 10^{-8}$ s. a) Find the magnitude of the electric field. b) Find the speed of the electron when it strikes the second plate.

## Section 22–6 Electric-Field Calculations

**22–21** A point charge $q_1 = -4.00$ nC is at the origin, and a second point charge $q_2 = +6.00$ nC is on the $x$-axis at $x = 0.800$ m. Find the electric field (magnitude and direction) at each of the following points on the $x$-axis: a) $x = 0.200$ m; b) $x = 1.20$ m; c) $x = -0.200$ m.

**22–22** Two particles having charges $q_1 = 1.00$ nC and $q_2 = 2.00$ nC are separated by a distance of 1.80 m. At what point along the line connecting the two charges is the total electric field due to the two charges equal to zero?

**22–23** In a rectangular coordinate system a positive point charge with a magnitude of $4.00 \times 10^{-8}$ C is placed at the point $x = +0.100$ m, $y = 0$, and an identical point charge is placed at $x = -0.100$ m, $y = 0$. Find the magnitude and direction of the electric field at the following points: a) the origin; b) $x = 0.200$ m, $y = 0$; c) $x = 0.100$ m, $y = 0.150$ m; d) $x = 0, y = 0.100$ m.

**22–24** A point charge $q_1 = +6.00$ nC is at the point $x = 0.800$ m, $y = 0.600$ m, and a second point charge $q_2 = -4.00$ nC is at the point $x = 0.800$ m, $y = 0$. Calculate the magnitude and direction of the resultant electric field at the origin due to these two point charges.

**22–25** Repeat Exercise 22–23 for the case in which the point charge at $x = +0.100$ m, $y = 0$ is positive and the other is negative.

**22–26** A long, straight wire has charge per unit length $3.00 \times 10^{-10}$ C/m. At what distance from the wire is the electric field magnitude equal to 0.600 N/C?

**22–27** What is the charge per unit area, in C/m$^2$, of an infinite plane sheet of charge if the electric field produced by the sheet of charge has magnitude 3.00 N/C?

## Section22–8 Electric Dipoles

***22–28** The potassium chloride molecule (KCl) has a dipole moment of $8.9 \times 10^{-30}$ C · m. a) If this dipole moment arises from two charges $\pm 1.6 \times 10^{-19}$ C separated by distance $l$, calculate $l$. b) What is the maximum magnitude of the torque that a uniform electric field with magnitude $6.0 \times 10^4$ N/C can exert on a KCl molecule? Show in a sketch the relative orientations of the electric dipole moment $\boldsymbol{p}$ and the electric field $\boldsymbol{E}$ when the torque is a maximum.

***22–29** The ammonia molecule ($NH_3$) has a dipole moment of $5.0 \times 10^{-30}$ C · m. Ammonia molecules in the gas phase are placed in a uniform electric field with magnitude $E = 3.0 \times 10^5$ N/C. a) What is the change in potential energy when the dipole moment $\boldsymbol{p}$ of the molecule changes its orientation with respect to the electric field from parallel to perpendicular? b) At what temperature $T$ is the average translational kinetic energy $\frac{3}{2}kT$ of the molecule equal to the change in potential energy calculated in part (a)? (Above this temperature, thermal agitation will prevent the dipoles from aligning with the field.)

## Section 22–9 Field Maps: A Case Study in Computer Analysis

**22–30** a) Verify that the distance between the two points specified in Eq. (22–25) is $c$. b) Verify that the line connecting the two points with the coordinate given in Eq. (22–25) lies along the direction of the resultant field $\boldsymbol{E}_r$ at a point $P$ that has coordinates $(x, y)$.

**22–31** Write a computer code to implement the procedure outlined in the 22 steps given in the text. a) Run your code for the case of two charges with equal magnitude but opposite sign. Compare your results to Fig. 22–27. b) Run your code for the case of two charges with equal magnitude and the same sign.

**22–32** In your computer program from Exercise 22–31, make the modification discussed in the text to let the lines drawn be proportional to the field magnitude at each point. Repeat the calculations for the two cases considered in Exercise 22–31.

**22–33** Modify the procedure of Exercise 22–31 in the way described in the text so that actual field lines will be drawn. Run the code you write for the two cases considered in Exercise 22–31. Compare your results to Fig. 22–20.

## PROBLEMS

**22–34** Three point charges are arranged along the $x$-axis. Point charge $q_1 = 6.00$ nC is located at $x = 0.300$ m, and point charge $q_2 = -4.00$ nC is at $x = -0.200$ m. A positive point charge $q_3$ is located at the origin. a) What must be the magnitude of $q_3$ for the resultant force on it to have magnitude $6.00 \times 10^{-4}$ N? b) What is the direction of the resultant force on $q_3$?

**22–35** Two small spheres, each with a mass of 16.0 g, are attached to silk threads 1.00 m long and hung from a common point. When the spheres are given equal quantities of negative charge, each thread makes an angle of 20.0° with the vertical. a) Draw a diagram showing all of the forces on each sphere. b) Find the magnitude of the charge on each sphere. c) The two threads are now shortened to length $l = 0.500$ m, while the charge on the spheres is held fixed. What will be the new angle that the threads each make with the vertical? (*Hint:* This part of the problem can be solved numerically by using trial values for $\theta$ and adjusting the values of $\theta$ until a self-consistent answer is obtained.)

**22–36** a) Suppose all the electrons in 20.0 g of hydrogen atoms could be located at the North Pole of the earth and all the protons at the South Pole. What would be the total force of attraction exerted on each group of charges by the other? The atomic mass of hydrogen is 1.01 g/mol. b) What would be the magnitude and direction of the force exerted by the charges in part (a) on a third charge that is positive, equal in magnitude to the total charge at one of the poles, and located at a point on the surface of the earth at the equator? Draw a diagram.

**22–37** Point charges of 6.00 nC are situated at each of three corners of a square whose side is 0.200 m. What are the magnitude and direction of the resultant force on a point charge of $-2.00$ nC if it is placed a) at the center of the square; b) at the vacant corner of the square?

**22–38** A charge $q_1 = -3.00$ nC is placed at the origin of an $xy$-coordinate system, and a charge $q_2 = 2.00$ nC is placed on the positive $y$-axis at $y = 4.00$ cm. a) If a third charge $q_3 = 4.00$ nC is now placed at the point $x = 3.00$ cm, $y = 4.00$ cm, find the $x$- and $y$-components of the total force exerted on this charge by the other two. b) Find the magnitude and direction of this force.

**22–39** Positive charge $Q$ is distributed uniformly along the positive $y$-axis between $y = 0$ and $y = a$. A negative point charge $-q$ lies on the positive $x$-axis, a distance $x$ from the origin (Fig. 22–29). Calculate the $x$- and $y$-components of the force that the charge distribution $Q$ exerts on $q$.

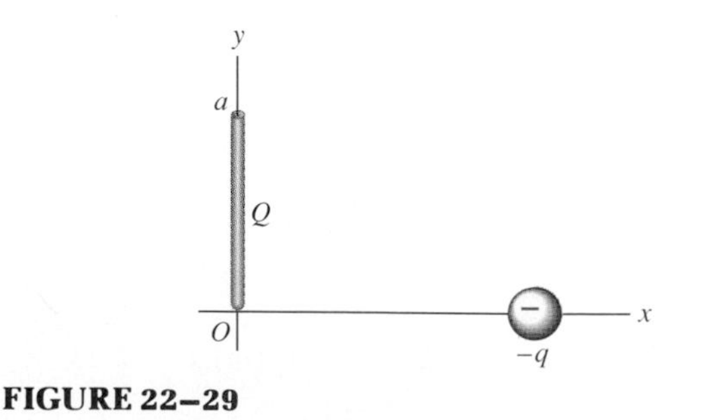


**FIGURE 22–29**

**22–40** A positive electric charge $Q$ is distributed uniformly along the positive $x$-axis from $x = 0$ to $x = a$. A positive point charge $q$ is located on the $x$-axis at $x = a + r$ so that it is a distance $r$ to the right of the end of $Q$ (Fig. 22–30). Calculate the magnitude and direction of the force that the charge distribution $Q$ exerts on $q$.

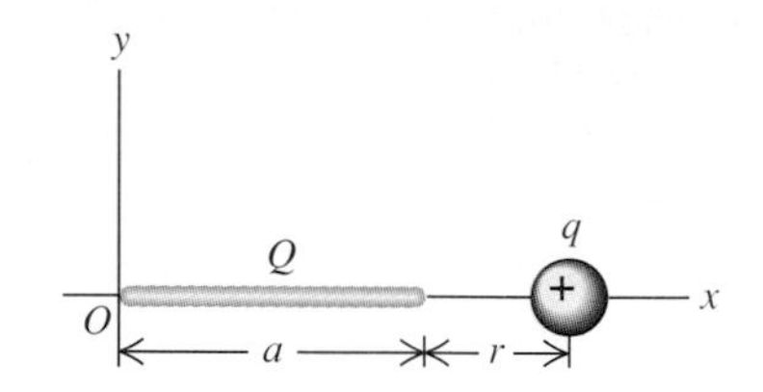


**FIGURE 22–30**

**22–41 Charge on Toner Particles.** Just outside the surface of a positively charged photocopier imaging drum the electric field has a magnitude of $2.00 \times 10^5$ N/C. What must be the magnitude of the negative charge on a toner particle that has mass $4.0 \times 10^{-12}$ kg if it is to be attracted to the drum with a force that is ten times its weight?

**22–42** Positive charge $+Q$ is distributed uniformly along the $+x$-axis from $x = 0$ to $x = a$. Negative charge $-Q$ is distributed uniformly along the $-x$-axis from $x = 0$ to $x = -a$. a) A positive point charge $q$ lies on the positive $y$-axis, a distance $y$ from the origin. Find the magnitude and direction of the force that the charge distribution exerts on $q$. Show that this force is proportional to $y^{-3}$ as $y$ becomes large. b) Suppose instead that the positive point charge $q$ lies on the positive $x$-axis, a distance $x > a$ from the origin. Find the magnitude and direction of the force that the charge distribution exerts on $q$. Show that this force is proportional to $x^{-3}$ as $x$ becomes large.

**22–43** An electron is projected into a uniform electric field that has a magnitude of 500 N/C. The direction of the field is vertically upward. The initial velocity of the electron has a magnitude of $4.00 \times 10^6$ m/s, and its direction is at an angle of 30.0° above the horizontal. a) Find the maximum distance the electron rises vertically above its initial elevation. b) After what horizontal distance does the electron return to its original elevation? c) Sketch the trajectory of the electron.

**22–44** A small sphere whose mass is 0.400 g carries a charge of $3.00 \times 10^{-10}$ C and is attached to one end of a silk fiber 8.00 cm long. The other end of the fiber is attached to a large vertical insulating sheet that has a surface charge density equal to $25.0 \times 10^{-6}$ C/m$^2$. Find the angle that the fiber makes with the vertical sheet when the sphere is in equilibrium.

**22–45** A negative point charge $q_1 = -5.00$ nC is on the $x$-axis at $x = 1.20$ m. A second point charge $q_2$ is on the $x$-axis at $x = -0.60$ m. What must be the sign and magnitude of $q_2$ for the resultant electric field at the origin to be a) 60.0 N/C in the $+x$-direction; b) 60.0 N/C in the $-x$-direction?

**22–46 Operation of an Inkjet Printer.** In an inkjet printer, letters are printed by squirting drops of ink at the paper

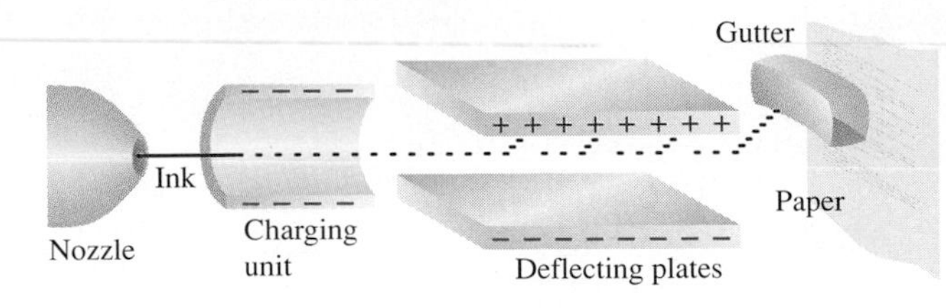


**FIGURE 22–31**

from a rapidly moving nozzle. The pattern on the paper is controlled by an electrostatic valve that determines at each nozzle position whether ink is squirted onto the paper or not. One scheme for doing this is illustrated in Fig. 22–31. The ink drops, $15 \mu\text{m}$ in radius, leave the nozzle and travel toward the paper at 20 m/s. The drops pass through a charging unit that gives each drop a negative charge $-q$ when the drop acquires a number of excess electrons. The drops then pass between parallel deflecting plates where there is a uniform vertical electric field with magnitude $8.0 \times 10^4$ N/C. If the drop is deflected more than 0.40 mm by the time it reaches the end of the deflection plate, it strikes a gutter and is returned to the ink supply. The charging unit can be rapidly turned off and on by a computerized controller. If drops are to pass through to the paper only when the charging unit is turned off, what minimum magnitude of charge must be given to each drop when the charging unit is on? (Assume that the density of the ink drop is the same as that of water, 1000 kg/m$^3$.)

**22–47** A charge of 16.0 nC is fixed at the origin of coordinates; a second charge of unknown magnitude is at $x = 3.00$ m, $y = 0$; and a third charge of 12.0 nC is at $x = 6.00$ m, $y = 0$. What are the sign and magnitude of the unknown charge if the resultant field at $x = 8.00$ m, $y = 0$ has a magnitude of 18.0 N/C and is in the $+x$-direction?

**22–48** Positive electric charge is uniformly distributed around a semicircle of radius $a$ with total charge $Q$ (Fig. 22–32). What is the electric field (magnitude and direction) at the center of curvature (point $P$)?

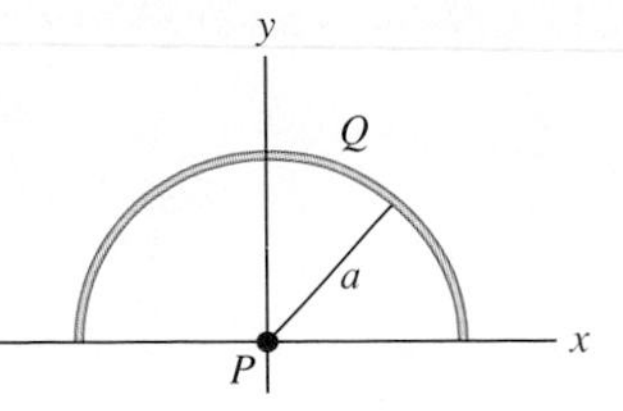


**FIGURE 22–32**

**22–49** Negative electric charge is distributed uniformly around a quarter-circle of radius $a$ with total charge $-Q$. The quarter-circle is in the first quadrant, with its center of curvature at the origin. What are the $x$- and $y$-components of the resultant electric field at the origin?

**22–50** Electric charge is distributed uniformly along each side of a square. Two adjacent sides have positive charge with total charge $+Q$ on each. a) If the other two sides have negative charge with total charge $-Q$ on each (Fig. 22–33), what are the $x$- and $y$-components of the resultant electric field at the center of the square? Each side of the square has length $a$. b) Repeat the calculation of part (a) if all four sides have positive charge $+Q$.

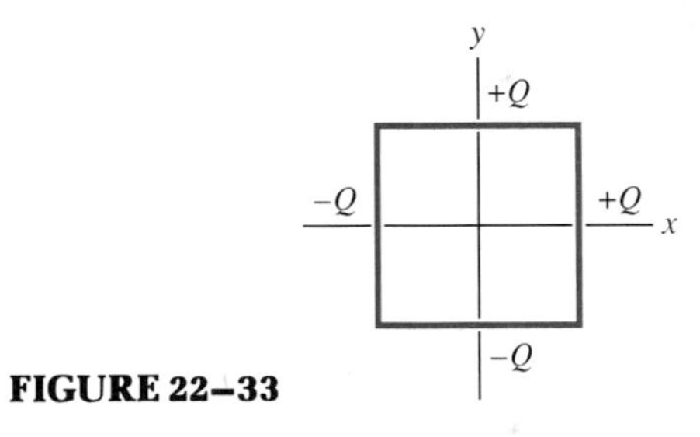


**FIGURE 22–33**

## CHALLENGE PROBLEMS

**22–51** Three charges are placed as shown in Fig. 22–34. It is known that the magnitude of $q_1$ is 4.00 $\mu$C, but its sign and the value of the charge $q_2$ are not known. The charge $q_3$ equals $+2.00$ $\mu$C, and the resultant force $\boldsymbol{F}$ on $q_3$ is measured to be entirely in the negative $x$-direction. a) Considering the different possible signs of $q_1$ and $q_2$, there are four possible force diagrams representing the forces $\boldsymbol{F}_1$ and $\boldsymbol{F}_2$ that $q_1$ and $q_2$ exert on $q_3$. Sketch these four possible force configurations. b) Using the sketches from part (a) and the fact that the net force on $q_3$ has no $y$-component and a negative $x$-component, deduce the signs of the charges $q_1$ and $q_2$. c) Calculate the magnitude of $q_2$. d) Determine $F$, the magnitude of the resultant force on $q_3$.

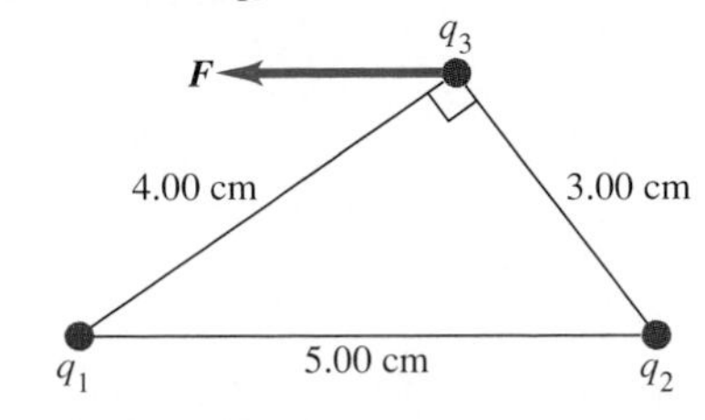


**FIGURE 22–34**

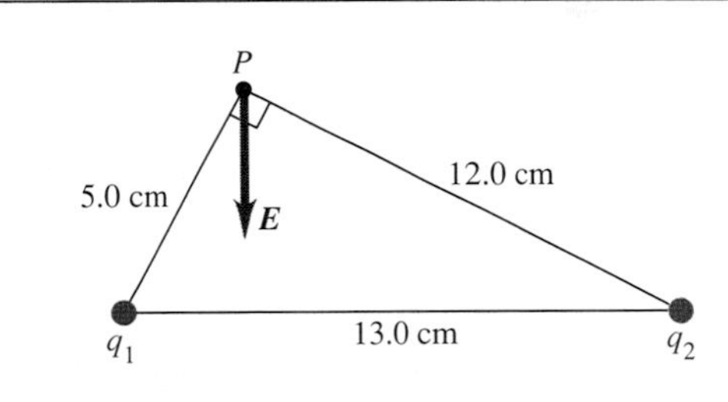

**FIGURE 22–35**

**22–52** Two charges are placed as shown in Fig. 22–35. It is known that the magnitude of $q_1$ is 4.00 $\mu$C, but its sign and the value of the other charge, $q_2$, are not known. The direction of the resultant electric field $\boldsymbol{E}$ at point $P$ is entirely in the negative $y$-direction. a) Considering the different possible signs of $q_1$ and $q_2$, there are four possible diagrams that could represent the electric fields $\boldsymbol{E}_1$ and $\boldsymbol{E}_2$ produced by $q_1$ and $q_2$. Sketch the four possible electric field configurations. b) Using the sketches from part (a) and the fact that the net electric field at $P$ has no $x$-component and a negative $y$-component, deduce the signs of $q_1$ and $q_2$. c) Determine the magnitude of the resultant field $\boldsymbol{E}$.

CHAPTER 23

# Gauss's Law

Gauss's law can help us determine the charge that produces an electric field, provided the field has spatial symmetry. Three important cases are fields with spherical symmetry (say, due to a point charge) ...

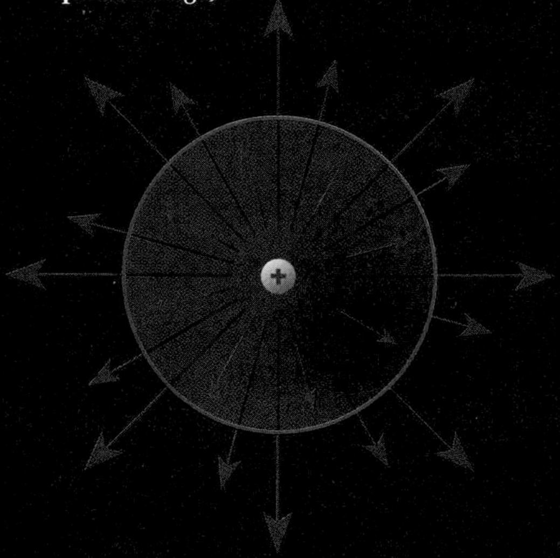

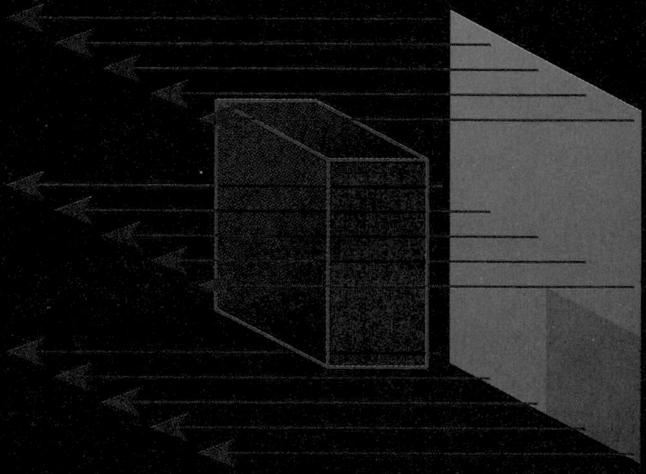

...or uniform symmetry in space (say, due to a large sheet of charge)...

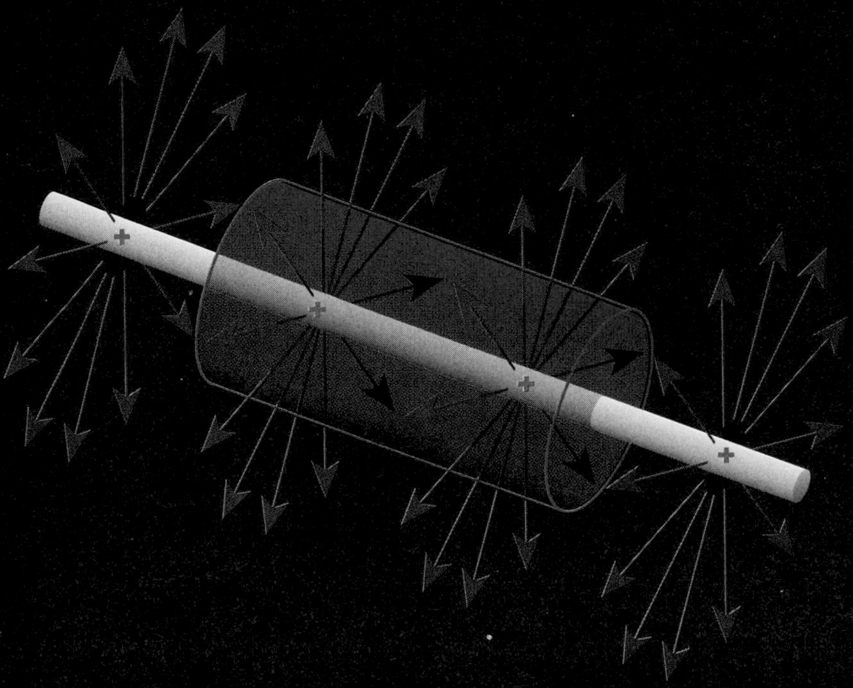

...or radial symmetry (say, due to a long line of charge) .

## KEY CONCEPTS

**• Electric flux through a surface is the product of the surface area and the component of electric field perpendicular to the surface.**

**• Gauss's law states that the total electric flux out of a closed surface is proportional to the total electric charge enclosed within the surface. This law is useful for calculating fields caused by charge distributions that have various symmetry properties.**

**• When a conductor has a net charge that is at rest, the charge lies entirely on the conductor's surface, and the electric field is zero everywhere within the conductor.**

**• Gauss's law is used to analyze experiments that test the validity of Coulomb's law with great precision.**

## INTRODUCTION

Often there's an easy way and a hard way to do a job; the easy way sometimes involves nothing more than having the right tools to use. In physics, one of the important tools used to simplify problems is to make use of *symmetry properties* of systems. Many physical systems have symmetry; for example, a cylindrical body doesn't look any different after you've rotated it on its axis, and a charged metal sphere looks just the same after you've turned it about any axis through its center. Calculations with a system that has symmetry properties can nearly always be simplified if we can make use of the symmetry.

Gauss's law is part of the key to using symmetry considerations in electric-field calculations. We'll find that many field calculations can be done much more simply with Gauss's law than with the methods that we used in Chapter 22. For example, the field of a straight-line or a plane-sheet charge distribution, which we derived in Section 22–6 using some fairly strenuous integrations, can be obtained in a few lines with the help of Gauss's law. This law can be derived from Coulomb's law, but it is so powerful that it can be used to solve some field problems that would be extremely complex if approached with Coulomb's law alone.

## 23–1 ELECTRIC FLUX

Gauss's law and Coulomb's law are two alternative formulations of the same basic relationship between a charge distribution and the electric field it creates. Coulomb's law describes the field at a *point* $P$ caused by a single *point* charge $q$. To calculate fields from an *extended* charge distribution, we have to represent that distribution as an assembly of point charges and use the superposition principle.

Gauss's law takes a more global view. Given any general distribution of charge, we surround it with an imaginary surface that encloses the charge. Then we look at the electric field at various points on this imaginary surface. Gauss's law is a relation between the field at *all* the points on the surface and the total charge enclosed within the surface. This may sound like a rather indirect way of expressing things, but it turns out to be a very powerful relationship.

In formulating Gauss's law we will use the concept of **electric flux,** also called *flux of the electric field*. We'll define this concept first, and then we'll discuss an analogy with fluid flow that will help you to develop intuition about it.

The definition of electric flux involves an area $A$ and the electric field at various points on the area. The area needn't be the surface of a real body; in fact, it will usually be an imaginary area in space. Consider first a small, flat area $A$ perpendicular to a uniform electric field $\boldsymbol{E}$ (Fig. 23–1a). We define the electric flux $\Phi_E$ through the area $A$ to be the product of the field magnitude $E$ and the area $A$:

$$\Phi_E = EA.$$

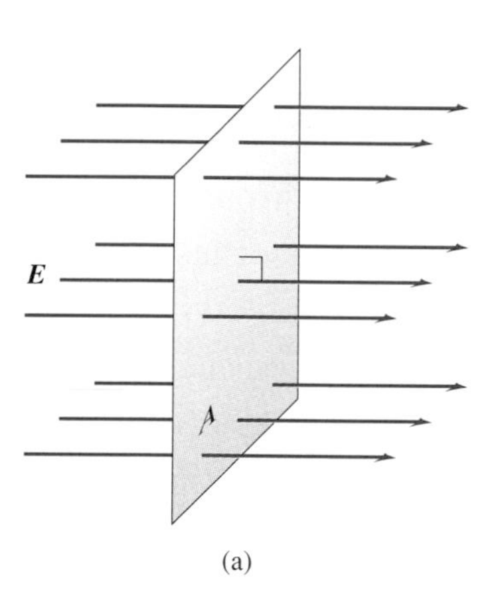


Roughly speaking, we can picture $\Phi_E$ in terms of the field lines passing through $A$. More area means more lines through the area, and stronger field means more closely spaced lines and therefore more lines per unit area.

If the area element $A$ is flat but not perpendicular to the field $\boldsymbol{E}$, then fewer field lines pass through it. In this case, what counts is the silhouette area of $A$ as we look along the direction of $\boldsymbol{E}$; this is the area $A_\perp$ in Fig. 23–1b, the *projection* of the area $A$ onto a surface perpendicular to $\boldsymbol{E}$. Two sides of the projected rectangle have the same length as the original one, but the other two are foreshortened by a factor $\cos\phi$; so the projected area $A_\perp$ is equal to $A\cos\phi$. We generalize our definition of electric flux for a uniform electric field to

$$\Phi_E = EA\cos\phi. \tag{23–1}$$

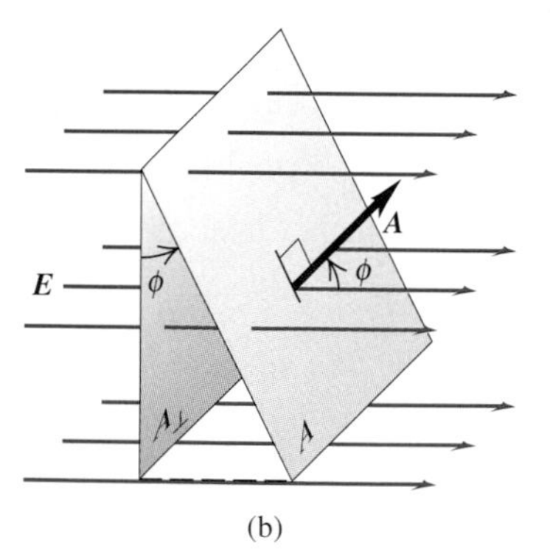


**23–1** A flat surface in a uniform electric field. (a) The electric flux through the surface equals $EA$. (b) When the area vector makes an angle $\phi$ with $\boldsymbol{E}$, $A_\perp = A\cos\phi$. The flux is zero when $\phi = 90°$.

Also, $E\cos\phi$ is the component of the vector $\boldsymbol{E}$ perpendicular to the area. Calling this component $E_\perp$, we can rewrite Eq. (23–1) as

$$\Phi_E = E_\perp A. \tag{23–2}$$

We can express the definition of electric flux more compactly by using the concept of *vector area* $\boldsymbol{A}$, a vector quantity with magnitude $A$ and a direction perpendicular to the area we are describing. The vector area $\boldsymbol{A}$ describes both the magnitude of an area and its orientation in space. In terms of $\boldsymbol{A}$, Eq. (23–1) becomes

$$\Phi_E = \boldsymbol{E}\cdot\boldsymbol{A}. \tag{23–3}$$

Equations (23–1), (23–2), and (23–3) express electric flux in different but equivalent ways. The SI unit for electric flux is $1\ \mathrm{N\cdot m^2/C}$.

Finally, what happens if the electric field $\boldsymbol{E}$ isn't uniform but varies from point to point over the area $A$? Or what if $A$ is part of a curved surface? Then we divide $A$ into many small elements $dA$, calculate the electric flux through each one, and integrate the results to obtain the total flux:

$$\Phi_E = \int E \cos\phi \, dA = \int E_\perp \, dA = \int \mathbf{E} \cdot d\mathbf{A}. \tag{23–4}$$

We call this integral the **surface integral** of the component $E_\perp$ over the area, or the surface integral of $\mathbf{E} \cdot d\mathbf{A}$.

The word *flux* comes from a Latin word meaning "flow." Even though an electric field is *not* a flow, an analogy with fluid flow will help to develop your intuition about electric flux. For water in a pipe the volume flow rate $dV/dt$ (cubic meters per second, for example) is the cross-section area $A$ multiplied by the flow speed $v$. More generally, we can consider the volume flow rate through *any* area in a flowing fluid, such as the flow through the wire rectangle with area $dA$ in Figure 23–2. When the area is perpendicular to the flow velocity $\mathbf{v}$ (Fig. 23–2a), $dV/dt = v\,dA$. When the rectangle is tilted at an angle $\phi$ (Fig. 23–2b), the area that counts is the silhouette area that we see when looking in the direction of $\mathbf{v}$. This area is $dA \cos\phi$, as shown, and the volume flow rate through $dA$ is

$$\frac{dV}{dt} = v\,dA\cos\phi.$$

In terms of the vector area element $d\mathbf{A}$ the volume flow rate is equal to $\mathbf{v}\cdot d\mathbf{A}$. If $\phi = 90°$, then $\cos\phi = 0$, and *no* fluid passes through the rectangle. Also, $v\cos\phi$ is equal to the component $v_\perp$ of $\mathbf{v}$ perpendicular to $d\mathbf{A}$, so $dV/dt = v_\perp\,dA$. Thus we can express the volume flow rate through $d\mathbf{A}$ in any of these forms:

$$\frac{dV}{dt} = v\,dA\cos\phi = v_\perp\,dA = \mathbf{v}\cdot d\mathbf{A}. \tag{23–5}$$

This quantity is called the *flux* of $\mathbf{v}$ through $d\mathbf{A}$; this is a natural term because it represents the volume rate of flow of fluid through the area. In the electric-field situation, *nothing is flowing,* but the analogy to fluid flow may help you to visualize the concept.

We can represent the direction of a vector area element by using a *unit vector* $\mathbf{n}$ perpendicular to the area; $\mathbf{n}$ stands for "normal." Then

$$d\mathbf{A} = \mathbf{n}\,dA.$$

A surface element has two sides, so there are two possible directions for $\mathbf{n}$ and $d\mathbf{A}$. We must always specify which direction we have chosen. With Gauss's law we will always work with the *total* flux through a *closed* surface that has an inside and an outside. We will always choose the direction of $\mathbf{n}$ to be *outward* from the surface, and we will speak of the flux *out of* the surface.

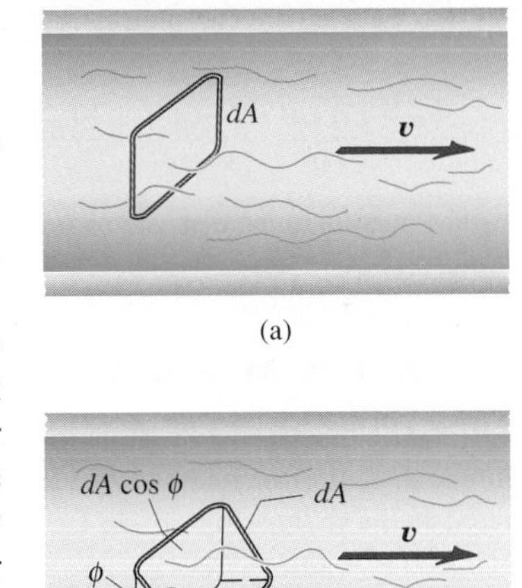


(a)

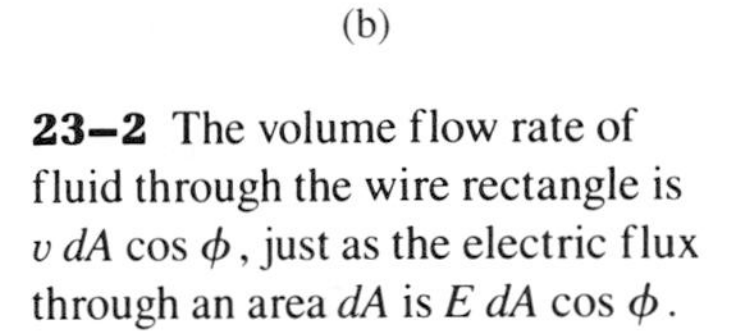


(b)

**23–2** The volume flow rate of fluid through the wire rectangle is $v\,dA\cos\phi$, just as the electric flux through an area $dA$ is $E\,dA\cos\phi$.

## EXAMPLE 23–1

**Electric flux through a disk** A disk with radius 0.10 m is oriented with its normal unit vector $\mathbf{n}$ at an angle of 30° to a uniform electric field $\mathbf{E}$ with magnitude $2.0 \times 10^3$ N/C (Fig. 23–3). a) What is the total electric flux through the disk? b) What is the total flux through the disk if it is turned so that its plane is parallel to $\mathbf{E}$? c) What is the total flux through the disk if its normal is parallel to $\mathbf{E}$?

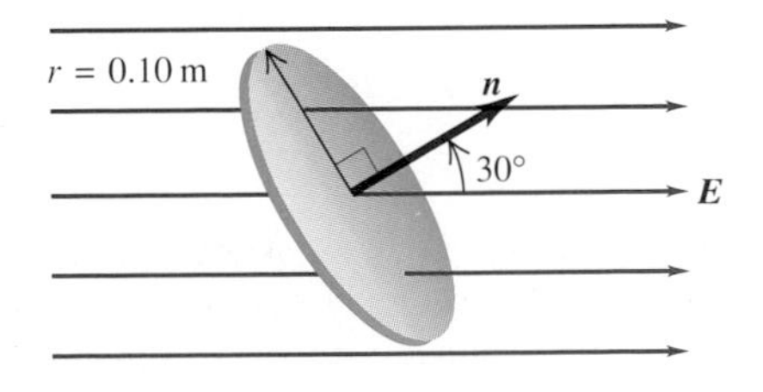


**23–3** The electric flux $\Phi_E$ through a disk depends on the angle between its normal $\mathbf{n}$ and the electric field $\mathbf{E}$.

**SOLUTION** a) The area is $A = \pi\,(0.10\text{ m})^2 = 0.0314\text{ m}^2$. From Eq. (23–1),

$$\Phi_E = EA\cos\phi = (2.0\times10^3\text{ N/C})(0.0314\text{ m}^2)(\cos 30°) = 54\text{ N}\cdot\text{m}^2/\text{C}.$$

b) The normal to the disk is now perpendicular to $\boldsymbol{E}$, so $\phi = 90°$, $\cos\phi = 0$, and $\Phi_E = 0$.

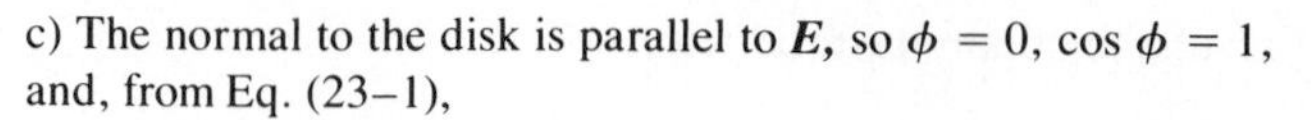

c) The normal to the disk is parallel to $\boldsymbol{E}$, so $\phi = 0$, $\cos\phi = 1$, and, from Eq. (23–1),

$$\Phi_E = EA\cos\phi = (2.0\times10^3\text{ N/C})\,(0.0314\text{ m}^2)\,(1) = 63\text{ N}\cdot\text{m}^2/\text{C}.$$

## EXAMPLE 23–2

**Electric flux through a sphere** A positive charge of magnitude 3.0 μC is surrounded by a sphere with radius 0.20 m centered on the charge (Fig. 23–4). Find the electric flux through the sphere due to this charge.

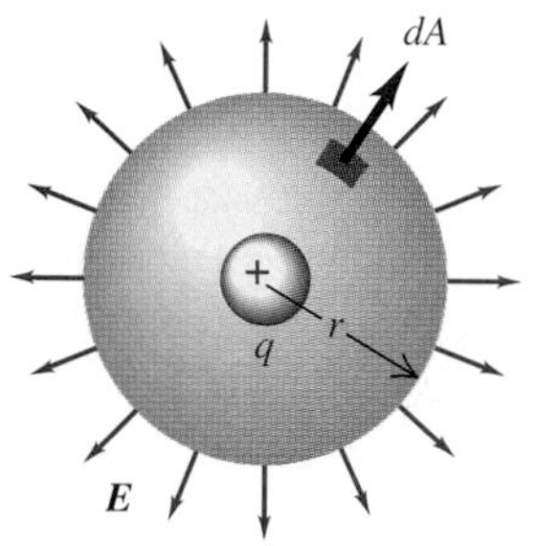


**23–4** Electric flux through a sphere centered on a charge.

**SOLUTION** At any point on the sphere the magnitude of $\boldsymbol{E}$ is

$$E = \frac{1}{4\pi\epsilon_0}\frac{q}{r^2} = (9.0\times10^9\text{ N}\cdot\text{m}^2/\text{C}^2)\,\frac{3.0\times10^{-6}\text{ C}}{(0.20\text{ m})^2} = 6.75\times10^5\text{ N/C}.$$

From symmetry the field is perpendicular to the spherical surface at every point. We take the positive direction for $\boldsymbol{n}$ and $E_\perp$ to be outward, so $E_\perp = E$, and the flux through a surface element $dA$ is $E\,dA$. In Eq. (23–4), $E$ is the same at every point and can be taken outside the integral; what remains is the integral $\int dA$, which is just the total area $4\pi r^2$ of the spherical surface. Thus the total flux out of the sphere is

$$\Phi_E = EA = (6.75\times10^5\text{ N/C})(4\pi)(0.20\text{ m})^2 = 3.4\times10^5\text{ N}\cdot\text{m}^2/\text{C}.$$

Note that the radius of the sphere cancelled out of this calculation; we would have obtained the same result with a sphere of radius 2.0 m or 200 m. There's a good reason for this, as we'll soon see. ■

## 23–2 GAUSS'S LAW

**Gauss's law** is an alternative to Coulomb's law for expressing the relationship between electric charge and electric field. It was formulated by Karl Friedrich Gauss (1777–1855), one of the greatest mathematicians of all time. Many areas of mathematics, from number theory and geometry to the theory of differential equations, bear the mark of his influence, and he made equally significant contributions to theoretical physics.

Gauss's law states that the total electric flux out of any closed surface (that is, a surface enclosing a definite volume) is proportional to the total (net) electric charge inside the surface. To develop this relationship, we'll start with the field of a single positive point charge $q$. The field lines radiate out equally in all directions. We place this charge at the center of an imaginary spherical surface with radius $R$. The magnitude $E$ of the electric field at every point on the surface is given by

$$E = \frac{1}{4\pi\epsilon_0}\frac{q}{R^2}.$$

At each point on the surface, $\boldsymbol{E}$ is perpendicular to the surface, and its magnitude is the same at every point, just as in Example 23–2 (Section 23–1). The total electric flux is just the product of the field magnitude $E$ and the total area $A = 4\pi R^2$ of the sphere:

$$\Phi_E = EA = \frac{1}{4\pi\epsilon_0}\frac{q}{R^2}(4\pi R^2) = \frac{q}{\epsilon_0}. \tag{23–6}$$

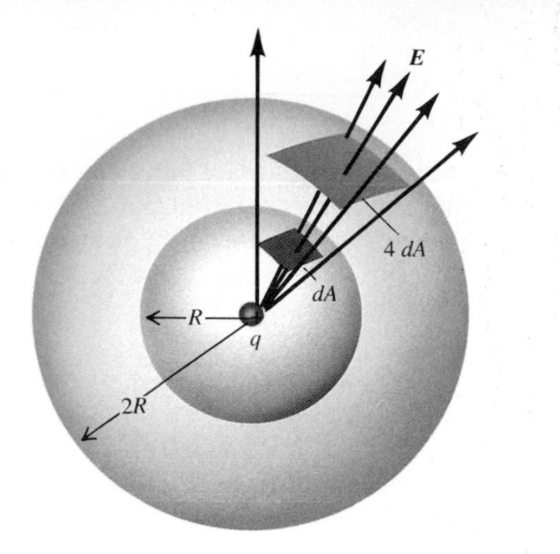


**23–5** Projection of an element of area $dA$ of a sphere of radius $R$ onto a concentric sphere of radius $2R$. The projection multiplies each linear dimension by two, so the area element on the larger sphere is $4\,dA$. The same number of lines and the same flux pass through each area element.

*The flux is independent of the radius R of the sphere.* It depends only on the charge $q$ enclosed by the sphere.

We can also interpret this result in terms of field lines. We consider two spheres with radii $R$ and $2R$, respectively (Fig. 23–5). According to Coulomb's law, the field magnitude is $\frac{1}{4}$ as great on the larger sphere as on the smaller, so the number of field lines per unit area should be $\frac{1}{4}$ as great. But the area of the larger sphere is four times as great, so the *total* number of field lines passing through is the same for both spheres.

What is true of the entire sphere is also true of any portion of its surface. In Fig. 23–5 an area $dA$ is outlined on a sphere of radius $R$ and projected onto the sphere of radius $2R$ by lines drawn from the center through points on the boundary of $dA$. The area projected onto the larger sphere is clearly $4\,dA$. The electric flux $E\,dA$ is the same for both areas and is independent of the radius of the sphere.

This projection technique shows us how to extend this discussion to nonspherical surfaces. Instead of a second sphere, let us surround the sphere of radius $R$ by a surface of irregular shape, as in Fig. 23–6a. Consider a small element of area $dA$ on the irregular surface; we note that this area is *larger* than the corresponding element on a spherical surface at the same distance from $q$. If a normal to the irregular surface makes an angle $\phi$ with a radial line from $q$, two sides of the area projected onto the spherical surface are foreshortened by a factor $\cos\phi$ (Fig. 23–6b). The other two sides are unchanged. Thus the quantity corresponding to $E\,dA$ for the spherical surface is $E\,dA\cos\phi$ for the irregular surface. But the *electric flux* through this element is the same as the flux through the corresponding element on the sphere.

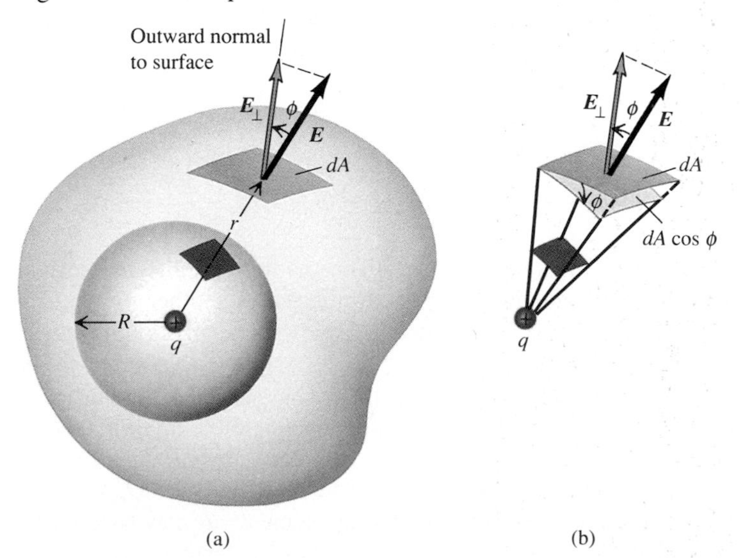


**23–6** (a) The outward normal to the surface makes an angle $\phi$ with the direction of $\boldsymbol{E}$. (b) The projection of the area element $dA$ onto the spherical surface is $dA\cos\phi$.

We can divide the entire irregular surface into elements $dA$, compute the electric flux $E\,dA\cos\phi$ for each, and sum the results by integrating, as in Eq. (23–4). Each of the area elements projects onto a corresponding element on the sphere. Thus the *total* electric flux through the irregular surface, given by any of the forms of Eq. (23–4), must be the same as the total flux through the sphere, which Eq. (23–6) shows is equal to $q/\epsilon_0$. Thus for the irregular surface,

$$\Phi_E = \oint \boldsymbol{E}\cdot d\boldsymbol{A} = \frac{q}{\epsilon_0}. \qquad (23\text{–}7)$$

This equation holds for *any* shape of surface, provided only that it is a *closed* surface enclosing the charge $q$. The circle on the integral sign reminds us that the integral is always taken over a *closed* surface.

The area elements $d\boldsymbol{A}$ and the corresponding unit vectors $\boldsymbol{n}$ always point *out of* the volume enclosed by the surface. The electric flux is then positive in areas where the electric field points out of the surface and negative where it is inward. Also, $E_\perp$ is positive at points where $\boldsymbol{E}$ points out of the surface and negative when $\boldsymbol{E}$ points into the surface.

For a closed surface enclosing *no* charge,

$$\Phi_E = \oint \boldsymbol{E}\cdot d\boldsymbol{A} = 0.$$

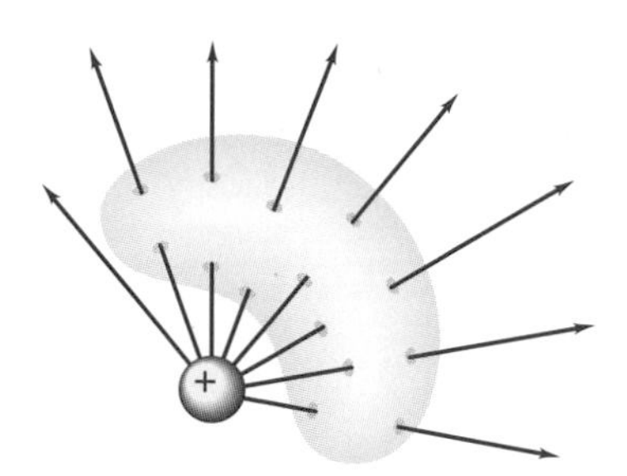


**23–7** Every electric field line from an *external* charge that enters any closed surface at one point leaves at another.

This is a mathematical statement of the fact that when a region contains no charge, any field lines (caused by charges *outside* the region) that enter on one side must leave again on the other side. Figure 23–7 illustrates this point. **Field lines can begin or end inside a region of space only when there is charge in that region.**

If the point charge in Fig. 23–6 is negative, the $\boldsymbol{E}$ field is directed radially *inward;* the angle $\phi$ is then greater than 90°, its cosine is negative, and the integral in Eq. (23–7) is negative. But since $q$ is also negative, Eq. (23–7) still holds.

Now comes the final step in obtaining the general form of Gauss's law. Suppose the surface encloses not just one point charge $q$ but several charges $q_1, q_2, q_3, \ldots$. The total (resultant) electric field $\boldsymbol{E}$ at any point is the vector sum of the $\boldsymbol{E}$ fields of the individual charges. Let $Q_{\text{encl}}$ be the *total* charge enclosed by the surface: $Q_{\text{encl}} = q_1 + q_2 + q_3 + \cdots$. Also let $\boldsymbol{E}$ be the *total* field at $d\boldsymbol{A}$ and $dE_\perp$ its component perpendicular to $d\boldsymbol{A}$. Then we can write an equation like Eq. (23–7) for each charge and its corresponding field, and add the results. When we do, we obtain the general statement of Gauss's law:

$$\oint \boldsymbol{E}\cdot d\boldsymbol{A} = \frac{Q_{\text{encl}}}{\epsilon_0}. \qquad (23\text{–}8)$$

Using the definition of $Q_{\text{encl}}$ and the various forms of Eq. (23–4), we can express this in the following equivalent forms:

$$\oint E\cos\phi\, dA = \oint E_\perp\, dA = \oint \boldsymbol{E}\cdot d\boldsymbol{A} = \frac{\Sigma q_i}{\epsilon_0} = \frac{Q_{\text{encl}}}{\epsilon_0}.$$

The various forms of the integral all say the same thing in different terms. In specific problems, one form is sometimes more convenient than another.

In Eq. (23–8), $Q_{\text{encl}}$ is always the algebraic sum of all the (positive and negative) charges enclosed by the surface and $\boldsymbol{E}$ is the *total* field at each point on the surface. Also note that this field is in general caused partly by charges inside the surface and partly by charges outside. The outside charges don't contribute to the total (net) flux through the surface, so Eq. (23–8) is still correct even when there are additional charges outside the surface that contribute to the electric field at the surface. When $Q_{\text{encl}} = 0$, the total flux through the surface must be zero, even though some areas may have positive flux and others negative.

It may look as though evaluating the integral in Eq. (23–8) is a hopeless task. Sometimes it is, but other times it is surprisingly easy. We'll work out several examples in the next section.

## EXAMPLE 23–3

Figure 23–8 shows the field produced by two equal and opposite point charges (an electric dipole). The electric flux through a surface is proportional to the number of electric field lines passing through it. Surface $A$ encloses only the positive charge, and 18 lines cross it in an outward direction. Surface $B$ encloses only the negative charge; it is also crossed by 18 lines, but in an inward direction. Surface $C$ encloses *both* charges. It is intersected by lines at 16 points; at 8 intersections the lines are outward, and at 8 they are inward. The *net* number of lines crossing in an outward direction is zero, and the net charge inside the surface is also zero. Surface $D$ is intersected at 6 points; at 3 the lines are outward, and at the other 3 they are inward. The net number of lines crossing in an outward direction, and the total charge enclosed, are both zero. Note that there are points on the surfaces where $\boldsymbol{E}$ is not perpendicular to the surface, but this does not affect the counting of the field lines.

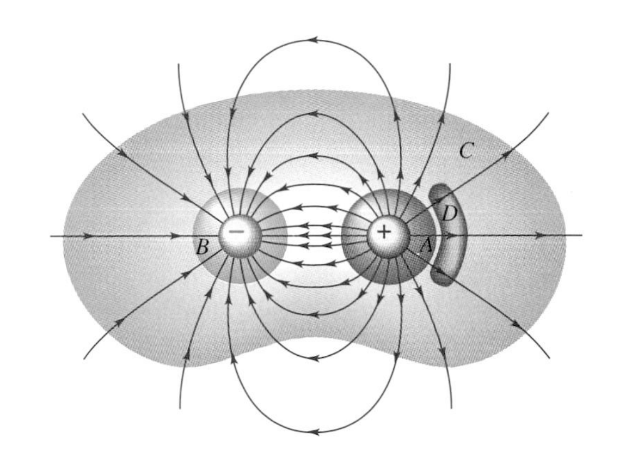


**23–8** The net number of field lines leaving a closed surface is proportional to the total charge enclosed.

## 23–3 APPLICATIONS OF GAUSS'S LAW

In principle, Gauss's law is valid for *any* distribution of charges and for *any* surface. In practical problems it is a useful relationship only when the system has recognizable symmetry properties that can be used to evaluate the flux integrals. Gauss's law is a two-way street. If we know the charge distribution and if it has enough symmetry to let us evaluate the integral in Gauss's law, we can find the field. Or if we know the field, we can use Gauss's law to find the charge distribution, such as charges on conducting surfaces. Following are examples of both kinds of applications. As you study them, watch for the role played by the symmetry properties of each system. The electric fields caused by several simple charge distributions are collected in a table in the chapter summary.

## PROBLEM-SOLVING STRATEGY

### Gauss's law

**1.** The first step is to select the surface (which we often call a *Gaussian surface*) that you are going to use with Gauss's law. If you are trying to find the field at a particular point, then that point must lie on your Gaussian surface.

**2.** The Gaussian surface does not have to be a real physical surface, such as a surface of a solid body. Often the appropriate surface is an imaginary geometric surface; it may be in empty space, embedded in a solid body, or partly both.

**3.** The Gaussian surface and the charge distribution must have some symmetry property, so that it is possible to actually evaluate the integral in Gauss's law. If the charge distribution has cylindrical or spherical symmetry, the Gaussian surface will usually be a coaxial cylinder or a concentric sphere, respectively.

**4.** Often you can think of the closed Gaussian surface as being made up of several separate areas, such as the sides and ends of a cylinder. The integral $\int E_{\perp}\, dA$ over the entire closed surface is always equal to the sum of the integrals over all the separate areas. Some of these integrals may be zero, as in points 6 and 7 below.

**5.** If $E$ is perpendicular (normal) to a surface with area $A$ at every point, and if it also has the same *magnitude* at every point on the surface, then $E_\perp = E =$ constant, and $\int E_\perp \, dA$ over that surface is equal to $EA$.

**6.** If $E$ is *tangent* to a surface at every point, then $E_\perp = 0$, and the integral over that surface is zero.

**7.** If $\mathbf{E} = \mathbf{0}$ at every point on a surface, the integral is zero.

**8.** Finally, in the integral $\int E_\perp \, dA$, $E_\perp$ is always the perpendicular component of the *total* electric field at each point on the surface. In general this field may be caused partly by charges within the volume and partly by charges outside. Even when there is *no* charge within the volume, the field at points on the surface is not necessarily zero. In that case, however, the *integral* over the closed surface is always zero.

## EXAMPLE 23–4

**Location of excess charge on a solid conductor** When excess charge is placed on a solid conductor and comes to rest, it resides entirely on the *surface*, not in the interior of the material. Here's the proof. We know from the discussion of Section 22–5 that in any electrostatic situation (charges at rest) the electric field $\mathbf{E}$ at every point in the interior of a conducting material is zero. If $\mathbf{E}$ were *not* zero, the charges would move. Suppose we construct a Gaussian surface inside the conductor, such as surface $A$ in Fig. 23–9. Because $\mathbf{E} = \mathbf{0}$ everywhere on this surface, Gauss's law requires that the net charge inside the surface is zero. Now imagine shrinking the surface down like a collapsing balloon until it encloses a region so small that we may consider it a point; then the charge at that point must be zero. We can do this anywhere inside the conductor, so *there can be no net charge at any point within the conductor.* Thus any excess charge at rest on a solid conductor must be located only on its surface, as shown.

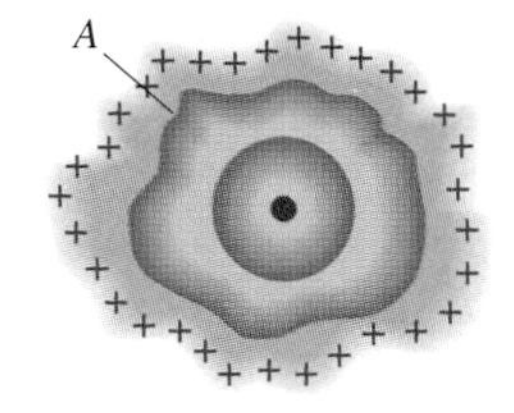


**23–9** Under electrostatic conditions, any excess charge resides entirely on the surface of a solid conductor.

## EXAMPLE 23–5

**Field of a charged conducting sphere** We place a charge $q$ on a solid conducting sphere with radius $R$. Find the electric field at any point outside the sphere.

**SOLUTION** From Example 23–4 we know that all the charge is on the surface of the sphere. We can use the property of spherical symmetry to show that the charge must be distributed *uniformly* over the surface and that the direction of the electric field at any point $P$ outside the sphere must be along a *radial* line between the center and point $P$.

The role of symmetry deserves careful discussion. When we say that the system is spherically symmetric, we mean that we can rotate it through any angle about any axis through the center. The system after rotation is indistinguishable from the original unrotated system. There is nothing about the system that distinguishes one direction or orientation in space from another. The charge is free to move on the conductor, and there is nothing about the conductor that would make it tend to concentrate more in some regions than others. If it were nonuniform, then when we rotated the system, the sphere would look the same but the charge distribution would look different. There is no property of the sphere that can make this happen. So we conclude that the surface charge distribution must be uniform.

A similar argument shows that the field must be radial. If we again rotate the system, the field pattern of the rotated system must be identical to that of the original system. If the field had a component at some point that was perpendicular to the radial direction, that component would have to be different after at least some rotations. Thus there can't be such a component, and the field must be *radial*. For the same reason the magnitude $E$ of the field depends only on the distance $r$ from the center. Thus the magnitude $E$ is the same at all points on a spherical surface with radius $r$, concentric with the conductor.

We take as our Gaussian surface an imaginary sphere with radius $r$ greater than the radius $R$ of the conducting sphere. The area of the Gaussian surface is $4\pi r^2$; $E$ is uniform over the sphere and perpendicular to it at each point. The integral in Gauss's law yields $E\,(4\pi r^2)$, and Eq. (23–8) then gives

$$4\pi r^2 E = \frac{q}{\epsilon_0} \quad \text{and} \quad E = \frac{1}{4\pi\epsilon_0}\frac{q}{r^2}. \tag{23–9}$$

This expression for the field at any point *outside* the sphere is the same as for a point charge. This shows that the field caused by the charged sphere is the same as though the entire charge were concentrated at its center. Just outside the surface of the sphere, where $r = R$,

$$E = \frac{1}{4\pi\epsilon_0}\frac{q}{R^2}.$$

Inside the sphere, as with any solid conductor when the charges are at rest, the field is zero. Thus when $r$ is less than $R$, $E = 0$. Figure 23–10 shows $E$ as a function of the distance $r$ from the center of the sphere. Also note that in the limit as $R \rightarrow 0$ the sphere becomes a point charge. Thus we have deduced Coulomb's law from Gauss's law. (In the derivation of Section 23–2 we deduced Gauss's law from Coulomb's law, so this completes the demonstration of their logical equivalence.)

We can also use this method for a conducting, hollow, spherical *shell* (a spherical conductor with a concentric spherical hole in the center) if there is no charge inside the hole. We take a spherical Gaussian surface with radius $r$ less than the radius of the hole. If there *were* a field inside the hole, it would have to be spherically symmetric (radial) as before, so $E = Q_{\text{encl}}/4\pi\epsilon_0 r^2$. But this time, $Q_{\text{encl}} = 0$, so $E$ must also be zero.

Can you use this same technique to find the electric field in the interspace between a charged sphere and a concentric hollow conducting sphere that surrounds it?

Because gravitational forces also have a $1/r^2$ dependence, there is a Gauss's law for gravitation. Reasoning similar to this discussion can be used to prove that the *gravitational* interaction of any spherically symmetric mass distribution, at any point outside the distribution, is the same as though all the mass were concentrated at the center. This is why we can treat spherically symmetric bodies as points when we calculate gravitational interactions. We proved this in Section 12–7 using some fairly strenuous analysis; the proof using Gauss's law for gravitation is almost trivial.

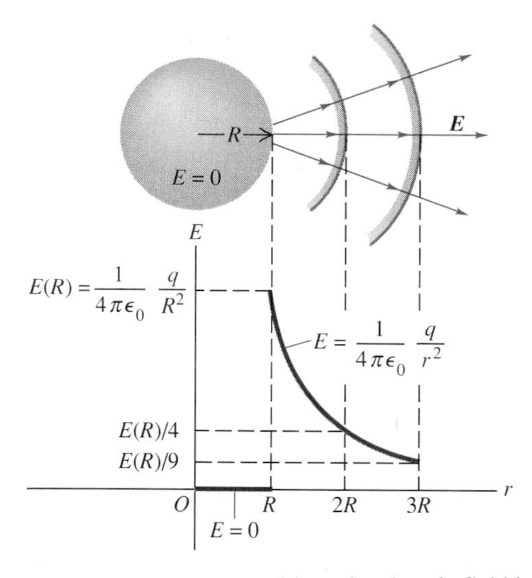


**23–10** Under electrostatic conditions the electric field inside a solid conducting sphere is zero. Outside the sphere, the electric field drops off as $1/r^2$, as though all the excess charges were concentrated at the center.

## EXAMPLE 23–6

**Field of a line charge** Electric charge is distributed uniformly along a long thin wire; the charge *per unit length* is $\lambda$. What is the electric field?

**SOLUTION** What is the symmetry? We can rotate the system through any angle about its axis, and we can shift it by any amount along the axis; in each case the resulting system is indistinguishable from the original. Using the same argument as in Example 23–5, we conclude that the electric field at each point does not change when either of these operations is carried out. The field cannot have any component parallel to the wire; if it did, there would have to be something to distinguish one end of the wire from the other; and there isn't. Also, the field cannot have any component tangent to a circle in a plane perpendicular to the wire with its center on the wire. If it did, we would have to explain why the component pointed in one direction around the wire rather than the other. What's left? Only a component radially outward from the wire at each point. So if the wire is very long and we are not too near either end, then the field lines outside the wire are *radial* and lie in planes perpendicular to the wire. The field *magnitude* must depend only on the radial distance from the wire.

These symmetry properties of the field suggest that we use as a Gaussian surface a *cylinder* with arbitrary radius $r$ and arbitrary length $l$ with its ends perpendicular to the wire (Fig. 23–11). The total charge inside the Gaussian surface is $Q_{\text{encl}} = \lambda l$. We break the surface integral into an integral over each end and one over the side. At the ends, $\boldsymbol{E}$ and $\boldsymbol{dA}$ are perpendicular, so they make no contribution to the integral. From symmetry, $\boldsymbol{E}$ is perpendicular to the side surface and parallel to $\boldsymbol{dA}$ at each point, so $E = E_\perp$ is the same everywhere on the side surface. The area of this surface is $2\pi rl$. (To make a paper cylinder with radius $r$ and height $l$, you need a paper rectangle with width $2\pi r$, height $l$, and area $2\pi rl$.) From Eq. (23–8),

$$(E)(2\pi rl) = \frac{\lambda l}{\epsilon_0} \quad \text{and} \quad E = \frac{1}{2\pi\epsilon_0}\frac{\lambda}{r}. \qquad (23\text{–}10)$$

This is the same result that we found in Example 22–11 (Section 22–6) by much more laborious means.

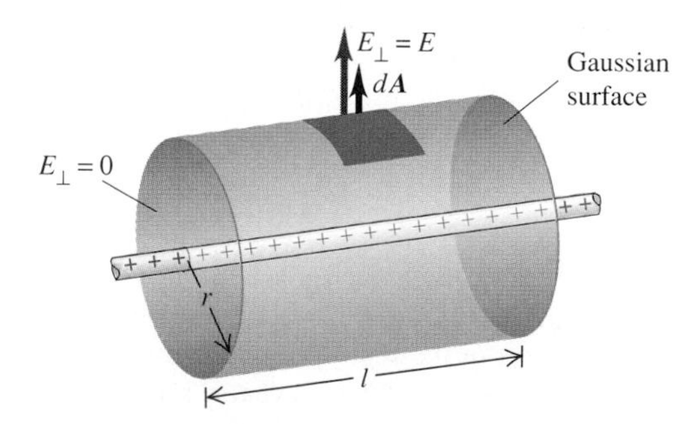


**23–11** The coaxial cylindrical Gaussian surface is used to find the electric field outside a long charged wire.

Note that although the *entire* charge on the wire contributes to the field $\boldsymbol{E}$, only the part of the total charge that is within the Gaussian surface is considered when we apply Gauss's law. This may seem strange; it looks as though we have somehow obtained the right answer by ignoring part of the charge and that the field of a *short* wire of length $l$ would be the same as that of a very long wire. But we *do* include the entire charge on the wire when we make use of the *symmetry* of the problem. If the wire is short, the symmetry with respect to shifts along the axis is not present, and the field is not uniform in magnitude over our Gaussian surface. Gauss's law is no longer useful, and the problem is best handled by using Coulomb's law, as we showed in Example 22–11.

We can use this method to show that the field at points outside a long uniformly charged cylinder is the same as though all the charge were concentrated on a line along its axis. We can also calculate the electric field in the space between a charged cylinder and a coaxial hollow conducting cylinder surrounding it. We leave these calculations as problems.

## EXAMPLE 23–7

**Field of an infinite plane sheet of charge** Find the electric field caused by a large, uniformly charged flat sheet if the charge per unit area is $\sigma$.

**SOLUTION** We use the Gaussian surface shown in Fig. 23–12, a cylinder with its axis perpendicular to the sheet of charge, with ends of area $A$. The charge distribution doesn't change if we slide it in any direction parallel to the plane. From this we conclude that at each point, $\boldsymbol{E}$ is perpendicular to the plane. From symmetry the field has the same magnitude $E$ at a given distance on either side of the surface and is directed normally away from the sheet of charge (if $\sigma$ is positive). Because $\boldsymbol{E}$ is perpendicular to the plane, it is parallel to the *side* walls of the cylinder, and $E_\perp$ at these walls is zero. At each end of the cylinder, $E_\perp$ is equal to $E$. The integral in Gauss's law becomes simply $2EA$. The net charge within the Gaussian surface is $\sigma A$; Gauss's law, Eq. (23–8), gives

$$2EA = \frac{\sigma A}{\epsilon_0}, \quad \text{and} \quad E = \frac{\sigma}{2\epsilon_0}. \tag{23–11}$$

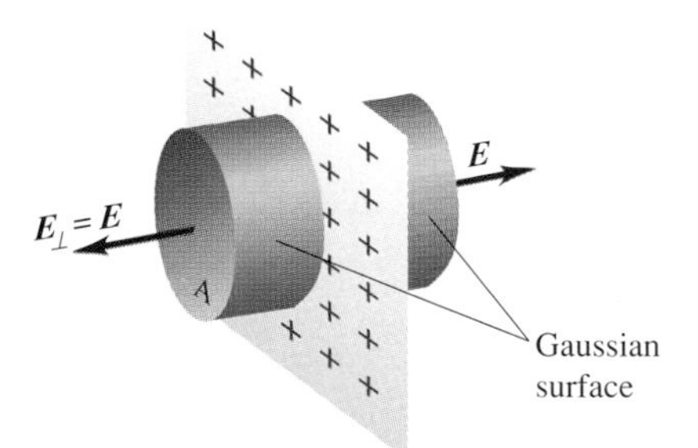


**23–12** Gaussian surface in the form of a cylinder for finding the field of an infinite plane sheet of charge.

This is the same result that we found in Example 22–12 (Section 22–6) using a much more complex calculation. The field is uniform and perpendicular to the plane. Its magnitude is *independent* of the distance from the sheet. The field lines are therefore straight, parallel, and uniformly spaced.

The assumptions that the sheet is infinitely large and has zero thickness are idealizations; nothing in nature is really infinitely large or thin. But Eq. (23–11) is a good approximation for points that are close to the sheet (compared to the dimensions) and not too near its edges. At such points the field is very nearly uniform and perpendicular to the plane.

## EXAMPLE 23–8

**Field between oppositely charged parallel conducting plates** Two large plane parallel conducting plates are given equal and opposite charges; the charge per unit area is $+\sigma$ for one and $-\sigma$ for the other. Find the electric field in the region between the plates.

**SOLUTION** The field between and around the plates is approximately as shown in Fig. 23–13a. Because opposite charges attract, most of the charge accumulates at the opposing faces of the plates. A small amount of charge resides on the *outer* surfaces of the plates, and there is some spreading, or "fringing," of the field at the edges. If the plates are very large in comparison to the distance between them, the fringing becomes negligible except near the edges. In this case we can assume that the field is uniform in the interior region between plates, as in Fig. 23–13b, and that the charges are distributed uniformly over the opposing surfaces.

We can use Eq. (23–11) for each plate. The electric field at any point is the resultant of the fields due to two sheets of charge with opposite sign. At points $a$ and $c$ in Fig. 23–13b, each of the components $\boldsymbol{E}_1$ and $\boldsymbol{E}_2$ has magnitude $\sigma/2\epsilon_0$, but they have opposite directions, and their resultant is zero. This is also true at every point within the material of each plate, consistent with the requirement that with charges at rest there can be no field within a conductor. At any point $b$ between the plates the components have the same direction; their resultant is

$$E = \frac{\sigma}{\epsilon_0}. \tag{23–12}$$

We can also get this result by applying Gauss's law to the surfaces shown by purple lines. We leave this as a problem.

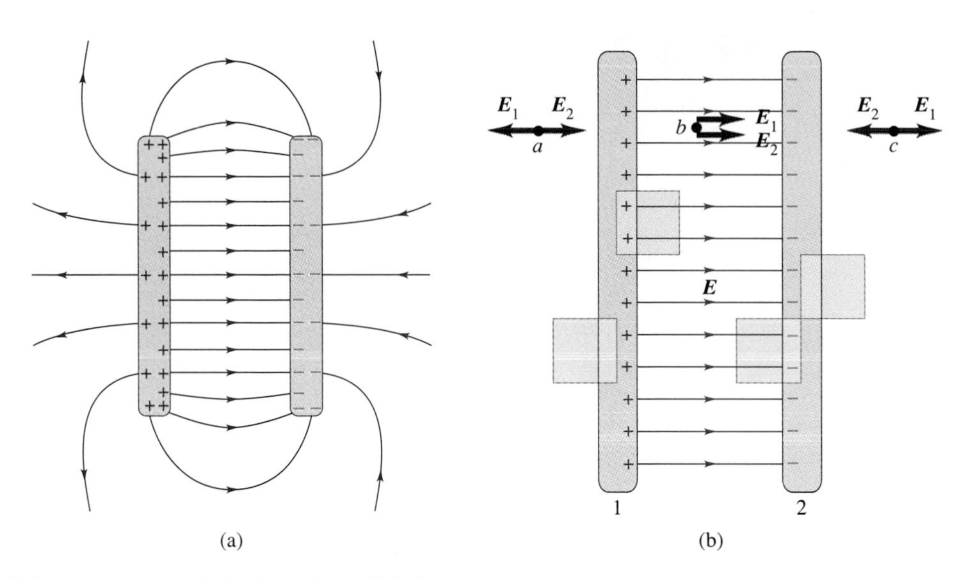


**23–13** Electric field between oppositely charged parallel plates.

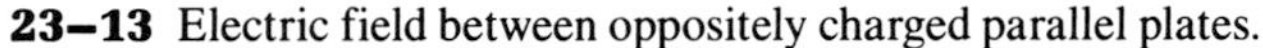

## EXAMPLE 23–9

**Uniformly charged sphere** Electric charge is distributed uniformly *throughout the volume* of an insulating sphere with radius $R$; the total charge is $Q$. Find the electric-field magnitude at a point $P$ inside the sphere at a distance $r$ from the center.

**SOLUTION** We choose as our Gaussian surface a sphere with radius $r$, concentric with the charge distribution. The volume charge density $\rho$ (charge per unit volume) is

$$\rho = \frac{Q}{4\pi R^3/3}.$$

The volume $V'$ enclosed by the Gaussian surface is $\frac{4}{3}\pi r^3$, so the total charge $Q_{\text{encl}}$ enclosed by that surface is

$$Q_{\text{encl}} = \rho V' = \frac{Q}{4\pi R^3/3}\frac{4}{3}\pi r^3 = Q\frac{r^3}{R^3}.$$

From symmetry the electric-field magnitude has the same value $E$ at every point on the Gaussian surface, and its direction at every point is radially outward. The total area of the surface is $4\pi r^2$, so the value of the sum in Gauss's law is simply $4\pi r^2 E$. We equate this to $Q_{\text{encl}}/\epsilon_0$, with $Q_{\text{encl}}$ given by the above equation. We find

$$4\pi r^2 E = \frac{Qr^3}{\epsilon_0 R^3}, \quad \text{or} \quad E = \frac{1}{4\pi\epsilon_0}\frac{Qr}{R^3}. \tag{23–13}$$

The field magnitude is proportional to the distance $r$ of the field point from the center of the sphere. At the center ($r = 0$), $E = 0$, as we should expect from symmetry. At the surface of the sphere ($r = R$) the field magnitude is

$$E = \frac{1}{4\pi\epsilon_0}\frac{Q}{R^2}.$$

This shows that at the surface the field has the same magnitude as though all the charge were concentrated at the center. As we have learned, this is also the case at any field point farther from the center than $R$.

Figure 23–14 is a graph of $E$ as a function of $r$ for this problem. For $r < R$, $E$ is directly proportional to $r$; and for $r > R$, $E$ varies as $1/r^2$. We remarked earlier that there is a Gauss's law for gravitational interactions. The result of this example is directly applicable to the gravitational field inside the earth. If we could drill a hole through the earth to its center, and if the density were uniform, we would find that the magnitude of the gravitational field varies with $r$ in the same way as the magnitude of the $\boldsymbol{E}$ field in Fig. 23–14. The *direction* of the gravitational field is toward the center; gravitational forces are always attractive.

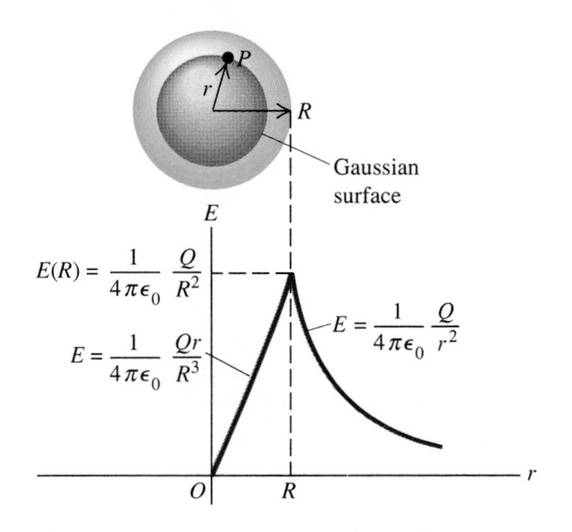


**23–14** The magnitude of the electric field of a uniformly charged insulating sphere. Compare this with Fig. 23–10, the field for a conducting sphere.

## 23–4 CHARGES ON CONDUCTORS

We have learned that in any electrostatic problem (where there is no net motion of charge) the electric field at every point within a conductor is zero and that charge on a solid conductor is located entirely on its surface (Fig. 23–15a). But what if there is a cavity inside the conductor (Fig. 23–15b)? If there is no charge in the cavity, we can use a Gaussian surface such as $A$ to show that the net charge on the surface *of the cavity* must be zero because $\boldsymbol{E} = \boldsymbol{0}$ everywhere on the Gaussian surface. In fact, we can prove in this situation not only that the *total* charge on the cavity surface is zero, but also that there can't be any charge *anywhere* on the cavity surface. We will postpone detailed proof of this statement until Chapter 24.

Suppose we place a small body with a charge $q$ inside a cavity in a conductor, insulated from it (Fig. 23–15c). Again $\boldsymbol{E} = \boldsymbol{0}$ everywhere on surface $A$, so according to Gauss's law the *total* charge inside this surface must be zero. Therefore there must be a total charge $-q$ on the cavity surface. The *total* charge on the conductor must remain zero, so a charge $+q$ must appear either on its outer surface or inside the material.

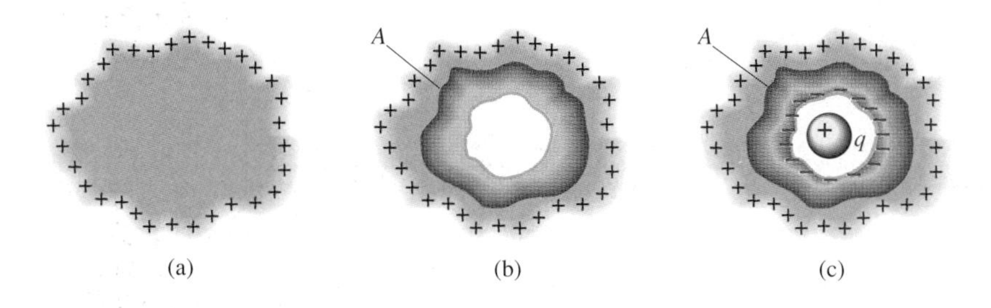


**23–15** (a) Charge on a solid conductor resides entirely on its outer surface. (b) If there is no charge inside the conductor's cavity, the net charge on the surface of the cavity is zero. (c) If there is a charge $q$ inside the cavity, the total charge on the cavity surface is $-q$.

To see that this charge can't be in the material, imagine first shrinking surface $A$ so that it's just barely bigger than the cavity. The field everywhere on $A$ is zero, so according to Gauss's law the total charge inside $A$ is zero. Now let surface $A$ expand until it is just inside the outer surface of the conductor. The field is still zero everywhere on surface $A$, so the total charge enclosed is still zero. We have not enclosed any additional charge by expanding surface $A$, and therefore there must be no charge in the interior of the material. We conclude that the charge $+q$ must appear on the outer surface. By the same reasoning, if the conductor originally had a charge $q'$, then the total charge on the outer surface after the charge $q$ is inserted into the cavity must be $q + q'$.

## EXAMPLE 23–10

The conductor shown in cross section in Fig. 23–16 carries a total charge of 7 nC. The charge within the cavity, insulated from the conductor, is $-5$ nC. How much charge is on each surface (inner and outer) of the conductor?

**SOLUTION** If the charge in the cavity is $q$, the charge on the cavity surface must be $-q$. In this case the charge on the cavity surface is $-(-5 \text{ nC}) = +5$ nC. The conductor carries a *total* charge of 7 nC, none of which is in the interior of the material. If 5 nC is on the inner surface (the surface of the cavity), then the remaining 2 nC must be on the outer surface.

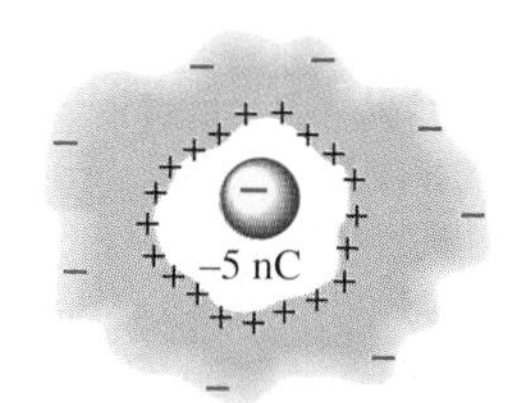


**23–16** There is no excess charge in the bulk material of this conductor. Charge is located only on the inner and outer surfaces. The innermost charge is represented by a single minus sign, instead of twenty.

We can now consider a historic experiment, shown in Fig. 23–17. We mount a conducting container, such as a metal pail with a lid, on an insulating stand. The container is initially uncharged. Then we hang a charged metal ball from an insulating thread, lower

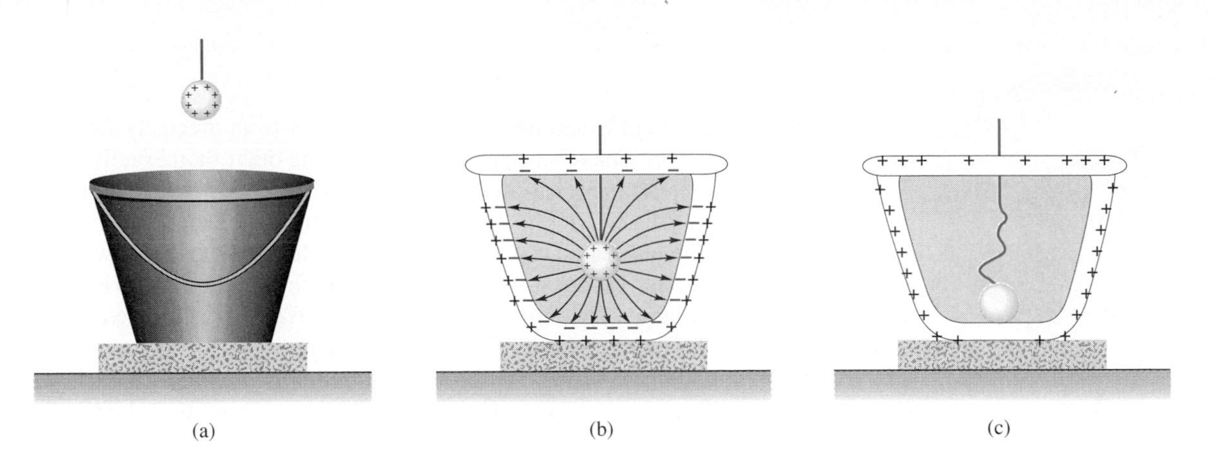


23–17 (a) A charged conducting ball suspended by an insulating thread outside a conducting container on an insulating stand. (b) The ball is lowered into the container, and the lid is put on. Charges are induced on the walls of the container. (c) When the ball is touched to the inner surface of the container, all its charge is transferred to the container and appears on the container's outer surface.

it into the pail, and put the lid on (Fig. 23–17b). Charges are induced on the walls of the container, as shown. But now we let the ball *touch* the inner wall (Fig. 23–17c). The surface of the ball becomes, in effect, part of the cavity surface. The situation is now the same as Fig. 23–15b; if Gauss's law is correct, the net charge on this surface must be zero. Thus the ball must lose all its charge. Finally, we pull the ball out; we find that it has indeed lost all its charge.

This experiment was performed by Faraday, using a metal icepail with a lid, and it is called **Faraday's icepail experiment.** (Similar experiments had been carried out earlier by Benjamin Franklin and Joseph Priestley, although with much less precision.) The result confirms the validity of Gauss's law and therefore of Coulomb's law. Nevertheless, Faraday's result was significant because Coulomb's experimental method, using a torsion balance and dividing of charges, was not very precise. It is very difficult to confirm the $1/r^2$ dependence of the electrostatic force with great precision by direct force measurements. Faraday's experiment tests the validity of Gauss's law and therefore of Coulomb's law with potentially much greater precision.

A contemporary version of this experiment is shown in Fig. 23–18. The details of the box labeled "power supply" aren't important; its job is to place charge on the outer sphere and remove it, on demand. The inner box with a dial is a sensitive electrometer, an instrument that can detect motion of extremely small amounts of charge between the outer and inner spheres. If Gauss's law is correct, there can never be any charge on the inner surface of the outer sphere. If so, there should be no flow of charge between spheres while the outer sphere is being charged and discharged. The fact that no flow is actually observed is a very sensitive confirmation of Gauss's law and therefore of Coulomb's law. The precision of the experiment is limited mainly by the electrometer, which can be astonishingly sensitive. The most recent (1986) experiments have shown that the

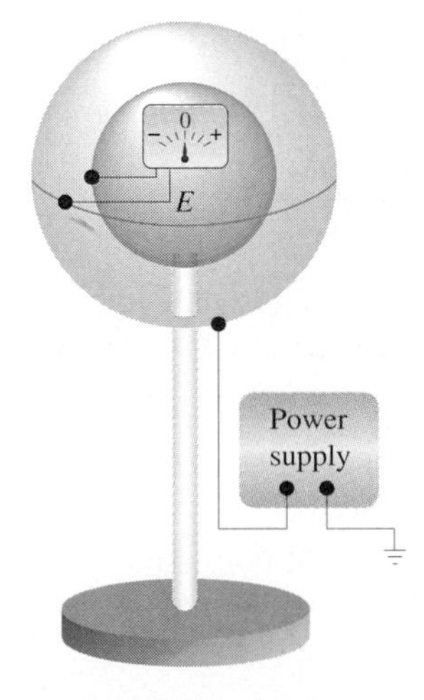


23–18 The outer spherical shell can be alternately charged and discharged by the power supply to which it is connected. If there is any flow of charge between the inner and outer shells, it is detected by the electrometer inside the inner shell.

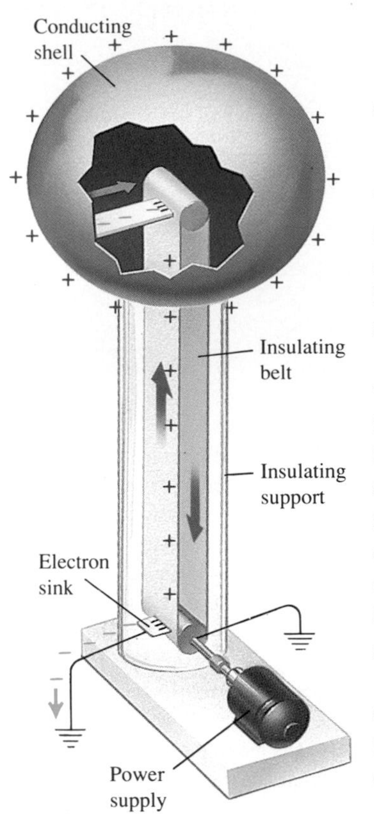


**23–19** Cutaway view of a Van de Graaff electrostatic generator. An electron source may be used instead of an electron sink to give the conducting shell a negative charge.

exponent 2 in the $1/r^2$ of Coulomb's law does not differ from precisely 2 by more than $10^{-16}$. So there is no reason to suspect that it is anything other than exactly 2.

Figure 23–19 shows the essential parts of a Van de Graaff electrostatic generator. The charged conducting ball of Fig. 23–17 is replaced by a charged belt that continuously carries charge to the inside of a conducting shell, only to have it carried away to the outside surface of the shell. As a result, the charge on the shell and the electric field around it can become very large very rapidly. The Van de Graaff generator is used as an accelerator of charged particles and for physics demonstrations.

This discussion also forms the basis for *electrostatic shielding.* Suppose we have a very sensitive electronic instrument that we want to protect from stray electric fields that might give erroneous measurements. We surround the instrument with a conducting box, or we line the walls, floor, and ceiling of the room with a conducting material such as sheet copper. The external electric field redistributes the free electrons in the conductor, leaving a net positive charge on the outer surface in some regions and a net negative charge in others (Fig. 23–20). This charge distribution causes an additional electric field such that the *total* field at every point inside the box is zero, as Gauss's law says it must be. The charge distribution on the box also alters the shapes of the field lines near the box, as the figure shows. Such a setup is often called a *Faraday cage*.

Finally, we note that there is a direct relation between the $\boldsymbol{E}$ field at a point just outside any conductor and the surface charge density $\sigma$ at that point. In general, $\sigma$ varies from point to point on the surface. We will show in Chapter 24 that at any such point the direction of $\boldsymbol{E}$ is always perpendicular to the surface.

To find a relation between $\sigma$ at any point on the surface and $E$ at that point, we construct a Gaussian surface in the form of a small cylinder (Fig. 23–21). One end face, with area $A$, lies within the conductor, and the other lies just outside. The charge within the Gaussian surface is $\sigma A$. The electric field is zero at all points within the conductor.

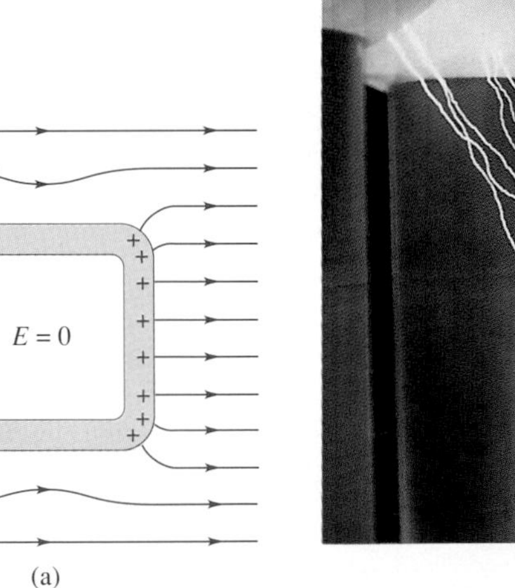


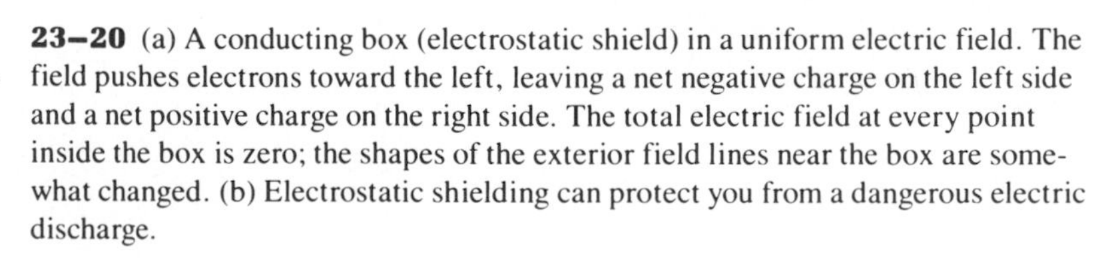


**23–20** (a) A conducting box (electrostatic shield) in a uniform electric field. The field pushes electrons toward the left, leaving a net negative charge on the left side and a net positive charge on the right side. The total electric field at every point inside the box is zero; the shapes of the exterior field lines near the box are somewhat changed. (b) Electrostatic shielding can protect you from a dangerous electric discharge.

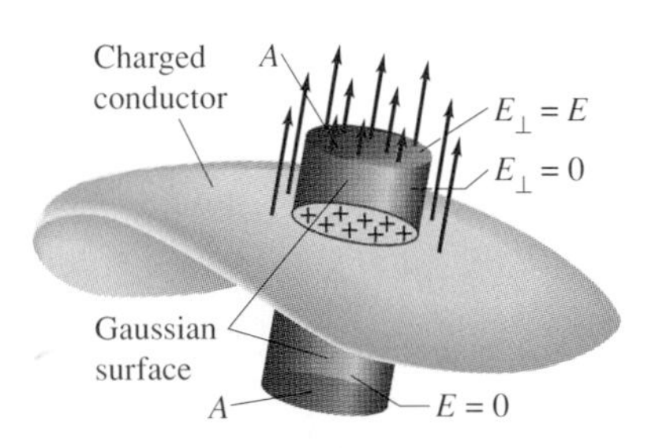


**23–21** The field just outside a charged conductor is perpendicular to the surface, and its magnitude is equal to $\sigma/\epsilon_0$.

Outside the conductor, the component of $\boldsymbol{E}$ perpendicular to the side walls of the cylinder is zero, and over the end face the perpendicular component is equal to $E$. Therefore, from Gauss's law,

$$EA = \frac{\sigma A}{\epsilon_0}, \qquad E = \frac{\sigma}{\epsilon_0}. \tag{23–14}$$

We can check this with the results that we have obtained for spherical, cylindrical, and plane surfaces.

We also showed in Example 23–8 that the field between two infinite flat oppositely charged conducting plates equals $\sigma/\epsilon_0$. In this case the field is the same at *all* distances from the plates, but in all other cases it decreases with increasing distance from the surface.

## EXAMPLE 23–11

Verify Eq. (23–14) for a conducting sphere with radius $R$ and total charge $q$.

**SOLUTION** In Example 23–5 (Section 23–3) we showed that the electric field just outside the surface is

$$E = \frac{1}{4\pi\epsilon_0}\frac{q}{R^2}.$$

The surface charge density is

$$\sigma = \frac{q}{4\pi R^2}.$$

Comparing these two expressions, we see that $E = \sigma/\epsilon_0$, as Eq. (23–14) states.

## EXAMPLE 23–12

**Electric field of the earth** The earth has a net electric charge. The resulting electric field near the surface can be measured with sensitive electronic instruments; it is about 150 N/C, directed toward the center of the earth. a) What is the corresponding surface charge density? b) What is the *total* surface charge of the earth?

**SOLUTION** a) We know from the direction of the field that $\sigma$ is negative (corresponding to a component $E_\perp$ *into* the surface). From Eq. (23–14),

$$\begin{aligned}\sigma = \epsilon_0 E_\perp &= (8.85 \times 10^{-12}\,\text{C}^2/\text{N}\cdot\text{m}^2)(-150\,\text{N/C})\\ &= -1.33 \times 10^{-9}\,\text{C/m}^2 = -1.33\,\text{nC/m}^2.\end{aligned}$$

b) The total charge $Q$ is the surface area $4\pi R_E^{\,2}$ times the charge density $\sigma$:

$$\begin{aligned}Q &= 4\pi(6.38 \times 10^6\,\text{m})^2(-1.33 \times 10^{-9}\,\text{C/m}^2)\\ &= -6.8 \times 10^5\,\text{C}.\end{aligned}$$

We could also use the result of Example 23–9. Solving for $Q$, we find

$$\begin{aligned}Q &= 4\pi\epsilon_0 R^2 E_\perp\\ &= (9.0 \times 10^9\,\text{N}\cdot\text{m}^2/\text{C}^2)^{-1}(6.38 \times 10^6\,\text{m})^2(-150\,\text{N/C})\\ &= -6.8 \times 10^5\,\text{C} = -680\,\text{kC}.\end{aligned}$$

# SUMMARY

- Electric flux is the product of an area element and the perpendicular component of $\boldsymbol{E}$, integrated over a surface:

$$\Phi_E = \int E\cos\phi\, dA = \int E_\perp\, dA = \int \boldsymbol{E}\cdot d\boldsymbol{A}. \tag{23–4}$$

- Gauss's law is logically equivalent to Coulomb's law. It states that the total electric flux out of any closed surface (the surface integral of the component of $\boldsymbol{E}$ normal to the surface) is proportional to the total charge $Q_{\text{encl}}$ enclosed by the surface:

$$\oint E\cos\phi\, dA = \oint E_\perp\, dA = \int \boldsymbol{E}\cdot d\boldsymbol{A} = \frac{\Sigma q_i}{\epsilon_0} = \frac{Q_{\text{encl}}}{\epsilon_0}. \tag{23–8}$$

### KEY TERMS

electric flux

surface integral

Gauss's law

Faraday's icepail experiment

• When excess charge is at rest on a conductor, it resides entirely on the surface, and $E = 0$ everywhere in the interior of the conductor.

• The table below lists electric fields caused by several charge distributions. In the table, $q$, $Q$, $\lambda$, and $\sigma$ refer to the *magnitudes* of the charge quantities.

| Charge distribution | Point in electric field | Electric field magnitude |
|---|---|---|
| Single point charge $q$ | Distance $r$ from $q$ | $E = \dfrac{1}{4\pi\epsilon_0}\dfrac{q}{r^2}$ |
| Charge $q$ on surface of conducting sphere with radius $R$ | Outside sphere, $r > R$<br>Inside sphere, $r < R$ | $E = \dfrac{1}{4\pi\epsilon_0}\dfrac{q}{r^2}$<br>$E = 0$ |
| Long wire, charge per unit length $\lambda$ | Distance $r$ from wire | $E = \dfrac{1}{2\pi\epsilon_0}\dfrac{\lambda}{r}$ |
| Long conducting cylinder with radius $R$, charge per unit length $\lambda$ | Outside cylinder, $r > R$<br>Inside cylinder, $r < R$ | $E = \dfrac{1}{2\pi\epsilon_0}\dfrac{\lambda}{r}$<br>$E = 0$ |
| Solid insulating sphere, charge $Q$ distributed uniformly throughout volume | Outside sphere, $r > R$<br>Inside sphere, $r < R$ | $E = \dfrac{1}{4\pi\epsilon_0}\dfrac{Q}{r^2}$<br>$E = \dfrac{1}{4\pi\epsilon_0}\dfrac{Qr}{R^3}$ |
| Two oppositely charged conducting plates with surface charge densities $+\sigma$ and $-\sigma$ | Any point between plates | $E = \dfrac{\sigma}{\epsilon_0}$ |

# EXERCISES

## Section 23–1 Electric Flux

**23–1** Example 22–11 showed that the electric field due to an infinite line of charge is perpendicular to the line and has magnitude $E = \lambda/2\pi\epsilon_0 r$. Consider a thin, hollow cylinder with radius $r = 0.160$ m and length $l = 0.400$ m. Running along the axis of the cylinder is an infinite line of positive charge, with charge per unit length $\lambda = 5.00\ \mu\text{C/m}$. a) What is the electric flux through the cylinder due to this infinite line of charge? b) What is the flux through the cylinder if the cylinder's radius is increased to $r = 0.320$ m? c) What is the flux through the cylinder if the cylinder's length is increased to $l = 0.800$ m?

**23–2 Electric Flux through a Cube.** Consider a uniform electric field in the $+x$-direction with magnitude $E = 6.00 \times 10^3$ N/C. a) What is the magnitude of the electric flux through one face of a cube that is 0.0800 m on a side and that is placed in this field with the plane of the face making an angle of 37.0° with the field direction? b) What is the total electric flux through all sides of the cube?

## Section 23–2 Gauss's Law

**23–3** A closed surface encloses a net charge of 5.20 μC. What is the net electric flux through the surface?

**23–4** The electric flux through a closed surface is found to be $3.60\ \text{N}\cdot\text{m}^2/\text{C}$. What quantity of charge is enclosed by the surface?

**23–5** A point charge $q = 3.00$ nC is at the center of a cube with sides of length 0.200 m. What is the electric flux through one of the six faces of the cube?

## Section 23–3 Applications of Gauss's Law

## Section 23–4 Charges on Conductors

**23–6 Photocopier Imaging Drum.** The cylindrical imaging drum of a photocopier (see Chapter 22) is to have an electric field just outside its surface of $2.00 \times 10^5$ N/C. a) If the drum has a surface area of 0.0610 m$^2$ (the area of an $8\frac{1}{2} \times 11$ in. sheet of paper), what total quantity of charge must reside on the surface of the drum? b) If the surface area of the drum is increased to 0.122 m$^2$ so that larger sheets of paper can be used, what total quantity of charge is required to produce the same $2.00 \times 10^5$ N/C electric field just above the surface?

**23–7** How many excess electrons must be added to an isolated spherical conductor 0.180 m in diameter to produce an electric field of 1300 N/C just outside the surface?

**23–8** The electric field in the region between a pair of oppositely charged plane parallel conducting plates, each 100 cm$^2$ in area, is $4.00 \times 10^4$ N/C. What is the charge on each plate? Neglect edge effects.

**23–9** Prove that the electric field outside an infinitely long cylindrical conductor with a uniform surface charge is the same as if all the charge were on the axis.

**23–10 A Sphere in a Sphere.** A conducting sphere carrying charge $q$ has radius $a$. It is inside a concentric hollow conducting sphere of inner radius $b$ and outer radius $c$. The hollow sphere has no net charge. a) Derive expressions for the electric field magnitude in terms of the distance $r$ from the center for the regions $r < a$, $a < r < b$, $b < r < c$, and $r > c$. b) Sketch a graph of the magnitude of the electric field as a function of $r$ from $r = 0$ to $r = 2c$. c) What is the charge on the inner surface of the hollow sphere? d) On the outer surface? e) Represent the charge of the small sphere by four plus signs. Sketch the field lines of the system within a spherical volume of radius $2c$.

**23–11** Apply Gauss's law to the purple Gaussian surfaces in Fig. 23–13b to calculate the electric field between and outside the plates.

# PROBLEMS

**23–12** The electric field $\boldsymbol{E}$ in Fig. 23–22 is everywhere parallel to the $x$-axis. The field has the same magnitude at all points in any given plane perpendicular to the $x$-axis (parallel to the $yz$-plane), but the magnitude is different for various planes. That is, $E_x$ depends on $x$ but not on $y$ and $z$, and $E_y$ and $E_z$ are zero. At points *in* the $yz$-plane (where $x = 0$), $E_x = 300$ N/C. (The volume shown could be a small section of a very large insulating slab 1.00 m thick, with its faces parallel to the $yz$-plane and with a charge distribution imbedded in it. The electric field is produced by charge outside the volume as well as by that within.) a) What is the electric flux through surface I in Fig. 23–22? b) What is the electric flux through surface II? c) If there is a total positive charge of 26.6 nC within the volume, what are the magnitude and direction of $\boldsymbol{E}$ at the face opposite surface I?

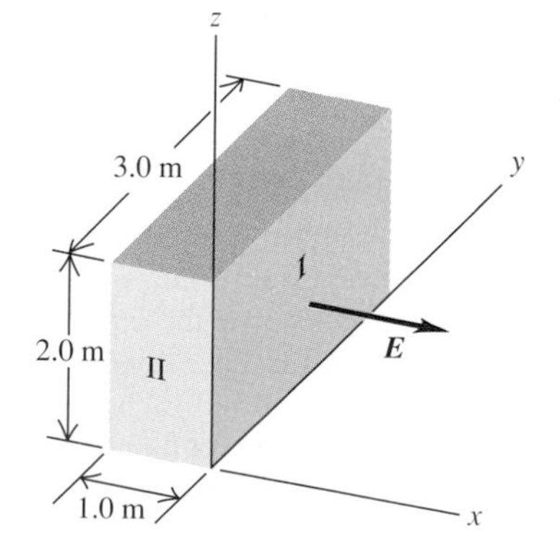


FIGURE 23–22

**23–13** A uniform electric field $\boldsymbol{E}_1$ is directed out of one face of a parallelepiped, and another uniform electric field $\boldsymbol{E}_2$ is directed into the opposite face (Fig. 23–23). $\boldsymbol{E}_1$ has a magnitude of $3.50 \times 10^4$ N/C, and $\boldsymbol{E}_2$ has a magnitude of $5.00 \times 10^4$ N/C.

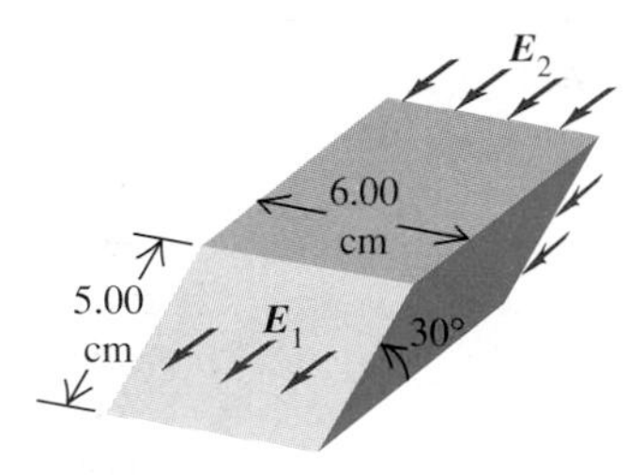


FIGURE 23–23

Assuming that there are no other electric field lines crossing the surfaces of the parallelepiped, determine the net charge contained within. (The electric field is produced by charges outside the parallelepiped as well as by those within.)

**23–14** A point charge $q_1 = 2.50$ nC is located at the origin, and a second point charge $q_2 = 5.00$ nC is on the $x$-axis at $x = 1.00$ m. What is the total electric flux due to these two point charges through a spherical surface with radius 0.500 m, centered at the origin?

**23–15** Positive charge $Q$ is distributed uniformly over the surface of a thin spherical insulating shell with radius $R$. Calculate the force (magnitude and direction) that the shell exerts on a positive point charge $q$ located a) a distance $r > R$ from the center of the shell (outside the shell); b) a distance $r < R$ from the center of the shell (inside the shell).

**23–16** A conducting spherical shell with inner radius $a$ and outer radius $b$ has a positive point charge $Q$ located at its center. The total charge on the shell is $-3Q$, and it is insulated from its surroundings (Fig. 23–24). a) Derive expressions for the electric field magnitude in terms of the distance $r$ from the center for the regions $r < a$, $a < r < b$, and $r > b$. b) What is the surface charge density on the inner surface of the conducting shell? c) What is the surface charge density on the outer surface of the conducting shell? d) Draw a sketch showing electric field lines and the location of all charges. e) Draw a graph of the electric field magnitude as a function of $r$.

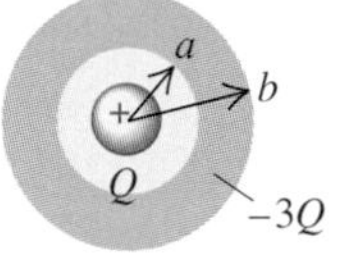


**FIGURE 23–24**

**23–17 Concentric Spherical Shells.** A small conducting spherical shell with inner radius $a$ and outer radius $b$ is concentric with a larger conducting spherical shell with inner radius $c$ and outer radius $d$ (Fig. 23–25). The inner shell has total charge $+2q$, and the outer shell has charge $+4q$. a) Calculate the electric field in terms of $q$ and the distance $r$ from the common center of the two shells for i) $r < a$; ii) $a < r < b$; iii) $b < r < c$; iv) $c < r < d$; v) $r > d$. Show your results in a sketch of $E(r)$ as a function of $r$. b) What is the total charge on the i) inner surface of the small shell; ii) outer surface of the small shell; iii) inner surface of the large shell; iv) outer surface of the large shell?

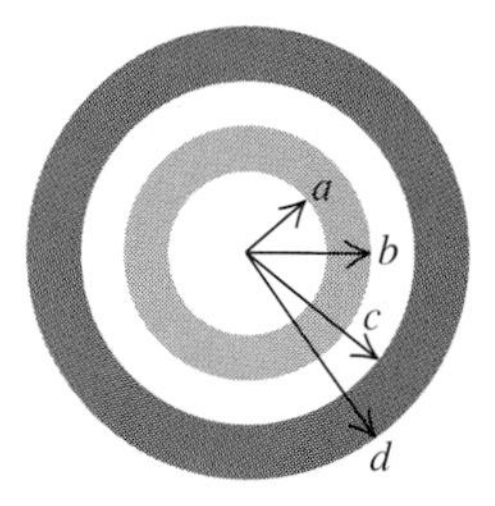


**FIGURE 23–25**

**23–18** Repeat Problem 23–17, but now let the outer shell have charge $-2q$. As in Problem 23–17, the inner shell has charge $+2q$.

**23–19** Repeat Problem 23–17, but now let the outer shell have charge $-4q$. As in Problem 23–17, the inner shell has charge $+2q$.

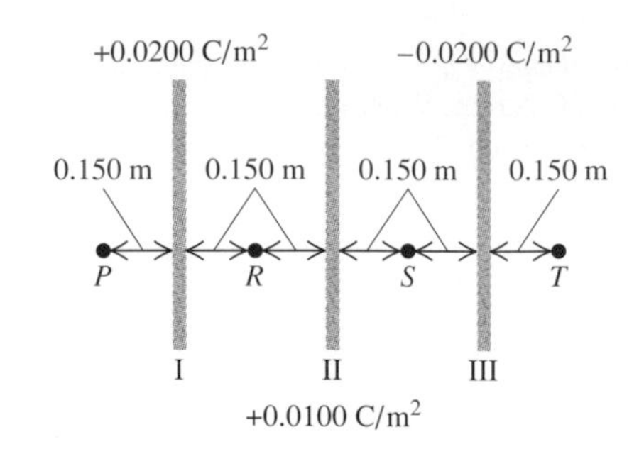


**FIGURE 23–26**

**23–20** Three large parallel insulating sheets have surface charge densities $+0.0200\ \mathrm{C/m^2}$, $+0.0100\ \mathrm{C/m^2}$, and $-0.0200\ \mathrm{C/m^2}$, respectively (Fig. 23–26). Adjacent sheets are a distance 0.300 m from each other. Calculate the resultant electric field (magnitude and direction) due to all three sheets at a) point $P$ (0.150 m to the left of sheet I); b) point $R$ (midway between sheets I and II); c) point $S$ (midway between sheets II and III); d) point $T$ (0.150 m to the right of sheet III).

**23–21 The Coaxial Cable.** A long coaxial cable consists of an inner cylindrical conductor with radius $a$ and an outer coaxial cylinder with inner radius $b$ and outer radius $c$. The outer cylinder is mounted on insulating supports and has no net charge. The inner cylinder has a uniform positive charge $\lambda$ per unit length. Calculate the electric field a) at any point between the cylinders, a distance $r$ from the axis; b) at any point outside the outer cylinder. c) Sketch a graph of the magnitude of the electric field as a function of the distance $r$ from the axis of the cable, from $r = 0$ to $r = 2c$. d) Find the charge per unit length on the inner surface and on the outer surface of the outer cylinder.

**23–22** A very long conducting tube (hollow cylinder) has inner radius $a$ and outer radius $b$. It carries charge per unit length $+\alpha$, where $\alpha$ is a positive constant with units of C/m. A line of charge lies along the axis of the tube and also has charge per unit length $+\alpha$. a) Calculate the electric field in terms of $\alpha$ and the distance $r$ from the axis of the tube for i) $r < a$; ii) $a < r < b$; iii) $r > b$. Show your results in a sketch of $E(r)$ as a function of $r$. b) What is the charge per unit length on i) the inner surface of the tube; ii) the outer surface of the tube?

**23–23** Repeat Problem 23–22, but now let the conducting tube have charge per unit length $-\alpha$. As in Problem 23–22, the line of charge has charge per unit length $+\alpha$.

**23–24** Suppose positive charge is uniformly distributed throughout a very long cylindrical volume with radius $R$ with charge per unit volume $\rho$. a) Derive the expression for the electric field inside the volume at a distance $r$ from the axis of the cylinder in terms of the charge density $\rho$. b) What is the electric field at a point outside the volume in terms of the charge per unit length $\lambda$ in the cylinder? c) Compare the answers to parts (a) and (b) when $r = R$. d) Sketch a graph of the magnitude of the electric field as a function of $r$ from $r = 0$ to $r = 3R$.

**23–25** A nonuniform but spherically symmetric distribution of charge has a charge density $\rho$ given as follows:

$$\rho = \rho_0(1 - r/R) \quad \text{for } r \leq R,$$

$$\rho = 0 \quad \text{for } r \geq R,$$

where $\rho_0 = 3Q/\pi R^3$ is a constant. a) Show that the total charge contained in the charge distribution is $Q$. b) Show that, for the region defined by $r \geq R$, the electric field is identical to that produced by a point charge $Q$. c) Obtain an expression for the electric field in the region $r \leq R$. d) Compare your results in parts (b) and (c) for $r = R$.

## CHALLENGE PROBLEMS

**23–26** Positive charge $Q$ is distributed uniformly over each of two spherical volumes of radius $R$. One sphere of charge is centered at the origin and the other at $x = 2R$ (Fig. 23–27). Find the magnitude and direction of the resultant electric field due to these two distributions of charge at the following points on the $x$-axis: a) $x = 0$; b) $x = R/2$; c) $x = R$; d) $x = 3R$.

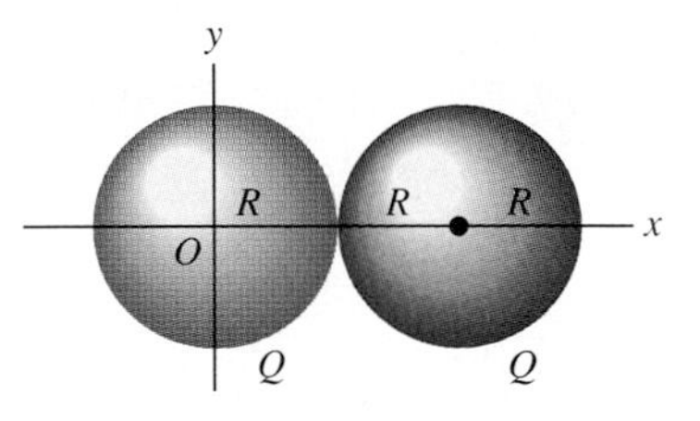


**FIGURE 23–27**

**23–27** A region in space contains charge that is distributed spherically such that the volume charge density $\rho$ is given by

$$\rho = \alpha \quad \text{for } r \leq R/2,$$

$$\rho = 2\alpha(1 - r/R) \quad \text{for } R/2 \leq r \leq R,$$

$$\rho = 0 \quad \text{for } r \geq R.$$

The total charge $Q$ is $3.00 \times 10^{-17}$ C, the radius $R$ of the spherical charge distribution is $2.00 \times 10^{-14}$ m, and $\alpha$ is a constant having units of C/m$^3$. a) Determine $\alpha$ in terms of $Q$ and $R$, and also determine its numerical value. b) Using Gauss's law, derive an expression for the magnitude of the electric field as a function of the distance $r$ from the center of the distribution. Do this separately for all three regions. Be sure to check that your results agree on the boundaries of the regions. c) What fraction of the total charge is contained within the region where $r \leq R/2$? d) If an electron with charge $q' = -e$ is oscillating back and forth about $r = 0$ (the center of the distribution) with an amplitude less than $R/2$, show that the motion is simple harmonic. [*Hint:* Review the definition of simple harmonic motion, as defined by Eq. (13–1). If it can be shown that the net force on the electron is of this form, then it follows that the motion is simple harmonic. Conversely, if the net force on the electron does not follow this form, the motion is not simple harmonic.] e) For the motion in part (d), what is the period? (Calculate a numerical value.) f) If the amplitude of the motion described in part (e) is greater than $R/2$, explain why the motion is no longer simple harmonic.

**23–28** A region in space contains charge that is distributed spherically in such a way that the volume charge density $\rho$ is given by

$$\rho = 3\alpha r/(2R) \quad \text{for } r \leq R/2,$$

$$\rho = \alpha[1 - (r/R)^2] \quad \text{for } R/2 \leq r \leq R,$$

$$\rho = 0 \quad \text{for } r \geq R.$$

The total charge is $Q$, the radius $R$ of the spherical charge distribution is $5.00 \times 10^{-10}$ m, and $\alpha = 3.00 \times 10^{11}$ C/m$^3$ and is constant. a) Determine $Q$ in terms of $\alpha$ and $R$, and also determine its numerical value. b) Using Gauss's law, derive an expression for the magnitude of the electric field as a function of the distance $r$ from the center of the distribution. Do this separately for all three regions. c) What fraction of the total charge is contained within the region where $R/2 \leq r \leq R$? d) What is the magnitude of the electric field at $r = R/2$? e) If an electron with charge $q' = -e$ is released from rest at any point in any of the three regions, the resulting motion will be oscillatory, although not simple harmonic. Why? (See Challenge Problem 23–27.)

CHAPTER 24

# Electric Potential Voltage

An oscilloscope is used to observe and measure rapidly varying potential differences (voltages) in electronic circuits. The voltage to be measured is connected to the deflection plates within the cathode ray tube of the oscilloscope. The electron beam's vertical deflection is then proportional to the magnitude of the applied voltage. The horizontal deflection is proportional to time, with a repeating sweep pattern; the beam traces out a graph of voltage as a function of time.

## KEY CONCEPTS

**• The force that acts on a charge moving in a static electric field is a conservative force, and there is an associated potential energy.**

**• Electric potential, a scalar quantity, is the potential energy per unit charge for the interaction of a charge with an electric field. It is usually called simply potential.**

**• The potential at a point, associated with a particular electric field, can be computed from the charges that cause the field or directly from the field itself.**

**• An equipotential surface is a surface on which the potential has the same value at every point. Equipotential surfaces are a useful addition to electric-field maps.**

**• If the potential is known as a function of position for every point in a region, the electric field at every point in the region can be determined.**

**• The Millikan oil-drop experiment provides a direct measurement of the magnitude of the electron charge.**

**• The concept of potential is useful in the analysis of many practical devices such as the cathode-ray tube.**

## INTRODUCTION

This chapter is about energy associated with electrical interactions. Every time you turn on a light or an electric motor, you are making use of electrical energy, a familiar part of everyday life and an indispensable ingredient of our technological society. In Chapters 6 and 7 we introduced the concepts of *work* and *energy* in a mechanical context; now we combine these concepts with what we have learned about electric charge, Coulomb's law, and electric fields.

When a charged particle moves in an electric field, the electric-field force does *work* on the particle. This work can always be expressed in terms of a potential energy, which in turn is associated with a new concept called *electric potential,* or simply *potential*. In circuits, potential is often called *voltage*. The practical applications of this concept cover a wide range, including electric circuits, electron beams in TV picture tubes, high-energy particle accelerators, and many other areas.

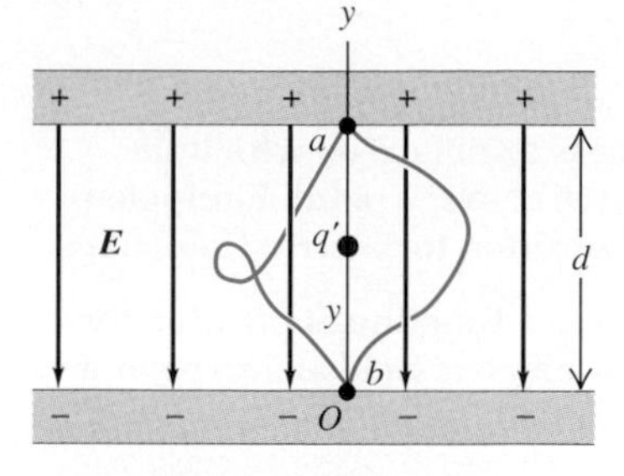


**24–1** A test charge $q'$ moving from $a$ to $b$ experiences a force of magnitude $q'E$; the work done by this force is $W_{a\to b} = q'Ed$ and is independent of the particle's path.

## 24–1 ELECTRIC POTENTIAL ENERGY

The opening sections of this chapter are about work, potential energy, and conservation of energy. Let's begin by reviewing several essential points from Chapters 6 and 7. First, when a force $\boldsymbol{F}$ acts on a particle that moves from point $a$ to point $b$, the work $W_{a\to b}$ done by the force is

$$W_{a\to b} = \int_a^b \boldsymbol{F}\cdot d\boldsymbol{l} = \int_a^b F\cos\phi\, dl, \tag{24–1}$$

where $d\boldsymbol{l}$ is an element of the particle's path and $\phi$ is the angle between $\boldsymbol{F}$ and $d\boldsymbol{l}$ at each point along the path.

Second, if the force field is *conservative*, as we defined the term in Section 7–4, this work can always be expressed in terms of a **potential energy** $U$. When the particle moves from a point where its potential energy is $U_a$ to a point where it is $U_b$, the work $W_{a\to b}$ done by the force is

$$W_{a\to b} = U_a - U_b. \tag{24–2}$$

When $W_{a\to b}$ is positive, $U_a$ is greater than $U_b$, and the potential energy *decreases*. That's what happens when a baseball falls from a high point ($a$) to a lower point ($b$) under the action of the earth's gravity. The force of gravity does positive work, and the gravitational potential energy decreases. When a ball is thrown upward, the gravitational force does negative work during the ascent, and the potential energy increases.

Third, the work-energy theorem says that the change in kinetic energy ($K_b - K_a$) during any displacement is equal to the total work done on the particle. So if Eq. (24–2) gives the *total* work, then $K_b - K_a = U_a - U_b$, which we usually write as

$$K_a + U_a = K_b + U_b. \tag{24–3}$$

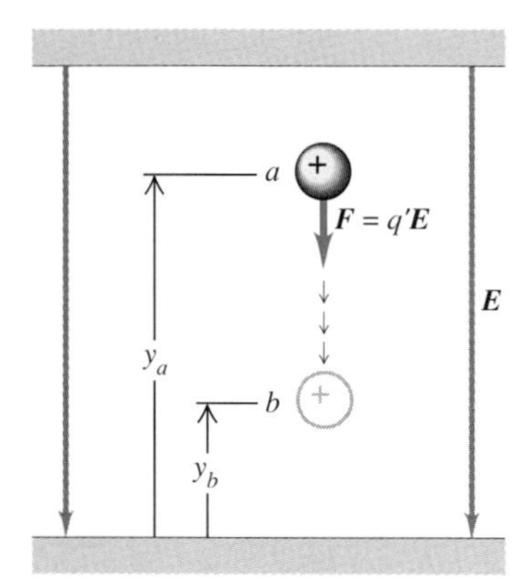


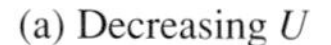
(a) Decreasing $U$

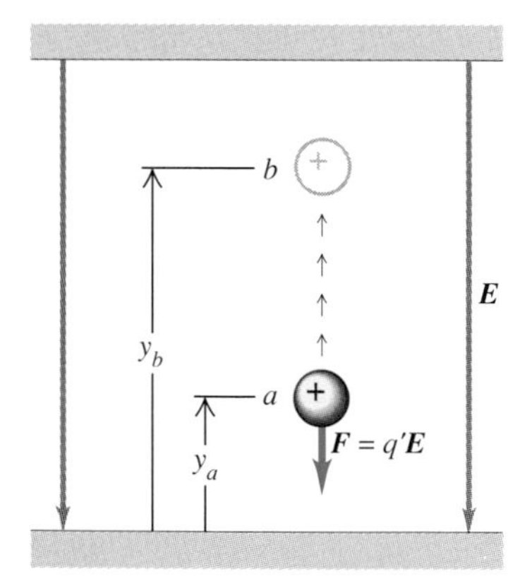


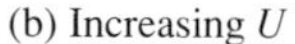
(b) Increasing $U$

**24–2** (a) When a positive charge moves in the direction of an electric field, the field does positive work and the potential energy decreases. (b) When a positive charge moves in the direction opposite to an electric field, the field does negative work and the potential energy increases.

Let's look at an electrical example of these basic concepts. In Fig. 24–1 a pair of charged parallel metal plates sets up a uniform electric field with magnitude $E$. The field exerts a downward force with magnitude $F = q'E$ on a positive test charge $q'$ as the charge moves a distance $d$ from point $a$ to point $b$. The force on the test charge is constant, independent of its location, so the work done by the electric field is

$$W_{a\to b} = Fd = q'Ed. \tag{24–4}$$

We can represent this work with a *potential-energy* function $U$, just as we did for gravitational potential energy in Section 7–2. The $y$-component of force, $F_y = -q'E$, is constant, and there is no $x$- or $z$-component, so the work is independent of the path the particle takes from $a$ to $b$. Just as the potential energy for the gravitational force $F_y = -mg$ was $U = mgy$, the potential energy for the electric-field force $F_y = -q'E$ is

$$U = q'Ey. \tag{24–5}$$

When the test charge moves from height $y_a$ to height $y_b$, the work done on the charge by the field is given by

$$W_{a\to b} = U_a - U_b = q'Ey_a - q'Ey_b = q'E(y_a - y_b). \tag{24–6}$$

When $y_a$ is greater than $y_b$ (Fig. 24–2a), the particle moves in the same direction as the $\boldsymbol{E}$ field, $U$ decreases, and the field does positive work. When $y_a$ is less than $y_b$ (Fig. 24–2b), the particle moves in the opposite direction to $\boldsymbol{E}$, $U$ increases, and the field does negative work. In particular, if $y_a = d$ and $y_b = 0$, then Eq. (24–6) gives $W_{a\to b} = q'Ed$, in agreement with Eq. (24–4).

If the test charge $q'$ is negative, the potential energy increases when it moves with the field and decreases when it moves against the field (Fig. 24–3).

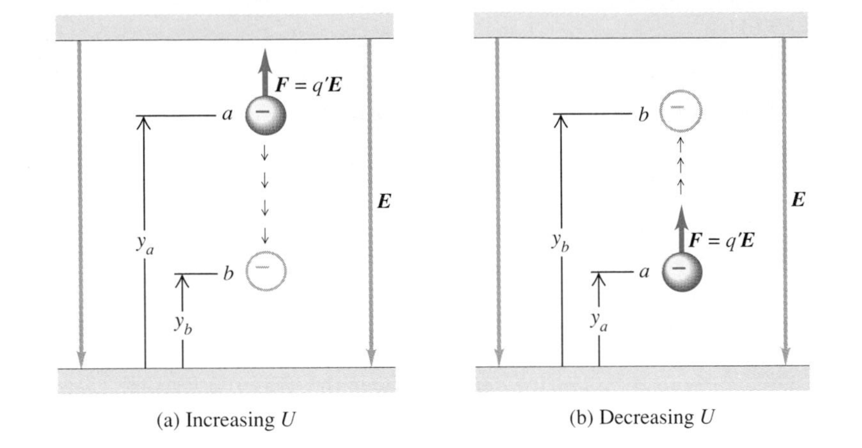


**24–3** When a negative charge moves in the direction of an electric field, the field does negative work and the potential energy increases. (b) When a negative charge moves in the direction opposite to an electric field, the field does positive work and the potential energy decreases.

We can represent any charge distribution as a collection of point charges. Therefore it's useful to calculate the work done on a test charge $q'$ moving in the electric field caused by a single *stationary* point charge $q$. We'll consider first a displacement along the *radial* line in Fig. 24–4, from point $a$ to point $b$. The force is *not* constant, and we have to integrate to calculate the work done on $q'$. The force on $q'$ is given by Coulomb's law, and its radial component is

$$F_r = \frac{1}{4\pi\epsilon_0}\frac{qq'}{r^2}. \tag{24–7}$$

If $q$ or $q'$ is negative, $F_r$ is also negative; if both $q$ and $q'$ are negative, $F_r$ is positive. The work $W_{a\to b}$ done on $q'$ by force $F_r$ as $q'$ moves from $a$ to $b$ is

$$W_{a\to b} = \int_{r_a}^{r_b} F_r\,dr = \int_{r_a}^{r_b} \frac{1}{4\pi\epsilon_0}\frac{qq'}{r^2}\,dr = \frac{qq'}{4\pi\epsilon_0}\left(\frac{1}{r_a} - \frac{1}{r_b}\right). \tag{24–8}$$

Thus the work for this particular path depends only on the end points.

In fact, the work is the same for *all possible* paths from $a$ to $b$. To prove this, we consider a more general displacement (Fig. 24–5) in which $a$ and $b$ *do not* lie on the same

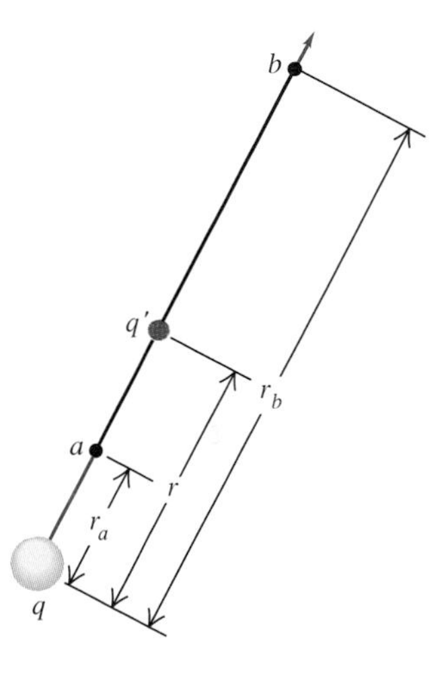


**24–4** Charge $q'$ moves along a straight line extending radially from charge $q$. As it moves from $a$ to $b$, the distance varies from $r_a$ to $r_b$.

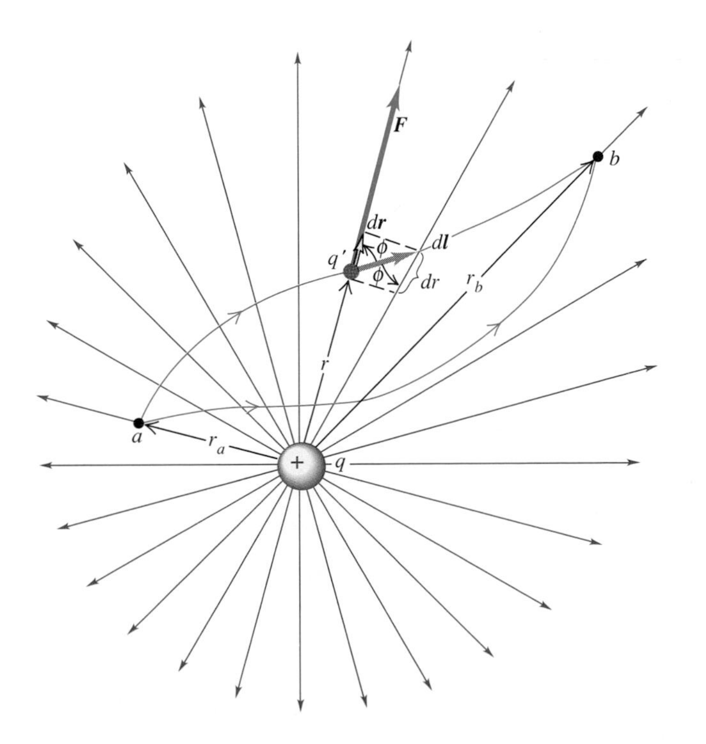


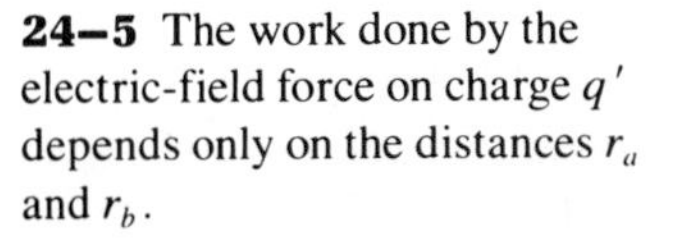

**24–5** The work done by the electric-field force on charge $q'$ depends only on the distances $r_a$ and $r_b$.

radial line. The work done on $q'$ during this displacement is given by

$$W_{a\to b} = \int_{r_a}^{r_b} F\cos\phi\, dl.$$

But the figure shows that $\cos\phi\, dl = dr$. That is, the work done during a small displacement $dl$ depends only on the change $dr$ in the distance $r$ between the charges, which is the *radial component* of the displacement. Thus Eq. (24–8) gives the work done by the field even for this more general displacement.

Equation (24–8) shows that the work done on $q'$ by the $\boldsymbol{E}$ field produced by $q$ depends only on $r_a$ and $r_b$, not on the details of the path. Also, if $q'$ returns to its starting point $a$ by a different path, the total work done in the round-trip displacement is zero. These are the needed characteristics for a conservative force field, as we defined it in Section 7–4. Thus thc force on $q'$ is a *conservative* force field.

Comparing Eqs. (24–2) and (24–8), we see that they are consistent if we define $qq'/4\pi\epsilon_0 r_a$ to be the potential energy $U_a$ when $q'$ is at point $a$, a distance $r_a$ from $q$, and $qq'/4\pi\epsilon_0 r_b$ to be the potential energy $U_b$ when $q'$ is at point $b$, a distance $r_b$ from $q$. Thus the potential energy $U$ when the test charge $q'$ is at *any* distance $r$ from charge $q$ is

$$U = \frac{1}{4\pi\epsilon_0}\frac{qq'}{r}. \tag{24–9}$$

Note that we have *not* assumed anything about the signs of $q$ and $q'$; Eq. (24–9) is valid for any combination of signs.

Gauss's law tells us that the electric field outside any spherically symmetric charge distribution is the same as though all the charge were concentrated at the center. Therefore Eq. (24–9) also holds if the test charge $q'$ is outside any spherically symmetric charge distribution with total charge $q$, at a distance $r$ from the center.

Potential energy is always defined relative to some reference point where $U = 0$. In Eq. (24–9), $U$ is zero when $q$ and $q'$ are infinitely far apart, or $r = \infty$. Therefore $U$ represents the work done on the test charge $q'$ by the field of $q$ when $q'$ moves from an initial distance $r$ to infinity. If $q$ and $q'$ have the same sign, the interaction is repulsive, this work is positive, and $U$ is positive at any finite separation. If they have opposite sign, the interaction is attractive and $U$ is negative.

## EXAMPLE 24–1

The positron (the antiparticle of the electron) has a mass of $9.11 \times 10^{-31}$ kg and a charge of $+1.60 \times 10^{-19}$ C. Suppose a positron moves in the vicinity of an alpha particle, which has a charge of $3.20 \times 10^{-19}$ C. The alpha particle is over 7000 times as massive as the positron, so we assume that it is at rest in some inertial frame of reference. When the positron is $1.00 \times 10^{-10}$ m from the alpha particle, it is moving directly away from it at a speed of $3.00 \times 10^{6}$ m/s. a) What is the positron's speed when the two particles are $3.00 \times 10^{-10}$ m apart? b) How would the situation change if the moving particle were an electron (same mass as the positron, but opposite charge)?

**SOLUTION** a) The electric force is conservative, so energy (kinetic plus potential) is conserved. Recalling the problem-solving strategy of Section 7–2, we list the positron's initial and final kinetic and potential energies, $K_a$, $K_b$, $U_a$, and $U_b$:

$$K_a = \tfrac{1}{2}mv_a^2 = \tfrac{1}{2}(9.11 \times 10^{-31}\text{ kg})(3.00 \times 10^{6}\text{ m/s})^2 = 4.10 \times 10^{-18}\text{ J},$$

$$K_b = \tfrac{1}{2}mv_b^2 = \tfrac{1}{2}(9.11 \times 10^{-31}\text{ kg})v_b^2,$$

$$U_a = \frac{1}{4\pi\epsilon_0}\frac{qq'}{r_a}$$

$$= (9.0 \times 10^{9}\text{ N}\cdot\text{m}^2/\text{C}^2)\frac{(3.20 \times 10^{-19}\text{ C})(1.60 \times 10^{-19)}\text{ C})}{1.00 \times 10^{-10}\text{ m}}$$

$$= 4.61 \times 10^{-18}\text{ J},$$

$$U_b = (9.0 \times 10^{9}\text{ N}\cdot\text{m}^2/\text{C}^2)\frac{(3.20 \times 10^{-19}\text{ C})(1.60 \times 10^{-19}\text{ C})}{3.00 \times 10^{-10}\text{ m}}$$

$$= 1.54 \times 10^{-18}\text{ J}.$$

From conservation of energy,

$$K_a + U_a = K_b + U_b,$$

$$4.10 \times 10^{-18}\text{ J} + 4.61 \times 10^{-18}\text{ J} = \tfrac{1}{2}(9.11 \times 10^{-31}\text{ kg})v_b^2 + 1.54 \times 10^{-18}\text{ J},$$

and finally,

$$v_b = 4.0 \times 10^{6}\text{ m/s}.$$

The force is repulsive, so the positron speeds up as it moves away from the stationary alpha particle.

b) If the moving charge is negative, the force on it is attractive rather than repulsive, and we expect it to slow down rather than speed up. The only difference in the above calculations is that both potential-energy quantities are negative.

The conservation-of-energy equation is

$$4.10 \times 10^{-18}\ \text{J} - 4.61 \times 10^{-18}\ \text{J} = \tfrac{1}{2}(9.11 \times 10^{-31}\ \text{kg})v_b^2 - 1.54 \times 10^{-18}\ \text{J},$$

$$v_b = 1.5 \times 10^6\ \text{m/s}.$$

---

We can generalize Eq. (24–9) for situations in which the $\boldsymbol{E}$ field in which charge $q'$ moves is caused by *several* point charges $q_1, q_2, q_3, \cdots$ at distances $r_1, r_2, r_3, \cdots$ from $q'$. The total electric field at each point is the *vector sum* of the fields due to the individual charges, and the total work done on $q'$ during any displacement is the sum of the contributions from the individual charges. We conclude that the potential energy associated with a test charge $q'$ at point $a$ in Fig. 24–6, due to a collection of charges $q_1, q_2, q_3, \cdots$ at distances $r_1, r_2, r_3, \cdots$ from the test charge $q'$ at point $a$, is the *algebraic* sum (*not* a vector sum)

$$U = \frac{q'}{4\pi\epsilon_0}\left(\frac{q_1}{r_1} + \frac{q_2}{r_2} + \frac{q_3}{r_3} + \cdots\right) = \frac{q'}{4\pi\epsilon_0}\sum_i \frac{q_i}{r_i}. \qquad (24\text{–}10)$$

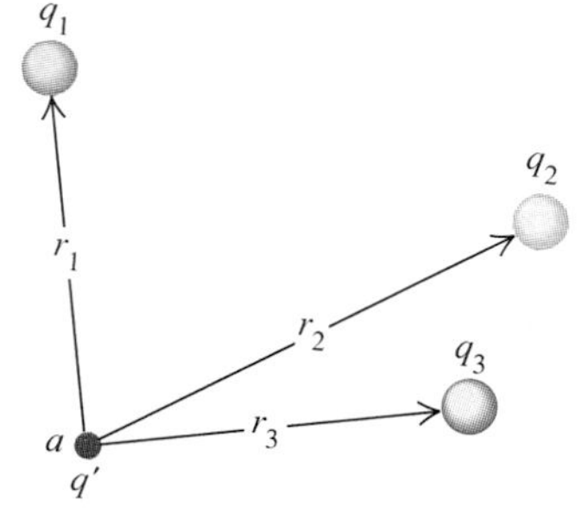


**24–6** Potential energy associated with a charge $q'$ at point $a$ depends on charges $q_1$, $q_2$, and $q_3$ and on their distances $r_1$, $r_2$, and $r_3$ from point $a$.

When $q'$ is at a different point $b$, the potential energy is given by the same expression, but $r_1, r_2, \cdots$ are the distances from $q_1, q_2, \cdots$ to point $b$. The work done on charge $q'$ when it moves from $a$ to $b$ along any path is equal to the difference $U_a - U_b$ between the potential energies when $q'$ is at $a$ and at $b$.

We can represent *any* charge distribution as a collection of point charges, so Eq. (24–10) shows that we can always find a potential-energy function for *any* static electric field. It follows that **every electric field due to a static charge distribution is a conservative force field.**

Equations (24–9) and (24–10) define $U$ to be zero when all the distances $r_1, r_2, \cdots$ are *infinite,* that is, when the test charge $q'$ is very far away from all the charges that produce the field. This position is in a sense equidistant from all the charges $q_i$ (except when the charge distribution itself extends to infinity). As with any potential-energy function, the reference point is arbitrary; we can always add a constant to make $U$ equal zero at any point we choose. Making $U = 0$ at infinity is a convenient reference level for electrostatic problems, but in circuit analysis other reference levels are often more convenient.

Equation (24–10) gives the potential energy associated with the presence of the test charge $q'$ in the $\boldsymbol{E}$ field produced by $q_1, q_2, q_3, \cdots$. But there is also potential energy involved in assembling these charges. If we start with charges $q_1, q_2, q_3, \cdots$ all separated from each other by infinite distances and then bring them together so that the distance between $q_i$ and $q_j$ is $r_{ij}$, the *total* potential energy is the sum of the energies of the pair interactions; we can write it as

$$U = \frac{1}{4\pi\epsilon_0}\sum_{ij}\frac{q_i q_j}{r_{ij}}. \qquad (24\text{–}11)$$

This sum extends over all *pairs* of charges; we don't let $i = j$ (because that would be an interaction of a charge with itself), and we count each pair only once. If we include a term with $i = 3$ and $j = 4$, we don't also include a term with $j = 3$ and $i = 4$. A neat way to express this limitation is to sum over all $i$'s and over the $j$'s for which $i > j$.

Here's a final comment about electric potential energy. We have defined it in terms of the work done *by the electric-field force* on a charged particle moving in the field, just as in Chapter 7 we defined potential energy in terms of the work done by a gravitational field or by an elastic force. When a particle moves from point $a$ to point $b$, the work done on it by the electric field is $W_{a \to b} = U_a - U_b$. When $U_a$ is greater than $U_b$, the field does positive work on the particle as it "falls" from a point of higher potential energy ($a$) to a point of lower potential energy ($b$).

An alternative but equivalent viewpoint is that in order to "raise" a particle from a point $b$ where the potential energy is $U_b$ to a point $a$ where it has a greater value $U_a$ (pushing two positive charges closer together, for example), we would need to apply an additional force $\boldsymbol{F}_{\text{ext}}$ that is equal and opposite to the electric-field force and does positive work. The potential-energy difference $U_a - U_b$ is then defined as the work done by $\boldsymbol{F}_{\text{ext}}$ during the *reverse* displacement from $b$ to $a$. Because $\boldsymbol{F}_{\text{ext}}$ is the negative of the electric-field force and the displacement is in the opposite direction, the definition of the potential difference $U_a - U_b$ is the same as before. We won't usually use this alternative viewpoint because the hypothetical additional force $\boldsymbol{F}_{\text{ext}}$ may or may not actually be present in a particular problem. We'll compare these two viewpoints further in the next section.

## 24–2 POTENTIAL

In the first section we looked at the potential energy $U$ associated with a test charge $q'$ in an electric field. Now we want to describe this potential energy on a "per unit charge" basis, just as electric field describes the force on a charged particle in the field on a "force per unit charge" basis. This leads us to the concept of electric potential, often called simply potential. This concept is very useful in calculations involving energies of charged particles. It also facilitates many electric-field calculations because it is closely related to the $\boldsymbol{E}$ field. When we need to calculate an electric field, it is often easier to calculate the potential first and then find the field from it.

**Potential** is *potential energy per unit charge.* We define the potential $V$ at any point in an electric field as *the potential energy $U$ per unit charge* associated with a test charge $q'$ at that point.

$$V = \frac{U}{q'}, \qquad \text{or} \qquad U = q'V. \tag{24–12}$$

Potential energy and charge are both scalars, so potential is a scalar quantity. From Eq. (24–12) its units are energy divided by charge. The SI unit of potential, 1 J/C, is called one **volt** (1 V) in honor of the Italian scientist Alessandro Volta (1745–1827).

$$1\ \text{V} = 1\ \text{volt} = 1\ \text{J/C} = 1\ \text{joule/coulomb}.$$

In circuit analysis, potential is often called voltage.

To put Eq. (24–2) on a "work per unit charge" basis, we divide both sides by $q'$, obtaining

$$\frac{W_{a \to b}}{q'} = \frac{U_a}{q'} - \frac{U_b}{q'} = V_a - V_b, \tag{24–13}$$

where $V_a = U_a/q'$ is the potential energy per unit charge at point $a$ and $V_b$ is that at $b$. We call $V_a$ and $V_b$ the *potential at point a* and *potential at point b,* respectively. To find the potential $V$ at a point due to any collection of point charges, we divide Eq. (24–10) by $q'$:

$$V = \frac{U}{q'} = \frac{1}{4\pi\epsilon_0} \sum_i \frac{q_i}{r_i}. \tag{24–14}$$

When we have a continuous distribution of charge along a line, over a surface, or through a volume, we divide the charge into elements $dq$, and the sum becomes an integral:

$$V = \frac{1}{4\pi\epsilon_0}\int \frac{dq}{r}, \qquad (24\text{–}15)$$

where $r$ is the distance from the charge element $dq$ to the field point where we are finding $V$. We'll work out several examples of such cases.

In deriving Eq. (24–15) we have used an expression for the potential of a point charge that is zero at an infinite distance from the charge; thus the $V$ defined by Eq. (24–15) is zero at points infinitely far away from *all* the charges. Later we'll encounter cases in which the charge distribution itself extends to infinity, and we won't be able to use this equation directly.

When we are given a collection of point charges, Eq. (24–14) is usually the easiest way to calculate the potential $V$. But in some problems in which the $\boldsymbol{E}$ field is known or can be found easily, it is easier to work directly with the field. The force $\boldsymbol{F}$ on the test charge $q'$ can be written as $\boldsymbol{F} = q'\boldsymbol{E}$, so

$$W_{a\to b} = \int_a^b \boldsymbol{F}\cdot d\boldsymbol{l} = \int_a^b q'\boldsymbol{E}\cdot d\boldsymbol{l}.$$

When we combine this with Eq. (24–13), the test charge $q'$ cancels, and we get

$$V_a - V_b = \int_a^b \boldsymbol{E}\cdot d\boldsymbol{l} = \int_a^b E\cos\phi\, dl. \qquad (24\text{–}16)$$

When the $\boldsymbol{E}$ field does positive work on a positive test charge that moves from $a$ to $b$, the potential must be greater at $a$ than at $b$. Or, using the alternative viewpoint mentioned at the end of Section 24–1, we say that to move a test charge from $b$ to $a$ would require an external force per unit charge equal to $-\boldsymbol{E}$. The work done by this external force is the potential difference $V_a - V_b$. That is,

$$V_a - V_b = -\int_b^a \boldsymbol{E}\cdot d\boldsymbol{l}.$$

Compared to Eq. (24–16), this has a negative sign and the limits are reversed. Therefore the two expressions are equal.

The difference $V_a - V_b$ is called the *potential of a with respect to b;* we sometimes abbreviate this difference as $V_{ab} = V_a - V_b$. This is often called the potential difference between $a$ and $b$, but that's ambiguous unless we specify which is the reference point. Note that potential, like electric field, is independent of the test charge $q'$ that we use to define it. When a positive test charge moves from higher to lower potential (that is, $V_a > V_b$), the electric field does positive work on it. A positive charge tends to "fall" from a high-potential region to a lower-potential one. The opposite is true for a negative charge.

An instrument that measures the difference of potential between two points is called a *voltmeter.* The principle of the common type of moving-coil voltmeter will be described later. There are also much more sensitive potential-measuring devices that use electronic amplification. Instruments that can measure a potential difference of 1 $\mu$V are common, and sensitivities down to $10^{-12}$ V can be attained.

## EXAMPLE 24–2

A proton (charge $1.60\times10^{-19}$ C) moves from point $a$ to point $b$ in a linear accelerator, along a straight line, a total distance $d = 0.50$ m. The electric field is uniform along this line in the direction from $a$ to $b$, with magnitude $E = 1.5\times10^{7}$ N/C. Determine a) the force on the proton; b) the work done on it by the field; and c) the potential difference $V_a - V_b$.

**SOLUTION** a) The force is in the same direction as the electric field, and its magnitude is

$$F = qE = (1.6 \times 10^{-19}\ \text{C})(1.5 \times 10^{7}\ \text{N/C}) = 2.4 \times 10^{-12}\ \text{N}.$$

b) The work done by this force is

$$W = Fd = (2.4 \times 10^{-12}\ \text{N})(0.50\ \text{m}) = 1.2 \times 10^{-12}\ \text{J}.$$

c) The potential difference is the work per unit charge, which is

$$V_a - V_b = \frac{W}{q} = \frac{1.2 \times 10^{-12}\ \text{J}}{1.6 \times 10^{-19}\ \text{C}} = 7.5 \times 10^{6}\ \text{J/C} = 7.5 \times 10^{6}\ \text{V} = 7.5\ \text{MV}.$$

Alternatively, $E$ is force per unit charge, so we can obtain the work per unit charge by multiplying $E$ by the distance $d$:

$$V_a - V_b = Ed = (1.5 \times 10^{7}\ \text{N/C})(0.50\ \text{m}) = 7.5 \times 10^{6}\ \text{J/C} = 7.5 \times 10^{6}\ \text{V} = 7.5\ \text{MV}.$$

## EXAMPLE 24–3

An electric dipole consists of two point charges, +12 nC and −12 nC, placed 10 cm apart (Fig. 24–7). Compute the potentials at points $a$, $b$, and $c$.

**SOLUTION** This is the same arrangement of charges as in Example 22–9 (Section 22–6). This time we have to evaluate the *algebraic* sum

$$\frac{1}{4\pi\epsilon_0}\sum_i \frac{q_i}{r_i}$$

at each point. At point $a$ the potential due to the positive charge is

$$\frac{1}{4\pi\epsilon_0}\frac{q_1}{r_1} = (9.0 \times 10^{9}\ \text{N}\cdot\text{m}^2/\text{C}^2)\frac{12 \times 10^{-9}\ \text{C}}{0.060\ \text{m}} = 1800\ \text{N}\cdot\text{m/C} = 1800\ \text{J/C} = 1800\ \text{V},$$

and the potential due to the negative charge is

$$\frac{1}{4\pi\epsilon_0}\frac{q_2}{r_2} = (9.0 \times 10^{9}\ \text{N}\cdot\text{m}^2/\text{C}^2)\frac{-12 \times 10^{-9}\ \text{C}}{0.040\ \text{m}} = -2700\ \text{N}\cdot\text{m/C} = -2700\ \text{J/C} = -2700\ \text{V}.$$

$V_a$ is the sum of these:

$$V_a = 1800\ \text{V} + (-2700\ \text{V}) = -900\ \text{V}.$$

At point $b$ the potential due to the positive charge is +2700 V, the potential due to the negative charge is −770 V, and

$$V_b = 2700\ \text{V} + (-770\ \text{V}) = 1930\ \text{V}.$$

At point $c$ the potential due to the positive charge is

$$\frac{1}{4\pi\epsilon_0}\frac{q_3}{r_3} = (9.0 \times 10^{9}\ \text{N}\cdot\text{m}^2/\text{C}^2)\frac{12 \times 10^{-9}\ \text{C}}{0.13\ \text{m}} = 830\ \text{V},$$

the potential due to the negative charge is −830 V, and the total potential is zero:

$$V_c = 830\ \text{V} + (-830\ \text{V}) = 0.$$

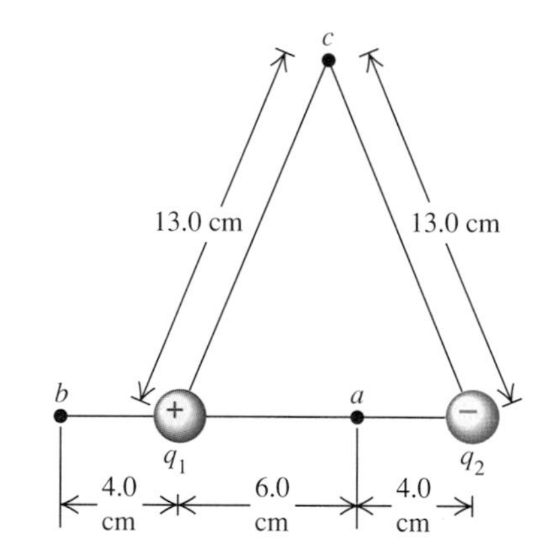


**24–7** What are the potentials at points $a$, $b$, and $c$ for this electric dipole?

## EXAMPLE 24–4

Compute the potential energies associated with a point charge of +4.0 nC if it is placed at points $a$, $b$, and $c$ in Fig. 24–7.

**SOLUTION** For any point charge $q$, $U = qV$. We use the values of $V$ from Example 24–3. At point $a$,

$$U_a = qV_a = (4.0 \times 10^{-9}\ \text{C})(-900\ \text{J/C}) = -3.6 \times 10^{-6}\ \text{J}.$$

At point $b$,

$$U_b = qV_b = (4.0 \times 10^{-9}\ \text{C})(1930\ \text{J/C}) = 7.7 \times 10^{-6}\ \text{J}.$$

At point $c$,

$$U_c = qV_c = 0.$$

(All these values correspond to $U$ and $V$ being zero at infinity.) Note that for point $c$, *no* work is done on the 4-nC charge if it moves from its original location to infinity *by any path*. In particular, if it moves along the perpendicular bisector of the line joining the other two charges, the field at every point on the bisector is perpendicular to it, so no work is done in any displacement along it.

## EXAMPLE 24–5

In Fig. 24–8 a particle with mass $m = 5.0$ g and charge $q' = 2.0$ nC starts from rest at point $a$ and moves in a straight line to point $b$. What is its speed $v$ at point $b$?

**SOLUTION** From conservation of energy,

$$K_a + U_a = K_b + U_b.$$

For this situation, $K_a = 0$ and $K_b = \frac{1}{2}mv^2$. We get the potential energies ($U$) from the potentials ($V$) using Eq. (24–12): $U_a = q'V_a$ and $U_b = q'V_b$. We then rewrite the energy equation as

$$0 + q'V_a = \tfrac{1}{2}mv^2 + q'V_b.$$

Solving this for $v$, we find

$$v = \sqrt{\frac{2q'(V_a - V_b)}{m}}.$$

We calculate the potentials using Eq. (24–14), just as we did in Example 24–3:

$$V_a = (9.0 \times 10^9\,\mathrm{N \cdot m^2/C^2})\left(\frac{3.0 \times 10^{-9}\,\mathrm{C}}{0.010\,\mathrm{m}} + \frac{-3.0 \times 10^{-9}\,\mathrm{C}}{0.020\,\mathrm{m}}\right)$$
$$= 1350\,\mathrm{V},$$

$$V_b = (9.0 \times 10^9\,\mathrm{N \cdot m^2/C^2})\left(\frac{3.0 \times 10^{-9}\,\mathrm{C}}{0.020\,\mathrm{m}} + \frac{-3.0 \times 10^{-9}\,\mathrm{C}}{0.010\,\mathrm{m}}\right)$$
$$= -1350\,\mathrm{V},$$

$$V_a - V_b = (1350\,\mathrm{V}) - (-1350\,\mathrm{V}) = 2700\,\mathrm{V}.$$

Finally,

$$v = \sqrt{\frac{2(2.0 \times 10^{-9}\,\mathrm{C})(2700\,\mathrm{V})}{5.0 \times 10^{-3}\,\mathrm{kg}}}$$
$$= 4.6 \times 10^{-2}\,\mathrm{m/s} = 4.6\,\mathrm{cm/s}.$$

We can check unit consistency by noting that 1 V = 1 J/C, so the numerator under the radical has units of J or kg · $\mathrm{m^2/s^2}$.

We can use exactly this same method to find the speed of an electron accelerated across a potential difference of 500 V in an oscilloscope tube or a potential difference of 20 kV in a TV picture tube. The end-of-chapter problems include several examples of such calculations.

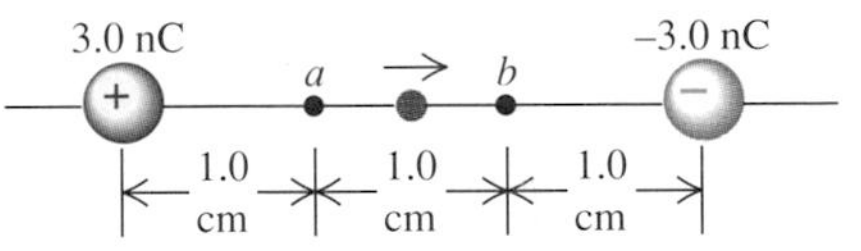


**24–8** The particle moves from point $a$ to point $b$; its acceleration is not constant.

# 24–3 CALCULATION OF POTENTIALS

Potential calculations usually follow one of two routes. If we know the charge distribution, we can use Eq. (24–14) or (24–15). Or if we know the electric field, we can use Eq. (24–16), defining the potential to be zero at some convenient place. Some problems require a combination of these approaches.

## PROBLEM-SOLVING STRATEGY

### Calculation of potential

**1.** Remember that potential is simply *potential energy per unit charge*. Understanding this simple statement can get you a long way.

**2.** To find the potential due to a collection of point charges, use Eq. (24–14). If you are given a continuous charge distribution, devise a way to divide it into infinitesimal elements identified by coordinates and coordinate differentials. Then use Eq. (24–15), expressing the sum as an integral. Carry out the integration, using appropriate limits to include the entire charge distribution. Be careful about which geometric quantities in the integral vary and which are constant.

**3.** If you are given the electric field or if you can find it using any of the methods of Chapter 22 or 23, it may be easier to calculate the work done on a test charge during a displacement from point $a$ to point $b$ and then use Eq. (24–16). When it's appropriate, make use of your freedom to define $V$ to be zero at some convenient place. For point charges this will usually be at infinity, but for other distributions of charge (especially those that extend to infinity themselves) it may be convenient or necessary to define $V$ to be zero at some finite distance from the charge distribution, say, at point $b$. This is just like defining $U$ to be zero at ground level in gravitational problems. Then the potential at any other point, say, $a$, can by found from Eq. (24–16), with $V_b = 0$.

**4.** Remember that potential is a *scalar* quantity, not a *vector.* It doesn't have components, and it would be wrong to try to use components of potential. However, you may have to use components of the vectors $\boldsymbol{E}$ and $d\boldsymbol{l}$ when you use Eq. (24–16).

## EXAMPLE 24–6

**Charged spherical conductor** A solid conducting sphere of radius $R$ has a total charge $q$. Find the potential everywhere, both outside and inside the sphere.

**SOLUTION** We used Gauss's law in Example 23–5 (Section 23–3) to show that at all points *outside* the sphere the field is the same as that of a point charge $q$ at the center of the sphere. *Inside* the sphere the field is zero everywhere; otherwise, charge would move within the sphere. If we take $V = 0$ at infinity, as we did for a point charge, then for a point outside the sphere, at a distance $r$ from its center, the potential is the same as for a point charge $q$ at the center, namely,

$$V = \frac{1}{4\pi\epsilon_0}\frac{q}{r}.$$

The potential at the surface of the sphere is

$$V = \frac{1}{4\pi\epsilon_0}\frac{q}{R}. \qquad (24\text{–}17)$$

Inside the sphere the field is zero everywhere, and no work is done on a test charge that moves from any point to any other point in this region. Thus the potential is the same at every point inside the sphere and is equal to its value $q/4\pi\epsilon_0 R$ at the surface. The field and potential are shown as functions of $r$ in Fig. 24–9 for a positive charge $q$. The electric field $\boldsymbol{E}$ at the surface has magnitude

$$E = \frac{1}{4\pi\epsilon_0}\frac{|q|}{R^2}. \qquad (24\text{–}18)$$

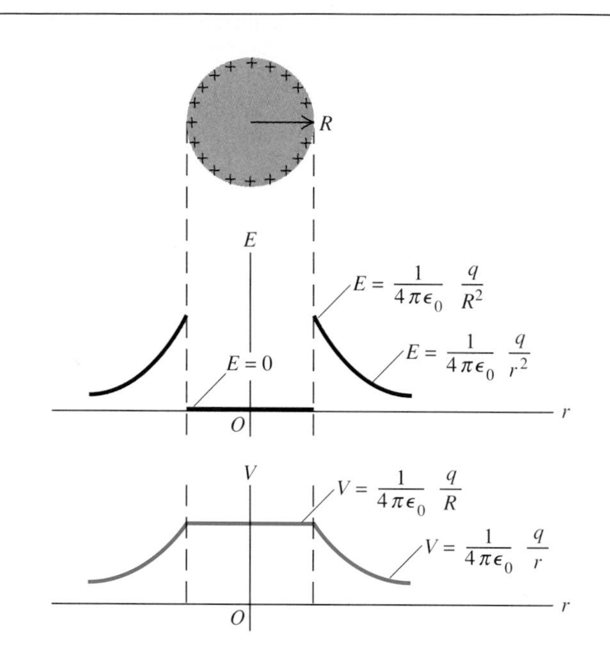


**24–9** Electric-field magnitude $E$ and potential $V$ at points inside and outside a charged spherical conductor.

Here is a practical application of the result of Example 24–6. The maximum potential to which a conductor in air can be raised is limited by the fact that air molecules become *ionized,* and air becomes a conductor, at an electric field magnitude of about $3 \times 10^6$ N/C. Assume for the moment that $q$ is positive. Comparing Eqs. (24–17) and (24–18), we note that at the surface of a conducting sphere the field and potential are related by $V = ER$. Thus if $E_m$ represents the electric field magnitude at which air becomes conductive (known as the *dielectric strength* of air), the maximum potential $V_m$ to which a spherical conductor can be raised is

$$V_m = RE_m.$$

For a sphere 1 cm in radius, in air,

$$V_m = (10^{-2}\text{ m})(3 \times 10^6\text{ N/C}) = 30{,}000\text{ V}.$$

No amount of "charging" could raise the potential of a sphere of this size, in air, higher than about 30,000 V. This is the reason that large spherical terminals are used on high-voltage machines such as Van de Graaff generators (Fig. 24–10). If we make $R = 2$ m, then

$$V_m = (2 \text{ m})(3 \times 10^6 \text{ N/C}) = 6 \times 10^6 \text{ V} = 6 \text{ MV}.$$

Such machines are sometimes placed in pressure tanks filled with a gas, such as sulfur hexafluoride ($SF_6$), that can withstand larger fields without becoming conductive.

At the other extreme is the effect produced by a surface of very *small* radius of curvature, such as a sharp point or thin wire. Since the maximum potential is proportional to the radius, even relatively small potentials applied to sharp points in air produce sufficiently high fields just outside the point to ionize the surrounding air, making it a conductor. The resulting current and its associated glow (visible in a dark room) is called *corona*. Copying machines use corona from fine wires to charge the imaging drum, and some paper-handling machines use corona to help drain off static charges that would otherwise cause paper jams. Along with the ionization comes ozone, which is unhealthful in large quantities. Car radio antennas have a ball on the end to help *prevent* corona that would cause static.

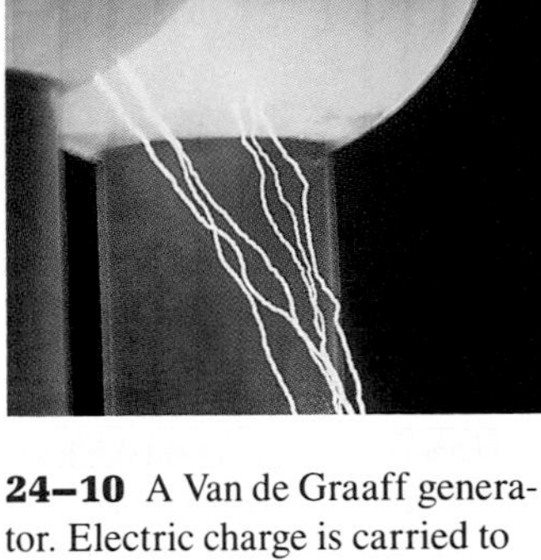

**24–10** A Van de Graaff generator. Electric charge is carried to the dome at the top by a moving belt inside the vertical column. Potential differences as large as $10^6$V can be developed.

## EXAMPLE 24–7

**Parallel plates** Find the potential at any height $y$ between the two charged parallel plates discussed at the beginning of Section 24–1.

**SOLUTION** The potential energy $U$ for a test charge $q'$ at a distance $y$ above the bottom plate (Fig. 24–11) is given by Eq. (24–5), $U = q'Ey$. The potential $V_y$ at point $y$ is the potential energy per unit charge, $V_y = U/q'$:

$$V = Ey.$$

We have chosen $U$, and therefore $V$, to be zero at point $b$, where $y = 0$. Even if we choose some different reference level, it is still true that

$$V_y - V_b = Ey.$$

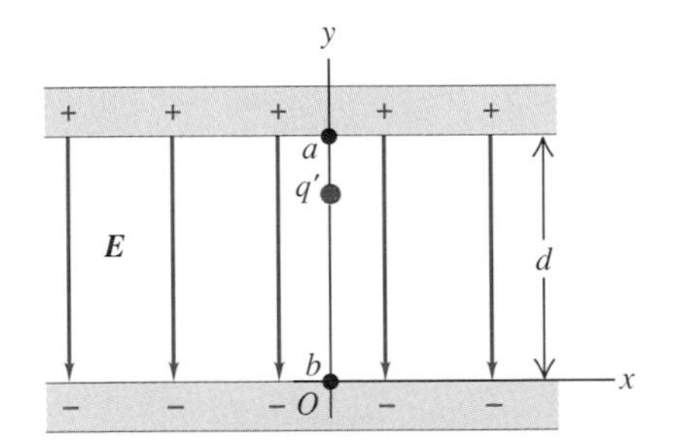


**24–11** The charged parallel plates from Fig. 24–1.

The potential decreases linearly with $y$ as we move from the upper to the lower plate. At point $a$, where $y = d$ and $V_y = V_a$,

$$V_a - V_b = Ed,$$

$$E = \frac{V_a - V_b}{d} = \frac{V_{ab}}{d}. \qquad (24\text{–}19)$$

That is, the electric field equals the potential difference between the plates divided by the distance between them. (*Caution!* This relation holds only for the planar geometry that we have described; it *doesn't* work for situations such as concentric cylinders or spheres, in which the $\boldsymbol{E}$ field isn't uniform.)

In Example 23–8 (Section 23–3) we derived the expression $E = \sigma/\epsilon_0$ for the electric field $E$ between two conducting plates in terms of the surface charge density $\sigma$ on a plate. Equation (24–19) is more generally useful than this expression because the potential difference $V_{ab}$ can be measured easily with a voltmeter, but there are no instruments that read surface charge density directly.

Equation (24–19) also shows that the unit of electric field can be expressed as 1 *volt per meter* (1 V/m), as well as 1 N/C:

$$1 \text{ V/m} = 1 \text{ N/C}.$$

In practice, the volt per meter is the usual unit of $E$.

## EXAMPLE 24–8

**Line charge and charged conducting cylinder** Find the potential at a distance $r$ from a line charge with linear charge density (charge per unit length) $\lambda$.

**SOLUTION** We found in Example 23–6 (Section 23–3) that the *field* $\boldsymbol{E}$ at a distance $r$ from a long straight-line charge or outside a long charged conducting cylinder has a radial component

$$E_r = \frac{1}{2\pi\epsilon_0}\frac{\lambda}{r}.$$

From Eq. (24–16) the potential of any point $a$ with respect to any other point $b$, at radial distances $r_a$ and $r_b$ from the line of charge

(Fig. 24–12a), is

$$V_a - V_b = \int_a^b \boldsymbol{E} \cdot d\boldsymbol{l} = \frac{\lambda}{2\pi\epsilon_0}\int_{r_a}^{r_b} \frac{dr}{r} = \frac{\lambda}{2\pi\epsilon_0} \ln \frac{r_b}{r_a}. \qquad (24\text{–}20)$$

If we take point $b$ at infinity and set $V_b = 0$, we find, for the potential $V_a$,

$$V_a = \frac{\lambda}{2\pi\epsilon_0} \ln \frac{\infty}{r_a} = \infty.$$

This shows that if we try to define $V$ to be zero at infinity, then it must be infinite at any finite distance from the line charge. This is therefore not a useful way to define $V$ for this problem. The difficulty, as we mentioned earlier, is that the charge distribution itself extends to infinity.

The way around this difficulty is the fact that we can define $V$ to be zero at any point we like. We set $V_b = 0$ at point $b$ with arbitrary radius $r_0$. Then at any other point $a$, with radius $r$, the potential $V$ is $V_a - V_b = V_a - 0$, or

$$V = \frac{\lambda}{2\pi\epsilon_0} \ln \frac{r_0}{r}. \qquad (24\text{–}21)$$

Equations (24–20) and (24–21) give the potential in the field of a cylinder only for values of $r$ equal to or greater than the radius $R$ of the cylinder. If we choose $r_0$ to be the cylinder radius $R$, so that $V = 0$ when $r = R$ (Fig. 24–12b), then at any point for which $r > R$,

$$V = \frac{\lambda}{2\pi\epsilon_0} \ln \frac{R}{r}, \qquad (24\text{–}22)$$

where $r$ is the distance from the axis of the cylinder.

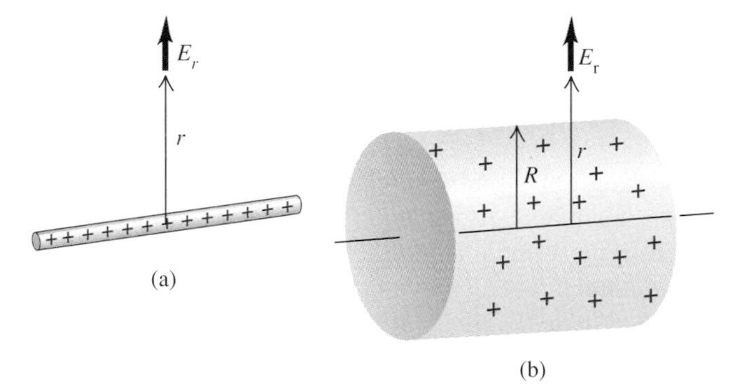


**24–12** Electric field outside (a) a long charged wire; (b) a long charged cylinder.

## EXAMPLE 24–9

**Charged circular ring** Electric charge is distributed uniformly around a thin ring of radius $a$, with total charge $Q$ (Fig. 24–13). Find the potential at a point $P$ on the ring's axis at a distance $x$ from the center of the ring.

**SOLUTION** We have seen this ring several times before, most recently in Example 22–10 (Section 22–6). Referring to that example, we note that the entire charge is at a distance $r = (x^2 + a^2)^{1/2}$ from point $P$. We conclude immediately that the potential at point $P$, which is a function of $x$, is

$$V(x) = \frac{1}{4\pi\epsilon_0} \frac{Q}{\sqrt{x^2 + a^2}}. \qquad (24\text{–}23)$$

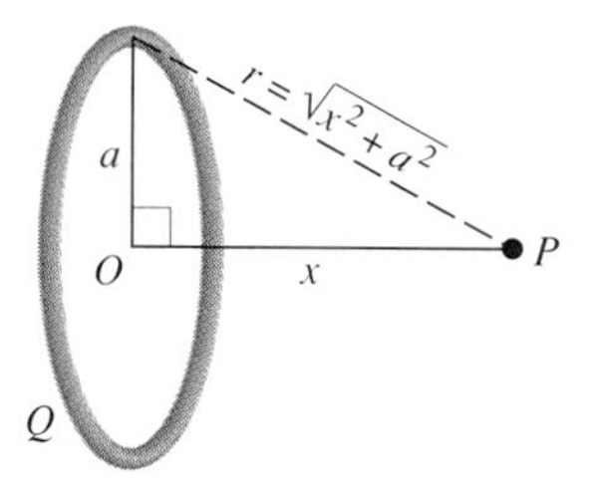


**24–13** All of the charge $Q$ is at the same distance $r$ from $P$.

Potential is a *scalar* quantity; there is no need to consider components of vectors in this calculation, as we had to do when we found the electric field at $P$, so the potential calculation is a lot simpler than the field calculation. When $x$ is much larger than $a$, Eq. (24–23) becomes approximately equal to

$$V(x) = \frac{1}{4\pi\epsilon_0} \frac{Q}{x}.$$

This corresponds to the potential of a point charge $Q$ at distance $x$. When we are very far away from a charged ring, it looks like a point charge.

## EXAMPLE 24–10

**Charged thin rod** Electric charge is distributed uniformly along a thin rod of length $2a$, with total charge $Q$. Find the potential at a point $P$ along the perpendicular bisector of the rod at a distance $x$ from its center.

**SOLUTION** This is the same situation as in Example 22–11 (Section 22–6). In Fig. 24–14 the element of charge $dQ$ corresponding to an element of length $dy$ on the rod is again given by $dQ = (Q/2a)dy$. The distance from $dQ$ to $P$ is $(x^2 + y^2)^{1/2}$, and the contribution $dV$ that it makes to the potential at $P$ is

$$dV = \frac{1}{4\pi\epsilon_0} \frac{Q}{2a} \frac{dy}{\sqrt{x^2 + y^2}}.$$

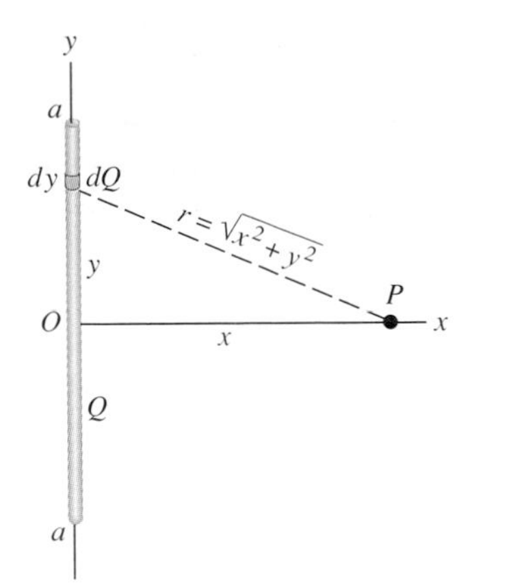


**24–14** Finding the electric potential on the perpendicular bisector of a uniformly charged rod of length $2a$.

To get the potential at $P$, we integrate this over the length of the rod, from $y = -a$ to $y = a$:

$$V(x) = \frac{1}{4\pi\epsilon_0}\frac{Q}{2a}\int_{-a}^{a}\frac{dy}{\sqrt{x^2 + y^2}}.$$

You can look up the integral in a table; the final result is

$$V(x) = \frac{1}{4\pi\epsilon_0}\frac{Q}{2a}\ln\frac{\sqrt{a^2 + x^2} + a}{\sqrt{a^2 + x^2} - a}.$$

When $x$ is very large, we expect $V$ to approach zero; we invite you to verify that it does so.

Again note that this problem is simpler than the calculation of $\boldsymbol{E}$ at point $P$ because potential is a scalar quantity and no vector calculations are involved.

## 24–4 EQUIPOTENTIAL SURFACES

Field lines (Section 22–7) help us visualize electric fields. In a similar way the potential at various points in an electric field can be represented graphically by **equipotential surfaces.** An equipotential surface is defined as a surface on which the potential is the same at every point. In a region where an electric field is present, we can construct an equipotential surface through any point. In diagrams we usually show only a few representative equipotentials, often with equal potential differences between adjacent surfaces. No point can be at two different potentials, so equipotential surfaces for different potentials can never touch or intersect.

The potential energy for a test charge is the same at every point on a given equipotential surface, so the $\boldsymbol{E}$ field does no work on a test charge when it moves from point to point on such a surface. It follows that the $\boldsymbol{E}$ field can never have a component tangent to the surface; such a component would do work on a charge moving on the surface. Therefore $\boldsymbol{E}$ must be perpendicular to the surface at every point. **Field lines and equipotential surfaces are always mutually perpendicular.** In general, field lines are curves and equipotentials are curved surfaces. For the special case of a *uniform* field, in which the field lines are straight, parallel, and equally spaced, the equipotentials are parallel *planes* perpendicular to the field lines.

Figure 24–15 shows several arrangements of charges. The field lines in the plane of the charges are represented by red lines, and the intersections of the equipotential

**24–15** Equipotential surfaces (blue lines) and electric field lines (red lines) for assemblies of point charges. (a) A single isolated positive charge. (b) An electric dipole. (c) Two equal positive charges. How would the diagrams change if the signs were reversed?

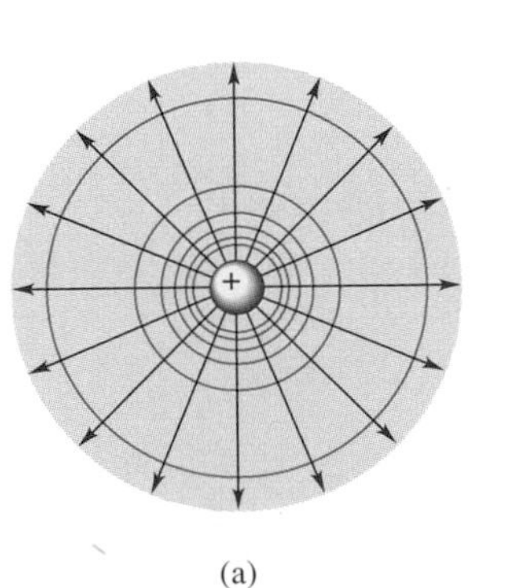
(a)

0 V

(b)

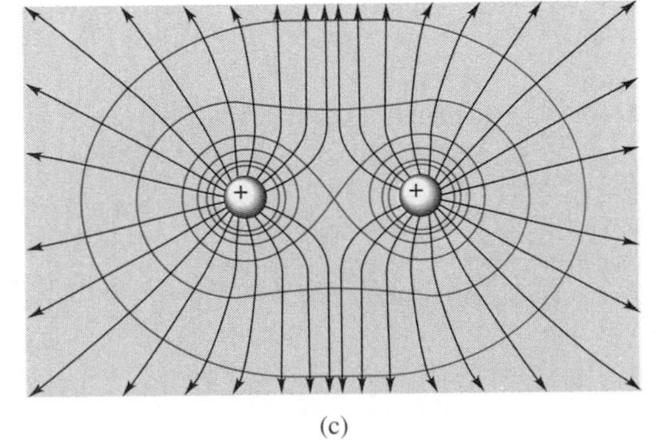
(c)

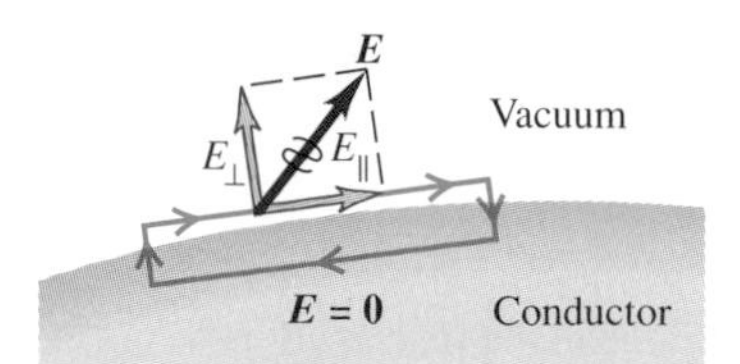


**24–16** Inside the conductor, $E = 0$. If $E$ just outside the conductor had a component $E_\parallel$ parallel to the conductor surface, then a test charge moving around the rectangular loop and returning to the starting point would have nonzero total work done on it by the field. Because the $E$ field is conservative, this is impossible. Therefore $E_\parallel$ must be zero, and $E$ just outside the surface is perpendicular to it.

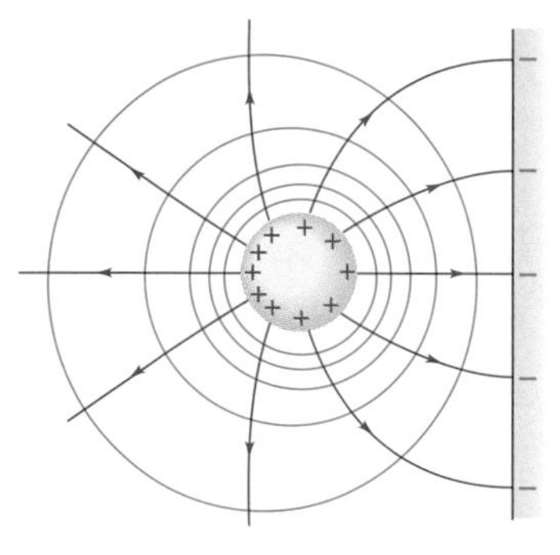

**24–17** When charges are at rest, a conducting surface is always an equipotential surface. Field lines are perpendicular to a conducting surface.

surfaces with this plane (that is, cross sections of these surfaces) are shown as blue lines. The actual field and equipotential surfaces are three-dimensional. At each crossing of an equipotential and a field line, the two are perpendicular.

We can prove that when all charges are at rest, **the electric field just outside a conductor must be perpendicular to the surface at every point.** We know that $E = 0$ everywhere inside the conductor; otherwise, charges would move. In particular, the component of $E$ tangent to the surface, just inside the surface at any point, is zero. It follows that the tangential component of $E$ is also zero just *outside* the surface. If it were not, a charge could move around a rectangular path partly inside and partly outside (Fig. 24–16) and return to its starting point with a net amount of work having been done on it. This would violate the conservative nature of electrostatic fields, so the tangential component of $E$ just outside the surface must be zero at every point on the surface. Thus $E$ is perpendicular to the surface at each point (Fig. 24–17). It follows that in an electrostatic situation **a conducting surface is always an equipotential surface.**

We can draw equipotentials so that adjacent surfaces have equal potential differences. Then in regions where the magnitude of $E$ is large the equipotential surfaces are close together because the field does a relatively large amount of work on a test charge in a relatively small displacement. Conversely, in regions where the field is weaker the equipotential surfaces are farther apart.

Finally, we can now prove a theorem that we quoted without proof in Section 23–4. The theorem is as follows: In an electrostatic situation, if a conductor contains a cavity, and if no charge is present inside the cavity, then there can be no net charge *anywhere* on the surface of the cavity. To prove this theorem, we first prove that *every point in the cavity is at the same potential*. In Fig. 24–18 the surface $A$ of the cavity is an equipotential surface, as we have just proved. Suppose point $P$ in the cavity is at a different potential; then we can construct a different equipotential surface $B$ including point $P$.

Now consider a Gaussian surface, shown as a purple line, between the two equipotential surfaces. Because of the relation between $E$ and the equipotentials, we know that the field at every point between the equipotentials is from $A$ toward $B$, or else at every point it is from $B$ toward $A$, depending on which equipotential surface is at higher potential. In either case the flux through this Gaussian surface is certainly not zero. Then Gauss's law says that the charge enclosed by the Gaussian surface cannot be zero. This contradicts our initial assumption that there is *no* charge in the cavity. So the potential at $P$ *cannot* be different from that at the cavity wall.

The entire region of the cavity must therefore be at the same potential. But for this to be true the electric field inside the cavity must be zero everywhere. Finally, Gauss's law shows that the electric field at any point on the surface of a conductor is proportional to the surface charge density $\sigma$ at that point. We conclude that *the surface charge density on the wall of the cavity is zero at every point.* This chain of reasoning may seem tortuous, but it is worth careful study.

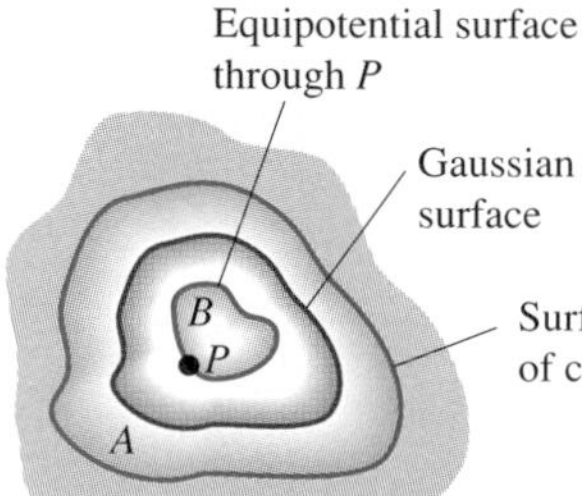


**24–18** A cavity in a conductor. If the cavity contains no charge, every point in the cavity is at the same potential, the electric field is zero everywhere, and there is no charge anywhere on the surface of the cavity.

# 24–5 POTENTIAL GRADIENT

Electric field and potential are closely related. Equation (24–16) expresses one aspect of that relationship:

$$V_a - V_b = \int_a^b \boldsymbol{E} \cdot d\boldsymbol{l} = -\int_b^a \boldsymbol{E} \cdot d\boldsymbol{l}. \tag{24–24}$$

If we know $\boldsymbol{E}$ at various points, we can use this equation to calculate potential differences. We should be able to turn this around; if we know the potential $V$ at various points, we can use it to determine $\boldsymbol{E}$. Regarding $V$ as a function of the coordinates $(x, y, z)$ of a point in space, we will show that the components of $\boldsymbol{E}$ are directly related to the *partial derivatives* of $V$ with respect to $x$, $y$, and $z$.

In Eq. (24–24), $V_{ab} = V_a - V_b$ is the potential of $a$ with respect to $b$, that is, the change of potential when a point moves from $b$ to $a$. We can write this as

$$V_{ab} = \int_b^a dV = -\int_a^b dV,$$

where $dV$ is the infinitesimal change of potential accompanying an infinitesimal element $d\boldsymbol{l}$ of the path from $b$ to $a$. Let $E_l$ be the component of $\boldsymbol{E}$ parallel to $d\boldsymbol{l}$; then, using the above expression for $V_{ab}$, we may rewrite our first equation as

$$-\int_a^b dV = \int_a^b E_l \, dl.$$

These two integrals must be equal for *any* pair of limits $a$ and $b$, and for this to be true the *integrands* must be equal. Thus for *any* infinitesimal displacement $d\boldsymbol{l}$,

$$-dV = E_l \, dl,$$

or

$$E_l = -\frac{dV}{dl}.$$

The derivative $dV/dl$ is the rate of change of $V$ for a displacement in the direction of $d\boldsymbol{l}$. In particular, if $d\boldsymbol{l}$ is parallel to the $x$-axis, then the component of $\boldsymbol{E}$ parallel to $d\boldsymbol{l}$ is just the $x$-component of $\boldsymbol{E}$, that is, $E_x$. Thus $E_x = -dV/dx$. Because $V$ is, in general, also a function of $y$ and $z$, we use partial derivative notation; a derivative in which only $x$ varies is written $\partial V/\partial x$. The $y$- and $z$-components of $\boldsymbol{E}$ are related to the corresponding derivatives of $V$ in the same way, so we have

$$E_x = -\frac{\partial V}{\partial x}, \qquad E_y = -\frac{\partial V}{\partial y}, \qquad E_z = -\frac{\partial V}{\partial z}. \tag{24–25}$$

In terms of unit vectors we can write $\boldsymbol{E}$ as

$$\boldsymbol{E} = -\left(\boldsymbol{i}\frac{\partial V}{\partial x} + \boldsymbol{j}\frac{\partial V}{\partial y} + \boldsymbol{k}\frac{\partial V}{\partial z}\right)$$
$$= -\left(\boldsymbol{i}\frac{\partial}{\partial x} + \boldsymbol{j}\frac{\partial}{\partial y} + \boldsymbol{k}\frac{\partial}{\partial z}\right)V. \tag{24–26}$$

In vector notation the following operation is called the **gradient** of the function $f$:

$$\boldsymbol{\nabla} f = \left(\boldsymbol{i}\frac{\partial}{\partial x} + \boldsymbol{j}\frac{\partial}{\partial y} + \boldsymbol{k}\frac{\partial}{\partial z}\right)f \tag{24–27}$$

The operator denoted by the symbol $\nabla$ is called "del." Thus, in vector notation, Eqs. (24–25) are summarized compactly as

$$\boldsymbol{E} = -\nabla V. \tag{24–28}$$

This is read "$\boldsymbol{E}$ is the negative of the gradient of $V$," or "$\boldsymbol{E}$ equals negative del $V$."

This equation doesn't depend on any particular choice of the zero point for $V$. If we change the zero point, the effect is to change $V$ at every point by the same amount; the derivatives of $V$ will be the same.

In problems in which $\boldsymbol{E}$ is radial with respect to a point or an axis and $r$ is the distance from the point or the axis, the relation corresponding to Eqs. (24–25) is

$$E_r = -\frac{\partial V}{\partial r}. \tag{24–29}$$

We note again that the unit of electric field can be expressed in either of these equivalent forms:

$$1 \text{ V/m} = 1 \text{ N/C}.$$

## EXAMPLE 24–11

**Potential and field of a point charge** We have shown that the potential at a radial distance $r$ from a point charge $q$ is

$$V = \frac{1}{4\pi\epsilon_0}\frac{q}{r}.$$

By symmetry the electric field is in the radial direction, so

$$E = E_r = -\frac{\partial V}{\partial r} = -\frac{\partial}{\partial r}\left(\frac{1}{4\pi\epsilon_0}\frac{q}{r}\right) = \frac{1}{4\pi\epsilon_0}\frac{q}{r^2},$$

in agreement with Eq. (22–6).

## EXAMPLE 24–12

**Potential and field outside a charged conducting cylinder** In Example 24–8 we found that the potential outside a charged conducting cylinder with radius $R$ and charge per unit length $\lambda$ is

$$V = \frac{\lambda}{2\pi\epsilon_0}\ln\frac{R}{r} = \frac{\lambda}{2\pi\epsilon_0}(\ln R - \ln r).$$

The electric field has only a radial component $E_r$; it is

$$E_r = -\frac{\partial V}{\partial r} = \frac{\lambda}{2\pi\epsilon_0 r},$$

in agreement with our previous result.

## EXAMPLE 24–13

**Potential and field of a ring of charge** In Example 24–9 we found that for a ring of charge with radius $a$ and total charge $Q$ the potential at a point $P$, a distance $x$ from the center of the ring on a line through the center perpendicular to the plane of the ring, is

$$V(x) = \frac{1}{4\pi\epsilon_0}\frac{Q}{\sqrt{x^2 + a^2}}.$$

From Eq. (24–25),

$$E_x = -\frac{\partial V}{\partial x} = \frac{1}{4\pi\epsilon_0}\frac{Qx}{(x^2 + a^2)^{3/2}}.$$

This agrees with the result that we obtained in Example 22–10 (Section 22–6).

In this example, $V$ does not appear to be a function of $y$, but it would *not* be correct to conclude that $\partial V/\partial y = 0$ and $E_y = 0$ everywhere. The reason is that our expression for $V$ is valid *only for points on the x-axis,* where $y = 0$. If we had the complete form of $V$, valid for *all* points in space, then we could use it to find $E_y = -\partial V/\partial y$ at any point, and so on.

## EXAMPLE 24–14

**Potential and field of a line charge** A charge $Q$ is uniformly distributed along a rod with length $2a$. In Example 24–10 (Section 24–3) we derived an expression for the potential at a point on the perpendicular bisector of such a rod at a distance $x$

from its center:

$$V(x) = \frac{1}{4\pi\epsilon_0}\frac{Q}{2a}\ln\frac{\sqrt{a^2+x^2}+a}{\sqrt{a^2+x^2}-a}.$$

The negative of the derivative of this expression with respect to $x$ is $E_x$. Carry out the differentiation and show that the result is the expression found in Example 22–11 (Section 22–6) by direct integration. (*Note:* Writing the ln of the fraction as the difference of two ln's simplifies the calculation.)

Several of these examples illustrate an important point. Often we can compute the electric field caused by a charge distribution in either of two ways: directly, by adding the $\boldsymbol{E}$ fields of point charges, or by first calculating the potential and then taking its negative gradient to find the field. The second method is often easier because potential is a *scalar* quantity, requiring at worst the integration of a scalar function. Electric field is a *vector* quantity, requiring computation of components for each element of charge and a separate integration for each component. Thus, quite apart from its fundamental significance, potential offers a very useful computational technique in field calculations.

Another important point is the relation of the *direction* of $\boldsymbol{E}$ to the behavior of $V$. At each point, $\boldsymbol{E}$ is always perpendicular to the equipotential surface through the point, and its direction is the direction in which $V$ *decreases* most rapidly. In all these examples the potential decreases as we move away from the charge distribution (assuming that the charge is positive), and the electric field points away from it. If the charge is negative, the potential increases algebraically (becoming less negative) as we move away from the charge, and $\boldsymbol{E}$ points *toward* the charge. The situation is completely analogous to gravitational potential energy (Sections 7–2 and 12–4). Near the surface of the earth, for example, the gravitational field points toward the center of the earth; this is the direction of most rapid decrease of gravitational potential energy. Heat flow provides another analogy; the direction of heat flow is the direction of the most rapid decrease of temperature $T$, that is, the direction of $-\boldsymbol{\nabla}T$.

Finally, we stress once more that if we know $\boldsymbol{E}$ as a function of position, we can calculate $V$, using Eq. (24–16) or (24–24), and if we know $V$ as a function of position, we can calculate $\boldsymbol{E}$ using Eq. (24–25), (24–26), or (24–29). Deriving $V$ from $\boldsymbol{E}$ is an integral operation, and deriving $\boldsymbol{E}$ from $V$ is a differential operation.

## 24–6 THE MILLIKAN OIL-DROP EXPERIMENT

In Section 22–3 we talked a little about *quantization* of charge. Have you ever wondered how the charge of an individual electron can be measured? The first solution to this formidable experimental problem was the **Millikan oil-drop experiment,** a brilliant piece of work carried out at the University of Chicago during the years 1909–1913 by Robert Andrews Millikan. We've postponed discussing this experiment until now so that we can use the concept of potential in our discussion.

Millikan's apparatus is shown schematically in Fig. 24–19a. Two parallel horizontal metal plates, $A$ and $B$, are insulated from each other and separated by a few millimeters. Oil is sprayed in very fine drops (diameter around $10^{-4}$ mm) from an atomizer above the upper plate, and a few drops are allowed to fall through a small hole in this plate. A beam of light is directed horizontally between the plates, and a telescope is set up with its axis at right angles to the light beam. The oil drops, illuminated by the light beam and viewed through the telescope, appear like tiny bright stars. A scale in the telescope permits precise measurements of the vertical positions of the drops, so their speeds can also be measured.

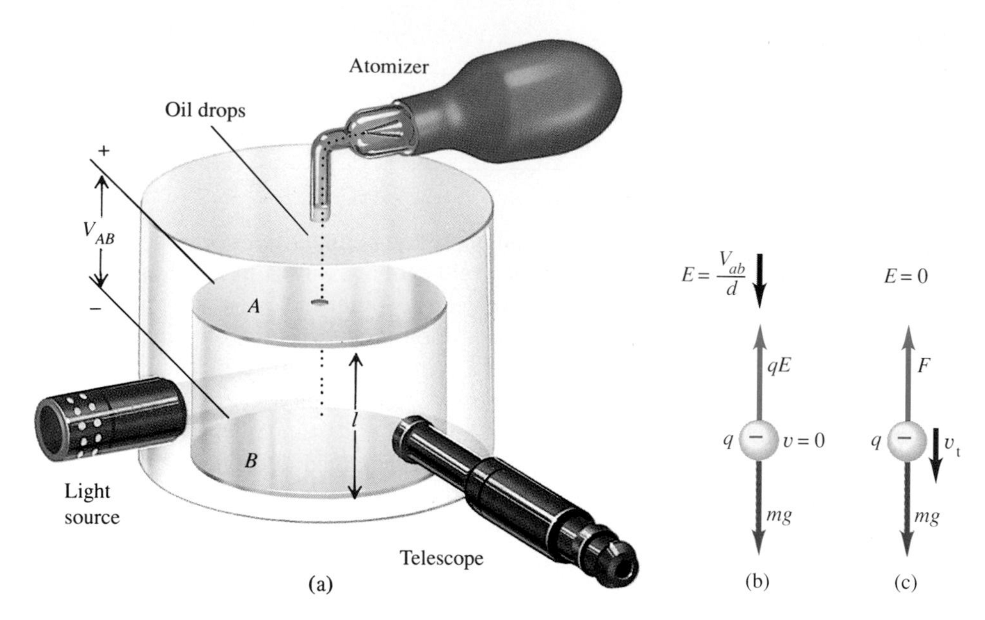


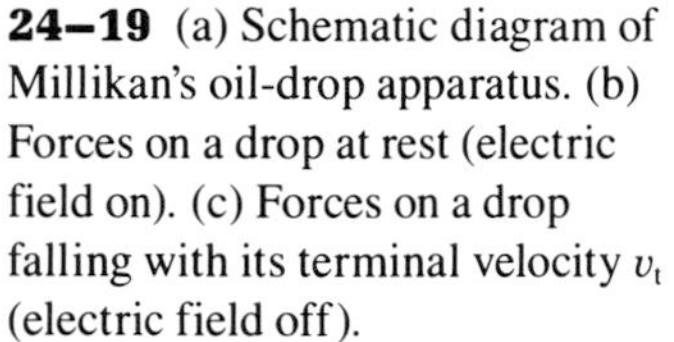

**24–19** (a) Schematic diagram of Millikan's oil-drop apparatus. (b) Forces on a drop at rest (electric field on). (c) Forces on a drop falling with its terminal velocity $v_t$ (electric field off).

Some of the oil drops are electrically charged because of frictional effects or because of ionization of the surrounding air by X-rays or radioactivity. The drops are usually negatively charged, but occasionally one with a positive charge is found.

Here's how we can measure the charge on a drop. Suppose a drop has a negative charge with absolute value $q$ and the plates are maintained at a potential difference such that there is a downward electric field of magnitude $E$ between them. The forces on the drop are then its weight $mg$ and the upward force $qE$. By adjusting the field $E$ we can make $qE$ equal to $mg$ (Fig. 24–19b). The drop is then in static equilibrium, and

$$q = \frac{mg}{E}. \tag{24–30}$$

The electric-field magnitude $E$ is the potential difference $V_{AB}$ divided by the distance $d$ between the plates, as we found in Example 24–7. We can find the mass $m$ of the drop if we know its radius $r$ because the mass equals the product of the density $\rho$ and the volume $4\pi r^3/3$. So we can rewrite Eq. (24–30) as

$$q = \frac{4\pi}{3}\frac{\rho r^3 g d}{V_{AB}}. \tag{24–31}$$

Everything on the right side of Eq. (24–31) is easy to measure except for the drop radius $r$, which is much too small to measure directly with any degree of precision. But here comes an example of Millikan's genius. He determined the drop's radius by cutting off the electric field and measuring the *terminal speed* $v_t$ of the drop as it fell. You may want to review the concept of terminal speed in Example 5–19 (Section 5–3). At terminal speed the weight $mg$ is just balanced by the viscous air-resistance force $F$ (Fig. 24–19c).

Here's how Millikan found the drop's radius from $v_t$. The viscous force $F$ on a sphere of radius $r$ moving with speed $v$ through a fluid with viscosity $\eta$ is given by Stokes' law, Eq. (14–31): $F = 6\pi\eta r v$. When the drop is falling at terminal speed, this just balances the weight $w = mg$ of the drop, which we can express in terms of the density $\rho$ and radius $r$ of the drop: $w = 4\pi\rho g r^3/3$. Equating this to the Stokes'-law expression for $F$, we find

$$\tfrac{4}{3}\pi r^3 \rho g = 6\pi\eta r v_t,$$

and

$$r = 3\sqrt{\frac{\eta v_t}{2\rho g}}. \tag{24–32}$$

Combining this expression for $r$ with Eq. (24–31), we find

$$q = 18\pi \frac{d}{V_{AB}}\sqrt{\frac{\eta^3 v_t^3}{2\rho g}}. \tag{24–33}$$

This expresses the absolute value $q$ of the charge in terms of measurable quantities.

The actual experiment was a little more complicated. Millikan had to correct for the buoyant force of air on the falling drop by replacing the density $\rho$ of the oil with $\rho - \rho_{air}$. He also had to use a more elaborate version of Stokes' law that takes into account the molecular nature of air because the intermolecular distances turn out to be in the same range as the sizes of the drops.

Millikan and his co-workers measured the charges of thousands of drops. Within the limits of their experimental error, every drop had a charge equal to some small integer multiple of a basic charge $e$. That is, they found drops with charges of $\pm 2e$, $\pm 5e$, and so on, but never with a value such as $0.76e$ or $2.49e$. A drop with charge $-e$ has acquired one extra electron; if its charge is $-2e$, it has acquired two extra electrons, and so on.

As we stated in Section 22–4, the best experimental value of the absolute value $e$ of the electron charge is

$$e = 1.60217733(49) \times 10^{-19}\,\text{C},$$

where the (49) indicates the uncertainty in the last two digits, 33.

The quark model of fundamental particle structure includes particles called *quarks* with fractional charges $\pm e/3$ and $\pm 2e/3$. But quarks always occur in combinations with a total charge that is an integer multiple of $e$, and the charge of a single quark is probably not observable. So any electric charge that can be directly observed is an integer multiple of $e$.

The magnitude $e$ of the electron charge can be used to define a unit of energy, the *electronvolt,* that is useful in many calculations with atomic and nuclear systems. When a particle with charge $q$ moves from a point where the potential is $V_a$ to a point where it is $V_b$, the change $\Delta U$ in the potential energy $U$ is

$$\Delta U = q(V_b - V_a) = qV_{ba}.$$

If the charge $q$ equals the magnitude $e$ of the electron charge, $1.602 \times 10^{-19}$ C, and the potential difference is $V_{ba} = 1$ V, the change in energy is

$$U = (1.602 \times 10^{-19}\,\text{C})(1\,\text{V}) = 1.602 \times 10^{-19}\,\text{J}.$$

This quantity of energy is defined to be 1 **electronvolt** (1 eV):

$$1\,\text{eV} = 1.602 \times 10^{-19}\,\text{J}.$$

The multiples meV, keV, MeV, GeV, and TeV are often used.

When a particle with charge $e$ moves through a potential difference of 1 V, its change in potential energy is 1 eV. If the charge is a multiple of $e$ such as $Ne$, the change in potential energy in electronvolts is $N$ times the potential difference in volts. For example, when an alpha particle, which has charge $2e$, moves between two points with a potential difference of 1000 V, the change in its potential energy is 2(1000 eV) = 2000 eV. To confirm this, we write

$$\Delta U = qV_{ba} = (2)(1.602 \times 10^{-19}\ \text{C})(1000\ \text{V})$$
$$= 3.204 \times 10^{-16}\ \text{J} = 2000\ \text{eV}.$$

Although we have defined the electronvolt in terms of *potential* energy, we can use it for *any* form of energy, such as the kinetic energy of a moving particle. When we speak of a "one-million-volt electron," we mean an electron with a kinetic energy of one million electronvolts (1 MeV), equal to $(10^6)(1.602 \times 10^{-19}\ \text{J}) = 1.602 \times 10^{-13}$ J.

## 24–7 THE CATHODE-RAY TUBE

Now let's look at how the concept of potential is applied to a class of devices called **cathode-ray tubes.** Such devices are found in oscilloscopes, and similar devices are used in TV picture tubes and computer displays. Figure 24–20 is a schematic diagram of the principal elements of a cathode-ray tube. The name goes back to the early 1900s. Cathode-ray tubes use an electron beam; before the basic nature of the beam was understood, it was called a cathode ray because it emanated from the cathode (negative electrode) of a vacuum tube.

The interior of a cathode-ray tube has a high vacuum, with a residual pressure of around 0.01 Pa ($10^{-7}$ atm) or less. At any greater pressure, collisions of electrons with air molecules would scatter the electron beam excessively. The *cathode,* at the left end in the figure, is raised to a high temperature by the *heater,* and electrons evaporate from the surface of the cathode. The *accelerating anode,* with a small hole at its center, is maintained at a high positive potential $V_1$, of the order of 1 to 20 kV, relative to the cathode. This potential causes an electric field, directed from right to left in the figure, in the region between the accelerating anode and the cathode. Electrons passing through the hole in the anode form a narrow beam and travel with constant horizontal velocity from the anode to the *fluorescent screen*. The area where the electrons strike the screen glows brightly.

The function of the *control grid* is to regulate the number of electrons that reach the anode and thus the brightness of the spot on the screen. The *focusing anode* ensures that electrons leaving the cathode in slightly different directions are focused down to a narrow beam and all arrive at the same spot on the screen. We won't need to worry about

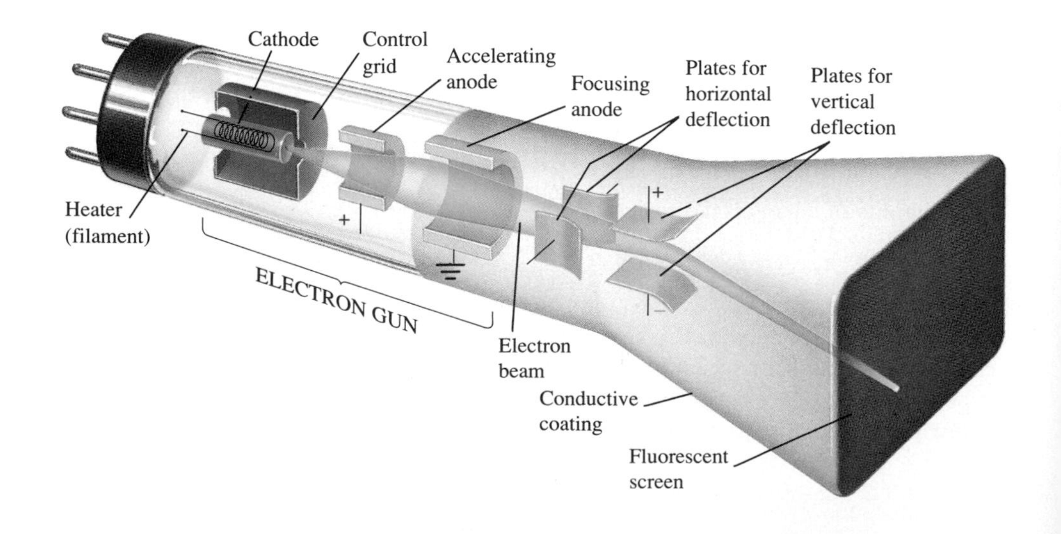


**24–20** Basic elements of a cathode-ray tube.

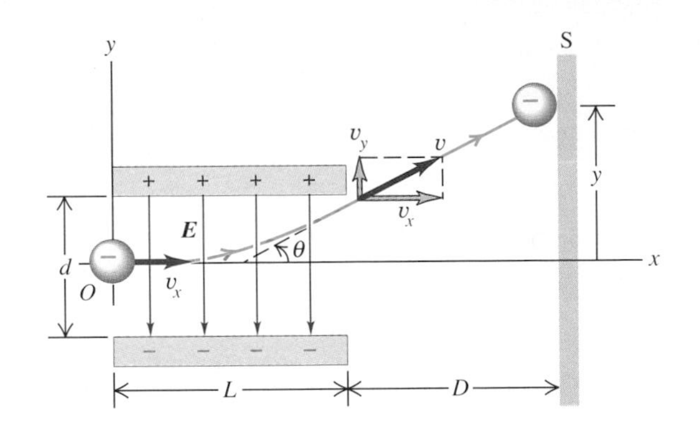


**24–21** Electrostatic deflection of an electron beam in a cathode-ray tube.

these two electrodes in the following analysis. The assembly of cathode, control grid, accelerating anode, and focusing electrode is called the *electron gun*.

The beam of electrons passes between two pairs of *deflecting plates.* An electric field between the first pair of plates deflects the electrons horizontally, and an electric field between the second pair deflects them vertically. If no deflecting fields are present, the electrons travel in a straight line from the hole in the accelerating anode to the center of the screen, where they produce a bright spot.

To analyze the electron motion, let's first calculate the speed $v$ of the electrons as they leave the electron gun. We can use the same method as in Example 24–5 (Section 24–2). The initial speeds of the electrons as they are emitted from the cathode are very small in comparison to their final speeds, so we assume that the initial speeds are zero. Then the speed $v_x$ of the electrons as they leave the electron gun is given by

$$v_x = \sqrt{\frac{2eV_1}{m}}. \tag{24–34}$$

As a numerical example, if $V_1 = 2000$ V,

$$v_x = \sqrt{\frac{2(1.60 \times 10^{-19}\ \text{C})(2.00 \times 10^3\ \text{V})}{9.11 \times 10^{-31}\ \text{kg}}} = 2.65 \times 10^7\ \text{m/s}.$$

The kinetic energy of an electron leaving the anode depends only on the *potential difference* between anode and cathode, not on the details of the fields or the electron trajectories within the electron gun.

If there is no electric field between the horizontal-deflection plates, the electrons enter the region between the vertical-deflection plates, shown in Fig. 24–21, with speed $v_x$. If there is a potential difference $V_2$ between these plates, with the upper plate positive (at higher potential), there is a *downward* electric field with magnitude $E = V_2/d$ between the plates. A constant upward force with magnitude $eE$ then acts on the electrons, and their upward ($y$-component) acceleration is

$$a_y = \frac{eE}{m} = \frac{eV_2}{md}. \tag{24–35}$$

The *horizontal* component of velocity $v_x$ is constant. The path of the electrons in the region between the plates is a parabolic trajectory, like the electron trajectory in Example 22–8 (Section 22–5) and the ballistic trajectories of Section 3–3. In all these cases the particle moves with constant $x$-velocity and constant $y$-acceleration. After the electrons emerge from this region, their paths again become straight lines, and they strike the screen at a point a distance $y$ above its center. We are going to prove that this distance is *directly proportional* to the deflecting potential $V_2$.

Proceeding one step at a time, we first note that the time $t$ required for the electrons to travel the length $L$ of the plates is

$$t = \frac{L}{v_x}. \tag{24–36}$$

During this time they acquire an upward velocity component $v_y$ given by

$$v_y = a_y t. \tag{24–37}$$

Combining this with Eqs. (24–35) and (24–36), we find

$$v_y = \frac{eV_2}{md}\frac{L}{v_x}. \tag{24–38}$$

When the electrons emerge from the deflecting field, their velocity $\boldsymbol{v}$ makes an angle $\theta$ with the $x$-axis given by

$$\tan\theta = \frac{v_y}{v_x}. \tag{24–39}$$

Ordinarily, the length $L$ of the deflection plates is much smaller than the distance $D$ from the plates to the screen. In this case the angle $\theta$ is also given approximately by $\tan\theta = y/D$. Combining this with Eq. (24–39), we find

$$\frac{y}{D} = \frac{v_y}{v_x}, \tag{24–40}$$

or, using Eq. (24–38) to eliminate $v_y$,

$$y = \frac{DeV_2L}{mdv_x^2}. \tag{24–41}$$

Finally, we substitute the expression for $v_x$ given by Eq. (24–34) into Eq. (24–41); the result is

$$y = \left(\frac{LD}{2d}\right)\frac{V_2}{V_1}. \tag{24–42}$$

The factor in parentheses depends only on the dimensions of the system, which are all constants. So we have proved that the deflection $y$ is proportional to the deflecting voltage $V_2$, as claimed. It is also *inversely* proportional to the accelerating voltage $V_1$. This isn't surprising; the faster the electrons are going, the less they are deflected by the deflecting voltage.

If there is also a field between the *horizontal* deflecting plates, the beam is also deflected in the horizontal direction, perpendicular to the plane of Fig. 24–21. The coordinates of the luminous spot on the screen are then proportional to the horizontal and vertical deflecting voltages, respectively. This is the principle of the cathode-ray oscilloscope. If the horizontal deflection voltage sweeps the beam from left to right at a uniform rate, the beam traces out a graph of the vertical voltage as a function of time. Oscilloscopes are extremely useful laboratory instruments in many areas of pure and applied science.

The picture tube in a television set is similar, but the beam is deflected by *magnetic* fields (to be discussed in later chapters) rather than electric fields. For the system that is currently used in the United States, the electron beam traces out the area of the picture 30 times per second in an array of 525 horizontal lines, and the intensity of the beam is varied to make bright and dark areas on the screen. (In a color set the screen is an array of dots of phosphors with three different colors.) The accelerating voltage in TV picture tubes ($V_1$ in the above discussion) is typically about 20 kV. Computer displays and monitors operate on the same principle, using a magnetically deflected electron beam to trace out images on a fluorescent screen. In this context the device is called a CRT (cathode-ray tube) display or a VDT (video display terminal).

## 24–8 EQUIPOTENTIAL LINES: A Case Study in Computer Analysis

In Section 22–9 we described a simple computer program that produces a field map for the electric field caused by an array of point charges in a plane. With a simple modification this same technique can be used to map the equipotential lines in the plane (the cross section in the $xy$-plane of the equipotential *surfaces*).

We use the fact that when an equipotential crosses a field line, the two are perpendicular. To map the equipotentials, we draw a line in each box of the array in a direction perpendicular to the field line in the box. The easiest way to do this is to use the result from analytic geometry that when two lines in a plane are perpendicular, their slopes are negative reciprocals of each other. The slope of the field line at each point is $E_{yr}/E_{xr}$, so the slope of the equipotential at the same point is $-E_{xr}/E_{yr}$. Referring to Eq. (22–25), which we used to draw lines with length $c$ parallel to the field, we see that we need only replace $E_{xr}$ by $E_{yr}$ and $E_{yr}$ by $-E_{xr}$ in this expression to get a line perpendicular to $\boldsymbol{E}$ at the point. That is, in Step 19 of the procedure in Section 22–9 we use instead of Eq. (22–25) the following:

$$\left(x - \frac{cE_{yr}}{2E_r}, y + \frac{cE_{xr}}{2E_r}\right) \quad \text{and} \quad \left(x + \frac{cE_{yr}}{2E_r}, y - \frac{cE_{xr}}{2E_r}\right). \tag{24–43}$$

Figure 24–22 shows the result of such a calculation for the case of two charges with equal magnitude but opposite sign.

This procedure doesn't plot actual equipotential curves; it plots segments of various curves. But these give us a good idea of the shapes of the equipotentials. As with the field lines in Section 22–9, we could write a more refined program that plots actual equipotentials.

For more complex geometries it is usually easier to take a completely different approach based on the fact that the potential function has to satisfy a certain differential equation called Laplace's equation. There are computational methods for finding solutions of this equation that are consistent with the conditions of a specific problem (for example, an electrostatic conducting surface is always equipotential, and so on). Once the potential is known as a function of position, we can do a numerical differentiation to find the components of field, using Eqs. (24–25). It is nearly always easier to compute the potential first, and then the field, than to compute the field directly.

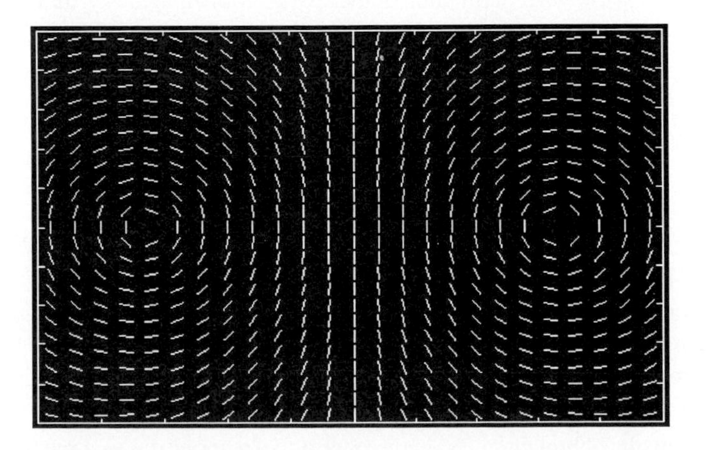

**24–22** Equipotential line segments for two charges with equal magnitude and opposite sign, the same charge arrangement as in Fig. 22–27.

## SUMMARY

### KEY TERMS

potential energy

potential

volt

equipotential surface

gradient

Millikan oil-drop experiment

electronvolt

cathode-ray tube

• The electric field caused by any collection of charges at rest is a conservative force field. The work $W$ done by the electric-field force on a charged particle moving in a field can be represented by a potential-energy function $U$:

$$W_{a\to b} = U_a - U_b. \tag{24-2}$$

The potential energy of a charge $q'$ in the electric field of a collection of charges $q_i$ is given by

$$U = \frac{q'}{4\pi\epsilon_0}\left(\frac{q_1}{r_1} + \frac{q_2}{r_2} + \frac{q_3}{r_3} + \cdots\right) = \frac{q'}{4\pi\epsilon_0}\sum_i \frac{q_i}{r_i}, \tag{24-10}$$

where $r_i$ is the distance from $q_i$ to $q'$. At points that are an infinite distance from all the charges, $U = 0$.

• Potential, denoted by $V$, is potential energy per unit charge. The potential at any point due to a collection of charges $q_i$ is

$$V = \frac{U}{q'} = \frac{1}{4\pi\epsilon_0}\sum_i \frac{q_i}{r_i}. \tag{24-14}$$

• The potential difference between two points $a$ and $b$, also called the potential of $a$ with respect to $b$, is given by the line integral of $\boldsymbol{E}$:

$$V_a - V_b = \int_a^b \boldsymbol{E}\cdot d\boldsymbol{l} = \int_a^b E\cos\phi\, dl. \tag{24-16}$$

Potentials can be calculated either by evaluating the sum in Eq. (24–14) or by first finding $\boldsymbol{E}$ and then using Eq. (24–16).

• An equipotential surface is a surface on which the potential has the same value at every point. At a point where a field line crosses an equipotential surface, the two are perpendicular. When all charges are at rest, the surface of a conductor is always an equipotential surface, and all points in the interior of a conductor are at the same potential. When a cavity within a conductor contains no charge, the entire cavity is an equipotential region, and there is no surface charge anywhere on the surface of the cavity.

• If the potential is known as a function of the spatial coordinates $x$, $y$, and $z$, the electric field $\boldsymbol{E}$ at any point is given by

$$\boldsymbol{E} = -\left(\boldsymbol{i}\frac{\partial V}{\partial x} + \boldsymbol{j}\frac{\partial V}{\partial y} + \boldsymbol{k}\frac{\partial V}{\partial z}\right)$$

$$= -\left(\boldsymbol{i}\frac{\partial}{\partial x} + \boldsymbol{j}\frac{\partial}{\partial y} + \boldsymbol{k}\frac{\partial}{\partial z}\right)V. \tag{24-26}$$

• Two equivalent sets of units for electric-field magnitude are volts per meter (V/m) and newtons per coulomb (N/C). One volt is one joule per coulomb (1 V = 1 J/C).

• The Millikan oil-drop experiment measured the electric charge of individual electrons by measuring the motion of electrically charged oil drops in an electric field. The size of the drop is determined by measuring its speed of free fall under gravity and the viscous force of air resistance given by Stokes' law.

• The electronvolt, abbreviated eV, is the energy corresponding to a particle with a charge equal to that of the electron moving through a potential difference of one volt.

The conversion factor is

$$1 \text{ eV} = 1.602 \times 10^{-19} \text{ J}.$$

• The cathode-ray tube uses an electron beam created by a set of electrodes called an electron gun. The beam is deflected by two sets of deflecting plates and then strikes a fluorescent screen and forms an image on it. Such tubes are used in oscilloscopes; similar tubes are found in television sets and video display terminals (VDT) for computers.

# EXERCISES

## Section 24–1 Electric Potential Energy

**24–1** A point charge $Q = +8.00\ \mu\text{C}$ is held fixed at the origin. A second point charge with a charge of $q = -0.500\ \mu\text{C}$ and a mass of $3.00 \times 10^{-4}$ kg is placed on the $x$-axis, 0.800 m from the origin. a) What is the electric potential energy of the pair of charges? (Take the potential energy to be zero when the charges have infinite separation.) b) The second point charge is released at rest. What is its speed when it is 0.200 m from the origin?

**24–2** A point charge $q_1 = +3.20\ \mu\text{C}$ is held stationary at the origin. How far away must a second point charge $q_2 = +5.20\ \mu\text{C}$ be placed for the electric potential energy $U$ of the pair of charges to be 0.600 J? (Take $U$ to be zero when the charges have an infinite separation.)

**24–3** A point charge $q_1 = -6.00\ \mu\text{C}$ is held stationary at the origin. A second point charge $q_2 = +4.00\ \mu\text{C}$ moves from the point $x = 0.160$ m, $y = 0$ to the point $x = 0.280$ m, $y = 0$. How much work is done by the electrical force on $q_2$?

**24–4** A point charge $q_1$ is held stationary at the origin. A second charge $q_2$ is placed at point $a$, and the electric potential energy of the pair of charges is $-4.60 \times 10^{-8}$ J. When the second charge is moved to point $b$, the electrical force on the charge does $2.40 \times 10^{-8}$ J of work. What is the electric potential energy of the pair of charges when the second charge is at point $b$?

**24–5 Two Approaching Positive Charges.** A small metal sphere, carrying a net charge of $q_1 = +6.00\ \mu\text{C}$, is held in a stationary position by insulating supports. A second small metal sphere, with a net charge of $q_2 = +2.50\ \mu\text{C}$ and mass 2.00 g, is projected toward $q_1$. When the two spheres are 0.800 m apart, $q_2$ is moving toward $q_1$ with a speed of 22.0 m/s (Fig. 24–23). Assume that the two spheres can be treated as point charges. Neglect the force of gravity. a) What is the speed of $q_2$ when the spheres are 0.500 m apart? b) How close does $q_2$ get to $q_1$?

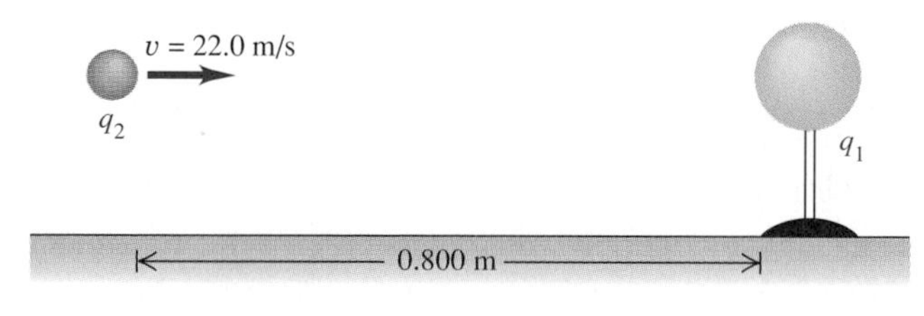


**FIGURE 24–23**

**24–6** Three equal point charges $q = 6.00 \times 10^{-7}$ C are placed at the corners of an equilateral triangle whose side is 1.00 m. What is the potential energy of the system? (Take as zero the potential energy of the three charges when they are infinitely far apart.)

## Section 24–2 Potential

**24–7** A point charge has a charge of $6.00 \times 10^{-11}$ C. At what distance from the point charge is the electric potential a) 12.0 V; b) 24.0 V?

**24–8** The potential at a distance of 0.600 m from a very small charged sphere is 48.0 V. If the sphere is treated as a point charge, what is its charge?

**24–9** A uniform electric field has magnitude $E$ and is directed in the positive $x$-direction. Consider point $a$ at $x = 0.80$ m and point $b$ at $x = 1.20$ m. The potential difference between these two points is 500 V. a) Which point, $a$ or $b$, is at the higher potential? b) Calculate the magnitude $E$ of the electric field. c) A negative point charge $q = -0.200\ \mu\text{C}$ is moved from $b$ to $a$. Calculate the work done on the point charge by the electric field.

**24–10** A particle with a charge of +3.00 nC is in a uniform electric field directed to the left. It is released from rest and moves to the left a distance of 5.00 cm; after this its kinetic energy is found to be $+2.50 \times 10^{-6}$ J. a) What work was done by the electrical force? b) What is the potential of the starting point with respect to the end point? c) What is the magnitude of the electric field?

**24–11** A charge of $3.00 \times 10^{-8}$ C is placed in a uniform electric field that is directed vertically upward and has a magnitude of $5.00 \times 10^4$ N/C. What work is done by the electrical force when the charge moves a) 0.450 m to the right; b) 0.800 m downward; c) 2.60 m at an angle of 45.0° upward from the horizontal?

**24–12** The potential at a certain distance from a point charge is 600 V, and the electric field is 300 N/C. a) What is the distance to the point charge? b) What is the magnitude of the charge?

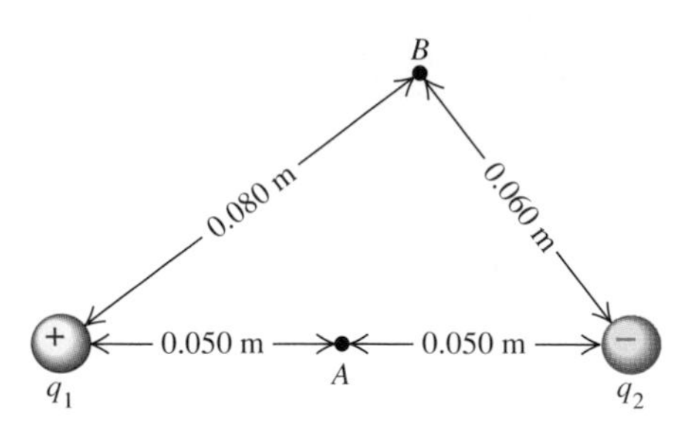


**FIGURE 24–24**

**24–13** Two point charges $q_1 = +4.00$ nC and $q_2 = -3.00$ nC are 0.100 m apart. Point $A$ is midway between them; point $B$ is 0.080 m from $q_1$ and 0.060 m from $q_2$ (Fig. 24–24). Find a) the potential at point $A$; b) the potential at point $B$; c) the work done by the electric field on a charge of 2.50 nC that travels from point $B$ to point $A$.

**24–14** Two stationary point charges $+2.80 \times 10^{-10}$ C and $-1.20 \times 10^{-10}$ C are separated by a distance of 5.00 cm. An electron is released from rest between the two charges, 1.00 cm from the negative charge, and moves along the line connecting the two charges. What is its speed when it is 1.00 cm from the positive charge?

**24–15 Potential Midway between Two Positive Charges.** Two positive point charges, each of magnitude $q$, are fixed on the $y$-axis at the points $y = +a$ and $y = -a$. a) Draw a diagram showing the positions of the charges. b) What is the potential $V_0$ at the origin? c) Show that the potential at any point on the $x$-axis is

$$V = \frac{1}{4\pi\epsilon_0}\frac{2q}{\sqrt{a^2 + x^2}}.$$

d) Sketch a graph of the potential on the $x$-axis as a function of $x$ over the range from $x = -4a$ to $x = +4a$. e) At what value of $x$ is the potential one-half that at the origin?

**24–16 Potential of Two Opposite Charges.** A positive charge $+q$ is located at the point $x = 0$, $y = -a$, and an equal negative charge $-q$ is located at the point $x = 0$, $y = +a$. a) Draw a diagram showing the positions of the charges. b) What is the potential at a point on the $x$-axis a distance $x$ from the origin? c) Derive an expression for the potential at a point on the $y$-axis a distance $y$ from the origin. d) Sketch a graph of the potential at points on the $y$-axis as a function of $y$ in the range from $y = -4a$ to $y = +4a$. Plot positive potentials upward, negative potentials downward.

**24–17** A simple type of vacuum tube known as a *diode* consists essentially of two electrodes within a highly evacuated enclosure. One electrode, the *cathode,* is maintained at a high temperature and emits electrons from its surface. A potential difference of a few hundred volts is maintained between the cathode and the other electrode, known as the *anode,* with the anode at the higher potential. Suppose a diode consists of a cylindrical cathode with radius 0.050 cm, mounted coaxially within a cylindrical anode 0.450 cm in radius. The potential of the anode is 360 V higher than that of the cathode. An electron leaves the surface of the cathode with zero initial speed. Find its speed when it strikes the anode.

## Section 24–3 Calculation of Potentials

**24–18** Two large parallel metal sheets carrying equal and opposite electric charges are separated by a distance of 0.0500 m. The electric field between them is uniform and has magnitude 500 N/C. a) What is the potential difference between the sheets? b) Which sheet is at higher potential, the one with positive charge or the one with negative charge?

**24–19** Two large parallel metal plates carry opposite charges. They are separated by 0.0800 m, and the potential difference between them is 500 V. a) What is the magnitude of the electric field (assumed to be uniform) in the region between the plates? b) What is the magnitude of the force this field exerts on a particle with charge +2.00 nC? c) Use the results of part (b) to compute the work done by the field on the particle as it moves from the higher-potential plate to the lower. d) Compare the result of part (c) to the change of potential energy of the same charge, computed from the electric potential.

**24–20** A total electric charge of 6.20 nC is distributed uniformly over the surface of a metal sphere with a radius of 0.200 m. If the potential is zero at a point at infinity, what is the value of the potential a) at a point on the surface of the sphere; b) at a point inside the sphere, 0.100 m from the center?

**24–21** A potential difference of 2.50 kV is established between parallel plates in air. If the air becomes electrically conducting when the electric field exceeds $3.00 \times 10^6$ N/C, what is the minimum separation of the plates?

**24–22** a) From the expression for $E$ obtained in Example 23–9 (Section 23–3), find the expression for the potential $V$ as a function of $r$, both inside and outside the sphere, relative to a point at infinity. b) Sketch graphs of $V$ and $E$ as functions of $r$ from $r = 0$ to $r = 3R$, and compare with Fig. 24–9.

## Section 24–4 Equipotential Surfaces

## Section 24–5 Potential Gradient

**24–23** A potential difference of 400 V is established between large parallel metal plates. Let the potential of one plate be 400 V and the other be 0 V. The plates are separated by $d = 1.20$ cm. a) On a sketch, show the equipotential surfaces that correspond to 0, 100, 200, 300, and 400 V. b) On your sketch, show the electric field lines. Does your sketch confirm that the field lines and equipotential surfaces are mutually perpendicular?

**24–24 A Sphere within a Shell.** A metal sphere with radius $r_a$ is supported on an insulating stand at the center of a hollow metal spherical shell with radius $r_b$. There is a charge $+q$ on the inner sphere and a charge $-q$ on the outer spherical shell. a) Calculate the potential $V(r)$ for i) $r < r_a$, ii) $r_a < r < r_b$; iii) $r > r_b$. Use the fact that the net potential is the sum of the potentials due to the individual spheres. Take $V(r)$ to be zero when $r$ is infinite. b) Show that the potential of the inner sphere with respect to the outer is

$$V_{ab} = \frac{q}{4\pi\epsilon_0}\left(\frac{1}{r_a} - \frac{1}{r_b}\right).$$

c) Use Eq. (24–29) and the result from part (a) to show that the electric field at any point between the spheres has the magnitude

$$E = \frac{V_{ab}}{(1/r_a - 1/r_b)}\frac{1}{r^2}.$$

d) Use Eq. (24–29) and the result from part (a) to find the electric field at a point outside the larger sphere at a distance $r$ from the center, where $r > r_b$. e) Suppose the charge on the outer sphere is not $-q$ but a negative charge of different magnitude, $-Q$. Show that the answers for parts (b) and (c) are the same as before but the answer for part (d) is different.

**24–25** A metal sphere with radius $r_a = 1.00$ cm is supported on an insulating stand at the center of a hollow metal spherical shell with radius $r_b = 10.0$ cm. Charge $+q$ is put on the inner sphere and charge $-q$ on the outer spherical shell. The magnitude of $q$ is chosen to make the potential difference between the spheres 500 V, with the inner sphere at higher potential. a) Use the result of Exercise 24–24(a) to calculate $q$. b) With the help of the result of Exercise 24–24(a), draw a sketch showing the equipotential surfaces that correspond to 500, 400, 300, 200, 100, and 0 V. c) On your sketch, show the electric field lines. Are the electric field lines and equipotential surfaces mutually perpendicular? Are the equipotential surfaces closer together when the magnitude of $\boldsymbol{E}$ is largest?

**24–26** Evaluate $-\partial V/\partial x$ for the $V(x)$ given in Example 24–14 (Section 24–5). Show that this gives an $E_x$ that agrees with that calculated in Example 22–11 (Section 22–6).

## Section 24–6
## The Millikan Oil-Drop Experiment

**24–27** An oil droplet with a mass of $3.00 \times 10^{-14}$ kg and a radius of $2.00 \times 10^{-6}$ m carries eight excess electrons. What is its terminal speed a) when falling in a region in which there is no electric field; b) when falling in an electric field of magnitude $4.00 \times 10^4$ N/C directed downward? (The viscosity of air is $180 \times 10^{-7}$ N·s/m$^2$. Neglect the buoyant force of the air.)

**24–28** In an apparatus for measuring the electronic charge $e$ by Millikan's method, an electric field of $5.86 \times 10^4$ V/m is required to maintain a certain charged oil drop at rest. If the plates are 1.50 cm apart, what potential difference between them is required?

**24–29** Find the potential energy of the interaction of two protons at a center-to-center distance of $2.40 \times 10^{-15}$ m, typical of the dimensions in atomic nuclei. Express your result in MeV.

## Section 24–7
## The Cathode-Ray Tube

**24–30** What is the final speed of an electron accelerated through a potential difference of 94.0 V if it has an initial speed of $3.00 \times 10^6$ m/s?

**24–31 Deflection in a CRT.** In Fig. 24–25 an electron is projected along the axis midway between the deflection plates of a cathode-ray tube with an initial speed of $8.00 \times 10^6$ m/s. The uniform electric field between the plates has a magnitude of $6.00 \times 10^3$ N/C and is upward. a) How far below the axis has the electron moved when it reaches the end of the plates? b) At what angle with the axis is it moving as it leaves the plates? c) How far below the axis will it strike the fluorescent screen S?

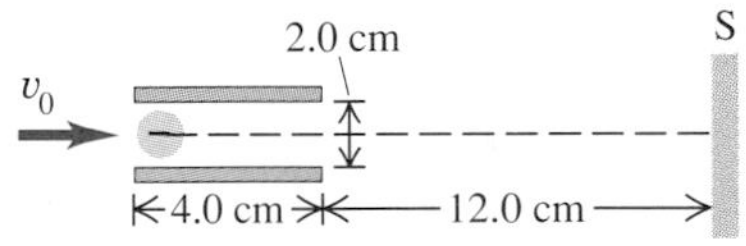

**FIGURE 24–25**

**24–32** The electric field in the region between the deflecting plates of a certain cathode-ray oscilloscope is $2.50 \times 10^4$ N/C. a) What is the magnitude of the force on an electron in this region? b) What is the magnitude of the acceleration of an electron when acted on by this force?

## Section 24–8
## Equipotential Lines: A Case Study in Computer Analysis

**24–33** Write a computer code to implement the procedure described in Section 24–8. Run your code for the case of two charges with equal magnitude but opposite sign. Compare the figure that your code generates to Fig. 24–22.

**24–34** Run your computer code from Exercise 24–33 for the following cases: a) a single positive point charge; b) two charges with equal magnitude and the same sign. In each case, compare your results to Fig. 24–15.

# PROBLEMS

**24–35** Physicists have succeeded in removing all 92 electrons from a uranium atom. Assume that the positive charge of the uranium nucleus is distributed uniformly throughout a spherical volume with a radius of $7.40 \times 10^{-15}$ m. a) If the potential is zero at a point at infinity, what is the value of the potential at the surface of the bare uranium nucleus? b) What is the electric potential energy of a pair of uranium nuclei whose surfaces are separated by a distance of 10.0 fm? (Express your answer in eV.)

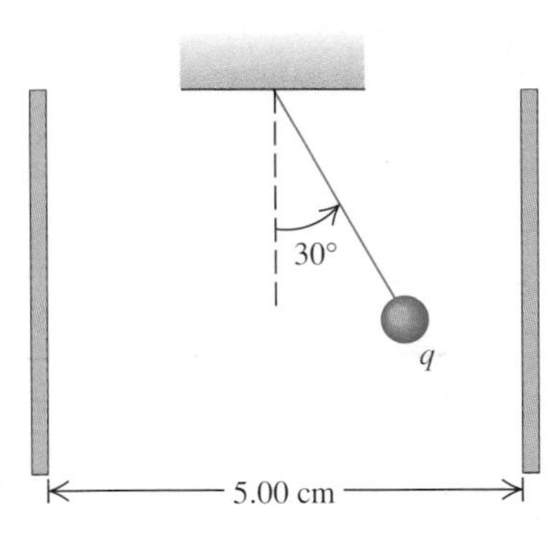


**FIGURE 24–26**

**24–36** A small sphere with a mass of 3.00 g hangs by a thread between two parallel vertical plates 5.00 cm apart. The charge on the sphere is $q = 6.00 \times 10^{-6}$ C. What potential difference between the plates will cause the thread to assume an angle of 30.0° with the vertical (Fig. 24–26)?

**24–37** Suppose another force in addition to the electrical force acts on the particle in Exercise 24–10 so that when it is released from rest, it moves to the right. After it has moved 5.00 cm, the additional force has done $9.00 \times 10^{-5}$ J of work, and the particle has $7.00 \times 10^{-5}$ J of kinetic energy. a) What work was done by the electrical force? b) What is the potential of the starting point with respect to the end point? c) What is the magnitude of the electric field?

**24–38** Consider the same distribution of charge as in Exercise 24–16. Show that the potential at a point on the $y$-axis is given by $V = -(1/4\pi\epsilon_0)2qa/y^2$ when $y >> a$.

**24–39** A potential difference of 1200 V is established between two parallel plates 4.00 cm apart. An electron is released from the negative plate at the same instant that a proton is released from the positive plate. a) How far from the positive plate will they pass each other? b) What is the ratio of their speeds just before they strike the opposite plates? c) What is the ratio of their kinetic energies just before they strike the opposite plates?

**24–40 Bohr Model of Hydrogen.** In the Bohr model of the hydrogen atom a single electron revolves around a single proton in a circle of radius $r$. Assume that the proton remains at rest. a) By equating the electrical force to the electron mass times its acceleration, derive an expression for the electron's speed. b) Obtain an expression for the electron's kinetic energy, and show that its magnitude is just half that of the electrical potential energy. c) Obtain an expression for the total energy, and evaluate it using $r = 5.29 \times 10^{-11}$ m. Give your numerical result in joules and in electronvolts.

**24–41** A vacuum tube diode was described in Exercise 24–17. Because of the accumulation of charge near the cathode, the electrical potential between the electrodes is not a linear function of the position, even with planar geometry, but is given by

$$V = Cx^{4/3},$$

where, for given operating conditions, $C$ is a constant, characteristic of a particular diode and operating conditions, and $x$ is the distance from the cathode (negative electrode). Assume that the distance between the cathode and anode (positive electrode) is 12.0 mm and the potential difference between electrodes is 160 V. a) Determine the value of $C$. b) Obtain a formula for the electric field between the electrodes as a function of $x$. c) Determine the force on an electron when the electron is halfway between the electrodes.

**24–42** Refer to Problem 22–38. a) Calculate the potential at the point $x = 3.00$ cm, $y = 0$ and at the point $x = 3.00$ cm, $y = 4.00$ cm due to the first two charges. b) If the third charge moves from the point $x = 3.00$ cm, $y = 0$ to the point $x = 3.00$ cm, $y = 4.00$ cm, calculate the work done on it by the field of the first two charges. Comment on the *sign* of this work. Is your result reasonable?

**24–43 Coaxial Cylinders.** A long metal cylinder with radius $a$ is supported on an insulating stand on the axis of a long, hollow metal tube with radius $b$. The positive charge per unit length on the inner cylinder is $\lambda$, and there is an equal negative charge per unit length on the outer cylinder. a) Calculate the potential $V(r)$ for i) $r < a$; ii) $a < r < b$; iii) $r > b$. Use the fact that the net potential is the sum of the potentials due to the individual conductors. Take $V = 0$ at the same $r_0$ for both conductors. b) Show that the potential of the inner cylinder with respect to the outer is

$$V_{ab} = \frac{\lambda}{2\pi\epsilon_0} \ln \frac{b}{a}.$$

c) Use Eq. (24–29) and the result from part (a) to show that the electric field at any point between the cylinders has magnitude

$$E = \frac{V_{ab}}{\ln(b/a)} \frac{1}{r}.$$

d) What is the potential difference between the two cylinders if the outer cylinder has no net charge?

**24–44 Geiger Counter.** A Geiger counter is used to detect radiation such as alpha particles by using the fact that the radiation ionizes the air along its path. The device consists of a thin wire passing along the axis of a hollow metal cylinder and insulated from it (Fig. 24–27). A large potential difference is established between the wire and the outer cylinder, with the wire at higher potential, and this sets up a strong radial electric field directed

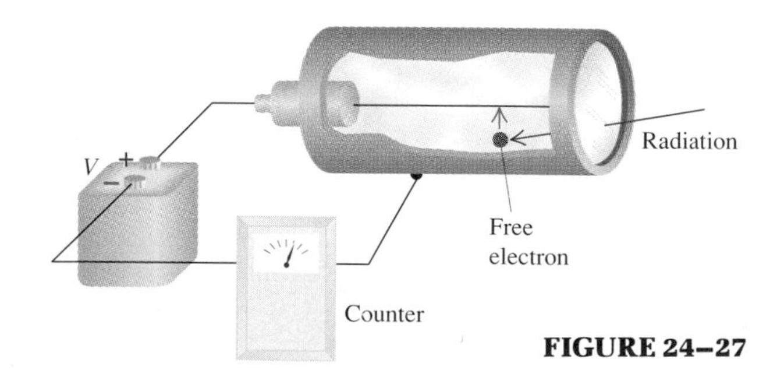


**FIGURE 24–27**

outward. When ionizing radiation enters the device, it ionizes a few air molecules. The free electrons produced are accelerated by the electric field toward the wire and, on the way there, ionize many more air molecules. Thus a current pulse is produced that can be detected by appropriate electronic circuitry and even converted to an audible "click." Suppose the radius of the central wire is 50.0 $\mu$m and the radius of the hollow cylinder is 2.00 cm. What potential difference between the wire and the cylinder is required to produce an electric field of $6.00 \times 10^4$ N/C at a distance of 1.00 cm from the wire? (Assume that the wire and cylinder are both very long in comparison to the cylinder radius, so the results of Problem 24–43 apply.)

**24–45 Electrostatic Precipitators.** Electrostatic precipitators use electrical forces to remove pollutant particles from smoke, in particular in the smokestacks of coal-burning power plants. One form of precipitator consists of a vertical hollow metal cylinder with a thin wire, insulated from the cylinder, running along its axis (Fig. 24–28). A large potential difference is established between the wire and the outer cylinder, with the wire at lower potential. This sets up a strong radial electric field directed inward. The field produces a region of ionized air near the wire. Smoke enters the precipitator at the bottom, ash and dust in it pick up electrons, and the charged pollutants are accelerated toward the outer cylinder wall by the electric field. Suppose the radius of the central wire is 80.0 $\mu$m, the radius of the cylinder is 12.0 cm, and a potential difference of 60.0 kV is established between the wire and the cylinder. Also assume that the wire and cylinder are both very long in comparison to the cylinder radius, so the results of Problem 24–43 apply. a) What is the magnitude of the electric field midway between the wire and the cylinder wall? b) What magnitude of charge must a 30.0-$\mu$g ash particle have if the electric field computed in part (a) is to exert a force ten times the weight of the particle?

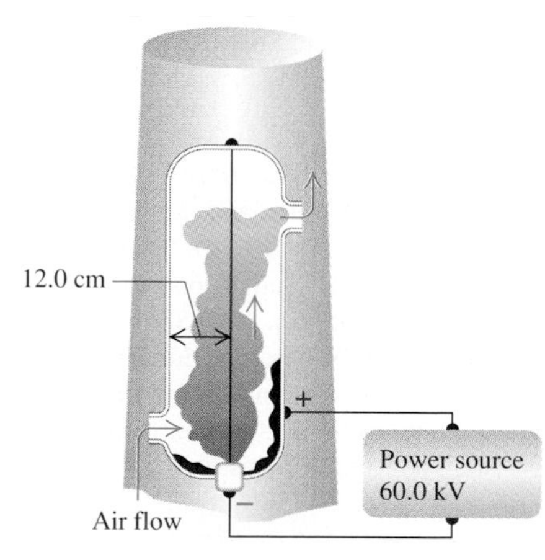


**FIGURE 24–28**

**24–46** Four line segments of charge are arranged to form a square with sides of length $a$. Calculate the potential at the center of the square, if the potential is zero at infinity, a) if two opposite sides are positively charged with charge $+Q$ each and the other two sides are negatively charged with charge $-Q$ each; b) if each side has positive charge $+Q$. (*Hint:* Use the result of Example 24–10.)

**24–47** a) From the expression for $E$ obtained in Problem 23–24, find the expressions for the potential $V$ as a function of $r$, both inside and outside the cylinder. Let $V = 0$ at the surface of the cylinder. In each case, express your result in terms of the charge per unit length $\lambda$ of the charge distribution. b) Sketch graphs of $V$ and $E$ as functions of $r$, from $r = 0$ to $r = 3R$.

**24–48** Electric charge is distributed uniformly along a semicircle of radius $a$ with total charge $Q$. Calculate the potential at the center of curvature if the potential is assumed to be zero at infinity.

**24–49** Electric charge is distributed uniformly along a thin rod of length $a$, with total charge $Q$. Take the potential to be zero at infinity. Find the potential at the following points (see Fig. 24–29): a) point $P$, a distance $x$ to the right of the rod; b) point $R$, a distance $y$ above the right-hand end of the rod. c) In part (a) and part (b), what does your result reduce to as $x$ or $y$ becomes large?

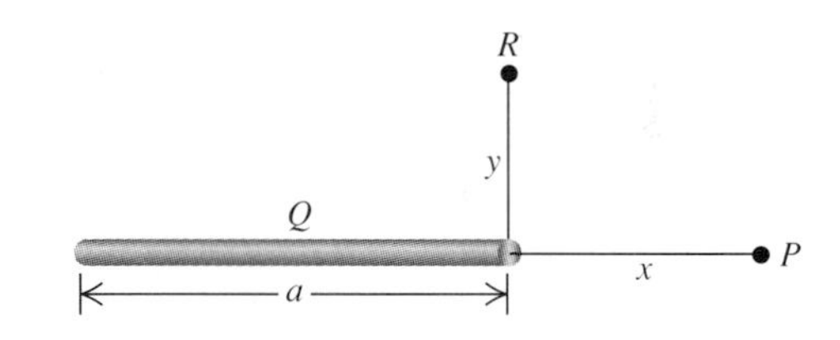


**FIGURE 24–29**

**24–50 Charged Raindrops.** a) A spherical raindrop has a radius of 0.400 mm. If it carries a negative charge of 1.60 pC, what is the potential at its surface? (Take the potential to be zero at an infinite distance from the raindrop.) b) Two identical raindrops, each with radius and charge as specified in part (a), collide and merge into one larger raindrop. What is the radius of this larger drop, and what is the potential at its surface?

**24–51** A charged oil drop in a Millikan oil-drop apparatus is observed to fall at constant speed 1.00 mm in a time of 27.4 s in the absence of any external field. The same drop can be held stationary in a field of $1.58 \times 10^4$ N/C. How many excess electrons has the drop acquired? The viscosity of air is $180 \times 10^{-7}\ \mathrm{N \cdot s/m^2}$. The density of the oil is 824 kg/m$^3$, and the density of air is 1.20 kg/m$^3$.

**24–52** An alpha particle with kinetic energy 10 MeV makes a head-on collision with a gold nucleus at rest. What is the distance of closest approach of the two particles? (Assume that the gold nucleus remains stationary and that it may be treated as a point charge. The atomic number of gold is 79. The alpha particle is a helium nucleus, with atomic number 2.)

**24–53** Two metal spheres of different sizes are charged such that the electrical potential is the same at the surface of each. Sphere $A$ has a radius three times that of sphere $B$. Let $Q_A$ and $Q_B$ be the charges on each sphere and $E_A$ and $E_B$ be the electric-field magnitudes at the surface of each sphere. What is a) the ratio $Q_B/Q_A$; b) the ratio $E_B/E_A$?

**24–54** A metal sphere of radius $R_1 = 0.160$ m has a charge $Q_1 = 3.20 \times 10^{-8}$ C. a) What are the electric field and electric potential at the surface of the sphere? This sphere is now connected by a long, thin conducting wire to another sphere, of radius $R_2 = 0.040$ m, that is several meters distant from the first sphere. Before the connection is made, this second sphere is uncharged. After electrostatic equilibrium has been reached, what is b) the total charge on each sphere; c) the electric potential at the surface of each sphere; d) the electric field at the surface of each sphere? Assume that the amount of charge on the wire is much less than the charge on each sphere.

**24–55** Use the charge distribution and the electric field calculated in Problem 23–25. a) Show that for $r \geq R$ the potential is identical to that produced by a point charge $Q$. (Take the potential to be zero at infinity.) b) Obtain an expression for the electric potential valid in the region $r \leq R$.

## CHALLENGE PROBLEMS

**24–56** The electric potential $V$ in a region of space is given by

$$V = ax^2 + ay^2 - 2az^2,$$

where $a$ is a constant. a) Derive an expression for the electric field $\boldsymbol{E}$ that is valid at all points in this region. b) The work done by the field when a 2.00-$\mu$C test charge moves from the point $(x, y, z) = (0, 0, 0.100 \text{ m})$ to the origin is measured to be $-5.00 \times 10^{-5}$ J. Determine $a$. c) Determine the electric field at the point $(0, 0, 0.100 \text{ m})$. d) Show that in every plane parallel to the $xy$-plane the equipotential lines are circles. e) What is the radius of the equipotential line corresponding to $V = 5000$ V and $z = \sqrt{2}$ m?

**24–57** Two point charges are moving toward the right along the $x$-axis. Point charge 1 has charge $q_1 = 2.00\ \mu$C, mass $m_1 = 6.00 \times 10^{-5}$ kg, and speed $v_1$. Point charge 2 is to the right of $q_1$ and has charge $q_2 = -5.00\ \mu$C, mass $m_2 = 3.00 \times 10^{-5}$ kg, and speed $v_2$. At a particular instant the charges are separated by a distance of 9.00 mm and have speeds $v_1 = 400$ m/s and $v_2 = 1200$ m/s. The only forces on the particles are the forces that they exert on each other. a) Determine the speed $v_{cm}$ of the center of mass of the system. b) The *relative energy* $E_r$ of the system is defined as the total energy minus the kinetic energy contributed by the motion of the center of mass:

$$E_r = E - \tfrac{1}{2}(m_1 + m_2)v_{cm}^2,$$

where $E = \frac{1}{2} m_1 v_1^2 + \frac{1}{2} m_2 v_2^2 + q_1 q_2/4\pi\epsilon_0 r$ is the total energy of the system and $r$ is the distance between the charges. Show that $E_r = \frac{1}{2}\mu v^2 + q_1 q_2/4\pi\epsilon_0 r$, where $\mu$ is called the *reduced mass* of the system and is equal to $m_1 m_2/(m_1 + m_2)$ and $v = v_2 - v_1$ is the relative speed of the moving particles. c) For the numerical values given above, calculate the numerical value of $E_r$. d) The value of $E_r$ can be used to determine whether the particles will "escape" from one another. For the conditions given above, will the particles escape from one another? Explain. e) If the particles do escape, what will be their final relative speed? That is, what will their relative speed be when $r \to \infty$? If the particles do not escape, what will be their distance of maximum separation? That is, what will be the value of $r$ when $v = 0$? f) Repeat parts (c) through (e) for $v_1 = 400$ m/s and $v_2 = 1800$ m/s when the separation is 9.00 mm.

**24–58 Charge Distribution in Spherical Coordinates.** In a certain region a charge distribution exists that is spherical but nonuniform. That is, the volume charge density $\rho(r)$ depends on the distance $r$ from the center of the distribution but not on the spherical polar angles $\theta$ and $\phi$. The electric potential $V(r)$ due to this charge distribution is given by

$$V(r) = \frac{\rho_0 a^2}{18\epsilon_0}\left[1 - 3\left(\frac{r}{a}\right)^2 + 2\left(\frac{r}{a}\right)^3\right] \quad \text{for } r \leq a,$$

$$V = 0 \quad \text{for } r \geq a,$$

where $\rho_0$ is a constant having units of C/m$^3$ and $a$ is a constant having units of meters. a) Derive expressions for the electric field for the regions $r \leq a$ and $r \geq a$. (*Hint:* Use the gradient operator for spherical polar coordinates,

$$\boldsymbol{\nabla} = \boldsymbol{i}_r \frac{\partial}{\partial r} + \boldsymbol{i}_\theta \frac{1}{r}\frac{\partial}{\partial \theta} + \boldsymbol{i}_\phi \frac{1}{r \sin\theta}\frac{\partial}{\partial \phi},$$

where $\boldsymbol{i}_r$, $\boldsymbol{i}_\theta$, and $\boldsymbol{i}_\phi$ are unit vectors in the $r$-, $\theta$-, and $\phi$-directions.) b) Derive an expression for $\rho(r)$ in each of the two regions $r \leq a$ and $r \geq a$. (*Hint:* Use Gauss's law for two spherical shells, one of radius $r$ and the other of radius $r + dr$. Then use the fact that $dq = 4\pi r^2 \rho dr$, where $dq$ is the charge contained in the infinitesimal spherical shell of thickness $dr$.) c) Show that the net charge contained in the volume of a sphere of radius greater than or equal to $a$ is zero. [*Hint:* Integrate the expressions derived in part (b) for $\rho$ over a spherical volume of radius greater than or equal to $a$.] Is this result consistent with the electric field for $r > a$ that you calculated in part (a)?

# CHAPTER 25

# Capacitance and Dielectrics

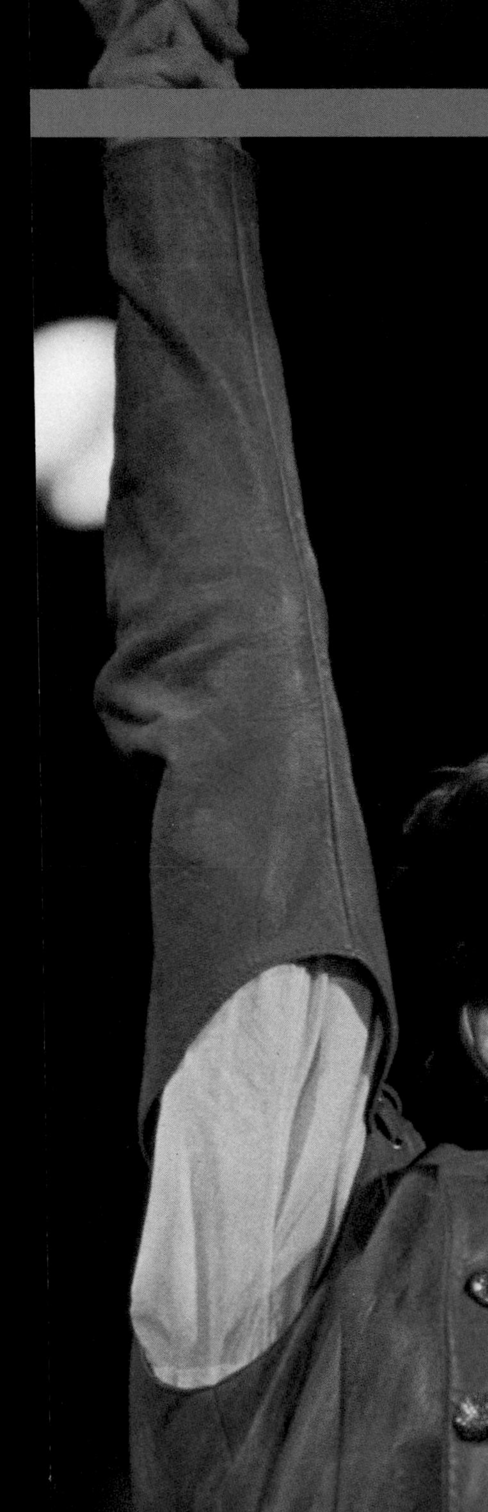

(a) In a condenser (capacitor) microphone, a flexible metal diaphragm forms one side of a parallel-plate capacitor.

(b) During the compression part of a sound wave, the diaphragm moves in, increasing the capacitance of the capacitor. Charge moves onto the plates, generating a positive output signal.

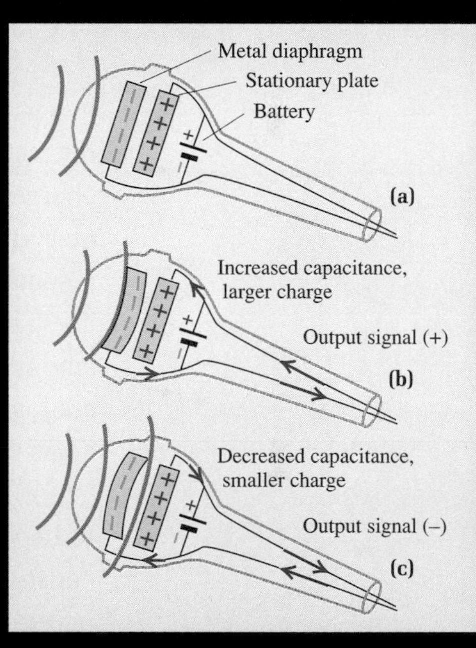


(c) During the rarefaction part of a sound wave, the diaphragm moves out, decreasing the capacitance. Charge moves off the plates, generating a negative output signal.

## KEY CONCEPTS

**• A capacitor consists of two conductors separated by vacuum or an insulator. When equal and opposite charges are placed on the conductors, the charge magnitude is proportional to the potential difference. The proportionality factor is called the capacitance. Capacitance depends on the size, shape, and separation of the conductors and on the material separating them.**

**• The behavior of capacitors connected in series or in parallel can be described in terms of an equivalent capacitance for the combination.**

**• The energy associated with a charged capacitor can be associated with the electric field produced by the charges and expressed in terms of an energy density (energy per unit volume).**

**• The effect of a dielectric (insulator) between the conductors of a capacitor can be understood on the basis of redistribution of charge within the dielectric, an effect called polarization.**

**• Gauss's law can be reformulated to include both the charges placed on the conductors and the polarization charge in the dielectric.**

## INTRODUCTION

When you set an old-fashioned spring mousetrap, pull back the string of a bow, or stretch a doorspring, you are storing mechanical energy as elastic potential energy. A capacitor is a device that stores *electric* potential energy and electric charge. The energy and charge can be recovered quickly, as in an electronic photoflash unit, or periodically, as in filters that smooth out a pulsating current or a resonant circuit that selects certain frequencies in the tuner of a radio or TV. The filter is the electric analog of a car suspension system that smooths out the bumps, and the behavior of the resonant circuit is analogous to normal modes of oscillation of mechanical systems such as a harmonic oscillator.

In these and many other applications, capacitors are among the indispensible elements of modern electronics. In principle, a capacitor consists of any two conductors separated by vacuum or an insulating material. When charges with equal magnitude and opposite sign are placed on the conductors, an electric field is established in the region between the conductors, and there is a potential difference between them. For any particular capacitor the ratio of charge to potential difference is a constant, called the *capacitance*. The capacitance depends on the sizes and shapes of the conductors and on the material between them. The energy stored in a charged capacitor is directly related to the electric field in the space between the conductors. When an insulating material (a *dielectric*) is present, the capacitance is increased because of redistribution of charge, called *polarization,* within the material. We'll learn about all these relationships in this chapter.

# 25–1 CAPACITORS

Any two conductors separated by an insulator form a **capacitor** (Fig. 25–1). In most practical applications the conductors have charges with equal magnitude and opposite sign, and the *net* charge on the capacitor as a whole is zero. We will assume throughout this chapter that this is the case. When we say that a capacitor has charge $Q$, we mean that the conductor at higher potential has charge $Q$ and the conductor at lower potential has charge $-Q$ (assuming that $Q$ is positive). Keep this in mind in the following discussion and examples.

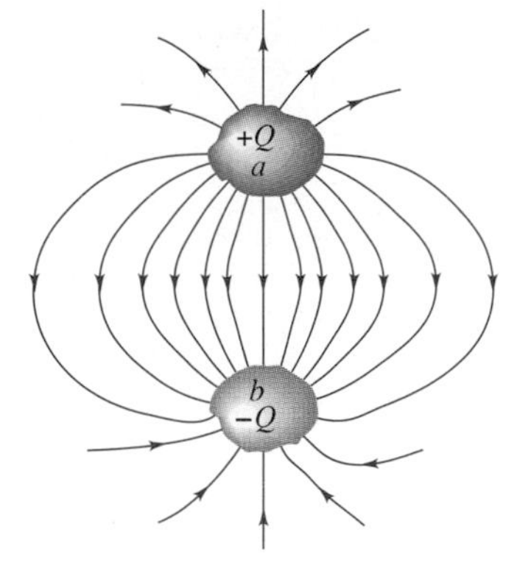


**25–1** Any two conductors insulated from one another form a capacitor.

The electric field at any point in the region between the conductors is proportional to the magnitude $Q$ of charge on each conductor. It follows that the *potential difference* $V_{ab}$ between the conductors (that is, the potential of the positively charged conductor $a$ with respect to the negatively charged conductor $b$) is also proportional to $Q$. If we double the magnitude of charge on each conductor, the charge density at each point doubles, the electric field at each point doubles, and the potential difference between conductors doubles; but the *ratio* of charge to potential difference does not change.

We define the **capacitance** $\boldsymbol{C}$ of a capacitor as the ratio of the magnitude of the charge $Q$ on *either* conductor to the magnitude of the potential difference $V_{ab}$ between the conductors:

$$C = \frac{Q}{V_{ab}}. \tag{25–1}$$

The SI unit of capacitance is called one **farad** (1 F), in honor of Michael Faraday. From Eq. (25–1), one farad is equal to one *coulomb per volt* (1 C/V):

$$1\ \text{F} = 1\ \text{C/V}.$$

In circuit diagrams a capacitor is represented by either of these symbols:

Figure 25–2 shows several practical capacitors.

Capacitors have thousands of practical uses, and contemporary electronics could not exist without them. In addition to the uses cited at the beginning of this chapter, capacitors are used in energy-storage units for pulsed lasers, in computer chips that store information, in circuits that improve the efficiency of ac power transmission lines, and in many other areas. The study of capacitors will also help us to develop insight into the behavior of electric fields and their interactions with matter.

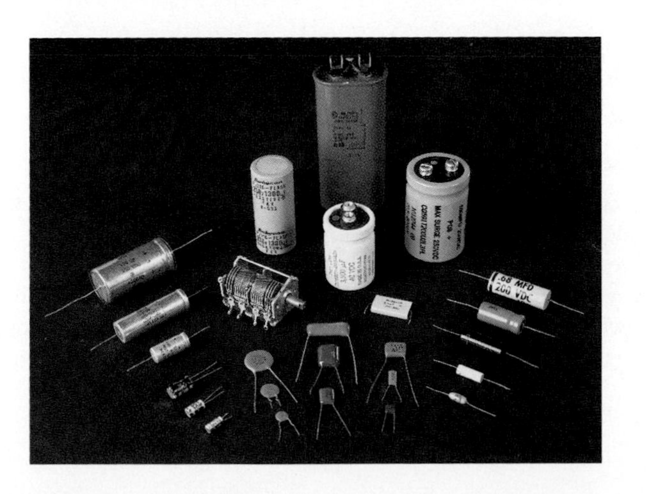


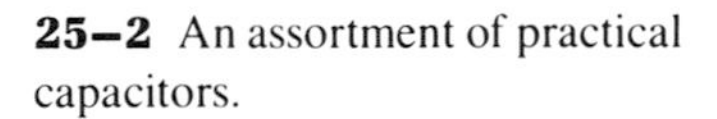

**25–2** An assortment of practical capacitors.

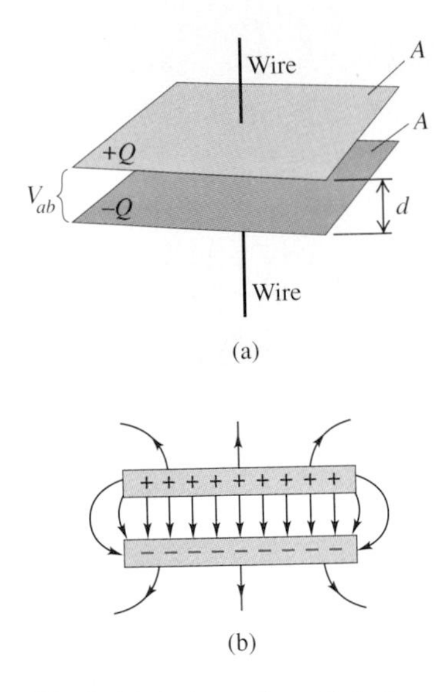


**25–3** (a) A charged parallel-plate capacitor. (b) When the separation of the plates is small in comparison to their size, the fringing of the ***E*** field at the edges is slight.

## 25–2 CALCULATING CAPACITANCE

The most common form of capacitor consists of two parallel conducting plates, each with area $A$, separated by a distance $d$ that is small in comparison with their dimensions (Fig. 25–3). Nearly all the field of such a capacitor is localized in the region between the plates, as shown. There is some "fringing" of the field at the edges, as shown in the figure, but if the distance between plates is small in comparison to their size, we can neglect this. The field between the plates is then *uniform,* and the charges on the plates are uniformly distributed over their opposing surfaces. We call this arrangement a **parallel-plate capacitor.**

We worked out the electric-field magnitude for this setup in Example 23–8 (Section 23–3), and it would be a good idea to review that discussion. We found that $E = \sigma/\epsilon_0$, where $\sigma$ is the magnitude of surface charge density on each plate. This is equal to the magnitude of total charge $Q$ on each plate divided by the area $A$ of the plate, or $\sigma = Q/A$, so the electric field $E$ can be expressed as

$$E = \frac{\sigma}{\epsilon_0} = \frac{Q}{\epsilon_0 A}.$$

The field is uniform; if the distance between the plates is $d$, then the potential difference ("voltage") between them is

$$V_{ab} = Ed = \frac{1}{\epsilon_0}\frac{Qd}{A}.$$

From this we see that the capacitance $C$ of a parallel-plate capacitor in vacuum is

$$C = \frac{Q}{V_{ab}} = \epsilon_0 \frac{A}{d}. \qquad (25\text{–}2)$$

The quantities $\epsilon_0$, $A$, and $d$ are constants for a given capacitor. The capacitance $C$ is therefore a constant, independent of the charge on the capacitor. It is directly proportional to the area $A$ of each plate and inversely proportional to their separation $d$.

In Eq. (25–2), if $A$ is in square meters and $d$ in meters, $C$ is in farads. The units of $\epsilon_0$ are $C^2/N \cdot m^2$, so we see that

$$1\ F = 1\ C^2/N \cdot m = 1\ C^2/J.$$

Because $1\ V = 1\ J/C$ (energy per unit charge), this is consistent with our definition $1\ F = 1\ C/V$. Finally, the units of $\epsilon_0$ can be expressed as

$$1\ C^2/N \cdot m^2 = 1\ F/m.$$

This relation is useful in capacitance calculations, and it also helps us to verify that Eq. (25–2) is dimensionally consistent.

We have assumed that there is only vacuum in the space between the plates. When matter is present, things are somewhat different. We will return to this topic in Section 25–5. Meanwhile, we remark that if the space contains air at atmospheric pressure instead of vacuum, the capacitance differs from the prediction of Eq. (25–2) by less than 0.06%.

## EXAMPLE 25–1

**Size of a 1-F capacitor** A parallel-plate capacitor has a capacitance of 1.0 F, and the plates are 1.0 mm apart. What is the area of the plates?

**SOLUTION** Solving Eq. (25–2) for the area $A$, we get

$$A = \frac{Cd}{\epsilon_0} = \frac{(1.0\ \text{F})(1.0 \times 10^{-3}\ \text{m})}{8.85 \times 10^{-12}\ \text{F/m}}$$
$$= 1.1 \times 10^{8}\ \text{m}^2.$$

This corresponds to a square about 10 km (about 6 miles) on a side! This area is about a third larger than Manhattan Island.

It used to be considered a good joke to send the newly graduated engineer to the stockroom for a 1-F capacitor. That's not as funny as it used to be; recently developed technology makes it possible to make 1-F capacitors that are a few centimeters on a side. One type uses activated carbon granules, of which 1 gram has a surface area of about 1000 $\text{m}^2$.

In many applications the most convenient units of capacitance are the *microfarad* ($1\ \mu\text{F} = 10^{-6}\ \text{F}$) and the *picofarad* ($1\ \text{pF} = 10^{-12}\ \text{F}$). For example, the power supply of an ac-powered AM radio contains several capacitors with capacitances of the order of 10 or more microfarads, and the capacitances in the tuning circuits are of the order of 10 to 100 picofarads.

## EXAMPLE 25–2

The plates of a parallel-plate capacitor are 5.00 mm apart and 2.00 $\text{m}^2$ in area. A potential difference of 10,000 V (10.0 kV) is applied across the capacitor. Compute a) the capacitance; b) the charge on each plate; and c) the magnitude of electric field in the space between them.

**SOLUTION** a) From Eq. (25–2),

$$C = \epsilon_0 \frac{A}{d} = \frac{(8.85 \times 10^{-12}\ \text{F/m})(2.00\ \text{m}^2)}{5.00 \times 10^{-3}\ \text{m}}$$
$$= 3.54 \times 10^{-9}\ \text{F} = 0.00354\ \mu\text{F}.$$

b) The charge on the capacitor is

$$Q = CV_{ab} = (3.54 \times 10^{-9}\ \text{C/V})(1.00 \times 10^{4}\ \text{V})$$
$$= 3.54 \times 10^{-5}\ \text{C} = 35.4\ \mu\text{C}.$$

That is, the plate at higher potential has charge $+35.4\ \mu\text{C}$, and the other plate has charge $-35.4\ \mu\text{C}$.

c) The electric field magnitude is

$$E = \frac{\sigma}{\epsilon_0} = \frac{Q}{\epsilon_0 A} = \frac{3.54 \times 10^{-5}\ \text{C}}{(8.85 \times 10^{-12}\ \text{C}^2/\text{N} \cdot \text{m}^2)(2.00\ \text{m}^2)}$$
$$= 2.00 \times 10^{6}\ \text{N/C}.$$

Alternatively, because the electric field equals the potential gradient,

$$E = \frac{V_{ab}}{d} = \frac{1.00 \times 10^{4}\ \text{V}}{5.00 \times 10^{-3}\ \text{m}} = 2.00 \times 10^{6}\ \text{V/m}.$$

(Remember that the newton per coulomb and the volt per meter are equivalent units.)

## EXAMPLE 25–3

**A cylindrical capacitor** A long cylindrical conductor has a radius $r_a$ and a linear charge density $+\lambda$. It is surrounded by a coaxial cylindrical conducting shell with inner radius $r_b$ and linear charge density $-\lambda$ (Fig. 25–4). Calculate the capacitance per unit length for this capacitor, assuming that there is vacuum in the space between cylinders.

**SOLUTION** This isn't a parallel-plate capacitor, so we can't use Eq. (25–2). Instead, we go back to the fundamental definition of capacitance, Eq. (25–1). To find the potential difference between the cylinders, we use a result that we worked out in Example 24–8 (Section 24–3). Equation (24–22) shows that if we take $V = 0$ at the inner surface, the potential at any distance is given by

$$V = \frac{\lambda}{2\pi\epsilon_0} \ln \frac{R}{r}.$$

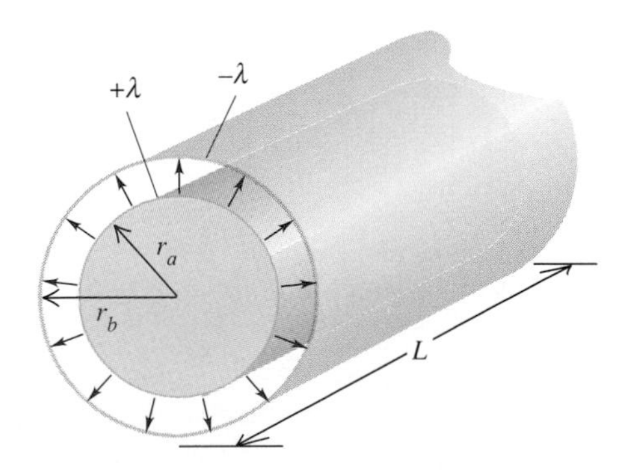


**25–4** A long cylindrical capacitor. The magnitude of charge in a length $L$ of either cylinder is $\lambda L$.

We can use this result for the present problem because, according to Gauss's law, the charge on the outer cylinder doesn't contribute to the field between cylinders. In our case the radius $R$ of the inner surface is $r_a$, and the potential of the outer surface (where $r = r_b$) is given by

$$V_{ba} = \frac{\lambda}{2\pi\epsilon_0} \ln \frac{r_a}{r_b}.$$

This potential difference is negative (assuming that $\lambda$ is positive) because the inner cylinder is at higher potential than the outer. So we can rewrite this as

$$V_{ab} = \frac{\lambda}{2\pi\epsilon_0} \ln \frac{r_b}{r_a}.$$

Capacitance is charge $Q$ divided by potential difference $V_{ab}$. The total charge $Q$ in a length $L$ is $Q = \lambda L$, so the capacitance $C$ of a length $L$ is

$$C = \frac{Q}{V_{ab}} = \frac{\lambda L}{\dfrac{\lambda}{2\pi\epsilon_0} \ln \dfrac{r_b}{r_a}}.$$

The capacitance per unit length is

$$\frac{C}{L} = \frac{2\pi\epsilon_0}{\ln (r_b/r_a)}. \tag{25-3}$$

Substituting $\epsilon_0 = 8.85 \times 10^{-12}$ F/m, we get

$$\frac{C}{L} = \frac{55.6 \text{ pF/m}}{\ln (r_b/r_a)}.$$

We see that the capacitance of the coaxial cylinders is determined entirely by the dimensions, just as for the parallel-plate case. Ordinary coaxial cables are made like this but with an insulating material instead of vacuum between the inner and outer conductors. A typical cable for TV antennas and VCR connections has a capacitance per unit length of 69 pF/m. ■

For more complex conductor shapes, the relation of capacitance to the shapes and dimensions of the conductors is more complicated than Eq. (25–2) or (25–3). In all cases, though, the capacitance of any pair of conductors is always determined completely by their geometric properties. When there is a material in the space between conductors, its properties also play a role, as we will see in Section 25–5.

## 25–3 CAPACITORS IN SERIES AND IN PARALLEL

Capacitors are manufactured with certain standard capacitances and working voltages. However, these standard values may not be the ones you actually need in a particular circuit. You can obtain the values you need by combining capacitors; the simplest combinations are a series connection and a parallel connection.

Figure 25–5a is a schematic diagram of a **series connection.** Two capacitors are connected in series (one after the other) between points $a$ and $b$, and a constant potential difference $V_{ab}$ is maintained. The capacitors are both initially uncharged. When a positive potential difference $V_{ab}$ is applied between points $a$ and $b$, the top plate of $C_1$ acquires a positive charge $Q$. The electric field of this positive charge pulls negative charge up to the bottom plate of $C_1$ until all of the field lines end on the bottom plate (Fig. 25–5). This requires that the lower plate have charge $-Q$. These negative charges had to come from the top plate of $C_2$, which becomes positively charged with charge $+Q$. This positive charge then pulls negative charge $-Q$ from the connection at point $b$ onto the bottom plate of $C_2$. The total charge on the lower plate of $C_1$ and the upper plate of $C_2$ must always

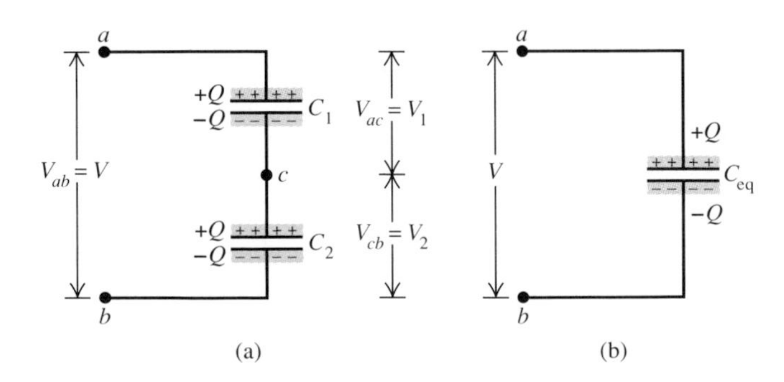


**25–5** (a) Two capacitors in series and (b) the equivalent capacitor.

be zero because these plates aren't connected to anything except each other. *In a series connection the magnitude of charge on all plates is the same.*

Referring again to Fig. 25–5a, we have

$$V_{ac} = V_1 = \frac{Q}{C_1}, \qquad V_{cb} = V_2 = \frac{Q}{C_2},$$

$$V_{ab} = V = V_1 + V_2 = Q\left(\frac{1}{C_1} + \frac{1}{C_2}\right),$$

$$\frac{V}{Q} = \frac{1}{C_1} + \frac{1}{C_2}. \tag{25–4}$$

The **equivalent capacitance** $C_{eq}$ of the series combination is defined as the capacitance of a *single* capacitor for which the charge $Q$ is the same as for the combination, when the potential difference $V$ is the same. For such a capacitor, shown in Fig. 25–5b,

$$C_{eq} = \frac{Q}{V}, \qquad \text{or} \qquad \frac{1}{C_{eq}} = \frac{V}{Q}. \tag{25–5}$$

Combining Eqs. (25–4) and (25–5), we find

$$\frac{1}{C_{eq}} = \frac{1}{C_1} + \frac{1}{C_2}.$$

We can extend this analysis to any number of capacitors in series; we find

$$\frac{1}{C_{eq}} = \frac{1}{C_1} + \frac{1}{C_2} + \frac{1}{C_3} + \cdots \quad \text{(series)}. \tag{25–6}$$

**The reciprocal of the equivalent capacitance of a series combination equals the sum of the reciprocals of the individual capacitances.** The magnitude of charge is the same on all plates of all the capacitors, but the potential differences of the individual capacitors are in general different.

The arrangement shown in Fig. 25–6a is called a **parallel connection.** Two capacitors are connected in parallel between points $a$ and $b$. In this case the upper plates of the two capacitors are connected together to form an equipotential surface, and the lower plates form another. *The potential difference is the same for both capacitors* and is equal to $V_{ab} = V$. The charges $Q_1$ and $Q_2$, which are not necessarily equal, are given by

$$Q_1 = C_1 V, \qquad Q_2 = C_2 V.$$

The *total* charge $Q$ of the combination, and thus on the equivalent capacitor, is

$$Q = Q_1 + Q_2 = V(C_1 + C_2),$$

so

$$\frac{Q}{V} = C_1 + C_2. \tag{25–7}$$

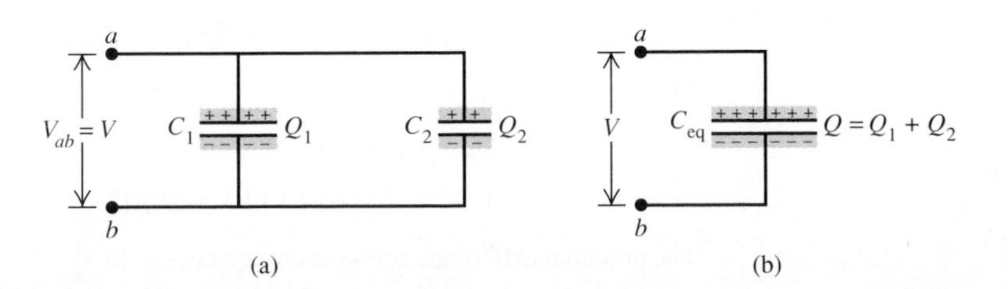


**25–6** (a) Two capacitors in parallel and (b) the equivalent capacitor.

The *equivalent* capacitance $C_{eq}$ of the parallel combination is defined as the capacitance of a single capacitor (Fig. 25–6b) for which the total charge is the same as in Fig. 25–6a. For this capacitor, $Q/V = C_{eq}$, so

$$C_{eq} = C_1 + C_2.$$

In the same way we can show that for any number of capacitors in parallel,

$$C_{eq} = C_1 + C_2 + C_3 + \cdots \quad \text{(parallel).} \tag{25–8}$$

**The equivalent capacitance of a parallel combination equals the *sum* of the individual capacitances.** The potential differences are the same for all capacitors, but their charges are in general different.

In a parallel connection the equivalent capacitance is always *greater than* any individual capacitance. In a series connection it is always *less than* any individual capacitance.

## PROBLEM-SOLVING STRATEGY

### Equivalent capacitance

**1.** Keep in mind that when we say that a capacitor has charge $Q$, we always mean that the plate at higher potential has charge $+Q$ and the other plate has charge $-Q$.

**2.** When capacitors are connected in series, as in Fig. 25–5a, they always have the same charge, assuming that they were uncharged before they were connected. The potential differences are *not* equal unless the capacitances are equal. The total potential difference across the combination is the sum of the individual potential differences.

**3.** When capacitors are connected in parallel, as in Fig. 25–6a, the potential difference $V$ is always the same for both. The charges on the two are *not* equal unless the capacitances are equal. The total charge on the combination is the sum of the individual charges.

**4.** For more complicated combinations you can sometimes identify parts that are simple series or parallel connections and replace these by their equivalent capacitances in a step-by-step reduction. Then, if you need to find the charge or potential difference for an individual capacitor, you may have to retrace your path to the original capacitors.

## EXAMPLE 25–4

In Figs. 25–5 and 25–6, let $C_1 = 6.0\ \mu\text{F}$, $C_2 = 3.0\ \mu\text{F}$, and $V_{ab} = 18$ V. Find the equivalent capacitance, and find the charge and potential difference for each capacitor when the two capacitors are connected a) in series; b) in parallel.

**SOLUTION** a) The equivalent capacitance of the series combination (Fig. 25–5a) is given by Eq. (25–6):

$$\frac{1}{C_{eq}} = \frac{1}{C_1} + \frac{1}{C_2} = \frac{1}{6.0\ \mu\text{F}} + \frac{1}{3.0\ \mu\text{F}}, \qquad C_{eq} = 2.0\ \mu\text{F}.$$

The charge $Q$ is

$$Q = C_{eq}V = (2.0\ \mu\text{F})(18\ \text{V}) = 36\ \mu\text{C}.$$

The potential differences across the capacitors are

$$V_{ac} = V_1 = \frac{Q}{C_1} = \frac{36\ \mu\text{C}}{6.0\ \mu\text{F}} = 6.0\ \text{V},$$

$$V_{cb} = V_2 = \frac{Q}{C_2} = \frac{36\ \mu\text{C}}{3.0\ \mu\text{F}} = 12\ \text{V}.$$

The *larger* potential difference appears across the *smaller* capacitor.

b) The equivalent capacitance of the parallel combination (Fig. 25–6a) is given by Eq. (25–8):

$$C_{eq} = C_1 + C_2 = 6.0\ \mu\text{F} + 3.0\ \mu\text{F} = 9.0\ \mu\text{F}.$$

The charges $Q_1$ and $Q_2$ are

$$Q_1 = C_1V = (6.0\ \mu\text{F})(18\ \text{V}) = 108\ \mu\text{C},$$

$$Q_2 = C_2V = (3.0\ \mu\text{F})(18\ \text{V}) = 54\ \mu\text{C}.$$

The potential difference across each capacitor is 18 V.

## EXAMPLE 25–5

Find the equivalent capacitance of the combination shown in Fig. 25–7a.

**SOLUTION** We first replace the 12-$\mu$F and 6-$\mu$F series combination by its equivalent capacitance; calling that $C'$, we use Eq. (25–6):

$$\frac{1}{C'} = \frac{1}{12\ \mu\text{F}} + \frac{1}{6\ \mu\text{F}}, \qquad C' = 4\ \mu\text{F}.$$

This gives us the equivalent combination shown in Fig. 25–7b. Next we find the equivalent capacitance of the three capacitors in parallel, using Eq. (25–8). Calling their equivalent capacitance $C''$, we have

$$C'' = 3\ \mu\text{F} + 11\ \mu\text{F} + 4\ \mu\text{F} = 18\ \mu\text{F}.$$

This gives us the equivalent combination shown in Fig. 25–7c. Finally, we find the equivalent capacitance $C_{eq}$ of these two capacitors in series:

$$\frac{1}{C_{eq}} = \frac{1}{18\ \mu\text{F}} + \frac{1}{9\ \mu\text{F}}, \qquad C_{eq} = 6\ \mu\text{F}.$$

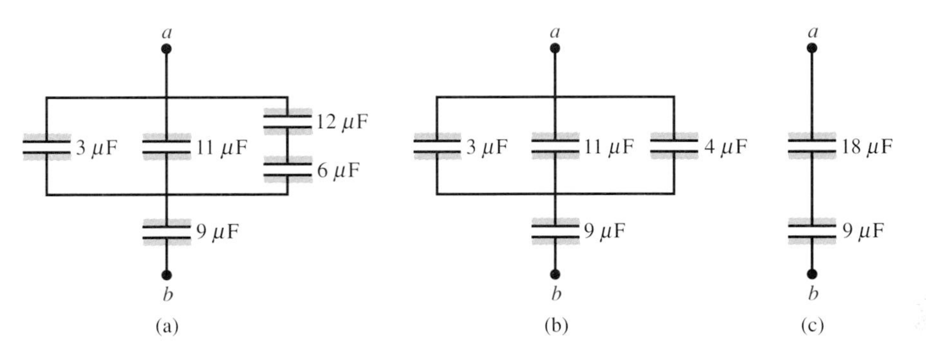


**25–7** (a) Finding the equivalent capacitance between points $a$ and $b$: (b) shows the equivalent capacitance of two capacitors in series; (c) shows the equivalent capacitance of three capacitors in parallel.

## 25–4 ELECTRIC-FIELD ENERGY

Many of the most important applications of capacitors depend on their ability to store energy. The opposite charges on the plates, separated and attracted toward each other, are analogous to a stretched spring or a body lifted in the earth's gravitational field. The potential energy corresponds to the energy input required to charge the capacitor and to the work done by the electrical forces when it becomes discharged. This work is analogous to the work done by a spring or the earth's gravity when the system returns from its displaced position to the reference position.

One way to calculate the potential energy $U$ of a charged capacitor is to calculate the work $W$ required to charge it. The final charge $Q$ and the final potential difference $V$ are related by

$$Q = CV.$$

Let $v$ and $q$ be the varying potential difference and charge during the charging process; then $v = q/C$. The work $dW$ required to transfer an additional element of charge $dq$ is

$$dW = v\,dq = \frac{q\,dq}{C}.$$

The total work $W$ needed to increase the charge $q$ from zero to a final value $Q$ is

$$W = \int_0^W dW = \frac{1}{C}\int_0^Q q\,dq = \frac{Q^2}{2C}.$$

This is also equal to the total work done by the electric field on the charge when $q$

decreases from an initial value $Q$ to zero, as the elements of charge $dq$ "fall" through potential differences $v$ that vary from $V$ down to zero.

If we define the potential energy of an *uncharged* capacitor to be zero, then $W$ is equal to the potential energy $U$ of the charged capacitor. The final potential difference $V$ between the plates is $V = Q/C$, so we can express $U$ (which is equal to $W$) as

$$U = \frac{Q^2}{2C} = \tfrac{1}{2}CV^2 = \tfrac{1}{2}QV. \tag{25–9}$$

When $Q$ is in coulombs, $C$ in farads (coulombs per volt), and $V$ in volts (joules per coulomb), $U$ is in joules.

The last form of Eq. (25–9) also shows that the total work $W$ is equal to the total charge $Q$ transferred, times the *average* potential difference $V/2$ during charging.

A charged capacitor is the electrical analog of a stretched spring with elastic potential energy $U = \frac{1}{2}kx^2$. The charge $Q$ is analogous to the elongation $x$, and the *reciprocal* of the capacitance, $1/C$, is analogous to the force constant $k$. The energy supplied to a capacitor in the charging process is analogous to the work we do on the spring when we stretch it.

The energy stored in a capacitor is related directly to the electric field between the capacitor plates. In fact, we can think of the energy as stored *in the field* in the region between the plates. To develop this relation, let's find the energy *per unit volume* in the space between the plates of a parallel-plate capacitor with plate area $A$ and separation $d$. We call this the **energy density,** denoted by $u$. From Eq. (25–9) the total energy is $\frac{1}{2}CV^2$, and the volume between the plates is simply $Ad$. The energy density is

$$u = \text{Energy density} = \frac{\frac{1}{2}CV^2}{Ad}. \tag{25–10}$$

From Eq. (25–2) the capacitance $C$ is given by $C = \epsilon_0 A/d$. The potential difference $V$ is related to the electric field magnitude $E$ by $V = Ed$. Using these expressions in Eq. (25–10), we find

$$u = \tfrac{1}{2}\epsilon_0 E^2. \tag{25–11}$$

Although we have derived this relation only for one specific kind of capacitor, it turns out to be valid for any capacitor in vacuum and indeed *for any electric field configuration in vacuum.* This result has an interesting implication. We think of vacuum as space with no matter in it, but vacuum can nevertheless have electric fields and therefore energy. It isn't necessarily just empty space. We will use Eq. (25–11) in Chapter 33 in connection with the energy transported by electromagnetic waves.

## EXAMPLE 25–6

In Fig. 25–8 we charge a capacitor $C_1$ by connecting it to a source of potential difference $V_0$ (not shown in the figure).

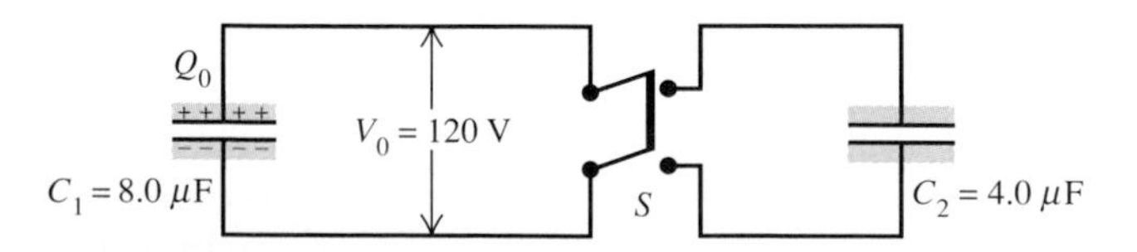


**25–8** When the switch $S$ is closed, the charged capacitor $C_1$ is connected to an uncharged capacitor $C_2$.

Let $C_1 = 8.0\ \mu$F and $V_0 = 120$ V. The charge $Q_0$ is

$$Q_0 = C_1V_0 = 960\ \mu\text{C},$$

and the energy stored in the capacitor is

$$U = \tfrac{1}{2}Q_0V_0 = \tfrac{1}{2}(960 \times 10^{-6}\ \text{C})(120\ \text{V}) = 0.058\ \text{J}.$$

a) What is the charge on each capacitor after we close switch $S$?
b) What is the total energy of the system after we close switch $S$?

**SOLUTION** a) The positive charge $Q_0$ becomes distributed over the upper plates of both capacitors, and the negative charge $-Q_0$ is distributed over the lower plates of both. Let $Q_1$ and $Q_2$ be the magnitudes of the final charges on the two capacitors. From conservation of charge,

$$Q_1 + Q_2 = Q_0.$$

In the final state, when the charges are no longer moving, both upper plates are at the same potential; they are connected by a wire and so form a single equipotential surface. Both lower plates are also at the same potential, different from that of the upper plates. The final potential difference between the plates, $V$, is therefore the same for both capacitors, and

$$Q_1 = C_1V, \qquad Q_2 = C_2V.$$

When we combine this with the preceding equation, we find

$$V = \frac{Q_0}{C_1 + C_2} = \frac{960\ \mu\text{C}}{12\ \mu\text{F}} = 80\ \text{V},$$

$$Q_1 = 640\ \mu\text{C}, \qquad Q_2 = 320\ \mu\text{C}.$$

b) The final energy of the system is

$$\tfrac{1}{2}Q_1V + \tfrac{1}{2}Q_2V = \tfrac{1}{2}Q_0V$$
$$= \tfrac{1}{2}(960 \times 10^{-6}\ \text{C})(80\ \text{V}) = 0.038\ \text{J}.$$

This is less than the original energy of 0.058 J; the difference has been converted to energy of some other form. The conductors become a little warmer because of their resistance, and some energy is radiated as electromagnetic waves. We'll study circuit behavior of capacitors in detail in Chapters 27 and 32.

## EXAMPLE 25–7

Suppose you want to store exactly one joule of electric potential energy in a volume of one cubic meter in vacuum. a) What is the magnitude of the required uniform electric field? b) If the field magnitude is ten times larger, how much energy is stored per cubic meter?

**SOLUTION** a) We solve Eq. (25–11) for $E$:

$$E = \sqrt{\frac{2u}{\epsilon_0}}$$

We subsitute numerical values:

$$E = \sqrt{\frac{2(1\ \text{J/m}^3)}{8.85 \times 10^{-12}\ \text{N}\cdot\text{m}^2/\text{C}^2}}$$
$$= 4.75 \times 10^5\ \text{N/C} = 4.75 \times 10^5\ \text{V/m}.$$

This is a sizable field, a potential difference of nearly a half million volts over a distance of one meter.

b) Equation (25–11) shows that $u$ is proportional to $E^2$. If $E$ increases by a factor of 10, $u$ increases by a factor of $10^2 = 100$, and the energy density is 100 J/m$^3$.

# 25–5 DIELECTRICS

Most capacitors have a nonconducting material, or **dielectric,** between their plates. A common type of capacitor uses long strips of metal foil for the plates, separated by strips of plastic sheet such as Mylar (Fig. 25–9). A sandwich of these materials is rolled up, forming a unit that can provide a capacitance of several microfarads in a compact package.

*Electrolytic* capacitors use as their dielectric an extremely thin layer of nonconducting oxide between a metal plate and a conducting solution. From Eq. (25–2) the capacitance is inversely proportional to the distance $d$ between the plates. Because of the thinness of the dielectric, electrolytic capacitors with relatively small dimensions may have a capacitance of the order of 100 or 1000 $\mu$F.

Placing a solid dielectric between the plates of a capacitor serves three functions. First, it solves the mechanical problem of maintaining two large metal sheets at a very small separation without actual contact.

Second, any dielectric material, when subjected to a sufficiently large electric field, experiences **dielectric breakdown,** a partial ionization that permits conduction through a material that is supposed to insulate. Many insulating materials can tolerate stronger electric fields without breakdown than can air.

Third, the capacitance of a capacitor of given dimensions is *greater* when there is a dielectric material between the plates than with air or vacuum. We can demonstrate this

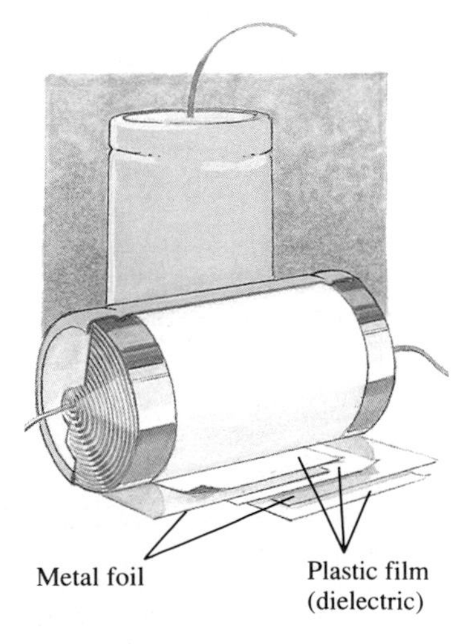


**25–9** A common type of capacitor is made from a rolled-up sandwich of metal foil and plastic film encased in a plastic shell.

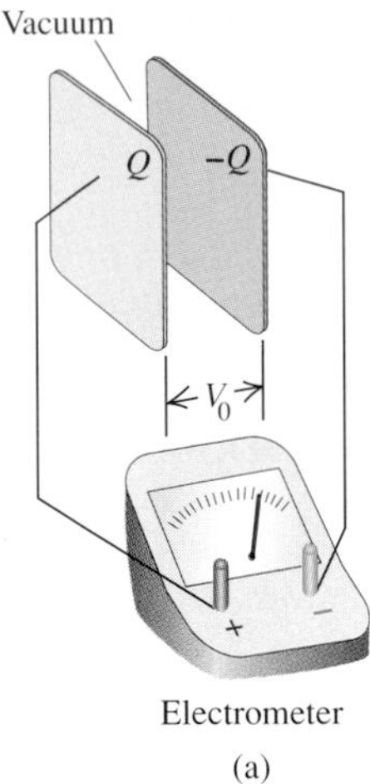

Electrometer

(a)

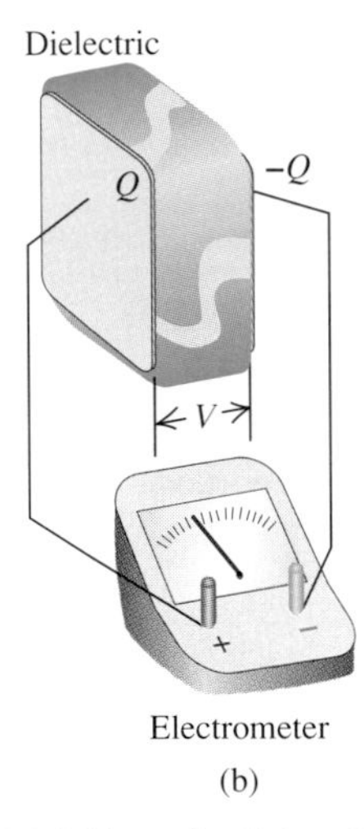


Electrometer

(b)

**25–10** Effect of a dielectric between the plates of a parallel-plate capacitor. The electrometer measures potential difference. (a) With a given charge, the potential difference is $V_0$. (b) With the same charge but with a dielectric between the plates, the potential difference $V$ is smaller than $V_0$.

effect with the aid of a sensitive electrometer, a device that measures the potential difference between two conductors without letting any appreciable charge flow from one to the other. Figure 25–10a shows an electrometer connected across a charged capacitor, with charge of magnitude $Q$ on each plate and potential difference $V_0$. When we insert a sheet of dielectric, such as glass, paraffin, or polystyrene, between the plates, the potential difference *decreases* to a smaller value $V$ (Fig. 25–10b). When we remove the dielectric, the potential difference returns to its original value $V_0$, showing that the original charges on the plates have not changed.

The original capacitance $C_0$ is given by $C_0 = Q/V_0$, and the capacitance $C$ with the dielectric present is $C = Q/V$. The charge $Q$ is the same in both cases, and $V$ is less than $V_0$, so we conclude that the capacitance $C$ with the dielectric present is *greater* than $C_0$. When the space between plates is completely filled by the dielectric, the ratio of $C$ to $C_0$ (equal to the ratio of $V_0$ to $V$) is called the **dielectric constant** of the material, $K$:

$$K = \frac{C}{C_0}. \tag{25–12}$$

When the charge is constant, the potential is *reduced* by a factor $K$:

$$K = \frac{V_0}{V}. \tag{25–13}$$

The dielectric constant $K$ is a pure number. Because $C$ is always greater than $C_0$, $K$ is always greater than unity. A few representative values of $K$ are given in Table 25–1. For vacuum, $K = 1$ by definition. For air at ordinary temperatures and pressures, $K$ is about 1.0006; this is so nearly equal to 1 that for most purposes an air capacitor is equivalent to one in vacuum.

When a dielectric material is inserted between the plates of a capacitor while the charge is kept constant, the potential difference between the plates decreases by a factor $K$. Therefore the electric field between the plates must decrease by the same factor. If $E_0$ is the vacuum value and $E$ the value with the dielectric, then

$$E = \frac{E_0}{K}. \tag{25–14}$$

The fact that $E$ is smaller when the dielectric is present means that the surface charge density is smaller. The surface charge on the conducting plates does not change, but an *induced* charge of the opposite sign appears on each surface of the dielectric (Fig. 25–11). These induced surface charges are a result of redistribution of charge within the dielectric material, a phenomenon called **polarization.** We will assume in this discus-

**TABLE 25–1 Values of Dielectric Constant $K$ at 20°C**

| Material | $K$ | Material | $K$ |
|---|---|---|---|
| Vacuum | 1 | Polyvinyl chloride | 3.18 |
| Air (1 atm) | 1.00059 | Plexiglas | 3.40 |
| Air (100 atm) | 1.0548 | Glass | 5–10 |
| Teflon | 2.1 | Neoprene | 6.70 |
| Polyethylene | 2.25 | Germanium | 16 |
| Benzene | 2.28 | Glycerin | 42.5 |
| Mica | 3–6 | Water | 80.4 |
| Mylar | 3.1 | Strontium titanate | 310 |

sion that the induced surface charge is *directly proportional* to the electric field in the material (like a spring that obeys Hooke's law). In that case, $K$ is a constant for any particular material. When very strong fields or anisotropic materials are present, the relationships become more complex, but we won't consider such cases here.

We can derive a relation between this induced surface charge and the charge on the plates. Let's denote the magnitude of the charge per unit area induced on the surfaces of the dielectric (the induced surface charge density) by $\sigma_i$. The magnitude of the surface charge density on the capacitor plates is $\sigma$, as usual. Then the *net* surface charge on each side of the capacitor has magnitude $(\sigma - \sigma_i)$, as shown in Fig. 25–11b. The field between the plates is related to the net surface charge density by Eq. (23–12), $E = \sigma_{net}/\epsilon_0$. Without and with the dielectric, respectively, we have

$$E_0 = \frac{\sigma}{\epsilon_0}, \qquad E = \frac{\sigma - \sigma_i}{\epsilon_0}.$$

Using these expressions in Eq. (25–14) and rearranging the result, we find

$$\sigma_i = \sigma\left(1 - \frac{1}{K}\right). \qquad (25\text{–}15)$$

This equation shows that when $K$ is very large, $\sigma_i$ is nearly as large as $\sigma$. In this case, $\sigma_i$ nearly cancels $\sigma$, and the field and potential difference are much smaller than their values in vacuum.

The product $K\epsilon_0$ is called the **permittivity** of the dielectric, denoted by $\epsilon$:

$$\epsilon = K\epsilon_0. \qquad (25\text{–}16)$$

In terms of $\epsilon$ we can express the electric field within the dielectric as

$$E = \frac{\sigma}{\epsilon}. \qquad (25\text{–}17)$$

The capacitance when the dielectric is present is given by

$$C = KC_0 = K\epsilon_0\frac{A}{d} = \epsilon\frac{A}{d}. \qquad (25\text{–}18)$$

In empty space, where $K = 1$, $\epsilon = \epsilon_0$. For this reason, $\epsilon_0$ is sometimes called the permittivity of free space, or the permittivity of vacuum. Because $K$ is a pure number, $\epsilon$ and $\epsilon_0$ have the same units, $C^2/N \cdot m^2$ or F/m.

We can repeat the derivation of Eq. (25–11) for the energy density $u$ in an electric field for the case in which a dielectric is present. The result is

$$u = \tfrac{1}{2}K\epsilon_0 E^2 = \tfrac{1}{2}\epsilon E^2. \qquad (25\text{–}19)$$

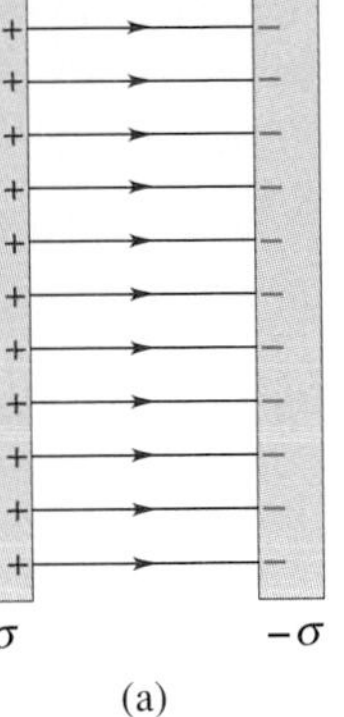


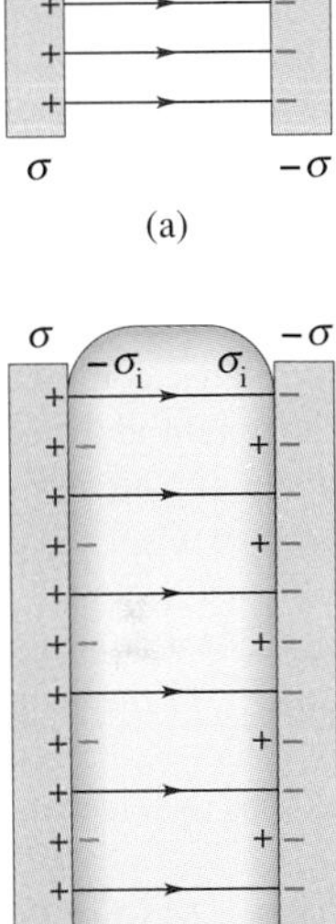


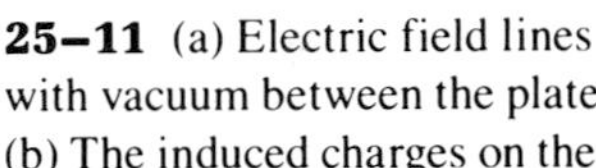
**25–11** (a) Electric field lines with vacuum between the plates. (b) The induced charges on the faces of the dielectric decrease the electric field.

## PROBLEM-SOLVING STRATEGY

### Dielectrics

**1.** As usual, be careful with units. Distances must be in meters; remember that a microfarad is $10^{-6}$ farads, and so on. Don't confuse the numerical value of $\epsilon_0$ with the value of $1/4\pi\epsilon_0$. There are several alternative sets of units for electric field magnitude, including N/C and V/m. The units of $\epsilon_0$ are $C^2/N \cdot m^2$ or F/m. Always check for consistency of units. It's a bit more complex with electrical quantities than it was in mechanics, but it's worth it!

**2.** In problems such as the following example it is easy to get lost in a blizzard of formulas. Ask yourself at each step what kind of quantity each symbol represents. For example, distinguish clearly between charges and charge densities and between electric fields and potentials. When you check numerical values, remember that the capacitance with a dielectric present is always greater than without one and that the induced surface charge density $\sigma_i$ on the dielectric is always less than the charge density $\sigma$ on the capacitor plates. With a given charge on a capacitor the electric field and potential difference are less with a dielectric present than without it.

## EXAMPLE 25–8

**Properties of dielectrics** Suppose the parallel plates in Fig. 25–11 have an area of 2000 cm$^2$ ($2.00 \times 10^{-1}$ m$^2$) and are 1.00 cm ($1.00 \times 10^{-2}$ m) apart. The capacitor is connected to a power supply and charged to a potential difference $V_0 = 3000$ V = 3.00 kV. Then it is disconnected from the power supply, and a sheet of insulating plastic material is inserted between the plates, completely filling the space between them. We find that the potential difference decreases to 1000 V. (The charge on each capacitor plate remains constant.) Compute a) the original capacitance $C_0$; b) the magnitude of charge $Q$ on each plate; c) the capacitance $C$ after the dielectric is inserted; d) the dielectric constant $K$ of the dielectric; e) the permittivity $\epsilon$ of the dielectric; f) the magnitude of induced charge $Q_i$ on each face of the dielectric; g) the original electric field $E_0$ between the plates; and h) the electric field $E$ after the dielectric is inserted.

**SOLUTION** Most of these quantities can be obtained in several different ways. Here is a representative sample; try to think of others and compare them.

a) $$C_0 = \epsilon_0 \frac{A}{d} = (8.85 \times 10^{-12}\ \text{F/m})\frac{2.00 \times 10^{-1}\ \text{m}^2}{1.00 \times 10^{-2}\ \text{m}} = 1.77 \times 10^{-10}\ \text{F} = 177\ \text{pF}.$$

b) $$Q = C_0V_0 = (1.77 \times 10^{-10}\ \text{F})(3.00 \times 10^3\ V) = 5.31 \times 10^{-7}\ \text{C} = 0.531\ \mu\text{C}.$$

c) $$C = \frac{Q}{V} = \frac{5.31 \times 10^{-7}\ \text{C}}{1.00 \times 10^3\ \text{V}} = 5.31 \times 10^{-10}\ \text{F} = 531\ \text{pF}.$$

d) $$K = \frac{C}{C_0} = \frac{5.31 \times 10^{-10}\ \text{F}}{1.77 \times 10^{-10}\ \text{F}} = \frac{531\ \text{pF}}{177\ \text{pF}} = 3.00.$$

Or, from Eq. (25–13),

$$K = \frac{V_0}{V} = \frac{3000\ \text{V}}{1000\ \text{V}} = 3.00.$$

e) $$\epsilon = K\epsilon_0 = (3.00)(8.85 \times 10^{-12}\ \text{C}^2/\text{N}\cdot\text{m}^2) = 2.66 \times 10^{-11}\ \text{C}^2/\text{N}\cdot\text{m}^2.$$

f) From Eq. (25–15),

$$Q_i = Q\left(1 - \frac{1}{K}\right) = (5.31 \times 10^{-7}\ \text{C})\left(1 - \frac{1}{3.00}\right) = 3.54 \times 10^{-7}\ \text{C}.$$

g) $$E_0 = \frac{V_0}{d} = \frac{3000\ \text{V}}{1.00 \times 10^{-2}\ \text{m}} = 3.00 \times 10^5\ \text{V/m}.$$

h) $$E = \frac{V}{d} = \frac{1000\ \text{V}}{1.00 \times 10^{-2}\ \text{m}} = 1.00 \times 10^5\ \text{V/m},$$

or

$$E = \frac{\sigma}{\epsilon} = \frac{Q}{\epsilon A} = \frac{5.31 \times 10^{-7}\ \text{C}}{(2.66 \times 10^{-11}\ \text{C}^2/\text{N}\cdot\text{m}^2)(2.00 \times 10^{-1}\ \text{m}^2)} = 1.00 \times 10^5\ \text{V/m},$$

or

$$E = \frac{\sigma - \sigma_i}{\epsilon_0} = \frac{Q - Q_i}{\epsilon_0 A} = \frac{(5.31 - 3.54) \times 10^{-7}\ \text{C}}{(2.00 \times 10^{-1}\ \text{m}^2)(8.85 \times 10^{-12}\ \text{C}^2/\text{N}\cdot\text{m}^2)} = 1.00 \times 10^5\ \text{V/m},$$

or, from Eq. (25–14),

$$E = \frac{E_0}{K} = \frac{3.00 \times 10^5\ \text{V/m}}{3.00} = 1.00 \times 10^5\ \text{V/m}.$$

## EXAMPLE 25–9

Find the total energy stored in the electric field of the capacitor in Example 25–8 and the energy density, both before and after the dielectric sheet is inserted.

**SOLUTION** Let the original energy be $U_0$, and let the energy with the dielectric in place be $U$. From Eq. (25–9),

$$U_0 = \tfrac{1}{2}C_0V_0^2 = \tfrac{1}{2}(1.77 \times 10^{-10}\ \text{F})(3000\ \text{V})^2 = 8.0 \times 10^{-4}\ \text{J},$$

$$U = \tfrac{1}{2}CV^2 = \tfrac{1}{2}(5.31 \times 10^{-10}\ \text{F})(1000\ \text{V})^2 = 2.7 \times 10^{-4}\ \text{J}.$$

The volume is $(0.200\ \text{m}^2)(0.0100\ \text{m}) = 0.00200\ \text{m}^3$. The original energy density (energy per unit volume) is

$$u_0 = \frac{8.0 \times 10^{-4}\ \text{J}}{0.00200\ \text{m}^3} = 0.40\ \text{J/m}^3$$

Alternatively,

$$u_0 = \tfrac{1}{2}\epsilon_0 E_0{}^2 = \tfrac{1}{2}(8.85 \times 10^{-12}\ \text{C}^2/\text{N}\cdot\text{m}^2)(3.0 \times 10^5\ \text{N/C})^2 = 0.40\ \text{J/m}^3.$$

With the dielectric in place,

$$u = \frac{2.7 \times 10^{-4}\ \text{J}}{0.00200\ \text{m}^3} = 0.13\ \text{J/m}^3.$$

This is one-third of the original energy density.

# Capacitors

Capacitors in electric circuits are analogous to springs in mechanical systems. They can store energy until it is needed or they can smooth out voltage fluctuations, as in power supplies.

# BUILDING PHYSICAL INTUITION

In many computer keyboards, each key is connected to a movable side of a parallel-plate capacitor. Pressing the key causes a change of capacitance, which is detected by the rest of the circuitry.

Microprocessor receives signal

Motion of charge detected by circuitry

Change in capacitance creates motion of charge

The bright blue or orange disks at the ends of these integrated circuit packages are thin-film capacitors. They serve to prevent a sudden large charge from damaging the circuit.

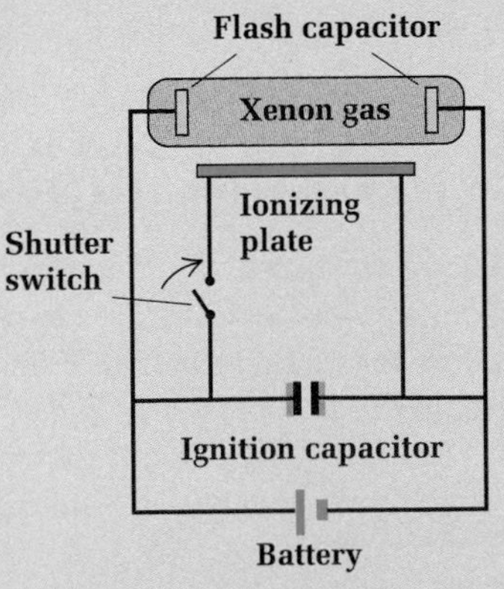


A photoflash unit uses two capacitors. Depressing the shutter button discharges the small ignition capacitor, which transmits charge to an ionizing plate outside the flash tube. The charged plate ionizes xenon gas within the tube, which provides a conducting path to discharge the main capacitor. When this capacitor discharges, the xenon gas produces a bright flash of light. Typically, the flash lasts for 0.0004 s and recycles in 0.3 s.

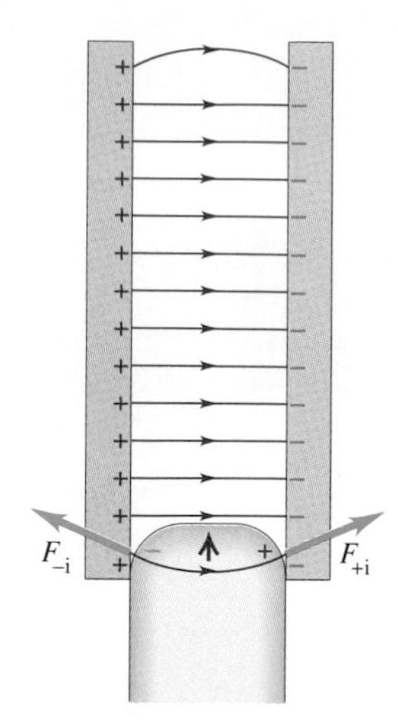


**25–12** The fringing field at the edges of the capacitor exerts forces on the induced surface charge of a dielectric, pulling the dielectric into the capacitor.

We can generalize the results of Example 25–9. When a dielectric is inserted into a capacitor while the charge on each plate remains the same, the value of $\epsilon$ increases by a factor of $K$, the electric field decreases by a factor of $1/K$, and the energy density $u = \frac{1}{2}\epsilon E^2$ decreases by a factor of $1/K$. Where did the energy go? The answer lies in the fringing field at the edges of a real parallel-plate capacitor. As Fig. 25–12 shows, that field exerts an attractive force on the dielectric, tending to pull the dielectric into the space between the plates and doing work on it as it does so. We could attach a spring to the bottom of the sheet in Fig. 25–12 and use this force to stretch the spring.

We mentioned earlier that when any dielectric material is subjected to a sufficiently strong electric field, it becomes a conductor. This phenomenon is called *dielectric breakdown*. Conduction occurs when the electric field is so strong that electrons are ripped loose from their molecules and crash into other molecules, liberating even more electrons. This avalanche of moving charge, forming a spark or arc discharge, often starts quite suddenly. Figure 25–13 shows a beautiful example of dielectric breakdown.

Capacitors always have maximum voltage ratings. When a capacitor is subjected to excessive voltage, an arc may form through a layer of dielectric, burning or melting a hole in it. This provides a conducting path between the conductors, creating a short circuit. If this path remains after the arc is extinguished, the device is rendered permanently useless as a capacitor.

The maximum electric field a material can withstand without the occurrence of breakdown is called its **dielectric strength.** This quantity is affected significantly by temperature, trace impurities, small irregularities in the metal electrodes, and other factors that are difficult to control. For this reason we can give only approximate figures for dielectric strengths. The dielectric strength of dry air is about $3 \times 10^6$ V/m. A few typical values of dielectric strength for common insulating materials are given in Table 25–2. Note that the values are all substantially greater than that for air. For example, a layer of polycarbonate that is 0.01 mm thick, which is about the smallest practical thickness, can withstand a maximum voltage of about $(3 \times 10^7 \text{ V/m})(1.0 \times 10^{-5} \text{ m}) = 300$ V.

**25–13** Dendritic pattern formed in a Plexiglas block by dielectric breakdown under the action of a very strong electric field.

**TABLE 25–2 Dielectric Constant and Dielectric Strength of Some Insulating Materials**

| Material | Dielectric constant, $K$ | Dielectric strength (V/m) |
|---|---|---|
| Polycarbonate | 2.8 | $3 \times 10^7$ |
| Polyester | 3.3 | $6 \times 10^7$ |
| Polypropylene | 2.2 | $7 \times 10^7$ |
| Polystyrene | 2.6 | $2 \times 10^7$ |
| Pyrex glass | 4.7 | $1 \times 10^7$ |

## *25–6 MOLECULAR MODEL OF INDUCED CHARGE

In Section 25–5 we discussed induced surface charges on a dielectric in an electric field. Now let's look at how these surface charges can come about. If the material were a *conductor,* the answer would be simple. Conductors contain charge that is free to move, and when an electric field is present, some of the charge redistributes itself on the surface so that there is no electric field inside the conductor. But an ideal dielectric has *no* charges that are free to move, so how can a surface charge occur?

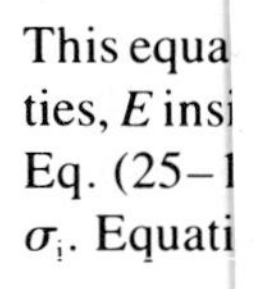

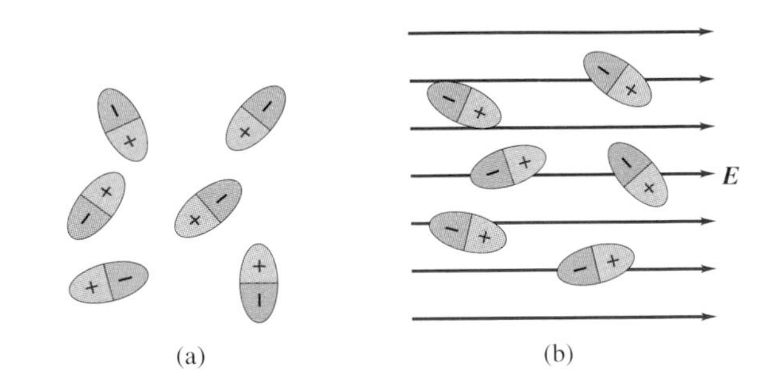


**25–14** (a) Polar molecules have random orientations when there is no applied electric field. (b) They tend to line up with an applied field **E**.

To understand this, we have to look at rearrangement of charge at the *molecular* level. Some molecules, such as $H_2O$ and $N_2O$, have equal amounts of positive and negative charge but a lopsided distribution, with excess positive charge concentrated on one side of the molecule and negative charge on the other. This arrangement is called an *electric dipole,* and the molecule is called a *polar molecule*. When no electric field is present in a gas or liquid with polar molecules, the molecules are oriented randomly (Fig. 25–14a). When they are placed in an electric field, however, they tend to orient themselves as in Fig. 25–14b, as a result of the electric-field torques.

Even a molecule that is *not* ordinarily polar *becomes* a dipole when it is placed in an electric field because the field pushes the positive charges in the molecules in the direction of the field and pushes the negative charges in the opposite direction. This causes redistribution of charge within the molecule (Fig. 25–15). Such dipoles are called *induced* dipoles. With either polar or nonpolar molecules the redistribution of charge caused by the field leads to the formation of a layer of charge on each surface of the dielectric material (Fig. 25–16). These layers are the surface charges described in Section 25–5; their surface charge density is denoted by $\sigma_i$. The charges are not free to move indefinitely, as they would be in a conductor, because each charge is bound to a molecule. They are in fact called **bound charges** to distinguish them from the **free charges** that are added to and removed from the conducting capacitor plates. In the interior of the material the net charge per unit volume remains zero. This redistribution of charge is called *polarization,* and we say that the material is *polarized.*

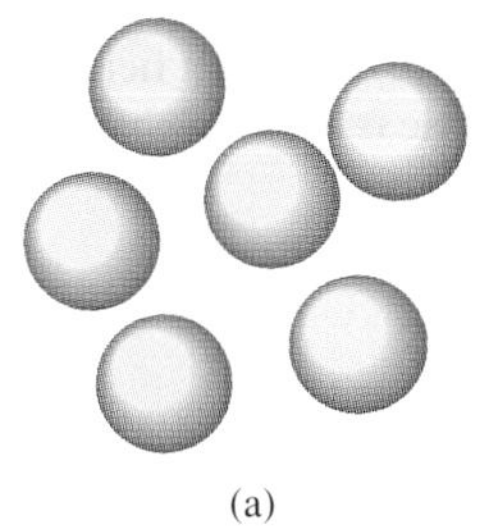


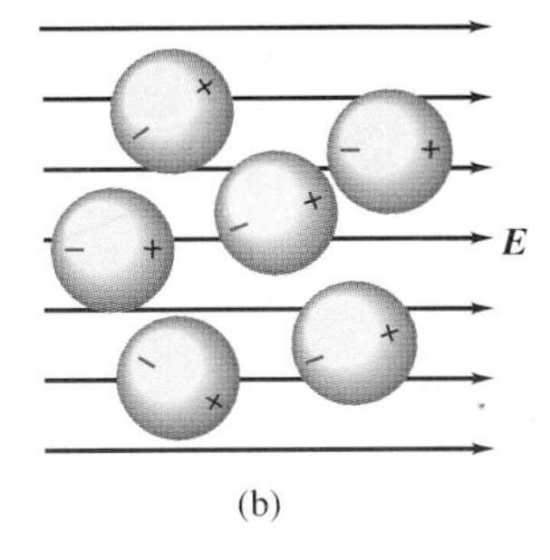


**25–15** (a) Nonpolar molecules have their positive and negative charge centers at the same point. (b) These centers become separated slightly by the applied electric field **E**.

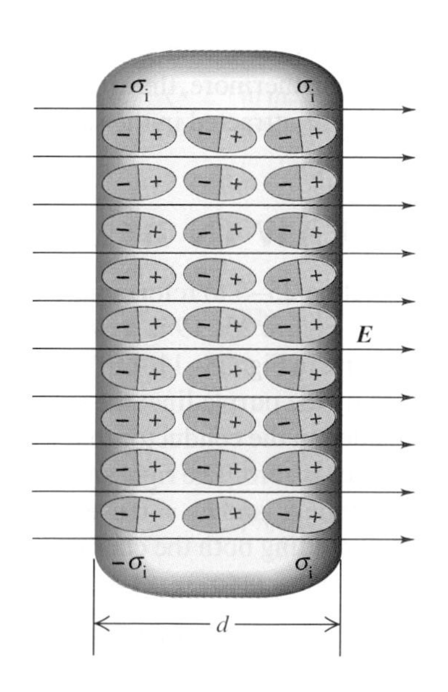


**25–16** Polarization of a dielectric in an electric field gives rise to thin layers of bound charges on the surfaces, creating surface charge densities $\sigma_i$ and $-\sigma_i$.

shows again that the effect of the dielectric is to increase the capacitance by a factor of $K$, as we already knew.

We can also relate Eq. (25–24) to the expression $C = \epsilon A/d$ for a parallel-plate capacitor. The quantity $4\pi r_a r_b$ is intermediate between the areas $4\pi r_a^2$ and $4\pi r_b^2$ of the two spheres and is in fact the *geometric mean* of these two areas, which we can denote by $A_{gm}$. The distance between spheres is $d = r_b - r_a$, so we can rewrite Eq. (25–24) as $C = \epsilon A_{gm}/d$. This is exactly the same form as for parallel plates. The point is that if the distance between spheres is very small in comparison to their radii, they behave like parallel plates with the same area and spacing.

**25–17**
two char
tion of a
induced
ner lines
the field.
a dielect
plates.

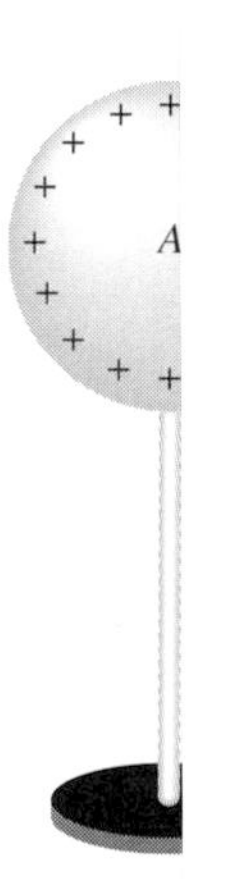

**25–18**
radial ele
charged s
charge be

## SUMMARY

### KEY TERMS

- capacitor
- capacitance
- farad
- parallel-plate capacitor
- series connection
- equivalent capacitance
- parallel connection
- energy density
- dielectric
- dielectric breakdown
- dielectric constant
- polarization
- permittivity
- dielectric strength
- bound charge
- free charge

• A capacitor is any pair of conductors separated by an insulating material. When the capacitor is charged, with charges of equal magnitude $Q$ and opposite sign on the two conductors, the potential $V_{ab}$ of the positively charged conductor with respect to the negatively charged conductor is proportional to $Q$. The capacitance $C$ is defined as

$$C = \frac{Q}{V_{ab}}. \qquad (25\text{–}1)$$

• A parallel-plate capacitor is made with two parallel plates, each with area $A$, separated by a distance $d$. If they are separated by vacuum, the capacitance is

$$C = \frac{Q}{V_{ab}} = \epsilon_0 \frac{A}{d}. \qquad (25\text{–}2)$$

The SI unit of capacitance is the farad, abbreviated F. One farad is one coulomb per volt (1 F = 1 C/V). Alternative units are

$$1\ \text{F} = 1\ \text{C}^2/\text{N}\cdot\text{m} = 1\ \text{C}^2/\text{J}.$$

The microfarad ($1\ \mu\text{F} = 10^{-6}$ F) and the picofarad ($1\ \text{pF} = 10^{-12}$ F) are commonly used.

• When capacitors with capacitances $C_1, C_2, C_3, \ldots$ are connected in series, the equivalent capacitance $C_{eq}$ is

$$\frac{1}{C_{eq}} = \frac{1}{C_1} + \frac{1}{C_2} + \frac{1}{C_3} + \cdots \quad \text{(series)}. \qquad (25\text{–}6)$$

When they are connected in parallel, the equivalent capacitance $C_{eq}$ is

$$C_{eq} = C_1 + C_2 + C_3 + \cdots \quad \text{(parallel)}. \qquad (25\text{–}8)$$

• The energy $U$ required to charge a capacitor $C$ to a potential difference $V$ and a charge $Q$ is equal to the energy stored in the capacitor and is given by

$$U = \frac{Q^2}{2C} = \tfrac{1}{2}CV^2 = \tfrac{1}{2}QV. \qquad (25\text{–}9)$$

This energy can be thought of as residing in the electric field between the conductors; the energy density $u$ (energy per unit volume) is

$$u = \tfrac{1}{2}\epsilon_0 E^2. \qquad (25\text{–}11)$$

• When the space between the conductors is filled with a dielectric material, the capacitance increases by a factor $K$, called the dielectric constant of the material. Surface charges induced on the surface of the dielectric decrease the electric field and potential difference between conductors by a factor $K$. The surface charge results from polarization, a microscopic rearrangement of charge in the dielectric due to the reorientation of polar molecules in an applied $\boldsymbol{E}$ field or the creation of induced dipole moments in a nonpolar material. Under sufficiently strong fields, dielectrics become conductors; this is called dielectric breakdown. The maximum field that a material can withstand without breakdown is called its dielectric strength.

- The energy density $u$ in an electric field in a dielectric is

$$u = \tfrac{1}{2} K\epsilon_0 E^2 = \tfrac{1}{2}\epsilon E^2. \qquad (25\text{–}19)$$

- Gauss's law can be reformulated for dielectrics as follows:

$$\oint K\mathbf{E} \cdot d\mathbf{A} = \frac{Q_{encl}}{\epsilon_0}. \qquad (25\text{–}22)$$

where $Q_{encl}$ includes only the free charge (not bound charge or polarization charge) enclosed by the Gaussian surface.

## EXERCISES

### Section 25–2 Calculating Capacitance

**25–1** A parallel-plate air capacitor has a capacitance of 500 pF and a charge of magnitude 0.200 $\mu$C on each plate. The plates are 0.400 mm apart. a) What is the potential difference between the plates? b) What is the area of each plate? c) What is the electric-field magnitude between the plates? d) What is the surface-charge density on each plate?

**25–2** The plates of a parallel-plate capacitor are 4.00 mm apart, and each carries a charge of magnitude $5.00 \times 10^{-8}$ C. The plates are in vacuum. The electric field between the plates has a magnitude of $4.00 \times 10^6$ V/m. a) What is the potential difference between the plates? b) What is the area of each plate? c) What is the capacitance?

**25–3** A capacitor has a capacitance of 6.40 $\mu$F. How much charge must be removed to lower the potential difference of its plates by 50.0 V?

### Section 25–3 Capacitors in Series and in Parallel

**25–4** In Fig. 25–5a, let $C_1 = 4.00$ $\mu$F, $C_2 = 6.00$ $\mu$F, and $V_{ab} = +54.0$ V. Calculate a) the charge on each capacitor; b) the potential difference across each capacitor.

**25–5** In Fig. 25–6a, let $C_1 = 4.00$ $\mu$F, $C_2 = 6.00$ $\mu$F, and $V_{ab} = +54.0$ V. Calculate a) the charge on each capacitor; b) the potential difference across each capacitor.

**25–6** In the circuit shown in Fig. 25–21, $C_1 = 2.00$ $\mu$F, $C_2 = 4.00$ $\mu$F, and $C_3 = 9.00$ $\mu$F. The applied potential is $V_{ab} = +48.0$ V. Calculate a) the charge on each capacitor; b) the potential difference across each capacitor; c) the potential difference between points $a$ and $d$.

**25–7** In Fig. 25–22, each capacitor has $C = 2.00$ $\mu$F and $V_{ab} = +36.0$ V. Calculate a) the charge on each capacitor; b) the potential difference across each capacitor; c) the potential difference between points $a$ and $d$.

**25–8** Two parallel-plate capacitors have plate spacings $d_1$ and $d_2$ and equal plate areas $A$. Show that when the capacitors are connected in series, the equivalent capacitance is the same as for a single capacitor with plate area $A$ and spacing $d_1 + d_2$.

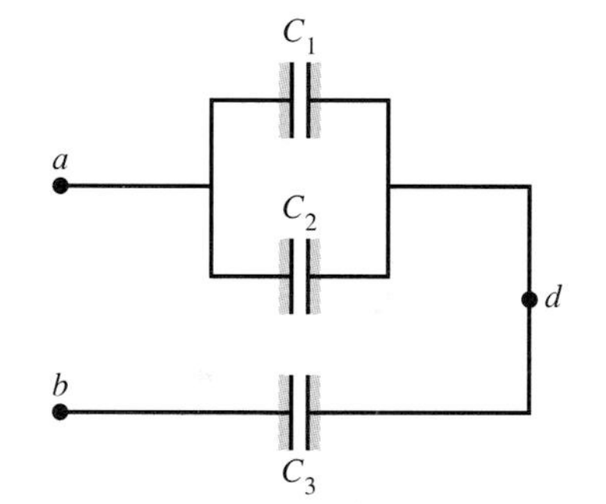


FIGURE 25–21

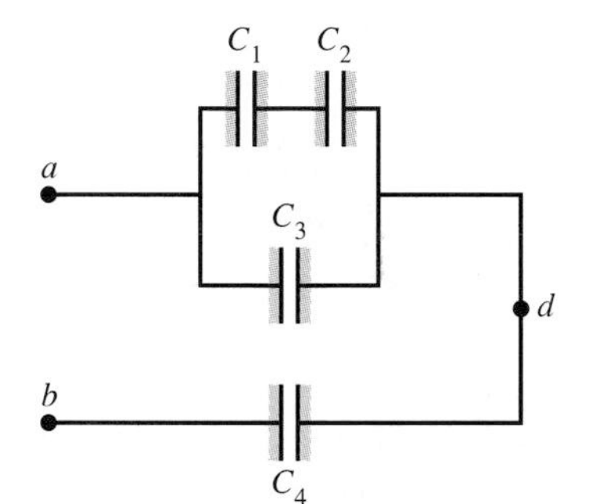


FIGURE 25–22

**25–9** Two parallel-plate capacitors have areas $A_1$ and $A_2$ and equal plate spacings $d$. Show that when the capacitors are connected in parallel, the equivalent capacitance is the same as for a single capacitor with plate area $A_1 + A_2$ and spacing $d$.

### Section 25–4 Electric-Field Energy

**25–10** An air capacitor is made from two flat parallel plates 1.20 mm apart. The magnitude of charge on each plate is 0.0150 $\mu$C when the potential difference is 200 V. a) What is the capacitance? b) What is the area of each plate? c) What maximum voltage can be applied without dielectric breakdown? (Dielectric breakdown for air occurs at an electric-field strength of $3.0 \times 10^6$ V/m.) d) When the charge is 0.0150 $\mu$C, what total energy is stored?

**25–11** A 300-$\mu$F capacitor is charged to 240 V. Then a wire is connected between the plates. How many joules of heat are produced as the capacitor discharges if all of the energy that was stored goes into heating the wire?

**25–12** An air capacitor consisting of two closely spaced parallel plates has a capacitance of 1000 pF. The charge on each plate is 5.00 $\mu$C. a) What is the potential difference between the plates? b) If the charge is kept constant, what will be the potential difference between the plates if the separation is doubled? c) How much work is required to double the separation?

**25–13** An 8.00-$\mu$F parallel-plate capacitor has a plate separation of 4.00 mm and is charged to a potential difference of 400 V. Calculate the energy density in the region between the plates, in units of J/m$^3$.

**25–14** A 20.0-$\mu$F capacitor is charged to a potential difference of 600 V. The terminals of the charged capacitor are then connected to those of an uncharged 10.0-$\mu$F capacitor. Compute a) the original charge of the system; b) the final potential difference across each capacitor; c) the final energy of the system; d) the decrease in energy when the capacitors are connected.

**25–15 Force on a Capacitor Plate.** A parallel-plate capacitor with plate area $A$ and separation $x$ is charged, with a charge of magnitude $q$ on each plate. a) What is the total energy stored in the capacitor? b) The plates are pulled apart an additional distance $dx$; now what is the total energy? c) If $F$ is the force with which the plates attract each other, then the difference in the two energies above must equal the work $dW = F\,dx$ done in pulling the plates apart. Show that $F = q^2/2\epsilon_0 A$. d) Explain why $F$ is not equal to $qE$, where $E$ is the electric field between the plates.

## Section 25–5 Dielectrics

**25–16** Show that Eq. (25–19) holds for a parallel-plate capacitor with a dielectric material between the plates. Use a derivation analogous to that used for Eq. (25–11).

**25–17** A parallel-plate capacitor is to be constructed using as a dielectric rubber that has a dielectric constant of 3.40 and a dielectric strength of $2.00 \times 10^7$ V/m. The capacitor is to have a capacitance of 1.50 nF and must be able to withstand a maximum potential difference of 6000 V. What is the minimum area the plates of the capacitor may have?

**25–18** Two oppositely charged conducting plates, with equal magnitude of charge per unit area, are separated by a dielectric 3.00 mm thick, with a dielectric constant of 4.50. The resultant electric field in the dielectric is $1.60 \times 10^6$ V/m. Compute a) the charge per unit area on the conducting plates; b) the charge per unit area on the surfaces of the dielectric.

**25–19** Two parallel plates have equal and opposite charges. When the space between the plates is evacuated, the electric field is $3.60 \times 10^5$ V/m. When the space is filled with dielectric, the electric field is $1.20 \times 10^5$ V/m. a) What is the charge density on the surface of the dielectric? b) What is the dielectric constant?

**25–20** Two parallel plates, each with an area of 40.0 cm$^2$, are given equal and opposite charges of magnitude $1.80 \times 10^{-7}$ C. The space between the plates is filled with a dielectric material, and the electric field within the dielectric is $3.30 \times 10^5$ V/m. a) What is the dielectric constant of the dielectric? b) What is the total induced charge on either face of the dielectric?

## Section 25–7 Gauss's Law in Dielectrics

***25–21** A parallel-plate capacitor has the volume between its plates filled with plastic with dielectric constant $K$. The magnitude of the charge on each plate is $Q$. Each plate has area $A$, and the distance between the plates is $d$. a) Use Gauss's law as stated in Eq. (25–22) to calculate the magnitude of the electric field in the dielectric. b) Use the electric field determined in part (a) to calculate the potential difference between the two plates. c) Use the result of part (b) to determine the capacitance of the capacitor. Compare your result to Eq. (25–12).

# PROBLEMS

**25–22 A Computer Keyboard.** In one type of computer keyboard, each key is connected to a small metal plate that serves as one plate of a parallel-plate air-filled capacitor (Fig. 25–23). When the key is depressed, the plate separation decreases and the capacitance increases. Electronic circuitry is used to detect the change in capacitance and thus to detect that the key has been pressed. Let the area of each metal plate be 50.0 mm$^2$ and the separation between the plates be 0.600 mm before the key is depressed. If the circuitry can detect a change in capacitance of 0.250 pF, how far must the key be depressed before the circuitry detects its depression?

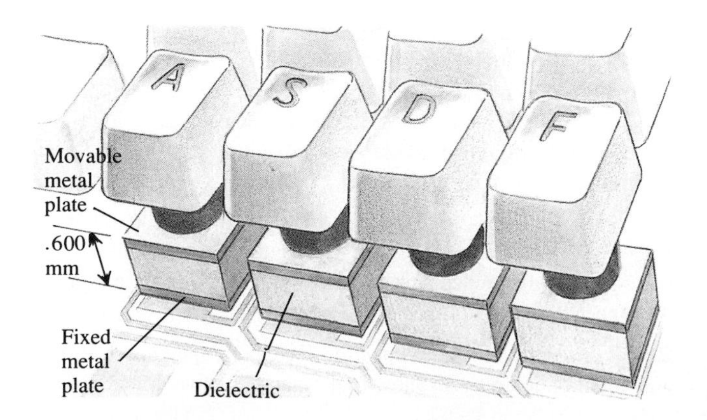


FIGURE 25–23

**25–23 An Electronic Photoflash Unit.** Electronic flash attachments for cameras contain a capacitor for storing the energy used to produce the flash. In one such unit the flash lasts for $\frac{1}{100}$ s with an average light power output of 600 W. a) If the conversion of electrical energy to light is 95% efficient (the rest of the energy goes to heat), how much energy must be stored in the capacitor for one flash? b) If the capacitance of the capacitor in the unit is 0.800 mF, what is the potential difference between its plates when the capacitor has stored in it the amount of energy calculated in part (a)?

**25–24** A parallel-plate air capacitor is made using two plates 0.200 m square, spaced 0.600 cm apart. It is connected to a 50.0-V battery. a) What is the capacitance? b) What is the charge on each plate? c) What is the electric field between the plates? d) What is the energy stored in the capacitor? e) If the battery is disconnected and then the plates are pulled apart to a separation of 1.20 cm, what are the answers to parts (a), (b), (c), and (d)?

**25–25** Suppose the battery in Problem 25–24 remains connected while the plates are pulled apart. What are the answers to parts (a), (b), (c), and (d) after the plates have been pulled apart?

**25–26** Several 0.500-$\mu$F capacitors are available. The voltage across each must not exceed 350 V. You need to make a capacitor with capacitance 0.500 $\mu$F to be connected across a potential difference of 600 V. a) Show in a diagram how an equivalent capacitor having the desired properties can be obtained. b) No dielectric is a perfect insulator, with infinite resistance. Suppose the dielectric in one of the capacitors in your diagram is a moderately good conductor. What will happen?

**25–27** In Fig. 25–24, $C_1 = C_5 = 3.00$ $\mu$F and $C_2 = C_3 = C_4 = 2.00$ $\mu$F. The applied potential is $V_{ab} = 600$ V. a) What is the equivalent capacitance of the network between points $a$ and $b$? b) Calculate the charge on each capacitor and the potential difference across each capacitor.

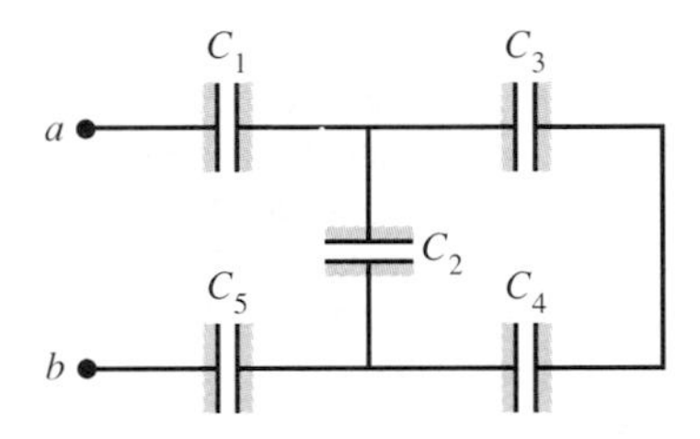


**FIGURE 25–24**

**25–28** A 2.00-$\mu$F capacitor and a 3.00-$\mu$F capacitor are connected in series across an 800-V supply line. a) Find the charge on each capacitor and the voltage across each. b) The charged capacitors are disconnected from the line and from each other and then reconnected, with terminals of like sign together. Find the final charge on each and the voltage across each.

**25–29** In Fig. 25–25, each capacitance $C_1$ is 9.00 $\mu$F, and each capacitance $C_2$ is 6.00 $\mu$F. a) Compute the equivalent capacitance of the network between points $a$ and $b$. b) Compute the charge on each of three capacitors nearest $a$ and $b$ when $V_{ab} = 900$ V. c) With 900 V across $a$ and $b$, compute $V_{cd}$.

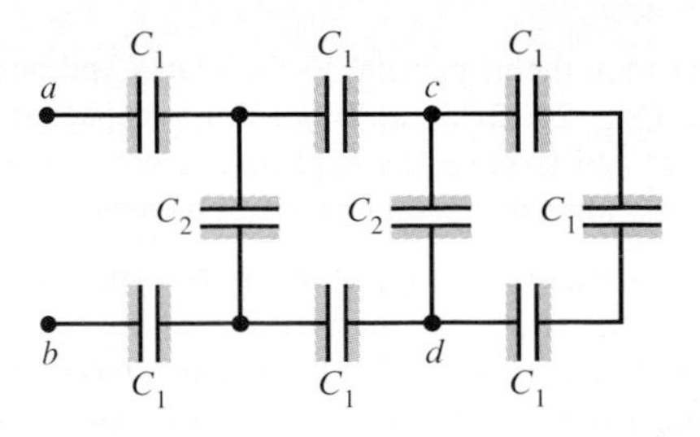


**FIGURE 25–25**

**25–30** In Fig. 25–6a, let $C_1 = 6.00$ $\mu$F, $C_2 = 3.00$ $\mu$F, and $V_{ab} = 36.0$ V. Suppose the charged capacitors are disconnected from the source and from each other and then reconnected with plates of *opposite* sign together. By how much does the energy of the system decrease?

**25–31** Three capacitors having capacitances of 8.00, 8.00, and 4.00 $\mu$F are connected in series across a 36.0-V line. a) What is the charge on the 4.00-$\mu$F capacitor? b) What is the total energy of all three capacitors? c) The capacitors are disconnected from the line and reconnected in parallel with each other, with the positively charged plates connected together. What is the voltage across the parallel combination? d) What is the total energy now stored in the capacitors?

**25–32** The capacitors in Fig. 25–26 are initially uncharged and are connected as in the diagram with switch $S$ open. The applied potential difference is $V_{ab} = +400$ V. a) What is the potential difference $V_{cd}$? b) What is the potential difference across each capacitor after switch $S$ is closed? c) How much charge flowed through the switch when it was closed?

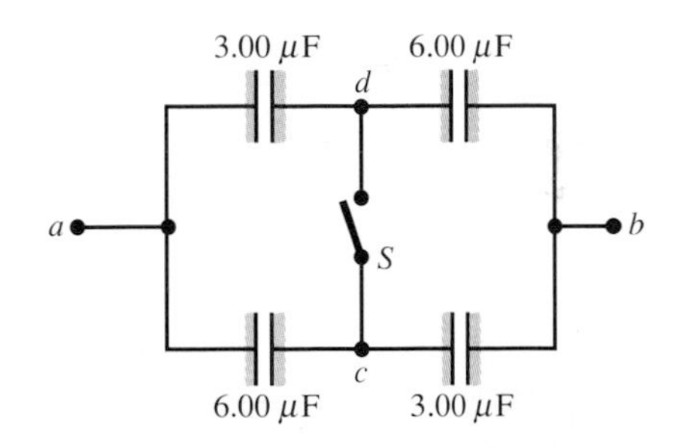


**FIGURE 25–26**

**25–33 Potential Difference across a Cell Wall.** Some cell walls in the human body have a double layer of surface charge, with a layer of negative charge inside and a layer of positive charge of equal magnitude on the outside. Consider a model for such a cell in which the surface charge densities are $\pm 0.50 \times 10^{-3}$ C/m$^2$ and the cell wall is $5.0 \times 10^{-9}$ m thick. Assume that the cell-wall material has a dielectric constant of $K = 6.00$. a) Find the electric-field magnitude in the wall between the two charge layers. b) Find the potential difference between inside and outside the cell. Which is at higher potential?

**25–34** An air capacitor is made using two flat plates, each with area $A$, separated by a distance $d$. Then a metal slab having thickness $a$ (less than $d$) and the same shape and size as the plates is inserted between them, parallel to the plates and not touching either plate (Fig. 25–27). a) What is the capacitance of this arrangement? b) Express the capacitance as a multiple of the capacitance $C_0$ when the metal slab is not present.

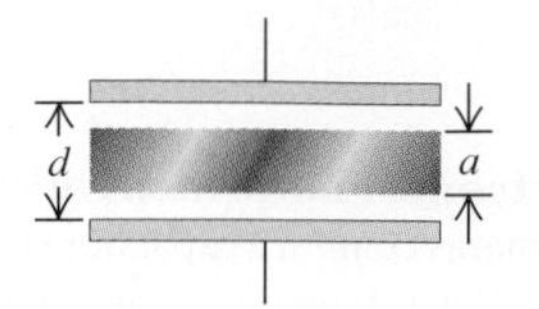


FIGURE 25–27

**25–35** A parallel-plate capacitor has the space between the plates filled with two slabs of dielectric, one with constant $K_1$ and one with constant $K_2$ (Fig. 25–28). Each slab has thickness $d/2$, where $d$ is the plate separation. Show that the capacitance is

$$C = \frac{2\epsilon_0 A}{d}\left(\frac{K_1 K_2}{K_1 + K_2}\right).$$

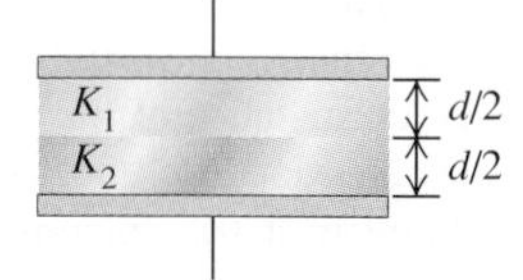

FIGURE 25–28

## CHALLENGE PROBLEMS

**25–36 Measuring Fuel in a Tank.** A fuel gauge uses a capacitor to determine the height of the fuel in a tank. The effective dielectric constant $K_{\text{eff}}$ changes from a value of 1 when the tank is empty to a value of $K$, the dielectric constant of the fuel, when the tank is full. The appropriate electronic circuitry can determine the effective dielectric constant of the combined air and fuel between the capacitor plates. Each of the two rectangular plates has a width $w$ and a length $L$ (Fig. 25–29). The height of the fuel between the plates is $h$. Neglect any fringing effects. a) Derive an expression for $K_{\text{eff}}$ as a function of $h$. b) What is the effective dielectric constant for a tank one-quarter full, one-half full, and three-quarters full if the fuel is gasoline ($K = 1.95$)? c) Repeat part (b) for methanol ($K = 33.0$). d) For which fuel is this fuel gauge more practical?

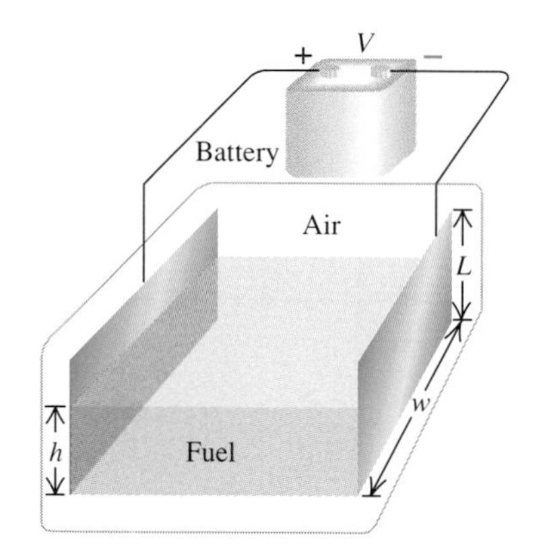


FIGURE 25–29

**25–37** Three square metal plates $A$, $B$, and $C$, each 8.00 cm on a side and 3.00 mm thick, are arranged as in Fig. 25–30. The plates are separated by sheets of paper 0.500 mm thick and with dielectric constant 4.00. The outer plates are connected together and connected to point $b$. The inner plate is connected to point $a$. a) Copy the diagram and show by plus and minus signs the charge distribution on the plates when point $a$ is maintained at a positive potential relative to point $b$. b) What is the capacitance between points $a$ and $b$?

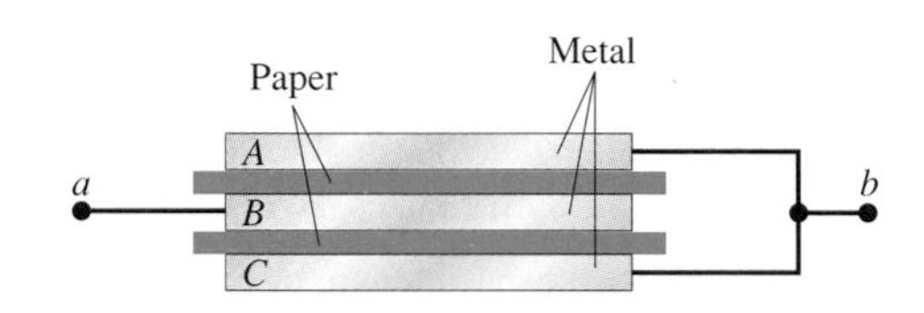


FIGURE 25–30

**25–38 A Capacitor on Springs.** A parallel-plate capacitor consists of two horizontal conducting plates of equal area $A$. The bottom plate is resting on a fixed support, and the top plate is suspended by four springs with spring constant $k$, positioned at each of the four corners of the top plate (Fig. 25–31). The plates, when uncharged, are separated by a distance $z_0$. A battery is connected to the plates and produces a potential difference $V$ between them. This causes the plate separation to decrease to $z$. Neglect any fringing effects. a) Show that the electrostatic force between the charged plates has a magnitude $\epsilon_0 AV^2/2z^2$. (*Note:* See Exercise 25–15.) b) Obtain an expression that relates the plate separation $z$ to the potential difference $V$ between the plates. The resulting equation will be cubic in $z$. c) Given the values $A = 0.250\ \text{m}^2$, $z_0 = 1.00\ \text{mm}$, $k = 25.0\ \text{N/m}$, and $V = 100\ \text{V}$, find the two values of $z$ for which the top plate will be in equilibrium. (*Hint:* You can solve the cubic equation that results by plugging a trial value of $z$ into the equation and then adjusting your guess until the equation

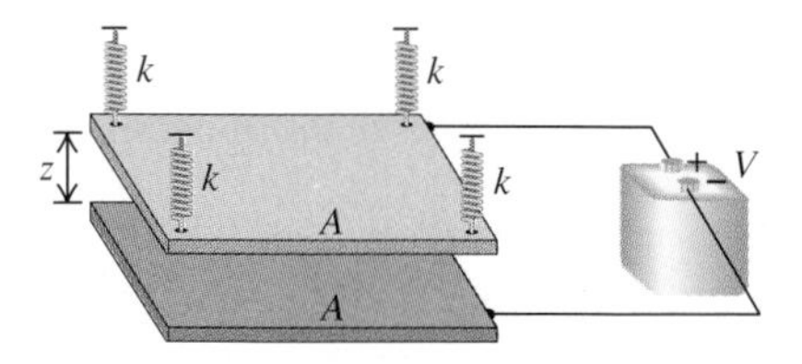

FIGURE 25–31

is satisfied to three significant figures. Locating the roots of the cubic equation graphically can help you in picking starting values of $z$ for this trial-and-error procedure. One root of the cubic equation has a nonphysical negative value.) d) For each of the two values of $z$ found in part (c), is the equilibrium stable or unstable? For stable equilibrium a small displacement of the object in question will give rise to a net force tending to return the object to the equilibrium position. For unstable equilibrium a small displacement gives rise to a net force that takes the object farther away from equilibrium.

**25–39** It is not always possible to combine capacitors in a simple series or parallel relationship. Consider the capacitors $C_x$, $C_y$, and $C_z$ in the network depicted in Fig. 25–32a. Such a configuration of capacitors, referred to as a delta network, cannot be transformed into a single equivalent capacitor because three terminals $a$, $b$, and $c$ exist in that network. It can be shown that as far as any effect on the external circuit is concerned, a delta network can be transformed into what is called a Y network. For example, the delta network of Fig. 25–32a can be replaced by the Y network of Fig. 25–32b. a) Show that the transformation equations that give $C_1$, $C_2$, and $C_3$ in terms of $C_x$, $C_y$, and $C_z$ are

$$C_1 = (C_xC_y + C_yC_z + C_zC_x)/C_x,$$
$$C_2 = (C_xC_y + C_yC_z + C_zC_x)/C_y,$$
$$C_3 = (C_xC_y + C_yC_z + C_zC_x)/C_z.$$

[*Hint:* The potential difference $V_{ac}$ must be the same in both circuits, as $V_{bc}$ must be. Furthermore, the charge $q_1$ that flows from point $a$ along the wire as indicated must be the same in both circuits, as must $q_2$. Obtain a relationship for $V_{ac}$ as a function of $q_1$ and $q_2$ (and the capacitances) for each network and a separate relationship for $V_{bc}$ as a function of the charges for each network. The coefficients of corresponding charges in corresponding equations must be the same for both networks.] b) Determine the equivalent capacitance of the network of capacitors between the terminals at the left end of the network for the network shown in Fig. 25–32c. [*Hint:* Use the delta-Y transformation derived in part (a). Use points $a$, $b$, and $c$ to form the delta, and transform the delta into a Y. The capacitors can then be easily combined by using the relationships for series and parallel combinations of capacitors.] c) Determine the charges of the 72.0-μF, 27.0-μF, 18.0-μF, 6.0-μF, 28.0-μF, and 21.0-μF capacitors and the potential differences across those capacitors.

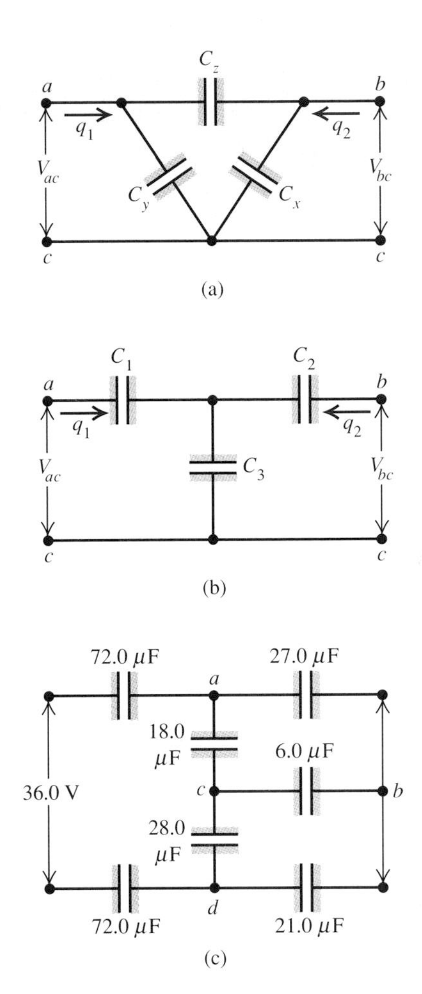


**FIGURE 25–32**

Modern personal computers can process a variety of visual information, including print, photographs, and drawings. In fact, the chapter openings of this book were put together on the computer terminal shown here.

CHAPTER 26

# Current, Resistance and Electromotive Force

Each binary digit makes up a bit. Eight bits, such as those that encode the letter A, comprise one byte. The file for this page alone registers 408,000 bytes of data.

The key to processing different kinds of data is to represent them in digital form, so that each dot of the image, and even the color of that dot, can be represented as a series of zeros and ones. In electronic terms, a pulse of current that doesn't change in a certain time interval can represent a "one." A current pulse that does change in the time interval can represent a "zero." Thus the computer's circuitry can manipulate data by manipulating pulses of current.

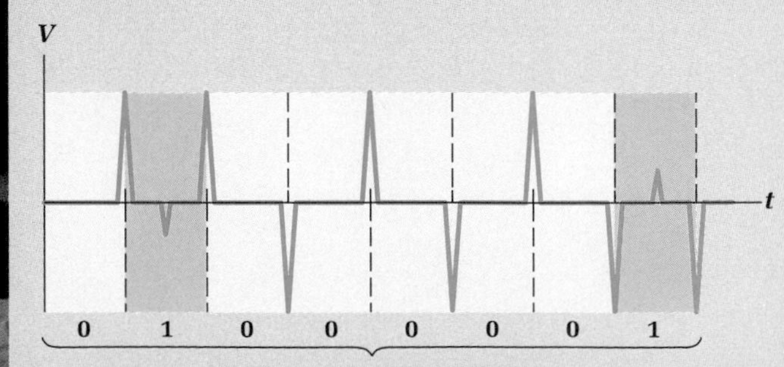

The letter A in binary code

## KEY CONCEPTS

**• Current is rate of flow of electric charge from one region to another. In a conductor, current depends on the drift velocity of the charged particles, their concentration, and their charges. Current density is current per unit area.**

**• In a material that obeys Ohm's law the ratio of electric field to current density is a constant, called resistivity, that depends on temperature. For a specific device that obeys Ohm's law, the ratio of potential difference to current is a constant called resistance, which also depends on temperature.**

**• A complete circuit carrying a steady current must include a source of electromotive force (emf), such as a battery or generator, that delivers energy to the circuit and in which charges move from regions of lower to higher potential energy. The sum of the potential differences and emf's around a closed circuit must be zero.**

**• The power input or output to any circuit device is the product of the current and the terminal potential difference.**

## INTRODUCTION

Electric circuits are at the heart of all radio and television transmitters and receivers, CD players, household and industrial power distribution systems, flashlights, computers, and the nervous systems of animals. In the past four chapters we have studied the interactions of electric charges *at rest;* now we're ready to study charges *in motion*. An *electric current* consists of motion of charge from one region to another. When this motion takes place within a conductor that forms a closed path, the path is called an electric circuit.

In this chapter we will study the basic properties of conductors and how they depend on temperature. We'll learn why a short, fat, cold copper wire is a better conductor than a long, skinny, hot steel wire. We'll study the properties of batteries and how they cause current and energy transfer in a circuit. In this analysis we will use the concepts of current, potential difference (or "voltage"), resistance, and electromotive force. Finally, we'll look at electric current in a material from a microscopic viewpoint.

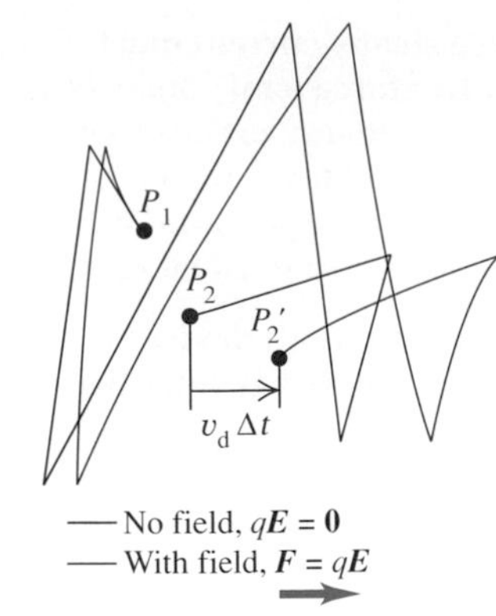


**26–1** The electric-field force imposes a small drift (greatly exaggerated here) on an electron's random motion. Without the field the electron's motion from point $P_1$ brings it to point $P_2$ after a time $\Delta t$. With an applied field the electron ends up at point $P_2'$, a distance $v_d \Delta t$ from $P_2$ in the direction of the field.

## 26–1 CURRENT

A **current** is any motion of charge from one region to another. In this section we'll discuss currents in conducting materials. In the *electrostatic* situations that we discussed in the past four chapters, there are *no* charges in motion, and the electric field everywhere within a conductor is zero. Now we're ready to let the charges move. To maintain a steady flow of charge in a conductor, we have to maintain a steady force on the mobile charges in the conductor, either with an electrostatic field or by other means that we'll consider later. For now, let's assume that there is an electric field $\boldsymbol{E}$ within the conductor, so a particle with charge $q$ experiences a force $\boldsymbol{F} = q\boldsymbol{E}$.

When a charged particle such as an electron moves in an electric field *in vacuum,* it accelerates continuously. However, the motion of an electron inside a conducting material such as a metal is very different because of frequent collisions with the atoms of the material. In a metal the free electrons have a lot of random motion, somewhat like the molecules in a gas but with much greater speeds, of the order of $10^6$ m/s.

When an electric field is applied to the metal, the forces that it exerts on the electrons lead to a small net motion or *drift* in the direction of the force, in addition to this random motion (Fig. 26–1). The electric field does work on the moving charges, and the resulting kinetic energy is transferred to the material by means of inelastic collisions with the ion cores, which vibrate about their equilibrium positions in a crystal lattice. This energy transfer increases their average vibrational energy and therefore the temperature of the material. Thus the motion of the charged particles consists of random motion with very large average speed and a much slower drift speed (often of the order of $10^{-4}$ m/s) in the direction of the electric-field force.

Current is defined to be the amount of charge transferred per unit time. In metals, the moving charges are always (negative) electrons, but in an ionized gas (plasma) or an ionic solution, both electrons and positively charged ions are moving. In a semiconductor material such as germanium or silicon, conduction is partly by electrons and partly by motion of *vacancies,* also known as *holes;* these are sites of missing electrons and act like positive charges.

Figure 26–2 shows a cross section of a conductor in which *positive* charges are moving. We define the current through the area of this cross section as *the net charge flowing through the area per unit time*. Thus, if a net charge $dQ$ flows through an area in

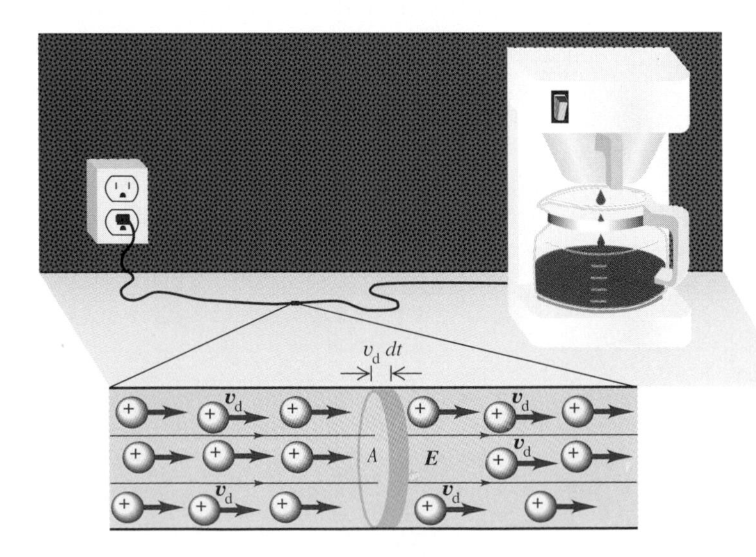


**26–2** The current through the cross-section area $A$ is the time rate of charge transfer through $A$. The random motion of each positive particle averages to zero, and the drift velocity is parallel to $\boldsymbol{E}$.

a time $dt$, the current through the area, denoted by $I$, is

$$I = \frac{dQ}{dt}. \tag{26–1}$$

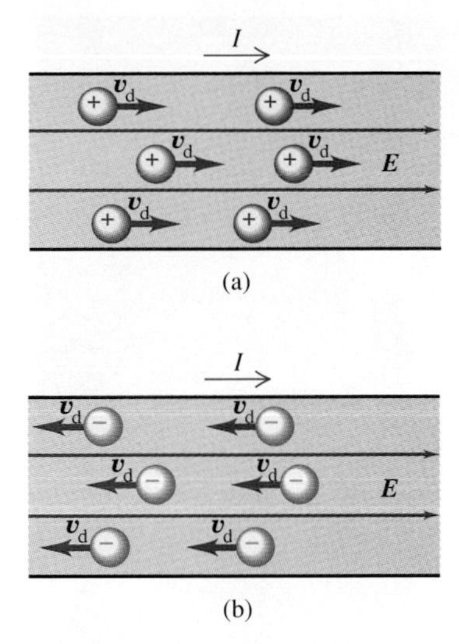


**26–3** Positive charges moving in the direction of the field produce the same positive current as negative charges moving in the direction opposite to the field.

Figure 26–3 compares the motion of positive and negative charges. In Fig. 26–3a the moving charges are positive and move in the direction of the electric field. In Fig. 26–3b they are negative and move oppositely to the electric field. In both cases the result is a net transfer of positive charge from left to right.

Current is a *scalar* quantity. The SI unit of current is the **ampere;** one ampere is defined to be *one coulomb per second* (1 A = 1 C/s). This unit is named in honor of the French scientist André Marie Ampère (1775–1836). The current in an ordinary flashlight (D-cell size) is about 0.5 to 1 A; the current in the starter motor of a car engine is around 200 A. Currents in radio and television circuits are usually expressed in *milliamperes* (1 mA $= 10^{-3}$ A) or *microamperes* (1 $\mu$A $= 10^{-6}$ A), and currents in computer circuits are expressed in *nanoamperes* (1 nA $= 10^{-9}$ A) or *picoamperes* (1 pA $= 10^{-12}$ A).

We can express the current through an area in terms of the **drift velocity** $\boldsymbol{v}_d$ of the moving charges. Let's consider again the situation of Fig. 26–2, a conductor with cross-section area $A$ and an electric field $\boldsymbol{E}$ directed from left to right. We'll discuss the case in which the free charges in the conductor are positive; then the electric force is in the same direction as the field. We'll also assume that the material of the conductor is homogeneous and isotropic, that is, that it has the same properties at all points and that all directions are equivalent.

Suppose there are $n$ charged particles per unit volume. We call $n$ the **concentration** of particles; its SI unit is $\mathrm{m}^{-3}$. Assume that all the particles move with a drift velocity with magnitude $v_d$. In a time $dt$, each particle moves a distance $v_d\,dt$. The particles that flow out of the right end of the shaded cylinder with length $v_d\,dt$ during $dt$ are the particles that were within this cylinder at the beginning of $dt$. The volume of the cylinder is $Av_d\,dt$, and the number of particles within it is $nAv_d\,dt$. If each particle has a charge $q$, the charge $dQ$ flowing out of the end of the cylinder in time $dt$ is

$$dQ = q\,(nAv_d\,dt) = nqv_dA\,dt,$$

and the current is

$$I = \frac{dQ}{dt} = nqv_dA. \tag{26–2}$$

The current *per unit cross-section area* is called the **current density** $J$. From Eq. (26–2),

$$J = \frac{I}{A} = nqv_d. \tag{26–3}$$

If the moving charges are negative rather than positive, the electric force is opposite to $\boldsymbol{E}$, and the drift velocity is right to left, as in Fig. 26–3b. But the *current* is still left to right; negative charge moving right to left and positive charge moving left to right would both increase the positive charge at the right of the section. In either case, particles flowing out at an end of the cylindrical section are continuously replaced by particles flowing *in* at the opposite end.

In general, a conductor may contain several different kinds of charged particles having charges $q_1, q_2, \ldots$, concentrations $n_1, n_2, \ldots$, and drift velocities with magnitudes

**26–4** A simple laboratory demonstration of electrolytic conductivity. The sodium chloride solution in the beaker is part of the circuit, and the bulb glows brightly. The current is carried by both positive and negative charges.

$v_{d1}, v_{d2}, \ldots$ . An example is conduction in an ionic solution (Fig. 26–4). The total current is then

$$I = A(n_1 q_1 v_{d1} + n_2 q_2 v_{d2} + \cdots). \tag{26–4}$$

When more than one kind of charge is moving, the total current density $J = I/A$ is given by

$$J = n_1 q_1 v_{d1} + n_2 q_2 v_{d2} + \cdots. \tag{26–5}$$

We can also define a vector current density $\boldsymbol{J}$ that includes the directions of the drift velocities:

$$\boldsymbol{J} = n_1 q_1 \boldsymbol{v}_{d1} + n_2 q_2 \boldsymbol{v}_{d2} + \cdots. \tag{26–6}$$

The direction of the drift velocity $\boldsymbol{v}_d$ is the same as that of the electric field $\boldsymbol{E}$ for a positive charge and opposite to $\boldsymbol{E}$ for a negative charge. In both cases the product $q\boldsymbol{v}_d$ is in the direction of $\boldsymbol{E}$. *The vector current density* $\boldsymbol{J}$ *always has the same direction as the field* $\boldsymbol{E}$*,* even in a metallic conductor, in which the moving charges are (negative) electrons with $\boldsymbol{v}_d$ in the direction *opposite* to $\boldsymbol{E}$.

In circuit analysis we will always describe currents *as though* they consisted entirely of positive charge flow, even in cases in which we know the actual current is due to electrons. In Chapter 28, when we study the effect of a *magnetic* field on a moving charge, we will consider a phenomenon called the Hall effect, in which the sign of the moving charges *is* important.

When there is a steady current in a closed loop (a "complete circuit"), the total charge in every segment of the conductor is constant. From the principle of conservation of charge (Section 22–3) the rate of flow of charge *out* at one end of a segment at any instant equals the rate of flow of charge *in* at the other end of the segment, and *the current is the same at all cross sections.* Current is *not* something that squirts out of the positive terminal of a battery and is consumed or used up by the time it reaches the negative terminal.

## EXAMPLE 26–1

An 18-gauge copper wire (the size usually used for lamp cords) has a nominal diameter of 1.02 mm. This wire carries a constant current of 1.67 A to a 200-W lamp. The density of free electrons is $8.5 \times 10^{28}$ electrons per cubic meter. Find the magnitudes of a) the current density and b) the drift velocity.

**SOLUTION** a) The cross-section area is

$$A = \frac{\pi d^2}{4} = \frac{\pi(1.02 \times 10^{-3}\text{ m})^2}{4}$$
$$= 8.2 \times 10^{-7}\text{ m}^2.$$

The magnitude of the current density is

$$J = \frac{I}{A} = \frac{1.67\text{ A}}{8.2 \times 10^{-7}\text{ m}^2}$$
$$= 2.0 \times 10^6\text{ A/m}^2.$$

b) Solving Eq. (26–3) for the drift velocity magnitude $v_d$, we find

$$v_d = \frac{J}{nq} = \frac{2.0 \times 10^6\text{ A/m}^2}{(8.5 \times 10^{28}\text{ m}^{-3})(1.6 \times 10^{-19}\text{ C})}$$
$$= 1.5 \times 10^{-4}\text{ m/s} \quad \text{(about 0.15 mm/s)}.$$

At this speed an electron would require 6700 s, or about 1 hr 50 min, to travel the length of a wire 1 m long. The speeds of random motion of the electrons are of the order of $10^6$ m/s. So in this example the drift speed is around $10^{10}$ times slower than the speed of random motion. Picture the electrons as bouncing around frantically, with a very slow and sluggish drift!

Since the electrons move this slowly, you may wonder why the light comes on right away when you turn on the switch. The reason is that the electric field is set up in the wire with a speed approaching the speed of light, and electrons start to move all along the wire at very nearly the same time. The time it takes for any individual electron to get from the switch to the light bulb isn't really relevant.

## 26–2 RESISTIVITY

The current density $\boldsymbol{J}$ in a conductor depends on the electric field $\boldsymbol{E}$ and on the properties of the material. In general this dependence can be quite complex. For some materials, especially metals, $\boldsymbol{J}$ is nearly *directly proportional* to $\boldsymbol{E}$; in that case the ratio of the magnitudes $E$ and $J$ is *constant*. In the following discussion we will assume that this simple proportion is valid, even though there are many situations in which it is not. This assumption is comparable to our representation of the behavior of friction forces as a direct proportion of friction and normal forces even though we knew that this was at best an approximate description.

We define the **resistivity** $\rho$ of a material as the ratio of the magnitudes of electric field and current density:

$$\rho = \frac{E}{J}. \tag{26–7}$$

The greater the resistivity, the greater the field needed to cause a given current density, or the smaller the current density caused by a given field. From Eq. (26–7) the units of $\rho$ are $(\text{V/m})/(\text{A/m}^2) = \text{V} \cdot \text{m/A}$. As we will discuss in the next section, 1 V/A is called one ohm (1 $\Omega$), so alternative units for $\rho$ are $\Omega \cdot \text{m}$ (ohm-meters). Representative values of resistivity are given in Table 26–1. A "perfect" conductor would have zero resistivity, and a "perfect" insulator would have an infinite resistivity. Metals and alloys have the smallest resistivities and are the best conductors. The resistivities of insulators are greater than those of the metals by an enormous factor, of the order of $10^{22}$.

Comparing Table 26–1 with Table 15–6 (Thermal Conductivities), we note that good electrical conductors, such as metals, are usually also good conductors of heat. Poor electrical conductors, such as ceramic and plastic materials, are also poor thermal conductors. In a metal the free electrons that carry charge in electrical conduction also provide the principal mechanism for heat conduction, so we should expect a correlation between electrical and thermal conductivity. Because of the enormous difference between resistivities of electrical conductors and insulators, it is easy to confine electric currents to well-defined paths or circuits. The variation in *thermal* conductivities is much

**TABLE 26–1 Resistivities at Room Temperature**

| | Substance | $\rho\ (\Omega \cdot \text{m})$ | Substance | $\rho\ (\Omega \cdot \text{m})$ |
|---|---|---|---|---|
| Conductors | | | Semiconductors | |
| Metals | Silver | $1.47 \times 10^{-8}$ | Pure Carbon | $3.5 \times 10^{-5}$ |
| | Copper | $1.72 \times 10^{-8}$ | Pure Germanium | 0.60 |
| | Gold | $2.44 \times 10^{-8}$ | Pure Silicon | 2300 |
| | Aluminum | $2.63 \times 10^{-8}$ | Insulators | |
| | Tungsten | $5.51 \times 10^{-8}$ | Amber | $5 \times 10^{14}$ |
| | Steel | $20 \times 10^{-8}$ | Glass | $10^{10} - 10^{14}$ |
| | Lead | $22 \times 10^{-8}$ | Lucite | $> 10^{13}$ |
| | Mercury | $95 \times 10^{-8}$ | Mica | $10^{11} - 10^{15}$ |
| Alloys | Manganin (Cu 84, Mn 12, Ni 4) | $44 \times 10^{-8}$ | Quartz (fused) | $75 \times 10^{16}$ |
| | Constantan (Cu 60, Ni 40) | $49 \times 10^{-8}$ | Sulfur | $10^{15}$ |
| | Nichrome | $100 \times 10^{-8}$ | Teflon | $> 10^{13}$ |
| | | | Wood | $10^{8} - 10^{11}$ |

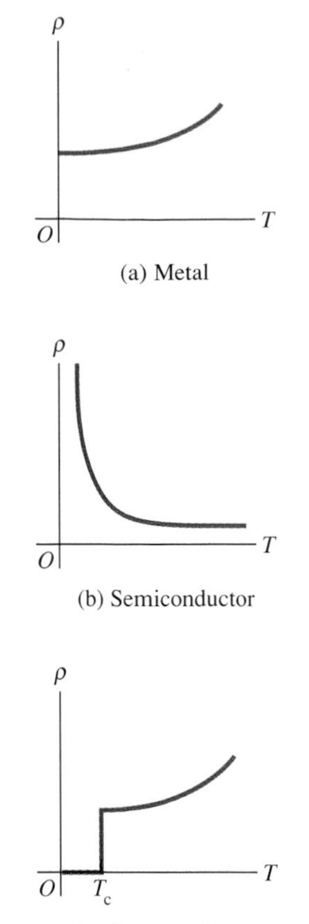


**26–5** Variation of resistivity with absolute temperature for (a) a normal metal; (b) a semiconductor; (c) a superconductor.

less, only a factor of $10^3$ or so, and it is usually impossible to confine heat currents to that extent.

*Semiconductors* have resistivities that are intermediate between those of metals and those of insulators. These materials are important not primarily because of their resistivities, but because of the way their resistivities are affected by temperature and by small amounts of impurities.

The proportionality of $J$ to $E$ for a metallic conductor at constant temperature was discovered by G. S. Ohm (1787–1854) and is called **Ohm's law.** A material that obeys Ohm's law reasonably well is called an *ohmic* conductor, or a *linear* conductor. Many materials show substantial departures from Ohm's-law behavior; they are *nonohmic,* or *nonlinear.* Ohm's law, like the ideal gas equation and Hooke's law, is an *idealized model* that describes the behavior of some materials quite well but is not a general description of *all* matter.

The reciprocal of resistivity is **conductivity.** Its units are $(\Omega \cdot \text{m})^{-1}$. Good conductors of electricity have larger conductivities than insulators do. Conductivity is the direct electrical analog of thermal conductivity.

Analogies with fluid flow can be a big help in developing intuition about electric current and circuits. For example, in wine making, the product is sometimes filtered to remove sediments. A pump forces the wine through the filter under pressure; if the flow rate is proportional to the pressure difference between the upstream and downstream sides, the behavior is analogous to Ohm's law. Maple syrup is also usually pressure-filtered.

The resistivity of a *metallic* conductor nearly always increases with increasing temperature (Fig. 26–5a). Over a small temperature range (up to 100 C° or so), the resistivity of a metal can be represented approximately by the equation

$$\rho_T = \rho_0[1 + \alpha(T - T_0)], \tag{26–8}$$

where $\rho_0$ is the resistivity at a reference temperature $T_0$ (often taken as 0°C or 20°C) and $\rho_T$ is the resistivity at temperature $T$. The factor $\alpha$ is called the **temperature coefficient of resistivity.** Some representative values are given in Table 26–2. The resistivity of carbon (a nonmetal) *decreases* with increasing temperature, and its temperature coefficient of resistivity is negative. The resistivity of the alloy manganin is practically independent of temperature.

The resistivity of a *semiconductor* decreases rapidly with increasing temperature (Fig. 26–5b). A small semiconductor crystal called a *thermistor* can be used to make a sensitive electronic thermometer. Its resistivity is used as a thermometric property. Other materials, such as platinum and carbon, are also used for electronic thermometers.

Some materials, including several metallic alloys and oxides, show a phenomenon called *superconductivity.* As the temperature decreases, the resistivity at first decreases

**TABLE 26–2 Temperature Coefficients of Resistivity (Approximate Values near Room Temperature)**

| Material | $\alpha$ [(C°)$^{-1}$] | Material | $\alpha$ [(C°)$^{-1}$] |
|---|---|---|---|
| Aluminum | 0.0039 | Lead | 0.0043 |
| Brass | 0.0020 | Manganin | 0.000000 |
| Carbon | −0.0005 | Mercury | 0.00088 |
| Constantan | 0.000002 | Nichrome | 0.0004 |
| Copper | 0.0039 | Silver | 0.0038 |
| Iron | 0.0050 | Tungsten | 0.0045 |

smoothly, like that of any metal. But then at a certain critical transition temperature $T_c$ a phase transition occurs, and the resistivity suddenly drops to zero, as shown in Fig. 26–5c. Once a current has been established in a superconducting ring, it continues indefinitely without the presence of any driving field.

Superconductivity was discovered in 1911 in the laboratory of H. Kamerlingh-Onnes. He had just discovered how to liquefy helium, which has a boiling temperature of 4.2 K at atmospheric pressure. Measurements of the resistivity of mercury at very low temperatures showed that below 4.2 K its resistance suddenly dropped to zero. For the next 75 years the highest $T_c$ attained was about 20 K. This meant that superconductivity occurred only when the material was cooled using (expensive and somewhat rare) liquid helium or (explosive) liquid hydrogen. But in January 1986 Karl Muller and Johannes Bednorz discovered an oxide of barium, lanthanum, and copper with a $T_c$ of nearly 40 K, and the race was on to develop "high-temperature" superconducting materials.

By 1987 a complex oxide of yttrium, copper, and barium had been found that has a $T_c$ well above the 77-K boiling temperature of (inexpensive and safe) liquid nitrogen. The current (1991) record for $T_c$ is about 125 K, and materials that are superconductors at room temperature may well become a reality. The implications of these discoveries for power-distribution systems, computer design, and transportation are enormous. Meanwhile, superconducting electromagnets cooled by liquid helium are used in particle accelerators and some experimental magnetic-levitation trains.

## 26–3 RESISTANCE

For a conductor that obeys Ohm's law, with resistivity $\rho$, the current density $\boldsymbol{J}$ at a point where the electric field is $\boldsymbol{E}$ is given by Eq. (26–7), which we can write as

$$\boldsymbol{E} = \rho\boldsymbol{J}. \qquad (26\text{–}9)$$

When Ohm's law is obeyed, $\rho$ is constant and $\boldsymbol{E}$ is directly proportional to $\boldsymbol{J}$.

Often we are more interested in the total current $I$ in a conductor and the potential difference $V$ between its ends. For example, suppose our conductor is a wire with uniform cross-section area $A$ and length $L$, as shown in Fig. 26–6. If the magnitudes of the current density $\boldsymbol{J}$ and the electric field $\boldsymbol{E}$ are uniform throughout the conductor, the total current $I$ is given by $I = JA$, and the potential difference $V$ between the ends is $V = EL$. When we solve these equations for $J$ and $E$, respectively, and substitute the results in Eq. (26–9), we obtain

$$\frac{V}{L} = \frac{\rho I}{A} \quad \text{or} \quad V = \frac{\rho L}{A}I. \qquad (26\text{–}10)$$

This shows that when $\rho$ is constant, the total current $I$ is proportional to the potential difference $V$.

The ratio of $V$ to $I$ for a particular conductor is called its **resistance** $R$:

$$R = \frac{V}{I}. \qquad (26\text{–}11)$$

Comparing this definition of $R$ to Eq. (26–10), we see that the resistance $R$ of a particular conductor is related to the resistivity $\rho$ of its material by

$$R = \frac{\rho L}{A}. \qquad (26\text{–}12)$$

If $\rho$ is constant, then so is $R$.

**26–6** A homogeneous conductor with uniform cross section. The current density is uniform over any cross section, and the electric field is constant along the length.

The equation

$$V = IR \tag{26–13}$$

is often called Ohm's law, but it is important to understand that the real content of Ohm's law is the direct proportionality (for some materials) of $V$ to $I$ or of $J$ to $E$. Equation (26–11) or (26–13) *defines* resistance $R$ for *any* conductor, whether or not it obeys Ohm's law, but only when $R$ is constant can we correctly call this relationship Ohm's law.

Equation (26–12) shows that the resistance of a wire or other conductor of uniform cross section is directly proportional to its length and inversely proportional to its cross-section area. It is also proportional to the resistivity of the material of which the conductor is made.

The flowing-fluid analogy is again useful. In analogy to Eq. (26–12) a thin water hose offers more resistance to flow than a fat one, and a long hose more resistance than a short one. We can increase the resistance to flow by stuffing the hose with cotton or sand; this corresponds to increasing the resistivity. The flow rate is approximately proportional to the pressure difference between the ends. Flow rate is analogous to current and pressure difference to potential difference ("voltage"). Let's not stretch this analogy too far, though; the flow rate of water in a pipe is usually *not* proportional to its cross-section area, as Eq. (14–29) shows.

The SI unit of resistance is the **ohm,** equal to one *volt per ampere* ($1\,\Omega = 1$ V/A). The *kilohm* ($1\,\text{k}\Omega = 10^3\,\Omega$) and the *megohm* ($1\,\text{M}\Omega = 10^6\,\Omega$) are also in common use. A 100-W, 120-V light bulb has a resistance (at operating temperature) of 140 $\Omega$. A 100-m length of 12-gauge copper wire (the size usually used in household wiring) has a resistance at room temperature of about 0.5 $\Omega$. Resistors in the range 0.01 $\Omega$ to $10^7\,\Omega$ can be bought off the shelf. Because the resistivity of a material varies with temperature, the resistance of a specific conductor also varies with temperature. For temperature ranges that are not too great we can describe this variation as approximately a linear relation, analogous to Eq. (26–8):

$$R_T = R_0[1 + \alpha(T - T_0)]. \tag{26–14}$$

In this equation, $R_T$ is the resistance at temperature $T$ and $R_0$ is the resistance at temperature $T_0$, often taken to be 20°C or 0°C. The temperature coefficient of resistivity $\alpha$ is the same constant that appears in Eq. (26–8). Within the limits of validity of Eq. (26–14) the *change* in resistance resulting from a temperature change $T - T_0$ is given by $R_0\alpha(T - T_0)$.

## EXAMPLE 26–2

The 18-gauge copper wire in Example 26–1 (Section 26–1) has a diameter of 1.02 mm and a cross-section area of $8.2 \times 10^{-7}\,\text{m}^2$. It carries a current of 1.67 A, and the current density is $2.0 \times 10^6$ A/m$^2$. Find a) the electric field; b) the potential difference between two points 50 m apart; c) the resistance of a 50-m length of this copper conductor.

**SOLUTION** a) From Table 26–1 the resistivity of copper is $1.72 \times 10^{-8}\,\Omega\cdot\text{m}$. From Eq. (26–7) the electric field $E$ is

$$E = \rho J = (1.72 \times 10^{-8}\,\Omega\cdot\text{m})(2.0 \times 10^6\,\text{A/m}^2) = 0.034\,\text{V/m}.$$

b) The potential difference is given by

$$V = EL = (0.034\,\text{V/m})(50\,\text{m}) = 1.7\,\text{V}.$$

c) The resistance of a 50-m length of this wire is

$$R = \frac{V}{I} = \frac{1.7\,\text{V}}{1.67\,\text{A}} = 1.0\,\Omega.$$

We can also obtain this result directly from Eq. (26–12):

$$R = \frac{\rho L}{A} = \frac{(1.72 \times 10^{-8}\,\Omega\cdot\text{m})\,(50\,\text{m})}{8.2 \times 10^{-7}\,\text{m}^2} = 1.0\,\Omega.$$

## EXAMPLE 26–3

Suppose the resistance of the wire in Example 26–2 is 1.05 Ω at a temperature of 20°C. Find the resistance at 0°C and at 100°C.

**SOLUTION** We use Eq. (26–14), with $T_0 = 20°\text{C}$ and $R_0 = 1.05\ \Omega$. From Table 26–2 the temperature coefficient of resistivity of copper is $\alpha = 0.0039\ (\text{C}°)^{-1}$. At $T = 0°\text{C}$,

$$\begin{aligned} R &= R_0[1 + \alpha(T - T_0)] \\ &= (1.05\ \Omega)(1 + [0.0039\ (\text{C}°)^{-1}][0°\text{C} - 20°\text{C}]) \\ &= 0.97\ \Omega. \end{aligned}$$

At $T = 100°\text{C}$,

$$\begin{aligned} R &= (1.05\ \Omega)(1 + [0.0039\ (\text{C}°)^{-1}][100°\text{C} - 20°\text{C}]) \\ &= 1.38\ \Omega. \end{aligned}$$

## EXAMPLE 26–4

The hollow cylinder shown in Fig. 26–7 has length $L$ and inner and outer radii $a$ and $b$. It is made of a material with resistivity $\rho$. What is the resistance between the inner and outer surfaces of the cylinder? (Each surface is an equipotential surface.)

**SOLUTION** We can't use Eq. (26–12) directly because the cross section through which the charge travels varies from $2\pi aL$ at the inner surface to $2\pi bL$ at the outer surface. Instead, we consider a thin cylindrical shell of inner radius $r$ and thickness $dr$. The area $A$ is then $2\pi rL$, and the length of the current path through the shell is $dr$. The resistance $dR$ of this shell, between inner and outer surfaces, is that of a conductor with length $dr$ and area $2\pi rL$, that is,

$$dR = \frac{\rho dr}{2\pi rL}.$$

The current has to pass successively through all such shells between the inner and outer radii $a$ and $b$, so the total resistance is the sum of the resistances of these shells. If the area $2\pi rL$ were constant, we could just integrate $dr$ from $r = a$ to $r = b$ to get the total length of the current path. But as the current passes through one shell after another, with increasing $r$, the area increases, so we have to integrate the above expression for $dR$. The total resistance is thus given by

$$R = \frac{\rho}{2\pi L}\int_a^b \frac{dr}{r} = \frac{\rho}{2\pi L}\ln\frac{b}{a}.$$

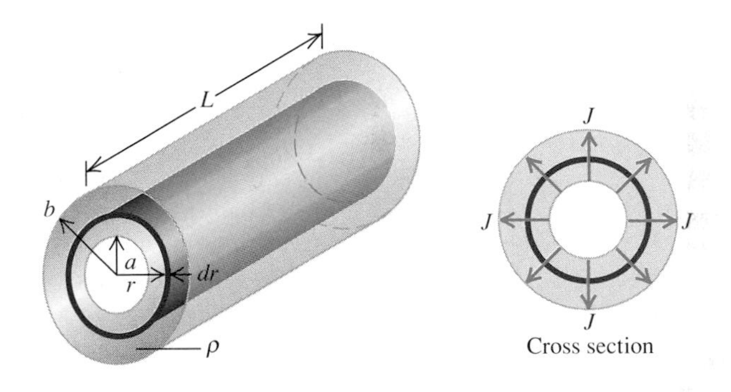


**26–7** Finding the resistance for radial current flow.

A circuit device made to have a specific value of resistance is called a **resistor.** Individual resistors used in electronic circuitry are often cylindrical in shape, a few millimeters in diameter and length, with wires coming out the ends. The resistance may be marked with a standard code using three or four color bands near one end (Fig. 26–8), according to the scheme shown in Table 26–3. The first two bands (starting with the band

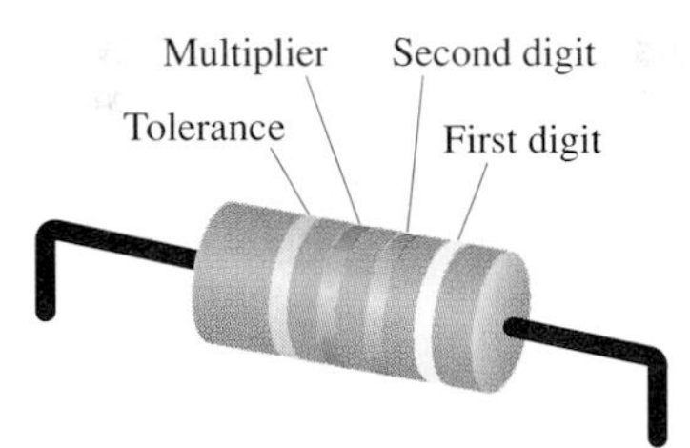


**26–8** This resistor has a value of 47 kΩ ± 10%.

**TABLE 26–3 Color Code for Resistors**

| Color | Value as digit | Value as multiplier |
|---|---|---|
| Black | 0 | 1 |
| Brown | 1 | 10 |
| Red | 2 | $10^2$ |
| Orange | 3 | $10^3$ |
| Yellow | 4 | $10^4$ |
| Green | 5 | $10^5$ |
| Blue | 6 | $10^6$ |
| Violet | 7 | $10^7$ |
| Gray | 8 | $10^8$ |
| White | 9 | $10^9$ |

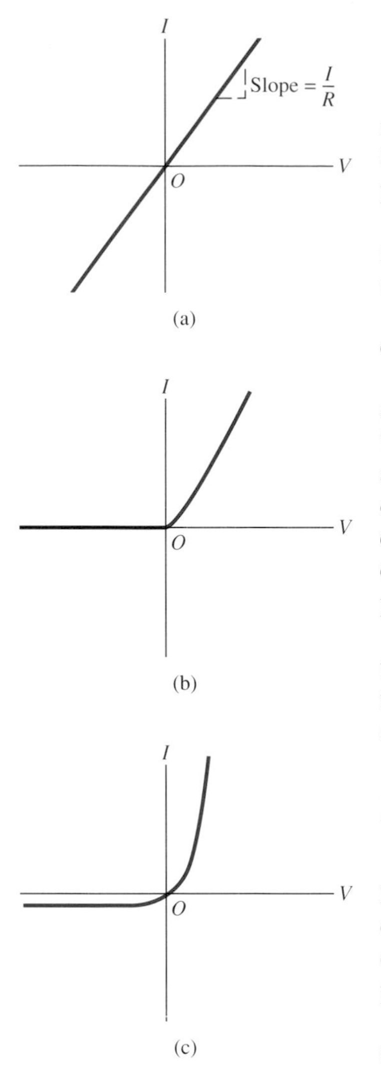


**26–9** Current-voltage relations for (a) a resistor that obeys Ohm's law; (b) a vacuum diode; (c) a semiconductor. Only for the resistor is $I$ proportional to $V$.

nearest an end) are digits, and the third is a power-of-10 multiplier, as shown in Fig. 26–8. For example, yellow-violet-orange means $47 \times 10^3\ \Omega$, or 47 k$\Omega$. The fourth band, if present, indicates the precision of the value; no band means $\pm 20\%$, a silver band $\pm 10\%$, and a gold band $\pm 5\%$. Another important characteristic of a resistor is the maximum *power* that it can dissipate without damage. We'll return to this point in Section 26–5.

For a resistor that obeys Ohm's law a graph of current as a function of potential difference (voltage) is a straight line (Fig. 26–9a). The slope of the line is $1/R$. As we have mentioned, not all devices obey Ohm's law. The relation of voltage to current may not be a direct proportion, and it may be different for the two directions of current. Figure 26–9b shows the behavior of a vacuum diode, a vacuum tube used to convert high-voltage alternating current to direct current. For positive potentials of anode with respect to cathode, $I$ is approximately proportional to $V^{3/2}$; for negative potentials the current is extremely small. The behavior of semiconductor diodes (Fig. 26–9c) is somewhat different but still strongly asymmetric, like a one-way valve in a circuit. Diodes are used to perform a wide variety of logic functions in computer circuitry.

Current-voltage relations are often temperature-dependent. At low temperatures the curve in Fig. 26–9c rises more steeply for positive $V$ than it does at higher temperatures, and at successively higher temperatures the asymmetry in the curve becomes less and less pronounced.

## 26–4 ELECTROMOTIVE FORCE AND CIRCUITS

For a conductor to have a steady current it must be part of a path that forms a closed loop, or **complete circuit.** But the path cannot consist entirely of resistance. In a resistor, charge always moves in the direction of decreasing potential energy. There must be some part of the circuit where the potential energy *increases.*

The problem is analogous to an ornamental water fountain that recycles its water (Fig. 26–10a). The water pours out of openings at the top, cascades down over the ter-

**26–10** (a) A pump is needed to keep the water circulating in this fountain. (b) A current in an ordinary circuit with no source of emf is analogous to this famous impossible scene by Escher.

races and spouts, and collects in a basin in the bottom. A pump then lifts it back to the top for another trip. Without the pump the water would just fall to the bottom and stay there. The situation of Fig. 26–10b doesn't occur in the real world.

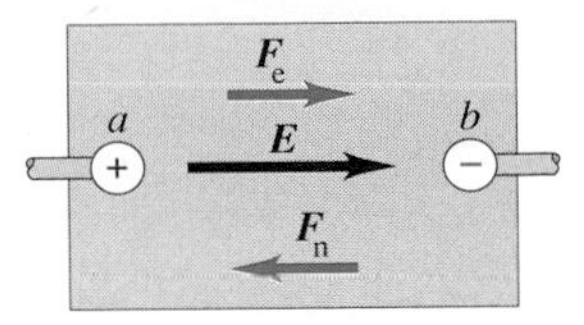


26–11 Schematic diagram of a source of emf in an "open-circuit" situation. The electric-field force $F_e = qE$ and the non-electrostatic force $F_n$ on a charge $q$ are shown. The work done by $F_n$ on a charge $q$ moving from $b$ to $a$ is equal to $q\mathcal{E}$, where $\mathcal{E}$ is the electromotive force. In the open-circuit situation, $F_e$ and $F_n$ have equal magnitude.

## Electromotive Force

In an electric circuit there must be a device somewhere in the loop where a charge travels "uphill," from lower to higher potential, despite the fact that the electrostatic force is trying to push it from higher to lower potential. The influence that makes charge move from lower to higher potential is called **electromotive force** (abbreviated **emf**). Every complete circuit with a steady current must include some device that provides emf (pronounced "ee-em-eff"). This is a poor term because emf is *not* a force but an energy-per-unit-charge quantity, like potential. The SI unit of emf is the same as the unit for potential, the volt (1 V = 1 J/C). A battery with an emf of 1.5 V does 1.5 J of work on every coulomb of charge that passes through it. We'll use the symbol $\mathcal{E}$ for emf.

Batteries, electric generators, solar cells, thermocouples, and fuel cells are all examples of sources of emf. All such devices convert energy of some form (mechanical, chemical, thermal, and so on) into electrical energy and transfer it into the circuit where the device is connected. An ideal source of emf maintains a constant potential difference between its terminals, independent of the current through it. We define emf quantitatively as the magnitude of this potential difference. As we will see, such an ideal source is a mythical beast, like the unicorn, the frictionless plane, and the free lunch. We will discuss later how real-life sources of emf differ in their behavior from this idealized model.

Figure 26–11 is a schematic diagram of a source of emf that maintains a potential difference between conductors $a$ and $b$, called the *terminals* of the device. Terminal $a$, marked +, is maintained at *higher* potential than terminal $b$, marked −. Associated with this potential difference is an electric field $\boldsymbol{E}$ in the region around the terminals, both inside and outside the source. The electric field inside the device is directed from $a$ to $b$, as shown. A charge $q$ within the source experiences an electric force $\boldsymbol{F}_e = q\boldsymbol{E}$. The source has to provide some additional influence, which we represent as a force $\boldsymbol{F}_n$ that pushes charge from $b$ to $a$ inside the device (opposite to the direction of $\boldsymbol{F}_e$) and maintains the potential difference. The origin of this additional influence depends on what kind of source we are talking about. In a generator the additional influence results from magnetic-field forces on moving charges. In a battery or fuel cell it is associated with diffusion processes and varying electrolyte concentrations resulting from chemical reactions. In an electrostatic machine such as a Van de Graaff generator an actual mechanical force is applied by a moving belt or wheel.

The potential $V_{ab}$ of point $a$ with respect to point $b$ is defined, as always, as the work per unit charge done by the electrostatic force $\boldsymbol{F}_e = q\boldsymbol{E}$ on a charge $q$ that moves from $a$ to $b$. The emf $\mathcal{E}$ of the source is the energy per unit charge supplied by the source during the "uphill" displacement from $b$ to $a$. For the ideal source of emf that we have described, the potential difference $V_{ab}$ is equal to the electromotive force $\mathcal{E}$:

$$V_{ab} = \mathcal{E} \quad \text{(no complete circuit).} \qquad (26\text{–}15)$$

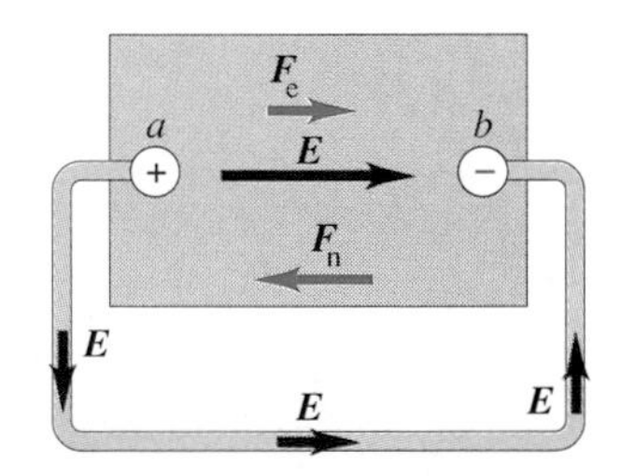


26–12 Schematic diagram of a source with a complete circuit. The vectors $F_n$ and $F_e$ are the forces on a charge $q$. The current is in the direction from $a$ to $b$ in the external circuit and from $b$ to $a$ within the source.

Now let's make a complete circuit by connecting a wire with resistance $R$ to the terminals of a source (Fig. 26–12). The charged terminals $a$ and $b$ of the source set up an electric field in the wire, and this causes a current in the wire from $a$ toward $b$. From $V = IR$ the current $I$ in the circuit is determined by

$$\mathcal{E} = V_{ab} = IR \quad \text{(ideal source of emf).} \qquad (26\text{–}16)$$

That is, when a charge $q$ flows around the circuit, the potential rise $\mathcal{E}$ as it passes through the source is numerically equal to the potential drop $V_{ab} = IR$ as it passes through the resistor. Once $\mathcal{E}$ and $R$ are known, this relation determines the current in the circuit. The current is the same at every point in the circuit. This follows from conservation of charge and from the fact that charge cannot accumulate in the circuit devices that we have described. (Otherwise, the potential differences would change with time.)

## Internal Resistance

Real sources don't behave exactly the way we have described because charge moving through the material of any real source encounters *resistance*. We call this the **internal resistance** of the source, denoted by $r$. If this resistance behaves according to Ohm's law, $r$ is constant. The current through $r$ has an associated drop in potential equal to $Ir$. The terminal potential difference $V_{ab}$ under complete-circuit conditions is then

$$V_{ab} = \mathcal{E} - Ir \quad \text{(source with internal resistance).} \tag{26–17}$$

The potential $V_{ab}$, called the **terminal voltage,** is less than the emf $\mathcal{E}$ because of the term $Ir$ representing the potential drop across the internal resistance $r$.

The current in the external circuit is still determined by $V = IR$. Combining this with Eq. (26–17), we find

$$\mathcal{E} - Ir = IR,$$

or

$$I = \frac{\mathcal{E}}{R + r} \quad \text{(source with internal resistance).} \tag{26–18}$$

That is, the current equals the source emf divided by the *total* circuit resistance $(R + r)$. Thus we can describe the behavior of a source in terms of two properties: an emf $\mathcal{E}$, which supplies a constant potential difference independent of current, in series with an internal resistance $r$.

To summarize, a circuit is a closed conducting path containing resistors, sources of emf, and possibly other circuit elements. Equation (26–17) shows that the algebraic sum of the potential differences and emf's around the path is zero. Also, the current in a simple loop is the same at every point. Charge is conserved; if the current were different at different points, there would be a continuing accumulation of charge at some points, and the current couldn't be constant.

Table 26–4 shows the usual symbols used in schematic circuit diagrams. We will use these symbols in most of the circuit analysis in the rest of this chapter and the next.

**TABLE 26–4 Symbols for Circuit Diagrams**

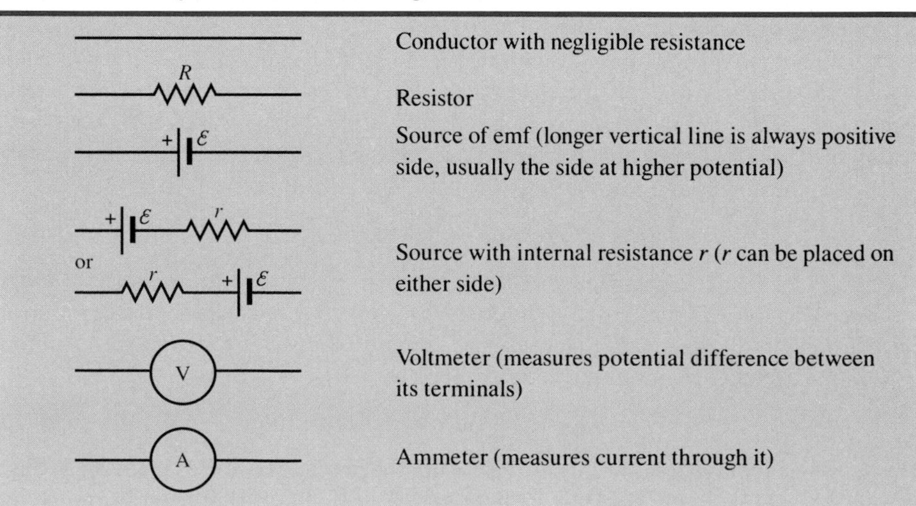

R
+
ℰ
r
or
V
A
Conductor with negligible resistance
Resistor
Source of emf (longer vertical line is always positive side, usually the side at higher potential)
Source with internal resistance r (r can be placed on either side)
Voltmeter (measures potential difference between its terminals)
Ammeter (measures current through it)

We'll discuss voltmeters and ammeters in Chapter 27. Idealized meters do not disturb the circuit in which they are connected. An idealized voltmeter has infinitely large resistance and measures potential difference without having any current diverted through it; an idealized ammeter has zero resistance and measures the current through it without having any potential difference between its terminals.

## EXAMPLE 26–5

**A source on open circuit** Figure 26–13 shows a source with an emf $\mathcal{E}$ of 12 V and an internal resistance $r$ of 2 Ω. (For comparison the internal resistance of a commercial 12-V lead storage battery is only a few thousandths of an ohm.) Determine the readings of meters V and A.

**SOLUTION** There is no current because there is no complete circuit. (There is no current through an idealized voltmeter.) The ammeter A reads $I = 0$. Because there is no current through the battery, there is no potential difference across its internal resistance. The potential difference $V_{ab}$ across its terminals is equal to the emf, $V_{ab} = \mathcal{E} = 12$ V, and the voltmeter reads $V = 12$ V.

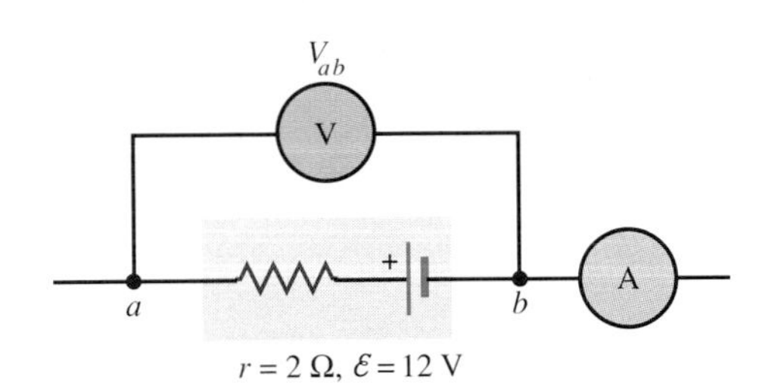


**26–13** A source on open circuit.

## EXAMPLE 26–6

**A source in a complete circuit** Using the battery in Example 26–5, we add a 4-Ω resistor to form the complete circuit shown in Fig. 26–14. What are the voltmeter and ammeter readings now?

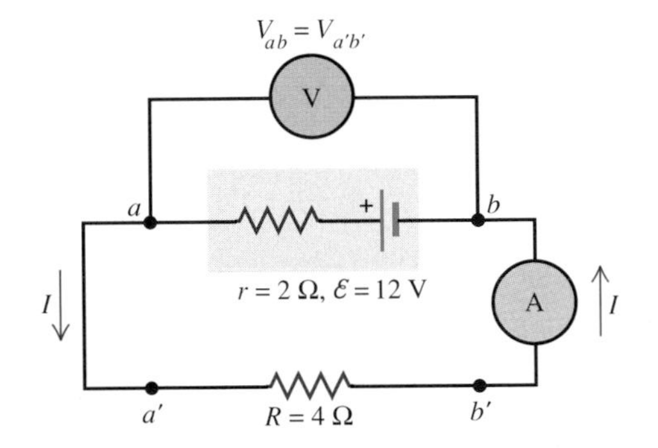


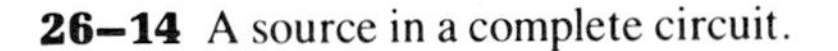

**26–14** A source in a complete circuit.

**SOLUTION** We now have a complete circuit and a current $I$ through the resistor $R$, determined by Eq. (26–18):

$$I = \frac{\mathcal{E}}{R + r} = \frac{12\ \text{V}}{4\ \Omega + 2\ \Omega} = 2\ \text{A}.$$

The ammeter A reads $I = 2$ A.

Our idealized conducting wires have zero resistance, so there is no potential difference between points $a$ and $a'$ or between $b$ and $b'$. That is, $V_{ab} = V_{a'b'}$. We can find $V_{ab}$ by considering $a$ and $b$ either as the terminals of the resistor or as the terminals of the source. Considering them as terminals of the resistor, we use Ohm's law ($V = IR$):

$$V_{a'b'} = IR = (2\ \text{A})\,(4\ \Omega) = 8\ \text{V}.$$

Considering them as terminals of the source, we have

$$V_{ab} = \mathcal{E} - Ir = 12\ \text{V} - (2\ \text{A})(2\ \Omega) = 8\ \text{V}.$$

Either way, we conclude that the voltmeter reads $V_{ab} = 8$ V.

## EXAMPLE 26–7

**A source with a short circuit** Using the same battery as in the last two examples, we now replace the 4-Ω resistor with a zero-resistance conductor, as shown in Fig. 26–15. What are the meter readings now?

**SOLUTION** Now we have a zero-resistance path between points $a$ and $b$, so we must have $V_{ab} = 0$, no matter what the current. Knowing this, we can find the current $I$ from the relation

$$V_{ab} = \mathcal{E} - Ir = 12\ \text{V} - I\,(2\ \Omega) = 0,$$
$$I = 6\ \text{A}.$$

The ammeter reads $I = 6$ A, and the voltmeter reads $V_{ab} = 0$.

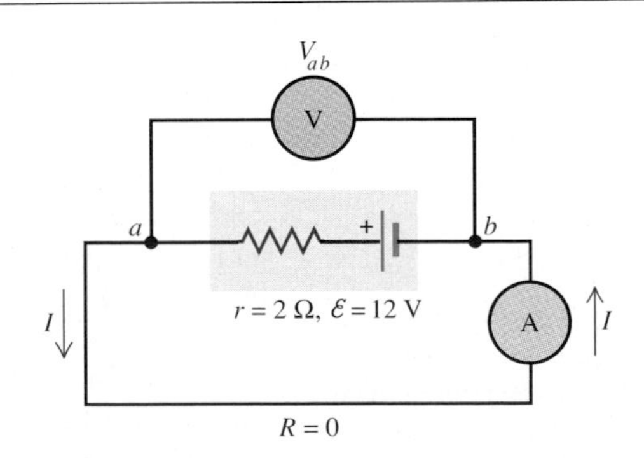


**26–15** A source on short circuit. (*Caution!* See the text for the dangers of this arrangement.)

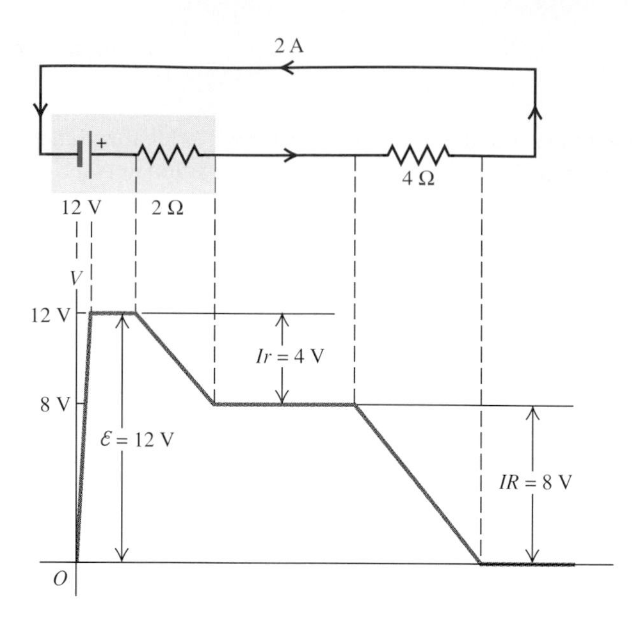


**26–16** Potential rises and drops in the circuit.

In all three examples we call $V_{ab}$ the *terminal voltage*. The situation in Example 26–7 is called a *short circuit*. The short-circuit current is equal to the emf $\mathcal{E}$ divided by the internal resistance $r$. *Caution*! A short circuit can be an extremely dangerous situation. Dont't try it! An automobile battery or a household power line has very small internal resistance (much less than in these examples), and the short-circuit current can be great enough to cause a small wire or a storage battery actually to explode.

Figure 26–16 is a graph showing how the potential varies around the complete circuit of Fig. 26–14. The horizontal axis doesn't necessarily represent actual distances but rather various points in the loop. If we take the potential to be zero at the negative terminal of the battery, then we have a rise $\mathcal{E}$ and a drop $Ir$ in the battery and an additional drop $IR$ in the external resistor, and as we finish our trip around the loop, the potential is back where it started.

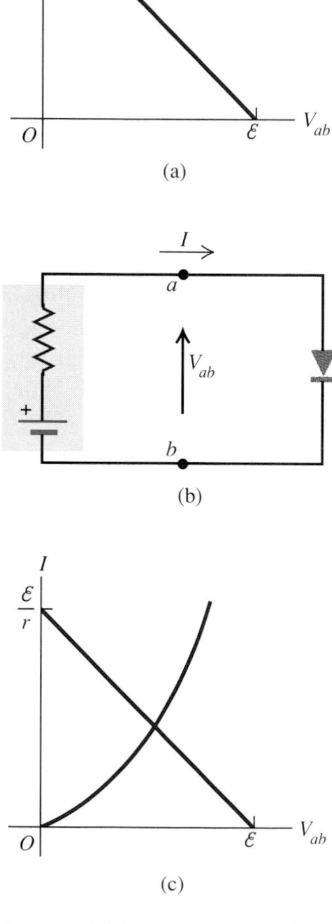


**26–17** (a) Current-voltage relation for a source with emf $\mathcal{E}$ and internal resistance $r$. (b) A circuit containing a source and a diode (a nonlinear element). (c) Simultaneous solution of $I$-$V_{ab}$ equations for this circuit.

We have mentioned graphical representation of current-voltage relations for circuit elements that don't obey Ohm's law; Fig. 26–9 shows two examples. The current-voltage relation for a *source* may also be represented graphically. For a source represented by Eq. (26–17), that is, for which

$$V_{ab} = \mathcal{E} - Ir,$$

the graph appears as in Fig. 26–17a. The intercept on the $V$-axis, corresponding to the open-circuit condition ($I = 0$), is at $V_{ab} = \mathcal{E}$, and the intercept on the $I$-axis, corresponding to a short-circuit situation ($V_{ab} = 0$), is at $I = \mathcal{E}/r$.

We can use this graph to find the current in a circuit containing a nonlinear device, such as a diode (Fig. 26–17b). The current-voltage relation for a semiconductor is shown in Fig. 26–17c. Equation (26–17), the current-voltage relation for the source, is also plotted on this graph. Each of the curves represents a current-voltage relation that must be satisfied, so the *intersection* of the two curves represents the only possible values of $V_{ab}$ and $I$. This amounts to a graphical solution of two simultaneous equations for $V_{ab}$ and $I$, one of which is nonlinear.

Finally, we remark that Eq. (26–17) is not always an adequate representation of the behavior of a source. The emf may not be constant, and what we have described as an internal resistance may actually be a more complex voltage-current relation that doesn't obey Ohm's law. Nevertheless, the concept of internal resistance frequently provides an adequate description of batteries, generators, and other energy converters. The difference between a fresh flashlight battery and an old one is not in the emf, which decreases only slightly with use, but mostly in the internal resistance, which may increase from a few ohms when fresh to as much as 1000 Ω or more after long use. Similarly, a car battery can deliver less current to the starter motor on a cold morning than when the battery is warm, not because the emf is appreciably less but because the internal resistance is temperature-dependent, increasing with decreasing temperature. Residents of northern Iowa have been known to soak their car batteries in warm water to provide greater starting power on very cold mornings!

## 26–5 ENERGY AND POWER IN ELECTRIC CIRCUITS

Let's now look at some energy and power relations in electric circuits. The box in Fig. 26–18 represents a circuit element with potential difference $V_a - V_b = V_{ab}$ between its terminals and current $I$ passing through it in the direction from $a$ toward $b$. This element might be a resistor, a battery, or something else; the details don't matter. As charge passes through the circuit element, the electric field does work on the charge. In a source of emf, additional work is done by the force $\boldsymbol{F}_n$ that we mentioned in Section 26–4.

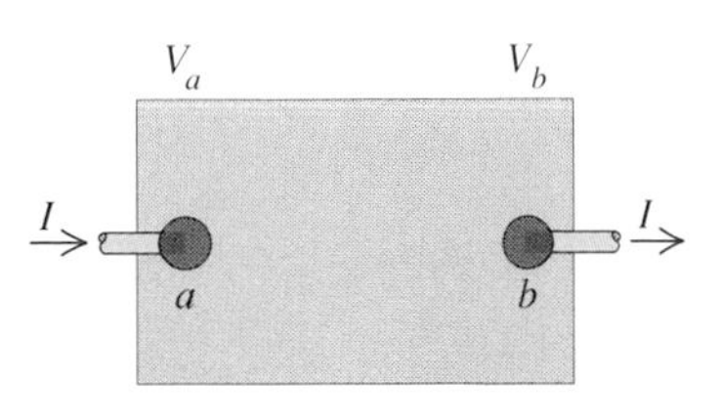


**26–18** The power input $P$ to the portion of the circuit between $a$ and $b$ is $P = V_{ab}I$.

The total work done on a charge $q$ passing through the circuit element is equal to the product of $q$ and the potential difference $V_{ab}$ (work per unit charge). When $V_{ab}$ is positive, the electric force (and possibly a nonelectric force) does a positive amount of work $qV_{ab}$ on the charge as it "falls" from potential $V_a$ to lower potential $V_b$. If the current is $I$, then in a time interval $dt$ an amount of charge $dQ = I\,dt$ passes through. The work $dW$ done on this amount of charge is

$$dW = V_{ab}\,dQ = V_{ab}I\,dt.$$

This work represents electrical energy transferred *into* this circuit element. The time rate of energy transfer is *power*, denoted by $P$. Dividing the above equation by $dt$, we obtain the *rate* at which the rest of the circuit delivers electrical energy to this circuit element:

$$\frac{dW}{dt} = P = V_{ab}I. \tag{26–19}$$

It may happen that the potential at $b$ is higher than at $a$; then $V_{ab}$ is negative, and there is a net transfer of energy *out of* the circuit element. The element is then acting as a source, delivering electrical energy into the circuit in which it is connected. This is the usual situation for a battery, which converts chemical energy into electrical energy and delivers it to the external circuit. Similarly, a generator converts mechanical energy to electrical energy and delivers it to the external circuit.

The unit of $V_{ab}$ is one volt, or one joule per coulomb, and the unit of $I$ is one ampere, or one coulomb per second. We can confirm that the SI unit of power is one watt:

$$(1\ \text{J/C})(1\ \text{C/s}) = 1\ \text{J/s} = 1\ \text{W}.$$

Let's consider a few special cases.

### Pure Resistance

If the circuit element in Fig. 26–18 is a resistor that obeys Ohm's law, the potential difference is $V_{ab} = IR$. From Eq. (26–19) the electrical power delivered to the resistor by the circuit is

$$P = V_{ab}I = I^2R = \frac{{V_{ab}}^2}{R}. \tag{26–20}$$

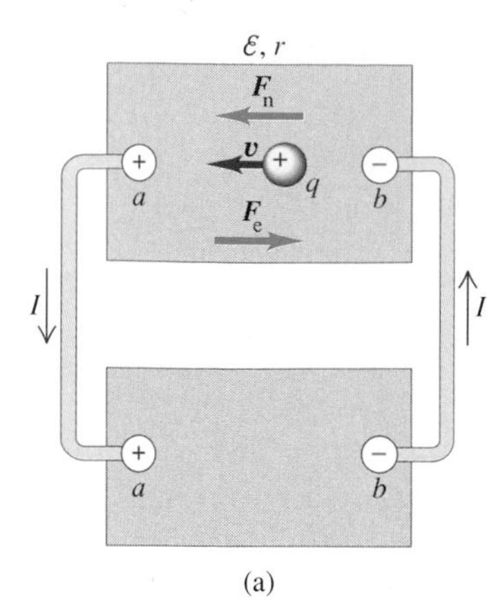


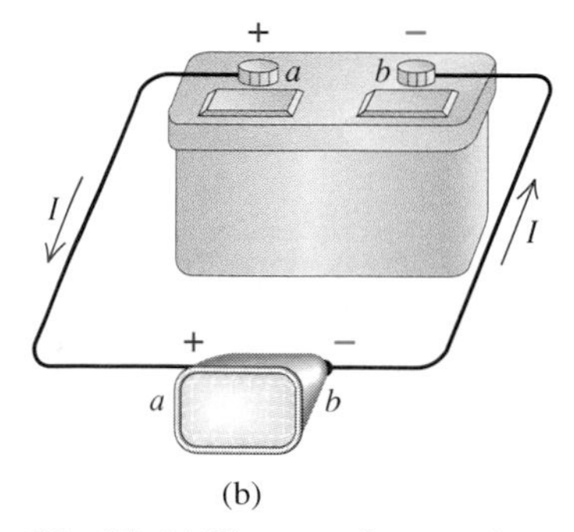


**26–19** (a) The rate of conversion of non-electrical to electrical energy in the source equals $\mathcal{E}I$. The rate of energy dissipation in the source is $I^2r$. The difference $\mathcal{E}I - I^2r$ is the power output of the source. (b) A familiar example of the circuit in part (a).

In this case the potential at $a$ is always higher than at $b$. Current enters the higher-potential terminal of the device, and Eq. (26–20) represents the rate of transfer of electrical energy *into* the circuit element.

What becomes of this energy? The moving charges collide with atoms in the resistor and transfer some of their energy to these atoms, increasing the internal energy of the material. Either the temperature of the resistor increases or there is a flow of heat out of it, or both. We say that energy is *dissipated* in the resistor at a rate $I^2R$. Too high a temperature can change the resistance unpredictably; the resistor may melt or even explode. Of course, some devices, such as electric heaters, are designed to get hot and transfer heat to their surroundings. But every resistor has a *power rating,* the maximum power that the device can dissipate without becoming overheated and damaged. In practical applications the power rating of a resistor is often just as important a characteristic as its resistance value.

## Power Output of a Source

The upper rectangle in Fig. 26–19a represents a source with emf $\mathcal{E}$ and internal resistance $r$, connected by ideal (resistanceless) conductors to an external circuit represented by the lower box. This could describe a car battery connected to the car's headlights (Fig. 26–19b). Point $a$ is at higher potential than point $b$: $V_a > V_b$. But now the current $I$ is *leaving* the device at the higher-potential terminal (rather than entering there). Energy is being delivered to the external circuit, and the rate of its delivery to the circuit is given by Eq. (26–19):

$$P = V_{ab}I.$$

For a source that can be described by an emf $\mathcal{E}$ and an internal resistance $r$ we may use Eq. (26–17):

$$V_{ab} = \mathcal{E} - Ir.$$

Multiplying this equation by $I$, we find

$$P = V_{ab}I = \mathcal{E}I - I^2r. \tag{26–21}$$

What do the terms $\mathcal{E}I$ and $I^2r$ mean? In Section 26–4 we defined the emf $\mathcal{E}$ as the work per unit charge performed on the charges by the non-electrostatic force as the charges are pushed "uphill" from $b$ to $a$ in the source. When a charge $dQ$ flows through the source, the work done on it by this non-electrostatic force is $\mathcal{E}\,dQ = \mathcal{E}I\,dt$. Thus $\mathcal{E}I$ is the *rate* at which work is done on the circulating charges by whatever agency causes the non-electrostatic force in the source. This term represents the rate of conversion of non-electrical energy to electrical energy within the source. The term $I^2r$ is the rate at which electrical energy is *dissipated* in the internal resistance of the source. The difference $\mathcal{E}I - I^2r$ is the *net* electrical power output of the source, that is, the rate at which the source delivers electrical energy to the remainder of the circuit.

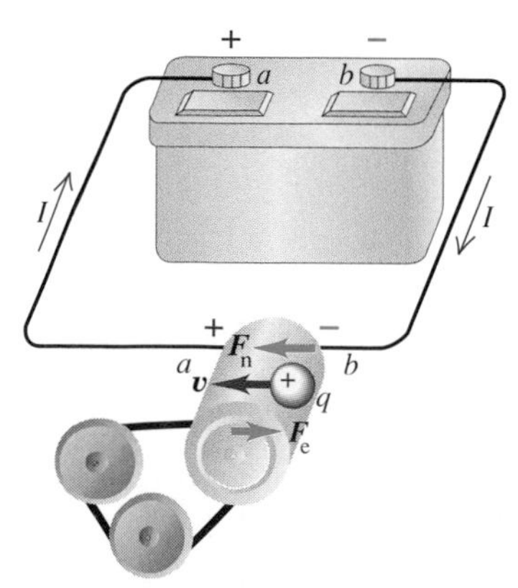


**26–20** When two sources are connected in a circuit, the source with the larger emf delivers energy to the other source.

## Power Input to a Source

Suppose the lower rectangle in Fig. 26–19a is itself a source, with an emf *larger* than that of the upper source and opposite to that of the upper source. Figure 26–20 shows a practical example, an automobile battery (the upper circuit element) being charged by the alternator (the lower element). The current $I$ in the circuit is then *opposite* to that shown in Fig. 26–19; the lower source is pushing current backward through the upper source. Instead of Eq. (26–17), we have for the upper source

$$V_{ab} = \mathcal{E} + Ir,$$

# Resistors

**All circuit elements, including wires, have some electrical resistance. Resistors can be used for specific purposes, such as providing large amounts of heat or light.**

## BUILDING PHYSICAL INTUITION

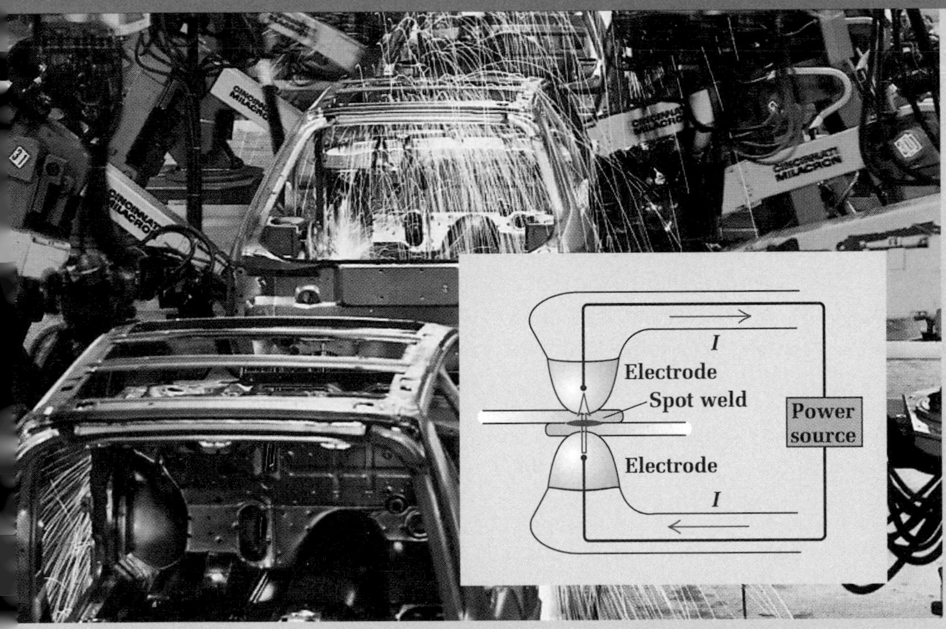


Resistance welding is a very common industrial process. Spot welding, shown here, joins two sheet metals through the heat and pressure applied by two electrodes. The electrode resistance is usually low (typically 100 mW) and the current can range from 3000 to 40,000 A, depending on the materials being welded and their thickness. A car may have as many as 10,000 spot welds.

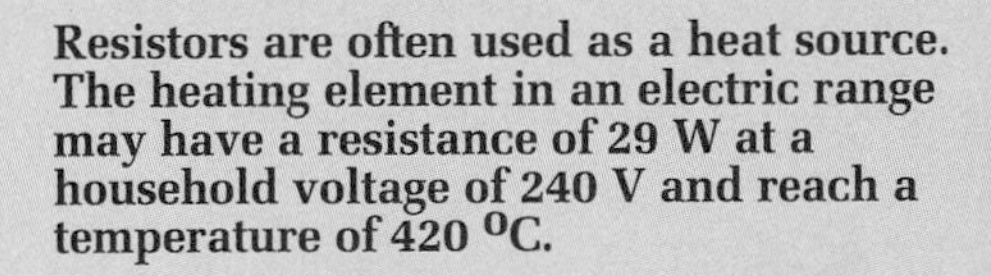

Resistors are often used as a heat source. The heating element in an electric range may have a resistance of 29 W at a household voltage of 240 V and reach a temperature of 420 $^{0}$C.

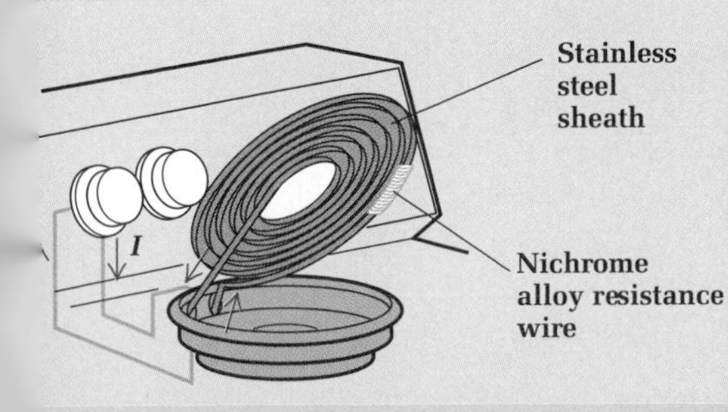


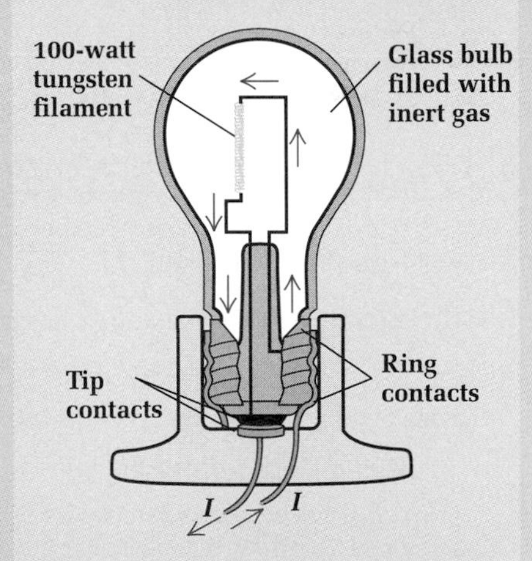


The filament in a 100-W light bulb has a resistance of 144 W at normal 120-V household voltage. The filament, usually made of tungsten, reaches a temperature of about 2500 $^{0}$C as current passes through it. Tungsten does not melt at this temperature, but it does glow white-hot, acting as a source of light.

and instead of Eq. (26–21), we have

$$P = V_{ab}I = \mathcal{E}I + I^2r. \qquad (26\text{–}22)$$

Work is being done *on*, rather than *by*, the agent that causes the non-electrostatic force. There is a conversion of electrical energy into non-electrical energy in the upper source, at a rate $\mathcal{E}I$. The term $I^2r$ is again the rate of dissipation of energy in the internal resistance of the upper source, and the sum $\mathcal{E}I + I^2r$ is the total electrical power *input* to the upper source. This is what happens when a rechargeable battery (a storage battery) is connected to a charger. The charger supplies electrical energy to the battery; part of it is converted to chemical energy, to be reconverted later, and the remainder is dissipated (wasted) in the battery's internal resistance, either warming the battery or causing a heat flow out of it. If you have a power tool with a rechargeable battery, you may have noticed that it gets warm while it is charging.

## PROBLEM-SOLVING STRATEGY

### Power and energy in circuits

**1.** A set of sign rules is useful in calculating the energy balance in a circuit. A source of emf $\mathcal{E}$ puts *positive* power $\mathcal{E}I$ into a circuit when the current $I$ runs through it from $-$ to $+$; the energy is converted from chemical energy in a battery, from mechanical energy in a generator, or whatever. But when current passes through a source in the direction from $+$ to $-$, the source is taking power out of the circuit (as in charging a storage battery, when electrical energy is converted back to chemical energy). In that case we count $\mathcal{E}I$ as negative. The internal resistance $r$ of a source always removes energy from the circuit, converting it into heat at a rate $I^2r$, irrespective of the direction of the current. This represents a *positive* power input to the device and a *negative* power input to the circuit. Similarly, a resistor $R$ always removes energy from the circuit at a rate given by $VI = I^2R = V^2/R$.

**2.** The energy balance can be expressed in one of two forms: either "net power input = net power output" or "the algebraic sum of the power inputs to the circuit is zero."

## EXAMPLE 26–8

Figure 26–21 shows the situation that we analyzed in Example 26–6. Find the rate of energy conversion (chemical to electrical) and the rate of dissipation of energy (conversion to heat) in the battery and its net power output.

**SOLUTION** The rate of energy conversion in the battery is

$$\mathcal{E}I = (12\ \text{V})(2\ \text{A}) = 24\ \text{W}.$$

The rate of dissipation of energy in the battery is

$$I^2r = (2\ \text{A})^2(2\ \Omega) = 8\ \text{W}.$$

The electrical power *output* of the source is the difference between these values, or 16 W. The terminal voltage is $V_{ab} = 8$ V. The power output is also given by

$$V_{ab}I = (8\ \text{V})(2\ \text{A}) = 16\ \text{W}.$$

The electrical power input to the resistor is

$$V_{a'b'}I = (8\ \text{V})(2\ \text{A}) = 16\ \text{W}.$$

This equals the rate of dissipation of electrical energy in the resistor:

$$I^2R = (2\ \text{A})^2(4\ \Omega) = 16\ \text{W}.$$

We can compare these numbers with Eq. (26–21):

$$V_{ab}I = \mathcal{E}I - I^2r.$$

In this example we have

$$16\ \text{W} = 24\ \text{W} - 8\ \text{W},$$

which verifies the consistency of the various power quantities.

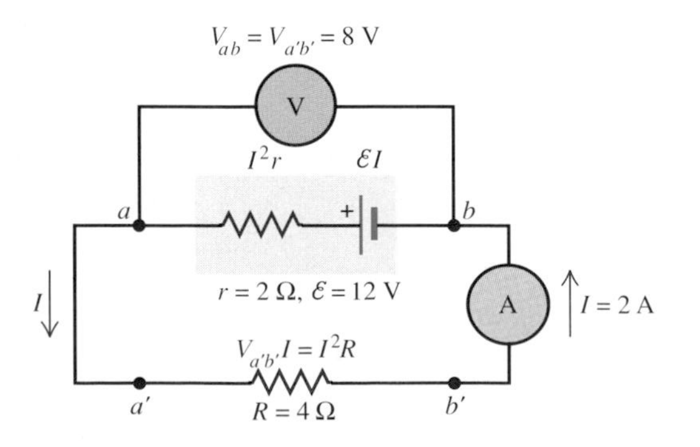


**26–21** Power relationships in a simple circuit.

## EXAMPLE 26–9

Figure 26–22 shows the short-circuited battery that we analyzed in Example 26–7. Find the rates of energy conversion and energy dissipation in the battery and its net power output.

**SOLUTION** The rate of energy conversion (chemical to electrical) in the battery is

$$\mathcal{E}I = (12\ \text{V})(6\ \text{A}) = 72\ \text{W}.$$

The rate of dissipation of energy in the battery is

$$I^2r = (6\ \text{A})^2(2\ \Omega) = 72\ \text{W}.$$

The net power output of the source, given by $V_{ab}I$, is zero because the terminal voltage $V_{ab}$ is zero. *All* of the converted energy is dissipated within the source. This is why a short-circuited battery is quickly ruined and in some cases may even explode.

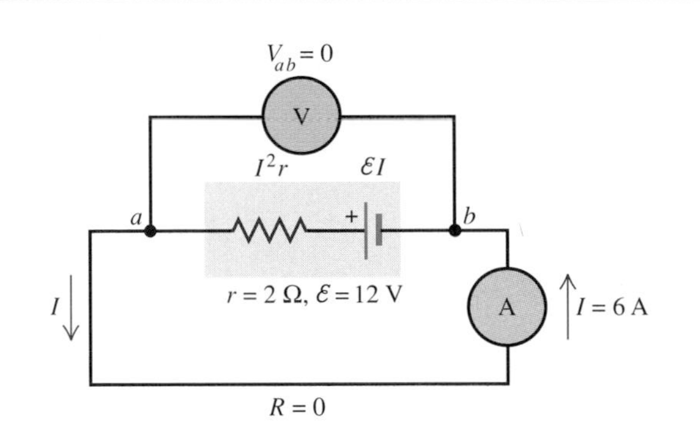


**26–22** Power relationships in a short circuit. With ideal wires and ammeter ($R = 0$), no electrical power leaves the battery.

# *26–6 THEORY OF METALLIC CONDUCTION

We can gain additional insight into electrical conduction by looking at the microscopic mechanisms of conductivity. We will consider only a very simple model that treats the electrons as classical particles and ignores their quantum-mechanical, wavelike behavior in solids. Using this model, we will derive an expression for the resistivity of a metal. Even though this model is not entirely correct conceptually, it will still help you to develop an intuitive idea of the microscopic basis of conduction.

In the simplest microscopic model of conduction in a metal, each atom in the crystal lattice gives up one or more of its outer electrons. These electrons are then free to move through the crystal lattice, colliding at intervals with the stationary positive ions. The motion of the electrons is analogous to molecules of a gas moving through a porous bed of sand, and they are often referred to as an "electron gas."

If there is no electric field, the electrons move in straight lines between collisions, the directions of their velocities are random, and on average they never get anywhere (Fig. 26–23a). But if an electric field is present, the paths curve slightly because of the acceleration caused by electric-field forces. Figure 26–23b shows a few free paths of an

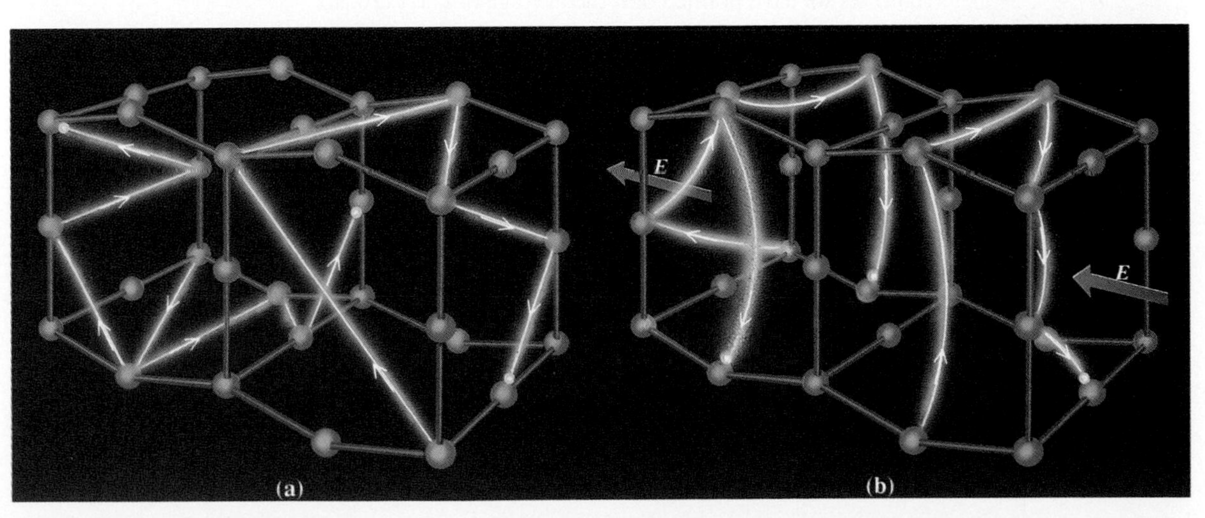


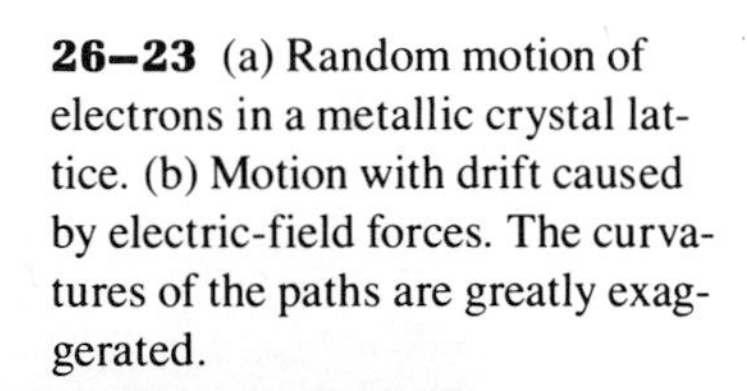

**26–23** (a) Random motion of electrons in a metallic crystal lattice. (b) Motion with drift caused by electric-field forces. The curvatures of the paths are greatly exaggerated.

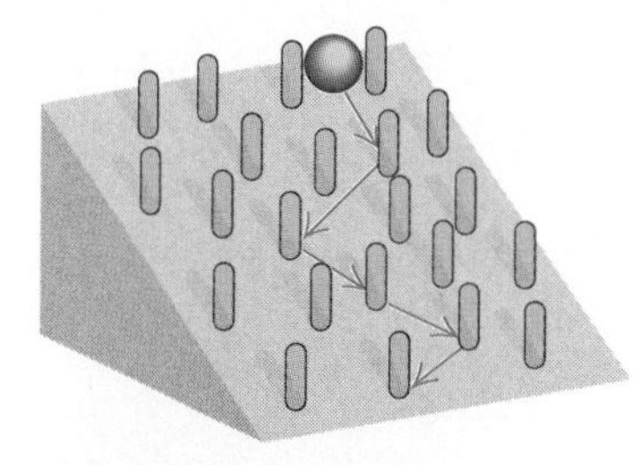

**26–24** The motion of a ball rolling down an inclined plane and bouncing off pegs in its path is analogous to the motion of an electron in a metallic conductor with an electric field present.

electron in an electric field directed from right to left. As we mentioned in Section 26–1, the average speed of random motion is of the order of $10^6$ m/s, and the average drift speed is *very much* smaller, of the order of $10^{-4}$ m/s. The average time between collisions is called the **mean free time,** denoted by $\tau$. Figure 26–24 shows a mechanical model that is analogous to this electron motion.

We would like to derive from this model an expression for the resistivity $\rho$ of a material, defined by Eq. (26–7):

$$\rho = \frac{E}{J}, \tag{26–23}$$

where $E$ and $J$ are the magnitudes of electric field and current density. The current density $\boldsymbol{J}$ is in turn given by Eq. (26–3):

$$\boldsymbol{J} = nq\boldsymbol{v}_{\mathrm{d}}, \tag{26–24}$$

where $n$ is the number of free electrons per unit volume, $q$ is the charge of each, and $\boldsymbol{v}_{\mathrm{d}}$ is their average drift velocity. (We also know that $q = -e$; we'll use that fact later.)

We need to relate the drift velocity $\boldsymbol{v}_{\mathrm{d}}$ to the electric field $\boldsymbol{E}$. The value of $\boldsymbol{v}_{\mathrm{d}}$ is determined by a steady-state condition in which, on the average, the velocity *gains* of the charges due to the force of the $\boldsymbol{E}$ field are just balanced by the velocity *losses* due to collisions.

To clarify this process, let's imagine turning on the two effects one at a time. Suppose that before time $t = 0$ there is no field. The electron motion is then completely random. A typical electron has velocity $\boldsymbol{v}_0$ at time $t = 0$, and the average value of $\boldsymbol{v}_0$ is zero, $(\boldsymbol{v}_0)_{\mathrm{av}} = \mathbf{0}$.

Then at time $t = 0$ we turn on a constant electric field $\boldsymbol{E}$. The field exerts a force $\boldsymbol{F} = q\boldsymbol{E}$ on each charge, and this causes an acceleration $\boldsymbol{a}$ in the direction of the force, given by

$$\boldsymbol{a} = \frac{\boldsymbol{F}}{m} = \frac{q\boldsymbol{E}}{m}, \tag{26–25}$$

where $m$ is the electron mass. *Every* electron has this acceleration.

We wait for a time $\tau$, the average time between collisions, and then we turn on the collisions. An electron that has velocity $\boldsymbol{v}_0$ at time $t = 0$ has a velocity at time $t = \tau$ equal to

$$\boldsymbol{v} = \boldsymbol{v}_0 + \boldsymbol{a}\tau.$$

The average value $\boldsymbol{v}_{\mathrm{av}}$ of $\boldsymbol{v}$ at this time is the sum of the averages of the two terms on the right. As we have pointed out, the average of $\boldsymbol{v}_0$ is zero, so

$$\boldsymbol{v}_{\mathrm{av}} = \boldsymbol{a}\tau = \frac{q\tau}{m}\boldsymbol{E}. \tag{26–26}$$

After time $t = \tau$ the tendency of the collisions to decrease the average velocity (by means of randomizing collisions) just balances the tendency of the $\boldsymbol{E}$ field to increase the average velocity. Thus the average velocity given by Eq. (26–26) is maintained and is equal to the drift velocity $\boldsymbol{v}_{\mathrm{d}}$:

$$\boldsymbol{v}_{\mathrm{d}} = \frac{q\tau}{m}\boldsymbol{E}. \tag{26–27}$$

Now we substitute this expression for the drift velocity $\boldsymbol{v}_{\mathrm{d}}$ into Eq. (26–24):

$$\boldsymbol{J} = nq\boldsymbol{v}_{\mathrm{d}} = \frac{nq^2\tau}{m}\boldsymbol{E}.$$

Comparing this with Eq. (26–23), which we can rewrite as $\boldsymbol{J} = \boldsymbol{E}/\rho$, and substituting $q = -e$, we see that the resistivity $\rho$ is given by

$$\rho = \frac{m}{ne^2\tau}. \tag{26–28}$$

If $n$ and $\tau$ are independent of $\boldsymbol{E}$, then the resistivity is independent of $\boldsymbol{E}$, and we are describing a material that obeys Ohm's law.

Turning the interactions on one at a time may seem artificial. But a little thought shows that the derivation would come out the same if each electron had its own clock and the $t = 0$ times were different for different electrons. If $\tau$ is the average time between collisions, then $v_d$ is still the average drift velocity, even though the motions of the various electrons aren't actually correlated in the way that we postulated.

What about the temperature dependence of resistivity? In an ideal crystal lattice with no atoms out of place, a correct quantum-mechanical analysis would let the free electrons move through the lattice with no collisions at all. But the atoms vibrate about their equilibrium positions. As the temperature increases, the amplitudes of these vibrations increase, the crystal becomes less ideal, collisions become more frequent, and the mean free time $\tau$ decreases. So this theory predicts that the resistivity of a metal increases with temperature. In a superconductor, roughly speaking, there are no inelastic collisions, $\tau$ is infinite, and the resistivity $\rho$ is zero.

In a pure semiconductor such as silicon or germanium, the number $n$ of charge carriers per unit volume is not constant but increases very rapidly with increasing temperature. This increase in $n$ far outweighs the decrease in the mean free time, and in a semiconductor the resistivity always decreases rapidly with increasing temperature. At low temperatures, $n$ is very small, and the resistivity becomes so large that the material can be considered an insulator.

## EXAMPLE 26–10

**Mean free time in copper** Calculate the mean free time between collisions in copper at room temperature.

**SOLUTION** From Example 26–1 and Table 26–1, $n = 8.5 \times 10^{28}/\text{m}^3$ and $\rho = 1.72 \times 10^{-8}\ \Omega \cdot \text{m}$. Also, for electrons, $m = 9.11 \times 10^{-31}$ kg and $e = 1.60 \times 10^{-19}$ C. Rearranging Eq. (26–28), we get

$$\tau = \frac{m}{ne^2\rho} = \frac{9.11 \times 10^{-31}\,\text{kg}}{(8.5 \times 10^{28}/\text{m}^3)(1.60 \times 10^{-19}\,\text{C})^2(1.72 \times 10^{-8}\,\Omega\cdot\text{m})}$$
$$= 2.4 \times 10^{-14}\,\text{s}.$$

Taking the reciprocal of this time, we find that each electron averages $42 \times 10^{12}$ collisions every second!

Electrons gain energy between collisions through the work done on them by the electric field. During collisions they transfer some of this energy to the atoms of the material of the conductor. This leads to an increase in the internal energy and temperature of the material. That's why wires carrying current get warm. When they are overloaded, they get too hot and can start a fire. The filament in an ordinary light bulb glows white-hot when current passes through it.

If the electric field in the material is large enough, an electron can gain enough energy between collisions to knock off electrons that are normally bound to atoms in the material. The electrons knocked off can then knock off more electrons, and so on, possibly leading to an avalanche of current. This is the microscopic basis of dielectric breakdown in insulators.

## *26–7 PHYSIOLOGICAL EFFECTS OF CURRENTS

Electrical potential differences and currents play a vital role in the nervous systems of animals. Conduction of nerve impulses is basically an electrical process, although the mechanism of conduction is much more complex than in simple materials such as metals. A nerve fiber, or *axon,* along which an electrical impulse can travel, includes a cylindrical membrane with one conducting fluid (electrolyte) inside and another outside (Fig. 26–25a). Chemical systems similar to those in batteries maintain a potential difference of the order of 0.1 V between these fluids.

When an electrical pulse is initiated, the nerve membrane temporarily becomes more permeable to the ions in the fluids, leading to a local drop in potential (Fig. 26–25b). As the pulse propagates, with a speed of about 30 m/s, the membrane recovers and the potential returns to its initial value (Fig. 26–25c).

The basically electrical nature of nerve-impulse conduction is responsible for the great sensitivity of the body to externally supplied electrical currents. Currents through the body as small as 0.1 A, which are much too small to produce significant heating, can be fatal because they interfere with nerve processes that are essential for vital functions such as heartbeat. The resistance of the human body is highly variable. Body fluids are usually quite good conductors because of the substantial ion concentrations, but the resistance of skin is relatively high, ranging from 500 kΩ for very dry skin to 1000 Ω or so when wet, depending also on the area of contact. If $R = 1000\ \Omega$, a current of 0.1 A requires a voltage of $V = IR = (0.1\ \text{A}) \cdot (1000\ \Omega) = 100\ \text{V}$.

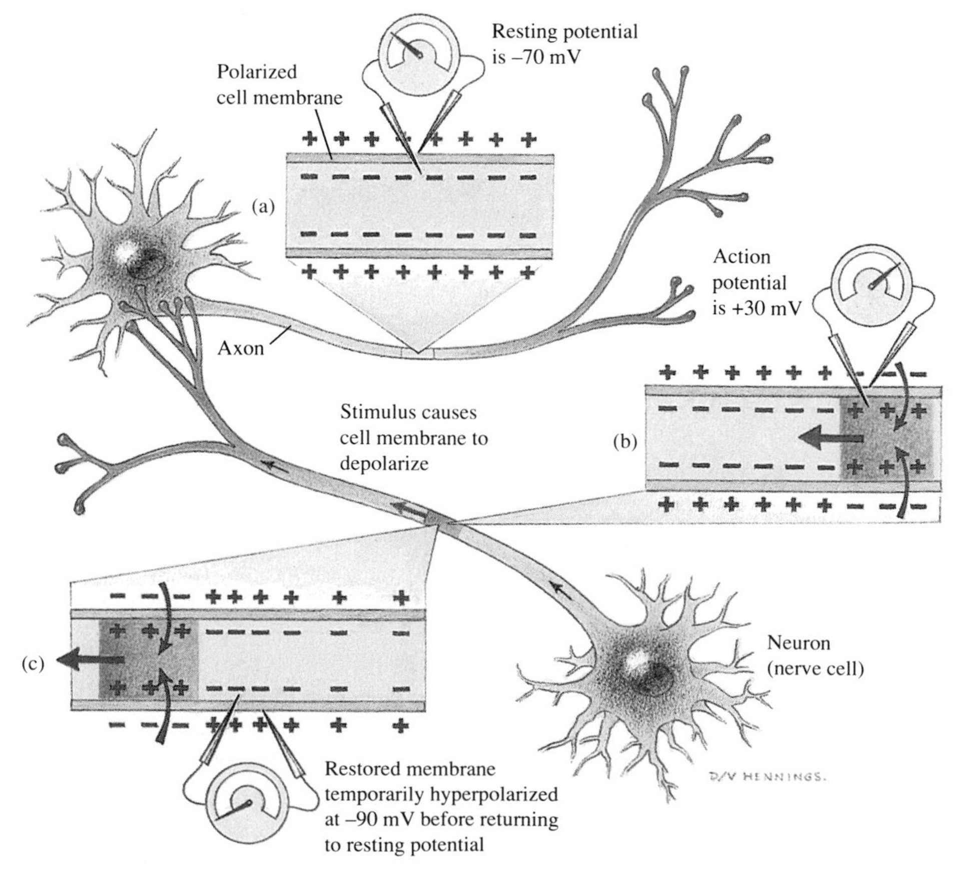


**26–25** (a) The cell membrane around a nerve fiber maintains a potential difference of about 0.1 V between inner and outer fluids. (b) An electrical stimulus depolarizes the membrane, and the potential difference drops. (c) The drop in potential difference propagates along the nerve fiber, which recovers its original value of potential difference after the pulse passes.

Even smaller currents can be very dangerous. A current of 0.01 A through an arm or leg causes strong, convulsive muscle action and considerable pain, and with a current of 0.02 A a person who is holding the conductor that is inflicting the shock is typically unable to release it. Currents of this magnitude through the chest can cause ventricular fibrillation, a disorganized twitching of heart muscles that pumps very little blood. Surprisingly, very large currents (over 0.1 A) are somewhat *less* likely to cause fatal fibrillation because the heart muscle is "clamped" in one position. The heart actually stops beating and is more likely to resume normal beating when the current is removed. The electric defibrillators used for medical emergencies apply a large current pulse to stop the heart (and the fibrillation) to give the heart a chance to restore normal rhythm.

Thus electric current poses three different kinds of hazards: interference with the nervous system, injury caused by convulsive muscle action, and burns from $I^2R$ heating.

The moral of this rather morbid story, if there is one, is that under certain conditions, voltages as small as 10 V can be dangerous. All electrical circuits and equipment should always be approached with respect and caution.

On the positive side, rapidly alternating currents can have beneficial effects. Alternating currents with frequencies of the order of $10^6$ Hz do not interfere appreciably with nerve processes and can be used for therapeutic heating for arthritic conditions, sinusitis, and a variety of other disorders. If one electrode is made very small, the resulting concentrated heating can be used for local destruction of tissue such as tumors or for cutting tissue in certain surgical procedures.

Study of particular nerve impulses is also an important *diagnostic* tool in medicine. The most familiar examples are electrocardiography (EKG) and electroencephalography (EEG). Electrocardiograms, obtained by attaching electrodes to the chest and back and recording the regularly varying potential differences, are used to study heart function. Similarly, electrodes attached to the scalp permit study of potentials in the brain, and the resulting patterns can be helpful in diagnosing epilepsy, brain tumors, and other disorders.

## SUMMARY

- Current is the amount of charge flowing through a specified area per unit time. The SI unit of current is the ampere, equal to one coulomb per second (1 A = 1 C/s). In terms of the charges $q_i$ and drift velocities $\boldsymbol{v}_{di}$ of the charge carriers in a material, the current $I$ through an area $A$ is

$$I = A(n_1 q_1 v_{d1} + n_2 q_2 v_{d2} + \cdots). \tag{26–4}$$

Current density $\boldsymbol{J}$ is current per unit cross-section area. In terms of the above quantities,

$$\boldsymbol{J} = n_1 q_1 \boldsymbol{v}_{d1} + n_2 q_2 \boldsymbol{v}_{d2} + \cdots. \tag{26–6}$$

Current is usually described in terms of a flow of positive charge, even when the actual charge carriers are negative or of both signs.

- Ohm's law, obeyed approximately by many materials, states that current density $\boldsymbol{J}$ is proportional to electric field $\boldsymbol{E}$. For such materials, resistivity $\rho$ is defined by

$$\rho = \frac{E}{J}. \tag{26–7}$$

The SI unit of resistivity is the ohm-meter ($1\ \Omega \cdot \text{m}$). Good conductors have small resistivity; good insulators have large resistivity. Resistivity usually increases with temperature; for small temperature changes, this variation can be represented approximately by

$$\rho_T = \rho_0[1 + \alpha(T - T_0)], \tag{26–8}$$

where $\alpha$ is the temperature coefficient of resistivity.

• For materials obeying Ohm's law the potential difference $V$ across a particular sample is proportional to the current $I$ through the sample:

$$V = IR, \tag{26–13}$$

where $R$ is the resistance of the sample. In terms of resistivity $\rho$, length $L$, and cross-section area $A$,

$$R = \frac{\rho L}{A}. \tag{26–12}$$

The SI unit of resistance is the ohm ($1\ \Omega = 1\ \text{V/A}$).

• A complete circuit is a conductor in the form of a loop, providing a continuous current-carrying path. A complete circuit carrying a steady current must contain a source of electromotive force (emf), symbolized by $\mathcal{E}$. The SI unit of electromotive force is the volt (1 V). An ideal source of emf maintains a constant potential difference, independent of current through the device, but every real source of emf has some internal resistance $r$. The terminal potential difference $V_{ab}$ then depends on current:

$$V_{ab} = \mathcal{E} - Ir \quad \text{(source with internal resistance).} \tag{26–17}$$

• Many devices do not obey Ohm's law but show a more complicated voltage-current relationship; often this must be represented graphically.

• A circuit element with a potential difference $V$ and a current $I$ puts energy into a circuit if the current direction is from lower to higher potential in the device and takes energy out of the circuit if the current is opposite. The power $P$ (rate of energy transfer) is $P = VI$. A resistor $R$ always takes energy out of a circuit, converting it to heat at a rate given by

$$P = V_{ab}I = I^2R = \frac{V_{ab}^{\ 2}}{R}. \tag{26–20}$$

• The microscopic basis of conduction in metals is the motion of electrons that move freely through the crystal lattice sites, bumping into ion cores in the lattice. In a crude classical model of this motion the resistivity of the material can be related to the electron mass, charge, speed of random motion, density, and mean free time between collisions.

## KEY TERMS

current
ampere
drift velocity
concentration
current density
resistivity
Ohm's law
conductivity
temperature coefficient of resistivity
resistance
ohm
resistor
complete circuit
electromotive force (emf)
internal resistance
terminal voltage
mean free time

# EXERCISES

## Section 26–1 Current

**26–1** A current of 5.00 A flows through an automobile headlight. How many coulombs of charge flow through the headlight in 2.00 h?

**26–2** A glass tube filled with gas has electrodes at each end. When a sufficiently high potential difference is applied between the two electrodes, the gas ionizes; electrons move toward the positive electrode, and positive ions move toward the negative electrode. a) What is the current in a hydrogen discharge if, in each second, $3.60 \times 10^{18}$ electrons and $1.15 \times 10^{18}$ protons move in opposite directions through a cross section of the tube? b) What is the direction of the current?

**26–3** The current in a wire varies with time according to the relation $I = 4.00\ \text{A} + (0.600\ \text{A/s}^2)t^2$. a) How many coulombs of charge pass a cross section of the wire in the time interval between $t = 0$ and $t = 10.0$ s? b) What constant current would transport the same charge in the same time interval?

**26–4** A silver wire 1.00 mm in diameter transfers a charge of 72.0 C in 1 h, 15.0 min. Silver contains $5.80 \times 10^{28}$ free electrons per cubic meter. a) What is the current in the wire? b) What is the magnitude of the drift velocity of the electrons in the wire?

## Section 26–2 Resistivity

## Section 26–3 Resistance

**26–5** In household wiring, a copper wire commonly known as 12-gauge is often used. Its diameter is 2.05 mm. Find the resistance of a 30.0-m length of this wire.

**26–6** A copper wire has a square cross section 2.00 mm on a side. It is 4.00 m long and carries a current of 5.00 A. The density of free electrons is $8.5 \times 10^{28}/\text{m}^3$. a) What is the current density in the wire? b) What is the electric field? c) How much time is required for an electron to travel the length of the wire?

**26–7** An aluminum wire carrying a current has diameter 0.800 mm. The electric field in the wire is 0.520 V/m. What is a) the current carried by the wire; b) the potential difference between two points in the wire 12.0 m apart; c) the resistance of a 12.0-m length of this wire?

**26–8** What length of copper wire 0.600 mm in diameter has a resistance of 1.00 Ω?

**26–9** The potential difference between points in a wire 8.00 m apart is 7.20 V when the current density is $3.40 \times 10^7\ \text{A/m}^2$. What is a) the electric field in the wire; b) the resistivity of the material of which the wire is made?

**26–10** What diameter must an aluminum wire have if its resistance is to be the same as that of an equal length of copper wire with diameter 1.80 mm?

**26–11** a) What is the resistance of a Nichrome wire at 0.0°C if its resistance is 100.00 Ω at 14.0°C? b) What is the resistance of a carbon rod at 30.0°C if its resistance is 0.0190 Ω at 0.0°C?

**26–12** A certain resistor has a resistance of 150.4 Ω at 20.0°C and a resistance of 162.4 Ω at 36.0°C. What is its temperature coefficient of resistivity?

**26–13** A carbon resistor is to be used as a thermometer. On a winter day when the temperature is 4.0°C, the resistance of the carbon resistor is 217.3 Ω. What is the temperature on a hot summer day when the resistance is 213.6 Ω?

## Section 26–4 Electromotive Force and Circuits

**26–14** A complete circuit consists of a 12.0-V automobile battery, a 4.20-Ω resistor (a headlight filament), and a switch. The internal resistance of the battery is 0.30 Ω. The switch is opened. What does an ideal voltmeter read when placed a) across the terminals of the battery; b) across the resistor; c) across the switch? d) Repeat parts (a), (b), and (c) for the case when the switch is closed.

**26–15** When switch $S$ in Fig. 26–26 is open, the voltmeter V connected across the terminals of the battery reads 1.48 V. When the switch is closed, the voltmeter reading drops to 1.37 V, and the ammeter A reads 1.30 A. Find the emf and internal resistance of the battery. Assume that the two meters are ideal, so they don't affect the circuit.

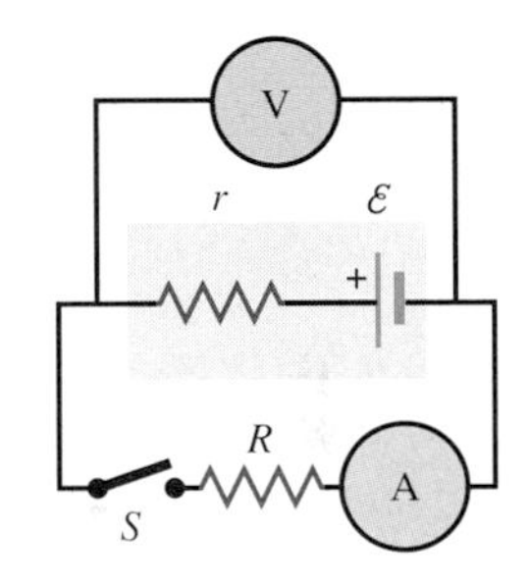


**FIGURE 26–26**

**26–16** Consider the circuit shown in Fig. 26–27. The terminal voltage of the 24.0-V battery is 21.6 V, and the current in the circuit is 4.00 A. What is a) the internal resistance $r$ of the battery; b) the resistance $R$ of the circuit resistor?

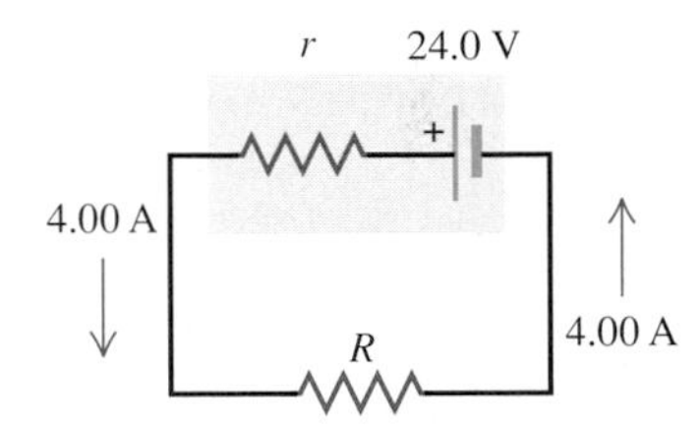


**FIGURE 26–27**

**26–17 A Battery Tester.** The internal resistance of a flashlight battery increases gradually with age, even though the battery is not used. The emf, however, remains fairly constant at about 1.5 V. A battery may be tested for age at the time of purchase by connecting an ammeter directly across the terminals of the battery and reading the current. The resistance of the ammeter is so small that the battery is practically short-circuited. a) The short-circuit current of a fresh flashlight battery (1.50-V emf) is 15.0 A. What is the internal resistance? b) What is the internal resistance if the short-circuit current is only 5.00 A? c) The short-circuit current of a 12.0-V car battery may be as great as 1000 A. What is its internal resistance?

**26–18** The circuit shown in Fig. 26–28 contains two batteries, each with an emf and an internal resistance, and two resistors. Find a) the current in the circuit; b) the terminal voltage $V_{ab}$ of the 16.0-V battery; c) the potential difference $V_{ac}$ of point $a$ with respect to point $c$.

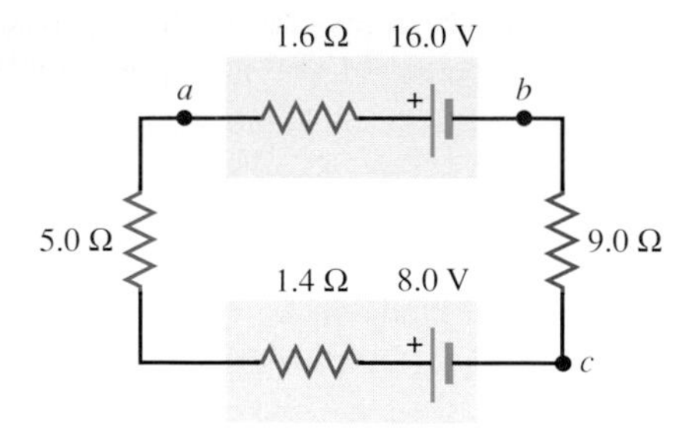


**FIGURE 26–28**

**26–19** The following measurements of current and potential difference were made on a resistor constructed of Nichrome wire:

| $I$ (A) | 0.50 | 1.00 | 2.00 | 4.00 |
|---|---|---|---|---|
| $V_{ab}$ (V) | 2.18 | 4.36 | 8.72 | 17.44 |

a) Make a graph of $V_{ab}$ as a function of $I$. b) Does Nichrome obey Ohm's law? c) What is the resistance of the resistor in ohms?

**26–20** The following measurements were made on a Thyrite resistor:

| $I$ (A) | 0.50 | 1.00 | 2.00 | 4.00 |
|---|---|---|---|---|
| $V_{ab}$ (V) | 4.76 | 5.81 | 7.05 | 8.56 |

a) Make a graph of $V_{ab}$ as a function of $I$. Does Thyrite obey Ohm's law? b) Construct a graph of the resistance $R = V_{ab}/I$ as a function of $I$.

## Section 26–5
## Energy and Power in Electric Circuits

**26–21** Consider a resistor with length $L$, uniform cross-section area $A$, and uniform resistivity $\rho$ that is carrying a current with uniform current density $J$. Use Eq. (26–20) to find the electrical power dissipated per unit volume, $p$. Express your result in terms of a) $E$ and $J$; b) $J$ and $\rho$; c) $E$ and $\rho$.

**26–22** A resistor develops heat at the rate of 360 W when the potential difference across its ends is 180 V. What is its resistance?

**26–23** A "480-W" electric heater is designed to operate from 120-V lines. a) What is its resistance? b) What current does it draw? c) If the line voltage drops to 110 V, what power does the heater take? (Assume that the resistance is constant. Actually, it will change because of the change in temperature.)

**26–24** A car radio operating on 12.0 V draws a current of 0.250 A. How much electrical energy is consumed by the radio in 4.00 h?

**26–25 Charging a Car Battery.** The capacity of a storage battery, such as those used in automobile electrical systems, is rated in ampere-hours (A · h). A 50-A · h battery can supply a current of 50 A for 1.0 h, or 25 A for 2.0 h, and so on. a) What total energy is stored in a 12-V, 50-A · h battery if its internal resistance is negligible? b) What volume (in liters) of gasoline has a total heat of combustion equal to the energy obtained in part (a)? (See Section 15–6. The density of gasoline is 900 kg/m$^3$.) c) If a windmill-powered generator has an average electrical power output of 400 W, how much time is required for it to charge the battery fully?

**26–26 The Household Flashlight.** A typical small flashlight contains two batteries, each having an emf of 1.50 V, connected in series with a bulb having resistance 16.0 Ω. a) If the internal resistance of the batteries is negligible, what power is delivered to the bulb? b) If the batteries last for 5 hours, what is the total energy delivered to the bulb? c) The resistance of real batteries increases as they run down. If the initial internal resistance is negligible, what is the combined internal resistance of both batteries when the power to the bulb has decreased to half its initial value? (Assume that the resistance of the bulb is constant. Actually, it will change somewhat as the temperature of the filament changes when the current through it changes.)

**26–27** In the circuit in Fig. 26–29, find a) the rate of conversion of internal (chemical) energy to electrical energy within the battery; b) the rate of dissipation of electrical energy in the battery; c) the rate of dissipation of electrical energy in the external resistor.

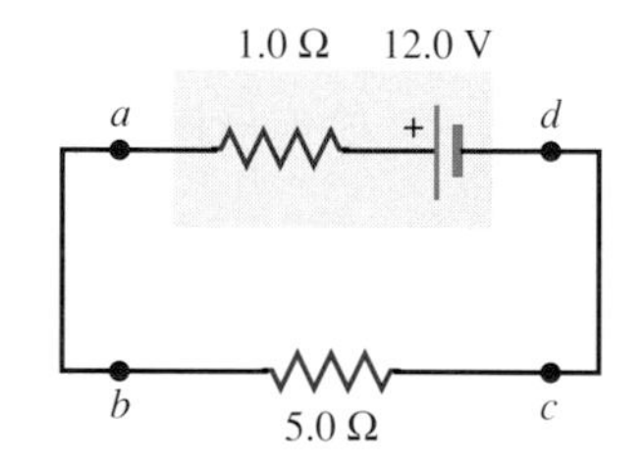


**FIGURE 26–29**

**26–28** Consider the circuit of Fig. 26–28. a) What is the total rate at which electrical energy is being dissipated in the 5.0-Ω and 9.0-Ω resistors? b) What is the power output of the 16.0-V battery? c) At what rate is electrical energy being converted to other forms in the 8.0-V battery? d) Show that the overall rate of conversion of non-electrical (chemical) energy to electrical energy equals the overall rate of dissipation of electrical energy in the circuit.

## Section 26–6
## Theory of Metallic Conduction

***26–29** For silver, the density of free electrons is $n = 5.80 \times 10^{28}/\text{m}^{-3}$. At room temperature, what mean free time $\tau$ gives a value of the resistivity that agrees with the value given in Table 26–1?

## Section 26–7 Physiological Effects of Currents

***26–30** The average bulk resistivity of the human body (apart from surface resistance of the skin) is about 5.0 Ω · m. The conducting path between the hands can be represented approximately as a cylinder 1.6 m long and 0.10 m in diameter. The skin resistance can be made negligible by soaking the hands in salt water (or seawater). a) What is the resistance between the hands if the skin resistance is negligible? b) What potential difference between the hands is needed for a lethal shock current of 100 mA? (Note that your result shows that small potential differences produce dangerous currents when the skin is damp.) c) With the current in part (b), what power is dissipated in the body?

***26–31** A person with body resistance between his hands of 10 kΩ accidently grasps the terminals of a 20-kV power supply. a) If the internal resistance of the power supply is 2000 Ω, what is the current through the person's body? b) What is the power dissipated in his body? c) If the power supply is to be made safe by increasing its internal resistance, what should the internal resistance be in order for the maximum current in the above situation to be 1.00 mA or less?

# PROBLEMS

**26–32** a) What is the potential difference $V_{ad}$ in the circuit of Fig. 26–30? b) What is the terminal voltage of the 4.00-V battery? c) A battery with emf 12.00 V and internal resistance 0.50 Ω is inserted in the circuit at $d$, with its negative terminal connected to the negative terminal of the 8.00-V battery. Now what is the difference of potential $V_{bc}$ between the terminals of the 4.00-V battery?

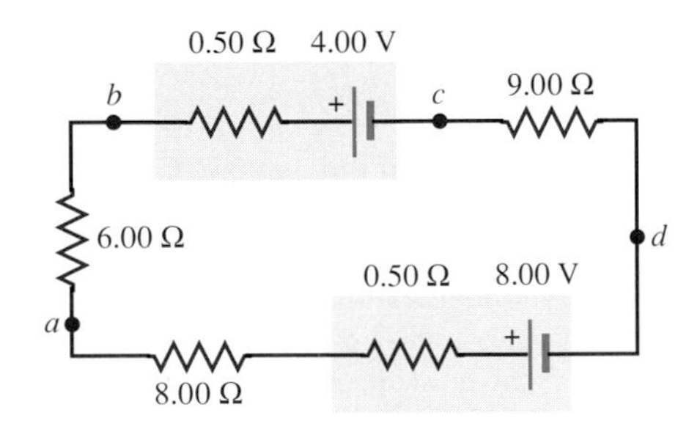


**FIGURE 26–30**

**26–33** An electrical conductor designed to carry large currents has a square cross section 2.00 mm on a side and is 12.0 m long. The resistance between its ends is 0.0640 Ω. a) What is the resistivity of the material? b) If the electric-field magnitude in the conductor is 1.20 V/m, what is the total current? c) If the material has $8.5 \times 10^{28}$ free electrons per cubic meter, find the average drift speed under the conditions of part (b).

**26–34** The region between two concentric conducting spheres with radii $a$ and $b$ is filled with a conducting material with resistivity $\rho$. a) Show that the resistance between the spheres is given by

$$R = \frac{\rho}{4\pi}\left(\frac{1}{a} - \frac{1}{b}\right).$$

b) Derive an expression for the current density as a function of radius, if the potential difference between the spheres is $V_{ab}$. c) Show that the result in part (a) reduces to Eq. (26–12) when the separation $L = b - a$ between the spheres is small.

**26–35 Leakage in a Dielectric.** Two parallel plates of a capacitor have equal and opposite charges $Q$. The dielectric has a dielectric constant $K$ and a resistivity $\rho$. Show that the "leakage" current $I$ carried by the dielectric is given by $I = Q/K\epsilon_0\rho$.

**26–36** A toaster using a Nichrome heating element operates on 120 V. When it is switched on at 20°C, the heating element carries an initial current of 1.46 A. A few seconds later the current reaches the steady value of 1.32 A. What is the final temperature of the element? The average value of the temperature coefficient of resistivity for Nichrome over the temperature range is $4.50 \times 10^{-4}\,(\mathrm{C}°)^{-1}$.

**26–37** The potential difference across the terminals of a battery is 9.2 V when there is a current of 3.00 A in the battery from the negative to the positive terminal. When the current is 2.00 A in the reverse direction, the potential difference becomes 11.2 V. a) What is the internal resistance of the battery? b) What is the emf of the battery?

**26–38** A semiconductor device in a personal computer does not obey Ohm's law but instead has the current-voltage relation $V = \alpha I + \beta I^2$, with $\alpha = 2.0$ Ω and $\beta = 0.500$ Ω/A. a) If the device is connected across a potential difference of 3.00 V, what is the current through the device? b) What potential difference is required to produce a current through the device twice that calculated in part (a)?

**26–39** A 12.0-V car battery with negligible internal resistance is connected to a 6.00-Ω resistor that obeys Ohm's law in series with a thermistor that does not obey Ohm's law but instead has a current-voltage relation $V = \alpha I + \beta I^2$, with $\alpha = 4.00$ Ω and $\beta = 1.20$ Ω/A. What is the current through the 6.00-Ω resistor?

**26–40** The open-circuit terminal voltage of a source is 9.00 V, and its short-circuit current is 4.00 A. a) What is the current when a resistor with resistance 2.00 Ω is connected to the terminals of the source? The resistor obeys Ohm's law. b) What is the current in the Thyrite resistor of Exercise 26–20 when connected across the terminals of this source? c) What is the terminal voltage of the source with the current calculated in part (b)?

**26–41** A 12.0-V automobile storage battery has a capacity of 60.0 A · h. (See Exercise 26–25.) Its internal resistance is 0.300 Ω. The battery is charged by passing a 15.0-A current through it for 4.00 h. a) What is the terminal voltage during charging? b) What total electrical energy is supplied to the battery during charging? c) What electrical energy is dissipated in the internal resistance during charging? d) The battery is now completely discharged through a resistor, again with a constant current of 15.0 A. What is the external circuit resistance? e) What total electrical energy is supplied to the external resistor? f) What total electrical energy is dissipated in the internal resistance? g) Why are the answers to parts (b) and (e) not equal?

**26–42** Repeat Problem 26–41 with charge and discharge currents of 30.0 A. The charging and discharging times will now be 2.00 h rather than 4.00 h. What differences in performance do you see?

**26–43** In the circuit of Fig. 26–31, find a) the current through the 8.0-Ω resistor; b) the total rate of dissipation of electrical energy in the 8.0-Ω resistor and in the internal resistance of the batteries. c) In one of the batteries, chemical energy is being converted into electrical energy. In which one is this happening, and at what rate? d) In one of the batteries, electrical energy is being converted into chemical energy. In which one is this happening, and at what rate? e) Show that the overall rate of production of electrical energy equals the overall rate of consumption of electrical energy in the circuit.

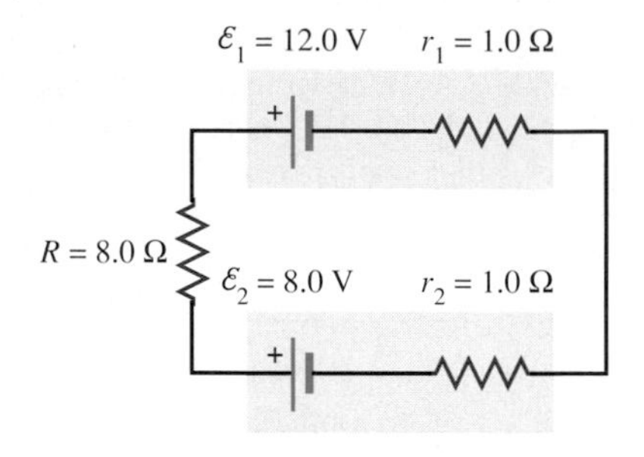


**FIGURE 26–31**

## CHALLENGE PROBLEMS

**26–44 Maximum Power of a Source.** A source with emf $\mathcal{E}$ and internal resistance $r$ is connected to an external circuit. a) Show that the power output of the source is maximum when the current in the circuit is one-half the short-circuit current of the source. b) If the external circuit consists of a resistance $R$, show that the power output is maximum when $R = r$, and that the maximum power is $\mathcal{E}^2/4r$.

**26–45** The definition of $\alpha$, the temperature coefficient of resistivity, is

$$\alpha = \frac{1}{\rho}\frac{d\rho}{dT},$$

where $\rho$ is the resistivity at the temperature $T$. Eq. (26–8) then follows if $\alpha$ is assumed to be constant and much smaller than $(T - T_0)^{-1}$. a) If $\alpha$ is not constant but is given by $\alpha = -n/T$, where $T$ is the Kelvin temperature and $n$ is a constant, show that the resistivity $\rho$ is given by $\rho = a/T^n$, where $a$ is a constant. b) From Fig. 26–9c, you can see that such a relation might be used as a rough approximation for a semiconductor. Using the values of $\rho$ and $\alpha$ for carbon from Table 26–1 and Table 26–2, determine $a$ and and $n$. (In Table 26–1, assume that "room temperature" means 293 K.) c) Using your result from part (b), determine the resistivity of carbon at $-196°$C and at 300°C. (Remember to express $T$ in kelvins.)

**26–46 Voltage across a Diode.** A semiconductor diode is a nonlinear device whose current-voltage relationship is described by

$$I = I_0[\exp(eV/kT) - 1],$$

where $I$ and $V$ are the current through and the voltage across the diode, respectively, $I_0$ is a constant characteristic of the device, $e$ is the magnitude of the electron charge, $k$ is Boltzmann's constant, and $T$ is the Kelvin temperature. Such a diode is connected in series with a resistor with $R = 1.00\ \Omega$ and a battery with $\mathcal{E} = 2.00$ V. The polarity of the battery is such that the current through the diode is in the forward direction (Fig. 26–32). The battery has negligible internal resistance. a) Obtain an equation for $V$. Note that you cannot solve for $V$ algebraically. b) Since $V$ cannot be solved algebraically, the value of $V$ must be obtained by using a numerical method. One approach is to try a value of $V$, see how the left- and right-hand sides of the equation compare for this $V$, and use this to refine your guess for $V$. Using $I_0 = 1.50$ mA and $T = 293$ K, obtain a solution (accurate to three significant figures) for the voltage drop $V$ across the diode and the current $I$ through it.

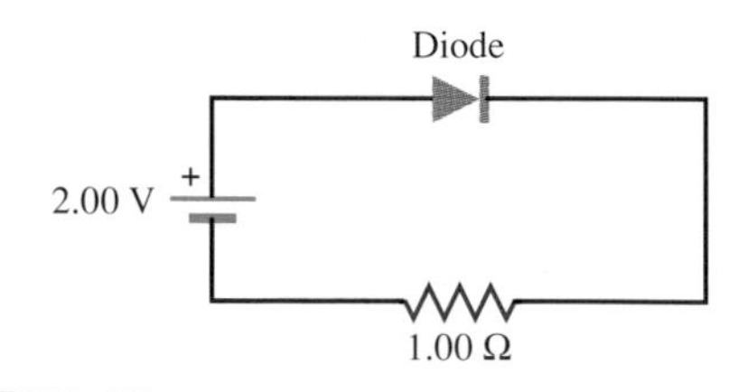


**FIGURE 26–32**

Modern electronic equipment, such as a compact disk player, contains an enormous number of complex circuits, each performing a different function. The circuits are integrated into devices, such as microprocessor chips, and these devices are wired into circuit boards.

CHAPTER 27

# Direct Current Circuits

The design of integrated circuits is an art in itself. Circuits are drawn, often by computer, and the drawings are used in a series of photographic processes to etch or deposit the proper conducting paths directly on a semiconductor chip.

## KEY CONCEPTS

**• When several resistors are connected in series or in parallel, there is an equivalent resistance for the combination. Resistors in series add directly; resistors in parallel add reciprocally.**

**• Kirchhoff's rules provide a general method for analyzing circuit networks. The current rule states that the algebraic sum of currents into any branch point is zero. The voltage rule states that the algebraic sum of potential differences around a closed loop is zero.**

**• A voltmeter measures the potential difference between two points; an ideal voltmeter has infinite resistance and zero current. An ammeter measures the current through it; an ideal ammeter has zero resistance.**

**• When a capacitor is charged or discharged by current through a resistor, the current and charge are exponential functions of time, approaching their final values asymptotically after a long time.**

## INTRODUCTION

If you look inside your TV, your computer, or your stereo receiver or under the hood of a car, you will find circuits of much greater complexity than the simple circuits that we studied in Chapter 26. Whether connected by wires or integrated in a semiconductor chip, these circuits often include several sources, resistors, and other circuit elements such as capacitors, transformers, and motors, interconnected in a *network*.

In this chapter we study general methods for analyzing networks, including finding unknown voltages, currents, and properties of circuit elements. We'll learn how to find the equivalent resistance for several resistors connected in series or in parallel. For more general networks we need two rules called *Kirchhoff's rules.* One is based on the principle of conservation of charge applied to a junction; the other is derived from energy conservation for a charge moving around a closed loop. We also discuss instruments for measuring various electrical quantities. We look at a circuit containing resistance and capacitance, in which the current varies with time. Finally, we look at how these principles are related to household wiring systems. ■

## 27–1 RESISTORS IN SERIES AND IN PARALLEL

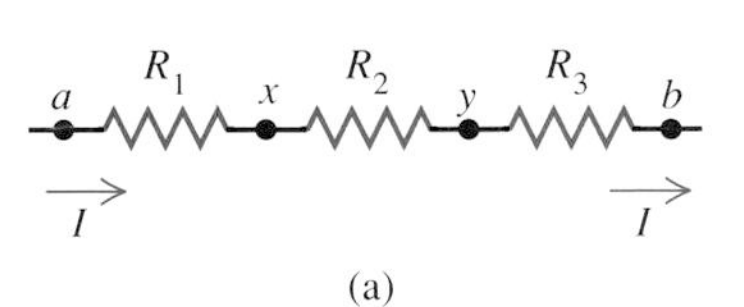

(a)

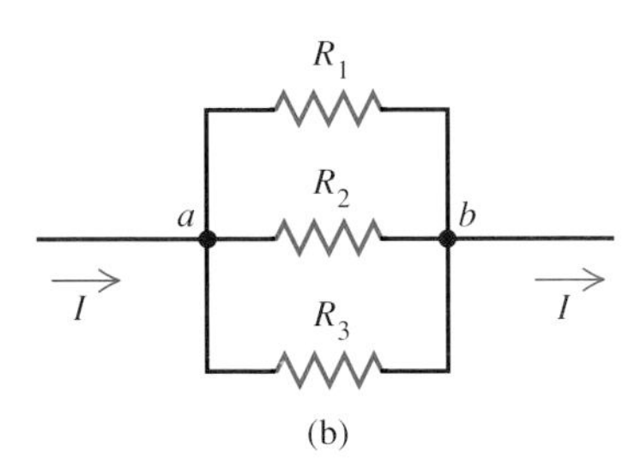

(b)

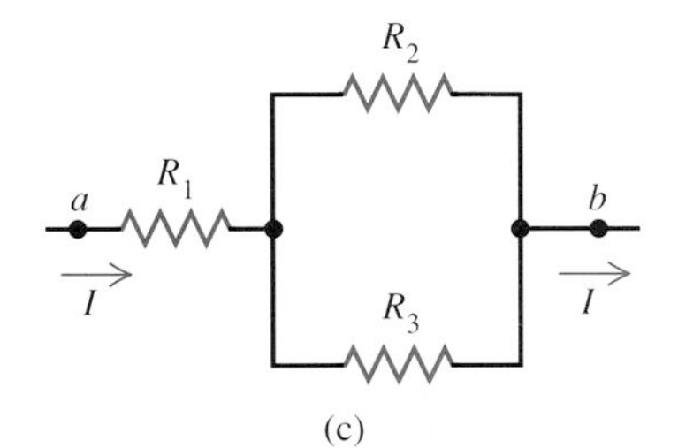

(c)

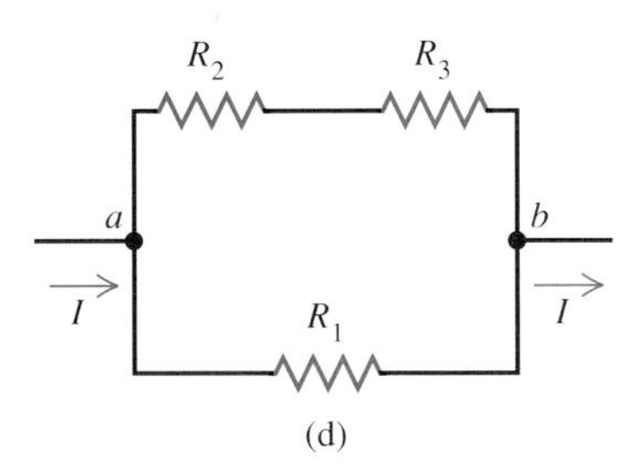

(d)

**27–1** Four different ways of connecting three resistors.

Resistors turn up in all kinds of circuits, ranging from hair dryers and space heaters to circuits that limit or divide current or reduce or divide a voltage. Suppose we have three resistors with resistances $R_1$, $R_2$, and $R_3$. Figure 27–1 shows four different ways in which they might be connected between points $a$ and $b$. In Fig. 27–1a the resistors provide only a single path between these points. When several circuit elements such as resistors, batteries, and motors are connected in sequence as in Fig. 27–1a, with only a single current path between the points, we say that they are connected in **series.** We studied *capacitors* in series in Section 25–3; we found that, because of conservation of charge, capacitors in series all have the same charge if they are initially uncharged. In circuits we're often more interested in the *current,* which is charge flow per unit time. Conservation of charge requires that when circuit elements are connected in series, the *current* is the same in each element.

The resistors in Fig. 27–1b are said to be connected in **parallel** between points $a$ and $b$. Each resistor provides an alternative path between the points. For circuit elements that are connected in parallel, the *potential difference* is the same across each element. We studied capacitors in parallel in Section 25–3.

In Fig. 27–1c, resistors $R_2$ and $R_3$ are in parallel, and this combination is in series with $R_1$. In Fig. 27–1d, $R_2$ and $R_3$ are in series, and this combination is in parallel with $R_1$.

For any combination of resistors that obey Ohm's law we can always find a single resistor that could replace the combination and result in the same total current and potential difference. The resistance of this single resistor is called the **equivalent resistance** of the combination. If any one of the networks in Fig. 27–1 were replaced by its equivalent resistance $R_{eq}$, we could write

$$V_{ab} = IR_{eq}, \qquad \text{or} \qquad R_{eq} = \frac{V_{ab}}{I},$$

where $V_{ab}$ is the potential difference between terminals $a$ and $b$ of the network and $I$ is the current at point $a$ or $b$. To compute an equivalent resistance, we assume a potential difference $V_{ab}$ across the actual network, compute the corresponding current $I$, and take the ratio $V_{ab}/I$.

For series and parallel combinations we can derive general equations for the equivalent resistance. If the resistors are in *series,* as in Fig. 27–1a, the current $I$ must be the same in all of them. (Otherwise, charge would be piling up at the junctions.) Applying $V = IR$ to each, we have

$$V_{ax} = IR_1, \qquad V_{xy} = IR_2, \qquad V_{yb} = IR_3.$$

The potential difference $V_{ab}$ is the sum of these three quantities:

$$V_{ab} = V_{ax} + V_{xy} + V_{yb} - I(R_1 + R_2 + R_3),$$

or

$$\frac{V_{ab}}{I} = R_1 + R_2 + R_3.$$

But $V_{ab}/I$ is, by definition, the equivalent resistance $R_{eq}$. Therefore

$$R_{eq} = R_1 + R_2 + R_3.$$

rizes these sign conventions. In each part of the figure "travel" is the direction in which we are going around a loop using Kirchhoff's loop law, not necessarily the direction of current.

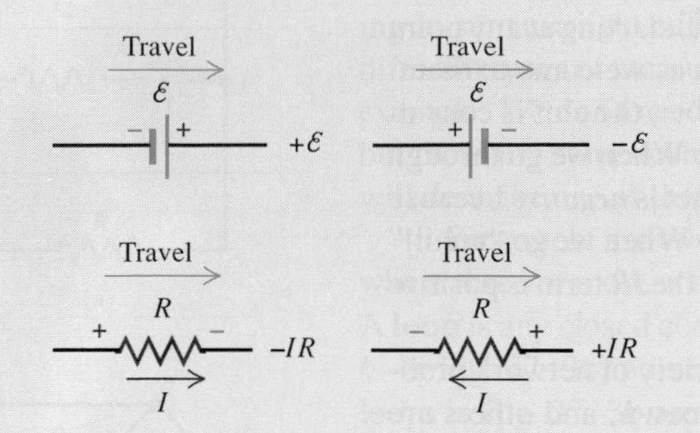


**27–6** Sign conventions to use in traveling around a circuit loop when using Kirchhoff's rules.

**5.** Equate the sum in Step 4 to zero.

**6.** If necessary, choose another loop to get a different relation among the unknowns, and continue until you have as many independent equations as unknowns or until every circuit element has been included in at least one of the chosen loops.

**7.** Finally, solve the equations simultaneously to determine the unknowns. This step is algebra, not physics, but it can be fairly complex. Be careful with algebraic manipulations; one sign error is fatal to the entire solution. Symbolic computer programs such as Mathematica® are very useful for this step.

**8.** You can use this same bookkeeping system to find the potential $V_{ab}$ of any point $a$ with respect to any other point $b$. Start at $b$ and add the potential changes that you encounter in going from $b$ to $a$, using the same sign rules as in Step 4. The algebraic sum of these changes is $V_{ab} = V_a - V_b$.

## EXAMPLE 27–2

**A one-loop circuit** The circuit shown in Fig. 27–7 contains two batteries, each with an emf and an internal resistance, and two resistors. Find the current in the circuit and the potential difference $V_{ab}$.

**SOLUTION** There is only one loop, so we don't need Kirchhoff's junction rule. To use the loop rule, we first assume a direction for the current, as shown. Then, starting at $a$ and going counterclockwise, we add potential increases and decreases and equate the sum to zero. The resulting equation is

$$-I(4\ \Omega) - 4\ \text{V} - I(7\ \Omega) + 12\ \text{V} - I(2\ \Omega) - I(3\ \Omega) = 0.$$

Collecting terms containing $I$ and solving for $I$, we find

$$8\ \text{V} = I(16\ \Omega), \qquad \text{and} \qquad I = 0.5\ \text{A}.$$

The result for $I$ is positive, showing that our assumed current direction is correct. For an exercise, try assuming the opposite direction for $I$; you should then get $I = -0.5$ A, indicating that the actual current is opposite to this assumption.

To find $V_{ab}$, the potential at $a$ with respect to $b$, we start at $b$ and go toward $a$, adding potential changes. There are two possible paths from $b$ to $a$; taking the lower one first, we find

$$V_{ab} = (0.5\ \text{A})(7\ \Omega) + 4\ \text{V} + (0.5\ \text{A})(4\ \Omega) = 9.5\ \text{V}.$$

Point $a$ is at 9.5 V higher potential than $b$. All the terms in this sum are positive because each represents an *increase* in potential as we go from $b$ toward $a$. If instead we use the upper path, the resulting equation is

$$V_{ab} = 12\ \text{V} - (0.5\ \text{A})(2\ \Omega) - (0.5\ \text{A})(3\ \Omega) = 9.5\ \text{V}.$$

Here the $IR$ terms are negative because our path goes in the direction of the current, with potential decreases through the resistors. The result is the same as that for the lower path, as it must be for the total potential change around the complete loop to be zero. In each case, potential rises are taken as positive and drops as negative.

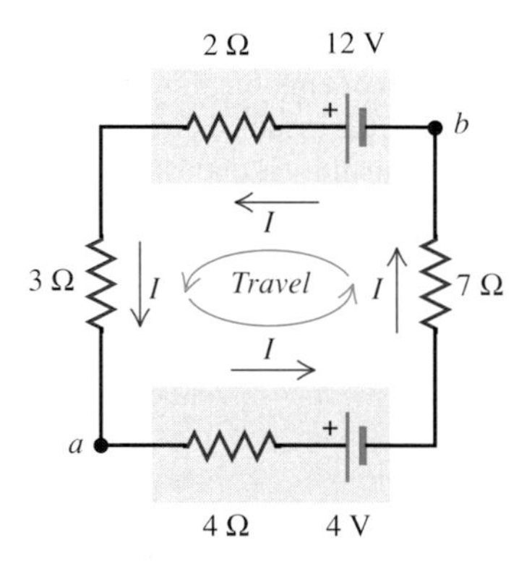


**27–7** In this example we travel around the loop in the same direction as the assumed current, so all the $IR$ terms are negative. The potential decreases as we travel from + to − through the bottom emf but increases as we travel from − to + through the top emf.

## EXAMPLE 27–3

**Charging a battery** In the circuit shown in Fig. 27–8 a 12-V power supply with unknown internal resistance $r$ (represented as a battery) is connected to a run-down rechargeable battery with unknown emf $\mathcal{E}$ and internal resistance 1 Ω and to a bulb with a resistance 3 Ω carrying a current of 2 A. Find the unknown current $I$, the internal resistance $r$, and the emf $\mathcal{E}$.

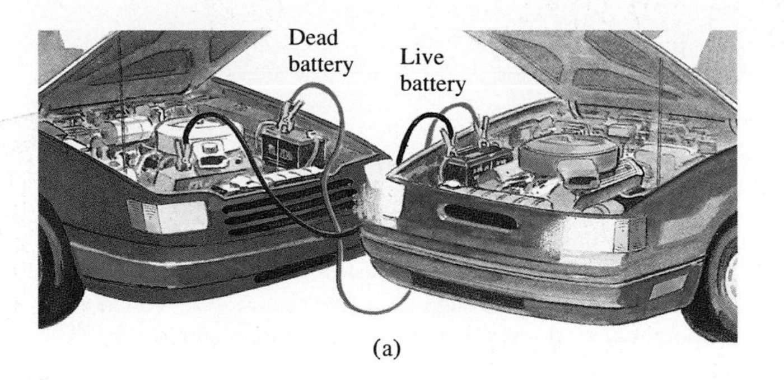


(a)

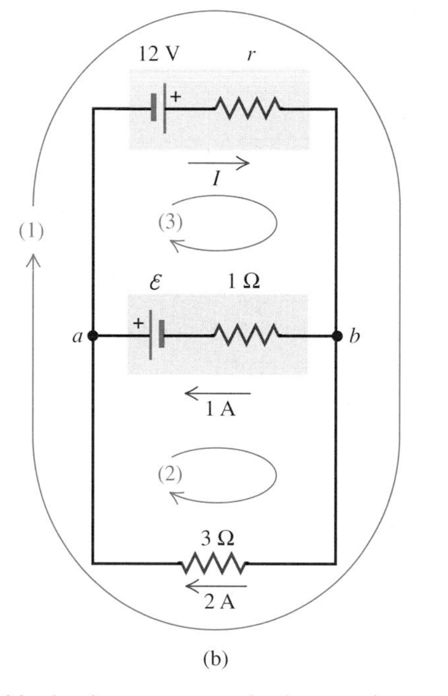


(b)

**27–8** In this circuit, a power supply charges a battery and lights a bulb.

**SOLUTION** First we apply the junction rule to point $a$, obtaining

$$-I + 1\text{ A} + 2\text{ A} = 0, \qquad I = 3\text{ A}.$$

To determine $r$, we apply the loop rule to the outer loop labeled (1); we find

$$12\text{ V} - (3\text{ A})\,r - (2\text{ A})(3\ \Omega) = 0,$$
$$r = 2\ \Omega.$$

The terms containing the resistances $r$ and $3\ \Omega$ are negative because we traverse our loop in the same direction as the current and thus find potential *drops*. If we had chosen to traverse loop (1) in the opposite direction, every term would have had the opposite sign, and the result for $r$ would have been the same.

To determine $\mathcal{E}$, we apply the loop rule to loop (2):

$$-\mathcal{E} + (1\text{ A})(1\ \Omega) - (2\text{ A})(3\ \Omega) = 0,$$
$$\mathcal{E} = -5\text{ V}.$$

The term for the 1-Ω resistor is positive because in traversing it in the direction opposite to the current, we find a potential *rise*. The negative value for $\mathcal{E}$ shows that the actual polarity of this emf is opposite to the assumption made in the figure; the positive terminal of this source is really on the right side. Alternatively, we could use loop (3), obtaining

$$12\text{ V} - (3\text{ A})(2\ \Omega) - (1\text{ A})(1\ \Omega) + \mathcal{E} = 0,$$
$$\mathcal{E} = -5\text{ V}.$$

As an additional consistency check, we note that $V_{ba} = V_b - V_a$ equals the voltage across the 3-Ω resistance, which is 6 V. Going from $a$ to $b$ by the top branch, we encounter potential differences $+12\text{ V} - (3\text{ A})(2\ \Omega) = +6\text{ V}$; by the middle branch, we find $-(-5\text{ V}) + (1\text{ A})(1\ \Omega) = +6\text{ V}$. The three ways of getting $V_{ba}$ give the same results. Make sure that you understand all the signs in these calculations.

## EXAMPLE 27–4

In Fig. 27–8 (in Example 27–3), find the power delivered by the 12-V power supply and by the battery and the power dissipated in each resistor.

**SOLUTION** The power output from the emf of the power supply is

$$P = \mathcal{E}I = (12\text{ V})(3\text{ A}) = 36\text{ W}.$$

The power dissipated in the power supply's internal resistance $r$ is

$$P = I^2r = (3\text{ A})^2(2\ \Omega) = 18\text{ W},$$

so the net power output of the power supply is 36 W − 18 W = 18 W. Alternatively, from Example 27–3 the terminal voltage of the battery is 6 V, so the net power output is

$$P = V_{ab}I = (6\text{ V})(3\text{ A}) = 18\text{ W}.$$

The power output of the battery's emf is

$$P = \mathcal{E}I = (-5\text{ V})(1\text{ A}) = -5\text{ W}.$$

This is negative because the 1-A current actually runs from the higher-potential side of the battery to the lower-potential side. We are storing energy in the battery as we charge it. Additional power is dissipated in the battery's internal resistance; this power is

$$P = I^2R = (1\text{ A})^2(1\ \Omega) = 1\text{ W}.$$

The total power input to the battery is thus $1\text{ W} + |-5\text{W}| = 6\text{ W}$. Of this, 5 W represent useful energy stored in the battery; the remainder is wasted in its internal resistance.

The power dissipated in the bulb is

$$P = I^2R = (2\text{ A})^2(3\ \Omega) = 12\text{ W}.$$

The net power from the power supply is 18 W.

## EXAMPLE 27–5

In Fig. 27–9, find the current in each resistor and the equivalent resistance of the network.

**SOLUTION** As we pointed out at the beginning of this section, this network cannot be represented in terms of series and parallel combinations. There are five different currents to determine, but by applying the junction rule to junctions *a* and *b* we can represent them in terms of three unknown currents, as shown in the figure. The current in the battery is $I_1 + I_2$.

We apply the loop rule to the three loops shown, obtaining the following three equations:

$$13\text{ V} - I_1(1\,\Omega) - (I_1 - I_3)(1\,\Omega) = 0; \qquad (1)$$

$$-I_2(1\,\Omega) - (I_2 + I_3)(2\,\Omega) + 13\text{ V} = 0; \qquad (2)$$

$$-I_1(1\,\Omega) - I_3(1\,\Omega) + I_2(1\,\Omega) = 0. \qquad (3)$$

This is a set of three simultaneous equations for the three unknown currents. They may be solved by various methods; one straightforward procedure is to solve the third equation for $I_2$, obtaining $I_2 = I_1 + I_3$, and then substitute this expression into the first two equations to eliminate $I_2$. When this is done, we are left with the two equations

$$13\text{ V} = I_1(2\,\Omega) - I_3(1\,\Omega), \qquad (1')$$

$$13\text{ V} = I_1(3\,\Omega) + I_3(5\,\Omega). \qquad (2')$$

Now we can eliminate $I_3$ by multiplying Eq. (1′) by 5 and adding the two equations.

We obtain

$$78\text{ V} = I_1(13\,\Omega), \qquad I_1 = 6\text{ A}.$$

We substitute this result back into Eq. (1′) to obtain $I_3 = -1$ A, and finally from Eq. (3) we find $I_2 = 5$ A. We note that the direction of $I_3$ is opposite to our initial assumption.

The total current through the network is $I_1 + I_2 = 11$ A, and the potential drop across it is equal to the battery emf, or 13 V. The equivalent resistance of the network is

$$R_{eq} = \frac{13\text{ V}}{11\text{ A}} = 1.2\,\Omega.$$

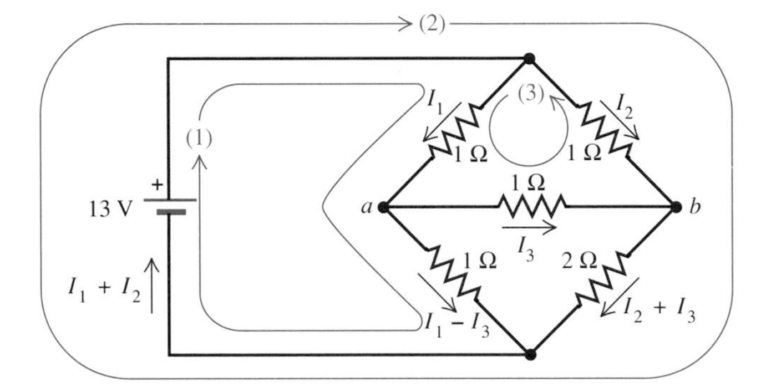


**27–9** A network circuit with several resistors.

## EXAMPLE 27–6

In Fig. 27–9 (in Example 27–5), find the potential difference $V_{ab}$.

**SOLUTION** To find $V_{ab} = V_a - V_b$, we start at point *b* and follow a path to point *a*, adding potential rises and drops as we go. The simplest path is through the central 1-Ω resistor. We have found $I_3 = -1$ A, showing that the actual current direction in this branch is from right to left. Thus as we go from *b* to *a*, there is a *drop* of potential with magnitude $IR = (1\text{ A})(1\,\Omega) = 1$ V, and $V_{ab} = -1$ V.

Alternatively, we may go through the lower two resistors. Then we have

$$I_2 + I_3 = 5\text{ A} + (-1\text{ A}) = 4\text{ A},$$

$$I_1 - I_3 = 6\text{ A} - (-1\text{ A}) = 7\text{ A},$$

and

$$V_{ab} = -(4\text{ A})(2\,\Omega) + (7\text{ A})(1\,\Omega) = -1\text{ V}.$$

We suggest that you try some other paths from *b* to *a* to verify that they also give this result.

# 27–3 ELECTRICAL MEASURING INSTRUMENTS

We've been talking about current and potential difference for two chapters; it's about time we said something about how to *measure* these quantities. Many common devices, including car instrument panels, battery chargers, and inexpensive electrical instruments, measure potential difference (voltage), current, or resistance using a device called a **d'Arsonval galvanometer.** In the following discussion we'll often just call it a *meter.* A pivoted coil of fine wire is placed in the magnetic field of a permanent magnet (Fig. 27–10). Attached to the coil is a spring, similar to the hairspring on the balance wheel of a watch. In the equilibrium position, with no current in the coil, the pointer is at

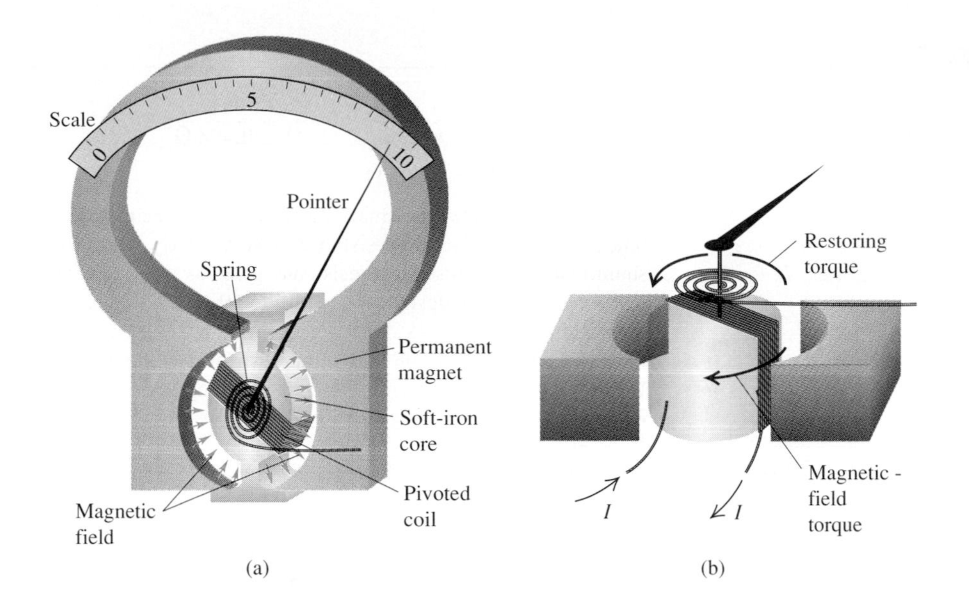


**27–10** (a) A d'Arsonval galvanometer, showing a pivoted coil with attached pointer, a permanent magnet supplying a magnetic field that is uniform in magnitude, and a spring to provide restoring torque, which opposes magnetic-field torque. (b) Pivoted coil around soft-iron core, with supports removed.

zero. When there is a current in the coil, the magnetic field exerts a torque on the coil that is proportional to the current. (We'll discuss this magnetic interaction in detail in Chapter 28.) As the coil turns, the spring exerts a restoring torque that is proportional to the angular displacement.

Thus the angular deflection of the coil and pointer is directly proportional to the coil current, and the device can be calibrated to measure current. The maximum deflection, typically 90° to 120°, is called *full-scale deflection*. The essential electrical characteristics of the meter are the current $I_{fs}$ required for full-scale deflection (typically of the order of 10 $\mu$A to 10 mA) and the resistance $R_c$ of the coil (typically of the order of 10 to 1000 $\Omega$).

The meter deflection is proportional to the *current* in the coil, but if the coil obeys Ohm's law, the current is proportional to the *potential difference* between the coil's terminals. Thus the deflection is also proportional to this potential difference. For example, consider a meter whose coil has a resistance $R_c$ of 20.0 $\Omega$ and that has full-scale deflection when the current in the coil is $I_{fs} = 1.00$ mA. The corresponding potential difference is

$$V = I_{fs}R_c = (1.00 \times 10^{-3}\ \text{A})(20.0\ \Omega) = 0.0200\ \text{V}.$$

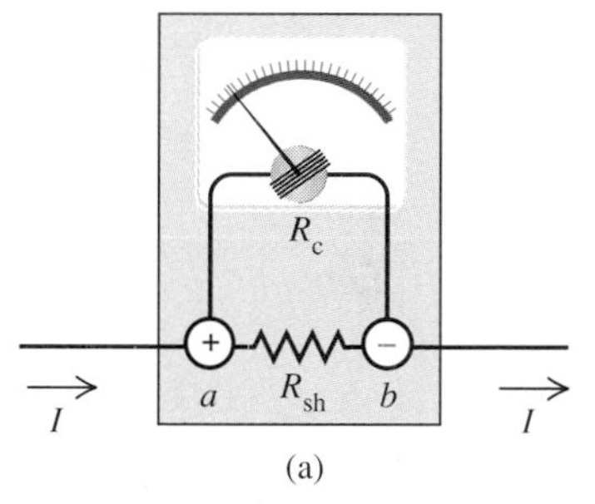


## Ammeters

A current-measuring instrument is usually called an **ammeter** (or milliammeter, microammeter, and so forth, depending on the range). *An ammeter always measures the current passing through it.* We can adapt any meter to measure currents larger than its full-scale reading by connecting a resistor in parallel with it (Fig. 27–11a) so that some of the current bypasses the meter coil. The parallel resistor is called a **shunt resistor,** or simply a *shunt,* and denoted as $R_{sh}$.

Suppose we want to make a meter with full-scale current $I_{fs}$ and coil resistance $R_c$ into an ammeter with full-scale reading $I_a$. To determine the shunt resistance $R_{sh}$ needed, we note that at full-scale deflection the total current through the parallel combination is $I_a$, the current through the meter is $I_{fs}$, and the current through the shunt is the difference $I_a - I_{fs}$. The potential difference $V_{ab}$ is the same for both paths, so

$$I_{fs}R_c = (I_a - I_{fs})R_{sh} \quad \text{(for an ammeter).} \tag{27–7}$$

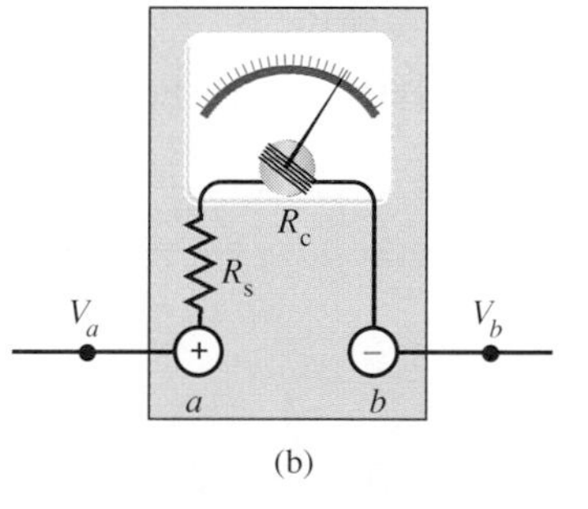


**27–11** (a) Internal connections of a moving-coil ammeter. (b) Internal connections of a moving-coil voltmeter.

## EXAMPLE 27–7

**Designing an ammeter** What shunt resistance is required to make the 1-mA, 20-Ω meter described above into an ammeter with a range of 0 to 50.0 mA?

**SOLUTION** We have $I_{fs} = 1.00\text{ mA} = 1.00 \times 10^{-3}\text{ A}$, $I_a = 50.0 \times 10^{-3}\text{ A}$, and $R_c = 20.0\ \Omega$. Solving Eq. (27–7) for $R_{sh}$, we find

$$R_{sh} = \frac{I_{fs} R_c}{I_a - I_{fs}} = \frac{(1.00 \times 10^{-3}\text{ A})(20.0\ \Omega)}{50.0 \times 10^{-3}\text{ A} - 1.00 \times 10^{-3}\text{ A}}$$
$$= 0.408\ \Omega.$$

The equivalent resistance $R_{eq}$ of the instrument is given by

$$\frac{1}{R_{eq}} = \frac{1}{R_c} + \frac{1}{R_{sh}} = \frac{1}{20\ \Omega} + \frac{1}{0.408\ \Omega},$$
$$R_{eq} = 0.400\ \Omega.$$

The shunt resistance is so small in comparison to the meter resistance that the equivalent resistance is very nearly equal to the shunt resistance. This shunt resistance gives us a low-resistance instrument with the desired range of 0 to 50.0 mA. At full-scale deflection, $I = 50.0$ mA, the current through the meter is 1.00 mA, the current through the shunt resistor is 49.0 mA, and $V_{ab} = 0.0200$ V. If the current $I$ is *less than* 50.0 mA, the coil current and the deflection are correspondingly less, but the resistance $R_{eq}$ is still 0.400 Ω.

An *ideal* ammeter would have *zero* resistance so that including it in a branch of a circuit would not affect the current in that branch. All real ammeters have some finite resistance, but it is always desirable for an ammeter to have as little resistance as possible.

## Voltmeters

The same basic meter may also be used to measure potential difference, or *voltage*. A voltage-measuring device is called a **voltmeter** (or millivoltmeter, and so on, depending on the range). A voltmeter always measures the potential difference between two points, and its terminals must be connected to these points. In the earlier example, the voltage across the meter coil at full-scale deflection was only $(1.00 \times 10^{-3}\text{ A})(20.0\ \Omega) = 0.0200$ V. We can extend this range by connecting a resistor $R_s$ in *series* with the coil (Fig. 27–11b). Then only a fraction of the total potential difference appears across the coil itself, and the remainder appears across $R_s$. For a voltmeter with full-scale reading $V_v$ we need a series resistor $R_s$ in Fig. 27–11b such that

$$V_v = I_{fs}(R_c + R_s) \quad \text{(for a voltmeter).} \tag{27–8}$$

## EXAMPLE 27–8

**Designing a voltmeter** How can we make a galvanometer with $R_c = 20.0\ \Omega$ and $I_{fs} = 0.00100$ A into a voltmeter with a maximum range of 10.0 V?

**SOLUTION** Solving Eq. (27–8) for $R_s$, we have

$$R_s = \frac{V_v}{I_{fs}} - R_c = \frac{10.0\text{ V}}{0.00100\text{ A}} - 20.0\ \Omega = 9980\ \Omega.$$

At full-scale deflection, $V_{ab} = 10.0$ V, the voltage across the meter is 0.0200 V, the voltage across $R_s$ is 9.98 V, and the current is 0.00100 A. In this case, most of the voltage appears across the series resistor. The equivalent voltmeter resistance is $R_{eq} = 20.0\ \Omega + 9980\ \Omega = 10{,}000\ \Omega$. Such a voltmeter is described as a "1,000 ohms-per-volt meter," referring to the ratio of resistance to full-scale deflection.

An ideal voltmeter would have *infinite* resistance so that connecting it between two points in a circuit would not alter any of the currents. Real voltmeters always have finite resistance, but a voltmeter should have large enough resistance that connecting it in a circuit does not change the other currents appreciably. Fairly elaborate meters with electronic amplification often have constant internal resistance of about 10 MΩ.

A voltmeter and an ammeter can be used together to measure *resistance* and *power*. The resistance $R$ of a resistor is the potential difference $V_{ab}$ between its terminals, divided by the current $I$: $R = V_{ab}/I$. The power input $P$ to any circuit element is the product of

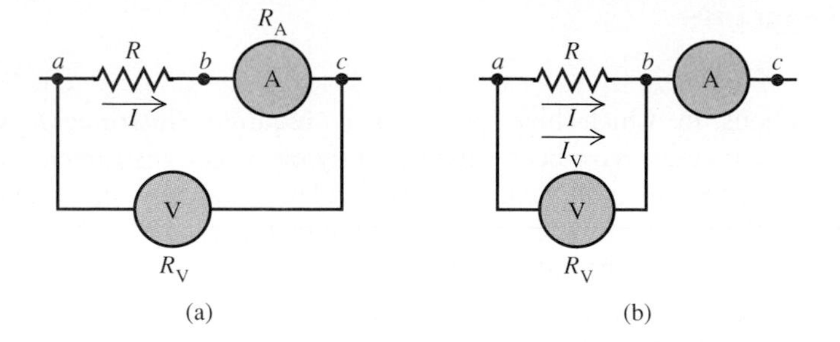


**27–12** Ammeter-voltmeter method for measuring resistance.

the potential difference across it and the current through it: $P = V_{ab}I$. In principle, the most straightforward way to measure $R$ or $P$ is to measure $V_{ab}$ and $I$ simultaneously.

With practical ammeters and voltmeters this isn't quite as simple as it seems. In Fig. 27–12a, ammeter A reads the current $I$ in the resistor $R$. Voltmeter V, however, reads the *sum* of the potential difference $V_{ab}$ across the resistor and the potential difference $V_{bc}$ across the ammeter. If we transfer the voltmeter terminal from $c$ to $b$, as in Fig. 27–12b, then the voltmeter reads the potential difference $V_{ab}$ correctly, but the ammeter now reads the *sum* of the current $I$ in the resistor and the current $I_V$ in the voltmeter. Either way, we have to correct the reading of one instrument or the other unless the corrections are small enough to be negligible.

## EXAMPLE 27–9

**Measuring resistance** Suppose we want to measure an unknown resistance $R$ using the circuit of Fig. 27–12a. The meter resistances are $R_V = 10{,}000\ \Omega$ and $R_A = 2.00\ \Omega$. If the voltmeter reads 12.0 V and the ammeter reads 0.100 A, what is the true resistance?

**SOLUTION** If the meters were ideal (that is, if $R_V = \infty$ and $R_A = 0$), the resistance would be simply $R = V/I = (12.0\ \text{V})/(0.100\ \text{A}) = 120\ \Omega$. But the voltmeter reading includes the voltage $V_{bc}$ across the ammeter as well as the voltage $V_{ab}$ across the resistor. We have $V_{bc} = IR_A = (0.100\ \text{A})(2.00\ \Omega) = 0.200\ \text{V}$, so the potential drop $V_{ab}$ across the resistor is $12.0\ \text{V} - 0.200\ \text{V} = 11.8\ \text{V}$, and the resistance is

$$R = \frac{V_{ab}}{I} = \frac{11.8\ \text{V}}{0.10\ \text{A}} = 118\ \Omega.$$

## EXAMPLE 27–10

Suppose the meters of Example 27–9 are connected to a different resistor, in the circuit shown in Fig. 27–12b, and the same readings are obtained. What is the true resistance?

**SOLUTION** In this case the voltmeter measures the potential across the resistor correctly; the difficulty is that the ammeter measures the voltmeter current $I_V$ as well as the current $I$ in the resistor. We have $I_V = V/R_V = (12.0\ \text{V})/(10{,}000\ \Omega) = 1.20\ \text{mA}$. The actual current $I$ in the resistor is $I = 0.100\ \text{A} - 0.0012\ \text{A} = 0.0988\ \text{A}$, and the resistance is

$$R = \frac{V_{ab}}{I} = \frac{12.0\ \text{V}}{0.0988\ \text{A}} = 121\ \Omega.$$

## Ohmmeters

An alternative method for measuring resistance is to use a d'Arsonval galvanometer in an arrangement called an **ohmmeter.** It consists of a meter, a resistor, and a source (often a flashlight cell) connected in series (Fig. 27–13). The resistance $R$ to be measured is connected between terminals $x$ and $y$.

The series resistance $R_s$ is adjusted so that when terminals $x$ and $y$ are short-circuited (that is, when $R = 0$), the meter deflects full-scale. When the circuit between $x$ and $y$ is *open* (that is, when $R = \infty$), there is no current and no deflection. For any intermediate value of $R$ the meter deflection depends on the value of $R$, and the meter scale can be calibrated to read the resistance $R$ directly. Larger currents correspond to smaller resistances, so this scale reads backward in comparison to the current scale.

You have probably seen multiple-range meters, or multimeters, that use d'Arsonval galvanometers. Such a meter uses a single-range, moving-coil meter; various ranges are provided by switching different resistances in parallel and series with the meter coil.

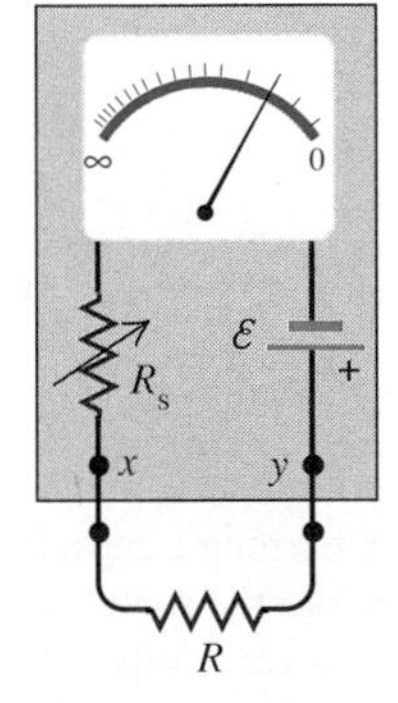


**27–13** Ohmmeter circuit. First connect $x$ directly to $y$ and adjust $R_s$ until the meter reads zero. Then connect $x$ and $y$ across the resistor $R$ and read the scale.

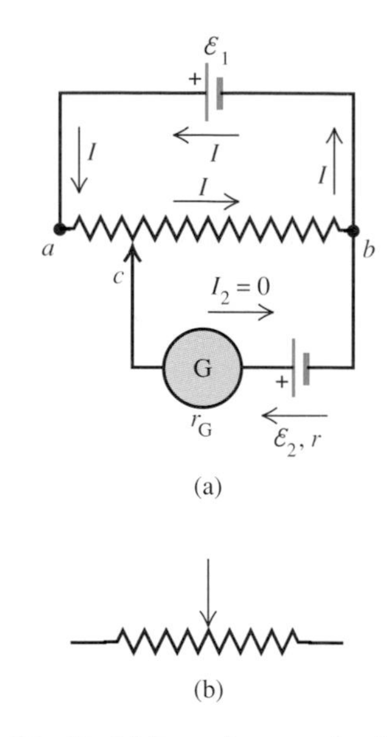


27–14 (a) Potentiometer circuit. (b) Circuit symbol for potentiometer (variable resistor).

In situations in which high precision is required, instruments containing d'Arsonval galvanometers have been supplanted by electronic instruments with direct digital readouts. These are more precise, stable, and mechanically rugged and are often more expensive than d'Arsonval devices. Digital voltmeters can be made with extremely high internal resistance, of the order of 100 MΩ.

## The Potentiometer

The *potentiometer* is an instrument that can be used to measure the emf of a source without drawing any current from the source; it also has a number of other useful applications. Essentially, it balances an unknown potential difference against an adjustable, measurable potential difference.

The principle of the potentiometer is shown schematically in Fig. 27–14. A resistance wire *ab* is permanently connected to the terminals of a source with emf $\mathcal{E}_1$. A sliding contact *c* is connected through the galvanometer G to a second source whose emf $\mathcal{E}_2$ is to be measured. Contact *c* is moved along the wire until a position is found at which the galvanometer shows no deflection. (Note that $V_{ab}$ must be greater than $\mathcal{E}_2$.) With $I_2 = 0$, Kirchhoff's loop rule gives

$$\mathcal{E}_2 = IR_{cb}.$$

If the resistance wire is uniform, $R_{cb}$ is proportional to the length of wire between *c* and *b*. We calibrate the device using a known voltage, and then any unknown emf can be measured by measuring the length of wire *cb*.

The term *potentiometer* is also used for any variable resistor, usually having a circular resistance element and a sliding contact controlled by a rotating shaft and knob. The circuit symbol for a potentiometer is shown in Fig. 27–14b.

## 27–4 RESISTANCE-CAPACITANCE CIRCUITS

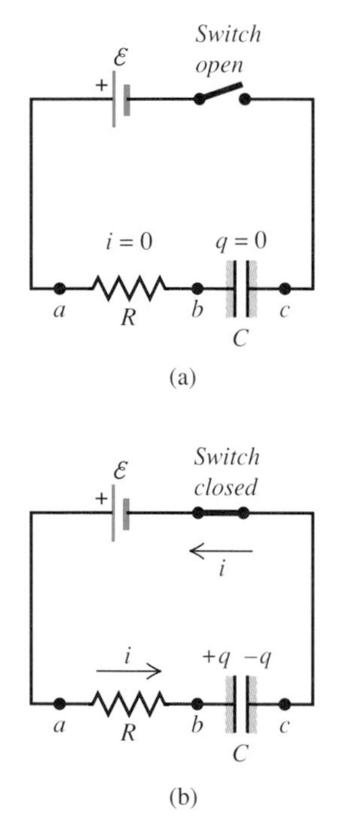


27–15 Charging a capacitor. (a) Just before the switch is closed, the charge is zero. When the switch closes (at $t = 0$), the current jumps from zero to $\mathcal{E}/R$. (b) At some later time *t*, as $t \to \infty$, $q \to Q_f$ and $i \to 0$.

In the circuits that we have analyzed up to this point we have assumed that all the emf's and resistances are *constant* (time-independent), so all the potentials, currents, and powers have also been independent of time. But in the simple act of charging a capacitor we find a situation in which the currents, voltages and powers *do* change with time. Figure 27–15 shows a simple circuit for charging a capacitor. We idealize the battery (or power supply) to have a constant emf $\mathcal{E}$ and zero internal resistance *r,* and we neglect the resistance of all the connecting conductors.

We begin with the capacitor initially uncharged; then at some initial time $t = 0$ we close the switch, completing the circuit and permitting current around the loop to begin charging the capacitor. The current begins at the same instant in every part of the circuit, and at each instant the current is the same in every part.

To distinguish between quantities that vary with time and those that are constant, we will use lowercase letters for time-varying voltages, currents, and charges and capital letters for constants. Because the capacitor is initially uncharged, the potential difference $v_{bc}$ across it is initially zero. At this time, from Kirchhoff's loop rule, the voltage $v_{ab}$ across the resistor *R* is equal to the battery emf $\mathcal{E}$. The initial current through the resistor, which we will call $I_0$, is given by Ohm's law: $I_0 = v_{ab}/R = \mathcal{E}/R$.

As the capacitor charges, its voltage $v_{bc}$ increases, and the potential difference $v_{ab}$ across the resistor decreases, corresponding to a decrease in current. The sum of these two voltages is constant and equal to $\mathcal{E}$. After a long time the capacitor becomes fully charged, the current decreases to zero, and the potential difference $v_{ab}$ across the resistor becomes zero. Then the entire battery emf $\mathcal{E}$ appears across the capacitor, $v_{bc} = \mathcal{E}$.

Let $q$ represent the charge on the capacitor and $i$ the current in the circuit at some time $t$ after the switch has been closed. The instantaneous potential differences $v_{ab}$ and $v_{bc}$ are

$$v_{ab} = iR, \qquad v_{bc} = \frac{q}{C}. \tag{27–9}$$

From Kirchhoff's loop rule,

$$\mathcal{E} - iR - \frac{q}{C} = 0. \tag{27–10}$$

Solving this equation for $i$, we find

$$i = \frac{\mathcal{E}}{R} - \frac{q}{RC}. \tag{27–11}$$

At time $t = 0$, when the switch is first closed, the capacitor is uncharged, so $q = 0$. Substituting $q = 0$ into Eq. (27–11), we find that the *initial* current $I_0$ is given by $I_0 = \mathcal{E}/R$, as we have already noted. If the capacitor were not in the circuit, the last term in Eq. (27–11) would not be present; then the current would be *constant* and equal to $\mathcal{E}/R$.

As the charge $q$ increases, the term $q/RC$ becomes larger, and the capacitor charge approaches its final value, which we will call $Q_f$. The current decreases and eventually becomes zero. When $i = 0$, Eq. (27–11) gives

$$\frac{\mathcal{E}}{R} = \frac{Q_f}{RC}, \qquad Q_f = C\mathcal{E}. \tag{27–12}$$

Note that the final charge $Q_f$ does not depend on $R$.

The current and the capacitor charge are shown as functions of time in Fig. 27–16. At the instant the switch is closed ($t = 0$), the current jumps from zero to its initial value $I_0 = \mathcal{E}/R$; after that it gradually approaches zero. The capacitor charge starts at zero and gradually approaches the final value $Q_f = C\mathcal{E}$.

We can derive general expressions for the charge $q$ and current $i$ as functions of time. In Eq. (27–11) we first replace $i$ by $dq/dt$:

$$\frac{dq}{dt} = \frac{\mathcal{E}}{R} - \frac{q}{RC} = -\frac{1}{RC}(q - C\mathcal{E}).$$

We can rearrange this to

$$\frac{dq}{q - C\mathcal{E}} = -\frac{dt}{RC}$$

and then integrate both sides. We change the integration variables to $q'$ and $t'$ so that we can use $q$ and $t$ for the upper limits. The lower limits are $q' = 0$ and $t' = 0$:

$$\int_0^q \frac{dq'}{q' - C\mathcal{E}} = -\int_0^t \frac{dt'}{RC}.$$

When we carry out the integration and rearrange the result, we get

$$\ln \frac{q - C\mathcal{E}}{-C\mathcal{E}} = -\frac{t}{RC},$$

$$\frac{q - C\mathcal{E}}{-C\mathcal{E}} = e^{-t/RC},$$

$$q = C\mathcal{E}(1 - e^{-t/RC}) = Q_f(1 - e^{-t/RC}). \tag{27–13}$$

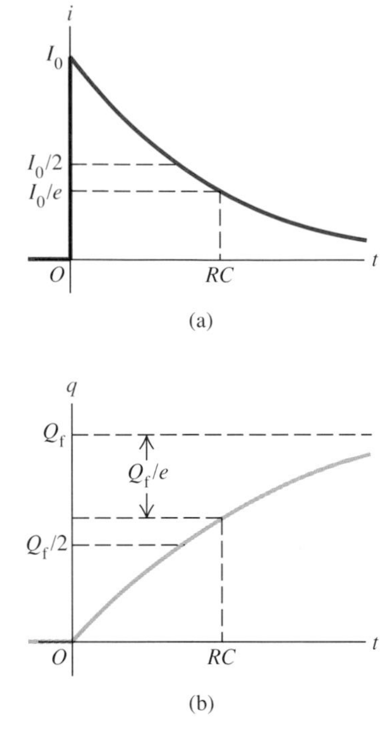


**27–16** Current and capacitor charge as functions of time for the circuit of Fig. 27–15. The initial current is $I_0$, and the final capacitor charge is $Q_f$.

The instantaneous current $i$ is just the time derivative of Eq. (27–13):

$$i = \frac{dq}{dt} = \frac{\mathcal{E}}{R}e^{-t/RC} = I_0 e^{-t/RC}. \tag{27–14}$$

The charge and current are both *exponential* functions of time. Fig. 27–16a is a graph of Eq. (27–14), and Fig. 27–16b is a graph of Eq. (27–13).

## Time Constant

After a time equal to $RC$ the current has decreased to $1/e$ (about 0.368) of its initial value. At this time the capacitor charge has reached $(1 - 1/e) = 0.632$ of its final value $Q_f = CV$. The product $RC$ is therefore a measure of how quickly the capacitor charges. (We invite you to verify that the product $RC$ has units of time.) We call $RC$ the **time constant,** or the **relaxation time,** of the circuit, denoted by $\tau$:

$$\tau = RC. \tag{27–15}$$

When $\tau$ is small, the capacitor charges quickly; when it is larger, the charging takes more time. A capacitor charges more quickly through a small resistor; it's easier for the charge to get through.

In Fig. 27–16a the horizontal axis is an *asymptote* for the curve. Strictly speaking, $i$ never becomes precisely zero. But the longer we wait, the closer it gets. After a time equal to $10RC$ the current has decreased to 0.000045 of its initial value. Similarly, the curve in Fig. 27–16b approaches the horizontal broken line labeled $Q_f$ as an asymptote. The charge $q$ never attains precisely this value, but after a time equal to $10RC$ the difference between $q$ and $Q_f$ is only 45 millionths of $Q_f$.

### EXAMPLE 27–11

A resistor with resistance $R = 10\ \mathrm{M\Omega}$ is connected in series with a capacitor with capacitance $1\ \mu\mathrm{F}$ and a battery with emf 12.0 V. At time $t = 0$ the capacitor is uncharged. a) What is the time constant? b) What fraction of the final charge is on the plates at time $t = 46$ s? c) What fraction of the initial current remains at $t = 46$ s?

**SOLUTION** a) The time constant is

$$\tau = RC = (10 \times 10^6\ \Omega)(10^{-6}\ \mathrm{F}) = 10\ \mathrm{s}$$

b) The fraction of the final charge is $q/Q_f$; from Eq. (27–13),

$$\frac{q}{Q_f} = 1 - e^{-t/RC} = 1 - e^{-(46\ \mathrm{s})/(10\ \mathrm{s})} = 0.99.$$

The capacitor is 99% charged after a time equal to $4.6RC$, or 4.6 time constants.

c) From Eq. (27–14),

$$\frac{i}{I_0} = e^{-4.6} = 0.010.$$

After 4.6 time constants the current has decreased to 1.0% of its initial value.

## Discharging a Capacitor

Now suppose that after the capacitor in Fig. 27–15b has acquired a charge $Q_0$, we remove the battery from the circuit and connect points $a$ and $c$ (Fig. 27–17). We reset our stopwatch so that the connection is made at time $t = 0$; at that time, $q = Q_0$. The capacitor then *discharges* through the resistor, and its charge eventually decreases to zero.

Again we let $i$ and $q$ represent the time-varying current and charge at some instant after the connection is made. Kirchhoff's loop rule now gives Eq. (27–11) but with $\mathcal{E} = 0$; that is,

$$i = \frac{dq}{dt} = -\frac{q}{RC}. \tag{27–16}$$

The current $i$ is now negative (opposite to the direction shown in Fig. 27–17b). At time $t = 0$, when $q = Q_0$, the initial current $I_0$ is $I_0 = -Q_0/RC$.

To find $q$ as a function of time, we rearrange Eq. (27–16), again change the names of the variables to $q'$ and $t'$, and integrate. This time the limits for $q'$ are $Q_0$ to $q$. We get

$$\int_{Q_0}^{q} \frac{dq'}{q'} = -\frac{1}{RC}\int_0^t dt',$$

$$\ln\frac{q}{Q_0} = -\frac{t}{RC},$$

$$q = Q_0 e^{-t/RC}. \tag{27–17}$$

The instantaneous current $i$ is the derivative of this:

$$i = \frac{dq}{dt} = -\frac{Q_0}{RC}e^{-t/RC} = I_0 e^{-t/RC}. \tag{27–18}$$

Both the current and the charge decrease exponentially with time. Comparing these results with Eqs. (27–13) and (27–14), we note that the expressions for the current are identical, except for the sign of $I_0$. The capacitor charge approaches zero asymptotically in Eq. (27–17), but the *difference* between $q$ and $Q_f$ approaches zero asymptotically in Eq. (27–13).

Energy considerations give us additional insight into the behavior of an $RC$ circuit. While the capacitor is charging, the instantaneous rate at which the battery delivers energy to the circuit is $P = \mathcal{E}i$. The instantaneous rate at which electrical energy is dissipated in the resistor is $i^2R$, and the rate at which energy is stored in the capacitor is $iv_{bc} = iq/C$. Multiplying Eq. (27–10) by $i$, we find

$$\mathcal{E}i = i^2R + iq/C. \tag{27–19}$$

This means that of the power $\mathcal{E}i$ supplied by the battery, part ($i^2R$) is dissipated in the resistor and part ($iq/C$) is stored in the capacitor.

The *total* energy supplied by the battery during charging of the capacitor equals the battery emf $\mathcal{E}$ multiplied by the total charge $Q_f$, or $\mathcal{E}Q_f$. The total energy stored in the capacitor, from Eq. (25–9), is $Q_f\,\mathcal{E}/2$. Thus of the energy supplied by the battery, *exactly half* is stored in the capacitor, and the other half is dissipated in the resistor. It is a little surprising that this half-and-half division of energy doesn't depend on $C$, $R$, or $\mathcal{E}$. This result can also be verified in detail by taking the integral over time of each of the power quantities mentioned above. We leave this calculation for your amusement.

Switch open

$i = 0$ $+Q_0$ $-Q_0$

a R b C c

(a)

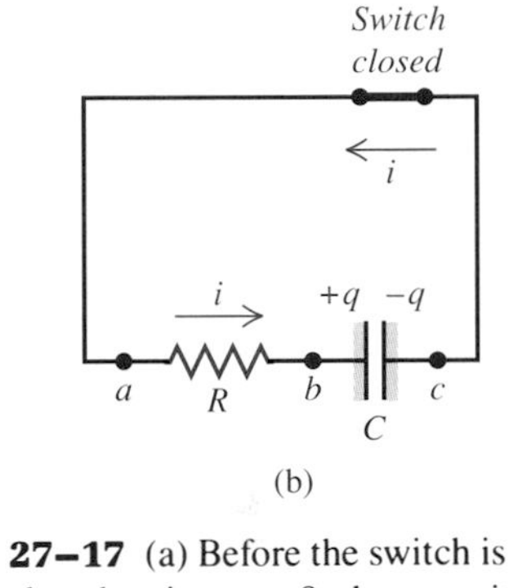


**27–17** (a) Before the switch is closed at time $t = 0$, the capacitor charge is $Q_0$ and the current is zero. (b) At time $t$ after the switch is closed, the capacitor charge is $q$ and the current is $i$. The actual current direction is opposite to the direction shown; $i$ is negative. After a long time, $q$ and $i$ both approach zero.

# 27–5 POWER DISTRIBUTION SYSTEMS: A Case Study in Circuit Analysis

We conclude this chapter with a brief discussion of practical household and automotive electric-power distribution systems. Automobiles use direct-current (dc) systems, and nearly all household, commercial, and industrial systems use alternating current (ac) because of the ease of stepping voltage up and down with transformers. Most of the same basic wiring concepts apply to both. We'll talk about alternating-current circuits in greater detail in Chapter 32.

The various lamps, motors, and other appliances to be operated are always connected *in parallel* to the power source (the wires from the power company for houses or

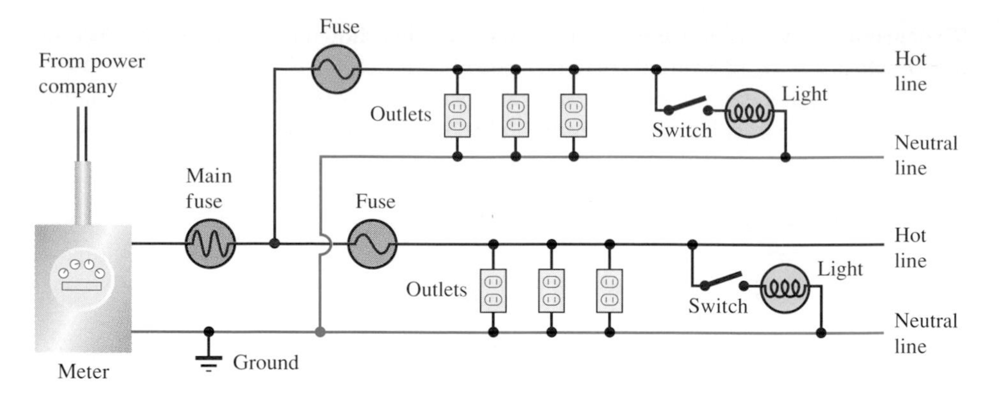


**27–18** Schematic diagram of part of a house wiring system. Only two branch circuits are shown; an actual system might have four to thirty branch circuits. Lamps and appliances may be plugged into the outlets. The grounding conductors, which normally carry no current, are not shown. Modern household systems usually have two "hot" lines with opposite polarity with respect to neutral and a voltage between them of 240 V. Such a system is shown in Fig. 27–20. (Note that actual wires have a different color coding system.)

the battery and alternator for a car). The basic idea of house wiring is shown in Fig. 27–18. One side of the "line," as the pair of conductors is called, is called the *neutral* side; it is always connected to "ground" at the entrance panel. For houses, ground is an actual electrode driven into the earth (usually a good conductor) and also sometimes connected to the household water pipes. Electricians speak of the "hot" side and the "neutral" side of the line. Most modern house wiring systems have *two* hot lines with opposite polarity with respect to the neutral; the voltage between them is twice the voltage between either hot line and the neutral line. We'll return to this detail later.

Household voltage is nominally 120 V in the United States and Canada, often 240 V in Europe. (For alternating current, which varies sinusoidally with time, these numbers represent the *root-mean-square* voltage, which is $1/\sqrt{2}$ times the peak voltage. We'll discuss this further in Section 32–4.) The power input $P$ to a device is given by Eq. (26–20): $P = VI$. For example, the current in a 100-W light bulb is

$$I = \frac{P}{V} = \frac{100\ \text{W}}{120\ \text{V}} = 0.83\ \text{A}.$$

The resistance of this bulb at operating temperature is

$$R = \frac{V}{I} = \frac{120\ \text{V}}{0.83\ \text{A}} = 144\ \Omega, \qquad \text{or} \qquad R = \frac{V^2}{P} = \frac{(120\ \text{V})^2}{100\ \text{W}} = 144\ \Omega.$$

Similarly, a 1500-W waffle iron draws a current of (1500 W)/(120 V) = 12.5 A and has a resistance, at operating temperature, of 9.6 Ω. Because of the temperature dependence of resistivity, the resistances of these devices are considerably less when they are cold. If you measure the resistance of a light bulb with an ohmmeter (whose small current causes very little temperature rise), you will probably get a value of about 10 Ω. When a light bulb is turned on, there is an initial surge of current as the filament heats up. That's why, when a light bulb gets ready to burn out, it nearly always happens when you turn it on.

The maximum current available from an individual circuit is limited by the resistance of the wires. The $I^2R$ power loss in the wires causes them to become hot, and in extreme cases this can cause a fire or melt the wires. Ordinary lighting and outlet wiring in houses usually uses 12-gauge wire. This has a diameter of 2.05 mm and can carry a maximum current of 20 A safely (without overheating). Larger sizes such as 8-gauge (3.26 mm) or 6-gauge (4.11 mm) are used for high-current appliances such as ranges and clothes dryers, and 2-gauge (6.54 mm) or larger is used for the main power lines entering a house.

Protection against overloading and overheating of circuits is provided by fuses or circuit breakers. A *fuse* contains a link of lead-tin alloy with a very low melting temperature; the link melts and breaks the circuit when its rated current is exceeded. A *circuit breaker* is an electromechanical device that performs the same function, using an electromagnet or a bimetallic strip to "trip" the breaker and interrupt the circuit when the current exceeds a specified value. Circuit breakers have the advantage that they can be reset after they are tripped. A blown fuse must be replaced, but fuses are somewhat more reliable in operation than circuit breakers are.

If your system has fuses and you plug too many high-current appliances into the same outlet, the fuse blows. *Do not* replace the fuse with one of larger rating; if you do, you risk overheating the wires and starting a fire. The only safe solution is to distribute the appliances among several circuits. Modern kitchens often have three or four separate 20-A circuits.

Contact between the hot and neutral sides of the line causes a *short circuit.* Such a situation, which can be caused by faulty insulation or by any of a variety of mechanical malfunctions, provides a very low-resistance current path, permitting a very large current that would quickly melt the wires and ignite their insulation if the current were not interrupted by a fuse or circuit breaker. An equally dangerous situation is a broken wire that interrupts the current path, creating an *open circuit.* This is hazardous because of the sparking that can occur at the point of intermittent contact.

In approved wiring practice, a fuse or breaker is placed *only* in the hot side of the line, never in the neutral side. Otherwise, if a short circuit should develop because of faulty insulation or other malfunction, the ground-side fuse could blow. The hot side would still be live and would pose a shock hazard if you touched the live conductor and a grounded object such as a water pipe. For similar reasons the wall switch for a light fixture is always in the hot side of the line, never the neutral side.

Further protection against shock hazard is provided by a third conductor called the *grounding wire,* included in all present-day wiring. This conductor corresponds to the long round or U-shaped prong of the three-prong connector plug on an appliance or power tool. It is connected to the neutral side of the line at the entrance panel, where the meter is. It normally carries no current, but it connects the metal case or frame of the device to ground. If a conductor on the hot side of the line accidentally contacts the frame or case, the grounding conductor provides a current path, and the fuse blows. Without the ground wire the frame could become "live," that is, at a potential 120 V above ground. Then if you touched it and a water pipe (or even a damp basement floor) at the same time, you could get a dangerous shock (Fig. 27–19). In some situations, especially for outlets located outdoors or near a sink or water pipes, a special kind of circuit breaker called a

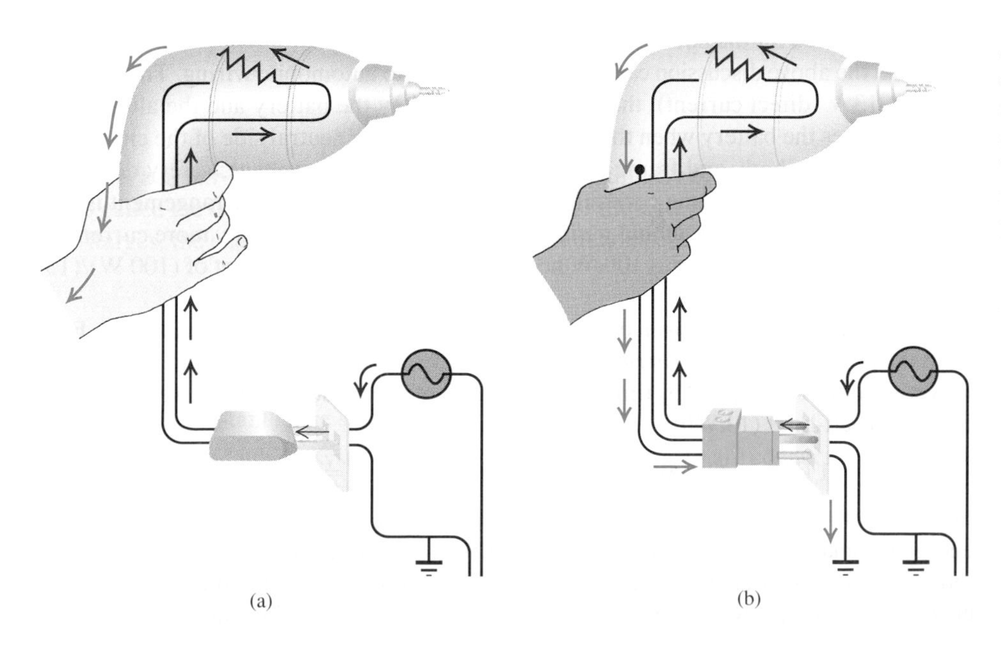


**27–19** (a) If a malfunctioning electric drill is connected to a wall socket via a two-prong plug, a person may receive a shock. (b) When the drill malfunctions when connected via a three-prong plug, a person touching it receives no shock, since electric charge flows through the third prong and into the ground rather than into the person's body. If the ground current is appreciable, the fuse blows.

## EXERCISES

### Section 27–1 Resistors in Series and in Parallel

**27–1** A 40.0-Ω resistor and a 60.0-Ω resistor are connected in parallel, and the combination is connected across a 120-V dc line. a) What is the resistance of the parallel combination? b) What is the total current through the parallel combination? c) What is the current through each resistor?

**27–2** Three resistors having resistances of 2.00 Ω, 3.00 Ω, and 4.00 Ω are connected in series to a 24.0-V battery that has negligible internal resistance. Find a) the equivalent resistance of the combination; b) the current in each resistor; c) the total current through the battery; d) the voltage across each resistor; e) the power dissipated in each resistor.

**27–3** The three resistors of Exercise 27–2 are connected in parallel to the same battery. Answer the same questions for this situation.

**27–4** Compute the equivalent resistance of the network in Fig. 27–21, and find the current in each resistor. The battery has negligible internal resistance.

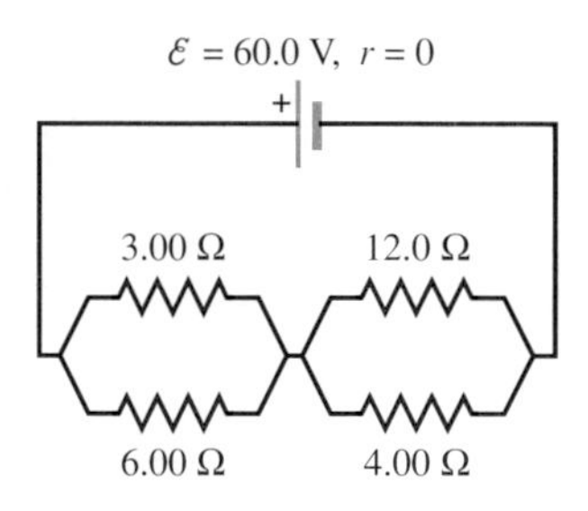


**FIGURE 27–21**

**27–5** Compute the equivalent resistance of the network in Fig. 27–22, and find the current in each resistor. The battery has negligible internal resistance.

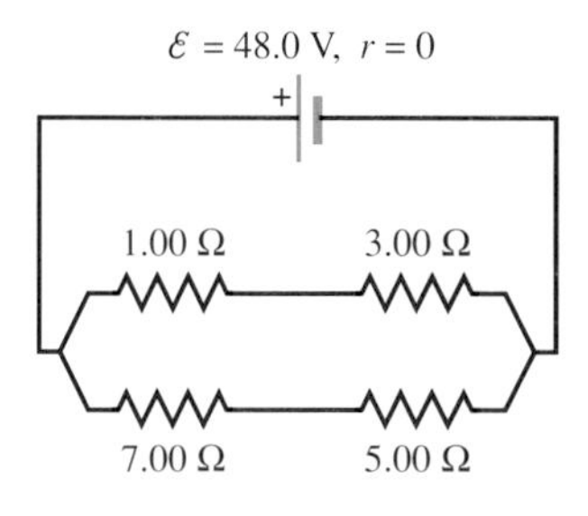


**FIGURE 27–22**

**27–6 Light Bulbs in Series.** A 25-W, 120-V light bulb and a 150-W, 120-V light bulb are connected in series across a 240-V line. Assume that the resistance of each bulb does not vary with current. (*Note:* This description of a light bulb gives the power it dissipates when connected to the stated potential difference; that is, a 25-W, 120-V light bulb dissipates 25 W when connected to a 120-V line.) a) Find the current through the bulbs. b) Find the power dissipated in each bulb. c) One bulb burns out very quickly. Which one? Why?

**27–7** a) The power rating of a 10,000-Ω resistor is 4.00 W. (The power rating is the maximum power that the resistor can safely dissipate without too great a rise in temperature.) What is the maximum allowable potential difference across the terminals of the resistor? b) We need a 20,000-Ω resistor, to be connected across a potential difference of 220 V. What power rating is required?

### Section 27–2 Kirchhoff's Rules

**27–8** Find the emf's $\mathcal{E}_1$ and $\mathcal{E}_2$ in the circuit of Fig. 27–23 and the potential difference of point $a$ relative to point $b$.

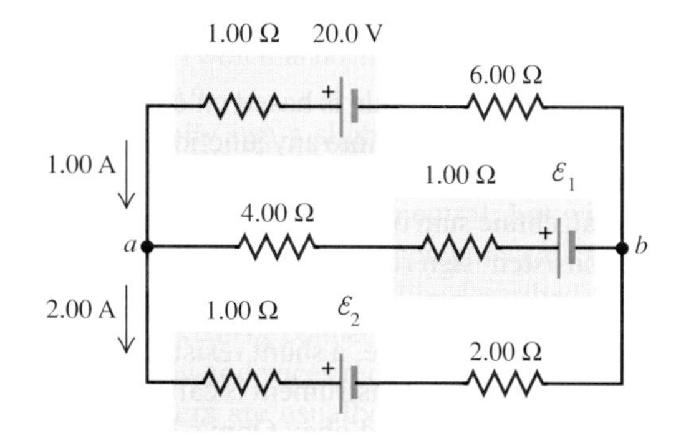


**FIGURE 27–23**

**27–9** In the circuit shown in Fig. 27–24, find a) the current in resistor $R$; b) the resistance $R$; c) the unknown emf $\mathcal{E}$. d) If the circuit is broken at point $x$, what is the current in the 28.0-V battery?

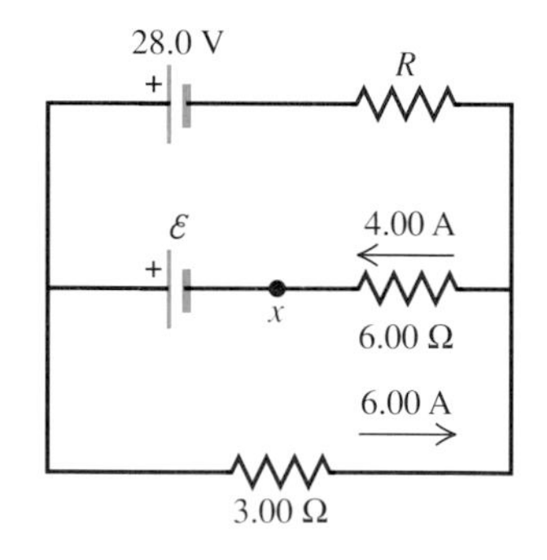


**FIGURE 27–24**

**27–10** In the circuit shown in Fig. 27–25, find a) the current in each branch; b) the potential difference $V_{ab}$ of point $a$ relative to point $b$.

**27–11** In the circuit shown in Fig. 27–26, find a) the current in the 3.00-Ω resistor; b) the unknown emf's $\mathcal{E}_1$ and $\mathcal{E}_2$; c) the resistance $R$. Note that three currents are given.

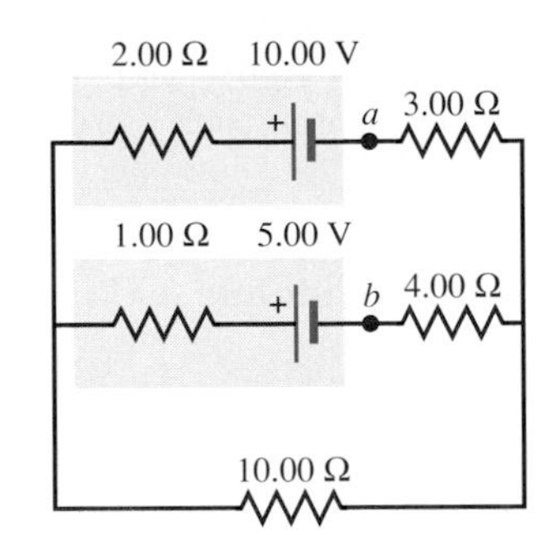


**FIGURE 27–25**

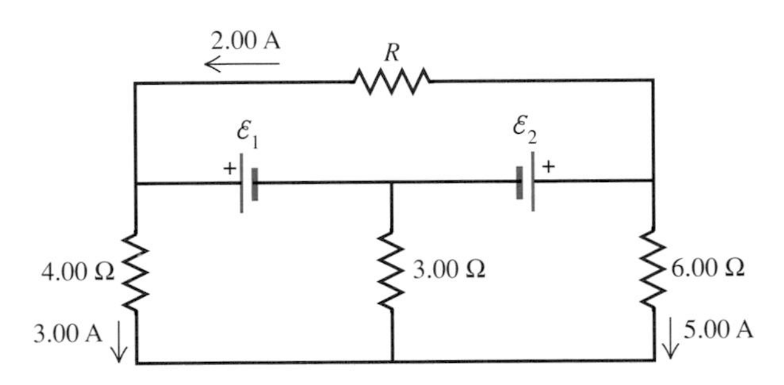


**FIGURE 27–26**

## Section 27–3 Electrical Measuring Instruments

**27–12** The resistance of a galvanometer coil is 50.0 Ω, and the current required for full-scale deflection is 300 $\mu$A. a) Show in a diagram how to convert the galvanometer to an ammeter reading 5.00 A full-scale, and compute the shunt resistance. b) Show how to convert the galvanometer to a voltmeter reading 150 V full-scale, and compute the series resistance.

**27–13** The resistance of the coil of a pivoted-coil galvanometer is 8.00 Ω, and a current of 0.0200 A causes it to deflect full-scale. We want to convert this galvanometer to an ammeter reading 10.0 A full-scale. The only shunt resistor available has a resistance of 0.0400 Ω. What resistance $R$ must be connected in series with the coil? (See Fig. 27–27.)

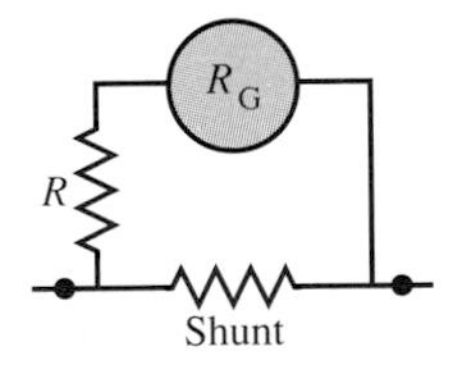


**FIGURE 27–27**

**27–14 A Multirange Ammeter.** The resistance of the moving coil of the galvanometer G in Fig. 27–28 is 32.0 Ω, and the galvanometer deflects full-scale with a current of 0.0100 A. Find the magnitudes of the resistances $R_1$, $R_2$, and $R_3$ required to convert the galvanometer to a multirange ammeter deflecting full-scale with currents of 10.0 A, 1.00 A, and 0.100 A. (When the meter is connected to the circuit being measured, one connection is made to the post marked + and the other to the post marked with the desired current range.)

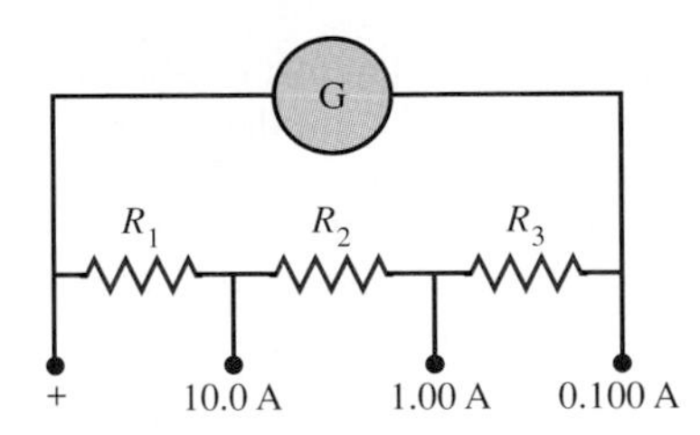


**FIGURE 27–28**

**27–15** Figure 27–29 shows the internal wiring of a "three-scale" voltmeter whose binding posts are marked +, 3.00 V, 15.0 V, and 150 V. The resistance of the moving coil, $R_G$, is 25.0 Ω, and a current of 1.00 mA in the coil causes it to deflect full-scale. Find the resistances $R_1$, $R_2$, and $R_3$ and the overall resistance of the meter on each of its ranges. (When the meter is connected to the circuit being measured, one connection is made to the post marked + and the other to the post marked with the desired voltage range.)

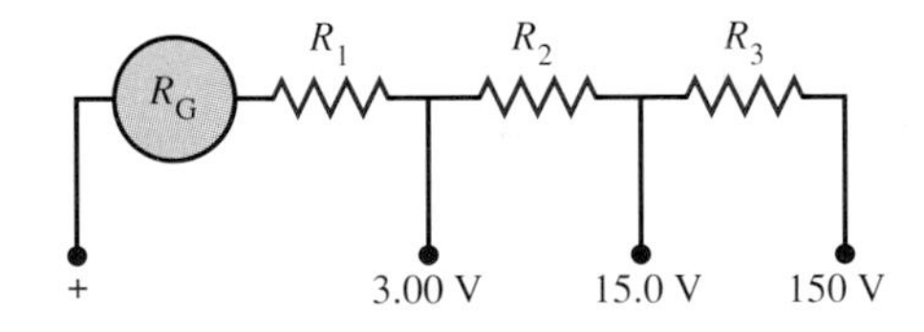


**FIGURE 27–29**

**27–16** A 100-V battery has an internal resistance of $r = 6.00$ Ω. a) What is the reading of a voltmeter having a resistance of $R_V = 500$ Ω when placed across the terminals of the battery? b) What maximum value may the ratio $r/R_V$ have if the error in the reading of the emf of a battery is not to exceed 5.0%?

**27–17** Consider the potentiometer circuit of Fig. 27–14. The resistor between $a$ and $b$ is a uniform wire with length $l$, with a sliding contact $c$ at a distance $x$ from $b$. An unknown emf $\mathcal{E}_2$ is measured by sliding the contact until the galvanometer G reads zero. a) Show that under this condition the unknown emf is given by $\mathcal{E}_2 = (x/l)\mathcal{E}_1$. b) Why is the internal resistance of the galvanometer not important? c) Suppose $\mathcal{E}_1 = 12.00$ V and $l = 1.000$ m. The galvanometer G reads zero when $x = 0.793$ m. What is the emf $\mathcal{E}_2$?

**27–18** In the ohmmeter in Fig. 27–30, M is a 1.00-mA meter having a resistance of 60.0 Ω. (A 1.00-mA meter deflects full-scale when the current through it is 1.00 mA.) The battery $B$ has an

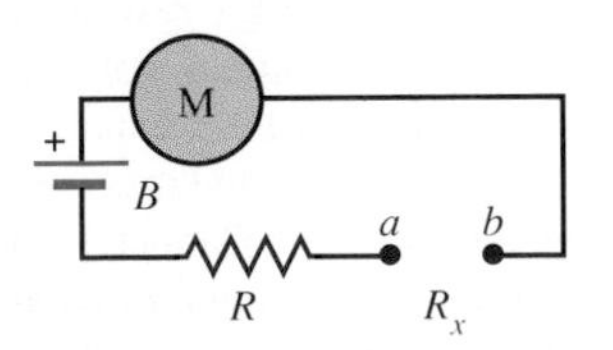

**FIGURE 27–30**

emf of 3.00 V and negligible internal resistance. $R$ is chosen so that when the terminals $a$ and $b$ are shorted ($R_x = 0$), the meter reads full-scale. When $a$ and $b$ are open ($R_x = \infty$), the meter reads zero. a) What is the resistance of the resistor $R$? b) What current indicates a resistance $R_x$ of 600 Ω? c) What resistances correspond to meter deflections of $\frac{1}{4}$, $\frac{1}{2}$, and $\frac{3}{4}$ of full scale if the deflection is proportional to the current through the galvanometer?

## Section 27–4 Resistance-Capacitance Circuits

**27–19** Verify that the product $RC$ has units of time.

**27–20** A 6.00-μF capacitor that is initially uncharged is connected in series with a $4.50 \times 10^3 \Omega$ resistor and an emf source with $\mathcal{E} = 300$ V and negligible internal resistance. Just after the circuit is completed, what is a) the voltage drop across the capacitor; b) the voltage drop across the resistor; c) the charge on the capacitor; d) the current through the resistor? e) A long time after the circuit is completed (after many time constants), what are the values of the preceding four quantities?

**27–21** A capacitor with capacitance $C = 2.50 \times 10^{-10}$ F is charged, with charge of magnitude $6.00 \times 10^{-8}$ C on each plate. The capacitor is then connected to a voltmeter that has internal resistance $4.00 \times 10^5 \Omega$. a) What is the current through the voltmeter just after the connection is made? b) What is the time constant of this $RC$ circuit?

**27–22** A capacitor is charged to a potential of 15.0 V and is then connected to a voltmeter having an internal resistance of 1.00 MΩ. After a time of 5.00 s the voltmeter reads 5.0 V. What is the capacitance?

**27–23** A 10.0-μF capacitor is connected through a 1.00-MΩ resistor to a constant potential difference of 60.0 V. a) Compute the charge on the capacitor at the following times after the connections are made: 0, 5.0 s, 10.0 s, 20.0 s, and 100.0 s. b) Compute the charging current at the same instants. c) Construct graphs of the results of parts (a) and (b) for a time interval of 20 s.

## Section 27–5 Power Distribution Systems: A Case Study in Circuit Analysis

**27–24** An electric dryer is rated at 4.00 kW when connected to a 240-V line. a) What is the current in the dryer? Is 12-gauge wire large enough to supply this current? b) What is the resistance of the dryer's heating element at its operating temperature? c) At 10.0 cents per kWh, how much does it cost per hour to operate the dryer?

**27–25** An 1800-W toaster, a 1400-W electric frying pan, and a 75-W lamp are plugged into the same outlet in a 20-A, 120-V circuit. (*Note:* See the note in Exercise 27–6. When plugged into the same outlet, the three devices are in parallel, so the voltage across each is 120 V, and the total current in the circuit is the sum of the currents through each device.) a) What current is drawn by each device? b) Will this combination blow the fuse?

**27–26** How many 60-W light bulbs can be connected across a 20-A, 120-V circuit without tripping the circuit breaker? (See the note in Exercise 27–6.)

**27–27** The heating element of an electric stove consists of a heater wire embedded within an electrically insulating material, which in turn is inside a metal casing. The heater wire has a resistance of 22.0 Ω at room temperature (23.0°C) and a temperature coefficient of resistivity $\alpha = 3.00 \times 10^{-3}$ (C°)$^{-1}$. The heating element operates from a 120-V line. a) When the heater element is first turned on, what current does it draw and what electrical power does it dissipate? b) When the heater element has reached an operating temperature of 280°C (536°F), what current does it draw and what electrical power does it dissipate?

# PROBLEMS

**27–28** Prove that when two resistors are connected in parallel, the equivalent resistance of the combination is always smaller than that of either resistor.

**27–29** A 1000-Ω, 2.00-W resistor is needed, but only several 1000-Ω, 1.00-W resistors are available. (See Exercise 27–7.) a) How can the required resistance and power rating be obtained by a combination of the available units? b) What power is dissipated in each resistor when 2.00 W is dissipated by the combination?

**27–30** a) A resistance $R_2$ is connected in parallel with a resistance $R_1$. Derive an expression for the resistance $R_3$ that must be connected in series with the combination of $R_1$ and $R_2$ so that the equivalent resistance is equal to the resistance $R_1$. Draw a diagram. b) A resistance $R_2$ is connected in series with a resistance $R_1$. Derive an expression for the resistance $R_3$ that must be connected in parallel with the combination of $R_1$ and $R_2$ so that the equivalent resistance is equal to $R_1$. Draw a diagram.

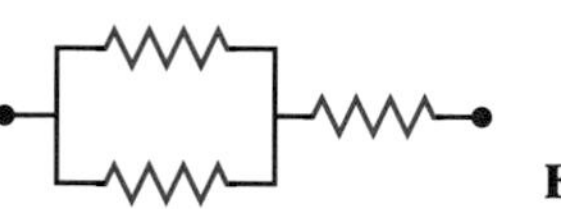

**FIGURE 27–31**

**27–31** Each of three resistors in Fig. 27–31 has a resistance of 2.00 Ω and can dissipate a maximum of 24.0 W without becoming excessively heated. What is the maximum power the circuit can dissipate?

**27–32** a) Calculate the equivalent resistance of the circuit of Fig. 27–32 between $x$ and $y$. b) What is the potential of point $a$ relative to point $x$ if the current in the 8.0-Ω resistor is 0.800 A in the direction from left to right in the figure?

**27–33** a) Find the potential of point $a$ with respect to point $b$ in Fig. 27–33. b) If points $a$ and $b$ are connected by a wire with negligible resistance, find the current in the 12.0-V battery.

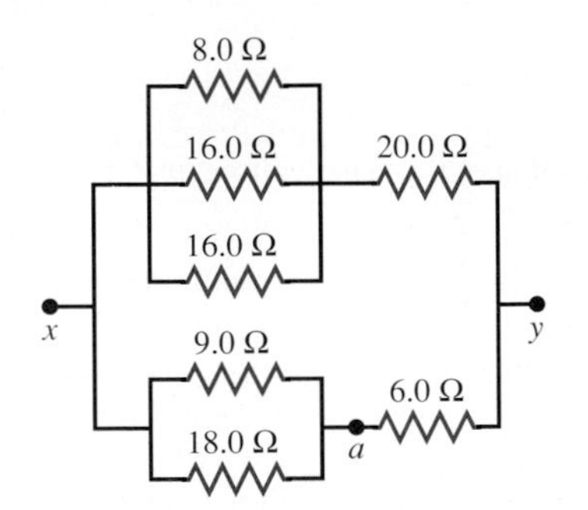


**FIGURE 27–32**

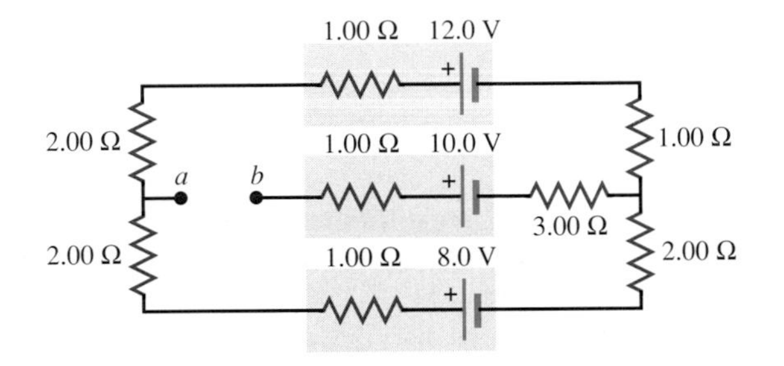


**FIGURE 27–33**

**27–34** Three identical resistors are connected in series. When a certain potential difference is applied across the combination, the total power dissipated is 45.0 W. What power would be dissipated if the three resistors were connected in parallel across the same potential difference?

**27–35** Calculate the three currents indicated in the circuit diagram of Fig. 27–34.

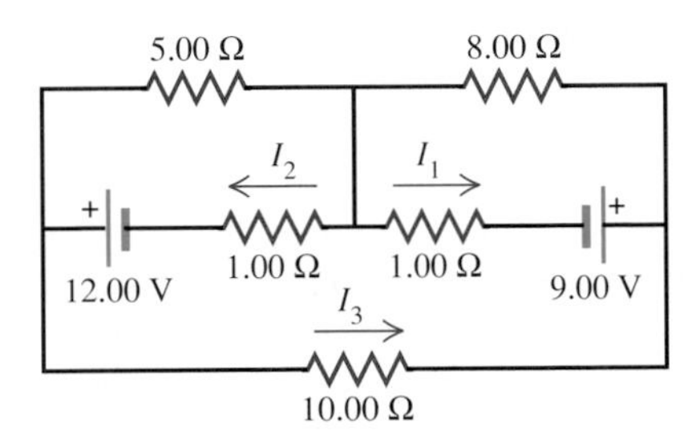


**FIGURE 27–34**

**27–36** What must the emf $\mathcal{E}$ in Fig. 27–35 be in order for the current through the 7.00-Ω resistor to be 3.00 A? Each emf source has negligible internal resistance.

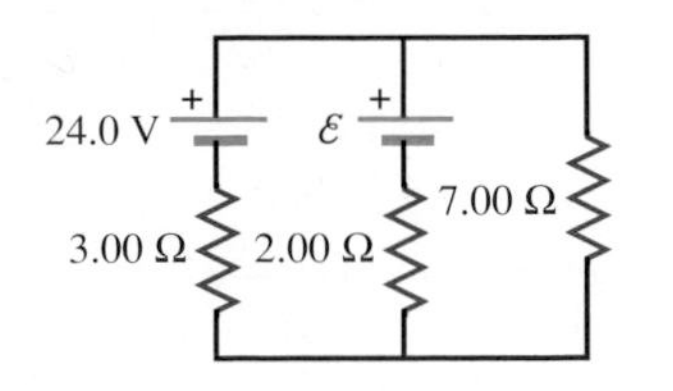


**FIGURE 27–35**

**27–37** Find the current through each of the three resistors of the circuit of Fig. 27–36. The emf sources have negligible internal resistance.

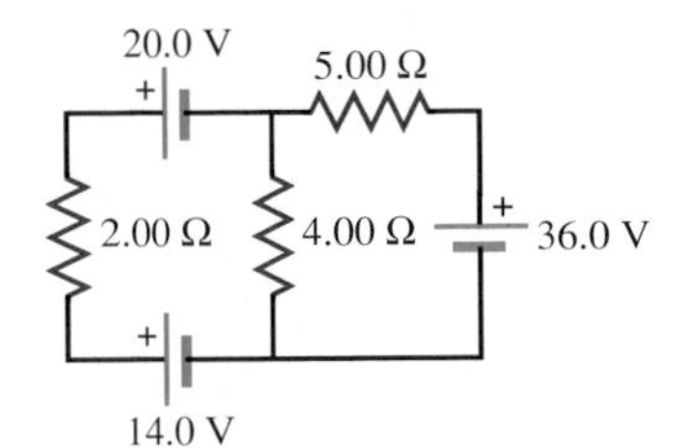


**FIGURE 27–36**

**27–38** Find the current through the battery and each resistor in the circuit shown in Fig. 27–37.

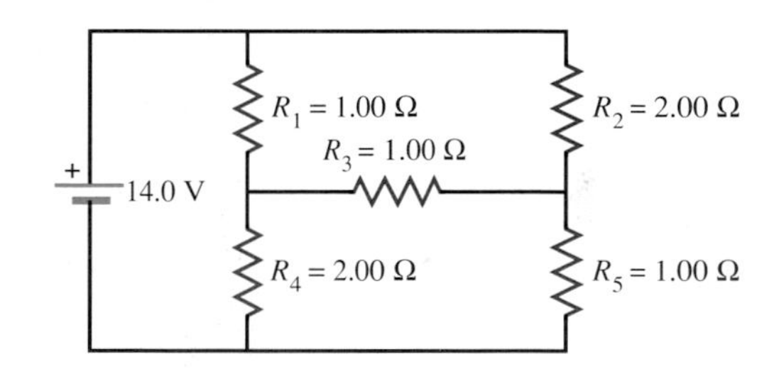


**FIGURE 27–37**

**27–39** Figure 27–38 employs a convention that is often used in circuit diagrams. The battery (or other power supply) is not shown explicitly. It is understood that the point at the top, labeled "36.0 V," is connected to the positive terminal of a 36.0-V battery having negligible internal resistance and that the "ground" symbol at the bottom is connected to its negative terminal. The circuit is completed through the battery, even though it is not shown on the diagram. a) In Fig. 27–38, what is the potential difference $V_{ab}$, the potential of point $a$ relative to point $b$, when the switch S is open? b) What is the current through switch S when it is closed? c) What is the equivalent resistance when switch S is closed?

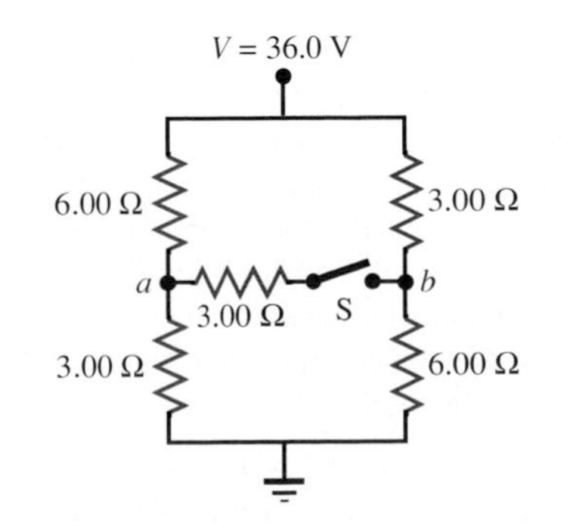

**FIGURE 27–38**

**27–40** (See Problem 27–39.) a) What is the potential of point $a$ with respect to point $b$ in Fig. 27–39 when switch S is open? b) Which point, $a$ or $b$, is at the higher potential? c) What is the final potential of point $b$ with respect to ground when switch S is closed? d) How much does the charge on each capacitor change when S is closed?

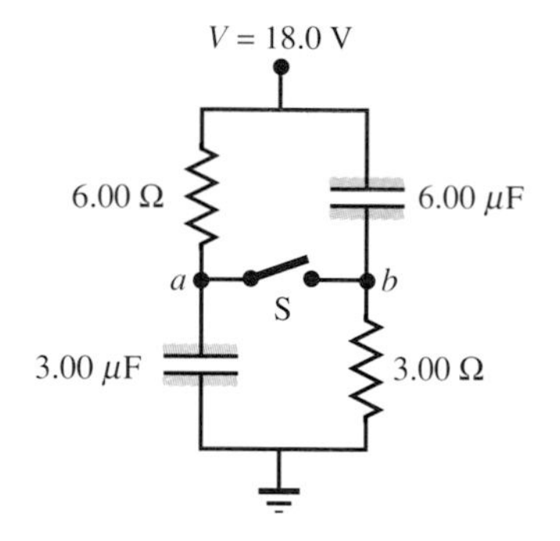


**FIGURE 27–39**

**27–41** (See Problem 27–39.) a) What is the potential of point $a$ with respect to point $b$ in Fig. 27–40 when switch S is open? b) Which point, $a$ or $b$, is at the higher potential? c) What is the final potential of point $b$ with respect to ground when switch S is closed? d) How much charge flows through switch S when it is closed?

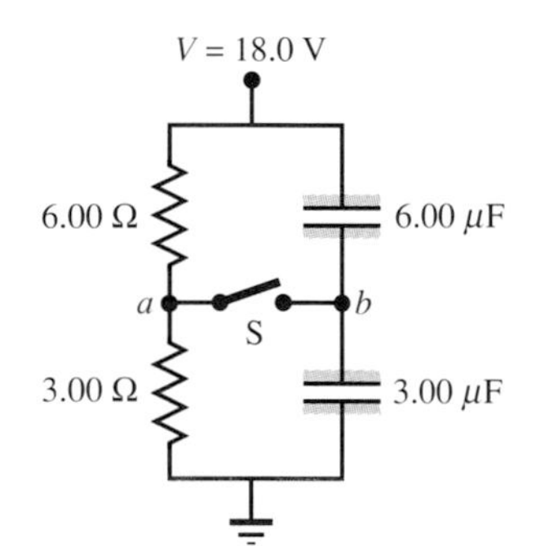


**FIGURE 27–40**

**27–42** A certain galvanometer has a resistance of 200 Ω and deflects full-scale with a current of 1.00 mA in its coil. This is to be replaced with a second galvanometer that has a resistance of 40.0 Ω and deflects full-scale with a current of 50.0 μA in its coil. Devise a circuit incorporating the second galvanometer such that the equivalent resistance of the circuit equals the resistance of the first galvanometer and the second galvanometer deflects full-scale when the current through the circuit equals the full-scale current of the first galvanometer.

**27–43** A 600-Ω resistor and a 400-Ω resistor are connected in series across a 90.0-V line. A voltmeter connected across the 600-Ω resistor reads 60.0 V. a) Find the voltmeter resistance. b) Find the reading of the same voltmeter if it is connected across the 400-Ω resistor.

**27–44** Point $a$ in Fig. 27–41 is maintained at a constant potential of 400 V above ground. (See Problem 27–39.) a) What is the reading of a voltmeter with the proper range and with resistance $3.00 \times 10^4$ Ω when connected between point $b$ and ground? b) What is the reading of a voltmeter with resistance $3.00 \times 10^6$ Ω? c) What is the reading of a voltmeter with infinite resistance?

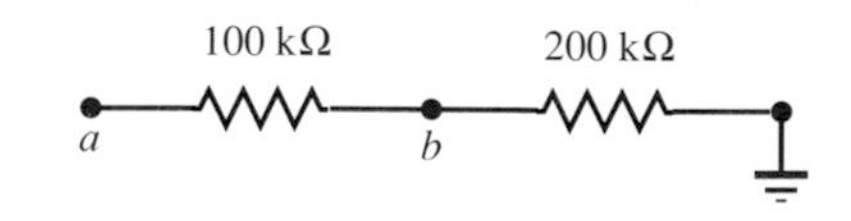


**FIGURE 27–41**

**27–45 Measuring Large Resistances.** A 150-V voltmeter has a resistance of 20,000 Ω. When connected in series with a large resistance $R$ across a 110-V line, the meter reads 60.0 V. Find the resistance $R$.

**27–46** Let $V$ and $I$ represent, respectively, the readings of the voltmeter and ammeter shown in Fig. 27–12, and let $R_V$ and $R_A$ be their equivalent resistances. a) When the circuit is connected as in Fig. 27–12a, show that

$$R = \frac{V}{I} - R_A.$$

b) When the connections are as in Fig. 27–12b, show that

$$R = \frac{V}{I - (V/R_V)}.$$

c) Show that the power delivered to the resistor in part (a) is $IV - I^2R_A$ and that in part (b) is $IV - (V^2/R_V)$.

**27–47 The Wheatstone Bridge.** The circuit shown in Fig. 27–42, called a *Wheatstone bridge*, is used to determine the value of an unknown resistor $X$ by comparison with three resistors $M$, $N$, and $P$ whose resistance can be varied. For each setting, the resistance of each resistor is precisely known. With switches $K_1$ and $K_2$ closed, these resistors are varied until the current in the galvanometer G is zero; the bridge is then said to be *balanced.*

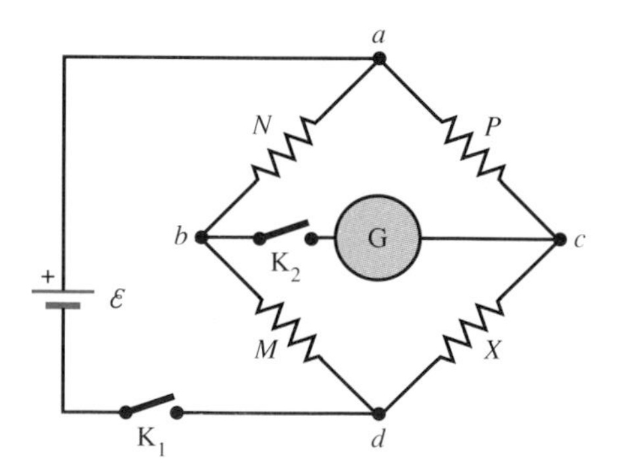


**FIGURE 27–42**

a) Show that under this condition the unknown resistance is given by $X = MP/N$. (This method permits very high precision in comparing resistors.) b) If the galvanometer G shows zero deflection when $M = 1000\ \Omega$, $N = 10.00\ \Omega$, and $P = 25.28\ \Omega$, what is the unknown resistance $X$?

**27–48** A $4.60 \times 10^3\ \Omega$ resistor is connected to the plates of a charged capacitor that has capacitance $C = 8.00 \times 10^{-10}$ F. The initial current through the resistor, just after the connection is made, is 0.300 A. What magnitude of charge was initially on each plate of the capacitor?

**27–49** A capacitor that is initially uncharged is connected in series with a resistor and an emf source with $\mathcal{E} = 200$ V and negligible internal resistance. Just after the circuit is completed, the current through the resistor is $8.00 \times 10^{-4}$ A, and the time constant for the circuit is 6.00 s. What are the resistance of the resistor and the capacitance of the capacitor?

**27–50** a) Using Eq. (27–18) for the current in a discharging capacitor, derive an expression for the instantaneous power $P = i^2R$ dissipated in the resistor. b) Integrate the expression for $P$ to find the total energy dissipated in the resistor, and show that this is equal to the total energy initially stored in the capacitor.

**27–51** The current in a charging capacitor is given by Eq. (27–14). a) The instantaneous power supplied by the battery is $\mathcal{E}i$. Integrate this to find the total energy supplied by the battery. b) The instantaneous power dissipated in the resistor is $i^2R$. Integrate this to find the total energy dissipated in the resistor. c) Find the final energy stored in the capacitor, and show that this equals the total energy supplied by the battery less the energy dissipated in the resistor, as obtained in parts (a) and (b). d) What fraction of the energy supplied by the battery is stored in the capacitor? How does this fraction depend on $R$?

**27–52** Two capacitors are charged in series by a 12.0-V battery that has an internal resistance of 1.00 Ω. There is a 5.00-Ω resistor in series between the capacitors (Fig. 27–43). a) What is the time constant of the charging circuit? b) After the switch has been closed for the time determined in part (a), what is the voltage across the 6.00-μF capacitor?

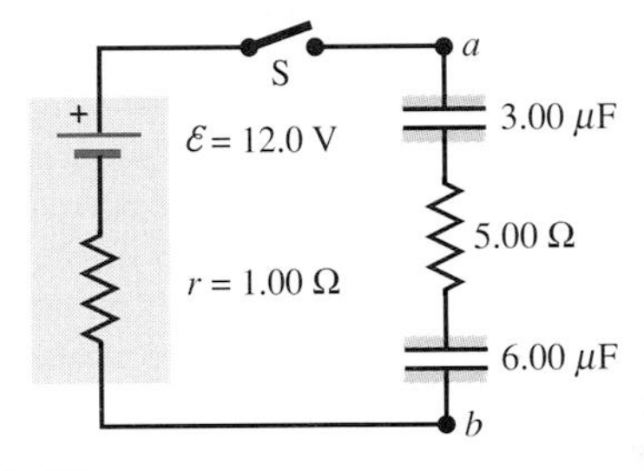

**FIGURE 27–43**

## CHALLENGE PROBLEMS

**27–53 An Infinite Network.** Prove that the resistance of the infinite network shown in Fig. 27–44 is equal to $(1 + \sqrt{3})r$.

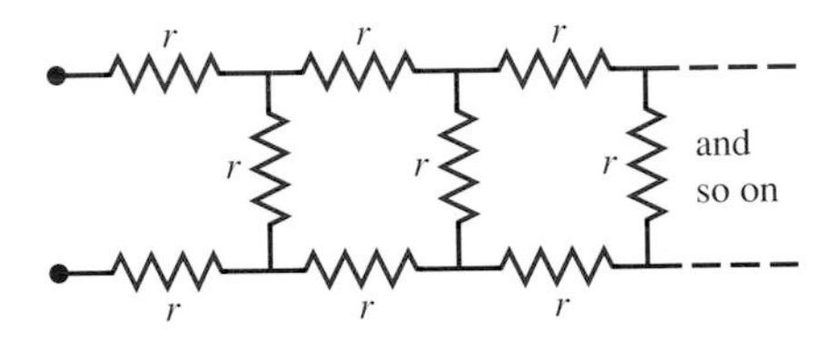


**FIGURE 27–44**

**27–54** Suppose a resistor $R$ lies along each edge of a cube (twelve resistors in all) with connections at the corners. Find the equivalent resistance between two diagonally opposite corners of the cube (points $a$ and $b$ in Fig. 27–45).

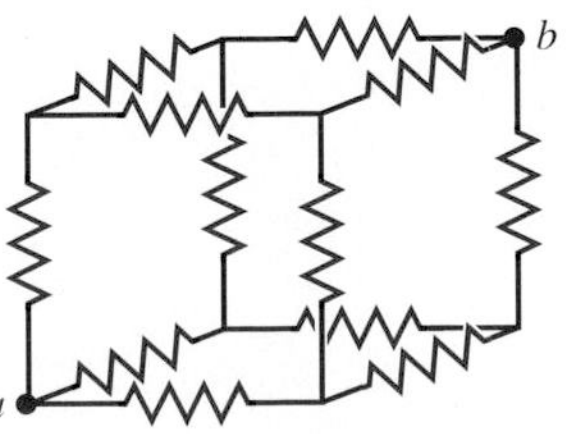

**FIGURE 27–45**

**27–55** According to the theorem of superposition, the response (current) in a circuit is proportional to the stimulus (voltage) that causes it. This is true even if there are multiple sources in a circuit. This theorem can be used to analyze a circuit without resorting to Kirchhoff's rules by considering the currents in the circuit to be the superposition of currents caused by each source independently. In this way the circuit can be analyzed by computing equivalent resistances rather than by using the (sometimes) more cumbersome method of Kirchhoff's rules. Furthermore, by using the superposition theorem it is possible to examine how the modification of a source in one part of the circuit will affect the currents in all parts of the circuit without having to use Kirchhoff's rules to recalculate all of the currents. Consider the circuit shown in Fig. 27–46. If the circuit were redrawn with the 55.0-V and 57.0-V

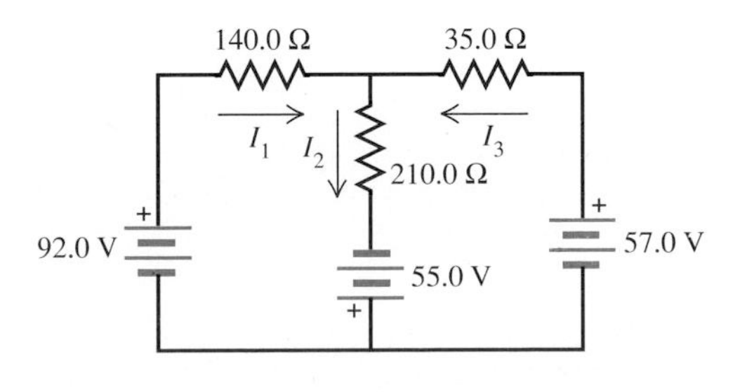


**FIGURE 27–46**

sources replaced by short circuits, the circuit could be analyzed by the method of equivalent resistances without resorting to Kirchhoff's rules, and the current in each branch could be found in a simple manner. Similarly, by redrawing the circuit with the 92.0-V and the 55.0-V sources replaced by short circuits, the circuit could again be analyzed in a simple manner. Finally, by replacing the 92.0-V and the 57.0-V sources with a short circuit the circuit could once again be analyzed simply. By superimposing the respective currents found in each of the branches by using the three simplified circuits, the actual current in each branch can be found. a) Using Kirchhoff's rules, find the branch currents in the 140.0-Ω, 210.0-Ω, and 35.0-Ω resistors. b) Using a circuit similar to the circuit of Fig. 27–46, but with the 55.0-V and 57.0-V sources replaced by a short circuit, determine the currents in each resistance. c) Repeat part (b) by replacing the 92.0-V and 55.0-V sources by short circuits, leaving the 57.0-V source intact. d) Repeat part (b) by replacing the 92.0-V and 57.0-V sources by short circuits, leaving the 55.0-V source intact. e) Verify the superposition theorem by taking the currents calculated in parts (b), (c), and (d) and comparing them with the currents calculated in part (a). f) If the 57.0-V source is replaced by a 90.0-V source, what will be the new currents in all branches of the circuit? [*Hint:* Using the superposition theorem, recalculate the partial currents calculated in part (c) using the fact that those currents are proportional to the source that is being replaced. Then superpose the new partial currents with those found in parts (b) and (d).]

**27–56 A Capacitor Burglar Alarm.** The capacitance of a capacitor can be affected by dielectric material that, although not inside the capacitor, is near enough to the capacitor to be polarized by the fringing electric field that exists near a charged capacitor. This effect is usually of the order of picofarads (pF), but it can be used with appropriate electronic circuitry to detect a change in the dielectric material surrounding the capacitor. Such a dielectric material might be the human body, and the effect described above might be used in the design of a burglar alarm. Consider the simplified circuit shown in Fig. 27–47. The voltage source has emf $\mathcal{E} = 1000$ V, and the capacitor has capacitance $C = 10.0$ pF. The electronic circuitry for detecting the current, represented as an ammeter in the diagram, has negligible resistance and is capable of detecting a current which persists at a level of at least 1.00 $\mu$A for at least 200 $\mu$s after the capacitance has changed abruptly from $C$ to $C'$. The burglar alarm is designed to be activated if the capacitance changes by 10%. a) Determine the charge on the 10.0-pF capacitor when it is fully charged. b) If the capacitor is fully charged before the intruder is detected, assuming that the time taken for the capacitance to change by 10% is small enough to be neglected, derive an equation that expresses the current through the resistor $R$ as a function of the time $t$ since the capacitance has changed. c) Determine the range of values of the resistance $R$ that will meet the design specifications of the burglar alarm. What happens if $R$ is too small? Too large? (*Hint:* You will not be able to solve this part analytically but must use numerical methods. $R$ can be expressed as a logarithmic function of $R$ plus known quantities. Use a trial value of $R$ and calculate from the expression a new value. Continue to do this until the input and output values of $R$ agree to within three significant figures.)

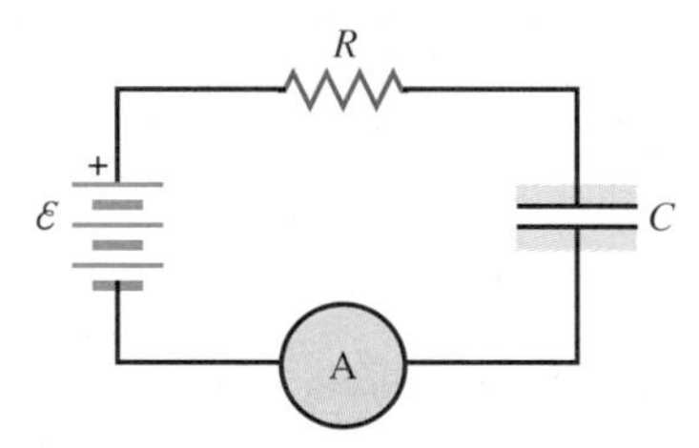


**FIGURE 27–47**

The Aurora Borealis, or Northern Lights, appear when atoms and molecules in the earth's upper atmosphere are hit by high-speed electrons from the sun's "solar wind." These particles collide under the concentrating effect of the earth's magnetic field near the poles.

# CHAPTER 28

# Magnetic Field and Magnetic Forces

This photo, taken from the space shuttle, shows a sheet of solar particles following the path of the earth's magnetic field as they bombard the upper atmosphere.

## KEY CONCEPTS

**• A magnetic field exerts a force on a moving charge. The magnitude of the force is proportional to the charge, to its speed, and to the magnitude of the magnetic field. The direction of the force is perpendicular to both the magnetic field and the particle's velocity.**

**• Magnetic fields can be represented by field maps using magnetic field lines. Magnetic flux is the product of surface area and the component of magnetic field perpendicular to the surface. The total magnetic flux out of any closed surface is always zero.**

**• Motion of charged particles in magnetic fields can be analyzed by using the magnetic force expression with Newton's second law. Such motion plays an important role in many fundamental experiments and practical devices.**

**• Magnetic fields exert forces on conductors that are carrying currents. These forces are essential for the operation of electric motors, moving-coil galvanometers, and many other devices.**

**• The Hall effect provides a direct demonstration of magnetic forces on moving charges in a conductor and a means of determining the signs of these charges.**

## INTRODUCTION

Everybody uses magnetic forces. Without them there would be no TV picture tubes, no electric motors, no microwave ovens, no loudspeakers, and no computer printers or disk drives. The most familiar aspects of magnetism are those associated with permanent magnets, which attract unmagnetized iron objects and can also attract or repel other magnets. A compass needle interacting with the earth's magnetism is an example of this interaction. But the fundamental nature of magnetism is the interaction of moving electric charges.

A magnetic field is established by a permanent magnet, by an electric current in a conductor, or by other moving charges. This magnetic field, in turn, exerts forces on other moving charges and current-carrying conductors. Magnetic forces are an essential aspect of the interactions of electrically charged particles. In this chapter we study magnetic forces and torques, and in Chapter 29 we examine the ways in which magnetic fields are *produced* by moving charges and currents.

## 28–1 MAGNETISM

Magnetic phenomena were first observed at least 2500 years ago in fragments of magnetized iron ore found near the ancient city of Magnesia (now Manisa, in western Turkey). It was discovered that when an iron rod is brought in contact with a natural magnet, the rod also becomes magnetized. When such a rod is suspended by a string from its center, it tends to line itself up in a north-south direction, like a compass needle. Magnets have been used for navigation at least since the eleventh century. Figure 28–1 shows a more contemporary example of magnetic forces of a permanent magnet.

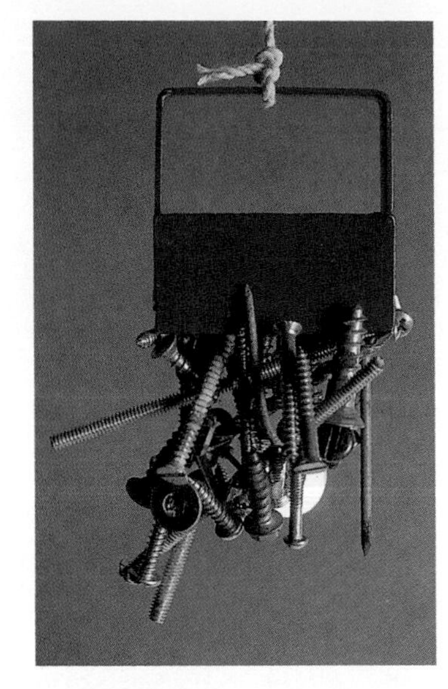

**28–1** Screws made of steel (a ferromagnetic material) are attracted to a permanent magnet. Screws made of brass or aluminum (nonferromagnetic materials) would not be attracted.

Before the relation of magnetic interactions to moving charges was understood, the interactions of bar magnets and compass needles were described in terms of *magnetic poles*. The end of a bar magnet that points north is called a *north pole*, or *N-pole*, and the other end is a *south pole*, or *S-pole*. Two opposite poles attract each other, and two like poles repel each other. The concept of magnetic poles is of limited usefulness and is somewhat misleading. There is no evidence that a single isolated magnetic pole exists; poles always appear in pairs. If a bar magnet is broken in two, each broken end becomes a pole. The existence of an isolated magnetic pole, or **magnetic monopole,** would have sweeping implications for theoretical physics. Extensive searches for magnetic monopoles have been carried out, so far without success.

A compass needle points north because the earth is a magnet; its geographical north pole is close to a magnetic *south* pole. The earth's magnetic axis is not quite parallel to its geographic axis (the axis of rotation), so a compass reading deviates somewhat from geographic north; this deviation, which varies with location, is called *magnetic declination*. Also, the magnetic field is not horizontal at most points on the earth's surface; its inclination up or down is described by the *angle of dip.*

A sketch of the earth's magnetic field is shown in Fig. 28–2. The lines show the direction in which a compass would point at each location. These lines are *magnetic*

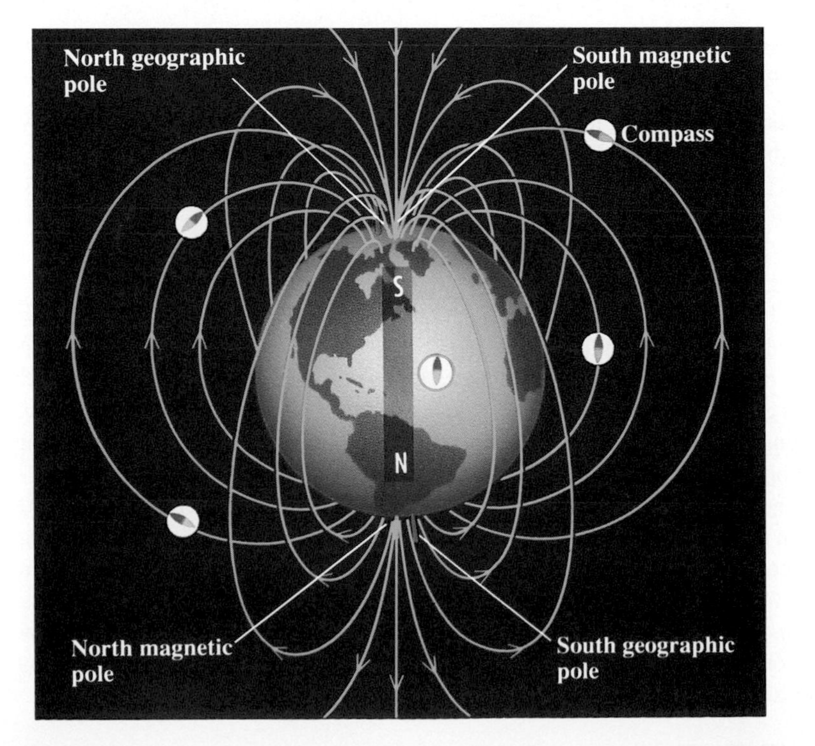


**28–2** A compass placed at any point in the earth's magnetic field will point in the direction of the field line at that point. Representing the earth's field as that of a tilted bar magnet is only a crude approximation of its fairly complex configuration.

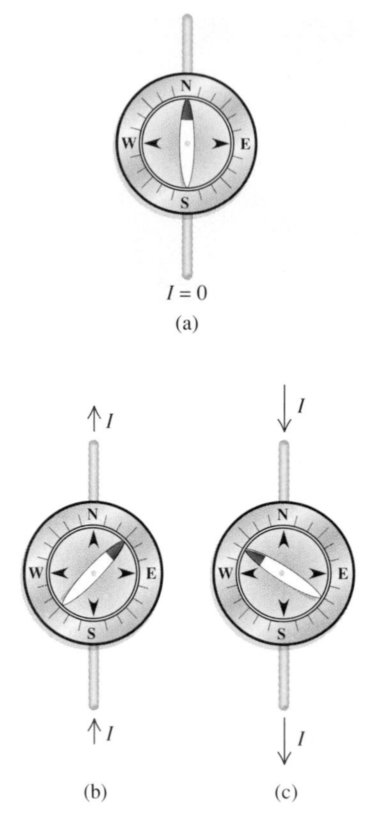


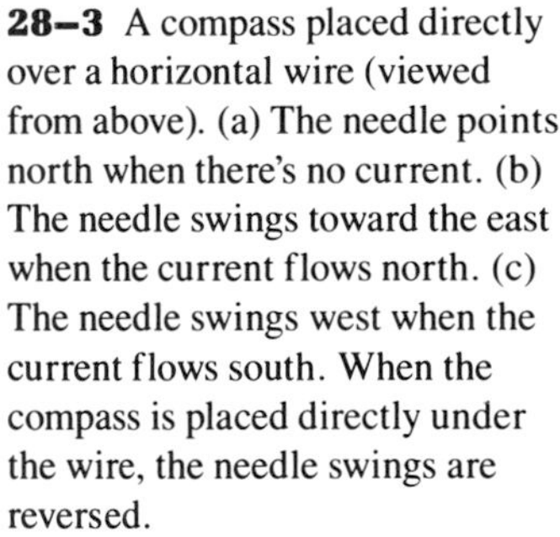

**28–3** A compass placed directly over a horizontal wire (viewed from above). (a) The needle points north when there's no current. (b) The needle swings toward the east when the current flows north. (c) The needle swings west when the current flows south. When the compass is placed directly under the wire, the needle swings are reversed.

*field lines,* discussed in detail in Section 28–3. The direction of the field at any point can be defined as the direction of the force that the field would exert on a magnetic north pole. In Section 28–2 we will introduce a more fundamental way to define magnetic field.

In 1819 the Danish scientist Hans Christian Oersted discovered that a compass needle was deflected by a current-carrying wire, as shown in Fig. 28–3. Similar investigations were carried out in France by André Ampère. A few years later, Michael Faraday in England and Joseph Henry in the United States discovered that moving a magnet near a conducting loop can cause a current in the loop and that a changing current in one conducting loop can cause a current in a separate loop. These observations were the first evidence of the relationship of magnetism to moving charges. We now know that electric and magnetic interactions are intimately intertwined. Over the next several chapters we will develop the unifying principles of electromagnetism, culminating in the expression of these principles in *Maxwell's equations.* These equations represent the synthesis of electromagnetism, just as Newton's laws of motion are the synthesis of mechanics, and like Newton's laws, they represent a towering achievement of the human intellect.

## 28–2 MAGNETIC FIELD

To introduce the concept of magnetic field, let's review our formulation of *electrical* interactions in Chapter 22, where we introduced the concept of *electric* field. We represented electrical interactions in two steps:

1. A distribution of electric charge at rest creates an electric field $\boldsymbol{E}$ in the surrounding space.
2. This field exerts a force $\boldsymbol{F} = q\boldsymbol{E}$ on any other charge $q$ that is present in the field.

We can describe magnetic interactions in the same way:

1. A moving charge or a current creates a **magnetic field** in the surrounding space (in addition to its *electric* field).
2. The magnetic field exerts a force $\boldsymbol{F}$ on any other moving charge or current that is present in the field.

Like electric field, magnetic field is a *vector field,* that is, a vector quantity associated with each point in space. We will use the symbol $\boldsymbol{B}$ for magnetic field.

In this chapter we'll concentrate on the *second* aspect of the interaction: Given the presence of a magnetic field, what force does it exert on a moving charge or a current? In Chapter 29 we will come back to the problem of how magnetic fields are *created* by moving charges and currents.

What are the characteristics of the magnetic force on a moving charge? First, its magnitude is proportional to the charge. If a 1-$\mu$C charge and a 2-$\mu$C charge move through a given magnetic field with the same velocity, the force on the 2-$\mu$C charge is twice as great as that on the 1-$\mu$C charge. The force is also proportional to the magnitude, or "strength," of the field; if we double the magnitude of the field without changing the charge or its velocity, the force doubles.

The magnetic force is also proportional to the particle's speed. This is quite different from the electric-field force, which is the same whether the charge is moving or not. A charged particle at rest experiences *no* magnetic force. Furthermore, the magnetic force $\boldsymbol{F}$ *does not* have the same direction as the magnetic field $\boldsymbol{B}$ but instead is always *perpendicular* to both $\boldsymbol{B}$ and $\boldsymbol{v}$. The magnitude $F$ of the force is found to be proportional to the component of $\boldsymbol{v}$ perpendicular to the field; when that component is zero (that is, when $\boldsymbol{v}$ and $\boldsymbol{B}$ are parallel or antiparallel), the force is zero.

Figure 28–4 shows these relationships. The direction of $\boldsymbol{F}$ is always perpendicular to the plane containing $\boldsymbol{v}$ and $\boldsymbol{B}$. Its magnitude is given by

$$F = |q|v_{\perp}B = |q|vB \sin\phi, \qquad (28\text{–}1)$$

where $|q|$ is the magnitude of the charge and $\phi$ is the angle measured from the direction of $\boldsymbol{v}$ to the direction of $\boldsymbol{B}$, as shown in the figure.

This description does not specify the direction of $\boldsymbol{F}$ completely; there are always two directions, opposite to each other and both perpendicular to the plane of $\boldsymbol{v}$ and $\boldsymbol{B}$. To complete the description, we use the same right-hand rule that we used to define the vector product in Section 1–10. (It would be a good idea to review that section before you go on.) Draw the vectors $\boldsymbol{v}$ and $\boldsymbol{B}$ with their tails together. Imagine turning $\boldsymbol{v}$ until it points in the direction of $\boldsymbol{B}$ (turning through the smaller of the two possible angles). Think of wrapping the fingers of your right hand around the line perpendicular to the plane of $\boldsymbol{v}$ and $\boldsymbol{B}$, as in Fig. 28–4, so that they curl around with this sense of rotation from $\boldsymbol{v}$ to $\boldsymbol{B}$; your thumb then points in the direction of the force $\boldsymbol{F}$ on a *positive* charge. (Alternatively, the direction of the force $\boldsymbol{F}$ on a positive charge is the direction in which a right-hand-thread screw would advance if turned the same way.)

This discussion shows that the force on a charge $q$ moving with velocity $\boldsymbol{v}$ in a magnetic field $\boldsymbol{B}$ is given, both in magnitude and in direction, by

$$\boldsymbol{F} = q\boldsymbol{v} \times \boldsymbol{B}. \qquad (28\text{–}2)$$

This is the first of several vector products that we will encounter in our study of magnetic-field relationships.

Equation (28–2) is valid for both positive and negative charges. When $q$ is negative, the direction of the force $\boldsymbol{F}$ is opposite to that of $\boldsymbol{v} \times \boldsymbol{B}$. If two charges with equal magnitude and opposite sign move in the same magnetic field $\boldsymbol{B}$ with the same velocity (Fig. 28–5), the forces have equal magnitude and opposite direction. Figures 28–4 and 28–5 show several examples of the relationships of the directions of $\boldsymbol{F}$, $\boldsymbol{v}$, and $\boldsymbol{B}$ for both positive and negative charges. Be sure that you understand these relationships and can verify these figures for yourself.

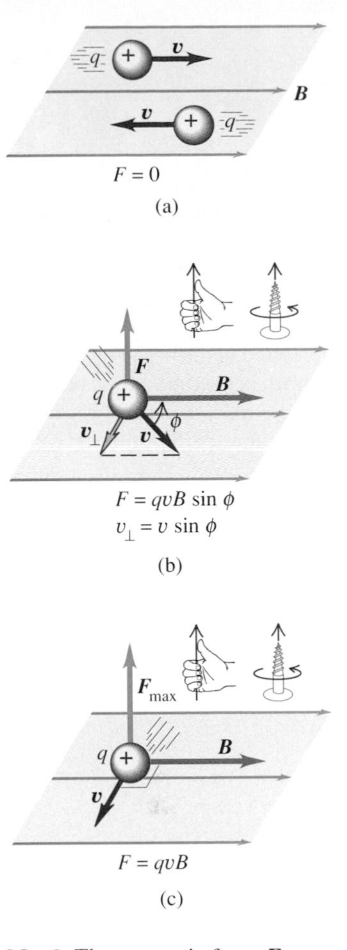


**28–4** The magnetic force $\boldsymbol{F}$ acting on a positive charge $q$ moving with velocity $\boldsymbol{v}$ is perpendicular to both the magnetic field $\boldsymbol{B}$ and $\boldsymbol{v}$. (a) The magnetic force is zero when $\boldsymbol{v}$ is parallel or antiparallel to $\boldsymbol{B}$. (b) When $\boldsymbol{v}$ is at an angle $\phi$ to $\boldsymbol{B}$, then $F = qvB \sin\phi$. (c) When $\boldsymbol{v}$ is perpendicular to $\boldsymbol{B}$, then $F = qvB$.

Equation (28–1) can be interpreted in a different but equivalent way. Recalling that $\phi$ is the angle between the directions of vectors $\boldsymbol{v}$ and $\boldsymbol{B}$, we may interpret $B \sin\phi$ as the component of $\boldsymbol{B}$ perpendicular to $\boldsymbol{v}$, that is, $B_{\perp}$. With this notation the force expression becomes

$$F = |q|vB_{\perp}. \qquad (28\text{–}3)$$

This form is equivalent to Eq. (28–1), but is sometimes more convenient to use, especially in problems involving *currents* rather than individual particles. We will discuss forces on currents later in this chapter.

From Eq. (28–1) the *units* of $B$ must be the same as the units of $F/qv$. Therefore the SI unit of $B$ is equivalent to $1\ \text{N}\cdot\text{s/C}\cdot\text{m}$, or, since one ampere is one coulomb per second ($1\ \text{A} = 1\ \text{C/s}$), $1\ \text{N/A}\cdot\text{m}$. This unit is called the **tesla** (abbreviated T), in honor of Nikola Tesla (1857–1943), the prominent Serbian-American scientist and inventor:

$$1\ \text{tesla} = 1\ \text{T} = 1\ \text{N/A}\cdot\text{m}.$$

The cgs unit of $B$, the **gauss** ($1\ \text{G} = 10^{-4}\ \text{T}$), is also in common use. Instruments for measuring magnetic field are sometimes called gaussmeters.

The magnetic field of the earth is of the order of $10^{-4}$ T, or 1 G. Magnetic fields of the order of 10 T occur in the interior of atoms and are important in the analysis of atomic spectra. The largest values of steady magnetic field that have been achieved in the laboratory are of the order of 30 T. Some pulsed-current electromagnets can produce fields of

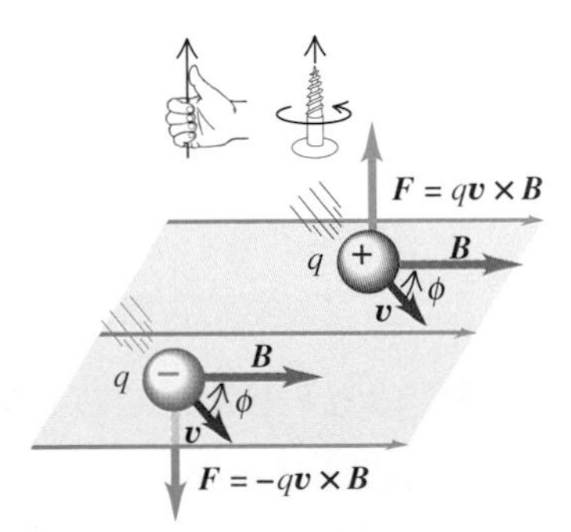


**28–5** In a magnetic field the magnetic forces are equal in magnitude but opposite in direction because the charges are equal in magnitude but opposite in sign.

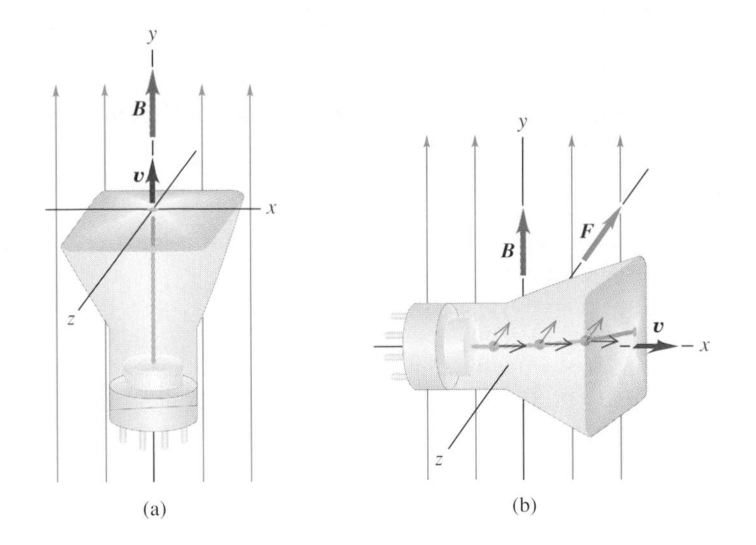


**28–6** (a) If the electron beam of a cathode-ray tube is undeflected when the beam is parallel to the $y$-axis, the $\boldsymbol{B}$ vector points either up or down. (b) If the beam is deflected in the negative $z$-direction when the tube axis is parallel to the $x$-axis, then the $\boldsymbol{B}$ vector points upward. The force $\boldsymbol{F}$ on the electrons points along the negative $z$-axis, opposite to the rule of Fig. 28–4, because $q$ is negative.

the order of 120 T for short time intervals of the order of a millisecond. The magnetic field at the surface of a neutron star is believed to be of the order of $10^8$ T.

To explore an unknown magnetic field, we can measure the magnitude and direction of the force on a *moving* test charge. The electron beam in a cathode-ray tube, discussed in Section 24–7, is a convenient device for making such measurements. The electron gun shoots out a narrow beam of electrons at a known speed. If there is no force to deflect the beam, it strikes the center of the screen. In principle, an old TV set with the deflection coils disconnected could be used for the same purpose.

In general, when a magnetic field is present, the electron beam is deflected. However, if the beam is parallel or antiparallel to the field, then in Eq. (28–1), $\phi = 0$ or $\pi$ and $F = 0$; there is no force and no deflection. If we find that the electron beam is undeflected when its direction is parallel to the $y$-axis, as in Fig. 28–6a, the $\boldsymbol{B}$ vector must point either up or down.

When we turn the tube 90°, so that its axis is along the $x$-axis (Fig. 28–6b), the beam is deflected in a direction corresponding to a force perpendicular to the plane of $\boldsymbol{B}$ and $\boldsymbol{v}$. We can perform additional experiments in which the angle between $\boldsymbol{B}$ and $\boldsymbol{v}$ is between zero and 90° to confirm Eq. (28–1) or (28–3) and the accompanying discussion. We note that the electron has a negative charge; the force in Fig. 28–6 is opposite in direction to the force on a positive charge.

When a charged particle moves through a region of space where *both* electric and magnetic fields are present, both fields exert forces on the particle. The total force $\boldsymbol{F}$ is the vector sum of the electric and magnetic forces:

$$\boldsymbol{F} = q(\boldsymbol{E} + \boldsymbol{v} \times \boldsymbol{B}). \tag{28–4}$$

## PROBLEM-SOLVING STRATEGY

### Magnetic forces

**1.** The biggest difficulty in determining magnetic forces is in relating the directions of the vector quantities. In evaluating $\boldsymbol{v} \times \boldsymbol{B}$, draw the two vectors $\boldsymbol{v}$ and $\boldsymbol{B}$ with their tails together so that you can visualize and draw the plane in which they lie. This also helps you to identify the angle $\phi$ between the two vectors and to avoid getting its complement or some other erroneous angle. Then remember that $\boldsymbol{F}$ is always perpendicular to this plane. The direction of $\boldsymbol{v} \times \boldsymbol{B}$ is determined by the right-hand rule; keep referring to Figs. 28–4 and 28–5 until you're sure you understand this. If $q$ is negative, the force is *opposite* to $\boldsymbol{v} \times \boldsymbol{B}$.

**2.** Whenever you can, do the problem in two ways. Do it directly from the geometric definition of the vector product. Then find the components of the vectors in some convenient axis system and calculate the vector product algebraically from the components. Check that the results agree.

## EXAMPLE 28–1

A proton beam moves through a uniform magnetic field with magnitude 2.0 T, directed along the positive $z$-axis, as in Fig. 28–7. The protons have velocity of magnitude $3.0 \times 10^5$ m/s in the $xz$-plane at an angle of 30° to the positive $z$-axis. Find the force on a proton (for which $q = 1.6 \times 10^{-19}$ C).

**SOLUTION** The right-hand rule shows that the direction of the force is along the negative $y$-axis. The magnitude of the force is calculated using Eq. (28–1).

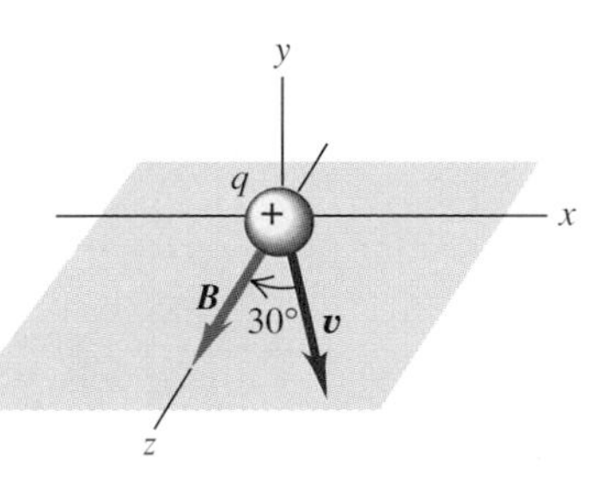


**28–7** Directions of $\boldsymbol{v}$ and $\boldsymbol{B}$ for a proton beam.

Thus we have

$$
\begin{aligned}
F &= qvB \sin \phi \\
&= (1.6 \times 10^{-19}\ \text{C})(3.0 \times 10^5\ \text{m/s})(2.0\ \text{T})(\sin 30°) \\
&= 4.8 \times 10^{-14}\ \text{N}.
\end{aligned}
$$

Alternatively, in vector language, with Eq. (28–2),

$$\boldsymbol{v} = (3 \times 10^5\ \text{m/s})[(\sin 30°)\boldsymbol{i} + (\cos 30°)\boldsymbol{k}],$$

$$\boldsymbol{B} = (2.0\ \text{T})\,\boldsymbol{k},$$

$$
\begin{aligned}
\boldsymbol{F} &= q\boldsymbol{v} \times \boldsymbol{B} \\
&= (1.6 \times 10^{-19}\ \text{C})(3 \times 10^5\ \text{m/s})[(\sin 30°)\,\boldsymbol{i} + (\cos 30°)\boldsymbol{k}] \times (2.0\ \text{T})\,\boldsymbol{k} \\
&= (-4.8 \times 10^{-14}\ \text{N})\boldsymbol{j}
\end{aligned}
$$

(because $\boldsymbol{i} \times \boldsymbol{k} = -\boldsymbol{j}$ and $\boldsymbol{k} \times \boldsymbol{k} = \boldsymbol{0}$).

If the beam consists of *electrons* rather than protons, the charge is negative ($q = -1.6 \times 10^{-19}$ C), and the direction of the force is reversed. The force is now directed along the *positive* $y$-axis; the magnitude is the same as before, $F = 4.8 \times 10^{-14}$ N.

# 28–3 MAGNETIC FIELD LINES AND MAGNETIC FLUX

We can represent any magnetic field by *lines,* just as we did for the earth's magnetic field in Fig. 28–2. The idea is the same as for the electric field lines that we introduced in Section 22–7. We draw the lines so that the line through any point is tangent to the magnetic field vector $\boldsymbol{B}$ at that point. Also, we draw the *number* of lines per unit area (perpendicular to the lines at a given point) to be proportional to the magnitude of the field at that point.

We call these lines **magnetic field lines.** They are sometimes called magnetic lines of force, but that's not a good name for them because, unlike electric field lines, they *do not* point in the direction of the force on a charge. Magnetic field lines *do* have the direction that a compass needle would point at each location; this may help you to visualize them. Just as with electric field lines, we draw only a few representative lines; otherwise, the lines would fill up all of space. Also, because the direction of $\boldsymbol{B}$ at each point is unique, field lines never intersect.

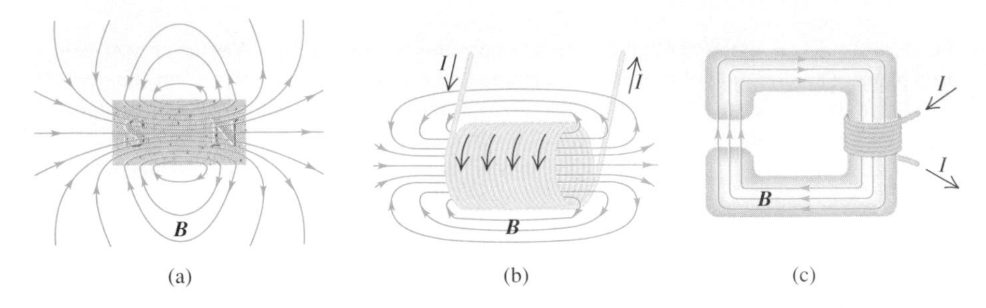


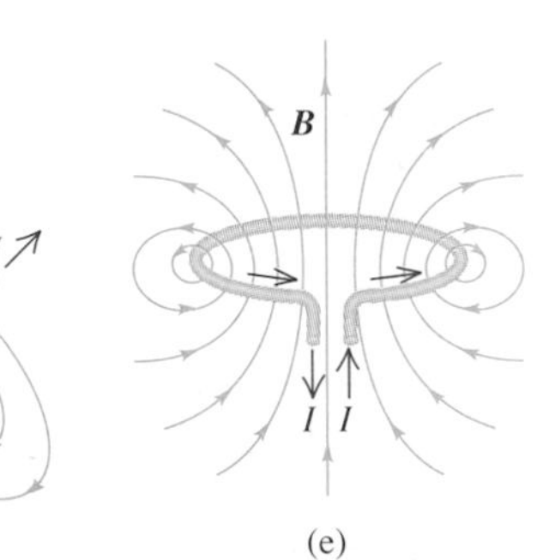


**28–8** Magnetic field lines in a plane through the center of (a) a permanent magnet, (b) a cylindrical coil, (c) an iron-core electromagnet. (d) Magnetic field lines in a plane perpendicular to a long, straight, current-carrying wire. (e) Magnetic field lines in a plane containing the axis of a circular loop.

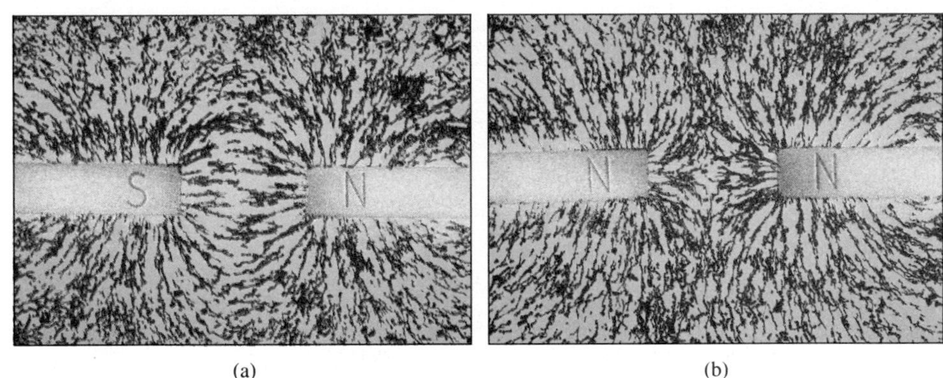


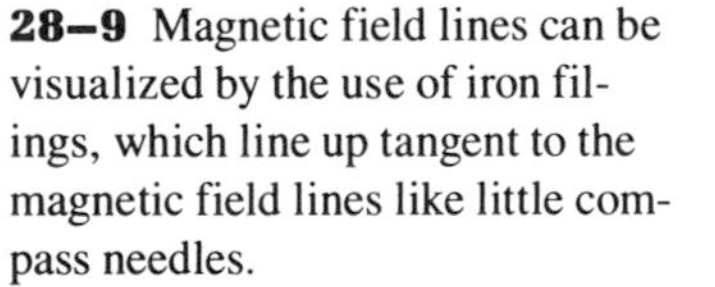

**28–9** Magnetic field lines can be visualized by the use of iron filings, which line up tangent to the magnetic field lines like little compass needles.

Magnetic field lines produced by several common sources of magnetic field are shown in Fig. 28–8. In the air space between the poles of the electromagnet shown in Fig. 28–8c, the field lines are approximately straight, parallel, and equally spaced, showing that the magnetic field in this region is approximately *uniform* (constant in magnitude and direction). Figure 28–9 shows the familiar technique of visualizing magnetic field lines with iron filings.

The magnetic field of the earth (Fig. 28–2) resembles the field of a bar magnet (Fig. 28–8a), with the magnet's axis tilted with respect to the earth's axis of rotation. The earth's field, thought to be caused by currents in its molten core, changes with time. There is geologic evidence that it has actually changed direction several times during the past 100 million years.

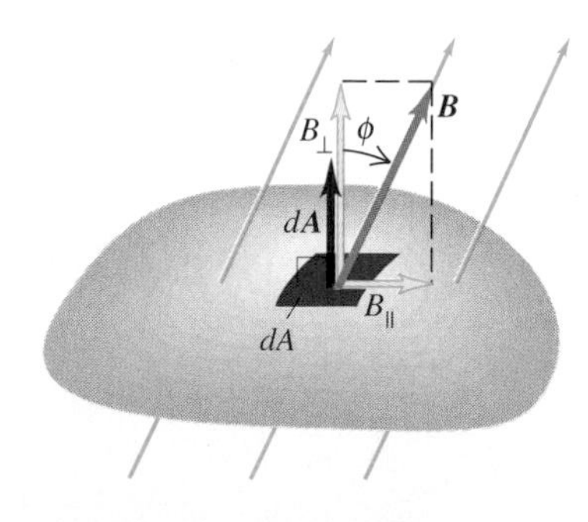


**28–10** The magnetic flux through an area element $dA$ is defined to be $d\Phi_B = B_\perp\, dA$.

We define the **magnetic flux** $\Phi_B$ through a surface just as we defined electric flux in connection with Gauss's law in Section 23–1. We can divide any surface into elements of area $dA$ (Fig. 28–10). For each element we determine $B_\perp$, the component of $\boldsymbol{B}$ normal to the surface at the position of that element, as shown. From the figure, $B_\perp = B \cos \phi$, where $\phi$ is the angle between the direction of $\boldsymbol{B}$ and a line perpendicular to the surface. (Be careful not to confuse $\phi$ with $\Phi_B$.) In general, this component varies from point to point on the surface. We define the magnetic flux $d\Phi_B$ through area $dA$ as

$$d\Phi_B = B_\perp\, dA = B \cos \phi\, dA = \boldsymbol{B} \cdot d\boldsymbol{A}. \qquad (28\text{–}5)$$

The *total* magnetic flux through the surface is the sum of the contributions from the individual area elements:

$$\Phi_B = \int B_\perp \, dA = \int B \cos\phi \, dA = \int \boldsymbol{B} \cdot d\boldsymbol{A}. \qquad (28\text{–}6)$$

(We have used the concepts of vector area and surface integral that we introduced in Section 23–1; you may want to review that discussion.) When we draw magnetic field lines, the number of lines passing through any surface is proportional to the magnetic flux through that surface.

Magnetic flux is a *scalar* quantity. In the special case in which $\boldsymbol{B}$ is uniform over a plane surface with total area $A$, $B_\perp$ and $\phi$ are the same at all points on the surface, and

$$\Phi_B = B_\perp A = BA \cos\phi. \qquad (28\text{–}7)$$

If $\boldsymbol{B}$ happens to be perpendicular to the surface, $\cos\phi = 1$, and Eq. (28–7) reduces to $\Phi_B = BA$. We will use the concept of magnetic flux extensively during our study of electromagnetic induction in Chapter 30.

According to Gauss's law, the total *electric* flux out of a closed surface is proportional to the total electric charge enclosed and to the number of electric field lines coming out of the surface. Electric field lines begin and end on electric charges. For example, if the surface encloses an electric dipole, the total flux is zero because the total charge is zero. If there were such a thing as a single magnetic charge (magnetic monopole), the total *magnetic* flux out of a closed surface would be proportional to the total magnetic charge strength enclosed. We have mentioned that no magnetic monopole has ever been observed, despite intensive searches. We conclude that **the total magnetic flux out of a closed surface is always zero.** Symbolically,

$$\oint \boldsymbol{B} \cdot d\boldsymbol{A} = 0 \quad \text{(closed surface).} \qquad (28\text{–}8)$$

This equation is sometimes called *Gauss's law for magnetism*. Electric monopoles (single electric charges) exist, but as far as we know, magnetic monopoles do not. It also follows from Eq. (28–8) that magnetic field lines are always continuous. They never have end points; such a point would indicate the existence of a monopole.

Our general definition of magnetic flux through an open surface, Eq. (28–6), has an ambiguity of sign because of the two possible choices of direction for the vector area element $d\boldsymbol{A}$. For Gauss's law, which always deals with *closed* surfaces, $d\boldsymbol{A}$ always points *out of* the surface. However, some applications of *magnetic* flux involve an *open* surface with a boundary line. In these cases we choose one of the possible sides of the surface to be the "positive" side and use that choice consistently.

The SI unit of magnetic flux is equal to the unit of magnetic field (1 T) times the unit of area (1 m$^2$). This unit is called the **weber** (1 Wb), in honor of Wilhelm Weber (1804–1891):

$$1 \text{ Wb} = 1 \text{ T} \cdot \text{m}^2.$$

Also, 1 T = 1 N/A · m, so

$$1 \text{ Wb} = 1 \text{ T} \cdot \text{m}^2 = 1 \text{ N} \cdot \text{m/A}.$$

If the element of area $dA$ in Eq. (28–5) is at right angles to the field lines, then $B_\perp = B$; calling this area $dA_\perp$, we have

$$B = \frac{d\Phi_B}{dA_\perp}. \qquad (28\text{–}9)$$

That is, the magnitude of magnetic field is equal to *flux per unit area* across an area at right angles to the magnetic field. For this reason, magnetic field $\boldsymbol{B}$ is sometimes called **magnetic flux density.** We can picture the total flux through a surface as proportional to the number of field lines passing through the surface, and the field (the flux density) as the number of lines *per unit area.*

## EXAMPLE 28–2

Figure 28–11 shows a flat surface with area 3.0 cm$^2$ in a uniform magnetic field. If the magnetic flux through this area is 0.90 mWb, calculate the magnitude of the magnetic field, and find the direction of the area vector.

**SOLUTION** Because $B$ and $\phi$ are the same at all points on the surface, we can use Eq. (28–7): $\Phi_B = BA\cos\phi$. The area $A$ is $3.0 \times 10^{-4}\ \text{m}^2$; the direction of $\boldsymbol{A}$ is perpendicular to the surface, so $\phi$ could be either 60° or 120°. But $\Phi_B$, $B$, and $A$ are all positive, so $\cos\phi$ must also be positive. This rules out 120°, so $\phi = 60°$, and we find

$$B = \frac{\Phi_B}{A\cos\phi} = \frac{0.90 \times 10^{-3}\ \text{Wb}}{(3.0 \times 10^{-4}\ \text{m}^2)(\cos 60°)} = 6.0\ \text{T}.$$

The area vector is perpendicular to the area in the direction shown in Fig. 28–11b.

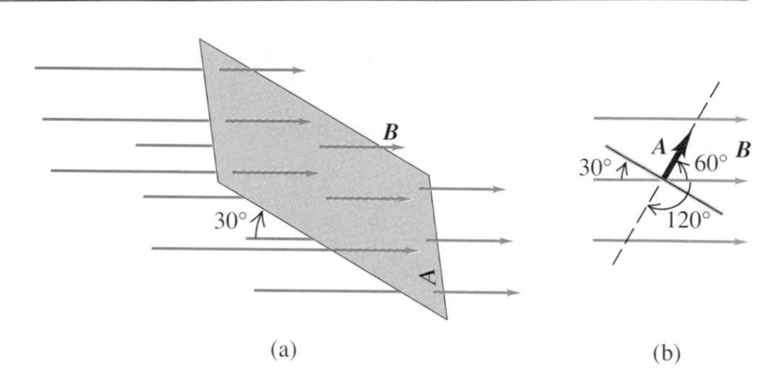


**28–11** (a) A flat area $A$ in a uniform magnetic field $\boldsymbol{B}$. (b) The angle $\phi$ could be $180° - (30° + 90°) = 60°$ or $180° - 60° = 120°$. The area vector $\boldsymbol{A}$ makes a 60° angle with $\boldsymbol{B}$. If the area rotates to $\phi = 90°$, $\Phi_B$ becomes zero; but if the area rotates so that $\boldsymbol{A}$ and $\boldsymbol{B}$ are parallel, $\Phi_B$ has its maximum value.

## 28–4 MOTION OF CHARGED PARTICLES IN A MAGNETIC FIELD

When a charged particle moves in a magnetic field, the motion is determined by Newton's laws, with the magnetic force given by Eq. (28–2). Figure 28–12 shows a simple example; a particle with positive charge $q$ is at point $O$, moving with velocity $\boldsymbol{v}$ in a uniform magnetic field $\boldsymbol{B}$ directed into the plane of the figure. The vectors $\boldsymbol{v}$ and $\boldsymbol{B}$ are perpendicular, so the magnetic force $\boldsymbol{F} = q\boldsymbol{v} \times \boldsymbol{B}$ has magnitude $F = qvB$, and its direction is as shown in the figure. The force is *always* perpendicular to $\boldsymbol{v}$, so it cannot change the *magnitude* of the velocity, only its direction. (To put it differently, the force can't do work on the particle, so the force can't change the particle's kinetic energy.) Thus the magnitudes of both $\boldsymbol{F}$ and $\boldsymbol{v}$ are constant. At points such as $P$ and $S$ in Fig. 28–12a the directions of force and velocity have changed as shown, but their magnitudes are the same. The parti-

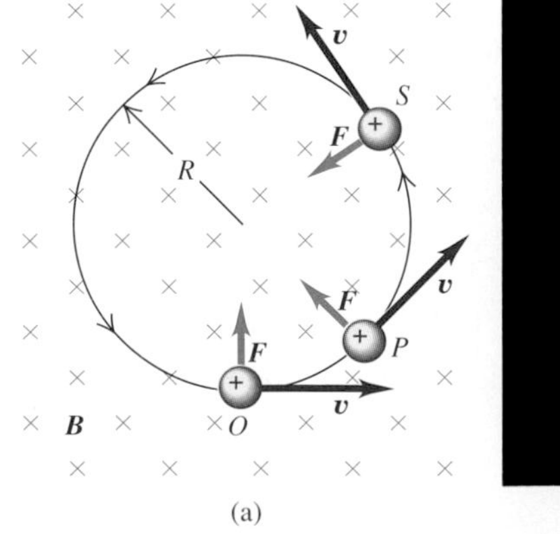

**28–12** (a) The orbit of a charged particle in a uniform magnetic field $\boldsymbol{B}$ is a circle when the initial velocity is perpendicular to the field. The crosses represent a uniform magnetic field directed straight into the page. (b) An electron beam curving in a magnetic field.

cle therefore moves under the influence of a constant-magnitude force that is always at right angles to the velocity of the particle. Comparing these conditions with the discussion of circular motion in Sections 3–4 and 5–4, we see that the particle's path is a *circle,* traced out with constant speed $v$ (Fig. 28–12b). The centripetal acceleration is $v^2/R$, and, from Newton's second law,

$$F = |q|vB = m\frac{v^2}{R}, \tag{28-10}$$

where $m$ is the mass of the particle. The radius $R$ of the circular path is

$$R = \frac{mv}{|q|B}. \tag{28-11}$$

In terms of the magnitude of the particle's momentum $p = mv$, we can also write this as $R = p/|q|B$. If the charge $q$ is negative, the particle moves *clockwise* around the orbit in Fig. 28–12a.

The angular velocity $\omega$ of the particle is given by Eq. (9–13): $\omega = v/R$. Combining this with Eq. (28–11), we get

$$\omega = \frac{v}{R} = v\frac{|q|B}{mv} = \frac{|q|B}{m}. \tag{28-12}$$

The number of revolutions per unit time is $\omega/2\pi$. This frequency is independent of the radius $R$ of the path. It is called the **cyclotron frequency;** in a particle accelerator called a *cyclotron,* particles moving in nearly circular paths are given a boost twice each revolution, increasing their energy and their orbital radii but not their angular velocity. Similarly, a *magnetron,* a common source of microwave radiation for microwave ovens and radar systems, emits radiation with a frequency equal to the frequency of circular motion of electrons in a vacuum chamber between the poles of a magnet.

If the direction of the initial velocity is *not* perpendicular to the field, the velocity *component* parallel to the field is constant (because there is no force parallel to the field), and the particle moves in a helix (Fig. 28–13). The radius of the helix is given by Eq. (28–11), where $v$ is now the component of velocity perpendicular to the $\boldsymbol{B}$ field.

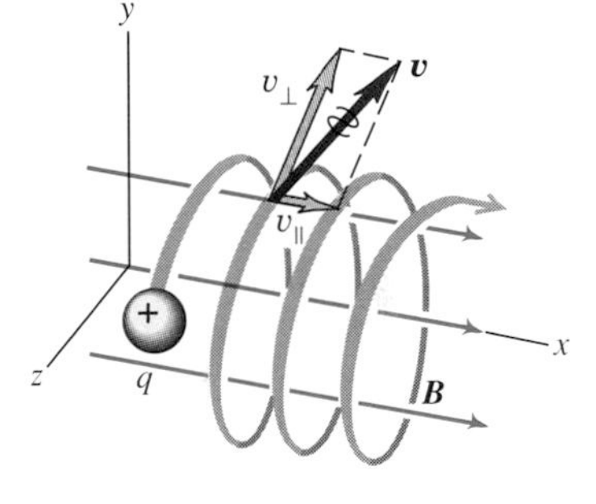


**28–13** When a charged particle with constant kinetic energy has velocity components both perpendicular and parallel to a uniform magnetic field, the particle moves in a helical path.

Motion of a charged particle in a non-uniform magnetic field is more complex. Figure 28–14 shows a field produced by two circular coils separated by some distance. Particles near either coil experience a magnetic force toward the center of the region; particles with appropriate speeds spiral repeatedly from one end of the region to the other and back. Because charged particles can be trapped in such a magnetic field, it is called a *magnetic bottle.* This technique is used to confine very hot plasmas having temperatures of the order of $10^6$ K. In a similar way the earth's non-uniform magnetic field traps charged particles coming from the sun in doughnut-shaped regions around the earth, as

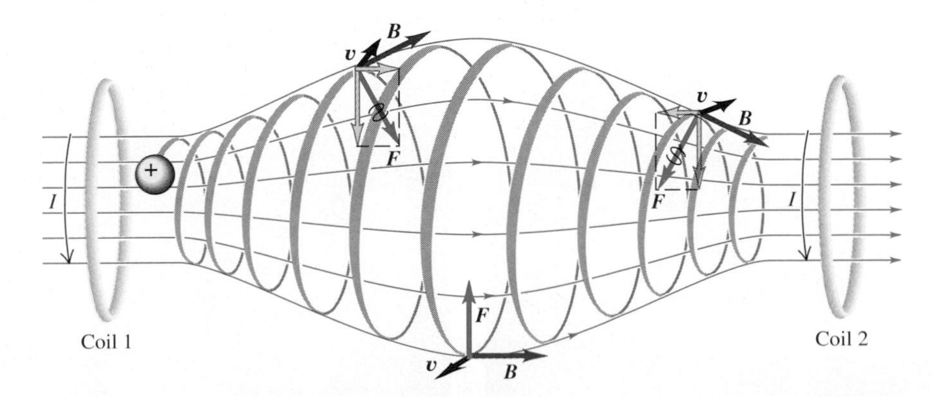


**28–14** A magnetic bottle: Particles near either end of the region experience a magnetic force toward the center of the region. This is one way of containing an ionized gas that has a temperature of the order of $10^6 K$, which would melt any material container.

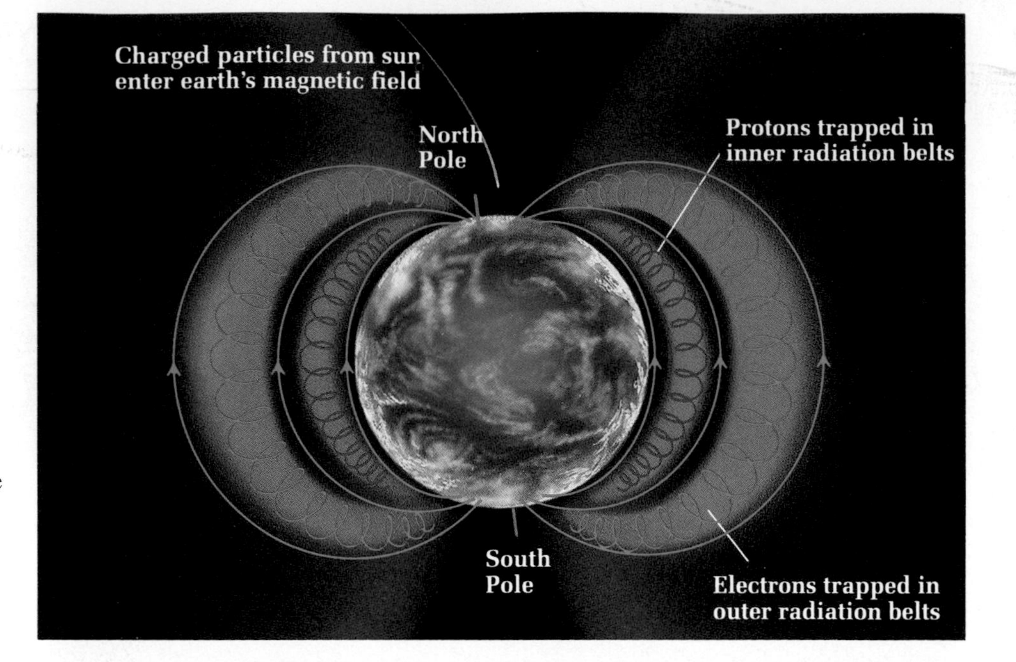


**28–15** The Van Allen radiation belts around the earth. Near the poles, charged particles can escape from these belts into the atmosphere, producing the Aurora Borealis (Northern Lights) or Aurora Australis (Southern Lights).

shown in Fig. 28–15. These regions, called the *Van Allen radiation belts,* were discovered in 1958 through observations made by instruments aboard the Explorer I satellite.

The magnetic force acting on a charged particle can never do *work* because at every instant the force is perpendicular to the velocity. A magnetic force can change the *direction* of motion, but it can never increase or decrease the *magnitude* of the velocity. **Motion of a charged particle under the action of a magnetic field alone is always motion with constant speed.**

Figure 28–16a is a photograph of the track made in a liquid-hydrogen bubble chamber by a high-energy electron moving in a magnetic field perpendicular to the plane of the page. There are other forces acting in addition to the magnetic-field force; as the electron plows through the liquid hydrogen, it collides with charged particles, losing energy (and speed). As a result, the radius of curvature decreases according to Eq. (28–11). Figure 28–16b shows tracks due to electron-positron pair production, also in a bubble chamber. Similar experiments using a cloud chamber provided the first experimental evidence in 1932 for the existence of the *positron,* or positive electron.

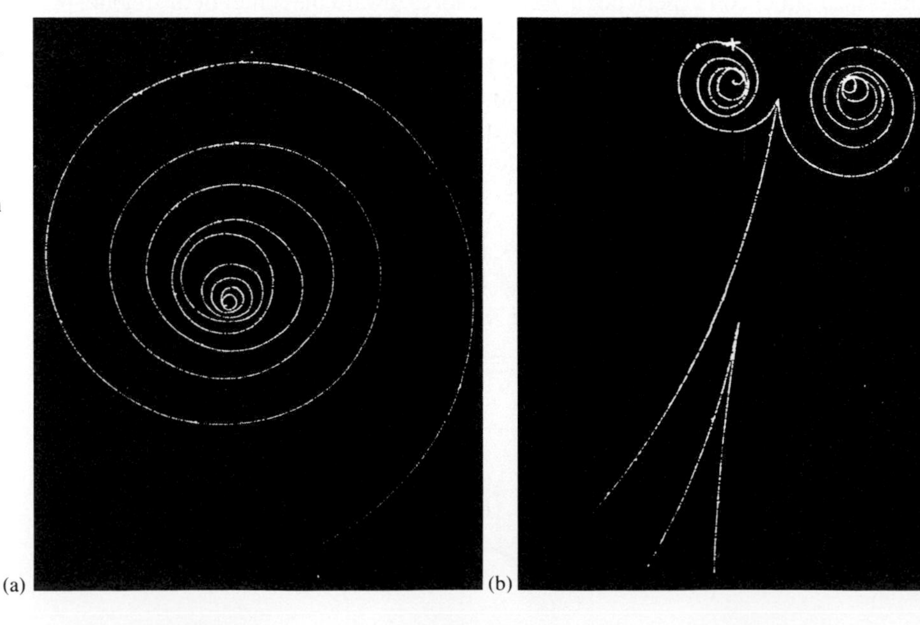


**28–16** (a) A track in a liquid-hydrogen bubble chamber made by an electron with initial kinetic energy of about 27 MeV. The magnetic field strength is 1.8 T. As the particle loses energy, the radius of curvature decreases; the maximum radius is about 5 cm. (b) Electron-positron pair production in a liquid-hydrogen bubble chamber. A high-energy gamma ray coming in from above scatters on an atomic electron. The paths of the recoil electron and of the electron-positron pair can be seen; the directions of curvature in the magnetic field show the signs of the charges.

## PROBLEM-SOLVING STRATEGY

### Motion in magnetic fields

**1.** In analyzing the motion of a charged particle in electric and magnetic fields, you will apply Newton's second law of motion, $\Sigma \boldsymbol{F} = m\boldsymbol{a}$, with $\Sigma \boldsymbol{F}$ given by $\Sigma \boldsymbol{F} = q(\boldsymbol{E} + \boldsymbol{v} \times \boldsymbol{B})$. Many of the problems are similar to the trajectory and circular-motion problems in Sections 3–3, 3–4, and 5–4, and it would be a good idea to review those sections.

**2.** Often the use of components is the most efficient approach. Choose a coordinate system, and then express all the vector quantities (including $\boldsymbol{E}$, $\boldsymbol{B}$, $\boldsymbol{v}$, $\boldsymbol{F}$, and $\boldsymbol{a}$) in terms of their components in this system. Then use $\Sigma \boldsymbol{F} = m\boldsymbol{a}$ in component form: $\Sigma F_x = ma_x$, and so forth. This approach is particularly useful when both electric and magnetic fields are present.

### EXAMPLE 28–3

**Electron motion in a microwave oven** A magnetron in a microwave oven emits microwaves with frequency $f = 2450$ MHz. What magnetic field strength is required for electrons to move in circular paths with this frequency?

**SOLUTION** The corresponding angular velocity is $\omega = 2\pi f = (2\pi)(2450 \times 10^6\ \text{s}^{-1}) = 1.54 \times 10^{10}\ \text{s}^{-1}$. From Eq. (28–12),

$$B = \frac{m\omega}{|q|} = \frac{(9.11 \times 10^{-31}\ \text{kg})(1.54 \times 10^{10}\ \text{s}^{-1})}{1.60 \times 10^{-19}\ \text{C}}$$
$$= 0.0877\ \text{T}.$$

This is a moderate field strength, easily produced with a permanent magnet. The frequency, incidentally, is one that is strongly absorbed by water molecules, so it is particularly useful for heating and cooking food.

## 28–5 APPLICATION OF MOTION OF CHARGED PARTICLES

This section describes several applications of the principles introduced in this chapter. Study them carefully, watching for applications of the problem-solving strategy outlined in the preceding section.

### Velocity Selector

A charged particle with mass $m$, charge $q$, and speed $v$ enters a region of space where the electric and magnetic fields are perpendicular to the particle's velocity and to each other, as shown in Fig. 28–17. The electric field $\boldsymbol{E}$ is vertically downward, and the magnetic field $\boldsymbol{B}$ is into the plane of the figure. If $q$ is positive, the electric force is downward, with magnitude $qE$, and the magnetic force is upward, with magnitude $qvB$. By adjusting the magnitudes $E$ and $B$, we can make these forces equal in magnitude; the total force is then zero, and the particle travels in a straight line with constant velocity. For zero total force, we need

$$\Sigma F_y = 0, \qquad -qE + qvB = 0, \qquad \text{and} \qquad v = \frac{E}{B}. \tag{28–13}$$

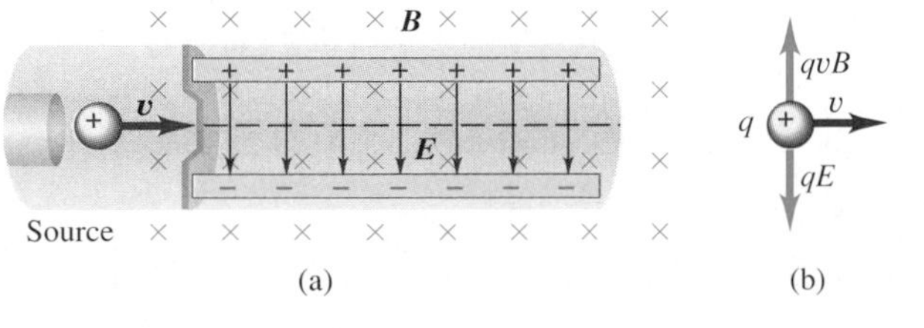


**28–17** Velocity selector: (a) Only charged particles with $v = E/B$ move through undeflected. (b) The electric and magnetic forces on a positive charge. The forces are reversed if the charge is negative.

Only particles with speeds equal to $E/B$ can pass through without being deflected by the fields. By adjusting $E$ and $B$ appropriately we can select particles having a particular speed for use in other experiments. Because $q$ divides out, Eq. (28–13) also works for electrons or other negatively charged particles.

## Thomson's *e/m* Experiment

In one of the landmark experiments in physics at the turn of the century, Sir J. J. Thomson used the idea just described to measure the ratio of charge to mass for the electron. For this experiment, carried out at the Cavendish Laboratory in Cambridge, England, in 1897, Thomson used the apparatus shown in Fig. 28–18. It is very similar in principle to the cathode-ray tube discussed in Section 24–7; you may want to review that section. In a highly evacuated glass container, electrons from the hot cathode are accelerated and formed into a beam by a potential difference $V$ between the two anodes A and A′. The speed $v$ of the electrons is determined by the accelerating potential $V$, just as in the derivation of Eq. (24–34). The kinetic energy $\frac{1}{2}mv^2$ equals the loss of electric potential energy $eV$, where $e$ is the magnitude of the electron charge:

$$\tfrac{1}{2}mv^2 = eV, \qquad \text{or} \qquad v = \sqrt{\frac{2eV}{m}}. \tag{28–14}$$

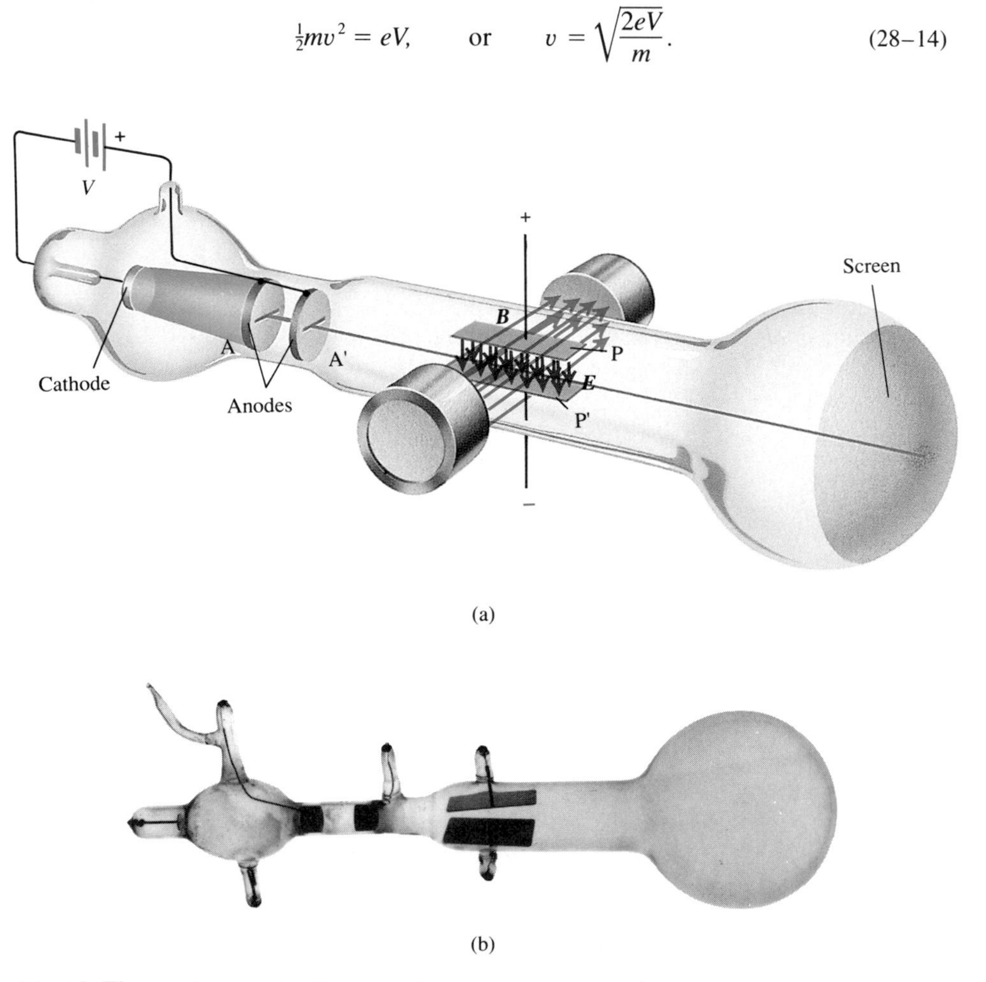


**28–18** Thomson's apparatus for measuring the ratio $e/m$ for cathode rays. (a) A sketch showing the crossed $E$ and $B$ fields. (b) A photo of the tube used by Thomson.

The electrons pass between the plates P and P′ and strike the screen at the end of the tube, which is coated with a material that fluoresces (glows) at the point of impact. The electrons pass straight through when Eq. (28–13) is satisfied; combining this with Eq. (28–14), we get

$$\frac{E}{B} = \sqrt{\frac{2eV}{m}}, \quad \text{and} \quad \frac{e}{m} = \frac{E^2}{2VB^2}. \tag{28–15}$$

The quantities $E$, $V$, and $B$ can be measured, so the ratio $e/m$ of charge to mass can be determined. It is *not* possible to measure $e$ or $m$ separately by this method, only their ratio.

The most significant aspect of Thomson's $e/m$ measurements was that he found a *single value* for this quantity. It did not depend on the cathode material, the residual gas in the tube, or anything else about the experiment. This independence showed that the particles in the beam, which we now call electrons, are a common constituent of all matter. Thus Thomson is credited with discovery of the first subatomic particle, the electron. He also found that the *speed* of the electrons in the beam was about one-tenth the speed of light, much larger than any previously measured material particle speed.

The most precise value of $e/m$ available, as of 1990, is

$$e/m = 1.75881962(53) \times 10^{11}\ \text{C/kg}.$$

In this expression, (53) indicates the likely uncertainty in the last two digits, 62.

Fifteen years after Thomson's experiments, Millikan succeeded in measuring the charge of the electron precisely with his famous oil-drop experiment, described in Section 24–6. This value, together with the value of $e/m$, enables us to determine the *mass* of the electron. The most precise value available at present is

$$m = 9.1093897(54) \times 10^{-31}\ \text{kg}.$$

## Mass Spectrometers

Techniques similar to Thomson's $e/m$ experiment can be used to measure masses of positive ions and thus measure atomic and molecular masses. In 1919, Francis Aston (1877–1945), a student of Thomson's, built the first of a family of instruments called **mass spectrometers.** A variation built by Bainbridge is shown in Fig. 28–19. Ions from a source pass through the slits $S_1$ and $S_2$, forming a narrow beam. Then the ions pass through a *velocity selector* with crossed $\boldsymbol{E}$ and $\boldsymbol{B}$ fields, as we have described, to block all ions except those with speeds $v$ equal to $E/B$. Finally, the ions pass into a region with a magnetic field $\boldsymbol{B}'$ perpendicular to the figure, where they move in circular arcs with radius $R$ determined by Eq. (28–11): $R = mv/qB'$. Particles with different masses strike the photographic plate at different points, and the values of $R$ can be measured. We assume that each ion has lost one electron, so the net charge of each ion is just $e$. With everything known in this equation except $m$, we can compute the mass $m$ of the ion.

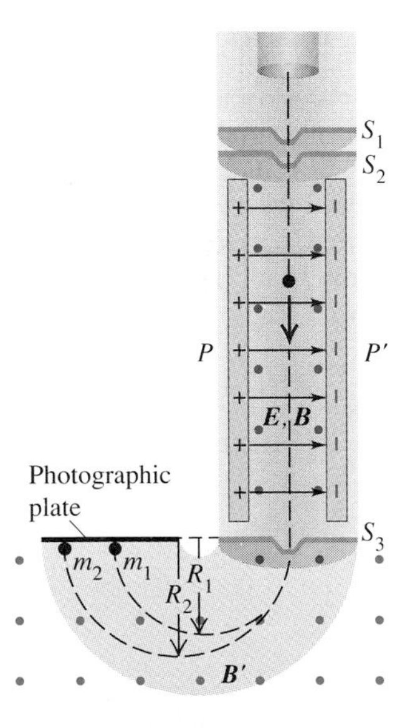


**28–19** Bainbridge's mass spectrometer utilizes a velocity selector to produce particles with uniform speed $v$. In the region of magnetic field $B'$, particles with larger mass travel in paths with larger radius $R$.

One of the earliest results from this work was the discovery that neon has two species of atoms, with atomic masses 20 and 22 g/mol. We now call these species **isotopes** of the element. Later experiments have shown that many elements have several isotopes, atoms that are identical in their chemical behavior but different in mass. The mass differences are due to differing numbers of neutrons in the *nuclei* of the atoms.

Present-day mass spectrometers can measure masses routinely with a precision of better than 1 part in 10,000. The photographic plate is replaced by a more sophisticated

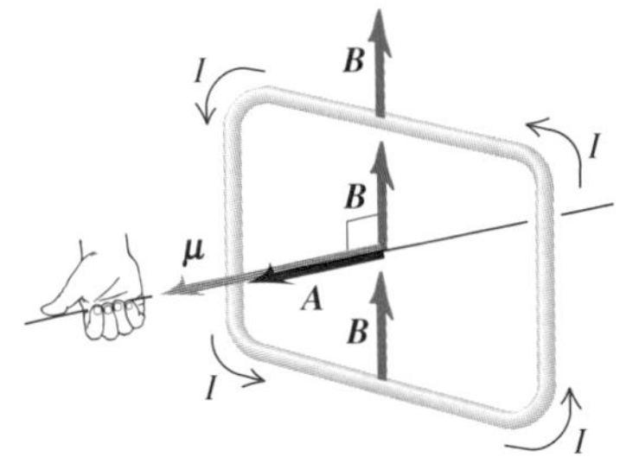


**28–28** The right-hand rule determines the direction of the magnetic moment of a current-carrying loop. This is also the direction of the loop's area vector $\boldsymbol{A}$; $\boldsymbol{\mu} = I\boldsymbol{A}$ is a vector equation.

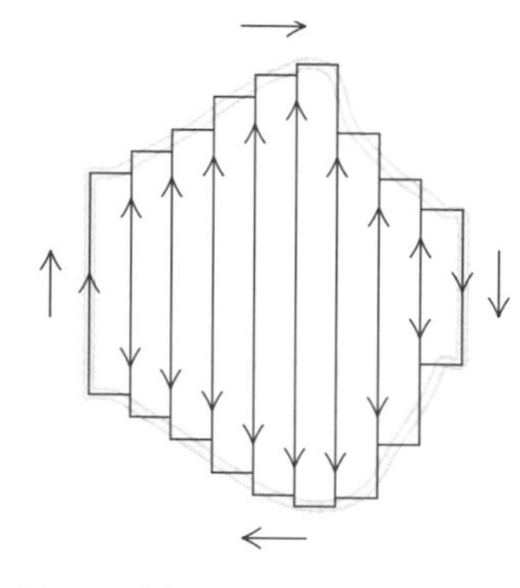

**28–29** The collection of rectangles exactly matches the irregular plane loop in the limit as the number of rectangles approaches infinity and the width of each rectangle approaches zero.

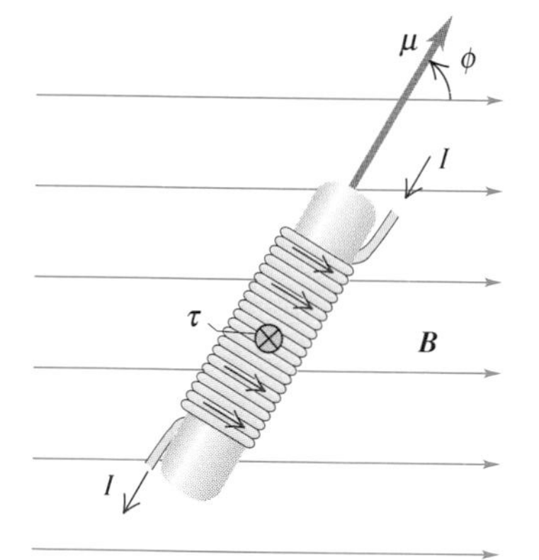


**28–30** The torque $\boldsymbol{\tau}$ on this solenoid in a uniform magnetic field is clockwise; that is, in the direction of $\boldsymbol{\mu} \times \boldsymbol{B}$ (straight into the page). It tends to align the magnetic moment and the axis of the solenoid with the magnetic field. An actual solenoid has many more turns, wrapped closely together.

where $\phi$ is the angle between the normal to the loop (the direction of the vector area $\boldsymbol{A}$) and $\boldsymbol{B}$. The torque $\tau$ tends to rotate the loop in the direction of *decreasing* $\phi$, that is, toward its stable equilibrium position, in which the loop lies in the $xy$-plane, perpendicular to the direction of the field $\boldsymbol{B}$ (Fig. 28–27c).

We can also define a vector magnetic moment $\boldsymbol{\mu}$ with magnitude $IA$. The direction of $\boldsymbol{\mu}$ is defined to be perpendicular to the plane of the loop, with a sense determined by the right-hand rule, as shown in Fig. 28–28: Wrap the fingers of your right hand around the perimeter of the coil in the direction of the current. Then extend your thumb so that it is perpendicular to the plane of the coil; its direction is the direction of $\boldsymbol{\mu}$ (and of the vector area $\boldsymbol{A}$ of the loop). The torque is greatest when $\boldsymbol{\mu}$ and $\boldsymbol{B}$ are perpendicular and zero when they are parallel or antiparallel. In the stable equilibrium position, $\boldsymbol{\mu}$ and $\boldsymbol{B}$ are parallel.

Finally, we can express this interaction in terms of vector torque $\boldsymbol{\tau}$, which we introduced in Section 10–1 and used for *electric*-dipole interactions in Section 22–8. From Eq. (28–24) the magnitude of $\boldsymbol{\tau}$ is equal to the magnitude of $\boldsymbol{\mu} \times \boldsymbol{B}$, and reference to Fig. 28–27 shows that the directions are also the same. So we have

$$\boldsymbol{\tau} = \boldsymbol{\mu} \times \boldsymbol{B}. \tag{28–25}$$

When a magnetic moment (magnetic dipole) changes orientation in a magnetic field, the field does work on it. For an infinitesimal angular displacement $d\phi$ the work $dW$ is given by $\tau\, d\phi$, and there is a corresponding change in potential energy. As the above discussion suggests, the potential energy is least when $\mu$ and $\boldsymbol{B}$ are parallel and greatest when they are antiparallel. To find an expression for the potential energy $U$ as a function of orientation, we can make use of the beautiful symmetry between the electric and magnetic dipole interactions. In Section 22–8 we found that the torque $\boldsymbol{\tau}$ on an *electric* dipole in an *electric* field is $\boldsymbol{\tau} = \boldsymbol{p} \times \boldsymbol{E}$ and that the corresponding potential energy is $U = -\boldsymbol{p} \cdot \boldsymbol{E}$. The torque on a *magnetic* dipole in a *magnetic* field is $\boldsymbol{\tau} = \boldsymbol{\mu} \times \boldsymbol{B}$, so we can conclude immediately that the corresponding potential energy is

$$U = -\boldsymbol{\mu} \cdot \boldsymbol{B} = -\mu B \cos \phi. \tag{28–26}$$

With this definition, $U$ is zero when the magnetic dipole is perpendicular to the magnetic field.

Although we have derived Eqs. (28–21) through (28–26) for a rectangular current loop, all these relations are valid for a plane loop of any shape. Any planar loop may be approximated as closely as we wish by a very large number of rectangular loops, as shown in Fig. 28–29. If these loops all carry equal currents in the same sense, then the forces on the sides of two loops adjacent to each other cancel, and the only forces that do not cancel are around the boundary. Thus all the above relations are valid for a plane current loop of any shape, with the magnetic moment $\boldsymbol{\mu}$ given by $\boldsymbol{\mu} = I\boldsymbol{A}$. We can also generalize this whole formulation to a coil consisting of $N$ planar loops close together; the effect is simply to multiply each force, the magnetic moment, the torque, and the potential energy by a factor of $N$.

An arrangement of particular interest is the **solenoid,** a helical winding of wire, such as a coil wound on a circular cylinder (Fig. 28–30). If the windings are closely spaced, the solenoid can be approximated by a number of circular loops lying in planes at right angles to its long axis. The total torque on a solenoid in a magnetic field is simply the sum of the torques on the individual turns. For a solenoid with $N$ turns in a uniform field $B$,

$$\tau = NIAB \sin \phi,$$

where $\phi$ is the angle between the axis of the solenoid and the direction of the field. The magnetic moment vector is along the axis. The torque is greatest when the solenoid axis is perpendicular to the magnetic field and zero when axis and field are parallel. The effect of this torque is to tend to rotate the solenoid into a position where its axis is parallel to the field.

The behavior of a solenoid in a magnetic field resembles that of a bar magnet or compass needle; both the solenoid and the magnet, if free to turn, orient themselves with their axes parallel to a magnetic field. We could use the torque on a solenoid for an alternative *definition* of magnetic field. The behavior of a bar magnet or compass needle is sometimes described in terms of magnetic forces on "poles" at its ends. For the solenoid, no such concept is needed. In fact, the moving electrons in a bar of magnetized iron play exactly the same role as the current in the windings of a solenoid, and the fundamental cause of the torque is the same in both cases. We will return to this subject later.

## EXAMPLE 28–8

**Torque on a circular coil** A circular coil of wire 0.0500 m in radius, with 30 turns, lies in a horizontal plane (Fig. 28–31). It carries a current of 5.00 A in a counterclockwise sense when viewed from above. The coil is in a uniform magnetic field directed toward the right, with magnitude 1.20 T. Find the magnetic moment and the torque on the coil.

**SOLUTION** The area of the coil is

$$A = \pi r^2 = \pi(0.0500 \text{ m})^2 = 7.85 \times 10^{-3} \text{ m}^2.$$

The magnetic moment of each turn of the coil is

$$\mu = IA = (5.00 \text{ A})(7.85 \times 10^{-3} \text{ m}^2) = 3.93 \times 10^{-2} \text{ A} \cdot \text{m}^2,$$

and the total magnetic moment of all 30 turns is

$$\mu_{\text{tot}} = (30)(3.93 \times 10^{-2} \text{ A} \cdot \text{m}^2) = 1.18 \text{ A} \cdot \text{m}^2.$$

The angle $\phi$ between the direction of $\boldsymbol{B}$ and the normal to the plane of the coil is 90°. From Eq. (28–22) the torque on each turn of the coil is

$$\tau = IBA \sin\phi = (5.00 \text{ A})(1.20 \text{ T})(7.85 \times 10^{-3} \text{ m}^2)(\sin 90°) = 0.0471 \text{ N} \cdot \text{m},$$

and the total torque on the coil is

$$\tau = (30)(0.0471 \text{ N} \cdot \text{m}) = 1.41 \text{ N} \cdot \text{m}.$$

Alternatively, from Eq. (28–24),

$$\tau = \mu_{\text{tot}} B \sin\phi = (1.18 \text{ A} \cdot \text{m}^2)(1.20 \text{ T})(\sin 90°) = 1.41 \text{ N} \cdot \text{m}.$$

The torque tends to rotate the right side of the coil down and the left side up, into a position where the normal to its plane is parallel to $\boldsymbol{B}$.

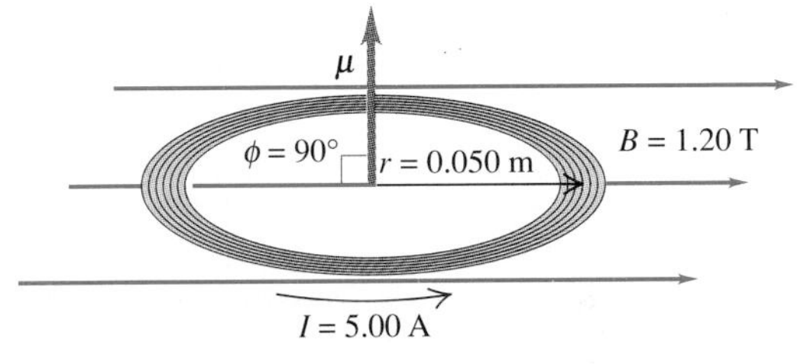


**28–31** A circular coil of current-carrying wire in a uniform magnetic field.

## EXAMPLE 28–9

If the coil in Example 28–8 rotates from its initial position to a position where its magnetic moment is parallel to $\boldsymbol{B}$, what is the change in potential energy?

**SOLUTION** From Eq. (28–26) the initial potential energy $U_1$ is

$$U_1 = -\mu B \cos\phi_1 = -(1.18 \text{ A} \cdot \text{m}^2)(1.20 \text{ T})(\cos 90°) = 0,$$

and the final potential energy $U_2$ is

$$U_2 = -\mu B \cos\phi_2 = -(1.18 \text{ A} \cdot \text{m}^2)(1.20 \text{ T})(\cos 0°) = -1.41 \text{ J}.$$

The change in potential energy is $-1.41$ J.

## EXAMPLE 28–10

What vertical forces applied to the left and right edges of the coil of Fig. 28–31 would be required to maintain it in equilibrium in its initial position?

**SOLUTION** An upward force of magnitude $F$ at the right side and a downward force of equal magnitude on the left side would have a total torque of

$$\tau = 2rF = (2)(0.0500 \text{ m})F.$$

This applied torque must be equal in magnitude to the magnetic-field torque, or 1.41 N · m, and we find that the required forces have magnitude 14.1 N.

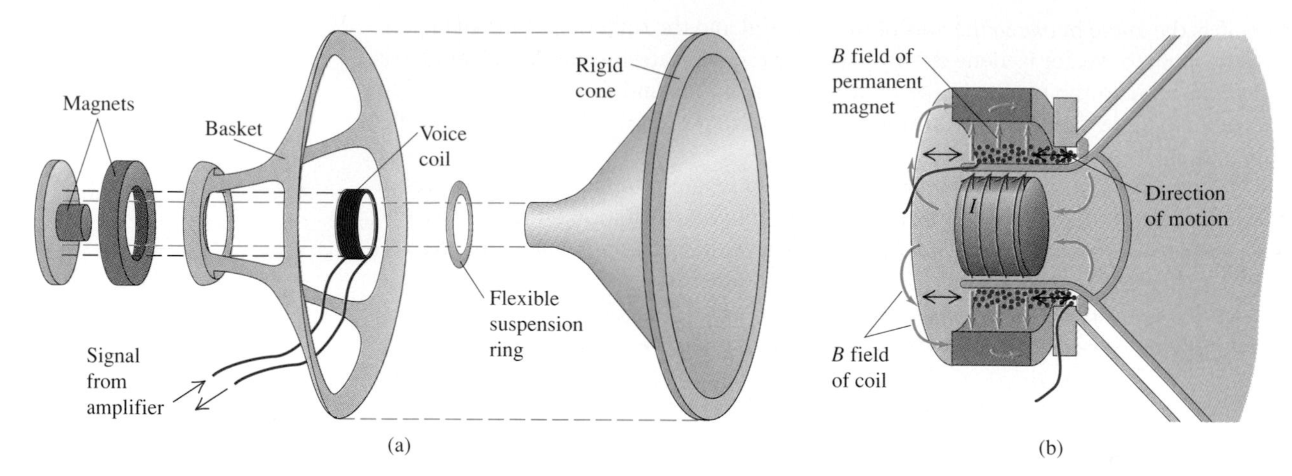


**28–32** (a) Components of a loudspeaker. (b) The permanent magnet creates a magnetic field that exerts axial forces on the current in the voice coil. The mechanical vibration of the speaker cone matches the electrical vibration of the current in the coil.

The d'Arsonval galvanometer, described in Section 27–3, makes use of a magnetic-field torque on a coil carrying a current. As Fig. 27–10 shows, the magnetic field is not uniform but is *radial,* so the side thrusts on the coil are always perpendicular to its plane. Thus the magnetic-field torque is directly proportional to the current, no matter what the orientation of the coil. A restoring torque proportional to the angular displacement of the coil is provided by two hairsprings, which also serve as current leads to the coil. When current is supplied to the coil, it rotates, along with its attached pointer, until the restoring spring torque just balances the magnetic-field torque. Thus the pointer deflection is proportional to the current. Such instruments can measure currents as small as $10^{-7}$ A, and modified versions can measure currents as small as $10^{-10}$ A. In situations requiring extreme sensitivity and mechanical ruggedness, however, d'Arsonval instruments have been mostly replaced by digital electronic instruments.

Figure 28–32 shows the construction of a loudspeaker. The magnetic field created by the permanent magnet exerts a force proportional to the current in the moving voice coil; this causes the speaker cone to vibrate with motion whose amplitude is proportional to the amplitude of the current in the coil.

## *28–8 THE DIRECT-CURRENT MOTOR

No one needs to be reminded of the importance of electric motors in contemporary society. Their operation depends on magnetic forces on current-carrying conductors. As an example, let's look at a simple type of direct-current motor, shown in Fig. 28–33. The center part is the *armature,* or *rotor;* it is a cylinder of soft steel that rotates about its axis (perpendicular to the plane of the figure).

Embedded in slots in the rotor surface (parallel to its axis) are insulated copper conductors. Current is led into and out of these conductors through graphite brushes making contact with a segmented cylinder called the *commutator,* shown in principle in Fig. 28–34. The commutator is an automatic switching arrangement that uses stationary sliding contacts called *brushes* to maintain the currents in the conductors in the directions shown in Fig. 28–33 as the rotor turns. The current in the field coils F and F′ sets up a magnetic field in the motor frame and in the gap between the pole pieces P and P′ and the rotor. (In some small motors the magnetic field is supplied by permanent magnets instead of electromagnets.) Some of the magnetic field lines are shown as blue lines. With the directions of field and rotor currents shown, the side thrust on each conductor in the rotor is such as to produce a *counterclockwise* torque on the rotor.

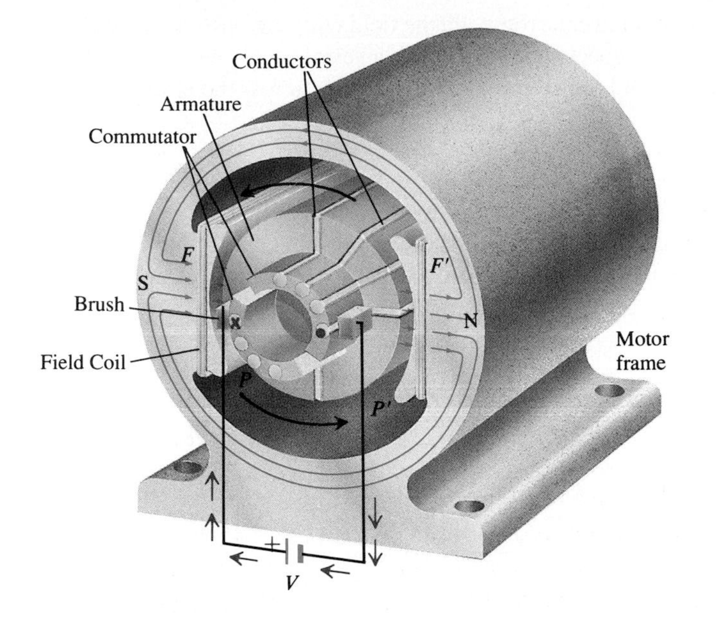


**28–33** Schematic diagram of a dc motor. The armature, or rotor, rotates on a shaft through its center, perpendicular to the plane of the figure.

A motor converts electrical energy to mechanical energy or work, and it requires electrical energy input. If the potential difference between its terminals is $V_{ab}$ and the current is $I$, then the power input is $P = V_{ab}I$. Even if the motor coils have negligible resistance, there must be a potential difference between the terminals if $P$ is to be different from zero. This potential difference results principally from magnetic forces exerted on the currents in the conductors of the rotor as they rotate through the magnetic field. The associated electromotive force $\mathcal{E}$ is called an *induced* emf, or a *back* emf, referring to the fact that its sense is opposite to that of the current. In Chapter 30 we will study induced emf's resulting from motion of conductors in magnetic fields.

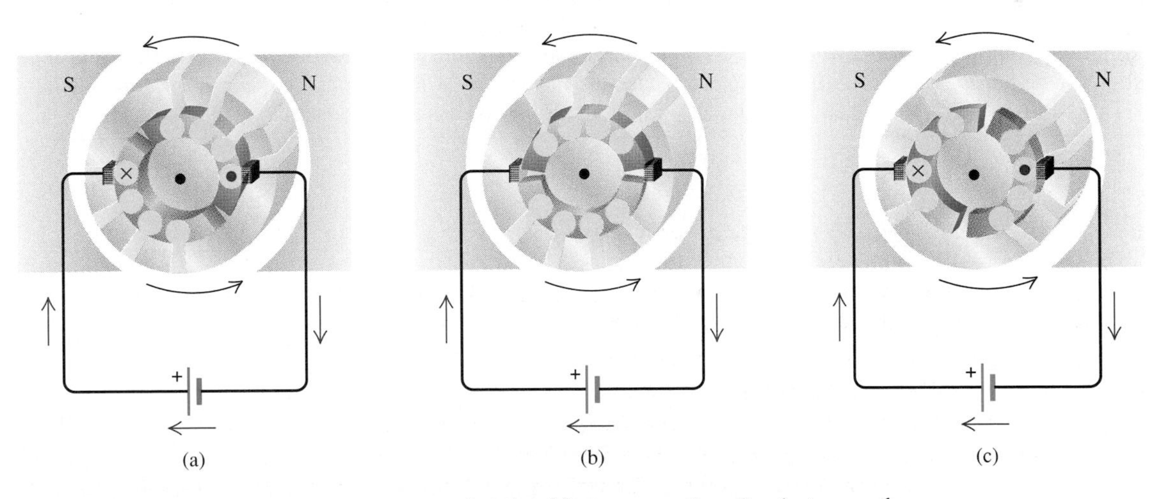


**28–34** The segments of this commutator are insulated from one another. Conductors on the rotor are shown in cross section; those with dots at their centers carry current out of the plane of the page, and those with crosses carry current into the plane of the page. (a) At the instant shown, the current $I$ goes through the armature coils shown, which are connected to the two segments in contact with the graphite brushes. (b) Instantaneously, the current reverses in the armature coils, just as the brushes contact different commutator segments. (c) One-half cycle after current flows into the coils, $I$ is reversed in the coils but is still the same in the external circuit.

In a *series* motor the rotor and the field windings are connected in series; in a *shunt* motor they are connected in parallel. In a series motor with internal resistance $r$, $V_{ab}$ is greater than $\mathcal{E}$, and the difference is the drop $Ir$ across the internal resistance; that is,

$$V_{ab} = \mathcal{E} + Ir. \qquad (28\text{–}28)$$

Because the magnetic-field force is proportional to velocity, $\mathcal{E}$ is *not* constant but is proportional to the speed of rotation of the rotor.

A dc motor is analogous to a battery being charged, which we discussed in Section 26–4. In a battery, electrical energy is converted to chemical rather than mechanical energy.

## EXAMPLE 28–11

**A series dc motor** A dc motor with its rotor and field coils connected in series has an internal resistance of 2.00 Ω. When running at full load on a 120-V line, it draws a current of 4.00 A. a) What is the emf in the rotor? b) What is the power delivered to the motor? c) What is the rate of dissipation of energy in the resistance of the motor? d) What is the mechanical power developed? e) What is the efficiency of the motor? f) What happens if the machine that the motor is driving jams and the rotor suddenly stops turning?

**SOLUTION** a) From Eq. (28–28),

$$V_{ab} = \mathcal{E} + Ir,$$

$$120\text{ V} = \mathcal{E} + (4.0\text{ A})(2.0\ \Omega),$$

$$\mathcal{E} = 112\text{ V}.$$

b) The power input $P$ from the source is

$$P = V_{ab}I = (120\text{ V})(4.0\text{ A}) = 480\text{ W}.$$

c) The power $P$ dissipated in the resistance $r$ is

$$P = I^2r = (4.0\text{ A})^2\,(2.0\ \Omega) = 32\text{ W}.$$

d) The mechanical power output is the electrical power input minus the rate of dissipation of energy in the motor's resistance (assuming that there are no other power losses):

$$P = 480\text{ W} - 32\text{ W} = 448\text{ W}.$$

e) The efficiency $e$ is the ratio of mechanical power output to electrical power input:

$$e = \frac{448\text{ W}}{480\text{ W}} = 0.93 = 93\%.$$

f) With the rotor stalled, the back emf $\mathcal{E}$ (which is proportional to rotor speed) goes to zero. From Eq. (28–28) the current becomes

$$I = \frac{V_{ab}}{r} = \frac{120\text{ V}}{2.0} = 60\text{ A},$$

and the power dissipation $P$ in the resistance $r$ becomes

$$P = I^2r = (60\text{ A})^2(2\ \Omega) = 7200\text{ W}.$$

If this massive overload doesn't blow a fuse or trip a circuit breaker, the coils will quickly melt. When the motor is first turned on, there's a momentary surge of current until the motor picks up speed. This surge causes greater-than-usual voltage drops in the power lines supplying the current. Similar effects are responsible for the momentary dimming of lights in a house when an air-conditioner or dishwasher motor starts.

## *28–9 THE HALL EFFECT

The reality of the forces acting on the moving charges in a conductor in a magnetic field is strikingly demonstrated by the **Hall effect,** an effect that is analogous to the sideways deflection of an electron beam in a magnetic field in vacuum. To describe this effect, let's consider a conductor in the form of a flat strip, as shown in Fig. 28–35. The current is in the direction of the $+x$-axis, and there is a uniform magnetic field $\boldsymbol{B}$ perpendicular to the plane of the strip, in the $+y$-direction. The drift velocity of the moving charges (charge magnitude $q$) has magnitude $v_d$. Part (a) of the figure shows the case of negative charges, such as electrons in a metal, and part (b) shows positive charges. In both cases the magnetic force is upward, just as the magnetic force on a conductor is the same whether the moving charges are positive or negative. In either case a moving charge is driven toward the *upper* edge of the strip by the magnetic force $F_z = |q|v_dB$.

If the charge carriers are electrons, as in Fig. 28–35a, an excess negative charge accumulates at the upper edge of the strip, leaving an excess positive charge at its lower edge. This accumulation continues until the resulting transverse electrostatic field $\boldsymbol{E}_e$

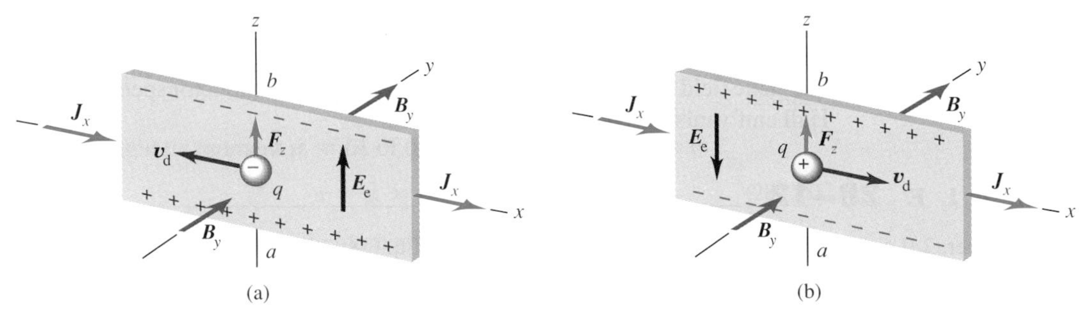


**28–35** Forces on charge carriers in a conductor in a magnetic field. (a) Negative current carriers (electrons) are pushed toward top of slab, leading to charge distribution as shown. Point $a$ is at higher potential than point $b$. (b) With positive current carriers, the polarity of the potential difference is opposite to that of part (a).

becomes large enough to cause a force (magnitude $|q|E_e$) that is equal and opposite to the magnetic force (magnitude $|q|v_dB$). After that, there is no longer any net sideways force to deflect the moving charges. This electric field causes a transverse potential difference between opposite edges of the strip, called the *Hall voltage,* or the *Hall emf.* The polarity depends on whether the moving charges are positive or negative. Experiments show that for metals the upper edge of the strip in Fig. 28–35a *does* become negatively charged, showing that the charge carriers in a metal are indeed negative electrons.

However, if the charge carriers are *positive,* as in Fig. 28–35b, then *positive* charge accumulates at the upper edge, and the potential difference is *opposite* to the situation with negative charges. Soon after the discovery of the Hall effect, in 1879, it was observed that some materials, particularly some *semiconductors,* show a Hall emf opposite to that of metals, as if their charge carriers were *positively* charged. We now know that these materials conduct by a process known as *hole conduction*. Within such a material there are locations, called *holes,* that would normally be occupied by an electron but are actually empty. A missing negative charge is equivalent to a positive charge. When an electron moves in one direction to fill a hole, it leaves another hole behind it. The hole, equivalent to a positive charge, migrates in the direction opposite to that of the electron movement.

In terms of the coordinate axes in Fig. 28–35b, the electrostatic field $\boldsymbol{E}_e$ for the positive-$q$ case is in the $-z$-direction; its $z$-component $E_z$ is negative. The magnetic field is in the $+y$-direction, and we write it as $B_y$. The magnetic-field force (in the $+z$-direction) is $qv_dB_y$. The current density $J_x$ is in the $+x$-direction. In the steady state, when the forces $qE_z$ and $qv_dB_y$ are equal in magnitude and opposite in direction,

$$qE_z + qv_dB_y = 0, \qquad \text{or} \qquad E_z = -v_dB_y.$$

This confirms that when $q$ is positive, $E_z$ is negative. The current density $J_x$ is

$$J_x = nqv_d.$$

Eliminating $v_d$ between these equations, we find

$$nq = \frac{-J_xB_y}{E_z}. \tag{28–29}$$

Note that this result (as well as the entire derivation) is valid for both positive and negative $q$. When $q$ is negative, $E_z$ is positive, and conversely.

We can measure $J_x$, $B_y$, and $E_z$, so we can compute the product $nq$. In both metals and semiconductors, $q$ is equal in magnitude to the electron charge, so the Hall effect permits a direct measurement of $n$, the concentration of current-carrying charges in the material. The *sign* of the charges is determined by the polarity of the Hall emf, as we have described.

The Hall effect can also be used for a direct measurement of electron drift speed $v_d$ in metals. As we have seen in Chapter 27, these speeds are very small, often of the order

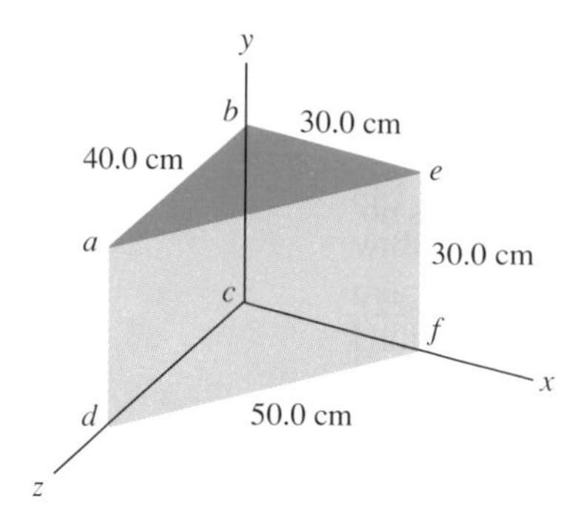


**FIGURE 28–37**

## Section 28–3
## Magnetic Field Lines and Magnetic Flux

**28–6** The magnetic field $\boldsymbol{B}$ in a certain region is 0.600 T, and its direction is that of the $+x$-axis in Fig. 28–37. a) What is the magnetic flux across the surface *abcd* in the figure? b) What is the magnetic flux across the surface *befc?* c) What is the magnetic flux across the surface *aefd?* d) What is the net flux through all five surfaces that enclose the shaded volume?

**28–7** A circular area with a radius of 0.400 m lies in the $xy$-plane. What is the magnetic flux through this circle due to a uniform magnetic field $B = 1.60$ T a) in the $+z$-direction; b) at an angle of 30.0° from the $+z$-direction; c) in the $+y$-direction?

## Section 28–4
## Motion of Charged Particles in a Magnetic Field

**28–8** A cyclotron is to accelerate protons to an energy of 3.00 MeV. The superconducting electromagnet of the cyclotron produces a magnetic field with magnitude 3.20 T. a) What is the radius of the circular orbit of the protons, and what is their angular velocity when they have achieved a kinetic energy of 1.50 MeV? b) Repeat part (a) when the protons have achieved their final kinetic energy of 3.00 MeV.

**28–9** An electron at point $A$ in Fig. 28–38 has a speed $v_0$ of $4.00 \times 10^6$ m/s. Find a) the magnitude and direction of the magnetic field that will cause the electron to follow the semicircular path from $A$ to $B$; b) the time required for the electron to move from $A$ to $B$.

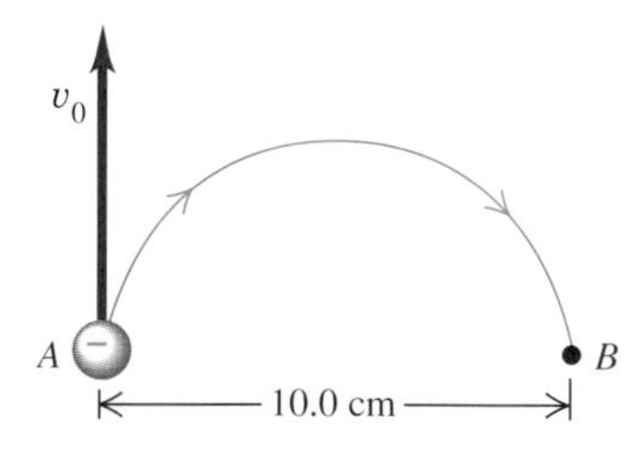


**FIGURE 28–38**

**28–10** Suppose the particle in Exercise 28–9 is a proton rather than an electron. Find the same quantities.

**28–11** A deuteron (the nucleus of an isotope of hydrogen) has a mass of $3.34 \times 10^{-27}$ kg and a charge of $+e$. The deuteron travels in a circular path with a radius of 0.0400 m in a magnetic field with magnitude 1.50 T. a) Find the speed of the deuteron. b) Find the time required for it to make one-half of a revolution. c) Through what potential difference would the deuteron have to be accelerated to acquire this speed?

**28–12** A singly charged $^7$Li ion has a mass of $1.16 \times 10^{-26}$ kg. It accelerates through a potential difference of 500 V and then enters a magnetic field with a magnitude of 0.300 T perpendicular to the path of the ion. What is the radius of the ion's path in the magnetic field?

**28–13 TV Picture Tube.** An electron in the beam of a television picture tube is accelerated by a potential difference of 20,000 V. Then it passes through a region of transverse magnetic field, where it moves in a circular arc with radius 0.150 m. What is the magnitude of the field?

## Section 28–5
## Application of Motion of Charged Particles

**28–14** a) What is the speed of a beam of electrons when the simultaneous influence of an electric field of $3.40 \times 10^5$ V/m and a magnetic field of $5.00 \times 10^{-2}$ T, with both fields normal to the beam and to each other, produces no deflection of the electrons? b) Show in a diagram the relative orientation of the vectors $\boldsymbol{v}$, $\boldsymbol{E}$, and $\boldsymbol{B}$. c) What is the radius of the electron orbit when the electric field is removed?

**28–15 Determining Mass of an Isotope.** The electric field between the plates of the velocity selector in a Bainbridge mass spectrometer (Fig. 28–19) is $1.20 \times 10^6$ V/m, and the magnetic field in both regions is 0.600 T. A stream of singly charged neon ions moves in a circular path with a radius of 0.728 m in the magnetic field. Determine the mass of one neon ion and the mass number of this isotope.

**28–16** In the Bainbridge mass spectrometer (Fig. 28–19), suppose the magnetic field magnitude $B$ in the velocity selector is 1.20 T, and ions having a speed of $4.00 \times 10^6$ m/s pass through undeflected. a) What is the electric field between the plates $P$ and $P'$? b) If the separation of the plates is 0.500 cm, what is the potential difference between plates?

## Section 28–6
## Magnetic Force on a Current-Carrying Conductor

**28–17** A horizontal rod 0.200 m long is mounted on a balance and carries a current. At the location of the rod is a uniform horizontal magnetic field with magnitude 0.0700 T and direction perpendicular to the rod. The magnetic force on the rod is measured by the balance and is found to be 0.240 N. What is the current?

**28–18** An electromagnet produces a magnetic field of 1.20 T in a cylindrical region of radius 5.00 cm between its poles. A straight wire carrying a current of 14.0 A passes through the center of this region and is perpendicular to it. What force is exerted on the wire?

**28–19** A wire along the $x$-axis carries a current of 7.00 A in the positive direction. Calculate the force (expressed in terms of unit vectors) exerted on a 1.00-cm section of the wire by these magnetic fields: a) $\boldsymbol{B} = -(0.600\ \mathrm{T})\boldsymbol{j}$; b) $\boldsymbol{B} = +(0.500\ \mathrm{T})\,\boldsymbol{k}$;

c) $\boldsymbol{B} = -(0.300 \text{ T})\boldsymbol{i}$; d) $\boldsymbol{B} = +(0.200 \text{ T})\boldsymbol{i} - (0.300 \text{ T})\boldsymbol{k}$; e) $\boldsymbol{B} = +(0.900 \text{ T})\boldsymbol{j} - (0.400 \text{ T})\boldsymbol{k}$.

**28–20** A straight, vertical wire carries a current of 8.00 A upward in a region between the poles of a large superconducting electromagnet, where the magnetic field has magnitude $B = 2.75$ T and is horizontal. What are the magnitude and direction of the magnetic force on a 1.00-cm section of the wire if the magnetic field direction is a) east; b) south; c) 30.0° south of west?

## Section 28–7
## Force and Torque on a Current Loop

**28–21** A circular coil of wire 8.00 cm in diameter has 12 turns and carries a current of 3.00 A. The coil is in a region where the magnetic field is 0.600 T. a) What is the maximum torque on the coil? b) In what position is the torque one-half as great as in part (a)?

**28–22** What is the maximum torque on a rectangular coil 5.00 cm × 12.0 cm and with 600 turns when carrying a current of 0.0700 A in a uniform field with magnitude 0.300 T?

**28–23** A coil with magnetic moment $\mu = 1.30 \text{ A} \cdot \text{m}^2$ is oriented initially with its magnetic moment parallel to a uniform magnetic field with $B = 0.750$ T. What is the change in potential energy of the coil when it is rotated 180° so that its magnetic moment is antiparallel to the field?

**28–24** The plane of a rectangular loop of wire 5.00 cm × 8.00 cm is parallel to a magnetic field with magnitude 0.150 T. The loop carries a current of 4.00 A. a) What torque acts on the loop? b) What is the magnetic moment of the loop? c) What is the maximum torque that can be obtained with the same total length of wire carrying the same current in this magnetic field?

**28–25** A circular coil with area $A$ and $N$ turns is free to rotate about a diameter that coincides with the $x$-axis. Current $I$ is circulating in the coil. There is a uniform magnetic field $\boldsymbol{B}$ in the positive $y$-direction. Calculate the magnitude and direction of the torque $\boldsymbol{\tau}$ and the value of the potential energy $U$, as given in Eq. (28–26), when the coil is oriented as shown in parts (a) through (d) of Fig. 28–39.

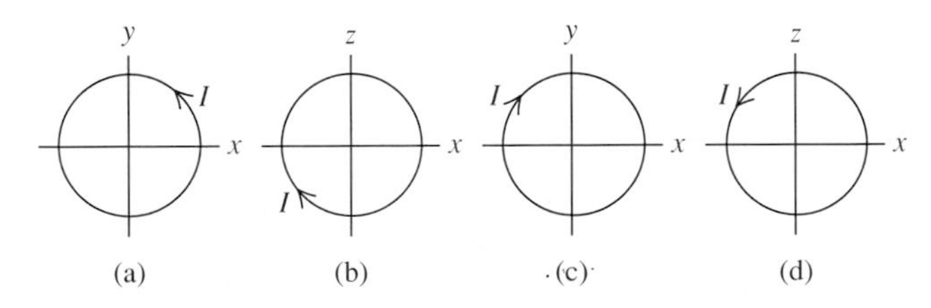


**FIGURE 28–39**

## Section 28–8
## The Direct-Current Motor

*__28–26__ A dc motor with its rotor and field coils connected in series has an internal resistance of 5.00 Ω. When running at full load on a 12.0-V line, the emf in the rotor is 105 V. a) What is the current drawn by the motor from the line? b) What is the power delivered to the motor? c) What is the mechanical power developed by the motor?

*__28–27__ In a shunt-wound dc motor (Fig. 28–40), the resistance $R_f$ of the field coils is 140 Ω, and the resistance $R_r$ of the rotor is 5.00 Ω. When a potential difference of 120 V is applied to the brushes and the motor is running at full speed delivering mechanical power, the current supplied to it is 4.50 A. a) What is the current in the field coils? b) What is the current in the rotor? c) What is the induced emf developed by the motor? d) How much mechanical power is developed by this motor?

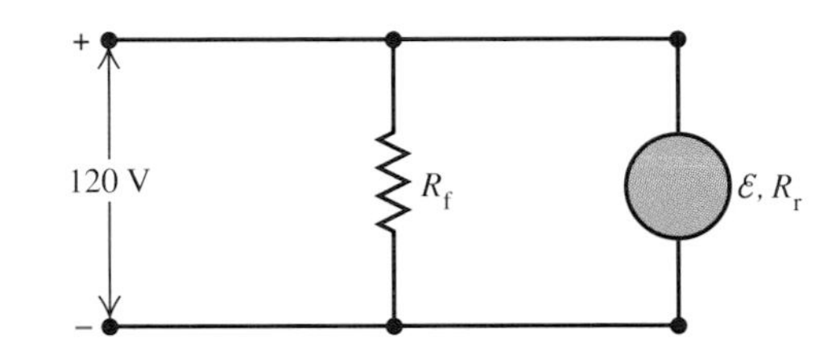


**FIGURE 28–40**

*__28–28__ A shunt-wound dc motor (Fig. 28–40) operates from a 120-V dc power line. The resistance of the field windings, $R_f$, is 240 Ω. The resistance of the rotor, $R_r$, is 4.00 Ω. When the motor is running, the rotor develops an emf $\mathcal{E}$. The motor draws a current of 4.50 A from the line. Friction losses amount to 50.0 W. Compute a) the field current; b) the rotor current; c) the emf $\mathcal{E}$; d) the rate of development of heat in the field windings; e) the rate of development of heat in the rotor; f) the power input to the motor; g) the efficiency of the motor.

## Section 28–9
## The Hall Effect

*__28–29__ Figure 28–41 shows a portion of a silver ribbon with $z_1 = 2.00$ cm and $y_1 = 1.00$ mm, carrying a current of 140 A in the $+x$-direction. The ribbon lies in a uniform magnetic field, in the $y$-direction, with magnitude 1.50 T. If there are $5.85 \times 10^{28}$ free electrons per cubic meter, find a) the magnitude of the drift velocity of the electrons in the $x$-direction; b) the magnitude and direction of the electric field in the $z$-direction due to the Hall effect; c) the Hall emf.

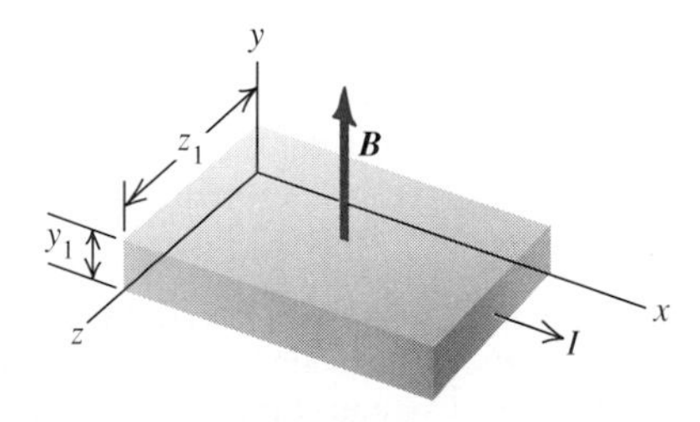


**FIGURE 28–41**

*__28–30__ Let Fig. 28–41 represent a strip of potassium of the same dimensions as those of the silver ribbon in Exercise 28–29. When the magnetic field is 5.00 T and the current is 100 A, the Hall emf is found to be 223 μV. What is the density of free electrons in the potassium?

## PROBLEMS

**28–31 Crossed *E* and *B* Fields.** A particle with initial velocity $\boldsymbol{v}_0 = (4.00 \times 10^3 \text{ m/s})\boldsymbol{i}$ enters a region of uniform electric and magnetic fields. The magnetic field in the region is $\boldsymbol{B} = -(0.600 \text{ T})\boldsymbol{j}$. Calculate the magnitude and direction of the electric field in the region if the particle is to pass through undeflected for a particle of charge a) $+0.400 \times 10^{-8}$ C; b) $-0.400 \times 10^{-8}$ C. Neglect the weight of the particle.

**28–32 Effect of the Earth's Magnetic Field on a TV Picture Tube.** Suppose the accelerating voltage in a TV picture tube is 8000 V. Calculate the approximate deflection of an electron beam over a distance of 0.40 m from the electron gun to the screen under the action of a transverse field with a magnitude of $5.0 \times 10^{-5}$ T (comparable to the magnitude of the earth's field), assuming that there are no other deflecting fields. Is this deflection significant?

**28–33** A particle carries a charge of 6.00 nC. When it moves with a velocity of $\boldsymbol{v}_1$ that has a magnitude of $3.00 \times 10^4$ m/s and is at 45.0° from the $+x$-axis in the $xy$-plane, a uniform magnetic field exerts a force $\boldsymbol{F}_1$ along the $-z$-axis (Fig. 28–42). When the particle moves with a velocity $\boldsymbol{v}_2$ that has a magnitude of $2.00 \times 10^4$ m/s and is along the $+z$-axis, there is a force $\boldsymbol{F}_2$ of magnitude $4.00 \times 10^{-5}$ N exerted on it along the $+x$-axis. What are the magnitude and direction of the magnetic field?

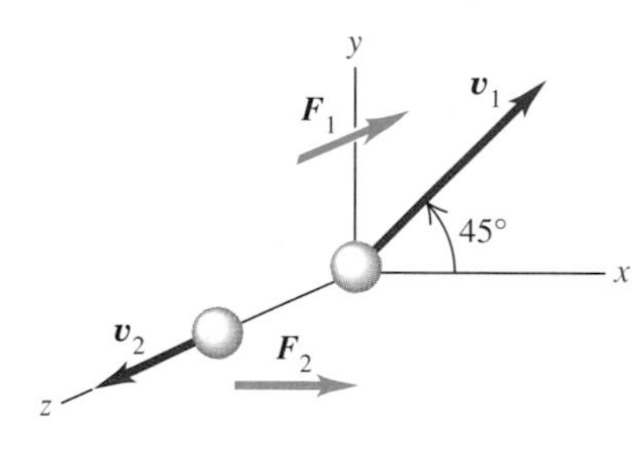


FIGURE 28–42

**28–34** The force on a charged particle moving in a magnetic field can be computed as the vector sum of the forces due to each separate component of the magnetic field. A particle with charge $3.50 \times 10^{-8}$ C has a velocity $v = 6.00 \times 10^5$ m/s in the $-x$-direction. It is moving in a uniform magnetic field with components $B_x = +0.200$ T, $B_y = -0.500$ T, and $B_z = +0.300$ T. What are the components of the force exerted on the particle by the magnetic field?

**28–35** An electron and an alpha particle (a doubly ionized helium atom) both move in a magnetic field in circular paths with the same tangential speed. Compute the ratio of the number of revolutions that the electron makes per second to the number made per second by the alpha particle. The mass of the alpha particle is $6.65 \times 10^{-27}$ kg.

**28–36** A particle having a charge $q = 2.00\ \mu$C is traveling with velocity $\boldsymbol{v} = (1.50 \times 10^3 \text{ m/s})\boldsymbol{j}$. The particle experiences force $\boldsymbol{F} = (2.00 \times 10^{-4} \text{ N})(3\boldsymbol{i} - 4\boldsymbol{k})$ due to a magnetic field $\boldsymbol{B}$. a) Determine $F$, the magnitude of $\boldsymbol{F}$. b) Determine $B_x$, $B_y$, and $B_z$, or at least as many of the three components as is possible from the information given. (See Problem 28–34.) c) If the magnitude of the magnetic field is 0.500 T, determine the remaining components of $\boldsymbol{B}$.

**28–37** Suppose the electric field between the plates $P$ and $P'$ in Fig. 28–19 is $1.50 \times 10^4$ V/m and the magnetic field in both regions is 0.600 T. If the source contains the three isotopes of magnesium, $^{24}$Mg, $^{25}$Mg, and $^{26}$Mg, and the ions are singly charged, find the distance between the lines formed by the three isotopes on the photographic plate. Assume that the atomic masses of the isotopes (in atomic mass units) are equal to their mass numbers. (One atomic mass unit = 1 amu = $1.66 \times 10^{-27}$ kg.)

**28–38** The force on a charged particle moving in a magnetic field can be computed as the vector sum of the forces due to each separate component of the particle's velocity. (See Problem 28–34.) A particle with charge $7.60 \times 10^{-8}$ C is moving in a region where there is a uniform magnetic field of 0.300 T in the $+x$-direction. At a particular instant of time the velocity of the particle has components $v_x = 2.50 \times 10^4$ m/s, $v_y = 9.00 \times 10^4$ m/s, and $v_z = -5.00 \times 10^4$ m/s. What are the components of the force on the particle at this time?

**28–39** A particle with positive charge $q$ and mass $m = 1.50 \times 10^{-15}$ kg is traveling through a region containing a uniform magnetic field $\boldsymbol{B} = -(0.220 \text{ T})\boldsymbol{k}$. At a particular instant of time the velocity of the particle is

$$\boldsymbol{v} = (1.00 \times 10^6 \text{ m/s})(4\boldsymbol{i} - 3\boldsymbol{j} + 12\boldsymbol{k}),$$

and the force $\boldsymbol{F}$ on the particle has a magnitude of 2.00 N. (See Problem 28–38.) a) Determine the charge $q$. b) Determine the acceleration $\boldsymbol{a}$ of the particle. c) Explain why the path of the particle is a helix, and determine the radius of curvature $R$ of the circular component of the helical path. d) Determine the cyclotron frequency of the particle. e) Although helical motion is not periodic in the full sense of the word, the $x$- and $y$-coordinates do vary in a periodic way. If the coordinates of the particle at $t = 0$ are $(x, y, z) = (R, 0, 0)$, determine its coordinates at a time $t = 2T$, where $T$ is the period of the motion in the $xy$-plane.

**28–40** An electron is moving in a circular path with radius $r = 4.00$ cm in the space between two concentric cylinders. The inner cylinder is a positively charged wire with radius $a = 1.00$ mm, and the outer cylinder is a negatively charged cylinder with radius $b = 5.00$ cm. The potential difference between the inner and outer cylinders is $V_{ab} = 120$ V, with the wire being at the higher potential. (See Fig. 28–43.) The electric field $\boldsymbol{E}$ in the region between the cylinders is radially outward and was shown in Problem 24–43 to have the magnitude $E = V_{ab}/[r \ln (b/a)]$. a) Determine the speed of the electron in order for it to maintain its circular orbit. Neglect both the gravitational and magnetic fields of the earth. b) Now include the effect of the earth's magnetic field. If the axis of symmetry of the cylinders is positioned parallel to the magnetic field of the earth, at what speed must the electron move to maintain the same circular orbit? Assume that the magnetic field of the earth has magnitude $1.00 \times 10^{-4}$ T and that its

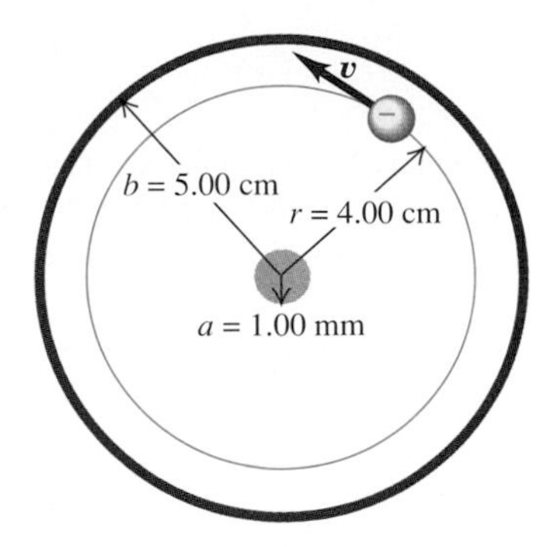

**FIGURE 28–43**

direction is out of the plane of the page in Fig. 28–43. c) Redo the calculation of part (b) for the case in which the magnetic field is in the opposite direction to that of part (b).

**28–41** The cube in Fig. 28–44, 0.500 m on a side, is in a uniform magnetic field of 0.200 T, parallel to the $x$-axis. The wire *abcdef* carries a current of 4.00 A in the direction indicated. a) Determine the magnitude and direction of the force acting on the segments *ab, bc, cd, de,* and *ef*. b) What are the magnitude and direction of the total force on the wire?

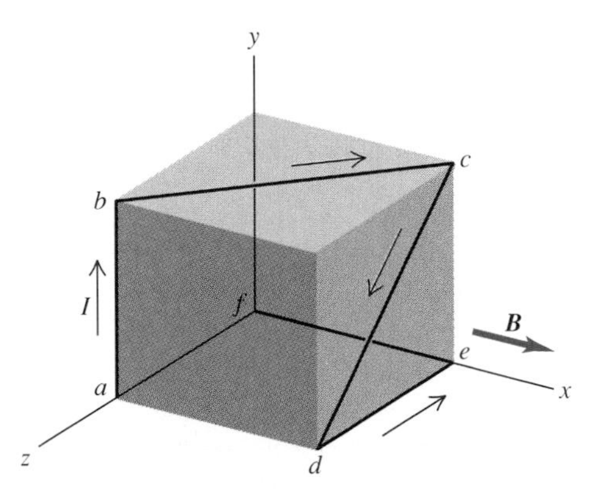

**FIGURE 28–44**

**28–42** A wire 0.150 m long lies along the $y$-axis and carries a current of 8.00 A in the $+y$-direction. The magnetic field is uniform and has components $B_x = 0.300$ T, $B_y = -1.20$ T, and $B_z = 0.500$ T. a) Find the components of force on the wire. (As in Problem 28–34, the resultant force is the vector sum of the forces due to each component of $\boldsymbol{B}$.) b) What is the magnitude of the total force on the wire?

**28–43 Torque on a Current Loop.** The rectangular loop of wire in Fig. 28–45 has a mass of 0.100 g per centimeter of length and is pivoted about side *ab* as a frictionless axis. The current in the wire is 8.00 A in the direction shown. Find the magnitude and direction of the magnetic field parallel to the $y$-axis that will cause the loop to swing up until its plane makes an angle of 30.0° with the $yz$-plane.

**28–44** The rectangular loop in Fig. 28–46 is pivoted about the $y$-axis and carries a current of 15.0 A in the direction indicated. a) If the loop is in a uniform magnetic field with magnitude 0.200 T, parallel to the $x$-axis, find the magnitude of the torque required to hold the loop in the position shown.) b) Repeat part (a) for the case in which the field is parallel to the $z$-axis. c) For each of the above magnetic fields, what magnitude of torque would be required if the loop were pivoted about an axis through its center and parallel to the $y$-axis?

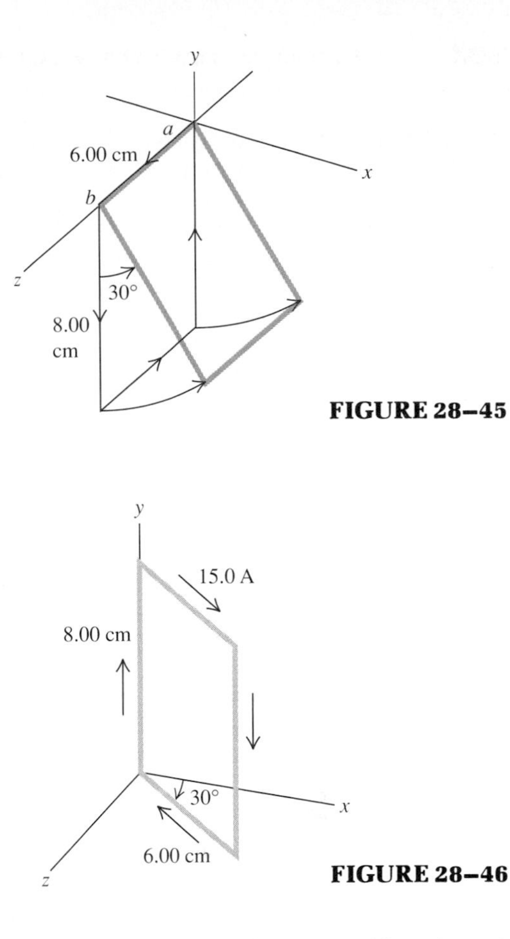


**28–45 A Voice Coil.** It was shown in Section 28–7 that the net force on a current loop in a *uniform* magnetic field is zero. The magnetic force on the voice coil of a loudspeaker (Fig. 28–32) is nonzero because the magnetic field at the coil is not uniform. A voice coil in a loudspeaker has 40 turns of wire and a diameter of 1.80 cm, and the current in the coil is 0.800 A. Assume that the magnetic field at each point of the wire of the coil has a constant magnitude of 0.200 T and is directed at an angle of 60.0° outward from the normal to the plane of the coil (Fig. 28–47). Let the axis of the coil be in the $y$-direction. The current in the coil is in the direction shown (counterclockwise as viewed from a point above the coil on the $y$-axis). Calculate the magnitude and direction of the resultant magnetic force on the coil.

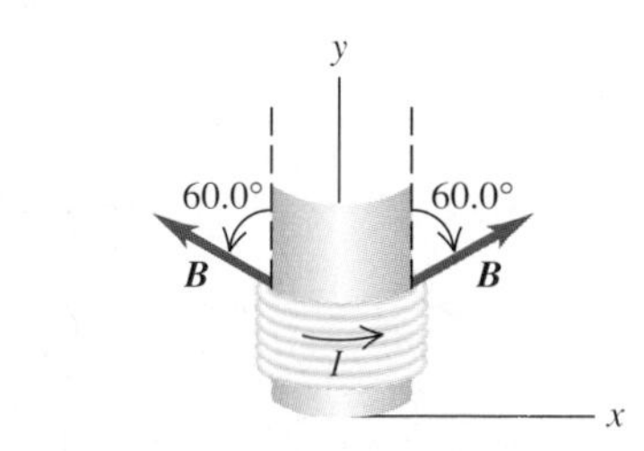


**FIGURE 28–47**

**28–46 Quark Model of the Neutron.** The neutron is a particle with zero charge but with a nonzero magnetic moment with magnitude $\mu = 9.66 \times 10^{-27}\ \text{A} \cdot \text{m}^2$. If the neutron is considered to be a fundamental entity with no internal structure, the two properties listed above seem to be contradictory. According to present theory in particle physics, a neutron is composed of three more fundamental particles called quarks. In this model the neutron consists of an "up quark" having a charge of $+2e/3$ and two "down quarks," each having a charge of $-e/3$. The combination of the three quarks produces a net charge of $2e/3 - e/3 - e/3 = 0$, as required, and if the quarks are in motion, they could also produce a nonzero magnetic moment. As a very simple model, suppose the up quark is moving in a counterclockwise circular path and the down quarks are moving in a clockwise circular path, all of radius $r$ and all with the same speed $v$ (Fig. 28–48). a) Obtain an expression for the current due to the circulation of the up quark. b) Obtain an expression for the magnitude $\mu_u$ of the magnetic moment due to the circulating up quark. c) Obtain an expression for the magnitude of the magnetic moment of the three-quark system. (Be careful to use the correct magnetic moment directions.) d) With what speed $v$ must the quarks move if this model is to reproduce the magnetic moment of the neutron? For the radius of the orbits, use $r = 1.20 \times 10^{-15}$ m, the radius of the neutron.

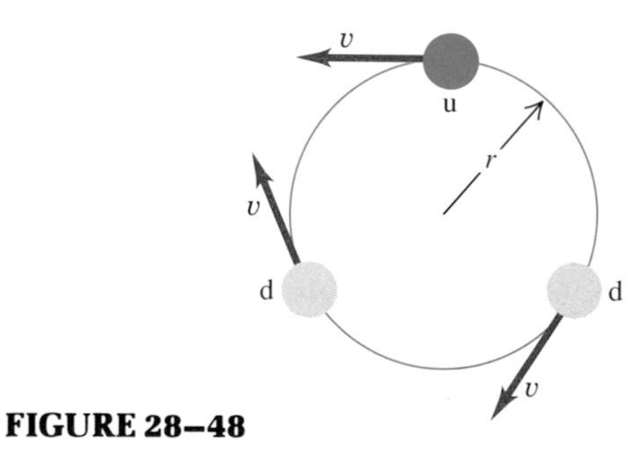


**FIGURE 28–48**

**28–47** An insulated wire with a length of 35.0 cm and mass $m = 9.79 \times 10^{-5}$ kg is bent into the shape of an inverted U such that the horizontal part has a length $l = 25.0$ cm. The bent ends of the wire, each 5.0 cm long, are completely immersed in two pools of mercury, and the entire structure is in a region containing a magnetic field with a magnitude of 0.0180 T and a direction into the page (Fig. 28–49). An electrical connection from the mercury pools is made through the ends of the wires. The mercury pools are connected to a 1.50-V battery and a switch S. Switch S is closed, and the wire jumps 0.700 m into the air, measured from its initial position. a) Determine the speed $v$ of the wire as it leaves the mercury. b) Assuming that the current $I$ through the wire was constant from the time the switch was closed until the wire left the mercury, determine $I$. c) Neglecting the resistance of the mercury and the circuit wires, determine the resistance of the moving wire.

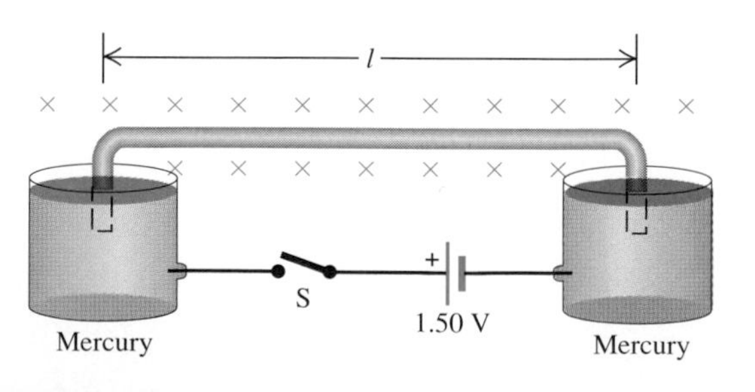


**FIGURE 28–49**

**28–48** It is fairly straightforward to derive Eq. (28–25) explicitly for a circular current loop. Consider a wire ring in the $xy$-plane with its center at the origin. The ring carries a counterclockwise current $I$ (Fig. 28–50). Let the magnetic field $\boldsymbol{B}$ be in the $+x$-direction, $\boldsymbol{B} = B_x\boldsymbol{i}$. (The result is easily extended to $\boldsymbol{B}$ in an arbitrary direction.) a) In Fig. 28–50, show that the element $d\boldsymbol{l} = R\, d\theta\,(-\sin\theta\boldsymbol{i} + \cos\theta\boldsymbol{j})$, and find $d\boldsymbol{F} = I\, d\boldsymbol{l} \times \boldsymbol{B}$. b) Integrate $d\boldsymbol{F}$ around the loop to show that the net force is zero. c) From part (a), find $d\boldsymbol{\tau} = \boldsymbol{r} \times d\boldsymbol{F}$, where $\boldsymbol{r} = R(\cos\theta\boldsymbol{i} + \sin\theta\boldsymbol{j})$. (Note that the $d\boldsymbol{l}$ is perpendicular to $\boldsymbol{r}$.) d) Integrate $d\boldsymbol{\tau}$ over the loop to find the total torque $\boldsymbol{\tau}$ on the loop. Show that the result can be written $\boldsymbol{\tau} = \boldsymbol{\mu} \times \boldsymbol{B}$, where $\mu = IA$. (*Note:* $\int \cos^2 x\, dx = \frac{1}{2}x + \frac{1}{4}\sin 2x$, $\int \sin^2 x\, dx = \frac{1}{2}x - \frac{1}{4}\sin 2x$, and $\int \sin x \cos x\, dx = \frac{1}{2}\sin^2 x$.)

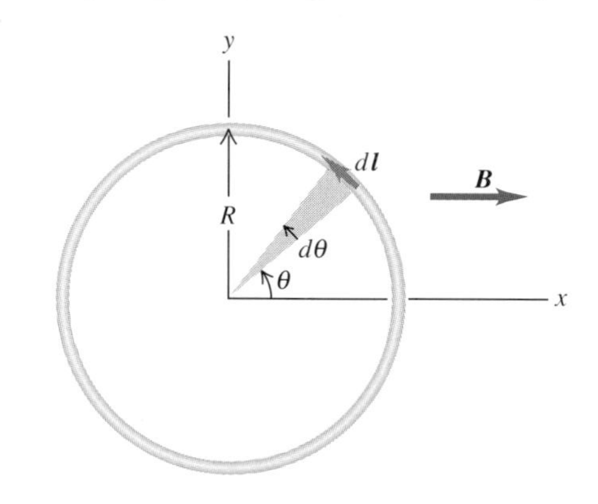


**FIGURE 28–50**

**28–49** A circular loop of wire with area 8.00 $\text{cm}^2$ carries a current of 15.0 A. The loop lies in the $xy$-plane. As viewed along the $z$-axis looking in toward the origin, the current is circulating counterclockwise. The torque produced by an external magnetic field $\boldsymbol{B}$ is given by $\boldsymbol{\tau} = (1.00 \times 10^{-3}\ \text{N} \cdot \text{m})(-6\boldsymbol{i} + 8\boldsymbol{j})$, and for this orientation of the loop the magnetic potential energy $U = -\boldsymbol{\mu} \cdot \boldsymbol{B}$ is negative. The magnitude of the magnetic field is 2.60 T. a) Determine the magnetic moment of the current loop. b) Determine the components $B_x$, $B_y$, and $B_z$ of $\boldsymbol{B}$.

**28–50** A circular ring with area 8.00 $\text{cm}^2$ and negligible mass is carrying a current of 50.0 A. The ring is free to rotate about a diameter. The ring, initially at rest, is immersed in a region of uniform magnetic field where $\boldsymbol{B}$ is given by $\boldsymbol{B} = (1.00 \times 10^{-2}\ \text{T})(3\boldsymbol{i} - 4\boldsymbol{j} - 12\boldsymbol{k})$. The ring is positioned initially such that its magnetic moment $\boldsymbol{\mu}_i$ is given by $\boldsymbol{\mu}_i = \mu(-0.600\boldsymbol{i} + 0.800\boldsymbol{j})$, where $\mu$ is the (positive) magnitude of the magnetic moment. The ring is released and turns through an angle of 90.0°, at which point its magnetic moment is given by $\boldsymbol{\mu}_f = -\mu\boldsymbol{k}$. a) Determine the decrease in potential energy of the ring. b) If the moment of inertia of the ring about a diameter is $1.70 \times 10^{-6}\ \mathrm{kg \cdot m^2}$, determine the angular velocity of the ring as it passes through the second position.

**28–51** Two positive ions having the same charge $q$ but different masses, $m_1$ and $m_2$, are accelerated horizontally from rest through a potential difference $V$. They then enter a region where there is a uniform magnetic field $\boldsymbol{B}$ normal to the plane of the trajectory. a) Show that if the beam entered the magnetic field along the $x$ axis, the value of the $y$-coordinate for each ion at any time $t$ is approximately

$$y = Bx^2\left(\frac{q}{8mV}\right)^{1/2},$$

provided that $y$ remains much smaller than $x$. b) Can this arrangement be used for isotope separation?

## CHALLENGE PROBLEMS

**28–52 A Cycloidal Path.** A particle with mass $m$ and charge $+q$ starts from rest at the origin in Fig. 28–51. There is a uniform electric field $\boldsymbol{E}$ in the $+y$-direction and a uniform magnetic field $\boldsymbol{B}$ directed out of the page. It is shown in more advanced books that the path is a *cycloid* whose radius of curvature at the top points is twice the $y$-coordinate at that level. a) Explain why the path has this general shape and why it is repetitive. b) Prove that the speed at any point is equal to $\sqrt{2qEy/m}$. (*Hint:* Use energy conservation.) c) Applying Newton's second law at the top point and taking as given that the radius of curvature here equals $2y$, prove that the speed at this point is $2E/B$.

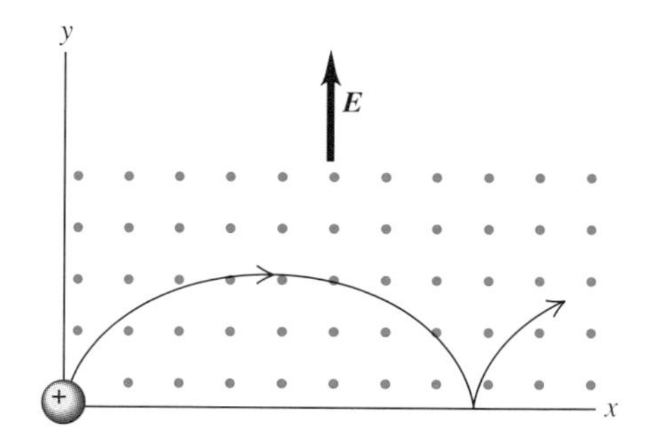


**FIGURE 28–51**

**28–53** A particle with charge $q = 4.00\ \mu\mathrm{C}$ and mass $m = 1.00 \times 10^{-11}$ kg is initially traveling in the $+y$-direction with a speed $v_0 = 3.00 \times 10^5$ m/s. It then enters a region containing a uniform magnetic field that is directed into and perpendicular to the page in Fig. 28–52. The magnitude of the field is 0.500 T. The region extends a distance of 25.0 cm along the initial direction of travel; 75.0 cm from the point of entry into the magnetic field region is a wall. The length of the field-free region is thus 50.0 cm. When the charged particle enters the magnetic field, it will follow a curved path whose radius of curvature is $R$. It then leaves the magnetic field after a time $t_1$, having been deflected a distance $\Delta x_1$. The particle then travels in the field-free region and strikes the wall after undergoing a total deflection $\Delta x$. a) Determine the radius $R$ of the curved part of the path. b) Determine $t_1$, the time the particle spends in the magnetic field. c) Determine $\Delta x_1$, the horizontal deflection at the point of exit from the field. d) Determine $\Delta x$, the total horizontal deflection.

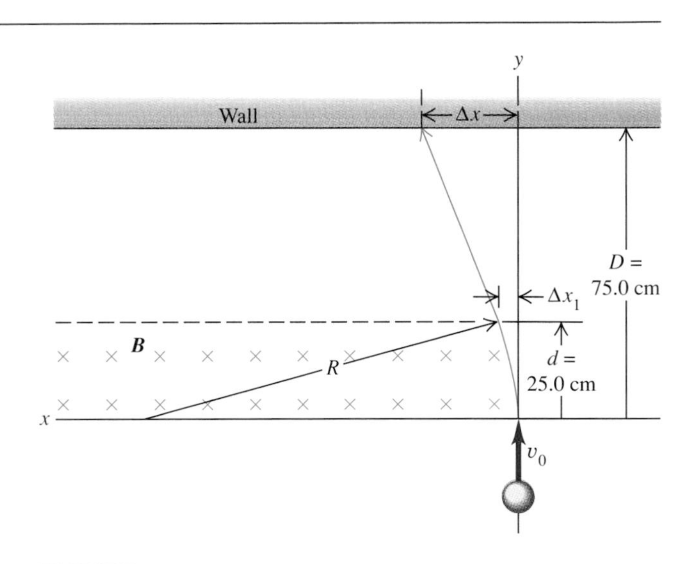


**FIGURE 28–52**

**28–54 The Electromagnetic Pump.** Magnetic forces acting on conducting fluids provide a convenient means of pumping these fluids. For example, such an electromagnetic pump can be used to pump blood without the damage to the cells that can be caused by a mechanical pump. A horizontal tube with rectangular cross section (height $h$, width $w$) is placed at right angles to a uniform magnetic field with magnitude $B$ so that a length $l$ of the tube is in the field (Fig. 28–53). The tube is filled with liquid sodium, and an electric current of density $J$ is maintained in the third mutually perpendicular direction. a) Show that the difference of pressure between a point in the liquid on a vertical plane through $ab$ and a point in the liquid on another vertical plane through $cd$, under conditions in which the liquid is prevented from flowing, is $\Delta p = JlB$. b) What current density is needed to provide a pressure difference of 1.00 atm between these two points if $B = 1.50$ T and $l = 0.0200$ m?

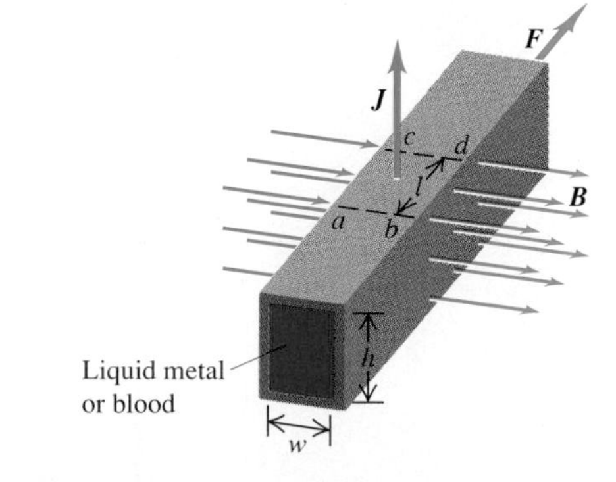

**FIGURE 28–53**

CHAPTER

# 29

# Sources of Magnetic Field

A quadrupole magnet has magnetic north poles in the upper right and lower left sections, looking along its axis; the upper left and lower right corners are magnetic south poles. The magnetic fields generated by these magnets are so precise that modern accelerators can operate for several days, with particles traveling a distance equivalent to 180 round trips between the earth and the sun, while staying within a path width of only a few millimeters.

A particle accelerator usually operates with two kinds of electromagnets. Dipole magnets steer the particle beam along its circular path. Quadrupole magnets, like the one shown here, focus the particles along the beam and prevent them from straying.

## KEY CONCEPTS

**• A moving electric charge creates a magnetic field in the space around it. The direction of the magnetic field at any point is perpendicular to the particle's velocity and to the vector from the particle to the point.**

**• The law of Biot and Savart gives the magnetic field created by a small element of a conductor carrying a current. It can be used to find the magnetic field created by any configuration of current-carrying conductors, such as a long, straight wire or a circular loop.**

**• Ampere's law provides an alternative formulation of the relation of magnetic fields to currents. It is analogous to Gauss's law in electrostatics.**

**• When a conductor is surrounded by a material, the magnetic field is modified by microscopic (atomic) currents in the material. Magnetization of magnetic materials is associated with these microscopic currents.**

**• A time-varying electric field can act as a source of magnetic field; this relationship is expressed by a fictitious current called displacement current associated with the varying electric field. The general form of Ampere's law requires the inclusion of displacement current.**

## INTRODUCTION

In Chapter 28 we studied the *forces* that are exerted on moving charges and on current-carrying conductors in a magnetic field. We didn't worry about how the magnetic field got there; we simply took its existence as a given fact. But how are magnetic fields *created?* We know that both permanent magnets and electric currents in electromagnets create magnetic fields. Now it's time to study these sources of magnetic field in detail.

We've learned that a charge at rest creates an electric field and that an electric field exerts a force on a charge at rest. But a *magnetic* field exerts a force only on a *moving* charge. Is it also true that a charge *creates* a magnetic field only when the charge is moving?

In a word, yes. Our analysis will begin with the magnetic field created by a single moving point charge. We can use this analysis to determine the field created by a small segment of a current-carrying conductor. Once we do that, we can in principle find the magnetic field produced by *any* shape of conductor, such as a long, straight wire, a circular loop, or a solenoid.

Then we will introduce Ampere's law, the magnetic analog of Gauss's law in electrostatics. Ampere's law lets us exploit symmetry properties in relating magnetic fields to their sources. We'll look at the role of magnetic materials in modifying magnetic fields. Finally, we will study the role of displacement current, which may involve no actual motion of charge, as a source of magnetic field. This relationship will be one of the key elements in our study of electromagnetic waves in Chapter 33.

## 29–1 MAGNETIC FIELD OF A MOVING CHARGE

Let's start with the basics, the magnetic field of a single moving point charge $q$. We call the location of the charge the **source point** and the point $P$ where we want to find the field the **field point.** In our study of *electric* fields in Chapter 22, we found that the $\boldsymbol{E}$ field of a point charge $q$, at a field point a distance $r$ from the charge, is proportional to $q$ and to $1/r^2$ and that the direction of this field (for positive $q$) is along the line from source point to field point. The corresponding relationship for the *magnetic* field $\boldsymbol{B}$ of a point charge $q$ moving with velocity $\boldsymbol{v}$ has some similarities to this relationship and some interesting differences.

Experiments show that the magnitude $B$ is also proportional to $q$ and to $1/r^2$. But there the similarity ends. The *direction* of $\boldsymbol{B}$ is *not* along the line from the source point (the moving charge) to the field point. Instead, $\boldsymbol{B}$ is perpendicular to the plane containing this line and the particle's velocity vector $\boldsymbol{v}$, as shown in Fig. 29–1. Furthermore, the field magnitude $B$ is proportional to the sine of the angle $\phi$ between these two directions. Finally, $B$ is proportional to the particle's speed $v$. Thus the magnitude $B$ of the magnetic field at point $P$ is given by

$$B = \frac{\mu_0}{4\pi}\frac{qv \sin\phi}{r^2}, \tag{29–1}$$

where $\mu_0/4\pi$ is a proportionality constant. The reason for writing the constant in this particular way will emerge shortly. We did something similar with Coulomb's law.

We can incorporate both the magnitude and direction of $\boldsymbol{B}$ into a single vector equation using the vector product. To avoid having to say "the direction from the source $q$ to the field point $P$" over and over, we introduce a unit vector $\hat{\boldsymbol{r}} = \boldsymbol{r}/r$ that points in the direction from charge $q$ to point $P$, that is, from the source point to the field point. Then the $\boldsymbol{B}$ field of a moving point charge is

$$\boldsymbol{B} = \frac{\mu_0}{4\pi}\frac{q\boldsymbol{v} \times \hat{\boldsymbol{r}}}{r^2}. \tag{29–2}$$

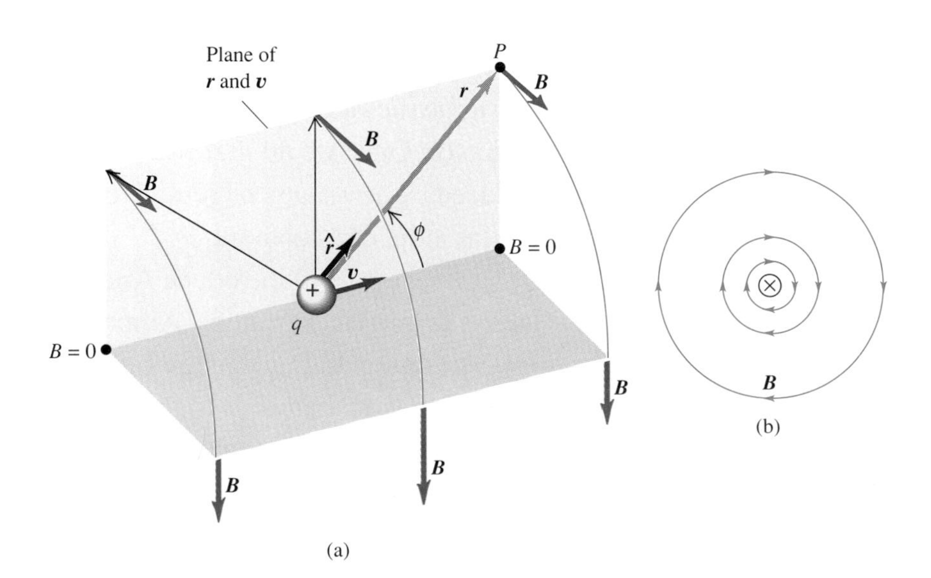


**29–1** (a) Magnetic field vectors due to a moving positive point charge $q$. At each point, $\boldsymbol{B}$ is perpendicular to the plane of $\boldsymbol{r}$ and $\boldsymbol{v}$, and its magnitude is proportional to the sine of the angle between them. (b) Magnetic field lines in a plane containing a moving positive charge. The charge is moving away from the reader, into the plane of the figure.

Figure 29–1 shows the relation of $\hat{\boldsymbol{r}}$ to $P$ and also shows the magnetic field $\boldsymbol{B}$ at several points in the vicinity of the charge. At all points along a line through the charge parallel to the velocity $\boldsymbol{v}$, the field is zero because $\sin\phi = 0$ at all such points. At any distance $r$ from $q$, $\boldsymbol{B}$ has its greatest magnitude at points lying in the plane perpendicular to $\boldsymbol{v}$ because at all such points, $\phi = 90°$ and $\sin\phi = 1$. The charge also produces an *electric* field in its vicinity; the electric-field vectors are not shown in the figure.

The field lines for the *electric* field of a positive point charge radiate outward from the charge. The *magnetic* field lines are completely different. The above discussion shows that for a point charge moving with velocity $\boldsymbol{v}$, the magnetic field lines are *circles* with centers along the line of $\boldsymbol{v}$ and lying in planes perpendicular to this line. The directions of these lines for a positive charge are given by a right-hand rule: Grasp the velocity vector $\boldsymbol{v}$ with your right hand so that your right thumb points in the direction of $\boldsymbol{v}$: your fingers then curl around the line of $\boldsymbol{v}$ in the same sense as the magnetic field lines. Figure 29–1a shows parts of a few field lines, and Fig. 29–1b shows some field lines in a plane through $q$, perpendicular to $\boldsymbol{v}$, as seen looking in the direction of $\boldsymbol{v}$.

As we discussed in Section 28–2, the unit of $B$ is one tesla (1 T):

$$1\ \text{T} = 1\ \text{N}\cdot\text{s/C}\cdot\text{m} = 1\ \text{N/A}\cdot\text{m}.$$

Using this with Eq. (29–1) or (29–2) and solving for $\mu_0$, we find that the units of the constant $\mu_0$ are

$$1\ \text{N}\cdot\text{s}^2/\text{C}^2 = 1\ \text{N/A}^2 = 1\ \text{Wb/A}\cdot\text{m} = 1\ \text{T}\cdot\text{m/A}.$$

In SI units the numerical value of $\mu_0$ is exactly $4\pi \times 10^{-7}$. Thus

$$\mu_0 = 4\pi \times 10^{-7}\ \text{N}\cdot\text{s}^2/\text{C}^2 = 4\pi \times 10^{-7}\ \text{Wb/A}\cdot\text{m} = 4\pi \times 10^{-7}\ \text{T}\cdot\text{m/A}. \tag{29–3}$$

This numerical value actually stems from the definition of the coulomb. We mentioned in Section 22–4 that the constant $1/4\pi\epsilon_0$ in Coulomb's law is related to the speed of light $c$:

$$\frac{1}{4\pi\epsilon_0} = (10^{-7}\ \text{N}\cdot\text{s}^2/\text{C}^2)c^2.$$

When we study electromagnetic waves in Chapter 33, we will find that their speed of propagation in vacuum, which is equal to the speed of light $c$, is given by

$$c^2 = \frac{1}{\epsilon_0\mu_0}. \tag{29–4}$$

If we solve the preceding equation for $\epsilon_0$, substitute the resulting expression into Eq. (29–4), and solve for $\mu_0$, we do get the value of $\mu_0$ stated above. This discussion is a little premature, but it may give you a hint about one of the nice unifying threads that runs through electromagnetic theory.

## EXAMPLE 29–1

**Forces between two moving protons** Two protons move with equal and opposite velocities (both small in magnitude compared to the speed of light $c$) parallel to the $x$-axis (Fig. 29–2, on page 808). At the instant shown, find the electric and magnetic forces on the upper proton, and determine their ratio.

**SOLUTION** The electric force is easy; Coulomb's law gives

$$F_E = \frac{1}{4\pi\epsilon_0}\frac{q^2}{r^2}.$$

The forces are repulsive, and the force on the upper proton is vertically upward (the $+y$-direction). To find the magnetic force, we first find the $\boldsymbol{B}$ field caused by the lower proton at the location of the upper one. From Eq. (29–1) and the right-hand rule we find that $\boldsymbol{B}$ is in the $+z$-direction, as shown in the figure, with magnitude

$$B = \frac{\mu_0}{4\pi}\frac{qv}{r^2}.$$

Alternatively, from Eq. (29–2),

$$\boldsymbol{B} = \frac{\mu_0}{4\pi}\frac{q(v\boldsymbol{i}) \times \boldsymbol{j}}{r^2} = \frac{\mu_0}{4\pi}\frac{qv}{r^2}\boldsymbol{k}.$$

The velocity of the upper proton is $-\boldsymbol{v}$, and the magnetic force on it is $\boldsymbol{F} = q(-\boldsymbol{v}) \times \boldsymbol{B}$. Combining this with the expressions for $\boldsymbol{B}$, we find

$$F_B = \frac{\mu_0}{4\pi}\frac{q^2v^2}{r^2},$$

or

$$\boldsymbol{F}_B = q(-\boldsymbol{v}) \times \boldsymbol{B} = q(-v\boldsymbol{i}) \times \frac{\mu_0}{4\pi}\frac{qv}{r^2}\boldsymbol{k} = \frac{\mu_0}{4\pi}\frac{q^2v^2}{r^2}\boldsymbol{j}.$$

The magnetic interaction is also repulsive. The ratio of the magnitudes of the two forces is

$$\frac{F_B}{F_E} = \frac{\mu_0}{4\pi}\frac{q^2v^2}{r^2}(4\pi\epsilon_0)\frac{r^2}{q^2} = \epsilon_0\mu_0 v^2.$$

Using the relationship $\epsilon_0\mu_0 = 1/c^2$, Eq. (29–4), we can express our result very simply as

$$\frac{F_B}{F_E} = \frac{v^2}{c^2}.$$

When $v$ is small in comparison to the speed of light $c$, the magnetic force is much smaller than the electric force.

Note that it is essential to use the same frame of reference in this entire calculation. We have described the velocities and the fields as they appear to an observer who is stationary in the coordinate system of Fig. 29–2. In a coordinate system moving with one of the charges, one of the velocities would be zero, so there would be *no* magnetic force. The explanation of this apparent paradox provided one of the paths that led to the special theory of relativity.

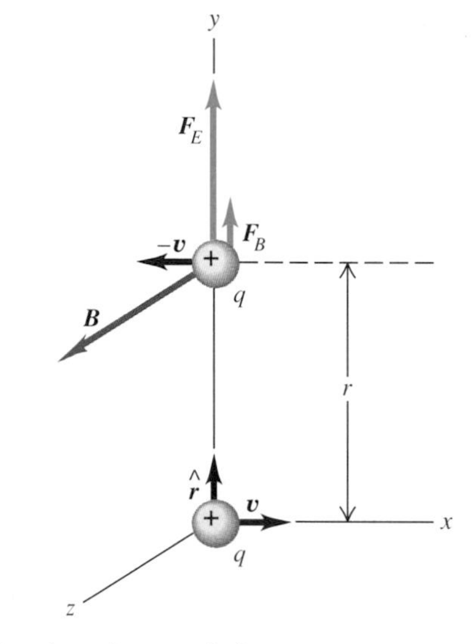


**29–2** Electric and magnetic forces on two protons.

The magnetic field, like the electric field, obeys the **superposition principle: The total magnetic field caused by several moving charges is the vector sum of the fields caused by the individual charges.** In the next section we'll use this fact to calculate the field caused by a current in a conductor.

## 29–2 MAGNETIC FIELD OF A CURRENT ELEMENT

We can use the principles of Section 29–1, including the superposition principle, to find the magnetic field at any field point $P$ produced by a current in a conductor. The total field is the vector sum of the fields due to all of the moving charges in the conductor.

We begin by calculating the magnetic field caused by a short segment $d\boldsymbol{l}$ of a current-carrying conductor, as shown in Fig. 29–3a. The volume of the segment is $A\,dl$, where $A$ is the cross-section area of the conductor. If we have $n$ charges $q$ per unit volume, then the total moving charge $dQ$ in the segment is

$$dQ = nqA\,dl.$$

The moving charges in this segment are equivalent to a single charge $dQ$, traveling with a velocity equal to the *drift* velocity, $\boldsymbol{v}_d$. (Fields due to the *random* motions of the carriers will, on the average, cancel out at every point.) From Eq. (29–1) the magnitude of the

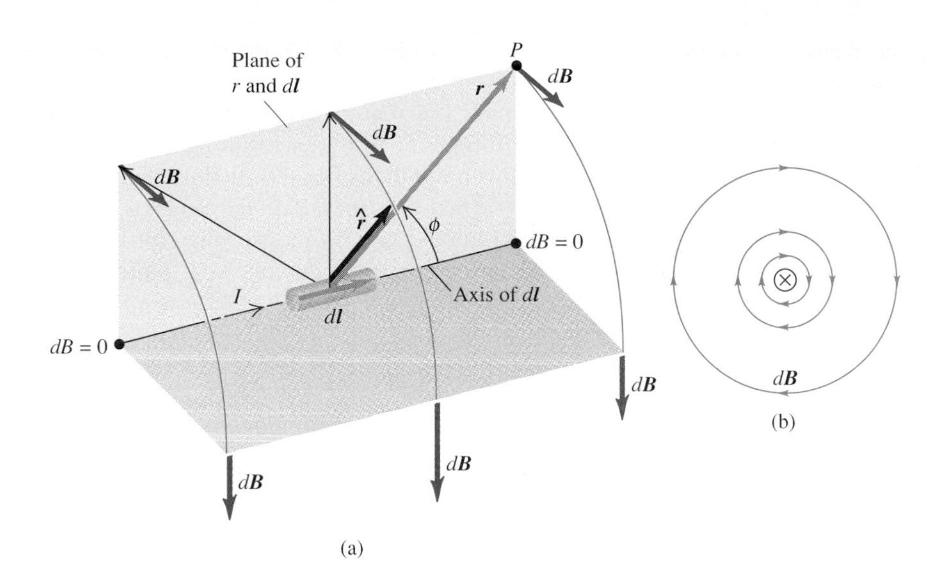


**29–3** (a) Magnetic-field vectors due to a current element ***dl***. (b) Magnetic field lines in a plane containing the current element ***dl***. The current is directed into the plane of the page.

resulting field ***dB*** at any point is

$$dB = \frac{\mu_0}{4\pi}\frac{dQv_d \sin\phi}{r^2} = \frac{\mu_0}{4\pi}\frac{nqv_d A\, dl \sin\phi}{r^2}.$$

But $nqv_d A$ equals the current $I$ in the element, so

$$dB = \frac{\mu_0}{4\pi}\frac{I\, dl \sin\phi}{r^2}, \qquad (29\text{–}5)$$

or in vector form,

$$\boldsymbol{dB} = \frac{\mu_0}{4\pi}\frac{I\, \boldsymbol{dl} \times \hat{\boldsymbol{r}}}{r^2}, \qquad (29\text{–}6)$$

where ***dl*** is a vector with length $dl$, in the same direction as the current in the conductor.

Equations (29–5) and (29–6) are called the **law of Biot and Savart.** To find the total magnetic field ***B*** at any point in space due to the current in a complete circuit, we have to integrate one of these expressions:

$$\boldsymbol{B} = \frac{\mu_0}{4\pi}\int \frac{I\, \boldsymbol{dl} \times \hat{\boldsymbol{r}}}{r^2}. \qquad (29\text{–}7)$$

In the following sections we will carry out this vector integration for several examples.

As shown in Fig. 29–3a, the field vectors ***dB*** and the magnetic-field lines are exactly like those set up by a positive charge $dQ$ moving in the direction of the drift velocity $\boldsymbol{v}_d$. The field lines are circles in planes perpendicular to ***dl*** and centered on the line of ***dl***. Their directions are given by the same right-hand rule that we introduced for point charges in Section 29–1. Figure 29–3b shows field lines in a plane containing the element ***dl*** and perpendicular to it; note the similarity to Fig. 29–1b for a moving point charge.

We can't verify Eq. (29–5) or (29–6) directly because we can never experiment with an isolated segment of a current-carrying circuit. What we measure experimentally is the *total* ***B*** for a complete circuit. But we can still verify these equations indirectly by calculating ***B*** for various current configurations and comparing the results with experimental measurements. We can also verify Eq. (29–1), the starting point of our derivation.

If matter is present in the space around a current-carrying conductor, the field at a field point $P$ in its vicinity will have an additional contribution resulting from the *magnetization* of the material. We'll return to this point in Section 29–8. However, unless the material is iron or some other ferromagnetic material, the additional field is so small that it is usually negligible. Additional complications arise if time-varying electric or magnetic fields are present or if the material is a superconductor; we'll return to these topics later.

## PROBLEM-SOLVING STRATEGY

### Magnetic-field calculations

**1.** Be careful about the directions of vector quantities. The current element $d\mathbf{l}$ always points in the direction of the current. The unit vector $\hat{\mathbf{r}}$ is always directed *from* the current element *toward* the point $P$ at which the field is to be determined, that is, from the source point toward the field point.

**2.** In some situations the $d\mathbf{B}$'s at point $P$ have the same direction for all the current elements. In that case the magnitude of the total field $\mathbf{B}$ is the sum of the magnitudes of the $d\mathbf{B}$'s. But often the $d\mathbf{B}$'s have different directions for different current elements. Then you have to set up a coordinate system and represent each $d\mathbf{B}$ in terms of its components. The integral for the total field $\mathbf{B}$ is then expressed in terms of an integral for each component. Sometimes you can use the symmetry of the situation to prove that one component must vanish. Always be alert for ways to use symmetry to simplify the problem.

**3.** Look for ways to use the superposition principle. If you know the fields produced by certain simple conductor shapes, and if you encounter a complex shape that can be represented as a combination of simple shapes, then you can use superposition to find the field of the complex shape. Examples are a rectangular loop and a semicircle with straight-line segments at both ends.

## EXAMPLE 29–2

A copper wire carries a steady current of 125 A to an electroplating tank. Find the magnetic field caused by a 1.0-cm segment of this wire, at a point 1.2 m away from it, if the point is a) straight out to the side of the segment; b) on a line at 30° to the segment, as shown in Fig. 29–4.

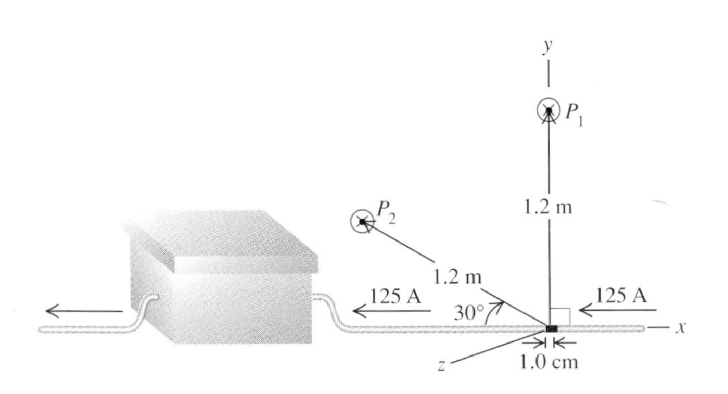


**29–4** Finding the magnetic field at two points due to a 1.0-cm segment of current-carrying wire.

**SOLUTION** a) From the right-hand rule the direction of $\mathbf{B}$ at each point is *into* the plane of the page. Or, using unit vectors, we note that $d\mathbf{l} = dl(-\mathbf{i})$. At point $P_1$, $\hat{\mathbf{r}} = \mathbf{j}$, so in Eq. (29–6),

$$d\mathbf{l} \times \hat{\mathbf{r}} = dl(-\mathbf{i}) \times \mathbf{j} = dl(-\mathbf{k}).$$

The negative $z$-direction is *into* the page.

To find the magnitude of $B$, we use Eq. (29–5). At point $P_1$,

$$B = \frac{\mu_0}{4\pi}\frac{I\,dl\,\sin\phi}{r^2}$$

$$= (10^{-7}\,\text{T}\cdot\text{m/A})\frac{(125\text{ A})(1.0\times10^{-2}\text{ m})(\sin 90°)}{(1.2\text{ m})^2}$$

$$= 8.7\times10^{-8}\text{ T}.$$

b) At point $P_2$,

$$B = (10^{-7}\,\text{T}\cdot\text{m/A})\frac{(125\text{ A})(1.0\times10^{-2}\text{ m})(\sin 30°)}{(1.2\text{ m})^2}$$

$$= 4.3\times10^{-8}\text{ T}.$$

Note that these magnetic-field magnitudes are very small in comparison to the typical values quoted in Section 28–2. Also note that the values are not the *total* fields at these points, but only the contributions from the short segment of conductor described.

# 29–3 MAGNETIC FIELD OF A STRAIGHT CONDUCTOR

An important application of the law of Biot and Savart is finding the magnetic field produced by a straight conductor with length $2a$ carrying a current $I$, at a point on its perpendicular bisector, at a distance $x$ from the conductor. The situation is shown in Fig. 29–5; the geometry is similar to the electric-field problem of Example 22–11 (Section 22–6), in which we found the electric field caused by a line of charge with length $2a$. The same integral appears in these two problems, but the behavior of the magnetic field is completely different from that of the electric field.

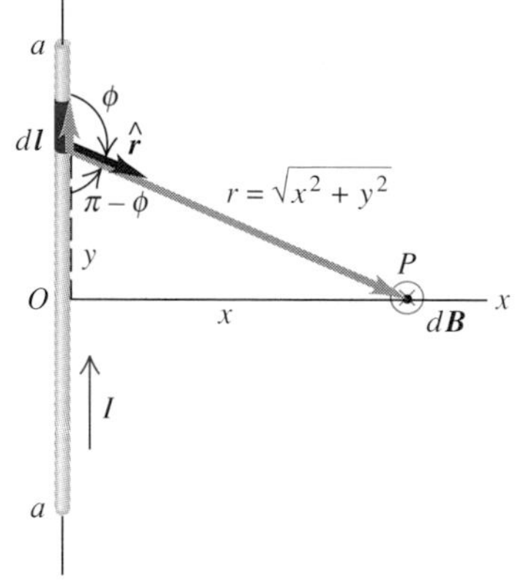


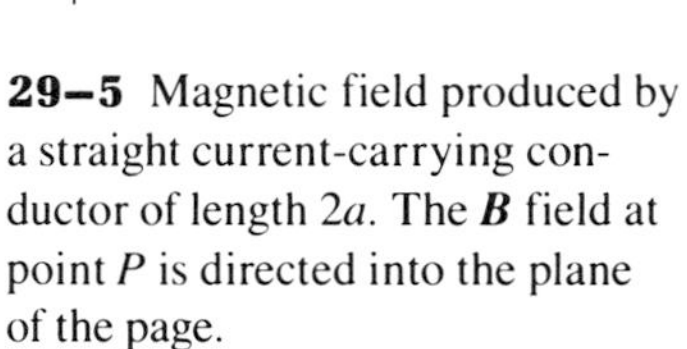

**29–5** Magnetic field produced by a straight current-carrying conductor of length $2a$. The $\boldsymbol{B}$ field at point $P$ is directed into the plane of the page.

We first use the law of Biot and Savart, Eq. (29–5), to find the field $d\boldsymbol{B}$ caused by the element of conductor $dl = dy$ shown in the figure. From the figure, $r = \sqrt{x^2 + y^2}$ and $\sin\phi = \sin(\pi - \phi) = x/\sqrt{x^2+y^2}$. From the right-hand rule or the vector product $d\boldsymbol{l} \times \boldsymbol{r}$, the *direction* of $d\boldsymbol{B}$ is perpendicular to the plane of the figure, into the plane. In this case the directions of the $d\boldsymbol{B}$'s from *all* elements of the conductor are the same. Thus, in integrating Eq. (29–5), we can just add the *magnitudes* of the $d\boldsymbol{B}$'s, a significant simplification.

Putting the pieces together, we find that the magnitude $B$ of the total $\boldsymbol{B}$ is

$$B = \frac{\mu_0 I}{4\pi}\int_{-a}^{a} \frac{x\,dy}{(x^2+y^2)^{3/2}}.$$

We can integrate this by trigonometric substitution or by using an integral table. The final result is

$$B = \frac{\mu_0 I}{4\pi}\frac{2a}{x\sqrt{x^2+a^2}}. \qquad (29\text{–}8)$$

When the length $(2a)$ of the conductor is very great in comparison to its distance $x$ from point $P$, we can consider it to be infinitely long. When $a$ is much larger than $x$, $\sqrt{x^2+a^2}$ is approximately equal to $a$, and so in the limit as $a \to \infty$, Eq. (29–8) becomes

$$B = \frac{\mu_0 I}{2\pi x}.$$

The physical situation has axial symmetry about the $y$-axis, so $\boldsymbol{B}$ must have the same *magnitude* at all points on a circle centered on the conductor and lying in a plane perpendicular to the conductor, and its *direction* is everywhere tangent to such a circle. Thus, at all points on a circle of radius $r$ around the conductor, the magnitude $B$ is given by

$$B = \frac{\mu_0 I}{2\pi r} \quad \text{(a long, straight wire).} \qquad (29\text{–}9)$$

## EXAMPLE 29–3

A long, straight conductor carries a current of 100 A. At what distance from the conductor is the magnetic field caused by the current equal in magnitude to the earth's magnetic field in Pittsburgh (about $0.5 \times 10^{-4}$ T)?

**SOLUTION** We use Eq. (29–9). Everything except $r$ is known, so we solve for $r$ and insert the appropriate numbers:

$$r = \frac{\mu_0 I}{2\pi B} = \frac{(4\pi \times 10^{-7}\,\mathrm{T\cdot m/A})(100\,\mathrm{A})}{(2\pi)(0.5 \times 10^{-4}\,\mathrm{T})} = 0.4\ \mathrm{m}.$$

At smaller distances the field becomes stronger; for example, when $r = 0.2$ m, $B = 1.0 \times 10^{-4}$ T, and so on.

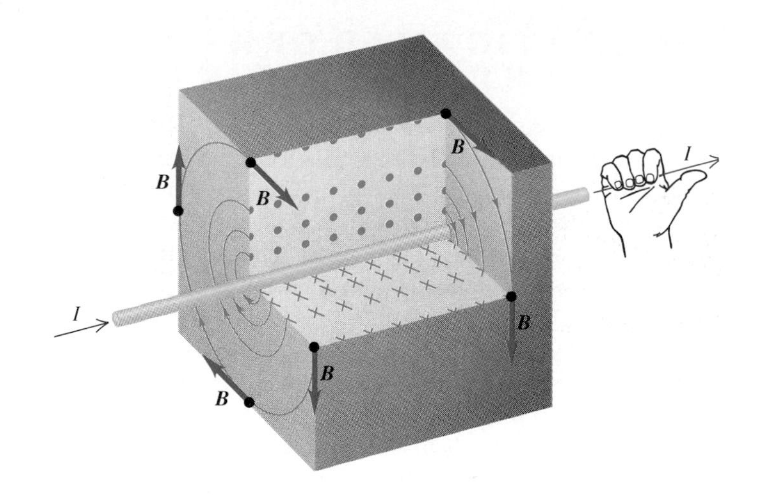


**29–6** Magnetic field around a long, straight conductor. The field lines are circles, with directions determined by the right-hand rule.

Part of the magnetic field around a long, straight conductor is shown in Fig. 29–6. The configuration of the magnetic field lines in this situation is completely different from that of the electric field lines in the analogous electrical situation. Electric field lines radiate outward from a positive line charge distribution (inward for negative charges). By contrast, these magnetic field lines *encircle* the current that acts as their source. Electric field lines begin and end at charges, but magnetic field lines are always continuous loops and *never* have end points, irrespective of the shape of the conductor that sets up the field.

We discussed this property of magnetic fields in Section 28–3, along with the fact that the total magnetic flux out of a closed surface is always zero. Gauss's law for magnetic fields,

$$\oint \mathbf{B} \cdot d\mathbf{A} = 0 \quad \text{(any closed surface)}, \tag{29–10}$$

expresses the fact that there are no isolated magnetic charges or magnetic monopoles. The number of magnetic field lines emerging from any closed surface must equal the number entering it.

## 29–4 FORCE BETWEEN PARALLEL CONDUCTORS

The interaction force between two long, current-carrying conductors is important in a variety of practical problems, and it also has fundamental significance in connection with the definition of the ampere. Figure 29–7 shows segments of two long, straight, parallel conductors separated by a distance $r$ and carrying currents $I$ and $I'$, respectively, in the same direction. Each conductor lies in the magnetic field set up by the other, so each experiences a force. The diagram shows some of the field lines set up by the current in the lower conductor.

From Eq. (29–9) the magnitude of the $\mathbf{B}$ vector at the upper conductor is

$$B = \frac{\mu_0 I}{2\pi r}.$$

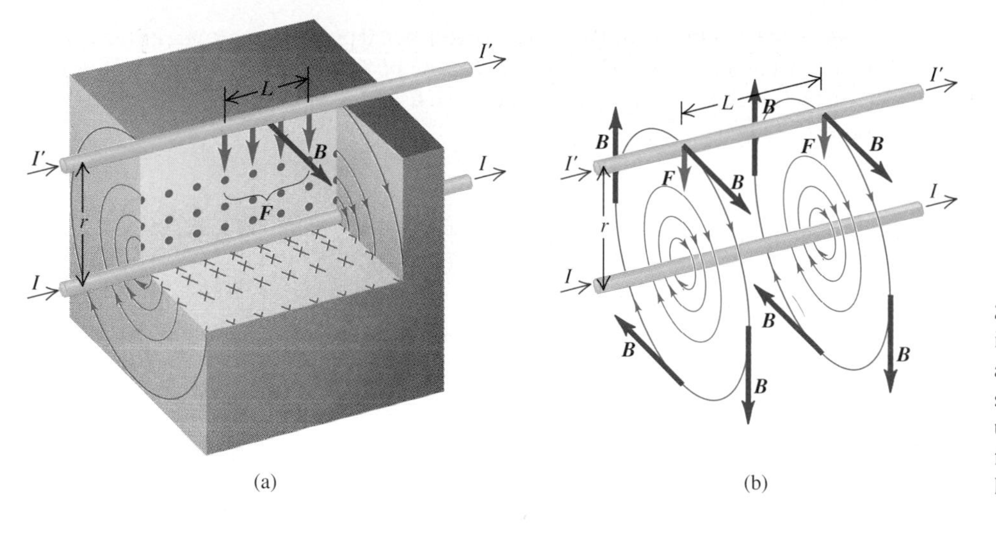


**29–7** Parallel conductors carrying currents in the same direction attract each other. The diagrams show the force exerted on the upper conductor by the magnetic field caused by the current in the lower conductor.

From Eq. (28–17) the force on a length $L$ of the upper conductor is

$$F = I'LB = \frac{\mu_0 II'L}{2\pi r},$$

and the force *per unit length* $F/L$ is

$$\frac{F}{L} = \frac{\mu_0 II'}{2\pi r}. \qquad (29\text{–}11)$$

The right-hand rule shows that the direction of the force on the upper conductor is *downward*.

The current in the upper conductor also sets up a field at the position of the lower one. Two successive applications of the right-hand rule show that the force on the lower conductor is *upward*. Thus two parallel conductors carrying current in the same direction *attract* each other. If the direction of either current is reversed, the forces also reverse. Parallel conductors carrying currents in *opposite* directions *repel* each other.

The fact that two straight, parallel conductors exert forces of attraction or repulsion on one another is the basis of the official SI definition of the **ampere:**

> **One ampere is that unvarying current which, if present in each of two parallel conductors of infinite length and one meter apart in empty space, causes each conductor to experience a force of exactly $2 \times 10^{-7}$ newtons per meter of length.**

This is an *operational definition;* it gives us an actual experimental procedure for measuring current and defining a unit of current. In principle, we could use this definition to calibrate an ammeter, using only a meter stick and a spring balance. For high-precision standardization of the ampere, coils of wire are used instead of straight wires, and their separation is made only a few centimeters. The complete instrument, which is capable of measuring currents with a high degree of precision, is called a *current balance*. This definition also forms the basis of the SI definition of the coulumb as the amount of charge transferred in one second by a current of one ampere.

Mutual forces of attraction exist not only between *wires* carrying currents in the same direction, but also between the longitudinal elements of a single current-carrying conductor. If the conductor is a liquid or an ionized gas (a plasma), these forces result in a constriction of the conductor, as if its surface were acted on by an external, inward pressure. The constriction of the conductor is called the *pinch effect*. The high temperature produced by the pinch effect in a plasma has been used in one technique to bring about nuclear fusion.

## EXAMPLE 29–4

Two straight, parallel, superconducting cables 4.5 mm apart carry equal currents of 15,000 A in opposite directions. Should we worry about the mechanical strength of these wires?

**SOLUTION** Because the currents are in opposite directions, the two conductors repel each other. From Eq. (29–11) the force per unit length is

$$\frac{F}{L} = \frac{\mu_0 II'}{2\pi r} = \frac{(4\pi \times 10^{-7}\,\mathrm{T\cdot m/A})(15{,}000\,\mathrm{A})^2}{(2\pi)(4.5 \times 10^{-3}\,\mathrm{m})}$$
$$= 1.0 \times 10^4\,\mathrm{N/m}.$$

This is a large force, something over one ton per meter, so mechanical strength of the conductors and insulating materials is certainly a significant consideration. Currents and separations of this magnitude are used in superconducting electromagnets in particle accelerators, and mechanical stress analysis is a crucial part of the design process.

## EXAMPLE 29–5

**Mechanical stress in a solenoid** Discuss the directions of the magnetic forces in a solenoid a) between adjacent turns of the coil; b) between opposite sides of the same turn.

**SOLUTION** a) Adjacent turns are carrying parallel currents and therefore attract one another, causing an axial, *compressive* stress.

b) Opposite sides of the same turn carry current in opposite directions and thus repel each other. Therefore the coil experiences forces that pull radially outward, creating *tensile* stress in the conductors. These forces are proportional to the square of the current; as mentioned above, they are an important consideration in electromagnet design.

## 29–5 MAGNETIC FIELD OF A CIRCULAR LOOP

If you look inside a doorbell, a relay, a transformer, or an electric motor, you will find coils of wire, often consisting of many circular loops. A current in such a coil is used to establish a magnetic field. So it is worthwhile to derive an expression for the magnetic field produced by a single circular conducting loop carrying a current or by $N$ closely spaced circular loops forming a coil.

Figure 29–8 shows a circular conductor with radius $a$, carrying a current $I$. The current is led into and out of the loop through two long, straight wires side by side; the

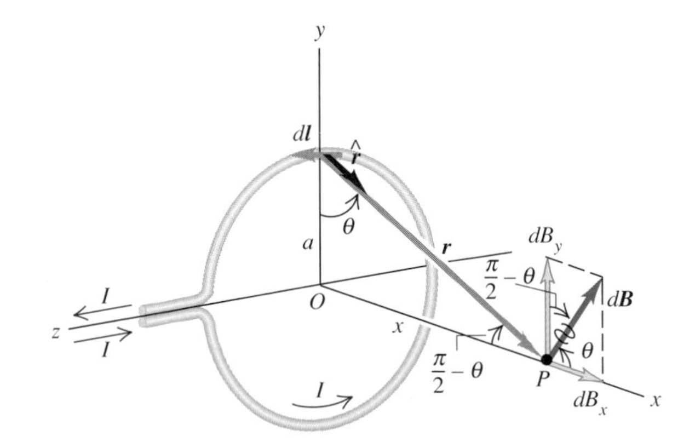


**29–8** Magnetic field of a circular loop. The segment $\boldsymbol{dl}$ causes the field $\boldsymbol{dB}$, lying in the $xy$-plane. Other $\boldsymbol{dl}$'s have different components perpendicular to the $x$-axis; these add to zero, and the $x$-components combine to give the total $\boldsymbol{B}$ field at point $P$.

currents in these straight wires are in opposite directions, and their magnetic fields very nearly cancel each other.

We can use the law of Biot and Savart, Eq. (29–5) or (29–6), to find the magnetic field at a point $P$ on the axis of the loop, at a distance $x$ from the center. As the figure shows, $\boldsymbol{dl}$ and $\hat{\boldsymbol{r}}$ are perpendicular, and the direction of the field $\boldsymbol{dB}$ caused by this particular element $\boldsymbol{dl}$ lies in the $xy$-plane. Also, $r^2 = x^2 + a^2$. The magnitude $dB$ of the field due to element $\boldsymbol{dl}$ is

$$dB = \frac{\mu_0 I}{4\pi} \frac{dl}{(x^2 + a^2)}. \tag{29–12}$$

The components of the vector $\boldsymbol{dB}$ are

$$dB_x = dB\cos\theta = \frac{\mu_0 I}{4\pi} \frac{dl}{(x^2 + a^2)} \frac{a}{(x^2 + a^2)^{1/2}}, \tag{29–13}$$

$$dB_y = dB\sin\theta = \frac{\mu_0 I}{4\pi} \frac{dl}{(x^2 + a^2)} \frac{x}{(x^2 + a^2)^{1/2}}. \tag{29–14}$$

Because the situation has rotational symmetry about the $x$-axis, there cannot be a component of $\boldsymbol{B}$ perpendicular to that axis. For every element $\boldsymbol{dl}$ there is a corresponding element on the opposite side of the loop with opposite direction. These two elements give equal contributions to the $x$-component of the field, given by Eq. (29–13), but *opposite* components perpendicular to the $x$-axis. Thus all the perpendicular components cancel, and only the $x$-components survive.

To obtain the *total* $x$-component, we integrate Eq. (29–13), including all the $\boldsymbol{dl}$'s around the loop. Everything in this expression except $dl$ is constant and can be taken outside the integral, so we have

$$B_x = \int \frac{\mu_0 I}{4\pi} \frac{a\,dl}{(x^2 + a^2)^{3/2}} = \frac{\mu_0 I a}{4\pi(x^2 + a^2)^{3/2}} \int dl.$$

The integral of $dl$ is just the circumference of the circle, $\int dl = 2\pi a$, and we finally get

$$B_x = \frac{\mu_0 I a^2}{2(x^2 + a^2)^{3/2}} \quad \text{(circular loop).} \tag{29–15}$$

Now suppose that instead of the single loop in Fig. 29–8 we have a coil consisting of $N$ closely spaced loops, all with the same radius. Each loop contributes equally to the field, and the total field is $N$ times the field of a single loop:

$$B_x = \frac{\mu_0 N I a^2}{2(x^2 + a^2)^{3/2}} \quad (N \text{ circular loops}). \tag{29–16}$$

At the center of the loop or loops, $x = 0$, and Eq. (29–16) reduces to

$$B_x = \frac{\mu_0 N I}{2a} \quad (\text{center of } N \text{ circular loops}). \tag{29–17}$$

As we go out along the axis, the field decreases in magnitude. Figure 29–9 shows a graph of $B_x$ as a function of $x$.

## EXAMPLE 29–6

A long coil consisting of 100 circular loops, with radius 0.60 m, carries a current of 5.0 A. a) Find the magnetic field at a point along the axis, 0.80 m from the center. b) At what distance from the center, along the axis, is the field magnitude one-eighth as great as at the center?

**SOLUTION** a) In Eq. (29–16) we have $I = 5.0$ A, $a = 0.60$ m, and $x = 0.80$ m. We find

$$B_x = \frac{(4\pi \times 10^{-7}\,\text{T}\cdot\text{m/A})(100)(5.0\,\text{A})(0.60\,\text{m})^2}{2[(0.80\,\text{m})^2 + (0.60\,\text{m})^2]^{3/2}} = 1.1 \times 10^{-4}\,\text{T}.$$

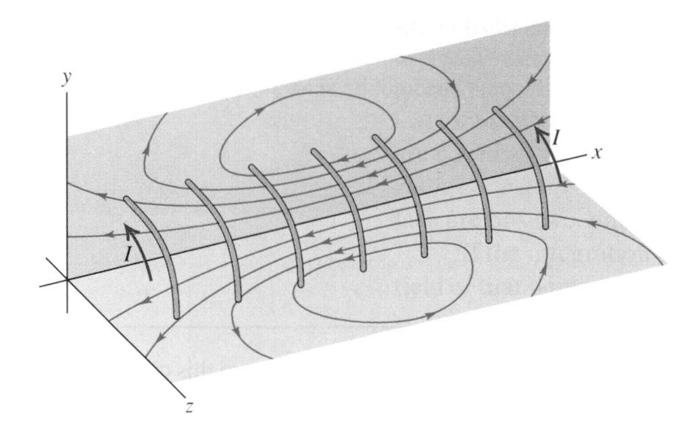


**29–14** Magnetic field lines produced by the current in a solenoid. For clarity, only a few turns are shown.

We can use Ampere's law to find the field at or near the center. We choose as our integration path the rectangle *abcd* in Fig. 29–15. Side *ab*, with length $L$, is parallel to the axis of the solenoid. Sides *bc* and *da* are taken to be very long, so side *cd* is far from the solenoid, and the field at side *cd* is negligibly small.

By symmetry the $\boldsymbol{B}$ field along side *ab* is parallel to this side and is constant, so for side *ab*, $B_{\parallel} = B$ and

$$\int \boldsymbol{B} \cdot d\boldsymbol{l} = BL.$$

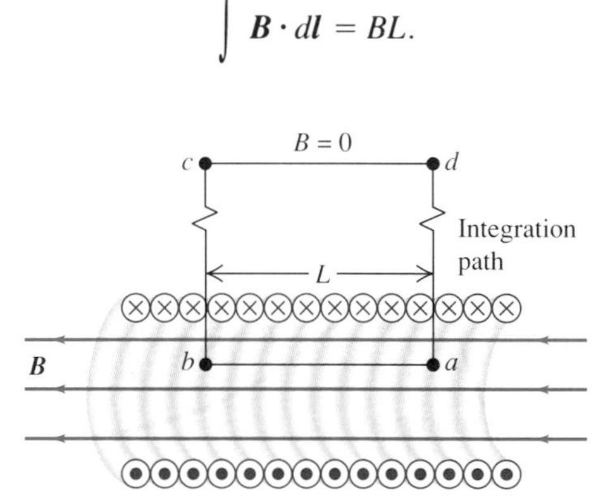


**29–15** Using Ampere's law to find the magnetic field near the center of a long, tightly wound solenoid.

Along sides *bc* and *da*, $B_{\parallel} = 0$ because $B$ is perpendicular to these sides; and along side *cd*, $B_{\parallel} = 0$ because $B = 0$. The integral $\oint \boldsymbol{B} \cdot d\boldsymbol{l}$ around the entire closed path therefore reduces to $BL$.

Let $n$ be the number of turns *per unit length* in the windings. The number of turns in length $L$ is then $nL$. Each of these turns passes once through the rectangle *abcd* and carries a current $I$, where $I$ is the current in the windings. The total current enclosed by the rectangle is then $I_{\text{encl}} = nLI$. From Ampere's law,

$$BL = \mu_0 nLI,$$

$$B = \mu_0 nI \quad \text{(solenoid).} \tag{29–21}$$

Side *ab* need not lie on the axis of the solenoid, so this calculation also proves that at the center the field is uniform over the entire cross section.

For points along the axis the field is strongest at the center of the solenoid and drops off near the ends. For a solenoid that is very long in comparison to its diameter, the field at each end is exactly half as strong as at the center. For a short, fat solenoid the relationship is more complicated. Figure 29–16 shows a graph of $B$ as a function of $x$ for points on the axis of a short solenoid.

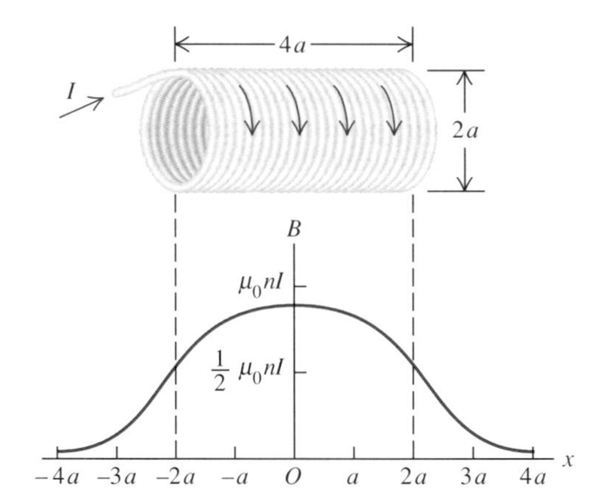


**29–16** Magnitude of the magnetic field at points along the axis of a solenoid with length $4a$, equal to four times its radius $a$. The field magnitude at each end is about half its value at the center. (Compare with Fig. 29–9 for the field of $N$ circular loops.)

## EXAMPLE 29–10

**Field of a toroidal solenoid** Figure 29–17a shows a *toroidal* (doughnut-shaped) *solenoid* wound with wire carrying a current $I$. In a practical version the turns would be more closely spaced. The black lines in Fig. 29–17b are integration paths that we want to use in applying Ampere's law. Consider path 1 first. By symmetry, if the toroidal solenoid produces any field at all in this region, it must be *tangent* to the path at all points, and $\oint \boldsymbol{B} \cdot d\boldsymbol{l}$ will equal the product of $B$ and the circumference $l = 2\pi r$ of the path. But the total current enclosed by the path is zero, so from Ampere's law the field $\boldsymbol{B}$ must be zero everywhere.

Similarly, if the toroidal solenoid produces any field at path 3, it must also be tangent to the path at all points. Each turn of the winding passes *twice* through the area bounded by this path, carrying equal currents in opposite directions. The *net* current $I_{\text{encl}}$ enclosed within this area is therefore zero, and hence $B = 0$ at all points of the path. Conclusion: *The field of the toroidal solenoid is confined completely to the space enclosed by the windings.* We can think of a toroidal solenoid as a solenoid that has been bent into a circle.

Finally, we consider path 2, a circle with radius $r$. Again by symmetry we expect the $\boldsymbol{B}$ field to be tangent to the path, and $\oint \boldsymbol{B} \cdot d\boldsymbol{l}$ equals $2\pi rB$. Each turn of the winding passes *once* through the area bounded by path 2. The total current enclosed by the path is $I_{\text{encl}} = NI$, where $N$ is the *total* number of turns in the winding. Then, from Ampere's law,

$$2\pi rB = \mu_0 NI,$$

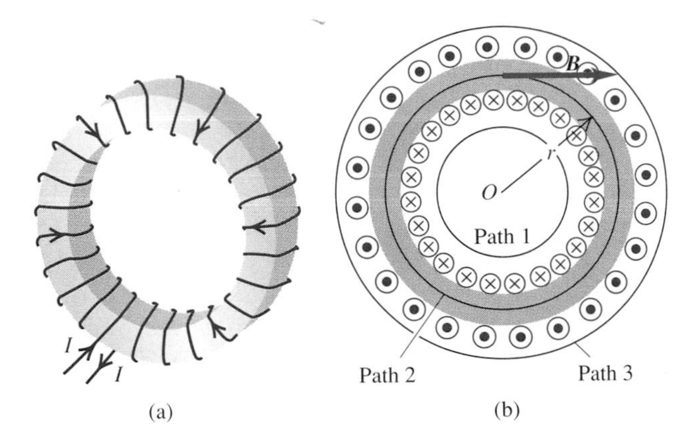


**29–17** (a) A toroidal solenoid. For clarity, only a few turns of the winding are shown. (b) Closed paths (black circles) used to compute the magnetic field $\boldsymbol{B}$ are set up by a current in a toroidal solenoid. The field is very nearly zero at all points except those within the space enclosed by the windings.

or

$$B = \frac{\mu_0 NI}{2\pi r} \quad \text{(toroidal solenoid).} \tag{29–22}$$

The magnetic field is *not* uniform over a cross section of the core because the radius $r$ is larger at the outer side of the section than at the inner side. However, if the radial thickness of the core is small in comparison to $r$, the field varies only slightly across a section. In that case, considering that $2\pi r$ is the circumferential length of the toroid and that $N/2\pi r$ is the number of turns per unit length $n$, the field may be written

$$B = \mu_0 nI,$$

just as at the center of a long, *straight* solenoid.

In a real toroidal solenoid the turns are not precisely circular loops but rather segments of a bent helix. As a result, the field outside is not strictly zero. To estimate its magnitude, we imagine Fig. 29–17a as roughly equivalent, for points outside the torus, to a circular loop with a single turn and radius $r$. Then we can show that the field at the *center* of the torus is smaller than the field inside by approximately a factor of $N/\pi$.

The equations derived above for the field in a closely wound straight or toroidal solenoid are strictly correct only for windings in *vacuum*. For most practical purposes, however, they can be used for windings in air or on a core of any nonferromagnetic, nonsuperconducting material. We will show in the next section how they are modified if the core is a ferromagnetic material. ■

## *29–8 MAGNETIC MATERIALS

In discussing how currents cause magnetic fields, we have assumed that the conductors are surrounded by vacuum. But transformers, motors, generators, and electromagnets nearly always have coils with iron cores to increase the magnetic field and confine it to desired regions. Permanent magnets, magnetic recording tapes, and computer disks depend directly on the magnetic properties of materials; when you store information on a computer disk, you are actually creating an array of microscopic permanent magnets on the disk. So it is worthwhile to examine some aspects of magnetic properties of materials.

If matter is present in the space surrounding a current-carrying conductor, the magnetic field is different from the case in which only vacuum surrounds the conductor. The atoms that make up all matter contain moving electrons, and these electrons form microscopic current loops that produce magnetic fields of their own. In many materials these currents are randomly oriented and cause no net magnetic field. But in some materials an external field (a field produced by currents outside the material) can cause these loops to become oriented preferentially with the field so that their magnetic fields *add* to the external field. We then say that the material is *magnetized.*

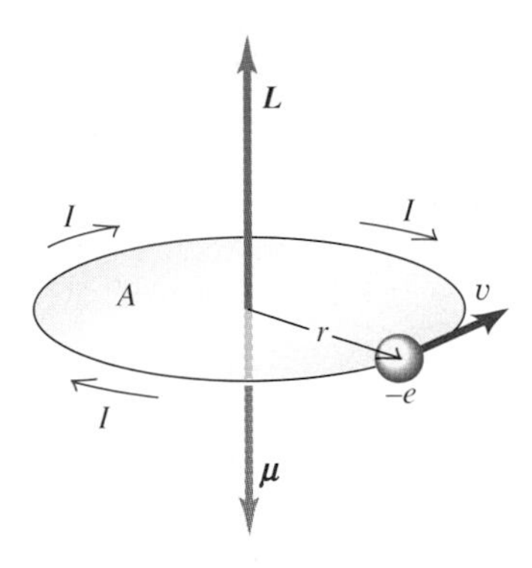


**29–18** An electron moving with speed $v$ in a circular orbit of radius $r$ has an angular momentum $\boldsymbol{L}$ and an oppositely directed orbital magnetic dipole moment $\boldsymbol{\mu}$. It also has a spin angular momentum and an oppositely directed spin magnetic dipole moment.

### The Bohr Magneton

Let's look at how these microscopic currents come about. Figure 29–18 shows a primitive classical model of an electron in an atom. We picture the electron (mass $m$, charge $-e$) as moving in a circular orbit with radius $r$ and speed $v$. This moving charge is equivalent to a current loop; in Section 28–7 we found that a current loop with area $A$ and current $I$ has a magnetic dipole moment $\mu$ given by $\mu = IA$. For the orbiting electron the area of the loop is $A = \pi r^2$. To find the current, we note that the number of revolutions that an electron makes per unit time is its speed divided by the circumference, $v/2\pi r$. So the total charge passing any point on the orbit per unit time is this quantity multiplied by the magnitude of the electron charge $e$. We identify this product as the equivalent current $I$ in the loop:

$$I = \frac{ev}{2\pi r}.$$

The magnetic moment $\mu = IA$ is then

$$\mu = \frac{ev}{2\pi r}(\pi r^2) = \frac{evr}{2}. \qquad (29\text{–}23)$$

It is useful to express $\mu$ in terms of the *angular momentum* $L$ of the electron. For a particle moving in a circular path the magnitude of angular momentum equals the magnitude of momentum $mv$ multiplied by the radius $r$; that is, $L = mvr$ (Section 10–5). Comparing this with Eq. (29–23), we can write

$$\mu = \frac{e}{2m}L. \qquad (29\text{–}24)$$

The reason Eq. (29–24) is useful in this discussion is that atomic angular momentum is *quantized;* its component in a particular direction is always an integer multiple of $h/2\pi$, where $h$ is a fundamental physical constant called *Planck's constant.* The numerical value of $h$ is

$$h = 6.626 \times 10^{-34}\ \text{J} \cdot \text{s}.$$

The quantity $h/2\pi$ represents a fundamental unit of angular momentum in atomic systems, just as $e$ is a fundamental unit of charge. Associated with quantization of $\boldsymbol{L}$ is a fundamental uncertainty in the *direction* of $\boldsymbol{L}$ and therefore of $\boldsymbol{\mu}$. In the following discussion, when we speak of the magnitude of a magnetic moment, a more precise statement would be "maximum component in a given direction." Similarly, to say that a magnetic moment $\boldsymbol{\mu}$ is aligned with a magnetic field $\boldsymbol{B}$ really means that $\boldsymbol{\mu}$ has its maximum possible component in the direction of $\boldsymbol{B}$; such components are always quantized.

Equation (29–24) shows that associated with the fundamental unit of angular momentum is a corresponding fundamental unit of magnetic moment. If $L = h/2\pi$, then

$$\mu = \frac{e}{2m}\left(\frac{h}{2\pi}\right) = \frac{eh}{4\pi m}. \qquad (29\text{–}25)$$

This quantity is called the **Bohr magneton,** denoted by $\mu_B$. Its numerical value is

$$\mu_B = 9.274 \times 10^{-24}\ \text{A} \cdot \text{m}^2 = 9.274 \times 10^{-24}\ \text{J/T}.$$

(We invite you to verify that these two sets of units are equivalent. The second set is useful when we compute the potential energy $U = -\boldsymbol{\mu} \cdot \boldsymbol{B}$ for a magnetic moment in a magnetic field.)

Electrons also have an intrinsic angular momentum, called *spin,* that is not related to orbital motion but that can be pictured in a classical model as spinning on an axis. This angular momentum also has an associated magnetic moment, and its magnitude turns out to be almost exactly one Bohr magneton. (Effects having to do with quantization of the electromagnetic field cause the spin magnetic moment to be about $1.001\mu_B$.)

## Paramagnetism

In an atom, most of the various orbital and spin magnetic moments of the electrons add up to zero. However, in some cases the atom has a net magnetic moment that is of the order of $\mu_B$. When such a material is placed in a magnetic field, the field exerts a torque on each magnetic moment, as given by Eq. (28–25): $\boldsymbol{\tau} = \boldsymbol{\mu} \times \boldsymbol{B}$. These torques tend to align the magnetic moments with the field, their position of minimum potential energy, as we discussed in Section 28–7. In this position the directions of the current loops are such as to *add* to the externally applied magnetic field.

It turns out that the additional $\boldsymbol{B}$ field produced by these microscopic electron current loops is proportional to the total magnetic moment $\boldsymbol{\mu}_{tot}$ per unit volume $V$ in the material. We call this vector quantity the **magnetization** of the material, denoted by $\boldsymbol{M}$:

$$\boldsymbol{M} = \frac{\boldsymbol{\mu}_{tot}}{V}. \tag{29–26}$$

The additional magnetic field due to magnetization of the material turns out to be equal simply to $\mu_0 \boldsymbol{M}$, where $\mu_0$ is the same constant that appears in the law of Biot and Savart and Ampere's law. When a material completely surrounds a conductor, the total magnetic field $\boldsymbol{B}$ in the material is

$$\boldsymbol{B} = \boldsymbol{B}_0 + \mu_0 \boldsymbol{M}, \tag{29–27}$$

where $\boldsymbol{B}_0$ is the field caused by the external current distribution.

A material showing the behavior just described is said to be **paramagnetic.** The result is that the magnetic field at any point in such a material is greater by a factor $K_m$, called the **relative permeability** of the material, than it would be if the material were replaced by vacuum. The value of $K_m$ is different for different materials; for common paramagnetic solids and liquids at room temperature, $K_m$ is typically 1.00001 to 1.003.

All the equations in this chapter that relate magnetic fields to their sources can be adapted to the situation in which the conductor is embedded in a paramagnetic material by simply replacing $\mu_0$ everywhere by $K_m \mu_0$. This product is usually denoted as $\mu$ and is called the **permeability** of the material:

$$\mu = K_m \mu_0. \tag{29–28}$$

This relation involves some really dangerous notation because we have also used $\mu$ for magnetic moment. It is customary to use $\mu$ for both quantities, but beware; from now on, every time you see a $\mu$, make sure you know whether it is permeability or magnetic moment. You can usually tell from the context.

The amount by which the relative permeability differs from unity is called the **magnetic susceptibility,** denoted by $\chi_m$:

$$\chi_m = K_m - 1. \tag{29–29}$$

Both $K_m$ and $\chi_m$ are dimensionless quantities. Values of magnetic susceptibility for several materials are given in Table 29–1.

The tendency of atomic magnetic moments to align themselves parallel to the magnetic field (where they have minimum potential energy) is opposed by random thermal motion, which tends to randomize their orientations. For this reason, paramagnetic susceptibility always decreases with increasing temperature. In many cases it is inversely proportional to the absolute temperature $T$, and the magnetization $M$ can be expressed as

$$M = C\frac{B}{T}. \tag{29–30}$$

This relation is called *Curie's law,* after its discoverer, Pierre Curie (1859–1906). The quantity $C$ is a constant, different for different materials, called the *Curie constant.*

**TABLE 29–1**

**Magnetic Susceptibilities of Paramagnetic and Diamagnetic Materials at $T$ = 20°C**

| Material | $\chi_m = K_m - 1$ ($\times 10^{-5}$) |
|---|---|
| Paramagnetic | |
| Iron ammonium alum | 66 |
| Uranium | 40 |
| Platinum | 26 |
| Aluminum | 2.2 |
| Sodium | 0.72 |
| Oxygen gas | 0.19 |
| Diamagnetic | |
| Bismuth | −16.6 |
| Mercury | −2.9 |
| Silver | −2.6 |
| Carbon (diamond) | −2.1 |
| Lead | −1.8 |
| Sodium chloride | −1.4 |
| Copper | −1.0 |

## EXAMPLE 29–11

For Eq. (29–27), show that the units are consistent.

**SOLUTION** Magnetization $\boldsymbol{M}$ is magnetic moment per unit volume. The units of magnetic moment are current times area ($A \cdot m^2$), so the units of magnetization are $(A \cdot m^2)/m^3 = A/m$. From Section 29–1 the units of the constant $\mu_0$ are $T \cdot m/A$. So the units of $\mu_0 \boldsymbol{M}$ are

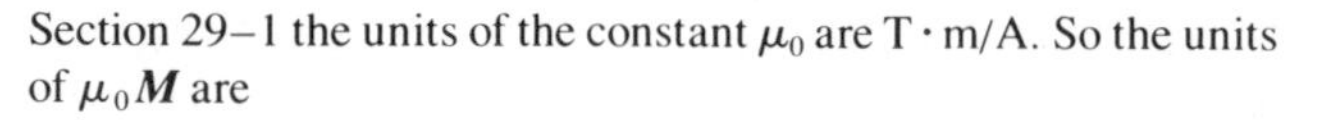

$$(T \cdot m/A)(A/m) = T,$$

also the unit of $B$.

## EXAMPLE 29–12

Nitric oxide (NO) is a paramagnetic compound whose molecules have a magnetic moment of one Bohr magneton each. In a magnetic field with magnitude $B = 1.5$ T, compare the interaction energy of the magnetic moments with the field to the average translational kinetic energy of the molecules at a temperature of 300 K.

**SOLUTION** In Section 28–7 we derived an expression for the potential energy of a magnetic moment in a $\boldsymbol{B}$ field: $U = -\boldsymbol{\mu} \cdot \boldsymbol{B} = -\mu B \cos\phi$. (This $\mu$ is magnetic moment.) In this case the maximum magnitude of this energy (occurring when $\cos\phi = \pm 1$) is

$$U_{\max} = \mu_B \cdot B = (9.27 \times 10^{-24}\ \text{J/T})(1.5\ \text{T}) = 1.4 \times 10^{-23}\ \text{J} = 8.7 \times 10^{-5}\ \text{eV}.$$

The average translational kinetic energy $K$ is given by the equipartition principle (Section 16–4):

$$K = \tfrac{3}{2}kT = \tfrac{3}{2}(1.38 \times 10^{-23}\ \text{J/K})(300\ \text{K}) = 6.2 \times 10^{-21}\ \text{J} = 0.039\ \text{eV}.$$

At a temperature of 300 K the magnetic interaction energy is much *smaller* than the random kinetic energy, so we expect only a slight degree of alignment. This is why paramagnetic susceptibilities at ordinary temperatures are usually very small.

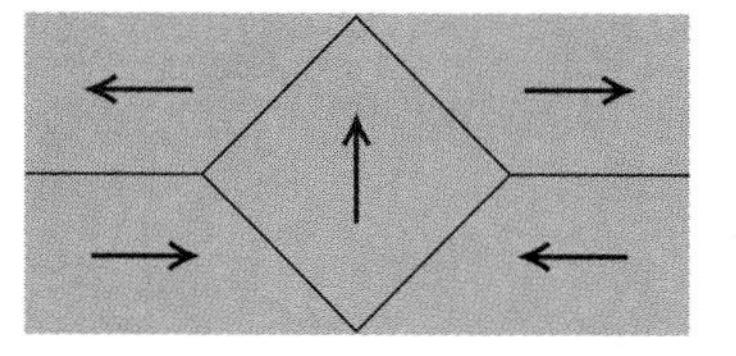

(a) No field

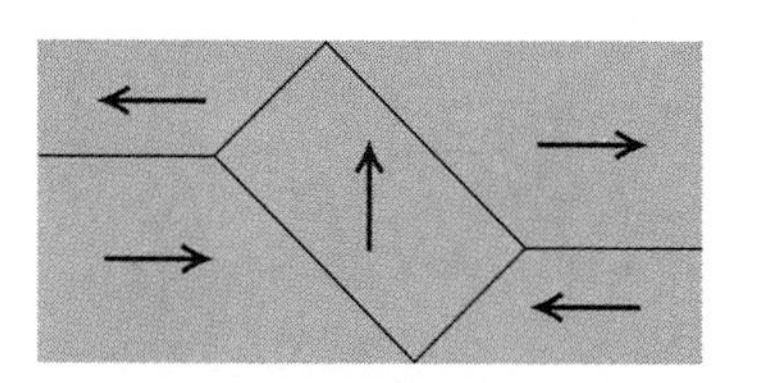

(b) Weak field

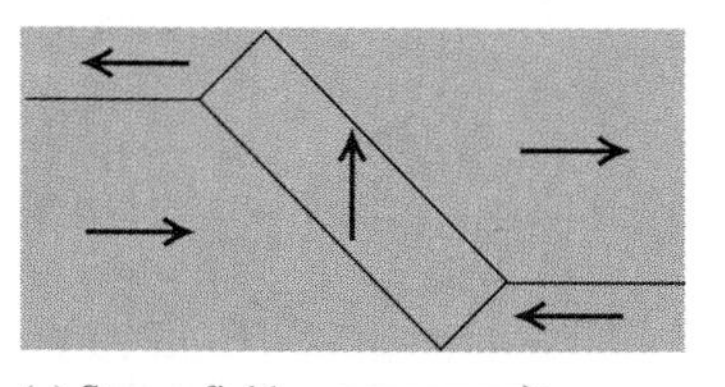

(c) Strong field

**29–19** In this drawing adapted from a magnified photo, the arrows show the directions of magnetization in the domains of a single crystal of nickel. Domains magnetized in the direction of an applied magnetic field grow larger.

### Diamagnetism

In some materials the total magnetic moment of all the atomic current loops is zero when there is no magnetic field present. But even these materials have magnetic effects because an external field alters electron motions within the atoms, causing additional current loops and induced magnetic moments comparable to the induced *electric* dipoles that we studied in Section 25–5. In this case the additional field caused by these current loops is always *opposite* in direction to that of the external field. (The reason for this is Faraday's law of induction, which we will study in Chapter 30. An induced current always tends to cancel the field change that caused it.)

Such materials are said to be **diamagnetic.** They always have negative susceptibility and relative permeability slightly *less than* unity, typically about 0.99990 to 0.99999 for solids and liquids. The susceptibility of a diamagnetic material is negative, as Table 29–1 shows. Diamagnetic susceptibilities are very nearly temperature-independent.

### Ferromagnetism

There is a third class of materials, called **ferromagnetic.** In these materials strong interactions between atomic magnetic moments cause them to line up parallel to each other in regions called **magnetic domains,** even when no external field is present. Figure 29–19 shows an example of magnetic domain structure. Within each domain, nearly all of the magnetic moments are parallel.

When there is no externally applied field, the orientations of the domain magnetizations are random, but when a field is present, they tend to orient themselves parallel to the field. The domain boundaries also shift; the domains magnetized in the field direction grow, and those magnetized in other directions shrink. Because the total magnetic moment of a domain may be many thousands of Bohr magnetons, there are much stronger aligning torques than with paramagnetic materials. The relative permeability is *much larger* than unity, typically of the order of 1000 to 100,000. Iron, cobalt, nickel, and many alloys containing these elements are ferromagnetic.

In ferromagnetic materials, as the external field is increased, a point is eventually reached at which nearly *all* the magnetic moments are aligned parallel to the external field. This condition is called *saturation magnetization;* after it is reached, further increase in the external field causes no increase in magnetization or in the additional field caused by the magnetization. As saturation is approached, the magnetization $\boldsymbol{M}$ is no longer proportional to the external magnetic field $\boldsymbol{B}_0$ (the field caused by the external currents).

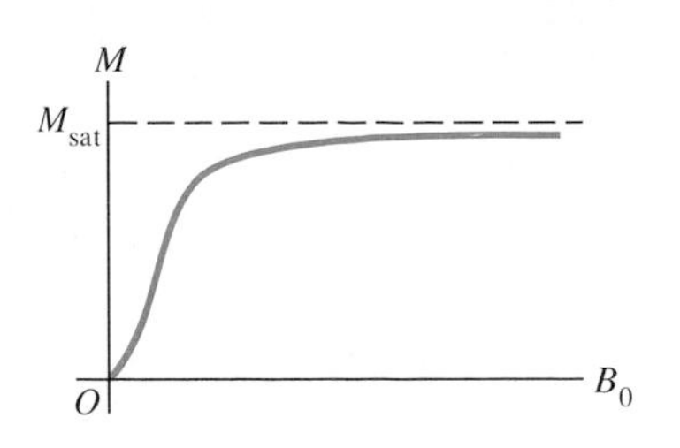


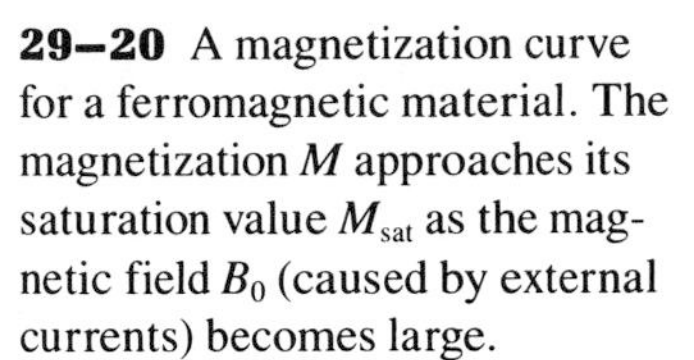

**29–20** A magnetization curve for a ferromagnetic material. The magnetization $M$ approaches its saturation value $M_{sat}$ as the magnetic field $B_0$ (caused by external currents) becomes large.

Figure 29–20 shows a "magnetization curve," a graph of magnetization $M$ as a function of external magnetic field $B_0$ for soft iron. An alternative description of this behavior is that $K_m$ is not constant but decreases as $B_0$ increases. In principle, paramagnetic materials also show saturation at sufficiently strong fields. In practice, the magnetic fields required are so large that departures from a linear relationship between $M$ and $B_0$ can be observed only at very low temperatures (1 K or so).

For many ferromagnetic materials the relation of magnetization to external magnetic field is different when the external field is increasing from when it is decreasing. Figure 29–21a shows a magnetization curve for such a material. When the material is magnetized to saturation and then the external field is reduced to zero, some magnetization remains. This behavior is characteristic of permanent magnets, which retain most of their saturation magnetization when the magnetizing field is removed. To reduce the magnetization to zero requires a magnetic field in the reverse direction.

This behavior is called **hysteresis,** and the curves in Fig. 29–21 are called *hysteresis loops.* Magnetizing and demagnetizing a material that has hysteresis involves the dissipation of energy, and the temperature of the material increases during such a process.

Ferromagnetic materials are widely used in electromagnets, transformer cores, and motors and generators, in which it is desirable to have as large a magnetic field as possible for a given current. Because hysteresis dissipates energy, materials used in these applications should usually have as narrow a hysteresis loop as possible. Soft iron is often used; it has high permeability without appreciable hysteresis (Fig. 29–21c). For permanent magnets a broad hysteresis loop, with large zero-field magnetization and large reverse field needed to demagnetize, is usually desirable (Fig. 29–21a). Many kinds of steel and many alloys, such as Alnico, are commonly used for permanent magnets. The remaining magnetic field in such a material, after it has been magnetized to near saturation, is typically of the order of 1 T, corresponding to a remaining magnetization $M = B/\mu_0$ of about 800,000 A/m.

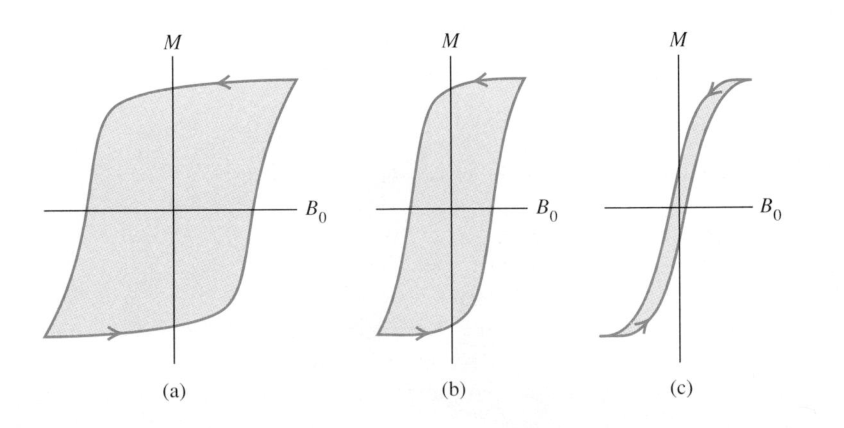


**29–21** Hysteresis loops. The materials of both (a) and (b) remain strongly magnetized when $B_0$ is reduced to zero. Since (a) is also hard to demagnetize, it would be good for permanent magnets. Since (b) magnetizes and demagnetizes more easily, it could be used as a computer memory material. The material of (c) would be useful for transformers and other alternating-current devices where zero hysteresis would be optimal.

## 29–9 DISPLACEMENT CURRENT

Ampere's law, in the form stated in Section 29–6, is incomplete. As we learn how it has to be modified for the most general situations, we will uncover some profound and fundamental aspects of the behavior of electric and magnetic fields.

To introduce the problem, let's consider the process of charging a capacitor (Fig. 29–22). Conductors lead current $i_C$ into one plate and out of the other; the charge $Q$ increases, and the electric field $\boldsymbol{E}$ between the plates increases. The notation $i_C$ indicates *conduction* current to distinguish it from another kind of current that we are about to encounter, called *displacement* current $i_D$. The lowercase $i$'s denote instantaneous values of currents that may vary with time.

Let's apply Ampere's law to the circular path shown. The integral $\oint \boldsymbol{B} \cdot d\boldsymbol{l}$ around this path equals $\mu_0 i_{encl}$. For the plane circular area bounded by the circle, $I_{encl}$ is just the current $i_C$ in the left conductor. But the surface that bulges out to the right is bounded by the same circle, and the current through that surface is zero. So $\oint \boldsymbol{B} \cdot d\boldsymbol{l}$ is equal to $\mu_0 i_C$, and at the same time it is equal to zero, a clear contradiction.

But something else is happening on the bulged-out surface. As the capacitor charges, the electric field $\boldsymbol{E}$ and the electric *flux* $\Phi_E$ through the surface are increasing. We can determine their rates of change in terms of the charge and current. The instantaneous charge $q$ is $q = Cv$, where $C$ is the capacitance and $v$ is the instantaneous potential difference. For a parallel-plate capacitor, $C = \epsilon_0 A/d$, where $A$ is the plate area and $d$ is the spacing. The potential difference $v$ between plates is $v = Ed$, where $E$ is the electric field magnitude between plates. (We neglect fringing and assume that $\boldsymbol{E}$ is uniform in the region between the plates.) If this region is filled with a material with permittivity $\epsilon$, we replace $\epsilon_0$ by $\epsilon$ everywhere; we use $\epsilon$ in the following discussion.

Substituting these expressions for $C$ and $v$ into $q = Cv$, we can express the capacitor charge $q$ as

$$q = Cv = \frac{\epsilon A}{d}(Ed) = \epsilon EA = \epsilon \Phi_E, \tag{29–31}$$

where $\Phi_E = EA$ is the electric flux through the surface.

As the capacitor charges, the rate of change of $q$ is the conduction current, $i_C = dq/dt$. Taking the derivative of Eq. (29–31), we get

$$i_C = \frac{dq}{dt} = \epsilon \frac{d\Phi_E}{dt}. \tag{29–32}$$

Now, stretching our imagination a little, we invent a fictitious current or pseudocurrent $i_D$ in the region between the plates, defined as

$$i_D = \epsilon \frac{d\Phi_E}{dt}. \tag{29–33}$$

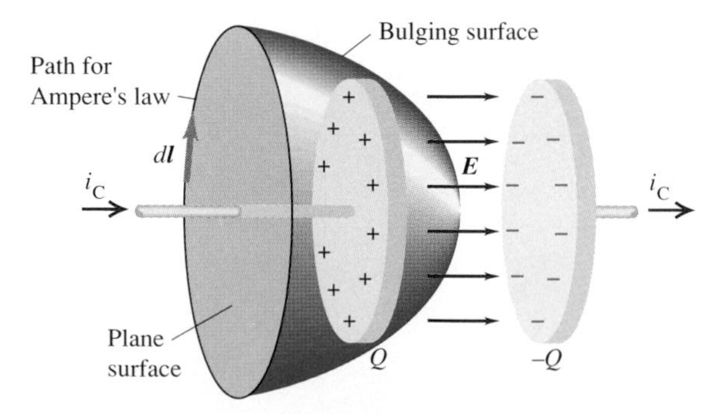


**29–22** Parallel-plate capacitor being charged. The conduction current through the plane surface is $i_C$, but there is no conductor current through the surface that bulges out to pass between the plates. The two surfaces have a common boundary, so this difference in $I_{encl}$ leads to an apparent contradiction in applying Ampere's law.

That is, we imagine that the changing flux through the curved surface is somehow equivalent, in Ampere's law, to a conduction current through that surface. We include this fictitious current, along with the real conduction current $i_C$, in Ampere's law:

$$\oint \boldsymbol{B} \cdot d\boldsymbol{l} = \mu_0(i_C + i_D). \tag{29–34}$$

In this form, Ampere's law is obeyed, no matter which surface we use in Fig. 29–22. For the flat surface, $i_D$ is zero; for the curved surface, $i_C$ is zero; and $i_C$ for the flat surface equals $i_D$ for the curved surface.

The fictitious current $i_D$ was invented in 1865 by James Clerk Maxwell (1831–1879), who called it **displacement current.** There is a corresponding displacement current density $j_D = i_D/A$; referring to Eq. (29–33) and using the fact that $E = \Phi_E/A$, we find

$$j_D = \epsilon \frac{dE}{dt}. \tag{29–35}$$

We have pulled the concept out of thin air, as Maxwell did, but we see that it enables us to save Ampere's law in situations such as Fig. 29–22.

Another benefit of displacement current is that it lets us generalize Kirchhoff's junction rule. Considering the left plate of the capacitor, we have conduction current into it but none out of it. But when we include the displacement current, we have conduction current coming in one side and an equal displacement current coming out the other side. With this generalized meaning of the term *current,* we can speak of current going *through* the capacitor.

You might well ask at this point whether displacement current has any real physical significance or whether it is just a ruse to satisfy Ampere's law and Kirchhoff's junction rule. Here's a fundamental experiment that helps to answer that question. We take a plane circular area between the capacitor plates, as shown in Fig. 29–23. If displacement current really plays the role in Ampere's law that we have claimed, then there ought to be a magnetic field in the region between the plates while the capacitor is charging. We can use our generalized Ampere's law, including displacement current, to predict what this field ought to be.

To be specific, let's picture round capacitor plates with radius $R$. To find the magnetic field at a point in the region between the plates, at a distance $r$ from the axis, we apply Ampere's law to a circle of radius $r$ passing through the point. This circle passes through points $a$ and $b$ in Fig. 29–23. The total current enclosed by the circle is equal to $j_D$ times its area, or $(i_D/\pi R^2)(\pi r^2)$. The integral $\oint \boldsymbol{B} \cdot d\boldsymbol{l}$ in Ampere's law is just $B$ times the circumference $2\pi r$ of the circle, and because $i_D = i_C$ for the charging capacitor,

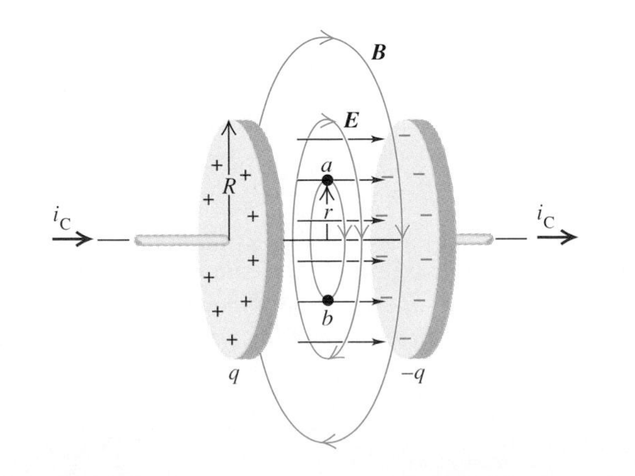


**29–23** A capacitor being charged by a current $i_C$ has a displacement current equal to $i_C$ with displacement-current density $j_D = \epsilon\, dE/dt$.

Ampere's law becomes

$$\oint \boldsymbol{B} \cdot d\boldsymbol{l} = 2\pi r B = \mu_0 \frac{r^2}{R^2} i_C,$$

or

$$B = \frac{\mu_0}{2\pi} \frac{r}{R^2} i_C. \tag{29–36}$$

This result predicts that in the region between the plates $\boldsymbol{B}$ is zero at the axis and increases linearly with distance from the axis. A similar calculation shows that *outside* the region between the plates, $\boldsymbol{B}$ is the same as though the wire were continuous and the plates not present at all.

When we *measure* the magnetic field in this region, we find that it really is there and that it behaves just as Eq. (29–36) predicts. This confirms directly the role of displacement current as a source of magnetic field. It is now established beyond a reasonable doubt that displacement current, far from being just a theoretical artifice, is a fundamental fact of nature. Maxwell's discovery was the bold step of an extraordinary genius. Indeed, as we will see later, displacement current was the missing link in the chain of electromagnetic theory that led Maxwell and others to the understanding of electromagnetic waves.

The calculation leading to Eq. (29–36) bears a strong resemblance to the one in Example 29–8 (Section 29–7), where we used Ampere's law to find the magnetic field inside a long, cylindrical conductor. In a certain sense we can think of the space between the conducting plates as a cylindrical conductor. There is no actual charge in motion to constitute a conduction current, but there is a displacement current. Thus it should not be too surprising that Eq. (29–36) is in fact identical to Eq. (29–20).

Two final comments: First, the generalized form of Ampere's law, Eq. (29–34), remains valid in a magnetic material, provided that magnetization is proportional to external field and we replace $\mu_0$ by $\mu$. Second, Ampere's law is valid even in empty space where there is no conduction current at all. This fact has profound implications; it means, among other things, that when $\boldsymbol{E}$ and $\boldsymbol{B}$ fields vary with time, they are interrelated. In particular, a changing electric field in a particular region of space induces a *magnetic* field in neighboring regions even when no conduction current and no matter are present. Conversely, we will learn in the next chapter that a changing *magnetic* field acts as a source of *electric* field. These relationships, first formulated in complete form by Maxwell in 1865, provide the key to a theoretical understanding of electromagnetic radiation and of light as a particular example of this radiation. We return to this topic in Chapter 33.

## SUMMARY

- The magnetic field $\boldsymbol{B}$ created by a charge $q$ moving with velocity $\boldsymbol{v}$ is

$$\boldsymbol{B} = \frac{\mu_0}{4\pi} \frac{q\boldsymbol{v} \times \hat{\boldsymbol{r}}}{r^2}, \tag{29–2}$$

where $\boldsymbol{r}$ is the vector from the source point (the location of $q$) to the field point $P$ and $\hat{\boldsymbol{r}}$ is a unit vector in that direction. Superposition principle: The total $\boldsymbol{B}$ field produced by several moving charges is the vector sum of the fields produced by the individual charges.

• The law of Biot and Savart: The magnetic field $\boldsymbol{dB}$ created by an element $\boldsymbol{dl}$ of a conductor with current $I$ is

$$d\boldsymbol{B} = \frac{\mu_0}{4\pi}\frac{I\,d\boldsymbol{l} \times \hat{\boldsymbol{r}}}{r^2}. \qquad (29\text{–}6)$$

The field created by a finite conductor is the integral of this expression over the length of the conductor.

• Magnetic field lines never have end points, and the surface integral of $\boldsymbol{B}$ over any closed surface is always zero:

$$\oint \boldsymbol{B} \cdot d\boldsymbol{A} = 0 \quad \text{(every closed surface).} \qquad (29\text{–}10)$$

This is the magnetic analog of Gauss's law and shows that there are no isolated magnetic poles.

• The magnetic field $\boldsymbol{B}$ at a distance $r$ from a long, straight wire carrying a current $I$ has magnitude

$$B = \frac{\mu_0 I}{2\pi r} \quad \text{(long, straight wire).} \qquad (29\text{–}9)$$

The magnetic field lines are circles coaxial with the wire, with directions given by the right-hand rule.

• The interaction force, per unit length, between two long, parallel conductors with currents $I$ and $I'$ has magnitude

$$\frac{F}{L} = \frac{\mu_0 I I'}{2\pi r}. \qquad (29\text{–}11)$$

The definition of the ampere is based on this relation.

• The magnetic field produced by a circular conducting loop with radius $a$, carrying current $I$, at a distance $x$ from its center, along its axis, has magnitude

$$B_x = \frac{\mu_0 I a^2}{2(x^2 + a^2)^{3/2}} \quad \text{(circular loop).} \qquad (29\text{–}15)$$

For $N$ loops, this expression is multiplied by $N$. At the center of the loop, where $x = 0$,

$$B_x = \frac{\mu_0 N I}{2a} \quad \text{(center of } N \text{ circular loops).} \qquad (29\text{–}17)$$

• Ampere's law states that the line integral of $\boldsymbol{B}$ around any closed path equals $\mu_0$ times the net current through the area enclosed by the path:

$$\oint \boldsymbol{B} \cdot d\boldsymbol{l} = \mu_0 I_{\text{encl}}. \qquad (29\text{–}19)$$

The positive sense of current is determined by a right-hand rule.

• The magnetic field in the interior of a long cylindrical conductor with radius $R$ carrying current $I$ has magnitude

$$B = \frac{\mu_0 I}{2\pi}\frac{r}{R^2}. \qquad (29\text{–}20)$$

• The magnetic field at the center of a long solenoid with $n$ turns per unit length, with current $I$, is

$$B = \mu_0 n I \quad \text{(solenoid).} \qquad (29\text{–}21)$$

## KEY TERMS

**source point**
**field point**
**superposition principle**
**law of Biot and Savart**
**ampere**
**Ampere's law**
**Bohr magneton**
**magnetization**
**paramagnetic**
**relative permeability**
**permeability**
**magnetic susceptibility**
**diamagnetic**
**ferromagnetic**
**magnetic domain**
**hysteresis**
**displacement current**

The magnetic field inside a toroidal solenoid with $N$ turns, with current $I$, at a distance $r$ from its axis, is

$$B = \frac{\mu_0 NI}{2\pi r} \quad \text{(toroidal solenoid).} \tag{29–22}$$

• For magnetic materials the magnetization of the material causes an additional contribution to $\boldsymbol{B}$. For paramagnetic and diamagnetic materials, $\mu_0$ is replaced in magnetic-field expressions by $\mu = K_m \mu_0$, where $\mu$ is the permeability of the material and $K_m$ is its relative permeability. The magnetic susceptibility $\chi_m$ is defined as $\chi_m = K_m - 1$. Magnetic susceptibilities for paramagnetic materials are small positive quantities; those for diamagnetic materials are small negative quantities. For ferromagnetic materials, $K_m$ is much larger than unity and is not constant. Some ferromagnetic materials are permanent magnets, retaining their magnetization even after the external magnetic field is removed.

• Displacement current $i_D$ acts as a source of magnetic field in exactly the same way as conduction current. It is defined as

$$i_D = \epsilon \frac{d\Phi_E}{dt}. \tag{29–33}$$

Ampere's law including displacement current is

$$\oint \boldsymbol{B} \cdot d\boldsymbol{l} = \mu_0(i_C + i_D). \tag{29–34}$$

Displacement current plays an essential role in the analysis of electromagnetic waves.

## EXERCISES

### Section 29–1 Magnetic Field of a Moving Charge

**29–1** A positive point charge with $q = 3.00\ \mu\text{C}$ has velocity $\boldsymbol{v} = (8.00 \times 10^6\ \text{m/s})\boldsymbol{i}$. At the instant when the point charge is at the origin, what is the magnetic field vector $\boldsymbol{B}$ that it produces at the following points: a) $x = 0.500$ m, $y = 0$, $z = 0$; b) $x = 0$, $y = -0.500$ m, $z = 0$; c) $x = 0$, $y = 0$, $z = +0.500$ m; d) $x = 0$, $y = -0.500$ m, $z = +0.500$ m?

**29–2** Two positive point charges $q$ and $q'$ are moving relative to an observer at point $P$ as shown in Fig. 29–24. a) What is the direction of the force that $q'$ exerts on $q$? b) What is the direction of the force that $q$ exerts on $q'$? c) If $v = v' = 7.00 \times 10^6$ m/s, what is the ratio of the magnitude of the magnetic force to that of the Coulomb force between the two charges?

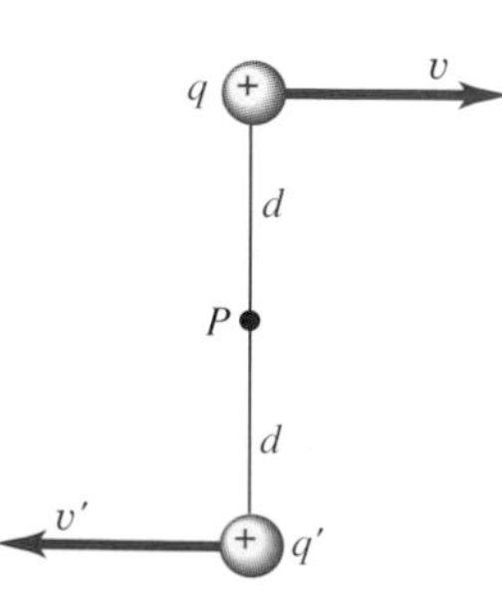

FIGURE 29–24

**29–3** A pair of point charges, $q = +5.00\ \mu\text{C}$ and $q' = -3.00\ \mu\text{C}$, are moving as shown in Fig. 29–25. At this instant, what are the magnitude and direction of the magnetic field produced at the origin? Take $v = v' = 6.00 \times 10^5$ m/s.

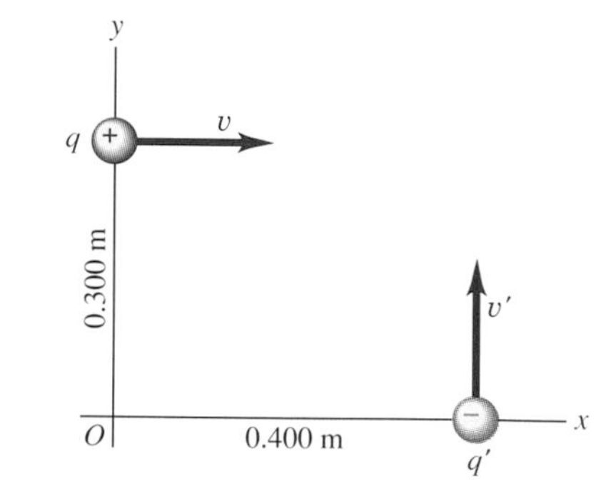

FIGURE 29–25

### Section 29–2 Magnetic Field of a Current Element

**29–4** A long, straight wire, carrying a current of 200 A, runs through a cubical wooden box, entering and leaving through holes in the centers of opposite faces (Fig. 29–26). The length of each side of the box is 20.0 cm. Consider an element of the wire 0.200 cm long at the center of the box. Compute the magnitude $dB$ of the

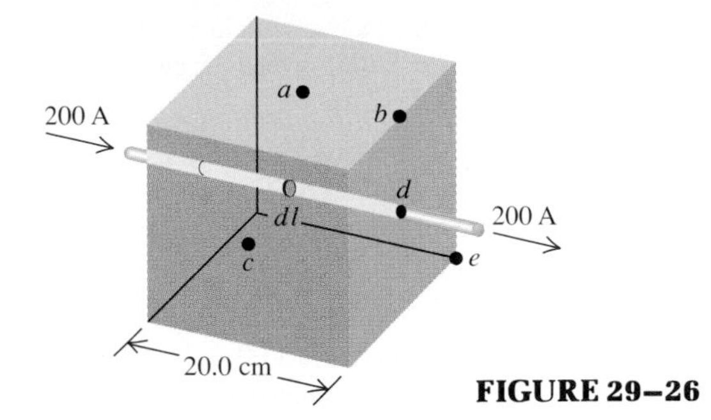

**FIGURE 29–26**

magnetic field produced by this element at the points $a$, $b$, $c$, $d$, and $e$ in Fig. 29–26. Points $a$, $c$, and $d$ are at the centers of the faces of the cube; point $b$ is at the midpoint of one edge; and point $e$ is at a corner. Copy the figure and show by vectors the directions and relative magnitudes of the field vectors. (*Note:* Assume that $dl$ is small in comparison to the distances from the current element to the points where the magnetic field is to be calculated.)

**29–5** Calculate the magnitude and direction of the magnetic field at point $P$ due to the current in the semicircular section of wire shown in Fig. 29–27. (Does the current in the long, straight section of the wire produce any field at $P$?)

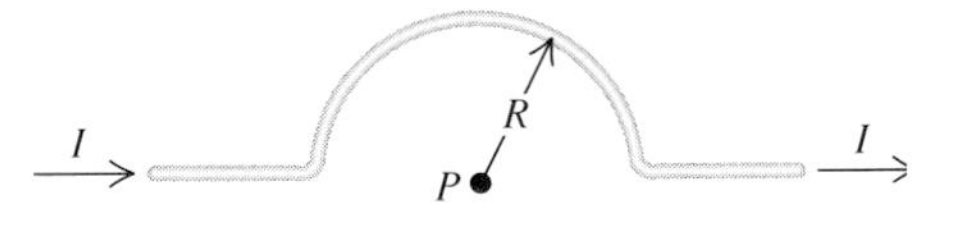

**FIGURE 29–27**

**29–6** Calculate the magnitude of the magnetic field at point $P$ of Fig. 29–28 in terms of $R$, $I_1$, and $I_2$. What does your expression give when $I_1 = I_2$?

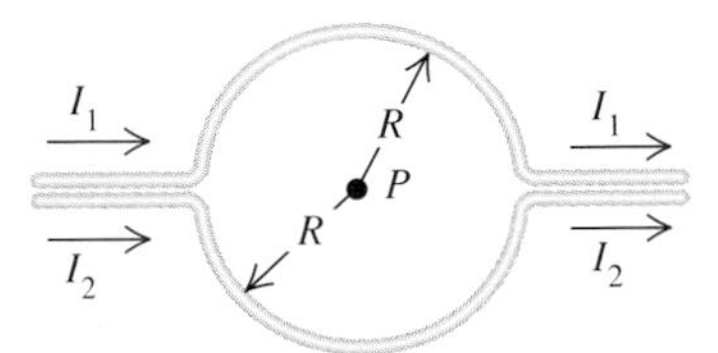

**FIGURE 29–28**

## Section 29–3
## Magnetic Field of a Straight Conductor

**29–7** Provide the details of the derivation of Eq. (29–8) from the equation that precedes it.

**29–8** You want to produce a magnetic field with a magnitude of $3.00 \times 10^{-4}$ T at a distance of 0.050 m from a long, straight wire. a) What current is required to produce this field? b) With the current found in part (a), what is the magnitude of the field at a distance of 0.100 m from the wire and at 0.200 m?

**29–9 Effect of Transmission Lines.** Two hikers are reading a compass under an overhead transmission line that is 5.00 m above the ground and carries a current of 800 A in a horizontal direction from east to west. a) Find the magnitude and direction of the magnetic field at a point on the ground directly under the conductor. b) One hiker suggests that they walk on another 50 m to avoid inaccurate compass readings caused by the current. Considering that the magnitude of the earth's field is of the order of $0.5 \times 10^{-4}$ T, is the current really a problem?

**29–10** A long, straight wire lies along the $y$-axis and carries a current of 5.00 A in the $-y$-direction (Fig. 29–29). In addition to the magnetic field due to the current in the wire, a uniform magnetic field $\boldsymbol{B}_0$ with magnitude $1.00 \times 10^{-6}$ T is in the $+x$-direction. What is the resultant field (magnitude and direction) at the following points in the $xz$-plane: a) $x = 0$, $z = 2.00$ m; b) $x = 2.00$ m, $z = 0$; c) $x = 0$, $z = -0.50$ m?

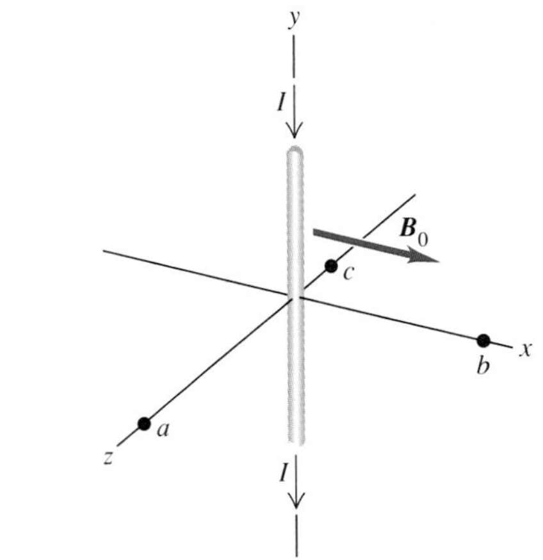

**FIGURE 29–29**

**29–11** Two long, straight, horizontal parallel wires, one above the other, are separated by a distance $2a$. If the wires carry equal currents with magnitude $I$ in opposite directions, what is the field magnitude in the plane of the wires at a point a) midway between them; b) at a distance $a$ above the upper wire? If the wires carry equal currents in the same direction, what is the field magnitude in the plane of the wires at a point c) midway between them; d) at a distance $a$ above the upper wire?

## Section 29–4
## Force between Parallel Conductors

**29–12** Two long, parallel wires are separated by a distance of 0.400 m (Fig. 29–30). The currents $I_1$ and $I_2$ have the directions shown. a) Calculate the magnitude of the force exerted by each wire on a 0.200-m length of the other. Is the force attractive or repulsive? b) If each current is doubled, so that $I_1$ becomes 10.0 A and $I_2$ becomes 4.00 A, what is the magnitude of the force that each wire exerts on a 2.00-m length of the other?

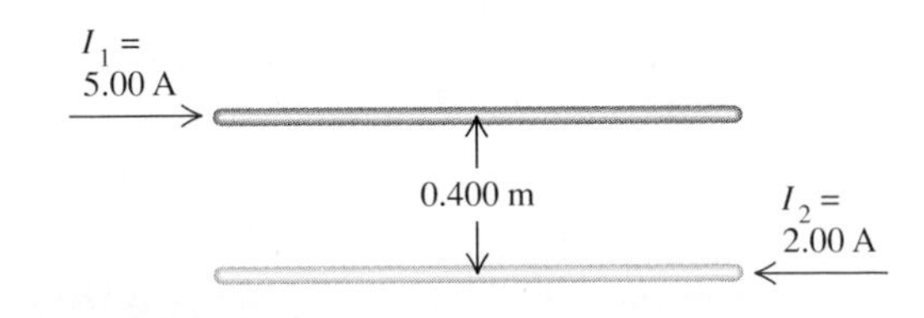

**FIGURE 29–30**

**29–13** Two long, parallel wires are separated by a distance of 0.100 m. The force per unit length each wire exerts on the other is $6.00 \times 10^{-5}$ N/m, and the wires repel each other. If the current in one wire is 2.00 A, a) what is the current in the second wire? b) Are the two currents in the same or in opposite directions?

**29–14** Three parallel wires each carry current $I$ in the directions shown in Fig. 29–31. If the separation between adjacent wires is $d$, calculate the magnitude and direction of the resultant magnetic force per unit length on each wire.

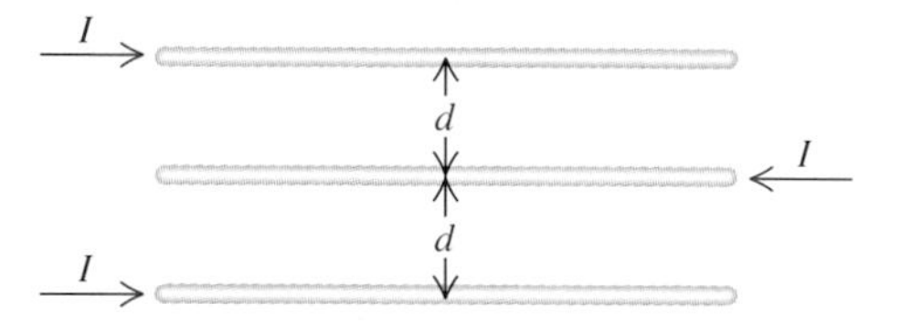


**FIGURE 29–31**

**29–15** A long, horizontal wire $AB$ rests on the surface of a table. Another wire $CD$ vertically above the first is 0.400 m long and free to slide up and down on the two vertical metal guides $C$ and $D$ (Fig. 29–32). The two wires are connected through the sliding contacts and carry a current of 40.0 A. The mass per unit length of the wire $CD$ is $5.00 \times 10^{-3}$ kg/m. To what equilibrium height $h$ will the wire $CD$ rise, assuming the magnetic force on it to be due wholly to the current in the wire $AB$?

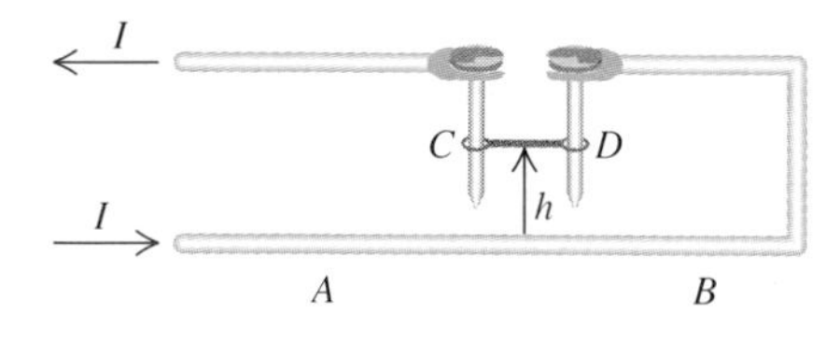

**FIGURE 29–32**

## Section 29–5 Magnetic Field of a Circular Loop

**29–16** A closely wound circular coil with a radius of 5.00 cm has 200 turns and carries a current of 0.300 A. What is the magnitude of the magnetic field a) at the center of the coil; b) at a point on the axis of the coil 10.0 cm from its center?

**29–17** A closely wound coil has a diameter of 18.0 cm and carries a current of 2.50 A. How many turns does it have if the magnetic field at the center of the coil is $4.19 \times 10^{-4}$ T?

## Section 29–6 Ampere's Law

**29–18** A closed curve encircles several conductors. The line integral $\oint \boldsymbol{B} \cdot d\boldsymbol{l}$ around this curve is $3.75 \times 10^{-5}$ T · m. What is the net current in the conductors?

## Section 29–7 Applications of Ampere's Law

**29–19** A wooden ring whose mean diameter is 0.180 m is wound with a closely spaced toroidal winding of 500 turns. Compute the magnitude of the magnetic field at a point at the center of the cross section of the windings when the current in the windings is 0.300 A.

**29–20** A solenoid with a length of 20.0 cm and a radius of 3.00 cm is closely wound with 200 turns of wire. The current in the windings is 6.00 A. Compute the magnetic field at a point near the center of the solenoid.

**29–21** A solenoid is designed to produce a magnetic field of 0.140 T at its center. The radius is 3.00 cm and the length 50.0 cm, and the wire can carry a maximum current of 10.0 A. a) What is the minimum number of turns per unit length the solenoid must have? b) What total length of wire is required?

**29–22** A toroidal solenoid (Fig. 29–17) has inner radius $r_1 = 0.200$ m and outer radius $r_2 = 0.280$ m. The solenoid has 300 turns and carries a current of 3.50 A. What is the magnitude of the magnetic field at each of the following distances from the center of the torus: a) 0.150 m; b) 0.240 m; c) 0.350 m?

**29–23 A Coaxial Cable.** A solid conductor with radius $a$ is supported by insulating disks on the axis of a tube with inner radius $b$ and outer radius $c$ (Fig. 29–33). If the central conductor and tube carry equal currents $I$ in opposite directions, derive an expression for the magnitude of the magnetic field a) at points outside the central solid conductor but inside the tube ($a < r < b$); b) at points outside the tube ($r > c$).

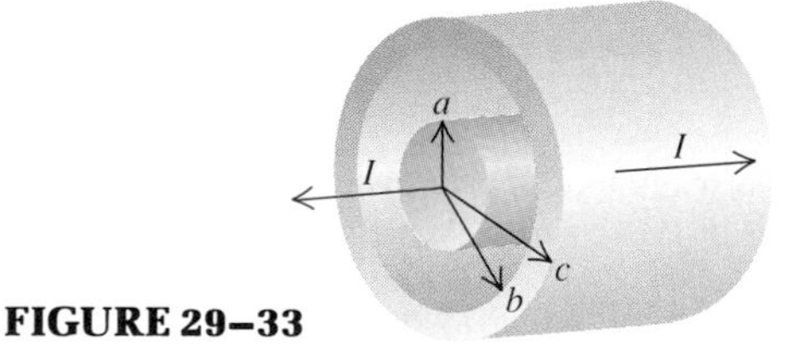


**FIGURE 29–33**

**29–24** Repeat Exercise 29–23 for the case in which the current in the central solid conductor is $I_1$ and that in the tube is $I_2$ and these currents are in the same direction rather than in opposite directions.

## Section 29–8 Magnetic Materials

***29–25 Curie's Law.** Experimental measurements of the magnetic susceptibility of iron ammonium alum are given below. Make a graph of $1/\chi_m$ against Kelvin temperature. Does the material obey Curie's law? If so, what is the Curie constant?

| $T$ (°C) | $\chi_m$ |
|---|---|
| −258.15 | $129 \times 10^{-4}$ |
| −173 | $19.4 \times 10^{-4}$ |
| −73 | $9.7 \times 10^{-4}$ |
| 27 | $6.5 \times 10^{-4}$ |

***29–26** A toroidal solenoid having 500 turns of wire and a mean radius of 0.0800 m carries a current of 0.400 A. The relative permeability of the core is 120. a) What is the magnetic field in the core? b) What part of the magnetic field is due to atomic currents?

***29–27** A toroidal solenoid with 400 turns is wound on a ring having a mean radius of 3.80 cm. Find the current in the winding that is required to set up a magnetic field of 0.150 T in the ring a) if the ring is made of annealed iron ($K_m = 1400$); b) if the ring is made of silicon steel ($K_m = 5200$).

***29–28** The current in the windings on a toroid is 2.00 A. There are 400 turns, and the mean radius is 0.400 m. The toroid is filled with a magnetic material. The magnetic field inside the windings is found to be 1.00 T. Calculate a) the relative permeability; b) the magnetic susceptibility of the material that fills the toroid.

### Section 29–9 Displacement Current

**29–29** In Fig. 29–23 the capacitor plates have area 4.00 cm$^2$ and separation 3.00 mm. The plates are in vacuum. The charging current $i_C$ has a *constant* value of 2.00 mA. At $t = 0$ the charge on the plates is zero. a) Calculate the charge on the plates, the electric field between the plates, and the potential difference between the plates when $t = 5.00\ \mu$s. b) Calculate $dE/dt$, the time rate of change of the electric field between the plates. Does $dE/dt$ vary with time? c) Calculate the displacement current density $j_D$ between the plates, and from this calculate the total displacement current $i_D$. How do $i_C$ and $i_D$ compare?

**29–30 Displacement Current in a Dielectric.** Suppose the parallel plates in Fig. 29–23 have an area of 2.00 cm$^2$ and are separated by a sheet of dielectric 1.00 mm thick, with dielectric constant 3.00. (Neglect fringing effects.) At a certain instant the potential difference between the plates is 140 V, and the conduction current $i_C$ equals 2.00 mA. a) What is the charge $q$ on each plate, at this instant? b) What is the rate of change of charge on the plates? c) What is the displacement current in the dielectric?

**29–31** A parallel-plate air-filled capacitor is being charged as in Fig. 29–23. The circular plates have radius 0.0500 m, and at a particular instant the conduction current in the wires is 0.600 A. a) What is the displacement current density $j_D$ in the air space between the plates? b) What is the rate at which the electric field between the plates is changing? c) What is the induced magnetic field between the plates at a distance of 0.0250 m from the axis? d) At 0.100 m from the axis?

**29–32** A copper wire with a circular cross-section area of 4.0 mm$^2$ carries a current of 40 A. The resistivity of the material is $2.0 \times 10^{-8}\ \Omega \cdot$m. a) What is the uniform electric field in the material? b) If the current is changing at the rate of 5000 A/s, at what rate is the electric field in the material changing? c) What is the displacement-current density in the material in part (b)? d) If the current is changing as in part (b), what is the magnitude of the magnetic field 5.0 cm from the center of the wire? Note that both the conduction current and the displacement current must be included in the calculation of $B$. Is the contribution from the displacement current significant?

## PROBLEMS

**29–33** A long, straight wire carries a current of 1.50 A. An electron is traveling in the vicinity of the wire. At the instant when the electron is traveling with a speed of $5.00 \times 10^4$ m/s parallel to the wire, 0.0800 m from it, and in the same direction as the current, what are the magnitude and direction of the force that the magnetic field of the current exerts on the moving electron?

**29–34** Figure 29–34 is an end view of two long, parallel wires perpendicular to the $xy$-plane, each carrying a current $I$ but in opposite directions. a) Copy the diagram, and show by vectors the $\boldsymbol{B}$ field of each wire and the resultant $\boldsymbol{B}$ field at point $P$. b) Derive the expression for the magnitude of $\boldsymbol{B}$ at any point on the $x$-axis in terms of the $x$-coordinate of the point. What is the direction of $\boldsymbol{B}$? c) Construct a graph of the magnitude of $\boldsymbol{B}$ at points on the $x$-axis. d) At what value of $x$ is the magnitude of $\boldsymbol{B}$ a maximum?

**29–35** Repeat Problem 29–34, but with the current in both wires directed into the plane of the figure.

**29–36** Suppose a third long, straight wire, parallel to the other two, passes through point $P$ in Fig. 29–34, and that each wire carries a current $I = 12.0$ A. Let $a = 0.300$ m and $x = 0.400$ m. Find the magnitude and direction of the force per unit length on the third wire a) if the current in it is directed into the plane of the figure; b) if current is directed out of the plane of the figure. (Use the results of Problem 29–34.)

**29–37** Two long, straight, parallel wires are 1.00 m apart (Fig. 29–35). The upper wire carries a current $I_1$ of 6.00 A into the plane of the paper. a) What must be the magnitude and direction of the current $I_2$ for the resultant field at point $P$ to be zero? b) What are

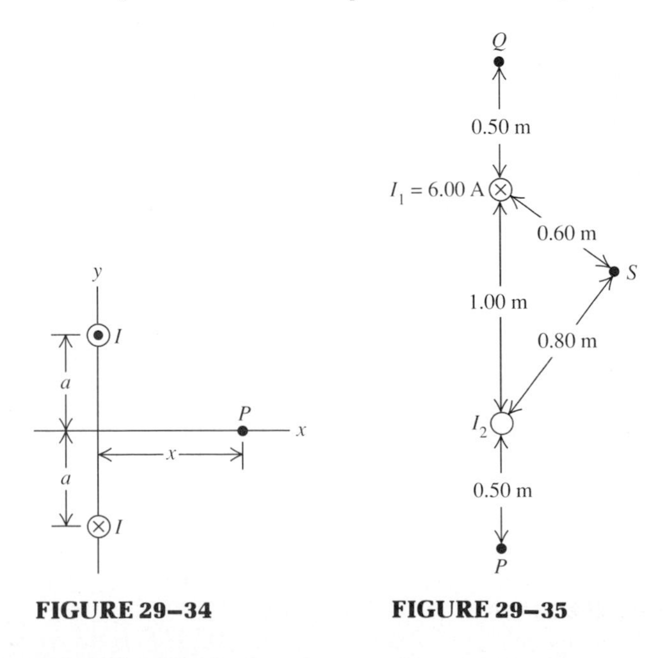


FIGURE 29–34

FIGURE 29–35

then the magnitude and direction of the resultant field at $Q$? c) What is then the magnitude of the resultant field at $S$?

**29–38** Two long, parallel wires are hung by cords 4.00 cm long from a common axis (Fig. 29–36). The wires have a mass per unit length of 0.0300 kg/m and carry the same current in opposite directions. What is the current in each wire if the cords hang at an angle of 6.00° with the vertical?

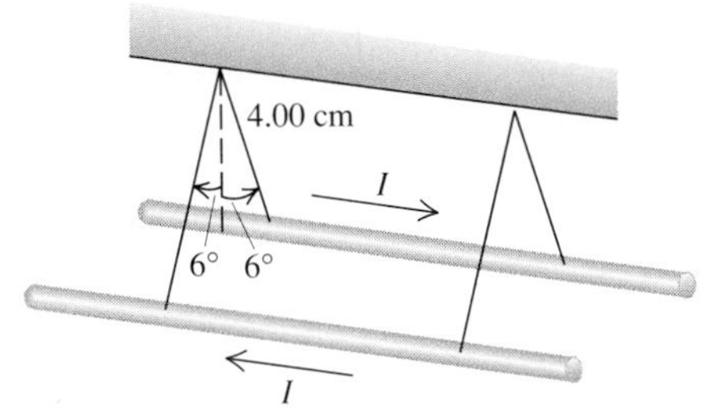

**FIGURE 29–36**

**29–39** The long, straight wire $AB$ in Fig. 29–37 carries a current of 14.0 A. The rectangular loop whose long edges are parallel to the wire carries a current of 5.00 A. Find the magnitude and direction of the resultant force exerted on the loop by the magnetic field of the wire.

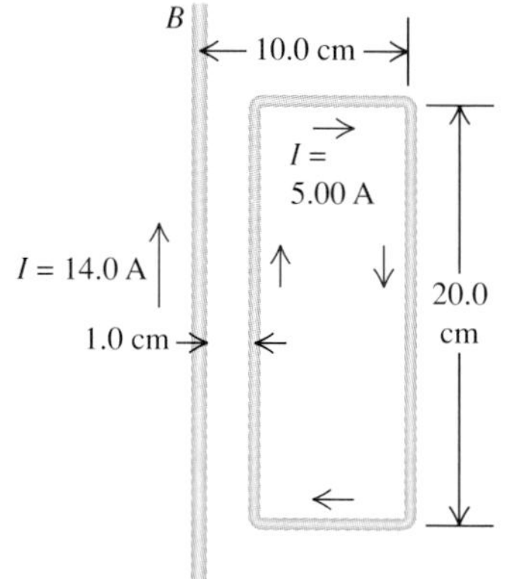


**FIGURE 29–37**

**29–40** Calculate the magnitude and direction of the magnetic field produced at point $P$ in Fig. 29–38 by current $I$ in the rectangular wire loop. (Point $P$ is at the center of the rectangle.) (*Hint:* The gap on the left-hand side where the wires enter and leave the rectangle is so small that this side of the rectangle can be taken to be a continuous wire with length $b$.)

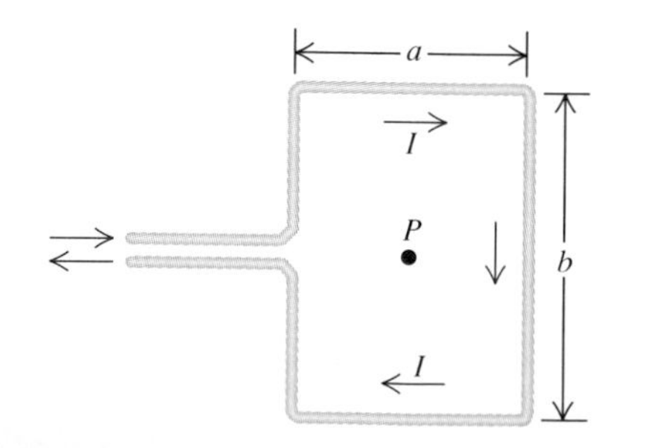

**FIGURE 29–38**

**29–41 Helmholtz Coils.** Figure 29–39 is a sectional view of two circular coils with radius $a$, each wound with $N$ turns of wire carrying a current $I$, circulating in the same direction in both coils. The coils are separated by a distance $a$ equal to their radii. In this configuration the coils are called Helmholtz coils; they produce a very uniform magnetic field in the region between them. a) Derive the expression for the magnitude $B$ of the magnetic field at a point on the axis a distance $x$ to the right of point $P$ that is midway between the coils. b) Sketch a graph of $B$ versus $x$ for $x = 0$ to $x = a/2$. Compare this sketch to one for the magnetic field due to the right-hand coil alone. c) From part (a), obtain an expression for the magnitude of the magnetic field at point $P$. d) Calculate the magnitude of the magnetic field at $P$ if $N = 200$ turns, $I = 5.00$ A, and $a = 0.300$ m.

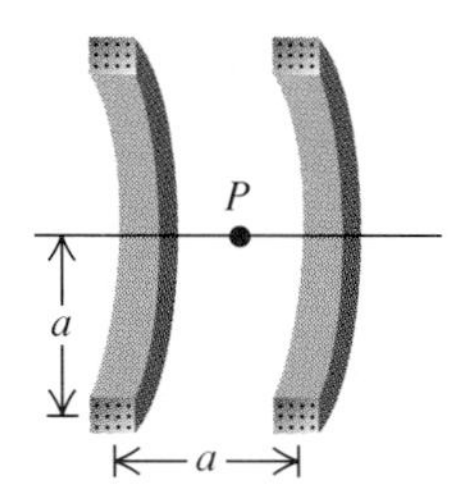


**FIGURE 29–39**

**29–42** The wire semicircles in Fig. 29–40 have radii $a$ and $b$. Calculate the resultant magnetic field (magnitude and direction) at point $P$.

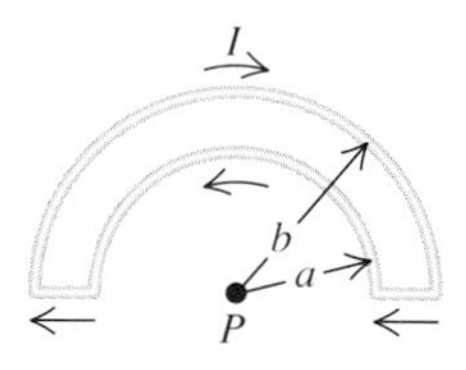


**FIGURE 29–40**

**29–43** The wire in Fig. 29–41 is infinitely long and carries current $I$. Calculate the magnitude and direction of the magnetic field that this current produces at point $P$.

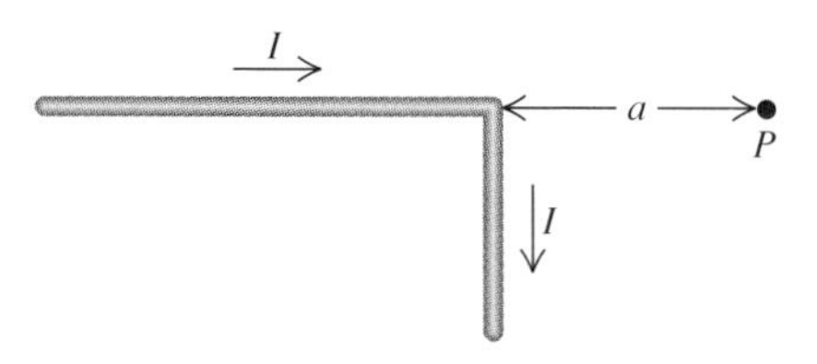


**FIGURE 29–41**

**29–44** The wire in Fig. 29–42 carries current $I$ in the direction shown. The wire consists of a very long, straight section, a quarter-circle with radius $R$, and another long, straight section. What are the magnitude and direction of the resultant magnetic field at the center of curvature of the quarter-circle section (point $P$)?

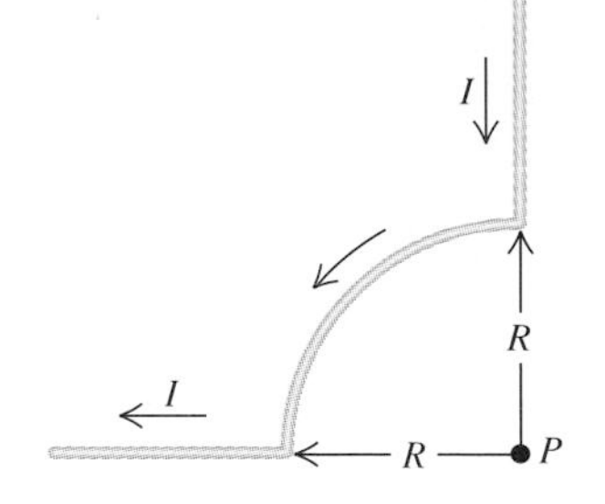


**FIGURE 29–42**

**29–45** A long, straight wire with circular cross section of radius $R$ carries current $I$. Assume that the current density is not constant across the cross section of the wire but rather varies as $J = \alpha r$, where $\alpha$ is a constant. a) Using the requirement that $J$ integrated over the cross section of the wire gives the total current $I$, calculate the constant $\alpha$ in terms of $I$ and $R$. b) Use Ampere's law to calculate the magnetic field $B(r)$ for i) $r \leq R$; ii) $r \geq R$.

**29–46** A conductor is made in the form of a hollow cylinder with inner and outer radii $a$ and $b$, respectively. It carries a current $I$, uniformly distributed over its cross section. Derive expressions for the magnitude of the magnetic field in the regions a) $r < a$; b) $a < r < b$; c) $r > b$.

**29–47 An Infinite Current Sheet.** Long, straight conductors with square cross section, each carrying current $I$, are laid side by side to form an infinite current sheet (Fig. 29–43). The conductors lie in the $xy$-plane, are parallel to the $y$-axis, and carry current in the $+y$-direction. There are $n$ conductors per meter of length measured along the $x$-axis. a) What are the magnitude and direction of the magnetic field a distance $a$ below the current sheet? b) What are the magnitude and direction of the magnetic field a distance $a$ above the current sheet?

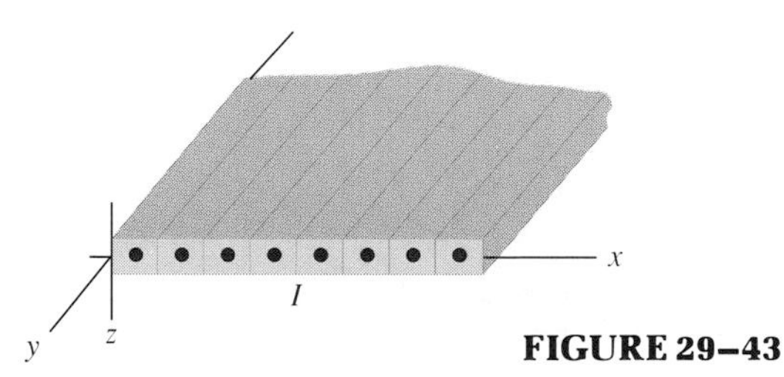


**FIGURE 29–43**

**29–48** Long, straight conductors with square cross section, each carrying current $I$, are laid side by side to form an infinite current sheet with current directed out of the plane of the page (Fig. 29–44). A second infinite current sheet is a distance $d$ below the first and is parallel to it. The second sheet carries current into the plane of the paper. Each sheet has $n$ conductors per meter of length. (Refer to Problem 29–47.) Calculate the magnitude and direction of the resultant magnetic field at a) point $P$ (above the upper sheet); b) point $R$ (midway between the two sheets); c) point $S$ (below the lower sheet).

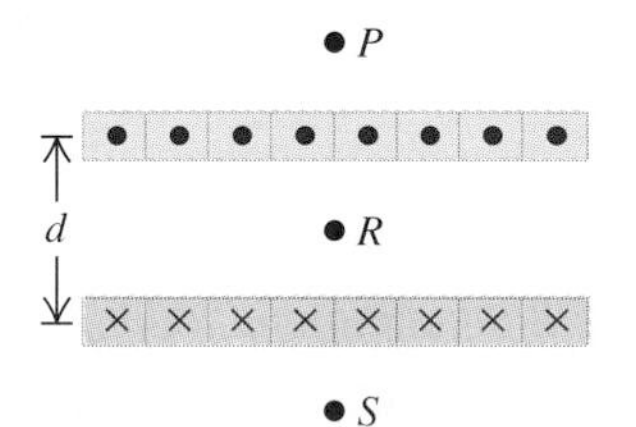


**FIGURE 29–44**

**29–49** A long, straight solid cylinder, oriented with its axis in the $z$-direction, carries a current whose current density is $\boldsymbol{J}$. The current density, although symmetrical about the cylinder axis, is not uniform but varies according to the relation

$$\boldsymbol{J} = \frac{2I_0}{\pi a^2}[1 - (r/a)^2]\boldsymbol{k} \quad \text{for } r \leq a,$$

$$\boldsymbol{J} = \boldsymbol{0} \quad \text{for } r \geq a,$$

where $a$ is the radius of the cylinder, $r$ is the radial distance from the cylinder axis, and $I_0$ is a constant having units of amperes. a) Show that $I_0$ is the total current passing through the entire cross section of the wire. b) Using Ampere's law, derive an expression for the magnitude of the magnetic field $\boldsymbol{B}$ in the region $r \geq a$. c) Obtain an expression for the current $I$ contained in a circular cross section of radius $r \leq a$ and centered at the cylinder axis. d) Using Ampere's law, derive an expression for the magnitude of the magnetic field $\boldsymbol{B}$ in the region $r \leq a$. How do your results in parts (b) and (d) compare for $r = a$?

**29–50** Consider the coaxial cable of Exercise 29–23. Assume the currents to be uniformly distributed over the cross sections of the solid conductor and the tube. Calculate the magnitude of the magnetic field at a) $r < a$ (inside the central conductor—does your answer give the expected result for $r = a$?); b) $b < r < c$ (inside the outer tube—does your answer give the expected results for $r = c$ and for $r = b$?).

**29–51** Integrate $B_x$ as given in Eq. (29–15) from $-\infty$ to $+\infty$. That is, calculate $\int_{-\infty}^{+\infty} B_x\,dx$. Explain the significance of your result.

**29–52** A rod of pure silicon (resistivity $\rho = 2300\ \Omega\cdot\text{m}$) is carrying a current. The electric field varies sinusoidally with time according to $E = E_0 \sin \omega t$, where $E_0 = 0.100$ V/m, $\omega = 2\pi f$, and the frequency is $f = 60$ Hz. a) Find the magnitude of the maximum conduction current density in the wire. b) Assuming that $\epsilon = \epsilon_0$, find the maximum displacement current density in the wire, and compare with the result of part (a). c) At what frequency $f$ would the maximum conduction and displacement densities become equal if $\epsilon = \epsilon_0$ (which is not actually the case)? d) At the frequency determined in part (b), what is the relative *phase* of the conduction and displacement currents?

**29–53** A capacitor has two parallel plates with area $A$ separated by a distance $d$. The space between the plates is filled with a material having dielectric constant $K$. The material is not a perfect insulator but has resistivity $\rho$. The capacitor is initially charged with charge of magnitude $Q_0$ on each plate, which gradually discharges by conduction through the dielectric. a) Calculate the conduction current density $j_C(t)$ in the dielectric. b) Show that at any instant the displacement current density in the dielectric is equal in magnitude to the conduction current density but opposite in direction, so that the *total* current density is zero at every instant.

## CHALLENGE PROBLEMS

**29–54** A wide, long insulating belt has a uniform positive charge per unit area $\sigma$ on its upper surface. Rollers at each end move the belt to the right at a constant speed $v$. Calculate the magnitude and direction of the magnetic field produced by the moving belt at a point just above its surface. [*Hint:* At points near the surface and far from its edges or ends, the moving belt can be considered to be an infinite current sheet (Problem 29–47).]

**29–55 A Charged Dielectric Disk.** A thin disk of dielectric material with radius $a$ has a total charge $+Q$ distributed uniformly over its surface. It rotates $n$ times per second about an axis perpendicular to the surface of the disk and passing through its center. Find the magnetic field at the center of the disk. (*Hint:* Divide the disk into concentric rings of infinitesimal width.)

**29–56** A wire in the shape of a semicircle with radius $a$ is oriented in the $zy$-plane with its center of curvature at the origin (Fig. 29–45). If the current in the wire is $I$, calculate the magnetic field components produced at point $P$, a distance $x$ out along the $x$-axis. (*Note:* Do not forget the contribution from the straight wire at the bottom of the semicircle that runs from $z = -a$ to $z = +a$. The fields of the two antiparallel currents at $z > a$ cancel.)

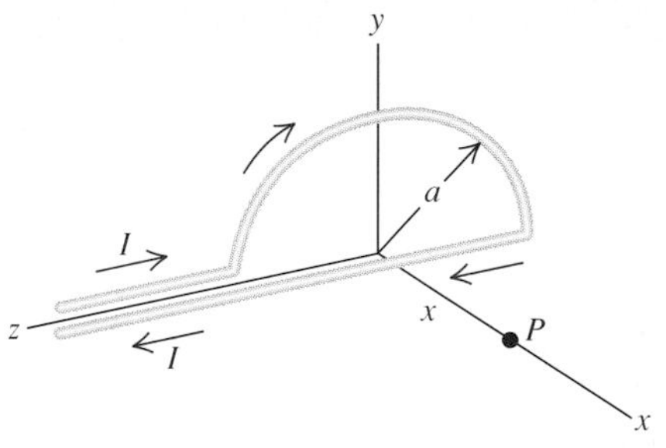

**FIGURE 29–45**

**29–57** Two long, straight conducting wires with linear mass density $\lambda = 1.75 \times 10^{-3}$ kg/m are suspended from cords so that they are horizontal, parallel to each other, and a distance $d = 2.00$ cm apart. The back ends of the wires are connected to one another by a slack low-resistance connecting wire. The positive side of a $1.00\text{-}\mu\text{F}$ capacitor that has been charged by a 1000-V source is connected to the front end of one of the wires, and the negative side of the capacitor is connected to the front end of the other wire (Fig. 29–46). When the connection is made, the wires are pushed aside by the repulsive force between the wires, and each wire has an initial horizontal velocity $v_0$. It can be assumed that the time for the capacitor to discharge is negligible in comparison to the time it takes for any appreciable displacement in the position of the wires to occur. a) Show that the initial velocity $v_0$ of either wire is given by

$$v_0 = \frac{\mu_0 Q_0^2}{4\pi\lambda RCd},$$

where $Q_0$ is the initial charge on the capacitor of capacitance $C$ and $R$ is the total resistance of the circuit. b) Determine $v_0$ numerically if $R = 0.0100\ \Omega$. c) To what height $h$ will each wire rise as a result of the circuit connection?

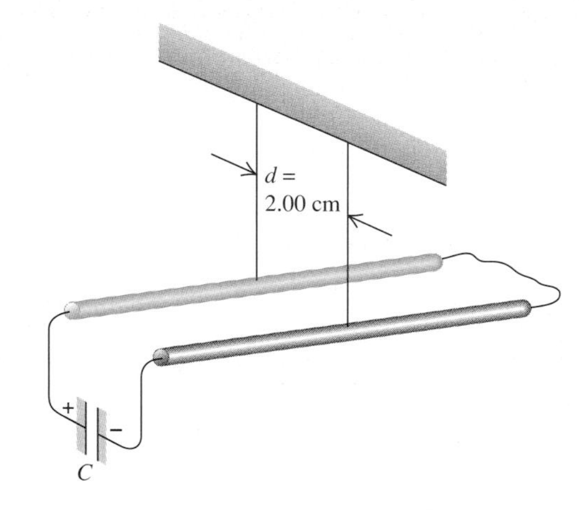


**FIGURE 29–46**

Recordings are made on magnetic tape by sending an electric signal from the microphone to a recording head, which is essentially a circular electromagnet with a gap in it. The electric signal produces a strong magnetic field across this gap. When a recording tape travels past the gap in the recording head, the field induces a magnetic pattern in the tape corresponding to the pattern of the original signal.

CHAPTER 30

# Electromagnetic Induction

Signal (sound) input
RECORDING HEAD
B
S
N
Unmagnetized tape
Signal recorded

The playback head of a tape deck reverses the recording process. As the magnetized tape travels past the gap in the playback head, the magnetized areas induce a magnetic field in the iron core of the playback head. The field changes according to the pattern recorded onto the tape, creating a changing magnetic flux in the coil wrapped around the core. This induces an emf in the coil, which becomes the electric signal that goes to the speakers.

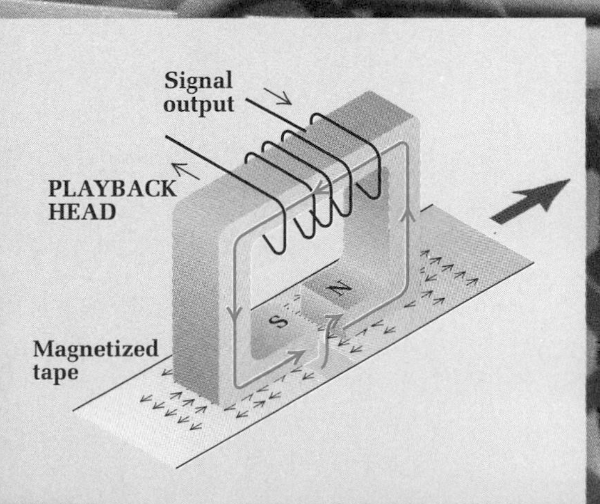

## KEY CONCEPTS

**• When the magnetic flux through a complete circuit changes, an electromotive force and a current are induced. Faraday's law states that the emf is equal in magnitude to the time rate of change of magnetic flux.**

**• When a conductor moves in a magnetic field, the induced emf can be understood on the basis of magnetic forces acting on the mobile charges in the conductor.**

**• Lenz's law states that the direction of an induced current is always such as to oppose the change of magnetic flux that induced it.**

**• A changing magnetic field induces an electric field that is nonconservative and of a sort that cannot be produced by any static charge distribution.**

**• Changing magnetic fields induce circulating currents called eddy currents in conducting materials.**

**• The fundamental relations of electromagnetism can be summarized neatly in four equations called Maxwell's equations.**

## INTRODUCTION

When you put a tape into your cassette player and lean back to listen, have you ever wondered how the player reads the music off the tape? The information is stored in tiny magnetized regions in the tape (of the order of 0.001 mm or 1 $\mu$m). The motion of these little magnets past the playback head induces currents that are amplified, processed, and fed to your loudspeakers or earphones.

What goes on in the playback head is an example of electromagnetic induction: A changing magnetic field causes a current in a circuit, even though there's no battery or other obvious source of electromotive force (emf). The heads in a computer disk drive read magnetically stored information from a disk in the same way.

In this chapter we discuss electromotive force that results from *magnetic* interactions. Many of the components of present-day electric power systems, including generators, transformers, and motors, depend directly on magnetically induced emfs. These systems would not be possible if we had to depend on chemical sources of emf such as batteries.

The central principle in this chapter is *Faraday's law.* This law relates induced emf to changing magnetic flux in a loop, often a closed circuit. We also discuss *Lenz's law,* which helps us to predict the directions of induced emf's and currents. This chapter provides the principles that we need to understand electrical energy conversion devices, including motors, generators, and transformers. It also paves the way for the analysis of electromagnetic waves in Chapter 33.

# 30–1 INDUCTION EXPERIMENTS

During the 1830s, several pioneering experiments with magnetically induced emf were carried out by Michael Faraday in England and Joseph Henry (first director of the Smithsonian Institution) in the United States. Figure 30–1 shows several examples. In Fig. 30–1a a coil of wire is connected to a galvanometer. When the nearby magnet is stationary, the meter shows no current. This isn't surprising; there is no source of emf in the circuit. But when we *move* the magnet, either toward or away from the coil, the meter shows current in the circuit, but *only* while the magnet is moving. If we keep the magnet stationary and move the coil and meter, we again detect a current during the motion. We call this an **induced current,** and the corresponding emf required to cause this current is called an **induced emf.**

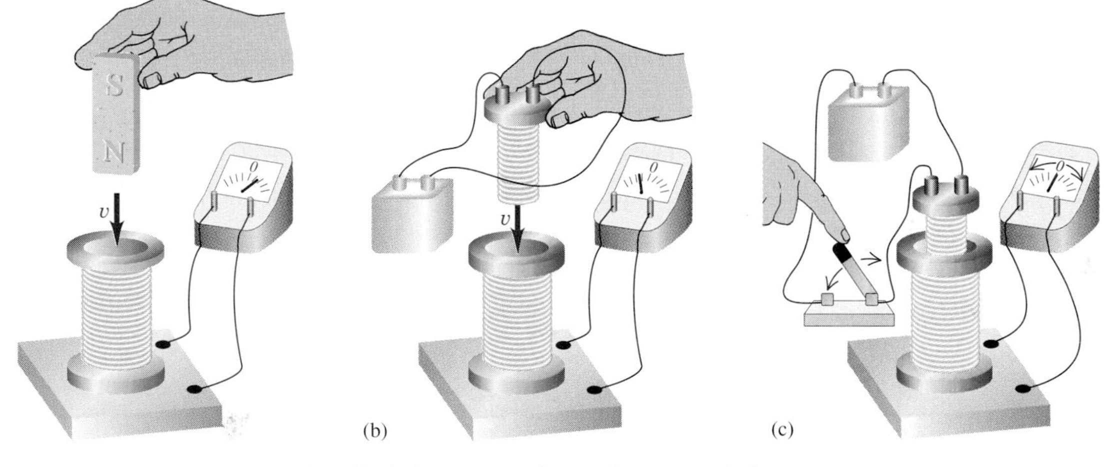


**30–1** (a) A magnet moving toward a coil of wire connected to a galvanometer induces a current in the coil. When magnet and coil move apart, the induced current is in the opposite direction. (b) A second coil carrying a current moves toward the coil connected to the galvanometer, inducing a current in it. When the coils move apart, the induced current is in the opposite direction. (c) When the switch is closed, a changing current in the inside coil induces a current through the galvanometer attached to the other coil.

In Fig. 30–1b we replace the magnet with a second coil connected to a battery. When the second coil is stationary, there is no current in the first. However, when we move the second coil toward or away from the first, or the first toward or away from the second, there is current in the first, but again *only* while one coil is moving relative to the other.

Finally, using the two-coil setup in Fig. 30–1c, we keep both coils stationary and vary the current in the second coil, either by opening and closing the switch or by changing the resistance $R$ (with the switch closed). We find that as we open or close the switch, there is a momentary current pulse in the first circuit. When we vary the resistance (and thus the current) in the second coil, there is an induced current in the first circuit, but only while the current in the second circuit is changing.

To explore further the common elements in such observations, let's consider a more detailed series of experiments with the situation shown in Fig. 30–2. We connect a coil of wire to a galvanometer and place it between the poles of an electromagnet whose magnetic field we can vary. Here's what we observe:

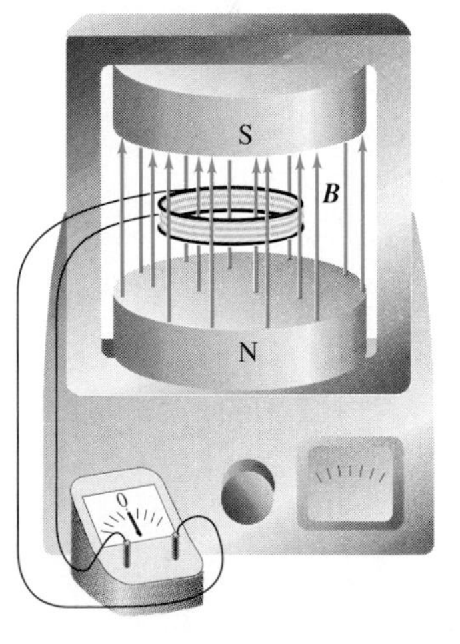


**30–2** When the $\boldsymbol{B}$ field is constant and the shape and location of the coil do not change, the induced current is zero. A current is induced when any of these factors change.

1. When there is no current in the electromagnet, $\boldsymbol{B} = \boldsymbol{0}$, and the galvanometer shows no current.
2. When the electromagnet is turned on, there is a momentary current through the meter as $B$ increases.

3. When $B$ levels off at a steady value, the current drops to zero, no matter how large $B$ is.
4. With the coil in a horizontal plane we deform it so as to decrease its area. The meter detects current *during* the deformation. When we return the coil to its original shape, there is current in the opposite direction while the area of the coil is increasing.
5. If we rotate the coil a few degrees about a horizontal axis, the meter detects current during the rotation in the same direction as when we decreased the area. When we rotate the coil back, there is a current in the opposite direction during this rotation.
6. If we jerk the coil out of the magnetic field, there is a current during the motion in the same direction as when we decreased the area.
7. If we decrease the number of turns in the coil by unwinding one or more turns, there is a current during the unwinding in the same direction as when we decreased the area. If we wind on more turns, there is a current in the opposite direction during the winding.
8. When the magnet is turned off, there is a momentary current in the direction opposite to the current when it was turned on.
9. The faster we carry out any of these changes, the greater the current.
10. If all these experiments are repeated with a coil with the same shape but different material and different resistance, the current in each case is inversely proportional to the total circuit resistance. This shows that the induced emf's that are causing the current do not depend on the material of the coil but only on its shape and the magnetic field.

The common element in all these experiments is changing *magnetic flux* $\Phi_B$ through the coil. Check back through the list to verify this statement. Faraday's law of induction, the subject of the next section, states that in all of these situations the induced emf is proportional to the *rate of change* of magnetic flux $\Phi_B$ through the circuit and that the direction of the induced emf depends on whether the flux is increasing or decreasing. For a coil with $N$ identical turns, the emf for a single turn is multiplied by a factor of $N$.

## 30–2 FARADAY'S LAW

The common element in all induction effects is changing magnetic flux through a circuit. *Faraday's law* states that the induced emf in a circuit is *directly proportional* to the time rate of change of magnetic flux through the circuit. Let's first review the concept of magnetic flux $\Phi_B$ (Section 28–3): For an area element $dA$ in a magnetic field $\boldsymbol{B}$ the magnetic flux $d\Phi_B$ through the area is

$$d\Phi_B = \boldsymbol{B} \cdot d\boldsymbol{A} = B\, dA \cos \phi,$$

where $\phi$ is the angle between $\boldsymbol{B}$ and $d\boldsymbol{A}$. (Be careful to distinguish between $\phi$ and $\Phi_B$.) The total magnetic flux $\Phi_B$ through a finite area is the integral of this expression over the area:

$$\Phi_B = \int \boldsymbol{B} \cdot d\boldsymbol{A} = \int B\, dA \cos \phi. \tag{30–1}$$

If $\boldsymbol{B}$ is uniform over a flat area $\boldsymbol{A}$, then

$$\Phi_B = \boldsymbol{B} \cdot \boldsymbol{A} = BA \cos \phi. \tag{30–2}$$

In Eqs. (30–1) and (30–2) we have to be careful to define the direction of the vector area $dA$ or $A$ unambiguously. There are always two directions perpendicular to any given area, and the sign of the magnetic flux through the area depends on which one we choose to be positive.

**Faraday's law of induction** states:

**The induced emf in a circuit equals the negative of the time rate of change of magnetic flux through the circuit.**

In symbols, Faraday's law is

$$\mathcal{E} = -\frac{d\Phi_B}{dt}. \tag{30–3}$$

To understand the negative sign, we have to introduce a sign convention for the induced emf $\mathcal{E}$. But first let's look at a simple sample of this law in action.

## EXAMPLE 30–1

**Current induced in a loop** In Figure 30–3 the magnetic field between the poles of the electromagnet is uniform at any time but is increasing at the rate of 0.020 T/s. The area of the conducting loop in the field is 120 cm$^2$, and the total circuit resistance, including the meter and the resistor, is 5.0 Ω. Find the induced emf and the induced current in the circuit.

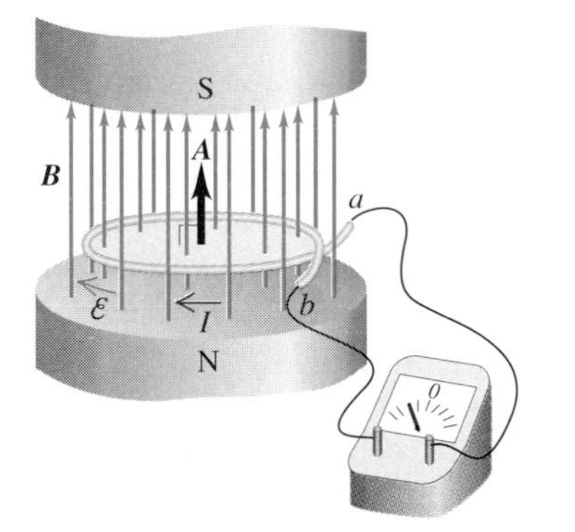


**30–3** A stationary conducting loop in an increasing magnetic field.

**SOLUTION** The area $A = 0.012\ \text{m}^2$ is constant, so the rate of change of magnetic flux is $A$ multiplied by the rate of change of magnetic field $B$:

$$\frac{d\Phi_B}{dt} = \frac{dB}{dt}A = (0.020\ \text{T/s})(0.012\ \text{m}^2)$$
$$= 2.4 \times 10^{-4}\ \text{V} = 0.24\ \text{mV}.$$

This, apart from a sign that we haven't discussed yet, is the induced emf $\mathcal{E}$.

It's worthwhile to verify unit consistency in this calculation. There are many ways to do this; one is to note that because of the magnetic force relation $\boldsymbol{F} = q\boldsymbol{v} \times \boldsymbol{B}$, 1 T = (1 N)/(1 C · m/s). The units of magnetic flux can thus be expressed as (1 T)(1 m$^2$) = 1 N · s · m/C, and the rate of change of magnetic flux as 1 N · m/C = 1 J/C = 1 V. So the unit of $d\Phi_B/dt$ is the volt, as required by Eq. (30–3). Also recall that the unit of magnetic flux is 1 T · m$^2$ = 1 Wb, so 1 V = 1 Wb/s.

Finally, the induced current $I$ in the circuit is

$$I = \frac{\mathcal{E}}{R} = \frac{2.4 \times 10^{-4}\ \text{V}}{5.0\ \Omega} = 4.8 \times 10^{-5}\ \text{A} = 0.048\ \text{mA}.$$

We haven't found the *direction* of the current yet. Doing that is our next task.

## Direction of Induced emf

We can find the direction of an induced emf or current by using Eq. (30–3), together with a set of sign rules, as follows.

1. Define a positive direction for the vector area $\boldsymbol{A}$.
2. The directions of $\boldsymbol{A}$ and the magnetic field $\boldsymbol{B}$ determine the sign of the magnetic flux $\Phi_B$ and its rate of change $d\Phi_B/dt$. Figure 30–4 (on page 842) shows several examples.
3. Curl the fingers of your right hand around the $\boldsymbol{A}$ vector with your right thumb in the direction of $\boldsymbol{A}$. An emf or current in the circuit that has the same direction as your curled fingers is positive, and one in the opposite direction is negative.

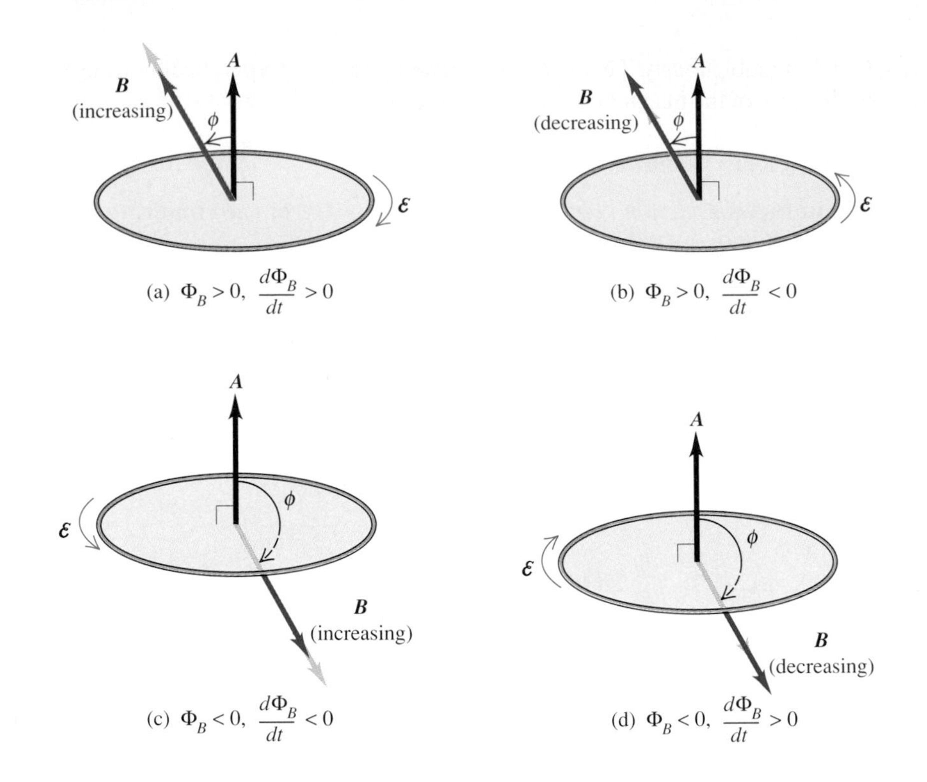


**30–4** The magnetic flux is becoming (a) more positive, (b) less positive, (c) more negative, and (d) less negative. Therefore $\Phi_B$ is increasing in (a) and (d) and decreasing in (b) and (c). In (a) and (d) the emf's are negative (clockwise as seen from above), and in (b) and (c) the emf's are positive (counterclockwise).

In Example 30–1, in which $\boldsymbol{A}$ is upward, a positive $\mathcal{E}$ would be directed counterclockwise around the loop, as seen from above. In this example, $\boldsymbol{A}$ and $\boldsymbol{B}$ are both upward, so $\Phi_B$ is positive. The magnitude $B$ is increasing, so $d\Phi_B/dt$ is positive. According to Eq. (30–3), $\mathcal{E}$ must be negative; its actual direction is clockwise around the loop, as shown in Fig. 30–3. If the loop were a battery, we would call point $a$ the + terminal and point $b$ the − terminal.

The induced current resulting from this emf is also clockwise. This current produces an additional magnetic field through the loop, and the right-hand rule shows that this field is *opposite* in direction to the field produced by the electromagnet. This is an example of a general rule called Lenz's law, which we'll study in Section 30–4. Lenz's law says that any induction effect tends to oppose the change that caused it; in this case the change is the increasing field through the loop. Lenz's law and the negative sign in Eq. (30–3) are both directly related to energy conservation. Think what would happen if the induced current were in the opposite direction. Then its field would *add* to the field already present, causing a further increase in flux and a further increase in current. We would have a runaway situation in which the current would increase (and energy would accumulate in the magnetic field) without limit, at least until the conductor melted. This would certainly violate energy conservation, and it doesn't happen in nature.

We invite you to check out the signs of the induced emf's and currents for the list of experiments cited at the end of Section 30–1. For example, when the loop is in a constant field and we tilt it or squeeze it to *decrease* the flux through it, the induced emf and current are counterclockwise, as seen from above.

If we have a coil with $N$ identical turns and the flux varies at the same rate through each turn, the induced emf's in the turns are all equal, are in *series,* and must be added. The total emf is

$$\mathcal{E} = -N\frac{d\Phi_B}{dt}. \tag{30–4}$$

## PROBLEM-SOLVING STRATEGY

### Faraday's law

**1.** To calculate the rate of change of magnetic flux, you first have to understand what is making the flux change. Is the conductor moving? Is it changing orientation? Is the magnetic field changing? Remember that it's not the flux itself that counts, but its *rate of change*.

**2.** Remember the sign rule for the positive directions of magnetic flux and emf, and use it consistently when you implement Eq. (30–3) or (30–4). If your conductor has $N$ turns in a coil, don't forget to multiply by $N$.

**3.** Use Faraday's law to obtain the emf. Interpret the sign of your result with reference to the sign rules to determine the direction of the induced current. If the circuit resistance is known, you can then calculate the current.

## EXAMPLE 30–2

A coil of wire containing 500 circular loops with radius 4.00 cm is placed between the poles of a large electromagnet, where the magnetic field is uniform, at an angle of 60° with the plane of the coil (Fig. 30–5). The field decreases at a rate of 0.200 T/s. What is the magnitude of the resulting induced emf?

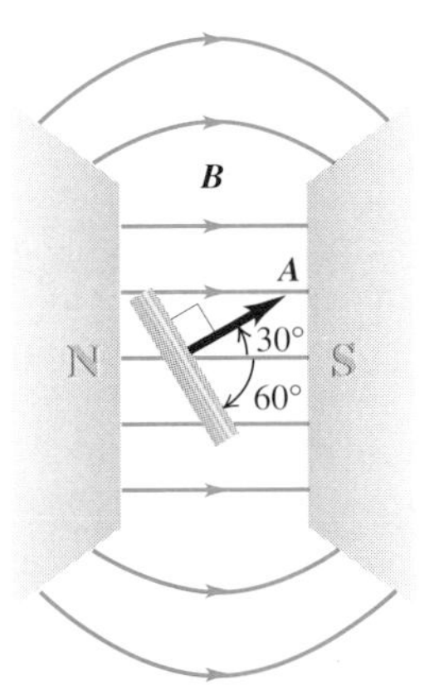


**30–5** The magnitude of $\boldsymbol{B}$ is decreasing. With the direction of $\boldsymbol{A}$ shown, the flux through the coil is decreasing, and $\mathcal{E}$ is positive. This corresponds to clockwise emf and current as you look in from the left along the direction of $\boldsymbol{A}$. The additional $\boldsymbol{B}$ field caused by the induced current tends to compensate for the decrease in flux.

**SOLUTION** Choose the direction for $\boldsymbol{A}$ shown in Fig. 30–5. Then $\phi$ is the angle between $\boldsymbol{A}$ and $\boldsymbol{B}$. The flux $\Phi_B$ at any time is given by $\Phi_B = BA\cos\phi$, and the rate of change of flux by $d\Phi_B/dt = (dB/dt)A\cos\phi$. In this case, $dB/dt = -0.200$ T/s, $A = \pi(0.0400\text{ m})^2 = 0.00503\text{ m}^2$, $\phi = 30°$ (*not* 60°), and

$$\begin{aligned}\frac{d\Phi_B}{dt} &= \frac{dB}{dt}A\cos 30° \\ &= (-0.200\text{ T/s})(0.00503\text{ m}^2)(0.866) \\ &= -0.000871\text{ T}\cdot\text{m}^2/\text{s} = -0.000871\text{ Wb/s}.\end{aligned}$$

From Eq. (30–4) the induced emf is

$$\begin{aligned}\mathcal{E} &= -N\frac{d\Phi_B}{dt} \\ &= -(500)(-0.000871\text{ Wb/s}) \\ &= 0.435\text{ V}.\end{aligned}$$

When we look in from the left, in the direction of the area vector (30° above the magnetic field $\boldsymbol{B}$) the positive direction for $\mathcal{E}$ is clockwise, according to the right-hand rule. The emf is positive and thus is also clockwise. If the ends of the wire are connected to a resistor, the direction of current in the coil is clockwise. A clockwise current gives added magnetic flux through the coil in the same direction as the flux from the electromagnet and therefore tends to oppose the decrease in total flux.

## EXAMPLE 30–3

**A simple alternator** Figure 30–6 (on page 844) shows a rectangular loop that rotates with constant angular velocity $\omega$ about the axis shown. The magnetic field $\boldsymbol{B}$ is uniform and constant. At time $t = 0$, $\phi = 0$. Determine the induced emf.

**SOLUTION** The flux $\Phi_B$ through the loop equals its area $A$ multiplied by the component of $\boldsymbol{B}$ perpendicular to the area; that is, $B\cos\phi$.

$$\Phi_B = BA\cos\phi = BA\cos\omega t.$$

Then

$$\mathcal{E} = -\frac{d\Phi_B}{dt} = \omega BA\sin\omega t.$$

The induced emf $\mathcal{E}$ varies sinusoidally with time (Fig. 30–6b).

Because $\mathcal{E}$ is directly proportional to $\omega$ and $B$, we could use the emf in a rotating coil to measure rotational speed or magnetic field. (This principle is used in some practical instruments to measure magnetic field.) When the plane of the loop is perpendicular to $\boldsymbol{B}$ ($\phi = 0$ or 180°), $\Phi_B$ reaches its maximum and minimum values. At these times its instantaneous rate of change is zero, and $\mathcal{E}$ is zero. Also, $\mathcal{E}$ is greatest when the plane of the loop is parallel to $\boldsymbol{B}$ ($\phi = 90°$ or 270°) and $\Phi_B$ is changing most rapidly. Finally, we note that the induced emf does not depend on the *shape* of the loop, but only on its area.

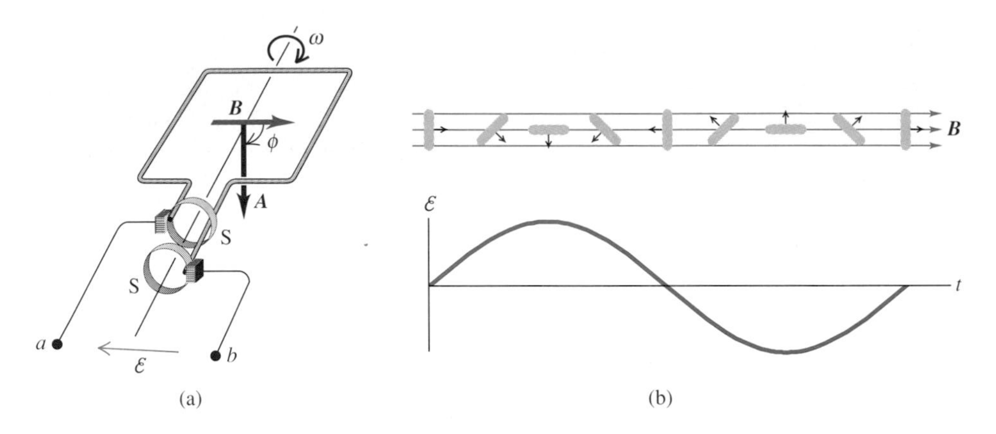


**30–6** (a) Schematic diagram of a simple alternator, using a conducting loop rotating in a magnetic field. Connections from each end of the loop to the external circuit are made by means of that end's slip ring S. The system is shown at the time when $\omega t = 90°$. (b) Graph of the resulting emf at terminals *ab*, along with corresponding positions of the loop during one complete rotation.

The rotating loop is the prototype of one kind of alternating-current generator, or *alternator;* it develops a sinusoidally varying emf. We can use it as a source of emf in an external circuit by use of two *slip rings* S, which rotate with the loop, as shown in Fig. 30–6a. Stationary contacts called *brushes* slide on the rings and are connected to the output terminals $a$ and $b$.

We can use a similar scheme to obtain an emf that always has the same sign. The arrangement shown in Fig. 30–7a is called a *commutator;* it reverses the connections to the external circuit at angular positions where the emf reverses. The resulting emf is shown in Fig. 30–7b. This device is the prototype of a dc generator. Commercial dc generators have a large number of coils and commutator segments; this arrangement smooths out the bumps in the emf, so the terminal voltage is not only one-directional, but also practically constant. This brush-and-commutator arrangement was also used in the direct-current motor that we discussed in Section 28–8. The motor's back emf is just the emf induced by the changing magnetic flux through its rotating coils.

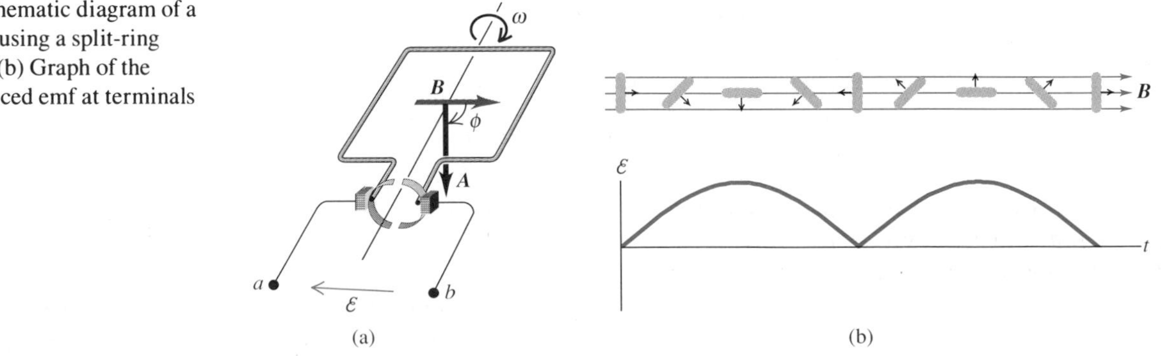


**30–7** (a) Schematic diagram of a dc generator, using a split-ring commutator. (b) Graph of the resulting induced emf at terminals *ab*.

## EXAMPLE 30–4

**The Faraday disk dynamo** A disk with radius $R$, shown in Fig. 30–8, lies in the $xy$-plane and rotates with constant angular velocity $\omega$ about the $z$-axis. The disk is in a uniform, constant $\boldsymbol{B}$ field parallel to the $z$-axis. Find the induced emf between the center and the rim of the disk.

**SOLUTION** We take as our circuit the outline of the red area in Fig. 30–8. There is no flux through the rectangular portion in the $yz$-plane because $\boldsymbol{B}$ is parallel to that plane. The red part of the disk in the $xy$-plane is a sector; its area is $\frac{1}{2}R^2\theta$, and the flux $\Phi_B$ through it is

$$\Phi_B = \tfrac{1}{2}BR^2\theta.$$

As the disk rotates, the red-shaded area increases. In a time $dt$ the angle $\theta$ increases by $d\theta = \omega dt$, and the flux increases by

$$d\Phi_B = \tfrac{1}{2}BR^2\,d\theta = \tfrac{1}{2}BR^2\omega\,dt.$$

The induced emf is

$$\mathcal{E} = -\frac{d\Phi_B}{dt} = -\tfrac{1}{2}BR^2\omega.$$

We can use this device as a source of emf in a circuit by completing the circuit through sliding contacts, or *brushes* (b in the figure). The emf in such a disk was studied by Faraday; the device is a called a *Faraday disk dynamo,* or a *homopolar generator.*

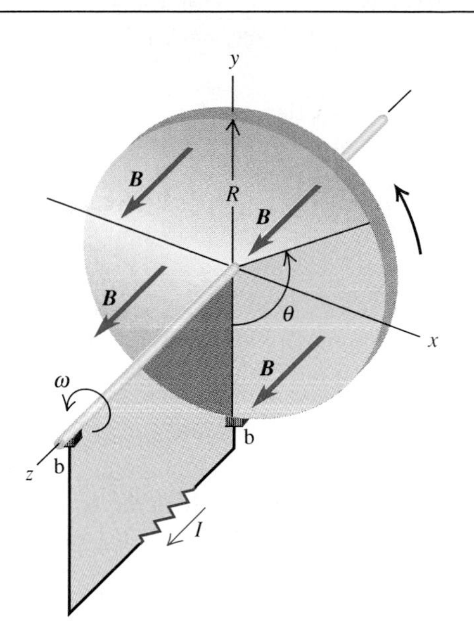


**30–8** A Faraday disk dynamo. The magnetic flux increases because the red-shaded area increases.

## EXAMPLE 30–5

**The search coil** One practical way to measure magnetic field strength uses a small, closely wound coil with $N$ turns called a *search coil*. If the area enclosed by the coil is $A$ and the area vector is initially aligned with a magnetic field with magnitude $B$, the flux $\Phi_B$ through the coil is $\Phi_B = BA$. Now if we quickly rotate the coil a quarter-turn about a diameter or snatch the coil out of the field, the flux decreases rapidly from $BA$ to zero. While the flux is decreasing, there is a momentary induced emf; and a momentary induced current occurs in the external circuit connected to the coil. The rate of change of flux through the coil is proportional to the current, or rate of flow of charge, so it is easy to show that the *total* flux change is proportional to the total charge that flows around the circuit. We can build an instrument that measures this total charge, and from this we can compute $B$. We leave the details as a problem. Strictly speaking, this method gives only the *average* field over the area of the coil. But if the area is small, this is very nearly equal to the field at the center of the coil.

## EXAMPLE 30–6

Figure 30–9 shows a U-shaped conductor in a uniform magnetic field $\boldsymbol{B}$ perpendicular to the plane of the figure, directed *into* the page. We lay a metal rod with length $L$ across the two ends of the conductor, forming a loop, and move the rod to the right with constant velocity $\boldsymbol{v}$. Find the magnitude and direction of the resulting induced emf.

**SOLUTION** The magnetic flux through the loop is changing because the area is increasing. In a time $dt$ the sliding rod moves a distance $v\,dt$ and the area increases by $dA = Lv\,dt$. Take the positive direction for area to be into the plane, parallel to $\boldsymbol{B}$. Then the magnetic flux through the loop is positive, and in time $dt$ it increases by an amount

$$d\Phi_B = B\,dA = BLv\,dt.$$

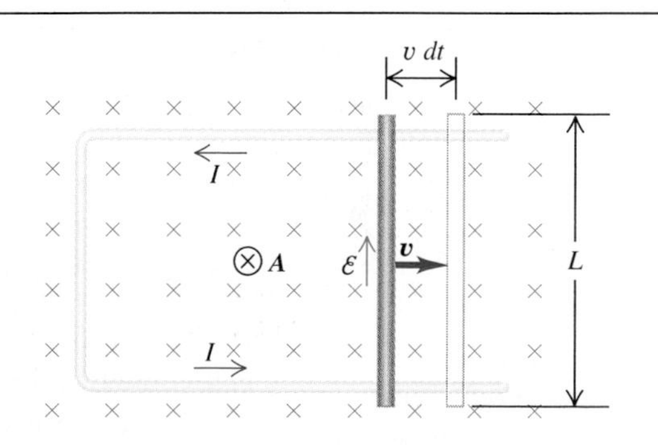


**30–9** The magnetic field $\boldsymbol{B}$ and the vector area $\boldsymbol{A}$ are both directed *into* the figure. The increase in magnetic flux (caused by an increase in area) induces the emf and current shown.

The induced emf is

$$\mathcal{E} = -\frac{d\Phi_B}{dt} = -BLv.$$

To interpret the negative sign, recall our sign rule: Point the right thumb in the direction of the vector area $\boldsymbol{A}$; the fingers curl in the direction of the positive sense of emf. In our case this is the clockwise direction around the loop in Fig. 30–9. The negative sign means that the actual emf is opposite to this, or counterclockwise, as shown in the figure. The induced current is also counterclockwise.

## 30–3 MOTIONAL ELECTROMOTIVE FORCE

When a conductor moves in a magnetic field, we can gain added insight into the resulting induced emf by considering the magnetic forces on charges in the conductor. Figure 30–10a shows the same moving rod that we just discussed in Example 30–6, separated for the moment from the U-shaped conductor. The magnetic field $\boldsymbol{B}$ is uniform and directed into the page, and we give the rod a constant velocity $\boldsymbol{v}$ to the right. A charged particle $q$ in the rod then experiences a magnetic force $\boldsymbol{F} = q\boldsymbol{v} \times \boldsymbol{B}$ with magnitude $F = |q|vB$. We'll assume in the following discussion that $q$ is positive; in that case the direction of this force is *upward* along the rod, from $b$ toward $a$. (We'll see later that the emf and current are the same whether $q$ is positive or negative.)

This magnetic force causes the free charges in the rod to move, creating an excess of positive charge at the upper end $a$ and negative charge at the lower end $b$. This, in turn, creates an electric field $\boldsymbol{E}$ in the direction from $a$ to $b$, which exerts a downward force $q\boldsymbol{E}$ on the charges. The accumulation of charge at the ends of the rod continues until $\boldsymbol{E}$ is large enough for the downward electric force (magnitude $qE$) to cancel exactly the upward magnetic force (magnitude $qvB$). Then $qE = qvB$ and the charges are in equilibrium, with point $a$ at higher potential than point $b$.

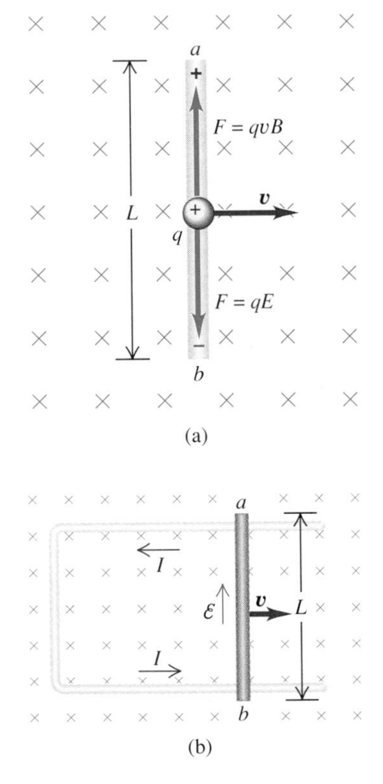


**30–10** A conducting rod moving in a uniform magnetic field. (a) The rod, the velocity, and the field are mutually perpendicular. (b) Direction of induced current in the circuit.

What is the magnitude of the potential difference $V_{ab}$? It is equal to the electric-field magnitude $E$ multiplied by the length $L$ of the rod. From the above discussion, $E = vB$, so

$$V_{ab} = EL = vBL, \qquad (30\text{–}5)$$

with point $a$ at higher potential than point $b$.

Now suppose the moving rod slides along a stationary U-shaped conductor, forming a complete circuit (Fig. 30–10b). No *magnetic* force acts on the charges in the stationary conductor, but there is an *electric* field caused by the charge accumulations at $a$ and $b$. Under the action of this field a current is established in the counterclockwise sense around this complete circuit. The moving rod has become a source of electromotive force; within it, charge moves from lower to higher potential, and in the remainder of the circuit, charge moves from higher to lower potential. We call this a **motional electromotive force,** denoted by $\mathcal{E}$; we can write

$$\mathcal{E} = vBL. \qquad (30\text{–}6)$$

This is the same result that we obtained in Section 30–2 using Faraday's law.

The emf associated with the moving rod in Fig. 30–10 is analogous to that of a battery with its positive terminal at $a$ and its negative terminal at $b$, although the origins of the two emf's are completely different. In each case a non-electrostatic force acts on the charges in the device in the direction from $b$ to $a$, and the emf is the work per unit charge done by this force when a charge moves from $b$ to $a$ in the device. When the device is connected to an external circuit, the direction of current is from $b$ to $a$ in the device and from $a$ to $b$ in the external circuit (in this case, the U-shaped conductor).

If we express $v$ in meters per second, $B$ in teslas, and $L$ in meters, $\mathcal{E}$ is in joules per coulomb or volts. We invite you to verify this.

## EXAMPLE 30–7

Suppose the length $L$ in Fig. 30–10 is 0.10 m, the velocity $v$ is 2.5 m/s, the total resistance of the loop is 0.030 Ω, and $B$ is 0.60 T. Find $\mathcal{E}$, the induced current, the force acting on the rod, and the mechanical power needed to keep the rod moving.

**SOLUTION** From Eq. (30–6) the emf $\mathcal{E}$ is

$$\mathcal{E} = vBL = (2.5\ \text{m/s})(0.60\ \text{T})(0.10\ \text{m}) = 0.15\ \text{V}.$$

The current $I$ in the loop is

$$I = \frac{\mathcal{E}}{R} = \frac{0.15\ \text{V}}{0.030\ \Omega} = 5.0\ \text{A}.$$

Because of this current, a force $\boldsymbol{F}$ acts on the rod in the direction *opposite* to its motion. From Eq. (28–16) this force has magnitude

$$F = ILB = (5.0\ \text{A})(0.10\ \text{m})(0.60\ \text{T}) = 0.30\ \text{N}.$$

To maintain the bar's constant-velocity motion, despite this resisting force, requires the application of an equal and opposite additional force. The rate at which this additional force does work is the *power P* needed to keep the rod moving:

$$P = Fv = (0.30\ \text{N})(2.5\ \text{m/s}) = 0.75\ \text{W}.$$

The rate at which the induced emf delivers electrical energy to the circuit is the product $\mathcal{E}I$:

$$P = \mathcal{E}I = (0.15\ \text{V})(5.0\ \text{A}) = 0.75\ \text{W}.$$

This is equal to the mechanical power input, $Fv$, as we should expect. The system is converting mechanical energy (work) into electrical energy. Finally, the rate of *dissipation* of electrical energy in the circuit resistance is $P = I^2R = (5.0\ \text{A})^2(0.030\ \Omega) = 0.75\ \text{W}$; this is also to be expected.

## EXAMPLE 30–8

**The Faraday dynamo, revisited** Consider again the rotating disk in Fig. 30–11, which we analyzed in Example 30–4 (Section 30–2). The conditions are the same as in that example. Find the induced emf between the center and the rim of the disk, using the concept of motional emf.

**SOLUTION** Parts of the disk at different distances from the center move at different speeds, so we can't just multiply $vB$ by the radius. Let's take as the moving conductor an infinitesimal radial length $dr$ at radius $r$. The speed $v$ of $dr$ is $v = \omega r$, and the motional emf $d\mathcal{E}$ induced in $dr$ is

$$d\mathcal{E} = vB\,dr = \omega Br\,dr.$$

The total emf between center and rim is the sum of all such contributions:

$$|\mathcal{E}| = \int_0^R \omega Br\,dr = \tfrac{1}{2}\omega BR^2.$$

This is the same result that we obtained using Faraday's law.

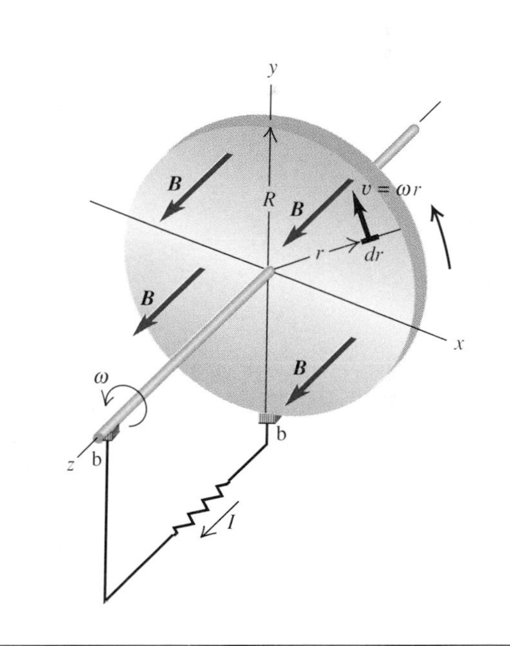


**30–11** The rotating disk again. The emf is induced along radial lines of the rotating disk and is connected to an external circuit through the two sliding contacts labeled b.

We can generalize the concept of motional emf for a conductor with any shape, moving in any magnetic field, uniform or not. For an element $d\boldsymbol{l}$ of conductor the contribution $d\mathcal{E}$ to the emf is the magnitude $dl$ multiplied by the component of $\boldsymbol{v} \times \boldsymbol{B}$ parallel to $dl$; that is,

$$d\mathcal{E} = \boldsymbol{v} \times \boldsymbol{B} \cdot d\boldsymbol{l}.$$

For any two points $a$ and $b$ the motional emf in the direction from $b$ to $a$ is

$$\mathcal{E} = \int_b^a \boldsymbol{v} \times \boldsymbol{B} \cdot d\boldsymbol{l}. \tag{30–7}$$

### EXAMPLE 30–9

An SR-71a Blackbird is flying north with a velocity $\boldsymbol{v} = (1200 \text{ m/s})\boldsymbol{j}$ (Fig. 30–12) in a region where the earth's magnetic field is

$$\boldsymbol{B} = (25\boldsymbol{j} - 43\boldsymbol{k}) \times 10^{-6}\text{ T}.$$

In the figure, $\boldsymbol{i}$ is toward the east, $\boldsymbol{j}$ is toward the (magnetic) north, and $\boldsymbol{k}$ is up. The plane is about 33 m long, with a 17-m wingspan. a) Find the magnitude and direction of the field. b) What is the motional emf from back to front? c) What is the motional emf from left to right between the wing tips?

**SOLUTION** a) The magnitude of the field is $(25^2 + 43^2)^{1/2} \times 10^{-6}\text{ T} = 50 \times 10^{-6}\text{ T}$. The direction is at angle $\theta$ below the horizontal (angle of dip) given by $\theta = \arctan(-43/25) = -60°$.

b) We use Eq. (30–7); $\boldsymbol{v}$ and $\boldsymbol{B}$ are constant, so

$$\mathcal{E} = \int \boldsymbol{v} \times \boldsymbol{B} \cdot d\boldsymbol{l} = (\boldsymbol{v} \times \boldsymbol{B}) \cdot \int d\boldsymbol{l} = \boldsymbol{v} \times \boldsymbol{B} \cdot \boldsymbol{l}.$$

First we find $\boldsymbol{v} \times \boldsymbol{B}$:

$$\boldsymbol{v} \times \boldsymbol{B} = (1200 \text{ m/s})\boldsymbol{j} \times (25\boldsymbol{j} - 43\boldsymbol{k}) \times 10^{-6}\text{ T}$$
$$= (-0.052 \text{ T}\cdot\text{m/s})\boldsymbol{i}.$$

The back-to-front line is in the direction of $\boldsymbol{j}$; it is perpendicular to $\boldsymbol{v} \times \boldsymbol{B}$, and the induced emf is zero.

c) For the wing we integrate from left to right between wing tips; then $\boldsymbol{l} = (17 \text{ m})\boldsymbol{i}$, and

$$\mathcal{E} = \boldsymbol{v} \times \boldsymbol{B} \cdot \boldsymbol{l} = (-0.052 \text{ T}\cdot\text{m/s})\boldsymbol{i} \cdot (17 \text{ m})\boldsymbol{i} = -0.88 \text{ V}.$$

The negative sign is analogous to a drop in potential as we go from left tip to right tip. Viewed as a battery, the wing has its positive terminal on the left (west) side. Note that the pilot can't use this induced emf as an energy source because the remaining part of any closed circuit in the plane will have an exactly opposite emf. The total flux through any such closed circuit is constant, provided that $\boldsymbol{B}$ is constant.

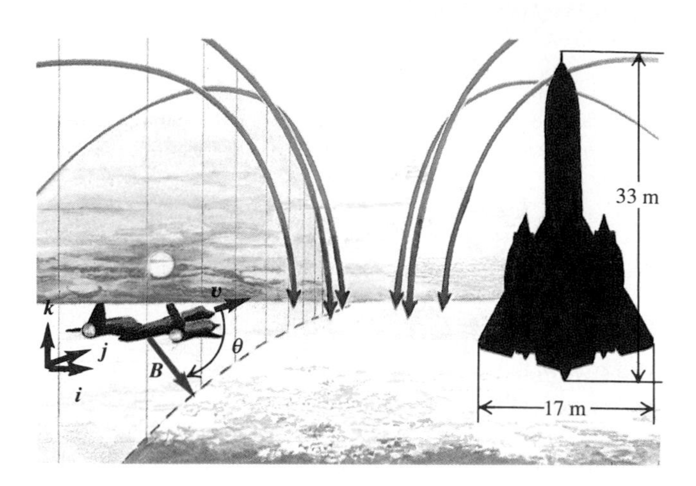


**30–12** Determining induced emf in a plane flying through the earth's magnetic field.

## 30–4 LENZ'S LAW

Lenz's law is a convenient alternative method for determining the sign or direction of an induced current or emf or the direction of the associated non-electrostatic field. Lenz's law is not an independent principle; it can be derived from Faraday's law. It always gives the same results as the sign rules that we have introduced in connection with Faraday's law, but it is often easier to use. Lenz's law also helps us to gain an intuitive understanding of various induction effects and of the role of energy conservation. H. F. E. Lenz (1804–1865) was a German scientist who duplicated independently many of the discoveries of Faraday and Henry. **Lenz's law** states:

> **The direction of any magnetic induction effect is such as to oppose the cause producing it.**

The "cause" may be motion of a conductor in a magnetic field, or it may be changing flux through a stationary circuit, or any combination. In the first case the direction of the induced current in the moving conductor is such that the direction of the magnetic-field force on the conductor is opposite in direction to its motion. The motion of the conductor, which caused the induced current, is opposed.

In the second case, where there is changing flux in a stationary circuit, the induced current sets up a magnetic field of its own. Within the area bounded by the circuit this field is *opposite* to the original field if the original field is *increasing* but is in the *same* direction as the original field if the latter is *decreasing.* That is, the induced current opposes the *change in flux* through the circuit (*not* the flux itself). In all these cases the induced current tries to preserve the *status quo* by opposing motion or a change of flux.

To have an induced current, we need a complete circuit. If a conductor does not form a complete circuit, then we can mentally complete the circuit between the ends of the conductor and use Lenz's law to determine the direction of the current. We can then deduce the polarity of the ends of the open-circuit conductor. The direction from the − end to the + end within the conductor is the direction the current would have if the circuit were complete.

## EXAMPLE 30–10

In Fig. 30–13, when the conducting rod moves to the right, a counterclockwise current is induced in the loop. The magnetic-field force exerted on the moving conductor as a result of this current is to the left, *opposing* the conductor's motion, in agreement with Lenz's law.

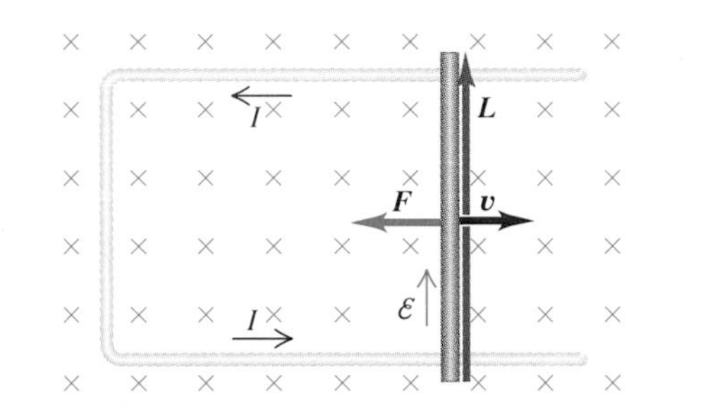


**30–13** The force $\boldsymbol{F} = I\boldsymbol{L} \times \boldsymbol{B}$ due to the induced current is to the left, opposite to $\boldsymbol{v}$.

## EXAMPLE 30–11

In the Faraday disk (or homopolar generator) shown in Fig. 30–14 the induced current in the disk is in the $-y$-direction, along a radius from the center of the disk to the contact ($b$) at the bottom of the rim. The magnetic-field force on this current is in the $-x$-direction. This force provides a torque that opposes the rotation of the disk. If we reverse the direction of rotation, the induced current also reverses direction, and again the magnetic-field force opposes the motion, again in agreement with Lenz's law.

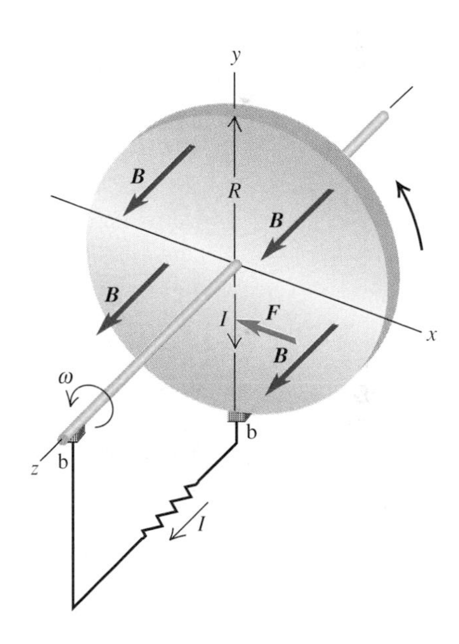


**30–14** The force on the current $I$ in the disk provides a torque that opposes the disk's rotation.

## EXAMPLE 30–12

In Fig. 30–15 the flux through the coil is increasing. The induced emf causes an induced current whose magnetic field inside the coil is downward, opposing the increase in flux. This situation is the same as in Example 30–1. Once again, Lenz's law is obeyed.

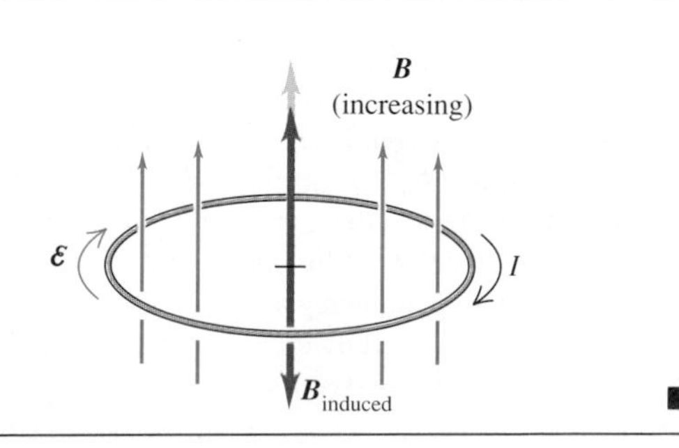


**30–15** The induced current is clockwise, as seen from above the loop. The added field that it causes is downward, opposing the increase in the upward field.

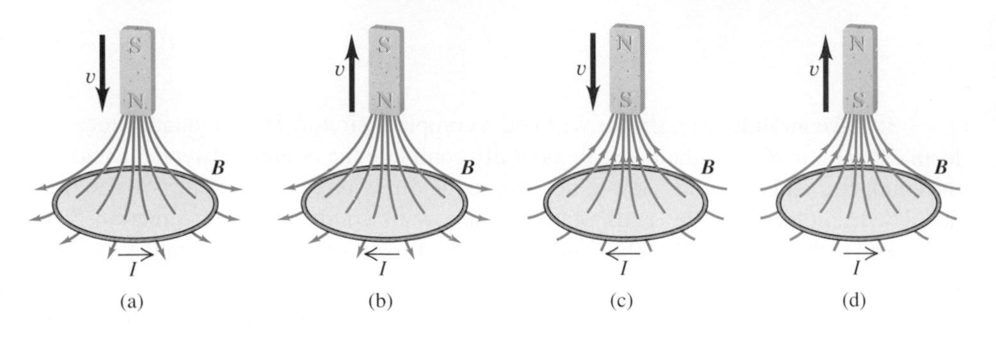


**30–16** Directions of induced currents as a bar magnet moves along the axis of a conducting loop. If $v = 0, I = 0$.

Figure 30–16 shows several applications of Lenz's law to a magnet moving near a conducting loop.

Lenz's law is also directly related to energy conservation. In Example 30–10, if the induced current were in the opposite direction, the resulting magnetic force on the rod would accelerate it to ever-increasing speed with no external energy source, despite the fact that electrical energy is being dissipated in the loop; this would be a clear violation of energy conservation.

## 30–5 INDUCED ELECTRIC FIELDS

We have studied induced emf's that occur when a conductor moves in a magnetic field and those that result from changing flux through a stationary conductor. With a moving conductor we can understand the induced emf on the basis of magnetic forces on charges in the conductor. We need to think a little more about what's happening in the second type of situation. In particular, just what is it that pushes the charges around the circuit when we have a stationary conductor with changing flux?

As an example, let's consider the situation shown in Fig. 30–17. A long, thin solenoid with cross-section area $A$ and $n$ turns per unit length is encircled at its center by a circular conducting loop. The galvanometer G measures the current in the loop. A current $I$ in the winding of the solenoid sets up a magnetic field $\boldsymbol{B}$ along the solenoid axis, as shown, with magnitude $B$ given by Eq. (29–21): $B = \mu_0 nI$, where $n$ is the number of turns per unit length. If we neglect the small field outside the solenoid, then the magnetic flux $\Phi_B$ through the loop is

$$\Phi_B = BA = \mu_0 nIA.$$

When the solenoid current $I$ changes with time, the magnetic flux $\Phi_B$ also changes, and according to Faraday's law, the induced emf in the loop is given by

$$\mathcal{E} = -\frac{d\Phi_B}{dt} = -\mu_0 nA\frac{dI}{dt}.$$

If the total resistance of the loop is $R$, the induced current in the loop, which we may call $I'$, is $I' = \mathcal{E}/R$.

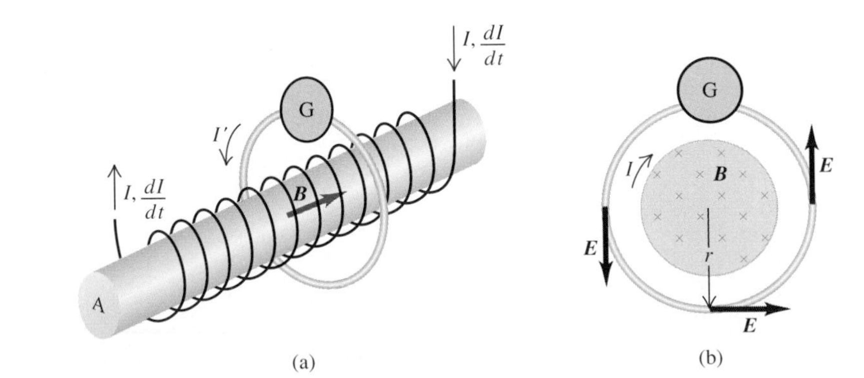


**30–17** (a) The windings of a long solenoid carry a current $I$ that is increasing at a rate $dI/dt$. The magnetic flux in the solenoid is increasing at a rate $d\Phi_B/dt$, and this changing flux passes through a wire loop. An emf $\mathcal{E} = -d\Phi_B/dt$ is induced in the loop, inducing a current $I'$ that is measured by the galvanometer G. (b) Cross-section view.

## EXAMPLE 30–13

Suppose the long solenoid in Fig. 30–17a is wound with 1000 turns per meter, and the current in its windings is increasing at the rate of 100 A/s. The cross-section area of the solenoid is 4.0 $\text{cm}^2 = 4.0 \times 10^{-4}\ \text{m}^2$. Find the magnitude of the induced emf.

**SOLUTION** We can use the result that we just worked out. The induced emf is

$$\mathcal{E} = -\frac{d\Phi_B}{dt} = -\mu_0 n A \frac{dI}{dt}$$

$$= -(4\pi \times 10^{-7}\ \text{Wb/A}\cdot\text{m})(1000\ \text{turns/m}) \times (4.0 \times 10^{-4}\ \text{m}^2)(100\ \text{A/s})$$

$$= -50 \times 10^{-6}\ \text{Wb/s} = -50 \times 10^{-6}\ \text{V} = -50\ \mu\text{V}.$$

In Fig. 30–17b the flux *into* the plane of the figure is increasing; according to our right-hand rule, the negative sign of $\mathcal{E}$ shows that the emf is in the counterclockwise direction. We invite you to verify that Lenz's law also predicts this direction.

---

But what *force* makes the charges move around the loop? It can't be a magnetic force because the conductor isn't moving in a magnetic field and in fact isn't even *in* a magnetic field. We are forced to conclude that there has to be an **induced electric field** in the conductor, *caused by the changing magnetic flux*. This may be a little jarring; we are accustomed to thinking about electric field as caused by electric charges, and now we are saying that a changing magnetic field somehow acts as a source of electric field. Furthermore, it's a strange sort of electric field. When a charge $q$ goes once around the loop, the total work done on it by the electric field must be equal to $q$ times the emf $\mathcal{E}$. That is, the electric field in the loop *is not conservative,* as we used the term in Chapter 24, because the line integral of $\mathbf{E}$ around a closed path is not zero. Indeed, this line integral, representing the work done by the induced $\mathbf{E}$ field per unit charge, is equal to the induced emf $\mathcal{E}$:

$$\oint \mathbf{E} \cdot d\mathbf{l} = \mathcal{E}. \qquad (30\text{–}8)$$

From Faraday's law the emf $\mathcal{E}$ is also the negative of the rate of change of magnetic flux through the loop. Thus for this case we can restate Faraday's law as

$$\oint \mathbf{E} \cdot d\mathbf{l} = -\frac{d\Phi_B}{dt}. \qquad (30\text{–}9)$$

As an example, suppose the circular loop in Fig. 30–17b has radius $r$. Because of axial symmetry, the electric field $\mathbf{E}$ has the same magnitude at every point on the circle and is tangent to it at each point. (Symmetry would also permit the field to be *radial,* but then Gauss's law would require the presence of charge inside the circle, and there is none.) The line integral in Eq. (30–9) becomes simply the magnitude $E$ times the circumference $2\pi r$ of the loop, $\oint \mathbf{E} \cdot d\mathbf{l} = 2\pi r E,$ and Eq. (30–9) gives

$$E = \frac{1}{2\pi r}\left|\frac{d\Phi_B}{dt}\right|. \qquad (30\text{–}10)$$

The directions of $\mathbf{E}$ at several points on the loop are shown in the figure. We know that $\mathbf{E}$ has to have this direction because $\oint \mathbf{E} \cdot d\mathbf{l}$ has to be negative when $d\Phi_B/dt$ is positive.

Now let's try to put the pieces together. Faraday's law, Eq. (30–3), is valid for two rather different situations. In one, an emf is induced by magnetic forces on charges when a conductor moves through a magnetic field. In the other a time-varying magnetic field induces an electric field in a stationary conductor and thus induces an emf. In the second case the $\mathbf{E}$ field is induced even when no conductor is present. This $\mathbf{E}$ field differs from an electro*static* field in an important way. It is *nonconservative;* the line integral $\oint \mathbf{E} \cdot d\mathbf{l}$ around a closed path is not zero, and when a charge moves around a closed path, the field does a nonzero amount of work on it. It follows that for such a field the concept of *potential* has no meaning. We call such a field a **non-electrostatic field.** In contrast, an electro*static* field is *always* conservative, as we discussed in Section 24–1, and always has an associated potential function. Despite this difference, the fundamental effect of *any* electric field is to exert a force $\boldsymbol{F} = q\boldsymbol{E}$ on a charge $q$. This relation is valid whether $\boldsymbol{E}$ is a conservative field produced by a charge distribution or a nonconservative field caused by changing magnetic flux.

So a changing magnetic field acts as a source of electric field of a sort that we *cannot* produce with any static charge distribution. This may seem strange, but it's the way nature behaves. Furthermore, this development is analogous to our generalized Ampere's law with displacement current, which shows that a changing *electric* field acts as a source of *magnetic* field. We'll explore this symmetry between the two fields in greater detail in Section 30–7 and in our study of electromagnetic waves in Chapter 33.

If there's any remaining doubt in your mind about the reality of magnetically induced electric fields, consider a few of the many practical applications. In the playback head of a tape deck, currents are induced in a stationary coil by the motion of variously magnetized regions of the tape past it. Computer disk drives operate on the same principle. Pickups in electric guitars use currents induced in stationary pickup coils by the vibration of nearby ferromagnetic strings. Alternators found in most cars use rotating permanent magnets to induce currents in stationary coils. The list goes on and on; whether we realize it or not, magnetically induced electric fields play an important role in everyday life.

## *30–6 EDDY CURRENTS

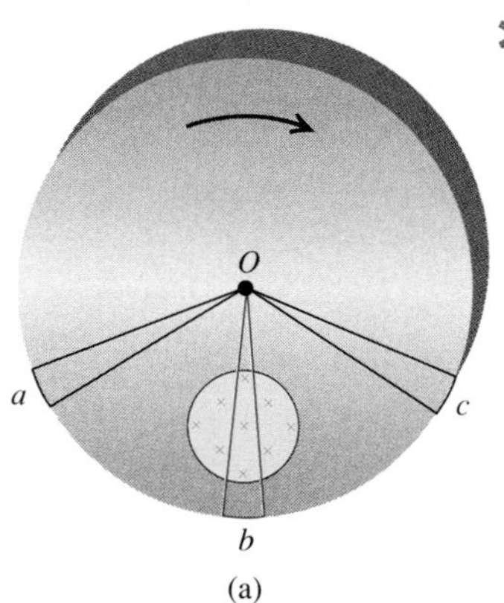

(a)

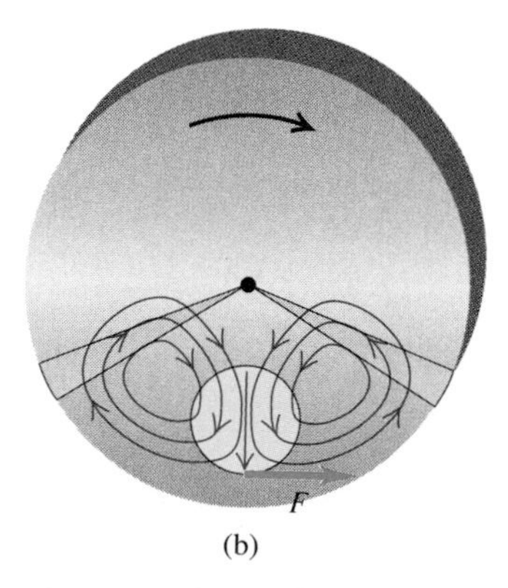

(b)

**30–18** (a) A metallic disk rotating through a perpendicular magnetic field. (b) Eddy currents formed by the induced emf.

In the examples of induction effects that we have studied, the induced currents have been confined to well-defined paths in conductors and other components forming a *circuit.* However, many pieces of electrical equipment contain masses of metal moving in magnetic fields or located in changing magnetic fields. In situations like these we can have induced currents that circulate throughout the volume of a material. Because their flow patterns resemble swirling eddies in a river, we call these **eddy currents.**

As an example, consider a metallic disk rotating in a magnetic field perpendicular to the plane of the disk but confined to a limited portion of the disk's area, as shown in Fig. 30–18a. Sector *Ob* is moving across the field and has an emf induced in it. Sectors *Oa* and *Oc* are not in the field, but they provide conducting paths for charges displaced along *Ob* to return from *b* to *O.* The result is a circulation of eddy currents in the disk, somewhat as sketched in Fig. 30–18b.

The downward current in the neighborhood of sector *Ob* experiences a sideways magnetic-field force toward the right that *opposes* the rotation of the disk, as Lenz's law predicts. The return currents lie outside the field, so they do not experience such forces. The interaction between the eddy currents and the field causes a braking action on the disk. Such effects can be used to stop the rotation of a circular saw quickly when the power is turned off. Some sensitive balances use this effect to damp out vibrations. Eddy current braking is also used on some electrically powered rapid transit vehicles. Electromagnets mounted in the cars induce eddy currents in the rails; the resulting magnetic fields cause braking forces on the electromagnets and thus on the cars.

As a second example, consider the core of an alternating-current transformer, shown in Fig. 30–19a. The alternating current in the primary winding P sets up an alternating flux within the core. This continuously changing flux causes an induced emf in the secondary winding S. The iron core, however, is also a conductor, and any section such as *AA* can be pictured as several conducting circuits, one within the other (Fig. 30–19b). The flux through each of these circuits is continually changing, so eddy currents circulate in the entire volume of the core, with lines of flow that form planes perpendicular to

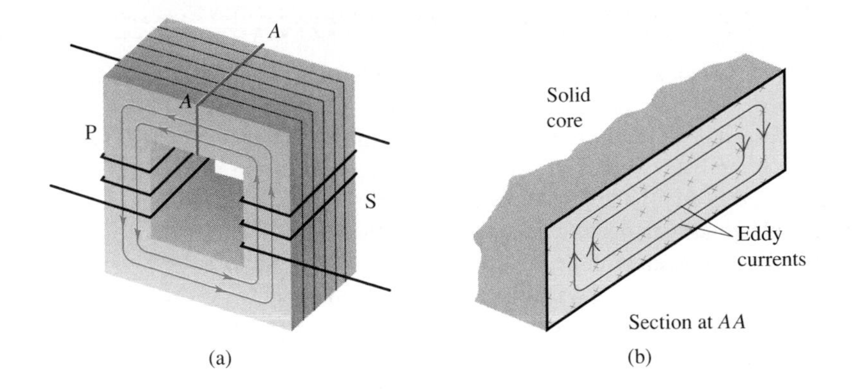


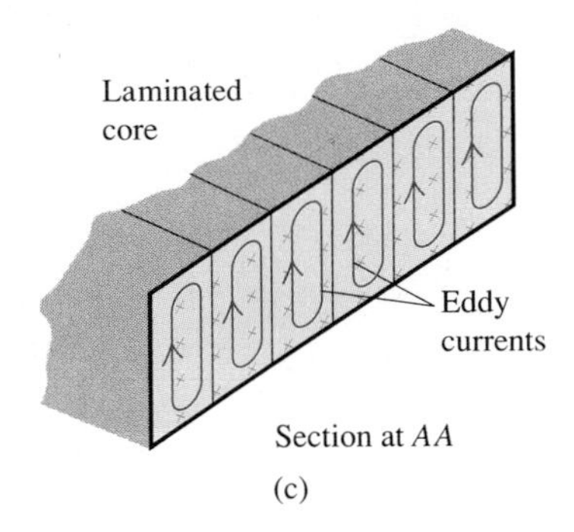


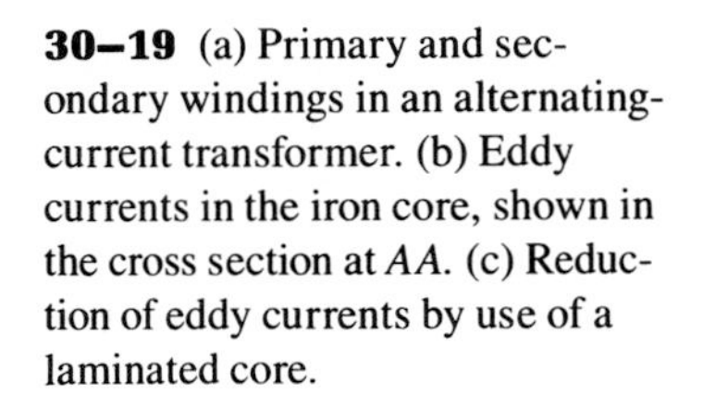
**30–19** (a) Primary and secondary windings in an alternating-current transformer. (b) Eddy currents in the iron core, shown in the cross section at $AA$. (c) Reduction of eddy currents by use of a laminated core.

the flux. These eddy currents are very undesirable, both because they waste energy through $I^2R$ heating and because of the opposing flux they themselves set up.

In actual transformers the eddy currents can be greatly reduced by the use of a *laminated* core, that is, one built up of thin sheets, or laminae. The large electrical surface resistance of each lamina, due to either a natural coating of oxide or an insulating varnish, effectively confines the eddy currents to individual laminae (Fig. 30–19c). The possible eddy-current paths are narrower, the induced emf in each path is smaller, and the eddy currents are greatly reduced.

In small transformers in which it is important to keep eddy-current losses to an absolute minimum, the cores are sometimes made of *ferrites,* which are complex oxides of iron and other metals. These materials are ferromagnetic, but they have much greater resistivity than pure iron.

Eddy currents have many practical uses. We have mentioned the use of eddy-current braking in balances, rotating machines, and rapid-transit trains. The shiny metal disk in the electric meter outside a house rotates as a result of eddy currents. These currents are induced in the disk by magnetic fields caused by sinusoidally varying currents in a coil. In induction furnaces, eddy currents are used to heat materials in completely sealed containers for processes in which it is essential to avoid the slightest contamination of the materials. Finally, the familiar metal detectors seen at security checkpoints in airports (Fig. 30–20a) operate by detecting eddy currents induced in metallic objects. Similar devices (Fig. 30–20b) are used to locate buried treasure such as bottlecaps and lost pennies.

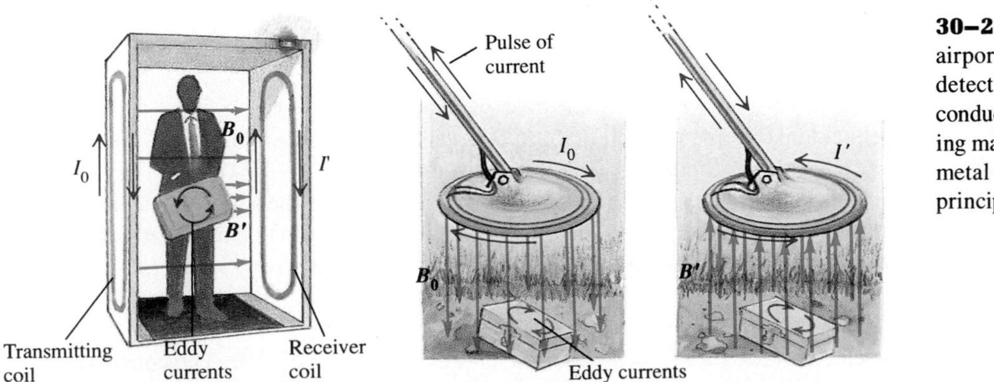


**30–20** (a) A metal detector at an airport security check point detects eddy currents induced in conducting objects by an alternating magnetic field. (b) Portable metal detectors work on the same principle.

## 30–7 MAXWELL'S EQUATIONS

We are now in a position to wrap up in a wonderfully neat package all the relationships between electric and magnetic fields and their sources that we have studied in the past several chapters. The package consists of four equations, called **Maxwell's equations.** You may remember Maxwell as the discoverer of the concept of displacement current, which we studied in Section 29–9. Maxwell didn't discover all of these equations single-handedly, but he put them together and recognized their significance, particularly in predicting the existence of electromagnetic waves.

For now we'll state Maxwell's equations in their simplest form, for the case in which there are charges and currents in otherwise empty space. Later we'll learn how to modify them for situations in which a dielectric or a magnetic material is present.

Two of Maxwell's equations involve an integral of $\boldsymbol{E}$ or $\boldsymbol{B}$ over a closed surface. The first is simply Gauss's law, Eq. (23–8). This law states that the surface integral of $E_{\perp}$ over any closed surface equals $1/\epsilon_0$ times the total charge $Q_{\text{encl}}$ enclosed within the surface:

$$\oint \boldsymbol{E} \cdot d\boldsymbol{A} = \frac{Q_{\text{encl}}}{\epsilon_0}. \qquad (30\text{–}11)$$

The second of Maxwell's equations is the analogous relation for *magnetic* fields, Eq. (28–8), which states that the surface integral of $B_{\perp}$ over any closed surface is always zero:

$$\oint \boldsymbol{B} \cdot d\boldsymbol{A} = 0. \qquad (30\text{–}12)$$

This statement is often called Gauss's law for magnetic fields. It means, among other things, that there are no magnetic monopoles (single magnetic charges) to act as sources of magnetic field.

The third equation is Ampere's law, including displacement current, and the fourth is Faraday's law. Ampere's law, as stated in Section 29–9, states that both conduction current $I_C$ and displacement current $\epsilon_0 \, d\Phi_E/dt$, where $\Phi_E$ is electric flux, act as sources of magnetic field:

$$\oint \boldsymbol{B} \cdot d\boldsymbol{l} = \mu_0\left(I_C + \epsilon_0 \frac{d\Phi_E}{dt}\right). \qquad (30\text{–}13)$$

Faraday's law, which we have studied in this chapter, states that a changing magnetic field or magnetic flux induces an electric field:

$$\oint \boldsymbol{E} \cdot d\boldsymbol{l} = -\frac{d\Phi_B}{dt}. \qquad (30\text{–}14)$$

The fact that the line integral in Eq. (30–14) is not zero shows that the $\boldsymbol{E}$ field produced by a changing magnetic flux is not conservative.

In general, the total $\boldsymbol{E}$ field at a point in space can be the superposition of a magnetically induced field $\boldsymbol{E}_{\text{n}}$ and a field $\boldsymbol{E}_{\text{e}}$ caused by a distribution of charges at rest. (The subscripts stand for non-electrostatic and electrostatic, respectively.) That is,

$$\boldsymbol{E} = \boldsymbol{E}_{\text{e}} + \boldsymbol{E}_{\text{n}}.$$

The electrostatic part $\boldsymbol{E}_{\text{e}}$ is *always* conservative, so $\oint \boldsymbol{E}_{\text{e}} \cdot d\boldsymbol{l} = 0$. The conservative part of the field does not contribute to the integral in Faraday's law, and we can take $\boldsymbol{E}$ in Eq. (30–14) to be the *total* electric field $\boldsymbol{E}$, including both the part due to charges $\boldsymbol{E}_{\text{e}}$ and the magnetically induced part $\boldsymbol{E}_n$. Also, because the nonconservative part $\boldsymbol{E}_n$ of the $\boldsymbol{E}$ field is not caused by static charges, $\oint \boldsymbol{E}_n \cdot d\boldsymbol{A}$ is always zero; the nonconservative field $\boldsymbol{E}_n$ does not contribute to the integral in Gauss's law. We conclude that in all of Maxwell's equations, $\boldsymbol{E}$ is the *total* electric field; these equations don't distinguish between conservative and nonconservative fields.

Comparing Eqs. (30–13) and (30–14), we see a remarkable symmetry. Equation (30–13) says that a changing electric flux creates a magnetic field, and Eq. (30–14) says that a changing magnetic flux creates an electric field. In fact, in empty space, where there is no conduction current and $I_C = 0$, the two equations have the same form, apart from a numerical constant and a negative sign, with the roles of $\boldsymbol{E}$ and $\boldsymbol{B}$ reversed. This same symmetry is also evident in the first two equations; in empty space, where there is no charge, the equations are identical in form, one containing $\boldsymbol{E}$, the other $\boldsymbol{B}$.

We can rewrite Eqs. (30–13) and (30–14) in different but equivalent forms by introducing the definitions of electric and magnetic flux, $\Phi_E = \int \boldsymbol{E} \cdot d\boldsymbol{A}$ and $\Phi_B = \int \boldsymbol{B} \cdot d\boldsymbol{A}$, respectively. In empty space, where there is no charge or current, $I_C = 0$ and $Q_{encl} = 0$, and we have

$$\oint \boldsymbol{B} \cdot d\boldsymbol{l} = \epsilon_0 \mu_0 \frac{d}{dt} \int \boldsymbol{E} \cdot d\boldsymbol{A}, \qquad (30\text{–}15)$$

$$\oint \boldsymbol{E} \cdot d\boldsymbol{l} = -\frac{d}{dt} \int \boldsymbol{B} \cdot d\boldsymbol{A}. \qquad (30\text{–}16)$$

Again we notice the symmetry for the roles of the $\boldsymbol{E}$ and $\boldsymbol{B}$ fields in these expressions.

The most remarkable feature of these equations is that a time-varying field of *either* kind induces a field of the other kind in neighboring regions of space. Maxwell recognized that these relationships predict the existence of electromagnetic disturbances consisting of time-varying electric and magnetic fields that travel, or *propagate,* from one region of space to another, even if no matter is present in the intervening space. Such disturbances, called *electromagnetic waves,* provide the physical basis for light, radio and television waves, infrared, ultraviolet, x rays, and the rest of the electromagnetic spectrum. We will return to this vitally important topic in Chapter 33.

Although it may not be obvious, *all* the basic relations between electric and magnetic fields and their sources are contained in Maxwell's equations. We can derive Coulomb's law from Gauss's law, we can derive the law of Biot and Savart from Ampere's law, and so on. When we add the equation that defines the $\boldsymbol{E}$ and $\boldsymbol{B}$ fields in terms of the forces they exert on a charge $q$,

$$\boldsymbol{F} = q(\boldsymbol{E} + \boldsymbol{v} \times \boldsymbol{B}), \qquad (30\text{–}17)$$

we have all the fundamental relations of electromagnetism!

Finally, we note that Eqs. (30–11) through (30–14) would have even greater symmetry between the $\boldsymbol{E}$ and $\boldsymbol{B}$ fields if single magnetic charges (magnetic monopoles) existed. The right side of Eq. (30–12) would contain the total *magnetic* charge enclosed by the surface, and the right side of Eq. (30–14) would include a magnetic monopole current term. Perhaps you can begin to see why many physicists wish that magnetic monopoles existed; they would help to perfect the meter of the mathematical poetry of Maxwell's equations.

The fact that electromagnetism can be wrapped up so neatly and elegantly was a very satisfying discovery. In conciseness and generality, Maxwell's equations are in the same league as Newton's laws of motion and the laws of thermodynamics. Indeed, that's really what science is all about—learning how to express very broad and general relationships in a concise and compact form. Maxwell's synthesis of electromagnetism stands as a towering intellectual achievement, comparable to the Newtonian synthesis described at the end of Section 12–6 and to the twentieth-century development of relativity, quantum mechanics, and the understanding of DNA. All are beautiful, and all are monuments to the achievements of which the human intellect is capable!

## 30–8 SUPERCONDUCTIVITY: A Case Study in Magnetic Properties

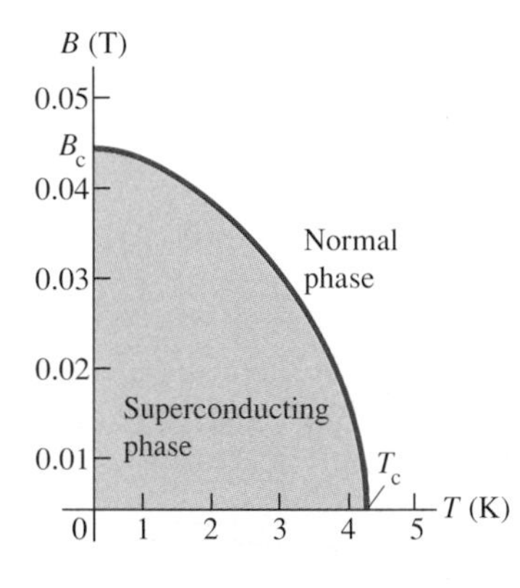


**30–21** Phase diagram for pure mercury, showing the superconducting transition critical temperature $T_c$ and its dependence on magnetic field. Other superconducting materials have similar curves but with different scales.

The most familiar property of a superconductor is the sudden disappearance of all electrical resistance when the material is cooled below a temperature called the *critical temperature,* denoted by $T_c$. We discussed this behavior and the circumstances of its discovery in Section 26–2. But superconductivity is far more than just the absence of measurable resistance. Superconductors also have extraordinary *magnetic* properties. We'll explore some of these properties in this section.

The first hint of these unusual magnetic properties was the discovery that for any superconducting material the critical temperature $T_c$ changes when the material is placed in an externally produced magnetic field $\boldsymbol{B}_0$. Figure 30–21 shows this dependence for mercury, the first element in which superconductivity was observed. As the external field magnitude $B_0$ increases, the superconducting transition occurs at lower and lower temperature. When $B_0$ is greater than about 0.04 T, *no* superconducting transition occurs. The minimum magnitude of magnetic field needed to eliminate superconductivity is called the *critical field,* denoted by $B_c$.

Figure 30–21 is a *phase diagram;* it shows which phase (normal or superconducting) occurs for any given conditions of temperature and magnetic field. It is analogous to the $p$-$T$ phase diagrams that we studied in Section 16–6, which show the phase of a substance (solid, liquid, or gas) that will exist for any given conditions of temperature and pressure.

Another important aspect of superconductivity is the complete absence of magnetic field inside a superconductor. When a homogeneous sphere of a normal diamagnetic or paramagnetic material (Fig. 30–22a) is placed in a magnetic field (Fig. 30–22b), the field inside the material is nearly the same as that outside. (We choose the spherical shape because the field inside the material then turns out to be uniform, avoiding complications associated with the shape of the specimen.) The relative permeability $K_m$ of such

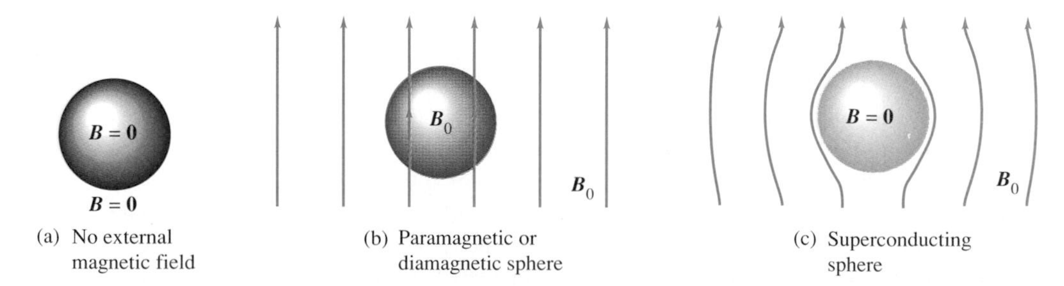


**30–22** (a) With no external field, $\boldsymbol{B}$ is zero everywhere. (b) When an external field $\boldsymbol{B}_0$ is applied to a normal paramagnetic or diamagnetic substance, the field inside the material is very nearly equal to $\boldsymbol{B}_0$. (c) When an external field $\boldsymbol{B}_0$ is applied to a superconducting substance, the field inside the material is zero.

materials (Section 29–8) usually differs from unity by less than 0.1%. But when a superconductor is placed in a magnetic field, we find that there is *no* field inside the material (provided that the external field magnitude is less than $B_c$). The field lines are distorted as shown in Fig. 30–22c. To confirm the absence of magnetic field inside the material, we can wrap a coil of wire around the specimen and then, with the material in the superconducting phase, turn on the external field. We find that no emf is induced in the coil; this shows that there is no flux change through the coil and therefore no $\boldsymbol{B}$ field in the specimen when the external field is turned on.

The absence of magnetic field inside a superconductor can be understood qualitatively on the basis of eddy currents induced in the material. Lenz's law shows that these currents always oppose the change that caused them; in this case the cause is increasing magnetic flux. In a superconductor, which offers no resistance at all to current, the induced eddy currents oppose the flux change so successfully that there is in fact *no* net change in flux.

Our discussion of magnetic materials in Section 29–8 shows that when a current-carrying conductor is completely surrounded by a magnetic material, the magnetic field $\boldsymbol{B}$ within the material differs from the field $\boldsymbol{B}_0$ that would be caused by the same current in vacuum because of the magnetization $\boldsymbol{M}$ (magnetic moment per unit volume) in the material. Specifically,

$$\boldsymbol{B} = \boldsymbol{B}_0 + \mu_0 \boldsymbol{M}. \tag{30–18}$$

This relation also holds for a toroidal solenoid wound on a core of magnetic material when there is no magnetic field outside the core. (In more general situations in which there is magnetic field both inside and outside the magnetic material the relationships are more complex.)

In a paramagnetic material the magnetization $\boldsymbol{M}$ results from preferential orientation of permanent magnetic dipoles associated with orbital and spin motion of electrons in the material. The dipoles tend to align themselves with the field direction; $\boldsymbol{B}_0$ and $\boldsymbol{M}$ have the same direction, and $\boldsymbol{B}$ is greater in magnitude than $\boldsymbol{B}_0$. In a diamagnetic material the magnetization results from *induced* magnetic dipoles; $\boldsymbol{M}$ is opposite in direction to $\boldsymbol{B}_0$, and $\boldsymbol{B}$ is less than $\boldsymbol{B}_0$. This behavior is in accordance with Lenz's law.

If the magnetization $\boldsymbol{M}$ is proportional to the external field $\boldsymbol{B}_0$, it is also proportional to the field $\boldsymbol{B}$ in the material; this relation can be expressed in terms of the magnetic susceptibility $\chi_m$ (defined in Section 29–8) as

$$\mu_0 \boldsymbol{M} = \frac{\chi_m}{1 + \chi_m} \boldsymbol{B}. \tag{30–19}$$

Paramagnetic materials have positive susceptibilities; diamagnetic materials have negative ones. For most paramagnetic materials, $\chi_m$ is very small, of the order of $10^{-3}$ or less. Diamagnetic susceptibilities are smaller still, usually $-10^{-4}$ or less. In both cases, $\mu_0 \boldsymbol{M}$ is usually much smaller in magnitude than $\boldsymbol{B}$.

In a superconductor, on the other hand, the induced currents and associated magnetization are large enough to reduce the magnetic field $\boldsymbol{B}$ in the material to zero, corresponding to a diamagnetic susceptibility $\chi_m = -1$ and a relative permeability $K_m = 0$. A superconductor is sometimes called a "perfect diamagnetic material."

Here's another aspect of the magnetic behavior of superconductors. We place a homogeneous sphere of a superconducting material in a uniform applied magnetic field $\boldsymbol{B}_0$, at a temperature $T$ greater than $T_c$. The material is then in the normal, not the superconducting phase. The field is as shown in Fig. 30–23a. Now we lower the temperature until the superconducting transition occurs. (We assume that $B_0$ is not so large as to prevent the phase transition.) What happens to the field?

Measurements of the field outside the sphere show that the field lines become distorted as in Fig. 30–23b. There is no longer any field inside the material, except possibly

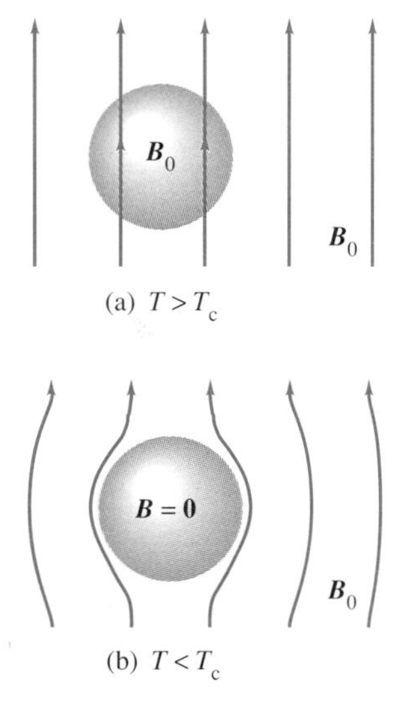


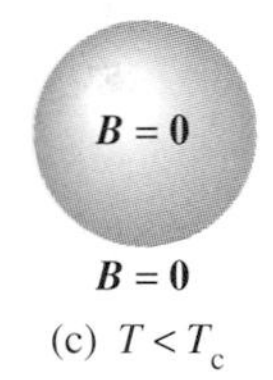


**30–23** When a superconducting material above the critical temperature is placed in an external magnetic field $\boldsymbol{B}_0$, the field inside the material is very nearly equal to $\boldsymbol{B}_0$. (b) As the material is cooled through the critical temperature $T_c$ and becomes superconducting, the magnetic flux is expelled from the material, and the field inside it becomes zero. (c) When the external field is turned off with the material in the superconducting phase, the field is zero everywhere; there is no change of magnetic flux in the material.

**30–24** A superconductor is supported by the repulsive force of a magnetic field.

in a very thin surface layer a hundred or so atoms thick. If a coil is wrapped around the sphere, the emf induced in the coil shows that during the superconducting transition the magnetic flux through the coil decreases from its initial value to zero; this is consistent with the absence of field inside the material. Finally, if the field is turned off while the material is still in its superconducting phase (Fig. 30–23c), no emf is induced in the coil, and measurements show no field outside the sphere.

We conclude that during a superconducting transition in the presence of the field $\boldsymbol{B}_0$, all of the magnetic flux is expelled from the bulk of the sphere, and the magnetic flux $\Phi_B$ through the coil becomes zero. This expulsion of magnetic flux is called the *Meissner effect.* As shown in Fig. 30–23b, this expulsion crowds the magnetic field lines closer together to the side of the sphere, increasing $\boldsymbol{B}$ there.

The diamagnetic nature of a superconductor has some interesting *mechanical* consequences. A paramagnetic or ferromagnetic material is attracted by a permanent magnet because the oriented magnetic dipoles in the material experience a net attractive force in the presence of the inhomogeneous magnetic field of the permanent magnet. For a diamagnetic material the magnetization is in the opposite sense, and a diamagnetic material is *repelled* by a permanent magnet. Figure 30–24 shows a specimen of a high-temperature superconductor being supported ("levitated") by this repulsive magnetic force.

The behavior that we have described is characteristic of what are called *type-I superconductors.* There is another class of superconducting materials called *type-II superconductors.* When such a material in the superconducting phase is placed in a magnetic field, the bulk of the material is superconducting, but there may be thin filaments of material, running parallel to the field, that remain in the normal phase. Currents circulate around the boundaries of these filaments, and there *is* magnetic flux inside them. Type-II superconductors have *two* critical magnetic fields; the first is $B_{c1}$, the field at which magnetic flux begins to enter the material, forming the filaments just described, and the second is $B_{c2}$, at which the material becomes normal. Type-II superconductors are used for electromagnets because they usually have much larger values of $B_{c2}$ than the $B_c$ of type-I materials, permitting much larger magnetic fields without destroying the superconducting state.

Until 1987 the record high value of $T_c$ was for a niobium-germanium alloy with $T_c = 23.2$ K. In that year, several complex compounds of copper, barium, oxygen and yttrium (and other elements) were found to be superconductors, with critical temperatures as high as 91 K. This is very significant because it is above the temperature of liquid nitrogen (about 77 K) and so is comparatively easy to attain. The dream of room-temperature superconductors, though not yet close to realization, no longer seems hopeless.

Many important and exciting applications of superconductors are under development. Superconducting electromagnets have been used in research laboratories for several years. Their advantages compared to conventional electromagnets include greater efficiency, compactness, and greater field magnitudes. Once a current is established in the coil of a superconducting electromagnet, no additional power input is required because there is no resistive energy loss. The coils can also be made more compact because there is no need to provide channels for the circulation of cooling fluids. Superconducting magnets routinely attain steady fields of the order of 10 T, much larger than the maximum fields available with conventional electromagnets.

The technological potential of superconductors, especially room-temperature superconductors, is breathtaking. For example, magnetic levitation of railroad cars will soon become a practical reality. Superconductors are attractive for long-distance electric power transmission and for energy-conversion devices, including generators, motors, and transformers. Superconducting devices can measure changes of magnetic flux of the order of $10^{-15}$ Wb. These have important medical applications in imaging devices. Development of practical applications of superconductor science promises to be an exciting chapter in contemporary technology.

## SUMMARY

• Faraday's law states that the induced emf in a circuit equals the negative of the time rate of change of magnetic flux through the circuit:

$$\mathcal{E} = -\frac{d\Phi_B}{dt}. \tag{30–3}$$

This relation is valid whether the flux change is caused by a changing magnetic field, motion of the conductor, or both.

• When a conductor with length $L$ moves with speed $v$ perpendicular to a uniform magnetic field with magnitude $B$, the induced emf is

$$\mathcal{E} = vBL. \tag{30–6}$$

More generally, when a conductor moves in a magnetic field $\boldsymbol{B}$, the induced emf in the direction from $b$ to $a$ is

$$\mathcal{E} = \int_b^a \boldsymbol{v} \times \boldsymbol{B} \cdot d\boldsymbol{l}. \tag{30–7}$$

• Lenz's law states that an induced current or emf always acts to oppose or cancel out the change that caused it. Lenz's law can be derived from Faraday's law, and is often easier to use.

• When an emf is induced by a changing magnetic flux through a stationary conductor, there is an induced electric field $\boldsymbol{E}$ of non-electrostatic origin such that

$$\oint \boldsymbol{E} \cdot d\boldsymbol{l} = -\frac{d\Phi_B}{dt}. \tag{30–9}$$

This field is nonconservative and cannot be associated with a potential.

• When a bulk piece of conducting material, such as a metal, is in a changing magnetic field or moves through a non-uniform field, currents called eddy currents are induced in the volume of the material.

### KEY TERMS

induced current
induced emf
Faraday's law of induction
motional electromotive force
Lenz's law
induced electric field
non-electrostatic field
eddy current
Maxwell's equations

• The relationships between electric and magnetic fields and their sources can be stated compactly in four equations called Maxwell's equations. They are

$$\oint \boldsymbol{E} \cdot d\boldsymbol{A} = \frac{Q_{\text{encl}}}{\epsilon_0}, \tag{30–11}$$

$$\oint \boldsymbol{B} \cdot d\boldsymbol{A} = 0, \tag{30–12}$$

$$\oint \boldsymbol{B} \cdot d\boldsymbol{l} = \mu_0\left(I_C + \epsilon_0 \frac{d\Phi_E}{dt}\right), \tag{30–13}$$

$$\oint \boldsymbol{E} \cdot d\boldsymbol{l} = -\frac{d\Phi_B}{dt}. \tag{30–14}$$

The first equation is Gauss's law of electrostatics; the second is Gauss's law for magnetic fields, stating the nonexistence of magnetic monopoles. The third is Ampere's law, generalized by Maxwell to include displacement current; and the fourth is Faraday's law. Together they form a complete basis for the relation of $\boldsymbol{E}$ and $\boldsymbol{B}$ fields to their sources.

• The force exerted on a charge $q$ by electric and magnetic fields $\boldsymbol{E}$ and $\boldsymbol{B}$ is

$$\boldsymbol{F} = q(\boldsymbol{E} + \boldsymbol{v} \times \boldsymbol{B}). \tag{30–17}$$

# EXERCISES

## Section 30–2 Faraday's Law

**30–1** A single loop of wire, with an enclosed area of 0.0900 m$^2$, is in a region of uniform magnetic field between the poles of a large superconducting electromagnet, with the field perpendicular to the plane of the loop. The magnetic field has an initial value of 3.80 T and is decreasing at a constant rate of 0.190 T/s. If the loop has a resistance of 0.400 Ω, what is the current induced in the loop?

**30–2** A coil of wire with 200 circular turns of radius 3.00 cm is in a uniform magnetic field that is perpendicular to the plane of the coil. The coil has a total resistance of 40.0 Ω. At what rate, in teslas per second, must the magnetic field be changing to induce a current of 0.240 A in the coil?

**30–3** A closely wound rectangular coil of 50 turns has dimensions 12.0 cm × 25.0 cm. In time $t = 0.0800$ s the plane of the coil is rotated from a position where it makes an angle of 45.0° with a magnetic field of 1.40 T to a position perpendicular to the field. What is the average emf induced in the coil?

**30–4** In a physics laboratory experiment a coil with 400 turns enclosing an area of 20 cm$^2$ is rotated in 0.020 s from a position where its plane is perpendicular to the earth's magnetic field to one where its plane is parallel to the field. What average emf is induced if the earth's magnetic field is $6.0 \times 10^{-5}$ T?

**30–5 A Faraday Disk Dynamo.** A large electromagnet requires 200 A at 5.00 V. The current is supplied by a disk dynamo 0.600 m in radius that turns in a magnetic field of 1.20 T, perpendicular to the plane of the disk. a) How many revolutions per second must the disk turn? b) What torque is required to turn the disk, assuming that all the mechanical energy is dissipated as heat in the large electromagnet?

**30–6** The armature of a small generator consists of a flat, square coil with 80 turns and sides with a length of 0.120 m. The coil rotates in a magnetic field of 0.0250 T. What is the angular velocity of the coil if the maximum emf produced is 20.0 mV?

**30–7 A Search Coil.** Derive the equation relating the total charge $Q$ that flows through a search coil to the magnetic field magnitude $B$. The search coil has $N$ turns, each with area $A$, and the flux through the coil is decreased from its initial maximum value to zero in a time $\Delta t$. The resistance of the coil is $R$, and the total charge is $Q = I\,\Delta t$, where $I$ is the average current induced by the change in flux.

**30–8** The cross-section area of a closely wound search coil (Exercise 30–7) having 20 turns is 1.50 cm$^2$, and its resistance is 9.00 Ω. The coil is connected through leads of negligible resistance to a charge-measuring instrument having an internal resistance of 16.0 Ω. Find the quantity of charge displaced when the coil is pulled quickly out of a region where $B = 1.80$ T to a point where the magnetic field is zero. The plane of the coil, when in the field, makes an angle of 90° with the magnetic field.

**30–9** A closely wound search coil (Exercise 30–7) has an area of 4.00 cm$^2$, 160 turns, and a resistance of 50.0 Ω. It is connected to a charge-measuring instrument whose resistance is 30.0 Ω. When the coil is rotated quickly from a position parallel to a uniform magnetic field to one perpendicular to the field, the instrument indicates a charge of $6.00 \times 10^{-5}$ C. What is the magnitude of the field?

**30–10** A coil 2.00 cm in radius, containing 500 turns, is placed in a magnetic field that varies with time according to $B = (0.0100 \text{ T/s})t + (2.00 \times 10^{-4} \text{ T/s}^3)t^3$. The coil is connected to a 500-Ω resistor, and its plane is perpendicular to the magnetic field. a) Find the magnitude of the induced emf in the coil as a function of time. b) What is the current in the resistor at time $t = 10.0$ s? (The resistance of the coil can be neglected.)

## Section 30–3 Motional Electromotive Force

**30–11** For Eq. (30–6), show that if $v$ is in meters per second, $B$ in teslas, and $L$ in meters, then the units of the right-hand side of the equation are joules per coulomb or volts (the correct SI units for $\mathcal{E}$).

**30–12** In Fig. 30–10a a rod with length $L = 0.250$ m moves with constant speed 6.00 m/s in the direction shown. The induced emf is 2.00 V. a) What is the magnitude of the magnetic field? b) Which point is at higher potential, $a$ or $b$?

**30–13** In Fig. 30–10b a rod with length $L = 0.400$ m moves in a magnetic field with magnitude $B = 1.20$ T. The emf induced in the moving rod is 3.20 V. a) What is the speed of the rod? b) If the total circuit resistance is 0.800 Ω, what is the induced current? c) What force (magnitude and direction) does the field exert on the rod as a result of this current?

**30–14** In Fig. 30–25 a rod with length $L = 0.150$ m moves in a magnetic field $\boldsymbol{B}$ directed into the plane of the figure. $B = 0.500$ T, and the rod moves with velocity $v = 6.00$ m/s in the direction shown. a) What is the motional emf induced in the rod? b) What is the potential difference between the ends of the rod? c) Which point, $a$ or $b$, is at higher potential?

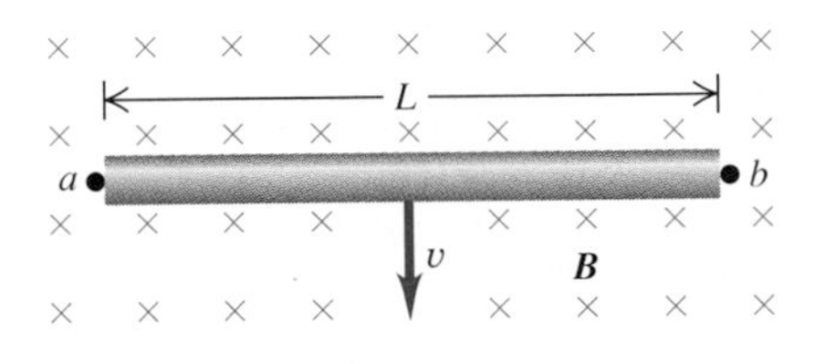


**FIGURE 30–25**

**30–15** A square loop of wire with resistance $R$ is moved at constant speed $v$ across a uniform magnetic field confined to a square region whose sides are twice the length of those of the square loop (Fig. 30–26). a) Sketch a graph of the external force $F$ needed to move the loop at constant speed, as a function of the coordinate $x$, from $x = -2L$ to $x = +2L$. (The coordinate $x$ is measured from the center of the magnetic-field region to the center of the loop. It is negative when the center of the loop is to the left of the center of the magnetic-field region. Take positive force to be to the right.) b) Sketch a graph of the induced current in the loop as a function of $x$. Take counterclockwise currents to be positive.

**30–16** Conducting rod $AB$ in Fig. 30–27 makes contact with metal rails $CA$ and $DB$. The apparatus is in a uniform magnetic field 0.400 T, perpendicular to the plane of the figure. a) Find the magnitude of the emf induced in the rod when it is moving toward the right with a speed 6.00 m/s. b) In what direction does

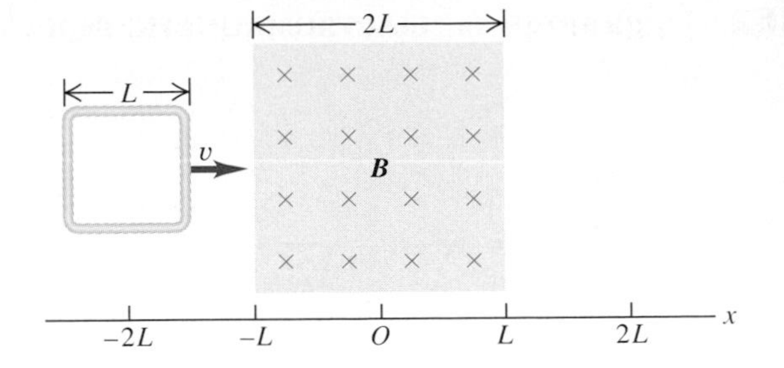


**FIGURE 30–26**

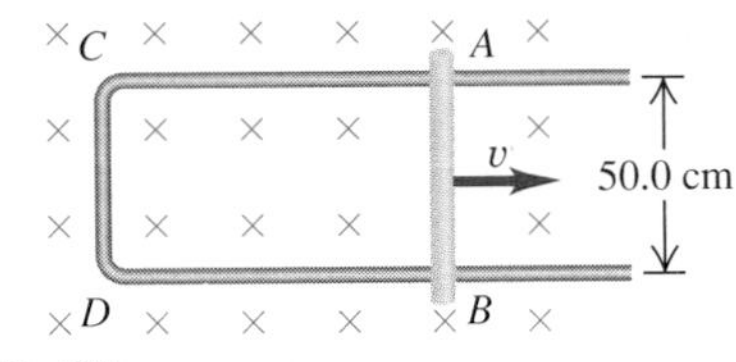


**FIGURE 30–27**

the current flow in the rod? c) If the resistance of the circuit $ABDC$ is 2.00 Ω (assumed to be constant), find the force (magnitude and direction) required to keep the rod moving to the right with a constant speed of 6.00 m/s. Neglect friction. d) Compare the rate at which mechanical work is done by the force ($Fv$) with the rate of development of heat in the circuit ($I^2R$).

## Section 30–4 Lenz's Law

**30–17** A circular loop of wire is in a region of spatially uniform magnetic field, as shown in Fig. 30–28. The magnetic field is directed into the plane of the figure. Calculate the direction (clockwise or counterclockwise) of the induced current in the loop when a) $B$ is increasing; b) $B$ is decreasing; c) $B$ is constant with value $B_0$.

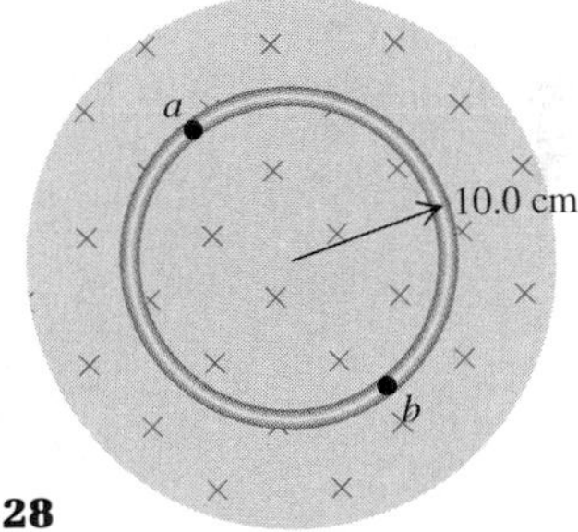


**FIGURE 30–28**

**30–18** Using Lenz's law, determine the direction of the current in resistor $ab$ of Fig. 30–29 when a) switch S is opened after having been closed for several minutes; b) coil $B$ is brought closer to coil $A$ with the switch closed; c) the resistance of $R$ is decreased while the switch remains closed.

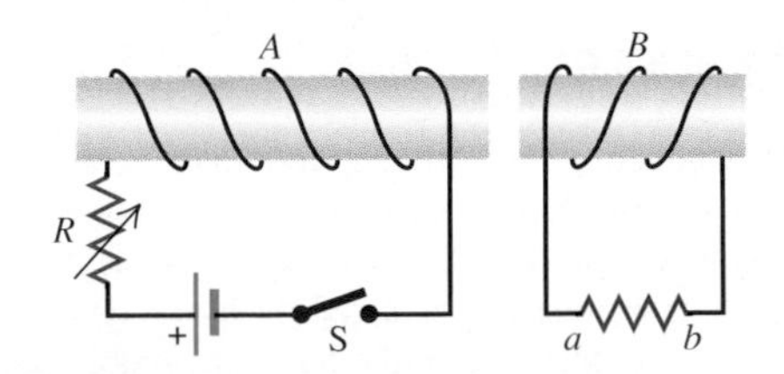


**FIGURE 30–29**

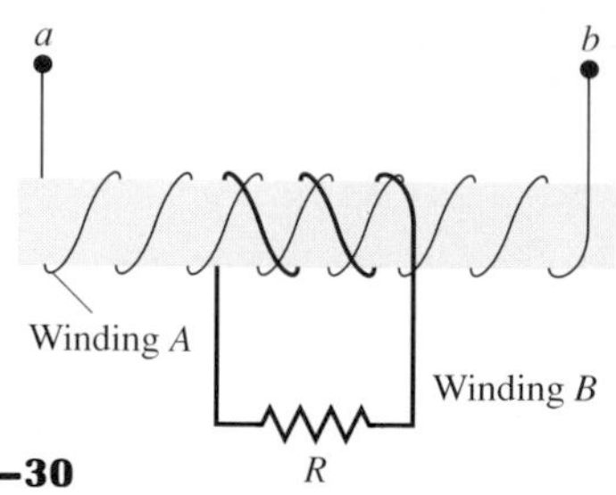


**FIGURE 30–30**

**30–19** A cardboard tube is wrapped with two windings of insulated wire wound in opposite directions, as in Fig. 30–30. Terminals $a$ and $b$ of winding $A$ may be connected to a battery through a reversing switch. State whether the induced current in the resistor $R$ is from left to right or from right to left in the following circumstances: a) the current in winding $A$ is from $a$ to $b$ and is increasing; b) the current is from $b$ to $a$ and is decreasing; c) the current is from $b$ to $a$ and is increasing.

## Section 30–5
## Induced Electric Fields

**30–20** In a simple transformer a long, straight solenoid with a cross-section area of 6.00 $cm^2$ is wound with 10 turns of wire per centimeter, and the windings carry a current of 0.250 A. A secondary winding of 8 turns encircles the solenoid at its center. When the primary circuit is opened, the magnetic field of the solenoid becomes zero in 0.0500 s. What is the average induced emf in the secondary?

**30–21** The magnetic field $\boldsymbol{B}$ at all points within the colored circle of Fig. 30–28 has an initial magnitude of 0.800 T. (The circle could represent approximately the space between the poles of a large electromagnet.) The magnetic field is directed into the plane of the figure and is decreasing at the rate of $-0.0600$ T/s. a) What is the shape of the field lines of the induced electric field in Fig. 30–28 within the colored circle? b) What are the magnitude and direction of this field at any point on the circular conducting ring with radius 0.100 m? c) What is the current in the ring if its resistance is 2.00 Ω? d) What is the potential difference between points $a$ and $b$ on the ring? e) If the ring is cut at some point and the ends are separated slightly, what will be the potential difference between the ends?

**30–22** The magnetic field within a long, straight solenoid with a circular cross section and a radius $R$ is increasing at a rate of $dB/dt$. a) What is the rate of change of flux through a circle with radius $r_1$ inside the solenoid, normal to the axis of the solenoid, and with center on the solenoid axis? b) Find the induced electric field inside the solenoid at a distance $r_1$ from its axis. Show the direction of this field in a diagram. c) What is the induced electric field *outside* the solenoid at a distance $r_2$ from the axis? d) Sketch a graph of the magnitude of the induced electric field as a function of the distance $r$ from the axis from $r = 0$ to $r = 2R$. e) What is the induced emf in a circular turn of radius $R/2$ that has its center on the solenoid axis? f) What is the induced emf if the radius in part (e) is $R$? g) What is the induced emf if the radius in part (e) is $2R$?

## Section 30–8
## Superconductivity: A Case Study in Magnetic Properties

**30–23** At temperatures near absolute zero, $B_c$ approaches 0.142 T for vanadium, a type-I superconductor. The normal phase of vanadium has a magnetic susceptibility close to zero. Consider a long, thin vanadium cylinder with its axis parallel to an external magnetic field $\boldsymbol{B}_0$, which is in the $+x$-direction. At points far from the ends of the cylinder, by symmetry, all the magnetic vectors are parallel to the $x$-axis. At temperatures near absolute zero, what are the resultant magnetic field $\boldsymbol{B}$ and the magnetization $\boldsymbol{M}$ inside and outside the cylinder (far from the ends) for a) $\boldsymbol{B}_0 = (0.126\text{ T})\boldsymbol{i}$; b) $\boldsymbol{B}_0 = (0.252\text{ T})\boldsymbol{i}$?

**30–24** The compound $SiV_3$ is a type-II superconductor. At temperatures near absolute zero the two critical fields are $B_{c1} = 55.0$ mT and $B_{c2} = 15.0$ T. The normal phase of $SiV_3$ has a magnetic susceptibility close to zero. Consider a long, thin $SiV_3$ cylinder with its axis parallel to the external magnetic field $\boldsymbol{B}_0$, which is in the $+x$-direction. At points far from the ends of the cylinder, by symmetry, all the magnetic vectors are parallel to the $x$-axis. At a temperature near absolute zero the external magnetic field is slowly increased from zero. What are the resultant magnetic field $\boldsymbol{B}$ and the magnetization $\boldsymbol{M}$ inside the cylinder at points far from its ends a) just before the magnetic flux begins to penetrate the material; b) just after the material becomes completely normal?

**30–25** A long, straight wire made of a type-I superconductor carries a constant current $I$. Show that the current cannot be uniformly spread over the wire's cross section but instead must all be at the surface.

**30–26** A type-II superconductor in an external field between $B_{c1}$ and $B_{c2}$ has regions that contain magnetic flux and have resistance and also has superconducting regions. What is the resistance of a long, thin cylinder of such material?

# PROBLEMS

**30–27** A conducting rod with length $L$, mass $m$, and resistance $R$ moves without friction on metal rails as shown in Fig. 30–9. A uniform magnetic field $\boldsymbol{B}$ is directed into the plane of the figure. The rod starts from rest and is acted on by a constant force $\boldsymbol{F}$ directed to the right. The rails have negligible resistance. a) Sketch a graph of the speed of the rod as a function of time. b) Find an expression for the terminal speed (the speed when the acceleration of the rod is zero).

**30–28** The cube in Fig. 30–31, 1.00 m on a side, is in a uniform magnetic field of 0.300 T, directed along the positive $y$-axis. Wires $A$, $C$, and $D$ move in the directions indicated, each with a speed of

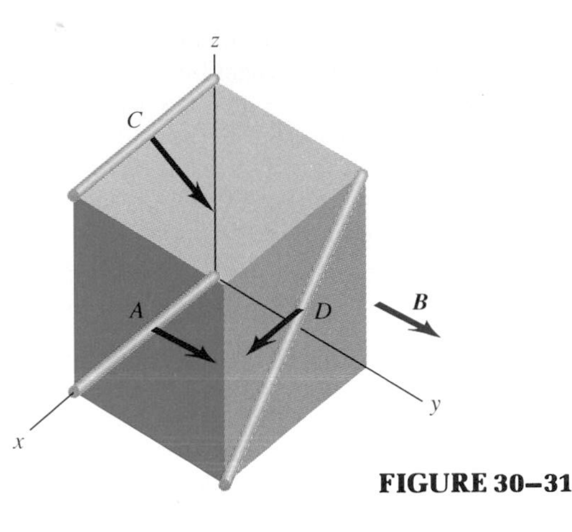


FIGURE 30–31

0.50 m/s. (Wire $A$ moves parallel to the $xy$-plane, $C$ moves at an angle of 45° below the $xy$-plane, and $D$ moves parallel to the $xz$-plane.) What is the potential difference between the ends of each wire?

**30–29** A slender rod 0.620 m long rotates about an axis through one end and perpendicular to the rod with an angular velocity of 8.00 rad/s. The plane of rotation of the rod is perpendicular to a uniform magnetic field with a magnitude of 0.500 T. a) What is the induced emf in the rod? b) What is the potential difference between its ends?

**30–30** Bar $ab$ moves without friction on conducting rails as shown in Fig. 30–32. There is a uniform magnetic field with magnitude $B = 0.60$ T directed into the plane of the page. You want to make an exercise machine out of this apparatus, in which the person exercises by pushing the bar back and forth with an average speed of 4.0 m/s. What should be the circuit resistance $R$ if the person moving the bar is to do work at an average rate of 200 W?

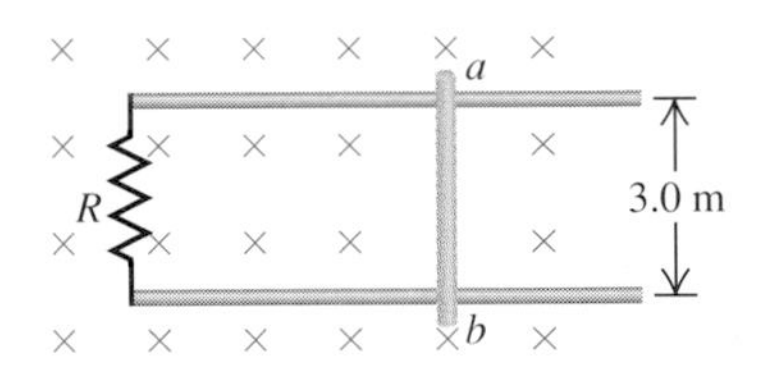


FIGURE 30–32

**30–31 A Rotating Loop.** The rectangular loop in Fig. 30–33, with area $A$ and resistance $R$, rotates at uniform angular velocity $\omega$ about the $y$-axis. The loop lies in a uniform magnetic field $\boldsymbol{B}$ in the direction of the $x$-axis. Let $t = 0$ in the position

FIGURE 30–33

shown in Fig. 30–33. Sketch the graphs of the following quantities as functions of time: a) the flux $\Phi_B$ through the loop; b) the rate of change of flux $d\Phi_B/dt$; c) the induced emf in the loop; d) the torque $\tau$ needed to keep the loop rotating at constant angular velocity. e) Show that the work of the external torque in one revolution is equal to the electrical energy dissipated (converted to heat) in the loop during one revolution.

**30–32** Suppose the loop in Fig. 30–33 is a) rotated about the $z$-axis; b) rotated about the $x$-axis; c) rotated about an edge parallel to the $y$-axis. What is the maximum induced emf in each case if $A = 400$ cm$^2$, $R = 2.00$ Ω, $\omega = 25.0$ rad/s, and $B = 0.500$ T?

**30–33** A flexible circular loop 0.100 m in diameter lies in a magnetic field with magnitude 1.50 T, directed into the plane of the page in Fig. 30–34. The loop is pulled at the points indicated by the arrows, forming a loop of zero area in 0.200 s. a) Find the average induced emf in the circuit. b) What is the direction of the current in $R$, from $a$ to $b$ or from $b$ to $a$?

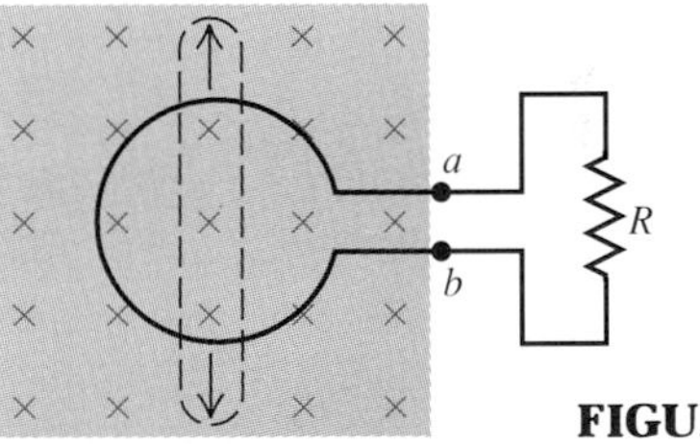


FIGURE 30–34

**30–34** The current in the infinite wire $AB$ of Fig. 30–35 is upward and is increasing steadily at a rate $di/dt$. a) At an instant when the current is $i$, what are the magnitude and direction of the field $\boldsymbol{B}$ at a distance $r$ from the wire? b) What is the flux $d\Phi_B$ through the narrow shaded strip? c) What is the total flux through the loop? d) What is the induced emf in the loop? e) Evaluate the numerical value of the induced emf if $a = 0.100$ m, $b = 0.300$ m, $L = 0.200$ m, and $di/dt = 1.20$ A/s.

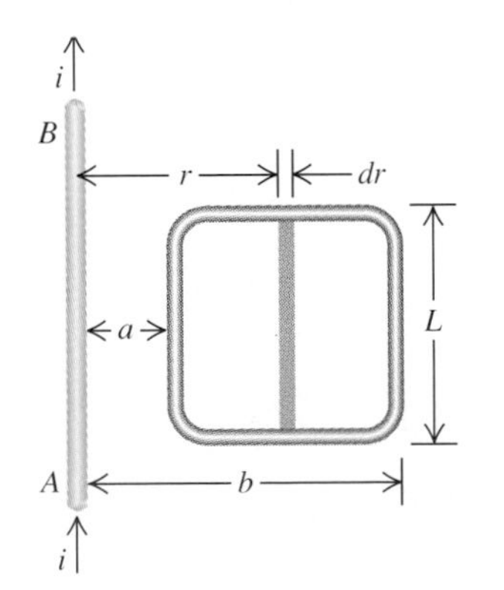


FIGURE 30–35

**30–35** The magnetic field $\boldsymbol{B}$ at all points within a circular region of radius $R$ is uniform in space and directed into the plane of

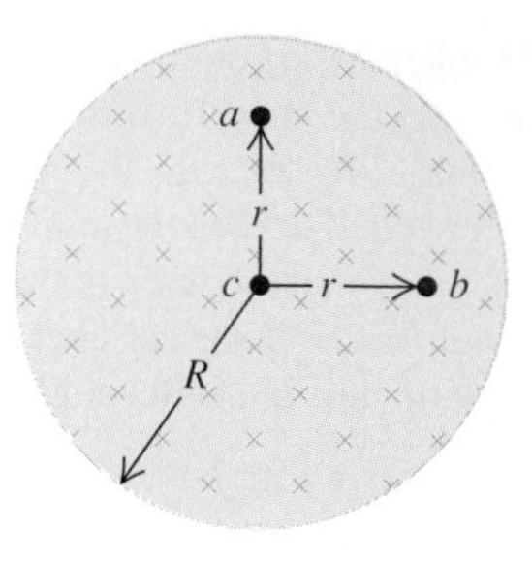


FIGURE 30–36

the page in Fig. 30–36. (The region could be a cross section inside the windings of a long, straight solenoid.) If the magnetic field is increasing at a rate $dB/dt$, what are the magnitude and direction of the force on a stationary positive point charge $q$ located at point $a$, $b$, and $c$? (Point $a$ is a distance $r$ above the center of the region, point $b$ is a distance $r$ to the right of the center, and point $c$ is at the center of the region.)

**30–36** A circular conducting ring with radius $r_0 = 0.0500$ m is oriented in the $xy$-plane and is in a region of magnetic field $\boldsymbol{B}$, where

$$\boldsymbol{B} = B_0 [1 - 3(t/t_0)^2 + 2(t/t_0)^3]\boldsymbol{k}.$$

Assume that $t_0 = 0.0100$ s and is constant, $r$ is the distance of an arbitrary point from the center of the ring, $t$ is the time, and $\boldsymbol{k}$ is a unit vector in the $+z$-direction. Also, $B_0 = 0.0800$ T and is constant. At points $a$ and $b$ (see Fig. 30–37) is a small gap in the ring with wires leading to an external circuit of resistance $R = 12.0\ \Omega$. There is no magnetic field at the location of the external circuit. a) Derive an expression, as a function of time, for the total magnetic flux $\Phi_B$ enclosed by the ring. b) Determine the emf induced in the ring at time $t = 5.00 \times 10^{-3}$ s. What is the polarity? c) Because of the internal resistance of the ring, the current through $R$ at the time given in part (b) is only 3.00 mA. Determine the internal resistance of the ring. d) Determine the emf in the ring at a time $t = 1.21 \times 10^{-2}$ s. What is the polarity? e) Determine the time at which the current through $R$ reverses its direction.

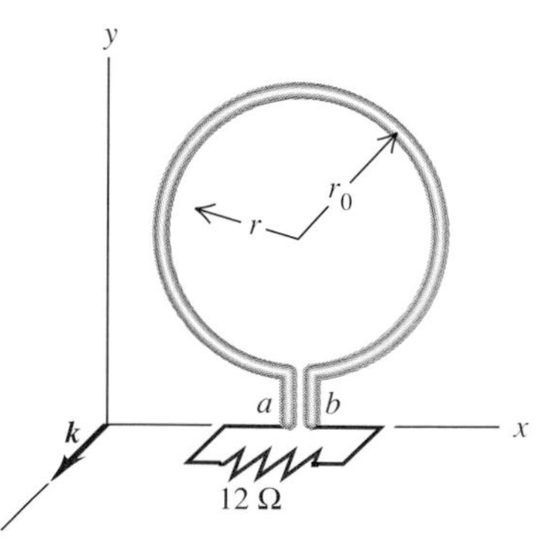


FIGURE 30–37

**30–37** The long, straight wire in Fig. 30–38a carries constant current $I$. A metal bar with length $L$ is moving at constant velocity $\boldsymbol{v}$, as shown in the figure. Point $a$ is a distance $d$ from the wire. a) Calculate the emf induced in the rod. b) Which point, $a$ or $b$, is at higher potential? c) If the bar is replaced by a rectangular wire loop of resistance $R$ (Fig. 30–38b), what is the magnitude of the current induced in the loop?

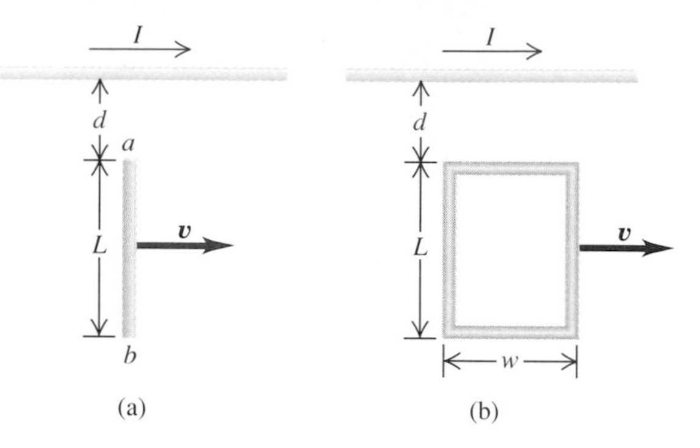


FIGURE 30–38

**30–38** A solenoid 0.500 m long and 0.0800 m in diameter is wound with 500 turns. A closely wound coil of 20 turns of insulated wire surrounds the solenoid at its midpoint, and the terminals of the coil are connected to a charge-measuring instrument. The total circuit resistance is 30.0 Ω. a) Find the quantity of charge displaced through the instrument when the current in the solenoid is quickly decreased from 3.00 A to 1.00 A. b) Draw a sketch of the apparatus, showing clearly the directions of the windings of the solenoid and coil and of the current in the solenoid. Show on your sketch the direction of the current in the coil when the solenoid current is decreasing.

**30–39 A Slide Wire.** A rectangular loop with width $a$ and a slide wire with mass $m$ are as shown in Fig. 30–39. A uniform magnetic field $\boldsymbol{B}$ is directed perpendicular to the plane of the loop into the plane of the figure. The slide wire is given an initial speed of $v_0$ and then released. Assume that there is no friction between the slide wire and the loop and that the resistance of the loop is negligible in comparison to the resistance $R$ of the slide wire. a) Obtain an expression for $F$, the magnitude of the force exerted on the wire while it is moving at speed $v$. b) Show that the distance $x$ that the wire moves before coming to rest is given by $x = mv_0 R/a^2B^2$.

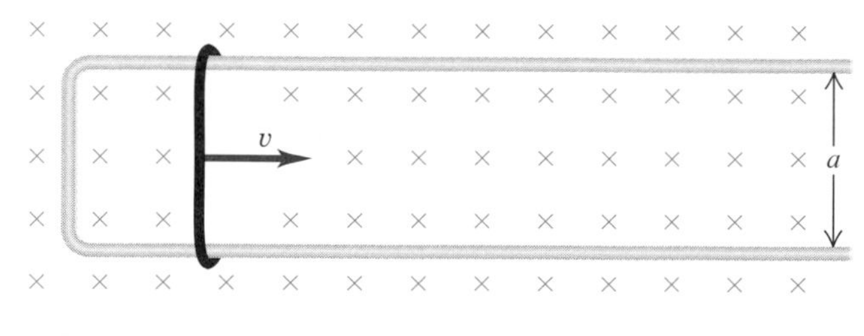


FIGURE 30–39

# CHALLENGE PROBLEMS

**30–40** A square conducting loop, 20.0 cm on a side, is placed in the same magnetic field as in Exercise 30–21. (See Fig. 30–40. The center of the square loop is at the center of the magnetic-field region.) a) Copy Fig. 30–40, and show by vectors the direction and relative magnitude of the induced electric field $\boldsymbol{E}$ at points $a$, $b$, and $c$. b) Prove that the component of $\boldsymbol{E}$ along the loop has the same value at every point of the loop and is equal to that of the ring of Fig. 30–28 (Exercise 30–21). c) What is the current induced in the loop if its resistance is 2.00 Ω? d) What is the potential difference between points $a$ and $b$?

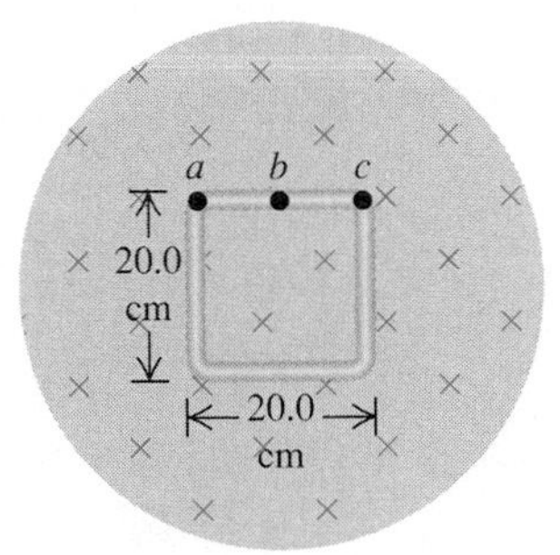


FIGURE 30–40

**30–41** A uniform square conducting loop, 20.0 cm on a side, is placed in the same magnetic field as in Exercise 30–21, with side $ac$ along a diameter and with point $b$ at the center of the field. (Fig. 30–41). a) Copy Fig. 30–41, and show by vectors the direction and relative magnitude of the induced electric field $\boldsymbol{E}$ at the lettered points. b) What is the induced emf in side $ac$? c) What is the induced emf in the loop? d) What is the current in the loop if its resistance is 2.00 Ω? e) What is the potential difference between points $a$ and $c$? Which is at higher potential?

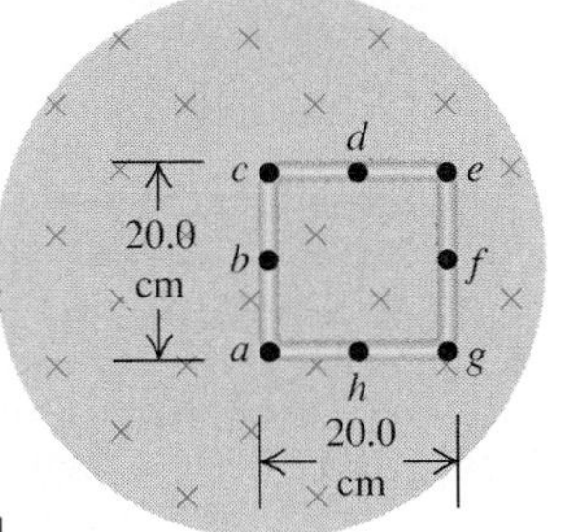


FIGURE 30–41

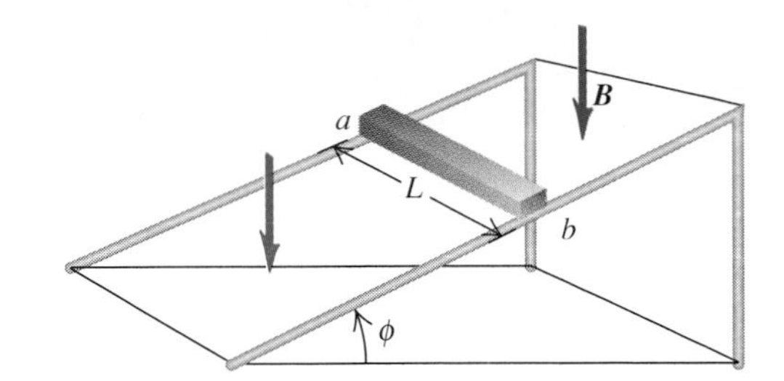


FIGURE 30–42

**30–42** A metal bar with length $L$, mass $m$, and resistance $R$ is placed on frictionless metal rails that are inclined at an angle $\phi$ above the horizontal. The rails have negligible resistance. There is a uniform magnetic field of magnitude $B$ directed downward in Fig. 30–42. The bar is released from rest and slides down the rails. a) Is the direction of the current induced in the bar from $a$ to $b$ or from $b$ to $a$? b) What is the terminal speed of the bar? c) What is the induced current in the bar when the terminal speed has been reached? d) After the terminal speed has been reached, at what rate is electrical energy being dissipated in the resistance of the bar? e) After the terminal speed has been reached, at what rate is work being done on the bar by gravity? Compare your answer to that in part (d).

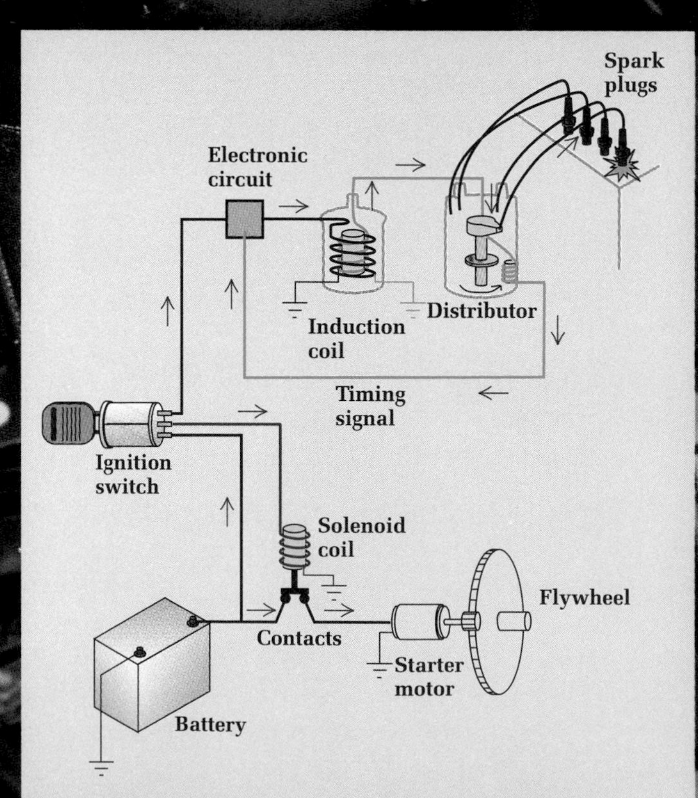

When the ignition switch is turned, it allows a small current from the battery to pass through a solenoid coil. Current in the coil induces a magnetic field. The resulting magnetic force on an iron plunger moves it within the coil, closing metal contacts that allow a much larger current from the battery to pass to the starter motor. Magnetic forces on the rotor cause it to turn; a gear on the rotor shaft turns the engine's flywheel and crankshaft.

An electronic circuit interrupts a current in the primary winding of the induction coil, and the magnetic field associated with this current collapses. The rapid change of this magnetic field induces a very high-voltage current in the secondary winding of the induction coil. This causes sparks in the spark plugs, which ignite the fuel-air mixture in the cylinders.

## KEY CONCEPTS

**• A changing current in one circuit causes a changing magnetic flux and an induced emf in a neighboring circuit, proportional to the rate of change of current. The proportionality factor is called the mutual inductance.**

**• A changing current in a circuit induces an emf in the same circuit, proportional to the rate of change of current. The proportionality factor is called the inductance. A device designed to provide inductance in a circuit is called an inductor.**

**• The energy needed to establish a current in an inductor is proportional to the square of the current. This energy is associated with the magnetic field; the energy per unit volume is proportional to the square of the field magnitude.**

**• Inductors have important and useful circuit properties. A circuit containing an inductor and a capacitor can have an *oscillating* current.**

## INTRODUCTION

How can a 12-volt car battery provide the thousands of volts needed to produce sparks across the gaps of the spark plugs in the engine? Electrical transmission lines often operate at 500,000 volts or more. Applied directly to your household wiring, this voltage would incinerate everything in sight. How can it be reduced to the relatively tame 120 or 240 volts required by familiar electrical appliances? The solutions of both of these problems, and many others concerned with varying currents in circuits, involve the *induction* effects studied in Chapter 30.

A changing current in a coil induces an emf in an adjacent coil. (This is the operating principle of the *transformer.*) The coupling between the coils is described by their *mutual inductance.* A changing current in a coil also causes an induced emf in that same coil. Such a coil is called an *inductor;* the relationship of current to emf is described by the *inductance* of the coil. In this chapter we will study energy relationships involving inductors, including the way a coil supplies energy to a spark plug, energy relations in a transformer, and energy storage in magnetic fields.

Finally, we'll study the circuit behavior of inductors. Inductors play an essential role in modern electronics, including communication equipment, power supplies, analog computers, fluorescent lights, and many other devices.

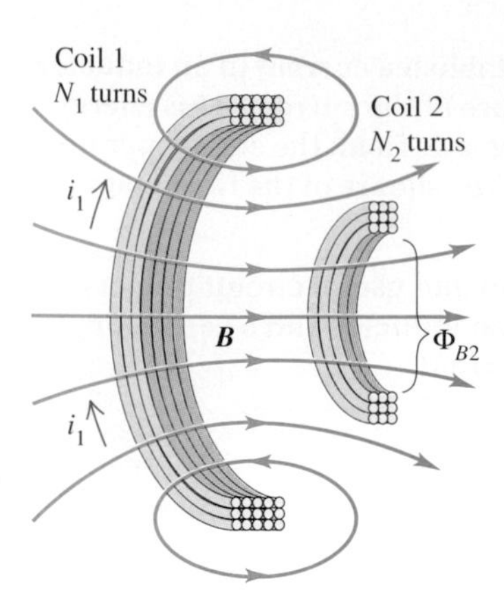


**31–1** In mutual inductance a portion of the magnetic flux set up by a current in coil 1 links with coil 2.

## 31–1 MUTUAL INDUCTANCE

According to Faraday's law, an emf is induced in a stationary circuit whenever the magnetic flux through the circuit varies with time. If this flux variation is caused by a varying current in a second circuit, as happens in a transformer, we can express the induced emf in terms of the varying *current,* rather than in terms of the varying *flux*. In this analysis we will use lowercase letters to represent quantities that vary with time; for example, a time-varying current is $i$, often with a subscript to identify the circuit.

Figure 31–1 is a cross-section view of two coils of wire. A current $i_1$ in coil 1 sets up a magnetic field, as indicated by the blue lines, and some of these field lines pass through coil 2. We denote the magnetic flux through each turn of coil 2, caused by the current $i_1$ in coil 1, as $\Phi_{B2}$. The magnetic field is proportional to $i_1$, so $\Phi_{B2}$ is also proportional to $i_1$. When $i_1$ changes, $\Phi_{B2}$ changes; this changing flux induces an emf $\mathcal{E}_2$ in coil 2 given by

$$\mathcal{E}_2 = -N_2\frac{d\Phi_{B2}}{dt}. \tag{31–1}$$

We could represent the proportionality of $\Phi_{B2}$ and $i_1$ in the form $\Phi_{B2} = (\text{constant})\, i_1$, but instead it is more convenient to include the number of turns $N_2$ in the relation. Introducing a proportionality constant $M_{21}$, we write

$$N_2\Phi_{B2} = M_{21}i_1. \tag{31–2}$$

From this,

$$N_2\frac{d\Phi_{B2}}{dt} = M_{21}\frac{di_1}{dt},$$

and we can rewrite Eq. (31–1) as

$$\mathcal{E}_2 = -M_{21}\frac{di_1}{dt}. \tag{31–3}$$

The constant $M_{21}$ is called the **mutual inductance** of the coils. It is defined by Eq. (31–2), which we may also write as

$$M_{21} = \frac{N_2\Phi_{B2}}{i_1}. \tag{31–4}$$

If the coils are in vacuum, $M_{21}$ depends only on the geometry of the two coils. If a magnetic material is present, $M_{21}$ also depends on the magnetic properties of the material. If the material has nonlinear magnetic properties (that is, if $K_m$ is not constant and magnetization is not proportional to magnetic field), then the flux $\Phi_{B2}$ is no longer directly proportional to the current $i_1$. In that case the mutual inductance is not constant. From now on we will assume that any magnetic material present has constant $K_m$ and that flux *is* directly proportional to current.

We can repeat our discussion for the opposite case, in which a changing current $i_2$ in coil 2 causes a changing flux $\Phi_{B1}$ and an emf $\mathcal{E}_1$ in coil 1. We might expect that the corresponding constant $M_{12}$ would be different from $M_{21}$ because in general the two coils are not identical and the flux through them is not the same. It turns out, however, that $M_{12}$ is *always* equal to $M_{21}$, even when the two coils are not symmetric. We call this common value the mutual inductance $M$; it characterizes completely the induced-emf interaction of two coils, and we can write

$$\mathcal{E}_2 = -M\frac{di_1}{dt} \quad \text{and} \quad \mathcal{E}_1 = -M\frac{di_2}{dt}. \tag{31–5}$$

The negative signs show that the direction of the emf induced in each coil is opposite to the rate of change of current in the other coil.

The SI unit of mutual inductance is called the **henry** (1 H), in honor of Joseph Henry (1797–1878), one of the discoverers of electromagnetic induction. From Eq. (31–4), one henry is equal to *one weber per ampere*. Other equivalent units, obtained by reference to Eq. (31–3), are *one volt-second per ampere* and *one ohm-second:*

$$1\text{ H} = 1\text{ Wb/A} = 1\text{ V}\cdot\text{s/A} = 1\,\Omega\cdot\text{s}.$$

## EXAMPLE 31–1

**The Tesla coil** In one form of Tesla coil (a high-voltage generator that you may have seen in a science museum), a long solenoid with length $L$ and cross-section area $A$ is closely wound with $N_1$ turns of wire. A coil with $N_2$ turns surrounds it at its center (Fig. 31–2). Find the mutual inductance.

**SOLUTION** A current $i_1$ in the solenoid sets up a magnetic field $\boldsymbol{B}_1$ at its center; from Eq. (29–21) the magnitude of $\boldsymbol{B}_1$ is

$$B_1 = \mu_0 n i_1 = \frac{\mu_0 N_1 i_1}{L}.$$

The flux through the central section equals $B_1 A$, and all of this flux links with the small coil. From Eq. (31–4) the mutual inductance $M$ is

$$M = \frac{N_2 \Phi_{B2}}{i_1} = \frac{N_2}{i_1}\frac{\mu_0 N_1 i_1}{L}A = \frac{\mu_0 A N_1 N_2}{L}.$$

Here's a numerical example to give you an idea of magnitudes. Suppose $L = 0.50$ m, $A = 10\text{ cm}^2 = 1.0 \times 10^{-3}\text{ m}^2$, $N_1 = 1000$ turns, and $N_2 = 10$ turns. Then

$$M = \frac{(4\pi \times 10^{-7}\text{ Wb/A}\cdot\text{m})(1.0 \times 10^{-3}\text{ m}^2)(1000)(10)}{0.50\text{ m}}$$
$$= 25 \times 10^{-6}\text{ Wb/A} = 25 \times 10^{-6}\text{ H} = 25\ \mu\text{H}.$$

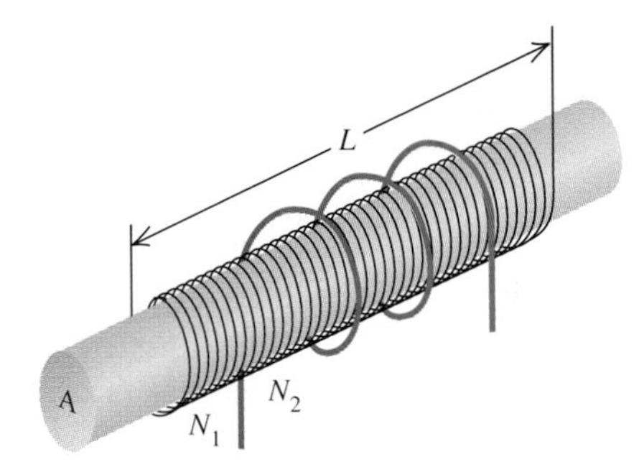


**31–2** One form of Tesla coil, a long solenoid with cross-section area $A$ and $N_1$ turns surrounded at its center by a small coil with $N_2$ turns.

## EXAMPLE 31–2

Suppose the current $i_2$ in the smaller coil in Example 31–1 is given by $i_2 = (2.0\text{ A/s})\,t$. a) At time $t = 3.0$ s, what is the average magnetic flux through each turn of the solenoid caused by the current in the smaller coil? b) What is the induced emf in the solenoid?

**SOLUTION** a) At time $t = 3.0$ s the current in coil 2 is $i_2 = (2.0\text{ A/s})(3.0\text{ s}) = 6.0$ A. To find the flux in the solenoid, we use Eq. (31–4), with the roles of coils 1 and 2 interchanged:

$$N_1 \Phi_{B1} = M i_2,$$
$$(\Phi_{B1})_{\text{av}} = \frac{M i_2}{N_1} = \frac{(25 \times 10^{-6}\text{ H})(6.0\text{ A})}{1000}$$
$$= 0.15 \times 10^{-6}\text{ Wb}.$$

This is an average value; the flux may vary somewhat from one turn to another.

b) The induced emf $\mathcal{E}_1$ is given by Eq. (31–5):

$$\mathcal{E}_1 = -M\frac{di_2}{dt}$$
$$= -(25 \times 10^{-6}\text{ H})\frac{d}{dt}(2.0\text{ A/s})\,t$$
$$= -(25 \times 10^{-6}\text{ H})(2.0\text{ A/s})$$
$$= -5.0 \times 10^{-5}\text{ V}.$$

This is a very small voltage in response to a very leisurely rate of change of current. In an operating Tesla coil, $di_2/dt$ would be alternating rapidly, and its magnitude would be billions of times as large as in this example.

# 31–2 INDUCTANCE WITHIN A CIRCUIT

In our discussion of mutual inductance we assumed that one circuit acted as the source of magnetic field. The emf under consideration was induced in a separate, independent circuit that linked some of the magnetic flux created by the first circuit. However, when a current is present in *any* circuit, this current sets up a magnetic field that links with *the*

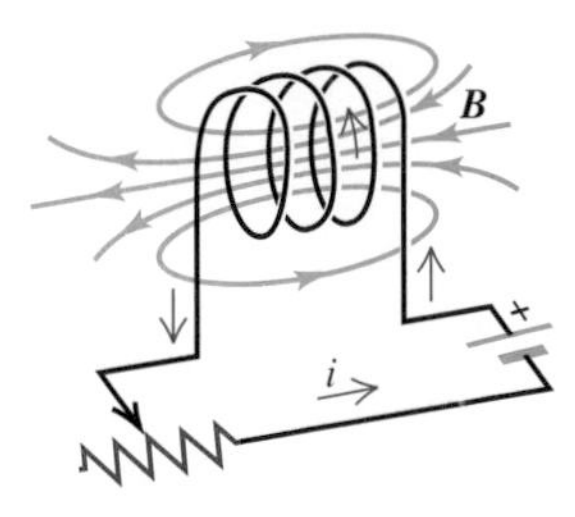


**31–3** A flux $\Phi_B$ linking a coil of $N$ turns. When the current in the circuit changes, the flux changes also, and a self-induced emf appears in the circuit.

*same* circuit and changes when the current changes. Any circuit that carries a varying current has an induced emf in it resulting from the variation in *its own* magnetic field. Such an emf is called a **self-induced emf.**

As an example, consider a coil with $N$ turns of wire, carrying a current $i$ (Fig. 31–3). As a result of this current, a magnetic flux $\Phi_B$ passes through each turn. In analogy to Eq. (31–4) we define the **inductance** $L$ of the circuit (sometimes called **self-inductance**):

$$L = \frac{N\Phi_B}{i}, \qquad \text{or} \qquad N\Phi_B = Li. \tag{31–6}$$

If $\Phi_B$ and $i$ change with time, then

$$N\frac{d\Phi_B}{dt} = L\frac{di}{dt}.$$

From Faraday's law, Eq. (30–4), the self-induced emf $\mathcal{E}$ is $-N\,d\Phi_B/dt$, so it follows that

$$\mathcal{E} = -L\frac{di}{dt}. \tag{31–7}$$

**The self-inductance of a circuit is the magnitude of the self-induced emf per unit rate of change of current.** From the definition the units of self-inductance are the same as those of mutual inductance; the SI unit of self-inductance is *one henry.* We will explore the significance of the negative sign in Eq. (31–7) below.

A circuit, or part of a circuit, that is designed to have a particular inductance is called an **inductor,** or a *choke*. Like resistors and capacitors, inductors are among the indispensible circuit elements of modern electronics. In the following sections we will explore the circuit behavior of inductors. The usual circuit symbol for an inductor is

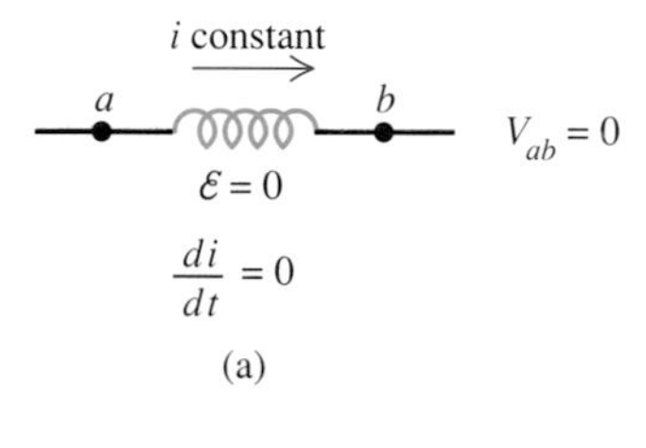


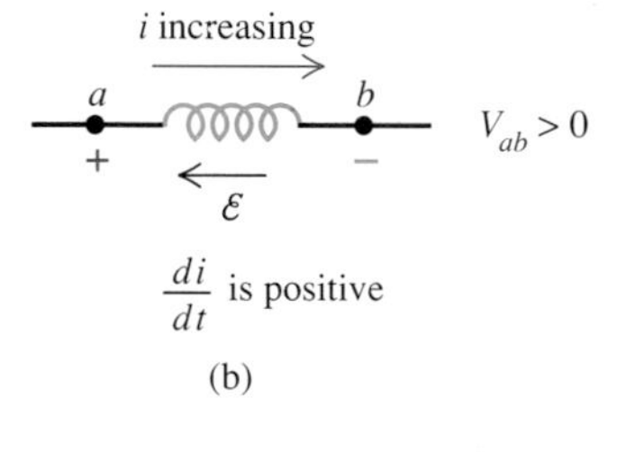


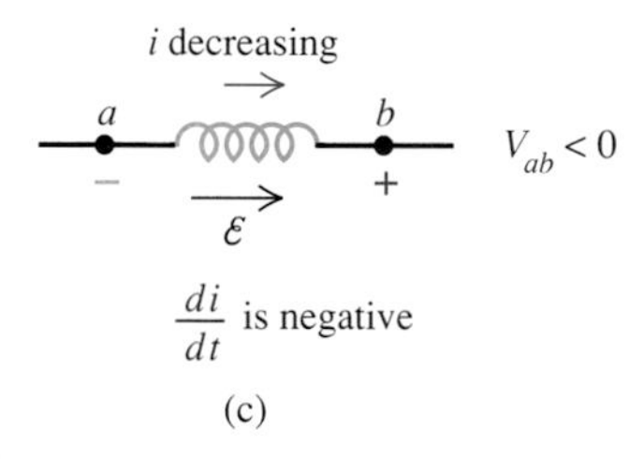


**31–4** An inductor with negligible resistance. When the inductor is traversed in the same direction as the current, the voltage $V_{ab}$ depends on the time rate of change of the current.

We can find the direction of the self-induced emf and the associated non-electrostatic field $\boldsymbol{E}$ from Lenz's law. The cause of the induced emf and the field is the *changing current* in the conductor, and the emf always acts to oppose this change. Figure 31–4 shows three cases. In Fig. 31–4a the current is constant, and $V_{ab} = 0$. In Fig. 31–4b the current is increasing, and $di/dt$ is positive. According to Lenz's law, the induced emf must oppose the increasing current. The emf therefore must be in the sense from $b$ to $a$; $a$ becomes the higher-potential terminal, and $V_{ab}$ is *positive,* as shown. The emf opposes the current increase caused by the external circuit. The direction of the emf is analogous to a battery with $a$ as its + terminal.

In Fig. 31–4c the situation is opposite. The current is decreasing, and $di/dt$ is negative. The induced emf $\mathcal{E}$ opposes this decrease, and $V_{ab}$ is negative. In both cases the induced emf opposes not the current itself but a *change* $di/dt$ in current. Thus its circuit behavior is quite different from that of a resistor. Figure 31–5 shows the comparison and summarizes the sign relations.

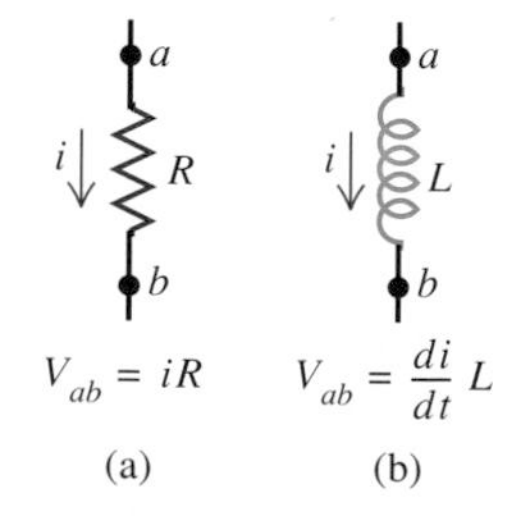


**31–5** When the positive direction of current is from $a$ to $b$; (a) for a resistor, $V_{ab}$ is *always* positive; (b) for a pure inductor, $V_{ab}$ is positive for increasing current, negative for decreasing current, and zero for constant current.

## EXAMPLE 31–3

An air-core toroidal solenoid with cross-section area $A$ and mean radius $r$ is closely wound with $N$ turns of wire (Fig. 31–6). Determine its self-inductance $L$. In calculating the flux, assume that $B$ is uniform across a cross section; neglect the variation of $B$ with distance from the toroid axis.

**SOLUTION** From Eq. (31–6), which defines inductance,

$$L = \frac{N\Phi_B}{i}.$$

To find $\Phi_B$, we first find the field magnitude $B$. From Eq. (29–22), at distance $r$ from the toroid axis, $B = \mu_0 Ni/2\pi r$. If we assume that the field has this magnitude over the entire cross-section area $A$, then the total flux through the cross section is

$$\Phi_B = BA = \frac{\mu_0 NiA}{2\pi r}.$$

*All* the flux links with each turn, and the self-inductance $L$ is

$$L = \frac{N\Phi_B}{i} = \frac{\mu_0 N^2 A}{2\pi r}.$$

Suppose $N = 200$ turns, $A = 5.0\ \text{cm}^2 = 5.0 \times 10^{-4}\ \text{m}^2$, and $r = 0.10$ m; then

$$L = \frac{(4\pi \times 10^{-7}\ \text{Wb/A}\cdot\text{m})(200)^2(5.0 \times 10^{-4}\ \text{m}^2)}{2\pi(0.10\ \text{m})}$$
$$= 40 \times 10^{-6}\ \text{H} = 40\ \mu\text{H}.$$

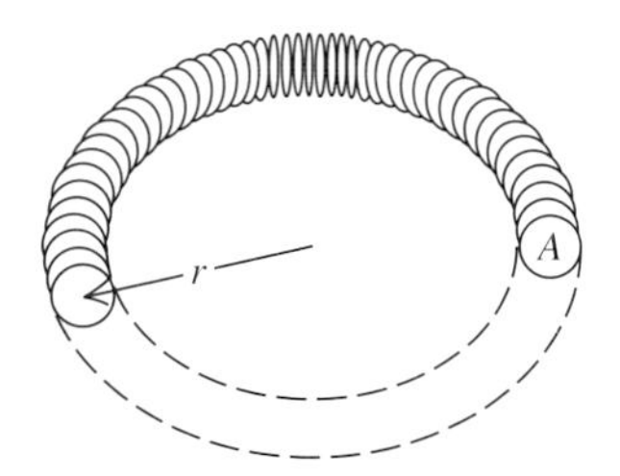


**31–6** Determining the self-inductance of a closely wound toroid.

## EXAMPLE 31–4

If the current in the coil in Example 31–3 increases uniformly from zero to 6.0 A in 0.30 s, find the magnitude and direction of the self-induced emf.

**SOLUTION** From Eq. (31–7),

$$|\mathcal{E}| = L\frac{di}{dt} = (40 \times 10^{-6}\ \text{H})\frac{6.0\ \text{A}}{0.30\ \text{s}}$$
$$= 8.0 \times 10^{-4}\ \text{V}.$$

The current is increasing, so according to Lenz's law, the direction of the emf is opposite to that of the current. This corresponds to Fig. 31–4b; the emf is in the direction from $b$ to $a$, like a battery with $a$ as the + terminal and $b$ as the − terminal, tending to oppose the current increase from the external circuit.

The self-inductance of a circuit depends on its size, shape, and number of turns. For $N$ turns close together it is always proportional to $N^2$. It also depends on the magnetic properties of the material enclosed by the circuit. In the above examples we assumed that the conductor is surrounded by vacuum. If the flux is concentrated in a region containing a magnetic material with permeability $\mu$, then in the expression for $B$ we must replace $\mu_0$ (the permeability of vacuum) by $\mu = K_m \mu_0$, as discussed in Section 29–8. If the material is diamagnetic or paramagnetic, this makes very little difference. If the material is *ferromagnetic,* however, the difference is of crucial importance. An inductor wound on a soft iron core having $K_m = 5000$ has an inductance approximately 5000 times as great as the same coil with an air core. Iron-core and ferrite-core inductors are very widely used in a variety of electronic and electric-power applications.

An added complication is that with ferromagnetic materials the magnetization is not always a linear function of magnetizing current, especially as saturation is approached. As a result, the inductance is not constant but can depend on current in a fairly complicated way. In our discussion we will ignore this complication and assume always that the inductance is constant. This is a reasonable assumption even for a ferromagnetic material if the magnetization remains well below the saturation level.

## 31–4 THE *R-L* CIRCUIT

An inductor is primarily a circuit device. Let's look at some examples of the circuit behavior of an inductor. One thing is clear already; we aren't going to see any sudden changes in the current through an inductor. Equation (31–7) shows that the greater the rate of change of current $di/dt$, the greater must be the potential difference between the inductor terminals. This equation, together with Kirchhoff's rules, gives us the principles that we need to analyze circuits containing inductors.

### PROBLEM-SOLVING STRATEGY

#### Inductors in circuits

**1.** When an inductor is used as a *circuit* device, all the voltages, currents, and capacitor charges are in general functions of time, not constants as they have been in most of our previous circuit analysis. But Kirchhoff's rules, which we studied in Chapter 27, are still valid. When the voltages and currents vary with time, Kirchhoff's rules hold at each instant of time.

**2.** As in all circuit analysis, getting the signs right is sometimes more challenging than understanding the principles. We suggest that you review the strategy in Section 27–2. In addition, study carefully the sign rule described with Eq. (31–7) and Figs. 31–4 and 31–5. In Kirchhoff's loop rule, when we go through an inductor in the *same* direction as the assumed current, we encounter a voltage *drop* equal to $L\,di/dt$, so the corresponding term in the loop equation is $-L\,di/dt$.

### Current Growth in an *R-L* Circuit

We can learn several basic things about inductor behavior by analyzing the circuit of Fig. 31–7. The resistor $R$ may be a separate circuit element, or it may be the resistance of the inductor windings; every real-life inductor has some resistance unless it is made of superconducting wire. By closing switch $S_1$ we can connect the $R$-$L$ combination to a source with constant emf $\mathcal{E}$. (We assume that the source has zero internal resistance, so the terminal voltage equals the emf.) Suppose both switches are initially open, and then at some initial time $t = 0$ we close switch $S_1$. As we have mentioned, the current cannot change suddenly from zero to some final value because of the infinite induced emf that would be involved. Instead, it begins to grow at a definite rate that depends on the value of $L$ in the circuit.

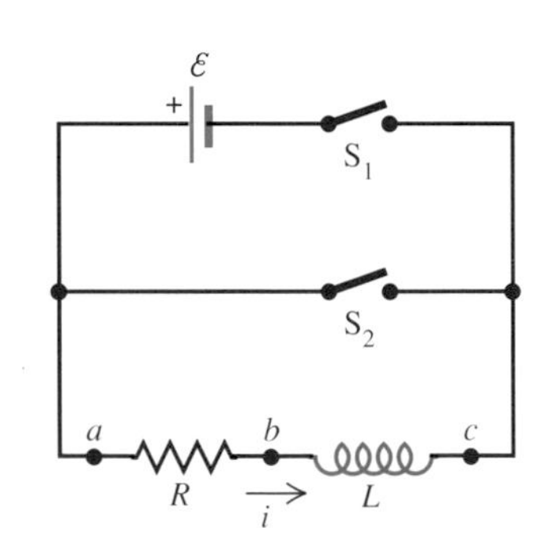


**31–7** An $R$-$L$ series circuit. Closing switch $S_1$ connects the $R$-$L$ combination to a source of emf $\mathcal{E}$; closing switch $S_2$ disconnects the combination from the source.

Let $i$ be the current at some time $t$ after switch $S_1$ is closed, and let $di/dt$ be its rate of change at that time. The potential difference $v_{bc}$ across the inductor at that time is

$$v_{bc} = L\frac{di}{dt},$$

and the potential difference $v_{ab}$ across the resistor is

$$v_{ab} = iR.$$

We apply Kirchhoff's voltage rule, starting at the negative terminal and proceeding counterclockwise around the loop:

$$\mathcal{E} - iR - L\frac{di}{dt} = 0. \qquad (31\text{–}11)$$

Solving this for $di/dt$, we find that the rate of increase of current is

$$\frac{di}{dt} = \frac{\mathcal{E} - iR}{L} = \frac{\mathcal{E}}{L} - \frac{R}{L}i. \qquad (31\text{–}12)$$

At the instant switch $S_1$ is first closed, $i = 0$ and the potential drop across $R$ is zero. The initial rate of change of current is

$$\left(\frac{di}{dt}\right)_{\text{initial}} = \frac{\mathcal{E}}{L}.$$

The greater the inductance $L$, the more slowly the current increases.

As the current increases, the term $(R/L)i$ in Eq. (31–12) also increases, and the *rate* of increase of current becomes smaller and smaller. When the current reaches its final *steady-state* value $I$, its rate of increase is zero. Then Eq. (31–12) becomes

$$\left(\frac{di}{dt}\right)_{\text{final}} = 0 = \frac{\mathcal{E}}{L} - \frac{R}{L}I,$$

and

$$I = \frac{\mathcal{E}}{R}.$$

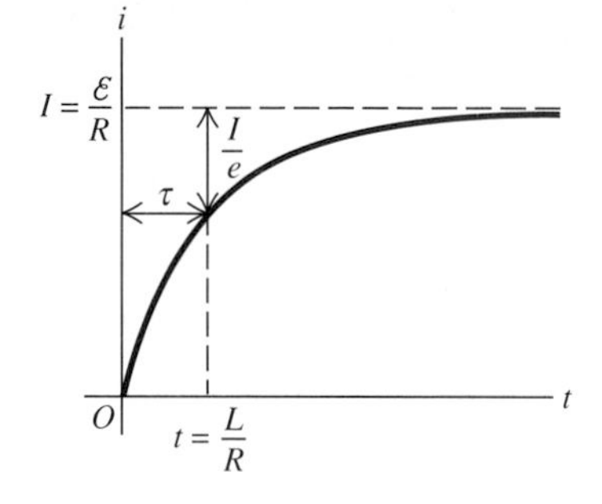


**31–8** Graph of $i$ versus $t$ for growth of current in an $R$-$L$ series circuit. The final current is $I = \mathcal{E}/R$; after one time constant, the current is $1 - 1/e$ of this value.

That is, the *final* current $I$ does not depend on the inductance $L$; it is the same as it would be if the resistance $R$ alone were connected to the source with emf $\mathcal{E}$.

The behavior of the current as a function of time is shown by the graph of Fig. 31–8. To derive the equation for this curve (an expression for current as a function of time), we proceed just as we did for the problem of the charging capacitor in Section 27–4. First we rearrange Eq. (31–12) to the form

$$\frac{di}{i - (\mathcal{E}/R)} = -\frac{R}{L}\,dt.$$

This separates the variables, with $i$ on the left side and $t$ on the right. Then we integrate both sides, renaming the integration variables $i'$ and $t'$ so that we can use $i$ and $t$ as the upper limits. (The lower limit for each is zero, corresponding to zero current at the initial time $t = 0$.) We get

$$\int_0^i \frac{di'}{i' - (\mathcal{E}/R)} = -\int_0^t \frac{R}{L}\,dt',$$

$$\ln\left(\frac{i - \mathcal{E}/R}{-\mathcal{E}/R}\right) = -\frac{R}{L}t.$$

Now we take exponentials of both sides and solve for $i$. We leave the details for you to work out; the final result is

$$i = \frac{\mathcal{E}}{R}\left(1 - e^{-(R/L)t}\right). \qquad (31\text{–}13)$$

This is the equation of the curve in Fig. 31–8. Taking the derivative of Eq. (31–13), we find

$$\frac{di}{dt} = \frac{\mathcal{E}}{L}e^{-(R/L)t}. \qquad (31\text{–}14)$$

At time $t = 0$, $i = 0$ and $di/dt = \mathcal{E}/L$. As $t \to \infty$, $i \to \mathcal{E}/R$ and $di/dt \to 0$, as we predicted.

As Fig. 31–8 shows, the instantaneous current $i$ first rises rapidly, then increases more slowly and approaches the final value $I = \mathcal{E}/R$ asymptotically. At a time $\tau$ equal to

$L/R$, the current has risen to $1 - 1/e$, or about 0.63 of its final value. The quantity $\tau = L/R$ is called the **time constant** for the circuit:

$$\tau = \frac{L}{R}. \tag{31–15}$$

In a time equal to $2\tau$ the current reaches 86% of its final value; in $5\tau$, 99.3%; and in $10\tau$, 99.995%.

The graphs of $i$ versus $t$ have the same general shape for all values of $L$. For a given value of $R$ the time constant $\tau$ is greater for greater values of $L$. When $L$ is small, the current rises rapidly to its final value; when $L$ is large, it rises more slowly. For example, if $R = 100\ \Omega$ and $L = 10$ H,

$$\tau = \frac{L}{R} = \frac{10\ \text{H}}{100\ \Omega} = 0.10\ \text{s},$$

and the current increases to about 63% of its final value in 0.10 s. But if $L = 0.010$ H, $\tau = 1.0 \times 10^{-4}\ \text{s} = 0.10$ ms, and the rise is much more rapid.

## EXAMPLE 31–7

In an $R$-$L$ circuit in a laboratory demonstration, measurements made with a digital multimeter and an oscilloscope indicate that the source emf is 6.3 V, the circuit resistance is 175 Ω, and the current takes 58 μs to build up from zero to 4.9 mA. a) What is the final current? b) What is the inductance? c) What is the time constant?

**SOLUTION** a) The final current $I$, which does not depend on the inductance $L$, is

$$I = \frac{\mathcal{E}}{R} = \frac{6.3\ \text{V}}{175\ \Omega} = 0.036\ \text{A} = 36\ \text{mA}.$$

b) To find $L$, we solve Eq. (31–13) for $L$. First we multiply through by $-R/\mathcal{E}$ and then add 1 to both sides to obtain

$$1 - \frac{iR}{\mathcal{E}} = e^{-(R/L)t}.$$

Then we take natural logs of both sides, solve for $L$, and insert the numbers:

$$\begin{aligned} L &= \frac{-Rt}{\ln(1 - iR/\mathcal{E})} \\ &= \frac{-(175\ \Omega)(58 \times 10^{-6}\ \text{s})}{\ln[1 - (4.9 \times 10^{-3}\ \text{A})(175\ \Omega)/(6.3\ \text{V})]} \\ &= 69\ \text{mH}. \end{aligned}$$

c) The time constant $\tau$ is given by Eq. (31–15):

$$\begin{aligned} \tau &= \frac{L}{R} = \frac{69 \times 10^{-3}\ \text{H}}{175\ \Omega} \\ &= 3.9 \times 10^{-4}\ \text{s} = 390\ \mu\text{s}. \end{aligned}$$

We note that 58 μs is much less than the time constant. The current has time to build up only from zero to 4.9 mA, a small fraction of its final value of 36 mA.

---

Energy considerations offer us additional insight into the behavior of an $R$-$L$ circuit. The instantaneous rate at which the source delivers energy to the circuit is $P = \mathcal{E}i$. The instantaneous rate at which energy is dissipated in the resistor is $i^2R$, and the rate at which energy is stored in the inductor is $iv_{bc} = Li\,di/dt$. When we multiply Eq. (31–11) by $i$ and rearrange, we find

$$\mathcal{E}i = Li\frac{di}{dt} + i^2R. \tag{31–16}$$

This shows that part of the power $\mathcal{E}i$ supplied by the source is dissipated ($i^2R$) in the resistor, and part is stored ($Li\,di/dt$) in the inductor. This discussion is completely analogous to our power analysis for a charging capacitor at the end of Section 27–4.

## Current Decay in an *R-L* Circuit

Now suppose switch $S_1$ in the circuit of Fig. 31–7 has been closed for a while and that the current has reached the value $I_0$. Resetting our stopwatch to redefine the initial time, we close switch $S_2$ at time $t = 0$, bypassing the battery. (At the same time we should open $S_1$ to save the battery from ruin.) The current through $R$ and $L$ does not instantaneously go to zero but decays smoothly, as shown in Fig. 31–9. The Kirchhoff's-rule loop equation is obtained from Eq. (31–11) by simply omitting the $\mathcal{E}$ term. We challenge you to retrace the steps in the above analysis and show that the current $i$ varies with time according to

$$i = I_0 e^{-(R/L)t}, \tag{31–17}$$

where $I_0$ is the initial current at time $t = 0$. The time constant, $\tau = L/R$, is the time for current to decrease to $1/e$, or about 37%, of its original value. In time $2\tau$ it has dropped to 13.5%, in time $5\tau$ to 0.67%, and in $10\tau$ to 0.0045%.

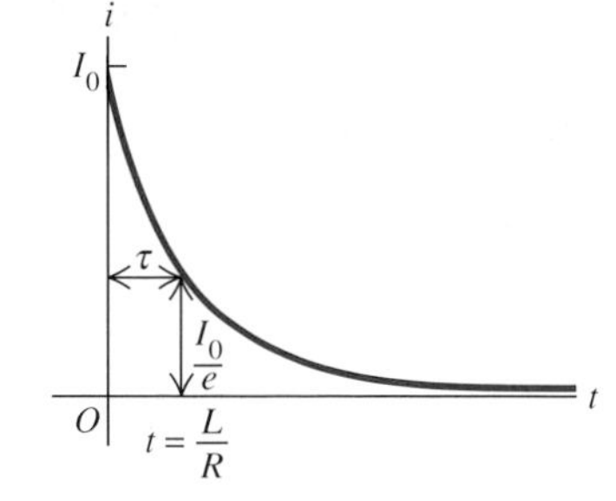


**31–9** Graph of $i$ versus $t$ for decay of current in an $R$-$L$ series circuit.

The energy needed to maintain the current during this decay is provided by the energy stored in the magnetic field of the inductor. The detailed energy analysis is simpler this time. In place of Eq. (31–16) we have

$$0 = Li\frac{di}{dt} + i^2R. \tag{31–18}$$

In this case, $Li\,di/dt$ is negative; this equation shows that the energy stored in the inductor *decreases* at a rate equal to the rate of dissipation of energy $i^2R$ in the resistor.

This entire discussion should look familiar; the situation is very similar to that of a charging and discharging capacitor, analyzed in Section 27–4. It would be a good idea to compare that section with our discussion of the $L$-$R$ circuit.

## EXAMPLE 31–8

**Energy in an $R$-$L$ Circuit** When the current in an $R$-$L$ circuit is decaying, what fraction of the original energy stored in the inductor has been dissipated after 2.3 time constants?

**SOLUTION** From Eq. (31–17) the current $i$ at any time $t$ is

$$i = I_0 e^{-(R/L)t}.$$

The energy $U$ in the inductor at *any* time is obtained by substituting this expression into $U = \frac{1}{2}Li^2$. We obtain

$$U = \tfrac{1}{2}LI_0^2 e^{-2(R/L)t} = U_0 e^{-2(R/L)t}$$

where $U_0 = \frac{1}{2}LI_0^2$ is the initial energy at time $t = 0$. When $t = 2.3\,\tau = 2.3L/R$, we have

$$U = U_0 e^{-2(2.3)} = U_0 e^{-4.6} = 0.010\,U_0.$$

That is, only 0.010, or 1.0%, of the energy remains, so 99.0% has been dissipated.

# 31–5 THE *L-C* CIRCUIT

A circuit containing an inductor and a capacitor shows an entirely new mode of behavior, characterized by *oscillating* current and charge. This is in sharp contrast to the *exponential* approach to a steady-state situation that we have seen with both $R$-$C$ and $R$-$L$ circuits. In the ***L*-*C* circuit** in Fig. 31–10 (on page 878), we charge the capacitor to a potential difference $V_m$ and initial charge $Q = CV_m$, as shown in Fig. 31–10a, and then close the switch. What happens?

The capacitor begins to discharge through the inductor. Because of the induced emf in the inductor, the current cannot change instantaneously; it starts at zero and eventually builds up to a maximum value $I_m$. During this buildup, the capacitor is discharging. At each instant the capacitor potential equals the induced emf, so as the capacitor discharges, the *rate of change* of current decreases. When the capacitor potential becomes zero, the induced emf is also zero, and the current has leveled off at its maximum value $I_m$. Figure 31–10b shows this situation; the capacitor has completely discharged. The potential difference between its terminals (and those of the inductor) has decreased to zero, and the current has reached its maximum value $I_m$.

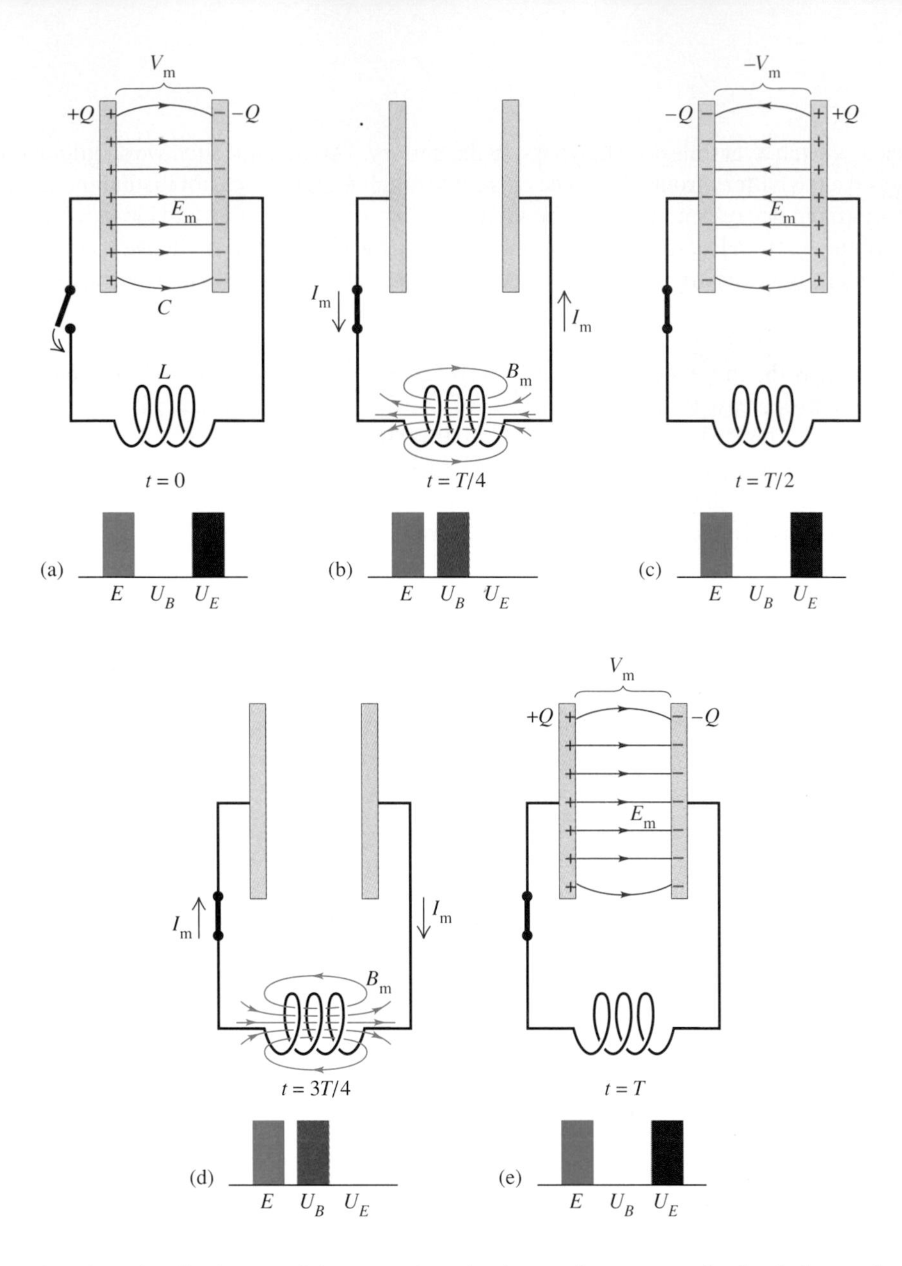


**31–10** Energy transfer between electric and magnetic fields in an oscillating $L$-$C$ series circuit. The switch is closed at time $t = 0$. As in simple harmonic motion (Fig. 13–3), the total energy $E$ remains constant.

During the discharge of the capacitor the increasing current in the inductor has established a magnetic field in the space around it, and the energy that was initially stored in the capacitor's electric field is now stored in the inductor's magnetic field.

The current cannot change instantaneously; as it persists, the capacitor begins to charge with polarity opposite to that in the initial state. As the current decreases, the magnetic field also decreases, inducing an emf in the inductor in the same direction as the current. Eventually, the current and the magnetic field reach zero, and the capacitor has been charged in the *opposite* sense to its initial polarity (Fig. 31–10c), with a potential difference $-V_m$ and charge $-Q$.

The process now repeats in the reverse direction; a little later, the capacitor has again discharged, and there is a current in the inductor in the opposite direction (Fig. 31–10d). Still later, the capacitor charge returns to its original value (Fig. 31–10e), and the whole process repeats. If there are no energy losses, the charges on the capacitor continue to oscillate back and forth indefinitely. This process is called an **electrical oscillation.**

From an energy standpoint the oscillations of an electrical circuit transfer energy from the capacitor's electric field to the inductor's magnetic field and back. The *total* energy associated with the circuit is constant. This is analogous to the transfer of energy in an oscillating mechanical system from potential to kinetic and back, with constant total energy. As we will see, this analogy goes much further.

To study the flow of charge in detail, we proceed just as we did for the $R$-$L$ circuit; we begin with Kirchhoff's loop rule. Figure 31–11 shows our definitions of $q$ and $i$. Note that if the capacitor is initially charged and begins to discharge, the initial current $i$ is negative, opposite to the (positive) direction shown. Starting at the lower right corner of the circuit in Fig. 31–11 and adding voltages as we go clockwise around the loop, we obtain

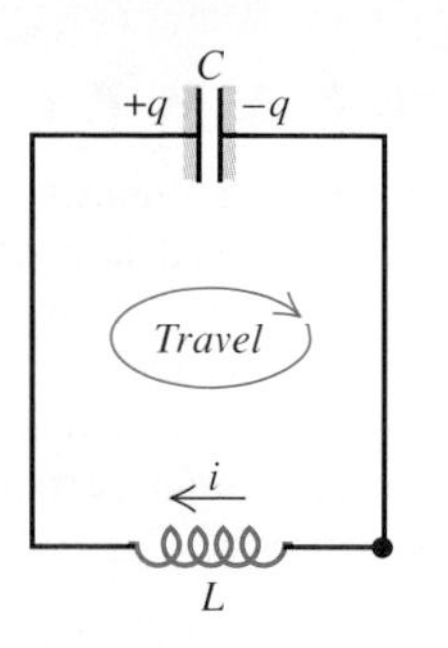


**31–11** Applying Kirchhoff's loop rule to the $L$-$C$ series circuit. The direction of travel around the loop in the loop equation is shown. Just after the switch is closed, the current is negative (opposite to the direction shown).

$$-L\frac{di}{dt}-\frac{q}{C}=0.$$

We also know that at each instant the current $i$ must be equal to the rate of change of capacitor charge $dq/dt$: $i = dq/dt$ and $di/dt = d^2q/dt^2$. We substitute this expression into the above equation and divide by $-L$ to obtain

$$\frac{d^2q}{dt^2}+\frac{1}{LC}q=0. \tag{31–19}$$

The form of this equation is exactly the same as the equation that we derived for simple harmonic motion in Section 13–3, Eq. (13–17):

$$\frac{d^2x}{dt^2}+\frac{k}{m}x=0.$$

(We suggest that you review that section before going on with this discussion.) In the $L$-$C$ circuit the capacitor charge $q$ plays the role of the displacement $x$, and the current $i = dq/dt$ is analogous to the particle's velocity $v = dx/dt$. The inductance $L$ is analogous to the mass $m$, and the reciprocal of the capacitance, $1/C$, is analogous to the force constant $k$.

Pursuing this analogy, we recall that the angular frequency $\omega = 2\pi f$ of the harmonic oscillator is equal to $(k/m)^{1/2}$, and the position is given as a function of time by

$$x = A\cos(\omega t+\phi), \tag{31–20}$$

where the amplitude $A$ and the phase angle $\phi$ depend on the initial conditions. In the analogous electrical situation the capacitor charge $q$ is given by

$$q = Q\cos(\omega t+\phi), \tag{31–21}$$

and the angular frequency $\omega$ of oscillation is given by

$$\omega=\sqrt{\frac{1}{LC}}. \tag{31–22}$$

We invite you to verify that Eq. (31–21) satisfies the loop equation, Eq. (31–19), when $\omega$ has the value given by Eq. (31–22). During this process you will find that the instantaneous current $i = dq/dt$ is given by

$$i = -\omega Q\sin(\omega t+\phi). \tag{31–23}$$

Thus the charge and current in the $L$-$C$ circuit oscillate sinusoidally with time with an angular frequency determined by the values of $L$ and $C$. The ordinary frequency $f$, the number of cycles per second, is equal to $\omega/2\pi$, as always. The constants $Q$ and $\phi$ are determined by the initial conditions. If at time $t = 0$ the capacitor has its maximum charge $Q$ and the current $i$ is zero, then $\phi = 0$. If at time $t = 0$, $q = 0$, then $\phi = \pm\pi/2$ rad.

## EXAMPLE 31–9

**An oscillating circuit** A 300-V dc power supply is used to charge a 25-$\mu$F capacitor. After the capacitor is fully charged, it is disconnected from the power supply and connected across a 10-mH inductor. The resistance in the circuit is negligible. a) Find the frequency of oscillation of the circuit. b) Find the capacitor charge and the circuit current 1.2 ms after the inductor and capacitor are connected.

**SOLUTION** a) The natural *angular* frequency is

$$\omega = \sqrt{\frac{1}{LC}} = \sqrt{\frac{1}{(10 \times 10^{-3}\ \text{H})(25 \times 10^{-6}\ \text{F})}}$$
$$= 2.0 \times 10^3\ \text{rad/s}.$$

The frequency $f$ is $1/2\pi$ times this:

$$f = \frac{\omega}{2\pi} = \frac{2.0 \times 10^3\ \text{rad/s}}{2\pi\ \text{rad/cycle}} = 320\ \text{Hz}.$$

b) We use Eq. (31–21) to find the capacitor charge as a function of time. The charge is maximum at $t = 0$, so $\phi = 0$ and $Q = CV_{\text{m}} = (25 \times 10^{-6}\ \text{F})(300\ \text{V}) = 7.5 \times 10^{-3}\ \text{C}$. The charge $q$ at any time is

$$q = (7.5 \times 10^{-3}\ \text{C}) \cos \omega t.$$

At time $t = 1.2 \times 10^{-3}$ s,

$$\omega t = (2.0 \times 10^3\ \text{rad/s})(1.2 \times 10^{-3}\ \text{s}) = 2.4\ \text{rad},$$
$$q = (7.5 \times 10^{-3}\ \text{C}) \cos (2.4\ \text{rad}) = -5.5 \times 10^{-3}\ \text{C}.$$

The current $i$ at any time is

$$i = -\omega Q \sin \omega t.$$

At time $t = 1.2 \times 10^{-3}$ s,

$$i = -(2.0 \times 10^3\ \text{rad/s})(7.5 \times 10^{-3}\ \text{C}) \sin (2.4\ \text{rad})$$
$$= -10\ \text{A}.$$

The negative sign shows that at this time the current is opposite to the direction defined as positive in Fig. 31–11.

---

We can also analyze the $L$-$C$ circuit using an energy approach. The analogy to simple harmonic motion is equally useful here. In the mechanical problem a body with mass $m$ is attached to a spring with force constant $k$. Suppose we displace the body a distance $A$ from its equilibrium position and release it from rest at time $t = 0$. The kinetic energy of the system at any later time is $\frac{1}{2}mv^2$, and its elastic potential energy is $\frac{1}{2}kx^2$. Because the system is conservative, the sum of these energies equals the initial energy of the system, $\frac{1}{2}kA^2$. We find the velocity $v$ at any position $x$ just as we did in Section 13–2, Eq. (13–7):

$$v = \pm\sqrt{\frac{k}{m}}\sqrt{A^2 - x^2}. \tag{31–24}$$

We integrated this to get $x$ as a function of $t$, and the result was Eq. (31–20).

The $L$-$C$ circuit is also a conservative system. Again let $Q$ be the maximum capacitor charge. The magnetic-field energy $\frac{1}{2}Li^2$ in the inductor at any time corresponds to the kinetic energy $\frac{1}{2}mv^2$ of the vibrating body. The electric-field energy $q^2/2C$ in the capacitor corresponds to the elastic potential energy $\frac{1}{2}kx^2$ of the spring. The sum of these energies equals the total energy $Q^2/2C$ of the system:

$$\tfrac{1}{2}Li^2 + \frac{q^2}{2C} = \frac{Q^2}{2C}.$$

Solving for $i$, we find that when the charge on the capacitor is $q$, the current $i$ is

$$i = \pm\sqrt{\frac{1}{LC}}\sqrt{Q^2 - q^2}. \tag{31–25}$$

Comparing this with Eq. (31–24), we see that current $i = dq/dt$ and charge $q$ are related in the same way as velocity $v = dx/dt$ and position $x$ in the mechanical problem. We could integrate Eq. (31–25) to obtain Eq. (31–21), as we did in the mechanical case.

The analogies between simple harmonic motion and oscillation of the $L$-$C$ circuit are summarized in Table 31–1. This striking parallel between the mechanical and electrical systems is only one of many such examples in physics. The parallel between electrical and mechanical (and acoustical) systems is so close that we can solve complicated mechanical and acoustical problems by setting up analogous electrical circuits and mea-

**TABLE 31–1 Oscillation of a Mass-Spring System Compared with Electrical Oscillation in an *L-C* Circuit**

| Mass-spring system | Inductor-capacitor circuit |
|---|---|
| Kinetic energy $= \frac{1}{2}mv^2$ | Magnetic energy $= \frac{1}{2}Li^2$ |
| Potential energy $= \frac{1}{2}kx^2$ | Electric energy $= \dfrac{q^2}{2C}$ |
| $\frac{1}{2}mv^2 + \frac{1}{2}kx^2 = \frac{1}{2}kA^2$ | $\frac{1}{2}Li^2 + \dfrac{q^2}{2C} = \dfrac{Q^2}{2C}$ |
| $v = \pm\sqrt{k/m}\sqrt{A^2 - x^2}$ | $i = \pm\sqrt{1/LC}\sqrt{Q^2 - q^2}$ |
| $v = \dfrac{dx}{dt}$ | $i = \dfrac{dq}{dt}$ |
| $\omega = \sqrt{\dfrac{k}{m}}$ | $\omega = \sqrt{\dfrac{1}{LC}}$ |
| $x = A\cos(\omega t + \phi)$ | $q = Q\cos(\omega t + \phi)$ |

suring the currents and voltages that correspond to the mechanical and acoustical quantities to be determined. This is the basic principle of one kind of analog computer. We see that in our comparison between the harmonic oscillator and the $L$-$C$ circuit, $m$ corresponds to $L$, $k$ to $(1/C)$, $x$ to $q$, and $v$ to $i$. This analogy can be extended to *damped oscillations,* which we consider in the next section.

## 31–6 THE *L-R-C* CIRCUIT

In our discussion of the $L$-$C$ circuit we assumed that there was no *resistance* in the circuit. This is an idealization, of course; every real inductor has resistance in its windings, and there may also be resistance in the connecting wires. The effect of resistance is to dissipate the electromagnetic energy in the circuit and convert it to other forms, such as internal energy in the circuit materials. Resistance in an electric circuit is analogous to friction in a mechanical system.

Suppose an inductor with inductance $L$ and a resistor of resistance $R$ are connected in series across the terminals of a charged capacitor. As before, the capacitor starts to discharge as soon as the circuit is completed. But because of $i^2R$ losses in the resistor, the magnetic-field energy acquired by the inductor when the capacitor is completely discharged is *less* than the original electric-field energy of the capacitor. In the same way, the energy of the capacitor when the magnetic field has decreased to zero is still smaller, and so on.

If the resistance $R$ is relatively small, the circuit still oscillates, but with **damped harmonic motion** (Fig. 31–12a), and we say that the circuit is **underdamped.** If we increase $R$, the oscillations die out more rapidly. When $R$ reaches a certain value, the

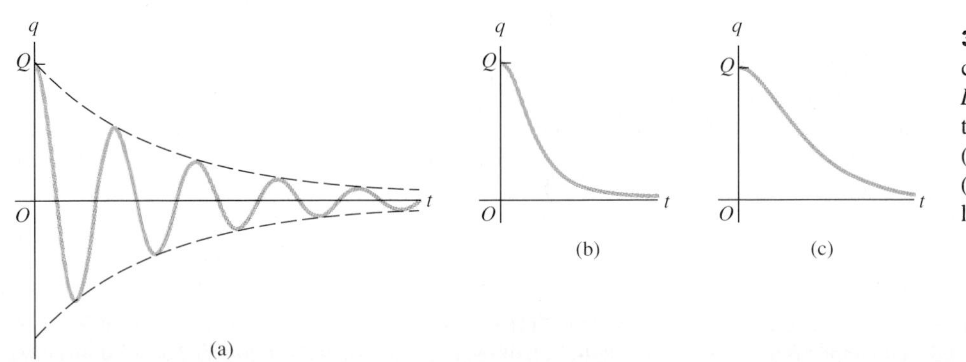


**31–12** Graphs of capacitor charge as a function of time in an $L$-$R$-$C$ series circuit with zero initial current: (a) underdamped (small $R$); (b) critically damped (larger $R$); (c) overdamped (very large $R$).

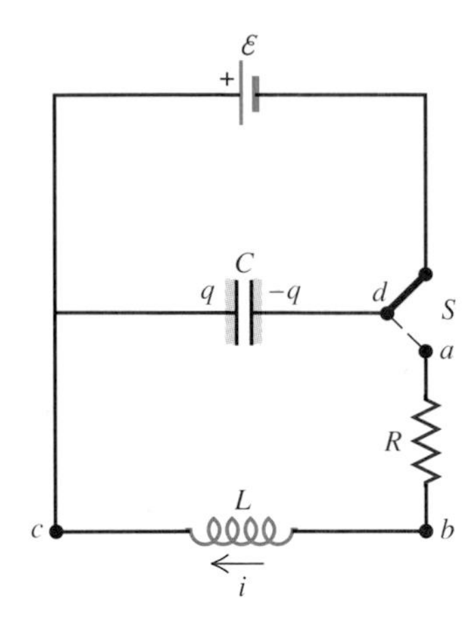


**31–13** An $L$-$R$-$C$ series circuit.

circuit no longer oscillates, and we say that it is **critically damped** (Fig. 31–12b). For still larger values of $R$, the circuit is **overdamped** (Fig. 31–12c), and the capacitor charge approaches zero more slowly. Note that we are using the same terms that we used to describe the behavior of the analogous mechanical system, the damped harmonic oscillator, in Section 13–7.

To analyze $L$-$R$-$C$ circuit behavior in detail, we consider the circuit shown in Fig. 31–13. It is like the $L$-$C$ circuit of Fig. 31–11 except for the added resistor $R$; we have also shown explicitly the source that charges the capacitor initially. The labeling of the positive senses of $q$ and $i$ are the same as for the $L$-$C$ circuit.

The analysis for finding the current $i$ and the capacitor charge $q$ as functions of time is given below. First we close the switch in the upward position, connecting the capacitor to a source with terminal voltage $V_m$ for a long enough time to ensure that the capacitor acquires its final charge $Q = CV_m$ and any initial oscillations have died out. Then at time $t = 0$ we flip the switch to the downward position. Note that the initial current is negative, opposite in direction to the direction of $i$ shown in the figure.

To find how $q$ and $i$ vary with time, we apply Kirchhoff's loop rule. Starting at point $a$ and going around the loop in the direction $abcda$, we obtain the equation

$$-iR - L\frac{di}{dt} - \frac{q}{C} = 0.$$

Replacing $i$ with $dq/dt$ and rearranging, we get

$$\frac{d^2q}{dt^2} + \frac{R}{L}\frac{dq}{dt} + \frac{1}{LC}q = 0. \tag{31–26}$$

Note that when $R = 0$, this reduces to Eq. (31–19).

There are general methods for obtaining solutions of Eq. (31–26). The form of the solution is different for the underdamped (small $R$) and overdamped (large $R$) cases. When $R^2$ is less than $4L/C$, the solution has the form

$$q = Qe^{-(R/2L)t}\cos\left(\sqrt{\frac{1}{LC} - \frac{R^2}{4L^2}}\,t + \phi\right). \tag{31–27}$$

We invite you to take the second derivative of this function and show by direct substitution that it does satisfy Eq. (31–26).

This solution corresponds to the *underdamped* behavior shown in Fig. 31–12a; the function represents a sinusoidal oscillation with an exponentially decaying amplitude. When $R = 0$, it reduces to Eq. (31–21). In general, though, the angular frequency of the oscillation is no longer $1/(LC)^{1/2}$ but is *less* than this because of the term containing $R$. The angular frequency $\omega'$ of the damped oscillations is given by

$$\omega' = \sqrt{\frac{1}{LC} - \frac{R^2}{4L^2}}. \tag{31–28}$$

When $R = 0$, this reduces to Eq. (31–22), $\omega = 1/(LC)^{1/2}$. As $R$ increases, $\omega'$ becomes smaller and smaller. When $R^2 = 4L/C$, the quantity under the radical becomes zero. The system no longer oscillates, and the case of *critical damping* (Fig. 31–12b) has been reached. For still larger values of $R$ the system behaves as in Fig. 31–12c. In this case the circuit is *overdamped,* and $q$ is given as a function of time by the sum of two decreasing exponential functions.

## EXAMPLE 31–10

What resistance $R$ is required (in terms of $L$ and $C$) to give an $L$-$R$-$C$ circuit a frequency that is one-half the undamped frequency?

**SOLUTION** We want the angular frequency $\omega'$ of Eq. (31–28) to be half the undamped angular frequency $\omega$ of Eq. (31–22).

## Inductors

**Inductors are important components of many ac circuits because they can regulate the amount of current in the circuit. Their role is similar to that of a variable mass in a mechanical mass-spring system, regulating the overall motion of the spring.**

# BUILDING PHYSICAL INTUITION

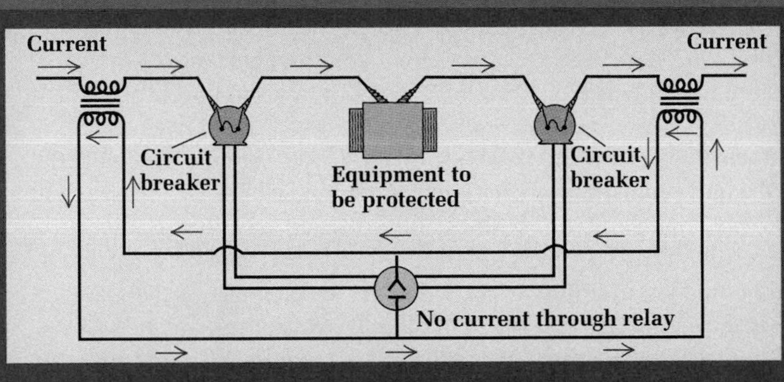


Inductors are a key component of electric protective devices, designed to detect unacceptable conditions in a circuit (for example, a voltage surge or short circuit). In a relay system, inductors are placed in the circuit on each side of the equipment to be protected. Normal current through the equipment produces a small secondary induced current through a circuit connecting the two inductors.

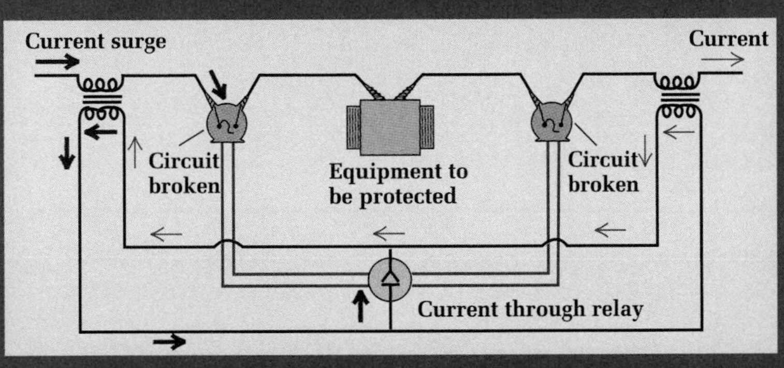


If a fault occurs in the main circuit line, it creates a change in the secondary current connecting the two inductors. This change in current is detected by a relay device that activates circuit breakers (shown in photo), isolating the protected piece of equipment from the dangerous circuit condition.

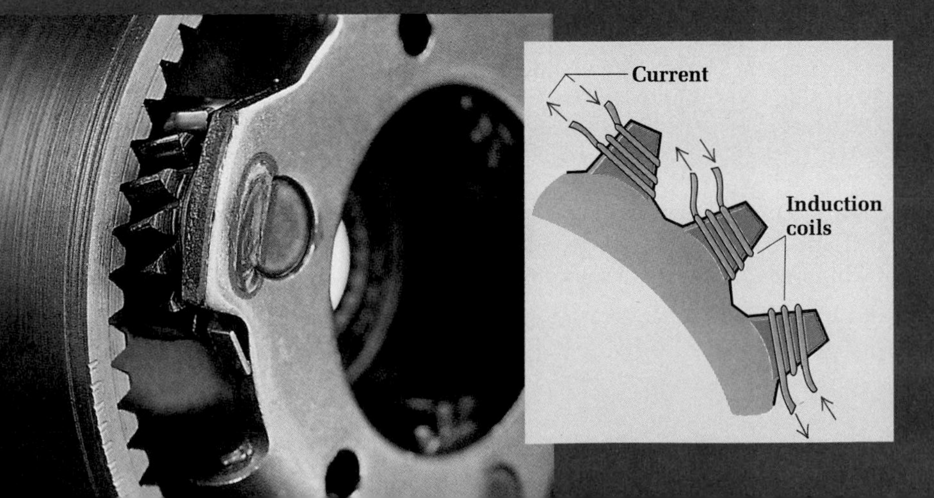


Induction heating is a common industrial process for hardening metals. The metal part is inserted into an induction coil or next to a flat coil, and is heated by eddy currents induced by the inductor's changing magnetic field. Then the metal is rapidly cooled, usually by quenching in water, which hardens the metal. Localized induction heating is often used to harden gear teeth.

Thus we have

$$\sqrt{\frac{1}{LC}-\frac{R^2}{4L^2}}=\frac{1}{2}\sqrt{\frac{1}{LC}}.$$

When we square both sides and solve for $R$, we get

$$R=\sqrt{\frac{3L}{C}}.$$

For example, adding 35 Ω to the circuit of Example 31–9 would reduce the frequency from 320 Hz to 160 Hz.

In the *underdamped* case the phase constant $\phi$ in the cosine function of Eq. (31–27) provides for the possibility of both an initial charge and an initial current at time $t = 0$ (analogous to an underdamped harmonic oscillator given both an initial displacement and an initial velocity).

We emphasize once more that the behavior of the $L$-$R$-$C$ circuit is completely analogous to that of the damped harmonic oscillator studied in Section 13–7. We invite you to verify, for example, that if you start with Eq. (13–38) and substitute $L$ for $m$, $1/C$ for $k$, and $R$ for $b$, the result is Eq. (31–26). Similarly, the crossover point between underdamping and overdamping occurs at $b^2 = 4km$ for the mechanical system and at $R^2 = 4L/C$ for the electrical one. Can you find still other aspects of this analogy?

With appropriate electronic circuitry, energy can be fed *into* an $L$-$R$-$C$ circuit at the same rate that it is dissipated by $i^2R$ losses. The behavior then is as though we had inserted a *negative resistance* into the circuit to make the *total* circuit resistance exactly zero. In this case the circuit oscillates with sustained oscillations of constant amplitude, just like the idealized $L$-$C$ circuit with no resistance.

Still more interesting aspects of this circuit's behavior emerge when we include a sinusoidally varying source of emf in the circuit. This is analogous to the *forced oscillations* that we discussed in Section 13–8, and there are analogous *resonance* effects. Such a circuit is called an *alternating-current* (ac) circuit; the analysis of ac circuits is the principal topic of the next chapter.

## SUMMARY

- When changing magnetic flux created by a changing current $i_1$ in one circuit links a second circuit, an emf $\mathcal{E}_2$ is induced in the second circuit. Changing current $i_2$ in the second circuit induces an emf $\mathcal{E}_1$ in the first circuit:

$$\mathcal{E}_2 = -M\frac{di_1}{dt}, \quad \text{and} \quad \mathcal{E}_1 = -M\frac{di_2}{dt}, \tag{31–5}$$

where $M$ is a constant called the mutual inductance. The value of $M$ depends on the geometry of the two coils and on the material between them. The SI unit of mutual inductance is the henry, abbreviated H. Equivalent units are

$$1\ \text{H} = 1\ \text{Wb/A} = \text{V}\cdot\text{s/A} = 1\ \Omega\cdot\text{s}.$$

- A changing current $i$ in any circuit induces an emf $\mathcal{E}$ in that same circuit, called a self-induced emf:

$$\mathcal{E} = -L\frac{di}{dt}, \tag{31–7}$$

where $L$ is a constant, depending on the geometry of the circuit and the material surrounding it, called inductance or self-inductance. A circuit device, usually including a coil of wire, intended to have a substantial inductance is called an inductor.

- An inductor with inductance $L$ carrying current $I$ has energy

$$U = \tfrac{1}{2}LI^2. \tag{31–8}$$

This energy is associated with the magnetic field of the inductor, with an energy density $u$ (energy per unit volume) given by

$$u = \frac{B^2}{2\mu_0} \tag{31-9}$$

if the field is in vacuum or

$$u = \frac{B^2}{2\mu} \tag{31-10}$$

in a material with magnetic permeability $\mu$.

• In a circuit containing a resistor $R$, an inductor $L$, and a source of emf $\mathcal{E}$, the growth and decay of current are exponential, with a characteristic time $\tau$ called the time constant:

$$\tau = \frac{L}{R}. \tag{31-15}$$

This is the time required for the increasing current to approach within a fraction $1 - (1/e)$ of its final value.

• A circuit containing inductance $L$ and capacitance $C$ undergoes electrical oscillations, with angular frequency $\omega$:

$$\omega = \sqrt{\frac{1}{LC}}. \tag{31-22}$$

Such a circuit is analogous to a mechanical harmonic oscillator, with inductance $L$ analogous to mass, the reciprocal of capacitance $1/C$ to force constant $k$, charge $q$ to displacement $x$, and current $i$ to velocity $v$.

• A series circuit containing inductance, resistance, and capacitance undergoes damped oscillations for sufficiently small resistance. The frequency $\omega'$ of damped oscillations is

$$\omega' = \sqrt{\frac{1}{LC} - \frac{R^2}{4L^2}}. \tag{31-28}$$

As $R$ increases, the damping increases; at a certain value of $R$ the behavior becomes overdamped and no longer oscillates. The crossover between underdamping and overdamping occurs when

$$R^2 = 4L/C.$$

There is a direct analogy between every aspect of the behavior of the $L$-$R$-$C$ circuit and the mechanical damped harmonic oscillator. This analogy and similar ones are widely used in analog computers.

**KEY TERMS**

- mutual inductance
- henry
- self-induced electromotive force
- inductance (self-inductance)
- inductor
- energy density
- time constant
- *L-C* circuit
- electrical oscillation
- damped harmonic motion
- underdamping
- critical damping
- overdamping

# EXERCISES

## Section 31–1 Mutual Inductance

**31–1** From Eq. (31–4), 1 H = 1 Wb/A, and from Eq. (31–3), 1 H = 1 Ω · s. Show that these two definitions are equivalent.

**31–2** Two coils are wound on the same form so that the flux from one coil links the turns of the second coil. When the current in the first coil is decreasing at a rate of −0.0850 A/s, the induced emf in the second coil has magnitude $3.00 \times 10^{-3}$ V. a) What is the mutual inductance of the pair of coils? b) If the second coil has five turns, what is the flux through each turn when the current in the first coil equals 1.60 A? c) If the current in the second coil increases at a rate of 0.0500 A/s, what is the induced emf in the first coil?

**31–3** Two coils have mutual inductance $M = 0.0300$ H. The current $i_1$ in the first coil increases at a uniform rate of 0.0500 A/s. a) What is the induced emf in the second coil? Is it constant? b) Suppose the current described is in the second coil rather than the first. What is the induced emf in the first coil?

**31–4** A toroidal solenoid has a mean radius $r$ and a cross-section area $A$ and is wound uniformly with $N_1$ turns. A second toroidal solenoid with $N_2$ turns is wound uniformly on top of the first. The two coils are wound in the same direction. What is their mutual inductance? (Neglect the variation of the magnetic field across the cross section of the toroid.)

## Section 31–2 Inductance within a Circuit

**31–5** At the instant when the current in an inductor is increasing at a rate of 0.0600 A/s, the self-induced emf is 0.0500 V. a) What is the inductance of the inductor? b) If the inductor is a solenoid with 300 turns, what is the magnetic flux through each turn when the current is 0.800 A?

**31–6** Show that the two expressions for inductance, $N\Phi_B/i$ and $-\mathcal{E}/(di/dt)$, have the same units.

**31–7** The inductor in Fig. 31–14 has inductance 0.600 H and carries a current in the direction shown that is decreasing at a uniform rate, $di/dt = -0.0300$ A/s. a) Find the self-induced emf. b) Which end of the inductor, $a$ or $b$, is at higher potential?

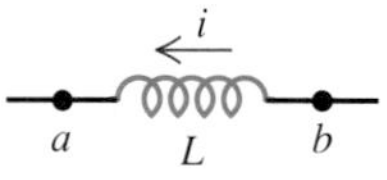

$i$

$a$ $L$ $b$

**FIGURE 31–14**

**31–8** An inductor with $L = 40.0$ H carries a current $i$ that varies with time according to $i = (0.180 \text{ A}) \sin([120\ \pi/\text{s}]t)$. Find an expression for the induced emf. What is the phase of $\mathcal{E}$ relative to $i$?

**31–9 Inductance of a Solenoid.** A long, straight solenoid has $N$ turns, uniform cross-section area $A$, and length $l$. Derive an expression for the inductance of the solenoid. Assume that the magnetic field is uniform inside the solenoid and zero outside. (This is a reasonable approximation if the solenoid is long and the turns are closely wound.)

**31–10** A toroidal solenoid has a cross-section area of 0.600 cm$^2$, a mean radius of 9.00 cm, and 2000 turns. It is filled with a core with a relative permeability 600. Calculate the inductance of the solenoid. (Neglect the variation of the magnetic field across the cross section of the toroid.)

## Section 31–3 Magnetic-Field Energy

**31–11** A toroidal solenoid has a mean radius of 0.120 m and a cross-section area of $4.00 \times 10^{-4}$ m$^2$. When the current is 25.0 A, the energy stored is 0.350 J. How many turns does the winding have?

**31–12** It has been proposed to use large inductors as energy storage devices. a) How much electrical energy is converted to light and heat by a 100-W light bulb in one day? b) If the amount of energy calculated in part (a) is stored in an inductor in which the current is 40.0 A, what is the inductance?

**31–13** It is proposed to store 1.00 kWh $= 3.60 \times 10^6$ J of electrical energy in a uniform magnetic field with magnitude 0.500 T. a) What volume (in vacuum) must the magnetic field occupy to store this amount of energy? b) If instead this amount of energy is to be stored in a volume (still in vacuum) of 0.125 m$^3$ (a cube 50.0 cm on a side), what magnetic field is required?

**31–14** An inductor used in a dc power supply has an inductance of 16.0 H and a resistance of 200 Ω and carries a current of 0.150 A. a) What is the energy stored in the magnetic field? b) At what rate is electrical energy dissipated in the resistor?

**31–15** Derive in detail Eq. (31–10) for the energy density in a toroidal solenoid filled with a magnetic material.

**31–16** A magnetic field with magnitude $B = 0.500$ T is uniform throughout a volume of 0.0200 m$^3$. Calculate the total magnetic energy in the volume if a) the volume is free space; b) the volume is filled with material with relative permeability 600.

## Section 31–4 The *R-L* Circuit

**31–17** Show that $L/R$ has units of time.

**31–18** Write an equation corresponding to Eq. (31–12) for the current in Fig. 31–7 just after switch $S_2$ is closed and $S_1$ is opened if the initial current is $I_0$. Solve the resulting differential equation and verify Eq. (31–17).

**31–19** An inductor with an inductance of 3.00 H and a resistance of 9.00 Ω is connected to the terminals of a battery with an emf of 12.0 V and negligible internal resistance. Find a) the initial rate of increase of current in the circuit; b) the rate of increase of current at the instant when the current is 1.00 A; c) the current 0.200 s after the circuit is closed; d) the final steady-state current.

**31–20** In Fig. 31–7, let $\mathcal{E}_s = 90.0$ V, $R = 400$ Ω, and $L = 0.200$ H. Initially, there is no current in the circuit. With switch $S_2$ open, switch $S_1$ is closed. a) Just after $S_1$ is closed, what are the potential differences $v_{ab}$ and $v_{bc}$? b) A long time (many time constants) after $S_1$ is closed, what are $v_{ab}$ and $v_{bc}$? c) What are $v_{ab}$ and $v_{bc}$ at an intermediate time when $i = 0.0500$ A?

**31–21** In Fig. 31–7, let $\mathcal{E}_s = 120$ V, $R = 500$ Ω, and $L = 0.200$ H. With switch $S_2$ open, switch $S_1$ is closed and left until a constant current is established. Then $S_2$ is closed and $S_1$ is opened, taking the battery out of the circuit. a) What is the initial current in the resistor just after $S_2$ is closed and $S_1$ is opened? b) What is the current in the resistor at $t = 2.00 \times 10^{-4}$ s? c) What is the potential difference between points $b$ and $c$ at $t = 2.00 \times 10^{-4}$ s? Which point is at higher potential? d) How long does it take the current to decrease to half its initial value?

**31–22** Refer to Exercise 31–19. a) What is the power input to the inductor from the battery as a function of time if the circuit is completed at $t = 0$? b) What is the rate of dissipation of energy in the resistance of the inductor as a function of time? c) What is the rate at which the energy of the magnetic field in the inductor is increasing as a function of time? d) Compare the results of parts (a), (b), and (c).

### Section 31–5 The *L-C* Circuit

**31–23** Show that $\sqrt{LC}$ has units of time.

**31–24 A Radio Tuning Circuit.** The maximum capacitance of a variable capacitor in a radio is 40.0 pF. a) What is the inductance of a coil connected to this capacitor if the natural frequency of the *L-C* circuit is $540 \times 10^3$ Hz, corresponding to one end of the AM radio broadcast band, when the capacitor is set to its maximum capacitance? b) The frequency at the other end of the broadcast band is $1600 \times 10^3$ Hz. What is the minimum capacitance of the capacitor if the natural frequency is adjustable over the range of the broadcast band?

**31–25 An Electrical Oscillator.** A capacitor with capacitance $6.00 \times 10^{-4}$ F is charged by connecting it to a 50.0-V battery. The capacitor is disconnected from the battery and connected across an inductor with $L = 3.00$ H. a) What are the angular frequency $\omega$ of the electrical oscillations and the period of these oscillations (the time for one oscillation)? b) What is the initial charge on the capacitor? c) How much energy is initially stored in the capacitor? d) What is the charge on the capacitor 0.0444 s after the connection to the inductor is made? e) At the time given in part (d), what is the current in the inductor? f) At the time given in part (d), how much electrical energy is stored in the capacitor and how much is stored in the inductor?

**31–26** Show that the differential equation of Eq. (31–19) is satisfied by the function $q = Q \cos \omega t$, with $\omega$ given by $1/(LC)^{1/2}$.

### Section 31–6 The *L-R-C* Circuit

**31–27** An *L-R-C* circuit has $L = 0.500$ H, $C = 2.00 \times 10^{-4}$ F, and resistance $R$. a) What is the natural angular frequency of the circuit when $R = 0$? b) What value must $R$ have to give a 10% decrease in natural frequency compared to the value calculated in part (a)?

**31–28** Show that the quantity $(L/C)^{1/2}$ has units of resistance (ohms).

## PROBLEMS

**31–29** The current in a coil of wire is initially zero but increases at a constant rate; after 10.0 s it is 50.0 A. The changing current induces an emf of 40.0 V in the coil. a) Determine the inductance of the coil. b) Determine the total magnetic flux through the coil when the current is 50.0 A. c) If the resistance of the coil is 25.0 Ω, determine the ratio of the rate at which energy is being stored in the magnetic field to the rate at which electrical energy is being dissipated by the resistance at the instant when the current is 50.0 A.

**31–30** A solenoid has length $l_1$, radius $r_1$, and number of turns $N_1$. A second, smaller solenoid with length $l_2$, radius $r_2$ ($r_2 < r_1$), and number of turns $N_2$ is placed at the center of the first solenoid so that their axes coincide. Assume that the magnetic field of the first solenoid at the location of the second is uniform and has a magnitude given by Eq. (29–21). a) What is the mutual inductance of the pair of solenoids? b) If the current in the large solenoid is increasing at the rate $di_1/dt$, what is the magnitude of the emf induced in the small solenoid? c) If the current in the smaller solenoid is increasing at the rate $di_2/dt$, what is the magnitude of the emf induced in the larger solenoid?

**31–31 A Differentiating Circuit.** The current in a resistanceless inductor is caused to vary with time as in the graph of Fig. 31–15. a) Sketch the pattern that would be observed on the screen of an oscilloscope connected to the terminals of the inductor. (The oscilloscope spot sweeps horizontally across the screen at a constant speed, and its vertical deflection is proportional to the potential difference between the inductor terminals.) b) Explain why the inductor can be described as a "differentiating circuit."

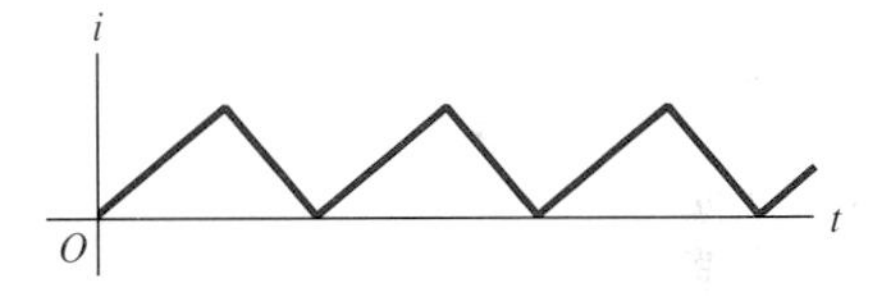


**FIGURE 31–15**

**31–32** Consider the toroidal solenoid of Exercise 31–4. a) Derive an expression for the inductance $L_1$ when only the first coil is used and an expression for $L_2$ when only the second coil is used. b) Show that $M^2 = L_1L_2$. This result is valid whenever all the flux linked by one coil is also linked by the other.

**31–33** We have neglected the variation of the magnetic field across the cross section of a toroidal solenoid. We will now examine the validity of that approximation. A certain toroidal solenoid has a rectangular cross section (Fig. 31–16, on page 888). It has $N$ uniformly spaced turns, with air inside. The magnetic field at a point inside the toroid is given by Eq. (29–22). *Do not* assume the field to be uniform over the cross section. a) Show that the magnetic flux through a cross section of the toroid is

$$\Phi_B = \frac{\mu_0 Nih}{2\pi} \ln\left(\frac{b}{a}\right).$$

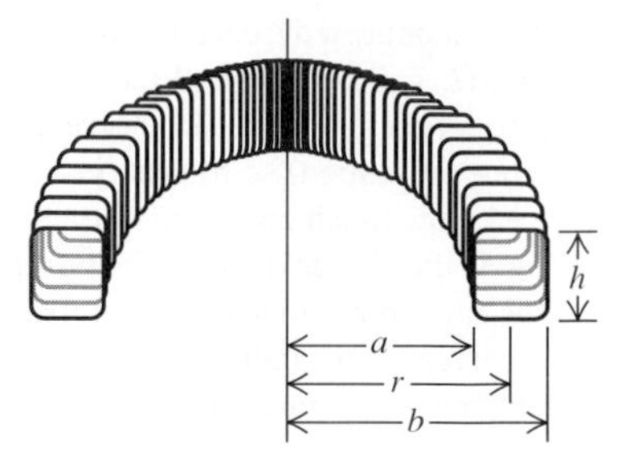


**FIGURE 31–16**

b) Show that the inductance of the toroidal solenoid is given by

$$L = \frac{\mu_0 N^2 h}{2\pi} \ln\left(\frac{b}{a}\right).$$

c) The fraction $b/a$ may be written as

$$\frac{b}{a} = \frac{a + b - a}{a} = 1 + \frac{b - a}{a}.$$

The power series expansion for $\ln(1 + x)$ is $\ln(1 + x) = x + x^2/2 + \cdots$. Show that when $b - a$ is much less than $a$, the inductance is approximately equal to

$$L = \frac{\mu_0 N^2 h(b - a)}{2\pi a}.$$

Compare this result with the result obtained in Example 31–3.

**31–34 A Coaxial Cable.** A small solid conductor with radius $a$ is supported by insulating disks on the axis of a thin-walled tube with inner radius $b$. The inner and outer conductors carry equal currents in opposite directions. a) Use Ampere's law to find the magnetic field at any point in the space between the conductors. b) Write the expression for the flux $d\Phi_B$ through a narrow strip of length $l$ parallel to the axis, of width $dr$, at a distance $r$ from the axis of the cable and lying in a plane containing the axis. c) Integrate your expression from part (b) to find the total flux linking a current $i$ in the central conductor. d) Show that the inductance of a length $l$ of the cable is

$$L = l\frac{\mu_0}{2\pi} \ln\left(\frac{b}{a}\right).$$

e) Use Eq. (31–8) to calculate the energy stored in the magnetic field for a length $l$ of the cable.

**31–35** Consider the coaxial cable of Problem 31–34. The conductors carry equal currents in opposite directions. a) Use Ampere's law to find the magnetic field at any point in the space between the conductors. b) Use the energy density for a magnetic field, Eq. (31–9), to calculate the energy stored in a thin cylindrical shell between the two conductors. Let the cylindrical shell have inner radius $r$, outer radius $r + dr$, and length $l$. c) Integrate your result in part (b) to find the total energy stored in the magnetic field for a length $l$ of the cable. d) Use the result of part (c) and Eq. (31–8) to calculate the inductance $L$ of a length $l$ of the cable. Compare your result to $L$ calculated from the expression in part (d) of Problem 31–34.

**31–36** A toroidal solenoid with $N$ turns and mean radius $r$ carries a current $i$. Neglect the variation of the magnetic field over the cross section of the toroid. a) What is the energy density in the magnetic field? b) What is the total magnetic-field energy? Find this energy using Eq. (31–8) and also by multiplying the energy density from part (a) by the volume of the toroid, which is $2\pi rA$; compare the two results.

**31–37** Uniform electric and magnetic fields $\boldsymbol{E}$ and $\boldsymbol{B}$ occupy the same region of free space. If $E = 500$ V/m, what is $B$ if the energy densities in the electric and magnetic fields are equal?

**31–38** You are asked to design an $L$-$C$ circuit in which the stored energy is $2.00 \times 10^{-4}$ J and the natural frequency is $\omega = 8.00 \times 10^4$ rad/s. The maximum voltage across the capacitor is to be 60.0 V. What values of $C$ and $L$ are required?

**31–39 Continuation of Exercise 31–19.** a) How much energy is stored in the magnetic field of the inductor one time constant after the battery has been connected? Compute this both by integrating the expression in Exercise 31–22(c) and by using Eq. (31–8), and compare the results. b) Integrate the expression obtained in Exercise 31–22(a) to find the *total* energy supplied by the battery during the time interval considered in part (a). c) Integrate the expression obtained in Exercise 31–22(b) to find the *total* energy dissipated in the resistance of the inductor during the same time period. d) Compare the results obtained in parts (a), (b), and (c).

**31–40 Continuation of Exercise 31–21.** a) What is the total energy initially stored in the inductor? b) At $t = 2.00 \times 10^{-4}$ s, at what rate is the energy stored in the inductor decreasing? c) At $t = 2.00 \times 10^{-4}$ s, at what rate is electrical energy being converted into heat in the resistor? d) Obtain an expression for the rate at which electrical energy is being converted into heat in the resistor as a function of time. Integrate this expression from $t = 0$ to $t = \infty$ to obtain the total electrical energy dissipated in the resistor. Compare your result to that of part (a).

**31–41** The equation preceding Eq. (31–26) may be converted into an energy relation. Multiply both sides of this equation by $-i = -dq/dt$. The first term then becomes $i^2R$. Show that the second term can be written $d(\frac{1}{2}Li^2)/dt$ and the third can be written $d(q^2/2C)/dt$. What does the resulting equation say about energy conservation in the circuit?

**31–42** An $L$-$C$ circuit consists of an inductor with $L = 0.800$ H and a capacitor of $C = 4.00 \times 10^{-4}$ F. The initial charge on the capacitor is 6.00 $\mu$C, and the initial current in the inductor is zero. a) What is the maximum voltage across the capacitor? b) What is the maximum current in the inductor? c) What is the maximum energy stored in the inductor? d) When the current in the inductor has half its maximum value, what is the charge on the capacitor, and what is the energy stored in the inductor?

**31–43** Consider the circuit of Fig. 31–17. Let $\mathcal{E} = 60.0$ V, $R_0 = 50.0\ \Omega$, $R = 150\ \Omega$, and $L = 5.00$ H. a) Switch $S_1$ is closed, and $S_2$ is left open. Just after $S_1$ is closed, what are the current $i_0$ through $R_0$ and the potential differences $v_{ac}$ and $v_{cb}$?

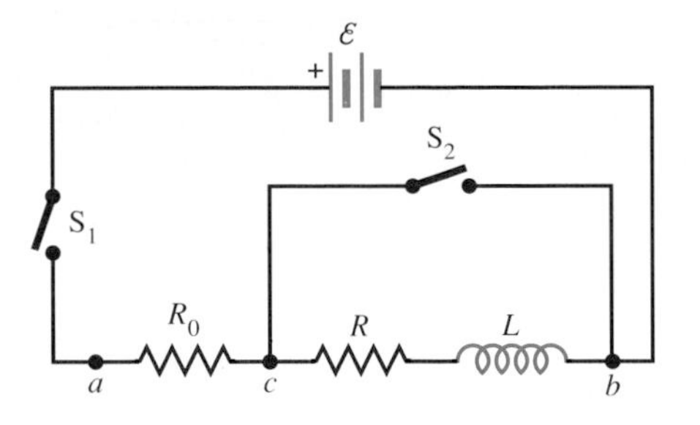


**FIGURE 31–17**

b) After $S_1$ has been closed a long time ($S_2$ is still open), so the current has reached its final steady value, what are $i_0$, $v_{ac}$, and $v_{cb}$? c) Find the expressions for $i_0$, $v_{ac}$, and $v_{cb}$ as functions of the time $t$ since $S_1$ was closed. Your results should agree with part (a) when $t = 0$ and with part (b) when $t \to \infty$. Sketch graphs of $i_0$, $v_{ac}$, and $v_{cb}$ versus time.

**31–44** After the current in the circuit of Fig. 31–17 has reached its final steady value with switch $S_1$ closed and $S_2$ open, $S_2$ is closed, thus short-circuiting the inductor. (Switch $S_1$ remains closed.) a) Just after $S_2$ is closed, what are $v_{ac}$ and $v_{cb}$, and what are the currents through $R_0$, $R$, and $S_2$? b) A long time after $S_2$ is closed, what are $v_{ac}$ and $v_{cb}$, and what are the currents through $R_0$, $R$, and $S_2$? c) Derive expressions for the currents through $R_0$, $R$, and $S_2$ as functions of the time $t$ that has elapsed since $S_2$ was closed. Your results should agree with part (a) when $t = 0$ and with part (b) when $t \to \infty$. Sketch graphs of these three currents as functions of time.

**31–45 Demonstrating Inductance.** A popular demonstration of inductance that is often used in physics courses employs a circuit such as the one shown in Fig. 31–18. Switch S is closed, and the light bulb, represented by resistance $R_1$, is seen to just barely glow. After a period of time, switch S is opened, and the bulb is then seen to light up brightly for a short period of time. This effect is easy to understand if you think of an inductor as a device that imparts an "inertia" to the current, preventing a discontinuous change in the current through it. a) Derive, as explicit functions of time, expressions for $i_1$ and $i_2$, the currents through the light bulb and inductor, after the switch S is closed. b) After a long period of time, steady-state conditions can be assumed. Obtain expressions for the steady-state currents in the bulb and the inductor. c) Switch S is now opened. Obtain an expression for the current through the inductor and light bulb as an explicit function of time. d) You have been asked to design demonstration apparatus using the circuit shown in Fig. 31–18 with a 48.0-H inductor and a 60.0-W light bulb. You are to connect a resistor in series with the inductor, and $R_2$ represents the sum of that resistance plus the internal resistance of the inductor. When switch S is opened, a transient current is to be set up that starts at 0.900 A and is not to fall below 0.300 A until after 0.200 s. For simplicity, assume that the resistance of the light bulb is constant and equals the resistance that the bulb must have to dissipate 60.0 W at 120 V. Determine $R_2$ and $\mathcal{E}$ for the given design considerations. e) With the numerical values determined in part (d), what is the current through the light bulb just before the switch is opened? Does this result confirm the qualitative description of what is observed in the demonstration?

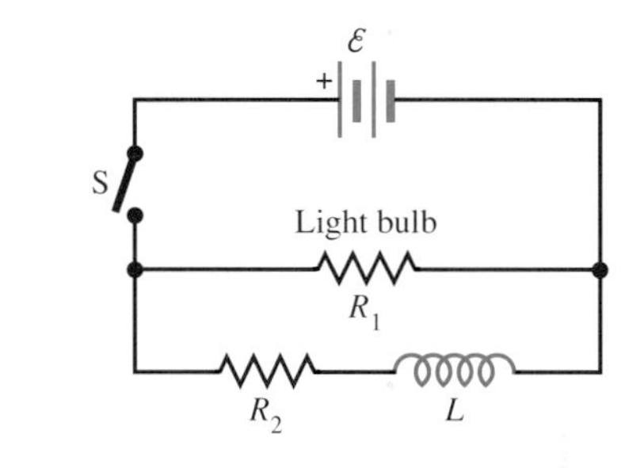


**FIGURE 31–18**

## CHALLENGE PROBLEMS

**31–46** An inductor is made with two coils wound close together on a form, so all the flux linking one coil also links the other. The number of turns is the same in each. If the inductance of one coil is $L$, what is the inductance when the two coils are connected a) in series; b) in parallel? In each case the current travels in the same sense around each coil. c) If an $L$-$C$ circuit with this inductor has natural frequency $\omega$ using one coil, what is the natural frequency when the two coils are used in series?

**31–47** Consider the circuit shown in Fig. 31–19. The values of the circuit elements are as follows: $\mathcal{E} = 60.0$ V, $L = 10.0$ H, $C = 20.0\ \mu$F, $R_1 = 25.0\ \Omega$, and $R_2 = 5000\ \Omega$. Switch S is closed at time $t = 0$, causing a current $i_1$ through the inductive branch and a current $i_2$ through the capacitive branch. The initial charge on the capacitor is zero, and the charge at time $t$ is $q_2$. a) Derive expressions for $i_1$, $i_2$, and $q_2$ as explicit functions of time. b) What is the initial current through the inductive branch? What is the initial current through the capacitive branch? c) What are the currents through the inductive and capacitive branches a long time after the switch has been closed? How long is a "long time"? Explain.

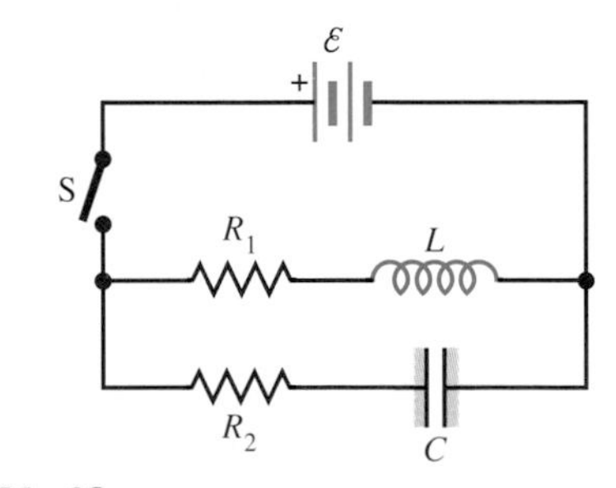


**FIGURE 31–19**

d) At what time $t_1$ will the currents $i_1$ and $i_2$ be equal? (*Hint:* You might consider using series expansions for the exponentials.) e) For the conditions given in part (d), determine $i_1$. f) The total current through the battery is $i = i_1 + i_2$. At what time $t_2$ will $i$ equal one-half of its final value? [*Hint:* As in part (d), the numerical work is greatly simplified if you make suitable approximations.]

**31–48** Consider the circuit shown in Fig. 31–20. The circuit elements are as follows: $\mathcal{E} = 40.0$ V, $L = 20.0$ H, $C = 6.25\ \mu$F, and $R = 1000\ \Omega$. At time $t = 0$, switch S is closed. The current through the inductance is $i_1$, the current through the capacitor branch is $i_2$, and the charge on the capacitor is $q_2$. a) Using Kirchhoff's rules, verify these circuit equations:

$$R(i_1 + i_2) + L\left(\frac{di_1}{dt}\right) = \mathcal{E},$$

$$R(i_1 + i_2) + \frac{q_2}{C} = \mathcal{E}.$$

b) What are the initial values of $i_1$, $i_2$, and $q_2$? c) Show by direct substitution that the following solutions for $i_1$ and $q_2$ satisfy the circuit equations from part (a). Also, show that they satisfy the initial conditions.

$$i_1 = (\mathcal{E}/R)(1 - [(2\omega RC)^{-1}\sin(\omega t) + \cos(\omega t)]\,e^{-\beta t}),$$

$$q_2 = \left(\frac{\mathcal{E}}{\omega R}\right)e^{-\beta t}\sin(\omega t),$$

where $\beta = (2RC)^{-1}$ and $\omega = [(LC)^{-1} - (2RC)^{-2}]^{1/2}$.
d) Determine the time $t_1$ at which $i_2$ first becomes zero.

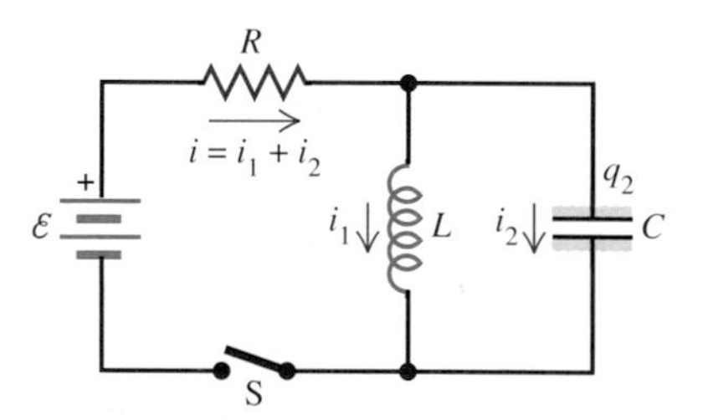


**FIGURE 31–20**

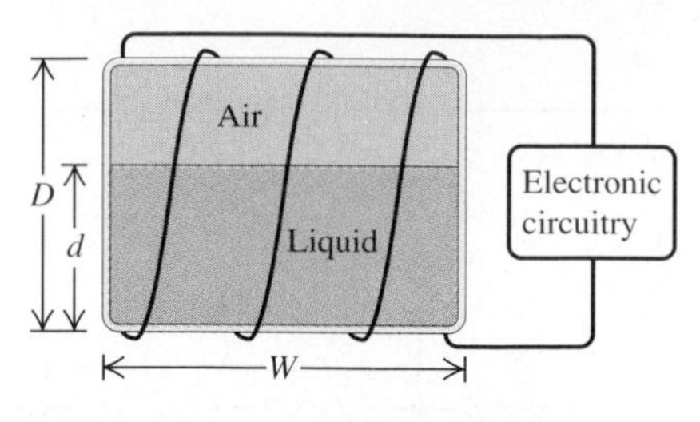


**FIGURE 31–21**

**31–49 A Volume Gauge.** A tank containing a liquid has turns of wire wrapped around it, causing it to act as an inductor. The liquid content of the tank can be measured by using its inductance to determine the height of the liquid in the tank. The inductance of the tank changes linearly from a value of $L_0$ corresponding to a relative permeability of 1 when the tank is empty to a value of $L_f$ corresponding to a relative permeability of $K_m$ (the relative permeability of the liquid) when the tank is full. The appropriate electronic circuitry can determine the inductance to five significant figures and thus the effective relative permeability of the combined air and liquid within the rectangular cavity of the tank. Each of the four sides of the tank has a width $W$ and a height $D$ (Fig. 31–21). The height of the liquid in the tank is $d$. Neglect any fringing effects and assume that the relative permeability of the material of which the tank is made can be neglected. a) Derive an expression for $d$ as a function of $L$, the inductance corresponding to a certain fluid height, and $L_0$ and $L_f$. b) What is the inductance (to five significant figures) for a tank that is one-quarter full, one-half full, three-quarters full, and completely full if the tank contains liquid oxygen? Take $L_0 = 0.75000$ H. The magnetic susceptibility of liquid oxygen is $\chi_m = 1.52 \times 10^{-3}$. c) Repeat part (b) for mercury. The magnetic susceptibility of mercury is given in Table 29–1. d) For which material is the gauge more practical?

CHAPTER 32

# Alternating Current

Contemporary distribution of electric power is made possible by transformers, which convert alternating-current electricity from low generating voltages to high voltages for transmission, and then back to lower voltages for consumption. Modern transformers are immersed in tanks of oil for insulation and cooling. The high-voltage coil, low-voltage coil, and connecting core are close together to minimize electrical losses and cost of materials.

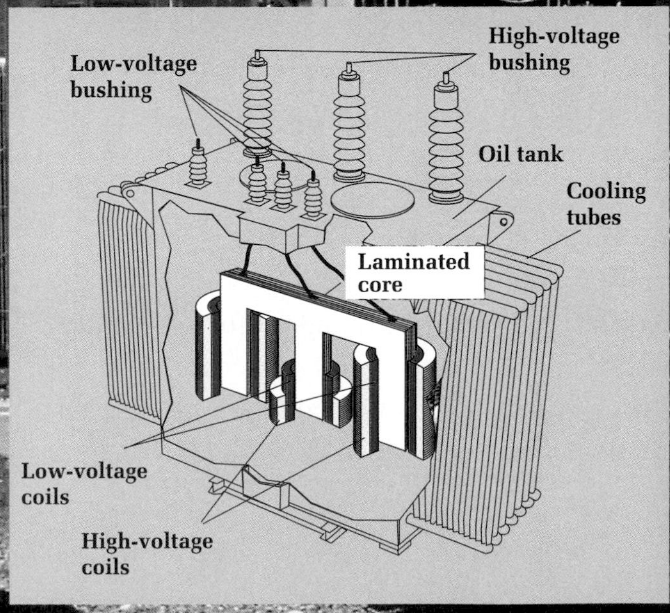

## KEY CONCEPTS

**• When voltages and currents in a circuit vary sinusoidally with time, the voltage and current amplitudes for an individual circuit element may be proportional, but in general there is a phase difference between voltage and current. The ratio of voltage and current amplitudes is called impedance; it depends on frequency. Sinusoidally varying voltages and currents can be represented by rotating vectors called phasors.**

**• Sinusoidally varying voltages and currents can be described by rectified average or root-mean-square values.**

**• Power in an ac circuit depends on the voltage and current amplitudes and on the phase difference between voltage and current.**

**• When inductors and capacitors are combined in a circuit, the effects of their reactances may cancel out at particular frequencies, leading to a maximum or minimum in voltage or current. This effect is called resonance.**

**• A transformer is used to step voltage up or down; the ratio of voltages in the two windings is proportional to the numbers of turns in the windings.**

## INTRODUCTION

During the 1880s there was a heated and acrimonious debate over the best method of electric-power distribution. Thomas Edison favored direct current (dc), that is, steady current that does not vary with time. George Westinghouse favored alternating current (ac), with sinusoidally varying voltages and currents. He argued that transformers (which we will study in this chapter) could be used to step the voltage up and down with ac but not with dc. Edison claimed that dc was inherently safer.

Eventually, the arguments of Westinghouse prevailed. Most present-day household and industrial power-distribution systems operate with alternating current. Any appliance that you plug into a wall outlet uses ac, and many battery-powered devices such as radios and cordless telephones make use of the dc supplied by the battery to create or amplify alternating currents. Circuits in modern communications equipment, including radios and televisions, make extensive use of ac.

In this chapter we will learn how resistors, inductors, and capacitors behave in circuits with sinusoidally varying voltages and currents. Many of the same principles that we found useful in Chapters 26, 29, and 31 are applicable, and there are several new concepts related to the circuit behavior of inductors and capacitors. A key idea in this discussion is the concept of *resonance,* which we studied in Chapter 13 for mechanical systems.

## 32–1 PHASORS AND ALTERNATING CURRENTS

We have already studied several sources of alternating emf (or voltage) that can supply an **alternating current** to a circuit. A coil of wire rotating with constant angular velocity in a magnetic field (Section 30–2) develops a sinusoidally alternating emf and is the prototype of the commercial alternating-current generator, or *alternator.* An $L$-$C$ circuit (Section 31–5) carries an alternating current with a frequency that may range from a few hertz to many millions.

We will use the term **ac source** for any device that supplies sinusoidally varying potential difference $v$ or current $i$. A sinusoidal voltage might be described by a function such as

$$v = V \cos \omega t. \tag{32–1}$$

In this expression, $V$ is the maximum potential difference, which we call the **voltage amplitude;** $v$ is the *instantaneous* potential difference: and $\omega$ is the *angular frequency,* equal to $2\pi$ times the frequency $f$. In the United States, commercial electric-power distribution systems always use a frequency $f = 60$ Hz, corresponding to an angular frequency $\omega = (2\pi \text{ rad})(60 \text{ s}^{-1}) = 377$ rad/s. Similarly, a sinusoidal current $i$ might be described by

$$i = I \cos \omega t, \tag{32–2}$$

where $I$ is the maximum current, or **current amplitude.** In the next section we'll look at the behavior of individual circuit elements when they carry a sinusoidal current. The usual circuit-diagram symbol for an ac source is

To represent sinusoidally varying voltages and currents, we will use vector diagrams similar to those we used in the study of harmonic motion in Section 13–4. In these diagrams, the instantaneous value of a quantity that varies sinusoidally with time is represented by the *projection* onto a horizontal axis of a vector with a length equal to the amplitude of the quantity. The vector rotates counterclockwise with constant angular velocity $\omega$. These rotating vectors are called **phasors,** and diagrams containing them are called **phasor diagrams.** Figure 32–1 shows a phasor diagram for the sinusoidal current described by Eq. (32–2). The projection of the phasor onto the horizontal axis at time $t$ is $I \cos \omega t$; this is why we chose to use the cosine function rather than the sine in Eq. (32–2).

A phasor is not a real physical quantity with a direction in space, such as velocity, momentum, or electric field. Rather, it is a *geometric* entity that provides a language for describing and analyzing physical quantities that vary sinusoidally with time. In Section 13–4 we used a single phasor to represent the position and motion of a point mass undergoing simple harmonic motion. In this chapter we will use phasors to *add* sinusoidal voltages and currents. Combining sinusoidal quantities with phase differences then becomes a matter of vector addition. We will find a similar use for phasors in Chapters 37 and 38, in our study of interference effects with light.

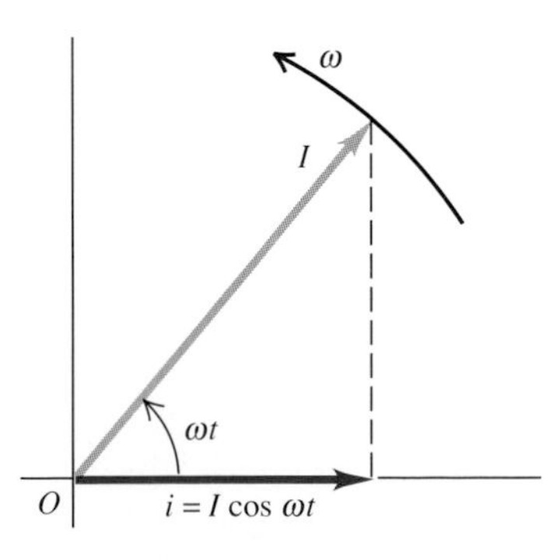


**32–1** Phasor diagram. The projection of the rotating vector onto the horizontal axis represents the instantaneous current.

How do we measure a sinusoidally varying current? In Section 27–3 we used a d'Arsonval galvanometer to measure steady currents. But if we pass a *sinusoidal* current through a d'Arsonval meter, the torque on the moving coil varies sinusoidally, being in one direction half the time and the opposite direction the other half. The *average* torque is zero. The needle may wiggle a little if the frequency is low enough, but its average deflection is zero. The *average* value of a sinusoidal current is always zero, so it isn't a very interesting quantity.

To get a one-way current through the galvanometer, which we can measure, we can use *diodes,* which we described in Section 26–3. A diode (or rectifier) is a device that

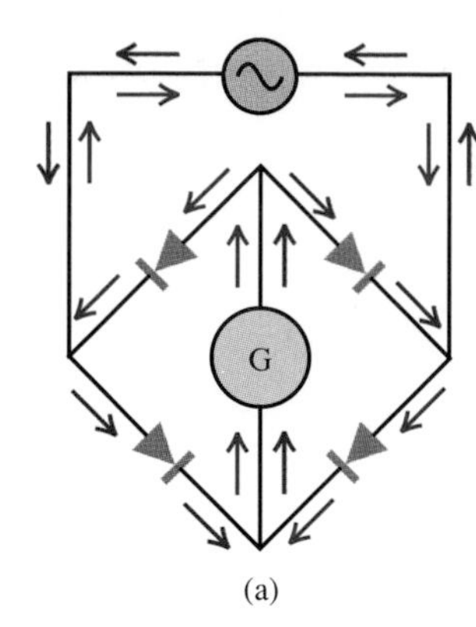


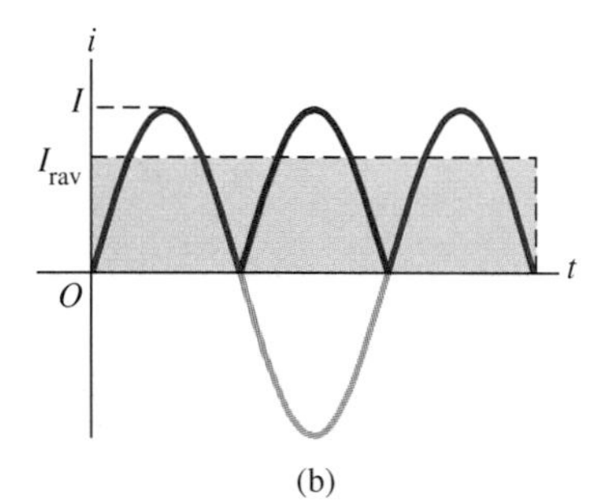


**32–2** (a) A full-wave rectifier circuit. (b) Graph of a full-wave rectified current and its average value.

conducts better in one direction than the other. An ideal diode has zero resistance for one direction of current and infinite resistance for the other. One possible arrangement is shown in Fig. 32–2a. The current through the galvanometer G is always upward, regardless of the direction of the current from the ac source (that is, regardless of which part of the cycle the source is in). The current through G is shown by the graph in Fig. 32–2b. It pulsates but always has the same direction, and the meter deflection is *not* zero. This arrangement of diodes is called a *full-wave rectifier circuit.*

The **rectified average current** $I_{rav}$ is defined so that during any whole number of cycles the total charge that flows is the same as though the current were constant with a value equal to the average value. In Fig. 32–2b the area under the curve must equal the rectangular area with height $I_{rav}$. We see that $I_{rav}$ is about two-thirds of the maximum current $I$; in fact, it turns out that the two are related by

$$I_{rav} = \frac{2}{\pi} I = 0.637I. \tag{32–3}$$

The galvanometer deflection is proportional to $I_{rav}$, and the scale can be calibrated to read $I$, $I_{rav}$, or $I_{rms}$ (discussed below). The notation $I_{rav}$ and the name *rectified average* current emphasize that this is *not* the average of the original current.

A more useful way to describe the average value of a quantity that can be either positive or negative is to use the concept of *root-mean-square (rms) value.* We encountered this concept in Section 16–3 in connection with the speeds of molecules in a gas. We *square* the instantaneous current $i$, take the *average* value of $i^2$, and finally take the *square root* of that average. This procedure defines the **root-mean-square (rms) current,** denoted as $I_{rms}$. Even when $i$ is negative, $i^2$ is always positive, so $I_{rms}$ is never zero (unless $i$ is zero at every instant).

Here's how we obtain $I_{rms}$. If the instantaneous current $i$ is given by $I \cos \omega t$, then

$$i^2 = I^2 \cos^2 \omega t.$$

We use a double-angle formula from trigonometry:

$$\cos^2 A = \tfrac{1}{2}(1 + \cos 2A).$$

We find

$$\begin{aligned} i^2 &= I^2 \tfrac{1}{2}(1 + \cos 2\omega t) \\ &= \tfrac{1}{2}I^2 + \tfrac{1}{2}I^2 \cos 2\omega t. \end{aligned}$$

The average of $\cos 2\omega t$ is zero because it is positive half the time and negative half the time. Thus the average of $i^2$ is simply $I^2/2$; the square root of this is $I_{rms}$:

$$I_{rms} = \frac{I}{\sqrt{2}}. \tag{32–4}$$

In the same way the root-mean-square value of a sinusoidal voltage with amplitude (maximum value) $V$ is

$$V_{rms} = \frac{V}{\sqrt{2}}. \tag{32–5}$$

We can convert a rectifying ammeter into a voltmeter by adding a series resistor, just as for the dc case that we discussed in Section 27–3. Meters that are used for ac voltage and current measurements are nearly always calibrated to read rms values, not maximum or rectified average values. Voltages and currents in power-distribution systems are always described in terms of their rms values. The usual household power supply, designated 120-volt ac, has an rms voltage of 120 V. The voltage amplitude is

$$V = \sqrt{2}V_{\text{rms}} = \sqrt{2}(120\text{ V}) = 170\text{ V}.$$

**EXAMPLE 32–1**

**Current in a personal computer** The plate on the back of a personal computer says that it draws 2.7 A from a 120-V, 60-Hz line. For this PC, what are a) the average current, b) the average of the square of the current, and c) the current amplitude?

**SOLUTION** a) The average of any sinusoidal alternating current over any whole number of cycles is zero.

b) The current given, 2.7 A, is the rms value, which is the *square root* of the *mean* (average) of the *square* of the current. That is,

$$I_{\text{rms}} = \sqrt{(i^2)_{\text{av}}}, \quad \text{or}$$

$$(i^2)_{\text{av}} = (I_{\text{rms}})^2 = (2.7\text{ A})^2 = 7.3\text{ A}^2.$$

c) From Eq. (32–4) the current amplitude $I$ is

$$I = \sqrt{2}I_{\text{rms}} = \sqrt{2}(2.7\text{ A}) = 3.8\text{ A}.$$

## 32–2 RESISTANCE AND REACTANCE

In this section we will derive voltage-current relations for an individual resistor, inductor, or capacitor carrying a sinusoidal current.

### Resistor in an ac Circuit

First let's consider a resistor with resistance $R$ with a sinusoidal current given by Eq. (32–2): $i = I\cos\omega t$, as in Fig. 32–3a. The current amplitude (maximum current) is $I$. From Ohm's law the instantaneous potential $v$ of point $a$ with respect to point $b$ is

$$v = iR = IR\cos\omega t. \tag{32–6}$$

The maximum voltage $V$, the *voltage amplitude*, is the coefficient of the cosine function, that is,

$$V = IR, \tag{32–7}$$

so we can also write

$$v = V\cos\omega t. \tag{32–8}$$

The current $i$ and voltage $v$ are both proportional to $\cos\omega t$, so the current is *in phase* with the voltage. Equation (32–7) shows that the current and voltage amplitudes are related in the same way as in a dc circuit.

Figure 32–3b shows graphs of $i$ and $v$ as functions of time. The vertical scales for $i$ and $v$ may be different, so the relative heights of the two curves are not significant. The corresponding phasor diagram is shown in Fig. 32–3c. Because $i$ and $v$ are *in phase* and have the same frequency, the current and voltage phasors rotate together; they are parallel at each instant. Their projections on the horizontal axis represent the instantaneous current and voltage, respectively.

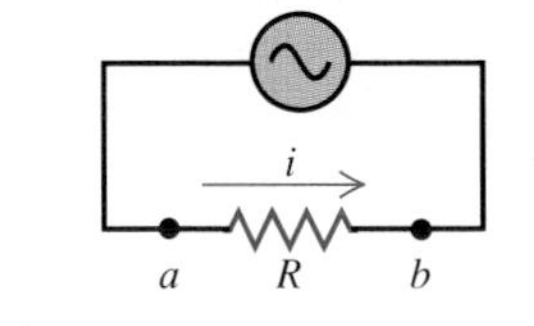


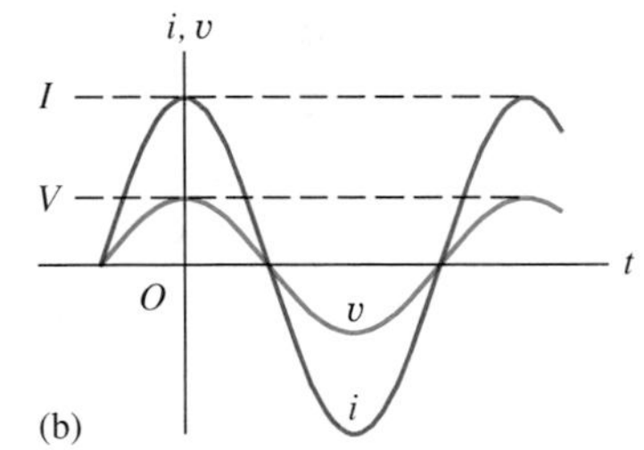


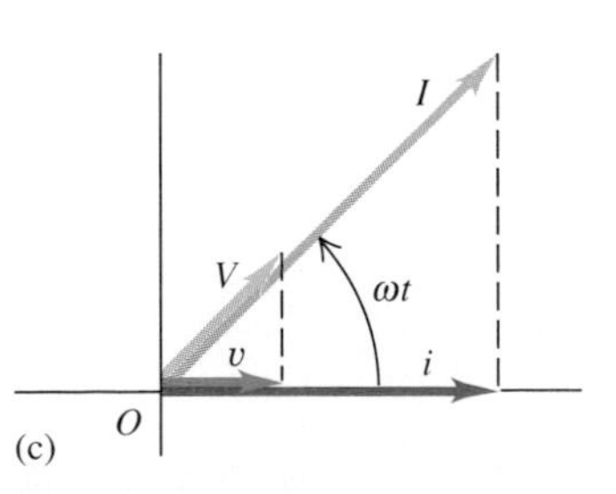


**32–3** (a) Resistance $R$ connected across an ac source. (b) Graphs of instantaneous voltage and current. (c) Phasor diagram; current and voltage are in phase.

### Inductor in an ac Circuit

Next, suppose we replace the resistor in Fig. 32–3 with a pure inductor with self-inductance $L$ and zero resistance (Fig. 32–4a). Again we assume that the current is $i = I\cos\omega t$. The induced emf in the direction of $i$ is given by Eq. (31–7): $\mathcal{E} = -L\,di/dt$. This

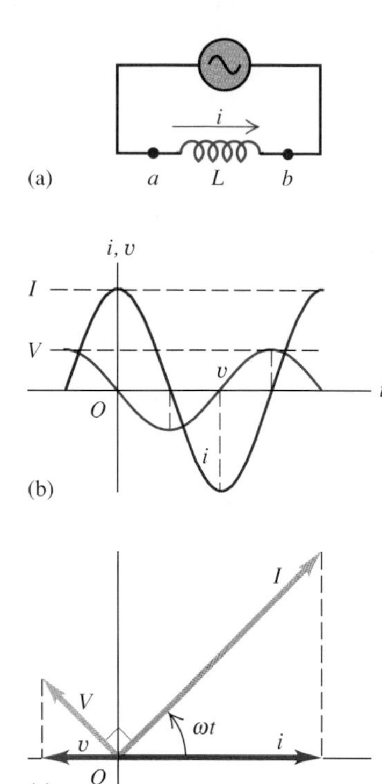


**32–4** (a) Inductance $L$ connected across an ac source. (b) Graphs of instantaneous voltage and current. (c) Phasor diagram; voltage *leads* current by 90°.

corresponds to the potential of point $b$ with respect to point $a$; the potential $v$ of point $a$ with respect to point $b$ is the negative of this, $v = L\,di/dt$, and we have

$$v = L\frac{di}{dt} = L\frac{d}{dt}(I\cos\omega t) = -I\omega L\sin\omega t. \tag{32–9}$$

The voltage across the inductor at any instant is proportional to the *rate of change* of the current. The points of maximum voltage on the graph correspond to maximum steepness of the current curve, and the points of zero voltage are the points where the current curve instantaneously levels off at its maximum and minimum values (Fig. 32–4b). The voltage and current are "out of step," or *out of phase,* by a quarter-cycle; the voltage peaks occur a quarter-cycle earlier than the current peaks, and we say that the voltage *leads* the current by 90°. The phasor diagram in Fig. 32–4c also shows this relationship; the voltage phasor is ahead of the current phasor by 90°.

We can also obtain this phase relationship by rewriting Eq. (32–9) using the identity $\cos(A + 90°) = -\sin A$:

$$v = I\omega L\cos(\omega t + 90°). \tag{32–10}$$

This result and the phasor diagram show that the voltage can be viewed as a cosine function with a "head start" of 90°.

For consistency in later discussions we will usually follow this pattern, describing the phase of the *voltage* relative to the *current,* not the reverse. Thus if the current $i$ in a circuit is

$$i = I\cos\omega t,$$

and the voltage $v$ of one point with respect to another is

$$v = V\cos(\omega t + \phi),$$

we call $\phi$ the **phase angle,** understanding that it gives the phase of the *voltage* relative to the *current.* For a pure inductor, $\phi = 90°$, and the voltage *leads* the current by 90°.

From Eq. (32–9) or (32–10) the voltage amplitude $V$ is

$$V = I\omega L. \tag{32–11}$$

We define the **inductive reactance** $X_L$ of an inductor as

$$X_L = \omega L. \tag{32–12}$$

Using $X_L$, we can write Eq. (32–11) in the same form as for a resistor ($V = IR$):

$$V = IX_L. \tag{32–13}$$

Because $X_L$ is the ratio of a voltage and a current, its unit is the same as for resistance, the ohm.

The inductive reactance of an inductor is directly proportional both to its inductance $L$ and to the angular frequency $\omega$; the greater the inductance and the higher the frequency, the *larger* the reactance. In some circuit applications, such as power supplies and radio-interference filters, inductors are used to block high frequencies while permitting lower frequencies or dc to pass through.

## EXAMPLE 32–2

The current amplitude in an inductor in a radio receiver is to be 250 $\mu$A when the voltage amplitude is 3.60 V at a frequency of 1.60 MHz (corresponding to the upper end of the AM broadcast band). What inductive reactance is needed? What inductance?

**SOLUTION** From Eq. (32–13),

$$X_L = \frac{V}{I} = \frac{3.60\text{ V}}{250 \times 10^{-6}\text{A}} = 1.44 \times 10^4\ \Omega = 14.4\text{ k}\Omega.$$

From Eq. (32–12), with $\omega = 2\pi f$, we find

$$L = \frac{X_L}{2\pi f} = \frac{1.44 \times 10^4\ \Omega}{2\pi(1.60 \times 10^6\text{ Hz})} = 1.43 \times 10^{-3}\text{ H} = 1.43\text{ mH}.$$

## Capacitor in an ac Circuit

Finally, suppose we connect a capacitor with capacitance $C$ to the source, as in Fig. 32–5a, producing a current $i = I \cos \omega t$ through the capacitor. You may object that charge can't really move *through* the capacitor because its two plates are insulated from each other. True enough, but at each instant, as the capacitor charges and discharges, we have a current $i$ into one plate, an equal current out of the other plate, and an equal *displacement* current between the plates, just as though the charge were being conducted through the capacitor. For this reason we often speak about alternating current *through* a capacitor.

To find the voltage $v$ of point $a$ with respect to point $b$, we first note that

$$i = \frac{dq}{dt} = I \cos \omega t.$$

Integrating this, we get

$$q = \frac{I}{\omega} \sin \omega t. \tag{32–14}$$

Also, the charge $q$ and voltage $v$ for a capacitor are related by Eq. (25–1), $q = Cv$. Using this in Eq. (32–14), we find

$$v = \frac{I}{\omega C} \sin \omega t. \tag{32–15}$$

In this case the instantaneous current $i$ is equal to the *rate of change* of the capacitor charge $q$, and because $q = Cv$, $i$ is also proportional to the rate of change of voltage. Figure 32–5b shows $v$ and $i$ as functions of $t$. Because $i = dq/dt = C\,dv/dt$, the current is greatest when the $v$ curve is rising or falling most steeply and zero when the $v$ curve instantaneously levels off at its maximum and minimum values.

The capacitor voltage and current are out of phase by a quarter-cycle. The peaks of voltage occur a quarter-cycle *after* the corresponding current peaks, and we say that the voltage *lags* the current by 90°. The phasor diagram in Fig. 32–5c shows this relationship; the voltage phasor is behind the current phasor by a quarter-cycle, or 90°.

We can also derive this phase difference by rewriting Eq. (32–15), using the identity $\cos(A - 90°) = \sin A$:

$$v = \frac{I}{\omega C} \cos(\omega t - 90°). \tag{32–16}$$

This cosine function has a "late start" of 90° compared with that for the current.

Equations (32–15) and (32–16) show that the *maximum* voltage $V$ (the voltage amplitude) is

$$V = \frac{I}{\omega C}. \tag{32–17}$$

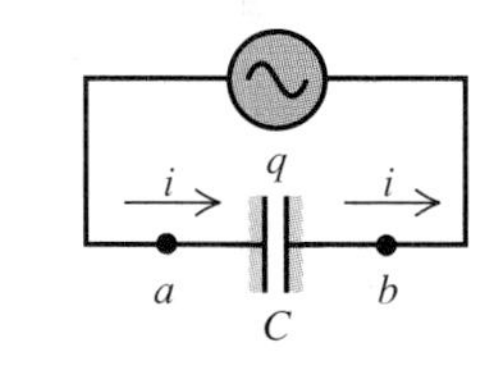


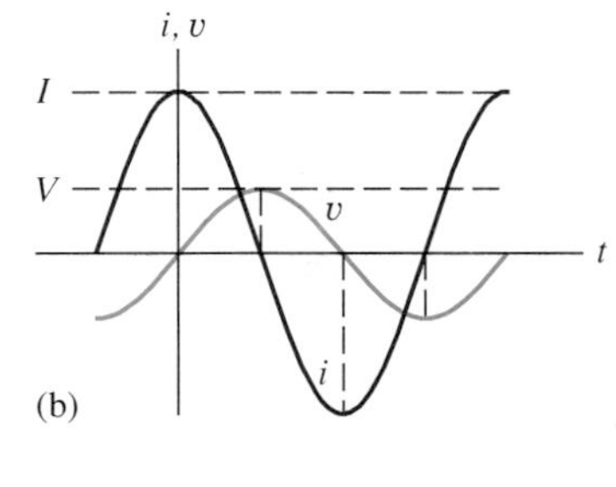


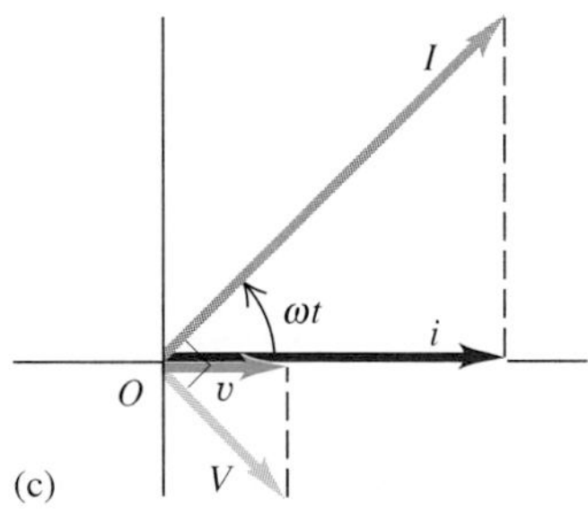


**32–5** (a) Capacitor $C$ connected across an ac source. (b) Graphs of instantaneous voltage and current. (c) Phasor diagram; voltage *lags* current by 90°.

To put this expression in the same form as for a resistor ($V = IR$), we define a quantity $X_C$, called the **capacitive reactance** of the capacitor, as

$$X_C = \frac{1}{\omega C}. \quad (32\text{–}18)$$

Then

$$V = IX_C. \quad (32\text{–}19)$$

Because $X_C$ is the ratio of a voltage and a current, its unit is the same as for resistance and inductive reactance, the ohm.

The capacitive reactance of a capacitor is inversely proportional both to the capacitance $C$ and to the angular frequency $\omega$; the greater the capacitance and the higher the frequency, the *smaller* the reactance $X_C$. Capacitors tend to pass high-frequency current and to block low-frequency and dc currents, just the opposite of inductors.

## EXAMPLE 32–3

A 300-Ω resistor is connected in series with a 5.0-μF capacitor. The voltage across the resistor is $v_R = (1.20\text{ V})\cos(2500\text{ rad/s})t$. a) Derive an expression for the circuit current. b) Determine the capacitive reactance of the capacitor. c) Derive an expression for the voltage $v$ across the capacitor.

**SOLUTION** a) Circuit elements in series have the same current; using $v = iR$, we find that the current $i$ is

$$i = \frac{v}{R} = \frac{(1.20\text{ V})\cos(2500\text{ rad/s})t}{300\ \Omega} = (4.0 \times 10^{-3}\text{ A})\cos(2500\text{ rad/s})t.$$

b) To find the capacitive reactance $X_C$, we use Eq. (32–18), with $\omega = 2500$ rad/s:

$$X_C = \frac{1}{\omega C} = \frac{1}{(2500\text{ rad/s})(5.0 \times 10^{-6}\text{ F})} = 80\ \Omega.$$

c) From Eq. (32–19) the voltage amplitude $V$ is

$$V = IX_C = (4.0 \times 10^{-3}\text{ A})(80\ \Omega) = 0.32\text{ V}.$$

The instantaneous capacitor voltage $v$ is given by Eq. (32–16):

$$v = V\cos(\omega t - 90°) = (0.32\text{ V})\cos[(2500\text{ rad/s})t - \pi/2\text{ rad}].$$

We converted 90° to $\pi/2$ rad, so all the angular quantities had the same units. In ac circuit analysis, phase angles are often given in degrees, so be careful to convert to radians when necessary.

**TABLE 32–1 Circuit Elements with Alternating Current**

| Circuit element | Amplitude relation | Circuit quantity | Phase of $v$ |
|---|---|---|---|
| Resistor | $V = IR$ | $R$ | In phase with $i$ |
| Inductor | $V = IX_L$ | $X_L = \omega L$ | Leads $i$ by 90° |
| Capacitor | $V = IX_C$ | $X_C = 1/\omega C$ | Lags $i$ by 90° |

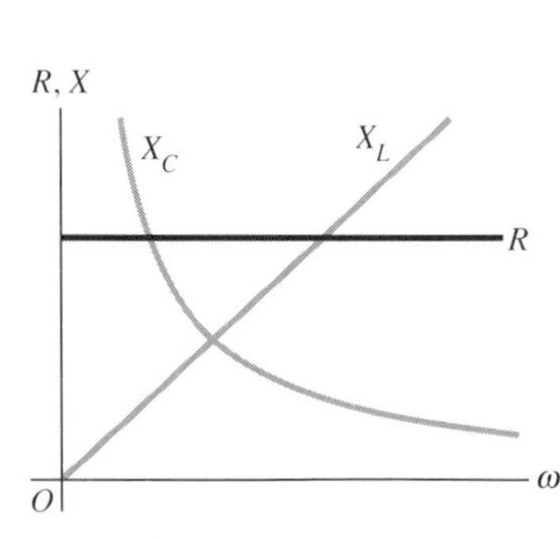

**32–6** Graphs of $R$, $X_L$, and $X_C$ as functions of angular frequency $\omega$.

Table 32–1 summarizes the relations of voltage and current amplitudes for the three circuit elements that we have discussed.

The graphs in Fig. 32–6 show how the resistance of a resistor and the reactances of an inductor and a capacitor vary with angular frequency. As the frequency approaches infinity, the reactance of the inductor approaches infinity, and that of the capacitor approaches zero. As the frequency approaches zero, the inductive reactance approaches zero, and the capacitive reactance approaches infinity. The limiting case of zero frequency corresponds to a dc circuit; in that case there is *no* current through a capacitor because $X_C \rightarrow \infty$, and there is no inductive effect because $X_L \rightarrow 0$.

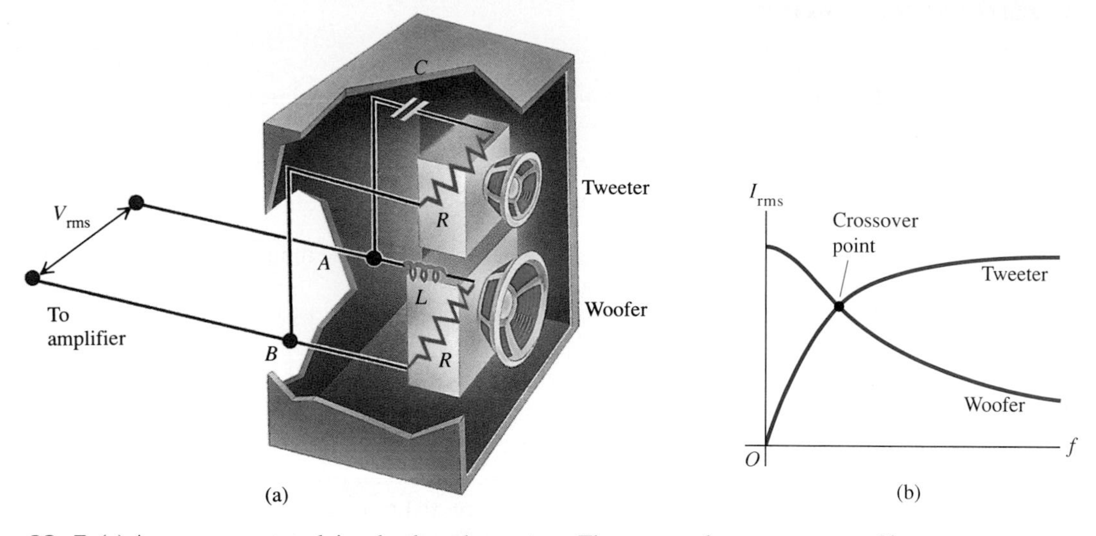


**32–7** (a) A crossover network in a loudspeaker system. The two speakers are connected in parallel to the amplifier; the inductance and capacitance feed the lower frequencies predominantly to the woofer, the higher frequencies to the tweeter. (b) Graphs of rms current in the tweeter and woofer as functions of frequency for a given rms amplifier voltage. The point where the two curves cross is called the crossover point.

Figure 32–7 shows an application of the above discussion to a loudspeaker system. The woofer and tweeter are connected in parallel across the amplifier output. The capacitor in the tweeter branch blocks the low-frequency components of sound but passes the higher frequencies; the inductor in the woofer branch does the opposite.

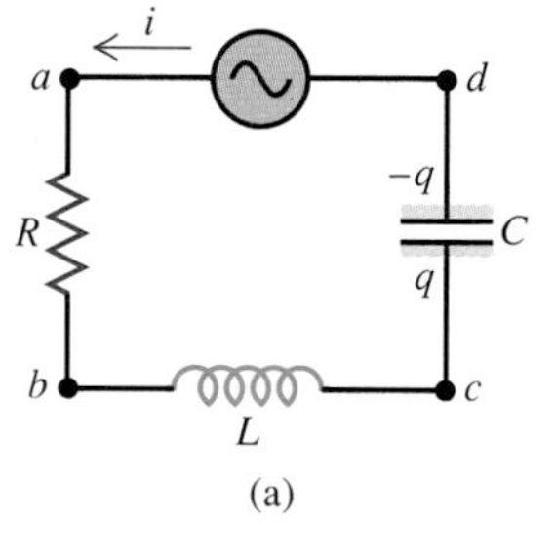


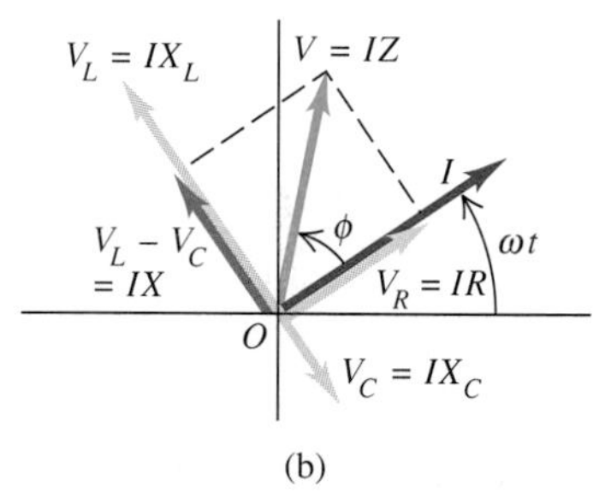


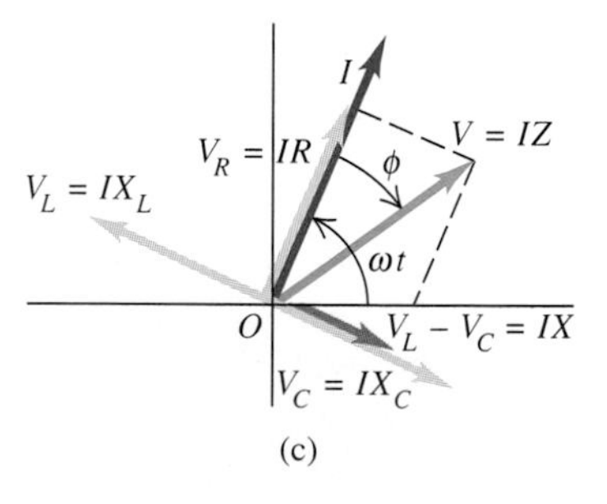


**32–8** (a) A series *L-R-C* circuit. (b) Phasor diagram for the case $X_L > X_C$. (c) Phasor diagram for the case $X_C > X_L$.

## 32–3 THE *L-R-C* SERIES CIRCUIT

Many ac circuits that are used in practical electronic systems involve resistance, inductive reactance, and capacitive reactance. A series circuit containing a resistor, an inductor, and a capacitor is shown in Fig. 32–8a. To analyze this and similar circuits, we will use a phasor diagram that includes the voltage and current phasors for each of the components. In this circuit, because of Kirchhoff's loop rule, the instantaneous *total* voltage $v_{ad}$ across all three components is equal to the source voltage at that instant. We will show that the phasor representing this total voltage is the *vector sum* of the phasors for the individual voltages. The complete phasor diagram for this circuit is shown in Fig. 32–8b. This may appear complex, but we will explain it one step at a time.

Let's assume that the source supplies a current $i$ given by $i = I \cos \omega t$. Because the circuit elements are connected in series, the current $i$ at any instant is the same at every point in the circuit. Thus a *single phasor* $I$, with length proportional to the current amplitude, represents the current in *all* circuit elements.

We use the symbols $v_R$, $v_L$, and $v_C$ for the instantaneous voltages across $R$, $L$, and $C$, and we use $V_R$, $V_L$, and $V_C$ for their maximum values. We denote the instantaneous and maximum *source* voltages by $v$ and $V$. Then $v = v_{ad}$, $v_R = v_{ab}$, $v_L = v_{bc}$, and $v_C = v_{cd}$.

We have shown that the potential difference between the terminals of a resistor is *in phase* with the current in the resistor and that its maximum value $V_R$ is

$$V_R = IR.$$

The phasor $V_R$ in Fig. 32–8b, in phase with the current phasor $I$, represents the voltage across the resistor. Its projection onto the horizontal axis at any instant gives the instantaneous potential difference $v_R$.

The voltage across an inductor *leads* the current by 90°. Its voltage amplitude is

$$V_L = IX_L.$$

The phasor $V_L$ in Fig. 32–8b represents the voltage across the inductor, and its projection onto the horizontal axis at any instant equals $v_L$.

The voltage across a capacitor *lags* the current by 90°. Its voltage amplitude is

$$V_C = IX_C.$$

The phasor $V_C$ in Fig. 32–8b represents the voltage across the capacitor, and its projection onto the horizontal axis at any instant equals $v_C$.

The instantaneous potential difference $v$ between terminals $a$ and $d$ is equal at every instant to the (algebraic) sum of the potential differences $v_R$, $v_L$, and $v_C$. That is, it equals the sum of the *projections* of the phasors $V_R$, $V_L$, and $V_C$. But the sum of the projections of these phasors is equal to the *projection* of their *vector* sum. So the vector sum $V$ must be the phasor that represents the source voltage $v$ and the instantaneous total voltage $v_{ad}$ across the series of elements.

To form this vector sum, we first subtract the phasor $V_C$ from the phasor $V_L$. (These two phasors always lie along the same line and have opposite directions.) This gives the phasor $V_L - V_C$. This is always at right angles to the phasor $V_R$, so from the Pythagorean theorem the magnitude of the phasor $V$ is

$$\begin{aligned} V &= \sqrt{V_R^2 + (V_L - V_C)^2} = \sqrt{(IR)^2 + (IX_L - IX_C)^2} \\ &= I\sqrt{R^2 + (X_L - X_C)^2}. \end{aligned} \tag{32–20}$$

The quantity $X_L - X_C$ is called the **reactance** of the circuit, denoted by $X$:

$$X = X_L - X_C. \tag{32–21}$$

Finally, we define the **impedance** $Z$ of the circuit as

$$Z = \sqrt{R^2 + (X_L - X_C)^2} = \sqrt{R^2 + X^2}, \tag{32–22}$$

so we can rewrite Eq. (32–20) as

$$V = IZ. \tag{32–23}$$

Equation (32–23) again has the same form as $V = IR$, with impedence $Z$ playing the same role as resistance $R$ in a dc circuit. Note, however, that the impedance is actually a function of $R$, $L$, and $C$, as well as of the angular frequency $\omega$. The complete expression for $Z$ for a series circuit is

$$\begin{aligned} Z &= \sqrt{R^2 + X^2} \\ &= \sqrt{R^2 + (X_L - X_C)^2} \\ &= \sqrt{R^2 + [\omega L - (1/\omega C)]^2}. \end{aligned} \tag{32–24}$$

Impedance is always a ratio of a voltage and a current; the SI unit of impedance is one ohm.

Equation (32–24) gives the impedance $Z$ only for a *series* $L$-$R$-$C$ circuit. But we can *define* the impedance of *any* network, using Eq. (32–23), as the ratio of the voltage amplitude to the current amplitude.

The angle $\phi$ shown in Fig. 32–8b is the phase angle of the source voltage $v$ with respect to the current $i$. From the diagram,

$$\tan\phi = \frac{V_L - V_C}{V_R} = \frac{I(X_L - X_C)}{IR} = \frac{X_L - X_C}{R} = \frac{X}{R}$$
$$= \frac{\omega L - 1/\omega C}{R}. \qquad (32\text{–}25)$$

The source voltage leads the current by an angle $\phi$. If the current is $i = I\cos\omega t$, then the source voltage $v$ is

$$v = V\cos(\omega t + \phi).$$

Figure 32–8b shows the behavior of a circuit in which $X_L > X_C$. If $X_L < X_C$, as in Fig. 32–8c, vector $V$ lies on the opposite side of the current vector $I$, and the voltage *lags* the current. In this case, $X = X_L - X_C$ is a *negative* quantity, $\tan\phi$ is negative, and $\phi$ is a negative angle between 0 and $-90°$.

Finally, we note that all the relations that we have developed for an *L-R-C* series circuit are still valid if one of the circuit elements is missing. If the resistor is missing, we set $R = 0$; if the inductor is missing, we set $L = 0$. But if the capacitor is missing, we set $C = \infty$, corresponding to the absence of any potential difference ($v = q/C$) or any capacitive reactance ($X_C = 1/\omega C$).

## PROBLEM-SOLVING STRATEGY

### Alternating-current circuits

**1.** In ac circuit problems it is nearly always easiest to work with angular frequency $\omega$. But you may be given the ordinary frequency $f$, expressed in hertz. Don't forget to convert, using $\omega = 2\pi f$.

**2.** Keep in mind a few basic facts about phase relationships. For a resistor, voltage and current are always *in phase*, and the two corresponding phasors in a phasor diagram always have the same direction. For an inductor the voltage always *leads* the current by 90° (that is, $\phi = +90°$), and the voltage phasor is always turned 90° counterclockwise from the current phasor. For a capacitor the voltage always *lags* the current by 90° (that is, $\phi = -90°$), and the voltage phasor is always turned 90° clockwise from the current phasor.

**3.** Remember that Kirchhoff's rules are applicable to ac circuits. All the voltages and currents are sinusoidal functions of time instead of being constant, but Kirchhoff's rules hold at each instant. Thus in a series circuit the instantaneous current is the same in all circuit elements; in a parallel circuit the instantaneous potential difference is the same across all circuit elements.

**4.** Reactance and impedance are analogous to resistance; each represents the ratio of voltage amplitude $V$ to current amplitude $I$ in a circuit element or combination of elements. But keep in mind that phase relations play an essential role; resistance and reactance have to be combined by *vector* addition of the corresponding phasors. When you have several circuit elements in series, for example, you can't just *add* all the numerical values of resistance and reactance; that would ignore the phase relations.

## EXAMPLE 32–4

In the series circuit of Fig. 32–8a, suppose $R = 300\ \Omega$, $L = 60$ mH, $C = 0.50\ \mu$F, $V = 50$ V, and $\omega = 10{,}000$ rad/s. Find the reactances $X_L$ and $X_C$, the impedance $Z$, the current amplitude $I$, the phase angle $\phi$, and the voltage amplitude across each circuit element.

**SOLUTION** From Eqs. (32–12) and (32–18),

$$X_L = \omega L = (10{,}000\ \text{rad/s})(60\ \text{mH}) = 600\ \Omega,$$
$$X_C = \frac{1}{\omega C} = \frac{1}{(10{,}000\ \text{rad/s})(0.50\times 10^{-6}\ \text{F})} = 200\ \Omega.$$

The reactance $X$ of the circuit is

$$X = X_L - X_C = 600\ \Omega - 200\ \Omega = 400\ \Omega,$$

and the impedance $Z$ is

$$Z = \sqrt{R^2 + X^2} = \sqrt{(300\ \Omega)^2 + (400\ \Omega)^2} = 500\ \Omega.$$

With source voltage amplitude $V = 50$ V, the current amplitude is

$$I = \frac{V}{Z} = \frac{50\ \text{V}}{500\ \Omega} = 0.10\ \text{A}.$$

The phase angle $\phi$ is

$$\phi = \arctan\frac{X_L - X_C}{R} = \arctan\frac{400\ \Omega}{300\ \Omega} = 53°.$$

Because $\phi$ is positive, we know that the voltage *leads* the current by 53°. From Eq. (32–7) the voltage amplitude $V_R$ across the resistor is

$$V_R = IR = (0.10\ \text{A})(300\ \Omega) = 30\ \text{V}.$$

From Eq. (32–13) the voltage amplitude $V_L$ across the inductor is

$$V_L = IX_L = (0.10\ \text{A})(600\ \Omega) = 60\ \text{V}.$$

From Eq. (32–19) the voltage amplitude $V_C$ across the capacitor is

$$V_C = IX_C = (0.10\ \text{A})(200\ \Omega) = 20\ \text{V}.$$

Note that the source voltage amplitude $V = 50$ V is *not* equal to the sum of the voltage amplitudes across the separate circuit elements. Make sure you understand why not!

---

In this entire discussion we have described magnitudes of voltages and currents in terms of their *maximum* values, the voltage and current *amplitudes.* But we remarked at the end of Section 32–1 that these quantities are usually described not in terms of their amplitudes but in terms of rms values. For any sinusoidally varying quantity the rms value is always $1/\sqrt{2}$ times the amplitude. All the relations between voltage and current that we have derived in this and the preceding sections are still valid if we use rms quantities throughout instead of amplitudes. For example, if we divide Eq. (32–23) by $\sqrt{2}$, we get

$$\frac{V}{\sqrt{2}} = \frac{I}{\sqrt{2}}Z,$$

which we can rewrite as

$$V_{\text{rms}} = I_{\text{rms}}Z. \tag{32–26}$$

We can translate Eqs. (32–7), (32–13), and (32–19) in exactly the same way.

Finally, we remark that what we have been describing throughout this section is the *steady-state* condition of a circuit, the state that exists after the circuit has been connected to the source for a long time. When the source is first connected, there may be additional voltages and currents, called *transients,* whose nature depends on the time in the cycle when the circuit is initially completed. A detailed analysis of transients is beyond our scope. They always die out after a sufficiently long time, and they do not affect the steady-state behavior of the circuit. But they can cause dangerous and damaging surges in power lines, and delicate electronic systems such as computers are often provided with power-line surge protectors.

## 32–4 POWER IN ALTERNATING-CURRENT CIRCUITS

Alternating currents play a central role in systems for distributing, converting, and using electrical energy, so it's important to look at power relationships in ac circuits. When a source with voltage amplitude $V$ and instantaneous potential difference $v$ supplies an instantaneous current $i$ (current amplitude $I$) to an ac circuit, the instantaneous power $p$ that it supplies is

$$p = vi.$$

Let's first see what this means for individual circuit elements.

We'll assume in each case that $i = I\cos\omega t$. Suppose first that the circuit consists of a *pure resistance R,* as in Fig. 32–3; then $i$ and $v$ are *in phase*. We obtain the graph repre-

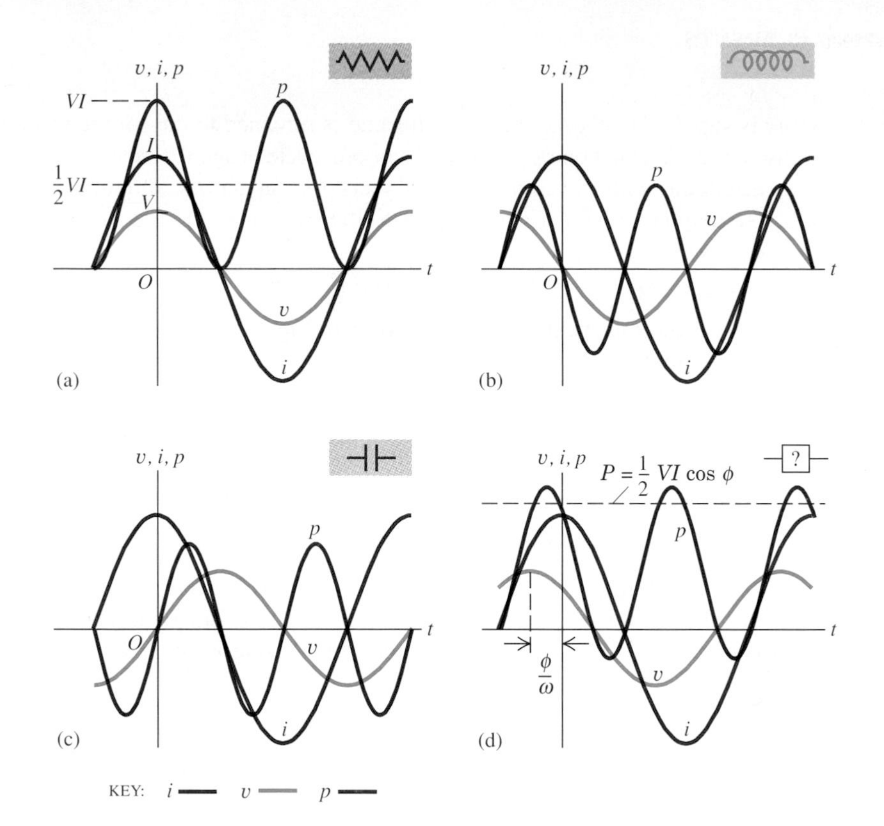


**32–9** Graphs of voltage, current, and power as functions of time for various circuits. (a) Instantaneous power input to a pure resistor. The average power is $\frac{1}{2}VI = V_{\text{rms}}I_{\text{rms}}$. (b) Instantaneous power input to a pure inductor. The average power is zero. (c) Instantaneous power input to a pure capacitor. The average power is zero. (d) Instantaneous power input to an arbitrary ac circuit. The average power is $\frac{1}{2}VI\cos\phi = V_{\text{rms}}I_{\text{rms}}\cos\phi$.

senting $p$ by multiplying the heights of the graphs of $v$ and $i$ in Fig. 32–3b at each instant. This graph is shown as the black curve in Fig. 32–9a. The product $vi$ is always positive because $v$ and $i$ are always either both positive or both negative. Energy is supplied to the resistor at every instant for both directions of $i$, although the power is not constant.

The power curve is symmetrical about a value equal to one-half of its maximum value $VI$, so the *average power* $P$ is

$$P = \tfrac{1}{2}VI. \tag{32–27}$$

An equivalent expression is

$$P = \frac{V}{\sqrt{2}}\frac{I}{\sqrt{2}} = V_{\text{rms}}I_{\text{rms}}. \tag{32–28}$$

Also, $V_{\text{rms}} = I_{\text{rms}}R$, so we can express $P$ in any of these equivalent forms:

$$P = I_{\text{rms}}{}^2R = \frac{V_{\text{rms}}{}^2}{R} = V_{\text{rms}}I_{\text{rms}}. \tag{32–29}$$

Note that these expressions have the same form as the corresponding relations for a dc circuit, Eq. (26–20). Also note that they are valid only for pure resistors, not for more complicated combinations of circuit elements.

Next we connect the source to a pure inductor $L$, as in Fig. 32–4. The voltage leads the current by 90°. When we multiply the curves of $v$ and $i$, the product $vi$ is *negative* during the half of the cycle when $v$ and $i$ have *opposite* signs. We get the power curve in Fig. 32–9b, which is symmetrical about the horizontal axis. It is positive half the time and negative the other half, and the average power is zero. When $p$ is positive, energy is being supplied to set up the magnetic field in the inductor; when $p$ is negative, the field is collapsing, and the inductor is returning energy to the source. The net energy transfer over one cycle is zero.

Finally, we connect the source to a pure capacitor $C$, as in Fig. 32–5. The voltage lags the current by 90°. Figure 32–9c shows the power curve; the average power is again zero. Energy is supplied to charge the capacitor and is returned to the source when the capacitor discharges. The net energy transfer over one cycle is again zero.

In *any* ac circuit, with any combination of resistors, capacitors, and inductors, the voltage $v$ has some phase angle $\phi$ with respect to the current $i$, and the instantaneous power $p$ is given by

$$p = vi = [V\cos(\omega t + \phi)]\,[I\cos\omega t]. \tag{32-30}$$

The instantaneous power curve has the form shown in Fig. 32–9d. The area under the positive loops is greater than that under the negative loops, and the average power is positive.

To derive an expression for the *average* power, which we will denote with a capital $P$, we apply the identity for the cosine of the sum of two angles to Eq. 32–30:

$$\begin{aligned} p &= [V(\cos\omega t\cos\phi - \sin\omega t\sin\phi)]\,[I\cos\omega t] \\ &= VI\cos\phi\cos^2\omega t - VI\sin\phi\cos\omega t\sin\omega t. \end{aligned}$$

From the discussion leading to Eq. (32–4) in Section 32–1 we see that the average value of $\cos^2\omega t$ (over one cycle) is $\frac{1}{2}$. The average value of $\cos\omega t\sin\omega t$ is zero because this product is equal to $\frac{1}{2}\sin 2\omega t$, whose average over a cycle is zero. So the average power $P$ is

$$P = \tfrac{1}{2}VI\cos\phi = V_{\text{rms}}I_{\text{rms}}\cos\phi. \tag{32-31}$$

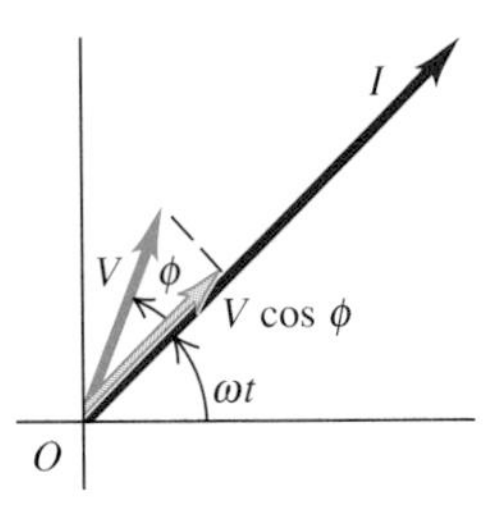


**32–10** The average power is half the product of $I$ and the component of $V$ in phase with $I$.

When $v$ and $i$ are *in phase*, the average power equals $\frac{1}{2}VI = V_{\text{rms}}I_{\text{rms}}$; when $v$ and $i$ are 90° *out of phase*, the average power is zero. In the general case, when $v$ has phase angle $\phi$ with respect to $i$, the average power equals $\frac{1}{2}I$ multiplied by $V\cos\phi$, the component of $V$ that is *in phase* with $I$. The relationship of the current and voltage phasors for this case is shown in Fig. 32–10. For the $L$-$R$-$C$ series circuit, $V\cos\phi$ is the voltage amplitude for the resistor, and Eq. (32–31) is the power dissipated in the resistor. The power dissipation in the inductor and capacitor is zero.

The factor $\cos\phi$ is called the **power factor** of the circuit. For a pure resistance, $\phi = 0$, $\cos\phi = 1$, and $P = V_{\text{rms}}I_{\text{rms}}$. For a pure (resistanceless) capacitor or inductor, $\phi = \pm 90°$, $\cos\phi = 0$, and $P = 0$. For an $L$-$R$-$C$ series circuit the power factor is equal to $R/Z$; we leave the proof of this statement as a problem.

## EXAMPLE 32–5

An electric hair dryer (Fig. 32–11) is rated at 1500 W at 120 V. Calculate a) the resistance, b) the rms current, and c) the maximum instantaneous power. Assume pure resistance.

**SOLUTION** a) We solve Eq. (32–29) for $R$ and substitute given values:

$$R = \frac{V_{\text{rms}}^2}{P} = \frac{(120\text{ V})^2}{1500\text{ W}} = 9.6\ \Omega.$$

b) From Eq. (32–28),

$$I_{\text{rms}} = \frac{P}{V_{\text{rms}}} = \frac{1500\text{ W}}{120\text{ V}} = 12.5\text{ A}.$$

Or, from Eq. (32–7),

$$I_{\text{rms}} = \frac{V_{\text{rms}}}{R} = \frac{120\text{ V}}{9.6\ \Omega} = 12.5\text{ A}.$$

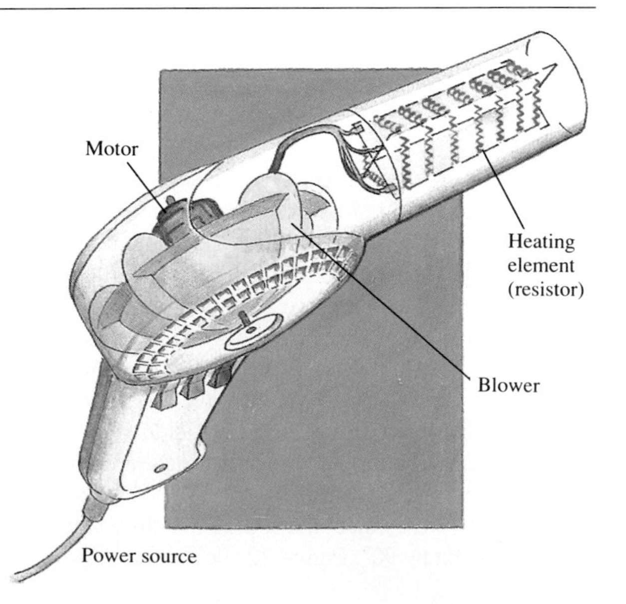


**32–11** The average power delivered to this hair dryer is equal to $V_{\text{rms}}^2/R$.

c) The maximum instantaneous power is $VI$; from Eq. (32–27),

$$VI = 2P$$
$$= 2(1500\ \text{W}) = 3000\ \text{W}.$$

To mislead the unwary consumer, some manufacturers of stereo amplifiers state power outputs in terms of the peak value rather than the lower average value.

## EXAMPLE 32–6

For the $L$-$R$-$C$ series circuit of Example 32–4, calculate a) the power factor and b) the average power to the entire circuit and to each circuit element.

**SOLUTION** a) The power factor is $\cos\phi = \cos 53° = 0.60$.

b) From Eq. (32–31) the average power to the circuit is

$$P = \tfrac{1}{2}VI\cos\phi = \tfrac{1}{2}(50\ \text{V})(0.10\ \text{A})(0.60) = 1.5\ \text{W}.$$

All of this power is dissipated in the resistor; the average power to a pure inductor or capacitor is always zero.

A low power factor (large angle of lag or lead) is usually undesirable in power circuits because, for a given potential difference, a large current is needed to supply a given amount of power. This results in large $I^2R$ losses in the transmission lines. Your electric power company may charge a higher rate to a client with a low power factor. Many types of ac machinery draw a lagging current; the power factor can be corrected by connecting a capacitor in parallel with the load. The leading current drawn by the capacitor compensates for the lagging current in the other branch of the circuit. The capacitor itself absorbs no net power from the line.

## 32–5 SERIES RESONANCE

The impedance of an $L$-$R$-$C$ series circuit depends on the frequency, as Eq. (32–24) shows. Figure 32–12a shows graphs of $R$, $X_L$, $X_C$, and $Z$ as functions of $\omega$. We have used a logarithmic angular frequency scale so that we can cover a wide range of frequencies. Because $X_L$ increases and $X_C$ decreases with increasing frequency, there is always one particular frequency at which $X_L$ and $X_C$ are equal and $X = X_L - X_C$ is zero. At this frequency the impedance $Z$ has its *smallest* value, equal to just the resistance $R$.

Suppose we connect an ac voltage source with constant voltage amplitude $V$ but variable angular frequency $\omega$ across an $L$-$R$-$C$ series circuit. As we vary $\omega$, the current amplitude $I$ varies with frequency as shown in Fig. 32–12b; its *maximum* value occurs at the frequency at which the impedance $Z$ is *minimum*. This peaking of the current amplitude at a certain frequency is called **resonance.** The angular frequency $\omega_0$ at which the resonance peak occurs is called the **resonance angular frequency.** This is the angular frequency at which the inductive and capacitive reactances are equal, so

$$X_L = X_C, \qquad \omega_0 L = \frac{1}{\omega_0 C}, \qquad \omega_0 = \frac{1}{\sqrt{LC}}. \tag{32–32}$$

Note that this is equal to the natural angular frequency of oscillation of an $L$-$C$ circuit, which we derived in Section 31–5, Eq. (31–22). The **resonance frequency** $f_0$ is $\omega_0/2\pi$.

Now let's look at what happens to the *voltages* in an $L$-$R$-$C$ series circuit at resonance. The current at any instant is the same in $L$ and $C$. The voltage across an inductor always *leads* the current by 90°, or a quarter-cycle, and the voltage across a capacitor always *lags* the current by 90°. Therefore the instantaneous voltages across $L$ and $C$ always differ in phase by 180°, or a half-cycle; they have opposite signs at each instant. If the *amplitudes* of these two voltages are equal, then they add to zero at each instant, and the *total* voltage $v_{bd}$ across the $L$-$C$ combination in Fig. 32–8a is exactly zero! This occurs only at the resonance frequency. Depending on the numerical values of $R$, $L$, and $C$, the voltages across $L$ and $C$ individually can be larger than that across $R$. At

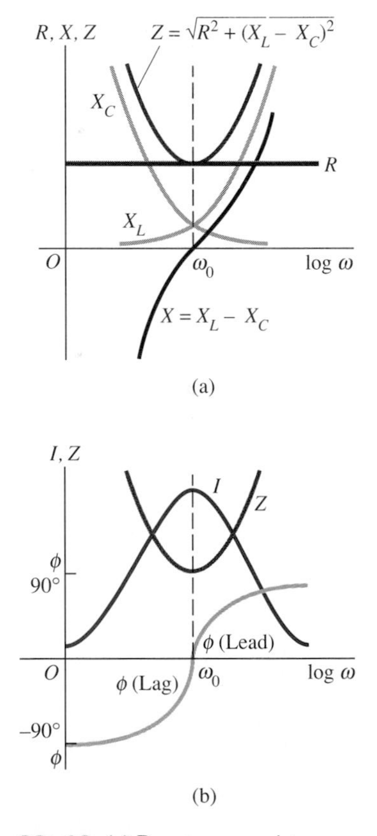


**32–12** (a) Reactance, resistance, and impedance as functions of angular frequency (logarithmic angular frequency scale).
(b) Impedance, current, and phase angle as functions of angular frequency (logarithmic angular frequency scale).

frequencies close to resonance the voltages across $L$ and $C$ individually can be *much larger* than the source voltage!

The *phase* of the voltage relative to the current is given by Eq. (32–25). At frequencies below resonance, $X_C$ is greater than $X_L$; the capacitive reactance dominates, the voltage *lags* the current, and the phase angle $\phi$ is between zero and $-90°$. Above resonance, the inductive reactance dominates; the voltage *leads* the current, and the phase angle is between zero and $+90°$. This variation of $\phi$ with angular frequency is shown in Fig. 32–12b.

If we can vary the inductance $L$ or the capacitance $C$ of a circuit, we can also vary the resonance frequency. This is how some radio or television receiving sets are "tuned" to receive particular stations. In the early days of radio this was accomplished by use of capacitors with movable metal plates whose overlap could be varied to change $C$. Nowadays it is more common to vary $L$ by using a coil with a ferrite core that slides in or out.

In a series $L$-$R$-$C$ circuit the impedance reaches its minimum value and the current reaches its maximum value at the resonance frequency. Figure 32–13 shows a graph of rms current as a function of frequency for such a circuit, with $V = 100$ V, $R = 500\ \Omega$, $L = 2.0$ H, and $C = 0.50\ \mu$F. This curve is called a *response curve,* or a *resonance curve*. The resonance angular frequency is $\omega_0 = (LC)^{-1/2} = 1000$ rad/s. As we expect, the curve has a peak at $\omega = 1000$ rad/s, the resonance angular frequency.

The resonance frequency is determined by $L$ and $C$; what happens when we change $R$? The figure also shows graphs of $I$ as a function of $\omega$ for $R = 200\ \Omega$ and for $R = 2000\ \Omega$. The curves are all similar for frequencies far away from resonance, where the impedance is dominated by $X_L$ or $X_C$. But near resonance, where $X_L$ and $X_C$ nearly cancel each other, the curve is higher and more sharply peaked for small values of $R$ than for larger values. The maximum height of the curve is in fact inversely proportional to $R$. A small $R$ gives a sharply peaked response curve, and a large value of $R$ gives a broad, flat curve.

The shape of the response curve is important in the design of radio and television receiving circuits. The sharply peaked curve is what makes it possible to discriminate between two stations broadcasting on adjacent frequency bands. But if the peak is *too* sharp, some of the information in the received signal is lost, such as the high-frequency sounds in music. The shape of the resonance curve is also related to the overdamped and underdamped oscillations that we described in Section 31–6. A sharply peaked resonance curve corresponds to a small value of $R$ and a lightly damped oscillating system; a broad flat curve goes with a large value of $R$ and a heavily damped system.

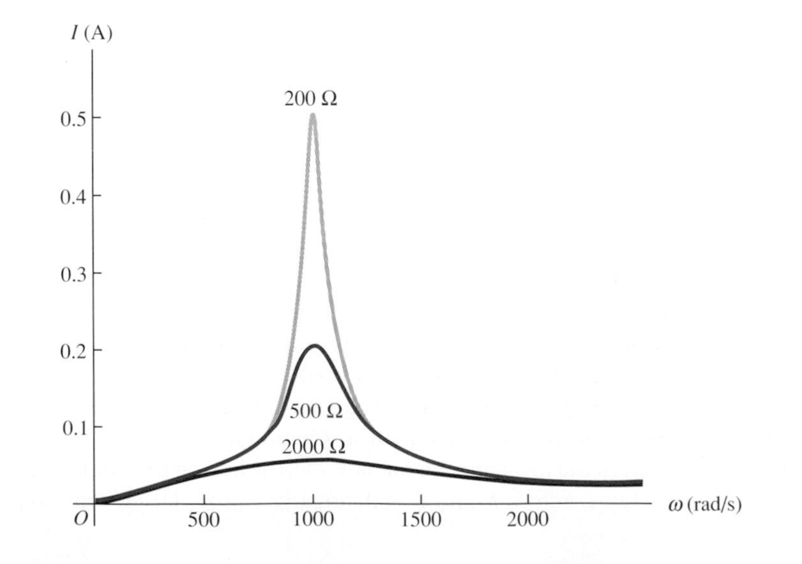


**32–13** Graph of rms current $I$ as a function of angular frequency $\omega$ (the red curve). The other curves show the relationship for different values of the resistance $R$, 2000 Ω (dark red curve) and 200 Ω (light red curve).

## EXAMPLE 32–7

**Tuning a radio** The series circuit in Fig. 32–14 is similar to arrangements that are sometimes used in radio tuning circuits. This circuit is connected to the terminals of an ac source with a constant rms terminal voltage of 1.0 V and a variable frequency. Find a) the resonance frequency; b) the inductive reactance, the capacitive reactance, the reactance, and the impedance at the resonance frequency; c) the rms current at resonance; and d) the rms voltage across each circuit element at resonance.

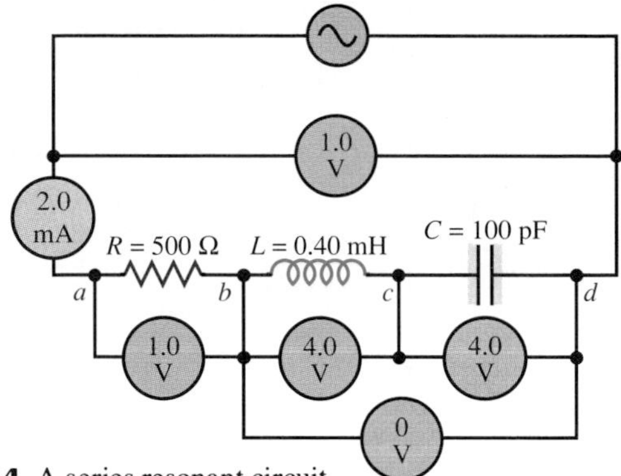


**32–14** A series resonant circuit.

**SOLUTION** a) The resonance angular frequency is

$$\omega_0 = \frac{1}{\sqrt{LC}} = \frac{1}{\sqrt{(0.40 \times 10^{-3}\ \text{H})(100 \times 10^{-12}\ \text{F})}} = 5.0 \times 10^6\ \text{rad/s}$$

The corresponding frequency $f = \omega/2\pi$ is

$$f = 8.0 \times 10^5\ \text{Hz} = 800\ \text{kHz}.$$

This corresponds to the lower part of the AM radio band.

b) At this frequency,

$$X_L = \omega L = (5.0 \times 10^6\ \text{rad/s})(0.40 \times 10^{-3}\ \text{H}) = 2000\ \Omega,$$

$$X_C = \frac{1}{\omega C} = \frac{1}{(5.0 \times 10^6\ \text{rad/s})(100 \times 10^{-12}\ \text{F})} = 2000\ \Omega,$$

$$X = X_L - X_C = 0.$$

From Eq. (32–22) the impedance $Z$ at resonance is equal to the resistance: $Z = R = 500\ \Omega$.

c) At resonance the rms current is

$$I = \frac{V}{Z} = \frac{V}{R} = \frac{1.0\ \text{V}}{500\ \Omega} = 0.0020\ \text{A} = 2.0\ \text{mA}.$$

d) The rms potential difference across the resistor is

$$V_R = IR = (0.0020\ \text{A})(500\ \Omega) = 1.0\ \text{V}.$$

The rms potential differences across the inductor and capacitor are, respectively,

$$V_L = IX_L = (0.0020\ \text{A})(2000\ \Omega) = 4.0\ \text{V},$$

$$V_C = IX_C = (0.0020\ \text{A})(2000\ \Omega) = 4.0\ \text{V}.$$

The rms potential difference across the inductor-capacitor combination ($V_{bd}$) is

$$V = IX = I(X_L - X_C) = 0.$$

The instantaneous potential differences across the inductor and the capacitor have equal amplitudes but are 180° out of phase and so add to zero at each instant. Note also that, at resonance, $V_R$ is equal to the source voltage $V$, but in this example, $V_L$ and $V_C$ are both considerably *larger* than $V$.

Resonance phenomena occur in all areas of physics; we have already seen one example in the forced oscillation of the harmonic oscillator (Section 13–8). In that case the amplitude of a mechanical oscillation peaked at a driving-force frequency close to the natural frequency of the system. The behavior of the *L-R-C* circuit is analogous to this. We suggest that you review that discussion now, looking for the analogies. Other important examples of resonance occur in acoustics, in atomic and nuclear physics, and in the study of fundamental particles (high-energy physics).

## 32–6 PARALLEL RESONANCE

A different kind of resonance occurs when a resistor, an inductor, and a capacitor are connected in parallel, as shown in Fig. 32–15a (on the next page). This circuit has resonance behavior similar to the *L-R-C* series circuit that we analyzed in Sections 32–3 and 32–5, but the roles of voltage and current are reversed. This time the instantaneous potential difference $v$ is the same for all three circuit elements and is equal to the source voltage, but the current is different in each of the three elements.

Figure 32–15b shows a phasor diagram; the phasor $V$ represents the common voltage across all elements. There are three current phasors, one for each circuit element. The phasor $I_R$, with amplitude $V/R$, is in phase with $V$; it represents the current in the resistor. Phasor $I_L$, with amplitude $V/X_L = V/\omega L$ and lagging $V$ by 90°, represents the

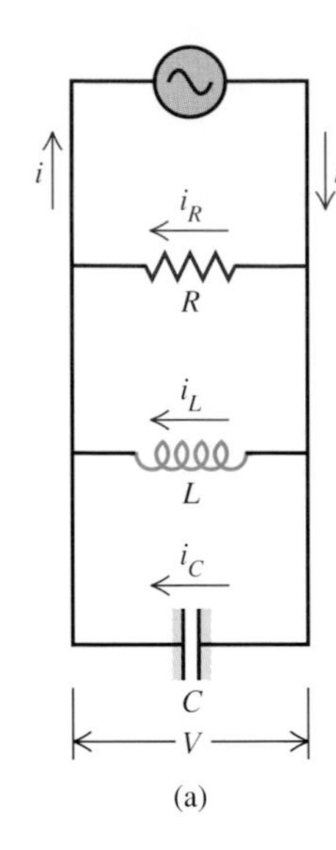


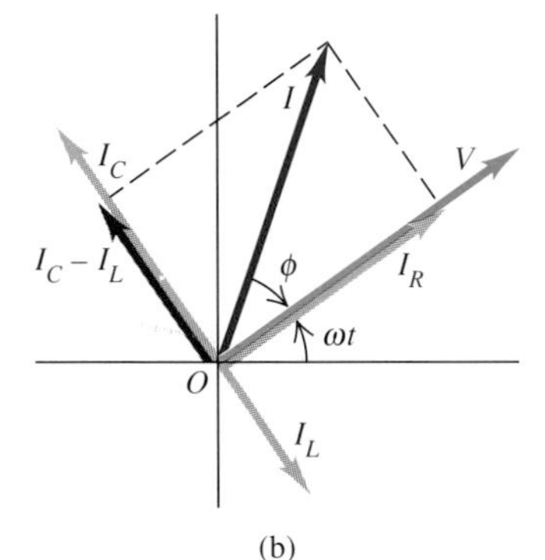


**32–15** (a) A parallel *L-R-C* circuit. (b) Phasor diagram showing current phasors for the three branches. The single voltage phasor *V* represents the voltage across all three branches.

current in the inductor. Phasor $I_C$, with amplitude $V/X_C = V\omega C$ and leading $V$ by 90°, represents the current in the capacitor. Thus the inductor current *lags* the voltage by 90°, the capacitor current *leads* the voltage by 90°, and the two currents differ in phase by 180° (a half-cycle).

From Kirchhoff's junction rule, the instantaneous current $i$ through the source equals the (algebraic) sum of the instantaneous currents $i_R$, $i_L$, and $i_C$. It is represented by the phasor $I$, the *vector* sum of phasors $I_R$, $I_L$, and $I_C$. Angle $\phi$ is the phase angle of the source voltage with respect to the source current, as usual.

From Fig. 32–15 the magnitude $I$ of the current phasor (representing the total current through the source) is

$$I = \sqrt{I_R{}^2 + (I_C - I_L)^2} = \sqrt{\left(\frac{V}{R}\right)^2 + \left(\omega CV - \frac{V}{\omega L}\right)^2}$$
$$= V\sqrt{\frac{1}{R^2} + \left(\omega C - \frac{1}{\omega L}\right)^2}. \qquad (32\text{–}33)$$

The source current amplitude $I$ is frequency-dependent, as expected. It is *minimum* when the second factor in the radical is zero; this occurs when the two reactances have equal magnitudes, at the resonance angular frequency $\omega_0$ given by Eq. (32–32).

Comparing this equation with Eq. (32–23), $V = IZ$, we see that the *impedance* $Z$ of this parallel combination is given by

$$\frac{1}{Z} = \sqrt{\frac{1}{R^2} + \left(\omega C - \frac{1}{\omega L}\right)^2}. \qquad (32\text{–}34)$$

At resonance, $1/Z$ has its smallest value, equal to $1/R$, so $Z$ itself reaches its *maximum* value when $\omega = \omega_0 = (1/LC)^{1/2}$. At this angular frequency, $Z = R$.

For a constant $V$, at resonance, the total current in the *parallel L-R-C* circuit is a *minimum,* but the *series L-R-C* circuit has *maximum* current. In the parallel circuit the currents in $L$ and $C$ are *always* exactly a half-cycle out of phase. When they also have equal *magnitudes,* they cancel each other completely, and the *total* current is simply the current through $R$. Indeed, when $\omega C = 1/\omega L$, Eq. (32–33) becomes simply $I = V/R$. This does *not* mean that there is *no* current in $L$ or $C$ at resonance, but only that the two currents cancel. If $R$ is large, the impedance $Z$ of the circuit near resonance is much *larger* than the individual reactances $X_L$ and $X_C$.

## EXAMPLE 32–8

In the parallel circuit of Fig. 32–15, suppose the elements, applied voltage, and angular frequency have the same values as in Example 32–4, where $R = 300\ \Omega$, $X_L = 600\ \Omega$, and $X_C = 200\ \Omega$. Determine a) the impedance of the parallel combination, b) the current amplitude for each element, and c) the total current amplitude.

**SOLUTION** a) Substituting the values from Example 32–4 into Eq. (32–34), we get

$$\frac{1}{Z} = \sqrt{\frac{1}{(300\ \Omega)^2} + \left(\frac{1}{200\ \Omega} - \frac{1}{600\ \Omega}\right)^2},$$
$$Z = 212\ \Omega.$$

b) For the resistor,

$$I_R = \frac{V}{R} = \frac{50\ \text{V}}{300\ \Omega} = 0.167\ \text{A}.$$

For the inductor,

$$I_L = \frac{V}{X_L} = \frac{50\ \text{V}}{600\ \Omega} = 0.083\ \text{A}.$$

For the capacitor,

$$I_C = \frac{V}{X_C} = \frac{50\ \text{V}}{200\ \Omega} = 0.25\ \text{A}.$$

c) The amplitude of the total current is

$$I = \frac{V}{Z} = \frac{50\ \text{V}}{212\ \Omega} = 0.24\ \text{A},$$

or

$$I = \sqrt{I_R{}^2 + (I_L - I_C)^2} = \sqrt{(0.167\ \text{A})^2 + (0.083\ \text{A} - 0.25\ \text{A})^2}$$
$$= 0.24\ \text{A}.$$

Make sure you understand that you *cannot* simply add the individual current amplitudes to get the amplitude of the total current because of the phase differences of the individual currents.

## 32–7 TRANSFORMERS

One of the great advantages of ac over dc for electric-power distribution is that it is much easier to step voltage levels up and down with ac than with dc. For long-distance power transmission it is desirable to use as high a voltage and as small a current as possible; this reduces $I^2R$ losses in the transmission lines, and smaller wires can be used, saving on material costs. Present-day transmission lines routinely operate at rms voltages of the order of 500 kV. On the other hand, safety considerations and insulation requirements dictate relatively low voltages in generating equipment and in household and industrial power distribution. The standard voltage for household wiring is 120 V in the United States and Canada and 240 V in most of Western Europe. The necessary voltage conversion is accomplished by use of **transformers.**

A transformer consists of two coils, electrically insulated from each other but wound on the same core, so they have mutual inductance, as we discussed in Section 31–1. The winding to which power is supplied is called the **primary;** the winding from which power is delivered is called the **secondary.** Transformers used in power-distribution systems have soft iron cores. The circuit symbol for an iron-core transformer is

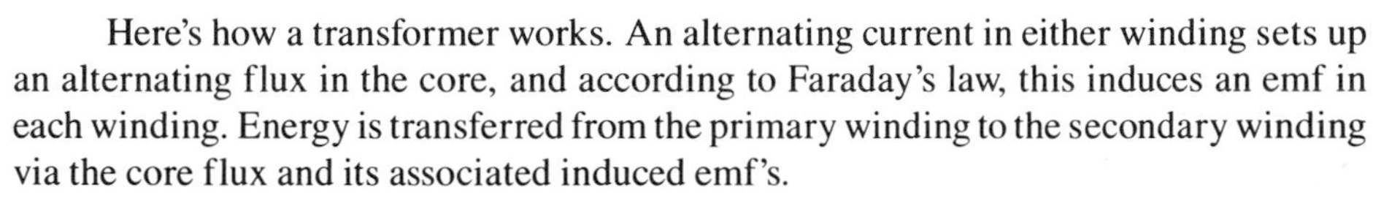

Here's how a transformer works. An alternating current in either winding sets up an alternating flux in the core, and according to Faraday's law, this induces an emf in each winding. Energy is transferred from the primary winding to the secondary winding via the core flux and its associated induced emf's.

An idealized transformer is shown in Fig. 32–16. We assume that all the flux is confined to the iron core, so at any instant the magnetic flux $\Phi_B$ is the same in the primary and secondary coils. We also neglect the resistance of the windings. The primary winding has $N_1$ turns, and the secondary winding has $N_2$ turns. When the magnetic flux

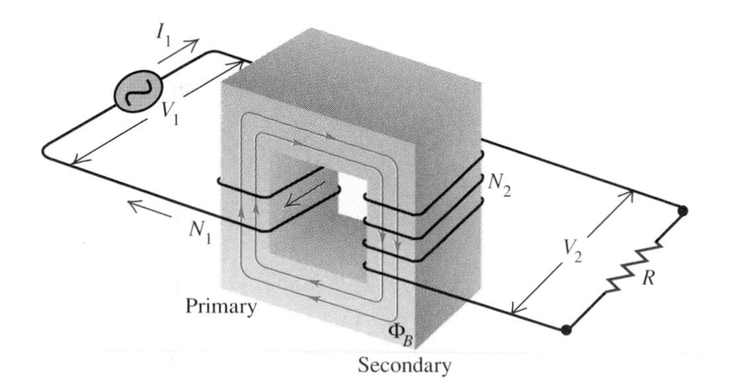


**32–16** Schematic diagram of an idealized step-up transformer with secondary connected to resistor $R$.

changes because of changing currents in the two coils, the resulting induced emf's are

$$\mathcal{E}_1 = -N_1\frac{d\Phi_B}{dt}, \quad \text{and} \quad \mathcal{E}_2 = -N_2\frac{d\Phi_B}{dt}. \tag{32-35}$$

Because the same flux links both primary and secondary, these expressions show that the induced emf *per turn* is the same in each. The ratio of the primary emf $\mathcal{E}_1$ to the secondary emf $\mathcal{E}_2$ is therefore equal to the ratio of primary to secondary turns:

$$\frac{\mathcal{E}_1}{\mathcal{E}_2} = \frac{N_1}{N_2}. \tag{32-36}$$

If the windings have zero resistance, the induced emf's $\mathcal{E}_1$ and $\mathcal{E}_2$ are equal to the corresponding terminal voltages $V_1$ and $V_2$, and

$$\frac{V_2}{V_1} = \frac{N_2}{N_1}. \tag{32-37}$$

(These two $V$'s can both be voltage amplitudes or both rms values.) By choosing the appropriate turns ratio $N_2/N_1$ we may obtain any desired secondary voltage from a given primary voltage. If $V_2 > V_1$, we have a *step-up* transformer; if $V_2 < V_1$, we have a *step-down* transformer.

If the secondary circuit is completed by a resistance $R$, then $I_2 = V_2/R$. From energy considerations the power delivered to the primary equals that taken out of the secondary, so

$$V_1I_1 = V_2I_2. \tag{32-38}$$

We can combine Eqs. (32–37) and (32–38) and the relation $I_2 = V_2/R$ to eliminate $V_2$ and $I_2$:

$$\frac{V_1}{I_1} = \frac{R}{(N_2/N_1)^2}. \tag{32-39}$$

This shows that when the secondary circuit is completed through a resistance $R$, the result is the same as if the *source* had been connected directly to a resistance equal to $R$ divided by the square of the turns ratio, $(N_2/N_1)^2$. In other words, the transformer "transforms" not only voltages and currents, but resistances (more generally, impedances) as well.

## EXAMPLE 32–9

A friend brings back from Europe a device that she claims to be the world's greatest coffee maker. Unfortunately, it was designed to operate from a 240-V line to obtain the 960 W power that it needs. a) What can she do to operate it at 120 V? b) What current will it draw from the 120-V line? c) What is its resistance? (The voltages are rms values.)

**SOLUTION** a) To get $V_2 = 240$ V with $V_1 = 120$ V, our friend needs a step-up transformer with a turns ratio of (240 V)/(120 V) = 2.

b) Assuming an ideal transformer, the power is 960 W = $I_1$(120 V), and $I_1 = 8.0$ A. The secondary current is then 4.0 A.

c) We have $V_1 = 120$ V, $I_1 = 8.0$ A, and $N_2/N_1 = 2$, so

$$\frac{V_1}{I_1} = \frac{120\text{ V}}{8.0\text{ A}} = 15\ \Omega.$$

From Eq. (32–39),

$$R = 2^2(15\ \Omega) = 60\ \Omega.$$

As a check, (240 V)/(60 Ω) = 4.0 A = $I_2$, and $V_2I_2$ = (240 V) × (4.0 A) = 960 W.

Equation (32–39) has many practical consequences. The power supplied by a source to a resistor depends on its resistance and the internal resistance of the source. It can be shown that the power transfer is greatest when the two resistances are *equal*. The same principle applies in both dc and ac circuits. When a high-impedance ac source must be connected to a low-impedance circuit, such as an audio amplifier connected to a loudspeaker, the source impedance can be *matched* to that of the circuit by use of a transformer with an appropriate turns ratio.

Real transformers always have some energy losses. The windings have some resistance, leading to $i^2R$ losses, although superconducting transformers may appear on the horizon in the next few years. There are also energy losses through hysteresis (Section 29–8) and eddy currents (Section 30–6) in the core. Hysteresis losses are minimized by the use of soft iron with a narrow hysteresis loop; and eddy currents are minimized by laminating the core. In spite of these losses, transformer efficiencies are usually well over 90%; in large installations they may reach 99%.

## SUMMARY

- An alternator or ac source produces an emf that varies sinusoidally with time. A sinusoidal voltage or current can be represented by a phasor, a vector that rotates counterclockwise with constant angular velocity $\omega$ equal to the angular frequency of the sinusoidal quantity. Its projection on the horizontal axis at any instant represents the instantaneous value of the quantity.
- For a sinusoidal current the rectified average and rms currents are related to the current amplitude $I$ by

$$I_{rav} = \frac{2}{\pi}I = 0.637I, \qquad (32–3)$$

$$I_{rms} = \frac{I}{\sqrt{2}}. \qquad (32–4)$$

- If the current $i$ in an ac circuit is

$$i = I \cos \omega t$$

and the voltage $v$ between two points is

$$v = V \cos (\omega t + \phi),$$

then $\phi$ is called the phase angle of the voltage relative to the current.

- The voltage across a resistor $R$ is in phase with the current, and the voltage and current amplitudes are related by

$$V = IR. \qquad (32–7)$$

The voltage across an inductor $L$ leads the current by 90°; the voltage and current amplitudes are related by

$$V = IX_L, \qquad (32–13)$$

where $X_L = \omega L$ is the inductive reactance of the inductor. The voltage across a capacitor $C$ lags the current by 90°; the voltage and current amplitudes are related by

$$V = IX_C, \qquad (32–19)$$

where $X_C = 1/\omega C$ is the capacitive reactance of the capacitor.

### KEY TERMS

alternating current
ac source
voltage amplitude
current amplitude
phasor
phasor diagram
rectified average current
root-mean-square (rms) current
phase angle
inductive reactance
capacitive reactance
reactance
impedance
power factor
resonance
resonance angular frequency
resonance frequency
transformer
primary
secondary

• In an $L$-$R$-$C$ series circuit the voltage and current amplitudes are related by

$$V = IZ, \tag{32–23}$$

where $Z$ is the impedance of the circuit:

$$Z = \sqrt{R^2 + [\omega L - (1/\omega C)]^2}. \tag{32–24}$$

The phase angle $\phi$ of the voltage relative to the current is

$$\tan \phi = \frac{\omega L - 1/\omega C}{R}. \tag{32–25}$$

• In an $L$-$R$-$C$ parallel circuit the voltage and current amplitudes are related by

$$V = IZ, \tag{32–23}$$

where $Z$ is the impedance of the circuit:

$$\frac{1}{Z} = \sqrt{\frac{1}{R^2} + \left(\omega C - \frac{1}{\omega L}\right)^2}. \tag{32–34}$$

• The average power input $P$ to an ac circuit is

$$P = \tfrac{1}{2}VI \cos \phi = V_{\text{rms}} I_{\text{rms}} \cos \phi, \tag{32–31}$$

where $\phi$ is the phase angle of voltage with respect to current. The quantity $\cos \phi$ is called the power factor.

• The current in an $L$-$R$-$C$ series circuit becomes maximum, and the impedance becomes minimum, at an angular frequency $\omega_0 = 1/(LC)^{1/2}$ called the resonance angular frequency. This phenomenon is called resonance. At resonance the voltage and current are in phase, and the impedance $Z$ is equal to the resistance $R$. The current in an $L$-$R$-$C$ parallel circuit becomes minimum, and the impedance becomes maximum, at the resonance angular frequency $\omega_0$.

• A transformer is used to transform the voltage and current levels in an ac circuit. In an ideal transformer with no energy losses, if the primary winding has $N_1$ turns and the secondary winding has $N_2$ turns, the two voltages are related by

$$\frac{V_2}{V_1} = \frac{N_2}{N_1}. \tag{32–37}$$

The primary and secondary voltages and currents are related by

$$V_1 I_1 = V_2 I_2. \tag{32–38}$$

# EXERCISES

## Section 32–1 Phasors and Alternating Currents

**32–1** The voltage across the terminals of an ac power supply varies with time according to Eq. (32–1). The voltage amplitude is $V = 50.0$ V. What is a) the average potential difference $V_{\text{av}}$ between the two terminals of the power supply; b) the root-mean-square potential difference $V_{\text{rms}}$?

## Section 32–2 Resistance and Reactance

**32–2** a) Compute the reactance of a 2.00-H inductor at frequencies of 60.0 Hz and 600 Hz. b) Compute the reactance of a 2.00-$\mu$F capacitor at the same frequencies. c) At what frequency is the reactance of a 2.00-H inductor equal to that of a 2.00-$\mu$F capacitor?

**32–3** a) What is the reactance of a 1.00-H inductor at a frequency of 50.0 Hz? b) What is the inductance of an inductor whose reactance is 1.00 Ω at 50.0 Hz? c) What is the reactance of a 1.00-$\mu$F capacitor at a frequency of 50.0 Hz? d) What is the capacitance of a capacitor whose reactance is 1.00 Ω at 50.0 Hz?

**32–4** A 4.00-$\mu$F capacitor is connected across an ac source whose voltage amplitude is kept constant at 50.0 V but whose frequency can be varied. Find the current amplitude when the angular frequency is a) 100 rad/s; b) 1000 rad/s; c) 10,000 rad/s. d) Show the results of parts (a) through (c) in a plot of log $I$ versus log $\omega$.

**32–5** An inductor with a self-inductance of 4.00 H and with negligible resistance is connected across the source in Exercise 32–4. Find the current amplitude when the angular frequency is a) 100 rad/s; b) 1000 rad/s; c) 10,000 rad/s. d) Show the results of parts (a) through (c) in a plot of log $I$ versus log $\omega$.

**32–6 A Radio Inductor.** You are designing a circuit for a radio receiver. In part of the circuit you want the current amplitude through a 0.600-H inductor to be 4.00 mA when a sinusoidal voltage with amplitude $V = 12.0$ V is applied across the inductor. What angular frequency is required?

**32–7 Kitchen Capacitance.** The wiring for a refrigerator contains a starter capacitor. A voltage amplitude of 170 V at a frequency of $f = 60.0$ Hz applied across the capacitor is to produce a current of 1.20 A through the capacitor. What capacitance $C$ is required?

## Section 32–3
## The *L-R-C* Series Circuit

**32–8** You have a resistor $R = 300\ \Omega$, an inductor $L = 0.600$ H, and a capacitor $C = 4.00\ \mu$F. Suppose you take the resistor and inductor and make a series circuit with a voltage source that has voltage amplitude 50.0 V and angular frequency 100 rad/s. a) What is the impedance of the circuit? b) What is the current amplitude? c) What are the voltage amplitudes across the resistor and across the inductor? d) What is the phase angle $\phi$ of the source voltage with respect to the current? Does the source voltage lag or lead the current? e) Construct the phasor diagram.

**32–9** Repeat Exercise 32–8 with the circuit consisting of only the resistor and the capacitor in series. For part (c), calculate the voltage amplitudes across the resistor and across the capacitor.

**32–10** Repeat Exercise 32–8 with the circuit consisting of only the capacitor and the inductor in series. For part (c), calculate the voltage amplitudes across the capacitor and across the inductor.

**32–11** A 300-Ω resistor is in series with a 0.100-H inductor and a 0.500-$\mu$F capacitor. Compute the impedance of the circuit and draw the phasor diagram a) at a frequency of 500 Hz; b) at a frequency of 1000 Hz. Compute, in each case, the phase angle of the source voltage with respect to the current, and state whether the source voltage lags or leads the current.

**32–12** a) Compute the impedance of an *L-R-C* series circuit at angular frequencies of 1000, 750, and 500 rad/s. Take $R = 400\ \Omega$, $L = 0.900$ H, and $C = 2.00\ \mu$F. b) Describe how the current amplitude varies as the frequency of the source is slowly reduced from 1000 rad/s to 500 rad/s. c) What is the phase angle of the source voltage with respect to the current when $\omega = 1000$ rad/s? d) Construct the phasor diagram when $\omega = 1000$ rad/s. e) Repeat parts (c) and (d) for $\omega = 500$ rad/s.

## Section 32–4
## Power in Alternating-Current Circuits

**32–13** The circuit in Exercise 32–11 carries an rms current of 0.250 A with a frequency of 1000 Hz. a) What average power is delivered by the source? b) What is the average rate at which electrical energy is dissipated in the resistor? c) What is the average rate at which electrical energy is dissipated (converted to other forms) in the capacitor? d) In the inductor? e) Compare the result of (a) to the sum of those of (b), (c), and (d).

**32–14 Power to a Large Inductor.** A large electromagnet coil with resistance $R = 300\ \Omega$ and inductance $L = 5.00$ H is connected across the terminals of a source that has voltage amplitude $V = 180$ V and angular frequency $\omega = 50.0$ rad/s. a) What is the power factor? b) What is the average power delivered by the source?

**32–15** a) Show that for an *L-R-C* series circuit the power factor is equal to $R/Z$. (*Hint:* Use the phasor diagram.) b) Show that for any ac circuit, not just one containing pure resistance only, the average power delivered by the voltage source is given by $P = I_{\mathrm{rms}}^2 R$.

**32–16** An *L-R-C* series circuit has a resistance of 90.0 Ω and an impedance of 150 Ω. The circuit is connected to a voltage source that has $V_{\mathrm{rms}} = 160$ V. What average power is delivered to the circuit by the source?

## Section 32–5
## Series Resonance

**32–17** Consider the *L-R-C* series circuit of Exercise 32–12. The voltage amplitude of the source is 110 V. a) At what angular frequency is the circuit in resonance? b) Sketch the phasor diagram at the resonance frequency. c) What is the reading of each voltmeter in Fig. 32–17 when the source frequency equals the

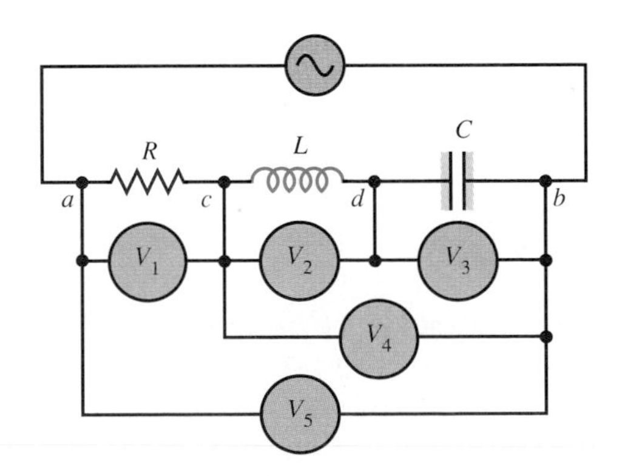

**FIGURE 32–17**

onance frequency? The voltmeters are calibrated to read rms voltages. d) What is the resonance angular frequency if the resistance is reduced to 100 Ω? e) What is the rms current at resonance?

**32–18** In an *L-R-C* series circuit, $L = 0.400$ H and $C = 8.00 \times 10^{-5}$ F. The voltage amplitude of the source is 240 V. a) What is the resonance angular frequency of the circuit? b) When the source operates at the resonance angular frequency, the current amplitude in the circuit is 0.500 A. What is the resistance $R$ of the resistor? c) At the resonance frequency, what are the peak voltages across the inductor, the capacitor, and the resistor?

**32–19** In an *L-R-C* series circuit, $R = 200\ \Omega$, $L = 0.800$ H, and $C = 0.0400\ \mu$F. The source has voltage amplitude $V = 300$ V and a frequency equal to the resonance frequency of the circuit. a) What is the power factor? b) What is the average power delivered by the source? c) The capacitor is replaced by one with $C = 0.0800\ \mu$F, and the source frequency is adjusted to the new resonance value. What is the average power delivered by the source?

**32–20** In an *L-R-C* series circuit, $R = 250\ \Omega$, $L = 0.400$ H, and $C = 0.0200\ \mu$F. a) What is the resonance angular frequency of the circuit? b) The capacitor can withstand a peak voltage of 350 V. If the voltage source operates at the resonance frequency, what maximum voltage amplitude can it have if the maximum capacitor voltage is not exceeded?

## Section 32–6 Parallel Resonance

**32–21** For the circuit of Fig. 32–15a, $R = 400\ \Omega$, $L = 0.500$ H, and $C = 0.600\ \mu$F. The voltage amplitude of the source is 120 V. a) What is the resonance frequency of the circuit? b) Sketch the phasor diagram at the resonance frequency. c) At the resonance frequency, what is the current amplitude through the source? d) At the resonance frequency, what is the current amplitude through the resistor, through the inductor, and through the branch containing the capacitor?

**32–22** For the circuit of Fig. 32–15a, $R = 200\ \Omega$, $L = 0.800$ H, and $C = 5.00\ \mu$F. When the source is operated at the resonance frequency, the current amplitude in the inductor is 0.600 A. Determine a) the current amplitude in the branch containing the capacitor; b) the current amplitude through the resistor.

**32–23** For the circuit of Fig. 32–15a, $R = 500\ \Omega$, $L = 0.200$ H, and $C = 0.200\ \mu$F. The source has voltage amplitude $V = 300$ V. a) If the source is operated at an angular frequency of 500 rad/s, what are the impedance of the circuit and the current amplitude through the source? b) If the source is operated at the resonance angular frequency, what are the impedance of the circuit and the current amplitude through the source?

**32–24** Consider the circuit of Fig. 32–15a, with the same numerical values as in Exercise 32–21. At resonance, determine a) the average rate at which electrical energy is being delivered by the source; b) the average rate at which electrical energy is being converted to heat in the resistor. Compare this to the result of part (a). c) Is the current through the inductor, and thus the energy stored in its magnetic field, zero at all times? If not, how can the result obtained in part (b) be explained? d) Calculate the maximum energy stored in the inductor at the resonance frequency. e) Calculate the maximum energy stored in the capacitor at the resonance frequency.

## Section 32–7 Transformers

**32–25 A Step-Down Transformer.** A transformer connected to a 120-V (rms) ac line is to supply 8.00 V (rms) to a low-voltage lighting system for a model-railroad village. The total equivalent resistance of the system is 4.00 Ω. a) What should be the ratio of primary to secondary turns of the transformer? b) What rms current must the secondary supply? c) What average power is delivered to the load? d) What resistance connected directly across the 120-V line would draw the same power as the transformer? Show that this is equal to 4.00 Ω times the square of the ratio of primary to secondary turns.

**32–26 A Step-Up Transformer.** A transformer connected to a 120-V (rms) ac line is to supply 16,800 V (rms) for a neon sign. To reduce shock hazard, a fuse is to be inserted in the primary circuit; the fuse is to blow when the current in the secondary circuit exceeds 10.0 mA. a) What is the ratio of secondary to primary turns of the transformer? b) What power must be supplied to the transformer when the secondary current is 10.0 mA? c) What current rating should the fuse in the primary circuit have?

**32–27** The internal resistance of an ac source is 9000 Ω. a) What should be the ratio of primary to secondary turns of a transformer to match the source to a load with a resistance of 10.0 Ω? (Matching means that the effective load resistance equals the internal resistance of the source. Refer to Exercise 32–25.) b) If the voltage amplitude of the source is 100 V, what is the voltage amplitude in the secondary circuit under open circuit conditions?

# PROBLEMS

**32–28** At a frequency $\omega_1$ the reactance of a certain capacitor equals that of a certain inductor. a) If the frequency is changed to $\omega_2 = 3\omega_1$, what is the ratio of the reactance of the inductor to that of the capacitor? Which reactance is larger? b) If the frequency is changed to $\omega_3 = \omega_1/2$, what is the ratio of the reactance of the inductor to that of the capacitor? Which reactance is larger?

**32–29** A coil has a resistance of 40.0 Ω. At a frequency of 100 Hz the voltage across the coil leads the current in it by 30.0°. Determine the inductance of the coil.

**32–30** Five infinite-impedance voltmeters, calibrated to read rms values, are connected as shown in Fig. 32–17. Take *R, L, C,* and $V$ as given in Exercise 32–8. What is the reading of each voltmeter if a) $\omega = 500$ rad/s; b) $\omega = 1000$ rad/s?

**32–31** Consider the circuit sketched in Fig. 32–17. The source has a voltage amplitude of 240 V, $R = 150\ \Omega$, and the reactance of the capacitor is 600 Ω. The voltage amplitude across the capacitor is 720 V. a) What is the current amplitude in the circuit? b) What is the impedance? c) What two values can the reactance of the inductor have?

**32–32** A large electromagnet coil is connected to an ac source that has frequency $f = 60.0$ Hz. The coil has resistance $R = 500$ Ω, and at this source frequency the coil has inductive reactance $X_L = 300\ \Omega$. What must be the rms voltage of the source if the coil is to consume an average electrical power of 400 W?

**32–33** An inductor having a reactance of 25.0 Ω and resistance $R$ gives off heat at the rate of 12.0 J/s when it carries a current of 0.500 A (rms). What is the impedance of the inductor?

**32–34** A circuit draws 280 W from a 110-V (rms), 60.0-Hz ac line. The power factor is 0.600, and the source voltage leads the current. a) What is the net resistance $R$ of the circuit? b) Find the capacitance of the series capacitor that will result in a power factor of unity when it is added to the original circuit. c) What power will then be drawn from the supply line?

**32–35** A series circuit has an impedance of 50.0 Ω and a power factor of 0.400 at 60.0 Hz. The source voltage lags the current. a) Should an inductor or a capacitor be placed in series with the circuit to raise its power factor? b) What size element will raise the power factor to unity?

**32–36** In an $L$-$R$-$C$ series circuit, $R = 300\ \Omega$, $X_C = 400\ \Omega$, and $X_L = 500\ \Omega$. The average power consumed in the resistor is 60.0 W. a) What is the power factor of the circuit? b) What is the rms voltage of the source?

**32–37** In an $L$-$R$-$C$ series circuit the phase angle is 40.0°, with the source voltage leading the current. The reactance of the capacitor is $X_C = 400\ \Omega$, and the resistor resistance is $R = 200\ \Omega$. The average power delivered by the source is 210 W. Find a) the reactance of the inductor; b) the rms current; c) the rms voltage of the source.

**32–38** A resistor with a resistance of 500 Ω and a capacitor with a capacitance of 2.00 μF are connected in parallel to an ac generator that supplies an rms voltage of 260 V at an angular frequency of 377 rad/s. Find a) the current amplitude in the resistor; b) the current amplitude in the capacitor; c) the phase angle (sketch the phasor diagram for the current phasors); d) the amplitude of the current through the generator.

**32–39 A High-Pass Filter.** One application of $L$-$R$-$C$ series circuits is to high-pass or high-pass filters, which filter out either the low- or high-frequency components of a signal. A high-pass filter is shown in Fig. 32–18, where the output voltage is taken across the $L$-$R$ combination. (The $L$-$R$ combination represents an inductive coil that also has resistance due to the large length of wire in the coil.) Derive an expression for $V_{out}/V_s$, the ratio of the output and source voltage amplitudes, as a function of the angular frequency $\omega$ of the source. Show that when $\omega$ is small, this ratio is proportional to $\omega$ and thus is small and that the ratio approaches unity in the limit of large frequency.

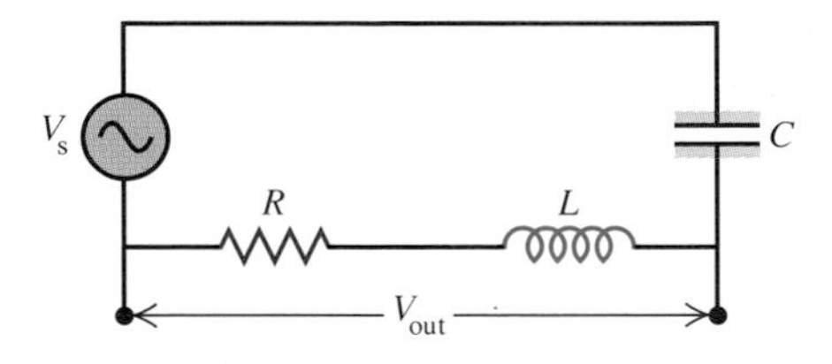


**FIGURE 32–18**

**32–40 A Low-Pass Filter.** Figure 32–19 shows a low-pass filter (see Problem 32–39); the output voltage is taken across the capacitor in an $L$-$R$-$C$ series circuit. Derive an expression for $V_{out}/V_s$, the ratio of the output and source voltage amplitudes, as a function of the angular frequency $\omega$ of the source. Show that when $\omega$ is large this ratio is proportional to $\omega$ and thus is very small and that the ratio approaches unity in the limit of small frequency.

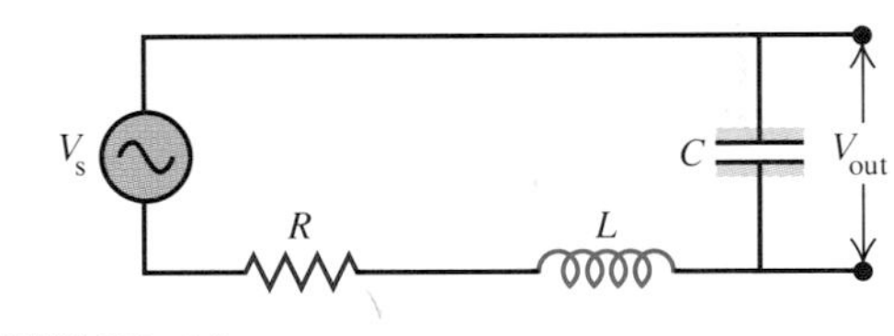


**FIGURE 32–19**

**32–41** A 100-Ω resistor, a 0.100-μF capacitor, and a 0.100-H inductor are connected in *parallel* to a voltage source with an amplitude of 160 V. a) What is the resonance angular frequency? b) What is the maximum current through the source at the resonance frequency? c) What is the maximum current in the resistor at resonance? d) What is the maximum current in the inductor at resonance? e) What is the maximum current in the branch containing the capacitor at resonance? f) What is the maximum energy stored in the inductor at resonance? In the capacitor?

**32–42** The same three components as in Problem 32–41 are connected in *series* to a voltage source with an amplitude of 160 V. a) What is the resonance angular frequency? b) What is the maximum current in the resistor at resonance? c) What is the maximum voltage across the capacitor at resonance? d) What is the maximum voltage across the inductor at resonance? e) What is the maximum energy stored in the capacitor at resonance and in the inductor at resonance?

**32–43** Consider the same circuit as in Problem 32–41 with the source operated at an angular frequency of 400 rad/s. a) What is the maximum current through the source? b) What is the maximum current in the resistor? c) What is the maximum current in the inductor? d) What is the maximum current in the branch containing the capacitor? e) What is the maximum energy stored in the inductor and in the capacitor?

**32–44** Consider the same circuit as in Problem 32–42 with the source operated at an angular frequency of 400 rad/s. a) What is the maximum current in the resistor? b) What is the maximum voltage across the capacitor? c) What is the maximum voltage across the inductor? d) What is the maximum energy stored in the capacitor and in the inductor?

**32–45** The current in a certain circuit varies with time as shown in Fig. 32–20. Find the average current and the rms current in terms of $I_0$.

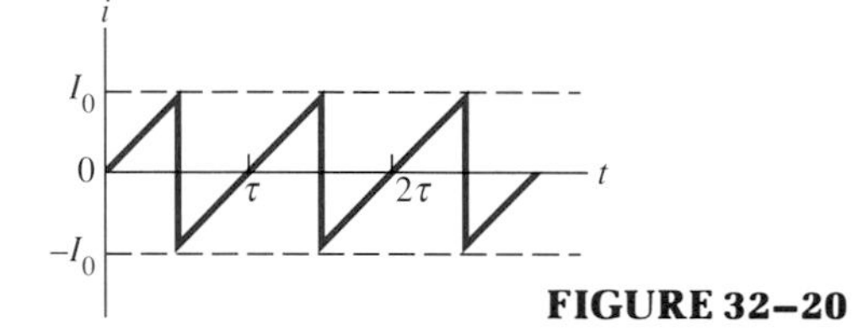


**FIGURE 32–20**

**32–46 Resonance Width.** Consider a series circuit with a source with terminal rms voltage $V_{\text{rms}} = 120$ V and frequency $\omega$. The inductor has $L = 1.80$ H, the capacitor has $C = 0.800$ $\mu$F, and the resistor has $R = 400$ $\Omega$. a) What is the resonance angular frequency $\omega_0$ of the circuit? b) What is the rms current through the source at resonance? c) For what two values of the angular frequency, $\omega_1$ and $\omega_2$, is the current half of the resonance value? d) The quantity $|\omega_1 - \omega_2|$ defines the *width* of the resonance. Calculate the resonance width for $R = 4.00$ $\Omega$, 40.0 $\Omega$, and 400 $\Omega$.

## CHALLENGE PROBLEMS

**32–47** a) At what angular frequency is the voltage amplitude across the *resistor* in an *L-R-C* series circuit at maximum value? b) At what angular frequency is the voltage amplitude across the *inductor* at maximum value? c) At what angular frequency is the voltage amplitude across the *capacitor* at maximum value?

**32–48 Complex Numbers in a Circuit.** The voltage drop across a circuit element in an ac circuit is not necessarily in phase with the current through that circuit element. Therefore the voltage amplitudes across the circuit elements in a branch in an ac circuit do not add algebraically. A method that is commonly employed to simplify the analysis of an ac circuit driven by a sinusoidal source is to represent the impedance $Z$ as a complex number rather than as a real number. The resistance $R$ is taken to be the real part of the impedance, and the reactance $X = X_L - X_C$ is taken to be the imaginary part. Thus, for a branch containing a resistor, inductor, and capacitor in series, the complex impedance would be given by $Z_{\text{cpx}} = R + iX$, where $i^2 = -1$. If the voltage amplitude across the branch is $V_{\text{cpx}}$ and the current amplitude is $I_{\text{cpx}}$, these quantities are related by $I_{\text{cpx}} = V_{\text{cpx}}/Z_{\text{cpx}}$. The actual current amplitude would be computed by taking the absolute value of the complex representation of the current amplitude, that is, $I = (I_{\text{cpx}}{}^* I_{\text{cpx}})^{1/2}$. The phase angle $\phi$ of the current with respect to the source voltage is given by $\tan(\phi) = \text{Im}(I_{\text{cpx}})/\text{Re}(I_{\text{cpx}})$. The voltage amplitudes $V_{R\text{cpx}}$, $V_{L\text{cpx}}$, and $V_{C\text{cpx}}$ across the resistance, inductance, and capacitance, respectively, are found by multiplying $I_{\text{cpx}}$ by $R$, $iX_L$, and $-iX_C$, respectively. If we use the complex representation for the voltage amplitudes, the voltage drop across a branch can be found by algebraic addition of the voltage drops across each circuit element, that is, $V_{\text{cpx}} = V_{R\text{cpx}} + V_{L\text{cpx}} + V_{C\text{cpx}}$. When we want to find the actual value of any current amplitude or voltage amplitude, the absolute value of the appropriate quantity is used.

Consider the series *L-R-C* circuit shown in Fig. 32–21. The angular frequency $\omega$ is 1000 rad/s, and the voltage amplitude of the source is 200 V. The values of the circuit elements are as shown. Use the phasor diagram techniques presented in Section 32–3 to solve for a) the current amplitude; b) the phase angle $\phi$ of the current with respect to the source voltage. (Note that this angle is the negative of the phase angle defined in Fig. 32–8.) Now apply the procedure outlined above to the same circuit. c) Determine the impedance of the circuit, $Z_{\text{cpx}}$, as represented by a complex number. Take the absolute value to obtain $Z$, the actual impedance of the circuit. d) Take the voltage amplitude of the source, $V_{\text{cpx}}$, to be real, and find the complex current amplitude $I_{\text{cpx}}$. Find the actual current amplitude by taking the absolute value of $I_{\text{cpx}}$. e) Find the phase angle $\phi$ of the current with respect to the source voltage by using the real and imaginary parts of $I_{\text{cpx}}$, as explained above. f) Find the complex representations of the voltages across the resistance, the inductance, and the capacitance. g) Adding the answers found in part (f), verify that the sum of these complex numbers is real and equal to 200 V, the voltage of the source.

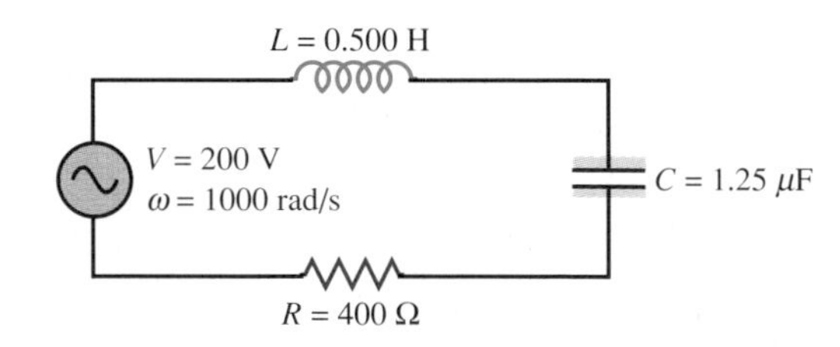


**FIGURE 32–21**

CHAPTER 33

# Electromagnetic Waves

The control tower at an airport relies on radar to indicate the exact position of aircraft and on radio to communicate with pilots. Both radar and radio are forms of electromagnetic waves.

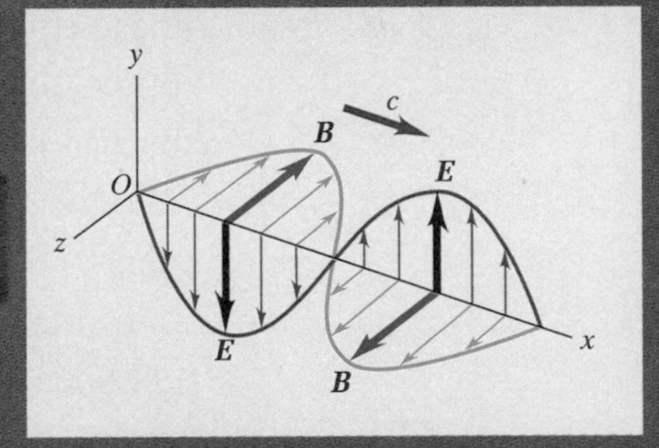


Electromagnetic waves consist of electric and magnetic fields propagating together through space. At points far from the source, the fields are perpendicular to each other and to the direction of propagation. Different kinds of electromagnetic waves have different wavelengths (or frequencies), but they all consist of perpendicular $E$ and $B$ fields and they all travel in vacuum with the same speed—the speed of light, $c$.

## KEY CONCEPTS

• **Maxwell's equations show that electric and magnetic fields that vary with time are interrelated. These equations predict the existence of electromagnetic waves that travel through space with definite speed, equal to the speed of light.**

• **In a plane electromagnetic wave, the electric and magnetic fields at each instant are uniform over any plane perpendicular to the direction of propagation. The simplest plane waves are sinusoidal waves, in which the *E* and *B* fields are sinusoidal functions of the spatial coordinates and time.**

• **Electromagnetic waves transmit energy and momentum from one region to another. The intensity of a wave is the average rate of energy transfer per unit area.**

• **The speed of electromagnetic waves in matter depends on the permittivity and permeability of the material.**

• **The superposition of two electromagnetic waves traveling in opposite directions can produce a standing wave, with nodes and antinodes, analogous to standing waves in mechanical systems.**

## INTRODUCTION

Energy from the sun, an essential requirement for life on earth, reaches us by means of electromagnetic waves that travel through 93 million miles of (nearly) empty space. What happens in that space as the waves pass through? What do these waves have in common with electromagnetic waves from other familiar sources such as TV and radio stations, microwave oscillators for ovens and radar, light bulbs, x-ray machines, and radioactive nuclei? Electromagnetic waves occur in an astonishing variety of physical situations, so it is important for us to make a careful study of their properties and behavior.

We have seen the role of Maxwell's equations in unifying electricity and magnetism into the single discipline of *electromagnetism.* These equations show that a time-varying magnetic field acts as a source of electric field and that a time-varying electric field acts as a source of magnetic field. Maxwell's equations give us the theoretical basis for understanding electromagnetic waves. These waves carry energy and momentum and have the property of polarization. In sinusoidal electromagnetic waves, the ***E*** and ***B*** fields are sinusoidal functions of time. The spectrum of electromagnetic waves covers an extremely broad range of frequency and wavelength. Light consists of electromagnetic waves, and our study of optics in the following chapters will be based in part on the electromagnetic nature of light.

# 33–1 INTRODUCTION TO ELECTROMAGNETIC WAVES

In the last several chapters we have studied various aspects of electric and magnetic fields. We've learned that when the fields don't vary with time, such as an electric field produced by charges at rest or the magnetic field of a steady current, we can analyze the electric and magnetic fields independently without considering interactions between them. But when the fields vary with time, they are no longer independent. Faraday's law (Section 30–2) tells us that a time-varying magnetic field acts as a source of electric field, as shown by induced emf's in inductances and transformers. And Ampere's law, including the displacement current discovered by James Clerk Maxwell (Section 29–9), shows that a time-varying electric field acts as a source of magnetic field. This mutual interaction between the two fields is summarized neatly in Maxwell's equations, presented in Section 30–7.

Thus, when *either* an electric or a magnetic field is changing with time, a field of the other kind is induced in adjacent regions of space. We are led (as Maxwell was) to consider the possibility of an electromagnetic disturbance, consisting of time-varying electric and magnetic fields, that can propagate through space from one region to another, even when there is no matter in the intervening region. Such a disturbance, if it exists, will have the properties of a *wave,* and an appropriate term is **electromagnetic wave.**

Such waves do exist; radio and television transmission, light, x rays, and many other kinds of radiation are examples of electromagnetic waves. In this chapter we will show how the existence of such waves is related to the principles of electromagnetism that we have studied thus far, and we will examine the properties of these waves.

As so often happens in the development of science, the theoretical understanding of electromagnetic waves originally took a considerably more devious path than the one just outlined. In the early days of electromagnetic theory (the early nineteenth century), two different units of electric charge were used, one for electrostatics and the other for magnetic phenomena involving currents. In the system of units that was used at that time, these two units of charge had different physical dimensions. Their *ratio* had units of velocity, and measurements showed that the ratio had a numerical value that was precisely equal to the speed of light, $3.00 \times 10^8$ m/s. At the time, physicists regarded this as an extraordinary coincidence and had no idea how to explain it.

During his search for understanding of this result, Maxwell proved in 1865 that an electromagnetic disturbance should propagate in free space with a speed equal to that of light and that light waves were therefore very likely to be electromagnetic in nature. At the same time he discovered that the basic principles of electromagnetism can be expressed in terms of the four equations that we now call **Maxwell's equations.** To review our discussion of Section 30–7, these four relationships are (1) Gauss's law for electric fields; (2) Gauss's law for magnetic fields, showing the absence of magnetic monopoles; (3) Ampere's law, including displacement current; and (4) Faraday's law. Maxwell's equations, in the integral form used in this text, are

$$\oint \boldsymbol{E} \cdot d\boldsymbol{A} = \frac{Q_{\text{encl}}}{\epsilon_0} \quad \text{(Gauss's law)}, \tag{23–8}$$

$$\oint \boldsymbol{B} \cdot d\boldsymbol{A} = 0 \quad \text{(Gauss's law for magnetism)}, \tag{28–8}$$

$$\oint \boldsymbol{B} \cdot d\boldsymbol{l} = \mu_0\left(I_C + \epsilon_0 \frac{d\Phi_E}{dt}\right) \quad \text{(Ampere's law)}, \tag{29–34}$$

$$\oint \boldsymbol{E} \cdot d\boldsymbol{l} = -\frac{d\Phi_B}{dt} \quad \text{(Faraday's law)}. \tag{30–9}$$

These equations apply to electric and magnetic fields *in vacuum.* If a material is present, the permittivity $\epsilon_0$ and permeability $\mu_0$ of free space are replaced by the permittivity $\epsilon$ and permeability $\mu$ of the material. If the regions of integration include several materials with different values of $\epsilon$ and $\mu$, then $\epsilon$ and $\mu$ have to be transferred to the left sides of Eqs. (23–8) and (29–34) and placed inside the integrals. The $\epsilon$ in Eq. 29–34 has to be included in the integral that gives $d\Phi_E/dt$.

In 1887, Heinrich Hertz produced in the laboratory, for the first time, electromagnetic waves with macroscopic wavelengths. To produce the waves, he used oscillating *L-C* circuits of the sort discussed in Section 31–5; he detected the waves with other circuits tuned to the same frequency. Hertz also produced electromagnetic *standing waves* and measured the distance between adjacent nodes (one half-wavelength) to determine the wavelength. Knowing the resonant frequency of his circuits, he then found the speed of the waves from the wavelength-frequency relation $v = \lambda f$. He established that their speed was the same as that of light; this verified Maxwell's theoretical prediction directly. The SI unit of frequency, one cycle per second, is named the *hertz* in honor of Hertz.

The possible use of electromagnetic waves for long-distance communication does not seem to have occurred to Hertz. It remained for the enthusiasm and energy of Marconi and others to make radio communication a familiar household experience.

## 33–2 SPEED OF AN ELECTROMAGNETIC WAVE

We are now ready to develop the basic ideas of electromagnetic waves and their relation to the principles of electromagnetism as summarized in Maxwell's equations. Our procedure will be to postulate a simple field configuration that has wavelike behavior. We'll assume an electric field that has only a $y$-component and a magnetic field with only a $z$-component, and we'll assume that both fields move together in the $+x$-direction with a speed $c$ that is initially unknown. Then we will test whether these fields are physically possible by asking whether they are consistent with Maxwell's equations, particularly Ampere's law and Faraday's law. We'll also show that the *wave equation,* which we encountered during our study of mechanical waves in Chapter 19, can be derived from Maxwell's equations.

### A Simple Plane Electromagnetic Wave

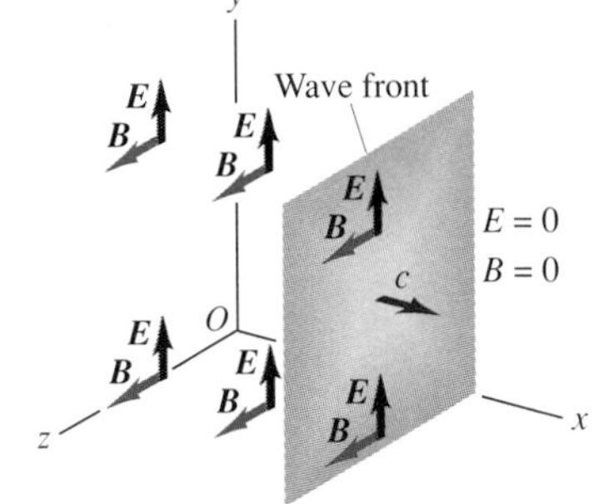


**33–1** An electromagnetic wave front. The ***E*** and ***B*** fields are uniform over the region to the left of the plane but are zero everywhere to the right of it. The plane representing the wave front moves to the right with speed $c$.

Using an $xyz$-coordinate system (Fig. 33–1), we imagine that all space is divided into two regions by using a plane perpendicular to the $x$-axis (parallel to the $yz$-plane). At every point to the left of this plane there are a uniform electric field $\boldsymbol{E}$ in the $+y$-direction and a uniform magnetic field $\boldsymbol{B}$ in the $+z$-direction, as shown. Furthermore, we suppose that the boundary plane, which we call the *wave front,* moves to the right with a constant speed $c$, as yet unknown. Thus the $\boldsymbol{E}$ and $\boldsymbol{B}$ fields travel to the right into previously field-free regions with a definite speed. The situation, in short, describes a rudimentary electromagnetic wave.

We won't concern ourselves with the problem of actually *producing* such a field configuration. Instead, we simply ask whether it is consistent with the laws of electromagnetism, that is, with Maxwell's equations. First we apply Faraday's law,

$$\oint \boldsymbol{E} \cdot d\boldsymbol{l} = -\frac{d\Phi_B}{dt}, \tag{33–1}$$

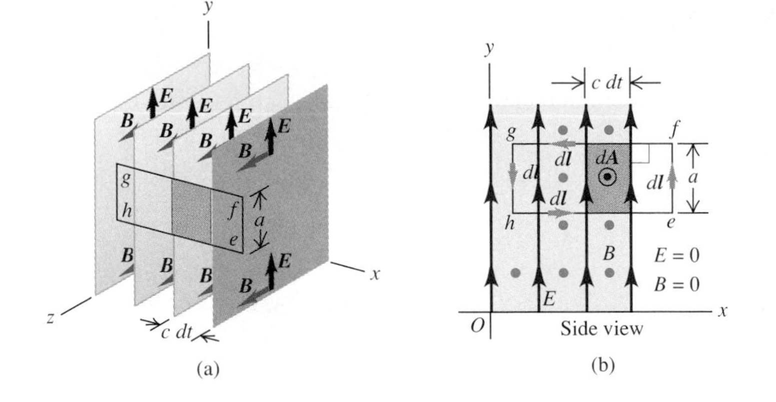


**33–2** In time $dt$ the wave front moves to the right a distance $c\,dt$. The magnetic flux through the rectangle in the $xy$-plane increases by an amount $d\Phi_B$ equal to the flux through the shaded rectangle with area $ac\,dt$; that is, $d\Phi_B = Bac\,dt$.

to the rectangle $efgh$ in the $xy$-plane (Fig. 33–2a). Figure 33–2b shows a cross section in the $xy$-plane; the magnetic field is perpendicular to this plane. The rectangle has height $a$ and width $l$. At the time shown, the wave front has progressed partway through the rectangle, and $\boldsymbol{E}$ is zero along the side $ef$. During the time interval $dt$ the wave front moves a distance $c\,dt$ to the right, sweeping out an area $ac\,dt$ of the rectangle. In this time the magnetic flux $\Phi_B$ through the rectangle increases by $d\Phi_B = B(ac\,dt)$, so the rate of change of magnetic flux is

$$\frac{d\Phi_B}{dt} = Bac. \tag{33–2}$$

In applying Faraday's law we take the vector area $d\boldsymbol{A}$ of rectangle $efgh$ to be in the $+z$-direction. With this choice, the change in magnetic flux is positive, and the right-hand rule requires that we integrate $\boldsymbol{E} \cdot d\boldsymbol{l}$ *counterclockwise* around the rectangle. At every point on side $ef$, $\boldsymbol{E}$ is zero. At every point on sides $fg$ and $he$, $\boldsymbol{E}$ is either zero or perpendicular to $d\boldsymbol{l}$. Only side $gh$ contributes to the integral. On this side, $\boldsymbol{E}$ and $d\boldsymbol{l}$ are opposite, and we obtain

$$\oint \boldsymbol{E} \cdot d\boldsymbol{l} = -Ea. \tag{33–3}$$

Now we substitute Eqs. (33–2) and (33–3) into Faraday's law, Eq. (33–1); we get

$$-Ea = -Bac,$$

$$E = cB. \tag{33–4}$$

This shows that the wave that we have assumed is possible (that is, consistent with Faraday's law) only if the wave speed $c$ and the magnitudes of the perpendicular vectors $\boldsymbol{E}$ and $\boldsymbol{B}$ are related as in Eq. (33–4).

Next we carry out a similar calculation using Ampere's law. There is no conduction current ($I_C = 0$), and Ampere's law is

$$\oint \boldsymbol{B} \cdot d\boldsymbol{l} = \mu_0 \epsilon_0 \frac{d\Phi_E}{dt}. \tag{33–5}$$

We move our rectangle so that it lies parallel to the $xz$-plane, as shown in Fig. 33–3, and we again look at the situation at a time when the wave front has traveled partway through the rectangle. In a time interval $dt$ the electric flux $\Phi_E$ through the rectangle increases by $d\Phi_E = Eac\,dt$. We take the vector area $d\boldsymbol{A}$ in the $+y$-direction, so this flux change is positive. The rate of change of electric flux is

$$\frac{d\Phi_E}{dt} = Eac. \tag{33–6}$$

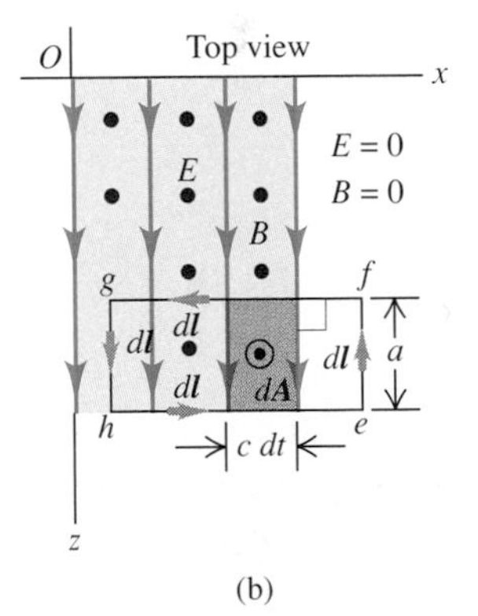


**33–3** In time $dt$ the electric flux through the rectangle in the $xz$-plane increases by an amount equal to $E$ times the area $ac\,dt$ of the shaded rectangle; that is, $d\Phi_E = Eac\,dt$. Thus, $d\Phi_E/dt = Eac$.

The right-hand rule requires that we integrate $\boldsymbol{B} \cdot d\boldsymbol{l}$ counterclockwise around the rectangle. The $\boldsymbol{B}$ field is zero at every point along side $ef$, and at each point on sides $fg$ and $he$ it is either zero or perpendicular to $d\boldsymbol{l}$. Only side $gh$, where $\boldsymbol{B}$ and $d\boldsymbol{l}$ are parallel, contributes to the integral, and we find

$$\oint \boldsymbol{B} \cdot d\boldsymbol{l} = Ba. \tag{33–7}$$

Substituting Eqs. (33–6) and (33–7) into Ampere's law, Eq. (33–5), we find

$$Ba = \epsilon_0 \mu_0 Eac,$$

$$B = \epsilon_0 \mu_0 cE. \tag{33–8}$$

Thus our assumed wave obeys Ampere's law only if $B$, $c$, and $E$ are related as in Eq. (33–8).

For *both* Ampere's law and Faraday's law to be obeyed at the same time, Eqs. (33–4) and (33–8) must both be satisfied. This can happen only when $\epsilon_0 \mu_0 c = 1/c$, or

$$c = \frac{1}{\sqrt{\epsilon_0 \mu_0}}. \tag{33–9}$$

Inserting the numerical values of these quantities, we find

$$c = \frac{1}{\sqrt{(8.85 \times 10^{-12}\ \mathrm{C^2/N \cdot m^2})(4\pi \times 10^{-7}\ \mathrm{N/A^2})}}$$
$$= 3.00 \times 10^8\ \mathrm{m/s}.$$

Our assumed wave is consistent with Maxwell's third and fourth equations (Ampere's law and Faraday's law), provided that the wave front moves with the speed given above. We recognize this as the speed of light!

We must also show that our wave satisfies Maxwell's first and second equations, that is, Gauss's laws for electric and magnetic fields. To do this, we take a box with rectangular sides parallel to the $xy$-, $xz$-, and $yz$-coordinate planes (Fig. 33–4). The box encloses no electric charge, and you should be able to show that the total electric flux and magnetic flux out of the box are both zero.

We have chosen a simple and primitive wave for our study in order to avoid mathematical complications, but this special case illustrates several important features of *all* electromagnetic waves:

1. The wave is **transverse;** both $\boldsymbol{E}$ and $\boldsymbol{B}$ are perpendicular to the direction of propagation of the wave and to each other. The direction of propagation is the direction of the vector product $\boldsymbol{E} \times \boldsymbol{B}$.
2. There is a definite ratio between the magnitudes of $\boldsymbol{E}$ and $\boldsymbol{B}$: $E = cB$.
3. The wave travels in vacuum with a definite and unchanging speed.
4. Unlike mechanical waves, which need the oscillating particles of a medium such as water or air to be transmitted, electromagnetic waves require no medium. What's "waving" in an electromagnetic wave are the electric and magnetic fields.

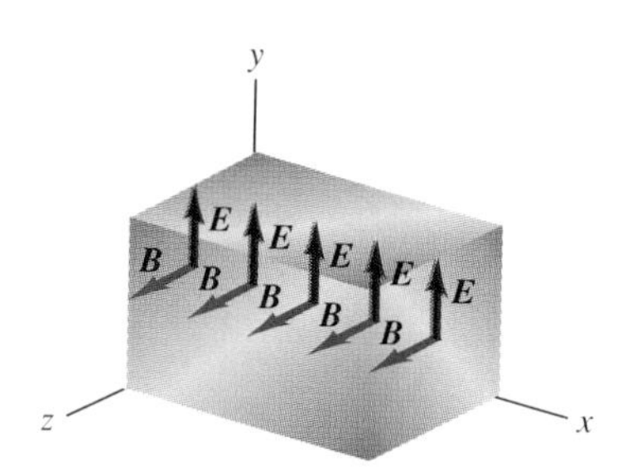


**33–4** Gaussian surface for a plane electromagnetic wave. The total electric flux and the total magnetic flux out of the surface are both zero.

We can generalize this discussion to a more realistic situation. Suppose we have several wave fronts in the form of parallel planes perpendicular to the $x$-axis, all moving to the right with speed $c$. Suppose that, within a single region between two planes, the $\boldsymbol{E}$ and $\boldsymbol{B}$ fields are the same at all points in the region but that they differ from region to region. Such a wave could be constructed by superposing several of the simple step waves that we described. The $\boldsymbol{E}$ and $\boldsymbol{B}$ fields, in waves as elsewhere, obey the superposition principle. When two waves are superposed, the total $\boldsymbol{E}$ field at each point is the vector sum of the $\boldsymbol{E}$ fields of the individual waves, and similarly for the total $\boldsymbol{B}$ field.

We can extend the above development to show that a wave with fields that vary in steps is also consistent with Ampere's and Faraday's laws, provided that the wave fronts all move with the speed $c$ given by Eq. (33–9). In the limit, when we make the individual steps smaller and smaller, we have a wave in which the $\boldsymbol{E}$ and $\boldsymbol{B}$ fields at any instant vary *continuously* along the $x$-axis. The entire field pattern moves to the right with speed $c$. In the next section we will consider waves in which $\boldsymbol{E}$ and $\boldsymbol{B}$ are *sinusoidal* functions of $x$ and $t$. Because at each point the magnitudes of $\boldsymbol{E}$ and $\boldsymbol{B}$ are related by $E = cB$, the periodic variations of the two fields in any periodic traveling wave must be *in phase*.

Electromagnetic waves have the property of *polarization*. We introduced this concept at the end of Section 19–4 in the context of transverse waves on a string. In the above discussion the choice of the $y$-direction for $\boldsymbol{E}$ was arbitrary. We could just as well have specified the $z$-axis for $\boldsymbol{E}$; then $\boldsymbol{B}$ would have been in the $-y$-direction. A wave in which $\boldsymbol{E}$ always lies along a certain axis is said to be *linearly polarized* along that axis. More generally, *any* wave traveling in the $x$-direction can be represented as a superposition of waves that are linearly polarized in the $y$- and $z$-directions. We will study polarization in greater detail, with special emphasis on polarization of light, in Chapter 34.

## *Derivation of the Wave Equation

Here is an alternate derivation of Eq. (33–9) for the speed of electromagnetic waves. It is more mathematical than our other treatment, but it includes an actual derivation of the wave equation. This part of the section can be omitted without loss of continuity in the chapter.

In Section 19–3 we showed that a function $y(x, t)$ that represents the displacement of any point in a mechanical wave traveling along the $x$-axis must satisfy a differential equation, Eq. (19–12):

$$\frac{\partial^2 y}{\partial x^2} = \frac{1}{v^2}\frac{\partial^2 y}{\partial t^2}. \tag{33–10}$$

This equation is called the **wave equation,** and $v$ is the speed of propagation of the wave.

To derive the corresponding equation for an electromagnetic wave, we consider again a *plane wave*. We assume that at each instant, $E_y$ and $B_z$ are uniform over any plane perpendicular to the $x$-axis, the direction of propagation. But now we let $E_y$ and $B_z$ vary continuously as we go along the $x$-axis; then each is a function of $x$ and $t$. We consider the values of $E_y$ and $B_z$ on two planes perpendicular to the $x$-axis, one at $x$ and one at $x + \Delta x$.

Following the same procedure as previously, we apply Faraday's law to a rectangle parallel to the $xy$-plane, Fig. 33–5. This figure is similar to Fig. 33–2. Let the left end $gh$ of the rectangle be at position $x$ and the right end $ef$ be at position $x + \Delta x$. We assume that $\Delta x$ is small, as shown. The values of $E_y$ on these two sides are then $E_y(x, t)$ and $E_y(x + \Delta x, t)$, respectively. When we apply Faraday's law to this rectangle, we find that instead of $\oint \boldsymbol{E} \cdot d\boldsymbol{l} = -Ea$ as before, we have

$$\oint \boldsymbol{E} \cdot d\boldsymbol{l} = -E_y(x, t)\, a + E_y(x + \Delta x, t)\, a$$
$$= a\,[E_y(x + \Delta x, t) - E_y(x, t)]. \tag{33–11}$$

To find the magnetic flux $\Phi_B$ through this rectangle, we assume that $\Delta x$ is small enough that $B_z$ is nearly uniform over the rectangle. In that case, $\Phi_B = B_z A = B_z a\, \Delta x$, and

$$\frac{d\Phi_B}{dt} = \frac{\partial B_z}{\partial t} a\, \Delta x.$$

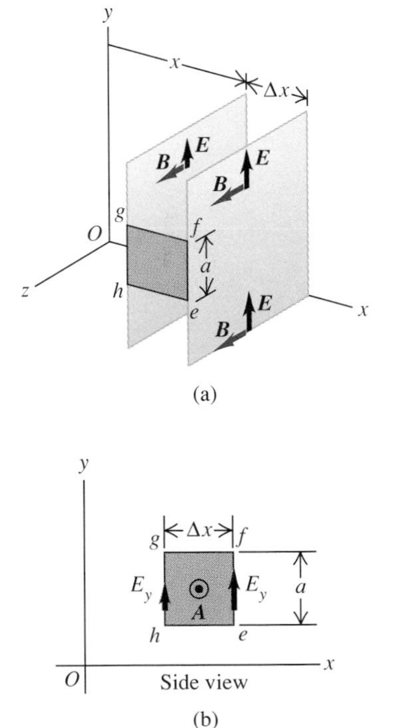


**33–5** Faraday's law applied to a rectangle with height $a$ and width $\Delta x$ parallel to the $xy$-plane.

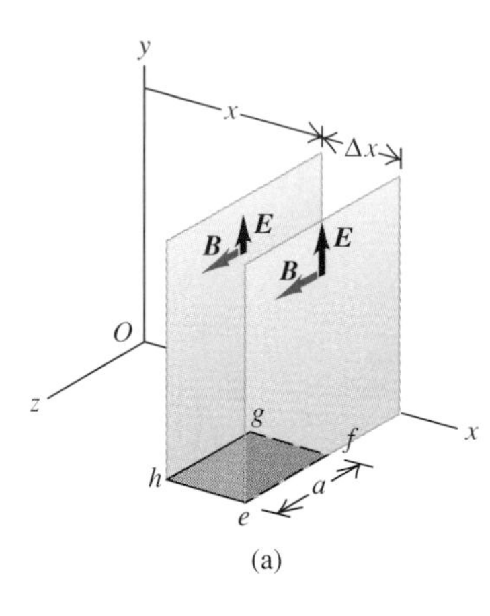


(a)

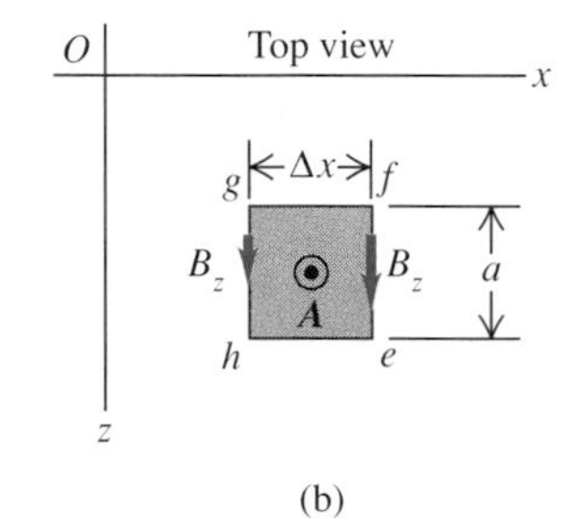


(b)

**33–6** Ampere's law applied to a rectangle with height $a$ and width $\Delta x$ parallel to the $xz$-plane.

We use partial-derivative notation as a reminder that $B_z$ is a function of both $x$ and $t$. When we substitute this expression and Eq. (33–11) into Faraday's law, Eq. (33–1), we get

$$a[E_y(x+\Delta x, t) - E_y(x, t)] = -\frac{\partial B_z}{\partial t} a\,\Delta x,$$

$$\frac{E_y(x+\Delta x, t) - E_y(x, t)}{\Delta x} = -\frac{\partial B_z}{\partial t}.$$

Finally, imagine shrinking the rectangle down to a sliver, as $\Delta x$ approaches zero. When we take the limit of this equation as $\Delta x \rightarrow 0$, we get

$$\frac{\partial E_y}{\partial x} = -\frac{\partial B_z}{\partial t}. \qquad (33\text{–}12)$$

This equation shows that if there is a time-varying component $B_z$ of magnetic field, there must also be a component $E_y$ of electric field that varies with $x$, and conversely. We put this relation on the shelf for now; we'll return to it soon.

Next we apply Ampere's law to the rectangle shown in Fig. 33–6. The line integral $\oint \boldsymbol{B} \cdot d\boldsymbol{l}$ becomes

$$\oint \boldsymbol{B} \cdot d\boldsymbol{l} = -B_z(x+\Delta x, t)\,a + B_z(x, t)\,a. \qquad (33\text{–}13)$$

Again assuming that the rectangle is narrow, we approximate the electric flux $\Phi_E$ through it as $\Phi_E = E_y A = E_y a\,\Delta x$. Its rate of change, which we need for Ampere's law, is then

$$\frac{d\Phi_E}{dt} = \frac{\partial E_y}{\partial t} a\,\Delta x.$$

When we substitute this expression and Eq. (33–13) into Ampere's law, Eq. (33–5), we get

$$-B_z(x+\Delta x, t)\,a + B_z(x, t)\,a = \epsilon_0 \mu_0 \frac{\partial E_y}{\partial t} a\,\Delta x.$$

Again we divide both sides by $a\,\Delta x$ and take the limit as $\Delta x \rightarrow 0$; we find

$$-\frac{\partial B_z}{\partial x} = \epsilon_0 \mu_0 \frac{\partial E_y}{\partial t}. \qquad (33\text{–}14)$$

Now comes the final step. We take the partial derivatives with respect to $x$ of both sides of Eq. (33–12), and we take the partial derivatives with respect to $t$ of both sides of Eq. (33–14). The results are

$$-\frac{\partial^2 E_y}{\partial x^2} = \frac{\partial^2 B_z}{\partial x \partial t},$$

$$-\frac{\partial^2 B_z}{\partial x \partial t} = \epsilon_0 \mu_0 \frac{\partial^2 E_y}{\partial t^2}.$$

Combining these two equations to eliminate $B_z$, we finally find

$$\frac{\partial^2 E_y}{\partial x^2} = \epsilon_0 \mu_0 \frac{\partial^2 E_y}{\partial t^2}. \qquad (33\text{–}15)$$

This is the *wave equation;* it has the same form as Eq. (33–10). The fact that the electric field $E_y$ must satisfy this equation shows that it behaves as a wave, with a pattern that travels through space with a definite speed. Furthermore, comparison of Eqs. (33–15) and (33–10) shows that the wave speed $v$ is given by

$$\frac{1}{v^2} = \epsilon_0\mu_0, \qquad \text{or} \qquad v = \frac{1}{\sqrt{\epsilon_0\mu_0}}.$$

This agrees with our previous result, Eq. (33–9), for the speed $c$ of electromagnetic waves.

We can show that $B_z$ must also satisfy Eq. (33–15). To prove this, we take the partial derivative of Eq. (33–12) with respect to $t$ and the partial derivative of Eq. (33–14) with respect to $x$ and combine the results. We leave this derivation as a problem (see Problem 33–21).

## 33–3 SINUSOIDAL WAVES

Sinusoidal electromagnetic waves are directly analogous to sinusoidal transverse mechanical waves on a stretched string. We studied these in Section 19–4; we suggest that you review that discussion. In a sinusoidal electromagnetic wave, the $\boldsymbol{E}$ and $\boldsymbol{B}$ fields at any point in space are sinusoidal functions of time, and at any instant of time the *spatial* variation of the fields is also sinusoidal.

Some sinusoidal electromagnetic waves share with the waves described in Section 33–2 the property that at any instant the fields are uniform over any plane perpendicular to the direction of propagation (as shown in Fig. 33–2). Such a wave is called a **plane wave.** The entire pattern travels in the direction of propagation with speed $c$. The directions of $\boldsymbol{E}$ and $\boldsymbol{B}$ are perpendicular to the direction of propagation (and to each other), so the wave is *transverse*.

The frequency $f$, the wavelength $\lambda$, and the speed of propagation $c$ of any periodic wave are related by the usual wavelength-frequency relation $c = \lambda f$. If the frequency $f$ is the power-line frequency of 60 Hz, the wavelength is

$$\lambda = \frac{c}{f} = \frac{3 \times 10^8\ \text{m/s}}{60\ \text{Hz}} = 5 \times 10^6\ \text{m} = 5000\ \text{km},$$

which is of the order of the earth's radius! For a wave with this frequency, even a distance of many miles includes only a small fraction of a wavelength. But if the frequency is $10^8$ Hz (100 MHz), which is typical of commercial FM radio stations, the wavelength is

$$\lambda = \frac{3 \times 10^8\ \text{m/s}}{10^8\ \text{Hz}} = 3\ \text{m},$$

and a moderate distance can include many complete waves.

Figure 33–7 shows a sinusoidal electromagnetic wave traveling in the $+x$-direction. The $\boldsymbol{E}$ and $\boldsymbol{B}$ vectors are shown only for a few points on the positive side of the $x$-axis. Imagine a plane perpendicular to the $x$-axis at a particular point and a particular time; the fields have the same values at all points in that plane. Of course, the values are

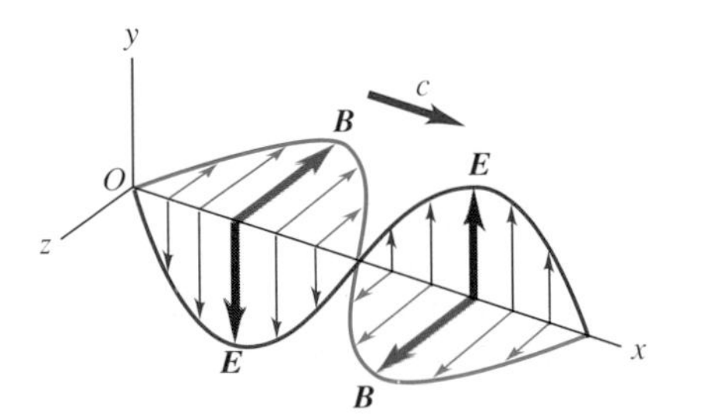


**33–7** Representation of the electric and magnetic fields as functions of $x$ for a linearly polarized sinusoidal plane electromagnetic wave. One wavelength of the wave is shown at time $t = 0$. The direction of propagation is the direction of $\boldsymbol{E} \times \boldsymbol{B}$. The fields are shown only for points along the $x$-axis.

different on different planes. In the planes where the $\boldsymbol{E}$ vector is in the $+y$-direction, $\boldsymbol{B}$ is in the $+z$-direction; where $\boldsymbol{E}$ is in the $-y$-direction, $\boldsymbol{B}$ is in the $-z$-direction. These directions illustrate the right-hand rule described in the first item in the list of characteristics in Section 33–2.

We can describe electromagnetic waves by means of *wave functions,* just as we did in Section 19–4 for waves on a string. One form of the wave function for a transverse wave traveling to the right along a stretched string is Eq. (19–7):

$$y = A \sin(\omega t - kx),$$

where $y$ is the transverse displacement from its equilibrium position at time $t$ of a point with coordinate $x$ on the string. The quantity $A$ is the maximum displacement, or *amplitude,* of the wave; $\omega$ is its *angular frequency,* equal to $2\pi$ times the frequency $f$; and $k$ is the *wave number,* or *propagation constant,* equal to $2\pi/\lambda$, where $\lambda$ is the wavelength.

Let $E$ and $B$ represent the instantaneous values and $E_{max}$ and $B_{max}$ represent the maximum values, or *amplitudes,* of the electric and magnetic fields in Fig. 33–7. The wave functions for the wave are then

$$E = E_{max} \sin(\omega t - kx),$$
$$B = B_{max} \sin(\omega t - kx). \quad (33\text{–}16)$$

We can also write these functions in vector form:

$$\boldsymbol{E} = E_{max}\boldsymbol{j} \sin(\omega t - kx),$$
$$\boldsymbol{B} = B_{max}\boldsymbol{k} \sin(\omega t - kx). \quad (33\text{–}17)$$

(Note the two different $k$'s, the unit vector $\boldsymbol{k}$ in the $z$-direction and the wave number $k$. Be careful!)

The sine curves in Fig. 33–7 represent instantaneous values of $E$ and $B$ as functions of $x$ at time $t = 0$. The wave travels to the right with speed $c$. Equations (33–16) show that at any point the sinusoidal oscillations of $\boldsymbol{E}$ and $\boldsymbol{B}$ are *in phase.* From Eq. (33–4) the amplitudes must be related by

$$E_{max} = cB_{max}. \quad (33\text{–}18)$$

These amplitude and phase relations are also required in order for $E$ and $B$ to satisfy Eqs. (33–12) and (33–14). Verification of this statement is left as a problem (see Problem 33–22).

Figure 33–8 shows the electric and magnetic fields of a wave traveling in the *negative* $x$-direction. At points where $\boldsymbol{E}$ is in the positive $y$-direction, $\boldsymbol{B}$ is in the *negative* $z$-direction; where $\boldsymbol{E}$ is in the negative $y$-direction, $\boldsymbol{B}$ is in the *positive* $z$-direction. We note that the direction of propagation is the direction of $\boldsymbol{E} \times \boldsymbol{B}$, as mentioned in Section 33–2

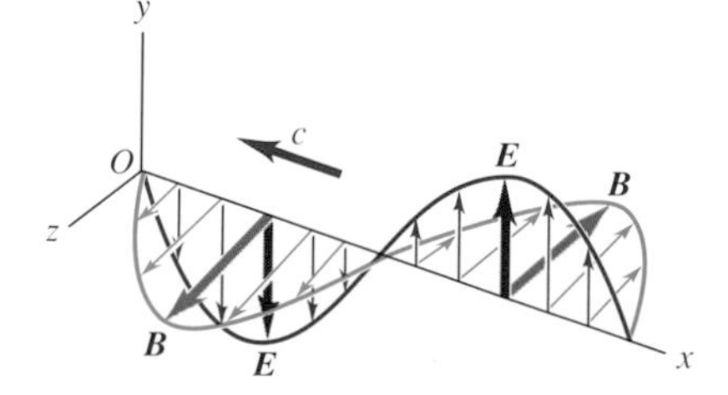


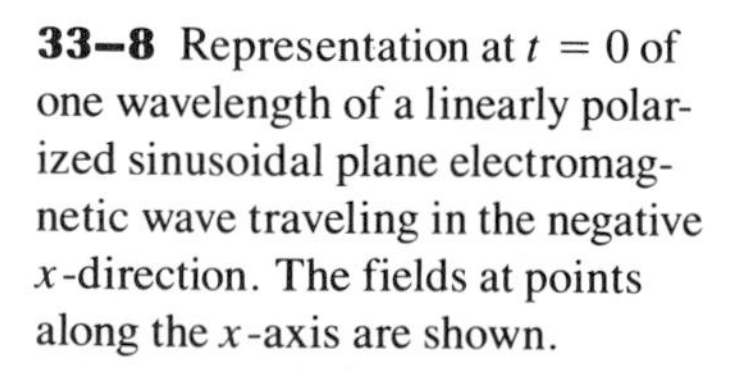
**33–8** Representation at $t = 0$ of one wavelength of a linearly polarized sinusoidal plane electromagnetic wave traveling in the negative $x$-direction. The fields at points along the $x$-axis are shown.

in the first item in the list of characteristics of electromagnetic waves. The wave functions for this wave are

$$E = -E_{\max} \sin(\omega t + kx),$$
$$B = B_{\max} \sin(\omega t + kx). \tag{33–19}$$

As with the wave traveling in the $+x$-direction, the sinusoidal oscillations of the $\boldsymbol{E}$ and $\boldsymbol{B}$ fields at any point are *in phase*.

Both of these sinusoidal waves are linearly polarized in the $y$-direction (because the $\boldsymbol{E}$ field always lies along the $y$-axis). Example 33–1 shows a wave that is linearly polarized in the $z$-direction.

## PROBLEM-SOLVING STRATEGY

### Electromagnetic waves

**1.** For the problems of this chapter the most important advice that we can give is to concentrate on basic relationships, such as the relation of $\boldsymbol{E}$ to $\boldsymbol{B}$ (both magnitude and direction), how the wave speed is determined, the transverse nature of the waves, and so on. Don't get sidetracked by mathematical details.

**2.** In the discussions of sinusoidal waves, both traveling and standing, you need to use the language of sinusoidal waves from Chapters 19 and 20. Don't hesitate to go back and review that material, including the problem-solving strategies presented in those chapters. Keep in mind the basic relationships for periodic waves, $v = \lambda f$ and $\omega = vk$. For electromagnetic waves in vacuum, $v = c$. Be careful to distinguish between ordinary frequency $f$, usually expressed in hertz, and angular frequency $\omega = 2\pi f$, expressed in rad/s. Also remember that the wave number $k$ is $k = 2\pi/\lambda$.

**3.** In the discussion of standing waves in Section 33–6, make sure you know what you mean by nodes and antinodes and which field you are talking about. In a standing wave, nodes of $\boldsymbol{E}$ coincide with antinodes of $\boldsymbol{B}$, and conversely. Compare this situation to the distinction between pressure nodes and displacement nodes in Section 20–4.

## EXAMPLE 33–1

A carbon dioxide laser emits a sinusoidal electromagnetic wave that travels in vacuum in the negative $x$-direction (Fig. 33–9). The wavelength is 10.6 $\mu$m, and the $\boldsymbol{E}$ field is along the $z$-axis, with maximum magnitude of 1.5 MV/m. Write vector equations for $\boldsymbol{E}$ and $\boldsymbol{B}$ as functions of time and position.

**SOLUTION** Equations (33–19) describe a wave traveling in the negative $x$-direction with $\boldsymbol{E}$ along the $y$-axis. That wave is linearly polarized along the $y$-axis. The wave in this example is linearly polarized along the $z$-axis. At points where $\boldsymbol{E}$ is in the positive $z$-direction, $\boldsymbol{B}$ must be in the positive $y$-direction in order

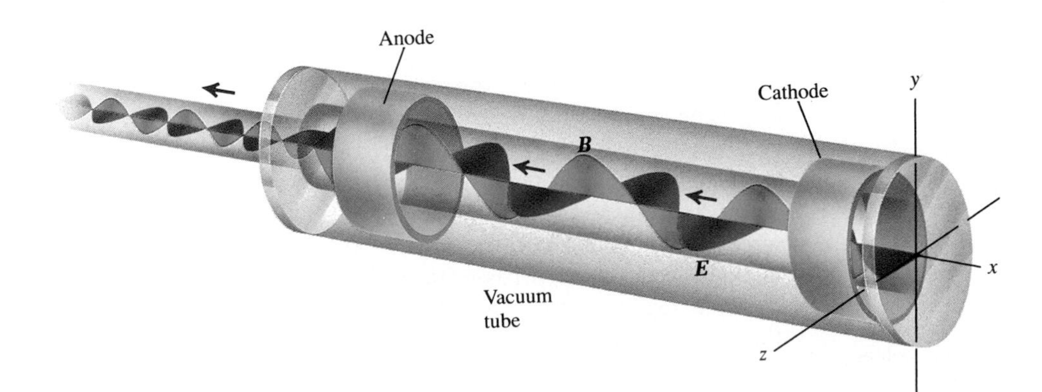


**33–9** A carbon dioxide laser emits a sinusoidal electromagnetic wave.

for the vector product $\boldsymbol{E} \times \boldsymbol{B}$ to be in the negative $x$-direction. A possible pair of wave functions to satisfy these requirements is

$$\boldsymbol{E} = E_{\max} \boldsymbol{k} \sin(\omega t + kx), \qquad \boldsymbol{B} = B_{\max} \boldsymbol{j} \sin(\omega t + kx).$$

Faraday's law requires $E_{\max} = cB_{\max}$, so

$$B_{\max} = \frac{E_{\max}}{c} = \frac{1.5 \times 10^6 \text{ V/m}}{3.0 \times 10^8 \text{ m/s}} = 5.0 \times 10^{-3} \text{ T}.$$

To check unit consistency, note that 1 V = 1 Wb/s and 1 Wb/m$^2$ = 1 T.

We have a wavelength $\lambda = 10.6 \times 10^{-6}$ m, so $k = 2\pi/\lambda = (2\pi \text{ rad})/(10.6 \times 10^{-6}\text{m}) = 5.93 \times 10^5$ rad/m. Also $\omega = ck = (3.00 \times 10^8 \text{ m/s})(5.93 \times 10^5 \text{ rad/m}) = 1.78 \times 10^{14}$ rad/s. Substituting these values into the above wave functions, we get

$$\boldsymbol{E} = (1.5 \times 10^6 \text{ V/m})\boldsymbol{k} \sin [(1.78 \times 10^{14} \text{ rad/s})t + (5.9 \times 10^5 \text{ rad/m})x],$$

$$\boldsymbol{B} = (5.0 \times 10^{-3} \text{ T})\boldsymbol{j} \sin [(1.78 \times 10^{14} \text{ rad/s})t + (5.9 \times 10^5 \text{ rad/m})x].$$

With these equations we can find the fields at any particular position and time by substituting specific values of $x$ and $t$.

## 33–4 ENERGY IN ELECTROMAGNETIC WAVES

It is a familiar fact that energy is associated with electromagnetic waves. Think of the sun's radiation and the radiation in microwave ovens. To derive detailed relationships for the energy in an electromagnetic wave, we begin with the expressions derived in Sections 25–4 and 31–3 for the **energy densities** in electric and magnetic fields; we suggest that you review those derivations now. Specifically, Eqs. (25–11) and (31–9) show that the total energy density $u$ in a region of space where $\boldsymbol{E}$ and $\boldsymbol{B}$ fields are present is given by

$$u = \frac{1}{2}\epsilon_0 E^2 + \frac{1}{2\mu_0}B^2. \tag{33–20}$$

For the simple electromagnetic waves studied so far, $\boldsymbol{E}$ and $\boldsymbol{B}$ are related by

$$B = \frac{E}{c} = \sqrt{\epsilon_0\mu_0}E. \tag{33–21}$$

Combining this with Eq. (33–20), we can also express the energy density $u$ as

$$u = \frac{1}{2}\epsilon_0 E^2 + \frac{1}{2\mu_0}\left(\sqrt{\epsilon_0\mu_0}E\right)^2 = \epsilon_0 E^2. \tag{33–22}$$

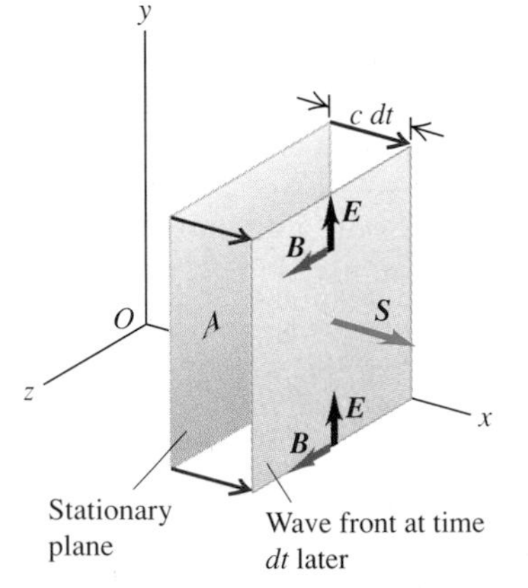


**33–10** Wave front at a time $dt$ after it passes through the stationary plane with area $A$. The volume between the plane and the wave front contains an amount of electromagnetic energy $uAc\,dt$.

This shows that the energy density associated with the $\boldsymbol{E}$ field in our simple wave is equal to the energy density of the $\boldsymbol{B}$ field.

In the wave described in Section 33–2 the $\boldsymbol{E}$ and $\boldsymbol{B}$ fields advance with time into regions where originally no fields were present, so it is clear that the wave transports energy from one region to another. We can describe this energy transfer in terms of *energy transferred per unit time per unit cross-section area*, or *power per unit area*, for an area perpendicular to the direction of wave travel.

To see how the energy flow is related to the fields, consider a stationary plane, perpendicular to the $x$-axis, that coincides with the wave front at a certain time. In a time $dt$ after this the wave front moves a distance $dx = c\,dt$ to the right of the plane. Considering an area $A$ on the stationary plane (Fig. 33–10), we note that the energy in the space to the right of this area must have passed through it to reach the new location. The volume $dV$ of the relevant region is the base area $A$ times the length $c\,dt$, and the energy $dU$ in this region is the energy density $u$ times this volume:

$$dU = u\,dV = (\epsilon_0 E^2)(Ac\,dt). \tag{33–23}$$

This energy passes through the area $A$ in time $dt$. The energy flow per unit time per unit area, which we will call $S$, is

$$S = \frac{1}{A}\frac{dU}{dt} = \epsilon_0 c E^2. \tag{33–24}$$

Using Eqs. (33–9) and (33–21), we can derive the alternative forms

$$S = \frac{\epsilon_0}{\sqrt{\epsilon_0 \mu_0}} E^2 = \sqrt{\frac{\epsilon_0}{\mu_0}} E^2 = \frac{EB}{\mu_0}. \tag{33–25}$$

The derivation is left as a problem (see Exercise 33–7).

The units of $S$ are energy per unit time per unit area, or power per unit area. The SI unit of $S$ is 1 J/s · m$^2$ or 1 W/m$^2$.

## Poynting Vector

We can define a *vector* quantity that describes both the magnitude and direction of the energy flow rate:

$$\mathbf{S} = \frac{1}{\mu_0} \mathbf{E} \times \mathbf{B}. \tag{33–26}$$

The vector $\mathbf{S}$ is called the **Poynting vector;** its magnitude is given by Eq. (33–24) or (33–25), and its direction is the direction of propagation of the wave. The magnitude $EB/\mu_0$ gives the flow of energy through a cross-section area perpendicular to the propagation direction per unit area and per unit time. The total energy flow per unit time (power, $P$) out of any closed surface is given by the integral over the surface,

$$P = \oint \mathbf{S} \cdot d\mathbf{A}.$$

### EXAMPLE 33–2

For the wave described in Section 33–2, suppose

$$E = 100 \text{ V/m} = 100 \text{ N/C}.$$

Find the value of $B$, the energy density, and the rate of energy flow per unit area $S$.

**SOLUTION** From Eq. (33–4),

$$B = \frac{E}{c} = \frac{100 \text{ V/m}}{3.00 \times 10^8 \text{ m/s}} = 3.33 \times 10^{-7} \text{ T}.$$

From Eq. (33–22),

$$u = \epsilon_0 E^2 = (8.85 \times 10^{-12} \text{ C}^2/\text{N} \cdot \text{m}^2)(100 \text{ N/C})^2$$
$$= 8.85 \times 10^{-8} \text{ N/m}^2 = 8.85 \times 10^{-8} \text{ J/m}^3.$$

The magnitude of the Poynting vector is

$$S = \frac{EB}{\mu_0} = \frac{(100 \text{ V/m})(3.33 \times 10^{-7} \text{ T})}{4\pi \times 10^{-7} \text{ Wb/A} \cdot \text{m}}$$
$$= 26.5 \text{ V} \cdot \text{A/m}^2 = 26.5 \text{ W/m}^2.$$

Alternatively,

$$S = \epsilon_0 c E^2$$
$$= (8.85 \times 10^{-12} \text{ C}^2/\text{N} \cdot \text{m}^2)(3.00 \times 10^8 \text{ m/s})(100 \text{ N/C})^2$$
$$= 26.5 \text{ W/m}^2.$$

For the sinusoidal waves studied in Section 33–3, and for other more complex waves, the electric and magnetic fields at any point vary with time, so the Poynting vector at any point is also a function of time. The *average* value of the magnitude of the Poynting vector at a point is called the **intensity** of the radiation at that point. The SI unit of intensity is the watt per square meter (W/m$^2$).

Let's work out the intensity of the sinusoidal wave described by Eqs. (33–16). We first substitute these expressions into Eq. (33–25):

$$S = \frac{EB}{\mu_0} = \frac{E_{max}B_{max}}{\mu_0}\sin^2(\omega t - kx)$$
$$= \frac{E_{max}B_{max}}{2\mu_0}[1 - \cos 2(\omega t - kx)].$$

The time-average value of $\cos 2(\omega t - kx)$ is zero because at any point it is positive during half a cycle and negative during the other half. So the average value $S_{av}$ of the Poynting vector magnitude over a full cycle is

$$S_{av} = \frac{E_{max}B_{max}}{2\mu_0}.$$

That is, the average value of $S$ for a sinusoidal wave (the intensity $I$ of the wave) is one-half of the maximum value. By using the relations $E_{max} = B_{max}c$ and $\epsilon_0\mu_0 = 1/c^2$ we can express the intensity in several equivalent forms:

$$I = S_{av} = \frac{E_{max}B_{max}}{2\mu_0} = \frac{E_{max}{}^2}{2\mu_0 c} = \frac{1}{2}\sqrt{\frac{\epsilon_0}{\mu_0}}E_{max}{}^2 = \frac{1}{2}\epsilon_0 c E_{max}{}^2. \quad (33\text{–}27)$$

We invite you to verify that these expressions are all equivalent.

For a wave traveling in the $-x$-direction, represented by Eqs. (33–19), the Poynting vector is in the $-x$-direction at every point, but the expressions for its magnitude, Eq. (33–27), are the same as for a wave traveling in the $+x$-direction. Verifying these statements is left as a problem (see Exercise 33–8).

## EXAMPLE 33–3

A radio station on the surface of the earth radiates a sinusoidal wave with an average total power of 50 kW (Fig. 33–11). Assuming that the transmitter radiates equally in all directions above the ground (which is unlikely in real situations), find the amplitudes $E_{max}$ and $B_{max}$ at a distance of 100 km from the antenna.

**SOLUTION** The first step is to find the magnitude $S_{av}$ of the average Poynting vector. Then we can find $E_{max}$ from Eq. (33–27) and $B_{max}$ from Eq. (33–4). We surround the antenna with an imaginary sphere with radius 100 km $= 1.00 \times 10^5$ m. The upper half of this sphere has area

$$A = 2\pi R^2 = 2\pi(1.00 \times 10^5\ \text{m})^2 = 6.28 \times 10^{10}\ \text{m}^2.$$

All the power radiated passes through this surface, so the power per unit area is

$$S_{av} = \frac{P}{A} = \frac{P}{2\pi R^2} = \frac{5.00 \times 10^4\ \text{W}}{6.28 \times 10^{10}\ \text{m}^2} = 7.96 \times 10^{-7}\ \text{W/m}^2.$$

From Eq. (33–27),

$$S_{av} = \frac{E_{max}{}^2}{2\mu_0 c}, \quad \text{and} \quad E_{max} = \sqrt{2\mu_0 c S_{av}},$$

so

$$E_{max} = \sqrt{2(4\pi \times 10^{-7}\ \text{Wb/A}\cdot\text{m})(3.00 \times 10^8\ \text{m/s})(7.96 \times 10^{-7}\ \text{W/m}^2)}$$
$$= 2.45 \times 10^{-2}\ \text{V/m}.$$

From Eq. (33–4),

$$B_{max} = \frac{E_{max}}{c} = 8.17 \times 10^{-11}\ \text{T}.$$

Note that the magnitude of $E_{max}$ is comparable to fields that are commonly seen in the laboratory but that $B_{max}$ is extremely *small* in comparison to ***B*** fields that we have seen in previous chapters. For this reason, most detectors of electromagnetic radiation respond to the effect of the electric field, not the magnetic field. Loop radio antennas are an exception.

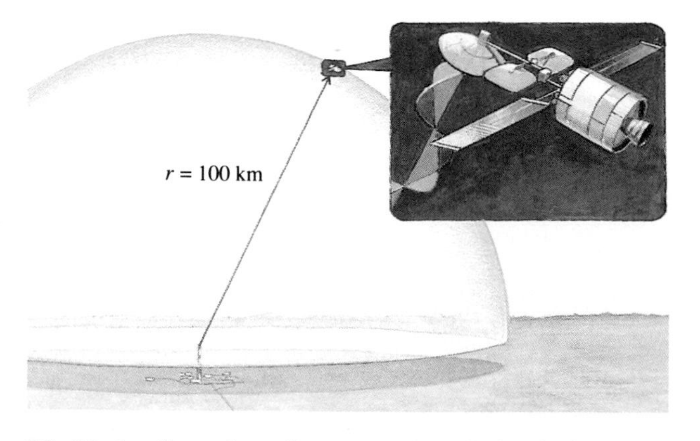


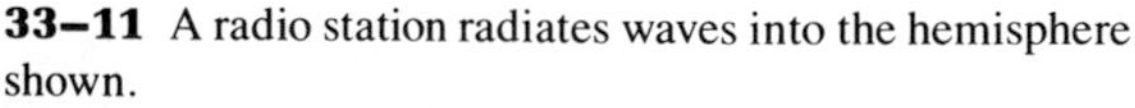

**33–11** A radio station radiates waves into the hemisphere shown.

## Radiation Pressure

As we have seen, the fact that electromagnetic waves transport energy follows directly from the fact that energy is required to establish electric and magnetic fields. It can also be shown that electromagnetic waves carry *momentum p,* with a corresponding momentum density (momentum $dp$ per volume $dV$) of magnitude

$$\frac{dp}{dV} = \frac{EB}{\mu_0 c^2} = \frac{S}{c^2}. \qquad (33\text{–}28)$$

This momentum is a property of the field; it is not associated with the mass of a moving particle in the usual sense.

There is also a corresponding momentum flow rate. The volume $dV$ passing through an area $A$ in time $dt$ is $dV = Ac\,dt$. When we substitute this into Eq. (33–28) and rearrange, we find that the momentum flow rate per unit area is

$$\frac{1}{A}\frac{dp}{dt} = \frac{S}{c} = \frac{EB}{\mu_0 c}. \qquad (33\text{–}29)$$

This is the momentum transferred per unit surface area per unit time. We obtain the *average* rate of momentum transfer by replacing $S$ by $S_{av} = I$.

This momentum is responsible for the phenomenon of **radiation pressure.** When an electromagnetic wave is completely absorbed by a surface perpendicular to the propagation direction, the time rate of change of momentum equals the *force* on the surface. Thus the average force per unit area, or pressure, is equal to $I/c$. If the wave is totally reflected, the momentum change is twice as great, and the pressure is $2I/c$. For example, the value of $I$ (or $S_{av}$) for direct sunlight before it passes through the earth's atmosphere is about 1.4 kW/m$^2$. The corresponding average pressure on a completely absorbing surface is

$$\frac{I}{c} = \frac{1.4 \times 10^3\ \text{W/m}^2}{3.0 \times 10^8\ \text{m/s}} = 4.7 \times 10^{-6}\ \text{Pa}.$$

The average pressure on a totally *reflecting* surface is twice this, $2I/c$, or $9.4 \times 10^{-6}$ Pa. These are very small pressures, of the order of $10^{-10}$ atmospheres, but they can be measured with sensitive instruments.

### EXAMPLE 33–4

A communications satellite has solar-energy-collecting panels with a total area of 4.0 m$^2$ (Fig. 33–12). Assuming that the sun's radiation is perpendicular to the panels and is completely absorbed, find a) the total solar power absorbed; b) the total force associated with radiation pressure.

**SOLUTION** From the above discussion the intensity $I$ (power per unit area) is $1.4 \times 10^3$ W/m$^2$. The total power is the intensity $I$ times the area $A$:

$$P = IA = (1.4 \times 10^3\ \text{W/m}^2)(4.0\ \text{m}^2) = 5.6 \times 10^3\ \text{W} = 5.6\ \text{kW}.$$

The radiation pressure $p$ (force per unit area) is $4.7 \times 10^{-6}$ Pa = $4.7 \times 10^{-6}$ N/m$^2$. The total force $F$ is the pressure $p$ times the area $A$:

$$F = pA = (4.7 \times 10^{-6}\ \text{N/m}^2)(4.0\ \text{m}) = 1.9 \times 10^{-5}\ \text{N}.$$

This is comparable to the weight (on earth) of a single grain of salt.

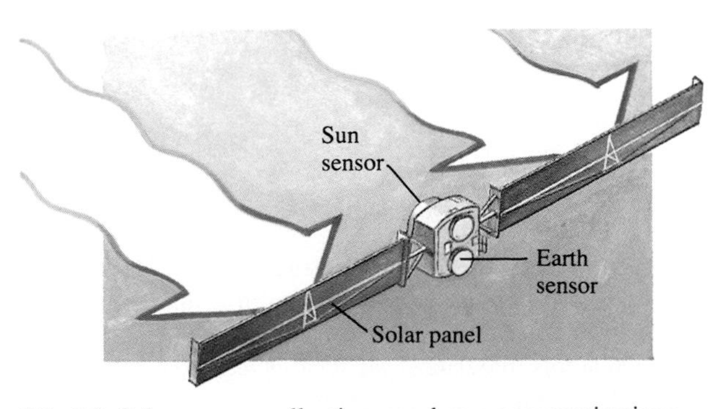


**33–12** Solar-energy-collecting panel on a communications satellite.

**33–13** The pressure of the sun's radiation helps push the tail of this comet away from the sun.

The pressure of the sun's radiation is partially responsible for pushing the tail of a comet away from the sun (Fig. 33–13). Radiation pressure is important in the structure of stars, in which it is often very much larger than in the above examples. Gravitational attractions tend to shrink a star, but this tendency is balanced by radiation pressure, so the size of the star is maintained through most stages of its evolution.

## *33–5 ELECTROMAGNETIC WAVES IN MATTER

Our initial discussion of electromagnetic waves has been restricted to waves *in vacuum,* but we can extend the analysis to electromagnetic waves in dielectrics. The wave speed is not the same as in vacuum, and we denote it by $v$ instead of $c$. Faraday's law is unaltered, but Eq. (33–4), derived from Faraday's law, is replaced by $E = vB$. In Ampere's law the displacement current density is given not by $\epsilon_0 dE/dt$ but by $\epsilon\, dE/dt = K\epsilon_0\, dE/dt$. Also, the constant $\mu_0$ in Ampere's law must be replaced by $\mu = K_m\mu_0$. The result is that Eq. (33–8) is replaced by

$$B = \epsilon\mu\, vE.$$

Following the same procedure as before, we find that the wave speed $v$ is

$$v = \frac{1}{\sqrt{\epsilon\mu}} = \frac{1}{\sqrt{KK_m}}\frac{1}{\sqrt{\epsilon_0\mu_0}} = \frac{c}{\sqrt{KK_m}}. \tag{33–30}$$

For most dielectrics (except for insulating ferromagnetic materials) the relative permeability $K_m$ is very nearly equal to unity. When $K_m \cong 1$,

$$v = \frac{1}{\sqrt{K}}\frac{1}{\sqrt{\epsilon_0\mu_0}} = \frac{c}{\sqrt{K}}.$$

Because $K$ is always greater than unity, the speed $v$ of electromagnetic waves in a dielectric is always *less* than the speed $c$ in vacuum by a factor of $1/\sqrt{K}$. The ratio of the speed $c$ in vacuum to the speed $v$ in a material is known in optics as the **index of refraction** $n$ of the material. When $K_m \cong 1$,

$$\frac{c}{v} = n = \sqrt{KK_m} \cong \sqrt{K}. \tag{33–31}$$

Usually, we can't use values of $K$ such as those in Table 25–1 in this equation because those values are measured using *constant* electric fields. When the fields vary rapidly with time, there is usually not time for the re-orienting of electric dipoles that occurs with steady fields. Values of $K$ with rapidly varying fields are usually much *smaller* than the values in the table. For example, $K$ for water is 80.4 for steady fields but only about 1.77 in the frequency range of visible light.

### EXAMPLE 33–5

An electromagnetic wave with a frequency of 100 MHz (the middle of the FM radio broadcast band) travels in an insulating ferrite (ferromagnetic) material with the properties $K = 10$ and $K_m = 1000$ at this frequency. a) What is the speed of propagation? b) What is the wavelength of the wave?

**SOLUTION** a) From Eq. (33–30),

$$v = \frac{c}{\sqrt{KK_m}} = \frac{3.00 \times 10^8\ \text{m/s}}{\sqrt{(10)(1000)}} = 3.00 \times 10^6\ \text{m/s}.$$

b) The wavelength is

$$\lambda = \frac{v}{f} = \frac{3.00 \times 10^6\ \text{m/s}}{100 \times 10^6\ \text{s}^{-1}} = 0.0300\ \text{m} = 3.00\ \text{cm}.$$

The wavelength in air for this frequency is

$$\lambda = \frac{c}{f} = \frac{3.00 \times 10^8\ \text{m/s}}{100 \times 10^6\ \text{s}^{-1}} = 3.00\ \text{m}.$$

When a dielectric is present, we need to modify the expressions for the energy density and the Poynting vector. The energy density is then

$$u = \frac{1}{2}\epsilon E^2 + \frac{1}{2\mu}B^2 = \epsilon E^2. \tag{33–32}$$

The energy densities in the $\boldsymbol{E}$ and $\boldsymbol{B}$ fields are still equal. The Poynting vector becomes

$$\boldsymbol{S} = \frac{1}{\mu}\boldsymbol{E} \times \boldsymbol{B}, \tag{33–33}$$

and its magnitude $S$ is

$$S = \frac{EB}{\mu} = \sqrt{\frac{\epsilon}{\mu}}E^2. \tag{33–34}$$

The intensity $I$ of a sinusoidal wave is obtained by simply replacing $\epsilon_0$ with $\epsilon$ and $\mu_0$ with $\mu$ in Eqs. (33–27):

$$I = \frac{E_{\text{max}}B_{\text{max}}}{2\mu} = \frac{E_{\text{max}}^2}{2\mu v} = \frac{1}{2}\sqrt{\frac{\epsilon}{\mu}}E_{\text{max}}^2 = \frac{1}{2}\epsilon v E_{\text{max}}^2. \tag{33–35}$$

Electromagnetic waves cannot propagate any appreciable distance in a *conducting* material because the $\boldsymbol{E}$ and $\boldsymbol{B}$ fields lead to currents that provide a mechanism for dissipating and reflecting the energy of the wave. For an ideal conductor with zero resistivity, $\boldsymbol{E}$ must be zero everywhere inside the material. When an electromagnetic wave strikes such a material, the wave is totally reflected. Real conductors with finite resistivity permit some penetration of the wave into the material, with partial reflection. A very thin metal film is partially transparent, but a thick layer is not. A polished metal surface such as silver is usually an excellent *reflector* of electromagnetic waves.

## 33–6 STANDING WAVES

Electromagnetic waves can be *reflected;* a conducting surface can serve as a reflector. The superposition principle holds for electromagnetic waves just as for all electric and magnetic fields. The superposition of an incident wave and a perfectly reflected wave forms a **standing wave.** The situation is analogous to standing waves on a stretched string, discussed in Section 20–2; we suggest that you review that discussion.

Suppose a sheet of an ideal conductor (zero resistivity) is placed in the $yz$-plane of Fig. 33–14 and a linearly polarized electromagnetic wave, traveling in the negative $x$-direction, strikes it. The surface of the conductor is an equipotential surface, as dis-

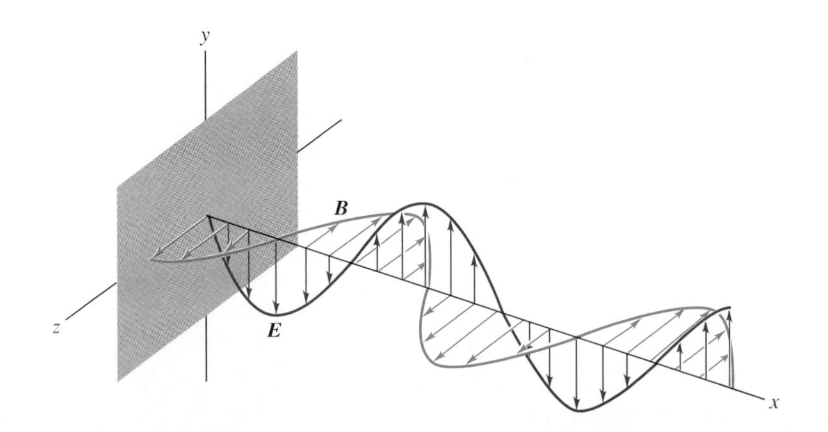


**33–14** Representation of the electric and magnetic fields of a linearly polarized electromagnetic standing wave when $\omega t = \pi/4$ rad. The pattern doesn't move along the $x$-axis, but at any point the $\boldsymbol{E}$ vectors diminish and the $\boldsymbol{B}$ vectors grow until $\omega t = \pi/2$ rad. In any plane perpendicular to the $x$-axis, $E$ is maximum where $B$ is zero, and conversely.

cussed in Section 24–4. We showed there that $\boldsymbol{E}$ can't have a component parallel to such a surface. Therefore, in the present situation, $\boldsymbol{E}$ must be zero everywhere in the $yz$-plane.

The $\boldsymbol{E}$ field of the incident wave induces sinusoidal currents in the conductor that keep $\boldsymbol{E}$ zero everywhere inside it and on its surface. These induced currents produce a *reflected* wave, traveling out from the plane in the $+x$-direction. From the superposition principle the total $\boldsymbol{E}$ field at any point to the right of the plane is the vector sum of the $\boldsymbol{E}$ fields of the incident and reflected waves; the same is true for the total $\boldsymbol{B}$ field.

Suppose the incident wave is described by the wave functions of Eqs. (33–19) and the reflected wave by Eqs. (33–16). (Compare these with Eqs. (19–7) and (19–8) for transverse waves on a string.) The superposition principle states that the total $\boldsymbol{E}$ field at any point is the sum of the $\boldsymbol{E}$ fields of the separate waves, and similarly for the $\boldsymbol{B}$ field. Therefore the wave functions for the superposition of the two waves are

$$E = E_{max}[-\sin(\omega t + kx) + \sin(\omega t - kx)],$$

$$B = B_{max}[\sin(\omega t + kx) + \sin(\omega t - kx)].$$

We can expand and simplify these expressions, using the identities

$$\sin(A \pm B) = \sin A \cos B \pm \cos A \sin B.$$

The results are

$$E = -2E_{max} \cos \omega t \sin kx, \tag{33–36}$$

$$B = 2B_{max} \sin \omega t \cos kx. \tag{33–37}$$

Equation (33–36) is analogous to Eq. (20–1) for a stretched string. We see that at $x = 0$, $E$ is *always* zero; this is required by the nature of the ideal conductor, which plays the same role as a fixed point at the end of the string. Furthermore, $E$ is zero at *all* times at points in those planes perpendicular to the $x$-axis for which $\sin kx = 0$; that is, $kx = 0$, $\pi, 2\pi, \ldots$, or

$$x = 0, \frac{\lambda}{2}, \lambda, \frac{3\lambda}{2}, \ldots. \tag{33–38}$$

These are called the **nodal planes** of the $\boldsymbol{E}$ field.

The total magnetic field is zero at all times at points in planes on which $\cos kx = 0$. This occurs where

$$x = \frac{\lambda}{4}, \frac{3\lambda}{4}, \frac{5\lambda}{4}, \ldots. \tag{33–39}$$

These are the nodal planes of the $\boldsymbol{B}$ field.

Figure 33–14 shows a standing-wave pattern at one instant of time. The magnetic field is *not* zero at the conducting surface ($x = 0$), and there is no reason it should be. The surface currents that must be present to make $\boldsymbol{E}$ exactly zero at the surface cause magnetic fields at the surface. The nodal planes of one field are midway between those of the other, and the nodal planes of each field are separated by one half-wavelength.

The total electric field is a *cosine* function of $t$, and the total magnetic field is a *sine* function of $t$. The sinusoidal variations of the two fields are therefore 90° out of phase at each point. At times when $\cos \omega t = 0$, the electric field is zero *everywhere* and the magnetic field is maximum. When $\sin \omega t = 0$, the magnetic field is zero everywhere and the electric field is maximum. This is in contrast to a wave traveling in one direction, as described by Eqs. (33–16) or (33–19) separately, when the sinusoidal variations of $\boldsymbol{E}$ and $\boldsymbol{B}$ at any particular point are *in phase*. It is interesting to check that Eqs. (33–36) and (33–37) satisfy the wave equation, Eq. (33–15). They also satisfy Eqs. (33–12) and (33–14); we leave the proofs of these statements as problems.

Pursuing the stretched-string analogy, we may now insert a second conducting plane, parallel to the first and a distance $L$ from it, along the $+x$-axis. This is analogous to the stretched string held at the points $x = 0$ and $x = L$. Both conducting planes must be nodal planes for $\boldsymbol{E}$; a standing wave can exist only when the second plane is placed at one of the $E = 0$ positions. That is, for a standing wave to exist, $L$ must be an integer multiple of $\lambda/2$. The possible wavelengths are

$$\lambda_n = \frac{2L}{n} \quad (n = 1, 2, 3, \ldots). \tag{33–40}$$

The corresponding frequencies are

$$f_n = \frac{c}{\lambda_n} = n\frac{c}{2L} \quad (n = 1, 2, 3, \ldots). \tag{33–41}$$

Thus there is a set of *normal modes*, each with a characteristic frequency, wave shape, and node pattern. By measuring the node positions we can measure the wavelength. If the frequency is known, the wave speed can be determined. This technique was, in fact, used by Heinrich Hertz in his pioneering investigations of electromagnetic waves.

Conducting surfaces are not the only reflectors of electromagnetic waves. Reflections also occur at an interface between two insulating materials with different dielectric or magnetic properties. The mechanical analog is a junction of two strings with equal tension but different linear mass density. In general, a wave incident on such a boundary surface is partly transmitted into the second material and partly reflected back into the first. The partial transmission and reflection of light at a glass surface is a familiar phenomenon; light is transmitted through a glass window, but its surfaces also reflect light.

## EXAMPLE 33–6

Prove that the intensity of the standing wave discussed in this section is zero.

**SOLUTION** The intensity of the wave is the average value $S_{av}$ of the Poynting vector. Let's first find the instantaneous value of $S$. Using the wave functions of Eqs. (33–36) and (33–37) with Eq. (33–25), which defines $S$, we find

$$S = \frac{EB}{\mu_0} = \frac{(-2E_{max}\cos\omega t \sin kx)(2B_{max}\sin\omega t\cos kx)}{\mu_0}.$$

Using the identity $\sin 2A = 2\sin A\cos A$, we can rewrite this as

$$S = -\frac{E_{max}B_{max}\sin 2\omega t\sin 2kx}{\mu_0}.$$

The average value of a sine function over any whole number of cycles is zero. Thus *the time average of $S$ at any point is zero*: $I = S_{av} = 0$. This is just what we should expect. We formed our standing wave by superposing two waves with the same frequency and amplitude traveling in opposite directions. All the energy transferred by one wave is completely cancelled by an equal amount transferred in the opposite direction by the other wave. When we use waves to transmit power, it is important to avoid standing waves.

## EXAMPLE 33–7

Electromagnetic standing waves are set up in a cavity used for electron spin resonance studies. The cavity has two parallel, highly conducting walls separated by 1.50 cm. a) Calculate the longest wavelength and lowest frequency of electromagnetic standing waves between the walls. b) Where in the cavity is the maximum magnitude of the electric field and of the magnetic field?

**SOLUTION** a) The longest possible wavelength corresponds to $n = 1$ in Eq. (33–40):

$$\lambda_1 = \frac{2L}{1} = \frac{2(1.50\text{ cm})}{1} = 3.00\text{ cm}.$$

The corresponding frequency is

$$f = \frac{c}{\lambda} = \frac{3.00\times10^8\text{ m/s}}{3.00\times10^{-2}\text{ m}} = 1.00\times10^{10}\text{ Hz} = 10\text{ GHz}.$$

Or, from Eq. (33–41) with $n = 1$,

$$f = 1\frac{c}{2L} = \frac{3.00\times10^8\text{ m/s}}{2(1.50\times10^{-2}\text{ m})} = 1.00\times10^{10}\text{ Hz}.$$

b) With $n = 1$ there is a single half-wavelength between the walls. The electric field has nodes at the walls ($E = 0$) and an antinode (where the maximum magnitude of $\boldsymbol{E}$ occurs) midway between them. The magnetic field has *antinodes* at the walls and a node midway between them.

## 33–7 THE ELECTROMAGNETIC SPECTRUM

As has been indicated at several points in this chapter, electromagnetic waves cover an extremely broad spectrum of wavelength and frequency. Radio and TV transmission, visible light, infrared and ultraviolet radiation, x rays, and gamma rays all form parts of the **electromagnetic spectrum.** The extent of this spectrum is shown in Fig. 33–15, which gives approximate wavelength and frequency ranges for the various segments. Despite vast differences in their uses and means of production, all these are electromagnetic waves. They all have the general characteristics described in preceding sections, including the common propagation speed (in vacuum) $c = 299{,}792{,}458$ m/s. All are the same in principle; they differ in frequency $f$ and wavelength $\lambda$, but the relation $c = \lambda f$ holds for each.

We can detect only a very small segment of this spectrum directly through our sense of sight. We call this range **visible light.** Its wavelengths range from about 400 to 700 nm (400 to $700 \times 10^{-9}$ m), with corresponding frequencies from about 750 to 430 THz (7.5 to $4.3 \times 10^{14}$ Hz). Different parts of the visible spectrum evoke in humans the sensations of different colors. Wavelengths for colors in the visible spectrum are (very approximately) as follows:

| | |
|---|---|
| 400 to 440 nm | Violet |
| 440 to 480 nm | Blue |
| 480 to 560 nm | Green |
| 560 to 590 nm | Yellow |
| 590 to 630 nm | Orange |
| 630 to 700 nm | Red |

By using special sources or filters we can select a narrow band of wavelengths with a range of, say, from 1 to 10 nm. Such light is approximately *monochromatic* (single-color) light. Absolutely monochromatic light with only a single wavelength is an unattainable idealization. When we use the expression "monochromatic light with wavelength 550 nm" with reference to a laboratory experiment, we really mean a small band of wavelengths *around* 550 nm. One distinguishing characteristic of light from a *laser* is that it is much more nearly monochromatic than light produced in any other way.

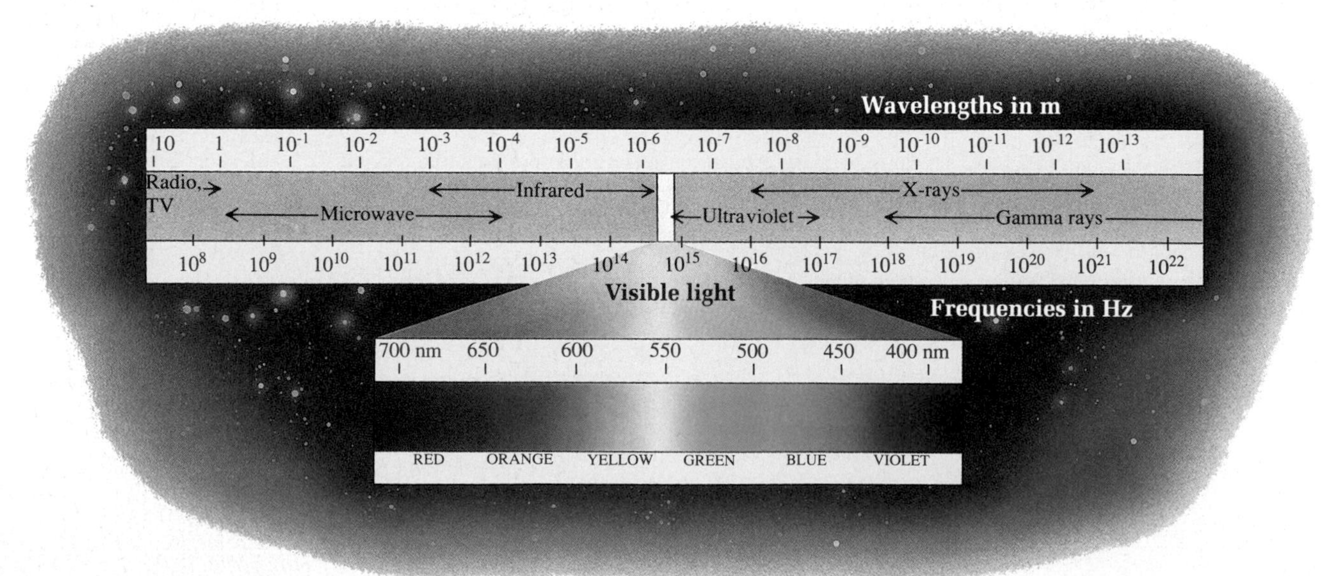


**33–15** The electromagnetic spectrum. The frequencies and wavelengths found in nature extend over such a wide range that we have to use a logarithmic scale to show all important bands. The boundaries between bands are somewhat arbitrary.

# The Electromagnetic Spectrum

The Andromeda galaxy seen in visible light. The spiral arms of the galaxy contain relatively young hot, blue stars, embedded in clouds of dust and gas. The central bulge contains older yellow and red giant stars.

# BUILDING PHYSICAL INTUITION

Visible light is just a small part of the electromagnetic spectrum. If we could see other forms of electromagnetic radiation, such as infrared or ultraviolet, the world would seem a very different place. Astronomers obtain information about astronomical objects according to the full range of radiation they emit, with often surprising results.

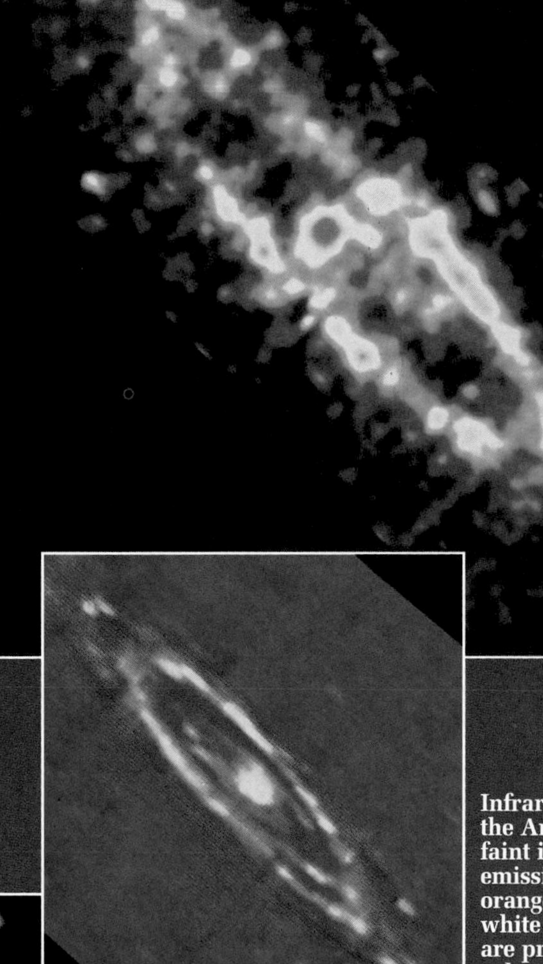

Radio-wave map of the Andromeda galaxy. The ring-like structure shows radio waves emitted by the spiral arms. The center of the galaxy is a nearly spherical source of intense radio waves. The orange and yellow patches in the ring are primarily thermal emissions; the central area is mostly (80%) due to electrons moving in magnetic fields.

Infrared image, processed by computer, of the Andromeda galaxy. Blue areas indicate faint infrared radiation; more intense emissions are indicated by green, red, orange, and yellow areas. The yellow and white areas show regions where young stars are probably forming. The center of the galaxy is a strong emitter at infrared wavelengths.

X-ray view of the Andromeda galaxy. Over 100 separate X-ray sources have been identified in this galaxy. The central area of the galaxy is a strong X-ray source, with about 1000 times the X-ray emission from the center of our own galaxy.

## *33–8 RADIATION FROM AN ANTENNA

Our discussion of electromagnetic waves in this chapter has centered mostly on *plane waves*, which propagate in a single direction (often chosen as the $x$-axis of our coordinate system). In any plane perpendicular to the direction of propagation of the wave, the $\boldsymbol{E}$ and $\boldsymbol{B}$ fields are uniform at any one instant of time. Though easy to describe, plane waves are by no means the simplest to produce experimentally. Any charge or current distribution that oscillates sinusoidally with time produces sinusoidal electromagnetic waves, but in general there is no reason to expect them to be plane waves.

A simple example of an oscillating charge distribution is an **oscillating electric dipole,** a pair of electric charges with equal magnitude and opposite sign, with the charge magnitude varying sinusoidally with time. Such an oscillating dipole can be constructed in various ways, depending on frequency, and we won't worry about the details. The radiation pattern is fairly complex, but at points far away from the dipole (in comparison to its dimensions and the wavelength of the radiation) it becomes fairly simple. We'll confine our description to this far region.

The radiation from an oscillating dipole is *not* a plane wave, but a wave that travels out radially in all directions from the source. The wave fronts are not planes; in the far region they are expanding concentric spheres centered at the source.

Figure 33–16 shows an oscillating dipole with maximum dipole moment $p_0$, aligned with the $z$-axis. The $\boldsymbol{E}$ and $\boldsymbol{B}$ fields at a point described by the spherical coordinates $(r, \theta, \phi)$ have the directions shown during half the cycle and the opposite direction during the other half. Their magnitudes in the far region are

$$E = \frac{p_0 k^2}{4\pi\epsilon_0} \frac{\sin\theta}{r} \sin(\omega t - kr),$$
$$B = \frac{p_0 k^2}{4\pi\epsilon_0 c} \frac{\sin\theta}{r} \sin(\omega t - kr). \quad (33\text{–}42)$$

Figure 33–17 shows a cross section of the radiation pattern at one instant. The circles are cross sections of the spherical wave fronts. At each point, $\boldsymbol{E}$ is in the plane of the section, and $\boldsymbol{B}$ is perpendicular to that plane. Both fields are perpendicular to the radial direction at each point, so the wave is *transverse*. The field magnitudes are greatest in the directions perpendicular to the dipole, where $\theta = \pi/2$, and there is *no* radiation in the direction of the dipole, where $\theta = 0$.

The direction of the Poynting vector $\boldsymbol{S}$ at each point is radially outward from the source. Because each field magnitude is proportional to $1/r$, the intensity $I$ (the average value of the Poynting vector magnitude) is proportional to $1/r^2$. It is also proportional to $\sin^2\theta$.

In the oscillating electric dipole the charges continuously *accelerate* as they vibrate back and forth with simple harmonic motion. In fact, it's a general result of Maxwell's

**33–16** One cycle in the production of an electromagnetic wave by an oscillating electric dipole. Only the electric field is indicated. The figure is not to scale.

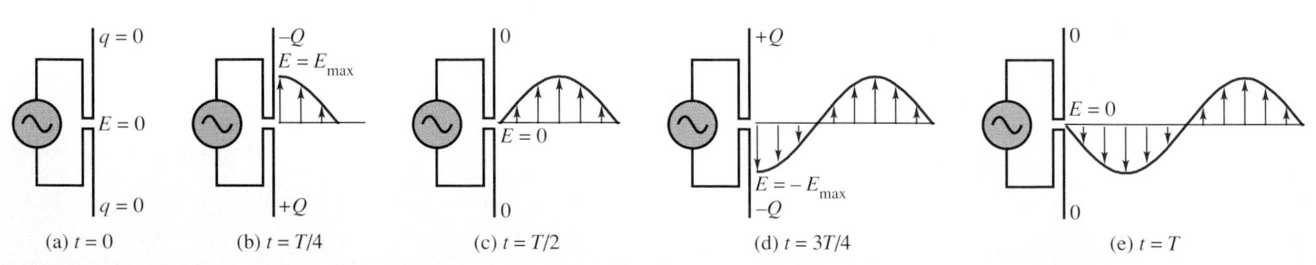

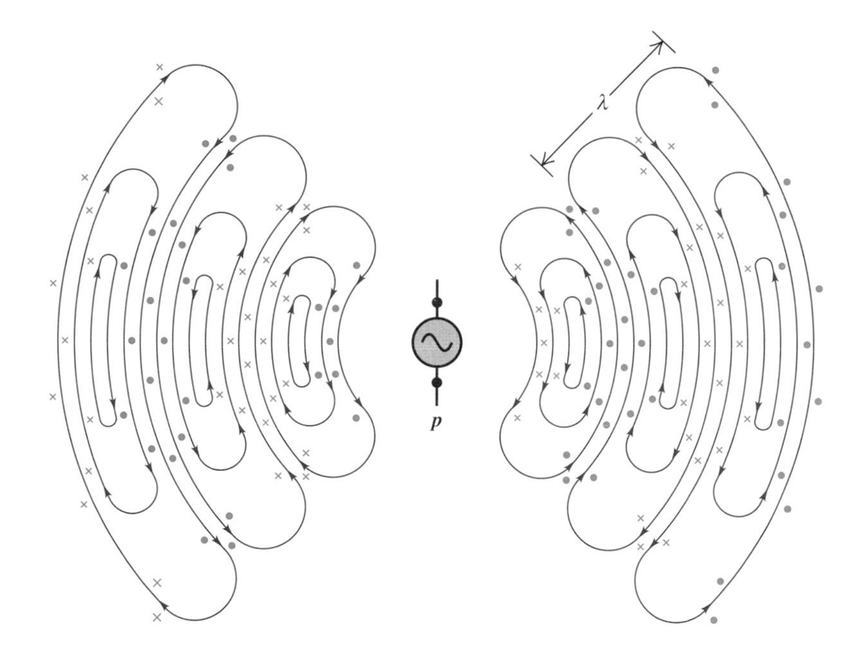


**33–17** Representation of the electric field (red lines) and the magnetic field (blue dots and crosses) in a plane containing an oscillating electric dipole. During one period the inner loop moves out and expands to become the outer loop. You can use Eq. (33–26) and the right-hand rule to show that the Poynting vector $\boldsymbol{S}$ is radially outward at each point within the pattern. No energy is radiated along the axis of the dipole.

equations that *every* accelerated charge radiates electromagnetic energy. This radiation is the reason for the shielding required around high-energy particle accelerators and high-voltage power supplies in TV sets. Oscillating charges are also present in magnetic-dipole sources, such as loop antennas, that use a sinusoidal current in a coil.

We have seen that electromagnetic waves can be *reflected* by conducting surfaces. When the surface is large in comparison to the wavelength of the radiation, the reflection behaves like reflection of light rays from a mirror, which we will study in Chapter 35. Large parabolic mirrors several meters in diameter are used as both transmitting and receiving antennas for microwave communications signals; typical wavelengths are a few centimeters. In transmission a small dipole located at the focus of a paraboloidal reflector emits spherical waves that are reflected and focused into a narrow, well-defined beam, just as the reflector in a car headlight focuses the multidirectional light from the glowing filament into a beam. The receiving reflector gathers wave energy over its whole area and reflects it to the focus of the parabola, where a detecting device is placed. Figure 33–18 shows an antenna installation using a large parabolic reflector.

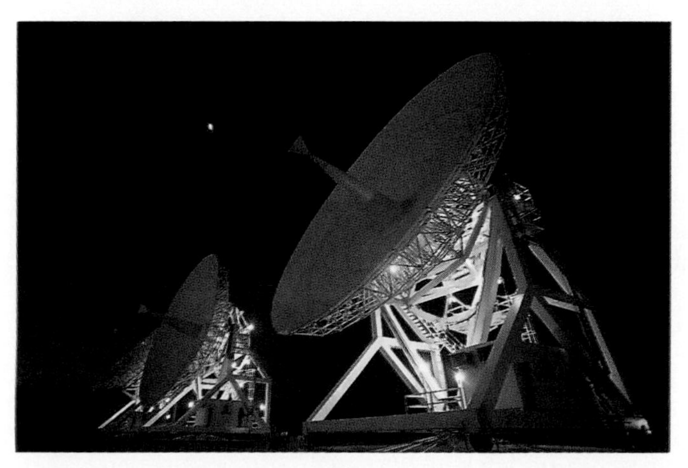

(a)

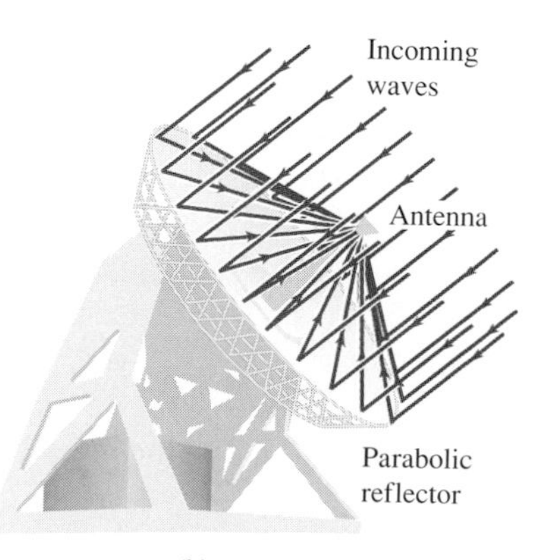


(b)

**33–18** (a) A microwave-receiving antenna in White Sands, New Mexico. (b) The parabolic shape of the antenna reflects incoming waves to the detecting device at the focus of the paraboloid.

## SUMMARY

### KEY TERMS

electromagnetic wave
Maxwell's equations
transverse wave
wave equation
plane wave
energy density
Poynting vector
intensity
radiation pressure
index of refraction
standing wave
nodal plane
electromagnetic spectrum
visible light
oscillating electric dipole

• Maxwell's equations predict the existence of electromagnetic waves that propagate in vacuum with a speed equal to the measured value of the speed of light. In a plane wave, ***E*** and ***B*** are uniform over any plane perpendicular to the propagation direction. Faraday's law requires that

$$E = cB, \tag{33–4}$$

Ampere's law requires that

$$B = \mu_0 \epsilon_0 c E, \tag{33–8}$$

where $c$ is the propagation speed. For both of these requirements to be satisfied,

$$c = \frac{1}{\sqrt{\epsilon_0 \mu_0}}. \tag{33–9}$$

• Electromagnetic waves are transverse; the ***E*** and ***B*** fields are perpendicular to the direction of propagation and to each other. The direction of propagation is the direction of $\mathbf{E} \times \mathbf{B}$.

• For a sinusoidal plane wave traveling in the $+x$-direction,

$$E = E_{max} \sin(\omega t - kx),$$
$$B = B_{max} \sin(\omega t - kx), \tag{33–16}$$

$$E_{max} = cB_{max}. \tag{33–18}$$

• The energy flow rate (power per unit area) in an electromagnetic wave is given by the Poynting vector ***S***:

$$\mathbf{S} = \frac{1}{\mu_0} \mathbf{E} \times \mathbf{B}. \tag{33–26}$$

The time-average value of the magnitude $EB/\mu_0$ of the Poynting vector is called the intensity $I$ of the wave. For a sinusoidal wave,

$$I = S_{av} = \frac{E_{max} B_{max}}{2\mu_0} = \frac{E_{max}^2}{2\mu_0 c} = \frac{1}{2}\sqrt{\frac{\epsilon_0}{\mu_0}} E_{max}^2 = \frac{1}{2}\epsilon_0 c E_{max}^2. \tag{33–27}$$

Electromagnetic waves also carry momentum; the rate of transfer of momentum per unit cross-section area is

$$\frac{1}{A}\frac{dp}{dt} = \frac{S}{c} = \frac{EB}{\mu_0 c}. \tag{33–29}$$

• When an electromagnetic wave travels through a dielectric, the wave speed $v$ is

$$v = \frac{1}{\sqrt{\epsilon \mu}} = \frac{1}{\sqrt{KK_m}} \frac{1}{\sqrt{\epsilon_0 \mu_0}} = \frac{c}{\sqrt{KK_m}}. \tag{33–30}$$

In a dielectric the Poynting vector is

$$\mathbf{S} = \frac{1}{\mu} \mathbf{E} \times \mathbf{B}, \tag{33–33}$$

and the intensity $I$ of a sinusoidal wave is

$$I = \frac{E_{max} B_{max}}{2\mu} = \frac{E_{max}^2}{2\mu v} = \frac{1}{2}\sqrt{\frac{\epsilon}{\mu}} E_{max}^2 = \frac{1}{2}\epsilon v E_{max}^2. \tag{33–35}$$

• If a reflecting surface is placed at $x = 0$, the incident and reflected waves form a standing wave. Nodal planes for $\boldsymbol{E}$ occur at $kx = 0, \pi, 2\pi, \ldots$, and nodal planes for $\boldsymbol{B}$ at $kx = \pi/2, 3\pi/2, 5\pi/2, \ldots$. At each point the sinusoidal variations of $E$ and $B$ are 90° out of phase; the nodes of $\boldsymbol{B}$ coincide with the antinodes of $\boldsymbol{E}$, and conversely.

• The electromagnetic spectrum covers a range of frequencies from at least 1 to $10^{24}$ Hz and a correspondingly broad range of wavelengths. Visible light is a very small part of this spectrum, with wavelengths of 400 to 700 nm.

• An oscillating dipole produces a wave that radiates out in all directions. At points far from the source, $\boldsymbol{E}$ and $\boldsymbol{B}$ are perpendicular to each other and to the radial direction, so this wave is transverse. The intensity depends on direction; it is zero along the dipole axis and greatest in the plane perpendicular to that axis.

## EXERCISES

### Section 33–2 Speed of an Electromagnetic Wave

**33–1 TV Ghosting.** In a TV picture, ghost images are formed when the signal from the transmitter travels directly to the receiver and indirectly after reflection from a building or other large metallic mass. In a 25-in. set, the ghost is about 1.0 cm to the right of the principal image if the reflected signal arrives 0.60 $\mu$s after the principal signal. In this case, what is the difference in path lengths for the two signals?

**33–2** a) How much time does it take light to travel from the sun to the earth, a distance of $1.49 \times 10^{11}$ m? b) The star Alpha Centauri is 4.3 light-years from the earth. (One light-year is the distance light travels in one year.) What is this distance in kilometers?

**33–3** Radio station WBUR in Boston broadcasts at a frequency of 90.9 MHz. At a point some distance from the transmitter, the maximum magnetic field of the electromagnetic wave that it emits is $2.40 \times 10^{-11}$ T. a) What is the wavelength of the wave? b) What is the maximum electric field?

**33–4** The maximum electric field in the vicinity of a certain radio transmitter is $6.00 \times 10^{-3}$ V/m. What is the maximum magnitude of the $\boldsymbol{B}$ field? How does this compare in magnitude with the earth's field?

### Section 33–3 Sinusoidal Waves

**33–5** A tunable dye laser emits a sinusoidal electromagnetic wave that travels in the $+y$-direction. The frequency is $f = 6.00 \times 10^{13}$ Hz, and the $\boldsymbol{B}$-field is along the $x$-axis with an amplitude of $4.00 \times 10^{-4}$ T. Write vector equations for $\boldsymbol{E}$ and $\boldsymbol{B}$ as functions of time and position $y$.

### Section 33–4 Energy in Electromagnetic Waves

**33–6** Consider each of the electric- and magnetic-field orientations given below. In each case, what is the direction of propagation of the wave? a) $\boldsymbol{E} = E\boldsymbol{i}$, $\boldsymbol{B} = B\boldsymbol{j}$; b) $\boldsymbol{E} = -E\boldsymbol{j}$, $\boldsymbol{B} = B\boldsymbol{i}$; c) $\boldsymbol{E} = E\boldsymbol{k}$, $\boldsymbol{B} = -B\boldsymbol{i}$; d) $\boldsymbol{E} = E\boldsymbol{j}$, $\boldsymbol{B} = -B\boldsymbol{k}$.

**33–7** Verify that all the expressions in Eq. (33–25) are equivalent.

**33–8** For an electromagnetic wave traveling in the $-x$-direction, as represented by Eqs. (33–19), show that the Poynting vector a) is in the $-x$-direction; b) has average magnitude given by Eqs. (33–27).

**33–9** At a distance of 50.0 km from a radio station antenna, the electric-field amplitude is $E_{max} = 0.0600$ V/m. a) What is the magnetic-field amplitude $B_{max}$ at this point? b) Assuming that the antenna radiates equally in all directions above the ground (which is probably not the case), what is the total power output of the station? c) At what distance from the antenna is $E_{max} = 0.0300$ V/m, half the above value?

**33–10 A Car Phone.** In a cellular phone communication network, a sinusoidal electromagnetic wave emitted by a microwave antenna has a wavelength of 4.00 cm and an electric field amplitude of $4.50 \times 10^{-2}$ V/m at a distance of 4.00 km from the antenna. a) What is the frequency of the wave? b) What is the magnetic field amplitude? c) What is the intensity (average power per unit area) of the wave?

**33–11** Assume that 10% of the power input to a 100-W lamp is radiated uniformly in all directions as light of wavelength 500 nm. Calculate $E_{max}$ and $B_{max}$ for the 500-nm light at a distance of 3.00 m from the source.

**33–12** At the floor of a room the intensity of light from bright overhead lights is 600 W/m$^2$. Find a) the average momentum density (momentum per unit volume); b) the average radiation pressure on a totally absorbing section of the floor.

**33–13** If the intensity of direct sunlight at a point on the earth's surface is 0.60 kW/m$^2$, find a) the average momentum density (momentum per unit volume); b) the average momentum flow rate (momentum carried through a surface area $A$ in unit time) in the sunlight.

**33–14** The intensity from a bright light source is 800 W/m$^2$. Find the average radiation pressure (in pascals) on a) a totally absorbing surface; b) a totally reflecting surface. Also express your results in atmospheres.

## Section 33–5 Electromagnetic Waves in Matter

*33–15** An electromagnetic wave with a frequency of 20.0 MHz propagates with a speed of $2.20 \times 10^8$ m/s in a certain piece of glass. Find a) the wavelength of the wave in the glass; b) the wavelength of a wave of the same frequency propagating in air; c) the refractive index $n$ of the glass for an electromagnetic wave with this frequency; d) the dielectric constant for glass at this frequency, assuming that the relative permeability is unity.

## Section 33–6 Standing Waves

**33–16** An electromagnetic standing wave in air has frequency $4.00 \times 10^{14}$ Hz. a) What is the distance between nodal planes of the $\boldsymbol{E}$ field? b) What is the distance between a nodal plane of $\boldsymbol{E}$ and the closest nodal plane of $\boldsymbol{B}$?

**33–17** An electromagnetic wave propagating in a certain material has frequency $2.00 \times 10^{10}$ Hz. The nodal planes of the $\boldsymbol{B}$ field are 5.00 mm apart. What are a) the wavelength of the wave in this material; b) the distance between adjacent nodal planes of the $\boldsymbol{E}$ field; c) the speed of propagation of the wave?

*33–18** Show that the electric and magnetic fields for a standing wave given by Eqs. (33–36) and (33–37) a) satisfy the wave equation, Eq. (33–15); b) satisfy Eqs. (33–12) and (33–14).

## Section 33–7 The Electromagnetic Spectrum

**33–19** For waves propagating in air, what is the wavelength in meters and nanometers of a) x rays with a frequency of $3.00 \times 10^{17}$ Hz; b) orange light with a frequency of $5.00 \times 10^{14}$ Hz?

**33–20** For an electromagnetic wave propagating in air, determine the frequency of a wave with a wavelength of a) 1.0 km; b) 1.0 m; c) 1.0 $\mu$m; d) 1.0 nm.

# PROBLEMS

*33–21** Show that the magnetic field $B_z(x, t)$ in a plane electromagnetic wave must satisfy Eq. (33–15). Do this by taking the partial derivative of Eq. (33–12) with respect to $t$ and the partial derivative of Eq. (33–14) with respect to $x$ and then combining the results.

*33–22** Consider a sinusoidal electromagnetic wave with electric and magnetic fields $\boldsymbol{E} = E_{\max}\boldsymbol{j} \sin(\omega t - kx)$ and $\boldsymbol{B} = B_{\max}\boldsymbol{k} \sin(\omega t - kx + \phi)$, with $\phi < 2\pi$. Show that if $\boldsymbol{E}$ and $\boldsymbol{B}$ are to satisfy Eqs. (33–12) and (33–14), then $E_{\max} = cB_{\max}$ and $\phi = 0$. (The result $\phi = 0$ means that the $\boldsymbol{E}$ and $\boldsymbol{B}$ fields oscillate in phase.)

**33–23** For a sinusoidal electromagnetic wave in vacuum, such as that described by Eqs. (33–16), show that the average density of energy in the electric field is the same as that in the magnetic field.

**33–24 Power from the Sun.** The energy flow to the earth associated with sunlight is about 1.4 kW/m$^2$. a) Find the maximum values of $E$ and $B$ for a sinusoidal wave with this intensity. b) The distance from the earth to the sun is about $1.5 \times 10^{11}$ m. Find the total average power radiated by the sun.

**33–25** A small mirror with area $A = 6.00$ cm$^2$ faces a light bulb that is 4.00 m away. At the mirror the electric-field amplitude of the light from the bulb is 15.0 V/m. a) How much energy is incident on the mirror in 1 s? b) What is the average radiation pressure exerted by the light from the bulb on the mirror? c) What is the total average radiant power output of the light bulb if the bulb is assumed to radiate uniformly in all directions?

**33–26** A 75.0-W light bulb radiates uniformly in all directions. At what distance from the bulb would the electric-field amplitude of the light be 8.00 V/m if it had a sinusoidal waveform?

**33–27** A small helium-neon laser (that emits red visible light) emits light with a power of 6.00 mW in a beam that has a diameter of 4.00 mm. a) What are the amplitudes of the electric and magnetic fields of the light? b) What are the average energy densities associated with the electric field and with the magnetic field? c) What is the total energy contained in a 0.500-m length of the beam?

**33–28** A plane sinusoidal electromagnetic wave in air has a wavelength of 6.00 cm and an $\boldsymbol{E}$-field amplitude of 50.0 V/m. a) What is the frequency? b) What is the $\boldsymbol{B}$ field amplitude? c) What is the intensity? d) What average force does this radiation exert on a totally absorbing surface with area 0.500 m$^2$ perpendicular to the direction of propagation?

**33–29 Poynting Vector in a Solenoid.** A very long solenoid with $n$ turns per unit length and radius $a$ carries a current $i$ that is increasing at a constant rate of $di/dt$. a) Calculate the induced electric field at a point inside the solenoid at a distance $r$ from the solenoid axis. b) Compute the magnitude and direction of the Poynting vector at this point. (The direction of $\boldsymbol{S}$ corresponds to flow of energy into the volume where energy is being stored in the magnetic field.)

**33–30** Consider the standing wave given by Eqs. (33–36) and (33–37). Let $\boldsymbol{E}$ be parallel to the $y$-axis and $\boldsymbol{B}$ be parallel to the $z$-axis; the equations then give the components of $\boldsymbol{E}$ and $\boldsymbol{B}$ along each axis. a) Plot the energy density as a function of $x$, $0 < x < \pi/k$, for the times $t = 0$, $\pi/4\omega$, $\pi/2\omega$, $3\pi/4\omega$, and $\pi/\omega$. b) Find the direction of $\boldsymbol{S}$ in the regions $0 < x < \pi/2k$ and $\pi/2k < x < \pi/k$ at the times $t = \pi/4\omega$, $3\pi/4\omega$. c) Use the results of part (b) to explain the plots obtained in part (a).

**33–31** A cylindrical conductor with circular cross section has radius $a$ and resistivity $\rho$ and carries constant current $I$. a) What are the magnitude and direction of the electric field vector $\boldsymbol{E}$ at a point just inside the wire at distance $a$ from the axis? b) What are the magnitude and direction of the magnetic field vector $\boldsymbol{B}$ at the same point? c) What are the magnitude and direction of the Poynting vector $\boldsymbol{S}$ at the same point? (The direction of $\boldsymbol{S}$ corresponds to energy entering the volume of the conductor, the energy that is dissipated in the conductor.) d) Use the result of part (c) to find the rate of flow of energy into the volume occupied by a length $l$ of the conductor. (*Hint:* Integrate the Poynting vector over the surface of this volume.) Compare your result to the rate of dissipation of energy in the same volume.

**33–32** A capacitor consists of two circular plates with radius $r$ separated by a distance $l$. Neglecting fringing, show that while the capacitor is being charged, the rate at which energy flows into the space between the plates is equal to the rate at which the electrostatic energy stored in the capacitor increases. (*Hint:* Integrate the Poynting vector over the surface of the space between the plates.)

**33–33** A circular loop of wire can be used as a radio frequency antenna. If a 0.400-m diameter antenna is located 500 m from a 10.0-MHz source with total power 1.00 MW, what is the maximum emf induced in the loop? (Assume that the plane of the antenna loop is perpendicular to the direction of the radiation's magnetic field and that the source radiates uniformly in all directions.)

**33–34 Microwave Power.** It has been proposed that to aid in satisfying the energy needs of the United States, solar-power-collecting satellites could be placed in earth orbit, and the power they collect could be beamed down to earth as microwave radiation. For a microwave beam with a cross-section area of 40.0 $m^2$ and a total power of 1.00 kW at the earth's surface, what is the amplitude of the electric field of the beam at the earth's surface?

**33–35** A space-walking astronaut has run out of fuel for her jet-pack and is floating 16.0 m from the space shuttle with zero relative velocity. The astronaut and all her equipment have a total mass of 200 kg. If she uses her 100-W flashlight as a light rocket, how long will it take her to reach the shuttle?

## CHALLENGE PROBLEMS

**33–36** Electromagnetic radiation is emitted by accelerating charges. The rate of energy emission from an accelerating charge that has charge $q$ and acceleration $a$ is given by

$$\frac{dE}{dt} = \frac{q^2a^2}{6\pi\epsilon_0 c^3},$$

where $c$ is the speed of light. a) Verify that this equation is dimensionally correct. b) If a proton with a kinetic energy of 5.0 MeV is traveling in a particle accelerator in a circular orbit of radius 1.0 m, what fraction of its energy does it radiate per second? c) Consider an electron orbiting with the same speed and radius. What fraction of its energy does it radiate per second?

**33–37 The Classical Hydrogen Atom.** The electron in a hydrogen atom can be considered to be in a circular orbit with a radius of 0.0529 nm and a kinetic energy of 13.6 eV. If the electron behaved classically, how much energy would it radiate per second? (See Challenge Problem 33–36.) What does this tell you about the use of classical physics in describing the atom?

**33–38 Solar Sailing.** The concept of solar sailing has appeared in science fiction and in proposals to NASA. A solar sailcraft uses a large, low-mass sail and the energy and momentum of sunlight for propulsion. The total power output of the sun is $3.9 \times 10^{26}$ W. a) Should the sail be absorbing or reflective? Why? b) How large a sail is necessary to propel a $2.0 \times 10^4$ kg spacecraft against the gravitational force of the sun? (Express your result in square miles.) c) Explain why your answer to part (b) is independent of the distance from the sun.

CHAPTER 34

# Nature and Propagation of Light

A rainbow appears over Isaac Newton's family home in Woolsthorpe, England. A rainbow is formed by refraction, reflection, and dispersion of light in spherical drops of water. Red bends most and violet least, with all other colors in between.

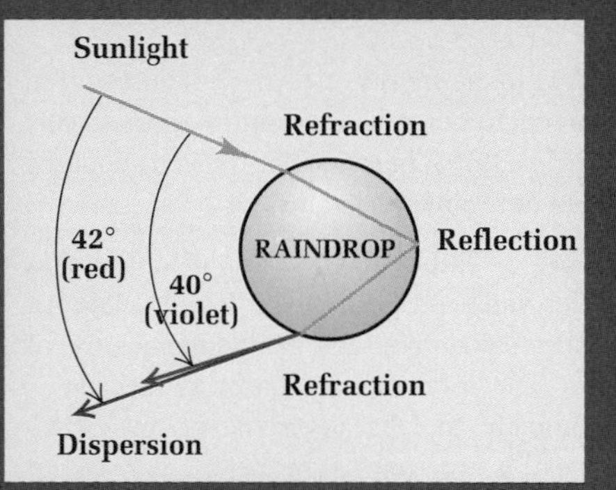


Your eye sees different colors of light corresponding to the angle at which you view drops at different heights. In the primary rainbow you see red light from the highest drops, which disperse red light most; you see violet light from the lowest drops, which disperse violet light least. (Dispersion angles are greatly exaggerated in this drawing.)

## KEY CONCEPTS

**• Light is an electromagnetic wave. Its propagation can be described by wave fronts or rays. Light also has particle (quantum) aspects.**

**• In geometric optics, light is described by rays that travel in straight lines in a homogeneous material. At an interface between two materials, a ray is partly reflected and partly transmitted, according to simple rules. In special cases a ray may be totally reflected.**

**• The optical properties of a material are described by its index of refraction. The dependence of index of refraction on wavelength is called dispersion.**

**• Light is a transverse wave and has polarization. Any light wave can be described as a superposition of waves polarized in two perpendicular directions. A polarizing filter passes only waves with a specific direction of polarization.**

**• The scattering of light by a gas depends on wavelength, and scattered light is partly polarized.**

**• Huygens' principle, used in analysis of wave behavior of light, states that every point in a wave front can be considered as the source of secondary waves that spread out from the wave front with definite speed.**

## INTRODUCTION

With this chapter we enter the area of optics, the study of the behavior of light and other electromagnetic waves. Not too many years ago, optics was considered a rather "ho-hum" area of physics. An understanding of mirrors and lenses was certainly important for the design of telescopes, microscopes, cameras, and eyeglasses, but there didn't seem to be much new on the horizon. But that was before the development of the laser, optical fibers, holograms, and optoelectronic devices such as LED's, solar cells, charge-coupled devices, photodiodes, and phototransistors, not to mention optical computers and new techniques in medical imaging.

To understand thoroughly the optical behavior of these exciting new devices and techniques and others yet to be developed, we need the basic principles in this chapter and the following ones. We begin with a study of the laws of reflection and refraction and the concepts of dispersion, polarization, and scattering of light. Along the way we compare the various possible descriptions of light in terms of *rays* and by *waves,* and we look at Huygens' principle, an important connecting link between these two viewpoints.

## 34–1 NATURE OF LIGHT

Until the time of Isaac Newton (1642–1727), most scientists thought that light consisted of streams of particles (called *corpuscles*) emitted by light sources. Galileo and others tried (unsuccessfully) to measure the speed of light. Around 1665, evidence of *wave* properties of light began to be discovered. By the early nineteenth century, evidence that light is a wave had grown very persuasive. We will study several optical wave phenomena in Chapters 37 and 38.

In 1873, James Clerk Maxwell predicted the existence of electromagnetic waves and calculated their speed of propagation, as we learned in Chapter 33. This development, along with the experimental work of Heinrich Hertz starting in 1887, showed conclusively that light is indeed an electromagnetic wave.

The wave picture of light is not the whole story, however. Several effects associated with the emission and absorption of light reveal that it also has a particle aspect, in that the energy carried by light waves is packaged in discrete bundles called *photons* or *quanta*. These apparently contradictory wave and particle properties have been reconciled since 1930 with the development of quantum electrodynamics, a comprehensive theory that includes *both* wave and particle properties. The *propagation* of light is best described by a wave model, but understanding emission and absorption requires a particle approach.

The fundamental sources of all electromagnetic radiation are electric charges in accelerated motion. All bodies emit electromagnetic radiation as a result of thermal motion of their molecules; this radiation, called *thermal radiation,* is a mixture of different wavelengths. At sufficiently high temperatures, all matter emits enough visible light to be self-luminous; a very hot body appears "red hot" or even "white hot." Thus hot matter in any form is a light source. Familiar examples are a candle flame, hot coals in a campfire, the coils in an electric room heater, and an incandescent lamp filament (which usually operates at a temperature of about 3000°C).

Light is also produced during electrical discharges through ionized gases. The bluish light of mercury-arc lamps, the orange-yellow of sodium-vapor lamps, and the various colors of "neon" signs are familiar. A variation of the mercury-arc lamp is the *fluorescent* lamp. This light source uses a material called a *phosphor* to convert the ultraviolet radiation from a mercury arc into visible light. This conversion makes fluorescent lamps more efficient than incandescent lamps in converting electrical energy into light.

A special light source that has attained prominence in the last 30 years is the *laser.* It can produce a very narrow beam of enormously intense radiation. High-intensity lasers have been used to cut through steel, to fuse high-melting-point materials, for microsurgery, and in many other applications. A significant characteristic of laser light is that it is much more nearly *monochromatic,* or of a single frequency, than light from any other source.

The speed of light in vacuum is a fundamental constant of nature. The first approximate measurement of the speed of light was made in 1676 by the Danish astronomer Olaf Roemer, from observations of the motion of one of Jupiter's satellites. The first successful *terrestrial* measurement was made by the French scientist Armand Fizeau in 1849, using a reflected light beam interrupted by a notched rotating disk. More refined versions of this experiment were carried out by Jean Foucault and by the American Physicist Albert A. Michelson (1852–1931).

From analysis of all measurements up to 1983, the most probable value for the speed of light in vacuum was

$$c = 2.99792458 \times 10^8 \text{ m/s}.$$

As we explained in Section 1–3, the definition of the second based on the cesium clock is precise to within one part in 10 trillion ($10^{13}$). Until 1983, the definition of the meter was much less precise, to about four parts in a billion ($10^9$). Any attempt to measure the speed of light with greater precision foundered on this limitation. For this reason, in November 1983 the General Conference of Weights and Measures redefined the meter by *defining* the speed of light in vacuum to be precisely 299,792,458 m/s. One meter is now defined to be the distance traveled by light in vacuum in a time of 1/299,792,458 s, with the second defined by the cesium clock.

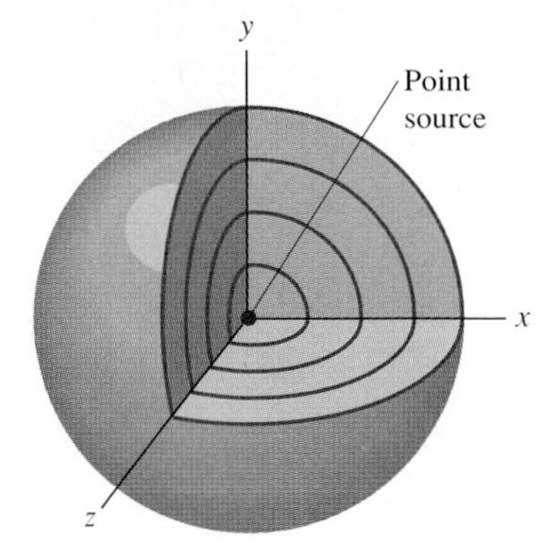


**34–1** Spherical wave fronts spread out uniformly in all directions from a point source in a homogeneous isotropic motionless medium such as still air. Electromagnetic waves do not require a medium for propagation.

We often use the concept of a **wave front** to describe wave propagation. We define a wave front as *the locus of all points at which the phase of vibration of a physical quantity associated with the wave is the same.* That is, at any instant, all points on a wave front are at the same part of the cycle of their periodic variation. A familiar example is a crest of a water wave; when we drop a pebble in a calm pool, the expanding circles formed by the wave crests are wave fronts. Similarly, when sound waves spread out in still air from a pointlike source, any spherical surface concentric with the source is a wave front, as shown in Fig. 34–1. The "pressure crests," the surfaces over which the pressure is maximum, form sets of expanding spheres as the wave travels outward from the source. In diagrams of wave motion we usually draw only parts of a few wave fronts, often choosing consecutive wave fronts that have the same phase, such as crests of a water wave. Such wave crests are separated from each other by one wavelength.

For a light wave (or any other electromagnetic wave) the quantity that corresponds to the displacement of the surface in a water wave or the pressure in a sound wave is the electric or magnetic field. We will often use diagrams that show the shapes of the wave fronts or their cross sections in some reference plane. For example, when electromagnetic waves are radiated by a small light source, we can represent the wave fronts as spherical surfaces concentric with the source or, as in Fig. 34–2a, by the intersections of these surfaces with the plane of the diagram. Far away from the source, where the radii of the spheres have become very large, a section of a spherical surface can be considered as a plane, and we have a *plane* wave (Fig. 34–2b).

It's often convenient to represent a light wave by **rays** rather than by wave fronts. Rays were used to describe light long before its wave nature was firmly established, and in a particle theory of light, rays are the paths of the particles. From the wave viewpoint *a ray is an imaginary line along the direction of travel of the wave.* In Fig. 34–2a the rays are the radii of the spherical wave fronts, and in Fig. 34–2b they are straight lines perpendicular to the wave fronts. When waves travel in a homogeneous isotropic material (a material with the same properties in all regions and in all directions), the rays are always straight lines normal to the wave fronts. At a boundary surface between two materials, such as the surface of a glass plate in air, the wave speed and the direction of a ray usually change, but the ray segments in the air and in the glass are straight lines.

The next several chapters will give you many opportunities to see the interplay of the ray, wave, and particle descriptions of light. The branch of optics for which the ray description is adequate is called **geometric optics;** the branch dealing specifically with wave behavior is called **physical optics.** This chapter and the two following ones are concerned mostly with geometric optics; in Chapters 37 and 38 we will study wave phenomena and physical optics.

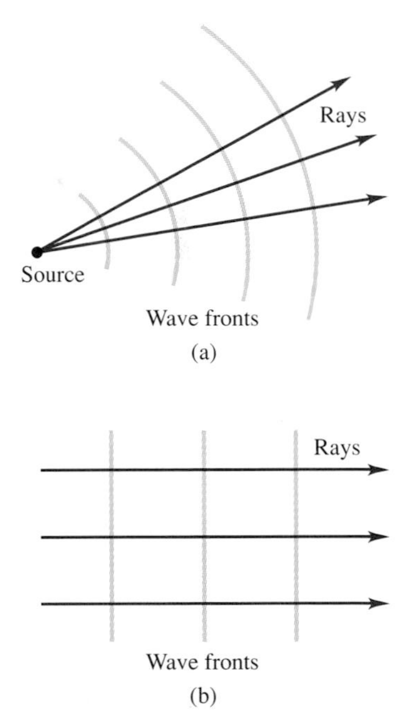


**34–2** Wave fronts and rays. (a) When the wave fronts are spherical, the rays radiate out from the center of the spheres. (b) When the wave fronts are planes, the rays are parallel.

## 34–2 REFLECTION AND REFRACTION

In this section we'll explore the basic elements of the *ray* model of light. When a light wave strikes a smooth interface (a surface separating two transparent materials such as air and glass or water and glass), the wave is in general partly reflected and partly

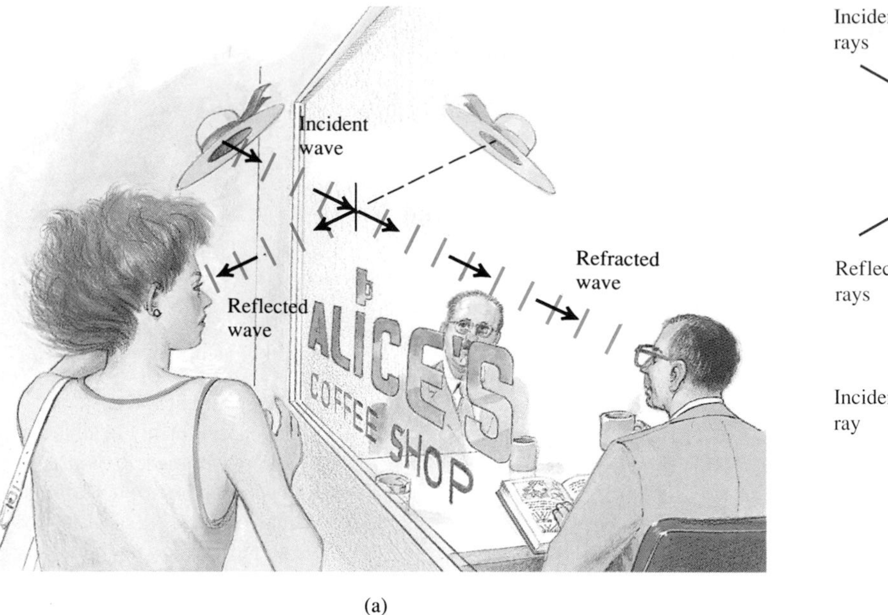


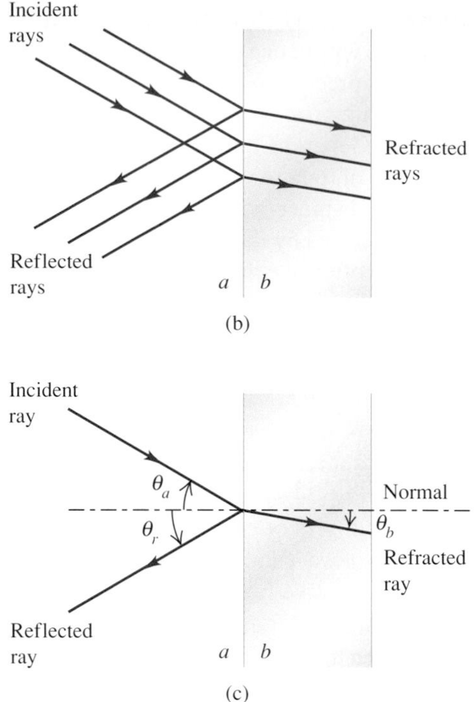


**34–3** (a) A plane wave is in part reflected and in part refracted at the boundary between two media (air and glass). (b) The waves in (a) are represented by rays. (c) For simplicity, only one example of an incident ray, a reflected ray, and a refracted ray is drawn.

*refracted* (transmitted) into the second material, as shown in Fig. 34–3a. For example, when you look into a store window from the street and see a reflection of the street scene, a person inside the store can look out *through* the window at the same scene, as light reaches him by refraction.

The segments of plane waves shown in Fig. 34–3a can be represented by bundles of rays forming *beams* of light (Fig. 34–3b). For simplicity we often draw only one ray in each beam (Fig. 34–3c). Representing these waves in terms of rays is the basis of *geometric optics.* We begin our study with the behavior of an individual ray.

We describe the directions of the incident, reflected, and refracted (transmitted) rays at a smooth interface between two optical materials in terms of the angles they make with the *normal* to the surface at the point of incidence, as shown in Fig. 34–3c. If the interface is rough, both the transmitted light and the reflected light are scattered in various directions, and there is no single angle of transmission or reflection. Reflection at a definite angle from a very smooth surface is called *specular reflection;* scattered reflection from a rough surface is called *diffuse reflection*. This distinction is illustrated in Fig. 34–4. Specular reflection also occurs at a very smooth opaque surface such as highly polished metal or plastic.

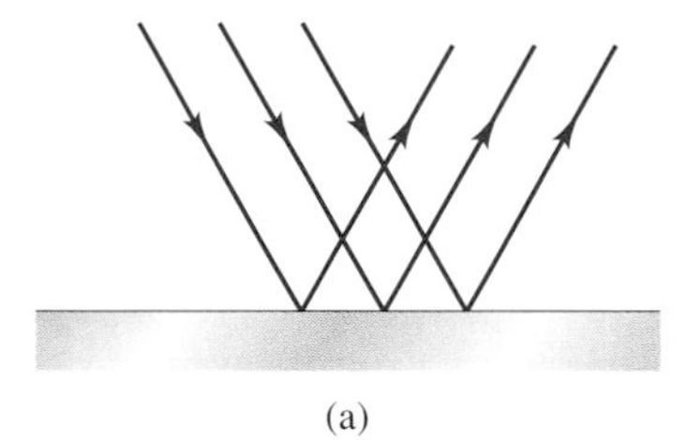


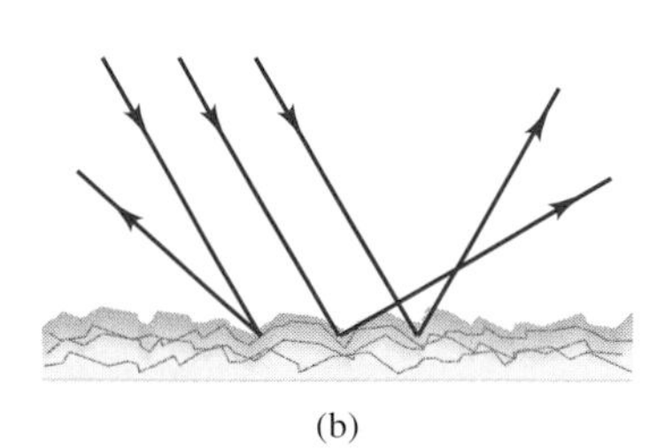


**34–4** (a) Specular reflection. (b) Diffuse reflection.

The **index of refraction** of an optical material (also called the *refractive index*), denoted by $n$, plays a central role in geometric optics. It is the ratio of the speed of light $c$ in vacuum to the speed $v$ in the material.

$$n = \frac{c}{v}. \tag{34–1}$$

Light always travels *more slowly* in a material than in vacuum, so $n$ for any material is always greater than unity. For vacuum, $n = 1$.

Experimental studies of the directions of the incident, reflected, and refracted rays at an interface between two optical materials lead to the following conclusions:

1. **The incident, reflected, and refracted rays and the normal to the surface all lie in the same plane.** If the incident ray is in the plane of the diagram and the boundary surface between the two materials is perpendicular to this plane, then the reflected and refracted rays are in the plane of the diagram.
2. **The angle of reflection $\theta_r$ is equal to the angle of incidence $\theta_a$ for all wavelengths and for any pair of substances.** That is, in Fig. 34–3c,

$$\theta_r = \theta_a. \tag{34–2}$$

   This relation, together with the fact that the incident and reflected rays and the normal all lie in the same plane, is called the **law of reflection.**
3. For monochromatic light and for a given pair of substances, $a$ and $b$, on opposite sides of the interface, **the ratio of the sines of the angles $\theta_a$ and $\theta_b$, where both angles are measured from the normal to the surface, is equal to the inverse ratio of the two indexes of refraction:**

$$\frac{\sin \theta_a}{\sin \theta_b} = \frac{n_b}{n_a}, \tag{34–3}$$

   or

$$n_a \sin \theta_a = n_b \sin \theta_b. \tag{34–4}$$

   This experimental result, together with the fact that the incident and refracted rays and the normal to the surface all lie in the same plane, is called the **law of refraction,** or **Snell's law** after Willebrord Snell (1591–1626). There is some doubt that Snell actually discovered it. The discovery that $n = c/v$ came much later.

Equations (34–3) and (34–4) show that when a ray passes from one material ($a$) into another material ($b$) having a larger index of refraction and smaller wave speed ($n_b > n_a$), the angle $\theta_b$ with the normal is *smaller* in the second material than the angle $\theta_a$ in the first, and the ray is bent *toward* the normal (Fig. 34–3c). When the second index is *less than* the first ($n_b < n_a$), the ray is bent *away from* the normal. The index of refraction of vacuum is unity by definition. When a ray passes from vacuum into a material, it is always bent *toward* the normal; when passing from a material into vacuum, it is always bent *away from* the normal. When the incident ray is perpendicular to the interface, $\theta_a = 0$, $\sin \theta_a = 0$, and the transmitted ray is not bent at all.

When a ray of light approaches the interface from the opposite side (from *below* in Fig. 34–3c), there are again reflected and refracted rays; these two rays, the incident ray, and the normal to the surface again lie in the same plane. The laws of reflection and refraction apply whether the incident ray is in material $a$ or $b$ in the figure. The path of a refracted ray is *reversible;* it follows the same path when going from $b$ to $a$ as when going from $a$ to $b$. The path of a ray *reflected* from any surface is also reversible.

The *intensities* of the reflected and refracted rays depend on the angle of incidence, the two indexes of refraction, and the polarization of the incident ray. For unpolarized light, the fraction reflected is smallest at *normal* incidence (0°), where it is about 4% for an air-glass interface, and it increases with increasing angle of incidence to 100% at grazing incidence, when $\theta_a = 90°$.

**SOLUTION** For mirror 1 the angle of incidence is $\theta_1$, and this equals the angle of reflection. The sum of interior angles in the triangle shown in the figure is 180°, so we see that the angles of incidence and reflection for mirror 2 are $90° - \theta_1$. The total change in direction of the ray after both reflections is therefore

$$(180° - 2\theta_1) + 2\theta_1 = 180°.$$

That is, the ray's final direction is opposite to its original direction.

An alternative viewpoint is that specular reflection reverses the sign of the component of light velocity perpendicular to the surface but leaves the other components unchanged. We invite you to verify this in detail and to use this result to show that when a ray of light is successively reflected by three mirrors forming a corner of a cube (a "corner reflector"), its final direction is again opposite to its original direction. This principle is widely used in tail-light lenses and highway signs to improve their night-time visibility. Apollo astronauts placed arrays of corner reflectors on the moon. By use of laser beams reflected from these arrays, the earth-moon distance has been measured to within 0.15 m.

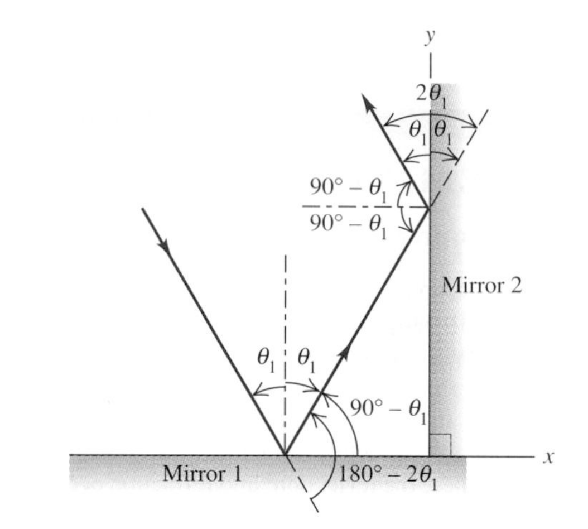


**34–6** A ray moving in the $xy$-plane. The first reflection changes the sign of the $y$-component of its velocity, and the second reflection changes the sign of the $x$-component. For a different ray with a $z$-component of velocity, a third mirror (perpendicular to the two shown) could be used to change the sign of that component.

# 34–3 TOTAL INTERNAL REFLECTION

Figure 34–7a shows several rays diverging from a point source $P$ in a material $a$ with index of refraction $n_a$. The rays strike the surface of a second material $b$ with index $n_b$, where $n_a > n_b$. From Snell's law,

$$\sin\theta_b = \frac{n_a}{n_b}\sin\theta_a.$$

Because $n_a/n_b$ is greater than unity, $\sin\theta_b$ is larger than $\sin\theta_a$; the ray is bent *away from* the normal. Thus there must be some value of $\theta_a$ *less than* 90° for which Snell's law gives $\sin\theta_b = 1$ and $\theta_b = 90°$. This is shown by ray 3 in the diagram, which emerges just grazing the surface, at an angle of refraction of 90°.

The angle of incidence for which the refracted ray emerges tangent to the surface is called the **critical angle,** denoted by $\theta_{\text{crit}}$. (A more detailed analysis using Maxwell's equations shows that as the incident angle approaches the critical angle, the transmitted intensity approaches zero.) If the angle of incidence is *greater than* the critical angle, the

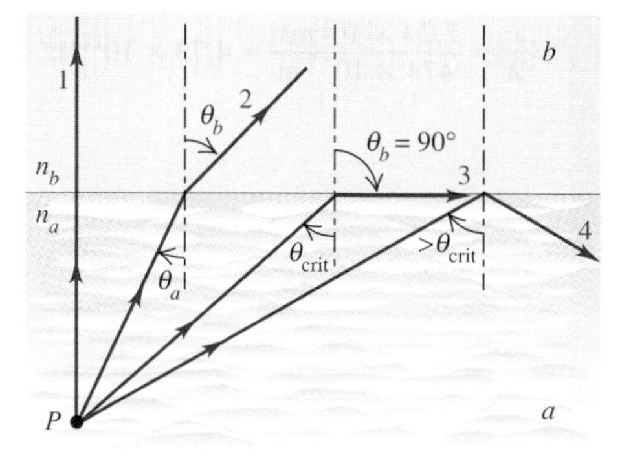


(a)

(b)

**34–7** (a) Total internal reflection. The angle of incidence $\theta_a$, for which the angle of refraction is 90°, is called the critical angle. The reflected portions of rays 1, 2, and 3 are omitted for clarity. (b) Rays of laser light enter the water in the fishbowl from above; they are reflected at the bottom by mirrors tilted at slightly different angles, and one ray undergoes total internal reflection at the air-water interface.

Infrared and visible-light waves have much higher frequencies than radio waves, so a modulated laser beam can transmit an enormous amount of information through a single fiber-optic cable. For example, the Carnegie-Mellon University computer system, which includes several thousand networked personal computers and workstations, is linked partly by fiber-optic cables of the type shown in Fig. 34–10b. Many telephone systems are connected by fiber optics.

Another advantage of fiber-optic cables is that they are electrical insulators. They are immune to electrical interference from lightning and other sources, and they don't allow unwanted currents between source and receiver. They are secure and difficult to "bug," but they are also difficult to splice and tap into.

## *34–4 DISPERSION

Ordinarily, white light is a superposition of waves with wavelengths extending throughout the visible spectrum. The speed of light *in vacuum* is the same for all wavelengths, but the speed in a material substance is different for different wavelengths. Therefore the index of refraction of a material depends on wavelength. The dependence of wave speed and index of refraction on wavelength is called **dispersion.** Figure 34–11 shows the variation of index of refraction with wavelength for a few common optical materials. The value of $n$ usually *decreases* with increasing wavelength and thus *increases* with increasing frequency. Light of longer wavelength usually has greater speed in a material than light of shorter wavelength.

Figure 34–12 shows a ray of white light incident on a prism. The deviation (change of direction) produced by the prism increases with increasing index of refraction and frequency and decreasing wavelength. Violet light is deviated most and red least, and other colors show intermediate deviations. When it comes out of the prism, the light is spread out into a fan-shaped beam, as shown. The light is said to be *dispersed* into a spectrum. The amount of dispersion depends on the *difference* between the refractive indexes for violet light and for red light. From Fig. 34–11 we can see that for a substance such as fluorite, whose refractive index for yellow light is small, the difference between the indexes for red and violet is also small. For silicate flint glass, both the index for yellow light and the difference between extreme indexes are larger.

The brilliance of diamond is due in part to its large dispersion and in part to its unusually large refractive index. Crystals of rutile and of strontium titanate, which can be produced synthetically, have about eight times the dispersion of diamond.

When you experience the beauty of a rainbow, you are seeing the combined effects of dispersion and total internal reflection. The light comes from behind you, is refracted into a water droplet, undergoes total internal reflection from the back surface of the

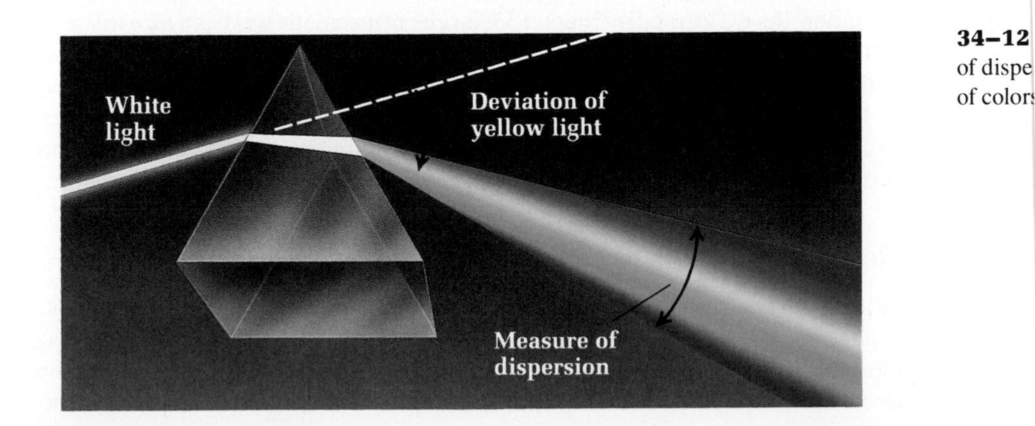


Index c
1.7
1.6
1.5
1.4
400

**34–11** refracti

**34–12** of dispe of color



(a)

(b)

**34–8** (a) Total internal reflection in a Porro prism. (b) A combination of two Porro prisms in binoculars.

sine of the angle of refraction, as computed by Snell's law, has to be greater than unity, which is impossible. Beyond the critical angle the ray *cannot* pass into the upper material; it is trapped in the lower material and is completely reflected internally at the boundary surface. This situation, called **total internal reflection,** occurs only when a ray is incident on an interface with a second material whose index of refraction is *smaller* than that of the material in which the ray is traveling.

We can find the critical angle for two given materials by setting $\theta_b = 90°$ ($\sin\theta_b = 1$) in Snell's law. We then have

$$\sin\theta_{\text{crit}} = \frac{n_b}{n_a}. \tag{34–6}$$

For a glass-air surface, with $n = 1.52$ for the glass,

$$\sin\theta_{\text{crit}} = \frac{1}{1.52} = 0.658, \qquad \theta_{\text{crit}} = 41.1°.$$

The fact that this critical angle is slightly less than 45° makes it possible to use a triangular prism with angles of 45°, 45°, and 90° as a totally reflecting surface. As reflectors, totally reflecting prisms have some advantages over metallic surfaces such as ordinary coated-glass mirrors. Light is *totally* reflected by a prism, but no metallic surface reflects 100% of the light incident on it. Also, the reflecting properties are permanent and not affected by tarnishing.

A 45°-45°-90° prism, used as in Fig. 34–8a, is called a *Porro prism.* Light enters and leaves at right angles to the hypotenuse and is totally reflected at each of the shorter faces. The total change of direction of the rays is 180°. Binoculars often use combinations of two Porro prisms, as shown in Fig. 34–8b.

### EXAMPLE 34–4

**A leaky periscope** A periscope uses two totally reflecting 45°-45°-90° prisms with total internal reflection on the sides opposite the 90° angles. It springs a leak, and the bottom prism is covered with water. Explain why the periscope no longer works.

**SOLUTION** The critical angle for water ($n_b = 1.33$) on glass ($n_a = 1.52$) is

$$\theta_{\text{crit}} = \arcsin\frac{1.33}{1.52} = 61.0°.$$

The 45° angle of incidence is *less than* the 61° critical angle for a totally reflecting prism, so total internal reflection does not occur at the glass-water boundary. Most of the light is transmitted into the water, and very little is reflected back into the prism.

**34–9** A light ray is "trapped" by internal reflections, provided that the angles shown exceed the critical angle.

When a beam of light enters at one end of a transparent r
totally reflected internally and is "trapped" within the rod eve
vided that the curvature is not too great. Such a rod is somet
bundle of fine glass or plastic fibers behaves in the same way
being flexible. A bundle may consist of thousands of individu
of 0.002–0.01 mm in diameter. If the fibers are assembled in t
tive positions of the ends are the same (or mirror images) at t
transmit an image, as shown in Fig. 34–10.

Fiber-optic devices have found a wide range of medical a
called *endoscopes,* which can be inserted directly into the brc
the colon, and so on for direct visual examination. A bundle o
a hypodermic needle for study of tissues and blood vessels fai

Fiber optics are also finding applications in communicat
are used to transmit a modulated laser beam. The amount o
transmitted by a wave (light, radio, or whatever) in a given t
frequency. To see qualitatively why this is so, consider modula
by chopping off some of the wave crests. Think of each crest
digit, with a chopped-off crest representing a zero and an un
number of binary digits that can be transmitted per unit time
frequency of the wave.

**34–10** (a) Image transmission by a bundle of fibers. (b) A fiber-optic cable, used to transmit a modulated laser beam for communication purposes, compared to a larger copper cable that has equal information-transmitting capacity.



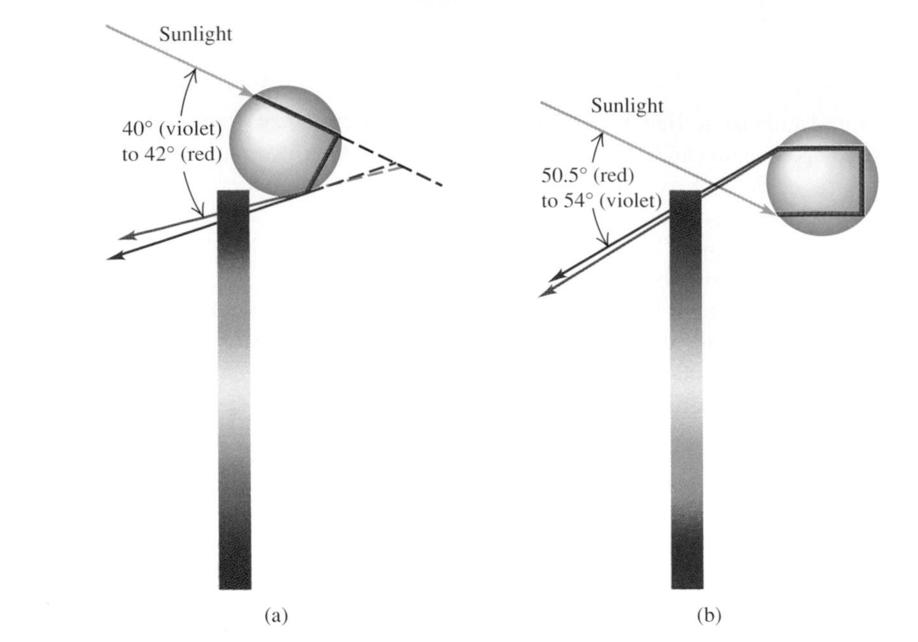


**34–13** Refraction in water drops, forming a rainbow. (a) The outside of the primary bow is red. (b) The inside of the fainter secondary bow is red.

droplet, and is reflected back to you (Fig. 34–13a). Dispersion causes different colors to be refracted at different angles. When you see a second, slightly larger rainbow with its colors reversed, you are seeing the results of dispersion and *two* total internal reflections (Fig. 34–13b).

## 34–5 POLARIZATION

Polarization occurs with all transverse waves. This chapter is about light, but to introduce basic polarization concepts, let's go back to some of the ideas presented in Chapter 19 about transverse waves on a string. For a string whose equilibrium position is along the $x$-axis, the displacements may be along the $y$-direction, as in Fig. 34–14a. In this case the string always lies in the $xy$-plane. But the displacements might instead be along the $z$-axis, as in Fig. 34–14b; then the string lies in the $xz$-plane.

When a wave has only $y$-displacements, we say that it is **linearly polarized** in the $y$-direction; a wave with only $z$-displacements is linearly polarized in the $z$-direction. For mechanical waves we can build a **polarizing filter** that permits only waves with a certain polarization direction to pass. In Fig. 34–14c the string can slide vertically in the slot without friction, but no horizontal motion is possible. This filter passes waves polarized in the $y$-direction but blocks those polarized in the $z$-direction.

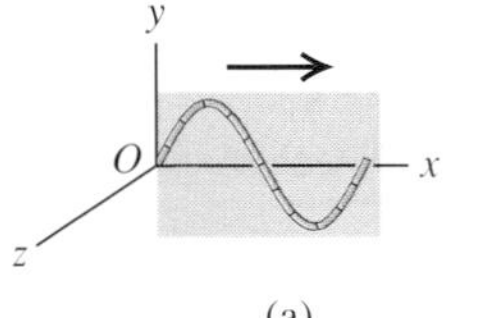


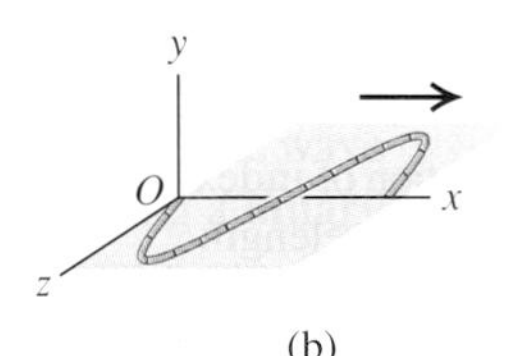


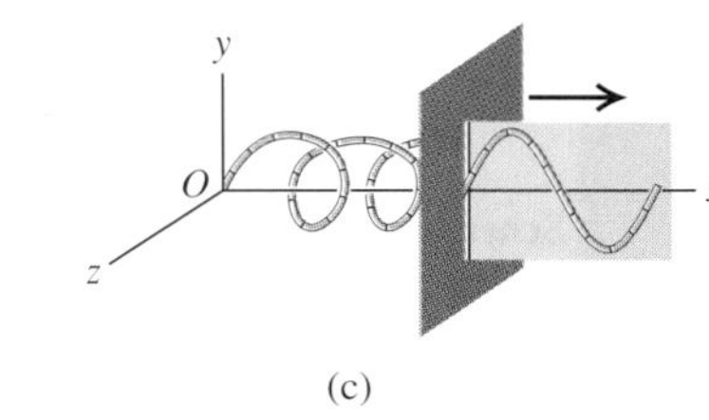


**34–14** (a) Transverse wave on a string, polarized in the $y$-direction. (b) Wave polarized in the $z$-direction. (c) A barrier with a frictionless vertical slot passes components polarized in the $y$-direction but blocks those polarized in the $z$-direction, acting as a polarizing filter.

This same language can be applied to electromagnetic waves, which also have polarization. As we learned in Chapter 33, an electromagnetic wave is a *transverse* wave; the fluctuating electric and magnetic fields are perpendicular to each other and to the direction of propagation. We always define the direction of polarization of an electromagnetic wave to be the direction of the *electric*-field vector, not the magnetic field, because most common electromagnetic-wave detectors (including the human eye) respond to the electric forces on electrons in materials, not the magnetic forces.

### Polarizing Filters

Polarizing filters can be made for electromagnetic waves; the details of construction depend on the wavelength. For microwaves with a wavelength of a few centimeters, a grid of closely spaced, parallel conducting wires insulated from each other will pass waves whose $\boldsymbol{E}$ fields are perpendicular to the wires but not those with $\boldsymbol{E}$ fields parallel to the

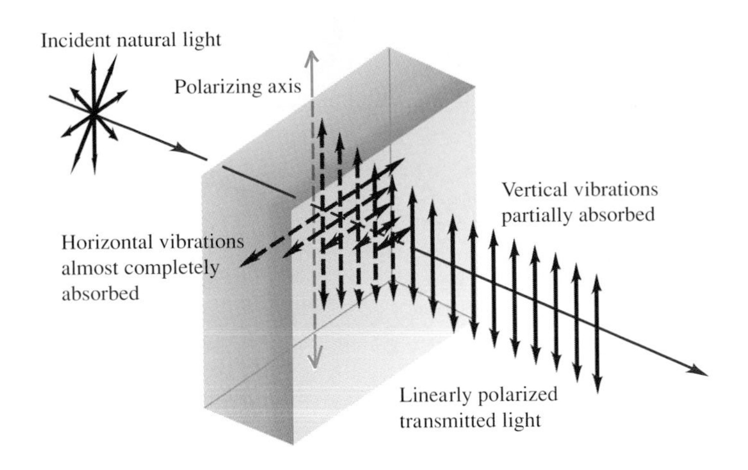


**34–15** Linearly polarized light transmitted by a polarizing filter. The components perpendicular to the polarizing axis are absorbed.

wires. The most common polarizing filter for light is a material known by the trade name Polaroid, widely used for sunglasses and polarizing filters for camera lenses. This material, developed originally by Edwin H. Land, incorporates substances that exhibit **dichroism,** the selective absorption of one of the polarized components much more strongly than the other (Fig. 34–15). A Polaroid filter transmits 80% or more of the intensity of waves polarized parallel to a certain axis in the material (called the **polarizing axis**), but only 1% or less of waves polarized perpendicular to this axis.

Waves emitted by a radio transmitter are usually linearly polarized. The vertical rod antennas used for CB radios emit waves that, in a horizontal plane around the antenna, are polarized in the vertical direction (parallel to the antenna). Rooftop TV antennas have horizontal elements in the United States and vertical elements in Great Britain because the transmitted waves have different polarizations.

Light from ordinary sources is *not* polarized. The "antennas" that radiate light waves are the molecules that make up the sources. The waves emitted by any one molecule may be linearly polarized, like those from a radio antenna. But any actual light source contains a tremendous number of molecules with random orientations, so the light emitted is a random mixture of waves that are linearly polarized in all possible transverse directions.

An ideal polarizing filter, or **polarizer,** passes 100% of the incident light polarized in the direction of the filter's *polarizing axis* but blocks completely all light polarized perpendicular to this axis. Such a device is an unattainable idealization, but the concept is useful in clarifying the basic ideas. In the following discussion we'll assume that all polarizing filters are ideal. In Fig. 34–16 unpolarized light (a random mixture of all polariza-

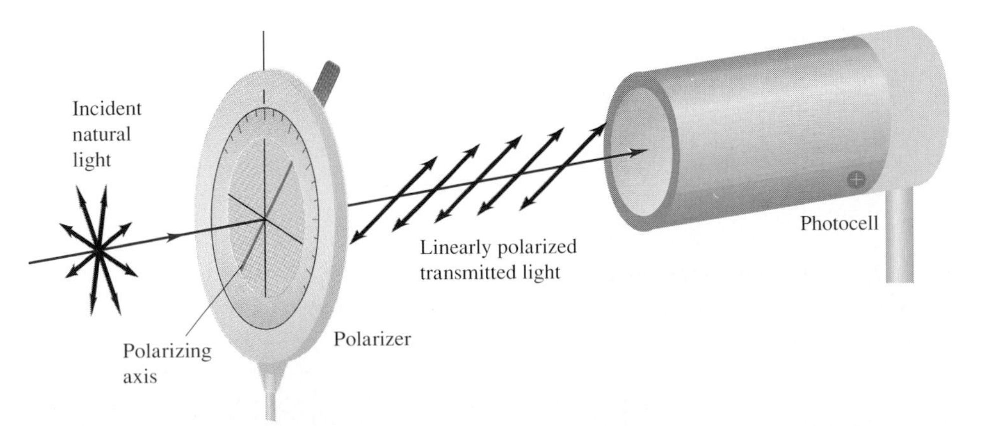


**34–16** The intensity of the transmitted linearly polarized light, measured by the photocell, is the same for all orientations of the polarizing filter. For an ideal polarizing filter, the transmitted intensity is half the incident intensity.

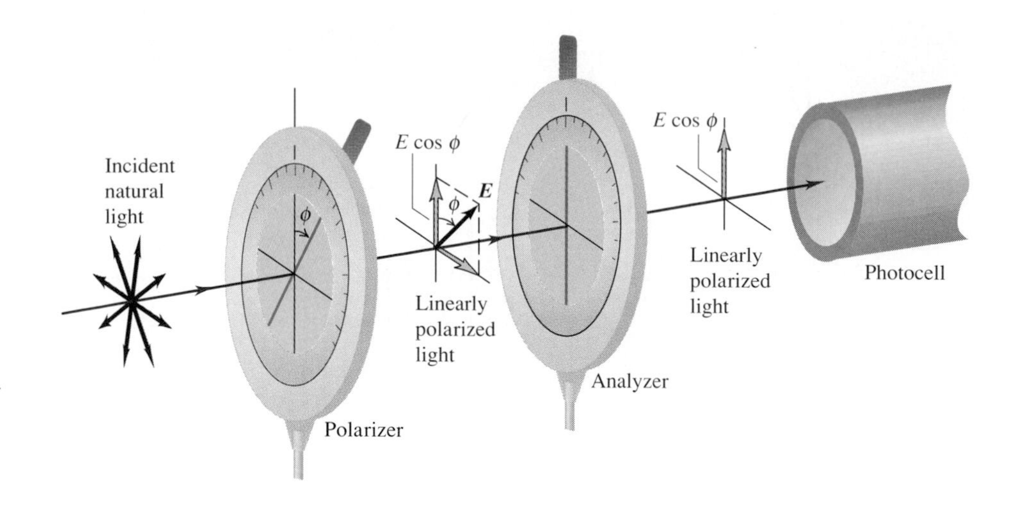


**34–17** The ideal analyzer transmits only the component parallel to its transmission direction, or polarizing axis.

tion states) is incident on a polarizer in the form of a flat plate. The polarizing axis is represented by the blue line. The $\boldsymbol{E}$ vector of the incident wave can be represented in terms of components parallel and perpendicular to the polarizing axis. The polarizer transmits only the components of $\boldsymbol{E}$ parallel to that axis. The light emerging from the polarizer is linearly polarized parallel to the polarizing axis.

When we measure the intensity (power per unit area) of the light transmitted through an ideal polarizer, using the photocell in Fig. 34–16, we find that it is exactly half that of the incident light, no matter how the polarizing axis is oriented. Here's why: We can resolve the $\boldsymbol{E}$ field of the incident wave into a component parallel to the polarizing axis and a component perpendicular to it. Because the incident light is a random mixture of all states of polarization, these two components are, on average, equal. The ideal polarizer transmits only the component parallel to the polarizing axis, so half the incident intensity is transmitted.

Now suppose we insert a second polarizer between the first polarizer and the photocell (Fig. 34–17). The polarizing axis of the second polarizer, or *analyzer,* is vertical, and the axis of the first polarizer makes an angle $\phi$ with the vertical. That is, $\phi$ is the angle between the polarizing axes of the two polarizers. We can resolve the linearly polarized light transmitted by the first polarizer into two components, as shown in Fig. 34–17, one parallel and the other perpendicular to the axis of the analyzer. Only the parallel component, with amplitude $E \cos \phi$, is transmitted by the analyzer. The transmitted intensity is greatest when $\phi = 0$; it is zero when $\phi = 90°$, that is, when polarizer and analyzer are *crossed*.

To find the transmitted intensity at intermediate angles, we recall from our discussion in Chapter 33 that the intensity of an electromagnetic wave is proportional to the *square* of the amplitude of the wave. The ratio of transmitted to incident *amplitude* is $\cos \phi$, so the ratio of transmitted to incident *intensity* is $\cos^2 \phi$. Thus we have

$$I = I_{\max} \cos^2 \phi, \qquad (34\text{–}7)$$

where $I_{\max}$ is the maximum intensity of light transmitted (at $\phi = 0$) and $I$ is the amount transmitted at angle $\phi$. This relation, discovered experimentally by Etienne Louis Malus in 1809, is called **Malus's law.**

## PROBLEM-SOLVING STRATEGY

### Linear polarization

**1.** Remember that in light waves or any other electromagnetic waves the $\boldsymbol{E}$ field is perpendicular to the propagation direction and is the direction of polarization. The polarization direction can be thought of as a two-headed arrow. When working with polarizing filters, you are really dealing with components of $\boldsymbol{E}$ parallel and perpendicular to the polarizing axis. Everything you know about components of vectors is applicable here.

**2.** The intensity (average power per unit area) of a wave is proportional to the *square* of its amplitude. If you find that two waves differ in amplitude by a certain factor, their intensities differ by the square of that factor.

**3.** Unpolarized light is a random mixture of all possible polarization states, so on average it has equal components in any two perpendicular directions. Partially linearly polarized light is a superposition of linearly polarized and unpolarized light.

### EXAMPLE 34–5

In Fig. 34–17 the incident unpolarized light has intensity $I_0$. Find the intensity transmitted by the first polarizer and the second if the angle between the axes of the two filters is 30°.

**SOLUTION** As explained above, the intensity after the first filter is $I_0/2$. According to Eq. (34–7) with $\phi = 30°$, the second filter reduces the intensity by a factor of $\cos^2 30° = \frac{3}{4}$. Thus the intensity transmitted by the second polarizer is $(I_0/2)\left(\frac{3}{4}\right) = \frac{3}{8}I_0$.

## Polarization by Reflection

Unpolarized light can be partially polarized by *reflection*. When unpolarized light strikes a reflecting surface between two optical materials, preferential reflection occurs for those waves in which the electric-field vector is parallel to the reflecting surface. In Fig. 34–18 the plane containing the incident and reflected rays and the normal to the surface is called the **plane of incidence.** At one particular angle of incidence, called the

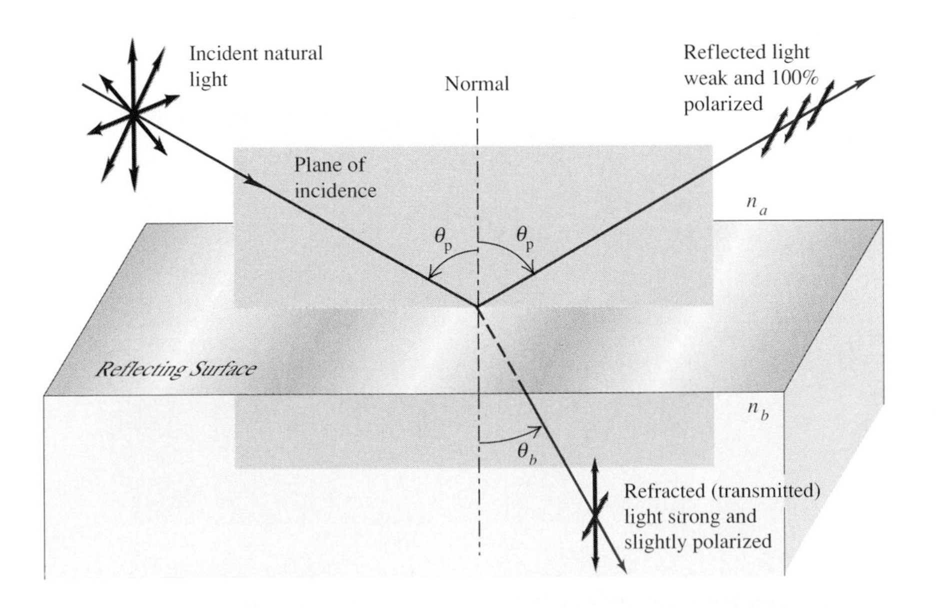


**34–18** When light is incident at the polarizing angle, the reflected light is linearly polarized.

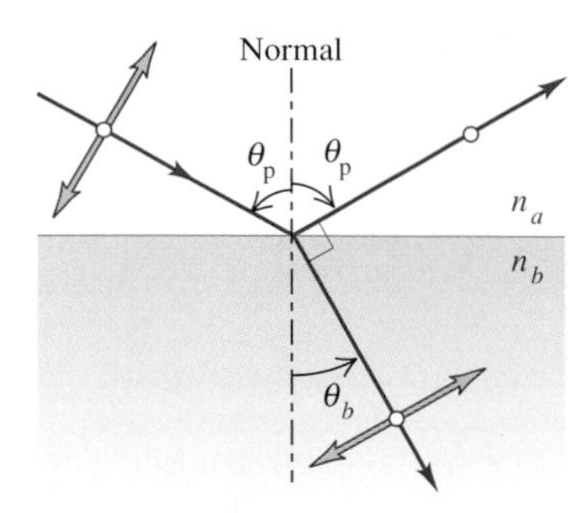


**34–19** When light is incident at the polarizing angle, the reflected and refracted rays are perpendicular to each other. The circles represent an $E$-component perpendicular to the plane of the figure.

**polarizing angle** $\theta_p$, only the light for which the $\boldsymbol{E}$ vector is perpendicular to the plane of incidence (parallel to the reflecting surface) is reflected. The reflected light is therefore linearly polarized perpendicular to the plane of incidence (parallel to the reflecting surface), as shown in Fig. 34–18.

When light is incident at the polarizing angle, *none* of the $\boldsymbol{E}$ field component *parallel* to the plane of incidence is reflected; this component is transmitted 100% in the *refracted* beam. So the *reflected* light is *completely* polarized. The *refracted* light is a mixture of the component parallel to the plane of incidence, all of which is refracted, and the remainder of the perpendicular component; it is therefore *partially* polarized.

In 1812, Sir David Brewster noticed that when the angle of incidence is equal to the polarizing angle $\theta_p$, the reflected ray and the refracted ray are perpendicular to each other, as shown in Fig. 34–19. In this case the angle of refraction $\theta_b$ becomes the complement of $\theta_p$, so $\sin\theta_b = \cos\theta_p$. From the law of refraction,

$$n_a \sin\theta_p = n_b \sin\theta_b,$$

so we find

$$n_a \sin\theta_p = n_b \cos\theta_p,$$

$$\tan\theta_p = \frac{n_b}{n_a}. \tag{34–8}$$

This relation is known as **Brewster's law.** Although it was discovered experimentally, it can also be *derived* from a wave model using Maxwell's equations.

## EXAMPLE 34–6

**Reflection from a swimming pool's surface** Sunlight reflects off the smooth surface of an unoccupied swimming pool. a) At what angle of reflection is the light completely polarized? b) What is the corresponding angle of refraction for the light that is transmitted (refracted) into the water? c) At night an underwater floodlight is turned on in the pool. Repeat parts (a) and (b) for rays from the floodlight that strike the smooth surface from below.

**SOLUTION** a) We're looking for the polarizing angle for light that is first in air, then in water, so $n_a = 1$ (air) and $n_b = 1.33$ (water). From Eq. (34–8),

$$\theta_p = \arctan\frac{n_b}{n_a} = \arctan\frac{1.33}{1.00} = 53.1°.$$

The angles are shown in Fig. 34–20a.

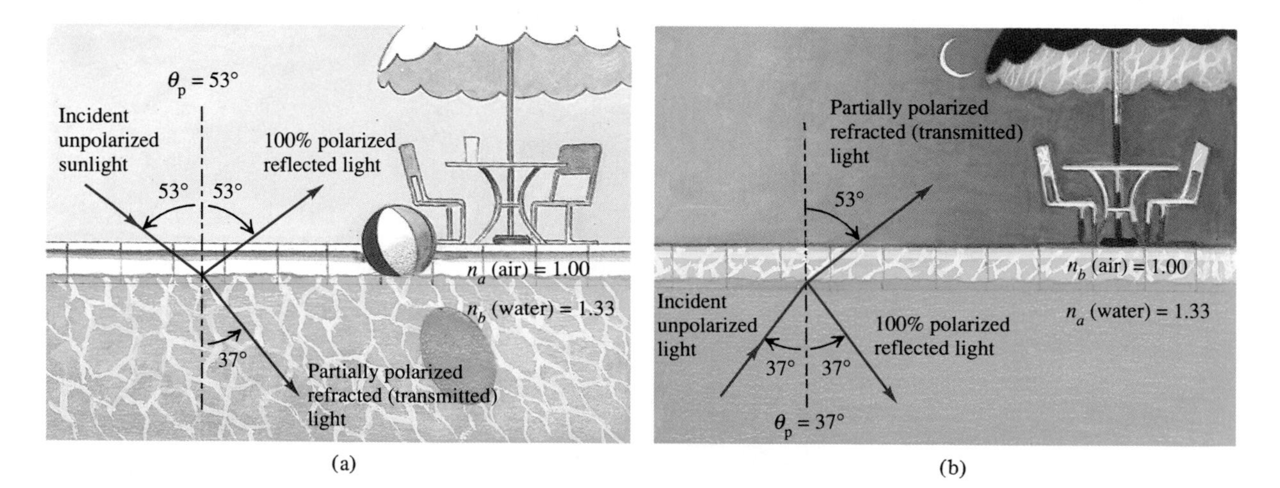


**34–20** Light striking the water-air interface at the polarizing angle (a) from the air side; (b) from the water side.

b) The reflected and refracted rays are perpendicular, so

$$\theta_p + \theta_b = 90°,$$

$$\theta_b = 90° - 53.1° = 36.9°.$$

c) Now the light is *first* in the water, then in the air, so $n_a = 1.33$ and $n_b = 1.00$. Again using Eq. (34–8), we have

$$\theta_p = \arctan\frac{1.00}{1.33} = 36.9°,$$

$$\theta_b = 90° - 36.9° = 53.1°.$$

The angles are shown in Fig. 34–20b. We see that the two polarizing angles for this interface add to 90°. This is *not* an accident. Do you see why?

---

Polarizing filters are widely used in sunglasses. When sunlight is reflected from a horizontal surface, the plane of incidence is vertical, and the reflected light contains a preponderance of light polarized in the horizontal direction. When the reflection occurs at a smooth asphalt road surface or the surface of a lake, it causes unwanted glare. Vision can be improved by eliminating this glare. The manufacturer makes the polarizing axis of the lens material vertical, so very little of the horizontally polarized light is transmitted to the eyes. The glasses also reduce the overall intensity in the transmitted light to somewhat less than 50% of the intensity of the unpolarized incident light.

## Circular and Elliptical Polarization

Light and other electromagnetic radiation can also have *circular* or *elliptical* polarization. To introduce these concepts, let's return once more to mechanical waves on a stretched string. In Fig. 34–14, suppose the two linearly polarized waves in parts (a) and (b) are in phase and have equal amplitude. When they are superposed, each point in the string has simultaneous $y$- and $z$-displacements of equal magnitudes. A little thought shows that the resultant wave lies in a plane oriented at 45° to the $y$- and $z$-axes (that is, in a plane making a 45°-angle with the $xy$- and $xz$-planes). The amplitude of the resultant wave is larger by a factor of $\sqrt{2}$ than that of either component wave, and the resultant wave is linearly polarized.

Now suppose the two equal-amplitude waves differ in phase by a quarter-cycle. Then the resultant motion of each point corresponds to a superposition of two simple harmonic motions at right angles, with a quarter-cycle phase difference. The $y$-displacement at a point is greatest at times when the $z$-displacement is zero, and vice versa. The motion of the string then no longer takes place in a single plane. It can be shown that each point on the string moves in a *circle* in a plane parallel to the $yz$-plane. Successive points on the string have successive phase differences, and the overall motion of the string has the appearance of a rotating helix. This particular superposition of two linearly polarized waves is called **circular polarization.** By convention the wave is said to be *right circularly polarized* when the sense of motion of a particle of the string, to an observer looking *backward* along the direction of propagation, is *clockwise;* the wave is said to be *left circularly polarized* if the opposite is the case.

Figure 34–21 (on the next page) shows the analogous situation for an electromagnetic wave. Two waves, polarized in the $y$- and $z$-directions, with a quarter-cycle phase difference, are superposed. The result is a wave in which the $\boldsymbol{E}$ vector at each point *rotates* clockwise. We call this a right circularly polarized electromagnetic wave.

If the phase difference between the two component waves is something other than a quarter-cycle, or if the two component waves have different amplitudes, then each point on the string traces out not a circle but an *ellipse*. The resulting wave is said to be **elliptically polarized.**

For electromagnetic waves with radio frequencies, circular or elliptical polarization can be produced by using two antennas at right angles, fed from the same transmitter but with a phase-shifting network that introduces the appropriate phase difference. For

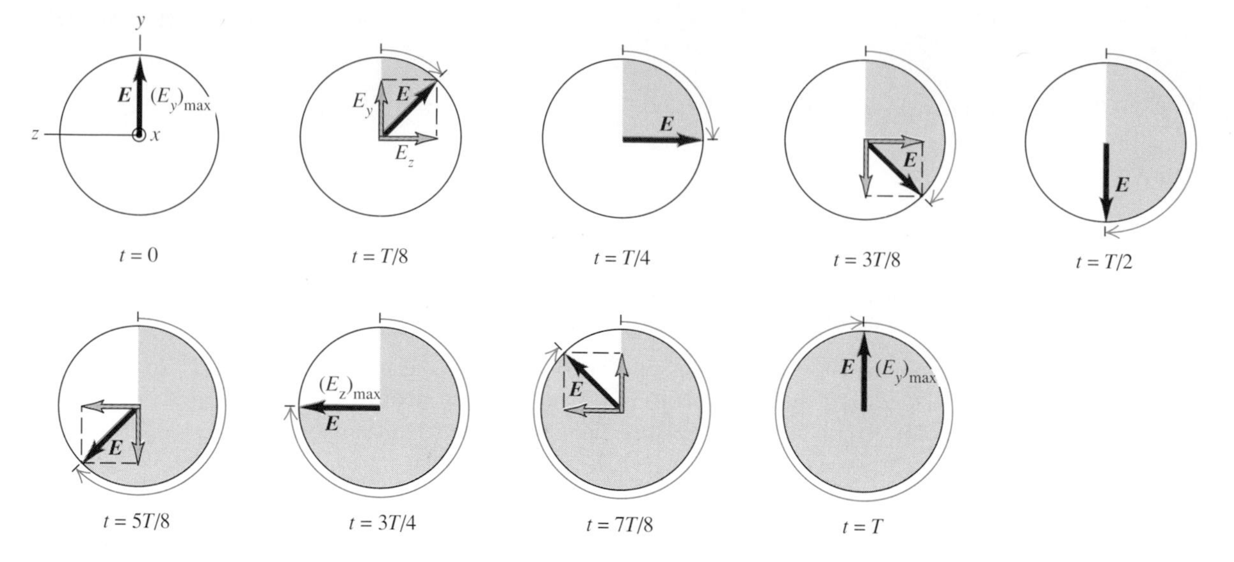


**34–21** Circular polarization. The $y$-component of $\boldsymbol{E}$ lags the $z$-component by a quarter-cycle. This phase difference results in right circular polarization if the wave is moving toward you (in the positive $x$-direction).

light the phase shift can be introduced by use of a material that exhibits **birefringence,** that is, has different indexes of refraction for different directions of polarization. A common example is calcite; when a calcite crystal is oriented appropriately in a beam of unpolarized light, its refractive index (for $\lambda = 589$ nm) is 1.658 for one direction of polarization and 1.486 for the perpendicular direction. When two waves with perpendicular directions of polarization enter such a material, they travel with different speeds. If they are in phase when they enter the material, then in general they are no longer in phase when they emerge. If the crystal is just thick enough to introduce a quarter-cycle phase difference, then the crystal converts linearly polarized light to circularly polarized light. Such a crystal is called a **quarter-wave plate.** Such a plate also converts circularly polarized light to linearly polarized light. Can you prove this?

## Photoelasticity

Some optical materials that are not normally birefringent become so when they are subjected to mechanical stress. This is the basis of the science of **photoelasticity.** Stresses in girders, boiler plates, gear teeth, and cathedral pillars can be analyzed by constructing a transparent model of the object, usually of a plastic material, subjecting it to stress, and examining it between a polarizer and an analyzer in the crossed position. Very complicated stress distributions can be studied by these optical methods. Figure 34–22 shows two photographs of photoelastic models under stress.

(a)

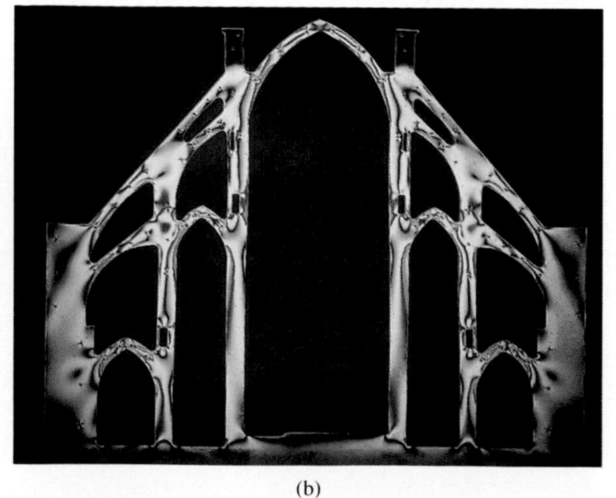

(b)

**34–22** (a) Photoelastic stress analysis of a plastic model of a machine part. (b) Stress analysis of a model of a cross section of a Gothic cathedral. The masonry construction used for this kind of building had great strength in compression but very little in tension. Inadequate buttressing and high winds sometimes caused tensile stresses in normally compressed structural elements, leading to some spectacular collapses.

## *34–6 SCATTERING OF LIGHT

The sky is blue. Skylight is partially polarized; you can see this by looking at the sky directly overhead through a polarizing filter. Sunsets are red. It turns out that one phenomenon is responsible for all three of these effects.

In Fig. 34–23 sunlight (unpolarized) comes from the left along the $x$-axis and passes over an observer looking vertically upward along the $y$-axis. (We are viewing the situation from the side.) Molecules of the earth's atmosphere are located at point $O$. The electric field in the beam of sunlight sets the electric charges in the molecules into vibration. Light is a transverse wave; the direction of the electric field in any component of the sunlight lies in the $yz$-plane, and the motion of the charges takes place in this plane. There is no field, and therefore no vibration, in the direction of the $x$-axis.

An incident light wave with its $\boldsymbol{E}$ field at an angle $\theta$ with the $z$-axis sets the electric charges in the molecules vibrating along the line of $\boldsymbol{E}$, as indicated by the double-ended arrow through point $O$. We can resolve this vibration into two components, one along the $y$-axis and the other along the $z$-axis. Each component in the incident light produces the equivalent of two molecular "antennas," oscillating with the same frequency as the incident light and lying along the $y$- and $z$-axes.

We mentioned in Section 33–8 that such an antenna does not radiate in the direction of its own length. The antenna along the $y$-axis does not send any light to the observer directly below it, although it does emit light in other directions. Therefore the only light reaching this observer comes from the other antenna, corresponding to the component of vibration along the $z$-axis. This light is linearly polarized, with its electric field parallel to the antenna. The vectors on the $y$-axis below point $O$ show the direction of polarization of the light reaching the observer.

The process that we have just described is called **scattering.** The energy of the scattered light is removed from the original beam, reducing its intensity. Detailed analysis of the scattering process shows that the intensity of the light scattered from air molecules increases in proportion to the fourth power of the frequency (inversely to the fourth power of the wavelength). Thus the intensity ratio for the two ends of the visible spectrum is $(700\ \text{nm}/400\ \text{nm})^4 = 9.4$. Roughly speaking, scattered light contains nine times as much blue light as red, and that's why the sky is blue.

Because skylight is partially polarized, polarizers are useful in photography. The sky in a photograph can be darkened by appropriate orientation of the polarizer axis. The

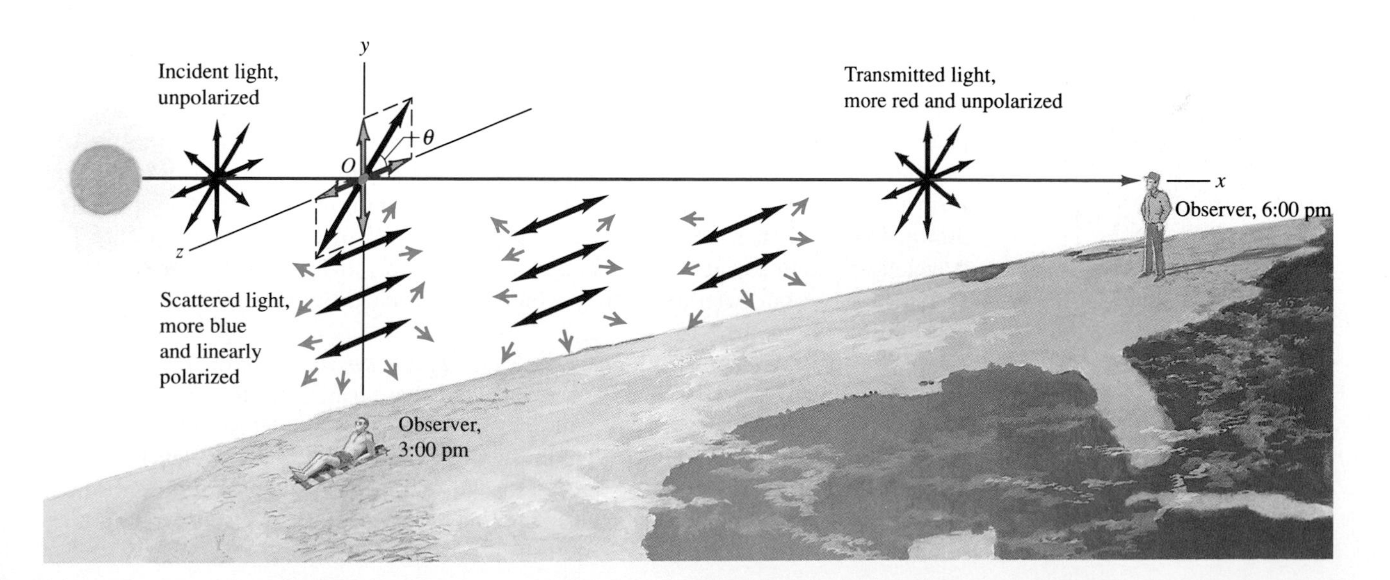


**34–23** Scattered light is linearly polarized and contains predominantly light from the blue end of the spectrum. The initial white light loses this blue light as it travels through the atmosphere, and the transmitted light contains predominantly light from the red end of the spectrum.

effect of atmospheric haze can be reduced in exactly the same way, and unwanted reflections can be controlled just as with polarizing sunglasses, discussed in Section 34–5.

Toward evening, when sunlight has to travel a long distance through the earth's atmosphere, a substantial fraction of the blue light is removed by scattering. White light minus blue light appears yellow or red. Thus, when sunlight with the blue component removed is incident on a cloud, the light reflected from the cloud to the observer has the yellow or red hue so often observed at sunset. If the earth had no atmosphere, we would receive *no* skylight at the earth's surface, and the sky would appear as black in the daytime as it does at night. To an astronaut in a spacecraft or on the moon, the sky appears black, not blue.

## 34–7 HUYGENS' PRINCIPLE

The laws of reflection and refraction of light rays that we introduced in Section 34–2 were discovered experimentally long before the wave nature of light was firmly established. However, we can *derive* these laws from wave considerations and show that they are consistent with the wave nature of light.

We begin with a principle called **Huygens' principle,** stated originally by Christiaan Huygens in 1678. This offers a geometrical method for finding, from the known shape of a wave front at some instant, the shape of the wave front at some later time. Huygens assumed that **every point of a wave front may be considered the source of secondary wavelets that spread out in all directions with a speed equal to the speed of propagation of the wave.** The new wave front at a later time is then found by constructing a surface *tangent* to the secondary wavelets, which is called the *envelope* of the wavelets. All the results that we obtain from Huygens' principle can also be obtained from Maxwell's equations. Thus it is not an independent principle, but it is often very convenient for calculations with wave phenomena.

Huygens' principle is illustrated in Fig. 34–24. The original wave front $AA'$ is traveling outward from a source, as indicated by the small arrows. We want to find the shape of the wave front after a time interval $t$. Let $v$ be the speed of propagation of the wave; then in time $t$ it travels a distance $vt$. We construct several circles (traces of spherical wavelets) with radius $r = vt$, centered at points along $AA'$. The trace of the envelope of these wavelets, which is the new wave front, is the curve $BB'$. We are assuming that the speed $v$ is the same at all points and in all directions.

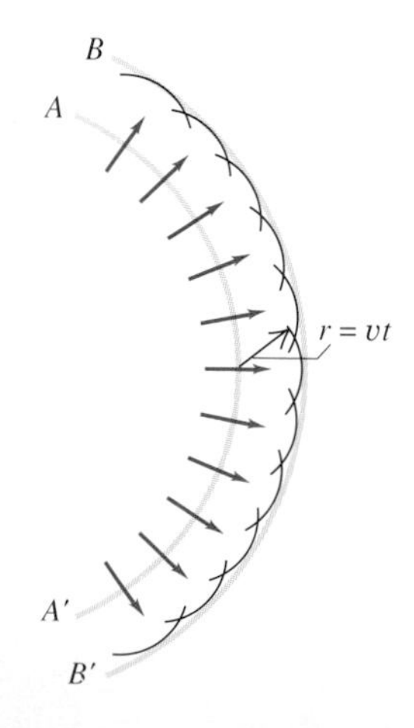

**34–24** Applying Huygens' principle to wave front $AA'$ to construct a new wave front $BB'$.

To derive the law of reflection from Huygens' principle, we consider a plane wave approaching a plane reflecting surface. In Fig. 34–25a the lines $AA'$, $BB'$, and $CC'$ represent successive positions of a wave front approaching the surface $MM'$. Point $A$ on the wave front $AA'$ has just arrived at the reflecting surface. We can use Huygens' principle to find the position of the wave front after a time interval $t$. With points on $AA'$ as centers we draw several secondary wavelets with radius $vt$, where $v$ is the speed of propagation of the wave. The wavelets that originate near the upper end of $AA'$ spread out unhindered, and their envelope gives the portion $OB'$ of the new wave front. The wavelets originating near the lower end of $AA'$, however, strike the reflecting surface. If the surface had not been there, they would have reached the positions shown by the broken circular arcs.

The effect of the reflecting surface is to *change the direction* of travel of those wavelets that strike it, so part of a wavelet that would have penetrated the surface actually lies to the left of it, as shown by the full lines. The first such wavelet is centered at point $A$ and is the wavelet nearest point $B$. The envelope of all such reflected wavelets is then the portion $OB$ of the wave front. The trace of the entire wave front at this instant is the bent line $BOB'$. A similar construction gives the line $CNC'$ for the wave front after another interval $t$.

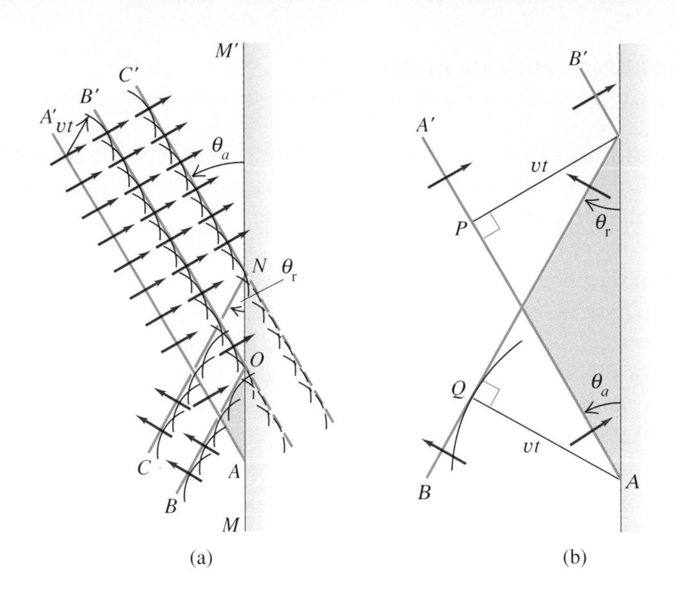


**34–25** (a) Successive positions of a plane wave $AA'$ as it is reflected from a plane surface. (b) Magnified portion of part (a).

From plane geometry the angle $\theta_a$ between the incident *wave front* and the *surface* is the same as that between the incident *ray* and the *normal* to the surface, and is therefore the angle of incidence. Similarly, $\theta_r$ is the angle of reflection. To find the relation between these angles, we consider Fig. 34–25b. From $O$ we draw $OP = vt$, perpendicular to $AA'$. Now $OB$, by construction, is tangent to a circle of radius $vt$ with center at $A$. If we draw $AQ$ from $A$ to the point of tangency, the triangles $APO$ and $OQA$ are equal because they are right triangles with the side $AO$ in common and with $AQ = OP = vt$. The angle $\theta_a$ therefore equals the angle $\theta_r$, and we have the law of reflection.

We can derive the law of *refraction* by a similar procedure. In Fig. 34–26a we consider a wave front, represented by line $AA'$, for which point $A$ has just arrived at the boundary surface $SS'$ between two transparent materials $a$ and $b$, with indexes of

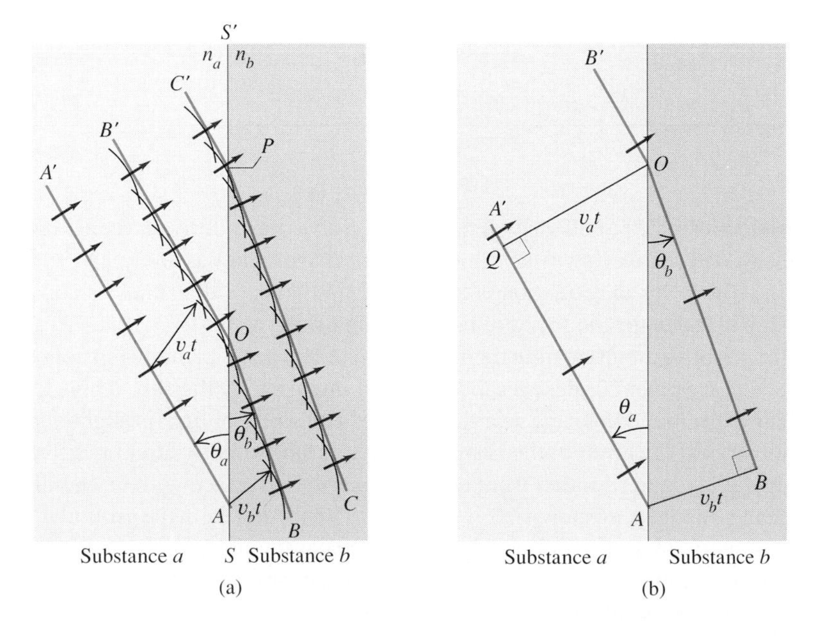


**34–26** (a) Successive positions of a plane wave front $AA'$ as it is refracted by a plane surface. (b) Magnified portion of part (a). The case $v_b < v_a$ is shown.

refraction $n_a$ and $n_b$ and wave speeds $v_a$ and $v_b$. (The *reflected* waves are not shown in the figure; they proceed exactly as in Fig. 34–25.) We can apply Huygens' principle to find the position of the refracted wave fronts after a time $t$.

With points on $AA'$ as centers we draw several secondary wavelets. Those originating near the upper end of $AA'$ travel with speed $v_a$ and, after a time interval $t$, are spherical surfaces of radius $v_a t$. The wavelet originating at point $A$, however, is traveling in the second material $b$ with speed $v_b$ and at time $t$ is a spherical surface of radius $v_b t$. The envelope of the wavelets from the original wave front is the plane whose trace is the broken line $BOB'$. A similar construction leads to the trace $CPC'$ after a second interval $t$.

The angles $\theta_a$ and $\theta_b$ between the surface and the incident and refracted wave fronts are the angle of incidence and the angle of refraction, respectively. To find the relation between these angles, refer to Fig. 34–26b. Draw $OQ = v_a t$, perpendicular to $AQ$, and draw $AB = v_b t$, perpendicular to $BO$. From the right triangle $AOQ$,

$$\sin \theta_a = \frac{v_a t}{AO},$$

and from the right triangle $AOB$,

$$\sin \theta_b = \frac{v_b t}{AO}.$$

Combining these, we have

$$\frac{\sin \theta_a}{\sin \theta_b} = \frac{v_a}{v_b}. \tag{34–9}$$

We have defined the index of refraction $n$ of a material as the ratio of the speed of light $c$ in vacuum to its speed $v$ in the material: $n_a = c/v_a$ and $n_b = c/v_b$. Thus

$$\frac{n_b}{n_a} = \frac{c/v_b}{c/v_a} = \frac{v_a}{v_b},$$

and we can rewrite Eq. (34–9) as

$$\frac{\sin \theta_a}{\sin \theta_b} = \frac{n_b}{n_a},$$

or

$$n_a \sin \theta_a = n_b \sin \theta_b,$$

which we recognize as Snell's law, Eq. (34–4). So we have derived Snell's law from a wave theory. Alternatively, we may choose to regard Snell's law as an experimental result that defines the index of refraction of a material; in that case this analysis confirms the relationship $v = c/n$ for the speed of light in a material.

Mirages offer an interesting demonstration of Huygens' principle in action. When the surface of pavement or desert sand is heated intensely by the sun, a hot, less dense layer of air with smaller $n$ forms near the surface. The speed of light is slightly greater in this hotter air near the ground, the Huygens wavelets have slightly larger radii, the wavefronts tilt slightly, and rays that were headed toward the surface with an incident angle near 90° can be bent up, as shown in Fig. 34–27. Light farther from the ground is bent less and travels nearly in a straight line. The observer sees the object in its natural position, with an inverted image below it, as though seen in a horizontal reflecting surface. Even when the turbulence of the heated air prevents a clear inverted image from being formed, the mind of the thirsty traveler can interpret the apparent reflecting surface as a sheet of water.

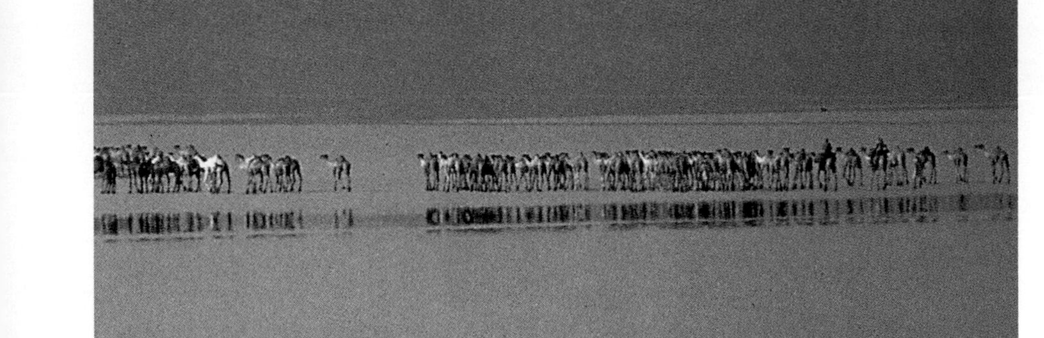

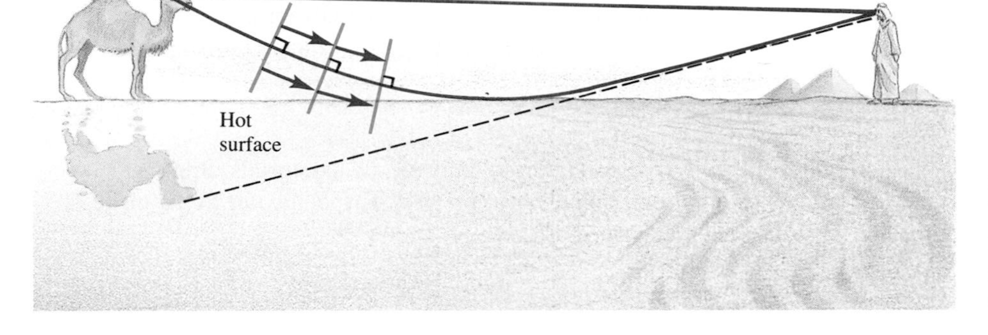


**34–27** A mirage is observed because wavelets near the hot surface have slightly greater radii $vt$, gradually tilting the wave fronts and bending the rays.

It is important to keep in mind that Maxwell's equations are the fundamental relations for electromagnetic wave propagation. But it is a remarkable fact that Huygens' principle anticipated Maxwell's analysis by two centuries. Maxwell provided the theoretical underpinning for Huygens' principle. Every point in an electromagnetic wave, with its time-varying electric and magnetic fields, acts as a source of the continuing wave, as predicted by Ampere's and Faraday's laws.

## SUMMARY

- Light is an electromagnetic wave. When emitted or absorbed, it also shows particle properties. It is emitted by accelerated electric charges that have been given excess energy by heat or electrical discharge. The speed of light is a fundamental physical constant.
- A wave front is a surface of constant phase; wave fronts move with a speed equal to the propagation speed of the wave. A ray is a line along the direction of propagation, perpendicular to the wave fronts. Representation of light by rays is the basis of geometrical optics.
- The index of refraction $n$ of a material is the ratio of the speed of light in vacuum $c$ to the speed $v$ in the material: $n = c/v$. The incident, reflected, and refracted rays and the normal to the interface all lie in a single plane called the plane of incidence. The law of reflection states that the angles of incidence and reflection are equal. The law of refraction is

$$n_a \sin\theta_a = n_b \sin\theta_b. \tag{34–4}$$

## KEY TERMS

wave front
ray
geometrical optics
physical optics
index of refraction (refractive index)
law of reflection
law of refraction (Snell's law)
critical angle
total internal reflection
dispersion
linear polarization
polarizing filter (polarizer)
dichroism
polarizing axis
Malus's law
plane of incidence
polarizing angle
Brewster's law
circular polarization
elliptical polarization
birefringence
quarter-wave plate
photoelasticity
scattering
Huygens' principle

Angles of incidence, reflection, and refraction are always measured from the normal to the surface.

• When a ray travels within a material of greater index of refraction $n_a$ toward an interface with one of smaller index $n_b$, total internal reflection occurs when the angle of incidence exceeds a critical value $\theta_{crit}$ given by

$$\sin\theta_{crit} = \frac{n_b}{n_a}. \tag{34–6}$$

• The variation of index of refraction $n$ with wavelength $\lambda$ is called dispersion. Usually, $n$ decreases with increasing $\lambda$.

• The direction of polarization of a linearly polarized electromagnetic wave is the direction of the $\boldsymbol{E}$ field. A polarizing filter passes radiation that is linearly polarized along its polarizing axis and blocks radiation polarized perpendicularly to that axis. When polarized light of intensity $I_{max}$ is incident on an analyzer and $\phi$ is the angle between the polarizing axes of the polarizer and analyzer, Malus's law states that the transmitted intensity $I$ is

$$I = I_{max}\cos^2\phi. \tag{34–7}$$

• When unpolarized light strikes an interface between two materials, Brewster's law states that the reflected light is completely polarized perpendicular to the plane of incidence if the angle of incidence $\theta_p$ is given by

$$\tan\theta_p = \frac{n_b}{n_a}. \tag{34–8}$$

• When two linearly polarized waves with a phase difference are superposed, the result is circularly or elliptically polarized light. In this case the $\boldsymbol{E}$ vector is not confined to a plane containing the direction of propagation, but describes circles or ellipses in planes perpendicular to the propagation direction.

• A birefringent material has different indexes of refraction for two perpendicular directions of polarization. Materials that become birefringent under mechanical stress form the basis of photoelastic stress analysis. A dichroic material has preferential absorption for one polarization direction.

• Light is scattered by air molecules. The scattered light is partially polarized.

• Huygens' principle states that if the position of a wave front at one instant is known, the position of the front at a later time can be constructed by imagining the front as a source of secondary wavelets. Huygens' principle can be used to derive the laws of reflection and refraction.

# EXERCISES

## Section 34–2 Reflection and Refraction

**34–1** A ray of light is incident on a plane surface separating two sheets of glass with refractive indexes 1.80 and 1.52. The angle of incidence is 35.0°, and the ray originates in the glass that has the higher index. Compute the angle of refraction.

**34–2** A parallel-sided plate of glass having a refractive index of 1.60 is in contact with the surface of water in a tank. A ray coming from above in air makes an angle of incidence of 38.0° with the top surface of the glass. a) What angle does the ray refracted into the water make with the normal to the surface? b) What is the dependence of this angle on the refractive index of the glass?

**34–3** A parallel beam of light in air makes an angle of 30.0° with the surface of a glass plate having a refractive index of 1.52. a) What is the angle between the reflected part of the beam and the surface of the glass? b) What is the angle between the refracted beam and the surface of the glass?

**34–4** Light with a frequency of $5.00 \times 10^{14}$ Hz travels in a block of plastic that has an index of refraction of 2.00. What is the wavelength of the light while it is in the plastic and while it is in vacuum?

**34–5** The speed of light with wavelength 656 nm is $1.90 \times 10^8$ m/s in a flint glass. a) What is the index of refraction of this glass at this wavelength? b) If this same light travels through air, what is its wavelength there?

**34–6 Seeing the Setting Sun.** The density of the earth's atmosphere increases as the surface of the earth is approached. This increase in density is accompanied by a corresponding increase in refractive index. a) Draw a diagram showing how the rays of light from a star or planet bend as they pass through the atmosphere. Indicate the apparent position of the light source. b) Explain how you can see the sun after it has set. c) Explain why the setting sun appears flattened.

**34–7** A beam of light has a wavelength of 500 nm in vacuum. a) What is the speed of this light in a piece of glass whose index of refraction at this wavelength is 1.70? b) What is the wavelength of these waves in the glass?

**34–8** Light of a certain frequency has a wavelength of 442 nm in water. What is the wavelength of this light in carbon disulfide?

**34–9** A parallel beam of light is incident on a prism, as shown in Fig. 34–28. Part of the light is reflected from one face and part from another. Show that the angle $\theta$ between the two reflected beams is twice the angle $A$ between the two reflecting surfaces.

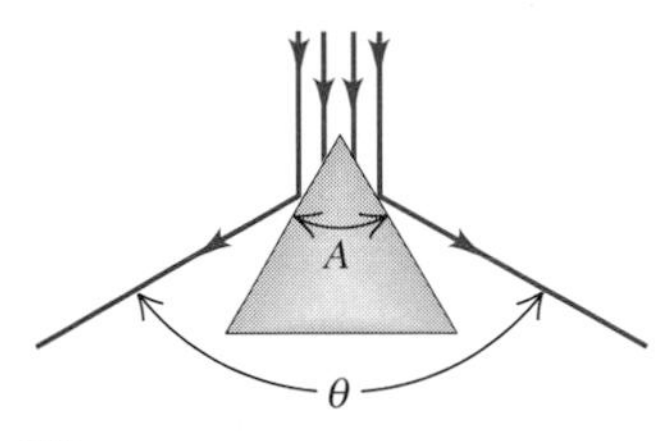


**FIGURE 34–28**

**34–10** Prove that a ray of light reflected from a plane mirror rotates through an angle of $2\theta$ when the mirror rotates through an angle $\theta$ about an axis perpendicular to the plane of incidence.

## Section 34–3 Total Internal Reflection

**34–11** The speed of a sound wave is 344 m/s in air and 1320 m/s in water. a) Which medium has the higher "index of refraction" for sound? b) What is the critical angle for a sound wave incident on the surface between air and water?

**34–12** A ray of light in glass with an index of refraction of 1.55 is incident on an interface with air. What is the *largest* angle the ray can make with the normal and not be totally reflected back into the glass?

**34–13** The critical angle for total internal reflection at a liquid-air interface is 37.0°. a) If a ray of light traveling in the liquid has an angle of incidence at the interface of 28.0°, what angle does the refracted ray in the air make with the normal? b) If a ray of light traveling in air has an angle of incidence at the interface of 28.0°, what angle does the refracted ray in the liquid make with the normal?

**34–14** A point source of light is 32.0 cm below the surface of a lake. Find the diameter of the largest circle at the surface through which light can emerge from the water.

## Section 34–5 Polarization

**34–15** Unpolarized light with intensity $I_0$ is incident on a polarizing filter. The emerging light strikes a second polarizing filter whose axis is at 60.0° to that of the first. Determine a) the intensity of the beam after it has passed through the second polarizer; b) its state of polarization.

**34–16** A polarizer and an analyzer are oriented so that the maximum amount of light is transmitted. To what fraction of its maximum value is the intensity of the transmitted light reduced when the analyzer is rotated through a) 30.0°; b) 45.0°; c) 60.0°?

**34–17 Three Polarizing Filters.** Three polarizing filters are stacked, with the polarizing axes of the second and third at 60.0° and 90.0°, respectively, with that of the first. a) If unpolarized light of intensity $I_0$ is incident on the stack, find the intensity and state of polarization of light emerging from each filter. b) If the second filter is removed, what is the intensity of the light emerging from each remaining filter?

**34–18** Light traveling in water strikes a glass plate at an angle of incidence of 50.0°; part of the beam is reflected and part is refracted. If the reflected and refracted portions make an angle of 90.0° with each other, what is the index of refraction of the glass?

**34–19** A parallel beam of unpolarized light in air is incident at an angle of 58.6° (with respect to the normal) on a plane glass surface. The reflected beam is completely linearly polarized. a) What is the refractive index of the glass? b) What is the angle of refraction of the transmitted beam?

**34–20** a) At what angle above the horizontal is the sun if sunlight reflected from the surface of a calm body of water is completely polarized? b) What is the plane of the $\boldsymbol{E}$ vector in the reflected light?

**34–21** Unpolarized light traveling in a liquid with refractive index $n$ is incident on the surface of the liquid, above which there is air. If the light is incident on the surface at an angle of 31.0° with respect to the normal, the light reflected back into the liquid is completely polarized. a) What is the refractive index $n$ of the liquid? b) What angle does the refracted light, traveling in the air, make with the normal to the surface?

## Section 34–6 Scattering of Light

*34–22** A beam of light, after passing through the Polaroid disk $P_1$ in Fig. 34–29, traverses a cell containing a scattering medium. The cell is observed at right angles through another Polaroid disk $P_2$. Originally, the disks are oriented so that the brightness of the field as seen by the observer is a maximum. a) Disk $P_2$ is rotated 90°. Is extinction produced? Explain. b) Disk $P_1$ is now rotated through 90°. Is the field bright or dark? Explain. c) Disk $P_2$ is then restored to its original position. Is the field bright or dark? Explain.

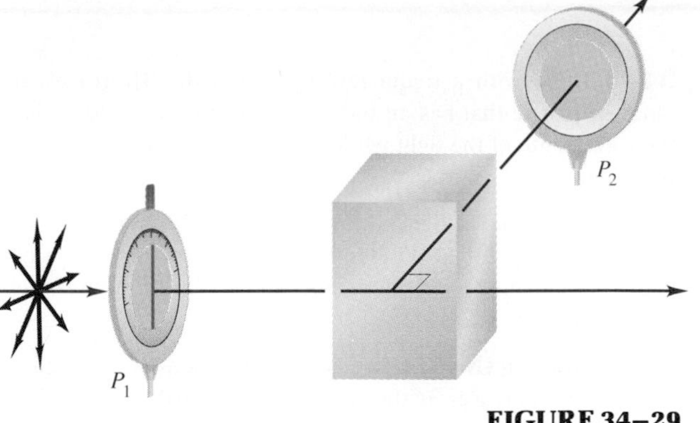


**FIGURE 34–29**

# PROBLEMS

**34–23 The Corner Reflector.** An inside corner of a cube is lined with mirrors to make a corner reflector (Example 34–3). A ray of light is reflected successively from each of the three mutually perpendicular mirrors; show that its final direction is always exactly opposite to its initial direction.

**34–24** A thin beam of light in air is incident on the surface of a glass plate having a refractive index of 1.50. What is the angle of incidence $\theta_a$ with this plate for which the angle of refraction is $\theta_a/2$, where both angles are measured relative to the normal?

**34–25** A ray of light is incident in air on a block of ice whose index of refraction is $n = 1.31$. What is the *largest* angle of incidence $\theta_a$ for which total internal reflection will occur at the vertical face (point $A$ in Fig. 34–30)?

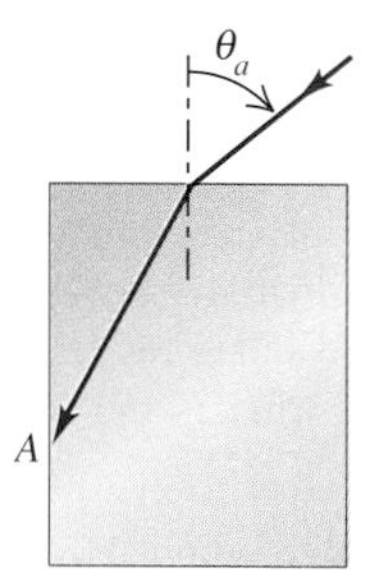


**FIGURE 34–30**

**34–26 Locating a Key in a Pool.** After a long day of driving, you take a late-night swim in a motel swimming pool. When you go to your room, you realize that you have lost your room key in the pool. You borrow a powerful flashlight and walk around the pool, shining the light into it. The light shines on the key, which is lying on the bottom of the pool, when the flashlight is held 1.2 m above the water surface and is directed at the surface a horizontal distance of 1.5 m from the edge (Fig. 34–31). If the water here is 4.0 m deep, how far is the key from the edge of the pool?

**34–27** You sight along the rim of a glass with vertical sides so that the top rim is lined up with the opposite edge of the bottom (Fig. 34–32a). The glass is a thin-walled hollow cylinder 16.0 cm high with top and bottom diameter of 8.0 cm. While you keep your eye in the same position, a friend fills the glass with a transparent liquid, and you then see a dime that is lying at the center of the bottom of the glass (Fig. 34–32b). What is the index of refraction of the liquid?

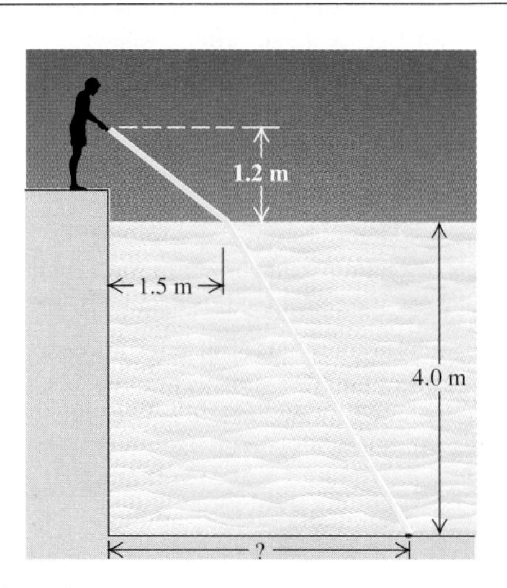

**FIGURE 34–31**

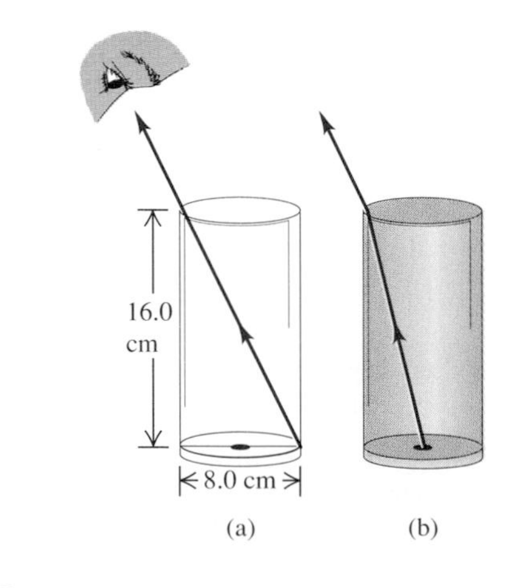

**FIGURE 34–32**

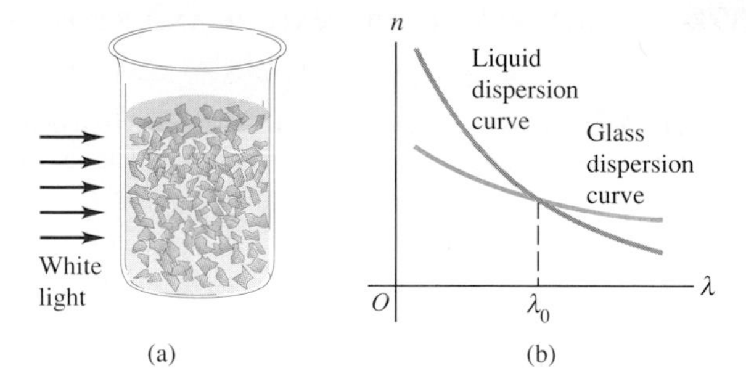

**FIGURE 34–35**

**34–28 Sonogram of the Heart.** Physicians use high-frequency ($f$ = 1–5 MHz) sound waves, called ultrasound, to image internal organs. The speed of these ultrasound waves is 1480 m/s in muscle and 344 m/s in air. a) At what angle from the normal does an ultrasound beam enter the heart if it leaves the lungs at an angle of 9.16° from the normal to the heart wall? (Assume that the speed of sound in the lungs is 344 m/s.) b) What is the critical angle for sound waves in air incident on muscle?

**34–29** Old photographic plates were made of glass with a light-sensitive emulsion on the front surface. This emulsion was somewhat transparent. When a bright point source is focused on the front of the plate, the developed photograph will show a halo around the image of the spot. If the glass plate is 3.00 mm thick and the halos have an inner radius of 5.24 mm, what is the index of refraction of the glass? (*Hint:* Light from the spot on the front surface is totally reflected at the back surface of the plate and comes back to the front surface.)

**34–30** A glass plate 3.00 mm thick, with an index of refraction of 1.50, is placed between a point source of light with wavelength 500 nm (in vacuum) and a screen. The distance from source to screen is 3.00 cm. How many wavelengths are there between the source and the screen?

**34–31** The prism of Fig. 34–33 has a refractive index of 1.48, and the angles $A$ are 30.0°. Two light rays $m$ and $n$ are parallel as they enter the prism. What is the angle between them after they emerge?

**34–32** A 45°-45°-90° prism is immersed in water. A ray of light is incident normally on one of its shorter faces. What is the minimum index of refraction that the prism must have if this ray is to be totally reflected within the glass at the long face of the prism?

**34–33** Light is incident normally on the short face of a 30°-60°-90° prism (Fig. 34–34). A drop of liquid is placed on the hypotenuse of the prism. If the index of the prism is 1.50, find the maximum index that the liquid may have if the light is to be totally reflected.

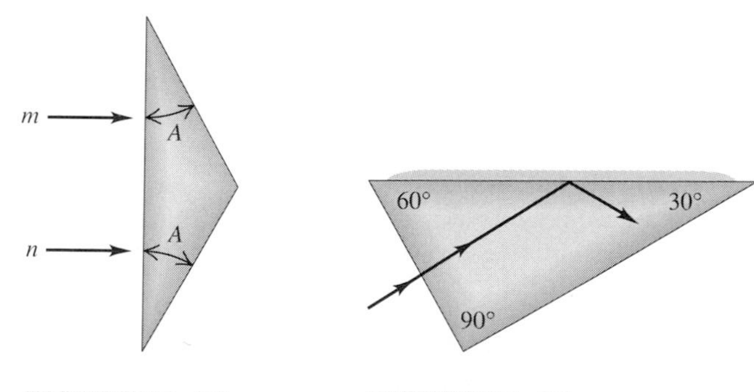


**FIGURE 34–33** **FIGURE 34–34**

**34–34 The Christiansen Filter.** The glass vessel shown in Fig. 34–35a contains a large number of small, irregular pieces of glass and a liquid. The dispersion curves of the glass and of the liquid are shown in Fig. 34–35b. Explain the behavior of a parallel beam of white light as it traverses the vessel. (This is known as a *Christiansen filter.*)

**34–35** A ray of light traveling in a block of glass ($n$ = 1.60) is incident on the top surface at an angle of 55.0° with respect to the normal. If a layer of oil is placed on the top surface of the glass, the ray is totally reflected. What is the maximum possible index of refraction of the oil?

**34–36** A beaker filled with a liquid whose index of refraction is 1.80 has a mirrored bottom that reflects the light incident on it. A light beam strikes the top surface of the liquid at an angle of 50.0° from the normal. At what angle from the normal will the beam exit from the liquid after traveling down through the liquid, reflecting from the mirrored bottom, and returning to the surface?

**34–37** Three polarizing filters are stacked, with the polarizing axes of the second and third at angles $\theta$ and 90°, respectively, with that of the first. Unpolarized light with intensity $I_0$ is incident on the stack. a) Derive an expression for the intensity of light transmitted through the stack as a function of $I_0$ and $\theta$. b) For what value of $\theta$ does the maximum transmission occur?

**34–38** The refractive index of a certain crown glass is 1.52. For what incident angle is light reflected from the surface of this glass completely polarized if the glass is immersed in a) air; b) water?

**34–39** A beam of light traveling horizontally is made of an unpolarized component with intensity $I_0$ and a polarized component with intensity $I_p$. The plane of polarization of the polarized component is oriented at an angle of $\theta$ with respect to the vertical. The following data give the intensity measured through a polarizer with an orientation of $\phi$ with respect to the vertical:

| $\phi$ | $I_{total}$ (W/m$^2$) | $\phi$ | $I_{total}$ (W/m$^2$) |
|---|---|---|---|
| 0.0° | 18.4 | 100.0° | 8.6 |
| 10.0° | 21.4 | 110.0° | 6.3 |
| 20.0° | 23.7 | 120.0° | 5.2 |
| 30.0° | 24.8 | 130.0° | 5.2 |
| 40.0° | 24.8 | 140.0° | 6.3 |
| 50.0° | 23.7 | 150.0° | 8.6 |
| 60.0° | 21.4 | 160.0° | 11.6 |
| 70.0° | 18.4 | 170.0° | 15.0 |
| 80.0° | 15.0 | 180.0° | 18.4 |
| 90.0° | 11.6 | | |

a) What is the orientation of the polarized component? (That is, what is the angle $\theta$?) b) What are the values of $I_0$ and $I_p$?

**34–40** A quarter-wave plate converts linearly polarized light to circularly polarized light. Prove that a quarter-wave plate also converts circularly polarized light to linearly polarized light.

**34–41 Optical Activity.** Many biologically active molecules are also optically active. When plane-polarized light traverses a solution of these compounds, its plane of polarization is rotated. Some compounds rotate the polarization clockwise; others rotate the polarization counterclockwise. The amount of rotation depends on the amount of material in the light path. The following data give the amount of rotation through two amino acids over a path length of 100 cm. From the data, find the relationship between the concentration $C$ in grams per 100 mL and the rotation in degrees of the polarization of each amino acid.

| Rotation | | |
|---|---|---|
| *l*-leucine | *d*-glutamic acid | Concentration (g/100 ml) |
| − 0.11° | 0.124° | 1.0 |
| − 0.22° | 0.248° | 2.0 |
| − 0.55° | 0.620° | 5.0 |
| − 1.10° | 1.24° | 10.0 |
| − 2.20° | 2.48° | 20.0 |
| − 5.50° | 6.20° | 50.0 |
| − 11.0° | 12.4° | 100.0 |

**34–42** A certain birefringent material has indexes of refraction $n_1$ and $n_2$ for the two perpendicular components of linearly polarized light passing through it. The corresponding wavelengths are $\lambda_0/n_1$ and $\lambda_0/n_2$, where $\lambda_0$ is the wavelength in vacuum. a) If the crystal is to function as a quarter-wave plate, the number of wavelengths of each component within the material must differ by $\frac{1}{4}$. Show that the minimum thickness for a quarter-wave plate is

$$d = \frac{\lambda_0}{4(n_1 - n_2)}.$$

b) Find the minimum thickness of a quarter-wave plate made of calcite if the indexes of refraction are 1.658 and 1.486 and the wavelength in vacuum is $\lambda_0 = 520$ nm.

## CHALLENGE PROBLEMS

**34–43 Angle of Deviation.** The incident angle $\theta_a$ in Fig. 34–36 is chosen so that the light passes symmetrically through the prism, which has refractive index $n$ and apex angle $A$. a) Show that the angle of deviation $\delta$ (the angle between the initial and final directions of the ray) is given by

$$\sin\frac{A + \delta}{2} = n \sin\frac{A}{2}.$$

(When the light passes through symmetrically, as shown, the angle of deviation is a minimum.) b) Use the result of part (a) to find the angle of deviation for a ray of light passing symmetrically through a prism having three equal angles ($A = 60.0°$) and $n = 1.60$. c) A certain glass has a refractive index of 1.50 for red light (700 nm) and 1.52 for violet light (400 nm). If both colors pass through symmetrically, as described in part (a), and if $A = 60.0°$, find the difference between the angles of deviation for the two colors.

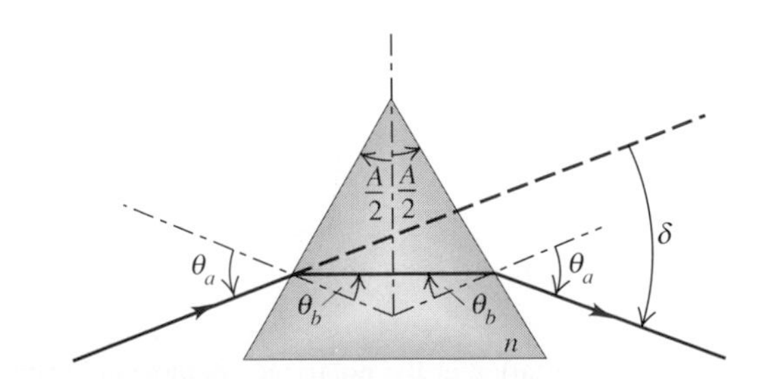


FIGURE 34–36

**34–44 Fermat's Principle of Least Time.** A ray of light traveling with speed $c$ leaves point 1 of Fig. 34–37 and is reflected to point 2. The ray strikes the reflecting surface a horizontal distance $x$ from point 1. a) Show that the time $t$ required for the light to travel from 1 to 2 is

$$t = \frac{\sqrt{y_1^2 + x^2} + \sqrt{y_2^2 + (l - x)^2}}{c}.$$

b) Take the derivative of $t$ with respect to $x$. Set the derivative equal to zero to show that this time reaches its *minimum* value when $\theta_1 = \theta_2$, which is the law of reflection and corresponds to the actual path taken by the light. This is an example of *Fermat's principle of least time,* which states that among all possible paths between two points, the one actually taken by a ray of light is that for which the time of travel is a minimum. (In fact, there are some cases in which the time is a maximum rather than a minimum.)

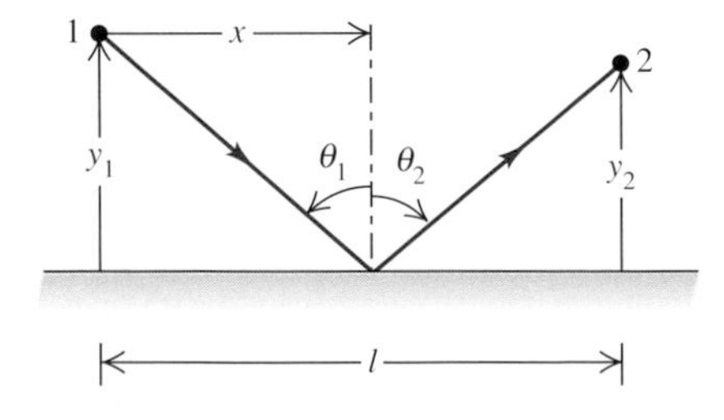

FIGURE 34–37

**34–45** A ray of light goes from point $A$ in a medium in which the speed of light is $v_1$ to point $B$ in a medium in which the speed is $v_2$ (Fig. 34–38). The ray strikes the interface a horizontal distance $x$ to the right of point $A$. a) Show that the time required for the light to go from $A$ to $B$ is

$$t = \frac{\sqrt{h_1^2 + x^2}}{v_1} + \frac{\sqrt{h_2^2 + (l - x)^2}}{v_2}.$$

b) Take the derivative of $t$ with respect to $x$. Set this derivative equal to zero to show that this time reaches its *minimum* value when $n_1 \sin\theta_1 = n_2 \sin\theta_2$. This is Snell's law, and it corresponds to the actual path taken by the light. This is another example of Fermat's principle of least time (see Challenge Problem 34–44).

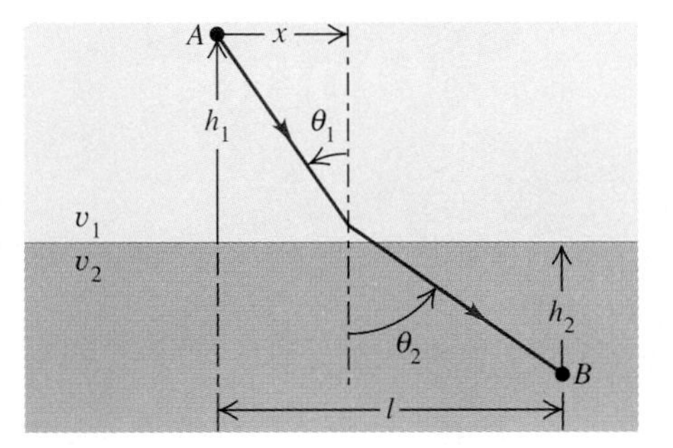


**FIGURE 34–38**

**34–46** Light is incident in air at an angle $\theta_a$ (Fig. 34–39) on the upper surface of a transparent plate, the surfaces of the plate being plane and parallel to each other. a) Prove that $\theta_a = \theta'_a$. b) Show that this is true for any number of different parallel plates. c) Prove that the lateral displacement $d$ of the emergent beam is given by the relation

$$d = t\frac{\sin(\theta_a - \theta_b)}{\cos\theta_b},$$

where $t$ is the thickness of the plate. d) A ray of light is incident at an angle of 60.0° on one surface of a glass plate 1.80 cm thick with an index of refraction 1.50. The medium on either side of the plate is air. Find the lateral displacement between the incident and emergent rays.

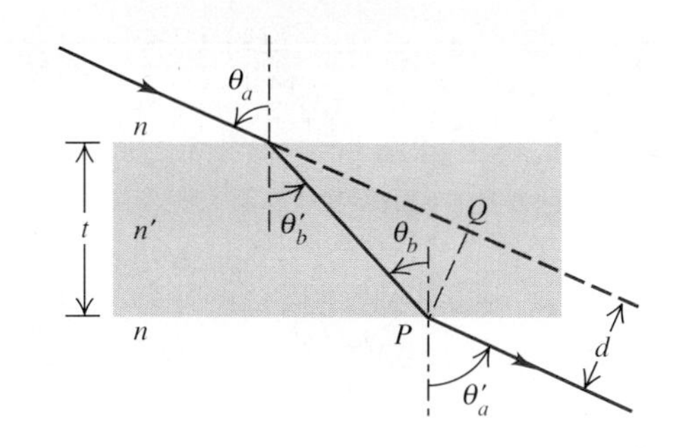


**FIGURE 34–39**

**34–47** Consider two vibrations of equal amplitude and frequency but differing in phase, one along the $x$-axis,

$$x = a\sin(\omega t - \alpha),$$

and the other along the $y$-axis,

$$y = a\sin(\omega t - \beta).$$

These can be written as follows:

$$\frac{x}{a} = \sin\omega t\cos\alpha - \cos\omega t\sin\alpha, \qquad (1)$$

$$\frac{y}{a} = \sin\omega t\cos\beta - \cos\omega t\sin\beta. \qquad (2)$$

a) Multiply Eq. (1) by $\sin\beta$ and Eq. (2) by $\sin\alpha$ and then subtract the resulting equations. b) Multiply Eq. (1) by $\cos\beta$ and Eq. (2) by $\cos\alpha$ and then subtract the resulting equations. c) Square and add the results of parts (a) and (b). d) Derive the equation $x^2 + y^2 - 2xy\cos\delta = a^2\sin^2\delta$, where $\delta = \alpha - \beta$. e) Use the above result to justify each of the diagrams in Fig. 34–40. In the figure, the angle given is the phase difference between two simple harmonic motions of the same frequency and amplitude, one horizontal (along the $x$-axis) and the other vertical (along the $y$-axis). The figure thus shows the resultant motion from the superposition of the two perpendicular harmonic motions.

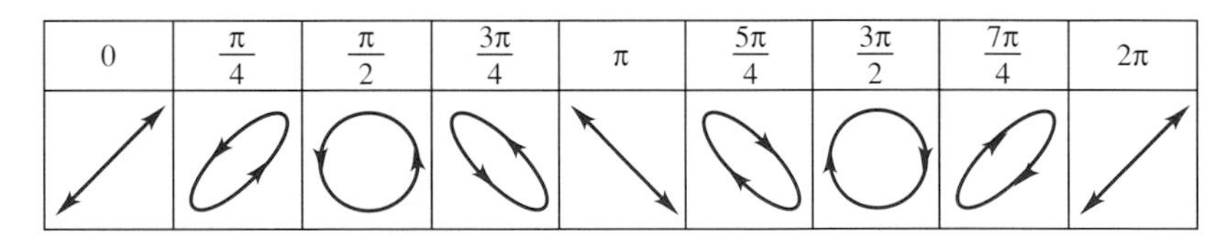


**FIGURE 34–40**

Kaleidoscopes contain three or more mirrors to produce multiple images of the object being viewed. The number of images and their symmetry about the central axis depends on the angle between the mirrors.

# CHAPTER 35

# Geometric Optics

Many common kaleidoscopes have two mirrors oriented at an angle of 60° to each other. This results in patterns of 6-fold symmetry, with one direct view of the object and five reflected images. The addition of a third mirror extends the pattern to fill the entire viewing plane.

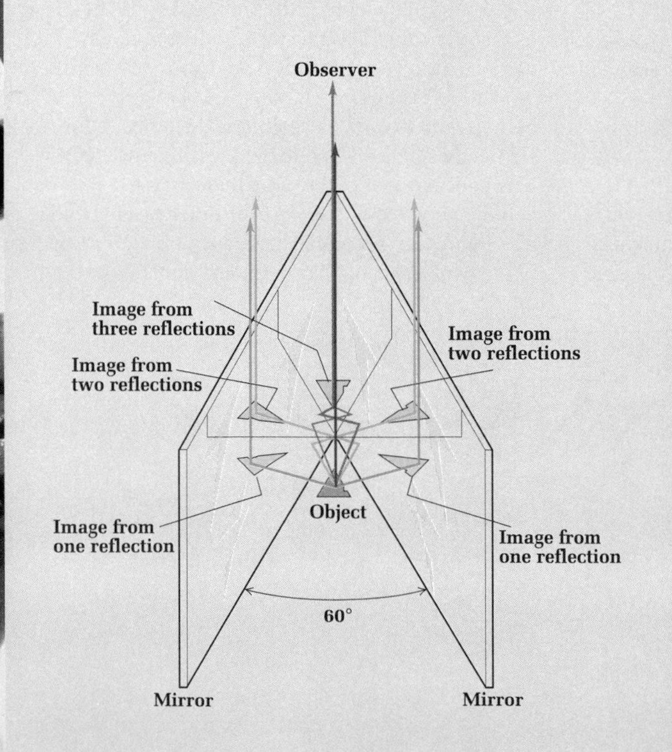


Three 4-inch mirrors

When this photograph was taken, a 50-cent coin was placed in the corner formed by two of the three mirrors. Its silhouette appears as a black circle in the repeated pattern of the kaleidoscope.

## KEY CONCEPTS

**• The concept of image plays a central role in geometric optics. When rays emerge from a point and are reflected or refracted, the outgoing rays converge to or appear to have diverged from an image point.**

**• A plane or spherical mirror forms an image of an object; the positions and sizes of object and image are related by simple equations. The ratio of image to object height is called the lateral magnification. A principal-ray diagram, showing a few particular rays, is useful in understanding the formation of images. Incoming rays parallel to the principal axis converge to or appear to diverge from a focal point.**

**• A spherical refracting surface or a lens with spherical surfaces forms an image; the sizes and positions of object and image are related by simple equations. The concepts of magnification and focal point have the same meaning as for spherical mirrors, and principal-ray diagrams are again helpful in understanding the formation of images. The focal length of a lens is determined by its index of refraction and the curvatures of its surfaces.**

## INTRODUCTION

When you look at yourself in a flat mirror, you appear actual size and right-side-up. But in a curved mirror you may appear larger or smaller, or even upside-down. Why is this? Mirrors, magnifying glasses, and telescopes are a familiar part of everyday life. We can use the ray model of light, introduced in the preceding chapter, to understand the behavior of these and similar instruments. All we need are the laws of reflection and refraction and some simple geometry and trigonometry.

In this chapter and the next, we'll make frequent use of the concept of *image*. When a politician worries about his image, he is thinking of how he looks to the voters. We will use the term in a related but more precise way, having to do with the behavior of a collection of rays that converge toward or appear to diverge from a point called an *image point.* In this chapter we analyze the formation of images by a single reflecting or refracting surface and by a thin lens. This discussion lays the foundation for analyses of many familiar optical instruments, including camera lenses, magnifiers, the human eye, microscopes, and telescopes. We will study these instruments in Chapter 36.

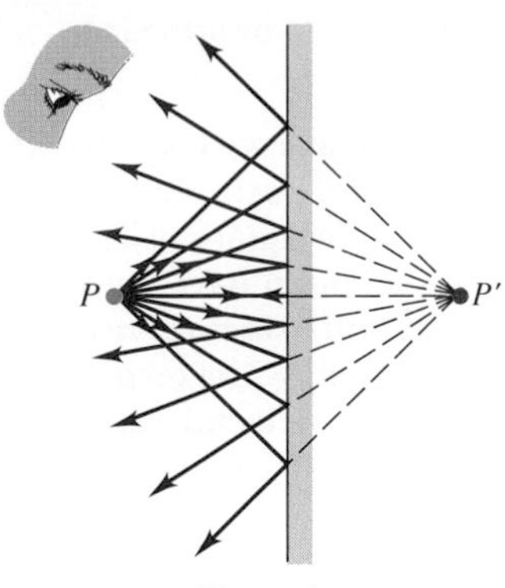


**35–1** The rays entering the eye after reflection from a plane mirror look as though they had come from point $P'$, the image point for object point $P$.

## 35–1 REFLECTION AT A PLANE SURFACE

The concept of **image** is the most important new idea in this chapter and the next. Consider Fig. 35–1. Several rays diverge from point $P$ and are reflected at the plane mirror, according to the law of reflection. After they are reflected, their final directions are the same as though they had come from point $P'$. We call point $P$ an *object point* and point $P'$ the corresponding *image point,* and we say that the mirror forms an *image* of point $P$. The outgoing rays (those going away from the mirror) don't really come from point $P'$, but their directions are the same as though they had come from that point.

Something similar happens at a plane refracting surface, as shown in Fig. 35–2. Rays coming from point $P$ are refracted at the interface between two optical materials. When the angles of incidence are small, the final directions of the rays after refraction are the same as though they had come from point $P'$, and again we call $P'$ an *image point.*

In both of these cases the rays spread out from point $P$, whether it is an actual point source of light or a point that scatters the light shining on it. An observer who can see only the rays spreading out from the surface (after they are reflected or refracted) *thinks* that they come from the image point $P'$. This image point is therefore a convenient way to describe the directions of the various reflected or refracted rays, just as the object point $P$ describes the directions of the rays arriving at the surface, *before* reflection or refraction.

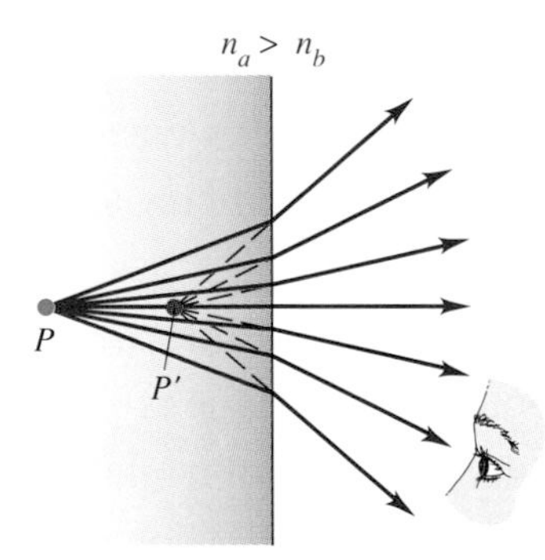


**35–2** The rays entering the eye after refraction at the plane interface look as though they had come from point $P'$, the image point for object $P$. When $n_b < n_a$, the image point $P'$ is closer to the surface than the object point $P$. The angles of incidence have been exaggerated for clarity.

To find the precise location of the image $P'$ that a plane mirror forms of an object point $P$, we use the construction shown in Fig. 35–3. The figure shows two rays diverging from an object point $P$ at a distance $s$ to the left of a plane mirror. We call $s$ the *object distance*. The ray $PV$ is incident normally on the mirror (that is, it is perpendicular to the mirror surface), and it returns along its original path.

The ray $PB$ makes an angle $\theta$ with $PV$. It strikes the mirror at an angle of incidence $\theta$ and is reflected at an equal angle with the normal. When we extend the two reflected rays backward, they intersect at point $P'$, at a distance $s'$ behind the mirror. We call $s'$ the *image distance*. The line between $P$ and $P'$ is perpendicular to the mirror. The two triangles are congruent, so $P$ and $P'$ are at equal distances from the mirror, and $s$ and $s'$ have equal magnitudes.

We can repeat the construction of Fig. 35–3 for each ray diverging from $P$. The directions of *all* the outgoing rays are the same *as though* they had originated at point $P'$, confirming that $P'$ is the *image* of $P$. The rays do not actually pass through point $P'$. In fact, if the mirror is opaque, there is no light at all on the back side. In cases like this, and like that of Fig. 35–2, the outgoing rays don't actually come from $P'$, and we call the image a **virtual image.** Later we will see cases in which the outgoing rays really *do* pass through an image point, and we will call the resulting image a **real image.** The images formed on a projection screen or the photographic film in a camera are real images.

Before we go further, let's introduce some general sign rules. These may seem unnecessarily complicated for the simple case of an image formed by a plane mirror, but

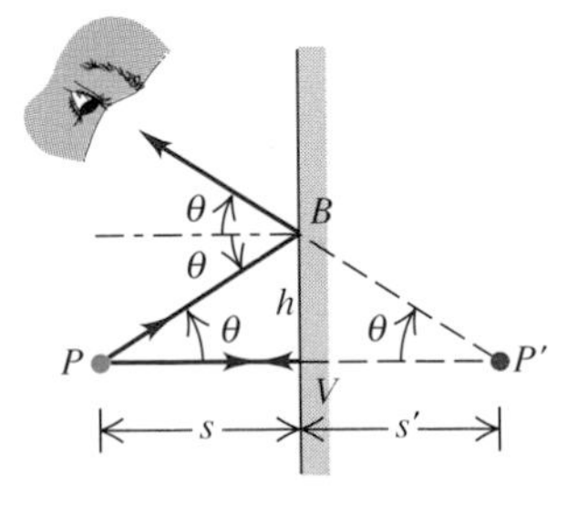


**35–3** After reflection at a plane surface, all rays originally diverging from the object point $P$ now diverge from the image point $P'$, although they do not *originate* at $P'$. Point $P'$ is called the *virtual image* of point $P$. The eye sees some of the outgoing rays and perceives them as having come from $P'$.

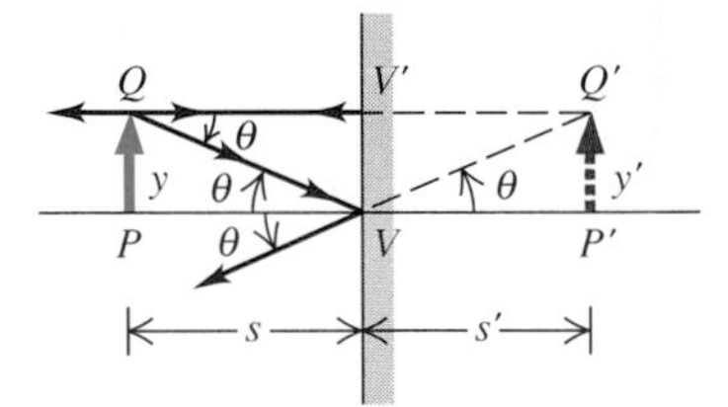


**35–4** Construction for determining the height of an image formed by reflection at a plane reflecting surface.

we want to state the rules in a form that will be applicable to *all* the situations that we will encounter. These will include image formation by a plane or spherical reflecting or refracting surface or a pair of refracting surfaces forming a lens. Here are the rules:

> When the object is on the same side of the reflecting surface as the incoming light, the object distance $s$ is positive; otherwise, it is negative.

> When the image is on the same side of the reflecting surface as the outgoing light, the image distance $s'$ is positive; otherwise, it is negative.

For a mirror the incoming and outgoing sides are always the same; in Fig. 35–3 they are both the left side.

In Fig. 35–3 the object distance $s$ is *positive* because the object point $P$ is on the incoming side (the left side) of the reflecting surface. The image distance $s'$ is *negative* because the image point $P'$ is *not* on the outgoing side (the left side) of the surface. The object and image distances $s$ and $s'$ are related simply by

$$s = -s' \quad \text{(plane mirror).} \tag{35–1}$$

Next we consider an object with finite size, parallel to the mirror, represented by the arrow $PQ$ with height $y$ in Fig. 35–4. Two of the rays from $Q$ are shown; *all* the rays from $Q$ appear to diverge from its image point $Q'$ after reflection. The image of the arrow is the line $P'Q'$, with height $y'$. Other points of the arrow $PQ$ have image points between $P'$ and $Q'$. The triangles $PQV$ and $P'Q'V$ are congruent, so the object $PQ$ and image $P'Q'$ have the same size and orientation, and $y = y'$.

The ratio of image to object height, $y'/y$, in *any* image-forming situation, is called the **lateral magnification** $m$; that is,

$$m = \frac{y'}{y}. \tag{35–2}$$

For a plane mirror the lateral magnification $m$ is unity. When you look at yourself in a plane mirror, your image is the same size as the real you.

We will often represent an object by an arrow. Its image may be an arrow pointing in the *same* direction as the object or in the *opposite* direction. When the directions are the same, as in Fig. 35–4, we say that the image is **erect;** when they are opposite, the image is **inverted.** The image formed by a plane mirror is always erect. A positive value of lateral magnification $m$ corresponds to an erect image; a negative value corresponds to an inverted image. That is, for an erect image $y$ and $y'$ always have the *same* sign, and for an inverted image they always have *opposite* signs.

Figure 35–5 shows a three-dimensional virtual image of a three-dimensional object formed by a plane mirror. The images $P'Q'$ and $P'S'$ are parallel to their objects,

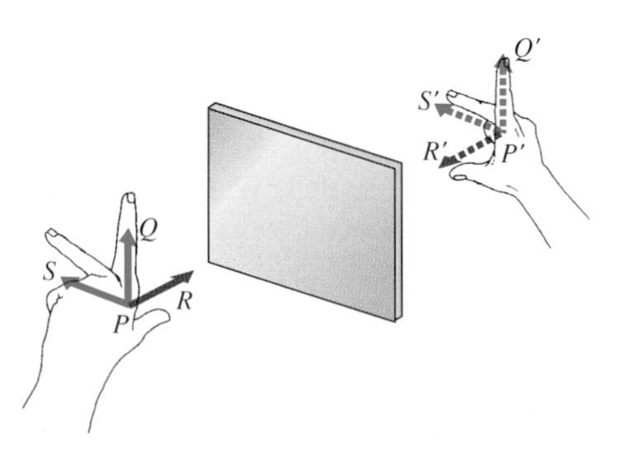


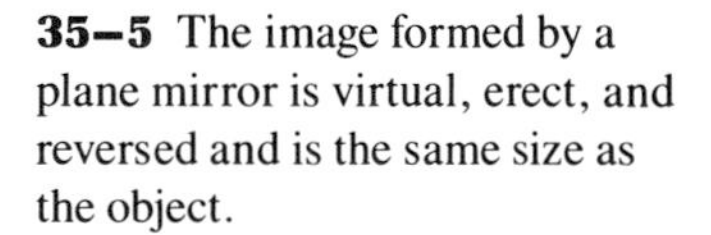

**35–5** The image formed by a plane mirror is virtual, erect, and reversed and is the same size as the object.

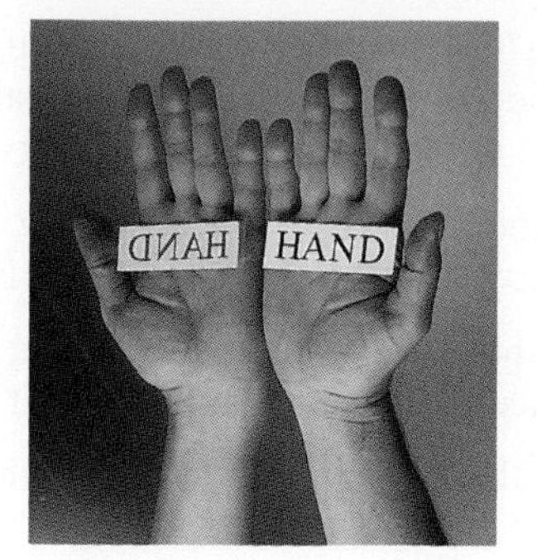


**35–6** The image formed by a plane mirror is reversed; the image of a right hand is a left hand, and so on. Are images of H's and A's reversed?

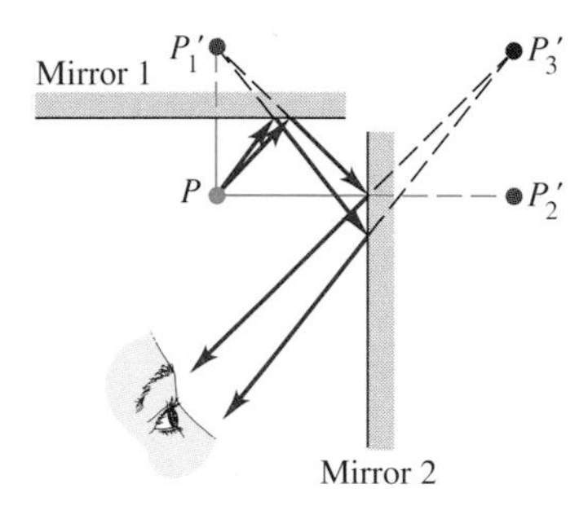


**35–7** Images $P_1'$ and $P_2'$ are formed by a single reflection of each ray. Image $P_3'$, located by treating either of the other images as an object, is formed by a double reflection of each ray.

but $P'R'$ is reversed relative to $PR$. The image of a three-dimensional object formed by a plane mirror is the same *size* as the object in all its dimensions, but the image and object are *not* identical. They are related in the same way as a right hand and a left hand, and indeed we speak of a pair of objects with this relationship as mirror-image objects. To verify this object-image relationship, point your two thumbs along $PR$ and $P'R'$, your forefingers along $PQ$ and $P'Q'$, and your middle fingers along $PS$ and $P'S'$. When an object and its image are related in this way, the image is said to be **reversed.** When the transverse dimensions of object and image are in the same direction, the image is erect. A plane mirror always forms an erect but reversed image. Figure 35–6 illustrates this point.

An important property of all images formed by reflecting or refracting surfaces is that an image formed by one surface or optical device can serve as the object for a second surface or device. Figure 35–7 shows a simple example. Mirror 1 forms an image $P_1'$ of the object point $P$, and mirror 2 forms another image $P_2'$, each in the way we have just discussed. But in addition the image $P_1'$ formed by mirror 1 can serve as an object for mirror 2, which then forms an image of this object at point $P_3'$ as shown. Similarly, mirror 1 can use the image $P_2'$ formed by mirror 2 as an object and form an image of it. We leave it to you to show that this image point is also at $P_3'$. Later in this chapter we will use this principle to locate the image formed by two successive curved-surface refractions in a lens.

## 35–2 REFLECTION AT A SPHERICAL SURFACE

Continuing our analysis of reflecting surfaces, we consider next the formation of an image by a *spherical* mirror. Figure 35–8a shows a spherical mirror with radius of curvature $R$, with its concave side facing the incident light. The **center of curvature** of the surface (the center of the sphere of which the surface is a part) is at $C$. Point $P$ is an object point; for the moment we assume that the distance from $P$ to $V$ is greater than $R$. The ray $PV$, passing through $C$, strikes the mirror normally and is reflected back on itself. Point $V$, at the center of the mirror surface, is called the **vertex** of the mirror, and the line $PCV$ is called the **optic axis.**

Ray $PB$, at an angle $\alpha$ with the axis, strikes the mirror at $B$, where the angles of incidence and reflection are $\theta$. The reflected ray intersects the axis at point $P'$. We will show that *all* rays from $P$ intersect the axis at the *same* point $P'$, as in Fig. 35–8b, no matter what $\alpha$ is, provided that it is a *small* angle. Point $P'$ is therefore the *image* of object point $P$. The object distance, measured from the vertex $V$, is $s$, and the image distance is $s'$. The object point $P$ is on the same side as the incident light, so according to the sign rule in Section 35–1, the object distance $s$ is positive. The image point $P'$ is on the same side as the reflected light, so the image distance $s'$ is also positive.

Unlike the reflected rays in Fig. 35–1, the reflected rays in Fig. 35–8b actually do intersect at point $P'$; then they diverge from $P'$ *as if* they had originated at this point. The image $P'$ is a *real* image.

A theorem from plane geometry states that an exterior angle of a triangle equals the sum of the two opposite interior angles. Using this theorem with triangles $PBC$ and $P'BC$ in Fig. 35–8a, we have

$$\phi = \alpha + \theta, \qquad \beta = \phi + \theta.$$

Eliminating $\theta$ between these equations gives

$$\alpha + \beta = 2\phi. \tag{35–3}$$

Now we need a sign rule for the radii of curvature of spherical surfaces:

When the center of curvature $C$ is on the same side as the outgoing (reflected) light, the radius of curvature is positive; otherwise, it is negative.

In Fig. 35–8, $R$ is positive because the center of curvature $C$ is on the same side of the mirror as the reflected light. This is always the case when reflection occurs at the *concave* side of a surface. For a *convex* surface the center of curvature is on the opposite side from the reflected light, and $R$ is negative.

We may now compute the image distance $s'$. Let $h$ represent the height of point $B$ above the axis, and $\delta$ the short distance from $V$ to the foot of this vertical line. We now write expressions for the tangents of $\alpha$, $\beta$, and $\phi$, remembering that $s$, $s'$, and $R$ are all positive quantities:

$$\tan\alpha = \frac{h}{s-\delta}, \qquad \tan\beta = \frac{h}{s'-\delta}, \qquad \tan\phi = \frac{h}{R-\delta}.$$

These trigonometric equations cannot be solved as simply as the corresponding algebraic equations for a plane mirror. However, *if the angle $\alpha$ is small,* the angles $\beta$ and $\phi$ are also small. The tangent of a small angle is nearly equal to the angle itself (in radians), so we can replace $\tan\alpha$ by $\alpha$, and so on, in the equations above. Also, if $\alpha$ is small, we can neglect the distance $\delta$ compared with $s'$, $s$, and $R$. So for small angles we have the approximate relations:

$$\alpha = \frac{h}{s}, \qquad \beta = \frac{h}{s'}, \qquad \phi = \frac{h}{R}.$$

Substituting these into Eq. (35–3) and dividing by $h$, we obtain a general relation among $s$, $s'$, and $R$:

$$\frac{1}{s} + \frac{1}{s'} = \frac{2}{R}. \tag{35–4}$$

This equation does not contain the angle $\alpha$. This means that *all* rays from $P$ that make sufficiently small angles with the optic axis intersect at $P'$ after they are reflected. Such rays, nearly parallel to the axis and close to it, are called **paraxial rays.**

Be sure you understand that Eq. (35–4), as well as many similar relations that we will derive later in this chapter and the next, is only *approximately* correct. It results from a calculation containing approximations, and it is valid only for paraxial rays. The term **paraxial approximation** is often used for the approximations that we have just described. As the angle $\alpha$ increases, the point $P'$ moves somewhat closer to the vertex; a spherical mirror, unlike a plane mirror, does not form a precise point image of a point object. This property of a spherical mirror is called *spherical aberration.* (The initially disappointing results from the Hubble space telescope resulted in part from errors in the corrections for spherical aberration.)

If $R = \infty$, the mirror becomes *plane,* and Eq. (35–4) reduces to Eq. (35–1), which we derived earlier for a plane reflecting surface.

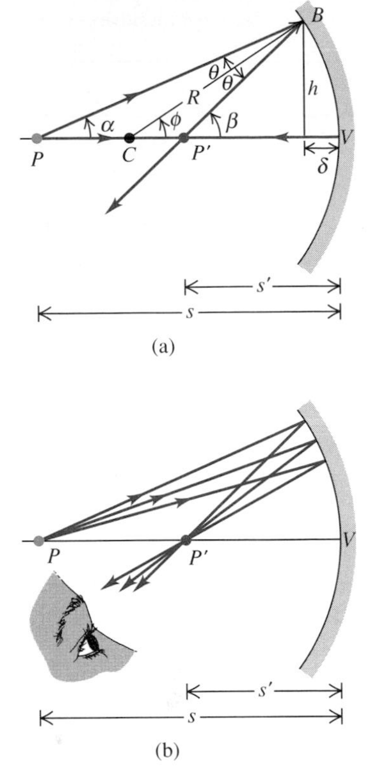


**35–8** (a) Construction for finding the position of the image $P'$ of a point object $P$, formed by a concave spherical mirror. (b) If the angle $\alpha$ is small, *all* rays from $P$ intersect at $P'$. The eye sees some of the outgoing rays and perceives them as having come from $P'$.

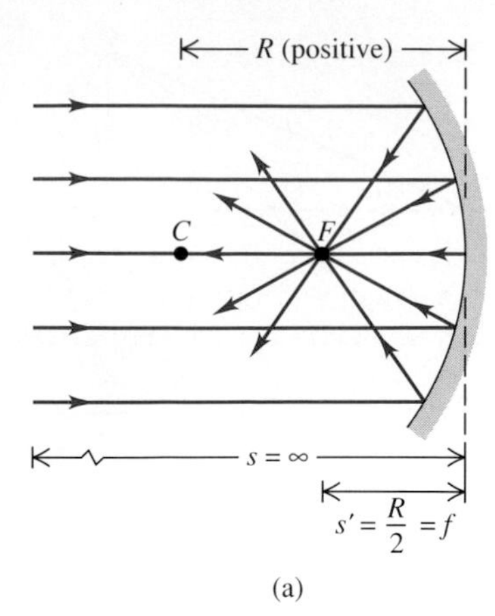


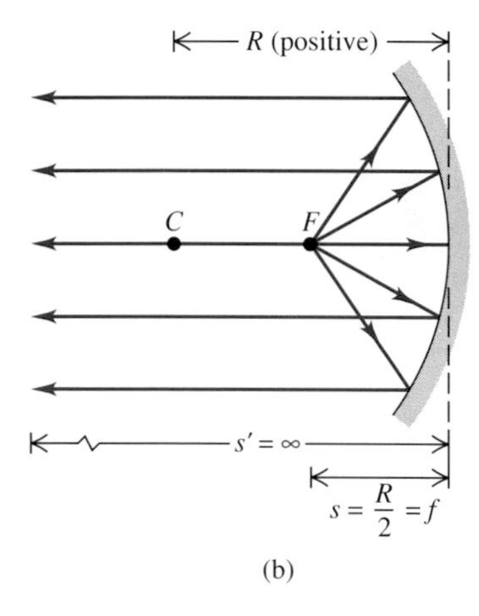


**35–9** (a) Incident rays parallel to the axis converge to the focal point $F$ of a concave mirror. (b) Rays diverging from the focal point $F$ of a concave mirror are parallel to the axis after reflection. The angles are exaggerated for clarity.

## Focal Point

When the object point $P$ is very far from the mirror ($s = \infty$), the incoming rays are parallel. From Eq. (37–4) the image distance $s'$ is given by

$$\frac{1}{\infty} + \frac{1}{s'} = \frac{2}{R}, \qquad s' = \frac{R}{2}.$$

The situation is shown in Fig. 35–9a. A beam of incident parallel rays converges, after reflection, to a point $F$ at a distance $R/2$ from the vertex of the mirror. Point $F$ is called the **focal point,** and its distance from the vertex, denoted by $f$, is called the **focal length.** We see that $f$ is related to the radius of curvature $R$ by

$$f = \frac{R}{2}. \tag{35–5}$$

We can discuss the opposite situation, shown in Fig. 35–9b. When the *image* distance $s'$ is very large, the outgoing rays are parallel to the optic axis. The object distance $s$ is then given by

$$\frac{1}{s} + \frac{1}{\infty} = \frac{2}{R}, \qquad s = \frac{R}{2}.$$

In Fig. 35–9b, rays coming to the mirror from the focal point are reflected parallel to the optic axis. Again we see that $f = R/2$.

Thus the focal point $F$ of a concave spherical mirror has the properties that (1) any incoming ray parallel to the optic axis is reflected through the focal point and (2) any incoming ray that passes through the focal point is reflected parallel to the optic axis. For spherical mirrors these statements are true only for paraxial rays; for parabolic mirrors they are *exactly* true.

We will usually express the relationship between object and image distances for a mirror, Eq. (35–4), in terms of the focal length $f$:

$$\frac{1}{s} + \frac{1}{s'} = \frac{1}{f}. \tag{35–6}$$

Now suppose we have an object with finite size, represented by the arrow $PQ$ in Fig. 35–10, perpendicular to the axis $PV$. The image of $P$ formed by paraxial rays is at $P'$. The object distance for point $Q$ is very nearly equal to that for point $P$, so the image $P'Q'$

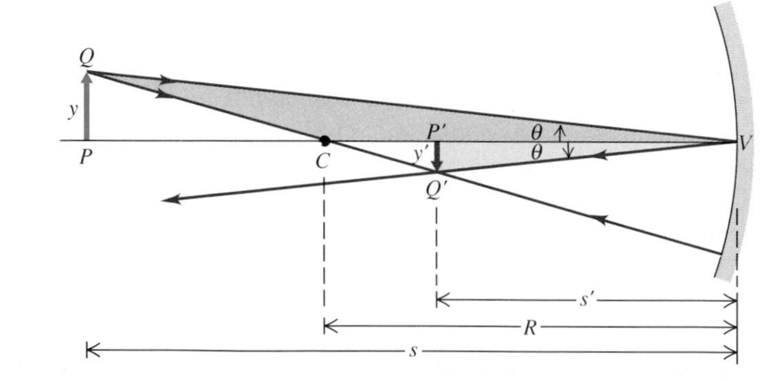


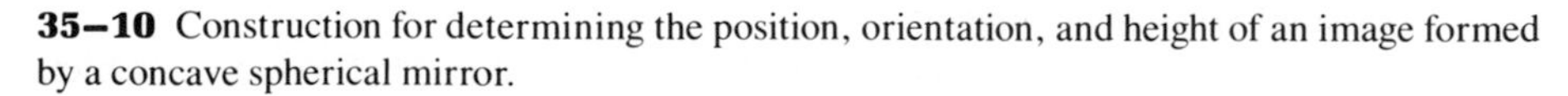
**35–10** Construction for determining the position, orientation, and height of an image formed by a concave spherical mirror.

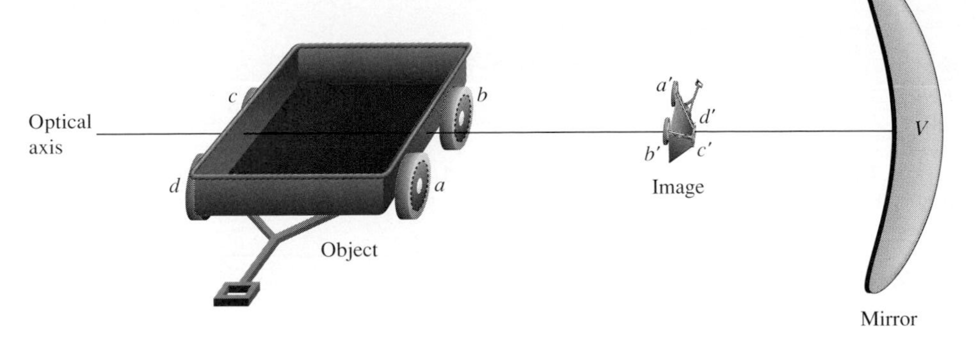


35–11 Schematic diagram of an object and its real, inverted, reduced image formed by a concave mirror. The face of the object closest to the mirror has the greatest magnitude of lateral magnification.

is nearly straight and is perpendicular to the axis. Note that object and image have different sizes, $y$ and $y'$, respectively, and that they have opposite orientations. We have defined the *lateral magnification* $m$ as the ratio of image size $y'$ to object size $y$:

$$m = \frac{y'}{y}.$$

Because triangles $PVQ$ and $P'VQ'$ in Fig. 35–10 are *similar,* we also have the relation $y/s = -y'/s'$. The negative sign is needed because object and image are on opposite sides of the optic axis; if $y$ is positive, $y'$ is negative. Therefore

$$m = \frac{y'}{y} = -\frac{s'}{s}. \qquad (35\text{–}7)$$

A negative value of $m$ indicates that the image is *inverted* relative to the object, as the figure shows. In cases that we will consider later, in which $m$ may be either positive or negative, a positive value always corresponds to an erect image and a negative value to an inverted one. For a *plane* mirror, $s = -s'$, so $y' = y$ and the image is erect, as we have already shown.

Although the ratio of image size to object size is called the *magnification,* the image formed by a mirror or lens may be either larger or smaller than the object. If it is smaller, then the magnification is less than unity in absolute value. The image formed by an astronomical telescope mirror or a camera lens is usually *much* smaller than the object. For three-dimensional objects the ratio of image distances to object distances measured *along* the optic axis is different from the ratio of *lateral* distances (the lateral magnification). In particular, if $m$ is a small fraction, the three-dimensional image of a three-dimensional object is reduced *longitudinally* much more than it is reduced *transversely.* Figure 35–11 shows this effect. Also, the image formed by a spherical mirror, like that of a plane mirror, is always reversed.

## EXAMPLE 35–1

A concave mirror forms an image, on a wall 3.0 m from the mirror, of the filament of a headlight lamp 10 cm in front of the mirror (Fig. 35–12). a) What is the radius of curvature of the mirror? b) What is the height of the image if the height of the object is 5.0 mm?

**SOLUTION** a) Both object distance and image distance are positive; we have $s = 10$ cm and $s' = 300$ cm. From Eq. (35–4),

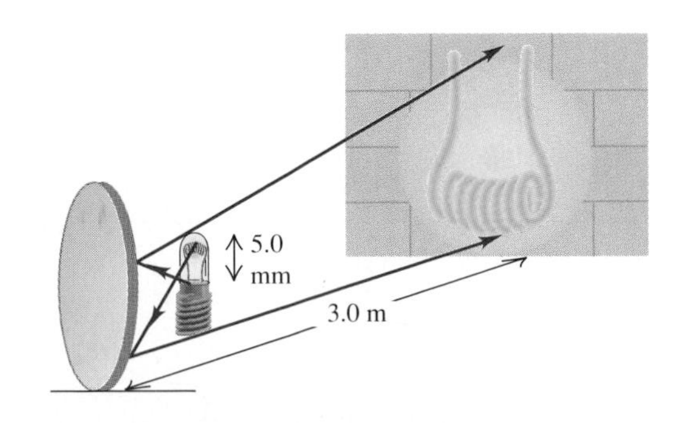


35–12 The mirror forms a real, enlarged, inverted image of the lamp filament.

$$\frac{1}{10\text{ cm}} + \frac{1}{300\text{ cm}} = \frac{2}{R},$$

$$R = 19.4\text{ cm}.$$

The focal length of the mirror is $f = R/2 = 9.7$ cm. In a headlight lamp the filament is usually placed close to the focal point, producing a beam of nearly parallel rays.

b) From Eq. (35–7),

$$m = \frac{y'}{y} = -\frac{s'}{s} = -\frac{300\text{ cm}}{10\text{ cm}} = -30.$$

The image is inverted (because $m$ is negative); its height is 30 times the height of the object, or (30)(5.0 mm) = 150 mm.

## Convex Mirrors

In Fig. 35–13a the *convex* side of a spherical mirror faces the incident light. The center of curvature is on the opposite side from the outgoing rays, so according to our sign rule, $R$ is negative. Ray $PB$ is reflected with the angles of incidence and reflection both equal to $\theta$. The reflected ray, projected backward, intersects the axis at $P'$. As with a concave mirror, *all* rays from $P$ that are reflected by the mirror diverge from the same point $P'$, provided that the angle $\alpha$ is small. Therefore $P'$ is the image of $P$. The object distance $s$ is positive, the image distance $s'$ is negative, and the radius of curvature $R$ is negative.

Figure 35–13b shows two rays diverging from the head of the arrow $PQ$ and the virtual image $P'Q'$ of this arrow. We leave it to you to show, by the same procedure that we used for a concave mirror, that

$$\frac{1}{s} + \frac{1}{s'} = \frac{2}{R}$$

and that the lateral magnification is

$$m = \frac{y'}{y} = -\frac{s'}{s}.$$

These expressions are exactly the same as those for a concave mirror. Thus when we use our sign rules consistently, Eqs. (35–4) and (35–7) are valid for both concave and convex mirrors.

When $R$ is negative (convex mirror), incoming rays parallel to the optic axis are not reflected through the focal point $F$. Instead, they diverge as though they had come from the point $F$ at a distance $f$ behind the mirror, as shown in Fig. 35–14a. In this case, $f$ is the focal length and $F$ is called a *virtual focal point.* The corresponding image distance $s'$ is negative, so both $f$ and $R$ are negative, and Eq. (35–5) holds for convex as well as concave mirrors. In Fig. 35–14b the incoming rays are converging as though they would meet at the virtual focal point $F$, and they are reflected parallel to the optic axis.

In summary, Eqs. (35–4) through (35–7), the basic relationships for image formation by a spherical mirror, are valid for both concave and convex mirrors, provided that we use the sign rules consistently.

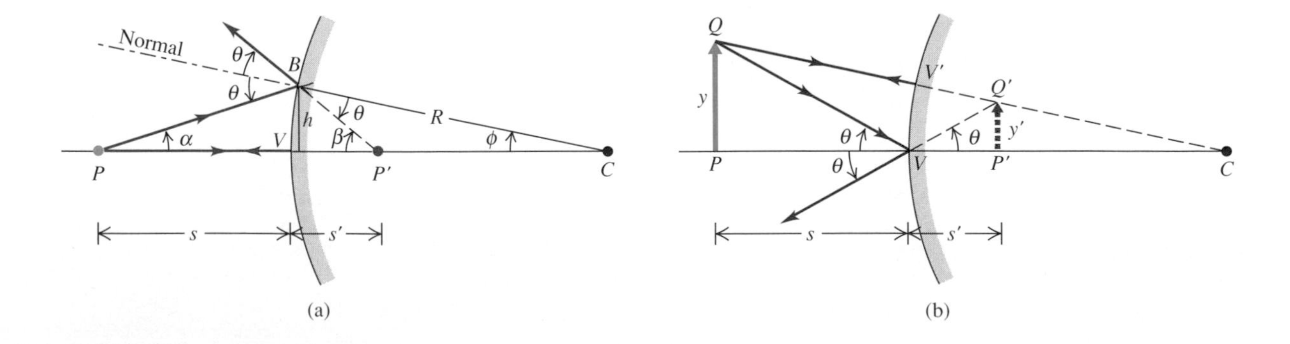


**35–13** Constructions for finding (a) the position and (b) the magnification of the image formed by a convex mirror.

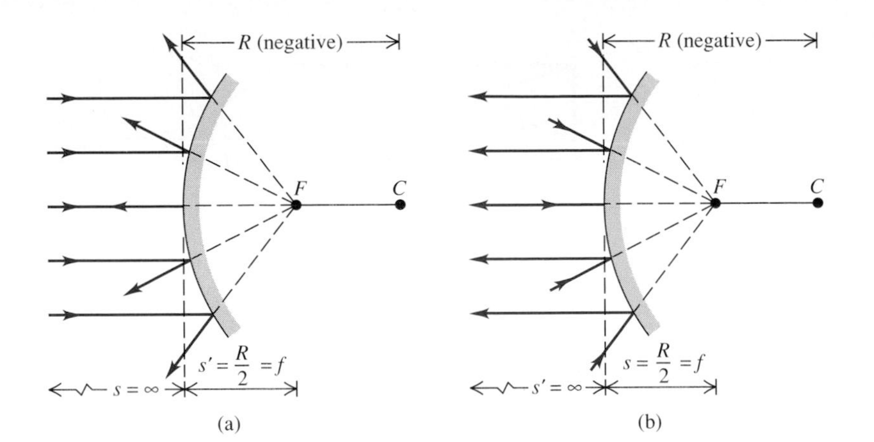


**35–14** (a) Incident rays parallel to the axis diverge as if from the virtual focal point $F$ of a convex mirror. (b) Rays aimed at the virtual focal point $F$ of a concave mirror are parallel to the axis after reflection. The angles are exaggerated for clarity.

### EXAMPLE 35–2

**Santa's image problem** Santa checks himself for soot, using his reflection in a shiny silvered Christmas tree ornament 0.750 m away. The diameter of the ornament is 7.20 cm. Standard reference works state that Santa is a "right jolly old elf," so we estimate his height as 1.6 m. Where and how tall is the image of Santa formed by the ornament? Is it erect or inverted?

**SOLUTION** The surface of the ornament closest to Santa acts as a convex mirror with radius $R = -(7.20\text{ cm})/2 = -3.60\text{ cm}$ and focal length $f = R/2 = -1.80\text{ cm}$. The object distance is $s = 0.75\text{ m} = 75\text{ cm}$. From Eq. (35–6),

$$\frac{1}{s'} = \frac{1}{f} - \frac{1}{s} = \frac{1}{-1.80\text{ cm}} - \frac{1}{75.0\text{ cm}},$$

$$s' = -1.76\text{ cm}.$$

Because $s'$ is negative, the image is behind the mirror, that is, opposite from the side of the outgoing light (Fig. 35–15), and it is virtual. The image is about halfway between the front surface of the ornament and its center. The lateral magnification $m$ is given by Eq. (35–7):

$$m = \frac{y'}{y} = -\frac{s'}{s} = -\frac{-1.76\text{ cm}}{75.0\text{ cm}} = 2.35 \times 10^{-2}.$$

Because $m$ is positive, the image is erect, and it is only about 0.0235 as tall as Santa himself:

$$y' = my = (0.0235)(1.6\text{ m}) = 3.8 \times 10^{-2}\text{ m} = 3.8\text{ cm}.$$

When the object distance $s$ is positive, a convex mirror *always* forms an erect, virtual, diminished, reversed image.

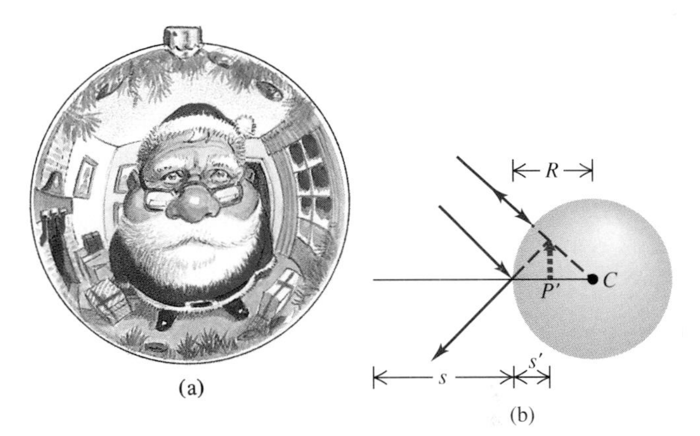


**35–15** (a) The ornament forms a virtual, reduced, erect image of Santa. (b) Two of the rays forming the image.

Spherical mirrors have many important uses. A concave mirror forms the light from a flashlight or headlight bulb into a parallel beam. Convex mirrors are used to give a wide-angled view to car and truck drivers, for shoplifting surveillance in stores, and at "blind" corners. A concave mirror with a long enough focal length so that your face is inside the focal point functions as a magnifier. Some solar-power plants use an array of plane mirrors to simulate an approximately spherical concave mirror. This is used to collect and direct the sun's radiation to the focal point, where a steam boiler is placed. You can probably think of other examples from your everyday experience.

## 35–3 GRAPHICAL METHODS FOR MIRRORS

We can find the position and size of the image formed by a mirror by a simple graphical method. This method consists of finding the point of intersection of a few particular rays

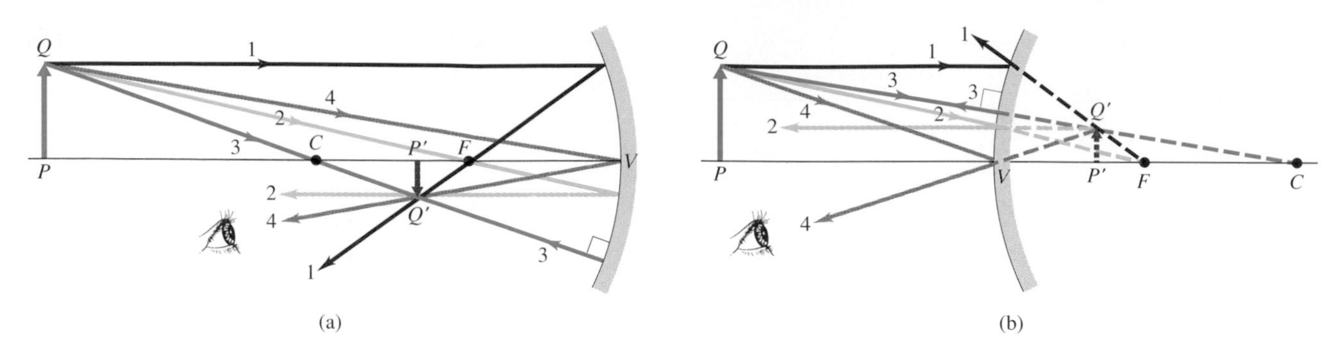


**35–16** Principal rays used in the graphical method of locating an image formed by a mirror.

that diverge from a point of the object (such as point $Q$ in Fig. 35–16) and are reflected by the mirror. Then (neglecting aberrations) *all* rays from this point that strike the mirror will intersect at the same point. For this construction we always choose an object point that is *not* on the optic axis. Four rays that we can usually draw easily are shown in Fig. 35–16. These are called **principal rays.**

1. *A ray parallel to the axis,* after reflection, passes through the focal point $F$ of a concave mirror or appears to come from the (virtual) focal point of a convex mirror.
2. *A ray through (or proceeding toward) the focal point $F$* is reflected parallel to the axis.
3. *A ray along the radius through or away from the center of curvature $C$* intersects the surface normally and is reflected back along its original path.
4. *A ray to the vertex $V$* is reflected forming equal angles with the optic axis.

Once we have found the position of the image point by means of the intersection of any two of these four principal rays, we can draw the path of any other ray from the object point to the same image point.

## PROBLEM-SOLVING STRATEGY

### Image formation by mirrors

**1.** The principal-ray diagram is to geometric optics what the free-body diagram is to mechanics! When you attack a problem involving image formation by a mirror, *always* draw a principal-ray diagram first if you have enough information. (The same advice should be applied to lenses in the following sections.) It is usually best to orient your diagrams consistently with the incoming rays traveling from left to right. Don't draw a lot of other rays at random; stick with the principal rays, the ones that you know something about.

**2.** If your principal rays don't converge at a real image point, you may have to extend them straight backward to locate a virtual image point. We recommend drawing the extensions with broken lines. Another useful aid is to color-code your principal rays, using red for 1, green for 2, blue for 3, and purple for 4 in the above list, or something similar.

**3.** Pay careful attention to signs on object and image distances, radii of curvature, and object and image heights. Make certain you understand that the same sign rules work for all four cases in this chapter: reflection and refraction from both plane and spherical surfaces. A negative sign on any one of the quantities mentioned above *always* has significance; use the equations and the sign rules carefully and consistently, and they will tell you the truth!

## EXAMPLE 35–3

A concave mirror has a radius of curvature with absolute value 20 cm. Find graphically the image of an object in the form of an arrow perpendicular to the axis of the mirror at each of the following object distances: a) 30 cm; b) 20 cm; c) 10 cm; and d) 5 cm. Check the construction by *computing* the size and magnification of each image.

**SOLUTION** The graphical constructions are shown in the four parts of Fig. 35–17. Study each of these diagrams carefully, comparing each numbered ray to the description above. Several points are worth noting. First, in Fig. 35–17b the object and image distances are equal. Ray 3 cannot be drawn in this case because a ray from $Q$ through the center of curvature $C$ does not strike the mirror. For the same reason, ray 2 cannot be drawn in Fig. 35–17c. In this case the outgoing rays are parallel, corresponding to an infinite image distance. In Fig. 35–17d the outgoing rays have no real intersection point; they must be extended backward to find the point from which they appear to diverge, that is, the *virtual image point* $Q'$.

Measurements of the figures, with appropriate scaling, give the following approximate image distances: (a) 15 cm, (b) 20 cm, (c) $\infty$ or $-\infty$ (because the outgoing rays are parallel and do not converge at any finite distance), (d) $-10$ cm. To *compute* these distances, we first note that $f = R/2 = 10$ cm; then we use Eq. (35–6):

a) $\dfrac{1}{30\text{ cm}} + \dfrac{1}{s'} = \dfrac{1}{10\text{ cm}}, \qquad s' = 15\text{ cm};$

b) $\dfrac{1}{20\text{ cm}} + \dfrac{1}{s'} = \dfrac{1}{10\text{ cm}}, \qquad s' = 20\text{ cm};$

c) $\dfrac{1}{10\text{ cm}} + \dfrac{1}{s'} = \dfrac{1}{10\text{ cm}}, \qquad s' = \infty \text{ (or } -\infty);$

d) $\dfrac{1}{5\text{ cm}} + \dfrac{1}{s'} = \dfrac{1}{10\text{ cm}}, \qquad s' = -10\text{ cm}.$

The lateral magnifications measured from the figures are approximately (a) $-1/2$, (b) $-1$, (c) $\infty$ or $-\infty$ (because the image distance is infinite), (d) $+2$. *Computing* the magnifications from Eq. (35–7), we find:

a) $m = -\dfrac{15\text{ cm}}{30\text{ cm}} = -\dfrac{1}{2};$

b) $m = -\dfrac{20\text{ cm}}{20\text{ cm}} = -1;$

c) $m = -\dfrac{\infty\text{ cm}}{10\text{ cm}} = -\infty;$

d) $m = -\dfrac{-10\text{ cm}}{5\text{ cm}} = +2.$

In (a) and (b) the image is inverted; in (d) it is erect.

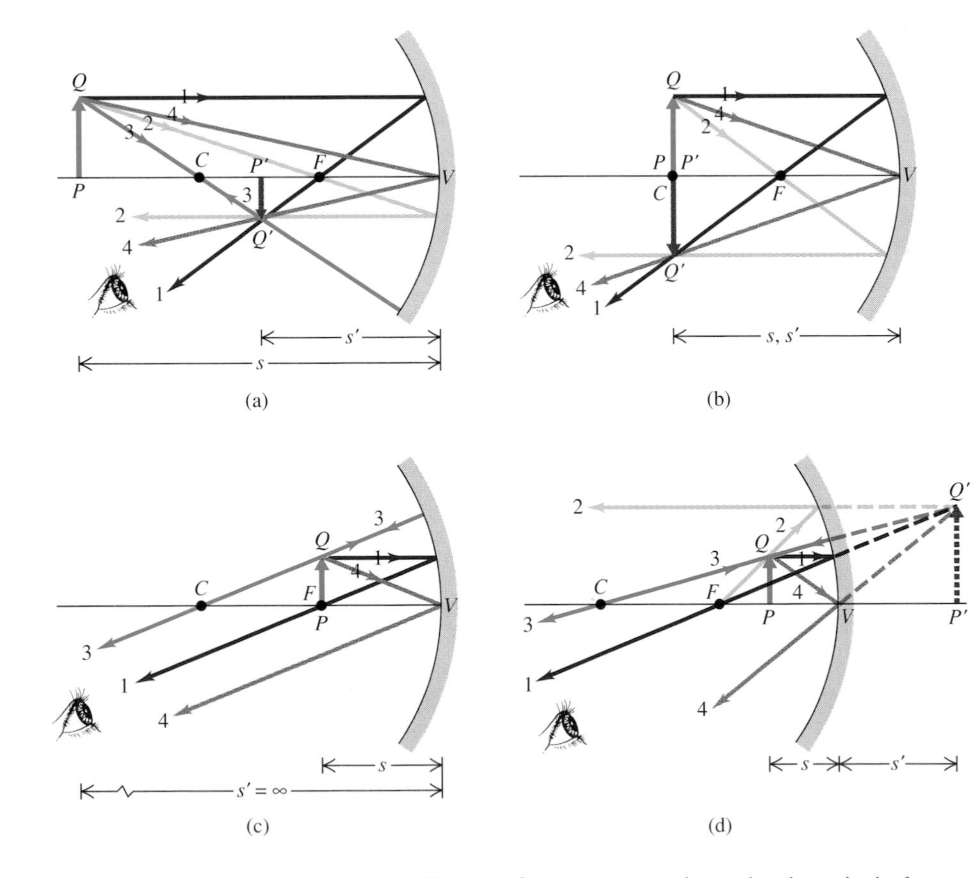


**35–17** Image of an object at various distances from a concave mirror, showing principal rays.

## 35–4 REFRACTION AT A SPHERICAL SURFACE

Our next topic is refraction at a spherical surface, that is, at a spherical interface between two optical materials with different indexes of refraction. This analysis is directly applicable to some real optical systems, such as the human eye. It also provides a steppingstone for the analysis of lenses, which usually have *two* spherical surfaces.

In Fig. 35–18 a spherical surface with radius $R$ forms an interface between two materials with indexes of refraction $n_a$ and $n_b$. The surface forms an image $P'$ of an object point $P$; we want to find how the object and image distances ($s$ and $s'$) are related. We will use the same sign rules that we used for spherical mirrors. The center of curvature $C$ is on the outgoing side of the surface, so $R$ is positive. Ray $PV$ strikes the vertex $V$ and is perpendicular to the surface (that is, to the tangent plane to the surface at the point of incidence $V$). It passes into the second material without deviation. Ray $PB$, making an angle $\alpha$ with the axis, is incident at an angle $\theta_a$ with the normal and is refracted at an angle $\theta_b$. These rays intersect at $P'$ at a distance $s'$ to the right of the vertex. The figure is drawn for the case where $n_b$ is greater than $n_a$. The object and image distances are both positive.

We are going to prove that if the angle $\alpha$ is small, *all* rays from $P$ intersect at the same point $P'$, so $P'$ is the *real image* of $P$. From the triangles $PBC$ and $P'BC$ we have

$$\theta_a = \alpha + \phi, \qquad \phi = \beta + \theta_b. \tag{35–8}$$

From the law of refraction,

$$n_a \sin\theta_a = n_b \sin\theta_b.$$

Also, the tangents of $\alpha$, $\beta$, and $\phi$ are

$$\tan\alpha = \frac{h}{s+\delta}, \qquad \tan\beta = \frac{h}{s'-\delta}, \qquad \tan\phi = \frac{h}{R-\delta}. \tag{35–9}$$

For paraxial rays we may approximate both the sine and tangent of an angle by the angle itself and neglect the small distance $\delta$. The law of refraction then gives

$$n_a\theta_a = n_b\theta_b.$$

Combining this with the first of Eqs. (35–8), we obtain

$$\theta_b = \frac{n_a}{n_b}(\alpha + \phi).$$

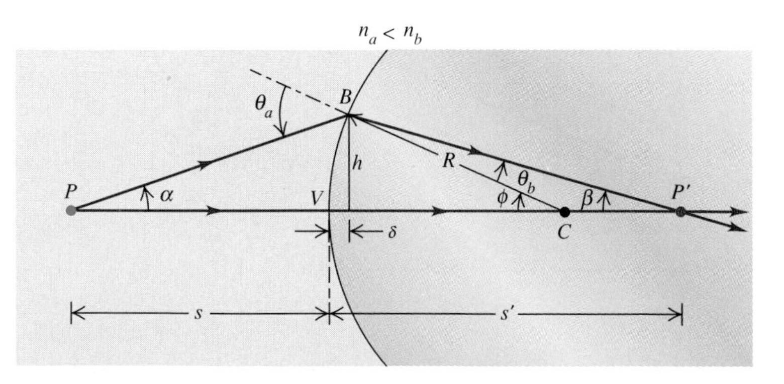


**35–18** Construction for finding the position of the image point $P'$ of a point object $P$ formed by refraction at a spherical surface.

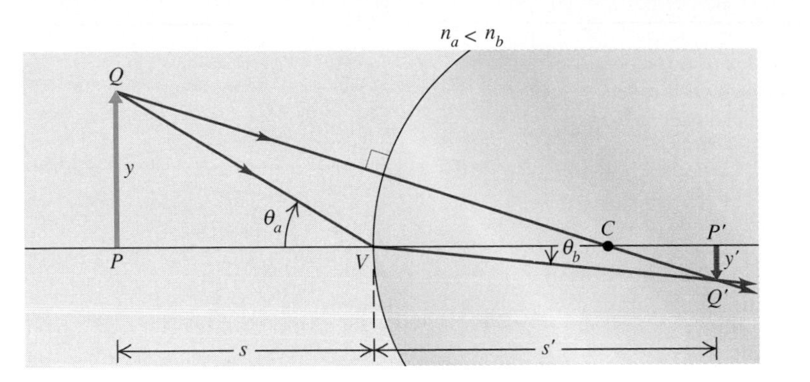


**35–19** Construction for determining the height of an image formed by refraction at a spherical surface.

When we substitute this into the second of Eqs. (35–8), we get

$$n_a\alpha + n_b\beta = (n_b - n_a)\phi. \qquad (35\text{–}10)$$

Now we use the approximations $\tan \alpha = \alpha$, and so on, in Eqs. (35–9) and neglect $\delta$; those equations then become

$$\alpha = \frac{h}{s}, \qquad \beta = \frac{h}{s'}, \qquad \phi = \frac{h}{R}.$$

Finally, we substitute these into Eq. (35–10) and divide out the common factor $h$. We obtain

$$\frac{n_a}{s} + \frac{n_b}{s'} = \frac{n_b - n_a}{R}. \qquad (35\text{–}11)$$

This equation does not contain the angle $\alpha$, so the image distance is the same for *all* paraxial rays from *P.*

To obtain the magnification for this situation, we use the construction in Fig. 35–19. We draw two rays from point *Q*, one through the center of curvature *C* and the other incident at the vertex *V.* From the triangles *PQV* and *P'Q'V,*

$$\tan \theta_a = \frac{y}{s}, \qquad \tan \theta_b = \frac{-y'}{s'},$$

and from the law of refraction,

$$n_a \sin \theta_a = n_b \sin \theta_b.$$

For small angles,

$$\tan \theta_a = \sin \theta_a, \qquad \tan \theta_b = \sin \theta_b,$$

so finally

$$\frac{n_a y}{s} = -\frac{n_b y'}{s'},$$

or

$$m = \frac{y'}{y} = -\frac{n_a s'}{n_b s}. \qquad (35\text{–}12)$$

## EXAMPLE 35–4

A cylindrical glass rod in air (Fig. 35–20) has index of refraction 1.52. One end is ground to a hemispherical surface with radius $R = 2.00$ cm. a) Find the image distance of a small object on the axis of the rod and 8.00 cm to the left of the vertex. b) Find the lateral magnification.

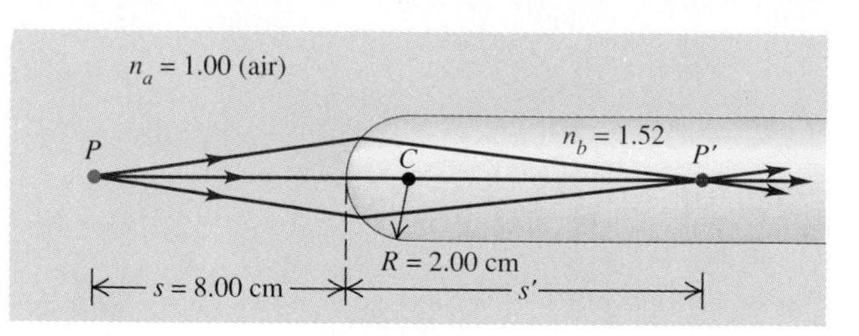


**35–20** The rod in air forms a real image.

**SOLUTION** We are given

$$n_a = 1.00, \qquad n_b = 1.52,$$

$$R = +2.00 \text{ cm}, \qquad s = +8.00 \text{ cm}.$$

a) From Eq. (35–11),

$$\frac{1.00}{8.00 \text{ cm}} + \frac{1.52}{s'} = \frac{1.52 - 1.00}{+2.00 \text{ cm}},$$

$$s' = +11.3 \text{ cm}.$$

The image is formed to the right of the vertex (because $s'$ is positive) at a distance of 11.3 cm from it.

b) From Eq. (35–12),

$$m = -\frac{n_a s'}{n_b s} = -\frac{(1.00)(11.3 \text{ cm})}{(1.52)(8.00 \text{ cm})} = -0.929.$$

The image is somewhat smaller than the object, and it is inverted. If the object is an arrow 1.00 mm high, pointing upward, the image is an arrow 0.929 mm high, pointing downward.

## EXAMPLE 35–5

The glass rod of Example 35–4 is immersed in water (index of refraction $n = 1.33$), as shown in Fig. 35–21. The other quantities have the same values as before. Find the image distance and lateral magnification.

**SOLUTION** From Eq. (35–11),

$$\frac{1.33}{8.00 \text{ cm}} + \frac{1.52}{s'} = \frac{1.52 - 1.33}{+2.00 \text{ cm}},$$

$$s' = -21.3 \text{ cm}.$$

The fact that $s'$ is negative means that after the rays are refracted by the surface, they are not converging but *appear* to diverge from a point 21.3 cm to the *left* of the vertex. We have seen a similar case in the refraction of spherical waves by a plane surface; we called the point a *virtual image*. In this example the surface forms a virtual image 21.3 cm to the left of the vertex.

The magnification in this case is

$$m = -\frac{(1.33)(-21.3 \text{ cm})}{(1.52)(8.00 \text{ cm})} = +2.33.$$

In this case the image is erect (because $m$ is positive) and 2.33 times as large as the object.

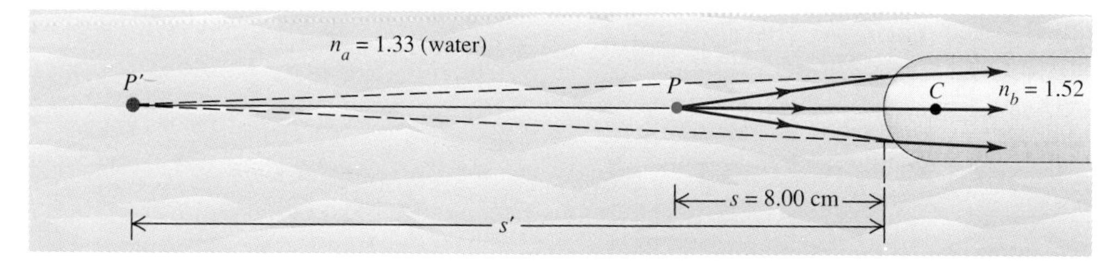


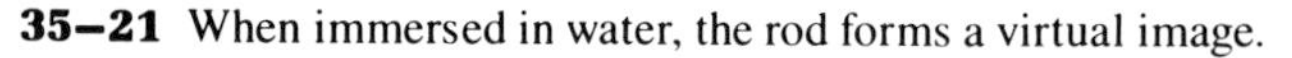

**35–21** When immersed in water, the rod forms a virtual image.

Equations (35–11) and (35–12) can be applied to both convex and concave refracting surfaces, provided that you use the sign rules consistently. It doesn't matter whether $n_b$ is greater or less than $n_a$. We suggest that you construct diagrams like Figs. 35–18 and 35–19, when $R$ is negative and when $n_b$ is less than $n_a$, and use them to derive Eqs. (35–11) and (35–12) for these cases.

Here's a final note on the sign rule for the radius of curvature $R$ of a surface. For the convex reflecting surface in Fig. 35–13 we considered $R$ negative, but in Fig. 35–18 the refracting surface with the same orientation has a *positive* value of $R$. This may seem inconsistent, but it isn't. Both cases are consistent with the rule that $R$ is positive when the center of curvature is on the outgoing side of the surface and negative when it is *not* on the outgoing side. When both reflection and refraction occur at a spherical surface, $R$ has one sign for the reflected light and the opposite sign for the refracted light.

An important special case of a spherical refracting surface is a *plane* surface between two optical materials. This corresponds to setting $R = \infty$ in Eq. (35–11). In this case,

$$\frac{n_a}{s} + \frac{n_b}{s'} = 0. \qquad (35\text{–}13)$$

To find the magnification $m$ for this case, we combine this equation with the general relation, Eq. (35–12), obtaining the simple result

$$m = 1. \qquad (35\text{–}14)$$

That is, the image formed by a *plane* refracting surface is always the same size as the object, and it is always erect.

Another example of image formation by a plane refracting surface is the appearance of a partly submerged drinking straw or canoe paddle (Fig. 35–22). When viewed from some angles, the object appears to have a sharp bend at the water surface because the submerged part appears to be only about three-quarters of its actual distance below the surface.

**35–22** An object partially submerged in water appears to bend at the surface, owing to the refraction of the light rays.

## EXAMPLE 35–6

Swimming pool owners know that the pool always looks shallower than it really is and that it is important to identify the deep parts conspicuously so that people who can't swim won't jump into water over their heads. If a nonswimmer looks straight down into water that is actually 2.00 m (about 6 ft, 7 in.) deep, how deep does it appear to be?

**SOLUTION** The situation is shown in Fig. 35–23. Suppose there is an arrow $PQ$ painted on the bottom of the pool at a depth $s = 2.00$ m. The refracting surface forms an image $P'Q'$ at a depth $s'$. To find $s'$, we use Eq. (35–13):

$$\frac{n_a}{s} + \frac{n_b}{s'} = 0,$$

$$\frac{1.33}{2.00\text{ m}} + \frac{1.00}{s'} = 0,$$

$$s' = -1.50\text{ m}.$$

The apparent depth is only about three-quarters of the actual depth, or about 4 ft, 11 in. A 6-ft nonswimmer who didn't allow for this effect would be in trouble. (The negative sign shows that the image is virtual and on the incoming side of the refracting surface.)

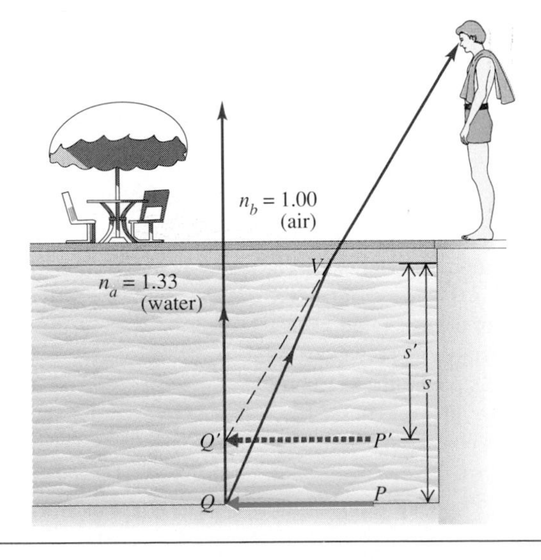


**35–23** Arrow $P'Q'$ is the virtual image of the underwater arrow $PQ$. The angles of the rays with the vertical are exaggerated for clarity.

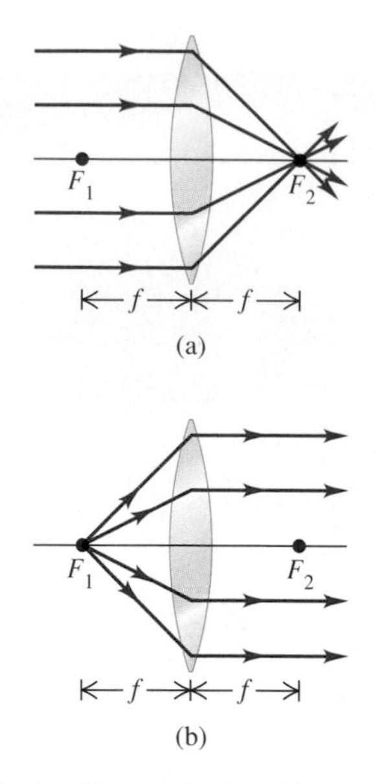


**35–24** First and second focal points of a converging thin lens. The numerical value of $f$ is positive.

## 35–5 THIN LENSES

The most familiar and widely used optical device (after the plane mirror) is the *lens.* A lens is an optical system with two refracting surfaces. The simplest lens has two *spherical* surfaces close enough together that we can neglect the distance between them (the thickness of the lens); we call this a **thin lens.** We can analyze this system in detail using the results of Section 35–4 for refraction by a single spherical surface. However, we postpone this analysis until later in this section so that we can first discuss the properties of thin lenses.

### Focal Points of a Lens

A lens of the type shown in Fig. 35–24 has the property that when a beam of parallel rays passes through it, the rays converge to a point $F_2$ (Fig. 35–24a). Similarly, rays passing through point $F_1$ emerge from the lens as a beam of parallel rays (Fig. 35–24b). The points $F_1$ and $F_2$ are called the first and second *focal points,* and the distance $f$ (measured from the center of the lens) is called the *focal length*. We have already used the concepts of focal point and focal length for spherical mirrors in Section 35–2. The two focal lengths in Fig. 35–24, both labeled $f$, *are always equal* for a thin lens, even when the two sides have different curvatures. We will derive this somewhat surprising result later in this section, when we derive the relationship of $f$ to the index of refraction of the lens and the radii of curvature of its surfaces. The central horizontal line is called the *optic axis,* as with spherical mirrors. The centers of curvature of the two spherical surfaces lie on and define the optic axis.

Figure 35–25 shows how we can determine the position of the image formed by a thin lens. Using the same notation and sign rules as before, we let $s$ and $s'$ be the object and image distances, respectively, and let $y$ and $y'$ be the object and image heights. Ray $QA$, parallel to the optic axis before refraction, passes through the second focal point $F_2$ after refraction. Ray $QOQ'$ passes undeflected straight through the center of the lens because at the center the two surfaces are parallel and (we have assumed) very close together. There is refraction where the ray enters and leaves the material but no net change in direction.

The two angles labeled $\alpha$ in Fig. 35–25 are equal. Therefore the two right triangles $PQO$ and $P'Q'O$ are *similar,* and ratios of corresponding sides are equal. Thus,

$$\frac{y}{s} = -\frac{y'}{s'}, \quad \text{or} \quad \frac{y'}{y} = -\frac{s'}{s}. \tag{35–15}$$

(The reason for the negative sign is that the image is below the optic axis and $y'$ is negative.) Also, the two angles labeled $\beta$ are equal, and the two right triangles $OAF_2$ and $P'Q'F_2$ are similar.

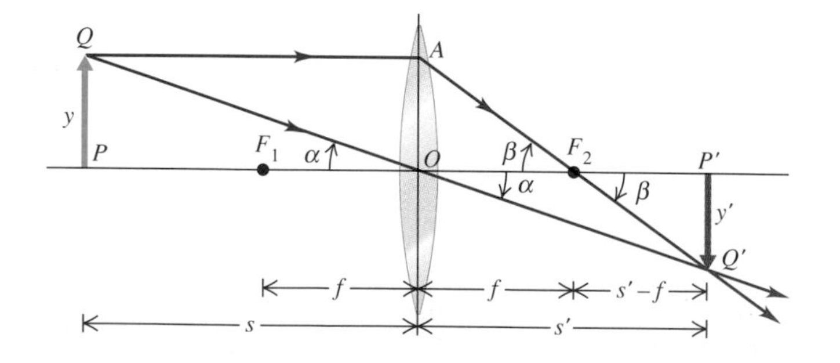


**35–25** Construction used to find image position for a thin lens. The ray $QAQ'$ is shown as bent at the midplane of the lens rather than at the two surfaces and ray $QOQ'$ is shown as a straight line to emphasize that the thickness of the lens is assumed to be very small.

Thus, we have

$$\frac{y}{f} = -\frac{y'}{s' - f},$$

or

$$\frac{y'}{y} = -\frac{s' - f}{f}. \tag{35–16}$$

We now equate Eqs. (35–15) and (35–16), divide by $s'$, and rearrange to obtain

$$\frac{1}{s} + \frac{1}{s'} = \frac{1}{f}. \tag{35–17}$$

This analysis also gives us the lateral magnification $m = y'/y$ of the system; from Eq. (35–15),

$$m = -\frac{s'}{s}. \tag{35–18}$$

The negative sign tells us that when $s$ and $s'$ are both positive, the image is *inverted,* and $y$ and $y'$ have opposite signs.

Equations (35–17) and (35–18) are the basic equations for thin lenses. It is pleasing to note that their *form* is exactly the same as that of the corresponding equations for spherical mirrors, Eqs. (35–6) and (35–7). As we will see, the same sign rules that we used for spherical mirrors are also applicable to lenses.

Figure 35–26 shows how a lens forms a three-dimensional image of a three-dimensional object. Point $R$ is nearer the lens than point $P$ is. From Eq. (35–17), image point $R'$ is farther from the lens than image point $P'$ is, and the image $P'R'$ points in the same direction as the object $PR$. Arrows $P'S'$ and $P'Q'$ are reversed in space relative to $PS$ and $PQ$.

Let's compare Fig. 35–26 with Fig. 35–5, which shows the image formed by a plane *mirror.* We note that the image formed by the lens is inverted, but it is *not* reversed. That is, if the object is a left hand, its image is also a left hand. You can verify this by pointing your left thumb along $PR$, your left forefinger along $PQ$, and your left middle finger along $PS$. Then rotate your hand 180°, using your thumb as an axis; this brings the fingers into coincidence with $P'Q'$ and $P'S'$. In other words, *inversion* of an image is equivalent to a rotation of 180° about the lens axis.

A bundle of parallel rays incident on the lens shown in Figure 35–24 converges to a real image after passing through the lens. This lens is called a **converging lens.** Its focal length is a positive quantity, and it is also called a *positive lens.*

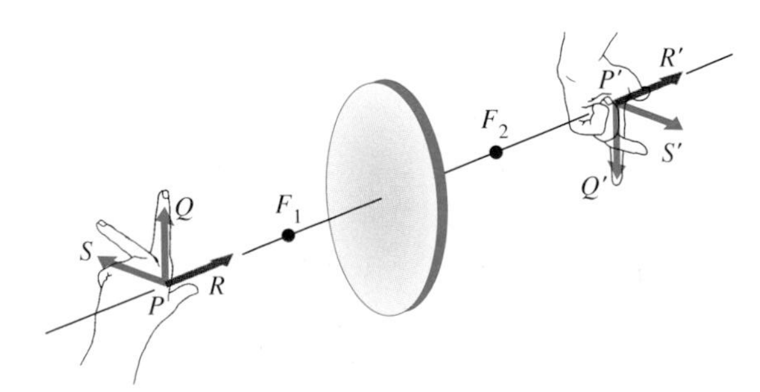


**35–26** The image of a three-dimensional object is *not* reversed by a lens.

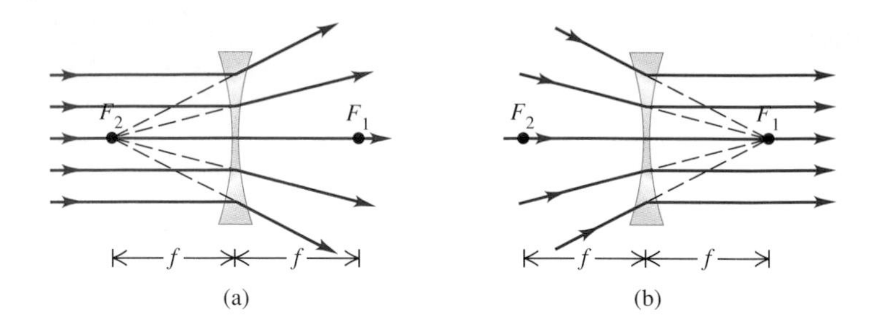


**35–27** Second and first focal points of a diverging thin lens. The numerical value of $f$ is negative.

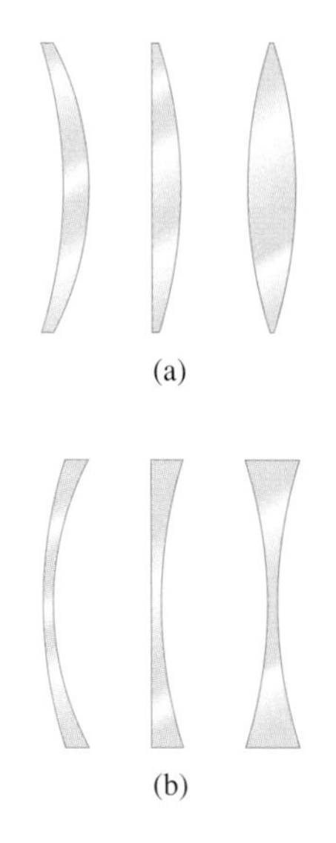


**35–28** (a) Meniscus, plano-convex, and double-convex converging lenses. (b) Meniscus, plano-concave, and double-concave diverging lenses. A converging lens is always thicker at the center than at its edges, and the reverse is true for a diverging lens. If the index of refraction of the surrounding material is greater than that of the lens, the lenses in (a) are diverging and those in (b) are converging.

A bundle of parallel rays incident on the lens in Fig. 35–27 *diverges* after refraction, and this lens is called a **diverging lens.** Its focal length is a negative quantity, and the lens is also called a *negative lens.* The focal points of a negative lens are reversed relative to those of a positive lens. The second focal point, $F_2$, of a negative lens is the point from which rays that are originally parallel to the axis *appear to diverge* after refraction, as in Fig. 35–27a. Incident rays converging toward the first focal point $F_1$, as in Fig. 35–27b, emerge from the lens parallel to its axis.

Equations (35–17) and (35–18) apply to *both* negative and positive lenses. Various types of lenses, both converging and diverging, are shown in Fig. 35–28. Any lens that is thicker in the center than at the edges is a converging lens with positive $f$, and any lens that is thicker at the edges than at the center is a diverging lens with negative $f$ (provided that the lens has greater index of refraction than the surrounding material). We can prove this using Eq. (35–20), which we have not yet derived.

## The Lensmaker's Equation

Now we proceed to derive Eq. (35–17) in more detail and at the same time derive the relationship between the focal length $f$ of the lens, its index of refraction $n$, and the radii of curvature $R_1$ and $R_2$ of its surfaces. We use the principle that an image formed by one reflecting or refracting surface can serve as the object for a second reflecting or refracting surface.

We begin with the somewhat more general problem of two spherical interfaces separating three materials with indexes of refraction $n_a$, $n_b$, and $n_c$, as shown in Fig. 35–29. The object and image distances for the first surface are $s_1$ and $s_1'$, and those for the second surface are $s_2$ and $s_2'$. We assume that the distance $t$ between the two surfaces is small enough to be neglected in comparison with the object and image distances. Then $s_2$ and $s_1'$ have the same magnitude but opposite sign. For example, if the first image is on the outgoing side of the first surface, $s_1'$ is positive. But when viewed as an object for the second surface, it is *not* on the incoming side of that surface. So we can say that $s_2 = -s_1'$.

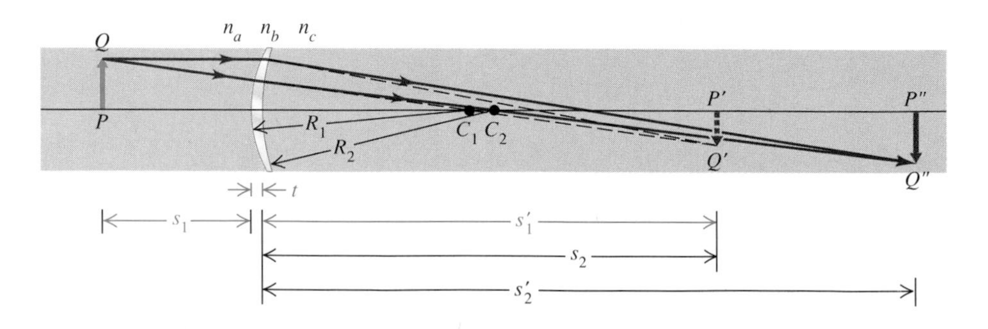


**35–29** The image formed by the first surface of a lens serves as the object for the second surface.

We need to use the single-surface equation, Eq. (35–11), twice, once for each surface. The two resulting equations are

$$\frac{n_a}{s_1} + \frac{n_b}{s_1'} = \frac{n_b - n_a}{R_1},$$

$$\frac{n_b}{s_2} + \frac{n_c}{s_2'} = \frac{n_c - n_b}{R_2}.$$

Ordinarily, the first and third materials are air or vacuum, so we set $n_a = n_c = 1$. The second index $n_b$ is that of the lens, which we now call simply $n$. Substituting these values and the relation $s_2 = -s_1'$ into the above equations, we get

$$\frac{1}{s_1} + \frac{n}{s_1'} = \frac{n-1}{R_1},$$

$$-\frac{n}{s_1'} + \frac{1}{s_2'} = \frac{1-n}{R_2}.$$

To get a relation between the initial object position $s_1$ and the final image position $s_2'$, we add these two equations. This eliminates the term $n/s_1'$, and we obtain

$$\frac{1}{s_1} + \frac{1}{s_2'} = (n-1)\left(\frac{1}{R_1} - \frac{1}{R_2}\right).$$

Finally, thinking of the lens as a single unit, we call the object distance simply $s$ instead of $s_1$ and the final image distance $s'$ instead of $s_2'$. Making these substitutions, we have

$$\frac{1}{s} + \frac{1}{s'} = (n-1)\left(\frac{1}{R_1} - \frac{1}{R_2}\right). \tag{35–19}$$

Now we compare this with the other thin-lens equation, Eq. (35–17). We see that the object and image distances $s$ and $s'$ appear in exactly the same places in both equations and that the focal length $f$ is given by

$$\frac{1}{f} = (n-1)\left(\frac{1}{R_1} - \frac{1}{R_2}\right). \tag{35–20}$$

This relation is called the *lensmaker's equation*. In the process of rederiving the thin-lens equation, we have also derived an expression for the focal length $f$ of a lens in terms of its index of refraction $n$ and the radii of curvature $R_1$ and $R_2$ of its surfaces.

We use all our previous sign conventions with Eqs. (35–19) and (35–20). For example, in Fig. 35–30, $s$, $s'$, and $R_1$ are positive, but $R_2$ is negative.

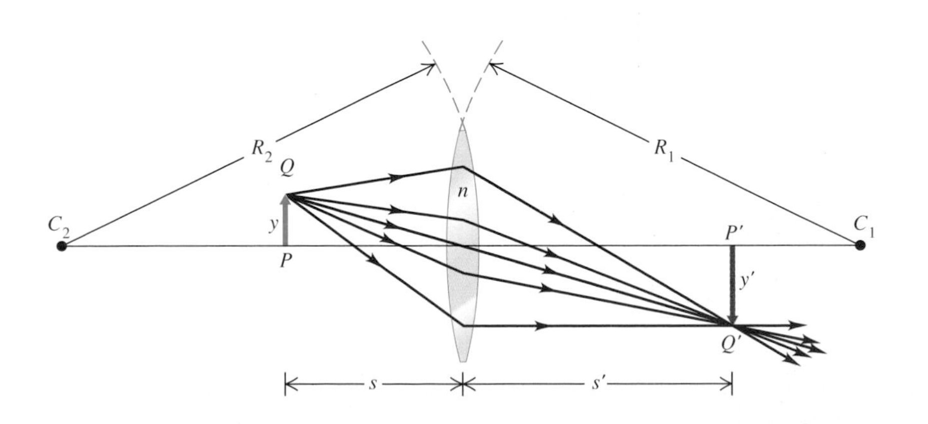


**35–30** A thin lens. The radius of curvature of the first surface is $R_1$, and that of the second surface is $R_2$. In this case, $R_1$ is positive and $R_2$ is negative. The focal length $f$ is positive, and the lens is a converging lens. Here, $s$ and $s'$ are also positive, and $m$ is therefore negative.

## EXAMPLE 35–7

Suppose the absolute values of the radii of curvature of the lens surfaces in Fig. 35–30 are 20 cm and 5 cm, respectively, and the index of refraction is $n = 1.52$. What is the focal length $f$ of the lens?

**SOLUTION** The center of curvature of the first surface is on the outgoing side, so $R_1$ is positive: $R_1 = +20$ cm. The center of curvature of the second surface is *not* on the outgoing side, so $R_2$ is negative: $R_2 = -5$ cm. From Eq. (35–20),

$$\frac{1}{f} = (1.52 - 1)\left(\frac{1}{20\text{ cm}} - \frac{1}{-5\text{ cm}}\right),$$

$$f = 7.7\text{ cm}.$$

It is not hard to generalize Eq. (35–20) to the situation in which the lens is immersed in a material with an index of refraction greater than unity. We invite you to work out the lensmaker's equation for this more general situation.

## 35–6 GRAPHICAL METHODS FOR LENSES

We can determine the position and size of an image formed by a thin lens by using a graphical method that is very similar to the one we used in Section 35–3 for spherical mirrors. Again we draw a few special rays called *principal rays* that diverge from a point of the object that is *not* on the optic axis. The intersection of these rays, after they pass through the lens, determines the position and size of the image. In using this graphical method we consider the entire deviation of a ray as occurring at the midplane of the lens, as shown in Fig. 35–31; this is consistent with the assumption that the distance between the lens surfaces is negligible.

The three principal rays whose paths are usually easy to trace for lenses are shown in Fig. 35–31. They are as follows:

1. *A ray parallel to the axis* after refraction by the lens passes through the second focal point $F_2$ of a converging lens or appears to come from the second focal point of a diverging lens.
2. *A ray through the center of the lens* is not appreciably deviated because at the center of the lens the two surfaces are parallel and close together.
3. *A ray through (or proceeding toward) the first focal point* $F_1$ emerges parallel to the axis.

When the image is real, the position of the image point is determined by the intersection of any two of the three principal rays (Fig. 35–31a). When the image is virtual, we extend the diverging outgoing rays backward to their intersection point (Fig. 35–31b). Once the image position is known, we can draw any other ray from the same point. Usually, nothing is gained by drawing a lot of additional rays.

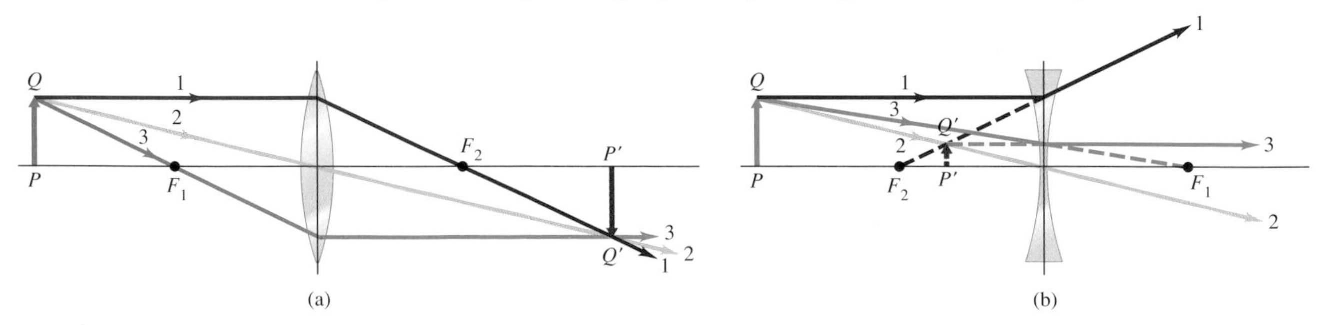


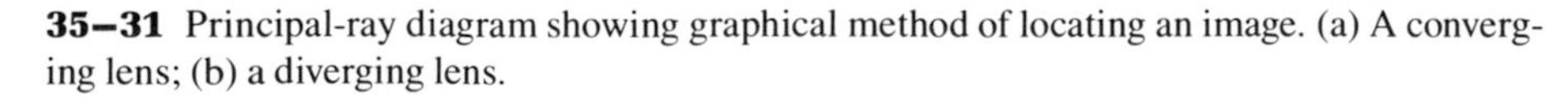
**35–31** Principal-ray diagram showing graphical method of locating an image. (a) A converging lens; (b) a diverging lens.

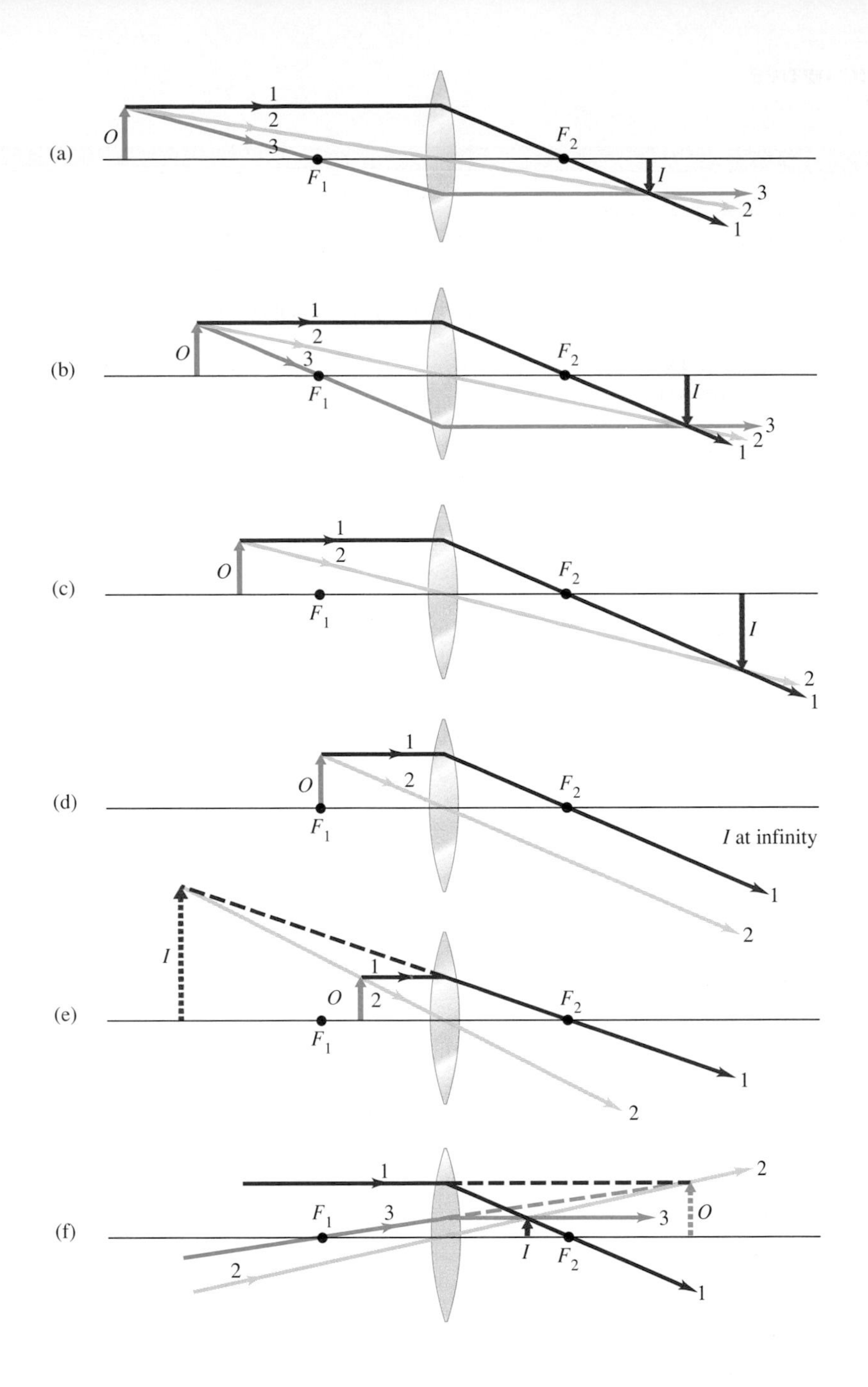


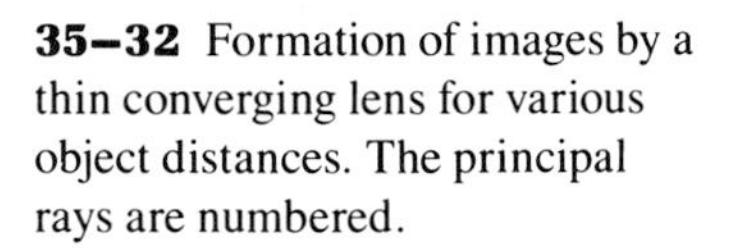
**35–32** Formation of images by a thin converging lens for various object distances. The principal rays are numbered.

Figure 35–32 shows several principal-ray diagrams for a converging lens for several object distances. We suggest that you study each of these diagrams very carefully, comparing each numbered ray with the above description. Several points are worth noting. In Fig. 35–32d the object is at the focal point; ray 3 cannot be drawn because it does not pass through the lens. In Fig. 35–32e the object distance is less than the focal length. The outgoing rays are divergent, and the *virtual image* is located by extending the outgoing rays backward. In this case the image distance $s'$ is negative. Figure 35–32f corresponds to a *virtual object.* The incoming rays do not diverge from a real object point $O$, but are *converging* as though they would meet at the virtual object point $O$ on the right side. The object distance $s$ is negative in this case. The image is real; the image distance $s'$ is positive and less than $f$.

## PROBLEM-SOLVING STRATEGY

### Image formation by a thin lens

**1.** The strategy outlined in Section 35–3 is equally applicable here, and we suggest that you review it now. Always begin with a principal-ray diagram if you have enough information. Orient your diagrams consistently so that light travels from left to right. For a lens there are only three principal rays, compared to four for a mirror. Don't just sketch these diagrams; draw the rays with a ruler and measure the distances carefully. Draw the rays so that they bend at the midplane of the lens, as shown in Fig. 35–31. Be sure to draw *all three* principal rays whenever possible. The intersection of any two determines the image, but if the third doesn't pass through the same intersection point, you know that you have made a mistake. Redundancy can be useful in spotting errors.

**2.** If the outgoing principal rays don't converge at a real image point, the image is virtual. Then you have to extend the outgoing rays backward to find the virtual image point, which lies on the *incoming* side of the lens.

**3.** The same sign rules that we have used for mirrors and single refracting surfaces are also applicable for thin lenses. Be extremely careful to get your signs right and to interpret the signs of results correctly.

**4.** Always determine the image position and size *both* graphically and by calculating. This gives an extremely valuable consistency check.

**5.** Remember that the *image* from one lens or mirror may serve as the *object* for another. In that case, be careful in finding the object and image *distances* for this intermediate image; be sure you include the distance between the two elements (lenses and/or mirrors) correctly.

## EXAMPLE 35–8

**Image formed by a converging lens** A converging lens has a focal length of 20 cm. Find graphically the image location for an object at each of the following distances from the lens: a) 50 cm; b) 20 cm; c) 15 cm; d) −40 cm. Determine the magnification in each case. Check your results by calculating the image position and magnification from Eqs. (35–17) and (35–18).

**SOLUTION** The principal-ray diagrams are shown in Figs. 35–32a, 35–32d, 35–32e, and 35–32f. The approximate image distances, from measurements of these diagrams, are 35 cm, $\infty$, −40 cm, and 15 cm, respectively, and the approximate magnifications are $-\frac{2}{3}$, $\infty$, | 3, and $+\frac{1}{3}$.

*Calculating* the image positions from Eq. (35–17), $1/s + 1/s' = 1/f$, we find

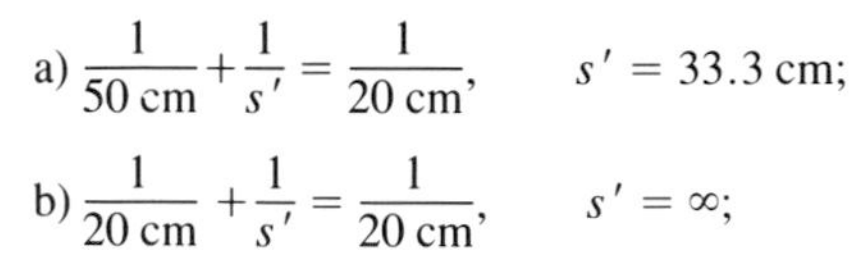

$$\text{a)}\ \frac{1}{50\text{ cm}} + \frac{1}{s'} = \frac{1}{20\text{ cm}}, \qquad s' = 33.3\text{ cm};$$

$$\text{b)}\ \frac{1}{20\text{ cm}} + \frac{1}{s'} = \frac{1}{20\text{ cm}}, \qquad s' = \infty;$$

$$\text{c)}\ \frac{1}{15\text{ cm}} + \frac{1}{s'} = \frac{1}{20\text{ cm}}, \qquad s' = -60\text{ cm};$$

$$\text{d)}\ \frac{1}{-40\text{ cm}} + \frac{1}{s'} = \frac{1}{20\text{ cm}}, \qquad s' = 13.3\text{ cm};$$

The graphical results are fairly close to these except for those from Fig. 35–32e, where the precision of the diagram is limited because the rays extended backward have nearly the same direction.

From Eq. (35–18), $m = -s/s'$, and the magnifications are

$$\text{a)}\ m = -\frac{33.3\text{ cm}}{50\text{ cm}} = -\frac{2}{3},$$

$$\text{b)}\ m = -\frac{\infty}{20\text{ cm}} = -\infty,$$

$$\text{c)}\ m = -\frac{-60\text{ cm}}{15\text{ cm}} = +4,$$

$$\text{d)}\ m = -\frac{13.3\text{ cm}}{-40\text{ cm}} = +\frac{1}{3}.$$

## EXAMPLE 35–9

**Image formed by a diverging lens** You are given a thin diverging lens. You find that a beam of parallel rays spreads out after passing through the lens, as though all the rays came from a point 20.0 cm from the center of the lens. You want to use this lens to form an erect virtual image that is one-third the height of the object. a) Where should the object be placed? b) Draw a principal-ray diagram.

**SOLUTION** a) The observation with incoming parallel rays shows that the focal length is $f = -20.0$ cm. We want the lateral magnification to be $m = +\frac{1}{3}$ (positive because the image is to be erect). From Eq. (35–18), $m = \frac{1}{3} = -s'/s$, so $s' = -s/3$. From Eq. (35–17),

$$\frac{1}{s} + \frac{1}{-s/3} = \frac{1}{-20.0\text{ cm}},$$

$$s = 40.0\text{ cm};$$
$$s' = -(40.0\text{ cm})/3 = -13.3\text{ cm}.$$

The image distance is negative, so the object and image are on the same side of the lens.

b) Figure 35–33 is a principal-ray diagram for this problem, with the rays numbered the same way as in Fig. 35–31b.

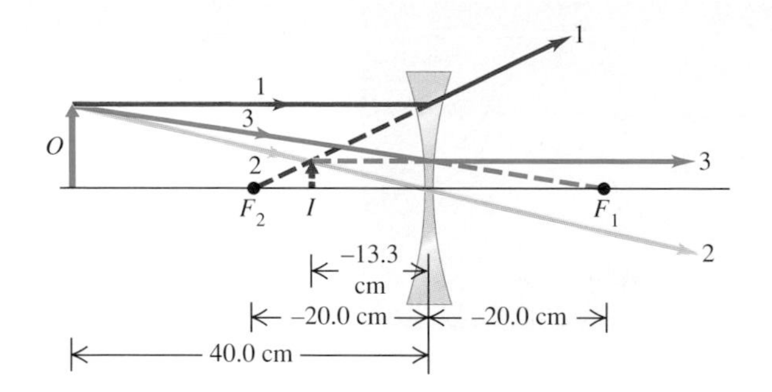


**35–33** Principal-ray diagram for an image formed by a thin diverging lens.

## SUMMARY

- When rays diverge from an object point $P$ and are reflected or refracted, the directions of the outgoing rays are the same as though they had diverged from a point $P'$, called the image point. If they actually converge at $P'$ and diverge again beyond it, $P'$ is a real image of $P$; if they only appear to have diverged from $P'$, it is a virtual image. Images can be either erect or inverted. An image formed by a plane or spherical mirror is always reversed; for example, the image of a right hand is a left hand.
- The lateral magnification $m$ in any reflecting or refracting situation is defined as the ratio of image height $y'$ to object height $y$:

$$m = \frac{y'}{y}. \qquad (35\text{–}2)$$

When $m$ is positive, the image is erect; when $m$ is negative, the image is inverted.

- Object-image relations derived in this chapter are valid only for rays that are close to and nearly parallel to the optic axis; these are called paraxial rays. Nonparaxial rays do not converge precisely to an image point; this effect is called spherical aberration.
- The focal point of a mirror is the point at which parallel rays converge after reflection from a concave mirror or the point from which they appear to diverge after reflection from a convex mirror. Rays diverging from the focal point of a concave mirror are parallel after reflection, and rays converging toward the focal point of a convex mirror are parallel after reflection. The distance from the focal point to the vertex is called the focal length, denoted as $f$. The focal points of a lens are defined similarly.
- The formulas for object distance $s$ and image distance $s'$ for plane and spherical mirrors and single refracting surfaces are summarized in the table below. The equation for a plane surface can be obtained from the corresponding equation for a spherical surface by setting $R = \infty$.

### KEY TERMS

image
virtual image
real image
lateral magnification
erect image
inverted image
reversed image
center of curvature
vertex
optic axis
paraxial rays
paraxial approximation

| | Plane mirror | Spherical mirror | Plane refracting surface | Spherical refracting surface |
|---|---|---|---|---|
| Object and image distances | $\frac{1}{s} + \frac{1}{s'} = 0$ | $\frac{1}{s} + \frac{1}{s'} = \frac{2}{R} = \frac{1}{f}$ | $\frac{n_a}{s} + \frac{n_b}{s'} = 0$ | $\frac{n_a}{s} + \frac{n_b}{s'} = \frac{n_b - n_a}{R}$ |
| Lateral magnification | $m = -\frac{s'}{s} = 1$ | $m = -\frac{s'}{s}$ | $m = -\frac{n_a s'}{n_b s} = 1$ | $m = -\frac{n_a s'}{n_b s}$ |

**focal point**
**focal length**
**principal ray**
**thin lens**
**converging lens**
**diverging lens**

The object-image relation for a thin lens is the same as for a spherical mirror:

$$\frac{1}{s} + \frac{1}{s'} = \frac{1}{f}. \quad (35\text{–}17)$$

The lensmaker's equation is

$$\frac{1}{f} = (n - 1)\left(\frac{1}{R_1} - \frac{1}{R_2}\right). \quad (35\text{–}20)$$

• The following sign rules are used with all plane and spherical reflecting and refracting surfaces:

$s$ is positive when the object is on the incoming side of the surface, negative otherwise.

$s'$ is positive when the image is on the outgoing side of the surface, negative otherwise.

$R$ is positive when the center of curvature is on the outgoing side of the surface, negative otherwise.

$m$ is positive when the image is erect, negative when it is inverted.

## EXERCISES

### Section 35–1 Reflection at a Plane Surface

**35–1** A candle 4.00 cm tall is 60.0 cm to the left of a plane mirror. Where is the image formed by the mirror, and what is the height of this image?

**35–2** The image of a tree just covers the height of a plane mirror 5.00 cm tall when the mirror is held 30.0 cm from the eye. The tree is 90.0 m from the mirror. What is its height?

### Section 35–2 Reflection at a Spherical Surface

### Section 35–3 Graphical Methods for Mirrors

**35–3 Telescopic Image of the Moon.** The diameter of the moon is 3480 km, and its distance from the earth is 386,000 km. Find the diameter of the image of the moon formed by a spherical concave telescope mirror with a focal length of 3.20 m.

**35–4** A spherical concave shaving mirror has a radius of curvature of 25.0 cm. a) What is the magnification of a person's face when it is 10.0 cm to the left of the vertex of the mirror? b) Where is the image? Is the image real or virtual? c) Draw a principal-ray diagram showing the formation of the image.

**35–5** An object 0.400 cm tall is placed 15.0 cm to the left of the vertex of a concave spherical mirror having a radius of curvature of 20.0 cm. a) Draw a principal-ray diagram showing the formation of the image. b) Determine the position, size, orientation, and nature (real or virtual) of the image.

**35–6** A concave mirror has a radius of curvature of 24.0 cm. a) What is its focal length? b) If the mirror is immersed in water (refractive index 1.33), what is its focal length?

**35–7** Prove that the image formed of a real object by a convex mirror is always virtual, regardless of the object's position.

**35–8** Repeat Exercise 35–5 for the case in which the mirror is convex.

**35–9** An object 1.50 cm tall is placed 8.00 cm to the left of the vertex of a convex spherical mirror whose radius of curvature has a magnitude of 24.0 cm. a) Draw a principal-ray diagram showing the formation of the image. b) Determine the position, size, orientation, and nature (real or virtual) of the image.

**35–10** Equations (35–6) and (35–7) were derived in the text for the case of a concave mirror. Repeat the similar derivation for a convex mirror, and show that the same equations result if you use the sign convention established in the text.

### Section 35–4 Refraction at a Spherical Surface

**35–11** Equations (35–11) and (35–12) were derived in the text for the case in which $R$ is positive and $n_a < n_b$. (See Figs. 35–18 and 35–19.) a) Carry through the derivation of these two equations for the case $R > 0$ and $n_a > n_b$. b) Carry through the derivation for $R < 0$ and $n_a < n_b$.

**35–12** The left end of a long glass rod 6.00 cm in diameter has a convex hemispherical surface 3.00 cm in radius. The refractive index of the glass is 1.50. Determine the position of the image if an object is placed in air on the axis of the rod at the following distances to the left of the vertex of the curved end: a) infinitely far; b) 16.0 cm; c) 4.00 cm.

**35–13** The rod of Exercise 35–12 is immersed in a liquid. An object 60.0 cm from the vertex of the left end of the rod and on its axis is imaged at a point 90.0 cm inside the rod. What is the refractive index of the liquid?

**35–14** The left end of a long glass rod 10.0 cm in diameter, with an index of refraction 1.50, is ground and polished to a convex hemispherical surface with a radius of 5.00 cm. An object in the form of an arrow 2.00 mm tall, at right angles to the axis of the rod, is located on the axis 25.0 cm to the left of the vertex of the convex surface. Find the position and height of the image of the arrow formed by paraxial rays incident on the convex surface. Is the image erect or inverted?

**35–15** Repeat Exercise 35–14 for the case in which the end of the rod is ground to a *concave* hemispherical surface with radius 5.00 cm.

**35–16 A Spherical Fish Bowl.** A small tropical fish is at the center of a spherical fish bowl 28.0 cm in diameter. a) Find the apparent position and magnification of the fish to an observer outside the bowl. The effect of the thin walls of the bowl may be neglected. b) A friend advised the owner of the bowl to keep it out of direct sunlight to avoid blinding the fish, which might swim into the focal point of the parallel rays from the sun. Is the focal point actually within the bowl?

**35–17** A speck of dirt is embedded 2.40 cm below the surface of a sheet of ice ($n = 1.309$). What is its apparent depth when viewed at normal incidence?

**35–18** A tank whose bottom is a mirror is filled with water to a depth of 24.0 cm. A small fish floats motionless 8.00 cm under the surface of the water. a) What is the apparent depth of the fish when viewed at normal incidence? b) What is the apparent depth of the image of the fish when viewed at normal incidence?

## Section 35–5 Thin Lenses

## Section 35–6 Graphical Methods for Lenses

**35–19** A converging lens has a focal length of 15.0 cm. For an object to the left of the lens, at distances of 20.0 cm and 5.00 cm, determine a) the image position; b) the magnification; c) whether the image is real or virtual; d) whether the image is erect or inverted. Draw a principal-ray diagram for each case.

**35–20** A converging lens with a focal length of 10.0 cm forms a real image 1.00 cm tall, 15.0 cm to the right of the lens. Determine the position and size of the object. Is the image erect or inverted? Draw a principal-ray diagram for this situation.

**35–21** Repeat Exercise 35–20 for the case in which the lens is diverging with a focal length of $-10.0$ cm.

**35–22** An object is 10.0 cm to the left of a lens. The lens forms an image 25.0 cm to the right of the lens. a) What is the focal length of the lens? Is the lens converging or diverging? b) If the object is 0.500 cm tall, how tall is the image? Is it erect or inverted? c) Draw a principal-ray diagram.

**35–23** A lens forms an image of an object. The object is 20.0 cm from the lens. The image is 6.00 cm from the lens on the same side as the object. a) What is the focal length of the lens? Is the lens converging or diverging? b) If the object is 0.400 cm tall, how tall is the image? Is it erect or inverted? c) Draw a principal-ray diagram.

**35–24** Prove that the image of a real object formed by a diverging lens is *always* virtual.

**35–25** A converging lens with a focal length of 6.00 cm forms an image of a 5.00-cm tall object that is to the left of the lens. The image is 40.0 cm tall and erect. Where are the object and image located? Is the image real or virtual?

**35–26** A converging lens with a focal length of 6.00 cm forms an image of a 40.0-cm tall object that is to the left of the lens. The image is 5.00 cm tall and inverted. Where are the object and image located in relation to the lens? Is the image real or virtual?

**35–27** A diverging meniscus lens (see Fig. 35–28b) with a refractive index of 1.48 has spherical surfaces whose radii are 4.00 cm and 2.50 cm. What is the position of the image if an object is placed 16.0 cm to the left of the lens?

**35–28** You are standing to the left of a lens that projects an image of you onto a wall 3.00 m to your right. The image is three times your height. a) How far are you from the lens? b) Is your image erect or inverted? c) What is the focal length of the lens? Is the lens converging or diverging?

**35–29 An Optical Glass Rod.** Both ends of a glass rod 10.0 cm in diameter, with index of refraction 1.50, are ground and polished to convex hemispherical surfaces. The radius of curvature at the left end is 5.00 cm, and the radius of curvature at the right end is 10.0 cm. The length of the rod between vertexes is 50.0 cm. An arrow 1.00 mm tall, at right angles to the axis and 20.0 cm to the left of the first vertex, constitutes the object for the first surface. a) What constitutes the object for the second surface? b) What is the object distance for the second surface? c) Is the object for the second surface real or virtual? d) What is the position of the image formed by the second surface?

**35–30** The rod of Exercise 35–29 is shortened to a distance of 15.0 cm between its vertexes; the curvatures of its ends remain the same. As in Exercise 35–29, an arrow 1.00 mm tall and 20.0 cm to the left of the first vertex constitutes the object for the first surface. a) What is the object distance for the second surface? b) Is the object for the second surface real or virtual? c) What is the position of the image formed by the second surface? d) Is the image real or virtual? Is it erect or inverted with respect to the original object? e) What is the height of the final image?

**35–31** Sketch the various possible thin lenses that can be obtained by combining two surfaces whose radii of curvature are 10.0 cm and 20.0 cm in absolute magnitude. Which are converging and which diverging? Find the focal length of each if the surfaces are made of glass with index of refraction 1.60.

## PROBLEMS

**35–32** What is the size of the smallest vertical plane mirror in which a man of height $h$ can see his full-length image?

**35–33** If you run toward a plane mirror at 3.00 m/s, at what speed is your image approaching you?

**35–34** An object is placed between two plane mirrors arranged at right angles to each other at a distance $d_1$ from the surface of one mirror and a distance $d_2$ from the other. a) How many images are formed? Show the location of the images in a diagram. b) Draw the paths of rays from the object to the eye of an observer.

**35–35** A concave mirror is to form an image of the filament of a headlight lamp on a screen 4.00 m from the mirror. The filament is 5.00 mm tall, and the image is to be 35.0 cm tall. a) How far in front of the vertex of the mirror should the filament be placed? b) What should be the radius of curvature of the mirror?

**35–36** Where must you place an object in front of a concave mirror with radius $R$ so that the image is real and one-third the size of the object? Where is the image?

**35–37** A luminous object is 5.00 m from a wall. You want to use a concave mirror to project an image of the object on the wall, with the image three times the size of the object. How far should the mirror be from the wall? What should its radius of curvature be?

**35–38 A Rear-View Car Mirror.** A mirror on the passenger side of your car is convex and has a radius of curvature with magnitude 20.0 cm. Another car is 8.0 m behind you and is seen in this side mirror. If this car is 1.5 m tall, what is the height of the image? (These mirrors usually have a warning attached that objects viewed in them are closer than they appear. Why is this so?)

**35–39** If the light incident from the left onto a convex mirror does not diverge from an object point but instead converges toward a point at a (negative) distance $s$ to the right of the mirror, this point is called a *virtual object.* a) For a convex mirror having a radius of curvature of 12.0 cm, for what range of virtual-object positions is a real image formed? b) What is the orientation of this real image? c) Draw a principal-ray diagram showing formation of such an image.

**35–40** An object is 24.0 cm from the center of a silvered spherical glass Christmas tree ornament 8.00 cm in diameter. What are the position and magnification of its image?

**35–41** What should the index of refraction of a transparent sphere be in order for paraxial rays from an infinitely distant object to be brought to a focus at the vertex of the surface opposite the point of incidence?

**35–42 A Method for Measuring Index of Refraction.** A microscope is focused on the upper surface of a glass plate. A second plate is then placed over the first. To focus on the bottom surface of the second plate, the microscope must be raised 0.800 mm. To focus on the upper surface, it must be raised 2.00 mm *farther.* Find the index of refraction of the second plate.

**35–43** A solid glass hemisphere having a radius of 10.0 cm and a refractive index of 1.50 is placed with its flat face downward on a table. A parallel beam of light with a circular cross section 0.400 cm in diameter travels straight down and enters the hemisphere at the center of its curved surface. a) What is the diameter of the circle of light formed on the table? b) How does your result depend on the radius of the hemisphere?

**35–44** A transparent rod 40.0 cm long is cut flat at one end and rounded to a hemispherical surface of radius 8.00 cm at the other end. A small object is embedded within the rod along its axis and halfway between its ends, 20.0 cm from the flat end and 20.0 cm from the vertex of the curved end. When viewed from the flat end of the rod, the apparent depth of the object is 12.5 cm from the flat end. What is its apparent depth when viewed from the curved end?

**35–45 A Three-Dimensional Image.** The longitudinal magnification is defined as $m' = ds'/ds$. It relates the longitudinal dimension of a small object to the longitudinal dimension of its image. a) For a spherical mirror, show that $m' = -m^2$. What is the significance of the fact that $m'$ is *always* negative? b) A wire frame in the form of a small cube 1.0 cm on a side is placed with its center on the axis of a concave mirror with radius of curvature 40.0 cm. All sides of the cube are either parallel or perpendicular to the axis. The cube face toward the mirror is 80.0 cm to the left of the mirror vertex. Find i) the location of the image of this face and of the opposite face of the cube; ii) the lateral and longitudinal magnifications; iii) the dimensions of each of the six cube faces of the image.

**35–46** Refer to Problem 35–45. Show that the longitudinal magnification $m'$ for refraction at a spherical surface is given by

$$m' = -\frac{n_b}{n_a}m^2.$$

**35–47** You are sitting in your parked car and notice a jogger approaching in the convex side mirror, which has a radius of curvature 2.00 m. If the jogger is running at a speed of 3.00 m/s, how fast does she appear to be moving when she is a) 10.0 m away; b) 2.0 m away?

**35–48** For refraction at a spherical surface, the first focal length $f$ is defined as the value of $s$ corresponding to $s' = \infty$, as shown in Fig. 35–34a. The second focal length $f'$ is defined as the value of $s'$ when $s = \infty$, as shown in Fig. 35–34b. a) Prove that $n_a/n_b = f/f'$. b) Prove that the general relation between object and image distance is

$$\frac{f}{s} + \frac{f'}{s'} = 1.$$

**35–49** A layer of benzene ($n = 1.50$) 2.00 cm deep floats on water ($n = 1.33$) that is 5.00 cm deep. What is the apparent distance from the upper benzene surface to the bottom of the water layer when it is viewed at normal incidence?

**35–50** A transparent rod 40.0 cm long and with a refractive index of 1.50 is cut flat at the right end and rounded to a hemispherical surface with a 12.0-cm radius at the left end. An object is placed on the axis of the rod 14.0 cm to the left of the vertex of the hemispherical end. a) What is the position of the final image? b) What is its magnification?

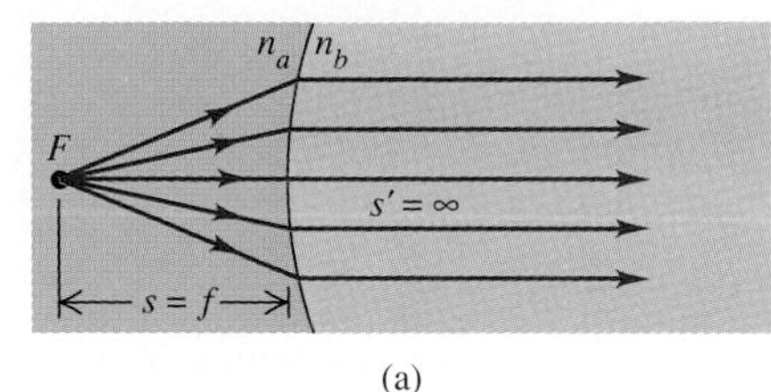

(a)

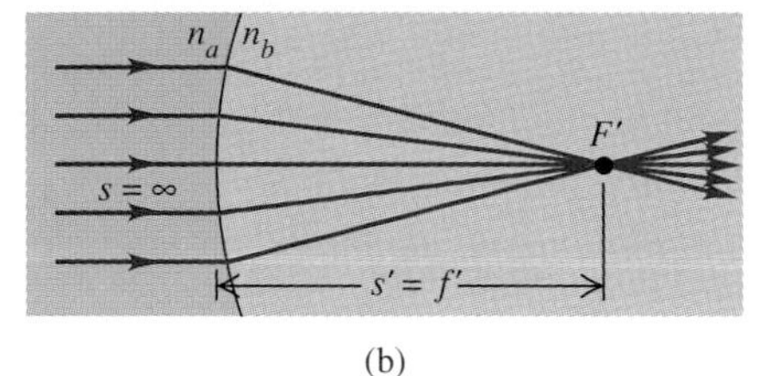


(b)

**FIGURE 35–34**

**35–51** A narrow beam of parallel rays enters a solid glass sphere in a radial direction. At what point outside the sphere are these rays brought to a focus? The radius of the sphere is 2.00 cm, and its index of refraction is 1.50.

**35–52** The radii of curvature of the surfaces of a thin converging meniscus lens are $R_1 = +10.0$ cm and $R_2 = +30.0$ cm. The index of refraction is 1.50. a) Compute the position and size of the image of an object in the form of an arrow 1.00 cm tall, perpendicular to the lens axis, 50.0 cm to the left of the lens. b) A second converging lens with the same focal length is placed 120 cm to the right of the first. Find the position and size of the final image. Is the final image erect or inverted with respect to the original object? c) Repeat part (b) with the second lens 40.0 cm to the right of the first. d) Repeat part (c) with the second lens having focal length $-30.0$ cm and being diverging.

**35–53** Three thin lenses, each with a focal length of 20.0 cm, are aligned on a common axis; adjacent lenses are separated by 30.0 cm. Find the position of the image of a small object on the axis, 40.0 cm to the left of the first lens.

**35–54** Two thin lenses with a focal length of magnitude 10.0 cm, the first converging and the second diverging, are placed 8.00 cm apart. An object 2.00 mm tall is placed 20.0 cm to the left of the first (converging) lens. a) How far from this first lens is the final image formed? b) Is the final image real or virtual? c) What is the height of the final image? Is the final image erect or inverted?

**35–55** An eyepiece consists of two converging thin lenses, each with a focal length 8.00 cm, separated by a distance of 3.00 cm. Where are the first and second focal points of the eyepiece?

**35–56** An object is placed 16.0 cm from a screen. a) At what two points between object and screen may a converging lens with a 3.00-cm focal length be placed to obtain an image on the screen? b) What is the magnification of the image for each position of the lens?

**35–57** An object to the left of a lens is imaged by the lens on a screen 14.0 cm to the right of the lens. When the lens is moved 2.00 cm to the right, the screen must be moved 2.00 cm to the left to refocus the image. Determine the focal length of the lens.

**35–58 Two Lenses in Contact.** a) Prove that when two thin lenses with focal lengths $f_1$ and $f_2$ are placed *in contact*, the focal length $f$ of the combination is given by

$$\frac{1}{f} = \frac{1}{f_1} + \frac{1}{f_2}.$$

b) A converging meniscus lens (Fig. 35–28a) has index of refraction 1.60 and radii of curvature for its surfaces of 4.00 and 8.00 cm. The concave surface is placed upward and filled with water. What is the focal length of the water-glass combination?

**35–59** When an object is placed at the proper distance to the left of a converging lens, the image is focused on a screen 25.0 cm to the right of the lens. A diverging lens is placed 12.5 cm to the right of the converging lens, and it is found that the screen must be moved 20.0 cm farther to the right to obtain a sharp image. What is the focal length of the diverging lens?

**35–60 A Lens in a Liquid.** A lens obeys Snell's law, bending light rays at each surface an amount determined by the index of refraction of the lens and the index of the medium in which the lens is located. a) Equation (35–20) assumes that the lens is surrounded by air. Consider instead a thin lens immersed in a liquid with refractive index $n_{\text{liq}}$. Prove that the focal length $f'$ is then given by Eq. (35–20) with $n$ replaced by $n/n_{\text{liq}}$. b) A thin lens with index $n$ has focal length $f$ in vacuum. Use the result of part (a) to show that when this lens is immersed in a liquid of index $n_{\text{liq}}$, it will have a new focal length given by

$$f' = \left[\frac{n_{\text{liq}}(n-1)}{n - n_{\text{liq}}}\right] f.$$

**35–61 A Light Source between Two Mirrors.** A convex mirror and a concave mirror are placed on the same optic axis, separated by a distance $L = 0.600$ m. The radius of curvature of each mirror has a magnitude of 0.400 m. A light sources is located a distance $x$ from the concave mirror, as shown in Fig. 35–35. a) What distance $x$ will result in the rays from the source returning to the source after reflecting first from the convex mirror and then from the concave mirror? b) Repeat part (a), but let the rays reflect first from the concave mirror and then from the convex one.

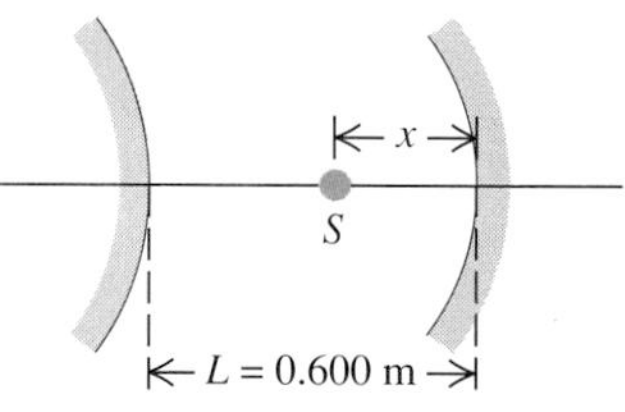


**FIGURE 35–35**

**35–62 A Light Source between a Mirror and a Lens.** In Fig. 35–36 (on the next page) the candle is at the center of curvature of the concave mirror, whose focal length is 10.0 cm. The converging lens has a focal length of 35.0 cm and is 85.0 cm to the right of the candle. The candle is viewed looking through the lens from the right. The lens forms two images of the candle. The first is formed by light passing directly through the lens. The second image is formed from the light that goes from the candle to the mirror, is reflected, and then passes through the lens.

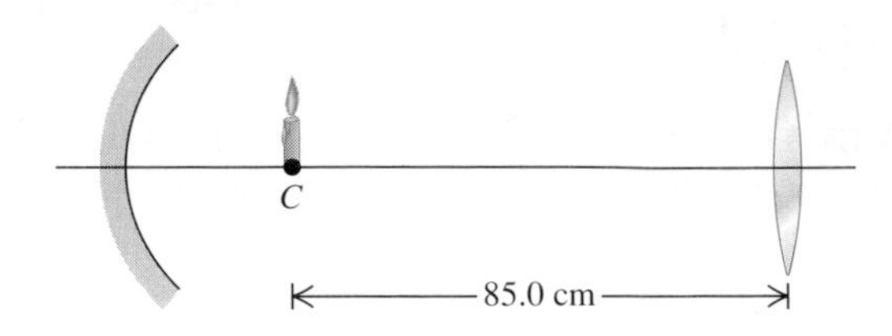


**FIGURE 35–36**

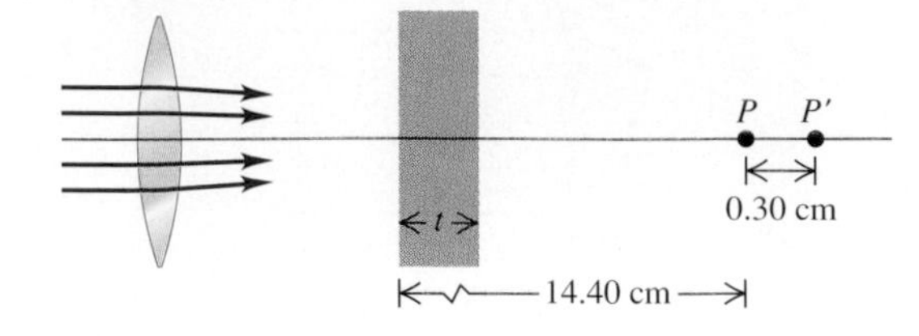


**FIGURE 35–37**

For *each* of these two images, answer the following questions: a) Draw a principal-ray diagram that locates the image. b) Where is the image? c) Is the image real or virtual? d) Is the image erect or inverted with respect to the original object?

**35–63** A thin-walled glass sphere with radius $R$ is filled with water. An object is placed a distance $3R$ from the surface of the sphere. Determine the position of the final image. The effect of the glass wall may be neglected. The refractive index of the water is $\frac{4}{3}$.

**35–64** A glass rod with a refractive index of 1.50 is ground and polished at both ends to hemispherical surfaces with radii of 5.00 cm. When an object is placed on the axis of the rod, 20.0 cm to the left of one end, the final image is formed 60.0 cm to the right of the opposite end. What is the length of the rod measured between the vertexes of the two hemispherical surfaces?

**35–65** A glass plate 3.00 cm thick, with an index of refraction of 1.50 and plane parallel faces, is held with its faces horizontal and its lower face 9.00 cm above a printed page. Find the position of the image of the page formed by rays making a small angle with the normal to the plate.

**35–66** A thick-walled wine goblet sitting on a table can be considered to be a glass sphere with an outer radius of 4.00 cm and a spherical cavity with a radius of 3.00 cm. The index of refraction of the goblet glass is 1.50. a) A beam of parallel light rays enters the side of the empty goblet along a horizontal radius. Where, if anywhere, will an image be formed? b) The goblet is filled with white wine ($n = 1.37$). Where is the image formed?

**35–67** Rays from a lens are converging toward a point image $P$ located to the right of the lens. What thickness $t$ of glass with an index of refraction 1.60 must be interposed between the lens and $P$ for the image to be formed at $P'$, located 0.30 cm to the right of $P$? The location of the piece of glass and of points $P$ and $P'$ are shown in Fig. 35–37.

## CHALLENGE PROBLEMS

**35–68** Two mirrors are placed together as shown in Fig. 35–38. a) Show that a point source in front of these mirrors and its two images lie on a circle. b) Find the center of the circle. c) Draw a diagram to show where an observer should stand in order to be able to see both images.

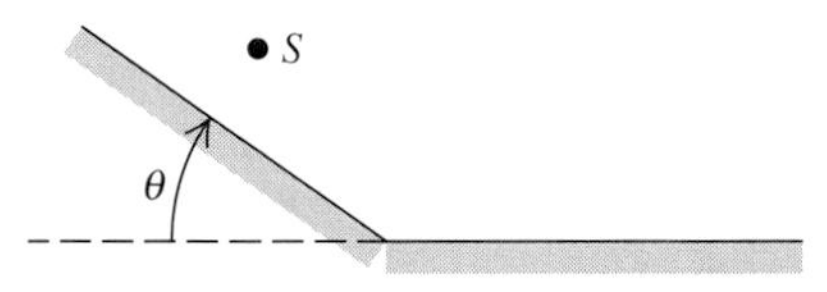


**FIGURE 35–38**

**35–69 An Object between Two Mirrors.** A convex spherical mirror with a focal length of magnitude 20.0 cm is placed 15.0 cm to the left of a plane mirror. An object 0.300 cm tall is placed midway between the surface of the plane mirror and the vertex of the spherical mirror. The spherical mirror forms multiple images of the object. Where are the two images of the object formed by the spherical mirror that are closest to the mirror, and how tall is each of them?

**35–70 An Object at an Angle.** A 20.0-cm-long pencil is placed at a 45° angle, with its center 15.0 cm above the optic axis and 45.0 cm from a lens with a 20.0-cm focal length, as shown in Fig. 35–39. Assume that the diameter of the lens is large enough for the paraxial approximation to be valid. a) Where is the image of the pencil? (Give the location of the images of the points $A$, $B$, and $C$ on the object, which are located at the eraser, point, and center of the pencil, respectively.) b) What is the length of the image, the distance between the images of points $A$ and $B$? c) Show the orientation of the image in a sketch.

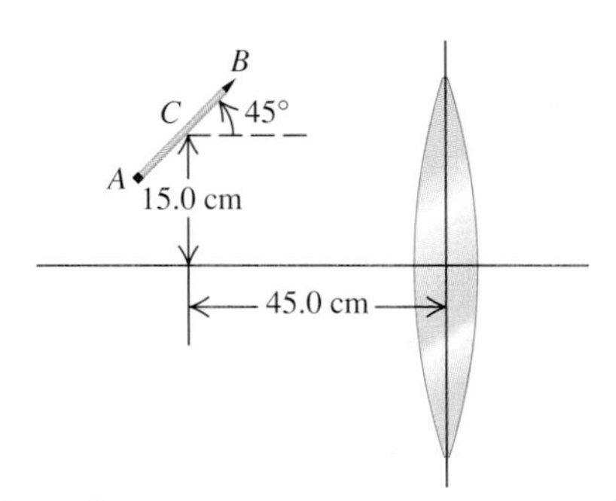


**FIGURE 35–39**

**35–71** a) For a lens with focal length $f$, find the smallest distance possible between the object and its real image. b) Sketch a graph of the distance between the object and its real image as a function of the distance of the object from the lens. Does your sketch agree with the result you found in part (a)?

**35–72** A solid glass sphere with radius $R$ and an index of refraction 1.60 is silvered over one hemisphere, as in Fig. 35–40. A small object is located on the axis of the sphere at a distance $2R$ to the left of the vertex of the unsilvered hemisphere. Find the position of the final image after all refractions and reflections have taken place.

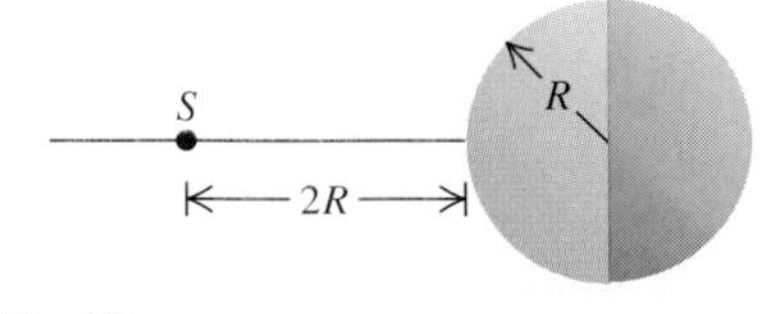

**FIGURE 35–40**

Most high-quality cameras are built to work with detachable lenses, so you can substitute a lens with a different focal length for the standard 35–mm focal-length lens. This series of pictures was taken with the same camera, using lenses with focal lengths of 28 mm, 105 mm, and 300 mm. The object distance remained the same for each photo, but the angle of view decreased from 75° to 25° to 8° for the larger lens. The result is a series of larger images that each fills the frame of the film.

# CHAPTER 36

# Optical Instruments

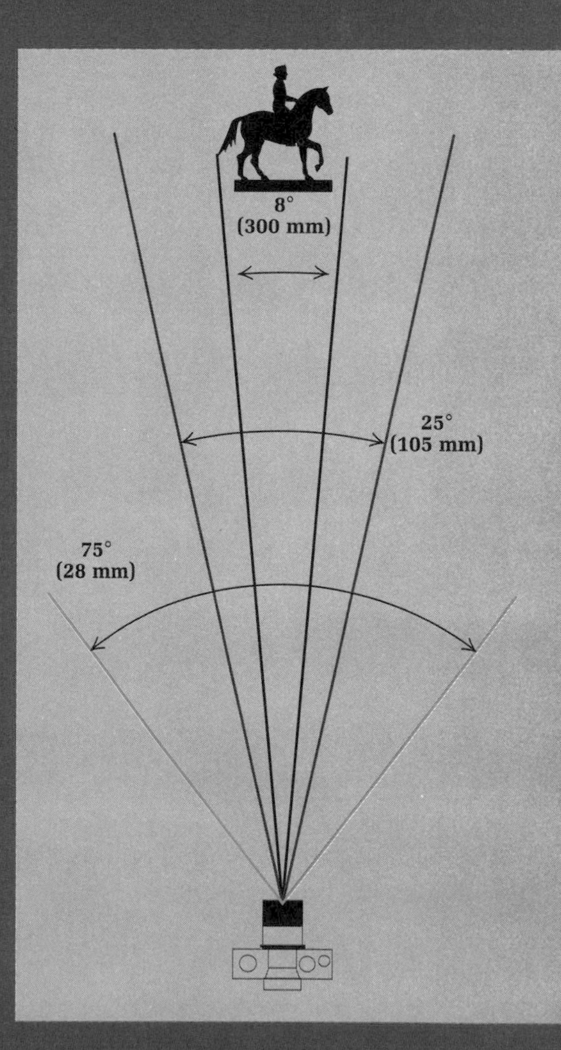

## KEY CONCEPTS

- **A camera uses a lens to form an inverted, usually reduced, real image of an object. The light energy reaching the film is determined by the exposure time and the lens diameter. A projector forms an inverted, enlarged, real image on a screen.**
- **In the eye the curved front surface of the cornea helps form an inverted, reduced, real image on the retina. Defects of vision can be corrected with appropriate lenses.**
- **A magnifier forms an erect, enlarged, virtual image with an angular size that is greater than when the object is viewed without a lens.**
- **The compound microscope uses two lens groups. The first (the objective lens) forms a real, enlarged image within the barrel of the instrument; the second (the ocular) forms a final enlarged virtual image.**
- **A telescope has an objective, either a lens or mirror, that forms a real image, which is then viewed by a second lens.**
- **Lens aberrations are departures from idealized behavior resulting from the failure of the paraxial approximation and from dependence of the index of refraction on wavelength.**

## INTRODUCTION

How does a camera resemble the human eye? What are the significant differences? What does a projector operator have to do to "focus" the picture on the screen? Why do large telescopes usually include curved mirrors as well as lenses? This chapter is concerned with these and other familiar optical systems. Some of these systems use several lenses or combinations of lenses and mirrors, but we can apply the basic principles of mirror and lens behavior that we studied in Chapter 35.

The concept of *image* plays an important role in the analysis of optical instruments. We continue to base our analysis on the *ray* model of light, so the content of this chapter comes under the general heading *geometric optics.*

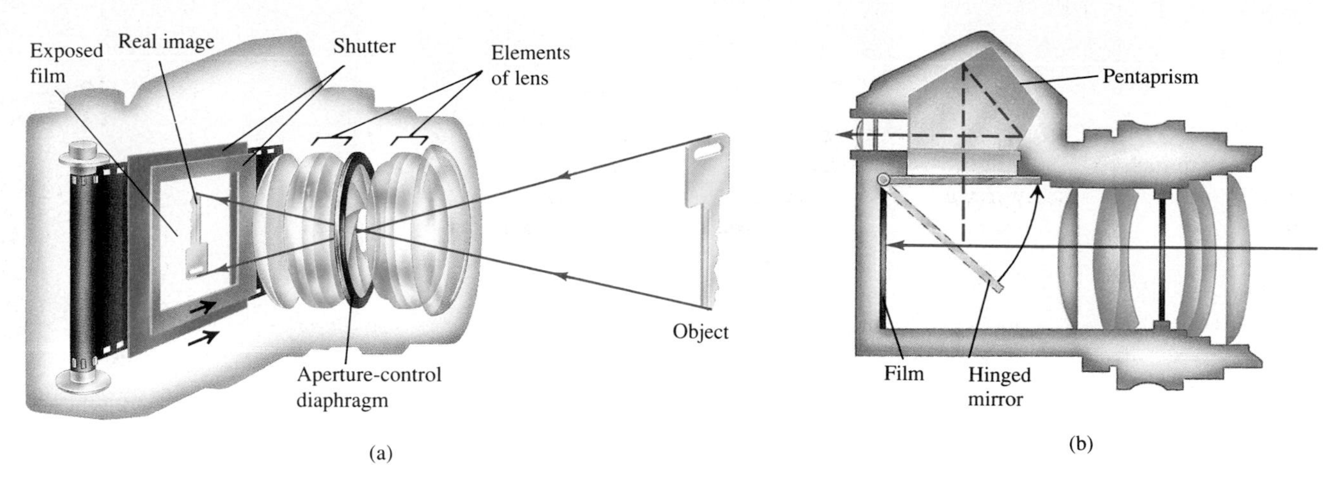


**36–1** Camera with aperture control. (a) The lens forms a real image of the object in the plane of the film. (b) The Zeiss Tessar lens, a typical compound lens, and the pentaprism for viewing through the lens. The hinged mirror swings up out of the way as the shutter opens.

## 36–1 THE CAMERA

The essential elements of a **camera** are a lens equipped with a shutter, a light-tight enclosure, and a light-sensitive film that records an image (Fig. 36–1a). The lens forms an inverted, usually reduced, real image on the film of the object being photographed. To provide proper image distances for various object distances, the lens is moved closer to or farther from the film, often by being turned in a threaded mount. Most lenses have several elements, permitting partial correction of various *aberrations,* including the dependence of index of refraction on wavelength and the limitations imposed by the paraxial approximation. (We'll discuss lens aberrations in Section 36–8). A classic lens design is the Zeiss "Tessar" design, shown in Fig. 36–1b.

In order for the film to record the image properly, the total light energy per unit area reaching the film (the exposure) must fall within certain limits. This is controlled by the *shutter* and the *lens aperture*. The shutter controls the time interval during which light enters the lens. This is usually adjustable in steps corresponding to factors of about two, often from 1 s to $\frac{1}{1000}$ s.

The intensity of light reaching the film is proportional to the effective area of the lens, which may be varied by means of an adjustable aperture, or *diaphragm,* a nearly circular hole with variable diameter $D$. The aperture size is usually described as a fraction of the focal length $f$ of the lens. A lens with a focal length $f = 50$ mm and an aperture diameter of 25 mm has an aperture of $f/2$; many photographers would call this an $f$-number of 2. In general,

$$f\text{-number} = \frac{\text{Focal length}}{\text{Aperture diameter}} = \frac{f}{D}. \tag{36–1}$$

Because the light intensity at the film is proportional to the area of the lens aperture and thus to the *square* of its diameter, changing the diameter by a factor of $\sqrt{2}$ changes the exposure by a factor of 2. Adjustable apertures (often called *f-stops*) usually have scales labeled with successive numbers related by factors of $\sqrt{2}$, such as

$$f/2,\quad f/2.8,\quad f/4,\quad f/5.6,\quad f/8,\quad f/11,\quad f/16,$$

and so on. The larger numbers represent smaller apertures and exposures, and each step corresponds to a factor of 2 in exposure (Fig. 36–2). The actual exposure is proportional to both the aperture area and the time of exposure. Thus $f/4$ and $\frac{1}{500}$ s, $f/5.6$ and $\frac{1}{250}$ s, and $f/8$ and $\frac{1}{125}$ s all correspond to the same exposure.

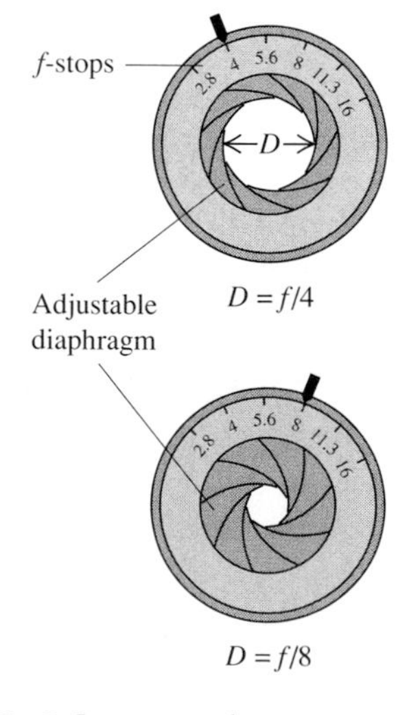


**36–2** Larger numbers mean smaller aperture diameters in a camera.

**36–3** As $f$ increases from 28 mm to 300 mm (a ratio of 1:10), the image size increases proportionally (see page 1003).

The choice of focal length for a camera lens depends on the film size and the desired angle of view, or *field*. The popular 35-mm camera has an image size on the film of 24 × 36 mm. The normal lens for such a camera usually has a focal length of about 50 mm; this permits an angle of view of about 45°. A lens with a longer focal length, often 135 mm or 200 mm, provides a *smaller* angle of view and a larger image of *part* of the object, compared with a normal lens. This gives the impression that the camera is *closer* than it really is, and such a lens is called a *telephoto lens.* A lens with shorter focal length, such as 35 mm or 28 mm, permits a wider angle of view and is called a *wide-angle lens.* An extreme wide-angle, or "fish-eye," lens may have a focal length as small as 6 mm. Figure 36–3 shows a scene photographed from the same point with lenses of various focal lengths.

### EXAMPLE 36–1

**Photographic exposures** A common telephoto lens for a 35-mm camera has a focal length of 200 mm and a range of $f$-stops from $f/5.6$ to $f/45$. a) What is the corresponding range of aperture diameters? b) What is the corresponding range of intensity of the film image?

**SOLUTION** a) From Eq. (36–1) the range of diameters is from

$$D = \frac{f}{f\text{-number}} = \frac{200\text{ mm}}{5.6} = 36\text{ mm}$$

to

$$D = \frac{200\text{ mm}}{45} = 4.4\text{ mm}.$$

b) Because the intensity is proportional to the square of the diameter, the ratio of the value of $f/5.6$ to the value at $f/45$ is approximately

$$\left(\frac{36\text{ mm}}{4.4\text{ mm}}\right)^2 \cong \left(\frac{45}{5.6}\right)^2 \cong 65 \quad (\text{about } 2^6).$$

If the correct exposure time at $f/5.6$ is $\frac{1}{1000}$ s, then at $f/45$ it is $(65)(\frac{1}{1000}\text{ s}) = \frac{1}{15}$ s.

---

The optical system for a television camera is the same in principle as that of an ordinary camera. The film is replaced by an electronic system that scans the image with a series of 525 parallel lines. The image brightness at points along these lines is translated into electrical impulses that can be broadcast, using electromagnetic waves with frequencies of the order of 100–400 MHz. The entire picture is scanned 30 times each second, so 30 × 525 = 15,750 lines are scanned each second. Some TV receivers emit a faint high-pitched sound at this scanning frequency (two octaves above the highest B on a piano).

## 36–2 THE PROJECTOR

A **projector** for viewing slides or motion pictures operates very much like a camera in reverse. The essential elements are shown in Fig. 36–4. Light from the source (an incandescent lamp bulb or, in large motion-picture projectors, a carbon-arc lamp) shines through the film, and the projection lens forms a real, enlarged, inverted image of the

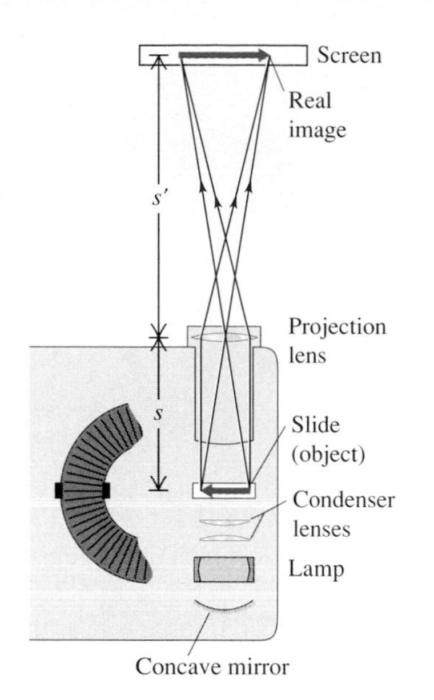


**36–4** A slide projector. The concave mirror and condenser lenses gather and direct the light from the lamp so that it will enter the projection lens after passing through the slide. The cooling fan, usually needed to take away excessive heat from the lamp, is not shown.

film on the projection screen. Additional lenses called *condenser lenses* are placed between lamp and film. Their function is to direct the light from the source so that most of it enters the projection lens after passing through the film. A concave mirror behind the lamp also helps direct the light. The condenser lenses must be large enough to cover the entire area of the film. The image on the screen is always real and inverted; this is why slides have to be put into a projector upside-down.

The position and size of the image projected on the screen are determined by the position and focal length of the projection lens.

## EXAMPLE 36–2

**A slide projector** An ordinary 35-mm color slide has a picture area 24 × 36 mm. What focal length would a projection lens have to have to project a 1.2 m × 1.8 m image of this picture on a screen 5.0 m from the lens?

**SOLUTION** We need a lateral magnification (disregarding sign) of (1.2 m)/(24 mm) = 50. From Eq. (35–18) the ratio $s'/s$ must also be 50. (The image is real, so $s'$ is positive.) We are given $s' = 5.0$ m, so $s = (5\text{ m})/50 = 0.10$ m.

Then, from Eq. (35–17),

$$\frac{1}{f} = \frac{1}{s} + \frac{1}{s'} = \frac{1}{0.10\text{ m}} + \frac{1}{5.0\text{ m}},$$

$$f = 0.098\text{ m} = 98\text{ mm}.$$

A popular focal length for home slide projectors is 100 mm; such lenses are readily available and would be the appropriate choice in this situation.

## *36–3 THE EYE

The optical behavior of the eye is similar to that of a camera. The essential parts of the human eye, considered as an optical system, are shown in Fig. 36–5 (on the next page). The eye is nearly spherical in shape and about 2.5 cm in diameter. The front portion is somewhat more sharply curved and is covered by a tough, transparent membrane called the *cornea*. The region behind the cornea contains a liquid called the *aqueous humor.* Next comes the *crystalline lens,* a capsule containing a fibrous jelly, hard at the center and progressively softer at the outer portions. The crystalline lens is held in place by ligaments that attach it to the ciliary muscle, which encircles it. Behind the lens the eye is filled with a thin, watery jelly called the *vitreous humor.* The indexes of refraction of both the aqueous humor and the vitreous humor are nearly equal to that of water, about 1.336. The crystalline lens, although not homogeneous, has an average index of refraction of 1.437. This is not very different from the indexes of the aqueous and vitreous humors; most of the refraction of light entering the eye occurs at the outer surface of the cornea.

**36–5** The eye. The ciliary muscle contracts to change the focal length of the crystalline lens in order to image close objects sharply.

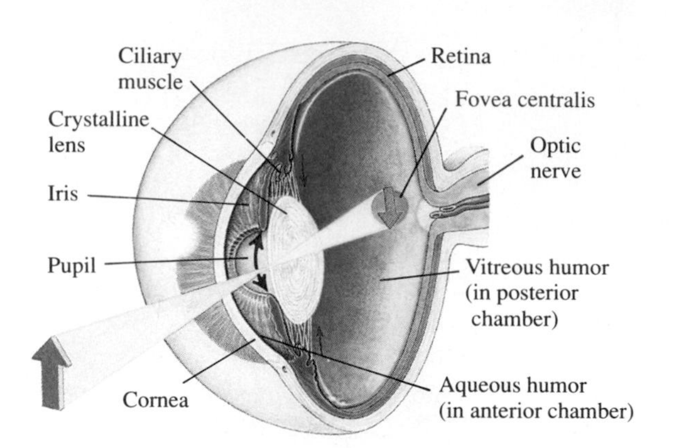


**TABLE 36–1**
**Receding of Near Point with Age**

| Age (years) | Near point (cm) |
|---|---|
| 10 | 7 |
| 20 | 10 |
| 30 | 14 |
| 40 | 22 |
| 50 | 40 |
| 60 | 200 |

Refraction at the cornea and the surfaces of the lens produces a *real image* of the object being viewed; the image is formed on the light-sensitive *retina* that lines the rear inner surface of the eye. The retina plays the same role as the film in a camera. The *rods* and *cones* in the retina act like an array of miniature photocells; they sense the image and transmit it via the *optic nerve* to the brain. Vision is most acute in a small central region called the *fovea centralis,* about 0.25 mm in diameter.

In front of the lens is the *iris.* It contains an aperture with variable diameter called the *pupil,* which opens and closes to adapt to changing light intensity. The receptors of the retina also have intensity adaptation mechanisms.

For an object to be seen sharply the image must be formed exactly on the retina. The lens-to-retina distance, corresponding to $s'$, does not change, but the eye accommodates to different object distances $s$ by changing the focal length of its lens. When the ciliary muscle surrounding the lens contracts, the lens bulges and the radii of curvature of its surfaces *decrease;* this decreases the focal length. For the normal eye an object at infinity is sharply focused when the ciliary muscle is relaxed. With increasing tension the focal length decreases to permit sharp imaging on the retina of closer objects. This process is called *accommodation*.

The extremes of the range over which distinct vision is possible are known as the *far point* and the *near point* of the eye. The far point of a normal eye is at infinity. The position of the near point depends on the amount by which the ciliary muscle can increase the curvature of the crystalline lens. The range of accommodation gradually diminishes with age as the crystalline lens loses its flexibility. For this reason, the near point gradually recedes as one grows older. This recession of the near point is called *presbyopia.* Table 36–1 shows the approximate position of the near point for an average person at various ages. For example, an average person 50 years of age cannot focus on an object closer than about 40 cm.

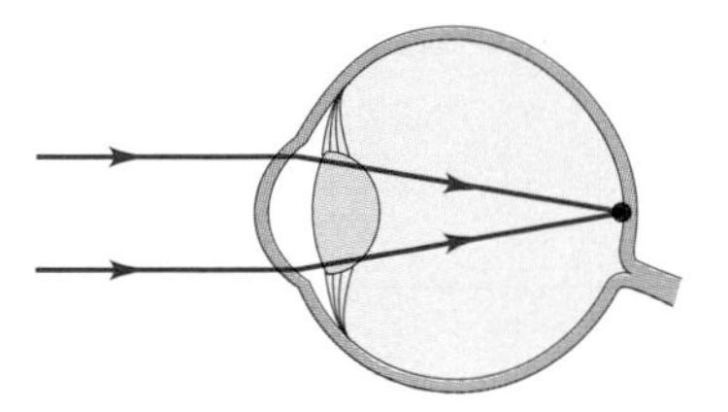

(a) Normal eye

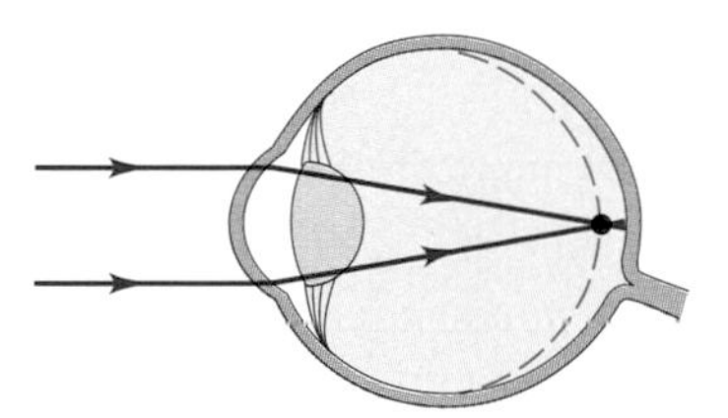

(b) Myopic eye

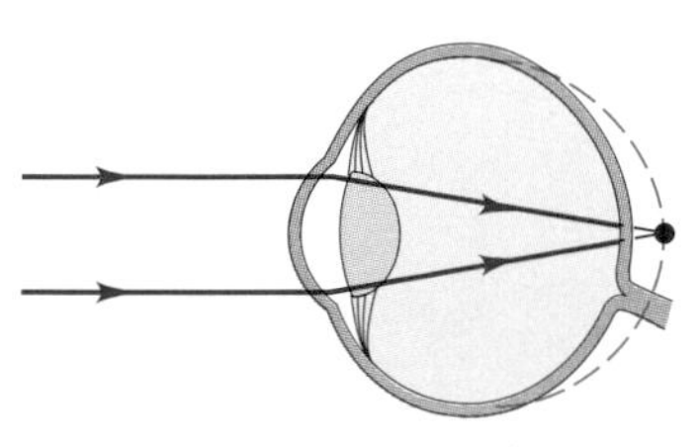

(c) Hyperopic eye

**36–6** Refractive errors for a myopic (nearsighted) and a hyperopic (farsighted) eye viewing a very distant object. The dashed blue curve indicates the correct position of the retina.

Several common defects of vision result from incorrect distance relations in the eye. A normal eye forms an image on the retina of an object at infinity when the eye is relaxed (Fig. 36–6a). In the *myopic* (nearsighted) eye the eyeball is too long from front to back in comparison with the radius of curvature of the cornea (or the cornea is too sharply curved), and rays from an object at infinity are focused in front of the retina (Fig. 36–6b). The most distant object for which an image can be formed on the retina is then nearer than infinity. In the *hyperopic* (farsighted) eye the eyeball is too short or the cornea is not curved enough, and the image of an infinitely distant object is behind the retina (Fig. 36–6c). The myopic eye produces *too much* convergence in a parallel bundle of rays for an image to be formed on the retina; the hyperopic eye produces *not enough* convergence.

*Astigmatism* refers to a defect in which the surface of the cornea is not spherical but is more sharply curved in one plane than another. As a result, horizontal lines may be imaged in a different plane from vertical lines (Fig. 36–7). Astigmatism may make it impossible, for example, to focus clearly on the horizontal and vertical bars of a window at the same time.

All these defects can be corrected by the use of corrective lenses (eyeglasses or contact lenses). The near point of either a presbyopic or a hyperopic eye is *farther* from the eye than normal. To see clearly an object at normal reading distance (often assumed to be 25 cm), such an eye needs a lens that forms a virtual image of the object at or beyond the near point. This can be accomplished by a converging (positive) lens, as shown in Fig. 36–8. In effect the lens moves the object farther away from the eye to a point where a sharp retinal image can be formed. Similarly, correcting the myopic eye involves using a diverging (negative) lens to move the image closer to the eye than the actual object, as shown in Fig. 36–9.

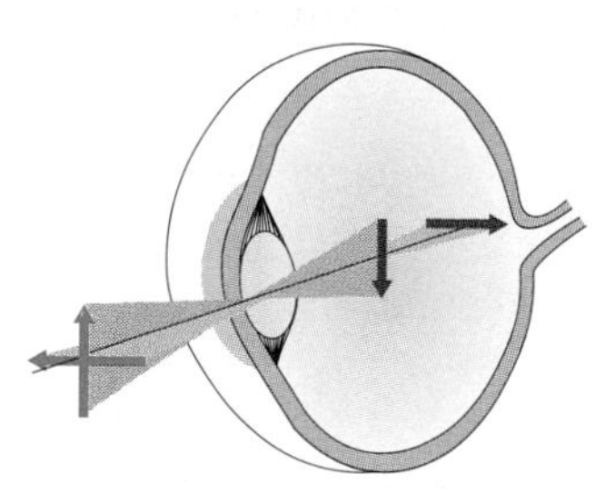

**36–7** Vertical lines are imaged in front of the retina by this astigmatic eye.

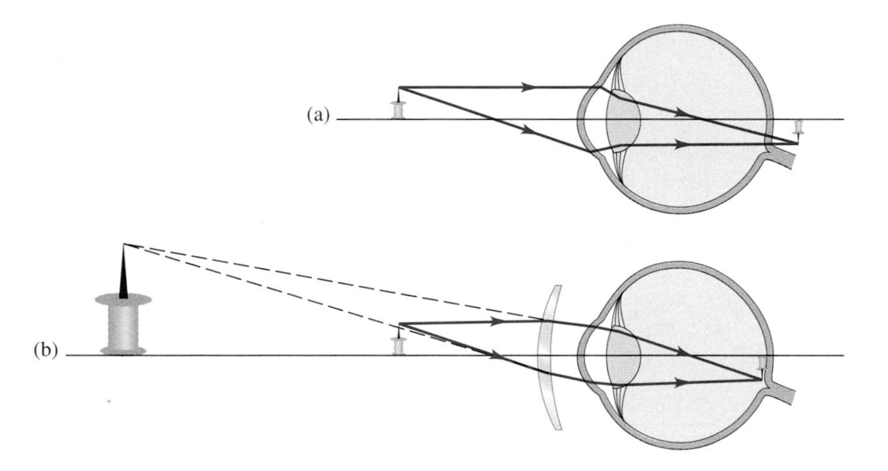


**36–8** (a) An uncorrected hyperopic (farsighted) eye. (b) A positive lens gives the extra convergence needed for a hyperopic eye to focus the image on the retina. The virtual image formed by the converging lens acts as an object at or past the near point.

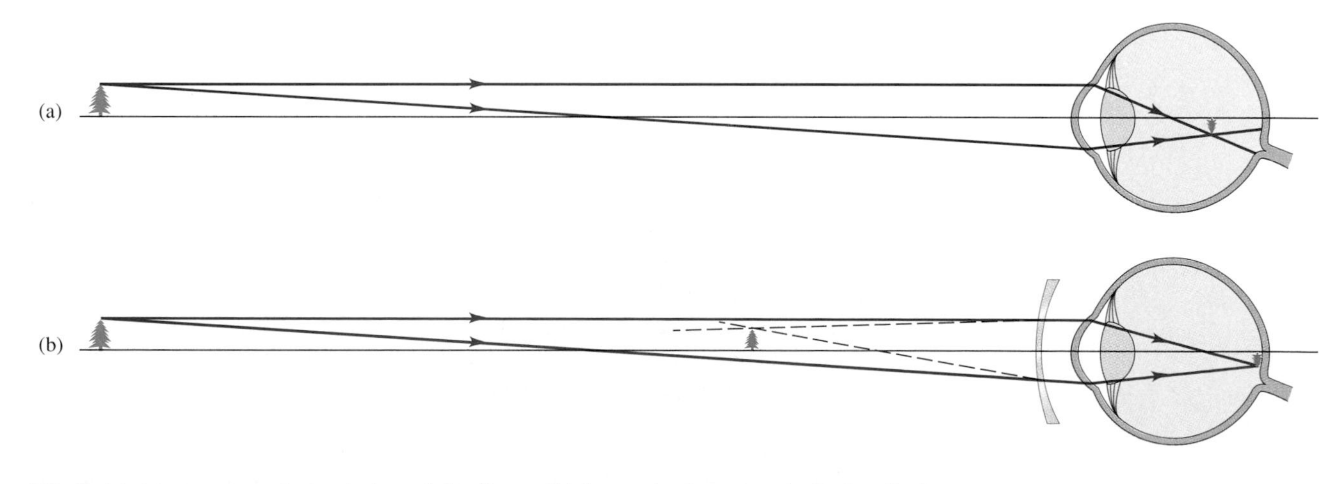


**36–9** (a) An uncorrected myopic (nearsighted) eye. (b) A negative lens spreads the rays farther apart to compensate for the excessive convergence of the myopic eye. The virtual image formed by the diverging lens acts as an object at or inside the far point.

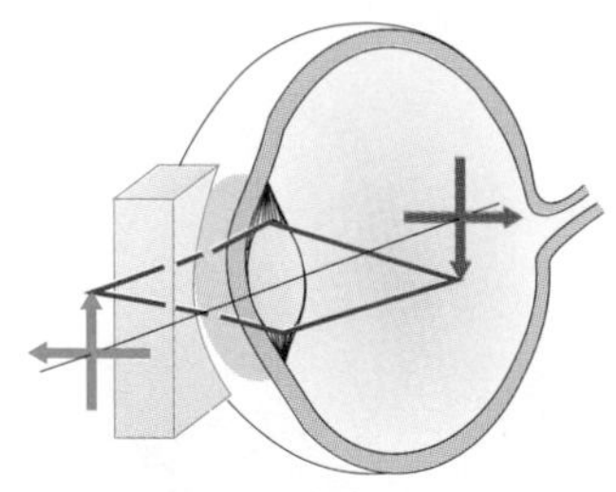

**36–10** The diverging cylindrical lens with horizontal axis separates rays farther apart vertically (but not horizontally) so that the vertical line is also imaged on the retina of this astigmatic eye.

Astigmatism is corrected by use of a lens with a *cylindrical* surface. For example, suppose the curvature of the cornea in a horizontal plane is correct to focus rays from infinity on the retina, but the curvature in the vertical plane is not great enough to form a sharp retinal image. When a cylindrical lens with its axis horizontal is placed before the eye, the rays in a horizontal plane are unaffected, but the additional divergence of the rays in a vertical plane causes these to be sharply imaged on the retina, as shown in Fig. 36–10.

Lenses for vision correction are usually described in terms of the **power,** defined as the reciprocal of the focal length expressed in meters. The unit of power is the **diopter.** Thus a lens with $f = 0.50$ m has a power of 2.0 diopters, $f = -0.25$ m corresponds to $-4.0$ diopters, and so on. The numbers on a prescription for glasses are usually powers expressed in diopters. When the correction involves both astigmatism and myopia or hyperopia, there are three numbers, one for the spherical power, one for the cylindrical power, and an angle to describe the orientation of the cylinder axis.

## EXAMPLE 36–3

**Correcting for farsightedness** The near point of a certain hyperopic eye is 100 cm in front of the eye. What lens should be used to see clearly an object 25 cm in front of the eye?

**SOLUTION** The lens should form a virtual image of the object at a location corresponding to the near point of the eye, 100 cm from it. That is, when $s = 25$ cm, we want $s'$ to be $-100$ cm. From the basic thin-lens equation, Eq. (35–17),

$$\frac{1}{f} = \frac{1}{s} + \frac{1}{s'} = \frac{1}{+25\text{ cm}} + \frac{1}{-100\text{ cm}},$$

$$f = +33\text{ cm}.$$

We need a converging lens with focal length $f = 33$ cm. The corresponding power is $1/0.33$ m, or $+3.0$ diopters.

## EXAMPLE 36–4

**Correcting for nearsightedness.** The far point of a certain myopic eye is 50 cm in front of the eye. What lens should be used to see clearly an object at infinity?

**SOLUTION** The far point of a *myopic* eye is nearer than infinity. To see clearly objects that are beyond the far point, such an eye needs a lens that forms a virtual image of such an object no farther from the eye than the far point. Assume that the virtual image is formed at the far point. Then when $s = \infty$, we want $s'$ to be $-50$ cm. From Eq. (35–17),

$$\frac{1}{f} = \frac{1}{s} + \frac{1}{s'} = \frac{1}{\infty} + \frac{1}{-50\text{ cm}},$$

$$f = -50\text{ cm}.$$

We need a *diverging* lens with focal length $-50$ cm $= -0.50$ m. The power is $-2.0$ diopter.

# 36–4 THE MAGNIFIER

The apparent size of an object is determined by the size of its image on the retina. If the eye is unaided, this size depends upon the *angle* $\theta$ subtended by the object at the eye, called its **angular size** (Fig. 36–11a).

To look closely at a small object, such as an insect or a crystal, you bring it close to your eye, making the subtended angle and the retinal image as large as possible. But your eye cannot focus sharply on objects closer than the near point, so the angular size of an object is greatest (subtends the largest possible viewing angle) when it is placed at the near point. In the following discussion we will assume an average viewer, for whom the near point is 25 cm from the eye.

A converging lens can be used to form a virtual image that is larger and farther from the eye than the object itself, as shown in Fig. 36–11b. Then the object can be moved closer to the eye, and the angular size of the image may be substantially larger than the angular size of the object at 25 cm without the lens. A lens used in this way is called a

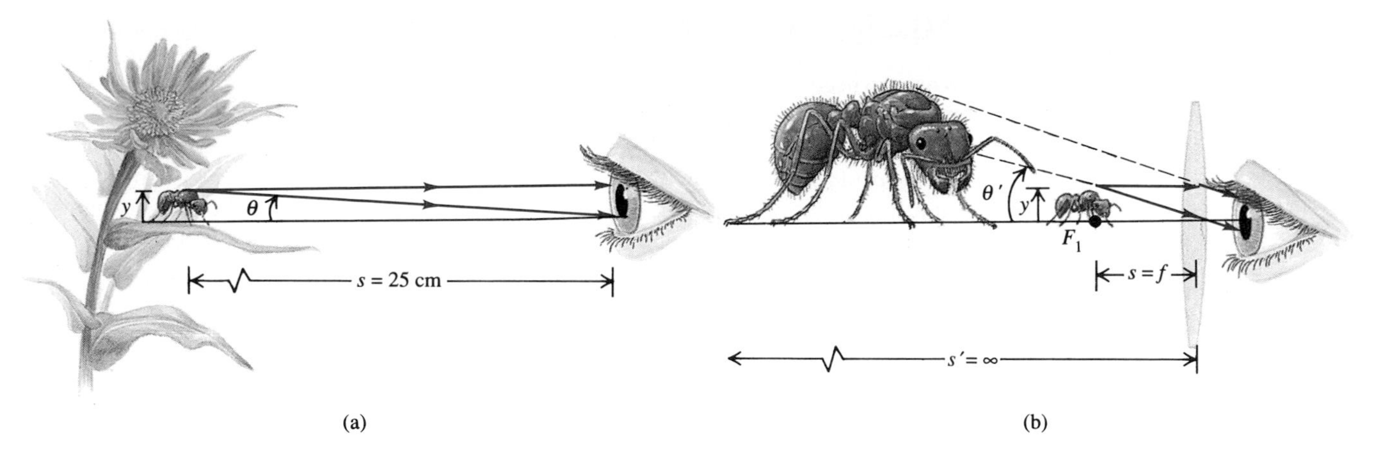


**36–11** (a) The subtended angle $\theta$ (angular size) is largest when the object is at the near point. (b) The simple magnifier gives a virtual image at infinity. This virtual image acts as a real object subtending a larger angle $\theta'$ for the eye.

*magnifying glass,* or simply a **magnifier.** The virtual image is most comfortable to view when it is placed at infinity, and in the following discussion we assume that this is done.

In Fig. 36–11a the object is at the near point, where it subtends an angle $\theta$ at the eye. In Fig. 36–11b a magnifier in front of the eye forms an image at infinity, and the angle subtended at the magnifier is $\theta'$. We define the **angular magnification** $M$ (not to be confused with the *lateral magnification* $m$) as the ratio of the angle $\theta'$ to the angle $\theta$:

$$M = \frac{\theta'}{\theta}. \tag{36–2}$$

To find the value of $M$, we first assume that the angles are small enough that each angle (in radians) is equal to its sine and its tangent. From Fig. 36–11, $\theta$ and $\theta'$ are given (in radians) by

$$\theta = \frac{y}{25\ \text{cm}},$$

$$\theta' = \frac{y}{f}.$$

Combining these expressions with Eq. (36–2), we find

$$M = \frac{\theta'}{\theta} = \frac{y/f}{y/25\ \text{cm}} = \frac{25\ \text{cm}}{f}. \tag{36–3}$$

It may seem that we can make the angular magnification as large as we like by decreasing the focal length $f$. In fact, the aberrations of a simple double-convex lens (to be discussed in Section 36–8) set an upper limit on $M$ of about $3\times$ to $4\times$. If these aberrations are corrected, the angular magnification may be made as great as $20\times$. When greater magnification than this is needed, we usually use a compound microscope, discussed in the next section.

## EXAMPLE 36–5

You have two plastic lenses, one double-convex and the other double-concave, each with a focal length with absolute value 10.0 cm. a) Which lens can you use as a simple magnifier? b) What is the angular magnification?

**SOLUTION** a) The virtual image formation shown in Fig. 36–11 requires a converging lens (positive focal length), so the double-convex lens is the one to use.

b) From Eq. (36–3) the angular magnification $M$ is

$$M = \frac{25\ \text{cm}}{10\ \text{cm}} = 2.5.$$

## 36–5 THE MICROSCOPE

When we need greater magnification than we can get with a simple magnifier, the instrument we usually use is the **microscope,** sometimes called a *compound microscope*. The essential elements of a microscope are shown in Fig. 36–12a. To analyze this system, we use the principle that an image formed by one optical element such as a lens or mirror can serve as the object for a second element. We used this principle in Section 35–5 when we derived the thin-lens equation by repeated application of the single-surface refraction equation.

The object $O$ to be viewed is placed just beyond the first focal point $F_1$ of the **objective,** a lens that forms a real and enlarged image $I$ (Fig. 36–12b). In a properly designed instrument this image lies just inside the first focal point $F_1'$ of the **eyepiece** (also called the *ocular*), which forms a final virtual image of $I$ at $I'$. The position of $I'$ may be anywhere between the near and far points of the eye. Both the objective and eyepiece of an actual microscope are highly corrected compound lenses, but for simplicity we show them here as simple thin lenses.

The objective forms an enlarged, real, inverted image that is viewed through the eyepiece. Thus the overall angular magnification $M$ of the compound microscope is the product of the *lateral* magnification $m_1$ of the objective and the *angular* magnification $M_2$ of the eyepiece. The first is given by

$$m_1 = -\frac{s_1'}{s_1},$$

where $s_1$ and $s_1'$ are the object and image distances, respectively, for the objective. Ordinarily, the object is very close to the focal point, and the resulting image distance $s_1'$ is very great in comparison to the focal length $f_1$ of the objective. Thus $s_1$ is approximately equal to $f_1$, and $m_1 = -s_1'/f_1$.

The eyepiece functions as a simple magnifier, as discussed in Section 36–4. From Eq. (36–3) its angular magnification is $M_2 = (25\text{ cm})/f_2$, where $f_2$ is the focal length of the eyepiece, considered as a simple lens. The overall magnification $M$ of the compound

Eyepiece
Real image
Objective lens
Object
Light source

(a)

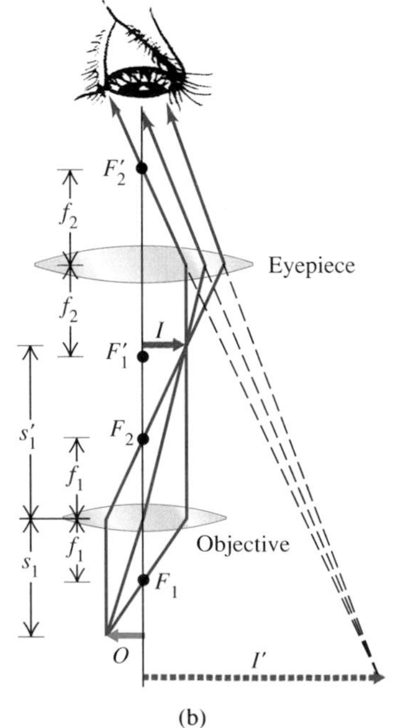


(b)

(c)

**36–12** (a) Elements of a microscope. (b) The objective forms a real, inverted image inside the first focal point $F_1'$ of the eyepiece. The eyepiece uses that image as an object and forms an enlarged virtual image that remains inverted. (c) Typically, optical microscopes can resolve details as small as 200 nm, comparable to the wavelength of light.

microscope (apart from a negative sign, which is customarily ignored) is the product of the two magnifications, that is,

$$M = m_1 M_2 = \frac{(25\text{ cm})s_1'}{f_1 f_2}, \qquad (36\text{–}4)$$

where $s_1'$, $f_1$, and $f_2$ are measured in centimeters. The final image is inverted with respect to the object (Fig. 36–12c). Microscope manufacturers usually specify the values of $m_1$ and $M_2$ for microscope components rather than the focal lengths of the objective and eyepiece.

## 36–6 TELESCOPES

The optical system of a refracting **telescope** is similar to that of a compound microscope. In both instruments the image formed by an objective is viewed through an eyepiece. The difference is that the telescope is used to view large objects at large distances and the microscope is used to view small objects close at hand.

An *astronomical telescope* is shown in Fig. 36–13. The objective forms a real, reduced image $I$ of the object, and the eyepiece forms a virtual, enlarged image of $I$. As with the microscope, the image $I'$ may be formed anywhere between the near and far points of the eye. Objects viewed with a telescope are usually so far away from the instrument that the first image $I$ is formed very nearly at the second focal point of the objective. This image is the object for the eyepiece. If the final image $I'$ formed by the eyepiece is at infinity (for most comfortable viewing by a normal eye), the first image must be at the first focal point of the eyepiece. The distance between objective and eyepiece, which is the length of the telescope, is therefore the *sum* of the focal lengths of objective and eyepiece, $f_1 + f_2$.

The angular magnification $M$ of a telescope is defined as the ratio of the angle subtended at the eye by the final image $I'$, to the angle subtended at the (unaided) eye by the object. We can express this ratio in terms of the focal lengths of objective and eyepiece. In Fig. 36–13 the ray passing through $F_1$, the first focal point of the objective, and through

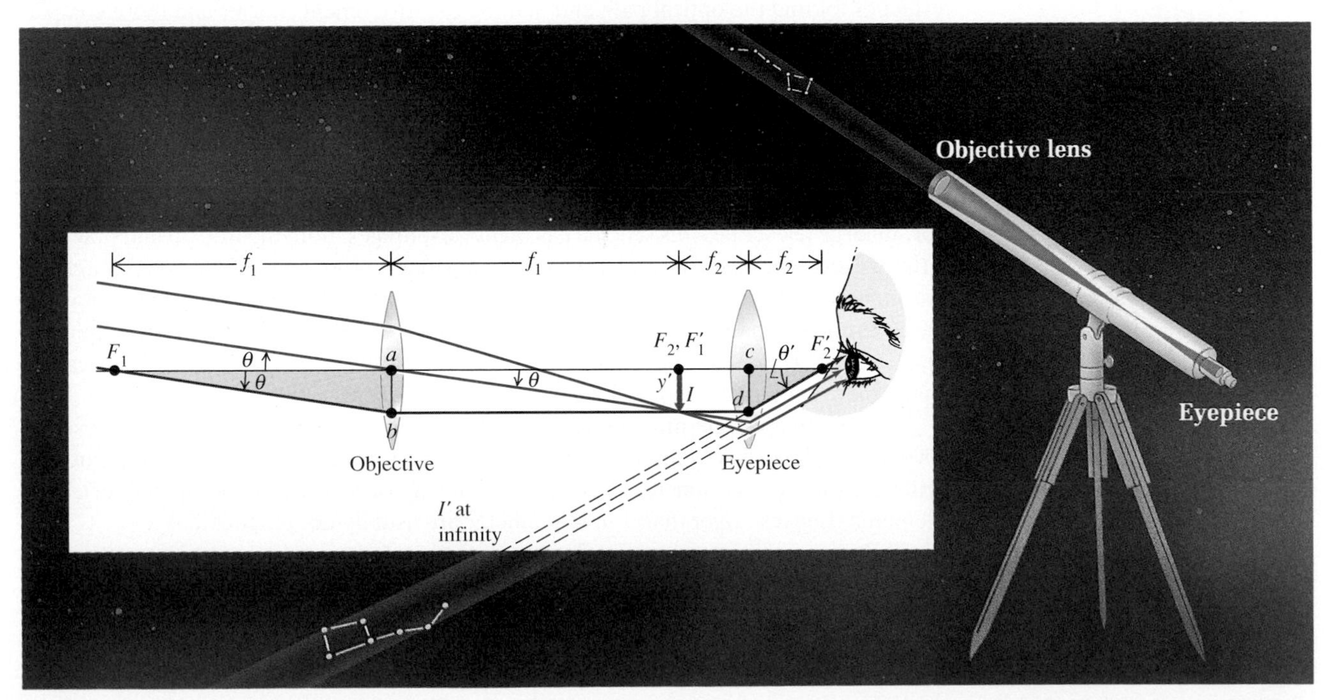


**36–13** Optical system of an astronomical telescope. The objective forms a real, inverted image of the distant object at its second focal point, which is also the first focal point of the eyepiece. The eyepiece uses that image as an object to form a magnified virtual image at infinity that remains inverted.

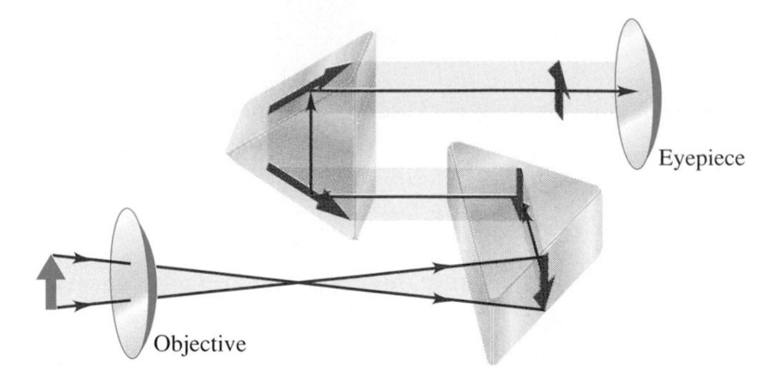


**36–14** Inversion of an image in prism binoculars.

$F_2'$, the second focal point of the eyepiece, has been emphasized. The object subtends an angle $\theta$ at the objective and would subtend essentially the same angle at the unaided eye. Also, since the observer's eye is placed just to the right of the focal point $F_2'$, the angle subtended at the eye by the final image is very nearly equal to the angle $\theta'$. Because $bd$ is parallel to the optic axis, the distances $ab$ and $cd$ are equal to each other and also to the height $y'$ of the image $I$. Because $\theta$ and $\theta'$ are small, they may be approximated by their tangents. From the right triangles $F_1ab$ and $F_2'cd$,

$$\theta = \frac{-y'}{f_1}, \qquad \theta' = \frac{y'}{f_2},$$

and the angular magnification $M$ is

$$M = \frac{\theta'}{\theta} = -\frac{y'/f_2}{y'/f_1} = -\frac{f_1}{f_2}. \tag{36–5}$$

The angular magnification $M$ of a telescope is equal to the ratio of the focal length of the objective to that of the eyepiece. The negative sign shows that the final image is inverted.

An inverted image is no particular disadvantage for astronomical observations. When we use a telescope or binoculars on earth, though, we want the image to be right-side-up. Inversion of the image is accomplished in *prism binoculars* by a pair of 45°-45°-90° totally reflecting prisms called *Porro prisms* (see Fig. 34–8). These are inserted between objective and eyepiece, as shown in Fig. 36–14. The image is inverted by the four internal reflections from the faces of the prisms at 45°. The prisms also have the effect of folding the optical path and making the instrument shorter and more compact than it would otherwise be. Binoculars are usually described by two numbers separated by a multiplication sign, such as $7 \times 50$. The first number is the angular magnification $M$, and the second is the diameter of the objective lenses (in millimeters). The diameter determines the light-gathering capacity of the objective lenses and thus the brightness of the image.

In the *reflecting telescope* (Fig. 36–15) the objective is replaced by a concave mirror. In large telescopes this scheme has many advantages, both theoretical and practical. Mirrors are inherently free of chromatic aberrations (dependence of focal length on wavelength), and spherical aberrations (associated with the paraxial approximation) are easier to correct than for a lens. The reflecting surface is sometimes parabolic rather than spherical. The material of the mirror need not be transparent, and the mirror can be made more rigid than a lens, which can be supported only at its edges.

The largest single-mirror reflecting telescope in the world, at the Crimea Observatory in the USSR, has a mirror 6 m in diameter. The Keck telescope, under construction, will have an overall diameter of 10 m; it is made up of 36 separate hexagonal reflecting elements. Lenses larger than 1 m in diameter are usually not practical.

Because the image in a reflecting telescope is formed in a region traversed by incoming rays, this image can be observed directly with an eyepiece only by blocking off part of the incoming beam (Fig. 36–15a); this is practical only for the very largest telescopes. Alternative schemes use a mirror to reflect the image out the side or through a hole in the mirror, as shown in Figs. 36–15b and 36–15c. This scheme is also used in

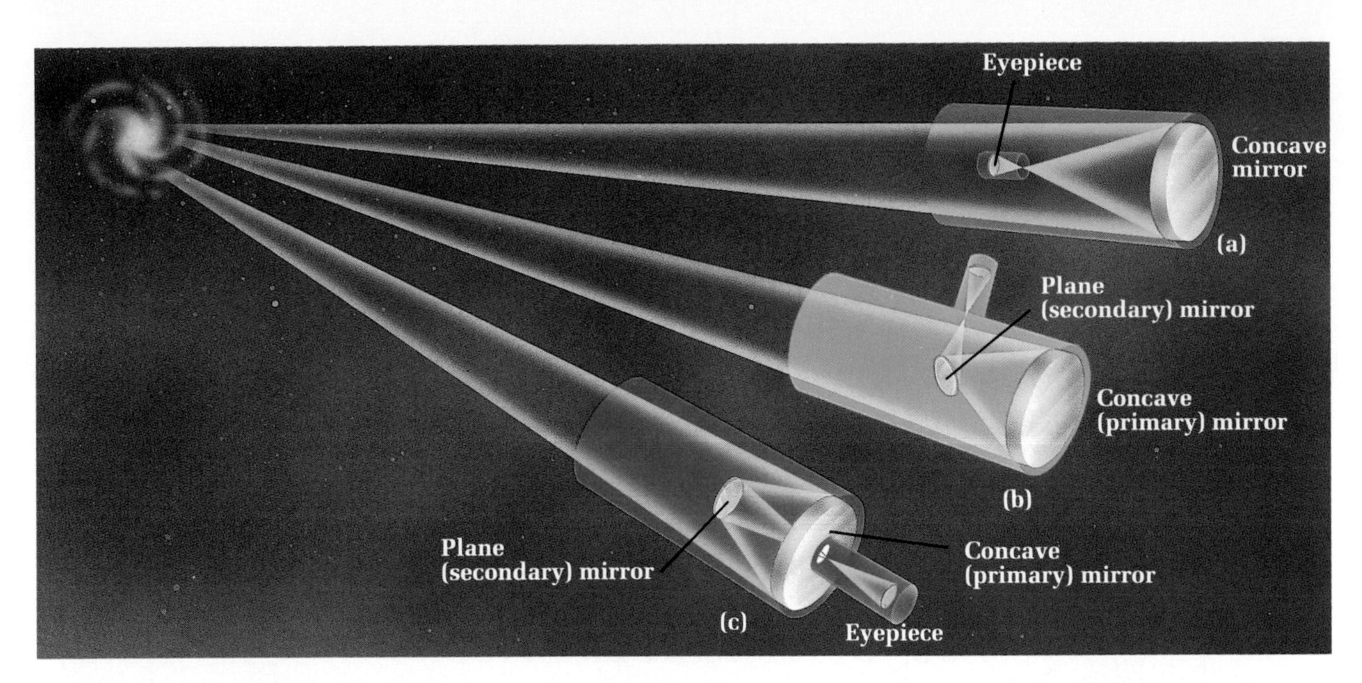


**36–15** Optical systems for reflecting telescopes. (a) The prime focus; (b) the Newtonian focus; (c) the Cassegrainian focus.

some long-focal-length telephoto lenses for cameras. In the context of photography, such a system is called a *catadioptric lens,* an optical system containing both reflecting and refraction elements.

The Hubble Space Telescope, launched on April 25, 1990, has a mirror 2.4 m in diameter, with an optical system similar to Fig. 36–15c. The initial disappointing performance of the Hubble telescope resulted from errors in correcting for spherical aberrations, which we will discuss in Section 36–8.

## *36–7 OTHER LENSES

In this section we mention briefly three special kinds of lenses. First is the *zoom lens,* which is not a single lens but a complex collection of several lens elements that give a continuously variable focal length, often over a range as great as from 10 to 1. Figure 36–16 shows a simple system with variable focal length and a typical zoom lens for a 35-mm camera. Zoom lenses are used with cameras, projectors, microscopes, and telescopes to give a range of image sizes of a given object. It is an enormously complex problem in optical design to keep the image in focus and maintain a constant $f$-number while the focal length changes. Typically, two groups of elements move within the lens, and a diaphragm opens and closes, as the focal length is varied.

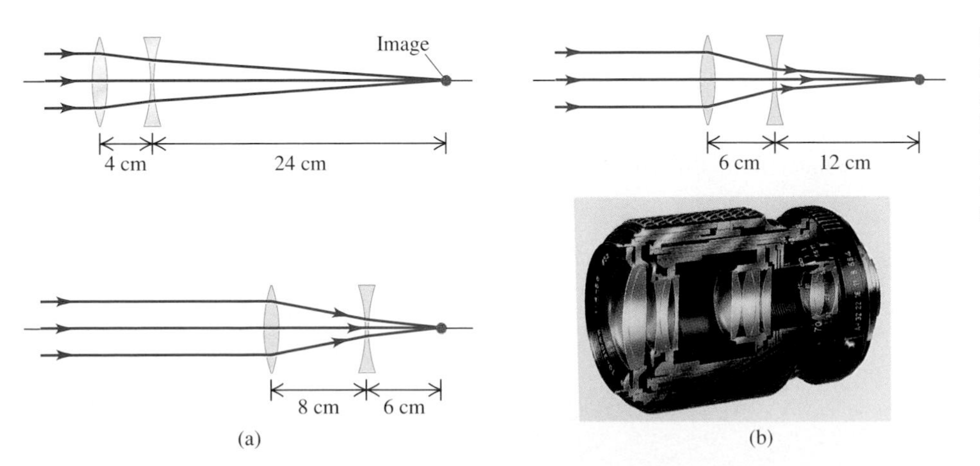


**36–16** (a) Principle of the zoom lens, using two elements with variable spacing. The effective focal length can be varied from 4 to 8 cm. (b) A typical zoom lens for a 35-mm camera, containing twelve elements arranged in four groups.

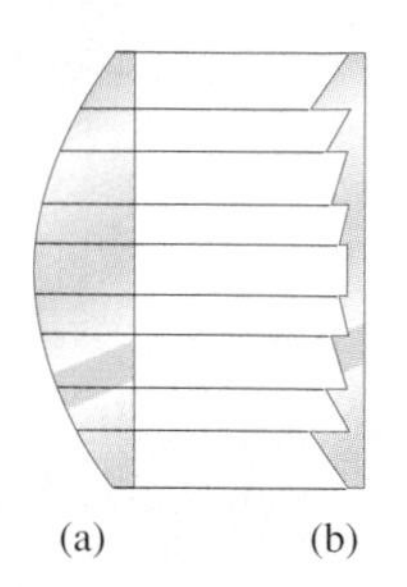


**36–17** Cross sections of (a) a converging ordinary lens; (b) a converging Fresnel lens.

Next is the *Fresnel lens.* The lenses that we've discussed so far work because of the bending of light rays at the surfaces. No refraction occurs in the interior of the lens. If we could eliminate some of the material between the surfaces, we could greatly reduce the weight of a large lens. This is what a Fresnel lens (Fig. 36–17) does. Each circular step copies the contour of the corresponding ring of an ordinary lens. Fresnel lenses don't usually have high optical quality, but they are very light and inexpensive for their size. They are used in traffic lights, overhead projectors, solar-cell light collectors, flat pocket magnifiers, and many other places.

Finally, a *graded-index lens* is a transparent disk with flat faces and a non-uniform index of refraction that is greatest at the center and decreases toward the rim (for a converging lens). Such a lens has a focusing action because Huygens' wavelets don't spread as fast at the center as at the edges. A plane wave front that arrives parallel to the first surface lags at the center as it passes through the disk, bending the rays toward the center (Fig. 36–18).

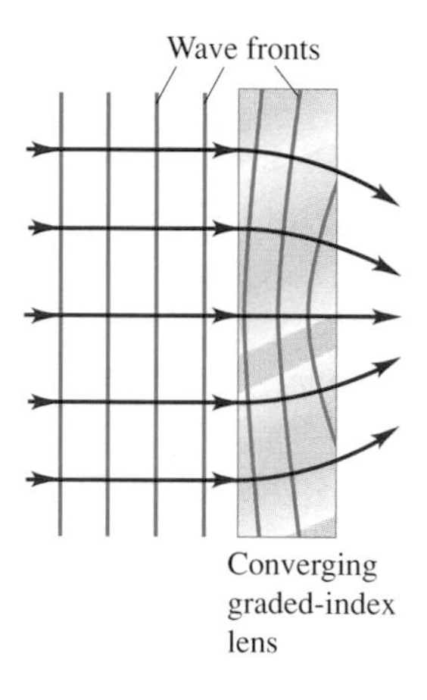


**36–18** For a converging graded-index lens, the index of refraction is greatest at the center.

## *36–8 LENS ABERRATIONS

An **aberration** is any failure of a mirror or lens to behave precisely according to the simple formulas that we have derived. Aberrations can be classified as **chromatic aberrations,** which involve wavelength-dependent imaging behavior, or **monochromatic aberrations,** which occur even with monochromatic (single-wavelength) light. Lens aberrations are not caused by faulty construction of the lens, such as irregularities in its surfaces, but are inevitable consequences of the laws of refraction at spherical surfaces.

Monochromatic aberrations are all related to the *paraxial approximation*. Our derivations of equations for object and image distances, focal lengths, and magnification have all been based upon this approximation. We have assumed that all rays are *paraxial,* that is, that they are very close to the optic axis and make very small angles with it. This condition is never obeyed exactly.

For any lens with an aperture of finite size, the cone of rays that forms the image of any point has a finite size. In general, nonparaxial rays that proceed from a given object point *do not* all intersect at precisely the same point after they are refracted by a lens. For this reason, the image formed by these rays is never perfectly sharp. *Spherical aberration* is the failure of rays from a point object on the optic axis to converge to a point image. Instead, the rays converge within a circle of minimum radius, called *the circle of least confusion,* and then diverge again, as shown in Fig. 36–19. The corresponding effect for points off the optic axis produces images that are comet-shaped figures rather than circles; this effect is called *coma.* Note that decreasing the aperture size cuts off the larger-angle rays, thus decreasing spherical aberration.

Spherical aberration is also present in spherical mirrors. Mirrors used in astronomical reflecting telescopes are usually paraboloidal rather than spherical; this shape completely eliminates spherical aberration for points on the axis. Paraboloidal shapes are much more difficult to fabricate precisely than spherical shapes. The initial

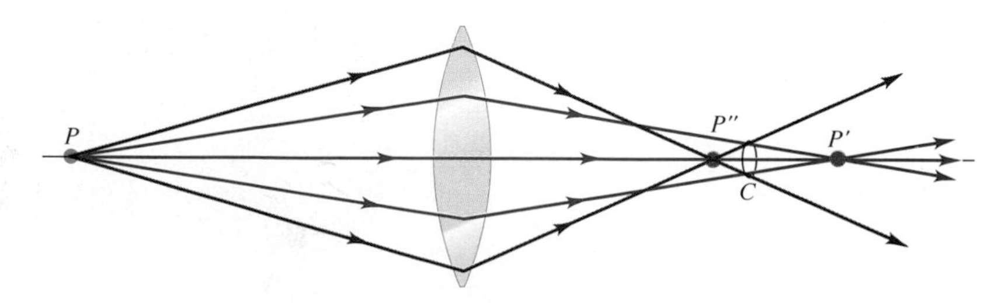


**36–19** Spherical aberration. The circle of least confusion is shown by $C$.

## Multiple-Mirror Telescopes

It is very difficult and costly to manufacture large mirrors and lenses with the accuracy needed in astronomical telescopes. The latest generation of telescopes use computer control to combine the light from several smaller mirrors, obtaining the effective light-gathering power of one much larger surface.

# BUILDING PHYSICAL INTUITION

The Multiple Mirror Telescope built in the late 1970s in Arizona uses six 1.8-m mirrors on a common mounting, collecting light equivalent to that from a single 4.5-m mirror.

The W. M. Keck telescope, presently under construction on Mauna Kea in Hawaii, will use 36 1.8-m hexagonal mirrors, forming the equivalent of a single 10-m mirror. Placement of each mirror will be controlled by computer to within one millionth of an inch. Initial testing in November, 1990, used nine mirrors, shown here, and achieved overall surface accuracy of 20-40 nm.

Radio telescopes can also operate with multiple antennas. Differences in path length from the radio-wave source to the various antennas creates an interference pattern that contains information about the location and features of the source. The Very Large Array in New Mexico consists of 27 parabolic antennas in a Y-shaped array, with each arm of the Y extending up to 21 km.

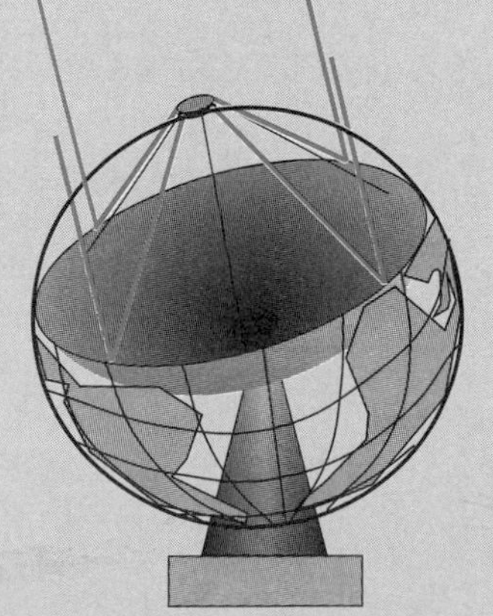

Plans for future arrays of radio antennas, using established radio telescopes around the world, would form an effective antenna almost the size of the earth.

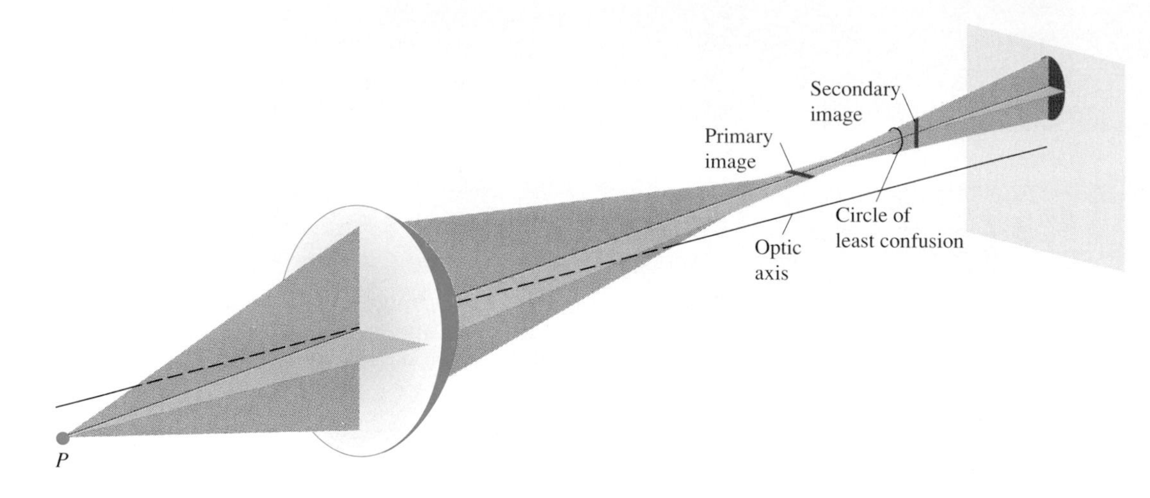


**36–20** Astigmatism of a lens for a point below the optic axis. The lens forms two images of the point, in planes perpendicular to each other.

disappointing results from the Hubble Space Telescope were associated with spherical aberration, the result of errors in measurement during the shaping process. These can be corrected by the use of additional lenses, but this will probably require a visit by a space shuttle.

*Astigmatism* is the imaging of a point off the axis as two perpendicular lines in different planes. In this aberration the rays from a point object converge, at a certain distance from the lens, to a line called the *primary image,* perpendicular to the plane defined by the optic axis and the object point. At a somewhat different distance from the lens they converge to a second line called the *secondary image,* which is *parallel* to that plane. This effect is shown in Fig. 36–20. The circle of least confusion (greatest convergence) appears between these two positions.

The location of the circle of least confusion depends on the object point's *transverse* distance from the axis as well as its *longitudinal* distance from the lens. As a result, object points lying in a plane are in general imaged not in a plane but in some curved surface; this effect is called *curvature of field.*

Finally, the image of a straight line that does not pass through the optic axis may be curved. As a result, the image of a square with the axis through its center may resemble a barrel (sides bent outward) or a pincushion (sides bent inward). This effect, called *distortion,* is not related to lack of sharpness of the image but results from a change in lateral magnification with distance from the axis.

*Chromatic aberrations* are a result of *dispersion,* the variation of index of refraction with wavelength that we discussed in Section 34–4. Dispersion causes the focal length of a lens to be somewhat different for different wavelengths. When an object is illuminated with light containing a mixture of wavelengths, different wavelengths are imaged at different points. The magnification of a lens also varies with wavelength; this effect is responsible for the rainbow-fringed images that are seen with inexpensive binoculars or telescopes. Reflectors are inherently free of chromatic aberrations; this is one of the reasons for their usefulness in large astronomical telescopes.

It is impossible to eliminate all these aberrations from a single lens, but in a compound lens with several optical elements the aberrations of one element may partially cancel those of another. Design of such lenses is an extremely complex problem, aided greatly in recent years by the use of computers. It is still impossible to eliminate all aberrations, but it *is* possible to decide which ones are most troublesome for a particular application and to design accordingly.

## EXAMPLE 36–6

**Chromatic aberration** A glass plano-convex lens has its flat side toward the object, and the other side has a radius of curvature of 30.0 cm. The index of refraction of the glass for violet light (wavelength 400 nm) is 1.537, and for red light (700 nm) it is 1.517. The color purple is a mixture of red and violet. A purple object is placed 80.0 cm from this lens. Where are the red and violet images formed?

**SOLUTION** We use the thin-lens equation in the form given by Eq. (35–19):

$$\frac{1}{s}+\frac{1}{s'}=(n-1)\left(\frac{1}{R_1}-\frac{1}{R_2}\right).$$

In this case, using the usual sign rules, we have $R_1=\infty$ and $R_2=-30.0$ cm. For violet light ($n=1.537$),

$$\frac{1}{80.0\text{ cm}}+\frac{1}{s'}=(1.537-1)\left(\frac{1}{\infty}-\frac{1}{-30.0\text{ cm}}\right),$$

$$s'=185\text{ cm}.$$

For red light ($n=1.517$) we find $s'=211$ cm. The violet light is refracted more than the red, and its image is formed closer to the lens. We see that a rather small variation in index of refraction causes a substantial displacement of the image.

## SUMMARY

- A camera forms a real, inverted, usually reduced image (on film) of the object being photographed. The amount of light striking the film is controlled by the shutter speed and the aperture. A projector is essentially a camera in reverse; a lens forms a real, inverted, enlarged image on a screen of the slide or motion-picture film.
- In the eye a real image is formed on the retina and transmitted to the optic nerve. Adjustment for various object distances is made by the ciliary muscle, which squeezes the crystalline lens, making it bulge and decreasing its focal length. A nearsighted eye is too long for its lens; a farsighted eye is too short. The power of a corrective lens, in diopters, is the reciprocal of the focal length, in meters.
- A simple magnifier creates a virtual image whose angular size is larger than that of the object itself at a distance of 25 cm, the nominal closest distance for comfortable viewing. The angular magnification is the ratio of the angular size of the virtual image to that of the object at this distance.
- In a compound microscope the objective forms a first image in the barrel of the instrument, and the eyepiece forms a final virtual image, often at infinity, of the first image. A telescope operates on the same principle, but the object is far away. In a reflecting telescope the objective is replaced by a concave mirror, which eliminates chromatic aberrations.
- Lens aberrations account for the failure of a lens to form a perfectly sharp image of an object. Monochromatic aberrations occur because of limitations of the paraxial approximation; chromatic aberrations result from the dependence of index of refraction on wavelength.

### KEY TERMS

camera
projector
power
diopter
angular size
magnifier
angular magnification
microscope
objective
eyepiece
telescope
aberration
chromatic aberration
monochromatic aberration

## EXERCISES

### Section 36–1 The Camera

**36–1** A camera lens has a focal length of 135 mm. How far from the lens should the subject for the photo be if the lens is 15.2 cm from the film?

**36–2 Photographing the Moon.** During a lunar eclipse, a picture of the moon (diameter $3.48\times10^6$ m, distance from earth $3.86\times10^8$ m) is taken with a camera whose lens has a focal length of 200 mm. What is the diameter of the image on the film?

**36–3 Choosing a Camera Lens.** The picture size on ordinary 35-mm camera film is 24 × 36 mm. Focal lengths of lenses available for 35-mm cameras typically include 28, 35, 50 (the "standard" lens), 85, 100, 135, 200, and 300 mm, among others. Which of these lenses should be used to photograph the following objects, assuming that the image is to fill most of the picture area? a) A cathedral 100 m tall and 150 m long at a distance of 150 m. b) An eagle with a wingspan of 2.0 m, at a distance of 12.0 m.

**36–4** When a camera is focused, the lens is moved away from or toward the film. If you take a picture of your friend, who is standing 6.00 m from the lens, using a camera with a lens with a 50-mm focal length, how far from the film is the lens? Will the whole image of your friend, who is 175 cm tall, fit on film that is 24 × 36 mm?

**36–5** The focal length of an $f/2.8$ camera lens is 90.0 mm. a) What is the aperture diameter of the lens? b) If the correct exposure of a certain scene is $\frac{1}{60}$ s at $f/2.8$, what is the correct exposure at $f/5.6$?

**36–6** Camera A, having an $f/8$ lens with an aperture diameter of 2.50 cm, photographs an object using the correct exposure time of $\frac{1}{40}$ s. What exposure time should be used with camera B in photographing the same object if this camera has an $f/4$ lens with an aperture diameter of 5.00 cm?

## Section 36–2 The Projector

**36–7** In Example 36–2 for a lens with a 98.0-mm focal length it was found that the slide needs to be placed 10.0 cm in front of the lens to focus the image on a screen 5.00 m from the lens. Now assume instead that a lens of 100-mm focal length is used. a) If the screen remains 5.00 m from the lens, how far in front of the lens should the slide be? b) Could the distance between the slide and lens be kept at 10.0 cm and the screen moved to achieve focus? If so, how far and in what direction would it have to be moved?

**36–8** The dimensions of the picture on a 35-mm color slide are 24 × 36 mm. An image of the slide is projected on a screen 6.00 m from the projector lens. The focal length of the projector lens is 120 mm. a) How far is the slide from the lens? b) What are the dimensions of the image on the screen?

## Section 36–3 The Eye

***36–9** a) Where is the near point of an eye for which a lens with a power of +3.00 diopters is prescribed? b) Where is the far point of an eye for which a lens with a power of −0.600 diopter is prescribed for distant vision?

***36–10 Curvature of the Cornea.** In a simplified model of the human eye, the aqueous and vitreous humors and the lens all have a refractive index of 1.40, and all the refraction occurs at the cornea, whose vertex is 2.60 cm from the retina. What should the radius of curvature of the cornea be in order to focus on the retina the image of an infinitely distant object?

***36–11** For the model of the eye described in Exercise 36–10, what should the radius of curvature of the cornea be in order to focus on the retina the image of an object that is a distance of 25.0 cm from the cornea's vertex?

***36–12** Determine the power of the corrective lenses required by a) a hyperopic eye whose near point is at 80.0 cm; b) a myopic eye whose far point is at 90.0 cm.

## Section 36–4 The Magnifier

**36–13** You are examining a flea with a magnifying lens that has focal length 3.00 cm. If the image of the flea is ten times the size of the flea, how far is the flea from the lens? Where, relative to the lens, is the image?

**36–14** A thin lens with a focal length of 8.00 cm is used as a simple magnifier. a) What maximum angular magnification is obtainable with this lens? b) When an object is examined through the lens, how close may it be brought to the lens? Assume that the image is at the near point, 25.0 cm from the eye.

**36–15** The focal length of a simple magnifier is 12.0 cm. Assume the magnifier to be a thin lens placed very close to the eye. a) How far in front of the magnifier should an object be placed if the image is formed at the observer's near point, 25.0 cm in front of her eye? b) If the object is 1.00 mm high, what is the height of its image formed by the magnifier?

## Section 36–5 The Microscope

**36–16** A certain microscope is provided with objectives that have focal lengths of 16 mm, 4 mm, and 1.9 mm and with eyepieces that have angular magnifications of 5 × and 10 × . Each objective forms an image 140 mm beyond its second focal point. Determine a) the largest overall magnification obtainable; b) the least overall magnification obtainable.

**36–17** The focal length of the eyepiece of a certain microscope is 2.50 cm. The focal length of the objective is 16.0 mm. The distance between objective and eyepiece is 21.4 cm. The final image formed by the eyepiece is at infinity. Treat all lenses as thin. a) What is the distance from the objective to the object being viewed? b) What is the magnitude of the linear magnification produced by the objective? c) What is the overall magnification of the microscope?

**36–18 Resolution of a Microscope.** The image formed by a microscope objective with a focal length of 4.00 mm is 180 mm from its second focal point. The eyepiece has a focal length of 28.0 mm. a) What is the magnification of the microscope? b) The unaided eye can distinguish two points at its near point as separate if they are about 0.10 mm apart. What is the minimum separation that can be resolved with this microscope?

## Section 36–6 Telescopes

**36–19** A telescope is made from two eyeglass lenses with focal lengths of 90.0 cm and 20.0 cm, the 90.0-cm lens being used as the objective. Both the object being viewed and the final image are at infinity. a) Find the angular magnification for the telescope. b) Find the height of the image formed by the objective of a building 80.0 m tall, 2.00 km away. c) What is the angular size of the final image as viewed by an eye very close to the eyepiece?

**36–20** The eyepiece of a refracting telescope (Fig. 36–13) has a focal length of 8.00 cm. The distance between objective and eyepiece is 2.20 m, and the final image is at infinity. What is the angular magnification of the telescope?

**36–21** Viewed from the earth the moon subtends an angle of approximately $\frac{1}{2}°$. What is the diameter of the image of the moon produced by the objective of the Lick Observatory telescope, a refractor having a focal length of 18.0 m?

**36–22** A reflecting telescope (Fig. 36–21a) is made by using a spherical mirror with a radius of curvature of 0.800 m and an eyepiece with a focal length of 1.00 cm. The final image is at infinity. a) What is the angular magnification? b) What is the distance between the eyepiece and the mirror's vertex if the object is taken to be at infinity?

**36–23** The Hubble Space Telescope uses an optical system similar to that shown in Fig. 36–21b (Cassegrain system). The image of a far distant object is focused on the detector through a hole in the large (primary) mirror. The primary mirror has a focal length of 2.4 m, the distance between the vertexes of the two mirrors is 1.5 m, and the distance from the vertex of the primary mirror to the detector is 0.25 m. Should the smaller mirror (the secondary) be concave or convex? What should its radius of curvature be?

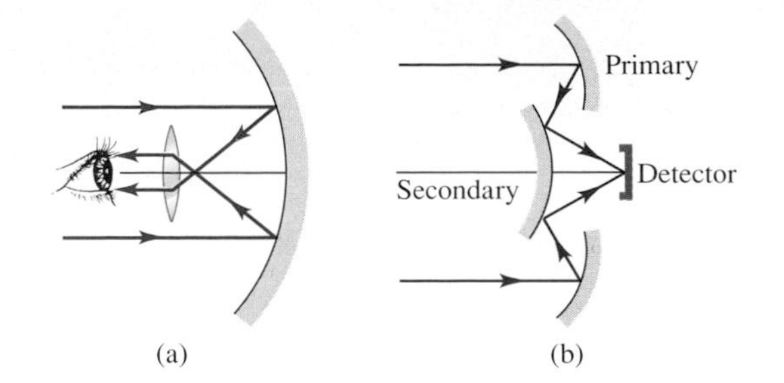


**FIGURE 36–21**

### Section 36–8 Lens Abberations

***36–24** A diverging double-concave lens (Fig. 35–28b) is ground from the glass of Example 36–6. The two radii of curvature each have magnitude 15.0 cm. The purple object of Example 36–6 is placed 40.0 cm to the left of the lens. Where are the red and violet images formed, and what is the distance between these two images?

## PROBLEMS

**36–25** Your camera has a lens with a 50.0-mm focal length and a viewfinder that is 20.0 mm (tall) by 30.0 mm (long). In taking a picture of a 3.20-m-long automobile you find that the image of the auto fills only two-thirds of the viewfinder. a) How far are you from the auto? b) How close should you stand if you want to fill the viewfinder frame with the auto's image?

**36–26 Resolution of a Camera.** The *resolution* of a camera lens can be defined as the maximum number of lines per millimeter in the image that can barely be distinguished as separate lines. A certain lens has a focal length of 50.0 mm and resolution of 100 lines/mm. What is the minimum separation of two lines 60.0 m away if they are to be visible in the image as separate lines?

***36–27** One form of cataract surgery is removal of the person's natural lens, which has become cloudy, and its replacement by an artificial one. The refracting properties of the replacement lens can be chosen so that the person's eye focuses distant objects. But there is no accommodation, and glasses or contact lenses are needed for close vision. What is the power, in diopters, of the corrective lenses that will enable a person who has had such surgery to focus on the page of a book at a distance of 36 cm?

***36–28** a) Show that when two thin lenses are placed in contact, the *power* of the combination in diopters, as defined in Section 36–3, is the sum of the powers of the separate lenses. Is this relation valid even when one lens has positive power and the other negative? b) Two thin converging lenses with 15.0-cm and 40.0-cm focal lengths are in contact. What is the power of the combination?

***36–29 A Nearsighted Eye.** A certain very nearsighted person cannot focus anything farther than 30.0 cm from the eye. Consider the simplified model of the eye described in Exercise 36–10. If the radius of curvature of the cornea is 0.75 cm when the eye is focusing on an object 30.0 cm from the cornea's vertex and the indexes of refraction are as described in Exercise 36–10, what is the distance from the cornea's vertex to the retina? What does this tell you about the shape of the nearsighted eye?

**36–30** A certain reflecting telescope, constructed as in Fig. 36–21a, has a mirror 10.0 cm in diameter with a radius of curvature of 0.800 m and an eyepiece with a focal length of 1.00 cm. If the angular magnification has a magnitude of 36 and the object is at infinity, find the position of the lens and the position and nature (real or virtual) of the final image. (*Note:* $|M|$ is *not* equal to $|f_1/f_2|$, so the image formed by the eyepiece is *not* at infinity.)

**36–31** A microscope with an objective that has a focal length of 9.00 mm and an eyepiece that has a focal length of 6.00 cm is used to project an image on a screen 1.00 m from the eyepiece. Let the image distance of the objective be 18.0 cm. a) What is the lateral magnification of the image? b) What is the distance between the objective and the eyepiece?

**36–32 The Galilean Telescope.** Figure 36–22 is a diagram of a *Galilean telescope,* or *opera glass,* with both the object and its final image at infinity. The image $I$ serves as a virtual object for the eyepiece. The final image is virtual and erect. a) Prove that the angular magnification is $M = -f_1/f_2$. b) A Galilean telescope is to be constructed with the same objective lens as in Exercise 36–19. What focal length should the eyepiece have if this telescope is to have the same magnification as the one in Exercise 36–19? c) Compare the lengths of the telescopes.

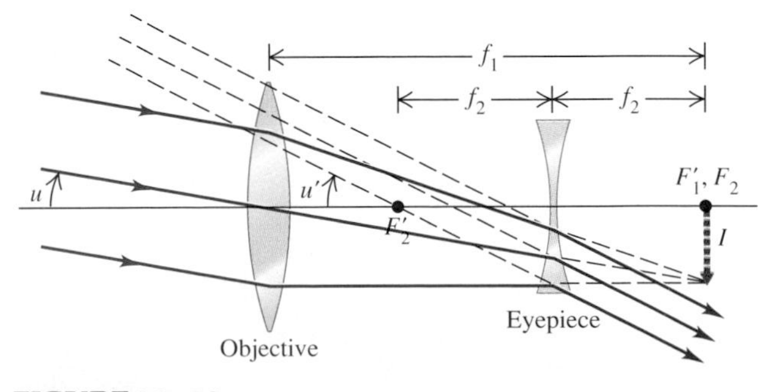


**FIGURE 36–22**

The colors reflected from a soap bubble result from interference of light reflected off the inner and outer surfaces of the bubble. What color you see in a particular spot is determined by the thickness of the bubble at that spot.

C H A P T E R 37

# Interference

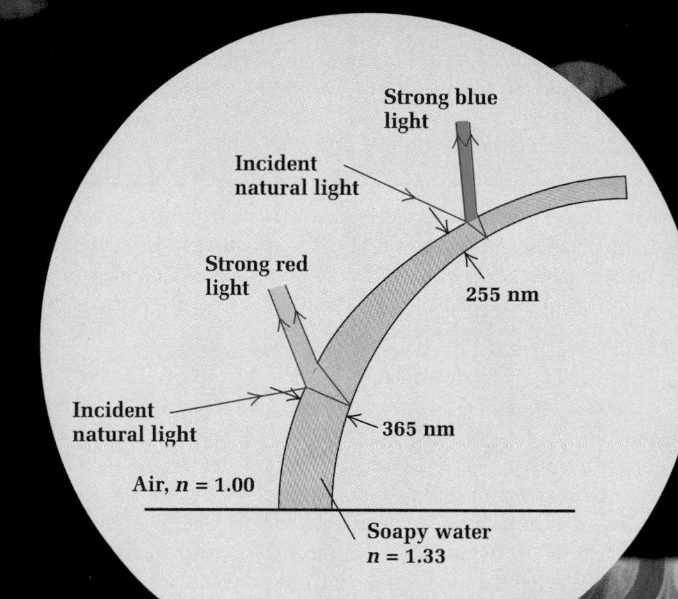


If we assume an index of refraction of 1.33 for soapy water, then a bubble thickness of 255 nm reflects blue light strongly and a bubble thickness of 365 nm reflects red light strongly. Thicknesses between these values give strong reflections for other colors. The result is a beautiful display of all the colors of the spectrum.

## KEY CONCEPTS

**• When waves from two identical sources overlap, the phase difference caused by the difference in path length causes the resultant wave amplitude to be greater in some directions and smaller in others than the amplitude of each wave. This effect is called interference. When the distance from each source is large in comparison to the distance between them, the directions of the maxima and minima are described by simple equations.**

**• The intensity in any direction in an interference pattern can be calculated by using an analysis based on phasor diagrams.**

**• Interference effects occur in reflections from thin films as a result of the path difference between waves reflected from the two surfaces.**

**• The Michelson interferometer permits high-precision measurements of wavelengths of light. It also played an important role in the experimental foundations of the special theory of relativity.**

**• In some situations, light shows particle properties. Its energy is carried in bundles called photons; the energy of each photon is proportional to the frequency of the light.**

## INTRODUCTION

If you've ever blown soap bubbles, you know that part of the fun is watching the multicolored reflections from the bubbles. An ugly black oil spot on the pavement can become a thing of beauty after a rain, when it reflects a rainbow of colors. These familiar sights give us a hint that there are aspects of light that we haven't yet explored. In our discussion of lenses, mirrors, and optical instruments, we have used the model of *geometric optics,* which represents light as *rays,* straight lines that are bent at a reflecting or refracting surface.

But many aspects of the behavior of light (including colors in soap bubbles and oil films) *can't* be understood on the basis of rays. We have already learned that light is fundamentally a *wave,* and in some situations we have to consider its wave properties explicitly. In this chapter and the next we will study *interference* and *diffraction* phenomena. When several waves overlap at a point, their total effect depends on their *phases* as well as their amplitudes. When light passes through apertures or around obstacles, the patterns that are formed are a result of the *wave* nature of light; they cannot be understood on the basis of rays. Such effects are grouped under the heading *physical optics.* We will look at several practical applications of physical optics, including diffraction gratings, x-ray diffraction, and holography. ■

## 37–1 INTERFERENCE AND COHERENT SOURCES

In our discussions of mechanical waves in Chapter 19 and electromagnetic waves in Chapter 33 we often talked about *sinusoidal* waves with a single frequency and a single wavelength. In optics, such a wave is characteristic of **monochromatic light** (light of a single color). Common sources of light, such as an incandescent light bulb or a flame, *do not* emit monochromatic light but rather a continuous distribution of wavelengths.

A precisely monochromatic light wave is an idealization, but monochromatic light can be *approximated* in the laboratory. For example, some filters block all but a very narrow range of wavelengths. Gas-discharge lamps, such as a mercury-vapor lamp, emit light with a discrete set of colors, each having a narrow band of wavelengths. The bright green line in the spectrum of a mercury-vapor lamp has a wavelength of about 546.1 nm, with a spread of the order of $\pm 0.001$ nm. By far the most nearly monochromatic source available at present is the *laser.* The familiar helium-neon laser, inexpensive and readily available, emits red light at 632.8 nm with a line width (wavelength range) of the order of $\pm 0.000001$ nm, or about 1 part in $10^9$. As we analyze interference and diffraction effects in this chapter and the next, we will often assume that we are working with monochromatic light.

The term **interference** refers to any situation in which two or more waves overlap in space. When this occurs, the total displacement at any point at any instant of time is governed by the **principle of linear superposition.** This is the most important principle in all of physical optics, so make sure you understand it well. It states: **When two or more waves overlap, the resultant displacement at any point and at any instant may be found by adding the instantaneous displacements that would be produced at the point by the individual waves if each were present alone.**

We use the term *displacement* in a general sense. For waves on the surface of a liquid, we mean the actual displacement of the surface above or below its normal level. For sound waves, the term refers to the excess or deficiency of pressure. For electromagnetic waves, we usually mean a specific component of electric or magnetic field.

To introduce the essential ideas of interference, let's first consider two identical sources of monochromatic waves separated in space by a certain distance. The two sources are permanently *in phase;* they vibrate in unison. They might be two agitators in a ripple tank, two loudspeakers driven by the same amplifier, two radio antennas powered by the same transmitter, or two small holes or slits in an opaque screen illuminated by the same monochromatic light source.

We position the sources at points $S_1$ and $S_2$ along the $y$-axis, equidistant from the origin, as shown in Fig. 37–1a. Let $P_0$ be any point on the $x$-axis. By symmetry the two distances from $S_1$ to $P_0$ and from $S_2$ to $P_0$ are *equal;* waves from the two sources thus require equal times to travel to $P_0$. Waves that leave $S_1$ and $S_2$ in phase arrive at $P_0$ in phase. The two waves add, and the total amplitude at $P_0$ is *twice* the amplitude of each individual wave.

Similarly, the distance from $S_2$ to point $P_1$ is exactly one wavelength *greater* than the distance from $S_1$ to $P_1$. A wave crest from $S_1$ arrives at $P_1$ exactly one cycle earlier than a crest emitted at the same time from $S_2$, and again the two waves arrive *in phase.* For point $P_2$ the path difference is *two* wavelengths, and again the two waves arrive in phase, and so on.

The addition of amplitudes that results when waves from two or more sources arrive at a point *in phase* is called **constructive interference,** or *reinforcement* (Fig. 37–1b). Let the distance from $S_1$ to any point $P$ be $r_1$ and the distance from $S_2$ to $P$ be $r_2$. Then the condition that must be satisfied for constructive interference to occur at $P$ is that

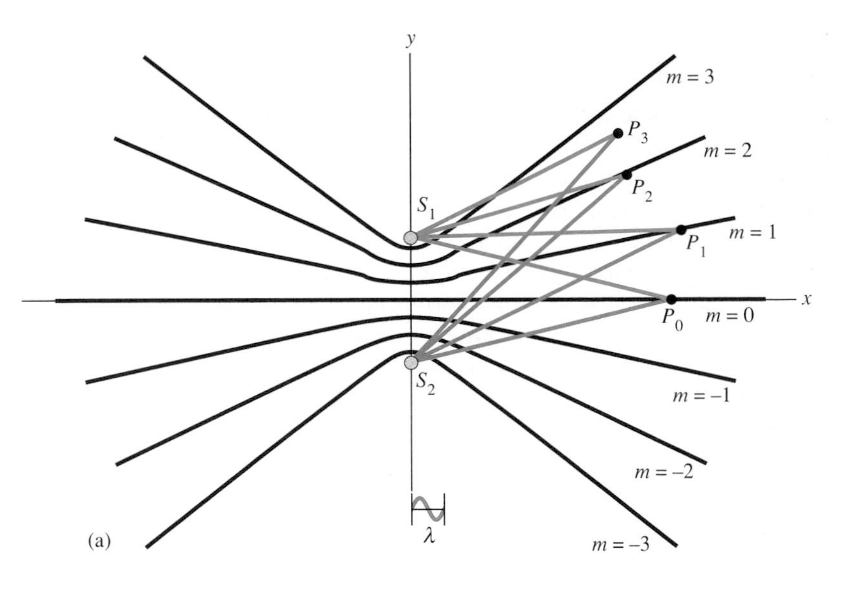


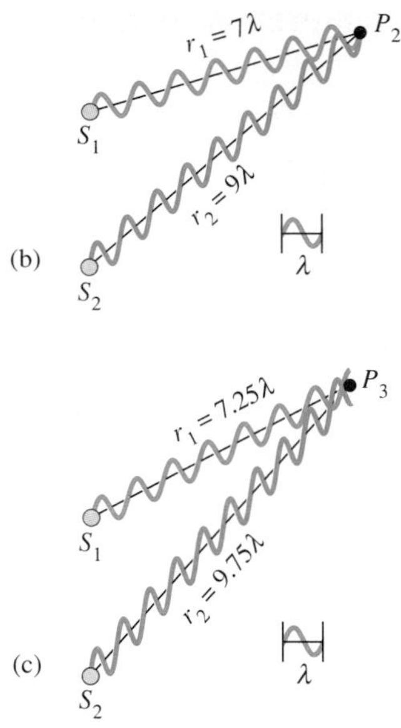


**37–1** Sources in phase. (a) Curves of maximum intensity (constructive interference) of waves from two monochromatic sources. In this example the distance between sources is $4\lambda$. (b) Constructive interference occurs when the path difference is a whole number of wavelengths, $m\lambda$. (c) Destructive interference occurs when the path difference is a half-integral number of wavelengths, $(m + \frac{1}{2})\lambda$.

the path difference $r_2 - r_1$ for the two sources must be an integral multiple of the wavelength $\lambda$:

$$r_2 - r_1 = m\lambda \quad (m = 0, \pm 1, \pm 2, \pm 3, \ldots). \tag{37–1}$$

For our example, all points satisfying this condition lie on the curves shown in Fig. 37–1a.

Intermediate between these curves is another set of curves for which the path difference for the two sources is a *half-integral* number of wavelengths. Waves from the two sources arrive at a point on one of these lines (such as point $P_3$ in Fig. 37–1a) exactly a half-cycle out of phase. A crest of one wave arrives at the same time as a crest in the opposite direction from the other wave (Fig. 37–1c), and the resultant amplitude is the *difference* between the two individual amplitudes. If the individual amplitudes are equal, then the *total* amplitude is zero! This condition is called **destructive interference,** or *cancellation*. For our example, the condition for destructive interference is

$$r_2 - r_1 = (m + \tfrac{1}{2})\lambda \quad (m = 0, \pm 1, \pm 2, \pm 3, \ldots). \tag{37–2}$$

Figures 37–1b and 37–1c suggest the phase relationships for constructive and destructive interference of two waves.

An example of this interference pattern is the familiar ripple-tank pattern shown in Fig. 37–2. The two wave sources are two agitators driven by the same vibrating mechanism. The regions of both maximum and zero amplitude are clearly visible. The superimposed color lines, like those in Fig. 37–1a, show lines of maximum amplitude.

A ripple tank is an inherently two-dimensional situation, but in some cases, such as with two loudspeakers or two radio-transmitter antennas, the pattern is three-dimensional. Think of rotating the color curves of Fig. 37–1a around the $y$-axis; then maximum constructive interference occurs at all points on the resulting surfaces of revolution.

For Eqs. (37–1) and (37–2) to hold, the two sources must *always* be in phase. When the wave is light, there is no practical way to achieve such a relationship with two separate sources because of the way light is emitted. In ordinary light sources, atoms gain excess energy by thermal agitation or by impact with accelerated electrons. An atom

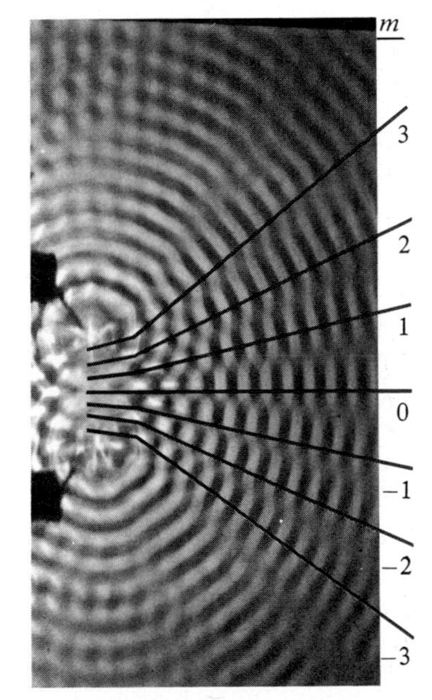


**37–2** Photograph of an interference pattern produced by water waves in a shallow ripple tank. The two wave sources are small balls moved up and down by the same vibrating mechanism. As waves move outward from the sources, they overlap and produce an interference pattern. The lines of maximum amplitude in the pattern are shown by superimposed red lines.

thus "excited" begins to radiate energy and continues until it has lost all the energy it can, typically in a time of the order of $10^{-8}$ s. The many atoms in a source ordinarily radiate in an unsynchronized and random phase relationship, and the light emitted from *two* such sources has no definite phase relation.

However, the light from a single source can be split so that parts of it emerge from two or more regions of space, forming two or more *secondary sources*. Then any random phase change in the source affects these secondary sources equally and does not change their *relative* phase. Light from two such sources, derived from a single primary source and with a definite, constant phase relation, is said to be **coherent.**

The distinguishing feature of light from a *laser* is that the emission of light from many atoms is *synchronized* in frequency and phase. As a result, the random phase changes mentioned above occur *much* less frequently. Definite phase relations are preserved over correspondingly much greater lengths in the beam, and laser light is much more *coherent* than ordinary light.

## 37–2 TWO-SOURCE INTERFERENCE

One of the earliest quantitative experiments with interference of light was performed in 1800 by the English scientist Thomas Young. His experiment involved interference of light from two sources, which we discussed in Section 37–1. Young's apparatus is shown in Fig. 37–3a. Monochromatic light emerging from a narrow slit $S_0$ (1.0 $\mu$m or so wide) falls on a screen with two other narrow slits $S_1$ and $S_2$, each 1.0 $\mu$m or so wide and a few micrometers apart. According to Huygens' principle, cylindrical wave fronts spread out from slit $S_0$ and reach slits $S_1$ and $S_2$ *in phase* because they travel equal distances from $S_0$. The waves emerging from slits $S_1$ and $S_2$ are therefore always in phase, so $S_1$ and $S_2$ are *coherent* sources (Fig. 37–3b). But the waves from these sources do not necessarily arrive at point $P$ in phase because of the path difference $(r_2 - r_1)$.

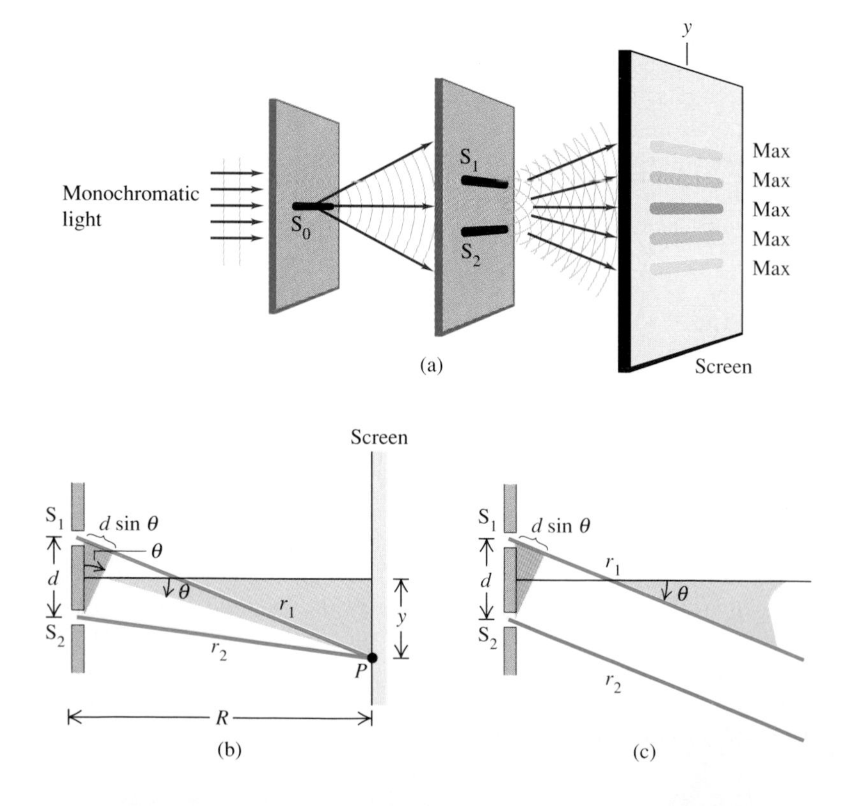


**37–3** (a) Interference of light waves passing through two slits. (b) Geometrical analysis of Young's experiment. The slits are horizontal and are seen from the side in cross section. (c) Approximate geometry when $R$ is much larger than $d$.

To simplify the following analysis, we assume that the distance $R$ from the slits to the screen is so large in comparison to the distance $d$ between the slits that the lines from $S_1$ and $S_2$ to $P$ are very nearly parallel, as shown in Fig. 37–3c. This is usually the case for experiments with light. The difference in path length is then given by

$$r_1 - r_2 = d \sin \theta . \qquad (37\text{–}3)$$

We found in Section 37–1 that constructive interference (reinforcement) occurs at a point $P$ (in a brightly illuminated region of the screen) when the path difference $d \sin \theta$ is an integral number of wavelengths, $m\lambda$, where $m = 0, \pm1, \pm2, \pm3, \ldots$. So constructive interference occurs at angles $\theta$ for which

$$d \sin \theta = m\lambda \quad (m = 0, \pm1, \pm2, \ldots). \qquad (37\text{–}4)$$

Similarly, destructive interference (cancellation) occurs, forming dark regions on the screen, at points for which the path difference is a half-integral number of wavelengths, $(m + \frac{1}{2})\lambda$:

$$d \sin \theta = (m + \tfrac{1}{2})\lambda \quad (m = 0, \pm1, \pm2, \ldots). \qquad (37\text{–}5)$$

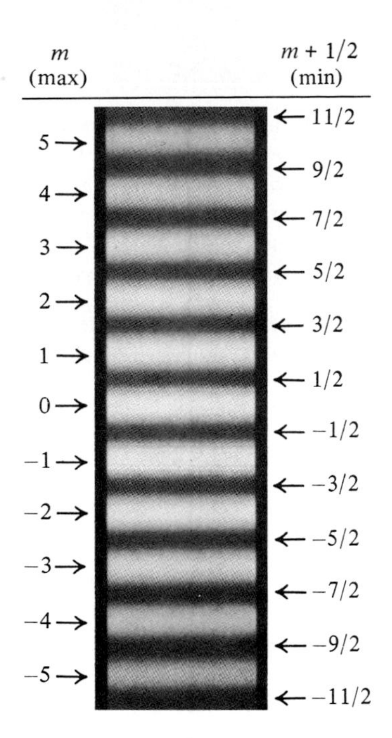


**37–4** Interference fringes produced by Young's double-slit experiment.

Thus the pattern on the screen of Fig. 37–3b is a succession of bright and dark bands. A photograph of such a pattern is shown in Fig. 37–4.

We can derive an expression for the positions of the centers of the bright bands on the screen. In Fig. 37–3b, $y$ is measured from the center of the pattern, corresponding to the distance from the center of Fig. 37–4. Let $y_m$ be the distance from the center of the pattern ($\theta = 0$) to the center of the $m$th bright band. We denote the corresponding value of $\theta$ as $\theta_m$; then

$$y_m = R \tan \theta_m .$$

In experiments such as this the $y_m$'s are often much smaller than $R$. This means that $\theta_m$ is very small, $\tan \theta_m$ is very nearly equal to $\sin \theta_m$, and

$$y_m = R \sin \theta_m .$$

Combining this with Eq. (37–4), we find

$$y_m = R\frac{m\lambda}{d}. \qquad (37\text{–}6)$$

We can measure $R$ and $d$, as well as the positions $y_m$ of the bright fringes, so this experiment provides a direct measurement of the wavelength $\lambda$. Young's experiment was in fact the first direct measurement of wavelengths of light.

Note that the distance between adjacent bright bands in the pattern is *inversely* proportional to the distance $d$ between the slits. The closer together the slits are, the more the pattern spreads out. When the slits are far apart, the bands in the pattern are closer together.

## EXAMPLE 37–1

In a two-slit interference experiment with the slits 0.20 mm apart and a screen at a distance of 1.0 m, the third bright fringe (not counting the central bright fringe straight ahead from the slits) is found to be displaced 7.5 mm from the central fringe. Find the wavelength of the light used.

**SOLUTION** From Eq. (37–6), with $m = 3$,

$$\lambda = \frac{y_m d}{mR} = \frac{(7.5 \times 10^{-3}\ \text{m})(0.20 \times 10^{-3}\ \text{m})}{(3)(1.0\ \text{m})}$$

$$= 500 \times 10^{-9}\ \text{m} = 500\ \text{nm}.$$

## EXAMPLE 37–2

**Broadcast pattern of a radio station** A radio station operating at a frequency of 1500 kHz = $1.5 \times 10^6$ Hz (near the top end of the AM broadcast band) has two identical vertical dipole antennas spaced 400 m apart, oscillating in phase. At distances much greater than 400 m, in what directions is the intensity greatest in the resulting radiation pattern?

**SOLUTION** The two antennas, seen from above in Fig. 37–5, correspond to sources $S_1$ and $S_2$ in Fig. 37–3. The wavelength is $\lambda = c/f = 200$ m. The directions of the intensity *maxima* are the values of $\theta$ for which the path difference is zero or an integral number of wavelengths, as given by Eq. (37–4). [Note that we can't use Eq. (37–6) because the angles $\theta_m$ aren't necessarily small.] Inserting the numerical values, with $m = 0, \pm 1, \pm 2$, we find

$$\sin\theta = \frac{m\lambda}{d} = \frac{m}{2}, \qquad \theta = 0, \pm 30°, \pm 90°.$$

In this example, values of $m$ greater than 2 or less than $-2$ give values of $\sin\theta$ greater than 1 or less than $-1$, which is impossible. There is *no* direction for which the path difference is three or more wavelengths. Thus values of $m$ of $\pm 3$ and beyond have no physical meaning in this example.

The angles for minimum intensity (destructive interference) are given by Eq. (37–5), with $m = -2, -1, 0, 1$:

$$\sin\theta = \frac{(m + \frac{1}{2})\lambda}{d} = \frac{m + \frac{1}{2}}{2}, \qquad \theta = \pm 14.5°, \pm 48.6°.$$

Other values of $m$ have no physical significance in this example.

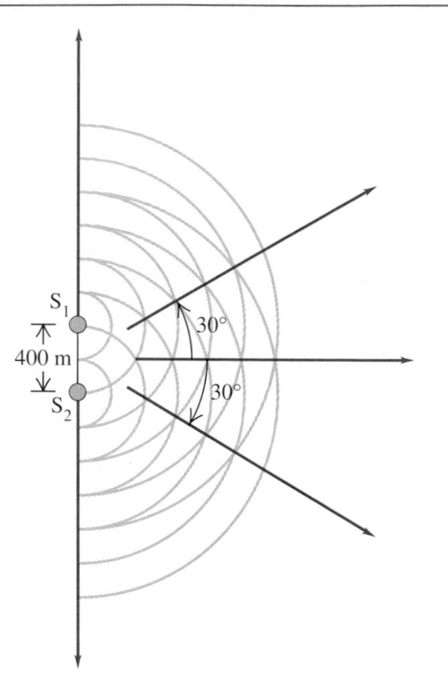


**37–5** Two radio antennas broadcasting in phase. Directions of maximum intensity are shown.

# 37–3 INTENSITY DISTRIBUTION IN INTERFERENCE PATTERNS

In Section 37–2 we found the directions of maximum and minimum intensity in a two-source interference pattern. We may also find the intensity at *any* point in the pattern. To do this, we have to combine the two sinusoidally varying fields (from the two sources) at a point $P$ in the radiation pattern, taking proper account of the phase difference of the two waves at point $P$, which results from the path difference. The intensity is then proportional to the square of the resultant electric-field amplitude, as we learned in Chapter 33.

Here is our program. We will assume that the two sinusoidal functions (corresponding to waves from the two sources) have equal amplitude $E$ and that the $\boldsymbol{E}$ fields lie along the same line (have the same polarization). This assumes that the sources are identical and neglects the slight amplitude difference caused by the unequal path lengths. From Eq. (33–27), each source by itself would give an intensity $\frac{1}{2}\epsilon_0 cE^2$ at point $P$. If the two sources are in phase, then the waves that arrive at $P$ differ in phase by an amount proportional to the difference in their path lengths, $(r_1 - r_2)$. If the phase angle between these arriving waves is $\phi$, then the two electric fields superposed at $P$ might be

$$E_1 = E\cos\omega t,$$

$$E_2 = E\cos(\omega t + \phi).$$

The superposition of the two fields at point $P$ is a sinusoidal function with some amplitude $E_P$ that depends on $E$ and the phase difference $\phi$. First we'll work on finding $E_P$ if $E$ and $\phi$ are known. Then we can find the intensity $I$ of the resultant wave, which is proportional to $E_P{}^2$. Finally, we'll derive the relation between the phase difference $\phi$ and the path difference, which is determined by the geometry of the situation.

To add the two sinusoidal functions with a phase difference, we use the *phasor* representation that we have previously used for simple harmonic motion (Section 13–4) and for voltages and currents in ac circuits (Section 32–1). We suggest that you review those sections so that phasors are fresh in your mind. Each sinusoidal function is represented by a rotating vector (phasor) whose projection on the horizontal axis at any instant represents the instantaneous value of the sinusoidal function.

In Fig. 37–6, $E_1$ is the horizontal component of the phasor representing the wave from source $S_1$, and $E_2$ is the phasor component for the wave from $S_2$. These two phasors have the same magnitude $E$, but $E_2$ is *ahead* of $E_1$ in phase by an angle $\phi$, as shown in the diagram. Both phasors rotate counterclockwise with constant angular velocity $\omega$, and the sum of the projections on the horizontal axis at any time gives the instantaneous value of the total $E$ field at point $P$. Thus the amplitude $E_P$ of the resultant sinusoidal wave at $P$ is the magnitude of the phasor labeled $E_P$ in the diagram, the *vector sum* of the other two phasors. To find $E_P$, we use the law of cosines, and the fact that $\cos(\pi - \phi) = -\cos\phi$:

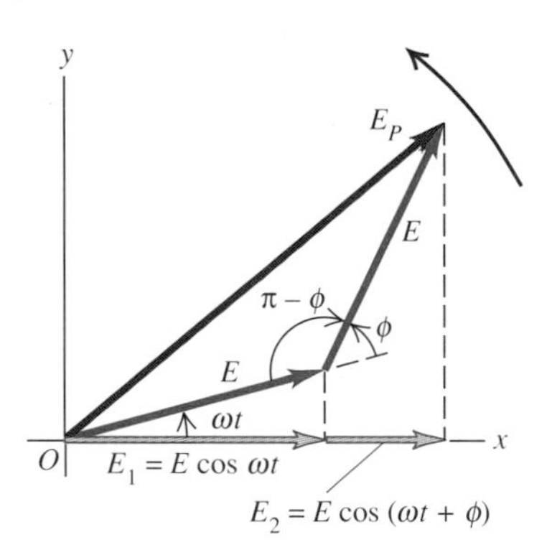


**37–6** Phasor diagram used to find the amplitude $E_P$ of the electric field at point $P$ where waves of equal amplitude $E$ with a phase difference $\phi$ are superposed.

$$E_P{}^2 = E^2 + E^2 - 2E^2\cos(\pi - \phi)$$
$$= E^2 + E^2 + 2E^2\cos\phi$$

Then, using the identity $1 + \cos\phi = 2\cos^2(\phi/2)$, we obtain

$$E_P{}^2 = 2E^2(1 + \cos\phi) = 4E^2\cos^2\left(\frac{\phi}{2}\right). \tag{37–7}$$

When the two waves are in phase, $\phi = 0$ and $E_P = 2E$. When they are exactly a half-cycle out of phase, $\phi = \pi$ rad $= 180°$, $\phi/2 = \pi/2$, $\cos^2(\phi/2) = 0$, and $E_P = 0$. Thus the superposition of two sinusoidal waves with the same amplitude and frequency but with a phase difference yields a sinusoidal wave with amplitude between zero and twice the individual amplitudes, depending on the phase difference.

To obtain the intensity $I$ at point $P$, we recall from Section 33–4 that $I$ is equal to the average magnitude of the Poynting vector, $S_{av}$. For a sinusoidal wave with $E$ field amplitude $E_P$, this is given by Eq. (33–27) with $E_{max}$ replaced by $E_P$. Thus we can express the intensity in any of the following equivalent forms:

$$I = S_{av} = \frac{E_P{}^2}{2\mu_0 c} = \frac{1}{2}\sqrt{\frac{\epsilon_0}{\mu_0}}\,E_P{}^2 = \frac{1}{2}\epsilon_0 c E_P{}^2. \tag{37–8}$$

The essential content of these expressions is that $I$ is proportional to $E_P{}^2$.

When we substitute Eq. (37–7) into the last expression in Eq. (37–8), we get

$$I = \frac{1}{2}\epsilon_0 c E_P{}^2 = 2\epsilon_0 c E^2\cos^2\frac{\phi}{2}. \tag{37–9}$$

In particular, the *maximum* intensity $I_0$, which occurs at points where the phase difference is zero ($\phi = 0$), is

$$I_0 = 2\epsilon_0 c E^2.$$

Note that the maximum intensity $I_0$ is *four times* (not twice) as great as the intensity $\frac{1}{2}\epsilon_0 c E^2$ from each individual source.

Substituting the expression for $I_0$ into Eq. (37–9), we can express the intensity $I$ at any point very simply, in terms of the maximum intensity $I_0$, as

$$I = I_0\cos^2\frac{\phi}{2}. \tag{37–10}$$

Note that if we average this over all possible phase differences, using the fact that the average of $\cos^2\phi$ is $\frac{1}{2}$, the result is $I_0/2 = \epsilon_0 c E^2$. This is just twice the intensity from each individual source, as we should expect. The total energy output from the two sources isn't changed by the interference effects, but the energy is redistributed. For some phase angles the intensity is four times as great as for an individual source, but for others it is zero. It averages out.

Now we have to find how the phase difference $\phi$ between the two fields at point $P$ is related to the geometry of the situation. We know that $\phi$ is proportional to the difference in path lengths from the two sources to point $P$. When the path difference is one wavelength, the phase difference is one cycle, and $\phi = 2\pi \text{ rad} = 360°$. When the path difference is $\lambda/2$, $\phi = \pi \text{ rad} = 180°$, and so on. In general, the ratio of $\phi$ to $2\pi$ is equal to the ratio of $r_1 - r_2$ to $\lambda$:

$$\frac{\phi}{2\pi} = \frac{r_1 - r_2}{\lambda}.$$

Thus a path difference $r_1 - r_2$ causes a phase difference given by

$$\phi = \frac{2\pi}{\lambda}(r_1 - r_2) = k(r_1 - r_2), \tag{37–11}$$

where $k = 2\pi/\lambda$ is the *wave number* introduced in Section 19–3.

If the material in the space between the sources and $P$ is anything other than vacuum, we must use the wavelength *in the material* in Eq. (37–11). If the material has index of refraction $n$, then

$$\lambda = \frac{\lambda_0}{n} \quad \text{and} \quad k = nk_0, \tag{37–12}$$

where $\lambda_0$ and $k_0$ are the values of $\lambda$ and $k$, respectively, in vacuum.

Finally, the path difference is given by Eq. (37–3):

$$r_1 - r_2 = d \sin\theta .$$

Combining this with Eq. (37–11), we find

$$\phi = k(r_1 - r_2) = kd \sin\theta = \frac{2\pi d}{\lambda} \sin\theta . \tag{37–13}$$

When we substitute this into Eq. (37–10), we find

$$\begin{aligned} I &= I_0 \cos^2 (\tfrac{1}{2}kd \sin\theta) \\ &= I_0 \cos^2 \left(\frac{\pi d}{\lambda} \sin\theta\right). \end{aligned} \tag{37–14}$$

The directions of *maximum* intensity occur when the cosine has the values $\pm 1$, that is, when

$$\frac{\pi d}{\lambda} \sin\theta = m\pi \quad (m = 0, \pm 1, \pm 2, \ldots),$$

or

$$d \sin\theta = m\lambda ,$$

in agreement with Eq. (37–4). We leave it as an exercise to show that Eq. (37–5) for the zero-intensity directions can also be derived from Eq. (37–14).

In Section 37–2 we noted that in experiments with light we can describe the positions of the bright fringes with the coordinate $y$, as in Eq. (37–6), and that ordinarily $y \ll R$. In that case, $\sin\theta$ is approximately equal to $y/R$, and we obtain the even simpler expressions

$$I = I_0 \cos^2 \left(\frac{kdy}{2R}\right) = I_0 \cos^2 \left(\frac{\pi dy}{\lambda R}\right). \tag{37–15}$$

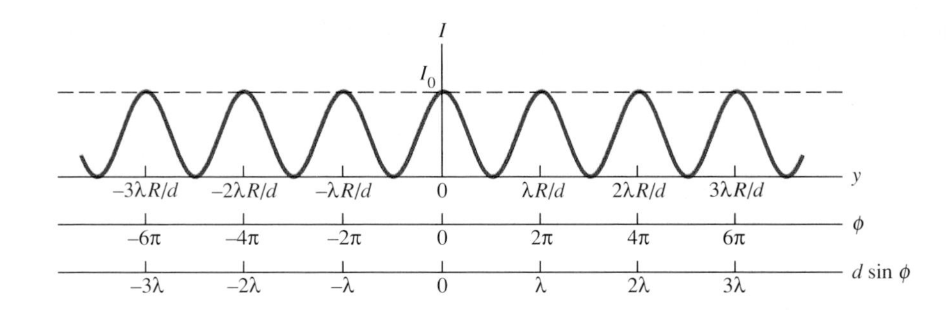


**37–7** Intensity distribution in the interference pattern from two identical slits. The three scales show the distance $y$ of a point in the pattern from the center ($y = 0$), the phase difference $\phi$ between the two waves at each point in the pattern, and the path difference, expressed as a multiple of the wavelength $\lambda$. The maxima occur at points where $\phi$ is a multiple of $2\pi$ and $d \sin \theta$ is a multiple of $\lambda$.

Figure 37–7 shows a graph of Eq. (37–15); we can compare this with the photographically recorded pattern of Fig. 37–4. The peaks in Fig. 37–7 all have the same intensity, but those in Fig. 37–4 fade off as we go away from the center. We'll explore the reasons for this variation in peak intensity in Chapter 38.

## EXAMPLE 37–3

**A directional transmitting antenna array** In Fig. 37–5, suppose the two sources are identical radio antennas 10 m apart, radiating waves with frequency $f = 60$ MHz in all directions. The intensity at a distance of 700 m in the $+x$-direction (corresponding to $\theta = 0$ in Fig. 37–3) is $I_0 = 0.020\ \text{W/m}^2$. a) What is the intensity in the direction $\theta = 4°$? b) In what direction near $\theta = 0$ is the intensity $I_0/2$? c) In what directions is the intensity zero?

**SOLUTION** Because the distance between the antennas is much less than 700 m, we assume that the amplitudes of the two waves are equal in all directions. We have to use Eq. (37–14); we can't use the approximate relation of Eq. (37–15) in this case because $\theta$ is not small. First we find the wavelength, using the relation $c = \lambda f$. We find

$$\lambda = \frac{c}{f} = \frac{3.0 \times 10^8\ \text{m/s}}{60 \times 10^6\ \text{s}^{-1}} = 5.0\ \text{m}.$$

The spacing between sources is $d = 10$ m, and Eq. (37–14) becomes

$$\begin{aligned} I &= I_0 \cos^2\left(\frac{\pi d}{\lambda} \sin \theta\right) \\ &= (0.020\ \text{W/m}^2) \cos^2\left[\frac{\pi(10\ \text{m})}{(5.0\ \text{m})} \sin \theta\right] \\ &= (0.020\ \text{W/m}^2) \cos^2 (2\pi\ \text{rad} \sin \theta). \end{aligned}$$

a) When $\theta = 4°$,

$$\begin{aligned} I &= (0.020\ \text{W/m}^2) \cos^2(2\pi\ \text{rad} \sin 4°) \\ &= 0.016\ \text{W/m}^2. \end{aligned}$$

This is about 82% of the intensity at $\theta = 0$.

b) The intensity $I$ equals $I_0/2$ when the cosine has the value $\pm 1/\sqrt{2}$; this occurs when $2\pi \sin \theta = \pm \pi/4$ rad, $\sin \theta = \pm \frac{1}{8}$, and $\theta = \pm 7.2°$. (This condition is also satisfied when $\sin \theta = \pm\frac{3}{8}$, $\pm\frac{5}{8}$, and $\pm\frac{7}{8}$. Can you verify this? These aren't the values the problem asked for. Why not?)

c) The intensity is zero when $\cos(2\pi \sin \theta) = 0$; this occurs when $2\pi \sin \theta = \pm\pi/2, \pm 3\pi/2, \pm 5\pi/2, \ldots,$

$$\sin \theta = \pm\tfrac{1}{4}, \pm\tfrac{3}{4}, \pm\tfrac{5}{4}, \ldots.$$

Values of $\sin \theta$ greater than 1 have no meaning, and we find

$$\theta = \pm 14.5°, \quad \pm 48.6°.$$

This is not just a hypothetical problem. It is often desirable to beam most of the radiated energy from a radio transmitter in particular directions rather than uniformly in all directions. Pairs or rows of antennas are often used to produce the desired radiation pattern.

## 37–4 INTERFERENCE IN THIN FILMS

You often see bright bands of color when light reflects from a soap bubble or from a thin layer of oil floating on water. These are the results of interference effects. Light waves are reflected from opposite surfaces of the thin films, and constructive interference between the two reflected waves (with different path lengths) occurs in different places

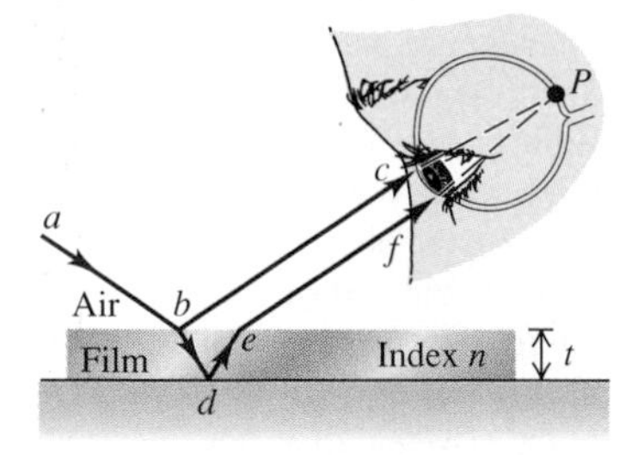


**37–8** Interference between rays reflected from the upper and lower surfaces of a thin film. The path differences and phase shifts at the surfaces cause some wavelengths of light to interfere constructively and others to interfere destructively.

for different wavelengths. The situation is shown schematically in Fig. 37–8. Light shining on the upper surface of a thin film with thickness $t$ is partly reflected at the upper surface (path *abc*). Light *transmitted* at the upper surface is partly reflected at the lower surface (path *abdef*). The two reflected waves come together at point $P$ on the retina of the eye. Depending on the phase relationship, they may interfere constructively or destructively. Different colors have different wavelengths, so the interference may be constructive for some colors and destructive for others. That's why you see colored rings or fringes.

Here is an example involving *monochromatic* light reflected from two nearly parallel surfaces at nearly normal incidence. Figure 37–9 shows two plates of glass separated by a thin wedge of air. We want to consider interference between the two light waves reflected from the surfaces adjacent to the air wedge, as shown. (Reflections also occur at the top surface of the upper plate and the bottom surface of the lower plate. To keep our discussion simple, we won't include those.) The situation is the same as in Fig. 37–8 except that the film thickness is not uniform. The path difference between the two waves is just twice the thickness $t$ of the air wedge at each point. At points where $2t$ is an integral number of wavelengths, we expect to see constructive interference and a bright area; where it is a half-integral number of wavelengths, we expect to see destructive interference and a dark area. Along the line where the plates are in contact, there is practically *no* path difference, and we expect a bright area.

When we carry out the experiment, the bright and dark fringes appear, but they are interchanged! Along the line of contact we find a *dark* fringe, not a bright one. This suggests that one or the other of the reflected waves has undergone a half-cycle phase shift during its reflection. In that case the two waves reflected at the line of contact are a half-cycle out of phase even though they have the same path length.

In fact, this phase shift can be predicted from Maxwell's equations and the electromagnetic nature of light. The details of the derivation are beyond our scope, but here is the result. Suppose a light wave with $\boldsymbol{E}$-field amplitude $E_{\mathrm{i}}$ is traveling in an insulating material with index of refraction $n_a$. It strikes, at normal incidence, an interface with another insulating material with index $n_b$. The amplitude $E_{\mathrm{r}}$ of the wave reflected from the interface is proportional to the amplitude $E_{\mathrm{i}}$ of the incident wave and is given by

$$E_{\mathrm{r}} = \frac{n_a - n_b}{n_a + n_b} E_{\mathrm{i}}. \qquad (37\text{–}16)$$

This result shows that the incident and reflected amplitudes have the same sign when $n_a$ is larger than $n_b$ and opposite sign when $n_b$ is larger than $n_a$. We can distinguish three cases, as shown in Fig. 37–10:

Figure 37–10a: When $n_a > n_b$, the first medium has greater optical density than the second. In this case, $E_{\mathrm{r}}$ and $E_{\mathrm{i}}$ have the same sign, and the phase shift of the reflected wave relative to the incident wave is zero. This is analogous to reflection of a transverse mechanical wave on a heavy rope at a point where it is tied to a lighter rope or a ring that can move vertically without friction.

Fig. 37–10b: When $n_a = n_b$, the amplitude $E_{\mathrm{r}}$ of the reflected wave is zero. The incident wave can't "see" the interface, and there *is no* reflected wave.

Fig. 37–10c: When $n_a < n_b$, the second material has greater optical density than the first. In this case, $E_{\mathrm{r}}$ and $E_{\mathrm{i}}$ have opposite signs, and the phase shift of the reflected wave relative to the incident wave is $\pi$ rad (180° or a half-cycle). This is analogous to reflection (with inversion) of a transverse mechanical wave on a light rope at a point where it is tied to a heavier rope or a rigid support.

**37–9** Interference between two light waves reflected from the two sides of an air wedge separating two glass plates. The path difference is $2t$.

Let's check this with the situation of Fig. 37–9. For the wave reflected from the upper surface of the air wedge, $n_a$ (glass) is greater than $n_b$, so this wave has zero phase

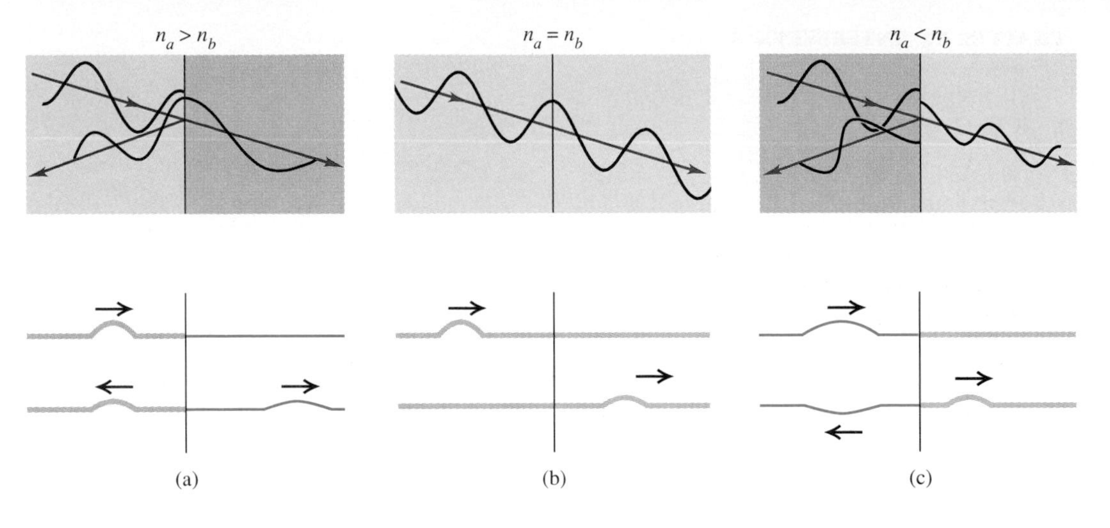


**37–10** Electromagnetic waves striking an insulator at normal incidence (shown as a small angle for clarity) and mechanical wave pulses on ropes. (a) There is no phase change upon reflection when $n_a > n_b$ and the transmitted wave moves faster than the incident wave. (b) There is no reflection when there is no difference in $n$ and no change in wave speed. (c) There is a half-cycle phase change upon reflection when $n_a < n_b$ and the transmitted wave moves more slowly than the incident wave.

shift. For the wave reflected from the lower surface, $n_a$ (air) is less than $n_b$ (glass), so this wave has a half-cycle phase shift. Waves reflected from the line of contact have no path difference to give additional phase shifts, and they interfere destructively; this is what we observe. We invite you to use the above principle to show that in Fig. 37–8 the wave reflected at point $b$ is shifted by a half-cycle, and the wave reflected at $d$ is not.

We can summarize this discussion symbolically: For a film with thickness $t$ and light at normal incidence, the reflected waves from the two surfaces interfere *constructively* if neither or both have a half-cycle reflection phase shift whenever the condition

$$2t = m\lambda \quad (m = 0, 1, 2, \ldots) \tag{37–17}$$

is satisfied, where $\lambda$ is the wavelength in the film. However, when *one* of the two waves has a half-cycle reflection phase shift, this equation is the condition for *destructive* interference.

Similarly, if neither wave, or both, has a half-cycle phase shift, the condition for *destructive* interference in the reflected waves is

$$2t = (m + \tfrac{1}{2})\lambda \quad (m = 0, 1, 2, \ldots). \tag{37–18}$$

However, if one wave has a half-cycle phase shift, this is the condition for *constructive* interference.

## EXAMPLE 37–4

Suppose the two glass plates in Fig. 37–9 are two microscope slides 10 cm long. At one end they are in contact; at the other end they are separated by a piece of paper 0.020 mm thick. What is the spacing of the interference fringes seen by reflection? Is the fringe at the line of contact bright or dark? Assume monochromatic light with $\lambda_0 = 500$ nm.

**SOLUTION** The wave reflected from the lower surface of the air wedge has a half-cycle phase shift, and the wave from the upper surface has none. Thus the fringe at the line of contact is dark. The condition for *destructive* interference (a dark fringe) is Eq. (37–17):

$$2t = m\lambda \quad (m = 0, 1, 2, \ldots).$$

From similar triangles in Fig. 37–9 the thickness $t$ at each point is proportional to the distance $x$ from the line of contact:

$$\frac{t}{x} = \frac{h}{l}.$$

Combining this with Eq. (37–17), we find

$$\frac{2xh}{l} = m\lambda,$$

$$x = m\frac{l\lambda}{2h} = m\frac{(0.10\text{ m})(500 \times 10^{-9}\text{m})}{(2)(0.020 \times 10^{-3}\text{m})} = m(1.25\text{ mm}).$$

Successive dark fringes, corresponding to successive integer values of $m$, are spaced 1.25 mm apart.

## EXAMPLE 37–5

Suppose the glass plates in Example 37–4 have $n = 1.52$ and the space between plates contains water ($n = 1.33$) instead of air. What happens now?

**SOLUTION** The phase changes are the same, and we again find a dark fringe at the line of contact. But the wavelength in water is

$$\lambda = \frac{\lambda_0}{n} = \frac{500 \text{ nm}}{1.33} = 376 \text{ nm}.$$

The fringe spacing is reduced by a factor of 1.33 and is equal to 0.94 mm.

## EXAMPLE 37–6

Suppose the top plate of the plates in Example 37–4 is a plastic material with $n = 1.40$, the wedge is filled with a silicone grease having $n = 1.50$, and the bottom plate is a dense flint glass with $n = 1.60$. Now what?

**SOLUTION** In this case there are half-cycle phase shifts at *both* surfaces of the wedge of grease, and the line of contact is at a *bright* fringe, not a dark one. The fringe spacing is again determined by the wavelength *in the film* (that is, in the silicone grease), $\lambda = (500 \text{ nm})/1.5 = 333$ nm; we invite you to show that the fringe spacing is 0.83 mm.

The results of Examples 37–4 and 37–5 show that the fringe spacing is proportional to the wavelength of the light used; the fringes are farther apart with red light (larger $\lambda$) than with blue light (smaller $\lambda$). If we use white light, the reflected light at any point is a mixture of wavelengths for which constructive interference occurs; the wavelengths that interfere destructively are weak or absent in the reflected light. But the colors that are weak in the *reflected* light are strong in the *transmitted* light. At any point, the color of the wedge as viewed by reflected light is the *complement* of its color as seen by transmitted light! Roughly speaking, the complement to any color is obtained by removing that color from white light (a mixture of all colors in the visible spectrum). For example, the complement of blue is yellow, and the complement of green is magenta.

## Newton's Rings

Figure 37–11a shows the convex surface of a lens in contact with a plane glass plate. A thin film of air is formed between the two surfaces. When you view the setup with monochromatic light, you see circular interference fringes (Fig. 37–11b). These were studied by Newton and are called *Newton's rings.* When viewed by reflected light, the center of the pattern is black. Can you see why this should be expected?

We can use interference fringes to compare the surfaces of two optical parts by placing the two in contact and observing the interference fringes. Figure 37–12 is a photograph made during the grinding of a telescope objective lens. The lower, larger-diameter, thicker disk is the correctly shaped master, and the smaller upper disk is the lens under test. The "contour lines" are Newton's interference fringes; each one indicates an

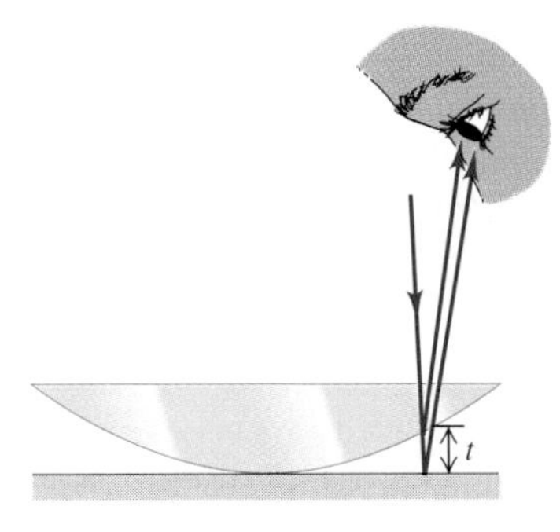


(a)

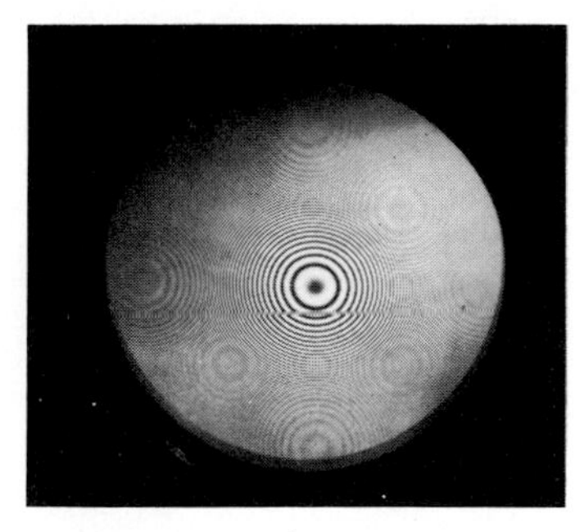

(b)

**37–11** (a) Air film between a convex lens and a plane surface. The thickness of the film $t$ increases from zero as we move out from the center, giving a series of alternating dark and bright rings for monochromatic light. (b) A photograph of Newton's rings.

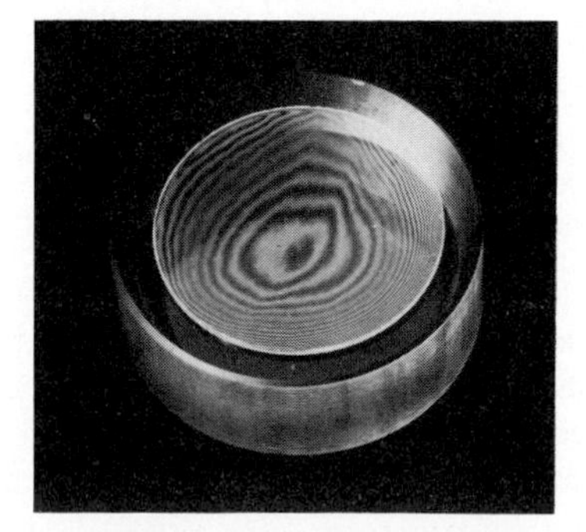

**37–12** The surface of a telescope objective under inspection during manufacture.

additional distance between the specimen and the master of a half-wavelength. At ten lines from the center spot the distance between the two surfaces is five wavelengths, or about 0.003 mm. This isn't very good; high-quality lenses are routinely ground with a precision of less than one wavelength. The surface of the primary mirror of the Hubble Space Telescope was ground to a precision of better than $\frac{1}{50}$ wavelength. Unfortunately, it was ground to incorrect specifications, creating one of the most precise errors in the history of optical technology.

"Nonreflecting" film

Air

Glass

$t = \frac{1}{4}\lambda$ (film)

**37–13** Destructive interference results when the film thickness is one-fourth of the wavelength in the film and the index of refraction of the film is intermediate between those of glass and air.

## Nonreflective Coatings

Nonreflective coatings for lens surfaces make use of thin-film interference. A thin layer or film of hard transparent material with an index of refraction smaller than that of the glass is deposited on the lens surface, as in Fig. 37–13. Light is reflected from both surfaces of the layer. In both reflections the light is reflected from a medium of greater index than that in which it is traveling, so the same phase change occurs in both reflections. If the film thickness is a quarter (one-fourth) of the wavelength *in the film* (assuming normal incidence), the total path difference is a half-wavelength. Light reflected from the first surface is then a half-cycle out of phase with light reflected from the second, and there is destructive interference.

The thickness of the nonreflective coating can be a quarter-wavelength for only one particular wavelength. This is usually chosen in the central yellow-green portion of the spectrum (550 nm), where the eye is most sensitive. Then there is somewhat more reflection at both longer and shorter wavelengths, and the reflected light has a purple hue. The overall reflection from a lens or prism surface can be reduced in this way from 4–5% to less than 1%. This treatment is particularly important in eliminating stray reflected light in highly corrected lenses with many air-glass surfaces. The same principle is used to minimize reflection from silicon photovoltaic solar cells ($n = 3.5$) by use of a thin surface layer of silicon monoxide (SiO, $n = 1.45$).

### EXAMPLE 37–7

A commonly used lens coating material is magnesium fluoride, $MgF_2$, with $n = 1.38$. What thickness should a nonreflecting coating have for 550-nm light if it is applied to glass with $n = 1.52$?

**SOLUTION** With this coating, the wavelength $\lambda$ of yellow-green light in the coating is

$$\lambda = \frac{\lambda_0}{n} = \frac{550 \text{ nm}}{1.38} = 400 \text{ nm}.$$

A nonreflecting film of $MgF_2$ should have a thickness of a quarter-wavelength, or 100 nm.

## Reflective Coatings

If a material with index of refraction *greater* than that of glass is deposited on glass with a thickness of a quarter-wavelength, then the reflectivity is *increased*. In this case there is a half-cycle phase shift at the air-film interface but none at the film-glass interface, and reflections from the two sides of the film interfere constructively. For example, a coating with index 2.5 allows 38% of the incident energy to be reflected, compared with 4% or so with no coating. By use of multiple-layer coatings, it is possible to achieve nearly 100% transmission or reflection for particular wavelengths. These coatings are used for infrared "heat reflectors" in motion-picture projectors, solar cells, and astronauts' visors and for color separation in color television cameras, to mention only a few applications.

We have talked mostly about interference in thin films using monochromatic light. When the incident light has a mixture of various wavelengths, such as white light, different wavelengths interfere constructively from different thicknesses of a film. The resulting interference fringes are rainbow-colored; Fig. 37–14 shows an example.

**37–14** Photo of a thin film of oil illuminated by white light.

## 37–5 THE MICHELSON INTERFEROMETER

The **Michelson interferometer** played an interesting role in the history of science during the latter part of the nineteenth century, and it has also had an important role in establishing high-precision length standards. In contrast to the Young two-slit experiment, which uses light from two very narrow sources, the Michelson interferometer uses light from a broad, spread-out source.

Figure 37–15 is a diagram of the interferometer's principal components. The figure shows the path of one ray from a point $A$ of an extended monochromatic source. This ray strikes a glass plate $C$ called a *beam splitter.* The right side of this plate has a thin coating of silver. Part of the light (ray 2) is reflected from the silvered surface at point $P$ to the mirror $M_2$ and back through $C$ to the observer's eye. The remainder of the light (ray 1) passes through the silvered surface and the compensator plate $D$ and is reflected from mirror $M_1$. It then returns through $D$ and is reflected from the silvered surface of $C$ to the observer. The compensator plate $D$ is cut from the same piece of glass as plate $C$, so its thickness is identical with that of $C$, within a fraction of a wavelength. Its purpose is to ensure that rays 1 and 2 pass through the same thickness of glass.

The whole apparatus is mounted on a very rigid frame, and a fine, very accurate micrometer screw is used to move mirror $M_2$. A simple interferometer is shown in Fig. 37–16.

If the distances $L_1$ and $L_2$ in Fig. 37–15 are exactly equal and the mirrors $M_1$ and $M_2$ are exactly at right angles, the virtual image of $M_1$ formed by reflection at the silvered surface of plate $C$ coincides with mirror $M_2$. If $L_1$ and $L_2$ are *not* exactly equal, the image of $M_1$ is displaced slightly from $M_2$; and if the mirrors are not exactly perpendicular, the image of $M_1$ makes a slight angle with $M_2$. Then the mirror $M_2$ and the virtual image of $M_1$ play the same roles as the two surfaces of a thin film (Section 37–4), and light reflected from these surfaces forms the same sort of interference fringes.

Suppose the angle between mirror $M_2$ and the virtual image of $M_1$ is just great enough that five or six vertical fringes are present in the field of view. If we now move the mirror $M_2$ slowly either backward or forward a distance $\lambda/2$, the effective film thickness changes by $\lambda$, and each fringe moves to the left or right a distance equal to the fringe spacing. If we observe the fringe positions through a telescope with a crosshair eyepiece,

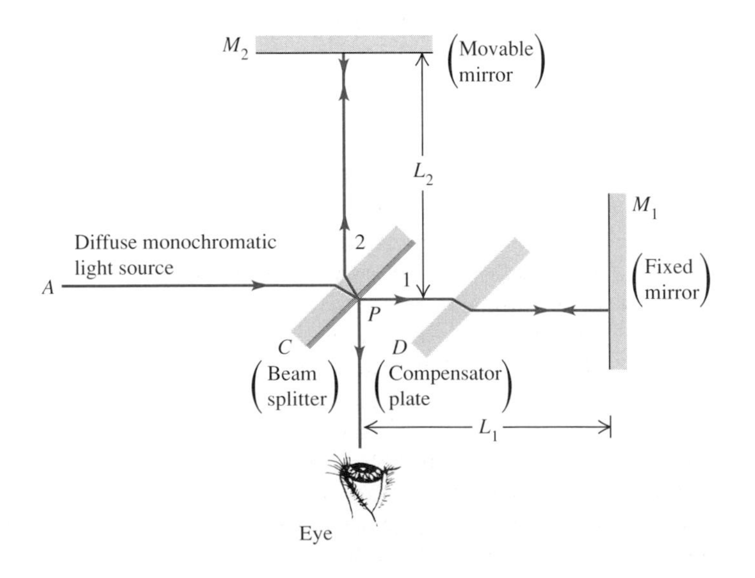


**37–15** The Michelson interferometer. The interference pattern seen by the eye results from the path difference for rays 1 and 2. The pattern shifts by one fringe when mirror $M_2$ moves a distance of one half-wavelength.

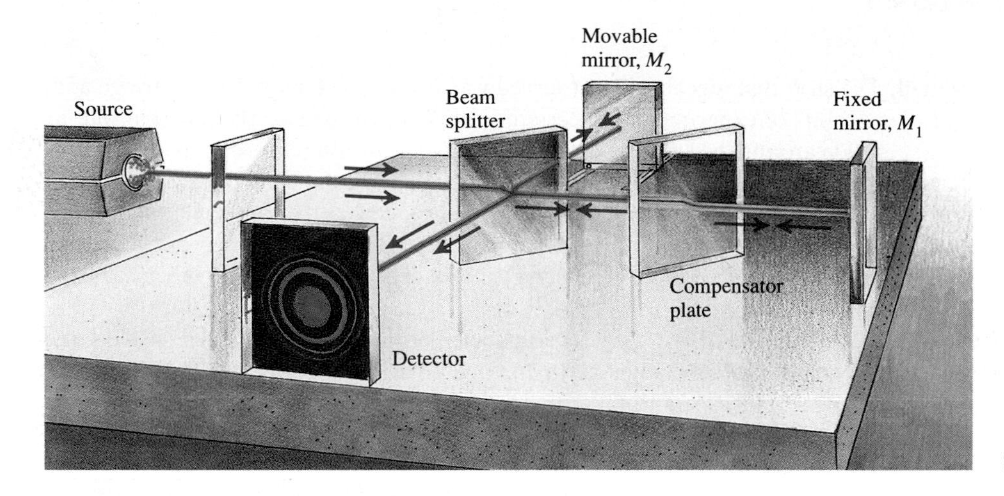


**37–16** Michelson interferometer.

and $m$ fringes cross the crosshair when we move the mirror a distance $x$, then

$$x = m\frac{\lambda}{2}, \quad \text{or} \quad \lambda = \frac{2x}{m}. \tag{37–19}$$

If $m$ is several thousand, the distance $x$ is large enough that it can be measured with good precision, and we can obtain a precise value for the wavelength $\lambda$.

Until 1983 the meter was defined as a length equal to 1,650,763.73 wavelengths of the orange-red light of krypton-86. Interferometers similar to the one we have described were used to measure distances as multiples of this wavelength by counting fringes and using Eq. (37–19). Such measurements were then used to calibrate other length-measuring instruments. As we mentioned in Section 1–3, this definition of the meter has been superseded by a length standard based on the unit of *time,* as defined by a cesium clock.

## EXAMPLE 37–8

If a Michelson interferometer is used with light of wavelength 605.78 nm, how many fringes pass the telescope crosshairs when mirror $M_2$ moves exactly one centimeter?

**SOLUTION** From Eq. (37–19) the number of fringes is

$$m = \frac{2x}{\lambda} = \frac{2(1.0000 \times 10^{-2}\text{m})}{605.78 \times 10^{-9}\text{m}} = 33{,}015.$$

That sounds like a lot of counting, but it is easily automated by use of a photocell and an electronic counter.

Another application of the Michelson interferometer with considerable historical interest is the **Michelson-Morley experiment.** To understand the purpose of this experiment, recall that before the electromagnetic theory of light and Einstein's special theory of relativity became established, most physicists believed that the propagation of light waves occurred in a medium called the **ether,** which was believed to permeate all space. In 1887, Michelson and Morley used the Michelson interferometer in an attempt to detect the motion of the earth through the ether. Suppose the interferometer in Fig. 37–15 is moving from left to right relative to the ether. According to nineteenth-century theory, this would lead to changes in the speed of light in the portions of the path shown as horizontal lines in the figure. There would be fringe shifts relative to the positions that the fringes *would have* if the instrument were at rest in the ether. Then when the entire instrument was rotated 90°, the other portions of the paths would be similarly affected, giving a fringe shift in the opposite direction.

Michelson and Morley expected that the motion of the earth through the ether would cause a fringe shift of about four-tenths of a fringe when the instrument was rotated. The shift that was actually observed was less than a hundredth of a fringe and, within the limits of experimental uncertainty, appeared to be exactly zero. Despite its orbital motion around the sun, the earth appeared to be *at rest* relative to the ether. This negative result baffled physicists until Einstein developed the special theory of relativity in 1905. Einstein postulated that the speed of a light wave has the same magnitude $c$ relative to *all* inertial reference frames, no matter what their velocity may be relative to each other. The presumed ether thus plays no role. The concept of the ether has been abandoned.

The theory of relativity is a well-established cornerstone of modern physics, and we will study it in detail in Chapter 39. In retrospect, the negative result of the Michelson-Morley experiment gives strong experimental support to the special theory of relativity, and it is often called the most significant "negative-result" experiment ever performed.

## 37–6 THE PHOTON: A Case Study in Quantum Physics

The interference effects discussed in this chapter, and similar effects to be discussed in the next chapter, show clearly the wave nature of light. But many other phenomena show a different aspect of the nature of light, in which it seems to behave as a stream of *particles.* For example, when a photograph of an interference pattern, such as Fig. 37–4, is made by using extremely weak monochromatic light and electronic image amplification, the pattern does not build up uniformly. Instead, first a light spot appears at one point, then another spot appears at another point, and so on. As the pattern emerges, the regions of maximum intensity have the greatest number of spots, the regions of zero intensity have none, and so forth.

This behavior suggests that the energy in a light wave is not a continuous stream, but rather a succession of little bundles of energy. These bundles are called *quanta,* or **photons.** We don't notice this effect in ordinary photography because the total number of spots is extremely large.

The concept of quantization of energy was introduced by Planck in 1900 as a computational technique in a calculation aimed at predicting the energy distribution among various wavelengths in the spectrum of radiation from hot objects (blackbody radiation, mentioned in Section 15–7). Einstein recognized in 1905 that quantization was far more than a computational trick, that instead it was a fundamental aspect of the nature of light. He used this concept to analyze the photoelectric effect, a process in which electrons are liberated from a conductive surface when light strikes it. Einstein assumed that an electron at the surface absorbs one photon and thus gains enough energy to escape from the surface.

Einstein assumed that the energy $E$ of an individual photon was proportional to the frequency $f$ of the light, with a proportionality constant $h$ that is now called **Planck's constant.** That is,

$$E = hf = \frac{hc}{\lambda}, \qquad (37\text{–}20)$$

where $c$ is the speed of light and $\lambda = c/f$ is the wavelength of the radiation in vacuum. Detailed measurements of the spectra of blackbody radiation and of the photoelectric effect confirmed the correctness of the photon concept and also enabled physcists to determine the numerical value of Planck's constant:

$$h = 6.626 \times 10^{-34}\ \text{J}\cdot\text{s}.$$

Thus light has a dual personality that includes both wavelike and particlelike properties. Light can form interference patterns similar to those of sound and water waves. But if the light in the pattern is examined closely enough, it is found to be made of particlelike photons. This duality is present not only in visible light but in the entire spectrum of electromagnetic radiation. In some experiments and with some frequency ranges, one aspect or the other may predominate, but fundamentally both are always present.

## EXAMPLE 37–9

A gamma-ray photon emitted during the decay of a radioactive cobalt-60 nucleus has an energy of $2.135 \times 10^{-13}$ J. What are the frequency and wavelength of this electromagnetic radiation?

**SOLUTION** From Eq. (37–20) we have

$$f = \frac{E}{h} = \frac{2.135 \times 10^{-13}\ \text{J}}{6.626 \times 10^{-34}\ \text{J}\cdot\text{s}} = 3.222 \times 10^{20}\ \text{Hz}.$$

This is roughly a million times larger than frequencies of visible light.

The wavelength $\lambda$ is

$$\lambda = \frac{c}{f} = \frac{3.00 \times 10^{8}\ \text{m/s}}{3.222 \times 10^{20}\ \text{Hz}} = 9.31 \times 10^{-13}\ \text{m}.$$

## EXAMPLE 37–10

The familiar red light emitted by a helium-neon laser has a wavelength of 632.8 nm. If the power output is 1.00 mW, how many photons of this light does the laser emit each second?

**SOLUTION** First we use Eq. (37–20) to find the energy of each photon:

$$E = \frac{hc}{\lambda} = \frac{(6.626 \times 10^{-34}\text{J}\cdot\text{s})(3.00 \times 10^{8}\text{m/s})}{632.8 \times 10^{-9}\text{m}} = 3.14 \times 10^{-19}\text{J}.$$

A power output of 1.00 mW means that the laser is emitting $1.00 \times 10^{-3}$ J of energy each second; the number of photons per second is

$$\frac{1.00 \times 10^{-3}\ \text{J/s}}{3.14 \times 10^{-19}\ \text{J/photon}} = 3.18 \times 10^{15}\ \text{photons/s}.$$

Example 37–10 shows that even a weak source of visible light, the 1.00-mW laser, emits an enormous number of photons each second, far too many to distinguish individually. Thus in many situations the quantization of visible light is not evident. In contrast, the energy and frequency of the photons of Example 37–9 are almost a million times greater than those of photons of visible light. With this much energy an individual gamma-ray photon is relatively easy to detect with any of several types of particle detectors, such as a Geiger counter or a solid-state detector. Conversely, the *wave* properties of visible light are easy to see in interference patterns, but interference effects with gamma rays of wavelength $9.30 \times 10^{-13}$ m are very difficult to detect.

Any material particle that has kinetic energy also has momentum, and photons also have momentum. Einstein showed, as part of his special theory of relativity (which we will study in Chapter 39), that a photon with energy $E$ has momentum $\boldsymbol{p}$ with magnitude $p$, and

$$E = pc. \tag{37–21}$$

Using the relation $c = \lambda f$ and rearranging, we find

$$p = \frac{E}{c} = \frac{hf}{c} = \frac{h}{\lambda}. \tag{37–22}$$

The *direction* of $\boldsymbol{p}$ is the direction in which the electromagnetic wave is traveling.

In Section 33–4 we discussed the phenomenon of *radiation pressure,* associated with the momentum carried by electromagnetic waves. Equation (37–22) shows us that this momentum is quantized. Just as the pressure that a gas exerts on a wall of its container results from the momentum change of gas molecules during impact, radiation pressure results from the momentum change of photons when they strike a surface and are absorbed or reflected.

We've seen that electromagnetic waves have particlelike properties, with the magnitude of the photon momentum given by $p = h/\lambda$. Nature is full of beautiful and sometimes surprising symmetries. In 1924, the French physicist Louis de Broglie suggested that particles as well as electromagnetic waves might have the dual wave-particle nature that we have described. If $p = h/\lambda$ holds for particles such as electrons as well as for photons, then a moving particle has a wavelength. Let's consider an electron (mass $m = 9.11 \times 10^{-31}$ kg) in the electron beam of a cathode-ray tube (described in Section 24–7). An electron accelerated through a potential difference of 100 V has a speed $v = 5.93 \times 10^{6}$ m/s. The momentum is

$$p = mv = (9.11 \times 10^{-31}\ \text{kg})(5.93 \times 10^{6}\ \text{m/s})$$
$$= 5.40 \times 10^{-24}\ \text{kg} \cdot \text{m/s}.$$

According to de Broglie's hypothesis, the wavelength $\lambda$ is

$$\lambda = \frac{h}{p} = \frac{6.626 \times 10^{-34}\ \text{J} \cdot \text{s}}{5.40 \times 10^{-24}\ \text{kg} \cdot \text{m/s}}$$
$$= 1.23 \times 10^{-10}\ \text{m} = 0.123\ \text{nm}.$$

This value is comparable to the spacing of atoms in a crystal lattice. In 1927, Davisson and Germer observed interference effects in the scattering of an electron beam from the surface of a crystal of nickel. They inferred correctly that their results demonstrated the wave properties of electrons. Measurements of the interference pattern enabled them to determine the wavelengths of the electrons and to confirm de Broglie's hypothesis directly.

Electron beams are now used routinely in electron microscopes. The small wavelengths make it possible to see details far smaller than the wavelengths of visible light and thus beyond the range of optical microscopes.

The number of photons in the universe is not a conserved quantity; photons are created and destroyed in many processes. This lack of permanence may seem to distinguish photons from other particles that we think of as having a permanent existence, such as electrons. But electrons (and all other particles) can also be created and destroyed, within the limitations imposed by various conservation laws including energy, momentum, angular momentum, electric charge, and others. Thus, ultimately, there is no fundamental distinction between photons and the other particles found in nature. Each particle has its individual characteristics and modes of interaction, but all are described with the same general language.

## SUMMARY

• Monochromatic light is light with a single frequency. Coherence is a definite, unchanging phase relationship between two waves. The overlap of waves from two coherent sources of monochromatic light forms an interference pattern. The principle of linear superposition states that the total wave disturbance at any point is the sum of the disturbances from the separate waves.

• When two light sources are in phase, constructive interference at a point occurs when the difference in path length from the two sources is zero or an integral number of wavelengths; destructive interference occurs when the path difference is a half-integral number of wavelengths. When the line from the sources to a point $P$ makes an angle $\theta$ with the line perpendicular to the line of the sources, and when the distance between sources is $d$, the condition for constructive interference is

$$d \sin \theta = m\lambda \quad (m = 0, \pm 1, \pm 2, \ldots). \tag{37–4}$$

The condition for destructive interference is

$$d \sin \theta = (m + \tfrac{1}{2})\lambda \quad (m = 0, \pm 1, \pm 2, \ldots). \tag{37–5}$$

When $\theta$ is very small, the position $y_m$ of the $m$th bright fringe is given by

$$y_m = R\frac{m\lambda}{d}. \tag{37–6}$$

• When two sinusoidal waves with equal amplitude $E$ and phase difference $\phi$ are superimposed, the resultant amplitude $E_P$ is

$$E_P{}^2 = 2E^2(1 + \cos \phi) = 4E^2 \cos^2 \frac{\phi}{2}, \tag{37–7}$$

and the intensity $I$ is

$$I = I_0 \cos^2 \frac{\phi}{2}. \tag{37–10}$$

When two sources emit waves in phase, the phase difference of the waves arriving at point $P$ is related to the difference in path length $(r_1 - r_2)$ by

$$\phi = \frac{2\pi}{\lambda}(r_1 - r_2) = k(r_1 - r_2). \tag{37–11}$$

• When light is reflected from both sides of a thin film of thickness $t$, constructive interference between the reflected waves occurs when

$$2t = m\lambda \quad (m = 0, 1, 2, \ldots), \tag{37–17}$$

unless a half-cycle phase shift occurs at one surface; then this is the condition for destructive interference. A half-cycle phase shift occurs during reflection whenever the index of refraction of the second material is greater than that of the first.

• The Michelson interferometer uses an extended monochromatic source and can be used for high-precision measurements of wavelengths. Its original purpose was to detect motion of the earth relative to a hypothetical ether, the supposed medium for electromagnetic waves. The concept of ether has been abandoned. The speed of light is the same relative to all observers. This is part of the foundation of the special theory of relativity.

• The energy in an electromagnetic wave is carried in bundles called photons. The energy $E$ of one photon is proportional to the frequency $f$:

$$E = hf = \frac{hc}{\lambda}, \tag{37–20}$$

where $h$ is a fundamental constant called Planck's constant.

## KEY TERMS

monochromatic light
interference
principle of linear superposition
constructive interference
destructive interference
coherent light
Michelson interferometer
Michelson-Morley experiment
ether
photon
Planck's constant

# EXERCISES

## Section 37–1 Interference and Coherent Sources

**37–1** Two coherent white light sources are in phase 1.80 $\mu$m apart and in line with the viewer. What visible wavelengths (400 nm to 700 nm) will the viewer see as brightest, owing to constructive interference?

**37–2 Radio Interference.** Two radio antennas $A$ and $B$ radiate in phase. Antenna $B$ is 110 m to the right of antenna $A$. Consider point $Q$ along the extension of the line connecting the antennas, a horizontal distance of 55 m to the right of antenna $B$ (Fig. 37–17). The frequency, and therefore the wavelength, of the emitted waves can be varied. a) What is the longest wavelength for which there will be destructive interference at point $Q$? b) What is the longest wavelength for which there will be constructive interference at point $Q$?

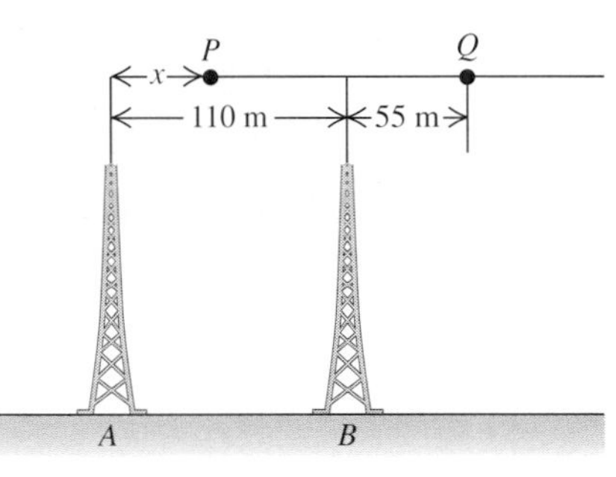


**FIGURE 37–17**

**37–3** A short-wave radio station operating at a frequency of 7.50 MHz has two identical antennas that radiate in phase. Antenna $B$ is 110 m to the right of antenna $A$ (Fig. 37–17). Consider point $P$ between the antennas and along the line connecting them, a horizontal distance $x$ to the right of antenna $A$. For what values of $x$ will constructive interference occur at point $P$?

## Section 37–2 Two-Source Interference

**37–4** Young's experiment is performed with sodium light ($\lambda$ = 589 nm). Fringes are measured carefully on a screen 1.20 m away from the double slit, and the center of the twentieth fringe (not counting the central bright fringe) is found to be 11.8 mm from the center of the central bright fringe. What is the separation of the two slits?

**37–5** Two slits are spaced 0.300 mm apart and are placed 60.0 cm from a screen. What is the distance between the second and third dark lines of the interference pattern when the slits are illuminated with coherent light with a wavelength of 600 nm?

**37–6** An FM radio station has a frequency of 100 MHz and uses two identical antennas mounted at the same elevation, 15.0 m apart. The antennas radiate in phase. The resulting radiation pattern has a maximum intensity along a horizontal line perpendicular to the line joining the antennas and midway between them. Assume that the intensity is observed at distances from the antennas that are much larger than 15.0 m. a) At what angles (measured from the line of maximum intensity) is the intensity maximum? b) At what angles is it zero?

**37–7** Coherent light from a mercury-arc lamp is passed through a filter that blocks everything except for one spectrum line in the green region of the spectrum. It then falls on two slits separated by 0.600 mm. In the resulting interference pattern on a screen 2.50 m away, adjacent bright fringes are separated by 2.27 mm. What is the wavelength?

## Section 37–3 Intensity Distribution in Interference Patterns

**37–8** Show that Eq. (37–14) gives zero-intensity directions that agree with Eq. (37–5).

**37–9** Two slits spaced 0.200 mm apart are placed 0.800 m from a screen and illuminated by coherent light that has a wavelength of 500 nm. The intensity at the center of the central maximum ($\theta = 0°$) is $6.00 \times 10^{-6}$ W/m$^2$. a) What is the distance on the screen from the center of the central maximum to the first minimum? b) What is the intensity at a point midway between the center of the central maximum and the first minimum?

**37–10 Intensity of Radio Signals.** An AM radio station has a frequency of 820 kHz; it uses two identical antennas at the same elevation, 150 m apart. The antennas radiate in phase, and the resulting interference pattern is observed at distances from the antennas that are much greater than 150 m. a) In what direction is the intensity maximum, considering points in a horizontal plane? b) Calling the maximum intensity in part (a) $I_0$, determine in terms of $I_0$ the intensity in directions making angles of 30°, 45°, 60°, and 90° to the direction of maximum intensity.

**37–11** Two slits spaced 0.300 mm apart are placed 0.600 m from a screen and illuminated by coherent light that has a wavelength of 540 nm. The intensity at the center of the central maximum ($\theta = 0°$) is $I_0$. a) What is the distance on the screen from the center of the central maximum to the first minimum? b) What is the distance on the screen from the center of the central maximum to the point where the intensity has fallen to $I_0/2$?

## Section 37–4 Interference in Thin Films

**37–12** Light with wavelength 500 nm is incident perpendicularly from air on a film $8.33 \times 10^{-5}$ cm thick and with refractive index 1.35. Part of the light is reflected from the first surface of the film, and part enters the film and is reflected back at the second surface, where the film is again in contact with air. a) How many waves are contained along the path of this second part of the light in the film? b) What is the phase difference between these two parts of the light as they leave the film?

**37–13** Suppose the top plate in Example 37–4 is plastic with $n = 1.40$, the wedge is filled with silicone grease having $n = 1.50$, and the bottom plate is glass with $n = 1.60$. Calculate the spacing between the dark fringes.

**37–14** A plate of glass 10.0 cm long is placed in contact with a second plate and is held at a small angle with it by a metal strip 0.100 mm thick placed under one end. The space between the plates is filled with air. The glass is illuminated from above with light having a wavelength of 520 nm. How many interference fringes are observed per centimeter in the reflected light?

**37–15** Two rectangular pieces of plate glass are laid one on the other on a table. A thin strip of paper is placed between them at one edge so that a very thin wedge of air is formed. The plates are illuminated by a beam of sodium light at normal incidence ($\lambda_0 = 589$ nm). Interference fringes are formed, with 20 per centimeter length of wedge measured normal to the edges in contact. Find the angle of the wedge.

**37–16** A thin film of polystyrene is used as a nonreflecting coating for fabulite. (See Table 34–1.) What is the minimum thickness of the film required? Assume that the wavelength of the light is 550 nm.

**37–17** What is the thinnest film of a coating with $n = 1.50$ on glass ($n = 1.60$) for which destructive interference of the green component (500 nm) of an incident white light beam in air can take place by reflection?

**37–18 Reflective Coating on a Car Window.** A plastic film with index of refraction 2.50 is put on the surface of a car window to increase the reflectivity and keep the interior of the car cooler. The window glass has index of refraction 1.60. a) What minimum thickness is required if light of wavelength 600 nm reflected from the two sides of the film is to interfere constructively? b) It is found to be difficult to manufacture and install coatings as thin as calculated in part (a). What is the next greatest thickness for which there will also be constructive interference?

## Section 37–5 The Michelson Interferometer

**37–19** How far must the mirror $M_2$ (Fig. 37–15) of the Michelson interferometer be moved so that 2000 fringes of krypton-86 light ($\lambda = 606$ nm) move across a line in the field of view?

**37–20** John first uses the 633-nm light from a helium-neon laser with a Michelson interferometer. He displaces the movable mirror away from him, counting 986 fringes moving across a line in his field of view. Then Linda replaces the laser with filtered 589-nm light from a sodium lamp and displaces the movable mirror toward her. She also counts 986 fringes, but they move across the line in her field of view opposite to the direction they moved for John. Assume that both John and Linda counted to 986 correctly. a) What distance did each person move the mirror? b) What is the resultant displacement of the mirror?

## Section 37–6 The Photon: A Case Study in Quantum Physics

**37–21** What is the range of photon energies for visible light, for which the wavelength range is 400 nm to 700 nm?

**37–22** An FM radio station broadcasts at 99.3 MHz with an output power of 50,000 W. How many photons does it emit during a 30-s commercial?

**37–23 Reaction Force on a Laser.** a) The photons leaving a laser exert a reaction force of magnitude $F$ on the laser. Show that $F = P/c$, where $P$ is the power output of the laser. b) An extraordinarily powerful laser used in controlled fusion experiments emits pulses of light with a peak power of $6.0 \times 10^{13}$ W. What is the magnitude of the reaction force at this peak power?

**37–24** How is the photon flux $\Phi$ (the number of photons per perpendicular area per time) related to the intensity $I$ and frequency $f$ of a beam of monoenergetic photons?

**37–25** An iron-57 nucleus has a mass of $9.454 \times 10^{-26}$ kg. It recoils with a kinetic energy of $3.132 \times 10^{-22}$ J after emitting a gamma ray photon. Find the magnitude of the momentum of the gamma ray, as well as its energy, frequency, and wavelength.

**37–26** A particle with mass $m$ has a speed $v$ and momentum with magnitude $mv$. a) In terms of these quantities, what is the de Broglie wavelength of the particle? b) What is the energy of a photon that has the same wavelength as the particle? c) A proton is traveling with speed $5.00 \times 10^4$ m/s. What is the energy of the photon that has the same wavelength as the proton? Express your answer in eV.

# PROBLEMS

**37–27 Two-Slit Interference in Water.** Suppose the entire apparatus (slits, screen, and space in between) of Exercise 37–5 is immersed in water. Then what is the distance between the second and third dark lines?

**37–28** Parallel light rays with wavelength $\lambda = 600$ nm fall on a double slit. On a screen 3.00 m away the distance between dark fringes is 3.60 mm. What is the separation of the slits?

**37–29** Two identical audio speakers connected to the same amplifier produce sound waves with frequency between 400 and 800 Hz. The speed of sound is 340 m/s. You find that where you are standing, you hear minimum-intensity sound. a) Explain why you hear minimum-intensity sound. b) To help you, a friend moves one of the speakers toward you. When the speaker has been moved 0.34 m, the sound that you hear has maximum intensity. What is the frequency of the sound? c) How much closer to you from the position in part (b) must the speaker be moved to the next position where you hear maximum intensity?

**37–30** Two radio antennas radiating in phase are located at points $A$ and $B$, 200 m apart. The radio waves have a frequency of 6.00 MHz. A radio receiver is moved out from point $B$ along a line perpendicular to the line connecting $A$ and $B$ (line $BC$ in Fig. 37–18, on page 1044). At what distances from $B$ will there be *destructive* interference? [*Note:* The distance of the receiver from the sources is not large in comparison to the separation of the sources, so Eq. (37–5) does not apply.]

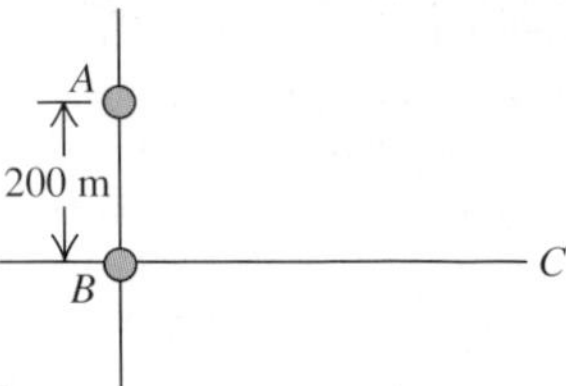


**FIGURE 37–18**

**37–31** Consider a two-slit interference pattern, for which the intensity distribution is given by Eq. (37–14). Let $\theta_m$ be the angular position of the $m$th bright fringe, where the intensity is $I_0$. Assume that $\theta_m$ is small, so $\sin\theta_m \cong \theta_m$. Let $\theta_m^{\,+}$ and $\theta_m^{\,-}$ be the two angles on either side of $\theta_m$ for which $I = \frac{1}{2}I_0$. The quantity $\Delta\theta_m = |\theta_m^{\,+} - \theta_m^{\,-}|$ is the half-width of the $m$th fringe. Calculate $\Delta\theta_m$. How does $\Delta\theta_m$ depend on $m$?

**37–32** What is the thinnest soap film that appears black when illuminated with light with a wavelength of 520 nm? The index of refraction of the film is 1.33, and there is air on both sides of the film.

**37–33** A glass plate that is 0.375 $\mu$m thick and surrounded by air is illuminated by a beam of white light normal to the plate. The index of refraction of the glass is 1.50. What wavelengths within the limits of the visible spectrum ($\lambda$ = 400 nm to 700 nm) are intensified in the reflected beam?

**37–34 Color of an Oil Spill.** An oil tanker spills a large amount of oil ($n = 1.40$) into the sea. Assume that the refractive index of sea water is 1.33. a) If you are overhead and look down onto the oil spill, what predominant color do you see at a point where the oil is 390 nm thick? b) If you swim under the slick and look up at the same place in the slick as in part (a), what visible wavelength (as measured in air) is predominant in the transmitted light?

**37–35** The radius of curvature of the convex surface of a plano-convex lens is 1.30 m. The lens is placed convex side down on a glass plate that is perfectly flat and illuminated from above with red light having a wavelength of 650 nm. Find the diameter of the third bright ring in the interference pattern.

**37–36** Newton's rings can be seen when a plano-convex lens is placed on a flat glass surface (Problem 37–35). If the lens has an index of refraction of $n = 1.50$ and the glass plate has an index of $n = 1.80$, the diameter of the second bright ring is 0.600 mm. If a liquid with an index of 1.30 is added to the space between the lens and the plate, what is the new diameter of this ring?

**37–37** In a Young's two-slit experiment a piece of glass with an index of refraction $n$ and a thickness $L$ is placed in front of the upper slit, as in Figure 37–3. a) Describe qualitatively what happens to the interference pattern. b) Derive an expression for the intensity $I$ of the light at points on a screen as a function of $n$, $L$, and $\theta$. Here $\theta$ is the usual angle measured from the center of the two slits. That is, determine the equation that is analogous to Eq. (37–14) but that also involves $L$ and $n$ for the glass plate. c) From your result in part (b), derive an expression for the values of $\theta$ that locate the maxima in the interference pattern [that is, derive an equation that is analogous to Eq. (37–4)].

**37–38 Index of Refraction of Gases.** A Michelson interferometer can be used to measure the index of refraction of a gas by placing an initially evacuated tube in one arm of the interferometer. The gas is then slowly added to the tube, and the number of fringes that cross the telescope crosshairs are counted. If the length of the tube is 4.00 cm and the light source is a sodium lamp (589 nm), what is the index of refraction of the gas if 40 fringes are seen to pass the view of the telescope? (*Note:* For gases it is convenient to give the value of $n - 1$ rather than $n$ itself, since the index differs little from unity.)

## CHALLENGE PROBLEMS

**37–39** The index of refraction of a glass rod is 1.48 at $T = 20.0$°C and varies linearly with temperature, with a coefficient of $2.50 \times 10^{-5}$/C°. The coefficient of linear expansion of the glass is $5.00 \times 10^{-6}$/C°. At 20.0°C the length of the rod is 2.00 cm. A Michelson interferometer has this glass rod in one arm, and the rod is being heated so that its temperature increases at a rate of 5.00 C°/min. The light source has wavelength $\lambda = 589$ nm, and the rod is initially at $T = 20.0$°C. How many fringes cross the field of view each minute?

**37–40** Figure 37–19 shows an interferometer known as *Fresnel's biprism*. The magnitude of the prism angle $A$ is extremely small. a) If $S_0$ is a very narrow source slit, show that the separation of the two virtual coherent sources $S_1$ and $S_2$ is given by $d = 2aA(n - 1)$, where $n$ is the index of refraction of the material of the prism. b) Calculate the spacing of the fringes of green light with wavelength 500 nm on a screen 2.00 m from the biprism. Take $a = 0.200$ m, $A = 5.00$ mrad, and $n = 1.50$.

**FIGURE 37–19**

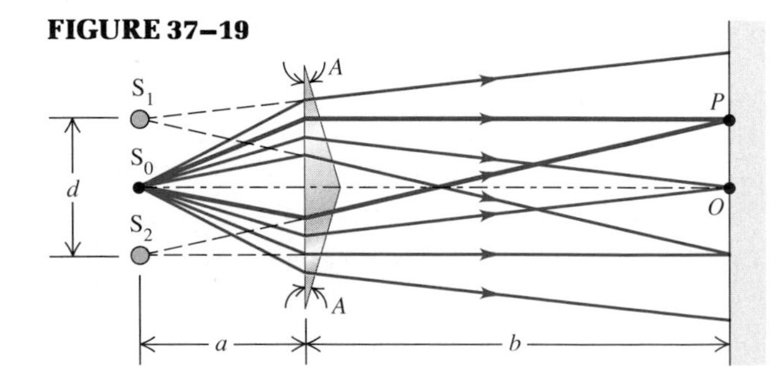

A hologram is an interference pattern formed by a reference beam of light and light reflected from an object and then recorded on photographic film. A three-dimensional image of the object can be reconstructed from this interference pattern, so that the image changes its appearance as you change your viewpoint. Because a high-quality interference pattern requires monochromatic, coherent light, holograms are normally made with laser light.

# 38

C H A P T E R

# Diffraction

Laser beam
Object beam
Beam splitter
Reference beam
Mirror
Mirror
Mirror
Photographic film

To form a useable hologram, the laser beam is split into two parts. The reference beam is directed right onto the photographic film, where it interferes with the second beam, which has been reflected off the object.

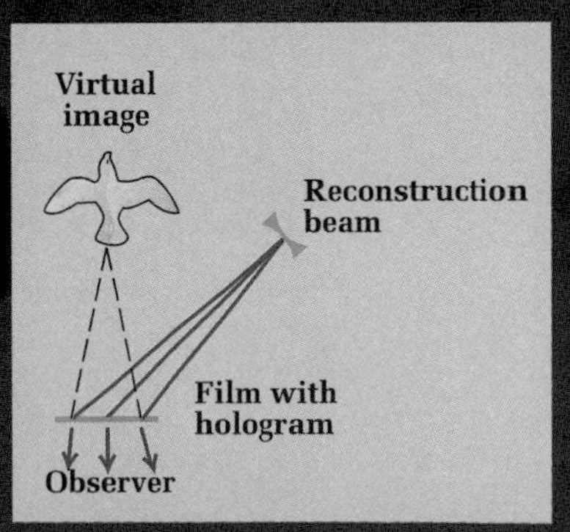


To view the holographic image, you look at light that has been transmitted through the developed film. The pattern of light waves appears to be coming from the original object, and you see a virtual image of it. In addition, part of the transmitted light forms a real image of the object (not shown).

## KEY CONCEPTS

**• Interference patterns formed when light passes through an aperture or around an edge are called diffraction patterns.**

**• Light emerging from a single rectangular slit forms a pattern similar but not identical to a two-source interference pattern. The intensity distribution can be calculated by vector addition of phasors, as with the two-source pattern.**

**• Diffraction patterns from multiple slits can be analyzed by the same methods that are used for a single slit. An array of many regularly spaced slits, called a diffraction grating, is useful for precise measurements of wavelengths of light.**

**• Interference patterns are formed when x rays are scattered from atoms in a crystal lattice. This effect, called x-ray diffraction, is an important research tool in investigating the structure of condensed matter.**

**• Diffraction patterns are formed when light passes through a circular aperture. Diffraction limits the resolution of optical instruments.**

**• A hologram is a special kind of interference pattern. A hologram forms a three-dimensional image that can be viewed from various directions.**

## INTRODUCTION

Holograms are turning up everywhere these days — on credit cards, postage stamps, and even cereal boxes. But what *is* a hologram? It's just a special kind of interference pattern, recorded on a photographic film and then reproduced. When properly illuminated, it forms a three-dimensional image of the original object.

We'll get to holograms later in this chapter; first we explore some simpler optical effects that involve the wave properties of light. When light from a point source falls on a straight edge, casting a shadow, the edge of the shadow is never perfectly sharp. Some light appears in the area that we expect to be in the shadow, and we find alternating bright and dark fringes in the illuminated area. More generally, light emerging from apertures doesn't behave precisely according to the predictions of the straight-line ray model of geometric optics. Light emerging from arrays of apertures forms interesting light-and-dark patterns, and by measuring these we can determine the wavelength precisely. Similar effects with x rays can be used to explore the atomic structure of solids and liquids. Waves do a lot of surprising things, and studying interference and diffraction effects will help us understand them. ■

## 38–1 FRESNEL AND FRAUNHOFER DIFFRACTION

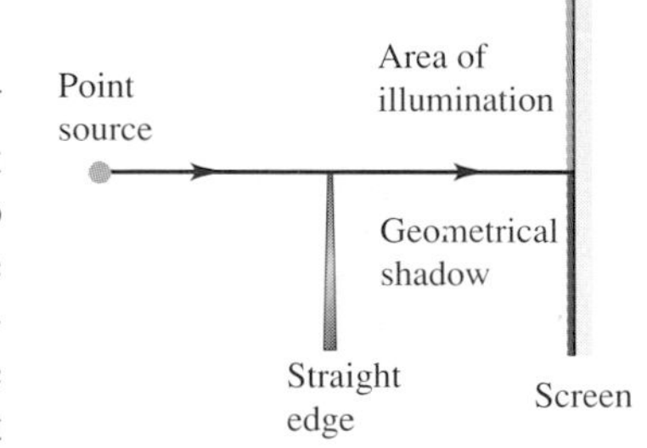


**38–1** Geometric optics predicts that a straight edge should give a shadow with a sharp boundary and a region of relatively uniform illumination above it.

According to *geometric* optics, when an opaque object is placed between a point light source and a screen, as in Fig. 38–1, the shadow of the object forms a perfectly sharp line. No light at all strikes the screen at points within the shadow, and the area outside the shadow is illuminated nearly uniformly. But we have already seen in the preceding chapter that the *wave* nature of light causes things to happen that can't be understood with the simple model of geometric optics. An important class of such effects occurs when light strikes a barrier with an aperture or an edge. The interference patterns formed in such a situation are grouped under the heading **diffraction.**

Here's an example. The photograph in Fig. 38–2a was made by placing a razor blade halfway between a pinhole, illuminated by monochromatic light, and a photographic film. The film recorded the shadow cast by the blade. Figure 38–2b is an enlargement of a region near the shadow of the left edge of the blade. The position of the *geometric* shadow line is indicated by arrows. The area outside the geometric shadow is bordered by alternating bright and dark bands. There is some light in the shadow region, although this is not very visible in the photograph. The first bright band, just outside the geometric shadow, is actually *brighter* than in the region of uniform illumination to the extreme left. This simple experiment gives us some idea of the richness and complexity of a seemingly simple phenomenon, the casting of a shadow by an opaque object.

We don't often observe diffraction patterns such as Fig. 38–2 in everyday life because most ordinary light sources are not monochromatic and are not point sources. If we use a frosted light bulb instead of a point source in Fig. 38–1, each wavelength of the light from every point of the bulb forms its own diffraction pattern, but the patterns overlap to such an extent that we can't see any individual pattern.

Diffraction is sometimes described as "the bending of light around an obstacle." But the process that causes diffraction effects is present in the propagation of *every* wave. When part of the wave is cut off by some obstacle, we observe diffraction effects that result from interference of the remaining parts of the wave fronts. Every optical instrument uses only a limited portion of a wave; for example, a telescope uses only the part of a wave admitted by its objective lens. Thus diffraction plays a role in nearly all optical phenomena.

Figure 38–3 (on page 1048) shows a diffraction pattern formed by a steel ball about 3 mm in diameter. Note the rings in the pattern, both outside and inside the geometric shadow area, and the bright spot at the very center of the shadow. The existence of this

**38–2** (a) Actual shadow of a razor blade illuminated by monochromatic light from a point source. (b) Enlarged shadow of the straight edge. The arrows show the position of the *geometric* shadow.

(a)

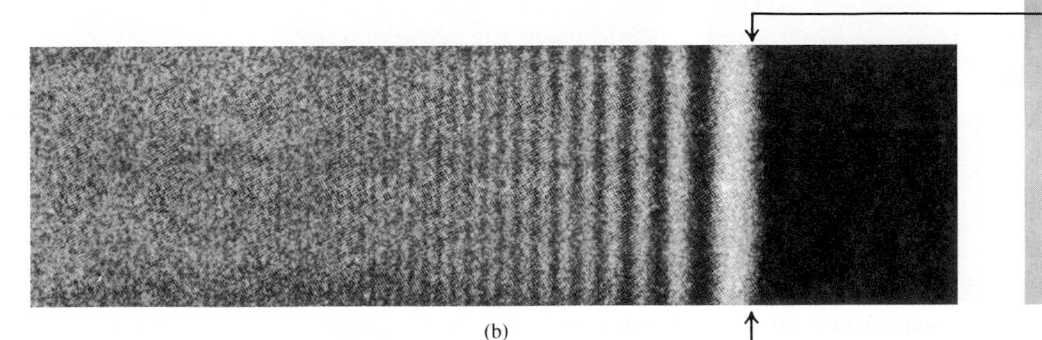

(b)

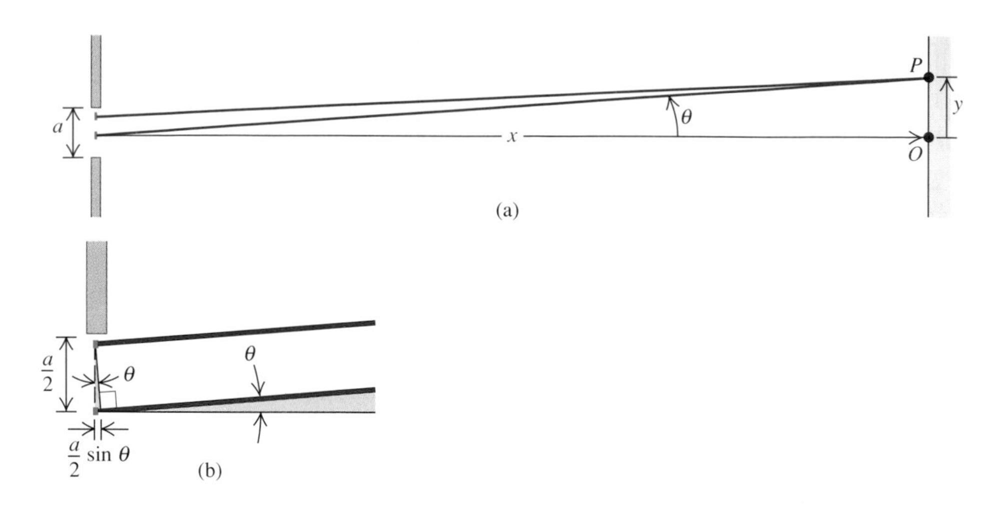


**38–6** Side view of a horizontal slit. (a) When the distance $x$ to the screen is much greater than the slit width $a$, the rays from a distance $a/2$ apart may be considered parallel. (b) Enlargement of half the slit. The ray from the middle of the slit travels a distance $(a/2)\sin\theta$ farther to point $P$ than the ray from the top edge of the slit.

pattern. That is, a dark fringe occurs whenever

$$\frac{a}{2}\sin\theta = \pm\frac{\lambda}{2}, \qquad \text{or} \qquad \sin\theta = \pm\frac{\lambda}{a}. \tag{38–1}$$

We may also divide the screen into quarters, sixths, and so on and use the above argument to show that a dark fringe occurs whenever $\sin\theta = 2\lambda/a,\ 3\lambda/a,$ and so on. Thus the condition for a *dark* fringe is

$$\sin\theta = \frac{m\lambda}{a} \quad (m = \pm 1, \pm 2, \pm 3, \ldots). \tag{38–2}$$

For example, if the slit width is equal to ten wavelengths ($a = 10\lambda$), dark fringes occur at $\sin\theta = \pm\frac{1}{10}, \pm\frac{2}{10}, \pm\frac{3}{10}, \ldots$. Between the dark fringes are bright fringes. We also note that $\sin\theta = 0$ is a *bright* band; in this case, light from the entire slit arrives at $P$ in phase. It would be wrong to put $m = 0$ in Eq. (38–2). The central bright fringe is wider than the others, as Fig. 38–4b shows. In the small-angle approximation we will use below, it is exactly *twice* as wide.

With light, the wavelength $\lambda$ is of the order of 500 nm $= 5 \times 10^{-7}$ m. This is often much smaller than the slit width $a$; a typical slit width is $10^{-2}$ cm $= 10^{-4}$ m. Therefore the values of $\theta$ in Eq. (38–2) are often so small that the approximation $\sin\theta = \theta$ is very good. In that case we can rewrite this equation as

$$\theta = \frac{m\lambda}{a}, \quad (m = \pm 1, \pm 2, \pm 3, \ldots).$$

Also, if the distance from slit to screen is $R$, and the vertical distance of the $m$th dark band from the center of the pattern is $y_m$, then $\tan\theta = y_m/R$. For small $\theta$ we may approximate $\tan\theta$ by $\theta$, and we then find

$$y_m = R\frac{m\lambda}{a}. \tag{38–3}$$

This equation has the same form as the equation for the two-slit pattern, Eq. (37–6), but here it gives the positions of the *dark* fringes in a *single-slit* pattern rather than the *bright* fringes in a *double-slit* pattern. Also, $m = 0$ is *not* a dark fringe. Be careful!

## EXAMPLE 38–1

You take a photograph of a straight black line drawn on a piece of white paper. The resulting negative has a transparent line surrounded by opaque black. Then you pass 633-nm laser light through the negative and observe the diffraction pattern on a screen 6.0 m away. You find that the distance between centers of the first minima beside the central bright fringe in the pattern is 27 mm. How wide is the clear line on the negative?

**SOLUTION** The angle $\theta$ in this situation is very small, so we can use the approximate relation of Eq. (38–3). Figure 38–6 shows the various distances. The distance $y_1$ from the central maximum to the first minimum on either side is half the distance between the two first minima, so $y_1 = (27 \text{ mm})/2$. Solving Eq. (38–3) for the slit width $a$ and substituting $m = 1$, we find

$$a = \frac{R\lambda}{y_1} = \frac{(6.0 \text{ m})(633 \times 10^{-9} \text{ m})}{(27 \times 10^{-3} \text{m})/2}$$
$$= 2.8 \times 10^{-4} \text{ m} = 0.28 \text{ mm}.$$

Can you show that the distance between the *second* minima on the two sides is 2(27 mm), and so on?

# 38–3 INTENSITY IN THE SINGLE-SLIT PATTERN

We can derive an expression for the intensity distribution for the single-slit pattern by the same method that we used to obtain Eq. (37–15) for the two-slit pattern. We again imagine a plane wave front at the slit subdivided into a large number of strips. We superpose the contributions of the Huygens' wavelets from all the strips at a point $P$ on a distant screen at an angle $\theta$ from the normal to the slit plane. To do this, we represent the sinusoidally varying $\boldsymbol{E}$ field from each strip with a phasor. Because of path differences, there are phase differences between the phasors. The magnitude of the vector sum of the phasors at each point $P$ is the amplitude $E_P$ of the total $\boldsymbol{E}$ field at that point.

At the point $O$ shown in Figure 38–6a, corresponding to the center of the pattern, where $\theta = 0$, there are *no* path differences; the phasors are all *in phase* (that is, have the same direction). In Fig. 38–7a we draw the phasors at time $t = 0$ and denote the resultant amplitude at $O$ by $E_0$. In this illustration we have divided the slit into 14 strips.

With this same subdivision, the wavelets at angle $\theta$ that arrive at $P$ from adjacent strips have phase differences between them because of the differences in path length, and the corresponding phasor diagram is shown in Fig. 38–7b. The vector sum of the phasors is now part of the perimeter of a many-sided polygon, and $E_P$, the amplitude of the resultant electric field at $P$, is the *chord*. The angle $\beta$ is the total phase difference between the wave from the top strip of Fig. 38–6 and the wave from the bottom strip, that is, the phase of the top strip with respect to the bottom strip.

We may imagine dividing the slit into narrower and narrower strips. In the limit, as the number of strips increases indefinitely, the phasor diagram becomes an *arc of a circle* (Fig. 38–7c), with arc length equal to the length $E_0$ in Fig. 38–7a. We can find the center $C$ of this arc by constructing perpendiculars at $A$ and $B$. From the definition of radian measure of angles, the radius of the arc is $E_0/\beta$, and the resultant amplitude $E_P$ (chord $AB$) is $2(E_0/\beta)\sin(\beta/2)$. We then have

$$E_P = E_0 \frac{\sin(\beta/2)}{\beta/2}. \tag{38–4}$$

The intensity at each point on the screen is proportional to the square of the amplitude, so if $I_0$ is the intensity in the straight-ahead direction for which $\theta = 0$ and $\beta = 0$, then the intensity $I$ at any point is

$$I = I_0 \left[\frac{\sin(\beta/2)}{(\beta/2)}\right]^2. \tag{38–5}$$

We can express the phase difference $\beta$ in terms of geometric quantities, as we did for the two-slit pattern. From Eq. (37–11) the phase difference is $2\pi/\lambda$ times the path difference. From Fig. 38–6 the path difference between the ray from the top of the slit

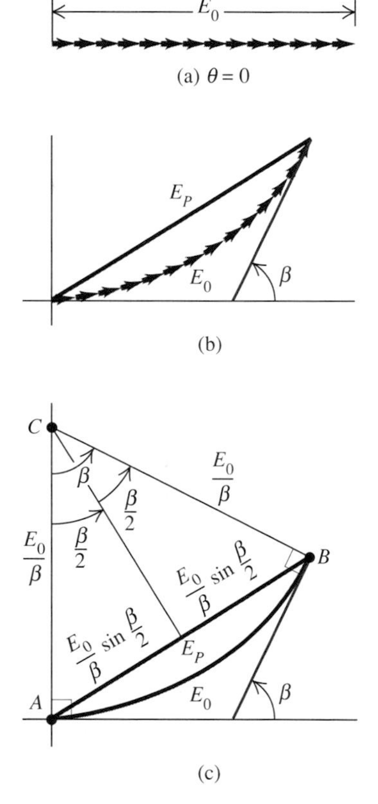


**38–7** (a) Phasor diagram when all elementary electric fields are in phase ($\theta = 0$, $\beta = 0$). (b) Phasor diagram when each elementary electric field differs in phase slightly from the preceding one. The total phase difference between the first and last phasors is $\beta$. (c) Limit reached by the phasor diagram when the slit is subdivided infinitely.

and that from the bottom is $a \sin\theta$. Therefore,

$$\beta = \frac{2\pi}{\lambda} a \sin\theta, \tag{38-6}$$

and Eq. (38–5) becomes

$$I = I_0 \left\{ \frac{\sin[\pi a(\sin\theta)/\lambda]}{\pi a(\sin\theta)/\lambda} \right\}^2. \tag{38-7}$$

This equation expresses the intensity directly in terms of the angle $\theta$. For actual calculations it is often easier first to calculate the phase angle $\beta$, using Eq. (38–6), and then to use Eq. (38–5).

Equation (38–7) is plotted in Fig. 38–8a, and a photograph of an actual diffraction pattern (turned 90° from the orientation of Fig. 38–4) is shown in Fig. 38–8b. Note that the central intensity peak is much larger than any of the others and that the peak intensities drop off rapidly as we go away from the center of the pattern.

The dark fringes in the pattern, the places where $I = 0$, occur when the numerator of Eq. (38–5) is zero. This happens when $\beta$ is a multiple of $2\pi$; from Eq. (38–6),

$$\frac{a\sin\theta}{\lambda} = m \quad (m = \pm 1, \pm 2, \ldots),$$

or

$$\sin\theta = \frac{m\lambda}{a} \quad (m = \pm 1, \pm 2, \ldots). \tag{38-8}$$

This agrees with our previous result, Eq. (38–2). Note again that $\beta = 0$ (corresponding to $\theta = 0$) is *not* a minimum. Equation (38–5) is indeterminate at $\beta = 0$, but we can evaluate the limit as $\beta \rightarrow 0$ using L'Hôpital's rule. We find that at $\beta = 0$, $I = I_0$, as expected.

We can also calculate the positions of the peaks and the peak intensities from Eq. (38–5). This is not quite as simple as it may appear. We might expect the peaks to occur where the sine function reaches the values $\pm 1$, namely, when $\beta = \pm\pi, \pm 3\pi, \pm 5\pi$:

$$\beta \cong \pm(2m+1)\pi \quad (m = 1, 2, 3, \ldots). \tag{38-9}$$

This is *approximately* correct, but because of the factor $(\beta/2)^2$ in the denominator of Eq. (38–5), the maxima don't occur precisely at these points. When we take the derivative of Eq. (38–5) with respect to $\beta$ and set it equal to zero, to try to find the maxima and minima, we get a transcendental equation that has to be solved numerically. There is *no* maximum near $\beta = \pm\pi$ because of the rapidly changing denominator. The first side maxima, near $\beta = \pm 3\pi$, actually occur at $\pm 2.860\pi$. The second maxima, near

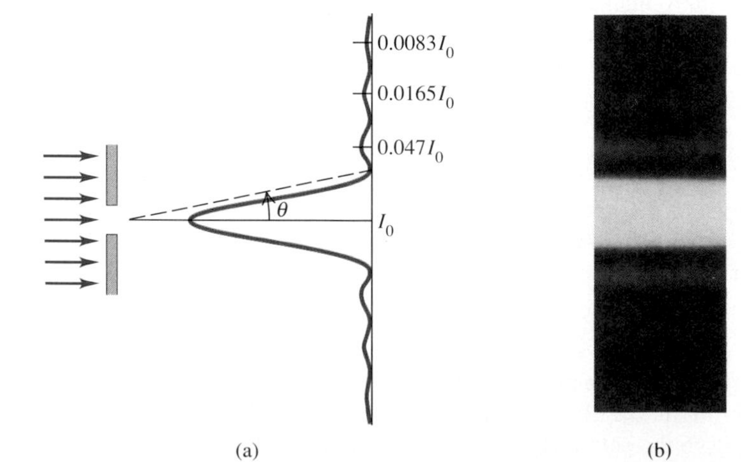


**38–8** (a) Intensity distribution for a single slit. (b) Photograph of the Fraunhofer diffraction pattern of a single slit.

$\beta = \pm 5\pi$, are actually at $\pm 4.918\pi$, and so on. The error in Eq. (38–9) vanishes in the limit of large $m$.

To find the intensities at the side maxima, we substitute these values of $\beta$ back into Eq. (38–5). Using the approximate expression in Eq. (38–9), we get

$$I_m \cong \frac{I_0}{(m + \frac{1}{2})^2 \pi^2}, \qquad (38\text{–}10)$$

where $I_m$ is the intensity of the $m$th side maximum and $I_0$ is the intensity of the central maximum. This formula gives the series of intensities

$$0.0450 I_0, \quad 0.0162 I_0, \quad 0.0083 I_0,$$

and so on. As we have pointed out, these equations are only approximately correct. The actual intensities of the side maxima turn out to be

$$0.0472 I_0, \quad 0.0165 I_0, \quad 0.0083 I_0, \quad \ldots.$$

Note that the intensities drop off very rapidly, as Fig. 38–8 also shows. Even the first side maxima have less than 5% of the intensity of the central maximum.

For small angles the angular spread of the diffraction pattern is inversely proportional to the slit width $a$, or, more precisely, to the ratio of $a$ to the wavelength $\lambda$. Figure 38–9 shows graphs of intensity $I$ as a function of $\theta$ for various values of $a/\lambda$.

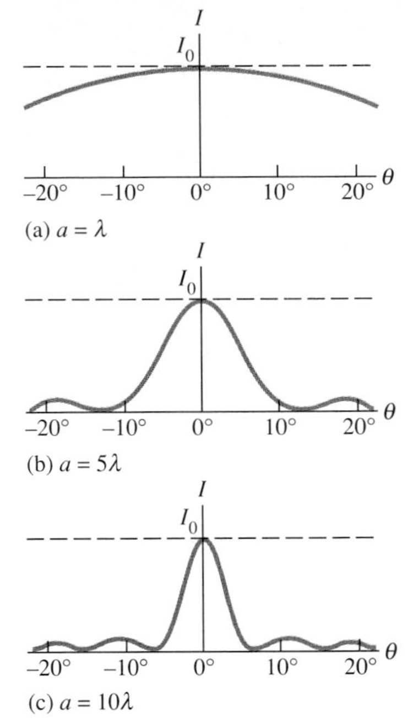


**38–9** (a) When the slit width $a$ is less than or equal to the wavelength $\lambda$, the intensity beyond the slit is spread over all angles. (b) Increasing the slit width decreases the angular width of the central maximum, as does (c) decreasing the wavelength.

With light, the wavelength $\lambda$ is ordinarily much smaller than the slit width $a$, and the values of $\theta$ in Eqs. (38–6) and (38–7) are so small that the approximation $\sin\theta = \theta$ is very good. With this approximation the position $\theta_1$ of the first minimum beside the central maximum, corresponding to $\beta/2 = \pi$, is, from Eq. (38–6),

$$\theta_1 = \frac{\lambda}{a}. \qquad (38\text{–}11)$$

This characterizes the width of the central maximum, and we see that it is *inversely* proportional to the slit width $a$. When the small-angle approximation is valid, the central maximum is exactly twice as wide as each side maximum. When $a$ is of the order of a centimeter or more, $\theta_1$ is so small that we can consider practically all the light to be concentrated at the geometric focus. But when $a$ is less than $\lambda$, the central maximum spreads over 180°, and the fringe pattern is not seen at all.

## EXAMPLE 38–2

a) In a single-slit diffraction pattern, what is the intensity at a point in the pattern where the total phase difference between wavelets from the top and bottom of the slit is 66 rad? b) If this point is 7.0° away from the central maximum, how many wavelengths wide is the slit?

**SOLUTION** a) We are given $\beta = 66$ rad, so $\beta/2 = 33$ rad; Eq. (38–7) becomes

$$I = I_0\left[\frac{\sin(33\text{ rad})}{33\text{ rad}}\right]^2 = (9.2 \times 10^{-4}) I_0.$$

This is the intensity of the *tenth* side maximum, which occurs at approximately $\beta = 21\pi$, approximately midway between the minima at $\beta = 20\pi$ and $22\pi$. (The actual location of this maximum is at $\beta = 20.98\pi$.)

b) We solve Eq. (38–6) for $a$:

$$a = \frac{\beta\lambda}{2\pi \sin\theta} = \frac{(66\text{ rad})\lambda}{(2\pi\text{ rad})\sin 7.0^\circ} = 86\lambda.$$

For example, for 550-nm light the slit width is (86)(550 nm) = $4.7 \times 10^{-5}$ m = 0.047 mm, or roughly $\frac{1}{20}$ mm.

## 38–4 MULTIPLE SLITS

In Sections 37–2 and 37–3 we analyzed interference from two point sources or from two very thin slits, and in Sections 38–2 and 38–3 we carried out a similar analysis for a single slit with finite width. When we have several slits or slits with finite width, additional interesting features occur.

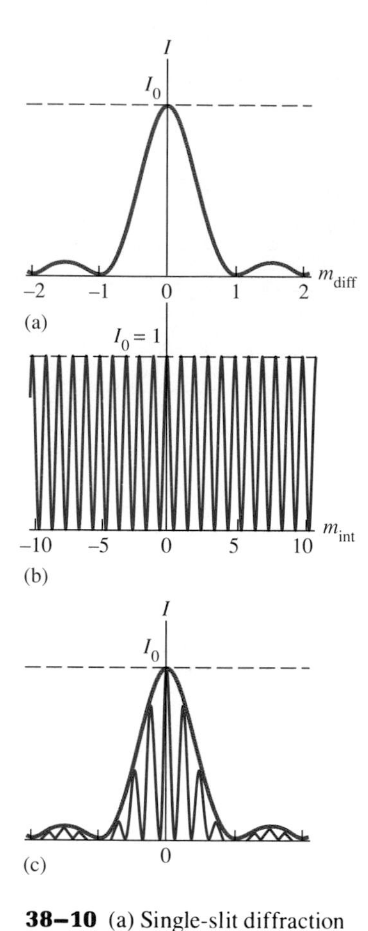


**38–10** (a) Single-slit diffraction pattern for a slit of width $a$. (b) Double-slit interference pattern for narrow slits whose separation $d$ is five times the slit width $a$. (c) Double-slit pattern for $d = 5a$. Every fifth interference maximum at the sides is missing because the interference maxima coincide with diffraction minima.

Let's take another look at the two-slit pattern. If the slits are very narrow (in comparison to the wavelength), as we assumed in Section 37–2, we can assume that light from each slit spreads out uniformly in all directions to the right of the slit. We used this assumption to calculate the interference pattern described by Eq. (37–10) or (37–15), consisting of a series of equally spaced, equally intense maxima. However, when the slits have finite width, the peaks in the two-slit interference pattern are modulated by the single-slit diffraction pattern that is characteristic of the width of each slit.

Figure 38–10a shows the intensity in a single-slit diffraction pattern with slit width $a$. Figure 38–10b shows the pattern formed by two very narrow slits with distance $d$ between slits, where $d$ is five times as great as the single-slit width $a$ in Fig. 38–10a. That is, $d = 5a$. We note that the minima in the single-slit pattern are five times as far apart as those in the two-slit pattern. Now suppose we widen each of the narrow slits to the same width $a$ as that of the single slit. Figure 38–10c shows the pattern from two slits with width $a$, separated by a distance (between centers) $d$. We see that the effect of the finite width of the slits is to superimpose the two patterns, that is, to multiply the two intensities at each point. The two-slit peaks are in the same positions as before, but their intensities are modulated by the single-slit pattern. The expression for the intensity shown in Fig. 38–10c is proportional to the product of the two-slit and single-slit expressions, Eqs. (37–10) and (38–5):

$$I = I_0 \cos^2 \frac{\phi}{2}\left[\frac{\sin(\beta/2)}{\beta/2}\right]^2, \qquad (38\text{–}12)$$

where, as before,

$$\phi = \frac{2\pi d}{\lambda}\sin\theta, \qquad \beta = \frac{2\pi a}{\lambda}\sin\theta.$$

Figure 38–11a shows a graph of Eq. (38–12) for the particular case $d = 4a$, and Fig. 38–11b is a photograph of an actual pattern. Shall we call this *interference* or *diffraction?* It is really both. The pattern results from superposition of waves coming from various parts of the two apertures, and this pattern shows that there is no fundamental distinction between interference and diffraction.

Next let's consider patterns produced by *several* very narrow slits. Assume that each slit is narrow in comparison to the wavelength, so its diffraction pattern spreads out nearly uniformly. Figure 38–12 shows an array of eight narrow slits, with distance $d$ between adjacent slits. Constructive interference occurs for rays at angle $\theta$ to the normal that arrive at point $P$ when the path difference between adjacent slits is an integral number of wavelengths,

$$d\sin\theta = m\lambda \quad (m = 0, 1, 2, \ldots).$$

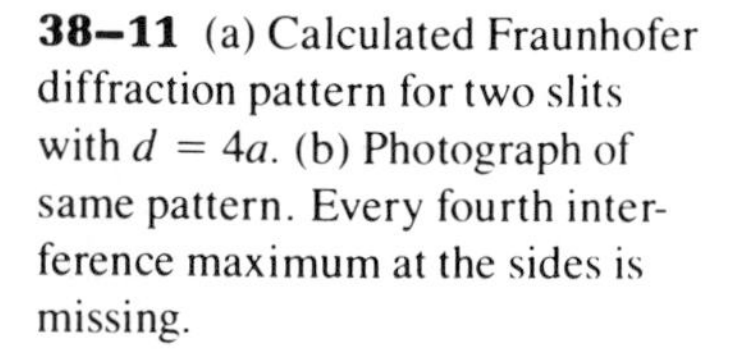

**38–11** (a) Calculated Fraunhofer diffraction pattern for two slits with $d = 4a$. (b) Photograph of same pattern. Every fourth interference maximum at the sides is missing.

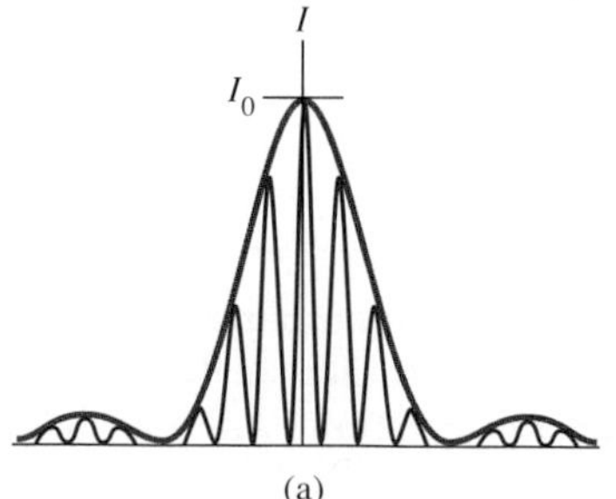

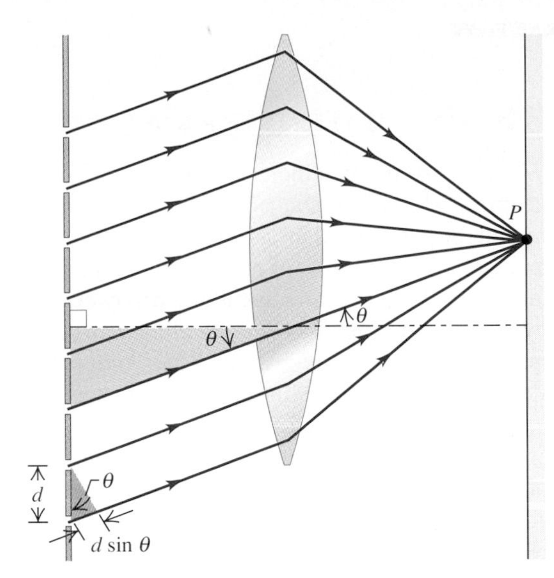


**38–12** In multiple-slit diffraction, rays from every slit arrive in phase to give a sharp maximum if the path difference between adjacent slits is a whole number of wavelengths.

Alternatively, reinforcement occurs when the phase difference $\phi$ at $P$ for light from adjacent slits is an integral multiple of $2\pi$. That is, the maxima in the pattern occur at the same positions as for a two-slit pattern with the same spacing. To this extent the pattern resembles the two-slit pattern.

But what happens *between* the maxima? In the two-slit pattern there is exactly one intensity minimum between each pair of maxima, corresponding to angles for which

$$d \sin \theta = (m + \tfrac{1}{2})\lambda \quad (m = 0, 1, 2, \ldots),$$

or for which the phase difference between waves from the two sources is $\pi$, $3\pi$, $5\pi$, and so on. In the eight-slit pattern these are also minima because the light from adjacent slits cancels out in pairs, corresponding to the phasor diagram in Fig. 38–13a. But in this case there are additional minima. For example, when the phase difference $\phi$ from adjacent sources is $\pi/4$, the phasor diagram is as shown in Fig. 38–13b, and the intensity is zero. When $\phi = \pi/2$, we get the phasor diagram of Fig. 38–13c, and again the intensity is zero. More generally, whenever $\phi$ is an integral multiple of $\pi/4$, the intensity is zero, except when it is a multiple of $2\pi$. Thus there are seven minima for every maximum. Detailed calculation shows that the interference pattern is as shown in Fig. 38–14. The maxima are in the same positions as for the two-slit pattern of Fig. 38–14a, but they are much narrower.

We invite you to show that when there are $N$ slits, a minimum occurs whenever $\phi = 2\pi/N$ and that there are $N - 1$ minima between each pair of large maxima. Thus the greater the value of $N$, the sharper the interference maxima become. From an energy standpoint the total intensity is proportional to $N$. The height of each principal maximum is proportional to $N^2$, so from energy conservation the width of each principal maximum must be proportional to $1/N$.

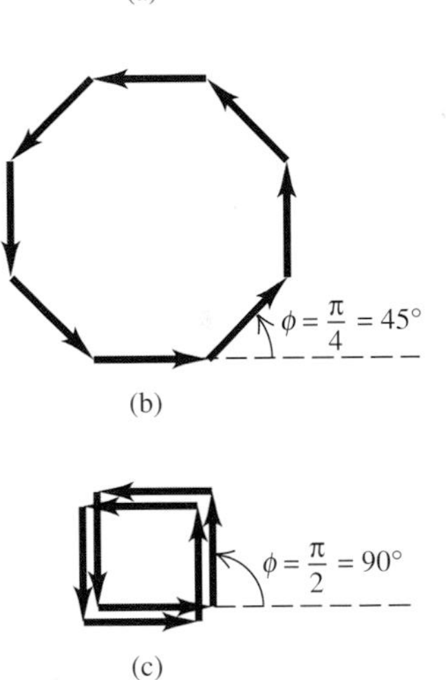


**38–13** Intensity maxima occur when the phase difference $\phi = 0$, $2\pi$, $4\pi$, .... Between the maxima at $\phi = 0$ and $\phi = 2\pi$ are seven minima, corresponding to $\phi = \pi/4$, $\pi/2$, $3\pi/4$, $\pi$, $5\pi/4$, $3\pi/2$, and $7\pi/4$. Phasor diagrams are shown for (a) $\phi = \pi$, (b) $\phi = \pi/4$, (c) $\phi = \pi/2$. Can you draw phasor diagrams for the other values?

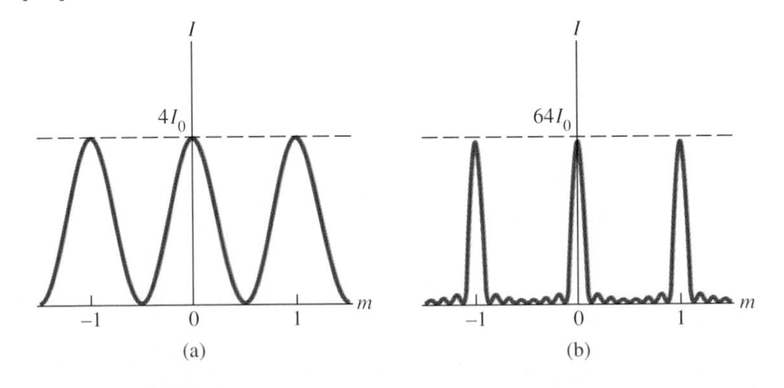


**38–14** Interference patterns for multiple narrow slits with spacing $d$ between adjacent slits. (a) Two slits. One minimum occurs between adjacent maxima. (b) Eight slits. Seven minima occur between adjacent pairs of the narrower large maxima. The vertical scales are different; if $I_0$ is the maximum intensity for a single slit, the maximum intensity for $N$ slits is $N^2I_0$.

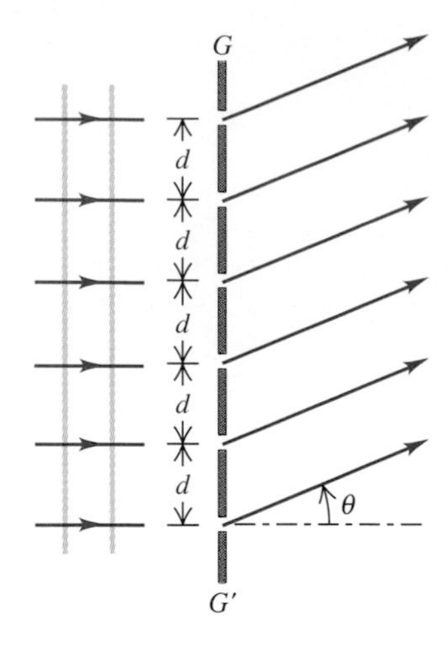


**38–15** A portion of a transmission diffraction grating.

## 38–5 THE DIFFRACTION GRATING

We have just seen that increasing the number of slits in an interference experiment while keeping the spacing of adjacent slits constant gives interference patterns with the maxima in the same positions as with two slits, but progressively sharper and narrower. An array of a large number of parallel slits, all with the same width $a$ and spaced equal distances $d$ between centers, is called a **diffraction grating.** The first one was constructed by Fraunhofer using fine wires. Gratings can be made by using a diamond point to scratch many equally spaced grooves on a glass or metal surface or by photographic reduction of a pattern of black and white stripes drawn on paper with a pen or a computer. For a grating, what we have been calling *slits* are often called *rulings* or *lines.*

In Fig. 38–15, $GG'$ is a cross section of a grating; the slits are perpendicular to the plane of the page. Only five slits are shown in the diagram, but an actual grating may contain several thousand. The spacing $d$ between centers of adjacent slits is called the *grating spacing;* it is typically about 0.002 mm. A plane wave of monochromatic light is incident normally on the grating from the left side. We assume far-field conditions; that is, the pattern is formed on a screen far enough away that all rays emerging from the grating in a particular direction are parallel.

We found in Section 38–4 that the principal intensity maxima occur in the same directions as for the two-slit pattern, directions for which the path difference for adjacent slits is an integer number of wavelengths. So the positions of the maxima are once again given by

$$d \sin\theta = m\lambda \quad (m = 0, \pm 1, \pm 2, \pm 3, \ldots). \tag{38–13}$$

Figure 38–16 shows intensity distributions for patterns from various numbers of slits, showing the progressive increase in sharpness as the number of slits increases.

When a grating containing hundreds or thousands of slits is illuminated by a parallel beam of monochromatic light, the pattern is a series of sharp lines at angles determined by Eq. (38–13). The $m = \pm 1$ lines are called the *first-order lines,* the $m = \pm 2$ lines the *second-order* lines, and so on. If the grating is illuminated by white light with a continuous distribution of wavelengths, each value of $m$ corresponds to a continuous spectrum in the pattern. The angle for each wavelength is determined by Eq. (38–13), which shows that for a given value of $m$, long wavelengths (the red end of the spectrum) lie at larger angles (are deviated more from the straight-ahead direction) than the shorter wavelengths at the violet end of the spectrum.

As Eq. (38–13) shows, the sines of the deviation angles of the maxima are proportional to the ratio $\lambda/d$. For substantial deviation to occur, the grating spacing $d$ should be of the same order of magnitude as the wavelength $\lambda$. Gratings for use in the visible spectrum usually have about 500 to 1500 slits per millimeter, so $d$ is of the order of 1000 nm.

**38–16** Interference patterns from $N$ equally spaced very narrow slits. As the number of slits increases, the peaks become narrower and higher. The width of each peak is proportional to $1/N$, and the height is proportional to $N^2$.

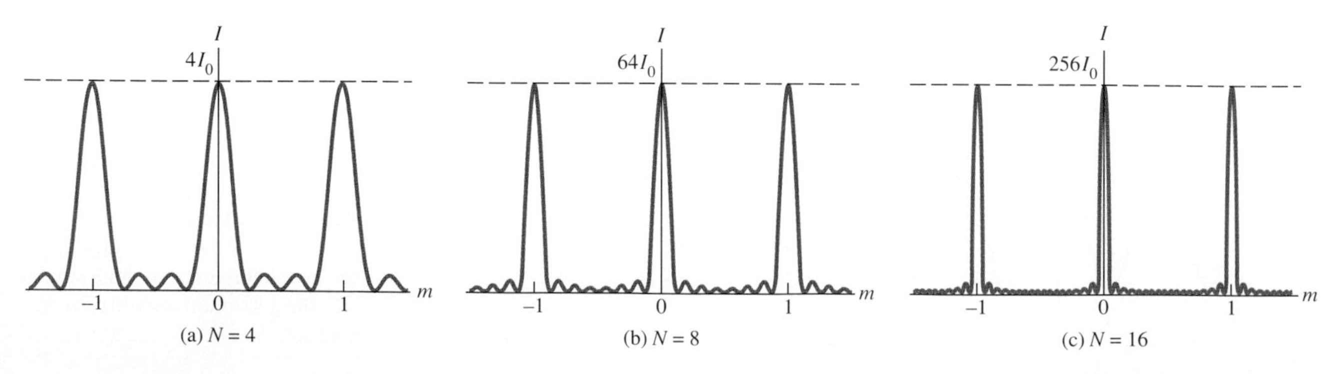

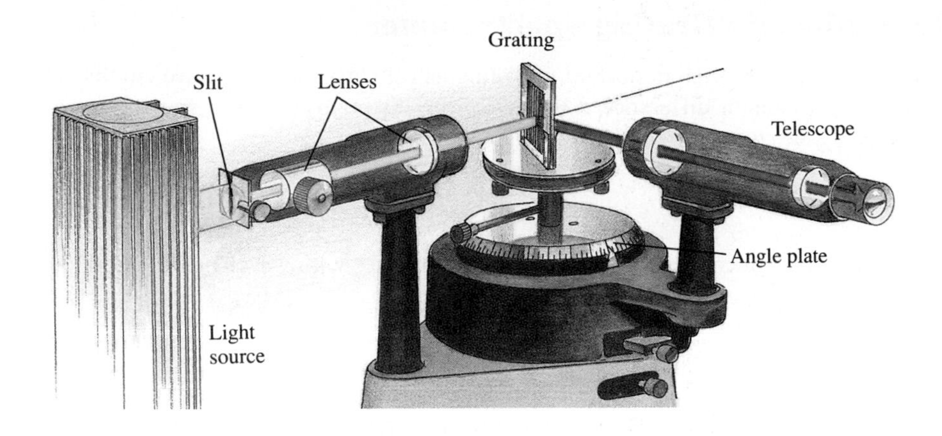


**38–17** A diffraction grating spectrometer. The lenses between source and grating form a beam of parallel rays incident on the grating; the beam is diffracted into various orders in directions that satisfy the equation $d \sin\theta = m\lambda$ ($m = 0, \pm 1, \pm 2 \ldots$). The diffracted beam is observed with a telescope with crosshairs, permitting precise measurements of the angle $\theta$.

## Grating Spectrometers

Diffraction gratings are widely used in spectrometry as a means of dispersing a light beam into spectra. If the grating spacing is known, then we can measure the angles of deviation and use Eq. (38–13) to compute the wavelength. A typical setup is shown in Fig. 38–17. A prism can also be used to disperse the various wavelengths through different angles because the index of refraction always varies with wavelength. But there is no simple relationship that describes this variation, so a spectrometer using a prism has to be calibrated with known wavelengths that are determined in some other way. Another difference is that a prism deviates red light the least and violet the most, and a grating does the opposite.

### EXAMPLE 38–3

**Width of a grating spectrum** The wavelengths of the visible spectrum are approximately 400 nm (violet) through 700 nm (red). Find the angular width of the first-order visible spectrum produced by a plane grating with 600 lines per millimeter when white light falls normally on the grating.

**SOLUTION** The first-order spectrum corresponds to $m = 1$. The grating spacing $d$ is

$$d = \frac{1}{600\ \text{slits/mm}} = 1.67 \times 10^{-6}\ \text{m}.$$

From Eq. (38–13), with $m = 1$, the angular deviation $\theta_v$ of the violet light (400 nm, or $400 \times 10^{-9}$m) is

$$\sin\theta_v = \frac{400 \times 10^{-9}\ \text{m}}{1.67 \times 10^{-6}\ \text{m}} = 0.240,$$

$$\theta_v = 13.9°.$$

The angular deviation $\theta_r$ of the red light (700 nm) is

$$\sin\theta_r = \frac{700 \times 10^{-9}\ \text{m}}{1.67 \times 10^{-6}\ \text{m}} = 0.419,$$

$$\theta_r = 24.8°.$$

So the first-order visible spectrum includes an angle of

$$24.8° - 13.9° = 10.9°.$$

### EXAMPLE 38–4

In the situation of Example 38–3, show that the violet end of the third-order spectrum overlaps the red end of the second-order spectrum.

**SOLUTION** From Eq. (38–13) the angular deviation of the third-order violet end ($m = 3$) is

$$\sin\theta_v = \frac{(3)(400 \times 10^{-9}\ \text{m})}{d} = \frac{1.20 \times 10^{-6}\ \text{m}}{d}.$$

The deviation of the second-order red end ($m = 2$) is

$$\sin\theta_r = \frac{(2)(700 \times 10^{-9}\ \text{m})}{d} = \frac{1.40 \times 10^{-6}\ \text{m}}{d}.$$

This shows that no matter what the grating spacing $d$ is, the largest angle (at the red end) for the second-order spectrum is always greater than the smallest angle (at the violet end) for the third-order spectrum, so the second and third orders *always* overlap.

## Resolution of a Grating Spectrometer

In spectroscopy it is often important to distinguish slightly differing wavelengths. The minimum wavelength difference $\Delta\lambda$ that can be distinguished by a spectrometer is described by the *chromatic resolving power R,* defined as

$$R = \frac{\lambda}{\Delta\lambda}. \tag{38–14}$$

A spectrometer that can barely distinguish the two lines with wavelengths 589.00 and 589.59 nm in the spectrum of sodium (called the *sodium doublet*) has a chromatic resolving power of 589/0.59, or 1000.

We can derive an expression for the resolving power of a diffraction grating when used as a spectrometer. Two different wavelengths give diffraction maxima at slightly different angles. As a reasonable (though arbitrary) criterion, let's assume that we can distinguish them as two separate peaks if the maximum of one coincides with the first minimum of the other.

From our discussion in Section 38–3 the $m$th-order maximum occurs when the phase difference $\phi$ for adjacent slits is $\phi = 2\pi m$. The first minimum beside that maximum occurs when $\phi = 2\pi m + 2\pi/N$, where $N$ is the number of slits. Since $\phi$ is also given by $\phi = (2\pi d \sin\theta)/\lambda$, the angular interval $d\theta$ corresponding to a small increment $d\phi$ in the phase shift can be obtained from the differential of this equation:

$$d\phi = \frac{2\pi d \cos\theta \, d\theta}{\lambda}.$$

When $d\phi = 2\pi/N$, this corresponds to the angular interval $d\theta$ between a maximum and the first adjacent minimum. Thus $d\theta$ is given by

$$\frac{2\pi}{N} = \frac{2\pi d \cos\theta \, d\theta}{\lambda}, \qquad \text{or} \qquad d\cos\theta \, d\theta = \frac{\lambda}{N}.$$

Now we need to find the angular spacing $d\theta$ between maxima for two slightly different wavelengths. This is easy; we have $d \sin\theta = m\lambda$, so the differential of this equation gives

$$d\cos\theta \, d\theta = m \, d\lambda.$$

According to our criterion, the limit of resolution is reached when these two angular spacings are equal. Equating the two expressions for the quantity $d\cos\theta \, d\theta$, we find

$$\frac{\lambda}{N} = m \, d\lambda.$$

If $\Delta\lambda$ is small, we can replace $d\lambda$ by $\Delta\lambda$, and the resolving power $R$ is given simply by

$$R = \frac{\lambda}{\Delta\lambda} = Nm. \tag{38–15}$$

The more lines ($N$), the better the resolution; and the higher-order maxima ($m$) in the diffraction pattern we use, the better the resolution.

### EXAMPLE 38–5

What minimum number of slits would be required in a grating to resolve the sodium doublet in the first order? In the fourth order?

**SOLUTION** As we found above, the resolving power needed is $R = 1000$. In the first order ($m = 1$) we need 1000 lines, but in the fourth order we need only 250 lines. These numbers are only approximate because of the arbitrary nature of our criterion for resolution and because real gratings always have slight imperfections in the line shapes and spacings.

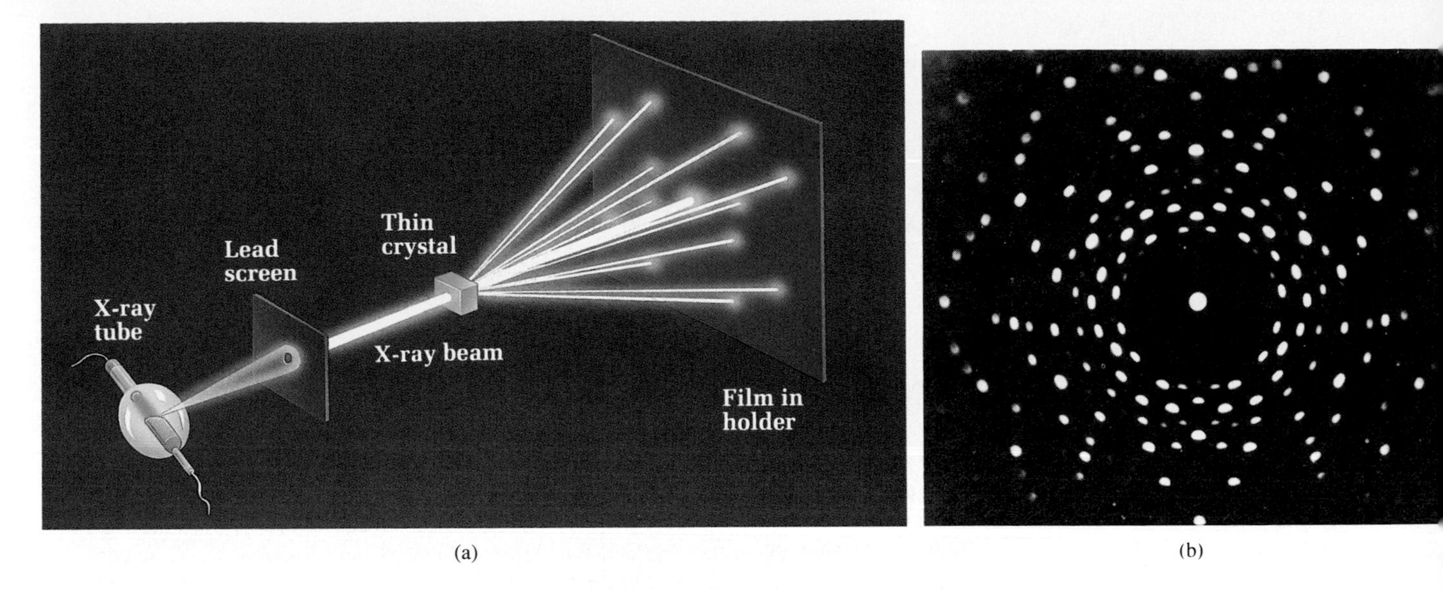


**38–18** (a) In an x-ray diffraction experiment, most x rays pass straight through the crystal, but some are scattered, forming an interference pattern that exposes the film in a pattern related to the atomic arrangement in the crystal. (b) Laue diffraction pattern formed by directing a beam of x rays at a thin section of quartz crystal.

## 38–6 X-RAY DIFFRACTION

X rays were discovered by Wilhelm Röntgen (1845–1923) in 1895, and early experiments suggested that they were electromagnetic waves with wavelengths of the order of $10^{-10}$ m. At about the same time the idea began to emerge that the atoms in a crystalline solid are arranged in a lattice in a regular repeating pattern, with spacing between adjacent atoms of the order of $10^{-10}$ m. Putting these two ideas together, Max von Laue (1879–1960) proposed in 1912 that a crystal might serve as a kind of three-dimensional diffraction grating for x rays. That is, a beam of x rays might be scattered (absorbed and re-emitted) by the individual atoms in a crystal, and the scattered waves might interfere just like waves from a diffraction grating.

The first experiments in **x-ray diffraction** were performed in 1912 by Friederich, Knipping, and von Laue, using the experimental setup sketched in Fig. 38–18a. The scattered x rays *did* form an interference pattern, which they recorded on photographic film. Figure 38–18b is a photograph of such a pattern. These experiments verified that x rays *are* waves, or at least have wavelike properties, and also that the atoms in a crystal *are* arranged in a regular pattern (Fig. 38–19). Since that time, x-ray diffraction has proved an invaluable research tool, both for measuring wavelengths of x rays and for the study of crystal structure.

To introduce the basic ideas, we consider first a two-dimensional scattering situation, as shown in Fig. 38–20a (on the next page), in which a plane wave is incident on a rectangular array of scattering centers. The situation might be a ripple tank with an array of small posts, or 3-cm microwaves striking an array of small conducting spheres, or x rays incident on an array of atoms. In the case of electromagnetic waves, the wave induces an oscillating electric dipole moment in each scatterer. These dipoles act like little antennas, emitting scattered waves. The resulting interference pattern is the superposition of all these scattered waves. The situation is different from that with a diffraction grating, in which the waves from all the slits are emitted *in phase* (for a plane wave at normal incidence). Here the scattered waves are *not* all in phase because their distances from the *source* are different. To compute the interference pattern, we have to consider the *total* path differences for the scattered waves, including both the distances from source to scatterer and from scatterer to observer.

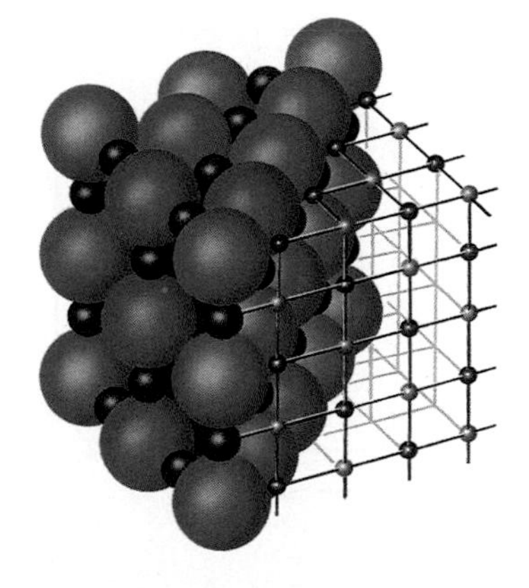

**38–19** Model of the arrangement of ions in a crystal of NaCl (table salt). Black spheres are Na; red spheres are Cl. The spacing of adjacent atoms is 0.282 nm. The electron clouds of the atoms actually overlap slightly, but the atoms are represented as small spheres for clarity.

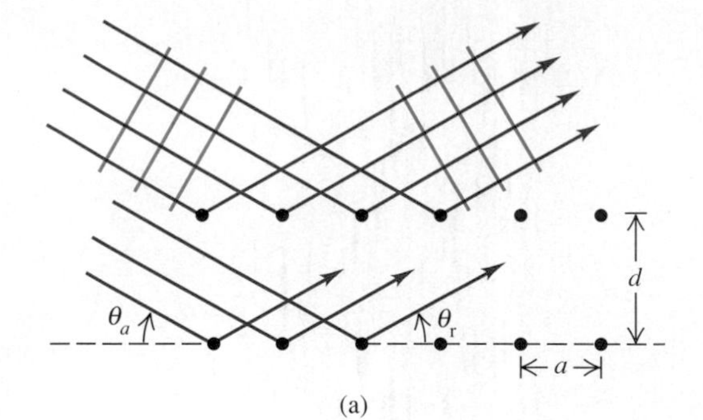


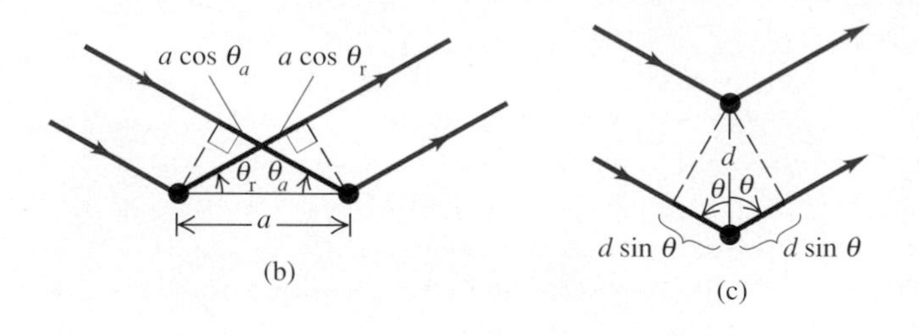


**38–20** (a) Scattering of waves from a rectangular array. (b) Interference of waves scattered from adjacent atoms in a row is constructive when $a \cos \theta_a = a \cos \theta_r$, that is, when the angle of incidence $\theta_a$ equals the angle of reflection $\theta_r$. (c) Interference from adjacent rows is also constructive when the path difference $2d \sin \theta$ equals an integral number of wavelengths, as in Eq. (38–16).

As Fig. 38–20b shows, the path length from source to observer is the same for all the scatterers in a single row if the angles $\theta_a$ and $\theta_r$ are equal, as shown. Scattered radiation from *adjacent* rows is *also* in phase if the path difference for adjacent rows is an integral number of wavelengths. Figure 38–20c shows that this path difference is $2d \sin \theta$. Therefore the conditions for radiation from the *entire array* to reach the observer in phase are (1) the angle of incidence must equal the angle of scattering and (2) the path difference for adjacent rows must equal $m\lambda$, where $m$ is an integer. We can express the second condition as

$$2d \sin \theta = m\lambda \quad (m = 1, 2, 3, \ldots). \tag{38–16}$$

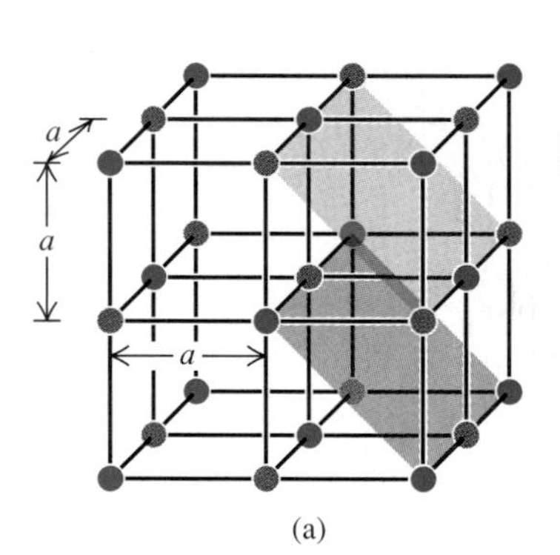


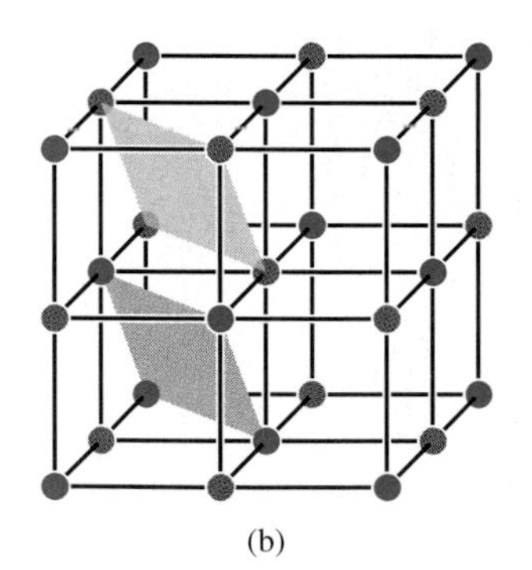


**38–21** Cubic crystal lattice, showing two different families of crystal planes. The spacing of the planes in (a) is $d = a/\sqrt{2}$; that of the planes in (b) is $a/\sqrt{3}$. There are also three sets of planes parallel to the cube faces, with spacing $a$.

In directions for which this condition is satisfied, we see a strong maximum in the interference pattern. We can describe this interference in terms of *reflections* of the wave from the horizontal rows of scatterers in Fig. 38–20a. Strong reflection (constructive interference) occurs at angles such that the incident and scattered angles are equal and Eq. (38–16) is satisfied.

Note that the angle $\theta$ is customarily measured with respect to the surface of the crystal. This is different from our usual usage, in which we measure $\theta$ with respect to the normal to the plane of an array of slits or a grating. Also, Eq. (38–16) is *not* the same as Eq. (38–13). Be careful!

We can extend this discussion to a three-dimensional array by considering *planes* of scatterers instead of *rows*. Figure 38–21 shows several different sets of parallel planes that pass through all the scatterers. Waves from all the scatterers in a given plane interfere constructively if the angles of incidence and scattering are equal. There is also constructive interference between planes when Eq. (38–16) is satisfied, where $d$ is now the distance between adjacent planes. Because there are many different sets of parallel planes, there are also many values of $d$ and many sets of angles that give constructive interference for the whole crystal lattice. This phenomenon is called **Bragg reflection,** and Eq. (38–16) is called the **Bragg condition,** in honor of Sir William Bragg and his son Laurence Bragg, two pioneers in x-ray analysis. *Caution:* Don't let the term *reflection* obscure the fact that we are dealing with an *interference* effect. In fact, the reflections from various planes are closely analogous to interference effects in thin films (Section 37–4).

As Fig. 38–18b shows, nearly complete cancellation occurs for all but certain very specific directions, where constructive interference occurs and forms bright spots. Such a pattern is usually called an x-ray *diffraction* pattern, although *interference* pattern might be more appropriate. This particular type of pattern is called a *Laue pattern*.

If the crystal lattice spacing is known, we can determine the wavelength of the x rays, just as we determined wavelengths of visible light by measuring diffraction patterns from slits or gratings. For example, we can determine the crystal lattice spacing for sodium chloride from its density and Avogadro's number. Then, once we know the x-ray wavelength, we can use x-ray diffraction to explore the structure and lattice spacing of crystals with unknown structure.

X-ray diffraction is by far the most important experimental tool in the investigation of crystal structure of solids. Atomic spacings in crystals can be measured precisely, and the lattice structure of complex crystals can be determined. X-ray diffraction also plays an important role in studies of the structures of liquids and of organic molecules. It has been one of the chief experimental techniques for working out the double-helix structure of DNA and subsequent advances in molecular genetics.

### EXAMPLE 38–6

You direct an x-ray beam with wavelength 0.154 nm at certain planes of a silicon crystal. As you increase the angle of incidence from zero, you find the first strong interference maximum from these planes when the beam makes an angle of 34.5° with the planes. a) How far apart are the planes? b) Will you find other interference maxima from these planes at larger angles?

**SOLUTION** To find the plane spacing $d$, we solve the Bragg equation, Eq. (38–16), for $d$ and set $m = 1$:

$$d = \frac{m\lambda}{2\sin\theta} = \frac{(1)(0.154\text{ nm})}{2\sin 34.5^\circ} = 0.136\text{ nm}.$$

This is the distance between adjacent planes.

b) To calculate other angles, we solve Eq. (38–16) for $\sin\theta$:

$$\sin\theta = \frac{m\lambda}{2d} = m\frac{0.154\text{ nm}}{2\,(0.136\text{ nm})} = m(0.566).$$

Values of $m$ of 2 or greater give values of $\sin\theta$ greater than unity, which is impossible. Therefore there are no other angles for interference maxima for this particular set of crystal planes.

## 38–7 CIRCULAR APERTURES AND RESOLVING POWER

We have studied in detail the diffraction patterns formed by long, thin slits or arrays of slits. But an aperture of *any* shape forms a diffraction pattern. The diffraction pattern formed by a *circular* aperture is of special interest because of its role in limiting the resolving power of optical instruments. In principle, we could compute the intensity at any point $P$ in the diffraction pattern by dividing the area of the aperture into small elements, finding the resulting wave amplitude and phase at $P$, and then integrating over the aperture area to find the resultant amplitude and intensity at $P$. In practice, the integration cannot be carried out in terms of elementary functions, but has to be done by numerical approximation. We will simply *describe* the pattern and quote a few relevant numbers.

The diffraction pattern formed by a circular aperture consists of a central bright spot surrounded by a series of bright and dark rings, as shown in Fig. 38–22. We can describe the pattern in terms of the angle $\theta$, representing the angular size of each ring. If the aperture diameter is $D$ and the wavelength is $\lambda$, the angular size $\theta_1$ of the first *dark* ring is given by

$$\sin\theta_1 = 1.22\frac{\lambda}{D}. \tag{38–17}$$

**38–22** Diffraction pattern formed by a circular aperture of diameter $D$, consisting of a central bright spot and alternating dark and bright rings. The angular size $\theta_2$ of the second dark ring is shown. (This diagram is not drawn to scale.)

The angular sizes of the next two dark rings are given by

$$\sin\theta_2 = 2.23\frac{\lambda}{D}, \qquad \sin\theta_3 = 3.24\frac{\lambda}{D}. \tag{38–18}$$

Between these are bright rings with angular sizes given by

$$\sin\theta = 1.63\frac{\lambda}{D}, \quad 2.68\frac{\lambda}{D}, \quad 3.70\frac{\lambda}{D}, \tag{38–19}$$

and so on. The central bright spot is called the **Airy disk,** in honor of Sir George Airy (1801–1892), Astronomer Royal of England, who first derived the expression for the intensity in the pattern. The angular size of the Airy disk is that of the first dark ring, given by Eq. (38–17).

The *intensities* in the bright rings drop off very quickly. When $D$ is much larger than the wavelength $\lambda$, the usual case for optical instruments, the peak intensity in the center of the first ring is only 1.7% of the value at the center of the Airy disk, and at the center of the second ring it is only 0.4%. Most (85%) of the total intensity falls within the Airy disk. Figure 38–23 shows a diffraction pattern from a circular aperture 1.0 mm in diameter.

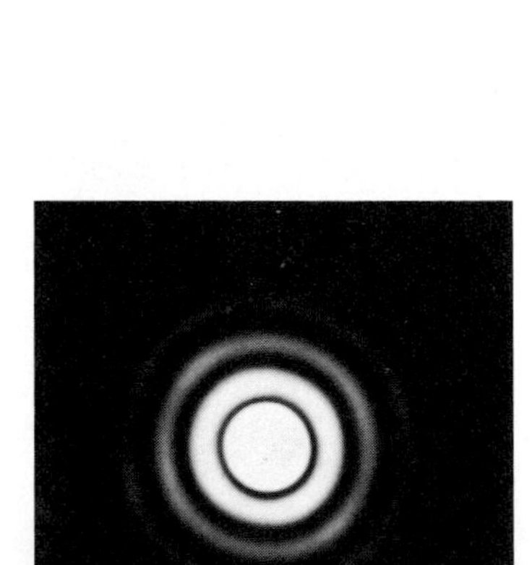

**38–23** Diffraction pattern formed by a circular aperture 1.0 mm in diameter. The Airy disk is overexposed so that other rings may be seen.

This analysis has far-reaching implications for image formation of lenses and mirrors. In our study of optical instruments in Chapter 36 we assumed that a lens with focal length $f$ focuses a parallel beam (plane wave) to a *point* at a distance $f$ from the lens. This assumption ignored diffraction effects. We now see that what we get is not a point but the diffraction pattern just described. If we have two point objects, their images are not two points but two diffraction patterns. When the objects are close together, their diffraction patterns overlap; if they are close enough, their patterns overlap almost completely and cannot be distinguished. The effect is shown in Fig. 38–24, which shows the patterns for four (very small) "point" objects. In Fig. 38–24a, the images of objects 1 and 2 are well separated, but the images of objects 3 and 4 on the right have merged together. In Fig. 38–24b, with a larger aperture diameter and resulting smaller Airy disks, images 3 and 4 are better resolved. In Fig. 38–24c, with a still larger aperture, they are well resolved.

A widely used criterion for resolution of two point objects, proposed by Lord Rayleigh (1887–1905) and called **Rayleigh's criterion,** is that the objects are just barely resolved (that is, distinguishable) if the center of one diffraction pattern coincides with the first minimum of the other. In that case the angular separation of the image centers is given by Eq. (38–17). The angular separation of the *objects* is the same as that of the *images,* so two point objects are barely resolved, according to Rayleigh's criterion, when their angular separation is given by Eq. (38–17).

The minimum separation of two objects that can just be resolved by an optical instrument is called the **limit of resolution** of the instrument. The smaller the limit of resolution, the greater the *resolution,* or **resolving power,** of the instrument. Diffraction

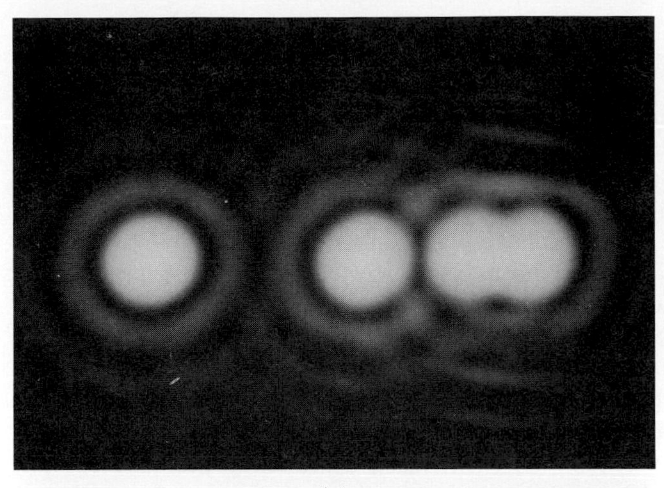

(a)

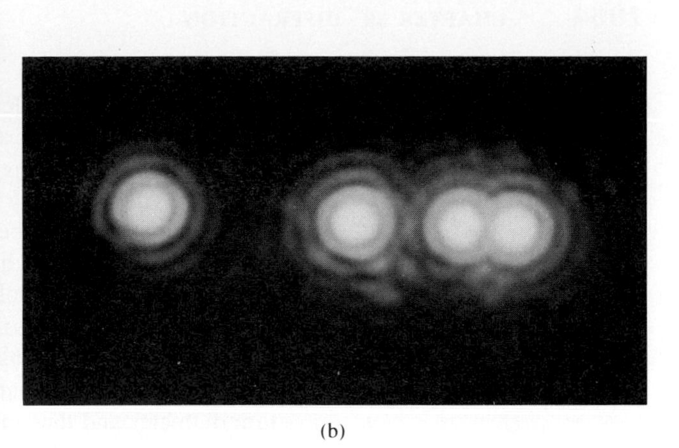

(b)

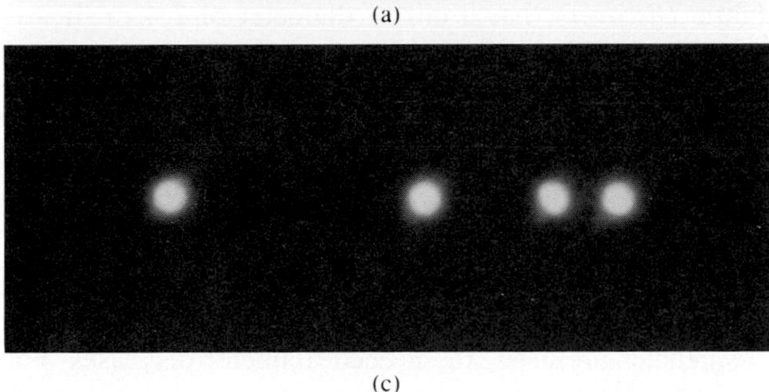

(c)

**38–24** Diffraction patterns of four "point" sources with a circular opening in front of the lens. (a) Here the opening is so small that the patterns of sources 3 and 4 overlap and are barely resolved by Rayleigh's criterion. Increasing the aperture decreases the size of the diffraction patterns, as shown in (b) and (c).

sets the ultimate limits on resolution of lenses. *Geometric* optics may make it seem that we can make images as large as we like. Eventually, though, we always reach a point at which the image becomes larger but does not gain in detail. The images in Fig. 38–24 would not become sharper with further enlargement.

## EXAMPLE 38–7

**Resolving power of a camera lens** A camera lens with focal length $f = 50$ mm and maximum aperture $f/2$ forms an image of an object 10 m away. a) If the resolution is limited by diffraction, what is the minimum distance between two points on the object that are barely resolved, and what is the corresponding distance between image points? b) How does the situation change if the lens is "stopped down" to $f/16$? Assume that $\lambda = 500$ nm in both cases.

**SOLUTION** a) The aperture diameter is $D = (50 \text{ mm})/2 = 25 \text{ mm} = 25 \times 10^{-3}$ m. From Eq. (38–17) the angular separation $\theta$ of two object points that are barely resolved is given by

$$\sin\theta \cong \theta = 1.22\frac{\lambda}{D} = 1.22\frac{500 \times 10^{-9}\text{ m}}{25 \times 10^{-3}\text{ m}} = 2.4 \times 10^{-5}\text{ rad}.$$

Let $y$ be the separation of the object points and $y'$ the separation of the corresponding image points. We know from our thin-lens analysis in Section 35–5 that, apart from sign, $y/s = y'/s'$. Thus the angular separations of the object points and the corresponding image points are both equal to $\theta$. Because $s >> f$, the image distance $s'$ is approximately equal to the focal length, $f = 50$ mm. Thus

$$\frac{y}{10\text{ m}} = 2.4 \times 10^{-5}, \qquad y = 2.4 \times 10^{-4}\text{ m} = 0.24\text{ mm};$$

$$\frac{y'}{50\text{ mm}} = 2.4 \times 10^{-5}, \qquad y' = 1.2 \times 10^{-3}\text{ mm} = 0.0012\text{ mm} \cong \tfrac{1}{800}\text{ mm}.$$

b) The aperture is now (50 mm)/16, or one-eighth, as large as before. The angular separation is eight times as great, and the values of $y$ and $y'$ are also eight times as great as before:

$$y = 1.9\text{ mm}, \qquad y' = 0.0096\text{ mm} = \tfrac{1}{100}\text{ mm}.$$

Setups to test the resolving power of a lens often use series of parallel lines with varying spacing, and the resolution in the image is often described in "lines per millimeter." The lens in Example 38–7 would be described as having a resolution of about 800 lines/mm when "wide open" and about 100 lines/mm when stopped down to $f/16$. Only the best-quality camera lenses approach this resolution. Photographers who always use the smallest possible aperture for maximum depth of field and (presumably) maximum sharpness should be aware that diffraction effects become more significant at small apertures. One cause of fuzzy images has to be balanced against another.

Another lesson to be learned is that resolution improves with shorter wavelengths. Ultraviolet microscopes have higher resolution than visible-light microscopes. In electron microscopes the resolution is limited by the wavelengths associated with the wavelike behavior of electrons (Section 37–6). These wavelengths can be made 100,000 times smaller than wavelengths of visible light, with a corresponding gain in resolution. Finally, one reason for building very large reflecting telescopes is to increase the aperture diameter and thus minimize diffraction effects. This also provides greater light-gathering area for viewing very faint stars. The Hubble Space Telescope, launched April 25, 1990 from the space shuttle *Discovery,* has a mirror diameter of 2.4 m. It was designed to resolve objects $2.8 \times 10^{-7}$ rad apart using 550-nm light. This would have been at least a factor of six better than earth-based telescopes, which are limited by atmospheric distortion. Because of errors in the grinding of the mirror, this resolving power has not yet been attained.

Diffraction is an important consideration for satellite "dishes," parabolic reflectors designed to receive satellite transmission. Satellite dishes have to be able to pick up transmission from two satellites only a few degrees apart transmitting at the same frequency, and the need to resolve two such transmissions determines the minimum diameter of the dish. As higher frequencies are used, the needed diameter decreases. For example, when two satellites 5.0° apart broadcast 7.5-cm microwaves, the minimum dish diameter to resolve them (by Rayleigh's criterion) is about 1.0 m.

The effective diameter of a telescope can be increased in some cases by using arrays of smaller telescopes. The Very-Long-Baseline Array (VLBA), a group of radio telescopes on earth, has a maximum separation of about 8000 km and can resolve radio signals to $10^{-8}$ rad. This astonishing resolution is comparable, in the optical realm, to seeing a parked car on the moon. The use of satellites may increase resolution even more in the future. Arrays of optical telescopes are under development, but none is yet in operation.

## 38–8 HOLOGRAPHY

**Holography** is a technique for recording and reproducing an image of an object without the use of lenses. Unlike the two-dimensional images recorded by an ordinary photograph or television system, a holographic image is truly three-dimensional. Such an image can be viewed from different directions to reveal different sides and from various distances to reveal changing perspective. If you had never seen a hologram, you wouldn't believe it was possible!

The basic procedure for making a hologram is shown in Fig. 38–25a. We illuminate the object to be holographed with monochromatic light, and we place a photographic film so that it is struck by scattered light from the object (object beam) and also by direct light from the source (reference beam). In practice, the source must be a laser, for reasons that we will discuss later. Interference between the direct and scattered light leads to the formation and recording of a complex interference pattern on the film.

To form the images, we simply project light (reconstruction beam) through the developed film, as shown in Fig. 38–25b. Two images are formed, a virtual image on the side of the film nearer the source, and a real image on the opposite side.

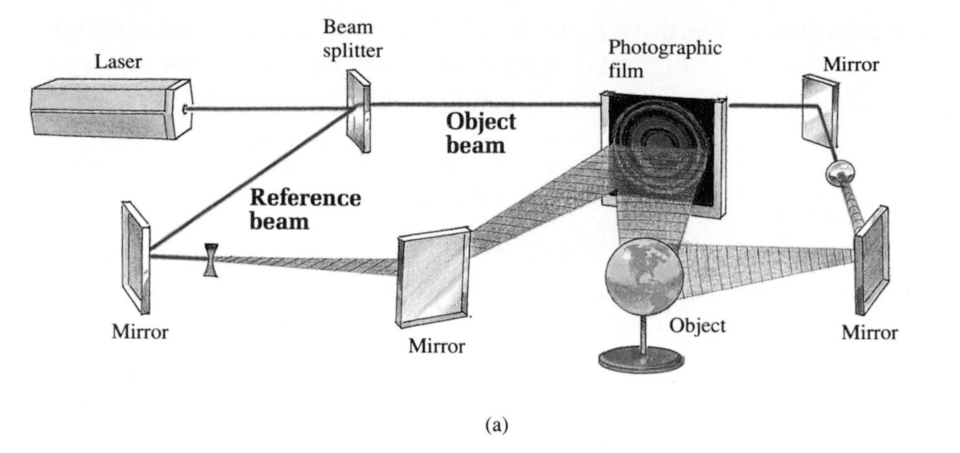


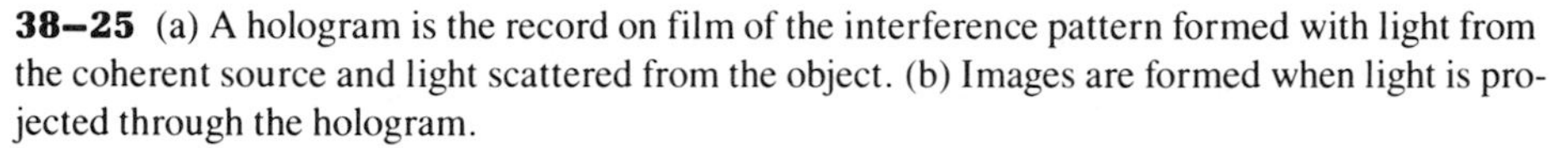
**38–25** (a) A hologram is the record on film of the interference pattern formed with light from the coherent source and light scattered from the object. (b) Images are formed when light is projected through the hologram.

A complete analysis of holography is beyond our scope, but we can gain some insight into the process by looking at how a single point is holographed and imaged. Consider the interference pattern formed on a photographic film by the superposition of an incident plane wave and a spherical wave, as shown in Fig. 38–26a. The spherical wave originates at a point source $P$ at a distance $b_0$ from the film; $P$ may in fact be a small object that scatters part of the incident plane wave. We assume that the two waves are monochromatic and coherent and that the phase relation is such that constructive interference occurs at point $O$ on the diagram. Then constructive interference will *also* occur

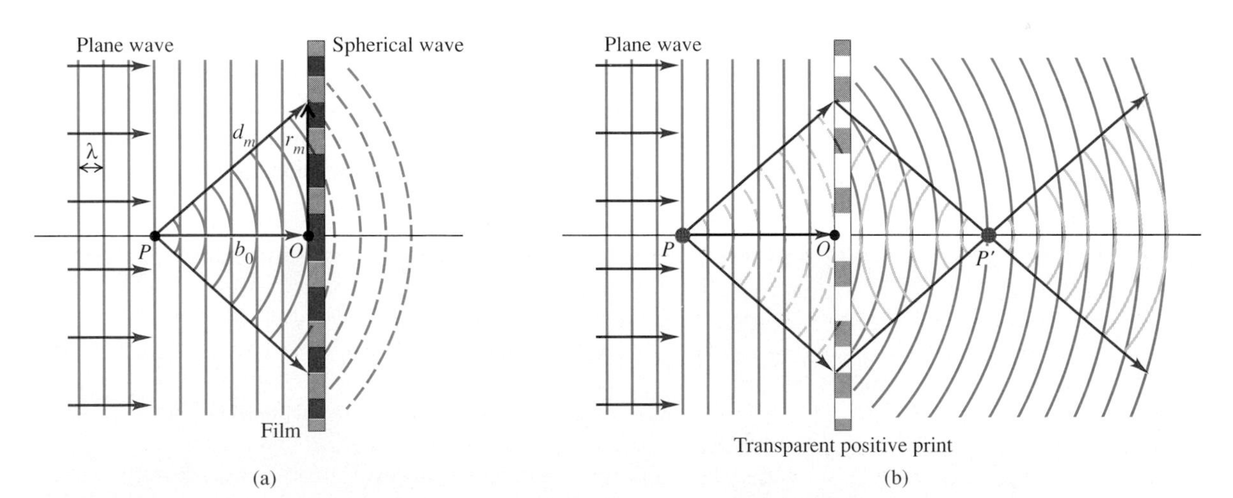


**38–26** (a) Constructive interference of the plane and spherical waves occurs in the plane of the film at every point $Q$ for which the distance $b_m$ from $P$ is greater than the distance $b_0$ from $P$ to $O$ by an integral number of wavelengths $m\lambda$. For the point shown, $m = 2$. (b) When a plane wave strikes the developed film, the diffracted wave consists of a wave converging to $P'$ and then diverging again and a diverging wave that appears to originate at $P$. These waves form the real and virtual images, respectively.

at any point $Q$ on the film that is farther from $P$ than $O$ is by an integral number of wavelengths. That is, if $b_m - b_0 = m\lambda$, where $m$ is an integer, then constructive interference occurs. The points where this condition is satisfied form circles centered at $O$, with radii $r_m$ given by

$$b_m - b_0 = \sqrt{b_0^2 + r_m^2} - b_0 = m\lambda \quad (m = 1, 2, 3, \ldots). \tag{38–20}$$

Solving this for $r_m^2$, we find

$$r_m^2 = \lambda(2mb_0 + m^2\lambda).$$

Ordinarily, $b_0$ is very much larger than $\lambda$, so we neglect the second term in parentheses, obtaining

$$r_m = \sqrt{2m\lambda b_0} \quad (m = 1, 2, 3, \ldots). \tag{38–21}$$

The interference pattern consists of a series of concentric bright circular fringes with radii given by Eq. (38–21). Between these bright fringes are dark fringes.

Now we develop the film and make a transparent positive print, so the bright-fringe areas have the greatest transparency on the film. Then we illuminate it with monochromatic plane-wave light of the same wavelength $\lambda$ that we used initially. In Fig. 38–26b, consider a point $P'$ at a distance $b_0$ along the axis from the film. The centers of successive bright fringes differ in their distances from $P'$ by an integral number of wavelengths, and therefore a strong *maximum* in the diffracted wave occurs at $P'$. That is, light converges to $P'$ and then diverges from it on the opposite side. Therefore $P'$ is a *real image* of point $P$.

This is not the entire diffracted wave, however. The interference of the wavelets that spread out from all the transparent areas form a second spherical wave that is diverging rather than converging. When traced back behind the film, this wave appears to be spreading out from point $P$. Thus the total diffracted wave from the hologram is a superposition of a spherical wave converging to form a real image at $P'$ and a spherical wave that diverges as though it had come from the virtual image point $P$.

Because of the principle of linear superposition, what is true for the imaging of a single point is also true for the imaging of any number of points. The film records the superposed interference pattern from the various points, and when light is projected through the film, the various image points are reproduced simultaneously. Thus the images of an extended object can be recorded and reproduced just as for a single point object. Figure 38–27 shows photographs of a holographic image from two different angles, showing the changing perspective in this three-dimensional image.

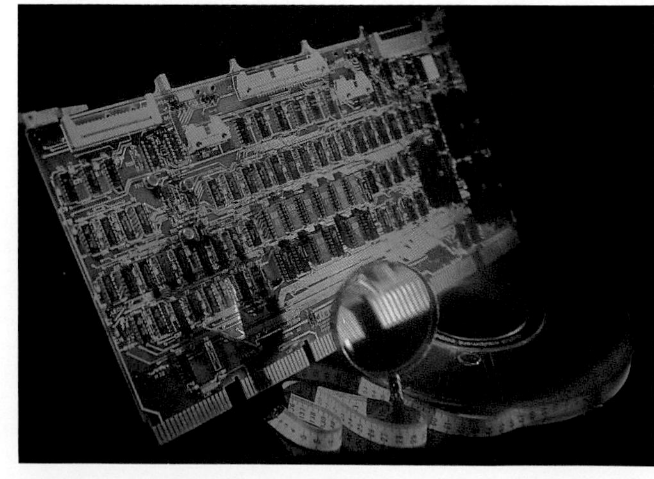

(a)

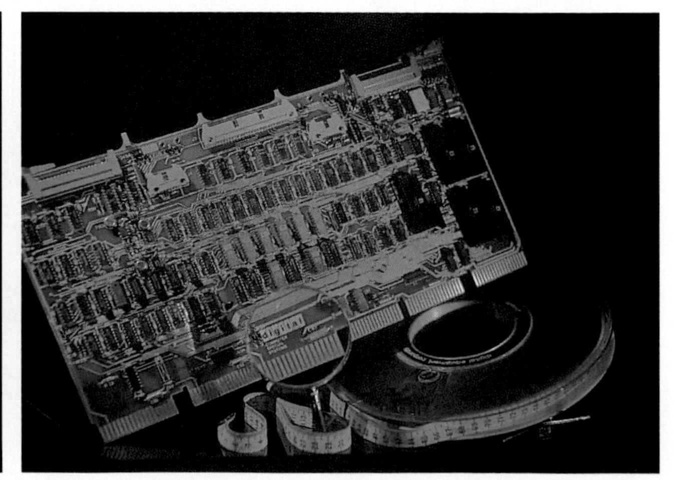


(b)

**38–27** Two views of a hologram. (a) From this angle you can look through the magnifying glass and see the row of connector tabs at the edge of the printed circuit board. (b) The angle of view has changed, so when you look through the glass, you see the manufacturer's name.

In making a hologram we have to overcome several practical problems. First, the light used must be *coherent* over distances that are large in comparison to the dimensions of the object and its distance from the film. Ordinary light sources *do not* satisfy this requirement, for reasons that we discussed in Section 37–1. Therefore laser light is essential for making a hologram. Second, extreme mechanical stability is needed. If any relative motion of source, object, or film occurs during exposure, even by as much as a wavelength, the interference pattern on the film is blurred enough to prevent satisfactory image formation. These obstacles are not insurmountable, however, and holography promises to become increasingly important in research, entertainment, and a wide variety of technological applications.

## SUMMARY

• Diffraction occurs when light passes through an aperture or around an edge. When source and observer are so far away from the obstructing surface that the outgoing rays can be considered parallel, it is called Fraunhofer diffraction; when the source or observer is at a finite distance, it is Fresnel diffraction.

• For a single narrow slit with width $a$ the condition for destructive interference (a dark fringe) at a point $P$ at angle $\theta$ is

$$\sin\theta = \frac{m\lambda}{a} \quad (m = \pm 1, \pm 2, \pm 3, \ldots). \tag{38–2}$$

The intensity $I$ at any angle $\theta$ is

$$I = I_0\left\{\frac{\sin[\pi a(\sin\theta)/\lambda]}{\pi a(\sin\theta)/\lambda}\right\}^2. \tag{38–7}$$

• A diffraction grating consists of a large number of thin parallel slits, spaced a distance $d$ apart. The condition for maximum intensity in the interference pattern is

$$d\sin\theta = m\lambda \quad (m = 0, \pm 1, \pm 2, \pm 3, \ldots). \tag{38–13}$$

This is the same condition for the two-source pattern, but for the grating the maxima are very sharp and narrow.

• A crystal serves as a three-dimensional diffraction grating for x rays with wavelengths of the same order of magnitude as the lattice spacing. For a set of crystal planes spaced a distance $d$ apart, constructive interference occurs when the angles of incidence and scattering (measured from the crystal planes) are equal and when

$$2d\sin\theta = m\lambda \quad (m = 1, 2, 3, \ldots). \tag{38–16}$$

This is called the Bragg condition.

• The diffraction pattern from a circular aperture of diameter $D$ consists of a central bright spot, called the Airy disk, and a series of concentric dark and bright rings. The angular size $\theta_1$ of the first dark ring, equal to the angular size of the Airy disk, is given by

$$\sin\theta_1 = 1.22\frac{\lambda}{D}. \tag{38–17}$$

Diffraction sets the ultimate limit on resolution (image sharpness) of optical instruments. According to Rayleigh's criterion, two point objects are just barely resolved when their angular separation $\theta$ is given by Eq. (38–17).

• A hologram is a photographic record of an interference pattern formed by light scattered from an object and light coming directly from the source. A hologram forms a true three-dimensional image of the object.

### KEY TERMS

diffraction
Fresnel diffraction
Fraunhofer diffraction
diffraction grating
x-ray diffraction
Bragg reflection
Bragg condition
Airy disk
Rayleigh's criterion
limit of resolution
resolving power
holography

## EXERCISES

### Section 38–2
### Diffraction from a Single Slit

**38–1** Monochromatic light from a distant source is incident on a slit 0.800 mm wide. On a screen 3.00 m away, the distance from the central maximum of the diffraction pattern to the first minimum is measured to be 1.80 mm. Calculate the wavelength of the light.

**38–2** Parallel rays of green mercury light with a wavelength of 546 nm pass through a slit covering a lens with a focal length of 40.0 cm. In the focal plane of the lens the distance from the central maximum to the first minimum is 12.0 mm. What is the width of the slit?

**38–3** Red light with a wavelength of 633 nm from a helium-neon laser passes through a slit 0.300 mm wide. The diffraction pattern is observed on a screen 4.0 m away. Define the width of a bright fringe as the distance between the minima on either side. a) What is the width of the central bright fringe? b) What is the width of the first bright fringe on either side of the central one?

**38–4** Light with a wavelength of 589 nm from a distant source is incident on a slit 0.850 mm wide, and the resulting diffraction pattern is observed on a screen 2.00 m away. What is the distance between the two dark fringes on either side of the central bright fringe?

### Section 38–3
### Intensity in the Single-Slit Pattern

**38–5** A slit 0.200 mm wide is illuminated by parallel rays of light that has a wavelength of 500 nm. The diffraction pattern is observed on a screen that is 4.00 m from the slit. The intensity at the center of the central maximum ($\theta = 0°$) is $5.00 \times 10^{-6}$ W/m$^2$. a) What is the distance on the screen from the center of the central maximum to the first minimum? b) What is the intensity at a point on the screen midway between the center of the central maximum and the first minimum?

**38–6** A diffraction pattern is formed by passing parallel rays of 500-nm light through a slit 0.250 mm wide. What is the phase angle $\beta$ (the phase difference between wavelets from the top and bottom of the slit) at a) the center of the central maximum; b) the third minimum out from the central maximum; c) 5.0° from the central maximum?

**38–7** In a single-slit diffraction pattern produced by a slit 0.400 mm wide, the phase angle $\beta$ (the phase difference between wavelets from the top and bottom of the slit) is $\pi/2$ rad at an angle of 4.0° from the central maximum. What is the wavelength of the light? (The light is not necessarily in the visible region of the spectrum.)

### Section 38–4
### Multiple Slits

**38–8** Consider the interference pattern produced by eight equally spaced narrow slits. There is an interference minimum when the phase difference $\phi$ between light from adjacent slits is $\pi/4$. The phasor diagram is given in Fig. 38–13b. For which pairs of slits is there totally destructive interference?

**38–9** Consider the interference pattern produced by four equally spaced narrow slits. By drawing the appropriate phasor diagram, show that there is an interference minimum when the phase difference $\phi$ from adjacent slits is (a) $\pi/2$; (b) $\pi$; (c) $3\pi/2$. In each case, for which pairs of slits is there totally destructive interference?

**38–10 Diffraction in an Interference Pattern.** Consider the interference pattern produced by two slits with width $a$ and separation $d$. Let $d = 4a$. a) Ignoring diffraction effects due to the slit width, at what angles $\theta$ from the central maximum will the next five maxima in the two-slit interference pattern occur? (Your answer will be in terms of $d$ and the wavelength $\lambda$ of the light.) b) Now include the effects of diffraction. If the intensity at $\theta = 0$ is $I_0$, what is the intensity at each of the angles calculated in part (a)? Compare your results to Fig. 38–11.

**38–11 Number of Fringes in a Diffraction Maximum.** In Fig. 38–11 the central diffraction maximum contains exactly seven interference fringes, and in this case, $d/a = 4$. a) What must the ratio $d/a$ be if the central maximum contains exactly five fringes? b) In the case considered in part (a), how many fringes are contained within the first diffraction maximum on one side of the central maximum?

### Section 38–5
### The Diffraction Grating

**38–12** a) What is the wavelength of light that is deviated in the first order through an angle of 16.0° by a transmission grating having 6000 lines/cm? b) What is the second-order deviation of this wavelength? Assume normal incidence.

**38–13** Plane monochromatic waves with wavelength 600 nm are incident normally on a plane transmission grating having 600 lines/mm. Find the angles of deviation in the first, second, and third orders.

**38–14** The wavelength range of the visible spectrum is approximately 400 nm to 700 nm. White light falls at normal incidence on a diffraction grating that has 200 lines/mm. Find the angular width of the visible spectrum in a) the first order; b) the third order. (*Note:* An advantage of working in higher orders is the greater angular spread and better resolution. A disadvantage is the overlapping of different orders, as shown in Example 38–4.)

**38–15** A plane transmission grating is ruled with 5000 lines/cm. Assume normal incidence. The $\alpha$ and $\delta$ lines of atomic hydrogen have wavelengths of 656 nm and 410 nm, respectively. Compute the angular separation in degrees between these lines in a) the first-order spectrum; b) the second-order spectrum.

**38–16 Identifying Isotopes by Spectra.** The light emitted by different isotopes of the same element is at slightly different wavelengths. A wavelength in the emission spectrum of a hydrogen atom is 656.45 nm; for deuterium the corresponding wavelength is 656.27 nm. What minimum number of slits is required to resolve these two wavelengths in the third order?

## Section 38–6 X-Ray Diffraction

**38–17** X rays with a wavelength of 0.0820 nm are scattered from the atoms of a crystal. The second-order maximum in the Bragg reflection occurs when the angle $\theta$ in Fig. 38–20 is 21.4°. What is the spacing between adjacent atomic planes in the crystal?

**38–18** Monochromatic x rays are incident on a crystal for which the spacing of the atomic planes is 0.400 nm. The first-order maximum in the Bragg reflection occurs when the incident and reflected x rays make an angle of 33.0° with the crystal planes. What is the wavelength of the x rays?

## Section 38–7 Circular Apertures and Resolving Power

**38–19** Monochromatic light with wavelength 480 nm passes through a circular aperture with diameter 2.40 $\mu$m. The resulting diffraction pattern is observed on a screen that is 5.00 m from the aperture. What is the diameter of the Airy disk on the screen?

**38–20** In Fig. 38–28, two point sources of light, $a$ and $b$, at a distance of 40.0 m from lens $L$ and 6.00 mm apart produce images at $c$ that are just resolved according to Rayleigh's criterion. The focal length of the lens is 20.0 cm. What is the diameter of the diffraction circles at $c$?

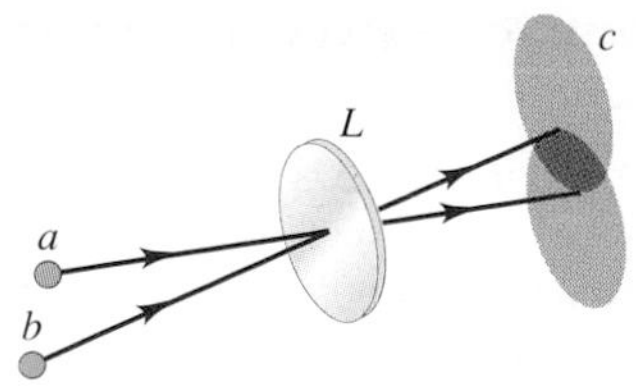

**FIGURE 38–28**

**38–21 A Telescope for the Sun.** You are asked to design a space telescope for earth orbit that can resolve (by Rayleigh's criterion) features on the sun that are 20.0 km apart. What minimum diameter lens is required? Assume a wavelength of 500 nm.

**38–22** A converging lens 8.00 cm in diameter has a focal length of 40.0 cm. If the resolution is diffraction-limited, how far away can an object be if points on it 6.00 mm apart are to be resolved (according to Rayleigh's criterion)? (Use $\lambda = 550$ nm.)

## Section 38–8 Holography

**38–23** If a hologram is made by using 600-nm light and then viewed with 500-nm light, how will the images look in comparison to those observed with 600-nm light?

**38–24** A hologram is made by using 600-nm light and is then viewed by using continuous-spectrum white light from an incandescent bulb. What will be seen?

**38–25 A Hologram Negative.** Ordinary photographic film reverses black and white, in the sense that the most brightly illuminated areas become blackest on development (thus the term *negative*). Suppose a hologram negative is viewed directly, without making a positive transparency. How will the resulting images differ from those obtained with the positive?

# PROBLEMS

**38–26** Suppose the entire apparatus (slits, screen, and space in between) in Exercise 38–4 is immersed in water ($n = 1.33$). Then what is the distance between the two dark fringes?

**38–27** A slit with width $a$ is placed in front of a lens with focal length 0.900 m. The slit is illuminated by parallel light with wavelength 600 nm, and the diffraction pattern of Fig. 38–8b is formed on a screen in the focal plane of the lens. If the photograph of Fig. 38–8b represents an enlargement to twice the actual size, what is the slit width?

**38–28** The intensity of light in the Fraunhofer diffraction pattern of a single slit is

$$I = I_0\left(\frac{\sin \delta}{\delta}\right)^2,$$

where

$$\delta = \frac{\pi a \sin \theta}{\lambda}.$$

a) Show that the equation for the values of $\delta$ at which $I$ is a maximum is $\tan \delta = \delta$. b) Determine the three smallest nonnegative values of $\delta$ that are solutions of this equation. (*Hint:* You can use a trial-and-error procedure. Guess a value for $\delta$ and adjust your guess to bring $\tan \delta$ closer to $\delta$. A graphical solution of the equation is very helpful in locating the solutions approximately, to get good initial guesses.)

**38–29** Consider a single-slit diffraction pattern. The center of the central maximum, where the intensity is $I_0$, is located at $\theta = 0$. a) Let $\theta_+$ and $\theta_-$ be the two angles on either side of $\theta = 0$ for which $I = \frac{1}{2}I_0$. The quantity $\Delta\theta = |\theta_+ - \theta_-|$ is called the full width at half maximum of the central diffraction maximum. Solve for $\Delta\theta$ when the ratio between the slit width $a$ and wavelength $\lambda$ is i) $a/\lambda = 2$; ii) $a/\lambda = 5$; iii) $a/\lambda = 10$. (*Hint:* Your equation for $\theta_+$ or $\theta_-$ cannot be solved analytically. You must use trial and error or solve it graphically.) b) The width of the central maximum can alternatively be defined as $2\theta_0$, where $\theta_0$ is the angle that locates the minimum on one side of the central maximum. Calculate $2\theta_0$ for each case considered in part (a), and compare to $\Delta\theta$.

**38–30** A slit 0.300 mm wide is illuminated by parallel rays of light that has a wavelength of 600 nm. The diffraction pattern is observed on a screen that is 0.900 m from the slit. The intensity at the center of the central maximum ($\theta = 0$) is $I_0$. a) What is the distance on the screen from the center of the central maximum to the first minimum? b) What is the distance on the screen from the center of the central maximum to the point where the intensity has fallen to $I_0/2$? (See Problem 38–28b for a hint about how to solve for the phase angle $\beta$.)

**38–31 Phasor Diagram for Eight Slits.** Consider the interference pattern produced by eight equally spaced narrow slits. In Section 38–4 it is claimed that there is an interference minimum when the phase difference $\phi$ between light from adjacent slits is $3\pi/4$. Draw the phasor diagram for this case. (*Note:* You may find use of different colored pencils helpful!) For which pairs of slits is there totally destructive interference?

**38–32** Consider the interference pattern produced by six equally spaced narrow slits. a) Draw the phasor diagram for the case for which the phase difference between light from adjacent slits is zero. In this case, if the electric field amplitude due to each individual slit is $E$, what is the resultant amplitude due to all six slits? If the intensity of light through each slit is $I$, what is the resultant intensity? b) Repeat part (a) for the case for which $\phi = 2\pi$. c) For what values of $\phi$ between 0 and $2\pi$ is there an interference minimum? For each $\phi$, draw the phasor diagram.

**38–33** What is the longest wavelength that can be observed in the fourth order for a transmission grating having 6000 lines/cm? Assume normal incidence.

**38–34** For Eq. (38–12), consider the case in which $d = a$. Show with a sketch that in this case the two slits reduce to a single slit with width $2a$. Then show that Eq. (38–12) reduces to Eq. (38–5) with slit width $2a$.

**38–35** A diffraction grating has 600 lines/mm. What is the highest order that contains the entire visible spectrum? (The wavelength range of the visible spectrum is approximately 400 nm to 700 nm.)

**38–36 X-Ray Diffraction of Salt.** X rays with a wavelength of 0.120 nm are scattered from a cubic array (of a sodium chloride crystal) for which the spacing of adjacent atoms is $a = 0.282$ nm. a) If diffraction from planes parallel to a cube face is considered, at what angles $\theta$ of the incoming beam relative to the crystal planes will maxima be observed? b) Repeat part (a) for diffraction produced by the planes shown in Fig. 38–21a, which are separated by $a/\sqrt{2}$.

**38–37** A telescope is used to observe two distant point sources 1.00 m apart with light of wavelength 500 nm. The objective of the telescope is covered with a slit having a width of 0.400 mm. What is the maximum distance in meters at which the two sources may be distinguished if the resolution is diffraction-limited and Rayleigh's criterion is used?

**38–38 Resolution of Radio and Optical Telescopes.** Diffraction occurs for circular antennas and circular mirrors, just as it does for circular apertures. What is the minimum angular separation $\theta$ of two sources of electromagnetic radiation that can be resolved (according to Rayleigh's criterion) by a) the 5.08-m-diameter (200-in.-diameter) reflecting telescope at Mt. Palomar observatory for a wavelength of 500 nm; b) the 305-m-diameter radio telescope at Arecibo, Puerto Rico for a wavelength of 80.0 m? (Note that this shows that radio telescopes must be much larger than optical ones to achieve the same diffraction-limited resolution.)

**38–39** An astronaut in the space shuttle can just resolve two point sources on earth that are 45.0 m apart. Assume that the resolution is diffraction-limited and use Rayleigh's criterion. What is the astronaut's altitude above the earth? Treat her eye as a circular aperture with a diameter of 4.00 mm (the diameter of her pupil), and take the wavelength of the light to be 550 nm.

## CHALLENGE PROBLEM

**38–40** The yellow sodium $D$ lines are a doublet with wavelengths of 589.0 and 589.6 nm and equal intensities. The sodium $D$ lines are used as a source in a Michelson interferometer. As the mirror is moved, it is noted that in addition to moving across the field of view, the interference fringes periodically appear and disappear. Why? How far is the mirror moved between disappearances of the fringes?

CHAPTER 39

# Relativity

This computer simulation shows a cubic array of balls connected by straight rods, viewed parallel to one of the edge directions. When the array is at rest, all rods appear straight and distances between adjacent balls are equal.

In this picture, the array is moving toward the viewer at a speed of 0.9c. The computer shows the appearance of a set of photons reaching the viewer simultaneously; however, not all these photons would have left the array at the same time. As a result, the rods seem to become curved and the edges of the array appear to recede.

Now the simulation shows the array moving toward the viewer at a speed of 0.99c. The image has become even more distorted.

## KEY CONCEPTS

**• The principle of relativity states that the laws of physics and the speed of light (in vacuum) are the same in all inertial frames of reference.**

**• Simultaneity is not an absolute concept; events that appear to be simultaneous in one frame of reference may not appear to be simultaneous in a second frame that is moving relative to the first. Distance and time intervals for two events look different when they are measured in frames of reference that are moving relative to each other. Each frame of reference has its own time scale.**

**• The Lorentz-transformation equations relate the position and time of an event observed in one frame of reference to its position and time in another frame moving relative to the first. They can be used to derive relative velocity relations for frames of reference moving relative to each other.**

**• For the principles of conservation of momentum and energy to be valid in all inertial frames of reference, the definitions of momentum and kinetic energy must be generalized. These generalizations lead to the concept of rest energy, which is energy associated with the mass of a body rather than with its motion.**

## INTRODUCTION

When the year 1905 began, Albert Einstein was an unknown 25-year-old clerk in the Swiss patent office. By the end of that year he had published three papers of extraordinary importance. One was an analysis of Brownian motion; a second (for which he was awarded the Nobel Prize) was on the photoelectric effect. In the third, Einstein introduced his *special theory of relativity,* proposing drastic revisions in the classical concepts of space and time.

The theory of relativity is based on just two simple postulates. One states that the laws of physics are the same in all inertial frames of reference, the other that the speed of light is the same in all inertial frames. These innocent-sounding propositions have far-reaching implications. Here are three: (1) When two observers moving relative to each other measure a time interval or a length, they may not get the same results. (2) Events that appear simultaneous to one observer may not appear so to another. (3) In order for the conservation principles for momentum and energy to be valid in all inertial systems, Newton's second law and the definitions of momentum and kinetic energy have to be revised.

Relativity has important consequences in *all* areas of physics, including thermodynamics, electromagnetism, optics, atomic and nuclear physics, and high-energy physics. Many of the results that we will derive in this chapter will run counter to your intuition, but the theory is on very solid ground in terms of its agreement with experimental observations.

## 39–1 INVARIANCE OF PHYSICAL LAWS

Einstein's **principle of relativity** states: **The laws of physics are the same in every inertial frame of reference.** Here are two examples. Suppose you watch two children playing catch with a ball while the three of you are aboard a train moving with constant velocity. No matter how carefully you study the motion of the ball, you can't tell how fast (or whether) the train is moving. This is because the laws of mechanics (Newton's laws) are the same in every inertial system.

Another example is the electromotive force (emf) induced in a coil of wire by a nearby moving permanent magnet. In the frame of reference in which the *coil* is stationary, the moving magnet causes a change of magnetic flux through the coil, and this induces an emf. In a different frame of reference in which the *magnet* is stationary, the motion of the coil through a magnetic field induces the emf. According to the principle of relativity, both of these points of view have equal validity, and both must predict the same induced emf. As we saw in Chapter 30, Faraday's law of induction can be applied to either description, and it does indeed satisfy this requirement. If the moving-magnet and moving-coil situations *did not* give the same results, we could use this experiment to distinguish one inertial frame from another. This would contradict the principle of relativity.

Equally significant is the prediction of the speed of electromagnetic radiation, derived from Maxwell's equations (Chapter 33). According to this analysis, light and all other electromagnetic waves travel in vacuum with a constant speed $c = 299{,}792{,}458$ m/s. (We will often use the approximate value $c = 3.00 \times 10^8$ m/s, which is within 1 part in 1000 of the exact value.) As we will see, the speed of light plays a central role in the theory of relativity.

During the nineteenth century, most physicists believed that light traveled through a hypothetical medium called the *ether,* just as sound waves travel through air. If so, the speed of light would depend on the motion of the observer relative to the ether and would therefore be different in different directions. The Michelson-Morley experiment, described in Section 37–5, was an effort to detect motion of the earth relative to the ether. Einstein's conceptual leap was to recognize that if Maxwell's equations are valid in all inertial frames, then the speed of light should also be the same in all frames and in all directions. In fact, Michelson and Morley detected *no* ether drift, and the ether concept was discarded. Although Einstein may not have known about this negative result, it supported his bold hypothesis that the speed of light is the same in all frames of reference.

Thus Einstein was led to his second postulate: **The speed of light is the same in all inertial frames of reference and is independent of the motion of the source.** Let's think about what this means. Suppose two observers measure the speed of light. One is at rest with respect to the light source, and the other is moving away from it. Both are in inertial frames of reference. According to the principle of relativity, the two observers must obtain the same result, despite the fact that one is moving with respect to the other.

If this seems too easy, consider the following situation. A spacecraft moving away from the earth at 1000 m/s fires a missile with a speed of 2000 m/s (relative to the spacecraft) in a direction directly away from the earth (Fig. 39–1, on the next page). What is the missile's speed relative to the earth? Simple, you say. An elementary problem in relative velocity (Section 2–7). The correct answer, according to Newtonian mechanics, is 3000 m/s. But now suppose the spacecraft turns on a searchlight, pointing in the same direction in which the missile was fired. An observer on the spacecraft measures the speed of light emitted by the searchlight and obtains the value $c$. According to our

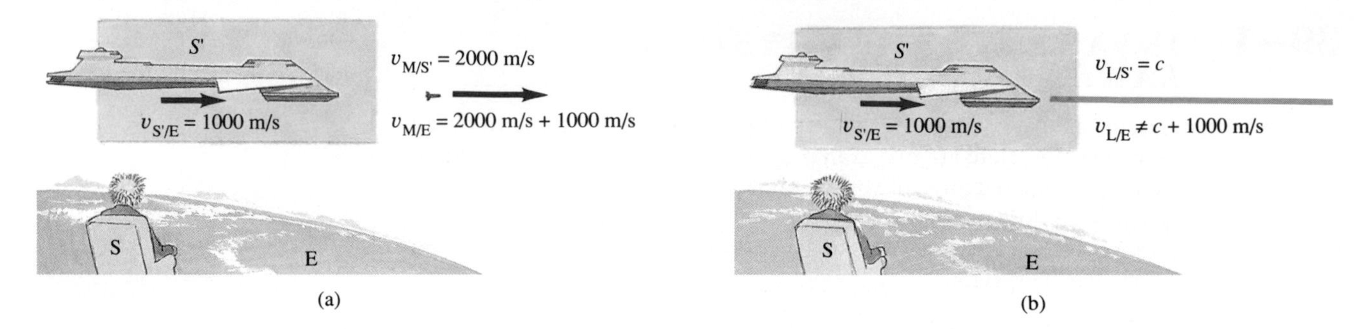


**39–1** A spaceship (S′) moves with speed $v_{S'/E} = 1000$ m/s relative to the earth (E). It fires a missile (M) with speed $v_{M/S'} = 2000$ m/s relative to the spaceship. (a) Newtonian mechanics tells us that the missile moves away from the earth at a speed of 3000 m/s. (b) Newtonian mechanics tells us that the light beam emitted by the spaceship moves away from the earth at a speed greater than $c$; this contradicts Einstein's second postulate.

previous discussion, the motion of the light after it has left the source cannot depend on the motion of the source. So the observer on earth who measures the speed of this same light must also obtain the value $c$, *not* $c + 1000$ m/s. This result contradicts our elementary notion of relative velocities, and it may not appear to agree with common sense. But "common sense" is intuition based on everyday experience, which does not usually include measurements of the speed of light.

Let's restate this argument symbolically, using the two inertial frames of reference, labeled $S$ for the observer on earth and $S'$ for the moving spacecraft, as shown in Fig. 39–2. To keep things as simple as possible, we have omitted the $z$-axes. The $x$-axes of the two frames lie along the same line, but the origin $O'$ of frame $S'$ moves relative to the origin $O$ of frame $S$ with constant velocity $u$ along the common $x$-$x'$-axis. We set our clocks so that the two origins coincide at time $t = 0$, so their separation at a later time $t$ is $ut$.

Now think about how we describe the motion of a particle $P$. This might be an exploratory vehicle launched from the spacecraft or a flash of light from a searchlight. We can describe the *position* of this point by using the earth coordinates $(x, y, z)$ in $S$ or the spacecraft coordinates $(x', y', z')$ in $S'$. The figure shows that these are related by

$$x = x' + ut, \qquad y = y', \qquad z = z'. \tag{39–1}$$

These equations, based on the familiar Newtonian notions of space and time, are called the **Galilean coordinate transformation.**

If point $P$ moves in the $x$-direction, its instantaneous velocity $v$ as measured by an observer stationary in $S$ is given by $v = dx/dt$. Its velocity $v'$ measured by an observer at rest in $S'$ is $v' = dx'/dt$. We can derive a relation between $v$ and $v'$ by taking the derivative with respect to $t$ of the first of Eqs. (39–1):

$$\frac{dx}{dt} = \frac{dx'}{dt} + u.$$

Now $dx/dt$ is the velocity $v$ measured in $S$, and $dx'/dt$ is the velocity $v'$ measured in $S'$.

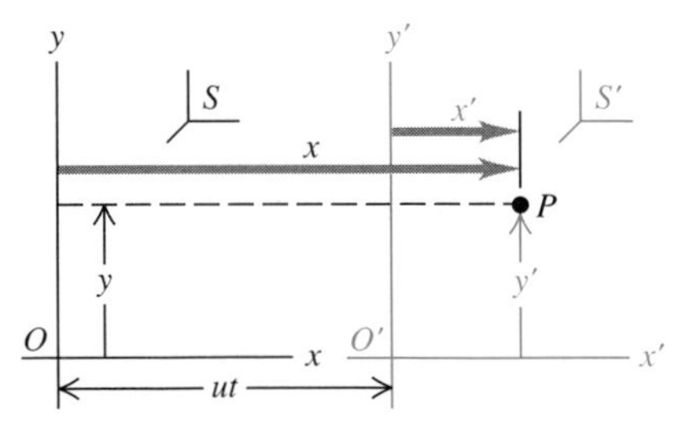


**39–2** The position of point $P$ can be described by the coordinates $x$ and $y$ in frame of reference $S$ or by $x'$ and $y'$ in $S'$. $S'$ moves relative to $S$ with constant velocity $u$ along the common $x$-$x'$-axis. The two origins $O$ and $O'$ coincide at time $t = t' = 0$.

Thus we get

$$v = v' + u. \tag{39–2}$$

This result agrees with our discussion of relative velocities in Sections 2–7 and 3–5.

Now here's the fundamental problem. Applied to the speed of *light,* Eq. (39–2) says $c = c' + u$. Einstein's second postulate, supported subsequently by a wealth of experimental evidence, says $c = c'$. This is a genuine inconsistency, not an illusion, and it demands resolution. If we accept this postulate, we are forced to conclude that Eqs. (39–1) and (39–2) *cannot* be precisely correct, despite our convincing derivation. They have to be modified to bring them into harmony with this principle.

The resolution involves some very fundamental modifications in our kinematic concepts. First is the seemingly obvious assumption that the observers in frames $S$ and $S'$ use the same *time scale.* We can state this formally by adding to Eqs. (39–1) a fourth equation:

$$t = t'.$$

Alas, we are about to show that the assumption $t = t'$ cannot be correct; the two observers *must* have different time scales. We must define the velocity $v'$ in frame $S'$ as $v' = dx'/dt'$, not as $dx'/dt$; the two quantities are not the same. The difficulty lies in the concept of *simultaneity,* which is our next topic. A careful analysis of simultaneity will help us to develop the appropriate modifications of our notions about space and time.

## 39–2 RELATIVE NATURE OF SIMULTANEITY

Measuring times and time intervals involves the concept of **simultaneity.** When you say that you awoke at seven o'clock, you mean that two *events* (your awakening and the arrival of the hour hand of your clock at the number seven) occurred *simultaneously.* The fundamental problem in measuring time intervals is that, in general, two events that are simultaneous in one frame of reference *are not* simultaneous in a second frame if it is moving relative to the first, even if both are inertial frames.

This may seem to be contrary to common sense. But here is a hypothetical experiment, devised by Einstein, that illustrates the point. Imagine a train moving with a speed comparable to $c$, with uniform velocity (Fig. 39–3a on page 1076). Two lightning bolts strike the train, one at each end. Each bolt leaves a mark on the train and one on the ground at the instant the bolt hits. The points on the ground are labeled $A$ and $B$ in the figure, and the corresponding points on the train are $A'$ and $B'$. An observer is standing on the ground at $O$, midway between $A$ and $B$. Another observer is at $O'$ at the middle of the train, midway between $A'$ and $B'$, moving with the train. Both observers see both light flashes emitted from the points where the lightning strikes.

Suppose the two light flashes reach the observer at $O$ simultaneously. He knows that he is the same distance from $A$ and $B$, so he concludes that the two bolts struck $A$ and $B$ simultaneously. But the observer at $O'$ is moving to the right with the train. She runs into the wave front from $B'$ before the wave front from $A'$ catches up to her. Because she is in the middle of the train, equidistant from $A'$ and $B'$, her observation is that both wave fronts took the same time to move the same distance at the same speed $c$. (Note that the speed of each wave front with respect to *either* observer is $c$.) Thus she concludes that the lightning bolt at the front of the train struck *earlier* than the one at the rear. The two events appear simultaneous to the observer at $O$ but not to the one at $O'$! **Whether or not two events at different locations are simultaneous depends on the state of motion of the observer.**

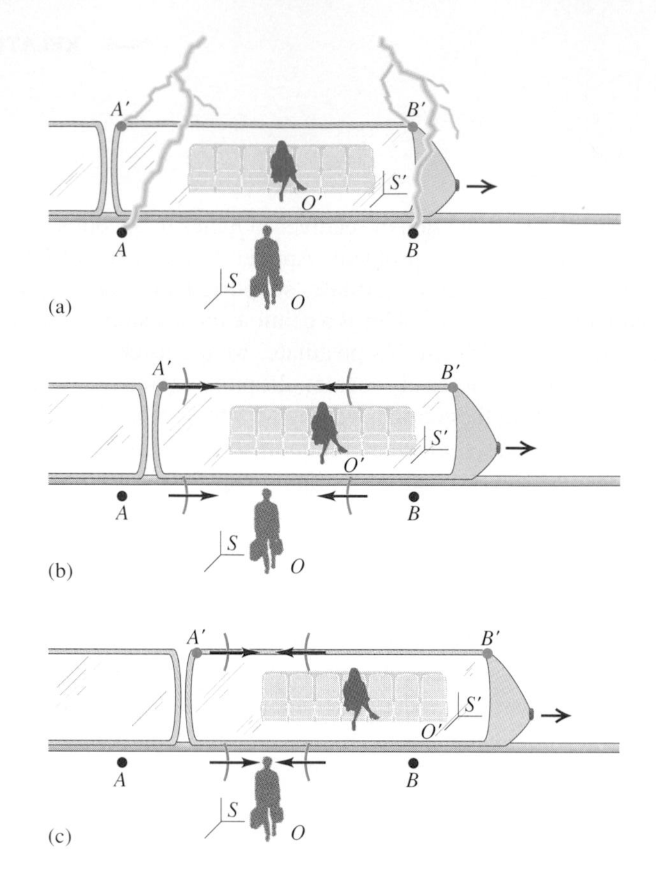


**39–3** (a) Lightning bolts strike the train and ground at each end. (b) The moving observer at point $O'$ sees the light from the front of the train first and concludes that the bolt at the front struck before the one at the rear. (c) The stationary observer at $O$ sees the light from the two bolts arrive at the same time and concludes that the lightning bolts struck simultaneously.

You may want to argue that in this example the lightning bolts really *are* simultaneous and that if the observer at $O'$ could communicate with the distant points without the time delay caused by the finite speed of light, she would realize this. But that would be erroneous; the finite speed of information transmission is not the real issue. If $O'$ is midway between $A'$ and $B'$, then in her frame of reference the time for a signal to travel from $A'$ to $O'$ is the same as the time to travel from $B'$ to $O'$. Two signals arrive simultaneously at $O'$ only if they were emitted simultaneously at $A'$ and $B'$. In this example they *do not* arrive simultaneously at $O'$, and so the observer at $O'$ must conclude that the events at $A'$ and $B'$ were *not* simultaneous.

Furthermore, there is no basis for saying that the observer at $O$ is right and the observer at $O'$ is wrong, or vice versa. According to the principle of relativity, no inertial frame of reference is preferred over any other in the formulation of physical laws. Each observer is correct *in his or her own frame of reference*. In other words, simultaneity is not an absolute concept. Whether two events are simultaneous depends on the frame of reference. Because of the essential role of simultaneity in measuring time intervals, it also follows that **the time interval between two events may be different in different frames of reference.** So our next task is to learn how to compare time intervals in different frames of reference.

## 39–3 RELATIVITY OF TIME

To derive a quantitative relation between time intervals in different coordinate systems, we consider another thought experiment. As before, a frame of reference $S'$ moves along the common $x$-$x'$-axis with constant speed $u$ relative to a frame $S$. For the present, let's assume that $u$ is always less than the speed of light $c$. An observer $O'$ in $S'$ with a light source directs a flash of light at a mirror a distance $d$ away, as shown in Fig. 39–4a. She measures the time interval $\Delta t_0$ for light to make the "round trip" to the mirror and back.

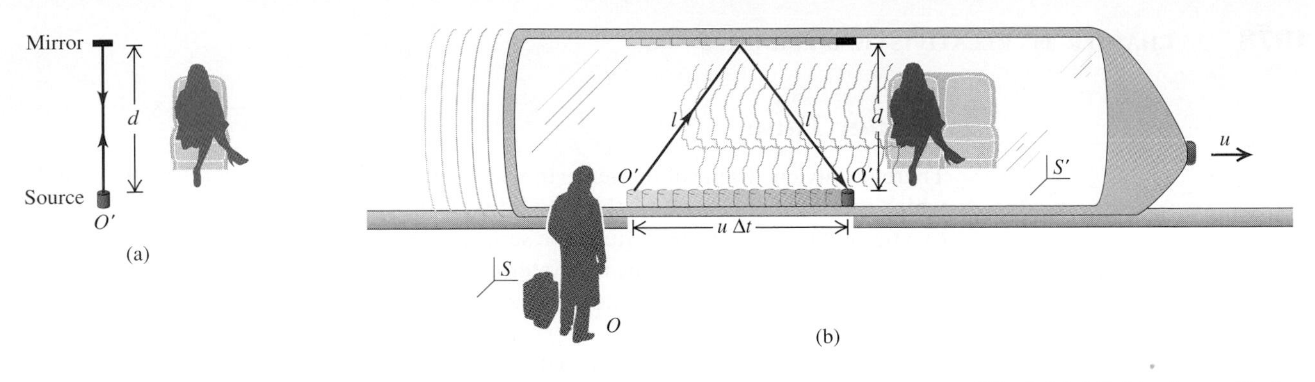


**39–4** (a) A light pulse emitted from a source at $O'$ and reflected back along the same line, as observed in $S'$. (b) Path of the same light pulse, as observed in $S$. The positions of $O'$ at the times of departure and return of the pulse are shown. The speed of the pulse is the same in $S$ as in $S'$, but the path is longer in $S$.

(We use the subscript zero as a reminder that the apparatus is at rest, with zero velocity, in frame $S'$.) The total distance is $2d$, so the time interval is

$$\Delta t_0 = \frac{2d}{c}. \tag{39–3}$$

The round-trip time measured by an observer in frame $S$ is a different interval $\Delta t$. During this time, the source moves relative to $S$ a distance $u\,\Delta t$. The total round-trip distance as seen in $S$ is not just $2d$ but $2l$, where

$$l = \sqrt{d^2 + \left(\frac{u\,\Delta t}{2}\right)^2}.$$

In writing this expression we have used the fact that the distance $d$ looks the same to both observers. This can (and indeed must) be justified by other thought experiments, but we won't go into this now. The speed of light is the same for both observers, so the relation in $S$ analogous to Eq. (39–3) is

$$\Delta t = \frac{2l}{c} = \frac{2}{c}\sqrt{d^2 + \left(\frac{u\,\Delta t}{2}\right)^2}. \tag{39–4}$$

We would like to have a relation between $\Delta t$ and $\Delta t_0$ that doesn't contain $d$. To get this, we solve Eq. (39–3) for $d$ and substitute the result into Eq. (39–4), obtaining

$$\Delta t = \frac{2}{c}\sqrt{\left(\frac{c\,\Delta t_0}{2}\right)^2 + \left(\frac{u\,\Delta t}{2}\right)^2}. \tag{39–5}$$

Now we square this and solve for $\Delta t$; the result is

$$\Delta t = \frac{\Delta t_0}{\sqrt{1 - u^2/c^2}}.$$

We may generalize this important result: If two events occur at the same point in space in a particular frame of reference, and if the time interval between them, as measured by an observer at rest in this frame (which we call the *rest frame* of this observer), is $\Delta t_0$, then an observer in a second frame moving with constant velocity $u$ relative to the first frame will measure the time interval to be $\Delta t$, where

$$\Delta t = \frac{\Delta t_0}{\sqrt{1 - u^2/c^2}}. \tag{39–6}$$

The denominator is always smaller than unity, so $\Delta t$ is always *larger* than $\Delta t_0$. Think of an old-fashioned pendulum clock that ticks once a second, as observed in its rest frame; this is $\Delta t_0$. When the time between ticks is measured by an observer in a frame moving with respect to the clock, the time interval $\Delta t$ that he observes is longer than one second, and he thinks that the clock is running slow. This effect is called **time dilation.** In brief, moving clocks appear to run slow. Note that this conclusion is a direct result of the fact that the speed of light is the same in both frames of reference.

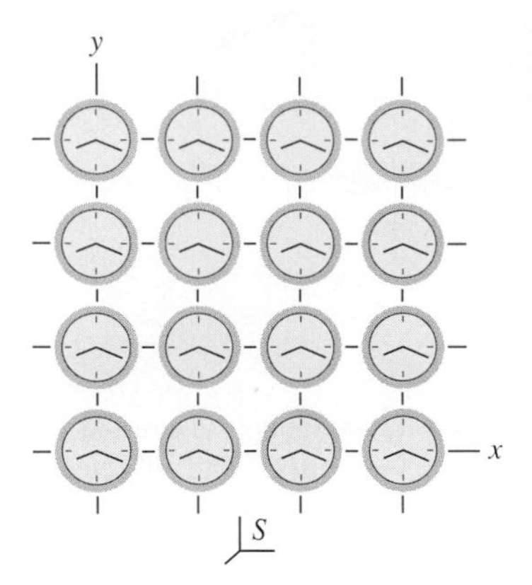


**39–5** A frame of reference pictured as a coordinate system with a set of synchronized clocks. For the third dimension, picture identical planes of these clocks parallel to the page, in front of and behind it, connected by grid lines perpendicular to the page.

## Proper Time

There is only one frame of reference in which the clock is at rest and infinitely many in which it is moving. Therefore the time interval measured between two events (such as two ticks of the clock) that occur at the same point (as viewed in a particular frame) is a more fundamental quantity than the interval between events at different points. We use the term **proper time** to describe the time interval $\Delta t_0$ between two events that occur *at the same point.*

We also note that Eq. (39–6) makes sense only when $u < c$; otherwise, the denominator is imaginary.

It is important to note that the time interval $\Delta t$ in Eq. (39–6) involves events that occur *at different space points* in the frame of reference $S$. Therefore it can't really be measured by a single observer at rest in $S$. But we can use *two* observers, one stationary at the location of the first event and the other at the second, each with his own clock. We can synchronize these two clocks without difficulty as long as they are at rest in the same frame of reference. For example, we could send a light pulse simultaneously to the two clocks from a point midway between them. When the pulses arrive, the observers set their clocks to a prearranged time. (But note that clocks that are synchronized in one frame of reference *are not* in general synchronized in any other frame.)

In thought experiments it's often helpful to imagine many observers with synchronized clocks at rest at various points in a particular frame of reference. We can picture a frame of reference as a coordinate grid with lots of synchronized clocks distributed around it, as suggested by Fig. 39–5. Only when a clock is moving relative to a given frame of reference do we have to watch for ambiguities of synchronization or simultaneity.

### EXAMPLE 39–1

**Time dilation at 0.99 $c$** A spaceship flies past the earth with a speed of $0.99c$ (about $2.97 \times 10^8$ m/s) relative to the earth. A high-intensity signal light (perhaps a pulsed laser) on the ship blinks on and off; each pulse lasts $2.20 \times 10^{-6}$ s, as measured on the spaceship. At a certain instant the ship is 1000 km above an observer on the earth and is traveling perpendicular to the observer's line of sight. What is the duration of each light pulse as measured by the observer on the earth, and how far does the ship travel relative to the earth during each pulse?

**SOLUTION** Let $S$ be the earth's frame of reference and let $S'$ be that of the spaceship. The time between laser pulses, measured by an observer on the spaceship (where the laser is at rest), is $2.20 \times 10^{-6}$ s. This is a proper time in $S'$, referring to two events, the starting and stopping of the pulse, that occur at the same point relative to $S'$. In the notation of Eq. (39–6), $\Delta t_0 = 2.20 \times 10^{-6}$ s. The corresponding interval $\Delta t$ on the earth ($S$) is given by Eq. (39–6):

$$\Delta t = \frac{\Delta t_0}{\sqrt{1 - u^2/c^2}} = \frac{2.20 \times 10^{-6}\ \text{s}}{\sqrt{1 - (0.99)^2}} = 15.6 \times 10^{-6}\ \text{s}.$$

Thus the time dilation in $S$ is about a factor of 7. During this interval the spaceship travels a distance $d$ relative to the earth, given by

$$d = u\,\Delta t = (0.99)(3.00 \times 10^8\ \text{m/s})(15.6 \times 10^{-6}\ \text{s})$$
$$= 4600\ \text{m} = 4.6\ \text{km}.$$

Note that the distance (1000 km) from spaceship to observer is not relevant; it is not involved in Eq. (39–6).

An experiment that is similar in principle to the spaceship example, though very different in detail, provided the first direct experimental confirmation of Eq. (39–6). The $\mu$ mesons, or *muons,* are unstable particles, first observed in cosmic rays. They decay with a mean lifetime of $2.2 \times 10^{-6}$ s, as measured in a frame of reference in which the particles are at rest. But in cosmic-ray showers, the particles are moving very fast. The mean lifetime of a muon with a speed of $0.99c$ is measured to be $15.6 \times 10^{-6}$ s. Note that the numbers are identical to those in the spaceship example; the duration of the light pulse is replaced by the mean lifetime of the muon. These measurements provide a direct experimental confirmation of Eq. (39–6).

### EXAMPLE 39–2

**Time dilation at jetliner speeds** An airplane flies from San Francisco to New York (about 4800 km, or $4.80 \times 10^6$ m) at a steady speed of 300 m/s (about 670 mi/h). How much time does the trip take as measured by an observer on the ground? By an observer in the plane?

**SOLUTION** The time measured by the ground observer corresponds to $\Delta t$ in Eq. (39–6); it is simply the distance divided by the speed:

$$\Delta t = \frac{4.80 \times 10^6\ \text{m}}{300\ \text{m/s}} = 1.60 \times 10^4\ \text{s} \quad (\text{or about } 4\tfrac{1}{2}\ \text{hours}).$$

We need two observers with synchronized clocks to measure this interval, one in San Francisco and one in New York, because the two events (takeoff and landing) occur at different space points in the ground frame of reference.

In the airplane's frame, these events occur at the *same* point. The time interval in the airplane, which can be measured by a single observer in the plane, is a proper time, corresponding to $\Delta t_0$ in Eq. (39–6). We have

$$\frac{u^2}{c^2} = \frac{(300\ \text{m/s})^2}{(3.00 \times 10^8\ \text{m/s})^2} = 10^{-12},$$

and from Eq. (39–6),

$$\Delta t_0 = (1.60 \times 10^4\ \text{s})\sqrt{1 - 10^{-12}}.$$

The radical can't be evaluated with adequate precision with an ordinary calculator. But we can expand it by using the binomial theorem (Appendix B):

$$(1 - 10^{-12})^{1/2} = 1 - (\tfrac{1}{2})10^{-12} + \cdots.$$

The remaining terms are of the order of $10^{-24}$ or smaller and can be discarded. The approximate result for $\Delta t_0$ is

$$\Delta t_0 = (1.60 \times 10^4\ \text{s})(1 - 0.500 \times 10^{-12}).$$

The proper time $\Delta t_0$, measured in the airplane, is very slightly less (by less than 1 part in $10^{12}$) than the time measured on the ground.

We don't notice such effects in everyday life. But as we mentioned in Section 1–3, present-day atomic clocks can attain a precision of about 1 part in $10^{13}$. A cesium clock traveling a long distance in a 747 has been used to measure this effect and thereby verify Eq. (39–6) directly. So this relationship, unlikely though it may seem, can be confirmed directly by experiment.

When the relative velocity $u$ of the two frames of reference $S$ and $S'$ is very small, the factor $1 - u^2/c^2$ is very nearly equal to unity, and Eq. (39–6) approaches the Newtonian relation $\Delta t = \Delta t_0$ (that is, the same time scale for all frames of reference). The quantity $1/(1 - u^2/c^2)^{1/2}$ appears so often in relativity that it is given the symbol $\gamma$:

$$\gamma = \frac{1}{\sqrt{1 - u^2/c^2}}. \tag{39–7}$$

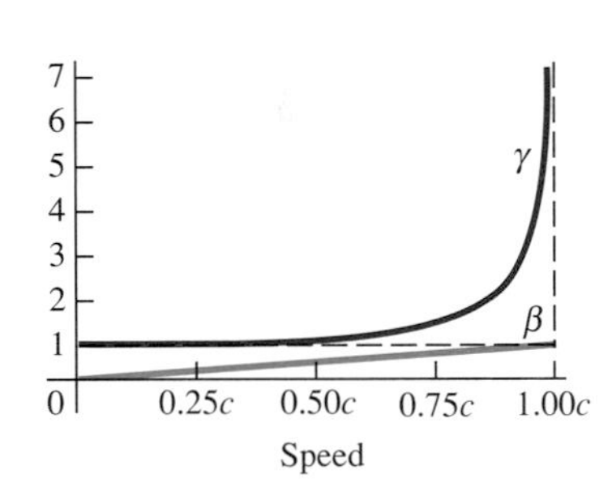


**39–6** Because $u$ is always less than or equal to $c$, $\gamma$ is always greater than or equal to unity. It approaches infinity as $u$ approaches $c$. On the other hand, $\beta$ is proportional to $u$; it is always less than or equal to unity, approaching unity as $u$ approaches $c$.

We will sometimes abbreviate this relationship by using the notation $\beta = u/c$; Eq. (39–7) then becomes

$$\gamma = \frac{1}{\sqrt{1 - \beta^2}}. \tag{39–8}$$

In terms of $\gamma$ we can express Eq. (39–6) simply as

$$\Delta t = \gamma\,\Delta t_0. \tag{39–9}$$

When the relative velocity $u$ of two frames of reference is very small in comparison to $c$, $\beta$ is much smaller than unity, and $\gamma$ is very nearly *equal* to unity. In that limit, Eq. (39–6) or (39–9) approaches the Newtonian relation $\Delta t = \Delta t_0$. Figure 39–6 shows graphs of $\gamma$ and $\beta$ as functions of $u$.

## The Twin Paradox

Equation (39–6) suggests an apparent paradox called the **twin paradox.** Consider identical-twin astronauts named Eartha and Astrid. Eartha remains on the earth, and her twin Astrid takes off on a high-speed trip through the galaxy. Because of time dilation, Eartha sees Astrid's heartbeat and all other life processes proceeding more slowly than her own. Thus Eartha thinks that Astrid ages more slowly, so when Astrid returns to the earth she is younger than Eartha.

Now here is the paradox: All inertial frames are equivalent. Can't Astrid make exactly the same arguments to conclude that Eartha is in fact the younger? Then each twin thinks that the other is younger, and that's a paradox.

To resolve the paradox, we recognize that the twins are *not* identical in all respects. If Eartha remains in an inertial frame at all times, Astrid must have an acceleration with respect to inertial frames during part of her trip in order to turn around and come back.

Eartha remains always at rest in the same inertial frame; Astrid does not. Thus there is a real physical difference between the circumstances of the two twins. Careful analysis shows that Eartha is correct; when Astrid returns, she *is* younger than Eartha.

## 39–4 RELATIVITY OF LENGTH

Just as the time interval between two events depends on the observer's frame of reference, the *distance* between two points may also depend on the observer's frame of reference. The concept of simultaneity is involved. Suppose you want to measure the length of a moving car. One way is to have two assistants make marks on the pavement *at the same time* at the positions of the front and rear bumpers. Then you measure the distance between the marks. If you mark the position of the front bumper at one time and that of the rear bumper half a second later, you won't get the car's true length. But we've learned that simultaneity isn't an absolute concept, so we have to proceed with caution.

To develop a relation between lengths in various coordinate systems, we consider another thought experiment. We attach a light source to one end of a ruler and a mirror to the other end (Fig. 39–7). The ruler is at rest in reference frame $S'$, and its length in this frame is $l_0$. Then the time $\Delta t_0$ required for a light pulse to make the round trip from source to mirror and back is

$$\Delta t_0 = \frac{2l_0}{c}. \tag{39–10}$$

This is a proper time interval because departure and return occur at the same point in $S'$.

In reference frame $S$ the ruler is moving to the right with speed $u$ during this travel of the light pulse. The length of the ruler in $S$ is $l$, and the time of travel from source to mirror, as measured in $S$, is $\Delta t_1$. During this interval the ruler, with source and mirror attached, moves a distance $u\,\Delta t_1$. The total length of path $d$ from source to mirror is not $l$ but

$$d = l + u\,\Delta t_1. \tag{39–11}$$

The light pulse travels with speed $c$, so it is also true that

$$d = c\,\Delta t_1. \tag{39–12}$$

Combining Eqs. (39–11) and (39–12) to eliminate $d$, we find

$$c\,\Delta t_1 = l + u\,\Delta t_1,$$

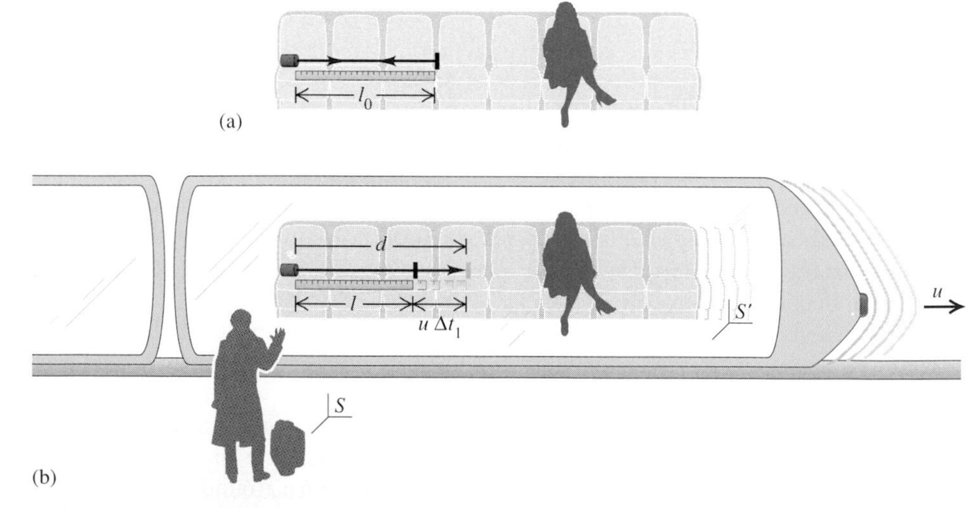


**39–7** (a) A light pulse is emitted from a source at one end of a ruler, reflected from a mirror at the opposite end, and returned to the source position. (b) Motion of the light pulse as seen by an observer in $S$. The distance traveled from source to mirror is greater, by the amount $u\,\Delta t_1$, than the length $l$ measured in $S$, as shown.

or

$$\Delta t_1 = \frac{l}{c - u}. \tag{39–13}$$

(This *does not* mean that light travels with speed $c - u$, but rather that the distance that the pulse travels is greater than $l$.)

In the same way we can show that the time $\Delta t_2$ for the return trip from mirror to source is

$$\Delta t_2 = \frac{l}{c + u}. \tag{39–14}$$

The *total* time $\Delta t = \Delta t_1 + \Delta t_2$ for the round trip, as measured in $S$, is

$$\Delta t = \frac{l}{c - u} + \frac{l}{c + u} = \frac{2l}{c(1 - u^2/c^2)}. \tag{39–15}$$

We also know that $\Delta t$ and $\Delta t_0$ are related by Eq. (39–6) because $\Delta t_0$ is proper time in $S'$. Thus Eq. (39–10) becomes

$$\Delta t \sqrt{1 - \frac{u^2}{c^2}} = \frac{2l_0}{c}. \tag{39–16}$$

Finally, combining this with Eq. (39–15) to eliminate $\Delta t$ and simplifying, we obtain

$$l = l_0 \sqrt{1 - \frac{u^2}{c^2}}. \tag{39–17}$$

Thus the length $l$ measured in $S$, in which the ruler is moving, is *shorter* than the length $l_0$ measured in its rest frame $S'$. (Note that if light traveled with speed $c \pm u$ in the moving frame, contrary to Einstein's second postulate, $l$ and $l_0$ would be equal.)

A length measured in the frame where the body is at rest (the rest frame of the body) is called a **proper length;** thus, $l_0$ is a proper length in $S'$, and the length measured in any other frame moving relative to $S'$ is *less than* $l_0$. This effect is called **length contraction.** In terms of the quantity $\gamma = (1 - u^2/c^2)^{-1/2}$ defined in Eq. (39–7), we can also express Eq. (39–17) as

$$l = \frac{l_0}{\gamma}. \tag{39–18}$$

## EXAMPLE 39–3

A crew member on the spaceship of Example 39–1 (Section 39–3) measures its length, obtaining the value 400 m. What is the length measured by an observer on the earth?

**SOLUTION** The length of the space ship in the frame in which it is at rest (400 m) corresponds to $l_0$ in Eq. (39–17), and we want to find the length $l$ measured by an observer on the earth. From Eq. (39–17),

$$l = l_0 \sqrt{1 - \frac{u^2}{c^2}}$$

$$= (400\ \text{m})\sqrt{1 - (0.99)^2}$$

$$= 56.4\ \text{m}.$$

To measure this quantity, we would have to observe the positions of the two ends of the spaceship simultaneously in the earth's reference frame; this could be done by two observers with synchronized clocks, as shown in Fig. 39–8. (These two observations will *not* appear simultaneous to an observer in the spaceship.)

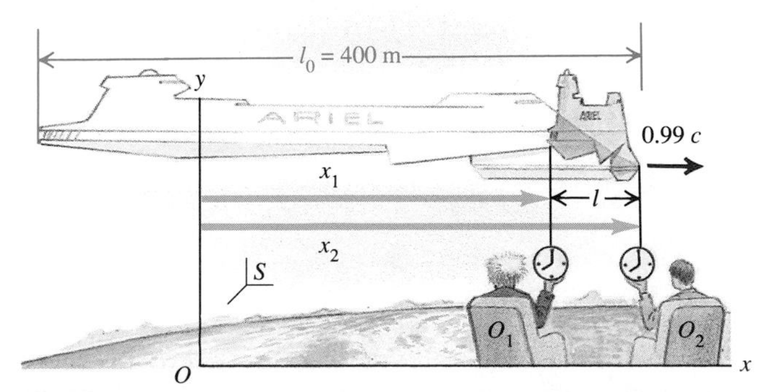


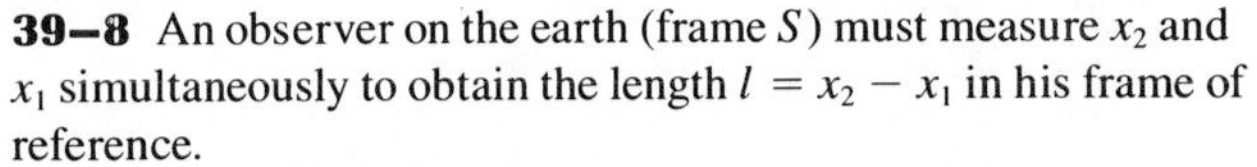

**39–8** An observer on the earth (frame $S$) must measure $x_2$ and $x_1$ simultaneously to obtain the length $l = x_2 - x_1$ in his frame of reference.

When $u$ is very small in comparison to $c$, the factor $\gamma$ in Eq. (39–18) approaches unity, and in the limit of small speeds we recover the Newtonian relation $l = l_0$. This and the corresponding result for time dilation show that Eqs. (39–1) are valid in the limit of speeds much smaller than $c$; only at speeds comparable to $c$ are changes needed.

We have derived Eq. (39–17) for lengths measured in the direction *parallel* to the relative motion of the two frames of reference. Lengths measured *perpendicular* to the direction of motion are *not* contracted. To prove this, consider two identical meter sticks. One stick is at rest in frame $S$ and lies along the positive $y$-axis with one end at $O$, the origin of $S$. The other is at rest in frame $S'$ and lies along the positive $y'$-axis with one end at $O'$, the origin of $S'$. An observer at rest in each frame stations himself at the 50-cm mark of his stick. At the instant the two origins coincide, the two sticks lie along the same line. At this instant, each observer makes a mark on the other's stick at the point that coincides with his own 50-cm mark.

Suppose for the sake of argument that $S$ sees the stick moving with $S'$ as longer than his own. Then the mark the observer in $S$ makes on the stick moving with $S'$ is *below* its center. In that case the observer in $S'$ will think that the stick in $S$ has become shorter, since half of its length coincides with *less* than half the length of the stick in $S'$. So $S'$ sees the moving sticks getting shorter, not longer. But this implies an asymmetry between the two frames. Any such asymmetry contradicts the basic postulate of relativity that all inertial frames are equivalent. We conclude that consistency with the postulates of relativity requires that both observers see the rulers as having the same length, even though to each observer one of them is stationary and one is moving. So *there is no length contraction perpendicular to the direction of relative motion of the coordinate systems.* In fact, we used this result in our derivation of Eq. (39–5) when we assumed that the distance $d$ is the same in both frames of reference.

Finally, let's think a little about the visual appearance of a moving three-dimensional body. If we could see the positions of all points of the body simultaneously, it would appear just to shrink in the direction of motion. But we *don't* see all the points simultaneously; light from points farther from us takes longer to get to us than light from points near to us, so we see the positions of farther points as they were at earlier times.

Suppose we have a cube with its faces parallel to the coordinate planes. When we look straight-on at the closest face of such a cube at rest, we see only that face. But when the cube is moving past us toward the right, we can also see the left side because of the effect just described. More generally, we can see some points that we couldn't see when the body was at rest because it moves out of the way of the light rays. Conversely, some light that can get to us when the body is at rest is blocked by the moving body. Because of all this, the cube appears rotated and distorted. Figure 39–9 shows a computer-generated image of a more complex body moving at a relativistic speed relative to the observer.

**39–9** Computer simulation of the appearance of an array of 25 rods with square cross section. The center rod is viewed end-on. (a) The array at rest; (b) the array moving to the right at $0.2c$; (c) the array moving to the right at $0.9c$.

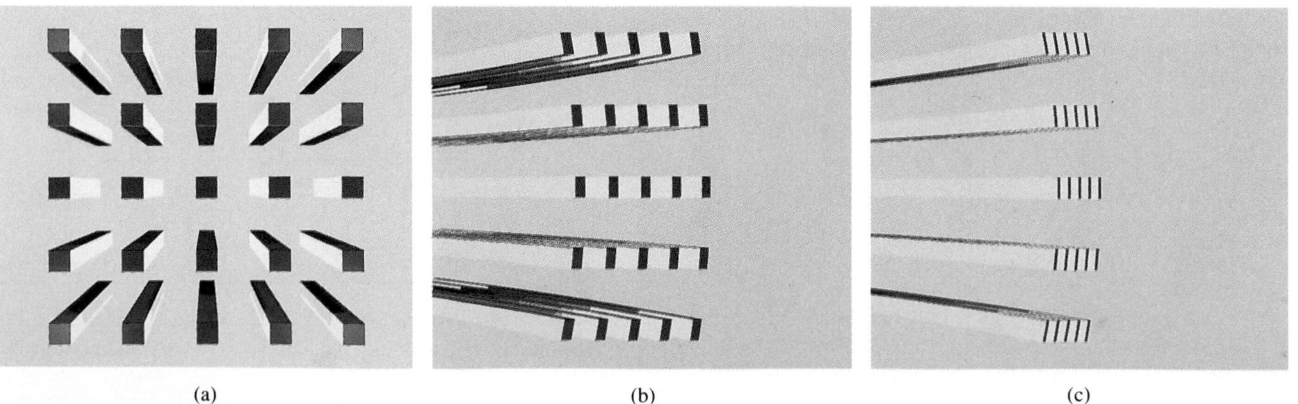

## 39–5 THE LORENTZ TRANSFORMATION

In Section 39–1 we discussed the Galilean coordinate transformation equations, Eqs. (39–1), which relate the coordinates $(x, y, z)$ of a point in frame of reference $S$ to the coordinates $(x', y', z')$ of the point in a second frame $S'$ moving with constant velocity $u$ relative to $S$ along the common $x$-$x'$-axis. Embedded in this formulation is the assumption that the time scale is the same in the two frames of reference, as expressed by the additional relation $t = t'$.

This transformation, as we have seen, is valid only in the limit when $u$ is much smaller than $c$. We are now ready to derive a more general transformation that is consistent with the principle of relativity. The more general relations are called the *Lorentz transformation*. In the limit of very small $u$ they reduce to the Galilean transformation, but they may also be used when $u$ is comparable to $c$.

The basic question is this: When an event occurs at point $(x, y, z)$ at time $t$, as observed in a frame of reference $S$, what are the coordinates $(x', y', z')$ and time $t'$ of the event as observed in a second frame $S'$ moving relative to $S$ with constant velocity $u$ along the $x$-direction?

To derive the transformation equations, we refer to Fig. 39–10, which is the same as Fig. 39–2. As before, we assume that the origins coincide at the initial time $t = t' = 0$. Then in $S$ the distance from $O$ to $O'$ at time $t$ is still $ut$. The coordinate $x'$ is a *proper length* in $S'$, so in $S$ it appears contracted by the factor $1/\gamma = (1 - u^2/c^2)^{1/2}$, as in Eq. (39–18). Thus the distance $x$ from $O$ to $P$, as seen in $S$, is not simply $x = ut + x'$ as in the Galilean transformation, but

$$x = ut + x'\sqrt{1 - \frac{u^2}{c^2}}. \tag{39–19}$$

Solving this equation for $x'$, we obtain

$$x' = \frac{x - ut}{\sqrt{1 - u^2/c^2}}. \tag{39–20}$$

This is half of the Lorentz transformation; the other half is the equation giving $t'$ in terms of $x$ and $t$. To obtain this, we note that the principle of relativity requires that the *form* of the transformation from $S$ to $S'$ be identical to that from $S'$ to $S$. The only difference is a change in the sign of the relative velocity $u$. Thus from Eq. (39–19) it must be true that

$$x' = -ut' + x\sqrt{1 - \frac{u^2}{c^2}}. \tag{39–21}$$

We now equate Eqs. (39–20) and (39–21) to eliminate $x'$. This gives us an equation for $t'$ in terms of $x$ and $t$. We will leave the algebraic details for you to work out; the result is

$$t' = \frac{t - ux/c^2}{\sqrt{1 - u^2/c^2}}. \tag{39–22}$$

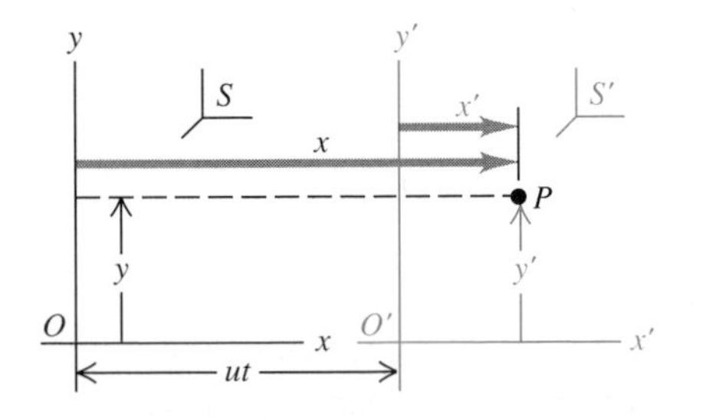


**39–10** As seen in reference frame $S$, $x'$ is contracted to $x'/\gamma$, so $x = ut + x'/\gamma$ and $x' = \gamma(x - ut)$.

As we discussed previously, lengths perpendicular to the direction of relative motion are not affected by the motion, so $y' = y$ and $z' = z$.

Collecting all the transformation equations, we have

$$
\begin{aligned}
x' &= \frac{x - ut}{\sqrt{1 - u^2/c^2}} = \gamma(x - ut), \\
y' &= y, \\
z' &= z, \\
t' &= \frac{t - ux/c^2}{\sqrt{1 - u^2/c^2}} = \gamma(t - ux/c^2).
\end{aligned}
\tag{39–23}
$$

These equations are the **Lorentz transformation,** the relativistic generalization of the Galilean transformation, Eqs. (39–1). When $u$ is much smaller than $c$, the radicals in the denominators approach unity, $\gamma$ approaches unity, and the second term in the numerator of the $t'$ equation approaches zero (because it contains the factor $u/c$). In this limit, Eqs. (39–23) become identical to Eqs. (39–1), and $t = t'$. In general, though, both the coordinates and time of an event in one frame depend on its coordinates and time in another frame. Space and time have become intertwined; we can no longer say that length and time have absolute meanings, independent of frame of reference.

We can use Eqs. (39–23) to derive the relativistic generalization of the Galilean velocity-transformation equation, Eq. (39–2). Suppose a body is moving along the $x$-axis; in a time $dt$ it moves a distance $dx$, as measured in frame $S$. We obtain the corresponding distance $dx'$ and time $dt'$ in $S'$ by taking differentials of Eqs. (39–23):

$$
\begin{aligned}
dx' &= \gamma(dx - u\,dt), \\
dt' &= \gamma(dt - u\,dx/c^2).
\end{aligned}
$$

We divide the first equation by the second and then divide numerator and denominator of the result by $dt$ to obtain

$$\frac{dx'}{dt'} = \frac{\dfrac{dx}{dt} - u}{1 - \dfrac{u}{c^2}\dfrac{dx}{dt}}.$$

Now $dx/dt$ is the velocity $v$ in $S$, and $dx'/dt'$ is the velocity $v'$ in $S'$, so we finally obtain

$$v' = \frac{v - u}{1 - uv/c^2}. \tag{39–24}$$

When $u$ and $v$ are much smaller than $c$, the denominator in Eq. (39–24) approaches unity, and we obtain the nonrelativistic result $v' = v - u$. The opposite extreme is the case $v = c$; then we find

$$v' = \frac{c - u}{1 - uc/c^2} = c.$$

This says that anything moving with speed $v = c$ relative to $S$ also has speed $v' = c$ relative to $S'$, despite the relative motion of the two frames. So Eq. (39–24) is consistent with our initial assumption that the speed of light is the same in all inertial frames.

Because there is no fundamental distinction between the two frames $S$ and $S'$, the expression for $v$ in terms of $v'$ must have the same form as Eq. (39–24), with $v$ and $v'$ interchanged and the sign of $u$ reversed. Carrying out these operations with Eq. (39–24), we find

$$v = \frac{v' + u}{1 + uv'/c^2}. \tag{39–25}$$

This can also be obtained algebraically by solving Eq. (39–24) for $v$.

## EXAMPLE 39–4

**Relative velocities** A spaceship moving away from earth with speed $0.90c$ fires a robot space probe in the same direction as its motion, with speed $0.70c$ relative to the spaceship. What is the probe's speed relative to the earth?

**SOLUTION** Let the earth's frame of reference be $S$ and let the spaceship's be $S'$ (Fig. 39–11). Then $u = 0.90c$ and $v' = 0.70c$. The nonrelativistic velocity addition formula would give a velocity relative to the earth of $1.60c$. The correct relativistic result, from Eq. (39–25), is

$$v = \frac{v' + u}{1 + uv'/c^2} = \frac{0.70c + 0.90c}{1 + (0.90c)(0.70c)/c^2} = 0.98c.$$

The incorrect classical value ($1.60c$) is too large by a factor of 1.63.

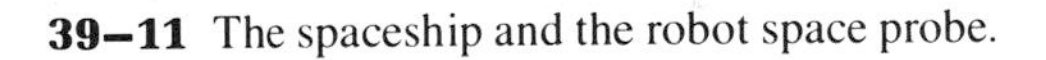


**39–11** The spaceship and the robot space probe.

## EXAMPLE 39–5

A scoutship from the earth tries to catch up with the spaceship of Example 39–4 by traveling at $0.95c$ relative to the earth. What is its speed relative to the spaceship?

**SOLUTION** Again we let the earth's frame of reference be $S$ and the spaceship's frame be $S'$. Again we have $u = 0.90c$, but now $v = 0.95c$. According to nonrelativistic velocity addition, the scoutship's velocity relative to the spaceship would be $0.05c$. We get the correct result from Eq. (39–24):

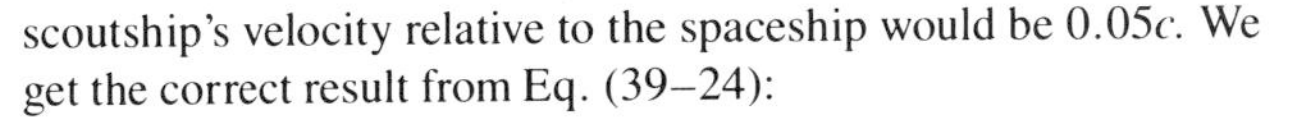

$$v' = \frac{v - u}{1 - uv/c^2} = \frac{0.95c - 0.90c}{1 - (0.90c)(0.95c)/c^2} = 0.34c.$$

The relativistically correct value of the relative velocity is nearly seven times as large as the incorrect classical value.

When $u$ is less than $c$, a body moving with a speed less than $c$ in one frame of reference always has a speed less than $c$ in *every other* frame of reference. This is one reason for thinking that no material body may travel with a speed greater than that of light, relative to *any* frame of reference. The relativistic generalizations of energy and momentum, which we will explore later, give further support to this hypothesis.

## *39–6 THE RELATIVISTIC DOPPLER EFFECT

An additional important consequence of relativistic kinematics is the Doppler effect for electromagnetic radiation. In our previous discussion of the Doppler effect (Section 21–4) we quoted without proof the formula, Eq. (21–20), for the frequency shift that results from motion of a source of electromagnetic waves relative to an observer. We can now derive that result.

Here's a statement of the problem. A source of light is moving with constant velocity $u$ toward an observer who is stationary in an inertial frame (Fig. 39–12, page 1086).

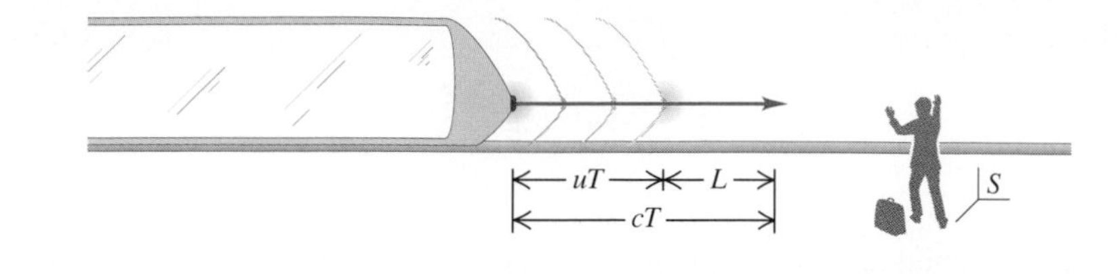


**39–12** The relativistic Doppler effect: A moving source emits a light pulse, then travels a distance $uT$ toward an observer and emits a second pulse. In the observer's reference frame, $S$, the second pulse is a distance $L$ behind the first, and the pulses arrive a time $L/c$ apart, where $L = (c - u)T$ and $T = \gamma T_0$.

The source emits light pulses with frequency $f_0$ and period $T_0 = 1/f_0$, as measured in its rest frame. What is the frequency $f$ at which these pulses arrive at the observer?

Let $T$ be the time interval between emission of successive pulses as observed in the rest frame of the observer. Note that this is *not* the interval between the *arrival* of successive pulses at the observer's position because the pulses are emitted at different points in the observer's frame. During a time $T$ the pulses move a distance $cT$, and the source moves a distance $uT$. The distance $L$ between successive pulses, as measured in the observer's frame, is thus $L = (c - u)T$. If one pulse arrives at the observer at a certain time, the next one arrives at a time later by $L/c$, which is the time a pulse takes to travel the distance $L$ at speed $c$. Thus the time between arrival of successive pulses at the observer's location is

$$\frac{L}{c} = \frac{(c - u)T}{c}.$$

The frequency $f$ with which these pulses are received is the reciprocal of this:

$$f = \frac{c}{L} = \frac{c}{(c - u)T}. \qquad (39\text{–}26)$$

So far we have followed the same pattern as for the Doppler effect for sound from a moving source (Section 21–4). In that discussion our next step was to equate $T$ to the time between emission of successive pulses by the source, which we have called $T_0$. But that would not be relativistically correct. The time $T_0$ is measured in the rest frame of the source, so it is a proper time. From Eq. (39–6), $T_0$ and $T$ are related by

$$T = \frac{T_0}{\sqrt{1 - u^2/c^2}} = \frac{cT_0}{\sqrt{c^2 - u^2}},$$

or, since $T_0 = 1/f_0$,

$$\frac{1}{T} = \frac{\sqrt{c^2 - u^2}}{cT_0} = \frac{\sqrt{c^2 - u^2}}{c} f_0.$$

When we substitute this expression for $1/T$ into Eq. (39–26), we get

$$f = \frac{c}{c - u}\frac{\sqrt{c^2 - u^2}}{c} f_0 = \sqrt{\frac{c + u}{c - u}} f_0. \qquad (39\text{–}27)$$

This shows that when the source moves *toward* the observer, the observed frequency is *greater* than $f_0$. The difference $f - f_0 = \Delta f$ is called the Doppler frequency shift. When $u/c$ is much smaller than unity, the fractional shift $\Delta f/f$ is approximately equal to $u/c$:

$$\frac{\Delta f}{f} \cong \frac{u}{c}.$$

This relationship was found in Problem 21–25, with some changes in notation.

When the source moves *away from* the observer, we change the sign of $u$ in Eq. (39–27) to get

$$f = \sqrt{\frac{c - u}{c + u}} f_0. \tag{39–28}$$

This agrees with Eq. (21–20), with minor notation changes.

With light, unlike sound, there is no distinction between motion of source and motion of observer; only the *relative* velocity of the two is significant. The last three paragraphs of Section 21–4 discuss several practical applications of the Doppler effect with light and other electromagnetic radiation; we suggest that you review those paragraphs now.

### EXAMPLE 39–6

We are observing the radiation from sodium atoms moving directly toward us. The frequency in the rest frame of the atoms is $5.08 \times 10^{14}$ Hz, but the frequency that we observe is $10.16 \times 10^{14}$ Hz. At what speed are the atoms moving toward us?

**SOLUTION** We are given $f/f_0 = 2.00$, and this is related to the relative velocity $u$ by Eq. (39–27). The simplest procedure is to solve this equation for $u$. That takes a little algebra; we'll leave it as an exercise for you to show that the result is

$$u = \frac{(f/f_0)^2 - 1}{(f/f_0)^2 + 1} c.$$

In this case we find

$$u = \frac{(2.00)^2 - 1}{(2.00)^2 + 1} c = 0.600c.$$

When the source moves toward us at $0.600c$, the frequency we measure is double the value measured in the rest frame of the source. We leave it as an exercise to show that when the source moves *away from* us at $0.600c$, the frequency we measure is *half* the value measured in the rest frame of the source.

## 39–7 RELATIVISTIC MOMENTUM

Newton's laws of motion have the same form in all inertial frames of reference. When we transform coordinates from one inertial frame to another, using the Galilean coordinate transformation, the laws are *invariant* (unchanging). But we have just learned that the principle of relativity forces us to replace the Galilean transformation with the more general Lorentz transformation. As we will see, this requires corresponding generalizations in the laws of motion and the definitions of momentum and energy.

The principle of conservation of momentum states that *when two bodies interact, the total momentum is constant,* provided that they are an *isolated* system (that is, they interact only with each other, not with anything else). If conservation of momentum is a valid physical law, it must be valid in *all* inertial frames of reference. Now, here's the problem: Suppose we look at a collision in one inertial coordinate system $S$ and find that momentum is conserved. Then we use the Lorentz transformation to obtain the velocities in a second inertial system $S'$. We find that if we use the Newtonian definition of momentum ($\mathbf{p} = m\mathbf{v}$), momentum is *not* conserved in the second system! If we are convinced that the principle of relativity and the Lorentz transformation are correct, the only way to save momentum conservation is to generalize the *definition* of momentum.

We won't derive the correct relativistic generalization of momentum, but here is the result. The **relativistic momentum $\mathbf{p}$** of a particle with mass $m$ moving with velocity $\mathbf{v}$ is

$$\mathbf{p} = \frac{m\mathbf{v}}{\sqrt{1 - v^2/c^2}}. \tag{39–29}$$

When the particle's speed $v$ is much less than $c$, this is approximately equal to the Newtonian expression $\mathbf{p} = m\mathbf{v}$, but in general the relativistic momentum is greater in magnitude than $mv$.

In Eq. (39–29), $m$ is a *constant* that is a fundamental characteristic of a particle and that describes its inertial properties. Because $\boldsymbol{p} = m\boldsymbol{v}$ is still valid in the limit of very small velocities, $m$ must be the same quantity that we used (and learned to measure) in our study of Newtonian mechanics. In relativistic mechanics, $m$ is often called the **rest mass** of a particle.

What about the relativistic generalization of Newton's second law? In Newtonian mechanics the most general form of the second law is

$$\boldsymbol{F} = \frac{d\boldsymbol{p}}{dt}. \tag{39–30}$$

That is, force equals time rate of change of momentum. Experiments show that this result is still valid in relativistic mechanics, provided that we use the relativistic momentum given by Eq. (39–29); that is, the relativistically correct generalization of Newton's second law is

$$\boldsymbol{F} = \frac{d}{dt}\frac{m\boldsymbol{v}}{\sqrt{1 - v^2/c^2}}. \tag{39–31}$$

Because momentum is no longer directly proportional to velocity, rate of change of momentum is no longer directly proportional to acceleration. As a result, *constant force does not cause constant acceleration*. For example, we can show that when force and velocity are both along the $x$-axis, the acceleration $a$ is given by

$$F = \frac{m}{(1 - v^2/c^2)^{3/2}}a, \qquad \text{or} \qquad a = \frac{F}{m}\left(1 - \frac{v^2}{c^2}\right)^{3/2}. \tag{39–32}$$

These equations are valid *only* for straight-line motion, where the vectors $\boldsymbol{F}$ and $\boldsymbol{v}$ always have the same or opposite direction.

As a particle's speed increases, the acceleration caused by a given force continuously *decreases*. As the speed approaches $c$, the acceleration approaches zero, no matter how great a force is applied. Thus it is impossible to accelerate a particle with nonzero rest mass to a speed equal to or greater than $c$, and the speed of light is sometimes called "the ultimate speed."

Equation (39–29) is sometimes interpreted to mean that a rapidly moving particle undergoes an increase in mass. If the mass at zero velocity (the rest mass) is denoted by $m$, then the "relativistic mass" $m_{\text{rel}}$ is given by

$$m_{\text{rel}} = \frac{m}{\sqrt{1 - v^2/c^2}}.$$

Indeed, when we consider the motion of a system of particles (such as moving gas molecules in a container), the total mass of the system is the sum of the relativistic masses of the particles, not the sum of their rest masses.

The concept of relativistic mass also has its pitfalls, however. As Eq. (39–32) shows, the relativistic generalization of Newton's second law is *not* $\boldsymbol{F} = m_{\text{rel}}\boldsymbol{a}$, and the relativistic kinetic energy of a particle is *not* $K = \frac{1}{2}m_{\text{rel}}v^2$. Thus this concept must be approached with great caution. For our discussion, dealing with individual particles, we will think of Eq. (39–29) as a generalized definition of momentum and retain the meaning of $m$ as a constant for each particle, independent of its state of motion. We will derive the correct relativistic kinetic-energy expression in Section 39–8.

We will again use the abbreviations $\beta = v/c$ and $\gamma = 1/\sqrt{1 - \beta^2}$. When we used these abbreviations in Section 39–3 with $u$ in place of $v$, $u$ was the relative velocity of two coordinate systems. Here $v$ is the speed of a particle in a particular coordinate system, that is, the speed of the particle's *rest frame* with respect to that system. In terms of $\gamma$, Eq. (39–29) becomes

$$\boldsymbol{p} = \gamma m\boldsymbol{v}, \tag{39–33}$$

and Eq. (39–32) is

$$F = \gamma^3 ma \quad \text{(for straight-line motion).} \tag{39–34}$$

### EXAMPLE 39–7

**Relativistic dynamics of a proton** A proton (rest mass $1.67 \times 10^{-27}$ kg, charge $1.60 \times 10^{-19}$ C) is moving parallel to an electric field with magnitude $E = 5.00 \times 10^5$ N/C. Find the magnitudes of momentum and acceleration at the instants when $v = 0.010c$, $0.90c$, and $0.99c$.

**SOLUTION** We can use Eqs. (39–33) and (39–34). First we find the values of $\gamma$ for the three speeds given; we find $\gamma = 1.00$, 2.29, 7.09. The values of momentum $p$ are

$$p_1 = (1.00)(1.67 \times 10^{-27}\ \text{kg})(0.010)(3.00 \times 10^8\ \text{m/s}) = 5.01 \times 10^{-21}\ \text{kg} \cdot \text{m/s};$$

$$p_2 = (2.29)(1.67 \times 10^{-27}\ \text{kg})(0.90)(3.00 \times 10^8\ \text{m/s}) = 1.03 \times 10^{-18}\ \text{kg} \cdot \text{m/s};$$

$$p_3 = (7.09)(1.67 \times 10^{-27}\ \text{kg})(0.99)(3.00 \times 10^8\ \text{m/s}) = 3.52 \times 10^{-18}\ \text{kg} \cdot \text{m/s}.$$

At higher speeds the relativistic values of momentum differ more and more from the nonrelativistic values computed from $p = mv$. The momentum at $0.99c$ is over three times as great as at $0.90c$ because of the increase in the factor $\gamma$.

The force on the proton is

$$F = qE = (1.60 \times 10^{-19}\text{C})(5.00 \times 10^5\text{N/C}) = 8.00 \times 10^{-14}\text{N}.$$

From Eq. (39–34),

$$a = \frac{F}{\gamma^3 m}.$$

When $v = 0.01c$ and $\gamma = 1.00$,

$$a_1 = \frac{8.00 \times 10^{-14}\ \text{N}}{(1)^3(1.67 \times 10^{-27}\ \text{kg})} = 4.79 \times 10^{13}\ \text{m/s}^2.$$

The other accelerations are smaller by a factor of $\gamma^3$:

$$a_2 = 3.99 \times 10^{12}\ \text{m/s}^2, \qquad a_3 = 1.34 \times 10^{11}\ \text{m/s}^2.$$

These are only 8.33% and 0.280% of the values predicted by nonrelativistic mechanics.

We note that as $v$ approaches $c$, the acceleration drops off very quickly. In the Stanford Linear Accelerator (Fig. 39–13) a path length of 3 km is needed to give electrons the speed that according to classical physics they could acquire in 1.5 cm.

**39–13** The Stanford Linear Accelerator (SLAC). Electrons are accelerated in a vacuum chamber in the form of a pipe 3 km long to an energy of about 50 GeV. This energy corresponds approximately to $\gamma = 100{,}000$ and $\beta = 0.99999999995$.

## 39–8 RELATIVISTIC WORK AND ENERGY

When we developed the relationship between work and kinetic energy in Chapter 6, we used Newton's laws of motion. When we generalize these laws according to the principle of relativity, we need a corresponding generalization of the definition of kinetic energy.

As Eq. (39–32) shows, a constant force on a body does not cause a constant acceleration (except when the speed of the body is very small). But we can still derive fairly easily the relativistic generalization of the work-energy principle. We begin with the definition of work. When the force and displacement are in the same direction, $W = \int F\, dx$. We substitute the expression for $F$ from Eq. (39–32), the generalization of Newton's second law:

$$W_{12} = \int_{x_1}^{x_2} F\, dx = \int_{x_1}^{x_2} \frac{ma\, dx}{(1 - v^2/c^2)^{3/2}}. \tag{39–35}$$

To derive the generalized expression for kinetic energy $K$ as a function of speed $v$, we would like to convert this to an integral on $v$. To do this, we first note that $dx$ and $dv$ are

**SOLUTION** a) The rest energy is

$$mc^2 = (9.09 \times 10^{-31}\ \text{kg})\,(2.998 \times 10^8\ \text{m/s})^2 = 8.187 \times 10^{-14}\ \text{J}.$$

From the definition of the electronvolt in Section 24–6 we have 1 eV = $1.602 \times 10^{-19}$ J. Using this, we find

$$\begin{aligned} mc^2 &= (8.187 \times 10^{-14}\ \text{J})(1\ \text{eV}/1.602 \times 10^{-19}\ \text{J}) \\ &= 5.11 \times 10^5\ \text{eV} = 0.511\ \text{MeV}. \end{aligned}$$

b) In calculations such as this it is often convenient to work with the quantity $\gamma$ defined by Eq. (39–7):

$$\gamma = \frac{1}{\sqrt{1 - v^2/c^2}}.$$

Solving this for $v$, we get

$$v = c\sqrt{1 - (1/\gamma)^2}.$$

The total energy $E$ of the accelerated electron includes its rest energy and the energy $eV$ that it gains from the electric field:

$$E = \gamma mc^2 = mc^2 + eV,$$

or

$$\gamma = 1 + \frac{eV}{mc^2}.$$

An electron accelerated through a potential difference of 20 kV gains an amount of energy 20 keV, so for this electron we have

$$\begin{aligned} \gamma &= 1 + \frac{20 \times 10^3\ \text{eV}}{0.511 \times 10^6\ \text{eV}} \\ &= 1.039, \end{aligned}$$

and

$$\begin{aligned} v &= c\sqrt{1 - (1/1.039)^2} = 0.272c \\ &= 8.15 \times 10^7\ \text{m/s}. \end{aligned}$$

The added energy is less than 4% of the rest energy, and the final speed is about a quarter of the speed of light.

Repeating the calculation for $V = 5.0$ MV, we find $eV/mc^2 = 9.78$, $\gamma = 10.78$, and $v = 0.996c$. The kinetic energy is much larger than the rest energy, and the speed is very close to $c$. Such a particle is said to be in the *extreme relativistic range.*

## EXAMPLE 39–9

Two protons ($M = 1.67 \times 10^{-27}$ kg) with equal speeds in opposite directions collide head-on, producing a $\pi^0$ meson ($m = 2.40 \times 10^{-28}$ kg). If all particles are at rest after the collision, find the initial speed of the protons.

**SOLUTION** Total energy (including rest energy) is conserved, so we have

$$2\gamma Mc^2 = 2Mc^2 + mc^2,$$

$$\gamma = 1 + \frac{m}{2M} = 1 + \frac{2.40 \times 10^{-28}\ \text{kg}}{2(1.67 \times 10^{-27}\ \text{kg})} = 1.072,$$

$$v = c\sqrt{1 - (1/\gamma)^2} = 0.360c.$$

The initial kinetic energy of each proton is $(\gamma - 1)Mc^2 = 0.072Mc^2$. The rest energy of a proton is 938 MeV, so its kinetic energy is (0.072)(938 MeV) = 67.5 MeV. We invite you to verify that the rest energy of the $\pi^0$ meson is twice this, or 135 MeV.

## 39–9 INVARIANCE

We have used the term **invariance** to describe the idea that, according to the principle of relativity, all the fundamental laws of physics must keep the same form, that is, must be *invariant,* when we shift our description from one inertial frame of reference to another. This word is also used in a more specific sense to describe *physical quantities* that have the same value in all inertial frames of reference. The speed of light is one obvious example, but there are others that are important.

For example, we can describe the position $(x, y, z)$ and time $t$ of an event in a frame of reference $S$ and the position $(x', y', z')$ and time $t'$ of that same event in a frame $S'$ moving with constant velocity relative to $S$. In general, $x$ is not equal to $x'$, nor is $t$ equal to $t'$. But we can prove that the quantity $x^2 + y^2 + z^2 - c^2t^2$ *does* have the same value in both systems, no matter what the event. That is, it is always true that the position and time of an event in $S$ are related to its position and time in $S'$ by

$$x^2 + y^2 + z^2 - c^2t^2 = x'^2 + y'^2 + z'^2 - c^2t'^2. \tag{39–43}$$

All we have to do to prove this is to substitute the Lorentz transformation equations, Eqs. (39–23), into the right side of Eq. (39–43) and simplify the result. We leave the details as a problem.

We say that the quantity $x^2 + y^2 + z^2 - c^2t^2$ is *invariant* under the Lorentz transformation. This invariance is directly related to the invariance of the speed of light. To see this, suppose a light pulse starts from the origin $O$ of frame $S$ at time $t = 0$; the square of its distance from the origin at any later time $t$ is $x^2 + y^2 + z^2$, and this must equal $(ct)^2$.

But when the same light pulse is observed in frame $S'$, an exactly similar equation must hold for the coordinates and time in $S'$. Both observers see a spherical wave front expanding at a rate $c$, as suggested by Fig. 39–15. Thus Eq. (39–43) is not so surprising. Indeed, it is an inevitable consequence of the invariance of the speed of light.

We can obtain another useful invariant quantity from Eq. (39–41). The rest mass $m$ of a particle is an inherent property of the particle and is therefore an invariant, independent of the frame of reference. Thus Eq. (39–41) tells us that the quantity $E^2 - p^2c^2$ is *invariant.* That is, if a particle has energy $E$ and momentum with magnitude $p$, computed from the velocity $\boldsymbol{v}$ measured in frame of reference $S$, and energy $E'$ and momentum with magnitude $p'$ computed from the velocity $\boldsymbol{v}'$ in $S'$, then

$$E^2 - p^2c^2 = E'^2 - p'^2c^2. \tag{39–44}$$

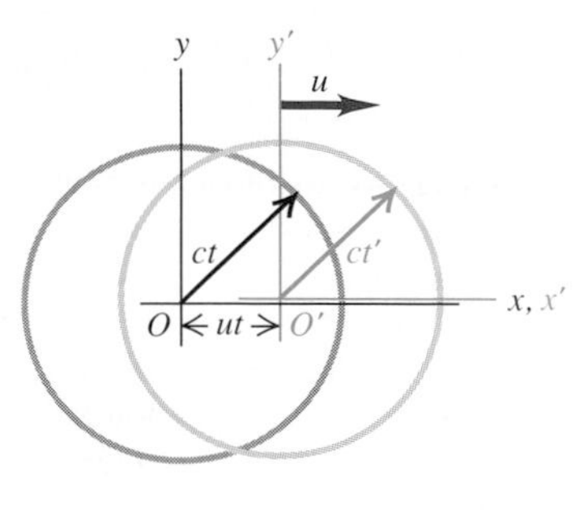


**39–15** According to Einstein's second postulate, observers in both frames of reference must see a spherical wave front expanding with speed $c$. For this to be true, $x - ct$ must equal $x' - ct'$.

It is not very hard to show that this relation is also valid for a collection of particles. That is, if we define the total energy (rest plus kinetic) of a set of particles as the sum of the energies of the individual particles and the total momentum as the vector sum of momenta of particles, then Eq. (39–44) also holds if we interpret $E$ as the *total* energy and $p$ as the magnitude of the *total* momentum. These relations are very useful in analyzing relativistic collision problems.

## 39–10 RELATIVITY AND NEWTONIAN MECHANICS

The sweeping changes required by the principle of relativity go to the very roots of Newtonian mechanics, including the concepts of length and time, the equations of motion, and the conservation principles. Thus it may appear that we have destroyed the foundations on which Newtonian mechanics is built. In one sense this is true, and yet the Newtonian formulation is still valid whenever speeds are small in comparison with the speed of light. In such cases, time dilation, length contraction, and the modifications of the laws of motion are so small that they are unobservable. In fact, every one of the principles of Newtonian mechanics survives as a special case of the more general relativistic formulation.

The laws of Newtonian mechanics are not *wrong;* they are *incomplete.* They are a limiting case of relativistic mechanics. They are *approximately* correct when all speeds are small in comparison to $c$, and in the limit when all speeds approach zero, they become exactly correct. Thus relativity does not completely destroy the laws of Newtonian mechanics but *generalizes* them. Newton's laws rest on a very solid base of experimental evidence, and it would be very strange to advance a new theory that is inconsistent with this evidence. This is a common pattern in the development of physical theory. Whenever a new theory is in partial conflict with an older, established theory, it still must yield the same predictions as the old in areas in which the old theory is supported by experimental evidence. Every new physical theory must pass this test, called the **correspondence principle.** This principle is a fundamental procedural rule in all of physics. There are many situations for which Newtonian mechanics is clearly inadequate, including all phenomena in which particle speeds are comparable to that of light or in which direct conversion of rest mass to energy occurs. But there is still a large area, including nearly all the behavior of macroscopic bodies in mechanical systems, in which Newtonian mechanics is still perfectly adequate.

At this point we may ask whether relativistic mechanics is the final word on this subject or whether *further* generalizations are possible or necessary. For example, inertial frames of reference have occupied a privileged position in our discussion. Can the principle of relativity be extended to noninertial frames as well?

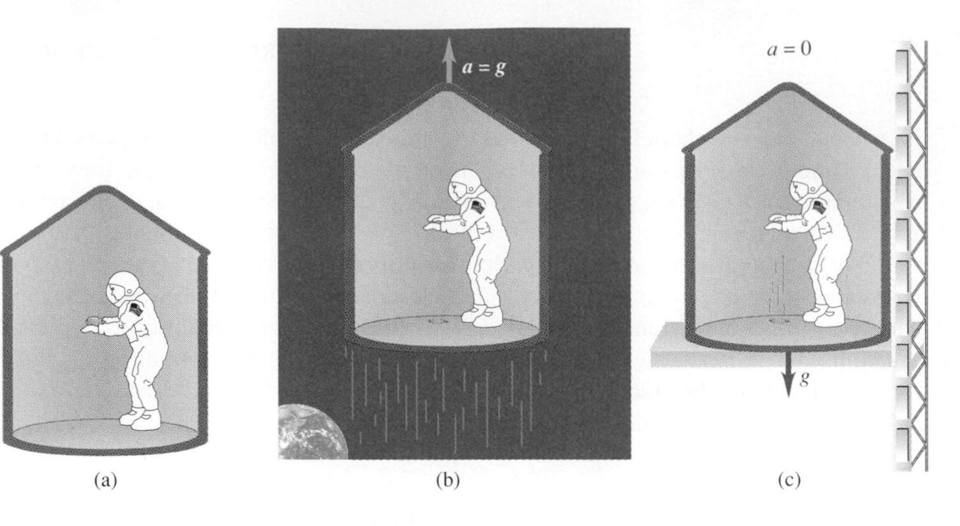


**39–16** (a) An astronaut drops his watch in the spaceship. (b) In gravity-free space the floor accelerates upward at $a = g$ and hits the watch. (c) On the planet's surface the watch accelerates downward at $a = g$ and hits the floor. From inside the spaceship the astronaut cannot distinguish situation (b) from situation (c).

Here's an example that illustrates some implications of this question. A student decides to go over Niagara Falls enclosed in a large wooden box. During her free fall she can perform experiments inside the box. An object released inside the box does not fall to the floor because both the box and the object are in free fall. But an alternative interpretation, from her point of view, is that the gravitational interaction with the earth, which causes her weight, has suddenly been turned off. As long as she remains in the box and it remains in free fall, she cannot tell whether she is indeed in free fall or whether the gravitational interaction has vanished.

A similar problem appears in a space station in orbit around the earth. Objects in the space station *seem* to be weightless, but without looking outside the station there is no way to determine whether gravity has been turned off or whether the station and all its contents are accelerating toward the center of the earth. Figure 39–16 makes a similar point for a spaceship accelerating relative to an inertial frame.

These considerations form the basis of Einstein's **general theory of relativity.** If we cannot distinguish experimentally between a uniform gravitational field at a particular location and an accelerated reference system, then there cannot be any real distinction between the two. Pursuing this concept, we may try to represent *any* gravitational field in terms of special characteristics of the coordinate system. This turns out to require even more sweeping revisions of our space-time concepts than did the special theory of relativity, and we find that, in general, the geometric properties of space are non-Euclidean (Fig. 39–17).

**39–17** A two-dimensional representation of a curved space. We imagine the space (a plane) as distorted as shown by a massive object (the sun). Light from a distant star (solid line) follows the distorted surface on its way to the earth. The broken line shows the direction from which the light *appears* to be coming.

The general theory of relativity has passed several experimental tests, including three proposed by Einstein. One test has to do with understanding the rotation of the axes of the planet Mercury's elliptical orbit, called the *precession of the perihelion*. (The perihelion is the point of closest approach to the sun.) Another test concerns the apparent bending of light rays from distant stars when they pass near the sun, and the third test is the *gravitational red shift,* the increase in wavelength of light proceeding outward from a massive source. Some details of the general theory are more difficult to test and remain speculative in nature, but this theory has played a central role in cosmological investigations of the structure of the universe, the formation and evolution of stars, and related matters. In recent years the theory has also been confirmed with several purely *terrestrial* experiments.

## SUMMARY

- All the fundamental laws of physics have the same form in all inertial frames of reference. The speed of light is the same in all inertial frames and is independent of the motion of the source. Simultaneity is not an absolute concept; two events that appear to be simultaneous in one frame do not appear to be simultaneous in a second frame moving relative to the first.

- If $\Delta t_0$ is the time interval between two events that occur at the same space point in a particular frame of reference, it is called a proper time. If the first frame moves with constant velocity $u$ relative to a second frame, the time interval $\Delta t$ between the events, as observed in the second frame, is

$$\Delta t = \frac{\Delta t_0}{\sqrt{1 - u^2/c^2}} = \gamma \Delta t_0, \qquad (39\text{–}6)$$

where

$$\gamma = \frac{1}{\sqrt{1 - u^2/c^2}}.$$

This effect is called time dilation.

- If $l_0$ is the distance between two points that are at rest in a particular frame of reference, it is called a proper length. If the first frame moves with constant velocity $u$ relative to a second frame, the distance $l$ between the points as measured in the second frame is

$$l = l_0\sqrt{1 - \frac{u^2}{c^2}} = \frac{l_0}{\gamma}. \qquad (39\text{–}17)$$

This effect is called length contraction.

- The Lorentz transformation relates the coordinates and time of an event in an inertial coordinate system $S$ to the coordinates and time of the same event as observed in a second system $S'$ moving with constant velocity $u$ relative to the first. The transformation equations are

$$\begin{aligned} x' &= \frac{x - ut}{\sqrt{1 - u^2/c^2}} = \gamma(x - ut), \\ y' &= y, \\ z' &= z, \\ t' &= \frac{t - ux/c^2}{\sqrt{1 - u^2/c^2}} = \gamma(t - ux/c^2). \end{aligned} \qquad (39\text{–}23)$$

### KEY TERMS

principle of relativity
Galilean coordinate transformation
simultaneity
time dilation
proper time
twin paradox
proper length
length contraction
Lorentz transformation
relativistic momentum
rest mass
rest energy
invariance
correspondence principle
general theory of relativity

If a particle's velocity in $S'$ is $v'$, its velocity $v$ in $S$ is

$$v = \frac{v' + u}{1 + uv'/c^2}. \tag{39–25}$$

• The Doppler effect is the apparent frequency shift in light from a moving source. For a source moving toward the observer with speed $u$, the apparent frequency $f$ is

$$f = \sqrt{\frac{c + u}{c - u}} f_0. \tag{39–27}$$

• In order for momentum conservation in collisions to hold in all coordinate systems, the definition of momentum must be generalized. For a particle of mass $m$ moving with velocity $\boldsymbol{v}$ the momentum $\boldsymbol{p}$ is defined as

$$\boldsymbol{p} = \frac{m\boldsymbol{v}}{\sqrt{1 - v^2/c^2}} = \gamma m\boldsymbol{v}. \tag{39–29}$$

• Generalizing the work-energy relation requires that we generalize the definition of kinetic energy $K$. For a particle of mass $m$ moving with speed $v$,

$$K = \frac{mc^2}{\sqrt{1 - v^2/c^2}} - mc^2 = (\gamma - 1)mc^2. \tag{39–38}$$

This form suggests assigning a rest energy $mc^2$ to a particle so that the total energy $E$, kinetic energy plus rest energy, is

$$E = K + mc^2 = \frac{mc^2}{\sqrt{1 - v^2/c^2}} = \gamma mc^2. \tag{39–40}$$

The total energy $E$ and magnitude of momentum $p$ for a particle with rest mass $m$ are related by

$$E^2 = (mc^2)^2 + (pc)^2. \tag{39–41}$$

• For an event observed in two different frames of reference,

$$x^2 + y^2 + z^2 - c^2t^2 = x'^2 + y'^2 + z'^2 - c^2t'^2. \tag{39–43}$$

The quantity $x^2 + y^2 + z^2 - c^2t^2$ is invariant. For a particle observed in two different frames of reference,

$$E^2 - p^2c^2 = E'^2 - p'^2c^2. \tag{39–44}$$

The quantity $E^2 - p^2c^2$ is invariant.

• The special theory of relativity is a generalization of Newtonian mechanics. All the principles of Newtonian mechanics are present as limiting cases when all the speeds are small in comparison to $c$. Further generalization to include accelerated frames of reference and their relation to gravitational fields leads to the general theory of relativity.

## EXERCISES

### Section 39–2 Relative Nature of Simultaneity

**39–1** Suppose the two lightning bolts appear to be simultaneous to an observer on the train in Section 39–2. Show that they *do not* appear to be simultaneous to an observer on the ground. Which appears to come first?

### Section 39–3 Relativity of Time

**39–2** A spaceship flies past Mars with a speed of $0.964c$ relative to the surface of the planet. When the spaceship is directly overhead at an altitude of 1500 km, a very bright signal light on the Martian surface blinks on and then off. An observer on Mars measures that the signal light was on for $3.00 \times 10^{-5}$ s. What is the duration of the light pulse measured by the pilot of the spaceship?

**39–3 Path of a Muon.** The $\mu^+$ meson (or positive muon) is an unstable particle with a lifetime of about $2.2 \times 10^{-6}$ s (measured in the rest frame of the muon). a) If the muon is made to travel at a very high speed relative to a laboratory, its lifetime is measured in the laboratory to be $1.9 \times 10^{-5}$ s. Calculate the speed of the muon expressed as a fraction of $c$. b) What distance, measured in the laboratory, does the particle travel during its lifetime?

**39–4** The $\pi^+$ meson (or positive pion), an unstable particle, lives on the average about $2.6 \times 10^{-8}$ s (measured in its own frame of reference) before decaying. a) If such a particle is moving with respect to the laboratory with a speed of $0.986c$, what lifetime is measured in the laboratory? b) What distance, measured in the laboratory, does the particle move before decaying?

**39–5** An advanced space probe in the twenty-first century travels from the earth with a speed of $5.00 \times 10^6$ m/s relative to the earth and then returns at the same speed. The probe carries an atomic clock that has been carefully synchronized with an identical clock that remains at rest on the earth. The probe returns after 1 year, as measured on the earth, has passed. What is the difference in the elapsed times on the two clocks? Which clock, the one in the probe or the one on the earth, shows the smallest elapsed time?

**39–6** An alien spacecraft is flying overhead at a great distance as you stand in your backyard. You see its searchlight blink on for $5.00 \times 10^{-3}$ s. The first officer on the spacecraft measures that the search light is on for $2.00 \times 10^{-5}$ s. a) Which of these two measured times is the proper time? b) What is the speed of the spacecraft relative to the earth expressed as a fraction of the speed of light $c$?

**39–7 Everyday Time Dilation.** Two atomic clocks are carefully synchronized. One remains in New York, and the other is loaded on a supersonic airplane that travels at an average speed of 400 m/s and then returns to New York. When the plane returns, the elapsed time on the clock that stayed behind is 5.00 h. By how much will the reading of the two clocks differ, and which clock will show the shorter elapsed time? (*Hint:* Use the fact that $u << c$ to simplify $\sqrt{1 - u^2/c^2}$ by a binomial expansion.)

## Section 39–4 Relativity of Length

**39–8** A meter stick moves past you at great speed. If you measure the length of the moving meter stick to be 0.760 m — for example, by comparing it to a meter stick that is at rest relative to you — what is the speed with which the meter stick is moving relative to you?

**39–9** In the year 2010 a spacecraft flies over Moon Station III at a speed of $0.800c$. A scientist on the moon measures the length of the moving spacecraft to be 160 m. The spacecraft later lands on the moon, and the same scientist measures the length of the now stationary spacecraft. What value does she get?

**39–10** An unstable particle is created in the upper atmosphere from a cosmic ray and travels straight down toward the surface ofthe earth with a speed of $0.99860c$ relative to the earth. A scientist in a laboratory on the surface of the earth measures that the particle is created at an altitude of 60.0 km. a) As measured by the scientist, how much time does it take the particle to travel the 60.0 km to the surface of the earth? b) Use the length contraction formula to calculate the distance from where the particle is created to the surface of the earth as measured in the particle's frame. c) In the particle's frame, how much time does it take the particle to travel from where it is created to the surface of the earth? Calculate this time both by the time dilation formula and also from the distance calculated in part (b). Do the two results agree?

**39–11** A muon created 20.0 km above the surface of the earth (as measured in the earth's frame) is traveling with a speed relative to the earth of $0.9954c$. The lifetime of the muon, measured in its own rest frame, is $2.2 \times 10^{-6}$ s. In the frame of the muon the earth is moving toward the muon with a speed of $0.9954c$. a) In the muon's frame, what is its height above the surface of the earth? b) In the muon's frame, how much closer does the earth get during the lifetime of the muon? What fraction is this of the muon's original height as measured in the muon's frame? c) In the earth's frame, what is the lifetime of the muon? In the earth's frame, how far does the muon travel during its lifetime? What fraction is this of the muon's original height in the earth's frame?

## Section 39–5 The Lorentz Transformation

**39–12** Solve Eqs. (39–23) to obtain $x$ and $t$ in terms of $x'$ and $t'$, and show that the resulting transformation has the same form as the original one except for a change of sign for $u$.

**39–13** Show in detail the derivation of Eq. (39–22) from Eqs. (39–20) and (39–21).

**39–14** Show the details of the derivation of Eq. (39–25) from (39–24).

**39–15** A spaceship moving relative to the earth at a large speed fires a rocket toward the earth with a speed of $0.840c$ relative to the spaceship. An earth-based observer measures that the rocket is approaching with a speed of $0.360c$. What is the speed of the spaceship relative to the earth? Is the spaceship moving toward or away from the earth?

**39–16 Relativistic Star Wars.** An enemy spaceship is moving toward your starfighter with a speed, as measured in your frame, of $0.600c$. The enemy ship fires a missile toward you at a speed of $0.900c$ relative to the enemy ship (Fig. 39–18). a) What is the speed of the missile relative to you? Express your answer in terms of the speed of light. b) If you measure that the enemy ship is $4.00 \times 10^6$ km away from you when the missile is fired, how

**FIGURE 39–18**

much time, measured in your frame, will it take the missile to reach you?

**39–17** A rebel fighter is in hot pursuit of a starfleet cruiser. As measured by an observer on the earth, the starfleet cruiser is traveling with a speed of $0.700c$, and the rebel fighter is traveling at $0.900c$ in the same direction (Fig. 39–19). What is the speed of the cruiser relative to the fighter?

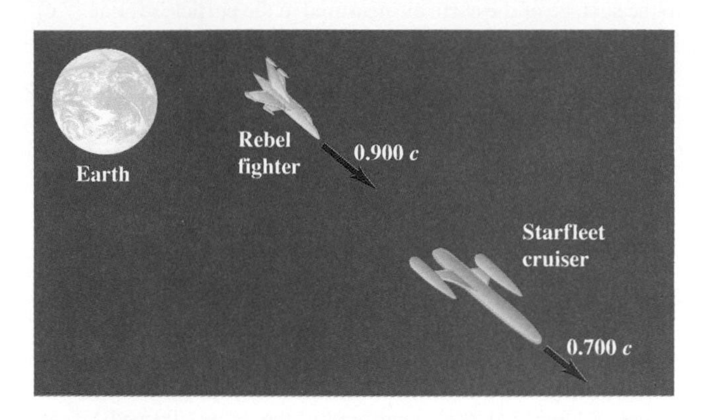


**FIGURE 39–19**

**39–18** Two particles emerge from a high-energy accelerator in opposite directions, each with a speed $0.750c$ as measured in the laboratory. What is the magnitude of the relative velocity of the particles?

**39–19** Two particles are created in a high-energy accelerator and move off in opposite directions. The speed of one of the particles, as measured in the laboratory, is $0.800c$, and the speed of each particle relative to the other is $0.900c$. What is the speed of the second particle, as measured in the laboratory?

## Section 39–6 The Relativistic Doppler Effect

***39–20** In terms of $c$, what relative velocity $u$ between source and observer produces a) a 1.0% decrease in frequency; b) a halving of the frequency of the observed light?

***39–21** How fast must you be approaching a red traffic light ($\lambda = 675$ nm) for it to appear green ($\lambda = 525$ nm)? If you used this as an excuse for not getting a ticket for running a red light, do you think you might get a speeding ticket instead?

## Section 39–7 Relativistic Momentum

**39–22** a) At what speed does the momentum of a particle differ by 1.0% from the value obtained by using the nonrelativistic expression $mv$? b) Is the correct relativistic value greater or less than that obtained from the nonrelativistic expression?

**39–23** At what speed is the momentum of a particle twice as great as the result obtained from the nonrelativistic expression $mv$?

**39–24 Relativistic Baseball.** Calculate the magnitude of the force required to give a 0.145-kg baseball an acceleration $a = 1.00$ m/s$^2$ in the direction of the baseball's initial velocity when this velocity has a magnitude of a) 10.0 m/s; b) $0.900c$; c) $0.990c$.

**39–25** A particle moves along a straight line under the action of a force lying along the same line. Show that the acceleration $a = dv/dt$ of the particle is given by Eq. (39–32).

## Section 39–8 Relativistic Work and Energy

**39–26** How much work must be done on a particle with mass $m$ to accelerate it a) from rest to a speed $0.090c$; b) from a speed $0.900c$ to a speed $0.990c$? (Express answers in terms of $mc^2$.)

**39–27** What is the speed of a particle whose kinetic energy is equal to a) its rest energy; b) five times its rest energy?

**39–28** In *positron annihilation* an electron and a positron (a positively charged electron) collide and disappear, producing electromagnetic radiation. If each particle has a mass of $9.11 \times 10^{-31}$ kg and they are at rest just before the annihilation, find the total energy of the radiation.

**39–29** Compute the kinetic energy of an electron using both the nonrelativistic and relativistic expressions, and compute the ratio of the two results (relativistic divided by nonrelativistic) for speeds of a) $5.00 \times 10^7$ m/s; b) $2.90 \times 10^8$ m/s.

**39–30 Energy of Fusion.** In a hypothetical nuclear-fusion reactor, two deuterium nuclei combine, or "fuse," to form one helium nucleus. The mass of a deuterium nucleus, expressed in atomic mass units (u), is 2.0136 u; that of a helium nucleus is 4.0015 u. (1 u $= 1.661 \times 10^{-27}$ kg.) a) How much energy is released when 1.0 kg of deuterium undergoes fusion? b) The annual consumption of electrical energy in the United States is of the order of $1.0 \times 10^{19}$ J. How much deuterium must react to produce this much energy?

**39–31** An electron in a certain x-ray tube is accelerated from rest through a potential difference of $1.40 \times 10^5$ V in going from the cathode to the anode. When it arrives at the anode, what is a) its kinetic energy in eV; b) its total energy in eV; c) its speed? d) What is the speed of the electron, calculated classically?

**39–32** Starting from Eq. (39–41), show that in the classical limit ($pc << mc^2$) the energy approaches the classical kinetic energy plus the rest mass energy.

**39–33** A particle has rest mass $3.32 \times 10^{-27}$ kg and momentum $9.65 \times 10^{-19}$ kg · m/s. a) What is the total energy (kinetic plus rest energy) of the particle? b) What is the kinetic energy of the particle? c) What is the ratio of the kinetic energy to the rest energy of the particle?

## Section 39–9 Invariance

**39–34** Use the Lorentz transformation, Eqs. (39–23), to prove Eq. (39–43).

## PROBLEMS

**39–35** The starships of the Solar Federation are marked with the symbol of the Federation, a circle, and starships of the Denabian Empire are marked with the Empire's symbol, an ellipse whose major axis is 1.50 times its minor axis ($a = 1.50b$ in Fig. 39–20). How fast relative to an observer does a Federation ship have to travel for its markings to be confused with those of an Empire ship?

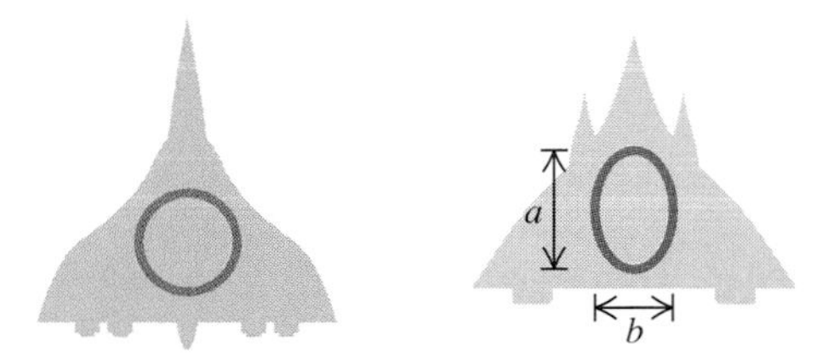


**FIGURE 39–20**

**39–36** A space probe is sent to the vicinity of the star Vega, which is 26.5 light-years from the earth. The probe travels with a speed of $0.9910c$. A young astronaut is 19 years old when the probe leaves the earth. What is her biological age when the probe reaches Vega?

**39–37 Fast Pions.** After being produced, a $\pi^+$ meson must travel down a 3.00-km-long tube to reach an experimental area. The $\pi^+$ meson has a lifetime of $2.6 \times 10^{-8}$ s. a) How fast must a $\pi^+$ meson travel if it is not to decay before it reaches the end of the tube? [Since $u$ will be very close to $c$, write $u = (1 - \Delta)c$ and give your answer in terms of $\Delta$ rather than $u$.] b) With a rest energy of 139.6 MeV, what is the $\pi^+$ meson's total energy at the speed calculated in part (a)?

**39–38** A cube of metal with sides of length $a$ sits at rest in a frame $S$ with one edge parallel to the $x$-axis. Therefore in $S$ the cube has volume $a^3$. Frame $S'$ moves along the $x$-axis with a speed $u$. To an observer in frame $S'$, what is the volume of the metal cube?

**39–39** By what minimum amount does the mass of 1.00 kg of ice increase when the ice melts and both water and ice are at 0°C?

**39–40** A photon with energy $E$ is emitted by an atom with mass $m$, which recoils in the opposite direction. a) Assuming that the motion of the atom can be treated nonrelativistically, compute the recoil speed of the atom. b) From the result of part (a), show that the recoil speed is much smaller than $c$ whenever $E$ is much smaller than the rest energy $mc^2$ of the atom.

**39–41** A particle is said to be in the *extreme relativistic range* when its kinetic energy is much larger than its rest energy. a) What is the speed of a particle (expressed as a fraction of $c$) such that the total energy is ten times the rest energy? b) What is the percent difference between the left and right sides of Eq. (39–41) if $(mc^2)^2$ is neglected for a particle with the speed calculated in part (a)?

**39–42** A nuclear bomb containing 12.0 kg of plutonium explodes. The sum of the rest masses of the products of the explosion is less than the original rest mass by 1 part in $10^4$. a) How much energy is released in the explosion? b) If the explosion takes place in 4.00 $\mu$s, what is the average power developed by the bomb? c) What mass of water could the released energy lift to a height of 1.00 km?

**39–43 Cerenkov Radiation.** The Soviet physicist P. A. Cerenkov discovered that a charged particle traveling in a solid with a speed exceeding the speed of light in that material radiates electromagnetic radiation. What is the minimum kinetic energy (in electronvolts) that an electron must have while traveling inside a slab of crown glass ($n = 1.52$) in order to create this Cerenkov radiation?

**39–44 Albert in Wonderland.** Einstein and Lorentz, being avid tennis players, play a fast-paced game on a court on which they stand 20.0 m from each other (Fig. 39–21). Being very skilled players, they play without a net. The tennis ball has a mass of 0.0580 kg. Neglect gravity and assume that the ball travels parallel to the ground as it travels between the two players. Unless specified otherwise, all measurements are made by the two men. a) Lorentz serves the ball at 100 m/s. What is the ball's kinetic energy? b) Einstein slams a return at $1.00 \times 10^8$ m/s. What is the ball's kinetic energy? c) During Einstein's return of the ball in part (b) a white rabbit runs beside the court in the direction from Einstein to Lorentz. The rabbit has a speed of $2.00 \times 10^8$ m/s relative to the two men. What is the speed of the rabbit relative to the ball? d) What does the rabbit measure as the length of the court, the distance from Einstein to Lorentz? e) How much time

**FIGURE 39–21**

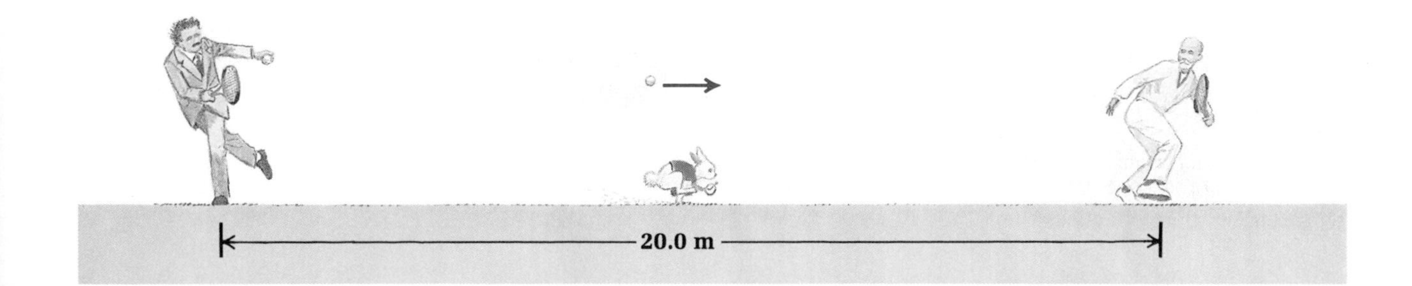

does it take for the rabbit to run the court, according to the players? f) The white rabbit carries a pocket watch. He uses this watch to measure the time (as he sees it) for the tennis court to pass by under him. What time does he measure?

***39–45** One of the wavelengths of light emitted by hydrogen atoms under normal laboratory conditions is $\lambda = 121.6$ nm. In the light emitted from a distant interstellar region this same spectral line is observed to be Doppler-shifted to $\lambda = 396.5$ nm. How fast are the emitting atoms moving relative to the earth? Are they approaching the earth or receding from it?

***39–46 Measuring Speed by Radar.** A highway patrolman measures the speed of cars approaching him with a device that sends out electromagnetic waves with frequency $f_0$ and then measures the shift in frequency $\Delta f$ of the waves reflected from the moving car. What fractional frequency shift $\Delta f/f_0$ is produced by a car speeding at 31.3 m/s (70 mph)? (See Problem 21–30.)

**39–47** A particle moves along the $x$-axis with speed $v$. A force in the $+y$-direction, having magnitude $F$, is applied. Show that the magnitude of the acceleration initially is

$$a = \frac{F}{m}(1 - v^2/c^2)^{1/2}.$$

This result and Eq. (39–32) were sometimes interpreted in the early days of relativity as meaning that a particle has a "longitudinal mass" given by $m/(1 - v^2/c^2)^{3/2}$ and a "transverse mass" given by $m/(1 - v^2/c^2)^{1/2}$. These terms are no longer in common use.

**39–48** Construct a right triangle in which one of the angles is $\alpha$, where $\sin \alpha = v/c$. ($v$ is the speed of a particle; $c$ is the speed of light.) If the base of the triangle (the side adjacent to $\alpha$) is the rest energy $mc^2$, show that a) the hypotenuse is the total energy; b) the side opposite $\alpha$ is $c$ times the relativistic momentum. c) Describe a simple graphical procedure for finding the kinetic energy $K$.

**39–49** A particle with mass $m$ accelerated by a constant force $F$ will, according to Newtonian mechanics, continue to accelerate without bound. That is, as $t \to \infty$, $v \to \infty$. Show that according to relativistic mechanics, the particle's speed approaches $c$ as $t \to \infty$. [*Note:* A standard integral is $\int (1 - x^2)^{-3/2}\, dx = x/\sqrt{1 - x^2}$.]

**39–50** Two events are observed in a frame of reference $S$ to occur at the same space point, the second occuring 2.00 s after the first. In a second frame $S'$ moving relative to $S$, the second event is observed to occur 2.75 s after the first. What is the difference between the positions of the two events as measured in $S'$?

## CHALLENGE PROBLEMS

**39–51 Pion Production.** In high-energy physics, new particles are created by collisions of fast-moving projectile particles with stationary particles. Some of the kinetic energy of the incident particle is used to create mass of the new particle. A proton-proton collision can result in the creation of 2 pions ($\pi^-$ and $\pi^+$):

$$p + p \to p + p + \pi^- + \pi^+.$$

a) Calculate the threshold kinetic energy of the incident proton that will allow this reaction to occur if the second proton is initially at rest. The rest mass of each pion is 139.6 MeV. (*Hint:* Working in the center-of-mass frame is useful here. See Problem 8–70, but here the Lorentz transformation must be used to relate the velocities in the laboratory and the center-of-mass frames.) b) How does this calculated threshold kinetic energy compare with the total rest mass energy of the created particles?

**39–52** The French physicist Armand Fizeau was the first to measure the speed of light accurately. He also found experimentally that the speed relative to the lab frame of light traveling in a tank of water that is itself moving at a speed $V$ relative to the lab frame is

$$v = \frac{c}{n} + kV.$$

Fizeau called $k$ the dragging coefficient and obtained an experimental value of $k = 0.44$. What value of $k$ would you calculate from relativistic transformations?

**39–53** Two events observed in a frame of reference $S$ have positions and times given by $(x_1, t_1)$ and $(x_2, t_2)$, respectively. a) Show that in a frame $S'$ moving along the $x$-axis just fast enough that the two events occur at the same point in $S'$, the time interval $\Delta t'$ between the two events is given by

$$\Delta t' = \sqrt{(\Delta t)^2 - \left(\frac{\Delta x}{c}\right)^2},$$

where $\Delta x = x_2 - x_1$ and $\Delta t = t_2 - t_1$. Show that therefore, if $\Delta x \geq c\,\Delta t$, there is *no* frame $S'$ in which the two events occur at the same point. The interval $\Delta t'$ is sometimes called the *proper time interval* for the events. Is this term appropriate? b) Show that if $\Delta x > c\,\Delta t$, there is a frame of reference $S'$ in which the two events occur *simultaneously*. Find the distance between the two events in $S'$. This distance is sometimes called a *proper length*. Is this term appropriate? c) Two events are observed in a frame of reference $S'$ to occur simultaneously at points separated by a distance of 1.00 m. In a second frame $S$ moving relative to $S'$ along the line joining the two points in $S'$, the two events appear to be separated by 2.50 m. What is the time interval between the events as measured in $S$? [*Hint:* Apply the result obtained in part (b).]

How do we study the physics of stars? Much of what we know comes from careful analysis of the light from the stars. Large telescopes, such as the Keck telescope shown here, collect light and reflect it through a diffraction grating, as we discussed in Ch. 38.

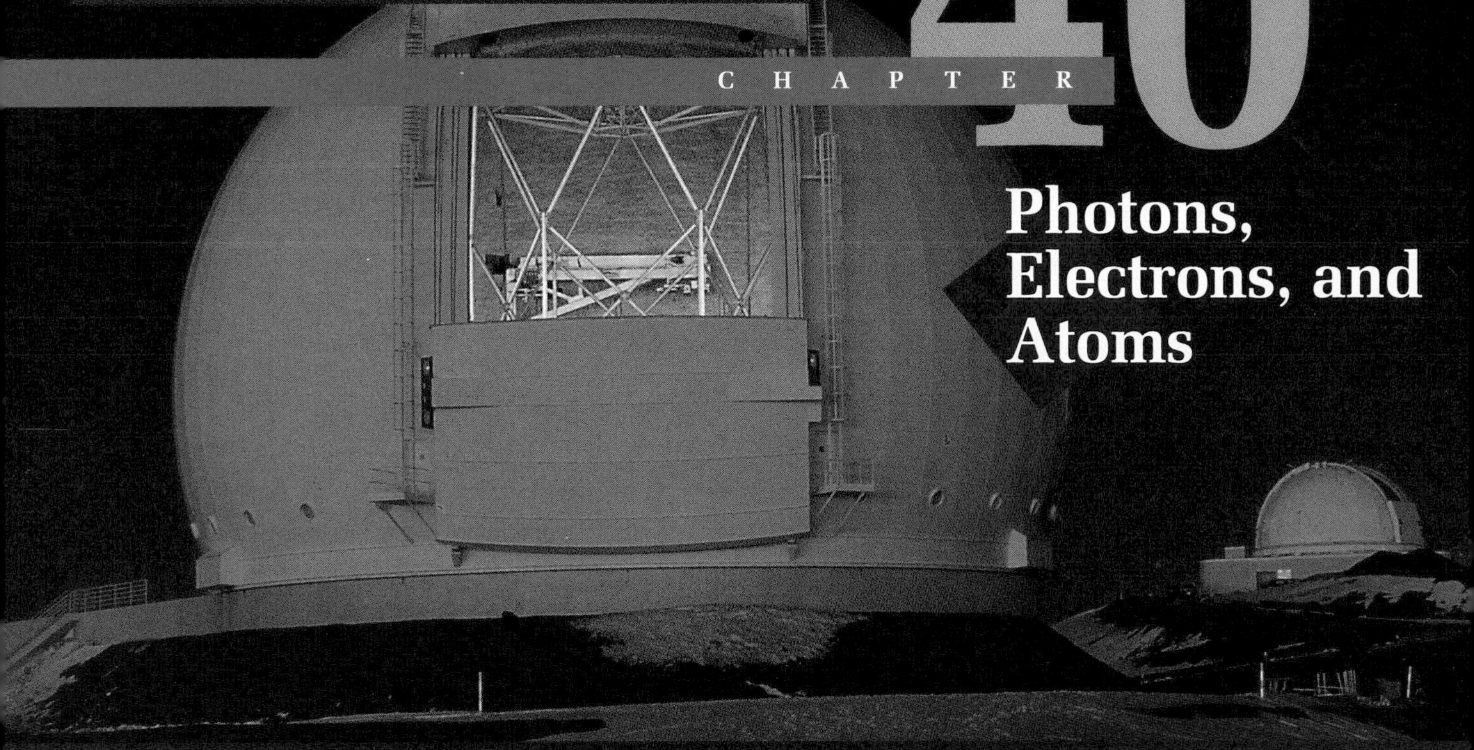

# CHAPTER 40

# Photons, Electrons, and Atoms

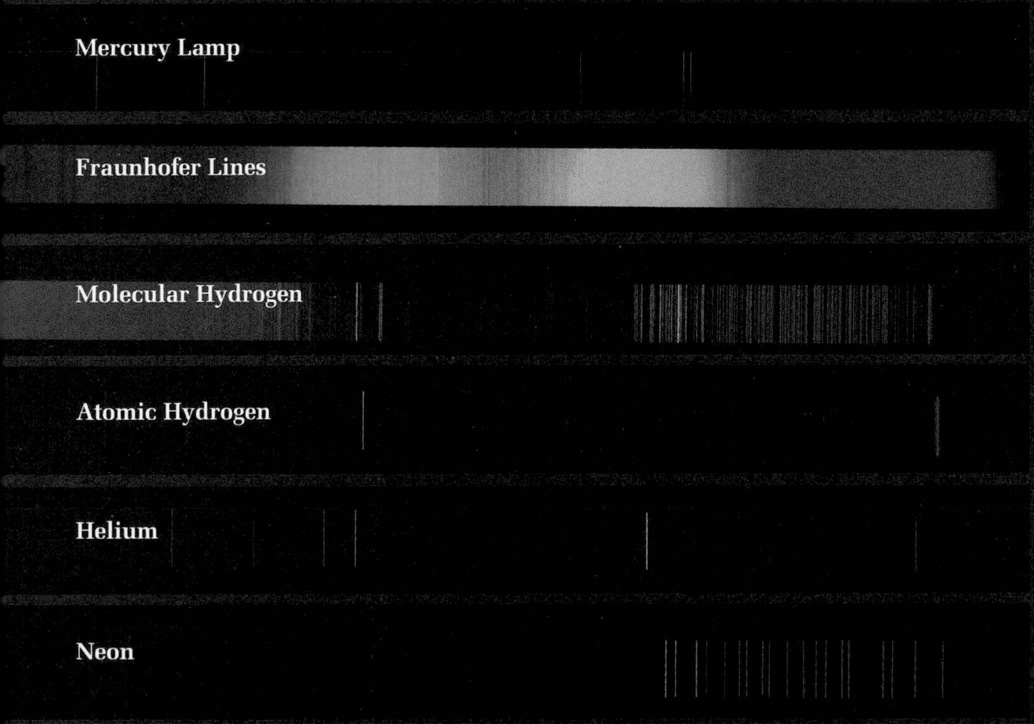

Mercury Lamp

Fraunhofer Lines

Molecular Hydrogen

Atomic Hydrogen

Helium

Neon

Spectra formed by a diffraction grating enable us to identify individual elements, as we will explain in this chapter. Astronomers have detected in starlight strong emission lines characteristic of hydrogen, indicating that stars consist mostly of hydrogen. Helium is also present in large amounts, with detectable amounts of metals such as iron, calcium, nickel, etc., depending on the star. Analysis of spectral lines enables us to measure such things as the star's temperature, rotational speed, density, and magnetic field strength, as well as its chemical composition.

## KEY CONCEPTS

**• Electromagnetic-wave energy is emitted and absorbed in packages of definite size called photons. The energy of a photon is proportional to the wave frequency. In the photoelectric effect, an electron at a conducting surface absorbs a photon and gains enough energy to escape from the surface.**

**• The energy associated with the internal motion of an atom can have only certain specific energy values called energy levels. A photon can be emitted or absorbed during a transition from one energy level to another. Line spectra, the operation of lasers, and the production and scattering of x rays can be understood on the basis of photons and energy levels.**

**• The positive charge and most of the mass of an atom are contained in a dense nucleus that is much smaller than the overall size of the atom.**

**• The Bohr model of the hydrogen atom is a mechanical model that correctly predicts the energy levels of this atom.**

**• Continuous-spectrum radiation from hot condensed matter can also be understood on the basis of photons.**

## INTRODUCTION

The work of Maxwell, Hertz, and others established firmly that light is an electromagnetic wave. Interference, diffraction, and polarization demonstrate the *wave nature* of light convincingly.

But many phenomena, particularly those concerned with emission and absorption of electromagnetic radiation, show a completely different aspect of its nature. The energy of an electromagnetic wave is emitted and absorbed in packages with a definite size, called *photons* or *quanta.* The energy of a single photon is proportional to the frequency of the radiation. In some experiments, light and other radiation behave like a stream of *particles.* We say that the energy is *quantized.*

The energy associated with internal motion in atoms is also quantized. Each kind of atom has a set of possible energy values called *energy levels.* It cannot have an energy between two levels.

These two basic ideas, photons and energy levels, take us a long way toward understanding a wide variety of otherwise puzzling observations. Among these are the photoelectric effect (the emission of electrons from a conducting surface when light strikes it), the *line spectra* emitted by gaseous elements, the operation of lasers, and the production and scattering of x rays. Analysis of these effects and their relation to atomic structure takes us to the threshold of *quantum mechanics,* which involves some radical changes in our views of the nature of radiation and of matter itself.

# 40–1 EMISSION AND ABSORPTION OF LIGHT

How is light produced? In Chapter 33 we discussed the experiments of Heinrich Hertz, who produced electromagnetic waves using oscillations in a resonant $L$-$C$ circuit similar to those we studied in Chapter 31. His frequencies were of the order of $10^8$ Hz. Frequencies of visible light are of the order of $10^{15}$ Hz, far higher than any frequency that can be attained with conventional electronic circuits. At the end of the nineteenth century, theorists speculated that waves in this frequency range might be produced by oscillating electric charges within individual atoms. Three great challenges facing theorists at the turn of the century were line spectra, the photoelectric effect, and the production of x rays. We'll describe each of these in turn.

## Line Spectra

A prism or a diffraction grating can be used to separate the various wavelengths in a beam of light into a *spectrum.* If the light source is a hot solid or liquid (such as a light bulb filament), the spectrum is *continuous;* light of all wavelengths is present (Fig. 40–1a). But if the source is a gas carrying an electrical discharge (as in a neon sign) or a volatile salt heated in a flame (as when table salt is thrown into a campfire), only a few colors appear, in the form of isolated sharp parallel lines (Fig. 40–1b). (Each "line" is an image of the spectrograph slit, deviated through an angle that depends on the frequency of the light forming the image.) A spectrum of this sort is called a **line spectrum.** Each line corresponds to a definite wavelength and frequency.

It was discovered early in the nineteenth century that each element has a certain set of wavelengths in its line spectrum. The spectrum of hydrogen always contains a certain set of wavelengths; sodium produces a different set; iron still another; and so on. The identification of elements by their spectra became a useful analytical technique. The characteristic spectrum of an atom was presumably related to its internal structure, but attempts to understand this relation on the basis of classical mechanics and electrodynamics were not successful.

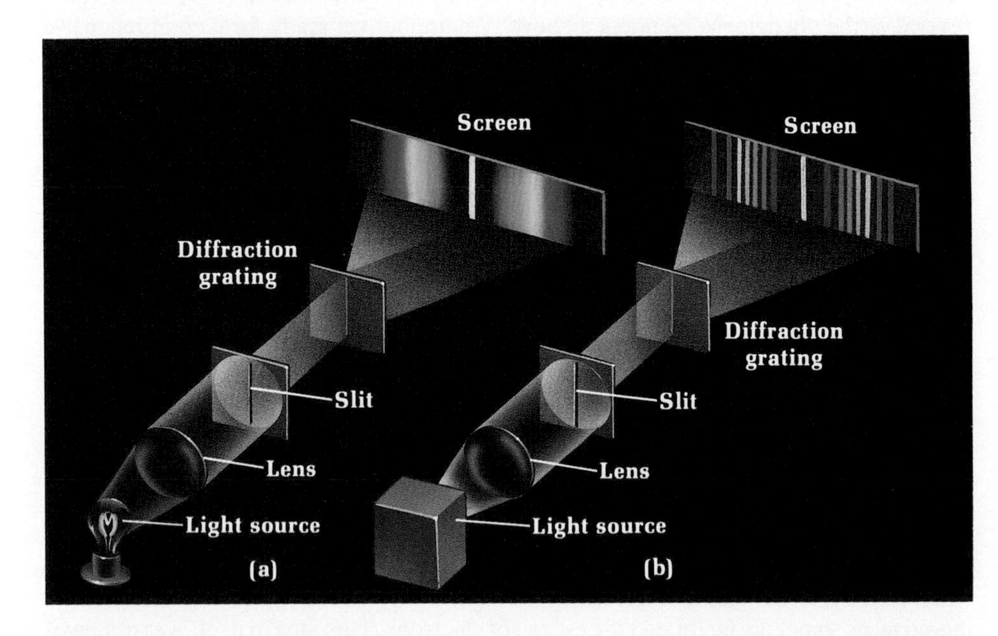


**40–1** (a) Continuous spectrum emitted by glowing light bulb filament. (b) Line spectrum emitted by mercury vapor lamp. The slit and diffraction grating are shown.

### Photoelectric Effect

There were also mysteries associated with *absorption* of light. In 1887, during his electromagnetic-wave experiments, Hertz discovered the *photoelectric effect.* When light strikes the surface of a conductor, some electrons near the surface absorb enough energy to surmount the potential-energy barrier at the surface and escape into the surrounding space. Detailed investigation of this effect revealed some puzzling features that couldn't be understood on the basis of classical optics. We will discuss these in the next section.

### X Rays

Still another area of unsolved problems in the emission and absorption of radiation centered around the production and scattering of *x rays,* discovered in 1895. These rays were produced in high-voltage glow discharge tubes, but no one understood how or why or what determined the wavelengths (which are much shorter than those of visible light). Even worse, when these rays collided with matter, the scattered rays sometimes had longer wavelengths than the original ones. This is analogous to a beam of blue light striking a mirror and reflecting back as red!

### Photons and Energy Levels

All these phenomena, and several others, pointed forcefully to the conclusion that classical optics, successful though it was in explaining lenses, mirrors, interference, and polarization, had its limitations. Understanding emission and absorption would require at least some extensions of classical theory. In fact, it has required something much more radical than that. All these phenomena result from the *quantum nature* of radiation. Electromagnetic radiation, along with its wave nature, has properties resembling those of *particles.* In particular, the *energy* in an electromagnetic wave is always emitted and absorbed in packages called *photons* or *quanta,* with energy proportional to the frequency of the radiation.

The two common threads woven through this chapter are the quantization of electromagnetic radiation and the existence of discrete energy levels in atoms. In the remainder of this chapter, we will show how these two concepts can contribute to understanding several of the phenomena mentioned above. We are not yet ready for a comprehensive theory of atomic structure; that will come in Chapters 42 and 43. But we'll look at the Bohr model of the hydrogen atom, one attempt to predict atomic energy levels on the basis of atomic structure.

## 40–2 THE PHOTOELECTRIC EFFECT

The **photoelectric effect** is the emission of electrons from the surface of a conductor when light strikes its surface. The liberated electrons absorb energy from the incident radiation and are thus able to overcome the potential-energy barrier that normally confines them inside the material. Think of this barrier as a curb separating a flat street from a raised sidewalk. A soccer ball rolling on the street can bounce off the curb and back into the street. If the ball is kicked hard enough, it can hop up onto the sidewalk, with a gain of potential energy (proportional to the height of the curb) equal to the loss of kinetic energy.

The photoelectric effect was first observed in 1887 by Hertz, quite by accident. He noticed that a spark would jump more readily between two electrically charged spheres when their surfaces were illuminated by the light from another spark. Light shining on the surfaces somehow facilitated the escape of electrons. This idea in itself was not revolutionary. The existence of the surface potential-energy barrier was already known. In 1883, Thomas Edison had discovered *thermionic emission,* in which the escape energy is supplied by heating the material to a very high temperature, liberating electrons by a process analogous to boiling a liquid. The minimum amount of energy an individual electron has to gain to escape from a particular surface is called the **work function** for that surface, denoted by $\phi$.

The photoelectric effect was investigated in detail by Wilhelm Hallwachs and Philipp Lenard during the years 1886–1900; their results were quite unexpected. We will describe their work in terms of a modern phototube, shown schematically in Fig. 40–2. Two conducting electrodes, the anode and the cathode, are enclosed in an evacuated glass tube. The battery, or other source of potential difference, creates an electric field in the direction from anode to cathode. Light (indicated by the arrows) falling on the photosensitive surface of the cathode causes a current in the external circuit; the current is measured by the galvanometer (G). Hallwachs and Lenard studied how this current varies with voltage and with the frequency and intensity of the light.

After the discovery of the electron in 1897, it became clear that the light causes electrons to be emitted from the cathode; once emitted, they are pushed by the electric field toward the anode. A high vacuum, with residual pressure of 0.01 Pa ($10^{-7}$ atm) or less, is needed to avoid collisions of electrons with gas molecules.

Here is what Hallwachs and Lenard found. When monochromatic light falls on the cathode, *no* photoelectrons at all are emitted unless the frequency of the light is greater than some minimum value called the **threshold frequency.** This critical frequency depends on the material of the cathode. For most metals it is in the ultraviolet (corresponding to wavelengths $\lambda$ of 200 to 300 nm), but for potassium and cesium oxides it is in the visible spectrum ($\lambda$ of 400 to 700 nm).

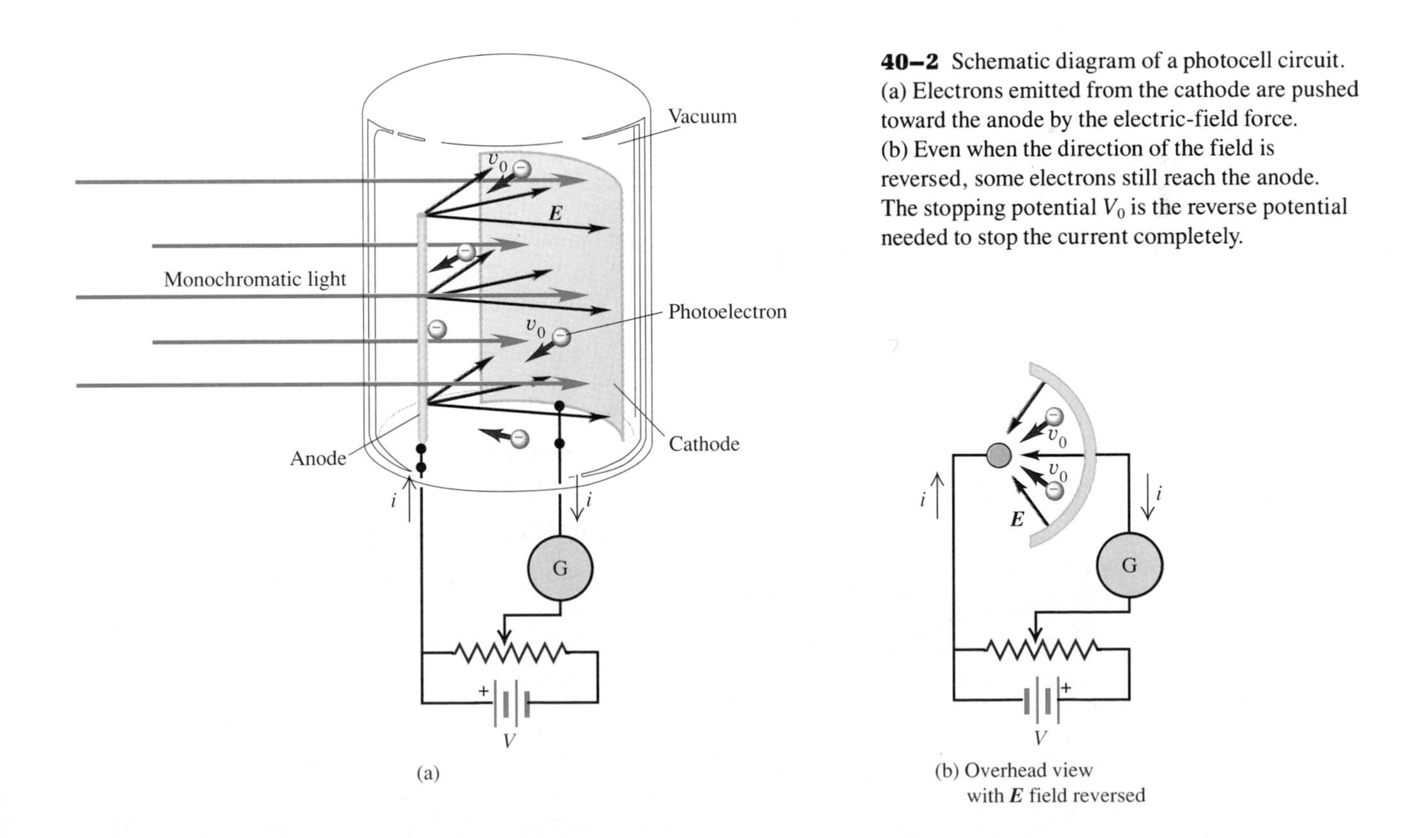


**40–2** Schematic diagram of a photocell circuit. (a) Electrons emitted from the cathode are pushed toward the anode by the electric-field force. (b) Even when the direction of the field is reversed, some electrons still reach the anode. The stopping potential $V_0$ is the reverse potential needed to stop the current completely.

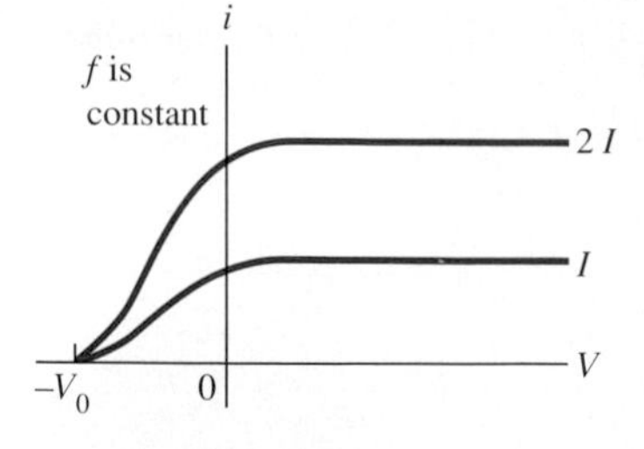


**40–3** Photocurrent $i$ as a function of the potential $V$ of anode with respect to cathode for a constant light frequency $f$. The stopping potential $V_0$ is independent of the intensity $I$, but the photocurrent for large positive $V$ is directly proportional to the intensity.

When the frequency $f$ is *greater than* the threshold value, some electrons are emitted from the cathode with substantial initial speeds. This is shown by the fact that, even with *no* electromotive force in the external circuit, a few electrons reach the collector, causing a small current in the external circuit. Indeed, even when the polarity of the potential difference $V$ is reversed (Fig. 40–2b) and the associated electric-field force on the electrons is directed back toward the cathode, some electrons still reach the anode. The electron flow stops completely only when the reversed potential $V$ is made large enough that the potential energy $eV$ is greater than the maximum kinetic energy $K_{max} = \frac{1}{2}mv_{max}^2$ of the emitted electrons. The reversed potential required to stop the flow completely is called the **stopping potential,** denoted by $V_0$:

$$K_{max} = \tfrac{1}{2}mv_{max}^2 = eV_0. \qquad (40\text{–}1)$$

Measuring the stopping potential $V_0$ therefore gives us a direct measurement of the maximum kinetic energy electrons have when they leave the cathode.

Figure 40–3 shows graphs of current as a function of battery potential for constant intensity and frequency of light. When the potential $V$ is sufficiently large and positive, the curve levels off, showing that *all* the emitted electrons are being collected by the anode. The reverse potential $V_0$ needed to reduce the current to zero is shown. The corresponding electron potential energy is $eV_0$.

When the intensity of light is increased, while its frequency is kept the same, the current levels off at a higher value, showing that more electrons are being emitted. But the stopping potential $V_0$ is found to be the same.

Finally, when the *frequency* of the incident monochromatic light is increased, the stopping potential $V_0$ increases. In fact, $V_0$ turns out to be a linear function of the *frequency* $f$. Figure 40–4 shows current as a function of voltage for two different frequencies, with the same intensity in each case.

These results are hard to understand on the basis of classical physics. When the intensity (energy per unit area per unit time) increases, electrons should be able to gain more energy, increasing the stopping potential $V_0$. But $V_0$ was found *not* to depend on intensity. Also, classical physics offers no explanation for the threshold effect; if we wait long enough, an electron should be able to acquire its needed escape energy from light with *any* frequency, but the experiments found a minimum threshold frequency $f_0$.

The correct analysis of the photoelectric effect was developed by Albert Einstein in 1905. Extending a proposal made four years earlier by Max Planck, Einstein postulated that a beam of light consists of small bundles of energy called **quanta** or **photons.** The energy $E$ of a photon is equal to a constant times its frequency $f$:

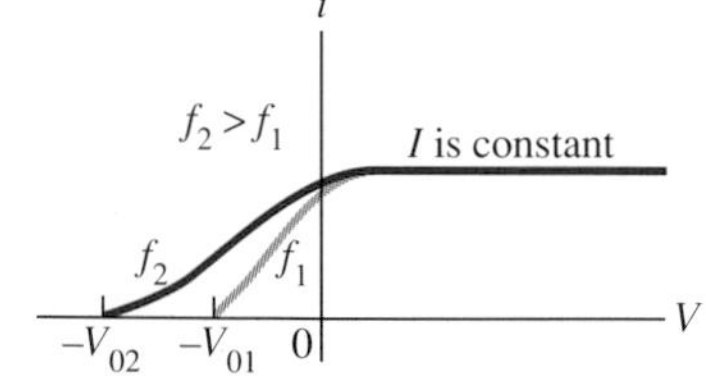


**40–4** Photocurrent $i$ as a function of anode potential $V$ with respect to cathode for two different light frequencies $f_1$ and $f_2$ with the same intensity. The stopping potential $V_0$ (and therefore the maximum kinetic energy of the photoelectrons) increases linearly with frequency.

$$E = hf, \qquad (40\text{–}2)$$

where $h$ is a universal constant called **Planck's constant.** The numerical value of this constant, to the accuracy known at present, is

$$h = 6.6260755(40) \times 10^{-34}\ \text{J} \cdot \text{s}.$$

A photon arriving at the surface of the conductor is absorbed by an electron. The energy transfer is an all-or-nothing process, in contrast to the continuous transfer of energy in the classical theory; the electron gets all the photon's energy or none at all. If this energy is greater than the surface potential-energy barrier or *work function* $\phi$, the electron can escape from the surface.

Then the *maximum* kinetic energy $K_{max} = \frac{1}{2}mv_{max}^2$ for an emitted electron is the energy $hf$ gained from a photon minus the work function $\phi$:

$$K_{max} = \tfrac{1}{2}mv_{max}^2 = hf - \phi. \qquad (40\text{–}3)$$

Combining this with Eq. (40–1), we find

$$eV_0 = hf - \phi. \qquad (40\text{–}4)$$

We can measure the stopping potential $V_0$ for each of several values of frequency $f$ for a given cathode material. A graph of $V_0$ as a function of $f$ turns out to be a straight line, verifying Eq. (40–4), and from such a graph we can determine both the work function $\phi$ for the material and the value of the quantity $h/e$. (Example 40–3 shows in detail how this can be done.) After the electron charge $e$ was measured by Robert Millikan in 1909 (Section 24–6), Planck's constant $h$ could also be determined from these measurements.

Electron energies and work functions are usually expressed in electronvolts (Section 24–6):

$$1 \text{ eV} = 1.602 \times 10^{-19} \text{ J}.$$

In terms of electronvolts, Planck's constant is

$$h = 4.136 \times 10^{-15} \text{ eV} \cdot \text{s}.$$

Table 40–1 lists a few typical work functions of elements. These values are approximate because they are very sensitive to surface impurities. For example, a thin layer of cesium oxide can reduce the work function of a metal to about 1 eV.

**TABLE 40–1 Work Functions of Elements**

| Element | Work function (eV) |
|---|---|
| Aluminum | 4.3 |
| Carbon | 5.0 |
| Copper | 4.7 |
| Gold | 5.1 |
| Nickel | 5.1 |
| Silicon | 4.8 |
| Silver | 4.3 |
| Sodium | 2.7 |

## EXAMPLE 40–1

Silicon films become better electrical conductors when illuminated by photons with energies of 1.14 eV or greater. (This behavior is called *photoconductivity*.) What is the corresponding wavelength?

**SOLUTION** The wavelength $\lambda$ is given by $\lambda = c/f$, so we can rewrite Eq. (40–2) as $E = hc/\lambda$, or

$$\lambda = \frac{hc}{E} = \frac{(4.136 \times 10^{-15} \text{ eV} \cdot \text{s})(3.00 \times 10^8 \text{ m/s})}{1.14 \text{ eV}}$$
$$= 1.09 \times 10^{-6} \text{ m} = 1090 \text{ nm}.$$

This is in the infrared region of the spectrum. The *minimum* energy of 1.14 eV corresponds to the *maximum* wavelength that causes photoconduction for silicon.

## EXAMPLE 40–2

In a photoelectric-effect experiment with light of a certain frequency, a reverse potential of 1.25 V is required to reduce the current to zero. Find a) the maximum kinetic energy and b) the maximum speed of the emitted photoelectrons.

**SOLUTION** a) From Eq. (40–1),

$$K_{max} = eV_0 = (1.60 \times 10^{-19} \text{ C})(1.25 \text{ V}) = 2.00 \times 10^{-19} \text{ J}.$$

To check unit consistency, recall that 1 V = 1 J/C. In terms of electronvolts,

$$K_{max} = (e)(1.25 \text{ V}) = 1.25 \text{ eV}.$$

b) From $K_{max} = \frac{1}{2} m v_{max}{}^2$, we get

$$v_{max} = \sqrt{\frac{2K_{max}}{m}} = \sqrt{\frac{2(2.00 \times 10^{-19} \text{ J})}{9.11 \times 10^{-31} \text{ kg}}}$$
$$= 6.63 \times 10^5 \text{ m/s}.$$

Note that this is much smaller than the speed of light $c$, so we are justified in using the nonrelativistic kinetic energy expression. An equivalent statement is that the kinetic energy is much smaller than the electron's rest energy $mc^2 = 0.511$ MeV.

## EXAMPLE 40–3

For a certain cathode material used in a photoelectric-effect experiment, a stopping potential of 3.0 V was required for light of wavelength 300 nm, 2.0 V for 400 nm, and 1.0 V for 600 nm. Determine the work function for this material and the value of Planck's constant.

**SOLUTION** According to Eq. (40–4), a graph of $V_0$ as a function of $f$ should be a straight line. We rewrite that equation as

$$V_0 = \frac{h}{e} f - \frac{\phi}{e}.$$

In this form we see that the *slope* of the line is $h/e$ and the *intercept* on the vertical axis (corresponding to $f = 0$) is at $-\phi/e$. The frequencies, obtained from $f = c/\lambda$ and $c = 3.00 \times 10^8$ m/s, are $0.5 \times 10^{15}$ Hz, $0.75 \times 10^{15}$ Hz, and $1.0 \times 10^{15}$ Hz, respectively. The graph is shown in Fig. 40–5 (on page 1108).

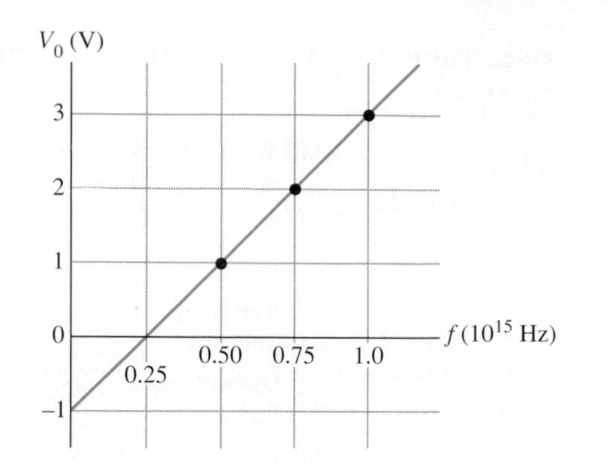


**40–5** Stopping potential as a function of frequency. For a different cathode material having a different work function, the line would be displaced up or down but would have the same slope, equal to $h/e$.

From the graph we find

$$-\frac{\phi}{e} = -1.0\text{ V},$$

$$\phi = 1.0\text{ eV} = 1.60 \times 10^{-19}\text{ J},$$

and

$$\text{Slope} = \frac{h}{e} = \frac{4.0\text{ V}}{1.00 \times 10^{15}\text{ s}^{-1}} = 4.0 \times 10^{-15}\text{ J}\cdot\text{s/C},$$

$$h = (4.0 \times 10^{-15}\text{ J}\cdot\text{s/C})(1.60 \times 10^{-19}\text{ C})$$
$$= 6.4 \times 10^{-34}\text{ J}\cdot\text{s}.$$

This experimental value differs by about 3% from the accepted value. The small value of $\phi$ suggests a composite surface, possibly cesium oxide, mentioned above.

We have discussed photons mostly in the context of light, but the concept is applicable to *all* regions of the electromagnetic spectrum, including radio waves, x rays, and so on. A photon of *any* electromagnetic radiation with frequency $f$ and wavelength $\lambda = c/f$ has energy $E$ given by

$$E = hf = \frac{hc}{\lambda}. \tag{40–5}$$

Furthermore, according to relativity theory, every particle that has energy must also have momentum, even if it has no rest mass. Photons have zero rest mass. According to Eq. (39–42), a photon with energy $E$ has momentum with magnitude $p$ given by $E = pc$. Thus the wavelength $\lambda$ of a photon and its momentum $p$ are related simply by

$$p = \frac{h}{\lambda}. \tag{40–6}$$

## EXAMPLE 40–4

The radio station WQED in Pittsburgh broadcasts at 89.3 MHz with a radiated power of 43.0 kW. How many photons does it emit each second?

**SOLUTION** The station sends out $43.0 \times 10^3$ joules each second. From Eq. (40–2) the energy of each photon emitted is

$$E = hf = (6.626 \times 10^{-34}\text{ J}\cdot\text{s})(89.3 \times 10^6\text{ Hz}) = 5.92 \times 10^{-26}\text{ J}.$$

The number of photons per second is therefore

$$\frac{43.0 \times 10^3\text{ J/s}}{5.92 \times 10^{-26}\text{ J/photon}} = 7.26 \times 10^{29}\text{ photons/s}.$$

With this huge number of photons leaving the station each second, the discreteness of the tiny individual bundles of energy isn't noticed; the radiated energy appears to be a continuous flow.

# 40–3 LINE SPECTRA AND ENERGY LEVELS

The origin of *line spectra,* which we described in Section 40–1, can be understood in general terms on the basis of two central ideas. One is the photon concept; the other is the concept of *energy levels* of atoms. These two ideas were combined by the Danish physicist Niels Bohr in 1913.

Bohr's hypothesis represented a bold breakaway from nineteenth-century ideas. His reasoning went like this. The line spectrum of an element results from the emission from the atoms of that element of photons with specific energies. During such an emission the internal energy of the atom changes by a definite amount. Therefore, said Bohr, each atom must be able to exist only with certain specific values of internal energy. Each atom has a set of possible **energy levels.** An atom can have an amount of internal energy equal to any one of these levels, but it *cannot* have an energy *intermediate* between two levels. All atoms of a given element have the same set of energy levels, but atoms of different elements have different sets.

According to Bohr, an atom can make a *transition* from one energy level to a lower level by emitting a photon with energy equal to the energy *difference* between the initial and final levels (Fig. 40–6). If $E_i$ is the initial energy of the atom, before such a transition, $E_f$ is its final energy, after the transition, and the photon's energy is $hf$, then

$$hf = E_i - E_f. \tag{40–7}$$

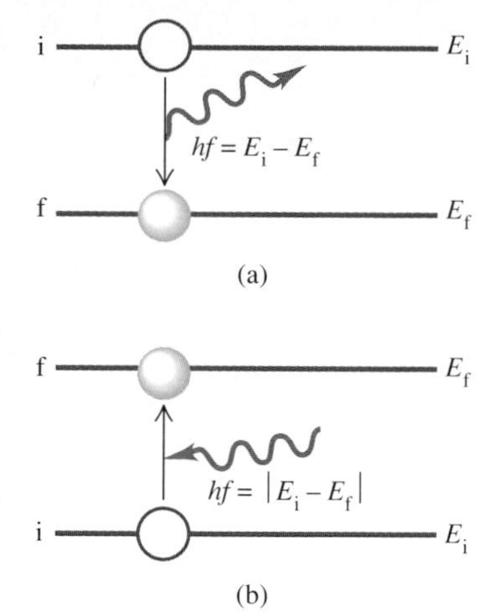


**40–6** (a) The atom drops from state i to a lower-energy state f as it emits a photon with energy equal to $E_i - E_f$. (b) The atom is raised from state i to a higher-energy state f as it absorbs a photon with energy equal to $E_f - E_i$.

For example, when an atom emits a photon of orange light with wavelength $\lambda = 600$ nm, the frequency $f$ is

$$f = \frac{c}{\lambda} = \frac{3.00 \times 10^8 \text{ m/s}}{600 \times 10^{-9} \text{ m}} = 5.00 \times 10^{14} \text{ Hz}.$$

The corresponding photon energy is

$$E = hf = (6.63 \times 10^{-34} \text{ J} \cdot \text{s})(5.00 \times 10^{14} \text{ Hz})$$
$$= 3.31 \times 10^{-19} \text{ J} = 2.07 \text{ eV}.$$

This photon is emitted during a transition between two states of the atom that differ in energy by 2.07 eV.

## Hydrogen Spectrum

By 1913 the spectrum of hydrogen, the least massive atom, had been studied intensively. Under proper conditions, atomic hydrogen emits the series of lines shown in Fig. 40–7. The line with the longest wavelength, or lowest frequency, in the red, is called $H_\alpha$; the next line, in the blue-green, is $H_\beta$; and so on. In 1885, Johann Balmer (1825–1898) had found (by trial and error) a formula that gives the wavelengths of these lines, which are now called the *Balmer series.* Balmer's formula is

$$\frac{1}{\lambda} = R\left(\frac{1}{2^2} - \frac{1}{n^2}\right), \tag{40–8}$$

where $\lambda$ is the wavelength, $R$ is an empirical constant called the *Rydberg constant* [chosen to make Eq. (40–8) fit the measured wavelengths], and $n$ may have the integer values 3, 4, 5, . . . . If $\lambda$ is in meters, the numerical value of $R$ is

$$R = 1.097 \times 10^7 \text{ m}^{-1}.$$

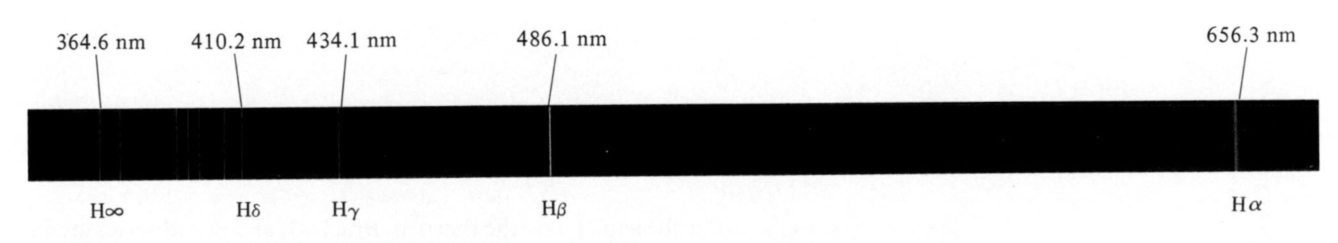


**40–7** The Balmer series of spectrum lines for atomic hydrogen.

Letting $n = 3$ in Eq. (40–8), we obtain the wavelength of the $H_\alpha$ line:

$$\frac{1}{\lambda} = (1.097 \times 10^7 \text{ m}^{-1})\left(\frac{1}{4} - \frac{1}{9}\right), \qquad \text{or} \qquad \lambda = 656.3 \text{ nm}.$$

For $n = 4$ we obtain the wavelength of the $H_\beta$ line, and so on. For $n = \infty$ we obtain the *smallest* wavelength in the series, $\lambda = 364.6$ nm.

Balmer's formula has a very direct relation to Bohr's hypothesis about energy levels. We can find the *photon energies* corresponding to the wavelengths of the Balmer series, using the relations $f = c/\lambda$ and $E = hf$. Multiplying Eq. (40–8) by $hc$, we find

$$\frac{hc}{\lambda} = hf = E = hcR\left(\frac{1}{2^2} - \frac{1}{n^2}\right) = \frac{hcR}{2^2} - \frac{hcR}{n^2}. \tag{40–9}$$

Now we compare Eqs. (40–7) and (40–9). The two agree if we identify $-hcR/n^2$ as the initial energy $E_i$ of the atom and $-hcR/2^2$ as its final energy $E_f$, in a transition in which a photon with energy $hf = E_i - E_f$ is emitted. Note that the negative signs in the energy expressions are needed for consistency with Eq. (40–7). The Balmer (and other) series suggests that the hydrogen atom has a series of energy levels, which we may call $E_n$, given by

$$E_n = -\frac{hcR}{n^2}, \quad n = 1, 2, 3, 4, \ldots. \tag{40–10}$$

(These energies are negative because we have chosen the zero for potential energy at infinite separation.) Each wavelength in the Balmer series corresponds to a transition from a state with $n = 3$ or greater to the state with $n = 2$.

The numerical value of the product $hcR$ is

$$\begin{aligned} hcR &= (6.626 \times 10^{-34} \text{ J} \cdot \text{s})(2.998 \times 10^8 \text{ m/s})(1.097 \times 10^7 \text{ m}^{-1}) \\ &= 2.179 \times 10^{-18} \text{ J} = 13.60 \text{ eV}. \end{aligned}$$

Thus the magnitudes of the energy levels given by Eq. (40–10) are approximately $-13.60$ eV, $-3.40$ eV, $-1.51$ eV, . . . .

Other spectral series for hydrogen have since been discovered. These are known, after their discoverers, as the Lyman, Paschen, Brackett, and Pfund series. Their wavelengths can be represented by formulas similar to Balmer's formula:

*Lyman series*

$$\frac{1}{\lambda} = R\left(\frac{1}{1^2} - \frac{1}{n^2}\right), \quad n = 2, 3, 4, \ldots;$$

*Paschen series*

$$\frac{1}{\lambda} = R\left(\frac{1}{3^2} - \frac{1}{n^2}\right), \quad n = 4, 5, 6, \ldots;$$

*Brackett series*

$$\frac{1}{\lambda} = R\left(\frac{1}{4^2} - \frac{1}{n^2}\right), \quad n = 5, 6, 7, \ldots;$$

*Pfund series*

$$\frac{1}{\lambda} = R\left(\frac{1}{5^2} - \frac{1}{n^2}\right), \quad n = 6, 7, 8, \ldots.$$

The Lyman series is in the ultraviolet, and the Paschen, Brackett, and Pfund series are in the infrared. We see that the Balmer series fits into the scheme between the Lyman and Paschen series.

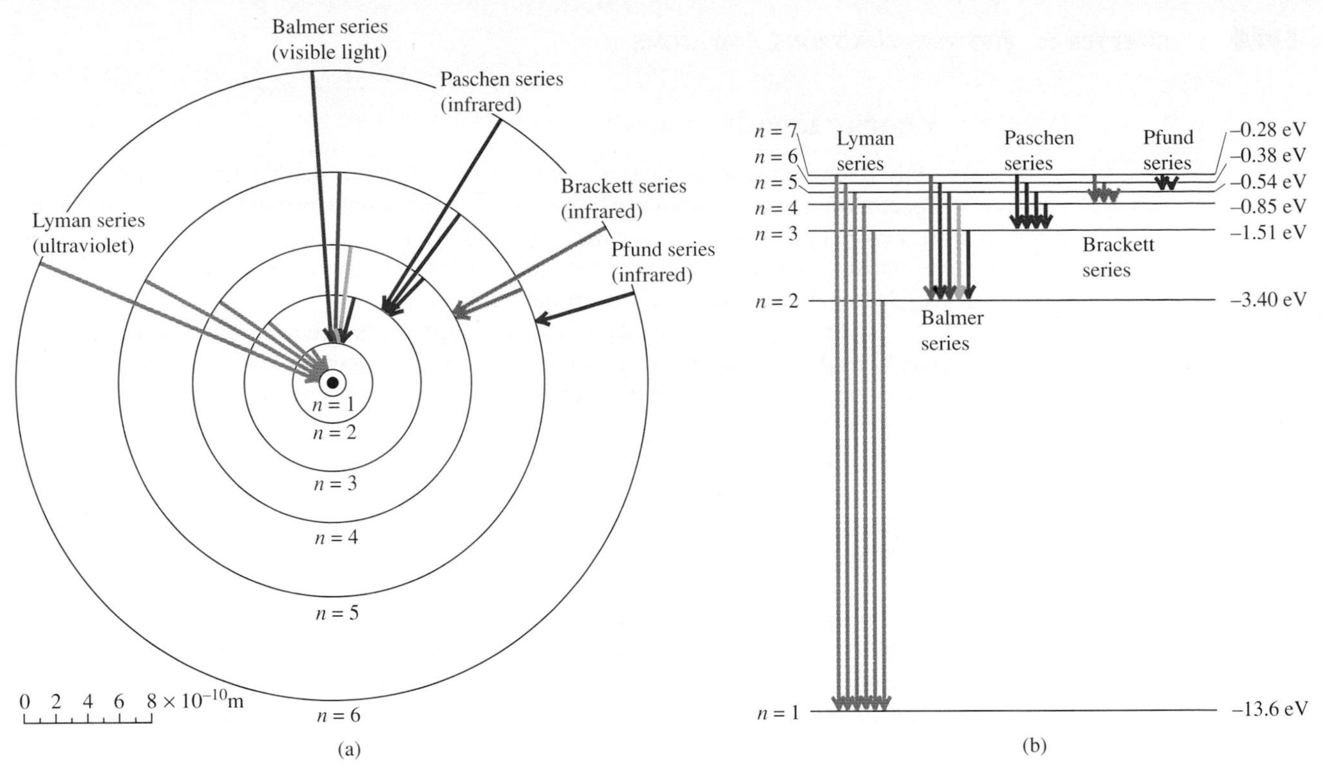


**40–8** (a) "Permitted" orbits of an electron in the Bohr model of a hydrogen atom. The transitions responsible for some of the lines of the various series are indicated by arrows. (b) Energy-level diagram, showing transitions corresponding to the various series.

Thus *all* the spectral series of hydrogen can be understood on the basis of Bohr's picture of transitions from one energy level (and corresponding electron orbit) to another, with the energy levels given by Eq. (40–10) with $n = 1, 2, 3, \ldots$. For the Lyman series the final state is always $n = 1$; for the Paschen series it is $n = 3$; and so on. The relation of the various spectral series to the energy levels and to electron orbits (which we'll discuss in Section 40–5) is shown in Fig. 40–8. Taken together, these spectral series give very strong support to Bohr's picture of energy levels in atoms.

We haven't yet discussed any way to *predict* what the energy levels for a particular atom should be. We haven't shown how to derive Eq. (40–10) from fundamental theory or to relate the Rydberg constant $R$ to other fundamental constants. We'll return to these problems later.

## Franck-Hertz Experiment

In 1914, James Franck and Gustav Hertz found even more direct experimental evidence for the existence of atomic energy levels. While studying the motion of electrons through mercury vapor under the action of an electric field, they found that when the electron kinetic energy was greater than 4.9 eV, the vapor emitted an ultraviolet spectrum line with wavelength 254 nm. Suppose mercury atoms have an energy level 4.9 eV above the lowest energy state. An atom can be raised to this level by collision with an electron; it later decays back to the lowest energy state by emitting a photon. According to Eq. (40–2), the energy of the photon should be

$$E = hf = \frac{hc}{\lambda} = \frac{(6.63 \times 10^{-34}\ \mathrm{J \cdot s})(3.00 \times 10^{8}\ \mathrm{m/s})}{254 \times 10^{-9}\ \mathrm{m}}$$
$$= 7.83 \times 10^{-19}\ \mathrm{J} = 4.9\ \mathrm{eV}.$$

This is equal to the measured electron energy, confirming the existence of the 4.9-eV energy level of the mercury atom.

## Energy Levels

Only a few elements (hydrogen, singly ionized helium, doubly ionized lithium) have spectra whose wavelengths can be represented by a simple formula such as Balmer's. But it is *always* possible to analyze the more complicated spectra of other elements in terms of transitions among various energy levels and to deduce the numerical values of these levels from the measured spectrum wavelengths.

Every atom has a lowest energy level, representing the *minimum* energy the atom can have. This state with lowest energy is called the **ground state,** and all states with higher energy are called **excited states.** A photon corresponding to a particular spectrum line is emitted when an atom makes a transition from an excited state to a lower excited state or to the ground state.

The energy levels for sodium are shown in Fig. 40–9. A sodium atom emits its characteristic yellow-orange light with wavelengths 589.0 and 589.6 nm when it makes transitions from the two closely spaced levels labeled *resonance levels* to the ground state.

A sodium atom in the ground state can also *absorb* a photon with wavelength 589.0 or 589.6 nm. To demonstrate this process, we pass a beam of light from a sodium-vapor lamp through a bulb containing sodium vapor. The atoms in the vapor absorb the 589-nm photons from the beam, reaching the resonance excited states; after a short time they return to the ground state, emitting photons in all directions and causing the sodium vapor to glow with the characteristic yellow light. The average time spent in the excited state is called the *lifetime* of the state; for the resonance levels of the sodium atom, the lifetime is about $1.6 \times 10^{-8}$ s.

More generally, a photon *emitted* when a sodium atom makes a transition from an excited state to the ground state can also be *absorbed* by another sodium atom that is initially in the ground state. If we pass white (continuous-spectrum) light through sodium

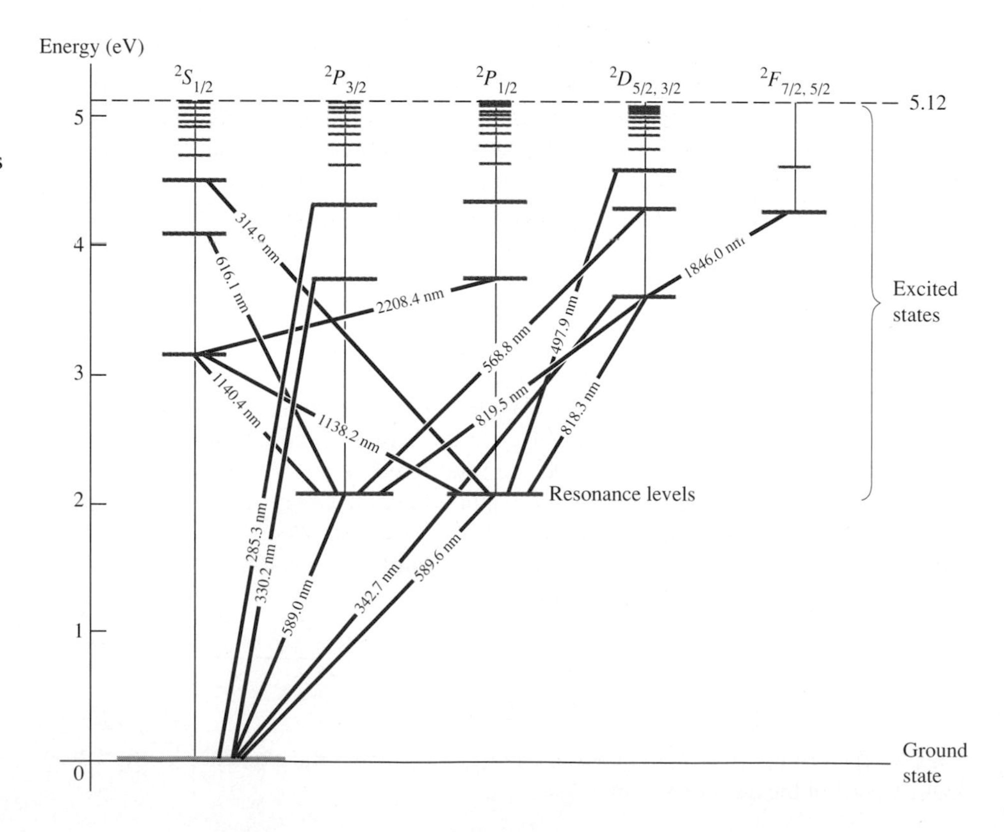


**40–9** Energy levels of the sodium atom. Numbers on the lines between levels are wavelengths. The column labels, such as $^2S_{1/2}$, refer to the quantum states of the valence electron (to be discussed in Chapter 43).

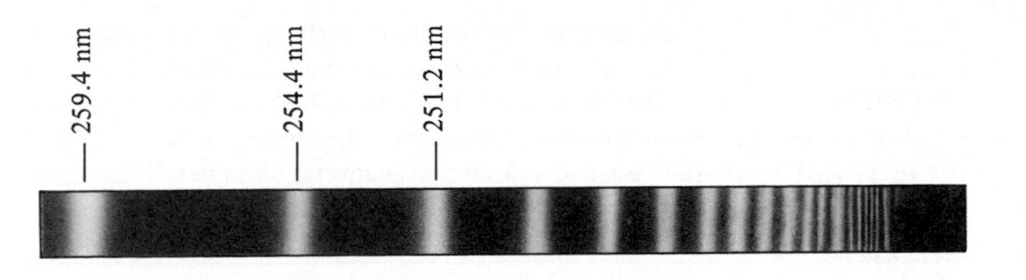


**40–10** Absorption spectrum of sodium. This is an image on photographic film. Black and white are reversed; the bright regions of the spectrum show as black, and the dark absorption lines appear as light lines.

vapor and look at the *transmitted* light with a spectrometer, we find series of dark lines, corresponding to the wavelengths that have been absorbed, as shown in Fig. 40–10. This is called an **absorption spectrum.**

A related phenomenon is *fluorescence*. An atom absorbs a photon (often in the ultraviolet region) to reach an excited state and then drops back to the ground state by successively emitting two or more photons with smaller energy and longer wavelength. For example, the ultraviolet radiation emitted from an electrical discharge in hot mercury vapor in a fluorescent tube is absorbed by the coating on the inside of the tube. The coating then re-emits light in the lower-frequency visible portion of the spectrum. Fluorescent lamps are more efficient than incandescent lamps in converting electrical energy to visible light because they do not waste as much energy producing (invisible) infrared photons.

## PROBLEM-SOLVING STRATEGY

### Photons and energy levels

**1.** Remember that photons are associated with electromagnetic waves. The wavelength $\lambda$ and frequency $f$ are related by $f = c/\lambda$. The energy $E$ of a photon can be expressed as $hf$ or $hc/\lambda$, whichever is more convenient for the problem at hand. Be careful with units; if $E$ is in joules, $h$ must be in J · s, $\lambda$ in meters, and $f$ in 1/seconds, or hertz. The magnitudes are in such unfamiliar ranges that common sense may not help if your calculation is wrong by a factor of $10^{10}$, so be careful with powers of 10.

**2.** It is often useful to measure energy in electronvolts and $h$ in eV · s. Keep these numbers handy: $1\text{ eV} = 1.602 \times 10^{-19}\text{ J}$ and $h = 4.136 \times 10^{-15}\text{ eV} \cdot \text{s}$.

**3.** Keep in mind that an electron moving through a potential difference of 1 volt gains or loses an amount of energy equal to 1 electronvolt. You will use the electronvolt a lot in this chapter and the next three, so it's important to get familiar with it now.

## EXAMPLE 40–5

A hypothetical atom has three energy levels: the ground-state level and 1.00 eV and 3.00 eV above the ground state. a) Find the frequencies and wavelengths of the spectrum lines for this atom. b) What wavelengths can be *absorbed* by this atom if it is initially in its ground state?

**SOLUTION** a) Figure 40–11 shows an energy-level diagram. The possible photon energies, corresponding to the transitions shown, are 1.00 eV, 2.00 eV, and 3.00 eV. For 1.00 eV we have, from Eq. (40–2),

$$f = \frac{E}{h} = \frac{1.00\text{ eV}}{4.136 \times 10^{-15}\text{ eV} \cdot \text{s}} = 2.42 \times 10^{14}\text{ Hz}.$$

For 2.00 eV and 3.00 eV, $f = 4.84 \times 10^{14}$ Hz and $7.25 \times 10^{14}$ Hz, respectively. We find the wavelengths from $\lambda = c/f$.

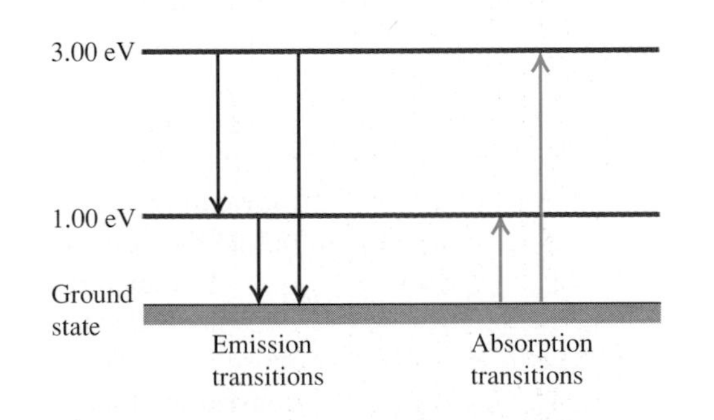


**40–11** Energy-level diagram, showing the possible transitions for emission from excited states and for absorption from the ground state.

For 1.00 eV,

$$\lambda = \frac{c}{f} = \frac{3.00 \times 10^8\ \text{m/s}}{2.42 \times 10^{14}\ \text{Hz}} = 1.24 \times 10^{-6}\ \text{m} = 1240\ \text{nm},$$

in the infrared region of the spectrum. For 2.00 eV and 3.00 eV, the wavelengths are 620 nm (red) and 414 nm (violet), respectively.

b) For an atom that is initially in the ground state, only a 1.00-eV or 3.00-eV photon can be absorbed; a 2.00-eV photon cannot be because there is no energy level 2.00 eV above the ground state. From the above calculations the corresponding wavelengths are 1240 nm and 414 nm, respectively; these two lines appear in the absorption spectrum for this atom.

The Bohr hypothesis established the relation of wavelengths to energy levels, but it provided no general principles for *predicting* the energy levels of a particular atom. Bohr provided a partial analysis for the hydrogen atom; we will discuss this in Section 40–5. A more general understanding of atomic structure and energy levels rests on the principles of *quantum mechanics,* which we will develop in Chapters 41 and 42. Quantum mechanics provides all the principles needed to calculate energy levels from fundamental theory. For many-electron atoms the calculations are so complex that they can be carried out only approximately.

## 40–4 THE NUCLEAR ATOM

Before we can make further progress in relating the energy levels of an atom to its internal structure, we need to have a better idea of what the inside of an atom looks like. We know that atoms are much smaller than wavelengths of visible light and that there is no hope of actually *seeing* an atom. But we can still describe how the mass and electric charge are distributed throughout the volume of the atom.

Here's where things stood in 1910. J. J. Thomson had discovered the electron and measured its charge-to-mass ratio ($e/m$) in 1897; and by 1909, Millikan had completed his first measurements of the electron charge $e$. These and other experiments showed that most of the mass of an atom had to be associated with the *positive* charge, not with the electrons. It was also known that the overall size of atoms is of the order of $10^{-10}$ m and that all atoms except hydrogen contain more than one electron. What was *not* known was how the mass and charge were distributed within the atom. Thomson had proposed a model in which the atom consisted of a sphere of positive charge, of the order of $10^{-10}$ m in diameter, with the electrons embedded in it like chocolate chips in a more or less spherical cookie.

The first experiments designed to probe the inner structure of the atom were the **Rutherford scattering experiments,** carried out in 1910–1911 by Sir Ernest Rutherford and two of his students, Hans Geiger and Ernest Marsden, at Cambridge, England. Each experiment consisted of projecting charged particles at the atoms under study and observing the deflections of the particles. The particle accelerators that are now in common use in high-energy physics laboratories had not yet been invented, and Rutherford's projectiles were *alpha particles* emitted from naturally radioactive elements. Alpha particles are identical with the nuclei of helium atoms: two protons and two neutrons bound together. They are ejected from unstable nuclei with speeds of the order of $10^7$ m/s, and they can travel several centimeters through air or 0.1 mm or so through solid matter before they are brought to rest by collisions.

Rutherford's experimental setup is shown schematically in Fig. 40–12. A radioactive material at the left emits alpha particles. Thick lead screens stop all particles except those in a narrow beam defined by small holes. The beam then passes through a target consisting of a thin gold, silver, or copper foil and strikes a screen coated with zinc sulfide, similar in principle to the screen of a TV picture tube. A momentary flash, or *scintillation,* can be seen on the screen whenever it is struck by an alpha particle. Rutherford counted the numbers of particles deflected through various angles.

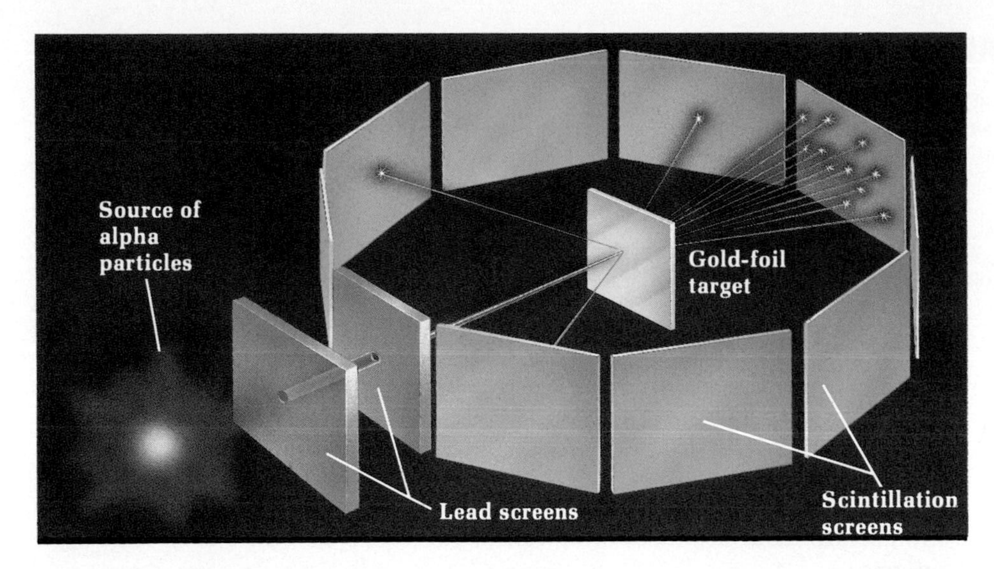


**40–12** The scattering of alpha particles by a thin metal foil. The source of alpha particles is a radioactive element such as radium. The two lead screens with small holes form a narrow beam of alpha particles, which are scattered by the gold foil. The directions of the scattered particles are determined from the scintillations on the screen.

Think of the atoms of the target material as packed together like marbles in a box. An alpha particle can pass through a thin sheet of metal foil, so the alpha particle must be able actually to pass through the interiors of atoms. The *total* electric charge of the atom is zero, so outside the atom there is no force on the alpha particle. Within the atom there are electrical forces caused by the electrons and by the positive charge. But the mass of an alpha particle is about 7300 times that of an electron. Momentum considerations show that the alpha particle can be scattered only a very small amount by its interaction with the much lighter electrons. It's like driving a car through a hailstorm; the hailstones don't deflect the car very much. Only interactions with the *positive* charge, which is tied to most of the mass of the atom, can deflect the alpha particle appreciably.

In the Thomson model, the positive charge is distributed through the whole atom. We can calculate the maximum deflection angle that an alpha particle can have in this situation. It turns out that the interaction potential energy is much *smaller* than the kinetic energy of the alpha particles and that the maximum deflection to be expected is only a few degrees (Fig. 40–13a).

The results were very different from this and were totally unexpected. Some alpha particles were scattered by nearly 180°, that is, almost straight backward (Fig. 40–13b). Rutherford wrote later:

> It was quite the most incredible event that ever happened to me in my life. It was almost as incredible as if you had fired a 15-inch shell at a piece of tissue paper and it came back and hit you.

Back to the drawing board! Suppose the positive charge, instead of being distributed through a sphere with atomic dimensions (of the order of $10^{-10}$ m), is all concentrated in a much *smaller* space. Then it would act like a point charge down to much smaller distances. The maximum repulsive force on the alpha particle would be much larger, and large-angle scattering, as in Fig. 40–13b, would be possible. Rutherford called this concentration of positive charge the **nucleus.** He again computed the numbers of particles expected to be scattered through various angles. Within the precision of his experiments, the computed and measured results agreed, down to distances of the order of $10^{-14}$ m. His experiments therefore established that the atom does have a nucleus, a very small, very dense structure, no larger than $10^{-14}$ m in diameter. The nucleus occupies only about $10^{-12}$ of the total volume of the atom, but it contains *all* the positive charge and at least 99.95% of the total mass of the atom.

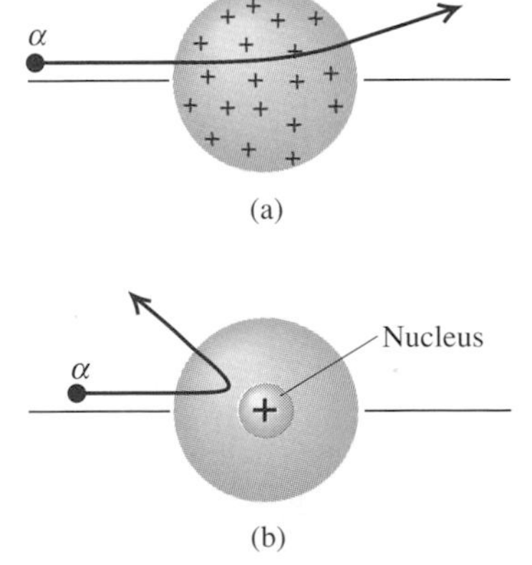


**40–13** (a) Alpha particle scattered through a small angle by the Thomson atom. (b) Alpha particle scattered through a large angle by the Rutherford nuclear atom.

## EXAMPLE 40–6

An alpha particle is aimed directly at a gold nucleus, which contains 79 protons. What minimum initial kinetic energy must the alpha particle have in order to approach within $4.0 \times 10^{-14}$ m of the center of the gold nucleus (assuming that its radius is less than this distance)?

**SOLUTION** We first find the potential energy of the system when the alpha particle is $4.0 \times 10^{-14}$ m from the center of the gold nucleus. From Eq. (24–9),

$$U = \frac{1}{4\pi\epsilon_0}\frac{qq'}{r} = (9.0 \times 10^9\,\mathrm{N \cdot m^2/C^2})\frac{(2)(79)(1.60 \times 10^{-19}\,\mathrm{C})^2}{4.0 \times 10^{-14}\,\mathrm{m}}$$

$$= 0.91 \times 10^{-12}\,\mathrm{J} = 5.7 \times 10^6\,\mathrm{eV} = 5.7\,\mathrm{MeV}.$$

If the alpha particle is to approach within $10^{-14}$ m of the gold nucleus before stopping, it must have at least 5.7 MeV of kinetic energy when it is far away from the nucleus. In fact, alpha particles emitted from naturally occurring radioactive elements typically have energies in the range 4 to 6 MeV. The common isotope of radium, $^{226}$Ra, emits an alpha particle with energy 4.78 MeV.

Figure 40–14 shows a computer simulation of the scattering of 5.0-MeV alpha particles from gold nuclei with radius $7.0 \times 10^{-15}$ m (the actual value) and from nuclei with a hypothetical radius ten times this great. In the second case there is *no* large-angle scattering.

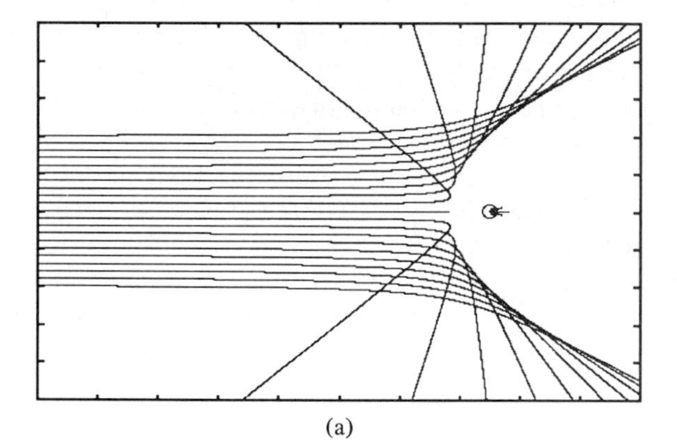

(a)

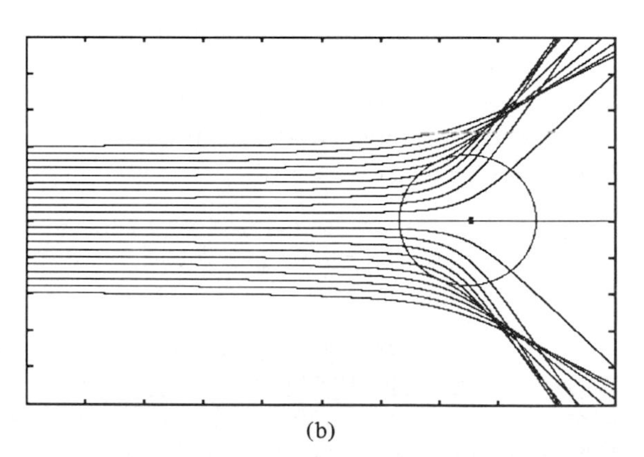

(b)

**40–14** Computer simulation of scattering of 5.0-MeV alpha particles from a gold nucleus. (a) A gold nucleus with radius $7.0 \times 10^{-15}$ m (its actual size); (b) a nucleus with radius $7.0 \times 10^{-14}$ m. In part (b) there is *no* large-angle scattering.

## 40–5 THE BOHR MODEL

At the same time (1913) that Bohr established the relationship between spectrum wavelengths and energy levels, he also proposed a mechanical model of the hydrogen atom. Using this model, now known as the **Bohr model,** he was able to *calculate* the energy levels of hydrogen and obtain agreement with values determined from spectra.

Rutherford's discovery of the atomic nucleus raised a serious question. What kept the negatively charged electrons at relatively large distances ($\sim 10^{-10}$ m) from the very small ($\sim 10^{-15}$ m), positively charged nucleus, despite their electrostatic attraction? Rutherford suggested that perhaps the electrons *revolve* in orbits about the nucleus, just as the planets revolve around the sun.

But according to classical electromagnetic theory, any accelerating electric charge (either oscillating or revolving) radiates electromagnetic waves (like the oscillating dipole in Section 33–8). The energy of a revolving electron should therefore decrease continuously, its orbit should become smaller and smaller, and it should eventually spiral into the nucleus (Fig. 40–15). Even worse, according to classical theory, the *frequency* of the electromagnetic waves emitted should equal the frequency of revolution. As the electrons radiated energy, their angular velocities would change continuously, and they would emit a *continuous* spectrum (a mixture of all frequencies), not the *line* spectrum actually observed.

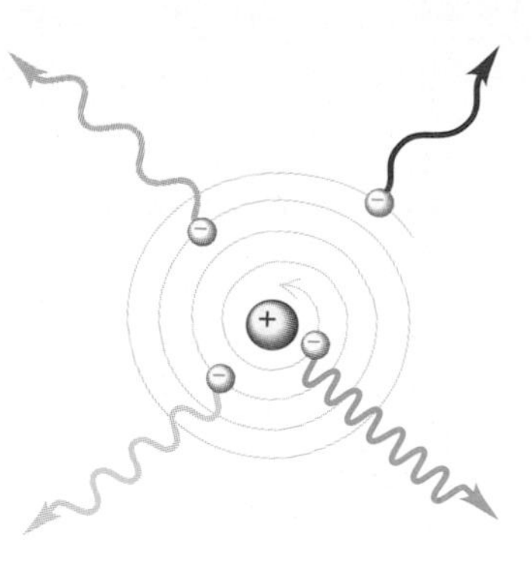

**40–15** Classical physics predicts that the electron should continuously radiate electromagnetic waves and spiral into the nucleus. The electron's angular velocity varies with its radius, so the frequency of the emitted radiation should change continuously.

## Stable Orbits

To solve this problem, Bohr made a revolutionary proposal. He postulated that an electron in an atom can revolve in certain circular *stable orbits,* with a definite energy associated with each orbit, *without* emitting radiation, contrary to the predictions of classical electromagnetic theory. According to Bohr, an atom radiates energy only when it makes a transition from one of these stable orbits to another, at the same time emitting (or absorbing) a photon with appropriate energy and frequency, as given by Eq. (40–7).

To determine the radii of the "permitted" orbits, Bohr introduced what we have to regard in hindsight as a brilliant intuitive guess. He postulated that the angular momentum is *quantized,* that the angular momentum of the electron must be an integral multiple of $h/2\pi$. (Note that the *units* of Planck's constant $h$, usually written as $\text{J}\cdot\text{s}$, are the same as the units of angular momentum, usually written as $\text{kg}\cdot\text{m}^2/\text{s}$.) From Section 10–5, Eq. (10–28), the angular momentum $L$ of a particle with mass $m$, moving with speed $v$ in a circle of radius $r$, is $L = mvr$. So Bohr's assumption may be stated as

$$mvr = n\frac{h}{2\pi},$$

where $n = 1, 2, 3, \ldots$. Each value of $n$ corresponds to a permitted value of the orbit radius, which we denote from now on by $r_n$, and a corresponding speed $v_n$. With this notation the above equation becomes

$$mv_n r_n = n\frac{h}{2\pi}. \tag{40–11}$$

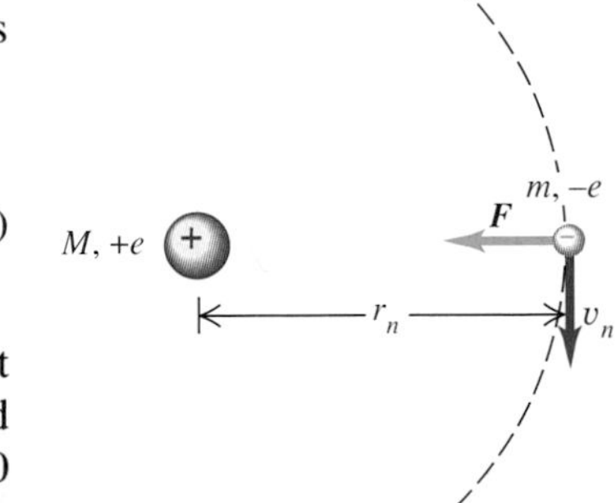


**40–16** The Bohr model of the hydrogen atom. The proton is assumed to be stationary; the electron revolves in a circle of radius $r_n$ with speed $v_n$. The electrostatic attraction provides the necessary centripetal acceleration.

Now let's consider a mechanical model of the hydrogen atom (Fig. 40–16) that incorporates this condition. This atom consists of a single electron with mass $m$ and charge $-e$, revolving around a single proton with charge $+e$. The proton is nearly 2000 times as massive as the electron, so we will assume that the proton does not move. We learned in Section 5–4 that when a particle with mass $m$ moves with speed $v_n$ in a circular orbit with radius $r_n$, its acceleration is $v_n^2/r_n$. According to Newton's second law, a force with magnitude $F = mv_n^2/r_n$ is needed to cause this acceleration. The force $F$ is provided by the electrical attraction between the two charges:

$$F = \frac{1}{4\pi\epsilon_0}\frac{e^2}{r_n^2},$$

so the $F = ma$ equation is

$$\frac{1}{4\pi\epsilon_0}\frac{e^2}{r_n^{\,2}} = \frac{mv_n^{\,2}}{r_n}. \tag{40–12}$$

When we solve Eqs. (40–11) and (40–12) simultaneously for $r_n$ and $v_n$, we get

$$r_n = \epsilon_0\frac{n^2h^2}{\pi me^2}, \tag{40–13}$$

$$v_n = \frac{1}{\epsilon_0}\frac{e^2}{2nh}. \tag{40–14}$$

Equation (40–13) shows that the orbit radius $r_n$ is proportional to $n^2$; the smallest orbit radius corresponds to $n = 1$. We'll denote this minimum radius, called the *Bohr radius,* as $a_0$:

$$a_0 = \epsilon_0\frac{h^2}{\pi me^2}. \tag{40–15}$$

Then Eq. (40–13) can be written as

$$r_n = n^2a_0. \tag{40–16}$$

The permitted, nonradiating orbits have radii $a_0$, $4a_0$, $9a_0$, and so on. The value of $n$ for each orbit is called the **quantum number** for the orbit.

The numerical values of the quantities on the right side of Eq. (40–15) are

$$\begin{aligned}\epsilon_0 &= 8.854 \times 10^{-12}\ \mathrm{C^2/N\cdot m^2},\\ h &= 6.626 \times 10^{-34}\ \mathrm{J\cdot s},\\ m &= 9.109 \times 10^{-31}\ \mathrm{kg},\\ e &= 1.602 \times 10^{-19}\ \mathrm{C}.\end{aligned}$$

Using these values in Eq. (40–15), we find that the radius $a_0$ of the smallest Bohr orbit is

$$\begin{aligned}a_0 &= \frac{(8.854 \times 10^{-12}\ \mathrm{C^2/N\cdot m^2})(6.626 \times 10^{-34}\ \mathrm{J\cdot s})^2}{(3.142)(9.109 \times 10^{-31}\ \mathrm{kg})(1.602 \times 10^{-19}\ \mathrm{C})^2}\\ &= 0.5292 \times 10^{-10}\ \mathrm{m}.\end{aligned}$$

This result is consistent with atomic dimensions estimated by other methods. We can use this value in Eq. (40–11) to find the orbit *speed* of the electron. We leave this calculation as an exercise; the result is that for the $n = 1$ state, $v_1 = 2.19 \times 10^6$ m/s. This is less than 1% of the speed of light, showing that relativistic considerations aren't significant.

## Energy Levels

We can use Eqs. (40–14) and (40–13) to find the kinetic and potential energies $K_n$ and $U_n$ for the electron in the orbit with quantum number $n$:

$$K_n = \tfrac{1}{2}mv_n^{\,2} = \frac{1}{\epsilon_0^2}\frac{me^4}{8n^2h^2},$$

$$U_n = -\frac{1}{4\pi\epsilon_0}\frac{e^2}{r_n} = -\frac{1}{\epsilon_0^2}\frac{me^4}{4n^2h^2}.$$

The total energy $E_n$ is the sum of the kinetic and potential energies:

$$E_n = K_n + U_n = -\frac{1}{\epsilon_0^2}\frac{me^4}{8n^2h^2}. \tag{40–17}$$

This expression has a negative sign because we have taken the potential energy to be zero when the electron is infinitely far from the nucleus. We are interested only in energy *differences,* so this doesn't matter.

In Fig. 40–8 the possible energy levels and orbits of the atom are labeled by values of the quantum number $n$. For each value of $n$ there are corresponding values of orbit radius $r_n$, speed $v_n$, angular momentum $L_n = nh/2\pi$, and total energy $E_n$. The energy of the atom is least when $n = 1$ and $E_n$ has its largest negative value. This is the *ground state* of the atom; it is the state with the smallest orbit, with radius $a_0$. For $n = 2, 3, 4, \ldots$, the absolute value of $E_n$ is smaller, and the energy is progressively larger (less negative). The orbit radius increases as $n^2$, as shown by Eq. (40–13).

Comparing the expression for $E_n$ in Eq. (40–17) with Eq. (40–10) (deduced from spectrum analysis), we see that they agree only if the coefficients are equal:

$$hcR = \frac{1}{\epsilon_0^2}\frac{me^4}{8h^2}, \qquad \text{or} \qquad R = \frac{me^4}{8\epsilon_0^2h^3c}. \tag{40–18}$$

This equation therefore shows us how to *calculate* the value of the Rydberg constant from the fundamental physical constants $m$, $c$, $e$, $h$, and $\epsilon_0$, all of which can be determined quite independently of the Bohr theory. When we substitute the numerical values of these quantities, we obtain the value $R = 1.097 \times 10^7\ \text{m}^{-1}$. This value is within 0.01% of the value determined from wavelength measurements; this agreement provides very strong and direct confirmation of Bohr's theory. We invite you to substitute numerical values into Eq. (40–18) and compute the value of $R$ to confirm these statements.

The *ionization energy* of the hydrogen atom is the energy required to remove the electron completely. Ionization corresponds to a transition from the ground state ($n = 1$) to an infinitely large orbit radius ($n = \infty$). The predicted energy is 13.61 eV. The ionization energy can also be measured directly; the result is 13.60 eV; the two values agree within 0.1%.

## EXAMPLE 40–7

Find the kinetic, potential, and total energies of the hydrogen atom in the $n = 2$ state, and find the wavelength of the photon emitted in the transition $n = 2 \rightarrow n = 1$.

**SOLUTION** We first note that the constant that appears in Eq. (40–17) is equal to $hcR$, which appears in Eq. (40–10) and which we have found to equal 13.6 eV:

$$\frac{me^4}{8\epsilon_0^2h^2} = hcR = 13.6\ \text{eV}.$$

Using this expression, we can rewrite Eq. (40–17) and the two preceding equations as

$$K_n = \frac{13.60\ \text{eV}}{n^2}, \qquad U_n = \frac{-27.20\ \text{eV}}{n^2}, \qquad E_n = \frac{-13.60\ \text{eV}}{n^2}.$$

For the $n = 2$ state we find $K_2 = 3.40$ eV, $U_2 = -6.80$ eV, and $E_2 = -3.40$ eV.

The energy of the emitted photon is $E_2 - E_1 = -3.40\ \text{eV} - (-13.6\ \text{eV}) = 10.2\ \text{eV} = 1.63 \times 10^{-18}$ J. This is equal to $hc/\lambda$, so we find

$$\lambda = \frac{hc}{E_2 - E_1} = \frac{(6.626 \times 10^{-34}\ \text{J}\cdot\text{s})(3.00 \times 10^8\ \text{m/s})}{1.63 \times 10^{-18}\ \text{J}}$$
$$= 122 \times 10^{-9}\ \text{m} = 122\ \text{nm}.$$

This is the wavelength of the Lyman alpha line, the longest-wavelength line in the Lyman series of ultraviolet lines in the hydrogen spectrum.

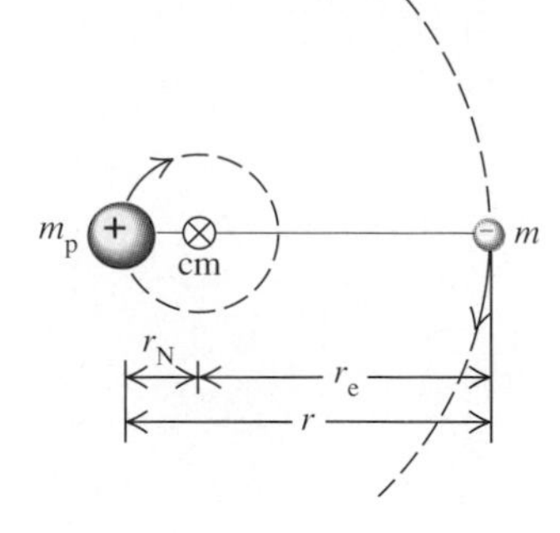


**40–17** The nucleus and the electron rotate about their common center of mass.

## Reduced Mass

The values of the Rydberg constant $R$ and the ionization energy of hydrogen predicted by Bohr's analysis are within 0.1% of the measured values. The agreement would be even better if we had not assumed that the nucleus (a proton) remains at rest. The proton and electron both revolve in circular orbits about their *center of mass* (Section 8–6), as shown in Fig. 40–17. It turns out that this motion can be taken into account very simply by using in Bohr's equations not the electron mass $m$ but a quantity called the **reduced mass** $m_r$ of the system, defined as

$$\frac{1}{m_r} = \frac{1}{m} + \frac{1}{m_p}, \qquad \text{or} \qquad m_r = \frac{mm_p}{m + m_p}, \tag{40–19}$$

where $m_p$ is the proton mass, $m_p = 1836.2m$. For the proton-electron system,

$$m_r = \frac{m(1836.2m)}{m + 1836.2m} = 0.99946m.$$

When this value is used instead of the electron mass $m$ in the Bohr equations, the predicted values are well within the limits of precision of the measured values.

Deuterium, also called *heavy hydrogen*, is composed of hydrogen atoms in which the nucleus is not a single proton but a proton and a neutron bound together to form a composite particle called the *deuteron*. The reduced mass of the deuterium atom turns out to be $0.99973m$. We should expect all the spectrum wavelengths of deuterium to be shifted relative to corresponding values for hydrogen by $0.99973m/0.99946m = 1.00027$. This is a small effect but well within the precision of modern spectrometers. This small wavelength shift led Harold Urey to the discovery of deuterium in 1932; he was subsequently awarded the Nobel prize in chemistry for this achievement.

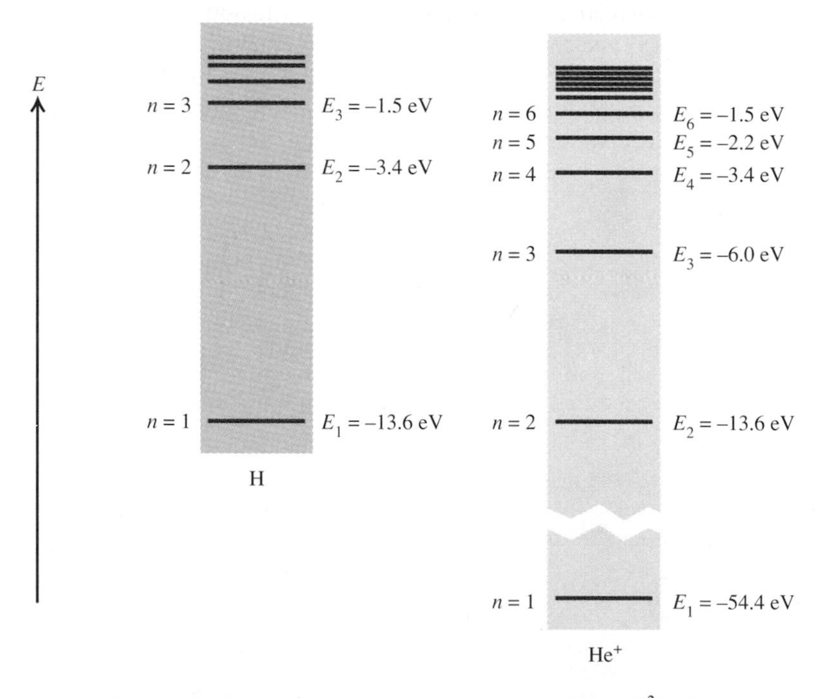


**40–18** Energy levels of H and $He^+$. Because of the additional factor $Z^2$ in the energy expression, Eq. (40–17), the energy of the $He^+$ ion with a given $n$ is almost exactly four times that of the H atom with the same $n$. There are small differences (of the order of 0.05%) because the reduced masses are slightly different.

The Bohr model can be extended to other one-electron atoms, such as singly ionized helium ($He^+$), doubly ionized lithium ($Li^{2+}$), and so on. Such atoms are called *hydrogenlike atoms*. Suppose the nuclear charge is not $e$ but $Ze$, where $Z$ is the *atomic number*, equal to the number of protons in the nucleus. The effect in the above analysis is to replace $e^2$ everywhere by $Ze^2$. In particular, the orbit radii $r_n$ given by Eq. (40–13) become smaller by a factor of $Z$, and the energy levels $E_n$ given by Eq. (40–17) are multiplied by $Z^2$. We invite you to verify these statements. The reduced-mass correction in these cases is less than 0.1%. Figure 40–18 compares the energy levels for H and for $He^+$, which has $Z = 2$.

Atoms of the *alkali metals* have one electron outside a core consisting of the nucleus and the inner electrons, with net core charge $+e$. These atoms are approximately hydrogenlike, especially in excited states. Even trapped electrons and electron vacancies in semiconducting solids act somewhat like hydrogen atoms.

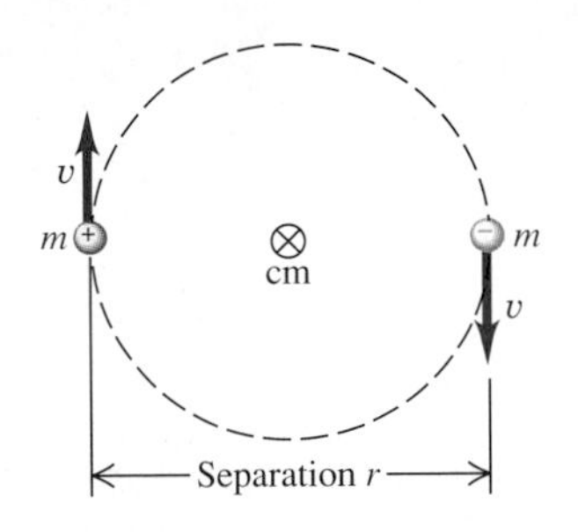


**40–19** In the Bohr model of positronium, the electron and the positron rotate about their common center of mass, which is located midway between them because they have equal mass. Their separation is twice either orbit radius.

The concept of reduced mass has other applications. A positronium "atom" (Fig. 40–19) consists of an electron and a positron, each with mass $m$, in orbit around their common center of mass. This structure lasts only about $10^{-6}$ s before the two particles disappear (converting their rest energy into electromagnetic radiation), but this is enough time to study the spectrum. The reduced mass is $m/2$, and the energy levels have exactly half the values of those in the simple Bohr model with infinite proton mass. Corresponding spectrum lines have been observed and measured, confirming the existence of the positronium atom.

Although the Bohr model predicted the energy levels of the hydrogen atom correctly, it raised as many questions as it answered. It combined elements of classical physics with new postulates that were inconsistent with classical ideas. There was no clear justification for restricting the angular momentum to multiples of $h/2\pi$, except that it led to the right answer. The model provided no insight into what happens *during* a transition from one orbit to another; the angular velocities of the electron motion were not the same as the angular frequencies of the emitted radiation, a result that is contrary to classical electrodynamics. And attempts to extend the model to atoms with two or more electrons were not successful. In Chapters 41 and 42 we will find that an even more radical departure from classical concepts was needed before the understanding of atomic structure could progress further.

## 40–6 THE LASER

A **laser** is a light source that produces a beam of highly coherent and very nearly monochromatic light as a result of cooperative emission from many atoms. The name *laser* is an acronym for "light amplification by stimulated emission of radiation." We can understand the principles of laser operation on the basis of photons and atomic energy levels. During the discussion we'll also introduce two new concepts, *stimulated emission* and *population inversion*.

If an atom has an excited state with energy level $E$ above the ground state, the atom in a ground state can absorb a photon with frequency $f$ given by $E = hf$. This process is shown schematically in Fig. 40–20a (on page 1122), which shows a gas in a transparent container. An atom A absorbs a photon, reaching an excited state denoted as $A^*$. Some time later, the excited atom returns to a ground state by emitting a photon with the same frequency as the one originally absorbed. This process is called *spontaneous emission;* the direction and phase of the emitted photon are random (Fig. 40–20b).

In **stimulated emission** (Fig. 40–20c) an incident photon encounters an excited atom. By a kind of resonance effect, the photon forces the atom to emit *another* photon with the same frequency, the same direction, the same phase, and the same polarization

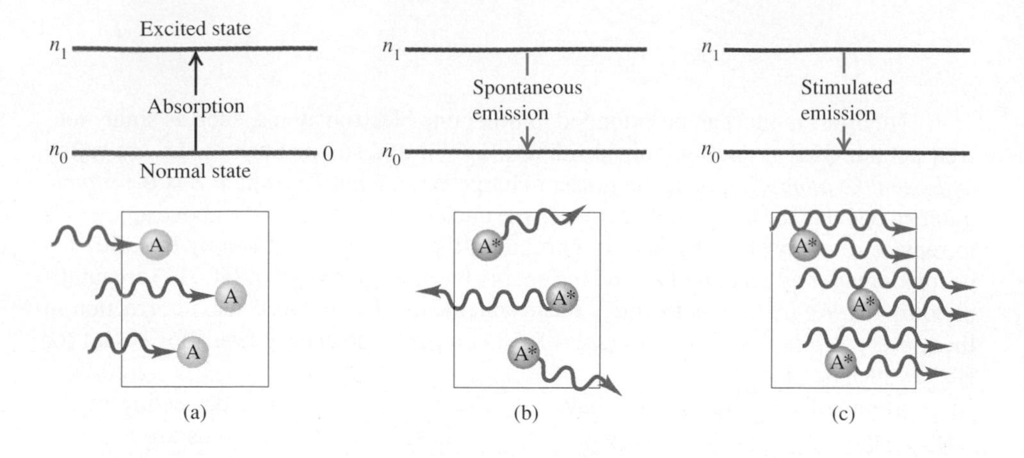


**40–20** Three interaction processes between an atom and radiation.

as the incident photon. The two photons thus have a definite phase relation and emerge together as *coherent* radiation. One photon goes in, and two come out—thus the name *light amplification*. The laser makes use of stimulated emission to produce a beam consisting of a large number of such coherent photons. Physicists sometimes describe stimulated emission by assigning personality traits to particles. Photons are gregarious; a photon likes to go where it can be with other identical photons. By contrast, electrons are hermits by nature, as we'll see in Section 43–4.

To discuss stimulated emission from atoms in excited states, we need to know something about how many atoms are in each of the various energy states. In a gas this is determined by the Maxwell-Boltzmann distribution function (Section 16–5). The function states that when a system is in thermal equilibrium at absolute temperature $T$, the number $n$ of atoms in a state with energy $E$ is proportional to $e^{-E/kT}$, where $k$ is Boltzmann's constant. (In Section 16–5, $E$ was the kinetic energy $\frac{1}{2}mv^2$ of a gas molecule; here we're talking about internal energy of an atom.) Because of the negative exponent, fewer atoms are in higher-energy states, as we should expect. If $E_0$ is the ground-state energy and $E_1$ the energy of an excited state, then the ratio of numbers of atoms in the two states is

$$\frac{n_1}{n_0} = \frac{e^{-E_1/kT}}{e^{-E_0/kT}} = e^{-(E_1 - E_0)/kT}. \tag{40–20}$$

For example, suppose $E_1 - E_0 = 2.0\ \text{eV} = 3.2 \times 10^{-19}\ \text{J}$, typical of photons of visible light. At $T = 3000\ \text{K}$ (the temperature of the filament in an incandescent light bulb),

$$\frac{E_1 - E_0}{kT} = \frac{3.2 \times 10^{-19}\ \text{J}}{(1.38 \times 10^{-23}\ \text{J/K})\,(3000\ \text{K})} = 7.73,$$

and

$$e^{-(E_1 - E_0)/kT} = e^{-7.73} = 0.00044.$$

That is, the fraction of atoms in a state 2 eV above the ground state is extremely small, even at this high temperature. The point is that at any reasonable temperature there aren't enough atoms in excited states for any appreciable amount of stimulated emission from these states to occur.

We could try to enhance the number of atoms in excited states by sending through the container a beam of radiation with frequency $f = E/h$ corresponding to the energy difference $E = E_1 - E_0$. Some of the atoms absorb photons of energy $E$ and are raised to an excited state, and the population ratio $n_1/n_0$ increases. But because $n_0$ is originally so much larger than $n_1$, an enormously intense beam of light would be required to increase $n_1$ to a value comparable to $n_0$. The rate at which energy is *absorbed* from the beam by the $n_0$ ground-state atoms far outweighs the rate at which energy is added to the beam by stimulated emission from the relatively rare ($n_1$) excited atoms.

We need to create a non-equilibrium situation in which $n_1$ is substantially increased in comparison to the normal equilibrium value given by the Maxwell-Boltzmann distribution, Eq. (40–20). Such a situation is called a **population inversion.** When $n_1$ is large enough, the rate of energy radiation by stimulated emission actually *exceeds* the rate of absorption. The system then acts as a *source* of radiation with photon energy $E$. Furthermore, because the photons are the result of stimulated emission, they all have the same frequency, phase, polarization, and direction. The resulting radiation is therefore very much more *coherent* than light from ordinary sources, in which the emissions of individual atoms are *not* coordinated. This coherent emission is exactly what happens in a laser.

The necessary population inversion can be achieved in a variety of ways. A familiar example is the helium-neon laser, a common and inexpensive laser that is available in many undergraduate laboratories. A mixture of helium and neon, each typically at a pressure of the order of $10^2$ Pa ($10^{-3}$ atm), is sealed in a glass enclosure provided with two electrodes. When a sufficiently high voltage is applied, a glow discharge occurs. Collisions between ionized atoms and electrons carrying the discharge current excite atoms to various energy states.

Figure 40–21a shows an energy-level diagram for the system. The notation used to label the various energy levels, such as $1s$, $3p$, and $5s$, refers to the states of the electrons in the atoms. (We'll discuss this notation in Chapter 43; we don't need detailed knowledge of the nature of the states to understand laser action.) Because of restrictions imposed by conservation of angular momentum, a helium atom with an electron excited to the $2s$ state cannot return to the ground ($1s$) state by emitting a 20.61-eV photon, as you might expect it to do. Such a state, in which single-photon emission is impossible, is called a *metastable state.* Helium atoms "pile up" in the metastable $2s$ state, creating a population inversion.

However, excited helium atoms *can* lose energy by energy-exchange collisions with neon atoms initially in the ground state. A $2s$ helium atom, with its internal energy of 20.61 eV and a little additional kinetic energy, can collide with a neon atom in the

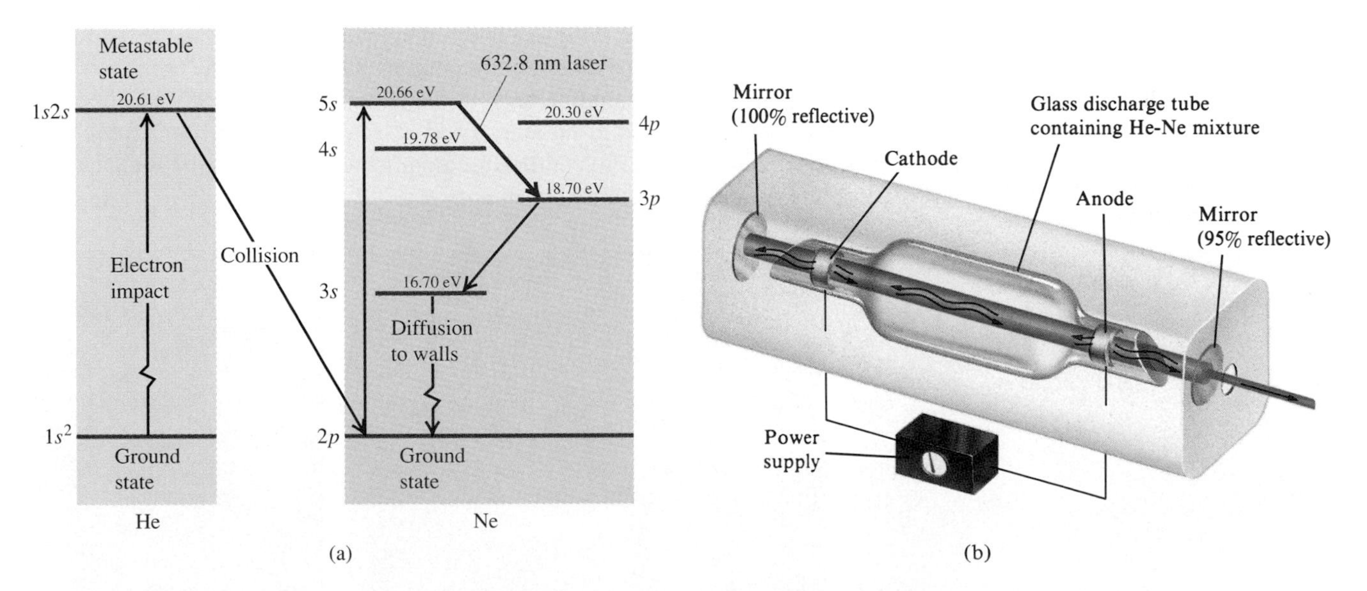


**40–21** (a) Energy-level diagram for a helium-neon laser. (b) Operation of a laser. The light emitted by electrons returning to lower-energy states is reflected between mirrors, so it continues to stimulate emission of more coherent light. One mirror is partially transmitting and allows the high-intensity light beam to leave the laser cavity.

ground state, exciting the *neon* atom to the 5*s* excited state at 20.66 eV and dropping the *helium* atom back to the 1*s* ground state. Thus, we have the necessary mechanism for a population inversion in neon, greatly enhancing the population in the 5*s* state compared to that in the 3*p* state.

Stimulated emissions during transitions from the 5*s* to the 3*p* state then result in the emission of highly coherent light at 632.8 nm, as shown in the diagram. In practice, the beam is sent back and forth through the gas many times by a pair of parallel mirrors (Fig. 40–21b), so as to stimulate emission from as many excited atoms as possible. One of the mirrors is partially transparent, so a portion of the beam emerges as an external beam. The net result of all these processes is a beam of light that is (1) very intense, (2) almost perfectly parallel, (3) highly monochromatic, and (4) spatially *coherent* at all points within a given cross section.

Other types of lasers use different processes to achieve the necessary population inversion. In a semiconductor laser the inversion is obtained by driving electrons and holes to a *p*-*n* junction (to be discussed in Section 44–8) with a steady electric field. In one type of *chemical* laser a chemical reaction forms molecules in metastable excited states. In a carbon dioxide gas dynamic laser the population inversion results from rapid expansion of the gas. There are also free-electron lasers and even x-ray lasers, with the population inversion caused by small nuclear explosions. A *maser* (acronym for "microwave amplification by stimulated emission of radiation") operates on the basis of population inversions in molecules, using energy levels that are much more closely spaced. The corresponding emitted radiation is in the microwave range.

In recent years, lasers have found a wide variety of practical applications (Fig. 40–22). The high intensity of a laser beam makes it a convenient drill. For example, a laser beam can drill a very small hole in a diamond for use as a die in drawing very small-diameter wire. Because the photons in a laser beam are strongly correlated in their directions, a laser beam can be very highly *collimated;* that is, it can travel long distances without appreciable spreading. Surveyors often use lasers, especially in situations requiring great precision, such as drilling a long tunnel from both ends.

Lasers are widely used in medicine. A laser can produce a very *narrow* beam that is intense enough to vaporize anything in its path. This property is used in the treatment of a detached retina; a short burst of radiation damages a small area of the retina, and the resulting scar tissue "welds" the retina back to the choroid from which it has become detached. Laser beams are used in surgery; blood vessels cut by the beam tend to seal themselves off, making it easier to control bleeding. Lasers are also used for selective destruction of tissue, as in the removal of tumors.

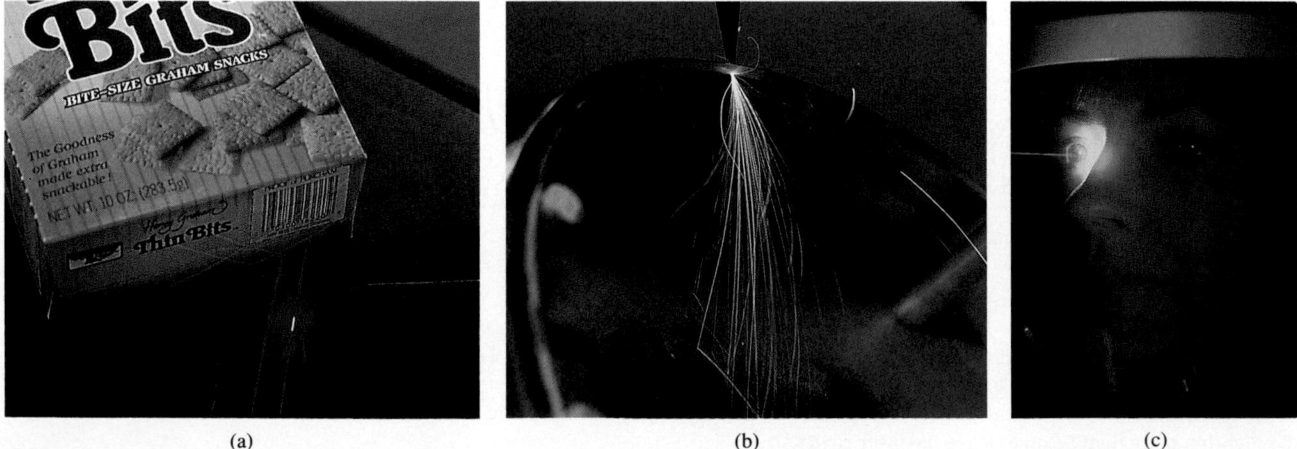


**40–22** Lasers have found many applications for consumer goods, industry, and medicine. (a) Laser beam scans a bar code at a supermarket check-out counter. (b) A carbon dioxide laser beam cuts precision holes in a jet engine component. (c) A laser beam used to treat a detached retina.

# 40–7 X-RAY PRODUCTION AND SCATTERING

The production and scattering of x rays provide additional examples of the quantum nature of electromagnetic radiation. X rays are produced when rapidly moving electrons that have been accelerated through a potential difference of the order of $10^3$ to $10^6$ V strike a metal target. They were first produced in 1895 by Wilhelm Röntgen (1845–1923), using an apparatus similar in principle to the setup shown in Fig. 40–23a. Electrons are "boiled off" from the heated cathode by thermionic emission and are accelerated toward the anode (the target) by a large potential difference $V$. The bulb is evacuated (residual pressure $10^{-7}$ atm or less) so that the electrons can travel from cathode to anode without colliding with air molecules. When $V$ is a few thousand volts or more, a very penetrating radiation is emitted from the anode surface. A typical x-ray machine is shown in Fig. 40–23b.

## X-Ray Photons

Because of their origin, it is clear that x rays are electromagnetic waves; like light, they are governed by quantum relations in their interaction with matter. Thus we can talk about x-ray photons or quanta, and the energy of an x-ray photon is related to its frequency and wavelength in the same way as for photons of light, $E = hf = hc/\lambda$. Typical x-ray wavelengths are 0.001 to 1 nm ($10^{-12}$ to $10^{-9}$ m). X-ray wavelengths can be measured quite precisely by crystal diffraction techniques, which we discussed in Section 38–6.

X-ray emission is the inverse of the photoelectric effect. In photoelectric emission the energy of a photon is transformed into kinetic energy of an electron; in x-ray production the kinetic energy of an electron is transformed into energy of a photon. The energy relation is exactly the same in both cases. For x-ray production we usually neglect the work function of the target because it is ordinarily very small in comparison to the other energies.

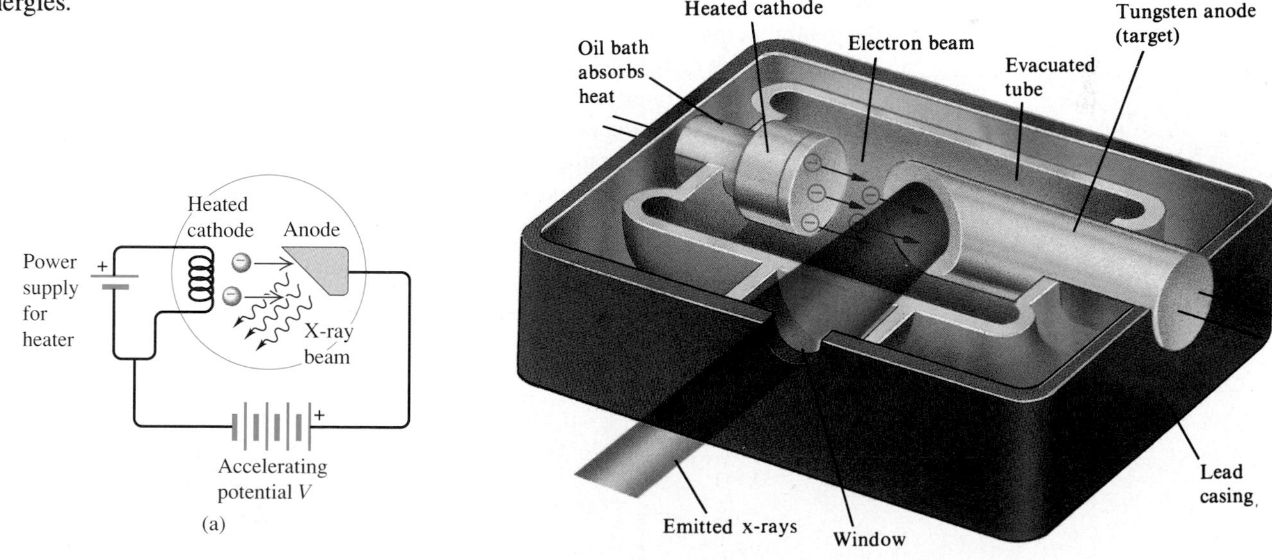


**40–23** (a) Apparatus used to produce x rays, similar to Röntgen's apparatus. Electrons are emitted thermionically from the heated cathode and are accelerated toward the anode; when they strike it, x rays are produced. (b) Cross section of an x-ray machine. The anode is liquid-cooled to prevent overheating.

Two distinct processes are involved in x-ray emission. Some electrons are slowed down or stopped by the target, and part or all of their kinetic energy is converted directly to a continuous spectrum of photons, including x rays. This process is called *bremsstrahlung* (German for "braking radiation"). Other electrons transfer their energy partly or completely to individual atoms within the target. These atoms are left in excited states; when they decay back to the ground state, they emit x-ray photons.

Each element has a set of atomic energy levels associated with x-ray photons and therefore also has a characteristic x-ray spectrum. The energy levels, called *x-ray energy levels,* are rather different in character from those associated with visible spectra. They correspond to vacancies in the inner electron configurations of complex atoms. The energy levels can be hundreds or thousands of electronvolts above the ground state, rather than a few electronvolts, as is typical for optical spectra. We will return to x-ray energy levels and spectra in Section 43–6.

## EXAMPLE 40–8

Electrons in an x-ray tube are accelerated by a potential difference of 10.0 kV. If an electron produces a photon on impact with the target, what is the minimum wavelength of the resulting x rays?

**SOLUTION** The *maximum* photon energy $hf_{max} = hc/\lambda_{min}$ occurs when *all* the initial kinetic energy $eV$ of the electron is used to produce a single photon:

$$eV = hf = \frac{hc}{\lambda}.$$

The corresponding *minimum* wavelength is

$$\lambda = \frac{hc}{eV} = \frac{(6.626 \times 10^{-34}\ \mathrm{J \cdot s})(3.00 \times 10^{8}\ \mathrm{m/s})}{(1.602 \times 10^{-19}\ \mathrm{C})(10.0 \times 10^{3}\ \mathrm{V})}$$
$$= 1.24 \times 10^{-10}\ \mathrm{m} = 0.124\ \mathrm{nm} = 124\ \mathrm{pm}.$$

We can measure the x-ray wavelength by crystal diffraction and confirm this prediction directly. The wavelength is smaller than that for typical visible-light photons by a factor of the order of 5000, and the photon energy is greater by the same factor.

## Compton Scattering

A phenomenon called **Compton scattering,** first observed in 1924 by A. H. Compton, provides additional direct confirmation of the quantum nature of x rays. When x rays strike matter, some of the radiation is *scattered,* just as visible light falling on a rough surface undergoes diffuse reflection. Compton discovered that some of the scattered radiation has smaller frequency (longer wavelength) than the incident radiation and that the change in wavelength depends on the angle through which the radiation is scattered. Specifically, if the scattered radiation emerges at an angle $\phi$ with respect to the incident direction (Fig. 40–24) and if $\lambda$ and $\lambda'$ are the wavelengths of the incident and scattered radiation, respectively, we find that

$$\lambda' - \lambda = \frac{h}{mc}(1 - \cos\phi), \qquad (40\text{–}21)$$

where $m$ is the electron mass.

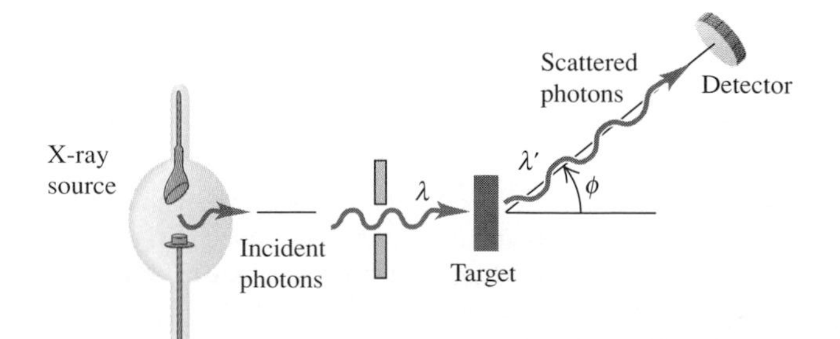


**40–24** A Compton-effect experiment.

Compton scattering cannot be understood on the basis of classical electromagnetic theory, which predicts that the scattered wave has the same wavelength as the incident wave. In contrast, the quantum theory provides a beautifully clear explanation. We imagine the scattering process as a collision of two *particles,* the incident photon and an electron that is initially at rest, as in Fig. 40–25a. The photon gives some of its energy and momentum to the electron, which recoils as a result of this impact. The final scattered photon has less energy, smaller frequency, and longer wavelength than the initial one (Fig. 40–25b).

We can derive Eq. (40–21) from the principles of conservation of energy and momentum. We outline the derivation below; we invite you to fill in the details. The electron energy may be in the relativistic range, so we have to use the relativistic energy-momentum relations, Eqs. (39–41) and (39–42). The initial photon has momentum with magnitude $p$ and energy $pc$; the final photon has momentum with magnitude $p'$ and energy $p'c$. The electron is initially at rest, so its initial momentum is zero and its initial energy is $mc^2$. The final electron momentum has magnitude $P$, and the final electron energy $E$ is given by $E^2 = (mc^2)^2 + (Pc)^2$. Then energy conservation gives us the relation

$$pc + mc^2 = p'c + E,$$

or

$$(pc - p'c + mc^2)^2 = E^2 = (mc^2)^2 + (Pc)^2. \tag{40–22}$$

We may eliminate the electron momentum $\boldsymbol{P}$ from this equation by using momentum conservation:

$$\boldsymbol{p} = \boldsymbol{p}' + \boldsymbol{P},$$

or

$$\boldsymbol{p} - \boldsymbol{p}' = \boldsymbol{P}. \tag{40–23}$$

By using the law of cosines with the vector diagram in Fig. 40–25c, we find

$$P^2 = p^2 + p'^2 - 2pp' \cos \phi. \tag{40–24}$$

We now substitute this expression for $P^2$ into Eq. (40–22) and multiply out the left side. We divide out a common factor $c^2$; several terms cancel, and when the resulting equation is divided through by $pp'$, the result is

$$\frac{mc}{p'} - \frac{mc}{p} = 1 - \cos \phi. \tag{40–25}$$

Finally, we substitute $p = h/\lambda$ and $p' = h/\lambda'$ and rearrange again to obtain Eq. (40–21).

The quantity $h/mc$ that appears in Eq. (40–21) has units of length, and it is called the *Compton wavelength* $\lambda_c$ for the electron. Its numerical value is

$$\lambda_c = \frac{h}{mc} = \frac{6.626 \times 10^{-34}\ \text{J} \cdot \text{s}}{(9.109 \times 10^{-31}\ \text{kg})(2.998 \times 10^8\ \text{m/s})}$$
$$= 2.426 \times 10^{-12}\ \text{m} = 2.426\ \text{pm}.$$

**40–25** Schematic diagram of Compton scattering showing (a) an electron initially at rest with an incident photon of wavelength $\lambda$ and momentum $\boldsymbol{p}$; (b) a scattered photon with longer wavelength $\lambda'$ and momentum $\boldsymbol{p}'$ and recoiling electron with momentum $\boldsymbol{P}$. The direction of the scattered photon makes an angle $\phi$ with that of the incident photon, and the angle between $\boldsymbol{p}$ and $\boldsymbol{p}'$ is also $\phi$. (c) Vector diagram for the conservation of momentum relation $\boldsymbol{p} = \boldsymbol{p}' + \boldsymbol{P}$. Applying the law of cosines to this triangle gives Eq. (40–24).

## EXAMPLE 40–9

For the x-ray photons in Example 40–8 ($\lambda = 0.124$ nm), at what angle do the Compton-scattered x rays have a wavelength that is 1.0% longer than that of the incident x rays? At what angle is it 0.05% longer?

**SOLUTION** In Eq. (40–21) we want $\Delta\lambda = \lambda' - \lambda$ to be 1.0% of 0.124 nm, or 1.24 pm. Using the value $h/mc = 2.426$ pm, we find

$$\Delta\lambda = \frac{h}{mc}(1 - \cos \phi),$$

$$1.24\ \text{pm} = (2.426\ \text{pm})(1 - \cos \phi),$$

$$\phi = 60.7°.$$

For $\Delta\lambda = 0.05\%$,

$$0.62 \text{ pm} = (2.426 \text{ pm})(1 - \cos\phi),$$

$$\phi = 41.9°.$$

Smaller angles give a smaller wavelength shift; in a grazing collision, the photon loses less energy than when the scattering angle is larger.

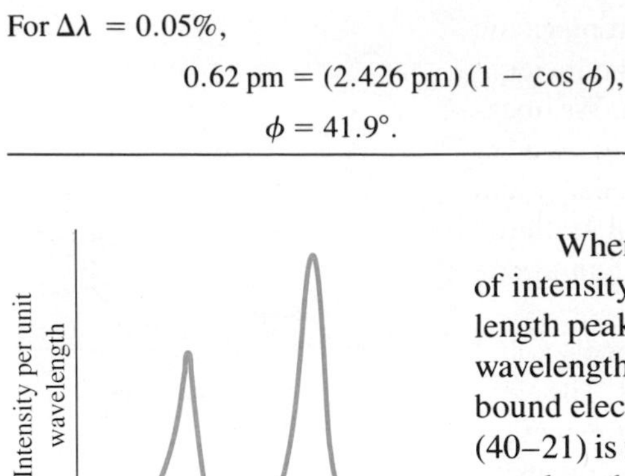


**40–26** Intensity as a function of wavelength for photons scattered at an angle of 135° in the Compton effect.

When the wavelengths of x rays scattered at a certain angle are measured, the curve of intensity as a function of wavelength has two peaks (Fig. 40–26). The longer-wavelength peak represents Compton scattering. The shorter-wavelength peak is at the same wavelength as that for the incident x rays; it corresponds to x-ray scattering from tightly bound electrons. In such scattering processes, the entire atom must recoil; the $m$ in Eq. (40–21) is then the mass of the entire atom rather than of a single electron. The resulting wavelength shifts are too small to measure.

X rays have many practical applications in medicine and industry. Because they can penetrate several centimeters of solid matter, they can be used to visualize the interiors of materials that are opaque to ordinary light, such as broken bones or defects in structural steel. The object to be visualized is placed between an x-ray source and a large sheet of photographic film; the darkening of the film is proportional to the radiation exposure. A crack or air bubble allows greater transmission and shows as a dark area. Bones appear lighter than the surrounding flesh because they contain greater proportions of elements with high atomic number (and greater absorption); in flesh, the light elements carbon, hydrogen, and oxygen predominate.

In the past decade, several vastly improved x-ray techniques have been developed. One widely used system is *computerized axial tomography;* the corresponding instrument is called a CAT scanner. The x-ray source produces a thin, fan-shaped beam that is detected on the opposite side of the subject by an array of several hundred detectors in a line. Each detector measures absorption along a thin line through the subject. The entire apparatus is rotated around the subject in the plane of the beam during a few seconds. The changing photon-counting rates of the detectors are recorded digitally; a computer processes this information and reconstructs a picture of density over an entire cross section of the subject. Density differences as small as 1% can be detected with CAT scans, and tumors and other anomalies much too small to be seen with older x-ray techniques can be detected.

X rays cause damage to living tissues. As x-ray photons are absorbed in tissues, they break molecular bonds and create highly reactive free radicals (such as neutral H and OH), which in turn can disturb the molecular structure of proteins and especially genetic material. Young and rapidly growing cells are particularly susceptible; thus x rays are useful for selective destruction of cancer cells. Conversely, however, a cell may be damaged by radiation but survive, continue dividing, and produce generations of defective cells; thus x rays can *cause* cancer.

Even when the organism itself shows no apparent damage, excessive radiation exposure can cause changes in the organism's reproductive system that will affect its offspring. The use of x rays in medical diagnosis has become an area of great concern in recent years; a careful assessment of the balance between risks and benefits of radiation exposure is essential in each individual case.

## 40–8 CONTINUOUS SPECTRA

Line spectra are emitted by matter in the gaseous state, in which the atoms are so far apart that interactions between them are negligible and each atom behaves as an isolated system. Hot matter in condensed states (solid or liquid) nearly always emits radiation with a *continuous* distribution of wavelengths rather than a line spectrum. Continuous-spectrum radiation emitted by an ideal surface that is completely nonreflecting is called **blackbody radiation.** By 1900 this radiation had been studied extensively, and several characteristics had been established.

First, the total rate of radiation of energy, per unit surface area, is proportional to the fourth power of the absolute temperature. We studied this relationship in Section 15–7 during our study of heat transfer mechanisms. The total intensity $I$ (power per unit area) emitted from the surface of an ideal radiator at absolute temperature $T$ is given by the **Stefan-Boltzmann law:**

$$I = \sigma T^4, \qquad (40\text{–}26)$$

where $\sigma$ is a fundamental physical constant called the *Stefan-Boltzmann constant.* In SI units,

$$\sigma = 5.6705 \times 10^{-8}\ \text{W/m}^2 \cdot \text{K}^4. \qquad (40\text{–}27)$$

Second, the power is not uniformly distributed over all wavelengths. Its distribution can be measured and described by a function $I(\lambda)$ with the property that $I(\lambda)\,d\lambda$ is the intensity corresponding to wavelengths in the interval $d\lambda$. The *total* intensity $I$, given by Eq. (40–26), is the *integral* of $I(\lambda)$ over all wavelengths, equal to the area under the $I(\lambda)$ curve:

$$I = \int_0^\infty I(\lambda)\,d\lambda. \qquad (40\text{–}28)$$

The function $I(\lambda)$ is also called the *spectral emittance.*

Measured distribution functions $I(\lambda)$ for three different temperatures are shown in Fig. 40–27. Each has a peak wavelength $\lambda_m$, where the emitted power is most concentrated. Experiment shows that $\lambda_m$ is inversely proportional to $T$. This relationship can be represented as

$$\lambda_m T = \text{constant} = 2.90 \times 10^{-3}\ \text{m} \cdot \text{K}. \qquad (40\text{–}29)$$

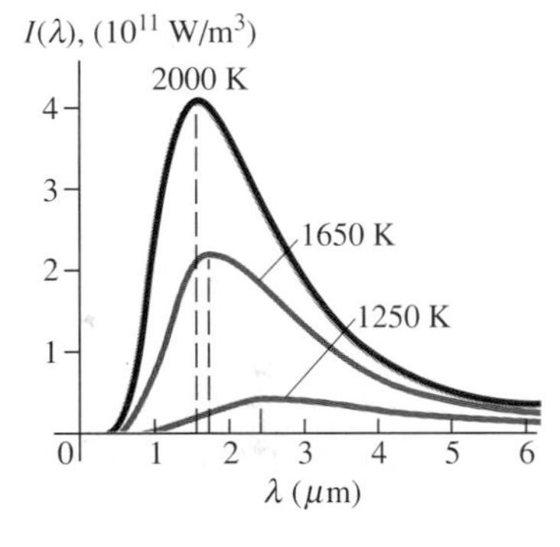


**40–27** Intensity distribution function $I(\lambda)$ for continuous-spectrum radiation from an ideal radiator. The power per unit area in a wavelength interval $d\lambda$ at wavelength $\lambda$ is $I(\lambda)\,d\lambda$. The total area under the curve at any temperature represents the total radiated power per unit area. Curves are plotted for three different temperatures, with darker color corresponding to higher temperature. The vertical lines show the value of $\lambda_m$ in Eq. (40–29) for each temperature. As the temperature increases, the peak grows larger and shifts to shorter wavelengths; the total area under the curve is proportional to $T^4$.

This rule is called the **Wien displacement law.** As the temperature rises, the peak of $I(\lambda)$ becomes higher and shifts to shorter wavelengths. This corresponds to the familiar fact that a body that glows yellow is hotter and brighter than one that glows red; yellow light has shorter wavelength than red. Finally, experiments show that the *shape* of the distribution function is the same for all temperatures; we can make a curve for one temperature fit any other temperature by simply changing the scales on the graph.

During the last decade of the nineteenth century, many attempts were made to *derive* these empirical results from basic principles. In one attempt, Rayleigh considered light enclosed in a rectangular box with perfectly reflecting sides. Such a box has a series of possible *normal modes* for the light waves, analogous to normal modes for a string held at both ends (Section 20–3). It seemed reasonable to assume that the distribution of energy among the various modes was given by the equipartition principle (Section 16–4), which had been successfully used in the analysis of heat capacities. A small hole in the box would behave as an ideal blackbody radiator.

Including both the electric- and magnetic-field energies, Rayleigh assumed that the total energy of each normal mode was equal to $kT$. Then by computing the *number* of normal modes corresponding to a wavelength interval $d\lambda$, Rayleigh could predict the distribution of wavelengths in the radiation within the box. Finally, he could compute the intensity distribution $I(\lambda)$ of the radiation emerging from a small hole in the box. His result was quite simple:

$$I(\lambda) = \frac{2\pi ckT}{\lambda^4}. \qquad (40\text{–}30)$$

At large wavelengths this formula agrees quite well with the experimental results shown in Fig. 40–27, but there is serious disagreement at small $\lambda$. The experimental curve falls to zero at small $\lambda$; Rayleigh's curve approaches infinity as $1/\lambda^4$, a result called in Rayleigh's time the "ultraviolet catastrophe." Even worse, the integral of Eq. (40–30) over all $\lambda$ is infinite, indicating an infinitely large *total* radiated intensity. Clearly, something is wrong.

Finally, in 1900, Planck succeeded in deriving a function, now called the **Planck radiation law,** that agreed very well with experimental intensity distribution curves. To do this, he made what seemed at the time to be a crazy assumption; he assumed that in Rayleigh's box a normal mode with frequency $f$ could gain or lose energy only in discrete steps with magnitude $hf$, where $h$ is the same constant that now bears Planck's name. This assumption was in sharp contrast to Rayleigh's point of view, which was that each normal mode could gain or lose energy in *any* increment.

Planck was not comfortable with this *quantum hypothesis;* he regarded it as a calculational trick rather than a fundamental principle. In a letter to a friend he called it "an act of desperation" into which he was forced because "a theoretical explanation had to be found at any cost, whatever the price." But five years later, Einstein extended this concept to explain the photoelectric effect (Section 40–2), and other evidence quickly mounted. By 1915 there was no longer any doubt about the validity of the quantum concept and the existence of photons. By discussing atomic spectra *before* discussing continuous spectra, we have departed from the historical order of things. The credit for inventing the concept of quantization goes to Planck, even though he didn't believe it at first.

We won't go into the details of Planck's derivation of the intensity distribution. Here is his result:

$$I(\lambda) = \frac{2\pi hc^2}{\lambda^5(e^{hc/\lambda kT} - 1)}, \tag{40–31}$$

where $h$ is Planck's constant, $c$ is the speed of light, $k$ is Boltzmann's constant, $T$ is the *absolute* temperature, and $\lambda$ is the wavelength. This function turns out to agree well with experimental intensity curves such as those in Fig. 40–27.

The Planck radiation law also contains the Wien displacement law and the Stefan-Boltzmann law as consequences. To derive the Wien law, we find the position of the maximum of the Planck law; we take the derivative of Eq. (40–31) and set it equal to zero to find the value of $\lambda$ at which $I(\lambda)$ is maximum. We leave the details as a problem; the result is

$$\lambda_{\mathrm{m}} = \frac{hc}{4.965kT}. \tag{40–32}$$

To obtain this result, you have to solve the equation

$$5 - x = 5e^{-x}. \tag{40–33}$$

The root of this equation, found by trial and error or more sophisticated means, is (approximately) 4.965. We invite you to evaluate the constant $hc/4.965k$ and show that it agrees with the empirical value given in Eq. (40–29).

We can obtain the Stefan-Boltzmann law by integrating Eq. (40–31) over all $\lambda$ to find the *total* radiated power per unit area. This calculation involves making a change of variable ($x = hc/\lambda kT$) and looking up an integral in a table. Again we leave the details as a problem; the result is

$$I = \int_0^\infty I(\lambda)\, d\lambda = \frac{2\pi^5 k^4}{15c^2h^3}T^4 = \sigma T^4. \qquad (40\text{–}34)$$

This is the Stefan-Boltzmann law, Eq. (40–26). This result also shows that the constant $\sigma$ in that law can be expressed in terms of a rather unlikely looking combination of other fundamental constants:

$$\sigma = \frac{2\pi^5 k^4}{15c^2h^3}. \qquad (40\text{–}35)$$

We invite you to plug in the values of $k$, $c$, and $h$, and verify that you get the value of $\sigma$ quoted above.

The general form of Eq. (40–32) is what we should expect from kinetic theory. If photon energies are typically of the order of $kT$, as suggested by the equipartition principle, then for a typical photon we would expect

$$E \cong kT \cong \frac{hc}{\lambda}, \qquad \text{and} \qquad \lambda \cong \frac{hc}{kT}. \qquad (40\text{–}36)$$

Indeed, a photon with a wavelength given by Eq. (40–32) has an energy $E = 4.965kT$.

Planck's equation, Eq. (40–31), looks so different from the unsuccessful Rayleigh expression, Eq. (40–30), that it may seem unlikely that they would agree at large values of $\lambda$. But when $\lambda$ is large, the exponent in the denominator of Eq. (40–31) is very small. We can then make a Taylor-series expansion of the exponential function, keeping only the first terms. We invite you to verify that when this is done, the result is Eq. (40–30), showing that the two expressions *do* agree in the limit of very large $\lambda$. We also note that the Rayleigh expression does not contain $h$. At very long wavelengths and correspondingly very small photon energies, quantum effects become unimportant.

## EXAMPLE 40–10

The temperature of the surface of the sun is approximately 6000 K. a) What is the peak-intensity wavelength $\lambda_m$? b) What is the total radiated power per unit area?

**SOLUTION** a) We use the Wien displacement law, Eq. (40–29):

$$\lambda_m = \frac{2.90 \times 10^{-3}\ \text{m}\cdot\text{K}}{T} = \frac{2.90 \times 10^{-3}\ \text{m}\cdot\text{K}}{6000\ \text{K}}$$
$$= 0.483 \times 10^{-6}\ \text{m} = 483\ \text{nm}.$$

This is in the blue-green region of the visible spectrum.

b) We get the total intensity from the Stefan-Boltzmann law, Eq. (40–26):

$$I = \sigma T^4 = (5.67 \times 10^{-8}\ \text{W}/\text{m}^2\cdot\text{K}^4)(6000\ \text{K})^4$$
$$= 7.35 \times 10^7\ \text{W/m}^2.$$

This is an enormous intensity.

## EXAMPLE 40–11

Find the intensity of light emitted from the surface of the sun in the wavelength range 600.0 to 605.0 nm.

**SOLUTION** For an exact result we should integrate Eq. (40–31) between the limits 600.0 and 605.0 nm, finding the area under the $I(\lambda)$ curve between these limits. This integral can't be evaluated in terms of familiar functions, so we *approximate* the area as the width of the interval (5.0 nm) multiplied by the height of the curve at $\lambda = 600$ nm, obtained by substituting this value in Eq. (40–31). First, at $\lambda = 600\ \text{nm} = 6.00 \times 10^{-7}$ m,

$$\frac{hc}{\lambda kT} = \frac{(6.626 \times 10^{-34}\ \text{J}\cdot\text{s})(3.00 \times 10^8\ \text{m/s})}{(6.00 \times 10^{-7}\ \text{m})(1.38 \times 10^{-23}\ \text{J/K})(6000\ \text{K})}$$
$$= 4.00.$$

$$I(\lambda) = \frac{2\pi(6.626 \times 10^{-34}\ \text{J}\cdot\text{s})(3.00 \times 10^8\ \text{m/s})^2}{(6.00 \times 10^{-7}\ \text{m})^5(e^{4.00} - 1)}$$
$$= 8.98 \times 10^{13}\ \text{W/m}^3.$$

The intensity in the 5.0-nm range from 600 to 605 nm is then

$$I(\lambda)\, d\lambda = (8.9 \times 10^{13}\ \text{W/m}^3)(5.0 \times 10^{-9}\ \text{m})$$
$$= 4.5 \times 10^5\ \text{W/m}^2.$$

## 40–9 WAVE-PARTICLE DUALITY

We have studied many examples of the behavior of light and other electromagnetic radiation. Some, including the interference and diffraction effects described in Chapters 37 and 38, demonstrate conclusively the *wave* nature of light. Others, the subject of the present chapter, point with equal force to the *particle* nature of light. At first glance, these two aspects seem to be in direct conflict. How can light be a wave and a particle at the same time?

The answer to this and similar questions will eventually take us to the heart of the conceptual basis of quantum mechanics. Let's start by considering again the diffraction pattern for a single slit, which we analyzed in Sections 38–2 and 38–3. Instead of recording the pattern on photographic film, we use a detector called a *photomultiplier* that can actually detect individual photons. Using the setup shown in Fig. 40–28, we place the photomultiplier at various positions for equal time intervals, count photons at each position, and plot out the intensity distribution.

We find that on the average the distribution of photons agrees with our predictions from Section 38–3. At points corresponding to the maxima of the pattern, we count a lot of photons; at minimum points, we count almost none; and so on. The graph of the counts at various points gives us the same diffraction pattern that we predicted with Eq. (38–7).

But suppose we now reduce the intensity to the point at which only a few photons per second pass through the slit. With only a few photons we can't expect to get the smooth diffraction curve that we found with very large numbers. In fact, there is no way to predict where any individual photon will go. To reconcile the wave and particle aspects of this pattern, we have to regard the pattern as a *statistical* distribution that tells us how many photons, on average, go various places, or the *probability* for an individual photon to land in each of several places. But we can't predict where an individual photon will go.

Now let's take a brief look at a quantum interpretation of a *two-slit* optical interference pattern, which we studied in Section 37–3. We can again trace out the pattern using a photomultiplier and a counter. We reduce the light intensity to a level of a few photons per second. Again we can't predict where an individual photon will go; the interference pattern is a *statistical distribution*.

It is tempting to think that each individual photon must pass through one slit or the other. We can test this viewpoint. Suppose we try to record the interference pattern on film by opening one slit for a while, then closing it and opening the other. This doesn't work; to form an interference pattern, the waves from the two slits have to be *coherent*. To reconcile this result with the photon picture, we have to assume that *every* photon goes

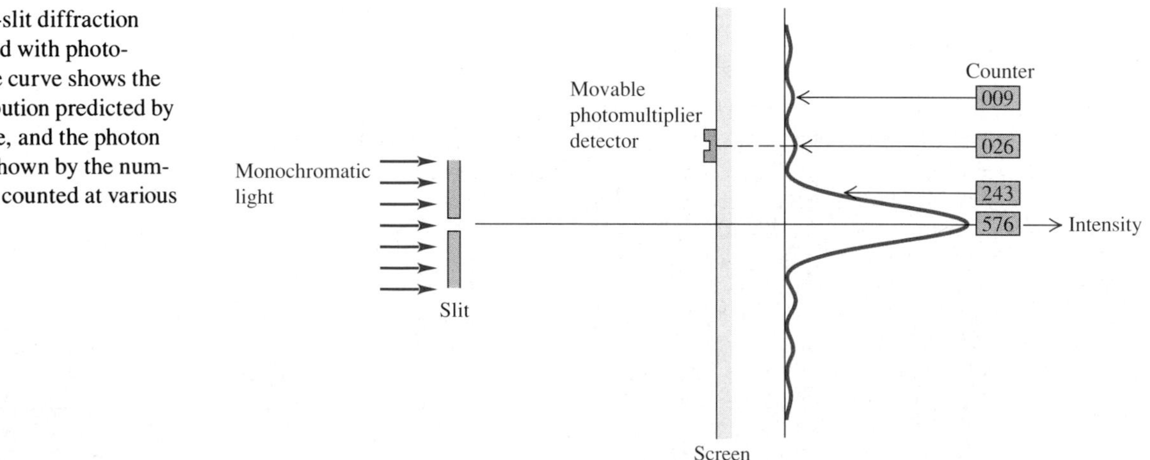


**40–28** Single-slit diffraction pattern observed with photomultipliers. The curve shows the intensity distribution predicted by the wave picture, and the photon distribution is shown by the numbers of photons counted at various positions.

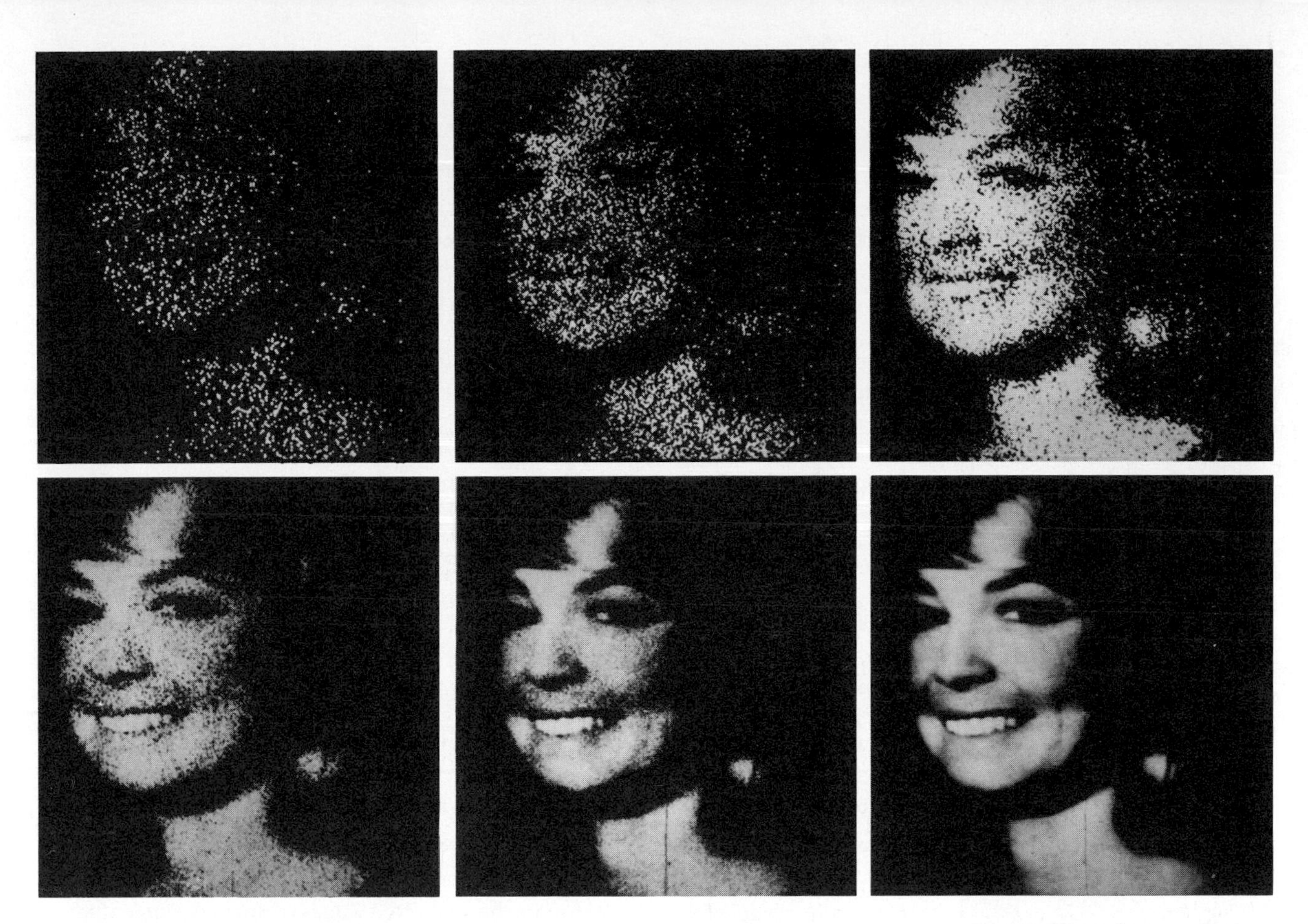

**40–29** These photographs are images made by small numbers of photons using electronic image amplification. In the upper pictures, each spot corresponds to one photon; under extremely faint light (few photons) the pattern seems almost random, but it emerges more distinctly as the light level increases (lower pictures).

through *both* slits and that *each photon interferes only with itself.* If this sounds like nonsense, remember that the photon is *not* localized at a particular point in space. It carries a definite amount of energy, but it still has wave properties that require extension in space. Most important, the photon is a conceptual framework to describe the quantization of energy in electromagnetic waves. There is nothing conceptually wrong with saying that every photon passes through both slits; indeed, this is the *only* consistent viewpoint.

Figure 40–29 shows a photograph reproduced by using increasing numbers of photons. The pattern emerges more clearly with increasing numbers of photons, just as the interference patterns discussed above are formed by large numbers of photons.

So is a photon a particle or a wave? There is no answer to that question. Electromagnetic radiation has *both* particle and wave aspects. Our goal must be to construct a theory that includes both these types of behavior. Such a comprehensive theory has been developed only in about the past 40 years, in the branch of physics called *quantum electrodynamics.* In this theory, the concept of energy levels of an atomic system is extended to electromagnetic fields. Just as an atom exists only in certain definite energy states, so the electromagnetic field has certain well-defined energy states, corresponding to the presence of various numbers of photons with various energies, momenta, and polarizations. This subject came to full flower 50 years after the conceptual birth of quantum mechanics with Planck's quantum hypothesis of 1900.

In the following chapters we will find that *particles,* such as electrons, also have a dual wave-particle personality. One of the great achievements of quantum mechanics has been to reconcile these apparently incompatible aspects of the behavior of photons, electrons, and other constituents of matter.

## SUMMARY

• The energy in an electromagnetic wave is carried in units called photons or quanta. The energy $E$ of one photon, for a wave with frequency $f$, is

$$E = hf = \frac{hc}{\lambda}. \tag{40–2}$$

• In the photoelectric effect, the stopping potential $V_0$ for electrons emitted by light with frequency $f$ is given by

$$eV_0 = hf - \phi. \tag{40–4}$$

• When an atom makes a transition from an energy state $E_i$ to a state $E_f$ by emitting a photon, the frequency $f$ of the photon is given by

$$hf = E_i - E_f. \tag{40–7}$$

The energy levels of the hydrogen atom are

$$E_n = -\frac{hcR}{n^2} = -\frac{13.60\text{ eV}}{n^2}, \quad n = 1, 2, 3, 4, \ldots, \tag{40–10}$$

where $R$ is a constant called the Rydberg constant. All the observed spectral series of hydrogen can be understood in terms of these levels.

• The Rutherford scattering experiments show that at the center of an atom is a dense nucleus, much smaller than the overall size of the atom but containing all of the positive charge and most of the mass.

• In the Bohr model of the hydrogen atom the permitted values of angular momentum are

$$mv_n r_n = n\frac{h}{2\pi}, \tag{40–11}$$

where $n = 1, 2, 3, \ldots$, and $n$ is called the quantum number for the state. The corresponding orbit radii $r_n$ and speeds $v_n$ are

$$r_n = \epsilon_0 \frac{n^2h^2}{\pi me^2}, \tag{40–13}$$

$$v_n = \frac{1}{\epsilon_0}\frac{e^2}{2nh}. \tag{40–14}$$

The energy levels are

$$E_n = K_n + U_n = -\frac{1}{\epsilon_0^2}\frac{me^4}{8n^2h^2} = -\frac{13.60\text{ eV}}{n^2} \tag{40–17}$$

• The laser operates on the principle of stimulated emission, in which many photons with identical wavelength and phase are emitted. Laser operation requires the creation of a non-equilibrium condition called a population inversion, in which there are more atoms in some higher-energy states than at thermal equilibrium.

• When x rays are emitted by electron impact on a target, the maximum frequency for electrons accelerated through a potential difference $V$ is given by

$$eV = hf_{\text{max}} = \frac{hc}{\lambda_{\text{min}}}.$$

### KEY TERMS

line spectrum
photoelectric effect
work function
threshold frequency
stopping potential
quantum
photon
Planck's constant
energy level
ground state
excited state
absorption spectrum
Rutherford scattering experiments
nucleus
Bohr model
quantum number
reduced mass
laser
stimulated emission
population inversion
Compton scattering
blackbody radiation
Stefan-Boltzmann law
Wien displacement law
Planck radiation law

Compton scattering is scattering of x-ray photons by electons. The wavelengths $\lambda$ and $\lambda'$ of incident and scattered photons are related to the scattering angle $\phi$ by

$$\lambda' - \lambda = \frac{h}{mc}(1 - \cos\phi). \tag{40–21}$$

• The Stefan-Boltzmann law states that the total radiated intensity $I$ from a blackbody surface is related to the absolute temperature $T$ by

$$I = \sigma T^4, \tag{40–26}$$

where $\sigma$ is a constant called the Stefan-Boltzmann constant. The wavelength $\lambda_m$ having maximum intensity in blackbody radiation is related to the absolute temperature $T$ by

$$\lambda_m T = \text{constant} = 2.90 \times 10^{-3}\,\text{m} \cdot \text{K}. \tag{40–29}$$

The Planck radiation gives the intensity distribution $I(\lambda)$ in blackbody radiation:

$$I(\lambda) = \frac{2\pi hc^2}{\lambda^5(e^{hc/\lambda kT} - 1)}. \tag{40–31}$$

• Electromagnetic radiation behaves as both waves and particles, and a comprehensive theory must include both these aspects of its behavior.

# EXERCISES

## Section 40–2 The Photoelectric Effect

**40–1** A nucleus in a transition from an excited state emits a gamma-ray photon with an energy of 2.50 MeV. a) What is the photon frequency? b) What is the photon wavelength? c) How does the wavelength compare with typical nuclear radii (of the order of $10^{-15}$ m)?

**40–2** A sodium-vapor lamp emits light with wavelength 589 nm. If the total power of the emitted light is 40.0 W, how many photons are emitted per second?

**40–3** A photon of green light has a wavelength of 510 nm. Find the photon's frequency, magnitude of momentum, and energy; express the energy in both joules and electronvolts.

**40–4 Medical Laser Photons.** A laser used to weld detached retinas emits light with a wavelength of 633 nm in pulses that are 20.0 ms in duration. The average power during each pulse is 0.600 W. a) How much energy is in each pulse in joules? In electronvolts? b) What is the energy of one photon in joules? In electronvolts? c) How many photons are in each pulse?

**40–5** A photon has momentum of magnitude $2.40 \times 10^{-27}$ kg · m/s. a) What is the energy of this photon? Give your answer in joules and in electronvolts. b) What is the wavelength of this photon? In what region of the electromagnetic spectrum does it lie?

**40–6** In the photoelectric effect, what is the relation between the threshold frequency $f_0$ and the work function $\phi$?

**40–7** A metal surface has a work function of 4.00 eV. What is the maximum speed of the photoelectrons emitted from the surface when it is exposed to light with a frequency of $3.40 \times 10^{15}$ Hz?

**40–8** The photoelectric threshold wavelength of tungsten is 272 nm. Calculate the maximum kinetic energy of the electrons ejected from a tungsten surface by ultraviolet radiation of wavelength 160 nm. (Express the answer in electronvolts.)

**40–9** When ultraviolet light with a wavelength of 254 nm from a mercury arc falls on a clean copper surface, the stopping potential necessary to stop emission of photoelectrons is 0.181 V. What is the photoelectric threshold wavelength for copper?

**40–10** The photoelectric work function of potassium is 2.30 eV. If light having a wavelength of 320 nm falls on potassium, find a) the stopping potential in volts; b) the kinetic energy in electronvolts of the most energetic electrons ejected; c) the speeds of these electrons.

## Section 40–3 Line Spectra and Energy Levels

**40–11** Calculate a) the frequency and b) the wavelength of the $H_\beta$ line of the Balmer series for hydrogen. This line is emitted in the transition from the $n = 4$ to $n = 2$ level.

**40–12** Find the longest and shortest wavelengths in the Lyman and Paschen series for hydrogen. In what region of the electromagnetic spectrum does each series lie?

**40–13** The silicon-silicon single bond that forms the basis of a (mythical) silicon-based creature, the Horta, has a bond strength of 3.60 eV. What wavelength photon would you need in a (mythical) phasor distintegration gun to destroy the Horta?

**40–14** The energy-level scheme for the hypothetical one-electron element Searsium is shown in Fig. 40–30. The potential energy of an electron is taken to be zero at an infinite distance from the nucleus. a) How much energy (in electronvolts) does it take to ionize an electron from the ground state? b) A 15-eV photon is absorbed by a Searsium atom. When the atom returns to its ground state, what possible energies can the emitted photons have? c) What will happen if a photon with an energy of 8 eV strikes a Searsium atom? Why? d) If photons emitted from Searsium transitions $n = 4 \rightarrow n = 2$ and $n = 2 \rightarrow n = 1$ will eject photoelectrons from an unknown metal, but the photon emitted from the transition $n = 3 \rightarrow n = 2$ will not, what are the limits (maximum and minimum possible values) of the work function of the metal?

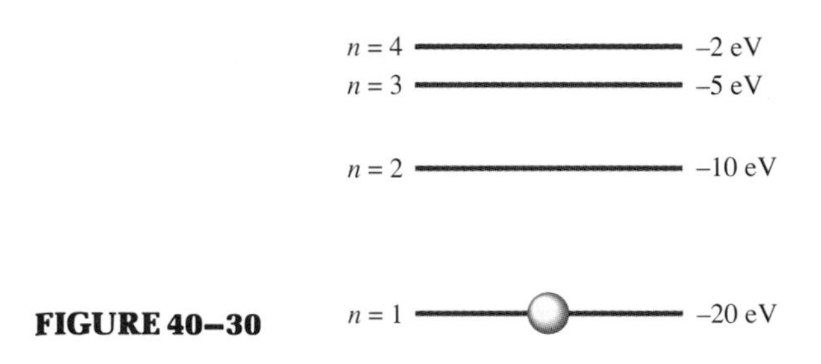


**FIGURE 40–30**

## Section 40–4 The Nuclear Atom

**40–15** A beam of alpha particles is incident on a gold nucleus. A particular alpha particle comes in "head-on" and stops $8.00 \times 10^{-14}$ m away from the center of the gold nucleus. Assume that the gold nucleus remains at rest. The mass of the alpha particle is $6.65 \times 10^{-27}$ kg. a) Calculate the electrostatic potential energy of the alpha particle when it has stopped. Express your result in joules and in MeV. b) What initial kinetic energy did the alpha particle have? Express in joules and in MeV. c) What was the initial speed of the alpha particle?

**40–16** A 4.50-MeV alpha particle from a $^{226}$Ra decay makes a head-on collision with a gold nucleus. a) What is the distance of closest approach of the alpha particle to the center of the nucleus? Assume that the gold nucleus remains at rest. b) What is the force on the alpha particle at the instant when it is at the distance of closest approach?

## Section 40–5 The Bohr Model

**40–17** a) Calculate the Bohr-model speed of the electron in a hydrogen atom in the $n = 1$, 2, and 3 levels. b) Calculate the orbital period in each of these levels. c) The average lifetime of the first excited level of a hydrogen atom is $1.0 \times 10^{-8}$ s. How many orbits does an electron in an excited atom complete before returning to the ground level?

**40–18 Rydberg Energy.** According to the Bohr model, the Rydberg constant $R$ is equal to $me^4/8\epsilon_0^2h^3c$. a) Calculate $R$ in $m^{-1}$ and compare with the experimental value. b) Calculate the energy (in electronvolts) of a photon whose wavelength equals $R^{-1}$. (This quantity is known as the Rydberg energy.)

**40–19** A singly ionized helium ion (a helium atom with one electron removed) behaves very much like a hydrogen atom, except that the nuclear charge is twice as great. a) How do the energy levels differ in magnitude from those of the hydrogen atom? b) Which spectral series for $He^+$ have lines in the visible spectrum? (Refer to Exercise 40–12.) c) For a given value of $n$, how does the radius of an orbit in $He^+$ relate to that for H?

**40–20** A hydrogen atom that is initially in the ground level absorbs a photon, which excites it to the $n = 4$ level. Determine the wavelength and frequency of the photon.

## Section 40–6 The Laser

**40–21** How many photons per second are emitted by a 1.50-mW helium-neon laser that has a wavelength of 633 nm?

**40–22** In Fig. 40–21, compute the energy difference for the 5$s$ to 3$p$ transition in neon; express your result in electronvolts and in joules. Compute the wavelength of a photon having this energy, and compare your result with the observed wavelength of the laser light.

**40–23** A large number of neon atoms are in thermal equilibrium at a temperature of 300 K. What is the ratio of the number of atoms in a 5$s$ state to the number in a 3$p$ state? The energies of these states are given in Fig. 40–21.

**40–24** A large number of hydrogen atoms are in thermal equilibrium at a temperature $T$. What must $T$ be for one atom to be in an $n = 2$ excited state for every 100 atoms in an $n = 1$ ground state?

## Section 40–7 X-Ray Production and Scattering

**40–25** The cathode-ray tubes that generated the picture in early color television sets were sources of x rays. If the acceleration voltage in a TV tube is 20.0 kV, what are the shortest-wavelength x rays produced? (Modern TV tubes contain shielding to stop these x rays.)

**40–26** a) What is the minimum potential difference between the filament and the target of an x-ray tube if the tube is to produce x rays with a wavelength of 0.150 nm? b) What is the shortest wavelength produced in an x-ray tube operated at $6.00 \times 10^4$ V?

**40–27** Complete the derivation of the Compton-scattering formula, Eq. (40–21), following the outline given in Eqs. (40–22) through (40–25).

**40–28** X rays are produced in a tube operating at 20.0 kV. After emerging from the tube, x rays with the minimum wavelength produced strike a target and are Compton-scattered through an angle of 60.0°. a) What is the original x-ray wavelength? b) What is the wavelength of the scattered x rays? c) What is the energy of the scattered x rays (in electronvolts)?

**40–29** X rays with initial wavelength 0.0600 nm undergo Compton scattering. What is the largest wavelength found in the scattered x rays?

**40–30** A beam of x rays with wavelength 50.0 pm are Compton-scattered by the electrons in a sample. At what angle from the incident beam should you look to find x rays with a wavelength of a) 53.1 pm? b) 50.0 pm?

### Section 40–8 Continuous Spectra

**40–31** Show that for large wavelength $\lambda$ the Planck distribution, Eq. (40–31), agrees with the Rayleigh distribution, Eq. (40–30).

**40–32** Determine $\lambda_m$, the wavelength at the peak of the Planck distribution, and the corresponding frequency $f$, at the following Kelvin temperatures: a) 4.00 K; b) 400 K; c) 4000 K.

**40–33** a) Show that the maximum in the Planck distribution, Eq. (40–31), occurs at a wavelength $\lambda_m$ given by $\lambda_m = hc/4.965kT$ (Eq. 40–32). As discussed in the text, 4.965 is the root of Eq. (40–33). b) Evaluate the constants in the expression derived in part (a) to show that $\lambda_m T$ has the numerical value given in the Wien displacement law, Eq. (40–29).

**40–34** The wavelength at the center of the visible region of the electromagnetic spectrum is 550 nm. What is the temperature of an ideal radiator whose radiation output peaks at this wavelength?

**40–35** Radiation has been detected from space that is characteristic of an ideal radiator at $T = 2.7$ K. (This radiation is usually interpreted as the result of the "big bang" beginning of the universe.) For this temperature, at what wavelength does the Planck distribution peak?

**40–36** An ideal radiator radiates with a total intensity of $I = 8.72$ kW/m$^2$. At what wavelength does the intensity distribution $I(\lambda)$ peak?

## PROBLEMS

**40–37** The directions of emission of photons from a source of radiation are random. According to the wave theory, the intensity of radiation from a point source varies inversely as the square of the distance from the source. Show that the number of photons from a point source passing out through a unit area is also given by an inverse-square law.

**40–38** a) If the average wavelength emitted by a 150-W light bulb is 600 nm, and 10% of the input power is emitted as visible light, approximately how many visible light photons are emitted per second? b) At what distance would this correspond to $1.00 \times 10^{12}$ photons per square centimeter per second if the light is emitted uniformly in all directions?

**40–39 Exposing Photographic Film.** The light-sensitive compound on most photographic films is silver bromide, AgBr. A film is "exposed" when the light energy absorbed dissociates this molecule into its atoms. (The actual process is more complex, but the quantitative result does not differ greatly.) The energy of dissociation of AgBr is $1.00 \times 10^5$ J/mol. For a photon that is just able to dissociate a molecule of silver bromide, find a) the photon energy in electronvolts, b) the wavelength of the photon, c) the frequency of the photon. d) What is the energy in electronvolts of a quantum of radiation having a frequency of 100 MHz? e) Explain the fact that light from a firefly can expose a photographic film, whereas the radiation from an FM station transmitting 50,000 W at 100 MHz cannot.

**40–40** From the kinetic-molecular theory of an ideal gas (Chapter 16), we know that the average translational kinetic energy of an atom is $\frac{3}{2}kT$. What is the wavelength of a photon that has this energy for $T = 300$ K (room temperature)?

**40–41** What will be the change in the stopping potential for photoelectrons emitted from a surface if the wavelength of the incident light is reduced from 400 nm to 320 nm?

**40–42** The photoelectric work functions for particular samples of certain metals are as follows: cesium, 2.1 eV; copper, 4.7 eV; potassium, 2.3 eV; and zinc, 4.3 eV. a) What is the threshold wavelength for each metal? b) Which of these metals *could not* emit photoelectrons when irradiated with visible light?

**40–43** When a certain photoelectric surface is illuminated with light of different wavelengths, the following stopping potentials are observed.

| Wavelength (nm) | Stopping potential (V) |
|---|---|
| 366 | 1.48 |
| 405 | 1.15 |
| 436 | 0.93 |
| 492 | 0.62 |
| 546 | 0.36 |
| 579 | 0.24 |

Plot the stopping potential as ordinate against the frequency of the light as abscissa. Determine a) the threshold frequency; b) the threshold wavelength; c) the photoelectric work function of the material (in electronvolts); d) the value of Planck's constant $h$ (assuming that the value of $e$ is known).

**40–44** An unknown element is found to have an absorption spectrum with lines at 3.0, 7.0, and 9.0 eV, and its ionization potential is 10.0 eV. a) Draw an energy-level diagram for this element. b) If a 9.0-eV photon is absorbed, what energies can the subsequently emitted photons have?

**40–45** A sample of hydrogen atoms is irradiated with light with a wavelength of 40.0 nm, and electrons are observed leaving the gas. If the hydrogen atoms are initially in a ground state, what is the maximum kinetic energy in electronvolts of these photoelectrons?

**40–46** Consider a hydrogenlike atom with nuclear charge $Z$. a) For what value of $Z$ (rounded to the nearest integer value) is the Bohr speed of the electron in the ground state equal to 10.0% of the speed of light? b) For what value of $Z$ (rounded to the nearest integer value) is the ionization energy of a ground state equal to 1.0% of the rest mass energy of the electron?

**40–47 Muonium.** The negative $\mu$-meson (or muon) has a charge equal to that of an electron but a mass 207 times as great. Consider a hydrogenlike atom consisting of a proton and a muon. a) What is the reduced mass of the atom? b) What is the ground-level energy (in electronvolts)? c) What is the separation between the muon and the proton in the $n = 1$ Bohr orbit? d) What is the wavelength of the radiation emitted in the transition from the $n = 2$ level to the $n = 1$ level?

**40–48** The mass of a deuteron is $3.34359 \times 10^{-27}$ kg, the mass of a proton is $1.672623 \times 10^{-27}$ kg, and the mass of an electron is $9.10939 \times 10^{-31}$ kg. a) Calculate the reduced mass of a deuterium atom and express your result in terms of the electron mass $m$. b) Consider the $n = 2$ to $n = 1$ transition in atomic hydrogen. Calculate the difference in the wavelength of light emitted in this transition by a hydrogen atom and by a deuterium atom.

**40–49** a) What is the least amount of energy in electronvolts that must be given to a hydrogen atom that is initially in its ground level so that it can emit the $H_\beta$ line (see Exercise 40–11) in the Balmer series? b) How many different possibilities of spectral line emissions are there for this atom when the electron starts in the $n = 4$ level and eventually ends up in the ground level? Calculate the wavelength of the emitted photon in each case.

**40–50** If hydrogen were monatomic, at what temperature would the average translational kinetic energy be equal to the energy required to raise a hydrogen atom from the ground level to the $n = 2$ excited level?

**40–51** If electrons in a metal had the same energy distribution as molecules in a gas at the same temperature (which is not actually the case), at what temperature would the average electron kinetic energy equal 1.0 eV, which is typical of work functions of metals? Comment on the relevance of your result to thermionic emission.

**40–52 Bohr Orbits of a Satellite.** A 20.0-kg satellite circles the earth once every 2.00 h in an orbit having a radius of 8060 km. a) Assuming that Bohr's angular-momentum postulate ($L = nh/2\pi$) applies to satellites just as it does to an electron in the hydrogen atom, find the quantum number $n$ of the orbit of the satellite. b) Show from Bohr's angular momentum postulate and Newton's law of gravitation that the radius of an earth-satellite orbit is directly proportional to the square of the quantum number, $r = kn^2$, where $k$ is the constant of proportionality. c) Using the result from part (b), find the distance between the orbit of the satellite in this problem and its next "allowed" orbit. (Calculate a numerical value.) d) Comment on the possibility of observing the separation of the two adjacent orbits. e) Do quantized and classical orbits correspond for this satellite? Which is the "correct" method for calculating the orbits?

**40–53** When a photon is emitted by an atom, the atom must recoil to conserve momentum. This means that the photon and the recoiling atom share the transition energy. For a hydrogen atom, calculate the correction due to recoil to the wavelength of the photon emitted when an electron in the $n = 5$ level returns to the ground level. (*Hint:* The correction is very small, so $|\Delta\lambda/\lambda| << 1$. Use this fact to help you obtain an approximate but very accurate expression for $\Delta\lambda$.)

**40–54 Target Material in an X-Ray Tube.** An x-ray tube is operating at 15.0 kV and 40.0 mA. a) If only 1.0% of the electric power supplied is converted into x rays, at what rate is the target being heated in joules per second? b) If the target has a mass of 0.300 kg and a specific heat of 147 J/kg · K, at what average rate would its temperature rise if there were no thermal losses? c) What must be the physical properties of a practical target material? What would be some suitable target elements?

**40–55** a) Calculate the maximum increase in x-ray wavelength that can occur during Compton scattering. b) What is the energy (in electronvolts) of the smallest-energy x-ray photon for which Compton scattering could result in doubling the original wavelength?

**40–56** A photon with a wavelength of 0.1300 nm is Compton-scattered through an angle of 180°. a) What is the wavelength of the scattered photon? b) How much energy is given to the electron? c) What is the recoil speed of the electron? Is it necessary to use the relativistic kinetic-energy relationship?

**40–57** A photon with $\lambda = 0.120$ nm collides with an electron that is initially at rest. After the collision the wavelength is 0.130 nm. a) What is the kinetic energy of the electron after the collision? b) If the electron is suddenly stopped (for example, in a solid target), it emits a photon. What is the wavelength of this photon?

**40–58** a) For Compton scattering of a photon from an electron, what is the largest possible wavelength shift? b) Repeat part (a) for Compton scattering of a photon from a proton.

**40–59** A beam of 2.00-MeV photons is Compton-scattered by the electrons in a target. At what angle from the incident beam direction will 0.500-MeV scattered photons be found?

**40–60 An Ideal Blackbody.** A large cavity with a very small hole and maintained at a temperature $T$ is a good approximation to an ideal radiator or blackbody. Radiation can pass into or out of the cavity only through the hole. The cavity is a perfect absorber, since any radiation incident on the hole becomes trapped inside the cavity. Such a cavity at 100°C has a hole with an area of 2.00 mm$^2$. How long does it take for the cavity to radiate 500 J of energy through the hole?

**40–61** a) Write the Planck distribution law in terms of the frequency $f$ rather than the wavelength $\lambda$, to obtain $I(f)$. b) Show that

$$\int_0^\infty I(\lambda)d\lambda = \frac{2\pi^5 k^4}{15c^2h^3}T^4,$$

where $I(\lambda)$ is the Planck distribution formula of Eq. (40–31). [*Hint:* Change the integration variable from $\lambda$ to $f$. You will need to use the following tabulated integral:

$$\int_0^\infty \frac{x^3}{e^{\alpha x}-1}dx = \frac{1}{240}\left(\frac{2\pi}{\alpha}\right)^4.]$$

c) The result of part (b) is $I$ and has the form of the Stefan-Boltzmann law, $I = \sigma T^4$ (Eq. 40–26). Evaluate the constants in (b) to show that $\sigma$ has the value given in Eq. (40–27).

## CHALLENGE PROBLEMS

**40–62** An x ray collides with an electron at rest. The final wavelength of the x ray is 0.00600 nm, and the final velocity of the struck electron is $8.60 \times 10^7$ m/s. a) What was the original wavelength of the x ray before the collision? b) Through what angle is the x ray scattered in the collision?

**40–63** a) Show that the frequency of revolution of an electron in its circular orbit in the Bohr model of the hydrogen atom is $f = me^4/4\epsilon_0^2 n^3 h^3$. b) Show that when $n$ is very large, the frequency of revolution equals the radiated frequency calculated from Eq. (40–7) for a transition from $n_1 = n + 1$ to $n_2 = n$. (This problem illustrates Bohr's *correspondence principle,* which is often used as a check on quantum calculations. When $n$ is small, quantum physics gives results that are very different from those of classical physics. When $n$ is large, the differences are not significant, and the two methods then "correspond." In fact, when Bohr first tackled the hydrogen atom problem, he sought to determine $f$ as a function of $n$ such that it would correspond to classical results for large $n$.)

**40–64** Consider a beam of monochromatic light with intensity $I$ incident on a perfectly absorbing surface oriented perpendicular to the beam. Use the photon concept to show that the radiation pressure exerted by the light on the surface is given by $I/c$.

CHAPTER

# 41

# The Wave Nature of Particles

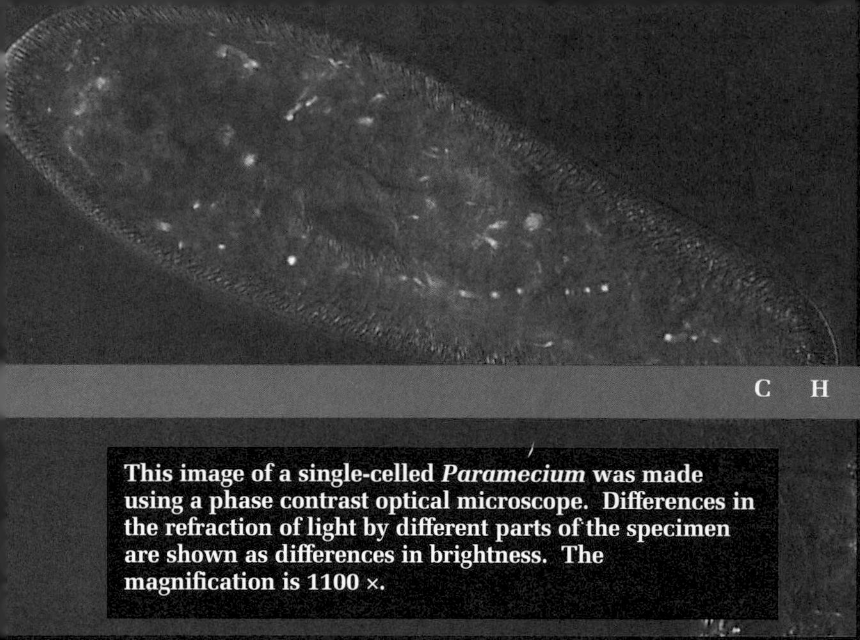

This image of a single-celled *Paramecium* was made using a phase contrast optical microscope. Differences in the refraction of light by different parts of the specimen are shown as differences in brightness. The magnification is 1100 ×.

A scanning electron microscope made this picture of a *Paramecium,* covered with hairlike cilia on its surface. The specimen was first covered with a thin metallic coating. The microscope's electron beam causes emission of electrons from this metal, and the pattern of emissions is analyzed by a computer to produce the image of the object. Only surface views of the specimen are possible, but they show great depth and realism. Magnification is 500 ×.

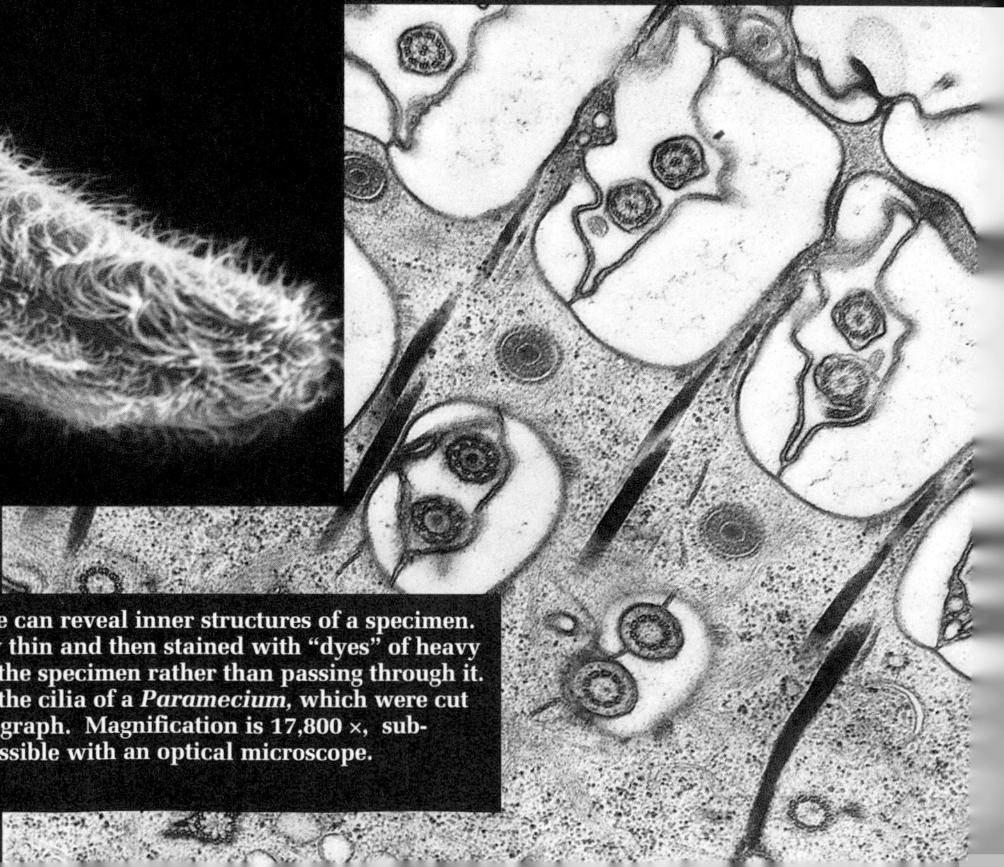

A transmission electron microscope can reveal inner structures of a specimen. The sample is first sliced extremely thin and then stained with "dyes" of heavy metal so the electrons scatter from the specimen rather than passing through it. This photo shows cross sections of the cilia of a *Paramecium,* which were cut through in order to make the photograph. Magnification is 17,800 ×, substantially greater than would be possible with an optical microscope.

## KEY CONCEPTS

**• Particles such as electrons behave like waves in some experiments; the wavelength depends on the particle's momentum.**

**• Particles can be diffracted by a crystal lattice in the same way as x rays are. Experiments with electron diffraction offer direct evidence for the wave nature of particles.**

**• It is impossible to determine a particle's position and momentum precisely at the same time. The Heisenberg uncertainty principle describes fundamental limitations on the precision with which position and momentum can be determined.**

**• The electron microscope uses an electron beam to form greatly enlarged images. Its resolution is limited by the electron wavelength, which can be much smaller than visible-light wavelengths.**

**• The dynamic state of a particle is described by a wave function, a function of the space coordinates and time. A wave packet has both wave and particle properties but is localized in space.**

## INTRODUCTION

In preceding chapters we've discussed the dual wave-particle nature of light and other electromagnetic radiation. Interference and diffraction demonstrate wave behavior; emission and absorption point to the photon concept. In particular, we can interpret atomic spectra in terms of photons and discrete energy levels in atoms.

A complete theory should also be able to *predict,* on theoretical grounds, the energy levels of any particular atom. The 1913 Bohr model of the hydrogen atom was a step in this direction. But it combined classical principles with new ideas that were inconsistent with classical theory, and it raised as many questions as it answered. More drastic departures from classical concepts were needed.

The theory that has emerged since about 1920 is called *quantum mechanics.* The concept of wave-particle duality has been extended to include particles as well as radiation. Particles show *wavelike* behavior in some situations. A particle is modeled as an inherently spread-out entity that can't be described as a point with a perfectly definite position and velocity.

Quantum mechanics is the key to understanding atoms and molecules, including their structures, spectra, chemical behavior, and many other properties. It has the happy effect of restoring unity and symmetry to our description of both particles and radiation.

## 41–1 DE BROGLIE WAVES

A major advance in the understanding of atomic structure began in 1924, about ten years after the Bohr model, with a proposition made by a French physicist, Louis de Broglie. His reasoning, freely paraphrased, went like this. Nature loves symmetry. Light is dualistic in nature, behaving in some situations like waves and in others like particles. If nature is symmetric, this duality should also hold for matter. Electrons and protons, which we usually think of as *particles,* may in some situations behave like *waves.*

If an electron acts like a wave, it should have a wavelength and a frequency. De Broglie postulated that a free electron with mass $m$, moving with speed $v$, should have a wavelength $\lambda$ related to its momentum $p = mv$ in exactly the same way as for a photon, as expressed by Eq. (40–6): $\lambda = h/p$. (In this discussion we assume that $v$ is much smaller than the speed of light, $c$.) The **de Broglie wavelength** of an electron is

$$\lambda = \frac{h}{p} = \frac{h}{mv}, \tag{41–1}$$

where $h$ is Planck's constant. The frequency $f$, according to de Broglie, is also related to the particle's energy $E$ in the same way as for a photon, namely,

$$E = hf. \tag{41–2}$$

Thus the relation of wavelength to momentum and the relation of frequency to energy, in de Broglie's hypothesis, are exactly the same for electrons as for photons.

To appreciate the enormous significance of de Broglie's proposal, we have to realize that at the time there was no direct experimental evidence that particles have wave characteristics. It is one thing to suggest a new hypothesis to explain experimental observations; it is quite another to propose such a radical departure from established concepts on theoretical grounds alone. But it was clear that a radical idea was needed. The dual nature of electromagnetic radiation had led to adoption of the photon concept, also a radical idea. The relatively complete lack of success in understanding atomic structure indicated that a similar revolution was needed in the mechanics of particles.

De Broglie's hypothesis was the beginning of that revolution. Within a few years after 1924 it was developed by Heisenberg, Schrödinger, Dirac, Born, and many others into a detailed theory called **quantum mechanics.** This development was well underway even before direct experimental evidence for the wave properties of particles was found.

Quantum mechanics involves sweeping revisions of our fundamental concepts of the description of matter. A particle is not a geometric point but an entity that is spread out in space. In some cases this spreading appears as a periodic pattern suggesting the *wavelike* properties proposed by de Broglie. The wave and particle aspects are not inconsistent; the particle model is a special case of a more general wave picture. The situation is comparable to the ray model of geometrical optics, which is a special case of the more general wave model of physical optics. Indeed, there is a very close analogy between optics and the description of the motion of particles.

The spatial distribution of a particle is defined by a function called a **wave function,** which is closely analogous to the wave functions that we have used for mechanical waves in Chapter 19 and for electromagnetic waves in Chapter 33. The wave function for a *free* electron with definite energy has a recurring wave pattern with definite wavelength and frequency.

The Bohr model pictured the energy levels of the hydrogen atom in terms of definite electron orbits, as shown in Fig. 40–7. This is an oversimplification and should not be taken literally. But the most important idea in Bohr's theory was the existence of discrete energy levels and their relation to the frequencies of emitted photons. The new quantum mechanics still assigns definite energy states to an atom but with a more general description of the electron motion in terms of wave functions. In the hydrogen atom the energy levels predicted by quantum mechanics turn out to be the same as those given by Bohr's theory. In more complicated atoms, for which the Bohr theory does not work, the quantum mechanical picture is in excellent agreement with observation.

The de Broglie wave hypothesis has an interesting relation to the Bohr model. We can use Eq. (41–1) to obtain the Bohr quantum condition that the angular momentum $L = mvr$ must be an integer multiple of $h$ divided by $2\pi$. The method is analogous to determining the normal-mode frequencies of a vibrating string held at both ends. We discussed this problem in Section 20–3; the central idea was to satisfy the **boundary conditions,** or the requirement that the displacement at each end of the string must always be zero. The ends are always nodes, and there may be additional nodes along the string. For the boundary conditions to be satisfied, the total length of the string must equal some *integral* number of half-wavelengths.

Now think of an electron as a wave wrapped around a circle in one of the Bohr orbits. In order for the wave to "come out even" and join onto itself smoothly, the circumference of this circle must include some *integral number* of wavelengths, as suggested by Fig. 41–1. For an orbit with radius $r$ and circumference $2\pi r$, we must have $2\pi r = n\lambda$, where $n = 1, 2, 3, \ldots$. According to the de Broglie relation, Eq. (41–1), the wavelength $\lambda$ of a particle with mass $m$, moving with speed $v$, is $\lambda = h/mv$. Combining these two equations, we find

$$2\pi r = n\frac{h}{mv}, \quad \text{and} \quad mvr = n\frac{h}{2\pi}. \tag{41–3}$$

We recognize this as identical to Eq. (40–11), Bohr's postulate that the angular momentum $L = mvr$ must equal an integer $n$ times $h/2\pi$. Thus the wave-mechanical picture leads naturally to Bohr's postulate.

To be sure, the idea of wrapping a wave around in a circular orbit is a rather vague notion. But the agreement of Eq. (41–3) with Bohr's hypothesis is much too remarkable to be a coincidence. It strongly suggests that the wave properties of electrons do indeed have something to do with atomic structure.

Later we will learn how wave functions for specific systems are determined by solving a wave equation called the Schrödinger equation. Boundary conditions play a central role in finding solutions of this equation and thus in determining possible energy levels, values of angular momentum, and other properties.

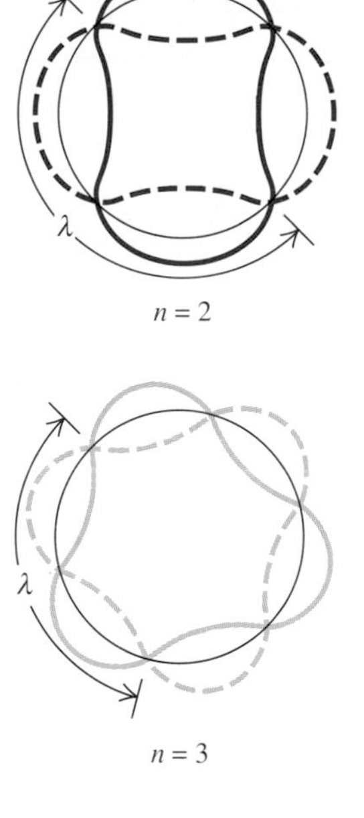


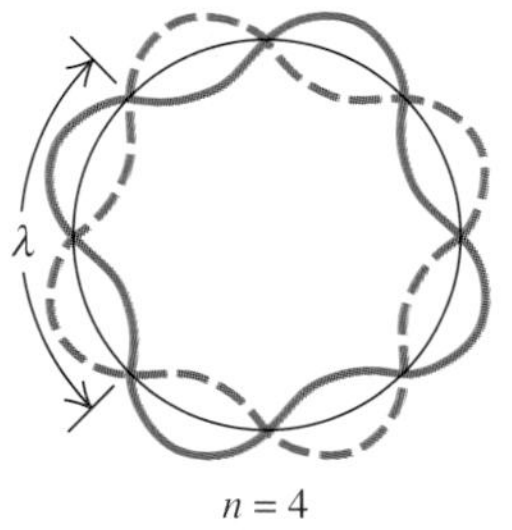


**41–1** Diagrams showing the idea of wrapping a standing wave around a circular orbit. For the wave to join onto itself smoothly, the circumference of the orbit must be an integral number $n$ of wavelengths. Examples are shown for $n = 2$, 3, and 4.

## PROBLEM-SOLVING STRATEGY

### Quantum mechanics

**1.** In atomic physics the orders of magnitude of physical quantities are so unfamiliar that often common sense isn't much help in judging the reasonableness of a result. It helps to remind yourself of some typical magnitudes of various quantities:

Size of an atom: $10^{-10}$ m
Mass of an atom: $10^{-26}$ kg
Mass of electron: $10^{-30}$ kg
Energy of an atomic state: $-1$ to $-10$ eV or $10^{-18}$ J for outer electrons (but some interaction energies are much smaller)
Speed of an electron in the Bohr atom: $10^6$ m/s
Electron charge: $10^{-19}$ C
$kT$ at room temperature: $\frac{1}{40}$ eV

You may want to add items to this list. These values will also help you in Chapter 45, when we have to deal with magnitudes characteristic of *nuclear* rather than atomic structure; these are often different by factors of $10^4$ to $10^6$. In working out problems, be very careful to handle powers of 10 properly. A gross error might not be obvious.

**2.** As in Chapter 40, energies may be expressed either in joules or in electronvolts. Be sure you use consistent units. Lengths, such as wavelengths, are always in meters if you use the other quantities consistently in SI units, such as $h = 6.626 \times 10^{-34}\ \mathrm{J \cdot s}$. If you want nanometers or something else, don't forget to convert. In some problems it's useful to express $h$ in eV · s: $h = 4.136 \times 10^{-15}\ \mathrm{eV \cdot s}$.

**3.** Kinetic energy can be expressed either as $K = \frac{1}{2}mv^2$ or (because $p = mv$) as $K = p^2/2m$. The latter form is often useful in calculations involving the de Broglie wavelength.

**4.** Aside from these calculational details, the main challenges of this chapter are conceptual, not computational. Try to keep an open mind when you encounter new and sometimes jarring ideas. Photons and electrons *can* have both wavelike and particlelike properties, but intuitive understanding of quantum mechanics takes time to develop.

## EXAMPLE 41–1

**Energy of a thermal neutron** Find the speed and kinetic energy of a neutron ($m = 1.675 \times 10^{-27}$ kg) that has a de Broglie wavelength $\lambda = 0.100$ nm, which is typical of atomic spacing in crystals. Compare the energy with the average kinetic energy of a gas molecule at room temperature ($T = 20°\mathrm{C}$).

**SOLUTION** From Eq. (41–1),

$$v = \frac{h}{\lambda m} = \frac{6.626 \times 10^{-34}\ \mathrm{J \cdot s}}{(0.100 \times 10^{-9}\ \mathrm{m})(1.675 \times 10^{-27}\ \mathrm{kg})} = 3.96 \times 10^3\ \mathrm{m/s};$$

$$K = \tfrac{1}{2}mv^2 = \tfrac{1}{2}(1.675 \times 10^{-27}\ \mathrm{kg})(3.96 \times 10^3\ \mathrm{m/s})^2 = 1.31 \times 10^{-20}\ \mathrm{J} = 0.0818\ \mathrm{eV}.$$

The average translational kinetic energy of a molecule of an ideal gas is given by Eq. (16–17):

$$K = \tfrac{3}{2}kT = \tfrac{3}{2}(1.38 \times 10^{-23}\ \mathrm{J/K})(293\ \mathrm{K}) = 6.07 \times 10^{-21}\ \mathrm{J} = 0.0379\ \mathrm{eV}.$$

The two energies are comparable in magnitude. In fact, a neutron with kinetic energy in this range is called a *thermal neutron*.

Diffraction of thermal neutrons, which we'll discuss in the next section, is used in the same way as x-ray diffraction to study crystal and molecular structure. Neutron diffraction has proved especially useful in the study of large organic molecules.

## 41–2 ELECTRON DIFFRACTION

De Broglie's wave hypothesis, radical though it seemed, almost immediately received direct experimental confirmation. The first direct evidence involved a diffraction experiment with electrons that was analogous to the x-ray diffraction experiments described in Section 38–6. In those experiments, atoms in a crystal act as a three-dimensional diffraction grating for x rays. An x-ray beam is strongly reflected when it strikes a crystal at an angle that gives constructive interference among the waves scattered from the various atoms in the crystal. We pointed out that these interference effects demonstrate the *wave* nature of x rays.

In 1927, Clinton Davisson and Lester Germer, working in the Bell Telephone Laboratories, were studying the surface of a piece of nickel by directing a beam of *electrons* at the surface and observing how many electrons bounced off at various angles. Their experimental setup is shown in Fig. 41–2a. The specimen was polycrystalline; like many ordinary metals, it consisted of many microscopic crystals bonded together with random orientations. The experimenters expected that even the smoothest surface attainable would still look rough to an electron and that the electron beam would be diffusely reflected, with a smooth distribution of intensity as a function of the angle $\theta$.

During the experiment, an accident occurred that permitted air to enter the vacuum chamber, and an oxide film formed on the metal surface. To remove this film, Davisson and Germer baked the specimen in a high-temperature oven, almost hot enough to melt it. Unknown to them, this had the effect of *annealing* the specimen, creating large single-crystal regions with crystal planes that were continuous over the width of the electron beam.

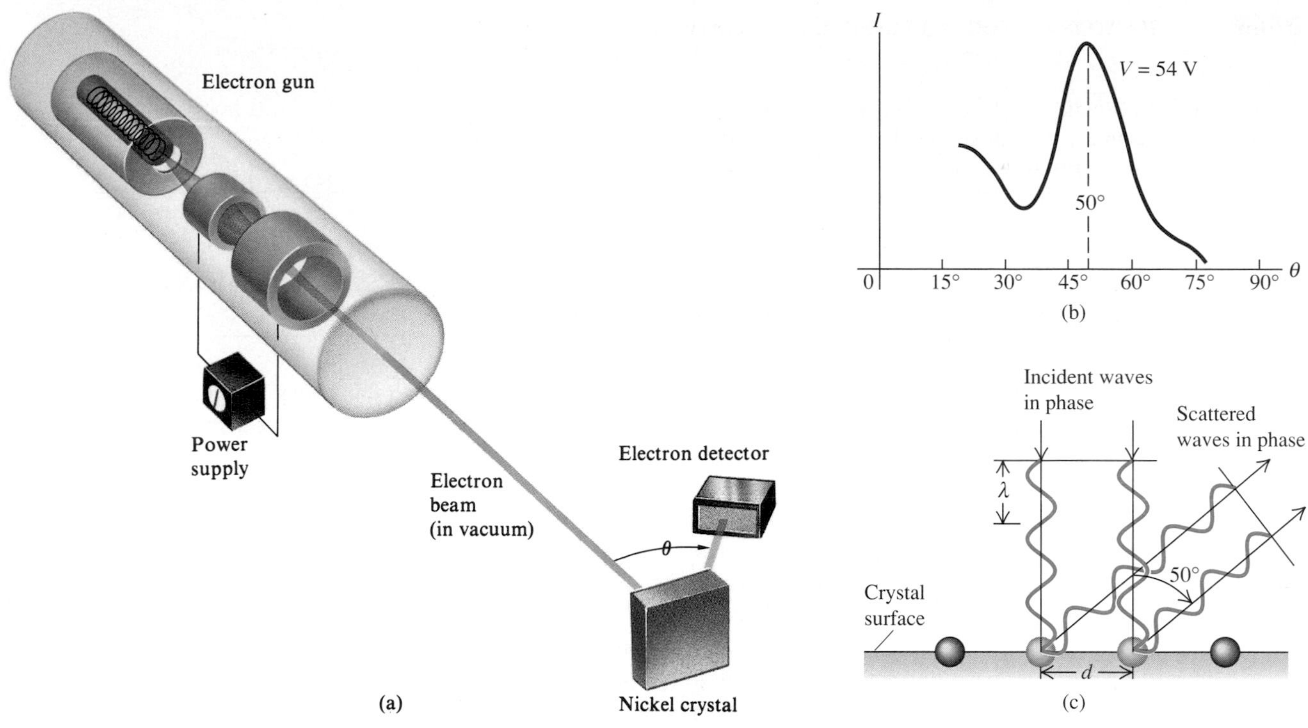


**41–2** (a) Apparatus for the Davisson-Germer experiment. Electrons emitted from the heated filament are accelerated by the electrodes in the electron gun and directed at the crystal, and the electrons in the scattered beam are observed by a detector. The orientation of the detector, described by the angle $\theta$, can be varied. The entire apparatus is enclosed in an evacuated chamber. (b) A graph of intensity of the scattered beam as a function of the angle $\theta$. The sharp peak at $\theta = 50°$ results from constructive interference between electron waves scattered from various atoms in the surface layer of the crystal. (c) Constructive interference of waves scattered from two adjacent atoms.

When the observations were repeated, the results were quite different. Sharp maxima in the intensity of the reflected electron beam occurred at specific angles (Fig. 41–2b), in contrast to the smooth variation of intensity with angle that Davisson and Germer had observed before the accident. The angular positions of the maxima depended on the accelerating voltage $V$ used to produce the electron beam. Davisson and Germer were familiar with de Broglie's hypothesis, and they noticed the similarity of this behavior to x-ray diffraction. This was not the effect they had been looking for, but they immediately recognized that the electron beam was being *diffracted*. They had discovered a very direct experimental confirmation of the wave hypothesis.

Davisson and Germer could determine the speeds of the electrons from the accelerating voltage $V$, so they could compute the de Broglie wavelength from Eq. (41–1). The electrons were scattered primarily by the plane of atoms at the surface of the crystal. Atoms in this surface plane are arranged in rows, with a distance $d$ between adjacent rows that can be measured by x-ray diffraction techniques. These rows act like a reflecting diffraction grating; the angles at which strong reflection occurs are the same as for a grating with spacing $d$ between adjacent lines. From Eq. (38–13), the angles of maximum intensity are given by

$$d \sin\theta = m\lambda \qquad (m = 1, 2, 3, \ldots), \tag{41–4}$$

where $\theta$ is the angle shown in Fig. 41–2a. The angles predicted by this equation, using the de Broglie wavelength, were found to agree with the observed values (Fig. 41–2b). Thus the accidental discovery of **electron diffraction** was the first direct evidence confirming de Broglie's hypothesis.

## EXAMPLE 41–2

In a particular electron-diffraction experiment using an accelerating voltage of 54 V, an intensity maximum occurs when the angle $\theta$ in Fig. 41–2a is 50°. The spacing of rows of atoms in the surface layer is found by x-ray diffraction to be $2.15 \times 10^{-10}$ m = 215 pm. Find the electron wavelength a) from the grating formula (assuming that $m = 1$); b) from the de Broglie formula. Compare your results.

**SOLUTION** a) From Eq. (41–4), with $m = 1$,

$$\lambda = d \sin\theta = (2.15 \times 10^{-10}\text{ m})(\sin 50°) = 1.65 \times 10^{-10}\text{ m}.$$

b) To find the de Broglie wavelength, we need the momentum $p = mv$ of the electrons. We can get this from energy considerations. The kinetic energy is $K = \frac{1}{2}mv^2 = p^2/2m$, and $K$ is also equal to $eV$, where $V$ is the accelerating *voltage:*

$$K = eV = \frac{p^2}{2m}, \qquad p = \sqrt{2meV}.$$

The de Broglie wavelength is

$$\lambda = \frac{h}{p} = \frac{h}{\sqrt{2meV}} \tag{41-5}$$

$$= \frac{6.626 \times 10^{-34}\ \mathrm{J \cdot s}}{\sqrt{2(9.109 \times 10^{-31}\ \mathrm{kg})(1.602 \times 10^{-19}\ \mathrm{C})(54\ \mathrm{V})}}$$

$$= 1.67 \times 10^{-10}\ \mathrm{m}.$$

The two numbers agree within the precision of the experimental results.

In 1928, just a year after the Davisson-Germer discovery, the English physicist G. P. Thomson carried out electron-diffraction experiments using a thin polycrystalline metallic foil as a target. Debye and Sherrer had used a similar technique several years earlier to study x-ray diffraction from polycrystalline specimens. Because of the random orientations of the individual microscopic crystals, such a diffraction pattern consists of intensity maxima forming rings around the direction of the incident beam. Thomson's results again confirmed the de Broglie relationship. Figure 41–3 shows both x-ray and electron diffraction patterns for a polycrystalline aluminum foil. (It is interesting to note that G. P. Thomson was the son of J. J. Thomson, who 31 years earlier had performed the definitive experiment to establish the *particle* nature of electrons.)

Additional experiments were soon carried out in many laboratories. In Germany, Estermann and Stern demonstrated diffraction of alpha particles. More recently, diffraction experiments have been performed with low-energy neutrons (see Example 41–1). Thus the wave nature of particles, so strange in 1924, became firmly established in the following years.

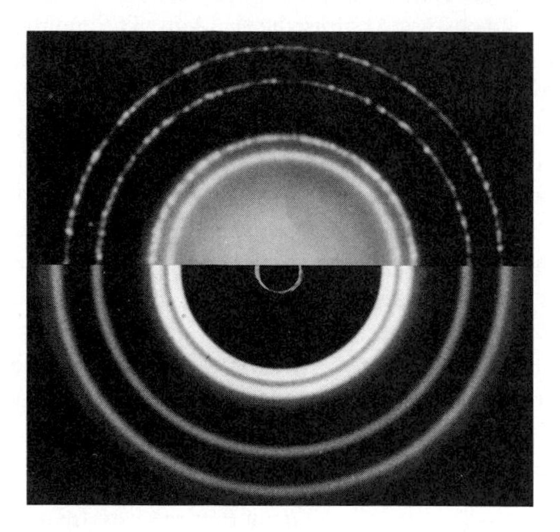

**41–3** X-ray and electron diffraction. The upper half of the photo shows the diffraction pattern for 71-pm x rays passing through aluminum foil. The lower half, with a different scale, shows the diffraction pattern for 600-eV electrons from aluminum.

## 41–3 PROBABILITY AND UNCERTAINTY

The discovery of the dual wave-particle nature of matter has forced us to reevaluate the kinematic language that we use to describe the position and motion of a particle. In classical Newtonian mechanics we think of a particle as a point. We can describe its state of motion at any instant with three spatial coordinates and three components of velocity. But, in general, such a specific description is not possible. When we look on a small enough scale, there are fundamental limitations on the precision with which we can determine the position and velocity of a particle. Many aspects of a particle's behavior can be stated only in terms of *probabilities.*

## Single-Slit Diffraction

To try to get some insight into the nature of the problem, let's review the optical single-slit diffraction experiment described in Section 38–2. Most (85%) of the intensity in the diffraction pattern is concentrated in the central maximum, bounded on either side by the first intensity *minimum*. We use $\theta_1$ to denote the angle between the central maximum and the first minimum. Using Eq. (38–2), with $m = 1$, we find that $\theta_1$ is given by $\sin\theta_1 = \lambda/a$, where $a$ is the slit width. If $\lambda$ is much smaller than $a$, then $\theta_1$ is very small, $\sin\theta_1$ is very nearly equal to $\theta_1$ (in radians), and

$$\theta_1 = \frac{\lambda}{a}. \tag{41–6}$$

Now we perform the same experiment again, but using a beam of *electrons* instead of a beam of monochromatic light (Fig. 41–4). We have to do the experiment in vacuum ($10^{-7}$ atm or less) so that the electrons don't bump into air molecules. We can produce the electron beam with a setup that is similar in principle to the electron gun in a cathode-ray tube. This produces a narrow beam of electrons, all of which have very nearly the same direction and speed and therefore also the same de Broglie *wavelength*.

The result of this experiment, recorded on photographic film or by use of more sophisticated detectors, is a diffraction pattern that is identical to the one shown in Fig. 38–8b. This gives us additional direct evidence of the *wave* nature of electrons. Most (85%) of the electrons strike the film in the region of the central maximum, but a few strike farther from the center, in the subsidiary maxima on both sides.

If we believe that electrons are waves, the wave behavior in this experiment isn't surprising. But if we try to interpret it in terms of *particles,* we run into very serious problems. First, the electrons don't all follow the same path, even though they all have the same initial state of motion. In fact, we can't predict the trajectory of an individual electron from knowledge of its initial state. The best we can do is to say that *most* of the electrons go to a certain region, *fewer* go to other regions, and so on. That is, we can describe only the *probability* for an individual electron to strike each of various areas on the film. This fundamental indeterminacy has no counterpart in Newtonian mechanics, in which the motion of a particle or a system can always be predicted if we know the initial position and motion with great enough precision.

Second, there are fundamental *uncertainties* in both position and momentum of an individual particle, and these two uncertainties are related inseparably. To clarify this point, let's go back to Fig. 41–4. An electron that strikes the film at the outer edge of the

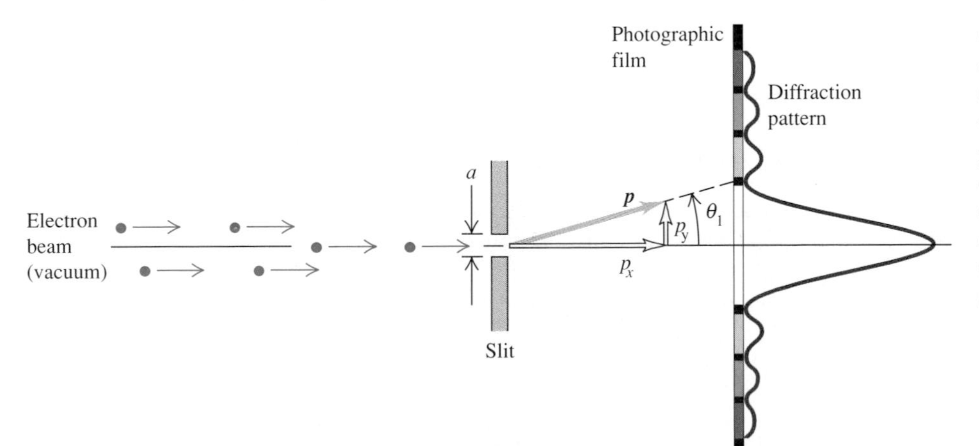


**41–4** An electron diffraction experiment. The graph at the right shows the degree of blackening of the film, which in any region is proportional to the number of electrons striking that region. The components of momentum of an electron striking the outer fringe of the central maximum, at angle $\theta_1$, are shown.

central maximum, at angle $\theta_1$, must have a component of momentum $p_y$ in the $y$-direction, as well as a component $p_x$ in the $x$-direction, despite the fact that the beam was initially directed along the $x$-axis. From the geometry of the situation, the two components are related by $p_y/p_x = \tan\theta_1$. If $\theta_1$ is small, we may use the approximation $\tan\theta_1 = \theta_1$, obtaining

$$p_y = p_x\theta_1. \tag{41-7}$$

Neglecting the 15% of the electrons that strike the film outside the central maximum (that is, at angles greater than $\lambda/a$), we see that the $y$-component of momentum may be as large as

$$p_y = p_x\frac{\lambda}{a}. \tag{41-8}$$

So the *uncertainty* $\Delta p_y$ in the $y$-component of momentum is at least as great as $p_x\lambda/a$:

$$\Delta p_y \geq p_x\frac{\lambda}{a}. \tag{41-9}$$

The narrower the slit width $a$, the broader is the diffraction pattern and the greater is the uncertainty in the $y$-component of momentum $p_y$.

The electron wavelength $\lambda$ is related to the momentum component $p_x = mv_x$ by the de Broglie relation, Eq. (41–1), which we can rewrite as $\lambda = h/p_x$. Using this relation in Eq. (41–9) and simplifying, we find

$$\Delta p_y \geq p_x\frac{h}{p_x a} = \frac{h}{a},$$

or

$$\Delta p_y a \geq h. \tag{41-10}$$

What does this result mean? The slit width $a$ represents the uncertainty in the *position* of an electron as it passes through the slit. We don't know exactly *where* in the slit each particle passes through. So both the $y$ position and the $y$-component of momentum have uncertainties, and the two uncertainties are related by Eq. (41–10). We can reduce the *momentum* uncertainty $\Delta p_y$ only by reducing the width of the diffraction pattern. To do this, we have to increase the slit width, which increases the *position* uncertainty. Conversely, when we *decrease* the position uncertainty by narrowing the slit, the diffraction pattern broadens, and the corresponding momentum uncertainty *increases.*

You may protest that it doesn't seem to be consistent with common sense for a particle not to have a definite position and momentum. We reply that what we call *common sense* is based on familiarity gained through experience. Our usual experience includes very little contact with the microscopic behavior of particles. Sometimes we have to accept conclusions that violate our intuition when we are dealing with areas that are far removed from everyday experience.

## The Uncertainty Principle

In more general discussions of uncertainty relations, the uncertainty of a quantity is usually described in terms of the statistical concept of *standard deviation,* which is a measure of the spread, or dispersion, of a set of numbers around their average value. If a coordinate $x$ has an uncertainty $\Delta x$ defined in this way, and if the corresponding momentum component $p_x$ has an uncertainty $\Delta p_x$, then the two uncertainties are found to be related in general by the inequality

$$\Delta x\,\Delta p_x \geq \frac{h}{2\pi}. \tag{41-11}$$

Equation (41–11) is one form of the **Heisenberg uncertainty principle.** It states that, in general, neither the momentum nor the position of a particle can be determined simultaneously with arbitrarily great precision, as classical physics would predict. Instead, the uncertainties in the two quantities play complementary roles, as we have described. Figure 41–5 shows the relationship between the two uncertainties.

It is tempting to suppose that we could get greater precision by using more sophisticated particle detectors in various areas of the slit. This turns out to be not possible. To detect a particle, the detector must *interact* with it, and this interaction unavoidably changes the state of motion of the particle, introducing uncertainty about its original state. A more detailed analysis of such hypothetical experiments shows that the uncertainties that we have described are fundamental and intrinsic. They *cannot* be circumvented *even in principle* by any experimental technique, no matter how sophisticated.

There is nothing special about the $x$-axis. In a three-dimensional situation with coordinates $(x, y, z)$ there is an uncertainty relation for each coordinate and its corresponding momentum component: $\Delta y\, \Delta p_y \geq h/2\pi$ and $\Delta z\, \Delta p_z \geq h/2\pi$. However, the uncertainty in one coordinate is *not* related to a different component of momentum. For example, $\Delta x$ is not related directly to $\Delta p_y$.

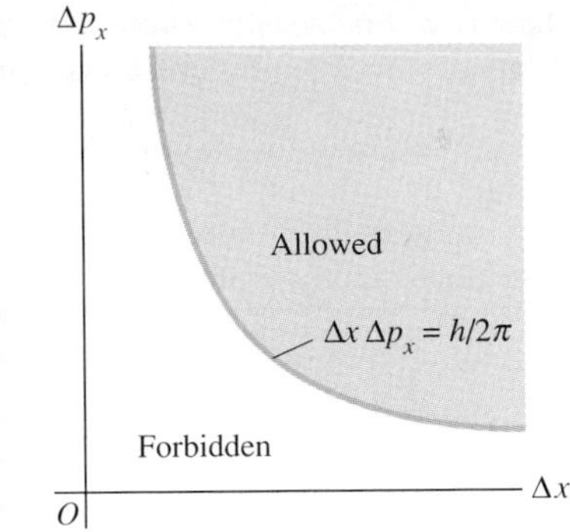


**41–5** The product $\Delta x \Delta p_x$ must be greater than or equal to $h/2\pi$.

## Uncertainty in Energy

There is also an uncertainty principle for *energy*. It turns out that the energy of a system also has inherent uncertainty. The uncertainty $\Delta E$ depends on the *time interval* $\Delta t$ during which the system remains in the given state. The relation is

$$\Delta E\, \Delta t \geq \frac{h}{2\pi}. \tag{41–12}$$

A system that remains in a certain state for a very long time (large $\Delta t$) can have a very well-defined energy (small $\Delta E$), but if it remains in that state for only a short time (small $\Delta t$), the uncertainty in energy must be correspondingly greater (large $\Delta E$). Figure 41–6 illustrates this idea.

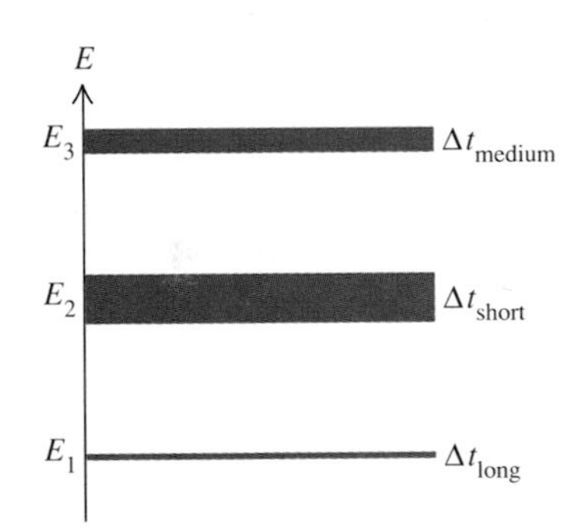


**41–6** The longer the lifetime of an energy state, the smaller is its spread in energy.

### EXAMPLE 41–3

An electron is confined inside a cubical box $1.0 \times 10^{-10}$ m on a side. a) Estimate the minimum uncertainty in the electron's momentum. b) If the electron has momentum with magnitude equal to the uncertainty found in part (a), what is its kinetic energy? Express the result in joules and in electronvolts.

**SOLUTION** a) Let's concentrate on $x$-components. In the uncertainty principle we have $\Delta x = 1.0 \times 10^{-10}$ m. From Eq. (41–11),

$$\Delta p_x = \frac{h}{2\pi \Delta x} = \frac{6.63 \times 10^{-34}\ \text{J}\cdot\text{s}}{2\pi(1.0 \times 10^{-10}\ \text{m})} = 1.1 \times 10^{-24}\ \text{kg}\cdot\text{m/s}.$$

b) An electron with this magnitude of momentum has kinetic energy

$$K = \frac{p_x^{\,2}}{2m} = \frac{(1.1 \times 10^{-24}\ \text{kg}\cdot\text{m/s})^2}{2(9.11 \times 10^{-31}\ \text{kg})}$$
$$= 6.1 \times 10^{-19}\ \text{J} = 3.8\ \text{eV}.$$

The box is roughly the same size as an atom, and the energy is of the same order of magnitude as typical electron energies in atoms. This is a very rough calculation, but it is reassuring that the magnitude of the energy is in a reasonable range of magnitude. If our result had differed from this by a factor of $10^6$, we would have had reason to be alarmed.

### EXAMPLE 41–4

A sodium atom in one of the "resonance levels" shown in Fig. 40–9 remains in that state for an average time of $1.6 \times 10^{-8}$ s before it makes a transition back to the ground state by emitting a photon with wavelength 589 nm and energy 2.109 eV. What is the uncertainty in energy of the resonance level? What is the wavelength spread of the corresponding spectrum line?

**SOLUTION** From Eq. (41–12),

$$\Delta E = \frac{h}{2\pi \Delta t} = \frac{6.626 \times 10^{-34}\ \text{J}\cdot\text{s}}{2\pi(1.6 \times 10^{-8}\ \text{s})}$$
$$= 6.6 \times 10^{-27}\ \text{J} = 4.1 \times 10^{-8}\ \text{eV}.$$

The atom remains an indefinitely long time in the ground state, so there is *no* fundamental uncertainty there. The fractional uncertainty of the resonance-level energy and of the photon energy is

$$\frac{4.1 \times 10^{-8}\ \text{eV}}{2.109\ \text{eV}} = 2 \times 10^{-8}.$$

The corresponding spread in wavelength, or "width," of the spectrum line is approximately

$$\Delta\lambda = (2 \times 10^{-8})(589\ \text{nm}) = 0.000012\ \text{nm}$$

This irreducible uncertainty is called the *natural line width* of this particular spectrum line. Though very small, it is within the limits of resolution of present-day spectrometers. Ordinarily, the natural line width is much smaller than line broadening from other causes such as collisions among atoms.

## Two-Slit Interference

Now let's take a brief look at a quantum interpretation of a *two-slit* optical interference pattern. We studied these patterns in detail for light in Sections 37–2 and 37–3, and in Section 40–9 we discussed their interpretation in terms of the probability for photons to strike various regions of the screen where the pattern is formed.

In view of our discussion of the wave properties of electrons, it is natural to ask what happens when we do a two-slit interference experiment with electrons. The answer is—exactly the same thing as with photons! We can again use photographic film (Fig. 41–7) or particle counters to trace out the interference pattern, as we did with photons. But we *cannot* predict where in the pattern any individual electron will land. We can't even ask which slit an individual electron passes through. Waves from the two slits have to be coherent in order for the two-slit interference pattern to be formed. To meet this requirement, we have to assume that *every electron goes through both slits!* If the idea of a *photon* passing through both slits made you uncomfortable, you'll really hate this. But remember that the electron is *not* a point; it is an inherently spread-out entity. There is nothing conceptually wrong with having it pass through both slits. As with photons, *each electron interferes only with itself.*

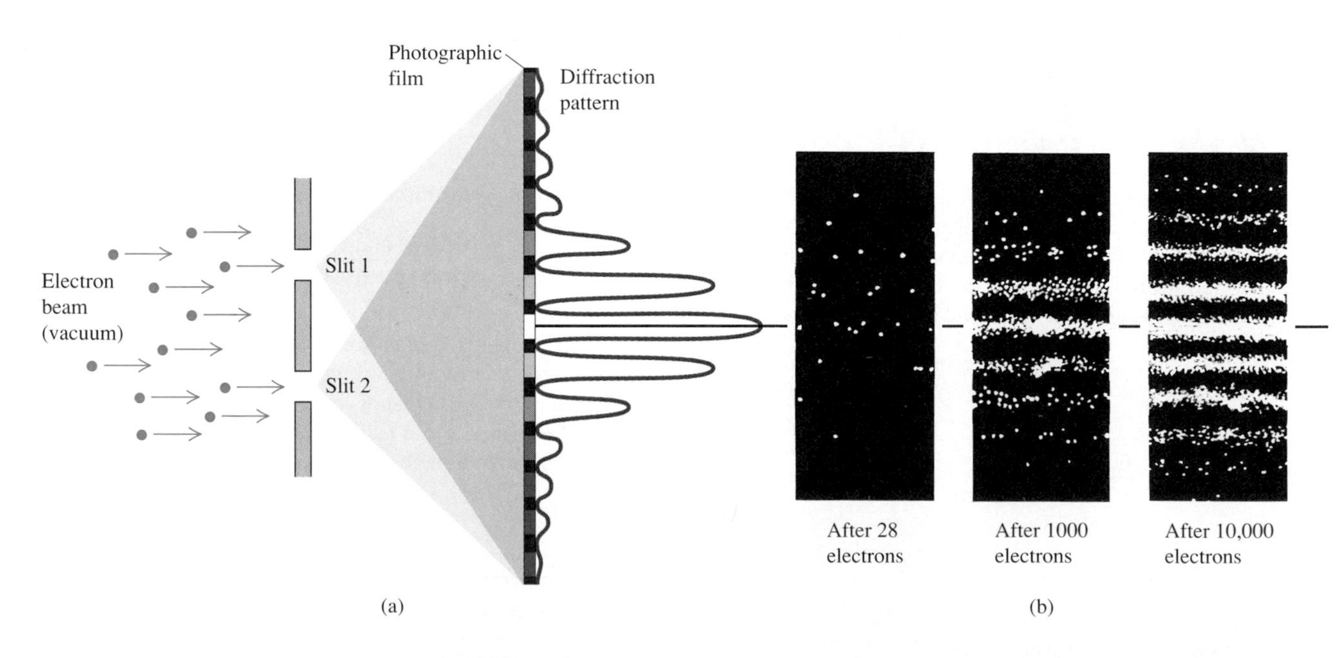


**41–7** (a) Formation of an interference pattern for electrons incident on two slits, (b) after 28 particles, after 1000 particles, and after 10,000 particles.

## 41–4 THE ELECTRON MICROSCOPE

The **electron microscope** offers an important and interesting example of the interplay of wave and particle properties of electrons. An electron beam can be used to form an image of an object in exactly the same way as a light beam. A ray of light can be bent by reflection or refraction, and an electron trajectory can be bent by an electric or magnetic field. Rays of light diverging from a point on an object can be brought to convergence by a converging lens, and electrons diverging from a small region can be brought to convergence by an electrostatic or magnetic lens. Figure 41–8a shows the construction of a simple electrostatic lens. The equipotential lines are shown in Fig. 41–8b, and an analogous optical system is shown in Fig. 41–8c. In each case the image can be made larger than the object, so both devices can act as magnifiers.

The analogy between light rays and electrons goes deeper. The *ray* model of geometric optics is an approximate representation of the more general *wave* model. Geometric optics (ray optics) is valid whenever interference and diffraction effects can be neglected. Similarly, the model of an electron as a point particle following a line trajectory is an approximate description of the actual behavior of the electron; this model is useful when we can neglect effects associated with the wave nature of electrons.

How is an electron microscope superior to an optical microscope? The *resolution* of an optical microscope is limited by diffraction effects, as we discussed in Section 38–7. Using wavelengths around 500 nm, an optical microscope can't resolve objects smaller than a few hundred nanometers, no matter how carefully its lenses are made. The resolution of an electron microscope is similarly limited by the wavelengths of electrons, but these wavelengths may be many thousands of times *smaller* than wavelengths of visible light. The useful magnification of an electron microscope can be thousands of times as great as that of an optical microscope.

Be sure you understand that the ability of the electron microscope to form an image *does not* depend on the wave properties of electrons. We can compute their trajectories by treating them as classical charged particles under the action of electric- and magnetic-field forces (in analogy to ray optics). Only when we talk about *resolution* do the wave properties become important.

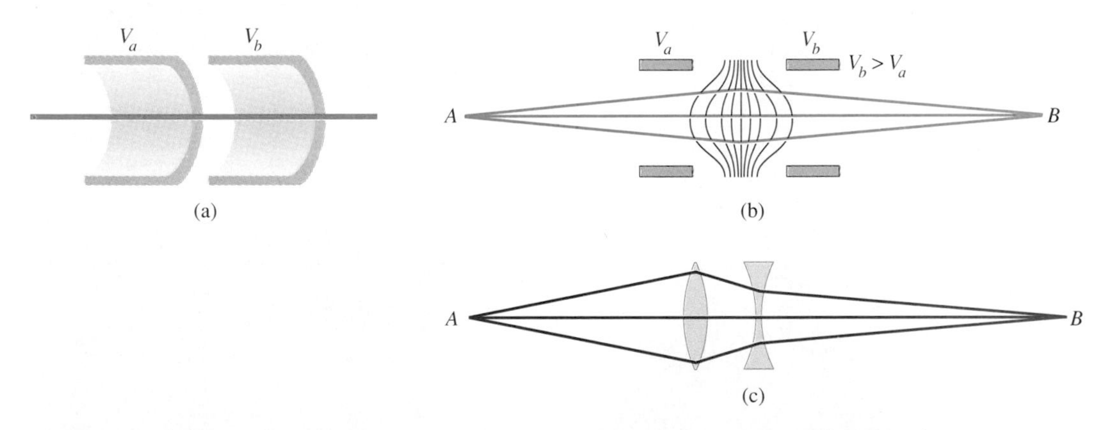


**41–8** (a) An electrostatic lens. The two cylinders are at different electrical potentials $V_a$ and $V_b$. (b) The red lines are cross sections of equipotential surfaces in the electric field between the cylinders. The electron trajectories are shown in blue; electrons diverging from point *A* are brought to a focus at point *B*. (c) Optical analog of the electrostatic lens in part (a).

## EXAMPLE 41–5

The electron beam in an electron microscope is formed by a setup similar to the electron gun in a cathode-ray tube (Section 24–7). What accelerating voltage is needed to produce electrons with wavelength 10 pm = 0.010 nm (roughly 50,000 times smaller than typical visible-light wavelengths)?

**SOLUTION** The wavelength is determined by Eq. (41–1), $\lambda = h/p$. The momentum $p$ is related to the kinetic energy, which is determined by the accelerating voltage $V$:

$$K = eV = \frac{p^2}{2m} = \frac{(h/\lambda)^2}{2m}.$$

Solving for $V$ and inserting the appropriate numbers, we find

$$\begin{aligned} V &= \frac{h^2}{2me\lambda^2} \\ &= \frac{(6.626 \times 10^{-34}\ \mathrm{J \cdot s})^2}{2(9.109 \times 10^{-31}\ \mathrm{kg})(1.602 \times 10^{-19}\ \mathrm{C})(10 \times 10^{-12}\ \mathrm{m})^2} \\ &= 1.5 \times 10^4\ \mathrm{V} = 15{,}000\ \mathrm{V}. \end{aligned}$$

This is approximately equal to the accelerating voltage for the electron beam in a TV picture tube. This example shows, incidentally, that the sharpness of a TV picture is *not* limited by electron-diffraction effects.

Most practical electron microscopes use magnetic rather than electrostatic lenses. A common setup includes three lenses in a compound-microscope arrangement, as shown in Fig. 41–9. Electrons are emitted from a hot cathode and accelerated by a potential difference, typically 10 to 100 kV. The electrons pass through a condensing lens and are formed into a parallel beam before passing through the specimen or object to be viewed. The objective lens then forms an intermediate image of this object, and the projection lens produces a final real image of that image. The latter two lenses play the

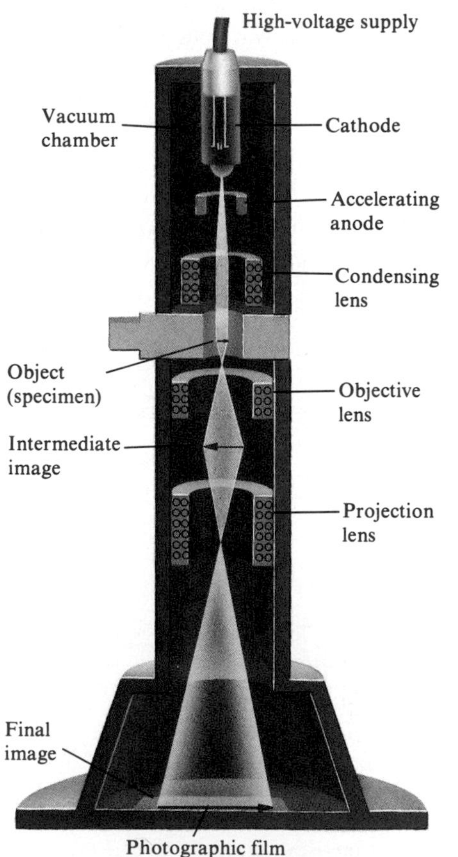


(a)

(b)

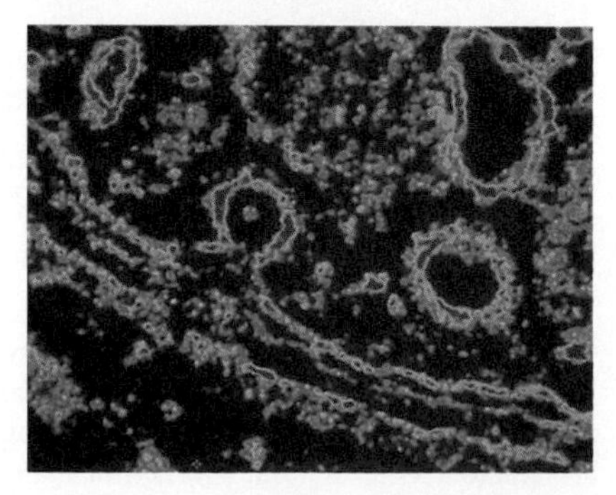

(c)

**41–9** (a) Schematic diagram of an electron microscope. The magnetic lenses, consisting of coils of wire carrying currents, are shown in cross section. The condensing lens forms a parallel beam of electrons that strikes the object. The objective lens forms an intermediate image that serves as the object for the final image formed by the projection lens. The final image is projected onto photographic film, a fluorescent screen, or the screen of a video camera. The magnification of each lens may be of the order of 100×, and the overall magnification of the order of 10,000×. The angles of the electron paths with the optic axis are greatly exaggerated; in actual instruments these angles are usually less than 0.01 rad (0.5°). The entire apparatus is enclosed in a vacuum chamber, not shown in the diagram. (b) An electron microscope. The electron-optical system is in the tall cylinder at the left; images can be viewed on the display screen on the console and can be recorded digitally. (c) Electron micrograph of a cell membrane. The image has been colorized by computer to produce a concentration profile: red indicates high concentration and blue low.

roles of the objective lens and eyepiece, respectively, of a compound optical microscope. The final image is recorded on photographic film or projected onto a fluorescent screen for viewing or photographing. The entire apparatus, including the specimen, must be enclosed in a vacuum container, just as with a cathode-ray tube; otherwise, electrons would collide with air molecules and muddle up the image. The specimen to be viewed is very thin, typically 10 to 100 nm, so the electrons are not slowed appreciably as they pass through.

We might think that when the electron wavelength is 0.01 nm (as in Example 41–5), the resolution would also be about 0.01 nm. In fact, it is seldom better than 0.5 nm, for several reasons. Large-aperture magnetic lenses have aberrations analogous to those of optical lenses, which we discussed in Section 36–8. The focal length of a magnetic lens depends on the current in the coil, which must be controlled precisely. The focal length also depends on the electron speed, which is never exactly the same for all electrons in the beam. This effect is analogous to chromatic aberration in an optical system.

An important variation is the *scanning electron microscope* (Fig. 41–10a). The electron beam is focused to a very fine line and is swept across the specimen, just as the electron beam in a TV picture tube traces out the picture. As the beam scans the specimen, electrons are knocked off and are collected by a collecting anode that is kept at a potential a few hundred volts positive with respect to the specimen. The current in the collecting anode is amplified and used to modulate the electron beam in a cathode-ray tube, which is swept in synchronization with the microscope beam. Thus the cathode-ray

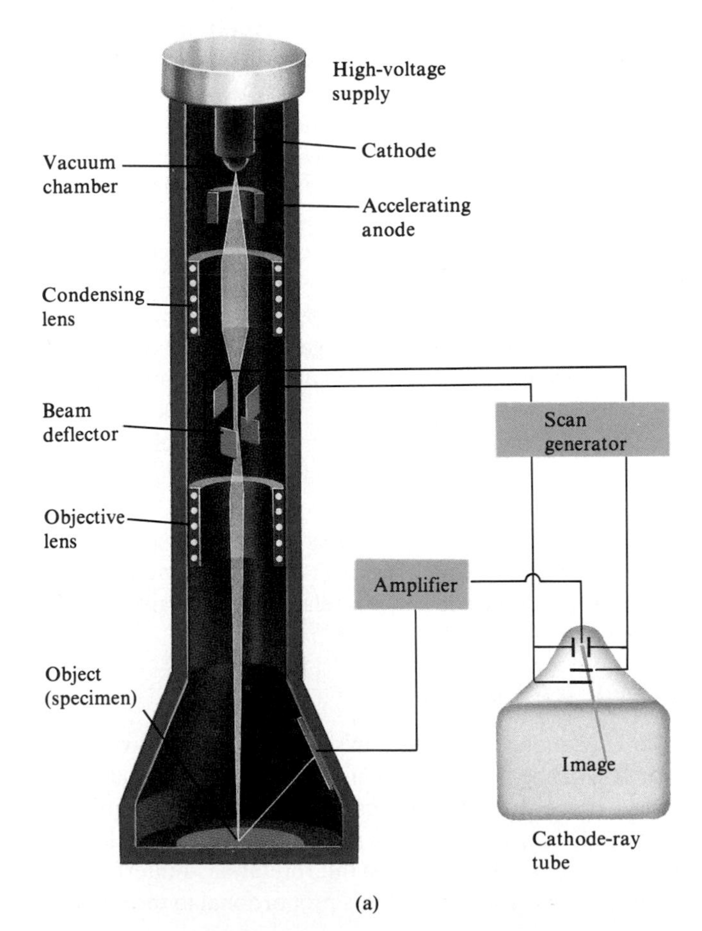


(a)

(b)

(c)

**41–10** (a) Schematic diagram of a scanning electron microscope. The beam deflector uses the signals from the scan generator to scan the beam across the specimen, and those signals simultaneously scan the electron beam in the video display. The signal received by the detector is used to modulate the video display beam, creating light and dark areas on the screen. (b) A scanning electron microscope. The electron-optical system is in the tall cylinder at the left; the console includes a display screen, the control system, and data-processing equipment. (c) A scanning electron micrograph of a sponge spicule.

tube traces out a greatly magnified image of the specimen. This scheme has several advantages. The specimen can be thick because the beam does not need to pass through it. Also, the knock-off electron production depends on the *angle* at which the beam strikes the surface. Thus scanning electron micrographs have a much greater three-dimensional appearance than conventional ones. The resolution is typically of the order of 10 nm, still much greater than the best optical microscopes. A scanning electron microscope is shown in Fig. 41–10b, and a photograph made with such an instrument is shown in Fig. 41–10c.

## 41–5 WAVE FUNCTIONS

We have seen persuasive evidence that on an atomic or subatomic scale a particle such as an electron can't be described as a point with three coordinates and three velocity components. In some situations a particle behaves like a wave, and we have spoken a few times about using a wave function to describe the state of a particle. Now we can describe more specifically the kinematic language that we have to use to replace the classical scheme of coordinates and velocity components.

Our new scheme for describing the state of a particle has a lot in common with the language of classical wave motion. In Chapter 19 we described transverse waves on a string by specifying the position of each point in the string at each instant of time by means of a *wave function* (Section 19–3). If $y$ represents the displacement from equilibrium, at time $t$, of a point on the string at a distance $x$ from the origin, then the function $y = f(x, t)$ or $y(x, t)$ represents the displacement of any point $x$ at any time $t$. Once we know the wave function for a particular wave motion, we know everything there is to know about the motion. We can find the position and velocity of any point on the string at any time, and so on. We worked out specific forms for these functions for *sinusoidal* waves, in which each particle undergoes simple harmonic motion.

We followed a similar pattern for sound waves in Chapter 21. The wave function $p(x, t)$ for a wave traveling along the $x$-direction represented the pressure variation at any point $x$ at any time $t$. We used this language once more in Section 33–3, where we used *two* wave functions to describe the electric and magnetic fields of *electromagnetic* waves at any point in space at any time.

Thus it is natural to use a wave function as the central element of our new language. The symbol usually used for this wave function is $\Psi$ or $\psi$. In general, $\Psi$ is a function of all the space coordinates and time. Just as the wave function $y(x, t)$ for mechanical waves on a string provides a complete description of the motion, the wave function $\Psi(x, y, z, t)$ for a particle contains all the information that can be known about the particle.

Mechanical waves travel through a medium; the stretched string is the medium for transverse waves on a string, and air is the medium for sound waves. But the wave function for a particle *does not* describe a wave that propagates through some material medium. The wave function describes the particle, but we can't define the function itself in terms of anything material. We can only describe how it is related to physically observable effects.

### Interpretation of the Wave Function

The wave function describes the distribution of a particle in space, just as the wave functions for an electromagnetic wave describe the distribution of the electric and magnetic fields. When we worked out interference and diffraction patterns in Chapters 37 and 38, we found that the intensity $I$ of the radiation at any point in a pattern is proportional to the square of the electric-field magnitude, or $E^2$. In the photon interpretation of interference and diffraction (Section 40–9), the intensity at each point is proportional to the number of photons striking near that point or, alternatively, to the *probability* for any individual photon to strike near the point.

In exactly the same way, the square of the wave function at each point represents the probability of finding the particle near that point. More precisely, we should say the square of the *absolute value* of the wave function, $|\Psi|^2$. This is necessary because, as we'll see later, $\Psi$ isn't necessarily a real quantity. It may be a *complex* quantity with real and imaginary parts. (The imaginary part of the function is a real function multiplied by the imaginary unit $i = \sqrt{-1}$.) A more precise statement is that for a particle moving in three dimensions the quantity $|\Psi|^2\,dV$ is the probability that the particle will be found within a volume $dV$ around the point at which $|\Psi|^2$ is evaluated. The particle is most likely to be found in regions where $|\Psi|^2$ is large, and so on. We have already used this interpretation in our discussion of electron-diffraction experiments. For a particle with charge, such as the electron, $|\Psi|^2$ is also proportional to the *charge density* at any point in space.

We have mentioned that for a particle in a particular state, the wave function $\Psi$ is a *complete* description of that state. That is, it contains within it all the information that can be known about the state. In the mathematical theory of quantum mechanics there are definite procedures for determining the average position of the particle, its average velocity, and dynamic quantities such as momentum, energy, and angular momentum. The required techniques are beyond the scope of this discussion, but they are well established and well supported by experimental results.

For a moving free particle the wave function is always a function of both the space coordinates and time. The value of $|\Psi|^2$ at a particular point also varies with time, reflecting the fact that the place where a moving particle is most likely to be found changes with time. But in some cases, such as an electron in an atom in a definite energy level, the value of $|\Psi|^2$ at each point is *constant,* independent of time. In such cases the probability distribution for the particle doesn't change with time. This is always true with states that have a definite energy, and we call such a state a **stationary state.** We will show later that such wave functions can be found by solving a differential equation that doesn't involve time explicitly. This is often a lot easier than solving a more general equation that involves both the space coordinates and time.

We have now reached a stage in our discussion comparable to the end of Chapter 3. At that point we had learned kinematic language for *describing* the motion of a particle but had not yet studied the *dynamic* principles (Newton's laws) that determine what motions are possible. We need a general dynamic principle, comparable to Newton's laws or Maxwell's equations, that will enable us to determine the wave function, or the possible wave functions, for specific physical situations. The needed principle is the *Schrödinger wave equation,* developed by Erwin Schrödinger in 1925. Physically possible states of a system are represented by wave functions that are solutions of this equation, just as physically possible waves on a string have wave functions that are solutions of the corresponding wave equation, Eq. (19–18). We'll study this equation in the next chapter. Among other things, we will learn that the solutions of the Schrödinger equation for any particular system yield a set of allowed *energy levels.* This discovery is of the utmost importance. Before the development of the Schrödinger equation there was no way to predict energy levels from any fundamental theory except the Bohr model for hydrogen, which had very limited success.

## Wave Packets

Finally, let's think once more about how an electron can be a particle and a wave at the same time. Armed with the idea of a wave function, we can be a little more specific about how to reconcile these seemingly incompatible aspects of particle behavior. First we note that a particle with a definite wavelength $\lambda$ also has a definite momentum $p,$ because of

**41–11** (a) Two sinusoidal waves with slightly different wavelengths, shown at one instant of time. (b) The superposition of these waves has a wavelength equal to the average of the two wavelengths but with varying amplitude, giving it a lumpy character not possessed by either individual wave.

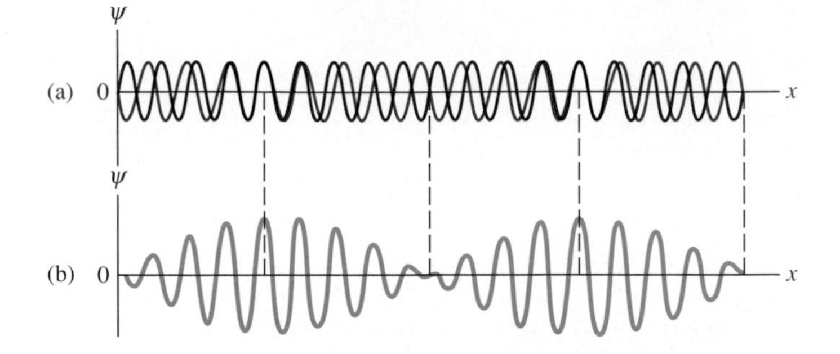


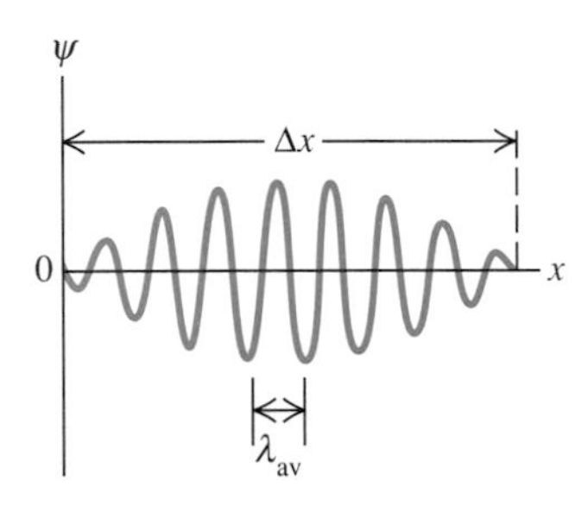


**41–12** Superposing a large number of sinusoidal waves with differing wavelengths and appropriate amplitudes can produce a wave pulse that has a periodic nature with an average wavelength $\lambda_{av}$ and yet is localized in a region of space $\Delta x$. Is this a particle or a wave?

the de Broglie relation, $\lambda = h/p$. For such a state there is *no* uncertainty in momentum: $\Delta p = 0$. The uncertainty principle, Eq. (41–11), says that $\Delta x\,\Delta p_x \geq h/2\pi$. If $\Delta p$ is zero, $\Delta x$ must be infinite. If we know the particle's momentum precisely, we have no idea at all *where* the particle is. Such a state is represented by a sinusoidal wave function with no beginning and no end.

We can superpose two or more sinusoidal functions to make a wave function that is more localized in space. To keep things simple, we'll imagine doing this only in one dimension and at one instant of time. Our wave functions are then functions of only the spatial coordinate $x$, and we denote them by $\psi$. In Section 21–3 we superposed two sinusoidal waves with slightly different frequencies (Fig. 21–4). The result was a wave that had a lumpy character not possessed by the individual waves. Imagine doing the same thing with two particle waves. Superposing two waves with slightly different wavelengths gives the wave shown in Fig. 41–11. A particle represented by this function is more likely to be found in some regions than in others, but the particle's momentum no longer has a definite value because there are two different wavelengths.

It's not hard to imagine superposing two additional sinusoidal waves with different wavelengths so as to reinforce alternate lumps in Fig. 41–11b and cancel out the in-between ones. Finally, if we superpose waves with a very large number of different wavelengths, we can construct a wave with only one lump (Fig. 41–12). Then, finally, we have something that begins to look like both a particle and a wave. It is a particle in the sense that it is localized in space; if we don't look too closely, it may look like a point. But it also has a periodic structure characteristic of a wave.

Such a wave is called a **wave packet.** We can represent such a superposition by an expression such as

$$\psi(x) = \int_0^\infty A(\lambda)\sin\frac{2\pi x}{\lambda}\,d\lambda, \tag{41–13}$$

where $\sin(2\pi x/\lambda)$ is a sinusoidal wave with wavelength $\lambda$. The integral represents a superposition in which we add a very large number of such waves with different values of $\lambda$, each with an amplitude $A(\lambda)$ that depends on $\lambda$.

It turns out that there is a very important relation between the two functions $\psi(x)$ and $A(\lambda)$. It is shown qualitatively in Fig. 41–13. If the function $A(\lambda)$ is sharply peaked,

**41–13** Two different functions $A(\lambda)$ and the corresponding functions $\psi(x)$. The width of $\psi(x)$ is inversely proportional to that of $A(\lambda)$.

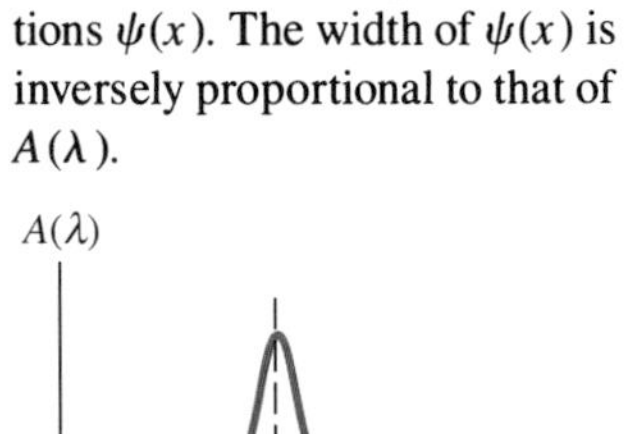


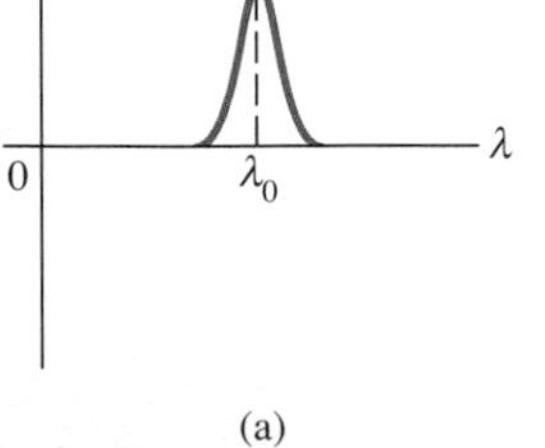


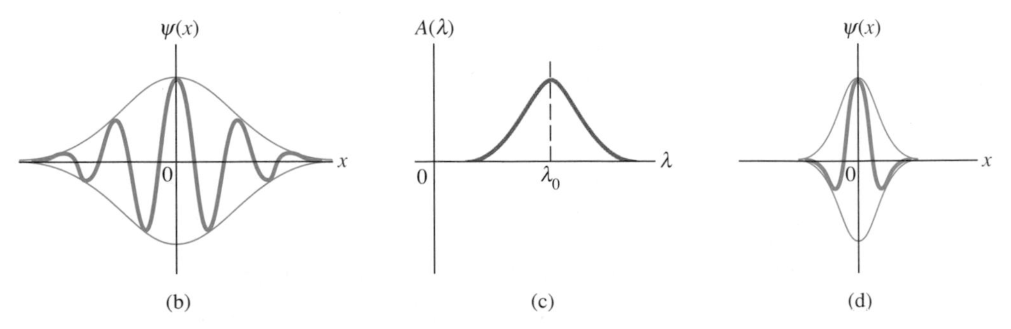

as in Fig. 41–13a, we are superposing only a narrow range of wavelengths. The resulting wave packet $\psi(x)$ is then relatively broad (Fig. 41–13b). But if we use a wide range of values of $\lambda$, so the function $A(\lambda)$ is broad (Fig. 41–13c), then the wave packet is more narrowly localized (Fig. 41–13d). You squeeze it in one place, and it bulges in another!

What we are seeing is the uncertainty principle in action. A narrow range of $\lambda$ means a narrow range of $p$ and thus a small $\Delta p$; the result is a relatively large $\Delta x$. A broad range of $\lambda$ corresponds to a large $\Delta p$, and the resulting $\Delta x$ is smaller. Thus we see that the uncertainty principle is an inevitable consequence of the de Broglie relation and the properties of integrals such as that in Eq. (41–13). These integrals are called *Fourier integrals;* they are a generalization of the concept of Fourier series, which we mentioned in Section 20–3. In both cases we are representing a complicated wave form as a superposition of sinusoidal functions. With Fourier series we use a sequence of frequencies (or values of $1/\lambda$) that are integer multiples of some basic value; with Fourier integrals we superpose functions with a *continuous* distribution of values of $\lambda$.

## SUMMARY

- Electrons and other particles have wave properties. The wavelength and frequency of the wave depend on the particle's momentum and energy in the same way as for photons:

$$\lambda = \frac{h}{p} = \frac{h}{mv}, \tag{41–1}$$

$$E = hf. \tag{41–2}$$

The state of a particle is described not by its coordinates and velocity components but by a wave function, a function of the space coordinates and time.

- A crystal lattice serves as a three-dimensional diffraction grating for both x rays and particles; diffraction of an electron beam from a metallic crystal provided the first direct confirmation of the wave nature of particles and of the correctness of Eq. (41–1). Diffraction experiments have also been carried out with alpha particles, neutrons, and other particles. The wavelength of an electron that has been accelerated through a potential difference $V$ is

$$\lambda = \frac{h}{p} = \frac{h}{\sqrt{2meV}}. \tag{41–5}$$

- It is impossible to make precise determinations of a coordinate of a particle and of the corresponding momentum component at the same time. The precision of such measurements is limited by the Heisenberg uncertainty principle, which is written as

$$\Delta x \, \Delta p_x \geq \frac{h}{2\pi} \tag{41–11}$$

for the $x$-component, with corresponding relations for the $y$- and $z$-components. The uncertainty in energy $\Delta E$ of a state that is occupied for a time $\Delta t$ is given by

$$\Delta E \, \Delta t \geq \frac{h}{2\pi}. \tag{41–12}$$

- The electron microscope uses an electron beam to produce greatly enlarged images of objects. The image formation does not depend on wave properties, but the image resolution is limited by the electron wavelengths, which can be thousands of times smaller than wavelengths for visible light.

**de Broglie wavelength**
**quantum mechanics**
**wave function**
**boundary condition**
**electron diffraction**
**Heisenberg uncertainty principle**
**electron microscope**
**stationary state**
**wave packet**

- The wave function for a system contains all the information that can be known about the physical state of the system. A wave function can be localized in a certain region of space and still have wave properties, giving it wave and particle aspects at the same time.

## EXERCISES

### Section 41–1 De Broglie Waves

**41–1** a) An electron moves with a speed of $5.00 \times 10^6$ m/s. What is its de Broglie wavelength? b) A proton moves with the same speed. Determine its de Broglie wavelength.

**41–2** For crystal diffraction experiments (discussed in Section 41–2), wavelengths of the order of 0.25 nm are often appropriate. Find the energy in electronvolts for a particle with this wavelength if the particle is a) a photon; b) an electron.

**41–3** a) What range of photon energies (in electronvolts) corresponds to the visible spectrum? b) What range of wavelengths would electrons in this energy range have?

**41–4 Wavelength of an Alpha Particle.** An alpha particle ($m = 6.64 \times 10^{-27}$ kg) emitted in the radioactive decay of radium has an energy of 4.60 MeV. What is its de Broglie wavelength?

**41–5 Wavelength of a Car.** Calculate the de Broglie wavelength of a 2000-kg car that is moving at 30.0 m/s. Will the car exhibit wavelike properties?

### Section 41–2 Electron Diffraction

**41–6** A beam of 1600-eV electrons scatters from a set of surface planes of a crystal. The $m = 1$ intensity maximum occurs when the angle $\theta$ in Fig. 41–2c is 34.0°. What is the spacing between adjacent crystal planes?

**41–7** A beam of 1.20-keV alpha particles ($m = 6.64 \times 10^{-27}$ kg) scatters from the surface planes of a crystal, which have a spacing of 72.0 pm. At what angle $\theta$ in Fig. 41–2c does the $m = 1$ intensity maximum occur?

**41–8** A beam of neutrons that all have the same energy scatters from the surface planes of a crystal that have a spacing of 96.0 pm. The $m = 1$ intensity maximum occurs when the angle $\theta$ in Fig. 41–2c is 28.0°. What is the kinetic energy of each neutron in the beam?

### Section 41–3 Probability and Uncertainty

**41–9** A scientist claims to have devised a new method of isolating individual particles that enables him to detect simultaneously their position along an axis to within 0.20 nm and their momentum component along this axis to within $5.0 \times 10^{-27}$ kg · m/s. Use the Heisenberg uncertainty principle to evaluate the validity of this claim.

**41–10** a) The uncertainty in the $y$-component of a proton's position is $2.0 \times 10^{-12}$ m. What is the minimum uncertainty in the $y$-component of the proton's velocity? b) The uncertainty in the $z$-component of an electron's velocity is 0.600 m/s. What is the minimum uncertainty in the $z$-coordinate of the electron?

**41–11** a) The $x$-coordinate of an electron is measured with an uncertainty of 0.20 mm. What is the $x$-component of the electron's velocity if the minimum percent uncertainty in this quantity is 1.0%? b) Repeat part (a) for a proton.

**41–12** An atom in a metastable state has a lifetime of 6.0 ms. What is the uncertainty in energy of the metastable state?

**41–13 Particle Lifetime.** An unstable particle produced in a high-energy collision has a mass that is five times that of a proton and an uncertainty in mass that is 2.0% of the particle's mass. Assuming that rest mass and energy are related by $E = mc^2$, estimate the lifetime of the particle.

### Section 41–4 The Electron Microscope

**41–14** a) In an electron microscope, what accelerating voltage is needed to produce electrons with wavelength 0.0400 nm? b) If protons are used instead of electrons, what accelerating voltage is needed to produce protons with wavelength 0.0400 nm?

**41–15** a) What is the de Broglie wavelength of an electron that has been accelerated through a potential difference of 400 V? b) What is the de Broglie wavelength of a proton accelerated through the same potential difference?

### Section 41–5 Wave Functions

**41–16** Consider the complex-valued function $f(x, y) = (1/\sqrt{2})(x + iy)$. Calculate $|f|^2$.

**41–17** Consider the complex-valued function $\psi(r) = Ae^{ikr}$, where $A$ and $k$ are real constants. Calculate $|\psi|^2$. How does $|\psi|^2$ depend on $r$?

## PROBLEMS

**41–18** A beam of 50-eV electrons traveling in the $+x$-direction passes through a slit that is parallel to the $y$-axis and 8.0 $\mu$m wide. The diffraction pattern is recorded on a screen 2.0 m from the slit. a) What is the de Broglie wavelength of the electrons? b) How much time does it take the electrons to travel from the slit to the screen? c) Use the width of the central maximum to calculate the uncertainty in the $y$-component of momentum of an electron just after it has passed through the slit. d) Use the result of part (c) and the Heisenberg uncertainty principle (Eq. 41–11) to estimate the minimum uncertainty in the $y$-coordinate of an electron just after it has passed through the slit. Compare your result to the width of the slit.

**41–19** What is the de Broglie wavelength of a red blood cell, with a mass of $1.00 \times 10^{-11}$ g, that is moving with a speed of 0.800 cm/s? Do we need to be concerned with the wave nature of blood cells when we describe the flow of blood in the body?

**41–20 Relativistic Matter Waves.** For relativistic particles the de Broglie relation $\lambda = h/p$ still holds, but the momentum is related to the total energy by $E^2 = (pc)^2 + (mc^2)^2$ (Eq. 39–41). Calculate the de Broglie wavelength for a) a proton with total energy 2.00 GeV; b) an electron with total energy 2.00 GeV.

**41–21** Suppose that the uncertainty in the position of a particle in a particular direction is of the order of its de Broglie wavelength. Show that in this case the minimum uncertainty in the particle's momentum component along the same direction is of the order of its momentum.

**41–22 Proton Energy in a Nucleus.** The radii of atomic nuclei are of the order of $1.0 \times 10^{-15}$ m. a) Estimate the minimum uncertainty in the momentum of a proton if it is confined within a nucleus. b) Take this uncertainty in momentum to be an estimate of the magnitude of the momentum. Use the relativistic expression of Eq. (39–41) to obtain an estimate of the kinetic energy of a proton confined within a nucleus.

**41–23 Electron Energy in a Nucleus.** The radii of atomic nuclei are of the order of $1.0 \times 10^{-15}$ m. a) Estimate the minimum uncertainty in the momentum of an electron if it is confined within a nucleus. b) Take this uncertainty in momentum to be an estimate of the magnitude of the momentum. Use the relativistic expression of Eq. (39–41) to obtain an estimate of the energy of an electron confined within a nucleus. c) Compare the energy calculated in part (b) to the magnitude of the Coulomb potential energy of a proton and an electron separated by $1.0 \times 10^{-15}$ m. On the basis of your result, could there be electrons within the nucleus? (*Note:* It is interesting to compare this result and that of Problem 41–22).

**41–24** Suppose that the uncertainty of position of an electron is equal to the radius of the $n = 1$ Bohr orbit for hydrogen. Estimate the uncertainty in the electron's momentum, and compare this with the magnitude of the momentum of the electron in the $n = 1$ Bohr orbit.

**41–25** In a TV picture tube the accelerating voltage is 1.20 kV, and the electron beam passes through an aperture 0.50 mm in diameter to a screen 0.300 m away. a) What is the uncertainty in position of the point where the electrons strike the screen? b) Does this uncertainty affect the clarity of the picture significantly? (Use nonrelativistic expressions for the motion of the electrons. This is fairly accurate and is certainly adequate for obtaining an estimate of uncertainty effects.)

**41–26** The average kinetic energy of a thermal neutron is $\frac{3}{2}kT$. What is the de Broglie wavelength associated with the neutrons in thermal equilibrium with matter at 300 K?

**41–27** The $\pi^0$ meson is an unstable particle produced in high-energy particle collisions. Its mass is about 264 times that of the electron, and it exists for an average lifetime of $0.87 \times 10^{-16}$ s before decaying into two gamma-ray photons. Assuming that the rest mass and energy of this particle are related by the Einstein relation $E = mc^2$, find the uncertainty in the mass of the particle and express it as a fraction of the mass.

**41–28 Atomic Spectra Uncertainties.** A certain atom has an energy level 2.80 eV above the ground level. When excited to this level, it remains on the average $2.0 \times 10^{-6}$ s before emitting a photon and returning to the ground level. a) What is the energy of the photon? What is its wavelength? b) What is the smallest possible uncertainty in energy of the photon? c) Show that $|\Delta E/E| = |\Delta\lambda/\lambda|$ when $|\Delta\lambda/\lambda|$ is small. Use this to calculate the magnitude of the smallest possible uncertainty in the wavelength of the photon.

**41–29 Doorway Diffraction.** If your wavelength were 1.0 m, you would undergo considerable diffraction in moving through a doorway. a) What must be your speed for you to have this wavelength? (Assume that your mass is 60.0 kg.) b) At the speed calculated in part (a), how many years would it take you to move 0.80 m (one step)? Would you notice diffraction effects as you walked through doorways?

**41–30** For x rays with wavelength 0.0200 nm, the $m = 1$ intensity maximum for a crystal occurs when the angle $\theta$ in Fig. 38–21 is 22.9°. At what angle $\theta$ does the $m = 1$ maximum occur when a beam of 9.00-keV electrons is used instead, if the electrons scatter from the surface planes of this same crystal?

**41–31** In another universe the value of Planck's constant is $6.63 \times 10^{-22}$ J · s. Assume that the physical laws and all other physical constants are the same as in our universe. In the other universe, an atom is in an excited state 16.0 eV above the ground state. The lifetime of this excited state (the average time the electron stays in this state) is $1.50 \times 10^{-3}$ s. What is the uncertainty in the energy of the photon emitted when the atom makes the transition from this excited state to the ground state?

**41–32 Zero-Point Energy.** Consider a particle of mass $m$ moving in a potential $U = \frac{1}{2}kx^2$, as in a mass-spring system. The total energy of the particle is $E = (p^2/2m) + \frac{1}{2}kx^2$. Assume that $p$ and $x$ are approximately related by the Heisenberg uncertainty principle, $px \cong h$. a) Calculate the minimum possible value of the energy $E$ and the value of $x$ that gives this minimum $E$. This lowest possible energy, which is not zero, is called the *zero-point energy.* b) For the $x$ calculated in part (a), what is the ratio of the kinetic energy to the potential energy of the particle?

**41–33** In another universe the value of Planck's constant is $0.0663\ \text{J}\cdot\text{s}$. Assume that the physical laws and all other physical constants are the same as in our universe. In the other universe, two physics students are playing catch with a baseball. They are 20 m apart, and one throws a 0.10-kg ball directly toward the other with a speed of 5.0 m/s. a) What is the uncertainty in the ball's horizontal momentum, in a direction perpendicular to that in which it is being thrown, if the student throwing the ball knows that it is located within a cube with volume $1000\ \text{cm}^3$ at the time she throws it? b) By what horizontal distance could the ball miss the second student?

**41–34** A particle has wave function

$$\psi(x, y, z) = Axe^{-\alpha x^2}e^{-\beta y^2}e^{-\gamma z^2},$$

where $A$, $\alpha$, $\beta$, and $\gamma$ are all constants. The probability that the particle will be found in the infinitesimal volume $dxdydz$ centered at the point $(x_0, y_0, z_0)$ is $|\psi(x_0, y_0, z_0)|^2\,dx\,dy\,dz$. a) At what value of $x_0$ is the particle most likely to be found? b) Are there values of $x_0$ for which the probability of finding the particle is zero? If so, at what $x_0$?

**41–35** Consider the wave packet defined in Eq. (41–13). It is often convenient to use the wave number $k = 2\pi/\lambda$ rather than the wavelength and the cosine rather than the sine, so the equation becomes

$$\psi(x) = \int_0^\infty B(k)\cos kx\,dk.$$

Let $B(k) = e^{-\alpha^2 k^2}$. a) The function $B(k)$ has its maximum value at $k = 0$. Let $k_h$ be the value of $k$ when $B(k)$ has fallen to half its maximum value and define the width of $B(k)$ as $w_k = k_h$. In terms of $\alpha$, what is $w_k$? b) Use integral tables to evaluate the integral that gives $\psi(x)$. For what value of $x$ is $\psi(x)$ maximum? c) Define the width of $\psi(x)$ as $w_x = x_h$, where $x_h$ is the positive value of $x$ when $\psi(x)$ has fallen to half its maximum value. Calculate $w_x$ in terms of $\alpha$. d) The momentum $p$ is equal to $hk/2\pi$, so the width of $B$ in momentum is $w_p = hw_k/2\pi$. Calculate the product $w_p w_x$ and compare to the Heisenberg uncertainty principle.

## CHALLENGE PROBLEMS

**41–36** The wave nature of particles results in the quantum mechanical situation that a particle confined in a box can assume only wavelengths that result in standing waves in the box, with nodes at the box walls. a) Show that an electron confined in a one-dimensional box of length $L$ will have energy levels given by

$$E_n = \frac{n^2h^2}{8mL^2}.$$

(*Hint:* Recall that the relation between the de Broglie wavelength and the speed of a particle is $mv = h/\lambda$. The energy of the particle is $\frac{1}{2}mv^2$.) b) If a hydrogen atom is modeled as a one-dimensional box with length equal to the Bohr radius, what is the energy (in electronvolts) of the ground level of the electron?

**41–37** You are entered in a contest to drop a marble with a mass of 30.0 g from the roof of a building onto a small target 30.0 m below. From uncertainty considerations, what is the typical distance by which you will miss the target, given that you aim with the highest possible precision? (*Hint:* The uncertainty $\Delta x_f$ in the $x$-coordinate of the marble when it reaches the ground comes in part from the uncertainty $\Delta x_i$ in the $x$-coordinate initially and in part from the initial uncertainty in $v_x$. The latter gives rise to an uncertainty in the horizontal motion of the marble as it falls. The uncertainties $\Delta x_i$ and $\Delta v_x$ are related by the uncertainty principle; a small $\Delta x_i$ gives rise to a large $\Delta v_x$, and vice versa. Find the $\Delta x_i$ that gives the smallest total uncertainty in $x$ at the ground. Ignore any effects of air resistance.)

A fire in the home can spread very quickly. These photographs show the spread of a fire under controlled laboratory conditions, one minute after combustion and 1.75 minutes after combustion. Smoke detectors are now required by law in most states to provide early warning of a fire. The two most common types of smoke detectors are photoelectric detectors and ionization detectors. Both are simple in operation, but both involve quantum-mechanical phenomena.

# CHAPTER 42

# Quantum Mechanics

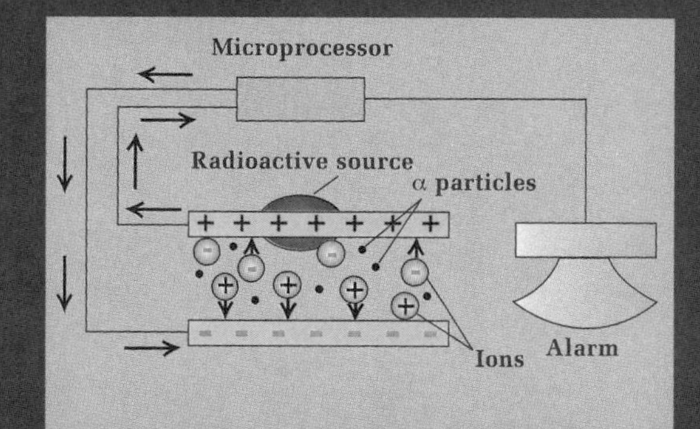


An ionization detector contains a small radioactive source of alpha particles. Usually the source is the element americium. The alpha particles ionize air molecules in the detector's chamber. The ionization allows a constant electric current between charged plates within the chamber. When smoke or other combustion products enter the chamber, they reduce the flow of ions between the plates, triggering the alarm.

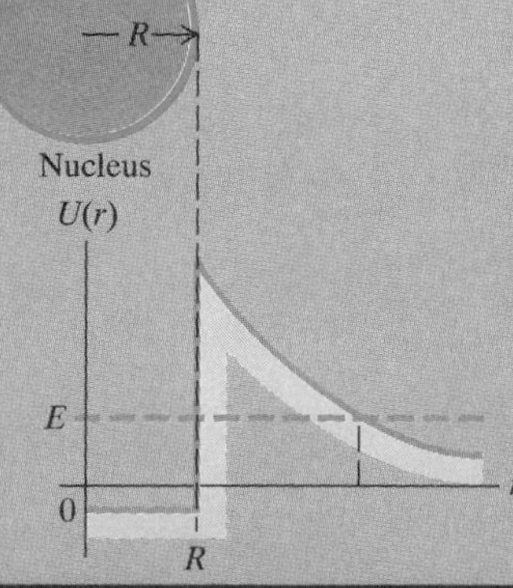


The emission of alpha particles from americium involves a phenomenon called tunneling. The alpha particle escapes from the nucleus of the atom even though it does not have enough energy to do so, according to classical physics.

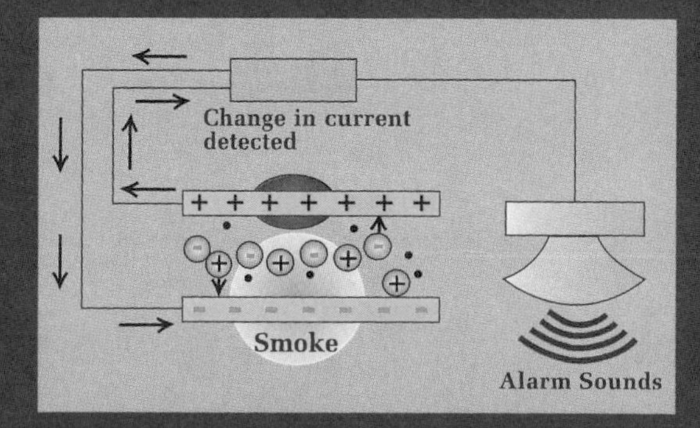

## KEY CONCEPTS

**• A particle confined between two rigid walls can be represented by a wave function. The requirement that the wave function must be zero at both walls leads to a set of possible energy levels for the particle.**

**• The Schrödinger equation determines the physically possible wave functions for a system. Solutions of the Schrödinger equation for a particular system also give the possible energy levels for that system.**

**• A particle in a potential well or near a barrier has a wave function that penetrates some distance into regions where in Newtonian mechanics the particle would have (impossible) negative kinetic energy. A particle can penetrate through a potential-energy barrier by a process called tunneling.**

**• The solutions of the Schrödinger equation for the harmonic oscillator yield a set of equally spaced energy levels.**

**• The Schrödinger equation can be generalized to three dimensions. The wave functions and potential-energy functions are then functions of three space coordinates. Solutions for the hydrogen atom provide the basis for analysis of more complex atoms.**

## INTRODUCTION

In the preceding chapter we discussed the wave properties of particles and the interpretation of the wave function for a system. We described a possible kind of wave function that has both particle and wave properties. Now we're ready for a systematic analysis of particles in bound states (such as electrons in atoms), including finding their possible wave functions and energy levels.

This analysis involves finding solutions of a fundamental equation called the Schrödinger equation. The wave functions for any system must be solutions of the Schrödinger equation for that system. Each solution corresponds to a definite energy, so solving the Schrödinger equation automatically gives the possible energy levels for a system. We'll discuss several simple applications of the Schrödinger equation.

Finally, we'll describe the generalization of the Schrödinger equation to three dimensions. This will pave the way for describing the wave functions for the hydrogen atom in Chapter 43. These in turn form the foundation for our analysis of more complex atoms, including the periodic table of the elements, the existence of x-ray energy levels, and other properties of atoms.

## 42–1 PARTICLE IN A BOX

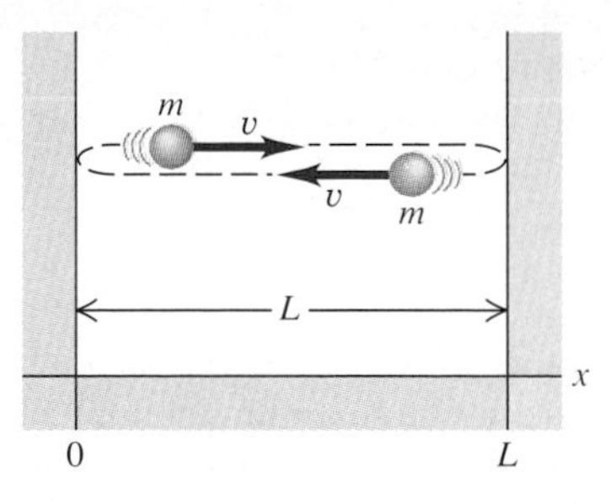


**42–1** A particle in a box. A particle with mass $m$ moves along a straight line, bouncing between two rigid walls that are a distance $L$ apart.

In this chapter we want to learn how to find wave functions and energy levels for various systems. As often happens, the simplest problems are not always the most interesting ones, but it's often helpful to begin exploring new principles with the simplest possible system.

Our first example fits that description. The system consists of a particle bouncing elastically back and forth between two rigid walls separated by a distance $L$ (Fig. 42–1). We'll make it a one-dimensional problem, with the particle always moving along the $x$-axis and the walls located at $x = 0$ and $x = L$. The collisions are perfectly elastic, so the particle never gains or loses energy; both its energy $E$ and the magnitude of its momentum $p$ are constant. This situation is often described succinctly as "a particle in a box."

Because the particle is confined to the region $0 \leq x \leq L$, we expect the wave function to be zero outside that region. Also, it seems physically reasonable that the wave function $\psi$ should be a *continuous* function of $x$. If it is, then it must be zero at $x = 0$ and at $x = L$. These two conditions are *boundary conditions* for the problem. These should look familiar (Fig. 42–2). They are the same conditions we used to find normal modes of a vibrating string in Section 20–3; you may want to review that discussion.

The magnitude of momentum is constant for the particle. Thus only one wavelength is involved, and it's reasonable to assume that $\psi(x)$ is a *sinusoidal* function of $x$. If so, a possible form is

$$\psi(x) = A \sin kx, \tag{42–1}$$

where $k$ is the **wave number** introduced in Section 19–3, Eq. (19–5):

$$k = \frac{2\pi}{\lambda}. \tag{42–2}$$

(We'll discuss the significance of the constant $A$ later.)

Equation (42–1) satisfies the requirement that $\psi(x)$ should be zero at $x = 0$. It is also zero at $x = L$ if we choose values of $k$ such that $kL = n\pi$ ($n = 1, 2, 3, \ldots$). The possible values of $k$ and $\lambda$ are therefore

$$k = \frac{n\pi}{L} \quad \text{and} \quad \lambda = \frac{2\pi}{k} = \frac{2L}{n}. \tag{42–3}$$

Now we are only two short steps from determining the energy levels. From $\lambda$ we can find the momentum $p$, using the de Broglie relation $p = h/\lambda$, and the energy $E$ is $p^2/2m$. For each $n$ there are corresponding values of $p$, $\lambda$, and $E$; let's call them $p_n$, $\lambda_n$, and $E_n$. Putting the pieces together, we get

$$p_n = \frac{h}{\lambda_n} = \frac{nh}{2L}, \tag{42–4}$$

$$E_n = \frac{p_n^{\,2}}{2m} = \frac{n^2h^2}{8mL^2}. \tag{42–5}$$

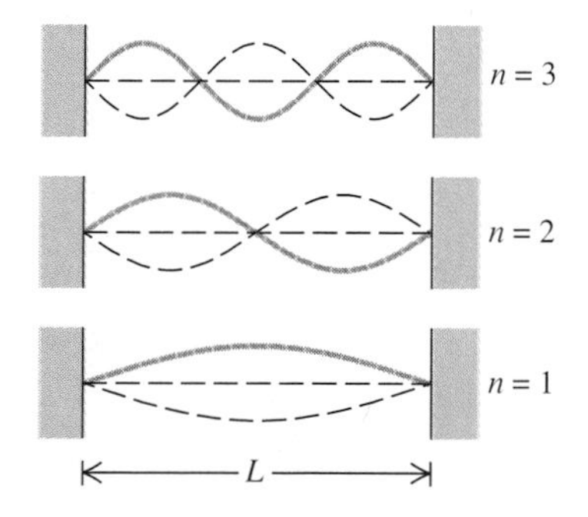


**42–2** Normal modes of vibration for a string with length $L$, held at both ends. Each end is a node, and there may be one or more additional nodes. The length is an integer number of half-wavelengths; $L = n\lambda/2$.

These are the possible energy levels for a particle in a box. Each value of $n$ corresponds to a wave function, which we denote by $\psi_n$. When we replace $k$ in Eq. (42–1) by the expression in Eq. (42–3), we find

$$\psi_n(x) = A \sin\frac{n\pi x}{L}. \tag{42–6}$$

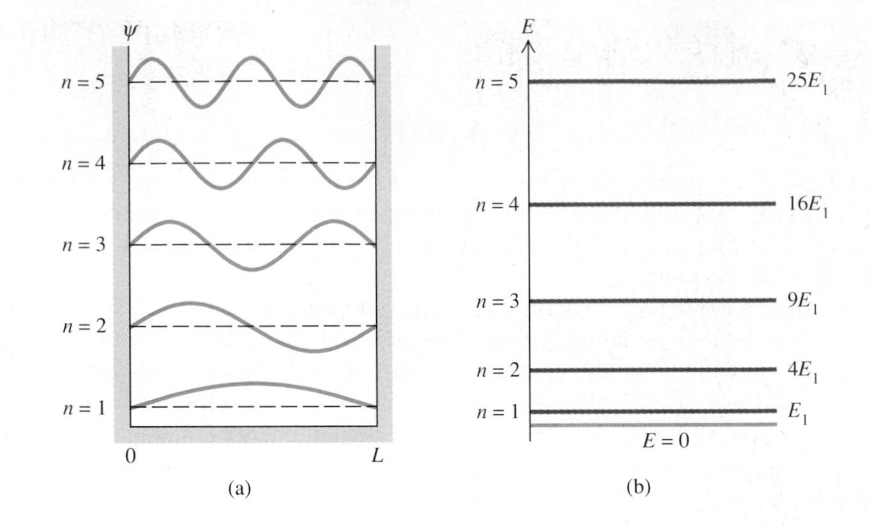


**42–3** (a) Wave functions for a particle in a box, with $n = 1, 2, 3, 4$, and 5. (b) Energy-level diagram for a particle in a box. Each energy is $n^2E_1$, where $E_1$ is the ground-state energy; $n$ is the quantum number for each state.

Figure 42–3a shows graphs of the wave functions for $n = 1, 2, 3, 4$, and 5, and Fig. 42–3b shows the energy-level diagram for this system. Successively higher levels, proportional to $n^2$, are spaced farther and farther apart, and there is an infinite number of levels.

## EXAMPLE 42–1

**Electron in an atom-size box** Find the lowest energy level for a particle in a box if the particle is an electron in a box $5.0 \times 10^{-10}$ m across, or a little bigger than an atom.

**SOLUTION** Using $n = 1$ in Eq. (42–5) to find the smallest $E$ (that is, the ground state), we get

$$E_1 = \frac{h^2}{8mL^2} = \frac{(6.626 \times 10^{-34}\ \mathrm{J \cdot s})^2}{8(9.109 \times 10^{-31}\ \mathrm{kg})(5.0 \times 10^{-10}\ \mathrm{m})^2}$$
$$= 2.4 \times 10^{-19}\ \mathrm{J} = 1.5\ \mathrm{eV}.$$

A particle trapped in a box is rather different from an electron trapped in an atom, but it is reassuring that this energy is of the same order of magnitude as actual atomic energy levels.

## EXAMPLE 42–2

**Neutron in a nucleus-size box** Repeat the calculation of Example 42–1 for a neutron in a box $5.0 \times 10^{-15}$ m across, or the size of a nucleus.

**SOLUTION** From Appendix F the mass of the neutron is $m = 1.675 \times 10^{-27}$ kg. We find

$$E_1 = \frac{h^2}{8mL^2} = \frac{(6.626 \times 10^{-34}\ \mathrm{J \cdot s})^2}{8(1.675 \times 10^{-27}\ \mathrm{kg})(5.0 \times 10^{-15}\ \mathrm{m})^2}$$
$$= 1.3 \times 10^{-12}\ \mathrm{J} = 8.2\ \mathrm{MeV}.$$

This result has the same order of magnitude as nuclear *binding energies,* expressed on a "per nucleon" basis, that is, the energy required to remove a single proton or neutron from the nucleus. Again, this agreement is reassuring.

You may want to repeat this calculation once more for a billiard ball ($m = 0.2$ kg) bouncing back and forth between the cushions of an idealized perfectly elastic billiard table ($L = 1.5$ m). You will find that the energy levels are somewhere around $10^{-67}$ J. This result shows that quantum effects won't have much effect on your billiard game.

Several additional points need to be discussed. First is the *probability* interpretation of the wave function $\psi(x)$, which we discussed in Section 41–5. In our one-dimensional situation, the quantity $|\psi|^2\,dx$ (with $\psi$ evaluated at a particular value of $x$) is the probability that the particle will be found in a small interval $dx$ near $x$. In our case,

$$|\psi|^2\,dx = A^2 \sin^2 \frac{n\pi x}{L}\,dx. \qquad (42\text{–}7)$$

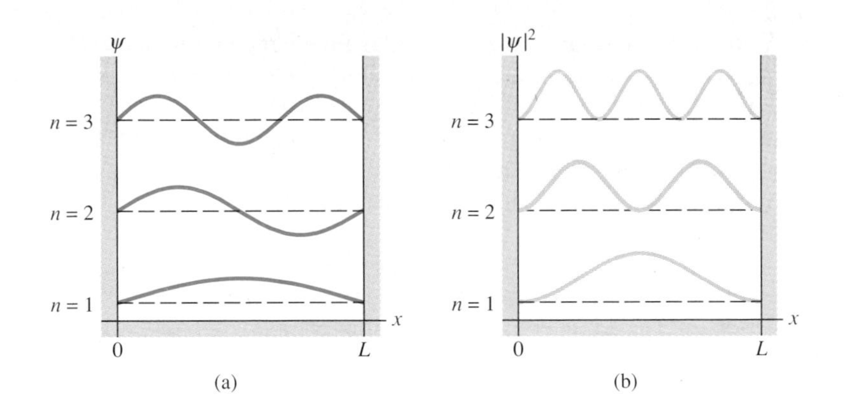


**42–4** Graphs of (a) $\psi$ and (b) $|\psi|^2$ for the first three wave functions ($n = 1, 2, 3$) for a particle in a box. The value of $|\psi|^2$ at each point is proportional to the probability of finding the particle in a small interval near that point.

The quantity $|\psi|^2$ is plotted in Fig. 42–4 for $n = 1, 2,$ and 3. We note that not all positions are equally likely. This is in contrast to the situation in classical mechanics, in which all positions between $x = 0$ and $x = L$ are equally likely.

We know that the particle *must* be somewhere in the interval $0 \leq x \leq L$. So the *sum* of the probabilities for all the $dx$'s in this interval (the *total* probability of finding the particle) must be unity. That is,

$$\int_0^L |\psi|^2 \, dx = \int_0^L A^2 \sin^2 \frac{n\pi x}{L} \, dx = 1. \tag{42–8}$$

Whether this equation is true depends on the value of the constant $A$. The value of the integral, which you can work out with the help of an integral table, is $A^2L/2$. Our probability interpretation of the wave function demands that $A^2L/2 = 1$, or $A = (2/L)^{1/2}$. (We note that the constant $A$ is *not* arbitrary; this is in contrast to the classical vibrating string problem, in which $A$ represented an amplitude that depended on initial conditions.)

A wave function with a factor such as $A$, chosen so that Eq. (42–8) is satisfied, is said to be *normalized*. The process of determining the constant is called **normalization.** Thus the normalized wave functions for a particle in a box are

$$\psi_n(x) = \sqrt{\frac{2}{L}} \sin \frac{n\pi x}{L}. \tag{42–9}$$

Wave functions used in quantum mechanics are nearly always normalized.

Next, let's check whether our results for the particle in a box are consistent with the uncertainty principle. The uncertainty in position is $\Delta x = L$ (the width of the box). From Eq. (42–4) the magnitude of momentum $p$ in state $n$ is $p = nh/2L$. A reasonable estimate of the momentum uncertainty is the difference in momentum of two states that differ by 1 in their $n$ values, that is, $\Delta p = h/2L$. Then the product $\Delta x \, \Delta p$ in the uncertainty principle, Eq. (41–11), is

$$\Delta x \, \Delta p = \frac{h}{2}.$$

This is similar to the limit imposed by the uncertainty principle. Note that the total energy *cannot* be zero; this would certainly violate the uncertainty principle. Do you see why?

Finally, we ask whether the wave functions for this problem have any *time* dependence. We recall that the wave functions for standing waves on a string, Eq. (20–1), contained a time factor cos $\omega t$. Is there a similar time dependence here? The answer, which we can't discuss in detail, is that there *is* a sinusoidal time dependence, but it can't have the form cos $\omega t$. If it did, all the probability expressions such as Eqs. (42–7) and (42–8) would also be time-dependent. If the time-dependent function $\Psi$ were normalized at one instant, it wouldn't be normalized at any other time.

It turns out that the time dependence is contained in a factor

$$e^{-i\omega t} = \cos \omega t - i \sin \omega t,$$

where $\omega$ is determined by the de Broglie energy-frequency relation, Eq. (41–2); that is, $\omega = 2\pi f = 2\pi E/h$. The exponential function is a complex quantity whose real and imaginary parts are cos $\omega t$ and $-i$ sin $\omega t$, respectively. This relation is called *Euler's formula;* if you haven't encountered it yet in your study of calculus, you probably will do so soon.

The *absolute value* of the exponential function is unity. If we multiply each $\psi$ by this factor, it doesn't change the value of $|\Psi|^2$. So in calculations with states that have definite energy we are justified in omitting the time factor. We pointed out in Section 41–5 that a state with definite energy is a *stationary state* with the property that, at each point, $|\Psi|^2$ is independent of time.

## 42–2 THE SCHRÖDINGER EQUATION

The Schrödinger equation is the basic relationship for determining wave functions and energy levels. We will apply it to several systems, including the particle in a box (which we've already discussed), the harmonic oscillator, and the hydrogen atom. We won't pretend that we can *derive* this equation from known principles. We can't; it is a new principle. But we can show how it is related to the de Broglie equations, and we can make it seem plausible. The ultimate test of the Schrödinger equation is to compare its predictions with experimental observations. As we will see, it passes such a test with flying colors.

We'll begin with a one-dimensional problem, a particle moving along the $x$-axis. The wave functions for a particle in a box have the general form

$$\psi(x) = A \sin kx = A \sin \frac{2\pi x}{\lambda}. \tag{42–10}$$

The energy $E$ of the corresponding state can be expressed, using the de Broglie relation, as

$$E = \frac{p^2}{2m} = \frac{h^2}{2\lambda^2 m} = \frac{h^2 k^2}{8\pi^2 m}. \tag{42–11}$$

Notice that the second derivative of Eq. (42–10) is $-k^2 A \sin kx$, that is, the original function multiplied by $-k^2$. So taking the second derivative and then multiplying by the factor $(-h^2/8\pi^2 m)$ has the same effect as multiplying by the factor $(-h^2/8\pi^2 m)(-k^2)$, which is equal to $E$. That is,

$$-\frac{h^2}{8\pi^2 m}\frac{d^2\psi}{dx^2} = E\psi. \tag{42–12}$$

We invite you to verify that Eq. (42–10) satisfies this equation, no matter what the value of $k$ is.

Equation (42–12) is the simplest form of the **Schrödinger equation.** We obtained it in a somewhat contrived way, but we can see that it is consistent with what we know about wave properties of particles and about the wave functions for a particle in a box.

We can avoid writing a lot of factors of $2\pi$ in later discussions by using the abbreviation $\hbar$ (read as "h-bar") for Planck's constant divided by $2\pi$:

$$\hbar = \frac{h}{2\pi} = 1.055 \times 10^{-34}\ \mathrm{J \cdot s}. \tag{42–13}$$

We can re-express the de Broglie relations in terms of $\hbar$, the wave number $k$, and the angular frequency $\omega$:

$$p = \hbar k, \qquad E = \hbar\omega. \tag{42–14}$$

In terms of $\hbar$, Eq. (42–12) is

$$-\frac{\hbar^2}{2m}\frac{d^2\psi}{dx^2} = E\psi. \tag{42–15}$$

So far we've been talking about a *free* particle, a particle that moves along the $x$-axis with no force acting on it (except for the impulsive forces that act when a particle in a box strikes a wall). Now suppose the particle is acted on by a force, with an $x$-component $F(x)$ that depends on $x$. We'll assume that the force is conservative, so there is a corresponding potential energy $U(x)$. In guessing how to generalize the Schrödinger equation for this more general problem, we start with the classical energy relation for motion under a conservative force: $K + U = E$, or

$$\frac{p^2}{2m} + U(x) = E.$$

Comparing this with Eq. (42–15), we see that a reasonable guess for the generalized one-dimensional Schrödinger equation is

$$-\frac{\hbar^2}{2m}\frac{d^2\psi(x)}{dx^2} + U(x)\psi(x) = E\psi(x). \tag{42–16}$$

(We have written $\psi(x)$ instead of just $\psi$ as a reminder that $\psi$ is indeed a function of $x$.) How do we know that this equation is really right? Because it works. Predictions made using this equation agree with experimental results and thus confirm that this is indeed the correct way to include interactions in the Schrödinger equation. We'll apply this equation to several problems in the following sections.

In working with this more general form of the Schrödinger equation, we will insist as we did before that the wave functions be *normalized:* $\int |\psi|^2\,dx = 1$. Note that if $\psi$ is a solution of Eq. (42–16), $A\psi$ is also a solution, where $A$ is any constant. We can always choose the value of $A$ to satisfy the normalization requirement.

We will also insist that the function $\psi(x)$ and its first derivative must be *continuous*. The basis for this requirement is less obvious than the need for normalization, although it can be justified by theoretical arguments. Here we merely point out that it is analogous to the requirement that the wave functions for a vibrating string, and their derivatives, should be continuous at points where the linear mass density or tension in the string changes. An exception occurs at any point where the potential becomes infinite; at such points only the function itself, not its derivative, needs to be continuous.

We can interpret the particle in a box in terms of Eq. (42–16) by representing the interaction with the walls in terms of the potential-energy function shown in Fig. 42–5. The potential energy is zero for $0 \le x \le L$ but infinitely great everywhere else. Then the wave function must be zero except where $0 \le x \le L$. We see that the continuity requirement then demands that $\psi(x)$ be zero at $x = 0$ and at $x = L$, as we assumed in our initial discussion of this problem.

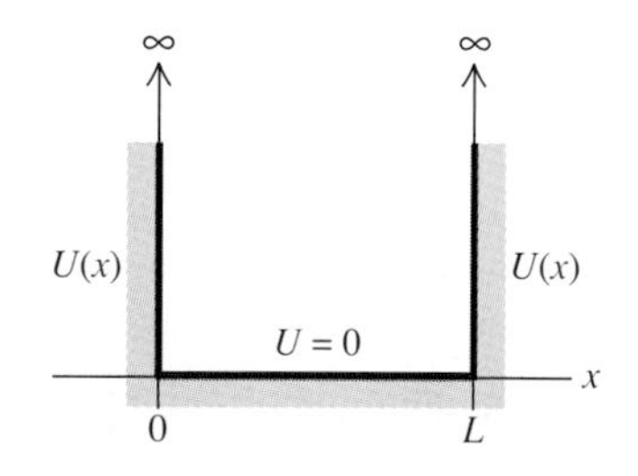


**42–5** The potential-energy function for a particle in a box. The potential energy $U$ is zero in the interval $0 \le x \le L$ and is infinite everywhere outside this interval.

To obtain the wave functions and energies for the particle in a box directly from Eq. (42–15), we note that $\psi(x)$ must satisfy this equation and the conditions $\psi(0) = 0$ and $\psi(L) = 0$. We rewrite Eq. (42–15) in the form

$$\frac{d^2\psi}{dx^2} = -\frac{2mE}{\hbar^2}\psi.$$

The solutions of this equation have the form sin $kx$ and cos $kx$, with $k = (2mE)^{1/2}/\hbar$. The condition $\psi(0) = 0$ is satisfied only by sin $kx$; and to satisfy $\psi(L) = 0$, we must set $kL = n\pi$. So both boundary conditions are satisfied only when

$$\frac{n\pi}{L} = \frac{\sqrt{2mE}}{\hbar}.$$

Solving this for $E$, we obtain

$$E = \frac{n^2\pi^2\hbar^2}{2mL^2}. \tag{42–17}$$

Substituting $\hbar = h/2\pi$, we see that this energy expression agrees with our earlier result, Eq. (42–5). The values of $k$ in the wave function are the same as we found previously, and the normalization is carried out just as before.

## 42–3 POTENTIAL WELLS

A **potential well** is a potential-energy function $U(x)$ that has a minimum. We introduced this term in Section 7–6, and we also used it in our discussion of periodic motion in Chapter 13. In Newtonian mechanics a particle trapped in a potential well can vibrate back and forth with periodic motion. Our first application of the Schrödinger equation, to the particle in a box, involved a rudimentary potential well, a function $U(x)$ that is zero within a certain interval and infinite everywhere else. This function corresponds approximately to some situations in nature, such as the surface of a crystal with electrons "trapped" inside.

A potential well that is more closely related to several actual physical situations is a well with straight sides but *finite* height. Figure 42–6 shows a potential-energy function that is zero in the interval $0 \leq x \leq L$ and has the value $U_0$ outside this interval. This function is often called a **square-well potential.** It could serve as a simple model of an electron within a metallic sheet with thickness $L$, moving perpendicular to the surfaces of the sheet. The electron can move freely inside the metal but has to climb a potential-energy barrier with height $U_0$ to escape from either surface of the metal. The quantity $U_0$ then corresponds to the *work function* we discussed in Section 40–2 in connection with the photoelectric effect. The motions of protons and neutrons within a nucleus can also be described approximately with a spherical three-dimensional version of such a potential well.

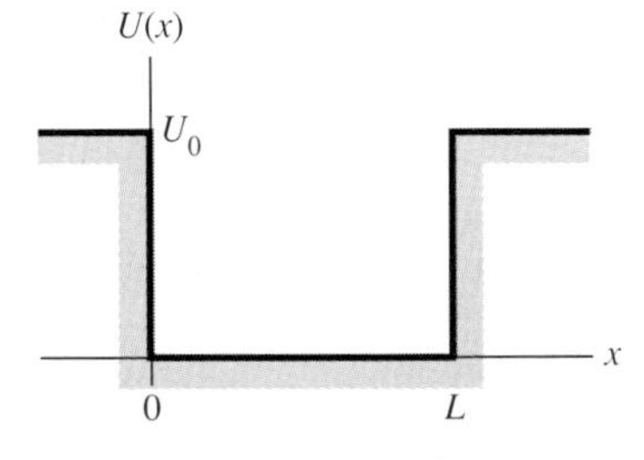


**42–6** A finite potential well, or square-well potential. The potential-energy function $U(x)$ is zero in the interval $0 \leq x \leq L$ and has the value $U_0$ everywhere outside this interval.

In Newtonian mechanics the particle is trapped (localized) in the well if the energy $E$ is less than $U_0$. In quantum mechanics, such a localized state is often called a **bound state.** If $E$ is greater than $U_0$, the particle is *not* bound. Let's consider bound-state solutions of the Schrödinger equation, corresponding to $E < U_0$. The easiest approach is to consider separately the regions where $U = 0$ and where $U = U_0$. Where $U = 0$, the Schrödinger equation is Eq. (42–16) with $U(x) = 0$. Rearranging this equation, we find

$$\frac{d^2\psi}{dx^2} = -\frac{2mE}{\hbar^2}\psi. \tag{42–18}$$

The solutions of this equation are sinusoidal, with the form

$$\psi = C_1 \sin\frac{\sqrt{2mE}}{\hbar}x + C_2 \cos\frac{\sqrt{2mE}}{\hbar}x, \tag{42–19}$$

where $C_1$ and $C_2$ are constants. So far, this looks a lot like the particle-in-a-box analysis in Section 42–2.

In the regions $x \leq 0$ and $x \geq L$ we again use Eq. (42–16) but with $U(x) = U_0$. Rearranging, we get

$$\frac{d^2\psi}{dx^2} = \frac{2m(U_0 - E)}{\hbar^2}\psi. \tag{42–20}$$

The quantity $U_0 - E$ is positive, so the solutions of this equation are exponential. Using the abbreviation $K = [2m(U_0 - E)]^{1/2}/\hbar$, we can write the solutions as

$$\psi = D_1 e^{Kx} + D_2 e^{-Kx}, \tag{42–21}$$

where $D_1$ and $D_2$ are constants.

The bound-state wave functions for this system are sinusoidal within the well and exponential outside. We have to use the positive exponent in the $x \leq 0$ region and the negative exponent in the $x \geq L$ region; otherwise, the wave function would go to infinity at large $|x|$. We also have to match the functions so that they satisfy the boundary conditions that we mentioned in Section 42–2: $\psi$ and $d\psi/dx$ must be continuous at the boundary points ($x = 0$ and $x = L$). If the wave function $\psi$ or the slope $d\psi/dx$ were to change discontinuously at a point, the second derivative $d^2\psi/dx^2$ would be *infinite* at that point. But that would violate the Schrödinger equation, which says that, at every point, $d^2\psi/dx^2$ is proportional to $U - E$. In our situation, $U - E$ is finite everywhere, so $d^2\psi/dx^2$ must also be finite everywhere.

Matching the sinusoidal and exponential functions at the boundary points so that they join smoothly is possible only for certain specific values of the total energy $E$, so this requirement determines the possible energy levels. There is no simple formula for the energy levels as there was for the infinitely deep well. Finding the levels is a fairly complex mathematical problem that requires solving a transcendental equation by numerical approximation; we won't go into the details. Figure 42–7 shows the general shape of a possible wave function. The most striking features of this function are the "tails" that extend outside the well into regions that are forbidden by Newtonian mechanics (because the particle would have to have negative kinetic energy). There is some probability for finding the particle *outside* the potential well, despite the fact that in classical mechanics this is impossible. This penetration into classically forbidden regions is a quantum effect that has no classical analog.

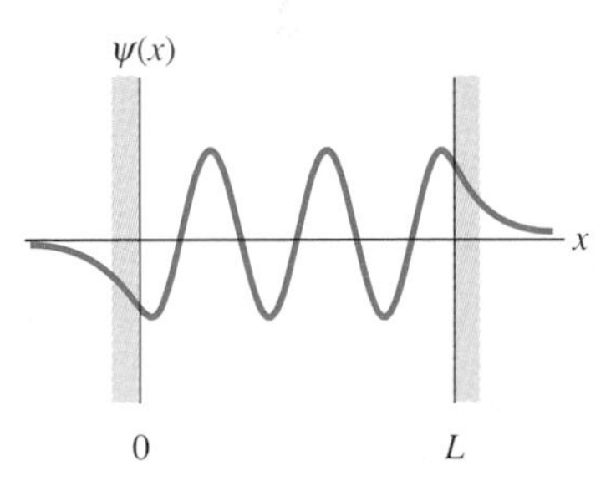

**42–7** A possible wave function for a particle in a finite potential well. The function is sinusoidal inside the well ($0 \leq x \leq L$) and exponential outside. It approaches zero asymptotically at large $|x|$. The functions must join smoothly at $x = 0$ and $x = L$; the function and its derivative must be continuous.

It is interesting to compare this situation with the infinitely deep well. First, because the wave functions for the finite well aren't zero at $x = 0$ and $x = L$, the wavelength of the sinusoidal part of each wave function is a little *longer* than it would be with an infinite well. This corresponds to reduced momentum and therefore reduced energy, so each energy level, including the ground state, is a little *lower* than for an infinitely deep well with the same width.

Second, a well with finite depth $U_0$ has only a *finite* number of bound states and corresponding energy levels, compared to the *infinite* number for an infinitely deep well. How many levels there are depends on the magnitude of $U_0$ in comparison with the ground-state energy for the infinite well, which we call $E_\infty$. From Eq. (42–17),

$$E_\infty = \frac{\pi^2\hbar^2}{2mL^2}. \tag{42–22}$$

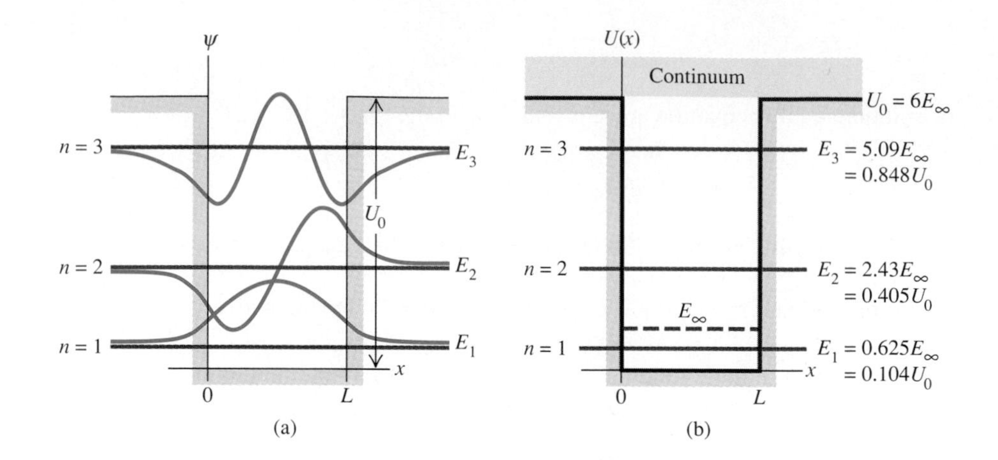


**42–8** (a) Wave functions for the three bound states for a particle in a finite potential well with depth $U_0$, for the case $U_0 = 6E_\infty$, where $E_\infty$ is the ground-state energy for a particle in an infinitely deep well with the same width. (b) Energy-level diagram for this system. The energies are expressed both as multiples of $E_\infty$ and as fractions of $U_0$. All energies greater than $U_0$ are possible; these states form a continuum.

When $U_0$ is much *larger* than $E_\infty$ (a very deep well), there are many levels, and the lowest few are nearly the same as for the infinitely deep well. When $U_0$ is only a few times as large as $E_\infty$, there are only a few bound states. (But there is always at least one bound state, no matter how shallow the well.) Finally, as with the infinitely deep well, there is *no* state with $E = 0$; such a state would violate the uncertainty principle.

Figure 42–8 shows the particular case for which $U_0 = 6E_\infty$; for this case there are three bound states. The energy levels are expressed both as fractions of the well depth $U_0$ and as multiples of $E_\infty$. Note that if the well were infinitely deep, the lowest three levels, as given by Eq. (42–17), would be $E_\infty$, $4E_\infty$, and $9E_\infty$. The wave functions for these three states are also shown.

It turns out that when $U_0$ is less than $E_\infty$, there is only one bound state. In the limit when $U_0$ is *much smaller* than $E_\infty$ (a very shallow or very narrow well), the energy of this single state is approximately $E = 0.68U_0$.

Figure 42–9 shows graphs of the probability distributions, that is, the values of $|\psi|^2$, for the wave functions shown in Fig. 42–8. As with the infinite well, not all positions are equally likely. We have already commented on the possibility of finding the particle outside the well, in the classically forbidden regions.

There are also states for which $E$ is *greater than* $U_0$. In this case the particle is not bound but is free to move through all values of $x$. Thus *any* energy $E$ greater than $U_0$ is possible. The free-particle states thus form a *continuum* rather than a discrete set of states with definite energy levels. The free-particle wave functions are sinusoidal both inside and outside the well. The wavelength is shorter inside the well than outside, corresponding to greater kinetic energy in the well than outside.

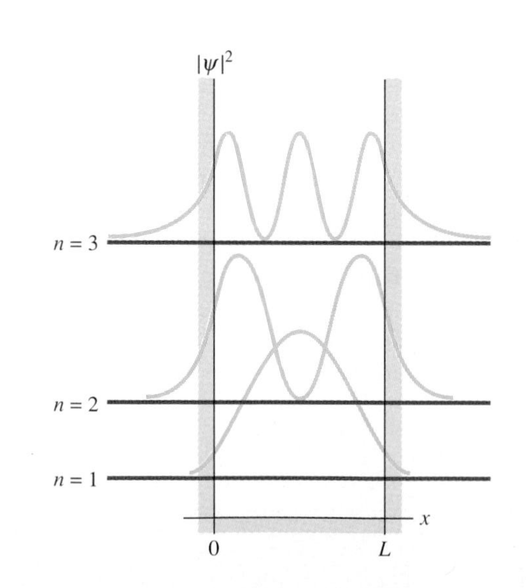


**42–9** Probability distributions (values of $|\psi|^2$) for the wave functions shown in Fig. 42–8 for a particle in a square-well potential.

## PROBLEM-SOLVING STRATEGY

### The Schrödinger equation

**1.** Be careful to distinguish between $h = 6.626 \times 10^{-34}\ \mathrm{J \cdot s}$ and $\hbar = h/2\pi = 1.055 \times 10^{-34}\ \mathrm{J \cdot s}$. Usually, $\hbar$ is more convenient to use in the Schrödinger equation, but in working with photon energies or wavelengths you may need $h$. Don't get the two confused.

**2.** When you have energies expressed in electronvolts, it's often convenient to use the value $h = 4.136 \times 10^{-15}\ \mathrm{eV \cdot s}$.

**3.** Think about the magnitudes of the various quantities. Atomic dimensions are typically 0.2 nm, typical energy levels in atoms are a few electronvolts, and so on. Visible-light photons have energies of a few electronvolts, wavelengths of a few hundred nanometers, and frequencies around $10^{15}$ Hz. If your calculations give you numbers with orders of magnitude that are wildly different from these, recheck your work. And always carry units through your calculations so that you can check unit consistency.

## EXAMPLE 42–3

**Electron in a square well** An electron is trapped in a square well with a width of 0.50 nm (comparable to a few atomic diameters). a) Find the ground-state energy if the well is infinitely deep. b) If the actual well depth is six times the result of part (a), find the energy levels. c) If the electron in the actual well makes a transition from the state with energy $E_2$ to the state with energy $E_1$ by emitting a photon, find the wavelength of the photon. In what region of the electromagnetic spectrum does the photon lie? d) If the electron is initially in its ground state and absorbs a photon, what is the minimum energy the photon must have to liberate the electron from the well? In what region of the spectrum does the photon lie?

**SOLUTION** a) The ground-state energy $E_\infty$ for an infinitely deep well is given by Eq. (42–22):

$$E_\infty = \frac{\pi^2\hbar^2}{2mL^2} = \frac{\pi^2(1.055 \times 10^{-34}\ \mathrm{J \cdot s})^2}{2(9.109 \times 10^{-31}\ \mathrm{kg})(0.50 \times 10^{-9}\ \mathrm{m})^2}$$
$$= 2.4 \times 10^{-19}\ \mathrm{J} = 1.5\ \mathrm{eV}.$$

b) We are given $U_0 = 6E_\infty$, so $U_0 = 6(1.5\ \mathrm{eV}) = 9.0\ \mathrm{eV}$. This is the same ratio as in the example shown in Fig. 42–8, so we can read off the energy levels. We find

$$E_1 = 0.104U_0 = 0.104(9.0\ \mathrm{eV}) = 0.94\ \mathrm{eV},$$
$$E_2 = 0.405U_0 = 0.405(9.0\ \mathrm{eV}) = 3.6\ \mathrm{eV},$$
$$E_3 = 0.848U_0 = 0.848(9.0\ \mathrm{eV}) = 7.6\ \mathrm{eV}.$$

Alternatively,

$$E_1 = 0.625E_\infty = 0.625(1.5\ \mathrm{eV}) = 0.94\ \mathrm{eV},$$
$$E_2 = 2.43E_\infty = 2.43(1.5\ \mathrm{eV}) = 3.6\ \mathrm{eV},$$
$$E_3 = 5.09E_\infty = 5.09(1.5\ \mathrm{eV}) = 7.6\ \mathrm{eV}.$$

Note that if the well had been infinitely deep, the energies would have been

$$E_1 = E_\infty = 1.5\ \mathrm{eV},$$
$$E_2 = 4E_\infty = 6.0\ \mathrm{eV},$$
$$E_3 = 9E_\infty = 13.5\ \mathrm{eV}.$$

As we mentioned earlier, the finite depth of the well *lowers* the energy levels in comparison to the values for an infinitely deep well.

c) The photon energy is

$$E_2 - E_1 = 3.6\ \mathrm{eV} - 0.94\ \mathrm{eV} = 2.7\ \mathrm{eV}$$
$$= 4.3 \times 10^{-19}\ \mathrm{J}.$$

We determine the wavelength from $E = hf = hc/\lambda$:

$$\lambda = \frac{hc}{E} = \frac{(6.626 \times 10^{-34}\ \mathrm{J \cdot s})(3.00 \times 10^8\ \mathrm{m/s})}{4.3 \times 10^{-19}\ \mathrm{J}}$$
$$= 4.6 \times 10^{-7}\ \mathrm{m} = 460\ \mathrm{nm}.$$

Alternatively,

$$\lambda = \frac{hc}{E} = \frac{(4.14 \times 10^{-15}\ \mathrm{eV \cdot s})(3.00 \times 10^8\ \mathrm{m/s})}{2.7\ \mathrm{eV}}$$
$$= 460\ \mathrm{nm}.$$

This photon is in the blue region of the visible spectrum.

d) The energy needed to lift the electron out of the well is the well depth (9.0 eV) minus the electron's initial energy (0.94 eV), or 8.1 eV $= 1.3 \times 10^{-18}$ J. We leave it to you to show that the corresponding photon wavelength is 150 nm, in the ultraviolet region of the spectrum.

The square-well potential described in this section has a number of practical applications. We mentioned earlier the example of an electron inside a metallic sheet. A three-dimensional version, in which $U$ is zero inside a spherical region with radius $R$ and has the value $U_0$ outside, provides a simple model to represent the interaction of a neutron with a nucleus in neutron-scattering experiments. In this context the model is called the *crystal-ball model* of the nucleus; this name refers to the fact that neutrons in such a potential are scattered in a way that's analogous to refraction of light by a crystal ball.

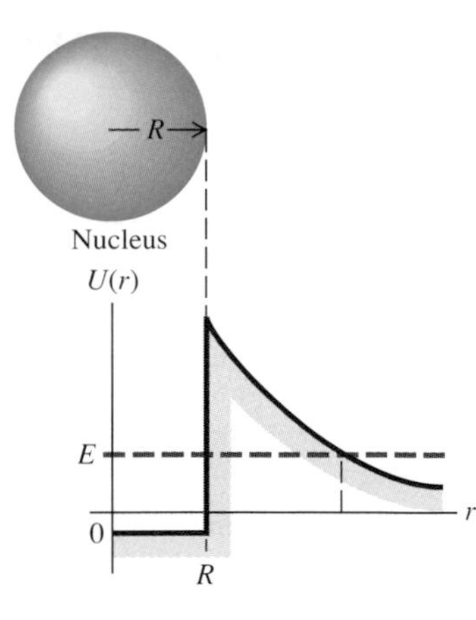


**42–14** Approximate potential-energy function for an alpha particle in a nucleus. The nucleus is represented as a sphere with radius $R$. When the alpha particle is inside the nucleus ($r < R$), the potential resembles a square well. When the alpha particle is outside, the potential is a $1/r$ function associated with the electrical repulsion. An alpha particle with energy $E$ greater than zero inside the well can tunnel through the barrier and escape from the nucleus.

the needle is at a positive potential with respect to the surface, electrons can tunnel through the surface potential-energy barrier and reach the needle. As Eq. (42–23) shows, the tunneling probability and thus the tunneling current depend critically on the width $L$ of the barrier (the distance between the surface and the needle tip). The needle is scanned across the surface and at the same time is moved perpendicular to the surface so as to maintain constant tunneling current. The needle motion is recorded, and after many parallel scans an image of the surface can be reconstructed. Extremely precise control of needle motion, including isolation from vibration, is essential. Figure 42–13c shows a scanning tunneling microscope image of iodine atoms adsorbed on the surface of a platinum crystal. The yellow spot indicates a missing atom.

In the realm of nuclear physics, a nuclear fusion reaction occurs when two nuclei tunnel through the barrier caused by their electrical repulsion and approach each other closely enough for the attractive nuclear force to cause them to fuse. Fusion reactions occur in all visible stars; without tunneling, the sun wouldn't shine. The emission of alpha particles from unstable nuclei also involves tunneling. An alpha particle at the surface of a nucleus encounters a potential barrier that results from the combined effect of the attractive nuclear force and the electrical repulsion of the remaining part of the nucleus (Fig. 42–14). The alpha particle tunnels through this barrier. Because the tunneling probability depends so critically on the barrier height and width, the lifetimes of alpha-emitting nuclei vary over an extremely wide range. We'll return to alpha decay in Chapter 45.

## 42–5 THE HARMONIC OSCILLATOR

In Newtonian mechanics the **harmonic oscillator** is a particle with mass $m$ acted on by a conservative force $F(x) = -kx$ (proportional to the particle's displacement $x$ from its equilibrium position, $x = 0$). The corresponding potential-energy function is $U(x) = \frac{1}{2}kx^2$ (Fig. 42–15). When the particle is displaced from equilibrium, it undergoes sinusoidal motion with angular frequency $\omega = (k/m)^{1/2}$. We studied this system in detail in Chapter 13; you may want to review that discussion.

A quantum-mechanical analysis of the harmonic oscillator using the Schrödinger equation is an interesting and useful project. The solutions provide insight into vibrations of molecules, the quantum theory of heat capacities, lattice vibrations in crystalline solids, and many other situations.

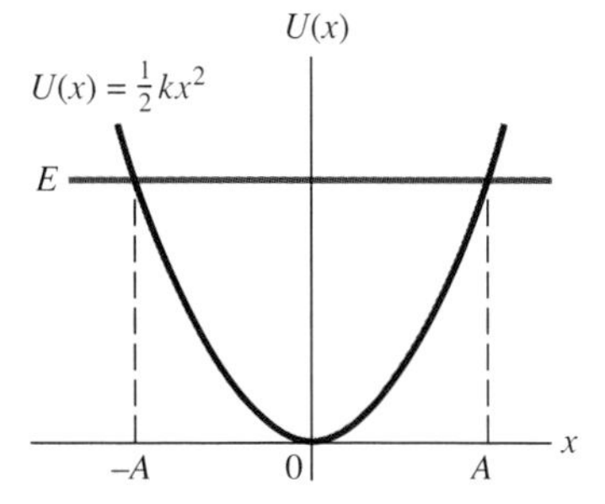


**42–15** Potential energy function for the harmonic oscillator, $U(x) = \frac{1}{2}kx^2$. The curve is a parabola. In Newtonian mechanics the amplitude $A$ is related to the total energy $E$ by $E = \frac{1}{2}kA^2$.

Before we get into the details, let's make an enlightened guess about the energy levels. The energy $E$ of a *photon* is related to its angular frequency $\omega$ by $E = \hbar\omega$. The harmonic oscillator has a characteristic angular frequency $\omega$, at least in Newtonian mechanics. So a reasonable guess would be that in the quantum-mechanical analysis the energy levels would be multiples of the quantity

$$\hbar\omega = \hbar\sqrt{\frac{k}{m}}.$$

This turns out to be a good guess; the energy levels are in fact half-integer ($\frac{1}{2}, \frac{3}{2}, \frac{5}{2}, \ldots$) multiples of $\hbar\omega$.

To get the Schrödinger equation for the harmonic oscillator, we write $\frac{1}{2}kx^2$ in place of $U(x)$ in Eq. (42–16):

$$-\frac{\hbar^2}{2m}\frac{d^2\psi}{dx^2} + \tfrac{1}{2}kx^2\psi = E\psi. \qquad (42\text{–}24)$$

The solutions of this equation are wave functions for the physically possible states of the system.

In the discussion of square-well potentials in Section 42–3 we found that the energy levels are determined by boundary conditions at the walls of the well. The harmonic-oscillator potential has no walls as such; what are the appropriate boundary conditions? The answer is that the wave functions must approach zero for very large $|x|$. We know that the particle is localized near the equilibrium position ($x = 0$), so the probability for finding it at large $x$ must grow very small as $x$ grows large.

Satisfying this requirement isn't as trivial as it may seem. Suppose we compute numerical solutions of Eq. (42–24), using the same method we used for our nonlinear oscillator analysis in Section 13–9. We start at some point $x$, choosing values of $\psi$ and $d\psi/dx$ at that point. Using the Schrödinger equation to evaluate $d^2\psi/dx^2$ at the point, we can compute $\psi$ and $d\psi/dx$ at a neighboring point $x + \Delta x$, and so on. By using a computer to iterate this process many times with sufficiently small steps $\Delta x$, we can compute the wave function with as great precision as we like.

There's no guarantee that a function obtained in this way will approach zero at large $x$. To see what kinds of trouble we can get into, let's rewrite Eq. (42–24) in the form

$$\frac{d^2\psi}{dx^2} = \frac{2m}{\hbar^2}\left(\tfrac{1}{2}kx^2 - E\right)\psi. \tag{42–25}$$

This equation shows that when $x$ is large enough (either positive or negative) to make the quantity $\frac{1}{2}kx^2 - E$ positive, the function $\psi$ and its second derivative have the same sign.

The second derivative of a function is the rate of change of *slope* of the function. Consider a point for which $\frac{1}{2}kx^2 - E > 0$. If $\psi$ is positive, the function is *concave upward.* Figure 42–16 shows several possible kinds of behavior. If the slope is positive, the function curves upward more and more steeply (curve $a$) and goes to infinity. If the slope is negative at the point, there are several possibilities. If the slope changes quickly enough (curve $b$), the curve bends up and again goes to infinity. If it doesn't change very quickly, the curve heads down and crosses the $x$-axis. After it has crossed, $\psi$ and $d^2\psi/dx^2$ are both negative (curve $c$), and the curve heads for *negative* infinity. Between these possibilities is the chance that the curve may bend just enough to glide in (curve $d$) and make a nice soft landing, asymptotic to the $x$-axis. In this case, $\psi$, $d\psi/dx$, and $d^2\psi/dx^2$ all approach zero at large $x$. In this possibility lies the only hope of satisfying the boundary condition that $\psi \rightarrow 0$ at large $x$, and it occurs only for certain very special values of the constant $E$.

This qualitative discussion offers some insight as to how the boundary conditions for this problem determine the possible energy levels. Equation (42–24) can also be solved exactly. The solutions, although not encountered in elementary calculus courses, are well known to mathematicians; they are called *Hermite functions.* Each one is an exponential function multiplied by a polynomial in $x$. The state with lowest energy (that is, the ground state) has the wave function

$$\psi(x) = Ce^{(-\sqrt{mk}x^2/2\hbar)}. \tag{42–26}$$

The constant $C$ is chosen to *normalize* the function; that is, to make $\int_{-\infty}^{\infty} |\psi|^2\, dx = 1$.

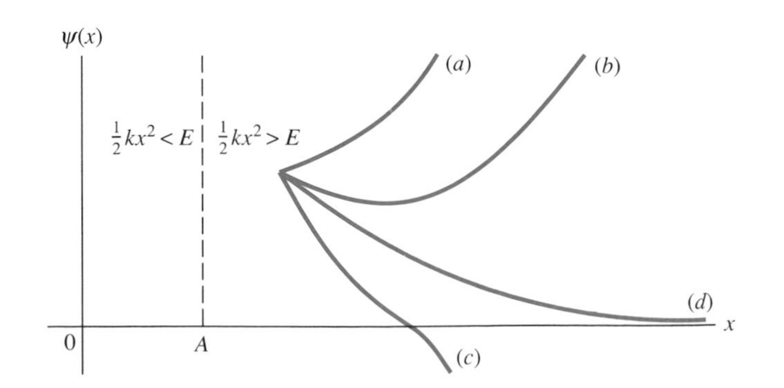


**42–16** Possible behaviors of harmonic-oscillator wave functions in the region $\frac{1}{2}kx^2 > E$. In this region, $\psi$ and $d^2\psi/dx^2$ have the same sign. The curve is concave upward when $\psi$ is positive and concave downward when $\psi$ is negative. Only curve ($d$) is an acceptable wave function for this system. This function approaches the $x$-axis asymptotically for large $x$.

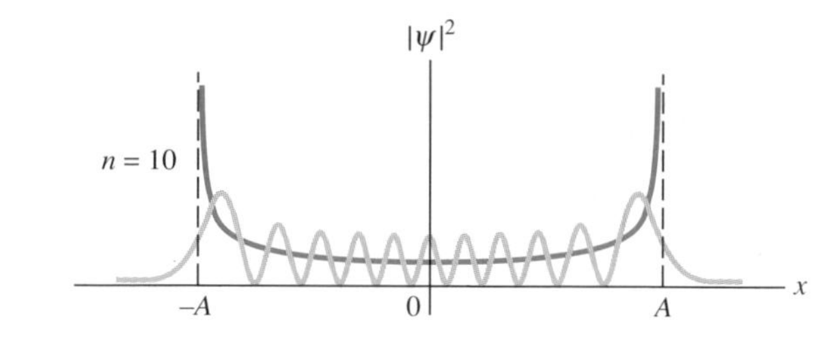


**42–20** Newtonian and quantum-mechanical probability distributions for a harmonic oscillator, for the state $n = 10$. The Newtonian amplitude $A$ is also shown.

This agreement improves with increasing $n$; Fig. 42–20 shows the classical and quantum-mechanical probability functions for $n = 10$.

In the Newtonian analysis of the harmonic oscillator, the minimum energy is zero, with the particle at rest at its equilibrium position. This is not possible in the quantum-mechanical analysis. There is no solution of the Schrödinger equation that has $E = 0$ and satisfies the boundary conditions. Furthermore, if there were such a state, it would violate the uncertainty principle because there would be no uncertainty in either position or momentum.

Indeed, the energy must be at least $\hbar\omega/2$ in order for the system to conform to the uncertainty principle. To see qualitatively why this is so, consider a Newtonian oscillator with total energy $\hbar\omega/2$. We can find the amplitude $A$ and the maximum velocity, just as we did in Section 13–2. The appropriate relations are Eqs. (13–6) and (13–8). Setting $E = \hbar\omega/2$, we find

$$E = \tfrac{1}{2}kA^2 = \tfrac{1}{2}\hbar\omega, \qquad A = \sqrt{\frac{\hbar}{\sqrt{km}}},$$

$$p_{\text{max}} = mv_{\text{max}} = m\sqrt{\frac{k}{m}}A = m\sqrt{\frac{k}{m}}\sqrt{\frac{\hbar}{\sqrt{km}}} = \sqrt{\hbar\sqrt{km}}.$$

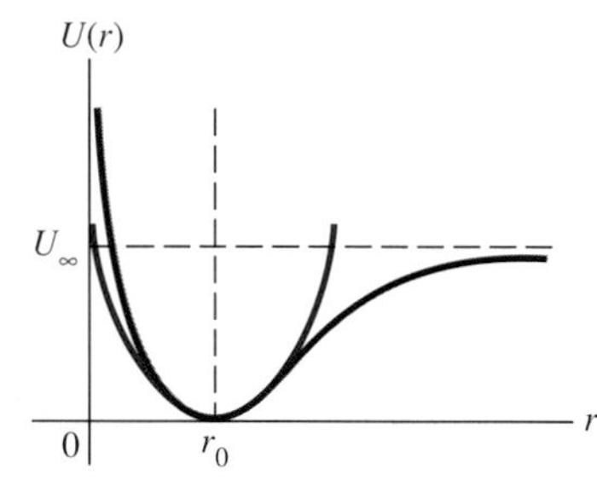


**42–21** A potential-energy function describing the interaction of two atoms in a diatomic molecule. At each point the force $F$ is $F = -dU/dr$. The equilibrium position is at $r = r_0$. When $r$ is greater than $r_0$, the force is attractive; when $r$ is less than $r_0$, it is repulsive. When $r$ is close to $r_0$, the curve is approximately parabolic (as shown by the bright red curve), and the motion is approximately simple harmonic. The energy needed to dissociate the molecule is $U_\infty$.

Finally, if we assume that $A$ represents the uncertainty $\Delta x$ in position and $p_{\text{max}}$ is the uncertainty $\Delta p$ in momentum, then the product of the two uncertainties is

$$\Delta x\, \Delta p = \sqrt{\frac{\hbar}{\sqrt{km}}}\sqrt{\hbar\sqrt{km}} = \hbar.$$

So the product comes out with the right magnitude to satisfy the uncertainty principle. If the energy had been less than $\hbar\omega/2$, the product $\Delta x\, \Delta p$ would have been less than $\hbar$, and the uncertainty principle would have been violated.

Even when a potential-energy function isn't precisely parabolic in shape, we may be able to approximate it by the harmonic-oscillator potential for sufficiently small displacements from equilibrium. Figure 42–21 shows a typical potential-energy function for an interatomic force in a molecule. At large separations it levels off, corresponding to the absence of force at great distances. But near the minimum point (the equilibrium position of the atoms) it is approximately parabolic. In this region the molecular vibration is approximately simple harmonic, with energy levels given by Eq. (42–28), as we assumed in Example 42–5.

## 42–6 THREE-DIMENSIONAL PROBLEMS

We have discussed the Schrödinger equation and its applications only for *one-dimensional* problems, the analog of a Newtonian particle moving along a straight line. The straight-line model is adequate for some applications; but to understand atomic structure, we need a three-dimensional generalization.

It's not difficult to guess what the three-dimensional Schrödinger equation should look like. First, the wave function $\psi$ is a function of all three space coordinates $(x, y, z)$. In general, the potential-energy function also depends on all three coordinates and can be written as $U(x, y, z)$. Next, recall that we obtained the term containing $d^2\psi/dx^2$ in the one-dimensional equation, Eq. (42–16), by a line of reasoning based on the relation of kinetic energy $K$ to momentum $p: K = p^2/2m$. If the particle's momentum has three components $(p_x, p_y, p_z)$, then the corresponding relation in three dimensions is

$$K = \frac{p_x{}^2}{2m} + \frac{p_y{}^2}{2m} + \frac{p_z{}^2}{2m}. \qquad (42\text{–}30)$$

These observations, taken together, suggest that the correct generalization of the Schrödinger equation to three dimensions is

$$-\frac{\hbar^2}{2m}\left(\frac{\partial^2\psi}{\partial x^2} + \frac{\partial^2\psi}{\partial y^2} + \frac{\partial^2\psi}{\partial z^2}\right) + U\psi = E\psi \qquad (42\text{–}31)$$

In this equation it is understood that $\psi$ and $U$ are functions of $x$, $y$, and $z$ (thus the partial-derivative notation).

We won't pretend that we have *derived* Eq. (42–31). As with the one-dimensional version, this equation has to be tested by comparison of its predictions with experimental results. As we will see in later chapters, Eq. (42–31) passes this test with flying colors, so we are confident that it *is* the correct equation.

In many practical problems, in atomic structure and elsewhere, the potential-energy function is *spherically symmetric;* that is, it depends only on the distance $r = (x^2 + y^2 + z^2)^{1/2}$ from the origin of coordinates. To exploit this symmetry, we use *spherical coordinates* $(r, \theta, \phi)$ (Fig. 42–22) instead of Cartesian coordinates $(x, y, z)$. Then a spherically symmetric potential-energy function is a function only of $r$, not of $\theta$ or $\phi$. This fact turns out to simplify greatly the problem of finding solutions of the Schrödinger equation, even though the derivatives are more complex when expressed in terms of spherical coordinates.

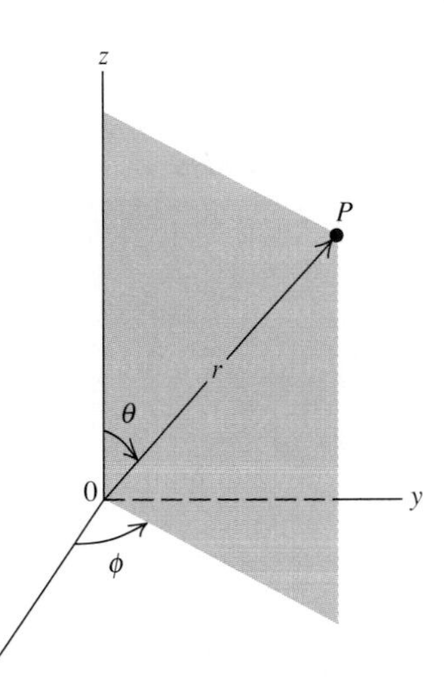


**42–22** The position of a point $P$ in space can be described by the rectangular coordinates $(x, y, z)$ or by the spherical coordinates $(r, \theta, \phi)$. Spherical coordinates are particularly useful in quantum mechanics in problems in which the potential energy depends only on $r$, the distance from the origin.

For the hydrogen atom the potential-energy function $U(r)$ is the familiar Coulomb's law function:

$$U(r) = -\frac{1}{4\pi\epsilon_0}\frac{e^2}{r}. \qquad (42\text{–}32)$$

We will find that for *all* spherically symmetric potential-energy functions $U(r)$, each possible wave function can be expressed as a product of three functions, one a function only of $r$, one only of $\theta$, and one only of $\phi$. Furthermore, the functions of $\theta$ and $\phi$ are *the same* for *every* spherically symmetric potential. This result is directly related to the problem of finding the possible values of *angular momentum* for the various states. We'll discuss these matters further in the next chapter.

## SUMMARY

- The energy levels for a particle in a box (an infinitely deep potential well with width $L$) are

$$E_n = \frac{p_n{}^2}{2m} = \frac{n^2h^2}{8mL^2}. \qquad (42\text{–}5)$$

The corresponding normalized wave functions are

$$\psi_n(x) = \sqrt{\frac{2}{L}}\sin\frac{n\pi x}{L}. \qquad (42\text{–}9)$$

- The Schrödinger equation for a particle moving on the $x$-axis with potential energy $U(x)$ is

$$-\frac{\hbar^2}{2m}\frac{d^2\psi(x)}{dx^2} + U(x)\psi(x) = E\psi(x). \qquad (42\text{–}16)$$

ately. What is the value of $\omega$ that makes $\Psi$ a solution? b) Show that $|\Psi|^2 = |\psi_n|^2$. (This says that the probability of finding the particle in any given region along the $x$-axis is independent of time.)

**42–27** An electron with kinetic energy 8.0 eV encounters a square barrier with height 10.0 eV. What is the width of the barrier if the electron has a 0.10% probability of tunneling through the barrier?

**42–28** An electron has a 1.0% probability of tunneling through a square barrier 15.0-eV high when the kinetic energy of the electron is 5.0 eV below the top of the barrier. What is the width of the barrier?

**42–29** Show that $\psi(x) = Cxe^{-m\omega x^2/2\hbar}$ is a solution to Eq. (42–24) with energy $E_1 = \frac{3}{2}\hbar\omega$.

**42–30** The wave function for the first excited state of a harmonic oscillator is given by $\psi(x) = Cxe^{-m\omega x^2/2\hbar}$, where $C$ is a normalization constant (Problem 42–29). a) At what values of $x$ is $|\psi|^2$ a maximum? b) At what values of $x$ is $|\psi|^2$ zero? Compare your results to what is shown in Fig. 42–19.

**42–31** For small amplitudes of swing the motion of a simple pendulum is harmonic with a period given by $T = 2\pi\sqrt{L/g}$, where $L$ is the length of the pendulum. For a pendulum with a period of 1.00 s, find the ground-state energy and the energy difference between adjacent energy levels. Express your results in joules and in electronvolts. Are these values detectable?

**42–32** A harmonic oscillator consists of a 0.015-kg mass on a spring. Its frequency is 2.00 Hz, and the mass has a speed of 0.40 m/s as it passes the equilibrium position. What is the value of the quantum number $n$ for its energy state?

**42–33 A Three-Dimensional Isotropic Harmonic Oscillator.** An oscillator has the potential energy function $U(x, y, z) = \frac{1}{2}k(x^2 + y^2 + z^2)$. (Isotropic means that the force constant $k$ is the same in all three coordinate directions.) a) Show that for this potential a solution to Eq. (42–31) is given by $\psi = \psi_{n_x}(x)\psi_{n_y}(y)\psi_{n_z}(z)$. Here $\psi_{n_x}$ is a solution to the one-dimensional Schrödinger equation, Eq. (42–24), with energy $E_{n_x} = (n_x + \frac{1}{2})\hbar\omega$. The functions $\psi_{n_y}(y)$ and $\psi_{n_z}(z)$ are analogous functions for oscillations in the $y$- and $z$-directions. Find the energy associated with this $\psi$. b) From your results in part (a), what are the ground level and first excited level energies of the three-dimensional isotropic oscillator?

## CHALLENGE PROBLEMS

**42–34** Section 42–1 considered a box with walls at $x = 0$ and $x = L$. Now consider a box with width $L$ but centered at $x = 0$, so that it extends from $x = -L/2$ to $x = +L/2$ (Fig. 42–24). Note that this box is symmetric about $x = 0$. a) Consider possible wave functions of the form $\psi(x) = A \sin kx$. Apply the boundary conditions at the wall to obtain the allowed energy levels. b) Another set of possible wave functions are functions of the form $\psi(x) = A \cos kx$. Apply the boundary conditions at the wall to obtain the allowed energy levels. c) Compare the energies obtained in parts (a) and (b) to the set of energies given in Eq. (42–5). d) An odd function $f$ satisfies the condition $f(x) = -f(-x)$, and an even function $g$ satisfies $g(x) = g(-x)$. Which of the wave functions from parts (a) and (b) are even and which are odd?

**42–35** Consider a potential well defined as follows: $U(x) = \infty$ for $x < 0$, $U(x) = 0$ for $0 < x < L$, and $U = U_0$ for $x > L$ (Fig. 42–25). Consider a particle with mass $m$ and kinetic energy $E < U_0$ that is trapped in the well. a) The boundary condition at the infinite wall ($x = 0$) is $\psi(0) = 0$. What must $\psi(x)$ be for $0 < x < L$ to satisfy both the Schrödinger equation and this boundary condition? b) The wave function must remain finite as $x \rightarrow \infty$. What must $\psi(x)$ be for $x > L$ to satisfy both the Schrödinger equation and this boundary condition at infinity? c) Impose the boundary conditions that $\psi$ and $d\psi/dx$ are continuous at $x = L$ and show that the energies of the allowed states are solutions of the equation $\xi \cot \xi = -\eta$, where $\xi = L\sqrt{2mE}/\hbar$ and $\eta = L\sqrt{2m(U_0 - E)}/\hbar$.

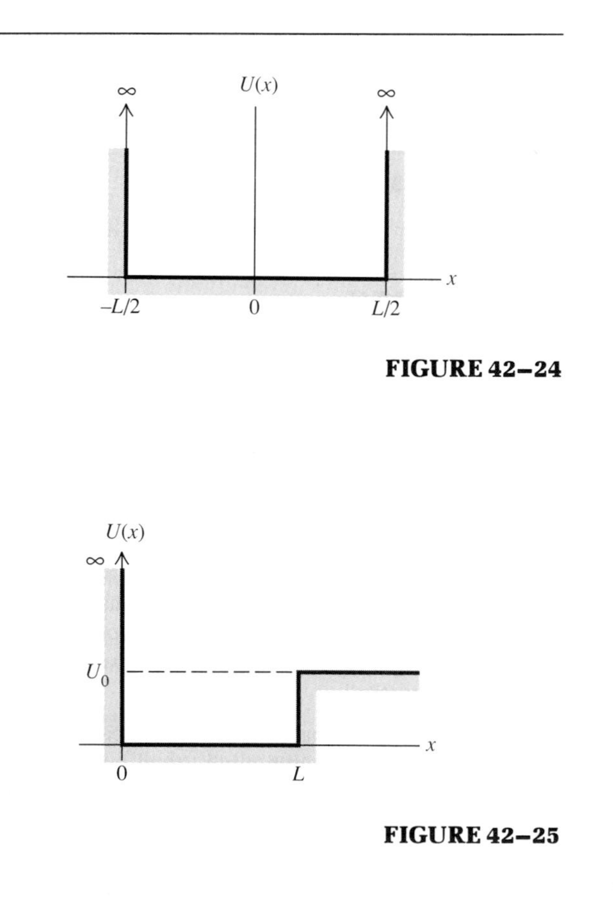


FIGURE 42–24

FIGURE 42–25

The ways atoms combine to form molecules or crystals are determined by the arrangements of electrons within the atoms. Two crystalline forms of carbon — graphite and diamond — are familiar. A third class of structures — the fullerenes — has recently been found. Among them is an especially stable molecule with 60 atoms arranged at the corners of a polyhedron made with 12 pentagons and 20 hexagons, just like a soccer ball. This structure is called a "buckyball," or "buckminsterfullerene," after the builder of geodesic domes. This chapter introduces the atomic physics needed to understand such structures.

CHAPTER 43

# Atomic Structure

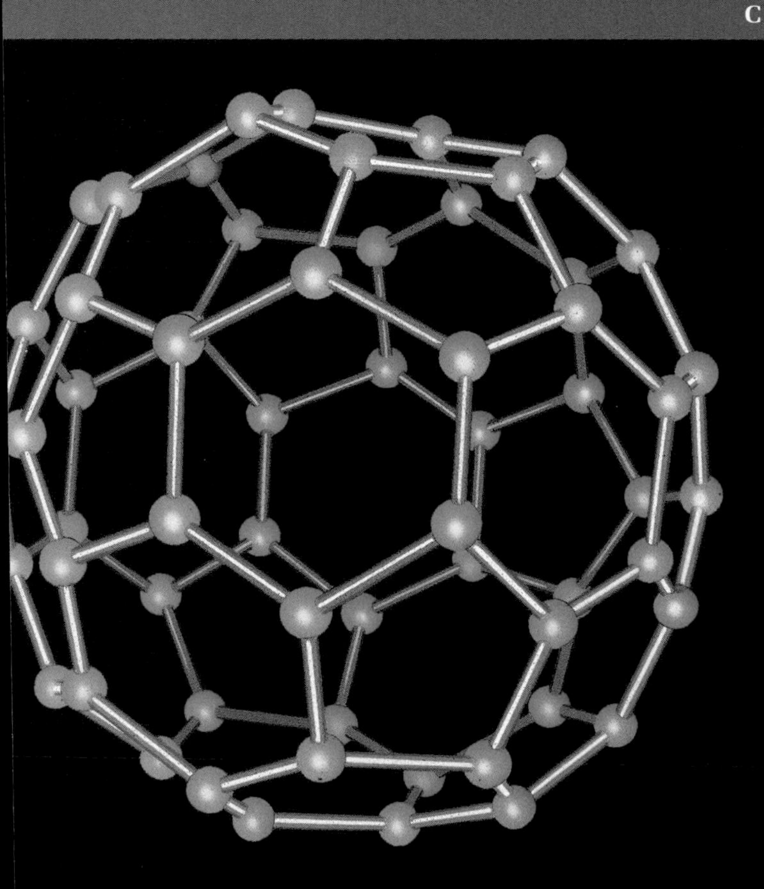

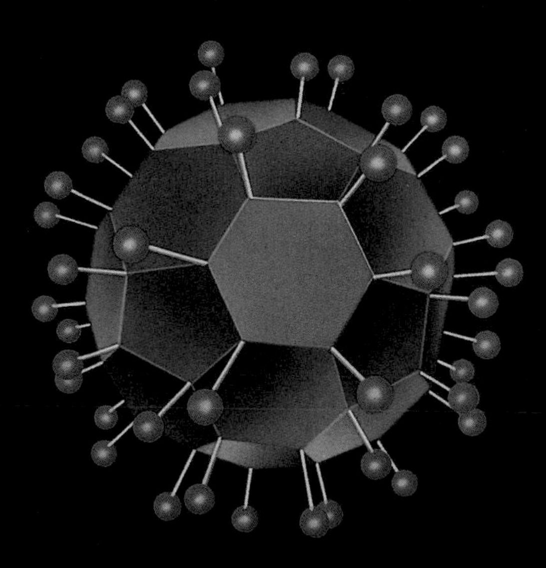

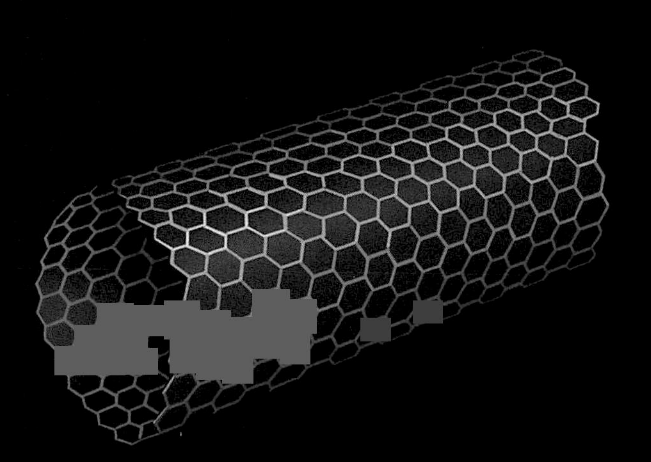

Research into the properties of fullerenes has been intensive and fast-moving. They are extremely resilient. Compounds with fluorine, such as $C_{60}F_{60}$ (called a teflon ball), appear to be excellent lubricants. Other fulleride compounds function as conductors, in- sulators, semiconductors, and even superconductors. Some evidence suggests that fullerene complexes may exhibit ferromagnetism without the presence of metals, a phenomenon never before observed. The trapping of atoms within the three-dimensional cage of a fullerene is a continuing area of research, promising new possibilities of chemical reactivity.

## KEY CONCEPTS

**• The Schrödinger equation can be solved exactly for the hydrogen atom. The wave functions are identified by a set of three quantum numbers. The energy levels are the same as those from the Bohr model, and the angular momentum is quantized.**

**• When an atom is placed in a magnetic field, the resulting interaction causes shifts in the energy levels and the spectrum wavelengths.**

**• Electrons have an intrinsic angular momentum called spin angular momentum in addition to their orbital angular momentum. Spin angular momentum is quantized.**

**• In the central-field approximation, each electron in a many-electron atom is assumed to move in the electric field of the nucleus and an averaged-out, spherically symmetric charge distribution representing all the other electrons. The periodic table of the elements can be understood on the basis of this model.**

**• An x-ray energy level is a state of an atom in which one of the inner electrons has been removed. Characteristic x-ray spectra are associated with transitions among such levels.**

## INTRODUCTION

Some physicists claim that all of chemistry is contained in the Schrödinger equation. This is an exaggeration, of course, but this equation can teach us a lot about the chemical behavior of elements and the nature of chemical bonds. We can gain insight into the periodic table of the elements and the microscopic basis of magnetism. It's surprising how much we can learn about the structure and properties of atoms by studying their magnetic interactions.

We begin with the solutions of the Schrödinger equation for the hydrogen atom. These solutions have quantized values of orbital angular momentum; we don't need a separate assumption about quantization, as we did with the Bohr model. We label the states with a set of quantum numbers, which we'll also use later with many-electron atoms. The electron also has an intrinsic spin angular momentum in addition to the orbital angular momentum associated with its motion.

The key to understanding many-electron atoms is the Pauli exclusion principle, a kind of microscopic zoning ordinance. This principle says that no two electrons in an atom can have the same quantum-mechanical state, just as two cars can't occupy the same parking space at the same time. Finally, we use the principles of this chapter to look at characteristic x-ray spectra.

# 43–1 THE HYDROGEN ATOM

Let's continue the discussion of the hydrogen atom that we began in Chapter 40. In the Bohr model, electrons moved in circular orbits like Newtonian particles, and we needed an *ad hoc* assumption about the allowed values of angular momentum. This model gave the correct energy levels of the hydrogen atom, as deduced from spectra. But it had many conceptual difficulties. It mixed classical physics with new and seemingly contradictory assumptions. It provided no insight into the process by which photons are emitted and absorbed. It could not be generalized to atoms with more than one electron. And perhaps most important, its picture of the electron as a point particle was inconsistent with the more general view we developed in Chapters 41 and 42.

Now let's apply the Schrödinger equation to the hydrogen atom. We discussed the three-dimensional version of this equation in Section 42–6. The hydrogen-atom problem is best formulated in spherical coordinates $(r, \theta, \phi)$; the potential energy is then simply

$$U(r) = -\frac{1}{4\pi\epsilon_0}\frac{e^2}{r}. \tag{43–1}$$

The Schrödinger equation with this potential-energy function can be solved exactly; the solutions are combinations of familiar functions. We won't go into a lot of detail, but we can describe the most important features of the procedure and the results.

First, the solutions are obtained by a method called *separation of variables,* in which we express the wave function $\psi(r, \theta, \phi)$ as a product of three functions, each one a function of only one of the three coordinates:

$$\psi(r, \theta, \phi) = R(r)\,\Theta(\theta)\,\Phi(\phi). \tag{43–2}$$

The function $R(r)$ depends *only* on $r$, and similarly for $\Theta(\theta)$ and $\Phi(\phi)$. When Eq. (43–2) is substituted into the Schrödinger equation, we get three separate equations, each containing only one of the coordinates. This is an enormous simplification because it reduces the problem of solving a fairly complex *partial* differential equation with three independent variables to the much simpler problem of solving three separate *ordinary* differential equations with one independent variable each.

The physically acceptable solutions of these equations are determined by *boundary conditions.* The radial function $R(r)$ must approach zero at large $r$ because we are describing *bound states* of the electron that are localized near the nucleus. This is analogous to the requirement that the harmonic-oscillator wave functions (Section 42–5) must approach zero at large $x$. The angular functions must be *periodic* functions. For example, the points $(r, \theta, \phi)$ and $(r, \theta, \phi + 2\pi)$ are the same point, so $\Phi(\phi + 2\pi)$ must equal $\Phi(\phi)$. That is, $\Phi(\phi)$ must be *periodic,* with period $2\pi$. Also, the angular functions must be *finite* for all relevant values of the angles. For example, there are solutions of the equation for $\Theta$ that blow up at $\theta = 0$ and $\theta = \pi$; these are unacceptable.

The radial functions $R(r)$ turn out to have the form $e^{-r/na_0}$ multiplied by a polynomial in $r$. The functions $\Theta(\theta)$ are polynomials containing various powers of $\sin\theta$ and $\cos\theta$, and the functions $\Phi(\phi)$ are simply $e^{im_l\phi}$, where $m_l$ is an integer whose significance we will discuss later.

While finding solutions that satisfy the boundary conditions, we also find the corresponding energy levels. These levels, which we denote by $E_n$ $(n = 1, 2, 3, \ldots)$, turn out to be identical to those from the Bohr model, as given by Eq. (40–17). Rewriting that expression using $\hbar$, we have

$$E_n = -\frac{1}{(4\pi\epsilon_0)^2}\frac{me^4}{2n^2\hbar^2} = -\frac{13.60\ \text{eV}}{n^2}. \tag{43–3}$$

We have called $n$ the *quantum number* for the state $E_n$. Now, anticipating the appearance of additional quantum numbers, we call $n$ the **principal quantum number.**

The fact that Eq. (43–3) can be obtained from the Schrödinger equation is an astonishing and critically important result. The Schrödinger analysis is completely different from the Bohr model, both formally and conceptually, yet they yield the same energy-level scheme; both agree with measured energies determined from spectra. Thus this result is a very significant confirmation of the validity of the Schrödinger approach.

## Quantization of Angular Momentum

The solutions that satisfy the boundary conditions mentioned above also have quantized values of *angular momentum.* Only certain discrete values of the magnitude and components of angular momentum are permitted. In discussing the Bohr model (Section 40–5), we mentioned that quantization of angular momentum had to be put in the model as an *ad hoc* assumption with no fundamental justification. With the Schrödinger equation it appears automatically!

The possible values of the magnitude $L$ of angular momentum are determined by the requirement that the $\Theta$ function must be finite at $\theta = 0$ and $\theta = \pi$. In a state with energy $E_n$ and principal quantum number $n$, the possible values of $L$ are

$$L = \sqrt{l(l+1)}\hbar \quad (l = 0, 1, 2, \ldots, n-1), \tag{43–4}$$

where $l$ is zero or a positive integer no larger than $n - 1$. We call $l$ the **angular-momentum quantum number.**

The permitted values of the *component* of $L$ in a given direction—for example, the $z$-component $L_z$—are determined by the requirement that the $\Phi$ function must be periodic with period $2\pi$. The possible values of $L_z$ are

$$L_z = m_l\hbar \quad (m_l = 0, \pm 1, \pm 2, \ldots, \pm l), \tag{43–5}$$

where $m_l$ can be zero or a positive or negative integer up to but no larger in magnitude than $l$. That is, $|m_l| \le l$. For reasons that will emerge later, we call $m_l$ the **magnetic quantum number.** We stress again that Eqs. (43–4) and (43–5) emerge as a consequence of the solutions of the Schrödinger equation; quantization of angular momentum doesn't have to be added as a separate assumption.

Note that the component $L_z$ can never be quite as large as $L$. For example, when $l = 2$ and $m_l = 2$, Eqs. (43–4) and (43–5) give

$$L = \sqrt{2(2+1)}\hbar = \sqrt{6}\hbar = 2.45\hbar,$$
$$L_z = 2\hbar.$$

Figure 43–1 shows the situation. For $l = 2$ and $m_l = 2$, the angle $\theta$ between the vector $\boldsymbol{L}$ and the $z$-axis is given by

$$\cos\theta = \frac{L_z}{L} = \frac{2}{2.45}, \qquad \theta = 35°.$$

We can't know the precise direction of $\boldsymbol{L}$; it can have any direction on the cone in the figure at an angle of $\theta = 35°$ with the $z$-axis.

That $|L_z|$ is always less than $L$ is also required by the uncertainty principle. Suppose we could know the precise *direction* of the angular momentum vector. For example, suppose $L_z$ were equal to $L$. Then we would know that the vector $\boldsymbol{L}$ is in the $z$-direction. This corresponds to a particle moving in the $xy$-plane, in which case the $z$-component of

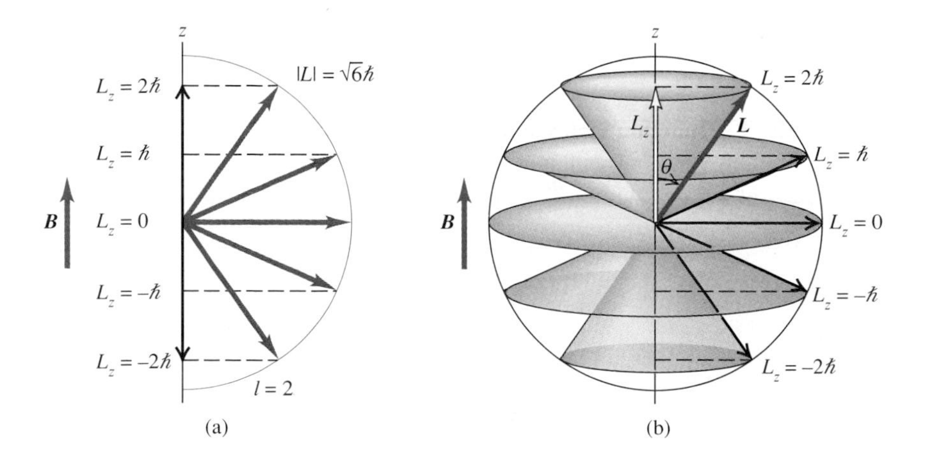


**43–1** (a) When $l = 2$, the magnitude of $L$ is $\sqrt{2(2+1)}\hbar = \sqrt{6}\hbar$, but the maximum $z$-component is $\pm 2\hbar$. The direction of $L$ is not definite; in this semiclassical vector picture, $L$ makes an angle of 35° with the $z$-axis. (b) Similar direction relations for other values of $L_z$.

momentum must be zero. But according to the uncertainty principle, this requires complete uncertainty in the coordinate $z$. This is impossible for a localized state; we conclude that we can't know the direction of $\boldsymbol{L}$ precisely. Thus, as we've already stated, the component of $\boldsymbol{L}$ in a given direction can never be quite as large as the magnitude $L$ (except when both $L$ and $L_z$ are zero).

You may wonder why we have singled out the $z$-axis for special attention. There's no fundamental reason for this; the atom certainly doesn't care what coordinate system we use. The point is that we can't determine all three components of angular momentum with certainty; we arbitrarily pick one as the component that we want to measure. Later we will discuss interactions of the atom with a magnetic field, and we will usually orient the field along the $z$-axis.

Another interesting feature of Eqs. (43–4) and (43–5) is that there are states for which $l = 0$ and the angular momentum is *zero.* This result has no analog in the Bohr model, where the electron always moved in an orbit and $L$ was never zero. The wave functions for $l = 0$ states are spherically symmetric; when there is no angular-momentum vector, there is nothing to favor any direction over any other. The $l = 0$ wave functions depend only on $r$; the $\theta$ and $\phi$ functions for these states are constants.

## Quantum Numbers

The wave functions for the hydrogen atom are labeled with values of three quantum numbers $n$, $l$, and $m_l$. The energy $E_n$ is determined by the principal quantum number $n$, according to Eq. (43–3). The magnitude of orbital angular momentum is determined by the angular-momentum quantum number $l$, as in Eq. (43–4). The component of angular momentum in a specified axis direction (customarily the $z$-axis) is determined by the magnetic quantum number $m_l$, as in Eq. (43–5). For each energy level $E_n$, there are several distinct states having the same energy but different values of $l$ and $m_l$. The only exception is the ground state ($n = 1$), for which the only possibility is $l = 0$, $m_l = 0$. The existence of several distinct states with the same energy is called **degeneracy;** it has no counterpart in the Bohr model.

States with various values of the angular-momentum quantum number $l$ are often labeled with letters, according to the following scheme:

$l = 0$: *s* state,
$l = 1$: *p* state,
$l = 2$: *d* state,
$l = 3$: *f* state,
$l = 4$: *g* state,
etc.

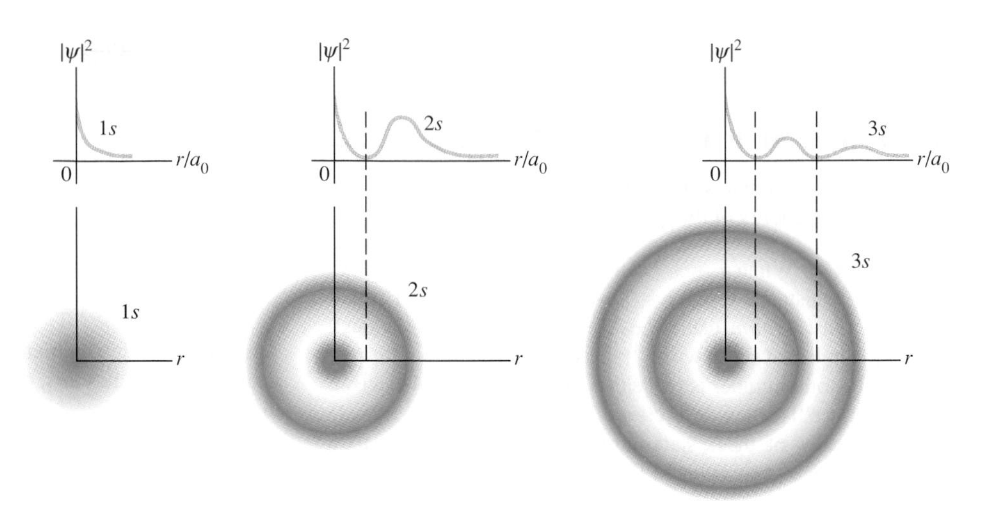


**43–3** Radial probability distribution for the spherically symmetric 1*s* wave function for the hydrogen atom.

The $r$ scales are labeled in multiples of the Bohr radius $a_0$, given by Eq. (40–15):

$$a_0 = \frac{\epsilon_0 h^2}{\pi m e^2} = \frac{4\pi\epsilon_0 \hbar^2}{m e^2} = 0.529 \times 10^{-10}\ \text{m}. \qquad (43\text{–}8)$$

Note that for the states having the largest possible $l$ for each $n$ (1*s*, 2*p*, 3*d*, and 4*f* states), the maximum value of $P(r)$ occurs at $n^2 a_0$. For these states, the electron is most likely to be found at a radius equal to that of the corresponding Bohr orbit. For *all* values of $l$, the probability for states with a given $n$ is concentrated near a radius of approximately $n^2 a_0$, confirming the "shell" picture described above. The *integral* of each $P(r)$ over all values of $r$ from 0 to $\infty$ is unity; the particle *must* be found at *some* distance from the nucleus.

Figure 43–3 shows the spherically symmetric probability distribution of the 1*s* wave function (the ground state) for the hydrogen atom, and Fig. 43–4 shows sketches of the three-dimensional probability distributions for a few hydrogen-atom wave functions.

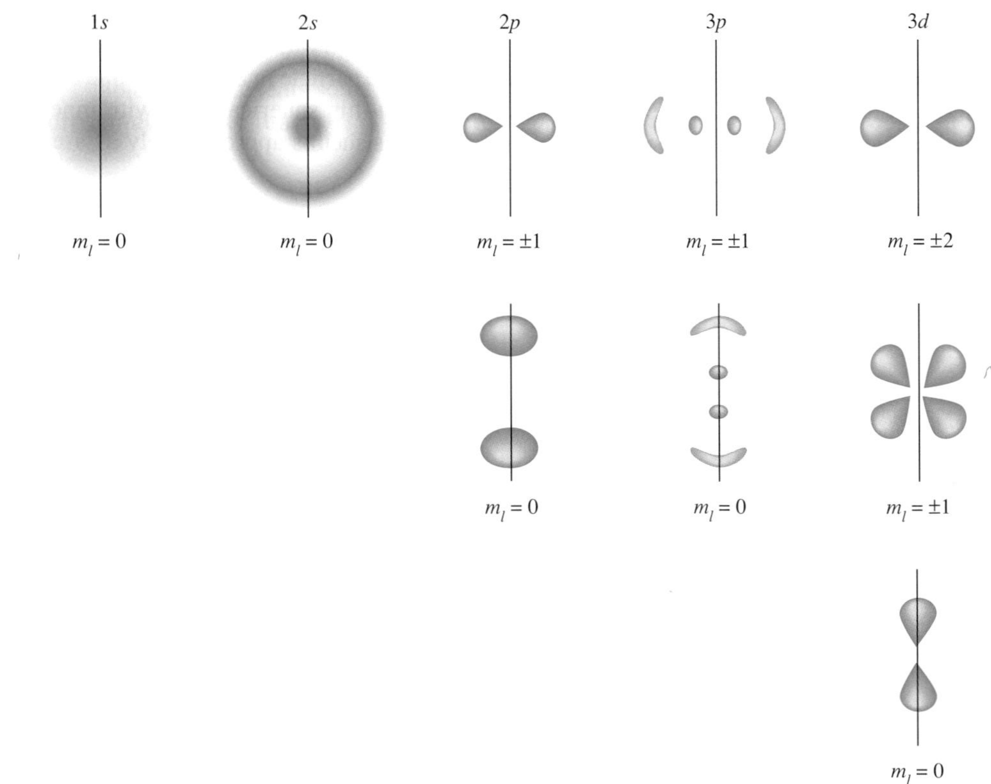


**43–4** Sketches of cross sections of three-dimensional probability distributions for a few quantum states of the hydrogen atom.

## EXAMPLE 43–3

The ground-state wave function for hydrogen (the $1s$ state) is

$$\psi_{1s}(r) = \frac{1}{\sqrt{\pi a_0^3}} e^{-r/a_0}.$$

a) Verify that this function is normalized. b) What is the probability that the electron will be found at a distance less than $a_0$ from the nucleus?

**SOLUTION** a) The normalization condition is $\int_0^\infty |\psi|^2\, dV = 1$. Using the volume element given by Eq. (43–6), we find

$$\int_0^\infty |\psi_{1s}|^2\, dV = \int_0^\infty \frac{1}{\pi a_0^3} e^{-2r/a_0} (4\pi r^2\, dr) = \frac{4}{a_0^3} \int_0^\infty r^2 e^{-2r/a_0}\, dr.$$

The following indefinite integral can be found in a table of integrals:

$$\int r^2 e^{-2r/a_0} dr = \left( -\frac{a_0 r^2}{2} - \frac{a_0^2 r}{2} - \frac{a_0^3}{4} \right) e^{-2r/a_0}.$$

Evaluating this between the limits 0 and $\infty$ is simple; it is zero at $\infty$ because of the exponential factor, and at 0 only the last term survives. The value of the integral is $a_0^3/4$. When we multiply by $4/a_0^3$, we find

$$\int_0^\infty |\psi_{1s}|^2\, dV = \frac{4}{a_0^3} \frac{a_0^3}{4} = 1.$$

This verifies that the wave function is normalized.

b) To find the probability that the electron is found at a distance less than $a_0$, we carry out the same integration but with the limits 0 and $a_0$. We'll leave the details as an exercise. From the upper limit we get $-5e^{-2}a_0^3/4$; the final result is

$$P(r < a_0) = \int_0^{a_0} |\psi_{1s}|^2\, dV = \frac{4}{a_0^3} \left( -\frac{5a_0^3 e^{-2}}{4} + \frac{a_0^3}{4} \right)$$
$$= 1 - 5e^{-2} = 0.323.$$

Thus in the ground state we expect to find the electron at a distance less than the Bohr radius from the nucleus about $\frac{1}{3}$ of the time and at a greater distance than this about $\frac{2}{3}$ of the time. In Fig. 43–2, about $\frac{2}{3}$ of the area under the $1s$ curve is at distances greater than $a_0$.

Finally, two generalizations we discussed with the Bohr model in Section 40–5 are equally valid in the Schrödinger analysis. First, if the nucleus is not infinitely massive, we replace the electron mass in Eqs. (43–3) and (43–8) by the *reduced mass* of the system, defined by Eq. (40–19). This correction is always less than 0.06% with ordinary atoms, but it is substantial with some exotic atoms, such as a hydrogen atom containing a $\mu$-meson instead of an electron or containing a positron instead of a proton (an atom of positronium). Second, our analysis is applicable to other single-electron atoms, such as $He^+$, $Li^{2+}$, and so on. In Eqs. (43–3) and (43–8), we replace $e^2$ by $Ze^2$, where $Z$ is the number of units of positive charge in the nucleus.

The Schrödinger analysis of the hydrogen atom is a lot more complex, both conceptually and mathematically, than Newtonian analysis of planetary motion or the semiclassical Bohr model. It deals with probabilities rather than certainties, and it predicts discrete rather than continuous behavior. But this analysis enables us to understand phenomena and to analyze problems for which classical mechanics and electromagnetism are inadequate. Added complexity is the price we pay for expanded understanding.

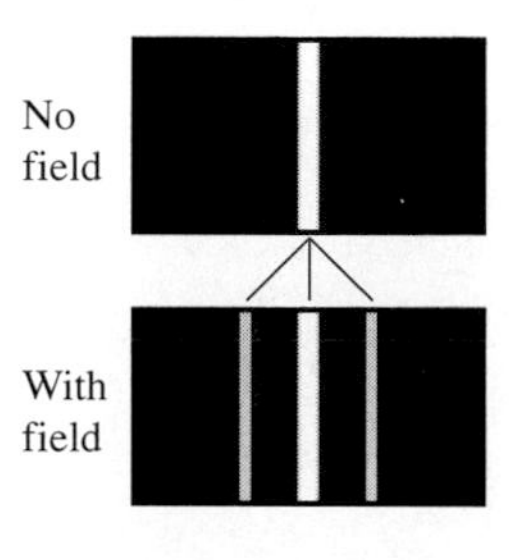


## 43–2 THE ZEEMAN EFFECT

The **Zeeman effect** is the shifting of atomic energy levels and the associated spectrum lines when the atoms are placed in a magnetic field. The Zeeman effect is a very direct experimental confirmation of the quantization of angular momentum. The following discussion also shows why we call $m_l$ the magnetic quantum number.

Atoms contain charges in motion, so it should not be surprising that the resulting magnetic-field forces cause changes in the motion and in the energy levels. As early as the middle of the nineteenth century, physicists speculated that the sources of visible light might be vibrating electric charge on an atomic level. In 1862, Michael Faraday placed light sources in a magnetic field in an attempt to observe changes in spectrum lines. His spectroscopic techniques were not refined enough to observe any effect. But in 1896 the Dutch physicist Pieter Zeeman, using improved instruments, found that in the presence of a magnetic field some spectrum lines were split into groups of closely spaced lines (Fig. 43–5). This effect now bears his name.

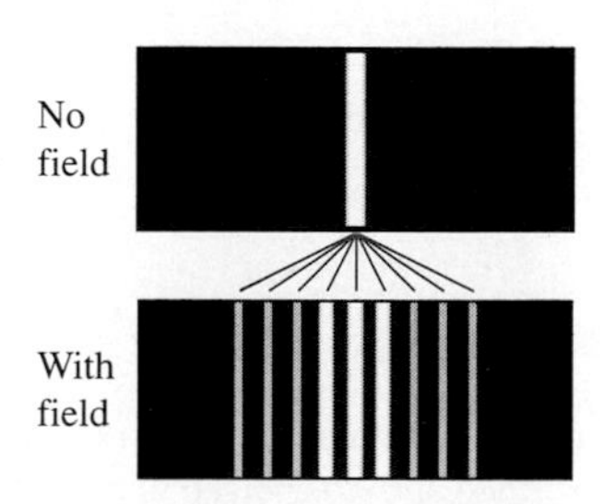


**43–5** Spectrum lines showing the Zeeman effect. When the source of radiation is placed in a magnetic field, the interactions of the atomic magnetic moment with the field split individual lines into sets of lines.

Let's begin our analysis of the Zeeman effect by reviewing the concept of **magnetic moment,** introduced in Section 28–7. A plane current loop with vector area $\boldsymbol{A}$ carrying current $I$ has a magnetic moment $\boldsymbol{\mu}$ in the direction of $\boldsymbol{A}$, with magnitude given by Eq. (28–23):

$$\mu = IA. \tag{43–9}$$

When a magnetic moment $\boldsymbol{\mu}$ is placed in a magnetic field $\boldsymbol{B}$, the field exerts a torque $\boldsymbol{\tau} = \boldsymbol{\mu} \times \boldsymbol{B}$. The potential energy $U$ associated with this interaction is given by Eq. (28–25):

$$U = -\boldsymbol{\mu} \cdot \boldsymbol{B}. \tag{43–10}$$

Now let's look at the interaction of a hydrogen atom with a magnetic field, using the Bohr model. The orbiting electron is equivalent to a current loop with radius $r$ and area $\pi r^2$. The average current $I$ is the average charge per unit time that passes a point of the orbit. This is equal to the charge $e$ divided by the time $T$ for one revolution, given by $T = 2\pi r/v$. Thus $I = ev/2\pi r$, and from Eq. (43–9) the magnitude of the magnetic moment $\boldsymbol{\mu}$ is

$$\mu = IA = \frac{ev}{2\pi r}\pi r^2 = \frac{evr}{2}. \tag{43–11}$$

We can also express this in terms of the magnitude $L$ of the angular momentum. From Eq. (10–28), $L = mvr$, so Eq. (43–11) becomes

$$\mu = \frac{e}{2m}L. \tag{43–12}$$

The ratio of magnetic moment to angular momentum, $e/2m$, is called the **gyromagnetic ratio.**

In the Bohr model, $L = nh/2\pi = n\hbar$, where $n = 1, 2, \ldots$. For the $n = 1$ state (the ground state) we find

$$\mu = \frac{e}{2m}\hbar = \tfrac{1}{2}(1.7588 \times 10^{11}\ \text{C/kg})(1.0546 \times 10^{-34}\ \text{J}\cdot\text{s})$$
$$= 9.274 \times 10^{-24}\ \text{A}\cdot\text{m}^2.$$

This quantity is a natural unit for magnetic moment; it is called one **Bohr magneton,** denoted by $\mu_B$:

$$\mu_B = \frac{e\hbar}{2m}. \tag{43–13}$$

We defined this quantity previously, in Section 29–8. Note that the units $\text{A}\cdot\text{m}^2$ and J/T are equivalent.

## EXAMPLE 43–4

Find the interaction potential energy when the hydrogen atom described above is placed in a magnetic field with magnitude 2.00 T and oriented perpendicular to the plane of the orbit.

**SOLUTION** From Eq. (43–10) the interaction energy $U$ when $\boldsymbol{B}$ and $\boldsymbol{\mu}$ are parallel is

$$U = -\boldsymbol{\mu}\cdot\boldsymbol{B} = -\mu B = -(9.274 \times 10^{-24}\ \text{A}\cdot\text{m}^2)(2.00\ \text{T})$$
$$= -1.85 \times 10^{-23}\ \text{J} = -1.16 \times 10^{-4}\ \text{eV}.$$

The magnetic-field interaction shifts the energy level of the state by this amount. When $\boldsymbol{\mu}$ and $\boldsymbol{B}$ are antiparallel, the energy is $+1.16 \times 10^{-4}$ eV. These energies are *smaller* than the energy levels of the atom by a factor of the order of $10^{-5}$. Note that because the electron charge is negative, the angular momentum and magnetic moment vectors are opposite; $\boldsymbol{B}$ and $\boldsymbol{\mu}$ are parallel when $\boldsymbol{B}$ and $\boldsymbol{L}$ are *antiparallel*.

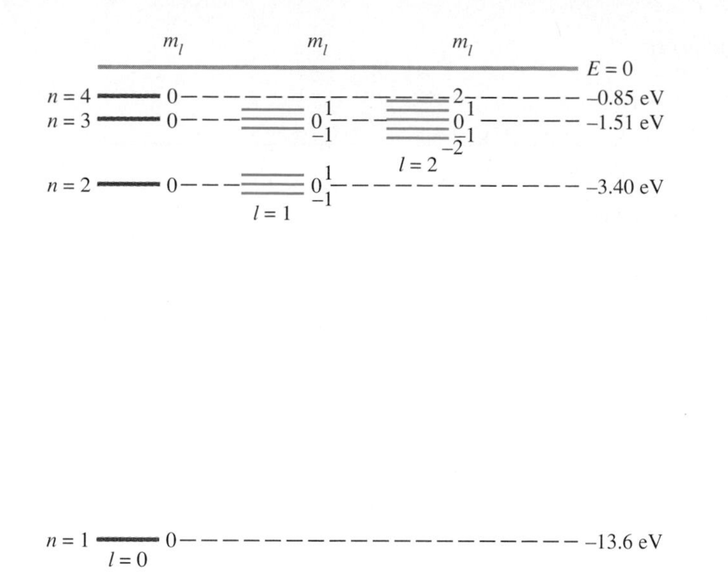


**43–6** Energy-level diagram for hydrogen, showing the splitting of energy levels resulting from the interaction of the magnetic moment of the electron's orbital motion with an external magnetic field. Values of $m_l$ are shown adjacent to the various levels. Relative magnitudes of splittings are exaggerated for clarity. Some of the $n = 4$ levels are not shown; can you draw them in?

We can generalize this discussion to states described by Schrödinger wave functions. It turns out that in the Schrödinger formulation, electrons have the same ratio of $\mu$ to $B$ (gyromagnetic ratio) as in the Bohr model, $e/2m$. Suppose the magnetic field $\boldsymbol{B}$ is directed along the $+z$-axis. From Eq. (43–10) the interaction energy $U$ of the atom's magnetic moment with the field is

$$U = -\mu_z B, \tag{43–14}$$

where $\mu_z$ is the $z$-component of the vector $\boldsymbol{\mu}$.

Now we use Eq. (43–12) to find $\mu_z$, recalling that $e$ is the *magnitude* of the electron charge and that the actual charge is $-e$. We find

$$\mu_z = -\frac{e}{2m}L_z. \tag{43–15}$$

For the Schrödinger wave functions, $L_z = m_l\hbar$, with $m_l = 0, \pm 1, \pm 2, \ldots, \pm l$, so

$$\mu_z = -\frac{e}{2m}L_z = -m_l\frac{e\hbar}{2m}. \tag{43–16}$$

(Be careful not to confuse the electron mass $m$ with the magnetic quantum number $m_l$.) Finally, we can express the interaction energy, Eq. (43–14), as

$$U = -\mu_z B = m_l\frac{e\hbar B}{2m} \quad (m_l = 0, \pm 1, \pm 2, \cdots, \pm l). \tag{43–17}$$

In terms of the Bohr magneton $\mu_B = e\hbar/2m$,

$$U = m_l\mu_B B. \tag{43–18}$$

The effect of the magnetic field is to shift each energy level by an amount $U$, given by Eq. (43–17). The magnitude of the shift depends on the value of the magnetic quantum number $m_l$, which is determined by the orientation of the angular-momentum vector.

An energy level with a particular value of the angular-momentum quantum number $l$ corresponds to $2l + 1$ different states, with values of $m_l$ ranging from $-l$ to $+l$. Without a magnetic field these states all have the same energy; that is, they are *degenerate*. In the presence of a magnetic field they are split into $2l + 1$ distinct energy levels; adjacent levels differ in energy by $e\hbar B/2m = \mu_B B$. The magnetic field removes the degeneracy. The effect on the energy levels of hydrogen is shown in Fig. 43–6. Spectrum lines corresponding to transitions involving these levels are correspondingly split and appear as a series of closely spaced spectrum lines instead of single lines.

## EXAMPLE 43–5

An atom in a state with $l = 1$ emits a photon with wavelength 600.000 nm as it decays to its ground state (with $l = 0$). Suppose the atom is placed in a magnetic field with magnitude $B = 2.00\ \text{T}$. Determine the shifts in the energy level and in the wavelength.

**SOLUTION** The energy of a 600-nm photon is

$$E = \frac{hc}{\lambda} = \frac{(4.14 \times 10^{-15}\ \text{eV} \cdot \text{s})(3.00 \times 10^{8}\ \text{m/s})}{600 \times 10^{-9}\ \text{m}} = 2.07\ \text{eV}.$$

The ground-state level has $l = 0$ and is not split by the field. The splitting of levels in the $l = 1$ state is given by Eq. (43–17):

$$U = m_l \frac{e\hbar B}{2m} = m_l \frac{(1.60 \times 10^{-19}\ \text{C})(1.054 \times 10^{-34}\ \text{J}\cdot\text{s})(2.00\ \text{T})}{2(9.11 \times 10^{-31}\ \text{kg})}$$

$$U = m_l(1.855 \times 10^{-23}\ \text{J}) = m_l(1.158 \times 10^{-4}\ \text{eV}).$$

When $l = 1$, the possible values of $m_l$ are $-1$, 0, and $+1$, and the three resulting levels are split by equal intervals of $1.158 \times 10^{-4}$ eV. This is a small fraction of the photon energy:

$$\frac{1.158 \times 10^{-4}/\text{eV}}{2.07\ \text{eV}} = 5.59 \times 10^{-5}.$$

The corresponding *wavelength* shifts are approximately $(5.59 \times 10^{-5})(600\ \text{nm}) = 0.034\ \text{nm}$. The original 600-nm line is split into a triplet with wavelengths 599.966 nm, 600.000 nm, and 600.034 nm. This splitting is well within the limit of resolution of modern spectrometers.

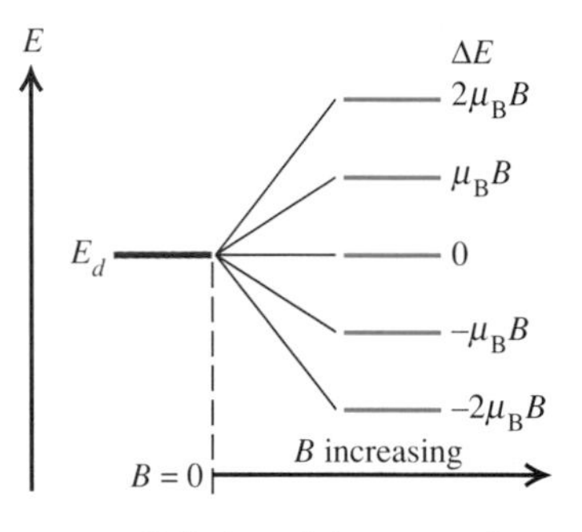


**43–7** Splitting of the energy levels of a $d$ state caused by an applied magnetic field, assuming only an orbital magnetic dipole moment.

Figure 43–7 shows what happens to a set of $d$ states ($l = 2$) of the hydrogen atom as the magnetic field increases. The fivefold degeneracy with zero field is removed as the levels diverge with increasing field. Figure 43–8 shows the splittings of the $3d$ and $2p$ states. Adjacent levels are separated by equal energies $e\hbar B/2m = \mu_B B$. In the absence of a magnetic field, a transition from a $3d$ to a $2p$ state would yield a single spectrum line. With the levels split as shown, it might seem that there are five possible photon energies.

In fact, there are only three possibilities. Not all combinations of initial and final levels are possible because of a restriction associated with conservation of angular momentum. The photon ordinarily carries off one unit ($\hbar$) of angular momentum. This fact leads to the requirements that in a transition, $l$ must change by 1 and $m_l$ must change by 0 or $\pm 1$. These requirements are called **selection rules.** Transitions that obey these rules are called *allowed transitions;* those that don't are *forbidden transitions.* The allowed transitions are solid lines in Fig. 43–8. We invite you to count the possible transition energies to convince yourself that there are three possibilities: the zero-field value, and that value plus or minus $e\hbar B/2m$. The corresponding spectrum lines are shown.

What we have described is called the *normal* Zeeman effect. This name distinguishes it from other cases in which spectroscopists have found Zeeman splitting into two or four lines, sometimes unequally spaced. Before this effect was understood, it was called the *anomalous* Zeeman effect. In fact, there's nothing particularly *normal* about

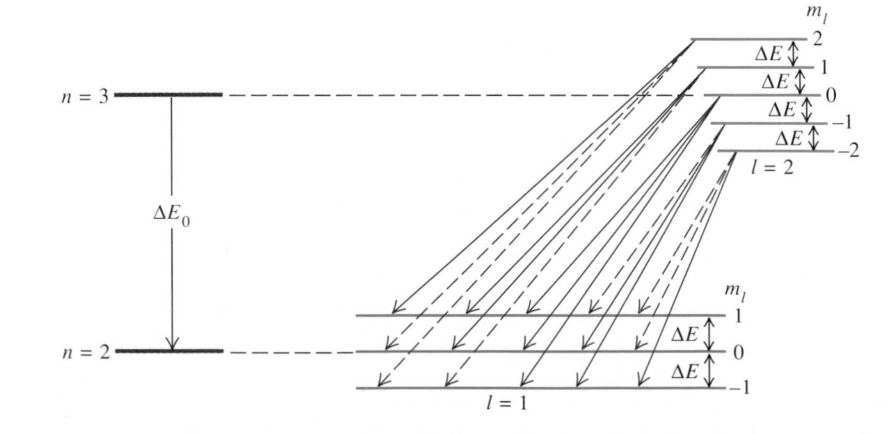


**43–8** The cause of the normal Zeeman effect. The magnetic field splits the levels, but selection rules allow transitions with only three different energy changes, giving three different frequencies (not to scale). Solid lines show allowed transitions; dashed lines show forbidden transitions.

what we have described. It is based entirely on the orbital angular momentum of the electron. It leaves out a very important consideration, the *spin* angular momentum, which we haven't yet discussed. That concept is the subject of the next section and provides the basis for understanding both the normal and anomalous Zeeman effects and many other observations as well.

## 43–3 ELECTRON SPIN

Despite the success of the Schrödinger equation in predicting the energy levels of the hydrogen atom, several experimental observations indicated that it didn't tell the whole story about the behavior of electrons in atoms. Some of the observed Zeeman-effect line splittings (Fig. 43–9) don't fit comfortably into the analysis we sketched in Section 43–2. And some energy levels show splittings that resemble the Zeeman effect even when there is *no* external magnetic field.

For example, when the lines in the hydrogen spectrum are examined with a high-resolution spectrometer, some lines are found to consist of sets of closely spaced lines called *multiplets.* Similarly, the famous orange-yellow line of sodium, corresponding to the $4p \rightarrow 3s$ transition of the valence electron, is found to be a doublet ($\lambda = 589.0, 589.6$ nm), suggesting that the $4p$ level might in fact be two closely spaced levels. The Schrödinger equation in its original form didn't predict any of this.

Similar anomalies appeared in 1922 in atomic-beam experiments performed by Otto Stern and Walter Gerlach. When they passed a beam of neutral atoms through a nonuniform magnetic field (Fig. 43–10, on page 1196), atoms were deflected according to the orientation of their magnetic moments with respect to the field. These experiments demonstrated the quantization of angular momentum in a very direct way; however, some atomic beams were split into an *even* number of components. This suggested a half-integer angular momentum that couldn't be understood on the basis of the Bohr model and similar pictures of atomic structure.

In 1925, two graduate students at the University of Leiden, Samuel Goudsmidt and George Uhlenbeck, proposed that the electron might have some additional motion. In a semiclassical picture, they suggested, the electron might behave like a spinning sphere of charge instead of a particle. If so, it would have additional angular momentum and magnetic moment. If these were quantized in the same way as orbital angular momentum and magnetic moment, they might help explain the observed energy-level anomalies.

To introduce the concept of **electron spin,** let's start with an analogy. The earth travels in a nearly circular orbit around the sun, and at the same time it *rotates* on its axis. Each motion has its associated angular momentum, which we call the *orbital* angular momentum and the *spin* angular momentum, respectively. The total angular momentum

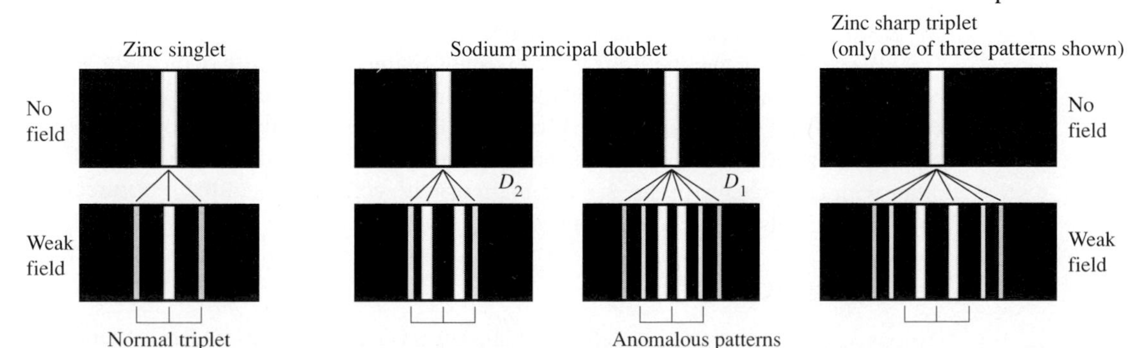


**43–9** Illustrations of the normal and anomalous Zeeman effects for several elements. The brackets under each illustration show the "normal" splitting predicted by neglecting the effect of electron spin.

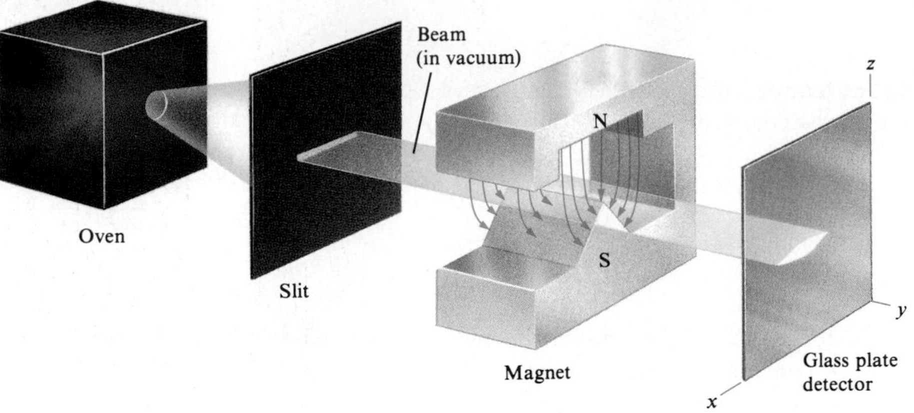


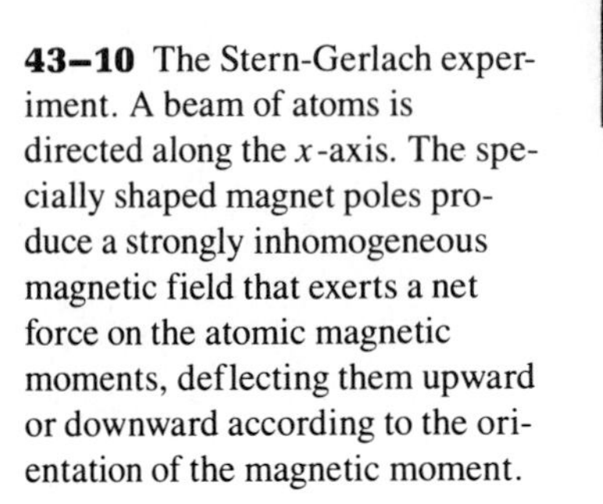

**43–10** The Stern-Gerlach experiment. A beam of atoms is directed along the $x$-axis. The specially shaped magnet poles produce a strongly inhomogeneous magnetic field that exerts a net force on the atomic magnetic moments, deflecting them upward or downward according to the orientation of the magnetic moment.

of the system is the vector sum of the two. If we were to model the earth as a single point, it could not have spin angular momentum. But when we model the earth as having finite size, spin angular momentum becomes possible.

In the Bohr model, suppose the electron is not just a point charge moving in an orbit, but a small spinning sphere moving in orbit. Then the electron has not only orbital angular momentum, but also spin angular momentum associated with its rotation on its axis. The sphere carries an electric charge, so the spinning motion leads to current loops and to a magnetic moment, as we discussed in Section 28–7. In a magnetic field the spin magnetic moment has an interaction energy in addition to that of the *orbital* magnetic moment (the Zeeman-effect interaction discussed in Section 43–2). We should see additional Zeeman shifts due to the spin magnetic moment.

As was mentioned above, such shifts *are* indeed observed in precise spectroscopic analysis. This and a variety of other experimental evidence have shown conclusively that the electron *does* have spin angular momentum and a spin magnetic moment that do not depend on its orbital motion but are intrinsic to the particle itself.

Like orbital angular momentum, spin angular momentum (usually denoted by $\boldsymbol{S}$) is found to be *quantized.* Suppose we have some apparatus that measures a particular component of $\boldsymbol{S}$, for example, the $z$-component, $S_z$. We find that the only possible values are

$$S_z = \pm\tfrac{1}{2}\hbar. \tag{43–19}$$

This relation is reminiscent of the expression $L_z = m_l\hbar$ for the $z$-component of orbital angular momentum, except that $S_z$ is *one-half* of $\hbar$ instead of an *integer* multiple. Equation (43–19) also suggests that the magnitude $S$ of the spin angular momentum is given by an expression that is analogous to Eq. (43–4) with $l$ replaced by the value $\frac{1}{2}$:

$$S = \sqrt{\tfrac{1}{2}(\tfrac{1}{2}+1)}\,\hbar = \sqrt{\tfrac{3}{4}}\,\hbar. \tag{43–20}$$

The electron is often called a "spin-$\frac{1}{2}$ particle."

We've learned that the Bohr model is an oversimplified picture of electron behavior. In quantum mechanics, in which the Bohr orbits are superseded by wave functions, we can't really *picture* electron spin. If we visualize a wave function as a cloud surrounding the nucleus, then we can imagine many tiny arrows distributed throughout the cloud, either with components in the $+z$-direction or with components in the $-z$-direction. But don't take this picture too seriously.

## Spin Quantum Number

The concept of electron spin is well established by a variety of experimental evidence. To label completely the state of the electron in a hydrogen atom, we now need a fourth quantum number $m_s$, the **spin quantum number,** to specify the electron spin orientation. The quantum number $m_s$ can take the value $+\frac{1}{2}$ or $-\frac{1}{2}$, so the $z$-component $S_z$ of spin angular momentum is

$$S_z = m_s\hbar \quad (m_s = \pm\tfrac{1}{2}). \tag{43–21}$$

The $z$-component ($\mu_z$) of the associated magnetic moment turns out to be related to $S_z$ by

$$\mu_z = -\frac{e}{m}S_z, \tag{43–22}$$

where $e$ and $m$ are (as usual) the charge and mass of the electron. When the atom is placed in a magnetic field, the interaction energy $-\boldsymbol{\mu}\cdot\boldsymbol{B}$ of the spin magnetic moment with the field causes further splittings in energy levels and in the corresponding spectrum lines.

Equation (43–22) shows that for electron spin, the gyromagnetic ratio $|\mu_z/S_z| = e/m$ is *twice* as great as the value $e/2m$ for *orbital* angular momentum and magnetic moment. This result has no classical analog; for a particle in a circular orbit, the gyromagnetic ratio is always $e/2m$.

## EXAMPLE 43–6

Calculate the interaction energy for a single electron in a magnetic field with magnitude 2.00 T.

**SOLUTION** The interaction energy is $U = -\boldsymbol{\mu}\cdot\boldsymbol{B}$. If $\boldsymbol{B}$ is along the $z$-axis, this is equal to $-\mu_z B$. From Eqs. (43–21) and (43–22),

$$\begin{aligned} U &= \frac{e}{m}S_z B = \pm\frac{e\hbar}{2m}B = \pm\mu_B B \\ &= \pm(9.274\times10^{-24}\ \text{J/T})(2.00\ \text{T}) = \pm1.85\times10^{-23}\ \text{J} \\ &= \pm1.16\times10^{-4}\ \text{eV}. \end{aligned}$$

The effect on energy levels is shown in Fig. 43–11. Note that the electron's spin magnetic moment is one Bohr magneton, even though the $z$-component of angular momentum is only $\pm\hbar/2$. The reason is that the gyromagnetic ratio is twice as great for spin as for orbital angular momentum.

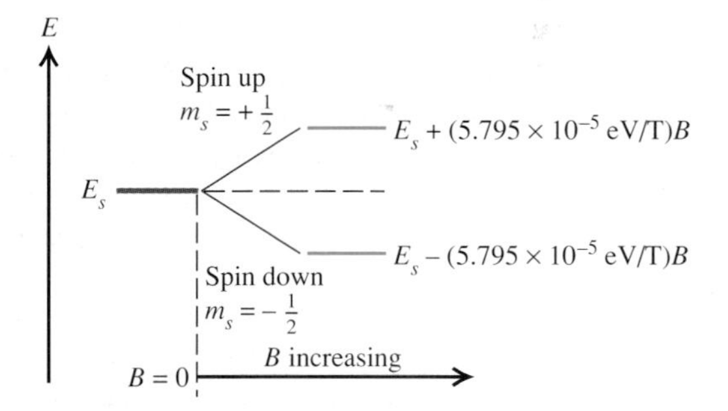


**43–11** An $s$ state of a single electron is split by interaction with a magnetic field.

## Spin-Orbit Coupling

The spin magnetic moment also causes splitting of some energy levels even when there is *no* external field. This effect is somewhat more subtle. In the Bohr model, an observer moving with the electron would see the positively charged nucleus revolving around him or her (just as to an observer on earth the sun seems to be orbiting the earth). This moving charge causes a magnetic field at the location of the electron (as seen in the electron's moving frame of reference). The resulting interaction energy with the spin magnetic moment causes a twofold splitting of this level, corresponding to the two possible orientations of electron spin.

Discussions based on the Bohr model can't be taken too seriously, but a similar result can be derived from the Schrödinger equation. The interaction energy $E$ can be expressed in terms of the scalar product of the angular-momentum vectors $\boldsymbol{L}$ and $\boldsymbol{S}$. This effect is called **spin-orbit coupling;** it is responsible for the small energy difference between the two closely spaced "resonance levels" of sodium shown in Fig. 40–9, and for the corresponding familiar doublet (589.0, 589.6 nm) in the spectrum of sodium.

### EXAMPLE 43–7

Calculate the effective magnetic field for the electron in the $3p$ states of the sodium atom.

**SOLUTION** The two lines in the sodium doublet result from transitions from the two $3p$ states, split by spin-orbit coupling, to the $3s$ state, which is *not* split because it has $L = 0$. The energies of the two photons are

$$E = \frac{hc}{\lambda} = \frac{(4.136 \times 10^{-15}\,\text{eV}\cdot\text{s})(2.998 \times 10^{8}\,\text{m/s})}{589.0 \times 10^{-9}\,\text{m}} = 2.1052\,\text{eV}$$

and

$$E = \frac{hc}{\lambda} = \frac{(4.136 \times 10^{-15}\,\text{eV}\cdot\text{s})(2.998 \times 10^{8}\,\text{m/s})}{589.6 \times 10^{-9}\,\text{m}} = 2.1031\,\text{eV}.$$

The energy difference is

$$2.1052\,\text{eV} - 2.1031\,\text{eV} = 0.0021\,\text{eV} = 3.4 \times 10^{-22}\,\text{J}.$$

This is the energy difference between the two $3p$ states. The difference is due to spin-orbit coupling, which raises one level and lowers the other, each by half of the energy difference, or $1.7 \times 10^{-22}$ J. From Example 43–6,

$$B = \frac{U}{\mu_B} = \frac{1.7 \times 10^{-22}\,\text{J}}{9.27 \times 10^{-24}\,\text{J/T}} = 18\,\text{T}.$$

This is a very strong field; to produce such a field in the laboratory would require superconducting electromagnets.

---

The orbital and spin angular momenta (**L** and **S**, respectively) can combine in various ways. We define the total angular momentum **J** as

$$\boldsymbol{J} = \boldsymbol{L} + \boldsymbol{S}. \tag{43–23}$$

The possible values of the magnitude $J$ are given in terms of a quantum number $j$ by

$$J = \sqrt{j(j+1)}\hbar. \tag{43–24}$$

We can then have states in which $j = l + \frac{1}{2}$, corresponding to the classical case in which **L** and **S** are parallel, and states in which $j = l - \frac{1}{2}$, where **L** and **S** are antiparallel. For example, when $l = 1$, $j$ can be $\frac{1}{2}$ or $\frac{3}{2}$. In spectroscopic notation these states are labeled $^2P_{1/2}$ and $^2P_{3/2}$, respectively. The superscript is the number of possible spin orientations, the letter $P$ (now capitalized) indicates states with $l = 1$, and the subscript is the value of $j$. This scheme is used to label the energy levels of the sodium atom (Fig. 40–9).

The various line splittings resulting from magnetic interactions are collectively called *fine structure*. There are also additional, much smaller splittings associated with the fact that the *nucleus* of the atom has a magnetic moment that interacts with the orbital and spin magnetic moments of the electrons. These effects are called *hyperfine structure*.

The fact that the gyromagnetic ratio for spin ($e/m$) is twice as great as the value for orbital angular momentum ($e/2m$) has to be put into our analysis as an *ad hoc* assumption; we use it because it works. In this respect it is subject to the same criticism as the quantization of angular momentum in the Bohr model. But in 1928, Paul Dirac developed a relativistic generalization of the Schrödinger equation for electrons. He showed that the spin gyromagnetic ratio $2(e/2m)$ is predicted by this equation automatically, without having to be put in as a separate assumption. So as time went on, more and more pieces of the atomic-structure puzzle dropped into place.

More recently, it has been found that the gyromagnetic ratio is not precisely twice as great for spin as for orbital angular momentum, but 2.0023193044 times as great. Believe it or not, this value has actually been measured with this precision and independently calculated from quantum electrodynamics. This is certainly one of the highest-precision comparisons of theory and experiment in the history of science.

## 43–4 MANY-ELECTRON ATOMS AND THE EXCLUSION PRINCIPLE

So far our analysis of atomic structure has concentrated on the hydrogen atom. That's natural; hydrogen, with only one electron, is the simplest atom. If we can't understand hydrogen, we certainly can't understand anything more complex. When the Bohr model was developed in 1915, attempts were made to apply it to the helium atom, which has two electrons. It's easy to picture two electrons moving on opposite sides of a circular orbit. But calculations with this model didn't agree at all with the observed properties or spectrum of helium. The Bohr model seemed to work only for hydrogen.

The general problem is an atom that in its normal (electrically neutral) state has $Z$ electrons; we call $Z$ the **atomic number.** The neutron has no charge; the proton and electron charges have the same magnitude but opposite sign. The total charge in the nucleus is $+Ze$, so the total electric charge in the normal atom is exactly zero.

We can apply the Schrödinger equation to this general problem. The complexity of the analysis increases very rapidly with increasing $Z$. Each electron interacts not only with the nucleus but also with every other electron. The wave functions and the potential energy are functions of $3Z$ coordinates, and the equation contains second derivatives with respect to all of them. The mathematical problem of finding solutions of such equations is so complex that it has not been solved exactly even for the helium atom.

Fortunately, various approximation schemes are available. The simplest approximation is to ignore all interactions between electrons and consider each electron as moving under the action only of the nucleus (considered to be a point charge). In this approximation the wave function for each electron is a hydrogenlike function with four quantum numbers $(n, l, m_l, m_s)$; the nuclear charge is $Ze$ instead of $e$. This requires replacement of every factor $e^2$ in the wave functions and the energy levels by $Ze^2$. In particular, the energy levels are given by Eq. (43–3) with $e^4$ replaced by $Z^2e^4$:

$$E_n = -\frac{1}{(4\pi\epsilon_0)^2}\frac{mZ^2e^4}{2n^2\hbar^2} = -\frac{Z^2}{n^2}(13.6\text{ eV}). \tag{43–25}$$

This approximation is fairly drastic; when there are many electrons, their interactions with each other are as important as the interaction of each with the nucleus. So this model isn't very useful for quantitative predictions.

## The Central-Field Approximation

A less drastic and more useful approximation is to think of all the electrons together as making up a charge cloud that is, on the average, *spherically symmetric.* We can then think of each individual electron as moving in the total electric field due to the nucleus and this averaged-out cloud of all the other electrons. There is a corresponding spherically symmetric potential-energy function $U(r)$. This picture is called the **central-field approximation;** it provides a useful starting point for the understanding of atomic structure. If you are disappointed that we have to make approximations at such an early stage in our discussion, keep in mind that we are dealing with problems that previously defied all attempts at analysis, with or without approximations.

In the central-field approximation we can again deal with one-electron wave functions. The Schrödinger equation differs from the equation for hydrogen only in that the $1/r$ potential-energy function is replaced by a different function $U(r)$. The angular functions are exactly the same as for hydrogen, so the angular-momentum *states* are also the same. The quantum numbers $l$, $m_l$, and $m_s$ have the same meaning as before, and the magnitude and $z$-component of angular momentum are again given by Eqs. (43–4) and (43–5).

The radial wave functions and probabilities are different because of the change in $U(r)$, so the energy levels are no longer given by Eq. (43–3). We can still label a state using the four quantum numbers $(n, l, m_l, m_s)$. In general, the energy of a state now depends on both $n$ and $l$, rather than just on $n$ as with hydrogen. The restrictions on values of the quantum numbers are the same as before:

$$l \le n-1, \qquad |m_l| \le l, \qquad m_s = \pm\tfrac{1}{2}. \tag{43–26}$$

$Z_{\text{eff}} = Z - 5$. To find the energy $E_L$ needed to remove an $L$ electron and create an $L$ x-ray energy level, we use Eq. (43–27) with $n = 2$ and $Z_{\text{eff}} = Z - 5$:

$$E_L = (Z - 5)^2(13.6 \text{ eV})/4. \tag{43–31}$$

When an $L$ electron drops into a vacancy in the $K$ shell, emitting an x-ray photon, the photon energy is the difference in the two energy levels:

$$E = E_K - E_L.$$

The corresponding photon frequency is $f = E/h$. To make a rough comparison with Moseley's law, first note that the $L$ levels are much smaller than the $K$ levels. Especially if $Z$ is substantial (20 or above), we make only a small error if we replace $(Z - 5)$ in Eq. (43–31) by $(Z - 1)$. When we do that, we get

$$\begin{aligned} E = E_K - E_L &= (Z - 1)^2(13.6 \text{ eV}) - (Z - 1)^2(13.6 \text{ eV})/4 \\ &= (\tfrac{3}{4})\,(Z - 1)^2\,(13.6 \text{ eV}), \\ f = \frac{E}{h} &= \frac{\frac{3}{4}(Z - 1)^2(13.6 \text{ eV})}{4.136 \times 10^{-15} \text{ eV} \cdot \text{s}} \\ &= (2.47 \times 10^{15} \text{ Hz})\,(Z - 1)^2. \end{aligned} \tag{43–32}$$

This agrees almost exactly with Moseley's law, Eq. (43–29). Indeed, considering the approximations we have made, the agreement is better than we have a right to expect. But our calculation does show how Moseley's law can be understood on the basis of transitions between x-ray energy levels.

The $K$ vacancy may also be filled by an electron falling from the $M$ or $N$ shell, assuming that these are occupied. If so, the x-ray spectrum shows a series of three lines, resulting from transitions in which the $K$-shell vacancy is filled by an $L$, $M$, or $N$ electron. This series is called a $K$ series. Figure 43–16 shows the $K$ series for copper ($Z = 29$), molybdenum ($Z = 42$), and tungsten ($Z = 74$). The three lines in each series are called the $K_\alpha$, $K_\beta$, and $K_\gamma$ lines. The $K_\alpha$ line is produced by the transition $K \rightarrow L$, the $K_\beta$ line by $K \rightarrow M$, and the $K_\gamma$ line by $K \rightarrow N$. (Note that $K \rightarrow L$ means that the *vacancy* moves from the $K$ to the $L$ shell; an *electron* moves from the $L$ shell to the $K$ shell.) There are other series of x-ray lines, called the $L$, $M$, and $N$ series, produced by the ejection of electrons from the $L$, $M$, and $N$ shells, respectively, rather than the $K$ shell. Electrons in these outer shells are farther away from the nucleus and are not held as tightly as those in the $K$ shell. Their removal requires less energy, and the x-ray photons emitted when these vacancies are filled have lower energy than those in the $K$ series.

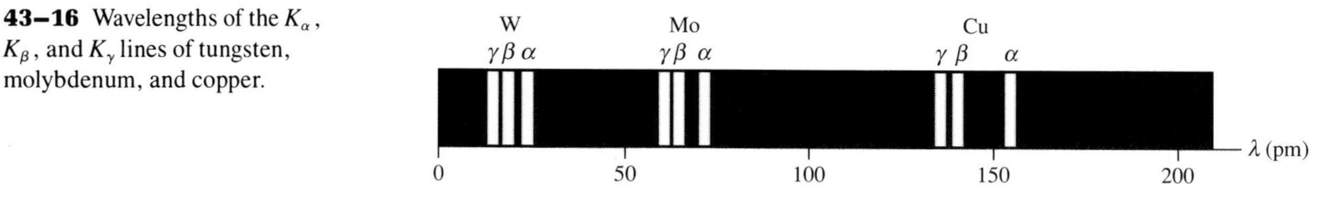


**43–16** Wavelengths of the $K_\alpha$, $K_\beta$, and $K_\gamma$ lines of tungsten, molybdenum, and copper.

## EXAMPLE 43–10

You measure the $K_\alpha$ wavelength for an unknown element, obtaining the value 154 pm. What is the element?

**SOLUTION** The corresponding frequency is

$$f = \frac{c}{\lambda} = \frac{3.00 \times 10^8 \text{ m/s}}{153 \times 10^{-12} \text{ m}} = 1.96 \times 10^{18} \text{ Hz}.$$

From Moseley's law, Eq. (43–29),

$$(2.48 \times 10^{15} \text{ Hz})\,(Z - 1)^2 = 1.96 \times 10^{18} \text{ Hz},$$
$$Z = 29.1$$

You know $Z$ has to be an integer; you conclude that $Z = 29$, corresponding to the element copper.

## EXAMPLE 43–11

What is the minimum wavelength of a characteristic x-ray line from copper? What is the corresponding photon energy?

**SOLUTION** The highest occupied shell in copper is the $N$ shell. The largest transition energy corresponds to the x-ray transition $K \rightarrow N$. The $N$ electrons, with $n = 4$ and $Z_{eff}$ equal to only about 2, are very loosely bound. From Eq. (43–27),

$$E_N = (2)^2(13.6\text{ eV})/(4)^2 = 3.4\text{ eV}.$$

This is so small in comparison to $E_K$ that we can ignore $E_N$. We assume the photon energy to be equal to $E_K$, given by Eq. (43–30):

$$E_K = (29 - 1)^2(13.6\text{ eV}) = 1.07 \times 10^4\text{ eV}.$$

The corresponding frequency is

$$f = \frac{E}{h} = \frac{1.07 \times 10^4\text{ eV}}{4.136 \times 10^{15}\text{ eV}\cdot\text{s}} = 2.58 \times 10^{18}\text{ Hz}.$$

The wavelength is

$$\lambda = \frac{c}{f} = \frac{3.00 \times 10^8\text{ m/s}}{2.58 \times 10^{18}\text{ Hz}} = 1.16 \times 10^{-10}\text{ m} = 116\text{ pm}.$$

This photon will be produced only if the accelerating voltage in the x-ray tube is greater than 10.7 kV; do you see why?

We can also observe x-ray *absorption* spectra, analogous to optical absorption spectra. Unlike optical spectra, the absorption wavelengths are not the same as those for emission. For example, the $K_\alpha$ emission line results from the transition $K \rightarrow L$. The reverse transition, $L \rightarrow K$, ordinarily doesn't occur because in the ground state there is no vacancy in the $L$ shell. To be absorbed, a photon must have enough energy to remove an electron from the atom completely, not just raise it to the $L$ shell. Experimentally, if we gradually increase the accelerating voltage and thus the maximum photon energy, we observe sudden increases in absorption when we reach the x-ray energies $E_K, E_L, \ldots$ . These sudden jumps of absorption are called *absorption edges* (Fig. 43–17).

X-ray energy levels and characteristic spectra provide a very useful analytical tool. Satellite-borne x-ray spectrometers are used to study x-ray emission lines from highly excited atoms in distant astronomical sources. X-ray spectra are also used in air-pollution monitoring and in studies of the abundance of various elements in rocks.

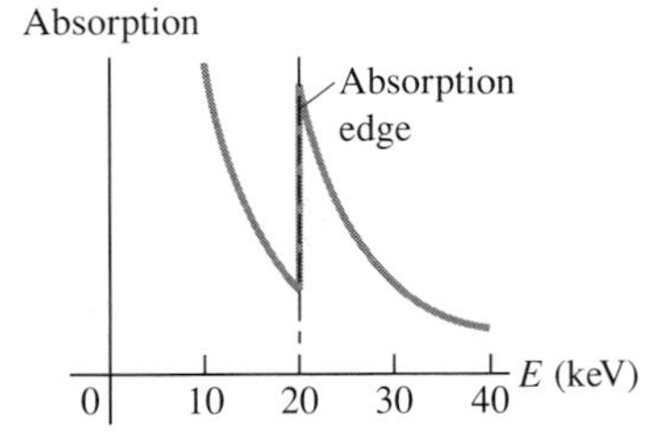


**43–17** X-ray absorption by molybdenum. A beam of x rays is passed through a slab of molybdenum, and the absorption is measured for various energies. A sharp increase in absorption occurs at the $K$ absorption edge, corresponding to excitation of the $K$ x-ray energy level at 20 keV.

## SUMMARY

- The Schrödinger equation for the hydrogen atom gives certain energy levels:

$$E_n = -\frac{1}{(4\pi\epsilon_0)^2}\frac{me^4}{2n^2\hbar^2} = -\frac{13.60\text{ eV}}{n^2}. \quad (43\text{–}3)$$

If the nucleus has charge $Ze$, there is also a factor $Z^2$. The possible values of angular momentum are

$$L = \sqrt{l(l+1)}\hbar \quad (l = 0, 1, 2, \ldots, n-1). \quad (43\text{–}4)$$

The possible values of the $z$-component of angular momentum are

$$L_z = m_l\hbar \quad (m_l = 0, \pm 1, \pm 2, \ldots, \pm l). \quad (43\text{–}5)$$

- The radial probability distribution function is

$$P(r)\,dr = |\psi|^2\,dV = 4\pi r^2|\psi|^2\,dr. \quad (43\text{–}7)$$

Atomic distances are often measured in units of the Bohr radius:

$$a_0 = \frac{\epsilon_0 h^2}{\pi m e^2} = \frac{4\pi\epsilon_0 \hbar^2}{m e^2} = 0.529 \times 10^{-10}\ \text{m}. \qquad (43\text{–}8)$$

• The magnetic moment corresponding to orbital angular momentum $\hbar$ is called one Bohr magneton $\mu_B$:

$$\mu_B = \frac{e\hbar}{2m}. \qquad (43\text{–}13)$$

The gyromagnetic ratio for orbital angular momentum is $e/2m$. The interaction energy of an electron with magnetic quantum number $m_l$ in a magnetic field $B$ along the $+z$-direction is

$$U = -\mu_z B = m_l \frac{e\hbar B}{2m} \quad (m_l = 0, \pm 1, \pm 2, \ldots, \pm l) \qquad (43\text{–}17)$$

or

$$U = m_l \mu_B B. \qquad (43\text{–}18)$$

• The spin angular momentum of an electron is

$$S = \sqrt{\tfrac{1}{2}(\tfrac{1}{2}+1)}\,\hbar = \sqrt{\tfrac{3}{4}}\,\hbar, \qquad (43\text{–}20)$$

with $z$-component

$$S_z = \pm \tfrac{1}{2}\hbar. \qquad (43\text{–}19)$$

The gyromagnetic ratio for electron spin is $e/m$.

• In the central-field approximation, each electron moves in the electric field of the nucleus and the averaged-out, spherically symmetric charge distribution of all the remaining electrons. The quantum numbers are the same as for hydrogen-atom states, but the energy levels depend on both $n$ and $l$ because of screening, the partial cancellation of the field of the nucleus by the inner electrons.

• If the effective charge seen by an electron is $Z_{\text{eff}}e$, the energy levels are approximately

$$E_n = -\frac{Z_{\text{eff}}^2}{n^2}(13.6\ \text{eV}). \qquad (43\text{–}27)$$

The value of $Z_{\text{eff}}$ for a particular electron depends on $n$ and $l$.

• The minimum x-ray wavelength for accelerating voltage $V$ is

$$\lambda_{\min} = \frac{hc}{eV}. \qquad (43\text{–}28)$$

• Moseley's law states that the shortest-wavelength x-ray from a target with atomic number $Z$ is

$$f = (2.48 \times 10^{15}\ \text{Hz})(Z-1)^2. \qquad (43\text{–}29)$$

X-ray energy levels are atomic states in which an electron is missing in one of the inner shells. Characteristic x-ray spectra result from transitions between x-ray energy levels.

## KEY TERMS

- principal quantum number
- angular-momentum quantum number
- magnetic quantum number
- degeneracy
- shell
- Zeeman effect
- magnetic moment
- gyromagnetic ratio
- Bohr magneton
- selection rules
- electron spin
- spin quantum number
- spin-orbit coupling
- atomic number
- central-field approximation
- exclusion principle
- periodic table of the elements
- screening
- x-ray energy level

# EXERCISES

## Section 43–1 The Hydrogen Atom

**43–1** Verify the claim made in step 2 of the Problem-Solving Strategy on page 1188 that the electric potential energy of a proton and an electron 0.10 nm apart has magnitude 14 eV.

**43–2** a) Make a chart showing all the possible sets of quantum numbers $l$ and $m_l$ for the states of the electron in the hydrogen atom when $n = 4$. How many combinations are there? b) What is the energy of these states?

**43–3** The orbital angular momentum of an electron has a magnitude of $2.583 \times 10^{-34}$ kg · m/s. What is the angular-momentum quantum number $l$ for the electron?

**43–4** Consider states with $l = 2$. a) In units of $\hbar$, what is the largest possible value of $L_z$? b) In units of $\hbar$, what is the value of $L$? Which is larger, $L$ or the maximum possible $L_z$? c) For each allowed value of $L_z$, what angle does the vector $\boldsymbol{L}$ make with the $+z$-axis? How does the angle for $m_l = l$ compare to that calculated for $l = 3$ in Example 43–2?

**43–5** Calculate, in units of $\hbar$, the magnitude of the maximum orbital angular momentum for an electron in a hydrogen atom for states with principal quantum numbers 1, 10, and 100. Compare each with the value of $n\hbar$ postulated in the Bohr model. What trend do you see?

**43–6** Show that $\Phi(\phi) = e^{im_l\phi}$ is periodic with period $2\pi$ if and only if $m_l$ is restricted to the values $0, \pm 1, \pm 2, \ldots$.

**43–7** In Example 43–3, fill in the missing details to show that $P(r < a_0) = 1 - 5e^{-2}$.

## Section 43–2 The Zeeman Effect

**43–8** a) In the Bohr model, what is the magnitude of the magnetic moment of a hydrogen atom in the $n = 2$ state? b) Find the interaction potential energy when a hydrogen atom in the $n = 2$ state is placed in a magnetic field with magnitude 3.00 T if the magnetic moment of the atom and the magnetic field are antiparallel.

**43–9** A hydrogen atom in a $3p$ excited state is placed in a uniform external magnetic field $\boldsymbol{B}$. What must $B$ be to split the $3p$ state into three levels with a splitting of $2.00 \times 10^{-5}$ eV between adjacent levels?

**43–10** Consider a hydrogen atom in an $l = 2$ state. In the absence of an external magnetic field the states with different $m_l$ have (approximately) the same energy. a) Calculate the splitting in electronvolts of the $m_l$ levels when the atom is put in a 0.400-T magnetic field that is in the $+z$-direction. b) Which $m_l$ level will have the lowest energy? c) Draw an energy-level diagram that shows the $l = 2$ levels with and without the external magnetic field.

**43–11** A hydrogen atom in the $4f$ state is placed in a magnetic field of 2.50 T that is in the $+z$-direction. a) Into how many levels is this state split by the interaction of the orbital angular momentum with the magnetic field? b) What is the energy separation between adjacent levels? c) What is the energy separation between the level of lowest energy and the level of highest energy?

## Section 43–3 Electron Spin

**43–12** A hydrogen atom in the $n = 1, m_s = \frac{1}{2}$ state is placed in a magnetic field with a magnitude of 1.50 T in the $+z$-direction. Find the interaction energy (in electronvolts) of the atom with the field due to the electron spin.

**43–13 Classical Electron Spin.** If you treat an electron as a classical uniform spherical particle with a radius of $1.0 \times 10^{-17}$ m, what is the angular velocity necessary to produce a spin angular momentum of $\hbar$?

**43–14** Calculate the energy difference between the $m_s = \frac{1}{2}$ and $m_s = -\frac{1}{2}$ levels of a hydrogen atom in the $n = 1, l = 0$ state when it is placed in a magnetic field of 0.600 T that is in the $+z$-direction. Which level, $m_s = \frac{1}{2}$ or $m_s = -\frac{1}{2}$, has the lower energy?

**43–15** A hydrogen atom in a particular orbital angular momentum state is found to have $j$ quantum numbers $\frac{5}{2}$ and $\frac{7}{2}$. What is the value of $l$ for the state?

**43–16** Give the different possible combinations of $l$ and $j$ for a hydrogen atom in the $n = 3$ level. In each case, give the spectroscopic notation of the state.

## Section 43–4 Many-Electron Atoms and the Exclusion Principle

**43–17 Neon Quantum Numbers.** Make a list of the four quantum numbers $n$, $l$, $m_l$, and $m_s$ for each of the ten electrons in the ground state of the neon atom.

**43–18** For germanium ($Z = 32$), make a list of the number of electrons in each state ($1s$, $2s$, $2p$, etc.).

**43–19** The $5s$ electron in Rb sees an effective charge of $2.771e$. Calculate the ionization energy of this electron.

**43–20** For magnesium the first ionization potential is 7.6 eV. The second ionization potential (additional energy required to remove a second electron) is almost twice this, 15 eV, and the third ionization potential is much larger, about 80 eV. How can these numbers be understood?

**43–21** Estimate the energy of the least penetrating $l$ level for a) the $n = 2$ states of $Be^+$; b) the $n = 4$ states of $Ca^+$.

**43–22** a) The energy of the 2$s$ state of lithium is $-5.391$ eV. Calculate the value of $Z_{\text{eff}}$ for this state. b) The energy of the 4$s$ state of potassium is $-4.339$ eV. Calculate the value of $Z_{\text{eff}}$ for this state. c) Compare $Z_{\text{eff}}$ for the 2$s$ state of lithium, the 3$s$ state of sodium (Example 43–8), and the 4$s$ state of potassium. What trend do you see?

### Section 43–5 X-Ray Spectra

**43–23** Calculate the energy, frequency, and wavelength of the $K_\alpha$ x ray for the elements a) Al ($Z = 13$); b) Mn ($Z = 25$); c) Rb ($Z = 37$).

**43–24** The energies of an electron in the $K$, $L$, and $M$ shells of the tungsten atom are $-69500$ eV, $-12000$ eV, and $-2200$ eV, respectively. Calculate the wavelengths of the $K_\alpha$ and $K_\beta$ x rays of tungsten.

**43–25** A $K_\alpha$ x ray emitted from a sample has an energy of 9.89 keV. What is the element?

## PROBLEMS

**43–26** For a hydrogen atom in an $l = 0$ state, the probability of finding the electron within a spherical shell with inner radius $r$ and outer radius $r + dr$ is $|\psi|^2\,dV$, with $dV$ given in Eq. (43–6). For a hydrogen atom in the 1$s$ ground state, at what value of $r$ does $r^2|\psi|^2$ have its maximum value? How does your result compare to the radius of the $n = 1$ orbit in the Bohr model?

**43–27** The wave function for a hydrogen atom in the 2$s$ state is

$$\psi_{2s}(r) = \frac{1}{\sqrt{32\pi a_0^3}}(2 - r/a_0)e^{-r/2a_0}.$$

a) Verify that this function is normalized. b) The Bohr radius for the $n = 2$ level is $4a_0$ (Eq. 40–16). Calculate the probability that an electron in the 2$s$ state will be found at a distance less than $4a_0$ from the nucleus.

**43–28** The normalized wave function for a hydrogen atom in the 2$s$ state is given in Problem 43–27. a) For a hydrogen atom in the 2$s$ excited state, at what value of $r$ does $r^2|\psi|^2$ have its largest value? (See Problem 43–26.) How does your result compare to the radius of the $n = 2$ orbit in the Bohr model (Eq. 40–16)? b) At what value of $r$ (other than $r = 0$ or $r = \infty$) is $r^2|\psi|^2$ equal to zero, so the probability of finding the electron at that separation from the nucleus is zero? Compare your results to Fig. 43–2.

**43–29 Classical Turning Point.** Consider a hydrogen atom in the 1$s$ state. a) For what value of $r$ is the total energy $E$ equal to the potential energy $U(r)$? Express your answer in terms of $a_0$. This value of $r$ is called the classical turning point. b) For $r$ greater than the classical turning point, $E < U(r)$, and classically the particle cannot be in this region, since the kinetic energy cannot be negative. Calculate the probability of the electron being found in this classically forbidden region.

**43–30** For an ion with nuclear charge $Z$ and a single electron, the electric potential energy is $-Ze^2/4\pi\epsilon_0 r$ and the expression for the energies of the states and for the normalized wave functions are obtained from those for hydrogen by replacing $e^2$ by $Ze^2$. Consider the $O^{7+}$ ion, with $Z = 8$ and one electron. a) What is the ground state energy in electronvolts? b) What is the ionization energy, the energy required to remove the electron from the ion if it is initially in the ground state? c) What is the Bohr radius $a_0$ for this ion? d) What is the wavelength of the photon emitted when the ion makes a transition from the $n = 2$ state to the $n = 1$ ground state?

**43–31** A hydrogen atom in an $n = 2$, $l = 1$, $m_l = -1$ state emits a photon when it decays to an $n = 1$, $l = 0$, $m_l = 0$ ground state. a) In the absence of an external magnetic field, what is the wavelength of this photon? b) If the atom is in a magnetic field in the $+z$-direction and with a magnitude of $B = 2.50$ T, what is the shift in the wavelength of the photon from the zero-field value? Does the magnetic field increase or decrease the wavelength? [Use the result of Problem 41–28(c). Disregard the effect of electron spin.]

**43–32 The Balmer $H_\alpha$ Line.** A hydrogen atom makes a transition from an $n = 3$ state to an $n = 2$ state (the Balmer $H_\alpha$ line) while in a magnetic field in the $+z$-direction and with magnitude 1.50 T. a) If the magnetic quantum number is $m_l = 2$ in the initial ($n = 3$) state and $m_l = 1$ in the final ($n = 2$) state, by how much is each energy level shifted from the zero-field value? b) By how much is the wavelength of the spectrum line shifted from the zero-field value? Is the wavelength increased or decreased? Disregard the effect of electron spin. [Use the result of Problem 41–28(c).]

**43–33** Consider the transition from a 3$d$ to a 2$p$ state of hydrogen in an external magnetic field. Assume that the effects of electron spin can be neglected (which is not actually the case) so the magnetic field interacts only with the orbital angular momentum. Identify each of the allowed transitions by the $m_l$ values of the initial and final states. For each of these allowed transitions, determine the shift of the transition energy from the zero-field value and show that there are three different transition energies.

**43–34 Electron Spin Resonance (ESR).** Electrons in the lower of two spin states in a magnetic field can absorb a photon of the right frequency and move to the higher state. Calculate the magnetic field magnitude required for this transition in a hydrogen atom with $n = 1$ and $l = 0$ to be induced by microwaves with wavelength 3.00 cm.

**43–35** A large number of hydrogen atoms in the $n = 1, l = 0$ state are placed in an external magnetic field that is in the $+z$-direction. Assume that the atoms are in thermal equilibrium at room temperature, $T = 300$ K. According to the Boltzmann distribution (Section 40–6), what is the ratio of the number of atoms in the $m_s = \frac{1}{2}$ state to the number in the $m_s = -\frac{1}{2}$ state when the magnetic field magnitude is a) $5.00 \times 10^{-5}$ T (approximately the earth's field); b) 0.100 T; c) 10.0 T?

**43–36** A lithium atom has three electrons, and the $^2S_{1/2}$ ground-state electron configuration is $1s^2 2s$. The $1s^2 2p$ excited state is split into two closely spaced levels, $^2P_{3/2}$ and $^2P_{1/2}$, by the spin-orbit interaction (Example 43–7). A photon with wavelength 67.09608 $\mu$m is emitted in the $^2P_{3/2} \rightarrow {}^2S_{1/2}$ transition, and a photon with wavelength 67.09761 $\mu$m is emitted in the $^2P_{1/2} \rightarrow {}^2S_{1/2}$ transition. Calculate the effective magnetic field seen by the electron in the $1s^2 2p$ state of the lithium atom. How does your result compare to that for the $3p$ level of sodium found in Example 43–7?

**43–37** Show that the total number of hydrogen-atom states (including different spin states) for a given value of the principal quantum number $n$ is $2n^2$. [*Hint:* The sum of the first $N$ integers $1 + 2 + 3 + \cdots + N$ is given by $N(N + 1)/2$.]

**43–38 Rydberg Atoms.** Rydberg atoms are atoms whose outermost electron is in an excited state with a very large principal quantum number. a) Why do all neutral Rydberg atoms with the same $n$ value have essentially the same ionization energy, independent of the total number of electrons in the atom? b) What is the ionization energy for a Rydberg atom with a principal quantum number of 290? What is the Bohr radius of the Rydberg electron? c) Repeat part (b) for $n = 732$. (Rydberg atoms with $n = 290$ have been produced in the laboratory, and atoms with $n = 732$ have been detected in interstellar space.)

**43–39** Estimate the minimum and maximum wavelengths of the characteristic x rays emitted by a) titanium ($Z = 22$); b) molybdenum ($Z = 42$).

**43–40** The energy needed to remove an electron in an $L$ shell is estimated in Eq. (43–31). In deriving Eq. (43–32) the estimate $E_L = (Z - 1)^2(13.6 \text{ eV})/4$ is used instead. Calculate the percent difference in the calculated $K_\alpha$ x-ray frequency obtained by using these two expressions for $E_L$ for a) Ca ($Z = 20$); b) Sn ($Z = 50$).

## CHALLENGE PROBLEM

**43–41** Repeat the calculation of Problem 43–29 for a one-electron ion with nuclear charge $Z$. (See Problem 43–30.) How does the probability of the electron being found in the classically forbidden region depend on $Z$?

electrons from one atom to another. As a result, many molecules having dissimilar atoms have electric dipole moments, that is, a preponderance of positive charge at one end and of negative charge at the other. Such molecules are called *polar molecules.* Water molecules have large electric dipole moments; these are responsible for the exceptionally large dielectric constant of liquid water.

## Van der Waals Bonds

Ionic and covalent bonds, with typical bond energies of 1 to 5 eV, are called *strong bonds.* There are also two types of *weaker* bonds with typical energies of 0.5 eV or less. One of these, the **van der Waals bond,** is an interaction between the electric dipole moments of two atoms or molecules. The bonding of water molecules in the liquid and solid states results partly from dipole-dipole interactions. The interaction potential energy drops off very quickly with distance $r$ between molecules, usually as $1/r^6$.

Even when an atom or molecule has no permanent dipole moment, fluctuating charge distributions can lead to fluctuating dipole moments; these in turn can induce dipole moments in neighboring structures. The resulting dipole-dipole interaction can be attractive and can lead to weak bonding of atoms or molecules. The low-temperature liquefaction and solidification of the inert gases and of molecules such as $H_2$, $O_2$, and $N_2$ is due to induced-dipole van der Waals interactions. Not much thermal-agitation energy is needed to break these weak bonds, so such substances usually exist in the liquid and solid states only at very low temperatures. Figure 44–4 shows a comparison of the van der Waals bond between two helium atoms and the covalent bond between two hydrogen atoms.

## Hydrogen Bonds

Another type of weak bond, the **hydrogen bond,** is analogous to the covalent bond, in which an electron pair binds two positively charged structures. In the hydrogen bond a proton ($H^+$ ion) gets between two atoms, polarizing them and attracting them by means of the induced dipoles. This bond is unique to hydrogen-containing compounds because only hydrogen has a singly ionized state with no remaining electron cloud; the hydrogen ion is a bare proton, much smaller than any other singly ionized atom. The bond energy

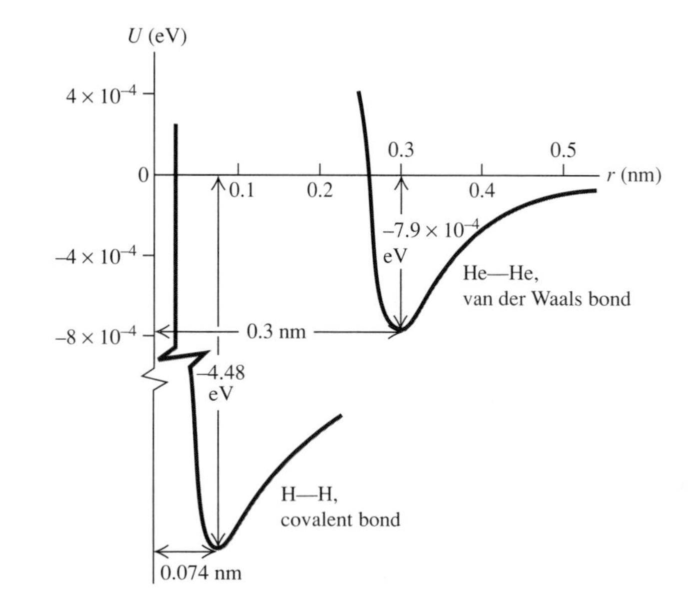


**44–4** Potential-energy curves for the covalent bond of two hydrogen atoms and the van der Waals bond between two helium atoms. The bond energy is over 5000 times as great for the covalent bond.

is usually less than 0.5 eV. Hydrogen bonding plays an essential role in many organic molecules, including the cross-linking of polymer chains such as polyethylene and cross-link bonding between the two strands of the double-helix DNA molecule. Hydrogen bonding also plays a role in the structure of ice.

All of these bond types play roles in the structures of *solids* as well as of molecules. Indeed, a solid is in many respects a giant molecule. Still another type of bonding, the *metallic bond,* comes into play in the structures of metallic solids. We'll return to this subject in Section 44–3.

# 44–2 MOLECULAR SPECTRA

Molecules have energy levels associated with rotational motion of the molecule as a whole and with vibrational motion of the atoms relative to each other. Transitions between these levels lead to *molecular spectra.*

## Rotational Energy Levels

In this discussion we'll concentrate mostly on *diatomic molecules,* to keep things as simple as possible. In Fig. 44–5 we picture a diatomic molecule as a rigid dumbbell (two masses $m_1$ and $m_2$ separated by a constant distance $r_0$) that can *rotate* about an axis through its center of mass. What are the energy levels associated with this motion?

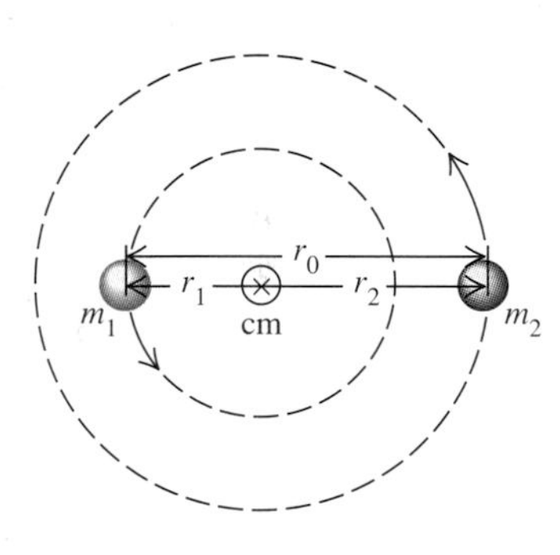


**44–5** Model of a diatomic molecule as two point masses $m_1$ and $m_2$ connected by a light rigid rod with length $r_0$. The distances of the masses from the center of mass are $r_1$ and $r_2$, and $r_1 + r_2 = r_0$.

We showed in Section 9–4 that when a rigid body with moment of inertia $I$ rotates with angular velocity $\omega$ about an axis through its center of mass, its kinetic energy is given by Eq. (9–17): $K = \frac{1}{2}I\,\omega^2$, where $I$ is the moment of inertia about an axis through the center of mass. We can also express the kinetic energy in terms of the magnitude $L$ of angular momentum, given by Eq. (10–30): $L = I\omega$. Combining these two equations, we find that $K = L^2/2I$. There is no potential energy, so the kinetic energy $K$ is equal to the total energy $E$:

$$E = \frac{L^2}{2I}. \qquad (44\text{–}1)$$

In a quantum-mechanical discussion of molecular rotation it is reasonable to assume that the angular momentum is quantized in the same way as for electrons in an atom, as in Eq. (43–4):

$$L^2 = l(l+1)\hbar^2 \qquad (l = 0, 1, 2, \ldots). \qquad (44\text{–}2)$$

Combining Eqs. (44–1) and (44–2), we obtain the *rotational energy levels:*

$$E = l(l+1)\frac{\hbar^2}{2I} \qquad (l = 0, 1, 2, \ldots). \qquad (44\text{–}3)$$

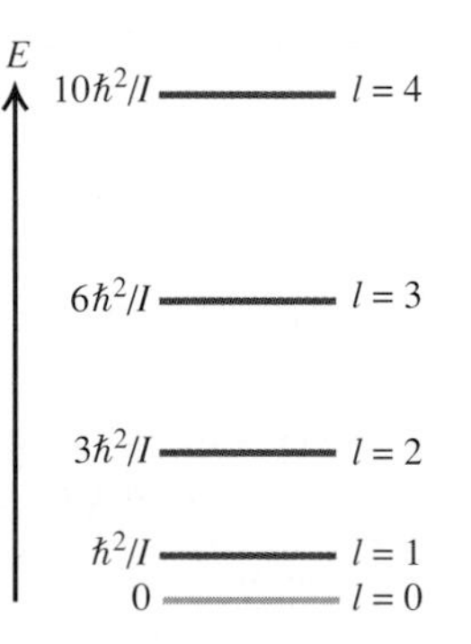


**44–6** The ground level and first four excited rotational energy levels for a diatomic molecule. The levels are not equally spaced.

Figure 44–6 is an energy-level diagram showing these rotational levels. Note that with increasing $l$, the spacing of adjacent levels increases. In fact, the energy difference between levels $l - 1$ and $l$ is $l\hbar^2/I$; can you prove this?

We can express the moment of inertia $I$ (about an axis through the center of mass and perpendicular to the molecule's axis) in terms of the *reduced mass* $m_r$ of the molecule:

$$m_r = \frac{m_1 m_2}{m_1 + m_2}. \qquad (44\text{–}4)$$

We introduced this quantity in Section 40–5 to accommodate a finite nuclear mass in the hydrogen atom. In Fig. 44–5 the distances $r_1$ and $r_2$ are the distances of the atoms from the center of mass. By definition of the center of mass, $m_1 r_1 = m_2 r_2$, and the figure also shows that $r_0 = r_1 + r_2$. Solving these equations for $r_1$ and $r_2$, we find

$$r_1 = \frac{m_2}{m_1 + m_2} r_0, \qquad r_2 = \frac{m_1}{m_1 + m_2} r_0. \tag{44–5}$$

The moment of inertia is $I = m_1 r_1^2 + m_2 r_2^2$; substituting Eq. (44–5), we find

$$I = m_1 \frac{m_2^2}{(m_1 + m_2)^2} r_0^2 + m_2 \frac{m_1^2}{(m_1 + m_2)^2} r_0^2 = \frac{m_1 m_2}{m_1 + m_2} r_0^2,$$

or

$$I = m_r r_0^2. \tag{44–6}$$

This is an elegant and useful result.

The reduced mass enables us to reduce this two-body problem to an equivalent one-body problem (a mass $m_r$ moving around a circle with radius $r_0$), just as we did with the hydrogen atom. Indeed, the only difference between this problem and the hydrogen atom is the difference in the radial forces. Thus, the angular-momentum states must be the same as for the hydrogen atom, and we are justified in using Eq. (44–2) for the angular momentum.

Let's look at some magnitudes: The moment of inertia of an oxygen ($O_2$) molecule is about $I = 2 \times 10^{-46}\ \text{kg} \cdot \text{m}^2$. We'll leave the derivation of this number as a problem. For this molecule the constant $\hbar^2/2I$ in Eq. (44–3) is approximately equal to

$$\frac{\hbar^2}{2I} = \frac{(1.05 \times 10^{-34}\ \text{J} \cdot \text{s})^2}{2(2 \times 10^{-46}\ \text{kg} \cdot \text{m}^2)}$$
$$= 3 \times 10^{-23}\ \text{J} = 2 \times 10^{-4}\ \text{eV}.$$

This amount of energy is much *smaller* than the atomic energy levels (a few electronvolts) that are typically associated with optical spectra. Photon energies for transitions among rotational levels are correspondingly small, and they fall in the *far infrared* region of the spectrum. The allowed transitions are determined by the same *selection rule* as for the hydrogen atom: $l$ must change by exactly one unit.

## EXAMPLE 44–2

**Hydrogen molecular spectrum** The two nuclei in the hydrogen molecule are 0.0741 nm apart. a) Find the energies of the three lowest rotational energy levels; express your results in electronvolts. b) Find the wavelength of the photon emitted in the transition from the $l = 2$ to the $l = 1$ state.

**SOLUTION** a) The mass of the hydrogen atom is 1.008 u $= 1.674 \times 10^{-27}$ kg. The reduced mass, from Eq. (44–4), is half of this: $m_r = 0.837 \times 10^{-27}$ kg. From Eq. (44–6),

$$I = m_r r^2 = (0.837 \times 10^{-27}\ \text{kg})(0.0741 \times 10^{-9}\ \text{m})^2$$
$$= 4.60 \times 10^{-48}\ \text{kg} \cdot \text{m}^2.$$

The rotational levels are given by Eq. (44–3):

$$E_l = l(l+1)\frac{\hbar^2}{2I} = l(l+1)\frac{(1.055 \times 10^{-34}\ \text{J} \cdot \text{s})^2}{2(4.60 \times 10^{-48}\ \text{kg} \cdot \text{m}^2)}$$
$$= l(l+1)(1.21 \times 10^{-21}\ \text{J}) = l(l+1)(0.00755\ \text{eV}).$$

(The quantity $\hbar^2/2I$ is larger than the value for the oxygen molecule because $I$ is smaller.) Substituting $l = 0, 1, 2$, we find

$$E_0 = 0,$$
$$E_1 = 0.0151\ \text{eV},$$
$$E_2 = 0.0453\ \text{eV}.$$

b) The photon energy is

$$E = E_2 - E_1$$
$$= 0.0302\ \text{eV} = 4.84 \times 10^{-21}\ \text{J}.$$

The photon wavelength is

$$\lambda = \frac{hc}{E} = \frac{(6.626 \times 10^{-34}\ \text{J} \cdot \text{s})(3.00 \times 10^8\ \text{m/s})}{4.84 \times 10^{-21}\ \text{J}}$$
$$= 4.11 \times 10^{-5}\ \text{m} = 41.1\ \mu\text{m} = 41{,}100\ \text{nm}.$$

This photon is in the far infrared region of the spectrum.

Rotational energy levels of water molecules play an important role in the operation of microwave ovens. The magnetron oscillator generates microwaves with a frequency of 2450 MHz; the corresponding photon energy is $1.01 \times 10^{-5}$ eV. The water molecule has a rotational energy level $1.01 \times 10^{-5}$ eV above the ground state, so the microwaves are strongly absorbed. Interaction of the rotating molecules with their surroundings then heats up the substance containing them.

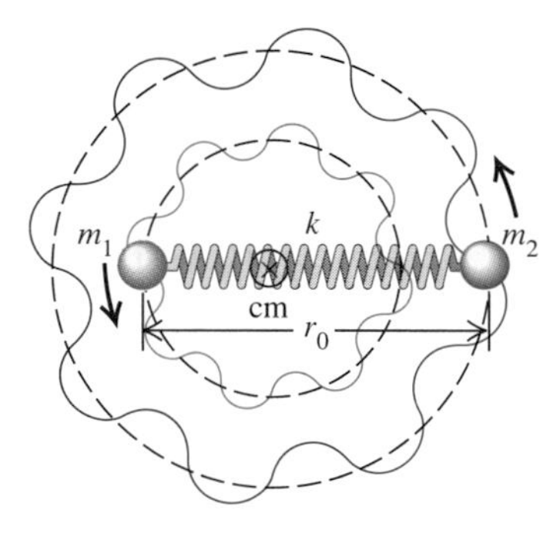


**44–7** Model of a diatomic molecule as two point masses $m_1$ and $m_2$ connected by a spring with force constant $k$.

## Vibrational Energy Levels

Molecules are never completely rigid. A more realistic model of a diatomic molecule represents the connection between atoms not as a rigid rod but as a *spring* (Fig. 44–7). Then in addition to rotating, the molecule can also *vibrate* along the line joining the atoms. If the force is proportional to the displacement from the equilibrium separation (a "spring" that obeys Hooke's law), the system is a harmonic oscillator. We discussed the quantum-mechanical harmonic oscillator in Section 42–5; the energy levels are given by Eq. (42–28), with the mass $m$ replaced with the reduced mass $m_r$:

$$E_n = \left(n + \tfrac{1}{2}\right)\hbar\omega = \left(n + \tfrac{1}{2}\right)\hbar\sqrt{\frac{k}{m_r}}. \tag{44–7}$$

This represents a series of equally spaced levels, with spacing $\Delta E$ between levels, where

$$\Delta E = \hbar\omega = \hbar\sqrt{\frac{k}{m_r}}. \tag{44–8}$$

Figure 44–8 is an energy-level diagram showing the vibrational states.

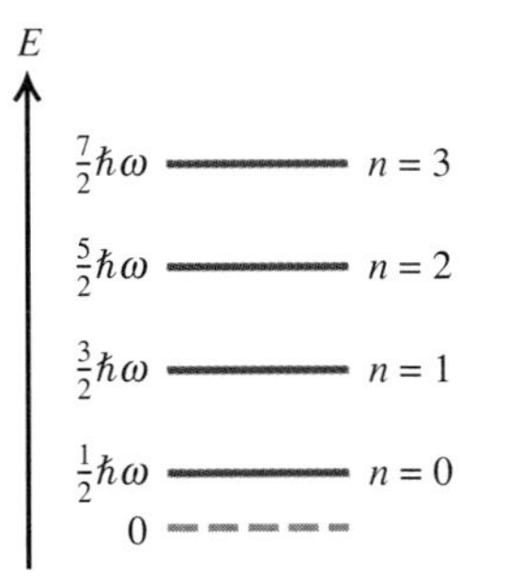


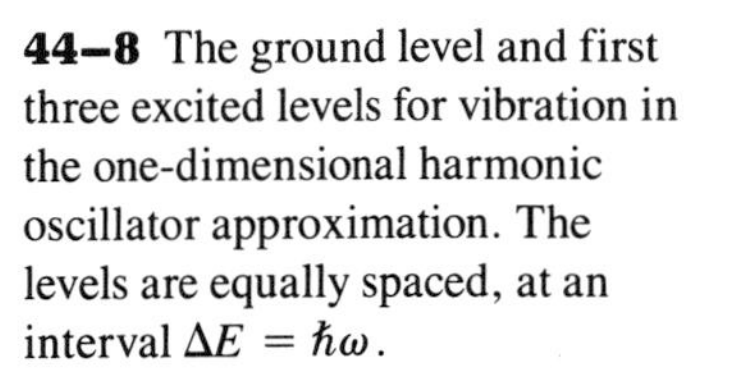

**44–8** The ground level and first three excited levels for vibration in the one-dimensional harmonic oscillator approximation. The levels are equally spaced, at an interval $\Delta E = \hbar\omega$.

## EXAMPLE 44–3

For the hydrogen molecule, $\Delta E$ is found to be 0.546 eV. a) Find the force constant $k$ for the interatomic force. b) Find the wavelength of a photon emitted during the transition from $n = 2$ to $n = 1$.

**SOLUTION** a) Solving Eq. (44–8) for $k$, we find

$$k = m_r\left(\frac{\Delta E}{\hbar}\right)^2 = (0.837 \times 10^{-27}\,\text{kg})\left(\frac{0.546\,\text{eV}}{6.583 \times 10^{-16}\,\text{eV}\cdot\text{s}}\right)^2$$
$$= 576\ \text{N/m}.$$

Force constants for diatomic molecules are typically 200 to 2000 N/m.

b) The photon energy is 0.546 eV; its wavelength is

$$\lambda = \frac{hc}{E} = \frac{(4.136 \times 10^{-15}\,\text{eV}\cdot\text{s})(3.00 \times 10^{8}\,\text{m/s})}{0.546\,\text{eV}}$$
$$= 2.27 \times 10^{-6}\,\text{m} = 2.27\,\mu\text{m} = 2270\,\text{nm}.$$

This is in the infrared region of the spectrum but closer to the visible region than the photon in Example 44–2. Vibrational energies, although usually much smaller than those of atomic spectra, are usually much *larger* than rotational energies.

We are given $\mu B = 9.27 \times 10^{-24}$ J, so

$$\frac{2\mu B}{kT} = \frac{2(9.27 \times 10^{-24}\ \text{J})}{(1.38 \times 10^{-23}\ \text{J/K})(300\ \text{K})} = 0.00448,$$

and

$$\frac{N_+}{N_-} = e^{0.00448} = 1.004.$$

At lower temperatures this ratio increases. We invite you to show that at 30 K it is 1.046 and that at 3 K it is 1.57. This partial alignment of spins parallel to the field causes *magnetization* of the material; this example shows one reason why paramagnetic susceptibilities always decrease with increasing temperature.

---

Once we know how many molecules are in each energy state $E_n$, we can compute the *total* energy $E_{\text{tot}}$ of the system, which is the number in each state multiplied by the energy of that state:

$$E_{\text{tot}} = \sum_n N(n)E_n. \tag{44–12}$$

The vibrational contribution to the molar heat capacity is $dE_{\text{tot}}/dT$ divided by the number of moles.

Now we'll describe how to implement this program, leaving out some of the computational details. First, the energy levels are given by Eq. (44–7):

$$E_n = (n + \tfrac{1}{2})\hbar\omega. \tag{44–13}$$

When this expression is substituted into Eq. (44–11), the sum is a geometric series that can be evaluated exactly. The result is

$$A = e^{\hbar\omega/2kT} - e^{-\hbar\omega/2kT}, \tag{44–14}$$

and the complete Maxwell-Boltzmann distribution for this system is

$$N(n) = N(e^{\hbar\omega/2kT} - e^{-\hbar\omega/2kT})\, e^{-(n+1/2)\hbar\omega/kT}. \tag{44–15}$$

Now we substitute this result and Eq. (44–13) into Eq. (44–12). Again the sum can be evaluated. We'll omit the details; the result is a surprisingly simple expression:

$$E_{\text{tot}} = \frac{N\hbar\omega}{2}\frac{e^{\hbar\omega/kT} + 1}{e^{\hbar\omega/kT} - 1}. \tag{44–16}$$

We can look at the extreme cases of very low and very high temperature. When $T$ is very small, the two exponentials become much larger than unity, and the fraction approaches unity. In this limit,

$$E_{\text{tot}} = \frac{N\hbar\omega}{2} \qquad (T \to 0). \tag{44–17}$$

Note that $\hbar\omega/2$ is the ground-state vibrational energy; no molecule can have less vibrational energy than this. So Eq. (44–17) says that at very low temperatures, *all* the molecules are in the ground state of vibrational motion, and the total energy is the ground-state energy $\frac{1}{2}\hbar\omega$ multiplied by the number $N$ of molecules. This is just what we should expect.

To look at the opposite limit, $T \to \infty$, note that in this limit, the exponents in Eq. (44–16) become very small, and we can use a Taylor series for each exponential: $e^x = 1 + x + x^2/2 + \cdots$. Keeping only the first two terms in each series, we get

$$E_n = \frac{N\hbar\omega}{2}\left(\frac{1 + \hbar\omega/kT + 1}{1 + \hbar\omega/kT - 1}\right) = \frac{N\hbar\omega}{2}\left(\frac{2 + \hbar\omega/kT}{\hbar\omega/kT}\right).$$

In the numerator of the last fraction, $\hbar\omega/kT$ is much smaller than 2 as $T \to \infty$, so we throw this term away. What remains can be simplified to

$$E_{\text{tot}} = NkT \qquad (T \to \infty). \tag{44–18}$$

This wonderfully simple result shows that in the high-temperature limit, where $kT$ is much larger than the steps between adjacent energy levels, the system obeys the classical equipartition principle, with $kT/2$ of kinetic and $kT/2$ of potential energy per molecule, for a total (for $N$ molecules) of $NkT$.

Finally, to get the general expression for the vibrational molar heat capacity, we set $N$ equal to Avogadro's number $N_A$ in Eq. (44–16) to get the energy of one mole. Then we take the derivative with respect to $T$ and replace the product $kN_A$ by the gas constant $R$ (recalling that $k$ is a gas constant on a "per molecule" basis). We leave the details as a problem; the final result is

$$C = R\left(\frac{\hbar\omega}{kT}\right)^2 \frac{e^{\hbar\omega/kT}}{(e^{\hbar\omega/kT} - 1)^2}. \qquad (44\text{–}19)$$

Again we can look at the limits of very high and very low temperatures. When $T$ is very large, we can make a Taylor expansion in the denominator as before; the denominator is approximately equal to $(\hbar\omega/kT)^2$. The exponential in the numerator is approximately unity, and the entire expression boils down to

$$C = R \qquad (T \to \infty). \qquad (44\text{–}20)$$

This again agrees with the equipartition theorem in the high-temperature limit. The low-temperature limit is harder to evaluate because Eq. (44–19) becomes indeterminate at $T = 0$, but it is not difficult to show that in the limit as $T$ approaches 0, $C$ approaches zero:

$$C = 0 \qquad (T \to 0). \qquad (44\text{–}21)$$

We leave the proof of this statement as a problem. Once more this result agrees with our expectations; at very low temperatures, all the molecules are in the ground state, and the total energy doesn't change appreciably with temperature.

Note the prominent role of the quantity $\hbar\omega/kT$ in Eqs. (44–16) and (44–19). The crucial quantity is the *ratio* of the energy $\hbar\omega$ to the characteristic thermal energy $kT$. Figure 44–11 shows graphs of $E_{av}$ and $C$ as functions of temperature, showing the low- and high-temperature limits that we have described.

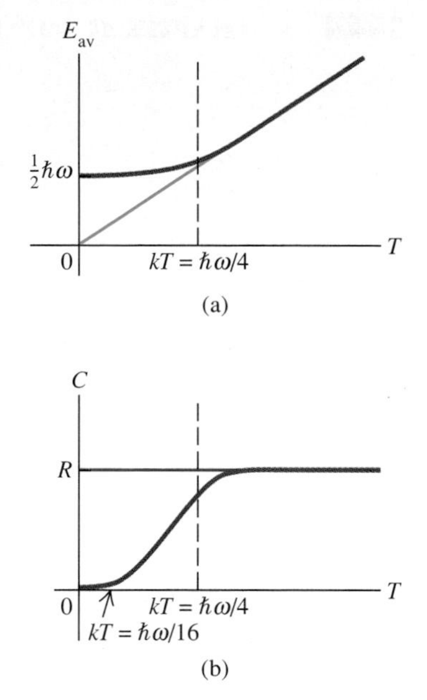


**44–11** (a) Average vibrational energy of a molecule as a function of temperature. For comparison, the prediction of the classical equipartition theorem is shown as a blue line. The temperature at which $kT = \hbar\omega/4$ is shown for reference. (b) Molar heat capacity as a function of temperature. For comparison, the prediction of the classical equipartition theorem is shown as a red line. The value of this function for a given value of $T$ is proportional to the slope of the $E_{av}$ versus $T$ graph at the same $T$.

## EXAMPLE 44–6

Derive expressions for the total vibrational energy and molar heat capacity for a diatomic gas at a temperature for which $kT = \hbar\omega$.

**SOLUTION** The problem refers to the "in-between" temperature range, the transition region where $kT$ is neither very large nor very small in comparison to $\hbar\omega$. Substituting $kT$ for $\hbar\omega$ in Eq. (44–16), we get

$$E_{tot} = \frac{NkT}{2}\left(\frac{e+1}{e-1}\right) = 1.082NkT.$$

The total energy is about 8% *greater* than the value $NkT$ predicted by the equipartition principle. This reflects the fact that the ground-state energy at $T = 0$ is not zero but $\hbar\omega/2$.

Substituting $kT$ for $\hbar\omega$ in Eq. (44–19), we get

$$C = R(1)\frac{e}{(e-1)^2} = 0.921R.$$

The molar heat capacity is somewhat *smaller* than the value $R$ predicted by the equipartition principle. When $\hbar\omega = kT$, the *slope* of the $E_{tot}$ curve in Fig. 44–11a is less than predicted by the equipartition principle.

A similar calculation can be carried out for the rotational levels. There are two complications. First, the levels aren't equally spaced, and the sum corresponding to Eq. (44–11) can't be evaluated exactly. Second, the energy levels are *degenerate;* each value of $l$ corresponds to $2l + 1$ levels with the same energy but different values of $m_l$. The degeneracy has to be included in the distribution $N(l)$ corresponding to Eq. (44–10).

Nevertheless, calculations based on this picture, using suitable approximations, agree well with the observed temperature dependence of heat capacities in temperature ranges where the rotational levels are comparable to $kT$.

This entire discussion can be adapted to the problem of heat capacities of solids, discussed briefly in Section 16–4. If each atom in a metallic crystal can be treated as a three-dimensional harmonic oscillator, the energy and heat capacity expressions are exactly the same as for the vibrational levels of diatomic molecules. There is an extra factor of 3 because each atom has three degrees of freedom. The high temperature limit, Eq. (16–30), becomes $C = 3R$, in agreement with the empirical *rule of Dulong and Petit.*

The assumption that the atoms in a crystal all vibrate independently with the same frequency is somewhat naive. A more sophisticated model proposed by Debye uses the frequencies of the *normal modes* of oscillation of the entire crystal; these have a spectrum of frequencies rather than a single frequency. We discussed normal modes of a crystal in Section 19–8, which you may want to review. A quantum of energy of a normal-mode vibration of the crystal is called a *phonon*. Calculations based on this model agree very well with measured values. During the 1920s the temperature dependence of heat capacities provided a very sensitive test, and ultimately a strong affirmation, of the "new" quantum mechanics.

## 44–4 STRUCTURE OF SOLIDS

The term *condensed matter* includes both solids and liquids. In both states, the interactions between atoms or molecules are strong enough to give the material a definite volume that changes relatively little with applied stress. The distances between adjacent atoms or molecules in condensed matter are of the same order of magnitude as the sizes of the atoms or molecules themselves, typically 0.1 to 0.5 nm.

Ordinarily, we think of a liquid as a material that can flow and of a solid as a material with a definite shape. On this basis there is no sharp division; glass flows at ordinary temperatures, behaving as an extremely viscous liquid. A more fundamental distinction between solid and liquid is one based on the arrangement of the atoms. A solid (sometimes called a *crystalline solid*) is characterized by *long-range order,* a recurring pattern of atomic positions that extends over many atoms. This pattern is called the *crystal structure.* Liquids have *short-range order* (correlations between neighboring atoms or molecules) but not long-range order. Again, there is no sharp distinction. Some solids have so many imperfections in their crystal structure that there is almost no long-range order; such solids are sometimes called *amorphous solids.* Conversely, some liquids have a fairly high degree of long-range order; a familiar example is the type of *liquid crystal* that is used for displays in digital watches, calculators, and laptop computers.

Crystalline solids usually have a definite melting temperature; examples are ice, salt, and lead. At the melting temperature a distinct phase transition occurs from a state with long-range order to one with only short-range order. Amorphous solids such as glass, butter, and tar have no definite melting point. They soften as the temperature rises, but there is no phase transition. In general, such materials have more in common with liquids than with crystalline solids, and physicists often think of them as liquids.

Nearly everything we know about crystal structures has been learned from diffraction experiments, initially with x rays, later with electrons and neutrons. A typical lattice dimension is 0.1 nm; we leave it as an exercise for you to show that 12.4-keV x rays, 150-eV electrons, and 0.0818-eV neutrons all have wavelengths of $\lambda = 0.1$ nm.

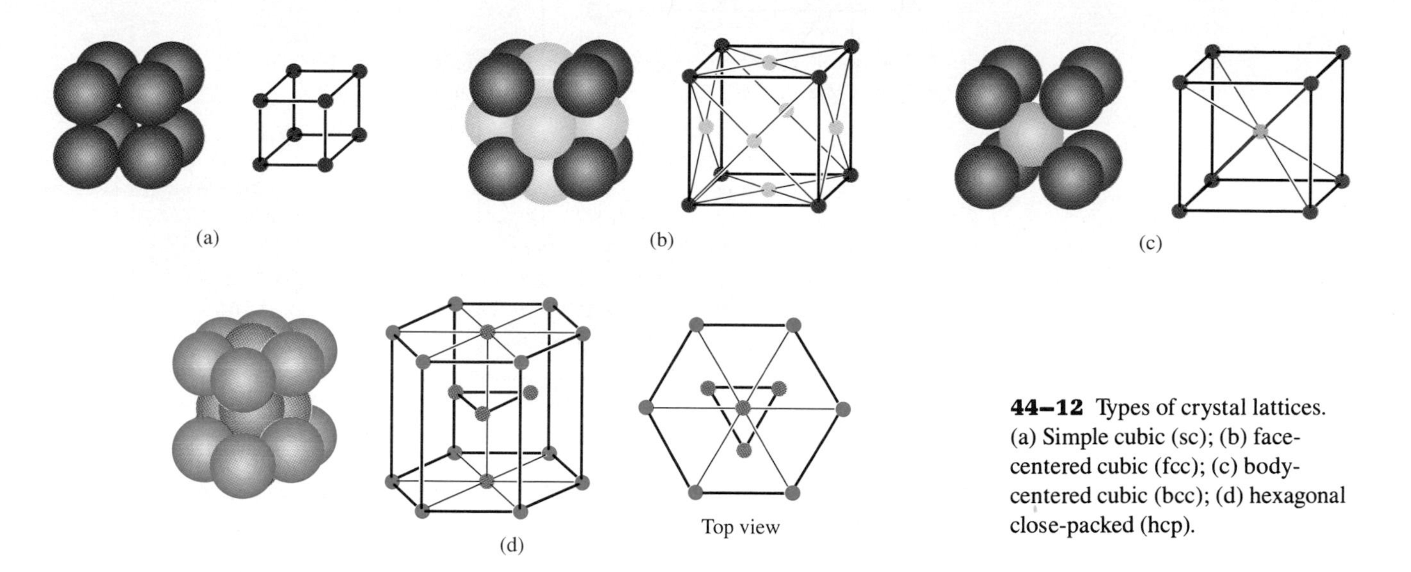


**44–12** Types of crystal lattices. (a) Simple cubic (sc); (b) face-centered cubic (fcc); (c) body-centered cubic (bcc); (d) hexagonal close-packed (hcp).

## Crystal Lattices

A crystal lattice is characterized by a repeating pattern that extends throughout the lattice. There are 14 general types of such patterns; here are a few common examples. The *simple cubic lattice* (sc) has a lattice site at each corner of a cubic array (Fig. 44–12a). The *face-centered cubic lattice* (fcc) is like the simple cubic but with an additional lattice site at the center of each cube face (Fig. 44–12b). The *body-centered cubic lattice* (bcc) is like the simple cubic but with an additional site at the center of each cube (Fig. 44–12c). The *hexagonal close-packed lattice* has layers of lattice sites in hexagonal patterns (Fig. 44–12d).

Each lattice site may be associated with a single atom or a group of atoms of either the same or different kinds. Such a group is called a *basis;* a complete description of a crystal structure includes both the lattice and the basis. The alkali metals have a bcc structure with one atom at each lattice site; each atom has eight nearest neighbors. The elements aluminum, calcium, copper, silver, and gold have fcc structures with one atom per site; each atom has 12 nearest neighbors.

Figure 44–13 shows the structure of sodium chloride ($NaCl$, ordinary salt). It may look like a simple cubic structure, but it isn't. The sodium and chloride ions each form an

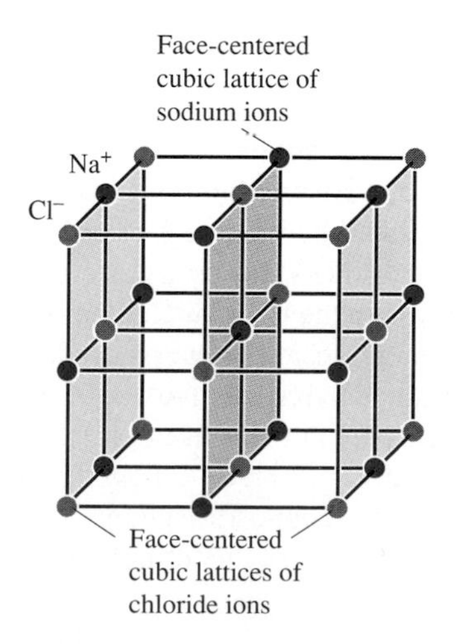


**44–13** Symbolic representation of a sodium chloride crystal with exaggerated distances between ions.

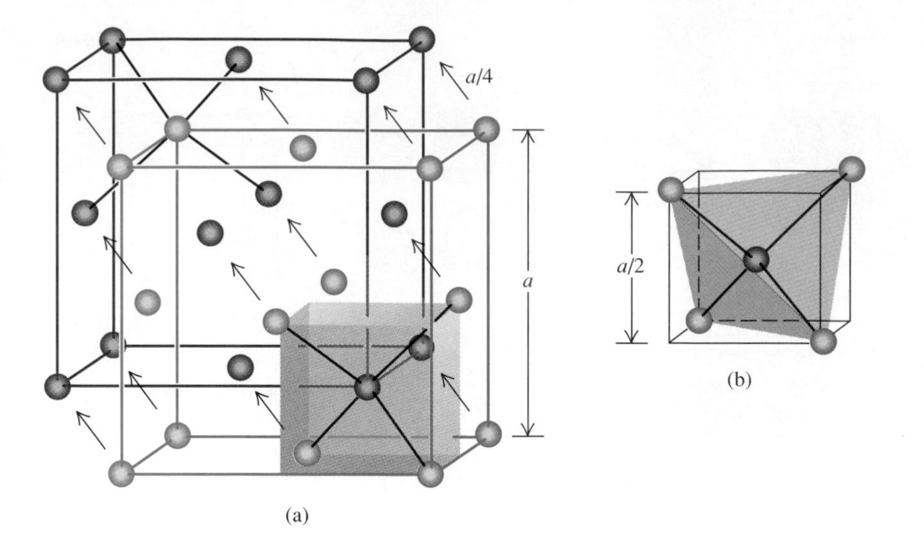


**44–14** (a) The diamond structure, shown as two interpenetrating face-centered cubic structures. Each purple atom is shifted up, back, and to the left, by a distance $a/4$, from the corresponding green atom. (b) The front lower right eighth of (a), showing four green sites at the corners of a tetrahedron inscribed in a cube with side length $a/2$, with a purple atom at the center.

fcc structure, so there are two interpenetrating fcc structures. Alternatively, we can say that each fcc lattice site is occupied by a sodium ion at the site and a chloride ion half a cube length above it. The basis consists of one sodium and one chloride ion.

Another example is the *diamond structure;* it's called that because it is the crystal structure of diamond, as well as of silicon and germanium (all Group IV elements in the periodic table). The diamond lattice is fcc; the basis consists of one atom at each site and a second atom displaced a quarter of a cube length in each of the three directions. Figure 44–14a will help you to visualize this. Fig. 44–14b shows the bottom right front eighth of the basic cube; the four atoms at alternate corners of this cube are at the corners of a regular tetrahedron, and there is an additional atom at the center. Thus each atom in the diamond structure is at the center of a regular tetrahedron with four nearest-neighbor atoms at the corners.

The structure of zinc sulfide is similar; the lattice is fcc, and the basis consists of one zinc and one sulfur atom, as in the diamond structure. Each zinc atom is at the center of a regular tetrahedron with four sulfur atoms at the corners, and vice versa. Gallium arsenide and other III-V compounds also have this structure.

## Bonding in Solids

The forces that are responsible for the regular arrangement of atoms in a crystal are the same as those involved in molecular bonds, and there is one additional type. *Ionic* and *covalent* molecular bonds are found in ionic and covalent crystals, respectively. The most familiar *ionic crystals* are the alkali halides, such as ordinary salt (NaCl). The positive sodium ions and the negative chloride ions occupy alternate positions in a cubic crystal lattice (Fig. 44–13). The forces are the familiar Coulomb's-law forces between charged particles. These forces have no preferred direction, and the arrangement in which the material crystallizes is determined by the relative size of the two ions. It's not difficult to prove that such a structure is *stable* in the sense that it has lower total energy than the separated ions. Example 44–7 gives a hint as to how the proof can be carried out. The negative potential energies of pairs of opposite charges are greater than the positive energies of pairs of like charges because the pairs of unlike charges are closer together, on average.

## EXAMPLE 44–7

**Potential energy of an ionic crystal** Consider a fictitious one-dimensional ionic crystal consisting of a very large number of alternating positive and negative ions with charges $e$ and $-e$, with equal spacing $a$ along a line. Prove that the total interaction potential energy is negative.

**SOLUTION** Let's pick an ion somewhere in the middle of the string and add up the potential energies of its interactions with all the ions to one side of it. We get the series

$$\begin{aligned} U &= -\frac{e^2}{4\pi\epsilon_0}\frac{1}{a} + \frac{e^2}{4\pi\epsilon_0}\frac{1}{2a} - \frac{e^2}{4\pi\epsilon_0}\frac{1}{3a} + \cdots \\ &= -\frac{e^2}{4\pi\epsilon_0 a}\left(1 - \frac{1}{2} + \frac{1}{3} - \frac{1}{4} + \cdots\right). \end{aligned}$$

You may notice the resemblance of the series in parentheses to the Taylor series for $\ln(1+x)$:

$$\ln(1+x) = x - \frac{x^2}{2} + \frac{x^3}{3} - \frac{x^4}{4} + \cdots.$$

When $x = 1$, we have the series in parentheses, and

$$U = -\frac{e^2}{4\pi\epsilon_0 a}\ln 2.$$

This is certainly a negative quantity. The atoms on the other side of the ion we're considering make an equal contribution to the potential energy. And if we include the potential energies of all pairs of atoms, the sum is certainly negative. So this structure has lower energy than the state in which all the ions are infinitely separated from each other.

---

The diamond structure of carbon, silicon, and germanium is a simple example of a *covalent crystal*. These elements are in Group IV of the periodic table, and each atom has four electrons in its outermost shell. Each atom forms a covalent bond with each of the four atoms at the corners of its tetrahedron. These bonds are strongly directional because of the asymmetrical electron distributions dictated by the exclusion principle, and the result is a tetrahedral structure.

A third crystal type, less directly related to the chemical bond than are ionic or covalent crystals, is the **metallic crystal.** In this structure, one or more of the outermost electrons in each atom become detached from the parent atom and are free to move through the crystal. These electrons are not localized near individual lattice sites. The corresponding electron wave functions extend over many atoms.

We can picture a metallic crystal as an array of positive ions (atoms from which one or more electrons have been removed) immersed in a sea of electrons whose attraction for the positive ions holds the crystal together (Fig. 44–15). This "sea" has many of the properties of a gas, and indeed we speak of the *electron-gas model* of metallic solids. The simplest version of this model is the *free-electron model,* in which interactions with the lattice are ignored completely. We'll return to this model in Section 44–6.

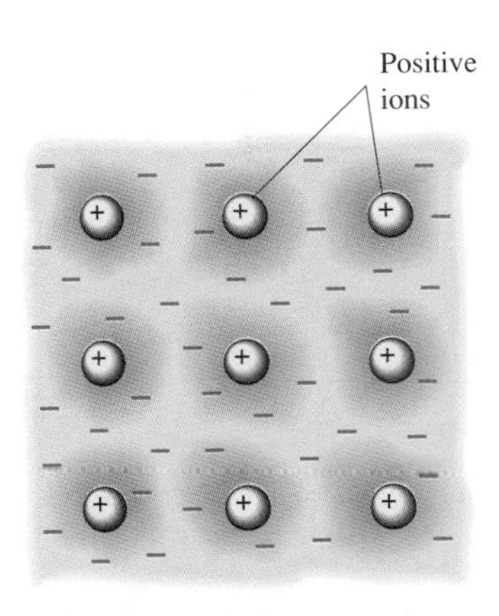


**44–15** Structure of a metallic solid. One or more electrons are detached from each atom and are free to wander around the crystal, forming the "electron gas." The wave functions for these electrons extend over many atoms. The positive ion cores vibrate around fixed locations in the crystal.

In a metallic crystal the atoms would like to form shared-electron bonds, but there aren't enough valence electrons available. Instead, electrons are shared among *many* atoms. This bonding is not strongly directional. The shape of the crystal lattice is determined primarily by considerations of **close packing,** that is, the maximum number of atoms that can fit into a given volume. The two most common metallic crystal lattices, the face-centered cubic and the hexagonal close-packed, are shown in Fig. 44–12. In each of these lattices, each atom has 12 nearest neighbors.

As we mentioned in Section 44–1, van der Waals interactions and hydrogen bonding also play a role in the structure of some solids. In polyethylene and similar polymers, covalent bonding of atoms forms long-chain molecules, and hydrogen bonding forms cross-links between adjacent chains. In solid water, both van der Waals forces and hydrogen bonds are significant, and together they determine the crystal structure of ice. Many other examples might be cited.

Our discussion has centered on *perfect crystals,* crystals in which the crystal structure extends uninterrupted throughout the entire material. Real crystals show a variety of departures from this idealized structure. Materials are often *polycrystalline,* composed of many small perfect single crystals bonded together at *grain boundaries.* Within

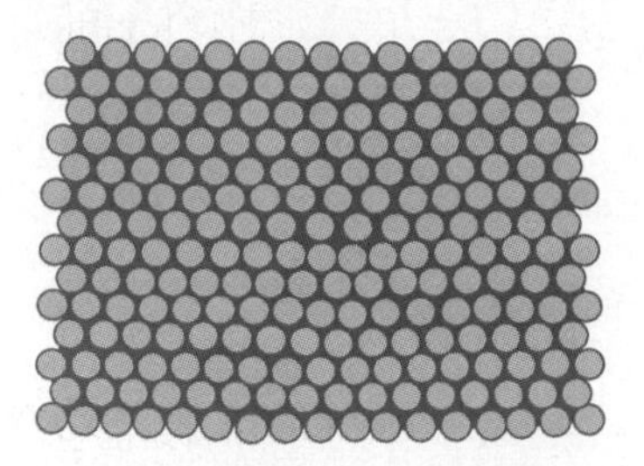

**44–16** A dislocation. The irregularity is seen most easily by viewing the figure from various directions at a grazing angle with the page.

a single crystal, *interstitial atoms* may occur in places where they do not belong, and there may be *vacancies,* lattice sites that should be occupied by an atom but are not. An imperfection of particular interest in semiconductors, which we will discuss in Section 44–7, is the *impurity atom,* a foreign atom occupying a regular lattice site (for example, arsenic in a silicon crystal).

A more complex kind of imperfection is the *dislocation,* shown schematically in Fig. 44–16, in which one plane of atoms slips relative to another. The mechanical properties of metallic crystals are influenced strongly by the presence of dislocations. The ductility and malleability of some metals depends on the presence of dislocations that can move through the lattice during plastic deformations.

## 44–5 ENERGY BANDS

The concept of **energy bands** in solids is a great help in understanding several properties of solids. To introduce the idea, suppose we have a large number $N$ of identical atoms, far enough apart that their interactions are negligible. Every atom has the same energy-level diagram. We can draw an energy-level diagram for the *entire system.* It looks just like the diagram for a single atom, but the exclusion principle, applied to the entire system, permits each state to be occupied by $N$ electrons instead of just one.

Now we begin to push the atoms uniformly closer together. Because of the electrical interactions and the exclusion principle, the wave functions begin to distort, especially those of the outer, or *valence,* electrons. The energy levels also shift, some upward and some downward, as the electron wave functions become less localized and extend over more and more atoms. Thus each valence electron state for the *system,* formerly a sharp energy level that can accommodate $N$ electrons, splits into a *band* containing $N$ closely spaced levels (Fig. 44–17). Ordinarily, $N$ is very large, of the order of Avogadro's number ($10^{23}$), so we can think of the levels as forming a *continuous* distribution of energies within a band. Between adjacent energy bands are gaps or forbidden regions where there are *no* allowed energy levels. The inner electrons in an atom are affected much less by nearby atoms than the valence electrons are, and their energy levels remain relatively sharp.

The nature of these energy bands determines whether the material is an electrical conductor or insulator. In insulators and semiconductors at absolute zero, the valence electrons completely fill the highest occupied band, called the **valence band.** The next higher band, called the **conduction band,** is completely empty. The energy gap separating the two may be of the order of 1 to 5 eV. This situation is shown in Figs. 44–18a and 44–18b. The electrons in the valence band are not free to

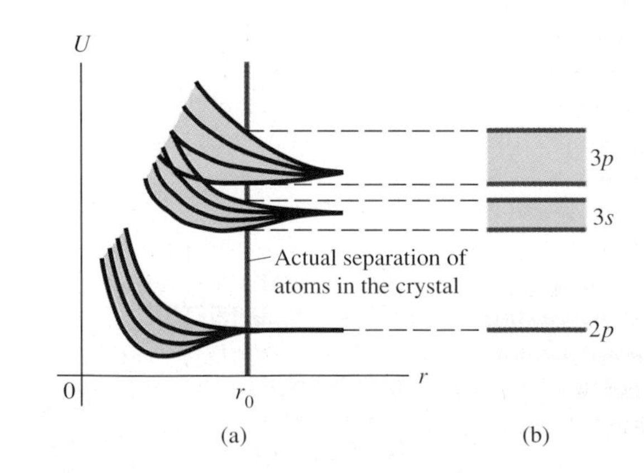


**44–17** Origin of energy bands in a solid. (a) As the atoms are pushed together, the energy levels spread into bands. The vertical line shows the actual atomic spacing in the crystal. (b) Symbolic representation of energy bands.

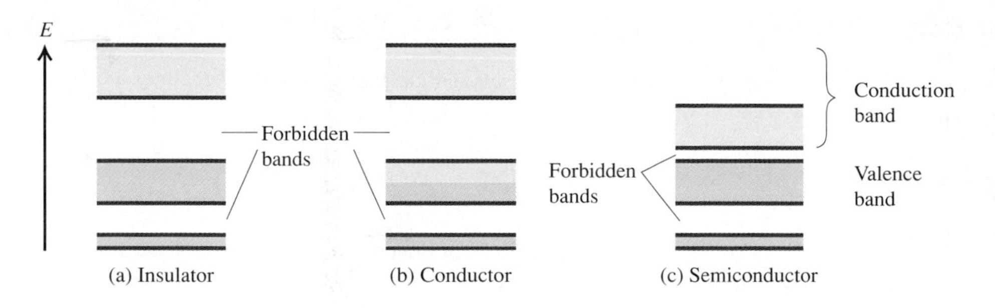


**44–18** Three types of energy-band structure. (a) An insulator; a completely full valence band is separated by a gap of several electronvolts from a completely empty conduction band, and electrons in the valence band cannot move. At finite temperatures very few electrons can reach the upper band. (b) A conductor; there is a partially filled valence band, and electrons in this band are free to move when an electric field is applied. (c) A semiconductor; a completely filled valence band is separated by a small gap of 1 eV or less from an empty conduction band; at finite temperatures, substantial numbers of electrons can reach the conduction band, where they are free to move.

move in response to an applied electric field. To move, an electron would have to go to a different quantum state with slightly different energy, but it can't do that because all the neighboring states are already occupied. The only way an electron can move is to jump into the conduction band. This would require an additional energy of a few electronvolts, and that much energy is not ordinarily available. The situation is like a completely filled floor in a parking garage; none of the cars can move because there is no place to go. If a car could jump up to the next empty floor, it could move!

However, at any temperature above absolute zero the atoms in the crystal have some vibrational motion, and there is some probability that an electron can gain enough energy from thermal motion to jump to the conduction band. Once in the conduction band, an electron is free to move in response to an applied electric field because plenty of nearby empty states are available to permit it to gain or lose energy in small increments. At any finite temperature there are always a few electrons in the conduction band, so no material is a perfect insulator. Furthermore, as the temperature increases, the population in the conduction band increases very rapidly. A doubling of the number of conduction electrons for a temperature rise of 10 C° is typical.

A distinguishing characteristic of all *metals* is that the valence band is only partly filled. The metal sodium is an example. The energy-level diagram for sodium in Fig. 44–17 shows that for an isolated atom the six $3p$ "resonance-level" states for the valence electron are about 2.1 eV above the two $3s$ ground states. But in the solid sodium crystal, the $3s$ and $3p$ *bands* spread out enough that they actually overlap, forming a single band that is only $\frac{1}{8}$ filled. The situation is similar to the one shown in Fig. 44–18b. Electrons in states near the top of the filled portion of the band have many adjacent unoccupied states available, and they can easily gain or lose small amounts of energy in response to an applied electric field. Therefore these electrons are mobile and can contribute to electrical and thermal conductivity. Metallic crystals always have partly filled bands. In the *ionic* NaCl crystal, on the other hand, there is no overlapping of bands. The $Na^+$ and $Cl^-$ ions both have filled-subshell structures corresponding to completely filled bands, and solid sodium chloride is an insulator.

When an insulating material is subjected to a large enough electric field, it becomes a conductor, an occurrence called *dielectric breakdown*. If the electric field in a material is so large that there is a potential difference of a few volts over a distance comparable to atomic sizes (that is, a field of the order of $10^{10}$ V/m), then the field can do enough work on a valence electron to boost it over the forbidden region and into the conduction band. Real insulators usually have dielectric strengths much *less* than this because of structural imperfections that provide some energy states in the forbidden region.

The concept of energy bands is very useful in understanding the properties of semiconductors, which we will study in a later section.

## EXAMPLE 44–8

**A germanium photocell** Pure germanium at low temperatures has a completely filled valence band separated by a gap of 0.67 eV from a completely empty conduction band. It is a poor electrical conductor, but its conductivity increases substantially when it is irradiated with electromagnetic waves of appropriate wavelength. Why? What wavelength is appropriate?

**SOLUTION** An electron in the valence band can absorb a photon with energy of 0.67 eV or greater and move into the conduction band, where it is a mobile charge. The wavelength of a 0.67-eV photon is

$$\lambda = \frac{hc}{E} = \frac{(4.136 \times 10^{-15}\ \text{eV} \cdot \text{s})(3.00 \times 10^{8}\ \text{m/s})}{0.67\ \text{eV}}$$
$$= 1.9 \times 10^{-6}\ \text{m} = 1900\ \text{nm}.$$

This wavelength is in the near infrared region of the spectrum. Germanium and silicon crystals are widely used as photocells, with variations to be discussed in Section 44–7.

## *44–6 FREE-ELECTRON MODEL OF METALS

Studying the energy states of electrons in metals can give us a lot of insight into their electrical and magnetic properties, the electron contributions to heat capacities, and other behavior. As we discussed in Section 44–4, one of the distinguishing features of a metal is that one or more valence electrons are detached from their home atom and can move freely around the lattice, with wave functions that extend over many lattice sites.

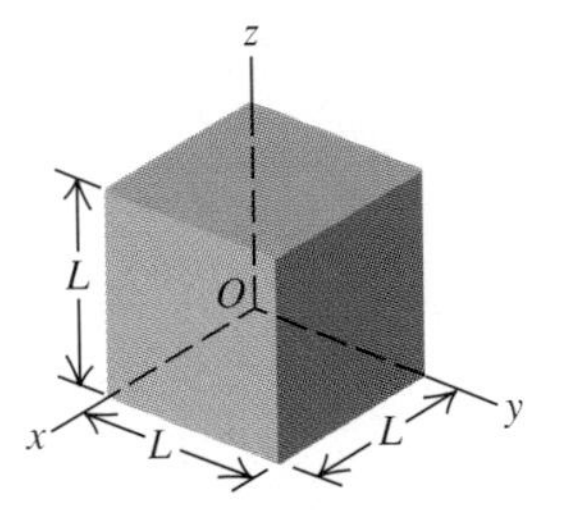


**44–19** A cubical box with rigid walls and side length $L$. This is the three-dimensional analog of the infinite square well discussed in Section 42–1. The energy levels for a particle in this box are given by Eq. (44–23).

In our simple model we'll assume that these electrons are completely free inside the material—they don't interact at all with the lattice or with each other—but that there are infinite potential-energy barriers at the surfaces. The wave functions and energy states are then the three-dimensional analogs of those for the (one-dimensional) particle in a box that we analyzed in Sections 42–1 and 42–2. Suppose the box is a cube with side length $L$ (Fig. 44–19). Then the possible wave functions, analogous to Eq. (42–6), are

$$\psi(x, y, z) = A \sin\frac{n_1 \pi x}{L} \sin\frac{n_2 \pi y}{L} \sin\frac{n_3 \pi z}{L}, \tag{44–22}$$

where $(n_1, n_2, n_3)$ is a set of three positive-integer quantum numbers that identify the state. We invite you to verify that these functions satisfy the boundary conditions. (They must be zero at the surfaces of the cube.) You can also substitute Eq. (44–22) into the Schrödinger equation, Eq. (42–31), to show that the energies of the states are

$$E = \frac{(n_1{}^2 + n_2{}^2 + n_3{}^2)\,\pi^2 \hbar^2}{2mL^2}. \tag{44–23}$$

This equation is the three-dimensional analog of Eq. (42–17).

### Density of States

Later we'll need to know the *number dN* of quantum states that have energies in a given range $dE$; $dN/dE$ is called the **density of states,** denoted by $g(E)$. We'll begin by working out an expression for $g(E)$. Think of a three-dimensional space with coordinates $(n_1, n_2, n_3)$. The distance $R$ from the origin to any point in that space is given by $n^2 = n_1{}^2 + n_2{}^2 + n_3{}^2$. Each point in that space with integer coordinates represents a quantum state, and each point corresponds to one unit of volume in the space. The total number $N$ of points with integer coordinates inside a sphere with radius $n_{\text{max}}$ is equal to the volume of the sphere, $4\pi n_{\text{max}}{}^3/3$. Because all our $n$'s are positive, we must take one *octant* of the sphere, with $\frac{1}{8}$ the total volume, or $\frac{1}{8}(4\pi n_{\text{max}}{}^3/3) = \pi n_{\text{max}}{}^3/6$. The particles are electrons, so each point corresponds to *two* states with opposite spin (spin quantum number $m_s = \pm\frac{1}{2}$). The total number $N$ of electron states corresponding to points inside the octant is

$$N = \frac{\pi n_{max}{}^3}{3}. \qquad (44\text{–}24)$$

The energy $E$ of each state can be expressed in terms of $n_{max}$. Equation (44–23) becomes

$$E = \frac{n_{max}{}^2 \pi^2 \hbar^2}{2mL^2}. \qquad (44\text{–}25)$$

We can combine Eqs. (44–24) and (44–25) to get a relation between $E$ and $N$ that doesn't contain $n_{max}$. We'll leave the details as an exercise; the result is

$$N = \frac{(2m)^{3/2} V E^{3/2}}{3\pi^2 \hbar^3}, \qquad (44\text{–}26)$$

where $V = L^3$ is the volume of the box. Equation (44–26) gives the total number of states with energies of $E$ or less.

To get the number of states $dN$ in an energy interval $dE$, we take differentials of both sides of Eq. (44–26). We get

$$dN = \frac{(2m)^{3/2} V E^{1/2}}{2\pi^2 \hbar^3} dE. \qquad (44\text{–}27)$$

The density of states $g(E)$ is equal to $dN/dE$, so from Eq. (44–27) we get

$$g(E) = \frac{(2m)^{3/2} V}{2\pi^2 \hbar^3} E^{1/2}. \qquad (44\text{–}28)$$

## Fermi-Dirac Distribution

Now we need to know how the electrons are distributed among the various quantum states at any given temperature. It wouldn't be right to use the Maxwell-Boltzmann distribution, for two very important reasons. The first reason is the *exclusion principle*. At very low temperatures the Maxwell-Boltzmann function predicts that nearly all the electrons would go into the lowest states, with small values of $n_1$, $n_2$, and $n_3$. But the exclusion principle allows only one electron in each state. At very low temperatures the electrons fill up the lowest *available* states, but there's not enough room for *all* of them to go into the lowest states. Thus a reasonable guess as to the shape of the distribution would be Fig. 44–20. At very low $T$ almost all the states are filled, up to some value $E_F$, and almost all states above this value are empty.

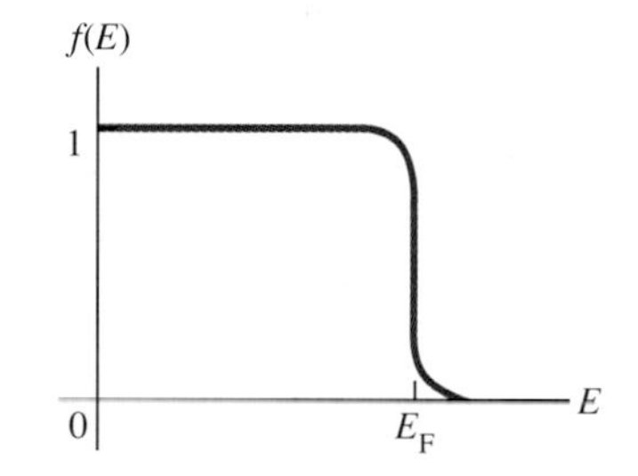


**44–20** A guess as to the probability distribution for occupation of electron energy states. Nearly all the states are occupied (probability 1) up to the value $E_F$, and nearly all those above $E_F$ are empty (occupation probability zero). As the temperature increases, we might expect more and more electrons to be in states with $E > E_F$.

The second reason we can't use the Maxwell-Boltzmann distribution is more subtle. It assumes that we are dealing with *distinguishable* particles. In principle we could put a tag on each molecule and know which is which. But overlapping electrons in a system such as a metal are *indistinguishable*. Suppose we have two electrons; a state in which the first is in an energy level $E_1$ and the second is in level $E_2$ is not distinguishable from a state in which the two electrons are reversed because we can't tell which electron is which.

The statistical distribution function that emerges from the exclusion principle and the indistinguishability requirement is called (after its inventors) the **Fermi-Dirac distribution.** The probability that a particular state with energy $E$ is occupied by an electron is given by the function

$$f(E) = \frac{1}{e^{(E - E_F)/kT} + 1}. \qquad (44\text{–}29)$$

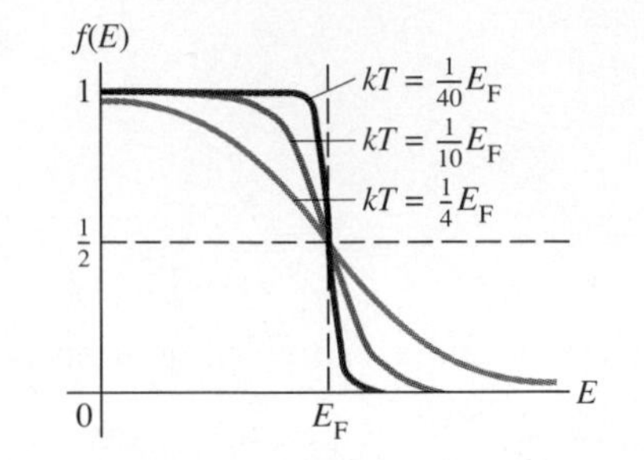


**44–21** Graphs of the Fermi-Dirac distribution function for various values of $kT$. When $kT$ is much less than $E_F$, nearly all the states up to $E_F$ are occupied, and nearly all states above $E_F$ are empty. As $T$ increases, more and more of the states above $E_F$ are occupied.

The constant $E_F$ is called the **Fermi level,** or the *Fermi energy;* we'll discuss its significance below.

Figure 44–21 shows graphs of Eq. (44–29) for several temperatures. The general shape of this function confirms our guess. When $E = E_F$, the exponent is zero, and $f(E) = \frac{1}{2}$. For $E < E_F$, the exponent is negative, and $f(E) > \frac{1}{2}$. For $E > E_F$, the exponent is positive, and $f(E) < \frac{1}{2}$. The shape depends on the ratio $E_F/kT$. At very small $T$ this ratio is very large. For $E < E_F$, the curve very quickly approaches 1, and for $E > E_F$, it quickly approaches zero. When $T$ is larger, the changes are more gradual. When $T$ equals zero, all the states up to the Fermi level $E_F$ are filled, and all above that level are empty.

## EXAMPLE 44–9

For free electrons in a solid, for what energy is the probability that a particular state is occupied 1% and 99%?

**SOLUTION** We solve Eq. (44–29) for $E$, obtaining

$$E = E_F + kT \ln\left(\frac{1}{f(E)} - 1\right).$$

When $f(E) = 0.01$,

$$E = E_F + kT \ln\left(\frac{1}{0.01} - 1\right) = E_F + 4.6kT.$$

A state that is $4.6kT$ above the Fermi level is occupied only 1% of the time.

When $f(E) = 0.99$,

$$E = E_F + kT \ln\left(\frac{1}{0.99} - 1\right) = E_F - 4.6kT.$$

A state that is $4.6kT$ below the Fermi level is occupied 99% of the time.

At very low temperatures these energy intervals become very small. Then levels very slightly below $E_F$ are nearly always full, and levels very slightly above $E_F$ are nearly always empty. In general, if the probability that a state at $\Delta E$ *above* $E_F$ is occupied is $P$, then the probability that a state $\Delta E$ *below* $E_F$ is occupied is $1 - P$. We leave the proof as a problem.

Equation (44–29) gives the probability that any specific state with energy $E$ is occupied at a temperature $T$. To get the actual number of electrons in any energy range $dE$, we have to multiply this probability by the density of states $g(E)$. Thus the number $dN$ of electrons with energies in the range $dE$ is

$$dN = g(E)f(E)\,dE = \frac{(2m)^{3/2}L^3E^{1/2}}{2\pi^2\hbar^3}\frac{1}{e^{(E-E_F)/kT}+1}dE. \tag{44–30}$$

The Fermi energy $E_F$ is determined by the total number $N$ of electrons; at any temperature the energy levels are filled up to a point at which all electrons are accommodated. At very low temperature there is a simple relation between $E_F$ and $N$. *All* levels below $E_F$ are filled; in Eq. (44–26) we set $N(E)$ equal to the total number of electrons $N$ and $E$ to the Fermi energy $E_F$:

$$N = \frac{(2m)^{3/2}VE_F^{3/2}}{3\pi^2\hbar^3}. \tag{44–31}$$

Solving for $E_F$, we get

$$E_F = \frac{3^{2/3}\pi^{4/3}\hbar^2}{2m}\left(\frac{N}{V}\right)^{2/3}. \tag{44–32}$$

The quantity $N/V$ is the number of valence electrons per unit volume. It is called the *electron concentration* and is usually denoted by $n$. With this notation, Eq. (44–32) becomes

$$E_F = \frac{3^{2/3}\pi^{4/3}\hbar^2 n^{2/3}}{2m}. \tag{44–33}$$

## EXAMPLE 44–10

At low temperatures, copper has a free-electron concentration of $8.45 \times 10^{28}\ \mathrm{m}^{-3}$. Using the free-electron model, find the Fermi energy at low temperatures.

**SOLUTION** We use Eq. (44–33):

$$E_F = \frac{3^{2/3}\pi^{4/3}(1.055 \times 10^{-34}\ \mathrm{J \cdot s})^2 (8.45 \times 10^{28}\ \mathrm{m}^{-3})^{2/3}}{2(9.11 \times 10^{-31}\ \mathrm{kg})}$$
$$= 11.3 \times 10^{-19}\ \mathrm{J} = 7.06\ \mathrm{eV}.$$

This is much *larger* than $kT$ at ordinary temperatures, so in determining $E_F$ it is a good approximation to take all the states below $E_F$ as completely full and all those above $E_F$ as completely empty.

We can also use Eq. (44–25) to find $n_{max}$ if $E$ and $L$ are known. We invite you to show that if $E = 5$ eV and $L = 1$ cm, $n_{max}$ is about $4 \times 10^7$. This number shows why we were justified in treating $n_{max}$ as a continuous variable in our density-of-states calculation.

## Average Electron Energy

We can calculate the *average* electron energy in a metal at very low temperature using the same assumptions that were used to find $E_F$. The number of electrons with energies in the range $dE$ is $g(E)\,dE$, with $g(E)$ given by Eq. (44–28). The energy of these electrons is $Eg(E)\,dE$, and the total energy of all the electrons is

$$E_{tot} = \int_0^{E_F} Eg(E)\,dE.$$

The simplest way to evaluate this expression is to compare Eqs. (44–28) and (44–32), noting that

$$g(E) = \frac{3NE^{1/2}}{2E_F^{3/2}}.$$

The integral then becomes

$$E_{tot} = \frac{3N}{2E_F^{3/2}}\int_0^{E_F} E^{3/2}\,dE = \tfrac{3}{5}NE_F. \qquad (44\text{–}34)$$

This result shows that at very low temperature the average electron energy $E_{av} = E_{tot}/N$ is equal to $\frac{3}{5}$ of the Fermi energy.

## EXAMPLE 44–11

a) Find the average energy of the electrons in copper (Example 44–10). b) If the equipartition principle were valid for this system, what would be the temperature of the electrons? c) What is the speed of an electron with kinetic energy equal to $E_F$?

**SOLUTION** a) The Fermi energy in copper is $11.3 \times 10^{-19}$ J = 7.06 eV. According to Eq. (44–34), the average energy is $\frac{3}{5}$ of this, or $6.78 \times 10^{-19}$ J = 4.24 eV.

b) If the equipartition principle were applicable (which it isn't), the average energy would equal $3kT/2$. In that case,

$$T = \frac{2E_{av}}{3k} = \frac{2(6.78 \times 10^{-19}\ \mathrm{J})}{3(1.38 \times 10^{-23}\ \mathrm{J/K})} = 3.28 \times 10^4\ \mathrm{K}.$$

c) We use $E_F = \frac{1}{2}mv_F{}^2$:

$$v_F = \sqrt{\frac{2E_F}{m}} = \sqrt{\frac{2(11.3 \times 10^{-19}\ \mathrm{J})}{9.11 \times 10^{-31}\ \mathrm{kg}}} = 1.57 \times 10^6\ \mathrm{m/s}.$$

This is called the *Fermi speed.* We invite to you show, for comparison, that electrons with average kinetic energy $3kT/2$ at room temperature would have an rms speed of $1.17 \times 10^5$ m/s, less than a tenth of the above result.

Example 44–11 shows that the average energies and rms speeds of electrons in metals are generally determined almost entirely by the exclusion principle; at ordinary temperature the behavior is very far from what the equipartition principle predicts. We could make a similar analysis to determine the electron contributions to heat capacities in a solid. If there is one conduction electron per atom, the equipartition principle would predict an additional molar heat capacity of $3R/2$ from electron kinetic energies. But when $E_F$ is much larger than $kT$, the usual situation, then only the few electrons near the Fermi energy change energies appreciably when the temperature changes. The number of such electrons is proportional to $kT/E_F$, so we should expect the electron heat capacity to be proportional to the product $(kT/E_F)(3R/2)$. A more detailed analysis shows that, in fact, the electron contribution to heat capacity of a metal is

$$C = \frac{\pi^2 kT}{2E_F} R,$$

which is in agreement with our prediction. We invite you to verify that if $T = 290$ K ($kT = \frac{1}{40}$ eV) and if $E_F = 5.0$ eV, then $C = 0.025R$, less than 2% of the prediction of the equipartition principle.

Fermi energies for metals typically fall in the range from 2 to 14 eV. The *Fermi temperature* $T_F$ is defined as $T_F = E_F/k$. Fermi temperatures for metals are typically in the range $2\text{–}16 \times 10^4$ K, and typical Fermi speeds are $0.8\text{–}2 \times 10^6$ m/s.

## 44–7 SEMICONDUCTORS

A **semiconductor** has an electrical resistivity that is intermediate between those of good conductors and those of good insulators. The tremendous importance of semiconductors in present-day electronics stems in part from the fact that their electrical properties are very sensitive to very small concentrations of impurities. We'll discuss the basic concepts, using the semiconductor elements silicon and germanium as examples.

Silicon and germanium are in Group IV of the periodic table. Each has four electrons in the outermost electron subshell (the 3*s* and 3*p* levels for Si, the 4*s* and 4*p* levels for Ge). Both crystallize in the diamond structure (Section 44–4) with covalent bonding; each atom lies at the center of a regular tetrahedron, forming a covalent bond with each of four nearest neighbors at the corners of the tetrahedron. All the valence electrons are involved in the bonding. The band structure (Section 44–5) has a completely filled valence band separated by an energy gap from an empty conduction band (Fig. 44–18c).

At very low temperatures these materials are insulators. Electrons in the valence band have no nearby levels available into which they can move in response to an applied electric field. However, the energy gap $E_g$ between the valence and conduction bands is unusually small: 1.14 eV for silicon and only 0.67 eV for germanium, compared to 5 eV or more for many insulators. Thus even at room temperature a substantial number of electrons can gain enough energy to jump the gap to the conduction band, where they are dissociated from their parent atoms and are free to move about the lattice. The number of these electrons increases rapidly with temperature.

### EXAMPLE 44–12

For a material with the band structure described above, find the probability that a state at the bottom of the conduction band is occupied, at a temperature of 300 K, if the band gap is a) 0.20 eV; b) 1.0 eV; c) 5.0 eV. Repeat the calculations for a temperature of 310 K. Assume that the Fermi energy is in the middle of the gap (Fig. 44–22).

**SOLUTION** a) We use the Fermi-Dirac distribution function, Eq. (44–29). For a band gap of 0.20 eV, $E - E_F = 0.1$ eV $= 1.60 \times 10^{-20}$ J,

$$\frac{E - E_F}{kT} = \frac{1.60 \times 10^{-20}\ \text{J}}{(1.38 \times 10^{-23}\ \text{J/K})(300\ \text{K})} = 3.86,$$

$$f(E) = \frac{1}{e^{3.86} + 1} = 0.0205.$$

With $E_g = 0.20$ eV and $T = 310$ K, the exponent is 3.74 and $f(E) = 0.0232$, a 13% increase for a temperature rise of only 10 K.

b) When $E_g = 1.0$ eV, both exponents are five times as large as before, 19.3 and 18.7; the values of $f(E)$ are $4.1 \times 10^{-9}$ and

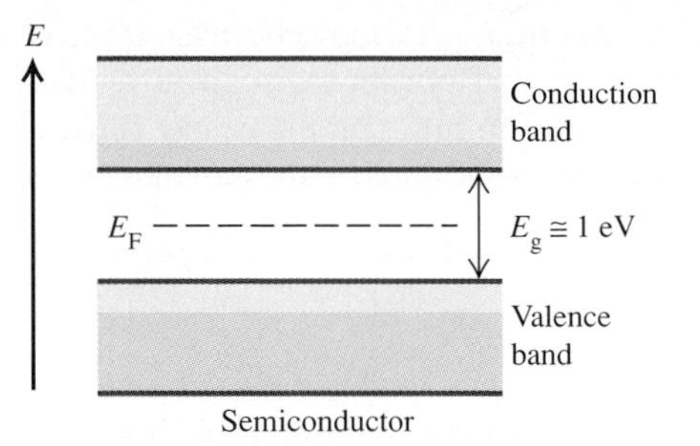


**44–22** Band structure of a semiconductor. At very low temperatures a completely filled valence band is separated by a narrow energy gap $E_g$ of 1 eV or less from a completely empty conduction band. Even at ordinary temperatures, many electrons are excited to the conduction band.

$7.6 \times 10^{-9}$. In this case the probability nearly doubles with a temperature rise of 10 K.

c) Finally, when $E_g = 5.0$ eV, the exponents are 96.6 and 93.5, and the values of $f(E)$ are $1.1 \times 10^{-42}$ and $2.5 \times 10^{-41}$. The probability increases by a factor of over 20 for a 10-K temperature rise, but it is still extremely small. A material with a 5-eV band gap has practically no electrons in the conduction gap and is a very good insulator.

This example illustrates two important points. First, the probability that a valence electron can reach the bottom of the conduction band is extremely sensitive to the width of the band gap. When the gap is 0.2 eV, the chance is about 2%, but when it is 1.0 eV, the chance is a few in a billion, and for a band gap of 5.0 eV it is essentially zero. Second, for any given band gap the probability is very temperature-dependent, more so for large gaps than for small.

In principle, we could continue the calculation in this example to find the actual density $n = N/V$ of electrons in the conduction band at any temperature. To do this, we would have to evaluate the integral $\int g(E)f(E)\, dE$ from the bottom of the conduction band to infinity. First we would need to know the density-of-states function $g(E)$. It wouldn't be correct to use Eq. (44–28) because the energy-level structure and the density of states for real solids are more complex than for the simple free-electron model. However, there are theoretical methods for predicting what $g(E)$ should be near the bottom of the conduction band, and such calculations have been carried out. Once we know $n$, we can determine the resistivity of the material (and its temperature dependence), using the analysis of Section 26–2, which you may want to review.

## Holes

When an electron is removed from a covalent bond, it leaves a vacancy where there would ordinarily be an electron. An electron from a neighboring atom can drop into this vacancy, leaving the neighbor with the vacancy. In this way the vacancy, called a **hole,** can travel through the lattice and serve as an additional current carrier. A hole behaves like a positively charged particle, even though the moving charges are electrons. It's like describing the motion of a bubble in a liquid. In a pure semiconductor, holes and electrons are always present in equal numbers. When an electric field is applied, they move in opposite directions (Fig. 44–23). The conductivity just described is called **intrinsic conductivity.** Another kind of conductivity, which we'll discuss later, is due to impurities.

The parking-garage analogy we mentioned in Section 44–5 helps to picture conduction in a semiconductor. A covalent crystal with no bonds broken is like a completely

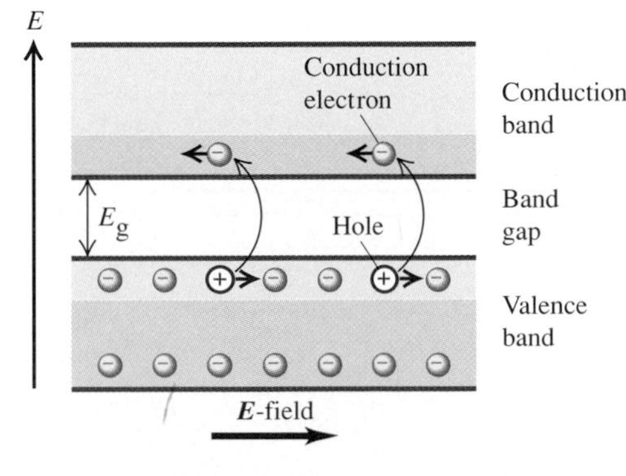


**44–23** Motion of electrons in the conduction band and of holes in the valence band of a semiconductor under the action of an applied electric field $\boldsymbol{E}$.

filled floor of a parking garage. No cars (electrons) can move because there is nowhere for them to go. But if one car is moved to the empty floor above, it can move freely, and the empty space it leaves behind also permits cars to move on the nearly filled floor. The motion of the vacant space corresponds to motion of a hole in the normally filled valence band.

## Impurities

Suppose we mix into melted germanium ($Z = 32$) a small amount of arsenic ($Z = 33$), the next element after germanium in the periodic table. Arsenic is in Group V; it has *five* valence electrons. When one of these electrons is removed, the remaining electron structure is essentially identical to that of germanium. The only difference is that it is smaller by a factor of $\frac{32}{33}$; the arsenic nucleus has a charge of $+33e$ rather than $+32e$, and it pulls the electrons in a little more. An arsenic atom can comfortably take the place of a germanium atom in the lattice. Four of its five valence electrons form the necessary nearest-neighbor covalent bonds.

The fifth valence electron is very loosely bound (Fig. 44–24a); it doesn't participate in the covalent bonds, and it is screened from the nuclear charge of $+33e$ by the 32 electrons, leaving a net effective charge of about $+e$. We might guess that the binding energy would be of the same order of magnitude as the energy of the $n = 4$ level in hydrogen, that is, $(\frac{1}{4})^2(13.6\text{ eV}) = 0.85\text{ eV}$. In fact, it is much smaller than this, only about 0.01 eV, because the electron's wave function extends over several atomic diameters, where neighboring atoms provide additional screening.

The energy level of this fifth electron corresponds in the band picture to an isolated energy level lying in the gap, 0.01 eV below the bottom of the conduction band (Fig. 44–24b). This level is called a *donor level,* and the impurity atom that is responsible for it is called a *donor impurity.* All Group V elements, including nitrogen, phosphorus, arsenic, antimony, and bismuth can serve as donor impurities. Even at ordinary temper-

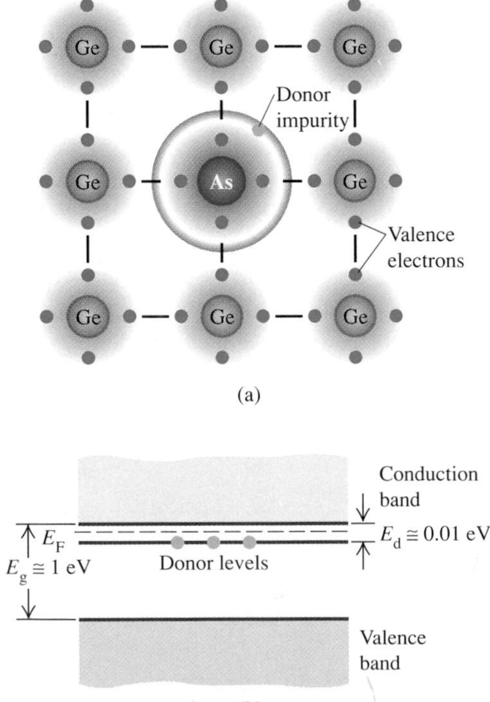

**44–24** (a) A donor ($n$-type) impurity has a fifth valence electron that does not participate in the covalent bonding and is very loosely bound. (b) Energy-band diagram for an $n$-type semiconductor. The donor levels lie just below the conduction band. At low temperatures the Fermi level is between the donor levels and the bottom of the conduction band.

atures, substantial numbers of electrons in donor levels can gain enough energy to climb into the conduction band, where they are free to wander through the lattice. We could calculate the probability for this event, using the same method as in Example 44–12. At $T = 300$ K the probability turns out to be about 0.4. (Note that the Fermi energy in this case is between the donor levels and the bottom of the conduction band.)

The corresponding positive charge is associated with the nuclear charge of the arsenic atom ($+33e$ instead of $+32e$). The remaining positive ion core is *not* free to move, in contrast to the situation with electrons and holes in pure germanium; it does not participate in conduction.

Example 44–12 shows that at ordinary temperatures and with a band gap of 1.0 eV, only a very small fraction (of the order of $10^{-9}$) of the valence electrons in pure germanium are able to escape their sites and participate in intrinsic conductivity. Thus we expect the conductivity of pure germanium to be less than that of good metallic conductors by a factor of the order of $10^{-9}$, and measurements bear out this prediction. However, a concentration of arsenic atoms as small as one part in $10^8$ can increase the conductivity so drastically that conduction due to impurities becomes by far the dominant mechanism. In this case the conductivity is due almost entirely to *negative* charge (electron) motion. We call the material an ***n*-type semiconductor,** with $n$-type impurities.

Adding atoms of an element in Group III (B, Al, Ga, In, Tl), with only *three* valence electrons, has an analogous effect. An example is gallium ($Z = 31$); placed in the germanium lattice, the gallium atom would like to form four covalent bonds, but it has only three outer electrons. It can, however, steal an electron from a neighboring germanium atom to complete the required four covalent bonds (Fig. 44–25a). The resulting atom has the same electron configuration as Ge but is larger by a factor of $\frac{32}{31}$ because the nuclear charge is $+31e$ instead of $+32e$.

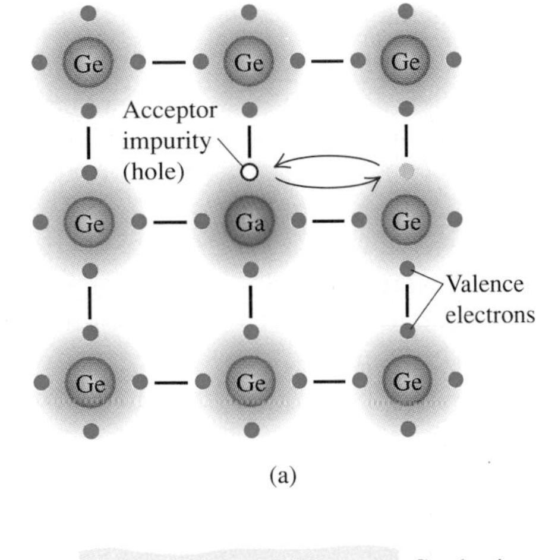


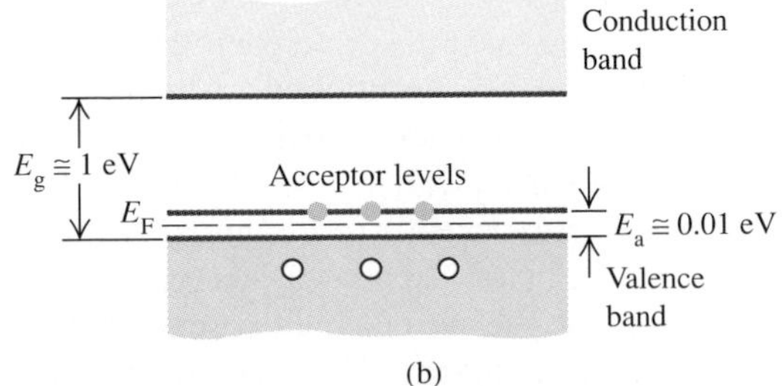


**44–25** (a) An acceptor ($p$-type) impurity has only three valence electrons. It can borrow an electron from a neighboring atom to complete four covalent bonds. The resulting hole is then free to move about the lattice. (b) Energy-band diagram for a $p$-type semiconductor. The acceptor levels lie just above the valence band. At low temperatures the Fermi level is between the top of the valence band and the acceptor levels.

This theft leaves the neighboring atom with a *hole,* or missing electron. The hole acts as a positive charge that can move through the lattice just as with intrinsic conductivity. The stolen electron is bound to the gallium atom in a level called an *acceptor level,* about 0.01 eV above the top of the valence band (Fig. 44–25b). The gallium atom thus completes the needed four covalent bonds, but it has a net charge of $-e$ (because there are 32 electrons and a nuclear charge of $31e$). This negative gallium ion is *not* free to move. A semiconductor with Group III impurities is called a ***p*-type semiconductor,** a material with $p$-type impurities.

The two types of impurities, $n$ and $p$, are also called *donors* and *acceptors,* respectively, and the deliberate addition of these impurity elements is called *doping.* Some semiconductors are doped with *both* $n$- and $p$-type impurities for reasons that we won't go into. Such materials are called *compensated* semiconductors.

We can verify the assertion that the current in $n$ and $p$ semiconductors really *is* carried by electrons and holes, respectively, by using the Hall effect (Section 28–9). The direction of the Hall emf is opposite in the two cases. Measurements of the Hall emf in various semiconductor materials confirm our analysis of the conduction mechanisms.

## 44–8 SEMICONDUCTOR DEVICES

Semiconductor devices play an indispensable role in contemporary electronics. In the early days of radio and television, transmitting and receiving equipment relied on vacuum tubes, but these have been almost completely replaced in the last three decades by solid-state devices, including transistors, diodes, integrated circuits, and other semiconductor devices. The only surviving vacuum tubes in radio and TV equipment are the picture tube in a TV receiver, imaging devices in TV cameras, and some microwave equipment. Equally significant are large-scale integrated circuits that incorporate the equivalent of many thousands of transistors, capacitors, resistors, and diodes on a silicon chip that is less than 1 cm square. Such chips form the heart of every pocket calculator, personal computer, and mainframe computer.

A thin slab of pure silicon or germanium can serve as a *photocell*. When the material is irradiated with light whose photons have at least as much energy as the band gap between the valence and conduction bands, an electron in the valence band can absorb a photon and jump to the conduction band, where it contributes to the conductivity. The conductivity therefore increases when the material is exposed to light. Detectors for charged particles operate on the same principle. A high-energy charged particle passing through the semiconductor material collides inelastically with valence electrons, exciting them from the valence to the conduction band and creating pairs of holes and conduction electrons. The conductivity increases momentarily, causing a pulse of current in the external circuit. Solid-state detectors are widely used in nuclear and high-energy physics research.

### The *p*-*n* Junction

In many semiconductor devices the essential principle is the fact that the conductivity of the material is controlled by impurity concentrations, which can be varied within wide limits and changed from one region of a device to another. An example is the ***p*-*n* junction,** a crystal of germanium or silicon with $p$-type impurities in one region and $n$-type impurities in another. The two regions meet in a boundary region called the *junction*. Originally, such junctions were produced by growing a crystal, usually by pulling a seed crystal very slowly away from the surface of a melted semiconductor, varying the concentration of impurities in the melt as the crystal was grown. The result was a crystal with two or more regions with different magnitudes of conductivity and conductivity types ($n$

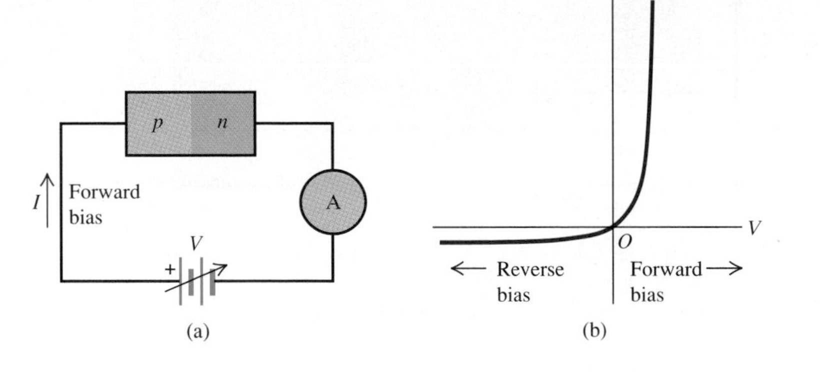


**44–26** (a) A semiconductor $p$-$n$ junction in a circuit; (b) graph showing the asymmetric voltage-current relationship. The curve is described by Eq. (44–35).

or $p$). There are much better ways to fabricate $p$-$n$ junctions now, but we don't need to go into the details.

When a $p$-$n$ junction is connected in an external circuit, as shown in Fig. 44–26a, and the potential $V$ across the device is varied, the current $I$ varies as shown in Fig. 44–26b. The device conducts much more readily in the direction from $p$ to $n$ than the reverse, in striking contrast to the symmetric behavior of materials that obey Ohm's law. Such a one-way device is called a **diode.** Later we'll discuss a simple model of $p$-$n$ junction behavior that predicts a voltage-current relation in the form

$$I = I_s(e^{eV/kT} - 1), \tag{44–35}$$

where $I_s$ is a constant characteristic of the device, $e$ is the electron charge, $k$ is Boltzmann's constant, and $T$ is absolute temperature. This expression is valid for both positive and negative values of $V$; note that $V$ and $I$ always have the same sign. As $V$ becomes very large and negative, $I$ approaches the value $-I_s$; the magnitude $I_s$ is called the *saturation current.*

## EXAMPLE 44–13

A certain $p$-$n$ junction diode has a saturation current $I_s = 1.00$ mA. At a temperature of 290 K, find the current when the voltage is 1.00 mV, $-$ 1.00 mV, 0.100 V, and $-$ 0.100 V.

**SOLUTION** At $T = 290$ K, $kT = \frac{1}{40}$ eV. When $V = 1.00$ mV $= 0.00100$ V, $eV/kT = e(0.00100\text{ V})/(\frac{1}{40}\text{ eV}) = 0.0400$. From Eq. (44–35) the current is

$$I = (1.00\text{ mA})(e^{0.0400} - 1) = 0.0408\text{ mA}.$$

When $V = -0.00100$ V,

$$I = (1.00\text{ mA})(e^{-0.0400} - 1) = -0.0392\text{ mA}.$$

The values of $I$ for the other two voltages are obtained in the same way; when $V = 0.100$ V, $I = 53.6$ mA, and when $V = -0.100$ V, $I = -0.98$ mA. Note that at $V = 1.00$ mV the current has nearly the same magnitude for both directions but that as the voltage increases, the directional asymmetry becomes more and more pronounced. At $|V| = 0.100$ V the negative current is nearly equal to the saturation value, and the positive current is over 50 times as great.

We can understand the behavior of a $p$-$n$ junction diode qualitatively on the basis of the conductivity mechanisms in the two regions. When the $p$ region is at higher potential than the $n$ region, corresponding to positive $V$ in Eq. (44–35), the resulting electric field is in the direction from $p$ to $n$. This is called the *forward direction,* and the potential is called *forward bias.* Holes in the $p$ region flow across the junction into the $n$ region, and electrons in the $n$ region move into the $p$ region; this flow constitutes a *forward current.* The opposite polarity is called *reverse bias;* in this case the field tends to push electrons from $p$ to $n$ and holes from $n$ to $p$. But there are very few mobile electrons in the $p$ region, only those associated with intrinsic conductivity and some that diffuse over from the $n$ region. Similarly, there are very few holes in the $n$ region. As a result, the current in the *reverse direction* is much smaller than that with the same potential in the forward direction.

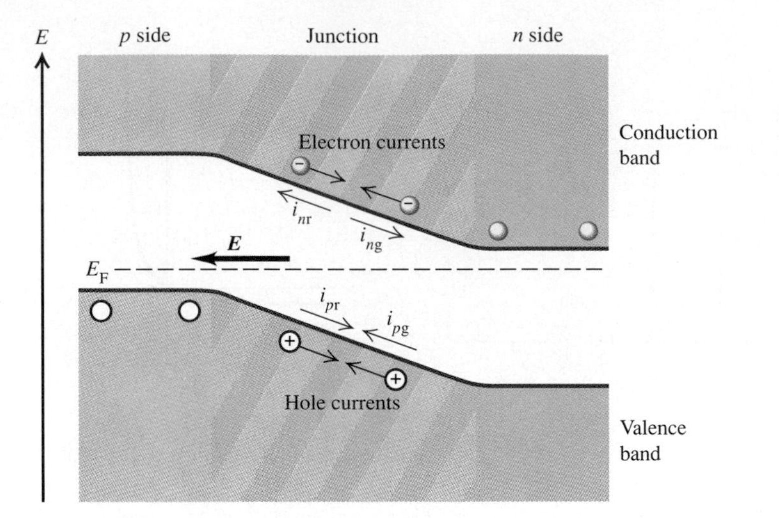


**44–27** A $p$-$n$ junction in equilibrium, with no externally applied field or potential difference. The generation and recombination currents exactly balance. The Fermi energy $E_F$ is the same on both sides of the junction; there is an excess of positive charge on the $n$ side and of negative charge on the $p$ side, resulting in an electric field in the direction shown. (The electron energy increases when the electric potential decreases because electrons have a negative charge.)

Let's take a more detailed look at the behavior of a $p$-$n$ junction. First, consider the equilibrium situation with no applied voltage (Fig. 44–27). There is a greater concentration of holes in the $p$ region than in the $n$ region, so holes tend to diffuse across the junction into the $n$ region. Once there, they recombine with excess electrons. Similarly, electrons diffuse from the $n$ region to the $p$ region and fall into holes there. The diffusion currents and the delays before recombination lead to a net positive charge in the $n$ region and a net negative charge in the $p$ region, causing an electric field in the direction from $n$ to $p$ at the junction. The potential energy associated with this field raises the electron energy levels in the $p$ region relative to the same levels in the $n$ region.

There are four currents across the junction, as shown in Fig. 44–27. The diffusion processes lead to *recombination currents* of holes and electrons, labeled $i_{pr}$ and $i_{nr}$. At the same time, electron-hole pairs are generated in the junction region by thermal excitation. The electric field described above sweeps these electrons and holes out of the junction in the directions shown. The corresponding currents, called *generation currents,* are labeled $i_{pg}$ and $i_{ng}$. At equilibrium the generation and recombination currents are equal:

$$i_{pg} = i_{pr} \qquad \text{and} \qquad i_{ng} = i_{nr}. \tag{44–36}$$

At thermal equilibrium the Fermi energy is the same at each point across the junction.

Now we apply a forward bias, a potential difference $V$ with the $p$ side at higher potential. This potential *decreases* the difference between the energy levels on the $p$ and $n$ sides (Fig. 44–28). It becomes easier for the electrons in the $n$ region to climb the poten-

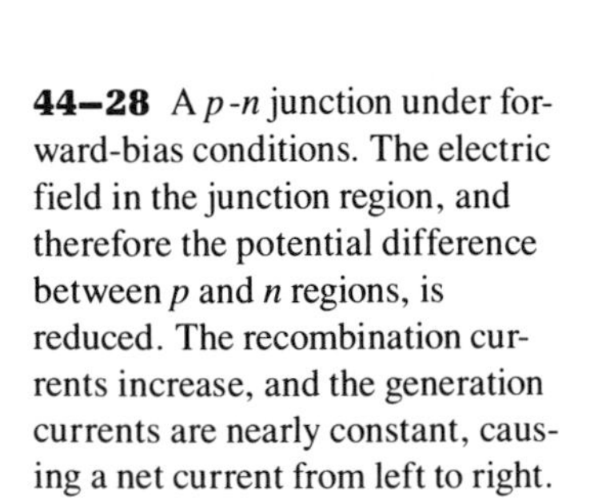

**44–28** A $p$-$n$ junction under forward-bias conditions. The electric field in the junction region, and therefore the potential difference between $p$ and $n$ regions, is reduced. The recombination currents increase, and the generation currents are nearly constant, causing a net current from left to right.

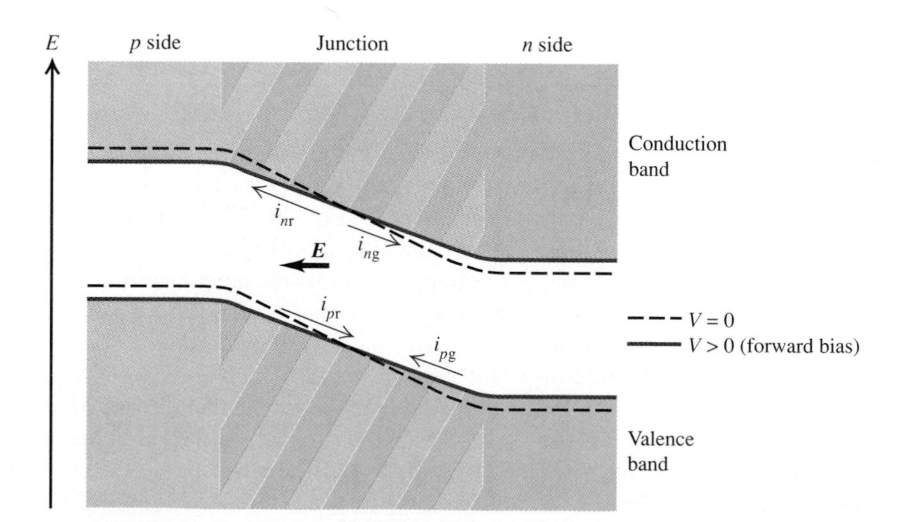

tial-energy hill and diffuse into the $p$ region and for the holes in the $p$ region to diffuse into the $n$ region. This effect increases both recombination currents by the Maxwell-Boltzmann factor $e^{\Delta E/kT} = e^{eV/kT}$. (We don't have to use the Fermi-Dirac distribution because most of the available states in the conduction band are empty and the exclusion principle has no effect.) The generation currents don't change appreciably, so the net hole current is

$$
\begin{aligned}
i_{\text{ptot}} &= i_{\text{pr}} - i_{\text{pg}} \\
&= i_{\text{pg}} e^{eV/kT} - i_{\text{pg}} \\
&= i_{\text{pg}}(e^{eV/kT} - 1).
\end{aligned} \tag{44–37}
$$

The net electron current is given by a similar expression, so the total current $I = i_{\text{ptot}} + i_{\text{ntot}}$ is

$$I = I_s\,(e^{eV/kT} - 1), \tag{44–38}$$

in agreement with Eq. (44–35). This entire discussion can be repeated for reverse bias (negative $V$ and $I$) with the same result. Therefore, Eq. (44–38) is valid for both positive and negative values.

Several effects make the behavior of practical $p$-$n$ junction diodes more complex than this simple analysis predicts. One effect, *avalanche breakdown,* occurs when the electric field in the junction becomes so strong that the carriers can gain enough energy between collisions to create electron-hole pairs during inelastic collisions. (A similar effect occurs in dielectric breakdown in insulators, discussed in Section 25–5.)

A second type of breakdown occurs when the reverse bias is large enough that the top of the valence band in the $p$ region is actually higher in energy than the bottom of the conduction band in the $n$ region (Fig. 44–29). If the junction region is thin enough, electrons can *tunnel* from the valence band of the $p$ region to the conduction band of the $n$ region. This process is called *Zener breakdown*. It occurs in Zener diodes, which are widely used for voltage regulation and protection against voltage surges.

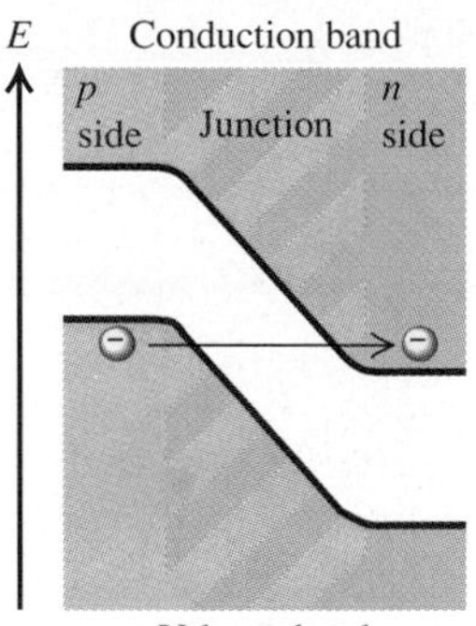


**44–29** Under reverse-bias conditions, the potential-energy difference between the $p$ and $n$ sides of a junction is greater than at equilibrium. If it is great enough, the bottom of the conduction band on the $n$ side may actually be below the top of the valence band on the $p$ side. If the junction is thin enough, electrons can tunnel through from the valence band to the conduction band; this process is called Zener breakdown.

A *light-emitting diode* (*LED*) is, as the name implies, a $p$-$n$ junction that emits light. When the junction is forward-biased, many holes are injected across the junction into the $n$ region, and electrons are injected into the $p$ region. When each minority carrier recombines with a majority carrier, the pair can emit a photon with energy approximately equal to the band gap. This energy (and therefore the photon wavelength and the color of the light) can be varied by using materials with different band gaps. Light-emitting diodes are widely used for digital displays in clocks, electronic equipment, automobile instrument panels, and many other applications.

The reverse process is called the *photovoltaic effect.* Here the material absorbs photons, and electron-hole pairs are created. Pairs that are created close to a $p$-$n$ junction migrate to it, where the electric field described above sweeps the electrons to the $n$ side and the holes to the $p$ side. We can connect this device to an external circuit, where it becomes a source of emf and power. Such a device is often called a *solar cell,* although sunlight isn't required. *Any* light with photon energies greater than the band gap will do. You might have a calculator powered by such cells. Production of low-cost photovoltaic cells for large-scale solar energy conversion is a very active field of research.

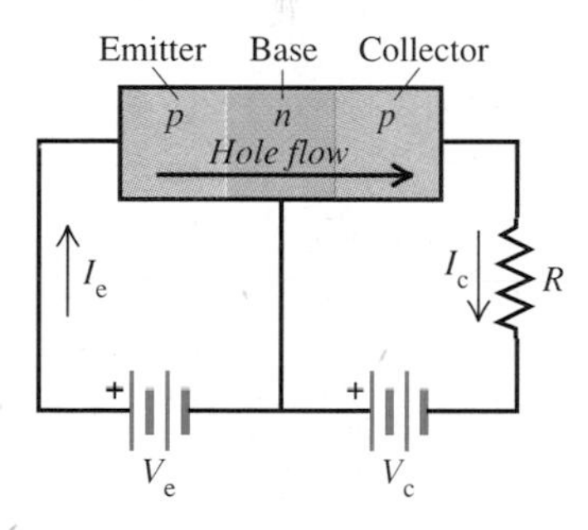


**44–30** Schematic diagram of a $p$-$n$-$p$ transistor and circuit. When $V_e = 0$, the current in the collector circuit is very small. When a potential $V_e$ is applied between emitter and base, holes travel from emitter to base, as shown; when $V_c$ is sufficiently large, most of the holes continue into the collector. The collector current $I_c$ is controlled by the emitter current $I_e$.

## Transistors

A *transistor* includes two $p$-$n$ junctions in a sandwich configuration, which may be either $p$-$n$-$p$ or $n$-$p$-$n$. A $p$-$n$-$p$ transistor is shown in Fig. 44–30. The three regions are called the emitter, base, and collector, as shown. When there is no current in the left loop of the circuit, there is only a very small current through the resistor $R$ because the voltage across the base-collector junction is in the reverse direction. But when a voltage is applied

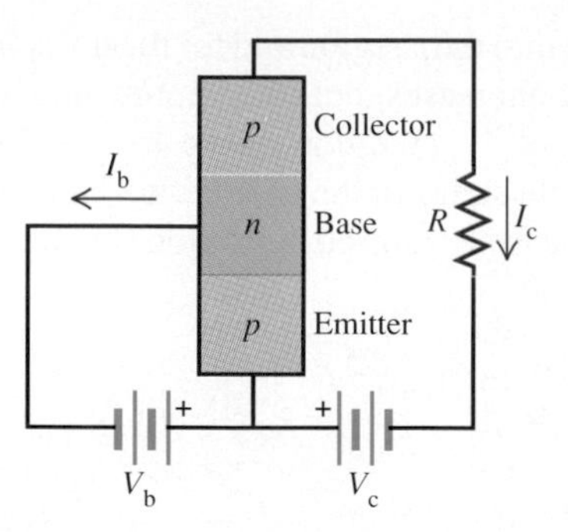


**44–31** A common-emitter circuit. When $V_b = 0$, $I_c$ is very small, and most of the voltage $V_c$ appears across the base-collector junction. As $V_b$ increases, the base-collector potential decreases, and more holes can diffuse into the collector; thus, $I_c$ increases. Ordinarily, $I_c$ is much larger than $I_b$.

between emitter and base, as shown, the holes traveling from emitter to base can travel *through* the base to the second junction, where they come under the influence of the collector-to-base potential difference and flow on through the resistor.

In this way the current in the collector circuit is *controlled* by the current in the emitter circuit. Furthermore, $V_c$ may be considerably larger than $V_e$, so the *power* dissipated in $R$ may be much larger than the power supplied to the emitter circuit by the battery $V_e$. Thus the device functions as a *power amplifier.* If the potential drop across $R$ is greater than $V_e$, it may also be a voltage amplifier.

In this configuration the *base* is the common element between the "input" and "output" sides of the circuit. Another widely used arrangement is the *common-emitter circuit,* shown in Fig. 44–31. In this circuit the current in the collector side of the circuit is much larger than in the base side, and the result is current amplification.

An important variation is the *field-effect transistor* (Fig. 44–32). A slab of $p$-type silicon is made with two $n$-type regions on the top, called the *source* and the *drain;* a metallic conductor is fastened to each. Separated from the slab by an insulating layer of silicon dioxide ($SiO_2$) is a third electrode called the *gate.* When there is no charge on the gate and a potential difference of either polarity is applied between the source and the drain, there is very little current because one of the $p$-$n$ junctions is reverse-biased.

Now we place a positive charge on the gate. There aren't many free electrons in the $p$-type material, but there are some, and they are attracted toward the gate. The resulting greatly enhanced concentration of electrons near the gate (and between the two junctions) permits electron current to flow between source and drain. The conductivity of the

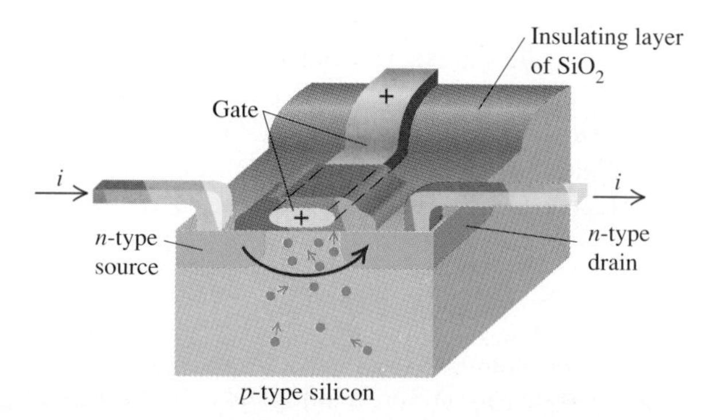


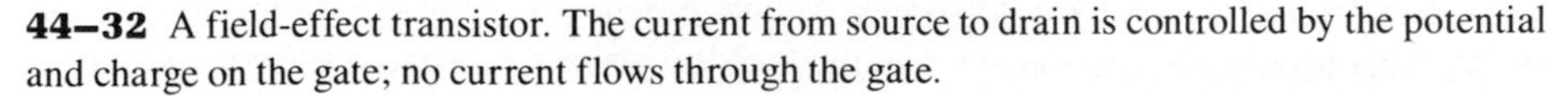
**44–32** A field-effect transistor. The current from source to drain is controlled by the potential and charge on the gate; no current flows through the gate.

region between source and drain depends critically on the electron concentration, so the current is very sensitive to the gate charge and potential, and the device functions as an amplifier. Note that there is very little current into or out of the gate. The device just described is called a MOSFET (metal-oxide-semiconductor field effect transistor).

## Integrated Circuits

A further refinement in semiconductor technology is the *integrated circuit.* By successively depositing layers of material and etching patterns to define current paths, we can combine the functions of several MOSFET transistors, capacitors, and resistors on a single square of semiconductor material that may be only a few millimeters on a side. An elaboration of this idea leads to *large-scale integrated circuits* and *very-large-scale integration* (VLSI).

Starting on a silicon chip base, various layers are built up, including evaporated metal layers for conducting paths and silicon-dioxide layers for insulators and for dielectric layers in capacitors. Appropriate patterns are etched into each layer by use of photosensitive etch-resistant materials onto which optically reduced patterns are projected. A circuit containing the functional equivalent of many thousands of transistors, diodes, resistors, and capacitors can be built up on a single chip. These so-called MOS (metal-oxide-semiconductor) chips are the heart of all pocket calculators and present-day computers, large and small. An example is shown in Fig. 44–33.

**44–33** Large-scale integrated circuits. A silicon chip only 1 cm on a side—about the size of a ladybug—can contain hundreds of thousands of transistors, providing as much computing power as a room-sized computer from 40 years ago.

The first semiconductor devices were invented in 1948. Since then, they have completely revolutionized the electronics industry, including applications in communications, computer systems, control systems, and many other areas.

# 44–9 SUPERCONDUCTIVITY

Superconductivity is the complete disappearance of all electrical resistance at low temperatures. We described at the end of Section 26–2 the circumstances of its discovery in 1911 by the Dutch physicist Heike Kamerlingh-Onnes. Since this discovery, many other superconducting metals, alloys, and other materials have been found. Each one has a characteristic superconducting transition temperature $T_c$ called its **critical temperature.** The highest critical temperature found so far for a *metallic* alloy is 23 K for a niobium-germanium alloy. Recently discovered *ceramic* materials have critical temperatures as high as 125 K (about $-150$°C or $-235$°F).

Below the superconducting transition temperature, the resistivity of a material is believed to be *exactly* zero. Experimental upper limits are of the order of $10^{-25}$ $\Omega \cdot$ m, compared with typical values of $10^{-8}$ $\Omega \cdot$ m for good conductors such as silver and copper at ordinary temperatures. When a current is magnetically induced in a superconducting ring, the current continues without measurable decrease for years!

A magnetic field can never exist inside a superconducting material. When we place a superconductor in a magnetic field, eddy currents are induced that exactly cancel the applied field everywhere inside the material (except for a surface layer a hundred or so atoms thick). Related to this behavior is the fact that an applied magnetic field lowers the critical temperature, as shown in Fig. 44–34. A sufficiently strong field eliminates the superconducting phase transition completely. Figure 44–34 is a *phase diagram* for the superconducting transition. The magnetic behavior of superconductors is discussed in greater detail in Section 30–8.

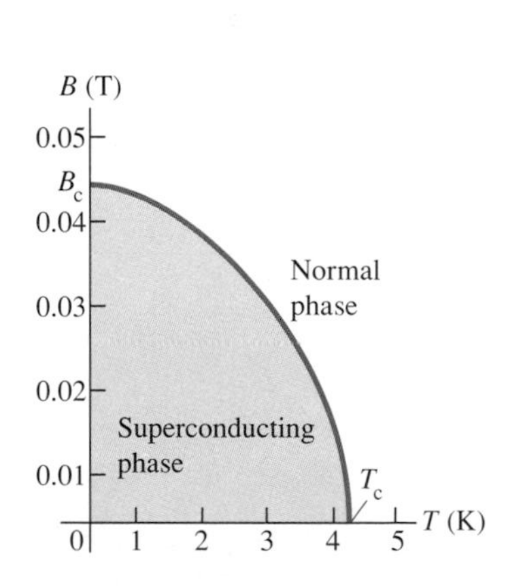


**44–34** Phase diagram for pure mercury, showing the superconducting transition critical temperature and its dependence on magnetic field. Other superconducting materials have similar curves but with different scales.

Although superconductivity was discovered in 1911, it was not well understood on a theoretical basis until 1957. In that year, John Bardeen, Leon Cooper, and Robert Schrieffer published the theory, now called the BCS theory, that was to earn them the Nobel prize in 1972. The key to the BCS theory is an interaction between *pairs* of conduction electrons, called *Cooper pairs,* caused by an interaction with the positive ion cores of the crystal lattice. Here's a rough qualitative picture of what happens. An electron exerts attractive forces on the positive ion cores, distorting the lattice slightly. The resulting slight concentration of positive charge then exerts an attractive force on another electron with momentum opposite to the first. At ordinary temperatures this electron-pair interaction is very small in comparison to energies of thermal motion, but at very low temperatures it becomes significant.

Bound together in this way, the pairs of electrons cannot *individually* gain or lose very small amounts of energy, as they would ordinarily be able to do in a partly filled conduction band. There is an energy gap in the allowed quantum states of the electron pairs, and at low temperatures there is not enough energy for a pair to jump this gap. Therefore the electrons can move freely through the lattice without any energy exchange through collisions.

Another class of materials is the *type II superconductors.* When such a material is placed in a magnetic field, the bulk of the material is superconducting, but there are thin filaments of material in the normal state, running parallel to the field. Currents circulate around the boundaries of these filaments, and there *is* magnetic flux inside the filaments. Type II superconductors are used for electromagnets because much larger magnetic fields can usually be present without destroying the superconducting state than with ordinary superconductors.

Still another type of material is the group of high-temperature superconductors, first discovered in 1987. These are not metals or alloys but *ceramic* materials that combine rare-earth metals, other metals, and nonmetals. An example is a compound of yttrium, barium, copper, and oxygen, with a critical temperature of 93 K. This temperature is very significant because it is above the boiling temperature of liquid nitrogen, 77 K. Until 1987, superconductivity had been achieved only at liquid helium temperatures, requiring sophisticated and expensive equipment. The critical temperatures of some of the new ceramics can be reached with liquid nitrogen, which is relatively cheap and easy to handle.

Many important and exciting applications of superconductors are under development. Superconducting electromagnets have been used in research laboratories for several years. Once a current is established in the coil of such a magnet, no additional power input is required because there is no resistive energy loss. The coils can also be made more compact because there is no need to provide channels for the circulation of cooling fluids. Thus superconducting magnets can attain very large fields much more easily and economically than conventional magnets; fields of the order of 10 T are fairly routine. These considerations also make superconductors attractive for long-distance electric-power transmissions, an active area of development.

One of the most glamorous applications of superconductors is in the field of magnetic levitation. Imagine a superconducting ring mounted on a railroad car that runs on a conducting guideway. The current induced in the guideway leads to a repulsive interaction with the ring, and levitation is possible. Magnetically levitated trains have been running on an experimental basis in Japan for several years, and similar development projects are also underway in Germany.

Superconducting materials with critical temperatures at or above room temperature no longer seem impossible. The potential implications of such developments for long-distance power transmission, magnetic levitation, and many other applications are fairly breathtaking. The high-temperature superconductors discovered so far are mechanically weak and brittle, like many ceramics, and are often chemically unstable. Fabricating conductors from them will pose very difficult technological problems.

## SUMMARY

• The principal types of molecular bonds are ionic, covalent, van der Waals, and hydrogen bonds.

• In a diatomic molecule the rotational energy levels are

$$E = l(l + 1)\frac{\hbar^2}{2I} \qquad (l = 0, 1, 2, \ldots). \tag{44–3}$$

The moment of inertia of a diatomic molecule is

$$I = m_r r_0^2, \tag{44–6}$$

where $m_r$ is the reduced mass, $m_r = \dfrac{m_1 m_2}{m_1 + m_2}$. (44–4)

The vibrational energy levels are

$$E_n = (n + \tfrac{1}{2})\hbar\,\omega = (n + \tfrac{1}{2})\hbar\sqrt{\frac{k}{m_r}}. \tag{44–7}$$

• The distribution of molecular energies is the Maxwell-Boltzmann distribution:

$$N(n) = NA\,e^{-E_n/kT}. \tag{44–10}$$

The vibrational contribution to the heat capacity approaches the value $R$ at high temperatures, but it approaches zero at low temperatures. At any temperature,

$$C = R\left(\frac{\hbar\,\omega}{kT}\right)^2 \frac{e^{\hbar\omega/kT}}{(e^{\hbar\omega/kT} - 1)^2}. \tag{44–19}$$

• Interatomic bonds in solids are of the same types as in molecules. There is one additional type—the metallic bond. There are several distinct types of crystal lattices.

• When atoms are bound together in a crystal, their energy levels spread out into bands. Insulators have a completely filled valence band separated by an energy gap from an empty conduction band. Conductors, including metals, have partially filled conduction bands.

• In a simple model of the behavior of metals, the electrons are treated as completely free particles inside the material. In this model the density of states is

$$g(E) = \frac{(2m)^{3/2}\,V}{2\pi^2\hbar^3}E^{1/2}. \tag{44–28}$$

The distribution of electrons in the various energy states is determined by the Fermi-Dirac distribution:

$$f(E) = \frac{1}{e^{(E - E_F)/kT} + 1}. \tag{44–29}$$

This distribution takes into account the exclusion principle and the indistinguishability of electrons.

• A semiconductor has covalent bonding but only a small energy gap between valence and conduction bands. Electrical properties are drastically changed by the addition of small concentrations of donor or acceptor impurities.

• Many semiconductor devices use one or more $p$-$n$ junctions, including diodes, transistors, and integrated circuits. The voltage-current relation for an ideal $p$-$n$ junction diode is

$$I = I_s\,(e^{eV/kT} - 1). \tag{44–35}$$

### KEY TERMS

- ionic bond
- covalent bond
- van der Waals bond
- hydrogen bond
- Maxwell-Boltzmann distribution
- metallic crystal
- energy band
- close packing
- valence band
- conduction band
- density of states
- Fermi-Dirac distribution
- Fermi level
- semiconductor
- hole
- intrinsic conductivity
- *n*-type semiconductor
- *p*-type semiconductor
- *p*-*n* junction
- diode
- critical temperature

# EXERCISES

## Section 44–1 Types of Molecular Bonds

**44–1** We know from Chapter 16 that the average kinetic energy of an ideal gas atom or molecule at Kelvin temperature $T$ is $\frac{3}{2}kT$. For what value of $T$ does this energy correspond to a) the bond energy of the van der Waals bond in $He_2$; b) the bond energy of the covalent bond in $H_2$? (The energy of the covalent bond in $H_2$ is 4.48 eV, and the bond energy for $He_2$ can be read from Fig. 44–4.)

**44–2 Binding Energy of KBr.** a) Calculate the electrical potential energy for a $K^+$ and a $Br^-$ ion separated by a distance of 0.29 nm, the equilibrium separation in the KBr molecule, if the ions are treated as point charges. b) The ionization energy of the potassium atom is 4.3 eV. Atomic bromine has an electron affinity of 3.5 eV. Use these data and the results of part (a) to estimate the binding energy of the KBr molecule. Do you expect the actual binding energy to be larger or smaller than your estimate?

**44–3** It is claimed in the text that the van der Waals bond between a pair of helium atoms is due to an induced dipole–induced dipole interaction and that this interaction has a potential energy that varies as $1/r^6$. The experimentally determined He–He interaction is shown in Fig. 44–4. Write $U(r)$ as $-C_6/r^6$ and calculate $C_6$ from the $U(r)$ given in the figure at $r = 0.50$ nm. Copy the figure and show on it the $-C_6/r^6$ interaction. What is the physical origin of the sharp repulsive interaction for $r < 0.3$ nm?

## Section 44–2 Molecular Spectra

**44–4** Show that the energy difference between rotational levels with angular momentum quantum numbers $l$ and $l-1$ is $l\hbar^2/I$.

**44–5** Show that the frequencies in a pure rotational spectrum (no change in vibrational level) are all integer multiples of the quantity $\hbar/2\pi I$. (Assume that the rotational quantum number $l$ changes by $\pm 1$ in the rotational transitions.)

**44–6 Moment of Inertia of Oxygen.** Assuming that the distance between nuclei in a diatomic oxygen molecule is 0.120 nm, calculate the moment of inertia about an axis through the center of mass of the two nuclei and perpendicular to the line joining them. The mass of an oxygen atom is $2.66 \times 10^{-26}$ kg.

**44–7** A lithium atom has mass $1.17 \times 10^{-26}$ kg, and a hydrogen atom has mass $1.67 \times 10^{-27}$ kg. If the equilibrium separation between the two nuclei in the LiH molecule is 0.159 nm, what is the energy separation between the $l = 1$ and $l = 2$ rotational levels?

**44–8** For the $I_2$ molecule the force constant is 172 N/m. If the mass of an iodine atom is $2.11 \times 10^{-25}$ kg, what is the energy separation in electronvolts between adjacent vibrational levels of the molecule?

**44–9** The vibration frequency for the molecule HCl is $8.97 \times 10^{13}$ Hz. The mass of a hydrogen atom is $1.67 \times 10^{-27}$ kg, and the mass of a chlorine atom is $5.81 \times 10^{-26}$ kg. a) What is the force constant $k$ for the interatomic force? b) What is the spacing between adjacent vibrational energy levels in joules and in electronvolts? c) What is the wavelength of a photon emitted in a transition between two adjacent vibrational levels? In what region of the spectrum does it lie?

## Section 44–3 Quantum Theory of Heat Capacities

***44–10** The spacing between the vibrational energy levels of the $N_2$ molecule is 0.292 eV, and the moment of inertia of the molecule for rotation about its center of mass is $I = 1.39 \times 10^{-46}$ kg. At what temperature is $kT$ equal to the excitation energy from the ground level to a) the first excited vibrational level; b) the first excited rotational level? Compare your results to those calculated in Example 44–4 for $H_2$.

***44–11** A large sample of hydrogen atoms in the $n = 1$, $l = 0$ ground level is placed in a magnetic field $\mathbf{B}$ in the $+z$-direction. At a temperature of 0.300 K, what must $B$ be for 0.10% of the atoms to be in the $m_s = +\frac{1}{2}$ spin state?

***44–12** Show that the vibrational contribution to the molar heat capacity goes to zero as $T \to 0$ (Eq. 44–21).

***44–13** For a hydrogen molecule the energy separation between adjacent vibrational levels is 0.54 eV. Use Eq. (44–19) to calculate the vibrational contribution to the specific heat capacity of $H_2$ for temperatures of a) 50 K; b) 300 K; c) 2000 K.

## Section 44–4 Structure of Solids

**44–14** Calculate the wavelength of a) a 12.4-keV x ray; b) a 150-eV electron; c) a 0.0818-eV neutron.

**44–15 Density of NaCl.** The spacing of adjacent atoms in a sodium-chloride crystal is 0.282 nm. The mass of a sodium atom is $3.82 \times 10^{-26}$ kg, and the mass of a chlorine atom is $5.89 \times 10^{-26}$ kg. Calculate the density of sodium chloride.

**44–16** Potassium bromide, KBr, has a density of $2.75 \times 10^3$ kg/m$^3$ and the same crystal structure as NaCl. The mass of a potassium atom is $6.49 \times 10^{-26}$ kg, and the mass of a bromine atom is $1.33 \times 10^{-25}$ kg. a) Calculate the average spacing between adjacent atoms in a KBr crystal. b) How does the value calculated in part (a) compare with the spacing in NaCl (Exercise 44–15)? Is the relation between the two values qualitatively what you would expect?

## Section 44–5 Energy Bands

**44–17 A Silicon Photocell.** Consider a photocell that uses pure silicon rather than germanium (Example 44–8). If the gap between valence and conduction bands in silicon is 1.14 eV, what maximum wavelength can be detected by a silicon photocell?

**44–18** The gap between valence and conduction bands in carbon is 7.0 eV. What is the maximum wavelength of a photon that can excite an electron from the top of the valence band into the conduction band? In what region of the electromagnetic spectrum does it lie?

### Section 44–6 Free-Electron Model of Metals

***44–19** a) Show that the wave function $\psi$ given in Eq. (44–22) is a solution of the three-dimensional Schrödinger equation (Eq. 42–31) with the energy as given by Eq. (44–23). b) What are the energies of the ground level and the lowest two excited levels, and what is the degeneracy of each level? (Include the factor of 2 in the degeneracy that is due to the two possible spin states.)

***44–20** Supply the details in the derivation of Eq. (44–26) from Eqs. (44–25) and (44–24).

***44–21** Calculate $n_{max}$ from Eq. (44–25) for $E = 5.0$ eV and $L = 1.0$ cm.

***44–22 Fermi Temperature of Potassium.** The Fermi energy of potassium is $E_F = 2.14$ eV. a) Find the average energy $E_{av}$ of the electrons. b) What is the speed of an electron that has energy $E_{av}$? c) At what Kelvin temperature $T$ is $kT$ equal to $E_F$? (This is the Fermi temperature for the metal.)

***44–23** Copper has a Fermi energy of 7.06 eV. Calculate, in terms of $R$, the electron contribution to the heat capacity at $T = 300$ K.

### Section 44–7 Semiconductors

**44–24** Germanium has a band gap of 0.67 eV. The Fermi energy is in the middle of the gap. a) For temperatures of 250 K, 300 K, and 350 K, calculate the probability $f(E)$ that a state at the bottom of the conduction band is occupied. b) At each of the temperatures in part (a), calculate the probability that a state at the top of the valence band is empty.

**44–25** Germanium has a band gap of 0.67 eV. Doping with arsenic adds filled donor levels in the gap 0.01 eV below the bottom of the conduction band. For temperatures of 250 K, 300 K, and 350 K, use Eq. (44–29) to calculate the probability $f(E)$ that a state at the bottom of the conduction band is occupied by thermal promotion of an electron from the donor level. (*Note:* Assume that the Fermi energy is midway between the donor levels and the bottom of the conduction band.)

### Section 44–8 Semiconductor Devices

**44–26** A $p$-$n$ junction has a saturation current $I_s = 5.00$ mA. a) At a temperature of 300 K, what voltage is needed to produce a positive current of 50.0 mA? b) For the voltage calculated in part (a), what is the negative current?

**44–27** a) A forward-bias voltage of 15.0 mV produces a positive current of 12.0 mA through a $p$-$n$ junction at 300 K. What does the positive current become if the forward-bias voltage is reduced to 10.0 mV? b) At each voltage of part (a), what is the reverse-bias negative current?

## PROBLEMS

**44–28** The binding energy of a KCl molecule is 4.43 eV. The ionization energy of a potassium atom is 4.3 eV, and the electron affinity of chlorine is 3.6 eV. Use these data to estimate the equilibrium separation between the two atoms in the KCl molecule.

**44–29** a) For the sodium chloride molecule (NaCl) discussed at the beginning of Section 44–1, what is the maximum separation of the ions for stability if they may be regarded as point charges? That is, what is the largest separation for which the energy of $Na^+ + Cl^-$, calculated in this model, is lower than the energy of the two separate atoms Na and Cl? b) Calculate this distance for the potassium bromide (KBr) molecule (Exercise 44–2).

**44–30** When a diatomic molecule undergoes a transition from the $l = 2$ to the $l = 1$ rotational state, a photon with wavelength 56.4 $\mu$m is emitted. What is the moment of inertia of the molecule for an axis through its center of mass?

**44–31** Use the data given in Example 44–3 to calculate the zero-point vibrational energy for $H_2$, or the vibrational energy the molecule has in the $n = 0$ ground vibrational level. How does this energy compare in magnitude with the $H_2$ bond energy of $-4.48$ eV?

**44–32** When an $O_2$ molecule undergoes a transition from the $n = 0$ to the $n = 1$ vibrational level, a photon with wavelength 6.32 $\mu$m is absorbed. Calculate the frequency of vibration $f$ and the force constant $k$ for the interatomic force. (The mass of an oxygen atom is $2.66 \times 10^{-26}$ kg.)

**44–33 Spectral Lines from Isotopes.** The equilibrium separation for NaCl is 0.2361 nm. The mass of a sodium atom is $3.8176 \times 10^{-26}$ kg. Chlorine has two stable isotopes, $^{35}Cl$ and $^{37}Cl$, that have different masses but identical chemical properties. The atomic mass of $^{35}Cl$ is $5.8068 \times 10^{-26}$ kg, and the atomic mass of $^{37}Cl$ is $6.1384 \times 10^{-26}$ kg. a) Calculate the wavelength of the photon emitted in the $l = 2 \rightarrow l = 1$ and $l = 1 \rightarrow l = 0$ transitions for $Na^{35}Cl$. b) Repeat part (a) for $Na^{37}Cl$. What are the differences in the wavelengths for the two isotopes?

**44–34** The rotational spectrum of HCl contains the following wavelengths:

$$60.4, \quad 69.0, \quad 80.4, \quad 96.4, \quad 120.4\ \mu\text{m}$$

Use this spectrum to find the moment of inertia of the HCl

The sun, like all normal stars, consists of gas, primarily hydrogen and helium. However, its inner structure is not uniform. This computer model shows one possible pattern in which pressure waves radiate out from the sun's center and reflect back from its surface. Outward-moving regions are colored blue, inward-moving regions are red.

CHAPTER 45

# Nuclear Physics

What is the sun's source of energy? The answer is thermonuclear fusion, converting $620 \times 10^9$ kg of hydrogen gas into $616 \times 10^9$ kg of helium gas every second. The missing $4 \times 10^9$ kg of rest mass is converted into energy. The primary sequence of nuclear reactions is the proton-proton chain, which operates at about $1.5 \times 10^7$ K. In hotter stars, up to $10^8$ K, the dominant energy source is the carbon cycle.

Proton-proton chain: Two protons fuse together and one decays into a neutron, forming a deuteron with the emission of a positron and a neutrino. A third proton fuses with the deuteron, forming helium-3 and emitting a gamma ray photon. Two helium-3 nuclei fuse, forming one nucleus of helium-4 and two protons. The protons repeat the cycle.

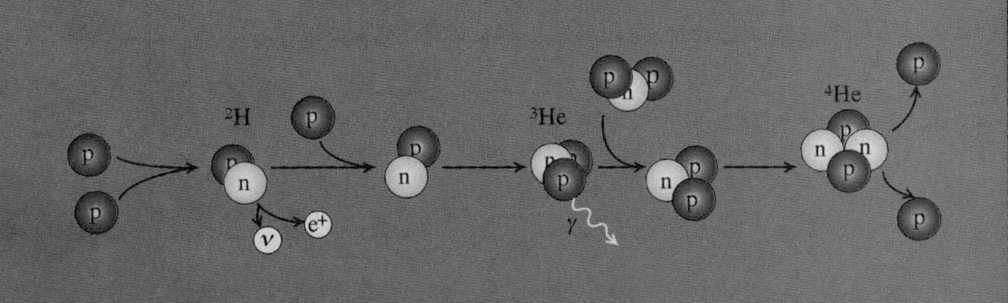

## KEY CONCEPTS

**• Nuclei are composed of protons and neutrons. All nuclei have about the same density. The mass of a nucleus is slightly less than the total mass of its constituents. Nuclei have angular momentum and magnetic moment.**

**• Nuclear stability is determined by the competition between the attractive nuclear force and the electrical repulsion of the protons. A nucleus is stable only if its neutron-proton ratio is within a certain range and the total number of nucleons is not too large.**

**• Unstable nuclei decay by emitting an alpha or beta particle, sometimes followed by a gamma photon. There may be a series of decays until a stable structure is reached. The rate of decay is described by the half-life or the decay constant.**

**• Radiation has a variety of effects on living tissue—some beneficial, some harmful.**

**• In a nuclear reaction, two nuclei or particles collide, yielding two new nuclei or particles.**

**• Nuclear fission is the splitting of a heavier nucleus into two smaller nuclei, which are always unstable beta emitters. Nuclear fusion is the joining of two light nuclei into a new larger nucleus.**

## INTRODUCTION

During this century, nuclear physics has had enormous effects on humankind, some beneficial, some catastrophic. Many people have strong opinions about the uses of nuclear physics. Ideally, those opinions should be based on understanding, not on prejudice or emotion, and we hope that this chapter will help in reaching that ideal.

Every atom contains at its center an extremely dense, positively charged *nucleus,* much smaller than the overall size of the atom but containing most of its total mass. We will look at several important general properties of nuclei and of the nuclear force that holds them together. The stability or instability of a particular nucleus is determined by the competition between the attractive nuclear force among the protons and neutrons and the repulsive electrical interactions among the protons. Unstable nuclei *decay,* transforming themselves spontaneously into other structures by a variety of decay processes. Structure-altering nuclear reactions can also be induced by impact on a nucleus of a particle or another nucleus. Two classes of reactions of special interest are *fission* and *fusion*.

## 45–1 PROPERTIES OF NUCLEI

The most obvious feature of the atomic nucleus is its size—20,000 to 200,000 times smaller than the atom itself. Since Rutherford's initial experiments (Section 40–4), many additional scattering experiments have been performed, using high-energy protons, electrons, and neutrons as well as alpha particles (helium nuclei). We can model the nucleus as a sphere; its radius $R$ depends on the mass, which in turn depends on the total number $A$ of nucleons (neutrons and protons) in the nucleus. The number $A$ is called the **mass number,** or the *nucleon number.* The radii of most nuclei are represented quite well by the equation

$$R = R_0 A^{1/3}, \tag{45–1}$$

where $R_0$ is an empirical constant:

$$R_0 = 1.2 \times 10^{-15}\ \text{m} = 1.2\ \text{fm}.$$

### Nuclear Density

The volume $V$ of a sphere is equal to $4\pi R^3/3$, so Eq. (45–1) shows that the *volume* of a nucleus is proportional to $A$ (that is, to the total mass). Therefore *the mass per unit volume* (proportional to $A/R^3$) is the same for all nuclei. That is, *all nuclei have approximately the same density.* This fact is of crucial importance in understanding nuclear structure.

### EXAMPLE 45–1

The most common isotope of iron has $A = 56$. Find the radius, approximate mass, and approximate density of the nucleus.

**SOLUTION** From Eq. (45–1) the radius is

$$R = R_0 A^{1/3} = (1.2 \times 10^{-15}\ \text{m})(56^{1/3}) = 4.6 \times 10^{-15}\ \text{m}.$$

Neglecting the mass difference between protons and neutrons and approximating each mass as $1.67 \times 10^{-27}$ kg, we find

$$M = (56)(1.67 \times 10^{-27}\ \text{kg}) = 9.4 \times 10^{-26}\ \text{kg}.$$

(This calculation also ignores the effect of the *binding energy* of the nucleus; we'll return to this concept later.) The volume is

$$V = \tfrac{4}{3}\pi R^3 = \tfrac{4}{3}\pi(4.6 \times 10^{-15}\ \text{m})^3 = 4.1 \times 10^{-43}\ \text{m}^3,$$

and the density $\rho$ is

$$\rho = \frac{m}{V} = \frac{9.4 \times 10^{-26}\ \text{kg}}{4.1 \times 10^{-43}\ \text{m}^3} = 2.3 \times 10^{17}\ \text{kg/m}^3.$$

The density of the *element* iron is about 8000 kg/m$^3$, so we see that the nucleus is of the order of $10^{13}$ times as dense as the bulk material. Densities of this magnitude are also found in so-called white dwarf stars, which are similar to gigantic nuclei. A 1-cm cube of material with this density would have a mass of $2.3 \times 10^{11}$ kg, or 230 million metric tons!

### Nuclear Masses

The basic building blocks of the nucleus are the *proton* and the *neutron*. The total number of protons, equal in a neutral atom to the number of electrons, is the **atomic number** $Z$. The number of neutrons, denoted by $N$, is called the **neutron number.** For any nucleus the mass number $A$ is the sum of the atomic number and the neutron number:

$$A = Z + N. \tag{45–2}$$

A single nuclear species (having specific values of both $Z$ and $N$) is called a **nuclide.** Table 45–1 lists values of $A$, $Z$, and $N$ for a few nuclides. The electron structure of an atom, which is responsible for its chemical properties, is determined by the charge $Ze$ of the nucleus. The table shows some nuclei with the same $Z$ but different $N$. These are nuclei of the same element, but they have different masses. *Isotopes* of an element are forms with different numbers of neutrons in their nuclei. A familiar example is chlorine (Cl, $Z = 17$). About 76% of chlorine nuclei have $N = 18$; the other 24% have $N = 20$.

**TABLE 45–1 Compositions of Some Common Nuclides**

| Nucleus | Mass number (total number of nuclear particles), $A$ | Atomic number (number of protons), $Z$ | Neutron number, $N = A - Z$ |
|---|---|---|---|
| ${}^{1}_{1}\mathrm{H}$ | 1 | 1 | 0 |
| ${}^{2}_{1}\mathrm{D}$ | 2 | 1 | 1 |
| ${}^{4}_{2}\mathrm{He}$ | 4 | 2 | 2 |
| ${}^{6}_{3}\mathrm{Li}$ | 6 | 3 | 3 |
| ${}^{7}_{3}\mathrm{Li}$ | 7 | 3 | 4 |
| ${}^{9}_{4}\mathrm{Be}$ | 9 | 4 | 5 |
| ${}^{10}_{5}\mathrm{B}$ | 10 | 5 | 5 |
| ${}^{11}_{5}\mathrm{B}$ | 11 | 5 | 6 |
| ${}^{12}_{6}\mathrm{C}$ | 12 | 6 | 6 |
| ${}^{13}_{6}\mathrm{C}$ | 13 | 6 | 7 |
| ${}^{14}_{7}\mathrm{N}$ | 14 | 7 | 7 |
| ${}^{16}_{8}\mathrm{O}$ | 16 | 8 | 8 |
| ${}^{23}_{11}\mathrm{Na}$ | 23 | 11 | 12 |
| ${}^{65}_{29}\mathrm{Cu}$ | 65 | 29 | 36 |
| ${}^{200}_{80}\mathrm{Hg}$ | 200 | 80 | 120 |
| ${}^{235}_{92}\mathrm{U}$ | 235 | 92 | 143 |
| ${}^{238}_{92}\mathrm{U}$ | 238 | 92 | 146 |

Different isotopes of an element have slightly different physical properties, such as melting and boiling temperatures and diffusion rates. The two common isotopes of uranium with $A = 235$ and $A = 238$ are separated industrially by taking advantage of the different diffusion rates of gaseous uranium hexafluoride ($UF_6$) containing the two isotopes.

Table 45–1 also shows the usual notation for individual nuclides: the symbol of the element, with a pre-subscript equal to $Z$ and a pre-superscript equal to the mass number $A$. The general format for an element El is ${}^{A}_{Z}\mathrm{El}$. The isotopes of chlorine mentioned above, with $A = 35$ and $A = 37$, respectively, are written ${}^{35}_{17}\mathrm{Cl}$ and ${}^{37}_{17}\mathrm{Cl}$. This notation is redundant because the name of the element determines the atomic number $Z$, but it is a useful aid to memory. The value of $Z$ is sometimes omitted.

The proton and neutron masses are

$$m_{\mathrm{p}} = 1.67262 \times 10^{-27}\ \mathrm{kg},$$

$$m_{\mathrm{n}} = 1.67493 \times 10^{-27}\ \mathrm{kg}.$$

These are nearly equal, so it is not surprising that many nuclear masses are approximately integer multiples of the proton or neutron mass. For precise measurements it is useful to define a new mass unit equal to $\frac{1}{12}$ of the mass of the neutral carbon atom with mass number $A = 12$. This mass is called one *atomic mass unit* (1 u):

$$1\ \mathrm{u} = 1.66054 \times 10^{-27}\ \mathrm{kg}.$$

In atomic mass units the masses of the proton, neutron, electron, and hydrogen atom are:

$$m_{\mathrm{p}} = 1.0072765\ \mathrm{u},$$

$$m_{\mathrm{n}} = 1.0086649\ \mathrm{u},$$

$$m_{\mathrm{e}} = 0.0005485799\ \mathrm{u},$$

$$m_{\mathrm{H}} = m_{\mathrm{p}} + m_{\mathrm{e}} = 1.0078250\ \mathrm{u}.$$

**TABLE 45–2 Atomic Masses of Light Elements**

| Element | Atomic number, $Z$ | Neutron number, $N$ | Atomic mass, u | Mass number, $A$ |
|---|---|---|---|---|
| Hydrogen (H) | 1 | 0 | 1.00783 | 1 |
| Deuterium (H) | 1 | 1 | 2.01410 | 2 |
| Helium (He) | 2 | 1 | 3.01603 | 3 |
| Helium (He) | 2 | 2 | 4.00260 | 4 |
| Lithium (Li) | 3 | 3 | 6.01512 | 6 |
| Lithium (Li) | 3 | 4 | 7.01600 | 7 |
| Beryllium (Be) | 4 | 5 | 9.01218 | 9 |
| Boron (B) | 5 | 5 | 10.01294 | 10 |
| Boron (B) | 5 | 6 | 11.00931 | 11 |
| Carbon (C) | 6 | 6 | 12.00000 | 12 |
| Carbon (C) | 6 | 7 | 13.00336 | 13 |
| Nitrogen (N) | 7 | 7 | 14.00307 | 14 |
| Nitrogen (N) | 7 | 8 | 15.00011 | 15 |
| Oxygen (O) | 8 | 8 | 15.99491 | 16 |
| Oxygen (O) | 8 | 9 | 16.99913 | 17 |
| Oxygen (O) | 8 | 10 | 17.99916 | 18 |

*Source: Encyclopedia of Physics,* Lerner and Trigg, eds. (Reading, Mass.: Addison-Wesley, 1981.)

These values are more precise than the u-to-kg conversion factor because atomic masses can be compared to each other with greater precision than they can be compared to the standard kilogram.

We can find the energy equivalent of 1 u from the relation $E = mc^2$:

$$E = (1.66054 \times 10^{-27}\ \text{kg})\,(2.99792 \times 10^{8}\ \text{m/s})^2$$
$$= 1.49242 \times 10^{-10}\ \text{J} = 931.494\ \text{MeV}.$$

The masses of some common atoms, including their electrons, are shown in Table 45–2. Such tables always give masses of *neutral* atoms (with $Z$ electrons) rather than masses of *bare* nuclei because it is much more difficult to measure masses of bare nuclei with high precision. To obtain the mass of a bare nucleus, subtract $Z$ times the electron mass.

## Nuclear Spin and Magnetic Moment

We discussed the intrinsic angular momentum of the electron in Section 43–3. It is a spin-$\frac{1}{2}$ particle; the magnitude $S$ of its spin angular momentum is

$$S = \sqrt{\tfrac{1}{2}(\tfrac{1}{2} + 1)}\hbar = \sqrt{\tfrac{3}{4}}\hbar, \tag{45–3}$$

and the $z$-component is

$$S_z = \pm\tfrac{1}{2}\hbar. \tag{45–4}$$

Protons and neutrons are also spin-$\frac{1}{2}$ particles, with spin angular momentum described by Eqs. (45–3) and (45–4). Thus we expect the nucleus to have angular momentum. In addition to the spin angular momentum, there may be *orbital* angular momentum associated with the motions of the nucleons within the nucleus. This is quantized in the same way as the orbital angular momentum of electrons in atoms.

The *total* nuclear angular momentum $\boldsymbol{J}$ has magnitude

$$J = \sqrt{j(j + 1)}\hbar \tag{45–5}$$

and $z$-component

$$J_z = m_j\hbar \quad (m_j = -j, -j+1, \ldots, j-1, j). \tag{45–6}$$

When the total number of nucleons $A$ is *even,* $j$ is an integer; when it is *odd,* $j$ is a half-integer. All nuclei for which both $Z$ and $N$ are even have $J = 0$, suggesting that pairing of particles with opposite spins may be an important consideration in nuclear structure. Nuclear angular momentum is usually called *nuclear spin,* even though in general it is a combination of orbital and spin angular momenta.

Associated with nuclear spin is a *magnetic moment.* When we discussed *electron* magnetic moments in Section 43–3, we introduced the Bohr magneton $\mu_B = e\hbar/2m_e$ as a natural unit of magnetic moment. We found that the electron-spin magnetic moment (more precisely, the magnitude of its component in a given direction, such as that of the $z$ axis) is almost exactly equal to $\mu_B$. In discussing *nuclear* magnetic moments we can define an analogous quantity, the **nuclear magneton** $\mu_n$:

$$\mu_n = \frac{e\hbar}{2m_p} = 5.05079 \times 10^{-27}\ \text{J/T}, \tag{45–7}$$

where $m_p$ is the proton mass. This unit is smaller than the Bohr magneton by a factor of $m_e/m_p = 1/1836$.

We might expect the spin magnetic moment of the proton to be approximately $\mu_n$. Instead, it turns out to be $2.7928\mu_n$. Even more surprising, the neutron, which has no charge, has a magnetic moment of $-1.9130\mu_n$. (The negative sign indicates that the magnetic moment is opposite in direction to the angular momentum.) These anomalous magnetic moments are related to the fact that the proton and neutron aren't really fundamental particles; they are made of particles called *quarks,* which we'll discuss in detail in the next chapter.

Nuclear magnetic moments are typically a few nuclear magnetons. When a nucleus is placed in a magnetic field $\boldsymbol{B}$, there is an interaction energy $U = -\boldsymbol{\mu} \cdot \boldsymbol{B}$ just as with atomic magnetic moments. The components of angular momentum and magnetic moment in the direction of the field are quantized, so a series of energy levels results from this interaction. The interactions of nuclear magnetic moments with electron magnetic moments in atoms cause additional splittings in atomic energy levels and spectra, an effect called *hyperfine structure.*

## EXAMPLE 45–2

Hydrogen atoms are placed in a magnetic field with magnitude 2.00 T. a) What is the energy difference between the states with nuclear spin component parallel and those with nuclear spin component antiparallel to the field (ignoring electron spin)? b) The protons can make transitions from one of these states to the other by emitting or absorbing a photon with energy equal to the energy difference of the two states. Find the frequency and wavelength of such a photon.

**SOLUTION** a) When the spin component is parallel to the field, the interaction energy is

$$U = -\mu_z B = -(2.7928)(5.05 \times 10^{-27}\ \text{J/T})(2.00\ \text{T})$$
$$= -2.82 \times 10^{-26}\ \text{J} = -1.76 \times 10^{-7}\ \text{eV}.$$

When the spin component is antiparallel to the field, the energy is $+2.82 \times 10^{-26}$ J, and the energy *difference* between the two states is

$$\Delta E = 2(2.82 \times 10^{-26}\ \text{J}) = 5.64 \times 10^{-26}\ \text{J}.$$

b) The corresponding photon frequency and wavelength are

$$f = \frac{E}{h} = \frac{5.64 \times 10^{-26}\ \text{J}}{6.626 \times 10^{-34}\ \text{J} \cdot \text{s}}$$
$$= 8.51 \times 10^{7}\ \text{Hz} = 85.1\ \text{MHz},$$

$$\lambda = \frac{c}{f} = \frac{3.00 \times 10^{8}\ \text{m/s}}{8.51 \times 10^{7}\ \text{s}^{-1}}$$
$$= 3.52\ \text{m}.$$

This frequency is just below the FM radio band. When a hydrogen specimen is placed in a 2.00-T magnetic field and then irradiated with radiation of this frequency, the flipping of the proton spins can be detected by the absorption of energy from the radiation.

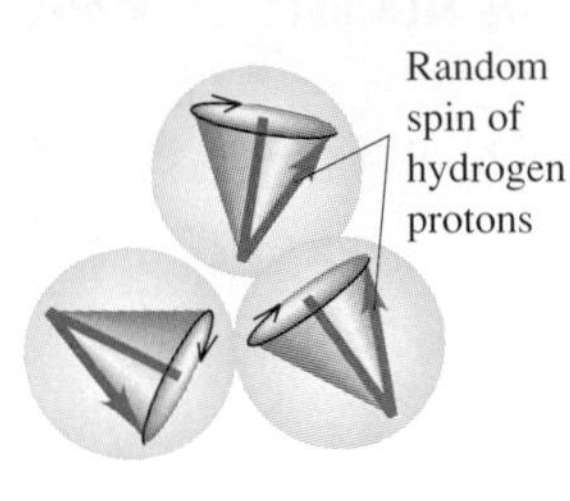


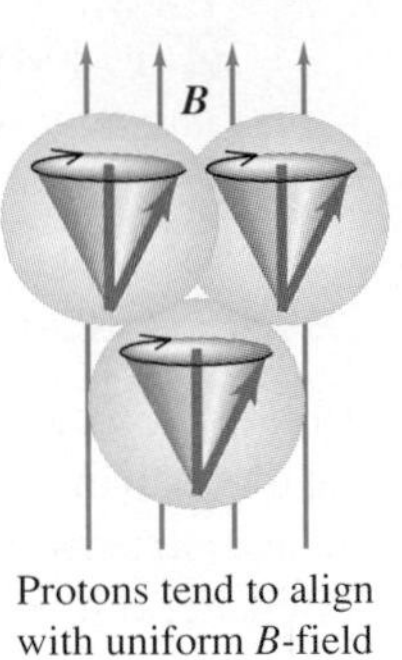


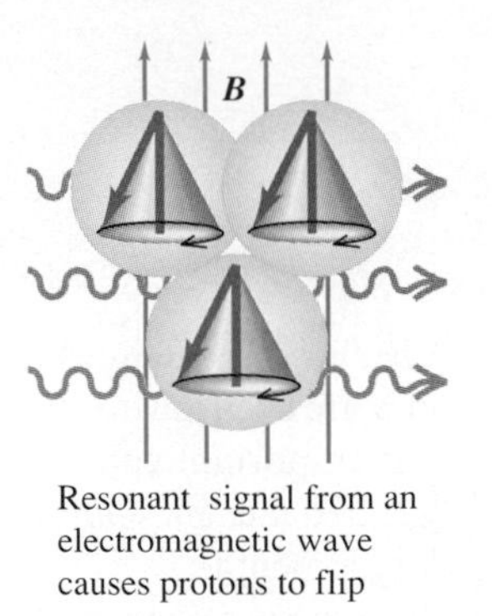


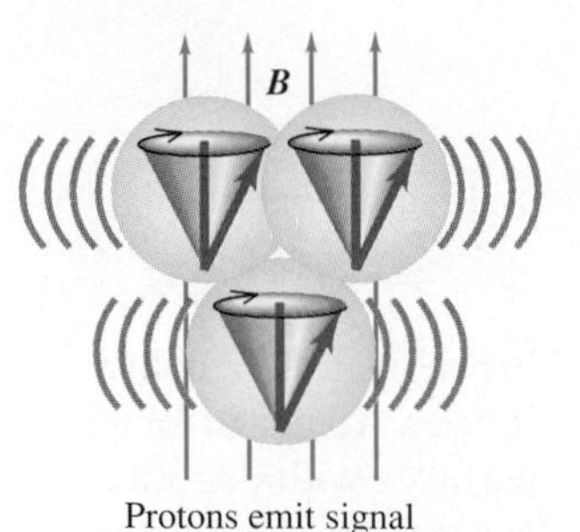


(a)

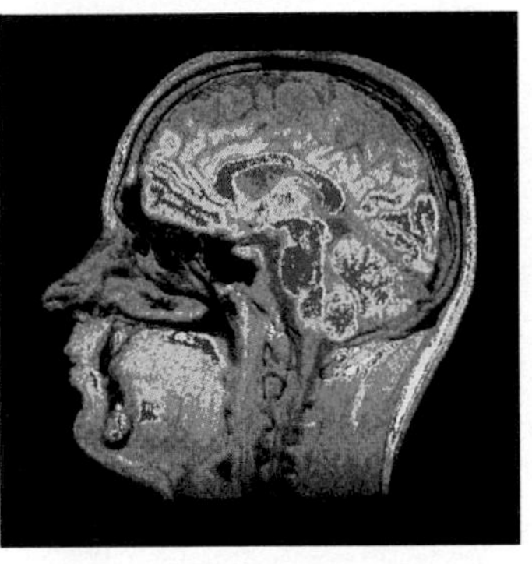

(b)

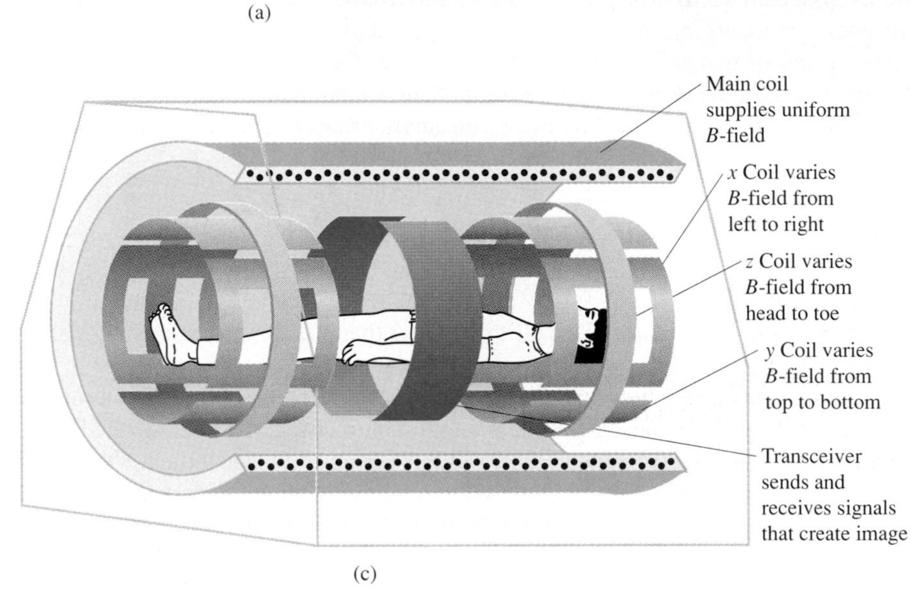


(c)

**45–1** Magnetic-resonance imaging (MRI). (a) Protons, the nuclei of hydrogen atoms in the tissue under study, normally have random spin orientations. In the presence of a strong magnetic field, they become aligned parallel to the field. A brief radio signal flips the spins; as they reorient themselves parallel to the field, they emit signals that are picked up by sensitive detectors. The differing magnetic environment in various regions permits reconstruction of an image showing the types of tissue present. (b) A color-enhanced MRI image showing a cross section through a patient's head. (c) An electromagnet used for MRI imaging.

Experiments of the sort referred to in Example 45–2 are called *nuclear magnetic resonance* (NMR). They have been carried out with many different nuclei. Frequencies and magnetic fields can be measured very precisely, so this technique permits precise measurements of nuclear magnetic moments. An elaboration of this basic idea leads to *magnetic resonance imaging* (MRI), a noninvasive imaging technique that discriminates among various body tissues on the basis of the differing magnetic environments of protons in the tissues. The principles of MRI are shown in Fig. 45–1.

## 45–2 BINDING ENERGY AND STABILITY

The total mass of a nucleus is always *less than* the total mass of its constituent parts because of the mass equivalent of the (negative) potential energy associated with the attractive forces that hold the parts together. The magnitude of this difference is called the **mass defect,** denoted by $\Delta M$. The magnitude of the total potential energy is called the **binding energy,** denoted by $E_B$. The mass defect $\Delta M$ for a nucleus with mass $M$ containing $Z$ protons and $N$ neutrons is defined as

$$\Delta M = Zm_p + Nm_n - M. \tag{45–8}$$

The binding energy is the mass defect multiplied by $c^2$; that is, $E_B = \Delta Mc^2$. With this definition, both $\Delta M$ and $E_B$ are positive, although the actual potential energy is always negative.

The simplest nucleus, that of hydrogen, is a single proton. Next comes the nucleus of $^2_1H$, the isotope of hydrogen with mass number 2, usually called *deuterium.* Its nucleus consists of a proton and a neutron bound together to form a particle called the *deuteron*. To find the binding energy of the deuteron, we calculate the mass defect $\Delta M$, using the values in Table 45–2:

$$\Delta M = m_H + m_n - m_D$$
$$= 1.00783\ u + 1.00866\ u - 2.01410\ u = 0.00239\ u.$$

(Note that $m_H$ and $m_D$ each include one electron.) The energy equivalent is

$$E = (0.00239\ u)(931.5\ MeV/u) = 2.23\ MeV.$$

This much energy would be required to pull the deuteron apart into a proton and a neutron. This value is unusually small; binding energies for most nuclei are about 8 MeV per nucleon.

## PROBLEM-SOLVING STRATEGY

### Nuclear structure

**1.** Familiarity with numerical magnitudes is helpful. The scale of things in nuclear structures is very different from that of atomic structures. The size of a nucleus is of the order of $10^{-15}$ m; the potential energy of interaction of two protons at this distance is $2.31 \times 10^{-13}$ J, or 1.44 MeV. Typical nuclear interaction energies are of the order of a few million electronvolts (MeV), rather than a few electronvolts (eV) as with atoms. Protons and neutrons are about 1840 times as massive as electrons. The binding energy per nucleon is roughly 1% of the rest energy of a nucleon. For comparison, the ionization energy of the hydrogen atom is only 0.003% of the electron's rest energy.

**2.** Angular momentum is of the same order of magnitude in both atoms and nuclei because it is determined by the value of Planck's constant $h$. But magnetic moments of nuclei are typically much *smaller* than those of electrons in atoms because the nuclear gyromagnetic ratio (the ratio of magnetic moment to angular momentum) is of the order of $e/2m_p$ instead of $e/2m_e$, almost 2000 times smaller than for orbital electron motion.

**3.** When doing energy calculations involving the mass defect, binding energies, and so on, note that mass tables nearly always list the masses of *neutral* atoms, including their full complements of electrons. To get the mass of a bare nucleus, you have to subtract the masses of these electrons. The binding energies of the electrons are much smaller, and we won't worry about them. Calculations of binding energy often involve subtracting two nearly equal quantities. To get enough precision in the difference, you often have to carry five or six significant figures, if that many are available. If not, you may have to be content with an approximate result.

**4.** Nuclear masses are usually measured in atomic mass units (u). A useful conversion factor is that the energy equivalent of 1 u is 931.5 MeV.

## EXAMPLE 45–3

**The carbon nucleus** Find the mass defect, the total binding energy, and the binding energy per nucleon for the common isotope of carbon, $^{12}_6C$.

**SOLUTION** The mass of the neutral carbon atom, including the nucleus and six electrons, is (by definition of the atomic mass unit) 12.00000 u. We obtain the mass of the bare nucleus by subtracting the mass of the six electrons:

$$m = 12.00000\ u - (6)(0.000549\ u) = 11.996706\ u.$$

The total mass of the six protons and six neutrons in the nucleus is

$$(6)(1.007276\ u) + (6)(1.008665\ u) = 12.095646\ u.$$

The mass defect is therefore

$$12.095646\ u - 11.996706\ u = 0.09894\ u.$$

The energy equivalent of this mass is

$$(0.09894\ u)(931.5\ MeV/u) = 92.16\ MeV.$$

The total binding energy for the 12 nucleons is 92.16 MeV. To pull the carbon nucleus completely apart into 12 separate nucleons would require a minimum of 92.16 MeV. The binding energy *per nucleon* is $\frac{1}{12}$ of this, or 7.68 MeV per nucleon.

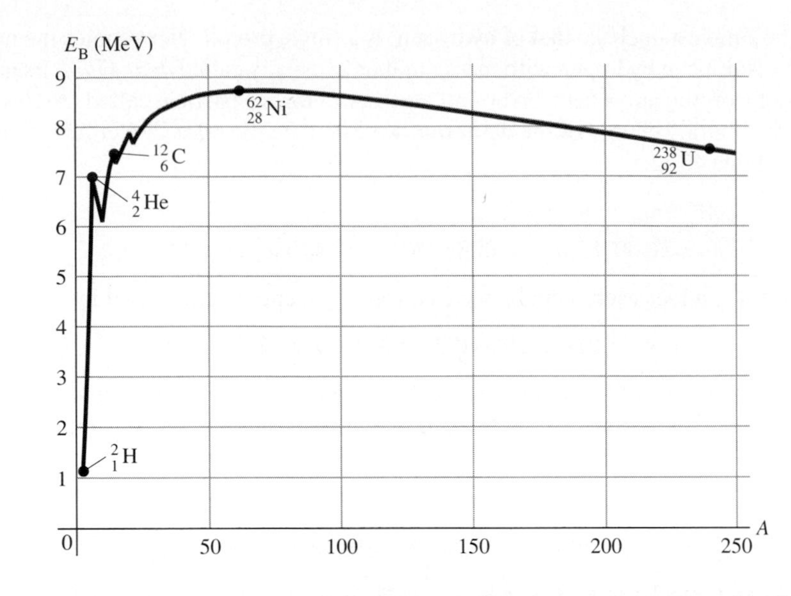


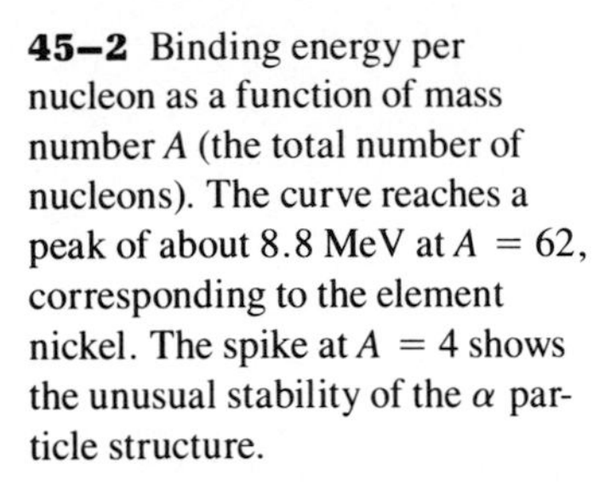
**45–2** Binding energy per nucleon as a function of mass number $A$ (the total number of nucleons). The curve reaches a peak of about 8.8 MeV at $A = 62$, corresponding to the element nickel. The spike at $A = 4$ shows the unusual stability of the $\alpha$ particle structure.

Nearly all stable nuclei, from the lightest to the most massive, have binding energies in the range of 6 to 9 MeV per nucleon; the largest values occur near $A = 60$. Figure 45–2 is a graph of binding energy per nucleon as a function of the mass number $A$. Note the spike at $A = 4$, showing the unusually large binding energy of the helium nucleus (alpha particle).

The energy of internal motion of a nucleus is quantized. Each nucleus has a set of allowed energy levels, including a *ground state* (state of lowest energy) and several *excited states.* Because of the great strength of nuclear interactions, excitation energies of nuclei are typically of the order of 1 MeV, compared with a few electronvolts for atomic energy levels. In ordinary physical and chemical transformations, the nucleus always remains in its ground state. When a nucleus is placed in an excited state, either by bombardment with high-energy particles or by a radioactive transformation, it can decay to the ground state by emission of one or more photons called **gamma rays,** or *gamma-ray photons,* with typical energies of the order of 1 MeV.

## The Nuclear Force

The forces that hold protons and neutrons together in the nucleus, despite the electrical *repulsion* of the protons, are an example of the *strong interaction* we mentioned in Section 5–6. In the context of nuclear structure this interaction is called the *nuclear force.* Here are some of its characteristics. First, it does not depend on charge; neutrons as well as protons are bound, and the binding is the same for both. Second, the nuclear force has short range, of the order of nuclear dimensions, that is, $10^{-15}$ m. (Otherwise, the nucleus would pull in additional protons and neutrons.) But within its range, the nuclear force is much stronger than electrical forces; otherwise, the nucleus could never be stable. Third, the nearly constant density of nuclear matter and the nearly constant binding energy per nucleon show that a particular nucleon cannot interact simultaneously with *all* the other nucleons in a nucleus, but only with those few in its immediate vicinity. This is different from electrical forces; *every* proton in the nucleus repels every other one. This limited number of interactions is called *saturation;* it is analogous to covalent bonding in molecules and solids. Finally, the nuclear force favors binding of *pairs* of protons or neutrons with opposite spins and of *pairs of pairs* (that is, a pair of protons and a pair of neutrons, each pair having opposite spins). For example, the alpha particle (two protons and two neutrons) is an exceptionally stable nucleus. We'll see other evidence for pairing effects later.

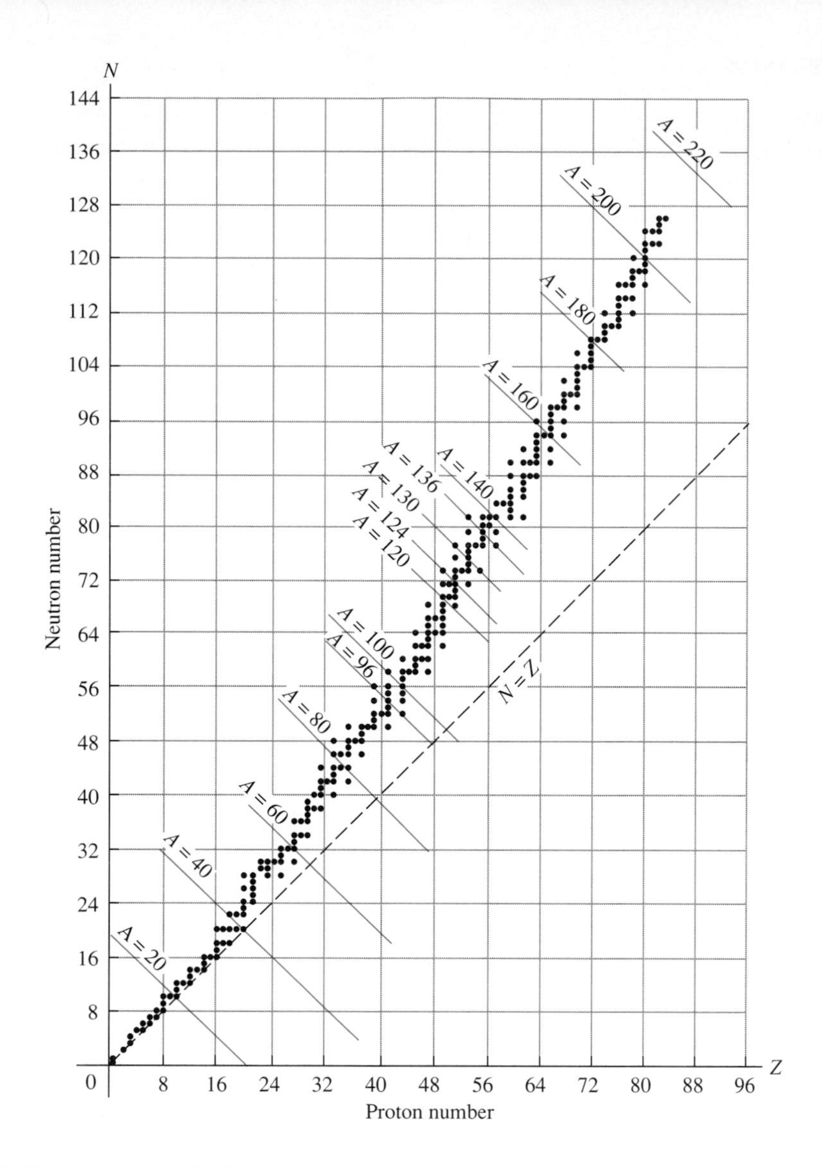


**45–3** Segrè chart, showing neutron number and proton number for stable nuclides. The number of neutrons exceeds the number of protons by an amount that increases with atomic number $Z$.

## Stability of Nuclei

Among about 2500 known nuclides, less than 300 are stable. The others are **radioactive;** they are unstable structures that decay to form other nuclides by emitting particles and electromagnetic radiation. The time scale of these decay processes ranges from a small fraction of a microsecond to billions of years. The *stable* nuclides are shown by dots on the graph in Fig. 45–3, where the neutron number $N$ and proton number (or atomic number) $Z$ for each nuclide are plotted. Such a chart is called a *Segrè chart,* after its inventor, Emilio Segrè.

Each blue line perpendicular to the line $N = Z$ represents a specific value of the mass number $A = Z + N$. Most lines of constant $A$ pass through only one or two stable nuclides; that is, there are usually only one or two stable nuclides with a given mass number. The lines at $A = 20$, $A = 40$, $A = 60$, and $A = 80$ are examples. In four cases these lines pass through *three* stable nuclides, namely, at $A = 96$, $A = 124$, $A = 130$, and $A = 136$. Only four stable nuclides have both odd $Z$ and odd $N$:

$$^{2}_{1}\text{H}, \qquad ^{6}_{3}\text{Li}, \qquad ^{10}_{5}\text{B}, \qquad ^{14}_{7}\text{N}.$$

These are called *odd-odd nuclides.* The absence of other odd-odd nuclei shows the influence of pairing. Also, there is *no* stable nuclide with $A = 5$ or $A = 8$. The $^{4}$He nucleus (alpha particle), with a pair of protons and a pair of neutrons, has no interest in accepting a fifth particle into its structure, and a collection of eight nucleons splits immediately into two $^4$He nuclei.

The points representing stable nuclides define a rather narrow stability region. For low mass numbers the numbers of protons and neutrons are approximately equal; that is, $N \cong Z$. The ratio $N/Z$ increases gradually with $A$, up to about 1.6 at large mass numbers. Points to the right of the stability region represent nuclides that have too many protons or not enough neutrons to be stable. To the left are nuclei with too many neutrons or not enough protons. The graph also shows that no nuclide with $A > 209$ or $Z > 83$ is stable. We also note that there is *no* stable nuclide with $Z = 43$ (technetium) or 61 (promethium). Figure 45–4, a three-dimensional version of the Segrè chart, shows the "valley of stability" for light nuclei (up to $Z = 22$).

The most important reason why some nuclei are stable and others are not is the competition between the attractive nuclear force and the repulsive electrical force. The nuclear force favors *pairs* of nucleons and *pairs of pairs*. If there were no electrical interactions, the most stable nuclei would be those with even and equal numbers of neutrons and protons, or $N = Z$. The electrical repulsion shifts the balance to favor greater numbers of neutrons, but a nucleus with *too many* neutrons is unstable because not enough of them are paired with protons. A nucleus with too many *protons* has too much repulsive electrical interaction, compared with the attractive nuclear interaction, to be stable.

As the number of nucleons increases, the total energy of electrical interaction increases faster than that of the nuclear interaction. To understand this, recall the discussion of electrostatic energy in Section 25–4. The energy of a capacitor with charge $Q$ is proportional to $Q^2$. The energy required to bring a total charge $Q$ together to form a spherical charge distribution with radius $R$ is proportional to $Q^2/R$. In the nucleus the (positive) electric potential energy increases approximately as $Z(Z-1)$, and the (negative) nuclear potential energy increases approximately as $A$, with corrections for surface and pairing effects.

As $A$ increases, the positive electric potential energy *per nucleon* grows faster than the negative nuclear potential energy per nucleon, until the point is reached at which stability is impossible. Thus the competition of electric and nuclear forces accounts for the fact that the neutron-proton ratio in stable nuclei increases with $Z$, and also for the fact that a nucleus cannot be stable if $A$ or $Z$ is too large. In later sections we'll describe various ways in which unstable nuclei decay.

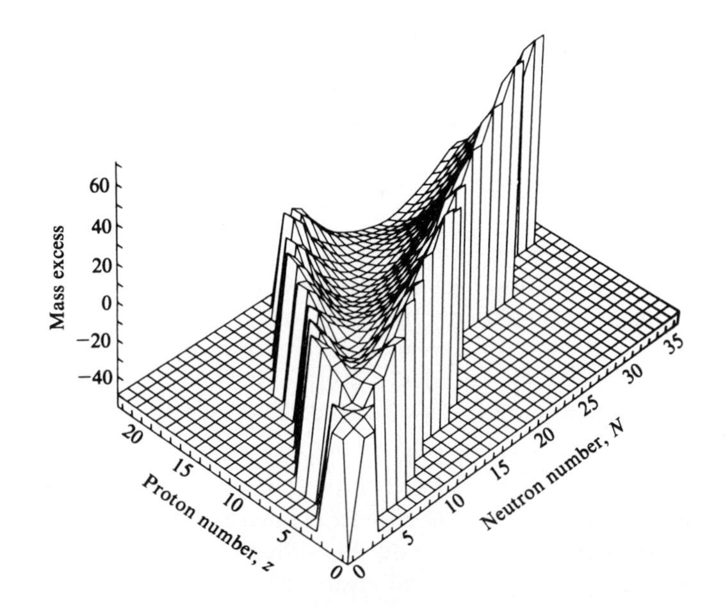


**45–4** A three-dimensional Segrè chart for light nuclides, up to $Z = 22$ (titanium). The quantity plotted on the third axis is the mass excess $(M - A)c^2$, where $M$ is the nuclide mass expressed in u. For light nuclides this is nearly equal to the negative of the binding energy, apart from an additive constant. The lowest points on the surface represent the most stable nuclides.

# 45–3 NUCLEAR STRUCTURE

The analysis of nuclear structure is more complex than the analysis of many-electron atoms. Two different kinds of interactions are involved (electrical and nuclear), and the nuclear force is not yet completely understood. Even so, we can gain some insight into nuclear structure by the use of simple models. We'll discuss briefly two rather different but successful models: the *liquid-drop model* and the *shell model*.

## The Liquid-Drop Model

The **liquid-drop model,** first proposed in 1936, is suggested by the fact that all nuclei have nearly the same density. The individual nucleons are analogous to molecules in a liquid, held together by short-range interactions and surface-tension effects. We can use this simple picture to derive a formula for the total binding energy of a nucleus. We'll include four contributions.

1. We've remarked that nuclear forces show *saturation;* an individual nucleon interacts only with a few of its nearest neighbors. As a first approximation, the *total* nuclear interaction energy should thus be proportional to the number of nucleons; we'll write this energy as $C_1A$, where $C_1$ is a constant to be determined.
2. The nucleons on the surface of the nucleus are less tightly bound than those in the interior because they have no neighbors outside the surface. So there should be a *negative* correction that is proportional to the surface area $4\pi R^2$. Because $R$ is proportional to $A^{1/3}$, this correction is proportional to $A^{2/3}$; we write it as $-C_2A^{2/3}$, where $C_2$ is another constant.
3. Every proton repels every other proton. The total repulsive electrostatic potential energy is proportional to $Z(Z-1)$ and inversely proportional to the radius $R$ and thus to $A^{1/3}$. This correction to the binding energy is negative because the nucleons are less tightly bound than they would be without the electrical repulsion. We write this correction as $-C_3Z(Z-1)/A^{1/3}$.
4. Finally, the nuclear force favors *pairing* of protons and neutrons. We need a negative correction to the nuclear interaction corresponding to the number of unpaired nucleons $|N-Z|$. The best agreement with observed binding energies is obtained by assuming this energy to be proportional to $(N-Z)^2/A$. Using $N = A - Z$ to express this in terms of $A$ and $Z$, we write the pairing correction as $-C_4(A-2Z)^2/A$.

The total binding energy $E_B$ is the sum of these four terms:

$$E_B = C_1A - C_2A^{2/3} - C_3\frac{Z(Z-1)}{A^{1/3}} - C_4\frac{(A-2Z)^2}{A}. \qquad (45\text{–}9)$$

The constants $C_1$, $C_2$, $C_3$, and $C_4$ are chosen to make this formula fit the observed binding energies of nuclei. The best values for these constants are

$$C_1 = 15.7 \text{ MeV},$$
$$C_2 = 17.8 \text{ MeV},$$
$$C_3 = 0.71 \text{ MeV},$$
$$C_4 = 23.7 \text{ MeV}.$$

The constant $C_1$ is the binding energy per nucleon due to the nuclear force, apart from surface and pairing corrections. This is almost 16 MeV, about double the *total* binding energy per nucleon in most nuclei.

## EXAMPLE 45–4

The common isotope of iron is $^{56}_{26}\mathrm{Fe}$. The measured mass of the neutral atom is 55.934939 u. a) Determine the mass defect, and from it find the total binding energy and the binding energy per nucleon. b) Calculate the binding energy from Eq. (45–9). Compare the two results.

**SOLUTION** a) We have $Z = 26$, $N = 30$, and $A = 56$. The mass defect is $26m_p + 30m_n - M_{\mathrm{Fe}}$, where $M_{\mathrm{Fe}}$ is the mass of the *bare* nucleus. This is easier to evaluate if we add and subtract 26 electrons; we then use the mass $m_H$ of the hydrogen atom in place of $m_p$ and the mass of the *neutral* iron atom in place of the mass of the bare nucleus. The mass defect is

$$(26)(1.00783\ \mathrm{u}) + (30)(1.008665\ \mathrm{u}) - 55.934939\ \mathrm{u} = 0.52859\ \mathrm{u}.$$

The energy equivalent of this is

$$(0.5286\ \mathrm{u})(931.5\ \mathrm{MeV/u}) = 492\ \mathrm{MeV}.$$

The binding energy per nucleon is (492 MeV)/56 = 8.79 MeV.

b) We substitute the given numbers into Eq. (45–9). The individual terms are:

1. $C_1A = (15.7\ \mathrm{MeV})(56) = 879.2\ \mathrm{MeV}$;
2. $-C_2A^{2/3} = -(17.8\ \mathrm{MeV})(56)^{2/3} = -260.5\ \mathrm{MeV}$;
3. $-C_3\dfrac{Z(Z-1)}{A^{1/3}} = -(0.71\ \mathrm{MeV})\dfrac{(26)(25)}{(56)^{1/3}} = -120.6\ \mathrm{MeV}$;
4. $-C_4\dfrac{(A-2Z)^2}{A} = -(23.7\ \mathrm{MeV})\dfrac{(56-52)^2}{56} = -6.8\ \mathrm{MeV}$.

We note that the pairing correction is small in comparison to the other terms. The total binding energy is the sum of these four terms, or 491 MeV; this agrees well with the value determined directly from the mass.

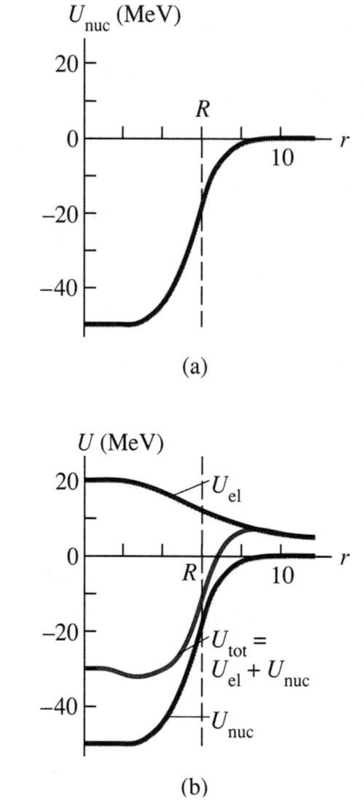

---

We can incorporate Eq. (45–9) into a formula for the mass of any neutral atom:

$$M = Zm_H + Nm_n - E_B. \tag{45–10}$$

Using $m_H$ rather than $m_p$ includes the masses of $Z$ electrons. Equation (45–10) is called the *semiempirical mass formula*. The name is apt; it is *empirical* in the sense that the $C$'s have to be determined empirically (experimentally), yet it does have a sound theoretical basis.

The liquid-drop model and the mass formula derived from it are quite successful in correlating nuclear masses, and we will see later that they are a great help in understanding decay processes of unstable nuclei. Some other aspects of nuclei, such as angular momentum and excited states, are better approached with different models.

## The Shell Model

The **shell model** of nuclear structure is analogous to the central-field approximation (Section 43–4). We picture each nucleon as moving in a potential that represents the averaged-out effect of all the other nucleons. This may not seem to be a very promising approach; the nuclear force is very strong, very short-range, and therefore strongly distance-dependent. However, in some respects, this model turns out to work fairly well.

The potential-energy function for the nuclear force is the same for protons as for neutrons. A reasonable assumption about the shape of this function is shown in Fig. 45–5a, a three-dimensional version of the square well that we discussed in Section 42–3. The corners are somewhat rounded, corresponding to the fact that the nucleus doesn't have a sharply defined surface. For protons there is an additional potential energy associated with electrical repulsion. We represent the average charge distribution as a sphere with uniform charge density, with radius $R$ and total charge $(Z-1)e$. Figure 45–5b shows the nuclear, electrical, and total potential energies for the proton.

In principle, we could solve the Schrödinger equation for a proton or neutron moving in such a potential. In any spherically symmetric potential, the angular-momentum states are the same as for the electrons in the central-field approximation in atomic physics. In particular, we can use the concept of *filled subshells* and their relation to stability. In atomic structure we found that the values $Z = 2, 10, 18, 36, 54, 86$ (the atomic numbers of the inert gases) correspond to particularly stable electron arrangements.



**45–5** Approximate potential-energy functions for nucleons in a nucleus. The approximate nuclear radius is $R$. (a) The potential energy due to the nuclear force. This is the same for protons and neutrons and is the *total* potential energy for neutrons. (b) The total potential energy $U_{\mathrm{tot}}$ for the proton is the sum of the nuclear ($U_{\mathrm{nuc}}$) and electrical ($U_{\mathrm{el}}$) potential energies.

A comparable effect occurs in nuclear structure. The numbers are different because the potential-energy function is different, and the subshells fill up in a different order from that for electrons in an atom. It is found that when the number of neutrons *or* the number of protons is 2, 8, 20, 28, 50, 82, or 126, the resulting structure is unusually stable, that is, has an unusually great binding energy. (Nuclei with $Z = 126$ have not been observed.) These numbers are called *magic numbers*. Nuclei in which $Z$ is a magic number tend to have more stable isotopes than average. There are several nuclei for which both $Z$ and $N$ are magic, including

$$^{4}_{2}\mathrm{He}, \quad ^{16}_{8}\mathrm{O}, \quad ^{40}_{20}\mathrm{Ca}, \quad ^{48}_{20}\mathrm{Ca}, \quad \text{and} \quad ^{208}_{82}\mathrm{Pb}.$$

All these nuclei have substantially larger binding energy per nucleon than do nuclei with neighboring values of $N$ or $Z$. They all have zero nuclear spin. The magic numbers correspond to filled-shell configurations of nucleon energy states.

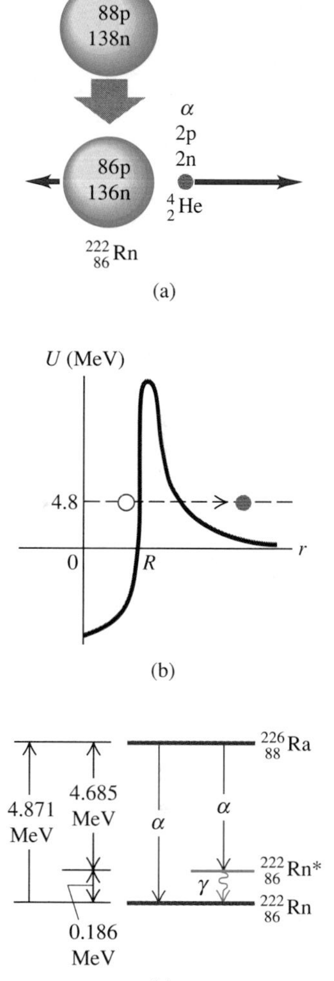


**45–6** (a) The nuclide $^{226}$Ra decays by $\alpha$ emission to $^{222}$Rn. (b) Potential-energy curve for $\alpha$ particle and $^{222}$Rn nucleus. The particle tunnels through the potential-energy barrier. (c) Energy-level diagram for the system, showing the excited state $^{222}$Rn$^*$ at an energy 0.186 MeV above the ground state. The system can decay from this state to the ground state $^{222}$Rn by emission of a $\gamma$ photon with energy 0.186 MeV.

## 45–4 RADIOACTIVITY

Nearly 90% of the 2500 known nuclides are *radioactive;* they are not stable, but *decay* into other nuclei. A nucleus is unstable if it is too big; no nucleus with $Z > 83$ or $A > 209$ is stable. This instability results from the competition between the attractive nuclear forces and the repulsive electrical forces. At sufficiently large $Z$ or $A$, repulsion always wins, and the nucleus comes apart. Another source of instability results from the pairing effects in nuclear forces, which favor pairs of protons paired with pairs of neutrons. If the neutron-proton ratio is wrong, a nucleus is unstable, even if $Z$ and $A$ don't exceed the limits mentioned above.

### Alpha and Beta Decay

The two most common decay modes are the emission of alpha ($\alpha$) and beta ($\beta$) particles. The **alpha particle** is identical to the $^4$He nucleus: two protons and two neutrons bound together, with zero total spin. Alpha emission occurs principally with nuclei that are too large to be stable. When a nucleus emits an $\alpha$ particle, its $N$ and $Z$ values each decrease by two, and $A$ decreases by four, moving it closer to stable territory on the Segrè chart.

A familiar example of an alpha emitter is radium, $^{226}$Ra (Fig. 45–6a). The speed of its $\alpha$ particle, determined from the curvature of its path in a transverse magnetic field, is about $1.5 \times 10^7$ m/s. The corresponding kinetic energy is

$$K = \tfrac{1}{2}(6.64 \times 10^{-27}\ \mathrm{kg})(1.5 \times 10^7\ \mathrm{m/s})^2$$
$$= 7.5 \times 10^{-13}\ \mathrm{J} = 4.7 \times 10^6\ \mathrm{eV} = 4.7\ \mathrm{MeV}.$$

This speed, although large, is only 5% of the speed of light, so we can use the nonrelativistic kinetic-energy expression. Alpha particles are always emitted with definite kinetic energy, determined by conservation of momentum and energy. They can travel several centimeters in air, or a few tenths or hundredths of a millimeter through solids, before they are brought to rest by collisions.

Heavy nuclei decay by emission of $\alpha$ particles rather than individual nucleons because alpha emission is energetically more favorable. The binding energy of an individual nucleon is of the order of 8 MeV, but the binding energy of the $\alpha$ particle taken as a unit may be much less. In alpha decay, the $\alpha$ particle tunnels through a potential-energy barrier, as shown in Fig. 45–6b. You may want to review the discussion of tunneling in Section 42–4.

A **beta-minus particle** is identical to an electron. It's not obvious how a nucleus can emit an electron if there aren't any electrons in the nucleus. Emission of a $\beta^-$ particle involves *transformation* of a neutron into a proton and an electron. We'll discuss such transformations in more detail in the next chapter; for now we'll just say that they happen. In fact, even a free neutron decays into a proton and an electron, with a mean lifetime of about 15 minutes.

Beta-minus decay usually occurs with nuclei for which the neutron-to-proton ratio $N/Z$ is too great for stability. In $\beta^-$ decay, $N$ decreases by one, $Z$ increases by one, and $A$ doesn't change. Note that both $\alpha$ and $\beta^-$ emission, by changing the $Z$ value of a nucleus, have the effect of changing one element into another.

Beta particles can be identified and their speeds measured with techniques similar to the Thomson experiments we described in Section 28–5. Their speeds range up to 0.9995 of the speed of light, so their motion is highly relativistic. They are emitted with a continuous spectrum of energies. This would not be possible if the only two particles were the $\beta^-$ particle and the recoiling nucleus; energy and momentum conservation would require a definite speed for the $\beta^-$ particle. We discussed this matter in Section 8–9; you may want to review that discussion. Thus a third particle must be involved; from charge conservation it must be neutral.

This particle is called a **neutrino** (symbolized by $\nu$); it has zero rest mass and zero charge and therefore produces very little observable effect when it passes through matter. It evaded detection until 1953, when Reines and Cowan succeeded in observing it directly. We now know that there are at least three varieties of neutrinos, the one associated with $\beta^-$ decay and two others associated with the decay of two unstable particles, the $\mu$ mesons and the $\tau$ particles. For reasons that we'll discuss in Chapter 46, the neutrino emitted in $\beta^-$ decay is called an *antineutrino,* denoted as $\bar{\nu}_e$. The basic process of decay can be represented as

$$n \longrightarrow p + \beta^- + \bar{\nu}_e. \tag{45–11}$$

After $\alpha$ or $\beta^-$ emission the remaining nucleus is sometimes left in an excited state (Fig. 45–6c). It can then decay to its ground state by emitting a photon, often called a *gamma ray photon* or simply a *gamma* ($\gamma$). Typical $\gamma$ energies are 10 keV to 5 MeV. For example, $\alpha$ particles emitted from radium have two possible kinetic energies, either 4.784 MeV or 4.602 MeV, corresponding to total released energy of 4.871 MeV or 4.685 MeV. When an $\alpha$ with the smaller energy is emitted, the resulting nucleus (which corresponds to the element *radon*) is left in an excited state. It then decays to its ground state by emitting a $\gamma$ photon with energy

$$(4.871 - 4.685)\ \text{MeV} = 0.186\ \text{MeV}.$$

A photon with this energy is observed during this decay.

## Natural Radioactivity

Several radioactive elements occur in nature. The study of natural radioactivity began in 1896, one year after Röntgen discovered x rays. Henri Becquerel discovered a radiation from uranium salts that seemed similar to x rays. Intensive investigation in the following two decades by Marie and Pierre Curie, Ernest Rutherford, and many others revealed that the emissions consist of positively and negatively charged and neutral particles; they were given the names $\alpha$, $\beta^-$, and $\gamma$ before their identities were firmly established.

The decaying nucleus is usually called the *parent nucleus;* the resulting nucleus is the *daughter nucleus.* When a radioactive nucleus decays, the daughter nucleus may also be unstable. In this case a *series* of successive decays occurs until a stable configuration is reached. Several such series are found in nature. The most abundant radioactive nucleus found on earth is uranium $^{238}U$, which undergoes a series of 14 decays, including eight $\alpha$ emissions and six $\beta^-$ emissions, terminating at the stable isotope of lead, $^{206}Pb$.

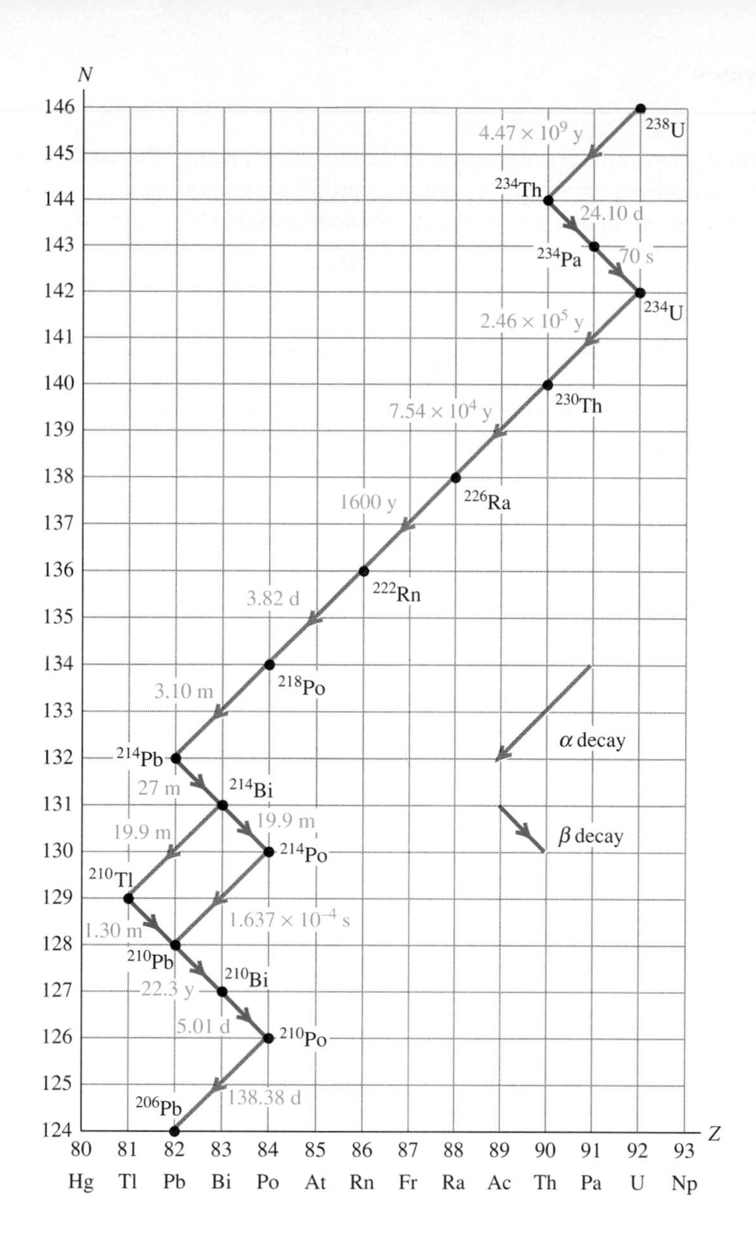


**45–7** Segrè chart showing the uranium $^{238}$U decay series, terminating with the stable nuclide $^{206}_{82}$Pb.

Radioactive decay series can be represented on a Segrè chart, as in Fig. 45–7. The neutron number $N$ is plotted vertically and the atomic number $Z$ horizontally. In $\alpha$ emission, both $N$ and $Z$ decrease by two. In $\beta^-$ emission, $N$ decreases by one, and $Z$ increases by one. The half-lives (discussed in the next section) are given in years (y), days (d), hours (h), minutes (m), or seconds (s).

Figure 45–7 shows the $^{238}$U decay series. The decays can also be represented in equation form; the first two decays in the series are written as

$$^{238}\text{U} \longrightarrow {}^{234}\text{Th} + \alpha,$$

$$^{234}\text{Th} \longrightarrow {}^{234}\text{Pa} + \beta^-,$$

or more briefly as

$$^{238}\text{U} \xrightarrow{\alpha} {}^{234}\text{Th},$$

$$^{234}\text{Th} \xrightarrow{\beta^-} {}^{234}\text{Pa}.$$

In the second process the $\beta^-$ decay leaves the daughter nucleus $^{234}$Pa in an excited state, from which it decays to the ground state by emitting a $\gamma$ photon with energy 92 keV. An excited state is denoted by an asterisk, so we can represent the $\gamma$ emission as

$$^{234}\text{Pa}^* \longrightarrow {}^{234}\text{Pa} + \gamma,$$

or

$$^{234}\text{Pa}^* \xrightarrow{\gamma} {}^{234}\text{Pa}.$$

An interesting feature of the $^{238}$U decay series is the branching that occurs at $^{214}$Bi. This nuclide decays to $^{210}$Pb by emission of an $\alpha$ and a $\beta^-$, which can occur in either order. We also note that the series includes unstable isotopes of several elements that also have stable isotopes, including thallium (Tl), lead (Pb), and bismuth (Bi). The unstable isotopes of these elements that occur in the $^{238}$U series all have too many neutrons to be stable.

Three other decay series are known. Two of these occur in nature, one starting with the uncommon isotope $^{235}$U and ending with $^{207}$Pb, the other starting with thorium ($^{232}$Th) and ending with $^{208}$Pb. The fourth series starts with neptunium ($^{237}$Np, an element that is not found in nature but produced in nuclear reactors) and ends with $^{209}$Bi.

Masses of nuclides can be measured very precisely with mass-spectrometer techniques of the sort described in Section 28–5. Armed with a table of nuclear masses, we can study energy relations in nuclear reactions; we can even predict which nuclides are stable and which are unstable. Here are two examples.

## EXAMPLE 45–5

**Radium decay** You are given the following masses:

$^{226}_{88}$Ra: 226.025403 u

$^{222}_{86}$Rn: 222.017571 u

$^{4}_{2}$He: 4.002603 u

Show that $\alpha$ emission is energetically possible, and find the kinetic energy of the emitted $\alpha$ particle.

**SOLUTION** First note that the masses given are those of the neutral atoms, including 88, 86, and 2 electrons, respectively, so we are including the same numbers of electrons in the initial and final states. This is essential.

Next, note that the $^{226}$Ra mass is greater than the sum of the $^{222}$Rn and $^{4}$He masses; the difference is

$$226.025403\text{ u} - (222.017571\text{ u} + 4.002603\text{ u}) = 0.005229\text{ u}.$$

The fact that this is positive shows that the decay is energetically possible.

The energy equivalent of 0.005229 u is

$$E = (0.005229\text{ u})(931.5\text{ MeV/u}) = 4.871\text{ MeV}.$$

Thus we expect the decay products to emerge with total kinetic energy 4.871 MeV. Momentum is also conserved; if the parent nucleus is at rest, the daughter nucleus and the $\alpha$ particle have equal and opposite momenta. Kinetic energy is $K = p^2/2m$, so the kinetic energy divides inversely as the masses of the two particles. The $\alpha$ gets 222/(222 + 4) of the total, or 4.78 MeV, equal to the observed $\alpha$ energy.

## EXAMPLE 45–6

The nuclide $^{60}$Co, an odd-odd unstable nucleus, is used in medical applications of radiation. Show that it is unstable against $\beta^-$ decay, and find the total energy of the decay products. The following masses are given:

$^{60}_{27}$Co: 59.933820 u

$^{60}_{28}$Ni: 59.930788 u

**SOLUTION** The $^{60}$Ni mass includes 28 electrons, one more than the $^{27}$Co mass. Even with that extra electron, its mass is less than that of $^{27}$Co by 0.003032 u, corresponding to an energy of 2.82 MeV. So the total energy of $\beta^-$, $\gamma$, and $\overline{\nu}_e$ emissions from $^{60}$Co is 2.82 MeV. Note that we don't have to add in the rest energy of the $\beta^-$; it's already included because the Ni mass includes one more electron than the Co mass.

Finally, we mention two additional decay modes. We have noted that $\beta^-$ decay occurs with nuclei with too large a neutron-to-proton ratio, $N/Z$. Nuclei for which $N/Z$ is *too small* for stability can emit a *positron,* a particle that is identical to the electron but with positive charge. (We'll discuss positrons in more detail in Section 46–1.) The basic process is

$$p \longrightarrow n + \beta^+ + \nu_e, \tag{45–12}$$

where $\beta^+$ is a positron. To avoid confusion, we'll usually use the notations $\beta^-$ and $\beta^+$ for electrons and positrons, respectively. The effect of $\beta^+$ emission is to decrease $Z$ by one and increase $N$ by one, moving the neutron-proton ratio toward a stable value.

A related process is *orbital-electron capture.* There are a few nuclei for which $\beta^+$ emission is not energetically possible but an orbital electron (usually in the $K$ shell) can be pulled into the nucleus. The basic process is

$$p + \beta^- \longrightarrow n + \nu_e. \tag{45–13}$$

Again the effect on thc nucleus is to decrease $Z$ by one and increase $N$ by one.

## 45–5 DECAY RATES

In any sample of a radioactive element, the number of radioactive nuclei decreases as the nuclei decay. This is a statistical process; there is no way to predict when any individual nucleus will decay. No change in physical or chemical environment, such as chemical reactions or heating or cooling, greatly affects the decay rate, but the rate varies over an extremely wide range for different nuclides.

Let $N$ be the number of radioactive nuclei in a sample at time $t$, and let $dN$ be the (negative) change in that number during a short time interval $dt$. The number of decays during $dt$ is $-dN$. The rate of change of $N$ is the negative quantity $dN/dt$; its absolute value is called the *decay rate,* or the **activity,** of the specimen. The larger the number of nuclei in the specimen, the more nuclei decay during any time interval. That is, the rate of change of $N$ is proportional to $N$ and is equal to a constant $\lambda$ multiplied by $N$:

$$\frac{dN}{dt} = -\lambda N. \tag{45–14}$$

The constant $\lambda$ is called the **decay constant,** and it has different values for different nuclides. The *probability* that any individual nucleus will decay during a time interval $dt$ is $\lambda\, dt$. A large value of $\lambda$ corresponds to rapid decay, a small value to slower decay.

The situation is reminiscent of a discharging capacitor, which we studied in Section 27–4. If $N_0$ is the number of nuclei at time $t = 0$, the solution of Eq. (45–14) is an exponential function:

$$N = N_0 e^{-\lambda t}. \tag{45–15}$$

Figure 45–8 is a graph of this function, showing the number of nuclei $N$ as a function of time.

The **half-life** $T_{1/2}$ is the time required for the number of radioactive nuclei to decrease to one-half of the original number $N_0$. Then half of those remaining decay during a second interval $T_{1/2}$, and so on. The numbers remaining after successive intervals of $T_{1/2}$ are $N_0/2, N_0/4, N_0/8, \ldots$.

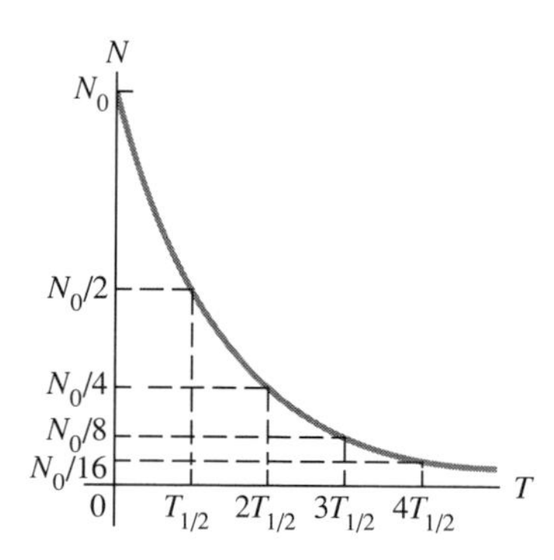


**45–8** Decay curve for a sample of a radioactive element.

### EXAMPLE 45–11

When lithium ($^7$Li) is bombarded by a proton, two $\alpha$ particles ($^4$He) are produced. Find the reaction energy.

**SOLUTION** The reaction can be written

$$^1_1\text{H} + {}^7_3\text{Li} \longrightarrow {}^4_2\text{He} + {}^4_2\text{He}.$$

Here are the initial and final masses:

| | | | |
|---|---|---|---|
| $A$: $^1_1$H | 1.00783 u | $C$: $^4_2$He | 4.00260 u |
| $B$: $^7_3$Li | 7.01600 u | $D$: $^4_2$He | 4.00260 u |
| | 8.02383 u | | 8.00520 u |

(Four electron masses are included on each side.) The mass decreases by 0.01863 u. The reaction energy is

$$(0.01863\ \text{u})\,(931.5\ \text{MeV/u}) = 17.4\ \text{MeV}.$$

The final total kinetic energy of the two separating $\alpha$ particles is 17.4 MeV *greater than* the initial total kinetic energy.

For a charged particle such as a proton or an $\alpha$ particle to penetrate the nucleus of another atom and cause a reaction, it must have enough initial kinetic energy to overcome the potential-energy barrier caused by the repulsive electrostatic forces. For example, in the reaction of Example 45–11, if the $^7$Li nucleus has a radius of $2.3 \times 10^{-15}$ m, as given by Eq. (45–1), the repulsive potential energy of the proton (charge $+e$) and the $^7$Li nucleus (charge $+3e$) at this distance is

$$U = \frac{1}{4\pi\epsilon_0}\frac{(e)(3e)}{r} = \frac{3(9.0 \times 10^9\ \text{N}\cdot\text{m}^2/\text{C}^2)\,(1.6 \times 10^{-19}\ \text{C})^2}{2.3 \times 10^{-15}\ \text{m}}$$
$$= 3.01 \times 10^{-13}\ \text{J} = 1.88 \times 10^6\ \text{eV} = 1.88\ \text{MeV}.$$

Even though the reaction is exothermic, the proton must have a minimum energy of about 1.9 MeV for the reaction to occur. (We haven't considered the possibility of *tunneling,* which can be significant.)

Absorption of *neutrons* by nuclei forms an important class of nuclear reactions. Heavy nuclei bombarded by neutrons in a nuclear reactor can undergo a series of neutron absorptions alternating with beta decays, in which the mass number $A$ increases by as much as 25. The *transuranic elements,* elements having $Z$ larger than 92, are produced in this way. These elements do not occur in nature. Seventeen transuranic elements, having $Z$ up to 109 and $A$ up to about 266, have been identified.

The analytical technique of *neutron activation analysis* uses similar reactions. When stable nuclei are bombarded by neutrons, some nuclei absorb neutrons and then undergo $\beta^-$ decay. The energies of the $\beta^-$ and $\gamma$ emissions depend on the parent nucleus and provide a means of identifying it. Quantities of elements that are far too small for conventional chemical analysis can be detected in this way, and neutron activation analysis is a powerful analytical technique.

## 45–8 NUCLEAR FISSION

**Nuclear fission** is a decay process in which an unstable nucleus splits into two fragments of comparable mass instead of emitting an $\alpha$ or $\beta$ particle. Fission was discovered in 1939 by Hahn and Strassman. Pursuing earlier work by Fermi, they bombarded uranium ($Z = 92$) with neutrons. The resulting radiation did not coincide with that of any known radioactive nuclide. Their meticulous chemical analysis led to the astonishing conclusion that they had found a radioactive isotope of barium ($Z = 56$). Later, radioactive krypton ($Z = 36$) was also found. Meitner and Frisch correctly interpreted these results as showing that uranium nuclei were splitting into two massive fragments called *fission fragments.* Two or three free neutrons usually appear along with the fission fragments.

Both the common (99.3%) isotope $^{238}$U and the uncommon (0.7%) isotope $^{235}$U (as well as several other nuclides) can be split by neutron bombardment: $^{235}$U by slow neutrons but $^{238}$U only by neutrons with a minimum of about 1 MeV of energy. Fission resulting from neutron absorption is called *induced fission*. Some nuclei can also undergo *spontaneous fission* without initial neutron absorption, but this is quite rare. When $^{235}$U absorbs a neutron, the resulting nuclide $^{236}$U$^*$ is in a highly excited state and splits into two fragments almost instantaneously. Strictly speaking, it is $^{236}$U$^*$, not $^{235}$U, that undergoes fission, but it's usual to speak of the fission of $^{235}$U.

Over 100 different nuclides, representing more than 20 different elements, have been found among the fission products. Figure 45–12 shows the distribution of mass numbers for fission fragments from the nuclide $^{235}$U. Most of the fragments have mass numbers from 90 to 100 and from 135 to 145; fission into two fragments with nearly equal mass is unlikely.

Here are two typical fission reactions:

$$^{235}_{92}\text{U} + {}^{1}_{0}\text{n} \longrightarrow {}^{236}_{92}\text{U}^* \longrightarrow {}^{144}_{56}\text{Ba} + {}^{89}_{36}\text{Kr} + 3\,{}^{1}_{0}\text{n},$$

$$^{235}_{92}\text{U} + {}^{1}_{0}\text{n} \longrightarrow {}^{236}_{92}\text{U}^* \longrightarrow {}^{140}_{54}\text{Xe} + {}^{94}_{38}\text{Sr} + 2\,{}^{1}_{0}\text{n}.$$

The total kinetic energy of the fission fragments is enormous, about 200 MeV (compared to typical $\alpha$ and $\beta$ energies of a few MeV). The reason for this is that nuclides at the high end of the mass spectrum (near $A = 240$) are less tightly bound than those nearer the middle ($A = 90$ to 145). Referring to Fig. 45–2, we see that the average binding energy per nucleon is about 7.6 MeV at $A = 240$ but about 8.5 MeV at $A = 120$. Therefore a rough estimate of the expected *increase* in binding energy during fission is about 8.5 MeV $-$ 7.6 MeV $=$ 0.9 MeV per nucleon, or a total of (235) (0.9 MeV) $=$ 210 MeV. (Keep in mind that binding energy represents *negative* potential energy; when binding energy increases, there is a corresponding *increase* in kinetic energy.)

Fission fragments always have too many neutrons to be stable. We noted in Section 45–2 that the neutron/proton ratio ($N/Z$) for stable nuclei is 1 for light nuclei but almost 1.6 for the heaviest nuclei because of the increasing influence of the electrical repulsion

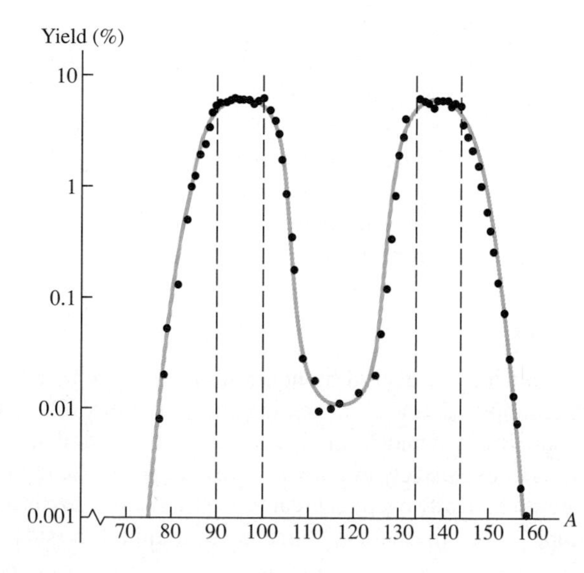


**45–12** Mass distribution of fission fragments from the fission of $^{236}$U* resulting from neutron absorption by $^{235}$U. The vertical scale is logarithmic.

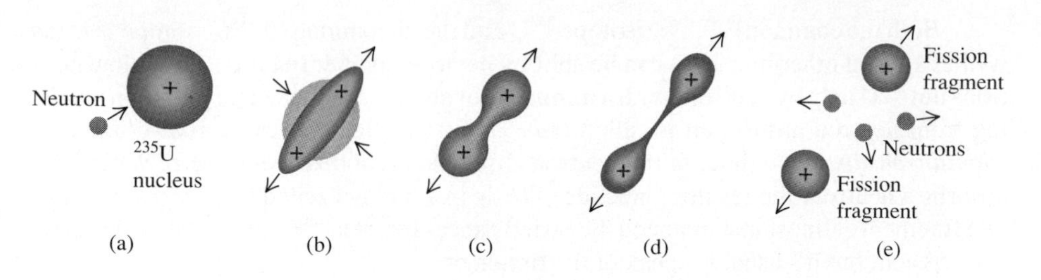


**45–13** (a) A $^{235}$U nucleus absorbs a neutron. (b) The resulting $^{236}U^*$ nucleus is in a highly excited state and oscillates strongly. (c) A neck develops, and electrical repulsion pushes the two lobes apart. (d) The two lobes separate, forming fission fragments. (e) Neutrons are emitted from the fission fragments at the time of fission or occasionally a few seconds later.

of the protons. The $N/Z$ value for stable nuclides is about 1.3 at $A = 100$ and 1.4 at $A = 150$. The fragments have about the same $N/Z$ as $^{235}$U, about 1.55. They respond to this surplus of neutrons by undergoing a series of $\beta^-$ decays (each of which increases $Z$ by one and decreases $N$ by one) until a stable value of $N/Z$ is reached. A typical example is

$$^{140}_{54}\text{Xe} \xrightarrow{\beta^-} {}^{140}_{55}\text{Cs} \xrightarrow{\beta^-} {}^{140}_{56}\text{Ba} \xrightarrow{\beta^-} {}^{140}_{57}\text{La} \xrightarrow{\beta^-} {}^{140}_{58}\text{Ce}.$$

The nuclide $^{140}$Ce is stable. This series of $\beta^-$ decays produces, on average, about 15 MeV of additional kinetic energy. This neutron excess also explains why two or three free neutrons are released during the fission.

## Liquid-Drop Model

We can understand fission qualitatively on the basis of the liquid-drop model of the nucleus (Section 45–3). The process is shown in Fig. 45–13 in terms of an electrically charged liquid drop. These sketches shouldn't be taken too literally, but they may help develop your intuition about fission. A $^{235}$U nucleus absorbs a neutron (Fig. 45–13a), becoming a $^{236}U^*$ nucleus with excess energy (Fig. 45–13b). This excess energy causes violent oscillations (Fig. 45–13c), during which a neck between two lobes develops. The electrical repulsion of these two lobes stretches the neck farther (Fig. 45–13d), and finally two smaller drops are formed (Fig. 45–13e).

This qualitative picture has been developed into a more quantitative theory to explain why some nuclei undergo fission and others don't. It is basically a tunneling phenomenon; Figure 45–14 shows a hypothetical potential-energy function for two possible fission fragments. If the neutron brings in an energy greater than the energy barrier height $E_B$, fission can occur. Even when there isn't quite enough energy to surmount the barrier, fission can take place by quantum-mechanical *tunneling,* discussed in Section 42–4. In fact, in principle, many stable heavy nuclei can fission by tunneling. But the probability depends very critically on the height and width of the barrier. For most nuclei this process is so unlikely that it is never observed.

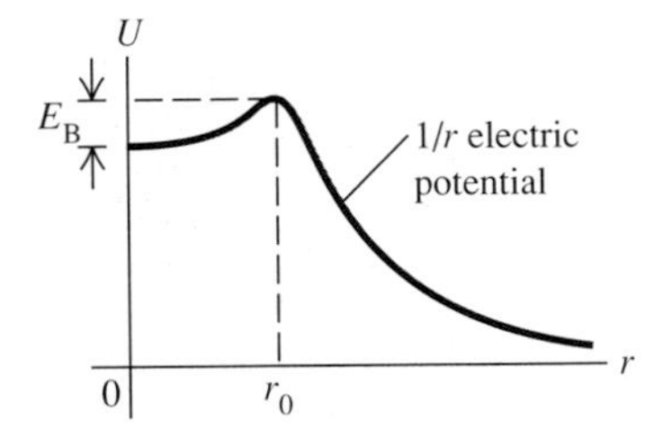


**45–14** Hypothetical potential-energy function for two fission fragments in a fissionable nucleus. At distances $r$ beyond the range of the nuclear force, the potential energy varies approximately as $1/r$. Fission occurs if there is an appreciable probability for tunneling through this potential-energy barrier.

## Chain Reactions

Fission of a uranium nucleus, triggered by neutron bombardment, releases other neutrons that can trigger more fissions, suggesting the possibility of a **chain reaction** (Fig. 45–15). The chain reaction may be made to proceed slowly and in a controlled manner in a nuclear reactor or explosively in a bomb. The energy release in a chain reaction is enormous, far greater than in any chemical reaction. (In a sense, *fire* is a chemical chain reaction.) For example, when uranium is "burned" to uranium dioxide in the chemical reaction

$$U + O_2 \longrightarrow UO_2,$$

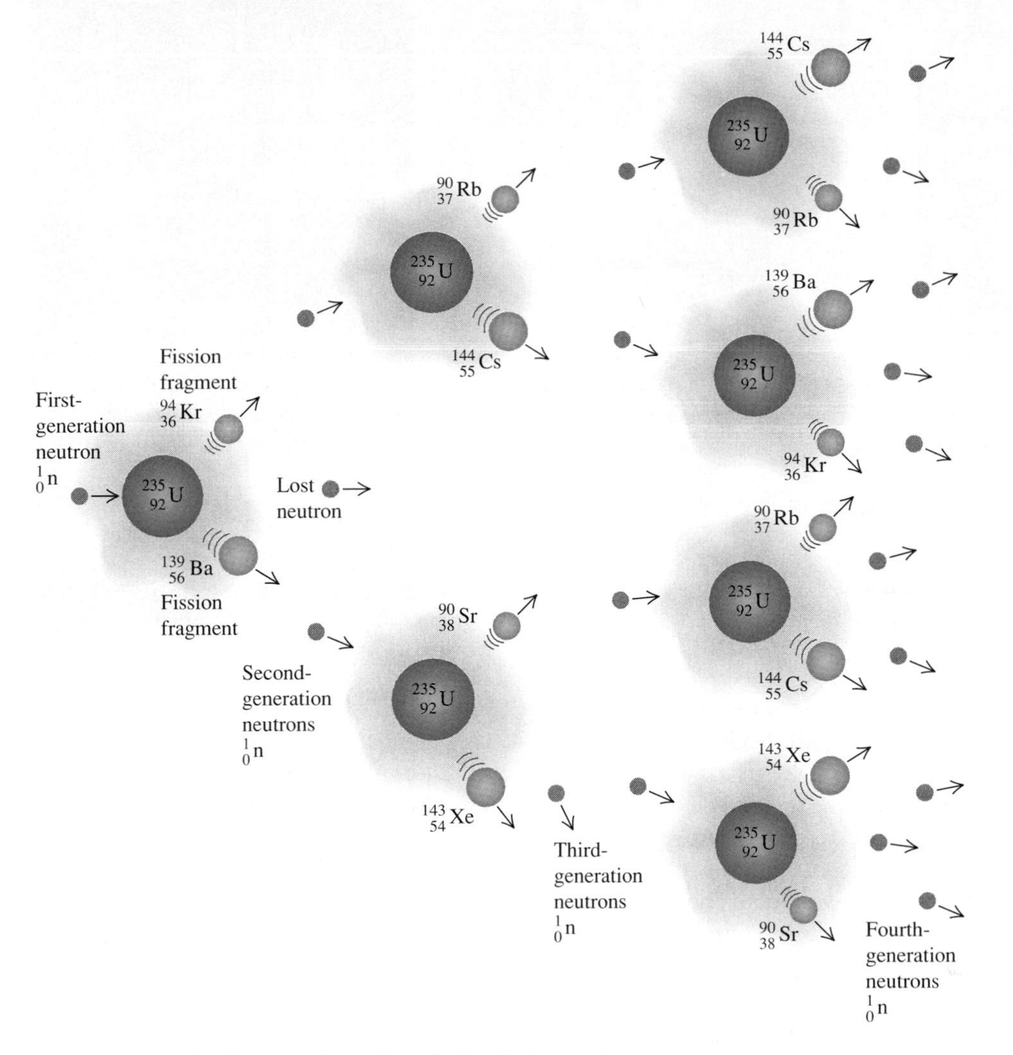


**45–15** Schematic diagram of a nuclear-fission chain reaction.

the heat of combustion is about 4500 J/g. Expressed as energy per atom, this is about 11 eV per uranium atom. By contrast, fission liberates about 200 MeV per atom, 20 million times as much energy.

## Nuclear Reactors

A *nuclear reactor* is a system in which a controlled nuclear chain reaction is used to liberate energy. In a nuclear power plant, this energy is used to generate steam, which operates a turbine and turns an electrical generator.

On average, each fission of a $^{235}$U nucleus produces about 2.5 free neutrons, so 40% of the neutrons are needed to sustain a chain reaction. The probability of neutron absorption by a nucleus is much larger for low-energy (less than 1 eV) neutrons than for the higher-energy (1 MeV or so) neutrons that are liberated during fission. In a nuclear reactor these neutrons are slowed down by collisions with nuclei in the surrounding material, called the *moderator,* so that they can cause further fissions. In nuclear power plants, the moderator is often water, occasionally graphite. The *rate* of the reaction is controlled by inserting or withdrawing *control rods* made of elements (often cadmium) whose nuclei

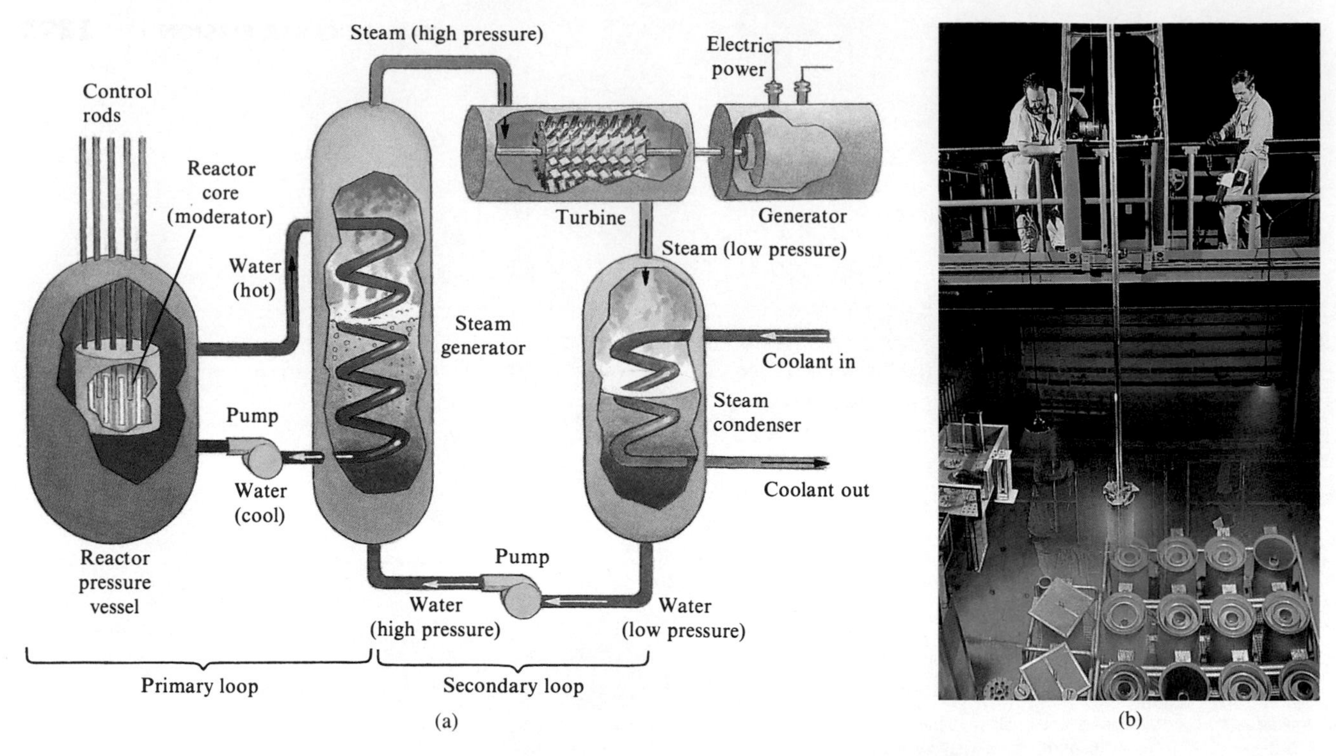


**45–16** (a) Schematic diagram of a nuclear power plant. (b) Technicians assembling components of a reactor core.

*absorb* neutrons without undergoing any additional reaction. The isotope $^{238}$U can also absorb neutrons, leading to $^{239}$U, but not with high enough probability for it to sustain a chain reaction by itself. Thus uranium that is used in reactors is often "enriched" by increasing the proportion of $^{235}$U above the natural value of 0.7%, typically to 3% or so, by isotope-separation processing.

The most familiar application of nuclear reactors is for the generation of electric power. As was noted above, the fission energy appears as kinetic energy of the fission fragments, and its immediate result is to heat the fuel elements and the surrounding water. This heat is used to generate steam to drive turbines, which drive the electrical generators. Figure 45–16 is a schematic diagram of a nuclear power plant. The energetic fission fragments heat the water surrounding the reactor core. The steam generator is a heat exchanger that takes heat from this highly radioactive water and generates nonradioactive steam to run the turbines.

A typical nuclear plant has an electric generating capacity of 1000 MW (or $10^9$ W). The turbines are heat engines and are subject to the efficiency limitations imposed by the second law of thermodynamics, discussed in Chapter 18. In modern nuclear plants the overall efficiency is about one-third, so 3000 MW of thermal power from the fission reaction are needed to generate 1000 MW of electrical power.

## EXAMPLE 45–12

What mass of uranium has to undergo fission per unit time to provide 3000 MW of thermal power?

**SOLUTION** Each second, we need 3000 MJ, or $3000 \times 10^6$ J. Each fission provides 200 MeV, which is

$$200\text{ MeV} = (200\text{ MeV})(1.6 \times 10^{-13}\text{ J/MeV}) = 3.2 \times 10^{-11}\text{ J}.$$

The number of fissions needed per second is

$$\frac{3000 \times 10^6\text{ J}}{3.2 \times 10^{-11}\text{ J}} = 0.94 \times 10^{20}.$$

Each uranium atom has a mass of about $(235)(1.66 \times 10^{-27}$ kg$) = 3.9 \times 10^{-25}$ kg, so the mass of uranium needed per second is

$$(0.94 \times 10^{20})(3.9 \times 10^{-25}\text{ kg}) = 3.7 \times 10^{-5}\text{ kg} = 37\text{ mg}.$$

In one day (86,400 s) the total consumption of uranium is

$$(3.7 \times 10^{-5}\text{ kg/s})(86{,}400\text{ s/d}) = 3.2\text{ kg/d}.$$

For comparison, note that the 1000-MW coal-fired power plant that we described in Section 18–9 burns 10,600 tons (about $10^7$ kg) of coal per day!

Nuclear fission reactors have many other practical uses. Among these are the production of artificial radioactive isotopes for medical and other research, production of high-intensity neutron beams for research in nuclear structure, and production of fissionable transuranic elements such as plutonium from the common isotope $^{238}$U. The last is the function of *breeder reactors.*

We mentioned above that about 15 MeV of the energy from fission of a $^{235}$U nucleus comes from the $\beta^-$ decays of the fission fragments rather than from the kinetic energy of the fragments themselves. This fact poses a serious problem with respect to control and safety of reactors. Even after the chain reaction has been completely stopped by insertion of control rods into the core, heat continues to be evolved by the $\beta$ decays, which cannot be stopped. For a 1000-MW reactor this heat power amounts to about 200 MW, which, in the event of total loss of cooling water, is more than enough to cause a catastrophic meltdown of the reactor core and possible penetration of the containment vessel. The difficulty in achieving a "cold shutdown" following an accident at the Three Mile Island nuclear power plant in Pennsylvania in March 1979 was a result of the continued evolution of heat due to $\beta^-$ decays.

The Chernobyl catastrophe of April 26, 1986 resulted from a combination of an inherently unstable design and several human errors that were committed during a test of the emergency core cooling system. Too many control rods were withdrawn to compensate for a decrease in power caused by a buildup of neutron absorbers such as $^{135}$Xe. The power level rose from 1% of normal to 100 times normal in 4 seconds; a steam explosion ruptured pipes in the core cooling system and blew the heavy concrete cover off the reactor. The graphite moderator caught fire and burned for several days. The total activity of the radioactive material released into the atmosphere has been estimated as about $10^8$ Ci.

Fission appears to set an upper limit on the production of transuranic nuclei, mentioned in Section 45–7. When a nucleus with $Z = 109$ is bombarded with neutrons, fission occurs essentially instantaneously; no $Z = 110$ nucleus is formed even for a short time. There are theoretical reasons to expect that nuclei in the vicinity of $Z = 114$, $N = 184$, might be stable with respect to spontaneous fission. In the shell model (Section 45–3) these numbers correspond to *filled shells* in the nuclear energy-level structure. Such nuclei, called *superheavy nuclei,* would still be unstable with respect to alpha emission, but they might live long enough to be identified. Attempts to produce superheavy nuclei in the laboratory have not been successful; whether they exist in nature is still an open question.

## 45–9 NUCLEAR FUSION

In **nuclear fusion** two or more small light nuclei come together, or *fuse,* to form a larger nucleus. Fusion reactions release energy for the same reason as fission reactions; the binding energy per nucleon after the reaction is greater than before. Referring to Fig. 45–2, we see that up to about $A = 60$ the binding energy per nucleon increases with $A$, so fusion of nearly any two light nuclei to make a nucleus with $A$ less than 60 is likely to be an exothermic reaction. In comparison to fission, we are moving toward the peak of this curve from the opposite side. Another way to express the energy relations is that the total rest mass of the products is less than that of the initial particles.

Here are three examples of energy-liberating fusion reactions:

$$ {}^1_1\mathrm{H} + {}^1_1\mathrm{H} \longrightarrow {}^2_1\mathrm{H} + \beta^+ + \nu_e, $$

$$ {}^2_1\mathrm{H} + {}^1_1\mathrm{H} \longrightarrow {}^3_2\mathrm{He} + \gamma, $$

$$ {}^3_2\mathrm{He} + {}^3_2\mathrm{He} \longrightarrow {}^4_2\mathrm{He} + {}^1_1\mathrm{H} + {}^1_1\mathrm{H}. $$

In the first reaction, two protons combine to form a deuteron ($^2$H) and a positron ($\beta^+$). In the second, a proton and a deuteron combine to form the light isotope of helium, $^3$He. In the third, two $^3$He nuclei fuse to form ordinary helium nuclei ($^4$He) and two protons.

The positrons that are produced during the first step of the proton-proton chain collide with electrons; mutual annihilation takes place, and their energy is converted into gamma radiation. The net effect of the chain, therefore, is the combination of four hydrogen nuclei into a helium nucleus and gamma and neutrino radiation. We can calculate the net energy release from the mass balance, taking care to include the same numbers of electrons in the initial and final states:

| | |
|---|---|
| Mass of four hydrogen atoms (including electrons) | 4.03132 u |
| Mass of one helium atom plus two electrons | 4.00370 u |
| Mass difference | 0.02762 u = 25.7 MeV. |

This series of fusion reactions is called the *proton-proton chain;* it is believed to take place in the interior of the sun and other stars. Each gram of the sun's mass contains about $4.5 \times 10^{23}$ protons. If all of these protons were fused into helium, the energy released would be about 130,000 kWh. If the sun were to continue to radiate at its present rate, it would exhaust its supply of protons in about 30 billion years.

## EXAMPLE 45–13

Two deuterons fuse to form a $^4$He nucleus. How much energy is liberated?

**SOLUTION** We find the relevant masses in Table 45–2. The mass of two deuterons plus two electrons (i.e., two neutral deuterium atoms) is 2(2.01410 u) = 4.02820 u. The mass of the neutral $^4$He atom, including two electrons, is 4.00260 u. The difference, 0.0256 u, is equivalent to (0.0256 u) (931.5 MeV/u) = 23.8 MeV. Thus 23.8 MeV must appear as kinetic energy or radiation.

For fusion of two nuclei to occur, they must come together to within the range of the nuclear force, typically of the order of $2 \times 10^{-15}$ m. To do this, they must overcome the electrical repulsion of their positive charges. For two protons at this distance the corresponding potential energy is about $1.1 \times 10^{-13}$ J or 0.7 MeV; this represents the initial *kinetic* energy that the fusion nuclei must have.

Atoms have this much energy only at extremely high temperatures. The discussion in Section 16–4 showed that the average translational kinetic energy of a gas molecule at temperature $T$ is $\frac{3}{2}kT$, where $k$ is Boltzmann's constant. The temperature at which this is equal to $E = 1.1 \times 10^{-13}$ J is determined by the relation

$$E = \tfrac{3}{2}kT,$$

$$T = \frac{2E}{3k} = \frac{2(1.1 \times 10^{-13}\ \text{J})}{3(1.38 \times 10^{-23}\ \text{J/K})} = 5.3 \times 10^9\ \text{K}.$$

Actually, temperatures of the order of $10^8$ K are sufficient for a fusion reaction because of the high-energy tail on the Maxwell-Boltzmann distribution function. These numbers show why fusion reactions are called *thermonuclear reactions.*

Intensive efforts are underway in many laboratories to achieve controlled fusion reactions, which potentially represent an enormous new resource of energy. At the temperatures mentioned, light atoms are fully ionized, and the resulting state of matter is called *plasma.* In one kind of experiment using *magnetic confinement,* a plasma is heated

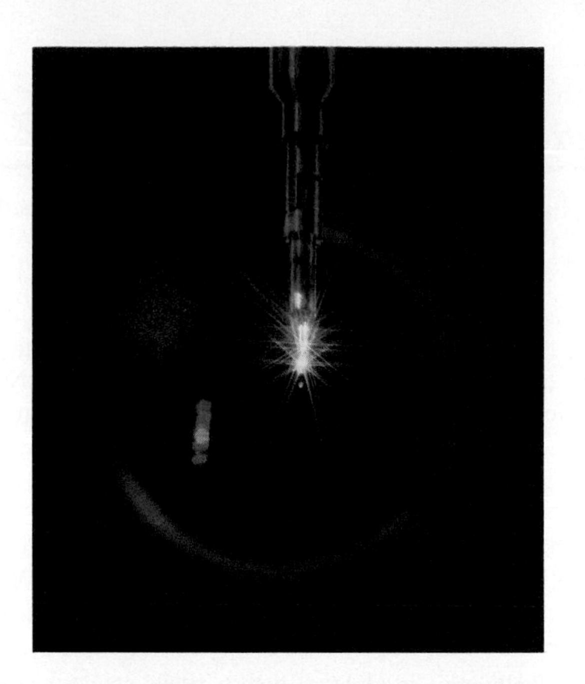

**45–17** The NOVA laser system at Lawrence Livermore National Laboratory, used for fusion research.

to an extremely high temperature by an electrical discharge, while being contained by appropriately shaped magnetic fields. In another, using *inertial confinement,* pellets of the material to be fused are heated by a high-intensity laser beam. One current laser experiment setup is shown in Fig. 45–17. Some of the reactions being studied are the following:

$$ {}^2_1\mathrm{H} + {}^2_1\mathrm{H} \longrightarrow {}^3_1\mathrm{H} + {}^1_1\mathrm{H} + 4\ \mathrm{MeV}, \tag{1} $$

$$ {}^3_1\mathrm{H} + {}^2_1\mathrm{H} \longrightarrow {}^4_2\mathrm{He} + {}^1_0\mathrm{n} + 17.6\ \mathrm{MeV}, \tag{2} $$

$$ {}^2_1\mathrm{H} + {}^2_1\mathrm{H} \longrightarrow {}^3_2\mathrm{He} + {}^1_0\mathrm{n} + 3.3\ \mathrm{MeV}, \tag{3} $$

$$ {}^3_2\mathrm{He} + {}^2_1\mathrm{H} \longrightarrow {}^4_2\mathrm{He} + {}^1_1\mathrm{H} + 18.3\ \mathrm{MeV}. \tag{4} $$

In the first reaction, two deuterons combine to form tritium (a proton plus two neutrons bound together, the nucleus of $^3$H) and a proton. In the second, the tritium nucleus combines with another deuteron to form helium and a neutron. The result of both of these reactions together is the conversion of three deuterons into a $^4$He nucleus, a proton, and a neutron, with the liberation of 21.6 MeV of energy. Reactions (3) and (4) together achieve the same conversion. In a plasma containing deuterium, the two pairs of reactions occur with roughly equal probability. As yet, no one has succeeded in producing these reactions under controlled conditions in such a way as to yield a net surplus of usable energy.

Methods of achieving fusion that don't require high temperatures are also being studied; these are called *cold fusion*. One possible scheme uses an unusual hydrogen molecule ion. The usual $H_2^+$ consists of two protons bound by one shared electron; the nuclear spacing is about 0.1 nm. If the protons are replaced by a deuteron ($^2$H) and a triton ($^3$H) and the electron by a *mu meson* (muon), which is 208 times as massive as the electron, the spacing is made smaller by a factor of 208. The probability for the two nuclei to tunnel through the repulsive potential-energy barrier and fuse then becomes appreciable. The prospect of making this process, called *muon-catalyzed fusion,* into a practical energy source is still distant.

## SUMMARY

### KEY TERMS

mass number
atomic number
neutron number
nuclide
nuclear magneton
mass defect
binding energy
gamma ray
radioactivity
liquid-drop model
shell model
alpha particle
beta-minus particle
neutrino
activity
decay constant
half-life
curie
relative biological effectiveness (RBE)
reaction energy
threshold energy
nuclear fission
chain reaction
nuclear fusion

• Nuclei are composed of protons and neutrons. All nuclei have about the same density. The radius of a nucleus with mass number $A$ is

$$R = R_0 A^{1/3}, \qquad (45\text{–}1)$$

where $R_0 = 1.2 \times 10^{-15}$ m. Isotopes are nuclei of the same element that have different numbers of neutrons. A single nuclear species is called a nuclide. Nuclear masses are measured in atomic mass units. Nuclei have angular momentum and magnetic moment.

• The mass of a nucleus is always less than the mass of the protons and neutrons in it. The difference is called the mass defect, and this quantity multiplied by $c^2$ is the binding energy. The nuclear force is short-range and favors pairs of particles. A nucleus is unstable if $A$ or $Z$ is too large or if the ratio $N/Z$ is wrong. Two widely used models of the nucleus are the liquid-drop model and the shell model; the latter is analogous to the central-field approximation in atomic structure.

• Unstable nuclei usually emit an alpha or beta particle, sometimes followed by a gamma photon. The rate of decay of an unstable nucleus is described by the half-life $T_{1/2}$ or the decay constant $\lambda$. If the number of nuclei at time $t = 0$ is $N_0$, then the number at time $t$ is

$$N = N_0 e^{-\lambda t}. \qquad (45\text{–}15)$$

A useful relation is

$$T_{\text{mean}} = \frac{1}{\lambda} = \frac{T_{1/2}}{\ln 2} = \frac{T_{1/2}}{0.693}. \qquad (45\text{–}17)$$

• The biological effect of any radiation depends on the product of the energy absorbed per unit mass and the relative biological effectiveness (RBE), which is different for different radiations.

• In a nuclear reaction, two nuclei or particles collide to produce two new nuclei or particles. Such reactions can be exothermic or endothermic. Several conservation laws, including charge, energy, momentum, angular momentum, and nucleon number, are obeyed.

• Nuclear fission is the splitting of a heavy nucleus into two lighter, always unstable, nuclei, with the release of energy. Nuclear fusion is the combining of two light nuclei into a heavier nucleus, with the release of energy.

## EXERCISES

### Section 45–1 Properties of Nuclei

**45–1** How many protons and how many neutrons are there in a nucleus of a) fluorine, $^{19}_{9}\text{F}$; b) nickel, $^{58}_{28}\text{Ni}$; c) lead, $^{208}_{82}\text{Pb}$?

**45–2** Consider the three nuclei of Exercise 45–1. a) Estimate the radius of each. b) Estimate the surface area of each. c) Estimate the volume of each. d) Determine the mass density (in kg/m$^3$) for each. (Assume that the mass is $A$ atomic mass units.) e) Determine the particle density (in particles/m$^3$) for each.

**45–3** Hydrogen atoms are placed in an external magnetic field. The protons can make transitions from states in which the nuclear spin is parallel and antiparallel to the field by absorbing a photon. What magnitude of magnetic field is required for this transition to be induced by photons with frequency 90.8 MHz?

### Section 45–2 Binding Energy and Stability

**45–4** Calculate the binding energy (in MeV) and the binding energy per nucleon of a) the lithium nucleus, ${}^{7}_{3}Li$; b) the helium nucleus, ${}^{4}_{2}He$. c) How do the results of parts (a) and (b) compare?

**45–5** What maximum wavelength $\gamma$ ray could break a deuteron up into a proton and a neutron? (This process is called photodisintegration.)

**45–6** Calculate for the common isotope of oxygen, ${}^{16}O$, a) the mass defect; b) the binding energy; c) the binding energy per nucleon.

### Section 45–3 Nuclear Structure

**45–7** What are the six known elements for which $Z$ is a magic number?

**45–8** The common isotope of calcium is ${}^{40}_{20}Ca$. The measured mass of the neutral atom is 39.962591 u. a) From the measured mass, determine the mass defect, and use it to find the total binding energy and the binding energy per nucleon. b) Calculate the binding energy from Eq. (45–9). Compare to the result you obtained in part (a). What is the percent difference?

**45–9** The common isotope of beryllium is ${}^{9}_{4}Be$. The measured mass of the neutral atom is 9.012182 u. a) Determine the total binding energy of ${}^{9}_{4}Be$ from the measured mass. b) Calculate this binding energy from Eq. (45–9). Compare to the result you obtained in part (a). What is the percent difference? Compare the accuracy of Eq. (45–9) for ${}^{9}Be$ to its accuracy for ${}^{56}Fe$ (Example 45–4).

### Section 45–4 Radioactivity

**45–10** What particle ($\alpha$ particle, electron, or positron) is emitted in the following radioactive decays: a) ${}^{222}_{89}Ac \rightarrow {}^{218}_{87}Fr$; b) ${}^{124}_{53}I \rightarrow {}^{124}_{52}Te$; c) ${}^{179}_{71}Lu \rightarrow {}^{179}_{72}Hf$?

**45–11** What nuclide is produced in the following radioactive decays: a) $\alpha$ decay of ${}^{238}_{94}Pu$; b) $\beta^-$ decay of ${}^{19}_{8}O$; c) $\beta^+$ decay of ${}^{25}_{13}Al$?

**45–12** The atomic mass of ${}^{56}_{26}Fe$ is 55.934939 u, and the atomic mass of ${}^{56}_{27}Co$ is 55.939841 u. a) Which of these nuclei will decay into the other? b) What type of decay will occur? c) How much kinetic energy will the products of the decay have?

**45–13 Tritium Decay.** Tritium, ${}^{3}_{1}H$, is an unstable isotope of hydrogen; its mass, including one electron, is 3.016049 u. a) Show that it must be unstable with respect to $\beta^-$ decay. b) Determine the total kinetic energy of the decay products, taking care to account for the electron masses correctly.

### Section 45–5 Decay Rates

**45–14** If you are of average mass, about 360 million nuclei in your body undergo radioactive decay each day. Express your activity in curies.

**45–15** The common (99.3%) isotope of uranium, ${}^{238}U$, has a half-life of $4.47 \times 10^9$ years, decaying to ${}^{234}Th$ by $\alpha$ emission. a) What is the decay constant? b) What mass of uranium would be required for an activity of 1.00 curie? c) How many $\alpha$ particles are emitted per second by 20.0 g of uranium?

**45–16** An unstable isotope of cobalt, ${}^{60}Co$, has one more neutron in its nucleus than the stable ${}^{59}Co$ and is a $\beta$ emitter with a half-life of 5.27 years. This isotope is widely used in medicine. A certain radiation source in a hospital contains 0.0300 mg of ${}^{60}Co$. a) What is the decay constant for this isotope? b) How many atoms are in the source? c) How many decays occur per second? d) What is the activity of the source in curies? How does this compare with the activity of an equal mass of radium (${}^{226}Ra$, half-life 1600 y)?

**45–17** A radioactive isotope has a half-life of 76.0 min. A sample is prepared that has an initial activity of $16.0 \times 10^{10}$ Bq. a) How many radioactive nuclei are initially present in the sample? b) How many are present after 76.0 min? What is the activity at this time? c) Repeat part (b) for a time of 152 min after the sample is first prepared.

**45–18 Radiocarbon Dating.** A sample from timbers at an archeological site containing 500 g of carbon provides 5010 decays per minute. What is the age of the sample?

**45–19** The unstable isotope ${}^{40}K$ is used for dating of rock samples. Its half-life is $1.28 \times 10^9$ years. a) How many decays occur per second in a sample containing $5.00 \times 10^{-6}$ g of ${}^{40}K$? b) What is the activity of the sample in curies?

### Section 45–6 Biological Effects of Radiation

**45–20** In an experimental radiation therapy, an equivalent dose of 0.600 rem is given by 0.800-MeV protons to a localized area of tissue with mass 0.150 kg. a) What is the absorbed dose in rads? b) How many protons are absorbed by the tissue? c) How many $\alpha$ particles of the same energy of 0.800 MeV are required to deliver the same equivalent dose of 0.600 rem?

**45–21** In a diagnostic x-ray procedure, $8.00 \times 10^{10}$ photons are absorbed by tissue with mass 0.600 kg. The x-ray wavelength is 0.0200 nm. a) What is the total energy absorbed by the tissue? b) What is the equivalent dose in rem?

**45–22** An amount of a radioactive source with a very long lifetime and activity 0.30 $\mu$Ci is ingested into a person's body. The radioactive material lodges in the lungs, where all of the 4.0-MeV $\alpha$ particles emitted are absorbed within a 0.50-kg mass of tissue. Calculate the absorbed dose and the equivalent dose for one year.

**45–23** A 50-kg person accidently ingests 2.0 Ci of tritium. Assume that the tritium spreads uniformly over the body and that each decay leads on average to the absorption of 5.0 keV of energy from the electrons emitted in the decay. The half-life of tritium is 12.3 y, and the RBE of the electrons is 1.0. Calculate the absorbed dose in rad and the equivalent dose in rem during one week.

## Section 45–7 Nuclear Reactions

## Section 45–8 Nuclear Fission

## Section 45–9 Nuclear Fusion

**45–24 Heat of Fission.** In the fission of one $^{238}$U nucleus, 200 MeV of energy is released. Express this energy in joules per mole, and compare with typical heats of combustion, which are of the order of $1.0 \times 10^5$ J/mol.

**45–25** Consider the nuclear reaction

$$^{2}_{1}\mathrm{H} + {}^{9}_{4}\mathrm{Be} \longrightarrow {}^{7}_{3}\mathrm{Li} + {}^{4}_{2}\mathrm{He}$$

a) How much energy is liberated? b) Estimate the threshold energy for this reaction.

**45–26** Consider the nuclear reaction

$$^{4}_{2}\mathrm{He} + {}^{7}_{3}\mathrm{Li} \longrightarrow {}^{10}_{5}\mathrm{B} + {}^{1}_{0}\mathrm{n}.$$

Is energy absorbed or liberated? How much?

**45–27** Consider the nuclear reaction

$$^{2}_{1}\mathrm{H} + {}^{14}_{7}\mathrm{N} \longrightarrow {}^{6}_{3}\mathrm{Li} + {}^{10}_{5}\mathrm{B}.$$

a) Calculate the $Q$-value for this reaction. b) If the $^{2}_{1}$H is incident on a stationary $^{14}_{7}$N, what minimum kinetic energy must it have for the reaction to occur?

**45–28 Energy from Nuclear Fusion.** Calculate the energy released in the fusion reaction

$$^{2}_{1}\mathrm{H} + {}^{3}_{1}\mathrm{H} \longrightarrow {}^{4}_{2}\mathrm{He} + {}^{1}_{0}\mathrm{n}.$$

The atomic mass of $^{3}_{1}$H (tritium) is 3.016049 u.

# PROBLEMS

**45–29** The experimentally determined mass of the nuclide $^{26}_{12}$Mg is 25.9760 u. (This is the nuclear mass and does not include the mass of the 12 electrons that would be present in the neutral magnesium atom.) Calculate the mass from the semiempirical mass formula (Eq. 45–10). What is the percent error of the result as compared to the experimental value? What percent error is made if the $E_B$ term is neglected entirely?

**45–30** Thorium $^{230}_{90}$Th decays to radium $^{226}_{88}$Ra by $\alpha$ emission. The masses of the neutral atoms are 230.033128 u and 226.025403 u. If the parent thorium nucleus is at rest, what is the kinetic energy of the emitted $\alpha$ particle? (Be sure to account for the recoil of the daughter nucleus.)

**45–31** Gold $^{198}_{79}$Au undergoes $\beta^-$ decay to an excited state of $^{198}_{80}$Hg. If the excited state decays by emission of a $\gamma$ photon with energy 0.412 MeV, what is the maximum kinetic energy of the electron emitted in the decay? (The recoil energy of the $^{198}_{80}$Hg can be neglected.) The masses of the neutral atoms in their ground states are 197.968217 u and 197.966743 u.

**45–32** The results of activity measurements on a radioactive sample are given below. a) Find the half-life. b) How many radioactive nuclei were present in the sample at $t = 0$? c) How many were present after 7.0 h?

| Time (h) | Counts/s |
|---|---|
| 0 | 20,000 |
| 0.5 | 14,800 |
| 1.0 | 11,000 |
| 1.5 | 8,130 |
| 2.0 | 6,020 |
| 2.5 | 4,460 |
| 3.0 | 3,300 |
| 4.0 | 1,810 |
| 5.0 | 1,000 |
| 6.0 | 550 |
| 7.0 | 300 |

**45–33** Measurements indicate that 27.83% of all rubidium atoms currently on earth are the radioactive $^{87}$Rb isotope. The rest are the stable $^{85}$Rb isotope. The half-life of $^{87}$Rb is $4.89 \times 10^{10}$ y. Assuming that no rubidium atoms have been formed since, what percentage of rubidium atoms were $^{87}$Rb when the solar system was formed $4.6 \times 10^9$ y ago?

**45–34** A 70.0-kg person experiences a whole-body exposure to $\alpha$ radiation with energy 4.77 MeV. A total of $2.00 \times 10^{12}$ $\alpha$ particles are absorbed. a) What is the absorbed dose in rads? b) What is the equivalent dose in rems? c) If the source is 0.0100 g of $^{226}$Ra (half-life 1600 years) somewhere in the body, what is the activity of this source? d) If all the $\alpha$ particles produced are absorbed, what time is required for this dose to be delivered?

**45–35** A $^{60}$Co source with activity 35.0 $\mu$Ci is embedded in a tumor that has a mass of 0.500 kg. The Co source emits $\gamma$ photons with average energy 1.25 MeV. Half the photons are absorbed in the tumor, and half escape. a) What energy is delivered to the tumor per second? b) What absorbed dose (in rads) is delivered per second? c) What equivalent dose (in rems) is delivered per second if the RBE for these $\gamma$ rays is 0.70? d) What exposure time is required for an equivalent dose of 200 rem?

**45–36** The nucleus ${}^{15}_{8}O$ has a half-life of 2.0 min; ${}^{19}_{8}O$ has a half-life of 0.45 min. a) If at some time a sample contains equal amounts of ${}^{15}_{8}O$ and ${}^{19}_{8}O$, what is the ratio of ${}^{15}_{8}O$ to ${}^{19}_{8}O$ after 2.0 min; b) after 10.0 min?

**45–37** A carbon specimen found in a cave believed to have been inhabited by early humans contained $\frac{1}{8}$ as much ${}^{14}C$ as an equal amount of carbon in living matter. Find the approximate age of the specimen.

**45–38** A patient is given a 2.0 $\mu$Ci dose of a radioactive isotope with a half-life of 2.0 h. Assuming that the entire dose remains in the body, how much time must elapse before the activity of the radioactive isotope is 8.5 counts/min, which is about three times the normal background of 3 counts/min?

**45–39 Neutron Decay Energy.** A free neutron at rest decays into a proton, an electron, and a neutrino, with a lifetime of about 15 min. Calculate the total kinetic energy of the decay products.

**45–40** a) What is the binding energy of the least strongly bound proton in ${}^{12}_{6}C$? b) The least strongly bound neutron in ${}^{13}_{6}C$?

**45–41** Consider the fusion reaction

$${}^{2}_{1}H + {}^{2}_{1}H \longrightarrow {}^{3}_{2}He + {}^{1}_{0}n$$

a) Estimate the barrier energy. b) Compute the energy liberated in this reaction in MeV and in joules. c) Compute the energy *per mole* of deuterium, remembering that the gas is diatomic, and compare with the heat of combustion of hydrogen, about $2.9 \times 10^5$ J/mol.

**45–42** About one-eighth of the ${}^{137}Cs$ present in the Chernobyl reactor was released. The isotope ${}^{137}Cs$ has a half-life for $\beta$ decay of 30.17 y and decays with the emission of a total of 1.17 MeV of energy per decay. Of this, 0.51 MeV goes to the emitted electron, and the remaining 0.66 MeV goes to a $\gamma$ ray. The radioactive ${}^{137}Cs$ is absorbed by plants, which are eaten by livestock and humans. How many ${}^{137}Cs$ atoms would need to be present in each kilogram of body tissue if an equivalent dose for one week is 3.0 Sv? Assume that all of the energy from the decay is deposited in that 1.0 kg of tissue and that the RBE of the electrons is 1.5.

**45–43 Natural Body Radioactivity.** On average, 0.21% of the mass of the human body is potassium, of which 0.012% is radioactive ${}^{40}K$, with a half-life of $1.25 \times 10^9$ y. From each decay, an average of 0.50 MeV of $\beta$ and $\gamma$ radiation is absorbed by the body. Take the RBE of the $\beta$ radiation to be 1.0. Calculate the absorbed dose in rads and the equivalent dose in rems for 70 y from the potassium in the body.

**45–44** A ${}^{186}Os$ nucleus at rest decays by the emission of a 2.76-MeV $\alpha$ particle. Calculate the atomic mass of the daughter nuclide produced by this decay, assuming that it is produced in its ground state. The atomic mass of ${}^{186}Os$ is 185.953830 u.

**45–45** Prove that when a particle with mass $m$ and kinetic energy $K$ collides with a stationary particle with mass $M$, the total kinetic energy $K_{cm}$ in the center-of-mass coordinate system (the energy available to cause reactions) is

$$K_{cm} = \left(\frac{M}{M + m}\right)K.$$

**45–46** Calculate the energy released in the fission reaction

$${}^{235}_{92}U + {}^{1}_{0}n \longrightarrow {}^{144}_{56}Ba + {}^{89}_{36}Kr + 3{}^{1}_{0}n.$$

Neglect the initial kinetic energy of the absorbed neutron. Use Eq. (45–9) to calculate the binding energies of the nuclei.

## CHALLENGE PROBLEMS

**45–47** The results of activity measurements on a mixed sample of radioactive elements is shown below. a) How many different nuclides are present in the mixture? b) What are their half-lives? c) How many nuclei of each type are initially present in the sample? d) How many of each type are present at $t = 6.0$ h?

| Time (h) | Counts/s |
|---|---|
| 0 | 7500 |
| 0.5 | 4120 |
| 1.0 | 2570 |
| 1.5 | 1790 |
| 2.0 | 1350 |
| 2.5 | 1070 |
| 3.0 | 872 |
| 4.0 | 596 |
| 5.0 | 414 |
| 6.0 | 288 |
| 7.0 | 201 |
| 8.0 | 140 |
| 9.0 | 98 |
| 10.0 | 68 |
| 12.0 | 33 |

**45–48 Industrial Radioactivity.** Radioisotopes are used in a variety of manufacturing and testing techniques. Wear measurements can be made using the following method. An automobile engine is produced with piston rings having a total mass of 100 g, which includes 10 $\mu$Ci of ${}^{59}Fe$ whose half-life is 45 days. The engine is test run for 1000 hours, after which the oil is drained and its activity is measured. If the activity of the engine oil is 70 decays/s, how much mass was worn from the piston rings per hour of operation?

**45–49** The isotope ${}^{128}I$ is created by the irradiation of ${}^{127}I$ with neutrons, using a neutron beam that creates $1.0 \times 10^6$ ${}^{128}I$ nuclei per second. The half-life of ${}^{128}I$ is 25 min. a) Sketch a graph of the number of ${}^{128}I$ nuclei present as a function of time. b) What is the activity of the sample 1, 10, 25, 50, 75, and 180 minutes after irradiation is begun? c) What is the maximum number of ${}^{128}I$ atoms that can be created in the sample after it is irradiated for a long time? (This steady-state situation is called *saturation.*) d) What is the maximum activity that can be produced?

C H A P T E R 46

# Particle Physics and Cosmology

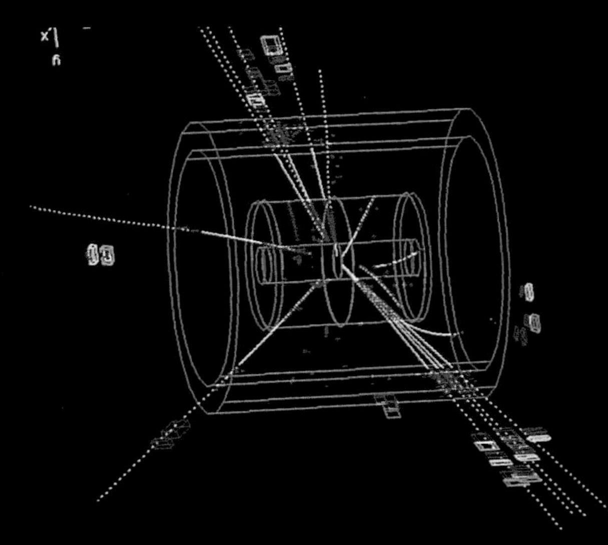

This computer-generated image from the Delphi particle detector at CERN shows a collision of fundamental particles. A $Z^0$ particle decays into two jets of hadrons and two isolated muons. The study of these high-energy particle interactions helps physicists understand the kinds of interactions that take place in the high-energy environments of galaxies and other astronomical objects.

Our home, the Milky Way Galaxy, as seen at near-infrared wavelengths by the Cosmic Background Explorer (COBE) satellite. Our solar system is about 28,000 light-years from the center of the Milky Way, which is a region of incredibly intense activity. Several million stars are packed closely together here, causing tremendous gravitational forces. Measurements of gamma radiation suggest that electron-positron annihilation takes place in the galaxy's center at a rate of $10^{13}$ kg of electrons per second.

## KEY CONCEPTS

**• Fundamental particles can be created and destroyed. Each particle has a corresponding antiparticle with the same mass and opposite charge. Interactions are mediated by exchange of particles.**

**• Particle accelerators provide beams of high-energy particles. Particle energies up to 1 TeV have been reached. With a stationary target, only part of the beam energy is available to cause reactions. Colliding-beam experiments avoid this limitation.**

**• Four classes of interactions are found in nature. Particles are classified in terms of their interactions, and interactions are classified in terms of conservation laws.**

**• Hadrons are composed of fundamental particles called quarks. In the standard model, the six leptons and six quarks are the truly fundamental particles.**

**• The electromagnetic and weak interactions are aspects of the same interaction. It may be possible to unify all four interactions. The early history of the universe is closely related to the behavior of fundamental particles.**

## INTRODUCTION

What is the world made of? What are the most fundamental constituents of matter? Philosophers and scientists have been asking this question for at least 2500 years. We still don't have anything that could be called a final answer, but we've come a long way. This final chapter is a progress report on where we are now and where we hope to go.

The chapter title, "Particle Physics and Cosmology," may seem strange. Fundamental particles are the *smallest* things in the universe, and cosmology deals with the *biggest* thing there is—the universe itself. Nonetheless, we'll see that there are very close ties between these two areas.

Fundamental particles, we'll find, are not permanent entities; they can be created and destroyed. The development of high-energy particle accelerators and associated detectors has played an essential role in our emerging understanding of particles. We can classify particles and their interactions in several ways in terms of conservation laws and symmetries, some of which are absolute and others obeyed only in certain kinds of interactions. We'll try in this final chapter to convey some idea of the remaining frontiers in this vital and exciting area of fundamental research.

## 46–1 FUNDAMENTAL PARTICLES

The idea that the world is made of *fundamental particles* has a long history. In about 400 B.C. the Greek philosophers Democritus and Leucippus suggested that matter is made of indivisible particles that they called *atoms,* a word derived from *a*- (not) and *tomos* (cut or divided). This idea lay dormant until about 1804, when John Dalton (1766–1844), often called the father of modern chemistry, discovered that many chemical phenomena could be explained on the basis of atoms of each element as the basic, indivisible building blocks of matter.

Toward the end of the nineteenth century it became clear that atoms are *not* indivisible. The existence of characteristic atomic spectra of elements suggested that atoms have internal structure, and J. J. Thomson's discovery of the electron in 1897 showed that atoms could be taken apart into charged particles. The hydrogen nucleus was identified as a proton, and in 1911 the sizes of nuclei were measured by Rutherford's experiments. Quantum mechanics, including the Schrödinger equation, blossomed during the next 15 years. Scientists were on their way to understanding the principles that underlie atomic structure, although many details remained to be worked out.

### The Neutron

The discovery of the neutron in 1930 was an important milestone. In that year, two German physicists, Bothe and Becker, observed that when beryllium, boron, or lithium was bombarded by $\alpha$ particles from the radioactive element polonium, the target material emitted a radiation with much greater penetrating power than the original $\alpha$ particles. Experiments by James Chadwick the following year showed that the emanation consisted of electrically neutral particles with mass approximately equal to that of the proton. Chadwick christened these particles *neutrons.* A typical reaction, using a beryllium target, is

$$^{4}_{2}\text{He} + {}^{9}_{4}\text{Be} \longrightarrow {}^{12}_{6}\text{C} + {}^{1}_{0}\text{n}. \qquad (46\text{–}1)$$

Because neutrons have no charge, they produce no ionization when they pass through matter, and they are not deflected by electric or magnetic fields. They interact only with nuclei; they can be slowed down during elastic scattering, and they can penetrate the nucleus. Slow neutrons can be detected by means of another nuclear reaction, the ejection of an $\alpha$ particle from a boron nucleus:

$$^{1}_{0}\text{n} + {}^{10}_{5}\text{B} \longrightarrow {}^{7}_{3}\text{Li} + {}^{4}_{2}\text{He}. \qquad (46\text{–}2)$$

The ejected $\alpha$ particle is easy to detect because of its charge.

The neutron was a welcome discovery because it cleared up a mystery about the composition of the nucleus. Before 1930, the mass of a nucleus was thought to be due only to protons, but no one understood why the charge-to-mass ratio was not the same for all nuclei. It soon became clear that all nuclei (except hydrogen) contain both protons and neutrons. In fact, the proton, the neutron, and the electron are the basic building blocks of atoms. One might think that would be the end of the story. On the contrary, it is barely the beginning.

### The Positron

The positive electron, or *positron,* was discovered by Carl D. Anderson in 1932, during an investigation of cosmic rays (very high-energy particles that come from outer space).

Figure 46–1 shows a historic photograph made with a *cloud chamber,* an instrument used to visualize the tracks of charged particles. The chamber contained a supercooled vapor; ions created by the passage of charged particles through the vapor served as nucleation centers, and liquid droplets formed around them, making a visible track.

The cloud chamber in Fig. 46–1 is in a magnetic field perpendicular to the plane of the chamber. The particle has passed through a thin lead plate in the chamber. The curvature of the track is greater above the plate than below it, showing that the speed was *less* above the plate than below. Therefore the particle had to be moving upward; it could not have *gained* energy passing through the lead. The density and curvature of the track suggested mass and charge equal to those of the electron. But the directions of the magnetic field and the velocity showed that the particle had *positive* charge. Anderson christened it the *positron*.

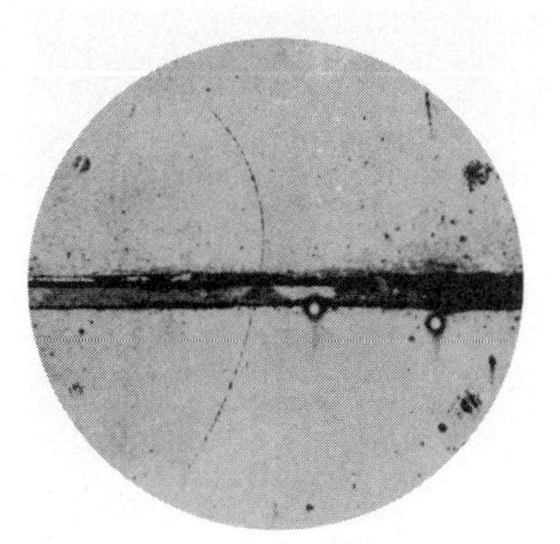

**46–1** Photograph of the cloud-chamber track made by the first positron ever observed. The photograph was made by Carl D. Anderson in 1932. The lead plate is 6 mm thick.

To theorists, the appearance of the positron was a welcome development. In 1928, P. A. M. Dirac had developed a relativistic generalization of the Schrödinger equation for the electron. One of its significant features was that it predicted the spin magnetic moment of the electron very well. We noted in Section 43–3 that the gyromagnetic ratio for electron spin is almost exactly twice as great ($e/m_e$) as the value for orbital angular momentum ($e/2m_e$). We had to put this ratio in as an *ad hoc* assumption, but it comes out of the Dirac equation automatically.

One of the puzzling features of the Dirac equation was that for free electrons it predicted not only a continuum of energy levels greater than $m_ec^2$ (including rest energy), as should be expected, but also a continuum of *negative* energy levels *less than* $-m_ec^2$ (Fig. 46–2a). That posed a problem; what was to prevent an electron from emitting a photon with energy $2m_ec^2$ or greater and hopping from a positive level to a negative level? It wasn't clear what these negative-energy levels meant, and there was no obvious way to get rid of them. Dirac's ingenious but somewhat implausible interpretation was that all the negative-energy states were filled with electrons that were for some reason unobservable. The exclusion principle would then forbid a transition to such an already occupied state.

A vacancy in a negative-energy level would act like a positive charge, just as a hole in the valence band of a semiconductor (Section 44–7) acts like a positive charge. Initially, Dirac tried to argue that such holes were protons. But after the discovery of the positron it became clear that the holes were observed physically as *positrons.* Furthermore, the Dirac energy-level picture provides a mechanism for the *creation* of positrons. When an electron in a negative-energy state absorbs a photon with energy greater than $2m_ec^2$, it goes to a positive state (Fig. 46–2b), in which it becomes observable. The hole that it leaves behind is observed as a positron; the result is creation of an electron-positron pair. Similarly, when an electron in a positive state falls into a hole, both disappear, and photons are emitted (Fig. 46–2c). Thus the Dirac theory leads naturally to the conclusion that, like photons, **electrons can be created and destroyed.** Photons can be created and destroyed singly, electrons only in pairs or in association with other particles. We'll see why later.

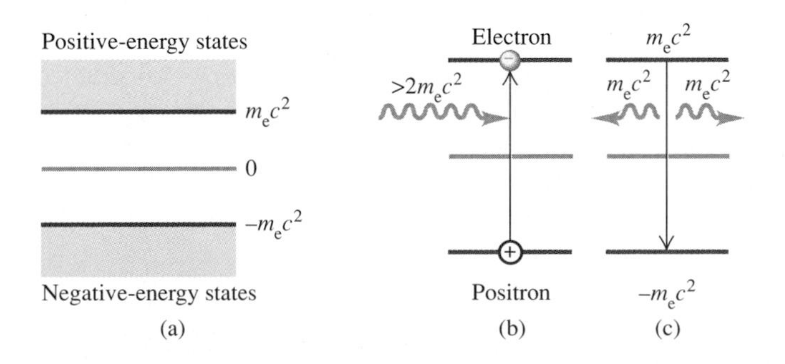


**46–2** (a) Energy levels for a free electron predicted by the Dirac equation. There is a continuum of levels with energies greater than $m_ec^2$ and another continuum of negative-energy states with energies less than $-m_ec^2$. (b) Raising of an electron from a negative- to a positive-energy state corresponds to electron-positron pair production. (c) An electron dropping from a positive-energy state to an unoccupied negative-energy state corresponds to positron annihilation.

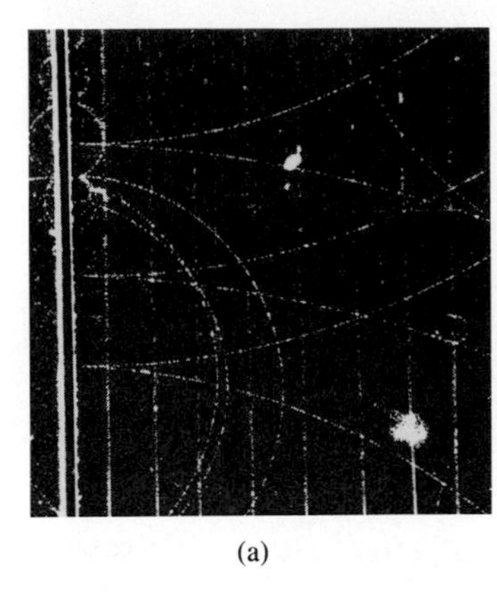

(a)

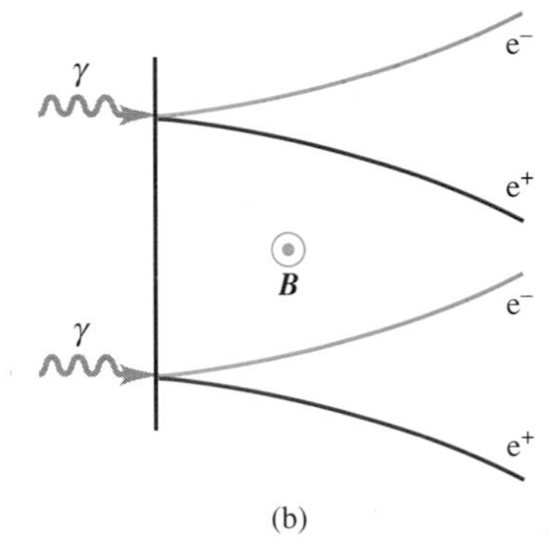


(b)

**46–3** (a) Photograph of bubble-chamber tracks of electron-positron pairs produced when 300-MeV $\gamma$ photons strike a lead sheet. A magnetic field perpendicular to the plane makes the electrons and positrons curve in opposite directions. (b) Diagram showing the pair-production process.

More recently, the Dirac theory has been reformulated to eliminate the infinite sea of negative-energy states and put electrons and positrons on the same footing. But the creation and destruction of electron-positron pairs remain. The Dirac theory is inherently a *many-particle* theory, and it provides the beginning of a theoretical framework for creation and destruction of all fundamental particles.

The masses of the positron and electron are identical; their charges are equal in magnitude but opposite in sign. The positron's spin angular momentum and magnetic moment have the same direction; they are opposite for the electron. Pairs of particles that are related to each other in this way are called **antiparticles.** We'll see several other examples of antiparticles later. In the following discussion we'll use the symbols $e^-$ for the electron and $e^+$ for the positron, and the generic term *electron* will often include both electrons and positrons. For other particles the antiparticle is often denoted by a bar over the symbol; for example, an antiproton is $\overline{p}$.

Positrons don't occur in ordinary matter. Electron-positron pairs are produced during high-energy collisions of charged particles or $\gamma$ rays with matter; the process is called *pair production* (Fig. 46–3). Electric charge is conserved, and enough energy $E$ must be available to account for the rest energy $2m_ec^2$ of the two particles. The minimum energy for pair production is

$$E = 2m_ec^2 = 2(9.11 \times 10^{-31}\text{ kg})(3.00 \times 10^8\text{ m/s})^2$$
$$= 1.64 \times 10^{-13}\text{ J} = 1.02\text{ MeV}.$$

The inverse process, $e^+e^-$ *annihilation,* occurs when a positron and an electron collide. Both particles disappear, and two or three $\gamma$ photons appear, with total energy of at least $2m_ec^2$. Decay into a *single* photon is impossible because such a process could not conserve both energy and momentum.

Positrons also occur in the decay of some unstable nuclei. Nuclei with *too many* neutrons for stability often emit a $\beta^-$ particle ($e^-$), converting a neutron to a proton and decreasing $N$ by one and increasing $Z$ by one. A nucleus with *too few* neutrons for stability may decay by converting a proton to a neutron, emitting a *positron,* increasing $N$ by one and decreasing $Z$ by one. Such nuclides don't occur in nature, but they can be produced artificially in nuclear reactors. An example is the unstable odd-odd nuclide $^{22}$Na, which has one less neutron than the stable $^{23}$Na. It decays with a half-life of 2.6 y by emitting a positron. The daughter nucleus is the stable even-even nuclide $^{22}$Ne. Positron emitters are used in positron emission tomography (PET), mentioned in Section 45–6.

## Particles as Force Mediators

In classical physics we describe the interaction of charged particles in terms of Coulomb's-law forces. In quantum mechanics we can describe this interaction in terms of emission and absorption of photons. Two electrons repel each other as one emits a photon and the other absorbs it, just as two skaters can push each other apart by tossing a ball back and forth between them (Fig. 46–4a). If the charges are opposite and the force is attractive, we imagine the skaters grabbing the ball away from each other (Fig. 46–4b). The electromagnetic interaction between two charged particles is *mediated,* or transmitted, by photons.

If charged-particle interactions are mediated by photons, where does the energy to create the photons come from? Recall our discussion of the uncertainty principle (Section 41–3). A state that exists for a short time $\Delta t$ has an uncertainty $\Delta E$ in its energy such that

$$\Delta E\,\Delta t \geq \hbar. \tag{46–3}$$

This uncertainty permits the creation of a photon with energy $\Delta E$, provided that it lives no longer than $\Delta t$ as given by Eq. (46–3). A photon that can exist for a short time because

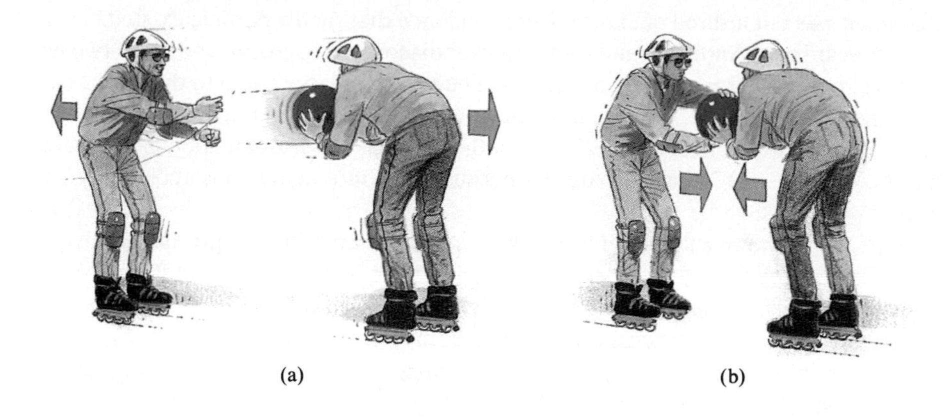


**46–4** (a) Two skaters exert repulsive forces on each other by tossing a ball back and forth. (b) Two skaters exert attractive forces on each other when one grabs the ball out of the other's hands.

of this energy uncertainty is called a *virtual photon*. It's as though there were an energy bank; you can borrow energy provided that you pay it back within the time limit. According to Eq. (46–3), the more you borrow, the sooner you have to pay it back. Later we'll discuss other virtual particles that live for a short time on "borrowed" energy.

## Mesons

Is there a particle that mediates the *nuclear* force? In 1935, the force between two nucleons appeared to be described by a potential energy $U(r)$ with the general form

$$U(r) = -f^2\frac{e^{-r/r_0}}{r} \quad \text{(nuclear potential).} \tag{46–4}$$

The constant $f$ characterizes the strength of the interaction, and $r_0$ describes its range. Figure 46–5 shows a graph of this function and compares it with the function $f^2/r$, which would be analogous to the *electrical* interaction of two electrons:

$$U(r) = \frac{1}{4\pi\epsilon_0}\frac{e^2}{r} \quad \text{(electrical potential).} \tag{46–5}$$

In 1935, the Japanese physicist Hideki Yukawa suggested that a hypothetical particle he called a **meson** might act as a mediator of the nuclear force. He showed that the range of the force was related to the mass of the particle. His argument went like this: The particle must live for a time $\Delta t$ that is long enough to travel a distance comparable to the range of the nuclear force, which was known from the sizes of nuclei and other information to be of the order of $r_0 = 1.5 \times 10^{-15}$ m $= 1.5$ fm. Assuming that the particle's speed is comparable to $c$, its lifetime $\Delta t$ must be of the order of

$$\Delta t = \frac{r_0}{c} = \frac{1.5 \times 10^{-15}\ \text{m}}{3.0 \times 10^{8}\ \text{m/s}} = 5.0 \times 10^{-24}\ \text{s}.$$

From Eq. (46–3) the necessary uncertainty $\Delta E$ in energy is

$$\Delta E = \frac{\hbar}{\Delta t} = \frac{1.05 \times 10^{-34}\ \text{J}\cdot\text{s}}{5.0 \times 10^{-24}\ \text{s}} = 2.1 \times 10^{-11}\ \text{J} = 130\ \text{MeV}.$$

The mass equivalent $\Delta m$ of this energy is

$$\Delta m = \frac{\Delta E}{c^2} = \frac{2.1 \times 10^{-11}\ \text{J}}{(3.00 \times 10^{8}\ \text{m/s})^2} = 2.3 \times 10^{-28}\ \text{kg}.$$

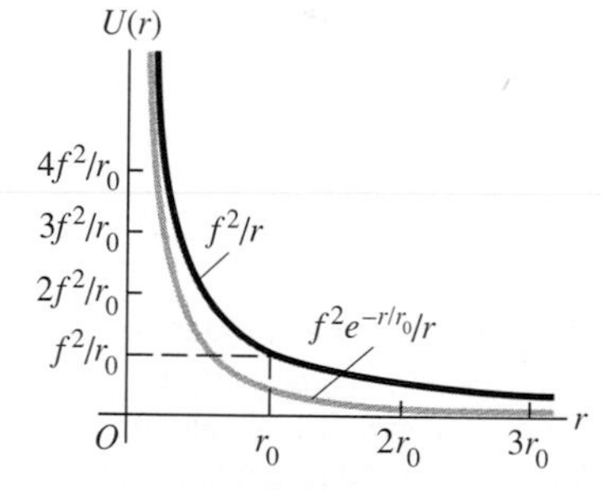


**46–5** Graph of the magnitude of the Yukawa potential-energy function for nuclear forces, $|U(r)| = f^2 e^{-r/r_0}/r$. The function approaches zero very rapidly for $r > r_0$. For comparison, the function $U(r) = f^2/r$, analogous to Coulomb's law, is also shown. The two functions are similar at small $r$, but the Yukawa potential drops off much more quickly at large $r$.

This is about 250 times the electron mass, and Yukawa postulated a particle with this mass to serve as the messenger for the nuclear force. This was a courageous act; at the time there was not a shred of experimental evidence that such a particle existed.

A year later, Anderson and Neddermeyer discovered in cosmic radiation two new particles, now called $\mu$ mesons, or **muons.** The $\mu^-$ has charge equal to that of the electron, and its antiparticle, the $\mu^+$, has a positive charge with equal magnitude. The two particles have equal mass, about 207 times the electron mass. But it soon became clear that muons were *not* Yukawa's particles because they interacted with nuclei only very weakly.

In 1947, *another* family of three mesons, called $\pi$ mesons, or **pions,** was discovered. Their charges are $+e$, $-e$, and zero, and their masses are about 270 times the electron mass. The pions interact strongly with nuclei, and they *are* the particles predicted by Yukawa. We'll discuss mesons further in later sections.

## 46–2 PARTICLE ACCELERATORS

Early nuclear physics experiments used $\alpha$ and $\beta$ particles from naturally occurring radioactive elements. Present-day accelerators can produce precisely controlled beams of high-energy particles such as protons and electrons. These accelerators use electric and magnetic fields to accelerate and guide beams of charged particles. A *linear accelerator* (LINAC) accelerates particles in a straight line. The earliest examples were the cathode-ray tubes of Thomson and the x-ray tubes of Coolidge. More sophisticated LINACs use a series of electrodes with gaps, to give the beam of particles a series of boosts. Most present-day high-energy linear accelerators use a traveling electromagnetic wave; the charged particles "ride" the wave more or less the way a surfer rides an incoming ocean wave. The highest-energy LINAC in the world today is the Stanford linear accelerator (SLAC), in which electrons and positrons are accelerated to 50 GeV in a tube 2 miles long.

### The Cyclotron

Many accelerators use magnets to deflect the charged particles into circular paths. The first of these was the *cyclotron,* invented in 1931 by Lawrence and Livingston at the University of California. In the cyclotron, shown schematically in Fig. 46–6a, particles with mass $m$ and charge $q$ move inside a vacuum chamber in a uniform magnetic field $\boldsymbol{B}$ per-

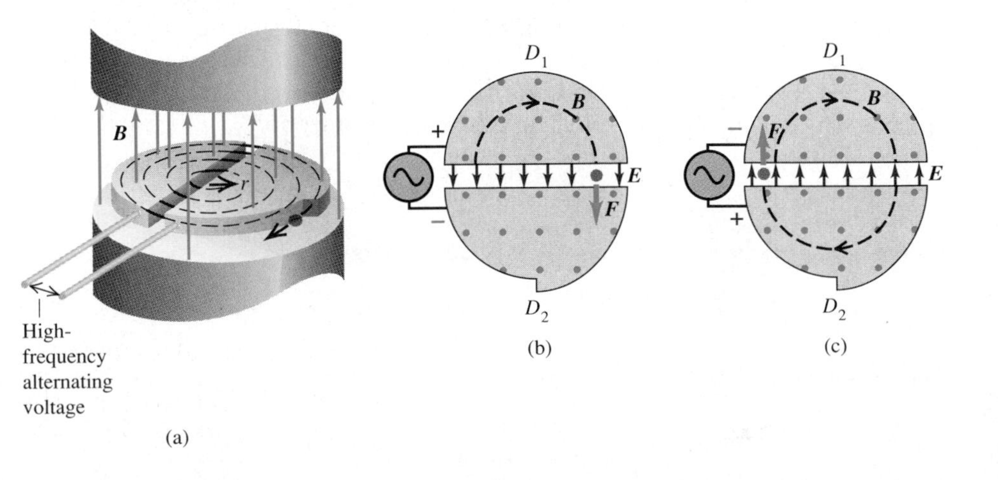


**46–6** (a) Schematic diagram of a cyclotron. (b) As the particle reaches the gap, it is accelerated by the electric-field force; the next semicircular orbit has somewhat larger radius. (c) By the time the particle reaches the gap again, the dee voltage has reversed, and the particle is again accelerated.

pendicular to the plane of their paths. The discussion in Section 28–4 showed that in such a field a particle with speed $v$ moves in a circular path with radius $r$ given by

$$r = \frac{mv}{|q|B} \tag{46–6}$$

and angular frequency

$$\omega = \frac{v}{r} = \frac{|q|B}{m}. \tag{46–7}$$

An alternating potential difference is applied between the two hollow electrodes $D_1$ and $D_2$ (called *dees*), creating an electric field in the gap between them. The potential and field change polarity twice each revolution (Figs. 46–6b and 46–6c), so the particles get a push each time they pass the gap. This boosts them into paths with larger radius, speed, and kinetic energy. The maximum speed $v_{max}$ and kinetic energy $K_{max}$ are determined by the radius $R$ of the electromagnet poles. Solving Eq. (46–6) for $v$, we find $v = |q|Br/m$, $v_{max} = |q|BR/m$, and

$$K_{max} = \tfrac{1}{2}mv_{max}{}^2 = \frac{q^2B^2R^2}{2m}. \tag{46–8}$$

## EXAMPLE 46–1

A cyclotron built during the 1930s has magnet poles with radius 0.50 m and a magnetic field with magnitude 1.50 T. Find the maximum particle energy if this cyclotron is used to accelerate protons.

**SOLUTION** For protons, $q = 1.60 \times 10^{-19}\,\text{C}$ and $m = 1.67 \times 10^{-27}$ kg. From Eq. (46–8) the maximum kinetic energy is

$$K_{max} = \frac{(1.60 \times 10^{-19}\,\text{C})^2 (1.50\,\text{T})^2 (0.50\,\text{m})^2}{2(1.67 \times 10^{-27}\,\text{kg})}.$$
$$= 4.31 \times 10^{-12}\,\text{J} = 2.7 \times 10^7\,\text{eV} = 27\,\text{MeV}.$$

This energy, which is much larger than energies available from natural radioactivity, can cause a variety of interesting nuclear reactions.

The maximum energy that can be attained with a cyclotron is limited by relativistic effects. The relativistic version of Eq. (46–7) is

$$\omega = \frac{|q|B}{m}\sqrt{1 - v^2/c^2}.$$

As the particles speed up, their angular frequency $\omega$ *decreases,* and their motion gets out of phase with the alternating dee voltage. In the *synchrocyclotron* the particles are accelerated in bursts. For each burst, the frequency of the alternating voltage is decreased as the particles speed up, maintaining the correct phase relation with the particles' motion.

Another limitation of the cyclotron is the difficulty of building very large electromagnets. The largest synchrocyclotron ever built has a vacuum chamber about 8 m in diameter and accelerates protons to energies of about 600 MeV.

## The Synchrotron

To attain higher energies, another type of machine, called the *synchrotron,* is more practical. Particles move in a vacuum chamber in the form of a thin doughnut, called the *accelerating ring.* The particle beam is bent around by a series of electromagnets placed around the ring. As the particles speed up, the magnetic field is increased so that the

(a) (b)

**46–7** The Tevatron, the 1000-GeV (1-TeV) accelerator at the Fermi National Accelerator Laboratory, Batavia, Illinois, the highest-energy accelerator in the world. (a) An aerial view of the main accelerator ring. (b) A section of the main tunnel. The original ring of conventional magnets (red and blue) and the new ring of superconducting magnets (yellow) can be seen. The conventional ring is used to inject protons and antiprotons into the superconducting ring. The two beams travel in opposite directions as they are accelerated to 1 TeV, and then they collide at interaction points.

particles retrace the same trajectory over and over. The Tevatron at the Fermi National Accelerator Laboratory (Fermilab) in Batavia, Illinois, is currently the world's highest-energy accelerator; it accelerates protons to a maximum energy of 1000 GeV. The accelerating ring is 2 km in diameter and uses superconducting electromagnets (Fig. 46–7). The accelerator and associated facilities cost about \$400 million to build. In each machine cycle, a few seconds long, it accelerates approximately $10^{13}$ protons.

## Available Energy

When a beam of high-energy particles collides with a stationary target, not all the kinetic energy of the incident particles is *available* to form new particle states. Because momentum must be conserved, the particles emerging from the collision must have some motion and thus some kinetic energy. Example 45–10 presented a nonrelativistic example of this principle. The maximum available energy is the kinetic energy in the frame of reference in which the total momentum is zero. We call this the *center-of-momentum system;* it is the relativistic generalization of the center-of-mass system we discussed in Section 8–7. In this system the total kinetic energy after the collision *can* be zero, so all the initial kinetic energy is available to cause the reaction being studied.

For a collision in which a target particle with mass $M$ is bombarded by a particle with mass $m$ and total energy (including rest energy) $E_m$, the total available energy $E_a$ in the center-of-momentum system (including rest energies of all the particles) is given by

$$E_a^2 = 2Mc^2E_m + (Mc^2)^2 + (mc^2)^2. \qquad (46\text{–}9)$$

When the masses of target and projectile particle are equal, this can be simplified to

$$E_a^2 = 2mc^2(E_m + mc^2). \qquad (46\text{–}10)$$

### EXAMPLE 46–2

**Threshold energy for pion production** A proton with kinetic energy $K$ collides with a proton at rest, producing a pion ($\pi^0$, rest energy 135 MeV). What minimum value of $K$ is needed?

**SOLUTION** The final state includes three particles: the two original protons (mass $m$) and the pion (mass $m_\pi$). The threshold energy corresponds to the case in which all three particles are at rest in the center-of-momentum system. The total available energy must be at least

$$E_a = (2m + m_\pi)c^2.$$

We substitute this expression into Eq. (46–10), simplify, and solve for $E_m$:

$$4m^2c^4 + 4mm_\pi c^4 + m_\pi{}^2c^4 = 2E_m mc^2 + 2(mc^2)^2,$$

$$E_m = mc^2 + m_\pi c^2\left(2 + \frac{m_\pi}{2m}\right).$$

The first term is the rest energy of the bombarding proton, and the remaining terms represent its kinetic energy. We see that the kinetic energy must be somewhat greater than twice the rest energy of the pion that we want to create. Using $mc^2 = 938$ MeV and $m_\pi c^2 = 135$ MeV in this expression, we find $m_\pi/2m = 0.072$ and

$$E_m = mc^2 + (135 \text{ MeV})(2 + 0.072)$$
$$= mc^2 + 280 \text{ MeV}.$$

To create a pion with a rest energy of 135 MeV requires a proton beam with kinetic energy of at least 280 MeV.

# Particle Detectors

In high-energy physics, particle accelerators and detectors are the primary experimental tools. The detectors, marvels of electronics and computer analysis, tell us what happens during a high-energy particle collision.

# BUILDING PHYSICAL INTUITION

The Collider Detector at Fermilab records over 100,000 proton-antiproton collisions per second. In this photo the arches pulled out to the sides contain parts of the calorimeter, which measures the energies of particles. The 5,000-ton detector slides into precise alignment with the Tevatron's particle beams so that collisions occur in the detector's center.

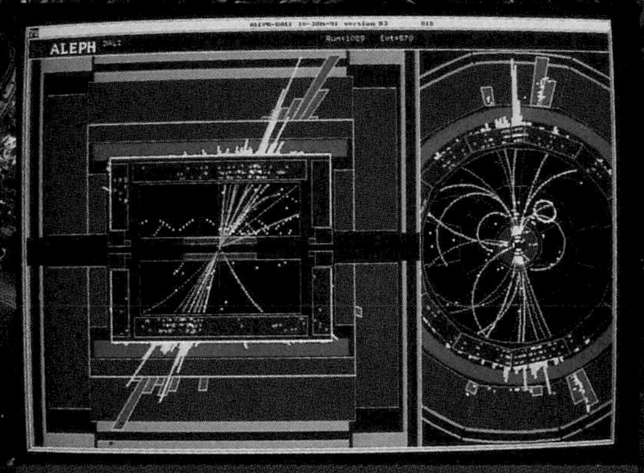


The Aleph detector at LEP shows side-on and end-on views of the same particle collision. Evidence gathered from analysis of these collisions has contributed greatly to our present understanding of fundamental particles and forces.

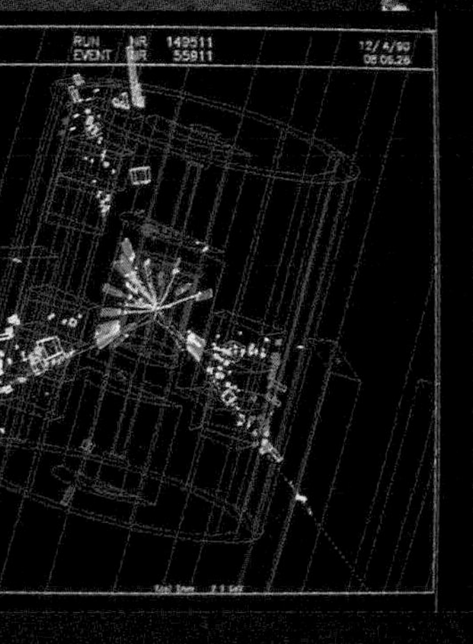


The L3 detector at the Large Electron-Positron (LEP) Collider at CERN (the European Laboratory for Particle Physics) shows graphic images of particle collisions based on the data collected. Red lines show the detector's structure; multicolored curves indicate particle trajectories; boxes show where particles hit the detector and how much energy was released. The detector records 500,000 numbers from each of 50,000 collisions per second: an astonishing $25 \times 10^9$ data values per second.

This computer-generated image shows a typical result of a proton-antiproton collision in the Tevatron as observed by the Collider Detector. Decay products of the collision include pions, kaons, muons, neutrinos, and others, each indicated by a separate color.

In the *extreme-relativistic range,* where the kinetic energy of the bombarding particle is much larger than its rest energy, available energy is a very severe limitation. Let's look again at Eq. (46–10), the case in which beam and target particles have equal masses. When $E_m$ is much greater than $mc^2$, we can neglect the second term on the right side. Then $E_a$ is

$$E_a = \sqrt{2mc^2E_m}. \tag{46–11}$$

## EXAMPLE 46–3

The Fermilab accelerator was originally designed for a proton-beam energy of 800 GeV ($800 \times 10^9$ eV). a) Find the available energy in a proton-proton collision. b) If the proton energy increases to 1000 GeV, what is the available energy?

**SOLUTION** For the proton, $mc^2 = 938\text{ MeV} = 0.938\text{ GeV}$.
a) When $E_m = 800$ GeV, then

$$E_a = \sqrt{2(0.938\text{ GeV})(800\text{ GeV})} = 38.7\text{ GeV}.$$

b) When $E_m = 1000$ GeV,

$$E_a = \sqrt{2(0.938\text{ GeV})(1000\text{ GeV})} = 43.3\text{ GeV}.$$

Increasing the beam energy by 200 GeV increases the available energy by only 4.6 GeV!

---

The limitation illustrated by Example 46–3 is circumvented in *colliding-beam experiments.* In these experiments, there is no stationary target; beams of particles and their antiparticles (such as electrons and positrons or protons and antiprotons) circulate in opposite directions in an arrangement called a *storage ring* or are accelerated simultaneously in a linear accelerator. In regions where the beams intersect, they are focused sharply onto one another, and collisions can occur. The total momentum in such a two-particle collision is zero; the laboratory system is the center-of-momentum system, and the available energy $E_a$ is the total energy of the two particles.

An example is the linear collider at the Stanford Linear Accelerator Center (SLAC) (Fig. 46–8). Electron and positron beams are accelerated alternately. At the end of the

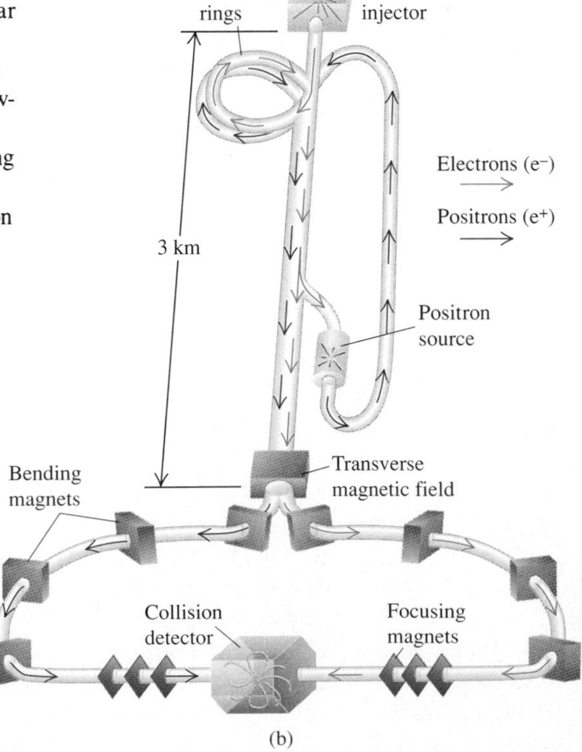


(a) (b)

**46–8** (a) The three-kilometer-long linear accelerator at the Stanford Linear Accelerator Center (SLAC) accelerates alternate bunches of positrons and electrons to 50 GeV; then magnets separate them and bend them around in the arcs (broken lines) to provide head-on $e^+e^-$ collisions. (b) Sketch showing the main components. The positrons are produced by pair production when part of the electron beam is diverted about two-thirds of the way along the accelerator and is deflected back to the start. The damping rings and focusing magnets help bunch and focus the beams, which have cross-section radii of about 1 $\mu$m ($\frac{1}{1000}$ mm).

LINAC, magnets separate the electrons and positrons and bend them around for head-on collisions, with total available energy of up to 100 GeV. Storage-ring facilities are located at DESY (German Electron Synchrotron) in Hamburg, Germany ($E_a$ up to 42 GeV total in electron-positron collisions), at the Cornell Electron Storage Ring Facility (CESR) ($E_a$ up to 16 GeV), and at the CERN (European Council for Nuclear Research) laboratory in Geneva, Switzerland (electron-positron storage ring, beam energy 85 GeV, $E_a = 170$ GeV).

The highest-energy accelerator project currently underway is the Superconducting Supercollider (SSC) (Fig. 46–9). In December 1988, the U.S. Department of Energy selected a site in Waxahachie, Texas, for this project. In two ring-shaped vacuum chambers in the collider's underground tunnel, 85 km (about 53 miles) in circumference, beams of protons will be steered in opposite directions by rings of superconducting magnets, accelerated to an energy of 20 TeV ($20 \times 10^{12}$ eV), and brought together for head-on collisions. Construction of this facility will require 8 to 10 years and cost about \$6 billion. When completed, it will have an operating budget of about \$250 million per year. This experimental facility will be a significant new tool for research into the most fundamental levels of the structure of matter. Funding for this project is a topic of continuing debate.

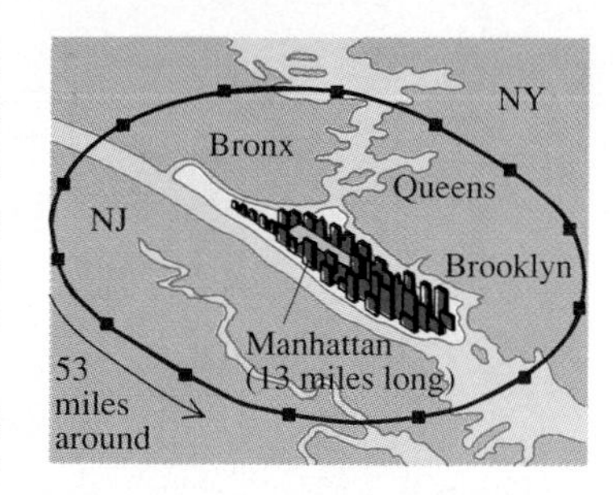


**46–9** The tunnel for the Superconducting Supercollider (SCC), to be built near Waxahachie, Texas, will be 85 km (53 mi) in circumference. The figure compares the size of the SSC to that of Manhattan Island.

# 46–3 PARTICLES AND INTERACTIONS

This section may seem like an expedition into the wilderness. From the security of 1931, when there were thought to be three permanent, unchanging fundamental particles, we enter the partly mapped territory of present-day particle physics. Particles are *not* permanent; they can be created and destroyed. Each particle has an *antiparticle*; for a few, the particle and its antiparticle are identical. Many particles are *unstable,* decaying spontaneously into other particles. Particles serve as mediators or transmitters of the various interactions; for example, photons mediate the electromagnetic interaction.

## The Four Forces

Let's review briefly where things stood in 1950. We'll start with the four categories of interactions (often called *forces*) that were known at that time. They are, in order of decreasing strength:

1. the strong interaction,
2. the electromagnetic interaction,
3. the weak interaction, and
4. the gravitational interaction.

The *electromagnetic* and *gravitational* forces are familiar from classical physics. Both are characterized by a $1/r^2$ dependence on distance. The gravitational force is very much *weaker* than the electromagnetic force. For example, the gravitational attraction of two protons is smaller than their electrical repulsion by a factor of about $0.8 \times 10^{-36}$. The gravitational force is of primary importance in the structure of stars and the large-scale behavior of the universe, but it is not believed to play a significant role in particle interactions at energies that are currently attainable.

The other two forces are less familiar. One, usually called the *strong interaction,* is responsible for the nuclear force and also for the production of pions and several other particles in high-energy collisions. Equation (46–4) is a possible potential-energy function for the nuclear force. The strength of the interaction is described by the constant $f^2$,

**TABLE 46–1 The Four Interactions of Nature**

| Interaction | Strength | Range | Mediating particle | | | |
|---|---|---|---|---|---|---|
| | | | Name | Rest mass | Charge | Spin |
| Strong | 1 | Short (~ 1 fm) | Gluon | 0 | 0 | 1 |
| Electromagnetic | $\frac{1}{137}$ | Long ($1/r^2$) | Photon | 0 | 0 | 1 |
| Weak | $10^{-9}$ | Short (~ 0.001 fm) | $W^{\pm}, Z^0$ | 81, 91 GeV/$c^2$ | $\pm e, 0$ | 1 |
| Gravitational | $10^{-38}$ | Long ($1/r^2$) | Graviton | 0 | 0 | 2 |

which has units of energy times distance. A better basis for comparison with other forces is the dimensionless ratio $f^2/\hbar c$, called the *coupling constant* for the interaction. (We invite you to verify that this ratio is a pure number and so must have the same value in all systems of units.) Observed behavior of nuclear forces suggests that $f^2/\hbar c \cong 1$. The dimensionless coupling constant for *electromagnetic* interactions is

$$\frac{1}{4\pi\epsilon_0}\frac{e^2}{\hbar c} = 7.297 \times 10^{-3} \cong \frac{1}{137}. \tag{46–12}$$

Thus the strong interaction is roughly 100 times as strong as the electromagnetic interaction; however, it drops off with distance more quickly than $1/r^2$.

The fourth interaction is called the *weak interaction*. It is responsible for beta decay, such as the conversion of a neutron into a proton, an electron, and an antineutrino. It is also responsible for the decay of many unstable particles (pions into muons, muons into electrons, $\Sigma$ particles into protons, and so on). It is a short-range interaction, like the strong interaction, but it is weaker by a factor of about $10^{-9}$.

Table 46–1 compares the main features of these four interactions. It also lists the mediating particle for each; we'll discuss these in the next section.

## More Particles

In Section 46–1 we mentioned the discovery of muons in 1937 and of pions in 1947, both in cosmic rays. The electric charges of the muons and the charged pions have the same magnitude $e$ as the electron charge. The two muons, $\mu^+$ and $\mu^-$, are antiparticles of each other; each has spin $\frac{1}{2}$ and a mass of about $106m_e = 106\ \text{MeV}/c^2$. They are unstable; each decays with a lifetime of $2.2 \times 10^{-6}$ s into an electron of the same sign and two neutrinos. (It's often convenient to represent particle masses in terms of the equivalent energy; for example, the muon rest mass is 106 MeV/$c^2$. We'll often use this notation in the following discussion.)

There are three kinds of pions, all with spin zero. The $\pi^{\pm}$ particles have masses of $273m_e$ (rest mass 140 MeV/$c^2$). They are unstable; each $\pi^{\pm}$ decays, with a lifetime of $2.6 \times 10^{-8}$ s, into a $\mu$ of the same sign and a neutrino. The $\pi^0$ is somewhat less massive, $264m_e$ (135 MeV/$c^2$); it decays with a lifetime of $0.87 \times 10^{-16}$ s into two $\gamma$ photons. The $\pi^+$ and $\pi^-$ are antiparticles; the $\pi^0$ is its own antiparticle. (That is, there is no distinction between particle and antiparticle.)

The existence of the *antiproton* (denoted by $\overline{\text{p}}$) had been suspected ever since the discovery of the positron. The $\overline{\text{p}}$ was found in 1955, when proton-antiproton (p-$\overline{\text{p}}$) pairs were created by use of a beam of 6-GeV protons from the Bevatron at the University of California at Berkeley. The *antineutron* was found soon afterward. Especially after 1960, as higher-energy accelerators and more sophisticated detectors were developed, a veritable blizzard of new unstable particles were identified. To describe and classify them, we need a small blizzard of new terms.

Initially, particles were classified by mass into three categories: (1) leptons ("light ones" such as electrons); (2) mesons ("intermediate ones" such as muons and pions); and (3) baryons ("heavy ones" such as nucleons and more massive particles). A more useful scheme is to classify particles in terms of their interactions. The two principal categories are *hadrons,* which have strong interactions, and *leptons,* which do not.

In the following discussion we will also distinguish between **bosons,** which have zero or integer spins, and **fermions,** which have half-integer spins. These names stem from the fact that all fermions obey the exclusion principle, on which the Fermi-Dirac distribution function is based. Bosons do not obey the exclusion principle and have a different distribution function, the Bose-Einstein distribution.

## Leptons

**Leptons** include the electrons ($e^{\pm}$), the muons ($\mu^{\pm}$), the tau particles ($\tau^{\pm}$), and three kinds of neutrinos ($\nu_e$, $\nu_\mu$, $\nu_\tau$), each with its corresponding antineutrino. The family of leptons is shown in Table 46–2. The taus, discovered in 1975, have spin $\frac{1}{2}$ and mass 1784 MeV/$c^2$. In all, there are six leptons and six antileptons. All leptons have spin $\frac{1}{2}$. Taus and muons are unstable; a tau can decay into a muon plus two neutrinos, and a muon decays into an electron plus two neutrinos. Until recently, all the neutrinos were believed to have zero rest mass; there is speculation and some experimental evidence that they may have small nonzero masses. We'll return to this point later. Note that $\tau$ particles are more massive than nucleons; they are classified as leptons because they have no strong interactions.

Leptons obey a *conservation principle.* Corresponding to the three kinds of leptons are three lepton numbers $L_e$, $L_\mu$, and $L_\tau$. The electron $e^-$ and the electron neutrino $\nu_e$ are assigned $L_e = 1$, and their antiparticles $e^+$ and $\overline{\nu}_e$ are given $L_e = -1$. Corresponding assignments of $L_\mu$ and $L_\tau$ are made for the $\mu$ and $\tau$ particles and their neutrinos. **In all interactions, each lepton number is separately conserved.** For example, in the decay of the $\mu^-$, the lepton numbers are

$$\begin{array}{ccccccc} \mu^- & \longrightarrow & e^- & + & \overline{\nu}_e & + & \nu_\mu, \\ L_\mu = 1 & & L_e = 1 & & L_e = -1 & & L_\mu = 1. \end{array}$$

**TABLE 46–2 The Leptons**

| Particle name | Symbol | Anti-particle | Rest mass (MeV/$c^2$) | $L_e$ | $L_\mu$ | $L_\tau$ | Lifetime (s) | Principal Decay Modes |
|---|---|---|---|---|---|---|---|---|
| Electron | $e^-$ | $e^+$ | 0.511 | +1 | 0 | 0 | Stable | |
| Neutrino (e) | $\nu_e$ | $\overline{\nu}_e$ | 0(?) | +1 | 0 | 0 | Stable | |
| Muon | $\mu^-$ | $\mu^+$ | 105.7 | 0 | +1 | 0 | $2.20 \times 10^{-6}$ | $e^-\overline{\nu}_e\nu_\mu$ |
| Neutrino ($\mu$) | $\nu_\mu$ | $\overline{\nu}_\mu$ | 0(?) | 0 | +1 | 0 | Stable | |
| Tau | $\tau^-$ | $\tau^+$ | 1784 | 0 | 0 | +1 | $<4 \times 10^{-13}$ | $\mu^-\overline{\nu}_\mu\nu_\tau$, $e^-\overline{\nu}_e\nu_\tau$ |
| Neutrino ($\tau$) | $\nu_\tau$ | $\overline{\nu}_\tau$ | 0(?) | 0 | 0 | +1 | Stable | |

### EXAMPLE 46–4

**Lepton reactions** Check conservation of lepton numbers for the following decay schemes:

a) $\mu^+ \rightarrow e^+ + \nu_e + \bar{\nu}_\mu$;

b) $\pi^- \rightarrow \mu^- + \bar{\nu}_\mu$;

c) $\pi^0 \rightarrow \mu^- + e^+ + \nu_e$.

**SOLUTION** a) The initial lepton numbers are $L_e = 0$ and $L_\mu = -1$. The final values are $L_e = -1 + 1 + 0 = 0$ and $L_\mu = 0 + 0 + (-1)$. Both lepton numbers are conserved. The number $L_\tau$ is not involved.

b) All lepton numbers are initially zero. The final values are $L_\mu = 1 + (-1) = 0$ and $L_e = 0$. Again, all lepton numbers are conserved.

c) The initial $L$'s are zero; in the final state, $L_e = -1 + 1 = 0$ and $L_\mu = 1$. Thus, $L_\mu$ is *not* conserved; this decay is forbidden by conservation of lepton number, and it does not occur in nature.

## Hadrons

**Hadrons,** the strongly interacting particles, are a more complex family than leptons. There are two subclasses: *mesons* and *baryons.* Mesons have spin 0 or 1, and baryons have spin $\frac{1}{2}$ or $\frac{3}{2}$. Therefore all mesons are bosons, and all baryons are fermions. Mesons include the pions, already mentioned, and several others including K mesons (or *kaons*), $\eta$ mesons, and other particles that we'll mention later. Baryons include the nucleons and several particles called *hyperons,* including the $\Lambda$, $\Sigma$, $\Xi$, and $\Omega$. These resemble the nucleons but are more massive. All the hyperons are unstable, decaying by various processes to other hyperons or to nucleons. *All* the mesons are unstable because all can decay to less massive particles, obeying all the conservation laws for such decays. Each hadron has an associated antiparticle, denoted with an overbar as with the antiproton $\bar{p}$.

Baryons obey the principle of **conservation of baryon number,** analogous to conservation of lepton numbers. We assign a baryon number $B = 1$ to each baryon (p, n, $\Lambda$, $\Sigma$, and so on) and $B = -1$ to each antibaryon ($\bar{p}$, $\bar{n}$, $\bar{\Lambda}$, $\bar{\Sigma}$, and so on). **In all interactions, the total baryon number is conserved.** This principle is the reason why the mass number $A$ must be conserved in all nuclear reactions.

### EXAMPLE 46–5

Which of the following reactions obey the law of conservation of baryon number?

a) $n + p \rightarrow n + p + p + \bar{p}$;

b) $n + p \rightarrow n + p + \bar{p}$.

**SOLUTION** In each case the initial baryon number is $1 + 1 = 2$.

a) The final baryon number for this reaction is $1 + 1 + 1 + (-1) = 2$, so baryon number is conserved.

b) The final baryon number for this reaction is $1 + 1 + (-1) = 1$. Baryon number is *not* conserved, and this reaction does not occur in nature.

### EXAMPLE 46–6

Antiprotons are produced by bombarding a stationary proton target (liquid hydrogen) with a beam of protons. Find the minimum beam energy needed for this reaction to occur.

**SOLUTION** Conservation of baryon number forbids the creation of single baryons; they must be created in particle-antiparticle pairs. The appropriate reaction is

$$p + p \rightarrow p + p + \bar{p} + p.$$

For this reaction to occur, the available energy $E_a$ in Eq. (46–10) must be at least $4mc^2$. With that substitution, Eq. (46–10) gives

$$(4mc^2)^2 = 2mc^2(E_m + mc^2),$$
$$E_m = 7mc^2.$$

The energy $E_m$ of the bombarding particle includes its rest energy $mc^2$, so the *kinetic* energy must be at least $6mc^2 = 6(938\ \text{MeV}) = 5.63$ GeV.

The search for the antiproton was one of the principal reasons for the construction of the Bevatron at the University of California (Berkeley), with beam energy of 6 GeV. The search was successful; in 1959, Segrè and Chamberlain were awarded the Nobel Prize for this discovery.

## Strangeness

The K mesons and the $\Lambda$ and $\Sigma$ hyperons appeared on the scene during the late 1950s. Because of their unusual behavior, they were called *strange particles.* They were produced in high-energy collisions such as $\pi^- + p$, and a K and a hyperon were always produced *together.* The abundance of the production process suggested that it was a *strong*-interaction process, but the relatively long lifetimes of these particles suggested that their decay was a *weak*-interaction process. The $K^0$ appeared to have *two* lifetimes, one (about $90 \times 10^{-12}$ s) characteristic of strong-interaction decays and another nearly 600 times longer. Were the K mesons hadrons or not?

The search for the answer to this question led physicists to introduce a new quantity called **strangeness.** The hyperons $\Lambda^0$ and $\Sigma^{\pm,0}$ were assigned a strangeness value $S = -1$, and the associated $K^0$ and $K^+$ mesons were assigned a value $S = +1$. The corresponding antiparticles had opposite strangeness, $S = +1$ for $\overline{\Lambda}{}^0$ and $\overline{\Sigma}{}^{\pm,0}$ and $S = -1$ for $\overline{K}{}^0$ and $K^-$. Strangeness was *conserved* in production processes such as

$$p + \pi^- \longrightarrow \overline{\Sigma}{}^- + K^-$$

and

$$p + \pi^- \longrightarrow \Lambda^0 + K^0.$$

The process

$$p + \pi^- \longrightarrow p + K^-$$

does not conserve strangeness, and it doesn't occur.

When strange particles decay individually, strangeness is usually *not* conserved. Typical processes include

$$\Sigma^+ \longrightarrow n + \pi^+,$$
$$\Lambda^0 \longrightarrow p + \pi^-,$$
$$K^- \longrightarrow \pi^+ + \pi^- + \pi^-.$$

In each of these the initial strangeness is 1 or $-1$, and the final value is zero. All observations of these particles are consistent with the conclusion that strangeness is conserved in strong interactions but that it can change by zero or one unit in weak interactions.

## Conservation Laws

The decay of strange particles provides our first example of a conditional conservation law, one that is obeyed in some interactions and not in others. Let's review *all* the conservation laws we know about and see what conclusions we can draw from them.

Several conservation laws are obeyed in *all* interactions. These include the familiar conservation laws: energy, momentum, angular momentum, and electric charge. These are called *absolute conservation laws.* Baryon number and the three lepton numbers are also conserved in all interactions. Strangeness is conserved in strong and electromagnetic interactions but *not* in all weak interactions.

Two other quantities are useful in classifying particles and their interactions. One is *isospin,* a quantity used to describe the charge independence of the strong interactions. The other is *parity,* which describes the comparative behavior of two systems that are mirror images of each other. Isospin is conserved in strong interactions, which are charge-independent, but not in electromagnetic or weak interactions. (The electromagnetic interaction is certainly *not* charge-independent.) Parity is conserved in strong and electromagnetic interactions but not in weak ones. Lee and Yang received the Nobel Prize in 1957 for laying the theoretical foundations for nonconservation of parity in weak interactions.

**TABLE 46–3 Some Known Hadrons and Their Properties**

| | Particle | Mass ($\mathrm{MeV}/c^2$) | Charge ratio, $Q/e$ | Spin | Baryon number, $B$ | Strangeness, $S$ | Mean lifetime (s) | Typical decay modes | Quark content |
|---|---|---|---|---|---|---|---|---|---|
| Mesons | $\pi^0$ | 135.0 | 0 | 0 | 0 | 0 | $0.87 \times 10^{-16}$ | $\gamma\gamma$ | $u\bar{u}, d\bar{d}$ |
| | $\pi^+$ | 139.6 | +1 | 0 | 0 | 0 | $2.6 \times 10^{-8}$ | $\mu^+\nu_\mu$ | $u\bar{d}$ |
| | $\pi^-$ | 139.6 | −1 | 0 | 0 | 0 | $2.6 \times 10^{-8}$ | $\mu^-\bar{\nu}_\mu$ | $\bar{u}d$ |
| | $\mathrm{K}^+$ | 493.7 | +1 | 0 | 0 | +1 | $1.24 \times 10^{-8}$ | $\mu^+\nu_\mu$ | $u\bar{s}$ |
| | $\mathrm{K}^-$ | 493.7 | −1 | 0 | 0 | −1 | $1.24 \times 10^{-8}$ | $\mu^-\bar{\nu}_\mu$ | $\bar{u}s$ |
| | $\eta^0$ | 548.8 | 0 | 0 | 0 | 0 | $\approx 10^{-18}$ | $\gamma\gamma$ | $u\bar{u}, d\bar{d}, s\bar{s}$ |
| Baryons | p | 938.3 | +1 | $\frac{1}{2}$ | 1 | 0 | Stable | — | $uud$ |
| | n | 939.6 | 0 | $\frac{1}{2}$ | 1 | 0 | 898 | $\mathrm{pe}^-\bar{\nu}_e$ | $udd$ |
| | $\Lambda^0$ | 1116 | 0 | $\frac{1}{2}$ | 1 | −1 | $2.63 \times 10^{-10}$ | $\mathrm{p}\pi^-$ or $\mathrm{n}\pi^0$ | $uds$ |
| | $\Sigma^+$ | 1189 | +1 | $\frac{1}{2}$ | 1 | −1 | $0.799 \times 10^{-10}$ | $\mathrm{p}\pi^0$ or $\mathrm{n}\pi^+$ | $uus$ |
| | $\Sigma^0$ | 1193 | 0 | $\frac{1}{2}$ | 1 | −1 | $7.4 \times 10^{-20}$ | $\Lambda^0\gamma$ | $uds$ |
| | $\Sigma^-$ | 1197 | −1 | $\frac{1}{2}$ | 1 | −1 | $1.48 \times 10^{-10}$ | $\eta\pi^-$ | $dds$ |
| | $\Xi^0$ | 1315 | 0 | $\frac{1}{2}$ | 1 | −2 | $2.90 \times 10^{-10}$ | $\Lambda^0\pi^0$ | $uss$ |
| | $\Xi^-$ | 1321 | −1 | $\frac{1}{2}$ | 1 | −2 | $1.64 \times 10^{-10}$ | $\Lambda^0\pi^-$ | $dss$ |
| | $\Delta^{++}$ | 1232 | +2 | $\frac{3}{2}$ | 1 | 0 | $10^{-23}$ | $\mathrm{p}\pi^+$ | $uuu$ |
| | $\Omega^-$ | 1672 | −1 | $\frac{3}{2}$ | 1 | −3 | $0.822 \times 10^{-10}$ | $\Lambda^0\mathrm{K}^-$ | $sss$ |
| | $\Lambda_c^+$ | 2285 | +1 | $\frac{1}{2}$ | 1 | 0 | $1.91 \times 10^{-13}$ | $\Sigma^+\pi\pi\pi$ | $udc$ |

Emerging from this discussion is the fact that conservation laws provide a basis for *classifying* particles and their interactions. Each conservation law is also associated with a *symmetry* property of the system. A familiar example is angular momentum. If a system is in an environment that has spherical symmetry, there can be no torque acting on it because the direction of the torque would violate the symmetry. In such a system, total angular momentum is *conserved*. When a conservation law is violated, the interaction is often described as a *symmetry-breaking interaction*.

Table 46–3 shows some of the hadrons that are currently known. We'll discuss the last column of the table (quark content) in the next section.

## 46–4 QUARKS

The leptons form a fairly neat package: three massive particles and three neutrinos, each with its antiparticle, and a conservation law relating their numbers. Physicists believe that leptons are genuinely fundamental particles. The hadron family, by contrast, is a mess. Table 46–3 shows only a sample of well over 100 hadrons that have been discovered since 1960, and it has become clear that these particles *do not* represent the most fundamental level of the structure of matter.

Our present understanding of the structure of hadrons is based on a proposal made initially in 1964 by Murray Gell-Mann and his collaborators. In this proposal, hadrons are *not* fundamental particles but are composite structures whose constituents are spin-$\frac{1}{2}$ fermions called **quarks.** (The name comes from the line "Three quarks for Muster Mark!" from *Finnegan's Wake* by James Joyce.) Each baryon is composed of three quarks ($qqq$), and each meson is a quark-antiquark pair ($q\bar{q}$). No other combinations seem to be necessary. This scheme requires that quarks have electric charges with magnitudes $\frac{1}{3}$ and $\frac{2}{3}$ of the electron charge $e$, which was previously thought to be the smallest unit of charge. Quarks also have fractional values of the baryon number $B$. Two quarks can combine with their spins parallel to form a particle with spin 1 or with their spins antiparallel to form a particle with spin 0. Similarly, three quarks can combine to form a particle with spin $\frac{1}{2}$ or $\frac{3}{2}$.

**TABLE 46–4 Properties of the Three Original Quarks**

| Symbol | $Q/e$ | Spin | Baryon number, $B$ | Strangeness, $S$ | Charm, $C$ | Bottomness, $B'$ | Topness, $T$ |
|---|---|---|---|---|---|---|---|
| $u$ | $\frac{2}{3}$ | $\frac{1}{2}$ | $\frac{1}{3}$ | 0 | 0 | 0 | 0 |
| $d$ | $-\frac{1}{3}$ | $\frac{1}{2}$ | $\frac{1}{3}$ | 0 | 0 | 0 | 0 |
| $s$ | $-\frac{1}{3}$ | $\frac{1}{2}$ | $\frac{1}{3}$ | $-1$ | 0 | 0 | 0 |

The first quark theory included three types (called *flavors*) of quarks, labeled ***u*** (up), ***d*** (down), and ***s*** (strange). Their principal properties are listed in Table 46–4. The corresponding antiquarks $\overline{u}$, $\overline{d}$, and $\overline{s}$ have opposite values of $Q$, $B$, and $S$. Protons, neutrons, $\pi$ and K mesons, and several hyperons can be constructed from these three quarks. For example, the proton quark content is ***uud***. Checking the table, we see that the values of $Q/e$ add to 1 and that the values of the baryon number $B$ also add to 1, as we should expect. The neutron is ***udd***, with total $Q = 0$ and $B = 1$. The $\pi^+$ meson is $u\overline{d}$, with $Q/e = 1$ and $B = 0$, and the $\mathrm{K}^+$ meson is $u\overline{s}$. The antiproton is $\overline{\mathrm{p}} = \overline{u}\overline{u}\overline{d}$, the negative pion is $\pi^- = \overline{u}d$, and so on. Figure 46–10 shows the quark content of several particles.

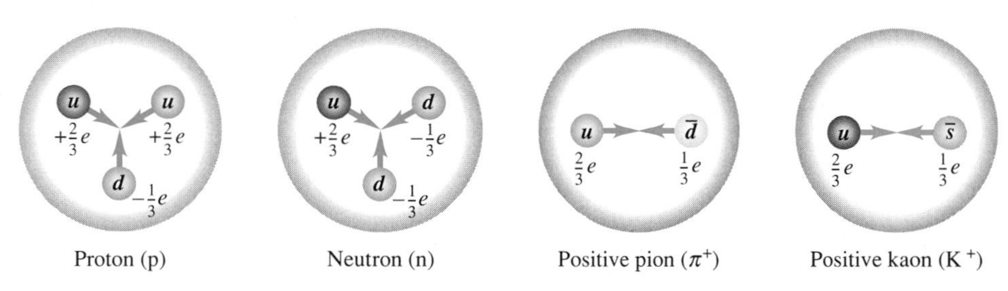


**46–10** Quark content of four hadrons. The various color combinations needed for color neutrality are not shown.

## EXAMPLE 46–7

Find the quark content of a) $\Sigma^+$; b) $\overline{\Lambda}^0$.

**SOLUTION** a) The $\Sigma^+$ has strangeness $S = -1$, so it must contain the $s$ quark. For total charge $+1$, the other two quarks must both be $u$, so $\Sigma^+$ is ***uus***.

b) First we find the quark content of $\Lambda^0$. It has $S = -1$, so it contains $s$. For zero total charge, the other two must be $u$ and $d$, so the $\Lambda^0$ is ***uds***. Then $\overline{\Lambda}^0$ is $\overline{u}\overline{d}\overline{s}$.

What holds quarks together? The attractive interactions among quarks are mediated by massless spin-1 bosons called **gluons** in exactly the same way that photons mediate the electromagnetic interaction and that pions mediate the nucleon-nucleon force in the old Yukawa theory.

Quarks, having spin $\frac{1}{2}$, are fermions and so are subject to the exclusion principle. This would seem to forbid a baryon's having two or three quarks with the same flavor. To avoid this difficulty, it is assumed that each quark comes in three varieties, whimsically called *colors*. Red, green, and blue are the usual choices. The exclusion principle applies separately to each color. A baryon always contains one red, one green, and one blue quark, so the baryon itself has no color. Each gluon has a color and an anticolor, and color is conserved during emission and absorption of a gluon by a quark. The gluon-exchange process changes the colors of the quarks in such a way that there is always one quark of each color in every baryon. But the color of an individual quark changes continually.

Similar processes occur in mesons such as pions. Gluon exchange binds the two quarks in a pion, which is always colorless. Suppose a pion initially consists of a blue

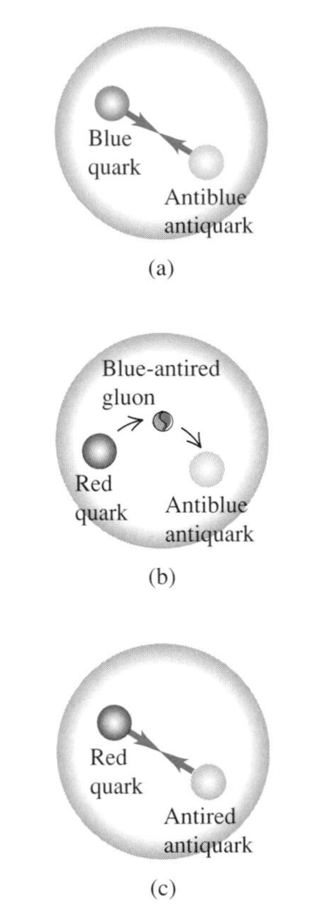


**46–11** (a) A pion containing a blue quark and an antiblue antiquark. (b) The blue quark emits a blue-antired gluon, changing to a red quark. (c) The gluon is absorbed by the antiblue antiquark, which becomes an antired antiquark. The pion now consists of a red-antired quark-antiquark pair. The actual quantum state of the pion is an equal superposition of red-antired, green-antigreen, and blue-antiblue pairs.

quark and an antiblue antiquark. The blue quark can emit a blue-antired virtual gluon, becoming a red quark. The gluon is absorbed by the antiblue antiquark, which then becomes an antired antiquark (Fig. 46–11). Color is conserved in each emission and absorption, but a blue-antiblue pair has been converted to a red-antired pair. So we have to think of a pion as a superposition of three quantum states: blue-antiblue, green-antigreen, and red-antired.

The theory of strong interactions is known as *quantum chromodynamics* (QCD). Individual free quarks have not been observed. In most QCD theories there are phenomena associated with the creation of quark-antiquark pairs that make it impossible to observe a single free, isolated quark. Nevertheless, an impressive body of experimental evidence supports the correctness of the quark structure of hadrons and the belief that quantum chromodynamics is the key to understanding the strong interactions.

Before the tau particles were discovered, there were four known leptons. This fact, together with some puzzling decay rates, led to the speculation that there might be a fourth quark flavor. This quark is labeled $\boldsymbol{c}$ (the *charmed* quark); it has $Q/e = \frac{2}{3}$, $B = \frac{1}{3}$, $S = 0$, and a new quantum number **charm** $C = +1$. This prediction was confirmed in 1974 at both the SLAC and Brookhaven accelerator laboratories by the observation of a meson with a mass of 3100 MeV/$c^2$. This meson, named $\Psi$ at SLAC and J at Brookhaven, was found to have several decay modes, decaying into $e^+e^-$, $\mu^+\mu^-$, or hadrons. The mean lifetime was found to be about $10^{-20}$ s. These results are consistent with J/$\Psi$ being a $\boldsymbol{c\bar{c}}$ system. Almost immediately after this, excited states of the $\boldsymbol{c\bar{c}}$ system, with higher energy, were observed. A few years later, individual mesons with the charm quantum number were also observed. These mesons, $D^0$ ($\boldsymbol{c\bar{u}}$) and $D^+$ ($\boldsymbol{c\bar{d}}$), and their excited states are now firmly established. A charmed baryon, $\Lambda_c^+$ ($\boldsymbol{udc}$), has also been observed.

In 1977 a meson with mass 9460 MeV/$c^2$, called upsilon ($\Upsilon$), was discovered at Brookhaven. Because it had properties similar to those of J/$\Psi$, it was conjectured that the meson was really the bound system of a new quark, $\boldsymbol{b}$ (the *bottom* quark), and its antiquark, $\boldsymbol{\bar{b}}$. The bottom quark has the value 1 of a new quantum number $B'$ (not to be confused with baryon number $B$) called *bottomness.* Excited states of $\Upsilon$ were soon observed, and the $B^+$ ($\boldsymbol{\bar{b}u}$) and $B^0$ ($\boldsymbol{\bar{b}d}$) mesons are now well established.

So far, we have five flavors of quarks ($\boldsymbol{u, d, s, c, b}$), and six flavors of leptons (e, $\mu$, $\tau$, $\nu_e$, $\nu_\mu$, and $\nu_\tau$). These are sufficient for the understanding of many aspects of the strong and weak interactions of hadrons and the weak interactions of leptons. But it is an appealing conjecture that nature is symmetric in its building blocks and that therefore there is a *sixth* quark. This quark, labeled $\boldsymbol{t}$ (top), should have $Q/e = \frac{2}{3}$, $B = \frac{1}{3}$, and a new quantum number, $T = 1$. No direct experimental evidence for the existence of the $\boldsymbol{t}$ quark has yet been found. Table 46–5 lists some properties of the six quarks. Each has a corresponding antiquark with opposite values of $Q$, $B$, $S$, $C$, $B'$, and $T$. (Some physicists insist that $\boldsymbol{t}$ and $\boldsymbol{b}$ stand for *truth* and *beauty,* not *top* and *bottom.* Beauty has been observed, but truth, as usual, is more elusive.)

**TABLE 46–5 Properties of Quarks**

| Symbol | $Q/e$ | Spin | Baryon number, $B$ | Strangeness, $S$ | Charm, $C$ | Bottomness, $B'$ | Topness, $T$ |
|---|---|---|---|---|---|---|---|
| $\boldsymbol{u}$ | $\frac{2}{3}$ | $\frac{1}{2}$ | $\frac{1}{3}$ | 0 | 0 | 0 | 0 |
| $\boldsymbol{d}$ | $-\frac{1}{3}$ | $\frac{1}{2}$ | $\frac{1}{3}$ | 0 | 0 | 0 | 0 |
| $\boldsymbol{s}$ | $-\frac{1}{3}$ | $\frac{1}{2}$ | $\frac{1}{3}$ | $-1$ | 0 | 0 | 0 |
| $\boldsymbol{c}$ | $\frac{2}{3}$ | $\frac{1}{2}$ | $\frac{1}{3}$ | 0 | $+1$ | 0 | 0 |
| $\boldsymbol{b}$ | $-\frac{1}{3}$ | $\frac{1}{2}$ | $\frac{1}{3}$ | 0 | 0 | $+1$ | 0 |
| $\boldsymbol{t}$ | $\frac{2}{3}$ | $\frac{1}{2}$ | $\frac{1}{3}$ | 0 | 0 | 0 | $+1$ |

## 46–5 THE STANDARD MODEL AND THE EIGHTFOLD WAY

The particles and interactions that we've discussed in this chapter provide a reasonably comprehensive picture of the fundamental building blocks of nature. There is enough confidence in the basic correctness of this picture that it is called the **standard model.**

The standard model includes three families of particles: (1) the six leptons, which have no strong interactions; (2) the six quarks, from which all hadrons are made; and (3) the particles that mediate the various interactions. We have already discussed gluons as the mediators of the strong interaction among quarks and photons as the mediators of the electromagnetic interactions among charged particles. Both gluons and photons are massless, spin-1 bosons.

The other two interactions are also mediated by particles. For the weak interactions these are the *weak bosons* $W^{\pm}$ and $Z^0$. They are also spin-1 bosons, but they are *not* massless. In fact, they have enormous masses, 81 GeV/$c^2$ for the W's and 91 GeV/$c^2$ for the $Z^0$. The existence of the weak bosons was confirmed by Carlo Rubbia, Simon van der Meer, and co-workers in 1983 in experiments at CERN, and they were awarded the Nobel Prize for this work in 1984.

There are persuasive theoretical reasons to believe that the gravitational interaction is mediated by a massless boson with spin 2 called the *graviton*. Because gravitational interactions are so weak, experimental investigation is extraordinarily difficult, and gravitons have not yet been observed.

Symmetry considerations play a very prominent role in all current work in particle theory. Here are two examples. Consider the eight spin-$\frac{1}{2}$ baryons we've mentioned: the familiar p and n; the strange $\Lambda^0$, $\Sigma^+$, $\Sigma^0$, and $\Sigma^-$; and the doubly strange $\Xi^0$ and $\Xi^-$. For each, we plot the values of charge $Q$ and strangeness $S$. The result is the hexagonal pattern shown in Fig. 46–12. A similar plot for the nine spin-0 mesons (six shown in Table 46–3 plus three others that are not included in that table) is shown in Fig. 46–13; the particles fall in exactly the same pattern!

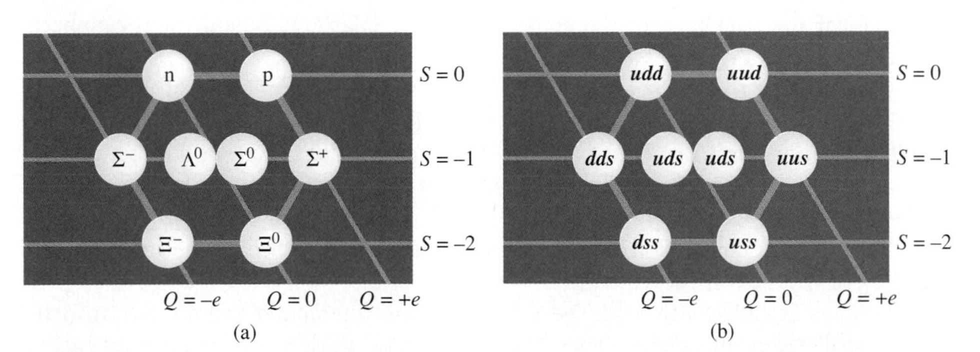


**46–12** (a) Plot of $S$ and $Q$ values for spin-$\frac{1}{2}$ baryons, showing the symmetry pattern of the eightfold way. (b) Quark content of each spin-$\frac{1}{2}$ baryon. The quark contents of the $\Sigma^0$ and $\Lambda^0$ are the same; the $\Sigma^0$ is an excited state of the $\Lambda^0$ and can decay into it by $\gamma$ emission.

**46–13** (a) Plot of $S$ and $Q$ values for nine spin-0 mesons, showing the symmetry pattern of the eightfold way. Each particle is directly opposite its antiparticle; each of the three particles in the center is its own antiparticle. (b) Quark content of each spin-0 meson.

The symmetries in quantum chromodynamics that lead to these and similar patterns are collectively called the **eightfold way.** They were discovered in 1961 by Murray Gell-Mann and independently by Yu val Ne'eman. (The name is a slightly irreverent reference to the Noble Eightfold Path, a set of principles for right living in Buddhism.) A similar pattern for the spin-$\frac{3}{2}$ baryons contains *ten* particles, arranged in a triangular pattern like pins in a bowling alley. When this pattern was first discovered, one of the particles was missing. But Gell-Mann gave it a name anyway ($\Omega^-$), predicted the properties it should have, and told experimenters what they should look for. Three years later the particle was found during an experiment at Brookhaven National Laboratory, a spectacular success for Gell-Mann's theory. The series of events is reminiscent of the way in which Mendeleev used gaps in the periodic table of the elements to predict properties of undiscovered elements and to guide chemists in their search for these elements.

## Unified Theories

It has long been a dream of particle theorists to be able to combine all the interactions of nature into a single unified theory. Einstein spent much of his later life trying to develop a unified field theory that would combine gravitation and electromagnetism. He was only partly successful.

In 1967, Sheldon Glashow, Abdus Salam, and Steven Weinberg proposed a theory that unifies the weak and electromagnetic forces. One outcome of the **electroweak theory** is a prediction of the weak-force mediator particles, the $Z^0$ and $W^{\pm}$ bosons, including their masses. The basic idea is that the mass difference between photons (zero mass) and the weak bosons makes the electromagnetic and weak interactions look quite different at low energies; but at sufficiently high energies (such as 1 TeV) the distinction disappears, and the two merge into a single interaction. This theory was verified experimentally in 1983 by two experimental groups working at the $p\bar{p}$ collider at CERN. The weak bosons were found, again with the help provided by the theoretical description, and their observed masses agreed with the predictions of the electroweak theory, a wonderful convergence of theory and experiment. The electroweak theory and quantum chromodynamics form the backbone of the standard model. Glashow, Salam, and Weinberg received the Nobel Prize in 1979.

A remaining difficulty with the electroweak theory is the fact that photons are massless but the weak bosons are very massive. To account for the broken symmetry among these interaction mediators, a particle called the Higgs boson has been proposed. Its mass is expected to be less than 1 TeV/$c^2$, but to produce it in the laboratory will require a center-of-mass energy of about 40 TeV. The search for the Higgs particle (or particles) is part of the mission of the SSC. Another goal is the direct experimental confirmation of the top (or truth) quark.

Perhaps at sufficiently high energies the strong interaction and the electroweak interaction have a similar convergence. If so, they can be unified to give a comprehensive theory of strong, weak, and electromagnetic interactions. Such schemes, called **grand unified theories** (GUTs), lean heavily on symmetry considerations, and they are still speculative in nature.

One interesting feature of some grand unified theories is that they predict the decay of the proton (in violation of conservation of baryon number), with an estimated lifetime of the order of $10^{30}$ to $10^{31}$ years. (For comparison, the age of the universe is estimated to be of the order of $10^{10}$ years.) Experiments are underway that should have detected the decay of the proton if its lifetime is $10^{30}$ years or less. Such decays have not been observed, but experimental work continues. Some GUTs also predict the existence of magnetic monopoles, which we mentioned in Chapter 28. At present there is no experimental evidence that magnetic monopoles exist, but the search goes on.

In the standard model the neutrinos have zero mass. Most experimental evidence supports this assumption. In most GUTs the neutrinos *must* have nonzero masses; values of a few MeV/$c^2$ have been suggested. If neutrinos do have mass, transitions called *neutrino oscillations* can occur in which one type of neutrino ($\nu_e$, $\nu_\mu$, or $\nu_\tau$) changes into another type. Experiments designed to detect neutrino oscillations are underway, but no positive results have yet been obtained. Recently (May 1991), several experimenters have independently reported observation of electron neutrinos ($\nu_e$) with the unexpectedly large rest mass of 17 keV/$c^2$. These observations await further confirmation.

Neutrino oscillations, if they do occur, may be able to clear up a mystery about the electron-neutrino flux from the sun. In an experiment that has been running for 20 years (Fig. 46–14), 380,000 L of perchloroethylene ($C_2Cl_4$) in a tank 4850 feet underground, in the Homestake gold mine in South Dakota, is used as a solar neutrino detector. The observed neutrino flux is only about a third of what should be expected on the basis of the fusion reactions that occur in the interior of the sun. Neutrino oscillations could play a role in resolving this discrepancy.

The ultimate dream of theorists is to unify all four fundamental interactions, including gravitation along with the strong and electroweak interactions that are included in GUTs. Such a unified theory is whimsically called a Theory of Everything (TOE). These theories range from the speculative to the fantastic. One popular ingredient is a space-time continuum with more than four dimensions, containing structures called *strings*. Another element is *supersymmetry*, which gives every boson a fermion "superpartner." Such concepts lead to the prediction of whole new families of particles, including sleptons, photinos, squarks bound together by gluinos, and even winos and wimps. Theorists are still very far away from a satisfactory TOE.

**46–14** Brookhaven National Laboratory's solar neutrino experiment, located 4850 feet underground in a gold mine in South Dakota to shield out cosmic rays and all other particles except neutrinos. The tank contains 100,000 gallons of perchloroethylene. Neutrinos from the interior of the sun are captured by $^{37}$Cl nuclei, converting them into $^{37}$Ar. The argon is then trapped and measured. The results of this long-term experiment indicate that the number of energetic neutrinos released from the sun is much lower than theory predicts. In an attempt to solve this neutrino puzzle, an international collaborative effort continues to collect data on the solar neutrino flux in a new underground laboratory in Italy.

## 46–6 THE EXPANDING UNIVERSE

In the last two sections of this chapter we'll explore briefly the connections between the early history of the universe and the interactions of fundamental particles. It is surprising that there are such close ties between the smallest-scale things we know about (the range of the weak interaction, of the order of $10^{-18}$ m) and the largest (the universe itself, of the order of at least $10^{26}$ m). Gravitational interactions play an essential role in the behavior of the universe. One of the great achievements of Newtonian mechanics, including the law of gravitation, was the understanding it brought to the motion of planets in the solar system. Astronomical evidence, such as observations of the motions of double stars around their common center of mass, shows that gravitational interactions also operate in larger astronomical systems, including stars, galaxies, and nebulae.

Until early in the twentieth century it was usually assumed that the universe was *static;* stars might move relative to each other, but there was not thought to be any general expansion or contraction. But stars have gravitational attractions. If everything is initially sitting still in the universe, why doesn't gravity just pull it all together into one big clump? Even Newton himself recognized the seriousness of this troubling question.

About 1930, astronomers began to find evidence that the universe is *not* static. The motions of distant galaxies relative to earth can be measured by observing the Doppler shifts in the wavelengths of their spectra. These shifts are always toward longer wavelength, showing that distant galaxies appear to be *receding* from us and from each other. We can derive a relation between the wavelength $\lambda$ of light from a source receding at speed $v$ and the wavelength $\lambda_0$ measured in the rest frame of the source. We invert Eq. (39–28) and use the relation $\lambda = c/f$; the result is

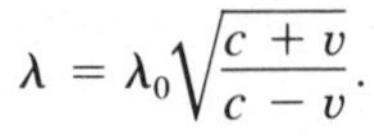

$$\lambda = \lambda_0\sqrt{\frac{c+v}{c-v}}. \qquad (46\text{–}13)$$

Wavelengths from receding sources are always shifted toward longer wavelengths, and these shifts are called the **red shift.**

We can solve Eq. (46–13) for $v$. We leave the details for a problem; the result is

$$v = \frac{(\lambda/\lambda_0)^2 - 1}{(\lambda/\lambda_0)^2 + 1}c. \tag{46–14}$$

## EXAMPLE 46–8

**Red shift from a distant galaxy** The spectrum lines of hydrogen are detected in light from a distant galaxy. The blue H$\gamma$ line ($\lambda_0 = 434.1$ nm) is observed with a wavelength $\lambda = 655$ nm, in the red end of the spectrum. At what speed is this galaxy receding from the earth?

**SOLUTION** First, we calculate $\lambda/\lambda_0 = (655 \text{ nm})/(434.1 \text{ nm}) = 1.51$. Then, from Eq. (46–14),

$$v = \frac{(1.51)^2 - 1}{(1.51)^2 + 1}c = 0.390c = 1.17 \times 10^8 \text{ m/s}.$$

The galaxy is receding from the earth with speed $0.206c$.

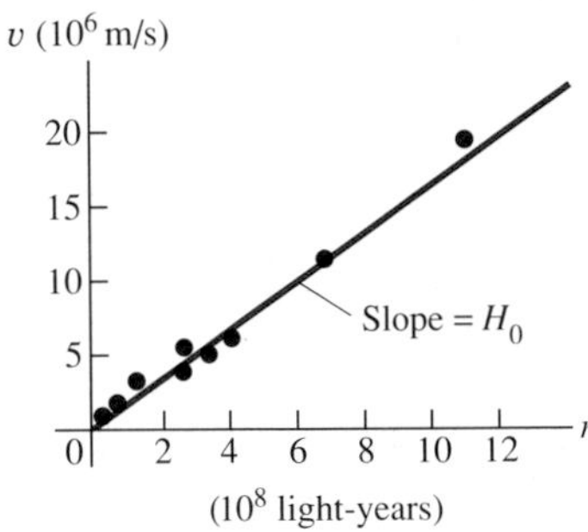


**46–15** Graph of recession speed as a function of distance for several galaxies. The graph is nearly a straight line, illustrating Hubble's law. The slope of the line is Hubble's constant, $H_0$.

The astronomer Edwin Hubble measured red shifts from many distant galaxies and came to an astonishing conclusion. The speed of recession $v$ of a galaxy is nearly proportional to its distance $r$ from us (Fig. 46–15). This relation is now called **Hubble's law;** its symbolic statement is

$$v = H_0 r, \tag{46–15}$$

where $H_0$ is an empirical constant called the *Hubble constant.* Its units are 1/time.

When distances are measured in meters and speeds in meters per second, the approximate value of $H_0$ is

$$H_0 = 1.8 \times 10^{-18} \text{ s}^{-1},$$

with an uncertainty of 25% or so. Astronomical distances are often measured in *light-years;* one light-year is the distance light travels in one year, and

$$1 \text{ light-year} = 9.46 \times 10^{15} \text{ m}.$$

The Hubble constant is usually expressed in the mixed units (m/s)/light-year:

$$H_0 = 17 \times 10^{-3} \text{ (m/s)/light-year}.$$

We don't know whether the Hubble constant is really *constant;* we've been observing it for only half a century, a mere blink of the eye on a cosmic time scale.

## EXAMPLE 46–9

In Example 46–8, find the distance of the galaxy from the earth, according to Hubble's law.

**SOLUTION** From Eq. (46–15),

$$r = \frac{v}{H_0} = \frac{1.17 \times 10^8 \text{ m/s}}{17 \times 10^{-3} \text{ (m/s)/light-year}}$$

$$r = 6.9 \times 10^9 \text{ light-year} = 6.5 \times 10^{25} \text{ m}.$$

This distance is so much greater than anything in human experience that it is very hard to develop any intuition about it.

Another aspect of Hubble's observations was that distant galaxies appeared to be receding *in all directions.* There is no particular reason to think that our galaxy is at the very center of the universe; if we were moving along with some other galaxy, everything else would still seem to be receding from us. That is, *the universe looks more or less the same, no matter where we are*. This important idea is called the **cosmological principle.** There are local fluctuations in density, but on average, the universe looks the same from all locations. And the laws of physics are the same everywhere.

## The Big Bang

An appealing hypothesis suggested by Hubble's law is that at some time in the past, all the matter in the universe was concentrated in a small space and was blown apart in an immense explosion called the **Big Bang,** giving all observable matter more or less the velocities we observe today. When did this happen? According to Hubble's law, matter at a distance $r$ away from us is traveling with speed $v = H_0 r$. The time $t$ needed to travel a distance $r$ is

$$t = \frac{r}{v} = \frac{r}{H_0 r} = \frac{1}{H_0} = 5.6 \times 10^{17}\ \text{s} = 1.8 \times 10^{10}\ \text{y}.$$

Thus we can understand Hubble's law if we assume that the Big Bang occurred about 18 billion years ago, the beginning of the universe. We have assumed that all speeds are *constant* after the Big Bang; this neglects any slowing down due to gravitational attraction. Whether there is appreciable slowing depends on the average density of matter. We'll return to this point later.

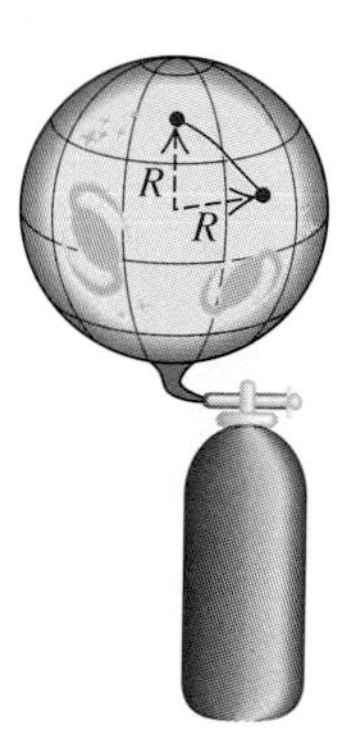


## Expanding Space

The general theory of relativity takes a somewhat different view of the expansion just described. According to this theory, what *appears* to us to be a Doppler shift caused by motion of galaxies away from us and away from each other is not really that but rather *the expansion of space itself.* This is not an easy concept to grasp, and if you haven't encountered it before, it may sound like doubletalk.

Here's an analogy that may help you to develop some intuition on this point. Suppose we are all bugs crawling around on a surface. We can't leave the surface, and we can see only ahead of us, not up or down. We are then living in a two-dimensional world; some writers have called such a world *flatland.* If the surface is a plane, we can locate our position with Cartesian coordinates $(x, y)$. If the plane extends indefinitely in both the $x$- and $y$-directions, we describe our space as having *infinite* extent or as being *unbounded.* No matter how far we go, we never reach an edge or a boundary.

An alternative habitat for us would be the surface of a sphere with radius $R$. The space would still seem infinite in the sense that we could crawl forever and never reach an edge or a boundary. Yet in this case the space is *finite* or *bounded.* To describe the location of a point in this space, we could use latitude and longitude (or the spherical coordinates $\theta$ and $\phi$).

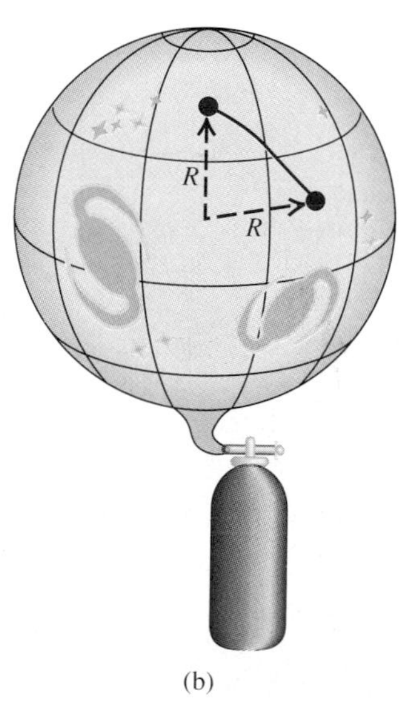


**46–16** (a) Points on the surface of a spherical balloon are described by their latitude and longitude coordinates. (b) The radius $R$ of the balloon has increased. The coordinates of the points are the same, but the distance between them has increased. The rate of recession for any two points is proportional to the original distance between them.

Now suppose the surface is a spherical balloon (Fig. 46–16). As we inflate the balloon more and more, increasing the parameter $R$, the coordinates of a point don't change, yet the distance between any two points gets larger and larger. Furthermore, as $R$ increases, the *rate of change* of distance between two points (or their recession speed) is proportional to their initial distance apart. *The recession speed is proportional to the distance,* just as with Hubble's law. For example, the distance from Pittsburgh to Miami is twice as great as the distance from Pittsburgh to Boston. If the earth were to begin to swell, Miami would recede from Pittsburgh twice as fast as Boston would.

We see that although the quantity $R$ isn't a coordinate in the usual sense, it nevertheless plays an essential role in any discussion of distance. It is the radius of curvature of our two-dimensional space, and it is also a varying *scale factor* that changes as this two-dimensional universe expands. Note that what we have called the Doppler effect is also a natural outgrowth of this viewpoint; the wavelength of light increases along with every other dimension as the universe expands.

Generalizing this picture to three dimensions isn't so easy. We have to think of our three-dimensional space as being embedded in a space with four or more dimensions, just as we visualized the two-dimensional spherical flatland as embedded in a three-dimensional Cartesian space. Our real three-space is *not Cartesian;* to describe its characteristics in any small region requires at least one additional parameter, the curvature of space, analogous to the radius $R$ of the sphere. In a sense this scale factor, which we'll also call $R$, describes the *size* of the universe, just as the radius of the sphere described the size of our two-dimensional spherical universe. Whether the real universe is *finite* is open to conjecture; we'll return to this question later.

We have spoken of the expansion of the universe with time, but we can also look backward to the initial time when everything was concentrated in a very small region, or perhaps a single point. This region (or point) *was* the universe at that time. It's important to understand that the Big Bang was not an expansion *in* space; it was an expansion *of* space.

## Critical Density

We've mentioned that the law of gravitation isn't consistent with a static universe. We need to look at the role of gravity in an *expanding* universe. Gravitational attractions should slow the initial expansion, but by how much? If they are strong enough, the universe should expand more and more slowly, eventually stop, and then begin to contract, perhaps all the way down to what's been called a *Big Crunch*. Some cosmological theories picture the universe this way, as a series of cataclysmic pulsations with a period of perhaps $10^{11}$ years. On the other hand, if gravitational forces are much weaker, they slow the expansion only a little, and the universe continues to expand forever.

The situation is analogous to the problem of escape velocity of a projectile launched from earth; we studied this problem in Section 12–4, and you may want to review that discussion. The total energy $E = K + U$ of a missile with speed $v$ at a distance $r$ from the center of the earth is

$$E = \tfrac{1}{2}mv^2 - \frac{Gmm_E}{r}. \qquad (46\text{–}16)$$

If $E$ is positive, the projectile has enough energy to move indefinitely far from earth ($r \to \infty$) and have some kinetic energy left over. If $E$ is negative, the kinetic energy $K = \tfrac{1}{2}mv^2$ becomes zero when $r = -Gmm_E/E$. No greater value of $r$ is possible, and the projectile can't escape the earth's gravity.

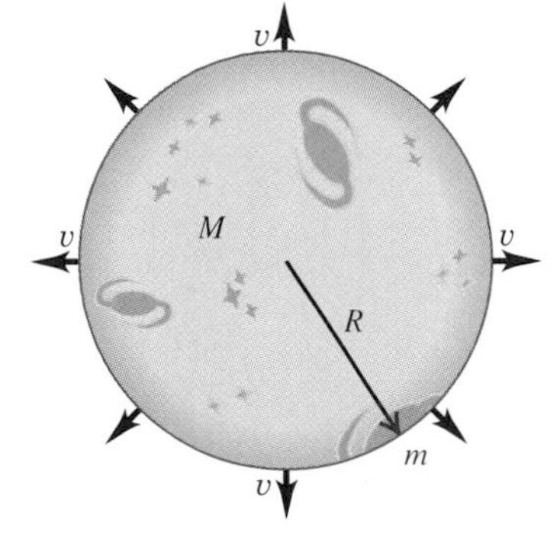


**46–17** The gravitational force on our galaxy (mass $m$) is the force exerted by the mass $M$ within the sphere with radius $R$. Because of the spherical symmetry, the mass outside this sphere has no effect.

We can carry out a similar analysis for the universe. Whether the universe continues to expand indefinitely depends on the average *density* of matter. If matter is relatively dense, there is a lot of gravitational attraction to slow and eventually stop the expansion and make the universe contract again. If not, the expansion continues indefinitely. We can derive an expression for the *critical density* $\rho_c$ needed to just barely stop the expansion.

Here's a calculation based on Newtonian mechanics; it isn't relativistically correct, but it illustrates the idea. Consider a large sphere with radius $R$, containing many galaxies (Fig. 46–17), with total mass $M$. Suppose our own galaxy has mass $m$ and is located at the surface of this sphere. According to the cosmological principle, the average distribution of matter around the sphere is spherically symmetric. We can apply Gauss's law to the mass distribution and the resulting gravitational field. The total gravitational force on our galaxy is just the force due to the mass $M$ inside the sphere. The force on our galaxy and potential energy $U$ are the same as though $m$ and $M$ were both points, so $U = -GmM/R$, just as in Section 12–4.

The total energy $E$ (kinetic plus potential) of our galaxy is

$$E = \tfrac{1}{2}mv^2 - \frac{GmM}{R}. \qquad (46\text{–}17)$$

If $E$ is *positive,* the galaxy has enough energy to escape from the gravitational attraction of the mass $M$ inside the sphere; in this case the universe keeps expanding forever. If $E$ is negative, the galaxy cannot escape, and the universe is eventually pulled back together. The cross-over between these two cases occurs when $E = 0$:

$$\tfrac{1}{2}mv^2 = \frac{GmM}{R}. \qquad (46\text{–}18)$$

The total mass $M$ inside the sphere is the density $\rho_c$ times the volume $4\pi R^3/3$:

$$M = \tfrac{4}{3}\pi R^3 \rho_c.$$

We'll assume that the speed $v$ of the galaxy is given by Hubble's law: $v = H_0 R$. Substituting these expressions for $M$ and $v$ into Eq. (46–18), we get

$$\tfrac{1}{2}m(H_0 R)^2 = \frac{Gm}{R}\left(\tfrac{4}{3}\pi R^3 \rho_c\right),$$

or

$$\rho_c = \frac{3H_0{}^2}{8\pi G}. \qquad (46\text{–}19)$$

This is the *critical density.* If the average density is less than $\rho_c$, the universe will continue to expand indefinitely; if it is greater, it will eventually stop expanding and begin to contract, possibly leading to the Big Crunch and then another Big Bang.

Putting numbers into Eq. (46–19), we find

$$\rho_c = \frac{3(1.8 \times 10^{-18}\ \text{s}^{-1})^2}{8\pi(6.67 \times 10^{-11}\ \text{m}^3/\text{kg}\cdot\text{s}^2)} = 5.8 \times 10^{-27}\ \text{kg/m}^3.$$

The mass of a hydrogen atom is $1.67 \times 10^{-27}$ kg, so this corresponds to about three hydrogen atoms per cubic meter.

## Dark Matter

Attempts have been made to estimate the actual average density of matter in the universe. We won't attempt to discuss the details; the estimates include both *luminous matter* (stars) and *nonluminous matter* (including black holes and interstellar gas). The total radiation in the universe is thought to have a mass equivalent of about 2% of the visible mass. All in all, the estimated total mass gives a density of the order of $\frac{1}{100}$ of $\rho_c$.

Some theorists believe that the universe must be closed and that the average density must be *equal* to $\rho_c$; then the expansion would approach zero after a long time. But for this to happen there would have to be a large amount of unseen *dark matter* in the universe. The nature of this missing matter is at present a mystery. Massive neutrinos are one possibility. The presence of dark matter is also suggested by recent measurements of the Doppler shift from opposite sides of rotating galaxies. The gravitational effect is 20 to 30 times as great as would be expected on the basis of the visible matter alone.

## 46–7 THE BEGINNING OF TIME

What an odd title for the very last section of a book! We will describe in general terms some of the current theories about the very early history of the universe and its relation to fundamental particle interactions. We'll find that an astonishing amount happened in the very first second. A lot of loose ends will be left untied, and many questions will be left unanswered. This is, after all, one of the frontiers of theoretical physics, and every day brings new theories, new discoveries, and new questions.

### Temperatures

The early universe was extremely dense and extremely hot, and the average particle energies were extremely large, all many orders of magnitude beyond anything that exists in the present universe. We can compare particle energy $E$ and absolute temperature $T$, using the equipartition principle

$$E = \tfrac{3}{2}kT, \qquad (46\text{–}20)$$

where $k$ is Boltzmann's constant, which we'll often express in eV/K:

$$k = 8.617 \times 10^{-5}\ \text{eV/K}.$$

When we're discussing orders of magnitude, we'll often drop the factor $\frac{3}{2}$ in Eq. (46–20) or approximate $3k/2$ as $10^{-4}$ eV/K.

### EXAMPLE 46–10

a) What is the average kinetic energy in electronvolts of particles at room temperature ($T = 300$ K) and at the surface of the sun ($T = 6000$ K)? b) What temperature corresponds to the ionization energy of the hydrogen atom, to the rest energy of the electron, and to the rest energy of the proton?

**SOLUTION** a) From Eq. (46–20),

$$E = \tfrac{3}{2}kT = \frac{3(8.617 \times 10^{-5}\ \text{eV/K})(300\ \text{K})}{2} = 0.0388\ \text{eV}.$$

The temperature of the sun's surface is larger by a factor of 6000 K/300 K = 20, so the average energy is 20(0.0388 eV) = 0.776 eV.

b) The ionization energy of hydrogen is 13.6 eV. From Eq. (46–20),

$$T = \frac{2E}{3k} = \frac{2(13.6\ \text{eV})}{3(8.617 \times 10^{-5}\ \text{eV/K})} = 1.05 \times 10^5\ \text{K}.$$

The rest energies of the electron and proton are 0.511 MeV and 938 MeV, respectively. Repeating the calculation for these values gives temperatures of $3.95 \times 10^9$ K, corresponding to the electron rest energy, and $7.26 \times 10^{12}$ K for the proton rest energy.

### Uncoupling of Interactions

The evolution of the universe has been characterized by a more or less linear growth of the scale factor $R$, which we can think of very roughly as characterizing the *size* of the universe, and by a corresponding decrease in average density. As the total gravitational potential energy increased during expansion, there were corresponding *decreases* in temperature and average particle energy. As this happened, the basic interactions became progressively uncoupled.

To understand this last point, recall that the unification of the electromagnetic and weak interactions occurs at energies large enough that the differences in mass among the various spin-1 bosons that mediate the interactions become insignificant by comparison. The electromagnetic interaction is mediated by the massless *photon,* the weak interaction by the weak bosons $W^{\pm}$ and $Z^0$ with masses of the order of 90 GeV/$c^2$. At energies much *less* than 90 GeV the two interactions seem quite different, but at energies much *greater* than 90 GeV they become part of a single interaction.

According to the grand unified theories (GUTs), a similar thing happens with the strong interaction. It becomes unified with the electroweak interaction at energies of the order of $10^{14}$ GeV, but at lower energies the two appear quite distinct. One of the reasons that GUTs are still very speculative is that there is no way to do controlled experiments in this energy range, larger by a factor of $10^{10}$ than even the proposed SSC energy.

Finally, at sufficiently high energies and short distances, it is assumed that gravitation becomes unified with the other three interactions. The distance at which this happens is thought to be of the order of $10^{-35}$ m. This distance, called the *Planck length* $l_P$, is determined by the speed of light $c$ and the fundamental constants of quantum mechanics and gravitation, $\hbar$ and $G$, respectively. The Planck length $l_P$ is defined as

$$l_P = \sqrt{\frac{\hbar G}{c^3}} \qquad (46\text{–}21)$$

$$= 1.616 \times 10^{-35}\ \text{m}.$$

We invite you to verify that this combination of constants does indeed have units of length. The *Planck time* $t_P = l_P/c$ is the time required for light to travel a distance $l_P$:

$$t_P = \frac{l_P}{c} = \sqrt{\frac{\hbar G}{c^5}} \qquad (46\text{–}22)$$

$$= 0.539 \times 10^{-43}\ \text{s}.$$

If we mentally go backward in time, we have to stop when we reach $t = 10^{-43}$ s because we have no adequate theory that unifies all four interactions. So as yet we have no way of knowing what happened or how the universe behaved at times earlier than the Planck time or when its size was less than the Planck length.

## The Standard Model

The description that follows is called the *standard model* of the history of the universe. The title may be slightly optimistic, but it does indicate that there are substantial areas of theory that rest on solid experimental foundations and are quite generally accepted. The figure on pages 1314–1315 presents a graphical description of this history, with the characteristic sizes, particle energies, and temperatures at various times. Referring to this chart frequently will help you understand the following discussion.

In the standard model of the Big Bang, the temperature of the universe at time $t = 10^{-43}$ s (the Planck time) was about $10^{32}$ K. Using Eq. (46–20), we find that the average energy per particle was of the order of

$$E = kT$$
$$= (10^{-4}\ \text{eV/K})\,(10^{32}\ \text{K}) \cong 10^{19}\ \text{GeV}.$$

In the unified theories this is about the energy below which gravity begins to behave as a separate interaction. This time therefore marks the end of any proposed TOE and the beginning of the GUT period.

During the GUT period, roughly $t = 10^{-43}$ to $10^{-35}$ s, the strong and electroweak forces were still unified, and the universe consisted of a soup of quarks and leptons transforming into each other so freely that there was no distinction between the two families of particles. There may also have been other, much more massive particles being freely created and destroyed. One important characteristic of GUTs is that at sufficiently high energies, baryon number is not conserved. (We mentioned earlier the proposed decay of the proton, which has not yet been observed.) Thus by the end of the GUT period the numbers of quarks and antiquarks may have been unequal. This point has important implications; we'll return to it at the end of the section.

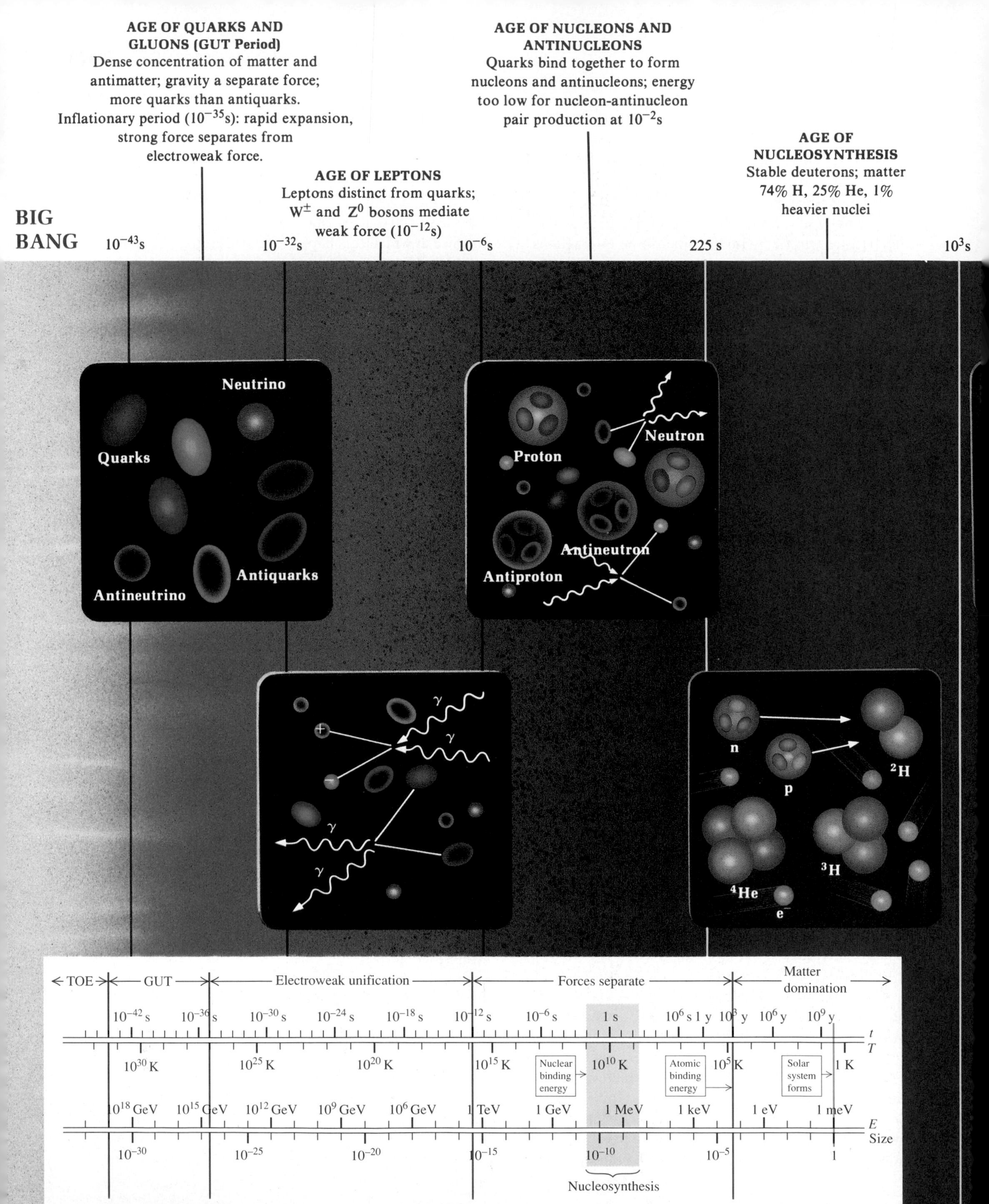
AGE OF QUARKS AND GLUONS (GUT Period)
Dense concentration of matter and antimatter; gravity a separate force; more quarks than antiquarks.
Inflationary period ($10^{-35}$s): rapid expansion, strong force separates from electroweak force.
AGE OF LEPTONS
Leptons distinct from quarks; $W^{\pm}$ and $Z^0$ bosons mediate weak force ($10^{-12}$s)
AGE OF NUCLEONS AND ANTINUCLEONS
Quarks bind together to form nucleons and antinucleons; energy too low for nucleon-antinucleon pair production at $10^{-2}$s
AGE OF NUCLEOSYNTHESIS
Stable deuterons; matter 74% H, 25% He, 1% heavier nuclei
BIG BANG
$10^{-43}$s
$10^{-32}$s
$10^{-6}$s
225 s
$10^3$s
Neutrino
Quarks
Antiquarks
Antineutrino
Proton
Neutron
Antineutron
Antiproton
γ
n
p
$^2$H
$^3$H
$^4$He
e$^-$
TOE
GUT
Electroweak unification
Forces separate
Matter domination
$10^{-42}$ s
$10^{-36}$ s
$10^{-30}$ s
$10^{-24}$ s
$10^{-18}$ s
$10^{-12}$ s
$10^{-6}$ s
1 s
$10^6$ s 1 y
$10^3$ y
$10^6$ y
$10^9$ y
t
T
$10^{30}$ K
$10^{25}$ K
$10^{20}$ K
$10^{15}$ K
Nuclear binding energy
$10^{10}$ K
Atomic binding energy
$10^5$ K
Solar system forms
1 K
$10^{18}$ GeV
$10^{15}$ GeV
$10^{12}$ GeV
$10^9$ GeV
$10^6$ GeV
1 TeV
1 GeV
1 MeV
1 keV
1 eV
1 meV
E
Size
$10^{-30}$
$10^{-25}$
$10^{-20}$
$10^{-15}$
$10^{-10}$
$10^{-5}$
1
Nucleosynthesis

Logarithmic scales show characteristic temperature, energy, and size of the universe as functions of time.

**AGE OF IONS**
Expanding, cooling gas of ionized H and He

# *A Brief History of the Universe*

**AGE OF ATOMS**
Neutral atoms form, pulled together by gravity; universe becomes transparent to most light

**AGE OF STARS AND GALAXIES**
Thermonuclear fusion begins in stars, forming heavier nuclei

$10^{13}$ s $\qquad$ $10^{15}$ s $\qquad$ **NOW**

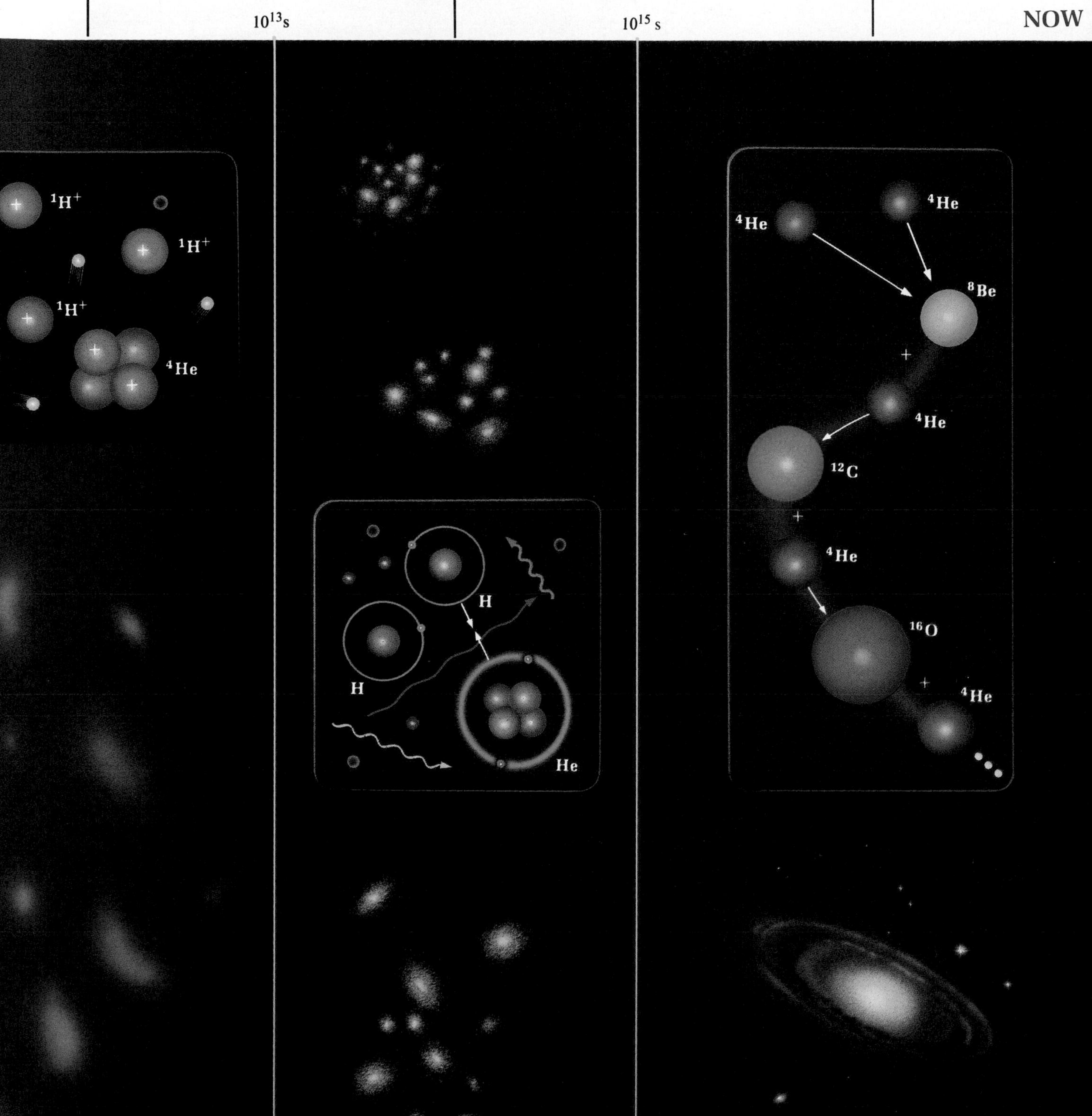

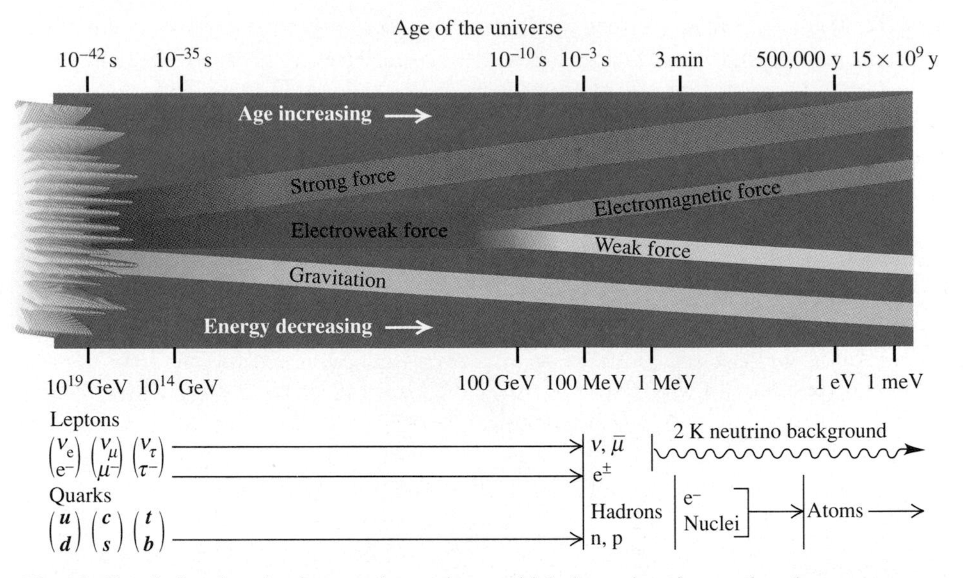


**46–18** Graph showing the times and energies at which the various interactions became uncoupled. The energy scale is backward because the energy decreased as the age of the universe increased.

By $t = 10^{-35}$ s the temperature had decreased to about $10^{27}$ K and the average energy to about $10^{14}$ GeV. At this energy the strong force separated from the electroweak force (Fig. 46–18), and baryon number and lepton numbers began to be separately conserved. In some models, called *inflationary models,* this separation of the strong force was analogous to a *phase change,* such as boiling of a liquid, with an associated heat of vaporization. Think of it as similar to boiling a heavy nucleus, pulling the particles apart beyond the short range of the nuclear force. As a result, the inflationary models predict, there was a very rapid expansion. In one model, the scale factor $R$ increased by a factor of $10^{50}$ in $10^{-32}$ s.

At $t = 10^{-32}$ s the universe was a mixture of quarks, leptons, and the mediating bosons (gluons, photons, and the weak bosons $W^\pm$ and $Z^0$). It continued to expand and cool from the inflationary period to $t = 10^{-6}$ s, when the temperature was about $10^{13}$ K and typical energies were about 1 GeV (comparable to the rest energy of a nucleon). At this time the quarks began to bind together to form nucleons and antinucleons. Up until $10^{-2}$ s there were still enough photons with energy of 1 GeV or more to cause enough photoproduction of nucleon-antinucleon pairs to balance the process of nucleon-antinucleon annihilation. By about $t = 10^{-2}$ s most photon energies fell below the threshold energy for such pair production. There was a slight excess of nucleons over antinucleons; as a result, nearly all of the antinucleons and many of the nucleons annihilated one another. A similar equilibrium occurred later between photoproduction and annihilation of electron-positron pairs. At about $t = 14$ s the average energy dropped to around 1 MeV, below the threshold for $e^+/e^-$ pair production. After pair production ceased, nearly all of the remaining positrons were annihilated, leaving the universe with many more protons and electrons than the antiparticles of each.

Up until about $t = 1$ s, neutrons could be produced in the endothermic reaction

$$e^- + p \longrightarrow n + \nu_e.$$

After this time, most electrons no longer had enough energy for this reaction. The average neutrino energy also decreased, and as the universe expanded, equilibrium reactions that involved *absorption* of neutrinos (which occurred with decreasing probability) became inoperative. At this time, in effect, the flux of neutrinos and antineutrinos throughout the universe "decoupled" from the rest of the universe. Because of the extraordinarily low cross section for neutrino absorption, most of this flux is still present today, although cooled greatly by expansion. The standard model predicts a present neutrino temperature of about 2 K, but no one has been able to devise an experiment to test this prediction.

## Nucleosynthesis

At about $t = 1$ s the ratio of protons to neutrons was determined by the Boltzmann distribution factor $e^{-\Delta E/kT}$, where $\Delta E$ is the difference between the neutron and proton rest energies, or 1.294 MeV. At a temperature of about $10^{10}$ K this gives about 4.5 times as many protons as neutrons. However, as we have discussed, free neutrons decay spontaneously to protons with a half-life of 616 s. This decay caused the proton-neutron ratio to increase until about $t = 225$ s. At this time the temperature was about $10^9$ K, and the average energy was well below 2 MeV.

This is a critical energy because the binding energy of the *deuteron* (a neutron and a proton bound together) is 2.22 MeV. A neutron bound in a deuteron does not decay spontaneously. As the average energy decreased, a proton and a neutron could combine to form a deuteron, and there were very few photons with enough energy (2.22 MeV) to dissociate the deuterons again. Therefore the combining of protons and neutrons into deuterons halted the decay of free neutrons.

The formation of deuterons starting at about $t = 225$ s marked the beginning of the period of formation of nuclei, or *nucleosynthesis*. At this time there were about seven protons for each neutron. The deuteron ($^2$H) can absorb a neutron and form a triton ($^3$H), or it can absorb a proton and form $^3$He. Then $^3$H can absorb a proton, and $^3$He can absorb a neutron, each yielding $^4$He (the $\alpha$ particle). A few $^7$Li nuclei may also have been formed by fusion of $^3$H and $^4$He nuclei. According to the theory, essentially all the $^1$H and $^4$He in the present universe was formed at this time. But then the building of nuclei almost ground to a halt. The reason is that there is *no* nuclide with mass number $A = 5$, stable or unstable. Alpha particles simply do not absorb neutrons or protons. The nuclide $^8$Be formed by fusion of two $^4$He nuclei is unstable, with an extremely short half-life, about $10^{-16}$ s. Note also that at this time the average energy was still much too large for electrons to be bound to nuclei; there were not yet any atoms.

### EXAMPLE 46–11

Nearly all of the protons and neutrons in the seven-to-one ratio at $t = 225$ s either formed $^4$He or remained as $^1$H. After this time, what was the relative abundance of $^1$H and $^4$He by mass?

**SOLUTION** The $^4$He nucleus contains two protons and two neutrons. For every two neutrons we have 14 protons. The two neutrons and two of the 14 protons make up one $^4$He nucleus, leaving 12 protons ($^1$H nuclei). So at this time there were 12 $^1$H nuclei for every $^4$He nucleus. The masses of $^1$H and $^4$He are about 1 u and 4 u, respectively, so there were 12 u of $^1$H for every 4 u of $^4$He. Therefore the mix was 75% $^1$H and 25% $^4$He by mass. This result agrees very well with estimates of the present H-He ratio in the universe, an important confirmation of this part of the theory.

Further nucleosynthesis did not occur until very much later, about $t = 2 \times 10^{13}$ s (about 700,000 y). At that time the temperature was about 3000 K and the average energy a few eV, and electrically neutral hydrogen and helium *atoms* could form. With the electrical repulsions of the nuclei cancelled out, gravitational attraction could pull the neutral atoms together to form galaxies and eventually stars. Thermonuclear reactions in stars are believed to have produced all of the more massive nuclei. In Section 45–9 we discussed one cycle of thermonuclear reactions in which $^1$H becomes $^4$He. These reactions are one of the sources of the energy radiated from stars.

As a star uses up its hydrogen, the inward gravitational pressure exceeds the outward radiation and gas pressure, and the star's core begins to contract. As it does so, the gravitational potential energy decreases, and the kinetic energies of its atoms increase. For stars with sufficient mass there is both enough energy and sufficient density to begin another process, *helium fusion*. First, two $^4$He nuclei fuse to form $^8$Be. The very short lifetime of this unstable nuclide is compensated by an usually large cross section for absorption of another $^4$He nucleus with a specific energy, a kind of resonance effect. Thus a substantial fraction of the $^8$Be nuclei fuse with $^4$He to form the stable nuclide $^{12}$C. The net result is the fusion of three $^4$He nuclei to form one $^{12}$C, the *triple-alpha process*. Then successive fusions with $^4$He give $^{16}$O, $^{20}$Ne, and $^{24}$Mg. All these reactions are exothermic. They release energy to heat up the star, and $^{12}$C and $^{16}$O can fuse to form elements with higher and higher atomic numbers.

The binding energy per nucleon for accessible nuclides peaks out at mass number $A = 56$ with the nuclide $^{56}$Fe, so exothermic fusion reactions stop with Fe. But successive neutron captures followed by beta decays can continue the synthesis of more massive nuclei. If the star is massive enough, it may eventually explode as a *supernova*, sending out into space the heavy elements produced by the earlier processes. In space, the debris and other interstellar matter can gravitationally bunch together to form new stars and planets. And that's how the stuff of our earth was formed.

## Background Radiation

In 1965, Arno Penzias and Robert Wilson discovered microwave radiation of unknown origin with no apparent preferred direction. Further research showed that the radiation had a frequency spectrum that fit Planck's blackbody radiation law, Eq. (40–31) (see Section 40–8). The wavelength of peak intensity was 1.1 mm (in the microwave region of the spectrum), with a corresponding absolute temperature $T = 2.7$ K. It was soon recognized that this was a remnant from the early evolution of the universe. We mentioned above that neutral atoms began to form at about $t = 700{,}000$ years. With far fewer charged particles than previously, the universe became transparent to electromagnetic radiation of long wavelength. The 3000-K blackbody radiation therefore survived, cooling to its present temperature of 2.7 K as the universe expanded. This *microwave background radiation* is among the most clear-cut experimental confirmations of the Big Bang.

### EXAMPLE 46–12

By approximately what factor has the universe expanded since $t = 700{,}000$ y?

**SOLUTION** Equation (40–29), the Wein displacement law, shows that the peak wavelength $\lambda_m$ in blackbody radiation is inversely proportional to absolute temperature:

$$\lambda_m T = 2.90 \times 10^{-3}\ \text{m} \cdot \text{K}.$$

As the universe expands, all wavelengths (including $\lambda_m$) increase in proportion to the scale factor $R$. The temperature decreased by a factor (2.7 K)/(3000 K), so the scale factor must have *increased* by this factor. Thus between $t = 700{,}000$ y and the present, the universe expanded by a factor of (3000 K)/(2.7 K), or about 1000.

## Matter and Antimatter

One of the most remarkable features of our universe is the asymmetry between matter and antimatter. One might think that the universe should have equal numbers of protons and antiprotons and of electrons and positrons, but this doesn't appear to be the case. There is no evidence for the existence of substantial amounts of antimatter (matter composed of antiprotons, antineutrons, and positrons) anywhere in the universe. Theories of the early universe must confront this imbalance.

We've mentioned that most GUTs include violation of conservation of baryon number at energies at which the strong and electroweak interactions have converged. If particle-antiparticle symmetry is also violated, we have a mechanism for making more quarks than antiquarks and eventually more nucleons than antinucleons. One serious problem is that any asymmetry created in this way during the GUT era might be wiped out by the electroweak interaction after the end of the GUT era. If so, there must be some mechanism that creates particle-antiparticle asymmetry at a much *later* time. The problem of the matter-antimatter asymmetry is still very much an open one.

We hope that this qualitative discussion has conveyed at least a hint as to the close connections between particle theory and cosmology. There are still lots of unanswered questions. Is the universe open or closed? There's not enough visible matter in sight to stop its expansion, but perhaps there is enough dark matter. There are so many neutrinos in the universe that even a 20-eV neutrino mass would provide enough additional matter, and a 17-keV neutrino mass would be far too much. The exotic particles postulated in connection with the unification theories (Section 46–5) may also be relevant for the dark matter problem.

We have no idea what happened during the first $10^{-43}$ s after the Big Bang because we have no quantum theory of gravity. The electroweak theory is on very firm ground, but there are many versions of grand unification theories from which to choose. And we are still very far from having a suitable theory that unifies all four interactions in nature. But this search for understanding of the physical world we live in continues to be one of the most exciting adventures of the human mind.

## SUMMARY

- Each particle has an antiparticle. Particles can be created and destroyed, some of them (including electrons and positrons) only in pairs or in conjunction with other particles.
- Particles serve as mediators for the fundamental interactions. The photon is the mediator of the electromagnetic interaction. Yukawa proposed the existence of mesons to mediate the nuclear interaction. Mediating particles that can exist only because of the uncertainty principle for energy are called virtual particles.
- Cyclotrons, linear accelerators, and synchrotrons are used to accelerate charged particles to high energies for experiments with particle interactions. Only part of the beam energy is available to cause reactions. This problem is avoided in colliding-beam experiments.
- Four fundamental interactions are found in nature: the strong, electromagnetic, weak, and gravitational interactions. Particles can be described in terms of their interactions and of quantities that are conserved in all or some of the interactions.
- Fermions have half-integer spins; bosons have integer spins. Leptons have no strong interactions. Strongly interacting particles are called hadrons. They include mesons, which are always bosons, and baryons, which are always fermions. There are conservation laws for three different lepton numbers and for baryon number.
- Additional quantum numbers, including strangeness and charm, are conserved in some interactions and not in others.

### KEY TERMS

antiparticle
meson
muon
pion
boson
fermion
lepton
hadron
strangeness
quark
gluon
charm

**standard model**
**eightfold way**
**electroweak theory**
**grand unified theory**
**red shift**
**Hubble's law**
**cosmological principle**
**Big Bang**

• Hadrons are composed of quarks. There are believed to be six types of quarks. The interaction between quarks is mediated by gluons. Quarks and gluons have an additional attribute called color.

• Symmetry considerations play an indispensable role in all fundamental-particle theories. The electromagnetic and weak interactions become unified at high energies into the electroweak interaction. In grand unified theories the strong interaction is also unified with these, but at much higher energies.

• According to the Big Bang model, the universe originated as a very small space, possibly a single point. Its expansion is described in terms of temperature, average particle energies, and distances. Whether or not it will continue to expand forever depends on the average density compared to a value called the critical density.

• The Big Bang model shows the close connections between particle interactions and the evolution of the universe.

# EXERCISES

## Section 46–1 Fundamental Particles

**46–1 "Maximum Power, Scotty!"** The starship *Enterprise,* of television and movie fame, is powered by the controlled combination of matter and antimatter. If the entire 200-kg antimatter fuel supply of the *Enterprise* combines with matter, how much energy is released?

**46–2** An electron and a positron, both with negligible kinetic energy, annihilate, producing two photons. a) Why must the momenta of the photons be equal and opposite? b) Find the energy, frequency, and wavelength of each photon.

**46–3** A neutral pion (mass 0.145 u) at rest decays into two $\gamma$ photons. Find the energy, frequency, and wavelength of each photon.

**46–4** Two equal-energy photons make a head-on collision and annihilate one another, producing a $\mu^+\mu^-$ pair. Calculate the maximum wavelength of the photons for this to occur. What happens if the photons have shorter wavelength than this maximum but still annihilate and produce a $\mu^+\mu^-$ pair?

**46–5** Consider the decay $^{22}\text{Na} \rightarrow {}^{22}\text{Ne} + \beta^+ + \nu_e$. a) Show that charge is conserved. b) Calculate the energy released in the decay. The atomic masses are 21.994435 u for $^{22}$Na and 21.991384 u for $^{22}$Ne.

## Section 46–2 Particle Accelerators

**46–6** The magnetic field in a cyclotron that is accelerating protons is 0.700 T. a) How many times per second should the potential across the dees reverse? (This is twice the frequency of the circulating protons.) b) The maximum radius of the cyclotron is 0.250 m. What is the maximum speed of the proton? c) Through what potential difference would the proton have to be accelerated to give it the speed calculated in part (b)?

**46–7** Deuterons in a cyclotron travel in a circle with radius 32.0 cm just before emerging from the dees. The frequency of the applied alternating voltage is 6.00 MHz. Find a) the magnetic field; b) the energy and speed of the deuterons upon emergence.

**46–8** Calculate the threshold kinetic energy for the production of a $\eta^0$ particle in the collision of a proton beam with a stationary proton target: $\text{p} + \text{p} \rightarrow \text{p} + \text{p} + \eta^0$. The rest mass of the $\eta^0$ is given by $\eta^0 c^2 = 549$ MeV.

**46–9** a) A high-energy beam of alpha particles collides with a stationary helium gas target. What must be the total energy of a beam particle if the available energy in the collision is 12.0 GeV? b) What must be the energy of each beam to produce the same available energy in a colliding-beam experiment?

**46–10** a) What is the speed of a proton that has total energy 1000 GeV, the maximum energy of the Tevatron at Fermilab? b) What is the angular frequency $\omega$ of a proton with the speed calculated in part (a) in a magnetic field of 4.00 T? Use both the nonrelativistic Eq. (46–7) and the correct relativistic expression, and compare the results.

## Section 46–3 Particles and Interactions

**46–11** Show that the coupling constant $f^2/\hbar c$ is dimensionless.

**46–12** If a $\tau$ lepton at rest decays into a muon at rest and two neutrinos, what is the total kinetic energy of the decay products?

**46–13** If a $\Sigma^+$ at rest decays into an n and a $\pi^+$, what is the total kinetic energy of the decay products?

**46–14** In each of the following decays, are the three lepton numbers conserved?

a) $\mu^- \rightarrow e^- + \nu_e + \overline{\nu}_\mu$
b) $\tau^- \rightarrow e^- + \overline{\nu}_e + \nu_\tau$
c) $\pi^+ \rightarrow e^+ + \gamma$
d) $\text{n} \rightarrow \text{p} + e^- + \overline{\nu}_e$

**46–15** In each of the following reactions or decays, is baryon number conserved?
a) $n \rightarrow p + e^- + \bar{\nu}_e$
b) $p + n \rightarrow p + \pi^0$
c) $p \rightarrow \pi^+ + \pi^0$
d) $p + p \rightarrow p + p + \pi^0$

**46–16** In each of the following reactions or decays, is strangeness conserved?
a) $K^+ \rightarrow \mu^+ + \nu_\mu$
b) $n + K^+ \rightarrow p + \pi^0$
c) $K^+ + K^- \rightarrow \pi^0$
d) $p + K^- \rightarrow \Lambda + \pi^0$

## Section 46–4 Quarks

**46–17** The quark content of the neutron is ***udd***. What is the quark content of the antineutron? Is the neutron its own antiparticle? What is the answer to this question for a $\pi^0$ particle?

**46–18** Use the method of Example 46–7 to deduce the quark content of a) a particle with charge $+2e$, baryon number 1, and strangeness 0; b) a particle with charge $+e$, baryon number $-1$, and strangeness $+2$; c) a particle with charge $-e$, baryon number 1, and strangeness $-3$.

**46–19** Determine the electric charge, baryon number, strangeness, and charm quantum numbers for the following quark combinations: a) $uus$; b) $c\bar{s}$; c) $\overline{ddu}$; d) $c\bar{b}$.

**46–20** What is the total kinetic energy of the decay products when a J/$\Psi$ particle at rest decays to $e^+ + e^-$?

## Section 46–5 The Standard Model and the Eightfold Way

**46–21** Nine of the spin-$\frac{3}{2}$ baryons are four $\Delta^*$ particles with strangeness 0, charges $+2e$, $+e$, 0, and $-e$, and each with mass 1232 MeV/$c^2$; three $\Sigma^*$ particles with strangeness $-1$, charges $+e$, 0, and $-e$, and each with mass 1385 MeV/$c^2$; and two $\Xi^*$ particles with strangeness $-2$, charges 0 and $-e$, and each with mass 1530 MeV/$c^2$. a) Place these particles on a plot of $S$ versus $Q$. Deduce the $Q$ and $S$ values of the missing $\Omega^-$ particle and place it on your diagram. Also label the particles with their masses. The mass of the $\Omega^-$ is 1672 MeV/$c^2$; is this value consistent with your diagram? b) Deduce the three-quark combinations (of $u$, $d$, and $s$) that comprise each of these ten particles. Redraw the plot of $S$ versus $Q$ from part (a) with each particle labeled by its quark content. What regularities do you see?

## Section 46–6 The Expanding Universe

**46–22** Derive Eq. (46–14) from (46–13).

**46–23 Measuring Distance to a Galaxy.** The spectrum of the sodium atom is detected in the light from a distant galaxy. a) If the 590.0 nm line is red-shifted to 1030.0 nm, at what speed is the galaxy receding from the earth? b) Use Hubble's law to calculate the distance of the galaxy from the earth.

## Section 46–7 The Beginning of Time

**46–24** a) Show that the expression for the Planck length, $\sqrt{\hbar G/c^3}$, has dimensions of length. b) Evaluate the numerical value of $\sqrt{\hbar G/c^3}$, and verify the value given in Eq. (46–21).

**46–25** Calculate the energy released in each of the following reactions: a) $p + {}^2H \rightarrow {}^3He$; b) $n + {}^3He \rightarrow {}^4He$.

**46–26** Calculate the energy released in the triple-alpha process $3\,{}^4He \rightarrow {}^{12}C$.

# PROBLEMS

**46–27 Superconducting Super Collider.** In the SSC, each proton will be accelerated to a total energy of 20 TeV. a) In the colliding beams, what is the available energy $E_a$ in a collision? b) What magnetic field magnitude $B$ is required if protons at this energy are to circulate in a circle with circumference 85 km? c) In a fixed target experiment in which a beam of protons is incident on a stationary proton target, what must be the total energy of the particles in the beam to produce the same available energy as in part (a)?

**46–28** A proton and an antiproton make a head-on collision with equal kinetic energies. Two gamma rays with wavelengths of 1.00 fm are produced. Calculate the kinetic energy of the incident proton.

**46–29** Calculate the threshold kinetic energy for the reaction

$$K^- + p \rightarrow \Lambda^0 + \eta^0$$

if a $K^-$ beam is incident on a stationary proton target.

**46–30** Calculate the threshold kinetic energy for the reaction

$$\pi^+ + p \rightarrow \Sigma^+ + K^+$$

if a $\pi^+$ beam is incident on a stationary proton target.

**46–31** Beams of $\pi^-$ mesons are being used experimentally in the treatment of cancer. What is the minimum total energy a pion can release in a tumor? (That is, what energy is released when a $\pi^-$ at rest decays to stable products? The actual energy deposited will be increased over this by the initial kinetic energy of the $\pi^-$.)

**46–32** A $K^+$ meson at rest decays into two $\pi$ mesons. a) What are the allowed combinations of $\pi^0$, $\pi^+$, and $\pi^-$ as decay products? b) Find the total kinetic energy of the $\pi$ mesons.

**46–33** Estimate the energy width (energy uncertainty) of the J/$\Psi$ if its mean lifetime is $1 \times 10^{-20}$ s. What fraction is this of its rest energy?

**46–34** The measured energy width of the $\phi$ meson is 4.2 MeV, and its mass is 1020 MeV/$c^2$. Using the uncertainty principle, Eq. (41–12), estimate the lifetime of the $\phi$ meson.

**46–35** A $\phi$ meson (Problem 46–34) at rest decays via $\phi \to K^+ + K^-$. a) Find the kinetic energy of the $K^+$ meson. (Assume that the two decay products share kinetic energy equally, since their masses are equal.) b) Suggest a reason why the decay $\phi \to K^+ + K^- + \pi^0$ has not been observed. c) Suggest reasons why the decays $\phi \to K^+ + \pi^-$ and $\phi \to K^+ + \mu^-$ have not been observed.

**46–36** Each of the reactions listed below is missing a single particle. Calculate the baryon number, charge, strangeness, and the three lepton numbers (where appropriate) of the missing particle, and from this identify the particle.

a) $p + p \to \Lambda^0 + K^+ + ?$

b) $\bar{\nu}_e + p \to n + ?$

c) $K^- + n \to \pi^- + ?$

d) $\mu^- \to e^- + \nu_\mu + ?$

**46–37** An $\Omega^-$ particle at rest decays into a $\Lambda^0$ and a $K^-$. a) Find the total kinetic energy of the decay products. b) What fraction of the energy is carried off by each particle? (For simplicity, use nonrelativistic momentum and kinetic-energy expressions.)

**46–38 Proton Decay.** Proton decay is a feature of some grand unification theories. One possible decay could be $p \to e^+ + \pi^0$, which violates both baryon and lepton number conservation, so the proton lifetime is expected to be very long. Suppose the proton half-life were $1.0 \times 10^{18}$ y. a) Calculate the energy deposited per kilogram of body tissue (in rad) due to the decay of the protons in your body in one year. Model your body as consisting entirely of water. Assume that the $\pi^0$ decays into two $\gamma$ rays, that the positron annihilates with an electron, and that all the energy produced in the primary and secondary decays remains in your body. b) Calculate the equivalent dose (in rem) assuming an RBE of 1.0 for all the radiation products and compare with the 0.1 rem due to the natural background and the 5.0 rem guideline for industrial workers. Based on your calculation, can the proton lifetime be as short as $1.0 \times 10^{18}$ y?

## CHALLENGE PROBLEMS

**46–39** A $\Lambda^0$ hyperon at rest decays into a proton and a $\pi^-$. a) Find the total kinetic energy of the decay products. b) What fraction of the energy is carried off by each particle? (Use relativistic momentum and kinetic energy expressions.)

**46–40** Consider a collision in which a stationary particle with mass $M$ is bombarded by a particle with mass $m$, speed $v_0$, and total energy (including rest energy) $E_m$. a) Use the Lorentz transformation to write the velocities $v_m$ and $v_M$ of particles $m$ and $M$ in terms of the speed $v_{cm}$ of the center of momentum. b) Use the fact that the total momentum in the center-of-momentum frame is zero to obtain an expression for $v_{cm}$ in terms of $m$, $M$, and $v_0$. c) Combine the results of parts (a) and (b) to obtain Eq. (46–9) for the total energy in the center-of-momentum frame.

APPENDIX A

# The International System of Units

The Système International d'Unités, abbreviated SI, is the system developed by the General Conference on Weights and Measures and adopted by nearly all the industrial nations of the world. It is based on the mksa (meter-kilogram-second-ampere) system. The following material is adapted from NBS Special Publication 330 (1981 edition) of the National Bureau of Standards.

| Quantity | Name of unit | Symbol | |
|---|---|---|---|
| | **SI base units** | | |
| length | meter | m | |
| mass | kilogram | kg | |
| time | second | s | |
| electric current | ampere | A | |
| thermodynamic temperature | kelvin | K | |
| luminous intensity | candela | cd | |
| amount of substance | mole | mol | |
| | **SI derived units** | | **Equivalent units** |
| area | square meter | $m^2$ | |
| volume | cubic meter | $m^3$ | |
| frequency | hertz | Hz | $s^{-1}$ |
| mass density (density) | kilogram per cubic meter | $kg/m^3$ | |
| speed, velocity | meter per second | m/s | |
| angular velocity | radian per second | rad/s | |
| acceleration | meter per second squared | $m/s^2$ | |
| angular acceleration | radian per second squared | $rad/s^2$ | |
| force | newton | N | $kg \cdot m/s^2$ |
| pressure (mechanical stress) | pascal | Pa | $N/m^2$ |
| kinematic viscosity | square meter per second | $m^2/s$ | |
| dynamic viscosity | newton-second per square meter | $N \cdot s/m^2$ | |
| work, energy, quantity of heat | joule | J | $N \cdot m$ |
| power | watt | W | J/s |
| quantity of electricity | coulomb | C | $A \cdot s$ |
| potential difference, electromotive force | volt | V | W/A, J/C |
| electric field strength | volt per meter | V/m | N/C |
| electric resistance | ohm | Ω | V/A |
| capacitance | farad | F | $A \cdot s/V$ |
| magnetic flux | weber | Wb | $V \cdot s$ |
| inductance | henry | H | $V \cdot s/A$ |
| magnetic flux density | tesla | T | $Wb/m^2$ |
| magnetic field strength | ampere per meter | A/m | |
| magnetomotive force | ampere | A | |
| luminous flux | lumen | lm | $cd \cdot sr$ |
| luminance | candela per square meter | $cd/m^2$ | |
| illuminance | lux | lx | $lm/m^2$ |
| wave number | 1 per meter | $m^{-1}$ | |
| entropy | joule per kelvin | J/K | |
| specific heat capacity | joule per kilogram kelvin | $J/kg \cdot K$ | |
| thermal conductivity | watt per meter kelvin | $W/m \cdot K$ | |

| Quantity | Name of unit | Symbol | Equivalent units |
|---|---|---|---|
| radiant intensity | watt per steradian | W/sr | |
| activity (of a radioactive source) | becquerel | Bq | $s^{-1}$ |
| radiation dose | gray | Gy | J/kg |
| radiation dose equivalent | sievert | Sv | J/kg |
| | **SI supplementary units** | | |
| plane angle | radian | rad | |
| solid angle | steradian | sr | |

## *DEFINITIONS OF SI UNITS*

**meter (m)** The *meter* is the length equal to the distance traveled by light, in vacuum, in a time of 1/299,792,458 second.

**kilogram (kg)** The *kilogram* is the unit of mass; it is equal to the mass of the international prototype of the kilogram. (The international prototype of the kilogram is a particular cylinder of platinum-iridium alloy that is preserved in a vault at Sèvres, France, by the International Bureau of Weights and Measures.)

**second (s)** The *second* is the duration of 9,192,631,770 periods of the radiation corresponding to the transition between the two hyperfine levels of the ground state of the cesium-133 atom.

**ampere (A)** The *ampere* is that constant current that, if maintained in two straight parallel conductors of infinite length, of negligible circular cross section, and placed 1 meter apart in vacuum, would produce between these conductors a force equal to $2 \times 10^{-7}$ newton per meter of length.

**kelvin (K)** The *kelvin,* unit of thermodynamic temperature, is the fraction 1/273.16 of the thermodynamic temperature of the triple point of water.

**ohm (Ω)** The *ohm* is the electric resistance between two points of a conductor when a constant difference of potential of 1 volt, applied between these two points, produces in this conductor a current of 1 ampere, this conductor not being the source of any electromotive force.

**coulomb (C)** The *coulomb* is the quantity of electricity transported in 1 second by a current of 1 ampere.

**candela (cd)** The *candela* is the luminous intensity, in a given direction, of a source that emits monochromatic radiation of frequency $540 \times 10^{12}$ hertz and that has a radiant intensity in that direction of 1/683 watt per steradian.

**mole (mol)** The *mole* is the amount of substance of a system that contains as many elementary entities as there are carbon atoms in 0.012 kg of carbon 12. The elementary entities must be specified and may be atoms, molecules, ions, electrons, other particles, or specified groups of such particles.

**newton (N)** The *newton* is that force that gives to a mass of 1 kilogram an acceleration of 1 meter per second per second.

**joule (J)** The *joule* is the work done when the point of application of 1 newton is displaced a distance of 1 meter in the direction of the force.

**watt (W)** The *watt* is the power that gives rise to the production of energy at the rate of 1 joule per second.

**volt (V)** The *volt* is the difference of electric potential between two points of a conducting wire carrying a constant current of 1 ampere, when the power dissipated between these points is equal to 1 watt.

**weber (Wb)** The *weber* is the magnetic flux that, linking a circuit of one turn, produces in it an electromotive force of 1 volt as it is reduced to zero at a uniform rate in 1 second.

**lumen (lm)** The *lumen* is the luminous flux emitted in a solid angle of 1 steradian by a uniform point source having an intensity of 1 candela.

**farad (F)** The *farad* is the capacitance of a capacitor between the plates of which there appears a difference of potential of 1 volt when it is charged by a quantity of electricity equal to 1 coulomb.

**henry (H)** The *henry* is the inductance of a closed circuit in which an electromotive force of 1 volt is produced when the electric current in the circuit varies uniformly at a rate of 1 ampere per second.

**radian (rad)** The *radian* is the plane angle between two radii of a circle that cut off on the circumference an arc equal in length to the radius.

**steradian (sr)** The *steradian* is the solid angle that, having its vertex in the center of a sphere, cuts off an area of the surface of the sphere equal to that of a square with sides of length equal to the radius of the sphere.

**SI Prefixes** The names of multiples and submultiples of SI units may be formed by application of the prefixes listed in Table 1–1, page 7.

APPENDIX B

# Useful Mathematical Relations

### *ALGEBRA*

$a^{-x} = \frac{1}{a^x} \qquad a^{(x+y)} = a^x a^y \qquad a^{(x-y)} = \frac{a^x}{a^y}$

***Logarithms:*** If $\log a = x$, then $a = 10^x$. $\log a + \log b = \log(ab)$ $\log a - \log b = \log(a/b)$ $\log(a^n) = n \log a$

If $\ln a = x$, then $a = e^x$. $\ln a + \ln b = \ln(ab)$ $\ln a - \ln b = \ln(a/b)$ $\ln(a^n) = n \ln a$

***Quadratic formula:*** If $ax^2 + bx + c = 0$, $\quad x = \dfrac{-b \pm \sqrt{b^2 - 4ac}}{2a}$.

### *BINOMIAL THEOREM*

$$(a + b)^n = a^n + na^{n-1}b + \frac{n(n-1)a^{n-2}b^2}{2!} + \frac{n(n-1)(n-2)a^{n-3}b^3}{3!} + \cdots$$

### *TRIGONOMETRY*

In the right triangle $ABC$, $x^2 + y^2 = r^2$.

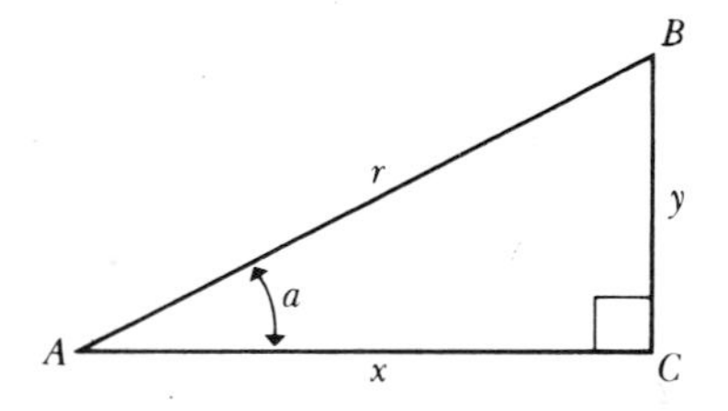


***Definitions of the trigonometric functions:*** $\sin a = y/r \qquad \cos a = x/r \qquad \tan a = y/x$

***Identities:***

$\sin^2 a + \cos^2 a = 1$

$\tan a = \dfrac{\sin a}{\cos a}$

$\sin 2a = 2 \sin a \cos a$

$\cos 2a = \cos^2 a - \sin^2 a = 2\cos^2 a - 1$

$\sin \frac{1}{2}a = \sqrt{\dfrac{1 - \cos a}{2}}$

$\cos \frac{1}{2}a = \sqrt{\dfrac{1 + \cos a}{2}}$

$\sin(-a) = -\sin a$

$\cos(-a) = \cos a$

$\sin(a \pm \pi/2) = \pm \cos a$

$\cos(a \pm \pi/2) = \pm \sin a$

$\sin(a \pm b) = \sin a \cos b \pm \cos a \sin b$

$\cos(a \pm b) = \cos a \cos b \pm \sin a \sin b$

$\sin a + \sin b = 2 \sin \frac{1}{2}(a + b) \cos \frac{1}{2}(a - b)$

$\cos a + \cos b = 2 \cos \frac{1}{2}(a + b) \cos \frac{1}{2}(a - b)$

### *GEOMETRY*

| | |
|---|---|
| Circumference of circle of radius *r*: | $C = 2\pi r$ |
| Area of circle of radius *r*: | $A = \pi r^2$ |
| Volume of sphere of radius *r*: | $V = 4\pi r^3/3$ |
| Surface area of sphere of radius *r*: | $A = 4\pi r^2$ |
| Volume of cylinder of radius *r* and height *h*: | $V = \pi r^2 h$ |

### *CALCULUS*

***Derivatives:***

$$\frac{d}{dx}x^n = nx^{n-1}$$

$$\frac{d}{dx}\sin ax = a\cos ax$$

$$\frac{d}{dx}\cos ax = -a\sin ax$$

$$\frac{d}{dx}e^{ax} = ae^{ax}$$

$$\frac{d}{dx}\ln ax = \frac{1}{x}$$

*Integrals:*

$$\int x^n\,dx = \frac{x^{n+1}}{n+1}$$

$$\int \frac{dx}{x} = \ln x$$

$$\int \sin ax\,dx = -\frac{1}{a}\cos ax$$

$$\int \cos ax\,dx = \frac{1}{a}\sin ax$$

$$\int e^{ax}dx = \frac{1}{a}e^{ax}$$

$$\int \frac{dx}{\sqrt{a^2 - x^2}} = \arcsin\frac{x}{a}$$

$$\int \frac{dx}{\sqrt{x^2 + a^2}} = \ln\left(x + \sqrt{x^2 + a^2}\right)$$

$$\int \frac{dx}{x^2 + a^2} = \frac{1}{a}\arctan\frac{x}{a}$$

$$\int \frac{dx}{(x^2 + a^2)^{3/2}} = \frac{1}{a^2}\frac{x}{\sqrt{x^2 + a^2}}$$

$$\int \frac{x\,dx}{(x^2 + a^2)^{3/2}} = -\frac{1}{\sqrt{x^2 + a^2}}$$

*Power series* (convergent for range of $x$ shown):

$$\sin x = x - \frac{x^3}{3!} + \frac{x^5}{5!} - \frac{x^7}{7!} + \cdots \qquad (\text{all } x)$$

$$\cos x = 1 - \frac{x^2}{2!} + \frac{x^4}{4!} - \frac{x^6}{6!} + \cdots \qquad (\text{all } x)$$

$$\tan x = x + \frac{x^3}{3} + \frac{2x^5}{15} + \frac{17x^7}{315} + \cdots \qquad (|x| < \pi/2)$$

$$e^x = 1 + x + \frac{x^2}{2!} + \frac{x^3}{3!} + \cdots \qquad (\text{all } x)$$

$$\ln(1 + x) = x - \frac{x^2}{2} + \frac{x^3}{3} - \frac{x^4}{4} + \cdots \qquad (|x| < 1)$$

APPENDIX C

# The Greek Alphabet

| Name | Capital | Lowercase | Name | Capital | Lowercase |
|---|---|---|---|---|---|
| Alpha | A | $\alpha$ | Nu | N | $\nu$ |
| Beta | B | $\beta$ | Xi | $\Xi$ | $\xi$ |
| Gamma | $\Gamma$ | $\gamma$ | Omicron | O | $o$ |
| Delta | $\Delta$ | $\delta$ | Pi | $\Pi$ | $\pi$ |
| Epsilon | E | $\epsilon$ | Rho | P | $\rho$ |
| Zeta | Z | $\zeta$ | Sigma | $\Sigma$ | $\sigma$ |
| Eta | H | $\eta$ | Tau | T | $\tau$ |
| Theta | $\Theta$ | $\theta$ | Upsilon | $\Upsilon$ | $\upsilon$ |
| Iota | I | $\iota$ | Phi | $\Phi$ | $\phi$ |
| Kappa | K | $\kappa$ | Chi | X | $\chi$ |
| Lambda | $\Lambda$ | $\lambda$ | Psi | $\Psi$ | $\psi$ |
| Mu | M | $\mu$ | Omega | $\Omega$ | $\omega$ |

# Periodic Table of the Elements

| Period | IA | IIA | IIIB | IVB | VB | VIB | VIIB | VIIIB | | VIIIB | IB | IIB | IIIA | IVA | VA | VIA | VIIA | Noble gases |
|---|---|---|---|---|---|---|---|---|---|---|---|---|---|---|---|---|---|---|
| 1 | 1<br>**H**<br>1.008 | | | | | | | | | | | | | | | | | 2<br>**He**<br>4.003 |
| 2 | 3<br>**Li**<br>6.941 | 4<br>**Be**<br>9.012 | | | | | | | | | | | 5<br>**B**<br>10.811 | 6<br>**C**<br>12.011 | 7<br>**N**<br>14.007 | 8<br>**O**<br>15.999 | 9<br>**F**<br>18.998 | 10<br>**Ne**<br>20.179 |
| 3 | 11<br>**Na**<br>22.990 | 12<br>**Mg**<br>24.305 | | | | | | | | | | | 13<br>**Al**<br>26.982 | 14<br>**Si**<br>28.086 | 15<br>**P**<br>30.974 | 16<br>**S**<br>32.064 | 17<br>**Cl**<br>35.453 | 18<br>**Ar**<br>39.948 |
| 4 | 19<br>**K**<br>39.098 | 20<br>**Ca**<br>40.08 | 21<br>**Sc**<br>44.956 | 22<br>**Ti**<br>47.90 | 23<br>**V**<br>50.942 | 24<br>**Cr**<br>51.996 | 25<br>**Mn**<br>54.938 | 26<br>**Fe**<br>55.847 | 27<br>**Co**<br>58.933 | 28<br>**Ni**<br>58.70 | 29<br>**Cu**<br>63.546 | 30<br>**Zn**<br>65.38 | 31<br>**Ga**<br>69.72 | 32<br>**Ge**<br>72.59 | 33<br>**As**<br>74.922 | 34<br>**Se**<br>78.96 | 35<br>**Br**<br>79.904 | 36<br>**Kr**<br>83.80 |
| 5 | 37<br>**Rb**<br>85.468 | 38<br>**Sr**<br>87.62 | 39<br>**Y**<br>88.906 | 40<br>**Zr**<br>91.22 | 41<br>**Nb**<br>92.906 | 42<br>**Mo**<br>95.94 | 43<br>**Tc**<br>(99) | 44<br>**Ru**<br>101.07 | 45<br>**Rh**<br>102.905 | 46<br>**Pd**<br>106.4 | 47<br>**Ag**<br>107.868 | 48<br>**Cd**<br>112.41 | 49<br>**In**<br>114.82 | 50<br>**Sn**<br>118.69 | 51<br>**Sb**<br>121.75 | 52<br>**Te**<br>127.60 | 53<br>**I**<br>126.905 | 54<br>**Xe**<br>131.30 |
| 6 | 55<br>**Cs**<br>132.905 | 56<br>**Ba**<br>137.33 | 57<br>**La**<br>138.905 | 72<br>**Hf**<br>178.49 | 73<br>**Ta**<br>180.948 | 74<br>**W**<br>183.85 | 75<br>**Re**<br>186.2 | 76<br>**Os**<br>190.2 | 77<br>**Ir**<br>192.22 | 78<br>**Pt**<br>195.09 | 79<br>**Au**<br>196.966 | 80<br>**Hg**<br>200.59 | 81<br>**Tl**<br>204.37 | 82<br>**Pb**<br>207.19 | 83<br>**Bi**<br>208.2 | 84<br>**Po**<br>(210) | 85<br>**At**<br>(210) | 86<br>**Rn**<br>(222) |
| 7 | 87<br>**Fr**<br>(223) | 88<br>**Ra**<br>(226) | 89<br>**Ac**<br>(227) | 104<br>**Rf(?)**<br>(261) | 105<br>**Ha(?)**<br>(262) | 106<br>(257) | 107<br>(260) | | | | | | | | | | | |

| | | | | | | | | | | | | | |
|---|---|---|---|---|---|---|---|---|---|---|---|---|---|
| 58<br>**Ce**<br>140.12 | 59<br>**Pr**<br>140.907 | 60<br>**Nd**<br>144.24 | 61<br>**Pm**<br>(145) | 62<br>**Sm**<br>150.35 | 63<br>**Eu**<br>151.96 | 64<br>**Gd**<br>157.25 | 65<br>**Tb**<br>158.925 | 66<br>**Dy**<br>162.50 | 67<br>**Ho**<br>164.930 | 68<br>**Er**<br>167.26 | 69<br>**Tm**<br>168.934 | 70<br>**Yb**<br>173.04 | 71<br>**Lu**<br>174.96 |
| 90<br>**Th**<br>(232) | 91<br>**Pa**<br>(231) | 92<br>**U**<br>(238) | 93<br>**Np**<br>(239) | 94<br>**Pu**<br>(239) | 95<br>**Am**<br>(240) | 96<br>**Cm**<br>(242) | 97<br>**Bk**<br>(245) | 98<br>**Cf**<br>(246) | 99<br>**Es**<br>(247) | 100<br>**Fm**<br>(249) | 101<br>**Md**<br>(256) | 102<br>**No**<br>(254) | 103<br>**Lr**<br>(257) |

For each element the average atomic mass of the mixture of isotopes occurring in nature is shown. For elements having no stable isotope, the approximate atomic mass of the most common isotope is shown in parentheses.

# Unit Conversion Factors

***LENGTH***

1 m = 100 cm = 1000 mm = $10^6$ $\mu$m = $10^9$ nm
1 km = 1000 m = 0.6214 mi
1 m = 3.281 ft = 39.37 in.
1 cm = 0.3937 in.
1 in. = 2.540 cm
1 ft = 30.48 cm
1 yd = 91.44 cm
1 mi = 5280 ft = 1.609 km
1 Å = $10^{-10}$ m = $10^{-8}$ cm = $10^{-1}$ nm
1 nautical mile = 6080 ft
1 light year = $9.461 \times 10^{15}$ m

***AREA***

1 $cm^2$ = 0.155 $in^2$
1 $m^2$ = $10^4$ $cm^2$ = 10.76 $ft^2$
1 $in^2$ = 6.452 $cm^2$
1 $ft^2$ = 144 $in^2$ = 0.0929 $m^2$

***VOLUME***

1 liter = 1000 $cm^3$ = $10^{-3}$ $m^3$ = 0.03531 $ft^3$ = 61.02 $in^3$
1 $ft^3$ = 0.02832 $m^3$ = 28.32 liters = 7.477 gallons
1 gallon = 3.788 liters

***TIME***

1 min = 60 s
1 h = 3600 s
1 d = 86,400 s
1 y = 365.24 d = $3.156 \times 10^7$ s

***ANGLE***

1 rad = 57.30° = 180°/$\pi$
1° = 0.01745 rad = $\pi$/180 rad
1 revolution = 360° = $2\pi$ rad
1 rev/min (rpm) = 0.1047 rad/s

***SPEED***

1 m/s = 3.281 ft/s
1 ft/s = 0.3048 m/s
1 mi/min = 60 mi/h = 88 ft/s
1 km/h = 0.2778 m/s = 0.6214 mi/h
1 mi/h = 1.466 ft/s = 0.4470 m/s = 1.609 km/h
1 furlong/fortnight = $1.662 \times 10^{-4}$ m/s

***ACCELERATION***

1 $m/s^2$ = 100 $cm/s^2$ = 3.281 $ft/s^2$
1 $cm/s^2$ = 0.01 $m/s^2$ = 0.03281 $ft/s^2$
1 $ft/s^2$ = 0.3048 $m/s^2$ = 30.48 $cm/s^2$
1 mi/h · s = 1.467 $ft/s^2$

***MASS***

1 kg = $10^3$ g = 0.0685 slug
1 g = $6.85 \times 10^{-5}$ slug
1 slug = 14.59 kg
1 u = $1.661 \times 10^{-27}$ kg
1 kg has a weight of 2.205 lb when $g$ = 9.80 $m/s^2$

***FORCE***

1 N = $10^5$ dyn = 0.2248 lb
1 lb = 4.448 N = $4.448 \times 10^5$ dyn

***PRESSURE***

1 Pa = 1 $N/m^2$ = $1.451 \times 10^{-4}$ $lb/in^2$ = 0.209 $lb/ft^2$
1 bar = $10^5$ Pa
1 $lb/in^2$ = 6891 Pa
1 $lb/ft^2$ = 47.85 Pa
1 atm = $1.013 \times 10^5$ Pa = 1.013 bar
= 14.7 $lb/in^2$ = 2117 $lb/ft^2$
1 mm Hg = 1 torr = 133.3 Pa

***ENERGY***

1 J = $10^7$ ergs = 0.239 cal
1 cal = 4.186 J (based on 15° calorie)
1 ft · lb = 1.356 J
1 Btu = 1055 J = 252 cal = 778 ft · lb
1 eV = $1.602 \times 10^{-19}$ J
1 kWh = $3.600 \times 10^6$ J

***MASS–ENERGY EQUIVALENCE***

1 kg ↔ $8.988 \times 10^{16}$ J
1 u ↔ 931.5 MeV
1 eV ↔ $1.073 \times 10^{-9}$ u

***POWER***

1 W = 1 J/s
1 hp = 746 W = 550 ft · lb/s
1 Btu/h = 0.293 W

# Numerical Constants

### *FUNDAMENTAL PHYSICAL CONSTANTS*

| *Name* | *Symbol* | *Value* |
|---|---|---|
| Speed of light | $c$ | $2.99792458 \times 10^{8}$ m/s |
| Charge of electron | $e$ | $1.602177 \times 10^{-19}$ C |
| Gravitational constant | $G$ | $6.67259 \times 10^{-11}$ N · m²/kg² |
| Planck's constant | $h$ | $6.6260755 \times 10^{-34}$ J · s |
| Boltzmann's constant | $k$ | $1.38066 \times 10^{-23}$ J/K |
| Avogadro's number | $N_A$ | $6.022 \times 10^{23}$ molecules/mol |
| Gas constant | $R$ | 8.314510 J/mol · K |
| Mass of electron | $m_e$ | $9.10939 \times 10^{-31}$ kg |
| Mass of neutron | $m_n$ | $1.67493 \times 10^{-27}$ kg |
| Mass of proton | $m_p$ | $1.67262 \times 10^{-27}$ kg |
| Permittivity of free space | $\epsilon_0$ | $8.854 \times 10^{-12}$ $C^2/N \cdot m^2$ |
| | $1/4\pi\epsilon_0$ | $8.987 \times 10^{9}$ $N \cdot m^2/C^2$ |
| Permeability of free space | $\mu_0$ | $4\pi \times 10^{-7}$ Wb/A · m |

### *OTHER USEFUL CONSTANTS*

| | | |
|---|---|---|
| Mechanical equivalent of heat | | 4.186 J/cal (15° calorie) |
| Standard atmospheric pressure | 1 atm | $1.013 \times 10^{5}$ Pa |
| Absolute zero | 0 K | − 273.15°C |
| Electronvolt | 1 eV | $1.602 \times 10^{-19}$ J |
| Atomic mass unit | 1 u | $1.66054 \times 10^{-27}$ kg |
| Electron rest energy | $mc^2$ | 0.511 MeV |
| Energy equivalent of 1 u | $Mc^2$ | 931.494 MeV |
| Volume of ideal gas (0°C and 1 atm) | $V$ | 22.4 liter/mol |
| Acceleration due to gravity (sea level, at equator) | $g$ | 9.78049 $m/s^2$ |

### *ASTRONOMICAL DATA*

| *Body* | *Mass, kg* | *Radius, m* | *Orbit radius, m* | *Orbit period* |
|---|---|---|---|---|
| Sun | $1.99 \times 10^{30}$ | $6.96 \times 10^{8}$ | — | — |
| Moon | $7.35 \times 10^{22}$ | $1.74 \times 10^{6}$ | $0.38 \times 10^{9}$ | 27.3 d |
| Mercury | $3.28 \times 10^{23}$ | $2.57 \times 10^{6}$ | $5.79 \times 10^{10}$ | 88.0 d |
| Venus | $4.82 \times 10^{24}$ | $6.31 \times 10^{6}$ | $1.08 \times 10^{11}$ | 224.7 d |
| Earth | $5.98 \times 10^{24}$ | $6.38 \times 10^{6}$ | $1.49 \times 10^{11}$ | 365.3 d |
| Mars | $6.42 \times 10^{23}$ | $3.38 \times 10^{6}$ | $2.28 \times 10^{11}$ | 687.0 d |
| Jupiter | $1.89 \times 10^{27}$ | $7.18 \times 10^{7}$ | $7.78 \times 10^{11}$ | 11.86 y |
| Saturn | $5.69 \times 10^{26}$ | $6.03 \times 10^{7}$ | $1.43 \times 10^{12}$ | 29.46 y |
| Uranus | $8.66 \times 10^{25}$ | $2.67 \times 10^{7}$ | $2.87 \times 10^{12}$ | 84.01 y |
| Neptune | $1.03 \times 10^{26}$ | $2.48 \times 10^{7}$ | $4.49 \times 10^{12}$ | 164.8 y |
| Pluto | $1.1 \times 10^{22}$ | $4 \times 10^{5}$ | $5.91 \times 10^{12}$ | 248.7 y |

# Answers to Odd-Numbered Problems

## CHAPTER 1

**1–1** 0.621 mi
**1–3** a) 0.403 mi/s b) 648 m/s
**1–5** 44.7 mi/gal
**1–7** 0.0056%
**1–9** a) 0.1% b) 0.01% c) 0.07%
**1–11** $5.50 \times 10^3$ kg/m$^3$
**1–13** 10,000
**1–15** $2 \times 10^5$ hairs
**1–17** \$1 $\times 10^8$
**1–19** $1 \times 10^6$ N ($2 \times 10^5$ lb); 1000 N (200 lb)
**1–21** 7.83 km, 37.7° N of E
**1–23** 66.2 m; 4° W of N
**1–25** $A_x = +7.22$ m, $A_y = +9.58$ m; $B_x = +11.5$ m, $B_y = -9.64$ m; $C_x = -3.0$ m, $C_y = -5.2$ m
**1–27** a) 11.1 m, 77.6° b) 28.5 m, 202°
**1–29** 5.03 km, 38.3° E of S
**1–31** a) 2.48 cm, 18.3° b) 4.09 cm, 83.7° c) 4.09 cm, 263.7°
**1–33** a) $A = 5.0$, $B = 5.1$ b) $3\boldsymbol{i} + 8\boldsymbol{j}$ c) 8.54, 69.4°
**1–35** $\boldsymbol{i}\cdot\boldsymbol{i} = 1$ $\quad \boldsymbol{j}\cdot\boldsymbol{i} = 0$ $\quad \boldsymbol{k}\cdot\boldsymbol{i} = 0$
$\boldsymbol{i}\cdot\boldsymbol{j} = 0$ $\quad \boldsymbol{j}\cdot\boldsymbol{j} = 1$ $\quad \boldsymbol{k}\cdot\boldsymbol{j} = 0$
$\boldsymbol{i}\cdot\boldsymbol{k} = 0$ $\quad \boldsymbol{j}\cdot\boldsymbol{k} = 0$ $\quad \boldsymbol{k}\cdot\boldsymbol{k} = 1$
**1–37** $-11$
**1–39** $\boldsymbol{i}\times\boldsymbol{i} = 0$ $\quad \boldsymbol{j}\times\boldsymbol{i} = -\boldsymbol{k}$ $\quad \boldsymbol{k}\times\boldsymbol{i} = \boldsymbol{j}$
$\boldsymbol{i}\times\boldsymbol{j} = \boldsymbol{k}$ $\quad \boldsymbol{j}\times\boldsymbol{j} = 0$ $\quad \boldsymbol{k}\times\boldsymbol{j} = -\boldsymbol{i}$
$\boldsymbol{i}\times\boldsymbol{k} = -\boldsymbol{j}$ $\quad \boldsymbol{j}\times\boldsymbol{k} = \boldsymbol{i}$ $\quad \boldsymbol{k}\times\boldsymbol{k} = 0$
**1–41** $-23\boldsymbol{k}$; 23
**1–43** a) $7.04 \times 10^{-10}$ s b) $5.11 \times 10^{12}$ cycles c) $2.06 \times 10^{26}$ cycles d) $4.60 \times 10^4$ s (about 13 h)
**1–45** a) $x$: 1.00 cm; $y$: $-1.17$ cm b) 1.54 cm, 310.5° counterclockwise from $+x$-axis
**1–47** 50.0 m, 42.7° W of N
**1–49** b) 3.75 km
**1–51** a) $A = 5.48$, $B = 4.12$ b) $-4\boldsymbol{i} - 3\boldsymbol{k}$ c) 5.00; yes
**1–53** a) 54.7° b) 35.3°
**1–55** 44.6 yd, 19.7° to right of straight downfield
**1–57** b) $\boldsymbol{C}\cdot(\boldsymbol{A}\times\boldsymbol{B})$

## CHAPTER 2

**2–1** a) 214 m/s b) 172 m/s
**2–3** 43 min
**2–5** a) 4.60 m/s b) 11.2 m/s c) 17.8 m/s
**2–7** 72.0 cm/s
**2–9** 1.66 s
**2–11** a) 0; 1.00 m/s$^2$; 1.50 m/s$^2$; 2.50 m/s$^2$; 2.50 m/s$^2$; 2.50 m/s$^2$; 1.00 m/s$^2$; 0 b) 2.5 m/s$^2$; 0.8 m/s$^2$; 0
**2–15** a) 11.7 m/s b) 0.56 m/s$^2$
**2–17** a) 0; 6.25 m/s$^2$; $-11.2$ m/s$^2$ b) 100 m; 230 m; 320 m
**2–19** a) 10 m b) 93 m
**2–21** a) 9000 m/s b) 98.6% c) 12.5 h
**2–23** a) 3.54 m/s b) 0.723 s
**2–25** a) 31.6 m/s b) 43.6 m c) 18.4 m/s
**2–27** a) 4.65 m/s, downward b) 9.80 m/s$^2$, downward
**2–29** a) 27.2 m/s b) 37.6 m c) 0 d) 9.80 m/s$^2$, downward
**2–31** a) 248 m/s$^2$ b) 25.3 c) 402 m d) $a = 20.6g$ if constant
**2–33** a) $x = (0.317\ \text{m/s}^3)t^3 - (0.010\ \text{m/s}^4)t^4$; $v = (0.95\ \text{m/s}^3)t^2 - (0.040\ \text{m/s}^4)t^3$ b) 79.4 m/s
**2–37** a) 26.7 s b) 80.0 s
**2–39** a) 4.4 m; $-22.4$ m/s; $-52.8$ m/s$^2$ c) 5.02 m/s; 0.784 s
**2–41** 3.5 m/s$^2$
**2–43** a) 85 km/h b) 32 km/h
**2–45** Hit the brake pedal
**2–47** a) 196 m b) 28.0 m/s
**2–49** a) 8.26 s b) 44.3 m c) Auto: 28.9 m/s; truck: 18.2 m/s
**2–51** 3.60 m from the point directly below you
**2–53** a) No b) Yes; 19.3 m/s; no
**2–55** a) 13.3 m b) 1.65 s
**2–57** a) 13.7 s b) 349 m c) 31.0 m/s
**2–59** 25.0 km
**2–63** a) 12.2 s; 73.5 m b) 2.20 m/s d) Time when bus overtakes her if she ran on past it; 9.80 m/s e) No f) 4.65 m/s; 25.8 s; 120 m

## CHAPTER 3

**3–1** a) $x$: $-1.8$ m/s; $y$: $+1.1$ m/s b) 2.1 m/s, 149°
**3–3** a) $x$: $-4.00$ m/s$^2$; $y$: $+9.00$ m/s$^2$ b) 9.85 m/s$^2$; 114°
**3–5** a) $\boldsymbol{v} = (-3.6\ \text{m/s})\boldsymbol{i} + (5.6\ \text{m/s}^2)t\boldsymbol{j}$; $\boldsymbol{a} = (5.6\ \text{m/s}^2)\boldsymbol{j}$ b) 17 m/s, 102°; 5.6 m/s$^2$, 90°
**3–7** a) 1.22 m b) 1.80 m c) $v_x = 3.60$ m/s, $v_y = -4.90$ m/s; 6.08 m/s, 53.7° below horizontal
**3–9** a) 0.16 m
**3–11** a) 0.777 s, 6.57 s b) 27.0 m/s, 28.4 m/s (up); 27.0 m/s, 28.4 m/s (down) c) 45.0 m/s, 53.1° below the horizontal
**3–13** a) 5500 m b) 34,000 m
**3–15** a) 54.5 m b) 64.7 m/s c) 377 m
**3–17** a) 1.5 m b) 0.89 m/s, downward
**3–19** 0.0337 m/s$^2$
**3–21** a) 5.79 m/s$^2$, upward b) 9.77 s
**3–23** a) 16.0° E of N b) 279 km/h
**3–25** a) 4.24 m/s at 34.4° N of E b) 286 s c) 686 m
**3–27** 38°
**3–29** 56 mph
**3–31** a) 16.0 m b) 3.49 m/s, 66.4° clockwise from $+x$-axis c) 1.60 m/s$^2$, $-y$-direction d) $t = 0$ e) $t = 0$: $x = 0$, $y = 19.0$ m; $t = 4.71$ s: $x = 6.60$ m, $y = 1.22$ m f) 6.82 m; 4.55 s
**3–33** a) $\boldsymbol{r} = (\alpha t - \frac{1}{3}\beta t^3)\boldsymbol{i} + (\frac{1}{2}\gamma t^2)\boldsymbol{j}$; $\boldsymbol{a} = (-2\beta t)\boldsymbol{i} + \gamma\boldsymbol{j}$ b) 4.4 m
**3–35** 204 m
**3–37** a) 42.8 m/s b) 42.0 m
**3–39** a) 184 m/s b) 835 m c) 139 m/s, 179 m/s (downward)
**3–41** a) 116 m b) 4.46 s c) 34.0 m/s
**3–43** 17.8 m/s
**3–45** a) 2.23 m above floor b) 3.84 m c) 8.66 m/s d) 3.09 m above floor; 0.62 m
**3–51** a) 113 km/h, 45° S of W b) 33.1° N of W
**3–53** a) 140 m b) $4.9 \times 10^{-4}$ c) Radius reduced
**3–55** a) $(2v_0^2 \cos^2(\phi + \theta)/g \cos\theta)(\tan[\phi + \theta] - \tan\theta) = (2v_0^2 \sin\phi/g\cos\theta)(\cos\phi - \tan\theta\sin\phi)$ b) $\phi = \pi/4 - \theta/2$
**3–57** $\Delta t = 0.5$ s: 11.51 m/s$^2$, 61.4° $\Delta t = 0.1$ s: 11.98 m/s$^2$, 84.3° $\Delta t = 0.05$ s: 12.00 m/s$^2$, 87.1°

## CHAPTER 4

**4–1** 69.3 N to the right; 40.0 N downward
**4–3** 482 N, 20.5°
**4–5** 5.65 kg
**4–7** 1.32 m/s$^2$
**4–9** 1200 N
**4–11** a) $x = 5.00$ m, 5.00 m/s b) $x = 35.0$ m, 10.0 m/s
**4–13** 6.53 kg; 24.2 N
**4–15** 6860 N
**4–17** a) Gravity force exerted on bottle by earth (downward) b) Gravity force exerted on earth by bottle (upward)
**4–19** 3.84 m/s$^2$
**4–21** 3.00 m/s$^2$
**4–23** a) 25.0 N b) 10.0 N
**4–25** $6mBt$
**4–27** 1700 N; 135° counterclockwise from $\boldsymbol{F}_1$
**4–29** 16.6 N; 90° clockwise from $+x$-axis
**4–31** 1990 N
**4–33** $4.74 \times 10^6$ N
**4–35** a) 4.20 m/s$^2$, upward b) 1.92 m/s$^2$, downward c) Yes, free fall
**4–37** a) $1.25 \times 10^5$ N b) $9.36 \times 10^4$ N c) $6.24 \times 10^4$ N d) $3.12 \times 10^4$ N e) Same magnitude, opposite direction
**4–39** a) 3.53 m/s$^2$ b) 11.7 m/s$^2$
**4–41** a) 3.53 m/s$^2$ b) 120 N c) 93.3 N
**4–43** $\boldsymbol{v} = (k_1 t/m)\boldsymbol{i} + (k_2 t^3/3m)\boldsymbol{j}$
**4–45** $\boldsymbol{r}(t) = (k_1 t^2/2m + k_2 k_3 t^5/120\, m^2)\boldsymbol{i} + (k_3 t^3/6m)\boldsymbol{j}$; $\boldsymbol{v}(t) = (k_1 t/m + k_2 k_3 t^4/24m^2)\boldsymbol{i} + (k_3 t^2/2m)\boldsymbol{j}$

## CHAPTER 5

**5–1** a) 15.0 N b) 30.0 N
**5–3** a) 1540 N b) 0.955°
**5–5** a) 296 N b) 23.1 kg
**5–7** a) 28.3 N, 28.3 N b) 28.3 N
**5–9** a) $w \sin\theta$ b) $2w \sin\theta$
**5–11** 588 N
**5–13** a) 2.00 m/s$^2$ b) 1.54 kg
**5–15** a) 2.96 m/s$^2$ b) 191 N
**5–17** a) Down incline, so rope is vertical b) No deflection in each stage
**5–19** a) 17 N b) 2.3 s
**5–21** a) 8.2 N b) 9.6 N c) Part (a): 1.4 N; part (b): 2.7 N
**5–25** a) Held back b) 506 N
**5–27** 39°
**5–29** a) $\mu_k(m_A + m_B)g$ b) $\mu_k m_A g$
**5–31** a) 1.63 m/s$^2$ b) 32.7 N
**5–33** a) $\mu_k mg/(\cos\theta - \mu_k \sin\theta)$ b) $1/\tan\theta$
**5–35** b) 8.0 N c) 26.4 N
**5–37** 16.3 m/s
**5–39** 10.0°
**5–41** 0.86 rev/min
**5–43** 2.80 m/s
**5–45** a) 82.0 m b) 6270 N
**5–47** $mg \tan\theta$
**5–49** $\mu_s/(1 + \mu_s)$
**5–51** a) 20.0 N b) 24.0 N
**5–53** a) 14.3 N b) 8.63 N
**5–55** 4.05 N
**5–57** a) 66.0 N eastward b) 58.8 N westward
**5–59** a) 18.7 kg b) $T_{AB} = 47.2$ N, $T_{BC} = 146.2$ N
**5–61** $a_1 = 2m_2 g/(4m_1 + m_2)$, $a_2 = m_2 g/(4m_1 + m_2)$
**5–63** 1.12 m above floor
**5–65** $g/\mu_s$
**5–67** a) Move up b) Remains constant c) Remains constant d) Stop
**5–69** a) 18.6 m/s b) 5.9 m/s
**5–71** b) 0.23 c) No
**5–73** a) To the right b) 83 m
**5–75** 160 N
**5–77** a) 81.1° b) No c) Bead sits at bottom of hoop ($\theta = 0°$)
**5–79** $T_{min} = 2\pi\sqrt{(h \sin\theta/g)(\sin\theta - \mu_s \cos\theta)/(\cos\theta + \mu_s \sin\theta)}$; $T_{max} = 2\pi\sqrt{(h \sin\theta/g)(\sin\theta + \mu_s \cos\theta)/(\cos\theta - \mu_s \sin\theta)}$
**5–81** a) $F = \mu_k w/(\cos\phi + \mu_k \sin\phi)$ c) 16.7°
**5–83** 128 m
**5–85** a) 6.67 m/s$^2$ b) 2.67 m/s$^2$ c) 4.50 m/s d) 5.00 m/s e) $x - x_0 = 6.51$ m, $v = 4.65$ m/s, $a = 0.463$ m/s$^2$ f) 1.73 s
**5–87** $(\cos\theta)^2$

## CHAPTER 6

**6–1** a) 3.00 J b) −0.600 J
**6–3** 360 J
**6–5** a) 116 N b) 500 J c) −500 J d) 0; 0
**6–7** a) 22.5 N; 45.0 N b) 1.12 J; 4.50 J
**6–9** 267 N
**6–11** a) 60.0 J b) −20.0 J
**6–13** 625 J
**6–15** 128 J
**6–17** a) 4.12 m/s b) 4.12 m/s (same)
**6–19** 0.140 m
**6–21** a) −28.4 J b) 18.4 m/s
**6–23** a) $v_0^2/(2\mu_k g)$ b) 95.7 m
**6–25** a) 1.37 m/s b) 1.10 m/s
**6–27** a) 0.0869 m b) 0.315 J
**6–29** 0.925 hp
**6–31** 470 W
**6–33** 2500 N
**6–35** b) 11.2 L c) 1.04 L/km, 0.044 gal/mi
**6–37** a) 322 N b) 20.8 kW (27.9 hp) c) 1.2 kW (1.6 hp) d) 2.74%
**6–39** a) 368 J b) −207 J c) 0 d) 161 J
**6–41** a) −8.00 N b) −64.0 N c) −30.0 J
**6–43** a) 0.080 N b) 0.640 N c) 0.048 J
**6–45** a) 48.0 m/s b) 5760 J
**6–47** a) −270 J b) 3810 J
**6–49** $5.3 \times 10^4$ N/m
**6–51** a) $1.20 \times 10^5$ J b) 1.39 W
**6–53** a) $1.18 \times 10^5$ J b) $1.60 \times 10^5$ J c) 4630 W
**6–55** $1.20 \times 10^3$ m$^3$
**6–57** a) 803 W b) 557 W c) 79.5 W

## CHAPTER 7

**7–1** $2.23 \times 10^6$ J
**7–3** a) 29.7 m/s b) 29.7 m/s (same as in part (a))
**7–5** −0.33 J
**7–7** a) 1920 J b) −376 J c) 1132 J d) 412 J e) 2.15 m/s$^2$; 8.29 m/s; 412 J (same)
**7–9** a) 70.0 J b) 17.5 J
**7–11** 0.076 m
**7–13** 1.7 m
**7–15** $F = 6C_6/x^7$, attractive
**7–17** $\boldsymbol{F} = -(2kx + k'y)\boldsymbol{i} - (2ky + k'x)\boldsymbol{j}$
**7–19** a) 212 N b) 956 N
**7–21** 0.408

**7–23** 4.43 m/s

**7–25** 48.2°

**7–27** a) 6.62 m/s b) 14.7 N

**7–29** a) 0.272 b) − 15.4 J

**7–31** 9.32 m/s

**7–33** a) $U(x) = \frac{1}{2}\alpha x^2 + \frac{1}{3}\beta x^3$ b) 5.45 m/s

**7–35** 134 J

**7–37** a) 3.87 m/s b) 0.10 m

**7–39** a) − 12.0 J b) − 16.0 J c) Nonconservative

## CHAPTER 8

**8–1** a) $2.00 \times 10^5$ kg · m/s b) (i) 40.0 m/s; (ii) 28.3 m/s

**8–3** 0.167 kg · m/s, $-x$-direction

**8–5** a) 3.82 m/s, in Gretzky's original direction b) − 6760 J

**8–7** a) 0.0597 m/s b) 0.115 m/s

**8–9** a) $v_B = m_A v_A / m_B$

**8–11** a) 1.50 m/s b) 1.50 J

**8–13** a) $A$: 29.3 m/s; $B$: 20.7 m/s b) 19.7%

**8–15** a) 1.33 m/s b) $-5.33 \times 10^4$ J c) 2.00 m/s

**8–17** 2.20 m/s

**8–19** 18.9 km/h, 45.0° S of E

**8–21** 496 m/s

**8–23** 0.300-kg: 1.08 m/s to left; 0.200-kg: 1.02 m/s to right

**8–25** a) 21.7 m/s, to right b) 18.3 m/s, to left

**8–27** 1120 N; no

**8–29** a) $At_2 + (B/3)t_2^3$ b) $(A/m)t_2 + (B/3m)t_2^3$

**8–31** $4.6 \times 10^6$ m from center of earth

**8–33** 50 kg

**8–35** a) 40.0 N b) Yes

**8–37** a) $7.2 \times 10^{-66}$ b) 0.223

**8–39** $3.6 \times 10^{-25}$ kg

**8–41** a) $5.69 \times 10^{-22}$ kg · m/s b) $4.62 \times 10^{-19}$ J

**8–43** 20

**8–45** a) 5.00 m/s, east b) 5.56 m/s, east c) 4.00 m/s, east

**8–47** a) 0.109 b) − 240 J c) 0.32 J

**8–49** 0.182 m

**8–51** a) 13.5 m/s at 124° from direction of final velocity of bullet b) No

**8–55** a) $Mv_{\text{cm}}^2/2$

**8–57** $A$: 26.0 m/s; $B$: 15.0 m/s; 90° from final velocity of $A$

**8–59** a) 32.5 m/s b) 18.2 m/s c) 22.5 m/s

**8–61** a) On bisector of angle $\alpha$, $(L/2)\cos(\alpha/2)$ below apex b) $L/3$ above center of bottom section c) On bisector of 90° angle, $0.177L$ from apex d) $0.289L$ above center of bottom section

**8–63** 0.600 m/s

**8–65** $1.74 \times 10^6$ m/s, $7.40 \times 10^5$ m/s; $v_{\text{Kr}} = 1.5 v_{\text{Ba}}$

**8–67** 10.0 N

**8–69** $\boldsymbol{v} = (0.312 \text{ m/s})\boldsymbol{i} - (1.75 \text{ m/s})\boldsymbol{j}$

**8–71** a) $1.34 v_{\text{ex}}$ b) $1.18 v_{\text{ex}}$ c) $2.10 v_{\text{ex}}$ d) 3.82 km/s

**8–73** b) $2L/3$

**8–75** a) $l^2 \lambda g/32$ b) $l^2 \lambda g/32$; same

**8–77** Yes

## CHAPTER 9

**9–1** a) 2.00 rad; 115° b) 1.20 m c) 47.7 cm

**9–3** a) $\omega = \gamma + 3\beta t^2$ b) 2.50 rad/s c) 5.50 rad/s, 3.50 rad/s

**9–5** $\alpha = 2b + 6ct$

**9–7** a) 15.4 s b) 9.79 rev

**9–9** a) − 1.67 rev/s², 54.2 rev b) 4.00 s

**9–11** 13.8 rad/s

**9–13** a) 5.50 m/s b) 127 rev/min

**9–15** 210 m/s

**9–17** a) 72.0 m/s² b) 6.00 m/s, 72.0 m/s²

**9–19** a) 0.180 m/s², 0, 0.180 m/s² b) 0.180 m/s², 0.754 m/s², 0.775 m/s² c) 0.180 m/s², 1.51 m/s², 1.52 m/s²

**9–21** a) 0.0640 kg · m² b) 0.0320 kg · m²

**9–23** 0.198 kg · m²

**9–25** 0.380 kg · m²

**9–27** $1.20 \times 10^5$ J

**9–29** 470 J

**9–31** $ML^2/3$

**9–33** a) $\gamma L^2/2$ b) $ML^2/2$; larger

**9–35** $2Ma^2/3$

**9–37** a) 0.12 kg · m² b) 0.21 kg · m² c) 0.33 kg · m²

**9–39** a) $\omega(t) = 2\gamma t - 3\beta t^2$ b) $\alpha(t) = 2\gamma - 6\beta t$ c) 5.21 rad/s, 2.08 s

**9–41** a) 1.5 m/s b) 75.4 rad/s

**9–43** a) 7.4 m/s b) 3.3 J c) 406 rad/s

**9–45** a) 211 rev/min b) 800 W

**9–47** a) − 1.47 J b) 5.42 rad/s c) 5.42 m/s d) 4.43 m/s

**9–49** $\sqrt{2gd(m_B - \mu_k m_A)/(m_A + m_B + I/R^2)}$

**9–51** a) $5.76 \times 10^{-3}$ kg · m² b) 3.40 m/s c) 4.95 m/s

**9–53** $3MR^2/5$

**9–55** $0.371MR^2$

## CHAPTER 10

**10–1** a) 80.0 N · m, counterclockwise b) 69.3 N · m, counterclockwise c) 40.0 N · m, counterclockwise d) 34.6 N · m, clockwise e) 0 f) 0

**10–3** 1.57 N · m, clockwise

**10–5** $-(0.50 \text{ N} \cdot \text{m})\boldsymbol{k}$

**10–7** a) 41.5 N, 7.50 N b) 0.0816 kg · m²

**10–9** 0.442

**10–11** a) 65.3 N b) 12.5 m/s c) 1.92 s d) 261 N

**10–13** a) 0.686 N b) 0.495 s c) 30.3 rad/s

**10–15** a) 1.88 N · m b) 157 rad c) 296 J d) 296 J

**10–17** a) 13.8 rad/s b) $1.51 \times 10^4$ J c) 756 W

**10–19** a) $7.16 \times 10^3$ N · m b) $2.86 \times 10^4$ N c) 62.8 m/s

**10–21** $4.36 \times 10^{-5}$ kg · m²/s

**10–23** 34.7 kg · m²/s

**10–25** a) 6.00 rad/s b) 0.00675 J c) 0.00675 J

**10–27** 0.119 rad/s; yes

**10–29** a) 0.562 rad/s b) Before: 270 J; after: 101 J; negative work done by friction

**10–31** a) 78.4 min b) $1.46 \times 10^5$ N · m

**10–33** a) 15.3 kg · m² b) − 1.60 N · m c) 90.0 rev

**10–35** 0.645 s

**10–37** $1.18 \times 10^3$ N

**10–39** a) 1.73 m/s² b) 13.0 N

**10–41** $F/2M$, $F/2$

**10–43** 2.45 m

**10–45** $g/3$

**10–47** 6000 J

**10–49** a) 5.10 rad/s b) 0.0177 m c) $1.01 \times 10^3$ m/s

**10–51** 4.98 rad/s

**10–53** a) $T = mv_1^2 r_1^2 / r^3$ b) $W = \frac{1}{2} m v_1^2 ([r_1/r_2]^2 - 1)$ c) Same

**10–55** a) $a = \mu_k g$; $\alpha = -2\mu_k g/R$ b) $\omega_0^2 R^2 / 18\mu_k g$ c) $-MR^2\omega_0^2/6$

## CHAPTER 11

**11–1** 0.108 m to left of center of second ball

**11–3** 1100 N; 0.727 m from 700-N force

**11–5** 3500 N

11–7 a) 2340 N b) 1360 N
11–11 a) $T = 2.60w$; pivot: $3.28w$ at 37.6° above horizontal b) $T = 4.10w$; pivot: $5.38w$ at 48.8° above horizontal
11–13 83.3 N each, in opposite directions
11–15 a) 5.42 N · m, counterclockwise b) 5.42 N · m, counterclockwise c) Same
11–17 $1.6 \times 10^{11}$ Pa
11–19 1.69 mm
11–21 a) $3.3 \times 10^{6}$ Pa b) $1.7 \times 10^{-5}$ c) $5.0 \times 10^{-5}$ m
11–23 a) Top: $2.4 \times 10^{-3}$; bottom: $1.6 \times 10^{-3}$ b) Top: $1.2 \times 10^{-3}$ m; bottom: $8.2 \times 10^{-4}$ m
11–25 $6.0 \times 10^{9}$ Pa; $1.7 \times 10^{-10}$ $Pa^{-1}$
11–27 $1.70 \times 10^{4}$ N
11–29 30.9 $m/s^2$
11–33 b) 2.32 m c) 2.75 m
11–35 a) 17.8 N, 52.2 N b) 33.5 N, 49.0 N
11–37 75.0 N, 100.0 N
11–39 a) 3.46 m b) 130 N
11–41 a) 60.0 N b) 53.1°
11–43 a) 268 N b) 232 N c) 366 N
11–45 Below: 1372 N; above: 588 N; above
11–47 a) Tips: 27°; slips: 17°; slips first b) Tips first
11–49 a) $A$: 70 N; $B$: 730 N b) 1.8 m
11–51 2.4 mm
11–53 a) 0.84 m from $A$ b) 0.76 m from $A$
11–55 a) 1.27 m b) Copper: $2.0 \times 10^{8}$ Pa; steel: $4.0 \times 10^{8}$ Pa b) Copper: $1.8 \times 10^{-3}$; steel: $2.0 \times 10^{-3}$
11–57 a) $F \cos^2\theta / A$ b) $F \sin\theta \cos\theta / A$ c) 0° d) 45°
11–59 a) 450 N b) 13,500 N c) 40°
11–61 $(xA^2 - k_0 V_0 F)/F v_0$
11–63 a) $9.8 \times 10^{-4}$ m b) 0.0294 J c) 0.0090 J d) −0.0384 J e) +0.0384 J

## CHAPTER 12

12–1 2.16
12–3 $2.58 \times 10^{8}$ m
12–5 $5.1 \times 10^{-8}$ $m/s^2$, downward along line midway between $A$ and $B$
12–7 $6.14 \times 10^{24}$ kg
12–9 $8.34 \times 10^{-11}$ $m/s^2$
12–11 $g_x = -22.5$ $m/s^2$; $g_y = 54.0$ $m/s^2$
12–13 $2.64 \times 10^{6}$ m
12–15 a) $2.38 \times 10^{3}$ m/s b) $3.53 \times 10^{4}$ m/s
12–17 7.3 m/s; yes (barely)
12–19 $7.26 \times 10^{3}$ m/s
12–21 a) 159 min b) 4.20 $m/s^2$
12–23 a) $U = -GMm/\sqrt{x^2 + a^2}$ b) $U = -GMm/x$ c) $F_x = -GMmx/(x^2 + a^2)^{3/2}$
12–25 a) 125 N b) 114 N
12–27 83.6 m/s
12–29 $4.25 \times 10^{-6}$
12–33 a) $g_x = +3.34 \times 10^{-10}$ $m/s^2$; $g_y = -8.01 \times 10^{-10}$ $m/s^2$ b) $8.68 \times 10^{-12}$ N; 67.4° clockwise from $+x$-axis
12–35 $3.59 \times 10^{7}$ m
12–37 $3.83 \times 10^{8}$ m
12–39 $2.20 \times 10^{-11}$ N; 73.1° clockwise from $+x$-axis
12–41 $0.010R_E = 6.4 \times 10^{4}$ m
12–43 $\sqrt{2Gm_E h / R_E(R_E + h)}$
12–45 $7.6 \times 10^{4}$ m/s
12–47 a) 50 kg: $1.49 \times 10^{-5}$ m/s; 100 kg: $7.46 \times 10^{-6}$ m/s b) $2.24 \times 10^{-5}$ m/s
12–49 $GM/d(L + d)$
12–51 $g_{\parallel} = 0$; $g_{\perp} = GM/a\sqrt{L^2 + a^2}$ (toward line)
12–53 2120 N

## CHAPTER 13

13–1 25.1 rad/s; 0.250 s
13–3 a) 300 N/m b) 90.0 N
13–5 a) 0.044 J b) 0.017 m c) 0.42 m/s
13–7 a) 0.475 s b) 2.10 Hz c) 13.2 rad/s
13–9 a) 0.250 s b) 25.1 rad/s c) 0.317 kg
13–13 a) 0.872 m b) 1.34 rad c) 57.0 J d) $x = (0.872$ m$)\cos[(7.07$ rad/s$)t - 1.34$ rad$]$
13–15 a) 114 $m/s^2$, 4.52 m/s b) −57 $m/s^2$, 3.9 m/s c) 0.0290 s
13–17 0.0397 m
13–19 a) $3.4 \times 10^{-8}$ kg · $m^2$ b) $5.3 \times 10^{-6}$ N · m
13–21 0.285 m
13–23 0.140 m
13–25 2.95 s
13–27 0.53 s
13–29 a) 4.19 Hz b) 21.9 kg/s
13–35 a) $3.43 \times 10^{3}$ $m/s^2$ b) $1.37 \times 10^{3}$ N c) 13.1 m/s
13–37 a) 1.57 s b) 0.0954 m c) 0.163
13–39 a) $AY/l_0$ b) $9.7 \times 10^{4}$ N/m
13–41 1.17 s
13–43 a) 0.349 m/s b) 0.609 $m/s^2$, downward c) 0.300 s d) 0.804 m
13–45 a) 2.52 m/s b) 0.184 m c) 0.444 s
13–47 0.898 m
13–49 a) $mg - m(2\pi f)^2 A \cos(2\pi f t + \phi)$ b) $(2\pi f_b)^2 A$
13–51 $2\pi\sqrt{3M/2k}$
13–53 0.882 m
13–55 a) 2.23 m b) Stick of length 0.50 m, pivoted 0.94 cm above its center
13–57 a) $\Delta T = -T\Delta g/2g$ b) $\Delta T/T = -\Delta g/2g$ c) 9.7977 $m/s^2$
13–59 a) Spring 1: 0.350 m; spring 2: 0.250 m b) 0.993 s
13–61 a) $T/T_0 = \sqrt{(1 + 3y^2/L^2)/(1 + 2y/L)}$ b) $y = 2L/3$

## CHAPTER 14

14–1 800 $kg/m^3$
14–3 0.0473 kg
14–5 a) $8.08 \times 10^{6}$ Pa b) $1.43 \times 10^{5}$ N
14–7 a) $1.06 \times 10^{3}$ Pa b) $4.00 \times 10^{3}$ Pa
14–9 a) $1.06 \times 10^{5}$ Pa b) $1.02 \times 10^{5}$ Pa c) $1.02 \times 10^{5}$ Pa d) $4.34 \times 10^{3}$ Pa
14–11 89.3%
14–13 0.600 $m^3$
14–15 a) 127 Pa b) 833 Pa c) 0.720 kg
14–17 6.67 Pa
14–19 a) 4.77 m/s b) 0.224 m
14–21 32.0 m/s
14–23 $4.08 \times 10^{-4}$ $m^2$
14–25 $2.45 \times 10^{5}$ Pa
14–27 1780 N
14–29 a) 1.95 m/s b) 0
14–31 a) 0.0934 $m^3/s$ b) $2.24 \times 10^{4}$ Pa c) 0.200 $m^3/s$
14–33 0.384 m/s
14–35 a) $1.10 \times 10^{8}$ Pa b) $1.08 \times 10^{3}$ $kg/m^3$; 5.3%
14–37 $2.61 \times 10^{4}$ N · m
14–39 a) $1.47 \times 10^{3}$ Pa b) 13.9 cm
14–41 a) $1.08 \times 10^{4}$ $m^3$ b) 11,200 kg
14–43 $9.18 \times 10^{6}$ kg; yes
14–45 a) 0.153 m b) 3.04 s

**14–47** $1.16 \times 10^{-4}$ m$^3$; 1.32 kg
**14–49** 0.0756 kg
**14–51** 11.5 N
**14–53** Rises, $5.30 \times 10^{-4}$ m
**14–55** a) Sink b) 178 N
**14–57** a) 36.8% b) 0.397 L
**14–59** 7.02 N · m
**14–61** 0.115 m
**14–63** 146 m/s
**14–65** $h_2 = 3h_1$
**14–67** a) $r_2 = \sqrt{v_1}r_1/(v_1^2 + 2g(y_1 - y_2))^{1/4}$ b) 1.96 m
**14–69** a) 1.16 cm/s b) 2.16 m/s
**14–71** a) 83.3 N
**14–73** a) $la/g$ b) $l^2\omega^2/2g$ c) Yes; yes
**14–75** a) $\sqrt{2gh}$ b) $p_a/\rho g - h$
**14–77** Air bubbles cause inaccuracies

## CHAPTER 15

**15–1** a) 104°F; yes b) 37.0°C
**15–3** $-40°$
**15–5** 77.34 K
**15–7** $-286$°C
**15–9** 79.96 m
**15–11** $1.8 \times 10^{-5}(\text{C}°)^{-1}$
**15–13** 1.405 cm
**15–15** $2.6 \times 10^{-5}(\text{C}°)^{-1}$
**15–17** 20 L
**15–19** a) $1.3 \times 10^{-5}(\text{C}°)^{-1}$ b) $1.3 \times 10^9$ Pa
**15–21** $1.79 \times 10^5$ J
**15–23** a) 44 J b) $3.2 \times 10^4$ J
**15–25** $8.01 \times 10^5$ J
**15–27** $3.92 \times 10^3$ J/kg · K
**15–29** $1.82 \times 10^4$ J; $4.34 \times 10^3$ cal; 17.2 Btu
**15–31** 357 m/s
**15–33** 3510 W; $1.20 \times 10^4$ Btu/h
**15–35** 24.7°C
**15–37** 3.11 kg
**15–39** 0.155 kg
**15–41** 0.156 kg
**15–45** a) $-7.7$°C b) 9.2 W
**15–47** 990 Btu; $1.04 \times 10^6$ J
**15–49** a) 500 K/m b) 23.1 W c) 70.0°
**15–51** 124 W
**15–53** 0.839 cm$^2$
**15–55** 8000 W/m$^2$
**15–57** 12°C
**15–59** a) $3.6 \times 10^{-6}$m$^3$ b) $-9.7$ kg/m$^3$
**15–61** a) 160°C b) $-190$°C
**15–63** 21.4 cm, 8.6 cm
**15–65** $3.5 \times 10^8$ Pa
**15–67** 45.7° C
**15–69** a) 44.9
**15–71** a) 566 J b) 5.66 J/mol · K c) 18.9 J/mol · K
**15–73** a) $8.40 \times 10^7$ J b) 21.4 C° c) 37.5 C°
**15–75** 4.19 g
**15–77** 45.4°C
**15–79** a) 60.0 W b) 25.8
**15–81** 69
**15–83** b) 480 h
**15–85** 171 min
**15–87** $1.71 \times 10^{-3}$ kg
**15–89** a) $2.4 \times 10^{-4}$ b) gains 10 s c) 1.9 C°; no
**15–91** a) $H = 4\pi kab(T_2 - T_1)/(b - a)$
b) $T = T_2 - b(r - a)(T_2 - T_1)/r(b - a)$
c) $H = 2\pi Lk(T_2 - T_1)/\ln(b/a)$
d) $T = T_2 - \ln(r/a)(T_2 - T_1)/\ln(b/a)$
**15–93** b) $T = 0$°C throughout d) $3.14 \times 10^3$ C°/m e) 121 W f) 0; 0 g) $1.11 \times 10^{-4}$ m$^2$/s h) $-10.9$ C°/s i) 9.14 s j) Decrease k) $-7.74$ C°/s
**15–95** 20.09 K

## CHAPTER 16

**16–1** 2.40 atm
**16–3** a) 70.0 mol b) $8.73 \times 10^6$ Pa; 86.2 atm
**16–5** $3.75 \times 10^4$ Pa
**16–7** 382° C
**16–9** 1.09 atm
**16–11** 0.206 L
**16–13** 108°C
**16–15** 13.9 mol; $8.36 \times 10^{24}$ molecules
**16–17** a) $18.0 \times 10^{-6}$ m$^3$ b) $3.10 \times 10^{-10}$ m c) About the same
**16–19** 3889° C
**16–21** 1.006
**16–23** a) $6.21 \times 10^{-21}$ J b) $1.87 \times 10^6$ m$^2$/s$^2$ c) $1.37 \times 10^3$ m/s d) $9.09 \times 10^{-24}$ kg · m/s e) $1.24 \times 10^{-19}$ N f) $1.24 \times 10^{-17}$ Pa g) $8.15 \times 10^{21}$ molecules h) $2.45 \times 10^{22}$ molecules i) $(v_x^2)_{av} = \frac{1}{3}(v^2)_{av}$
**16–25** a) 2078 J b) 1247 J
**16–27** $C_V = 24.9$ J/mol · K; $c_V = 1390$ J/kg · K; vibration plays a significant role
**16–31** Solid (ice) plus vapor
**16–33** 2.24 atm
**16–35** $9.27 \times 10^{-4}$ g
**16–37** a) $7.60 \times 10^4$ Pa b) 1.17 g
**16–39** a) 8.08 cylinders b) 5880 N c) 5450 N
**16–41** $6.0 \times 10^{27}$ atoms
**16–43** a) 517 m/s b) 299 m/s
**16–45** a) $1.24 \times 10^{-18}$ kg b) $4.16 \times 10^7$ molecules c) $1.37 \times 10^{-7}$ m; no
**16–47** b) Oxygen: $1.60 \times 10^5$ K; hydrogen: $1.01 \times 10^4$ K
**16–51** 23.0%
**16–53** a) 18.4 m/s b) 8.38 m/s; 4.48 m/s c) 1.40 m
**16–55** b) $T_c = 8a/27Rb$; $(V/n)_c = 3b$ c) $p_c = a/27b^2$ d) 2.67 e) $H_2$: 3.28; $N_2$: 3.44; $H_2O$: 4.34
**16–57** a) $F(r) = U_0[(12/r)(\sigma/r)^{12} - (6/r)(\sigma/r)^6]$ b) $r_1 = \sigma$; $r_2 = 1.12\sigma$; $(r_1/r_2) = 0.891$ c) $U_0/4$

## CHAPTER 17

**17–1** 0
**17–3** $3.74 \times 10^3$ J
**17–5** a) 1; 3 b) 1; absorbed
**17–7** a) Absorbs b) + 400 J c) Liberates; 400 J
**17–9** a) Yes b) No c) Negative
**17–11** 240 J
**17–13** a) $-8.00 \times 10^4$ J b) $-2.00 \times 10^5$ J c) No
**17–15** a) 519 J b) 0 c) Liberates 519 J
**17–17** $C_V = 25.0$ J/mol · K; $C_p = 33.3$ J/mol · K
**17–19** 182°C; 4.66 atm
**17–21** a) 250 J b) 0
**17–23** a) 30.0 J b) Liberates 65.0 J c) $ad$: 16.0 J; $db$: 14.0 J
**17–25** 0.384 m$^3$
**17–27** a) $3.68 \times 10^{-4}$ m$^3$ b) $-551$ J c) $2.30 \times 10^6$ J d) $2.30 \times 10^6$ J
**17–29** a) 0.154 m b) 212° C c) $1.15 \times 10^5$ J
**17–31** a) $Q = 600$ J; $\Delta U = 0$ b) $Q = 0$; $\Delta U = -600$ J

**17–33** b) $1.25 \times 10^4$ J c) 0 d) $1.25 \times 10^4$ J e) 0.226 $m^3$
**17–35** a) $W = 1250$ J; $Q = 4380$ J; $\Delta U = 3180$ J b) $W = 0$; $Q = -3130$ J; $\Delta U = -3130$ J c) 0
**17–37** $W = nRT \ln [(V_2 - nb)/(V_1 - nb)] + an^2[(V_1 - V_2)/V_1V_2]$

## CHAPTER 18

**18–1** a) 25.0% b) 6000 J c) 0.160 g d) $8.00 \times 10^4$ W; 107 hp
**18–3** a) 28.6% b) 500 MW
**18–5** a) 15,000 J b) 60.0%
**18–7** 370° C
**18–9** a) $1.50 \times 10^4$ J b) $4.50 \times 10^4$ J
**18–13** a) 279 K b) 30.2% c) 145 J
**18–15** a) $1.47 \times 10^7$ J b) $1.32 \times 10^6$ J
**18–17** $-4.84 \times 10^3$ J/K
**18–19** 22.6 J/K
**18–21** 15.2 J/K
**18–23** 167 $m^2$
**18–25** 57.3 L
**18–27** Water: 44.7 $m^3$; Glauber salt: 10.5 $m^3$
**18–29** a) $p_1 = 1.00$ atm, $V_1 = 7.39 \times 10^{-3}$ $m^3$; $p_2 = 2.00$ atm, $V_2 = 7.39 \times 10^{-3}$ $m^3$; $p_3 = 1.00$ atm, $V_3 = 1.12 \times 10^{-2}$ $m^3$ b) 1→2: $Q = 1120$ J, $W = 0$, $\Delta U = 1120$ J; 2→3: $Q = 0$, $W = 542$ J, $\Delta U = -542$ J; 3→1: $Q = -967$, $W = -387$ J, $\Delta U = -580$ J; c) 156 J d) 156 J e) 13.9%
**18–31** 10.6%
**18–33** a) $7.36 \times 10^5$ J b) $9.33 \times 10^5$ J c) $1.98 \times 10^5$ J d) 3.73
**18–35** +16.5 J/K

## CHAPTER 19

**19–1** a) 10.8 m b) 282 Hz
**19–3** a) 17.2 m, 0.0172 m b) 74.0 m, 0.0740 m
**19–5** a) 1.33 m/s b) 0.400 m
**19–7** a) $f = 50.0$ Hz, $T = 0.0200$ s, $k = 20.9$ $m^{-1}$ b) $y = (0.0800 \text{ m}) \sin [(314 \text{ rad/s})t - (20.9 \text{ m}^{-1})x]$ c) 0.0690 m
**19–9** b) $+x$-direction
**19–11** a) 52.2 m/s b) 0.217 m
**19–13** 0.325 s
**19–15** $1.15 \times 10^9$ Pa
**19–17** 94.7 m
**19–19** 34 m/s
**19–23** a) Slope b) $a\sqrt{(k'/m)} \cos (ka/2)$ c) $\lambda = 2a$ d) $\lambda >> a$
**19–25** $6.2 \times 10^{12}$ Hz, $3.9 \times 10^{13}$ rad/s
**19–27** a) 62.5 Hz, 393 rad/s, 3.93 $m^{-1}$ b) $y(x, t) = -(0.0800 \text{ m}) \sin [(393 \text{ rad/s})t - (3.93 \text{ m}^{-1})x]$ c) $y(x, 0) = -(0.0800 \text{ m}) \sin [(393 \text{ rad/s})t]$ d) $y(x, 1.20 \text{ m}) = -(0.0800 \text{ m}) \sin [(393 \text{ rad/s})t - 4.71 \text{ rad}]$ e) 31.4 m/s f) 0.0800 m; 31.4 m/s
**19–31** 0.256 s
**19–33** $Y/2500$
**19–35** a) 368 N b) 368 N + (9.6 N/m)$x$ c) 4.10 s

## CHAPTER 20

**20–3** a) 0, 0.60 m, 1.20 m, 1.80 m, 2.40 m b) 0.30 m, 0.90 m, 1.50 m, 2.10 m
**20–5** a) 48.0 m/s b) 192 N
**20–7** a) 50.0 cm b) No
**20–9** a) 3.20 m, 12.5 Hz b) 1.60 m, 25.0 Hz c) 1.07 m, 37.5 Hz
**20–11** a) Fundamental: 1.80 m; 1st overtone: 0.60 m, 1.80 m; 2nd overtone: 0.36 m, 1.08 m, 1.80 m b) Fundamental: 0; 1st overtone: 0, 1.20 m; 2nd overtone: 0, 0.72 m, 1.44 m
**20–13** a) 35.2 Hz b) 17.6 Hz
**20–17** a) 175 Hz, 350 Hz, 525 Hz, 700 Hz, . . . b) 88 Hz, 262 Hz, 438 Hz, 612 Hz, . . .
**20–19** a) 40.0 m/s b) Four times each second
**20–21** a) 4.30 m b) 6, 7
**20–23** a) 187 N b) 12%
**20–25** a) 375 m/s b) 1.39
**20–27** a) 0.657 m b) 56.2°C
**20–29** a) 99.8 N c) $-5.0$ Hz; raised

## CHAPTER 21

**21–1** a) 16.2 Pa; below pain threshold b) 648 Pa; far above pain threshold
**21–3** $1.09 \times 10^{-12}$ W/$m^2$
**21–7** 0.0469 W/$m^2$; 107 dB
**21–9** 17.5 dB
**21–11** 436.8 Hz; 443.2 Hz
**21–13** a) 0.785 m b) 0.935 m c) 432 Hz d) 368 Hz
**21–15** a) 457 Hz b) 352 Hz
**21–17** a) 30.0° b) 3.06 s
**21–19** a) $9.09 \times 10^{-3}$ Pa b) $7.96 \times 10^{-9}$ m c) 60.0 m
**21–21** a) 0.30 m/s b) 1.40 m
**21–23** 9.86 m/s
**21–25** c) 66.0 m/s
**21–27** a) 0.0493 m b) 280 Hz
**21–29** a) 180° or $\lambda/2$ b) $A$ only: $3.98 \times 10^{-6}$ W/$m^2$, 66.0 dB; $B$ only: $1.06 \times 10^{-5}$ W/$m^2$, 70.3 dB c) $1.59 \times 10^{-6}$ W/$m^2$, 62.0 dB
**21–31** a) $vt$ b) $(v - v_1)/f_0$ c) $f_0t(v + v_2)/(v - v_1)$ d) $v$ e) $(v - v_1)(v - v_2)/f_0(v + v_2)$ f) $f_0(v + v_1)(v + v_2)/(v - v_1)(v - v_2)$ g) 1646 Hz

## CHAPTER 22

**22–1** $1.93 \times 10^5$ C
**22–3** 0.225 N
**22–5** a) 2.11 $\mu$C, 2.11 $\mu$C b) 1.49 $\mu$C, 2.98 $\mu$C
**22–7** 5.08 m
**22–9** $5.31 \times 10^{-7}$ N, $+y$-direction
**22–11** 0.540 N, $+x$-direction
**22–13** 2.74 m
**22–15** a) 6.00 N/C, upward b) $9.61 \times 10^{-19}$ N, upward
**22–17** $5.57 \times 10^{-11}$ N/C
**22–19** $6.97 \times 10^3$ N/C
**22–21** a) 1050 N/C, $-x$-direction b) 312 N/C, $+x$-direction c) 845 N/C, $+x$-direction
**22–23** a) 0 b) $3.99 \times 10^4$ N/C, $+x$-direction c) $2.00 \times 10^4$ N/C, 76.7° counterclockwise from $+x$-axis d) $2.54 \times 10^4$ N/C, $+y$-direction
**22–25** a) $7.19 \times 10^4$ N/C, $-x$-direction b) $3.20 \times 10^4$ N/C, $+x$-direction c) $1.33 \times 10^4$ N/C, 110° counterclockwise from $+x$-axis d) $2.54 \times 10^4$ N/C, $-x$-direction
**22–27** $5.31 \times 10^{-11}$ C/$m^2$
**22–29** a) $1.50 \times 10^{-24}$ J b) 0.0724 K
**22–35** b) 1.72 $\mu$C c) 31.7°
**22–37** a) $5.39 \times 10^{-6}$ N, away from vacant corner b) $5.16 \times 10^{-6}$ N, toward opposite corner

**22–39** $F_x = -qQ/(4\pi\epsilon_0 x\sqrt{x^2+a^2})$; $F_y = (qQ/4\pi\epsilon_0 a)(1/x - 1/\sqrt{x^2+a^2})$
**22–41** $1.96 \times 10^{-15}$ C
**22–43** a) 2.27 cm b) 15.8 cm
**22–45** a) 1.15 nC b) −3.65 nC
**22–47** −31.2 nC
**22–49** $E_x = +Q/2\pi^2\epsilon_0 a^2$; $E_y = +Q/2\pi^2\epsilon_0 a^2$
**22–51** b) $q_1<0$, $q_2>0$ c) $|q_2| = 1.69\ \mu$C d) 56.2 N

## CHAPTER 23

**23–1** a) $2.26 \times 10^5$ N·m²/C b) $2.26 \times 10^5$ N·m²/C c) $4.52 \times 10^5$ N·m²/C
**23–3** $5.87 \times 10^5$ N·m²/C
**23–5** 56.5 N·m²/C
**23–7** $7.31 \times 10^9$
**23–11** $E = \sigma/\epsilon_0$ between plates; $E = 0$ outside plates
**23–13** $-1.99 \times 10^{-10}$ C
**23–15** a) $qQ/4\pi\epsilon_0 r^2$, away from center of shell b) 0
**23–17** a) (i) $E = 0$; (ii) $E = 0$; (iii) $E = 2q/4\pi\epsilon_0 r^2$, radially outward; (iv) $E = 0$; (v) $E = 6q/4\pi\epsilon_0 r^2$, radially outward b) (i) 0; (ii) $+2q$; (iii) $-2q$; (iv) $+6q$
**23–19** a) (i) $E = 0$; (ii) $E = 0$; (iii) $E = 2q/4\pi\epsilon_0 r^2$, radially outward; (iv) $E = 0$; (v) $E = 2q/4\pi\epsilon_0 r^2$, radially inward b) (i) 0; (ii) $+2q$; (iii) $-2q$; (iv) $-2q$
**23–21** a) $E = \lambda/2\pi\epsilon_0 r$, radially outward b) $E = \lambda/2\pi\epsilon_0 r$, radially outward d) Inner: $-\lambda$; outer: $+\lambda$
**23–23** a) (i) $E = \alpha/2\pi\epsilon_0 r$; (ii) $E = 0$; (iii) $E = 0$ b) (i) $-\alpha$; (ii) 0
**23–25** c) $(\rho_0 r/12\epsilon_0)(4 - 3(r/R))$ d) Same
**23–27** a) $\alpha = 8Q/5\pi R^3 = 1.91 \times 10^{24}$ C/m³ b) $r \le R/2$: $E = 8Qr/15\pi\epsilon_0 R^3$; $R/2 \le r \le R$: $E = (4Qr/15\pi\epsilon_0 R^3)(4 - 3(r/R)) - Q/60\pi\epsilon_0 r^2$; $r \ge R$: $E = Q/4\pi\epsilon_0 r^2$ c) 4/15 e) $5.59 \times 10^{-23}$ s f) $E$ not proportional to $r$

## CHAPTER 24

**24–1** a) −0.0449 J b) 30.0 m/s
**24–3** −0.578 J
**24–5** a) 19.6 m/s b) 0.207 m
**24–7** a) 0.0449 m b) 0.0225 m
**24–9** a) point $a$ b) 1250 V/m c) $1.00 \times 10^{-4}$ J
**24–11** a) 0 b) $-1.20 \times 10^{-3}$ J c) $2.76 \times 10^{-3}$ J
**24–13** a) 180 V b) 0 c) $-4.50 \times 10^{-7}$ J
**24–15** a) $2q/4\pi\epsilon_0 a$ b) $x = \pm\sqrt{3}a$
**24–17** $1.12 \times 10^7$ m/s
**24–19** a) 6250 V/m b) $1.25 \times 10^{-5}$ N c) $1.00 \times 10^{-6}$ J d) Same
**24–21** 0.833 mm
**24–25** a) 5.01 $\mu$C
**24–27** a) 0.433 mm/s b) 0.358 mm/s
**24–29** 0.600 MeV
**24–31** a) 1.32 cm b) 33.4° c) 9.23 cm
**24–35** a) $1.79 \times 10^7$ V b) $4.91 \times 10^8$ eV
**24–37** a) $-2.00 \times 10^5$ J b) $-6.67 \times 10^3$ V c) $1.33 \times 10^5$ V/m
**24–39** a) $2.18 \times 10^{-5}$ m b) 42.9 (electron's speed larger) c) 1
**24–41** a) $5.82 \times 10^4$ V/m$^{4/3}$ b) $-\frac{4}{3}Cx^{1/3}$ c) $2.26 \times 10^{-15}$ N, toward the anode
**24–43** a) (i) $V = (\lambda/2\pi\epsilon_0)\ln(b/a)$; (ii) $V = (\lambda/2\pi\epsilon_0)\ln(b/r)$; (iii) $V = 0$ d) $V_{ab} = (\lambda/2\pi\epsilon_0)\ln(b/a)$
**24–45** a) $1.37 \times 10^5$ V/m b) $2.15 \times 10^{-11}$ C
**24–47** a) $r \le R$: $V = (\lambda/4\pi\epsilon_0)(1 - (r/R)^2)$; $r \ge R$: $V = (\lambda/2\pi\epsilon_0)\ln(R/r)$
**24–49** a) $(Q/4\pi\epsilon_0 a)\ln(1 + a/x)$ b) $(Q/4\pi\epsilon_0 a)\ln[(a + \sqrt{a^2+y^2})/y]$ c) Point P: $V \to Q/4\pi\epsilon_0 x$; point R: $V \to Q/4\pi\epsilon_0 y$
**24–51** 3
**24–53** a) 1/3 b) 3
**24–55** b) $V(r) = (Q/2\pi\epsilon_0 R)(1 - (r/R)^2 + \frac{1}{3}(r/R)^3)$
**24–57** a) 667 m/s c) −3.59 J d) No e) 25.1 mm f) (c) 9.61 J; (d) yes; (e) 980 m/s

## CHAPTER 25

**25–1** a) 400 V b) 0.0226 m² c) $1.00 \times 10^6$ V/m d) $8.85 \times 10^{-6}$ C/m²
**25–3** $3.20 \times 10^{-4}$ C
**25–5** a) $Q_1 = 216\ \mu$C; $Q_2 = 324\ \mu$C b) $V_1 = V_2 = 54$ V
**25–7** a) $Q_1 = 14.4\ \mu$C; $Q_2 = 14.4\ \mu$C; $Q_3 = 28.8\ \mu$C; $Q_4 = 43.2\ \mu$C b) $V_1 = 7.2$ V; $V_2 = 7.2$ V; $V_3 = 14.4$ V; $V_4 = 21.6$ V c) 14.4 V
**25–11** 8.64 J
**25–13** 0.0443 J/m³
**25–15** a) $q^2x/2\epsilon_0 A$ b) $q^2(x + dx)/2\epsilon_0 A$
**25–17** 0.0150 m²
**25–19** a) $2.12 \times 10^{-6}$ C/m² b) 3.00
**25–21** a) $E = Q/\epsilon_0 AK$ b) $V = Qd/\epsilon_0 AK$ $C = K\epsilon_0 A/d$
**25–23** a) 6.32 J b) 126 V
**25–25** a) 29.5 pF b) 1.48 nC c) $4.17 \times 10^3$ V/m d) $3.69 \times 10^{-8}$ J
**25–27** a) 1.00 $\mu$F b) $Q_1 = 600\ \mu$C; $Q_2 = 400\ \mu$C; $Q_3 = 200\ \mu$C; $Q_4 = 200\ \mu$C; $Q_5 = 600\ \mu$C; $V_1 = 200$ V; $V_2 = 200$ V; $V_3 = 100$ V; $V_4 = 100$ V; $V_5 = 200$ V
**25–29** a) 3.00 $\mu$F b) $Q_1 = 2.70$ mC; $Q_2 = 1.80$ mC c) 100 V
**25–31** a) 72.0 $\mu$C b) $1.30 \times 10^{-3}$ J c) 10.8 V d) $1.17 \times 10^{-3}$ J
**25–33** a) $9.41 \times 10^6$ V/m b) 0.0471 V
**25–37** b) $9.07 \times 10^{-10}$ F
**25–39** b) 14.0 $\mu$F c) 72-$\mu$F: 504 $\mu$C, 7.0 V; 27-$\mu$F: 207 $\mu$C, 10.0 V; 18-$\mu$F: 234 $\mu$C, 13.0 V; 6-$\mu$F: 18 $\mu$C, 3.0 V; 28-$\mu$F: 252 $\mu$C, 9.0 V; 21-$\mu$F: 252 $\mu$C, 12.0 V

## CHAPTER 26

**26–1** $3.60 \times 10^4$ C
**26–3** a) 240 C b) 24.0 A
**26–5** 0.156 Ω
**26–7** a) 9.94 A b) 6.24 V c) 0.628 Ω
**26–9** a) 0.900 V/m b) $2.65 \times 10^{-8}$ Ω·m
**26–11** a) 99.4 Ω b) 0.0187 Ω
**26–13** 38°C
**26–15** 1.48 V, 0.085 Ω
**26–17** a) 0.100 Ω b) 0.300 Ω c) 0.0120 Ω
**26–19** b) Yes c) 4.36 A
**26–21** a) $EJ$ b) $\rho J^2$ c) $E^2/\rho$
**26–23** a) 30.0 Ω b) 4.00 A c) 403 W
**26–25** a) $2.16 \times 10^6$ J b) 0.0522 L c) 1.50 h
**26–27** a) 24.0 W b) 4.0 W c) 20.0 W
**26–29** $4.16 \times 10^{-14}$ s
**26–31** a) 1.67 A b) $2.78 \times 10^4$ W c) $2.00 \times 10^7$ Ω
**26–33** a) $2.13 \times 10^{-8}$ Ω·m b) 225 A c) 4.13 mm/s
**26–37** a) 0.40 Ω b) 10.4 V
**26–39** 1.06 A
**26–41** a) 16.5 V b) $3.56 \times 10^6$ J c) $9.72 \times 10^5$ J d) 0.500 Ω e) $1.62 \times 10^6$ J f) $9.72 \times 10^5$ J g) Electrical energy dissipated in the internal resistance during charging and discharging

**26–43** a) 0.40 A b) 1.6 W c) 12-V battery: 4.8 W d) 8-V battery: 3.2 W e) 4.8 W = 3.2 W + 1.6 W

**26–45** b) $a = 8.0 \times 10^{-5}\ \Omega \cdot m \cdot K^n$; $n = 0.15$ c) $-196°C$: $\rho = 4.3 \times 10^{-5}\ \Omega \cdot m$; 300°C: $\rho = 3.2 \times 10^{-5}\ \Omega \cdot m$

## CHAPTER 27

**27–1** a) 24.0 Ω b) 5.00 A c) 40-Ω: 3.00 A; 60-Ω: 2.00 A

**27–3** a) 0.923 Ω b) 2-Ω: 12.0 A; 3-Ω: 8.00 A; 4-Ω: 6.00 A c) 26.0 A d) 24.0 V across each e) 2-Ω: 288 W; 3-Ω: 192 W; 4-Ω: 144 W

**27–5** $R_{eq} = 3.00\ \Omega$; 1-Ω: 12.0 A; 3-Ω: 12.0 A; 7-Ω: 4.00 A; 5-Ω: 4.00 A

**27–7** a) 200 V b) 2.42 W

**27–9** a) 2.00 A b) 5.00 Ω c) 42.0 V d) 3.50 A

**27–11** a) 8.00 A b) $e_1 = 36.0$ V; $e_2 = 54.0$ V c) 9.00 Ω

**27–13** 12.0 Ω

**27–15** $R_1 = 2.98 \times 10^3\ \Omega$; $R_2 = 1.20 \times 10^4\ \Omega$; $R_3 = 1.35 \times 10^5\ \Omega$; 3 V: $3.00 \times 10^3\ \Omega$; 15 V: $1.50 \times 10^4\ \Omega$; 150 V: $1.50 \times 10^5\ \Omega$

**27–17** b) Measurement made with no current through it c) 9.52 V

**27–21** a) $6.00 \times 10^{-4}$ A b) $1.00 \times 10^{-4}$ s

**27–23** a) 0; $2.36 \times 10^{-4}$ C; $3.79 \times 10^{-4}$ C; $5.19 \times 10^{-4}$ C; $6.00 \times 10^{-4}$C b) $6.00 \times 10^{-5}$ A; $3.64 \times 10^{-5}$ A; $2.21 \times 10^{-5}$ A; $8.12 \times 10^{-6}$ A; $2.72 \times 10^{-9}$ A

**27–25** a) Toaster: 15.0 A; frying pan: 11.7 A; lamp: 0.625 A b) Yes ($I_{tot} = 27.3$ A)

**27–27** a) 5.45 A; 655 W b) 3.08 A; 370 W

**27–29** a) Two in series in parallel with two in series b) 0.500 W

**27–31** 36.0 W

**27–33** a) + 0.22 V b) 0.464 A

**27–35** $I_1 = 0.848$ A; $I_2 = 2.14$ A; $I_3 = 0.171$ A

**27–37** 2-Ω: 5.21 A; 4-Ω: 1.10 A; 5-Ω: 6.32 A

**27–39** a) − 12.0 V b) 1.71 A c) 4.20 Ω

**27–41** a) − 6.0 V b) Point *b* c) + 6.0 V d) + 54.0 μC, in direction of *b* to *a*

**27–43** a) 3000 Ω b) 33.3 V

**27–45** $1.67 \times 10^4\ \Omega$

**27–47** b) 2528 Ω

**27–49** $R = 2.50 \times 10^5\ \Omega$; $C = 2.40 \times 10^{-5}$F

**27–51** a) $e^2C$ b) $e^2C/2$ c) $e^2C/2$ d) 50%

**27–55** a) 140-Ω: 0.300 A; 210-Ω: 0.500 A; 35-Ω: 0.200 A b) 140-Ω: 0.541 A; 210-Ω: 0.077 A; 35-Ω: −0.464 A c) 140-Ω: −0.287 A; 210-Ω: 0.192 A; 35-Ω: 0.479 A d) 140-Ω: 0.046 A; 210-Ω: 0.231 A; 35-Ω: 0.185 A f) 140-Ω: 0.083 A; 210-Ω: 0.644 A; 35-Ω: 0.561 A

## CHAPTER 28

**28–1** Positive

**28–3** a) $(1.75 \times 10^{-3}\ N)\boldsymbol{k}$ b) $-(1.75 \times 10^{-3}\ N)\boldsymbol{i} - (1.05 \times 10^{-3}\ N)\boldsymbol{j}$

**28–5** $a$: $-qvB\boldsymbol{k}$; $b$: $qvB\boldsymbol{j}$; $c$: 0; $d$: $-(qvB/\sqrt{2})\boldsymbol{j}$; $e$: $-(qvB/\sqrt{2})(\boldsymbol{j} + \boldsymbol{k})$

**28–7** a) 0.804 Wb b) 0.697 Wb c) 0

**28–9** a) $4.55 \times 10^{-4}$ T, into paper b) $3.93 \times 10^{-8}$ s

**28–11** a) $2.88 \times 10^6$ m/s b) $4.37 \times 10^{-8}$ s c) 86.3 kV

**28–13** $3.18 \times 10^{-3}$ T

**28–15** $3.50 \times 10^{-26}$ kg; 21

**28–17** 17.1 A

**28–19** a) $-(0.0420\ N)\boldsymbol{k}$ b) $-(0.0350\ N)\boldsymbol{j}$ c) 0 d) $+(0.0210\ N)\boldsymbol{j}$ e) $+(0.0280\ N)\boldsymbol{j} + (0.0630\ N)\boldsymbol{k}$

**28–21** a) 0.109 N · m b) Angle of 30° between ***B*** and normal to coil

**28–23** + 1.95 J

**28–25** a) $\boldsymbol{\tau} = -(NIAB)\boldsymbol{i}$, $U = 0$ b) $\tau = 0$; $U = -NIAB$ c) $\boldsymbol{\tau} = +(NIAB)\boldsymbol{i}$, $U = 0$ d) $\tau = 0$; $U = +NIAB$

**28–27** a) 0.857 A b) 3.64 A c) 102 V d) 371 W

**28–29** a) $7.47 \times 10^{-4}$ m/s b) $1.12 \times 10^{-3}$ V/m c) $2.24 \times 10^{-5}$ V

**28–31** a) $(2400\ N/C)\boldsymbol{k}$ b) $(2400\ N/C)\boldsymbol{k}$

**28–33** $B = 0.333$ T, $-y$-direction

**28–35** $3.65 \times 10^3$

**28–37** 0.864 mm

**28–39** a) 1.82 μC b) $(2.67 \times 10^{14}\ m/s^2)(3\boldsymbol{i} + 4\boldsymbol{j})$ c) $R = 0.0188$ m d) $2.67 \times 10^8$ rad/s e) $x = R$, $y = 0$, $z = 0.565$ m

**28–41** a) *ab*: 0.400 N, $-z$-direction; *bc*: 0.400 N, $-y$-direction; *cd*: 0.566 N, midway between $+y$- and $+z$-axis; *de*: 0.400 N, $-y$-direction; *ef*: 0 b) 0.400 N, $-y$-direction

**28–43** 0.0165 T, $+y$-direction

**28–45** 0.313 N, $-y$-direction

**28–47** a) 3.57 m/s b) 2.98 A c) 0.503 Ω

**28–49** a) $0.0120\ A \cdot m^2$ b) $B_x = +0.667$ T; $B_y = +0.500$ T; $B_z = +2.46$ T

**28–51** b) Yes

**28–53** a) 1.50 m b) $8.37 \times 10^{-7}$ s c) 0.0210 m d) 0.106 m

## CHAPTER 29

**29–1** a) 0 b) $-(9.60 \times 10^{-6}\ T)\boldsymbol{k}$ c) $-(9.60 \times 10^{-6}\ T)\boldsymbol{j}$ d) $-(3.39 \times 10^{-6}\ T)(\boldsymbol{j} + \boldsymbol{k})$

**29–3** $4.46 \times 10^{-6}$ T, in $-z$-direction

**29–5** $\mu_0 I/4R$, into paper

**29–9** a) $3.20 \times 10^{-5}$ T, south b) Yes

**29–11** a) $\mu_0 I/\pi a$ b) $\mu_0 I/3\pi a$ c) 0 d) $2\mu_0 I/3\pi a$

**29–13** a) 15.0 A b) Opposite

**29–15** 6.53 mm

**29–17** 24

**29–19** $3.33 \times 10^{-4}$ T

**29–21** a) $1.11 \times 10^4$ turns/m b) $1.05 \times 10^3$ m

**29–23** a) $\mu_0 I/2\pi r$ b) 0

**29–25** Yes; 1540 A · K/T · m

**29–27** a) 0.0509 A b) 0.0137 A

**29–29** a) $q = 1.00 \times 10^{-8}$ C; $E = 2.82 \times 10^6$ V/m; $V = 8.41 \times 10^3$ V b) $5.65 \times 10^{11}$ V/m · s; no c) $j_D = 5.00\ A/m^2$; $i_D = 2.00$ mA; $i_C = i_D$

**29–31** a) $76.4\ A/m^2$ b) $8.63 \times 10^{12}$ V/m · s c) $1.20 \times 10^{-6}$ T d) $1.20 \times 10^{-6}$ T

**29–33** $3.00 \times 10^{-20}$ N, away from wire

**29–35** b) $B = \mu_0 Ix/\pi(a^2 + x^2)$, $-y$-direction d) $x = a$

**29–37** a) 2.00 A, out of plane b) $2.13 \times 10^{-6}$ T, to the right c) $2.06 \times 10^{-6}$ T

**29–39** $2.52 \times 10^{-4}$ N, toward wire

**29–41** a) $B = \mu_0 NIa^2/2([(\frac{a}{2} + x)^2 + a^2]^{-3/2} + [(\frac{a}{2} - x)^2 + a^2]^{-3/2})$ c) $B = 8\mu_0 NI/a5^{3/2}$ d) $3.00 \times 10^{-3}$ T

**29–43** $\mu_0 I/4\pi a$, out of page

**29–45** a) $\alpha = 3I/2\pi R^3$ b) (i) $B = \mu_0 Ir^2/2\pi R^3$; (ii) $B = \mu_0 I/2\pi r$

**29–47** a) $B = \mu_0 nI/2$, $+x$-direction b) $B = \mu_0 nI/2$, $-x$-direction

**29–49** b) $B = \mu_0 I/2\pi r$ c) $I = (I_0 r^2/a^2)(2 - r^2/a^2)$ d) $B = (\mu_0 Ir/2\pi a^2)(2 - r^2/a^2)$; same

**29–51** $\int_{-\infty}^{\infty} B_x\, dx = \mu_0 I$; agrees with Ampere's law

**29–53** a) $(Q_0/KA\epsilon_0\rho)\exp(-t/K\rho\epsilon_0)$

**29–55** $\mu_0 nQ/a$

**29–57** b) 0.286 m/s c) 4.16 mm

## CHAPTER 30

**30–1** 0.0428 A

**30–3** 7.69 V

**30–5** a) 3.68 rev/s b) 43.2 N · m

**30–7** $NBA/R$

**30–9** 0.0750 T

**30–13** a) 6.67 m/s b) 4.00 A c) 1.92 N, to the left

**30–17** a) Counterclockwise b) Clockwise c) $I = 0$

**30–19** a) Right to left b) Right to left c) Left to right

**30–21** a) Concentric circles b) $3.00 \times 10^{-3}$ V/m; tangent to ring and clockwise c) $9.42 \times 10^{-4}$ A d) 0 e) $1.88 \times 10^{-3}$ V

**30–23** a) Inside: $B = 0$, $\boldsymbol{M} = -(1.00 \times 10^5 \text{ A/m})\boldsymbol{i}$; outside: $\boldsymbol{B} = (0.126 \text{ T})\boldsymbol{i}$, $\boldsymbol{M} = 0$ b) Inside: $\boldsymbol{B} = (0.252 \text{ T})\boldsymbol{i}$, $\boldsymbol{M} = 0$; outside: $\boldsymbol{B} = (0.252 \text{ T})\boldsymbol{i}$, $\boldsymbol{M} = 0$

**30–27** b) $v_t = FR/L^2B^2$

**30–29** a) 0.769 V b) 0.769 V

**30–33** a) 0.0589 V b) $a$ to $b$

**30–35** Point $a$: $F = \frac{1}{2}qr\, dB/dt$, to left; point $b$: $F = \frac{1}{2}qr\, dB/dt$, toward top of page; point $c$: $F = 0$

**30–37** a) $(\mu_0 I v/2\pi)\ln(1 + L/d)$ b) Point $a$ c) 0

**30–39** a) $B^2a^2v/R$

**30–41** b) 0 c) $2.40 \times 10^{-3}$ V d) $1.20 \times 10^{-3}$ A e) $6.00 \times 10^{-4}$ V; point $a$

## CHAPTER 31

**31–3** a) $1.50 \times 10^{-3}$ V; yes b) $1.50 \times 10^{-3}$ V

**31–5** a) 0.833 H b) $2.22 \times 10^{-3}$ Wb

**31–7** a) $1.80 \times 10^{-2}$ V b) $a$

**31–9** $L = \mu_0 N^2 A/l$

**31–11** 1296

**31–13** a) 36.2 m$^3$ b) 8.51 T

**31–19** a) 4.00 A/s b) 1.00 A/s c) 0.602 A d) 1.33 A

**31–21** a) 0.240 A b) 0.146 A c) 72.8 V; point $c$ d) $2.77 \times 10^{-4}$ s

**31–25** a) 23.6 rad/s; 0.267 s b) 0.0300 C c) 0.750 J d) 0.0150 C e) −0.612 A f) Capacitor: 0.187 J; inductor: 0.563 J

**31–27** a) 100 rad/s b) 43.6 Ω

**31–29** a) 8.00 H b) 400 Wb c) 0.0320

**31–35** a) $B = \mu_0 i/2\pi r$ b) $dU = \mu_0 i^2 l\, dr/4\pi r$ c) $U = \mu_0 i^2 l \ln(b/a)/4\pi$ d) $L = \mu_0 l \ln(b/a)/2\pi$; same

**31–37** $1.67 \times 10^{-6}$ T

**31–39** a) 1.07 J b) 1.96 J c) 0.896 J d) $(b) = (a) + (c)$

**31–43** a) $i_0 = 0$; $v_{ac} = 0$; $v_{cb} = 60.0$ V b) $i_0 = 0.300$ A; $v_{ac} = 15.0$ V; $v_{cb} = 45.0$ V c) $i_0(t) = (0.300 \text{ A})(1 - e^{-t/0.025 \text{ s}})$; $v_{ac}(t) = (15.0 \text{ V})(1 - e^{-t/0.025 \text{ s}})$; $v_{cb}(t) = (15.0 \text{ V})(3.00 + e^{-t/0.025 \text{ s}})$

**31–45** a) $i_1 =$ ™$/R_1$; $i_2 = (\text{e}/R_2)(1 - e^{-R_2 t/L})$ b) $i_1 =$ ™$/R_1$; $i_2 =$ ™$/R_2$ c) $i = (\text{e}/R_2)e^{-(R_1 + R_2)t/L}$ d) $R_2 = 24\ \Omega$; e = 21 V e) 0.089 A

**31–47** a) $i_1 = (\text{e}/R_1)(1 - e^{-R_1 t/L})$; $i_2 = (\text{e}/R_2)(1 - e^{-t/R_2 C})$; $q_2 = (\text{e}C)(1 - e^{-t/R_2 C})$ b) $i_1 = 0$; $i_2 = 0.0120$ A c) $i_1 = 2.40$ A; $i_2 = 0$; $t \geq 4.0$ s d) $1.97 \times 10^{-3}$ s e) 0.0118 A f) 0.277 s

**31–49** a) $d = D(L - L_0)/(L_f - L_0)$ b) 0.75029 H; 0.75057 H; 0.75086 H; 0.75115 H c) 0.74999 H; 0.74999 H; 0.74998 H; 0.74998 H d) oxygen

## CHAPTER 32

**32–1** a) 0 b) 35.4 V

**32–3** a) 314 Ω b) $3.18 \times 10^{-3}$ H c) $3.18 \times 10^3\ \Omega$ d) $3.18 \times 10^{-3}$ F

**32–5** a) 0.125 A b) 0.0125 A c) 0.00125 A

**32–7** 18.7 $\mu$F

**32–9** a) 2520 Ω b) 0.0199 A c) 5.96 V; 49.6 V d) −83.2°; voltage lags

**32–11** a) 440 Ω; −47.1°; voltage lags b) 431 Ω; +45.9°; voltage leads

**32–13** a) 18.8 W b) 18.8 W c) 0 d) 0 e) $(a) = (b) + (c) + (d)$

**32–17** a) 745 rad/s c) $V_1 = 77.8$ V; $V_2 = 130$ V; $V_3 = 130$ V; $V_4 = 0$; $V_5 = 77.8$ V d) 745 rad/s e) 0.778 A

**32–19** a) 1.00 b) 225 W c) 225 W

**32–21** a) 291 Hz c) 0.300 A d) 0.300 A; 0.131 A; 0.131 A

**32–23** a) 99.0 Ω; 3.03 A b) 500 Ω; 0.600 A

**32–25** a) 15 b) 2.00 A c) 16.0 W d) 900 Ω

**32–27** a) 30 b) 3.33 V

**32–29** 0.0367 H

**32–31** a) 1.20 A b) 200 Ω c) 468 Ω or 732 Ω

**32–33** 54.1 Ω

**32–35** a) Inductor b) $L = 0.122$ H

**32–37** a) 568 Ω b) 1.02 A c) 268 V

**32–39** $\sqrt{(\omega^2L^2 + R^2)/(R^2 + [\omega L - 1/\omega C]^2)}$

**32–41** a) $1.00 \times 10^4$ rad/s b) 1.60 A c) 1.60 A d) 0.160 A e) 0.160 A f) 1.28 mJ; 1.28 mJ

**32–43** a) 4.30 A b) 1.60 A c) 4.00 A d) $6.40 \times 10^{-3}$ A e) 0.800 J; $1.28 \times 10^{-3}$ J

**32–45** a) 0 b) $I_0/\sqrt{3}$

**32–47** a) $1/\sqrt{LC}$ b) $1/\sqrt{LC - R^2C^2/2}$ c) $\sqrt{1/LC - R^2/2L^2}$

## CHAPTER 33

**33–1** 180 m

**33–3** a) 3.30 m b) $7.20 \times 10^{-3}$ V/m

**33–5** $\boldsymbol{E} = (1.20 \times 10^5 \text{ V/m})\boldsymbol{k} \sin[(3.77 \times 10^{14} \text{ rad/s})t - (1.26 \times 10^6 \text{ rad/m})x]$; $\boldsymbol{B} = (4.00 \times 10^{-4} \text{ T})\boldsymbol{i} \sin[(3.77 \times 10^{14} \text{ rad/s})t - (1.26 \times 10^6 \text{ rad/m})x]$

**33–9** a) $2.00 \times 10^{-10}$ T b) 75.0 kW c) 100.0 km

**33–11** 8.16 V/m; $2.72 \times 10^{-8}$ T

**33–13** a) $6.68 \times 10^{-15}$ kg/m$^2$ · s b) $2.00 \times 10^{-6}$ Pa

**33–15** a) 11.0 m b) 15.0 m c) 1.36 d) 1.86

**33–17** a) 10.0 mm b) 5.00 mm c) $2.00 \times 10^8$ m/s

**33–19** a) $9.99 \times 10^{-10}$ m = 0.999 nm b) $6.00 \times 10^{-7}$ m = 600 nm

**33–25** a) $1.79 \times 10^{-4}$ J b) $1.99 \times 10^{-9}$ Pa c) 60.0 W

**33–27** a) 600 V/m; $2.00 \times 10^{-6}$ T b) $7.96 \times 10^{-7}$ J/m$^3$; $7.96 \times 10^{-7}$ J/m$^3$ c) $1.00 \times 10^{-11}$ J

**33–29** a) $\frac{1}{2}\mu_0 nr\, di/dt$ b) $\frac{1}{2}\mu_0 n^2 ri\, di/dt$; radially inward

**33–31** a) $E = \rho I/\pi a^2$ in direction of current b) $B = \mu_0 I/2\pi a$, clockwise if current is directed out of page c) $S = \rho I^2/2\pi^2 a^3$, radially inward d) $dU/dt = \rho l I^2/\pi a^2$; same

**33–33** 0.816 V

**33–35** 38.5 h

**33–37** $2.91 \times 10^{11}$ eV

## CHAPTER 34

**34–1** 42.8°

**34–3** a) 30.0° b) 55.3°

**34–5** a) 1.58 b) 1035 nm

**34–7** a) $1.76 \times 10^8$ m/s b) 294 nm

**34–11** a) Air b) 15.1°

**34–13** a) 51.3° b) 16.4°

**34–15** a) $0.125I_0$ b) Linearly polarized along axis of second polarizer

**34–17** a) First: $0.500I_0$; second: $0.125I_0$; third: $0.0938I_0$; in each case linearly polarized along axis of polarization b) $0.500I_0$; 0

**34–19** a) 1.64 b) 31.4°
**34–21** a) 1.66 b) 59.0°
**34–25** 57.8°
**34–27** 1.84
**34–29** 1.52
**34–31** 35.5°
**34–33** 1.30
**34–35** 1.31
**34–37** a) $I_0(\sin 2\theta)^2/8$ b) 45°
**34–39** a) $\theta = 35°$ b) $I_0 = 10.1$ W/m$^2$; $I_p = 19.9$ W/m$^2$
**34–41** *l*-Leucine: $\theta = (-0.11°)C$; *d*-glutamic acid: $\theta = (0.124°)C$
**34–43** b) 46.3° c) 1.7°

## CHAPTER 35

**35–1** 60.0 cm to right of mirror; 4.00 cm
**35–3** 2.88 cm
**35–5** b) 30.0 cm to left of mirror vertex; 0.800 cm tall; inverted; real
**35–9** b) 4.80 cm to right of vertex; 0.90 cm; erect; virtual
**35–13** 1.38
**35–15** 10.7 cm to left of vertex; 0.571 mm tall; erect
**35–17** 1.83 cm
**35–19** $s = 20$ cm: (a) 60.0 cm to right of lens, (b) $-3.00$, (c) real, (d) inverted; $s = 5$ cm: (a) 7.50 cm to left of lens, (b) $+1.50$, (c) virtual, (d) erect
**35–21** Object is 6.00 cm to right of lens and is 0.400 cm tall; erect
**35–23** a) $-8.57$ cm; diverging b) 0.120 cm; erect
**35–25** Object: 5.25 cm to left of lens; image: 42.0 cm to left of lens; virtual
**35–27** 7.44 cm to left of lens
**35–29** a) Image formed by first surface b) $+20.0$ cm c) Real d) 40.0 cm to left of second vertex
**35–31** $\pm 33.3$ cm; $\pm 11.1$ cm
**35–33** 6.00 m/s
**35–35** a) 5.71 cm b) 11.3 cm
**35–37** 7.50 m; 3.75 m
**35–39** a) $|s| < 6.00$ cm b) Erect
**35–41** $n = 2.00$
**35–43** a) 0.267 cm b) Independent of hemisphere's radius
**35–45** b) (i) 26.7 cm; 26.6 cm; (ii) $-0.33$; $-0.11$; (iii) face toward mirror and opposite face: 0.33 cm × 0.33 cm; top and bottom faces: 0.33 cm × 0.11 cm; side faces: 0.33 cm × 0.11 cm
**35–47** a) 0.0248 m/s b) 0.333 m/s
**35–49** 5.09 cm
**35–51** 3.00 cm from center of sphere
**35–53** 140 cm to right of third lens
**35–55** 3.08 cm from either end of eyepiece
**35–57** 4.85 cm
**35–59** $-20.3$ cm
**35–61** a) 0.276 m b) 0.276 m
**35–63** $4R$ from center of sphere, on opposite side from object
**35–65** 1.00 cm above page
**35–67** 0.80 cm
**35–69** 5.45 cm to left of vertex of spherical mirror; 0.22 cm tall and 10.6 cm to left of vertex of spherical mirror; 0.14 cm tall
**35–71** a) $4f$

## CHAPTER 36

**36–1** 1.21 m
**36–3** a) 35 mm b) 200 mm
**36–5** a) 32.1 mm b) $\frac{1}{15}$ s
**36–7** a) 10.2 cm b) No
**36–9** a) 1.00 m b) 1.67 m
**36–11** 0.692 cm
**36–13** 2.70 cm; 27.0 cm from lens
**36–15** a) 8.11 cm b) 3.08 mm
**36–17** a) 17.5 mm b) 10.8 c) 108
**36–19** a) $-4.50$ b) 3.60 cm c) 0.180 rad
**36–21** 16 cm
**36–23** Convex; 3.7 m
**36–25** a) 8.05 m b) 5.38 m
**36–27** $+2.78$ diopters
**36–29** 2.80 cm; elongated
**36–31** a) 298 b) 24.4 cm

## CHAPTER 37

**37–1** 450 nm, 600 nm
**37–3** 15 m, 35 m, 55 m, 75 m, 95 m
**37–5** 1.20 mm
**37–7** 545 nm
**37–9** a) 1.00 mm b) $3.00 \times 10^{-6}$ W/m$^2$
**37–11** a) 0.540 mm b) 0.270 mm
**37–13** 0.833 mm
**37–15** 0.0337°
**37–17** 83.3 nm
**37–19** 0.606 mm
**37–21** $2.84 \times 10^{-19}$ J to $4.97 \times 10^{-19}$ J (1.77 eV to 3.10 eV)
**37–23** b) $2.00 \times 10^5$ N
**37–25** $p = 7.70 \times 10^{-24}$ kg·m/s; $E = 2.31 \times 10^{-15}$ J; $f = 3.48 \times 10^{18}$ Hz; 0.0861 nm
**37–27** 0.900 mm
**37–29** a) Destructive interference b) 500 Hz c) 0.68 m
**37–31** $\lambda/2d$, independent of $m$
**37–33** 450 nm
**37–35** 2.91 mm
**37–37** a) Interference pattern moves down on screen b) $I = I_0 \cos^2(\frac{\pi}{\lambda}[d \sin\theta + (n-1)L])$ c) $d \sin\theta = m\lambda - (n-1)L$
**37–39** 4.65 fringes/min

## CHAPTER 38

**38–1** 480 nm
**38–3** a) 16.9 mm b) 8.44 mm
**38–5** a) 1.00 cm b) $2.03 \times 10^{-6}$ W/m$^2$
**38–7** 112 $\mu$m
**38–9** a) 1–3, 2–4; b) 1–2, 3–4; c) 1–3, 2–4
**38–11** a) 3 b) 2
**38–13** 17.5°, 36.9°, 64.2°
**38–15** a) 7.32° b) 16.8°
**38–17** 0.225 nm
**38–19** 2.44 m
**38–21** 4.54 m
**38–27** 0.108 mm
**38–29** a) (i) 25.6°; (ii) 10.2°; (iii) 5.08° b) (i) 60.0°; (ii) 23.1°; (iii) 11.5°
**38–31** 1–5, 2–6, 3–7, 4–8
**38–33** 417 nm
**38–35** Second order
**38–37** 800 m
**38–39** 268 km

# Index

# Photo Credits

**Chapter 1: Opener**—Jupiter and its moons, NASA; Newton's drawing from the *Principia,* courtesy AIP Niels Bohr Library; space probe Galileo, courtesy Hughes Communications, Inc., all rights reserved; **1–1(a),** John G. Ross/Photo Researchers, Inc.; **1–1(b),** NASA; **1–2(a),** courtesy of Bibliothèque Nationale, Paris; **1–2(b),** collection E.S./EXPLORER; **1–2(c),** NASA; **1–4(a),(b),** courtesy of the National Institute of Standards and Technology, Boulder Laboratories, U.S. Department of Commerce; **Building physical intuition**—spiral galaxy, Jack Zehrt, FPG International; the sun, National Optical Astronomy Observatories; the earth, NASA; the Statue of Liberty, Comstock Inc./Jack Elness; students, Richard Laird, FPG International; computer chip, courtesy Texas Instruments; red blood cells, Howard Sochurek/The Stock Shop; atomic array, Srinivas Manne and Scot Gould, University of California, Santa Barbara, CA; **1–5,** © ND-Viollet/Roger-Viollet.

**Chapter 2: Opener**—race horse, Bill Strode, Woodfin Camp & Associates; racetrack diagram, courtesy of Churchill Downs; **2–6,** NASA; **2–18,** Fundamental Photographs, New York.

**Chapter 3: Opener**—aircraft carrier, U.S. Air Force; inset of jet, © George Hall/Check Six 1991; **3–11, 3–13,** Fundamental Photographs, New York; **3–15(b),** Ronald C. Modra/*Sports Illustrated;* **3–21,** Will McIntyre/Photo Researchers, Inc.

**Chapter 4: Opener**—moving train, Garnier; **4–1(a),** courtesy of Santa Fe Railway; **4–1(b),** Vandystadt/Photo Researchers, Inc.; **4–1(c),** Smithsonian Institution, photo no. 77-14453; **4–14,** NASA; **Building physical intuition—**sprinter, John McDonough/*Sports Illustrated;* basketball players, Joe Patronite/Allsport; underwater divers, © The Stock Market/Mark Lawrence.

**Chapter 5: Opener**—ice boat, Marshall Henrichs; **5–15(a), (b),** courtesy of Central Scientific Company; **5–22,** Vandystadt/Photo Researchers, Inc.; **5–28(a),** SPL/Photo Researchers, Inc.; **5–28(b),** NASA; **5–28(c),** National Optical Astronomy Observatories; **5–28(d),** Lawrence Berkeley Laboratory, University of California.

**Chapter 6: Opener**—bicyclist, Ronald C. Modra/*Sports Illustrated;* **6–1(a),** Fundamental Photographs, New York; **6–1(b),** courtesy of New York Power Authority; **6–1(c),** Lowell Georgia/Photo Researchers, Inc.; **6–5,** Mark Gerson, FPG International; **6–15,** Comstock Inc./Comstock Inc.

**Chapter 7: Opener**—high diver, Robert S. Beck/*Sports Illustrated;* **7–1,** copyright © 1964 Estate of Harold E. Edgerton, courtesy of Palm Press, Inc.; **7–3(a),** VS#200014/Allsport; **7–3(b),** FOCUS ON SPORTS; **7–14,** T. Zimmerman, FPG International; **7–19,** PHOTRI/W. Geiersperger; **Building physical intuition**—Saturn's moons, NASA; spark plug, © The Stock Market/Palmer/Kane, Inc.; crystal, Van Pelt, Photographer.

**Chapter 8: Opener**—bighorn rams, Jerome Wexler/Photo Researchers, Inc.; **8–2,** NASA; **8–17,** Mike Powell/Allsport; **8–21,** Fundamental Photographs, New York; **8–22(b),** John V. A. F. Neal/Photo Researchers, Inc.; **8–24,** NASA.

**Chapter 9: Opener**—Hurricane Gilbert, courtesy Aero-Status Center; news clippings, Marshall Henrichs; **9–3(b),** Gorcher.

**Chapter 10: Opener**—spiral galaxy, © The Stock Market/Photri, Inc.; **10–12(b),** Fundamental Photographs, New York.

**Chapter 11: Opener**—Notre Dame cathedral, Adam Woolfitt, Woodfin Camp & Associates; **11–2,** © The Exploratorium, photos by Nancy Rodger; **Building physical intuition**—arched bridge, suspension bridge, Marshall Henrichs.

**Chapter 12: Opener**—Neptune, NASA; **12–13(b),** courtesy Hughes Communications, Inc., all rights reserved; **12–15(a),** NASA; **12–15(b),** © The Stock Market/Frank P. Rossotto; **12–22,** NASA.

**Chapter 13: Opener**—EKG reading, Will McIntyre/Photo Researchers, Inc.; grandfather clock, Marshall Henrichs; **13–2,** M. Sudderth; **13–21(a),(b),** AP/Wide World Photos.

**Chapter 14: Opener**—albatross, Jeff Foott/TOM STACK & ASSOCIATES; **14–2,** courtesy of Central Scientific Company; **14–7,** Peter B. Kaplan/Photo Researchers, Inc.; **14–12,** M. Sudderth; **14–15,** © The Stock Market/Otto Rogge; **14–20,** from Sears et al., *College Physics,* 7th ed., © 1991 Addison-Wesley Publishing Co., Reading, MA, reprinted with permission; **Building physical intuition**—airplane in a wind tunnel, Gary S. Settles/Photo Researchers, Inc.; trout, Comstock Inc./George Lepp; Frisbee, courtesy Wham-O, Inc.; **14–28,** copyright © 1936 Estate of Harold E. Edgerton, courtesy of Palm Press, Inc.; **14–32,** from Sears et al., *College Physics,* 7th ed., © 1991 Addison-Wesley Publishing Co., Reading, MA, reprinted with permission; **14–33,** Fundamental Photographs, New York; **14–34,** courtesy of M. L. Salby and R. R. Garcia.

**Chapter 15: Opener**—molten steel, Comstock Inc./Jack Elness; craftsman working with metal, Jack Fields/Photo Researchers, Inc.; **15–4,** courtesy of Central Scientific Company; **15–5(a),** courtesy of Central Scientific Company; **15–9,** Marshall Henrichs; **15–11,** courtesy of Steven Brigance; **15–13,** Fundamental Photographs, New York; **15–17,** Lockheed Missiles & Space Company, Inc., Russ Underwood, photographer; **15–22(a),** John Raffo/Photo Researchers; **15–22(b),** courtesy Texas Instruments.

**Chapter 16: Opener**—fireworks over Capitol, George Chan/Photo Researchers, Inc.; **16–1,** Union Pacific Museum Collection; **16–4,** courtesy of Geri Murphy; **16–5,** from *Everest: The Testing Place* (New York: McGraw-Hill, 1985), reprinted with permission; **16–10,** SPL/Photo Researchers, Inc.; **16–11,** Fundamental Photographs, New York.

**Chapter 17: Opener**—race car pit stop, PHOTRI/J. Clark; **17–2(a),** PHOTRI/W. Geiersperger; **17–2(b),** AP/Wide World Photos; **17–3,** USGS Photographic Library.

**Chapter 18: Opener**—pool table and balls, Garnier; **Building physical intuition**—hands with playing cards, Marshall Henrichs; melting ice, courtesy of John Yurka; mixing dyes, Fundamental Photographs, New York; **18–13(a),** Comstock Inc./Marvin Koner; **18–13(b),** GM Corporation; **18–14,** courtesy of Paul Gipe & Assoc.

**Chapter 19: Opener**—earthquake damage in San Francisco, USGS Photographic Library; **19–1,** John V. A. F. Neal/Photo Researchers, Inc.; **19–18,** Comstock Inc./Ernest Braun.

**Chapter 20: Opener**—organ at Wellesley College, Marshall Henrichs; **20–1,** PSSC PHYSICS, 2nd edition (1965), D. C. Heath & Company with Education Development Center, Inc., Newton, MA; **20–5(a)-(d),** Fundamental Photographs, New York; **20–9,** courtesy Steinway & Sons; **20–18,** courtesy AT&T Archives; **20–19,** Comstock Inc./Russ Kinne.

**Chapter 21: Opener**—dolphins, © The Stock Market/Mark Lawrence; **21–2,** Lynn Goldsmith/LGI; **21–12(b),** © John Gaffney/Mach 1, Inc.; **21–14,** courtesy of Pamela K. Clark.

**Chapter 22: 22–21,** from PSSC PHYSICS, 2nd edition (1965), D. C. Heath and Company with Education Development Center, Inc., Newton, MA.

**Chapter 23: 23–20(b),** Comstock Inc./Russ Kinne.

**Chapter 24: Opener**—oscilloscope, Fundamental

photographs, New York; **24–10,** Comstock Inc./Russ Kinne.

**Chapter 25: Opener**—Mick Jagger, AP/Wide World Photos; **25–2,** Fundamental Photographs, New York; **Building physical intuition**—circuit board, Photo Researchers; computer keyboard, photo flash unit, Marshall Henrichs; **25–12,** courtesy of L. Williams and C. Cooke, High Voltage Research Laboratory, MIT.

**Chapter 26: Opener**—color monitor, Marshall Henrichs; **26–4,** Fundamental Photographs, New York; **26–10(a),** Bruce Byers, FPG International; **26–10(b),** © 1961 M. C. Escher/Cordon Art-Baarn-Holland; **Building physical intuition**—welding, photograph courtesy of Cincinnati Milacron Marketing Co.; light bulb, stove heating element, Marshall Henrichs.

**Chapter 27: Opener**—CD player, Garnier; **27–4,** Van D. Bucher/Photo Researchers.

**Chapter 28: Opener**—aurora borealis, Malcolm Lockwood Photograph; red-tipped auroral bands, courtesy of Robert Overmyer, Thomas Hallinan, Don Lind, and the Geophysical Institute, University of Alaska Fairbanks; **28–1,** Marshall Henrichs; **28–9,** Fundamental Photographs, New York; **28–12(a),** courtesy of Central Scientific Company; **28–16,** Lawrence Berkeley Laboratory, University of California; **28–18(b),** courtesy of Cavendish Laboratory, University of Cambridge; **28–20(a),** TSQ700 Mass Spectrometer System, photograph courtesy of Finnigan MAT, San Jose, CA; **Building physical intuition**—Fermilab Accelerator, Fermilab Visual Media Services; Stanford Linear Accelerator, courtesy of Stanford Linear Accelerator Center and U.S. Department of Energy; **28–25,** M. Bertinetti/Photo Researchers, Inc.

**Chapter 29: Opener**—magnet in a particle accelerator, Deutsches Elektronen-Synchrontron, Hamburg; quadrupole magnet, Lawrence Berkeley Laboratory, University of California.

**Chapter 30: Opener**—cassette player, Marshall Henrichs; **30–24,** Ken Gatherum–Boeing Computer Services Richland, Inc.

**Chapter 31: Opener**—car engine, Marshall Henrichs; **Building physical intuition**—circuit breaker, courtesy of New York Power Authority; gears, Marshall Henrichs.

**Chapter 32: Opener**—transformer, Marshall Henrichs.

**Chapter 33: Opener**—transmitting tower, Marshall Henrichs; **33–13,** NASA; **Building physical intuition**—road map of Andromeda galaxy, Max Planck Institut für Radioastronomie; optical photograph, National Optical Astronomy Observatories; infrared image, courtesy of JPL/IPAC; x-ray view, Smithsonian Institution Astrophysical Observatory; **33–18(a),** Farrell Grehan, FPG International.

**Chapter 34: Opener**—rainbow, courtesy of American Institute of Physics, Niels Bohr Library; **34–7,** © The Exploratorium, photo by Nancy Rodger; **34–10(a),** courtesy of Barry Blanchard; **34–10(b),** Corning Incorporated; **34–22(b),** Sepp Seitz, Woodfin Camp & Associates; **34–27(a),** K. Nomachi/Photo Researchers, Inc.

**Chapter 35: Opener**—kaleidoscope image, courtesy of Steven Brigance; **35–6,** Marshall Henrichs; **35–23,** Grant Heilman, Photography, Inc.

**Chapter 36: Opener**—Boston Common, Marshall Henrichs; **36–3,** Marshall Henrichs; **36–12,** Michael Abbey/Photo Researchers, Inc.; **36–16(b),** courtesy Tokina Optical, Inc.; **Building physical intuition**—Keck Telescope, California Association for Research in Astronomy; VLA, MMT, courtesy NRAO/AUI; Multiple Mirror Telescope (Smithsonian), Smithsonian Institution/University of Arizona.

**Chapter 37: Opener**—soap bubble, Marshall Henrichs; **37–2,** from PSSC PHYSICS, 2nd edition (1965), D. C. Heath and Company with Education Development Center, Inc., Newton, MA; **37–4,** from Sears et al., *College Physics,* 7th ed., © 1991 Addison-Wesley Publishing Co., Reading, MA, reprinted with permission; **37–11(b),** courtesy of Bausch & Lomb; **37–12,** courtesy of Bausch & Lomb; **37–14,** Tom Branch/Photo Researchers, Inc..

**Chapter 38: Opener**—hologram, Chuck O'Rear, Woodfin Camp & Associates; **38–2,** from Sears et al., *College Physics,* 7th ed., © 1991 Addison-Wesley Publishing Co., Reading, MA, reprinted with permission; **38–3,** from Hecht, *Optics,* 2nd ed., © 1987 Addison-Wesley Publishing Co., Reading, MA, reprinted with permission; **38–8(b), 38–11(b),** from Sears et al., *College Physics,* 7th ed., © 1991 Addison-Wesley Publishing Co., Reading, MA, reprinted with permission; **38–18(b),** courtesy Dr. B. E. Warren; **38–23, 38–24,** from Sears et al., *College Physics,* 7th ed., © 1991 Addison-Wesley Publishing Co., Reading, MA, reprinted with permission; **38–27,** courtesy of Media Interface, Ltd.

**Chapter 39: Opener**—realistic array, courtesy Dr. Ping Kang Hsiung; **39–9,** courtesy Dr. Ping Kang Hsiung; **39–14,** Fermilab Visual Media Services.

**Chapter 40: Opener**—Keck telescope, © 1992 Roger Ressmeyer–Starlight; atomic spectra, Bausch & Lomb; **40–7,** Bausch & Lomb; **40–22(a),** David R. Frazier/Photo Researchers; **40–22(b),** Bruce H. Frisch/Photo Researchers; **40–22(c),** Will McIntyre/Photo Researchers; **40–29,** courtesy of General Electric Company.

**Chapter 41: Opener**—paramecium, Brian Parker/Tom Stack & Associates; SEM of paramecium, Gary W. Grimes, Hofstra University; transmission EM of paramecium, Gary W. Grimes, Hofstra University; **41–3,** from PSSC PHYSICS, 2nd edition (1965), D.C. Heath and Company with Education Development Center, Inc., Newton, MA; **41–7(b),** from Elisha Higgins, *Physics I,* © 1968 W.A. Benjamin, Inc., reprinted with permission of Addison-Wesley Publishing Co.; **41–9(b),** courtesy of JEOL USA, Inc.; **41–9(c),** courtesy of Carl Zeiss, Inc., Thornwood, NY 10594; **41–10(b),** Lawrence Migdale/Photo Researchers; **41–10(c),** courtesy of Carl Zeiss, Inc., Thornwood, NY 10594.

**Chapter 42: Opener**—house fires, Fundamental Photographs, New York; **42–13(b),** IBM Research; **42–13(c),** Digital Instruments.

**Chapter 44: Opener**—LCD flat monitor, Sharp; lap-top computer, Toshiba America Information Systems, Inc.; **44–10,** R.C. Herman; **44–33,** courtesy of Hewlett-Packard Company.

**Chapter 45: Opener**—computer simulation of sun's pressure waves, John W. Harvey, National Solar Observatory; **45–1(b),** SPL/Photo Researchers; **45–9,** National Museum of the American Indian, Smithsonian Institution; **45–10(a),** Hank Morgan/Photo Researchers; **45–10(b),** Science Source/Photo Researchers; **45–16(b),** Oak Ridge National Laboratory; **45–17,** Lawrence Livermore National Laboratory.

**Chapter 46: Opener**—Milky Way galaxy, NASA; particle reaction, CERN photo; **46–1,** Lawrence Berkeley Laboratory, University of California; **46–3(a),** Lawrence Berkleley Laboratory, University of California; **46–7(a), (b)** Fermilab Visual Media Services; **Building physical intuition**—Collider Detector, Fermilab Visual Media Services; L3 detector image, CERN photo; Aleph detector image, CERN photo; computer image, Fermilab Visual Media Services; **46–8(a),** Stanford Linear Accelerator Center/U.S. Department of Energy; **46–14,** Brookhaven National Laboratory.

APPENDIX F

# Numerical Constants

### *FUNDAMENTAL PHYSICAL CONSTANTS*

| *Name* | *Symbol* | *Value* |
|---|---|---|
| Speed of light | $c$ | $2.99792458 \times 10^{8}$ m/s |
| Charge of electron | $e$ | $1.602177 \times 10^{-19}$ C |
| Gravitational constant | $G$ | $6.67259 \times 10^{-11}$ N · m²/kg² |
| Planck's constant | $h$ | $6.6260755 \times 10^{-34}$ J · s |
| Boltzmann's constant | $k$ | $1.38066 \times 10^{-23}$ J/K |
| Avogadro's number | $N_A$ | $6.022 \times 10^{23}$ molecules/mol |
| Gas constant | $R$ | 8.314510 J/mol · K |
| Mass of electron | $m_e$ | $9.10939 \times 10^{-31}$ kg |
| Mass of neutron | $m_n$ | $1.67493 \times 10^{-27}$ kg |
| Mass of proton | $m_p$ | $1.67262 \times 10^{-27}$ kg |
| Permittivity of free space | $\epsilon_0$ | $8.854 \times 10^{-12}$ $C^2/N \cdot m^2$ |
| | $1/4\pi\epsilon_0$ | $8.987 \times 10^{9}$ $N \cdot m^2/C^2$ |
| Permeability of free space | $\mu_0$ | $4\pi \times 10^{-7}$ Wb/A · m |

### *OTHER USEFUL CONSTANTS*

| | | |
|---|---|---|
| Mechanical equivalent of heat | | 4.186 J/cal (15° calorie) |
| Standard atmospheric pressure | 1 atm | $1.013 \times 10^{5}$ Pa |
| Absolute zero | 0 K | − 273.15°C |
| Electronvolt | 1 eV | $1.602 \times 10^{-19}$ J |
| Atomic mass unit | 1 u | $1.66054 \times 10^{-27}$ kg |
| Electron rest energy | $mc^2$ | 0.511 MeV |
| Energy equivalent of 1 u | $Mc^2$ | 931.494 MeV |
| Volume of ideal gas (0°C and 1 atm) | $V$ | 22.4 liter/mol |
| Acceleration due to gravity (sea level, at equator) | $g$ | $9.78049$ m/s² |

### *ASTRONOMICAL DATA*

| *Body* | *Mass, kg* | *Radius, m* | *Orbit radius, m* | *Orbit period* |
|---|---|---|---|---|
| Sun | $1.99 \times 10^{30}$ | $6.96 \times 10^{8}$ | — | — |
| Moon | $7.35 \times 10^{22}$ | $1.74 \times 10^{6}$ | $0.38 \times 10^{9}$ | 27.3 d |
| Mercury | $3.28 \times 10^{23}$ | $2.57 \times 10^{6}$ | $5.79 \times 10^{10}$ | 88.0 d |
| Venus | $4.82 \times 10^{24}$ | $6.31 \times 10^{6}$ | $1.08 \times 10^{11}$ | 224.7 d |
| Earth | $5.98 \times 10^{24}$ | $6.38 \times 10^{6}$ | $1.49 \times 10^{11}$ | 365.3 d |
| Mars | $6.42 \times 10^{23}$ | $3.38 \times 10^{6}$ | $2.28 \times 10^{11}$ | 687.0 d |
| Jupiter | $1.89 \times 10^{27}$ | $7.18 \times 10^{7}$ | $7.78 \times 10^{11}$ | 11.86 y |
| Saturn | $5.69 \times 10^{26}$ | $6.03 \times 10^{7}$ | $1.43 \times 10^{12}$ | 29.46 y |
| Uranus | $8.66 \times 10^{25}$ | $2.67 \times 10^{7}$ | $2.87 \times 10^{12}$ | 84.01 y |
| Neptune | $1.03 \times 10^{26}$ | $2.48 \times 10^{7}$ | $4.49 \times 10^{12}$ | 164.8 y |
| Pluto | $1.1 \times 10^{22}$ | $4 \times 10^{5}$ | $5.91 \times 10^{12}$ | 248.7 y |